ICU	intensive care unit
ID	intradermal
IF	intrinsic factor
IFN	interferon
Ig	immunoglobin
IM	intramuscular; infectious mononucleosis
IN	internist
IOP	intraocular pressure
IPPA	inspection, palpation, percussion, auscultation
IPSP	inhibitory postsynaptic potential
IUD	intrauterine device
IV	intravenous
IVC	inferior vena cava
IVF	in vitro fertilization
IVP	intravenous pyelogram
IVT	intravenous transfusion
JGA	juxtaglomerular apparatus
KS	Kaposi's sarcoma
KUB	kidneys, ureters, bladder
LBB	left breast biopsy
LDL	low-density lipoprotein
LFT	liver function test
LG	laryngectomy
LH	luteinizing hormone
LLQ	left lower quadrant
LMP	last menstrual period
LOC	loss of consciousness
LP	lumbar puncture
LPN	licensed practical nurse
LRI	lower respiratory infection
LUQ	left upper quadrant
LVAD	left ventricular assist device
MAb	monoclonal antibody
mEq/l	milliequivalents per liter
MG	myasthenia gravis
MI	myocardial infarction
MLT	medical laboratory technologist
mm^3	cubic millimeter
mm Hg	millimeters of mercury
MOA	medical office assistant
MRI	magnetic resonance imaging
MS	multiple sclerosis
MSH	melanocyte-stimulating hormone
MSOF	multisystem organ failure
MVP	mitral valve prolapse
MVR	minute volume of respiration
NBM	nothing by mouth
ND	natural death
NE	norepinephrine
NGU	nongonococcal urethritis
NLMC	nocturnal leg muscle cramping
NMG	neuromuscular junction
NPN	nonprotein nitrogen
NREM	nonrapid eye movement
NSAID	nonsteroidal antiinflammatory drug
NSU	nonspecific urethritis
NTG	nitroglycerin
NTP	normal temperature and pressure
OB/GYN	obstetrician-gynecologist; obstetrics-gynecology
OC	oral contraceptive
OD	overdose; right eye
OHS	open heart surgery
OI	opportunistic infection
OR	operating room
ORT	operating room technician
OT	oxytocin
OTC	over-the-counter
OV	office visit
P	pressure
PABA	para-aminobenzoic acid
PCP	*Pneumocystis carinii* pneumonia
PCV	packed cell volume
PD	Parkinson's disease
PE	pulmonary embolism; physical examination
PED	pediatrics; pediatrician
Peff	effective filtration pressure
PEG	pneumoencephalogram
PEMF	pulsating electromagnetic field
PET	positron emission tomography
PG	prostaglandin
pH	hydrogen-ion concentration
PID	pelvic inflammatory disease
PKU	phenylketonuria
PMH	past medical history
PMN	polymorphonuclear leukocyte
PMP	plasma membrane protein
PMS	premenstrual syndrome
PNS	peripheral nervous system
PRL	prolactin
PROG	progesterone
PT	prothrombin time; physical therapist
PTCA	percutaneous transluminal coronary angioplasty
PTD	permanent and total disability
PTH	parathyroid hormone
PTT	partial thromboplastin time
PTX	pneumothorax
PUBS	percutaneous umbilical blood sampling
PUL	percutaneous ultrasonic lithotripsy
Px	prognosis; pneumothorax
PX	physical examination
R	roentgen (unit of x radiation)
RA	rheumatoid arthritis
RAD	radiation absorbed dose
RAS	reticular activating system
RBB	right breast biopsy
RBC	red blood cell; red blood count
RBOW	rupture of bag of waters
RDA	recommended daily allowance
RDS	respiratory distress syndrome
REM	rapid eye movement
Rh	*Rhesus*
RHC	respirations have ceased
RIA	radioimmunoassay
RK	radial keratotomy
RLQ	right lower quadrant
RLX	relaxin
RM	radical mastectomy
RN	registered nurse
RNA	ribonucleic acid
ROS	review of symptoms
RR	respiratory rate
RRR	regular rate and rhythm (heart)
RS	Reye's syndrome
RT	radiotherapy; radiologic technologist
SCA	sickle-cell anemia
SCD	sudden cardiac death
SCID	severe combined immunodeficiency syndrome
SDS	same-day surgery
SF	synovial fluid
SG	skin graft; specific gravity
SH	social history
SIDS	sudden infant death syndrome
SIG	sigmoidoscopy; sigmoidoscope
SIW	self-inflicted wound
SLE	systemic lupus erythematosus
SMD	senile macular degeneration
SNS	somatic nervous system
SOB	shortness of breath
SPF	sun protection factor
S/S (Sx)	signs and symptoms
STD	sexually transmitted disease
SubQ or SQ	subcutaneous
SV	stroke volume
SVC	superior vena cava
T	temperature
TAH	total artificial heart
TB	tuberculosis
TBI	total body irradiation
TIA	transient ischemic attack
TLI	total lymphoid irradiation
Tm	transport maximum
TM	transcendental meditation
TMJ	temporomandibular joint
TND	transient neurologic deficit
TOP	termination of pregnancy
t-PA	tissue plasminogen activator
TPE	therapeutic plasma exchange
TPN	total parenteral nutrition
TPR	temperature, pulse, and respiration
TSH	thyroid-stimulating hormone
TSS	toxic shock syndrome
Tx	treatment
UA	urinalysis
UDO	undetermined origin
UG	urogenital
URI	upper respiratory infection
US	ultrasound; ultrasonography
UTI	urinary tract infection
UV	ultraviolet
VD	venereal disease
VDRL	venereal disease research laboratory test (blood test for syphilis)
VF	ventricular fibrillation
VPC	ventricular premature contraction
VS	vital signs
VT	ventricular tachycardia
VV	varicose veins; vulva and vagina
WBC	white blood cell; white blood count
WNL	within normal limits
X-match	cross-match
XRT	x-ray therapy

ALLIED HEALTH

Bastian
ILLUSTRATED REVIEW OF ANATOMY AND PHYSIOLOGY Series (1993)
BASIC CONCEPTS OF CHEMISTRY, THE CELL, AND TISSUES
THE MUSCULAR AND SKELETAL SYSTEMS
THE NERVOUS SYSTEM
THE ENDOCRINE SYSTEM
THE CARDIOVASCULAR SYSTEM
THE LYMPHATIC AND IMMUNE SYSTEMS
THE RESPIRATORY SYSTEM
THE DIGESTIVE SYSTEM
THE URINARY SYSTEM
THE REPRODUCTIVE SYSTEM

Kreier/Mortensen
PRINCIPLES OF INFECTION, RESISTANCE, AND IMMUNITY (1990)

Telford/Bridgman
INTRODUCTION TO FUNCTIONAL HISTOLOGY (1989)

Tortora
INTRODUCTION TO THE HUMAN BODY, Second Edition (1991)

Tortora
PRINCIPLES OF HUMAN ANATOMY, Sixth Edition (1992)

Tortora/Grabowski
PRINCIPLES OF ANATOMY AND PHYSIOLOGY, Seventh Edition (1993)

Volk
BASIC MICROBIOLOGY, Seventh Edition (1992)

LIFE SCIENCES

Beck/Liem/Simpson
LIFE: AN INTRODUCTION TO BIOLOGY, Third Edition (1991)

Harris
CONCEPTS IN ZOOLOGY (1992)

Herrmann
CELL BIOLOGY (1989)

Hill/Wyse
ANIMAL PHYSIOLOGY, Second Edition (1988)

Jenkins
HUMAN GENETICS, Second Edition (1989)

Kaufman
PLANTS: THEIR BIOLOGY AND IMPORTANCE (1990)

Kleinsmith/Kish
PRINCIPLES OF CELL BIOLOGY (1987)

Mix/Farber/King
BIOLOGY: THE NETWORK OF LIFE (1992)

Nickerson
GENETICS: A GUIDE TO BASIC CONCEPTS AND PROBLEM SOLVING (1989)

Nybakken
MARINE BIOLOGY: AN ECOLOGICAL APPROACH, Third Edition (1993)

Penchenik
A SHORT GUIDE TO WRITING ABOUT BIOLOGY, Second Edition (1993)

Rischer/Easton
FOCUS ON HUMAN BIOLOGY (1992)

Russell
GENETICS, Third Edition (1992)

Shostak
EMBRYOLOGY: AN INTRODUCTION TO DEVELOPMENTAL BIOLOGY (1991)

Wallace
BIOLOGY: THE WORLD OF LIFE, Sixth Edition (1992)

Wallace/Sanders/Ferl
BIOLOGY: THE SCIENCE OF LIFE, Third Edition (1991)

Webber/Thurman
MARINE BIOLOGY, Second Edition (1991)

ECOLOGY AND ENVIRONMENTAL STUDIES

Kaufman/Franz
BIOSPHERE 2000: PROTECTING OUR GLOBAL ENVIRONMENT (1993)

Krebs
ECOLOGY: THE EXPERIMENTAL ANALYSIS OF DISTRIBUTION AND ABUNDANCE, Third Edition (1984)

Krebs
ECOLOGICAL METHODOLOGY (1988)

Krebs
THE MESSAGE OF ECOLOGY (1987)

Pianka
EVOLUTIONARY ECOLOGY, Fourth Edition (1988)

Smith
ECOLOGY AND FIELD BIOLOGY, Fourth Edition (1991)

Smith
ELEMENTS OF ECOLOGY, Third Edition (1992)

Yodzis
INTRODUCTION TO THEORETICAL ECOLOGY (1989)

LABORATORY MANUALS

Beishir
MICROBIOLOGY IN PRACTICE: A SELF-INSTRUCTIONAL LABORATORY COURSE, Fifth Edition (1991)

Donnelly
LABORATORY MANUAL FOR HUMAN ANATOMY: WITH CAT DISSECTIONS, Second Edition (1993)

Donnelly/Wistreich
LABORATORY MANUAL FOR ANATOMY AND PHYSIOLOGY: WITH CAT DISSECTIONS, Fourth Edition (1993)

Donnelly/Wistreich
LABORATORY MANUAL FOR ANATOMY AND PHYSIOLOGY: WITH FETAL PIG DISSECTIONS (1993)

Eroschenko
LABORATORY MANUAL FOR HUMAN ANATOMY WITH CADAVERS (1990)

Tietjen
THE HARPERCOLLINS BIOLOGY LABORATORY MANUAL (1991)

Tietjen
LABORATORY MANUAL TO ACCOMPANY BIOLOGY: THE NETWORK OF LIFE (1992)

Tietjen/Harris
HARPERCOLLINS ZOOLOGY LABORATORY MANUAL (1992)

COLORING BOOKS

Diamond/Scheibel/Elson
THE HUMAN BRAIN COLORING BOOK (1985)

Elson
THE ZOOLOGY COLORING BOOK (1982)

Griffin
THE BIOLOGY COLORING BOOK (1987)

Kapit/Macey/Meisami
THE PHYSIOLOGY COLORING BOOK (1987)

Kapit/Elson
THE ANATOMY COLORING BOOK, Second Edition (1993)

Niesen
THE MARINE BIOLOGY COLORING BOOK (1982)

Young
THE BOTANY COLORING BOOK (1982)

PRINCIPLES
OF
ANATOMY
AND
PHYSIOLOGY

ABOUT
THE
AUTHORS

Gerard J. Tortora

Jerry Tortora is a professor of biology and teaches human anatomy and physiology and microbiology at Bergen Community College in Paramus, New Jersey. He is the Biology Coordinator and has just completed 33 years as a teacher, the past 27 at Bergen CC. He received his B.S. in biology from Fairleigh Dickinson University in 1962 and his M.A. in biology from Montclair State College in 1965. He has also taken graduate courses in education and science at Columbia University and Rutgers University. He belongs to numerous biology organizations, such as the Human Anatomy and Physiology Society (HAPS), the American Association for the Advancement of Science (AAAS), the American Association of Microbiology (ASM), and the Metropolitan Association of College and University Biologists (MACUB). Jerry is the author of a number of best-selling anatomy and physiology, anatomy, and microbiology textbooks and several laboratory manuals.

Sandra Reynolds Grabowski

Sandy Grabowski is an instructor in the Department of Biological Sciences at Purdue University in West Lafayette, Indiana. Since 1977 she has taught human anatomy and physiology to students in a wide range of academic programs. In 1992 students selected her as one of the top 10 teachers in the School of Science at Purdue. Sandy received her B.S. in biology and Ph.D. in neurophysiology from Purdue. She is an active member of the Human Anatomy and Physiology Society (HAPS), served as editor of *HAPS News* from 1990 through 1992, and was elected to serve a three-year term as President-Elect, President, and Past-President from 1992 to 1995. In addition, she is a member of the American Association for the Advancement of Science (AAAS), the Association for Women in Science (AWIS), the National Science Teachers Association (NSTA), the Society for College Science Teachers (SCST), and the Association of Biology Laboratory Educators (ABLE).

EIGHTH EDITION

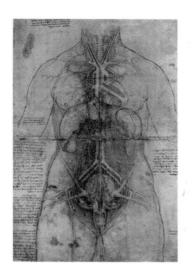

PRINCIPLES OF ANATOMY AND PHYSIOLOGY

Gerard J. Tortora
BIOLOGY COORDINATOR
BERGEN COMMUNITY COLLEGE

Sandra Reynolds Grabowski
PURDUE UNIVERSITY

HarperCollins*CollegePublishers*

Executive Editor: Bonnie Roesch
Senior Developmental Editor: Thom Moore
Project Coordination and Text Design: Electronic Publishing Services Inc.
Cover Designer: Mary McDonnell
Front Cover Illustration: Alinari/Art Resource, New York
Back Cover Illustration: Kevin Sommerville
Art Coordinator: Claudia Durrell
Photo Researcher: Mira Schachne
Electronic Production Manager: Mike Kemper
Manufacturing Manager: Willie Lane
Electronic Page Makeup: Electronic Publishing Services Inc.
Printer and Binder: RR Donnelley & Sons Company
Cover Printer: Coral Graphics

For permission to use copyrighted material, grateful acknowledgment is made to the copyright holders on pp. C-1–C-4, which are hereby made part of this copyright page.

***Principles of Anatomy and Physiology*, Eighth Edition**

Library of Congress Cataloging-in-Publication Data

Tortora, Gerard J.
    Principles of anatomy and physiology / Gerard J. Tortora, Sandra
Reynolds Grabowski. — 8th ed.
      p. cm.
    Includes index.
    ISBN 0-673-99355-8 (instructor edition)   0-673-99354-X
      (student edition)
    1. Human physiology. 2. Human anatomy. I. Grabowski,
Sandra Reynolds. II. Title.
    [DNLM: 1. Anatomy. 2. Physiology. QS 4 T712p 1996]
QP34.5.T67 1996
612—dc20
DNLM/DLC
for Library of Congress   95-23451
                             CIP

96 97 98 9 8 7 6 5 4 3 2

To Lynne Marie, Gerard Joseph, Jr., Kenneth Stephen, Christopher Andrew, and Anthony Gerard, who make it all worthwhile.

G.J.T.

To my husband Zbig, for his steadfast and loving support.

S.R.G.

ABOUT THE COVER

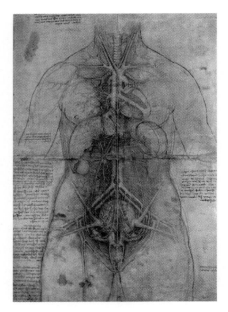

At first glance the elegant illustrations that grace the front and back cover of the eighth edition of Principles of Anatomy and Physiology *appear to be from the same era. Yet nearly 500 years of learning about human anatomy and physiology separate these drawings.*

The front cover is Leonardo da Vinci's so-called Great Lady Anatomy, *drawn about 1508. Although more widely known today for about a dozen paintings, including* Mona Lisa *and* The Last Supper, *Leonardo left a legacy of nearly a thousand biological sketches. Like many other sketches,* Great Lady Anatomy *shows marginal notes in which Leonardo, using his unique mirror-image script, included speculations about some aspects of biological function.*

Unfortunately, what this masterpiece achieves in artistic beauty, it fails to provide in anatomical and physiological accuracy. The trachea bifurcates into equally inclined primary bronchi, the liver is misshapen and confined to the right side, the heart has only two chambers, the aortic branches and iliac vessels are misrepresented, and the apparently pregnant uterus sprouts horn-like structures.

These inaccuracies are rather surprising in light of Leonardo's claim to have dissected ten cadavers. Most likely, he had access only to portions of cadavers over many years. Leonardo may have never even dissected a female; indeed, the figure outline of Great Lady Anatomy *looks masculine. Leonardo was also influenced by his extensive animal dissections. His representation of the uterus and the aortic arch is consistent with their appearance in other mammals. But perhaps the greatest impediment to accuracy was the adherence of Leonardo and others of his time to the philosophy of Galen, a second-century physician, whose writings engendered 1300 years of scientific dogma.*

On the back cover is a composition by Kevin Somerville, a well-known contemporary medical illustrator who also has created dozens of the drawings in Principles of Anatomy and Physiology, *Eighth Edition. This image is anatomically correct and integrates form and function in a superimposed fashion. Somerville attractively blends an aged appearance with modern-day science, serving as a tribute to generations of artists who have striven for accurate, beautiful portrayals of human anatomy and physiology.*

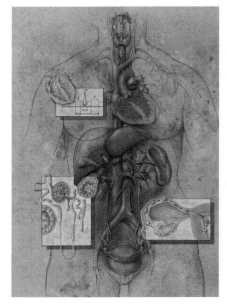

Kevin Petti
Associate Professor
Department of Science and Health
San Diego Miramar College

CONTENTS IN BRIEF

CONTENTS IN DETAIL

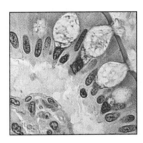

CLINICAL APPLICATIONS

CLINICAL APPLICATIONS

CLINICAL APPLICATIONS

UNIT 2   PRINCIPLES OF SUPPORT AND MOVEMENT   141

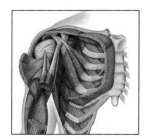

Chapter 6   Bone Tissue   142

Chapter 7   The Skeletal System: The Axial Skeleton   162

CLINICAL APPLICATIONS

Chapter 11   The Muscular
                    System   269

CLINICAL APPLICATIONS

UNIT 3   CONTROL SYSTEMS OF
         THE HUMAN BODY   330

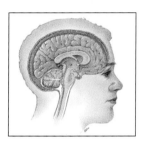

Chapter 12   Nervous Tissue   331

CLINICAL APPLICATIONS

Chapter 16 The Special Senses 453

CLINICAL APPLICATIONS

Chapter 17 The Autonomic Nervous System 487

UNIT 4   MAINTENANCE OF THE HUMAN BODY   551

Chapter 19   The Cardiovascular System: The Blood   552

CLINICAL APPLICATIONS

Chapter 20   The Cardiovascular System: The Heart   578

CLINICAL APPLICATIONS

CLINICAL APPLICATIONS

Chapter 23   The Respiratory System   707

CLINICAL APPLICATIONS

Chapter 24   The Digestive System   752

CLINICAL APPLICATIONS

Chapter 25   Metabolism   806

CLINICAL APPLICATIONS

Chapter 26   The Urinary System   847

CLINICAL APPLICATIONS

Chapter 27   Fluid, Electrolyte, and Acid–Base Homeostasis   890

CLINICAL APPLICATIONS

UNIT 5   CONTINUITY   907

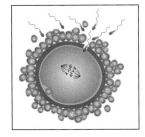

Chapter 28   The Reproductive Systems   908

CLINICAL APPLICATIONS

CLINICAL APPLICATIONS

PREFACE

Principles of Anatomy and Physiology is designed for an introductory course in anatomy and physiology and assumes no prior knowledge of the human body by the student. It is geared to students preparing for careers in health-related professions, such as nursing, occupational therapy, physical therapy, medical technology, medicine, and dentistry. Because of its scope, the text is also useful for students in the biological sciences, science technology, science education, and physical education programs.

This eighth edition of *Principles of Anatomy and Physiology* builds on the phenomenal and unprecedented success of the past seven editions. Previous editions have been so successful because of clear and concise readability, outstanding art programs, the introduction of so many unique and innovative pedagogical aids, and the refinement of tried and tested learning devices. In the eighth edition the themes and organization, the features just mentioned, and the accuracy of previous editions, upon which anatomy and physiology instructors and students have depended for over twenty years, have been retained. The text has been completely revised, however, and includes many new and revised figures, up-to-date physiology, some new and innovative learning devices, and interesting, new clinical applications.

Each instructor approaches the teaching of anatomy and physiology from a different perspective and background. The advantage of two authors with very different backgrounds working together proved practical and beneficial in the seventh edition. Once again, our collaboration has resulted in a text that presents a body of knowledge filtered and refined both by an anatomist and a physiologist. Thus the balance between anatomy and physiology continues to be fine-tuned. In addition, we have emphasized correlations between normal physiology and pathophysiology, normal anatomy and pathology, and homeostasis and homeostatic imbalances.

THEMES AND ORGANIZATION

Homeostasis and Homeostatic Imbalances

As in the past, the eighth edition has two underlying themes: homeostasis and homeostatic imbalances. **Homeostasis**, the condition in which the body's internal environment remains within certain physiological limits, is immediately introduced in Chapter 1 and continued throughout the book. The *negative feedback illustrations*, so well received in the last edition, have been modified and enhanced where appropriate. These diagrams help make potentially confusing concepts much easier to understand. They are used whenever applicable to help students grasp the dynamic counterbalancing act that systems must perform to maintain normal anatomy and physiology.

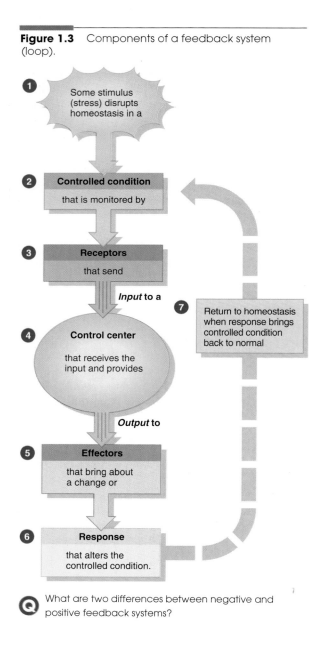

Figure 1.3 Components of a feedback system (loop).

What are two differences between negative and positive feedback systems?

Figure 5.7 Thermoregulation: homeostasis of body temperature by the skin.

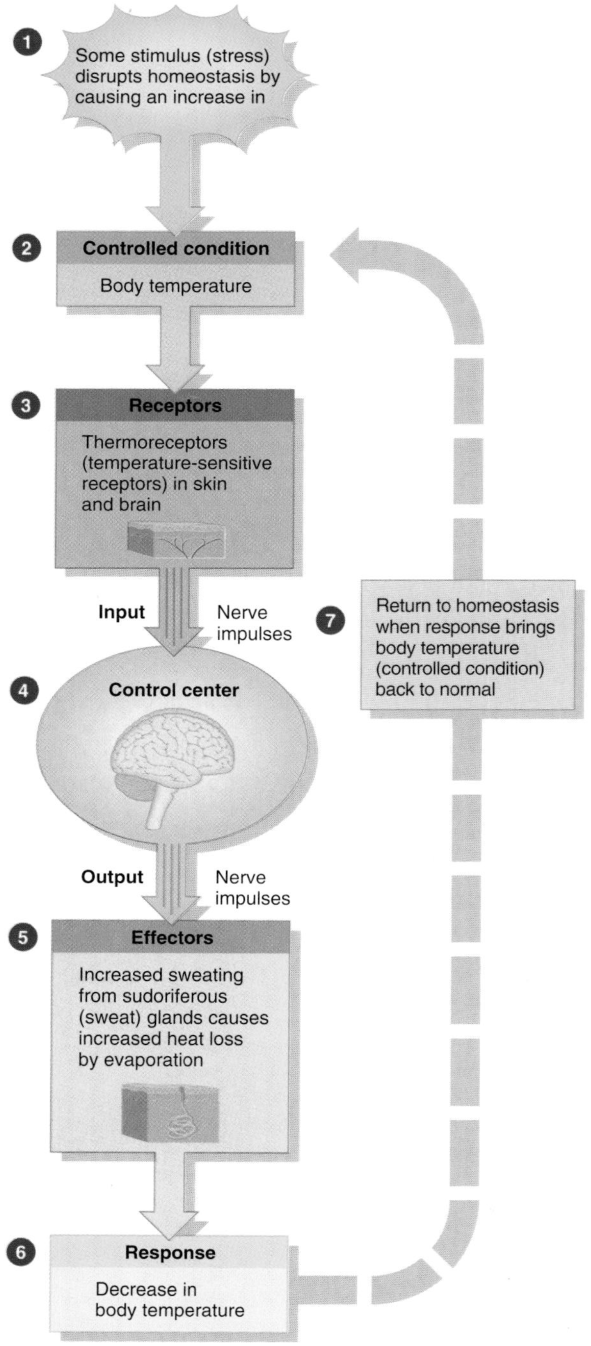

Why is this a negative (rather than a positive) feedback cycle?

Homeostatic Imbalances, the other theme of the textbook, represent disruptions in homeostasis that could result in illness or even death. Once students have been introduced to normal anatomy and physiology, they are provided with examples of homeostatic imbalances through clinical applications and disorders. These discussions further illustrate the relevance of normal mechanisms.

Clinical applications are in-text discussions that apply knowledge of normal body processes to clinical situations. These discussions are now readily identifiable by the use of a stethoscope icon.

Disorders: Homeostatic Imbalances at the end of most chapters provide additional explanations of important medical problems. These include answers to many of the questions that students ask about medical conditions and diseases.

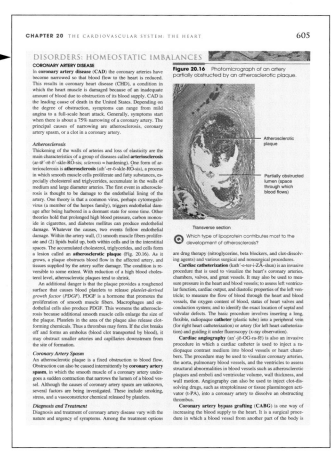

CHAPTER 20 THE CARDIOVASCULAR SYSTEM: THE HEART   605

DISORDERS: HOMEOSTATIC IMBALANCES

CORONARY ARTERY DISEASE

In **coronary artery disease (CAD)** the coronary arteries have become narrowed so that blood flow to the heart is reduced. This results in coronary heart disease (CHD), a condition in which the heart muscle is damaged because of an inadequate amount of blood due to obstruction of its blood supply. CAD is the leading cause of death in the United States. Depending on the degree of obstruction, symptoms can range from mild angina to a full-scale heart attack. Generally, symptoms start when there is about a 75% narrowing of a coronary artery. The principal causes of narrowing are atherosclerosis, coronary artery spasm, or a clot in a coronary artery.

Atherosclerosis

Thickening of the walls of arteries and loss of elasticity are the main characteristics of a group of diseases called **arteriosclerosis** (ar-tē'-rē-ō'-skle-RŌ-sis; *sclerosis* = hardening). One form of arteriosclerosis is **atherosclerosis** (ath'-er-ō-skle-RŌ-sis), a process in which smooth muscle cells proliferate and fatty substances, especially cholesterol and triglycerides, accumulate in the walls of medium and large diameter arteries. The first event in atherosclerosis is thought to be damage to the endothelial lining of the artery. One theory is that a common virus, perhaps cytomegalovirus (a member of the herpes family), triggers endothelial damage after being harbored in a dormant state for some time. Other theories hold that prolonged high blood pressure, carbon monoxide in cigarettes, and diabetes mellitus can produce endothelial damage. Whatever the causes, two events follow endothelial damage. Within the artery wall, (1) smooth muscle fibers proliferate and (2) lipids build up, both within cells and in the interstitial spaces. The accumulated cholesterol, triglycerides, and cells form a lesion called an **atherosclerotic plaque** (Fig. 20.16). As it grows, a plaque obstructs blood flow in the affected artery, and tissues supplied by the artery suffer damage. The condition is reversible to some extent. With reduction of a high blood cholesterol level, atherosclerotic plaques tend to shrink.

An additional danger is that the plaque provides a roughened surface that causes blood platelets to release *platelet-derived growth factor (PDGF)*. PDGF is a hormone that promotes the proliferation of smooth muscle fibers. Macrophages and endothelial cells also produce PDGF. This worsens the atherosclerosis because additional smooth muscle cells enlarge the size of the plaque. Platelets in the area of the plaque also release clot-forming chemicals. Thus a thrombus may form. If the clot breaks off and forms an embolus (blood clot transported by blood), it may obstruct smaller arteries and capillaries downstream from the site of formation.

Coronary Artery Spasm

An atherosclerotic plaque is a fixed obstruction to blood flow. Obstruction can also be caused intermittently by **coronary artery spasm**, in which the smooth muscle of a coronary artery undergoes a sudden contraction that narrows the lumen of a blood vessel. Although the causes of coronary artery spasm are unknown, several factors are being investigated. These include smoking, stress, and a vasoconstrictor chemical released by platelets.

Diagnosis and Treatment

Diagnosis and treatment of coronary artery disease vary with the nature and urgency of symptoms. Among the treatment options

Figure 20.16 Photomicrograph of an artery partially obstructed by an atherosclerotic plaque.

Atherosclerotic plaque

Partially obstructed lumen (space through which blood flows)

Transverse section

Which type of lipoprotein contributes most to the development of atherosclerosis?

are drug therapy (nitroglycerine, beta blockers, and clot-dissolving agents) and various surgical and nonsurgical procedures.

Cardiac catheterization (kath'-e-ter-i-ZA-shun) is an invasive procedure that is used to visualize the heart's coronary arteries, chambers, valves, and great vessels. It may also be used to measure pressure in the heart and blood vessels; to assess left ventricular function, cardiac output, and diastolic properties of the left ventricle; to measure the flow of blood through the heart and blood vessels, the oxygen content of blood, status of heart valves and conduction system; and to identify the exact location of septal and valvular defects. The basic procedure involves inserting a long, flexible, radiopaque **catheter** (plastic tube) into a peripheral vein (for right heart catheterization) or artery (for left heart catheterization) and guiding it under fluoroscopy (x-ray observation).

Cardiac angiography (an'-jē-OG-ra-fē) is also an invasive procedure in which a cardiac catheter is used to inject a radiopaque contrast medium into blood vessels or heart chambers. The procedure may be used to visualize coronary arteries, the aorta, pulmonary blood vessels, and the ventricles to assess structural abnormalities in blood vessels such as atherosclerotic plaques and emboli and ventricular volume, wall thickness, and wall motion. Angiography can also be used to inject clot-dissolving drugs, such as streptokinase or tissue plasminogen activator (t-PA), into a coronary artery to dissolve an obstructing thrombus.

Coronary artery bypass grafting (CABG) is one way of increasing the blood supply to the heart. It is a surgical procedure in which a blood vessel from another part of the body is

---

TREATING SPORTS INJURIES

Most sports injuries should be treated initially with **RICE** therapy, which stands for **R**est, **I**ce, **C**ompression, and **E**levation. Immediately apply ice and rest and elevate the injured part. Then apply an elastic bandage, if possible, to compress the injured tissue. Continue using RICE for 2–3 days, and resist the temptation to apply heat, which may worsen the swelling. Follow-up treatment may include alternating moist heat and ice massage to enhance blood flow in the injured area. Sometimes, nonsteroidal anti-inflammatory drugs (NSAIDs), such as aspirin or ibuprofen, or local injections of corticosteroids, such as cortisone, are needed. During the recovery period, it is important to keep active with an alternate fitness program. And careful exercise is needed to rehabilitate the injured area itself. ■

Organization

The book follows the same unit and topic sequence as its seven earlier editions. It is divided into five principal areas of concentration. Unit 1, "Organization of the Human Body," provides an understanding of the structural and functional levels of the body, from molecules to organ systems. Unit 2, "Principles of Support and Movement," analyzes the anatomy and physiology of the skeletal system, articulations, and the muscular system. Unit 3, "Control Systems of the Human Body," emphasizes the importance of neural communication in the immediate maintenance of homeostasis, the role of sensory receptors in providing information about the internal and external environment, and the significance of hormones in maintaining long-term homeostasis. Unit 4, "Maintenance of the Human Body," explains how body systems function to maintain homeostasis on a day-to-day basis through the processes of circulation, respiration, digestion, cellular metabolism, urinary functions, and buffer systems. Unit 5, "Continuity," covers the anatomy and physiology of the reproductive systems, development, and the basic concepts of genetics and inheritance.

THE ILLUSTRATION PROGRAM

The illustration program is one of the signature features of *Principles of Anatomy and Physiology*. Once again, it has been reviewed and refined to continue the standard of excellence expected. To emphasize structural and functional relations, colors are used in a consistent and meaningful manner throughout the text. For example, sensory structures, sensory neurons, and sensory regions of the brain are shades of blue ■, whereas motor structures are light red ■. Membrane phospholipids are gray ■ and orange ■, the cytosol is sand ■, and extracellular fluid is blue ■. Negative and positive feedback loops also use color cues to aid students in understanding and recognizing the concept. Stimulus and response are both orange ■ since they both alter the controlled condition. The controlled condition itself is green ■, the receptor is blue ■, the control center is purple ■, and the effector is red ■. Such color cues provide additional help for students who are trying to learn complex anatomical and physiological concepts.

Successful Illustration Features Maintained and Enhanced

Questions with figures are questions that accompany most figures. They are designed to help students interpret the figures and correlate them with the textual material. This very well-received feature has been refined and expanded in the eighth edition. Many of these questions require students to synthesize verbal and visual information, think critically, or draw conclusions. Answers are given at the end of each chapter.

Figure 12.15 Presynaptic facilitation and inhibition. An excitatory or inhibitory neurotransmitter released by neuron 1 can facilitate or inhibit release of neurotransmitter by neuron 2 at its synapse with neuron 3.

 Presynaptic facilitation increases the amount of neurotransmitter released whereas presynaptic inhibition decreases it.

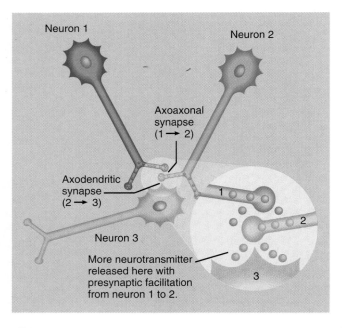

Q Suppose that neuron 2 transmits pain information. Will more or less pain be felt if neuron 1 provides presynaptic inhibition?

Orientation insets, introduced in the last edition, have been greatly expanded and modified in the eighth edition. In one type of inset, not only are planes used to indicate where certain sections are made, but the planes are now labeled so that the reader can more easily relate the planes to the sections that result when a part of the body is cut (a). Other insets contain a directional arrow to indicate the direction from which the body is viewed (superior, inferior, posterior, anterior, and so on) (b). Still other insets have arrows leading from or to them to direct attention to enlarged and detailed parts of illustrations (c). Added to Chapter 25 are new orientation diagrams for metabolic reactions, which are miniature versions of larger illustrations with a selected part highlighted (d). These are designed to help students relate the various steps of a process, such as the electron transport chain, to the overall process, in this case cellular respiration.

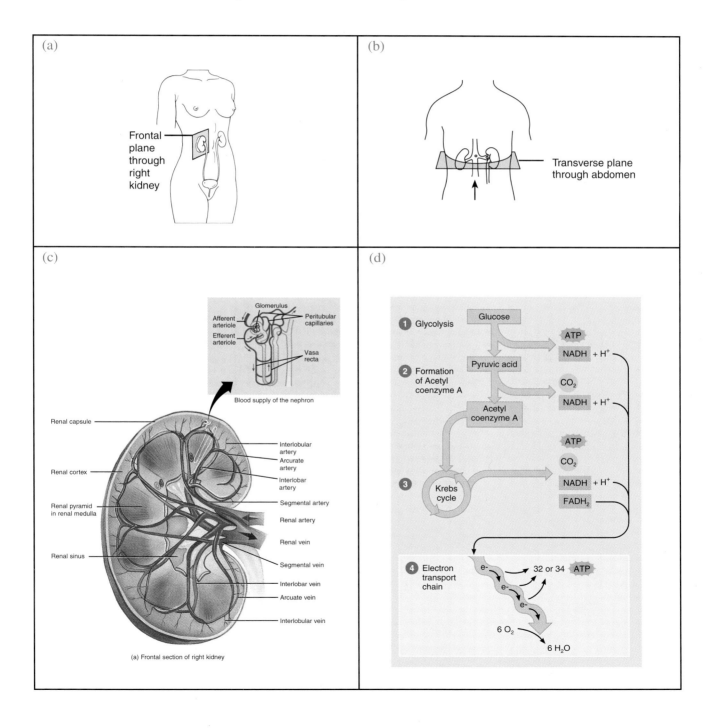

New Illustration Features
New cadaver photographs have replaced most of the older ones for this edition and several additional photos have been added. The dissections were prepared under the supervision of Mark Nielsen of the University of Utah and photographed by Borge Andersen specifically for use in this textbook.

New illustrations have been added throughout the book to amplify both anatomical and physiological concepts. Many illustrations have been revised for better clarity. These additions and enhancements are in keeping with the standard of excellence that *Principles of Anatomy and Physiology* always strives to exceed.

Overview of Functions is a unique new feature that juxtaposes the anatomical components and a brief functional overview for each body system. These function boxes accompany the first figure of most chapters, as well as selected other figures. They permit students to visually integrate the anatomy and physiology of a body system at the outset.

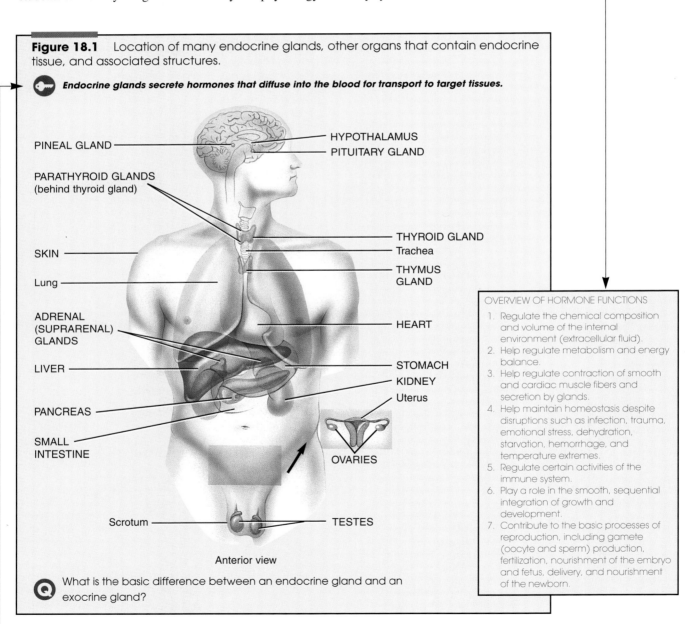

Figure 18.1 Location of many endocrine glands, other organs that contain endocrine tissue, and associated structures.

🔑 *Endocrine glands secrete hormones that diffuse into the blood for transport to target tissues.*

PINEAL GLAND
HYPOTHALAMUS
PITUITARY GLAND
PARATHYROID GLANDS (behind thyroid gland)
THYROID GLAND
Trachea
SKIN
THYMUS GLAND
Lung
ADRENAL (SUPRARENAL) GLANDS
HEART
LIVER
STOMACH
KIDNEY
Uterus
PANCREAS
SMALL INTESTINE
OVARIES
Scrotum
TESTES

Anterior view

OVERVIEW OF HORMONE FUNCTIONS
1. Regulate the chemical composition and volume of the internal environment (extracellular fluid).
2. Help regulate metabolism and energy balance.
3. Help regulate contraction of smooth and cardiac muscle fibers and secretion by glands.
4. Help maintain homeostasis despite disruptions such as infection, trauma, emotional stress, dehydration, starvation, hemorrhage, and temperature extremes.
5. Regulate certain activities of the immune system.
6. Play a role in the smooth, sequential integration of growth and development.
7. Contribute to the basic processes of reproduction, including gamete (oocyte and sperm) production, fertilization, nourishment of the embryo and fetus, delivery, and nourishment of the newborn.

❓ What is the basic difference between an endocrine gland and an exocrine gland?

Key Concept Statements are new and unique to anatomy and physiology textbooks. They are concise statements incorporated into most illustrations, symbolized by a 🔑, that capture the essence of a key concept that has been stated in the text and then amplified in the illustration.

Flow charts included in some figures permit students to concentrate on the movement of a substance from one area to another while still visualizing the anatomical components at the same time.

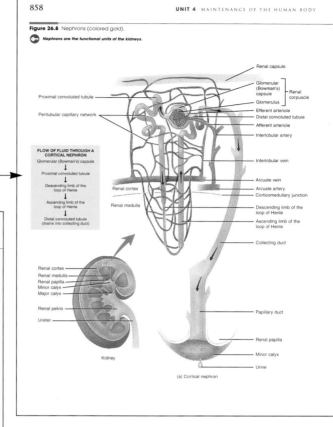

858   **UNIT 4** MAINTENANCE OF THE HUMAN BODY

Figure 26.8 Nephron (colored gold).

Nephrons are the functional units of the kidneys.

(a) Cortical nephron

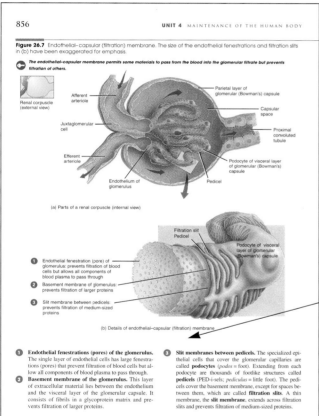

856   **UNIT 4** MAINTENANCE OF THE HUMAN BODY

Figure 26.7 Endothelial–capsular (filtration) membrane. The size of the endothelial fenestrations and filtration slits in (b) have been exaggerated for emphasis.

The endothelial–capsular membrane permits some materials to pass from the blood into the glomerular filtrate but prevents filtration of others.

(a) Parts of a renal corpuscle (internal view)

(b) Details of endothelial–capsular (filtration) membrane

① **Endothelial fenestrations (pores) of the glomerulus.** The single layer of endothelial cells has large fenestrations (pores) that prevent filtration of blood cells but allow all components of blood plasma to pass through.

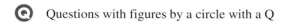
② **Basement membrane of the glomerulus.** This layer of extracellular material lies between the endothelium and the visceral layer of the glomerular capsule. It consists of fibrils in a glycoprotein matrix and prevents filtration of larger proteins.

③ **Slit membranes between pedicels.** The specialized epithelial cells that cover the glomerular capillaries are called **podocytes** (*podos* = foot). Extending from each podocyte are thousands of footlike structures called **pedicels** (PED-i-sels; *pediculus* = little foot). The pedicels cover the basement membrane, except for spaces between them, which are called **filtration slits**. A thin membrane, the **slit membrane**, extends across filtration slits and prevents filtration of medium-sized proteins.

Correlating sequential processes in text and art is achieved through the use of numbered lists in the text that correspond to numbered segments in the accompanying art. This is done extensively throughout the book and permits the reader to connect the text description with the illustration under consideration more easily.

NEW ICONS ARE USED THROUGHOUT THE BOOK.

 Key concept statements are identified by a key

Clinical applications by a stethoscope

 Questions with figures by a circle with a Q

Developmental anatomy descriptions by a fetus

EXCEPTIONAL PEDAGOGICAL STRUCTURE

The highly praised pedagogical features, introduced in this text over the course of its editions, continue to be refined and molded based on feedback from students who rely on their effectiveness. Each feature described below has been considered for its individual value as well as for how it works as a part of the entire pedagogical system. The variety of features ensures that students with differing learning styles have help in understanding and studying important concepts.

Student Objectives, on the opening page of each chapter, is a list of outcomes that can be expected after reading and understanding the chapter comments. These have been *page referenced* for the first time in the eighth edition.

Chapter Contents at a Glance, also on the opening page of each chapter, is an outline of the topics covered and chapter organization. Readers who wish to find a specific section or clinical application will appreciate the addition of page numbers, also new to this edition.

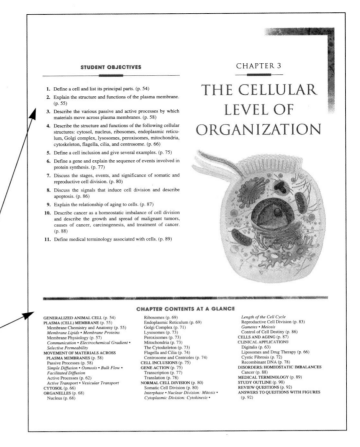

STUDENT OBJECTIVES

CHAPTER 3

THE CELLULAR LEVEL OF ORGANIZATION

1. Define a cell and list its principal parts. (p. 54)
2. Explain the structure and functions of the plasma membrane. (p. 55)
3. Describe the various passive and active processes by which materials move across plasma membranes. (p. 58)
4. Describe the structure and functions of the following cellular structures: cytosol, nucleus, ribosomes, endoplasmic reticulum, Golgi complex, lysosomes, peroxisomes, mitochondria, cytoskeleton, flagella, cilia, and centrosome. (p. 66)
5. Define a cell inclusion and give several examples. (p. 75)
6. Define a gene and explain the sequence of events involved in protein synthesis. (p. 77)
7. Discuss the stages, events, and significance of somatic and reproductive cell division. (p. 80)
8. Discuss the signals that induce cell division and describe apoptosis. (p. 86)
9. Explain the relationship of aging to cells. (p. 87)
10. Describe cancer as a homeostatic imbalance of cell division and describe the growth and spread of malignant tumors, causes of cancer, carcinogenesis, and treatment of cancer. (p. 88)
11. Define medical terminology associated with cells. (p. 89)

CHAPTER CONTENTS AT A GLANCE

GENERALIZED ANIMAL CELL (p. 54)
PLASMA (CELL) MEMBRANE (p. 55)
  Membrane Chemistry and Anatomy (p. 55)
  Membrane Lipids • Membrane Proteins
  Membrane Physiology (p. 57)
  Communication • Electrochemical Gradient •
  Selective Permeability
MOVEMENT OF MATERIALS ACROSS
  PLASMA MEMBRANES (p. 58)
  Passive Processes (p. 58)
  Simple Diffusion • Osmosis • Bulk Flow •
  Facilitated Diffusion
  Active Processes (p. 62)
  Active Transport • Vesicular Transport
CYTOSOL (p. 66)
ORGANELLES (p. 68)
  Nucleus (p. 68)

Ribosomes (p. 69)
Endoplasmic Reticulum (p. 69)
Golgi Complex (p. 71)
Lysosomes (p. 73)
Peroxisomes (p. 73)
Mitochondria (p. 73)
The Cytoskeleton (p. 73)
Flagella and Cilia (p. 74)
Centrosome and Centrioles (p. 74)
CELL INCLUSIONS (p. 75)
GENE ACTION (p. 75)
  Transcription (p. 77)
  Translation (p. 78)
NORMAL CELL DIVISION (p. 80)
  Somatic Cell Division (p. 80)
  Interphase • Nuclear Division: Mitosis •
  Cytoplasmic Division: Cytokinesis •

Length of the Cell Cycle
  Reproductive Cell Division (p. 83)
  Gametes • Meiosis
  Control of Cell Destiny (p. 86)
CELLS AND AGING (p. 87)
CLINICAL APPLICATIONS
  Digitalis (p. 63)
  Liposomes and Drug Therapy (p. 66)
  Cystic Fibrosis (p. 72)
  Recombinant DNA (p. 78)
DISORDERS: HOMEOSTATIC IMBALANCES
  Cancer (p. 88)
MEDICAL TERMINOLOGY (p. 89)
STUDY OUTLINE (p. 90)
REVIEW QUESTIONS (p. 92)
ANSWERS TO QUESTIONS WITH FIGURES
  (p. 92)

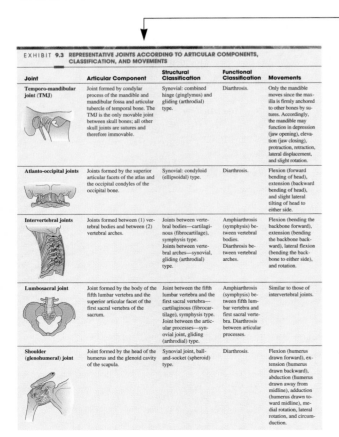

EXHIBIT 9.3 REPRESENTATIVE JOINTS ACCORDING TO ARTICULAR COMPONENTS, CLASSIFICATION, AND MOVEMENTS

Joint	Articular Component	Structural Classification	Functional Classification	Movements
Temporo-mandibular joint (TMJ)	Joint formed by condylar process of the mandible and mandibular fossa and articular tubercle of temporal bone. The TMJ is the only movable joint between skull bones; all other skull joints are sutures and therefore immovable.	Synovial: combined hinge (ginglymus) and gliding (arthrodial) type.	Diarthrosis.	Only the mandible moves since the maxilla is firmly anchored to other bones by sutures. Accordingly, the mandible may function in depression (jaw opening), elevation (jaw closing), protraction, retraction, lateral displacement, and slight rotation.
Atlanto-occipital joints	Joints formed by the superior articular facets of the atlas and the occipital condyles of the occipital bone.	Synovial: condyloid (ellipsoidal) type.	Diarthrosis.	Flexion (forward bending of head), extension (backward bending of head), and slight lateral tilting of head to either side.
Intervertebral joints	Joints formed between (1) vertebral bodies and between (2) vertebral arches.	Joints between vertebral bodies—cartilaginous (fibrocartilage), symphysis type. Joints between vertebral arches—synovial, gliding (arthrodial) type.	Amphiarthrosis (symphysis) between vertebral bodies. Diarthrosis between vertebral arches.	Flexion (bending the backbone forward), extension (bending the backbone backward), lateral flexion (bending the backbone to either side), and rotation.
Lumbosacral joint	Joint formed by the body of the fifth lumbar vertebra and the superior articular facet of the first sacral vertebra of the sacrum.	Joint between the fifth lumbar vertebra and the first sacral vertebra—cartilaginous (fibrocartilage), symphysis type. Joint between the articular processes—synovial joint, gliding (arthrodial) type.	Amphiarthrosis (symphysis) between fifth lumbar vertebra and first sacral vertebra. Diarthrosis between articular processes.	Similar to those of intervertebral joints.
Shoulder (glenohumeral) joint	Joint formed by the head of the humerus and the glenoid cavity of the scapula.	Synovial joint, ball-and-socket (spheroid) type.	Diarthrosis.	Flexion (humerus drawn forward), extension (humerus drawn backward), abduction (humerus drawn away from midline), adduction (humerus drawn toward midline), medial rotation, lateral rotation, and circumduction.

Exhibits are tables that summarize concepts or take material out of the narrative to consolidate information for review. Illustrations have been added to many more Exhibits in this edition.

Language aids are provided to help students with the difficult task of learning a new vocabulary at the same time that they are attempting to master the topical coverage. Three types of aid are presented:

- *Phonetic pronunciations*, presented in an easily remembered format (see Glossary page G-1), allow the reader to confidently and correctly pronounce terms that may be new. Pronunciations appear in parentheses immediately after introduction of a new term. This feature has been expanded considerably in the eighth edition.

- *Word roots* are derivations designed to provide an understanding of the meaning of new terms. They also appear in parentheses when a term is introduced. More than 200 new word roots were added for this edition.

- *Singular and plural forms* of a term are given when the term is first introduced unless the plural is formed merely by adding an "s" or "es."

Cross-References are references that guide the reader to previously studied concepts for reinforcement or to new material for a more complete discussion of the topic at hand. These references help students more thoroughly grasp the interrelatedness of the body's systems.

Medical Terminology, located at the end of most chapters, is a listing of relevant new terms, together with phonetic pronunciations, that helps students build a medical vocabulary.

Study Outlines are a concise summary of important topics covered in each chapter. Page numbers permit easy reference to text pages for amplification of details.

Review Questions are questions designed to help students achieve the learning objectives stated at the beginning of the chapter. Page numbers guide students to the appropriate pages for confirming answers.

---

484   UNIT 3 CONTROL SYSTEMS OF THE HUMAN BODY

MEDICAL TERMINOLOGY

Achromatopsia (a-krō'-ma-TOP-sē-a; *a* = without; *chrom* = color) Complete color blindness.

Ametropia (am'-e-TRŌ-pē-a; *ametro* = disproportionate; *ops* = eye) Refractive defect of the eye resulting in an inability to focus images properly on the retina.

Anopsia (an-OP-sē-a; *opsia* = vision) A defect of vision.

Blepharitis (blef-a-RI-tis; *blepharo* = eyelid; *itis* = inflammation of) An inflammation of the eyelid.

Conjunctivitis (pinkeye) An inflammation of the conjunctiva, caused by bacteria such as pneumococci, staphylococci, or *Hemophilus influenzae*, that is very contagious and more common in children. Conjunctivitis may also be caused by irritants, such as dust, smoke, or pollutants in the air, in which case it is not contagious.

Eustachitis (yoo'-stā-KĪ-tis) An inflammation or infection of the auditory (Eustachian) tube.

Exotropia (ek'-sō-TRŌ-pē-a; *ex* = out; *tropia* = turning) Turning outward of the eyes.

Keratitis (ker'-a-TĪ-tis; *kerato* = cornea) An inflammation or infection of the cornea.

Labyrinthitis (lab'-i-rin-THĪ-tis) An inflammation of the labyrinth (inner ear).

Mydriasis (mi-DRĪ-a-sis) Dilated pupil.

Myringitis (mir'-in-JĪ-tis; *myringa* = eardrum) An inflammation of the eardrum; also called **tympanitis**.

Nystagmus (nis-TAG-mus; *nystazein* = to nod) A rapid involuntary movement of the eyeballs, possibly caused by a disease of the central nervous system. It is associated with conditions that cause vertigo.

Otalgia (ō-TAL-jē-a; *oto* = ear; *algia* = pain) Earache.

Otosclerosis (ō'-tō-skle-RŌ-sis; *oto* = ear; *sclerosis* = harden-

ing) Pathological process that may be hereditary in which new bone is deposited around the oval window. The result may be immobilization of the stapes, leading to conduction deafness.

Photophobia (fō'-tō-FŌ-bē-a; *photo* = light; *phobia* = fear) Abnormal visual intolerance to light.

Ptosis (TŌ-sis; *ptosis* = fall) Falling or drooping of the eyelid. (This term is also used for the slipping of any organ below its normal position.)

Retinoblastoma (ret'-i-nō-blas-TŌ-ma; *blast* = bud; *oma* = tumor) A tumor arising from immature retinal cells and accounting for 2% of childhood malignancies.

Scotoma (skō-TŌ-ma; *scotoma* = darkness) An area of reduced or lost vision in the visual field. Also called a **blind spot** (other than the normal blind spot at the optic disc).

Strabismus (stra-BIZ-mus) An imbalance in the extrinsic eye muscles that produces a squint (formerly called "cross-eyes"). **Amblyopia** (am'-blē-Ō-pē-a) is the term used to describe the loss of vision in an otherwise normal eye that, because of muscle imbalance, cannot focus in synchrony with the other eye.

Tinnitus (ti-NĪ-tus) A ringing, roaring, or clicking in the ears.

Trachoma (tra-KŌ-ma) A serious form of conjunctivitis, the greatest single cause of blindness in the world. It is caused by a bacterium called *Chlamydia trachomatis*. The disease produces an excessive growth of subconjunctival tissue and invasion of blood vessels into the cornea, which progresses until the entire cornea is opaque, causing blindness.

Vertigo (VER-ti-gō; *vertex* = whorl) A sensation of spinning or movement in which the world is revolving or the person is revolving in space.

STUDY OUTLINE

OLFACTORY SENSATIONS: SMELL (p. 454)

1. The receptors for olfaction, which are bipolar neurons, are in the nasal epithelium.
2. In olfactory reception, a generator potential develops and triggers one or more nerve impulses.
3. Adaptation to odors occurs quickly, and the threshold of smell is low.
4. Axons of olfactory receptors form the olfactory (I) nerves, which convey nerve impulses to the olfactory bulbs, olfactory tracts, limbic system, and cerebral cortex (temporal and frontal lobes).

GUSTATORY SENSATIONS: TASTE (p. 455)

1. The receptors for gustation, the gustatory receptor cells, are located in taste buds.
2. Substances to be tasted must be in solution in saliva.
3. Receptor potentials developed in gustatory receptor cells cause the release of neurotransmitter, which gives rise to nerve impulses.
4. Adaptation to taste occurs quickly, and the threshold varies with the taste involved.

5. Gustatory receptor cells trigger nerve impulses in cranial nerves VII, IX, and X. Taste signals then pass to the medulla oblongata, thalamus, and cerebral cortex (parietal lobe).

VISUAL SENSATIONS (p. 457)

Accessory Structures of the Eye (p. 457)

1. Accessory structures of the eyes include the eyebrows, eyelids, eyelashes, and the lacrimal apparatus.
2. The lacrimal apparatus consists of structures that produce and drain tears.

Anatomy of the Eyeball (p. 459)

1. The eye is constructed of three coats: (a) fibrous tunic (sclera and cornea), (b) vascular tunic (choroid, ciliary body, and iris), and (c) retina (nervous tunic).
2. The retina consists of pigment epithelium and a neural portion (photoreceptor layer, bipolar cell layer, ganglion cell layer, horizontal cells, and amacrine cells).
3. The anterior cavity contains aqueous humor; the vitreous chamber contains the vitreous body.

CHAPTER 16 THE SPECIAL SENSES   485

Image Formation and Convergence (p. 464)

1. Image formation on the retina involves refraction of light rays by the cornea and lens, accommodation of the lens by an increase in its curvature for near vision, and constriction of the pupil to prevent light rays from entering the eye through the periphery of the lens.
2. In convergence, the eyeballs move medially so they are both directed toward an object being viewed.

Physiology of Vision (p. 467)

1. The first step in vision is the absorption of light by photopigments in rods and cones (photoreceptors) and isomerization of *cis*-retinal.
2. Once receptor potentials develop in rods and cones, they decrease the release of inhibitory neurotransmitter, which induces graded potentials in bipolar cells and horizontal cells.
3. Horizontal cells transmit inhibitory signals to bipolar cells; bipolar or amacrine cells transmit excitatory signals to ganglion cells, which depolarize and initiate nerve impulses.
4. Impulses from ganglion cells are conveyed into the optic (II) nerve, through the optic chiasm and optic tract, to the thalamus. From the thalamus visual signals pass to the cerebral cortex (occipital lobe).

AUDITORY SENSATIONS AND EQUILIBRIUM (p. 472)

1. The external or outer ear consists of the auricle, external auditory canal, and eardrum (tympanic membrane).

2. The middle ear consists of the auditory or Eustachian tube, ossicles, oval window, and round window.
3. The internal or inner ear consists of the bony labyrinth and membranous labyrinth. The internal ear contains the spiral organ (organ of Corti), the organ of hearing.
4. Sound waves enter the external auditory canal, strike the eardrum, pass through the ossicles, strike the oval window, set up waves in the perilymph, strike the vestibular membrane and scala tympani, increase pressure in the endolymph, strike the basilar membrane, and stimulate hairs on the spiral organ (organ of Corti).
5. Hair cells convert a mechanical force into a receptor potential.
6. Hair cells release neurotransmitter, which initiates nerve impulses in the first-order sensory neurons.
7. The cochlear branch of the vestibulocochlear (VIII) nerve terminates in the thalamus. From the thalamus auditory signals pass to the temporal lobes of the cerebral cortex.
8. Static equilibrium is the orientation of the body relative to the pull of gravity. The maculae of the utricle and saccule are the sense organs of static equilibrium.
9. Dynamic equilibrium is the maintenance of body position in response to movement. The cristae in the semicircular ducts are the sense organs of dynamic equilibrium.
10. Most vestibular branch fibers of the vestibulocochlear (VIII) nerve enter the brain stem and terminate in the pons; the remaining fibers enter the cerebellum.

REVIEW QUESTIONS

1. Describe the structure of olfactory receptors. What is the function of basal cells? (p. 454)
2. Describe adaptation to odors. What type of threshold does olfaction have? (p. 455)
3. Discuss the origin and path of a nerve impulse that results in olfaction. (p. 455)
4. Describe the structure of gustatory receptors. (p. 455)
5. How are gustatory receptors stimulated? (p. 456)
6. Describe adaptation to taste. What type of threshold does taste have? (p. 457)
7. Discuss how a nerve impulse for gustation travels from a taste bud to the brain. (p. 457)
8. Describe the structure and importance of the following accessory structures of the eye: eyelids, eyelashes, and eyebrows. (p. 457)
9. What is the function of the lacrimal apparatus? Explain how it operates. (p. 459)
10. By means of a labeled diagram, indicate the principal anatomical structures of the eye. (p. 460)
11. Describe the location and contents of the chambers of the eye. What is intraocular pressure (IOP)? How is the scleral venous sinus (canal of Schlemm) related to this pressure? (p. 463)
12. Describe the histology of the neural portion of the retina. (p. 461)
13. Explain how each of the following events is related to the physiology of vision: (a) refraction of light, (b) accommodation of the lens, and (c) constriction of the pupil. (p. 464)
14. Distinguish emmetropia, myopia, hypermetropia, and astigmatism by means of a diagram. (p. 465)

15. What is convergence? How does it occur? (p. 466)
16. Describe the structure of rods and cones. (p. 467)
17. Explain how photopigments respond to light and recover in darkness. (p. 467)
18. How do receptor potentials develop in photoreceptors? (p. 468)
19. Explain how the retina processes visual input. (p. 469)
20. Describe the path of a visual impulse from the optic (II) nerve to the brain. (p. 470)
21. Define visual field. Relate the visual field to image formation on the retina. (p. 470)
22. Diagram the principal parts of the external, middle, and internal ear. Describe the function of each part labeled. (p. 472)
23. What are sound waves? In what units are sound intensities measured? (p. 475)
24. Explain the events involved in the transmission of sound from the auricle to the spiral organ (organ of Corti). (p. 478)
25. What is the sensory pathway for sound impulses from the cochlear branch of the vestibulocochlear (VIII) nerve to the brain? (p. 479)
26. Compare the function of the maculae in the saccule and utricle in maintaining static equilibrium with the role of the cristae in the semicircular ducts in maintaining dynamic equilibrium. (p. 480)
27. Describe the path of a nerve impulse that results in static and dynamic equilibrium. (p. 482)
28. Define the following: corneal transplant (p. 459), detached retina (p. 461), cataract (p. 463), color blindness (p. 468), perforated eardrum (p. 473), otoacoustic emissions (p. 476), and cochlear implant (p. 479).

Appendices are the sources of various types of information. *Appendix A: Measurements* summarizes U.S., metric, and apothecary units of length, mass, volume, and time. *Appendix B: Normal Values for Selected Blood and Urine Tests* contains a listing of reference values, given in both conventional U.S. units and S.I. units, for the principal constituents of these fluids. *Appendix C: Periodic Table* is included for easy reference. In addition, the end papers of the book include lists of *Symbols and Medical Abbreviations*, *Eponyms Used in the Text*, and *Selected Terms Used in Writing Prescriptions.*

APPENDIX A

MEASUREMENTS

UNITS OF MEASUREMENT

When you measure something, you are comparing it with some standard scale to determine its *magnitude*. How long is it? How much does it weigh? How fast is it going? Some measurements are made directly by comparing the unknown quantity with the known unit of the same kind, for example, weighing a patient on a scale and taking the reading directly in pounds. Other measurements are indirect and are done by calculation, for example, counting a person's blood cells in a certain number of squares on a microscope slide and then calculating the total blood count.

Regardless of how a measurement is taken, it always requires two things: a *number* and a *unit*. When recording the weight of a patient, you would not just say 145. You have to give both the number (145) and the unit (pounds). When you count blood cells, you report the measurement as 10,000 (number) white blood cells per cubic millimeter of blood (unit).

All the units in use can be expressed in terms of one of three special units called **fundamental units**. These fundamental units have been established arbitrarily as length, mass, and time. Mass is perhaps an unfamiliar term to you. **Mass** is the amount of matter an object contains. The mass of this textbook is the same whether it is measured in a laboratory, under the sea, on top of a mountain, or even on the moon. No matter where you take it, it still has the same quantity of matter. **Weight**, on the other hand, is determined by the pull of gravity on an object. This textbook will not have the same weight on earth as on the moon because of the differences in gravitation. However, as long as we are dealing only with earthbound objects, weight and mass may be considered synonymous terms because the force of gravity on the surface of the earth is nearly constant. Thus, weight remains nearly the same regardless of where the measurements are taken.

All units other than the fundamental ones are **derived units**—they can always be written as some combination of the three fundamental units. For example, units of volume are derived from units of length (the volume of a cube = length × width × height). Units of speed are combinations of distance and time (miles per hour).

Units are grouped into systems of measurement. The two principal systems of measurement commonly used are the U.S. and the metric systems. The apothecary system is used by physicians and pharmacists.

U.S. SYSTEM

The *U.S. system* of measurement is used in everyday household work, industry, and some fields of engineering. The fundamental units in the U.S. system are the foot (length), the pound (mass), and the second (time).

The basic problem with the U.S. system is that there is no *uniform* progression from one unit to another. If you want to convert a measurement of 2½ yd to feet, you have to multiply it by 3 because there are 3 ft in a yard. If you want to convert the same length to inches, you have to multiply by 12 (or by 36) because there are 12 in. in a foot. In other words, to convert one unit of length to another, it is necessary to use *different* numbers each time. Conversions in the metric system are much easier since they are based on progressions of the number 10.

Exhibit A.1 lists U.S. units of measurement.

EXHIBIT A.1 U.S. UNITS OF MEASUREMENT	
Fundamental or Derived Unit	**Units and U.S. Equivalents**
Length	1 inch (in.) = 0.083 foot
	1 foot (ft) = 12 in.
	= 0.333 yard
	1 yard (yd) = 3 ft = 36 in.
	1 mile (mi) = 1,760 yd = 5,280 ft
Mass	1 grain (gr) = 0.002285 ounce
	1 dram (dr) = 27.34 gr
	= 0.063 ounce
	1 ounce (oz) = 16 dr = 437.5 gr
	1 pound (lb) = 16 oz = 7,000 gr
	1 ton = 2,000 lb
Time	second (sec) = 1/86,400 of a day
	1 minute (min) = 60 sec
	1 hour (hr) = 60 min = 3,600 sec
	1 day = 24 hr = 1,440 min
	= 86,400 sec
Volume	1 fluidram (fl dr) = 0.125 fluidounce
	1 fluidounce (fl oz) = 8 fl dr
	= 0.0625 quart
	= 0.008 gallon
	1 pint (pt) = 16 fl oz = 128 fl dr
	1 quart (qt) = 2 pt = 32 fl oz
	= 256 fl dr
	= 0.25 gallon
	1 gallon (gal) = 4 qt = 8 pt
	= 128 fl oz
	= 1,024 fl dr

GLOSSARY OF COMBINING FORMS, WORD ROOTS, PREFIXES, AND SUFFIXES

PRONUNCIATION KEY

1. The most strongly accented syllable appears in capital letters, for example, bilateral (bī-LAT-er-al) and diagnosis (dī-ag-NŌ-sis).

2. If there is a secondary accent, it is noted by a prime ('), for example, constitution (kon'-sti-TOO-shun) and physiology (fiz'-ē-OL-ō-jē). Any additional secondary accents are also noted by a prime, for example, decarboxylation (dē'-kar-bok'-si-LĀ-shun).

3. Vowels marked with a line above the letter are pronounced with the long sound as in the following common words:
   ā as in *māke*
   ē as in *bē*
   ī as in *īvy*
   ō as in *pōle*

4. Vowels not so marked are pronounced with the short sound as in the following words:
   a as in *above*
   e as in *bet*
   i as in *sip*
   o as in *not*
   u as in *bud*

5. Other phonetic symbols are used to indicate the following sounds:
   oo as in *sue*
   yoo as in *cute*
   oy as in *oil*

Many medical terms are "compound" words; that is, they are made up of one or more word roots or combining forms of word roots with prefixes or suffixes. For example, *leukocyte* (white blood cell) is a combination of *leuko*, the combining form for the word root meaning "white," and *cyt*, the word root meaning "cell." Learning the medical meanings of the fundamental word roots will enable you to analyze many long, complicated terms.

The following list includes the most commonly used combining forms, word roots, prefixes, and suffixes used in making medical terms and an example for each.

COMBINING FORMS AND WORD ROOTS

Acou-, Acu- hearing Acoustics (A-KOO-stiks), the science of sounds or hearing.

Acr-, Acro- extremity Acromegaly (ak'-rō-MEG-a-lē), hyperplasia of the nose, jaws, fingers, and toes.

Aden-, Adeno- gland Adenoma (ad-en'-Ō-ma), a tumor with a glandlike structure.

Alg-, Algia- pain Neuralgia (nyoo-RAL-ja), pain along the course of a nerve.

Angi- vessel Angiocardiography (an'-jē-ō-kard-ē-OG-ra-fē), x-ray of the great blood vessels and heart after intravenous injection of radiopaque fluid.

Arthr-, Arthro- joint Arthropathy (ar-THROP-a-thē), disease of a joint.

Aut-, Auto- self Autolysis (aw-TOL-i-sis), destruction of cells of the body by their own enzymes after death.

Bio- life, living Biopsy (BĪ-op-sē), examination of tissue removed from a living body.

Blast- germ, bud Blastocyte (BLAS-tō-sīt), an embryonic or undifferentiated cell.

Blephar- eyelid Blepharitis (blef-a-RĪT-is), inflammation of the eyelids.

Brachi- arm Brachialis (brā-kē-AL-is), muscle that flexes the forearm.

Bronch- trachea, windpipe Bronchoscopy (bron-KOS-kō-pē), direct visual examination of the bronchi.

Bucc- cheek Buccocervical (bū-kō-SER-vi-kal), pertaining to the cheek and neck.

Capit- head Decapitate (dē-KAP-i-tāt), to remove the head.

Carcin- cancer Carcinogenic (kar-sin-ō-JENK-ik), causing cancer.

Cardi-, Cardia-, Cardio- heart Cardiogram (KARD-ē-o-gram), a recording of the force and form of the heart's movements.

Cephal- head Hydrocephalus (hi-drō-SEF-a-lus), enlargement of the head due to an abnormal accumulation of fluid.

Glossaries appear at the end of the book. One glossary presents combining forms, word roots, prefixes, and suffixes. The other is a comprehensive glossary containing definitions and pronunciations.

H.E.A.R.T.

Principles of Anatomy and Physiology, Eighth Edition, is the centerpiece of an extensive, integrated system of support materials called **H.E.A.R.T.** (**H**uman Anatomy and Physiology **E**ducation **A**nd **R**esource **T**ools). H.E.A.R.T. includes materials for both professors and students, each carfully produced to enhance the learning environment. All teaching and learning styles—whether auditory, visual, or verbal—are supported.

For complete information on any of the following items, please call your local HarperCollins representative, call the toll-free HarperCollins Customer Communication Line at **1-800-8-HEART-1**, contact us through e-mail at **tortora@harpercollins.com** or write to:

Marketing Manager—Allied Health Sciences
HarperCollins College Publishers
10 East 53rd Street
New York, NY 10022-5299

● **Professors Resource Manual** Prepared by Ronald Dunn of Kapiolani Community College, the manual contains in each chapter lecture outlines, an extended list of objectives, lecture enhancements, teaching tips, suggested resources, and supplementary teaching materials such as reading lists and medical tests that can be reproduced for student use.

● **Electronic Lecture Guide** Prepared by Ronald Dunn of Kapiolani Community College, these expanded and enriched lecture outlines are made available to you on disk so that you can easily customize them to your particular course needs and use the content to create handouts for your students. Included on the disk are image reference numbers to correlating material on the HarperCollins Videodisc. Available for both PC and Macintosh environments.

● **Testbank** Prepared by Pamela Langley of New Hampshire Technical Community College, the testbank for the eighth edition is all new. It consists of approximately 4000 multiple choice, essay, short answer, matching, and true/false questions. Each question for the testbank has been reviewed and revised appropriately in order to provide you with the most useful testbank ever. It is available in printed form as well as on Testmaster for use with PCs or Macintosh computers. Quizmaster, a coordinating program to Testmaster, allows students to take timed or untimed tests on-line. Upon completing a test, a student can see his or her test score and view or print a diagnostic report that lists those topics or objectives that have been mastered and those that need to be restudied. When Quizmaster is installed on a network, student scores are saved on disk, and you, the professor, can use the Quizmaster utility program to view records and print reports for individual students, class sections, and entire courses.

● **Practice Tests** Prepared by Arthur Reed of the University of Hawaii, *Practice Tests* is an easy-to-use software package for student self-testing on-line. The program offers an introductory outline keyed to each chapter in the text and then a variety of multiple choice, true/false, and fill-in-the-blank study questions. Immediate feedback on correct answers is provided. Upon completing the quiz for each chapter, students can review their scores and print diagnostic guides which provide them with page references to the text for review and study. A flash-card feature aids students in learning difficult glossary terms. Available for both PC and Macintosh computers.

● **Transparencies** A set of over 550 transparency acetates—virtually every illustration in the text—is provided to adopters. These full-color illustrations have been painstakingly prepared with enlarged labels whenever possible in order to facilitate use in lecture halls. Transparencies are conveniently packaged in a box containing individual folders for each chapter in the book.

● **Histology Slides** Prepared by Victor P. Eroschenko of the University of Idaho, this set of 157 histology slides was prepared specifically to accompany the text. The set reflects all tissue types and is correlated to all the major body systems.

● **Laboratory Manuals** Three laboratory manuals are available to coordinate with your course. Patricia J. Donnelly and George A. Wistreich of East Los Angeles College have revised their successful manuals with cat or pig dissections. Victor Eroschenko of the University of Idaho has revised his manual, which focuses on the study of anatomy with the use of models or prosected cadavers. Each of the manuals coordinates well with the text, providing a visual match as a result of shared art and a consistency of use in terms. All of the manuals are accompanied by instructor's manuals.

● **A Brief Atlas of the Human Skeleton** New to the eighth edition, this brief photographic atlas of the components of the skeleton by Gerard J. Tortora has been carefully crafted to provide students with an effective supplement to the illustrations in the text, as well as a guide for learning the skeletal system in the laboratory. Photographs have been labeled and oriented, and reference is made to the corresponding text figure and discussion.

● **Art Notebook** Also new for the eighth edition, this notebook is the perfect tool for organized note taking in class. Left-hand pages present black and white representations of the same color illustration in the textbook and transparency package. These are offered without labels and leaders so that the students can fill these in during your lecture according to your expectations. The right-hand page is a blank sheet for taking notes on the topic during lecture.

● **Learning Guide** Written by Kathleen Schmidt Prezbindowski of the College of Mount St. Joseph with Jerry Tortora, this multifaceted student study guide emphasizes active learning. It includes numerous and varied exercises, labeling and coloring diagrams, and mastery tests. Each chapter begins with a framework that helps students visualize the relationships among key concepts and terms. Wordbytes, checkpoints, and clinical challenges all offer opportunites for students to enhance their understanding of the textual material.

● **Concept Maps** Revised for the eighth edition by Janice Smith of Tarrant County Junior College, the concept maps are visual maps of the written text, a tool to help students link important ideas together in an organized fashion. Each map correlates closely to the main topics and objectives introduced in each chapter. In addition, the maps have been prepared so that they can easily be reproduced as overheads for classroom use.

● **Bastian Illustrated Review Series** Written by Glenn Bastian, this series of ten workbooks serves as a quick and efficient study review. The highly functional format of the texts includes a series of labeled images accompanied on facing pages by descriptive text. The same images, without labels, are repeated in the last section of each workbook for self-testing. Each book covers a specific system and includes a system-specific glossary.

● **Coloring Books** Many students benefit from the interactive and enjoyable activity of studying with one or more of the very popular HarperCollins coloring books, including *The Anatomy Coloring Book*, by Kapit and Elson; *The Physiology Coloring Book*, by Kapit, Macey, and Meisami; and *The Human Brain Coloring Book*, by Diamond, Scheibel, and Elson.

● **Multimedia** A wide array of multimedia items are available to support and enhance the study of anatomy and physiology. All are closely linked to this text. Included in this category of resources are a new CD-ROM that will enable you and your students to examine, manipulate, and explore the systems of the human body from organismic to molecular levels of structure and function; *HarperCollins Clip Art of the Human Body*, based on the art from the Bastian series; *The HarperCollins Videodisc of Anatomy and Physiology*, a comprehensive and useful disc corresponding directly with the text; *HarperCollins Animated Tutorials for Physiology*, a CD-ROM (or disks) that combines animation with interactive quizzes on four of the toughest physiological concepts that students face; *The Tortora Pronunciation Dictionary*, which gives students needed help in learning a new language; *Physiofinder: Investigative Modules in Physiology*, a program of laboratory simulations; *Body Language*, a superior drill-and-practice program for learning name and location of anatomical structures; and *The Anatomist*, a CD-ROM providing helpful study tools based on *The Anatomy Coloring Book*. Demonstration versions of most packages are available to professors. Please contact us for full details on any of these packages or to receive a demo for review.

Acknowledgments

This book reflects the dedication and efforts of many people who have labored tirelessly to ensure that it is a work of the highest quality possible in college publishing. The guiding spirit throughout the creation of the eighth edition has been Bonnie Roesch, Executive Editor. We are greatly indebted to her for the vision she has for the continued success of a classic text and for her consistent encouragement and support throughout every phase of textbook development. Thom Moore, Senior Developmental Editor, shepherded the manuscript through each revision cycle and coordinated all interactions among members of the Tortora–Grabowski team. Editorial Assistant Sarah Schlegel put in long hours to prepare our first-ever digital art log, and even had time to recruit a truly outstanding group of reviewers. Cyndy Taylor, the Supplements Editor, has put together the finest ancillaries package we have yet seen. The production team included several individuals who expertly tracked the thousands of details that go into the making and manufacturing of this text. They include, from Electronic Publishing Services Inc., Jeff Chen, Tara Felitto, and Kurt Scherwatzky, and at HarperCollins, Mike Kemper, Paula Soloway, and Willie Lane. Our sincere thanks to all of you for your contributions.

The beautiful line drawings by medical illustrators Leonard Dank, Sharon Ellis, Lauren Keswick, Lynn O'Kelley, Biaggio John Melloni, Hilda Muinos, Nadine Sokol, Kevin Somerville, Beth Willert, Jared Schneidmen Design, and Page Two Associates continue to enhance every aspect of the text. Mira Schachne's work researching photographs for the text has once again been of the highest quality. Mark Nielsen of the University of Utah provided the beautiful new cadaver photographs for this edition. Mark was aided by Shawn Miller, a teaching assistant at Utah, who worked hard to dissect and prepare cadavers for photographing. Other aides to Mark included Matt Bingham, Chris Roach, Julie Maughan, and Doug Bardugan. The photography studio of Borge Andersen sent the very capable Jan Schou and Jim Frankoski to work with Mark on these new photographs. Our very special thanks to Claudia Durrell for her care and expertise in overseeing the all-important visual elements of this text. It is hard to imagine doing this text without her assistance. The elegant cover of this edition is a design of Mary McDonnell. The outstanding backcover illustration is the work of Kevin Sommerville.

We are grateful to both for their talent and creativity. Our thanks to Kevin Petti of San Diego Miramar College for writing the historical perspective about the art on the covers.

Michael C. Kennedy of Hahnemann University reviewed the entire illustration program for accuracy of drawing, labeling, and description—an especially valuable contribution. Randy McKee of the University of Wisconsin at Parkside carefully proofread the entire manuscript and art and made significant suggestions for improvement. We owe special recognition to Richard Welton of Southern Oregon University, who has maintained a steady line of communication with us throughout many editions. We thank him for his outstanding suggestions and his continual support.

We appreciate the contributions of Sandy's colleagues at Purdue University, who willingly provided perspectives in their areas of expertise, and the continued encouragement of Lou Sherman, her department head. During this revision Paul V. Malven, Patrick J. Daly, Sanford E. Ostroy, and Henry Weiner also provided greatly appreciated expert consultation on selected topics in their areas of specialty.

We are, as always, extremely grateful to the many professors who take the time to review manuscript drafts and offer suggestions for improvement or refinement. For the eighth edition, we are also pleased to have benefited from the input of two outstanding student focus groups. The contributions of both students and professors have had a significant impact on this revision.

Student Focus Group Participants

Paul Aguilla, Cuyahoga Community College
Eddie Baker, University of Texas, Southwest
Karen Barnes, Cuyahoga Community College
Lauren Brady, Northlake College
Debra Crocker, Cuyahoga Community College
Cleacia Foster, Richland Community College
Lauren Goulden, University of Texas, Southwest
Brenda Harper, Cuyahoga Community College
Leah Kaltreider, University of Texas, Southwest
Swarupa Koneru, University of Texas, Southwest
Lauren Scheidegger, Cuyahoga Community College
Lisa Tekavec, Cleveland State University
Belinda Trevino, Brookhaven Community College
Deanna Turosky, Cleveland State University
Carol Waker, Brookhaven Community College

Eighth Edition Reviewers

John Aliff, Dekalb College
Merlyn Anderberg, Spokane Falls Community College
Joan I. Barber, Delaware Technical and Community College
Charles J. Biggers, University of Memphis

Barbara Boss, St. Petersburg Junior College
Sandi Bushor-Gardner, South Suburban College
Joe Connell, Leeward Community College
John C. Conroy, University of Winnipeg at Manitoba
Lenda Cook, Florida Community College at Jacksonville
Lynita M. Cooksey, Arkansas State University
Rosemary Davenport, Gulf Coast Community College
William Dunscombe, Union County College
John Emes, British Columbia Institute of Technology
Victor P. Eroschenko, University of Idaho
James D. Fawcett, University of Nebraska at Omaha
Brian Feige, Charles S. Mott Community College
Mildred Fowler, Tidewater Community College
David Gantt, Georgia Southern University
John H. Green, Nicholls State University
Beverly Grundset, St. Petersburg Junior College
Hector Harima, Florida Community College
John Harling, Okanagan College
Richard D. Harrington, Riviere College
Patricia E. Hawker, St. Louis Community College at Forest Park
Mary Healy, Springfield College
Jean Helgeson, Collin County Community College
Sarah Caruthers Jackson, Florida Community College
Brian R. Johnson, Cuyahoga Community College
Gary Johnson, Madison Area Technical College
Wallis Jones, DeKalb College
Warren R. Jones, Loyola University
Beverly L. Ketcham, Hillsborough Community College
William C. Kleinelp, Jr., Middlesex Community College
Michael L. Kovacs, Broward Community College
Johanna Krontiris-Litowitz, Youngstown State University
Elden Martin, Bowling Green State University
Randall M. McKee, University of Wisconsin at Parkside
Eddie C. McNack, Houston Community College
Margaret S. Merkley, Delaware Technical and Community College
A. Kenneth Moore, Seattle Pacific University
Aubrey Morris, Pensacola Junior College
Holly Morris, Lehigh Carbon Community College
Richard Mostardi, University of Akron
Elizabeth A. Murray, College of Mount St. Joseph
Linda R. Nichols, Santa Fe Community College
Brian O'Connor, University of Massachusetts at Amherst
Roberta O'Dell-Smith, University of New Orleans
John Olson, Villanova University
Justicia Opoku-Edusei, University of Maryland at College Park
S. Arthur Reed, University of Hawaii at Manoa
Virginia Rivers, Truckee Meadows Community College
Don Rubbelke, Lakeland Community College

Henry Rushin, Humber College
Martha DePecol Sanner, Middlesex Technical
    Community College
P. George Simone, Eastern Michigan University
David S. Smith, San Antonio College
Francis J. Sullivan, Front Range Community College
Stuart Sumida, California State University at San
    Bernadino
Ralph W. Stevens, Old Dominion University
Clarence Thompson, University of North Dakota
Caryl Tickner, Stark Technical College
Steve Trautwein, Southeast Missouri State University
Itzeck Vatnick, Widener University
John Vaughan, St. Petersburg Junior College
Michael A. Vitale, Daytona Beach Community College
Robert C. Wall, Lake-Sumpter Community College
Jane Wallace, Chattanooga State Technical College
James Wallis, St. Petersburg Junior College
Clarence C. Wolfe, Northern Virginia Community
    College
Bonnie Wood, University of Maine at Orono

Finally, we would like to express deep appreciation to the thousands of instructors and nearly two million students who have used this text in previous editions. Over the past two decades, you have been "the heart of Tortora" because of the feedback—comments, criticisms, requests, and reviews—you have provided. We invite all readers and users of this eighth edition of our text to continue the tradition of sending your suggestions to us, so that we can include them in our plans for subsequent editions.

Gerard J. Tortora, Biology Coordinator
Natural Sciences and Mathematics S229
Bergen Community College
400 Paramus Road
Paramus, NJ 07652

Sandra Reynolds Grabowski
Department of Biological Sciences
1392 Lilly Hall of Life Sciences
Purdue University
West Lafayette, IN 47907-1392

FOR STUDENTS:
MAKING THE MOST OF THIS TEXT

A variety of special learning aids are designed into this book to help you build a strong foundation in anatomy and physiology. Each of these has been developed and refined based on feedback from students using previous editions of this text. With the help of their comments and suggestions, we have created a pedagogical plan for each chapter that will help you understand the content with greater ease and more enjoyment. They're described here so you can familiarize yourself with them before you begin.

Chapter Outlines and Student Objectives

Each chapter begins with **Chapter Contents at a Glance** and a listing of **Student Objectives**. Before you read the chapter, please review these carefully. Each objective states a skill or a piece of knowledge that you should acquire. To meet these objectives, you will have to perform several activities. Obviously, you must read the chapter carefully. If there are sections that you do not understand after one reading, you should reread those sections before continuing. As you read, pay particular attention to the figures and exhibits; they have been carefully coordinated with the textual narrative.

Illustrations

Studying the illustrations in this text is as important as reading the text. We have designed the illustrations so that they are as effective as possible in your study of the topic. In order to get the most out of the visual parts of this text, you should learn to integrate all of the additional tools we have added to the illustrations to aid in your comprehension. Each illustration starts with a clearly written **Legend**, describing what you are looking at in the drawing or photograph. Following the legend for many figures you will see a **Key Concept Statement**. These are concise statements that capture the essence of a key concept that has been stated in the text and is now reflected in the art. For many illustrations you will also find an **Orientation Inset** to help you understand the anatomical focus of the main piece of art. For selected pieces of art throughout, illustrations are accompanied by an **Overview of Functions** box. These help you visually integrate the anatomy and physiology being presented. Finally, at the bottom of most illustrations you will find **Questions with Figures**. If you try to answer these questions as you go along, they will serve as self-checks to help you understand the material. Often it will be possible to answer a question by examining the figure. Such questions reinforce a visual message by putting it into words. In other cases, a question will encourage you to integrate the knowledge you've gained by reading the associated text with the information presented in the figure. Some questions may prompt you to think critically about the topic at hand or predict a consequence in advance of its description in the text. If you find you can't immediately come up with an answer, first reexamine the verbal description of the figure in the text. Then, read the next section of the text to see if it offers a clue. We've provided answers at the end of each chapter so you can confirm that you're on the right track.

Summary Exhibits

These exhibits summarize several preceding pages of text material. Here, in a nutshell, are the highlights of most chapters in a format that is perfect for reviewing and organizing the concepts.

Cross-Referencing

Many cross-references have been added to help you relate concepts to previously learned material. In addition, some cross-references direct you ahead for a more complete discussion of topics under consideration.

End-of-Chapter Activities

At the end of each chapter are two—sometimes three—other learning guides that you may find useful. The first, a **Study Outline**, is a concise summary of important topics discussed in the chapter. This section is designed to consolidate the essential points covered in the chapter so that you may recall and relate them to one another. For convenience, page numbers are listed next to key topics so you can easily refer to specific passages in the text for clarification or amplification. **Review Questions** are next. These are a series of questions designed specifically to help you master the objectives. Again, page numbers help you locate the answers. After you have answered the review questions, you should return to the beginning of the chapter and reread the objectives to determine whether you achieved the goals. A third aid, **Medical Terminology**, appears in some chapters. This is a listing of terms designed to build your medical vocabulary. Every medical term is accompanied by a phonetic pronunciation.

Appendixes

There are three useful appendixes: **Measurements** summarizes U.S., metric, and apothecary units of length, mass, volume, and time; **Normal Values for Selected Blood and Urine Tests** contains a listing of reference values for the principal constituents of these fluids; and a **Periodic Table** is included for easy reference. In addition, the endpapers of the text include lists of **Symbols and Medical Abbreviations**, **Eponyms Used in the Text**, and **Selected Terms Used in Writing Prescriptions**.

Glossaries

Two glossaries appear at the end of the book. The first presents combining forms, word roots, prefixes, and suffixes. The second is a comprehensive glossary of terms.

Pronunciations

As a further aid, we have included pronunciations for many terms that may be new to you. These appear in parentheses immediately following new words, and most are repeated in the Glossary of Terms in the back of the book. Look at these words carefully and say them out loud several times. Learning to pronounce a new word will help you remember it and make it a useful part of your vocabulary. Take a few minutes now to read the following pronunciation key, so it will be familiar as you encounter new words. The key is repeated at the beginning of the Glossary of Combining Forms, Word Roots, Prefixes, and Suffixes.

Pronunciation Key

1. The most strongly accented syllable appears in capital letters, for example, bilateral (bī-LAT-er-al) and diagnosis (dī-ag-NŌ-sis).

2. If there is a secondary accent, it is noted by a prime ('), for example, constitution (kon′-sti-TOO-shun) and physiology (fiz′-ē-OL-ō-jē). Any additional secondary accents are also noted by a prime, for example, decarboxylation (dē′-kar-bok′-si-LĀ-shun).

3. Vowels marked by a line above the letter are pronounced with the long sound as in the following common words:

    ā as in *māke*
    ē as in *bē*
    ī as in *īvy*
    ō as in *pōle*

4. Vowels not so marked are pronounced with the short sound as in the following words:

    a as in *above*
    e as in *bet*
    i as in *sip*
    o as in *not*
    u as in *bud*

5. Other phonetic symbols are used to indicate the following sounds:

    oo as in *sue*
    yoo as in *cute*
    oy as in *oil*

We wish you a rewarding experience in your study of anatomy and physiology. It is our sincere hope that you find this text useful now and throughout your journey toward your career goal. Please feel free to write to us with questions, comments, or suggestions as many of the students who preceded you have done. *To your success!*

Gerard J. Tortora, Biology Coordinator
Natural Sciences and Mathematics S229
Bergen Community College
400 Paramus Road
Paramus, NJ 07652

Sandra Reynolds Grabowski
Department of Biological Sciences
1392 Lilly Hall of Life Sciences
Purdue University
West Lafayette, IN 47907-1392

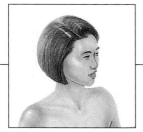

UNIT 1

ORGANIZATION OF THE HUMAN BODY

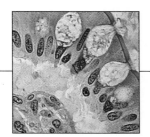

THIS UNIT EXPLAINS HOW THE HUMAN BODY IS organized at different levels. By studying the various regions and parts of the body, you will discover the importance of the chemicals that make it up. You will then find out how cells, tissues, and organs form the systems that keep humans alive and healthy.

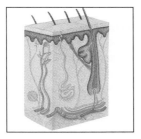

CHAPTER 1

AN INTRODUCTION TO THE HUMAN BODY

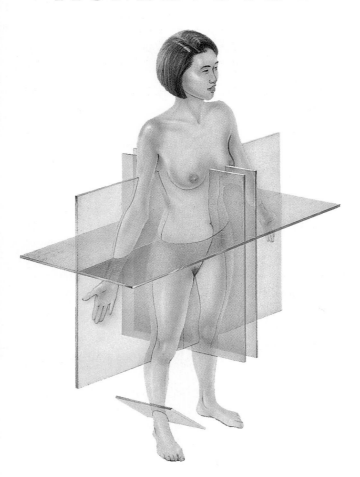

ou are beginning a study of the human body to learn its organization and functions. First, you will gain a basic understanding of how the body is organized, how its different parts normally work, and how various conditions affect its operation to maintain life and health. Then, you will be able to understand what happens when the body is injured, diseased, or placed under extreme stress. This chapter introduces the various systems that compose the human body. You will also discover that all body systems influence one another. As an example, consider how just two body systems—the integumentary and skeletal systems—cooperate.

The integumentary system, which includes the skin, hair, and nails, protects all other body systems, including the skeletal system, which includes all the bones and joints of the body. The skin serves as a barrier between the outside environment and internal tissues and organs. It also participates in the production of vitamin D, which is needed for proper deposition of calcium and other minerals into bone; these minerals are essential for normal bone growth and development. The skeletal system, in turn, provides support for the integumentary system, serves as a reservoir for calcium—hoarding it in times of plenty and releasing it for

other tissues in times of need—and generates cells that help the skin resist invasion by pathogens. As you study the body systems in more detail, you will discover how they work together to maintain health, provide protection from disease, and allow for reproduction of the species.

ANATOMY AND PHYSIOLOGY DEFINED

Two branches of science—anatomy and physiology—provide the foundation for understanding the body's parts and functions. **Anatomy** (a-NAT-ō-mē; *anatome* = to cut up) is the study of **structure** and the relationships among structures. Although anatomy was first studied by **dissection** (dis-SEK-shun; *dis* = apart; *secare* = to cut) of cadavers, many other techniques now contribute to the advance of anatomical knowledge (for examples, see Exhibit 1.4 on page 20). Whereas anatomy deals with structures of the body, **physiology** (fiz′-ē-OL-ō-jē) deals with **functions** of the body parts, that is, how they work.

Exhibit 1.1 describes several subdivisions of anatomy and physiology.

EXHIBIT **1.1** SELECTED SUBDIVISIONS OF ANATOMY AND PHYSIOLOGY

Subdivisions of Anatomy	Description	Subdivisions of Physiology	Description
Surface anatomy	Study of the form (morphology) and markings of the surface of the body.	**Cell physiology**	Study of the functions of cells.
Gross (macroscopic) anatomy	Study of structures that can be examined without using a microscope.	**Pathophysiology** (PATH-ō-fiz-ē-ol′-ō-jē; *patho* = disease)	Study of functional changes associated with disease and aging.
Systemic (systematic) anatomy	Study of specific systems of the body such as the nervous or respiratory system.	**Exercise physiology**	Study of changes in cell and organ functions during muscular activity.
Regional anatomy	Study of a specific region of the body such as the head or chest.	**Neurophysiology** (NOO-ro-fiz-ē-ol′-ō-jē; *neuro* = nerve or nervous system)	Study of functional characteristics of nerve cells.
Radiographic (rā′-dē-ō-GRAF-ik; *radio* = ray; *graph* = to write) **anatomy**	Study of the structure of the body that includes the use of x-rays.	**Endocrinology** (en′-dō-kri-NOL-ō-jē; *endo* = within; *crin* = to secrete)	Study of hormones (chemical regulators in the blood) and how they control body functions.
Developmental anatomy	Study of development from the fertilized egg to adult form.		
Embryology (em′-brē-OL-ō-jē; *embryon* = embryo; *logos* = study of)	Study of development from the fertilized egg through the eighth week in utero.	**Cardiovascular** (kar-dē-ō-VAS-kyoo-lar; *cardio* = heart; *vascular* = blood vessels) **physiology**	Study of functions of the heart and blood vessels.
Histology (hiss′-TOL-ō-jē; *histio* = tissue)	Microscopic study of the structure of tissues.	**Immunology** (im′-yoo-NOL-ō-jē; *immunis* = free)	Study of body defense mechanisms.
Cytology (sī-TOL-ō-jē; *cyto* = cell)	Chemical and microscopic study of the structure of cells.	**Respiratory** (re-SPĪ-ra-to′-rē; *respiratio* = to breathe) **physiology**	Study of functions of the air passageways and lungs.
Pathological (path′-ō-LOJ-i-kal; *patho* = disease) **anatomy**	Study of structural changes (from gross to microscopic) associated with disease.	**Renal** (RĒ-nal; *renes* = kidneys) **physiology**	Study of the functions of the kidneys.

Because possible functions reflect structural organization, you will learn about the human body by studying its anatomy and physiology together. You will see how the structure of a part of the body is adapted for performing certain functions. For example, the bones of the skull are tightly joined to protect the brain. The bones of the fingers, by contrast, are more loosely joined to allow various types of movements. The teeth have different shapes for biting, tearing, and grinding food. The air sacs in the lungs are so thin that they easily permit the movement of oxygen into the blood for use by body cells and the movement of carbon dioxide out of the blood to be exhaled.

LEVELS OF STRUCTURAL ORGANIZATION

The human body consists of several levels of structural organization that are associated with one another: chemical, cellular, tissue, organ, system, and organism (Fig. 1.1). The level of organization having the smallest components, the **chemical level**, includes all atoms and molecules in the body. Certain atoms, such as carbon (C), hydrogen (H), oxygen (O), nitrogen (N), and calcium (Ca), are essential for maintaining life. Atoms can combine to form molecules in the body. Some examples of molecules are proteins, carbohydrates, fats, and vitamins.

Molecules, in turn, combine to form structures at the next higher level of organization—the **cellular level**. **Cells** are the basic structural and functional units of an organism. Among the many kinds of cells in your body are muscle cells, nerve cells, and blood cells. Figure 1.1 shows four different types of cells from the lining of the stomach. As you will see on page 68, cells contain specialized structures called **organelles** (or-gan-ELZ), such as the nucleus, mitochondria, and lysosomes, that perform specific functions.

The next higher level of structural organization is the **tissue level**. **Tissues** are groups of cells (and the materials surrounding them) that usually arise from common ancestor cells and work together to perform a particular function. The four basic types of tissue in your body are *epithelial tissue, muscle tissue, connective tissue,* and *nervous tissue.* For example, the cells shown in Fig. 1.1 form an epithelial tissue that lines the stomach. Each type of cell in this particular tissue contributes to digestion in specific ways.

Throughout the body, different kinds of tissues combine to form the next higher level of organization—the **organ level**. **Organs** are structures that are composed of two or more different types of tissues, have specific functions, and usually have recognizable shapes. Examples of organs are the heart, liver, lungs, brain, and stomach. Figure 1.1 shows how several tissues make up one organ, the stomach.

1. The outer covering of the stomach is the *serosa,* a layer of epithelial tissue and connective tissue that protects the stomach and reduces friction when the stomach moves and rubs against other body organs.
2. The *muscle tissue layers* contract to churn and mix food and push it on to the next digestive organ (the small intestine).
3. The *epithelial tissue layer* lining the stomach contributes fluid and chemicals that aid digestion.

The next level of structural organization in the body is the **system level**. A **system** consists of several related organs that have a common function. For example, the organs of the digestive system, which break down and absorb food, include the mouth, pharynx (throat), esophagus, stomach, small intestine, and large intestine. Other organs—the salivary glands (which produce saliva), liver, and pancreas—are also part of the digestive system because they release materials needed for digestion. Sometimes an organ is part of more than one system. The pancreas, for example, is part of both the digestive system and the hormone-producing endocrine system.

Structurally, the largest level is the **organismic level**. All the parts of the body functioning with one another comprise the total **organism**—one living individual.

In the chapters that follow, you will study the anatomy and physiology of the major body systems. Exhibit 1.2 on page 6 introduces the components and functions of these systems in the order they are discussed in the book.

LIFE PROCESSES

All living forms carry on certain processes that distinguish them from nonliving things. Following are some of the important life processes of humans:

1. **Metabolism** (me-TAB-ō-lizm; *metabole* = change) is the sum of all the chemical processes that occur in the body. One phase of metabolism, called **catabolism** (ca-TAB-ō-lizm; *cata* = downward), involves breaking down large, complex molecules into smaller, simpler ones. An example is the splitting of proteins in food into amino acids, the building blocks of proteins. During catabolism, oxygen provided by the respiratory system and nutrients broken down in the digestive system are used to generate chemical energy in a form that can be used by body cells to carry out their activities. The other phase, called **anabolism** (a-NAB-ō-lizm; *ana* = upward), uses the energy from catabolism to build the body's structural and functional components. An example of anabolism is the synthesis of proteins that make up muscles and bones. Some metabolic processes create chemical wastes. These then leave the body (are excreted), mainly by way of the urinary system (in urine), the respiratory system (exhalation of carbon dioxide), and the digestive system (in feces).

FIGURE 1.1 Levels of structural organization in the human body.

The levels of structural organization are chemical, cellular, tissue, organ, system, and organismic.

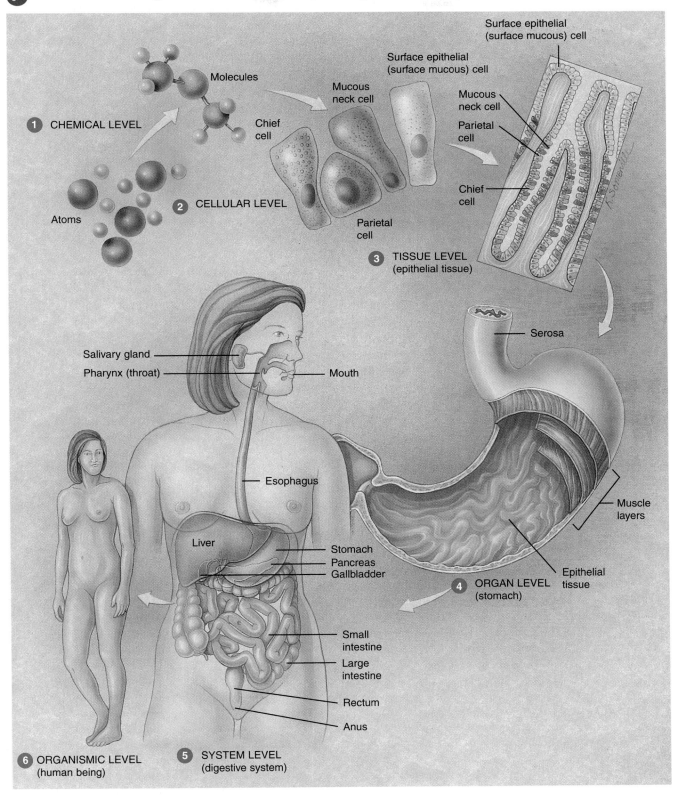

Which level of structural organization is composed of two or more different types of tissues and usually has a recognizable shape?

EXHIBIT 1.2 MAJOR SYSTEMS OF THE HUMAN BODY

1. Integumentary system
Components: The skin and structures derived from it, such as hair, nails, and sweat and oil glands.
Function: Helps regulate body temperature, protects the body, eliminates some wastes, helps produce vitamin D, and monitors certain stimuli such as changes in temperature and pressure.
Reference: See Fig. 5.1.

2. Skeletal system
Components: All the bones of the body, their associated cartilages, and the joints of the body.
Function: Supports and protects the body, assists in body movements, houses cells that give rise to blood cells, and stores minerals.
Reference: See Fig. 7.1.

3. Muscular system
Components: Specifically refers to skeletal muscle tissue, which is muscle usually attached to bones. Other muscle tissues are smooth and cardiac.
Function: Powers movements of the body, such as walking or throwing a ball, stabilizes body positions (posture), and generates heat.
Reference: See Fig. 11.3.

4. Nervous system
Components: Brain, spinal cord, nerves, and special sense organs, such as the eyes and ears.
Function: Regulates body activities through action potentials (nerve impulses) by detecting changes in the internal and external environments, interpreting the changes, and responding to the changes by inducing muscular contractions or glandular secretions.
Reference: See Figs. 13.2 and 14.1.

5. Endocrine system
Components: All hormone-producing glands and cells such as the pituitary gland, thyroid gland, and pancreas.
Function: Regulates body activities through hormones, chemicals transported in the blood to various target organs of the body.
Reference: See Fig. 18.1.

6. Cardiovascular system
Components: Blood, heart, and blood vessels.
Function: Distributes oxygen and nutrients to cells, carries carbon dioxide and wastes away from cells, helps maintain the acid–base balance of the body, protects against disease, prevents hemorrhage by forming blood clots, and helps regulate body temperature.
Reference: See Figs. 20.1, 21.20, and 21.25.

7. Lymphatic and immune systems
Components: Lymph, lymphatic vessels, and structures or organs containing lymphatic tissue (large numbers of white blood cells called lymphocytes), such as the spleen, thymus gland, lymph nodes, and tonsils.
Function: Returns proteins and plasma (liquid portion of blood) to the cardiovascular system, transports fats from the gastrointestinal tract to the cardiovascular system, serves as a site of maturation and proliferation of certain white blood cells, and helps protect against disease through the production of proteins called antibodies, as well as other responses.
Reference: See Fig. 22.1.

8. Respiratory system
Components: Lungs and a series of associated passageways leading into and out of them.
Function: Supplies oxygen, eliminates carbon dioxide, helps regulate the acid–base balance of the body, and helps produce vocal sounds.
Reference: See Fig. 23.1.

9. Digestive system
Components: A long tube called the gastrointestinal tract and associated organs that include the salivary glands, liver, gallbladder, and pancreas.
Function: Performs the physical and chemical breakdown of food and absorption of nutrients for use by cells and helps eliminate solid and other wastes.
Reference: See Fig. 24.1.

10. Urinary system
Components: Kidneys, ureters, urinary bladder, and urethra that together produce, store, and eliminate urine.
Function: Regulates the volume and chemical composition of blood, eliminates wastes, regulates fluid and electrolyte balance, helps maintain the acid–base and calcium balance of the body, and secretes a hormone that regulates red blood cell production.
Reference: See Fig. 26.1.

11. Reproductive system
Components: Organs (testes and ovaries) that produce reproductive cells or gametes (sperm and ova) and other organs such as the uterine (Fallopian) tubes and uterus in females and the epididymis, ductus (vas) deferens, and penis in males that transport and store reproductive cells.
Function: Produces gametes, which can unite to form a new organism, and hormones that help to regulate metabolism.
Reference: See Figs. 28.1 and 28.11.

---

2. **Responsiveness** is the ability to detect and respond to changes in the external or internal environment. Different cells detect different sorts of changes and respond in characteristic ways. Neurons (nerve cells) respond by generating electrical signals, known as action potentials (nerve impulses). Muscle cells respond by contracting, that is, becoming shorter, to move body parts. Endocrine cells in the pancreas respond to elevated blood glucose (sugar) level by secreting the hormone insulin. Other body cells respond to insulin by taking up glucose, which lowers the blood level to normal.

3. **Movement** includes motion of the whole body, individual organs, single cells, or even organelles inside cells. For example, the coordinated contraction of several leg

muscles moves your whole body from one place to another when you walk or run. After you eat a meal that contains fats, your gallbladder contracts and releases bile to help in the digestion of fats. When a body tissue is damaged or infected, certain white blood cells move from the blood into the tissue to help clean up and repair the area. And inside individual cells, various cell parts move from one position to another.

4. **Growth** is an increase in size that results from an increase in the number or size of cells or both. Sometimes, a tissue increases in size because the materials between cells increase in amount. In growing bone, for example, mineral deposits accumulate around the bone cells.

5. **Differentiation** is the change that a cell undergoes to develop from an unspecialized to a specialized state. Specialized cells have structural and functional characteristics that differ from their undifferentiated ancestor cells. For example, several different types of red and white blood cells differentiate from the same type of undifferentiated ancestor cell in red bone marrow (see Fig. 19.2). Also, through differentiation, a fertilized egg develops into an embryo, and then into a fetus, infant, child, and finally an adult.

6. **Reproduction** refers to either the formation of new cells for growth, repair, or replacement, or the production of a new individual. Through reproduction, life continues from one generation to the next.

HOMEOSTASIS: MAINTAINING PHYSIOLOGICAL LIMITS

As we have seen, the human body is composed of various systems and organs, each of which consists of millions of cells. These cells need relatively stable conditions to function effectively and contribute to the survival of the body as a whole. The maintenance of stable conditions for its cells is an essential function of every many-celled organism. Physiologists call such relative stability homeostasis, and it is one of the major themes of this textbook. **Homeostasis** (hō′-mē-ō-STĀ-sis; *homeo* = same; *stasis* = standing still) is a condition in which the body's internal environment remains within certain physiological limits.

An important aspect of homeostasis is regulation of the volume and composition of **body fluids**, which are dilute, watery solutions found inside cells and surrounding them (Fig. 1.2). Proper functioning of body cells depends on precise regulation of the composition of the surrounding fluids. Fluid within cells is called **intracellular** (*intra* = within, inside) **fluid** (**ICF**). Fluid outside body cells is called **extracellular** (*extra* = outside) **fluid** (**ECF**). The ECF that fills the narrow spaces between cells of tissues is called **interstitial** (in′-ter-STISH-al; *inter* = between) **fluid, inter-**

cellular fluid, or **tissue fluid**. The ECF in blood vessels is termed **plasma**. Among the substances dissolved in the water of ICF and ECF are oxygen, nutrients, and electrically charged chemical particles called *ions,* such as sodium ions (Na^+) and chloride ions (Cl^-), all needed to maintain life, as well as wastes.

The smallest blood vessels in the body are called **blood capillaries**. Capillary walls are so thin that water and solutes in plasma can filter out into the interstitial fluid. On a daily basis, about the same volume of fluid leaves the interstitial spaces and returns to capillaries. This movement in both directions across capillary walls is known as **capillary exchange**. It provides needed materials, such as nutrients, oxygen, ions, and so on, to tissue cells and removes wastes, for example, carbon dioxide. A small quantity of interstitial fluid flows into lymphatic vessels, which are present alongside blood vessels in most tissues, and forms another type of ECF called **lymph**. Eventually, lymph also returns to the bloodstream. Thus the volume of each of these extracellular fluids remains stable.

Since interstitial fluid surrounds all body cells, it is often called the body's **internal environment**. An organism is said to be in homeostasis when conditions in the internal environment are maintained within physiological limits. When homeostasis is disturbed, illness may result. If the body fluids are not eventually brought back into homeostasis, death may occur.

FIGURE 1.2 Intracellular and extracellular fluid. Intracellular fluid is the fluid within cells. Extracellular fluid (ECF) is found in blood vessels as plasma and between tissue cells as interstitial fluid.

 Interstitial fluid is the body's internal environment.

INTRACELLULAR FLUID (ICF)
(inside tissue cells and blood cells)

- Tissue cells
- Blood capillary
- Blood cells

Interstitial fluid   Plasma

EXTRACELLULAR FLUID (ECF)

Q What substances are present in ECF and ICF?

Stress and Homeostasis

Homeostasis in all organisms is continually disturbed by **stress**, any stimulus that tends to create an imbalance in the internal (within the body) environment. The stress may come from the external (outside the body) environment in the form of physical stimuli such as heat, cold, loud noises, or lack of oxygen. Or the stress may originate within the body in the form of stimuli such as low blood glucose level or increased acidity of the extracellular fluid. Stress also comes from psychological stimuli in our social environment—the demands of work and school, for example. Most stresses are mild and routine, and the responses of body cells quickly restore balance in the internal environment. Poisoning, overexposure to temperature extremes, severe infection, and death of one's spouse are examples of extreme stresses, situations in which homeostasis may fail.

Fortunately, the body has many regulating (homeostatic) devices that oppose the forces of stress and bring the internal environment back into balance. Some people live in deserts where the daytime temperatures easily reach 49°C (120°F). Others work outside all day in subzero weather. Yet everyone's internal body temperature remains near 37°C (98.6°F). Mountain climbers exercise strenuously at high altitudes, where the oxygen content of the air is low. But once they adjust to the new altitude, they usually do not suffer from oxygen shortage. The extremes in temperature and in oxygen content of the air are external stresses, and the exercise performed is an internal stress, yet the body compensates and remains in homeostasis.

Walter B. Cannon (1871–1945), an American physiologist, coined the term *homeostasis*. He noted that the heat produced by muscles during strenuous exercise would curdle and inactivate the body's proteins if the body did not dissipate heat quickly. Besides heat, muscles also produce lactic acid during exercise. If the body did not have a homeostatic mechanism for reducing the amount of acid, the extracellular fluid would become too acidic and destroy the cells. Each body structure, from the cellular to the systemic level, contributes in some way to keeping the internal environment within normal limits.

Regulation of Homeostasis by the Nervous and Endocrine Systems

Many homeostatic responses of the body are regulated by the nervous system and the endocrine system working together or independently. The nervous system regulates homeostasis by detecting deviations from the balanced state and then sending messages in the form of nerve impulses to the proper organs to counteract the stress. For instance, when muscle cells are active they remove a great deal of oxygen from the blood. They also give off a lot of carbon dioxide, which enters the blood. Certain nerve cells detect these chemical changes in the blood and send impulses to the brain. In response, the brain sends impulses to the heart, which cause it to pump blood more rapidly through the body. As a result, the blood gives up carbon dioxide and takes on oxygen more rapidly. Simultaneously, the brain sends nerve impulses to the muscles that control breathing to contract more often. The result is exhalation of more carbon dioxide and inhalation of more oxygen.

The endocrine system—a group of glands that secrete chemical regulators, called **hormones**, into the blood—also regulates homeostasis. For example, the hormone epinephrine (adrenalin) also helps to increase heart rate when carbon dioxide builds up in the blood. Whereas nerve impulses cause rapid changes, hormones usually work more slowly. Both means of regulation work toward the same end—achieving homeostasis.

Feedback Systems (Loops)

A **feedback system** (**loop**) involves a cycle of events in which information about the status of a condition is continually monitored and fed back (reported) to a central control region (Fig. 1.3). A feedback system consists of three basic components—control center, receptor, and effector.

1. The **control center** determines the value at which some aspect of the body, called a **controlled condition**, should be maintained. In the body, there are hundreds of controlled conditions. A few examples are heart rate, blood pressure, acidity of the blood, blood glucose level, body temperature, and breathing rate. The control center receives information about the status of a controlled condition from a receptor and then determines an appropriate course of action.
2. The **receptor** monitors changes in the controlled condition and then sends the information, called the *input,* to the control center. Any stress that changes a controlled condition is called a **stimulus**. For example, a stimulus such as exercise raises the body temperature (the controlled condition), and thermal (heat) receptors send input to the control center, which in this case is in the brain.
3. The **effector** receives information, called the *output,* from the control center and produces a **response** (effect). So while you are exercising, your brain (control center) signals for increased secretions by your sweat glands (effectors). As sweat evaporates from the skin, body temperature drops back to normal.

The change that occurs in the controlled condition is continually monitored by the receptor and fed back to the control center. If the response reverses the original stimulus, as in the example just described, the system is a **negative feedback system** (**loop**). If the response enhances or intensifies the original stimulus, the system is a **positive feedback system** (**loop**).

FIGURE 1.3 Components of a feedback system (loop).

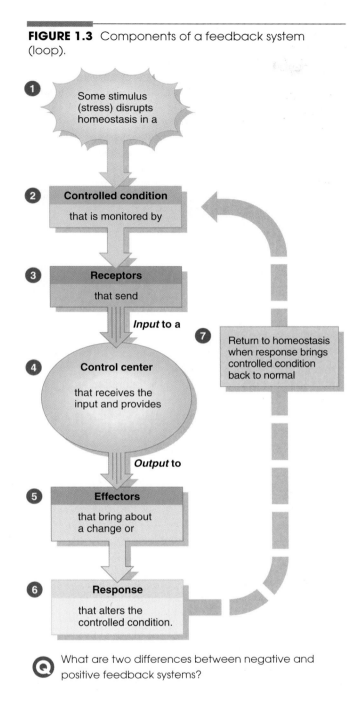

1. Some stimulus (stress) disrupts homeostasis in a

2. **Controlled condition**
   that is monitored by

3. **Receptors**
   that send

   Input to a

4. **Control center**
   that receives the input and provides

   Output to

5. **Effectors**
   that bring about a change or

6. **Response**
   that alters the controlled condition.

7. Return to homeostasis when response brings controlled condition back to normal

Q What are two differences between negative and positive feedback systems?

Negative feedback systems tend to maintain conditions that require frequent monitoring and adjustment within physiological limits. Examples of conditions regulated by negative feedback are body temperature and blood glucose level. Positive feedback systems, on the other hand, are important for conditions that do not occur often and do not require continual fine-tuning. Since positive feedback systems tend to intensify or amplify a controlled condition, they usually are shut off by some mechanism outside the system if they are part of a normal physiological response.

Given the characteristics of both negative and positive feedback systems, it is not surprising that most feedback systems in the body are negative. Positive feedback systems can be destructive and result in various disorders, yet some are normal and beneficial. For example, during blood clotting, which helps stop loss of blood from a cut, the initial signal is amplified until the blood clot forms and bleeding is under control. Then, other substances help turn off the clotting response. Positive feedback mechanisms also contribute during birth of a baby to strengthen labor contractions and during immune responses to provide defense against pathogens.

Having considered the components and operation of feedback systems in general, we can now look at the relationship of feedback systems to homeostasis in the body. As examples, we will describe the homeostasis of blood pressure, a negative feedback system, and labor contractions, a positive feedback system.

Homeostasis of Blood Pressure: Negative Feedback

Blood pressure (BP) is the force exerted by blood as it presses against the walls of the blood vessels. When the heart beats faster or harder, BP increases; when total blood volume increases, BP also rises.

If some stimulus (stress), either internal or external, causes blood pressure (controlled condition) to rise, the following sequence of events occurs (Fig. 1.4). The higher pressure is detected by pressure-sensitive nerve cells (the receptors) in the walls of certain arteries. They send nerve impulses (input) to the brain (control center), which interprets the impulses and responds by sending nerve impulses (output) to the heart (effector). Heart rate decreases and blood pressure drops (response). This returns blood pressure (controlled condition) to normal, and homeostasis is restored.

A second set of effectors also contributes to maintaining normal blood pressure. Small arteries, called *arterioles,* have muscular walls that can constrict (narrow) or dilate (widen) upon receiving appropriate signals from the brain. When a stimulus causes blood pressure to increase, pressure-sensitive nerve cells (receptors) in certain arteries send nerve impulses (input) to the brain (control center). The brain interprets the messages and responds by sending fewer nerve impulses (output) to the arterioles. This causes the arterioles (effectors) to dilate (response). Thus the blood flowing through the wider arterioles meets less resistance, blood pressure drops back to normal, and homeostasis is restored.

Homeostasis of Labor Contractions: Positive Feedback

The hormone oxytocin is produced in the brain and enhances muscular contraction (controlled condition) of the pregnant uterus (see Fig. 18.11). When labor begins, the uterus is stretched (stimulus) and pressure-sensitive nerve cells in the uterine wall (receptors) send nerve impulses (input) to the brain (control center). The brain responds by causing the release of oxytocin (output). Oxytocin enters the blood, is carried to the uterus, and stimulates the muscles of the uterus (effector) to contract more forcefully (response).

FIGURE 1.4 Homeostasis of blood pressure by a negative feedback system. Note that the response is fed back into the system, and the system continues to lower blood pressure until there is a return to homeostasis.

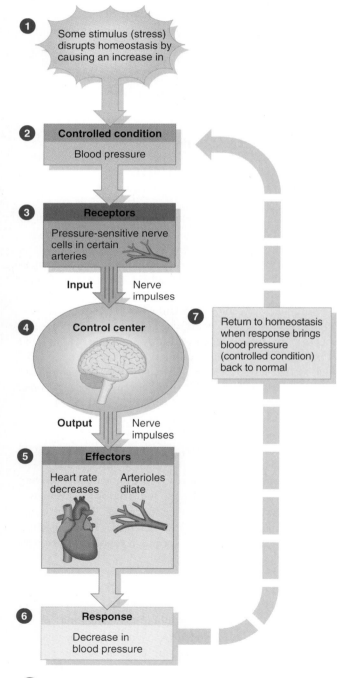

1 Some stimulus (stress) disrupts homeostasis by causing an increase in

2 **Controlled condition**

Blood pressure

3 **Receptors**

Pressure-sensitive nerve cells in certain arteries

Input   Nerve impulses

4 **Control center**

7 Return to homeostasis when response brings blood pressure (controlled condition) back to normal

Output   Nerve impulses

5 **Effectors**

Heart rate decreases   Arterioles dilate

6 **Response**

Decrease in blood pressure

Q What are three examples of positive feedback systems?

As the baby's head moves down into the birth canal, further stretching of the uterus occurs, causing the release of more oxytocin, and producing even more forceful uterine contractions. Thus this is a positive feedback cycle. The cycle is broken by the birth of the baby, which decreases uterine stretching and inhibits the release of oxytocin.

Disease: Homeostatic Imbalance

As long as the various body processes remain within normal physiological limits, body cells function efficiently, homeostasis is maintained, and the organism is healthy. When one or more components of the body lose their ability to contribute to homeostasis, however, body processes do not function efficiently. If the homeostatic imbalance is moderate, disease may result; if it is severe, death may result.

A **disease** is a pathological process with a definite set of characteristics in which part or all of the body is not carrying on its normal functions. A **local disease** is one that affects one part or a limited region of the body. A **systemic disease** affects either the entire body or several parts. Each disease alters body structures and functions in characteristic ways. A patient may experience certain **symptoms**, which are *subjective* changes in body functions that are not apparent to an observer, for example, headache or nausea. *Objective* changes that a clinician can observe and measure are called **signs**. Signs can be either anatomical or physiological changes, such as swelling, fever, rash, paralysis, and so forth.

DIAGNOSIS OF DISEASE

Diagnosis (dī′-ag-NŌ-sis; *dia* = through; *gnosis* = knowledge) is the art of distinguishing one disease from another or determining the nature of a disease. It is an early step in evaluating a disease, usually after a medical history is taken and a physical examination is given. Taking a **medical history** consists of collecting information about past events that might be related to a patient's illness (chief complaint, history of present illness, past medical problems, family medical problems, social history, and review of symptoms). A **physical examination** is a methodical evaluation that includes inspection (looking at or into a patient with various instruments), palpation (touching to feel irregularities), auscultation (listening), percussion (striking gently), measuring vital signs (temperature, pulse, respiratory rate, and blood pressure), and sometimes laboratory tests.

The science that deals with why, when, and where diseases occur and how they are transmitted in a human community is known as **epidemiology** (ep′-i-dē-mē-OL-ō-jē; *epi* = on or among; *demos* = people; *logos* = study of). The science that deals with the effects and uses of drugs in the treatment of disease is called **pharmacology** (far′-ma-KOL-ō-jē; *pharmakon* = drug or poison). ■

ANATOMICAL TERMS

In anatomy, descriptions of any region or part of the human body assume that the body is in a specific position called the **anatomical position**. In the anatomical position, the subject stands erect (upright position) facing the observer, with feet flat on the floor, arms placed at the sides, and palms turned forward (Fig. 1.5). Once the body is in the anatomical position, it is easier to visualize and understand how it

FIGURE 1.5 Anatomical position. The common names and anatomical terms, in parentheses, are indicated for many of the regions of the body. For example, the chest is the thoracic region.

In the anatomical position, the subject stands erect facing the observer, with feet flat on the floor, arms at sides, and palms turned forward.

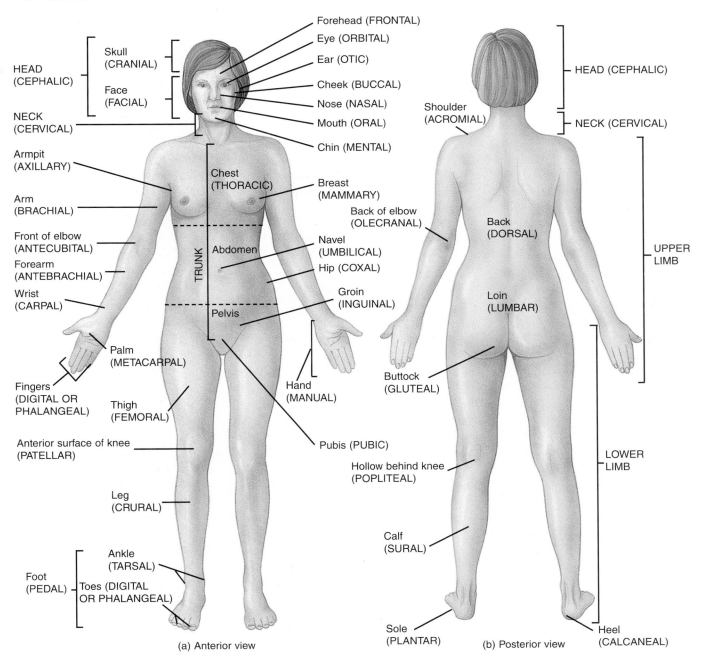

(a) Anterior view

(b) Posterior view

Q What is the usefulness of defining one standard anatomical position?

is organized into various regions. Having one standard anatomical position allows directional terms to be clear; any body part or region can be described relative to any other part. The common nouns and adjectives that designate the principal body regions are given in Fig. 1.5. These terms are used to refer to a particular region of the body, such as the arm (brachial region) or the nose (nasal region).

To locate various body structures in relation to one another, anatomists use certain **directional terms**. Such terms are precise and avoid the use of unnecessary words. There are many pairs of directional terms with opposite meanings, for example, superior and inferior. Important directional terms are defined in Exhibit 1.3, and the parts of the body referred to in the examples are labeled in Fig. 1.6. If you

EXHIBIT **1.3** **DIRECTIONAL TERMS**[a]

Term	Definition	Example
Superior (soo′-PEER-ē-or) **(cephalic or cranial)**	Toward the head or the upper part of a structure.	The heart is superior to the liver.
Inferior (in′-FEER-ē-or) **(caudal)**	Away from the head or toward the lower part of a structure.	The stomach is inferior to the lungs.
Anterior (an-TEER-ē-or) or **ventral**	Nearer to or at the front of the body. In the **prone** position, the body position is anterior side down. In the **supine** position, the body position is anterior side up.	The sternum (breastbone) is anterior to the heart.
Posterior (pos-TEER-ē-or) or **dorsal**	Nearer to or at the back of the body.	The esophagus is posterior to the trachea.
Medial (MĒ-dē-al) **(mesial)**	Nearer to the midline of the body or a structure. The midline is an imaginary vertical line that divides the body into equal left and right sides.	The ulna is on the medial side of the forearm.
Lateral (LAT-er-al)	Farther from the midline of the body or a structure.	The radius is on the lateral side of the forearm.
Intermediate (in′-ter-MĒ-dē-at)	Between two structures.	The ring finger is intermediate between the little and middle fingers.
Ipsilateral (ip-si-LAT-er-al)	On the same side of the body.	The gallbladder and ascending colon of the large intestine are ipsilateral.
Contralateral (con′-tra-LAT-er-al)	On the opposite side of the body.	The ascending and descending colons of the large intestine are contralateral.
Proximal (PROK-si-mal)	Nearer to the attachment of a limb to the trunk; nearer to the point of origin.	The humerus is proximal to the radius.
Distal (DIS-tal)	Farther from the attachment of a limb to the trunk; farther from the point of origin.	The phalanges are distal to the carpals .
Superficial (soo′-per-FISH-al)	Toward or on the surface of the body.	The muscles of the thoracic wall are superficial to the organs in the thoracic cavity. (See Fig. 1.10.)
Deep (DĒP)	Away from the surface of the body.	The ribs are deep to the skin of the chest. (See Fig. 1.10.)

[a] Study this exhibit with Figs. 1.5 and 1.6 to visualize the examples given.

FIGURE 1.6 Directional terms. Study Exhibit 1.3 with this figure to understand the directional terms: *superior, inferior, anterior, posterior, medial, lateral, intermediate, ipsilateral, contralateral, proximal,* and *distal.*

Directional terms are very precise terms that locate various parts of the body in relation to one another.

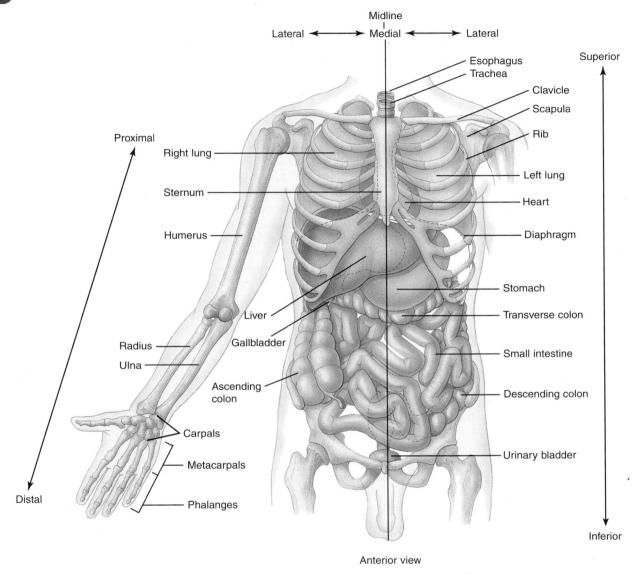

Anterior view

Is the radius proximal to the humerus? Is the esophagus anterior to the trachea? Are the ribs superficial to the lungs? Is the urinary bladder medial to the ascending colon? Is the sternum lateral to the descending colon?

study the exhibit and the figure together, the directional relations among various body parts will be made clear.

PLANES AND SECTIONS

You will also study parts of the body relative to **planes** (imaginary flat surfaces) that pass through it (Fig. 1.7). A

sagittal (SAJ-i-tal; *sagittalis* = arrow) **plane** is a vertical plane that divides the body or an organ into right and left sides. More specifically, if such a plane passes through the midline of the body or organ and divides it into *equal* right and left sides, it is called a **midsagittal (median) plane**. If the sagittal plane does not pass through the midline but instead divides the body or an organ into *unequal* right and left sides, it is called a **parasagittal** (*para* = near) **plane**. A **frontal** or **coronal** (kō-RŌ-nal; *corona* = crown) **plane**

FIGURE 1.7   Planes of the human body.

 The frontal, transverse, sagittal, and oblique planes divide the body in specific ways.

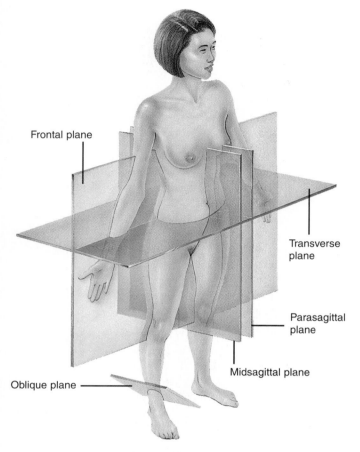

Frontal plane

Transverse plane

Parasagittal plane

Midsagittal plane

Oblique plane

Right anterolateral view

Q Which plane divides the brain into anterior and posterior portions?

frontal section, and a *midsagittal section*—provide different views of the brain.

BODY CAVITIES

Body cavities are confined spaces within the body that contain internal organs. The cavities help protect, separate, and support the organs. The various body cavities may be separated from each other by structures such as muscles, bones, or ligaments. Figure 1.9 shows the two principal ones, the dorsal and ventral body cavities. The **dorsal body cavity** is located near the dorsal (back) surface of the body. It is further subdivided into a **cranial cavity**, which is formed by the cranial (skull) bones and contains the brain, and a **vertebral (spinal) canal**, which is formed by the vertebrae of the backbone and contains the spinal cord and the beginnings (roots) of spinal nerves.

The other principal body cavity is the **ventral body cavity**. This cavity is located on the ventral (front) aspect of the body. A thin, slippery tissue called a **serous membrane** lines the wall of the ventral body cavity and covers the organs within it. The organs inside are called **viscera** (VIS-er-a). The ventral body cavity also has two principal subdivisions—a superior portion, called the **thoracic** (thor-AS-ik) **cavity** (or chest cavity), and an inferior portion, called the **abdominopelvic** (ab-dom′-i-nō-PEL-vik) **cavity**. The structure that divides the ventral body cavity into the thoracic and abdominopelvic cavities is the diaphragm (DĪ-a-fram; *diaphragma* = partition or wall), an important muscle for breathing.

The thoracic cavity has several compartments: two pleural cavities, the mediastinum, and one pericardial cavity. Each of the two **pleural** (PLOOR-al; *pleur* = rib or side) **cavities** surrounds a lung (Fig. 1.10). Each pleural cavity is a small, fluid-filled space between the part of the serous membrane that covers the lung and the part that lines the wall of the thoracic cavity. As you will see in Chapter 4, a serous membrane is a double-layered membrane that (1) *lines* a body cavity that does not open directly to the exterior and (2) *covers* the organs within that cavity. The serous membrane associated with the lungs is called the **pleura**. Between the pleural cavities is the **mediastinum** (mē′-dē-as-TĪ-num; *media* = middle; *stare* = to stand in), a broad, median partition—actually a mass of tissues—medial to the lungs that extends from the sternum (breastbone) to the vertebral column (backbone). It includes all the contents of the thoracic cavity except the lungs themselves (Fig. 1.10). Among the structures in the mediastinum are the heart, esophagus, trachea, thymus gland, and many large blood and lymphatic vessels. Within the mediastinum is the **pericardial** (per′-i-KAR-dē-al; *peri* = around; *cardi* = heart) **cavity**. It is a fluid-filled space

divides the body or an organ into anterior (front) and posterior (back) portions. A **transverse (cross-sectional** or **horizontal) plane** divides the body or an organ into superior (top) and inferior (bottom) portions. Sagittal, frontal, and transverse planes are all at right angles to one another. An **oblique plane**, on the other hand, passes through the body or an organ at an angle between the transverse plane and either a sagittal or frontal plane.

When you study a body region, you will often view it in **section**, meaning that you look at only one flat surface of the three-dimensional structure. It is important to know the plane of the section so you can understand the anatomical relationship of one part to another. Figure 1.8 indicates how three different sections—a *transverse (cross) section,* a

FIGURE 1.8 Planes and sections through different parts of the brain. The planes are shown in the diagrams on the left and the resulting sections are shown in the photographs on the right.

 Planes divide the body in various ways to produce sections.

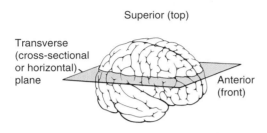

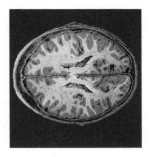

(a)

Transverse (cross) section

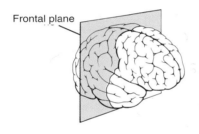

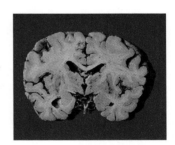

(b)

Frontal section

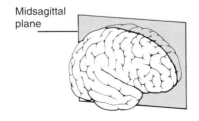

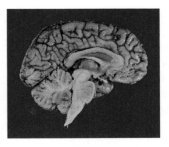

(c)

Midsagittal section

Which plane divides the brain into equal right and left sides?

FIGURE 1.9 Body cavities.

 The two principal cavities are the dorsal and ventral body cavities.

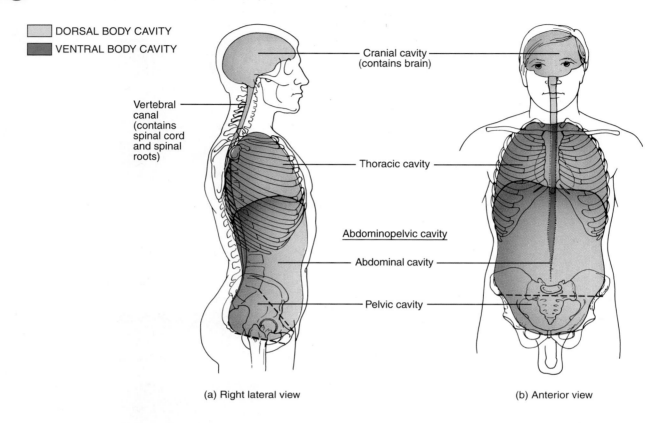

DORSAL BODY CAVITY
VENTRAL BODY CAVITY

Cranial cavity
(contains brain)

Vertebral
canal
(contains
spinal cord
and spinal
roots)

Thoracic cavity

Abdominopelvic cavity

Abdominal cavity

Pelvic cavity

(a) Right lateral view

(b) Anterior view

In which cavities are the following organs located: urinary bladder, stomach, heart, pancreas, small intestine, lungs, internal female reproductive organs, thymus gland, spleen, rectum, liver? Use the following symbols for your response: T = thoracic cavity, A = abdominal cavity, or P = pelvic cavity.

between the part of the serous membrane that covers the heart and the part that lines the thoracic cavity. The serous membrane associated with the heart is called the **pericardium**.

The **abdominopelvic cavity**, as the name suggests, is divided into two portions, although no wall separates them (see Fig. 1.9). The serous membrane that lines the abdominopelvic cavity and covers the organs within it is called the **peritoneum**. (Because the kidneys lie posterior to the peritoneum, between it and the posterior abdominal wall, they are not strictly within the abdominopelvic cavity.) The superior portion of the abdominopelvic cavity, the **abdominal** (*abdere* = to hide, because it hides the viscera) **cavity**, contains the stomach, spleen, liver, gallbladder, pancreas, small intestine, and most of the large intes-

tine. The inferior portion, the **pelvic cavity,** contains the urinary bladder, portions of the large intestine, and the internal organs of reproduction. The pelvic cavity is the region between two imaginary planes, shown by dashed lines in Fig. 1.9a.

ABDOMINOPELVIC REGIONS AND QUADRANTS

To describe the location of organs easily, the abdominopelvic cavity is divided into nine **abdominopelvic regions** (Fig. 1.11a). Note which organs and parts of organs are in the different regions by carefully examining Fig. 1.11b,c.

FIGURE 1.10 Thoracic cavity. The two pleural cavities surround the right and left lungs and the pericardial cavity surrounds the heart. The mediastinum is medial to the lungs and extends from the sternum to the backbone. *Note:* The arrow in the inset indicates the direction from which the section is viewed (superior). This aid is used throughout the book.

 The diaphragm separates the thoracic cavity from the abdominal cavity.

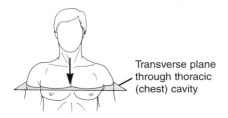

Transverse plane through thoracic (chest) cavity

ANTERIOR (front)

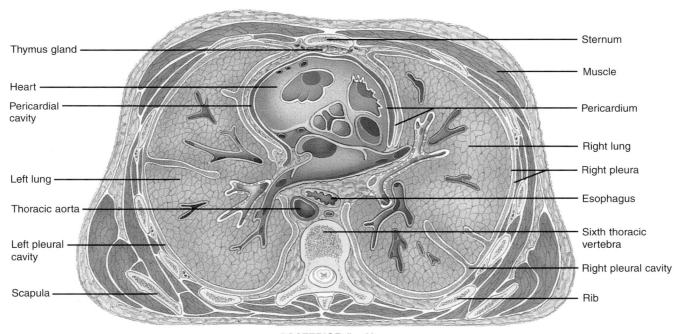

Thymus gland —
Heart —
Pericardial cavity —
Left lung —
Thoracic aorta —
Left pleural cavity —
Scapula —

— Sternum
— Muscle
— Pericardium
— Right lung
— Right pleura
— Esophagus
— Sixth thoracic vertebra
— Right pleural cavity
— Rib

POSTERIOR (back)

Superior view of transverse section

 Which of the following structures are contained in the mediastinum: thymus gland, right lung, heart, esophagus, thoracic aorta, rib, left pleural cavity?

Although some parts of the body in the illustrations may be unfamiliar to you at this point, they will be discussed in detail in later chapters.

The abdominopelvic cavity is divided more simply into **quadrants** (KWOD-rantz; *quad* = a one-fourth part). These are shown in Fig. 1.12. In this method, a horizontal line and a vertical line are passed through the **umbilicus** (um-bi-LĪ-kus; *umbo* = knob) or navel. Whereas the nine-region division is more widely used for anatomical studies, the quadrant division is more commonly used by clinical staff for locating the site of an abdominopelvic pain, tumor, or other abnormality.

FIGURE 1.11   Abdominopelvic cavity. (a) The nine regions. The subcostal (top horizontal) line is drawn just inferior to the rib cage, across the inferior portion of the stomach; the transtubercular (bottom horizontal) line is drawn just inferior to the tops of the hipbones. The left and right midclavicular (two vertical) lines are drawn through the midpoints of the clavicles (collar bones), just medial to the nipples. The four lines divide the abdominopelvic cavity into a larger middle section and smaller left and right sections. (b) The greater omentum has been removed. (c) Many anterior organs have been removed, exposing the more posterior structures. The internal reproductive organs in the pelvic cavity are shown in Figs. 28.1 and 28.11.

 The nine-region designation is used for anatomical studies.

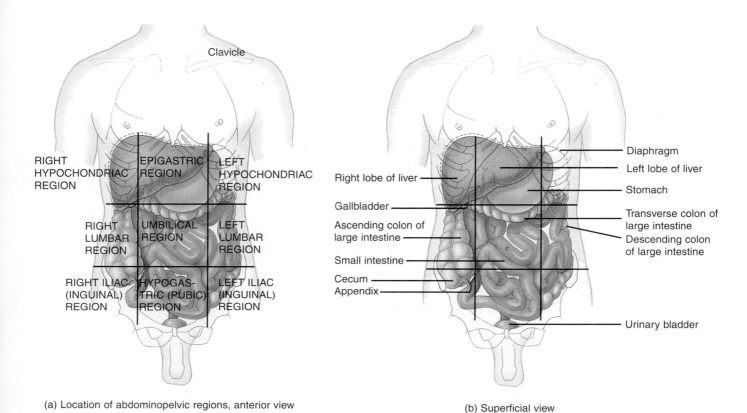

(a) Location of abdominopelvic regions, anterior view

(b) Superficial view

AUTOPSY

To determine the cause of death accurately, it is necessary to perform an **autopsy** (AW-top-sē; *auto* = self; *opsis* = to see with one's own eyes). An autopsy can uncover the existence of diseases not detected during life. It also may support the accuracy of diagnostic tests, establish the beneficial and adverse effects of drugs, reveal the impact of environmental influences on the body, and educate health-care students. Moreover, an autopsy can reveal conditions that may affect offspring or siblings (such as congenital heart defects). An autopsy may be needed in a criminal investigation, but it may also help resolve disputes between beneficiaries and insurance companies. ■

MEDICAL IMAGING

Various kinds of **medical imaging** procedures allow visualization of structures inside our bodies. The images provide clues to both abnormal anatomy and deviations from normal physiology. They are increasingly helpful for precise diagnosis of a wide range of disorders. The grandfather of all medical imaging techniques is conventional radiography, in therapeutic use since the late 1940s. The newer techniques not only contribute to diagnosis of disease, but they also are advancing our understanding of normal physiology. Exhibit 1.4 describes some commonly used medical imaging techniques. Other imaging methods, such as cardiac catheterization, will be discussed in later chapters.

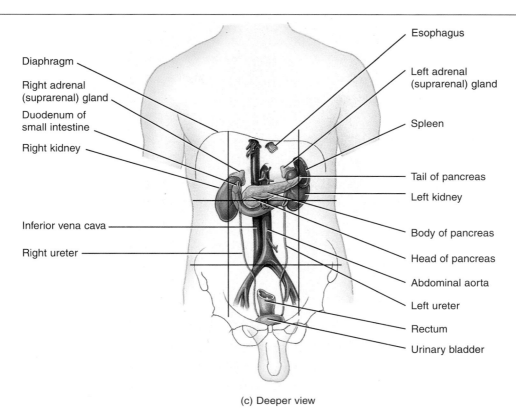

(c) Deeper view

Q In which abdominopelvic region is each of the following found: most of the liver, transverse colon, urinary bladder, spleen?

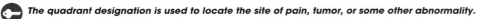

FIGURE 1.12 Quadrants of the abdominopelvic cavity. The two lines intersect at right angles at the umbilicus (navel).

🔑 *The quadrant designation is used to locate the site of pain, tumor, or some other abnormality.*

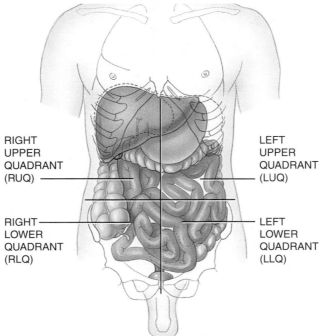

Anterior view

 In which quadrant would the pain from appendicitis (inflammation of the appendix) be felt?

EXHIBIT 1.4 SUMMARY OF MEDICAL IMAGING PROCEDURES

Procedure and Example	Description
Conventional radiography	A single barrage of x-rays passes through the body and produces a two-dimensional image of the interior of the body called a **radiograph** (RĀ-dē-ō-graf) or **x-ray**. *Comment:* Overlap of structures can make diagnosis difficult, and subtle differences in tissue density cannot always be discerned.

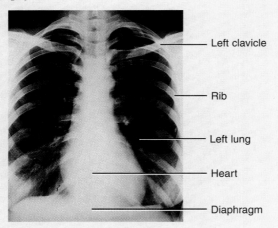

Left clavicle

Rib

Left lung

Heart

Diaphragm

Radiograph of thorax (chest) (anterior view)

Computed tomography (CT) scanning
[formerly called computerized axial tomography (CAT) scanning]

An x-ray beam moves in an arc around the body, producing an image of a transverse section called a **CT scan** on a video monitor hooked to a computer. By stacking images one on top of the other, the computer can construct three-dimensional images that can be rotated, for example, to plan plastic surgery.
Comment: Quick, painless CT scans have replaced exploratory surgery in many cases. Detailed images can reveal tumors, aneurysms (bulges in blood vessels), kidney stones, gallstones, infections, tissue damage, and deformities.

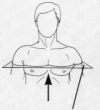

Transverse plane through thorax (chest)

Note: In keeping with radiographic convention, as in CT scans, the section is viewed from the inferior aspect. Thus the left side of the body appears on the right side of the photograph.

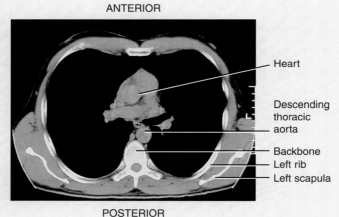

ANTERIOR

Heart

Descending thoracic aorta

Backbone
Left rib
Left scapula

POSTERIOR

CT scan of a transverse section of the thorax (chest) (inferior view)

Dynamic spatial reconstruction (DSR)

A highly sophisticated x-ray machine produces moving, three-dimensional, life-size images from any view.
Comment: By using a computer, an image can be rotated, tipped, "sliced open," enlarged, replayed, and viewed in slow motion or at high speeds; good procedure for imaging heart, lung, and blood vessels; measuring movements and volumes, and assessing tissue damage.

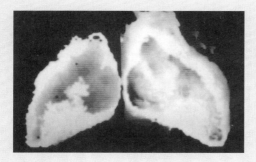

DSR image of "opened" heart

STUDENT OBJECTIVES

1. Distinguish between matter and energy and describe five forms of energy. (p. 26)

2. Identify by name and symbol the principal chemical elements of the human body. (p. 26)

3. Describe the structure of an atom. (p. 26)

4. Explain how ionic, covalent, and hydrogen bonds form. (p. 30)

5. Define a chemical reaction and explain the basic differences between synthesis, decomposition, exchange, reversible, oxidation–reduction, exergonic, and endergonic chemical reactions. (p. 32)

6. List and compare the properties of inorganic acids, bases, salts, and water. (p. 34)

7. Define pH and explain the role of buffer systems in homeostasis. (p. 36)

8. Compare the structure and functions of carbohydrates, lipids, proteins, deoxyribonucleic acid (DNA), ribonucleic acid (RNA), and adenosine triphosphate (ATP). (p. 38)

9. Describe the characteristics and importance of enzymes. (p. 46)

CHAPTER 2

THE CHEMICAL LEVEL OF ORGANIZATION

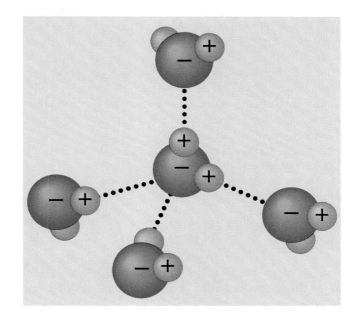

CHAPTER CONTENTS AT A GLANCE

You are what you eat! The common substances you eat and drink, for example, water, meat, vegetables, sugar, table salt, and cooking oil, provide matter and energy to sustain life. Since chemicals compose your body and all body activities are chemical in nature, it is important to become familiar with the language and fundamental ideas of chemistry.

MATTER AND ENERGY

All living and nonliving things consist of **matter**, which is anything that occupies space and has mass. Matter exists in one of three states: solid, liquid, or gas. Although the terms mass and weight often are used interchangeably, there is a distinction. Mass is the amount of matter that a substance contains. Weight is the force of gravity acting on a mass. Although your mass is the same, your weight depends on where you are—at sea level on a beach in Florida, at a high altitude in the Rocky Mountains, or in a space station orbiting the Earth. Objects weigh less when they are farther from the Earth's core because the pull of gravity is weaker. In outer space, weight is close to zero but mass remains the same as it was on Earth at sea level.

Energy is the capacity to do work, that is, to put mass into motion. Energy can be neither created nor destroyed. This principle is known as conservation of energy. But one form of energy can be converted into another. For example, when sunlight strikes your body and warms your skin, radiant energy is being converted into heat energy. The two principal kinds of energy are **potential energy** (inactive or stored energy) and **kinetic energy** (energy of motion). The energy stored in a battery, a tightly coiled spring, water behind a dam, or a person poised to jump down the steps is potential energy. Any object in motion, from a molecule, to a baseball, to a river, has kinetic energy. Potential and kinetic energy exist in several forms:

1. **Radiant energy**, for example, visible light, is energy that travels in waves. The waves may be spaced closer together (short wavelength) or farther apart (long wavelength) depending on the type of radiant energy. From longest to shortest wavelength, forms of radiant energy are: radio waves, microwaves (used in microwave ovens), infrared waves (cause heating), visible light waves, ultraviolet waves (cause sunburn and skin cancer), and x-rays and gamma rays (used for medical imaging).
2. **Electrical energy** results from the flow of charged particles. The electrical energy that powers your computer depends on the flow of electrons. In your body, electrical energy involves the movement of charged particles called ions. Action potentials (impulses) in nerve and muscle cells are examples of electrical energy.
3. **Heat energy** is the energy transferred from one thing to another because of a difference in temperature. A hot object, such as a flame, will transfer heat energy to a nearby cooler object, for example, a hamburger.
4. **Chemical energy** is a form of potential energy. During the breaking apart or forming of chemicals, chemical energy is absorbed or released in the form of radiant, heat, or electrical energy. The building processes of the body—the construction of bones, the growth of hair and nails, the replacement of injured cells—all require energy. When body cells break down nutrients, they store chemical energy in a form that can be used for building processes.
5. **Mechanical energy** is the energy due to either the position or the movement of a mass. It may be potential energy, if the mass is still, or kinetic energy, if the mass is moving. Many body processes involve the conversion of chemical energy into mechanical energy to perform movements, for example, beating of the heart to pump blood throughout the body.

Chemical Elements

All forms of matter are made up of a limited number of building blocks called **chemical elements**, which are substances that cannot be split into simpler substances by ordinary chemical reactions. Scientists now recognize 109 different elements, of which 92 occur in nature. (Physicists can artificially create the other 17 elements in experiments using powerful machines called accelerators, but their existence is brief.)

Elements are given letter abbreviations, usually derived from the first or first and second letters of the English or Latin name for the element. Such letter abbreviations are called **chemical symbols**. Examples of chemical symbols are H (hydrogen), C (carbon), O (oxygen), N (nitrogen), Ca (calcium), Na (*natrium* = sodium), K (*kalium* = potassium), Fe (*ferrum* = iron), and P (phosphorus). Twenty-six of the 92 naturally occurring elements are present in your body. Oxygen, carbon, hydrogen, and nitrogen make up about 96% of the body's mass. Calcium, phosphorus, potassium, sulfur, sodium, chlorine, magnesium, iodine, and iron make up about 3.9% of the body's mass. Thirteen other chemical elements are called **trace elements** because they are present in minute concentrations. They compose about 0.1% of total body mass. Exhibit 2.1 lists the major and trace elements.

Structure of Atoms

Each element is made up of **atoms**, the smallest units of matter that enter into chemical reactions. An **element** is simply a quantity of matter composed of atoms that are all of the same type. A sample of the element carbon, such as pure coal dust or a diamond, contains only carbon atoms. A tank of helium gas contains only helium atoms. The smallest atoms (hydrogen) are less than 0.00000001 centimeter (cm) or $\frac{1}{250,000,000}$ inch in diameter, and the largest atoms are

EXHIBIT **2.1** **CHEMICAL ELEMENTS PRESENT IN THE BODY**

Chemical Element (Symbol)	Percent[a]	Comments
Oxygen (O)	65.0	Constituent of water and organic molecules (carbon-containing molecules that are made by a living organism); needed for cellular respiration, which produces adenosine triphosphate (ATP), an energy-rich chemical in cells.
Carbon (C)	18.5	Found in every organic molecule.
Hydrogen (H)	9.5	Constituent of water, all foods, and most organic molecules; contributes to acidity when it is positively charged (H^+).
Nitrogen (N)	3.2	Component of all proteins and nucleic acids. The nucleic acids are deoxyribonucleic acid (DNA) and ribonucleic acid (RNA).
Calcium (Ca)	1.5	Contributes to hardness of bones and teeth; needed for many body processes, for example, blood clotting, movement of structures inside cells, release of hormones, and contraction of muscle.
Phosphorus (P)	1.0	Component of many proteins, nucleic acids, and ATP; required for normal bone and tooth structure.
Potassium (K)	0.4	The ionized form (K^+) is the most abundant cation (positively charged particle) inside cells and is important in conduction of nerve and muscle impulses and for muscle contraction.
Sulfur (S)	0.3	Component of some vitamins and many proteins, especially muscle proteins.
Sodium (Na)	0.2	The ionized form (Na^+) is the most plentiful cation outside cells, is essential in blood to maintain water balance, and is needed for conduction of nerve and muscle impulses and for muscle contraction.
Chlorine (Cl)	0.2	The ionized form (Cl^-) is the most plentiful anion (negatively charged particle) in extracellular fluid and is essential to maintain water balance.
Magnesium (Mg)	0.1	Component of bone; many enzymes (molecules that increase the speed of chemical reactions in living organisms) need the ionized form (Mg^{2+}) to function properly.
Iodine (I)	0.1	Needed for production of thyroid hormones by the thyroid gland.
Iron (Fe)	0.1	The ionized forms (Fe^{2+} and Fe^{3+}) are essential components of hemoglobin (oxygen-carrying protein in blood) and of some enzymes needed for ATP production.
Aluminum (Al) Boron (B) Chromium (Cr) Cobalt (Co) Copper (Cu) Fluorine (F) Manganese (Mn) Molybdenum (Mo) Selenium (Se) Silicon (Si) Tin (Sn) Vanadium (V) Zinc (Zn)		These elements are called **trace elements** because they are present in minute concentrations. Some have important functions (described in Exhibit 25.5 on page 838) whereas others are present but have no known function.

[a] Percent of total body mass.   Compose about 96% of total body mass.   Compose about 3.9% of total body mass.   Compose about 0.1% of total body mass.

only five times larger. If 50 million of the largest atoms were placed end to end, they would span approximately 2.5 cm (1 in.).

An atom consists of three major types of subatomic particles: neutrons, protons, and electrons (Fig. 2.1). Negatively charged **electrons** (e^-) orbit the dense central core or **nucleus**. Within the nucleus are positively charged particles called **protons** (p^+) and uncharged (neutral) particles called **neutrons** (n^0). The electron orbits or shells drawn in diagrams like Fig. 2.1 are very simplified, showing just the

Figure 2.1 Structure of an atom. In this simplified diagram of a carbon atom, note the central location of the nucleus. The nucleus contains six neutrons and six protons, although not all are visible in this view. The six electrons move about the nucleus in regions called electron shells, shown here as circles.

 An atom is the smallest unit of matter that enters into a chemical reaction.

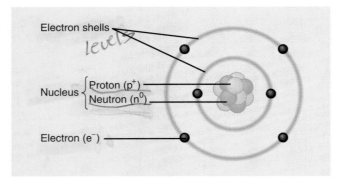

 What is the atomic number of this atom?

average (or most probable) distance between the nucleus and a particular group of orbiting electrons. Actually, the fast-moving electrons follow no defined path around the nucleus. You can picture the electrons as forming a cloud of negative charge around the nucleus.

The number of electrons in an atom of an element always equals the number of protons. Since each electron carries one negative charge, the negatively charged electrons and the positively charged protons balance each other. Thus each atom is electrically neutral; its total charge is zero.

The *number of protons* in the nucleus distinguishes the atoms of one element from those of another. Figure 2.2 shows that a hydrogen atom contains one proton. A helium atom contains two. A carbon atom has six, and so on. Each different kind of atom has a different number of protons in its nucleus. The number of protons in an atom is called the atom's **atomic number**. Therefore each kind of atom, or element, has a different atomic number. For example, oxygen has an atomic number of 8 because its nucleus has 8 protons, whereas sodium has an atomic number of 11 because its nucleus has 11 protons.

The **mass number** of an atom is the total number of protons and neutrons. For sodium, which has 12 neutrons, the mass number is 23 (11 protons plus 12 neutrons = 23). Atoms of one element may have different mass numbers because they have different numbers of neutrons. Different atoms of an element that have the same number of protons but different numbers of neutrons are called **isotopes**. All isotopes of an element have identical *chemical properties* because they have the same number of electrons. (As you will see shortly, the chemical properties of an atom are a function of its electrons.) In a sample of oxygen, for exam-

ple, most atoms have 8 neutrons, but a few have 9 or 10, even though all have 8 protons and 8 electrons. The stable isotopes of oxygen are designated ^{16}O, ^{17}O, and ^{18}O (or O-16, O-17, and O-18). The numbers indicate the mass number (total number of protons and neutrons) in each isotope.

Certain isotopes called **radioactive isotopes (radioisotopes)** are unstable; their nuclear structure decays or changes to a more stable configuration. Examples are H-3, C-14, O-15, and O-19. In decaying, these atoms emit radiation (alpha or beta particles or gamma rays) that can be monitored with radiation detectors. Such instruments can not only detect emissions from a radioactive isotope but with the aid of computers can also form an image of its distribution within the body. The decay of a particular radioisotope may be slow or fast, taking thousands of years or a fraction of a second. Each radioactive isotope has a characteristic **half-life**, which is the time required for half of the radioactive atoms in a sample to decay.

The standard unit for measuring the mass of atoms and their subatomic particles is a **dalton**. A neutron has a mass of 1.008 daltons, and a proton has a mass of 1.007 daltons. The mass of an electron is 0.0005 dalton, about 2000 times smaller than the mass of a neutron or proton. The **atomic mass (atomic weight)** is the average mass of all stable atoms of an element and reflects the relative proportion of atoms with different mass numbers. The atomic mass of chlorine, for example, is 35.453. While most chlorine atoms have 18 neutrons (mass number = 35), a few have 20 neutrons (mass number = 37). The atomic mass is slightly less than the total masses of all the neutrons, protons, and electrons in an element because energy is released and thus some mass (less than 1%) is lost when the atom's components bind together.

MEDICAL IMAGING USING RADIOACTIVE ISOTOPES

Cells don't distinguish between radioactive isotopes and stable isotopes of an element because all isotopes have the same number of electrons and thus the same chemical behavior. Radioactive isotopes can be used to study both the structure and function of particular tissues. For example, thallium-201 is used to image the heart (thallium imaging) and to assess adequacy of blood flow to the heart muscle. Various other radioactive isotopes are also used to study bones (bone scan), the brain (PET scan), the lungs (lung scan), the thyroid gland (thyroid scan), and the kidneys (kidney scan). ■

Electrons and Chemical Reactions

When atoms combine with or break apart from other atoms, a **chemical reaction** occurs. In the process, new products with different chemical properties are formed. Chemical reactions are the foundation of all life processes, and electron interactions are the basis of all chemical reactions.

In their motion around the nucleus, electrons tend to spend most of the time in specific atomic regions. Figure 2.2 depicts these most likely regions as circles lying at varying distances from the nucleus. Each circle represents one **electron shell**, which can hold a certain maximum number of electrons. For instance, the electron shell nearest the nucleus never holds more than two electrons, no matter what the element. This is the first shell. The second shell holds a maximum of eight electrons, while the third can hold up to 18 electrons. Higher shells (there are as many as seven) can contain many more electrons. Iodine, the most massive element present in the human body, has a total of 53 electrons, with 2 in the first shell, 8 in the second shell, 18 in the third shell, 18 in the fourth shell, and 7 in the fifth shell.

To achieve stability, atoms tend to either empty their outermost shell or fill it to the maximum. In so doing, they may give up, accept, or share electrons with other atoms—whichever is most favorable. The **valence** (combining capacity) is the number of extra or deficient electrons in the **valence shell**, which is the outermost shell. Take a look at the chlorine atom in Fig. 2.2. Its outermost shell, which happens to be the third shell, has seven electrons. Although the third shell can hold up to 18 electrons, one stable form is reached at 8 electrons. Thus chlorine, having 7 electrons, tends to pick up an electron that another atom has lost. Sodium, by contrast, has only one electron in its third (valence) shell. It is much easier for sodium to give up one electron than to fill the third shell by accepting seven more electrons. Atoms of a few elements, like helium and neon, have completely filled outer shells and do not tend to gain or lose electrons. Such elements are called **inert elements** and do not participate in chemical reactions.

Atoms with incompletely filled outer shells, such as sodium and chlorine, tend to combine with each other in chemical reactions. During a reaction the atoms can exchange or share valence electrons and thereby fill or empty their outer shells. When two or more atoms share electrons after a chemical reaction, the resulting combination is called a **molecule** (MOL-e-kyool). A molecule may contain two atoms of the same kind, as in hydrogen and oxygen molecules:

Figure 2.2 Atomic structures of several stable atoms that have important roles in living systems.

When atoms take part in chemical reactions, they lose, gain, or share electrons in their outermost (valence) shell.

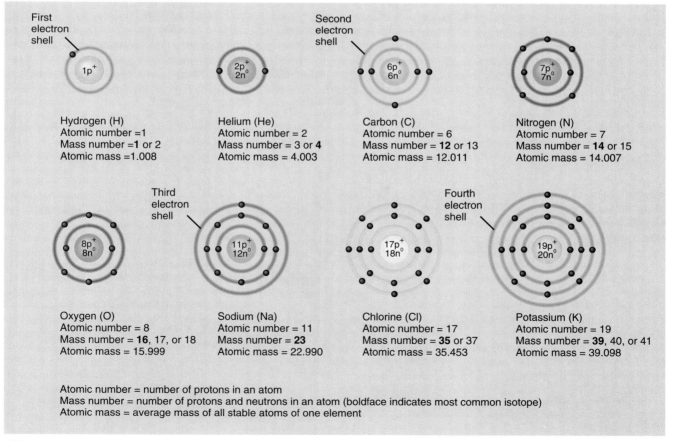

First electron shell

Second electron shell

Hydrogen (H)
Atomic number =1
Mass number =**1** or 2
Atomic mass =1.008

Helium (He)
Atomic number = 2
Mass number = 3 or **4**
Atomic mass = 4.003

Carbon (C)
Atomic number = 6
Mass number = **12** or 13
Atomic mass = 12.011

Nitrogen (N)
Atomic number = 7
Mass number = **14** or 15
Atomic mass = 14.007

Third electron shell

Fourth electron shell

Oxygen (O)
Atomic number = 8
Mass number = **16**, 17, or 18
Atomic mass = 15.999

Sodium (Na)
Atomic number = 11
Mass number = **23**
Atomic mass = 22.990

Chlorine (Cl)
Atomic number = 17
Mass number = **35** or 37
Atomic mass = 35.453

Potassium (K)
Atomic number = 19
Mass number = **39**, 40, or 41
Atomic mass = 39.098

Atomic number = number of protons in an atom
Mass number = number of protons and neutrons in an atom (boldface indicates most common isotope)
Atomic mass = average mass of all stable atoms of one element

Q Which of these elements is inert?

H_2 and O_2. The subscript 2 shows there are two atoms in the molecule. Two or more different kinds of atoms may also react to form a molecule, as in the hydrogen chloride molecule: HCl. Here one atom of hydrogen shares electrons with one atom of chlorine.

A **compound** is a chemical substance that can be broken down into two or more different elements by chemical means. A compound always contains atoms of two or more different elements. Hydrogen chloride (HCl), which dissolves in water to form hydrochloric acid in the digestive juices of the stomach, and NaCl (table salt) are compounds. A molecule of hydrogen (H_2) is not.

Chemical Bonds

The atoms of a molecule are held together by forces of attraction called **chemical bonds**, a form of potential energy. The types of bonds we will consider are ionic bonds, covalent bonds, and hydrogen bonds.

Ions and Ionic Bonds

Atoms are electrically neutral because the number of positively charged protons equals the number of negatively charged electrons. When an atom gains or loses electrons, however, this balance is upset. If the atom gains electrons, it acquires a negative charge; if it loses electrons, it is left with a positive charge. A particle with a negative or positive charge is called an **ion**. An ion is symbolized by writing the chemical symbol followed by the number of its positive (+) or negative (−) charges. A compound that dissociates into positive and negative ions in solution is called an **electrolyte** (e-LEK-trō-līt) because the solution can conduct an electric current. (The chemistry and importance of electrolytes are discussed in detail in Chapter 27.)

Consider the sodium ion (Fig. 2.3a). An atom of sodium has 11 protons and 11 electrons, with 1 electron in its valence (outermost) shell. When sodium gives up its valence electron, it is left with 11 protons and only 10 electrons. It is an **electron donor** because it gives up an electron. The atom now has a positive charge of one (+1) and is called a sodium ion (written Na^+). Similarly, a calcium atom can donate two electrons to form Ca^{2+}.

Another example is the formation of the chloride ion (Fig. 2.3b). Chlorine has a total of 17 electrons, 7 of them in its valence shell. Since this shell is stable with eight electrons, chlorine tends to pick up an electron that another atom has lost. Thus chlorine is an **electron acceptor**. Accepting an electron brings the total to 18 electrons. However, the chloride ion still has only 17 positively charged protons in its nucleus. It therefore has a negative charge of one (−1) and is written as Cl^-.

The positively charged sodium ion (Na^+) and the negatively charged chloride ion (Cl^-) attract each other; unlike charges attract. The attraction, called an **ionic bond**, holds the two ions together, and sodium chloride (NaCl), or table salt, is formed (Fig. 2.3c). Thus ionic bonds form when oppositely charged ions are attracted to one another. In the body, ionic bonds are found mainly in teeth and bones. Most ions in the body are dissolved in body fluids and thus are electrolytes.

Generally, if the outer electron shell is less than half-filled, an atom loses electrons and forms a positively charged ion called a **cation** (KAT-ī-on). Examples of cations are the potassium ion (K^+), sodium ion (Na^+), and calcium ion (Ca^{2+}). By contrast, if the outer shell is more than half-filled, an atom tends to accept electrons and form a negatively charged ion called an **anion** (AN-ī-on). Examples of anions include the iodide ion (I^-), chloride ion (Cl^-), and sulfide ion (S^{2-}).

Figure 2.3   Ions and ionic bond formation. (a) Sodium (Na) tends to give up the single electron in its valence shell; it is an electron donor. (b) Chlorine (Cl) tends to pick up one electron to completely fill its valence shell; it is an electron acceptor. (c) Ionic bonds hold oppositely charged ions together. Na^+ joins Cl^- to form sodium chloride, ordinary table salt.

 An ionic bond is an attraction that holds together ions with different charges.

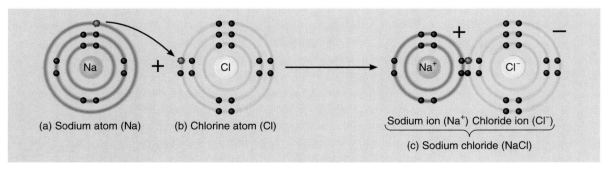

(a) Sodium atom (Na)   (b) Chlorine atom (Cl)   Sodium ion (Na^+) Chloride ion (Cl^-)

(c) Sodium chloride (NaCl)

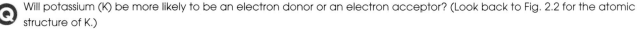 Will potassium (K) be more likely to be an electron donor or an electron acceptor? (Look back to Fig. 2.2 for the atomic structure of K.)

Covalent Bonds

The second type of chemical bond is the **covalent bond**. This kind of bond is much more common in the body than an ionic bond. When a covalent bond forms, neither of the combining atoms loses or gains electrons. Instead, the two atoms *share* one, two, or three pairs of electrons.

One way a hydrogen atom can fill its outer shell is to combine with another hydrogen atom to form the molecule H_2 (Fig. 2.4a). In H_2 molecules, the two atoms share a pair of electrons. Each hydrogen atom has its own electron plus a shared electron from the other atom. The two electrons orbit the nuclei (single protons) of both atoms, but they are most often in the region between the two nuclei.

When two atoms share one pair of electrons, as in the H_2 molecule, a **single covalent bond** is formed. A single covalent bond is written as a single line between the atoms

Figure 2.4   Covalent bond formation. To the right are simpler ways to represent these molecules. In structural formulas, each covalent bond is written as a straight line between the symbols for two atoms. In molecular formulas, the number of atoms in each molecule is noted by subscripts.

In a covalent bond, two atoms share one, two, or three pairs of valence electrons.

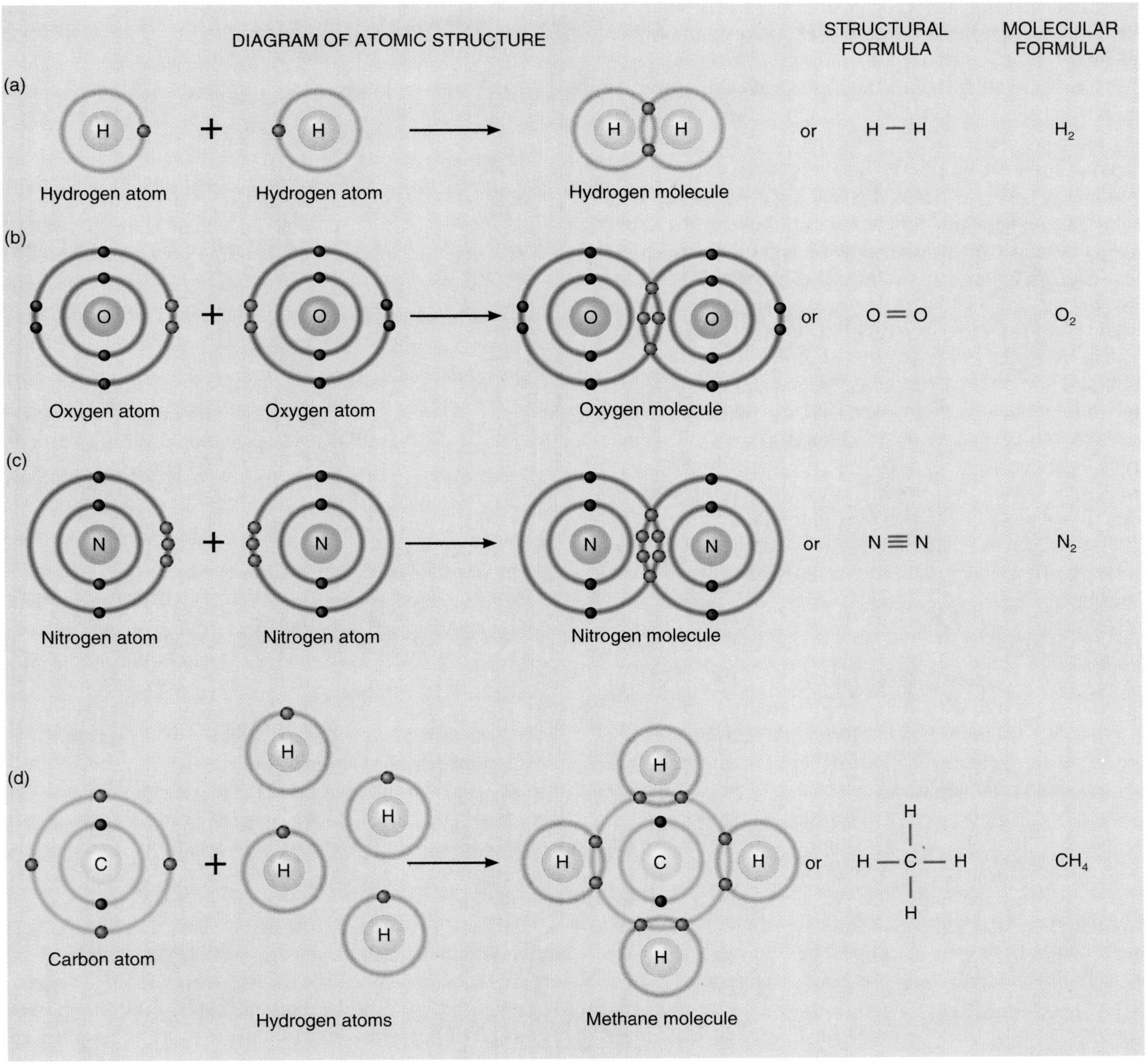

What is the principal difference between an ionic bond and a covalent bond?

(H—H). When two atoms share *two* pairs of electrons, the result is a **double covalent bond**, written as two parallel lines (O=O in Fig. 2.4b). A **triple covalent bond**, written as *three* parallel lines, occurs when three pairs of electrons are shared (N≡N in Fig. 2.4c).

The same principles that apply to covalent bonding between atoms of the same element also apply to atoms of different elements. Methane (CH_4), also known as marsh gas, is an example of covalent bonding between atoms of different elements (Fig. 2.4d). The outer shell of the carbon atom can hold eight electrons but has only four of its own. Each hydrogen atom can hold two electrons but has only one of its own. In the methane molecule the carbon atom shares four pairs of electrons. One pair is shared with each hydrogen atom. Each of the four carbon electrons orbits both the carbon nucleus and a hydrogen nucleus. Each hydrogen electron likewise orbits both its own nucleus and the carbon nucleus.

In some covalent bonds, atoms share the electrons equally; that is, one atom does not attract the shared electrons more strongly than the other atom. This is called a **nonpolar covalent bond**. The bonds between two identical atoms always are nonpolar covalent bonds (Fig. 2.4a–c). Another example of a nonpolar covalent bond is the single covalent bond between carbon and hydrogen (Fig. 2.4d).

In other covalent bonds, the sharing of electrons between atoms is unequal; that is, one atom attracts the shared electrons more strongly than the other. This type of covalent bond is a **polar covalent bond**. The most important example in living systems is the bond between oxygen and hydrogen in a molecule of water (see Fig. 2.6a). When polar covalent bonds form, the resulting molecule has a partial negative charge (written as δ^-) near the atom that attracts electrons more strongly and a partial positive charge (written as δ^+) near at least one other atom. As we will discuss in more detail later in the chapter, polar covalent bonds allow water to dissolve many molecules that are important to life.

Hydrogen Bonds

In a **hydrogen bond** two other atoms (usually oxygen or nitrogen) associate with a hydrogen atom. A hydrogen atom that is covalently bonded to one oxygen or nitrogen atom is also attracted to the partial negative charge of another covalently bonded nitrogen or oxygen atom. Because hydrogen bonds are weak, only about 5% as strong as covalent bonds, they do not bind atoms into molecules. However, they do serve as links between molecules, for example, between water molecules (see Fig. 2.6b) or between various parts of the same molecule (see Fig. 2.16). Even though single hydrogen bonds are weak, large molecules may contain hundreds of these bonds. Thus they confer considerable strength and stability. In addition, hydrogen bonds determine the three-dimensional shape of large molecules, such as proteins and nucleic acids. Usually, a specific molecular shape is needed for a particular molecule to carry out a given function.

Chemical Reactions

Chemical reactions involve the making and breaking of bonds between atoms. Through thousands of reactions, body structures are built and body functions are carried out. Atoms, ions, and molecules all are continuously moving and colliding with one another; they have kinetic energy. A sufficiently forceful collision disrupts electron interactions and can break an existing chemical bond or form a new one. Each chemical reaction requires a specific level of energy. The collision energy needed for a chemical reaction to occur is called the **activation energy**. This is the amount of energy needed to disrupt the stable electronic configuration of a specific molecule so that the electrons can be rearranged. Even if colliding particles have the minimum energy needed to react, no reaction will take place unless the particles are properly oriented toward each other. Two factors influence the chance that a collision will occur and cause a chemical reaction.

1. **Concentration.** The more particles present in a confined region, the greater the chance that they will collide.
2. **Temperature.** Up to a point, the higher the speed of particles, the greater the chance that their collision will result in a reaction. A higher temperature increases the particles' speeds. At the same temperature, smaller particles have higher speeds than larger ones.

After a chemical reaction, *the total number of atoms is the same*. Because the atoms are rearranged, however, there are new molecules with new properties. The term **metabolism** refers to all chemical reactions occurring in an organism. In this section, we will look at the main types of chemical reactions common to all living cells. Once you have learned them, you will be able to understand the chemical reactions discussed later.

Energy Balance in Chemical Reactions

In chemical reactions, breaking bonds requires energy and forming bonds releases energy. Because most chemical reactions involve both breaking old bonds and forming new bonds, the overall reaction may either release energy or require energy. Chemical reactions that *release energy* as they occur are termed **exergonic** (*exo* = outside; *ergon* = work) **reactions**. In an exergonic reaction, the energy released as new bonds form is *greater* than the energy needed to break apart old bonds. On the other hand, reactions that *require energy* to occur are termed **endergonic** (*endon* = inside) **reactions**. In an endergonic reaction, the energy released as new bonds form is *less* than the energy needed to break apart old bonds. In other words, energy must be provided. In general, the reactions that occur as nutrients, such as glucose, are being catabolized (broken down) are exergonic. Some of the energy given off is temporarily stored in a special molecule called **ATP** (**adenosine triphosphate**). If a molecule of glucose is completely broken down, the chemical energy in its bonds can be donated to as many as 38 molecules of ATP. This energy then can be used to drive reactions that lead to the building of body structures, such as your muscles and bones, which most often are endergonic reactions. The energy in ATP can also be used to do mechanical work, such as contraction (shortening) of muscle or movement of substances into or out of cells.

Synthesis Reactions—Anabolism

When two or more atoms, ions, or molecules combine to form new and larger molecules, the process is called a **synthesis reaction**. The word *synthesis* means "to put together." Synthesis reactions can be expressed in the following way:

$$A \quad + \quad B \quad \xrightarrow{\text{Combine to form}} \quad AB$$

Atom, ion, or molecule A   Atom, ion, or molecule B   New molecule AB

The combining substances, A and B, are called the **reactants**; the substance formed by the combination is the **product**. The arrow indicates the direction in which the reaction proceeds. An example of a synthesis reaction is:

$$N_2 \quad + \quad 3H_2 \quad \longrightarrow \quad 2NH_3$$

One nitrogen molecule   Three hydrogen molecules   Two ammonia molecules

All the synthesis reactions that occur in your body are collectively called anabolic reactions, or simply **anabolism** (a-NAB-ō-lizm). Overall, anabolic reactions are usually endergonic (energy-requiring). Combining amino acids to form proteins is an example of anabolism. Anabolism is one aspect of metabolism; catabolism (described next) is the other. The importance of anabolism and catabolism is considered in more detail in Chapter 25.

Decomposition Reactions—Catabolism

In a **decomposition reaction** a chemical is broken down into smaller parts. Large molecules are split up into smaller molecules, ions, or atoms. A decomposition reaction occurs in this way:

$$AB \quad \xrightarrow{\text{Breaks down into}} \quad A \quad + \quad B$$

Molecule AB   Atom, ion, or molecule A   Atom, ion, or molecule B

For example, under the proper conditions, methane can decompose into carbon and hydrogen molecules:

$$CH_4 \quad \longrightarrow \quad C \quad + \quad 2H_2$$

One methane molecule   One carbon atom   Two hydrogen molecules

All the decomposition reactions that occur in your body are collectively called catabolic reactions, or simply **catabolism** (ka-TAB-ō-lizm). Overall, catabolic reactions are usually exergonic (energy-releasing). The breakdown of glucose to pyruvic acid, with the net production of two molecules of ATP, is an example (see Fig. 25.4).

Exchange Reactions

Many reactions, such as **exchange reactions**, are partly synthesis and partly decomposition. One type of exchange reaction works like this:

$$AB \quad + \quad CD \quad \longrightarrow \quad AD \quad + \quad BC$$

The bonds between A and B and between C and D break (decomposition) and new bonds then form (synthesis) between A and D and between B and C. As you will see in Chapter 27, buffer reactions, which help maintain normal acid–base balance, are exchange reactions.

Reversible Reactions

Metabolic reactions may proceed in only one direction, as indicated by the previous right-pointing arrows, or they may be reversible. When a chemical reaction is reversible, the product can revert to the original reactants. A **reversible reaction** is indicated by two arrows:

$$AB \quad \underset{\text{Combine to form}}{\overset{\text{Breaks down into}}{\rightleftharpoons}} \quad A \quad + \quad B$$

Some reactions are reversible only under special conditions:

$$AB \quad \underset{\text{Heat}}{\overset{\text{Water}}{\rightleftharpoons}} \quad A \quad + \quad B$$

Whatever is written above or below the arrows indicates the condition needed for the reaction to occur. In this case, AB breaks down into A and B only when water is added, and A and B react to produce AB only when heat is applied. An example of a reversible reaction is the breakdown and reformation of ATP:

$$ATP \quad \underset{\text{Combine to form}}{\overset{\text{Breaks down into}}{\rightleftharpoons}} \quad ADP \quad + \quad \textcircled{P} \quad + \quad E$$

Adenosine triphosphate   Adenodine diphosphate   Phosphate group   Energy

Oxidation–Reduction Reactions

Oxidation is the *removal of electrons* from a molecule and results in a decrease in the energy content of the molecule. In many cellular oxidations, a hydrogen ion or proton (H^+, a hydrogen nucleus with no orbiting electrons) and a hydride ion (H^-, a hydrogen nucleus with two orbiting electrons) are removed at the same time; this is equivalent to the removal of two hydrogen atoms ($H^+ + H^- = 2H$). Because most biological oxidations involve the loss of hydrogen atoms, they are called *dehydrogenation reactions*. An example of an oxidation is the conversion of lactic acid into pyruvic acid.

$$\begin{array}{c} COOH \\ | \\ HC-OH \\ | \\ CH_3 \end{array} \quad \xrightarrow[\text{(oxidation)}]{\text{Remove 2H (}H^+ + H^-\text{)}} \quad \begin{array}{c} COOH \\ | \\ C=O \\ | \\ CH_3 \end{array}$$

Lactic acid   Pyruvic acid

Reduction is the opposite of oxidation. It is the *addition of electrons* to a molecule and results in an increase in the energy content of the molecule. An example of reduction is the conversion of pyruvic acid into lactic acid.

$$
\begin{array}{ccc}
\text{COOH} & & \text{COOH} \\
| & \xrightarrow[\text{(reduction)}]{\text{Add 2H (H}^+ + \text{H}^-)} & | \\
\text{C}=\text{O} & & \text{HC}-\text{OH} \\
| & & | \\
\text{CH}_3 & & \text{CH}_3 \\
\text{Pyruvic acid} & & \text{Lactic acid}
\end{array}
$$

Within a cell, oxidation and reduction reactions are always coupled; that is, whenever one substance is oxidized, another is almost simultaneously reduced. Such coupled reactions are referred to as **oxidation–reduction (redox) reactions**.

CHEMICAL COMPOUNDS AND LIFE PROCESSES

Most of the chemicals in your body exist in the form of compounds. Biologists and chemists divide these compounds into two principal classes: inorganic compounds and organic compounds. Usually, **inorganic compounds** are small and lack carbon; many contain ionic bonds. They include oxygen, carbon dioxide, water, and many salts, acids, and bases. **Organic compounds**, on the other hand, always contain carbon. Carbon has four electrons in its outer (valence) shell. It can combine with a variety of atoms, including other carbon atoms, to form rings and straight or branched chains. Carbon chains are the backbone for many substances of living cells. Organic compounds are held together mostly or entirely by covalent bonds. Important categories of organic compounds in the body include carbohydrates, lipids, proteins, nucleic acids, and adenosine triphosphate (ATP).

INORGANIC COMPOUNDS

Inorganic Acids, Bases, and Salts

When molecules of inorganic acids, bases, or salts dissolve in water, they undergo **ionization** (ī′-on-i-ZĀ-shun) or **dissociation** (dis′-sō-sē-Ā-shun); that is, they separate into ions.

An **acid** (Fig. 2.5a) may be defined as a substance that dissociates into one or more **hydrogen ions** (H^+) and one or more *anions* (negative ions). Since H^+ is a single proton with one positive charge, an acid may also be defined as a proton donor. A **base**, by contrast (Fig. 2.5b), dissociates into one or more **hydroxide ions** (OH^-) and one or more *cations* (positive ions). A base may also be viewed as a proton acceptor. Hydroxide ions have a strong attraction for protons.

Figure 2.5 Ionization of inorganic acids, bases, and salts.

 Ionization is the separation of inorganic acids, bases, and salts into ions in a solution.

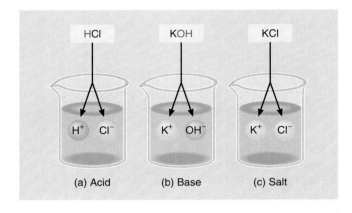

(a) Acid   (b) Base   (c) Salt

Q The compound $CaCO_3$ (calcium carbonate) dissociates into a calcium ion (Ca^{2+}) and a carbonate ion (CO_3^{2-}). Is it an acid, a base, or a salt? What about H_2SO_4, which dissociates into two H^+ and one SO_4^{2-}?

A **salt**, when dissolved in water, dissociates into cations and anions, neither of which is H^+ or OH^- (Fig. 2.5c). Acids and bases react with one another to form salts. For example, the combination of hydrochloric acid (HCl), an acid, and potassium hydroxide (KOH), a base, produces potassium chloride (KCl), a salt, and water (H_2O). This exchange reaction can be written:

$$
\underset{\text{Acid}}{\text{HCl}} + \underset{\text{Base}}{\text{KOH}} \longrightarrow \underset{\text{Salt}}{\text{KCl}} + \underset{\text{Water}}{H_2O}
$$

Solid forms of salts are present in bones and teeth, where they contribute to the great strength of these tissues. The ions of salts also provide many essential chemical elements in intracellular fluid and extracellular fluids such as lymph, blood, and the interstitial fluid of tissues.

Water and Solutions

Water is an important, and the most abundant, inorganic substance in all living systems. In body fluids the solvent is water. A **solvent** is a liquid or gas in which some other material (solid, liquid, or gas), called a **solute**, has been dissolved. The combination of solvent plus solute is called a **solution**. Common examples of solutions are milk and tears. Solutes, such as the calcium salts and proteins in milk, do not settle out and accumulate on the bottom of the container. In a **suspension**, by contrast, the suspended material may mix with the liquid or suspending medium for some time, but it will eventually settle out. An example of a suspension is blood. When freshly drawn from the body, blood has an even reddish color. After it sits for a while in a test tube, the top layer appears clear. This is the liquid portion of blood, called plasma. Red blood cells that have settled out of the suspension drift to the bottom of the tube, making this layer look red.

A few tissues, such as tooth enamel and bone, have very little water. In most tissues, water predominates—accounting for 60% of red blood cells, 75% of muscle tissue, and 92% of blood plasma. Giving a percentage is one way to express the **concentration** of a solution, which specifies the relative amounts of water and solutes. Another common way to express a solution's concentration is in chemical units called **moles per liter (mol/liter)**, which relate to the total number of molecules in a given volume of solution. Exhibit 2.2 describes these two ways of expressing concentration.

Water has many properties that explain why it is such a prevalent and important compound in living systems. These include:

1. **Water participates in chemical reactions.** During digestion, for example, water can be added to large nutrient molecules to break them down into smaller molecules (see Fig. 2.8a). This kind of breakdown (called hydrolysis and described shortly) is necessary if the

body is to use the energy in nutrients. On the other hand, when two smaller molecules join to form one larger one in a synthesis reaction (called dehydration synthesis and described shortly), a water molecule is removed. Such reactions occur in the production of proteins and other large molecules (see Fig. 2.13).

2. **Water has a high heat capacity.** In comparison to most substances, water has a high capacity for absorbing heat with only a modest increase in temperature. It can also give off a great deal of heat with only a small decrease in temperature. The high heat capacity of water is the reason that it is used in automobile radiators; it cools the engine by absorbing heat without its temperature rising to an unacceptably high level. The large amount of water in a living organism lessens the impact of environmental temperature changes and thereby helps to maintain the homeostasis of body temperature.

3. **Water requires a large amount of heat to change from a liquid to a gas.** When water (perspiration) evaporates from the skin, it takes with it large quantities of heat and provides an excellent cooling mechanism. Water's heat of vaporization is high.

4. **Water serves as a lubricant.** It is a major part of mucus and other lubricating fluids. Lubrication is especially necessary in the chest and abdomen, where internal organs touch and slide over one another. It is also needed at joints, where bones, ligaments, and tendons rub against one another. In the gastrointestinal tract, water in mucus moistens foods, which aids their smooth passage.

5. **Water is an excellent solvent and suspending medium.** Alchemists in medieval times tried to find a universal solvent, a substance that would dissolve all other materials. They found nothing that worked as well as water. Although it is the most versatile solvent known, it is not a "universal solvent." If it were, no container could hold it; the container would dissolve!

    The versatility of water as a solvent for ionized or polar substances is due to its polar covalent bonds, in which electrons are not shared equally between atoms, and its bent shape. In a molecule of water, there are both positive and negative areas (Fig. 2.6a). When two hydrogen atoms bond covalently to an oxygen atom, the shared electrons spend more time around oxygen than hydrogen. Since electrons have a negative charge, the unequal sharing causes the oxygen atom to have a slight negative charge (δ^-) and each hydrogen atom to have a slight positive charge (δ^+).

    To understand the dissolving capability of water, consider what happens when a crystal of a salt such as sodium chloride (NaCl) is placed in water. The sodium and chloride ions at the surface of the salt crystal are exposed to the polar water molecules. The negatively charged oxygen portion of water molecules is attracted to the sodium ion (Na^+) of the salt. And the positively charged hydrogen portions of water molecules are attracted to the chloride ions (Cl^-) of the salt (Fig. 2.6c).

E X H I B I T **2.2** **WAYS TO EXPRESS THE CONCENTRATION OF SOLUTIONS**

Definition	Example
Percent (weight per volume) Number of grams of a substance per 100 milliliters (ml) of solution.	To make a 10% NaCl solution, take 10 g of NaCl (common table salt) and add enough water to make a total of 100 ml of solution.
Moles 1 mole = weight, in grams, of combined atomic weights of atoms that make up a molecule or compound.	1 mole of NaCl = 22.9 g (atomic weight of Na) + 34.45 g (atomic weight of Cl) = 58.44 g of NaCl.
Moles per liter 1 mole per liter = 1 mol/liter. Dissolve 1 mole of a substance in enough water to make 1 liter total of solution.	Take 58.44 g NaCl ; add enough water to make a total of 1 liter of solution = 1 mol/liter NaCl.
Millimoles per liter 1 millimole/liter = 1 mmol/liter = $\frac{1}{1000}$ mol/liter = 0.001 mol/liter	Take 58.44 mg NaCl = 0.058 g NaCl; add enough water to make a total of 1 liter of solution = 1 mmol/liter NaCl.

Figure 2.6   Polar water molecules dissolve salts and polar substances. Because the oxygen nucleus attracts the shared electrons more strongly (a), the oxygen end of a water molecule has a partial negative charge (written δ^-) and the hydrogen ends have a partial positive charge (written δ^+). When a crystal of sodium chloride is dropped into water (c), the slightly negative oxygen end of water molecules is attracted to the positive sodium ions (Na^+); the slightly positive hydrogen portions of the water molecules are attracted to the negative chloride ions (Cl^-).

 Water is a versatile solvent due to its polar covalent bonds, in which electrons are shared unequally.

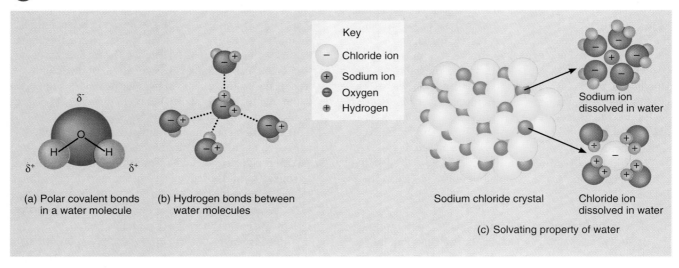

(a) Polar covalent bonds in a water molecule

(b) Hydrogen bonds between water molecules

Key
— Chloride ion
+ Sodium ion
⊖ Oxygen
⊕ Hydrogen

Sodium ion dissolved in water

Sodium chloride crystal

Chloride ion dissolved in water

(c) Solvating property of water

Table sugar (sucrose) easily dissolves in water but is not an electrolyte. Is it likely that all the covalent bonds between atoms in table sugar are nonpolar bonds?

Soon, water molecules surround some Na^+ and Cl^- and separate them from each other. In this way, water molecules dissolve and dissociate (take apart) the ions of the salt.

The solvating and suspending properties of water are essential to health and survival. Since water can dissolve or suspend so many different substances, it is an ideal medium for metabolic reactions. Because they are together in a common fluid, reactants and other necessary materials, such as enzymes, readily collide to form products. Water also dissolves waste products and flushes them out of the body in the urine.

Acid–Base Balance: The Concept of pH

To ensure homeostasis, intracellular and extracellular fluids must contain almost balanced quantities of acids and bases. The more hydrogen ions (H^+) dissolved in a solution, the more acidic the solution; conversely, the more hydroxide ions (OH^-), the more basic (alkaline) the solution. Chemical reactions are very sensitive to even small changes in the acidity or alkalinity. Any departure from the narrow limits of normal H^+ and OH^- concentrations greatly disrupts body functions.

A solution's acidity or alkalinity is expressed on the **pH scale**, which runs from 0 to 14 (Fig. 2.7). This scale is based on the concentration of H^+ in moles per liter. A pH of 7

means that a solution contains one ten-millionth (0.0000001) of a mole of hydrogen ions per liter. The number 0.0000001 is written as 10^{-7} in scientific notation (exponential form), which indicates that the number is one with the decimal point moved 7 places to the left. To convert this value to pH, the negative exponent (-7) is changed to a positive number (7). A solution with a H^+ concentration of 0.0001 (10^{-4}) moles per liter has a pH of 4; a solution with a H^+ concentration of 0.000000001 (10^{-9}) moles per liter has a pH of 9; and so on.

The midpoint in the scale is 7, where the concentrations of H^+ and OH^- are equal. A substance with a pH of 7, such as distilled (pure) water, is neutral. A solution that has more H^+ than OH^- is an **acidic solution** and has a pH below 7. A solution that has more OH^- than H^+ is a **basic (alkaline) solution** and has a pH above 7. A change of one whole number on the pH scale represents a tenfold change from the previous concentration. A pH of 1 denotes 10 times more H^+ than a pH of 2. A pH of 3 indicates 10 times fewer H^+ than a pH of 2 and 100 times fewer H^+ than a pH of 1.

Maintaining pH: Buffer Systems

Although the pH of body fluids may differ, the normal limits for the various fluids are generally quite specific and narrow. Exhibit 2.3 shows the pH values for certain body fluids

Figure 2.7 The pH scale. A pH below 7 indicates an acidic solution; that is, there are more H+ than OH−. The lower the numerical value of the pH, the more acidic is the solution because the H+ concentration becomes progressively greater. A pH above 7 indicates a basic (alkaline) solution; that is, there are more OH− than H+. The higher the pH, the more basic the solution.

🔑 *At pH 7 (neutrality), the concentrations of H+ and OH− are equal (10⁻⁷ mol/liter).*

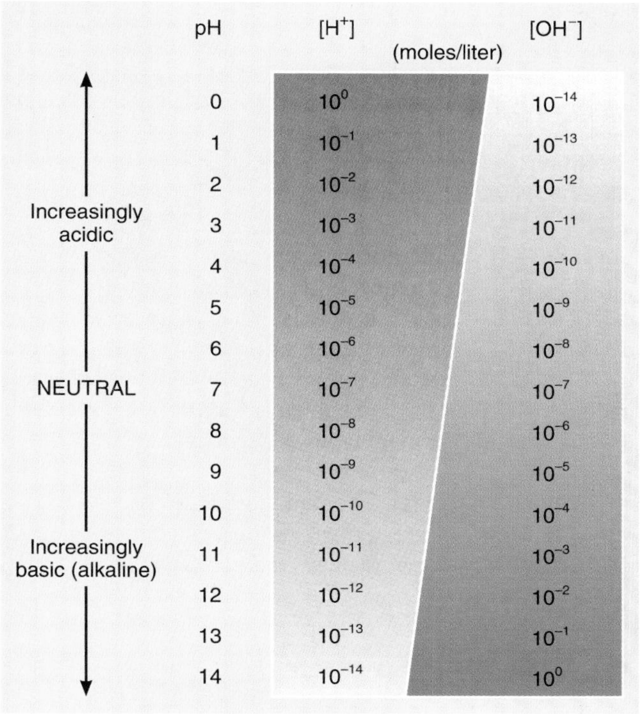

pH	$[H^+]$	$[OH^-]$
		(moles/liter)
0	10^0	10^{-14}
1	10^{-1}	10^{-13}
2	10^{-2}	10^{-12}
3	10^{-3}	10^{-11}
4	10^{-4}	10^{-10}
5	10^{-5}	10^{-9}
6	10^{-6}	10^{-8}
7	10^{-7}	10^{-7}
8	10^{-8}	10^{-6}
9	10^{-9}	10^{-5}
10	10^{-10}	10^{-4}
11	10^{-11}	10^{-3}
12	10^{-12}	10^{-2}
13	10^{-13}	10^{-1}
14	10^{-14}	10^0

Increasingly acidic

NEUTRAL

Increasingly basic (alkaline)

❓ What is the concentration of H+ and OH− at pH 6? Which pH is more acidic, 6.82 or 6.91? Which pH is closer to neutral, 8.41 or 5.59?

EXHIBIT 2.3 pH VALUES OF SELECTED SUBSTANCES

Substance	pH Value
Gastric juice (digestive juice of the stomach)[a]	1.2–3.0
Lemon juice	2.2–2.4
Grapefruit juice, vinegar, wine	3.0
Carbonated soft drink	3.0–3.5
Vaginal fluid[a]	3.5–4.5
Pineapple juice, orange juice	3.5
Tomato juice	4.2
Coffee	5.0
Urine[a]	4.6–8.0
Saliva[a]	6.35–6.85
Milk	6.6–6.9
Distilled (pure) water	7.0
Blood[a]	7.35–7.45
Semen (fluid containing sperm)[a]	7.20–7.60
Cerebrospinal fluid (fluid associated with nervous system)[a]	7.4
Pancreatic juice (digestive juice of the pancreas)[a]	7.1–8.2
Eggs	7.6–8.0
Bile (liver secretion that aids fat digestion)[a]	7.6–8.6
Milk of magnesia	10.0–11.0
Lye	14.0

[a] Substances in the human body.

compared with common substances. Homeostatic mechanisms maintain the pH of blood between 7.35 and 7.45, slightly more basic than pure water. Saliva is slightly acidic, and semen is slightly basic. Because the kidneys help remove excess acid from the body, urine can be quite acidic. Even though strong acids and bases are continually taken into and formed by the body, the pH of fluids inside and outside cells remains almost constant. One important reason is the presence of **buffer systems**.

The function of a buffer system is to convert strong acids or bases, which are relatively unstable and ionize easily, into weak acids or bases, which are relatively stable and do not ionize easily. Strong acids (or bases) ionize easily and contribute many H+ (or OH−) to a solution. They therefore change the pH drastically, which may cause considerable damage. Weak acids (or bases) do not dissociate as much. They contribute fewer H+ (or OH−) and have less effect on

the pH. The chemicals that replace strong acids or bases with weak ones are called **buffers**.

One important buffer system is the **carbonic acid–bicarbonate buffer system**. It includes the bicarbonate ion (HCO_3^-), which can act as a weak base, and carbonic acid, which can act as a weak acid. Thus the buffer system can compensate for either an excess or a shortage of H+. For example, if there is an excess of H+ (an acidic condition), HCO_3^- can function as a weak base and remove the excess H+ as follows:

$$H^+ \;+\; HCO_3^- \longrightarrow H_2CO_3 \longrightarrow H_2O \;+\; CO_2$$

Hydrogen ion   Bicarbonate ion (weak base)   Carbonic acid   Water   Carbon dioxide

On the other hand, if there is a shortage of H+ ions (an alkaline condition), H_2CO_3 can function as a weak acid and provide H+ as follows:

$$H_2CO_3 \longrightarrow H^+ \;+\; HCO_3^-$$

Carbonic acid (weak acid)   Hydrogen ion   Bicarbonate ion

Buffers and their roles in maintaining acid–base balance are described in more detail in Chapter 27 on page 900.

ORGANIC COMPOUNDS

As noted earlier, organic compounds always contain carbon, which can form four covalent bonds. Other elements most often found in organic compounds are hydrogen (one bond), oxygen (two bonds), and nitrogen (three bonds). Sulfur (two bonds) and phosphorus (five bonds) appear less often. Additional elements are present but only in a few organic compounds.

Carbon has several properties that make it particularly useful to living organisms. For one thing, it can react with one to thousands of other carbon atoms to form large molecules of many different shapes. This means that the body can build many compounds out of carbon, hydrogen, and oxygen, each of which has a particular structure or function. The large size of most carbon-containing molecules and the fact that some do not dissolve easily in water make them useful materials for building body structures.

Carbon compounds are mostly or entirely held together by covalent bonds and tend to decompose easily. This means that organic compounds are also a good source of energy. Ionic compounds are not good energy sources because they form new ionic bonds as soon as the old ones are broken.

Small organic molecules can combine into very large molecules called **macromolecules**. Macromolecules are usually **polymers**. A polymer is a large molecule formed by covalent bonding of many repeating small building-block molecules called **monomers** (**subunits**). When two monomers join together, the reaction usually involves the elimination of a molecule of water. This type of reaction is called **dehydration** (*dehydration* = loss of water) **synthesis** because a molecule of water is released. Macromolecules such as carbohydrates, lipids, proteins, and nucleic acids are assembled in the cell by dehydration synthesis. It is also possible for macromolecules to be broken down into monomers by the addition of water. This type of reaction is called **hydrolysis** (hī-DROL-i-sis), which means to split apart by using water. Molecules that have the same molecular formula but different structures are called **isomers** (Ī-so-merz). For example, the molecular formula for both glucose and fructose is $C_6H_{12}O_6$ but the individual atoms are positioned differently (see Fig. 2.8).

Carbohydrates

Carbohydrates include sugars, starches, glycogen, and cellulose. They are a large and diverse group of organic compounds and have several functions, although they represent only 2–3% of your total body weight. Plants store carbohydrate as starch and use the carbohydrate cellulose to build the cell wall. Cellulose is the most plentiful organic substance on earth. Although humans eat cellulose, they do not digest it. It does, however, create bulk, which helps to move food and wastes along the gastrointestinal tract. In animals the principal function of carbohydrates is to provide a readily available source of energy to drive metabolic reactions. Only a few carbohydrates form structural units in animals. One example is deoxyribose, a type of sugar. It is a building block of deoxyribonucleic acid (DNA), the molecule that carries hereditary information (see Fig. 2.16a).

Some carbohydrates are converted to other substances, which are used to build structures and to generate ATP. For example, the sugar in a donut you eat for breakfast could be used to power your walk to work. Other carbohydrates function as food reserves. The prime storage carbohydrate in animals is glycogen, which is stored in the liver and skeletal muscles.

Carbon, hydrogen, and oxygen are the elements found in carbohydrates. The ratio of hydrogen to oxygen atoms is usually 2:1, the same as in water. Although there are exceptions, the general rule for carbohydrates is one carbon atom (C) for each water molecule (H_2O), which is why they are called carbohydrates (= watered carbon). Carbohydrates are divided into three major groups based on size: monosaccharides, disaccharides, and polysaccharides. Monosaccharides and disaccharides also are known as **simple sugars**. Exhibit 2.4 summarizes the main types of carbohydrates.

1. **Monosaccharides.** The building blocks (monomers) of carbohydrates are termed **monosaccharides** (mon′-ō-SAK-a-rīds; *mono* = one; *sakcharon* = sugar). They

---

EXHIBIT **2.4**   SUMMARY OF CARBOHYDRATES

Type of Carbohydrate	Description
Monosaccharides	The building blocks (monomers) of carbohydrates; include *glucose* (the main energy-supplying molecule of the body), *fructose* (found in fruits), *galactose* (present in milk), *deoxyribose* (in DNA), and *ribose* (in RNA).
Disaccharides	Formed by joining together two monosaccharides in a dehydration synthesis reaction; most common in food are *sucrose* or table sugar (glucose + fructose) and *lactose* or milk sugar (glucose + galactose).
Polysaccharides	Large carbohydrates composed of many monosaccharide monomers; include *glycogen* (main stored form of carbohydrate in animals), *starch* (main carbohydrate in food), and *cellulose* (not digested by humans but adds bulk, which aids movement of food through intestines).

contain from three to seven carbon atoms. Those with three carbons are called trioses. The number of carbon atoms in the molecule is indicated by the prefix, *tri*. There are also tetroses (four-carbon sugars), pentoses (five-carbon sugars), hexoses (six-carbon sugars), and heptoses (seven-carbon sugars). Glucose, a hexose, is the main energy-supplying molecule of the body.

2. **Disaccharides.** Two monosaccharide molecules can combine by dehydration synthesis to form one **disaccharide** (dī-SAK-a-rid; *di* = two) molecule and a molecule of water (Fig. 2.8). For example, molecules of the monosaccharides glucose and fructose combine to form a molecule of the disaccharide sucrose (table sugar) as follows:

$$C_6H_{12}O_6 \ + \ C_6H_{12}O_6 \ \longrightarrow \ C_{12}H_{22}O_{11} \ + \ H_2O$$

Glucose   Fructose   Sucrose

(monosaccharide)   (monosaccharide)   (disaccharide)   Water

Although glucose and fructose have the same molecular formula, they are different monosaccharides because the relative positions of the oxygens and carbons are different. Look at Fig. 2.8a to appreciate this difference. Also, the formula for sucrose is $C_{12}H_{22}O_{11}$, not $C_{12}H_{24}O_{12}$, because a molecule of water is split out

as the two monosaccharides are joined. In every dehydration synthesis reaction, a molecule of water is lost.

Disaccharides can also be split into smaller, simpler molecules by adding water. This reverse chemical reaction is called hydrolysis. A molecule of sucrose, for example, may be hydrolyzed into its components of glucose and fructose by the addition of water. Figure 2.8 also shows this reaction.

Another important disaccharide is lactose, or milk sugar. It consists of the monosaccharides glucose and galactase. As you will discover in Chapter 24, some people cannot digest lactose, a condition called lactose intolerance.

3. **Polysaccharides.** The third major group of carbohydrates is the **polysaccharide** (pol'-ē-SAK-a-rid; *poly* = many) family. These large carbohydrates contain tens or hundreds of monosaccharides joined through dehydration synthesis reactions. The principal polysaccharide in the human body is glycogen, which is composed of glucose units linked to each other and is stored in the liver and skeletal muscles. Like disaccharides, polysaccharides can be broken down into monosaccharides through hydrolysis reactions. For example, when the blood glucose level falls, liver cells have the ability to

---

Figure 2.8 The monosaccharides glucose and fructose and the disaccharide sucrose. (a) In dehydration synthesis (read from left to right), two smaller molecules, glucose and fructose, are joined to form a larger molecule of sucrose. Note the loss of a water molecule. In hydrolysis (read from right to left), the larger sucrose molecule is broken down into two smaller molecules, glucose and fructose. Here, a molecule of water is added to sucrose for the reaction to occur.

🔑 *Monosaccharides are the building blocks (monomers) of carbohydrates.*

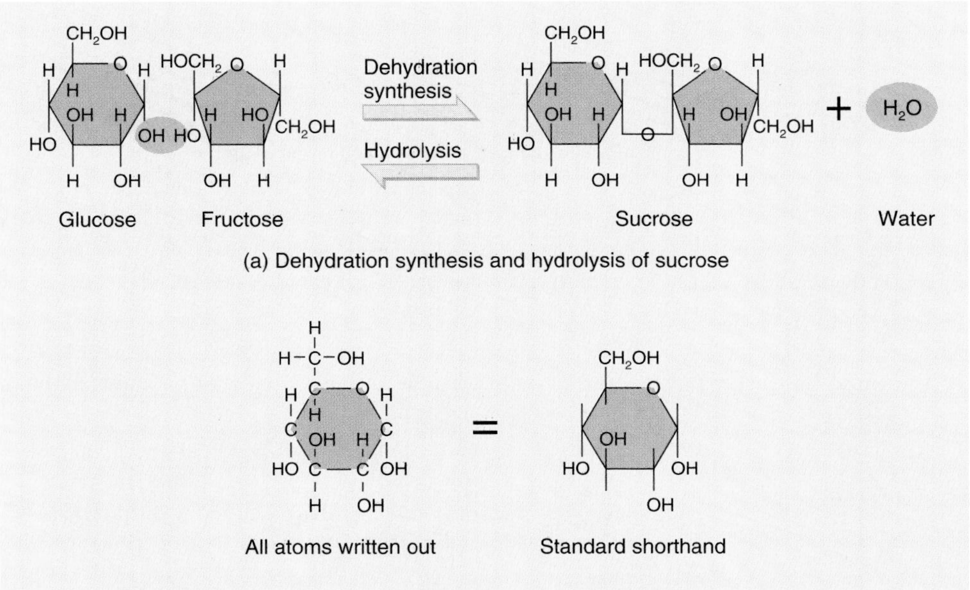

(a) Dehydration synthesis and hydrolysis of sucrose

(b) Alternate chemical structures of organic molecules (shown here is glucose)

❓ Is dehydration synthesis an anabolic or a catabolic reaction? How many carbons can you count in fructose? In sucrose?

break down glycogen into glucose and release it into the blood. In this way, glucose is made available to body cells, where it can be broken down to synthesize ATP. Unlike simple sugars such as fructose and sucrose, however, polysaccharides usually are not soluble in water and are not sweet.

Lipids

A second group of important organic compounds is lipids (*lipos* = fat). Lipids comprise 18–25% of body weight in lean adults. Like carbohydrates, lipids contain carbon, hydrogen, and oxygen. Unlike carbohydrates, they do not have a 2:1 ratio of hydrogen to oxygen. The amount of oxygen in lipids is usually less than that in carbohydrates, so there are fewer polar covalent bonds. As a result, most lipids are insoluble in polar solvents such as water; they are **hydrophobic** (*hydro* = water; *phobic* = fearing). Nonpolar solvents such as chloroform and ether, however, readily dissolve lipids. Because they are hydrophobic, only the smallest lipids (some fatty acids) can travel freely in the watery blood. For blood transport, most lipids are combined with proteins to form water-soluble **lipoproteins**.

The diverse lipid family includes triglycerides (fats and oils), phospholipids (lipids that contain phosphorus), steroids (lipids that contain rings of carbon atoms), eicosanoids (20-carbon lipids), fatty acids, fat-soluble vitamins (vitamins A, D, E, and K), and lipoproteins. Exhibit 2.5 summarizes the various types of lipids and highlights their roles in the human body.

1. **Triglycerides.** The most plentiful lipids in your body and in your diet are the **triglycerides** (tri-GLI-cer-īdes), also called neutral fats. At room temperature, triglycerides may be either solids (fats) or liquids (oils), and they are the body's most highly concentrated form of chemical energy. Triglycerides provide more than twice as much energy per gram as either carbohydrates or proteins. Our capacity to store triglycerides in adipose (fat) tissue is unlimited, for all practical purposes. Excess dietary carbohydrates, proteins, fats, and oils all have the same fate; they are deposited in adipose tissue as triglycerides.

   A triglyceride consists of two types of building blocks: a single glycerol and three fatty acid molecules. A three-carbon **glycerol** molecule forms the backbone of a triglyceride (Fig. 2.9). Three **fatty acids** are attached, by dehydration synthesis reactions, one to each

---

EXHIBIT 2.5 TYPES OF LIPIDS IN THE BODY

Type of Lipid	Function
Triglycerides (fats and oils)	Protection, insulation, major energy storage molecules in the body; can be broken down by hydrolysis into glycerol and three fatty acids, which then may be catabolized to provide ATP.
Phospholipids	Major lipid component of cell membranes; found in high concentrations in the nervous system.
Steroids	
Cholesterol	Constituent of all animal cell membranes; precursor of bile salts, vitamin D, and steroid hormones.
Bile salts	Substances that emulsify or suspend fats before their digestion and absorption; needed for absorption of fat-soluble vitamins (A, D, E, and K).
Vitamin D	Hormone that aids in regulation of calcium level in the body; necessary for bone growth and repair.
Adrenocortical hormones	Most important are cortisol, which helps regulate metabolism, resistance to stress, and inflammation, and aldosterone, which helps regulate salt and water balance in the body.
Sex hormones	Estrogens and progesterone (produced in large quantities by females) and testosterone (produced in large quantity by males) stimulate reproductive functions and sexual characteristics.
Eicosanoids	Membrane-associated lipids made by most cells in the body that have diverse effects on blood clotting, inflammation, immunity, stomach acid secretion, airway diameter, lipid breakdown, and smooth muscle contraction in the uterus and gastrointestinal tract.
Other lipids	
Fatty acids	Important energy-supplying molecules that can be catabolized to provide adenosine triphosphate (ATP) or used to synthesize triglycerides and phospholipids.
Carotenes	Carotenes are needed for synthesis of vitamin A, which is then used to make photopigments (visual pigments) in the retina of the eye.
Vitamin E	May promote wound healing, prevent scarring, and contribute to the normal structure and function of the nervous system; functions as an antioxidant, by preventing the oxidation of certain molecules in the body.
Vitamin K	Required in the synthesis of certain proteins needed for blood clotting.
Lipoproteins	Several types of lipid and protein particles in the blood that help transport lipids to the liver and to adipose (fat) tissue, carry cholesterol to tissues, and remove excess cholesterol from the blood.

Figure 2.9 Triglycerides. Structure and reactions of (a) glycerol and (b) a fatty acid. Each time a glycerol and a fatty acid are joined in dehydration synthesis, a molecule of water is lost. (c) Triglycerides consist of one molecule of glycerol joined to three molecules of fatty acids, which vary in length and the number and location of double bonds between carbon atoms (C=C). Shown here is a triglyceride molecule that contains two saturated fatty acids and a monounsaturated fatty acid. The kink (bend) in the oleic acid occurs at the double bond.

Glycerol and fatty acids are the building blocks of triglycerides.

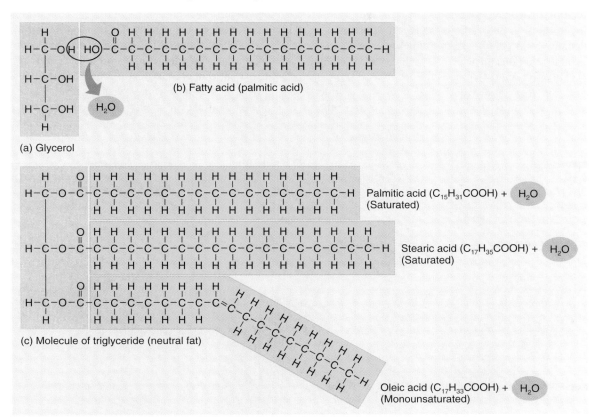

(a) Glycerol

(b) Fatty acid (palmitic acid)

Palmitic acid ($C_{15}H_{31}COOH$) + H_2O
(Saturated)

Stearic acid ($C_{17}H_{35}COOH$) + H_2O
(Saturated)

(c) Molecule of triglyceride (neutral fat)

Oleic acid ($C_{17}H_{33}COOH$) + H_2O
(Monounsaturated)

 During a dehydration synthesis reaction, a molecule of water is removed. Does the oxygen in the water come from the glycerol or from a fatty acid?

carbon of the glycerol backbone. A reverse reaction, hydrolysis, breaks down a single molecule of a triglyceride into three fatty acids and glycerol.

In later chapters, we will refer to saturated, monounsaturated, and polyunsaturated fats. **Saturated** fats are triglycerides that contain *only single* covalent bonds between fatty acid carbon atoms. They do not contain any double bonds between fatty acid carbon atoms and thus each carbon atom is *saturated with hydrogen atoms* (for example, palmitic acid and stearic acid in Fig. 2.9c). Triglycerides with many saturated fatty acids tend to be solid at room temperature and occur mostly in animal tissues. They also occur in a few plant products, such as cocoa butter, palm oil, and coconut oil.

Monounsaturated fats contain fatty acids with one double covalent bond between two fatty acid carbon atoms and thus are not completely saturated with hydrogen atoms (for example, oleic acid in Fig. 2.9c).

Olive oil and peanut oil are rich in triglycerides with monounsaturated fatty acids.

Polyunsaturated fats *contain more than one* double covalent bond between fatty acid carbon atoms. An example is linoleic acid. Corn oil, safflower oil, sunflower oil, cottonseed oil, sesame oil, and soybean oil contain a high percentage of polyunsaturated fatty acids.

SATURATED FATS AND ATHEROSCLEROSIS

Atherosclerosis is a progressively worsening disorder in which fatty plaques (deposits) form in the walls of arteries. Plaque buildup narrows the passage and severely limits blood flow in advanced cases. Although many factors contribute, people who eat a diet high in saturated fats run a greater risk of developing atherosclerosis than do people who eat a diet lower in saturated fats. Also, a diet high in saturated fats causes blood cholesterol (a steroid, described next) to rise, and blood cholesterol contributes to fatty plaque formation. To reduce dietary intake of saturated fats and cholesterol, eat less red meat (beef, pork, and lamb) and fat-laden dairy products (whole milk, butter, and cheese). ∎

2. **Phospholipids.** Like triglycerides, phospholipids have a glycerol backbone. Two fatty acid chains are attached to the first two carbons. In the third position, a phosphate group (PO_4^{3-}) links a small charged group that usually contains nitrogen (N) to the backbone (Fig. 2.10). This portion of the molecule (the "head") is polar and can form hydrogen bonds with water molecules. The two fatty acids (the "tails"), on the other

hand, are nonpolar and can interact only with other lipids. Molecules that have both polar and nonpolar portions are said to be **amphipathic** (am-fi-PATH-ic). Phospholipids, which have this unusual property, line up tails-to-tails in a double row to make up much of the membrane that surrounds each cell (Fig. 2.10c).

3. **Steroids.** Steroids such as cholesterol, sex hormones, cortisol, bile salts, and vitamin D, have four rings of carbon atoms (Fig. 2.11). Their structure differs considerably from that of the triglycerides, but they also are nonpolar, fat-soluble molecules. Although cholesterol can contribute to buildup of fatty plaques in atherosclerosis, it also serves important functions—as a component of animal cell membranes and as the starting mate-

Figure 2.10   Phospholipids. (a) In the synthesis of phospholipids, two fatty acids attach to the first two carbons of the glycerol backbone. A phosphate group links a small charged group to the third carbon in glycerol. In (b), the circle represents the polar head region and the two wavy lines represent the two nonpolar tails. Double bonds in the fatty acid hydrocarbon chain often form a kink in the tail.

🔑 *Phospholipids are amphipathic molecules, having both polar and nonpolar regions.*

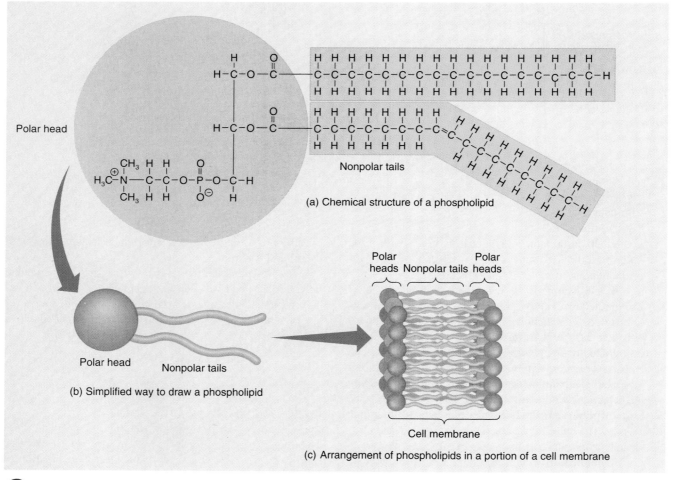

(a) Chemical structure of a phospholipid

(b) Simplified way to draw a phospholipid

(c) Arrangement of phospholipids in a portion of a cell membrane

Ⓠ Which portion of a phospholipid will be hydrophilic and which portion will be hydrophobic?

Figure 2.11 Steroids. All steroids have four rings of carbon atoms (labeled A–D).

Cholesterol is the starting material for synthesis of other steroids in the body.

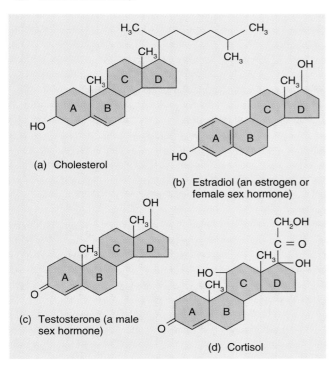

(a) Cholesterol

(b) Estradiol (an estrogen or female sex hormone)

(c) Testosterone (a male sex hormone)

(d) Cortisol

How is the structure of estradiol different from that of testosterone?

rial for synthesis of other steroids. These include bile salts, vitamin D, adrenocortical hormones (produced by the outer portion or cortex of the adrenal gland) and sex hormones (see Exhibit 2.5).

4. **Eicosanoids** (ī-KŌ-sa-noids). These lipids are derived from a 20-carbon fatty acid called arachidonic acid. The two principal subclasses of eicosanoids are the prostaglandins (pros′-ta-GLAN-dins) and leukotrienes (loo′-kō-TRĪ-ēns). Prostaglandins are involved in a wide variety of body functions. They modify responses to hormones, contribute to the inflammatory response (Chapter 22), prevent stomach ulcers, dilate (enlarge) airways to the lungs, regulate body temperature, and influence formation of blood clots, to name just a few effects. Leukotrienes participate in allergic and inflammatory responses.

5. **Other lipids.** Body lipids also include fatty acids (which can undergo either hydrolysis to provide ATP or dehydration synthesis to build triglycerides and phospholipids), carotenes (yellow-orange pigments in egg yolk, carrots, and tomatoes that are needed for vision), vitamins E and K, and lipoproteins.

Proteins

Proteins, a third principal group of organic compounds, are much more complex in structure and have a larger range of functions than carbohydrates or lipids. A normal, lean adult body is 12–18% protein. Some proteins have a structural role—they are cellular building materials. Other proteins have physiological roles. For example, proteins in the form of enzymes speed up most essential biochemical reactions. Other proteins provide the machinery for muscle contraction. Antibodies are proteins that defend against invading microbes. Some hormones that regulate homeostasis also are proteins. Exhibit 2.6 describes several functions of proteins.

Amino Acids and Polypeptides

Proteins always contain carbon, hydrogen, oxygen, and nitrogen. Many proteins also contain sulfur. Just as monosaccharides are the building blocks of polysaccharides, **amino acids** are the building blocks (monomers) of proteins. Each

EXHIBIT **2.6** **FUNCTIONS OF PROTEINS**

Type of Protein	Functions
Structural	Form the structural framework of various parts of the body. *Examples:* collagen in bone and other connective tissues and keratin in the skin, hair, and fingernails.
Regulatory	Function as hormones, regulate various physiological processes, control growth and development, and mediate responses of the nervous system. *Examples:* insulin, which regulates blood glucose level, and substance P, which mediates the sensation of pain in the nervous system.
Contractile	Allow shortening of muscle tissue, which produces movement. *Examples:* myosin and actin.
Immunological	Aid responses that protect the body against foreign substances and invading pathogens. *Examples:* antibodies and interleukins.
Transport	Carry vital substances throughout the body. *Example:* hemoglobin, which transports most oxygen and some carbon dioxide in the blood.
Catalytic	Act as enzymes and function in regulating biochemical reactions. *Examples:* salivary amylase, lipase, and lactase.

of the 20 different amino acids has three important groups attached to a central carbon atom (Fig. 2.12a): (1) an amino group ($-NH_2$), (2) a carboxyl (acid) group ($-COOH$), and (3) a side chain (R group). At the normal pH of body fluids, both the amino group and the carboxyl group are ionized (Fig. 2.12b). The distinctive side chain gives each amino acid its individual identity (Fig. 2.12c).

Synthesis of a protein takes place in stepwise fashion—one amino acid is joined to a second, a third is then added to the first two, and so on. The covalent bond between each pair of amino acids is called a **peptide bond**. It always forms between the carboxyl group ($-COOH$) of one amino acid and the amino group ($-NH_2$) of another. At the site of formation of a peptide bond, a molecule of water is removed (Fig. 2.13). Thus this is a dehydration synthesis reaction. When two amino acids combine, a **dipeptide** results. Adding another amino acid to a dipeptide produces a **tripeptide**. Further additions of amino acids result in the formation of a chainlike **peptide** (4–10 amino acids) or **polypeptide** (10–2000 or more amino acids). Small proteins contain as few as 50 amino acids. A protein may have only one polypeptide chain or several. For example, there are four polypeptide chains in hemoglobin—the molecule that transports oxygen in the blood (see Fig. 19.3b).

A great variety of proteins is possible because each variation in the number or sequence of amino acids can produce a different protein. The situation is similar to using an alphabet of 20 letters to form words. Each letter is like a different amino acid, and each word is like a different peptide, polypeptide, or protein.

Levels of Structural Organization

Proteins exhibit four levels of structural organization. The **primary structure** is the unique sequence of amino acids making up a polypeptide strand (Fig. 2.14a). It is genetically determined. A change in the primary structure can have serious consequences. In sickle cell anemia, for example, a nonpolar amino acid (valine) replaces a polar amino acid (glutamate) at just two locations in the protein hemoglobin.

Figure 2.12   Amino acids. (a) In keeping with their name, amino acids have an amino group and a carboxyl (acid) group. The side chain (R group) is different in each amino acid. (b) At pH close to 7, the amino group and the carboxyl group both are ionized. (c) Glycine is the simplest amino acid; the side chain is merely an H atom. Cysteine is one of two amino acids that contain sulfur (S). The side chain in tyrosine contains a six-carbon ring. Lysine has a second amino group at the end of the side chain.

 Each different amino acid has a unique side chain.

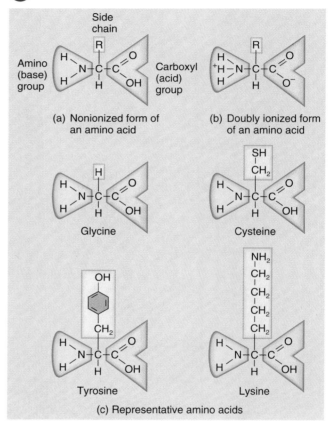

(c) Representative amino acids

In an amino acid, what is the minimum number of carbon atoms? Nitrogen atoms?

Figure 2.13   Formation of a peptide bond between two amino acids during dehydration synthesis. In this example, glycine is joined to alanine (forming a dipeptide).

 Amino acids are the building blocks (monomers) of proteins.

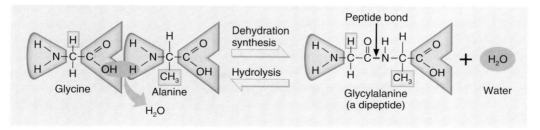

What type of reaction is used in protein catabolism?

Figure 2.14 Levels of structural organization in proteins. (a) The primary structure is the sequence of amino acids in the polypeptide. (b) Common secondary structures include helixes and pleated sheets. (c) The tertiary structure is the overall folding pattern that produces a distinctive, three-dimensional shape. (d) The quaternary structure in a protein is the arrangement of two or more polypeptide chains relative to one another.

 The unique shape of each protein permits it to carry out specific functions.

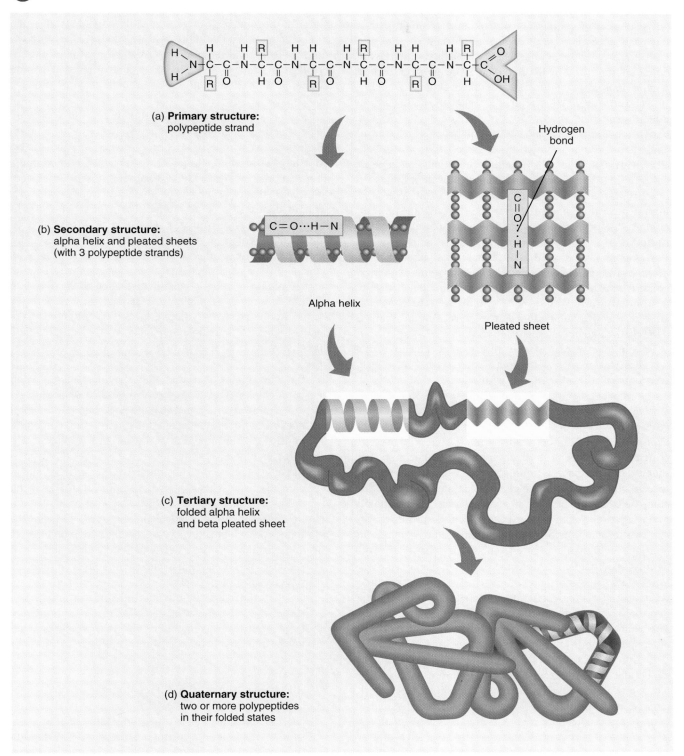

(a) **Primary structure:** polypeptide strand

(b) **Secondary structure:** alpha helix and pleated sheets (with 3 polypeptide strands)

Alpha helix

Hydrogen bond

Pleated sheet

(c) **Tertiary structure:** folded alpha helix and beta pleated sheet

(d) **Quaternary structure:** two or more polypeptides in their folded states

Q What types of bonds hold the side chains (R groups) of amino acids to one another to form the tertiary structure?

This substitution diminishes hemoglobin's water solubility. As a result, the altered hemoglobin tends to form crystals inside red blood cells, producing deformed, sickle-shaped cells that cannot squeeze properly through narrow blood vessels.

The **secondary structure** of a protein is the repeated twisting or folding of neighboring amino acids in the polypeptide chain (Fig. 2.14b). Common secondary structures are clockwise spirals called alpha (α) helixes (singular is helix) and pleated sheets. The secondary structure of a protein is stabilized by hydrogen bonds, which are found at regular intervals along the polypeptide backbone.

The **tertiary** (TUR-shē-er′-ē) **structure** refers to the three-dimensional shape of a polypeptide chain. The tertiary folding pattern may allow amino acids at opposite ends of the chain to be close neighbors (Fig. 2.14c). The side chains (R groups) of the amino acids may bond to each other by hydrogen bonds, ionic bonds, covalent bonds between sulfur atoms called disulfide bridges, and hydrophobic interactions. The tertiary structure is unique for each protein, and it determines how a particular protein will function.

When a protein contains more than one polypeptide chain, it also has a quaternary structure. The arrangement of the individual polypeptide chains relative to one another defines the **quaternary** (KWA-ter-ner′-ē) **structure**. The bonds that hold polypeptide chains together are basically the same as those that maintain the tertiary structure (Fig. 2.l4d).

Proteins vary tremendously in structure. Different proteins have different architectures and different three-dimensional shapes. This variation in structure and shape is directly related to their diverse functions. Homeostatic mechanisms maintain the temperature and chemical composition of body fluids, which allow body proteins to keep their proper three-dimensional shapes. When a cell makes a protein, the polypeptide chain folds to assume a certain shape. One reason for folding of the polypeptide is that some of its parts are attracted to water (hydrophilic), and other parts are repelled by it (hydrophobic). In practically every case, the function of a protein depends on its ability to recognize and bind to some other molecule, as a key fits in a lock. For example, an enzyme binds specifically with its substrate, that is, the molecule on which the enzyme acts (see Fig. 2.15). A hormone binds to a specific protein on a cell whose function it will alter. An antibody protein binds to a foreign substance (antigen) that has invaded the body. The unique shape of a protein permits it to interact with other specific molecules to carry out specific functions.

If a protein encounters a hostile environment in which temperature, pH, or electrolyte concentration is altered, it may unravel and lose its characteristic shape (secondary, tertiary, and quaternary structure). This process is called **denaturation**. Denatured proteins are no longer functional. A common example of denaturation is seen in frying an egg. In a raw egg the protein (albumin) is soluble and the egg white appears as a clear, viscous fluid. When heat is applied to the egg, however, the protein changes shape, becomes insoluble, and looks white.

Enzymes

As we have seen, chemical reactions occur when chemical bonds are made or broken as atoms, ions, or molecules collide with one another. Normal body temperature and pressure are too low for chemical reactions to occur at a rate rapid enough to maintain life. Although raising the temperature, pressure, and the number of reacting particles can increase the frequency of collisions and also increase the rate of chemical reactions, such changes could denature proteins and damage or kill cells.

Enzymes are the living cell's solution to this problem. They speed up chemical reactions by lowering the activation energy and properly orienting the colliding molecules. And they do this without increasing the temperature or pressure—in other words, without disrupting or killing the cell. Substances that can speed up chemical reactions in this way, without themselves being altered, are called **catalysts**. In living cells, **enzymes** function as catalysts. Most enzymes consist of a protein portion, called the **apoenzyme**, and a nonprotein portion, called a **cofactor**. The cofactor may be a metal ion, such as iron, magnesium, zinc, or calcium, or an organic molecule called a **coenzyme**. Coenzymes often are vitamin derivatives. Together, the apoenzyme and cofactor form a **holoenzyme**, or whole enzyme. Although enzymes catalyze selected reactions, they do so with great efficiency and with many built-in controls.

1. **Specificity.** Enzymes are highly specific catalysts. Each particular enzyme affects only specific **substrates**—molecules on which the enzyme acts. In some cases, a portion of the enzyme, called the **active site**, is thought to "fit" the substrate like a key fits in a lock (see Fig. 2.15). In other cases, the active site changes its shape to fit snugly around the substrate once the substrate enters the active site. This is known as an **induced fit**. Of the more than 1000 known enzymes, each has a characteristic three-dimensional shape with a specific surface configuration, which allows it to recognize and bond to certain substrates. When an enzyme is denatured, the active site loses its unique shape and can no longer fit together with its substrate.

   Not only does an enzyme select a particular substrate, it also catalyzes a specific reaction. From the large number of diverse molecules in a cell, an enzyme must "find" the correct substrate and then take it apart or merge it with another substrate to form one or more specific products.

2. **Efficiency.** Under optimal conditions, enzymes can catalyze reactions at rates that are from 100 million (10^8) to 10 billion (10^{10}) times more rapid than those of similar reactions occurring without enzymes. The **turnover number** (number of substrate molecules converted to product per enzyme molecule in one second)

is generally between 1 and 10,000 and can be as high as 600,000.

3. **Control.** Enzymes are subject to a variety of cellular controls. Their rate of synthesis and their concentration at any given time are under the control of a cell's genes. Substances within the cell may either enhance or inhibit activity of a given enzyme. Many enzymes occur in both active and inactive forms in cells. The rate at which the inactive form becomes active or vice versa is determined by the environment inside the cell. Some enzymes require cofactors or coenzymes for effective function.

Exactly how enzymes lower the activation energy of a reaction is not completely understood. However, an enzyme is thought to work as shown in Fig. 2.15.

1 The surface of the substrate makes contact with the active site on the surface of the enzyme molecule, forming a temporary intermediate compound called the **enzyme–substrate complex**.

2 The substrate molecule is transformed by rearrangement of existing atoms, breakdown of the substrate molecule, or combination of several substrate molecules. The transformed substrate molecules are called the **products** of the reaction.

3 After the reaction is completed and the products of the reaction move away from the enzyme, the unchanged enzyme is free to attach to another substrate molecule.

The names of enzymes usually end in the suffix **-ase**. All enzymes can be grouped according to the types of chemical reactions they catalyze. For example, *oxidases* add oxygen, *kinases* add phosphate, *dehydrogenases* remove hydrogen, *ATPases* split ATP, *anhydrases* remove water, *proteases* break down proteins, and *lipases* break down lipids.

GALACTOSEMIA

Galactosemia (ga-lak-tō-SĒ-mē-a) is an inherited disorder in which galactose cannot be converted to glucose because the needed enzyme is missing. An infant with the disorder will fail to thrive within a week after birth due to anorexia (loss of appetite), vomiting, and diarrhea unless galactose and lactose are removed from the diet. The disaccharide lactose (milk sugar) is broken down into galactose and glucose. If treatment is delayed, the infant remains physically small and becomes mentally retarded. Treatment consists of eliminating galactose and lactose from the diet. ■

Nucleic Acids: Deoxyribonucleic Acid (DNA) and Ribonucleic Acid (RNA)

Nucleic (noo-KLE-ic) **acids**, compounds named because they were first discovered in the nuclei of cells, are huge organic molecules that contain carbon, hydrogen, oxygen, nitrogen, and phosphorus. Nucleic acids are of two varieties.

Figure 2.15 How an enzyme works.

An enzyme speeds up a chemical reaction without being altered or consumed.

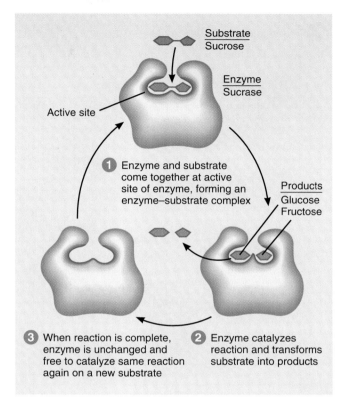

Substrate
Sucrose

Enzyme
Sucrase

Active site

1 Enzyme and substrate come together at active site of enzyme, forming an enzyme–substrate complex

Products
Glucose
Fructose

3 When reaction is complete, enzyme is unchanged and free to catalyze same reaction again on a new substrate

2 Enzyme catalyzes reaction and transforms substrate into products

What type of reaction is illustrated here? One product of this reaction is not illustrated. What is it?

Deoxyribonucleic (dē-ok′-sē-rī-bō-noo-KLĒ-ik) **acid** (**DNA**) forms the inherited genetic material inside each cell. Each **gene** is a segment of a DNA molecule. Our genes determine which traits we inherit, and by controlling protein synthesis, they regulate most of the activities that take place in our cells throughout a lifetime. When a cell divides, its hereditary information passes on to the next generation of cells. **Ribonucleic acid** (**RNA**), the second type of nucleic acid, relays instructions from the genes to guide each cell's assembly of amino acids into proteins.

The monomers of nucleic acids are called **nucleotides**. A nucleic acid is a chain composed of repeating nucleotide units. Each nucleotide of DNA consists of three parts (Fig. 2.16a):

1. **A base (nitrogenous base).** Four nitrogen-containing structures containing atoms of C, H, O, and N called **bases** or **nitrogenous bases** are present in DNA and RNA. In DNA they are adenine (A), thymine (T), cytosine (C), and guanine (G). Whereas adenine and guanine are larger, double-ring bases called **purines** (PYOO-rēns), thymine and cytosine are smaller, single-ring bases called **pyrimidines** (pī-RIM-id-ēns). The nucleotides are

Figure 2.16 DNA molecule. (a) A nucleotide consists of a base, a pentose sugar, and a phosphate group. (b) The paired bases project toward the center of the double helix. The structure is stabilized by hydrogen bonds (dotted lines) between each base pair. There are two hydrogen bonds between adenine and thymine and three between cytosine and guanine.

Nucleotides are the building blocks (monomers) of nucleic acids.

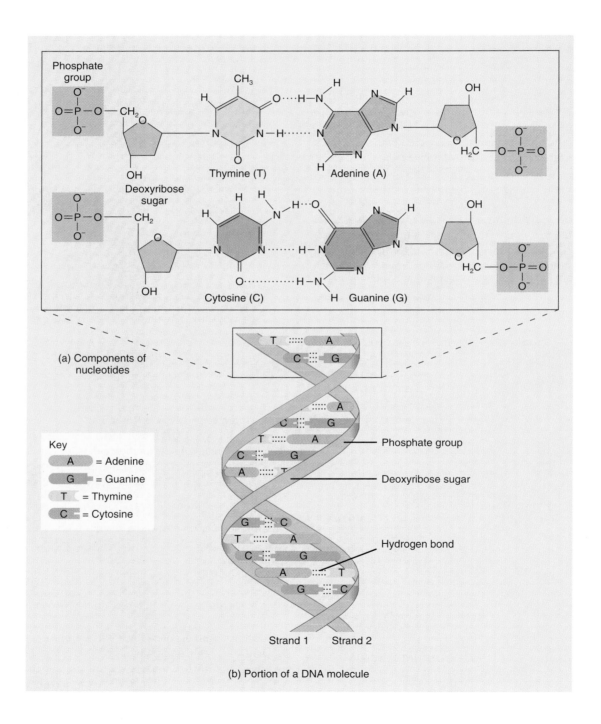

(a) Components of nucleotides

Key
A = Adenine
G = Guanine
T = Thymine
C = Cytosine

Phosphate group

Deoxyribose sugar

Hydrogen bond

Strand 1   Strand 2

(b) Portion of a DNA molecule

Which bases always pair with one another?

named according to the base that is present. Thus a nucleotide containing thymine is called a thymine nucleotide. One containing adenine is called an adenine nucleotide, and so on.

2. **A pentose sugar.** A five-carbon sugar called **deoxyribose** attaches to each base in DNA.

3. **A phosphate group.** Alternating phosphate groups (PO_4^{3-}) and pentoses form the backbone of a DNA strand, while the bases protrude from the backbone chain (Fig. 2.16b).

In 1953, F. H. C. Crick of Great Britain and J. D. Watson, a young American scientist, published a brief paper describing how these three components might be arranged in DNA. Their insights into data gathered by others led them to construct a model so elegant and simple that the scientific world immediately knew it was correct! In the Watson–Crick **double helix** model, DNA resembles a twisted rope ladder (Fig. 2.l6b). Two strands of alternating phosphate groups and deoxyriboses form the uprights of the ladder. Paired bases, held together by hydrogen bonds, form the rungs. Adenine always pairs with thymine, and cytosine always pairs with guanine. If you know the sequence of bases in one strand of DNA, you can predict the sequence on the complementary (second) strand. Each time DNA is copied, as an organism adds cells, the two strands unwind. Each strand serves as the template or mold on which to construct a new second strand (see Chapter 3).

DNA FINGERPRINTING

A technique called DNA fingerprinting is used in research and a variety of legal situations to help identify and convict criminals. In each person, certain DNA segments contain base sequences that are repeated several times. Both the number of repeat copies in one region and the number of regions subject to repeat are different from one person to another. DNA fingerprinting can be done with minute quantities of DNA, for example, from a single strand of hair, a drop of semen, or a spot of blood. It also can be used to determine the identity of a child's parents. ■

RNA, the second variety of nucleic acid, differs from DNA in several respects. In humans, RNA is single-stranded, the sugar in the RNA nucleotide is the pentose **ribose**, and RNA contains the pyrimidine base uracil (U) rather than thymine. Cells contain three different kinds of RNA called messenger RNA, ribosomal RNA, and transfer RNA. Each has a specific role to perform in carrying out the instructions coded in DNA (see page 77).

Adenosine Triphosphate

Adenosine (a-DEN-ō-sēn) **triphosphate** or **ATP** (Fig. 2.17) is the "energy currency" of living systems. It functions to temporarily store and then transfer the energy liber-

---

Figure 2.17 Structure of ATP and ADP. The two phosphate bonds that can be used to transfer energy are indicated by "squiggles" (∼). Most often energy transfer involves hydrolysis of the terminal phosphate bond.

 ATP stores chemical energy for various cellular activities.

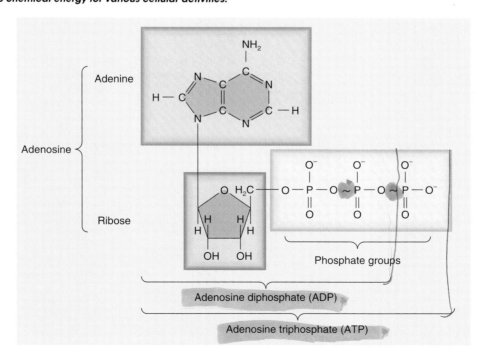

What are some cellular activities that depend on energy supplied by ATP?

ated in exergonic catabolic reactions to cellular activities that require energy (endergonic reactions). Among these cellular activities are muscular contractions, movement of chromosomes during cell division, movement of cytoplasm within cells, transporting substances across cell membranes, and synthesizing larger molecules from smaller ones (anabolism). Structurally, ATP consists of three phosphate groups attached to an adenosine unit composed of adenine and the five-carbon sugar ribose.

When the terminal phosphate group (PO_4^{3-}), symbolized by Ⓟ in the following discussion, is hydrolyzed by addition of a water molecule, the reaction liberates energy. This energy is used by the cell to power its activities. Removal of the terminal phosphate group leaves a molecule called **adenosine diphosphate** (**ADP**). This reaction may be represented as follows:

ATP ⟶ ADP + Ⓟ + E

Adenosine triphosphate　　　Adenosine diphosphate　Phosphate group　Energy

The energy supplied by the catabolism of ATP into ADP is constantly being used by the cell. Since the supply of ATP at any given time is limited, a mechanism exists to replenish it: a phosphate group is added to ADP to manufacture more ATP. The reaction may be represented as follows:

ADP + Ⓟ + E ⟶ ATP

Adenosine diphosphate　Phosphate group　Energy　　　Adenosine triphosphate

Of course, energy is required to produce ATP. The energy needed to attach a phosphate group to ADP is supplied mainly by the catabolism of glucose in a process called cellular respiration. Cellular respiration has two phases:

1. **Anaerobic.** In the absence of oxygen, glucose is partially broken down by a process called glycolysis into pyruvic acid. Each glucose that is converted into a pyruvic acid molecule yields two molecules of ATP.
2. **Aerobic.** In the presence of oxygen, glucose is completely broken down into carbon dioxide and water. These reactions generate heat and a large number of ATP molecules: 36–38 ATP from each glucose.

The details of cellular respiration will be discussed in Chapters 10 and 25.

STUDY OUTLINE

MATTER AND ENERGY (p. 26)

Introduction (p. 26)
1. Matter is anything that occupies space and has mass.
2. Energy is the capacity to do work and is of two principal kinds: potential (stored) and kinetic (energy of motion).
3. Forms of energy include mechanical, heat, chemical, radiant, and electrical.

Chemical Elements (p. 26)
1. All forms of matter are composed of chemical elements.
2. Oxygen, carbon, hydrogen, and nitrogen make up 96% of body weight.

Structure of Atoms (p. 26)
1. Each element is made up of small units called atoms.
2. Atoms consist of a nucleus, which contains protons and neutrons, and electrons that move about the nucleus in regions called electron shells.
3. The number of protons (the atomic number) distinguishes the atoms of one element from those of another.
4. Mass number is the total number of protons and neutrons.
5. Different atoms of an element that have the same number of protons but different numbers of neutrons are called isotopes.
6. The atomic mass (atomic weight) is the average mass of all stable atoms of an element.

Electrons and Chemical Reactions (p. 28)
1. When atoms combine or break apart from other atoms, a chemical reaction occurs.
2. Atoms always attempt to either empty their outermost (valence) shell or fill it to the maximum to achieve stability.

3. When two or more atoms share electrons in a chemical reaction, a molecule is formed.
4. A compound is a substance that can be broken down into two or more different elements by chemical means.

Chemical Bonds (p. 30)
1. The atoms of a molecule are held together by forces of attraction called chemical bonds.
2. In an ionic bond, valence electrons are transferred from one atom to another. The transfer forms ions, whose unlike charges attract each other and form ionic bonds.
3. In a covalent bond, there is a sharing of pairs of valence electrons. Covalent bonds may be nonpolar or polar.
4. In a hydrogen bond, two other atoms (usually oxygen or nitrogen) associate with a hydrogen atom.

Chemical Reactions (p. 32)
1. Several factors determine if a chemical reaction will occur: concentration of particles, speed of particles, activation energy, and proper orientation of the particles.
2. Endergonic reactions require energy, whereas exergonic reactions release energy.
3. Synthesis reactions involve the combination of reactants to produce larger molecules. The reactions are anabolic and usually endergonic.
4. In decomposition reactions, a substance breaks down into smaller molecules. The reactions are catabolic and usually exergonic.
5. Exchange reactions involve the replacement of one atom or atoms by another atom or atoms.
6. In reversible reactions, end products can revert to the original combining molecules.

7. In an oxidation–reduction (redox) reaction, one compound is oxidized (electrons are removed) while another compound is reduced (electrons are added).

CHEMICAL COMPOUNDS AND LIFE PROCESSES (p. 34)

1. Inorganic substances usually lack carbon and are small molecules.
2. Organic substances always contain carbon. Most organic substances contain covalent bonds.

INORGANIC COMPOUNDS (p. 34)

1. Inorganic acids, bases, and salts dissociate into ions in water. An acid ionizes into hydrogen ions (H^+); a base ionizes into hydroxide ions (OH^-). A salt ionizes into neither H^+ nor OH^- ions. Cations are positively charged ions; anions are negatively charged ions.
2. Two ways to express the concentration of a solution are percent and moles per liter.
3. Water is the most abundant substance in the body. It is an excellent solvent, lubricating, and suspending medium; participates in chemical reactions; and has a high heat capacity.
4. The pH of body fluids must remain fairly constant for the body to maintain homeostasis. On the pH scale, 7 represents neutrality. Values below 7 indicate acidic solutions, and values above 7 indicate alkaline solutions.
5. Buffer systems usually consist of a weak acid and a weak base. They soak up excess H^+ and excess OH^-, which helps maintain pH homeostasis.

ORGANIC COMPOUNDS (p. 38)

1. Carbohydrates are sugars or starches that provide most of the energy needed for life. They may be monosaccharides, disaccharides, or polysaccharides. Carbohydrates, and other organic molecules, are joined together to form larger molecules with the loss of water by a process called dehydration synthesis. In the reverse process, called hydrolysis, large molecules are broken down into smaller ones upon the addition of water.
2. Lipids are a diverse group of compounds that include triglycerides (fats and oils), phospholipids, steroids, carotenes, vitamins A, D, E, and K, and eicosanoids. Triglycerides protect, insulate, provide energy, and are stored. Phospholipids are important membrane components. Eicosanoids (prostaglandins and leukotrienes) modify hormone responses, contribute to inflammation, dilate airways, and regulate body temperature.
3. Proteins are constructed from amino acids. They give structure to the body, regulate processes, provide protection, help muscles to contract, transport substances, and serve as enzymes. Structural levels of organization among proteins include: primary, secondary, tertiary, and quaternary. Enzymes are catalysts that are highly specific in terms of substrates with which they react, efficient in terms of the number of substrate molecules with which they react, and subject to a variety of cellular controls.
4. Deoxyribonucleic acid (DNA) and ribonucleic acid (RNA) are nucleic acids consisting of nitrogenous bases, five-carbon sugars, sugar, and phosphate groups. DNA is a double helix and is the primary chemical in genes. RNA differs in structure and chemical composition from DNA and is mainly concerned with protein synthesis reactions.
5. Adenosine triphosphate (ATP) is the principal energy-transferring molecule in living systems. When it transfers energy to an endergonic reaction, it is decomposed to adenosine diphosphate (ADP) and Ⓟ. ATP is synthesized from ADP and Ⓟ using the energy supplied by various decomposition reactions, particularly of glucose.

REVIEW QUESTIONS

1. What is the relationship of matter and energy to the body? Distinguish the various forms of energy. (p. 26)
2. Define a chemical element. List the chemical symbols for 10 different chemical elements. Which chemical elements make up the bulk of the human organism? (p. 26)
3. What is an atom? Diagram the positions of the nucleus, protons, neutrons, and electrons in an atom of oxygen and in an atom of nitrogen. (p. 26)
4. What is an atomic number? Compare it to an atomic mass and mass number. (p. 28)
5. What is an isotope? A radioisotope? Describe several medical uses of radioisotopes. (p. 28)
6. What is an electron shell? How are electron shells related to chemical reactions? (p. 29)
7. How are chemical bonds formed? Distinguish between an ionic bond and a covalent bond. Give at least one example of each. (p. 30)
8. Can you determine how a molecule of $MgCl_2$ is ionically bonded? Magnesium has two electrons in its outer energy level. Construct a diagram to verify your answer. (p. 30)
9. Refer to Fig. 2.4b, c. See if you can determine why there is a double covalent bond between atoms in an oxygen molecule (O_2) and a triple covalent bond between atoms in a nitrogen molecule (N_2). (p. 31)
10. Define a hydrogen bond. Why are hydrogen bonds important? (p. 32)
11. What factors determine if a collision will occur to cause a chemical reaction? (p. 32)
12. How do endergonic reactions differ from exergonic reactions? What is the role of ATP in these reactions? (p. 32)
13. What are the principal kinds of chemical reactions? How are anabolism and catabolism related to synthesis and decomposition reactions, respectively? (p. 33)
14. Identify what kind of reaction each of the following represents (p. 33):

   a. $H_2 + Cl_2 \longrightarrow 2HCl$

   b. $3NaOH + H_3PO_4 \longrightarrow Na_3PO_4 + 3H_2O$

   c. $CaCO_3 + CO_2 + H_2O \longrightarrow Ca(HCO_3)_2$

   d. $HNO_3 \longrightarrow H^+ + NO_3^-$

   e. $NH_3 + H_2O \rightleftharpoons NH_4^+ + OH^-$

15. How do inorganic compounds differ from organic compounds? List and define the principal inorganic and organic compounds that are important to the human body. (p. 34)

16. Define an inorganic acid, a base, and a salt. How does the body acquire some of these substances? List some functions of the chemical elements furnished as ions of salts. (p. 34)

17. What are the essential functions of water in the body? Explain the solvating property of water. Describe two ways to express the concentration of a solution (p. 35)

18. What is pH? Why is it important to maintain a relatively constant pH? What is the pH scale? (p. 36)

19. List the normal pH values of some common fluids, biological solutions, and foods. Refer to Exhibit 2.3 and select the two substances whose pH values are closest to neutrality. Is the pH of milk or of cerebrospinal fluid closer to 7? Is the pH of bile or of urine farther from neutrality? (p. 37)

20. What are the components of a buffer system? What is the function of a buffer? How is buffering an example of homeostasis? (p. 37)

21. Define a carbohydrate. Why are carbohydrates essential to the body? How are carbohydrates classified? (p. 38)

22. Compare dehydration synthesis and hydrolysis. Why are they significant? (p. 38)

23. How do lipids differ from carbohydrates? Explain the importance of the following lipids to the body: triglycerides, phospholipids, steroids, lipoproteins, and eicosanoids. (p. 40)

24. Distinguish among saturated, monounsaturated, and polyunsaturated fats. How do saturated fats relate to atherosclerosis? (p. 41)

25. Define a protein. What is a peptide bond? Discuss the classification of proteins on the basis of function and levels of structural organization. (p. 43)

26. What is an enzyme? What are some characteristics of enzymes? (p. 46)

27. What is a nucleic acid? How do deoxyribonucleic acid (DNA) and ribonucleic acid (RNA) differ with regard to chemical composition, structure, and function? (p. 47)

28. What is adenosine triphosphate (ATP)? What is the essential function of ATP in the human body? How is this function accomplished? (p. 49)

29. What is galactosemia? (p. 47)

30. What is DNA fingerprinting? What is its value? (p. 49)

ANSWERS TO QUESTIONS WITH FIGURES

2.1 Six.

2.2 Helium (He), because it has a completely filled valence (outermost) shell.

2.3 K is an electron donor; when it ionizes, it becomes a cation, K^+.

2.4 An ionic bond involves the loss and gain of electrons; a covalent bond involves the sharing of pairs of electrons.

2.5 $CaCO_3$ is a salt, and H_2SO_4 is an acid.

2.6 Since it easily dissolves in a polar solvent (water), table sugar must have several polar covalent bonds.

2.7 $[H^+] = 10^{-6}$ mol/liter; $[OH^-] = 10^{-12}$ mol/liter. A pH of 6.82 is more acidic than a pH of 6.91. Both pH = 8.41 and pH = 5.59 are 1.41 pH units from neutral (pH = 7).

2.8 Anabolic; there are 6 carbons in fructose, 12 in sucrose.

2.9 The fatty acid.

2.10 The polar head region is hydrophilic (water-loving), and the nonpolar tail region is hydrophobic.

2.11 The only difference is the number of double bonds and the type of side group attached to ring A.

2.12 Two carbons; one nitrogen (see structure of glycine in Fig. 2.12c).

2.13 Hydrolysis.

2.14 Hydrogen bonds, ionic bonds, covalent bonds (disulfide bridges), and hydrophobic interactions.

2.15 Hydrolysis. Water.

2.16 Thymine always pairs with adenine and cytosine always pairs with guanine.

2.17 Muscular contractions, movement of chromosomes, transportation of substances across cell membranes, and synthesis (anabolic) reactions.

THE CELLULAR LEVEL OF ORGANIZATION

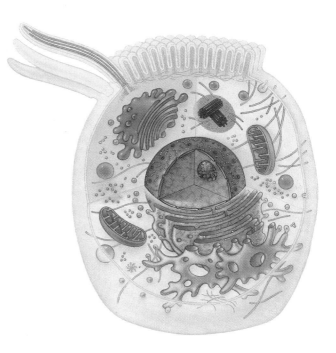

t is at the cellular level of organization that activities essential to life occur and disease processes originate. A **cell** is the basic, living, structural, and functional unit of the body. **Cytology** (sī-TOL-ō-jē; *cyt* = cell; *logos* = study of) is the branch of science concerned with the study of cells. This chapter focuses on the structure, functions, and reproduction of cells.

GENERALIZED ANIMAL CELL

The **generalized animal cell** illustrated in Fig. 3.1 is a composite of many different cells in the body. While most cells have many of the features shown in this diagram, few cells have all the illustrated features. For ease of study, we can divide a cell into four principal parts:

1. **Plasma (cell) membrane.** Outer, limiting membrane separating the cell's internal components from the extracellular materials and external environment. Extracellular materials, which are substances external to the cell surface, will be examined together with tissues in Chapter 4.
2. **Cytosol.** The thick, semifluid intracellular fluid is termed the **cytosol** (SĪ-tō-sol). The cytosol contains many dissolved proteins and enzymes, nutrients, ions, and other small molecules, which all participate in various phases

Figure 3.1   Generalized animal cell based on electron microscopic studies.

The cell is the basic, living, structural and functional unit of the body.

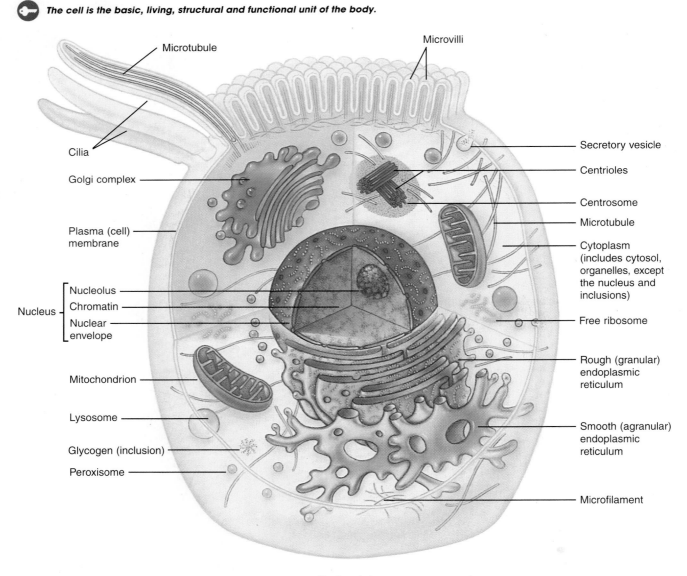

Sectional view

Q List the four principal parts of a cell.

of metabolism. Organelles and inclusions are suspended in the cytosol. The term **cytoplasm** includes cytosol, all organelles (except the nucleus), and inclusions.

3. **Organelles.** Highly organized structures with characteristic shapes that are specialized for specific cellular activities. The organelle with the largest volume is the **nucleus**, which contains most of a cell's genetic material.

4. **Inclusions.** Temporary structures in the cytoplasm that contain secretions and storage products of the cell.

PLASMA (CELL) MEMBRANE

The thin barrier that separates the internal components of a cell from extracellular materials is the **plasma membrane**, also called the **cell membrane** (Fig. 3.2). The plasma membrane is the gatekeeper that regulates passage of substances into and out of the cell.

The **fluid mosaic model** of membrane structure describes the molecular arrangement of the plasma membrane and other membranes in living organisms. A mosaic is a pattern of many small pieces fitted together. According to this model, the membrane is a mosaic of proteins floating like icebergs in a sea of lipids.

Membrane Chemistry and Anatomy

Plasma membranes of typical animal cells are about a 50:50 mix by weight of proteins and lipids that are held together by noncovalent interactions. Since proteins are larger and more massive than lipids, however, there are about 50 lipid molecules for each protein molecule.

Membrane Lipids

About 75% of the lipids are **phospholipids**, lipids that contain phosphorus. Present in smaller amounts are cholesterol (a steroid) and glycolipids, which are lipids with one or more sugar groups attached. The phospholipids line up in two parallel layers forming a **phospholipid bilayer** (Fig. 3.2). This arrangement occurs because the phospholipids are **amphipathic** (*amphi* = on both sides, double; *pathos* = feeling). Amphipathic molecules have a dual nature: they contain both polar and nonpolar regions (see Fig. 2.10). The polar part is the phosphate-containing "head," which is hydrophilic (= water loving). The nonpolar parts are the two

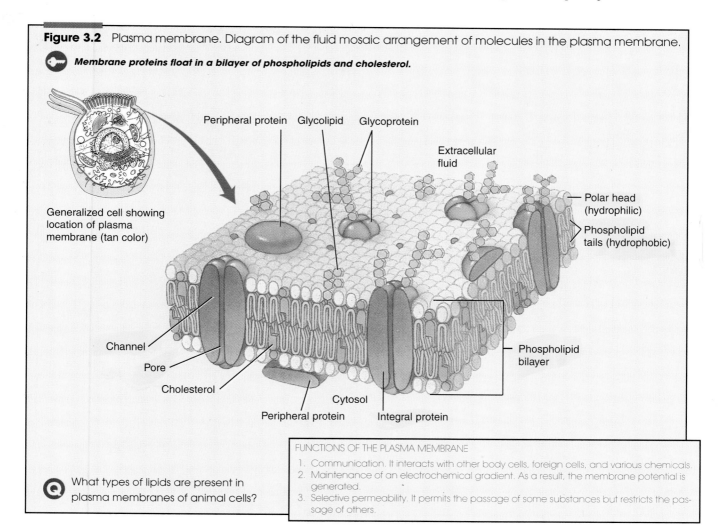

Figure 3.2 Plasma membrane. Diagram of the fluid mosaic arrangement of molecules in the plasma membrane.

🔑 **Membrane proteins float in a bilayer of phospholipids and cholesterol.**

Generalized cell showing location of plasma membrane (tan color)

Peripheral protein · Glycolipid · Glycoprotein

Extracellular fluid

Polar head (hydrophilic)

Phospholipid tails (hydrophobic)

Channel

Pore

Cholesterol

Peripheral protein

Cytosol

Integral protein

Phospholipid bilayer

FUNCTIONS OF THE PLASMA MEMBRANE
1. Communication. It interacts with other body cells, foreign cells, and various chemicals.
2. Maintenance of an electrochemical gradient. As a result, the membrane potential is generated.
3. Selective permeability. It permits the passage of some substances but restricts the passage of others.

❓ What types of lipids are present in plasma membranes of animal cells?

fatty acid "tails," which are hydrophobic (= water fearing). The molecules orient in the bilayer so that the heads face outward on either side, toward the watery cytosol and extracellular fluid (fluid outside cells). The tails face each other in the membrane's interior. The bilayer forms the basic framework of the plasma membrane.

Glycolipids, which account for about 5% of membrane lipids, are also amphipathic. They appear only in the layer that faces the extracellular fluid. Although membrane glycolipids are the target of certain bacterial toxins (poisons), their normal functions are still being unraveled. They are important for adhesion among cells, may mediate cell-to-cell recognition and communication, and contribute to regulation of cellular growth and development.

The remaining 20% of membrane lipids are **cholesterol** molecules, which are located among the phospholipids in animal cells. Plant cell membranes lack cholesterol. The stiff steroid rings of cholesterol strengthen the membrane of an animal cell but decrease its flexibility.

The phospholipid bilayer is dynamic because the lipids can move sideways and exchange places in their own layer. Flip-flop of molecules between layers, however, rarely occurs. The bilayer is also self-healing; if a needle is pushed through it and pulled out, the puncture site seals.

Membrane Proteins

Membrane proteins are of two types: integral and peripheral (Fig. 3.2). **Integral proteins** extend across the phospholipid bilayer among the fatty acid tails. Most (maybe all) integral proteins are **glycoproteins**, which are proteins with attached sugar groups. The sugar portion of a glycoprotein faces the extracellular fluid. **Peripheral proteins** do not extend across the phospholipid bilayer. They are loosely attached to the inner and outer surfaces of the membrane and are easily separated from it.

To a large extent, membrane proteins and glycoproteins determine what functions a cell can perform. In general, the types of lipids present vary only slightly from one membrane to another. On the other hand, the variety of proteins that can be included in a given membrane is enormous, and the membrane proteins have many roles (Fig. 3.3). Some integral proteins are **channels** that have a **pore** (hole) through which certain substances can flow into or out of the cell. Others act as **transporters** (carriers) to move a substance from one side of the membrane to the other (described on pages 61–62).

Integral proteins also serve as recognition sites called **receptors**. These are molecules that can identify and attach to a specific molecule, such as a hormone, a neurotransmitter, or a nutrient, that is important for some cellular function. A molecule that specifically binds to a receptor by forces other than covalent bonds is called a **ligand** (LI-gand; *ligare* = to bind) of that receptor.

Some integral and peripheral proteins are **enzymes**. Other peripheral proteins facing the cytosol serve as **cytoskeleton anchors**, forming an attachment between the plasma membrane and filaments of the cytoskeleton. Membrane glycoproteins and glycolipids often are **cell identity markers**.

Figure 3.3   Functions of membrane proteins.

 Membrane proteins and glycoproteins largely reflect the functions a cell can perform.

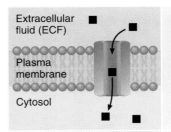

1 Channel
Allows specific substance (■) to move through water-filled passageway (pore). Most plasma membranes include specific channels for several ions, most commonly K^+ and Cl^-.

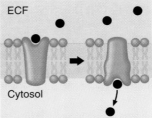

2 Transporter (Carrier)
Transports specific substance (●) across membrane by changing shape. For example, amino acids, needed to synthesize new proteins, enter body cells via transporters.

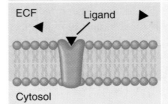

3 Receptor
Recognizes specific ligand (▲) and alters cell's function in some way. For example, antidiuretic hormone binds to receptors in the kidneys and changes the water permeability of certain plasma membranes.

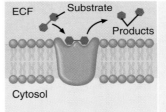

4 Enzyme
Catalyzes reaction inside or outside cell (depending on which direction the active site faces). For example, lactase protruding from epithelial cells lining your small intestine splits the disaccharide lactose in the milk you drink.

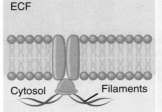

5 Cytoskeleton Anchor
Anchors filaments and tubules of the cytoskeleton inside cell to the membrane to provide structural stability and shape for the cell. May also participate in movement of the cell.

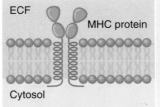

6 Cell Identity Marker
Distinguishes your cells from anyone else's (unless you are an identical twin). An important class of such markers are the major histocompatibility (MHC) proteins.

When a cell is stimulated by the hormone insulin, the insulin first binds to a protein in the plasma membrane. Which of the illustrations best represents this membrane protein?

They may enable a cell to recognize other cells of its own kind during tissue formation or to recognize and respond to potentially dangerous foreign cells. The ABO blood type markers are one example of cell identity markers. When you receive a transfusion, the blood must be of the same type as your own.

Membrane Physiology

Knowing the structure of the plasma membrane, we can now describe its several important physiological properties.

Communication

The plasma membrane functions in cellular communication. This includes interactions with other body cells, foreign cells, and ligands such as hormones, neurotransmitters, enzymes, nutrients, and antibodies in the extracellular fluid.

Electrochemical Gradient

The plasma membrane encloses the cellular contents and separates them from the extracellular fluid or, sometimes, from the external environment. The membrane maintains an electrical and chemical gradient (difference), called an **electrochemical gradient**, between the inside and outside of the cell. The chemical portion of the electrochemical gradient arises because the membrane maintains very different chemical compositions in the cytosol and the extracellular fluid. Although chemicals other than ions are present in different concentrations inside and outside living cells, for now we note only the differences in ion concentrations (Fig. 3.4). In extracellular fluid, the main cation (positively charged ion) is Na^+ and the main anion (negatively charged ion) is Cl^-. In cytosol, the main cation is K^+, while the two dominant anions are organic phosphates (PO_4^{3-} groups attached to organic molecules such as ATP) and negatively charged amino acids in proteins.

The electrical gradient arises because the inside surface of the membrane is more negatively charged than the outside surface in most cells. As a result, there is a voltage called the **membrane potential** across the membrane. *Voltage* is a form of potential energy that occurs when positive and negative charges are separated; examples of such charge separation are found in ordinary flashlight and car batteries. In living cells, a small separation of ions (charged particles) by the membrane gives rise to the membrane potential. This separation of charges occurs only near the plasma membrane. Overall, both the cytosol and extracellular fluid are electrically neutral; if you added up all the positive and negative charges in each fluid, they would balance except in a thin region adjacent to the membrane. The voltage across the plasma membrane of cells throughout your body usually is between −20 and −200 millivolts (mV). (One mV is one-thousandth of a volt = 0.001 V.) The negative sign in front of the number means that the inside is negative relative to the outside. Under certain conditions, some cells exhibit a positive membrane potential. The electro-

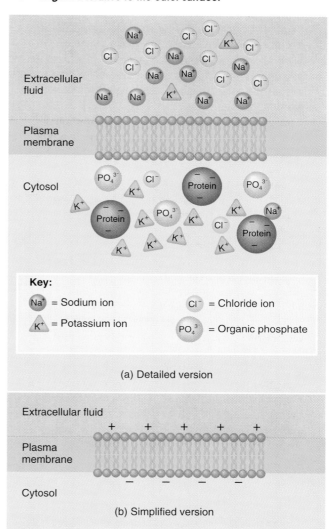

Figure 3.4 Electrochemical gradient.

 The inner surface of the plasma membrane of most cells is negative relative to the outer surface.

Key:

Na^+ = Sodium ion Cl^- = Chloride ion

K^+ = Potassium ion PO_4^{3-} = Organic phosphate

(a) Detailed version

(b) Simplified version

Q Which ions congregate closest to the membrane on the cytosolic (inner) side?

chemical gradient and the resulting membrane potential are important for the proper functioning of most cells. How the membrane potential arises is discussed on page 342.

Selective Permeability

The plasma membrane regulates entry and exit of materials. It permits passage of certain substances and restricts the passage of others. This property of membranes is called **selective permeability**.

A membrane is said to be *permeable* to a substance if it allows the free passage of that substance. Although plasma membranes are not completely permeable to any substance, they do permit some substances to pass more readily than others. Water, for example, usually passes more easily than most other substances. The selective permeability of a plasma membrane to different substances depends on several factors that relate to the structure of the membrane.

1. **Lipid solubility.** Substances that dissolve in lipids (nonpolar, hydrophobic molecules) pass easily through the phospholipid bilayer of the plasma membrane.
2. **Size.** Most large molecules, such as proteins, cannot pass through the plasma membrane. A few very small, uncharged polar molecules can pass through the phospholipid bilayer.
3. **Charge.** The phospholipid bilayer portion of the plasma membrane is impermeable to all charged molecules and ions. Some charged substances do pass through the membrane, however, by moving through the pore of a channel or by being carried from one side to the other by transporters. The negative membrane potential of most cells aids the inflow of cations and hinders the inflow of anions.
4. **Presence of channels and transporters.** Cell membranes are permeable to a variety of polar and charged substances, including ions, that cannot cross the phospholipid bilayer. Proteins in the membrane help many substances cross the membrane, much as a ferry boat transports people across a lake. These integral membrane proteins increase membrane permeability in two ways. Channel proteins form a water-filled **pore** (hole) that pierces the membrane. Other proteins act as **transporters**; they pick up a substance on one side of the membrane and shuttle it through to the other side before releasing it. Most channels and transporters are very selective, helping only a specific solute to cross the membrane.

MOVEMENT OF MATERIALS ACROSS PLASMA MEMBRANES

The mechanisms that enable substances to move across cell membranes are essential to the life of the cell. Certain substances, for example, must move into the cell to support needed metabolic reactions, while waste materials or harmful substances must be moved out. Mechanisms that move substances across a membrane without using cellular energy (released by splitting ATP) are **passive processes**. In **active processes**, the cell uses some of its own energy from the splitting of ATP to move the substance across the membrane.

Passive Processes

The passive processes that we will discuss are simple diffusion, osmosis, bulk flow, and facilitated diffusion. These transport processes depend on pressure or concentration differences and the process of diffusion.

Simple Diffusion

Because all substances have kinetic energy (energy of motion), molecules and ions are continually moving about, colliding with one another, and moving off in various directions. The random mixing of ions and molecules in a solution due to their kinetic energy is called **simple diffusion** (*diffus* = spreading) or **diffusion**.

If a particular ion or molecule is present in high concentration in one area and in low concentration in another area, the difference in concentration between the two areas forms a **concentration gradient**. When two such areas are connected, more particles diffuse from the region of high concentration to the region of low concentration than diffuse in the opposite direction. This difference in diffusion between two regions having different concentrations is called **net diffusion**. Substances undergoing net diffusion (from a high to a low concentration) are said to move *down* or *with* their concentration gradient. After a period of time, they become evenly distributed. The point at which there is no further net change in concentration is called **equilibrium**. At equilibrium, although simple diffusion continues due to the kinetic energy of the particles, there is no further *net* diffusion.

Because diffusion depends on the kinetic energy of the particles, diffusion occurs more rapidly when temperature increases. Also, a larger concentration gradient (difference) produces faster net diffusion. And smaller molecules diffuse more rapidly than larger ones.

As an example of diffusion, think what happens if you place a crystal of dye in a water-filled container (Fig. 3.5). Just next to the dye, the color is intense because the dye concentra-

---

Figure 3.5   Principle of diffusion. Crystals of dye in a cylinder of water dissolve (beginning) and there is net diffusion from the region of higher dye concentration to regions of lower dye concentration (intermediate).

 At equilibrium, net diffusion stops but simple diffusion continues.

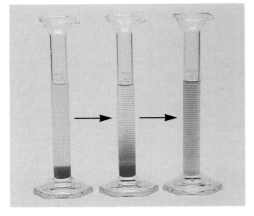

Beginning   Intermediate   Equilibrium

 How would having a fever affect body processes that involve diffusion?

tion is greatest there. At increasing distances, the color is lighter and lighter because the dye concentration is less. Some time later, at equilibrium, the solution of water and dye has a uniform color. Both the dye molecules and the water molecules have undergone net diffusion, down their concentration gradients, until they are evenly dispersed among one another.

In the examples just noted, no membrane was involved. Substances may also diffuse through a membrane, if the membrane is permeable to them. Lipid-soluble molecules, such as oxygen, carbon dioxide, nitrogen, steroids, fat-soluble vitamins (A, E, D, and K), glycerol, small alcohols, and ammonia diffuse through the phospholipid bilayer of the plasma membrane, into and out of cells (Fig. 3.6a). Water also diffuses easily through the phospholipid bilayer, as has been shown in many experiments using artificially made, pure phospholipid membranes. Diffusion is important in the movement of oxygen and carbon dioxide between blood and body cells and between blood and air within the lungs during breathing. It also influences the absorption of some nutrients and excretion of wastes by body cells, activities that contribute to homeostasis.

Small substances that are not lipid-soluble also may diffuse into or out of cells through small water-filled pores of channels formed by integral proteins (Fig. 3.6b). Important examples include sodium ions (Na^+), potassium ions (K^+), calcium ions (Ca^{2+}), chloride ions (Cl^-), bicarbonate ions (HCO_3^-), and urea. Water itself also diffuses into and out of cells through the pores of channels. Diffusion of substances through channels generally is much slower than diffusion through the phospholipid bilayer and is now considered to be one type of facilitated diffusion (described on page 61).

Osmosis

Another passive process is **osmosis** (oz-MŌ-sis). It is the net diffusion of a solvent, which is water in living systems, through a selectively permeable membrane. Water moves by osmosis across a membrane from an area of higher water concentration (where the solute concentration is lower) to an area of lower water concentration (where the solute concentration is higher). A simple apparatus may be used to demonstrate osmosis (Fig. 3.7). A sac made of cellophane, a permeable membrane that permits water but not sugar molecules to pass, is filled with a solution that is 20% sugar (sucrose) and 80% water. The upper portion of the cellophane sac is wrapped tightly about a stopper through which a glass tube is fitted. The sac is then placed into a beaker containing pure (100%) water.

The *water concentration* on the two sides of the cellophane membrane is different. There is a lower water concentration inside the sac because the addition of sugar molecules has decreased the water concentration. As a result, osmosis occurs. Water moves across the membrane down its concentration gradient, from the beaker into the cellophane sac. You can see that osmosis is occurring by observing the rising fluid level in the glass tube.

What happens to the concentration and volume of the solution inside the cellophane sac? There is no movement of sugar from the sac into the beaker because the cellophane is

Figure 3.6 Diffusion through the plasma membrane.

Lipid-soluble substances can diffuse through the phospholipid bilayer whereas water-soluble materials may pass through the water-filled pores of channels.

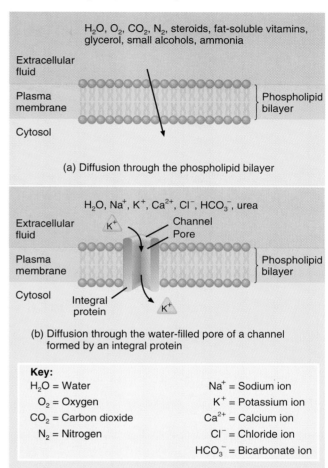

(a) Diffusion through the phospholipid bilayer

(b) Diffusion through the water-filled pore of a channel formed by an integral protein

Key:

H_2O = Water	Na^+ = Sodium ion
O_2 = Oxygen	K^+ = Potassium ion
CO_2 = Carbon dioxide	Ca^{2+} = Calcium ion
N_2 = Nitrogen	Cl^- = Chloride ion
	HCO_3^- = Bicarbonate ion

Which term better describes most molecules that can diffuse through the lipid bilayer: hydrophobic or hydrophilic?

impermeable to molecules of sugar. They are too large to pass through the pores of the membrane. As water moves into the cellophane sac, the sugar solution becomes more dilute, and the increased volume and pressure force the solution up the glass tubing. At equilibrium, the fluid level in the tube will not rise further because water molecules will be leaving and entering the cellophane sac at the same rate.

● **Osmotic Pressure** The pressure of the water column in the glass tube above the cellophane sac at equilibrium exactly balances the osmotic pressure of the sugar solution. **Osmotic pressure** is defined as the pressure required to stop movement of pure water into a solution containing solutes when the two are separated by a membrane permeable only to the water. The higher the concentration of solutes, the higher the osmotic pressure of a solution. A

Figure 3.7   Principle of osmosis. A sac made of cellophane (a selectively permeable membrane) contains a sugar (sucrose) solution and is immersed in a beaker of pure water. The arrows indicate movement of water molecules into and out of the sac. Sugar molecules are too large to pass out of the sac. (a) As the experiment starts, water molecules diffuse into the sac down the water concentration gradient. This is osmosis. (b) After some time, the volume of the sugar solution inside the sac has increased. At equilibrium, osmosis has stopped because the number of water molecules entering and leaving the cellophane sac is equal.

 Osmosis is the net diffusion of water molecules through a selectively permeable membrane.

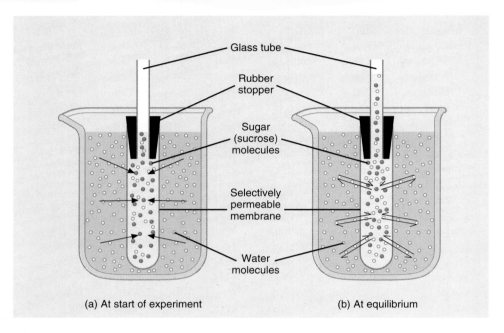

(a) At start of experiment   (b) At equilibrium

Will the fluid level in the tube rise until the sugar concentrations are the same in the sac and in the beaker?

solution having a higher osmotic pressure will tend to take up water from a solution that has a lower osmotic pressure if the two solutions are separated by a membrane that is permeable to water. Osmotic pressure is equal to the pressure in the water column but is exerted in the opposite direction.

Osmotic pressure is an important force in the movement of water between various compartments of the body. Normally, the osmotic pressure of intracellular fluid (inside cells) is the same as the osmotic pressure of the interstitial fluid outside cells. Because the osmotic pressure on both sides of the membrane is equal, cell volume remains relatively constant. But the osmotic pressure of blood plasma is greater than the osmotic pressure of interstitial fluid surrounding capillary walls. This difference tends to draw tissue water into blood capillaries. In conditions that decrease the blood osmotic pressure, as may occur in liver disease, excess water remains in interstitial fluid. Such buildup of interstitial fluid is termed edema. Blood osmotic pressure is discussed further in Chapter 21 on page 618.

● **Tonicity** Osmosis may also be understood by considering the effects of different water concentrations on cell shape. To maintain the normal shape of a cell, for example, a red blood

cell (RBC), the cell must be bathed in an **isotonic** (*iso* = same) **solution** (Fig. 3.8a). This is a solution in which the total concentrations of water molecules (solvent) and impermeable solute particles are the *same* on both sides. The concentrations of water and solute in the fluid outside the RBC must be the same as the concentration of the fluid inside the cell. Under ordinary circumstances, a 0.9% NaCl (salt) solution, called a **normal saline solution**, is isotonic for RBCs. Whereas the membrane of a RBC permits the water to move back and forth, it is nearly impermeable to Na^+ and Cl^-, the solutes. When RBCs are bathed in 0.9% NaCl, water molecules enter and exit the cells at the same rate, allowing the RBCs to maintain their normal shape and volume.

A different situation results if RBCs are placed in a solution that has a *lower* concentration of solutes and therefore a higher concentration of water. This is called a **hypotonic** (*hypo* = less than, under) **solution** (Fig. 3.8b). In this condition, water molecules enter the cells faster than they leave, causing the RBCs to swell and eventually burst. The rupture of RBCs in this manner is called **hemolysis** (hē-MOL-i-sis). Distilled water is strongly hypotonic.

A **hypertonic** (*hyper* = greater than, above) **solution** has a *higher* concentration of solutes and a lower concentration of

Figure 3.8 Tonicity and red blood cells.

An isotonic solution is one in which cell shape stays the same because there is no net water movement into or out of the cell.

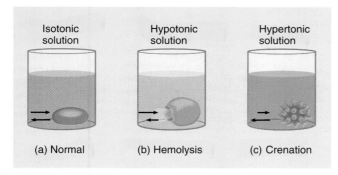

(a) Normal   (b) Hemolysis   (c) Crenation

Is a 2% NaCl solution hypotonic, hypertonic, or isotonic?

water than the RBCs (Fig. 3.8c). One example of a hypertonic solution is a 2% NaCl solution. In such a solution, water molecules move out of the cells faster than they enter. This causes the cells to shrink. The shrinkage of RBCs in this manner is called **crenation** (kri-NĀ-shun). RBCs and other body cells may be damaged or destroyed if exposed to hypertonic or hypotonic solutions. For this reason, most intravenous (IV) solutions, which are infused into the blood of patients, are isotonic.

Bulk Flow

The movement in the same direction of *large* numbers of ions, molecules, or particles that are dissolved or carried in a medium such as fluid or air is called **bulk flow**. The substances move in unison in response to forces that push them, such as differences in hydrostatic (water) pressure or air pressure, and they move at rates far greater than can be accounted for by simple diffusion or osmosis alone. Such movement is always from an area of higher pressure to an area of lower pressure and continues as long as a pressure difference exists. Two examples of bulk flow in the body are the flow of blood within vessels and the movement of air into and out of the lungs.

In another example of bulk flow, substances move across capillary walls from the blood into interstitial fluid. This process is called **filtration** and is driven by the pressure difference between blood and interstitial fluid. Blood pressure is a hydrostatic (water) pressure that is generated by the pumping action of the heart. Blood hydrostatic pressure forces most small- to medium-sized molecules such as water, nutrients, gases, ions, hormones, and vitamins through the capillary walls. Larger proteins remain in the blood because they are too big to be forced through the capillary walls. The greatest amount of filtration occurs in the kidneys. Blood hydrostatic pressure forces water and waste products such as creatinine through specialized filtering

units (glomeruli) in the kidneys. Molecules of many harmful substances and waste products are small enough to be filtered and then they can be eliminated in the urine.

Facilitated Diffusion

Many ions, urea, glucose, fructose, galactose, and certain vitamins, which are too lipid-insoluble to diffuse through the phospholipid bilayer, can cross the plasma membrane by **facilitated diffusion**. In this process, the substance moves *down its concentration gradient* from a region of higher concentration to a region of lower concentration with the help of specific integral proteins in the membrane that serve as water-filled **channels** or **transporters** (**carriers**) for each type of substance. The rate of facilitated diffusion is determined by the size of the concentration difference on the two sides of the membrane and the number of channels or transporters available.

Glucose is an important substance that enters many body cells by facilitated diffusion, as follows (Fig. 3.9):

1 First, glucose attaches to a transporter on the outside of the membrane. Different cells have different glucose transporters, which are called GluT1, GluT2, and so on—numbered in the order they were discovered.

2 Then the transporter changes shape.

3 Glucose passes through the membrane and is released inside the cell.

After glucose has entered a cell by facilitated diffusion, an enzyme called a kinase attaches a phosphate group to

Figure 3.9 Facilitated diffusion of glucose.

 Facilitated diffusion across a membrane requires a channel or transporter protein but does not use ATP.

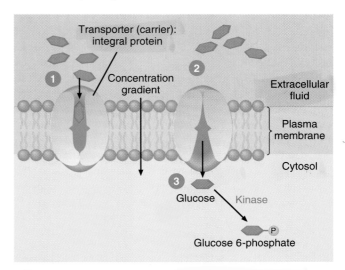

What determines the rate of facilitated diffusion?

produce glucose 6-phosphate. This reaction keeps the intracellular concentration of glucose itself very low so that the concentration gradient always favors facilitated diffusion of glucose into, not out of, cells.

Selective permeability of the plasma membrane is often regulated to achieve homeostasis. For example, facilitated diffusion of glucose into certain cells (those having GluT4) is greatly accelerated by the hormone insulin. Thus insulin lowers blood glucose level by speeding the movement of glucose from the blood into body cells.

Active Processes

Some substances that need to enter or leave body cells cannot move across plasma membranes passively because they need to move against the concentration gradient. Such substances can cross the membrane by **active processes**. These are the ATP-consuming processes of active transport and vesicular transport. In **active transport**, particular integral membrane proteins act as ATP-driven pumps to push certain ions and some smaller molecules across the membrane. **Vesicular transport** involves the formation of vesicles that either detach from the plasma membrane while bringing materials into the cell or merge with the plasma membrane to release materials from the cell. Large particles, such as whole bacteria and red blood cells, and large molecules, for example, polysaccharides and proteins, may enter and leave cells by vesicular transport. Like active transport, vesicular transport requires energy derived from splitting ATP.

Active Transport

The list of actively transported substances includes several ions such as Na^+, K^+, H^+, Ca^{2+}, I^-, Cl^-; amino acids; and monosaccharides. (Recall that these substances may also cross the membrane by facilitated diffusion if the proper channel proteins and transporters are present.) There are two types of **active transport**. In **primary active transport**, energy derived from splitting ATP *directly* moves or "pumps" a substance across the membrane. The cell uses energy from ATP to change the shape of transport proteins in the plasma membrane. A typical body cell expends about 40% of the ATP it generates for primary active transport. Drugs that turn off ATP production—for example, the poison cyanide—are lethal because they shut down active transport in cells throughout the body. In **secondary active transport**, the energy stored in an ionic concentration difference (gradient) is used to drive substances across the membrane. Because the ion concentration difference is established by primary active transport, secondary active transport *indirectly* uses energy obtained from splitting ATP.

● **Primary Active Transport: The Sodium Pump** The most prevalent primary active transport mechanism expels sodium ions (Na^+) from cells and brings potassium ions (K^+) in. These pumps maintain a low concentration of Na^+ in the cytosol by pumping it out against the concentration gradient. They also move K^+ into cells against its concentration gradient.

Because of the ions it moves, this primary active transport mechanism is called the **Na^+/K^+ pump** or, more simply, the **sodium pump**. All cells have hundreds of sodium pumps in each square micrometer (μm^2) of membrane surface. Since the pump is a protein that acts as an enzyme to split the needed ATP, it is also called **Na^+/K^+ ATPase**. Figure 3.10 shows our current understanding of the sodium pump, which is actually a group of similar pumps with slightly different properties. Many operate as diagrammed in Fig. 3.10, expelling three Na^+ each time they import two K^+. Others exchange one Na^+ for each K^+. The sodium pump must work continually because K^+ and Na^+ slowly leak across the plasma membrane through leakage channels.

● **Secondary Active Transport: Symporters and Antiporters** The sodium pump maintains a large concentration difference of Na^+ across the plasma membrane. The sodium ions have stored (potential) energy just as water behind a dam has potential energy. Accordingly, if there is a route for Na^+ to leak back in, some of the stored energy can be converted to kinetic energy (energy of motion) and used to transport other substances *against* their concentration differences. Normally, the plasma membrane is fairly impermeable to Na^+ because there are few leakage channels that admit Na^+. Secondary active transport proteins harness the energy in the Na^+ concentration difference by providing easy routes for Na^+ to leak into cells. This is like allowing water behind a dam to flow through an outlet pipe and turn a turbine to produce electrical power.

When two substances move in the same direction across a plasma membrane, the process is called **symport** or **cotransport** (Fig. 3.11a). For example, glucose, galactose, and amino acids enter cells lining the small intestine and the kidney tubules via Na^+ symporters. This is how dietary monosaccharides and amino acids are absorbed and nutrients filtered by the kidneys are returned to the bloodstream so they are not lost in the urine. Through **antiport** or **countertransport** two substances move in opposite directions across the plasma membrane (Fig. 3.11b). In most cells, Na^+/Ca^{2+} antiporters keep Ca^{2+} concentration low in the cytosol. Likewise, Na^+/H^+ antiporters help regulate the cytosol's pH (H^+ concentration) by using the Na^+ gradient to expel H^+.

The protein that forms a symporter or antiporter works by simultaneously binding to Na^+ and the other substance and then changing its shape so that both substances cross the membrane at the same time. The larger the concentration difference of Na^+ across the membrane, the faster the secondary active transport.

Figure 3.10 The sodium pump (Na⁺/K⁺ ATPase). Sodium ions (Na⁺) are expelled and potassium ions (K⁺) are imported. The pump will not work unless Na⁺ and ATP are present in the cytosol and K⁺ is present in the extracellular fluid.

🔑 *The sodium pump maintains a low intracellular concentration of sodium ions.*

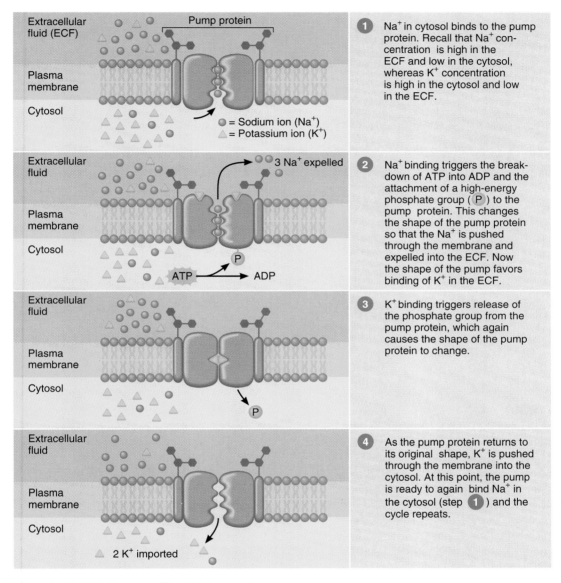

1 Na⁺ in cytosol binds to the pump protein. Recall that Na⁺ concentration is high in the ECF and low in the cytosol, whereas K⁺ concentration is high in the cytosol and low in the ECF.

2 Na⁺ binding triggers the breakdown of ATP into ADP and the attachment of a high-energy phosphate group (P) to the pump protein. This changes the shape of the pump protein so that the Na⁺ is pushed through the membrane and expelled into the ECF. Now the shape of the pump favors binding of K⁺ in the ECF.

3 K⁺ binding triggers release of the phosphate group from the pump protein, which again causes the shape of the pump protein to change.

4 As the pump protein returns to its original shape, K⁺ is pushed through the membrane into the cytosol. At this point, the pump is ready to again bind Na⁺ in the cytosol (step **1**) and the cycle repeats.

Q What is the role of ATP in the operation of this pump?

DIGITALIS

Because it strengthens the heartbeat, digitalis (Lanoxin®) often is given to patients with heart failure, a condition of weakened pumping action by the heart. Digitalis exerts its effect by slowing the sodium pump, which lets more Na⁺ accumulate inside heart muscle cells. The result is a smaller Na⁺ concentration differ-ence across the membrane. In turn, the Na⁺/Ca²⁺ antiporters slow down due to the smaller Na⁺ concentration difference. As a result, more Ca²⁺ remains inside heart muscle cells, and this increases the force of heart muscle contraction. As this example illustrates, the balance between the concentrations of Na⁺ and Ca²⁺ in the cytosol and extracellular fluid is crucial to the normal functioning of nerve and muscle cells. ■

Figure 3.11   Secondary active transport mechanisms. (a) Symporters carry two substances across the membrane in the same direction. (b) Antiporters carry two substances across the membrane in opposite directions.

Secondary active transport mechanisms use the energy stored in an ionic concentration difference (here, for Na⁺). Because this difference is maintained by primary active transport pumps that split ATP, secondary active transport consumes ATP indirectly.

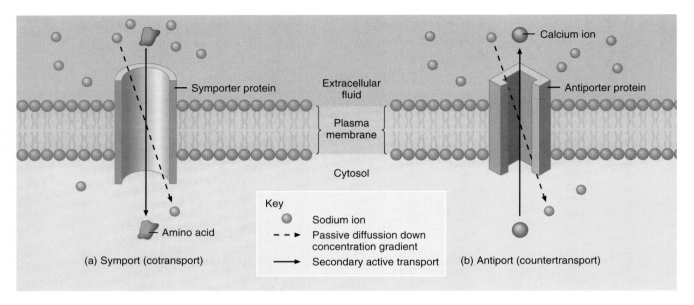

Q What is the main difference between secondary and primary active transport mechanisms?

Vesicular Transport

We will now examine how *larger* substances move across plasma membranes by several types of vesicular (bulk) transport: phagocytosis, pinocytosis, receptor-mediated endocytosis, and exocytosis.

Phagocytosis, pinocytosis, and receptor-mediated endocytosis are different types of **endocytosis** (*endo* = into; *cyt* = cell; *osis* = process); that is, they are processes that bring substances *into* cells. In endocytosis, a segment of the plasma membrane surrounds the substance to be taken in, encloses it, and brings it into the cell.

Exocytosis (*exo* = out of) is a reverse process; it discharges substances *from* cells. Endocytosis and exocytosis are important because molecules and particles of material that would normally be restricted from crossing the plasma membrane because of their large size can be brought in or removed from the cell.

● **Phagocytosis** In **phagocytosis** (fag′-ō-sī-TŌ-sis; *phagein* = to eat) or "cell eating," projections of the plasma membrane and cytoplasm, called **pseudopods** (SOO-dō-pods; *pseudo* = false; *pous* = foot), surround large solid particles outside the cell and then engulf them (Fig. 3.12a). Once the particle is sur-

rounded, the pseudopods fuse. In this way, the particle becomes enclosed within a sac of membrane and enters the cytoplasm. The membrane-bounded sac is called a **phagocytic vesicle** (**phagosome**). Solid material inside the vesicle is digested by enzymes provided by lysosomes (described on page 73).

Certain cells, called **phagocytes**, perform phagocytosis and thereby destroy bacteria and other foreign substances (Fig. 3.12b). Important phagocytes include neutrophils, a type of white blood cell, and macrophages, which are present in most body tissues. The process of phagocytosis is a vital defense mechanism that helps protect us from disease (described further on page 682).

● **Pinocytosis** In **pinocytosis** (pi′-nō-sī-TŌ-sis; *pinein* = to drink) or "cell drinking," the engulfed material is a tiny droplet of extracellular fluid rather than a solid particle. Moreover, no pseudopods are formed. Instead, the membrane folds inward, forming a **pinocytic vesicle** that allows the liquid to flow inward and then surrounds the liquid. The pinocytic vesicle next detaches from the rest of the intact membrane. Whereas only certain cells are capable of phagocytosis, most cells carry on pinocytosis.

● **Receptor-Mediated Endocytosis** Although similar to pinocytosis, **receptor-mediated endocytosis** is a highly selective process in which cells can take up specific molecules or particles. The extracellular fluid that bathes cells contains a large number of chemicals, most of which are present in low concentration. Some of these substances are ligands that bind to specific protein receptors in the plasma membrane. Ligands serve many functions. For example, cholesterol, iron, and vitamins are needed for metabolic reactions that sustain life. Other ligands include hormones that deliver messages to cells, so cells can respond in specific ways, and waste products that certain cells have the ability to break down.

Although receptor-mediated endocytosis normally imports *needed* materials, some viruses can sneak into and infect body cells by this process. For example, the human immunodeficiency virus (HIV), which causes acquired immune deficiency syndrome (AIDS), enters cells by attaching to a glycoprotein receptor called CD4. This receptor is present at the surface of certain types of body cells such as white blood cells called helper T cells. After binding to CD4, HIV enters the cell by receptor-mediated endocytosis.

Figure 3.12 Phagocytosis.

 Phagocytic cells destroy microbes and other foreign substances, an activity that helps protect against disease.

Step 1: Membrane of phagocytic cell begins to extend projections called pseudopods toward a large solid particle.

Step 2: Pseudopods lengthen, extending along the particle.

Step 4: Pseudopods fuse, creating a phagocytic vesicle inside the cell. The particle is then destroyed by digestive enzymes.

Step 3: Pseudopods nearly encircle the particle.

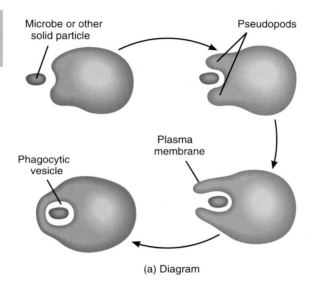

Microbe or other solid particle

Pseudopods

Plasma membrane

Phagocytic vesicle

(a) Diagram

Pseudopod

White blood cell

Pseudopod

Microbe

White blood cell engulfs microbe

White blood cell destroys microbe

(b) Color-enhanced electron micrographs showing phagocytosis

 Is phagocytosis an example of endocytosis or exocytosis?

Receptor-mediated endocytosis occurs as follows (Fig. 3.13):

1 On the extracellular side of the plasma membrane the ligand binds to its specific receptor, which is an integral membrane protein.

2 This interaction causes the membrane to fold inward, forming an **endocytic vesicle** that contains the ligands bound to their receptors.

3 After a vesicle moves inward from the plasma membrane, it merges with other endocytic vesicles to form a larger structure called an **endosome**.

4 Within the endosome, receptors separate from their ligands.

5 The portion of the endosome that contains the receptors pinches off and moves back to the plasma membrane. There the receptors are reinserted into the plasma membrane by exocytosis.

6 The remaining portion of the endosome, containing the ingested material, merges with a lysosome. The ingested materials are then degraded by digestive enzymes from the lysosome or released for use by the cell.

● **Exocytosis** During **exocytosis** membrane-enclosed structures called **secretory vesicles** that form inside the cell fuse with the plasma membrane and release their contents into the extracellular fluid (see Fig. 3.19). Exocytosis occurs in all cells but is especially important in two types of cells: (1) nerve cells, which release their neurotransmitter substances by this process, and (2) secretory cells, for example, cells that secrete digestive enzymes or protein hormones, such as insulin.

Exhibit 3.1 lists and summarizes the passive and active processes that move materials across plasma membranes.

LIPOSOMES AND DRUG THERAPY

Despite the impact of antibiotics and other drugs in the treatment of disease, they generally have disadvantages in that they are cleared from the blood rapidly and may be toxic to healthy cells as well as microorganisms and diseased cells. In an attempt to overcome these problems, scientists are experimenting with artificially made phospholipid vesicles called **liposomes** (LIP-ō-sōms). When mixed with a solution containing a drug, the phospholipids can form small vesicles that enclose a tiny droplet of the drug solution. Experiments are also underway to use liposomes for gene therapy, for example, to transfer good copies of the defective gene into people with cystic fibrosis. Liposomes can be injected, inhaled as an aerosol, or applied to the skin. The liposomes are taken into cells by diffusion or endocytosis. Liposomes release their contents slowly, which maintains a more even blood level and reduces toxicity.

Scientists are also experimenting with incorporating anticancer drugs into low-density lipoprotein (LDL) particles, which enter body cells by receptor-mediated endocytosis. LDL is a natural particle so it may escape being attacked by the body defense systems. Some artificially produced particles are thought to be attacked by defense mechanisms before they are able to deliver their drugs. ■

Figure 3.13 Receptor-mediated endocytosis.

🔑 *Normally, receptor-mediated endocytosis imports materials that are needed by cells.*

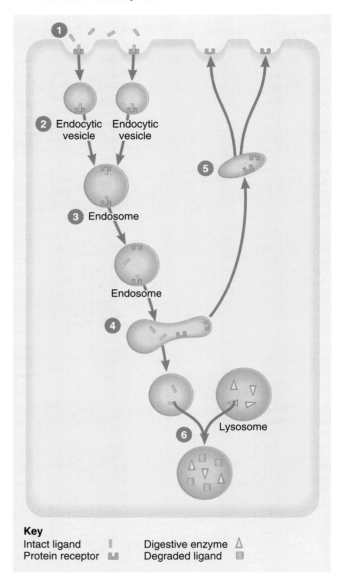

Key

Intact ligand	Digestive enzyme △
Protein receptor	Degraded ligand

Q How does receptor-mediated endocytosis differ from pinocytosis?

CYTOSOL

The intracellular fluid, in which organelles and inclusions are suspended and solutes are dissolved, is called the **cytosol** (SĪ-tō-sol). Physically, cytosol is a viscous, transparent, gel-like fluid containing suspended particles and a series of minute tubules and filaments that form a cytoskeleton (described on page 74). **Cytoplasm** (SĪ-tō-plazm′; *cyto* = cell; *plassein* = to form) includes cytosol, all organelles (except the nucleus), and inclusions.

EXHIBIT **3.1** **SUMMARY OF MOVEMENT OF MATERIALS ACROSS PLASMA MEMBRANES**

Process	Description	Examples
PASSIVE PROCESSES	Substances move down a concentration gradient from an area of higher to lower concentration or pressure; cell does not expend energy.	
Diffusion	**Simple diffusion** is the *random* movement of molecules or ions due to their kinetic energy. In **net diffusion**, a substance moves *down a concentration gradient* until it is equally distributed. At that point (**equilibrium**), net diffusion ceases but simple diffusion continues.	H_2O, O_2, CO_2, N_2, steroids, fat-soluble vitamins, glycerol, small alcohols, and ammonia (through the phospholipid bilayer).
Osmosis	Diffusion of water molecules across a selectively permeable membrane from an area of higher to lower concentration of water until an equilibrium is reached. The solute concentration produces osmotic pressure.	A solvent, usually water in living systems.
Bulk flow	Movement in the same direction of large numbers of ions, molecules, or particles that are dissolved or carried in a medium such as water or air as a result of a hydrostatic (water) or air pressure difference. **Filtration** is a type of bulk flow in which water and solutes pass through capillary walls from blood into interstitial fluid or urine.	Air flowing into and out of lungs; fluid flowing through capillary walls.
Facilitated diffusion	Movement of ions or molecules across a selectively permeable membrane aided by specific membrane proteins that serve as channels or transporters for particular substances.	Many ions, urea, glucose, fructose, galactose, and certain vitamins, as well as water.
ACTIVE PROCESSES	Cell expends energy by splitting ATP to move substance against a concentration gradient.	
Primary active transport	Movement of ions or molecules across a selectively permeable membrane, from a region of lower to higher concentration by pump proteins that use energy from the splitting of ATP.	Na^+, K^+, Ca^{2+}, H^+, I^-, Cl^-, and other ions.
Secondary active transport	Simultaneous movement of two substances across the membrane, one of which is Na^+. Uses energy supplied by the Na^+ concentration gradient, which is maintained by primary active transport pumps.[a] **Symporters** (cotransporters) move Na^+ and another substance in the same direction across the membrane. **Antiporters** (countertransporters) move Na^+ and another substance in opposite directions across the membrane.	Symport—Glucose, into cells lining the small intestine and the kidney tubules; amino acids into most body cells. Antiport—Ca^{2+}, H^+ out of many cells.
Vesicular transport	Processes in which small vesicles bud from the plasma membrane to bring material into or take it out of a cell.	
Phagocytosis	"Cell eating"; movement of solid particles through the plasma membrane. Pseudopods extend around the substance, enclose it, and bring it into the cell, forming a phagocytic vesicle (phagosome).	Solid materials such as bacteria, viruses, or aged red blood cells.
Pinocytosis	"Cell drinking"; movement of extracellular fluid droplets into cell by infolding of plasma membrane, forming pinocytic vesicle.	Solutes dissolved in extracellular fluid.
Receptor-mediated endocytosis	Mechanism for selected substances (ligands) to move into cells; ligand binds to receptor at extracellular surface of plasma membrane; membrane then folds inward, forming endocytic vesicle.	Low density lipoprotein (LDL) particles; some vitamins, minerals, and hormones.
Exocytosis	Process for exporting substances from the cell in which vesicles fuse with the plasma membrane and release their contents into the extracellular fluid.	Neurotransmitters, protein hormones, and digestive enzymes.

[a] Symporters and antiporters using ions other than Na^+ exist but are less common.

Chemically, cytosol is 75–90% water plus solid components. Proteins, carbohydrates, lipids, and inorganic substances comprise most of the solids. Inorganic substances and smaller organic substances, such as simple sugars and amino acids, are soluble in water and are present as solutes. Larger organic compounds, like proteins and the polysaccharide glycogen, are found as **colloids**—particles that remain suspended in the surrounding medium although they are not dissolved. The colloids bear electrical charges that repel each other and thus remain suspended and separated.

Functionally, cytosol is the medium in which many metabolic reactions occur. The cytosol receives raw materials from the external environment by way of extracellular fluid and obtains usable energy from them by decomposition reactions.

ORGANELLES

Despite many chemical activities occurring at the same time in the cell, there is little interference of one reaction with another. This is because the cell has many different compartments provided by its **organelles** ("little organs"). These are specialized structures that have characteristic appearances and specific roles in growth, maintenance, repair, and control. The numbers and types of organelles vary in different kinds of cells, depending on their functions.

Nucleus

The **nucleus** (NOO-klē-us; *nucleus* = kernel) is usually a spherical or oval organelle and is the largest structure in the cell (Fig. 3.14). Within the nucleus are most of the hereditary units of the cell, called **genes**, which control cellular structure and direct many cellular activities. (Mitochondria—described shortly—also contain a few genes.) The nuclear genes are arranged in single file along structures termed **chromosomes** (*chroma* = colored; *soma* = body). Human somatic (body) cells have 46 chromosomes, 23 inherited

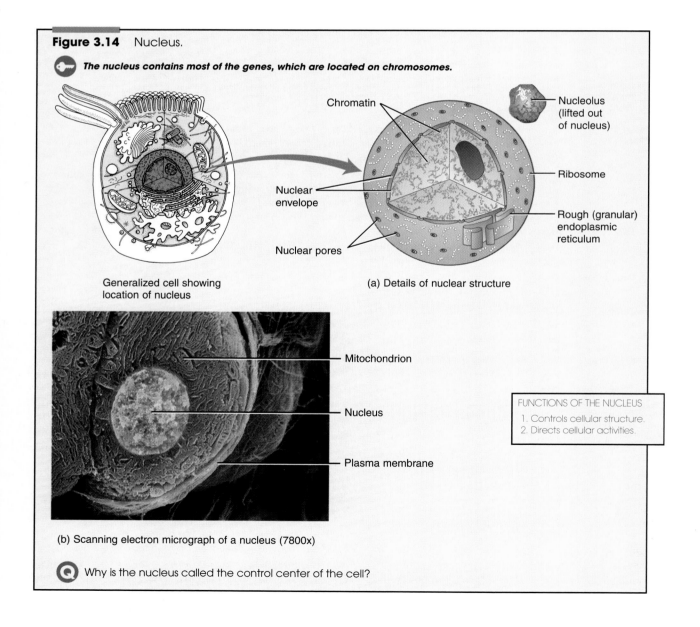

Figure 3.14   Nucleus.

The nucleus contains most of the genes, which are located on chromosomes.

Chromatin

Nucleolus (lifted out of nucleus)

Nuclear envelope

Ribosome

Nuclear pores

Rough (granular) endoplasmic reticulum

Generalized cell showing location of nucleus

(a) Details of nuclear structure

Mitochondrion

Nucleus

Plasma membrane

FUNCTIONS OF THE NUCLEUS
1. Controls cellular structure.
2. Directs cellular activities.

(b) Scanning electron micrograph of a nucleus (7800x)

Q   Why is the nucleus called the control center of the cell?

from each parent. Most body cells contain a nucleus, although some, such as mature red blood cells, do not. Skeletal muscle cells and a few other cells contain several nuclei.

A double membrane called the **nuclear envelope** separates the nucleus from the cytoplasm. Both layers of the nuclear envelope are phospholipid bilayers similar to the plasma membrane. Water-filled **nuclear pores** in the envelope allow most ions and water-soluble molecules to shuttle between the nucleus and the cytoplasm. Because nuclear pores are about ten times larger in diameter than the pores of channels in the plasma membrane, even large molecules such as RNA and various proteins can pass through.

One or more spherical bodies called **nucleoli** (noo-KLĒ-ō-lī′; singular is **nucleolus**) are present inside the nucleus. They are clusters of protein, DNA, and RNA that are not enclosed by a membrane. Nucleoli are the sites of assembly of ribosomes (described next) and contain a type of RNA called ribosomal RNA. Ribosomes play a key role in protein synthesis. Nucleoli disperse and disappear during cell division and reorganize once new cells are formed.

A chromosome is a very long DNA molecule that is coiled and packed into an amazingly compact structure together with several proteins. In a cell that is not dividing, the 46 chromosomes are a tangled mass of stringy threads, which is called **chromatin** and has a fuzzy appearance under the light microscope. Electron micrographs reveal that chromatin has a "beads-on-a-string" structure. Each "bead" is a **nucleosome** and consists of double-stranded DNA wrapped twice around a core of eight proteins called **histones** (Fig. 3.15). Histones help organize the coiling and folding of DNA. The "string" between "beads" is **linker DNA**, which holds adjacent nucleosomes together. Another histone promotes coiling of nucleosomes into a larger diameter **chromatin fiber**, which then folds into large **loops**. In cells that are not dividing, this is the extent of DNA packing. However, before cell division the DNA replicates (duplicates) and the loops condense even more into **chromatids**. During this stage of cell division, a pair of chromatids constitutes a chromosome that is easily seen as a rod-shaped structure under the light microscope.

Ribosomes

Ribosomes (RĪ-bō-sōms) are tiny granules that contain **ribosomal RNA (rRNA)** and many ribosomal proteins. The rRNA is synthesized by DNA in the nucleolus. Ribosomes were so named because of their high content of *ribo*nucleic acid. This particular type of RNA was subsequently named rRNA to distinguish it from the other types of RNA in the cell. Structurally, a ribosome consists of two subunits, one about half the size of the other (Fig. 3.16). Functionally, ribosomes are the sites of protein synthesis, which is discussed later in the chapter.

Some ribosomes, called **free ribosomes**, float in the cytosol; they have no attachments to other organelles (see Fig.

Figure 3.15 Packing of DNA into chromosomes.

A chromosome is a highly coiled and folded DNA molecule that is combined with protein molecules.

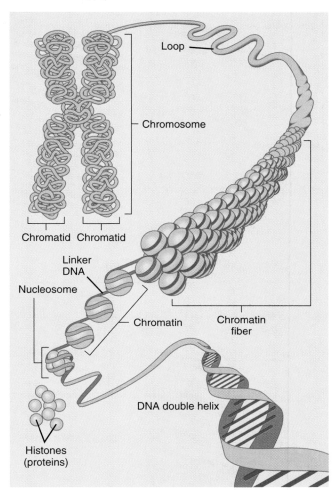

What are the components of a nucleosome?

3.1). The free ribosomes occur singly or in clusters, and they are concerned primarily with synthesizing proteins for use inside the cell. Other ribosomes attach to a cellular structure called the endoplasmic reticulum (see Fig. 3.17). These ribosomes are involved in the synthesis of proteins destined for insertion in the plasma membrane or for export from the cell.

Endoplasmic Reticulum

The **endoplasmic reticulum** (en′-dō-PLAS-mik rē-TIK-yoo-lum; *endo* = within; *plasmic* = cytoplasm; *reticulum* = network) or **ER** is a system of membrane-enclosed channels of varying shapes called **cisterns** (SIS-terns; *cistern* = cavity

Figure 3.16 Ribosomes.

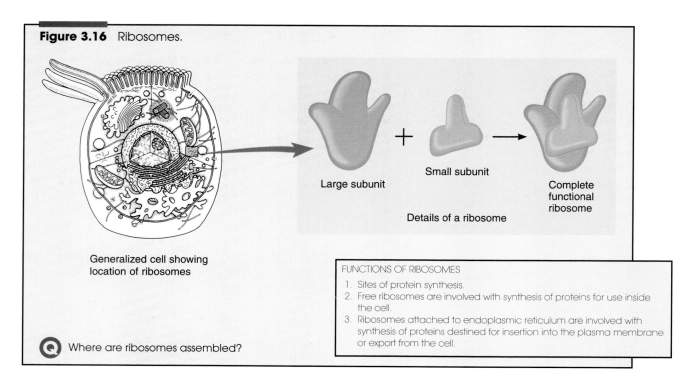

Details of a ribosome

Large subunit

Small subunit

Complete functional ribosome

Generalized cell showing location of ribosomes

Q Where are ribosomes assembled?

FUNCTIONS OF RIBOSOMES

1. Sites of protein synthesis.
2. Free ribosomes are involved with synthesis of proteins for use inside the cell.
3. Ribosomes attached to endoplasmic reticulum are involved with synthesis of proteins destined for insertion into the plasma membrane or export from the cell.

Figure 3.17 Endoplasmic reticulum.

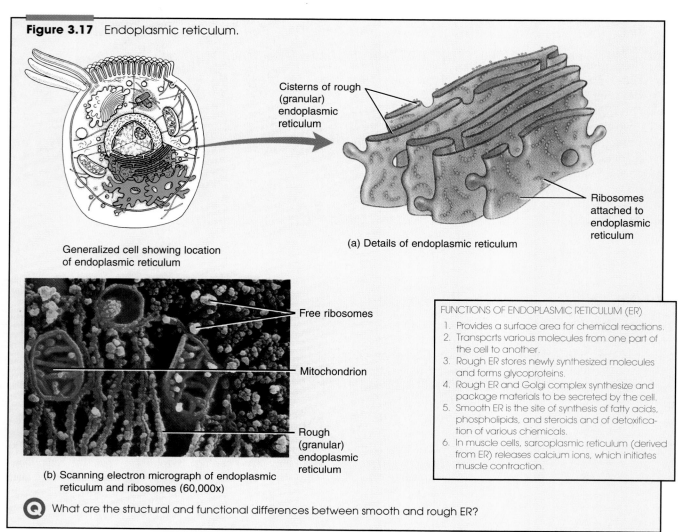

Generalized cell showing location of endoplasmic reticulum

Cisterns of rough (granular) endoplasmic reticulum

Ribosomes attached to endoplasmic reticulum

(a) Details of endoplasmic reticulum

Free ribosomes

Mitochondrion

Rough (granular) endoplasmic reticulum

(b) Scanning electron micrograph of endoplasmic reticulum and ribosomes (60,000x)

FUNCTIONS OF ENDOPLASMIC RETICULUM (ER)

1. Provides a surface area for chemical reactions.
2. Transports various molecules from one part of the cell to another.
3. Rough ER stores newly synthesized molecules and forms glycoproteins.
4. Rough ER and Golgi complex synthesize and package materials to be secreted by the cell.
5. Smooth ER is the site of synthesis of fatty acids, phospholipids, and steroids and of detoxification of various chemicals.
6. In muscle cells, sarcoplasmic reticulum (derived from ER) releases calcium ions, which initiates muscle contraction.

Q What are the structural and functional differences between smooth and rough ER?

or reservoir) or **cisternae** (Fig. 3.17). The ER is continuous with the nuclear envelope. The two types of ER are **rough (granular) ER**, which is studded with ribosomes, and **smooth (agranular) ER**, which has no ribosomes.

Ribosomes associated with rough ER synthesize proteins. The rough ER also serves as a temporary storage area for newly synthesized molecules and may add sugar groups to certain proteins, thus forming glycoproteins. Together, the rough ER and the Golgi complex (another organelle, described next) synthesize and package molecules that will be secreted from the cell.

Smooth ER is the site of fatty acid, phospholipid, and steroid synthesis. Also, in certain cells, enzymes within the smooth ER can inactivate or detoxify a variety of chemicals, including alcohol, pesticides, and carcinogens (cancer-causing agents). In muscle cells, calcium ions released from the sarcoplasmic reticulum, which resembles smooth ER, trigger contraction.

Golgi Complex

The **Golgi** (GOL-jē) **complex** (**apparatus**) is an organelle located near the nucleus. In cells with high secretory activity, the Golgi complex is extensive. It consists of flattened sacs called **cisterns** (**cisternae**), stacked upon each other like a pile of plates with expanded bulges at their edges (Fig. 3.18). Associated with the cisterns are small **Golgi vesicles**, which cluster along the expanded edges of the cisterns.

The Golgi complex processes, sorts, packages, and delivers proteins and lipids to the plasma membrane and forms lysosomes and secretory vesicles. All proteins destined for export from the cell follow a similar route: ribosomes (site of protein synthesis) → rough ER cistern → transport vesicles → Golgi complex → secretory vesicles → release to exterior of the cell by exocytosis. Proteins and lipids destined for inclusion in the plasma membrane or for use inside lysosomes also pass through the Golgi complex.

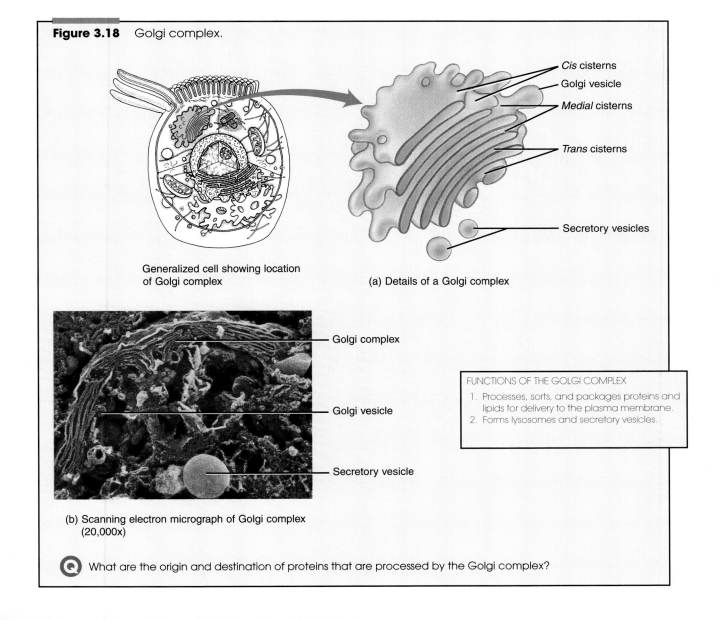

Figure 3.18   Golgi complex.

Cis cisterns
Golgi vesicle
Medial cisterns
Trans cisterns
Secretory vesicles

Generalized cell showing location of Golgi complex

(a) Details of a Golgi complex

Golgi complex
Golgi vesicle
Secretory vesicle

(b) Scanning electron micrograph of Golgi complex (20,000x)

FUNCTIONS OF THE GOLGI COMPLEX
1. Processes, sorts, and packages proteins and lipids for delivery to the plasma membrane.
2. Forms lysosomes and secretory vesicles.

Q What are the origin and destination of proteins that are processed by the Golgi complex?

A trip through a Golgi complex normally occurs as follows (Fig. 3.19):

1 Within the cistern of the rough ER, proteins (and glycoproteins) become surrounded by a piece of the ER membrane, which forms a **transport vesicle**.

2 The transport vesicle buds off the ER and moves to the Golgi complex.

3 Here, the vesicle fuses with the side of the Golgi stack closest to the ER, which is termed the *cis* or **entry cistern**. As a result of this fusion, the proteins enter the Golgi complex.

4 Once inside, they move through the *cis* cistern into a **medial (middle) cistern**. Their movement from one cistern to the next is probably by way of vesicles that bud from the edges of each cistern in the stack.

5 As the proteins pass through the Golgi cisterns, they are modified in various ways depending on their function and destination. After moving through one or more medial cisterns, the proteins enter the *trans* or **exit cistern**.

6 The *trans* cistern packages the modified proteins into vesicles.

7 Some vesicles become secretory vesicles, which undergo exocytosis to discharge their contents into the extracellular fluid. Certain cells of the pancreas secrete digestive enzymes in this way. The vesicle membrane prevents the enzymes within from digesting the contents of the pancreatic cell itself.

8 Other vesicles that leave the Golgi complex are loaded with digestive enzymes intended for use within the cell. They become organelles called lysosomes (described next).

CYSTIC FIBROSIS

Sometimes a disorder results from faulty instructions for routing of a molecule. Such is apparently the case in cystic fibrosis, a deadly inherited disease (see page 747). The protein produced by the mutated cystic fibrosis gene fails to reach the plasma membrane, where it should be inserted to help pump chloride ions (Cl^-) out of certain cells. Evidently, the pump protein becomes stuck in the endoplasmic reticulum or Golgi complex and never reaches its correct destination. The result is an imbalance in the transport of fluid and ions across the plasma membrane and excessive buildup of mucus outside certain types of cells. The mucus clogs the lining of the airways to the lungs, causing breathing difficulty, and prevents proper secretion of digestive enzymes by the pancreas. ■

Figure 3.19   Packaging of synthesized proteins into secretory vesicles and lysosomes by the Golgi complex.

All proteins exported from the cell are processed in the Golgi complex.

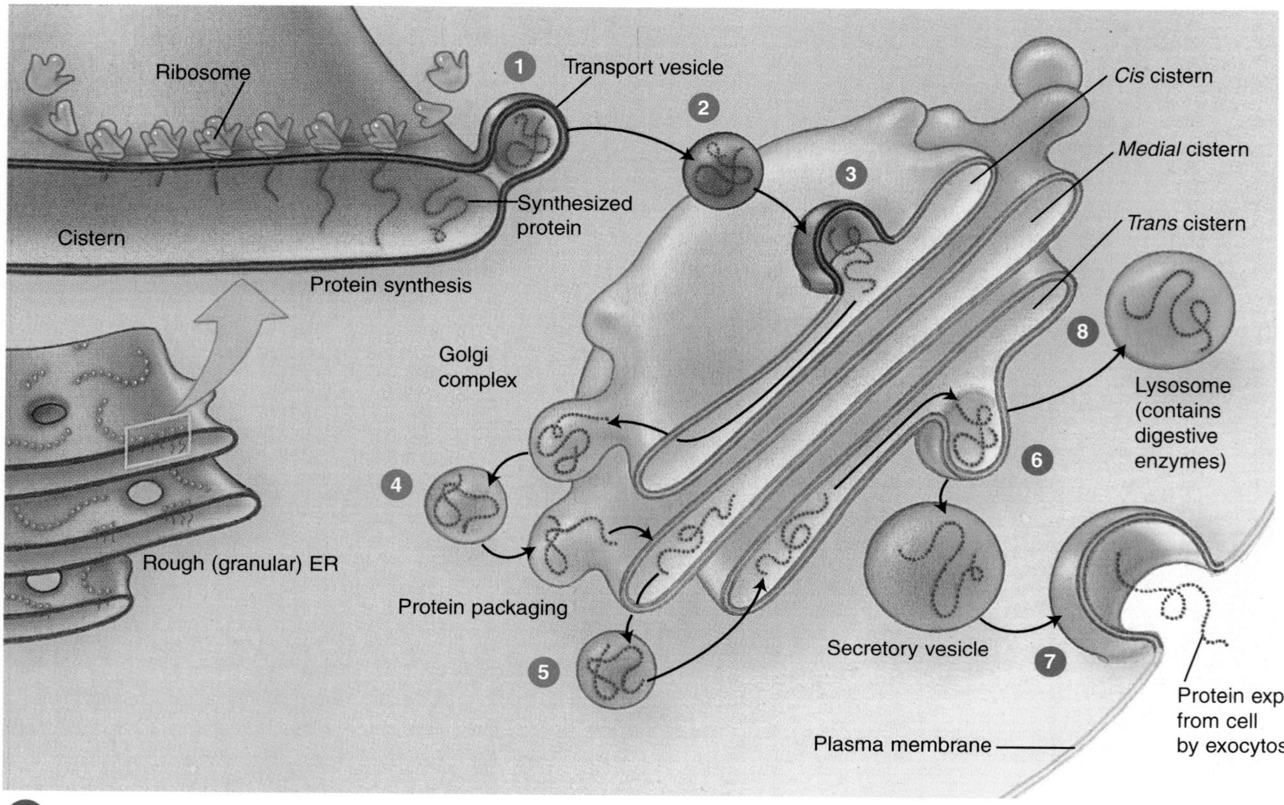

What is the advantage of packaging digestive enzymes into vesicles?

Lysosomes

Lysosomes (LĪ-sō-sōms; *lysis* = dissolution; *soma* = body) are membrane-enclosed vesicles (see Fig. 3.19) that form in the Golgi complex. Inside are as many as 40 kinds of powerful digestive (hydrolytic) enzymes capable of breaking down a wide variety of molecules. Some disorders are caused by faulty lysosomes. For example, Tay–Sachs disease is an inherited absence of a single lysosomal enzyme. This enzyme normally breaks down a membrane glycolipid (called ganglioside GM_2) that is especially prevalent in nerve cells. As GM_2 accumulates, the nerve cells function less efficiently. The child becomes blind, demented, and uncoordinated and dies, usually before the age of 5.

Lysosomal enzymes work best at an acidic pH. The lysosomal membrane includes active transport pumps that drive hydrogen ions (H^+) into the lysosomes. Thus the interior of a lysosome has a pH of 5, which is 100 times more acidic than the cytosolic pH of 7.

Lysosomal enzymes digest bacteria and other substances that enter the cell in phagocytic vesicles during phagocytosis, pinocytic vesicles during pinocytosis, or endosomes during receptor-mediated endocytosis. Eventually, the products of digestion are small enough to pass out of the lysosome into the cytosol. There, they are recycled to synthesize various molecules needed by the cell.

Lysosomal enzymes also help recycle the cell's own structures. A lysosome can engulf another organelle, digest it, and return the digested components to the cytosol for reuse. In this way, old organelles are continually replaced. The process by which worn-out organelles are digested is called **autophagy** (aw-TOF-a-jē; *auto* = self; *phagein* = to eat). A human liver cell recycles about half its contents every week. Lysosomal enzymes may also destroy their host cell, a process known as **autolysis** (aw-TOL-i-sis). Autolysis occurs after death and in some pathological conditions. Lysosomes also function in extracellular digestion. Lysosomal enzymes released at sites of injury help digest cellular debris, which prepares the injured area for effective repair.

Peroxisomes

Another group of organelles similar in structure to lysosomes, but smaller, are called **peroxisomes** (pe-ROKS-i-sōms; *perox* = peroxide; *soma* = body). See Fig. 3.1. They are so named because they usually contain one or more enzymes that use molecular oxygen to oxidize (remove hydrogen atoms from) various organic substances. Such reactions produce hydrogen peroxide (H_2O_2). One of the enzymes in peroxisomes, called *catalase,* uses the H_2O_2 generated by other enzymes to oxidize a variety of substances, including phenol, formic acid, formaldehyde, and alcohol. All of these are toxic substances that may enter the bloodstream. This type of oxidation is especially important in liver and kidney cells, where peroxisomes detoxify many potentially harmful substances.

Mitochondria

Known as the "powerhouses" of the cell, **mitochondria** (mī-tō-KON-drē-a; singular is **mitochondrion**; *mitos* = thread; *chondros* = granule) are the main sites for generation of ATP. A mitochondrion consists of two membranes, each of which is similar in structure to the plasma membrane (Fig. 3.20). The outer mitochondrial membrane is smooth, but the inner membrane is arranged in a series of folds called **cristae** (KRIS-tē; singular is **crista**; *crista* = ridge). The central cavity of a mitochondrion, enclosed by the inner membrane and cristae, is called the **matrix**.

The elaborate folds of the cristae provide an enormous surface area for a series of chemical reactions known as **cellular respiration**, which provides most of a cell's ATP supply (see page 815). Enzymes that catalyze these reactions are located in the matrix and on the cristae. Active cells, such as muscle, liver, and kidney tubule cells, have a large number of mitochondria and use ATP at a high rate.

Mitochondria self-replicate; they divide to increase in number. Their replication is controlled by genes within the mitochondria (mitochondrial DNA). Self-replication usually occurs in response to increased cellular need for ATP and at the time of cell division. Although each cell's nucleus contains genes from both your mother and father, mitochondrial genes usually are inherited only from your mother. The head of a sperm, which is the part that penetrates and fertilizes an egg, normally lacks mitochondria.

The Cytoskeleton

Cellular shape and the capability to carry out a variety of coordinated cellular movements depend on a complex internal network of filamentous proteins in the cytoplasm called the **cytoskeleton**. The cytoskeleton is responsible for movement of whole cells, such as phagocytes, and for movement of organelles and some chemicals within the cell. Three main types of protein filaments—microfilaments, microtubules, and intermediate filaments—comprise the cytoskeleton (see Fig. 3.1).

Microfilaments are rodlike structures of varying length that are formed from the protein actin. In muscle tissue, actin filaments (thin filaments) and myosin filaments (thick filaments) slide past one another to produce contraction (shortening) of muscle cells. In nonmuscle cells, actin microfilaments provide support and shape. They also assist in cell movement (for example, in phagocytes and cells of developing embryos) and movements within cells (secretion, phagocytosis, pinocytosis).

Microtubules are larger than microfilaments. They are relatively straight, slender, cylindrical structures that consist of a protein called **tubulin**. The major center for assembly of microtubules is an organizing region known as the centrosome

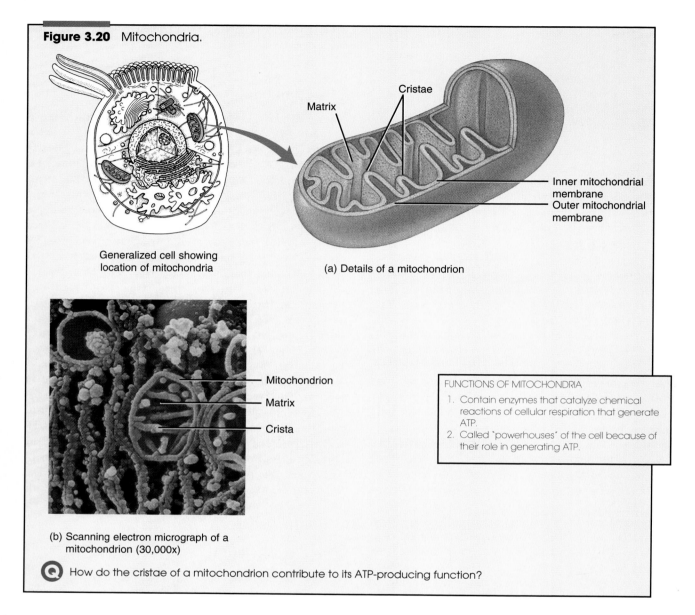

Figure 3.20 Mitochondria.

Generalized cell showing location of mitochondria

(a) Details of a mitochondrion

Cristae

Matrix

Inner mitochondrial membrane

Outer mitochondrial membrane

Mitochondrion

Matrix

Crista

FUNCTIONS OF MITOCHONDRIA

1. Contain enzymes that catalyze chemical reactions of cellular respiration that generate ATP.
2. Called "powerhouses" of the cell because of their role in generating ATP.

(b) Scanning electron micrograph of a mitochondrion (30,000x)

Q How do the cristae of a mitochondrion contribute to its ATP-producing function?

(see Fig. 3.21). Together with microfilaments, microtubules help support and shape cells. Microtubules also function like a conveyor belt to move various substances and organelles through the cytosol, and they assist in the movement of phagocyte pseudopods.

Intermediate filaments, so-named because their size is in between that of microfilaments and microtubules, are exceptionally strong and tough. Several proteins form intermediate filaments. They provide structural reinforcement inside cells, hold organelles (such as the nucleus) in place, and associate closely with microtubules to give shape to the cell.

Flagella and Cilia

Some body cells have projections for moving the entire cell or for moving substances along the external surface of the cell. If the projections are few and long in proportion to the size of the cell (occurring singly or in pairs), they are called

flagella (fla-JEL-a; singular is *flagellum* = whip). The only example of a flagellum in the human body is the sperm cell tail, which is used for locomotion (see Fig. 28.5). If the projections are numerous and short, resembling many hairs, they are called **cilia** (SIL-ē-a; *cilia* = eyelashes). Ciliated cells of the respiratory tract remove foreign particles trapped in mucus from the airways (see Fig. 23.6c). Their movement is paralyzed by the nicotine in cigarette smoke. For this reason, smokers must cough to remove foreign particles from their airways. Both flagella and cilia contain nine pairs of microtubules that form a ring around two single microtubules in the center. Both also contain cytosol and are covered by the plasma membrane.

Centrosome and Centrioles

Near the nucleus is a dense area of cytoplasmic material with radiating microtubules called a **centrosome**. Centrosomes serve as centers for organizing microtubules in nondividing cells and for forming the mitotic spindle during cell division (described on page 83). Within the centrosome is a pair of cylindrical structures called **centrioles** (Fig. 3.21). Each centriole is composed of nine clusters of three microtubules arranged in a circular pattern. Centrioles lack the two central single microtubules found in flagella and cilia. The long axis of one centriole is at a right angle to the long axis of the other. Centrioles play a role in the formation and regeneration of flagella and cilia.

CELL INCLUSIONS

Cell inclusions are a large and diverse group of chemical substances produced by cells. Although some have recognizable shapes, they are not bounded by a membrane. These products are principally organic molecules and may appear or disappear at various times in the life of a cell. Examples include melanin, glycogen, and triglycerides. **Melanin** is a pigment stored in certain cells of the skin, hair, and eyes. It protects the body by screening out harmful ultraviolet rays from the sun. **Glycogen** is a polysaccharide that is stored in liver, skeletal muscle, and the inner lining of the uterus and vagina. When the body requires quick energy, liver cells can break down the glycogen into glucose and release it into the blood. **Triglycerides**, which are stored in adipocytes (fat cells), may be broken down to synthesize ATP.

The major parts of a cell and their functions are summarized in Exhibit 3.2.

GENE ACTION

Although cells synthesize many chemicals to maintain homeostasis, much of the cellular machinery promotes protein production. Cells are basically protein factories that constantly synthesize large numbers of diverse proteins. The proteins, in turn, determine the physical and chemical characteristics of cells and therefore of organisms. Some proteins are structural, helping form plasma membranes, microfilaments, microtubules, centrioles, flagella, cilia, mitochondria, and other parts of cells. Other proteins serve as hormones, antibodies, and contractile elements in muscle tissue. Still other proteins are enzymes that regulate the rates of the myriad chemical reactions that occur in cells.

The instructions for making proteins are contained in DNA (deoxyribonucleic acid). Cells make proteins by translating the genetic information encoded in DNA into specific proteins. In this process, the information in a region of DNA is first *transcribed* (copied) to produce a specific molecule of RNA (ribonucleic acid). Then, the information contained in RNA is *translated* into a corresponding specific sequence of amino acids in a newly produced protein molecule. Let us take a look at how DNA directs protein synthesis by considering the two principal steps in production of proteins: transcription and translation.

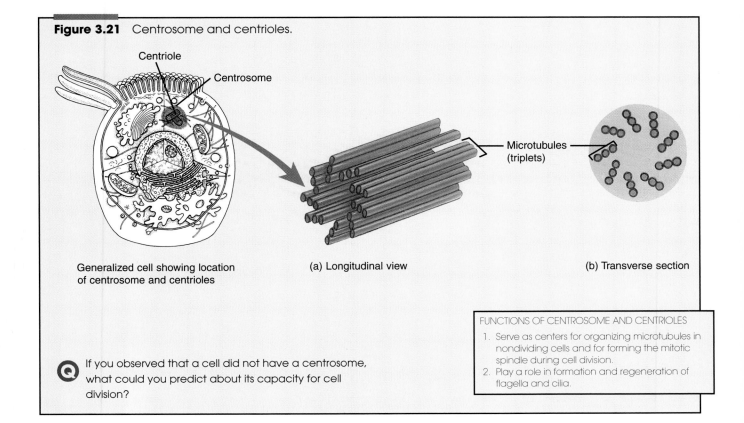

Figure 3.21 Centrosome and centrioles.

Centriole

Centrosome

Microtubules (triplets)

Generalized cell showing location of centrosome and centrioles

(a) Longitudinal view

(b) Transverse section

FUNCTIONS OF CENTROSOME AND CENTRIOLES

1. Serve as centers for organizing microtubules in nondividing cells and for forming the mitotic spindle during cell division.
2. Play a role in formation and regeneration of flagella and cilia.

Q If you observed that a cell did not have a centrosome, what could you predict about its capacity for cell division?

EXHIBIT **3.2   CELL PARTS AND THEIR FUNCTIONS**

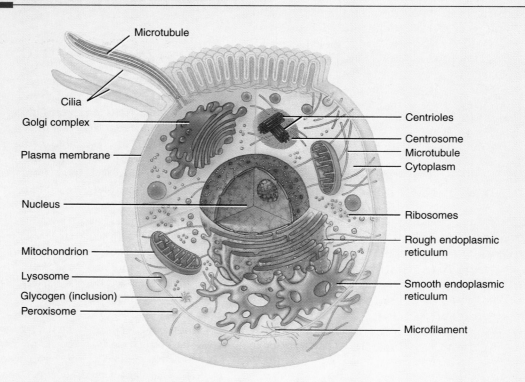

Part	Functions
PLASMA MEMBRANE	Protein-studded phospholipid bilayer that surrounds the cell; protects cellular contents; makes contact with other cells; provides receptors for hormones, enzymes, and antibodies; mediates the entrance and exit of materials.
CYTOSOL	Viscous, transparent, gel-like intracellular fluid containing water, ions, and many types of organic molecules such as enzymes; medium in which many of the cell's chemical reactions occur. Cytoplasm includes cytosol, all organelles (except the nucleus), and inclusions.
ORGANELLES	Organized compartments within the cytoplasm that have specific structures and functions.
Nucleus	Surrounded by the nuclear envelope; contains genes and controls cellular activities.
Ribosomes	Sites of protein synthesis; may be either free in the cytosol or attached to endoplasmic reticulum.
Endoplasmic reticulum (ER)	Membrane-bounded network of channels that provides surface area for many types of chemical reactions; ribosomes attached to rough ER synthesize proteins that will be secreted; smooth ER (no attached ribosomes) synthesizes lipids and detoxifies certain molecules.
Golgi complex	Stack of disc-shaped membrane-enclosed cisterns; packages proteins and lipids into secretory vesicles for export or insertion into the plasma membrane; forms lysosomes.
Lysosomes	Contain digestive enzymes that break down molecules and microbes.
Peroxisomes	Contain enzymes that use molecular oxygen to oxidize various organic substances.
Mitochondria	Sites of production of most ATP during cellular respiration.
Cytoskeleton	Includes three types of protein filaments (microfilaments, microtubules, and intermediate filaments) that give the cell shape and allow coordinated movements of organelles; in some cells the cytoskeleton is responsible for movement of the cell itself.
Flagella and cilia	Allow movement of entire cell (flagella) or movement of substances along surface of cell (cilia).
Centrosome and centrioles	Centrosome helps organize microtubules in nondividing cell and forms mitotic spindle during cell division; centrioles play a role in the formation and regeneration of flagella and cilia.
INCLUSIONS	Chemical substances produced by cells that are not surrounded by a membrane, for example, melanin, glycogen, and triglycerides (fats and oils).

Transcription

Transcription is the process by which the genetic information encoded in DNA is copied onto a strand of RNA. The genetic information stored in the sequence of bases in DNA nucleotides serves as a blueprint or template for rewriting the same information into a complementary sequence of bases in RNA nucleotides (Fig. 3.22). Three forms of RNA are made from the DNA template: **messenger RNA (mRNA)**, which directs synthesis of a polypeptide chain, **ribosomal RNA (rRNA)**, which comes together with ribosomal proteins to make up ribosomes, and **transfer RNA (tRNA)**, which binds to amino acids during translation. Each tRNA can bind specifically to one of the 20 different types of amino acids. One can thus define a **gene** as a sequence of DNA that codes for a particular mRNA, rRNA, or tRNA. Because no two people have the same DNA molecules, unless they are identical twins, the nucleotide sequence is the key to each person's uniqueness.

Transcription of DNA is catalyzed by the enzyme *RNA polymerase.* During transcription, bases pair in the same manner as they do in the DNA double helix (see Fig. 2.16). The base cytosine (C) in the DNA template dictates the base guanine (G) in the new RNA strand; a G in the DNA template dictates a C in the RNA strand; and a thymine (T) in the DNA template dictates an adenine (A) in the RNA. Since RNA contains uracil (U) instead of thymine, an A in the DNA template dictates a U in the RNA. As an example, if the template portion of DNA has the base sequence ATGCAT, the newly transcribed RNA strand would have the complementary base sequence UACGUA.

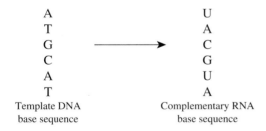

A	U
T	A
G	C
C	G
A	U
T	A
Template DNA base sequence	Complementary RNA base sequence

Only one of the two DNA strands serves as a template for RNA synthesis. This strand is referred to as the **sense strand**. The other strand, the one not transcribed, is the complement of the sense strand and is called the **antisense strand**.

Within DNA are regions, called **introns**, that *do not* code for synthesis of part of a protein. Introns are located between regions called **exons**, regions that *do* code for parts of proteins. Initially, an mRNA transcript includes both introns and exons. The RNA regions corresponding to DNA introns, however, are deleted (cut out) and the exons are spliced (rejoined) before mRNA leaves the nucleus and enters the cytoplasm to proceed to the next step of protein synthesis. This process, called **mRNA splicing**, deletes about three-fourths of the original RNA. A single gene may be able to code for more than one protein if the initial RNA transcript can be spliced in different ways to make different

Figure 3.22 Transcription. When RNA synthesis is complete, mRNA leaves the nucleus and enters the cytoplasm, where translation occurs.

During transcription, the genetic information in DNA is copied to RNA.

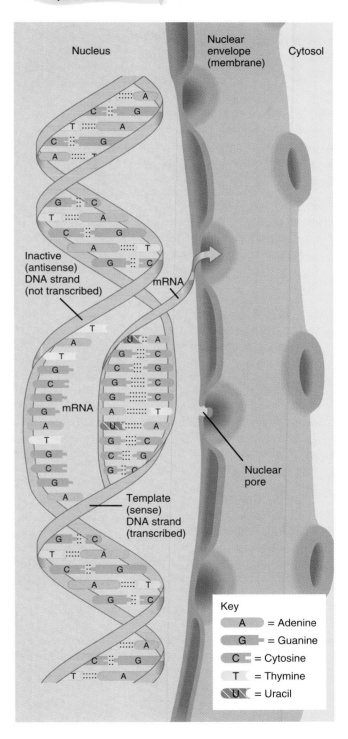

Key
A = Adenine
G = Guanine
C = Cytosine
T = Thymine
U = Uracil

 If the DNA template had the base sequence AGCT, what would be the mRNA base sequence? What enzyme catalyzes transcription of DNA?

mRNAs. During cell differentiation, for example, some splicing patterns may change to produce different proteins at particular stages of differentiation. Once synthesized, mRNA, rRNA (in ribosomes), and tRNA leave the nucleus of the cell. In the cytoplasm, they participate in the next principal step in protein synthesis—translation.

Translation

Just as a DNA molecule provides the template for making an mRNA, so mRNA provides a template for synthesizing a protein. **Translation** is the process whereby the *nucleotide sequence* in a molecule of mRNA specifies the *amino acid sequence* for a protein molecule. In the mRNA molecule, each set of three consecutive nucleotide bases is called a **codon** and specifies one amino acid. Most mRNA molecules contain 300–3000 nucleotides. Since each set of three nucleotides codes for one amino acid, most proteins contain between 100 and 1000 amino acids. The term **genetic code** refers to the set of rules that relates the base sequence of DNA, the corresponding codons of mRNA, and the amino acids that the codons represent.

As shown in Fig. 3.23, the key events of translation are as follows:

1 In the cytoplasm, the small ribosomal subunit binds to one end of the mRNA molecule and finds the **start codon**, a sequence where translation will begin. The large ribosomal subunit then joins in.

2 In the cytosol, each type of transfer RNA binds to one kind of amino acid and brings it to the ribosome. One end of the tRNA carries a specific amino acid, and another part of each tRNA has a triplet of nucleotides called an **anticodon**.

3 By base pairing, the tRNA anticodon recognizes and attaches to a complementary codon on mRNA. For example, if the mRNA codon is AUG, then a tRNA having the anticodon UAC would attach. In the process, the tRNA brings along the specific amino acid, methionine in this case.

4 Once the first tRNA has attached to mRNA, the ribosome moves exactly three nucleotides along the mRNA, and the next tRNA, carrying its amino acid, moves into position.

5 When an amino acid attaches to its tRNA, ATP is used to form a high-energy bond between the amino acid and the tRNA. Such an activated amino acid can form a peptide bond with a second amino acid, if the second amino acid is brought into correct position by its own tRNA. The larger ribosomal subunit contains the enzymes needed for peptide bond formation. As the peptide bond forms, tRNA detaches from the first amino acid.

6 The proper amino acids are brought into line, one by one, peptide bonds form between them, and the protein progressively lengthens. Each time the ribosome moves one codon along the mRNA, an "empty" tRNA

is ejected. The released tRNA can then pick up another molecule of the same amino acid.

7 When the specified protein is complete, synthesis is terminated by a special **stop codon**. The assembled protein is then released from the ribosome.

8 After protein synthesis, the large and small ribosomal subunits separate.

As each ribosome moves along an mRNA strand, it "reads" the information coded in mRNA and synthesizes a protein according to that information. Thus the ribosome synthesizes the protein by translating the nucleotide codon sequence into an amino acid sequence. Protein synthesis progresses at a rate of about 15 amino acids per second. As the ribosome moves along the mRNA and before it completes synthesis of the whole protein, another ribosome may attach behind it and begin translation of the same mRNA strand. In this way, several ribosomes may be attached to the same mRNA. This assembly is called a **polyribosome**. Several ribosomes moving together along the same mRNA molecule permit the translation of a one mRNA into several identical proteins almost simultaneously.

Remember that the base sequence of the gene determines the sequence of bases in mRNA, which, in turn, determines the order of amino acids in a given protein. Thus each gene is responsible for making a particular protein as follows:

$$ \text{DNA} \xrightarrow{\text{Transcription}} \text{mRNA} \xrightarrow{\text{Translation}} \text{Protein} $$

RECOMBINANT DNA

Different kinds of cells make different proteins following instructions encoded in the DNA of their genes. Since 1973, scientists have been able to alter those instructions in bacteria, viruses, and yeast cells by inserting genes from other organisms. This causes the host organism to produce proteins it normally does not synthesize. Organisms so altered are called **recombinants**, and their DNA, a combination of DNA from different sources, is called **recombinant DNA**. When recombinant DNA functions properly, the host will synthesize the protein specified by the new gene it has acquired. The new technology that has arisen from manipulating the genetic material is called **genetic engineering**.

The practical applications of recombinant DNA technology are enormous. Strains of recombinant bacteria are now producing many important therapeutic substances. These include *human growth hormone* (hGH), required for growth during childhood and important in metabolism of adults; *insulin,* a hormone that helps regulate blood sugar level and is used by diabetics; *interferon* (IFN), an antiviral (and possibly anticancer) substance; *factor VIII,* a blood clotting factor missing in people with hemophilia A; *erythropoietin,* a hormone that stimulates formation of red blood cells; *monoclonal antibodies* to diagnose and treat cancer and assist in AIDS research; and many other substances. Scientists are also using recombinant DNA techniques in attempts to develop vaccines against several viruses, including those that cause AIDS, herpes, and influenza. ∎

Figure 3.23 Translation. Note that during protein synthesis the ribosomal subunits join together. When protein synthesis ceases, they separate.

🔑 *During translation, information in mRNA specifies the amino acid sequence of a protein.*

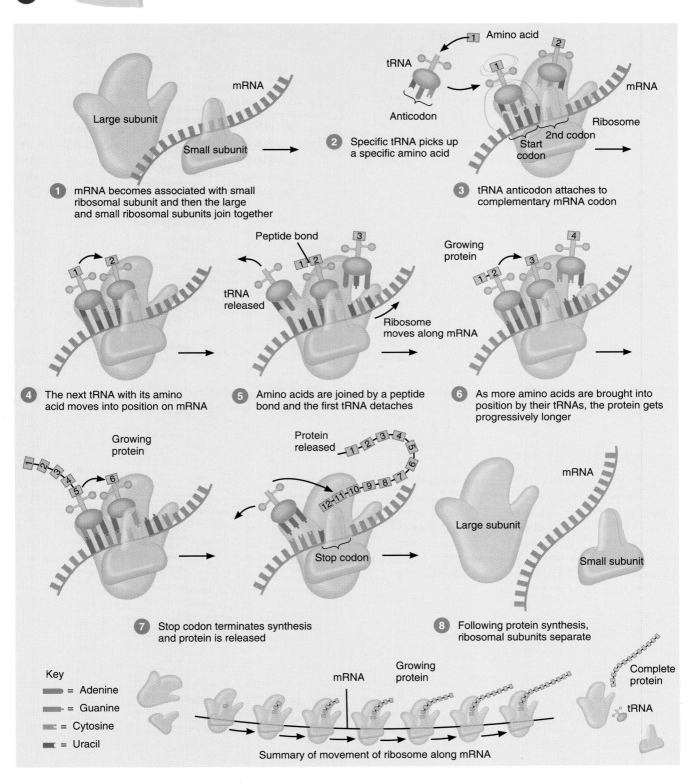

1 mRNA becomes associated with small ribosomal subunit and then the large and small ribosomal subunits join together

2 Specific tRNA picks up a specific amino acid

3 tRNA anticodon attaches to complementary mRNA codon

4 The next tRNA with its amino acid moves into position on mRNA

5 Amino acids are joined by a peptide bond and the first tRNA detaches

6 As more amino acids are brought into position by their tRNAs, the protein gets progressively longer

7 Stop codon terminates synthesis and protein is released

8 Following protein synthesis, ribosomal subunits separate

Key
▬ = Adenine
▬ = Guanine
▬ = Cytosine
▬ = Uracil

Summary of movement of ribosome along mRNA

❓ Once a specified protein is complete, how is its synthesis terminated?

NORMAL CELL DIVISION

Most of the cell activities mentioned thus far maintain the life of the cell on a day-to-day basis. However, somatic cells ultimately become damaged, diseased, or worn out. New cells must be produced by cell division as replacements. In a 24-hour period, the average adult loses billions of cells from different parts of the body. Cells that have a short life span, such as cells of the outer layer of skin, are completely replaced every few days. In addition, cell division produces new somatic cells for tissue growth and new germ cells, which are specialized to form sperm and egg cells.

Cell division is the process by which cells reproduce themselves. It consists of a nuclear division and a cytoplasmic division. Because *nuclear* division can be either one of two types, two kinds of cell division are recognized.

In the first kind of division, **somatic cell division**, a single starting cell called a **parent cell** divides to produce two identical cells called **daughter cells**. This process consists of a nuclear division called **mitosis** and a cytoplasmic division called **cytokinesis**. The process ensures that each daughter cell has the same *number* and *kind* of chromosomes as the original parent cell. This kind of cell division replaces dead or injured cells and adds new ones for tissue growth.

The second type of cell division, **reproductive cell division**, is the mechanism by which sperm and egg cells are produced. These are the cells needed to form a new organism. The process consists of a special nuclear division called **meiosis** (reduction division) followed by **cytokinesis**. We will discuss somatic cell division first.

Figure 3.24 The cell cycle.

 In a complete cell cycle, one parent cell divides into two daughter cells.

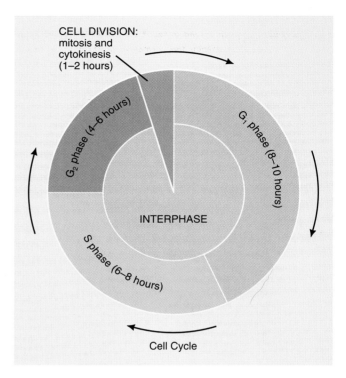

CELL DIVISION: mitosis and cytokinesis (1–2 hours)

G_2 phase (4–6 hours)

G_1 phase (8–10 hours)

INTERPHASE

S phase (6–8 hours)

Cell Cycle

During which phase of interphase do the chromosomes replicate?

Somatic Cell Division

Human cells, except for eggs and sperm, contain 23 pairs of chromosomes. Two chromosomes that belong to a pair, one contributed by the mother and one contributed by the father, are called **homologous chromosomes**. They have similar genes, which are usually arranged in the same order. When a cell reproduces, it must replicate (duplicate) all its chromosomes so that its hereditary traits may be passed on to the next generation of cells.

Interphase

The **cell cycle** is a series of activities through which a cell passes from the time it is formed until it reproduces. It is the growth and division of a single cell into two daughter cells. The cell cycle consists of two major activities: interphase and cell division (Fig. 3.24). When a cell is between divisions, it is said to be in **interphase**. During this stage the replication of DNA, centrosome, and centrioles occurs and the RNA and protein needed to produce structures required for doubling all cellular components are manufactured.

Interphase consists of three distinct phases: G_1, S, and G_2 (Fig. 3.24). During the G_1- (for gap or growth) **phase**, cells

are engaged in growth, metabolism, and the production of substances required for division. The period of interphase during which chromosomes replicate is called the **S-** (for synthesis) **phase**. After the S-phase, there is another growth phase called the G_2-**phase**. Since the G-phases are periods when there are no events related to chromosomal replication, they are thought of as gaps or interruptions in DNA synthesis. Cells that are destined never to divide again are permanently arrested in the G_1-phase, a state termed the G_0-phase. Most nerve cells are in this state. Once a cell enters the S-phase, it is committed to go through cell division.

When DNA replicates during the S-phase, its helical structure partially uncoils (Fig. 3.25) and the two strands separate at the points where hydrogen bonds connect base pairs. Each exposed base then picks up a complementary base (with its associated sugar and phosphate group). This uncoiling and complementary base pairing continues until each of the two original DNA strands is matched and joined with one newly formed DNA strand. The original DNA molecule has become two DNA molecules.

Figure 3.25 Replication of DNA. The two strands of the double helix separate by breaking the hydrogen bonds between nucleotides. New nucleotides attach at the proper sites, and a new strand of DNA is synthesized alongside each of the original strands.

Replication doubles the amount of DNA. Thus each daughter cell contains the normal amount after a somatic cell division.

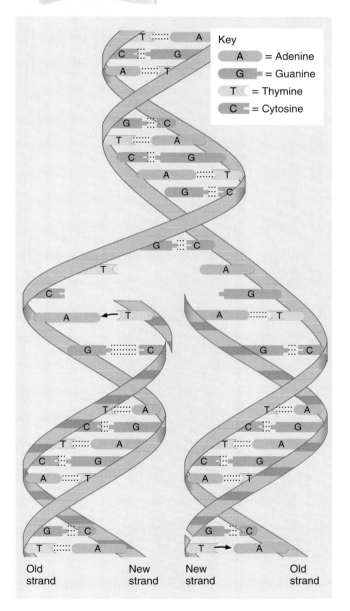

Key
A = Adenine
G = Guanine
T = Thymine
C = Cytosine

| Old strand | New strand | New strand | Old strand |

To which class of bases do adenine and guanine belong?

A microscopic view of a cell during interphase shows a clearly defined nuclear envelope, nucleoli, and chromatin (Fig. 3.26a). The absence of visible chromosomes is another physical characteristic of interphase. Once a cell completes its repli-

cation of DNA, centrosomes, and centrioles and its production of RNA and proteins during interphase, mitosis begins.

Nuclear Division: Mitosis

The events that take place during mitosis and cytokinesis are plainly visible under a microscope because chromatin condenses into chromosomes. The process called **mitosis** is the distribution of the two sets of chromosomes into two separate and equal nuclei. It results in the *exact* duplication of genetic information. For convenience, biologists divide the process into four stages: prophase, metaphase, anaphase, and telophase. Mitosis is a continuous process, however, with one stage merging imperceptibly into the next.

● **Prophase** The first stage of mitosis is called **prophase** (*pro* = before) (Fig. 3.26b). During early prophase, the chromatin fibers condense and shorten into chromosomes. Condensation may prevent entangling of the long DNA strands as they move during mitosis or meiosis. Since DNA replication took place during interphase, each prophase chromosome contains a pair of identical double-stranded chromatids. Each chromatid pair is held together by a small spherical body called a **centromere** that is required for the proper segregation of chromosomes. Attached to the outside of each centromere is a protein complex known as the **kinetochore** (ki-NET-ō-kor), whose function will be described shortly.

Later in prophase, the nucleoli disappear, during which time synthesis of RNA is temporarily halted, and the nuclear envelope breaks down and is absorbed into the cytosol. In addition, each centrosome and its centrioles move to opposite poles (ends) of the cell. As they do so, the centrosomes start to form the **mitotic spindle**, a football-shaped assembly of microtubules that are responsible for the movement of chromosomes. The lengthening of microtubules between centrosomes pushes the centrosomes to the poles of the cell so that the spindle extends from pole to pole. As the mitotic spindle continues to develop, three types of microtubules form: (1) **nonkinetochore microtubules** grow from centrosomes, extend inward, but do not bind to kinetochores; (2) **kinetochore microtubules** grow from centrosomes, extend inward, and attach to kinetochores; and (3) **aster microtubules** grow out of centrosomes but radiate outward from the mitotic spindles. The spindle is an attachment site for chromosomes, and it also distributes chromosomes to opposite poles of the cell.

● **Metaphase** During **metaphase** (*meta* = after), the second stage of mitosis, the centromeres of the chromatid pairs line up at the exact center of the mitotic spindle. This midpoint region is called the **metaphase plate** or **equatorial plane region** (Fig. 3.26c).

● **Anaphase** The third stage of mitosis, **anaphase** (*ana* = upward), is characterized by the splitting and separation of centromeres and the movement of the two sister chromatids of each pair toward opposite poles of the cell (Fig. 3.26d).

Figure 3.26 Cell division: mitosis and cytokinesis. Start looking at the sequence at (a) and read clockwise until you complete the process.

🔑 *In somatic cell division, a single diploid parent cell divides to produce two identical diploid daughter cells.*

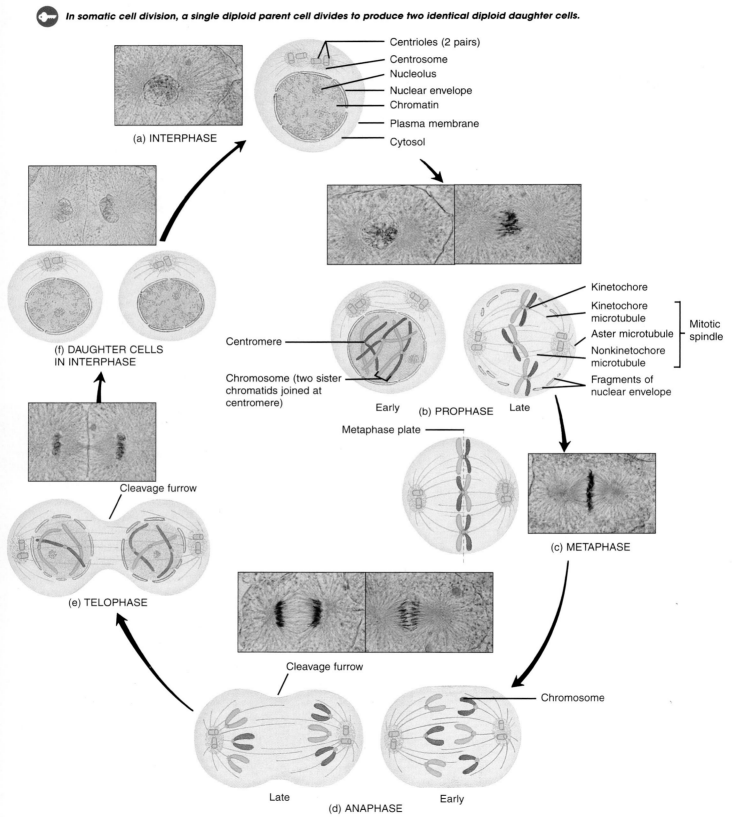

(a) INTERPHASE

Centrioles (2 pairs)
Centrosome
Nucleolus
Nuclear envelope
Chromatin
Plasma membrane
Cytosol

(f) DAUGHTER CELLS
IN INTERPHASE

Centromere

Chromosome (two sister chromatids joined at centromere)

Early   (b) PROPHASE   Late

Kinetochore
Kinetochore microtubule
Aster microtubule
Nonkinetochore microtubule
} Mitotic spindle
Fragments of nuclear envelope

Metaphase plate

(c) METAPHASE

Cleavage furrow

(e) TELOPHASE

Cleavage furrow

Chromosome

Late   (d) ANAPHASE   Early

❓ When does cytokinesis begin?

Once separated, the sister chromatids are referred to as daughter **chromosomes**. The movement of chromosomes is due to shortening of kinetochore microtubules and elongation of the nonkinetochore microtubules, processes that increase the distance between separated chromosomes. As the chromosomes are pulled by the microtubules during anaphase, they appear V-shaped as the centromeres lead the way and seem to drag the trailing parts of the chromosomes toward opposite poles of the cell.

• **Telophase** The final stage of mitosis, **telophase** (*telos* = far or end), begins as soon as chromosomal movement stops. Telophase is essentially the opposite of prophase. During telophase, the identical sets of chromosomes at opposite poles of the cell uncoil and revert to the threadlike chromatin form. A new nuclear envelope forms around each chromatin mass, new nucleoli reappear in the daughter nuclei, and eventually the mitotic spindle breaks up (Fig. 3.26e).

Cytoplasmic Division: Cytokinesis

Division of a parent cell's cytoplasm and organelles is called **cytokinesis** (sī-tō-ki-NĒ-sis; *cyto* = cell; *kinesis* = motion). It begins in late anaphase or early telophase with formation of a **cleavage furrow**, a slight indentation of the plasma membrane that extends around the center of the cell (Fig. 3.26d–f). The furrow gradually deepens until opposite surfaces of the cell make contact and the cell is split in two. When cytokinesis is complete, interphase begins. The result of cytokinesis is two separate daughter cells, each with separate portions of cytoplasm and organelles and its own set of identical chromosomes.

If we consider the cell cycle in its entirety, the sequence of events is G_1-phase → S-phase → G_2-phase → mitosis → cytokinesis (see Fig. 3.24). A summary of the events that occur during interphase and cell division is presented in Exhibit 3.3.

Length of the Cell Cycle

The time required for one cell cycle varies with the kind of cell, its location, and other factors such as temperature. Furthermore, the different stages of mitosis are not equal in duration. The G_1-phase is the most variable, ranging from almost nonexistent in rapidly dividing cells to days, weeks, years, or a lifetime. Mammalian cells studied in laboratory cultures often have the following time intevals during a cell cycle (see Fig. 3.24).

G_1-phase	8–10 hours
S-phase	6–8 hours
G_2-phase	4–6 hours
Mitosis and cytokinesis	1–2 hours

Within the mitosis and cytokinesis time interval, prophase is longest and anaphase is shortest. As you can see, mitosis and cytokinesis represent only a small part of the life cycle of a cell. Together, the various phases of the cell cycle require 18–24 hours in many cultured mammalian cells.

EXHIBIT **3.3** **SUMMARY OF EVENTS IN THE CELL CYCLE**

Phase	Activity
INTERPHASE	Cell is between divisions; chromosomes are not visible with a light microscope.
G_1-phase	Cell engages in growth, metabolism, and production of substances required for division; no chromosomal replication occurs.
S-phase	Chromosomes replicate (synthesis of new DNA and its associated proteins).
G_2-phase	Same as for G_1 period.
CELL DIVISION	Single parent cell produces two chromosomally identical daughter cells; chromosomes are visible with a light microscope.
Mitosis	Nuclear division—distribution of two sets of chromosomes into separate and equal nuclei.
Prophase	Chromatin fibers shorten and coil into visible chromosomes (paired chromatids), nucleoli and nuclear envelope disappear, each centrosome moves with its centrioles to opposite poles of cell, and centrosomes form mitotic spindle.
Metaphase	Centromeres of chromatid pairs line up at metaphase plate of cell.
Anaphase	Centromeres divide and identical sets of chromosomes move to opposite poles of cell.
Telophase	Nuclear envelope reappears and encloses chromosomes, chromosomes resume chromatin fiber form, nucleoli reappear, and mitotic spindle breaks up.
Cytokinesis	Cytoplasmic divison—Cleavage furrow forms around center of cell, progresses inward, and separates cytoplasm into two separate and usually equal portions.

Reproductive Cell Division

In sexual reproduction, each new organism is produced by the union and fusion of two different germ cells, one produced by each parent. The germ cells, called **gametes**, are the secondary oocytes, produced in the female gonads (ovaries), and the sperm, produced in the male gonads (testes). The union and fusion of gametes is called **fertilization**, and the cell thus produced is known as a **zygote** (*zygosis* = a joining). The zygote contains one set of chromosomes (DNA) from each parent and, through repeated mitotic divisions, develops into a new organism.

Gametes

Gametes differ from all other body cells (somatic cells) with respect to the number of chromosomes in their nuclei. Somatic cells, such as brain cells, stomach cells, kidney cells, and others, contain 23 pairs of chromosomes, or a total of 46 chromosomes. One member of each pair is inherited from each parent. A gamete, on the other hand, has only 23 chromosomes, just one member of each pair. The two chromosomes that make up a pair are called **homologous** (hō-MOL-ō-gus) **chromosomes**, or **homologs**. They contain similar genes arranged in the same or almost the same order. When examined under the light microscope (see Fig. 29.18), homologous chromosomes look very similar. The exception to this general rule is a pair of chromosomes called the **sex chromosomes**, designated X and Y. In females the homologous pair of sex chromosomes consists of two X chromosomes; in males the pair consists of an X and a Y chromosome. The other 22 pairs of chromosomes are called **autosomes**.

The symbol n is used to designate the number of different chromosomes. In humans, $n = 23$. Since somatic cells contain two sets of chromosomes, they are $2n$ and are called **diploid** (DIP-loyd; *diploos* = double; *eidos* = form) **cells**. Gametes, with a single set of chromosomes, are $1n$ and are called **haploid** (HAP-loyd; *haploos* = single) **cells**. If gametes had the same number of chromosomes as somatic cells, the zygote formed from their fusion would have double the normal number ($4n$). With each succeeding generation, the number of chromosomes would double. Chromosome number does not double with each generation, however, because of a special nuclear division called **meiosis** (mī-Ō-sis; *meio* = less). Meiosis occurs only in the development of gametes, and it results in the production of haploid cells that contain only 23 chromosomes. As you will see later, meiosis also accounts for genetic variation in daughter cells. Mitosis, by contrast, usually produces exact genetic copies.

Meiosis

The formation of haploid sperm cells in the testes of the male occurs in several phases as part of the process called **spermatogenesis** (sper'-mat-tō-JEN-e-sis). The formation of haploid secondary oocytes (potential ova) in the ovaries of the female occurs as part of **oogenesis** (ō'-ō-JEN-e-sis) and also involves several phases. Spermatogenesis and oogenesis are discussed on pages 911 and 926, respectively. At this point, we will examine only the essentials of meiosis, which is a common feature of both.

Meiosis occurs in two successive nuclear divisions referred to as **reduction division (meiosis I)** and **equatorial division (meiosis II)**. During the interphase that precedes reduction division of meiosis, the chromosomes replicate. This replication is similar to that in the interphase before mitosis of somatic cell division.

● **Reduction division (meiosis I)** Once chromosomal replication is complete, reduction division begins. It consists of

four phases: prophase I, metaphase I, anaphase I, and telophase I (Fig. 3.27a–d).

Prophase I is an extended phase in which the chromosomes shorten and thicken, the nuclear envelope and nucleoli disappear, and the mitotic spindle appears. Unlike prophase of mitosis, however, the chromosomes become arranged in homologous pairs.

In metaphase I, the homologous pairs of chromosomes line up along the metaphase plate of the cell, with the homologous chromosomes side by side. (Recall that there is no pairing of homologous chromosomes during metaphase of mitosis.) The centromeres of each chromatid pair form kinetochore microtubules that attach the centromeres to opposite poles of the cell.

During anaphase I the members of each homologous pair separate, with one member of each pair moving to an opposite pole of the cell. The centromeres do not split and the paired chromatids, held by a centromere, remain together. (Recall that during anaphase of mitosis, the centromeres split and the sister chromatids separate.)

Telophase I and cytokinesis are similar to telophase and cytokinesis of mitosis. The net effect of reduction division is that each resulting daughter cell contains the haploid number of chromosomes; each cell contains only one member of each pair of the original homologous chromosomes in the starting parent cell. The interphase between reduction division and equatorial division is either brief or lacking altogether.

● **Crossing-over** During prophase I of meiosis two events occur that are not seen in prophase I of mitosis (or prophase II of meiosis). First, the two chromatids of each pair of chromosomes join, an event called **synapsis**. The resulting four chromatids form a **tetrad**. Second, portions of one chromatid may be exchanged with portions of another (Fig. 3.28). When such an exchange occurs, it is termed **crossing-over**. This process, among others, permits an exchange of genes among homologous chromatids so that the resulting daughter cells are genetically unlike each other and unlike the parent cell that produced them. It results in **genetic recombination**—the formation of new combinations of genes—and accounts for part of the great genetic variation among humans and other organisms that form gametes by meiosis.

● **Equatorial division (meiosis II)** The second phase of meiosis, equatorial division, consists of four phases: prophase II, metaphase II, anaphase II, and telophase II (Fig. 3.27e–h). These phases are similar to those that occur during mitosis; the centromeres split and the sister chromatids separate and move toward opposite poles of the cell.

● **Summary of meiosis** Reduction division starts with a parent cell with the diploid number of chromosomes and ends up with two daughter cells, each with the haploid number. During equatorial division, each haploid cell formed during reduction division divides, and the net result is four haploid cells that are all genetically different. As you will see later, all

Figure 3.27 Meiosis. See text for details.

In reproductive cell division, a single diploid parent cell undergoes reduction division and equatorial division to produce four haploid gametes that are genetically different from the parent cell.

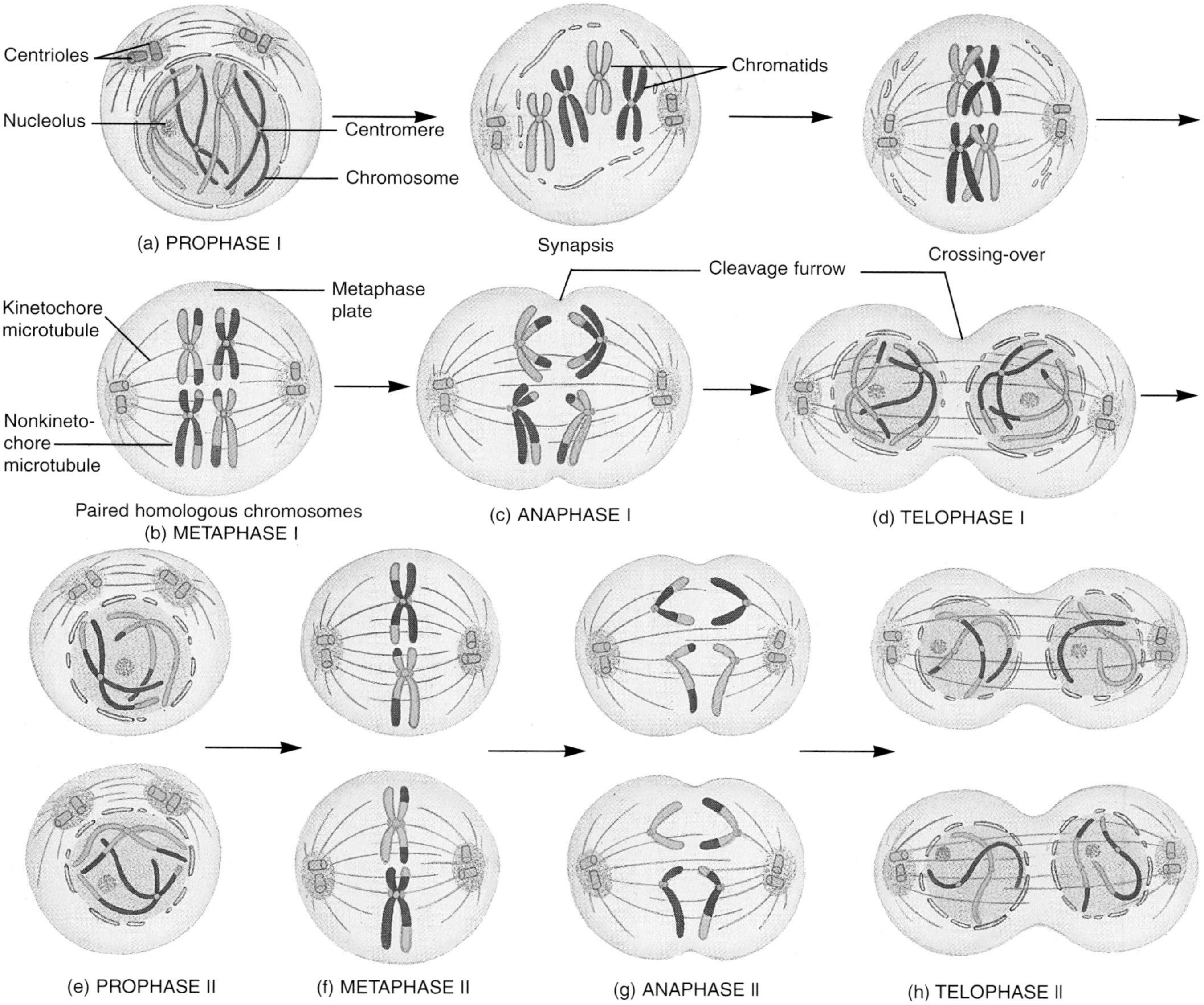

Centrioles
Nucleolus
Centromere
Chromosome
(a) PROPHASE I
Chromatids
Synapsis
Crossing-over
Cleavage furrow
Kinetochore microtubule
Metaphase plate
Nonkinetochore microtubule
Paired homologous chromosomes
(b) METAPHASE I
(c) ANAPHASE I
(d) TELOPHASE I
(e) PROPHASE II
(f) METAPHASE II
(g) ANAPHASE II
(h) TELOPHASE II

When does chromosomal replication occur in meiosis?

Figure 3.28 Crossing-over within a tetrad.

Crossing-over permits an exchange of genes between homologous chromosomes.

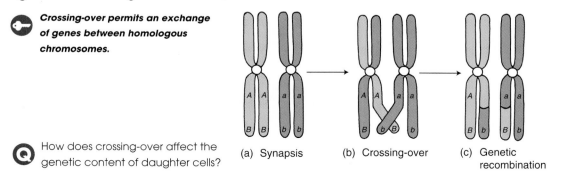

A A a a
B B b b
A A a a
B b B b
A a a
B b B b
(a) Synapsis
(b) Crossing-over
(c) Genetic recombination

How does crossing-over affect the genetic content of daughter cells?

four haploid cells develop into sperm cells in the testes of the male. Only one of the haploid cells has the potential to develop into an oocyte in the female. The others become structures called *polar bodies* that disintegrate. Although reduction division reduces the number of chromosomes, equatorial division is necessary to separate the four chromatids of each chromosome pair into four sperm or one oocyte and three polar bodies. Failure to do so results in genetic disorders such as Down syndrome (described on page 984).

A very simplified comparison of mitosis and meiosis is illustrated in Fig. 3.29.

Control of Cell Destiny

A cell has three possible destinies—to remain alive and functioning without dividing, to grow and divide, or to die. Homeostasis is maintained when there is a balance between cell proliferation and cell death. The signals that tell a cell when to exist in the G_0-phase, when to divide, and when to die have been the subjects of intense and fruitful research during the past decade. A key signal that induces cell division (both mitosis and meiosis) is a substance called **maturation promoting factor** (**MPF**). One component of MPF is a group of enzymes called **cdc2 proteins** because they participate in the cell division cycle (cdc). Another component of MPF is a protein called **cyclin**, so named because its level rises and falls during the cell cycle. Cyclin builds up in the cell during interphase and activates cdc2 proteins and thus MPF. As a result, the cell undergoes mitosis (or meiosis). By the end of mitosis, cyclin levels are low. Accordingly, cdc2 proteins are not activated, MPF is not activated, mitosis (or meiosis) stops, and the cell enters another interphase.

Cellular death also is a regulated process. Throughout the lifetime of an organism, certain cells undergo genetically programmed cell death. This type of cell death, termed **apoptosis** (ap-ō-TŌ-sis; *apoptosis* = a falling off, like dead leaves from a tree), occurs when the protein products of "cell-suicide" genes become activated. In a cell undergoing apoptosis, the DNA fragments, the nucleus condenses, mitochondria cease to function, and the cytoplasm shrinks, although the plasma membrane remains intact. Neighboring phagocytes rapidly engulf and degrade the cellular corpse. This normal type of cell death that occurs in cells scattered throughout a tissue contrasts with **necrosis** (ne-KRŌ-sis; *nekros* = death), a pathological type of cell death that results from tissue injury. In necrosis, many adjacent cells swell, burst, and spill their cytoplasm into the interstitial fluid. This mess of cellular debris usually stimulates an inflammatory response, which does not occur in apoptosis. Apoptosis functions to remove unneeded cells during embryological development, regulate the number of cells in a tissue, and eliminate potentially dangerous cells such as cancer cells.

Abnormalities in genes that participate in regulation of cell division or apoptosis are associated with many diseases. For example, some cancers are caused by loss of genes called **tumor-suppressor genes**. These genes produce proteins that normally inhibit cell division. Loss or alteration of a tumor-suppressor gene called *p53* on chromosome 17 is the most common genetic change in a wide variety of tu-

Figure 3.29   Comparison between (a) mitosis and (b) meiosis.

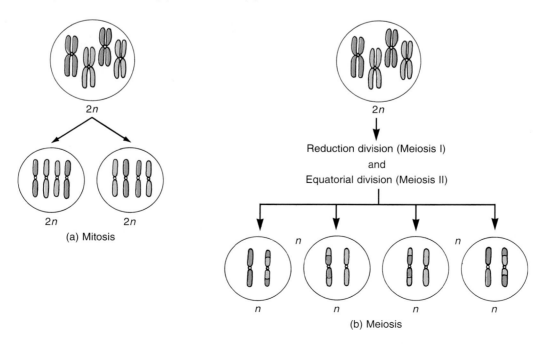

(a) Mitosis

(b) Meiosis

Reduction division (Meiosis I)
and
Equatorial division (Meiosis II)

Q   Where in the body does meiosis occur?

mors, including breast and colon cancers. The normal p53 protein arrests cells in the G_1-phase, which prevents cell division, and also is needed to initiate apoptosis. Another family of genes, called *BCL*, produce proteins that block apoptosis. In some tumors *BCL* is always "turned on," which inhibits cell death and decreases the effectiveness of some anticancer drugs.

Understanding the genetic regulators of cell division and apoptosis more fully is likely to lead to improved strategies for treating disorders as diverse as cancer, AIDS, stroke, Alzheimer's disease, heart attack, and viral infections. It may be possible to find new ways to speed up cell division to accelerate wound healing, stimulate division of nerve cells to replace ones lost to injury or disease, slow cell division or induce apoptosis to prevent the growth of cancer cells, or rescue needed cells from apoptosis.

CELLS AND AGING

Aging is a normal process accompanied by a progressive alteration of the body's homeostatic adaptive responses. It produces observable changes in structure and function and increases vulnerability to environmental stress and disease. The specialized branch of medicine that deals with the medical problems and care of elderly persons is called **geriatrics** (jer'-ē-AT-riks; *geras* = old age; *iatrike* = surgery, medicine).

The obvious characteristics of aging are well known: graying and loss of hair, loss of teeth, wrinkling of skin, decreased muscle mass, and increased fat deposits. The physiological signs of aging are gradual deterioration in function and capacity to respond to environmental stresses. Metabolism slows, as does the ability to maintain a constant internal environment (homeostasis) in response to changes in temperature, diet, and oxygen supply. These signs of aging are related to a net decrease in the number of cells in the body and to dysfunctions in cells that remain.

Glucose, the most abundant sugar in the body, may play a role in the aging process. Glucose is haphazardly added to proteins inside and outside cells, forming irreversible cross-links between adjacent protein molecules. With advancing age, more cross-links form, and this may contribute to the stiffening and loss of elasticity that occur in aging tissues.

Although many millions of new cells normally are produced each minute, several kinds of cells in the body—heart cells, skeletal muscle cells, nerve cells—are not replaced because they are arrested permanently in the G_0-phase. Experiments have shown that many other cell types have only a limited capability to divide. Cells grown outside the body divide only a certain number of times and then stop. These observations suggest that cessation of mitosis is a normal, genetically programmed event. According to this view, "aging" genes are part of the genetic blueprint at birth, and they turn on at preprogrammed times, slowing down or halting processes vital to life.

Another theory of aging is the **free radical theory**. Free radicals are electrically charged molecules that have an unpaired electron, for example, **superoxide** (Fig. 3.30). Such molecules are unstable and highly reactive. A free radical produces oxidative damage in a nearby lipid, protein, or nucleic acid by "stealing" an electron to accompany its unpaired electron. Some effects are wrinkled skin, stiff joints, and hardened arteries. Normal cellular metabolism—for example, cellular respiration in mitochondria—produces some free radicals. Others are present in air pollution, radiation, and certain foods we eat. Naturally occuring enzymes, such as *superoxide dismutase, peroxidase,* and *catalase* in peroxisomes and in the cytosol, normally dispose of free radicals. Dietary substances, such as vitamin E, vitamin C, beta-carotene, and selenium, are antioxidants that inhibit free radical formation. Besides accelerating the aging process, free radicals contribute to reperfusion damage after a stroke or heart attack (see page 589) and are probably one factor in the development of cancer.

Two discoveries bolster the free radical theory of aging. Strains of fruit flies bred for longevity produce larger-than-normal amounts of superoxide dismutase, which functions to neutralize free radicals. Also, injection of genes that lead to production of superoxide dismutase into fruit fly embryos prolongs their average lifetime.

Whereas some theories of aging explain the process at the cellular level, others concentrate on regulatory mechanisms operating within the entire organism. For example, the immune system, which manufactures antibodies against foreign invaders, may start to attack the body's own cells. This autoimmune response might be caused by changes in cell identity markers at the surface of cells, causing antibodies to attach and mark the cell for destruction. As surface changes in cells increase, the autoimmune response intensifies, producing the well-known signs of aging.

The effects of aging on the various body systems are discussed in their respective chapters.

Figure 3.30 Atomic structures of an oxygen molecule and a superoxide free radical.

 A free radical has an unpaired electron in the outer (valence) shell.

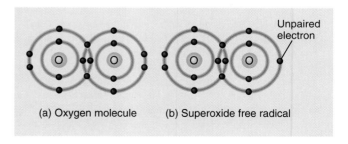

(a) Oxygen molecule   (b) Superoxide free radical

 Why are free radicals very reactive chemically?

DISORDERS: HOMEOSTATIC IMBALANCES

CANCER

When cells in some area of the body divide without control, the excess of tissue that develops is called a **tumor**, **growth**, or **neoplasm** (NĒ-ō-plazm). The study of tumors is called **oncology** (on-KOL-ō-jē; *onco* = swelling or mass; *logos* = study of), and a physician who specializes in this field is called an **oncologist**. Tumors may be cancerous and sometimes fatal, or they may be quite harmless. A cancerous growth is called a **malignant tumor** or **malignancy**. One property of a malignant tumor is its ability to undergo **metastasis** (me-TAS-ta-sis), the spread of cancerous cells to other parts of the body. A noncancerous growth is called a **benign tumor.** Benign tumors do not metastasize, but they may be removed if they interfere with a normal body function or are disfiguring.

Types of Cancer

Cancers are classified by their microscopic appearance and the body site from which they arise. The name of the cancer is derived from the type of tissue in which it develops. Most human cancers are **carcinomas** (kar-sin-ŌM-az; *carc* = cancer; *oma* = tumor), malignant tumors that arise from epithelial cells. **Melanomas** (mel-an-ŌM-az; *melano* = black), for example, are cancerous growths of melanocytes, skin cells that produce the pigment melanin. **Sarcoma** (sar-KŌ-ma) is a general term for any cancer arising from muscle cells or connective tissues. For example, **osteogenic sarcomas** (*osteo* = bone; *genic* = origin), the most frequent type of childhood cancer, destroy normal bone tissue, which is one type of connective tissue, and eventually spread to other areas of the body. **Leukemia** (loo-KĒ-mē-a) is a cancer of blood-forming organs characterized by rapid growth and distorted development of leukocytes (white blood cells) and their precursors. **Lymphoma** (lim-FŌ-ma) is a malignant disease of lymphatic tissue, for example, lymph nodes. An example is Hodgkin's disease.

Growth and Spread of Cancer

Cells of malignant tumors duplicate continually and very often quickly and without control. Such an increase in the number of cells due to an increase in the frequency of cell division is called **hyperplasia** (hi-per-PLĀ-zē-a). Initially, malignant cells invade surrounding tissues. As the cancer grows, it expands and begins to compete with normal tissues for space and nutrients. Eventually, the normal tissue decreases in size (atrophies) and dies.

Following the nearby invasion, some of the malignant cells may detach from the initial (primary) tumor. They may invade a body cavity or enter the blood or lymph, which can lead to widespread metastasis. Next, those malignant cells that survive in the blood or lymph invade other body tissues and establish secondary tumors.

Finally, the secondary tumors become vascularized. They undergo **angiogenesis**, which is the growth of new networks of blood vessels. Any new tissue, whether it results from repairing a wound, normal growth, or tumors, requires a blood supply to deliver nutrients and oxygen. Proteins that serve as chemical triggers for blood vessel growth in tumor tissue are called **tumor angiogenesis factors** (**TAFs**). In all stages of metastasis, the malignant cells resist the antitumor defenses of the body. The pain associated with cancer develops when the growth puts pressure on nerves or blocks a passageway so that secretions build up pressure.

Causes of Cancer

What triggers a normal cell to lose control and become abnormal? Several factors contribute. First, there are environmental agents: substances in the air we breathe, the water we drink, and the food we eat. A chemical or other environmental agent that produces cancer is called a **carcinogen** (car-SIN-ō-jen). The World Health Organization estimates that carcinogens may be associated with 60–90% of all human cancers. Examples of carcinogens are hydrocarbons found in cigarette tar, radon gas from the earth, and ultraviolet (UV) radiation in sunlight. Adopting a cancer-prevention life-style and avoiding environmental carcinogens can help decrease your risk of getting cancer.

Viruses are a second cause of cancer. These agents are tiny packages of nucleic acids, either DNA or RNA, that are capable of infecting cells and converting them to virus-producers. Although the link between viruses and cancer is strongly established for a variety of animal cancers, the relation in human cancers is less predictable. Clearly, the number of people infected with these viruses is much larger than the number who develop cancer. But evidence suggests that chronic viral infections are associated with up to one-fifth of all cancers. These include:

1. **Human T-cell leukemia–lymphoma virus-1** (**HTLV-1**), associated with leukemia, a malignant disease of blood-forming tissues, and lymphoma, a cancer of lymphatic tissue.
2. **Human immunodeficiency virus** (**HIV**), associated with Kaposi's sarcoma, a cancer of the inner lining of some blood vessels that is often seen in those who have acquired immune deficiency syndrome (AIDS).
3. **Epstein–Barr virus** (**EBV**), which causes infectious mononucleosis, associated with Burkitt's lymphoma, a cancer of white blood cells, nasopharyngeal carcinoma (common in Chinese males), and Hodgkin's disease, a lymphatic system cancer.
4. **Hepatitis B virus** (**HBV**), associated with liver cancer.
5. **Human papillomavirus** (**HPV**), which causes genital warts (benign growths), associated with cancer of the cervix, vagina, penis, and colon.
6. **Type 2 herpes simplex virus**, which causes genital herpes, implicated in cancer of the cervix of the uterus.

Intensive research efforts are now directed toward studying **oncogenes** (ON-kō-jēnz). These genes have the ability to transform a normal cell into a cancerous cell when they are *inappropriately activated*. Oncogenes develop from normal genes that regulate growth and development, called **proto-oncogenes**. These genes may undergo some change that either causes them to produce an abnormal product or disrupts their control so that they are expressed inappropriately, making their products in excessive amounts or at the wrong time. It is believed that some oncogenes cause excessive production of growth factors, chemicals that stimulate cell growth. Other oncogenes may cause changes in a surface receptor, causing it to send signals as though it were being activated by a growth factor. As a result, the growth pattern of the cell becomes abnormal.

Every cell contains proto-oncogenes, which carry out normal cellular functions until a malignant change occurs. It appears that some proto-oncogenes are activated to oncogenes by mutations in which the DNA of the proto-oncogene is altered. A **mutation** is a permanent structural change in a gene. Carcinogens may induce mutations. Other proto-oncogenes are activated by a rearrangement of the chromosomes so that segments of DNA are exchanged. Rearrangement activates proto-oncogenes by placing them near genes that enhance their activity. Viruses are believed to cause cancer by inserting their own oncogenes or proto-oncogenes into the host cell's DNA.

Carcinogenesis: A Multistep Process

Carcinogenesis (kar′-si-nō-JEN-e-sis), the process by which carcinomas develop, is a multistep process in which as many as ten distinct mutations may have to accumulate in a cell before it becomes cancerous. The progression of gene changes leading to cancer is best understood for colon (colorectal) cancer. Such cancers, as well as lung and breast cancer, take years or decades to develop. In colon cancer, the tumor begins as an area of increased cell proliferation that results from one mutation. This growth then progresses to abnormal, but noncancerous, growths called adenomas. As two or three additional mutations occur, the adenomas develop through successive stages of increasing severity. Finally, with mutation of p53, a carcinoma develops. The fact that so many mutations are needed for a cancer to develop indicates that cell growth is normally controlled with many sets of checks and balances.

The identification of mutations underlying cancers, such as those that affect the colon, lungs, and breasts, may aid scientists in preventing, diagnosing, and treating these malignancies. For example, if it is known that a particular cancer requires several mutations in a multistep process and if people could be identified after only one or two mutations had taken place, they could be counseled to avoid carcinogens that would produce future mutations—mutations that could drive cells into malignancy. Also, by understanding the steps involved in the progression of cancer, scientists may be able to design better treatments.

Treatment of Cancer

Treating cancer is difficult because it is not a single disease and because all the cells in a single tumor population do not behave in the same way. Although most cancers are thought to derive from a single abnormal cell, by the time a tumor reaches a clinically detectable size, the cancer may contain a diverse population of cells. For example, some cells metastasize and others do not. Some cells divide and others do not. Some are sensitive to drugs and some are resistant. Because of differences in drug resistance, a single chemotherapeutic drug may destroy susceptible cells but permit resistant cells to proliferate.

Certain tumor cells may be simultaneously resistant to completely unrelated drugs, a phenomenon called **multi-drug resistance**. This broad-based resistance to chemotherapeutic drugs is due to a plasma membrane protein called P-glycoprotein (PGP) that accumulates in resistant tumor cells. PGP actively transports drugs out, and the cells are not killed. Scientists think that if they can find ways to inhibit the action of PGP, then anticancer drugs can work more effectively to kill tumor cells.

Another stumbling block encountered by blood-borne anticancer drugs is the physical barrier that solid tumors, such as those that arise in the breasts, lungs, colon, and other organs, develop. Deep within such tumors are high-pressure areas that collapse blood vessels in the tumor. This makes it difficult, if not impossible, for blood-borne anticancer agents to penetrate the tumor. Besides chemotherapy, radiation therapy, surgery, cryothermia (freezing), hyperthermia (abnormally high temperature), and immunotherapy (bolstering the body's own defenses) may be used alone or in combination.

MEDICAL TERMINOLOGY

NOTE TO THE STUDENT

Each chapter in this text that discusses a major system of the body is followed by a glossary of **medical terminology**. Both normal and pathological conditions of the system are included in these glossaries. By becoming familiar with these terms, you will start to build your medical vocabulary.

Atrophy (AT-rō-fē; *a* = without; *tropho* = nourish) A decrease in the size of cells with subsequent decrease in the size of the affected tissue or organ; wasting away.

Biopsy (BĪ-op-sē; *bio* = life; *opsis* = vision) The removal and microscopic examination of tissue from the living body for diagnosis.

Dysplasia (dis-PLĀ-zē-a; *dys* = abnormal; *plas* = to grow) Alteration in the size, shape, and organization of cells due to chronic irritation or inflammation; may progress to a neoplasm or revert to normal if the stress is removed.

Hyperplasia (hī-per-PLĀ-zē-a; *hyper* = over) Increase in the number of cells of a tissue due to increased rate of cell division.

Hypertrophy (hī-PER-trō-fē) Increase in the size of cells without cell division.

Metaplasia (met′-a-PLĀ-zē-a; *meta* = change) The transformation of one type of cell into another.

Necrosis (ne-KRŌ-sis; *necros* = death; *osis* = condition) Death of a group of cells.

Neoplasm (NĒ-ō-plazm; *neo* = new) Any abnormal formation or growth, usually a malignant tumor.

STUDY OUTLINE

GENERALIZED ANIMAL CELL (p. 54)

1. A cell is the basic, living, structural and functional unit of the body.
2. A generalized cell is a composite that represents various cells of the body.
3. Cytology is the science concerned with the study of cells.
4. The principal parts of a cell are the plasma (cell) membrane, cytosol, organelles, and inclusions.

PLASMA (CELL) MEMBRANE (p. 55)

Membrane Chemistry and Anatomy (p. 55)

1. The plasma (cell) membrane surrounds the cell and separates it from other cells and the external environment.
2. It is composed primarily of phospholipids and proteins. The proteins are integral and peripheral.
3. According to the fluid mosaic model, the membrane is a mosaic of proteins floating like icebergs in a sea of lipids.

Membrane Physiology (p. 57)

1. The plasma membrane functions in cellular communication, establishment of an electrochemical gradient, and selective permeability.
2. The membrane's selectively permeable nature restricts the passage of certain substances. Substances can pass through the membrane depending on their lipid solubility, size, electrical charges, and the presence of channels and transporters.

MOVEMENT OF MATERIALS ACROSS PLASMA MEMBRANES (p. 58)

Passive Processes (p. 58)

1. Passive processes depend on the concentration of substances and their kinetic energy.
2. Net diffusion is the net movement of molecules or ions from an area of higher concentration to an area of lower concentration until an equilibrium is reached.
3. Osmosis is the movement of water through a selectively permeable membrane from an area of higher water concentration to an area of lower water concentration.
4. In an isotonic solution, red blood cells maintain their normal shape; in a hypotonic solution, they undergo hemolysis; in a hypertonic solution, they undergo crenation.
5. Bulk flow is the movement in the same direction of large numbers of ions, molecules, or particles that are dissolved or carried in a medium such as water or air as a result of a hydrostatic (water) pressure or air pressure difference.
6. Filtration is a type of bulk flow that involves the movement of water and dissolved substances across a capillary wall from blood into interstitial fluid.
7. In facilitated diffusion, certain substances, such as ions or glucose, move through the membrane with the help of a channel or transporter protein.

Active Processes (p. 62)

1. Active processes depend on the use of ATP by the cell. The two principal types are active transport and bulk transport.
2. Active transport is the movement of a substance across a cell membrane from lower to higher concentration using energy derived from ATP either directly (primary active transport) or indirectly (secondary active transport).
3. The most prevalent primary active transport pump is the sodium pump.

4. Secondary active transport mechanisms include both symporters and antiporters.
5. Phagocytosis is the ingestion of solid particles. It is an important process used by some white blood cells to destroy bacteria that enter the body.
6. Pinocytosis is the ingestion of fluid. In this process, the fluid becomes surrounded by a pinocytic vesicle.
7. Receptor-mediated endocytosis is the selective uptake of large molecules and particles (ligands) by cells.

CYTOSOL (p. 66)

1. The semifluid intracellular fluid is termed cytosol. Cytosol is composed mostly of water plus proteins, carbohydrates, lipids, and inorganic substances.
2. Cytoplasm includes the cytosol, all organelles (except the nucleus), and inclusions.

ORGANELLES (p. 68)

Introduction (p. 68)

1. Organelles are specialized structures in the cytosol that have characteristic appearances and functions.
2. They play specific roles in cellular growth, maintenance, repair, and control.

Nucleus (p. 68)

1. Usually the largest organelle, the nucleus, controls cellular activities and contains the genetic information.
2. Most body cells have a single nucleus; some (red blood cells) have none, whereas others (skeletal muscle cells) have several.
3. Parts of the nucleus include the nuclear envelope, nucleoli, and chromosomes.
4. Chromosomes consist of subunits called nucleosomes that are composed of DNA (genetic material) and histone proteins. The various levels of DNA packing are represented by nucleosomes, chromatin fibers, loops, chromatids, and chromosomes.

Ribosomes (p. 69)

1. Ribosomes are tiny granules consisting of ribosomal RNA and ribosomal proteins.
2. Free ribosomes float in the cytosol and have no attachments to other organelles; other ribosomes attach to endoplasmic reticulum.
3. Functionally, ribosomes are the sites of protein synthesis.

Endoplasmic Reticulum (p. 69)

1. The ER is a network of membrane-enclosed channels that connects with the nuclear membrane.
2. Rough (granular) ER has ribosomes attached to it. Smooth (agranular) ER does not contain ribosomes.
3. The ER transports substances, stores newly synthesized molecules, synthesizes and packages molecules, detoxifies chemicals, and releases calcium ions involved in muscle contraction.

Golgi Complex (p. 71)

1. The Golgi complex consists of several stacked, flattened membranous sacs (cisterns) referred to as *cis, medial,* and *trans.*

2. The principal functions of the Golgi complex are to process, sort, and deliver proteins and lipids to the plasma membrane, lysosomes, and secretory vesicles.

Lysosomes (p. 73)

1. Lysosomes are membrane-enclosed vesicles that are formed by the Golgi complex and contain digestive enzymes.
2. They are found in large numbers in white blood cells, which carry on phagocytosis.
3. Lysosomes function in intracellular digestion, digestion of worn-out organelles (autophagy), digestion of cellular contents (autolysis) during embryological development, and extracellular digestion.

Peroxisomes (p. 73)

1. Peroxisomes are similar to lysosomes but smaller.
2. They contain enzymes (for example, catalase) that use molecular oxygen to oxidize various organic substances.

Mitochondria (p. 73)

1. Mitochondria consist of a smooth outer membrane and a folded inner membrane surrounding the interior matrix. The inner folds are called cristae.
2. The mitochondria are called "powerhouses" of the cell because they produce most of a cell's ATP.

The Cytoskeleton (p. 73)

1. Together, microfilaments, microtubules, and intermediate filaments form the cytoskeleton.
2. The cytoskeleton provides organization for chemical reactions and assists in transporting chemicals and organelles through the cytosol.

Flagella and Cilia (p. 74)

1. These cellular projections have the same basic structure and are used in movement.
2. If projections are few (typically occurring singly or in pairs) and long, they are called flagella. If they are numerous and short, they are called cilia.
3. The flagellum on a sperm cell moves the entire cell. The cilia on cells of the respiratory tract move foreign matter trapped in mucus along the cell surfaces toward the throat for elimination.

Centrosome and Centrioles (p. 74)

1. The dense area of cytoplasm containing the centrioles is called a centrosome. Centrosomes serve as centers for organizing microtubules in interphase cells and the mitotic spindle during cell division.
2. Centrioles are paired cylinders arranged at right angles to one another that play a role in the formation and regeneration of flagella and cilia.

CELL INCLUSIONS (p. 75)

1. Cell inclusions are chemical substances produced by cells. They are usually organic and may have recognizable shapes.
2. Examples of cell inclusions are melanin, glycogen, and triglycerides.

GENE ACTION (p. 75)

1. Most of the cellular machinery is concerned with synthesizing proteins.
2. Cells make proteins by translating the genetic information encoded in DNA into specific proteins. This involves transcription and translation.
3. In transcription, genetic information encoded in DNA is copied onto a strand of messenger RNA (mRNA).

4. DNA also synthesizes ribosomal RNA (rRNA) and transfer RNA (tRNA).
5. The process of using the information in the base sequence of mRNA to dictate the amino acid sequence of a protein is known as translation. Each three sequential bases (a codon) of mRNA specifies one amino acid.
6. mRNA associates with ribosomes, which consist of rRNA and protein.
7. Specific amino acids attach to molecules of tRNA. Another portion of the tRNA has a triplet of bases called an anticodon.
8. The anticodon of a tRNA binds to a specific codon of mRNA, thus bringing a specific amino acid into position. The ribosome moves along an mRNA strand and amino acids are joined to form a growing polypeptide.

NORMAL CELL DIVISION (p. 80)

1. Cell division is the process by which cells reproduce themselves. It consists of nuclear division (mitosis or meiosis) and cytoplasmic division (cytokinesis).
2. Cell division that results in an increase in the number of body cells is called somatic cell division and involves a nuclear division called mitosis plus cytokinesis.
3. Cell division that results in the production of sperm and oocytes is called reproductive cell division and consists of a nuclear division called meiosis plus cytokinesis.

Somatic Cell Division (p. 80)

1. Before mitosis and cytokinesis, the DNA molecules, or chromosomes, replicate themselves so the same chromosomes can be passed on to the next generation of cells.
2. A cell between divisions carrying on every life process except division is said to be in interphase, which consists of three phases: G_1, S, and G_2.
3. Mitosis is the replication and distribution of two sets of chromosomes into separate and equal nuclei; it consists of prophase, metaphase, anaphase, and telophase.
4. Cytokinesis usually begins in late anaphase and ends in telophase.
5. A cleavage furrow forms at the cell's metaphase plate and progresses inward, cutting through the cell to form two separate portions of cytoplasm.

Reproductive Cell Division (p. 83)

1. Gametes contain the haploid (n) chromosome number and most somatic cells contain the diploid ($2n$) chromosome number.
2. Meiosis is the process that produces haploid gametes. It consists of two successive nuclear divisions called reduction division (meiosis I) and equatorial division (meiosis II).
3. During reduction division, homologous chromosomes undergo synapsis (pairing) and crossing-over; the net result is two haploid daughter cells that are genetically unlike each other and unlike the parent cell that produced them.
4. During equatorial division, the two haploid daughter cells divide to form four haploid cells.

Control of Cell Destiny (p. 86)

1. A cell can remain alive and functioning without dividing, grow and divide, or die.
2. Maturation promoting factor (MPF) induces cell division (both mitosis and meiosis). MPF consists of cyclin, which builds up during interphase, and cdc2 proteins, which are activated by cyclin.

3. Apoptosis is the term used for programmed cell death, a normal type of cell death that begins during embryological development and continues for the lifetime of an organism.

4. Certain genes regulate both cell division and apoptosis. Abnormalities in these genes are associated with a wide variety of diseases and disorders.

CELLS AND AGING (p. 87)

1. Aging is a normal process accompanied by progressive alteration of the body's homeostatic adaptive responses.

2. Many theories of aging have been proposed, including genetically programmed cessation of cell division, excessive immune responses, and buildup of free radicals.

REVIEW QUESTIONS

1. Define a cell. What are the four principal portions of a cell? What is meant by a generalized cell? (p. 54)
2. Discuss the chemistry of the plasma membrane. (p. 55)
3. Describe the various functions of the plasma membrane. What determines selective permeability? (p. 57)
4. What are the major differences between passive processes and active processes that move substances across plasma membranes? (p. 58)
5. Define and give an example of each of the following: simple diffusion, osmosis, filtration, facilitated diffusion, primary and secondary active transport, phagocytosis, pinocytosis, and receptor-mediated endocytosis. (p. 58)
6. Compare the effect of isotonic, hypertonic, and hypotonic solutions on red blood cells. What is osmotic pressure? (p. 60)
7. Discuss the chemical composition and physical nature of the cytosol. What is its function? (p. 67)
8. What is an organelle? By means of a labeled diagram, indicate the parts of a generalized animal cell. (p. 68)
9. Describe the structure and functions of the nucleus of a cell. Describe how DNA is packed into chromosomes. (p. 68)
10. Discuss the distribution and functions of ribosomes. (p. 69)
11. Distinguish between rough and smooth endoplasmic reticulum (ER). What are the functions of ER? (p. 69)
12. Describe the structure and functions of the Golgi complex. (p. 71)
13. List and describe the functions of lysosomes. (p. 73)
14. What is the importance of peroxisomes? (p. 73)
15. Why are mitochondria called "powerhouses" of the cell? (p. 73)
16. Contrast the structure and functions of microfilaments, microtubules, and intermediate filaments. (p. 73)
17. What are the structural and functional differences between flagella and cilia? (p. 74)
18. Describe the structure and functions of the centrosome. (p. 74)
19. Define a cell inclusion. Give examples. (p. 75)
20. Summarize the steps in transcription and translation. (p. 77)
21. Distinguish between the two types of cell division. Why is each important? (p. 80)
22. Define interphase. When does DNA replicate itself? (p. 80)
23. Describe the principal events of each stage of mitosis. (p. 81)
24. Distinguish between haploid (n) and diploid ($2n$) cells. (p. 84)
25. Define meiosis and contrast the principal events of reduction division and equatorial division. (p. 84)
26. How is cell destiny controlled? (p. 86)
27. What is aging? List some of the characteristics of aging. (p. 87)

ANSWERS TO QUESTIONS WITH FIGURES

3.1 Plasma membrane, cytosol, organelles, inclusions.
3.2 Phospholipids, cholesterol, and glycolipids.
3.3 ❸ Receptor.
3.4 The ions closest to the cytosolic face of the membrane are the anions: mainly proteins and organic phosphates.
3.5 Since a fever represents an increase in body temperature, all diffusion processes would be increased.
3.6 Hydrophobic.
3.7 No, that can never happen, since the beaker always contains pure water. Sugar cannot cross the cellophane membrane.
3.8 Hypertonic.
3.9 The size of the concentration difference on the two sides of the membrane and the number of available transporters.
3.10 ATP adds phosphate to the pump protein, which changes the pump's three-dimensional shape.
3.11 ATP is not used to drive the activity of symporter or antiporter proteins whereas it does directly power the pump protein in primary active transport.
3.12 Endocytosis, because it is a process that brings material into the cell.
3.13 Pinocytosis is a less specific process than receptor-mediated endocytosis.
3.14 It controls cellular activity by regulating which proteins will be synthesized.

3.15 Double-stranded DNA wrapped twice around a core of eight histones (proteins).
3.16 In the nucleoli inside the nucleus.
3.17 Rough ER has attached ribosomes whereas smooth ER does not. Rough ER synthesizes proteins that will be exported from the cell, while smooth ER is associated with lipid synthesis and other metabolic reactions.
3.18 The proteins originate in the ER and become encased within a secretory vesicle or lysosome.
3.19 The membrane of the vesicle holds the digestive enzymes inside so they cannot inappropriately break down molecules within the cytosol.
3.20 They increase surface area for chemical reactions.
3.21 It probably will not be able to undergo cell division.
3.22 UCGA. RNA polymerase.
3.23 A stop codon terminates synthesis.
3.24 S-phase.
3.25 Purines.
3.26 Usually starts in late anaphase or early telophase.
3.27 During the interphase that precedes reduction division.
3.28 Genetically, the daughter cells are unlike each other and unlike the parent cell that produced them.
3.29 In testes of males and ovaries of females.
3.30 They have an unpaired electron.

STUDENT OBJECTIVES

1. Define a tissue and classify the tissues of the body into four major types. (p. 94)

2. Describe the structure and functions of the three principal types of cell junctions. (p. 94)

3. Describe the general features of epithelial tissue. (p. 95)

4. List the structure, location, and function for the following types of epithelium: simple squamous, simple cuboidal, simple columnar (nonciliated and ciliated), stratified squamous, stratified cuboidal, stratified columnar, transitional, and pseudostratified columnar. (p. 96)

5. Define a gland and distinguish between exocrine and endocrine glands. (p. 103)

6. Describe the general features of connective tissue. (p. 104)

7. Discuss the cells, ground substance, and fibers that compose connective tissue. (p. 104)

8. List the structure, function, and location of mesenchyme; mucous connective tissue; areolar connective tissue; adipose tissue; reticular connective tissue; dense regular and irregular connective tissue; elastic connective tissue; cartilage; bone; and blood. (p. 113)

9. Define a membrane and describe the location and function of mucous, serous, and synovial membranes. (p. 115)

10. Contrast the three types of muscle tissue with regard to structure, location, and modes of control. (p. 116)

11. Describe the structural features and functions of nervous tissue. (p. 118)

12. Describe tissue repair in restoring homeostasis. (p. 118)

CHAPTER 4

THE TISSUE LEVEL OF ORGANIZATION

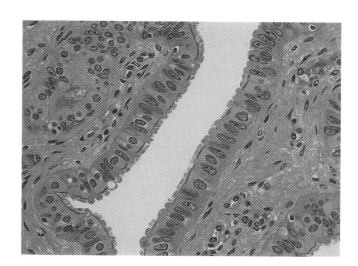

CHAPTER CONTENTS AT A GLANCE

ells are highly organized, but they do not function as isolated units in your body. They work together in groups called tissues. A **tissue** is a group of cells that usually have a common embryonic origin and function together to carry out specialized activities. The structure and properties of a specific tissue are influenced by such things as the nature of the extracellular material that surrounds the tissue cells and connections between the cells that compose the tissue. The science that deals with the study of tissues is **histology** (hiss′-TOL-ō-jē; *histio* = tissue; *logos* = study of).

TYPES OF TISSUES AND THEIR ORIGINS

Body tissues can be classified into four principal types according to their function and structure:

1. Epithelial tissue, which covers body surfaces; lines hollow organs, body cavities, and ducts; and forms glands.
2. Connective tissue, which protects and supports the body and its organs; binds organs together; stores energy reserves as fat; and provides immunity.
3. Muscle tissue, which is responsible for movement and generation of force.
4. Nervous tissue, which initiates and transmits action potentials (nerve impulses) that help coordinate body activities.

About 8 days after a sperm fertilizes a secondary oocyte (potential ovum), the mass of cells that results from several cell divisions embeds in the lining of the uterus and begins to form three primary **germ layers: ectoderm, endoderm,** and **mesoderm**. These are the embryonic tissues from which all tissues and organs of the body develop. Epithelial tissues develop from all three germ layers. Connective tissues and most muscle tissues all derive from mesoderm. Nervous tissue develops from ectoderm. (See Exhibit 29.1 on page 965 for a more detailed list of structures derived from the primary germ layers.)

Epithelial tissue and connective tissue, except for bone tissue and blood, are discussed in detail in this chapter. The general features of bone tissue and blood will be introduced here, but their detailed discussion occurs later in Chapters 6 and 19, respectively. Similarly, the structure and function of muscle tissue and nervous tissue are examined in detail in Chapters 10 and 12, respectively.

The cells of most epithelial tissues are continually replaced by mitosis of progenitor cells. Thus epithelial tissue is constantly being regenerated. Connective tissues with a good blood supply, such as bone, regenerate better than those with a poorer blood supply, such as cartilage and fibrous connective tissue. This can affect the rate at which some injuries will heal. Muscle cells and neurons (nerve cells) are highly differentiated (specialized) in adults and lack the capacity for mitosis. As a result, muscle tissue and nervous tissue cannot regenerate.

Surrounding all body cells is **extracellular fluid** (**ECF**), which provides a medium for dissolving and mixing solutes, transporting substances, and carrying out chemical reactions. The two major subdivisions of ECF are **interstitial (intercellular) fluid**, the fluid that fills the microscopic spaces (interstitial spaces) between cells in tissues, and **plasma**, the liquid portion of blood, found in blood vessels (see Fig. 1.2).

Normally, most cells within a tissue remain in place, anchored to other cells, basement membranes (described shortly), and connective tissues. A few cells, such as phagocytes, routinely move through the body, searching for invaders. Before birth, certain cells migrate extensively as part of the growth and development process.

MEDICAL STUDIES OF TISSUES

A **pathologist** (pa-THOL-ō-gist; *pathos* = disease) is a physician who specializes in laboratory studies of cells and tissues to help other physicians reach accurate diagnoses. One of the principal functions of a pathologist is to examine tissues for any changes that might indicate disease. Pathologists also conduct autopsies (see page 18) and supervise laboratory personnel who examine blood and other body fluids.

A **biopsy** (BĪ-op-sē) is the removal of a sample of living tissue for microscopic examination. It is used to help diagnose many disorders, especially cancer, and to discover the cause of unexplained infections and inflammations. Once the tissue sample is removed, it may be preserved, stained to highlight special properties, and cut into thin sections for microscopic observation. ■

CELL JUNCTIONS

Most epithelial cells, some muscle cells, and some nerve cells are tightly joined to form a close functional unit. The points of contact between adjacent plasma membranes are called **cell junctions**. Three principal types of cell junctions serve distinct functions: (1) **tight junctions** form fluid-tight seals between cells like the seal on a Ziploc® sandwich bag; (2) **anchoring junctions** fasten cells to one another or to the extracellular material; and (3) **gap junctions** permit electrical or chemical signals to pass from cell to cell (Fig. 4.1).

Tight junctions are common among epithelial cells that line the stomach, intestines, and urinary bladder. They prevent fluid in a cavity from leaking into the body by passing between cells.

Anchoring junctions are common in tissues subjected to friction and stretching. Examples include the outer layer of the skin, the muscle tissue of the heart, the neck of the uterus (which is greatly stretched during childbirth), and the epithelial lining of the gastrointestinal tract. The most common type of anchoring junction is called a **desmosome**. These structures form firm attachments between cells some-

Figure 4.1 Cell junctions.

 Most epithelial cells and some muscle and nerve cells contain cell junctions.

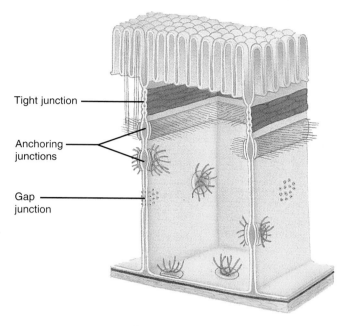

Tight junction

Anchoring junctions

Gap junction

Types and locations of cell junctions

 Which type of junction functions in communication between adjacent cells?

what like spot welds. On the cytoplasmic surface of a desmosome, intermediate filaments of the cytoskeleton attach to a dense plaque of proteins.

Gap junctions allow the rapid spread of action potentials from one cell to the next in some parts of the nervous system and in muscle of the heart and gastrointestinal tract. In a developing embryo, chemical and electrical signals that regulate growth and differentiation may travel by way of gap junctions. At a gap junction, adjacent plasma membranes approach each other, not fusing but leaving a gap of 2–3 nm (1 nm = $^{1}/_{1,000,000,000}$ m). Spanning the gap are proteins called **connexons** that form minute fluid-filled tunnels. Through the connexons, ions and small molecules, such as glucose and amino acids, can pass directly from the cytosol of one cell into the cytosol of the next. Note that cancer cells do not have gap junctions and therefore cannot communicate with each other. As a result, cell division is not coordinated and occurs in an uncontrolled manner.

EPITHELIAL TISSUE

Epithelial (ep-i-THĒ-lē-al) **tissue**, or **epithelium** (plural is **epithelia**), may be divided into two types: (1) **covering and lining epithelium** and (2) **glandular epithelium**. Covering

and lining epithelium forms the superficial layer of the skin and some internal organs. It forms the inner lining of blood vessels, ducts, body cavities, and the interiors of the respiratory, digestive, urinary, and reproductive systems. Glandular epithelium constitutes the secreting portion of glands, such as sweat glands and the thyroid gland. Epithelial tissue also combines with nervous tissue to make up special sense organs for smell, hearing, vision, and touch.

General Features of Epithelial Tissue

Following are the general features of epithelial tissue:

1. Epithelium consists largely or entirely of closely packed cells with little extracellular material between adjacent cells.
2. Epithelial cells are arranged in continuous sheets, in either single or multiple layers.
3. Epithelial cells have an **apical (free) surface**, which is exposed to a body cavity, lining of an internal organ, or the exterior of the body, and a **basal surface**, which is attached to the basement membrane (Fig. 4.2).
4. Cell junctions are plentiful, providing secure attachments among the cells.
5. Epithelia are **avascular** (*a* = without; *vasculum* = vessel). The blood vessels that supply nutrients and remove wastes are located in adjacent connective tissue. The exchange of materials between epithelium and connective tissue is by diffusion.

Figure 4.2 Surfaces of epithelial cells and the structure and location of the basement membrane.

 The basement membrane is found between epithelium and connective tissue.

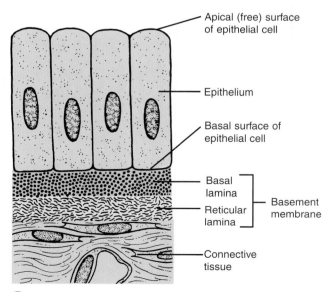

Apical (free) surface of epithelial cell

Epithelium

Basal surface of epithelial cell

Basal lamina

Reticular lamina

Basement membrane

Connective tissue

 What are the functions of the basement membrane?

6. Epithelia adhere firmly to nearby connective tissue, which holds the epithelium in position and prevents it from being torn. The attachment between the epithelium and the connective tissue is a thin extracellular layer called the **basement membrane** (Fig. 4.2). It consists of two layers. The **basal lamina** (*lamina* = thin, flat plate or layer) contains collagen, laminin, and proteoglycans secreted by the epithelium. Cells in the connective tissue secrete the second layer, the **reticular lamina**, which contains reticular fibers, fibronectin, and glycoproteins. The basement membrane provides physical support for epithelium, provides for cell attachment, serves as a filter in the kidneys, and guides cell migration during development and tissue repair.
7. Epithelia have a nerve supply.
8. Since epithelium is subject to a certain amount of wear and tear and injury, it has a high capacity for renewal (high mitotic rate).
9. Epithelia are diverse in origin. They are derived from all three primary germ layers (ectoderm, mesoderm, and endoderm).
10. Functions of epithelia include protection, filtration, lubrication, secretion, digestion, absorption, transportation, excretion, sensory reception, and reproduction.

Covering and Lining Epithelium

Arrangement of Layers

Epithelial cells tend to be arranged in one or more layers depending on the function of the particular body part. The terms used to refer to their classification by arrangement of layers include the following:

1. **Simple epithelium** is a single layer of cells found in areas where activities such as diffusion, osmosis, filtration, secretion, and absorption occur.
2. **Stratified epithelium** contains two or more layers of cells that protect underlying tissues in areas where there is considerable wear and tear.
3. **Pseudostratified epithelium** contains one layer of a mixture of cell shapes. The tissue has a multilayered (stratified) appearance because the nuclei lie at different levels and not all cells reach the surface. Those cells that do reach the surface are either ciliated or secrete mucus (see page 101, Pseudostratified columnar epithelium).

Cell Shapes

Epithelial tissue contains cells with four basic shapes:

1. **Squamous** (SKWĀ-mus; *squama* = flat) cells are flat and attach to each other like tiles. Their thinness allows for rapid movement of substances through them.
2. **Cuboidal** cells are thicker, being cube or hexagon shaped. They produce several important body **secretions** (fluids that are produced and released by cells, such as sweat and digestive enzymes). They may also function in **absorption** (intake) of fluids and other substances, such as digested foods in the intestines.
3. **Columnar** cells are tall and cylindrical, thereby protecting underlying tissues. Some may have cilia. They may also be specialized for secretion and absorption.
4. **Transitional** cells readily change in shape from flat to columnar and often change shape due to distention (stretching), expansion, or movement of body parts.

Classification

Considering layers and cell shapes in combination, we may classify covering and lining epithelium as follows:

I. **Simple**
   A. Squamous
   B. Cuboidal
   C. Columnar
II. **Stratified**
   A. Squamous [1]
   B. Cuboidal [1]
   C. Columnar [1]
   D. Transitional
III. **Pseudostratified columnar**

Each of the epithelial tissues described in the following sections is illustrated in Exhibit 4.1. Along with the illustrations are the descriptions, locations, and functions of the tissues.

Simple Epithelium

● **Simple Squamous Epithelium** This consists of a single layer of flat cells. Its surface resembles a tiled floor. The nucleus of each cell is oval or spherical. Since simple squamous epithelium has only one layer of cells, it is highly adapted for diffusion, osmosis, and filtration. Simple squamous epithelium is found in body parts that are subject to little wear and tear.

Simple squamous epithelium that lines the heart, blood vessels, and lymphatic vessels and forms the walls of capillaries is known as **endothelium** (*endo* = within; *thelium* = covering). Simple squamous epithelium that forms the epithelial layer of serous membranes is called **mesothelium** (*meso* = middle). Endothelium and mesothelium both derive from the embryonic mesoderm.

● **Simple Cuboidal Epithelium** The cuboidal shape of the cells is obvious only when the tissue is sectioned at right angles to the surface. The nuclei are usually round. Simple cuboidal epithelium performs the functions of secretion and absorption.

● **Simple Columnar Epithelium** When sectioned at right angles to the surface, the cells appear rectangular with oval nuclei. Simple columnar epithelium exists in two forms: **nonciliated simple columnar epithelium** and **ciliated simple columnar epithelium**. The nonciliated type contains

---

[1] This classification is based on the shape of the apical (free) surface cells.

EXHIBIT **4.1** **EPITHELIAL TISSUES**

COVERING AND LINING EPITHELIUM

Simple squamous epithelium   *Description:* Single layer of flat cells.
Location: Lines heart, blood vessels, lymphatic vessels, air sacs of lungs, glomerular (Bowman's) capsule of kidneys, and inner surface of the tympanic membrane (eardrum) of ear; forms epithelial layer of serous membranes.
Function: Filtration, diffusion, osmosis, and secretion in serous membranes.

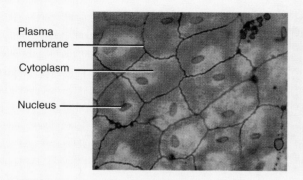

Plasma membrane

Cytoplasm

Nucleus

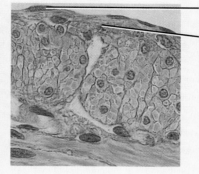

Nucleus of simple squamous cell

Connective tissue

Surface view of mesothelial lining of peritoneal cavity (243x)

Sectional view of intestinal serosa (245x)

Diagram of simple squamous epithelium

Simple cuboidal epithelium   *Description:* Single layer of cube-shaped cells.
Location: Covers surface of ovary, lines anterior surface of capsule of the lens of eye, forms the pigmented epithelium at the back of the eye, and lines kidney tubules and smaller ducts of many glands.
Function: Secretion and absorption.

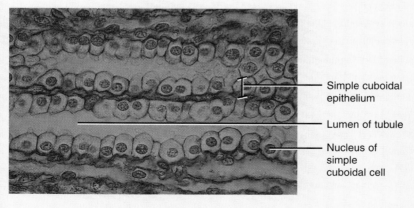

Simple cuboidal epithelium

Lumen of tubule

Nucleus of simple cuboidal cell

Sectional view of kidney tubules (575x)

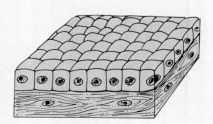

Diagram of simple cuboidal epithelium

Exhibit continues

EXHIBIT **4.1**   **EPITHELIAL TISSUES**   (**CONTINUED**)

COVERING AND LINING EPITHELIUM, *continued*

Nonciliated simple columnar epithelium

Description: Single layer of nonciliated rectangular cells; contains goblet cells and microvilli in some locations.

Location: Lines the gastrointestinal tract from the stomach to the anus, ducts of many glands, and gallbladder.

Function: Secretion and absorption.

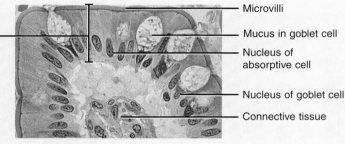

Nonciliated simple columnar epithelium

Microvilli

Mucus in goblet cell

Nucleus of absorptive cell

Nucleus of goblet cell

Connective tissue

Sectional view of epithelium of a villus from the lining of the small intestine (400x)

Diagram of nonciliated simple columnar epithelium

Ciliated simple columnar epithelium

Description: Single layer of ciliated rectangular cells; contains goblet cells in some locations.

Location: Lines a few portions of upper respiratory tract, uterine (Fallopian) tubes, uterus, some paranasal sinuses, and central canal of spinal cord.

Function: Moves fluids or particles along a passageway by ciliary action.

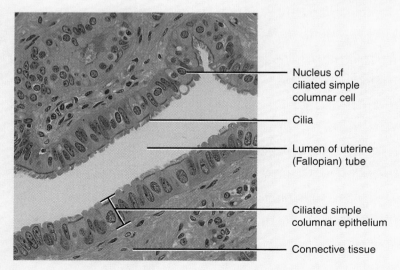

Nucleus of ciliated simple columnar cell

Cilia

Lumen of uterine (Fallopian) tube

Ciliated simple columnar epithelium

Connective tissue

Sectional view of uterine (Fallopian) tube (275x)

Diagram of ciliated simple columnar epithelium

Stratified squamous epithelium

Description: Several layers of cells; cuboidal to columnar shape in deep layers; squamous cells in superficial layers; basal cells replace surface cells as they are lost.

Location: Keratinized variety forms superficial layer of skin; nonkeratinized variety lines wet surfaces such as lining of the mouth, esophagus, part of epiglottis, and vagina and covers the tongue.

Function: Protection.

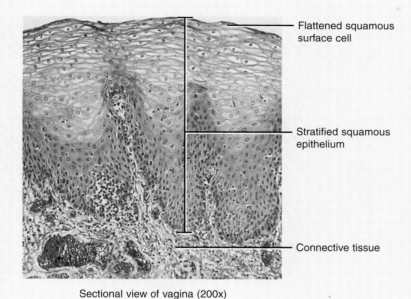

Sectional view of vagina (200x)

Diagram of stratified squamous epithelium

Stratified cuboidal epithelium

Description: Two or more layers of cells in which the superficial cells are cube-shaped.

Location: Ducts of adult sweat glands and part of male urethra.

Function: Protection.

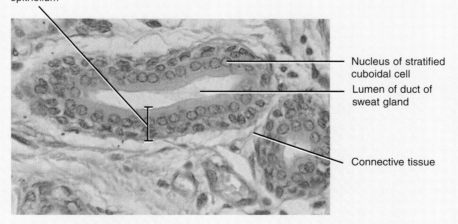

Sectional view of the duct of a sweat gland (450x)

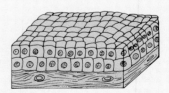

Diagram of stratified cuboidal epithelium

Exhibit continues

EXHIBIT **4.1**   **EPITHELIAL TISSUES**   (CONTINUED)

COVERING AND LINING EPITHELIUM, *continued*

Stratified columnar epithelium

Description: Several layers of polyhedral cells; columnar cells are only in the superficial layer.
Location: Lines part of urethra, large excretory ducts of some glands, small areas in anal mucous membrane, and portion of conjunctiva of eye.
Function: Protection and secretion.

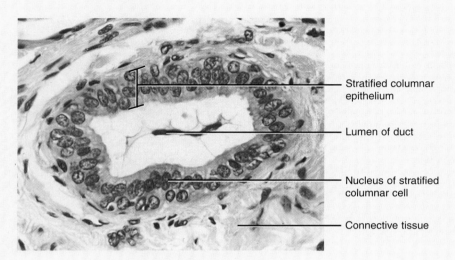

Stratified columnar epithelium

Lumen of duct

Nucleus of stratified columnar cell

Connective tissue

Sectional view of the duct of the submandibular salivary gland (495x)

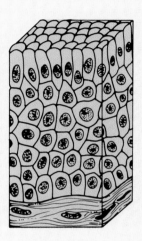

Diagram of stratified columnar epithelium

Transitional epithelium

Description: Appearance is variable (transitional); shape of superficial cells ranges from squamous to cuboidal depending on the degree of distention (stretching).
Location: Lines urinary bladder and portions of ureters and urethra.
Function: Permits distention.

Lumen of urinary bladder

Nucleus of transitional cell

Connective tissue

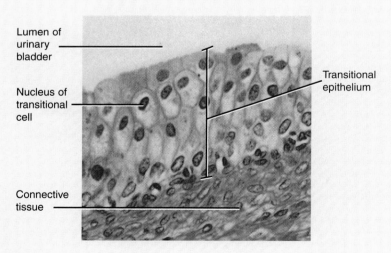

Transitional epithelium

Sectional view of urinary bladder in relaxed state (240x)

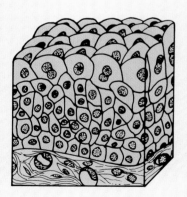

Diagram of relaxed transitional epithelium

Pseudostratified columnar epithelium

Description: Not a true stratified tissue; nuclei of cells are at different levels; all cells are attached to basement membrane, but not all reach the free surface.
Location: Pseudostratified ciliated columnar epithelium lines most of upper respiratory tract; pseudostratified nonciliated columnar epithelium lines larger ducts of many glands, epididymis, and part of male urethra.
Function: Secretion and movement of mucus by ciliary action.

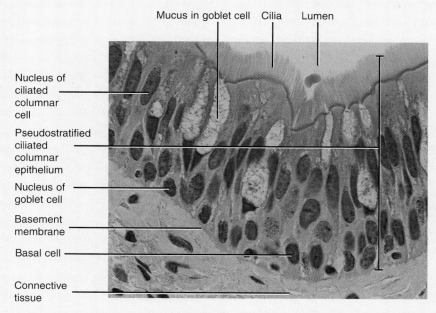
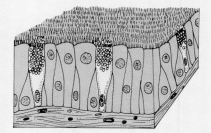

Mucus in goblet cell   Cilia   Lumen

Nucleus of ciliated columnar cell

Pseudostratified ciliated columnar epithelium

Nucleus of goblet cell

Basement membrane

Basal cell

Connective tissue

Diagram of pseudostratified columnar epithelium

Photomicrograph of enlarged aspect of tracheal epithelium (850x)

GLANDULAR EPITHELIUM

Exocrine glands

Description: Secretory products released into ducts.
Location: Sweat, oil, ear wax, and mammary glands of the skin; digestive glands such as salivary glands that secrete into mouth cavity and pancreas that secretes into the small intestine.
Function: Produce mucus, perspiration, oil, ear wax, milk, or digestive enzymes.

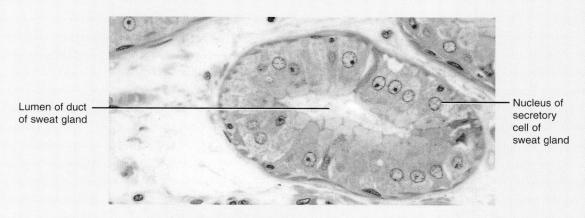

Lumen of duct of sweat gland

Nucleus of secretory cell of sweat gland

Sectional view of the secretory portion of a sweat gland (1532x)

Exhibit continues

EXHIBIT **4.1** **EPITHELIAL TISSUES** (CONTINUED)

Endocrine glands   *Description:* Secretory products (hormones) diffuse into blood after passing through extracellular fluid.
Location: Examples include pituitary gland at base of brain, pineal gland in brain, thyroid and parathyroid glands near larynx (voice box), adrenal glands above kidneys, pancreas near stomach, ovaries in pelvic cavity, testes in scrotum, and thymus gland in thoracic cavity.
Function: Produce hormones that regulate various body activities.

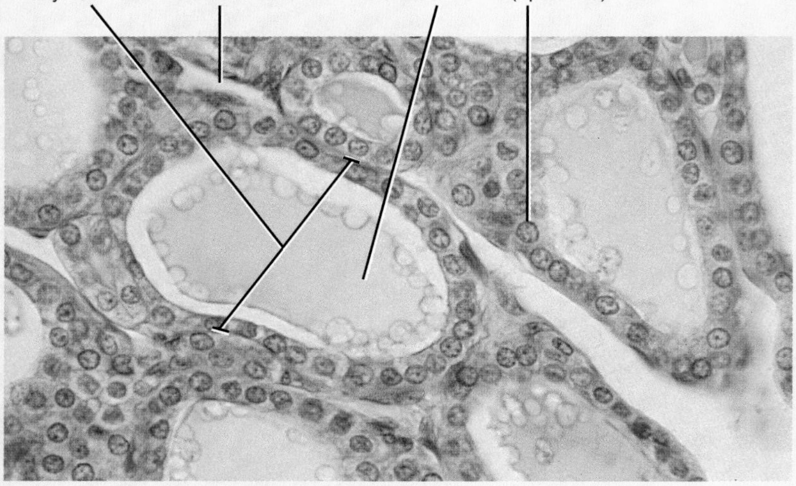

Sectional view of thyroid gland (500x)

microvilli and goblet cells. **Microvilli** (*micro* = small; *villus* = tuft of hair) are microscopic fingerlike cytoplasmic projections that serve to increase the surface area of the plasma membrane. This increases the amount of digested nutrients and fluids that can be absorbed into the cell and thus into the body. **Goblet cells** are modified columnar cells that secrete mucus, which is a slightly sticky fluid. Mucus accumulates toward the apical (free) surface of the cell, causing that area to bulge out. The whole cell then resembles a goblet or wine glass. The secreted mucus serves as a lubricant for the linings of the digestive, respiratory, reproductive, and most of the urinary tracts. Mucus also helps to trap dust that enters the body, especially in the respiratory tract. In the digestive tract, mucus prevents destruction of the epithelial lining by digestive enzymes.

Another modification of columnar epithelium is found in cells with hairlike processes called **cilia** (*ciliaris* = resembling an eyelash). In a few parts of the upper respiratory tract, ciliated columnar cells are interspersed with goblet cells. Mucus secreted by the goblet cells forms a film over the respiratory surface that traps foreign particles that are inhaled. The cilia wave in unison and move the mucus, with any foreign particles, toward the throat, where it can be swallowed or spit out. Cilia also function in the female reproductive tract where they help move a secondary oocyte down the uterine (Fallopian) tube for fertilization or a fertilized ovum down the uterine tube into the uterus for implantation.

Stratified Epithelium

In contrast to simple epithelium, stratified epithelium has at least two layers of cells. Thus it is more durable and can protect underlying tissues from the external environment and from wear and tear. Some cells of stratified epithelia also produce secretions. The name of the specific kind of stratified epithelium depends on the shape of the cells in the superficial layer.

● **Stratified Squamous Epithelium** In the superficial layers of this type of epithelium, the cells are flat, whereas in the deep layers, cells vary in shape from cuboidal to columnar. The basal (deepest) cells continually replicate by cell division. As new cells grow, the cells of the basal layer continually shift upward toward the surface. As they move farther from the deep layer and their blood supply (in the underlying connective tissue), they become dehydrated, shrunken, and harder. At the surface, the cells lose their cell junctions and are rubbed off. Old cells are sloughed off and replaced as new cells continually emerge. The Pap smear, a screening test for precancer and cancer of the uterus, cervix, and vagina (see page 930), consists of examining cells that are sloughed off.

Stratified squamous epithelium exists in two forms: keratinized and nonkeratinized. In **keratinized stratified squamous epithelium**, a tough layer of keratin is deposited in the surface cells. **Keratin** (*kerato* = horny) is a protein that is waterproof and resistant to friction and helps repel bacteria. **Nonkeratinized stratified squamous epithelium** does not contain keratin and remains moist.

• **Stratified Cuboidal Epithelium** This fairly rare type of epithelium sometimes consists of more than two layers of cells. Its function is mainly protective, as in the male urethra.

• **Stratified Columnar Epithelium** Like stratified cuboidal epithelium, this type of tissue also is uncommon in the body. Usually the basal layer or layers consist of shortened, irregular polyhedral cells. Only the superficial cells are columnar in form. It functions in protection and secretion.

• **Transitional Epithelium** This kind of epithelium is variable in appearance, depending on whether it is relaxed or distended (stretched). In the relaxed state, it looks similar to stratified cuboidal epithelium except that the superficial cells tend to be large and rounded. This allows the tissue to be distended without the outer cells breaking apart from one another. When stretched, they are drawn out into squamous-shaped cells, giving the appearance of stratified squamous epithelium. Because of this capability, transitional epithelium lines hollow structures that are subjected to expansion from within, such as the urinary bladder. Its function is to help prevent rupture of these organs.

Pseudostratified Columnar Epithelium

The third category of covering and lining epithelium is called pseudostratified columnar epithelium. The nuclei of the cells are at varying depths. Even though all the cells are attached to the basement membrane in a single layer, some cells do not reach the free surface. When the tissue is sectioned, these features give the false impression of a multilayered tissue, the reason for the name *pseudo*stratified (*pseudo* = false) epithelium. In pseudostratified ciliated columnar epithelium, the cells that reach the surface either secrete mucus or bear cilia that sweep away mucus and trapped foreign particles for eventual elimination from the body. Pseudostratified nonciliated columnar epithelium contains no cilia or goblet cells.

Glandular Epithelium

The function of glandular epithelium is secretion, accomplished by glandular cells that often lie in clusters deep to the covering and lining epithelium. A **gland** may consist of one cell or a group of highly specialized epithelial cells that secrete substances into ducts, onto a surface, or into the blood. All glands of the body are classified as exocrine or endocrine.

The secretions of **exocrine** (*exo* = outside; *krin* = to secrete) **glands** flow onto the free surface of covering and lining epithelia, usually by way of tubelike ducts. Thus exocrine secretions reach the skin surface or the lumen of a hollow organ. The secretions include mucus, perspiration, skin oil, ear wax, and digestive enzymes. Examples of exocrine glands are sweat glands, which secrete perspiration (sweat), and salivary glands, which secrete mucus and digestive enzymes.

The secretions of **endocrine** (*endo* = within) **glands** enter the extracellular fluid and then diffuse directly into the bloodstream without flowing through a duct. These secretions, called hormones, regulate many metabolic and physiological activities to maintain homeostasis. The pituitary, thyroid, and adrenal glands are examples of endocrine glands. Endocrine glands will be described in detail in Chapter 18.

Structural Classification of Exocrine Glands

Exocrine glands are classified into two structural types: multicellular and unicellular. Most glands are **multicellular glands**, which have many cells that form a distinctive microscopic structure or macroscopic organ. Examples are sweat, oil, and salivary glands.

Unicellular glands are single-celled. An important unicellular exocrine gland is the goblet cell (see page 98, Nonciliated simple columnar epithelium). Although goblet cells do not contain ducts, they are often classified as unicellular mucus-secreting exocrine glands. Goblet cells are present in the epithelial lining of the digestive, respiratory, urinary, and reproductive systems. They produce mucus, which lubricates the apical surface of these tissues.

Functional Classification of Exocrine Glands

The functional classification of exocrine glands is based on whether a secretion is a product of a cell or consists of entire or partial glandular cells themselves. **Holocrine** (HŌ-lō-krin; *holos* = entire; *krin* = to secrete) **glands** accumulate a secretory product in their cytosol. The cell then dies and is discharged with its contents as the glandular secretion (Fig. 4.3a). The discharged cell is replaced by a new cell. One example of a holocrine gland is a sebaceous (oil) gland of the skin. **Merocrine** (MER-ō-krin; *meros* = a part) **glands** simply form the secretory product and discharge it from the cell (Fig. 4.3b). Most exocrine glands of the body are merocrine. Examples of merocrine glands are the salivary glands. **Apocrine** (AP-ō-krin; *apo* = from) **glands** accumulate their secretory product at the apical surface of the secreting cell. That portion of the cell pinches off from the rest of the cell to form the secretion (Fig. 4.3c). The remaining part of the cell repairs itself and repeats the process. The mammary glands are apocrine glands.

CONNECTIVE TISSUE

The most abundant and most widely distributed tissue in the body is **connective tissue**. It binds together, supports, and strengthens other body tissues, protects and insulates internal organs, and compartmentalizes structures such as skeletal muscles. Blood, a fluid connective tissue, is the major transport system within the body, whereas adipose (fat) tissue is the major site of stored energy reserves.

Figure 4.3 Functional classification of multicellular exocrine glands.

 Functional classification is based on whether a secretion is a product of a cell or consists of an entire or partial glandular cell.

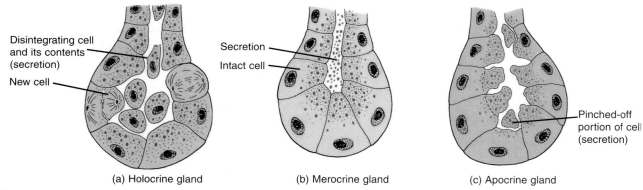

| (a) Holocrine gland | (b) Merocrine gland | (c) Apocrine gland |

Q What is an example of a holocrine gland, merocrine gland, and apocrine gland?

General Features of Connective Tissue

Following are the general features of connective tissue:

1. Connective tissue consists of three basic elements: cells, ground substance, and fibers. Together, the ground substance and fibers, which are outside the cells, form the **matrix**, which is noncellular. Unlike epithelial cells, connective tissue cells rarely touch one another; they are separated by a considerable amount of matrix.
2. In contrast to epithelia, connective tissues do not usually occur on free surfaces, such as the covering or lining of internal organs, lining of a body cavity, or the external surface of the body. However, joint cavities are lined by a type of connective tissue called areolar connective tissue.
3. Except for cartilage, connective tissue, like epithelium, has a nerve supply.
4. Unlike epithelium, connective tissue usually is highly vascular (has a rich blood supply). Exceptions include cartilage, which is avascular (has no blood supply), and tendons, which have a scanty blood supply.
5. The matrix of a connective tissue, which may be fluid, semifluid, gelatinous, fibrous, or calcified, is usually secreted by the connective tissue cells and adjacent cells and determines the tissue's qualities. In blood, the matrix, which is not secreted by blood cells, is liquid. In cartilage, it is firm but pliable. In bone, it is considerably harder and not pliable.

Connective Tissue Cells

The cells in connective tissue are derived from embryonic mesodermal cells called mesenchymal cells. The immature cells of each major type of connective tissue have names that end in *-blast,* which means to bud, sprout, or produce.

These are fibroblasts in loose and dense connective tissue, chondroblasts in cartilage, and osteoblasts in bone. These cells retain the capacity for mitosis and secrete the matrix (ground substance and fibers) that is characteristic of the tissue. In cartilage and bone, once the matrix is produced, the immature cells differentiate. The mature cells have names that end in *-cyte,* which means cell (chondrocytes and osteocytes). Mature cells have reduced capacity for cell division and matrix formation and are mostly involved in maintaining the matrix.

Following is a description of some of the cells that are found in various types of connective tissue. The specific tissues to which they belong are described on pages 113–114.

1. **Fibroblasts** (FĪ-brō-blasts; *fibra* = fiber) are large, flat, spindle-shaped cells with branching processes; they secrete the molecules that form the matrix.
2. **Macrophages** (MAK-rō-fā-jez; *macro* = large; *phagein* = to eat), or **histiocytes**, develop from **monocytes**, a type of white blood cell (see Fig. 19.2). Macrophages have an irregular shape with short branching projections and are capable of engulfing bacteria and cellular debris by phagocytosis. Thus they provide a vital defense for the body. Whereas wandering macrophages leave the blood and migrate to infected tissues, fixed macrophages remain in certain tissues and organs of the body.
3. **Plasma cells** are small and either round or irregular in shape. They develop from a type of white blood cell called a **B lymphocyte** (**B cell**). Plasma cells secrete antibodies and, accordingly, provide a defense mechanism through immunity. Although they are found in many places in the body, most reside in connective tissues, especially in the gastrointestinal tract and the mammary glands.
4. **Mast cells** are abundant alongside blood vessels. They produce histamine, a chemical that dilates small blood

vessels during inflammation. Mast cells also contain heparin. When produced by other cells of the body, heparin functions as an anticoagulant that prevents blood from clotting in blood vessels. However, heparin from mast cells is a poor anticoagulant. It is in the form of heparin proteoglycan and may serve to bind certain intracellular constituents of mast cells.

Connective Tissue Matrix

Each type of connective tissue has unique properties, due to accumulation of specific matrix materials between the cells. Matrix contains protein **fibers** embedded in a fluid, gel, or solid **ground substance**. The ground substance is amorphous, meaning it has no specific shape. Connective tissue cells usually produce the ground substance and deposit it in the space between the cells.

Ground Substance

Besides chemicals normally found in extracellular fluid, ground substance contains a diversity of large molecules. Some are proteoglycans and glycosaminoglycans, which are complex combinations of polysaccharides and proteins. Several examples are as follows. **Hyaluronic** (hī-a-loo-RON-ik) **acid** is a viscous, slippery substance that binds cells together, lubricates joints, and helps maintain the shape of the eyeballs. It also appears to play a role in helping phagocytes migrate through connective tissue during development and repair of a wound. **Chondroitin** (kon-DROY-tin) **sulfate** is a jellylike substance that provides support and adhesiveness in cartilage, bone, the skin, and blood vessels. The skin, tendons, blood vessels, and heart valves contain **dermatan sulfate**, while bone, cartilage, and the cornea of the eye contain **keratan sulfate**. Another class of molecules in matrix are **adhesion proteins**, for example, fibronectin, laminin, collagen, and fibrinogen. Adhesion proteins interact with receptors on plasma membranes to anchor cells in position and to provide traction for the movement of cells.

The ground substance supports cells, binds them together, and provides a medium through which substances are exchanged between the blood and cells. Until recently, it was thought to function mainly as an inert scaffolding to support tissues. Now it is clear that the ground substance is quite active in functions such as influencing tissue development, migration, proliferation, shape, and even metabolic functions.

Fibers

Fibers in the matrix are synthesized by fibroblasts and provide strength and support for tissues. Three types of fibers are embedded in the matrix between the cells of connective tissue: collagen, elastic, and reticular fibers.

Collagen (*kolla* = glue) **fibers**, of which there are at least five different types, are very tough and resistant to a pulling force, yet allow some flexibility in the tissue because they are not taut. These fibers often occur in bundles made up of many tiny fibrils lying parallel to one another. The bundle arrangement affords great strength. Chemically, collagen fibers consist of the protein *collagen*. This is the most abundant protein in your body, representing about 25% of the total protein. Collagen fibers are found in most types of connective tissues, especially bone, cartilage, tendons, and ligaments.

Elastic fibers are smaller in diameter than collagen fibers. They also branch and join together to form a network within a tissue. In the assembly of an elastic fiber, molecules of a protein called *elastin* form cross links with one another within a scaffolding of large glycoproteins. A particular glycoprotein, named *fibrillin*, is essential to the stability of an elastic fiber. Like collagen fibers, elastic fibers provide strength. In addition, they can be stretched up to 150% of their relaxed length without breaking. Elastic fibers are plentiful in skin, blood vessel walls, and lung tissue.

Reticular (*rete* = net) **fibers**, consisting of the protein *collagen* and a coating of *glycoprotein,* provide support in the walls of blood vessels and form a network around fat cells, nerve fibers, and skeletal and smooth muscle cells. They are much thinner than collagen fibers and form branching networks. Like collagen fibers, reticular fibers provide support and strength and also form the **stroma** (*stroma* = bed covering) or supporting framework of many soft organs, such as the spleen and lymph nodes. These fibers also help form the basement membrane.

Classification of Connective Tissues

Classification of connective tissues is difficult because of the diversity of cells, ground substance, and fibers and the differences in their relative proportions. Thus the separation of connective tissues into categories is not always clear-cut. We will classify them as follows:

I. **Embryonic connective tissue**
   A. Mesenchyme
   B. Mucous connective tissue
II. **Mature connective tissue**
   A. Loose connective tissue
      1. Areolar connective tissue
      2. Adipose tissue
      3. Reticular connective tissue
   B. Dense connective tissue
      1. Dense regular connective tissue
      2. Dense irregular connective tissue
      3. Elastic connective tissue
   C. Cartilage
      1. Hyaline cartilage
      2. Fibrocartilage
      3. Elastic cartilage
   D. Bone (osseous) tissue
   E. Blood (vascular tissue)

Each of the connective tissues described in the following sections is illustrated in Exhibit 4.2. Along with the illustrations are the descriptions, locations, and functions of the tissues.

EXHIBIT **4.2** **CONNECTIVE TISSUES**

EMBRYONIC CONNECTIVE TISSUE

MESENCHYME

Description: Consists of irregularly shaped mesenchymal cells embedded in a semifluid ground substance that contains delicate reticular fibers.
Location: Deep to skin and along developing bones of embryo; some mesenchymal cells found in adult connective tissue, especially along blood vessels.
Function: Forms all other kinds of connective tissue.

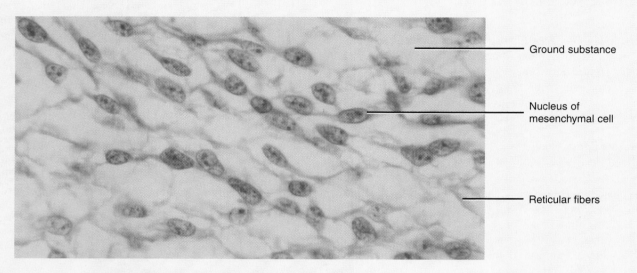

Ground substance

Nucleus of mesenchymal cell

Reticular fibers

Sectional view of mesenchyme from a developing fetus (800x)

MUCOUS CONNECTIVE TISSUE

Description: Consists of star-shaped cells embedded in a viscous, jellylike ground substance that contains fine collagen fibers.
Location: Umbilical cord of fetus.
Function: Support.

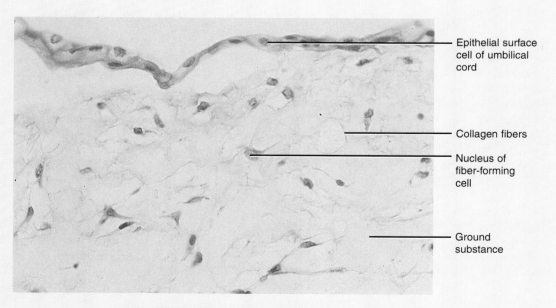

Epithelial surface cell of umbilical cord

Collagen fibers

Nucleus of fiber-forming cell

Ground substance

Sectional view of the umbilical cord (457x)

MATURE CONNECTIVE TISSUE

LOOSE CONNECTIVE TISSUE
Areolar connective tissue

Description: Consists of fibers (collagen, elastic, and reticular) and several kinds of cells (fibroblasts, macrophages, plasma cells, adipocytes, and mast cells) embedded in a semifluid ground substance.
Location: Subcutaneous layer of skin, papillary (superficial) region of dermis of skin, mucous membranes, blood vessels, nerves, and around body organs.
Function: Strength, elasticity, and support.

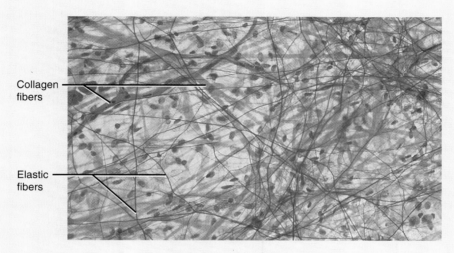

Collagen fibers

Elastic fibers

Sectional view of subcutaneous tissue (224x)

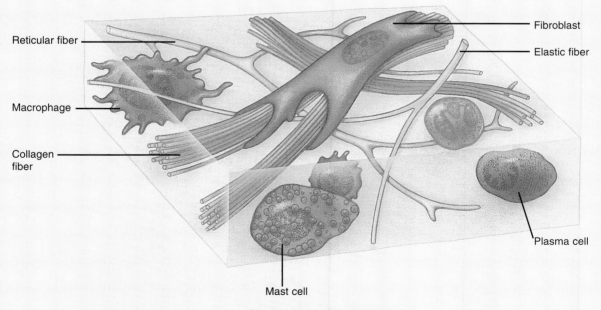

Reticular fiber

Macrophage

Collagen fiber

Fibroblast

Elastic fiber

Plasma cell

Mast cell

Diagram of areolar connective tissue

Exhibit continues

EXHIBIT **4.2** **CONNECTIVE TISSUES** (CONTINUED)

MATURE CONNECTIVE TISSUE, *continued*

LOOSE CONNECTIVE TISSUE, *continued*

Adipose tissue

Description: Consists of adipocytes, cells that are specialized to store triglycerides (fats and oils) in a large central area in the cytoplasm; nuclei are peripherally located.

Location: Subcutaneous layer of skin, around heart and kidneys, yellow bone marrow of long bones, and padding around joints and behind eyeball in eye socket.

Function: Reduces heat loss through skin, serves as an energy reserve, supports, and protects; in newborns, brown fat generates considerable heat (thermogenesis) that probably helps to maintain proper body temperature.

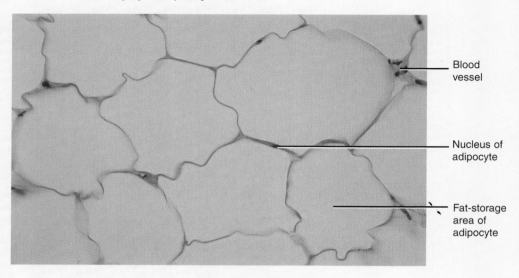

Blood vessel

Nucleus of adipocyte

Fat-storage area of adipocyte

Sectional view of adipocytes of white fat of the pancreas (350x)

Reticular connective tissue

Description: Consists of a network of interlacing reticular fibers and reticular cells.

Location: Stroma (framework) of liver, spleen, lymph nodes; red bone marrow that gives rise to blood cells, and reticular lamina of the basement membrane.

Function: Forms stroma of organs; binds together smooth muscle tissue cells.

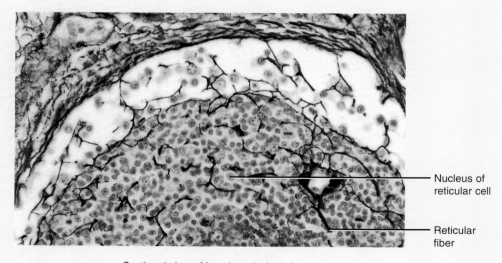

Nucleus of reticular cell

Reticular fiber

Sectional view of lymph node (496x)

DENSE CONNECTIVE TISSUE

Dense regular connective tissue

Description: Matrix looks shiny white; consists of predominantly collagen fibers arranged in parallel bundles; fibroblasts present in rows between bundles.
Location: Forms tendons (attach muscle to bone), most ligaments (attach bone to bone), and aponeuroses (sheetlike tendons that attach muscle to muscle or muscle to bone).
Function: Provides strong attachment between various structures.

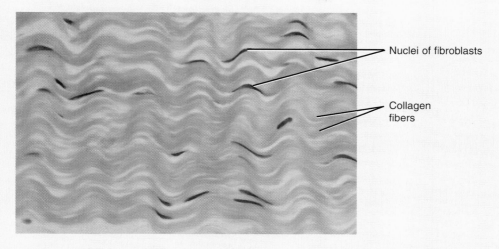

Nuclei of fibroblasts

Collagen fibers

Sectional view of a tendon (250x)

Dense irregular connective tissue

Description: Consists of predominantly collagen fibers, randomly arranged, and a few fibroblasts; usually forms a sheet.
Location: Fasciae, reticular (deeper) region of dermis of skin, perichondrium (membrane around cartilage), periosteum (membrane around bone), joint capsules, dura mater (outer membrane around brain and spinal cord), membrane capsules around various organs (kidneys, liver, testes, lymph nodes), and heart valves.
Function: Provides strength.

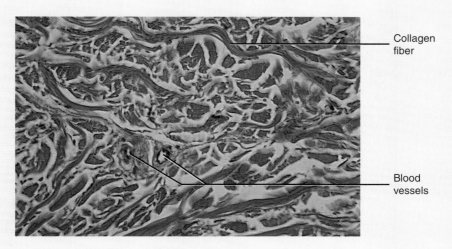

Collagen fiber

Blood vessels

Sectional view of dermis of skin (275x)

Exhibit continues

EXHIBIT **4.2 CONNECTIVE TISSUES** **(CONTINUED)**

MATURE CONNECTIVE TISSUE, *continued*

DENSE CONNECTIVE TISSUE, *continued*

Elastic connective tissue

Description: Consists of predominantly freely branching elastic fibers; fibroblasts present in spaces between fibers.
Location: Lung tissue, walls of elastic arteries, trachea, bronchial tubes, true vocal cords, suspensory ligament of penis, and ligamenta flava of vertebrae (ligaments between vertebrae).
Function: Allows stretching of various organs.

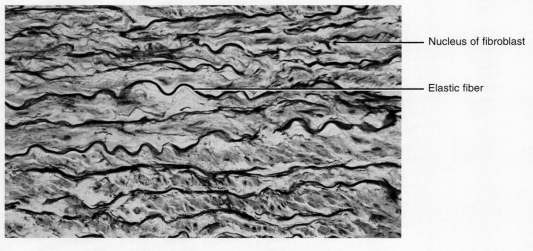

Nucleus of fibroblast

Elastic fiber

Sectional view of aorta (largest artery in the body) (335x)

CARTILAGE

Hyaline cartilage

Description: Consists of a bluish white and shiny ground substance with fine collagen fibers; contains numerous chondrocytes; most abundant type of cartilage.
Location: Ends of long bones, anterior ends of ribs, nose, parts of larynx, trachea, bronchi, bronchial tubes, and embryonic skeleton.
Function: Provides smooth surfaces for movement at joints, flexibility, and support.

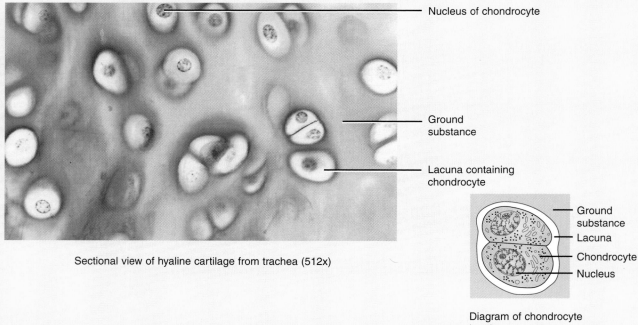

Nucleus of chondrocyte

Ground substance

Lacuna containing chondrocyte

Ground substance
Lacuna
Chondrocyte
Nucleus

Sectional view of hyaline cartilage from trachea (512x)

Diagram of chondrocyte in a lacuna

Fibrocartilage

Description: Consists of chondrocytes scattered among bundles of collagen fibers within the matrix.

Location: Pubic symphysis (point where hipbones join anteriorly), intervertebral discs (discs between vertebrae), and menisci (cartilage pads) of knee.

Function: Support and fusion.

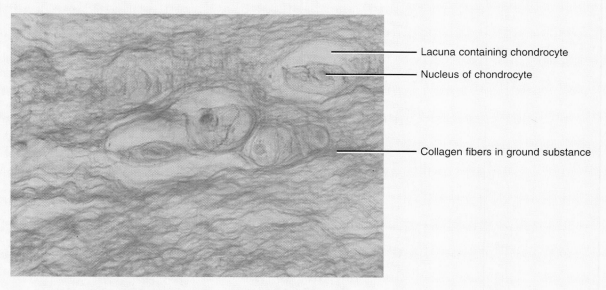

— Lacuna containing chondrocyte

— Nucleus of chondrocyte

— Collagen fibers in ground substance

Sectional view of fibrocartilage from patellar tendon insertion (742x)

Elastic cartilage

Description: Consists of chondrocytes located in a threadlike network of elastic fibers within the matrix.

Location: Epiglottis of larynx, external ear (auricle), and auditory (Eustachian) tubes.

Function: Gives support and maintains shape.

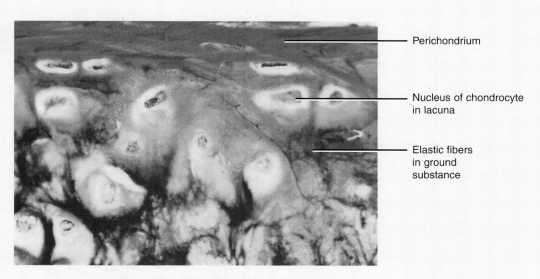

— Perichondrium

— Nucleus of chondrocyte in lacuna

— Elastic fibers in ground substance

Sectional view of elastic cartilage from auricle of ear (742x)

Exhibit continues

EXHIBIT **4.2** **CONNECTIVE TISSUES** (CONTINUED)

MATURE CONNECTIVE TISSUE, *continued*

BONE (OSSEOUS) TISSUE

Description: Compact bone consists of osteons (Haversian systems) that contain lamellae, lacunae, osteocytes, canaliculi, and central (Haversian) canals. See also Fig. 6.3a. Spongy bone consists of thin plates called trabeculae; spaces between trabeculae are filled with red bone marrow. See Fig. 6.3b,c.

Location: Both compact and spongy bone comprise the various parts of bones of the body.

Function: Support, protection, storage, houses blood-forming tissue, and serves as levers that act together with muscle tissue to provide movement.

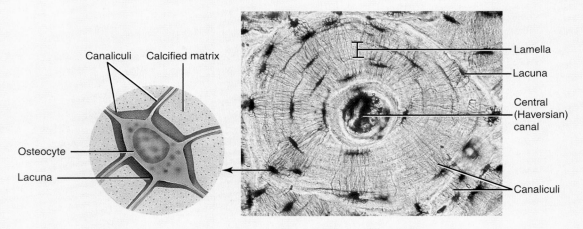

Sectional view of an osteon (Haversian system)
from the femur (thigh bone) (285x)

BLOOD (VASCULAR TISSUE)

Description: Consists of plasma and formed elements. The formed elements are erythrocytes (red blood cells), leukocytes (white blood cells), and platelets. See also Fig. 19.2.

Location: Within blood vessels (arteries, arterioles, capillaries, venules, and veins).

Function: Erythrocytes transport oxygen and carbon dioxide; leukocytes carry on phagocytosis and are involved in allergic reactions and immunity; platelets are essential for the clotting of blood.

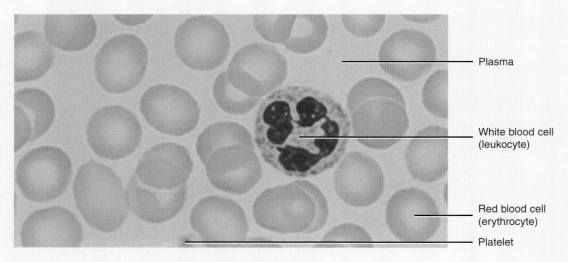

Blood smear (1000x)

Embryonic Connective Tissue

Connective tissue that is present primarily in the embryo or fetus is called **embryonic connective tissue**. The term *embryo* refers to a developing human from fertilization through the first two months of pregnancy; *fetus* refers to a developing human from the third month of pregnancy to birth.

One example of embryonic connective tissue found almost exclusively in the embryo is **mesenchyme** (MEZ-en-kīm)—the tissue from which all other connective tissues eventually arise. It is composed of irregularly shaped mesenchymal cells, a semifluid ground substance, and delicate reticular fibers. Mesenchymal cells in mature connective tissue differentiate into fibroblasts that assist in wound healing.

Another kind of embryonic connective tissue is **mucous connective tissue** (**Wharton's jelly**), found primarily in the umbilical cord of the fetus. It is a form of mesenchyme that contains star-shaped cells, a more viscous, jellylike ground substance, and collagen fibers.

Mature Connective Tissue

Mature connective tissue exists in the newborn, has cells produced from mesenchyme, and does not change after birth. It is subdivided into several kinds: loose connective tissue, dense connective tissue, cartilage, bone, and blood.

Loose Connective Tissue

In **loose connective tissue** the fibers are *loosely* woven, and there are many cells. The types are areolar connective tissue, adipose tissue, and reticular connective tissue.

● **Areolar Connective Tissue** One of the most widely distributed connective tissues in the body is **areolar** (a-RĒ-ō-lar; *areola* = a small space) **connective tissue**. It contains several kinds of cells, including fibroblasts, macrophages, plasma cells, mast cells, adipocytes, and a few white blood cells. All three types of fibers—collagen, elastic, and reticular—are present and randomly arranged. The fluid, semifluid, or gelatinous ground substance contains hyaluronic acid, chondroitin sulfate, dermatan sulfate, and keratan sulfate.

The matrix materials normally aid the passage of nutrients from the blood vessels of the connective tissue into adjacent cells and tissues. The thick consistency of hyaluronic acid, however, may impede the movement of some drugs. If an enzyme called **hyaluronidase** is injected into the tissue, the ground substance changes to a watery consistency. This feature is of clinical importance because the reduced viscosity hastens the absorption and diffusion of injected drugs and fluids through the tissue and thus can lessen tension and pain. White blood cells, sperm, and some bacteria produce hyaluronidase. Combined with adipose tissue, areolar connective tissue forms the **subcutaneous** (sub′-kyoo-TĀ-nē-us; *sub* = under; *cutis* = skin) **layer**—the layer of tissue that attaches the skin to underlying tissues and organs.

MARFAN SYNDROME

Marfan (MAR-fan) **syndrome** is an inherited disorder caused by a defective fibrillin gene. The result is abnormal elastic fibers. Tissues rich in elastic fibers are malformed or weakened. Structures affected most seriously are the covering layer of bones (periosteum), the ligament that suspends the lens of the eye, and the walls of the large arteries. People with Marfan syndrome tend to be tall and have disproportionately long arms, legs, fingers, and toes. A common sign is blurred vision caused by displacement of the lens of the eye. The most life-threatening complication of Marfan syndrome is weakening of the aorta (main artery that emerges from the heart), which can suddenly burst. ■

● **Adipose Tissue** Adipose tissue is a loose connective tissue in which the cells, called **adipocytes** (*adeps* = fat) are specialized for storage of triglycerides (fats and oils). Adipocytes are derived from fibroblasts. As the cell fills with a large triglyceride droplet, the cytoplasm and nucleus are pushed to the edge (periphery) of the cell. Adipose tissue is found wherever areolar connective tissue is located. Adipose tissue is a good insulator and can therefore reduce heat loss through the skin. It is a major energy reserve and generally supports and protects various organs.

Most of the fat in adults is **white fat**, the type just described. There is another type called **brown fat**. The brown color is due to a very rich blood supply and numerous mitochondria, which contain colored cytochrome pigments that participate in cellular respiration. Although brown fat is widespread in the fetus and infant, in adults only small amounts are present. Brown fat generates considerable heat (thermogenesis) and probably helps to maintain proper body temperature in the newborn. The heat generated by the many mitochondria is carried away by the extensive blood supply to other body tissues.

● **Reticular Connective Tissue** Reticular connective tissue consists of fine interlacing reticular fibers and reticular cells. Reticular connective tissue forms the framework of certain organs and helps to bind together certain cells.

Dense Connective Tissue

Dense connective tissue contains more numerous, thicker and *densely* packed fibers but considerably fewer cells than loose connective tissue. The types are dense regular connective tissue, dense irregular connective tissue, and elastic connective tissue.

● **Dense Regular Connective Tissue** In this tissue, bundles of collagen fibers have a regular (orderly), parallel arrangement that confers great strength. The tissue structure withstands pulling in one direction. Fibroblasts, which produce the fibers and ground substance, appear in rows between the fibers. The tissue is silvery white and tough, yet somewhat pliable.

● **Dense Irregular Connective Tissue** This tissue contains collagen fibers that are usually irregularly arranged. It is found in parts of the body where tensions are exerted in various directions. The tissue usually occurs in sheets. Heart valves, the perichondrium (a membrane around cartilage), and the periosteum (membrane around bone) are classified as dense irregular connective tissue although they have a fairly orderly arrangement of collagen fibers.

● **Elastic Connective Tissue** Elastic connective tissue has a predominance of freely branching elastic fibers. These fibers give the unstained tissue a yellowish color. Fibroblasts are present in the spaces between the fibers. Elastic connective tissue can be stretched and will snap back into shape (elasticity). It also provides strength.

Cartilage
Cartilage is capable of enduring considerably more stress than the connective tissues just discussed. Cartilage consists of a dense network of collagen fibers and elastic fibers firmly embedded in chondroitin sulfate, a rubbery component of the ground substance. Whereas the strength of cartilage is due to its collagen fibers, its resilience (ability to assume its original shape after deformation) is due to chondroitin sulfate.

The cells of mature cartilage, called **chondrocytes** (KON-drō-sīts; *chondros* = cartilage), occur singly or in groups within spaces called **lacunae** (la-KOO-nē-; singular is *lacuna* = little lake) in the matrix. The surface of cartilage is surrounded by a membrane of dense irregular connective tissue called the **perichondrium** (per'-i-KON-drē-um; *peri* = around). Unlike other connective tissues, cartilage has no blood vessels or nerves, except for those in the perichondrium. Three kinds of cartilage are recognized: hyaline cartilage, fibrocartilage, and elastic cartilage (see pages 110–111).

● **Hyaline Cartilage** This cartilage, also called **gristle**, contains a resilient gel as its ground substance and appears in the body as a bluish-white, shiny substance. The fine collagen fibers, although present, are not visible with ordinary staining techniques, and the prominent chondrocytes are found in lacunae. Hyaline cartilage is the most abundant cartilage in the body. It affords flexibility and support and reduces friction and absorbs shock at joints. Hyaline cartilage is the weakest of the three types of cartilage.

● **Fibrocartilage** Chondrocytes are scattered among clearly visible bundles of collagen fibers within the matrix of this type of cartilage. This tissue combines strength and rigidity and is the strongest of the three types of cartilage.

● **Elastic Cartilage** In this tissue, chondrocytes are located in a threadlike network of elastic fibers within the matrix. Elastic cartilage provides strength and elasticity and maintains the shape of certain organs.

● **Growth and Repair of Cartilage** Cartilage growth is slow; metabolically, it is an inactive tissue. When injured or inflamed, cartilage repair proceeds slowly, in large part because cartilage is avascular. Substances needed for repair and blood cells that participate in tissue repair must diffuse or migrate from the perichondrium into the cartilage. Bone, which has a rich blood supply, heals much more rapidly than cartilage.

The growth of cartilage follows two basic patterns. In **interstitial (endogenous) growth**, the cartilage increases rapidly in size through the division of existing chondrocytes and continuous deposition of increasing amounts of matrix by the chondrocytes. The formation of new chondrocytes and their production of new matrix cause the cartilage to expand from within—thus the term *inter*stitial growth. This growth pattern occurs while the cartilage is young and pliable—during childhood and adolescence.

In **appositional (exogenous) growth**, activity of cells in the inner chondrogenic layer of the perichondrium leads to growth. The deeper cells of the perichondrium, the fibroblasts, divide. Some differentiate into chondroblasts (immature cells) and then into chondrocytes. As differentiation occurs, the chondroblasts surround themselves with matrix and become chondrocytes. As a result, matrix accumulates on the surface of the cartilage, increasing its size. A new layer of cartilage forms beneath the perichondrium on the surface of the cartilage, causing it to grow in width. Appositional growth starts later than interstitial growth and continues through adolescence.

With aging, cartilage tends to calcify; that is, minerals deposit in it. Also, cartilage may ossify or change into bone (see page 147). As a result, chondrocytes die because they are poorly nourished.

CARTILAGE AND CANCER

When cartilage grows, it does not allow the growth of blood vessels into it. It was found that this is due to the production of chemicals by cartilage called **antiangiogenesis factors (AAFs)**. Since cancers are the result of uncontrolled cell division, which is supported by blood vessel growth, AAFs may have great potential for treating cancer. Some AAFs have been isolated and are now in clinical trials for treatment of various types of cancers. They are also being tested for the treatment of stomach ulcers. ■

Bone (Osseous) Tissue
Together, cartilage, joints, and **bone** or **osseous** (OS-ē-us) **tissue** comprise the skeletal system. In addition to bone tissue, other components of bone that are also connective tissues include the periosteum (covering around a bone), red and yellow bone marrow, and the endosteum (lining of a space in a bone that stores yellow bone marrow).

Bone tissue is classified as either compact (dense) or spongy (cancellous), depending on how the matrix and cells are organized. The basic unit of compact bone is called an **osteon** (**Haversian system**). Each osteon is composed of lamellae (*lamella* = little plate or layer), concentric rings of matrix that consist of mineral salts (mostly tricalcium phosphate and calcium carbonate) that give bone its hardness and collagen fibers that give bone its strength; **lacunae,** small spaces between lamellae that contain mature bone cells called **osteocytes; canaliculi** (*canaliculi* = small channels or canals), minute canals that project from lacunae and contain projections from osteocyts that provide multiple routes for nutrients to reach osteocytes and wastes to leave them; and a **central (Haversian) canal** that contains blood vessels and nerves. Spongy bone has no osteons; instead, it consists of plates of bone called **trabeculae** (*trabecula* = little beam), which contain lamellae, osteocytes, lacunae, and canaliculi. Spaces between lamellae are filled with red bone marrow. The details of the histology of bone begin on page 143.

The skeletal system has several functions. It supports soft tissues, protects delicate structures, and works with skeletal muscles to generate movement. Bone stores calcium and phosphorus, houses red bone marrow, which produces several kinds of blood cells, and houses yellow bone marrow, which contains triglycerides as an energy source.

Blood (Vascular Tissue)

Blood (**vascular tissue**) is a connective tissue with a liquid matrix called plasma. Suspended in the plasma are formed elements—cells and cell fragments. **Plasma** is a straw-colored liquid that consists mostly of water with a wide variety of dissolved substances (nutrients, wastes, enzymes, hormones, respiratory gases, and ions). The formed elements are red blood cells (erythrocytes), white blood cells (leukocytes), and platelets. See also Fig. 19.2. **Red blood cells** function in transporting oxygen to body cells and removing carbon dioxide from them. **White blood cells** are involved in phagocytosis, immunity, and allergic reactions. **Platelets** function in blood clotting.

The details of blood are considered in Chapter 19.

MEMBRANES

The combination of an epithelial layer and an underlying connective tissue layer constitutes an **epithelial membrane**. The principal epithelial membranes of the body are mucous membranes, serous membranes, and the cutaneous membrane, or skin. The skin is an organ of the integumentary system and is discussed in the next chapter. Another kind of membrane, a **synovial membrane**, contains only connective tissue and no epithelium.

Mucous Membranes

A **mucous membrane** lines body cavities that open directly to the exterior. Mucous membranes line the entire digestive, respiratory, and reproductive systems and much of the urinary system. They consist of a lining layer of epithelium and an underlying layer of connective tissue.

The epithelial layer of a mucous membrane is an important aspect of the body's defense mechanisms. It is a barrier that microbes and other pathogens have difficulty penetrating. Usually, tight junctions connect the cells, so materials cannot leak in between. Goblet and other cells of the epithelial layer of a mucous membrane secrete mucus. This slippery fluid prevents the cavities from drying out, traps particles in the respiratory passageways, and lubricates food as it moves through the gastrointestinal tract. In addition, the epithelial layer secretes some of the enzymes needed for digestion and is the site of food and fluid absorption in the gastrointestinal tract. As you will see later, the epithelium of a mucous membrane varies greatly in different parts of the body. For example, the epithelium of the mucous membrane of the small intestine is simple columnar, while that of the large airways to the lungs is ciliated pseudostratified columnar.

The connective tissue layer of a mucous membrane is called the **lamina propria** (LAM-i-na PRŌ-prē-a). The lamina propria is so named because it belongs to the mucous membrane (*proprius* = one's own). It binds the epithelium to the underlying structures and allows some flexibility of the membrane. It also holds the blood vessels in place and protects underlying muscles from abrasion or puncture. Oxygen and nutrients diffuse from the lamina propria to the epithelium covering it, while carbon dioxide and wastes diffuse in the opposite direction.

Serous Membranes

A **serous** (*serous* = watery) **membrane** lines body cavities that do not open directly to the exterior, and it covers the organs that lie within those cavities. Serous membranes consist of thin layers of areolar connective tissue covered by a layer of mesothelium (simple squamous epithelium), and they are composed of two layers. The layer attached to the cavity wall is called the **parietal** (pa-RĪ-e-tal; *paries* = wall) **layer**. The layer that covers and attaches to the organs inside these cavities is the **visceral** (*viscus* = body organ) **layer**. The serous membrane lining the thoracic cavity and covering the lungs is called the **pleura**. The serous membrane lining the heart cavity and covering the heart is the **pericardium** (*cardio* = heart). The serous membrane lining the abdominal cavity and covering the abdominal organs and some pelvic organs is called the **peritoneum**.

The epithelial layer of a serous membrane secretes a watery lubricating fluid, called **serous fluid**, that allows the organs to glide easily against one another or against the walls of the cavities. The connective tissue layer of the serous membrane consists of a thin layer of areolar connective tissue.

Synovial Membranes

Synovial (sin-Ō-vē-al) **membranes** line the cavities of the freely movable joints (see Fig. 9.1a). *Syn* means together or junction, here referring to a place where joints come together. Like serous membranes, synovial membranes line structures that do not open to the exterior. Unlike mucous, serous, and cutaneous membranes, they do not contain epithelium and are therefore not epithelial membranes. They are composed of areolar connective tissue with elastic fibers and varying amounts of fat. Synovial membranes secrete **synovial fluid**, which lubricates the cartilage at the ends of bones during their movements and nourishes the cartilage covering the bones at joints. These are articular synovial membranes. Other synovial membranes line cushioning sacs, called bursae, and tendon sheaths in our hands and feet that ease the movement of muscle tendons.

MUSCLE TISSUE

Muscle tissue consists of fibers (cells) that are beautifully constructed to generate force for contraction. As a result of this characteristic, muscle tissue provides motion, maintains posture, and generates heat. Based on location and certain structural and functional characteristics, muscle tissue is classified into three types: skeletal, cardiac, and smooth (Exhibit 4.3).

Skeletal muscle tissue is named for its location—attached to bones. It is also **striated**; that is, the fibers contain alternating light and dark bands (striations) that are perpendicular to the long axes of the fibers. The striations are visible under a microscope. A single skeletal muscle fiber is very long, roughly cylindrical in shape, and has many nuclei, which are located at the periphery of the cell. Within a whole muscle, the individual fibers are parallel to each other. Skeletal muscle is also **voluntary** because it can be made to contract or relax by conscious control.

Cardiac muscle tissue forms the bulk of the wall of the heart. Like skeletal muscle, it is striated. However, unlike skeletal muscle tissue, it is **involuntary**; its contraction is usually not under conscious control. Cardiac muscle fibers exhibit branching and are squarish in cross section. The fibers usually have only one nucleus that is centrally located; sometimes there are two nuclei. Cardiac muscle fibers attach end to end to each other by transverse thickenings of the plasma membrane called **intercalated** (*intercalare* = to insert between) **discs**, which contain both gap junctions and desmosomes. Intercalated discs are unique to cardiac muscle. Desmosomes form firm attachments between cells, like spot welds, and strengthen the tissue. Gap junctions provide a route for quick conduction of muscle action potentials throughout the heart.

EXHIBIT **4.3 MUSCLE TISSUE**

Skeletal muscle tissue

Description: Long, cylindrical, striated fibers with many peripherally located nuclei; voluntary control.

Location: Usually attached to bones by tendons.

Function: Motion, posture, heat production (thermogenesis).

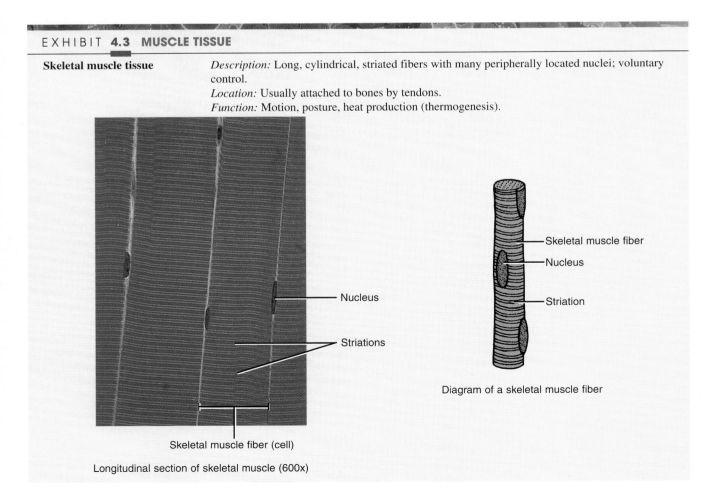

Nucleus

Striations

Skeletal muscle fiber (cell)

Longitudinal section of skeletal muscle (600x)

Skeletal muscle fiber

Nucleus

Striation

Diagram of a skeletal muscle fiber

Cardiac muscle tissue

Description: Branched striated fibers with one or two centrally located nuclei; contains intercalated discs; mainly involuntary control.
Location: Heart wall.
Function: Pumps blood to all parts of the body.

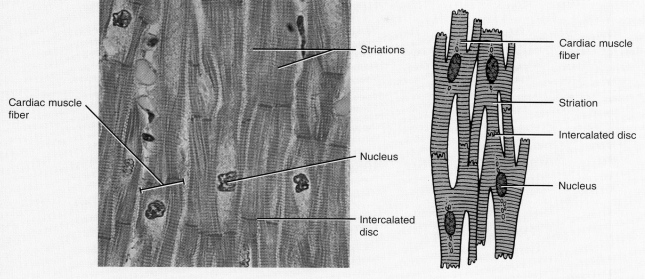

Longitudinal section of cardiac muscle (700x)

Diagram of cardiac muscle fibers

Smooth muscle tissue

Description: Spindle-shaped, nonstriated fibers with one centrally located nucleus; usually involuntary control.
Location: Walls of hollow internal structures such as blood vessels, airways to the lungs, stomach, intestines, gallbladder, urinary bladder, and uterus.
Function: Motion (constriction of blood vessels and airways, propulsion of foods through gastrointestinal tract, contraction of urinary bladder and gallbladder).

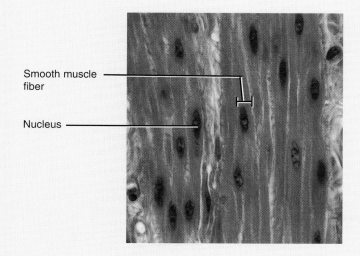

Longitudinal section of smooth muscle (840x)

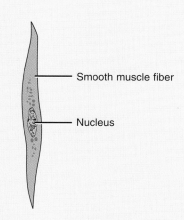

Diagram of a smooth muscle fiber

Smooth muscle tissue is located in the walls of hollow internal structures such as blood vessels, airways to the lungs, the stomach, intestines, gallbladder, and urinary bladder. Its contraction helps constrict (narrow) the lumen of blood vessels, break down food, move food and fluids through the body, and eliminate wastes. Smooth muscle fibers are usually **involuntary**, and they are **nonstriated** (**smooth**). A smooth muscle fiber is small and spindle-shaped (thickest in the middle with each end tapering to a point) and contains a single, centrally located nucleus. Many individual fibers in some smooth muscle tissue, for example, in the wall of the intestines, are connected by gap junctions and thus contract as a single unit. In other locations, for example, the iris of the eye, smooth muscle fibers contract individually, like skeletal muscle fibers, because gap junctions are absent.

Chapter 10 provides a more detailed discussion of muscle tissue.

NERVOUS TISSUE

Despite the awesome complexity of the nervous system, it consists of only two principal kinds of cells: neurons and neuroglia. **Neurons** (*neuro* = nerve) or nerve cells are sensitive to various stimuli, they convert stimuli into nerve impulses (nerve action potentials), and they conduct these impulses to other neurons, muscle tissue, or glands. Most neurons consist of three basic portions: a cell body and two kinds of processes—dendrites and axons (Exhibit 4.4). The **cell body** contains the nucleus and other typical organelles. **Dendrites** (*dendro* = tree) are tapering, highly branched, and usually short. They are the major receiving or input portion of a neuron. The **axon** (*axon* = axis) of a neuron is a single, thin, cylindrical process that may be very long. It is the output portion of a neuron, conducting nerve impulses toward another neuron or other tissue.

Neuroglia (*glia* = glue) do not generate or conduct nerve impulses. They do, however, have many important functions (see Exhibit 12.1 on page 334). They are of clinical interest because they are often the sites of tumors of the nervous system. The detailed structure and function of neurons and neuroglia are considered in Chapter 12.

TISSUE REPAIR: RESTORING HOMEOSTASIS

Epithelial tissues, which in some locations endure considerable wear and tear and even injury, and connective tissues generally have a continuous capacity for renewal. In some cases, immature, undifferentiated cells called **stem cells** divide to replace lost or damaged cells. For example, stem cells reside in protected positions in the epithelium of the skin and gastrointestinal tract to replenish cells sloughed from the apical (free) surface. And stem cells in bone mar-

EXHIBIT 4.4   NERVOUS TISSUE

Description: Consists of neurons (nerve cells) and neuroglia. Neurons (nerve cells) consist of a cell body and processes extending from the cell body called dendrites or axons. Neuroglia do not generate or conduct nerve impulses but have other important functions.
Location: Nervous system.
Function: Exhibits sensitivity to various types of stimuli, converts stimuli into nerve impulses, and conducts nerve impulses to other neurons, muscle fibers, or glands.

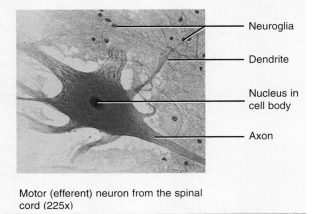

Motor (efferent) neuron from the spinal cord (225x)

row continually provide new red and white blood cells. In other cases, mature, differentiated cells can duplicate themselves by mitosis, for example, hepatocytes (liver cells) and endothelial cells in blood vessels.

Stem cells, known as satellite cells, are present in muscle tissue, but they do not divide rapidly enough to replace damaged skeletal or cardiac muscle fibers. And existing skeletal or cardiac muscle fibers rarely split to form new fibers. Smooth muscle fibers do proliferate to some extent but much more slowly than do the cells of epithelial or connective tissues. Although experiments recently have revealed the presence of some stem cells in the brain, they normally do not undergo mitosis to replace damaged neurons. Discovering why is a major goal of researchers who hope to find ways to encourage repair of damaged or diseased nervous tissue.

Tissue repair is the process by which tissues replace dead or damaged cells. New cells originate by cell division from the **stroma**, the supporting connective tissue, or from the **parenchyma** (pa-REN-ki-ma), cells that form the organ's functioning part. The restoration of an injured organ or tissue to normal structure and function depends entirely on which type of cell—parenchymal or stromal—is active in the repair.

If parenchymal cells accomplish the repair, tissue **regeneration** is possible. A perfect or near-perfect reconstruction of the injured tissue may occur. However, if fibroblasts of the stroma are active in the repair, the replacement tissue will be new connective tissue called scar tissue. In this process, fibroblasts synthesize collagen and other matrix materials that aggregate to form scar tissue. The process of scar tissue formation is known as **fibrosis**. Since scar tissue is not specialized to perform the functions of the parenchymal tissue, the function of the tissue is impaired.

The cardinal factor in tissue repair is the capacity of parenchymal tissue to regenerate. This capacity, in turn, depends on the ability of the parenchymal cells to replicate quickly.

ADHESIONS

Scar tissue can cause **adhesions**, abnormal joining of tissues. They commonly form in the abdomen around a site of previous inflammation such as an inflamed appendix, or they can develop after surgery. Although adhesions do not always cause problems, they can decrease tissue flexibility, cause obstruction (such as in the intestine), and make a later operation more difficult. Surgery may be required to release adhesions. ■

Repair Process

If injury to a tissue, for example the skin, is slight, drainage and reabsorption of pus (accumulation of leukocytes and fluid resulting from inflammation), followed by parenchymal regeneration, may lead to complete repair.

When tissue and cell damage are extensive and severe, as in large, open wounds, both the connective tissue stroma and the parenchymal cells are active in repair. This repair involves the rapid cell division of many fibroblasts and the manufacture of new collagen fibers to provide strength. Blood capillaries also sprout new buds to serve the healing tissue. All these processes create an actively growing connective tissue called **granulation tissue**. This new tissue forms across a wound or surgical incision to provide a framework (stroma). The framework supports the epithelial cells that migrate into the open area and fill it. The newly formed granulation tissue also secretes a fluid that kills bacteria.

Conditions Affecting Repair

Three factors affect tissue repair: nutrition, blood circulation, and age. Nutrition is vital since the healing process places a great demand on the body's store of nutrients. Adequate protein in the diet is important because most of the cell structure is made from proteins. Vitamins also play a direct role in wound healing. Among the vitamins involved and their roles in wound healing are the following:

1. Vitamin A is essential in the replacement of epithelial tissues, especially in the respiratory tract.
2. The B vitamins—thiamine, nicotinic acid, riboflavin—are coenzymes needed by many enzyme systems in cells. These vitamins may relieve pain and are necessary for division of the cells that accomplish repair.
3. Vitamin C directly affects the normal production and maintenance of matrix materials, especially collagen. Vitamin C also strengthens and promotes the formation of new blood vessels. With vitamin C deficiency, even superficial wounds fail to heal, and the walls of the blood vessels become fragile and are easily ruptured.
4. Vitamin D is necessary for the proper absorption of calcium from the intestine. Calcium gives bones their hardness and is necessary for the healing of fractures.
5. Vitamin E is believed to promote healing of injured tissues and may prevent scarring.
6. Vitamin K is needed for production of certain proteins that participate in blood clotting and thus prevent the injured person from bleeding to death.

In tissue repair, proper blood circulation is essential to transport oxygen, nutrients, antibodies, and many defensive cells to the injured site. The blood also plays an important role in the removal of tissue fluid, bacteria, foreign bodies, and debris. These elements would otherwise interfere with healing.

Generally, tissues heal faster and leave less obvious scars in the young than in the aged. In fact, surgery performed on fetuses leaves no scars. The younger body is generally in a better nutritional state, its tissues have a better blood supply, and its cells have a faster metabolic rate. Thus cells can synthesize needed materials and divide more quickly. The extracellular components of tissues also change with age. Collagen fibers, responsible for the strength in tendons, increase in number and change in quality with aging. These changes in the collagen of arterial walls are as much responsible for their loss of extensibility as are the deposits associated with atherosclerosis, the deposition of fatty materials in arterial walls. Elastin, another extracellular component, is responsible for the elasticity of blood vessels and skin. It thickens, fragments, and acquires a greater affinity for calcium with age—changes that may also be associated with the development of atherosclerosis.

STUDY OUTLINE

TYPES OF TISSUES AND THEIR ORIGINS (p. 94)

1. A tissue is a group of similar cells that usually have a similar embryological origin and are specialized for a particular function.
2. Depending on their function and structure, the various tissues of the body are classified into four principal types: epithelial, connective, muscular, and nervous.

3. Extracellular fluid is external to body cells. Examples are interstitial fluid and plasma.

CELL JUNCTIONS (p. 94)

1. Cell junctions are points of contact between adjacent plasma membranes.

2. Types are tight junctions (form fluid-tight seals between cells), anchoring junctions such as desmosomes (fasten cells to one another or the matrix), and gap junctions (permit electrical or chemical signals to pass between cells).

EPITHELIAL TISSUE (P. 95)

1. The subtypes of epithelium include covering and lining epithelium and glandular epithelium.
2. Some general characteristics of epithelium are: consists mostly of cells with little extracellular material, arranged in sheets, attached to a basement membrane, avascular, has a nerve supply, high capacity for renewal, and derived from all three primary germ layers.

Covering and Lining Epithelium (p. 96)

1. Epithelial layers are arranged as simple (one layer), stratified (several layers), and pseudostratified (one layer that appears as several); cell shapes include squamous (flat), cuboidal (cube-like), columnar (rectangular), and transitional (variable).
2. Simple squamous epithelium consists of a single layer of flat cells. It is adapted for diffusion and filtration and is found in lungs and kidneys. Endothelium lines the heart and blood vessels. Mesothelium lines the thoracic and abdominopelvic cavities and covers the organs within them.
3. Simple cuboidal epithelium consists of a single layer of cube-shaped cells. It is adapted for secretion and absorption. It is found covering ovaries, in kidneys and eyes, and lining some glandular ducts.
4. Nonciliated simple columnar epithelium consists of a single layer of nonciliated rectangular cells. It lines most of the gastrointestinal tract. Specialized cells containing microvilli perform absorption. Goblet cells secrete mucus. Ciliated simple columnar epithelium consists of a single layer of ciliated rectangular cells. It is found in a few portions of the upper respiratory tract where it moves foreign particles trapped in mucus out of the respiratory tract.
5. Stratified squamous epithelium consists of several layers of cells; cells of the superficial layer are flat. It is protective. Nonkeratinized variety lines the mouth; keratinized variety forms outer layer of skin.
6. Stratified cuboidal epithelium consists of several layers of cells; cells of the superficial layer are cube-shaped. It is found in adult sweat glands and a portion of male urethra.
7. Stratified columnar epithelium consists of several layers of cells; cells of the superficial layer are rectangular. It protects and secretes. It is found in a portion of male urethra and large excretory ducts of some glands.
8. Transitional epithelium consists of several layers of cells whose appearance varies with the degree of stretching. It lines the urinary bladder and is capable of stretching.
9. Pseudostratified columnar epithelium has only one layer but gives the appearance of many. Ciliated variety contains goblet cells and lines most of upper respiratory tract; nonciliated variety has no goblet cells and lines ducts of many glands, epididymis, and part of male urethra.

Glandular Epithelium (p. 103)

1. A gland is a single cell or a grouping of epithelial cells adapted for secretion.
2. Exocrine glands (sweat, oil, mammary, and digestive glands) secrete into ducts or directly onto a free surface.
3. Structural classification of exocrine glands includes unicellular and multicellular glands.

4. Functional classification of exocrine glands includes holocrine, merocrine, and apocrine glands.
5. Endocrine glands secrete hormones into the blood.

CONNECTIVE TISSUE (p. 103)

1. Connective tissue is the most abundant body tissue.
2. Some general characteristics of connective tissue are: It consists of cells, ground substance, and fibers; has abundant matrix with relatively few cells; does not usually occur on free surfaces; has a nerve supply (except for cartilage); and is highly vascular (except for cartilage, tendons, and ligaments).

Connective Tissue Cells (p. 104)

1. Cells in connective tissue are derived from mesenchyme.
2. Types include fibroblasts (secrete matrix), macrophages (phagocytes), plasma cells (secrete antibodies), mast cells (produce histamine), and adipocytes.

Connective Tissue Matrix (p. 105)

1. The ground substance and fibers comprise the matrix.
2. Substances found in the ground substance include hyaluronic acid, chondroitin sulfate, dermatan sulfate, and keratan sulfate.
3. The ground substance supports, binds, provides a medium for the exchange of materials, and is active in influencing cell functions.
4. The fibers provide strength and support and are of three types.
5. Collagen fibers (composed of collagen) are found in large amounts in bone, tendons, and ligaments; elastic fibers (composed of elastin and glycoproteins) are found in skin, blood vessels, and lungs; and reticular fibers (composed of collagen and glycoproteins) are found around fat cells, nerve fibers, and skeletal and smooth muscle cells.

Embryonic Connective Tissue (p. 113)

1. Mesenchyme forms all other connective tissues.
2. Mucous connective tissue is found in the umbilical cord of the fetus, where it gives support.

Mature Connective Tissue (p. 113)

1. Mature connective tissue is connective tissue that differentiates from mesenchyme and exists in the newborn and does not change after birth. It is subdivided into several kinds: connective tissue proper, cartilage, bone tissue, and blood. Subtypes include loose connective tissue, dense connective tissue, cartilage, bone, and blood.
2. Loose connective tissue includes areolar connective tissue, adipose tissue, and reticular connective tissue.
3. Areolar connective tissue consists of the three types of fibers, several cells, and a semifluid ground substance. It is found in the subcutaneous layer and mucous membranes and around blood vessels, nerves, and body organs.
4. Adipose tissue consists of adipocytes that store triglycerides. It is found in subcutaneous layer, around organs, and in the yellow bone marrow of long bones.
5. Reticular connective tissue consists of reticular fibers and reticular cells and is found in the liver, spleen, and lymph nodes.
6. Dense connective tissue includes dense regular connective tissue, dense irregular connective tissue, and elastic connective tissue.
7. Dense regular connective tissue consists of parallel bundles of collagen fibers and fibroblasts. It forms tendons, most ligaments, and aponeuroses.

8. Dense irregular connective tissue consists of usually randomly arranged collagen fibers and a few fibroblasts. It is found in fasciae, dermis of skin, and membrane capsules around organs.

9. Elastic connective tissue consists of elastic fibers and fibroblasts. It is found in the walls of large arteries, lungs, trachea, and bronchial tubes.

10. Cartilage contains chondrocytes and has a rubbery matrix (chondroitin sulfate) containing collagen and elastic fibers.

11. Hyaline cartilage is found in the embryonic skeleton, at the ends of bones, in the nose, and in respiratory structures. It is flexible, allows movement, and provides support.

12. Fibrocartilage is found in the pubic symphysis, intervertebral discs, and menisci (cartilage pads) of the knee joint.

13. Elastic cartilage maintains the shape of organs such as the epiglottis of the larynx, auditory (Eustachian) tubes, and external ear.

14. Cartilage enlarges by interstitial growth (from within) and appositional growth (from without).

15. Bone (osseous tissue) consists of mineral salts and collagen fibers that contribute to the hardness of bone and cells called osteocytes. It supports, protects, helps provide movement, stores minerals, and houses blood-forming tissue.

16. Blood (vascular tissue) consists of plasma and formed elements (red blood cells, white blood cells, and platelets). Functionally, its cells transport oxygen and carbon dioxide, carry on phagocytosis, participate in allergic reactions, provide immunity, and bring about blood clotting.

MEMBRANES (p. 115)

1. An epithelial membrane consists of an epithelial layer overlying a connective tissue layer. Examples are mucous, serous, and cutaneous membranes.

2. Mucous membranes line cavities that open to the exterior, such as the gastrointestinal tract.

3. Serous membranes (pleura, pericardium, peritoneum) line closed cavities and cover the organs in the cavities. These membranes consist of parietal and visceral layers.

4. The cutaneous membrane is the skin.

5. Synovial membranes line joint cavities, bursae, and tendon sheaths and consist of areolar connective tissue instead of epithelium.

MUSCLE TISSUE (p. 116)

1. Muscle tissue is modified for contraction. It provides motion, maintenance of posture, and heat production.

2. Skeletal muscle tissue is attached to bones, is striated, and is voluntary.

3. Cardiac muscle tissue forms most of the heart wall, is striated, and is usually involuntary.

4. Smooth muscle tissue is found in the walls of hollow internal structures (blood vessels and viscera), is nonstriated and is usually involuntary.

NERVOUS TISSUE (p. 118)

1. The nervous system is composed of neurons (nerve cells) and neuroglia (protective and supporting cells).

2. Most neurons consist of a cell body and two types of processes called dendrites and axons.

3. Neurons are sensitive to stimuli, convert stimuli into nerve impulses, and conduct nerve impulses.

TISSUE REPAIR: RESTORING HOMEOSTASIS (p. 118)

1. Tissue repair is the replacement of damaged or destroyed cells by healthy ones.

2. It begins during the active phase of inflammation and is not completed until after harmful substances in the inflamed area have been neutralized or removed.

Repair Process (p. 119)

1. If the injury is superficial, tissue repair involves pus removal (if pus is present) and parenchymal regeneration.

2. If damage is extensive, granulation tissue is involved.

Conditions Affecting Repair (p. 119)

1. Nutrition is important to tissue repair. Various vitamins (A, some B, C, D, E, and K) and a protein-rich diet are needed.

2. Adequate circulation of blood is needed.

3. The tissues of young people repair rapidly and efficiently; the process slows down with aging.

REVIEW QUESTIONS

1. Define a tissue. What are the four basic types of human tissue? (p. 94)

2. What is extracellular fluid (ECF)? Why is it important? (p. 94)

3. Describe the three basic types of cell junctions and the functions of each. (p. 94)

4. Describe the various layering arrangements and cell shapes of epithelium. (p. 96)

5. Distinguish covering and lining epithelium from glandular epithelium. What characteristics are common to all epithelial tissues (epithelia)? Describe the structure of the basement membrane. (p. 96)

6. How is epithelium classified? List the various types. (p. 96)

7. For each of the following kinds of epithelium, briefly describe the microscopic appearance, location in the body, and functions: simple squamous, simple cuboidal, simple columnar (nonciliated and ciliated), stratified squamous (keratinized and nonkeratinized), stratified cuboidal, stratified columnar, transitional, and ciliated and nonciliated pseudostratified columnar. (pp. 96–103)

8. Define the following terms: endothelium, mesothelium, secretion, absorption, goblet cell, and keratin. (p. 96)

9. What is a gland? Distinguish between endocrine and exocrine glands. (p. 103)

10. Describe the classification of exocrine glands according to structure and function and give at least one example of each. (p. 103)

11. Enumerate the ways in which connective tissue differs from epithelium. (p. 104)

12. Describe the cells, ground substance, and fibers that comprise connective tissue. (p. 104)

13. How are connective tissues classified? List the various types. (p. 105)

14. How are embryonic connective tissue and mature connective tissue distinguished? (p. 113)

15. Describe the following connective tissues with regard to microscopic appearance, location in the body, and function: areolar connective tissue, adipose tissue, reticular connective tissue, dense regular connective tissue, dense irregular connective tissue, elastic connective tissue, hyaline cartilage, fibrocartilage, elastic cartilage, bone (osseous) tissue, and blood (vascular tissue). (pp. 106–115)

16. Define the following terms: matrix, ground substance, hyaluronic acid, chondroitin sulfate, dermatan sulfate, keratan sulfate, collagen fiber, elastic fiber, reticular fiber, fibroblast, macrophage, plasma cell, mast cell, adipocyte, chondrocyte, lacuna, osteocyte, lamella, canaliculus, and osteon (Haversian system). (pp. 105–115)

17. Distinguish between the interstitial and appositional growth of cartilage. (p. 114)

18. Define the following kinds of membranes: mucous, serous, cutaneous, and synovial. Where is each located in the body? What are their functions? (p. 115)

19. How is muscle tissue classified? What are its functions? (p. 116)

20. Distinguish between neurons and neuroglia. Describe the structure and function of neurons. (p. 118)

21. Following are some descriptions of various tissues of the body. For each description, name the tissue described.
    a. An epithelium that permits distention (stretching).
    b. A single layer of flat cells concerned with filtration.
    c. Forms all other kinds of connective tissue.
    d. Specialized for fat storage.
    e. An epithelium with waterproofing qualities.
    f. Forms the framework of many organs.
    g. Produces perspiration, wax, oil, or digestive enzymes.
    h. Cartilage that shapes the external ear.
    i. Contains goblet cells and lines the intestine.
    j. Most widely distributed connective tissue.
    k. Forms tendons, ligaments, and aponeuroses.
    l. Specialized for the secretion of hormones.
    m. Provides support in the umbilical cord.
    n. Lines kidney tubules and is specialized for absorption and secretion.
    o. Permits extensibility of lung tissue.
    p. Stores red bone marrow, protects, supports.
    q. Nonstriated, usually involuntary muscle tissue.
    r. Consists of erythrocytes, leukocytes, and platelets.
    s. Composed of a cell body, dendrites, and axon.

22. What is meant by tissue repair? Distinguish between stromal and parenchymal repair. What is the importance of granulation tissue? (p. 118)

23. What conditions affect tissue repair? (p. 119)

24. Define the following: biopsy (p. 94), Marfan syndrome (p. 113), and adhesions (p. 119).

ANSWERS TO QUESTIONS WITH FIGURES

4.1 Gap junctions allow the spread of electrical and chemical signals from cell to cell.

4.2 Provides physical support for epithelium, provides for cell attachment, serves as a filter in the kidneys, and guides cell migration during development and tissue repair.

4.3 Mammary gland is an apocrine gland, sebaceous (oil) gland is a holocrine gland, and salivary gland is a merocrine gland.

STUDENT OBJECTIVES

1. Describe the anatomy and seven functions of the skin. (p. 124)

2. Explain the basis for skin color. (p. 128)

3. Compare the anatomy, distribution, and physiology of epidermal derivatives: hair, sebaceous (oil) glands, sudoriferous (sweat) glands, ceruminous glands, and nails. (p. 129)

4. Outline the steps involved in epidermal wound healing and deep wound healing. (p. 132)

5. Describe how the skin helps maintain normal body temperature. (p. 134)

6. Describe the effects of aging on the integumentary system. (p. 135)

7. Describe the development of the epidermis, its derivatives, and the dermis. (p. 136)

8. Describe the causes and effects for the following skin disorders: burns, skin cancer, acne, and pressure sores. (p. 136)

9. Define medical terminology associated with the integumentary system. (p. 138)

CHAPTER 5

THE INTEGUMENTARY SYSTEM

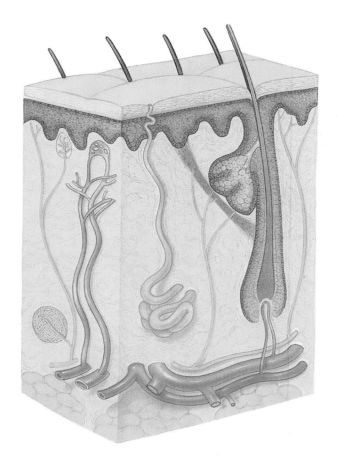

CHAPTER CONTENTS AT A GLANCE

f all the body's organs, none is more easily in-spected or more exposed to infection, disease, and injury than the skin. Because of its visibil-ity, skin reflects our emotions and some as-pects of normal physiology, as evidenced by frowning, blushing, and sweating. Changes in skin color may indicate homeostatic imbalances in the body; for example, a bluish skin color is one sign of heart failure. Abnormal skin erup-tions or rashes such as chickenpox, cold sores, or measles may reveal systemic infections or diseases of internal or-gans. Other disorders may involve just the skin itself, such as warts, age spots, or pimples. The skin's location makes it vulnerable to damage from trauma, sunlight, microbes, and pollutants in the environment.

Many interrelated factors affect both the appearance and health of the skin, including nutrition, hygiene, circulation, age, immunity, genetic traits, environmental stress, psycho-logical state, and drugs. So important is the skin to one's image that many people spend much time and money to re-store it to a more normal or youthful appearance.

A group of tissues that performs a specific function is an **organ**. The next higher level of organization is a **system**—a group of organs working together toward common goals. The organs that make up the **integumentary** (in′-teg-yoo-MEN-tar-ē; *integumentum* = covering) **system** of the body are the skin and its derivatives, such as hair, nails, glands, and nerve endings.

The developmental anatomy of the integumentary system is considered at the end of the chapter.

SKIN

The **skin** is an organ because it consists of different tissues that are joined to perform specific activities. It is one of the largest organs of the body in surface area and weight. In adults, the skin covers an area of about 2 square meters (22 square feet) and weighs 4½–5 kg (10–11 lb). Its thickness is 0.5–4.0 mm, depending on location. The skin is not just a simple, thin coat that keeps the body together and provides protection. It performs several essential functions that will be described shortly. **Dermatology** (der′-ma-TOL-ō-jē; *derm* = skin; *logos* = study of) is the medical specialty that deals with diagnosing and treating skin disorders.

Anatomy

Structurally, the skin consists of two principal parts (Fig. 5.1). The superficial, thinner portion, which is composed of *epithelial tissue*, is called the **epidermis**. The epidermis is attached to the deeper, thicker, *connective tissue* part called the **dermis**. Deep to the dermis is a **subcutaneous (subQ)**

layer. This layer, also called the **superficial fascia** or **hypo-dermis**, consists of areolar and adipose tissues. Fibers from the dermis extend into the subcutaneous layer and anchor the skin to it. The subcutaneous layer, in turn, attaches to underlying tissues and organs.

Physiology

Skin serves several functions, which are introduced in this chapter and explained in more detail in later chapters.

1. **Regulation of body temperature.** In response to high environmental temperature or strenuous exercise, the evaporation of sweat from the skin surface helps lower an elevated body temperature to normal (see Fig. 5.7). In response to low environmental temperature, produc-tion of sweat is decreased, which helps conserve heat. Changes in the flow of blood to the skin also help regu-late body temperature.
2. **Protection.** The skin covers the body and provides a physical barrier that protects underlying tissues from physical abrasion, bacterial invasion, dehydration, and ultraviolet (UV) radiation. Hair and nails also have pro-tective functions, as described shortly.
3. **Sensation.** The skin contains abundant nerve endings and receptors that detect stimuli related to temperature, touch, pressure, and pain.
4. **Excretion.** Besides removing heat and some water from the body, sweat also is the vehicle for loss of a small quantity of ions and several organic compounds.
5. **Immunity.** Certain cells of the epidermis are important components of the immune system, which fends off foreign invaders.
6. **Blood reservoir.** The dermis houses extensive net-works of blood vessels that carry 8–10% of the total blood flow in a resting adult. In moderate exercise, skin blood flow may increase, which helps dissipate heat from the body. During strenuous exercise, however, skin blood vessels constrict (narrow) somewhat, which allows more of the blood to circulate through contract-ing muscles.
7. **Synthesis of vitamin D.** What is commonly called **vitamin D** is actually a group of closely related com-pounds. Synthesis of vitamin D begins with activation of a precursor molecule in the skin by UV rays in sun-light. Enzymes in the liver and kidneys then modify the molecule, finally producing **calcitriol**, the *most active form* of vitamin D. Calcitriol contributes to the home-ostasis of body fluids by aiding absorption of calcium in foods from the digestive tract into the blood. Thus vitamin D is a hormone, since it is produced in one lo-cation in the body, transported by the blood, and then exerts its effect in another location. For this reason, the skin can be considered an endocrine organ.

Figure 5.1 Structure of the skin and underlying subcutaneous tissue. The stratum lucidum shown is not apparent on hairy skin but is included in this diagram of hairy skin so that you can see the relationship of all five epidermal strata.

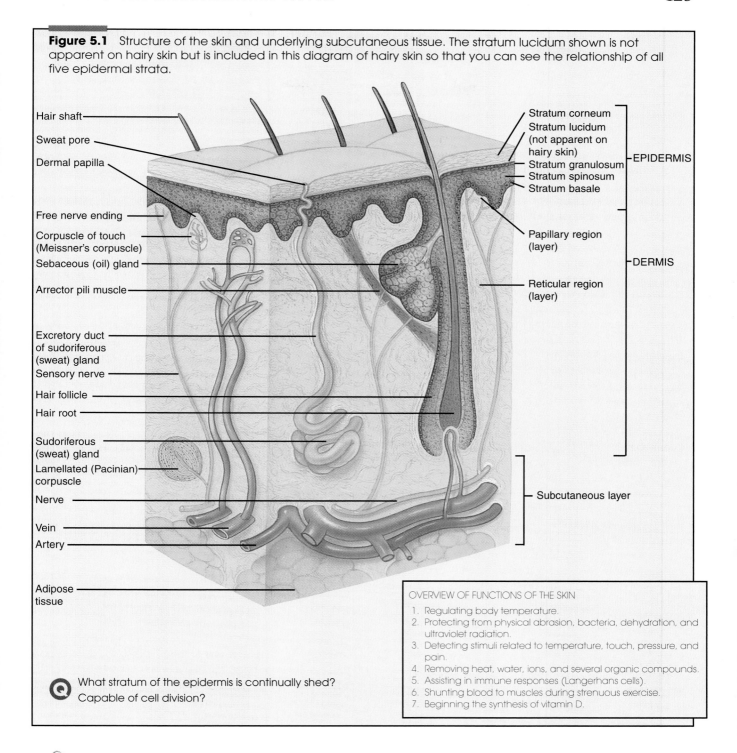

OVERVIEW OF FUNCTIONS OF THE SKIN
1. Regulating body temperature.
2. Protecting from physical abrasion, bacteria, dehydration, and ultraviolet radiation.
3. Detecting stimuli related to temperature, touch, pressure, and pain.
4. Removing heat, water, ions, and several organic compounds.
5. Assisting in immune responses (Langerhans cells).
6. Shunting blood to muscles during strenuous exercise.
7. Beginning the synthesis of vitamin D.

Q What stratum of the epidermis is continually shed? Capable of cell division?

VITAMIN D DEFICIENCY

During most of the year, an hour per week in the sunlight with the hands, arms, and face exposed meets the body's needs for activation of the vitamin D precursor. Additional sun exposure merely increases one's risk of skin cancer. Also, all milk sold in the United States is fortified with a D vitamin, so most people take in an adequate amount. Vitamin D deficiency can arise in those who completely cover their skin when outdoors (for example, Muslim women), in those who rarely venture outside, or in people who drink little or no milk. Elderly people living in nursing homes or confined to a hospital bed have double the risk for developing vitamin D deficiency. They may stay indoors for long periods, and they may drink little or no milk because they cannot digest it. The enzyme lactase, which is needed to split the milk sugar lactose, often declines as we age, producing milk intolerance. ■

Epidermis

The **epidermis** (*epi* = above) is composed of keratinized stratified squamous epithelium and contains four principal types of cells (Fig. 5.2). About 90% of the epidermal cells are **keratinocytes** (ker-a-TIN-ō-sits; *kerato* = horny), so named because they produce a protein called **keratin**. This substance helps waterproof and protect the skin and underlying tissues from light, heat, microbes, and many chemicals. Anchoring junctions called desmosomes (see Fig. 4.1) weld keratinocytes to one another.

Melanocytes (MEL-a-nō-sits; *melan* = black), which produce the pigment melanin, comprise about 8% of the epidermal cells. Their long, slender projections extend between and transfer granules of melanin to keratinocytes. **Melanin** is a brown-black pigment that contributes to skin color and absorbs UV light. Once inside keratinocytes, the melanin granules cluster to form a protective veil over the nucleus, on the side toward the skin surface. In this way they shield the genetic material from damaging UV light.

The third type of cell in the epidermis is known as a **Langerhans** (LANG-er-hans) **cell**. These cells arise from bone marrow and migrate to the epidermis. They interact with white blood cells called helper T cells in immune responses and are easily damaged by UV radiation.

A fourth type of cell found in the epidermis is called a **Merkel cell**. These cells are located in the deepest layer (stratum basale) of the epidermis of hairless skin, where they are attached to keratinocytes by desmosomes. Merkel cells make contact with the flattened portion of the ending of a sensory neuron (nerve cell), called a **tactile (Merkel) disc**, and are thought to function in the sensation of touch.

Four or five distinct layers of cells form the epidermis. In most regions of the body the epidermis is about 0.1 mm thick and has four recognizable layers. Where exposure to friction is greatest, such as in the palms and soles, the epidermis is thicker (1–2 mm) and has five layers (Fig. 5.2). Constant exposure of thin or thick skin to friction or pressure stimulates formation of a **callus**, an abnormal thickening of the epidermis.

The names of the five layers (strata) of the epidermis, from the deepest to the most superficial, are:

1. **Stratum basale** (*basale* = base). This single layer of cuboidal- to columnar-shaped cells contains stem cells, which are capable of continued cell division, and melanocytes. The stem cells multiply, producing keratinocytes, which push up toward the surface and become part of the more superficial layers. The nuclei of the keratinocytes degenerate, and the cells die. Eventually, the cell remnants are shed from the surface layer of the epidermis. During embryological development, other stem cells in the stratum basale migrated into the dermis and gave rise to sweat and oil glands and hair follicles. The stratum basale is sometimes referred to as the **stratum germinativum** (jer′-mi-na-TĒ-vum; *germ* = sprout) to indicate its role in germinating new cells. The stratum basale also contains tactile (Merkel) discs that are sensitive to touch.

2. **Stratum spinosum** (*spinosum* = thornlike or prickly). This stratum of the epidermis contains 8–10 layers of polyhedral (many-sided) cells that fit closely together. The cells here appear to be covered with prickly spines because the cells shrink apart when the tissue is prepared for microscopic examination. At each spinelike projection, filaments of the cytoskeleton insert into desmosomes, which tightly join the cells to one another. Long projections of the melanocytes extend among the keratinocytes, which take in melanin by phagocytosis of these melanocyte projections.

3. **Stratum granulosum** (*granulum* = little grain). This stratum of the epidermis consists of three to five layers of flattened cells that develop darkly staining granules of a substance called **keratohyalin** (ker′-a-tō-HĪ-a-lin). This compound is the precursor of **keratin**, a protein found in the superficial layer of the epidermis. Keratin molecules assemble into intermediate filaments that form a barrier to protect deeper layers from injury and microbial invasion and make the skin waterproof. The nuclei of the cells in the stratum granulosum are in various stages of degeneration. As their nuclei break down, the cells can no longer carry on vital metabolic reactions, and they die.

4. **Stratum lucidum** (*lucidus* = clear). Normally, this stratum is more apparent in the thick skin of the palms and soles. It consists of three to five layers of clear, flat, dead cells that contain droplets of an intermediate substance (eleidin) that is eventually transformed to keratin.

5. **Stratum corneum** (*corneum* = horny). This stratum consists of 25–30 layers of flat, dead cells completely filled with keratin. These cells are continuously shed and replaced by cells from deeper strata. The stratum corneum serves as an effective barrier against light, heat, bacteria, and many chemicals.

In the process of **keratinization**, cells newly formed in the basal layers undergo a developmental process as they are pushed to the surface. As the cells relocate, they accumulate keratin. At the same time the cytoplasm, nucleus, and other organelles disappear, and the cells die. Eventually, the keratinized cells slough off and are replaced by underlying cells that, in turn, become keratinized. The whole process by which a cell forms in the basal layer, rises to the surface, becomes keratinized, and sloughs off takes 2–4 weeks.

Epidermal growth factor (EGF) is a protein hormone that stimulates growth of epithelial and epidermal cells during tissue development, repair, and renewal. Certain oncogenes, genes that can turn normal cells into cancerous ones, cause tumors by permanently turning on EGF stimulation of cells, which then proliferate without control.

Dermis

The second principal part of the skin, the **dermis** (*derma* = skin), is composed of connective tissue containing collagen and elastic fibers (see Fig. 5.1). The few cells in the dermis in-

Figure 5.2 Cell types and layers in the epidermis of thick skin. Thin skin is similar to thick skin, but the stratum lucidum is often not noticed and the stratum corneum is thinner.

🗝 *The superficial layer of the skin—the epidermis—is keratinized stratified squamous epithelium.*

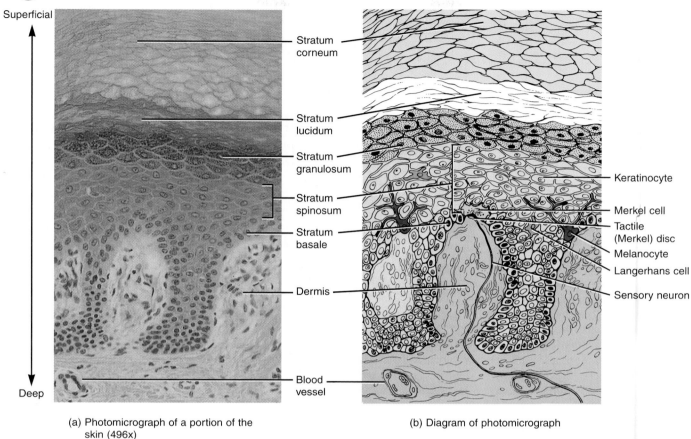

(a) Photomicrograph of a portion of the skin (496x)

(b) Diagram of photomicrograph

Ⓠ What hormone stimulates renewal of epidermal cells?

clude fibroblasts, macrophages, and adipocytes. The dermis is very thick in the palms and soles and very thin in areas such as the eyelids and scrotum. It also tends to be thicker on the posterior than the anterior aspect of the body and thicker on the lateral than the medial aspects of the limbs. Blood vessels, nerves, glands, and hair follicles are embedded in the dermis.

The superficial portion of the dermis, about one-fifth of the thickness of the total layer, is named the **papillary region** (**layer**). It consists of areolar connective tissue containing fine elastic fibers. Its surface area is greatly increased by small, fingerlike projections called **dermal papillae** (pa-PIL-ē; *papilla* = nipple). These nipple-shaped structures indent the epidermis, and many contain loops of capillaries. Some dermal papillae also contain tactile receptors called **corpuscles of touch** (**Meissner's corpuscles**), nerve endings that are sensitive to touch. Dermal papillae of thick skin cause ridges in the overlying epidermis. Secretions produced at the tops of these ridges leave fingerprints on objects that are handled.

The deeper portion of the dermis is called the **reticular** (*rete* = net) **region** (**layer**). It consists of dense, irregular

connective tissue containing interlacing bundles of collagen and some coarse elastic fibers. Spaces between the fibers are occupied by hair follicles, nerves, oil glands, the ducts of sweat glands, and a small quantity of adipose tissue. Varying thicknesses of the reticular region contribute to differences in the thickness of skin.

The combination of collagen and elastic fibers in the reticular region provides the skin with strength, extensibility, and elasticity. (**Extensibility** is the ability to stretch; **elasticity** is the ability to return to original shape after stretching.) The ability of the skin to stretch can readily be seen in pregnancy, obesity, and edema. Small tears that occur in the dermis during extreme stretching are initially red and remain visible afterward as silvery white streaks called **striae** (STRĪ-ē; *stria* = streak) or stretch marks.

The reticular region is attached to underlying organs, such as bone and muscle, by the subcutaneous layer, also called the **hypodermis** or **superficial fascia**. In addition to areolar connective tissue and adipose tissue, the subcutaneous layer also contains nerve endings called **lamellated** or **Pacinian** (pa-SIN-ē-an) **corpuscles** that are sensitive to pressure (see Fig. 15.1). Nerve endings sensitive to cold are found in and just deep to the dermis, while those sensitive to heat are located in the intermediate and superficial dermis.

LINES OF CLEAVAGE

In certain regions of the body, collagen fibers tend to orient more in one direction than another. **Lines of cleavage (tension lines)** in the skin indicate the predominant direction of underlying collagen fibers. The lines are especially evident on the palmar surfaces of the fingers, where they are arranged parallel to the long axis of the digits. Lines of cleavage are of particular interest to a surgeon because an incision running parallel to the collagen fibers will heal with only a fine scar. An incision made across the rows of fibers disrupts the collagen, and the wound tends to gape open and heal in a broad, thick scar. ■

Skin Color

Three pigments—melanin, carotene, and hemoglobin—give skin a wide variety of colors. Melanin is located mostly in the epidermis. The amount of **melanin** varies the skin color from pale yellow to black. Melanocytes are most plentiful in the mucous membranes, penis, nipples of the breasts and the area just around the nipples (areola), face, limbs, and choroid and iris of the eye. The *number* of melanocytes is about the same in all races. Differences in skin color are due mainly to the *amount of pigment* the melanocytes produce and disperse to keratinocytes. In some people, melanin tends to accumulate in patches called **freckles**. As one grows older, **liver (age) spots** may develop. These are flat skin patches that look like freckles and range in color from light brown to black. Like freckles, liver spots are accumulations of melanin, and neither one tends to become cancerous.

Melanocytes synthesize melanin from the amino acid *tyrosine* in the presence of an enzyme called *tyrosinase*. Synthesis occurs in an organelle called a **melanosome**. Exposure to UV radiation increases the enzymatic activity within melanosomes and leads to increased melanin production. Both the amount and the darkness of melanin increase, which tans the skin and further protects the body against UV radiation. Thus, within limits, melanin serves a protective function. However, as you will see later, excessive skin exposure to UV radiation can lead to skin cancer. A tan disappears when the excess melanin is lost in keratinocytes as they are shed.

An inherited inability of an individual of any race to produce melanin results in **albinism** (AL-bin-izm; *albus* = white). Most albinos (al-BĪ-nōs), individuals affected with albinism, have melanocytes but the cells are unable to synthesize tyrosinase. Melanin is thus absent in hair, eyes, and skin. In another condition, called **vitiligo** (vit-i-LĪ-gō), the partial or complete loss of melanocytes from patches of skin produces irregular white spots. It may be an autoimmune disease in which the body produces antibodies that destroy melanocytes.

Carotene (KAR-ō-tēn; *keraton* = carrot) is a yellowish orange pigment that gives carrots and egg yolks their color. It is the precursor of vitamin A, which is used to synthesize pigments needed for vision. Carotene is found in the stratum corneum and fatty areas of the dermis and subcutaneous layer.

When little melanin and carotene are present, the epidermis is translucent. The epidermis itself has no blood vessels, a characteristic of all epithelia. The skin of Caucasians appears pink to red, depending on the amount and quality of the blood moving through capillaries in the dermis. The red color is due to **hemoglobin**, the pigment that carries oxygen in red blood cells.

SKIN COLOR CLUES

The color of skin and mucous membranes can provide clues for diagnosing certain problems. When blood is not picking up an adequate amount of oxygen in the lungs, such as in a baby who has stopped breathing, mucous membranes, nail beds, and light-colored skin appear bluish or **cyanotic** (si-an-OT-ic; *kyanos* = blue). This is because hemoglobin that is depleted of oxygen looks deep, purplish blue. **Jaundice** (JON-dis; *jaune* = yellow), a yellowed appearance of the whites of the eyes and of light-colored skin, usually indicates liver disease. It is due to a buildup of bilirubin, a yellow pigment found in bile. In **Addison's disease** there is increased skin pigmentation due to excessive secretion of hormones produced by the anterior pituitary gland called melanocyte stimulating hormone (MSH) and adrenocorticotropic hormone (ACTH), which stimulate melanocytes. **Erythema** (er-e-THĒ-ma; *erythros* = red), redness of the skin, is caused by engorgement of capillaries in the dermis with blood. In people who have light-colored skin, exercise and embarrassment may cause noticeable erythema, especially of the face. Exercise and embarrassment produce the same, though often unobservable, reaction in dark-colored skin. Erythema also occurs with skin injury, infection, inflammation, or allergic reactions. ■

Epidermal Ridges

The superficial surface of the skin of the palms, fingers, soles, and toes is marked by a series of ridges. They appear either as straight lines or as a pattern of loops and whorls, as on the tips of the fingers and toes. **Epidermal ridges** develop during the third and fourth fetal months as the epidermis conforms to the contours of the underlying dermal papillae (see Fig. 5.1). The ridges increase the grip of the hand or foot by increasing friction and acting like tiny suction cups. Because the ducts of sweat glands open on the tops of the epidermal ridges as sweat pores, the sweat and ridges form fingerprints (or footprints) when a smooth object is touched. The ridge pattern, which is genetically determined, is unique for each individual. Normally, it does not change throughout life, except to enlarge, and thus can serve as the basis for identification through fingerprints or footprints.

SKIN GRAFTS

Sometimes, the germinal portion of epidermis is destroyed, and new skin cannot regenerate. Such wounds require **skin grafts**. The most successful type of skin graft involves the transplantation of a segment of skin from a donor site to a recipient site of the same individual (*autograft*) or an identical twin (*isograft*).

If skin loss is so extensive that conventional grafting is impossible, a self-donation procedure called **autologous** (aw-TOL-ō-gus) **skin transplantation** may be employed. In this procedure, used most often on severely burned patients, small amounts of an individual's epidermis are removed and grown in the laboratory to produce thin sheets of skin. Then, they are transplanted back to the patient where they adhere to burn wounds and generate a permanent skin to cover burned areas. The process normally takes 3–4 weeks. The obvious advantage to this procedure is that a lot of new epithelium can be grown from a very small skin sample. During the waiting period, wound dressing or skin from another person (*homograft* or *allograft*), such as a cadaver, or animal (*heterograft*) may be used to help protect the patient from potentially fatal fluid loss and infection. Although homografts and heterografts are temporary because they are ultimately rejected by the immune system, they prevent loss of fluid, electrolytes, and protein from burn sites while also decreasing pain and increasing the patient's mobility.

One type of synthetic skin that is used as part of autologous skin transplantation consists of a protective plastic layer that serves as the temporary epidermis and an underlying layer composed of collagen fibers and cartilage derived from animals that serves as the temporary dermis. After the synthetic skin is applied to the burned area, fibroblasts produce collagen fibers to replace those in the synthetic skin, which decompose. After a few months, when the reconstruction of the dermis is complete, sheets of the patient's newly grown epidermis are transplanted over the wound area. ■

EPIDERMAL DERIVATIVES

Organs that develop from the embryonic epidermis—hair, glands, nails—have a host of important functions. Hair and nails protect the body. The sweat glands help regulate body temperature. The enamel of teeth is also an epidermal derivative and is discussed on page 761.

Hair

Hairs, or **pili** (PI-lē), are growths of the epidermis variously distributed over the body. Their primary function is protection. Although the protection is limited, hair on the head guards the scalp from injury and the sun's rays. It also de-

creases heat loss. Eyebrows and eyelashes protect the eyes from foreign particles. Hair in the nostrils protects against inhaling insects and foreign particles. Hair serves a similar protective function in the external ear canal. Touch receptors associated with hair follicles (hair root plexuses) are activated whenever a hair is even slightly moved. Normal hair loss in an adult scalp is about 70–100 hairs per day. Both the rate of growth and the replacement cycle may be altered by illness, diet, high fever, surgery, blood loss, or severe emotional stress. Rapid weight-loss diets that severely restrict calories or protein increase hair loss. An increase in the rate of shedding can also occur for 3–4 months after childbirth and with certain drugs and radiation therapy for cancer.

Anatomy

A hair is composed of columns of dead, keratinized cells welded together. The **shaft** is the superficial portion of the hair, which projects from the surface of the skin (see Fig. 5.l). The shaft of straight hair is round in cross section, that of wavy hair is oval, and that of woolly hair is elliptical or kidney-shaped. The **root** is the portion of the hair deep to the surface that penetrates into the dermis and sometimes into the subcutaneous layer (see Fig. 5.1). The shaft and root both consist of three concentric layers (Fig. 5.3a). The inner **medulla** is composed of two or three rows of polyhedral cells containing pigment granules and air spaces. The middle **cortex** forms the major part of the shaft and consists of elongated cells that contain pigment granules in dark hair but mostly air in white hair. The **cuticle of the hair**, the outermost layer, consists of a single layer of thin, flat cells that are the most heavily keratinized. Cuticular cells are arranged like shingles on the side of a house, with their free edges pointing toward the end of the hair (Fig. 5.3b).

Surrounding the root of the hair is the **hair follicle**, which is made up of an external root sheath and an internal root sheath. The **external root sheath** is a downward continuation of the epidermis. Near the surface, it contains all the epidermal layers. At the base of the hair follicle, the external root sheath contains only the stratum basale. The **internal root sheath** forms a cellular tubular sheath between the external root sheath and the hair.

At the base of each hair follicle is an enlarged, layered structure, the **bulb**. This structure houses a nipple-shaped indentation, the **papilla of the hair**, which contains areolar connective tissue. The papilla of the hair contains many blood vessels and provides nourishment for the growing hair. The bulb also contains a ring of cells called the **matrix**, which is the germinal layer. The cells of the matrix derive from the stratum basale. They are responsible for the growth of existing hairs and produce new hairs by cell division when older hairs are shed. This replacement occurs within the same follicle. Matrix cells also give rise to the cells of the internal root sheath.

Sebaceous (oil) glands and a bundle of smooth muscle cells are also associated with hairs. Details of the sebaceous

Figure 5.3   Hair. The relationship of a hair to the epidermis, sebaceous (oil) glands, and arrector pili muscle can also be seen in Fig. 5.1.

Hairs are growths of epidermis composed of dead, keratinized cells.

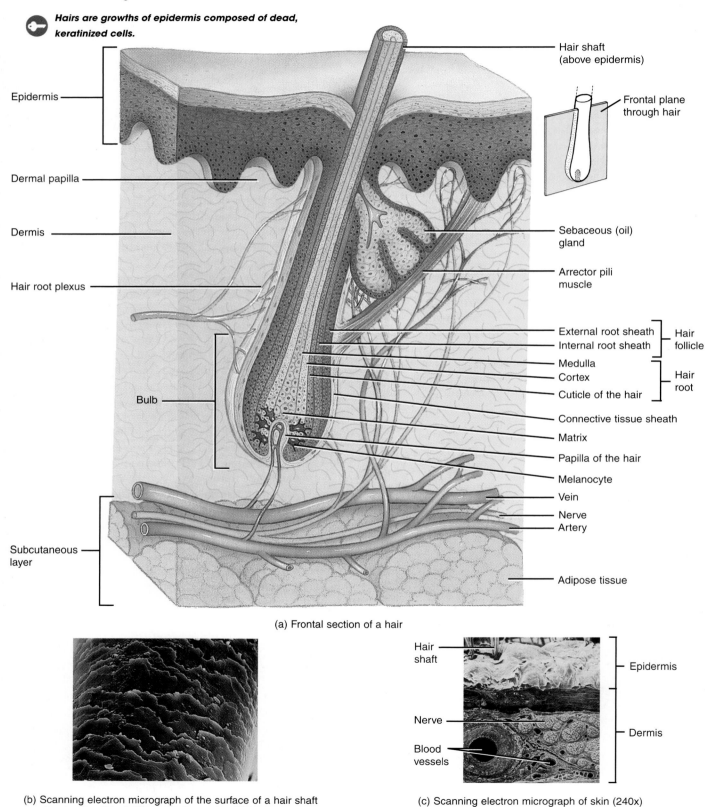

(a) Frontal section of a hair

(b) Scanning electron micrograph of the surface of a hair shaft showing the shingle-like cuticular scales (1720x)

(c) Scanning electron micrograph of skin (240x)

 Why does it hurt when you pluck a hair out but not when you have a haircut?

glands will be discussed shortly. The smooth muscle is called an **arrector** (*arrector* = to raise) **pili** (plural is **arrectores pilorum**); it extends from the superficial dermis of the skin to the side of the hair follicle. The arrector pili muscles contract under stresses of fright, cold, and emotions and pull the hairs into a vertical position. This response makes a furry animal look larger, and thus it may appear more threatening to an aggressor. In people, with little body hair, it serves no apparent purpose, merely causing "goose bumps" or "gooseflesh" because the skin around the shaft forms slight elevations.

Around each hair follicle are nerve endings, called **hair root plexuses**, that are sensitive to touch. They respond if a hair shaft is moved.

Growth

Each hair follicle goes through a **growth cycle**, which consists of a growth stage and a resting stage. During the *growth stage,* a hair is formed by cells of the matrix that differentiate, divide, become keratinized, and die. As new cells are added at the base of the hair root, the hair grows longer. In time, the growth of the hair stops and the resting stage begins. During the *resting stage,* the matrix is inactive and the hair follicle atrophies somewhat. After the resting stage, a new growth cycle begins in which a new hair replaces the old hair and the old hair is pushed out of the hair follicle. In general, scalp hair grows for about 3 years and rests for about 1–2 years.

Color

The color of hair is due primarily to melanin. It is synthesized by melanocytes scattered in the matrix of the bulb and passes into cells of the cortex and medulla (Fig. 5.3a). Dark-colored hair contains mostly true melanin. Blond and red hair contain variants of melanin in which there is iron and more sulfur. Gray hair occurs with a progressive decline in tyrosinase, an enzyme necessary for synthesis of melanin. White hair results from accumulation of air bubbles in the medullary shaft.

HAIR AND HORMONES

At puberty, when their testes begin secreting significant quantities of male sex hormones (androgens), males develop the typical male pattern of hair growth, including a beard and a hairy chest. In females, both the ovaries and the adrenal glands produce small quantities of androgens. Occasionally, a tumor of one of these glands oversecretes androgens and causes **hirsutism** (hur-SOO-tiz-um; *hirsutus* = shaggy). This is a condition of excessive hairiness of the upper lip, chin, chest, inner thighs, and abdomen in females or prepubertal males.

Surprisingly, androgens also must be present for the most common form of baldness, **male-pattern baldness**, to occur. In genetically predisposed males, androgens somehow inhibit hair growth. Minoxidil (Rogaine®) is a potent vasodilator, that is, a drug that widens blood vessels and increases circulation. When applied topically and used continuously, it does stimulate some hair regrowth in some persons with thinning hair due to male-pattern baldness. For many, however, the hair growth is meager, and it does not help people who already are bald. ■

Glands

Several kinds of glands are associated with the skin: sebaceous (oil) glands, sudoriferous (sweat) glands, ceruminous glands, and mammary glands. The anatomy and physiology of the mammary glands, which are modified sudoriferous glands, will be discussed on page 934 as components of the female reproductive system.

Sebaceous (Oil) Glands

Sebaceous (se-BĀ-shus; *sebaceus* = oily or fatty) **glands**, or **oil glands**, with few exceptions, are connected to hair follicles (see Figs. 5.1 and 5.3a). The secreting portions of the glands lie in the dermis and open into the necks of hair follicles or directly onto a skin surface (lips, glans penis, labia minora, and tarsal glands of the eyelids). Absent in the palms and soles, sebaceous glands vary in size and shape in other regions of the body. For example, they are small in most areas of the trunk and limbs, but large in the skin of the breasts, face, neck, and upper chest.

Sebaceous glands secrete an oily substance called **sebum** (SĒ-bum), which is a mixture of fats, cholesterol, proteins, and inorganic salts. Sebum helps keep hair from drying and becoming brittle, prevents excessive evaporation of water from the skin, keeps the skin soft and pliable, and inhibits the growth of certain bacteria. When sebaceous glands of the face become enlarged because of accumulated sebum, **blackheads** develop. The color of blackheads is due to melanin and oxidized oil, not dirt. Since sebum is nutritive to certain bacteria, **pimples** or **boils** often result.

Sudoriferous (Sweat) Glands

Three to four million **sudoriferous** (soo'-dor-IF-er-us; *sudor* = sweat; *ferre* = to bear) or **sweat glands** empty their secretions onto the skin surface (see Fig. 5.1). They are divided into two principal types, eccrine and apocrine, based on their structure, location, and type of secretion.

Eccrine (*ekkrinein* = to secrete) **sweat glands** are much more common than apocrine sweat glands. They are distributed throughout the skin except for the margins of the lips, nail beds of the fingers and toes, glans penis, glans clitoris, labia minora, and eardrums. Eccrine sweat glands are most numerous in the skin of the palms and the soles; their density can be as high as 450 per square centimeter (3000 per square inch) in the palms. The secretory portion of eccrine sweat glands is frequently located in the subcutaneous layer, and the excretory duct extends outward through the dermis and epidermis. It ends as a pore at the surface of the epidermis (see Fig. 5.1).

Apocrine (*apo* = from) **sweat glands** are found mainly in the skin of the axilla (armpit), pubic region, and areolae (pigmented areas around the nipples) of the breasts. The secretory portion of apocrine sweat glands is located in the dermis or subcutaneous layer, and the excretory duct opens into hair follicles. Apocrine sweat glands begin to function at puberty and produce a more viscous secretion than eccrine sweat glands. They are stimulated during emotional stresses and sexual excitement and the secretions are commonly known as a "cold sweat."

Sweat (**perspiration**) is the fluid produced by sweat glands. Most comes from eccrine sweat glands because they are so much more numerous. Sweat is a mixture of water, ions (mostly Na^+ and Cl^-), urea, uric acid, amino acids, ammonia, glucose, lactic acid, and ascorbic acid. Its principal function is to help regulate body temperature by providing a cooling mechanism (described shortly). Sweat also eliminates a small amount of wastes from the body.

Mammary glands, which are specialized sudoriferous glands, secrete milk. They will be discussed in Chapter 28 along with the female reproductive system.

Ceruminous Glands

In the ear, modified sweat glands called **ceruminous** (se-ROO-mi-nus; *cera* = wax) **glands** produce a waxy secretion. The secretory portions of the ceruminous glands lie in the subcutaneous layer, deep to sebaceous glands. Their excretory ducts open either directly onto the surface of the external auditory canal (ear canal) or into ducts of sebaceous glands. The combined secretion of the ceruminous and sebaceous glands is called **cerumen**. Cerumen, together with hairs in the external auditory canal, provides a sticky barrier that prevents the entrance of foreign bodies.

Nails

Nails are plates of tightly packed, hard, keratinized cells of the epidermis. The cells form a clear, solid covering over the dorsal surfaces of the terminal portions of the fingers and toes. Each nail (Fig. 5.4) consists of a nail body, a free edge, and a nail root. The **nail body** is the portion of the nail that is visible, the **free edge** is the part that may extend past the distal end of the digit, and the **nail root** is the portion that is buried in a fold of skin. Most of the nail body is pink because of blood flowing through underlying capillaries. The free edge appears white because there are no underlying capillaries. The whitish semilunar area of the proximal end of the body is called the **lunula** (LOO-nyoo-la; *lunula* = little moon). It appears whitish because the vascular tissue underneath does not show through owing to the thickened stratum basale in the area. The **eponychium** (ep′-ō-NIK-ē-um; *onyx* = nail) or **cuticle** is a narrow band of epidermis that extends from the margin of the nail wall (lateral border), adhering to it. It occupies the proximal border of the nail and consists of stratum corneum.

The epithelium deep to the nail root is known as the **nail matrix.** This is the region where nail growth occurs. Growth occurs by the transformation of superficial cells of the matrix into nail cells. In the process, the outer, harder layer is pushed forward over the stratum basale. The average growth in the length of fingernails is about 1 mm (0.04 in.) per week. The growth rate is somewhat slower in toenails. The longer the digit, the faster the nail grows. Nail growth is faster in the summer and on the most-used hand. Adding supplements such as gelatin to an otherwise healthy diet has no effect on making nails grow faster or stronger.

Functionally, nails help us grasp and manipulate small objects in various ways and provide protection against trauma to the ends of the digits, and allow us to scratch various parts of the body.

THE SKIN AND HOMEOSTASIS

The skin affords two excellent examples of how a part of the body contributes to homeostasis. One involves wound healing. When the skin is damaged by various stimuli (stresses), certain mechanisms go into operation to restore it to its normal or near-normal structure and functions. The other example is concerned with body temperature. The skin is one of the major organs that helps restore a normal body temperature should some stimulus raise or lower it.

Skin Wound Healing

Here we discuss the basic mechanisms by which skin wounds are repaired. First we consider wounds that affect primarily the epidermis. Then we describe wounds that extend into the dermis and subcutaneous layer.

Epidermal Wound Healing

The exposed location of the skin (and mucous membranes) makes it vulnerable to trauma as a result of physical and chemical stimuli (stresses). One common type of epidermal wound is an **abrasion** (a-BRĀ-shun; *abrasio* = scraped area), a portion of skin that has been scraped away, such as might be experienced in the form of a skinned knee or elbow. Another type is a first-degree or second-degree burn (see page 136). In an epidermal wound, the central portion of the wound may extend to the dermis, while the edges of the wound usually involve only slight damage to superficial epidermal cells.

In response to injury, basal epidermal cells in the area of the wound break their contacts with the basement membrane. These cells then enlarge and migrate across the wound (Fig. 5.5). The cells appear to migrate as a sheet until advancing cells from opposite sides of the wound meet. When epidermal cells encounter each other, their continued migration is stopped by **contact inhibition.** According to this phenomenon, when one epidermal cell encounters an-

Figure 5.4 Structure of nails. Shown Is a fingernail.

Nails cells arise by transformation of epithelial cells in the nail matrix.

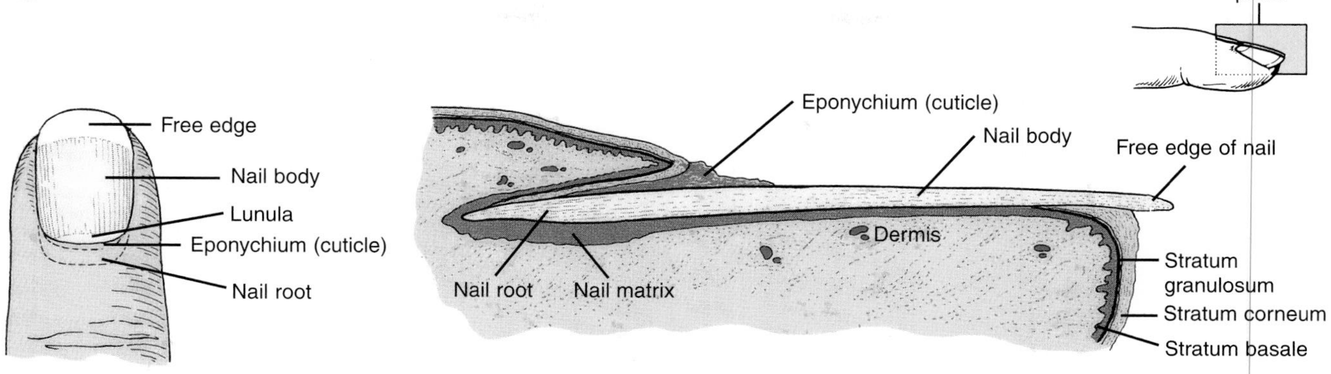

Sagittal plane

Free edge
Nail body
Lunula
Eponychium (cuticle)
Nail root

Eponychium (cuticle)
Nail body
Free edge of nail
Dermis
Nail root   Nail matrix
Stratum granulosum
Stratum corneum
Stratum basale

(a) Viewed from above

(b) Sagittal section

Q Why are nails so hard?

other, its direction of movement changes until it encounters another like cell, and so on. Continued migration of the epidermal cell stops when it is finally in contact on all sides with other epidermal cells. Contact inhibition appears to occur only among like cells; in other words, contact inhibition does not occur between epidermal cells and other types of cells. Malignant cells do not "obey the rules" of contact inhibition. Thus they have the ability to invade body tissues with few restrictions.

While some basal epidermal cells are migrating, epidermal growth factor is stimulating others to divide and replace the ones that have left. Migration continues until the wound is resurfaced. Following this, the migrated cells themselves divide to form new strata, thus thickening the new epidermis. The events involved in epidermal wound healing occur within 24–48 hours after wounding.

Deep Wound Healing

When an injury extends to tissues deep to the epidermis, the repair process is more complex than epidermal healing, and scar formation results.

The first step in deep wound healing involves inflammation, a vascular and cellular response that serves to dispose of microbes, foreign material, and dying tissue in preparation for repair. During the **inflammatory phase**, a blood clot forms in the wound and loosely unites the wound edges. Vasodilation and increased permeability of blood vessels enhance delivery of white blood cells called neutrophils and monocytes (macrophages) that phagocytize microbes, and mesenchymal cells, which develop into fibroblasts (Fig. 5.6a).

In the next phase, the **migratory phase**, the clot becomes a scab and epithelial cells migrate beneath the scab to bridge

Figure 5.5 Epidermal wound healing.

In an epidermal wound, the injury does not extend into the dermis.

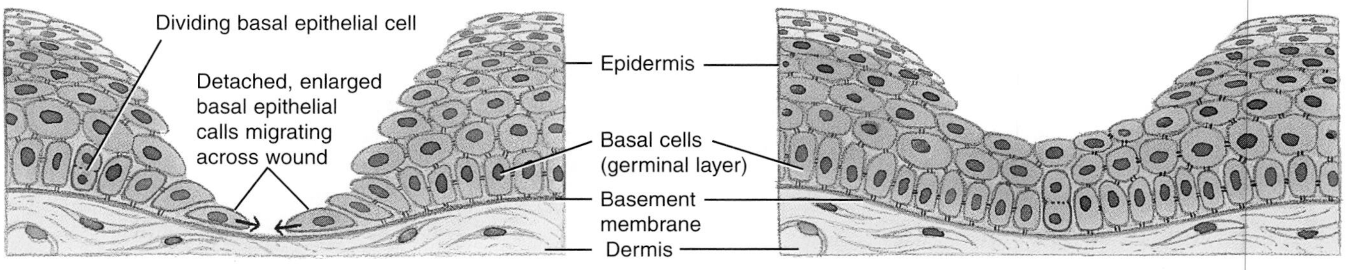

Dividing basal epithelial cell

Detached, enlarged basal epithelial calls migrating across wound

Epidermis

Basal cells (germinal layer)

Basement membrane

Dermis

(a) Division of basal cells and migration across wound

(b) Resurfacing of wound

Q Would you expect an epidermal wound to bleed? Why?

Figure 5.6 Deep wound healing.

 In a deep wound, the injury extends deep to the epidermis.

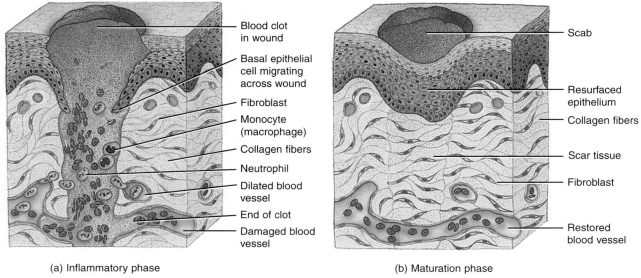

Blood clot in wound

Basal epithelial cell migrating across wound

Fibroblast

Monocyte (macrophage)

Collagen fibers

Neutrophil

Dilated blood vessel

End of clot

Damaged blood vessel

Scab

Resurfaced epithelium

Collagen fibers

Scar tissue

Fibroblast

Restored blood vessel

(a) Inflammatory phase

(b) Maturation phase

Why is the arrival of phagocytic white blood cells (neutrophils and monocytes) at the scene of a wound helpful?

the wound. Fibroblasts migrate along fibrin threads and begin synthesizing scar tissue (collagen fibers and glycoproteins), and damaged blood vessels begin to regrow. During this phase, tissue filling the wound is called **granulation tissue**.

The **proliferative phase** is characterized by extensive growth of epithelial cells beneath the scab, deposition by fibroblasts of collagen fibers in random patterns, and continued growth of blood vessels.

In the final phase, the **maturation phase**, the scab sloughs off once the epidermis is restored to normal thickness. Collagen fibers become more organized, fibroblasts decrease in number, and blood vessels are restored to normal (Fig. 5.6b).

As noted in Chapter 4, the process of scar tissue formation is called **fibrosis**. Sometimes, so much scar tissue is formed that a raised scar results, that is, one that is elevated above the normal epidermal surface. Such a scar may remain within the boundaries of the original wound (**hypertrophic scar**), or it may extend beyond the boundaries of the original wound into normal surrounding tissues (**keloid scar**). Scar tissue differs from normal skin in that its collagen fibers are more densely arranged. Also, it has fewer blood vessels and may not contain hair, skin glands, or sensory neurons.

Thermoregulation: Homeostasis of Body Temperature

The skin plays a major role in thermoregulation, that is, the homeostasis of body temperature. As warm-blooded animals, we are able to maintain our body temperature at a re-

markably constant 37°C (98.6°F) even though the environmental temperature varies greatly. Negative feedback systems (see Fig. 1.3) ensure that body temperature (a controlled condition) fluctuates very little.

Suppose you are in a place where the temperature is 38°C (101°F). Heat (the stimulus) continually flows from the environment to your body, raising body temperature. To counteract this change in a controlled condition, a sequence of events is set into operation (Fig. 5.7). In the skin and brain are temperature-sensitive receptors (nerve endings) called thermoreceptors that detect the stimulus and send nerve impulses (input) to your brain. A temperature control region in the brain (control center) then sends nerve impulses (output) to the sweat glands (effectors), which produce perspiration more rapidly. As the sweat evaporates from the surface of your skin, heat is lost, and your body temperature drops to normal (return to homeostasis). At this point, sweat glands are no longer stimulated. Also, when environmental temperature is low, sweat glands produce less perspiration.

Note that temperature regulation by the skin involves a *negative* feedback system because the response (cooling) is opposite to the stimulus (heating) that started the cycle. Also, the thermoreceptors continually monitor body temperature and feed back information to keep the brain informed. The brain, in turn, continues to send impulses to the sweat glands and blood vessels until the temperature returns to 37°C (98.6°F).

Regulating the rate of sweating is just one mechanism for adjusting body temperature. This and other mechanisms for thermoregulation are discussed in more detail in Chapter 25 on page 809.

Figure 5.7 Thermoregulation: homeostasis of body temperature by the skin.

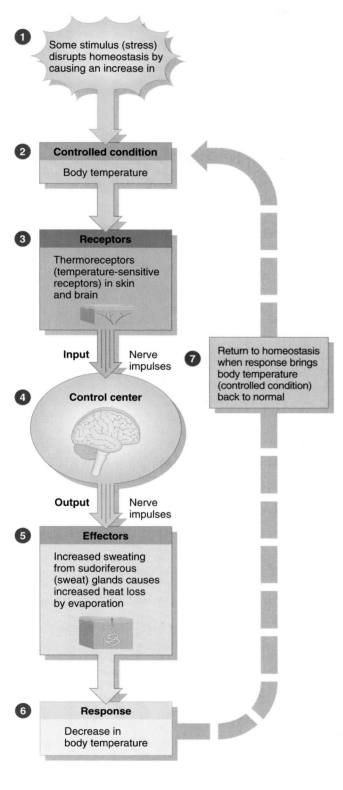

1 Some stimulus (stress) disrupts homeostasis by causing an increase in

2 **Controlled condition**

Body temperature

3 **Receptors**

Thermoreceptors (temperature-sensitive receptors) in skin and brain

Input   Nerve impulses

7 Return to homeostasis when response brings body temperature (controlled condition) back to normal

4 **Control center**

Output   Nerve impulses

5 **Effectors**

Increased sweating from sudoriferous (sweat) glands causes increased heat loss by evaporation

6 **Response**

Decrease in body temperature

Q Why is this a negative (rather than a positive) feedback cycle?

AGING AND THE INTEGUMENTARY SYSTEM

Although skin is constantly aging, pronounced effects do not become noticeable until a person reaches the late forties. Around that time, collagen fibers begin to decrease in number, stiffen, break apart, and form into a shapeless, matted tangle. Elastic fibers lose some of their elasticity, thicken into clumps, and fray. As a result, the skin forms crevices and furrows known as wrinkles. Fibroblasts, which produce both collagen and elastic fibers, decrease in number, and macrophages become less efficient phagocytes. Smoking greatly accelerates these changes so that the skin of a 40-year-old smoker may resemble that of a much older person.

With increased age, the hair and nails grow more slowly. Langerhans cells dwindle in number, thus decreasing the immune responsiveness of older skin. Decreased size of sebaceous (oil) glands leads to dry and broken skin that is more susceptible to infection. Production of sweat diminishes, which probably contributes to the increased incidence of heat stroke in the elderly. There is a decrease in the number of functioning melanocytes, resulting in gray hair and atypical skin pigmentation. An increase in the size of some melanocytes produces pigmented blotching (liver spots). Blood vessels in the dermis become thicker walled and less permeable, and subcutaneous fat is lost. In general, aged skin is thinner than young skin, especially the dermis, and migration of cells from the basal layer to the epidermal surface slows considerably. Aged skin also heals poorly and becomes more susceptible to pathological conditions such as skin cancer, itching, pressure sores, and shingles.

PHOTODAMAGE

Nearly everyone has experienced the effect of acute overexposure to sunlight—a sunburn. Acute overexposure to the ultraviolet (UV) rays in sunlight causes sunburn. Even if sunburn does not occur, the UV rays cause photodamage of the DNA in epidermal cells and the extracellular matrix materials, such as collagen and elastic fibers, in the dermis. Over the long term, chronic UV exposure accelerates the aging of skin (photoaging) and is an important factor in development of nearly all skin cancers (see page 138). Protect your skin from UV rays by wearing a hat that shades the face and clothing that covers the body. Use a sunblock with a SPF (sun protection factor) greater than 15 on skin areas that are exposed to sunlight.

In recent years, **tanning salons** have become very popular. Many salons claim to use longer, "safe" wavelengths of UV light (UVA). According to medical authorities, however, UVA is just as harmful as shorter UV wavelengths (UVB). Among the harmful effects of UV exposure are increased risk for skin cancer, premature aging and wrinkling of the skin, suppression of the immune system, and damage to the lens or retina of the eye. Another problem that can develop is chemical photosensitivity. In this condition, adverse skin reactions to certain chemicals, including some drugs, that are applied topically or taken internally occur at the same time a person is exposed to UV radiation.

One treatment for wrinkles and liver spots in photodamaged skin is the drug **tretinoin** (**Retin-A**). It is a derivative of vitamin A that has been used to treat acne since the 1960s. Among the reported benefits of using tretinoin are increased thickness of the epidermis, smoother stratum corneum, diminished number and size of melanocytes, increased production of collagen and elastin, dilation of blood vessels in the dermis, and regression of precancerous lesions. Dermatologists caution that tretinoin does not improve advanced changes associated with aging, and it may increase the risk of cancers induced by UV light exposure. ■

DEVELOPMENTAL ANATOMY OF THE INTEGUMENTARY SYSTEM

Throughout this book, the developmental anatomy of the body systems will be discussed, usually at the end of each appropriate chapter. The principal features of embryonic development are not treated in detail until Chapter 29. To help you understand the development of organ systems, we will explain and review a few terms at this point.

As part of the early development of a fertilized egg, a portion of the developing embryo differentiates into three layers of tissue called **primary germ layers**. On the basis of position, they are referred to as **ectoderm** (*ecto* = outside), **mesoderm** (*meso* = middle), and **endoderm** (*endo* = inside). They are the embryonic tissues from which all tissues and organs of the body will eventually develop (see Exhibit 29.1 on page 965).

The *epidermis* is derived from the **ectoderm**. At the beginning of the second month, the ectoderm consists of sim-

ple cuboidal epithelium. These cells become flattened and are known as the **periderm**. By the fourth month, all layers of the epidermis are formed and each layer assumes its characteristic structure.

Nails develop during the third month. Initially, they consist of a thick layer of epithelium called the **primary nail field**. The nail itself is keratinized epithelium and grows distally from its base. It is not until the ninth month that the nails reach the distal tips of the digits.

Hair follicles develop between the third and fourth months as ingrowths of the stratum basale of the epidermis into the deeper dermis. The ingrowths soon differentiate into the bulb, papilla of the hair, beginnings of the epithelial portions of sebaceous (oil) glands, and other structures associated with hair follicles. By the fifth or sixth month, the follicles produce delicate fetal hair called **lanugo** (la-NOO-gō), first on the head and then on other parts of the body. The lanugo is usually shed before birth.

The epithelial (secretory) portions of *sebaceous glands* develop from the sides of the hair follicles and remain connected to the follicles.

The epithelial portions of *sudoriferous (sweat) glands* are also derived from ingrowths of the stratum basale of the epidermis into the dermis. They appear during the fourth month on the palms and soles and a little later in other regions. The connective tissue and blood vessels associated with the glands develop from **mesoderm**.

The *dermis* is derived from **mesodermal cells**. The mesodermal cells in a zone deep to the ectoderm undergo changes into the connective tissues that form the dermis.

DISORDERS: HOMEOSTATIC IMBALANCES

BURNS

Tissue damage from excessive heat, electricity, radioactivity, or corrosive chemicals that destroy (denature) proteins in the exposed cells is called a **burn**. Burns disrupt homeostasis because they destroy the protection afforded by the skin. They permit microbial invasion and infection, loss of fluid, and loss of thermoregulation. The injury to tissues directly or indirectly in contact with the damaging agent, such as the skin or the linings of the respiratory and gastrointestinal tracts, is the local effect of a burn. Generally, however, the systemic effects of a burn are a greater threat to life. They may include (1) a large loss of water, plasma, and plasma proteins, which causes shock; (2) bacterial infection; (3) reduced circulation of blood; (4) decreased production of urine; and (5) diminished immune responses.

A **first-degree burn** involves only the surface epidermis. It is characterized by mild pain and erythema (redness) but no blisters. Skin functions remain intact. The pain and damage caused by a first-degree burn may be lessened by immediately flushing it with cold water. Generally, a first-degree burn will heal in about two to three days and may be accompanied by flaking or peeling. A typical sunburn is an example of a first-degree burn.

A **second-degree burn** destroys the entire epidermis and possibly parts of the dermis. Some skin functions are lost. In a second-degree burn, there is redness, blister formation, edema, and pain. (Blister formation is due to separation of the epidermis from the dermis with tissue fluid accumulation between.) Epidermal derivatives, such as hair follicles, sebaceous glands, and sweat glands, usually are not injured. If there is no infection, deep second-degree burns heal without grafting in about three to four weeks. Scarring may result. First- and second-degree burns are collectively referred to as **partial-thickness burns**.

A **third-degree burn** or **full-thickness burn** destroys the epidermis, dermis, and the epidermal derivatives. Skin functions are lost. Such burns vary in appearance from marble-white to mahogany colored to charred, dry wounds. There is marked edema, and the burned region is numb because sensory nerve endings have been destroyed. Regeneration is slow, and much granulation tissue forms before being covered by epithelium.

The seriousness of a burn is determined by its depth, extent, and area involved, as well as the person's age and general health. When the burn area exceeds 70%, more than half the victims die. A quick means for estimating the surface area affected by a burn in an adult is the **rule of nines** (Fig. 5.8a).

1. Count 9% if the anterior and posterior surfaces of the head and neck are affected.
2. Count 9% for the anterior and 9% for the posterior surfaces of each upper limb.
3. Count four times nine or 36% for the anterior and posterior surfaces of the trunk, including the buttocks.
4. Count 9% for the anterior and 9% for the posterior surfaces of each lower limb as far up as the buttocks (total of 36% for both lower limbs).

5. Count 1% for the perineum (per-i-NE-um), which includes the anal and urogenital regions.

A more accurate way to estimate the amount of surface area affected by a burn is the **Lund–Browder method**. This method estimates the extent by comparing the areas affected to the percentage of total surface area for body parts shown in Fig. 5.8b. For example, if the anterior of the head and neck and the whole right hand of an adult are affected, the burn covers 7% (3.5% for anterior head + 1% for anterior neck + 2.5% for anterior and posterior aspects of hand) of the body surface. Because the proportions of the body change with growth, the percentages

Figure 5.8 Methods for determining the extent of a burn. In (a) the numbers give approximate proportions whereas in (b) the numbers give the actual proportions of various body regions.

 Whereas the rule of nines (a) is a quicker way to determine the extent of a burn, the Lund–Browder method (b) is more accurate.

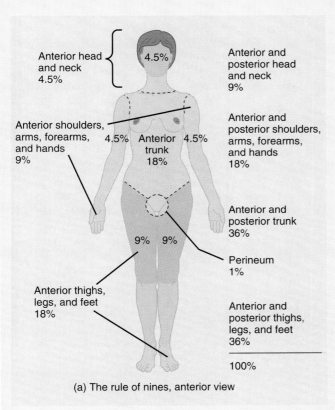

(a) The rule of nines, anterior view

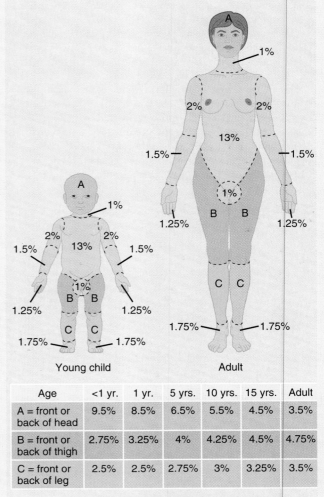

Young child   Adult

Age	<1 yr.	1 yr.	5 yrs.	10 yrs.	15 yrs.	Adult
A = front or back of head	9.5%	8.5%	6.5%	5.5%	4.5%	3.5%
B = front or back of thigh	2.75%	3.25%	4%	4.25%	4.5%	4.75%
C = front or back of leg	2.5%	2.5%	2.75%	3%	3.25%	3.5%

(b) The Lund–Browder method, anterior view

 Why would burns to the head of a child be more life threatening than similar burns in an adult?

DISORDERS: HOMEOSTATIC IMBALANCES CONTINUED

vary for different ages. Thus the extent of burn damage can be estimated fairly accurately for any age group.

SKIN CANCER

Excessive sun exposure can result in **skin cancer**, and everyone, regardless of skin pigmentation, is a potential victim of skin cancer if exposure to sunlight is sufficiently intense and prolonged. Natural skin pigment can never give complete protection. A **solar keratosis** (ker′-a-TŌ-sis; *kera* = horn) is a precancerous skin lesion induced by sunlight. These lesions commonly form on skin areas exposed to the sun and may be round or irregularly shaped, with a rough, scaly surface.

The three most common forms of skin cancer all are caused in part by chronic sun exposure. (1) **Basal cell carcinomas** account for over 75% of all skin cancers. The tumors arise from the epidermis and rarely metastasize (spread to other tissues). (2) **Squamous cell carcinomas** also arise from the epidermis, and they have a variable tendency to metastasize. Most arise from pre-existing lesions on sun-exposed skin. (3) **Malignant melanomas** (*melan* = black; *oma* = tumor) arise from melanocytes and are the most prevalent life-threatening cancer in young women. Malignant melanomas metastasize rapidly and can kill a person within months of diagnosis. Often, but not always, there is a 10- to 20-year interval between photodamage to the skin and diagnosis of skin cancer.

It is a good practice to examine skin moles periodically. Take note of those that enlarge, develop highly irregular borders, have uneven surfaces, or display a mixture of colors. These signs of changing appearance or bleeding may indicate a developing cancer. Fortunately, most skin cancers involve basal and squamous cells and can be treated by surgical excision.

Among the risk factors for skin cancer are:

1. **Skin type.** Persons with light-colored skin and red or blond hair who never tan but always burn are at higher risk.

2. **Sun exposure.** People who live in areas with many days of sunlight per year (near the equator) and at high altitudes have a higher incidence of skin cancer. Likewise, people engaged in outdoor occupations and those who have suffered three or more severe sunburns have a higher risk.

3. **Family history.** Skin cancer rates are higher in some families than in others.

4. **Age.** Older people are more prone to skin cancer owing to longer total exposure to sunlight.

5. **Immune system function.** Persons whose immune system is suppressed, either by disease or from drugs, have a higher incidence of skin cancer.

ACNE

Acne is an inflammation of sebaceous (oil) glands that usually begins at puberty when the sebaceous glands grow in size and increase production of sebum. Although testosterone, a male sex hormone, appears to be the most potent circulating hormone for sebaceous gland stimulation, adrenal and ovarian hormones stimulate sebaceous secretions in females. Acne occurs predominantly in sebaceous follicles that have been colonized by bacteria, which may thrive in the lipid-rich sebum. When this occurs, a cyst or sac of connective tissue cells can destroy and displace epidermal cells, resulting in permanent scarring, a condition called **cystic acne**.

PRESSURE SORES

Pressure sores, also known as **decubitus** (dē-KYOO-bi-tus) **ulcers** or **bedsores**, are caused by a constant deficiency of blood to tissues overlying a bony projection that has been subjected to prolonged pressure against an object such as a bed, cast, or splint. The deficiency results in tissue ulceration. Small breaks in the epidermis become infected, and the sensitive subcutaneous and deeper tissues are damaged. Eventually, the tissue is destroyed. Pressure sores are seen most often in patients who are bedridden for long periods of time.

MEDICAL TERMINOLOGY

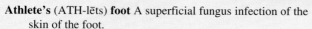

Athlete's (ATH-lēts) **foot** A superficial fungus infection of the skin of the foot.

Chickenpox Highly contagious disease that begins in the respiratory system and is caused by the varicella-zoster virus and characterized by vesicular eruptions on the skin that fill with pus, rupture, and form a scab before healing. Also called **varicella** (var′-i-SEL-a). Shingles is caused by reactivation of latent chickenpox viruses.

Cold sore (KŌLD sor) A lesion, usually in the oral mucous membrane, caused by herpes simplex virus type 1 (HSV-1), transmitted by oral or respiratory routes. Triggering factors include UV radiation, hormonal changes, and emotional stress. Also called a **fever blister**.

Contusion (kon-TOO-shun; *contundere* = to bruise) Condition in which tissue deep to the skin is damaged, but the epidermis is not broken.

Corn (KORN) A painful conical thickening of the stratum corneum of the epidermis found principally over toe joints and between the toes and often caused by pressure. It may be hard or soft, depending on the location. Hard corns are usually found over toe joints, and soft corns are usually found between the fourth and fifth toes.

Cyst (SIST; *cyst* = sac containing fluid) A sac with a distinct connective tissue wall, containing a fluid or other material.

German measles Highly contagious disease that begins in the respiratory system and is caused by the rubella virus and characterized by a rash of small red spots on the skin. Also called **rubella** (roo-BEL-a).

Hemangioma (hē-man′-jē-Ō-ma; *hemo* = blood; *angio* = blood vessel; *oma* = tumor) Localized tumor of the skin and subcutaneous layer that results from an abnormal increase in blood vessels; one type is a **portwine stain**, a flat, pink, red, or purple lesion present at birth, usually at the nape of the neck.

Hives (HĪVZ) Condition of the skin marked by reddened elevated patches that are often itchy. Most commonly caused by infections, physical trauma, medications, emotional stress, food additives, and certain foods. Also called **urticaria** (yoor-ti-KAR-ē-a).

Impetigo (im′-pe-TĪ-go) Superficial skin infection caused by staphylococci or streptococci; most common in children.

Intradermal (in′-tra-DER-mal; *intra* = within) Within the skin. Also called **intracutaneous.**

Laceration (las′-er-Ā-shun; *lacerare* = to tear) Wound or irregular tear of the skin.

Measles Highly contagious disease caused by the measles virus that begins in the respiratory system and is characterized by a rash on the skin. Also called **rubeola** (roo-bē-Ō-la).

Nevus (NE-vus) A round, pigmented, flat or raised skin area that may be present at birth or develop later. Varying in color from yellow-brown to black. Also called a **mole** or **birthmark.**

Pruritus (proo-RĪ-tus; *pruire* = to itch) Itching, one of the most common dermatological disorders. It may be caused by skin disorders (infections), systemic disorders (cancer, kidney failure), or psychogenic factors (emotional stress).

Topical (TOP-i-kal) Pertaining to a definite area; local. Also in reference to a medication, applied to the surface rather than ingested or injected.

Wart (WORT) Mass produced by uncontrolled growth of epithelial skin cells; caused by a virus (papillomavirus). Most warts are noncancerous.

STUDY OUTLINE

SKIN (p. 124)

1. The skin and the organs derived from it (hair, glands, and nails) constitute the integumentary system.
2. The skin is one of the larger organs of the body. The principal parts of the skin are the superficial epidermis and deep dermis. The dermis overlies the subcutaneous layer.
3. Following are the functions of the skin: regulating body temperature, protection, sensation, excretion, immunity, blood reservoir, and synthesis of vitamin D.
4. The epidermis consists of keratinocytes, melanocytes, Langerhans cells, and Merkel cells.
5. The epidermal layers, from deepest to most superficial, are the strata basale, spinosum, granulosum, lucidum, and corneum. The stratum basale undergoes continuous cell division and produces all other layers.
6. The dermis consists of a papillary region and a reticular region. The papillary region is composed of areolar connective tissue containing fine elastic fibers, dermal papillae, and corpuscles of touch (Meissner's corpuscles). The reticular region is composed of dense irregular connective tissue containing collagen fibers and some elastic fibers, adipose tissue, hair follicles, nerves, sebaceous (oil) glands, and ducts of sudoriferous (sweat) glands.
7. Lines of cleavage indicate the direction of collagen fiber bundles in the dermis and are considered during surgery.
8. The color of skin is due to melanin, carotene, and hemoglobin.
9. Epidermal ridges increase friction for better grasping ability and provide the basis for fingerprints and footprints.

EPIDERMAL DERIVATIVES (p. 129)

1. Epidermal derivatives are structures developed from the embryonic epidermis.
2. Among the epidermal derivatives are hair, skin glands (sebaceous, sudoriferous, and ceruminous), and nails.

Hair (p. 129)

1. Hairs are epidermal growths that function in protection.
2. A hair consists of a shaft superficial to the surface, a root that penetrates the dermis and subcutaneous layer, and a hair follicle.

3. The growth cycle of a hair follicle includes a growth stage and a resting stage.
4. Associated with each hair is a sebaceous (oil) gland, arrector pili muscle, and a hair root plexus.
5. New hairs develop from cell division of the matrix in the bulb; hair replacement and growth occur in a cyclical pattern. Male-pattern baldness is caused by androgens and heredity.

Glands (p. 131)

1. Sebaceous (oil) glands are usually connected to hair follicles; they are absent in the palms and soles. Sebaceous glands produce sebum, which moistens hairs and waterproofs the skin. Enlarged sebaceous glands may produce blackheads, pimples, and boils.
2. Sudoriferous (sweat) glands are divided into apocrine and eccrine. Apocrine sweat glands are limited in distribution to the skin of the axilla, pubis, and areolae; their ducts open into hair follicles. Eccrine sweat glands have an extensive distribution; their ducts terminate at pores at the surface of the epidermis. Sudoriferous glands produce perspiration (sweat), which carries small amounts of wastes to the surface and assists in maintaining body temperature.
3. Ceruminous glands are modified sudoriferous glands that secrete cerumen. They are found in the external auditory canal (ear canal).

Nails (p. 132)

1. Nails are hard, keratinized epidermal cells over the dorsal surfaces of the terminal portions of the fingers and toes.
2. The principal parts of a nail are the body, free edge, root, lunula, eponychium, and matrix. Cell division of the matrix cells produces new nails.

THE SKIN AND HOMEOSTASIS (p. 132)

Skin Wound Healing (p. 132)

1. In an epidermal wound, usually the central portion of the wound extends deep down to the dermis, whereas the wound edges involve only superficial damage to the epidermal cells.
2. Epidermal wounds are repaired by enlargement and migration of basal cells, contact inhibition, and division of migrating and stationary basal cells.

3. During the inflammatory phase, a blood clot unites the wound edges, epithelial cells migrate across the wound, vasodilation and increased permeability of blood vessels deliver phagocytes, and fibroblasts form.

4. During the migratory phase, epithelial cells beneath the scab bridge the wound, fibroblasts begin to synthesize scar tissue, and damaged blood vessels begin to regrow.

5. During the proliferative phase, the events of the migratory phase intensify, and the open wound tissue is called granulation tissue.

6. During the maturation phase, the scab sloughs off, the epidermis is restored to normal thickness, collagen fibers become more organized, fibroblasts begin to disappear, and blood vessels are restored to normal.

Thermoregulation: Homeostasis of Body Temperature (p. 134)

1. One of the functions of the skin is to help maintain a normal body temperature of 37°C (98.6°F).

2. If environmental temperature is high, skin and brain thermoreceptors sense the stimulus (heat) and generate impulses (input) that are transmitted to the brain (control center). The brain then sends impulses (output) to sweat glands (effectors) to produce perspiration. As the perspiration evaporates, the skin is cooled and body temperature returns to normal.

3. The skin-cooling response is a negative feedback mechanism.

AGING AND THE INTEGUMENTARY SYSTEM (p. 135)

1. Most effects of aging begin to occur when an individual reaches the late forties.

2. Among the effects of aging are wrinkling, loss of subcutaneous fat, atrophy of sebaceous glands, and decrease in the number of melanocytes and Langerhans cells.

DEVELOPMENTAL ANATOMY OF THE INTEGUMENTARY SYSTEM (p. 136)

1. The epidermis, hair, nails, and skin glands are epidermal derivatives.

2. The dermis is derived from wandering mesodermal cells.

REVIEW QUESTIONS

1. What structures comprise the integumentary system? (p. 124)
2. List the seven principal functions of the skin. (p. 124)
3. Compare the structure of epidermis and dermis. What is the subcutaneous layer? (p. 126)
4. Describe the various cells that comprise the epidermis. List and describe the epidermal layers from the deepest outward. What is the importance of each layer? (p. 126)
5. Compare the structure of the papillary and reticular regions of the dermis. (p. 127)
6. Explain the factors that produce skin color. What is an albino? (p. 128)
7. How are epidermal ridges formed? Why are they important? (p. 128)
8. List the receptors in the epidermis, dermis, and subcutaneous layer and indicate the location and role of each. (p. 127)
9. Describe the structure of a hair. How are hairs moistened? What produces "goose bumps" or "gooseflesh"? (p. 129)
10. Contrast the locations and functions of sebaceous (oil) glands, sudoriferous (sweat) glands, and ceruminous glands. (p. 131)
11. Compare the locations and secretions of eccrine and apocrine sweat glands. (p. 131)
12. Describe the principal parts of a nail. (p. 132)
13. Outline the steps involved in epidermal wound healing and deep wound healing. (p. 132)
14. Explain with a labeled diagram how the skin helps maintain normal body temperature. (p. 135)
15. Describe the effects of aging on the integumentary system. (p. 135)
16. Describe the origin of the epidermis, its derivatives, and the dermis. (p. 136)
17. Define the following: lines of cleavage (p. 128), freckle (p. 128), liver (age) spot (p. 128), cyanosis, jaundice, erythema (p. 128), skin graft (p. 129), male-pattern baldness (p. 131), and photodamage (p. 135).

ANSWERS TO QUESTIONS WITH FIGURES

5.1 Stratum corneum; stratum basale.

5.2 Epidermal growth factor (EGF).

5.3 Plucking a hair damages hair root plexuses in the dermis, which causes the sensation of pain. Because the cells of a hair shaft are already dead and there are no nerves in the hair shaft, cutting hair is not painful.

5.4 They are composed of tightly packed, hard, keratinized epidermal cells.

5.5 Epidermal wounds do not bleed because there are no blood vessels in the epidermis.

5.6 In addition to phagocytizing microbes, they can help clean up cellular debris that results from the wound.

5.7 The result of the effectors (lowering body temperature) is opposite to the initial stimulus (rising body temperature).

5.8 The head represents a larger percentage of the total body surface in children.

UNIT TWO

PRINCIPLES OF SUPPORT AND MOVEMENT

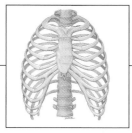

THIS UNIT CONSIDERS TWO PRIMARY THEMES: support and movement. You will study the various ways in which the body is supported and the different movements it can perform. Both support and movement are made possible by the cooperative effort of bones, joints, and muscles.

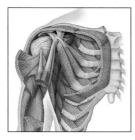

CHAPTER 6

BONE TISSUE

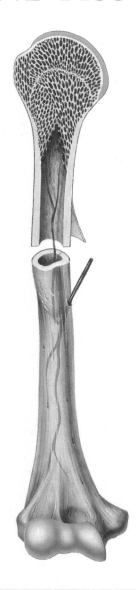

STUDENT OBJECTIVES

1. Discuss the functions of bone. (p. 143)

2. Identify the parts of a long bone. (p. 143)

3. Describe the histological features of compact and spongy bone tissue. (p. 145)

4. Contrast the steps involved in intramembranous and endochondral ossification. (p. 148)

5. Describe the processes involved in bone remodeling. (p. 152)

6. Define a fracture, describe several common kinds of fractures, and describe the sequence of events involved in fracture repair. (p. 153)

7. Describe the role of bone in calcium homeostasis. (p. 155)

8. Explain the effects of exercise and aging on the skeletal system. (p. 156)

9. Describe the development of the skeletal system. (p. 157)

10. Contrast the causes and clinical symptoms associated with osteoporosis and Paget's disease. (p. 159)

11. Define medical terminology associated with bone tissue. (p. 159)

CHAPTER CONTENTS AT A GLANCE

Bone tissue forms most of the skeleton, the framework that supports and protects our organs and allows us to move. Remarkably strong but lightweight, bone is a dynamic, ever-changing tissue. Throughout life, it is continually being broken down and reformed. **Osteology** (os-tē-OL-ō-jē; *osteon* = bone; *logos* = study of) is the study of bone structure and treatment of bone disorders.

The individual bones that form the skeleton and the joints that connect bones are the topics of Chapters 7–9. Here we focus on bone tissue itself. The developmental anatomy of bone is considered at the end of the chapter.

PHYSIOLOGY: FUNCTIONS OF BONE

Bone tissue and the skeletal system perform several basic functions.

1. **Support.** Bone provides a framework for the body by supporting soft tissues and providing points of attachment for many skeletal muscles.
2. **Protection.** Bones protect many internal organs from injury. For example, cranial bones protect the brain, vertebrae surround the spinal cord, the rib cage encloses the heart and lungs, and the hipbones guard internal reproductive organs.
3. **Assists in movement.** Skeletal muscles attach to bones. When muscles contract, they pull on bones and together they produce movement. Movements of muscles, bones, and joints are discussed in detail in later chapters.
4. **Mineral homeostasis.** Bone tissue stores several minerals, especially calcium and phosphorus, which are important in muscle contraction and nerve activity, among other functions. On demand, bone releases minerals into the blood to maintain critical mineral balances and to distribute them to other parts of the body.
5. **Site of blood cell production.** Within certain parts of bones is a connective tissue called **red bone marrow**, which produces red blood cells, white blood cells, and platelets. This process is called **hemopoiesis** (hēm-ō-poy-Ē-sis; *haimatos* = blood; *poiein* = to make). Besides blood cells in immature stages, red bone marrow contains adipose cells and macrophages.
6. **Storage of energy.** Lipids stored in cells of a second type of bone marrow, called yellow bone marrow, are an important chemical energy reserve. **Yellow bone marrow** consists primarily of adipose cells and a few scattered blood cells.

ANATOMY: STRUCTURE OF BONE

Structurally, the skeletal system consists of bone tissue, cartilage, red and yellow bone marrow, and periosteum, the membrane around bones. We described the microscopic structure of cartilage on page 114. Here the focus is the anatomy and histology of bone tissue.

The structure of bone may be analyzed by first considering the parts of a long bone such as the humerus, the arm bone (Fig. 6.1). A long bone is one that has greater length than width. A typical long bone consists of the following parts:

1. **Diaphysis** (dī-AF-i-sis; *dia* = through; *physis* = growth). The shaft or long, main portion of the bone.
2. **Epiphyses** (e-PIF-i-sēz; *epi* = above). The distal and proximal extremities or ends of the bone (singular is **epiphysis**).
3. **Metaphysis** (me-TAF-i-sis; *meta* = after). The region in a mature bone where the diaphysis joins the epiphysis. In a growing bone, it is the region that includes the epiphyseal plate where cartilage is replaced by bone. The epiphyseal plate is a layer of hyaline cartilage that allows the diaphysis of the bone to grow in length. This is described later in the chapter.
4. **Articular cartilage.** A thin layer of hyaline cartilage covering the epiphysis where the bone forms an articulation (joint) with another bone. The cartilage reduces friction and absorbs shock at freely movable joints.
5. **Periosteum** (per'-ē-OS-tē-um). The periosteum (*peri* = around; *osteo* = bone) is a membrane around the surface of the bone not covered by articular cartilage. It consists of two layers (see Fig. 6.3a). The outer **fibrous layer** is composed of dense, irregular connective tissue containing blood vessels, lymphatic vessels, and nerves that pass into the bone. The inner **osteogenic** (os'-tē-ō-JEN-ik) **layer** contains elastic fibers, blood vessels, and bone cells. The periosteum is essential for bone growth in diameter, repair, and nutrition. It also serves as a point of attachment for ligaments and tendons.
6. **Medullary** (MED-yoo-lar'-ē; *medulla* = central part of a structure) or **marrow cavity.** This is the space within the diaphysis that contains the fatty yellow bone marrow in adults.
7. **Endosteum** (end-OS-tē-um; *endo* = within). Lining the medullary cavity is the endosteum, a membrane that contains osteoprogenitor cells and osteoclasts (described shortly).

HISTOLOGY OF BONE TISSUE

Like other connective tissues, **bone**, or **osseous** (OS-ē-us), **tissue** contains an abundant matrix surrounding widely separated cells. The matrix is about 25% water, 25% protein

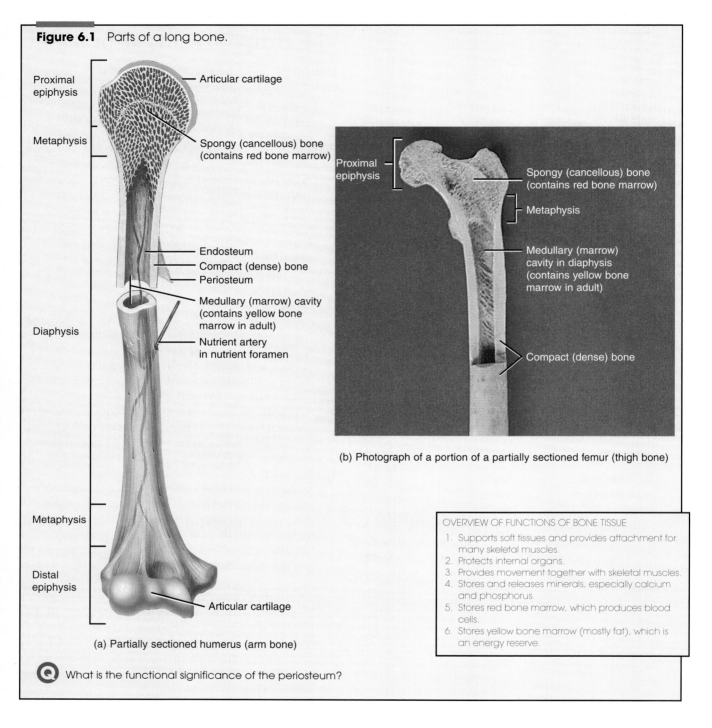

Figure 6.1 Parts of a long bone.

Proximal epiphysis

Articular cartilage

Metaphysis

Spongy (cancellous) bone (contains red bone marrow)

Endosteum

Compact (dense) bone

Periosteum

Medullary (marrow) cavity (contains yellow bone marrow in adult)

Diaphysis

Nutrient artery in nutrient foramen

Metaphysis

Distal epiphysis

Articular cartilage

(a) Partially sectioned humerus (arm bone)

Proximal epiphysis

Spongy (cancellous) bone (contains red bone marrow)

Metaphysis

Medullary (marrow) cavity in diaphysis (contains yellow bone marrow in adult)

Compact (dense) bone

(b) Photograph of a portion of a partially sectioned femur (thigh bone)

OVERVIEW OF FUNCTIONS OF BONE TISSUE

1. Supports soft tissues and provides attachment for many skeletal muscles.
2. Protects internal organs.
3. Provides movement together with skeletal muscles.
4. Stores and releases minerals, especially calcium and phosphorus.
5. Stores red bone marrow, which produces blood cells.
6. Stores yellow bone marrow (mostly fat), which is an energy reserve.

Q What is the functional significance of the periosteum?

fibers, and 50% mineral salts. There are four types of cells in bone tissue: osteoprogenitor (osteogenic) cells, osteoblasts, osteocytes, and osteoclasts (Fig. 6.2).

1. **Osteoprogenitor** (os′-tē-ō-prō-JEN-i-tor; *osteo* = bone; *pro* = precursor; *gen* = to produce) **cells** are unspecialized cells derived from mesenchyme, the tissue from which all connective tissues are derived. They can undergo mitosis and develop into osteoblasts. Osteoprogenitor cells are found in the inner portion of the pe-

riosteum, in the endosteum, and in canals in bone that contain blood vessels.

2. **Osteoblasts** (OS-tē-ō-blasts′; *blast* = germ or bud) are the cells that form bone, but they have lost the ability to divide by mitosis. They secrete collagen and other organic components needed to build bone tissue.

3. **Osteocytes** (OS-tē-ō-sīts′; *cyte* = cell) are mature bone cells that are derived from osteoblasts; they are the principal cells of bone tissue. Like osteoblasts, osteocytes have no mitotic potential. Osteoblasts are found

on the surfaces of bone, but as they surround themselves with matrix materials they become trapped in their secretions and become osteocytes. Osteocytes no longer secrete matrix materials. Whereas osteoblasts initially form bone tissue, osteocytes maintain daily cellular activities of bone tissue, such as the exchange of nutrients and wastes with the blood.

4. **Osteoclasts** (OS-tē-ō-clasts′; *clast* = to break) settle on the surfaces of bone and function in bone resorption (destruction of matrix), which is important in the development, growth, maintenance, and repair of bone.

The matrix of bone, unlike that of other connective tissues, contains abundant mineral salts, primarily a crystallized form of tricalcium phosphate $[Ca_3(PO_4)_2 \cdot (OH)_2]$ called **hydroxyapatite** and some calcium carbonate $(CaCO_3)$. In addition, there are small amounts of magnesium hydroxide, fluoride, and sulfate. As these salts are deposited in the framework formed by the collagen fibers of the matrix, crystallization occurs and the tissue hardens. This process is called **calcification** or **mineralization**.

Although the *hardness* of bone depends on the crystallized inorganic mineral salts, without the organic collagen fibers it would be very brittle. Like reinforcing metal rods in concrete, the collagen fibers and other organic molecules provide bone with *tensile strength,* which is resistance to being stretched or torn apart. Collagen fibers make bone less brittle than other calcium-based products, such as egg shells and oyster shells. If the inorganic minerals are removed, for example, by soaking a bone in weak acid such as vinegar, the result is a rubbery, flexible structure.

At one time, it was thought that calcification simply occurred when enough mineral salts were present to form crystals. Now, however, it is known that the process occurs only in the presence of collagen. Mineral salts accumulate in microscopic spaces between collagen fibers. There the salts crystallize and become hardened. Then, after the spaces are filled, mineral salts deposit around the collagen fibers, where the salts again crystallize and harden. The combination of crystallized salts and collagen is responsible for the hardness that is characteristic of bone.

Bone is not completely solid but has many small spaces (sometimes microscopic only) between its hard components. Some spaces provide channels for blood vessels that supply bone cells with nutrients. Other spaces are storage areas for bone marrow. Depending on the size and distribution of the spaces, the regions of a bone may be categorized as compact or spongy.

Compact Bone Tissue

Compact (**dense**) **bone tissue** contains few spaces. It forms the external layer of all bones of the body and the bulk of the diaphyses of long bones. Compact bone tissue provides protection and support and helps the long bones resist the stress of weight placed on them.

Compare the differences between spongy and compact bone tissues by looking at the highly magnified section in Fig. 6.3a. (See Fig. 6.1 also.) One main difference is that adult compact bone has a concentric-ring structure, whereas spongy bone appears as an irregular latticework. Blood vessels, lymphatic vessels, and nerves from the periosteum penetrate the compact bone through **perforating (Volkmann's) canals**. The blood vessels and nerves of these canals connect with those of the medullary cavity, periosteum, and **central (Haversian) canals**. The central canals run longitudinally through the bone. Around the canals are **concentric lamellae** (la-MEL-ē)—rings of hard, calcified matrix. Between the lamellae are small spaces called

---

Figure 6.2 Types of cells in bone tissue.

 Osteoprogenitor cells undergo mitosis and develop into osteoblasts, which secrete the matrix materials.

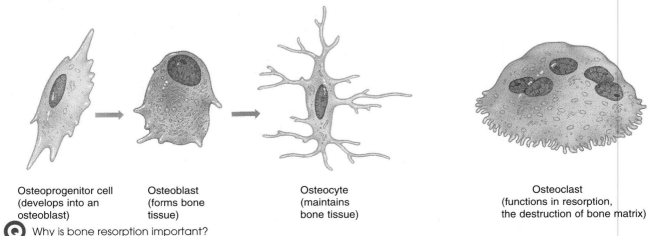

Osteoprogenitor cell (develops into an osteoblast)

Osteoblast (forms bone tissue)

Osteocyte (maintains bone tissue)

Osteoclast (functions in resorption, the destruction of bone matrix)

Q Why is bone resorption important?

Figure 6.3   Histology of bone.

Osteocytes lie in lacunae arranged in concentric circles around a central canal in compact bone and in lacunae arranged irregularly in trabeculae of spongy bone.

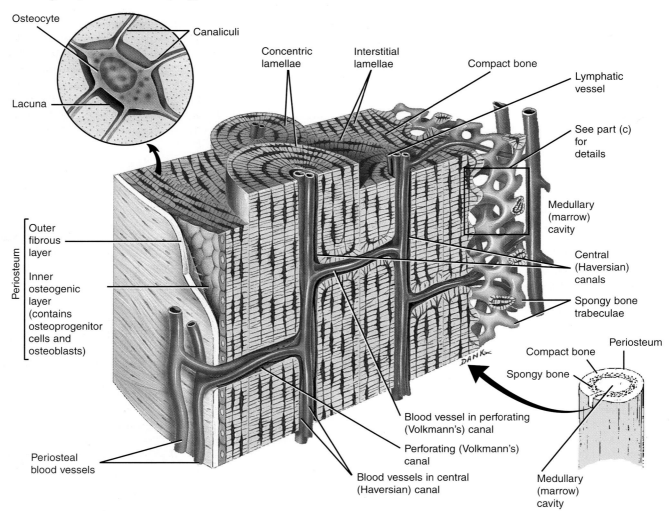

(a) Enlarged aspect of several osteons
(Haversian systems) in compact bone

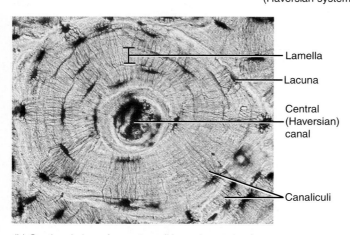

(b) Sectional view of an osteon (Haversian system)
from the femur (thigh bone) (285x)

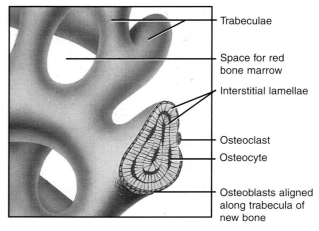

(c) Enlarged aspect of spongy bone trabeculae

  As people age, some central (Haversian) canals may become blocked. What effect would this have on osteocytes?

lacunae (la-KOO-nē; singular is *lacuna* = little lake), which contain osteocytes. **Osteocytes**, as noted earlier, are mature bone cells that no longer secrete matrix materials.

Radiating in all directions from the lacunae are tiny canals called **canaliculi** (kan′-a-LIK-yoo-lī; singular is *canaliculus* = small channel or canal), which are filled with extracellular fluid and slender fingerlike processes of osteocytes (Fig. 6.3a). The canaliculi connect lacunae with one another and, eventually, with the central canals. Thus there is an intricate, miniature canal system throughout the bone. Because materials cannot diffuse adequately through the calcified matrix, the branching canaliculi provide many routes for nutrients and oxygen to reach the osteocytes and wastes to diffuse away. Osteocytes from neighboring lacunae also form gap junctions with each other, which allows movement of materials from cell to cell.

Each central canal, with its surrounding lamellae, lacunae, osteocytes, and canaliculi, forms an **osteon (Haversian system)**. Adult compact bone tissue is the only connective tissue that has a basic structural unit associated with it, the osteon. The areas between osteons contain **interstitial lamellae**. These also have lacunae with osteocytes and canaliculi. Interstitial lamellae are fragments of older osteons that have been partially destroyed during bone rebuilding or growth.

Spongy Bone Tissue

In contrast to compact bone, spongy (cancellous) bone does not contain true osteons (Fig. 6.3b,c). See Fig. 6.1 also. It consists of lamellae arranged in an irregular latticework of thin plates of bone called **trabeculae** (tra-BEK-yoo-lē; singular is *trabecula* = little beam). The macroscopic spaces between the trabeculae of some bones are filled with red bone marrow, which produces blood cells. Within the trabeculae are osteocytes that lie in lacunae. Radiating from the lacunae are canaliculi. Osteocytes in the trabeculae receive nourishment directly from the blood circulating through the marrow cavities. Osteons are not necessary in spongy bone because osteocytes are not deeply buried (as in compact bone) and have access to nutrients directly from the blood.

Spongy bone makes up most of the bone tissue of short, flat, and irregularly shaped bones and most of the epiphyses of long bones. Spongy bone tissue in the hipbones, ribs, breastbone (sternum), backbones (vertebrae), skull, and ends of some long bones is the only site of red bone marrow storage and thus hemopoiesis in adults.

Blood and Nerve Supply

Bone is richly supplied with blood, and blood vessels are especially abundant in portions of bone containing red bone marrow. Blood vessels pass into bones from the periosteum. Here we shall consider the blood supply to a long bone only.

Because the artery to the diaphysis of a long bone is usually the largest, it is referred to as the **nutrient artery**. The artery first enters the bone during the early development and passes through an opening in the diaphysis, the **nutrient foramen** (see Figs. 6.1a and 6.5, number ③). On entering the medullary cavity, the nutrient artery divides into a proximal and a distal branch, each supplying most of the bone marrow, inner portion of compact bone of the diaphysis, and metaphysis. As the nutrient artery passes through compact bone on its way to the medullary cavity, it sends branches into the central (Haversian) canals. Branches of the **epiphyseal arteries** in the epiphysis supply the bone marrow, metaphysis, and bony tissue of the epiphysis (see Fig. 6.5, number ④). **Periosteal arteries**, accompanied by nerves, enter the diaphysis at many points through perforating (Volkman's) canals and supply the outer part of the compact bone of the diaphysis. Veins accompany the several types of arteries; the principal veins leave the bone by numerous openings at the epiphyses of the bone.

Although the nerve supply to bone is not extensive, some nerves accompany blood vessels and some occur in the periosteum. The nerves of the periosteum are primarily concerned with pain, as might be associated with a fracture or tumor.

BONE FORMATION: OSSIFICATION

Most people think of all bone as a very hard, rigid material. Yet the bones of an infant are quite soft and become rigid only after growth stops during late adolescence. Even then, bone is constantly broken down and rebuilt. It is a dynamic, living tissue. Let us now see how bones are formed and how they grow.

The process by which bone forms is called **ossification** (os′-i-fi-KĀ-shun; *facere* = to make). The "skeleton" of a human embryo is composed of fibrous connective tissue membranes formed by embryonic connective tissue (mesenchyme) and pieces of hyaline cartilage that are loosely shaped like bones. They provide the supporting structures for ossification. Ossification begins around the sixth or seventh week of embryonic life and continues throughout adulthood. Bone formation follows one of two patterns.

1. **Intramembranous** (in′-tra-MEM-bra-nus; *intra* = within; *membranous* = membrane) **ossification** refers to the formation of bone directly on or within loose fibrous connective tissue membranes. Such bones form *directly* from mesenchyme without first going through a cartilage stage. As you will see on page 168, the fontanels ("soft spots") of an infant's skull, which are composed of loose fibrous connective tissue membranes, are also eventually replaced by bone through intramembranous ossification.

2. **Endochondral** (en'-dō-KON-dral; *endo* = within; *chondro* = cartilage) **ossification** refers to the formation of bone in hyaline cartilage. In this process, mesenchyme is transformed into chondroblasts which produce a hyaline cartilage matrix that is gradually replaced by bone.

These two kinds of ossification do *not* lead to differences in the gross structure of mature bones. They are simply different methods of bone formation. Both mechanisms involve the replacement of a pre-existing connective tissue with bone.

The first stage in the development of bone is the migration of embryonic mesenchymal cells into the area where bone formation is about to begin. These cells increase in number and size and become osteoprogenitor cells. In some skeletal structures where capillaries are lacking, they become chondroblasts; in others where capillaries are present, they become osteoblasts. The **chondroblasts** are responsible for cartilage formation. Osteoblasts form bone tissue by intramembranous or endochondral ossification.

Intramembranous Ossification

The flat bones of the skull, mandible (lower jawbone), and clavicles (collarbones) develop directly on or within loose fibrous connective tissue membranes formed by mesenchymal cells. This process of **intramembranous ossification** occurs as follows (Fig. 6.4):

1   At the site where the bone will develop, mesenchymal cells in fibrous connective tissue membranes cluster and differentiate, first into osteoprogenitor cells and then into osteoblasts. The site of such a cluster is called a **center of ossification**. Osteoblasts secrete the organic matrix (osteoid) of bone until they are completely surrounded by it.

2   Then secretion of matrix stops and the cells, now called osteocytes, lie in lacunae and extend narrow cytoplasmic processes into canaliculi that radiate in all directions. Within a few days, calcium and other mineral salts are deposited and the matrix hardens or calcifies. Thus calcification is just one aspect of ossification.

3   As the bone matrix forms, it develops into **trabeculae** that fuse with one another to create the open latticework appearance of spongy bone. The spaces between trabeculae fill with vascularized connective tissue, which differentiates into red bone marrow. On the outside of the bone, vascularized mesenchyme condenses.

4   The condensed mesenchyme develops into the periosteum. Eventually, most surface layers of the spongy bone are replaced by compact bone, but spongy bone remains in the center of the bone. Much of this newly formed bone will be remodeled (destroyed and reformed) so the bone may reach its final adult size and shape.

Endochondral Ossification

The replacement of cartilage by bone is called **endochondral (intracartilaginous) ossification**. Most bones of the body are formed in this way, but this type of ossification is best observed in a long bone. It proceeds as follows (Fig. 6.5):

1   **Development of the cartilage model.** At the site where the bone is going to form, mesenchymal cells crowd together in the shape of the future bone. The mesenchymal cells differentiate into chondroblasts that produce cartilage matrix so the model consists of hyaline cartilage. In addition, a membrane called the **perichondrium** (per-i-KON-drē-um) develops around the cartilage model.

2   **Growth of the cartilage model.** The cartilage model grows in length by continual cell division of chondrocytes accompanied by further secretion of cartilage matrix by the daughter cells. This pattern of growth that results in an increase in length is called **interstitial** (in'-ter-STISH-al) **growth**, which means growth from within. Growth of the cartilage in thickness is mainly due to the addition of more matrix to its periphery by new chondroblasts that develop from the perichondrium. This growth pattern of cartilage in which matrix is deposited on its surface is called **appositional** (a-pō-ZISH-a-nal) **growth**.

As the cartilage model continues to grow, chondrocytes in its midregion hypertrophy (increase in size), probably because they accumulate glycogen for ATP production and produce enzymes to catalyze further chemical reactions. Some hypertrophied cells burst, releasing their contents, which change the pH of the matrix. These chemical changes trigger calcification. Once the cartilage becomes calcified, other chondrocytes die because nutrients no longer diffuse quickly enough through the matrix. The lacunae of the cells that have died are now empty, and the thin partitions between them break down forming small cavities.

In the meantime, a nutrient artery penetrates the perichondrium and then the bone through a hole in the bone (nutrient foramen). This occurs in the midregion of the model, stimulating osteoprogenitor cells in the perichondrium to differentiate into osteoblasts. The cells lay down a thin shell of compact bone under the perichondrium called the **periosteal bone collar**. Once the perichondrium starts to form bone, it is known as the **periosteum**.

3   **Development of the primary ossification center.** Near the middle of the model, periosteal capillaries grow into the disintegrating calcified cartilage. These vessels, and the associated osteoblasts, osteoclasts, and red bone marrow cells, are known as the **periosteal bud**. On growing into the cartilage model, the

Figure 6.4 Intramembranous ossification.

Intramembranous ossification involves the formation of bone directly on or within loose fibrous connective tissue membranes.

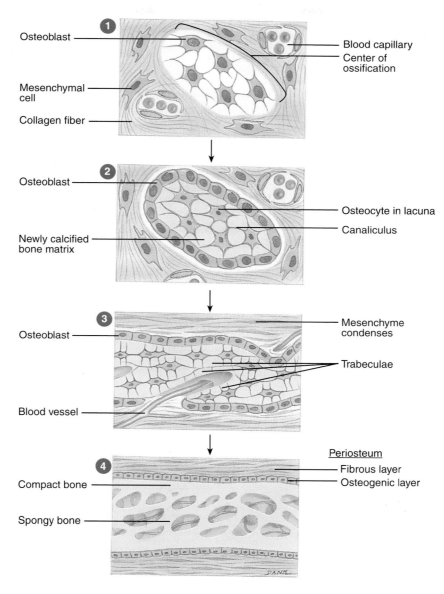

1
Osteoblast
Mesenchymal cell
Collagen fiber
Blood capillary
Center of ossification

2
Osteoblast
Newly calcified bone matrix
Osteocyte in lacuna
Canaliculus

3
Osteoblast
Blood vessel
Mesenchyme condenses
Trabeculae

4
Compact bone
Spongy bone
Periosteum
Fibrous layer
Osteogenic layer

Which bones of the body develop by intramembranous ossification?

capillaries induce growth of a **primary ossification center**, a region where bone tissue will replace most of the cartilage. Osteoblasts then begin to deposit bone matrix over the remnants of calcified cartilage, forming spongy bone trabeculae. As the ossification center enlarges toward the ends of the bone, osteoclasts break down the newly formed spongy bone trabeculae. This activity leaves a cavity, the medullary (marrow) cav-

ity, in the core of the model. The cavity then fills with red bone marrow. Primary ossification proceeds *inward* from the external surface of the bone.

4 **Development of the diaphysis and epiphysis.** The diaphysis (shaft), which was once a solid mass of hyaline cartilage, is replaced by compact bone, the core of which contains a red bone marrow-filled medullary cavity. When blood vessels (epiphyseal arteries) enter

Figure 6.5 Endochondral ossification of the tibia (shinbone).

🔑 **Endochondral ossification involves the replacement of cartilage by bone.**

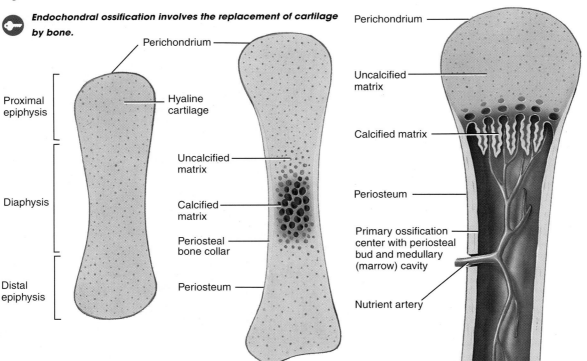

Proximal epiphysis

Diaphysis

Distal epiphysis

Perichondrium

Hyaline cartilage

Uncalcified matrix

Calcified matrix

Periosteal bone collar

Periosteum

Perichondrium

Uncalcified matrix

Calcified matrix

Periosteum

Primary ossification center with periosteal bud and medullary (marrow) cavity

Nutrient artery

① Mesenchymal cells differentiate into chondroblasts, which form the hyaline cartilage model

② Cartilage model grows by interstitial and appositional growth, chondrocytes in midregion calcify the matrix, vacated lacunae form small cavities, osteoblasts in perichondrium produce periosteal bone collar

③ With development of periosteal bud, primary ossification center and medullary cavity form

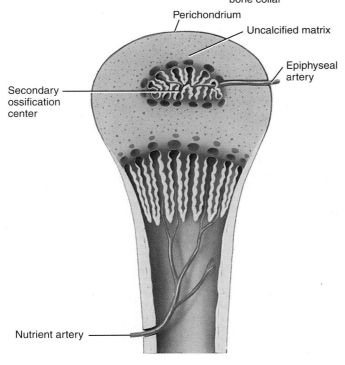

Perichondrium

Uncalcified matrix

Secondary ossification center

Epiphyseal artery

Nutrient artery

④ Development of secondary ossification center in epiphysis. A secondary ossification center also develops in the distal epiphysis of a long bone

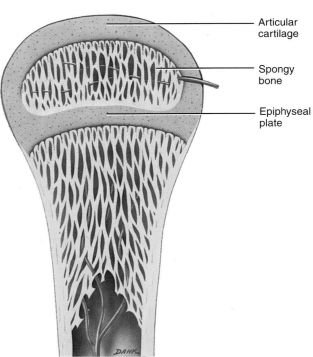

Articular cartilage

Spongy bone

Epiphyseal plate

⑤ Remnants of hyaline cartilage as articular cartilage and epiphyseal plate

❓ Radiographs of an 18-year-old basketball star do not reveal epiphyseal lines. Is she likely to grow taller?

the epiphyses, **secondary ossification centers** develop, usually around the time of birth.

 In the secondary ossification centers, bone formation is similar to that in the primary ossification centers. One difference, however, is that spongy bone remains in the interior of the epiphyses (no medullary cavities are formed in the epiphyses). Also, hyaline cartilage remains covering the epiphyses as the articular cartilage and between the diaphysis and epiphysis as the epiphyseal plate, which is responsible for the lengthwise growth of long bones. Secondary ossification proceeds *outward* from the center of the epiphysis toward the outer surface of the bone.

BONE GROWTH

During childhood, bones throughout the body enlarge by appositional growth and long bones lengthen by addition of bone material at the epiphyseal plate. Growth in length of bones normally ceases by age 25, although bones may continue to thicken.

Growth in Length

To understand how a bone grows in length, you will need to know some of the details of the structure of the epiphyseal plate.

The **epiphyseal** (ep′-i-FIZ-ē-al; *epiphyein* = to grow upon) **plate** is a layer of hyaline cartilage in the metaphysis of a growing bone that consists of four zones (Fig. 6.6). The **zone of resting cartilage** is nearest the epiphysis and consists of small, scattered chondrocytes. The cells do not function in bone growth (thus the term "resting"); they anchor the epiphyseal plate to the bone of the epiphysis.

The **zone of proliferating cartilage** consists of slightly larger chondrocytes arranged like stacks of coins. Chondrocytes divide to replace those that die at the diaphyseal surface of the epiphyseal plate.

The **zone of hypertrophic** (hi-per-TROF-ik) or **maturing cartilage** consists of even larger chondrocytes that are also arranged in columns. The movement of the epiphysis farther from the diaphysis is the result of cell divisions in the zone of proliferating cartilage and maturation of the cells in the zone of hypertrophic cartilage.

The **zone of calcified cartilage** is only a few cells thick and consists mostly of dead cells because the matrix around them has calcified. The calcified matrix is taken up by osteoclasts, and the area is invaded by osteoblasts and capillaries from the bone in the diaphysis. These cells lay down bone on the calcified cartilage that persists. As a result, the diaphyseal border of the epiphyseal plate is firmly cemented to the bone of the diaphysis.

The activity of the epiphyseal plate is the only mechanism by which the diaphysis can increase in length. The plate allows the diaphysis of the bone to increase in length until early adulthood. It also shapes the articular surfaces. As a child grows, cartilage cells are produced by mitosis on the epiphyseal side of the plate. They are then destroyed, and the cartilage is replaced by bone on the diaphyseal side of the plate. In this way, the thickness of the epiphyseal

Figure 6.6 Epiphyseal plate.

The epiphyseal plate allows the diaphysis of a bone to increase in length.

Radiograph of a portion of the tibia and fibula of a 10-year-old child

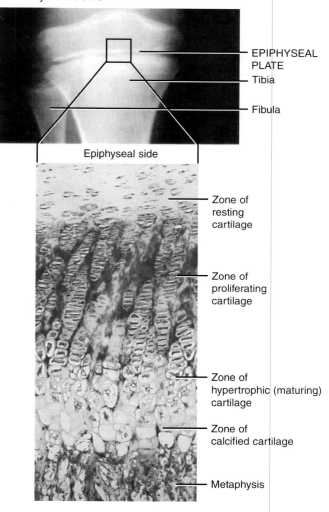

EPIPHYSEAL PLATE

Tibia

Fibula

Epiphyseal side

Zone of resting cartilage

Zone of proliferating cartilage

Zone of hypertrophic (maturing) cartilage

Zone of calcified cartilage

Metaphysis

Diaphyseal side

Photomicrograph of epiphyseal plate (360x)

 What accounts for the lengthwise expansion of the diaphysis?

plate remains almost constant, but the bone on the diaphyseal side increases in length. If a bone fracture also damages the epiphyseal plate, the fractured bone will be shorter than its normal counterpart once adult stature is reached. This is because damage to cartilage, which is avascular, accelerates closure of the epiphyseal plate, and growth in length of the bone is inhibited. A bone fracture that does not involve the epiphyseal plate usually heals quite well due to the rich blood supply of bone.

Eventually, the epiphyseal cartilage cells stop dividing and bone replaces the cartilage. The newly formed bony structure in the metaphysis is called the **epiphyseal line**, a remnant of the once active epiphyseal plate. With the appearance of the epiphyseal line, bone stops growing in length. The clavicle is the last bone to stop growing. In general, lengthwise growth of bones is completed earlier in females than in males.

Growth in Thickness

Unlike cartilage, which can grow by both interstitial and appositional growth, bone can grow in thickness or diameter only by appositional growth. It occurs as follows. First, the bone lining the medullary cavity is destroyed by osteoclasts in the endosteum so that the cavity increases in diameter. At the same time, osteoblasts from the periosteum add new bone tissue to the outer surface. Initially, diaphyseal and epiphyseal ossification produce only spongy bone. Later, the outer region of spongy bone is reorganized into compact bone.

Hormonal Regulation of Bone Growth

Before puberty, bone growth is stimulated mainly by human growth hormone (hGH), which is produced by the anterior pituitary gland, and insulin-like growth factors (IGFs), which are produced locally by bone and also by the liver in response to hGH stimulation. Oversecretion of hGH produces giantism, in which a person becomes much taller and heavier than normal, whereas undersecretion of hGH produces dwarfism (short stature). Thyroid hormones, from the thyroid gland, and insulin, from the pancreas, also stimulate normal bone growth.

At puberty, estrogens and testosterone, sex hormones produced by the ovaries and testes, start to be released in larger quantities. These hormones are responsible for the sudden growth or "growth spurt" that occurs during the teenage years. Estrogens (more than one form exists) also promote changes in the skeleton that are typical of females, for example, a wider pelvis, whereas testosterone promotes skeletal changes typical of males. (See Exhibit 8.1 on page 212 and Fig. 8.15 for a comparison of typical male and female pelvises.)

BONE HOMEOSTASIS

Bone, like skin, forms before birth but continually renews itself thereafter. **Remodeling** is the ongoing replacement of old bone tissue by new bone tissue. It takes place at different rates in various body regions. The distal portion of the thighbone (femur) is replaced about every four months. By contrast, bone in certain areas of the shaft will not be completely replaced during a lifetime. Bone constantly remodels and redistributes its matrix along lines of mechanical stress. Compact bone is formed from spongy bone. However, even after bones have reached their adult shapes and sizes, old bone is continually destroyed, and new bone tissue is formed in its place. Remodeling also removes worn and injured bone, replacing it with new tissue. Reservoirs of calcium needed for other tissues become available during remodeling. Several hormones continually regulate exchanges of calcium and phosphate ions between blood and bones.

Remodeling

Osteoclasts are responsible for bone resorption (destruction of matrix). A delicate homeostasis exists between the actions of osteoclasts in removing minerals and collagen and of bone-making osteoblasts in depositing minerals and collagen. Should too much new tissue be formed, the bones become abnormally thick and heavy. If too much mineral is deposited in the bone, the surplus may form thick bumps, or spurs, on the bone that interfere with movement at joints. A loss of too much calcium or inadequate formation of new tissue weakens the bones. They may break, as occurs in osteoporosis (see page 159), or become too flexible, as in rickets in children and osteomalacia in adults. Both rickets and osteomalacia are caused by insufficient calcium or vitamin D, which results in inadequate mineralization of bone. Abnormal acceleration of the remodeling process results in a condition called Paget's disease (see page 159).

In the process of resorption, osteoclasts put forth projections that secrete protein-digesting lysosomal enzymes and several acids (lactic, carbonic, and citric). The enzymes digest collagen and other organic substances, while the acids dissolve the bone minerals, mainly hydroxyapatite. Under a microscope, osteoclasts have a phagocytic appearance, and it is presumed that they also phagocytize whole fragments of collagen and bone minerals.

Normal bone growth in the young, bone remodeling in the adult, and repair of fractured bone depend on (1) adequate minerals, most importantly calcium, phosphorus, magnesium, boron, and manganese; (2) vitamins A, B_{12}, C, and D; (3) several hormones, most importantly human growth hormone, sex hormones (estrogens and testosterone), insulin, insulin-like growth factors, thyroid hormones, calcitonin, and parathyroid hormone; and (4) amount of exercise that places stress on bones (weight-bearing activities).

Fracture and Repair of Bone

A **fracture** is any break in a bone. Although bone has a generous blood supply, healing sometimes takes months. Sufficient calcium and phosphorus to strengthen and harden new bone is deposited only gradually. Bone cells generally also grow and reproduce slowly. Moreover, in a fractured bone the blood supply is decreased, which helps to explain the difficulty in the healing of an infected bone. Recall that cartilage injuries heal even more slowly, however, because capillaries are present only in the perichondrium of cartilage. The following steps occur in the repair of a bone fracture (Fig. 6.7):

1 As a result of the fracture, blood vessels crossing the fracture line are broken. These include vessels in the periosteum, osteons (Haversian systems), and medullary cavity. As blood pours from the torn ends of the vessels, it forms a clot in and about the site of the fracture. This clot, called a **fracture hematoma** (hē′-ma-TŌ-ma), usually occurs 6–8 hours after the injury. Since the circulation of blood ceases when the fracture hematoma forms, bone cells and periosteal cells at the fracture site die. The hematoma serves as a focus for the cellular invasion that follows. Swelling and inflammation occur after formation of the fracture hematoma, and there is a considerable amount of cell death and debris. Blood capillaries grow into the blood clot and phagocytes (neutrophils and macrophages) plus osteoclasts begin to remove the traumatized tissue in and around the fracture hematoma. This may take up to several weeks.

2 The infiltration of blood capillaries into the fracture hematoma helps organize it into granulation tissue (see page 119), now called a **procallus**. Next, fibroblasts from the periosteum and osteoprogenitor cells from the periosteum, endosteum, and marrow invade the procallus. The fibroblasts produce collagen fibers, which help connect the broken ends of the bones. Osteoprogenitor cells develop into chondroblasts in areas farther away from healthy bone tissue where the environment is avascular. Here chondroblasts begin to produce fibrocartilage and the procallus is transformed into a **fibrocartilaginous (soft) callus**. A **callus** is actually a mass of repair tissue that bridges the broken ends of the bones. The stage of the fibrocartilaginous callus lasts about 3 weeks.

3 In areas closer to healthy bone tissue where the environment is more vascular, osteoprogenitor cells develop into osteoblasts, which begin to produce spongy bone trabeculae. The trabeculae join living and dead portions of the original bone fragments. In time, the

---

Figure 6.7 Steps involved in repair of a bone fracture.

 Bone heals more rapidly than cartilage because its blood supply is more plentiful.

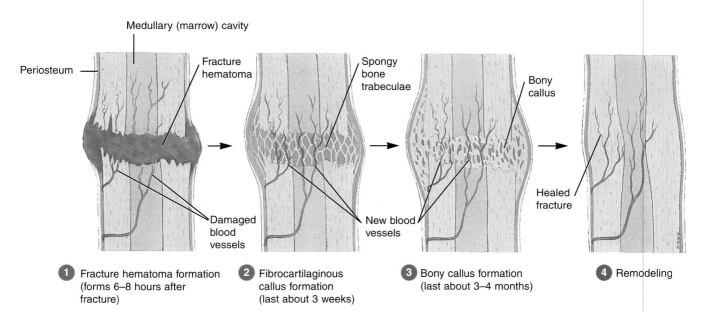

1 Fracture hematoma formation (forms 6–8 hours after fracture)

2 Fibrocartilaginous callus formation (last about 3 weeks)

3 Bony callus formation (last about 3–4 months)

4 Remodeling

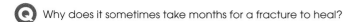 Why does it sometimes take months for a fracture to heal?

fibrocartilage is converted to spongy bone and the callus is then referred to as a **bony** (**hard**) **callus**. The stage of the bony callus lasts about 3–4 months.

4 The final phase of fracture repair is **remodeling** of the callus. Dead portions of the original fragments are gradually resorbed by osteoclasts. Compact bone replaces spongy bone around the periphery of the fracture. Sometimes, the healing is so complete that the fracture line is undetectable, even in a radiograph (x-ray). However, a thickened area on the surface of the bone usually remains as evidence of a healed fracture site.

Although fractures may be classified in several different ways, the following list gives common categories.

1. **Partial**: the break across the bone is incomplete.
2. **Complete**: the break across the bone is complete, so that the bone is broken into two or more pieces.
3. **Closed (simple)**: the bone does not break through the skin.
4. **Open (compound)**: the broken ends of the bone protrude through the skin (Fig. 6.8a and see Fig. 6.9).
5. **Comminuted** (KOM-i-nyoo-ted): the bone has splintered at the site of impact, and smaller fragments of bone lie between the two main fragments (Fig. 6.8b).
6. **Greenstick**: a partial fracture in which one side of the bone is broken and the other side bends; occurs only in children (Fig. 6.8c).
7. **Spiral**: the bone usually is twisted apart.
8. **Transverse**: a fracture at right angles to the long axis of the bone.
9. **Impacted**: one fragment is firmly driven into the other (Fig 6.8d).
10. **Displaced**: the anatomical alignment of the bone fragments is not preserved.
11. **Nondisplaced**: the anatomical alignment of the bone fragments is preserved.
12. **Stress**: microscopic fractures resulting from inability to withstand repeated stressful impact. Usually, they result from repeated, strenuous activities such as running, basketball, jumping, or aerobic dancing. About 25% of all stress fractures involve the shinbone (tibia).
13. **Pathologic**: weakening of a bone caused by disease processes such as osteogenic sarcoma (bone cancer), osteomyelitis (inflammation of a bone), osteoporosis (decreased bone mass), or osteomalacia.
14. **Pott's**: a fracture of the distal end of the lateral leg bone (fibula), with serous injury of the distal tibial articulation (Fig 6.8e).
15. **Colles'** (KOL-ez): a fracture of the distal end of the lateral forearm bone (radius) in which the distal fragment is displaced posteriorly (Fig. 6.8f).

Figure 6.8 illustrates some of these types of fractures.

Figure 6.8   Types of fractures.

 A fracture is any break in a bone.

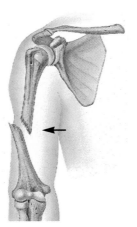
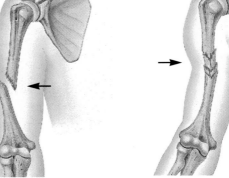

(a) Open fracture

(b) Comminuted fracture

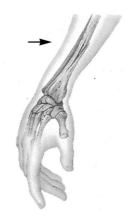
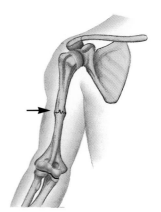

(c) Greenstick fracture

(d) Impacted fracture

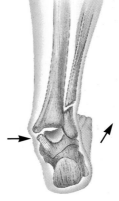
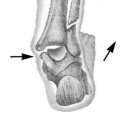
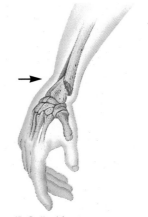

(e) Pott's fracture

(f) Colles' fracture

 What is a callus?

Diagnosis of the type of fracture that has occurred involves both visual inspection and examination of radiographs. Figure 6.9 shows a radiograph of an open (compound) fracture of the arm bone (humerus). Usually, the fractured ends of a bone can be reduced (aligned to their normal positions) by manipulation without surgery. This procedure of setting a fracture is called **closed reduction**. In other cases, the fracture must be exposed by surgery before the break is rejoined. This procedure is known as **open reduction**.

Bone's Role in Calcium Homeostasis

Bone is the major reservoir of calcium in the body, storing more than 99% of the total. The blood level of calcium ions (Ca^{2+}) is very closely regulated; even small changes in Ca^{2+} concentration are deadly. The heart may stop (cardiac arrest) if the concentration goes too high or breathing may cease (respiratory arrest) if the level falls too low. Most functions of nerve cells depend on just the right level of Ca^{2+}. Also, many enzymes require Ca^{2+} as a cofactor, and blood clotting requires Ca^{2+}. The role of bone in calcium homeostasis is to "buffer" blood calcium level, releasing Ca^{2+} to the blood when the blood level decreases and taking Ca^{2+} back when the level rises. Hormones regulate these exchanges.

The most important hormone that regulates Ca^{2+} exchange between bone and blood is **parathyroid hormone** (**PTH**), secreted by the parathyroid glands (see Fig. 18.18). PTH secretion is linked to several negative feedback systems that adjust blood Ca^{2+} concentration (controlled condition). Look at the negative feedback cycle in Fig. 6.10. If some stimulus causes blood Ca^{2+} level to fall, parathyroid gland cells (receptors) detect this change. The control center is the gene for PTH within the nucleus of a parathyroid gland cell. One input signal to the control center is increased level of a molecule known as cyclic AMP (adenosine

monophosphate) in the cytosol. Cyclic AMP accelerates reactions that "turn on" the PTH gene, PTH synthesis speeds up, and more PTH (output) is released into the blood. PTH increases the number and activity of osteoclasts (effectors),

---

Figure 6.10 Negative feedback system for the regulation of blood calcium ion (Ca^{2+}) concentration.

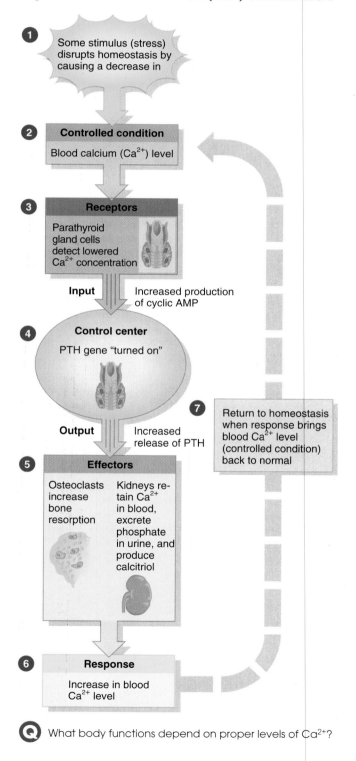

1 Some stimulus (stress) disrupts homeostasis by causing a decrease in

2 **Controlled condition**
Blood calcium (Ca^{2+}) level

3 **Receptors**
Parathyroid gland cells detect lowered Ca^{2+} concentration

Input Increased production of cyclic AMP

4 **Control center**
PTH gene "turned on"

Output Increased release of PTH

7 Return to homeostasis when response brings blood Ca^{2+} level (controlled condition) back to normal

5 **Effectors**
Osteoclasts increase bone resorption   Kidneys retain Ca^{2+} in blood, excrete phosphate in urine, and produce calcitriol

6 **Response**
Increase in blood Ca^{2+} level

Q What body functions depend on proper levels of Ca^{2+}?

Figure 6.9 Radiograph (x-ray) of an open (compound) fracture of the humerus.

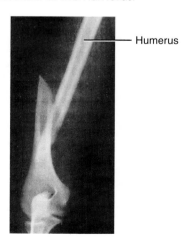

—— Humerus

which step up the pace of bone resorption. The resulting release of Ca^{2+} (and phosphate ions) from bone into blood (response) returns the blood Ca^{2+} level to normal.

PTH also affects the kidneys. It promotes (1) recovery of Ca^{2+} so it is not lost in the urine, (2) elimination of phosphate ions in the urine, and (3) formation of calcitriol, the active form of vitamin D. These kidney effects of PTH all augment the bone effect—they elevate blood Ca^{2+} concentration.

Another hormone also contributes to the homeostasis of blood Ca^{2+}. **Calcitonin (CT)** is secreted by certain thyroid gland cells (parafollicular cells) when blood Ca^{2+} rises above normal. It inhibits osteoclastic activity, speeds Ca^{2+} uptake by bone from blood, and accelerates Ca^{2+} deposit into bones. The net result is that calcitonin promotes bone formation and decreases blood Ca^{2+} level.

EXERCISE AND BONE TISSUE

Within limits, bone has the ability to alter its strength in response to mechanical stress. When placed under such stress, bone tissue becomes stronger with time, through increased deposition of mineral salts and production of collagen fibers. Another effect of stress is to increase the production of calcitonin, which inhibits bone resorption. Without mechanical stress, bone does not remodel normally since resorption outstrips bone formation. Removal of mechanical stress weakens bone through demineralization (loss of bone minerals) and collagen reduction. The main mechanical stresses on bone are those that result from the pull of skeletal muscles and the pull of gravity. If a person is bedridden or has a fractured bone in a cast, the strength of the unstressed bones diminishes. Astronauts subjected to the weightlessness of space also lose bone mass. In both cases, the bone loss can be dramatic, as much as 1% per week. Bones of athletes, which are repetitively and highly stressed, become notably thicker than those of nonathletes. Weight-bearing activities, such as walking or moderate weight lifting, help build and retain bone mass.

Exhibit 6.1 summarizes the factors that influence bone growth, remodeling, and repair.

AGING AND BONE TISSUE

Aging has two principal effects on bone tissue. The first is the loss of calcium and other minerals from bone matrix (demineralization). This loss usually begins after age 30 in

EXHIBIT **6.1** SUMMARY OF FACTORS THAT INFLUENCE GROWTH, REMODELING, AND REPAIR OF BONE	
Factor	**Comment**
MINERALS **Calcium** and **phosphorus**	Make bone matrix hard.
Magnesium	Deficiency inhibits osteoblasts.
Boron	May inhibit calcium loss and increase levels of estrogens.
Manganese	Inhibits formation of new bone tissue.
VITAMINS **Vitamin A**	Controls activity, distribution, and coordination of osteoblasts and osteoclasts; toxic in high doses.
Vitamin B$_{12}$	May inhibit osteoblast activity.
Vitamin C	Helps maintain bone matrix; deficiency leads to decreased collagen production, which inhibits bone growth and delays fracture repair.
Vitamin D (calcitriol)	Active form (calcitriol) is formed in the skin and kidneys from dietary precursor; helps build bone by increasing absorption of calcium from intestine into blood; may reduce the risk of osteoporosis but is toxic in high doses.

Factor	Comment
HORMONES	
Human growth hormone (hGH)	Secreted by the anterior pituitary gland; promotes general growth of all body tissues, including bone, mainly by stimulating production of insulin-like growth factors (IGFs).
Insulin-like growth factors (IGFs)	Stimulate uptake of amino acids and synthesis of proteins; promote tissue repair and bone growth.
Sex hormones (several estrogens and testosterone)	Estrogens secreted by ovaries and testosterone secreted by testes increase bone-building activity of osteoblasts to promote bone growth; responsible for characteristic feminine and masculine skeletal differences.
Insulin	Secreted by pancreas; promotes normal bone growth and maturity.
Thyroid hormones (thyroxine and tri-iodothyronine)	Secreted by thyroid gland; promote normal bone growth and maturity.
Calcitonin (CT)	Secreted by thyroid gland; promotes bone formation by inhibiting activity of osteoclasts, speeding up Ca^{2+} absorption from blood, and accelerating Ca^{2+} deposit in bones.
Parathyroid hormone (PTH)	Secreted by parathyroid glands; promotes bone resorption by increasing the number and activity of osteoclasts; enhances recovery of Ca^{2+} from urine; promotes formation of the active form of vitamin D (calcitriol).
EXERCISE	Weight-bearing activities help build thicker, stronger bones and retard the loss of bone mass that occurs as people age.

females, accelerates greatly around age 40–45 as levels of estrogens decrease, and continues until as much as 30% of the calcium in bones is lost by age 70. In males, calcium loss typically does not begin until after age 60. The loss of calcium from bones is one of the problems in a condition called osteoporosis (see page 159).

The second principal effect of aging on the skeletal system results from a decrease in the rate of protein synthesis. This causes a decreased ability to produce the organic portion of bone matrix, mainly collagen, which normally gives bone its tensile strength. The loss of tensile strength causes the bones to become very brittle and susceptible to fracture. In some elderly people protein synthesis slows, in part, because activity of human growth hormone decreases.

 DEVELOPMENTAL ANATOMY OF THE SKELETAL SYSTEM

As noted earlier, both intramembranous and endochondral ossification begin when **mesenchymal (mesodermal) cells** migrate into the area where bone formation will occur. In some skeletal structures, mesenchymal cells develop into **chondroblasts** that form *cartilage.* In other skeletal structures, mesenchymal cells develop into **osteoblasts** that form *bone tissue* by intramembranous or endochondral ossification (discussed on page 148).

Discussion of the development of the skeletal system provides us with an excellent opportunity to trace the development of the limbs. The *limbs* make their appearance about the fifth week as small elevations at the sides of the trunk called **limb buds** (Fig. 6.11a). They consist of masses of general **mesoderm** covered by **ectoderm**. At this point, a mesenchymal skeleton exists in the limbs; some of the masses of mesoderm surrounding the developing bones will become the skeletal muscles of the limbs.

By the sixth week, the limb buds develop a constriction around the middle portion. The constriction produces distal segments of the upper buds called **hand plates** and distal segments of the lower buds called **foot plates**. These plates represent the beginnings of the *hands* and *feet,* respectively. At this stage of limb development, a cartilaginous skeleton is present. By the seventh week (Fig. 6.11c), the *arm, forearm,* and *hand* are evident in the upper limb bud, and the *thigh, leg,* and *foot* appear in the lower limb bud. Endochondral ossification has begun. By the eighth week (Fig. 6.11d), the *shoulder, elbow,* and *wrist* areas become apparent.

Figure 6.11   External features of a developing human embryo at various stages of development. Many of the labeled structures are discussed in later chapters; they are indicated here to help you orient to the figure.

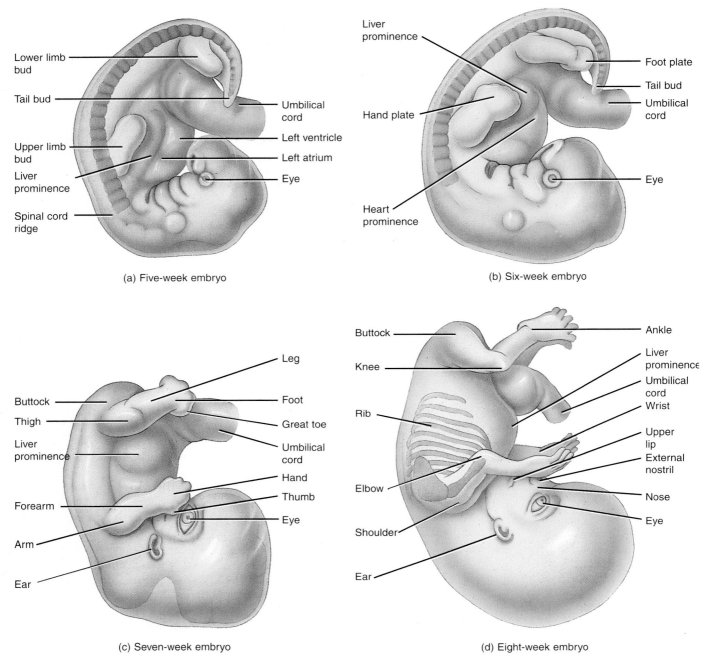

(a) Five-week embryo

(b) Six-week embryo

(c) Seven-week embryo

(d) Eight-week embryo

DISORDERS: HOMEOSTATIC IMBALANCES

OSTEOPOROSIS

Twenty million people in the United States suffer from **osteoporosis** (os′-tē-ō-pō-RŌ-sis), literally a condition of porous bones. It is characterized by decreased bone mass and increased susceptibility to fractures. The basic problem is that bone resorption outpaces bone formation. The disorder primarily affects middle-aged and elderly people—women more than men and whites more than blacks. Between puberty and mid-life, sex hormones (estrogens in women and testosterone in men) and other hormones maintain bone tissue by stimulating the osteoblasts to form new bone. Women produce much smaller amounts of estrogens after menopause, and men produce smaller amounts of testosterone as they age. As a result, the osteoblasts become less active, and there is a decrease in bone mass.

Although most common in women over age 50, osteoporosis can also occur in a variety of conditions. These include female runners and ballet dancers whose body fat drops so low that they stop menstruating and produce inadequate levels of estrogens, marathoners whose caloric intake is inadequate, teenagers on junk-food diets, individuals suffering from eating disorders, people allergic to dairy products, nursing mothers, and those exposed to prolonged treatment with cortisone. Often, the first symptom of osteoporosis is a pathological fracture. Bone mass becomes so depleted that the skeleton can no longer withstand the mechanical stresses of everyday living. For example, a hip fracture might result from sitting down too quickly. Osteoporosis causes more than 250,000 hip fractures a year, and complications from osteoporosis are the 12th leading cause of death in the United States. It is responsible for shrinkage of the backbone (vertebrae) and height loss, hunched backs, bone fractures, and considerable pain. Osteoporosis afflicts the entire skeletal system.

Risk factors for developing osteoporosis, besides race and gender, are (1) body build (short females are at greater risk since they have less total bone mass), (2) weight (thin females, anorectic females, and those who overdo exercise are at greater risk because adipose tissue is a source of an estrogen that retards bone loss), (3) smoking (smoking decreases blood levels of estrogens), (4) calcium deficiency or malabsorption, (5) vitamin D deficiency, (6) exercise (sedentary people are more likely to develop bone loss), (7) certain drugs (alcohol, some diuretics, cortisone, and tetracycline promote bone loss), (8) a family history of osteoporosis (daughters of women with osteoporosis have reduced bone mass), and (9) menopause (which may be premature due to excessive exercise).

Adequate diet and exercise are the mainstays for preventing osteoporosis. In postmenopausal women, estrogen replacement therapy (ERT; low doses of estrogen and sometimes progesterone, another sex hormone), calcium supplements, and weight-bearing exercise help prevent or retard the development of osteoporosis. The most important aspect of treatment is prevention. Adequate calcium intake and exercise in her early years may be more beneficial to a woman than ERT and calcium supplements when she is older.

PAGET'S DISEASE

Paget's disease is characterized by a greatly accelerated remodeling process in which osteoclastic resorption is massive and new bone formation by osteoblasts is extensive. As a result, there is an irregular thickening and softening of the bones and greatly increased vascularity, especially in bones of the skull, pelvis, and limbs.

MEDICAL TERMINOLOGY

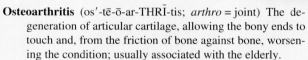

Osteoarthritis (os′-tē-ō-ar-THRĪ-tis; *arthro* = joint) The degeneration of articular cartilage, allowing the bony ends to touch and, from the friction of bone against bone, worsening the condition; usually associated with the elderly.

Osteogenic (os′-tē-Ō-JEN-ik) **sarcoma** (sar-KŌ-ma; *sarcoma* = connective tissue tumor) Bone cancer that primarily affects osteoblasts and occurs most often in the bones of teenagers during their growth spurt; most common sites are the metaphyses of the thighbone (femur), shinbone (tibia), and arm bone (humerus); metastases occur most often in lungs; treatment consists of multidrug chemotherapy and removal of the malignant growth or amputation of the limb.

Osteomyelitis (os′-tē-ō-mī-el-Ī-tis; *myelos* = marrow) Inflammation of a bone, especially the marrow, caused by a pathogenic organism, especially *Staphylococcus aureus*.

Osteopenia (os′-tē-ō-PĒ-nē-a; *osteo* = bone; *penia* = poverty) Reduced bone mass due to a decrease in the rate of bone synthesis to a level insufficient to compensate for normal bone breakdown; any decrease in bone mass below normal. An example is osteoporosis.

STUDY OUTLINE

PHYSIOLOGY: FUNCTIONS OF BONE (p. 143)

1. Bone (osseous) tissue and the skeletal system function in support, protection, movement, mineral homeostasis, red and white blood cell production, and storage of energy.

ANATOMY: STRUCTURE OF BONE (p. 143)

1. Structurally, the skeletal system contains bone, cartilage, bone marrow, and the periosteum.
2. Parts of a typical long bone are the diaphysis (shaft), proximal and distal epiphyses (ends), metaphyses, articular cartilage, periosteum, medullary (marrow) cavity, and endosteum.

HISTOLOGY OF BONE TISSUE (p. 143)

Introduction

1. Bone tissue consists of widely separated cells surrounded by large amounts of matrix.
2. The four principal types of cells are osteoprogenitor cells, osteoblasts, osteocytes, and osteoclasts.
3. The matrix of bone contains abundant mineral salts (mostly hydroxyapatite) and collagen fibers.

Compact Bone Tissue (p. 145)

1. Compact (dense) bone tissue consists of osteons (Haversian systems) with little space between them.
2. Compact bone lies over spongy bone and composes most of the bone tissue of the diaphysis. Functionally, compact bone protects, supports, and resists stress.

Spongy Bone Tissue (p. 147)

1. Spongy (cancellous) bone tissue does not contain osteons. It consists of trabeculae surrounding many red bone marrow-filled spaces.
2. It forms most of the structure of short, flat, and irregular bones, and the epiphyses of long bones.
3. Functionally, spongy bone stores red bone marrow and provides some support.

Blood and Nerve Supply (p. 147)

1. Long bones are supplied by nutrient, epiphyseal, and periosteal arteries; veins accompany the arteries.
2. Nerves accompany blood vessels in bone and are present in the periosteum.

BONE FORMATION: OSSIFICATION (p. 147)

1. Bone forms by a process called ossification (osteogenesis), which begins when mesenchymal cells become transformed into osteoprogenitor cells. These undergo cell division and give rise to cells that differentiate into osteoblasts and osteoclasts.
2. The process begins during the sixth or seventh week of embryonic life and continues throughout adulthood. The two types of ossification, intramembranous and endochondral, involve the replacement of a preexisting connective tissue with bone.
3. Intramembranous ossification occurs within loose fibrous connective tissue membranes of the embryo and the adult.
4. Endochondral ossification occurs within a hyaline cartilage model. The primary ossification center of a long bone is in the diaphysis. Cartilage degenerates, leaving cavities that merge to form the medullary cavity. Osteoblasts lay down bone. Next, ossification occurs in the epiphyses, where bone replaces cartilage, except for the epiphyseal plate.

BONE GROWTH (p. 151)

1. The anatomical zones of the epiphyseal plate are the zones of resting cartilage, proliferating cartilage, hypertrophic cartilage, and calcified cartilage.
2. Because of the activity of the epiphyseal plate, the diaphysis of a bone increases in length.
3. Bone grows in diameter as a result of the addition of new bone tissue by periosteal osteoblasts around the outer surface of the bone (appositional growth).
4. Growth hormone, insulin-like growth factors, estrogen, and testosterone stimulate bone growth.

BONE HOMEOSTASIS (p. 152)

Remodeling (p. 152)

1. Remodeling is the replacement of old bone tissue by new bone tissue.
2. Old bone is continually destroyed by osteoclasts, whereas new bone is constructed by osteoblasts.
3. Remodeling requires minerals (calcium, phosphorus, magnesium, boron, and manganese), vitamins (A, B_{12}, C, and D), and hormones (human growth hormone, sex hormones, insulin, insulin-like growth factors, thyroid hormones, parathyroid hormone, and calcitonin).

Fracture and Repair of Bone (p. 153)

1. A fracture is any break in a bone.
2. The types of fractures include partial, complete, closed (simple), open (compound), comminuted, greenstick, spiral, transverse, impacted, Pott's, Colles', displaced, nondisplaced, and stress.
3. Fracture repair involves formation of a fracture hematoma, fibrocartilaginous callus, bone callus, and remodeling.

Bone's Role in Calcium Homeostasis (p. 155)

1. Bone is the major reservoir for calcium (Ca^{2+}) in the body.
2. Parathyroid hormone (PTH) secreted by the parathyroid gland increases blood Ca^{2+} level.
3. Calcitonin (CT) secreted by the thyroid gland decreases blood Ca^{2+} level.

EXERCISE AND BONE TISSUE (p. 156)

1. Mechanical stress increases bone strength by increasing deposition of mineral salts and production of collagen fibers.
2. Removal of mechanical stress weakens bone through demineralization and collagen reduction.

AGING AND BONE TISSUE (p. 156)

1. The principal effect of aging is a loss of calcium from bones, which may result in osteoporosis.
2. Another effect is a decreased production of matrix proteins (mostly collagen), which makes bones more susceptible to fracture.

DEVELOPMENTAL ANATOMY OF THE SKELETAL SYSTEM (p. 157)

1. Bone forms from mesoderm by intramembranous or endochondral ossification.
2. Limbs develop from limb buds, which consist of mesoderm and ectoderm.

REVIEW QUESTIONS

1. List and describe the six principal functions of bone tissue. (p. 143)
2. Diagram the parts of a long bone and list the functions of each part. (p. 144)
3. Describe the four types of cells in bone tissue. (p. 144)
4. What is the composition of the matrix of bone tissue? (p. 145)
5. Distinguish between spongy and compact bone in terms of microscopic appearance, location, and function. (p. 145)
6. Diagram the microscopic appearance of compact bone and indicate the functions of the various components. (p. 145)
7. What is meant by ossification? Describe the initial events of ossification. (p. 147)
8. Outline the major events of intramembranous and endochondral ossification and explain the main differences. (p. 147)
9. Describe the histology of the various zones of the epiphyseal plate. How does the plate grow? What is the significance of the epiphyseal line? (p. 151)
10. Explain how bone grows in thickness or diameter. (p. 152)
11. How do hormones regulate bone growth? (p. 152)
12. Define remodeling and describe the factors that contribute to the process. (p. 152)
13. What is a fracture? Distinguish several principal kinds of fractures. Outline the three basic steps of fracture repair. (p. 153)
14. Describe the role of bone tissue in calcium homeostasis. (p. 155)
15. Explain the effects of exercise and aging on the skeletal system. (p. 156)
16. Describe the development of the skeletal system. (p. 157)
17. What are the principal symptoms and causes of osteoporosis and Paget's disease? (p. 159)

ANSWERS TO QUESTIONS WITH FIGURES

6.1 The periosteum is essential for growth in bone diameter, bone repair, and bone nutrition. It also serves as a point of attachment for ligaments and tendons.

6.2 It is necessary for the development, growth, maintenance, and repair of bone.

6.3 Because the central canals are the main blood supply to the osteocytes of an osteon, their blockage would lead to death of the osteocytes.

6.4 Flat bones of the skull, mandible (lower jawbone), and clavicles (collarbones).

6.5 Yes. The epiphyseal lines are indications of growth zones that have ceased to function. The absence of epiphyseal lines indicates that bone is still lengthening.

6.6 Cell divisions in the zone of proliferating cartilage and maturation of the cells in the zone of hypertrophic cartilage.

6.7 Calcium and phosphorus deposition is a slow process, and bone cells generally grow and reproduce slowly.

6.8 A mass of repair tissue that bridges the ends of broken bones.

6.10 Heartbeat, respiration, nerve cell functioning, enzyme functioning, and blood clotting, to name just a few.

CHAPTER 7

THE SKELETAL SYSTEM: THE AXIAL SKELETON

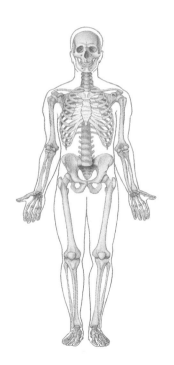

CHAPTER CONTENTS AT A GLANCE

ithout the skeletal system, you would be unable to perform movements such as walking or grasping. The slightest jar to your head or chest could damage the brain or heart. It would even be impossible to chew food. The framework of bones and cartilage that protects organs and allows movement is called the **skeletal** (*skeletos* = dried up) **system**. Since the skeletal system forms the framework of the body, a familiarity with the names, shapes, and positions of individual bones will help you understand some of the other organ systems. For example, the radial artery, the site where pulse is usually taken, is named for its proximity to the radius, the lateral bone of the forearm. The frontal lobe of the brain lies deep to the frontal (forehead) bone. The tibialis anterior muscle is located near the anterior surface of the tibia (shinbone). The ulnar nerve is named for its proximity to the ulna, the medial bone of the forearm.

Movements such as throwing a ball, biking, and walking require the coordinated use of bones and muscles. To understand how muscles produce different movements, you need to learn where on bones the muscles attach and the types of joints acted on by the contracting muscles. Many bones also serve as anatomical and surgical landmarks. For example, parts of certain bones serve to locate structures within the skull and to outline the lungs and heart, and abdominal and pelvic viscera. Blood vessels and nerves often run parallel to bones. These structures can be located more easily if the bone is identified first.

The specialized branch of medicine that deals with the preservation and restoration of the skeletal system, joints, and associated structures is called **orthopedics** (or′-thō-PĒ-diks; *ortho* = correct or straighten; *pais* = child).

TYPES OF BONES

Almost all the bones of the body may be classified into four principal types on the basis of shape: long, short, flat, and irregular. **Long bones** (see Fig. 6.1) have greater length than width and consist of a shaft and a variable number of extremities (ends). They are slightly curved for strength. A curved bone absorbs the stress of the body weight at several different points so the stress is evenly distributed. If such bones were straight, the weight of the body would be unevenly distributed and the bone would fracture easily. Long bones consist mostly of *compact bone tissue,* which is dense and has few spaces. They also contain considerable amounts of *spongy bone tissue,* which has larger spaces. Long bones include those in the thigh (femur), leg (tibia and fibula), toes (phalanges), arm (humerus), forearm (ulna and radius), and fingers (phalanges).

Short bones are somewhat cube-shaped and nearly equal in length and width. They are spongy bone except at the surface, where there is a thin layer of compact bone. Examples of short bones are the wrist (carpal) and ankle (tarsal) bones (see Figs. 8.6 and 8.14).

Flat bones are generally thin and composed of two nearly parallel plates of compact bone enclosing a layer of spongy bone. Flat bones afford considerable protection and provide extensive areas for muscle attachment. Flat bones include the cranial bones (which protect the brain) and the breastbone (sternum) and ribs (which protect organs in the thorax and are shown in Fig. 7.22), and the shoulder blades (scapulae).

Irregular bones have complex shapes and cannot be grouped into any of the three categories just described. They also vary in the amount of spongy and compact bone present. Such bones include the backbones (vertebrae), shown in Fig. 7.17, and certain facial bones.

Two additional types of bones are not included in this classification by shape but instead are classified by location. **Sutural** (SOO-chur-al; *sutura* = seam) or **Wormian** (named after a Dutch anatomist, O. Wormian) **bones** are small bones located within the joints (sutures) of certain cranial bones (see Fig. 7.5). Their number varies greatly from person to person. **Sesamoid** (*sesamoides* = shaped like a sesame seed) **bones** are small bones that are embedded in tendons where considerable pressure develops, for instance, in the thumb and great toe. These, like sutural bones, are also variable in number. Some sesamoid bones change the direction of pull of a tendon. Two sesamoid bones, the kneecaps (patellae), are present in everyone (see Fig. 8.12).

BONE SURFACE MARKINGS

The surfaces of bones have various structural features adapted to specific functions. These features are called **surface markings**. Long bones that bear a lot of weight have large, rounded ends that can form sturdy joints and provide adequate surface area for the attachment of ligaments and muscles. Other bones have depressions that receive the rounded ends. Rough areas serve as points of attachment for muscles, tendons, and ligaments. Grooves on the surfaces of bones provide for the passage of blood vessels. Openings occur where blood vessels and nerves pass into or through the bone. Exhibit 7.1 describes the different bone surface markings and their functions.

DIVISIONS OF THE SKELETAL SYSTEM

The adult human skeleton consists of 206 named bones grouped in two principal divisions: the **axial skeleton** and the **appendicular skeleton**. Refer to Fig. 7.1 to see how the two divisions join to form the complete skeleton. The bones of the axial skeleton are shown in blue. The longitudinal **axis**, or center, of the human body is a straight line that runs through the body's center of gravity. This imaginary line extends

EXHIBIT **7.1** **BONE SURFACE MARKINGS**

Marking	Description	Example
DEPRESSIONS AND OPENINGS		
Fissure (FISH-ur)	A narrow, cleftlike opening between adjacent parts of bones through which blood vessels or nerves pass.	Superior orbital fissure of the sphenoid bone (Fig. 7.14).
Fontanel (fon-ta-NEL; *fontenelle* = little fountain)	Space between skull bones at birth, filled with dense fibrous connective tissue.	Anterior fontanel between frontal and parietal bones (Fig. 7.6).
Foramen (fo-RĀ-men; *foramen* = hole)	An opening through which blood vessels, nerves, or ligaments pass.	Infraorbital foramen of the maxilla (Fig. 7.2).
Fossa (*fossa* = basinlike depression)	A depression in or on a bone.	Mandibular fossa of the temporal bone (Fig. 7.7).
Sulcus (*sulcus* = ditchlike groove)	A groove that accommodates a soft structure such as a blood vessel, nerve, or tendon.	Intertubercular sulcus of the humerus (Fig. 8.4).
Meatus (mē-Ā-tus; *meatus* = passageway) or **canal**	A tubelike passageway within a bone.	External auditory (acoustic) meatus of the temporal bone (Fig. 7.3).
Paranasal sinus (*sinus* = cavity or hollow place)	An air-filled cavity within a bone that connects to the nasal cavity.	Frontal sinus of the frontal bone (Fig. 7.12).
PROCESSES		
PROCESSES THAT FORM JOINTS		
Condyle (KON-dil; *condylus* = knucklelike process)	A large, rounded articular prominence.	Medial condyle of the femur (Fig. 8.10).
Facet	A smooth, flat surface.	Articular facet of a vertebra for the tubercle of rib (Fig. 7.19).
Head	A rounded articular projection supported on the constricted portion (neck) of a bone.	Head of the femur (Fig. 8.10).
PROCESSES TO WHICH TENDONS, LIGAMENTS, AND OTHER CONNECTIVE TISSUES ATTACH		
Crest	A prominent border or ridge.	Iliac crest of the hipbone (Fig. 8.7).
Epicondyle (*epi* = above)	A prominence above a condyle.	Medial epicondyle of the femur (Fig. 8.10).
Linea (line)	A less prominent ridge than a crest.	Linea aspera of the femur (Fig. 8.10).
Spinous process (**spine**)	A sharp, slender process.	Spinous process of a vertebra (Fig. 7.17).
Trochanter (trō-KAN-ter)	A very large projection found only on the femur.	Greater trochanter of the femur (Fig. 8.10).
Tubercle (TOO-ber-kul; *tube* = knob)	A small, rounded process.	Greater tubercle of the humerus (Fig. 8.4).
Tuberosity	A large, rounded, usually roughened process.	Ischial tuberosity of the hipbone (Fig. 8.7).

through the head and down to the space between the feet. The axial skeleton consists of the bones that lie around the axis: skull bones, hyoid bone, ribs, sternum (breastbone), and vertebrae (bones of the backbone). Although the auditory ossicles (ear bones) are not considered part of the axial or appendicular skeleton, but rather as a separate group of bones, they are placed with the axial skeleton for convenience. The middle portion of each ear contains three auditory ossicles held together by a series of ligaments. The auditory ossicles are exceedingly small bones named for their shapes. Their names are the malleus, incus, and stapes, commonly called the hammer, anvil, and stirrup, respectively. They vibrate in response

to sound waves that strike the eardrum and have a key role in the mechanism of hearing. This is described in detail in Chapter 16.

The appendicular skeleton contains the bones of the **upper** and **lower limbs** (**extremities**), plus the bones called **girdles** whose function is to connect the limbs to the axial skeleton. Exhibit 7.2 on page 166 presents the standard grouping of the 80 bones of the axial skeleton and the 126 bones of the appendicular skeleton.

We will study bones by examining the various regions of the body. First, we will look at the skull and see how its bones relate to each other. We will then move on to the vertebral

Figure 7.1 Divisions of the skeletal system. The axial skeleton is indicated in blue. Note the position of the hyoid bone (not shown here) in Fig. 7.3.

🔑 *The adult human skeleton consists of 206 bones grouped into axial and appendicular divisions.*

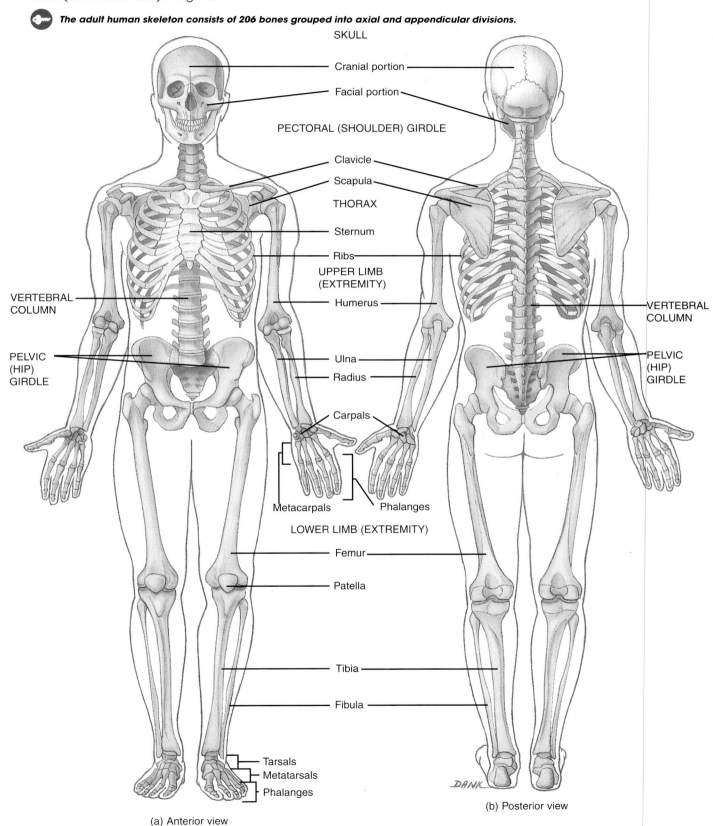

(a) Anterior view

(b) Posterior view

Q Which of the following bones are part of the axial or the appendicular skeleton: skull, clavicle, vertebral column, shoulder girdle, humerus, pelvic girdle, and femur?

EXHIBIT 7.2 DIVISIONS OF THE ADULT SKELETAL SYSTEM

Regions of the Skeleton	Number of Bones
AXIAL SKELETON	
Skull	
Cranium	8
Face	14
Hyoid	1
Auditory ossicles	6
Vertebral column	26
Thorax	
Sternum	1
Ribs	24
	Subtotal = 80
APPENDICULAR SKELETON	
Pectoral (shoulder) girdles	
Clavicle	2
Scapula	2
Upper limbs (extremities)	
Humerus	2
Ulna	2
Radius	2
Carpals	16
Metacarpals	10
Phalanges	28
Pelvic (hip) girdle	
Hip, pelvic, or coxal bone	2
Lower limbs (extremities)	
Femur	2
Fibula	2
Tibia	2
Patella	2
Tarsals	14
Metatarsals	10
Phalanges	28
	Subtotal = 126
	Total = 206

column (backbone) and the chest. In Chapter 8 we will examine the bones of the pectoral (shoulder) girdle, upper limbs, pelvic (hip) girdle, and lower limbs. This regional approach will allow you to see how the many bones of the body relate to one another.

SKULL

The **skull**, which contains 22 bones, rests on the superior end of the vertebral column. It includes two sets of bones: cranial bones and facial bones. The **cranial** (*cranium* = brain case) **bones** form the cranial cavity and enclose and protect the brain. The 8 cranial bones are the frontal bone, parietal bones (2), temporal bones (2), occipital bone, sphenoid bone, and ethmoid bone. There are 14 **facial bones** that form the face: nasal bones (2), maxillae or maxillas (2), zygomatic bones (2), mandible, lacrimal bones (2), palatine bones (2), inferior nasal conchae (2), and vomer. Be sure you can locate all the skull bones in the various views of the skull (Figs. 7.2–7.5 on pages 167–170).

Overview

Besides the large cranial cavity, the skull also contains several smaller cavities. These include the nasal cavity and orbits (eye sockets), which open to the exterior. Certain skull bones also contain paired cavities that are lined with mucous membranes and are called paranasal sinuses. The sinuses open into the nasal cavity. Also within the skull are small cavities that house the structures involved in hearing and equilibrium.

Other than the auditory ossicles, the mandible is the only movable bone of the skull. Most of the skull bones are held together by immovable joints called sutures. These are especially noticeable on the outer surfaces of the bones.

The skull has numerous surface markings such as foramina and fissures through which blood vessels and nerves pass. As various skull bones are described, you will learn the names of important surface markings.

Functions of Cranial and Facial Bones

In addition to protecting the brain, the cranial bones also have other functions. Their inner surfaces attach to membranes (meninges) that stabilize the position of the brain, blood vessels, and nerves. Their outer surfaces provide large areas of attachment for muscles that move various parts of the head.

Besides forming the framework of the face, the facial bones protect and provide support for the entrances to the digestive and respiratory systems. The bones also provide attachment for muscles that are involved in producing various facial expressions, such as frowning, fear, surprise, and happiness.

Together, the cranial and facial bones protect and support the delicate special sense organs for vision, taste, smell, hearing, and equilibrium (balance).

Figure 7.2 Skull.

🔑 *The skull consists of two sets of bones: cranial and facial.*

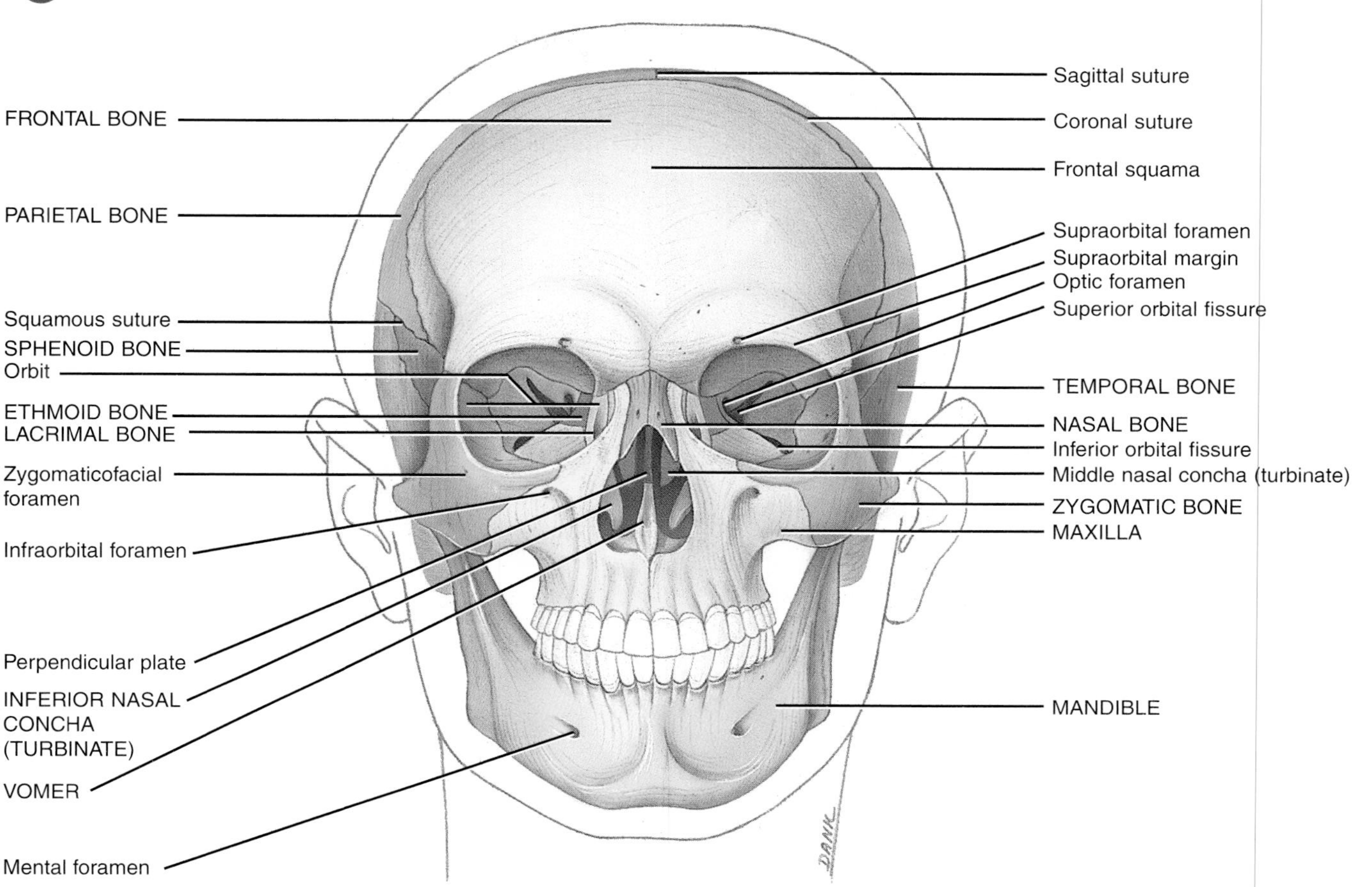

FRONTAL BONE

PARIETAL BONE

Squamous suture
SPHENOID BONE
Orbit

ETHMOID BONE
LACRIMAL BONE

Zygomaticofacial foramen

Infraorbital foramen

Perpendicular plate

INFERIOR NASAL CONCHA (TURBINATE)

VOMER

Mental foramen

Sagittal suture

Coronal suture

Frontal squama

Supraorbital foramen
Supraorbital margin
Optic foramen
Superior orbital fissure

TEMPORAL BONE

NASAL BONE
Inferior orbital fissure
Middle nasal concha (turbinate)
ZYGOMATIC BONE
MAXILLA

MANDIBLE

Anterior view

Ⓠ Which of the bones shown here are cranial bones?

Sutures

A **suture** (SOO-chur; *sutura* = seam) is an immovable joint that is found only between skull bones. Sutures hold all skull bones together. There are several types of sutures that can be distinguished according to how the margins of the bones unite. In some sutures, the margins of the bones are fairly smooth. In other sutures, the margins overlap. In still other sutures, the margins interlock in a jigsaw fashion. This latter arrangement provides sutures with added strength and decreases their chance of fracturing.

The names of many sutures reflect the bones that they unite. For example, the frontozygomatic suture is between the frontal bone and zygomatic bone. Similarly, the sphenoparietal suture is between the sphenoid bone and parietal bone. In other cases, however, the names of sutures are not so obvious.

Of the many sutures that are found in the skull, we will identify only four prominent ones (Figs. 7.2–7.6):

1. **Coronal** (kō-RŌ-nal; *corona* = crown) **suture.** This suture unites the frontal bone and two parietal bones. See Fig. 7.3.
2. **Sagittal** (SAJ-i-tal; *sagitta* = arrow) **suture.** This suture unites the two parietal bones. The sagittal suture is so named because in the infant, before the bones of the skull are firmly united, the suture and the fontanels (soft spots) associated with it somewhat resemble an arrow (see Fig. 7.6).
3. **Lambdoid** (LAM-doyd) **suture.** This suture unites the parietal bones and occipital bones. The suture is so named because of its resemblance to the Greek letter lambda (Λ), as can be seen in Fig. 7.5.

Figure 7.3   Skull. Although the hyoid bone is not part of the skull, it is included in the illustration for reference.

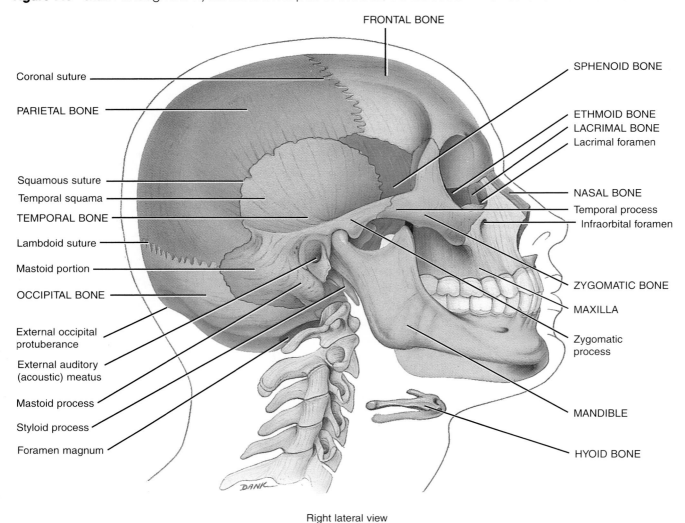

FRONTAL BONE

SPHENOID BONE

Coronal suture

ETHMOID BONE
LACRIMAL BONE
Lacrimal foramen

PARIETAL BONE

Squamous suture

NASAL BONE

Temporal squama

Temporal process

TEMPORAL BONE

Infraorbital foramen

Lambdoid suture

Mastoid portion

ZYGOMATIC BONE

OCCIPITAL BONE

MAXILLA

External occipital
protuberance

Zygomatic
process

External auditory
(acoustic) meatus

Mastoid process

MANDIBLE

Styloid process

Foramen magnum

HYOID BONE

Right lateral view

Q   What are the major bones on either side of the squamous suture, the lambdoid suture, and coronal suture?

4.   **Squamous** (SKWĀ-mus; *squama* = flat) **suture.** This suture unites the parietal bones and temporal bones. It is so named because the part of the temporal bone that unites with the parietal bone to form the suture is a thin, flat region called the temporal squama.

There may be sutural bones in the sagittal and lambdoid sutures.

Fontanels

The skeleton of a newly formed embryo consists of cartilage or fibrous connective tissue membrane structures shaped like bones. Gradually, bone replaces the cartilage or fibrous con-

nective tissue membrane in a process called ossification. At birth, membrane-filled spaces called **fontanels** (fon'-ta-NELZ; *fontenelle* = little fountain) still exist between cranial bones (Fig. 7.6 on page 170). These "soft spots" are areas of fibrous connective tissue membranes that will eventually be replaced with bone by intramembranous ossification and become sutures. Functionally, the fontanels enable the fetal skull to modify its size and shape as it passes through the birth canal and permit rapid growth of the brain during infancy. In addition, fontanels help a physician gauge the degree of brain development by their state of closure and serve as landmarks (anterior fontanel) for withdrawal of blood for analysis from the superior sagittal sinus (a large vein around the brain). Although an infant may have many fontanels at birth, the form and location of several are fairly constant (Exhibit 7.3 on page 171).

Figure 7.4 Skull.

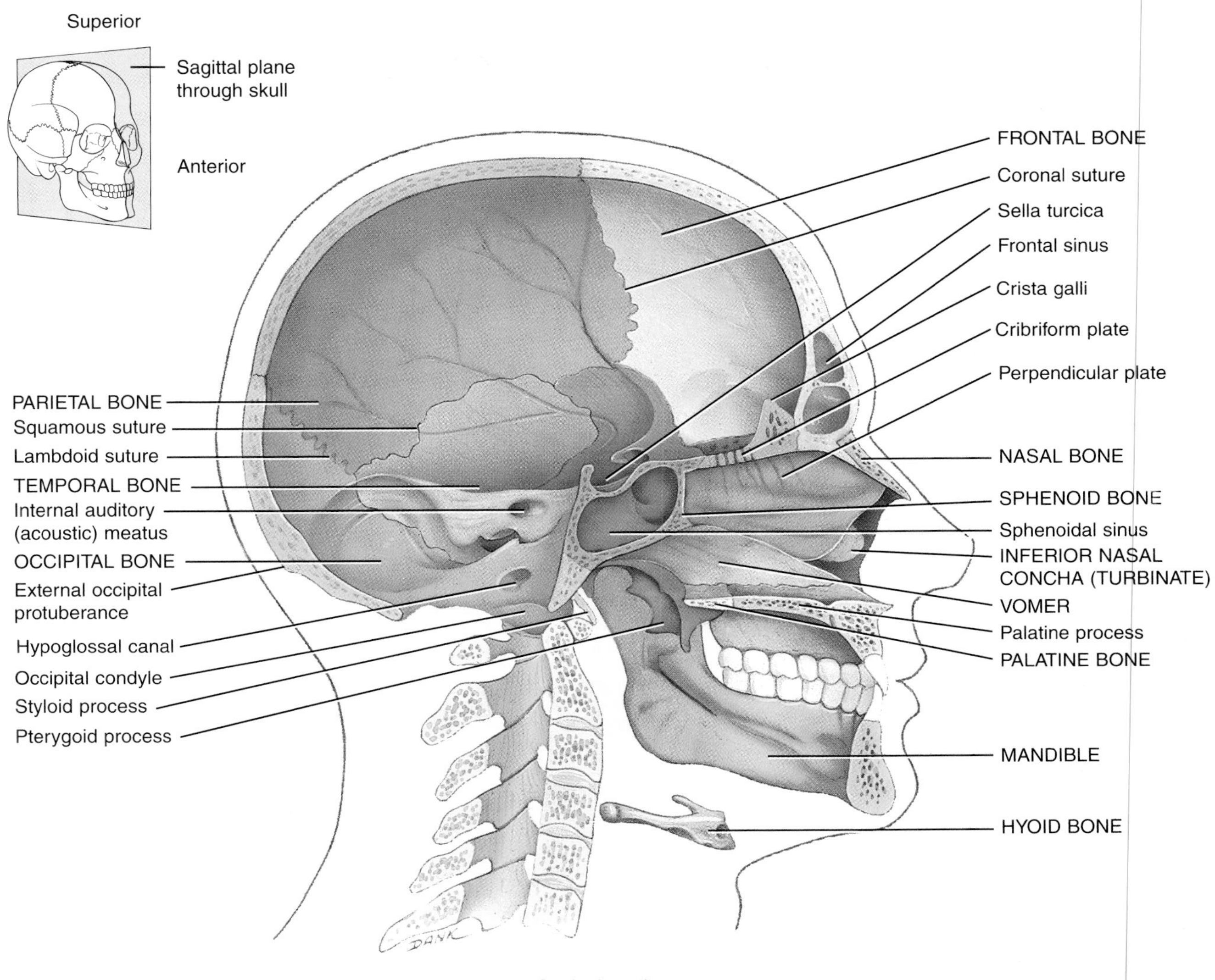

Superior

Sagittal plane through skull

Anterior

FRONTAL BONE
Coronal suture
Sella turcica
Frontal sinus
Crista galli
Cribriform plate
Perpendicular plate

NASAL BONE

SPHENOID BONE
Sphenoidal sinus
INFERIOR NASAL CONCHA (TURBINATE)
VOMER
Palatine process
PALATINE BONE

MANDIBLE

HYOID BONE

PARIETAL BONE
Squamous suture
Lambdoid suture
TEMPORAL BONE
Internal auditory (acoustic) meatus
OCCIPITAL BONE
External occipital protuberance
Hypoglossal canal
Occipital condyle
Styloid process
Pterygoid process

Sagittal section

Q Which of the bones shown here are facial bones?

Cranial Bones

Frontal Bone

The **frontal bone** forms the forehead (the anterior part of the cranium), the roofs of the **orbits** (eye sockets), and most of the anterior part of the cranial floor. Soon after birth, the left and right sides of the frontal bone are united by the **frontal suture** (see Fig. 7.6a), which usually disappears by age 6. If it persists throughout life, it is referred to as the **metopic suture**.

If you examine Fig. 7.2, you will note the **frontal squama**. This scalelike plate forms the forehead. It gradually slopes inferiorly from the coronal suture, then curves abruptly to become almost vertical. Superior to the orbits, the frontal bone thickens, forming the **supraorbital** (*supra* = above) **margin**. From this margin the frontal bone extends posteriorly to form the roof of the orbit and part of the floor of the cranial cavity. Within the supraorbital margin, slightly medial to its midpoint, is a hole called the **supraorbital foramen** (plural is **foramina**). As the various foramina associated with cranial bones are discussed, refer to Exhibit 7.4 on page 180 to note which structures pass through them. The **frontal sinuses** lie deep to the frontal squama. Among other functions, paranasal sinuses act as sound chambers that give the voice resonance.

Figure 7.5   Skull. The sutures are exaggerated for emphasis.

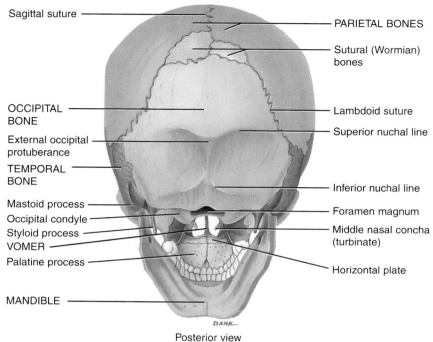

Posterior view

Which sutures may contain sutural bones?

Figure 7.6   Fontanels of the skull at birth.

Fontanels are membrane-filled spaces between cranial bones that are present at birth.

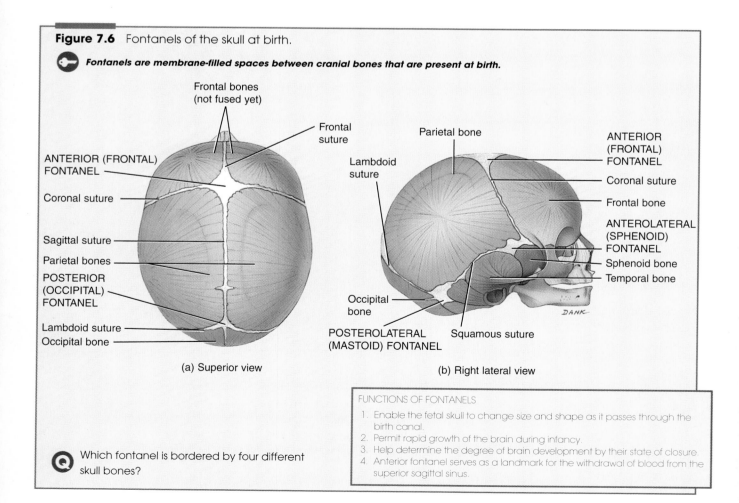

(a) Superior view

(b) Right lateral view

FUNCTIONS OF FONTANELS
1. Enable the fetal skull to change size and shape as it passes through the birth canal.
2. Permit rapid growth of the brain during infancy.
3. Help determine the degree of brain development by their state of closure.
4. Anterior fontanel serves as a landmark for the withdrawal of blood from the superior sagittal sinus.

Which fontanel is bordered by four different skull bones?

Figure 7.9 Ethmoid bone.

 The ethmoid bone is the major supporting structure of the nasal cavity.

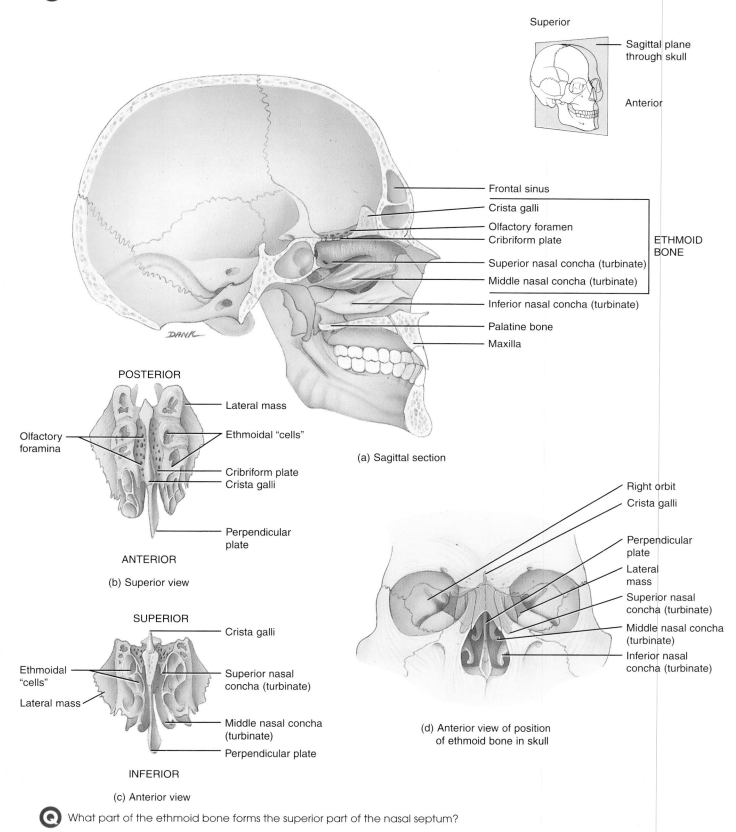

Superior

Sagittal plane through skull

Anterior

Frontal sinus

Crista galli

Olfactory foramen

Cribriform plate

ETHMOID BONE

Superior nasal concha (turbinate)

Middle nasal concha (turbinate)

Inferior nasal concha (turbinate)

Palatine bone

Maxilla

(a) Sagittal section

POSTERIOR

Lateral mass

Ethmoidal "cells"

Olfactory foramina

Cribriform plate

Crista galli

Perpendicular plate

ANTERIOR

(b) Superior view

SUPERIOR

Crista galli

Ethmoidal "cells"

Superior nasal concha (turbinate)

Lateral mass

Middle nasal concha (turbinate)

Perpendicular plate

INFERIOR

(c) Anterior view

Right orbit

Crista galli

Perpendicular plate

Lateral mass

Superior nasal concha (turbinate)

Middle nasal concha (turbinate)

Inferior nasal concha (turbinate)

(d) Anterior view of position of ethmoid bone in skull

Q What part of the ethmoid bone forms the superior part of the nasal septum?

Figure 7.10   Cranial fossae.

Cranial fossae contain depressions for brain convolutions, blood vessels, and foramina.

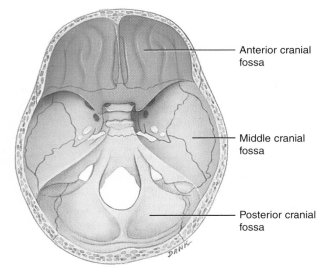

Anterior cranial fossa

Middle cranial fossa

Posterior cranial fossa

Q Which cranial fossa contains the cerebellum?

and has a small median portion and two expanded lateral positions. The median portion is formed by part of the body of the sphenoid bone, and the lateral portions are formed by the greater wings of the sphenoid bone, temporal squama, and parietal bone. The middle cranial fossa cradles the temporal lobes of the cerebral hemispheres. The last fossa, at the most inferior level, is the **posterior cranial fossa**, the largest of the fossae. It is formed largely by the occipital bone and the petrous and mastoid portions of the temporal bone. It is a very deep fossa that accomodates the cerebellum, pons, and medulla oblongata of the brain.

Facial Bones

The shape of the face changes dramatically during the first two years after birth. The brain and cranial bones expand, the teeth form and erupt, and the paranasal sinuses increase in size. Growth of the face ceases at about 16 years of age.

Nasal Bones

The paired **nasal bones** meet at the midline (see Fig. 7.2) and form part of the bridge of the nose. The major portion of the nose consists of cartilage.

Maxillae

The paired **maxillae** (mak-SIL-ē; *macerae* = to chew; singular is **maxilla**) unite to form the upper jawbone (Fig. 7.11) and articulate with every bone of the face except the mandible, or lower jawbone. They form part of the floors of

the orbits, part of the lateral walls and floor of the nasal cavity, and most of the hard palate. The hard palate is a bony partition formed by the maxillae and palatine bones and comprises the roof of the mouth.

Each maxilla contains a **maxillary sinus** that empties into the nasal cavity (see Fig. 7.12). The **alveolar** (al-VĒ-ō-lar; *alveolus* = hollow) **process** is an arch that contains the **alveoli** (sockets) for the maxillary (upper) teeth. The **palatine process** is a horizontal projection of the maxilla that forms the anterior three-quarters of the hard palate. The maxillary bones unite, and the fusion is normally completed before birth.

A fissure associated with the maxilla and sphenoid bone is the **inferior orbital fissure**. It separates the greater wing of the sphenoid and the maxilla (see Fig. 7.14).

CLEFT PALATE AND CLEFT LIP

Usually the palatine processes of the maxillary bones unite during weeks 10–12 of embryonic development. Failure to do so can result in a condition called **cleft palate**. The condition may also involve incomplete fusion of the horizontal plates of the palatine bones (see Fig. 7.7). Another form of this condition, called **cleft lip**, involves a split in the upper lip. Cleft lip and cleft palate often occur together. Depending on the extent and position of the cleft, suckling of an infant, speech, and swallowing may be affected. Facial and oral surgeons recommend closure of cleft lip in utero or during the first year of life, and surgical results are excellent. Repair of cleft palate is done between the first and second year of life, ideally before the child begins to talk. Orthodontic therapy may be needed to align the teeth. Here again, results usually are excellent. ■

Paranasal Sinuses

Although they are not cranial or facial bones, this is an appropriate point to discuss the **paranasal** (*para* = beside) **sinuses**. These are paired cavities in certain cranial and facial bones near the nasal cavity (Fig. 7.12). The paranasal sinuses are lined with mucous membranes that are continuous with the lining of the nasal cavity. Skull bones containing paranasal sinuses are the frontal, sphenoid, ethmoid, and maxillae. (The paranasal sinuses were described in the discussion of each of these bones.) Besides producing mucus, the paranasal sinuses serve as resonating chambers for sound as we speak or sing.

SINUSITIS

Secretions produced by the mucous membranes of the paranasal sinuses drain into the nasal cavity. An inflammation of the membranes due to an allergic reaction or infection is called **sinusitis**. If the membranes swell enough to block drainage into the nasal cavity, fluid pressure builds up in the paranasal sinuses, and a sinus headache results. ■

Figure 7.11 Maxillae.

🔑 *The maxillae articulate with every bone of the face, except the mandible.*

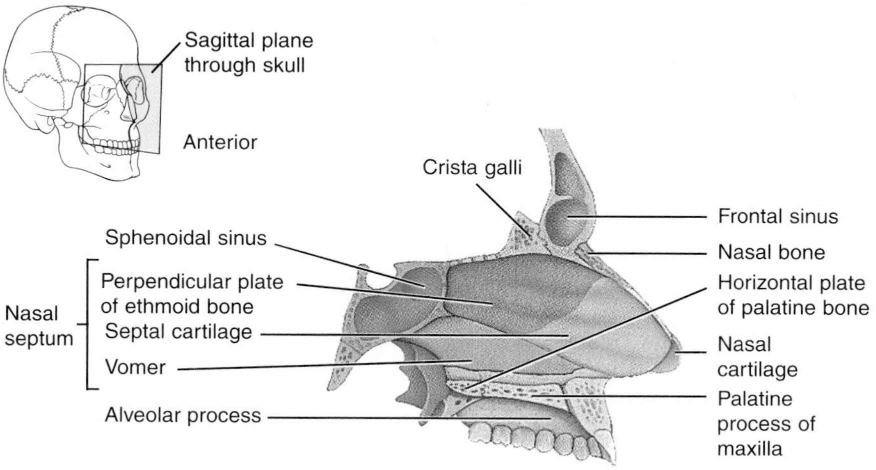

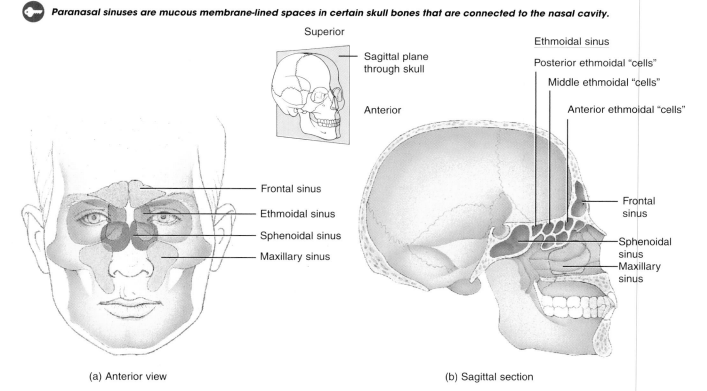

(a) Sagittal section

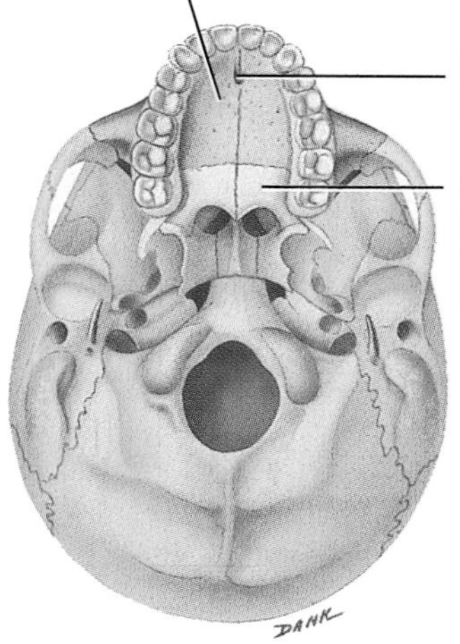

(b) Inferior view

❓ What bones form the hard palate?

Figure 7.12 Paranasal sinuses.

🔑 *Paranasal sinuses are mucous membrane-lined spaces in certain skull bones that are connected to the nasal cavity.*

(a) Anterior view

(b) Sagittal section

❓ What are the two main functions of the paranasal sinuses?

Zygomatic Bones

The two **zygomatic bones,** commonly called cheekbones, form the prominences of the cheeks and part of the lateral wall and floor of each orbit (see Fig. 7.14). They articulate with the frontal, maxilla, sphenoid, and temporal bones.

The **temporal process** of the zygomatic bone projects posteriorly and articulates with the zygomatic process of the temporal bone to form the **zygomatic arch** (see Fig. 7.7).

Mandible

The **mandible** (*mandere* = to chew), or lower jawbone, is the largest, strongest facial bone (Fig. 7.13). It is the only skull bone that moves (other than the auditory ossicles).

In the lateral view, you can see that the mandible consists of a curved, horizontal portion, the **body**, and two perpendicular portions, the **rami** (RĀ-mi; singular is **ramus** = branch). The **angle** of the mandible is the area where each ramus meets the body. Each ramus has a posterior **condylar** (KON-di-lar) **process** that articulates with the mandibular fossa and articular tubercle of the temporal bone to form the temporomandibular joint (TMJ). It also has an anterior **coronoid** (KOR-ō-noyd) **process** to which the temporalis muscle attaches. The depression between the coronoid and condylar processes is called the **mandibular notch**. The **alveolar process** is an arch containing the **alveoli** (sockets) for the mandibular (lower) teeth.

The **mental** (*mentum* = chin) **foramen** is approximately inferior to the second premolar tooth. It is near this foramen that dentists reach the mental nerve when injecting anesthetics. A second foramen associated with the mandible is the **mandibular foramen** on the medial surface of the ramus, another site often used by dentists to inject anesthetics. The

mandibular foramen is the beginning of the **mandibular canal**, which runs anteriorly in the ramus deep to the roots of the teeth.

TEMPOROMANDIBULAR JOINT SYNDROME

One problem associated with the temporomandibular joint (TMJ) is **TMJ syndrome**. It is characterized by dull pain around the ear, tenderness of the jaw muscles, a clicking or popping noise when opening or closing the mouth, limited or abnormal opening of the mouth, headache, tooth sensitivity, and abnormal wearing of the teeth. TMJ syndrome can be caused by improperly aligned teeth, grinding or clenching the teeth, trauma to the jaw, or arthritis. Treatment may involve application of moist heat or ice, a soft diet, taking aspirin, muscle retraining, adjusting or reshaping the teeth, orthodontic treatment, or surgery. ■

Lacrimal Bones

The paired **lacrimal** (LAK-ri-mal; *lacrima* = teardrop) **bones** are thin and roughly resemble a fingernail in size and shape. They are the smallest bones of the face. These bones are posterior and lateral to the nasal bones and they form a part of the medial wall of each orbit. The lacrimal bones each contain a **lacrimal foramen** through which a nasolacrimal or tear (TEER) duct passes (see Fig. 7.3). They can be seen in the anterior and lateral views of the skull in Figs. 7.2 and 7.3.

Palatine Bones

The two **palatine** (PAL-a-tīn) **bones** are L-shaped. They form the posterior portion of the hard palate, part of the

---

Figure 7.13 Mandible.

 The mandible is the largest and strongest facial bone.

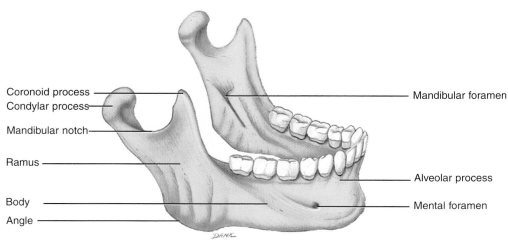

Coronoid process
Condylar process
Mandibular notch
Ramus
Body
Angle
Mandibular foramen
Alveolar process
Mental foramen

Lateral view

Q What is the distinctive feature of the mandible among all the skull bones?

floor and lateral wall of the nasal cavity, and a small portion of the floors of the orbits. The posterior portion of the hard palate, which separates the nasal cavity from the oral cavity, is formed by the **horizontal plates** of the palatine bones. These can be seen in Fig. 7.7.

Inferior Nasal Conchae

Refer to the views of the skull in Figs. 7.2 and 7.9a. The two **inferior nasal conchae** (KONG-kē) or **turbinates** are scroll-like bones that form a part of the lateral wall of the nasal cavity and project into the nasal cavity inferior to the superior and middle nasal conchae of the ethmoid bone. They serve the same function as the superior and middle nasal conchae of the ethmoid bone; that is, they promote turbulent circulation and filter and warm air before it passes into the lungs. The inferior nasal conchae are separate bones and not part of the ethmoid.

Vomer

The **vomer** (= plowshare) is a roughly triangular bone that forms the inferior and posterior part of the nasal septum. It is clearly seen in the anterior view of the skull in Fig. 7.2 and the inferior view in Fig. 7.7.

The vomer articulates with the septal cartilage of the nasal septum that divides the external nose into right and left sides. Its superior border articulates with the perpendicular plate of the ethmoid bone. The structures that form the **nasal septum** (*septum* = partition) are the perpendicular

plate of the ethmoid, septal cartilage, vomer, and parts of the palatine bones and maxillae (see Fig. 7.10a).

DEVIATED NASAL SEPTUM

A **deviated nasal septum** (**DNS**) is one that is deflected laterally from the midline of the nose. The deviation usually occurs at the junction of bone with the septal cartilage. A DNS may occur as a result of a developmental abnormality or trauma. If the deviation is severe, it may entirely block the nasal passageway. Even a partial blockage may lead to infection. If inflammation occurs, it may cause nasal congestion, blockage of the paranasal sinus openings, chronic sinusitis, headache, and nosebleeds. ■

Orbits

Each **orbit** (eye socket) is a pyramid-shaped space that contains the eyeball and associated structures. It is formed by seven bones of the skull (Fig. 7.14) and has four walls that converge posteriorly to form an apex (posterior end). The roof of the orbit consists of parts of the frontal and sphenoid bones. Portions of the zygomatic and sphenoid bones form the lateral wall. The floor of the orbit is formed by parts of the maxilla, zygomatic, and palatine bones. Portions of the maxilla, lacrimal, ethmoid, and sphenoid bones form the medial wall.

The structures that pass through the various openings of the orbit are indicated in Exhibit 7.4.

Figure 7.14   Details of the orbit (eye socket).

 The orbit is a pyramid-shaped space that contains the eyeball and associated structures.

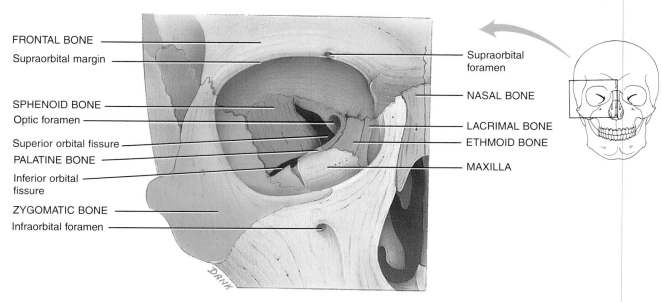

FRONTAL BONE
Supraorbital margin
SPHENOID BONE
Optic foramen
Superior orbital fissure
PALATINE BONE
Inferior orbital fissure
ZYGOMATIC BONE
Infraorbital foramen

Supraorbital foramen
NASAL BONE
LACRIMAL BONE
ETHMOID BONE
MAXILLA

Anterior view of right orbit

 Name the bones that form the orbit.

EXHIBIT **7.4** **SUMMARY OF FORAMINA OF THE SKULL**

Foramen	Location	Structures Passing Through
Carotid (relating to carotid artery in neck) (Fig. 7.7)	Petrous portion of temporal bone.	Internal carotid artery.
Greater palatine (*palatum* = palate) (Fig. 7.7)	Posterior angle of hard palate of palatine bones.	Greater palatine nerve and greater palatine vessels.
Hypoglossal (*hypo* = under; *glossus* = tongue) (Fig. 7.8)	Superior to base of occipital condyles.	Cranial nerve XII (hypoglossal) and branch of ascending pharyngeal artery.
Incisive (*incisive* = pertaining to incisor teeth) (Fig. 7.7)	Posterior to incisor teeth in maxilla.	Branches of greater palatine vessels and nasopalatine nerve.
Inferior orbital (*inferior* = below; *orbital* = orbits) (Fig. 7.14)	Between greater wing of sphenoid bone and maxilla.	Maxillary branch of cranial nerve V (trigeminal), zygomatic nerve, and infraorbital vessels.
Infraorbital (*infra* = below) (Fig. 7.2)	Inferior to orbit in maxilla.	Infraorbital nerve and artery.
Jugular (*jugular* = pertaining to jugular vein) (Fig. 7.7)	Posterior to carotid canal between petrous portion of temporal bone and occipital bone.	Internal jugular vein, cranial nerves IX (glossopharyngeal), X (vagus), and XI (accessory).
Lacerum (*lacerum* = lacerated) (Fig. 7.8a)	Bounded anteriorly by sphenoid bone, posteriorly by petrous portion of temporal bone, and medially by the sphenoid bone and occipital bone.	Branch of ascending pharyngeal artery.
Lacrimal (*lacrima* = pertaining to tears) (Fig. 7.3)	Lacrimal bone.	Lacrimal (tear) duct.
Lesser palatine (*palatum* = palate) (Fig. 7.7)	Posterior to greater palatine foramen in palatine bones.	Lesser palatine nerves and artery.
Magnum (*magnum* = large) (Fig. 7.7)	Occipital bone.	Medulla oblongata and its membranes (meninges), cranial nerve XI (accessory), and vertebral and spinal arteries.
Mandibular (*mandare* = to chew) (Fig. 7.13)	Medial surface of ramus of mandible.	Inferior alveolar nerve and vessels.
Mastoid (*mastoid* = breast-shaped) (Fig. 7.7)	Posterior border of mastoid process of temporal bone.	Emissary vein to transverse sinus and branch of occipital artery to dura mater.
Mental (*mentum* = chin) (Fig. 7.13)	Inferior to second premolar tooth in mandible.	Mental nerve and vessels.
Olfactory (*olfacere* = to smell) (Fig. 7.8a)	Cribriform plate of ethmoid.	Cranial nerve I (olfactory).
Optic (*optikas* = eye) (Fig. 7.8a)	Between superior and inferior portions of lesser wing of sphenoid bone.	Cranial nerve II (optic) and ophthalmic artery.
Ovale (*ovale* = oval) (Fig. 7.8a)	Greater wing of sphenoid bone.	Mandibular branch of cranial nerve V (trigeminal).
Rotundum (*rotundum* = round opening) (Fig. 7.8a)	Junction of anterior and medial parts of sphenoid bone.	Maxillary branch of cranial nerve V (trigeminal).
Spinosum (*spinosum* = spine-like) (Fig. 7.8a)	Posterior angle of sphenoid bone.	Middle meningeal vessels.
Stylomastoid (*stylo* = stake or pole) (Fig. 7.7)	Between styloid and mastoid processes of temporal bone.	Cranial nerve VII (facial) and stylomastoid artery.
Superior orbital (*superior* = above) (Fig. 7.14)	Between greater and lesser wings of sphenoid bone.	Cranial nerves III (oculomotor), IV (trochlear), ophthalmic branch of V (trigeminal) and VI (abducens).
Supraorbital (*supra* = above) (Fig. 7.2)	Supraorbital margin of orbit in frontal bone.	Supraorbital nerve and artery.
Zygomaticofacial (*zygoma* = cheek bone; *facies* = face) (Fig. 7.2)	Zygomatic bone.	Zygomaticofacial nerve and vessels.

Foramina

Many **foramina** (singular is **foramen**) associated with the skull were mentioned along with the descriptions of the cranial and facial bones they penetrate. As preparation for studying other systems of the body, especially the nervous and cardiovascular systems, these foramina, plus some additional ones, and the structures passing through them are listed in Exhibit 7.4. For your convenience and for future reference, the foramina are listed alphabetically.

HYOID BONE

The single **hyoid bone** (*hyoedes* = U-shaped) is a unique component of the axial skeleton because it does not articulate with any other bone (see Fig. 7.3). Rather, it is suspended from the styloid processes of the temporal bones by ligaments and muscles. The hyoid is located in the neck between the mandible and larynx. It supports the tongue and provides attachment for some of its muscles and for muscles of the neck and pharynx. The hyoid bone consists of a horizontal **body** and paired projections called the **lesser horns** or **cornua** and the **greater horns** or **cornua** (Fig. 7.15). **Cornu,** which is the singluar, means horn. Muscles and ligaments attach to these paired projections.

The hyoid bone, as well as the cartilage of the larynx (voice box) and trachea (windpipe), are often fractured during strangulation. As a result, they are carefully examined in an autopsy when strangulation is suspected.

VERTEBRAL COLUMN

Divisions

The **vertebral** or **spinal column (spine)**, together with the sternum and ribs, forms the skeleton of the **trunk** of the body. Whereas the spinal column consists of bone, the spinal cord consists of nervous tissue. The vertebral column makes up about two-fifths of the total height of the body and is composed of a series of bones called **vertebrae** (VER-te-brē; singular is **vertebra**). The length of the column is about 71 cm (28 in.) in an average adult male and about 61 cm (24 in.) in an average adult female. In effect, the vertebral column is a strong, flexible rod that bends anteriorly, posteriorly, and laterally and rotates. It encloses and protects the spinal cord, supports the head, and serves as a point of attachment for the ribs and the muscles of the back. Between vertebrae are openings called **intervertebral foramina**. The nerves that connect the spinal cord to various parts of the body pass through these openings.

The adult vertebral column typically contains 26 vertebrae (Fig. 7.16a,b). These are distributed as follows: 7 **cervical vertebrae** (*cervix* = neck) in the neck region; 12

Figure 7.15 Hyoid bone.

 The hyoid bone provides attachment for muscles of the tongue, neck, and pharynx.

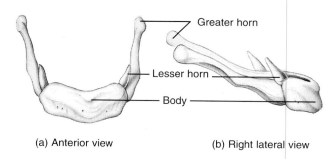

Greater horn

Lesser horn

Body

(a) Anterior view   (b) Right lateral view

How is the hyoid bone distinguished from all other bones of the axial skeleton?

thoracic vertebrae (*thorax* = chest) posterior to the thoracic cavity; 5 **lumbar vertebrae** (*lumbus* = loin) supporting the lower back; 5 **sacral vertebrae** fused into one bone called the **sacrum** (SĀ-krum; *sacrum* = sacred or holy); and usually 4 **coccygeal** (kok-SIJ-ē-al; *kokkyx* = resembling the bill of a cuckoo) **vertebrae** fused into one or two bones called the **coccyx** (KOK-six). Before fusion of the sacral and coccygeal vertebrae, the total number of vertebrae is 33. Whereas the cervical, thoracic, and lumbar vertebrae are movable, the sacrum and coccyx are immovable.

Between adjacent vertebrae from the second vertebra (axis) to the sacrum are **intervertebral discs**. Each disc has an outer fibrous ring consisting of fibrocartilage called the **annulus** (*annulus* = ring) **fibrosus** and an inner soft, pulpy, highly elastic structure called the **nucleus pulposus** (*pulposus* = pulp; Fig. 7.16d; see Fig. 7.24). The discs form strong joints, permit various movements of the vertebral column, and absorb vertical shock. Under compression, they flatten, broaden, and bulge from their intervertebral spaces.

Normal Curves

When viewed from the side, the vertebral column shows four **normal curves** (Fig. 7.16b). The **cervical** and **lumbar curves** are anteriorly convex (bulging out) while the **thoracic** and **sacral curves** are anteriorly concave (cupping in). The curves of the vertebral column are important because they increase its strength, help maintain balance in the upright position, absorb shock during walking and running, and help protect the column from fracture.

In the fetus, there is only a single anteriorly concave curve (Fig. 7.16c). At approximately the third month after birth, when an infant begins to hold its head erect, the cervical curve develops. Later, when the child sits up, stands, and walks, the lumbar curve develops, thus separating the thoracic and sacral

Figure 7.16   Vertebral column. In (d), the relative size of the disc has been enlarged for emphasis. A "window" has been cut in the annulus fibrosus so that the nucleus pulposus can be seen.

🔑 *The adult vertebral column typically contains 26 vertebrae.*

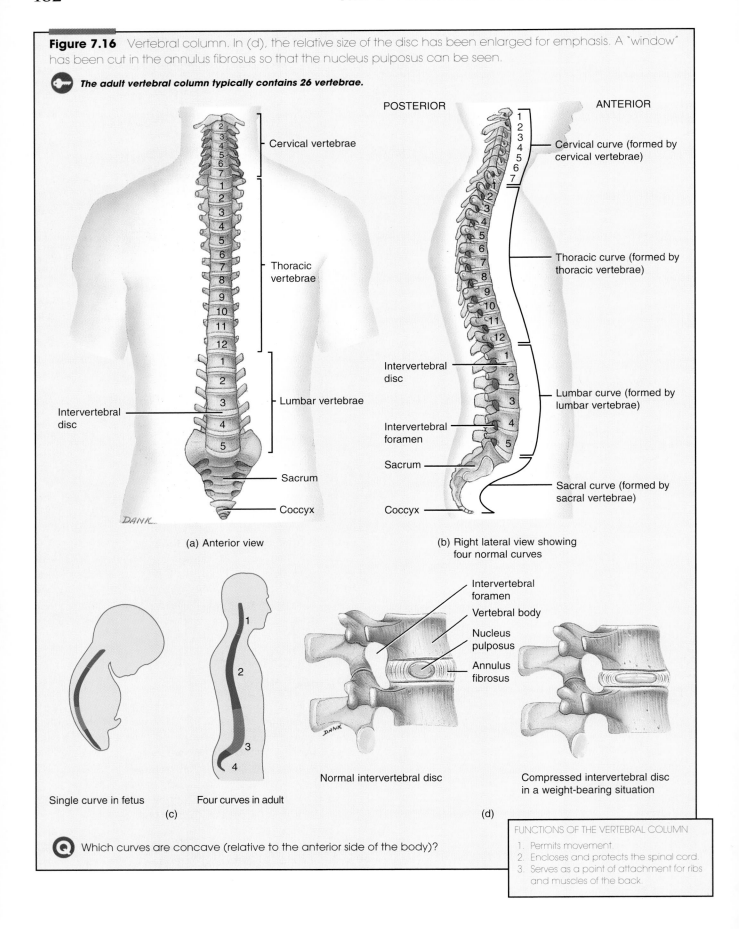

(a) Anterior view

(b) Right lateral view showing four normal curves

Single curve in fetus

Four curves in adult

(c)

Normal intervertebral disc

Compressed intervertebral disc in a weight-bearing situation

(d)

❓ Which curves are concave (relative to the anterior side of the body)?

FUNCTIONS OF THE VERTEBRAL COLUMN
1. Permits movement.
2. Encloses and protects the spinal cord.
3. Serves as a point of attachment for ribs and muscles of the back.

curves. The thoracic and sacral curves retain the anterior concavity of the fetus and thus are called **primary curves**. Because they are modifications of the original fetal curve, the cervical and lumbar curves are called **secondary curves**.

Typical Vertebra

Vertebrae in different regions of the vertebral column vary in size, shape, and detail, but they are similar enough that we can discuss the parts and functions of a typical vertebra (Fig. 7.17).

1. The **body** is the thick, disc-shaped anterior portion that is the weight-bearing part of a vertebra. Its superior and inferior surfaces are roughened for the attachment of the fibrocartilaginous intervertebral discs. The anterior and lateral surfaces contain nutrient foramina for blood vessels.
2. The **vertebral (neural) arch** extends posteriorly from the body of the vertebra. With the body of the vertebra, it

surrounds the spinal cord. It is formed by two short, thick processes, the **pedicles** (PED-i-kuls; *pediculus* = little feet), which project posteriorly from the body to unite with the laminae. The **laminae** (LAM-i-nē-; *lamina* = thin layer) are the flat parts that join to form the posterior portion of the vertebral arch. The space that lies between the vertebral arch and body contains the spinal cord, adipose and areolar connective tissues, and blood vessels. This space is known as the **vertebral foramen**. The vertebral foramina of all vertebrae together form the **vertebral (spinal) canal**. The pedicles contain superior and inferior indentations called **vertebral notches**. When they are stacked on top of one another, there is an opening between adjoining vertebrae on each side of the column. Each opening, called an **intervertebral foramen**, permits the passage of a single spinal nerve.
3. Seven **processes** arise from the vertebral arch. At the point where a lamina and pedicle join, a **transverse process** extends laterally on each side. A single **spinous process (spine)** projects posteriorly and inferiorly from

Figure 7.17   Typical vertebra as illustrated by a thoracic vertebra. In (b), only one spinal nerve has been included and it has been extended beyond the intervertebral foramen for clarity.

A vertebra consists of a body, vertebral arch, and several processes.

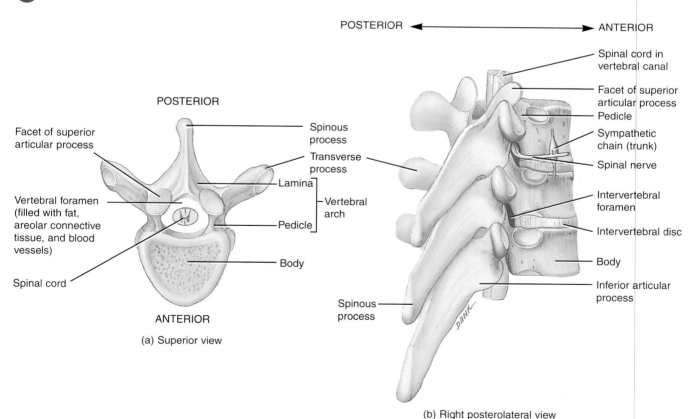

(a) Superior view

(b) Right posterolateral view

What are the functions of the vertebral and intervertebral foramina?

the junction of the laminae. These three processes serve as points of attachment for muscles. The remaining four processes form joints with other vertebrae above and below. The two **superior articular processes** of a vertebra articulate with the two inferior articular processes of the vertebra immediately superior to them. The two **inferior articular processes** of a vertebra articulate with two superior articular processes of the vertebra immediately below them. The articulating surfaces of the articular processes are referred to as **facets** (*facet* = little face).

Cervical Region

Vertebrae are numbered by region, in sequence from superior to inferior. When viewed from above, it can be seen that the bodies of the **cervical vertebrae** (C1–C7) are smaller than those of the thoracic vertebrae (Fig. 7.18). The cervical vertebral arches, however, are larger. The spinous processes of the second through sixth cervical vertebrae are often *bifid*, that is, split into two parts. All cervical vertebrae have three foramina: the vertebral foramen and two transverse foramina. The vertebral foramina of cervical vertebrae are the largest in the spinal column because they house the cervical enlargement of the spinal cord. Each cervical transverse process contains a **transverse foramen** through which the vertebral artery and its accompanying vein and nerve fibers pass.

The first two cervical vertebrae differ considerably from the others. The first cervical vertebra (C1), the **atlas**, like the mythological Atlas who supported the world on his shoulders, supports the head. The atlas is a ring of bone with **anterior** and **posterior arches** and large **lateral masses**. It lacks a body and a spinous process. The superior surfaces of the lateral masses, called **superior articular**

Figure 7.18   Cervical vertebrae.

🔑 *The cervical vertebrae are found in the neck region.*

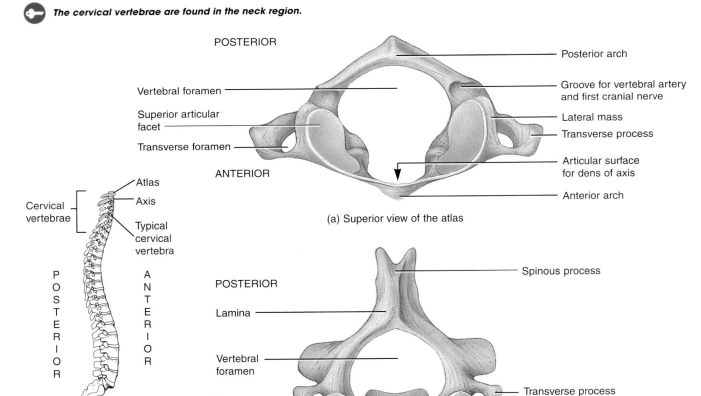

(a) Superior view of the atlas

(b) Superior view of the axis

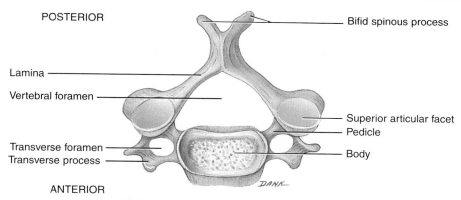

POSTERIOR

Bifid spinous process

Lamina

Vertebral foramen

Superior articular facet

Pedicle

Transverse foramen

Body

Transverse process

ANTERIOR

(c) Superior view of a typical cervical vertebra

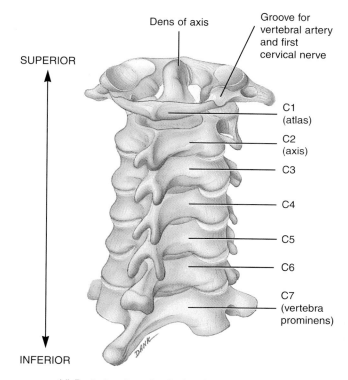

Dens of axis

Groove for
vertebral artery
and first
cervical nerve

SUPERIOR

C1
(atlas)

C2
(axis)

C3

C4

C5

C6

C7
(vertebra
prominens)

INFERIOR

(d) Posterior view of articulated cervical vertebrae

Q Which bones permit movement of the head to signify no?

facets, are concave and articulate with the occipital condyles of the occipital bone. This articulation permits the movement seen when moving the head to signify yes. The inferior surfaces of the lateral masses, the **inferior articular facets**, articulate with the second cervical vertebra. The transverse processes and transverse foramina of the atlas are quite large.

The second cervical vertebra (C2), the **axis**, does have a body. A peglike process called the **dens** (*dens* = tooth) or **odontoid** (*odontoid* = toothlike) **process** projects up through the anterior portion of the vertebral foramen of the atlas. The dens makes a pivot on which the atlas and head rotate as in moving the head to signify no. This arrangement permits side-to-side rotation of the head. In various instances of trauma, the dens of the axis may be driven into the medulla oblongata of the brain. When whiplash injuries result in death, this type of injury is the usual cause.

The third through sixth cervical vertebrae (C3–C6) correspond to the structural pattern of the typical cervical vertebra previously described.

The seventh cervical vertebra (C7), called the **vertebra prominens** (*prominens* = projection), is somewhat different. It is marked by a large, nonbifid spinous process that may be seen and felt at the base of the neck.

Thoracic Region

Thoracic vertebrae (T1–T12) are considerably larger and stronger than cervical vertebrae (Fig. 7.19). In addition, the spinous processes on T1–T10 are long, laterally flattened, and directed inferiorly. The spinous process of T11–T12 are shorter, broader, and directed less inferiorly. Compared to cervical vertebrae, thoracic vertebrae also have longer and heavier transverse processes.

The best distinguishing feature of the thoracic vertebrae is that they articulate with the ribs. The articulating surfaces of the vertebrae are called **facets** and **demifacets** (half-facets). Except for the eleventh and twelfth thoracic vertebrae, the transverse processes have facets for articulating with the *tubercles* of the ribs. The bodies of thoracic vertebrae also have facets or demifacets for articulation with the *heads* of the ribs. As you can see in Fig. 7.19c, T1 has a superior facet and an inferior demifacet on each side. T2–T8 have a superior and inferior demifacet on each side. T9 has a superior demifacet on each side, and T10–T12 have a superior facet on each side. Movements of the thoracic region are limited by thin intervertebral (*inter* = between) discs and the attachment of ribs to the sternum.

Lumbar Region

The **lumbar vertebrae** (L1–L5) are the largest and strongest in the vertebral column because the amount of body weight supported by the vertebrae increases toward the inferior end of the vertebral column (Fig. 7.20 on page 188). Their various projections are short and thick. The superior articular processes are directed medially instead of superiorly. The inferior articular processes are directed laterally instead of inferiorly. The spinous processes are quadrilateral in shape, thick, and broad and project nearly straight rather than inferiorly. The spinous processes are well-adapted for the attachment of the large back muscles.

Sacrum

The **sacrum** is a triangular bone formed by the union of five sacral vertebrae. These are indicated in Fig. 7.21 on page

189 as S1–S5. Fusion normally begins between 16 and 18 years of age and is usually completed by 30 years of age. The sacrum serves as a strong foundation for the pelvic girdle. It is positioned at the posterior portion of the pelvic cavity medial to the two hipbones.

The concave anterior side of the sacrum faces the pelvic cavity. It is smooth and contains four **transverse lines** (**ridges**) that mark the joining of the sacral vertebral bodies. At the ends of these lines are four pairs of **anterior (pelvic) sacral foramina**. The lateral portion of the superior surface contains a smooth surface called the **sacral ala** (wing), which is formed by the fused transverse processes of the first sacral vertebra (S1).

The convex, posterior surface of the sacrum contains a **median sacral crest**, the fused spinous processes of the upper sacral vertebrae; a **lateral sacral crest**, the transverse processes of the sacral vertebrae; and four pairs of **posterior (dorsal) sacral foramina**. These foramina communicate with the anterior sacral foramina through which nerves and blood vessels pass. The **sacral canal** is a continuation of the vertebral canal. The laminae of the fifth sacral vertebra, and sometimes the fourth, fail to meet. This leaves an inferior entrance to the vertebral canal called the **sacral hiatus** (hī-Ā-tus; *hiatus* = opening). On either side of the sacral hiatus are the **sacral cornua**, the inferior articular processes of the fifth sacral vertebra. They are connected by ligaments to the coccyx.

The superior border of the sacrum exhibits an anteriorly projecting border, the **sacral promontory** (PROM-on-tō-rē). It is an obstetrical landmark for measurements of the pelvis. On both lateral surfaces, the sacrum has a large **auricular surface** for articulating with the ilium of each hipbone. Posterior to the auricular surface is a roughened surface, the **sacral tuberosity**, that contains depressions for the attachment of ligaments. The sacral tuberosity is another surface of the sacrum that unites with the hipbone to form the sacroiliac joint. The **superior articular processes** of the sacrum articulate with the fifth lumbar vertebra.

Whereas the female sacrum is shorter, wider, and more curved between S2 and S3, the male sacrum is longer, narrower, and less curved.

Coccyx

The **coccyx** is also triangular in shape and is formed by the fusion of the usually four coccygeal vertebrae. These are indicated in Fig. 7.21 as Co1–Co4. Fusion generally occurs between 20 and 30 years of age. On the lateral surfaces of the coccyx are a series of **transverse processes**, the first pair being the largest. The coccyx articulates superiorly with the sacrum. In females, the coccyx points inferiorly; in males, it points anteriorly.

Figure 7.19 Thoracic vertebrae.

The thoracic vertebrae are found in the chest region and articulate with the ribs.

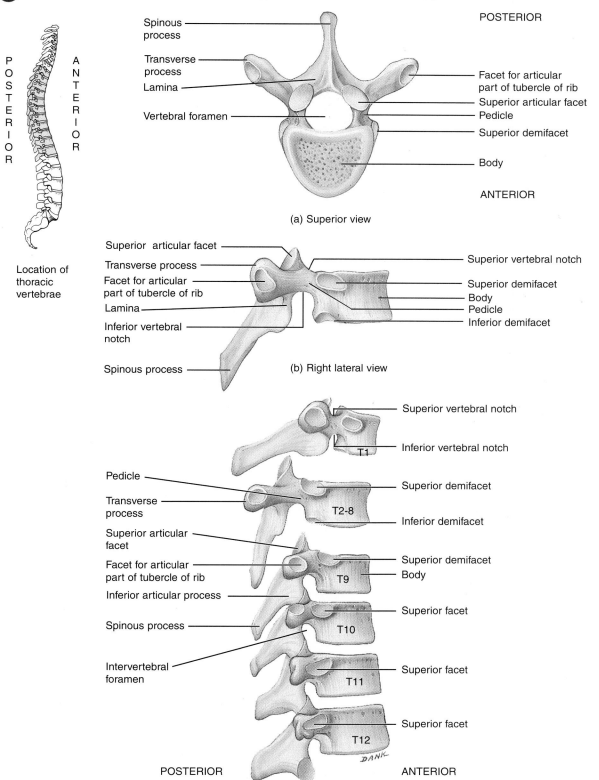

Location of thoracic vertebrae

(a) Superior view

- Spinous process
- Transverse process
- Lamina
- Vertebral foramen
- POSTERIOR
- Facet for articular part of tubercle of rib
- Superior articular facet
- Pedicle
- Superior demifacet
- Body
- ANTERIOR

(b) Right lateral view

- Superior articular facet
- Transverse process
- Facet for articular part of tubercle of rib
- Lamina
- Inferior vertebral notch
- Spinous process
- Superior vertebral notch
- Superior demifacet
- Body
- Pedicle
- Inferior demifacet

(c) Right lateral view of articulated thoracic vertebrae

- Superior vertebral notch
- Inferior vertebral notch
- T1
- Pedicle
- Transverse process
- Superior demifacet
- T2-8
- Inferior demifacet
- Superior articular facet
- Facet for articular part of tubercle of rib
- Superior demifacet
- Body
- T9
- Inferior articular process
- Superior facet
- Spinous process
- T10
- Intervertebral foramen
- Superior facet
- T11
- Superior facet
- T12
- POSTERIOR
- ANTERIOR
- DANK

Which parts of a thoracic vertebra articulate with a rib?

Figure 7.20　Lumbar vertebrae.

🔑 *Lumbar vertebrae are found in the lower back.*

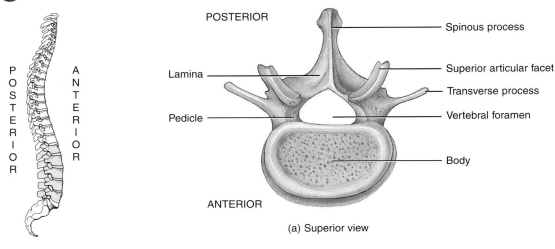

POSTERIOR

Lamina

Pedicle

Spinous process

Superior articular facet

Transverse process

Vertebral foramen

Body

ANTERIOR

(a) Superior view

Location of
lumbar vertebrae

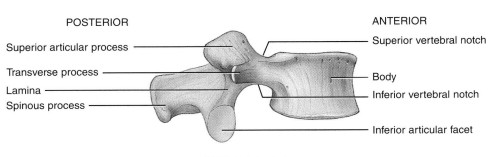

POSTERIOR　　　　　　　　　　　　ANTERIOR

Superior articular process

Transverse process

Lamina

Spinous process

Superior vertebral notch

Body

Inferior vertebral notch

Inferior articular facet

(b) Right lateral view

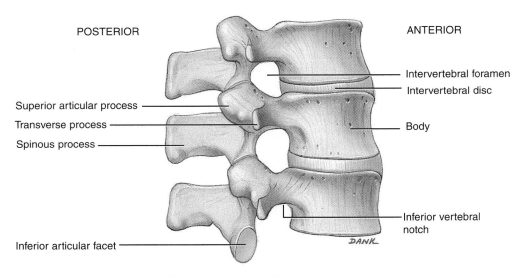

POSTERIOR　　　　　　　　　　　　ANTERIOR

Superior articular process

Transverse process

Spinous process

Inferior articular facet

Intervertebral foramen

Intervertebral disc

Body

Inferior vertebral
notch

DANK

(c) Right lateral view of articulated lumbar vertebrae

❓ Why are the lumbar vertebrae the largest and strongest in the vertebral column?

Figure 7.21 Sacrum and coccyx.

The sacrum is formed by the union of five sacral vertebrae and the coccyx is formed by the union of usually four coccygeal vertebrae.

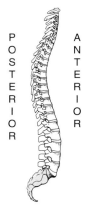

P O S T E R I O R   A N T E R I O R

Location of sacrum and coccyx

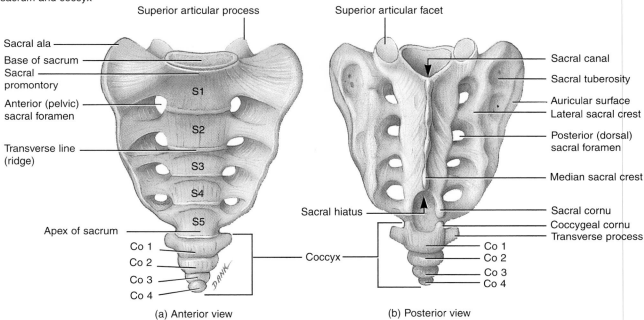

Superior articular process

Sacral ala
Base of sacrum
Sacral promontory
Anterior (pelvic) sacral foramen
Transverse line (ridge)

S1
S2
S3
S4
S5

Apex of sacrum
Co 1
Co 2
Co 3
Co 4

Coccyx

(a) Anterior view

Superior articular facet

Sacral canal
Sacral tuberosity
Auricular surface
Lateral sacral crest
Posterior (dorsal) sacral foramen
Median sacral crest
Sacral cornu
Coccygeal cornu
Transverse process

Sacral hiatus

Co 1
Co 2
Co 3
Co 4

(b) Posterior view

Q How many foramina pierce the sacrum and what is their function?

EPIDURAL ANESTHESIA

Anesthetic agents that act on the sacral and coccygeal nerves are sometimes injected through the sacral hiatus, a procedure called **epidural anesthesia** that is used most often in obstetrics. Since the sacral hiatus is between the sacral cornua, the cornua are important bony landmarks for locating the hiatus. Anesthetic agents may also be injected through the posterior (dorsal) sacral foramina. ■

THORAX

The term **thorax** refers to the entire chest. The skeletal portion of the thorax is a bony cage formed by the sternum, costal cartilages, ribs, and the bodies of the thoracic vertebrae (Fig. 7.22).

The thoracic cage is narrower at its superior end and broader at its inferior end. It is flattened from anterior to posterior. The thoracic cage encloses and protects the organs

Figure 7.22   Skeleton of the thorax.

 The bones of the thorax enclose and protect organs in the thoracic cavity and upper abdominal cavity.

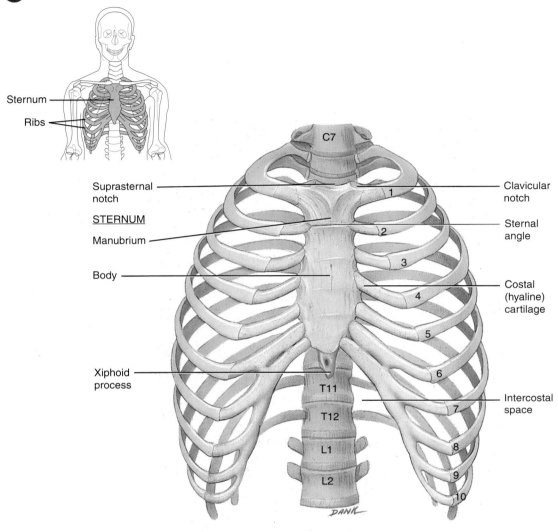

Anterior view

Which ribs are true ribs, false ribs, and floating ribs?

in the thoracic cavity and upper abdominal cavity. It also provides support for the bones of the shoulder girdle and upper limbs.

Sternum

The **sternum**, or breastbone, is a flat, narrow bone measuring about 15 cm (6 in.) in length. It is located in the middle of the anterior thoracic wall. During thoracic surgery, it may be split in the midsagittal plane to allow surgeons access to

structures in the thoracic cavity, such as the thymus gland, heart, and great vessels of the heart.

The sternum consists of three portions (Fig. 7.22): the **manubrium** (ma-NOO-brē-um; *manubrium* = handlelike), the superior portion; the **body**, the middle, largest portion; and the **xiphoid** (ZĪ-foyd; *xipho* = swordlike) **process**, the inferior, smallest portion. The junction of the manubrium and body forms the **sternal angle**. The manubrium has a depression on its superior surface called the **suprasternal (jugular) notch**. Lateral to the suprasternal notch are **clavicular notches** that articulate with the medial ends of the clavicles. The manubrium also articulates with the costal cartilages of the first and second ribs. The body of the sternum articulates directly or indirectly with the costal cartilages of the second through tenth ribs. The xiphoid process has no ribs attached to it but provides attachment for some abdominal muscles. The xiphoid process consists of hyaline cartilage during infancy and childhood and does not ossify completely until about age 40. If the hands of a rescuer are incorrectly positioned during cardiopulmonary resuscitation (CPR), there is danger of fracturing the xiphoid process, and driving it into internal organs.

STERNAL PUNCTURE

Since the sternum contains red bone marrow throughout life and because it is readily accessible and has thin compact bone, it is a common site for withdrawal of marrow for biopsy. Under a local anesthetic, a wide-bore needle is introduced into the marrow cavity of the sternum for aspiration of a sample of red bone marrow. This procedure is called a **sternal puncture**. ■

Ribs

Twelve pairs of **ribs** make up the sides of the thoracic cavity (Fig. 7.22). The ribs increase in length from the first through seventh, then decrease in length to the twelfth rib. Each articulates posteriorly with its corresponding thoracic vertebra.

The first through seventh pairs of ribs have a direct anterior attachment to the sternum by a strip of hyaline cartilage called **costal cartilage** (*costa* = rib). These ribs are called **true (vertebrosternal) ribs**. The remaining five pairs of ribs are referred to as **false ribs** because their costal cartilages either attach indirectly to the sternum or do not attach to the sternum at all. The cartilages of the eighth, ninth, and tenth pairs of ribs attach to each other and then to the cartilages of the seventh pair of ribs. These false ribs are called **vertebrochondral ribs**. The eleventh and twelfth pairs of ribs are false ribs designated as **floating (vertebral) ribs** because their anterior ends do not attach to the sternum. Instead they terminate in the abdominal muscles. They attach only posteriorly to the thoracic vertebrae.

We will examine the parts of a typical (third through ninth) rib (Fig. 7.23 on page 192). The **head** is a projection at the posterior end of the rib. It consists of one or two **facets** that articulate with facets on the bodies of adjacent thoracic vertebrae. The **neck** is a constricted portion just lateral to the head. A knoblike structure on the posterior surface where the neck joins the body is called a **tubercle** (TOO-ber-kul). It consists of (1) a **nonarticular part** that affords attachment to the ligament of the tubercle and (2) an **articular part** that articulates with the facet of a transverse process of the inferior of the two vertebrae to which the head of the rib is connected. The **body (shaft)** is the main part of the rib. A short distance beyond the tubercle, there is an abrupt change in the curvature of the shaft. This point is called the **costal angle**. The inner surface of the rib has a **costal groove** that houses blood vessels and a small nerve.

In summary, the posterior portion of the rib is connected to a thoracic vertebra by its head and articular part of a tubercle. The facet of the head fits into a facet on the body of one vertebra or the demifacets of two adjoining vertebrae. The articular part of the tubercle articulates with the facet of the transverse process of the vertebra.

Spaces between ribs, called **intercostal** (*inter* = between) **spaces**, are occupied by intercostal muscles, blood vessels, and nerves (see Fig. 7.22). Surgical access to the lungs or other structures in the thoracic cavity is commonly undertaken through an intercostal space. Special rib retractors are used to create a wide separation between ribs. The costal cartilages are sufficiently elastic in younger individuals to permit considerable bending without breaking.

If you examine Fig. 7.22, you will notice that the first rib is the shortest, broadest, and most sharply curved of the ribs. The first rib is an important landmark because of its close relationship to the nerves of the brachial plexus, which provides the entire nerve supply of the shoulder and upper limb, two major blood vessels, the subclavian artery and vein, and two skeletal muscles, the anterior and medial scalene muscles. The superior surface of the first rib has two shallow grooves, one for the subclavian vein and one for the subclavian artery and inferior trunk of the brachial plexus. The superior surface also serves as the point of attachment for the anterior and middle scalene muscles. The second rib is thinner, less curved, and considerably longer than the first. The tenth rib has a single articular facet on its head. The eleventh and twelfth ribs also have single articular facets on their heads, but no necks, tubercles, or costal angles.

RIB FRACTURES

Rib fractures represent the most common chest injuries and usually result from direct blows, most commonly from steering wheel impact, falls, and crushing injuries to the chest. Ribs tend to break at the point where greatest force is applied, but may also break at the weakest point (that is, the site of greatest curvature), which is just anterior to the costal angle. In children the ribs are highly elastic, and fractures are less frequent than in adults. Since the first two ribs are protected by the clavicle and pectoralis major muscle, and the last two ribs are mobile, they are the least commonly injured. The middle ribs are the ones most commonly fractured. In some cases, fractured ribs may cause damage to the heart, great vessels of the heart, lungs, trachea, bronchi, esophagus, spleen, liver, and kidneys. ■

Figure 7.23   Typical rib. The arrow in the inset in (c) indicates the direction from which the bones are viewed (superior).

 Each rib articulates posteriorly with its corresponding thoracic vertebra.

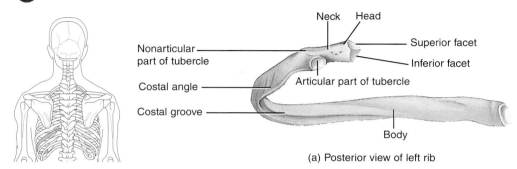

(a) Posterior view of left rib

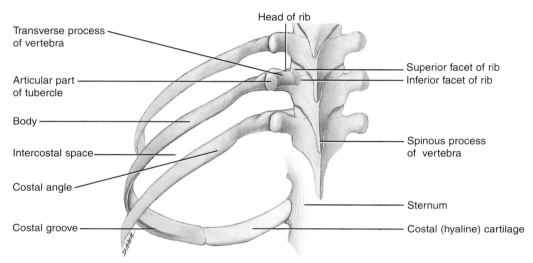

(b) Posterior view of left ribs articulated with thoracic vertebrae and the sternum

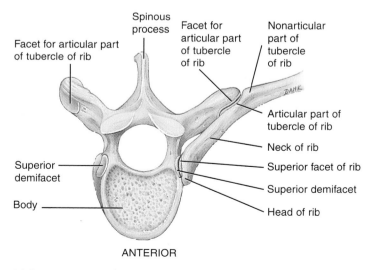

(c) Superior view of left rib articulated with a thoracic vertebra

How does a rib articulate with a thoracic vertebra?

DISORDERS: HOMEOSTATIC IMBALANCES

HERNIATED (SLIPPED) DISC

In their function as shock absorbers, intervertebral discs are constantly being compressed. If the anterior and posterior ligaments of the discs become injured or weakened, the pressure developed in the nucleus pulposus may be great enough to rupture the surrounding fibrocartilage (annulus fibrosus). If this occurs, the nucleus pulposus may herniate (protrude) posteriorly or into one of the adjacent vertebral bodies. This condition is called a **herniated (slipped) disc**.

Most often the nucleus pulposus slips posteriorly toward the spinal cord and spinal nerves (Fig. 7.24). This movement exerts pressure on the spinal nerves, causing considerable, sometimes very acute, pain. If the roots of the sciatic nerve, which passes from the spinal cord to the foot, are pressured, the pain radiates down the posterior aspect of the thigh, through the calf, and occasionally into the foot. If pressure is exerted on the spinal cord itself, some of its neurons may be destroyed.

Figure 7.24   Herniated (slipped) disc.

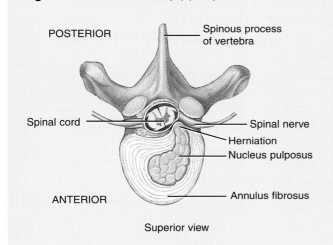

POSTERIOR

Spinous process
of vertebra

Spinal cord

Spinal nerve

Herniation

Nucleus pulposus

ANTERIOR

Annulus fibrosus

Superior view

ABNORMAL CURVES

Various conditions may exaggerate the normal curves of the vertebral column or the column may acquire a lateral bend, resulting in **abnormal curves** of the spine.

Scoliosis (skō′-lē-Ō-sis; *scolios* = twisted) is a lateral bending of the vertebral column, usually in the thoracic region. This is the most common of the abnormal curves. It may result from congenitally (present at birth) malformed vertebrae, chronic sciatica, paralysis of muscles on one side of the vertebral column, poor posture, or one leg being shorter than the other.

Kyphosis (kī-FŌ-sis; *kyphos* = a hump), also known as hunchback, is an exaggeration of the thoracic curve of the vertebral column. In tuberculosis of the spine, vertebral bodies may partially collapse, causing an acute angular bending of the vertebral column. In the elderly, degeneration of the intervertebral discs leads to kyphosis. Kyphosis may also be caused by rickets or poor posture. It is also common in females with advanced osteoporosis. The term *round-shouldered* is an expression for mild kyphosis.

Lordosis (lor-DŌ-sis; *lordos* = bent backward), also known as swayback, is an exaggeration of the lumbar curve of the vertebral column. It may result from increased weight of the abdomen as in pregnancy or extreme obesity, poor posture, rickets, and tuberculosis of the spine.

SPINA BIFIDA

Spina bifida (SPĪ-na BIF-i-da) is a congenital defect of the vertebral column in which laminae fail to unite at the midline. In serious cases, protrusion of the membranes (meninges) around the spinal cord or the spinal cord itself produces perilous problems, such as partial or complete paralysis, partial or complete loss of urinary bladder control, and the absence of reflexes. Spina bifida may be diagnosed prenatally by a test of the mother's blood, sonography, or amniocentesis (withdrawal of amniotic fluid for analysis).

STUDY OUTLINE

TYPES OF BONES (p. 163)

1. On the basis of shape, bones are classified as long, short, flat, or irregular.
2. Sutural (Wormian) bones are found within the sutures of certain cranial bones. Sesamoid bones develop in tendons or ligaments.

BONE SURFACE MARKINGS (p. 163)

1. Surface markings are structural features visible on the surfaces of bones.
2. Functions of markings include joint formation, muscle attachment, or passage of nerves and blood vessels.
3. Terms that describe markings include fissure, foramen, meatus, fossa, process, condyle, head, facet, tuberosity, crest, and spine.

DIVISIONS OF THE SKELETAL SYSTEM (p. 163)

1. The axial skeleton consists of bones arranged along the longitudinal axis. The parts of the axial skeleton are the skull, hyoid bone, vertebral column, sternum, and ribs.
2. The appendicular skeleton consists of the bones of the girdles and the upper and lower limbs (extremities). The parts of the appendicular skeleton are the pectoral (shoulder) girdles, bones of the upper limbs, pelvic (hip) girdle, and bones of the lower limbs.

SKULL (p. 166)

1. The skull consists of the cranium and the face (22 bones).
2. Sutures are immovable joints between bones of the skull. Examples are coronal, sagittal, lambdoid, and squamous sutures.
3. Fontanels are fibrous, connective tissue membrane-filled spaces between the cranial bones of fetuses and infants. The major fontanels are the anterior, posterior, anterolaterals, and posterolaterals. They fill in with bone and become sutures.
4. The 8 cranial bones include the frontal, parietal (2), temporal (2), occipital, sphenoid, and ethmoid.

5. The 14 facial bones are the nasal (2), maxillae (2), zygomatic (2), mandible, lacrimal (2), palatine (2), inferior nasal conchae (2), and vomer.

6. Paranasal sinuses are cavities in bones of the skull that communicate with the nasal cavity. They are lined by mucous membranes. The cranial bones containing the paranasal sinuses are the frontal, sphenoid, ethmoid, and maxillae.

7. The orbits (eye sockets) are formed by seven bones of the skull.

8. The foramina of the skull bones provide passages for nerves and blood vessels.

HYOID BONE (p. 181)

1. The hyoid bone is a U-shaped bone that does not articulate with any other bone.

2. It supports the tongue and provides attachment for some of its muscles as well as for some neck muscles and muscles of the pharynx.

VERTEBRAL COLUMN (p. 181)

1. The bones of the adult vertebral column are the cervical vertebrae (7), thoracic vertebrae (12), lumbar vertebrae (5), sacrum (5, fused), and the coccyx (4, fused).

2. The vertebral column contains four curves (cervical, thoracic, lumbar, and sacral) that give strength, support, and balance.

3. The vertebrae are similar in structure, each usually consisting of a body, vertebral arch, and seven processes. Vertebrae in the different regions of the column vary in size, shape, and detail.

THORAX (p. 189)

1. The thoracic skeleton consists of the sternum, ribs and costal cartilages, and thoracic vertebrae.

2. The thoracic cage protects vital organs in the chest area and upper abdomen.

REVIEW QUESTIONS

1. Describe the importance of the skeletal system to the body. (p. 163)

2. What are the four principal types of bones? Give an example of each. Distinguish between a sutural (Wormian) and a sesamoid bone. (p. 163)

3. What are bone surface markings? Describe and give an example of each. (p. 163)

4. Distinguish between the axial and appendicular skeletons. What subdivisions and bones are contained in each? (p. 163)

5. What are the bones that compose the skull? The cranium? The face? (p. 166)

6. Define a suture. What are the four prominent sutures of the skull? Where are they located? (p. 167)

7. What is a fontanel? Describe the location of the six fairly constant fontanels. (p. 168)

8. What is a paranasal sinus? What cranial bones contain paranasal sinuses? (p. 176)

9. Describe the components of each orbit. (p. 179)

10. What bones form the skeleton of the trunk? Distinguish between the number of nonfused vertebrae found in the adult vertebral column and that of a child. (p. 181)

11. What are the normal curves in the vertebral column? What are the functions of the curves? (p. 181)

12. What are the principal distinguishing characteristics of the bones of the various regions of the vertebral column? (p. 184)

13. What bones form the skeleton of the thorax? What are the functions of the thoracic skeleton? (p. 189)

14. How are ribs classified? (p. 191)

15. Define the following: black eye (p. 171), cleft palate and cleft lip (p. 176), sinusitis (p. 176), temporomandibular joint (TMJ) syndrome (p. 178), deviated nasal septum (DNS) (p. 179), epidural anesthesia (p. 189), and sternal puncture (p. 191).

ANSWERS TO QUESTIONS WITH FIGURES

7.1 Axial skeleton: skull, vertebral column. Appendicular skeleton: clavicle, shoulder girdle, humerus, pelvic girdle, femur.

7.2 Frontal, parietal, sphenoid, ethmoid, and temporal.

7.3 Squamous suture is between the parietal and temporal bones. Lambdoid suture is between the parietal and occipital bones. Coronal suture is between frontal and parietal bones.

7.4 Nasal, maxillae, mandible, palatine, inferior nasal conchae, and vomer.

7.5 Sagittal and lambdoid.

7.6 Anterolateral (sphenoid).

7.7 The temporal bone; the carotid foramen.

7.8 Crista galli, frontal, parietal, temporal, occipital, temporal, parietal, frontal, crista galli.

7.9 Perpendicular plate.

7.10 Posterior cranial fossa.

7.11 Maxillae and palatine bones.

7.12 They produce mucus and serve as resonating chambers for vocalization.

7.13 The mandible is the only movable skull bone, other than the auditory ossicles.

7.14 Frontal, sphenoid, zygomatic, maxilla, lacrimal, ethmoid, and palatine.

7.15 It does not articulate with any other bone.

7.16 The thoracic and sacral curves are concave.

7.17 The vertebral foramina enclose the spinal cord while the intervertebral foramina provide spaces for spinal nerves to exit the vertebral column.

7.18 Atlas and axis.

7.19 Facets and demifacets.

7.20 The body weight supported by vertebrae increases toward the inferior end of the vertebral column.

7.21 There are four pairs of foramina, for a total of eight. Each anterior sacral foramen joins a posterior sacral foramen at the intervertebral foramen. Nerves and blood vessels pass through these openings in the bone.

7.22 True ribs (pairs 1–7), false ribs (pairs 8–12), and floating ribs (pairs 11 and 12).

7.23 The facet of the head fits into a facet on the body of a vertebra and the articular part of the tubercle articulates with the facet of the transverse process of a vertebra.

STUDENT OBJECTIVES

1. Identify the bones of the pectoral (shoulder) girdle and their principal markings. (p. 196)

2. Identify the upper limb, its component bones, and their principal markings. (p. 197)

3. Identify the components of the pelvic (hip) girdle and their principal markings. (p. 202)

4. Identify the lower limb, its component bones, and their principal markings. (p. 206)

5. Define the structural features and importance of the arches of the foot. (p. 211)

6. Compare the principal structural differences between female and male skeletons, especially those that pertain to the pelvis. (p. 212)

THE SKELETAL SYSTEM: THE APPENDICULAR SKELETON

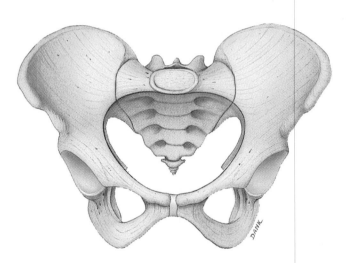

CHAPTER CONTENTS AT A GLANCE

his chapter discusses the bones of the appendicular skeleton, that is, the bones of the pectoral (shoulder) and pelvic (hip) girdles and upper and lower limbs. The differences between female and male skeletons are also described.

PECTORAL (SHOULDER) GIRDLE

Each of the two **pectoral** (PEK-tō-ral) or **shoulder girdles** (Fig. 8.1) consists of a **clavicle** (KLAV-i-kul; *clavis* = key)

Figure 8.1   Right pectoral (shoulder) girdle and upper limb.

 The shoulder girdle attaches the bones of the upper limb to the axial skeleton.

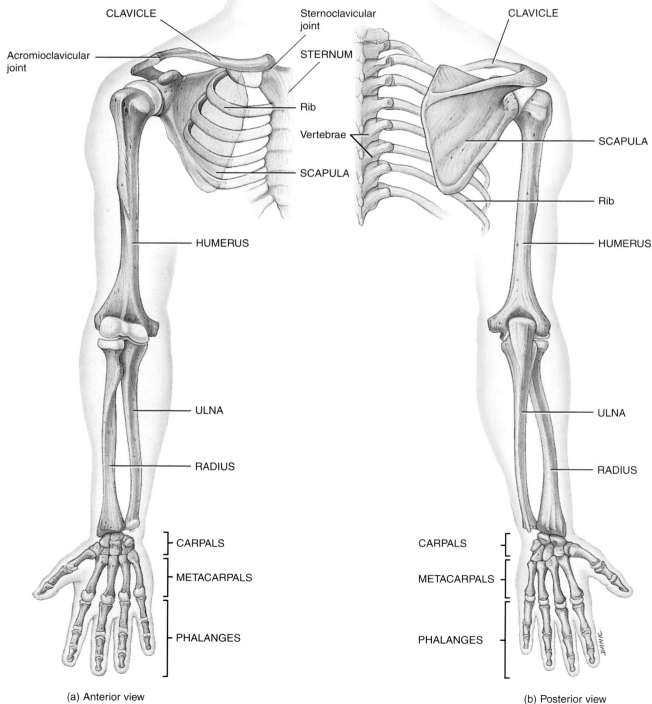

(a) Anterior view

(b) Posterior view

What bones make up the pectoral (shoulder) girdle?

and a **scapula** (SCAP-yoo-la), which attach the bones of the upper limbs (extremities) to the axial skeleton. The clavicle is the anterior component and articulates with the sternum at the sternoclavicular joint. The posterior component is the scapula, which is held in position by complex muscle attachments. It articulates with the clavicle and humerus. The pectoral girdles have no articulation with the vertebral column.

Clavicle

Each clavicle, or **collarbone**, is a long, slender S-shaped bone with two curves, one convex and one concave (Fig. 8.2). The medial one-third of the clavicle is convex anteriorly, whereas the lateral one-third is concave anteriorly. Since the junction of two curves is the weakest point of a structure, this is the most frequent site of clavicular fractures. The clavicles lie horizontally in the superior and anterior part of the thorax superior to the first rib.

The medial end of the clavicle, the **sternal extremity**, is rounded and articulates with the sternum to form the **sternoclavicular joint**. The broad, flat, lateral end, termed the **acromial** (a-KRŌ-mē-al) **extremity**, articulates with the acromion of the scapula. This joint is called the **acromioclavicular joint**. (Refer to Fig. 8.1 for a view of these articulations.) The **conoid** (KŌ-noyd; *konos* = cone) **tubercle** on the inferior surface of the lateral end of the bone serves as a point of attachment for the conoid ligament. The **costal tuberosity** on the inferior surface of the medial end also serves as a point of attachment for the costoclavicular ligament.

FRACTURED CLAVICLE

Because of its position, the clavicle transmits forces from the upper limb to the trunk. If such forces are excessive, as in falling on one's outstretched arm, a **fractured clavicle** may result. It is one of the most frequently broken bones in the body. ■

Scapula

Each scapula, or **shoulder blade**, is a large, triangular, flat bone situated in the posterior part of the thorax between the levels of the second and seventh ribs (Fig. 8.3). The medial borders of the scapulae (plural) lie about 5 cm (2 in.) from the vertebral column.

A sharp ridge, the **spine**, runs diagonally across the posterior surface of the flattened, triangular **body** of the scapula. The lateral end of the spine projects as a flattened, expanded process called the **acromion** (a-KRŌ-mē-on; *acro* = top or summit), easily felt as the high point of the shoulder. This process articulates with the clavicle. Inferior to the acromion is a depression called the **glenoid cavity (fossa)**. This cavity articulates with the head of the humerus (arm bone) to form the shoulder joint.

The thin edge of the bone near the vertebral column is the **medial (vertebral) border**. The thick edge closer to the arm is the **lateral (axillary) border**. The medial and lateral

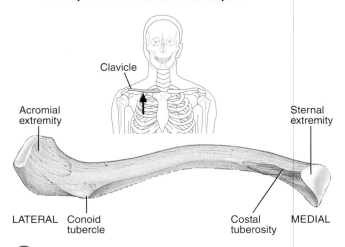

Figure 8.2   Right clavicle. The arrow in the inset indicates the direction from which the clavicle is viewed (inferior).

 The clavicle articulates medially with the sternum and laterally with the acromion of the scapula.

Clavicle

Acromial extremity

Sternal extremity

LATERAL   Conoid tubercle   Costal tuberosity   MEDIAL

❓ Which part of the clavicle is its weakest point?

borders join at the **inferior angle**. The superior edge of the scapula, called the **superior border**, joins the vertebral border at the **superior angle**. The **scapular notch** is a prominent indentation along the superior border through which the suprascapular nerve passes.

At the lateral end of the superior border is a projection of the anterior surface called the **coracoid** (KOR-a-koyd; *korakodes* = like a crow's beak) **process** to which muscles attach. Above and below the spine are two fossae: the **supraspinous** (soo′-pra-SPĪ-nus) **fossa** and the **infraspinous fossa**, respectively. Both serve as surfaces of attachment for shoulder muscles that have similar names (supraspinatus and infraspinatus). On the ventral (costal) surface is a slightly hollowed-out area called the **subscapular fossa**, also a surface of attachment for shoulder muscles.

UPPER LIMB (EXTREMITY)

The **upper limbs (extremities)** consist of 60 bones. Figure 8.1 shows the skeleton of the right upper limb. Each upper limb includes a humerus in the arm, ulna and radius in the forearm, and carpals (wrist bones), metacarpals (palm bones), and phalanges (finger bones) in the hand.

Humerus

The **humerus** (HYOO-mer-us), or arm bone, is the longest and largest bone of the upper limb. It articulates proximally with the scapula and distally at the elbow with both the ulna and radius (see Fig. 8.1).

Figure 8.3   Right scapula.

The glenoid cavity of the scapula articulates with the head of the humerus to form the shoulder joint.

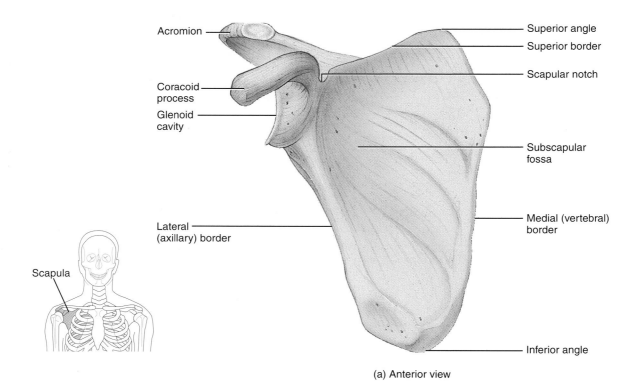

(a) Anterior view

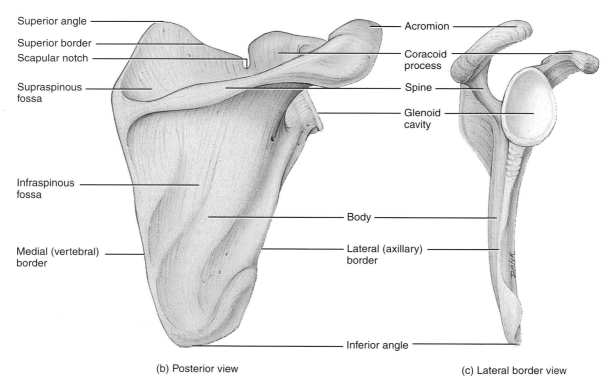

(b) Posterior view

(c) Lateral border view

Which part of the scapula forms the high point of the shoulder?

The proximal end of the humerus features a **head** that articulates with the glenoid cavity of the scapula (Fig. 8.4). It also has an **anatomical neck**, the former site of the epiphyseal plate, which is an oblique groove just distal to the head. The **greater tubercle** is a lateral projection distal to the neck. It is the most laterally palpable bony landmark of the shoulder region. The **lesser tubercle** is an anterior projection. Between these tubercles runs an **intertubercular sulcus (bicipital groove)**. The **surgical neck** is a constricted portion just distal to the tubercles and is so named because fractures often occur here.

The **body (shaft)** of the humerus is cylindrical at its proximal end. It gradually becomes triangular and is flattened and broad at its distal end. Laterally, at the middle portion of the shaft, there is a roughened, V-shaped area called the **deltoid tuberosity**. This area serves as a point of attachment for the deltoid muscle.

The following parts are found at the distal end of the humerus. The **capitulum** (ka-PIT-yoo-lum), meaning a small head, is a rounded knob that articulates with the head of the radius. The **radial fossa** is an anterior depression that receives the head of the radius when the forearm is flexed (bent). The **trochlea** (TRŌK-lē-a), located medial to the capitulum, is a spool-shaped surface that articulates with the ulna. The **coronoid** (KOR-ō-noyd; *korne* = crown-shaped) **fossa** is an anterior depression that receives the coronoid process of the ulna when the forearm is flexed. The **olecranon** (ō-LEK-ra-non) **fossa** is a posterior depression

Figure 8.4 Right humerus in relation to the scapula, ulna, and radius.

The humerus is the longest and largest bone of the upper limb.

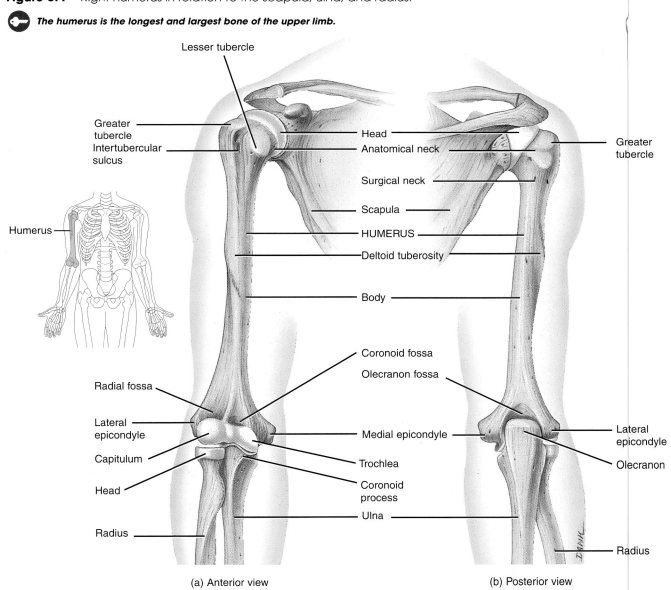

(a) Anterior view   (b) Posterior view

Which parts of the humerus articulate with the radius at the elbow? With the ulna at the elbow?

that receives the olecranon of the ulna when the forearm is extended (straightened). The **medial epicondyle** and **lateral epicondyle** are rough projections on either side of the distal end to which most muscles of the forearm are attached. The ulnar nerve lies on the posterior surface of the medial epicondyle and may easily be rolled between the finger and the medial epicondyle.

Ulna and Radius

The **ulna** is located on the medial aspect (little finger side) of the forearm and is longer than the radius (Fig. 8.5). At the proximal end of the ulna is the **olecranon** or **olecranon process**, which forms the prominence of the elbow. The **coronoid process** is an anterior projection that, together with

Figure 8.5   Right ulna and radius in relation to the humerus and carpals.

 The elbow joint is formed by the articulation of the trochlea of the humerus and the trochlear notch of the ulna and by the articulation of the head of the radius with the capitulum of the humerus and radial notch of the ulna.

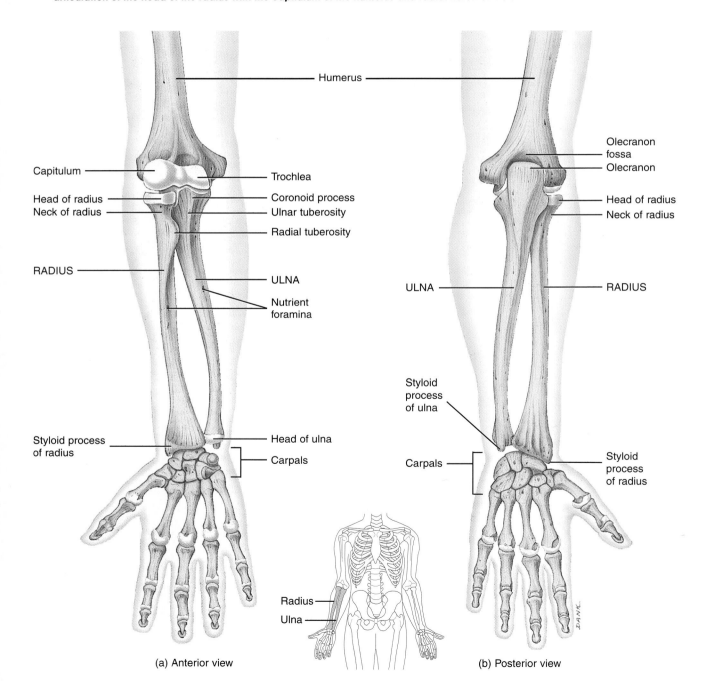

(a) Anterior view

(b) Posterior view

the olecranon, receives the trochlea of the humerus. Just inferior to the coronoid process is the **ulnar tuberosity**. The **trochlear (semilunar) notch** is a large curved area between the olecranon and the coronoid process. The trochlea of the humerus fits into this notch to form part of the elbow joint. The **radial notch** is a depression located laterally and inferiorly to the trochlear notch. It receives the head of the radius.

The distal end of the ulna consists of a **head** that is separated from the wrist by a fibrocartilage disc. A **styloid** (*stylo* = pillar) **process** is on the posterior side of the distal end.

The **radius** is located on the lateral aspect (thumb side) of the forearm. The proximal end of the radius has a disc-shaped **head** that articulates with the capitulum of the humerus and radial notch of the ulna to form another part of

The arrows in the insets in (c) and (d) indicate the direction from which the bones are viewed (inferior and lateral, respectively).

(c) Inferior view of distal ends of radius and ulna

(d) Lateral view of proximal end of ulna

(e) Medial view in relation to humerus

With which bones does the proximal head of the radius articulate?

the elbow joint. It also has a raised, roughened area on the medial side called the **radial tuberosity**. This is a point of attachment for the biceps brachii muscle. Inferior to the head is the constricted **neck**. The shaft of the radius widens distally to form a concave inferior surface that articulates with two bones of the wrist called the lunate and scaphoid bones. Also at the distal end is a **styloid process** on the lateral side and a medial, concave **ulnar notch** for articulation with the distal end of the ulna.

Carpals, Metacarpals, and Phalanges

The skeleton of the hand has three regions: (1) proximal carpus, (2) intermediate metacarpus, and (3) distal phalanges.

The **carpus** (wrist) consists of eight small bones, the **carpals**, joined to one another by ligaments (Fig. 8.6). The carpals are arranged in two transverse rows, with four bones in each row, and they are named for their shapes. In the anatomical position, the carpals in the proximal row, from the lateral to medial position, are the **scaphoid** or **navicular** (resembles a boat), **lunate** (resembles a crescent moon in its anteroposterior aspect), **triquetrum** (has three articular surfaces), and **pisiform** (pea-shaped). In about 70% of carpal fractures, only the scaphoid is broken. The carpals in the distal row, from the lateral to medial position, are the **trapezium** (four-sided), **trapezoid** (also four-sided), **capitate** (the largest carpal bone whose rounded projection, the head, articulates with the lunate), and **hamate** (named for a large hook-shaped projection on its anterior surface).

Together, the concavity formed by the pisiform and hamate (on the ulnar side) and the scaphoid and trapezium (on the radial side) plus the flexor retinaculum (deep fascia) constitute a space called the **carpal tunnel**. Through it pass the long flexor tendons of the digits and thumb and the median nerve. Narrowing of the carpal tunnel gives rise to a condition called carpal tunnel syndrome (see page 380).

The five bones of the **metacarpus** (*meta* = after or beyond), called **metacarpals**, constitute the palm of the hand. Each metacarpal bone consists of a proximal **base**, an intermediate **shaft**, and a distal **head**. The metacarpal bones are numbered I to V, starting with the one proximal to the thumb. The bases articulate with the distal row of carpal bones. The heads articulate with the proximal phalanges of the fingers. The heads of the metacarpals are commonly called the "knuckles" and are readily visible when the fist is clenched.

The **phalanges** (fa-LAN-jēz; *phalanx* = closely knit row), or bones of the fingers, number 14 in each hand. A single bone of the finger (or toe) is referred to as a **phalanx** (FA-lanks). Each phalanx consists of a proximal **base**, an intermediate **shaft**, and a distal **head**. There are two phalanges in the thumb (**pollex**) and three phalanges in each of the other fingers. In order from the thumb, they are commonly referred to as the index finger, middle finger, ring finger, and little finger. The first row of phalanges, the **proximal row**, articulates with the metacarpal bones and

second row of phalanges. The second row of phalanges, the **middle row**, articulates with the proximal row and the third row. The third row of phalanges, the **distal row**, articulates with the middle row. The thumb has no middle phalanx.

PELVIC (HIP) GIRDLE

The **pelvic (hip) girdle** consists of the two **hipbones** or **coxal** (KOK-sal; *coxa* = hip) **bones** (see Fig. 8.15), which attach the lower limbs to the trunk at the sacrum. The pelvic girdle provides a strong and stable support for the vertebral column and viscera. The hipbones are united to each other anteriorly at a joint called the **pubic symphysis** (PYOO-bik SIM-fi-sis). They unite posteriorly with the sacrum.

Each of the two hipbones of a newborn consists of three components: a superior **ilium**, an inferior and anterior **pubis**, and an inferior and posterior **ischium** (IS-kē-um) (Fig. 8.7). Eventually, the three separate bones fuse into one. The area of fusion is a deep, lateral fossa called the acetabulum, which serves as the socket for the head of the femur. Although the adult hipbones are both single bones, it is common to discuss the bones as if they still consisted of three portions.

Ilium

The ilium (meaning flank) is the largest of the three subdivisions of the hipbone. Its superior border, the **iliac crest**, ends anteriorly in the **anterior superior iliac spine**. Inferior to it is the **anterior inferior iliac spine**. Posteriorly, the iliac crest ends in the **posterior superior iliac spine**. Inferior to it is the **posterior inferior iliac spine**. The spines serve as points of attachment for muscles of the trunk, hip, and thighs. Inferior to the posterior inferior iliac spine is the **greater sciatic** (sī-AT-ik) **notch** through which the sciatic nerve, the longest nerve in the body, passes. The medial surface of the ilium contains the **iliac fossa**. It is a concavity where the iliacus muscle attaches. Posterior to this fossa is the **auricular** (*auricula* = little ear) **surface**, which articulates with the sacrum to form the **sacroiliac joint** (see Fig. 8.15).

Ischium

The ischium, meaning hip, is the inferior, posterior portion of the hipbone. It contains a prominent **ischial spine**, a **lesser sciatic notch** inferior to the spine, and an **ischial tuberosity**. This prominent tuberosity may hurt someone's thigh when you sit on their lap. The rest of the ischium, the **ramus** (meaning branch) joins with the pubis, and together they surround the **obturator** (OB-too-rā′ter) **foramen**, the largest foramen in the skeleton. The term obturator means closed up. The foramen is so named because, even though blood vessels and nerves pass through it, it is nearly completely closed by a fibrous membrane (obturator membrane).

Figure 8.6 Right wrist and hand in relation to the ulna and radius.

The skeleton of the hand consists of the proximal carpals, intermediate metacarpals, and distal phalanges.

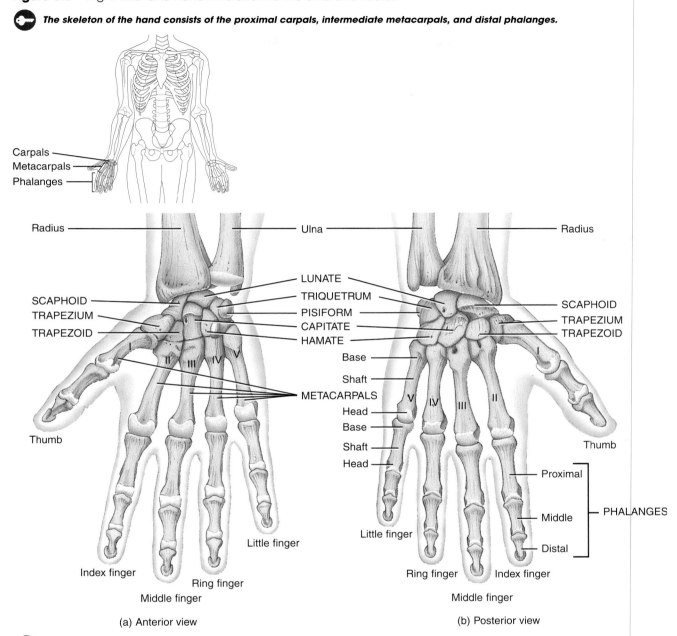

(a) Anterior view

(b) Posterior view

Which wrist bone is fractured most often?

Pubis

The pubis (*pubis* = pubic hair) is the anterior and inferior part of the hipbone. It consists of a **superior ramus**, an **inferior ramus**, and a **body** between the rami that contributes to the formation of the pubic symphysis. The anterior border of the body is known as the **pubic crest** and at its lateral end is a projection, the **pubic tubercle**.

As noted earlier, the pubic symphysis is the joint between the two hipbones (see Fig. 8.15). It consists of a pad of fibrocartilage. The **acetabulum** (as'-e-TAB-yoo-lum; *aceta-*

bulum = vinegar cup) is the fossa formed by the ilium, ischium, and pubis. It is the socket for the head of the femur.

Pelvis

Together with the sacrum and coccyx, the two hipbones of the pelvic girdle form the basinlike structure called the **pelvis** (plural is **pelvises** or **pelves**), as shown in Fig. 8.8. The superior and inferior portions of the pelvis are separated from each other by a plane that connects the sacral

Figure 8.7   Right hipbone. In the three divisions of the hipbone shown in (c), the lines of fusion of the ilium, ischium, and pubis are not always visible in an adult.

 The two hipbones form the pelvic girdle, which attaches the lower limbs to the axial skeleton and supports the vertebral column and viscera.

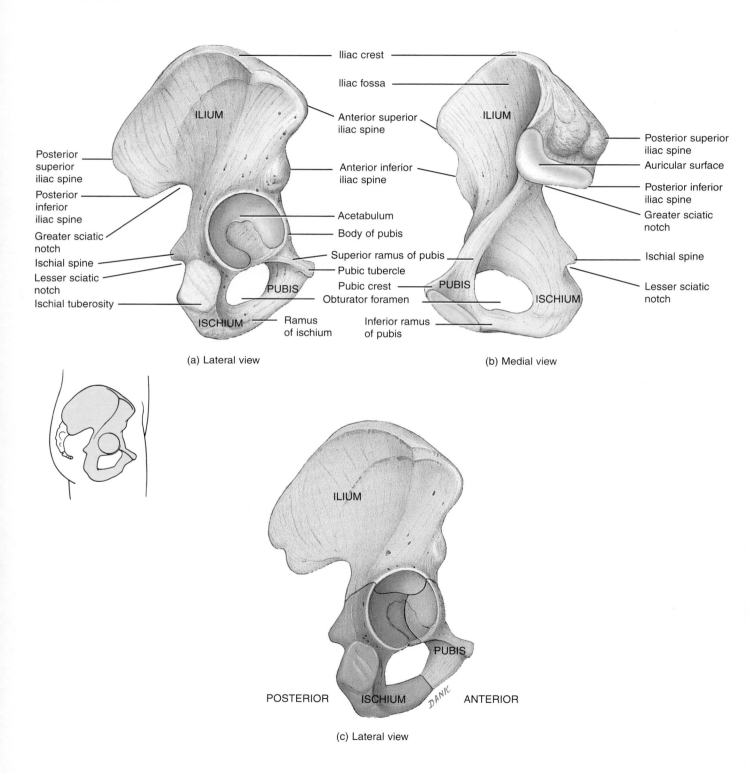

(a) Lateral view

(b) Medial view

(c) Lateral view

Which part of the hipbone articulates with the femur? With the sacrum?

Figure 8.8 Pelvis. Shown here is the female pelvis.

The hipbones are united anteriorly at the pubic symphysis and posteriorly at the sacrum.

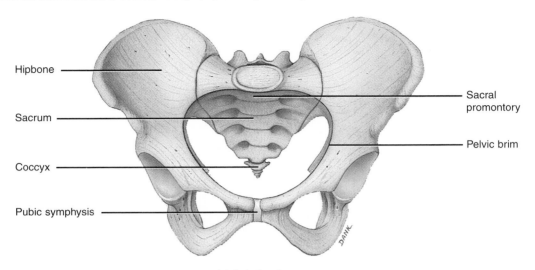

Hipbone

Sacrum

Coccyx

Pubic symphysis

Sacral promontory

Pelvic brim

(a) Anterior view

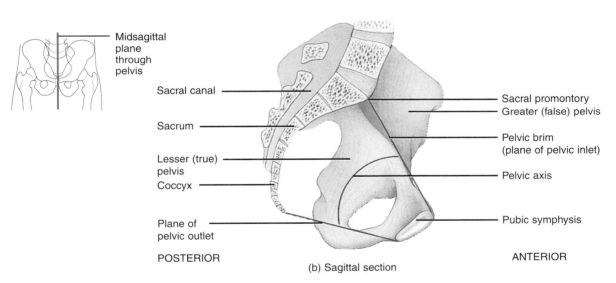

Midsagittal plane through pelvis

Sacral canal

Sacrum

Lesser (true) pelvis

Coccyx

Plane of pelvic outlet

POSTERIOR

Sacral promontory
Greater (false) pelvis

Pelvic brim (plane of pelvic inlet)

Pelvic axis

Pubic symphysis

ANTERIOR

(b) Sagittal section

Which part of the pelvis surrounds the pelvic viscera in the pelvic cavity?

promontory on the posterior side, pubic symphysis on the anterior side, and several points on the lateral side. The circumference of this oblique plane is called the **pelvic brim**.

The portion of the pelvis above the pelvic brim is called the **greater (false) pelvis**. It is bordered by the lumbar vertebrae (posteriorly), the upper portions of the hipbones (laterally), and the abdominal wall (anteriorly). The greater pelvis is actually part of the abdomen and does not contain any pelvic organs, except for the urinary bladder when it is full and the uterus during pregnancy.

The portion of the pelvis inferior to the pelvic brim is called the **lesser (true) pelvis**. It is bounded by the sacrum

and coccyx (posteriorly), inferior portions of the ilium and ischium (laterally), and pubic bones (anteriorly). The lesser pelvis surrounds the pelvic cavity (see Fig. 1.9). The superior opening of the lesser pelvis, actually the pelvic brim, is called the **pelvic inlet**; the inferior opening is called the **pelvic outlet** (Fig. 8.8b). Whereas the pelvic inlet in females is wide and oval-shaped, in males it is narrower and heart-shaped. The **pelvic axis** is an imaginary curved line passing through the lesser pelvis at right angles to the center of the planes of the pelvic inlet and outlet. During childbirth it is the course taken by the baby's head as it descends through the pelvis.

PELVIMETRY

Pelvimetry is the measurement of the size of the inlet and outlet of the birth canal. Measurement of the pelvic cavity in pregnant females is important because the fetus passes through the narrower opening of the lesser pelvis at birth. ■

LOWER LIMB (EXTREMITY)

The **lower limbs** (**extremities**) are composed of 60 bones (Fig. 8.9). Each lower limb includes the femur in the thigh, patella (kneecap), fibula and tibia in the leg, and tarsals (ankle bones), metatarsals, and phalanges (toes) in the foot.

Figure 8.9   Right pelvic (hip) girdle and lower limb.

🔑 *The pelvic girdle attaches the lower limb to the axial skeleton.*

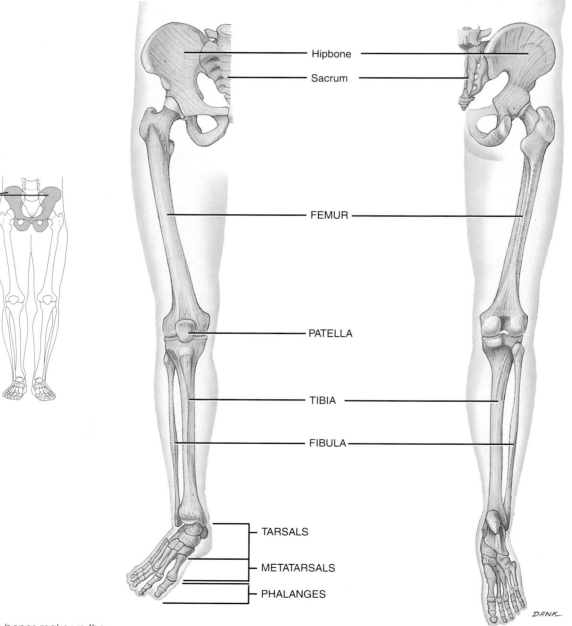

Q Which bones make up the pelvic (hip) girdle?

(a) Anterior view

(b) Posterior view

Femur

The **femur**, or thighbone, is the longest, heaviest, and strongest bone in the body (Fig. 8.10). Its proximal end articulates with the hipbone. Its distal end articulates with the tibia and patella. The **body** (**shaft**) of the femur angles me-

dially and, as a result, the knee joints are brought nearer at the midline. The degree of convergence is greater in females because the female pelvis is broader.

The proximal end of the femur consists of a rounded **head** that articulates with the acetabulum of the hipbone to form the hip joint. The head contains a small centered

---

Figure 8.10   Right femur in relation to the hipbone, patella, tibia, and fibula.

The acetabulum of the hipbone and head of the femur articulate to form the hip joint.

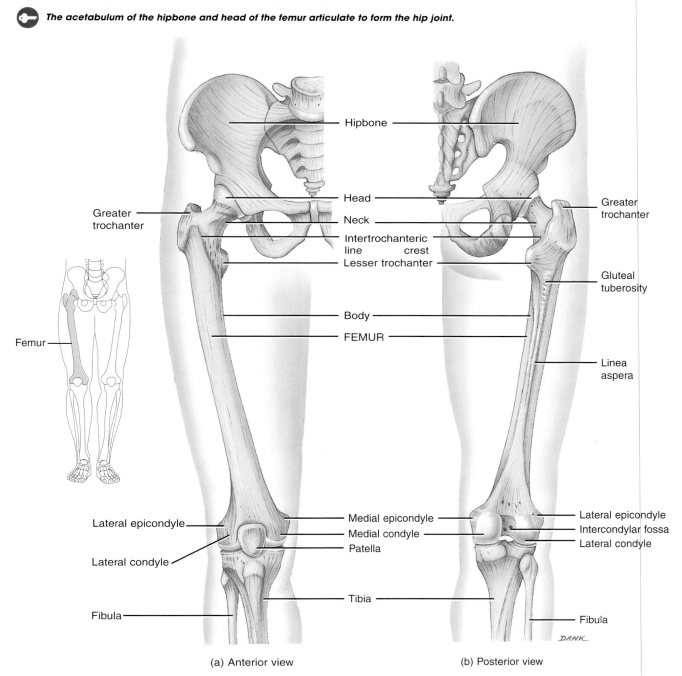

(a) Anterior view   (b) Posterior view

Figure continues

Figure 8.10   (continued) The arrow in the inset in (c) indicates the direction from which the femur is viewed (medial).

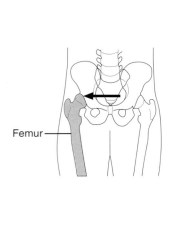

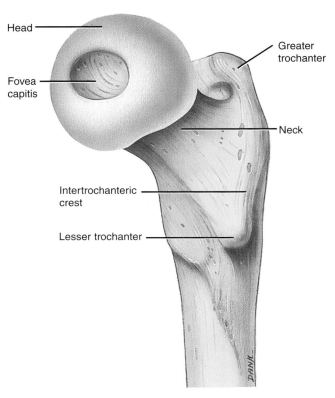

(c) Medial view of proximal end of femur

Changes in what part of the femur are related to "knock-knee" and "bowleg"?

depression (pit) called the **fovea capitis** (FŌ-vē-a CAP-i-tis). The ligament of the head of the femur connects the femur to the acetabulum of the hipbone. The **neck** of the femur is a constricted region distal to the head. A fairly common fracture in the elderly occurs at the neck of the femur, frequently as a result of osteoporosis. Apparently, the neck becomes so weak that it fails to support the weight of body. This condition is commonly known as a broken hip. The **greater trochanter** (trō-KAN-ter) and **lesser trochanter** are projections that serve as points of attachment for some of the thigh and buttock muscles. The greater trochanter is the prominence felt and seen anterior to the hollow on the side of the hip. The lesser trochanter is inferior and medial to the greater trochanter. Between the trochanters on the anterior surface is a narrow **intertrochanteric line**. Between the trochanters on the posterior surface is an **intertrochanteric crest**.

Inferior to the intertrochanteric crest on the **body (shaft)** of the femur is a vertical ridge called the **gluteal tuberosity**. It blends into another vertical ridge called the **linea aspera** (LIN-ē-a AS-per-a). Both ridges serve for the attachment of several thigh muscles.

The distal end of the femur is expanded and includes the **medial condyle** and **lateral condyle**. These articulate with the medial and lateral condyles of the tibia. Superior to the condyles are the **medial epicondyle** and **lateral epicondyle**. A depressed area between the condyles on the posterior surface is called the **intercondylar** (in′-ter-KON-di-lar) **fossa**. The **patellar surface** is located between the condyles on the anterior surface.

Pathologic changes in the angle of the neck of the femur result in abnormal posture of the lower limbs. A decreased angle produces "knock-knee" condition (**genu valgum**). An abnormally large angle produces "bowleg" condition (**genu varum**). Either condition places an abnormal strain on the knee joints.

Patella

The **patella** (*patella* = small plate) or kneecap, is a small, triangular bone located anterior to the knee joint (Fig. 8.11). It is a sesamoid bone that develops in the tendon of the quadriceps femoris muscle. The broad superior end of the patella is called the **base**. The pointed inferior end is

Figure 8.11 Right patella.

🔑 *The patella articulates with the lateral and medial condyles of the femur.*

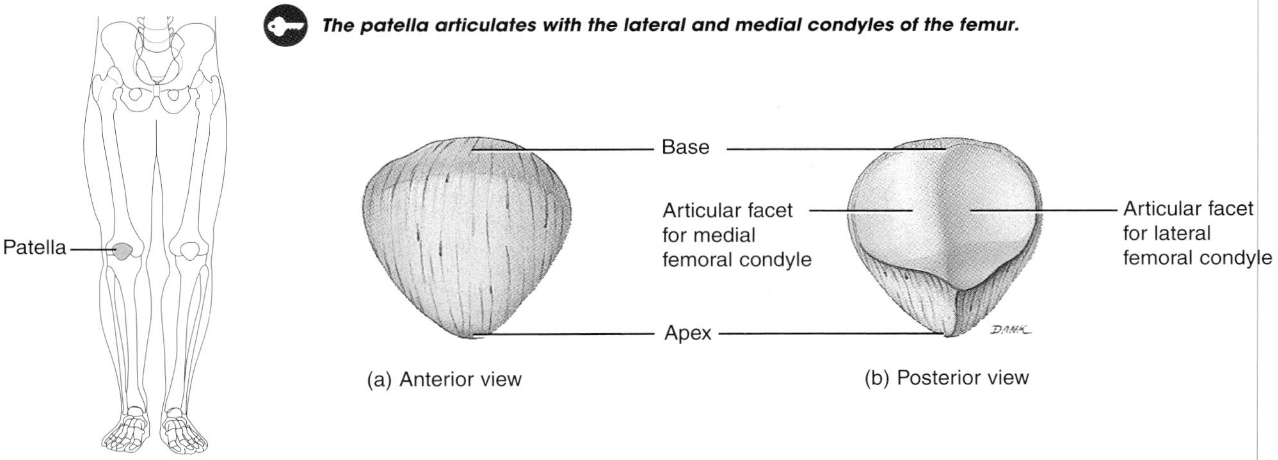

Patella

Base

Articular facet
for medial
femoral condyle

Articular facet
for lateral
femoral condyle

Apex

(a) Anterior view

(b) Posterior view

❓ Because the patella develops in the tendon of the quadriceps femoris muscle, it is classified as which type of bone?

the **apex**. The posterior surface contains two **articular facets**, one for the medial condyle and the other for the lateral condyle of the femur. The patellar ligament attaches the patella to the tibial tuberosity of the tibia. The function of the patella is to increase the leverage of the tendon of the quadriceps femoris muscle (see Fig. 11.19a), to maintain the position of the tendon when the knee is bent (flexed), and to protect the knee joint.

Tibia and Fibula

The **tibia**, or shinbone, is the larger, medial bone of the leg (Fig. 8.12). It bears the weight of the body. The tibia articulates at its proximal end with the femur and fibula and at its distal end with the fibula and talus bone of the ankle.

The proximal end of the tibia is expanded into a **lateral condyle** and a **medial condyle**. These articulate with the condyles of the femur. The inferior surface of the lateral condyle articulates with the head of the fibula. The slightly concave condyles are separated by a superior projection called the **intercondylar eminence.** The **tibial tuberosity** on the anterior surface is a point of attachment for the patellar ligament. Inferior to and continuous with the tibial tuberosity is a sharp ridge that can be felt deep to the skin and is known as the **anterior border (crest)**.

The medial surface of the distal end of the tibia forms the **medial malleolus** (mal-LĒ-ō-lus; *malleus* = little hammer). This structure articulates with the talus bone of the ankle and forms the prominence that can be felt on the medial surface of your ankle. The **fibular notch** articulates with the distal end of the fibula.

The **fibula** is parallel and lateral to the tibia. It is considerably smaller than the tibia. The **head** of the fibula, the

proximal end, articulates with the inferior surface of the lateral condyle of the tibia inferior to the level of the knee joint. The distal end has a projection called the **lateral malleolus** that articulates with the talus bone of the ankle. This forms the prominence on the lateral surface of the ankle. As noted, the fibula also articulates with the tibia at the fibular notch.

Tarsals, Metatarsals, and Phalanges

The skeleton of the foot has three regions: (1) the proximal tarsus, (2) the intermediate metatarsus, and (3) the distal phalanges.

The **tarsus** is a collective term for the seven **tarsal bones** of the ankle (Fig. 8.13). They include the **talus** (TĀ-lus; *talas* = ankle bone) and **calcaneus** (kal-KĀ-nē-us), or heel bone, located in the posterior part of the foot. The anterior tarsal bones are the **cuboid, navicular** (boat-shaped), and three **cuneiform** (*cuneiform* = wedge-shaped) **bones** called the **first (medial), second (intermediate),** and **third (lateral) cuneiforms.** The talus, the most superior tarsal bone, is the only bone of the foot that articulates with the fibula and tibia. It is bordered on one side by the medial malleolus of the tibia and on the other side by the lateral malleolus of the fibula. During walking, the talus initially bears the entire weight of the body. About half the weight is then transmitted to the calcaneus. The remainder is transmitted to the other tarsal bones. The calcaneus is the largest and strongest tarsal bone.

The **metatarsus** consists of five **metatarsal bones** numbered I to V from the medial to lateral position. Like the metacarpals of the palm of the hand, each metatarsal consists of a proximal **base**, an intermediate **shaft**, and a distal

Figure 8.12   Right tibia and fibula in relation to the femur, patella, and talus. The arrow in the inset in (c) indicates the direction from which the tibia is viewed (lateral).

The tibia articulates with the femur and fibula proximally and fibula and talus distally.

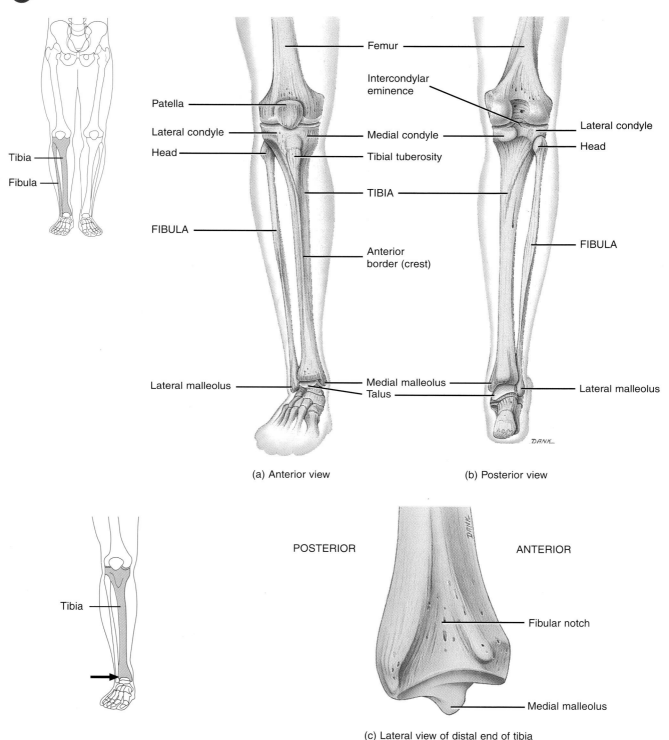

Tibia

Fibula

Femur

Patella

Lateral condyle

Head

FIBULA

Lateral malleolus

Intercondylar eminence

Medial condyle

Tibial tuberosity

TIBIA

Anterior border (crest)

Medial malleolus

Talus

Lateral condyle

Head

FIBULA

Lateral malleolus

(a) Anterior view

(b) Posterior view

Tibia

POSTERIOR

ANTERIOR

Fibular notch

Medial malleolus

(c) Lateral view of distal end of tibia

Q Which leg bone bears the weight of the body?

Figure 8.13   Right foot. The arrow in the inset indicates the direction from which the foot is viewed (superior).

🔑 *The skeleton of the foot consists of the proximal tarsals, intermediate metatarsals, and distal phalanges.*

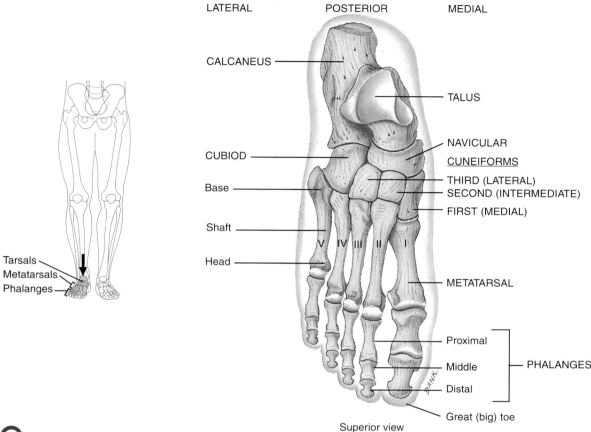

LATERAL   POSTERIOR   MEDIAL

CALCANEUS

TALUS

NAVICULAR

CUNEIFORMS

CUBIOD

THIRD (LATERAL)

Base

SECOND (INTERMEDIATE)

FIRST (MEDIAL)

Shaft

V IV III II I

Head

METATARSAL

Proximal

Middle — PHALANGES

Distal

Great (big) toe

Superior view

Tarsals
Metatarsals
Phalanges

❓ Which tarsal bone articulates with the tibia and fibula?

head. The metatarsals articulate proximally with the first, second, and third cuneiform bones and with the cuboid. Distally, they articulate with the proximal row of phalanges. The first metatarsal is thicker than the others because it bears more weight.

The **phalanges** of the foot resemble those of the hand in both number and arrangement. Each also consists of a proximal **base**, an intermediate **shaft**, and a distal **head.** The great or big toe (**hallux**), has two large, heavy phalanges called proximal and distal phalanges. The other four toes each have three phalanges—proximal, middle, and distal.

Arches of the Foot

The bones of the foot are arranged in two **arches** (Fig. 8.14). These arches enable the foot to support the weight of the body, provide an ideal distribution of body weight over the hard and soft tissues of the foot, and provide leverage while walking. The arches are not rigid. They yield as weight is applied and spring back when the weight is lifted,

thus helping to absorb shocks. Usually, the arches are fully developed by age 12 or 13.

The **longitudinal arch** has two parts. Both consist of tarsal and metatarsal bones arranged to form an arch from the anterior to the posterior part of the foot. The **medial** (inner) **part** of the longitudinal arch originates at the calcaneus. It rises to the talus and descends through the navicular, the three cuneiforms, and the heads of the three medial metatarsals. The **lateral** (outer) **part** of the longitudinal arch also begins at the calcaneus. It rises at the cuboid and descends to the heads of the two lateral metatarsals.

The **transverse arch** is formed by the navicular, three cuneiforms, and the bases of the five metatarsals.

FLATFOOT, CLAWFOOT, AND BUNIONS

The bones composing the arches are held in position by ligaments and tendons. If these ligaments and tendons are weakened, the

Figure 8.14   Arches of the right foot.

 Arches help the foot support and distribute the weight of the body and provide leverage while walking.

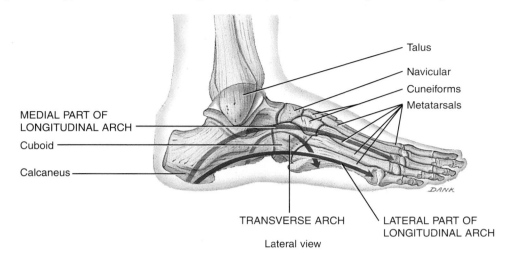

Lateral view

Q What structural aspect of the arches allows them to absorb shocks?

height of the medial longitudinal arch may decrease or "fall." The result is **flatfoot** and is characterized by lateral deviation of the foot. Causes include excessive weight, postural abnormalities, weakened supporting tissues, and genetic predisposition. A custom-designed arch support (orthotic) may be prescribed to treat flatfoot.

Clawfoot is a condition in which the medial longitudinal arch is abnormally elevated. It is often caused by muscle imbalance, such as may result from poliomyelitis.

A **bunion** (hallux valgus) is a deformity of the great toe. Although the condition may be inherited, it is typically caused by wearing tightly fitting shoes and is characterized by lateral deviation of the proximal phalanx of the great toe and medial displacement of metatarsal I. Arthritis of the first metatarsophalangeal joint may also be a predisposing factor. The condition produces inflammation of bursae (fluid-filled sacs at the joint), bone spurs, and calluses. ■

FEMALE AND MALE SKELETONS

The bones of the male are generally larger and heavier than those of the female. The articular ends are thicker in relation to the shafts. Also, since certain muscles of the male are generally larger than those of the female, the points of attachment—tuberosities, lines, ridges—are usually larger in the male skeleton.

Many significant structural differences between female and male skeletons occur in the pelvis. Most are structural adaptations for pregnancy and childbirth. The female pelvis is wider and shallower than the male pelvis. Because of these characteristics, the female pelvis provides more room in the lesser (true) pelvis, especially in the pelvic inlet and pelvic outlet, to accommodate the passage of the infant's head at birth. Typical differences are listed in Exhibit 8.1 and illustrated in Fig. 8.15.

EXHIBIT **8.1** **DIFFERENCES IN TYPICAL FEMALE AND MALE PELVISES**

Point of Comparison	Female	Male
Greater (false) pelvis (see Fig. 8.8b)	Shallow.	Deep.
Pelvic brim (pelvic inlet)	Larger and more oval.	Heart-shaped.
Pubic arch (arch formed by pubic rami)	Greater than a 90° angle.	Less than a 90° angle.
Ilium	Less vertical.	More vertical.
Iliac fossa (see Fig. 8.7b)	Shallow.	Deep.
Iliac crest	Less curved.	More curved.
Acetabulum	Small.	Large.
Obturator foramen	Oval.	Round.

Figure 8.15 Comparison of female and male pelvises.

 Many structural adaptations of the female pelvis are related to pregnancy and childbirth.

Right lateral view
of female pelvis

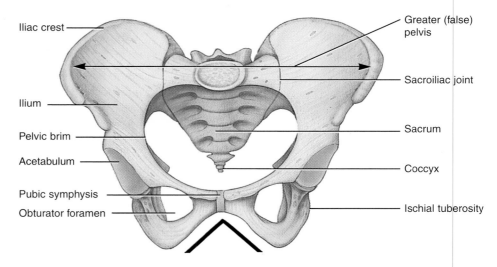

Iliac crest

Greater (false) pelvis

Ilium

Sacroiliac joint

Pelvic brim

Sacrum

Acetabulum

Coccyx

Pubic symphysis

Obturator foramen

Ischial tuberosity

Pubic arch (greater than 90°)

(a) Anterior view of female pelvis

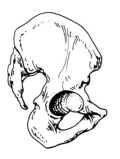

Right lateral view
of male pelvis

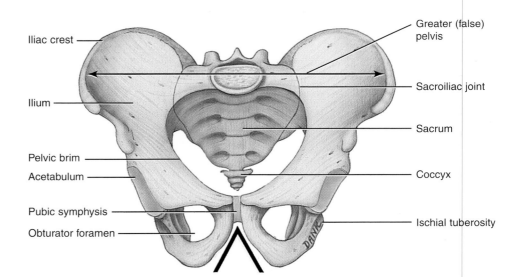

Iliac crest

Greater (false) pelvis

Ilium

Sacroiliac joint

Pelvic brim

Sacrum

Acetabulum

Coccyx

Pubic symphysis

Obturator foramen

Ischial tuberosity

Pubic arch (less than 90°)

(b) Anterior view of male pelvis

Q What is the main difference between the pelvic brim in females and males?

STUDY OUTLINE

PECTORAL (SHOULDER) GIRDLE (p. 196)

1. Each pectoral (shoulder) girdle consists of a clavicle and scapula.
2. Each attaches an upper limb (extremity) to the trunk.

UPPER LIMB (EXTREMITY) (p. 197)

1. There are 60 bones in the upper limbs (extremity).
2. The bones of each upper limb include the humerus, ulna, radius, carpals, metacarpals, and phalanges.

PELVIC (HIP) GIRDLE (p. 202)

1. The pelvic (hip) girdle consists of two hipbones.
2. It attaches the lower limbs to the trunk at the sacrum.
3. Each hipbone consists of three fused components—ilium, pubis, and ischium.

LOWER LIMB (EXTREMITY) (p. 206)

1. There are 60 bones in the lower limbs (extremities).
2. The bones of each lower limb include the femur, tibia, patella, fibula, tarsals, metatarsals, and phalanges.
3. The bones of the foot are arranged in two arches, the longitudinal arch and the transverse arch, to provide support and leverage.

FEMALE AND MALE SKELETONS (p. 212)

1. Male bones are generally larger and heavier than female bones and have more prominent markings for muscle attachment.
2. The female pelvis is adapted for pregnancy and childbirth. Differences in male and female pelvic structure are listed in Exhibit 8.1.

REVIEW QUESTIONS

1. What is the pectoral (shoulder) girdle? Why is it important? (p. 196)
2. What are the bones of the upper limb? (p. 197)
3. What is the pelvic (hip) girdle? Why is it important? (p. 202)
4. What is pelvimetry? What is the clinical importance of pelvimetry? (p. 206)
5. What are the bones of the lower limb? (p. 206)
6. In what ways do the upper limb and lower limb differ structurally? (pp. 197, 206)
7. Describe the structure of the longitudinal and transverse arches of the foot. What is the function of an arch? (p. 211)
8. What are the principal structural differences between typical female and male pelvises? Use Exhibit 8.1 as a guide in formulating your response. (p. 212)

ANSWERS TO QUESTIONS WITH FIGURES

8.1 Clavicle and scapula.
8.2 The junction of the two curves.
8.3 Acromion.
8.4 Radius: capitulum and radial fossa; ulna: trochlea, coronoid fossa, and olecranon fossa.
8.5 The capitulum of the humerus and the radial notch of the ulna.
8.6 Scaphoid.
8.7 Femur: acetabulum; sacrum: auricular surface.
8.8 Lesser (true) pelvis.
8.9 Hipbones.
8.10 The angle of the neck.
8.11 Sesamoid.
8.12 Tibia.
8.13 Talus.
8.14 They are not rigid, yielding when weight is applied and springing back when weight is lifted.
8.15 Larger and more oval in females and heart-shaped in males.

STUDENT OBJECTIVES

1. Define an articulation (joint) and identify the factors that determine the types and degree (range) of movement at a joint. (p. 216)

2. Classify joints on the basis of structure and function. (p. 216)

3. Contrast the structure, kind of movement, and location of immovable, slightly movable, and freely movable joints. (p. 216)

4. Describe the structure, types, and movements of freely movable joints. (p. 217)

5. Describe the shoulder and knee joints with respect to the bones that enter into their formation, structural classification, articular components, and movements. (p. 224)

6. Describe the causes and symptoms of common joint disorders, including rheumatism, rheumatoid arthritis (RA), osteoarthritis (OA), gouty arthritis, Lyme disease, bursitis, ankylosing spondylitis, sprain, and strain. (p. 234)

7. Define medical terminology associated with articulations. (p. 235)

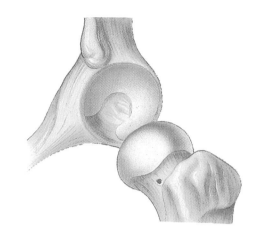

CHAPTER CONTENTS AT A GLANCE

ones are too rigid to bend without being damaged. Fortunately, flexible connective tissues form joints that hold bones together while still permitting some degree of movement in many cases. Since most body movements occur at joints, you can appreciate their importance if you imagine how a cast over the knee joint makes walking difficult or how a splint on a finger limits the ability to manipulate small objects. A few joints do not permit any movement at all but do provide a great degree of protection.

An **articulation** (**joint**) is a point of contact between bones, between cartilage and bones, or between teeth and bones. When we say that one bone *articulates* with another, we mean that one bone forms a joint with another bone. The scientific study of joints is called **arthrology** (ar-THROL-ō-jē; *arthro* = joint; *logos* = study of). **Kinesiology** (ki-nē-sē-OL-ō-jē; *kinesis* = movement) is the study of motion of the human body.

Generally, the closer the bones fit at the point of contact, the stronger the joint. Tightly fitted joints, however, restrict movement. The looser the fit, the greater the movement. Unfortunately, loosely fitted joints are prone to dislocation (displacement). Several other factors also affect joint movement: (1) the precise manner in which the articulating bones fit together; (2) the flexibility (tension or tautness) of the tissues that bind the bones together; and (3) the position of ligaments, muscles, and tendons (discussed in detail later).

CLASSIFICATION OF JOINTS

Joints may be categorized into structural classes, based on anatomical characteristics, or into functional classes, based on the type of movement they permit.

Structural Classification

The structural classification of joints is based on the presence or absence of a space between the articulating bones that is called a synovial (joint) cavity (described on page 217) and the type of connective tissue that binds the bones together. Structurally, a joint is classified as (1) **fibrous**, if there is no synovial (joint) cavity and the bones are held together by fibrous (collagenous) connective tissue; (2) **cartilaginous**, if there is no synovial cavity and the bones are held together by cartilage; or (3) **synovial**, if there is a synovial cavity and the bones forming the joint are united by a surrounding articular capsule and frequently by accessory ligaments (described in detail later).

Functional Classification

The functional classification of joints takes into account the degree of movement they permit. Functionally, a joint is classified as follows:

1. A **synarthrosis** (sin′-ar-THRŌ-sis; *syn* = together; *arthros* = joint) is an immovable joint; plural is **synarthroses**.
2. An **amphiarthrosis** (am′-fē-ar-THRŌ-sis; *amphi* = on both sides) is a slightly movable joint; plural is **amphiarthroses**.
3. A **diarthrosis** (dī-ar-THRŌ-sis; *diarthros* = movable joint) is a freely movable joint; plural is **diarthroses**.

SYNARTHROSES (IMMOVABLE JOINTS)

There are three types of synarthroses, or immovable joints: sutures, gomphoses, and synchondroses.

Suture

A **suture** (SOO-cher; *sutura* = seam) is a fibrous joint composed of a thin layer of dense fibrous connective tissue that unites bones of the skull. An example of a suture is the coronal suture between the frontal and parietal bones (see Fig. 7.3). The irregular, interlocking edges of sutures give them added strength and decrease their chance of fractures. Some sutures, although present during childhood, are replaced by bone in the adult. Such a suture is referred to as a **synostosis** (sin′-os-TŌ-sis; *syn* = together; *osteon* = bone), or bony joint—a joint in which there is a complete fusion of bone across the suture line. An example of a synostosis is the frontal suture between the left and right sides of the frontal bone that begins to fuse during infancy (see Fig. 7.6a).

Gomphosis

A **gomphosis** (gom-FŌ-sis; *gomphosis* = to bolt together) is a type of fibrous joint in which a cone-shaped peg fits into a socket. The substance between the two is the periodontal ligament. The only examples are the articulations of the roots of the teeth with the sockets of the alveolar processes of the maxillae and mandible (see Fig. 24.6).

Synchondrosis

A **synchondrosis** (sin′-kon-DRŌ-sis; *syn* = together; *chondros* = cartilage) is a cartilaginous joint in which the connecting material is hyaline cartilage. The most common type of synchondrosis is the epiphyseal plate (see Fig. 6.6). Such a joint connects the epiphysis and diaphysis of a growing bone. Since the hyaline cartilage is eventually replaced by bone or fibrocartilage when growth ceases, the joint is temporary. It is replaced by a synostosis or symphysis (described next).

AMPHIARTHROSES (SLIGHTLY MOVABLE JOINTS)

There are two types of amphiarthroses, or slightly movable joints: syndesmoses and symphyses.

Syndesmosis

A **syndesmosis** (sin′-dez-MŌ-sis; *syndesmo* = band or ligament) is a fibrous joint in which there is considerably more fibrous connective tissue than there is in a suture. As a result, the fit between the bones is not quite as tight. The fibrous connective tissue forms an interosseous membrane or ligament that permits some degree of flexibility and movement. An example of a syndesmosis is the distal articulation between the tibia and fibula (see Fig. 8.12).

Symphysis

A **symphysis** (SIM-fi-sis; *symphysis* = growing together) is a cartilaginous joint in which the connecting material is a broad, flat disc of fibrocartilage. This type of joint is found in intervertebral discs between the bodies of vertebrae (see Fig. 7.16). Recall that the outer portion of an intervertebral disc is fibrocartilaginous material (annulus fibrosus). The pubic symphysis between the anterior surfaces of the hipbones is another example of a symphysis (see Fig. 8.15).

DIARTHROSES (FREELY MOVABLE JOINTS)

Diarthroses, or freely movable joints, have a variety of shapes and permit several different types of movements. First, we discuss the general structure of a diarthrosis and then consider the different types.

Structure of Diarthroses

A distinguishing anatomical feature of diarthroses is the presence of a space, called a **synovial** (si-NŌ-vē-al) (**joint**) **cavity** (Fig. 9.1), that separates the articulating bones. For this reason, diarthroses are also referred to as **synovial joints**. Another characteristic of such joints is the presence of **articular cartilage**. Articular cartilage (which is the hyaline type) covers the surfaces of the articulating bones but does not bind the bones together. Articular cartilage reduces friction at the joint when the bones move and helps absorb shock.

A sleevelike **articular capsule** surrounds a diarthrosis, encloses the synovial cavity, and unites the articulating bones. The articular capsule is composed of two layers. The outer layer, called the **fibrous capsule**, usually consists of

dense, irregular connective tissue. It attaches to the periosteum of the articulating bones at a variable distance from the edge of the articular cartilage. The flexibility of the fibrous capsule permits considerable movement at a joint, whereas its great tensile strength resists dislocation. The fibers of some fibrous capsules are arranged in parallel bundles and are therefore highly adapted to resist recurrent strain. Such bundles of fibers are called **ligaments** (*ligare* = to bind) and are given special names. Examples of such ligaments include the coracohumeral and glenohumeral ligaments associated with the shoulder joint (see Fig. 9.7a) and the tibial and fibular collateral ligaments associated with the knee joint (see Fig. 9.8a). The strength of such ligaments is one of the principal factors in holding bone to bone. Diarthroses are freely movable joints because of the presence of the joint cavity and the arrangement of the articular capsule and ligaments.

The inner layer of the articular capsule is formed by a **synovial membrane**. The synovial membrane is composed of areolar connective tissue with elastic fibers and a variable amount of adipose tissue. It secretes **synovial fluid** (**SF**). This fluid, which fills the synovial cavity, has several functions. It lubricates and reduces friction in the joint. It also supplies nutrients to and removes metabolic wastes from the chondrocytes of the articular cartilage. (Recall that cartilage is avascular.) Synovial fluid also contains phagocytic cells that remove microbes and debris resulting from wear and tear in the joint. Synovial fluid consists of hyaluronic acid and an interstitial fluid formed from blood plasma. It is similar in appearance and consistency to uncooked egg white. When there is no joint movement, the fluid is quite viscous, but as joint movement increases, the fluid becomes less viscous. The amount of synovial fluid varies according to the size of the joint. For example, a large joint, such as the knee, may contain 3–4 ml (about ⅛ oz) of fluid. It forms a thin, viscous film over the surfaces within the articular capsule.

Many diarthroses also contain **accessory ligaments**, called extracapsular ligaments and intracapsular ligaments. **Extracapsular ligaments** lie *outside* the articular capsule. An example is the oblique popliteal ligament of the knee joint (see Fig. 9.8b). **Intracapsular ligaments** occur *within* the articular capsule but are excluded from the synovial cavity by folds of the synovial membrane. Examples are the anterior and posterior cruciate ligaments of the knee joint (see Fig. 9.8e).

Inside some synovial joints, such as the knee, are pads of fibrocartilage that lie between the articular surfaces of the bones. They are attached by their margins to the fibrous capsule. These pads are called **articular discs (menisci)**. In Fig. 9.8e you can see the lateral and medial menisci of the knee joint. The discs usually subdivide the synovial cavity into two separate spaces. Articular discs allow two bones of different shapes to fit tightly; they modify the shape of the joint surfaces of the articulating bones. Articular discs also help to maintain the stability of the joint and direct the flow of synovial fluid within the joint to areas of greatest friction.

Figure 9.1   Diarthrosis. Generalized structure.

 A diarthrosis is distinguished by a synovial (joint) cavity between articulating bones.

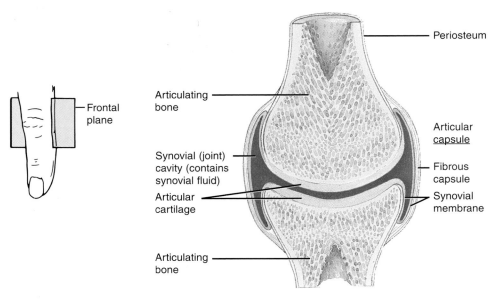

Periosteum

Articulating bone

Synovial (joint) cavity (contains synovial fluid)

Articular cartilage

Articulating bone

Articular capsule

Fibrous capsule

Synovial membrane

Frontal plane

(a) Frontal section of a generalized diarthrotic (synovial) joint

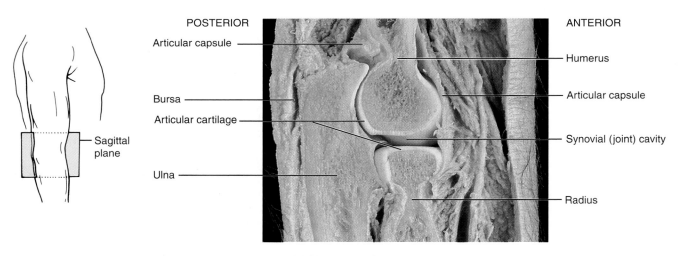

POSTERIOR

Articular capsule

Bursa

Articular cartilage

Ulna

Sagittal plane

ANTERIOR

Humerus

Articular capsule

Synovial (joint) cavity

Radius

(b) Photograph of sagittal section of elbow joint

Structurally, how is this type of joint classified?

TORN CARTILAGE

A tearing of articular discs in the knee is commonly called **torn cartilage**. The condition occurs frequently among athletes. Such damaged cartilage requires surgical removal (meniscectomy), or it will begin to wear and cause arthritis. At one time, knee joint surgery for torn cartilage required cutting through several layers of healthy tissue and removing much, if not all, of the cartilage. This procedure is usually painful and expensive and does not always provide full recovery. These problems have largely been overcome by arthroscopy (described on page 231). ■

The various motions of the body create friction between moving parts. Saclike structures called **bursae** (*bursa* = pouch or purse) are strategically situated to alleviate friction in some joints, such as the shoulder and knee joints. Bursae resemble joint capsules in that their walls consist of connective tissue lined by a synovial membrane. They are also filled with a fluid similar to synovial fluid. Bursae are located between the skin and bone in places where skin rubs over bone. They are also found between tendons and bones, muscles and bones, and ligaments and bones and within articular capsules (see Fig. 9.8c). Such fluid-filled sacs cushion the movement of one part of the body over another. Inflammation of a bursa is called **bursitis** (see page 235).

Factors Affecting Movement at Diarthroses

Several factors contribute to keeping the articular surfaces of diarthroses in contact. How the surfaces contact one another, in turn, determines the type and extent of motion that is possible.

1. First is the **structure or shape of the articulating bones**, which determines how they fit together. An interlocking shape is very obvious at the hip joint, where the head of the femur articulates with the acetabulum of the hipbone. This type of fit allows rotational movement. Other shapes permit a diversity of motions, described shortly.

2. A second factor is the **strength and tension (tautness) of the joint ligaments**. The different components of a fibrous capsule are tense only when the joint is in certain positions. Tense ligaments not only restrict the range of movement but also direct the movement of the articulating bones with respect to each other. In the knee joint, for example, the major ligaments have less tension when the knee is bent but more tension when the knee is straightened. Also, when the knee is straightened, the surfaces of the articulating bones are in fullest contact with each other.

3. A third factor that holds joints together but also restricts movement is the **arrangement and tension of the mus-**cles around the joint. Muscle tension reinforces the restraint placed on a joint by ligaments. A good example of the effect of muscle tension on a joint is seen at the hip joint. When the thigh is raised with the knee straight, the movement is restricted by the tension of the hamstring muscles on the posterior surface of the thigh. But if the knee is bent, the tension on the hamstring muscles is lessened, and the thigh can be raised further.

4. In a few joints, the **apposition** (coming together) **of soft parts** may limit mobility. For example, if you bend your arm at the elbow, it can move no further after the anterior surface of the forearm meets with and presses against the arm.

5. Joint flexibility may also be affected by **hormones**. For example, relaxin, a hormone produced by the placenta and ovaries, increases the flexibility of the pubic symphysis and ligaments between the sacrum and hipbone and sacrum and coccyx. This allows expansion of the pelvic outlet, which eases delivery of the baby.

Types of Diarthroses

Though all diarthroses have a generally similar structure, the shape of the articulating surfaces varies. Accordingly, diarthroses are divided into six subtypes: gliding, hinge, pivot, condyloid, saddle, and ball-and-socket joints.

Gliding Joint

The articulating surfaces of bones in a **gliding** or **arthrodial** (ar-THRŌ-dē-al) **joint** are usually flat. A gliding movement is the simplest kind that can occur at a joint. Only side-to-side and back-and-forth movements are permitted (Fig. 9.2a). Twisting and rotation are prevented at gliding joints, generally because ligaments or adjacent bones restrict the range of movement. Some joints that glide are those between the carpals and between the tarsals. The heads and tubercles of ribs glide on the bodies and transverse processes of vertebrae. Also, the clavicle glides on the sternum and the scapula.

Hinge Joint

In a **hinge** or **ginglymus** (JIN-gli-mus; *ginglymos* = hinge) **joint**, the convex surface of one bone fits into the concave surface of another bone. Hinge joints include the knee, elbow, ankle, and interphalangeal joints. Movement is primarily in a single plane, and the joint is therefore known as **monaxial** or **uniaxial** (Fig. 9.2b). The motion is similar to that of a hinged door. Movement is usually flexion and extension. **Flexion** decreases the angle between articulating bones. For example, it occurs when you bend your knee or elbow (Fig. 9.3). **Extension** increases the angle between articulating bones, often to restore a body part to its anatomical position after it has been flexed (Fig. 9.3). Some hinge joints are capable of **hyperextension** (Fig. 9.3a,b,d), continuation of extension beyond the anatomical position, such as when the head bends backward.

Figure 9.2   Subtypes of diarthroses. For each subtype shown, there is a drawing of the actual joint and a simplified diagram.

 Diarthroses are classified into subtypes on the basis of the shapes of the articulating bone surfaces.

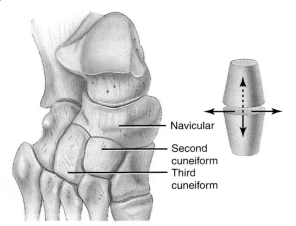

(a) Gliding joint between the navicular and second and third cuneiforms of the tarsus in the foot

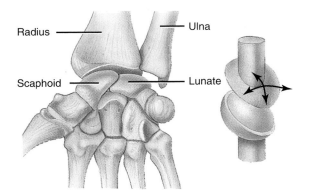

(d) Condyloid joint between radius and scaphoid and lunate bones of the carpus (wrist)

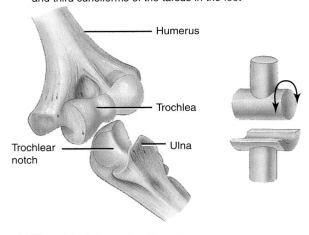

(b) Hinge joint between trochlea of humerus and trochlear notch of ulna at the elbow

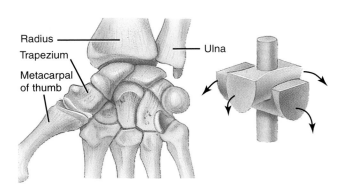

(e) Saddle joint between trapezium of carpus (wrist) and metacarpal of thumb

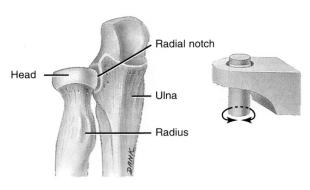

(c) Pivot joint between head of radius and radial notch of ulna

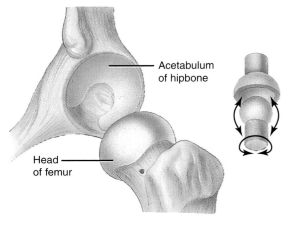

(f) Ball-and-socket joint between head of the femur and acetabulum of the hipbone

 Which joints shown are biaxial?

Figure 9.3 Flexion, extension, and hyperextension.

Whereas flexion decreases the angle between articulating bones, extension increases the angle.

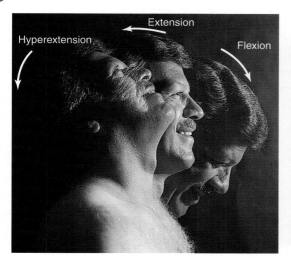

(a) Atlanto-occipital and cervical intervertebral joints

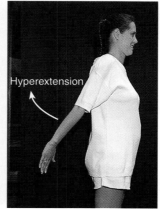
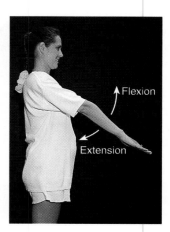

(b) Shoulder joint

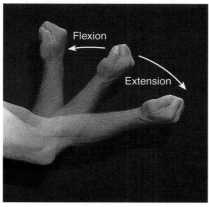

(c) Elbow joint

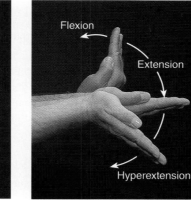

(d) Wrist joint

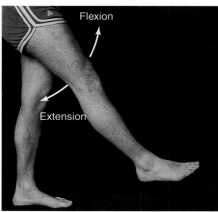

(e) Hip joint

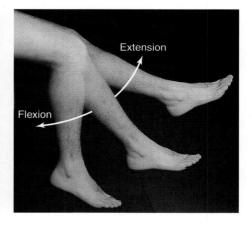

(f) Knee joint

 Why are hinge joints classified as monaxial?

Pivot Joint

In a **pivot** or **trochoid** (TRŌ-koyd; *trokhos* = wheel) **joint**, a rounded or pointed surface of one bone articulates within a ring formed partly by another bone and partly by a ligament. The primary movement permitted is **rotation**, where a bone moves in a single plane around its longitudinal axis. The joint is therefore monaxial. During rotation, no other motion is permitted. The atlas rotates around the dens of the axis when you turn your head from side to side to indicate "no" (Fig. 9.4). Another pivot joint is found between the proximal ends of the ulna and radius (see Fig. 9.2c) and it allows us to turn the palms forward (or upward) and backward (or downward). In **medial** (**internal**) **rotation**, the anterior surface of a bone or limb moves toward the midline. For example, turning the palm from anterior (as in the anatomical position) to posterior results in medial rotation of the forearm. In **lateral** (**external**) **rotation**, the anterior surface moves away from the midline. Turning the palm from posterior to anterior results in lateral rotation of the forearm.

Condyloid Joint

In a **condyloid** (KON-di-loyd; *kondylos* = knuckle) or **ellipsoidal joint**, an oval-shaped condyle of one bone fits into an elliptical cavity of another bone (see Fig. 9.2d). The joint at the wrist between the radius and carpals is condyloid. The movement permitted by such a joint is in two planes, side-to-side and back-and-forth, as when you flex and extend and abduct and adduct (described shortly) your wrist. Such a joint is said to be **biaxial**. At a condy-loid joint, it is possible to combine the movements of flexion–extension and abduction–adduction in succession to produce a movement called **circumduction**. In circumduction, the distal end of a part of the body moves in a circle. For example, you can circumduct your humerus by moving your arm in a circle (see Fig. 9.5d). Since circumduction is a combined movement, it is not considered to be a separate plane of movement.

Abduction (= taking away) usually refers to movement of a bone away from the midline of the body (Fig. 9.5a,b). An example is moving the arms upward from the sides of the trunk and away from the body. With the fingers and toes, however, the midline of the body is not used as the line of reference. Abduction of the fingers (not the thumb) is a movement away from an imaginary line drawn through the middle finger; in other words, it is spreading the fingers (Fig. 9.5c). Abduction of the thumb moves the thumb away from the plane of the palm at a right angle to the palm. Abduction of the toes is relative to an imaginary line drawn through the second toe. Instead of abduction, the term **lateral flexion** is used to refer to bending the trunk to the left or right away from the midline.

Adduction (= to move toward) usually refers to movement toward the midline of the body (Fig. 9.5a,b). An example of adduction is returning the arm to the side after abduction. As in abduction, adduction of the fingers is relative to the middle finger (Fig. 9.5c), and adduction of the toes is relative to the second toe. In adduction of the thumb, the thumb moves toward the plane of the palm at a right angle to the palm. Abduction and adduction are also angular movements.

Saddle Joint

In a **saddle** or **sellaris** (sel-A-ris; *sellar* = saddle) **joint**, the articular surface of one bone is saddle-shaped, and the articular surface of the other bone is shaped like the legs of a rider sitting in the saddle. The joint between the trapezium of the carpus and metacarpal of the thumb is an example of a saddle joint. Movements at a saddle joint are side-to-side and back-and-forth (see Fig. 9.2e). The saddle joint is a modified condyloid joint in which the movement is somewhat freer. Such joints are biaxial and also permit circumduction. In circumduction, the thumb is moved in a circle.

The saddle joint between the trapezius and metacarpal of the thumb permits opposition. This is the movement of the thumb so that the tip of the thumb can meet the tip of any other digits on the same hand. Opposition provides humans and other primates the ability to grasp and manipulate objects.

Ball-and-Socket Joint

A **ball-and-socket** or **spheroid** (SFĒ-royd; *spheroideum* = shaped like a sphere) **joint** consists of a ball-like surface of one bone fitted into a cuplike depression of another bone.

Figure 9.4   Rotation.

 In a pivot joint, the pointed surface of one bone fits into a ring formed by another bone and a ligament.

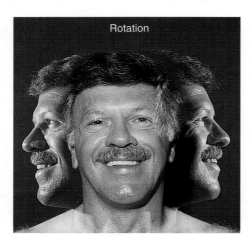

Rotation

Rotation of the head at the atlantoaxial joint

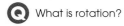 What is rotation?

Figure 9.5 Abduction, adduction, and circumduction.

 In circumduction, the distal end of a part of the body moves in a circle.

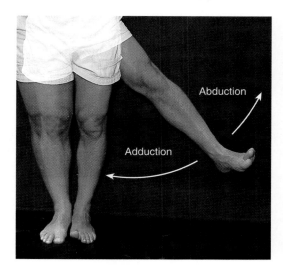

(a) Hip joint

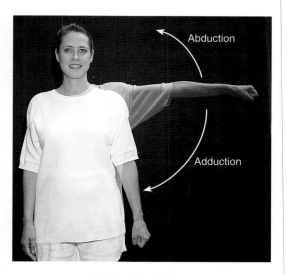

(b) Shoulder joint

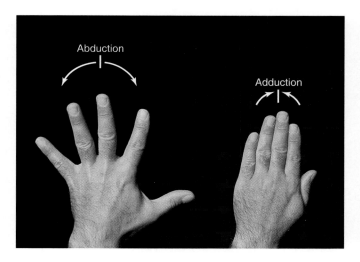

(c) Metacarpophalangeal joints

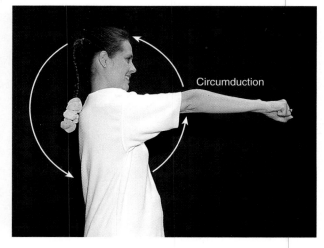

(d) Shoulder joint

 You can remember adduction as "*add*ing your limb to your trunk." Why is this an accurate learning device?

The only examples of ball-and-socket joints are the shoulder joint and hip joint (Fig. 9.2f). Such joints are said to be **triaxial** because they permit movement in three planes: flexion–extension, abduction–adduction, and rotation. The shoulder joint also permits circumduction, as when winding up to pitch a ball (Fig. 9.5d).

A summary of joints based on functional classification is presented in Exhibit 9.1.

EXHIBIT **9.1** **SUMMARY OF JOINTS BASED ON FUNCTIONAL CLASSIFICATION**

Type	Description	Examples
SYNARTHROSIS (IMMOVABLE)		
Suture	Found only between bones of the skull; articulating bones united by a thin layer of dense fibrous connective tissue.	Coronal suture between frontal and parietal bones.
Gomphosis	Cone-shaped peg fits into a socket; articulating bones united by periodontal ligament.	Roots of teeth in alveoli (sockets).
Synchondrosis	Connecting material is hyaline cartilage.	Temporary joint between the diaphysis and epiphysis of a long bone.
AMPHIARTHROSIS (SLIGHTLY MOVABLE)		
Syndesmosis	Articulating bones united by dense fibrous connective tissue.	Distal ends of tibia and fibula.
Symphysis	Connecting material is a broad, flat disc of fibrocartilage.	Intervertebral joints and pubic symphysis.
DIARTHROSIS (FREELY MOVABLE)		
Gliding	Articulating surfaces usually flat.	Intercarpal and intertarsal joints.
Hinge	Convex surface fits into a concave surface. Monaxial (flexion–extension).	Elbow, ankle, and interphalangeal joints.
Pivot	Rounded or pointed surface fits into a ring formed partly by bone and partly by a ligament. Monaxial (rotation).	Joint between atlas and axis and joint at proximal ends of radius and ulna.
Condyloid	Oval-shaped condyle fits into an elliptical cavity. Biaxial (flexion–extension; abduction–adduction). Circumduction also occurs.	Joint between radius and carpals.
Saddle	Articular surface of one bone is saddle-shaped, and the articular surface of the other bone is shaped like the legs of a rider sitting in the saddle. Biaxial (flexion–extension; abduction–adduction). Circumduction also occurs.	Joint between trapezium of carpus and metacarpal of thumb.
Ball-and-socket	Ball-like surface fits into a cuplike depression. Triaxial (flexion–extension; abduction–adduction; rotation). Circumduction also occurs.	Shoulder and hip joints.

Special Movements at Diarthroses

In addition to gliding movements, flexion, extension, hyperextension, rotation, abduction, adduction, and circumduction, several other movements also occur at synovial joints. These are called **special movements** and occur only at particular joints.

Elevation is an upward movement of a part of the body, and **depression** is a downward movement of a part of the body (Fig. 9.6a,b). For example, you elevate your mandible when you close your mouth and depress your mandible when you open your mouth. You also can elevate and depress your shoulders. **Protraction** is the movement of the mandible or shoulder girdle forward on a plane parallel to the ground. Thrusting the jaw outward is protraction of the mandible (Fig. 9.6c). Bringing your arms forward until the elbows touch requires protraction of the shoulder girdle. **Retraction** is the movement of a protracted part of the body backward on a plane parallel to the ground. Pulling the lower jaw back in line with the upper jaw is retraction of the mandible (Fig. 9.6d).

Six special movements relate specifically to the foot and hand. **Inversion** is the movement of the soles inward (medially) so that they face each other (Fig. 9.6e). **Eversion** is the movement of the soles outward (laterally) so that they face away from each other (Fig. 9.6f). Whereas **dorsiflexion** involves bending of the foot in the direction of the dorsum (upper surface), **plantar flexion** involves bending the foot in the direction of the plantar surface (sole) (Fig. 9.6g). **Supination** is a movement of the forearm in which the palm of the hand is turned anteriorly or superiorly; **pronation** is a movement of the forearm in which the palm is turned posteriorly or inferiorly (Fig. 9.6h).

A summary of movements that occur at synovial joints is presented in Exhibit 9.2.

Details of the Shoulder and Knee Joints

In this section we will examine some of the principal structural and functional features of two major diarthroses: the shoulder joint and the knee joint.

Figure 9.6 Special movements.

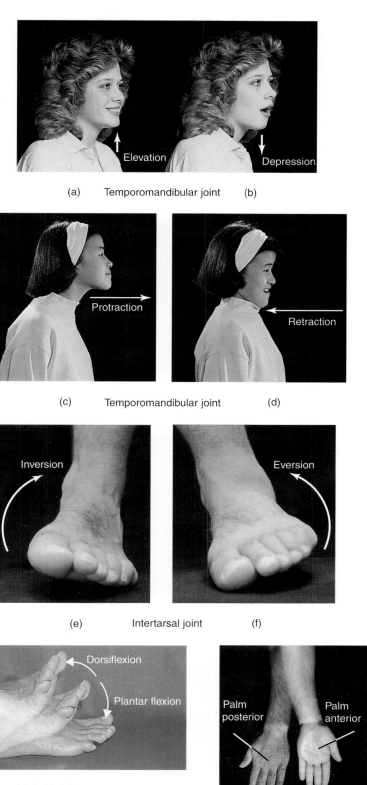

(a)   Temporomandibular joint   (b)

(c)   Temporomandibular joint   (d)

(e)   Intertarsal joint   (f)

(g) Ankle joint

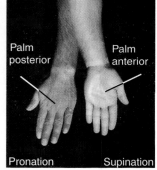

(h) Proximal radioulnar joint

Why are these movements called special movements?

EXHIBIT **9.2**   **SUMMARY OF MOVEMENTS AT SYNOVIAL JOINTS**

Movement	Definition
GLIDING	The surface of one bone moves back and forth and from side to side over another surface. During the movement there is no angular or rotary motion.
ANGULAR	There is an increase or decrease at the angle between bones.
Flexion	Involves a decrease in the angle between the surfaces of the articulating bones.
Extension	Involves an increase in the angle between the surfaces of the articulating bones.
Hyperextension	Continuation of extension beyond the anatomical position.
Abduction	Movement of a bone away from the midline.
Adduction	Movement of a bone toward the midline.
Circumduction	A combination of flexion–extension and abduction–adduction in succession, in which the distal end of a part of the body moves in a circle.
ROTATION	Movement of a bone around its longitudinal axis; may be medial (toward the midline) or lateral (away from the midline).
SPECIAL	Occur at specific joints.
Inversion	Movement of the soles inward so that they face each other.
Eversion	Movement of the soles outward so that they face away from each other.
Dorsiflexion	Bending the foot in the direction of the dorsum (upper surface).
Plantar flexion	Bending the foot in the direction of the plantar surface (sole).
Protraction	Movement of the mandible or shoulder girdle forward on a plane parallel to the ground.
Retraction	Movement of the mandible or shoulder girdle backward on a plane parallel to the ground.
Supination	Movement of the forearm in which the palm is turned anteriorly or superiorly.
Pronation	Movement of the forearm in which the palm is turned posteriorly or inferiorly.
Elevation	Movement of a part of the body upward.
Depression	Movement of a part of the body downward.

Shoulder (Glenohumeral) Joint

The **shoulder** (**glenohumeral**) **joint** is formed by the head of the humerus and glenoid cavity of the scapula. It is a ball-and-socket joint. The anatomical components of the shoulder joint are as follows (Fig. 9.7):

1. The **articular capsule** is a thin, loose sac that completely envelops the joint. It extends from the glenoid cavity to the anatomical neck of the humerus. The inferior part of the capsule is its weakest area.
2. The **coracohumeral ligament** is a strong, broad ligament that strengthens the superior part of the articular capsule and extends from the coracoid process of the scapula to the greater tubercle of the humerus (Fig. 9.7a,b).
3. The **glenohumeral ligaments** are three thickenings of the articular capsule over the anterior surface of the joint (Fig. 9.7a,b). They extend from the glenoid cavity to the lesser tubercle and anatomical neck of the humerus. These ligaments are often indistinct or absent and provide only minimal strength.
4. The **transverse humeral ligament** is a narrow sheet extending from the greater tubercle to the lesser tubercle of the humerus (Fig. 9.7a).
5. The **glenoid labrum** is a narrow rim of fibrocartilage around the edge of the glenoid cavity (Fig. 9.7b,c). It slightly deepens and enlarges the glenoid cavity.
6. Four **bursae** are associated with the shoulder joint. They are the **subscapular bursa**, **subdeltoid bursa**, **subacromial bursa**, and **subcoracoid bursa**.

The shoulder joint has more freedom of movement than any other joint of the body. Among the movements permitted are flexion (humerus drawn forward), extension (humerus drawn backward), abduction (humerus drawn away from midline), adduction (humerus drawn toward midline), medial rotation, lateral rotation, and circumduction. This freedom results from the looseness of the articular capsule and shallowness of the glenoid cavity in relation to the large size of the head of the humerus. The cavity receives little more than one-third of the head of the humerus. These same anatomical features that lead to freedom of movement also result in instability.

Although the ligaments of the shoulder joint strengthen it to some extent, most of the strength results from the muscles that surround the joint, especially the **rotator cuff muscles**. These muscles (supraspinatus, infraspinatus, teres minor, and subscapularis) join the scapula to the humerus (see Fig. 11.15). The tendons of the muscles, together called the **rotator cuff**, encircle the joint (except for the inferior portion) and fuse with the articular capsule. The rotator cuff muscles work as a group to hold the head of the humerus in the glenoid cavity.

Figure 9.7 Right shoulder (glenohumeral) joint. The arrow in the inset in (b) indicates the direction from which the shoulder joint is viewed (lateral).

Most of the stability of the shoulder joint results from the arrangement of the rotator cuff muscles.

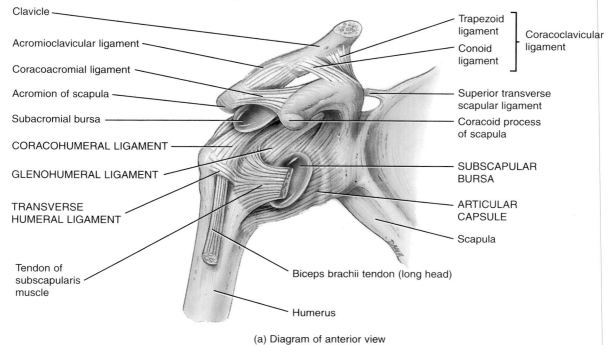

(a) Diagram of anterior view

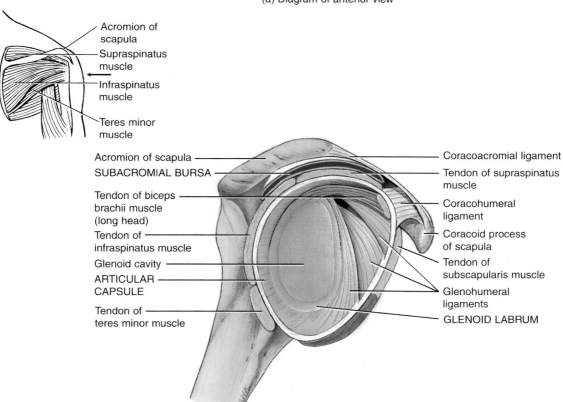

(b) Diagram of the lateral view (opened)

Figure continues

Figure 9.7   (continued)

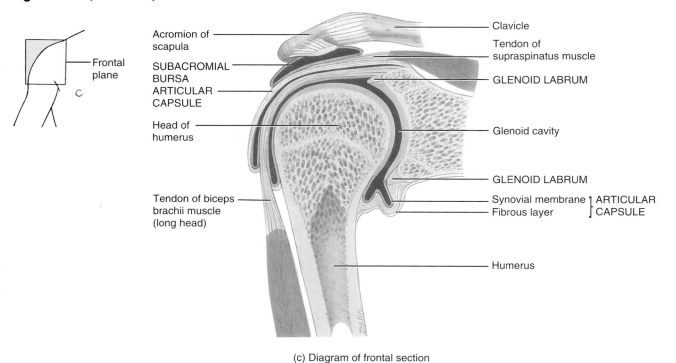

Frontal plane

Acromion of scapula

SUBACROMIAL BURSA
ARTICULAR CAPSULE

Head of humerus

Tendon of biceps brachii muscle (long head)

Clavicle

Tendon of supraspinatus muscle

GLENOID LABRUM

Glenoid cavity

GLENOID LABRUM

Synovial membrane ⎫ ARTICULAR
Fibrous layer   ⎬ CAPSULE

Humerus

(c) Diagram of frontal section

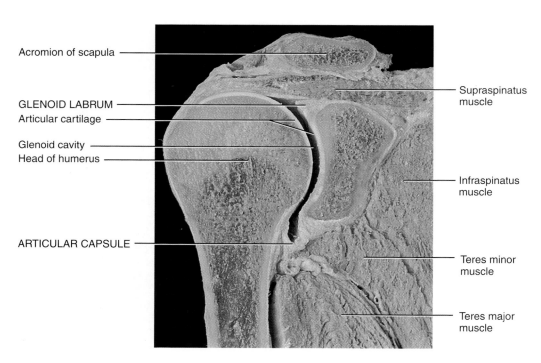

Acromion of scapula

GLENOID LABRUM
Articular cartilage

Glenoid cavity
Head of humerus

ARTICULAR CAPSULE

Supraspinatus muscle

Infraspinatus muscle

Teres minor muscle

Teres major muscle

(d) Photograph of frontal section

Q Why does the shoulder joint have more freedom of movement than any other joint of the body?

DISLOCATION AND ROTATOR CUFF INJURY

Dislocation, or **luxation** (luks-Ā-shun; *luxatio* = dislocation), is the displacement of a bone from a joint with tearing of ligaments, tendons, and articular capsules. A partial or incomplete dislocation is called a **subluxation**. A dislocation is usually caused by a blow or fall, although unusual physical effort may lead to this condition. The joint most commonly dislocated in adults is the shoulder joint because its socket is quite shallow. Usually, the head of the humerus becomes displaced inferiorly, where the articular capsule is least protected. Dislocations of the mandible, elbow, fingers, knee, or hip are less common.

Rotator cuff injury is a common injury among baseball pitchers, owing to shoulder movements that involve vigorous circumduction. Most often, there is tearing of the supraspinatus muscle tendon of the rotator cuff. This tendon is especially predisposed to wear-and-tear changes because of its location between the head of the humerus and acromion of the scapula, which compresses the tendon during shoulder movements. ■

Knee (Tibiofemoral) Joint

The **knee (tibiofemoral) joint** is the largest joint of the body. It actually consists of three joints: (1) an intermediate patellofemoral joint between the patella and the patellar surface of the femur; (2) a lateral tibiofemoral joint between the lateral condyle of the femur, lateral meniscus, and lateral condyle of the tibia; and (3) a medial tibiofemoral joint between the medial condyle of the femur, medial meniscus, and medial condyle of the tibia. The patellofemoral joint is a gliding joint, and the lateral and medial tibiofemoral joints are modified hinge joints.

The anatomical components of the knee joint are as follows (Fig. 9.8):

1. **Articular capsule.** No complete, independent capsule unites the bones. The ligamentous sheath surrounding the joint consists mostly of muscle tendons or expansions of them. There are, however, some capsular fibers connecting the articulating bones.
2. The **medial and lateral patellar retinacula** (singular is **retinaculum**) are fused tendons of insertion of the

---

Figure 9.8   Right knee (tibiofemoral) joint.

 The knee joint is the largest and most complex joint in the body.

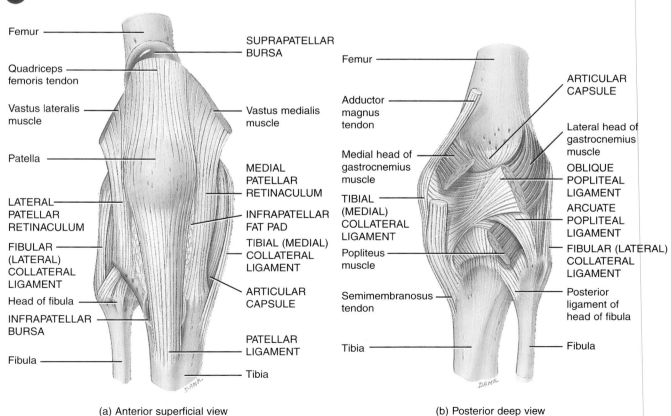

(a) Anterior superficial view   (b) Posterior deep view

Figure continues

Figure 9.8   (continued)

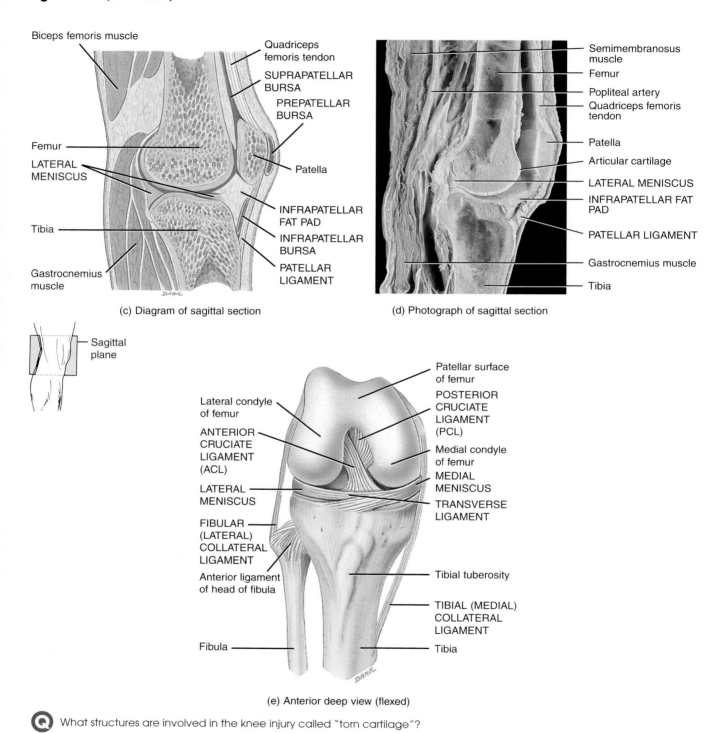

(c) Diagram of sagittal section

(d) Photograph of sagittal section

(e) Anterior deep view (flexed)

Q What structures are involved in the knee injury called "torn cartilage"?

quadriceps femoris muscle and the fascia lata (deep fascia of the thigh) that strengthen the anterior surface of the joint (Fig. 9.8a).

3. The **patellar ligament** is the continuation of the common tendon of insertion of the quadriceps femoris muscle. It extends from the patella to the tibial tuberosity and also strengthens the anterior surface of the joint. An **infrapatellar fat pad** separates the posterior surface of the ligament from the synovial membrane of the joint.

4. The **oblique popliteal ligament** is a broad, flat ligament that extends from the intercondylar fossa of the femur to the head of the tibia (Fig. 9.8b). The tendon of the semimembranosus muscle is superficial to the ligament and passes from the medial condyle of the tibia to the lateral condyle of the femur. The ligament and tendon afford strength for the posterior surface of the joint.

5. The **arcuate popliteal ligament** extends from the lateral condyle of the femur to the styloid process of the head of the fibula (Fig. 9.8b). It strengthens the lower lateral part of the posterior surface of the joint.

6. The **tibial (medial) collateral ligament** is a broad, flat ligament on the medial surface of the joint that extends from the medial condyle of the femur to the medial condyle of the tibia (Fig. 9.8b). The ligament is crossed by tendons of the sartorius, gracilis, and semitendinosus muscles, all of which strengthen the medial aspect of the joint.

7. The **fibular (lateral) collateral ligament** is a strong, rounded ligament on the lateral surface of the joint that extends from the lateral condyle of the femur to the lateral side of the head of the fibula (Fig. 9.8b). It strengthens the lateral aspect of the joint. The ligament is covered by the tendon of the biceps femoris muscle. The tendon of the popliteal muscle is deep to the ligament.

8. The **intracapsular ligaments** connect the tibia and femur within the capsule (Fig. 9.8e).
   a. The **anterior cruciate ligament (ACL)** extends posteriorly and laterally from the area anterior to the intercondylar eminence of the tibia to the posterior part of the medial surface of the lateral condyle of the femur. This ligament is stretched or torn in about 70% of all serious knee injuries.
   b. The **posterior cruciate ligament (PCL)** extends anteriorly and medially from a depression on the posterior intercondylar area of the tibia and lateral meniscus to the anterior part of the medial surface of the medial condyle of the femur.

9. The **articular discs (menisci)** are two fibrocartilage discs between the tibial and femoral condyles. They help to compensate for the irregular shapes of the articulating bones and circulate synovial fluid (Fig. 9.8c–e).
   a. The **medial meniscus** is a semicircular (C-shaped) piece of fibrocartilage. Its anterior end is attached to the area anterior to the intercondylar eminence of the tibia, anterior to the anterior cruciate ligament. Its posterior end is attached to the area posterior to the intercondylar eminence of the tibia between the attachments of the posterior cruciate ligament and lateral meniscus.
   b. The **lateral meniscus** is a nearly circular (an incomplete O-shaped) piece of fibrocartilage. Its anterior end is attached anterior to the intercondylar eminence of the tibia and lateral and posterior to the anterior cruciate ligament. Its posterior end is attached posterior to the intercondylar eminence of the tibia

and anterior to the posterior end of the medial meniscus. The medial and lateral menisci are connected to each other by the **transverse ligament** and to the margins of the head of the tibia by the **coronary ligaments**.

10. The more important **bursae** of the knee are the **prepatellar bursa**, **infrapatellar bursa**, and **suprapatellar bursa** (Fig. 9.8c).

KNEE INJURIES

The knee joint is the joint most vulnerable to damage because of the stresses to which it is subjected and because there is no interlocking of the articulating bones; reinforcement is strictly by ligaments and tendons. The most common type of **knee injury** in football is rupture of the tibial (medial) collateral ligament, often associated with tearing of the anterior cruciate ligament and medial meniscus (torn cartilage). Usually, a blow to the lateral side of the knee causes the injury. When a knee is examined for such an injury, the three Cs are kept in mind: collateral ligament, cruciate ligament, and cartilage.

A **swollen knee** may occur immediately or hours after an injury. Immediate swelling is due to escape of blood from damaged blood vessels adjacent to areas where there is rupture of the anterior cruciate ligament, torn menisci, fractures, or collateral ligament sprains. Delayed swelling is due to excessive production of synovial fluid as a result of irritation of the synovial membrane, a condition commonly referred to as "water on the knee."

A **dislocated knee** refers to the displacement of the tibia relative to the femur. Accordingly, such dislocations are classified as anterior, posterior, medial, lateral, or rotatory. The most common type is anterior dislocation, resulting from hyperextension of the knee. A frequent consequence of a dislocated knee is damage to the popliteal artery. ∎

Exhibit 9.3 lists selected joints of the body, relating their articular components, structural and functional classification, and movements.

Examining and Repairing Diarthroses

Arthroscopy

Arthroscopy (ar-THROS-kō-pē; *arthro* = joint; *skopein* = to view) is a procedure that involves examination of the interior of a joint, usually the knee, using an arthroscope, a lighted instrument the diameter of a pencil. It is used to determine the nature and extent of damage following knee injury; to remove torn cartilage and repair cruciate ligaments in the knee; to obtain tissue samples for analysis and to perform surgery on other joints, such as the shoulder, elbow, ankle, and wrist; and to monitor the progression of disease and the effects of therapy. Since arthroscopy requires only small incisions, recovery is more rapid than with conventional surgery. However, in some cases, a cast or splint may be worn for several days, depending on the extent of the procedure.

EXHIBIT **9.3** **REPRESENTATIVE JOINTS ACCORDING TO ARTICULAR COMPONENTS, CLASSIFICATION, AND MOVEMENTS**

Joint	Articular Component	Structural Classification	Functional Classification	Movements
Temporo-mandibular joint (TMJ)	Joint formed by condylar process of the mandible and mandibular fossa and articular tubercle of temporal bone. The TMJ is the only movable joint between skull bones; all other skull joints are sutures and therefore immovable.	Synovial: combined hinge (ginglymus) and gliding (arthrodial) type.	Diarthrosis.	Only the mandible moves since the maxilla is firmly anchored to other bones by sutures. Accordingly, the mandible may function in depression (jaw opening), elevation (jaw closing), protraction, retraction, lateral displacement, and slight rotation.
Atlanto-occipital joints	Joints formed by the superior articular facets of the atlas and the occipital condyles of the occipital bone.	Synovial: condyloid (ellipsoidal) type.	Diarthrosis.	Flexion (forward bending of head), extension (backward bending of head), and slight lateral tilting of head to either side.
Intervertebral joints	Joints formed between (1) vertebral bodies and between (2) vertebral arches.	Joints between vertebral bodies—cartilaginous (fibrocartilage), symphysis type. Joints between vertebral arches—synovial, gliding (arthrodial) type.	Amphiarthrosis (symphysis) between vertebral bodies. Diarthrosis between vertebral arches.	Flexion (bending the backbone forward), extension (bending the backbone backward), lateral flexion (bending the backbone to either side), and rotation.
Lumbosacral joint	Joint formed by the body of the fifth lumbar vertebra and the superior articular facet of the first sacral vertebra of the sacrum.	Joint between the fifth lumbar vertebra and the first sacral vertebra—cartilaginous (fibrocartilage), symphysis type. Joint between the articular processes—synovial joint, gliding (arthrodial) type.	Amphiarthrosis (symphysis) between fifth lumbar vertebra and first sacral vertebra. Diarthrosis between articular processes.	Similar to those of intervertebral joints.
Shoulder (glenohumeral) joint	Joint formed by the head of the humerus and the glenoid cavity of the scapula.	Synovial joint, ball-and-socket (spheroid) type.	Diarthrosis.	Flexion (humerus drawn forward), extension (humerus drawn backward), abduction (humerus drawn away from midline), adduction (humerus drawn toward midline), medial rotation, lateral rotation, and circumduction.

Joint	Articular Component	Structural Classification	Functional Classification	Movements
Elbow joint	Joint formed by the trochlea of the humerus, the trochlear notch of the ulna, and the head of the radius.	Synovial joint, hinge (ginglymus) type.	Diarthrosis.	Flexion and extension of forearm.
Wrist (radiocarpal) joint	Joint formed by the distal end of the radius, the distal surface of the articular disc separating the carpal and distal radioulnar joint, and the scaphoid, lunate, and triquetrum carpal bones.	Synovial joint, condyloid (ellipsoidal) type.	Diarthrosis.	Flexion, extension, abduction, adduction, and circumduction.
Hip (coxal) joint	Joint formed by the head of the femur and the acetabulum of the hipbone.	Synovial, ball-and-socket (spheroid) type.	Diarthrosis.	Flexion, extension, abduction, adduction, circumduction, and rotation.
Knee (tibiofemoral) joint	Formed by three joints: (1) a lateral tibiofemoral joint between the lateral condyle of the femur, lateral meniscus, and lateral condyle of the tibia; (2) an intermediate patellofemoral joint between the patella and the patellar surface of the femur; and (3) a medial tibiofemoral joint between the medial condyle of the femur, medial meniscus, and medial condyle of the tibia.	Lateral and medial tibiofemoral joints—synovial, hinge (ginglymus) type. Patellofemoral joint—partly synovial, gliding (arthrodial) type.	Diarthrosis.	Flexion, extension, slight medial rotation, and lateral rotation in flexed position.
Ankle (talocrural) joint	Joints between (1) the distal end of the tibia and its medial malleolus and the talus and (2) the lateral malleolus of the fibula and the talus.	Both joints—synovial, hinge (ginglymus) type.	Diarthrosis.	Dorsiflexion and plantar flexion.

Arthroplasty

Arthroplasty (AR-thrō-plas'-tē; *arthro* = joint; *plasty* = plastic repair of) refers to surgical replacement of joints. The most common joint replacements are those involving the hip and knee. Each year in the United States over 150,000 people undergo **total hip replacement** (**hip arthroplasty**), a procedure involving both the acetabulum of the hipbone and head of the femur. Thousands of partial hip replacements, involving only the femur, are also performed annually.

In total hip replacement, prefabricated prostheses (artificial devices) replace the damaged portions of the acetabulum plus the head and neck of the femur. The acetabular component consists of polyethylene, while the femoral component is made of cobalt–chrome. These materials are designed to withstand a high degree of stress. Once the appropriate components are selected, they are attached to the healthy portion of bone with acrylic cement and screws.

Total hip replacement may benefit individuals with osteoarthritis, rheumatoid arthritis, hip fractures, dislocations, metabolic bone diseases (osteoporosis and osteomalacia), and congenital and developmental deformities of the hip.

DISORDERS: HOMEOSTATIC IMBALANCES

RHEUMATISM

Rheumatism refers to any painful state of the supporting structures of the body—bones, ligaments, joints, tendons, or muscles. Arthritis is a form of rheumatism in which the joints have become inflamed.

ARTHRITIS

The term **arthritis** refers to many different diseases, most of which are characterized by inflammation of one or more joints. Inflammation, pain, and stiffness may also be present in adjacent parts of the body, such as the muscles near the joint.

Rheumatoid Arthritis (RA)

Rheumatoid (ROO-ma-toyd; *rheuma* = discharge) **arthritis** (**RA**) is an autoimmune disease in which the immune system of the body attacks its own tissues, in this case its own cartilage and joint linings. RA is characterized by inflammation of the joint, swelling, pain, and loss of function. Usually, this form occurs bilaterally: if your left wrist is affected, your right wrist is also likely to be affected, although usually not to the same degree.

The primary symptom of rheumatoid arthritis is inflammation of the synovial membrane. If untreated, the membrane thickens and synovial fluid accumulates. The resulting pressure causes pain and tenderness. The membrane then produces an abnormal granulation tissue, called *pannus,* that adheres to the surface of the articular cartilage and sometimes erodes the cartilage completely. When the cartilage is destroyed, fibrous tissue joins the exposed bone ends. The tissue ossifies and fuses the joint so that it is immovable—the ultimate crippling effect of rheumatoid arthritis (Fig. 9.9). The growth of the granulation tissue causes the distortion of the fingers that is so typical of hands that have been affected by this disease.

Osteoarthritis (OA)

Osteoarthritis (os'-tē-ō-ar-THRĪ-tis) is a degenerative joint disease that apparently results from a combination of aging, irritation of the joints, and wear and abrasion. It is commonly known as "wear-and-tear" arthritis.

Osteoarthritis is a noninflammatory, progressive disorder of movable joints, particularly weight-bearing joints. It is charac-

Figure 9.9   Rheumatoid arthritis. Radiograph (x-ray) of the right hand and wrist showing changes characteristic of rheumatoid arthritis. The arrows indicate fusion of the bones resulting in obliteration of the joint cavities.

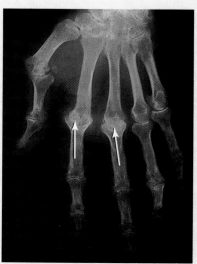

Anterior view

terized by the deterioration of articular cartilage and by the formation of new bone in the subchondral areas and at the margins of the joint. The cartilage slowly degenerates, and as the bone ends become exposed, small bumps, or **spurs**, of new osseous tissue are deposited on them. These spurs decrease the space of the joint cavity and restrict joint movement. Unlike rheumatoid arthritis, osteoarthritis usually affects only the articular cartilage. The synovial membrane is rarely destroyed, and other tissues are unaffected. A major distinction between osteoarthritis and rheumatoid arthritis is that the former strikes the larger joints (knees, hips) first, whereas the latter first strikes smaller joints.

lthough bones provide leverage and form the framework of the body, they cannot move the body by themselves. Motion results from alternating contraction (shortening) and relaxation of muscles, which constitute 40–50% of total body weight. Your muscular strength reflects the prime function of muscle—changing chemical energy (in the form of ATP) into mechanical energy to generate force, perform work, and produce movement. Muscle tissues also function to stabilize body positions, regulate organ volume, and generate heat. The scientific study of muscles is known as **myology** (mi-OL-ō-jē; *myo* = muscle; *logos* = study of).

The developmental anatomy of the muscular system is considered at the end of the chapter.

TYPES OF MUSCLE TISSUE

The three kinds of muscle tissue—skeletal, cardiac, and smooth—differ from one another in their microscopic anatomy, location, and control by the nervous and endocrine systems.

Skeletal muscle tissue is so-named because it is attached primarily to bones, and it moves parts of the skeleton. (Some skeletal muscles are also attached to skin, other muscles, or deep fascia.) Skeletal muscle tissue is said to be **striated** because alternating light and dark bands (striations) are visible when the tissue is examined under a microscope. It is a **voluntary** muscle tissue because it can be made to contract and relax by conscious control.

Cardiac muscle tissue forms most of the heart. It is also **striated** but is **involuntary**; that is, its contraction is usually not under conscious control. Cardiac muscle includes a pacemaker system that causes the heart to beat; this built-in intrinsic rhythm is called **autorhythmicity**. Several neurotransmitters and hormones adjust heart rate by speeding or slowing the pacemaker.

Smooth muscle tissue is located in the walls of hollow internal structures, such as blood vessels, the stomach, and the intestines, as well as most other abdominal organs. It is also found in the skin attached to hair follicles. Under the microscope, this tissue looks **nonstriated** or **smooth**. It is usually **involuntary** muscle tissue, often has autorhythmicity, and is also influenced by certain hormones and neurotransmitters. (Regulation of cardiac and smooth muscle by hormones and neurotransmitters is discussed in Chapter 17.)

Thus all muscle tissues are classified in the following way: (1) skeletal (striated, voluntary) muscle tissue; (2) cardiac (striated, involuntary) muscle tissue; and (3) smooth (nonstriated, involuntary) muscle tissue (summarized in Exhibit 10.3 on page 263).

FUNCTIONS OF MUSCLE TISSUE

Through sustained contraction or alternating contraction and relaxation, muscle tissue has four key functions: producing motion, moving substances within the body, providing stabilization, and generating heat.

1. **Motion**. Motion is obvious in movements such as walking and running, and in localized movements, such as grasping a pencil or nodding the head. These movements rely on the integrated functioning of bones, joints, and skeletal muscles.

2. **Movement of substances within the body**. Cardiac muscle contractions pump blood to all body tissues and help regulate blood pressure. Smooth muscle contractions aid in the movement of food through the gastrointestinal tract, substances that aid digestion such as bile and enzymes into the gastrointestinal tract, sperm and ova through the reproductive systems, and urine through the urinary system. Skeletal muscle contractions promote the flow of lymph and the return of venous blood to the heart.

3. **Stabilizing body positions and regulating organ volume**. Besides producing movements, skeletal muscle contractions maintain the body in stable positions, such as standing or sitting. Postural muscles display sustained contractions when a person is awake; for example, partially contracted neck muscles hold the head upright. In a similar manner, sustained contractions of smooth muscles (sphincters) may prevent outflow of the contents of a hollow organ. Temporary storage of food in the stomach or urine in the urinary bladder is possible because smooth muscles close off the exit route.

4. **Thermogenesis (generating heat)**. As skeletal muscle contracts to perform work, a by-product is heat. Much of the heat released by muscle is used to maintain normal body temperature. Muscle contractions are thought to generate as much as 85% of all body heat. This is why active cheering helps warm you up during a cold weather football game. Involuntary contractions of skeletal muscle, known as **shivering**, can increase thermogenesis by several hundred percent (see Fig. 10.11).

CHARACTERISTICS OF MUSCLE TISSUE

Muscle tissue has five principal characteristics that enable it to carry out its functions and thus contribute to homeostasis.

1. **Excitability (irritability)**, a property of both muscle cells and nerve cells (neurons), is the ability to respond to certain stimuli by producing electrical signals called action potentials (impulses). For muscle, the stimuli that trigger action potentials are chemicals—neurotransmitters, released by neurons, or hormones distributed by the blood.

2. **Conductivity** is the ability of a cell, especially a muscle cell or neuron, to propagate or conduct action potentials along the plasma membrane.

3. **Contractility** is the ability of muscle tissue to shorten and thicken (contract), thus generating force to do work. Muscle contracts in response to one or more muscle action potentials.

4. **Extensibility** means that muscle can be extended (stretched) without damaging the tissue. Most skeletal muscles are arranged in opposing pairs. While one is contracting, the other not only is relaxed but usually is being stretched.

5. **Elasticity** means that muscle tissue tends to return to its original shape after contraction or extension.

Skeletal muscle is the focus of much of this chapter. Cardiac muscle and smooth muscle are described briefly here, but in more detail later (with discussions of the heart in Chapter 20, the various organs containing smooth muscle, and the autonomic nervous system in Chapter 17).

ANATOMY AND INNERVATION OF SKELETAL MUSCLE TISSUE

To understand how skeletal muscle contraction produces movement, one needs some knowledge of its nerve and blood supply, connective tissue components, and microscopic anatomy.

Connective Tissue Components

Connective tissue surrounds and protects muscle tissue. The term **fascia** (FASH-ē-a; *fascia* = bandage) refers to a sheet or broad band of fibrous connective tissue deep to the skin or around muscles and other organs of the body. **Superficial fascia (subcutaneous layer)** is immediately deep to the skin (see Fig. 11.21). It is composed of areolar connective tissue and adipose tissue and has four important functions: (1) It stores water and fat. Much of the fat of an overweight person is in the superficial fascia. (2) It reduces the rate of heat loss. (3) It provides mechanical protection against traumatic blows. (4) It provides a framework for nerves and blood vessels to enter and exit muscles.

Deep fascia is dense, irregular connective tissue that lines the body wall and limbs and holds muscles together, separating them into functional groups. Deep fascia allows free movement of muscles, supports nerves, blood vessels, and lymphatic vessels, and fills spaces between muscles.

Three layers of dense, irregular connective tissue extend from the deep fascia to further protect and strengthen skeletal muscle (Figs. 10.1 and 10.4). The outermost layer, encir-

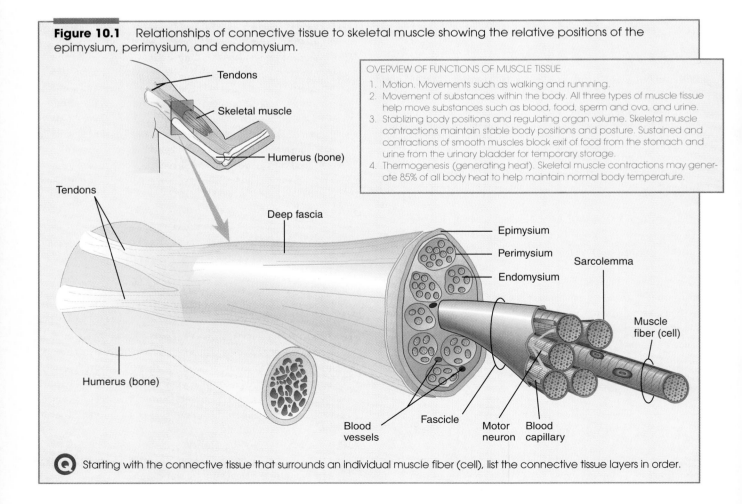

Figure 10.1 Relationships of connective tissue to skeletal muscle showing the relative positions of the epimysium, perimysium, and endomysium.

Tendons
Skeletal muscle
Humerus (bone)

OVERVIEW OF FUNCTIONS OF MUSCLE TISSUE
1. Motion. Movements such as walking and runnning.
2. Movement of substances within the body. All three types of muscle tissue help move substances such as blood, food, sperm and ova, and urine.
3. Stablizing body positions and regulating organ volume. Skeletal muscle contractions maintain stable body positions and posture. Sustained and contractions of smooth muscles block exit of food from the stomach and urine from the urinary bladder for temporary storage.
4. Thermogenesis (generating heat). Skeletal muscle contractions may generate 85% of all body heat to help maintain normal body temperature.

Tendons
Humerus (bone)
Deep fascia
Epimysium
Perimysium
Endomysium
Sarcolemma
Muscle fiber (cell)
Blood vessels
Fascicle
Motor neuron
Blood capillary

Starting with the connective tissue that surrounds an individual muscle fiber (cell), list the connective tissue layers in order.

cling the whole muscle, is the **epimysium** (ep-i-MĪZ-ē-um; *epi* = upon). **Perimysium** (per-i-MĪZ-ē-um; *peri* = around) surrounds bundles of 10–100 or more individual muscle fibers. These bundles are called **fascicles**. Penetrating into the interior of each fascicle and separating individual muscle fibers from one another is **endomysium** (en′-dō-MĪZ-ē-um; *endo* = within).

Epimysium, perimysium, and endomysium are all continuous with and contribute collagen fibers to the connective tissue that attaches the muscle to other structures, such as bone or other muscle. All three may extend beyond the muscle fibers as a **tendon** (*tendere* = to stretch out)—a cord of dense connective tissue that attaches a muscle to the periosteum of a bone. An example is the calcaneal (Achilles) tendon of the gastrocnemius (calf) muscle (see Fig. 11.21a). When the connective tissue elements extend as a broad, flat layer, the tendon is called an **aponeurosis** (*apo* = from; *neuron* = a tendon). This structure also attaches muscle to the coverings of a bone, another muscle, or the skin. An example of an aponeurosis is the galea aponeurotica on top of the skull (see Fig. 11.4). Certain tendons, especially those of the wrist and ankle, are enclosed by tubes of fibrous connective tissue called **tendon sheaths** (see Fig. 11.17). They are similar in structure to bursae and contain a film of synovial fluid. Tendon sheaths reduce friction as tendons slide back and forth.

TENOSYNOVITIS

Tenosynovitis (ten′-ō-sin-ō-VĪ-tis) is an inflammation of the tendons, tendon sheaths, and synovial membranes surrounding certain joints. The tendons most often affected are at the wrists, shoulders, elbows (producing tennis elbow), finger joints (producing trigger finger), ankles, and feet. The affected sheaths sometimes become visibly swollen because of fluid accumulation. Tenderness and pain are often associated with movement of the body part. The condition often follows trauma, strain, or excessive exercise. ■

Nerve and Blood Supply

Skeletal muscles are well supplied with nerves and blood vessels (see Fig. 11.16c). Neurons that stimulate muscle to contract are called **motor neurons**. When muscle contracts, it uses a good deal of adenosine triphosphate (ATP) and therefore needs large amounts of nutrients and oxygen for ATP production. Moreover, the waste products of the reactions that produce ATP must be eliminated. Thus prolonged muscle action depends on a rich blood supply to deliver nutrients and oxygen and remove wastes and heat. Capillaries, the microscopic blood vessels where substances can pass into or out of blood, are plentiful within the endomysium. Thus each muscle fiber (cell) is in close contact with one or more capillaries.

The Motor Unit

A motor neuron delivers the stimulus that ultimately causes a muscle fiber to contract. A motor neuron plus all the skeletal muscle fibers it stimulates is called a **motor unit** (Fig. 10.2).

Figure 10.2 Motor units. Shown are two motor neurons, one in red and one in green, each supplying the muscle fibers of its motor unit.

 A motor unit consists of a motor neuron plus the muscle fibers it stimulates.

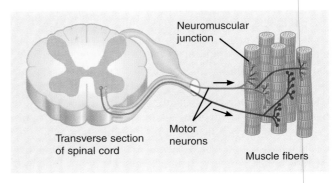

 How will the strength of contraction of a large motor unit compare with that of a small motor unit? (Assume each muscle fiber generates about the same amount of tension.)

A single motor neuron makes contact with an average of 150 muscle fibers. This means that activation of one neuron causes the simultaneous contraction of about 150 muscle fibers. All the muscle fibers of a motor unit contract and relax together. Muscles that control precise movements have many small motor units. For example, muscles of the larynx (voice box) that control voice production have as few as two or three muscle fibers per motor unit and muscles controlling eye movements may have 10–20 muscle fibers per motor unit. Muscles of the body that are responsible for powerful gross movements, such as the biceps brachii in the arm and gastrocnemius in the leg, have some motor units with as many as 2000 muscle fibers each. Stimulation of a motor neuron produces a contraction in all the muscle fibers of a particular motor unit. Accordingly, the total strength of a contraction depends, in part, on which motor units are activated.

The Neuromuscular Junction

Excitable cells (neurons and muscle fibers) communicate with one another and with other target cells, for example, endocrine or exocrine gland cells, at specialized regions called **synapses** (*synapsis* = connection). At most synapses a small gap, called the **synaptic cleft**, separates the two cells. Since the cells do not physically touch, the action potential from one cell cannot "jump the gap" to excite the next cell. Rather, the first cell communicates with the second by releasing a chemical called a **neurotransmitter**. The particular type of synapse formed between a motor neuron and a skeletal muscle fiber is called the **neuromuscular junction** (NMJ) or **myoneural junction** (Fig. 10.3a).

Each motor neuron has a threadlike **axon** that extends from the spinal cord to a group of skeletal muscle fibers.

Close to its target skeletal muscle fibers, the axon branches into several **axon terminals** (Fig. 10.3b). In motor neurons, the distal tip of each axon terminal expands into a cluster of **synaptic end bulbs** that contain many membrane-enclosed sacs called **synaptic vesicles** (Fig. 10.3c). Inside each synaptic vesicle are thousands of neurotransmitter molecules. Although many different neurotransmitters exist, the one released at the NMJ is **acetylcholine** (as´-ē-til-KŌ-lēn), abbreviated **ACh**. The region of the muscle fiber plasma membrane that is adjacent to the axon terminals is called the **motor end plate**. It contains **acetylcholine receptors** (Fig. 10.3d), which are integral membrane proteins that recognize and bind specifically to ACh. The term neuromuscular junction includes all motor neuron axon terminals, with their synaptic end bulbs, plus the motor end plate of the muscle fiber, which typically contains 30–40 million ACh receptors.

When a nerve impulse (nerve action potential) reaches the synaptic end bulbs, it triggers exocytosis of synaptic vesicles. In this process, the synaptic vesicles fuse with the plasma membrane and liberate ACh, which diffuses into the synaptic cleft between the motor neuron and the motor end plate. When ACh binds to its receptor, a channel that passes small cations, most importantly Na^+, opens. The inrush of Na^+ changes the resting membrane potential, which triggers a muscle action potential that travels along the muscle cell plasma membrane and initiates the events leading to muscle contraction. (Chapter 12 presents details of how action potentials arise.)

Since skeletal muscle fibers often are very long cells, the neuromuscular junction usually is located near the midpoint of the fiber. Muscle action potentials then spread from the center of the fiber toward both ends. This arrangement permits nearly simultaneous contraction of all parts of the fiber.

ELECTROMYOGRAM

The recording of electrical activity in resting and contracting muscles is called **electromyography** (e-lek´-trō-mi-OG-ra-fē; *electro* = electricity; *myo* = muscle; *graph* = to write). An **electromyogram** (**EMG**) is a record of skeletal muscle electrical activity. It is similar to the more familiar electrocardiogram (ECG), which is a record of the heart's electrical signals (see Fig. 20.10). To record EMGs, a flat metal plate is placed on the skin over the muscle to be tested. Then, a thin sterile needle attached by wires to a recording machine is inserted through the skin into the muscle. The electrical activity of the muscle is recorded at rest and during contraction. It is then displayed as electrical waves on an oscilloscope and amplified to produce sounds over an audio speaker.

EMGs may be used to determine the cause of muscular weakness or paralysis, to evaluate involuntary muscle twitching, to determine why abnormal levels of muscle enzymes (such as creatine phosphokinase) appear in blood, and to serve as a component of biofeedback studies.

Nerve conduction studies are sometimes done together with EMG testing. A nerve is stimulated electrically through the skin, while a recording device detects the muscle's response. Such studies can help determine if nerve damage is the cause of muscle weakness. ■

Microscopic Anatomy of Skeletal Muscle

Microscopic examination of a typical skeletal muscle reveals hundreds or thousands of very long, cylindrical cells called **muscle fibers** or **myofibers** (Fig. 10.4b). The muscle

---

Figure 10.3   Neuromuscular junction (NMJ).

 A neuromuscular junction is the synapse between a motor neuron and a skeletal muscle fiber.

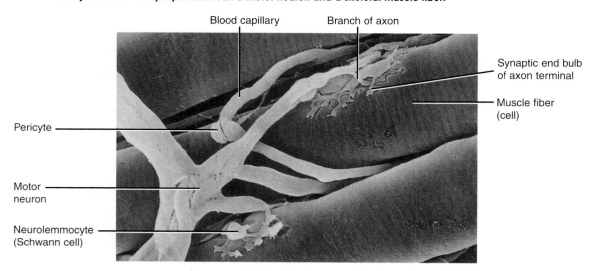

(a) Scanning electron micrograph (1650x)

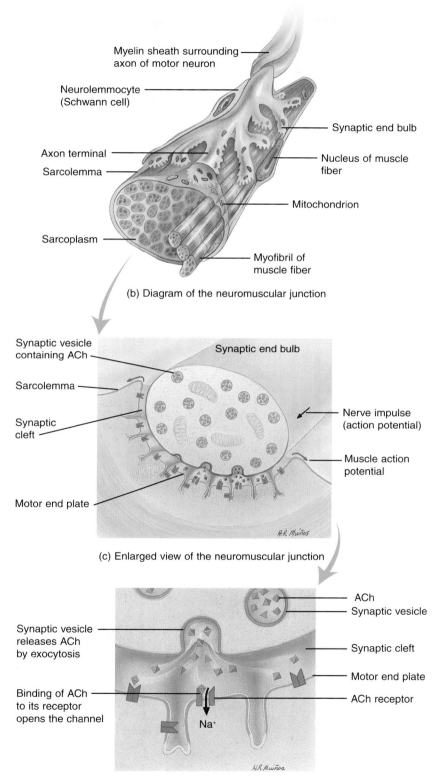

Myelin sheath surrounding axon of motor neuron

Neurolemmocyte (Schwann cell)

Axon terminal

Sarcolemma

Synaptic end bulb

Nucleus of muscle fiber

Mitochondrion

Sarcoplasm

Myofibril of muscle fiber

(b) Diagram of the neuromuscular junction

Synaptic vesicle containing ACh

Sarcolemma

Synaptic cleft

Motor end plate

Synaptic end bulb

Nerve impulse (action potential)

Muscle action potential

(c) Enlarged view of the neuromuscular junction

Synaptic vesicle releases ACh by exocytosis

Binding of ACh to its receptor opens the channel

Na⁺

ACh

Synaptic vesicle

Synaptic cleft

Motor end plate

ACh receptor

(d) Binding of acetylcholine to ACh receptors in the motor end plate

What is a motor end plate?

Figure 10.4 Organization of a skeletal muscle from gross to molecular levels.

The structural organization of a skeletal muscle from macroscopic to microscopic is: skeletal muscle, fascicle (bundle of muscle fibers), muscle fiber, myofibril, thin and thick filaments.

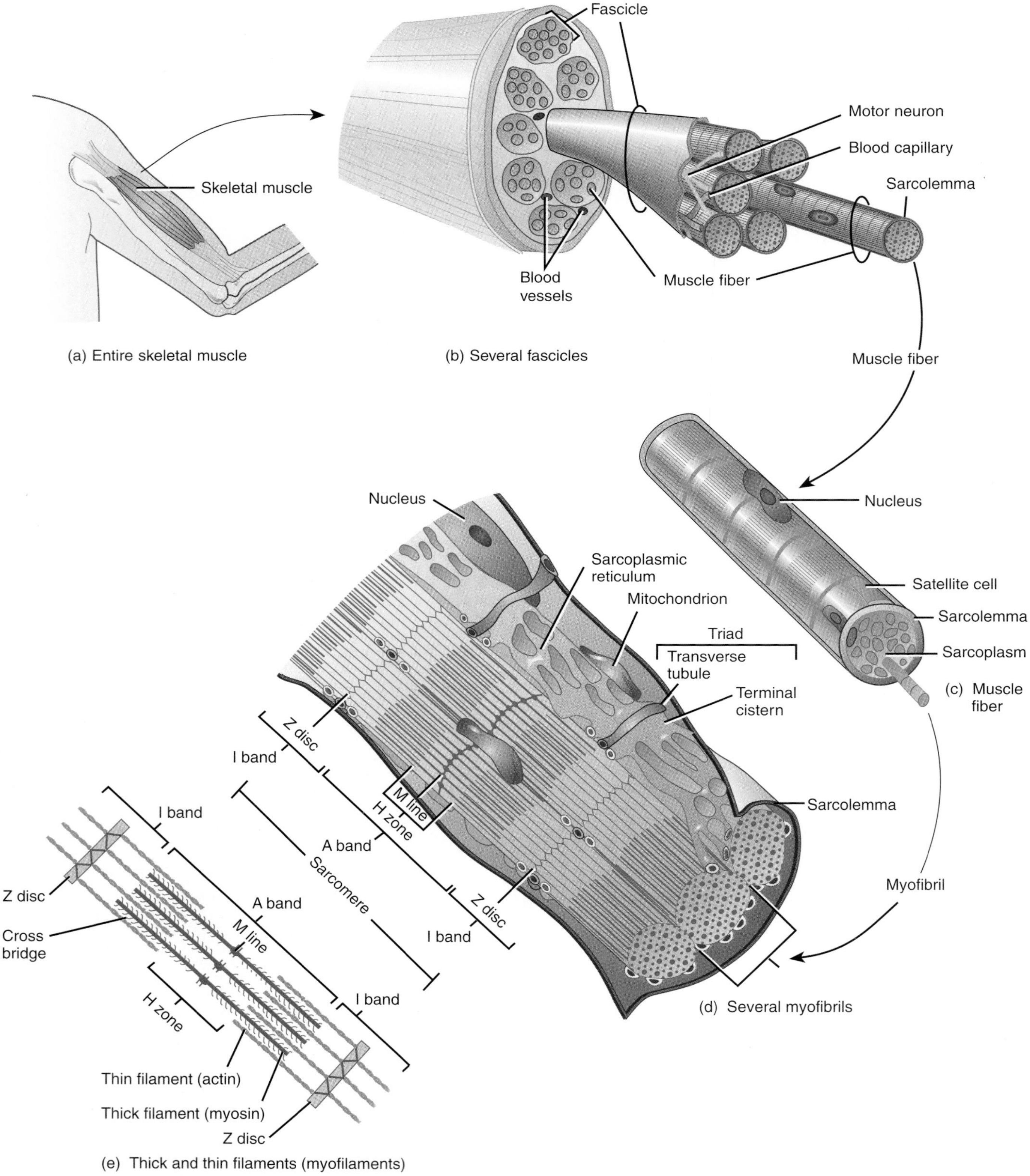

(a) Entire skeletal muscle

(b) Several fascicles

(c) Muscle fiber

(d) Several myofibrils

(e) Thick and thin filaments (myofilaments)

Q Which bands within a sarcomere appear darker? Lighter?

fibers lie parallel to one another and range from 10 to 100 μm in diameter.[1] While a typical length is 100 μm, some are up to 30 cm (12 in.) long. The plasma membrane of a muscle cell is termed the **sarcolemma** (*sarco* = flesh; *lemma* = sheath), and it surrounds the muscle fiber's cytoplasm, which is called **sarcoplasm** (Fig. 10.4c). Because skeletal muscle fibers arise from the fusion of many smaller cells (myoblasts) during embryonic development, each fiber has many nuclei. The nuclei are at the periphery of the cell, next to the sarcolemma, conveniently out of the way of the contractile elements (described shortly). The mitochondria lie in rows throughout the muscle fiber, strategically close to the muscle proteins that use ATP during contraction.

At high magnification the sarcoplasm appears stuffed with little threads. These small structures are the **myofibrils**. Although myofibrils extend lengthwise within the muscle fiber, their prominent alternating light and dark bands make the whole muscle cell look striated or striped (Fig. 10.4d). The bands are called **cross-striations**.

Myofibrils

Myofibrils are the contractile elements of skeletal muscle. They are 1–2 μm in diameter and contain three types of even smaller structures called **filaments (myofilaments)**. The diameter of the **thin filaments** is about 8 nm, whereas that of the **thick filaments** is about 16 nm (Fig. 10.4e).[2] Depending on whether the muscle is contracting or relaxing, the thick and thin filaments overlap one another to a greater or lesser extent. The pattern of their overlap causes the cross-striations seen both in single myofibrils and in whole muscle fibers. A third type of filament, the elastic filament, will be described shortly.

The filaments inside a myofibril do not extend the entire length of a muscle fiber. They are arranged in compartments called **sarcomeres** (*meros* = part), which are the basic functional units of striated muscle fibers (Figs. 10.5 and 10.6a). Narrow plate-shaped regions of dense material called **Z discs** (**lines**) separate one sarcomere from the next. Within each sarcomere is a darker area, called the **A band**. It consists mostly of the thick filaments and includes portions of the thin filaments where they overlap the thick filaments. A lighter, less dense area called the **I band** contains the rest of the thin filaments but no thick filaments. The Z disc passes through the center of each I band. The alternating darker A bands and lighter I bands give the muscle fiber its striated appearance. A narrow **H zone** in the center of each A band contains thick but not thin filaments. Dividing the H zone is the **M line**, formed by protein molecules that connect adjacent thick filaments.

The two contractile proteins in muscle are myosin and actin. About 200 molecules of the protein **myosin** form a

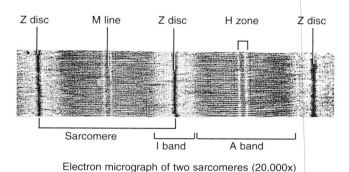

Figure 10.5 Transmission electron micrograph of mammalian skeletal muscle. The Z discs separate one sarcomere from another.

 The overlap of thick and thin filaments produces an alternating pattern of lighter and darker bands (striations).

Electron micrograph of two sarcomeres (20,000x)

Which filaments connect into the Z disc?

single thick filament (Fig. 10.6b). Each myosin molecule is shaped like two golf clubs twisted together. The **myosin tails** (golf club handles) point toward the M line in the center of the sarcomere. The projections, called **myosin heads** or **cross bridges**, extend out toward the thin filaments. Tails of neighboring myosin molecules lie parallel to one another, forming the shaft of the thick filament. The heads project from all around the shaft in a spiraling fashion.

Thin filaments extend from anchoring points within the Z discs. Their main component is **actin**. Also present in the thin filament are smaller amounts of two regulatory proteins, **tropomyosin** and **troponin** (Fig. 10.6c). Individual actin molecules have an irregular shape. They join to form an actin filament that is twisted into a helix. (For many years it was thought that two helical strands intertwined to form an actin filament. Recent evidence suggests there is just a single twisted chain of actin molecules.) On each actin molecule is a **myosin-binding site**, a location where a myosin head (cross bridge) can attach. In relaxed muscle, tropomyosin covers the myosin-binding sites on actin and thus blocks attachment of myosin heads to actin.

A more recently recognized component of the sarcomere is the **elastic filament** (Fig. 10.6a). It is composed of the protein titin (connectin), the third most plentiful protein in skeletal muscle (after actin and myosin). Titin anchors thick filaments to the Z discs and thereby helps stabilize the position of the thick filaments. It may also play a role in recovery of the resting sarcomere length when a muscle is stretched or during relaxation. The protein was named titin because it has a huge (titanic) molecular weight (or connectin because of its connecting function). Exhibit 10.1 reviews the types of filaments in skeletal muscle fibers.

Figure 10.6   Detailed structure of muscle filaments. (a) The relation of thick (myosin), thin (actin), and elastic (titin) filaments in a sarcomere. Note that actin filaments are anchored directly at the Z discs whereas myosin filaments are connected to the Z discs by titin (also known as connectin). (b) About 200 myosin molecules comprise a thick filament. The myosin tails all point toward the center of the sarcomere. (c) Thin filaments contain actin, troponin, and tropomyosin.

 Myofibrils contain three types of filaments: thin, thick, and elastic.

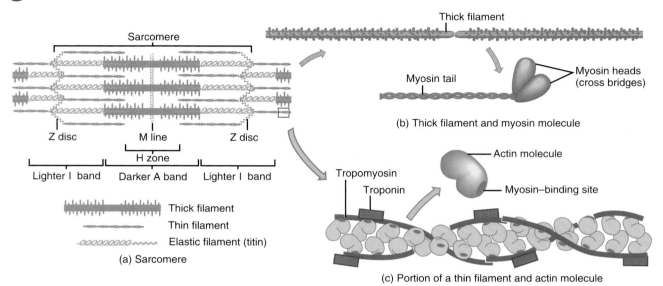

 Which proteins are present in the A band? In the I band?

Filament	Size and Protein Composition[a]	Functions
Thick filament	16 nm diameter; contains myosin (44%).	Myosin heads (cross bridges) move thin filaments toward center of sarcomere during contraction.
Thin filament	8 nm diameter; contains actin (22%), troponin (5%), and tropomyosin (5%).	Contains myosin head binding sites; slides along thick filament during contraction.
Elastic filament	Less than 1 nm diameter; contains titin, also called connectin (9%).	Anchors thick filaments to Z discs and stabilizes them during contraction and relaxation.

EXHIBIT **10.1** **TYPES OF FILAMENTS IN SKELETAL MUSCLE FIBERS**

[a]Percentage of total protein in muscle filaments.

EXERCISE-INDUCED MUSCLE DAMAGE

Electron micrographs of muscle tissue taken from athletes before and after a marathon race or other types of severe exercise reveal considerable damage, including torn sarcolemmas in some muscle fibers, damaged myofibrils, and disrupted Z discs. Microscopic muscle damage after exercise also is indicated by increases in blood levels of myoglobin and certain enzymes (lactic acid dehydrogenase and creatine phosphokinase, for example) that normally are confined inside muscle fibers. From 12 to 48 hours after a bout of strenuous exercise, skeletal muscles often become sore. Such **delayed onset muscle soreness** (**DOMS**) is accompanied by stiffness, tenderness, and swelling. Although the causes of DOMS are not completely understood, microscopic muscle damage appears to be a major contributing factor. ■

Sarcoplasmic Reticulum and Transverse Tubules

A fluid-filled system of cisterns (*cisterna* = reservoir or cavity) called the **sarcoplasmic reticulum** (**SR**) encircles each myofibril (see Fig. 10.4d). This elaborate tubular system is similar to smooth endoplasmic reticulum in nonmuscle cells. In a relaxed muscle fiber, the sarcoplasmic reticulum stores Ca^{2+}. Release of Ca^{2+} from the sarcoplasmic reticulum into the sarcoplasm around the thick and thin filaments triggers muscle contraction. The calcium ions leave the sarcoplasmic reticulum through channels in its membrane called **Ca^{2+} release channels**.

The **transverse tubules** (**T tubules**) are tunnel-like infoldings of the sarcolemma. They penetrate toward the center of the muscle fiber at right angles to the myofilaments. In mammalian skeletal muscle, there are two transverse tubules in each sarcomere, one at each A–I band junction. T tubules are open to the outside of the fiber and are filled with extracellular fluid. On both sides of a transverse tubule are dilated end sacs of the sarcoplasmic reticulum called **terminal cisterns**. The term **triad** (*tri* = three) refers to a transverse tubule and the terminal cisterns on either side of it.

coplasm starts filament sliding, while a decrease turns off the sliding process.

When a muscle fiber is relaxed (not contracting), the concentration of Ca^{2+} in its sarcoplasm is low (Fig. 10.8a). This is because the sarcoplasmic reticulum (SR) membrane contains **Ca^{2+} active transport pumps** that move Ca^{2+} from the sarcoplasm into the SR. Ca^{2+} is stored or sequestered inside the SR. As a muscle action potential travels along the sarcolemma and into the transverse tubule system, however, Ca^{2+} release channels open in the SR membrane (Fig. 10.8b). As a result, Ca^{2+} floods into the sarcoplasm around the thick and thin filaments. The Ca^{2+} released from the sarcoplasmic reticulum combine with troponin, causing it to change shape. This shape change moves the troponin–tropomyosin complex away from the myosin-binding sites on actin (Fig. 10.8b).

CONTRACTION OF MUSCLE

In the mid-1950s Jean Hanson and Hugh Huxley had a revolutionary insight into the mechanism of muscle contraction. Previously, scientists had imagined that muscle contraction must be a folding process, somewhat like closing an accordion. Hanson and Huxley proposed, however, that skeletal muscle shortens during contraction because the thick and thin filaments slide past one another. Their model is known as the **sliding filament mechanism** of muscle contraction.

Sliding Filament Mechanism

During muscle contraction, myosin heads pull on the thin filaments, causing them to slide inward toward the H zone at the center of the sarcomere (Fig. 10.7). The myosin cross bridges may even pull the thin filaments of each sarcomere so far inward that their ends overlap in the center of the sarcomere (Fig. 10.7c). As the thin filaments slide inward, the Z discs come toward each other, and the sarcomere shortens, but the lengths of the thick and thin filaments do not change. The sliding of the filaments and shortening of the sarcomeres cause shortening of the whole muscle fiber and ultimately the entire muscle.

Role of Calcium and Regulator Proteins

The sliding filament model explains the mechanism of contraction, but what starts and stops sliding of the filaments? An increase in Ca^{2+} concentration in the sar-

Figure 10.7 Sliding filament mechanism of muscle contraction. For simplicity, the elastic filaments are not illustrated.

 During muscle contraction, thin filaments move inward toward the H zone.

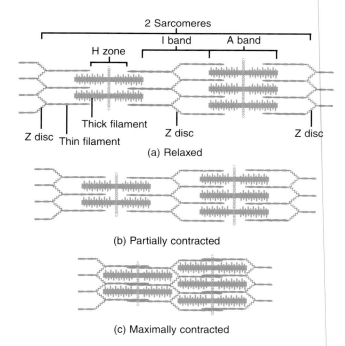

(a) Relaxed

(b) Partially contracted

(c) Maximally contracted

 What happens to the I band and H zone during contraction? Do the lengths of the thick and thin filaments change?

Figure 10.8   Regulation of contraction by troponin and tropomyosin. (a) The level of Ca^{2+} in the sarcoplasm is low during relaxation because it is pumped into the sarcoplasmic reticulum (SR) by Ca^{2+} active transport pumps. (b) A muscle action potential traveling along a transverse tubule opens calcium release channels in the SR and Ca^{2+} flows into the sarcoplasm. Note that contraction is occurring because the thin filaments are closer in the center of the sarcomere.

An increase in the level of Ca^{2+} in the sarcoplasm starts the movement of thin filaments; when the level of Ca^{2+} declines, movement stops.

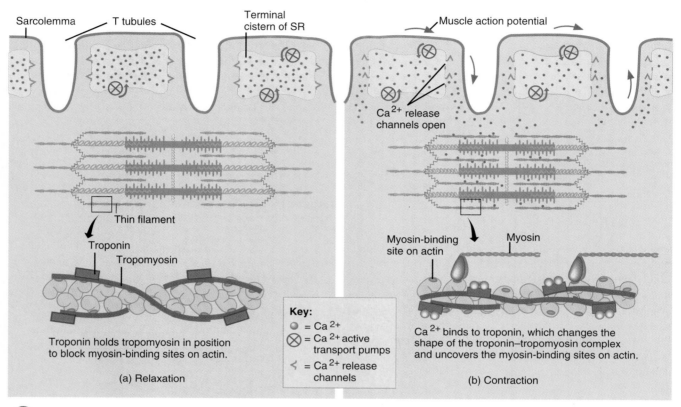

Which muscle protein binds Ca^{2+} that is released from the sarcoplasic reticulum?

The Power Stroke and the Role of ATP

As we have seen, muscle contraction requires Ca^{2+}. It also requires energy, in the form of ATP. Figure 10.9 shows the sequence of events during sliding of the filaments.

1   While the muscle is relaxed, ATP attaches to ATP-binding sites on the myosin cross bridges (heads). A portion of each myosin head acts as an ATPase, an enzyme that splits the ATP into ADP + Ⓟ (phosphate group) through a hydrolysis reaction. This reaction transfers energy from ATP to the myosin head, even before contraction begins. The myosin cross bridges are thus in an activated (energized) state.

2   When the sarcoplasmic reticulum releases Ca^{2+} and Ca^{2+} level rises in the sarcoplasm, tropomyosin moves away from its blocking position.

3   The activated myosin heads spontaneously bind to the myosin-binding sites on actin.

4   The shape change that occurs as myosin heads bind to actin produces the **power stroke** of contraction.

During the power stroke, the myosin heads swivel toward the center of the sarcomere, like the oars of a boat during rowing. This action draws the thin filaments past the thick filaments toward the H zone. As the myosin heads swivel, they release ADP.

5   Once the power stroke is complete, ATP again combines with the ATP-binding sites on the myosin heads. As ATP binds, the myosin head detaches from actin.

6   Again, the myosin ATPase splits ATP, transferring its energy to the myosin head, which returns to its original upright position.

7   The myosin head is then ready to combine with another myosin-binding site further along the thin filament.

The cycle of steps 3 through 7 repeats over and over as long as ATP is available and the Ca^{2+} level near the thin filament is high. The myosin heads keep rotating back and forth with each power stroke, pulling the thin filaments toward the H zone. At any one instant, about half of the myosin heads are bound to actin and are swiveling. The other half are detached and preparing to swivel again. Contraction is analogous to running on a nonmotorized treadmill. One foot (myosin head) strikes the belt (thin filament) and pushes it backward (toward the H zone). Then the other foot comes down and imparts a second push. The belt soon moves smoothly while the runner (thick filament) remains stationary. And like the legs of a runner, the myosin heads need a constant supply of energy to keep going!

This continual movement of myosin heads applies the force that draws the Z discs toward each other, and the sarcomere shortens. The myofibrils thus contract, and the whole muscle fiber shortens. During a maximal muscle contraction, the distance between Z discs can decrease to half the resting length. But the power stroke does not always result in shortening of the muscle fibers and the whole muscle. Contraction without shortening is called an **isometric contraction**, for example, in trying to lift a very heavy object. The myosin heads (cross bridges) swivel and generate force, but the thin filaments do not slide inward.

Relaxation

Two changes permit a muscle fiber to relax after it has contracted. First, acetylcholine is rapidly broken down by an enzyme called **acetylcholinesterase (AChE)**. AChE is attached to collagen fibers in the extracellular matrix of the synaptic cleft and is probably synthesized by the muscle fibers. When action potentials cease in the motor neuron, release of ACh stops, and AChE rapidly breaks down the ACh already present in the synaptic cleft. This ends the generation of muscle action potentials, and the Ca^{2+} release channels in the sarcoplasmic reticulum membrane close.

Second, Ca^{2+} active transport pumps rapidly remove Ca^{2+} from the sarcoplasm into the sarcoplasmic reticulum, where molecules of a calcium-binding protein, appropriately called **calsequestrin**, bind to the Ca^{2+}. This reaction takes Ca^{2+} out of solution and allows even more Ca^{2+} to be sequestered within the SR. The concentration of Ca^{2+} is 10,000 times lower in the sarcoplasm of a relaxed muscle fiber than

Figure 10.9   Role of ATP and the power stroke of muscle contraction. Sarcomeres shorten through repeated cycles in which the myosin heads (cross bridges) attach to actin, swivel, and detach.

During the power stroke of contraction, myosin heads swivel and move the thin filaments past the thick filaments toward the center of the sarcomere.

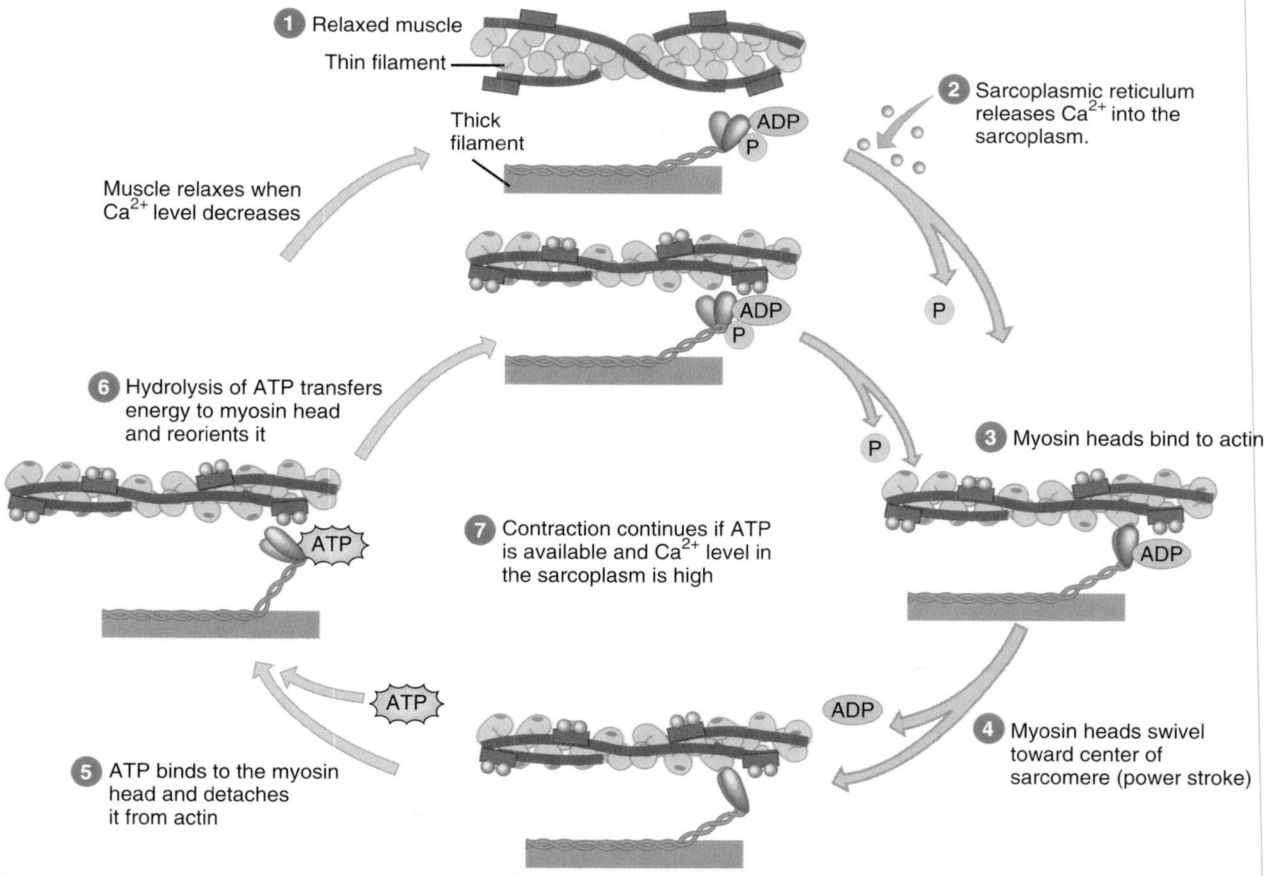

1. Relaxed muscle
Thin filament
Thick filament

Muscle relaxes when Ca^{2+} level decreases

2. Sarcoplasmic reticulum releases Ca^{2+} into the sarcoplasm.

ADP
P

6. Hydrolysis of ATP transfers energy to myosin head and reorients it

3. Myosin heads bind to actin

ADP

7. Contraction continues if ATP is available and Ca^{2+} level in the sarcoplasm is high

ATP

5. ATP binds to the myosin head and detaches it from actin

4. Myosin heads swivel toward center of sarcomere (power stroke)

ADP

What would happen if ATP were suddenly not available after the sarcomere had started to shorten?

inside the SR. As Ca²⁺ level drops in the sarcoplasm, the tropomyosin–troponin complex moves back over the myosin-binding sites on actin. This prevents further binding of myosin heads to actin, and the thin filaments slip back to their relaxed positions. Figure 10.10 summarizes the events associated with contraction and relaxation of a muscle fiber.

RIGOR MORTIS

After death, autolysis begins in muscle fibers and Ca²⁺ leaks out of the sarcoplasmic reticulum. The Ca²⁺ binds to troponin and triggers attachment of myosin cross bridges to actin. ATP

Figure 10.10   Summary of the events of contraction and relaxation.

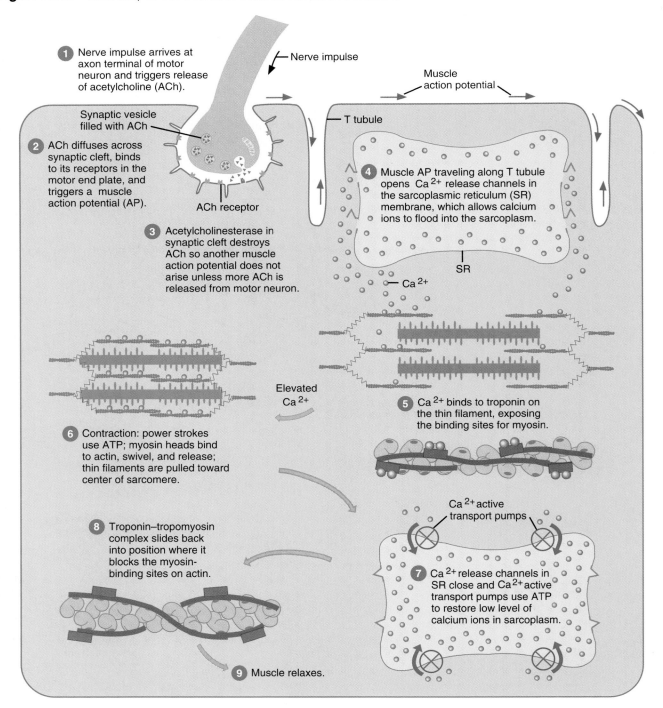

What are three functions of ATP in skeletal muscle contraction?

synthesis has ceased, however, so the myosin cross bridges cannot detach from actin. The resulting condition, in which muscles are in a state of rigidity (cannot contract or stretch), is called **rigor mortis** (rigidity of death). Rigor mortis lasts about 24 hours but disappears as tissues begin to disintegrate. ■

Homeostasis of Body Temperature

Both smooth and skeletal muscles play important roles in maintaining the body's thermal homeostasis. The main contribution of smooth muscle is in regulation of blood vessel diameter. When smooth muscle in the walls of skin arterioles relaxes, the arterioles dilate, and more blood flows to the skin. This permits greater transfer of heat from the warm blood through the skin to the environment. On the other hand, when heat conservation is needed, smooth muscle in the skin blood vessels contracts. As a result, the vessels constrict, less blood flows through the skin, and less heat is lost.

During contraction of skeletal muscles, only a small amount of the energy stored in body chemicals is used for mechanical work (movement). As much as 85% is released as heat (thermogenesis). A portion of the released heat helps maintain a normal body temperature. Excess heat is eliminated through the skin and lungs. If body temperature (controlled condition) decreases, one result is shivering, which causes involuntary thermogenesis. This increase in skeletal muscle tone can raise heat production by several hundred percent. Shivering is initiated by a part of the brain called the hypothalamus (control center). It acts via a negative feedback system to produce enough heat to raise body temperature back to normal (Fig. 10.11). Other mechanisms for regulation of body temperature are discussed on page 809.

ADJUSTING TENSION IN WHOLE MUSCLES

A single action potential in a motor neuron elicits a single contraction in all the muscle fibers of its motor unit. The contraction is said to be **all-or-none** because individual muscle fibers will contract to their fullest extent. In other words, muscle fibers do not contract partially. The force of their contraction can vary only slightly. Local chemical conditions (for example, nutrient and oxygen availability) and recent contraction of the muscle fibers influence the force of contraction to a small extent.

You know, however, that a whole muscle can have graded contractions (varying sizes) to perform different tasks. For example, your arm muscles do not contract to the same extent when you lift a textbook as when you lift a piece of paper. But they do contract in a smooth, graded fashion to lift either object. The amount of tension (force) that a skeletal muscle can develop depends on the frequency of stimulation of muscle fibers by motor neurons, the length of muscle fibers just before they contract, the number of

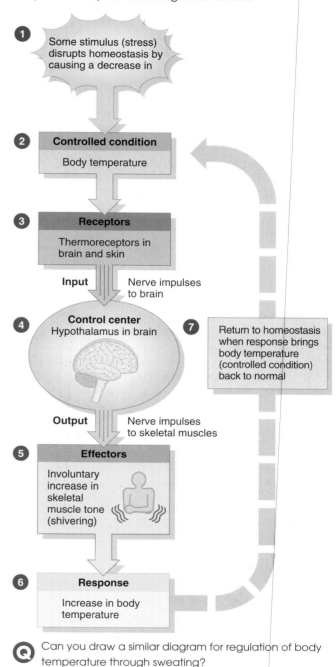

Figure 10.11 Negative feedback regulation of body temperature by the shivering mechanism.

1 Some stimulus (stress) disrupts homeostasis by causing a decrease in

2 **Controlled condition**
Body temperature

3 **Receptors**
Thermoreceptors in brain and skin

Input Nerve impulses to brain

4 **Control center**
Hypothalamus in brain

7 Return to homeostasis when response brings body temperature (controlled condition) back to normal

Output Nerve impulses to skeletal muscles

5 **Effectors**
Involuntary increase in skeletal muscle tone (shivering)

6 **Response**
Increase in body temperature

Q Can you draw a similar diagram for regulation of body temperature through sweating?

muscle fibers contracting (number of motor units recruited and size of individual motor units), and structural components of the muscle itself.

Twitch Contraction

A **twitch contraction** is a brief contraction of all the muscle fibers in a motor unit of a muscle in response to a single action potential in its motor neuron. In the laboratory, a twitch also can be produced by direct electrical stimulation of a motor neuron or its muscle fibers. Much of our knowledge of muscle contraction has come from experiments performed on isolated, excised muscle.

Figure 10.12 is a graph of a twitch contraction. The record of a muscle contraction is called a **myogram**. Note that a brief period exists between application of the stimulus and the beginning of contraction; this is the **latent period**. During this time, Ca^{2+} is being released from the sarcoplasmic reticulum, the filaments start to exert force (usually taking some slack out of the system), and finally shortening begins. The latent period lasts about 2 milliseconds (msec; 1 msec = $\frac{1}{1000}$ sec). The second phase, the **contraction period**, has a duration of 10–100 msec and is indicated by the upward tracing. The third phase, the **relaxation period**, also lasts about 10–100 msec and is indicated by the downward tracing. It is caused by the active transport of Ca^{2+} back into the sarcoplasmic reticulum, which results in relaxation. The duration of these periods depends on the type of muscle fiber. Some fibers, such as those that move the eyes, are fast-twitch fibers (described shortly) and have a contraction period as brief as 10 msec and an equally brief relaxation period. Others, such as those that move the legs, are slow-twitch fibers, with contraction and relaxation periods of 100 msec or so.

If two stimuli are applied, one immediately after the other, the muscle will respond to the first stimulus but not to the second. When a muscle fiber receives enough stimulation to contract, it temporarily loses its excitability and cannot respond again until its responsiveness is regained. This period of lost excitability is called the **refractory period** and is a characteristic of all nerve and muscle cells. The duration of the refractory period varies with the muscle involved. Skeletal muscle has a short refractory period of about 5 msec (0.005 sec). Cardiac muscle has a long refractory period of about 300 msec (0.30 sec).

Frequency of Stimulation

When two stimuli are applied and the second one is delayed until the refractory period is over, the skeletal muscle will respond to both stimuli. In fact, if the second stimulus is applied after the refractory period, but before the muscle fiber has finished relaxing, the second contraction will be stronger than the first. This phenomenon, in which stimuli arrive at different times and cause larger contractions, is called **wave (temporal) summation** (Fig. 10.13a).

Tetanus

When a skeletal muscle is stimulated at a rate of 20–30 times per second, it can only partly relax between stimuli. The result is a sustained contraction called **incomplete (unfused) tetanus** (Fig. 10.13b). Stimulation at an increased rate (80–100 stimuli per second) results in **complete (fused) tetanus**, a sustained contraction that lacks even partial relaxation between stimuli (Fig. 10.13c).

Both kinds of tetanus result from the addition of Ca^{2+} released from the sarcoplasmic reticulum by the second, and subsequent, stimuli to the Ca^{2+} still in the sarcoplasm from the first stimulus. Relaxation is either partial or does not occur at all. Most voluntary muscular contractions involve short-term tetanic contractions and are thus smooth, sustained contractions.

Staircase Effect (Treppe)

When a muscle has been relaxed for some time and then is stimulated to contract by several identical stimuli that are too far apart for wave summation to occur, each of the first few contractions is a little stronger than the last. This phenomenon is known as the **staircase effect** or **treppe** (TREP-eh; = staircase). After the first few contractions, the muscle reaches its peak of performance and can undergo its strongest contractions. The explanation for the staircase effect may be the same as for tetanus—a progressive buildup of Ca^{2+} in the sarcoplasm. Successive stimuli cause Ca^{2+} to flow out of the sarcoplasmic reticulum faster than the active transport pumps take them back in. Up to a certain point, as Ca^{2+} builds up and binds to troponin, more power strokes can occur, and filament sliding intensifies. Also, other internal conditions in the muscle, such as temperature, pH, and viscosity, have changed. A rise in temperature, for example, could provoke stronger contractions.

Length of Muscle Fibers

As you already know, a skeletal muscle fiber contracts when myosin cross bridges of thick filaments connect with actin on the thin filaments. A muscle fiber develops its greatest tension when there is optimal overlap between thick and thin filaments (Fig. 10.14). At the optimum sar-

Figure 10.12   Myogram of a twitch contraction. The arrow at 0 indicates the time at which the stimulus was applied.

🔑 *A myogram is a record of a muscle contraction.*

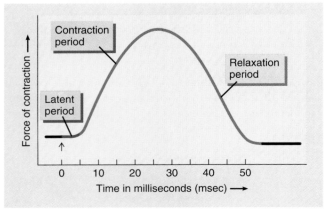

❓ What events occur during the latent period?

Figure 10.13 Myograms showing effect of different frequencies of stimulation. (a) When a second stimulus (longer arrow) is applied before the muscle has finished relaxing, the second contraction is stronger than the first. (The broken line indicates the force of contraction expected in a single twitch.) (b) In incomplete tetanus the curve looks jagged due to partial relaxation of the muscle between stimuli. (c) In complete tetanus, the contraction force is sustained until after the last stimulus.

 Most voluntary muscle contractions involve short-term tetanic contraction and are thus smooth, sustained contractions.

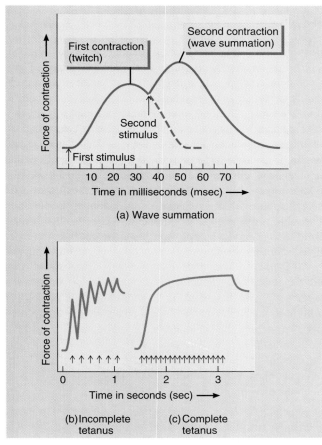

(a) Wave summation

(b) Incomplete tetanus

(c) Complete tetanus

In (a), would the peak force of the second contraction be larger or smaller if the second stimulus were applied at 45 msec rather than at 35 msec?

comere length, the number of myosin cross bridges making contact with thin filaments brings about a maximal force of contraction.

If the sarcomeres of a muscle fiber are stretched to a longer length, fewer myosin cross bridges can make contact with thin filaments, and the force of contraction decreases. If a skeletal muscle fiber is stretched to 175% of its optimal length, no myosin cross bridges can bind to thin filaments and no contraction occurs. At sarcomere lengths less than the optimum, the force of contraction also decreases. This is

because thick filaments crumple as they are compressed by the Z discs, resulting in fewer myosin cross bridge contacts with thin filaments. In the intact body, resting muscle fiber length is rarely less than 70% or more than 130% of the optimum. This is because skeletal muscles are firmly anchored to bones and other inelastic tissues.

Number of Muscle Fibers Contracting

The process of increasing the number of active motor units is called **recruitment** (**multiple motor unit summation**). The various motor neurons to a whole muscle fire asynchronously. While some motor units are active (contraction), others are inactive (relaxed). This pattern of activity of motor neurons prevents muscle fatigue while maintaining contraction by allowing a brief rest for the inactive motor units. The alternating motor units relieve one another so that the contraction can be sustained for long periods.

Recruitment also is one factor responsible for producing smooth movements rather than a series of jerky movements. As indicated before, the number of muscle fibers innervated by one motor neuron varies greatly. Precise movements require tiny changes in muscle contraction. Therefore, in such muscles, the motor units are small. In this way, when a motor unit is recruited or turned off, slight but controlled changes occur in muscle contraction. On the other hand, large motor units are active where maintaining a constant position or posture is important and precision is not.

Figure 10.14 Length–tension relationship in a skeletal muscle fiber. Maximum tension occurs at a sarcomere length of 2.2 μm (= 100%).

 A muscle fiber develops its greatest tension when there is optimal overlap of the thick and thin filaments.

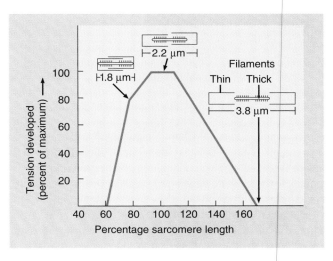

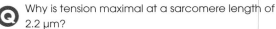

Why is tension maximal at a sarcomere length of 2.2 μm?

Muscle Tone

Involuntary activation of a small number of motor units causes sustained, small contractions that give a firmness to a relaxed skeletal muscle. This firmness is known as **muscle tone** (*tonos* = tension). At any instant, a few muscle fibers are contracted while most are relaxed. This small amount of contraction firms up a muscle without producing movement and is essential for maintaining posture. Asynchronous firing of motor units allows muscle tone to be sustained continuously. When the muscles in the back of the neck are in tonic contraction, for example, they keep the head upright and prevent it from slumping forward onto the chest, but they do not apply enough force to pull the head back into hyperextension.

HYPOTONIA AND HYPERTONIA

Hypotonia refers to decreased or lost muscle tone. Such muscles are said to be **flaccid** (FLAK-sid or FLAS-sid). Flaccid muscles are loose and appear flattened rather than rounded. Certain disorders of the nervous system may result in **flaccid paralysis**, which is characterized by loss of muscle tone, loss or reduction of tendon reflexes, and atrophy (wasting away) of muscles.

Hypertonia refers to increased muscle tone and is expressed in two ways: spasticity or rigidity. **Spasticity** is characterized by increased muscle tone (stiffness) associated with an increase in tendon reflexes and pathological reflexes (such as the Babinski sign, which is described on page 374). Certain disorders of the nervous system may result in **spastic paralysis**, partial paralysis in which the muscles exhibit spasticity. **Rigidity** refers to conditions of increased muscle tone in which reflexes are not affected. ■

Active and Passive Tension

As thin filaments start to slide past thick filaments, they pull on the Z discs, which, in turn, pull on neighboring sarcomeres. Eventually, whole muscle cells are pulling on the connective tissue wrappings. Some of these structural components are elastic: they stretch slightly before they start to transfer the force or tension being generated by the sliding filaments. The elastic components include the elastic filaments, connective tissue around the muscle fibers (epimysium, perimysium, and endomysium), and tendons that attach muscle to bone. The tension generated by contractile elements (thin and thick filaments) is called **active tension**. The tension generated by elastic elements is called **passive tension** and is not related to muscular contraction. It depends on the degree of muscle stretch; within limits, the more a muscle is stretched, the greater its passive tension. As a skeletal muscle starts to shorten, it first pulls on its connective tissue coverings and tendons. The coverings and tendons stretch, they become taut, and the tension passed through the tendons pulls on the bones to which they are attached. The result is movement of a part of the body.

The stretch of elastic elements is also related to wave summation and tetanus (see Fig. 10.13). During wave summation, elastic elements are not given much time to spring back between contractions, and thus they remain taut. While in this state, the elastic elements do not require very much stretching before the beginning of the next muscular contraction. The combination of the tautness of the elastic elements and the partially contracted state of the filaments enables the force of another contraction to be added more quickly to the one before.

Isotonic and Isometric Contractions

Isotonic (*iso* = equal; *tonos* = tension) **contractions** occur when you move a constant load through the range of motions possible at a joint. During such a contraction, the tension remains almost constant. There are two types of isotonic contractions. In a **concentric contraction**, the muscle shortens and pulls on another structure, such as a bone, to produce movement and to reduce the angle at a joint (Fig. 10.15a). Picking up a book involves concentric contractions of the biceps brachii muscle in the arm. When the overall length of a muscle increases during a contraction, it is called an **eccentric contraction** (Fig. 10.15b). As you lower the book to place it back on the table, the previously shortened biceps gradually lengthens while it continues to contract. For reasons that are not well understood, repeated eccentric contractions produce more muscle damage and more delayed onset muscle soreness than concentric contractions.

An **isometric contraction** occurs when the muscle does not or cannot shorten, but the tension on the muscle increases greatly (Fig. 10.15c). An example would be holding a book in a steady position. The book pulls the arm downward, stretching the shoulder and arm muscles. The isometric contraction of the shoulder and arm muscles counteracts the stretch. The two forces—contraction and stretching—applied in opposite directions create the tension. Isometric contractions are important because they stabilize some joints as others are moved. Although isometric contractions do not result in body movement, energy is still expended. Most activities include both isotonic and isometric contractions.

MUSCULAR ATROPHY AND HYPERTROPHY

Muscular atrophy (A-trō-fē) is a wasting away of muscles. Individual muscle fibers decrease in size as a result of progressive loss of myofibrils. Muscles atrophy if they are not used. This is termed **disuse atrophy**. Bedridden individuals and people with casts experience atrophy because the flow of nerve impulses to inactive muscle is greatly reduced. If the nerve supply to a muscle is cut, the muscle undergoes **denervation atrophy**. In about 6 months to 2 years, the muscle will be one-quarter its original size and the muscle fibers will be replaced by connective tissue. The transition to connective tissue, when complete, cannot be reversed.

Figure 10.15   Comparison between isotonic (concentric and eccentric) and isometric contractions.

In an isotonic contraction, tension remains constant as muscle length decreases or increases; in an isometric contraction, tension increases greatly without change in muscle length.

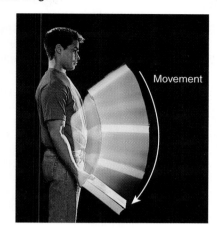

(a) Concentric contraction

(b) Eccentric contraction

(c) Isometric contraction

 What kind of contraction occurs in your neck muscles while you are reading your class assignment?

Muscular hypertrophy (hī-PER-trō-fē), the opposite of atrophy, refers to an increase in the diameter of muscle fibers owing to the production of more myofibrils, mitochondria, sarcoplasmic reticulum, and so forth. It results from very forceful, repetitive muscular activity, such as strength training. Hypertrophied muscles are capable of more forceful contractions. ■

MUSCLE METABOLISM

Contraction of muscle requires energy, but surprisingly little ATP is present inside muscle fibers; there is just enough to power contraction for a few seconds. Unlike most cells of the body, skeletal muscle fibers often function in a "stop and go" manner; they switch between virtual inaction and great activity. If strenuous exercise is to continue for more than a few seconds, additional ATP must be produced. On demand, the muscle fiber's metabolic machinery goes into high gear to step up ATP production.

Phosphagen System

Muscle fibers have a unique molecule called **creatine phosphate** (**phosphocreatine**) that can transfer its high-energy phosphate group to ADP (adenosine diphosphate), thus forming ATP (adenosine triphosphate) and creatine (Fig. 10.16a). Creatine phosphate is about three to five times more plentiful than ATP. Together, creatine phosphate and ATP constitute the **phosphagen system** and provide enough energy for muscles to contract maximally for about 15 seconds. This energy system is used for maximal short bursts of energy, for example, to run a 100-meter dash.

Glycogen–Lactic Acid System

When muscle activity continues and the supply of creatine phosphate is depleted, glucose is catabolized to generate ATP. Glucose easily passes into contracting muscle fibers from the blood (via facilitated diffusion) and also is produced by breakdown of glycogen within muscle fibers (Fig. 10.16b). A series of ten reactions known as **glycolysis** quickly splits each glucose molecule into two molecules of pyruvic acid. This process uses two ATP but forms four ATP for a net gain of two (the reactions are shown in Fig. 25.4). Since glycolysis does not require oxygen, it is said to be an **anaerobic process**.

Ordinarily, the pyruvic acid formed by glycolysis enters mitochondria. There, its oxidation produces a large amount of ATP from ADP. During some activities, however, there is not enough oxygen to completely break down pyruvic acid. When this happens, most of the pyruvic acid is converted to lactic acid. About 80% of the lactic acid produced in this way diffuses out of the skeletal muscle fibers into the blood. Heart muscle fibers, kidney cells, and liver cells can use lactic acid to produce ATP. Also, liver cells can convert some of the lactic acid back to glucose. This conversion has two benefits—providing new glucose molecules and reducing acidity. However, some lactic acid accumulates in blood and in muscle tissue. The glycogen–lactic acid system just described can provide enough energy for about 30–40 seconds of maximal muscle activity, for example, to run a 300-meter race. Eventually, muscle glycogen must also be restored. This is accomplished mainly through eating carbohydrate-rich foods and may take several days, depending on the intensity of exercise.

Aerobic System

Muscular activity that lasts longer than half a minute depends increasingly on **aerobic processes**, that is, reactions requiring oxygen. If sufficient oxygen is present, enzymes in the mitochondria can completely oxidize pyruvic acid to carbon dioxide, water, ATP, and heat (Fig. 10.16c). This process, called **cellular respiration** or **biological oxidation**, is described in more detail in Chapter 25. Although cellular respiration is slower than glycolysis, it yields much more ATP, about 36 molecules of ATP from each glucose molecule.

Muscle tissue has two sources of oxygen: (1) oxygen that diffuses into muscle fibers from the blood and (2) oxygen that is released by **myoglobin** inside muscle fibers. Both myoglobin and hemoglobin (in red blood cells) are oxygen-

binding proteins; they bind oxygen when it is plentiful and release it in times of scarcity.

The aerobic system will provide enough ATP for prolonged activity so long as sufficient oxygen and nutrients are available. Besides pyruvic acid obtained from glycolysis of glucose, these nutrients include fatty acids (from the breakdown of triglycerides in adipose cells) and amino acids (from the breakdown of proteins). In activities that last more than 10 minutes, the aerobic system provides more than 90% of the needed ATP. During a long-term event, such as a marathon race, close to 100% of the ATP is produced aerobically.

The maximal rate of oxygen consumption that is possible during the aerobic catabolism of pyruvic acid is called **maximal oxygen uptake**. It is influenced by gender (higher in males), age (highest at about age 20), and size (increases

Figure 10.16   Muscle metabolism during contraction. (a) Creatine phosphate, formed while the muscle is relaxed, transfers a high-energy phosphate group to ADP, forming ATP. (b) Breakdown of muscle glycogen into glucose and metabolism of glucose via glycolysis produce both ATP and lactic acid. Since no oxygen is needed, this is an anaerobic pathway. (c) Within mitochondria, pyruvic acid, amino acids, and fatty acids are used to produce ATP.

During a long-term event such as a marathon race, most ATP is produced aerobically.

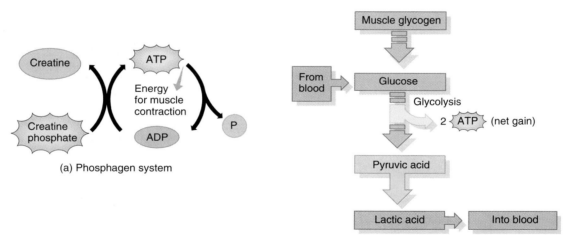

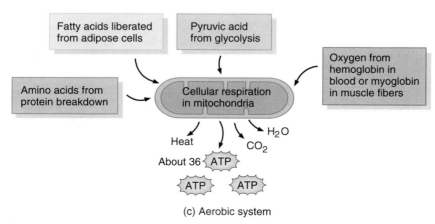

(c) Aerobic system

Where in the muscle fiber are the events shown here occurring?

with body size). Highly trained athletes can have maximal oxygen uptakes that are twice those of untrained people, owing to a combination of both heredity and training. As a result, they are capable of greater muscular strength and endurance than untrained people.

ENDURANCE VERSUS STRENGTH TRAINING

Regular aerobic workouts that feature many isotonic contractions, for example, jogging or aerobic dancing, increase the blood supply of skeletal muscles and thus improve blood flow. Activities that include many isometric contractions, for example, weight lifting, rely more on anaerobic production of ATP through glycolysis. Such activities also stimulate synthesis of muscle proteins, and the result, over a period of time, is an increase in muscle size (muscle hypertrophy). As a result, aerobic training builds endurance for prolonged activities whereas anaerobic training builds muscle strength for short-term feats. **Interval training** is a workout regimen that incorporates both types of training, for example, alternating sprints (anaerobic activity) with jogging (aerobic activity). ■

Oxygen Consumption After Exercise

During muscular exercise, blood vessels in muscles dilate, blood flow increases, and oxygen delivery increases. Up to a point, the available oxygen is sufficient to meet the energy needs of the contracting muscles. When muscular exertion is very great, however, oxygen cannot be supplied to muscle fibers fast enough and cellular respiration cannot produce enough ATP.

After exercise has stopped, heavy breathing continues for a period of time, and oxygen consumption is above the resting level. Depending on the intensity of the exercise, the recovery period may be just a few minutes or several hours. In 1922 A. V. Hill coined the term **oxygen debt** for the added oxygen that is taken into the body after exercise, over and above the resting oxygen consumption. He proposed that this extra oxygen was all used to "pay back" or restore metabolic conditions to the resting level. In this view, the extra oxygen is used to (1) convert lactic acid into glycogen stores in the liver, (2) resynthesize creatine phosphate and ATP, and (3) replace the oxygen removed from myoglobin.

The metabolic changes that occurred during exercise, however, account for only some of the extra oxygen used after exercise. Postexercise oxygen use also is boosted by ongoing changes. First, body temperature is elevated after strenuous exercise. This increases the pace of chemical reactions throughout the body. Faster reactions use ATP more rapidly, and more oxygen is needed to produce ATP. Second, the heart and muscles used in breathing are still working harder than they were at rest and thus consume more ATP. Third, tissue repair processes are occurring at an increased pace. For these reasons, a better term than oxygen debt for the elevated oxygen use after exercise is **recovery oxygen consumption**.

Muscle Fatigue

If a skeletal muscle or group of skeletal muscles is overstimulated, the strength of contraction becomes progressively weaker until the muscle no longer responds. The inability of a muscle to maintain its strength of contraction or tension is called **muscle fatigue**. It occurs when a muscle cannot produce enough ATP to meet its needs. Several factors appear to contribute to muscle fatigue, including insufficient oxygen, depletion of glycogen, buildup of lactic acid, failure of action potentials in the motor neuron to release enough acetylcholine, and unexplained fatigue mechanisms in the central nervous system.

TYPES OF SKELETAL MUSCLE FIBERS

Skeletal muscle fibers are not all identical in structure and function. For example, they vary in color depending on their content of myoglobin, which is a red-colored protein. Skeletal muscle fibers that have a high myoglobin content are termed **red muscle fibers**. Those that have a low content of myoglobin, on the other hand, are called **white muscle fibers**. Red muscle fibers also have more mitochondria and more blood capillaries than white muscle fibers.

As mentioned earlier, skeletal muscle fibers contract and relax with different velocities. Whether a fiber is slow-twitch or fast-twitch depends on how rapidly it splits ATP. Fast-twitch fibers split ATP more quickly. In addition, skeletal muscle fibers vary in the metabolic reactions they use to generate ATP and in how quickly they fatigue. Based on these structural and functional characteristics, skeletal muscle fibers are classified into three types (Exhibit 10.2):

1. **Slow oxidative (type I) fibers.** These fibers, which are the smallest in diameter, are also called **slow-twitch** or **fatigue-resistant fibers**. They contain large amounts of myoglobin, many mitochondria, and many blood capillaries. They thus look red and have a high capacity to generate ATP by the aerobic system, which is why they are called oxidative fibers. They split ATP at a slow rate and, as a result, have a slow contraction velocity. These fibers are very resistant to fatigue. Such fibers are found in large numbers in the postural muscles, for example, in neck muscles that hold the head upright.
2. **Fast oxidative (type IIA) fibers.** These fibers, which are intermediate in diameter, are also called **fast-twitch A** or **fatigue-resistant fibers**. They also contain large amounts of myoglobin, many mitochondria, and many blood capillaries. Consequently, they are red and have a high capacity for generating ATP by oxidative processes. Such fibers also split ATP at a very rapid rate, and as a result, contraction velocity is fast. Fast oxidative fibers are resistant to fatigue but not quite as much as

EXHIBIT **10.2** **CHARACTERISTICS OF THE THREE TYPES OF SKELETAL MUSCLE FIBERS**

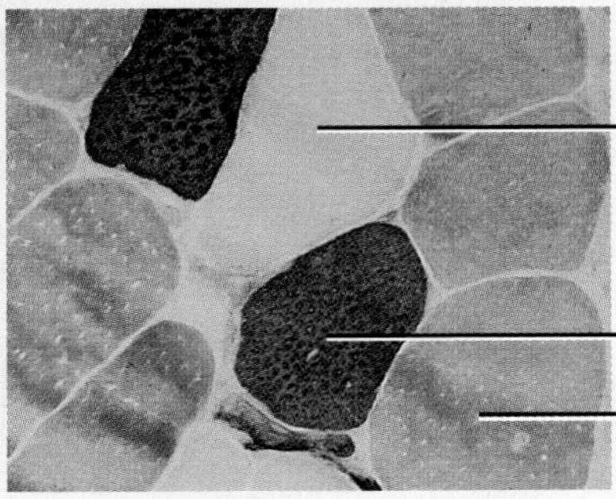

Three types of skeletal muscle fibers in transverse section (440x)

Structural Features	Slow Oxidative (Type I, Slow-Twitch, Fatigue-Resistant Fibers)	Fast Oxidative (Type IIA, Fast-Twitch A, Fatigue-Resistant Fibers)	Fast Glycolytic (Type IIB, Fast-Twitch B, Fatigable Fibers)
Diameter of fiber	Smallest.	Intermediate.	Largest.
Myoglobin content	Large amount.	Large amount.	Small amount.
Mitochondria	Many.	Many.	Few.
Capillaries	Many.	Many.	Few.
Color	Red.	Red to pink.	White (pale).
Functional Features			
Main method of ATP production	Aerobic (oxygen-requiring) processes.	Aerobic (oxygen-requiring) processes.	Glycolysis (anaerobic process).
Rate of ATP hydrolysis	Slow.	Fast.	Fast.
Velocity of contraction	Slow.	Fast.	Fast.
Resistance to fatigue	High (very fatigue-resistant).	Intermediate (moderately fatigue-resistant).	Low (not fatigue-resistant).
Glycogen stores	Low.	Intermediate.	High.
Order of recruitment	First.	Third.	Second.
Activities	Maintaining posture, endurance activities, such as running a marathon.	Walking, running, sprinting.	Rapid, intense movements of short duration, such as throwing a ball or weight lifting.

slow oxidative fibers. Sprinters tend to have a large proportion of fast oxidative fibers in their leg muscles.

3. **Fast glycolytic (type IIB) fibers.** These fibers, which are the largest in diameter, are also called **fast-twitch B** or **fatigable fibers.** They have a low myoglobin content, relatively few mitochondria, and relatively few blood capillaries. They do, however, contain large amounts of glycogen. Fast glycolytic fibers are white. They generate ATP by anaerobic processes (glycolysis) that are not able to supply skeletal muscle fibers continuously with sufficient ATP. Accordingly, these fibers fatigue easily. They are the largest diameter fibers, and they split ATP at a fast rate so that their contraction is strong and rapid. Muscles of the arms contain many of these fibers.

Most skeletal muscles of the body are a mixture of all three types of skeletal muscle fibers, but their proportion varies depending on the usual action of the muscle. For ex-ample, postural muscles of the neck, back, and legs have a high proportion of slow oxidative fibers. Muscles of the shoulders and arms are not constantly active but are used intermittently, usually for short periods of time, to produce large amounts of tension such as in lifting and throwing. These muscles have a high proportion of fast glycolytic fibers. Leg muscles not only support the body but are also used for walking and running. Such muscles have large numbers of both slow and fast oxidative fibers.

Even though most skeletal muscles are a mixture of all three types of skeletal muscle fibers, the skeletal muscle fibers of any one motor unit are all the same type. But, the different motor units in a muscle may be recruited at different times, depending on need. For example, if only a weak contraction is needed to perform a task, only slow oxidative motor units are activated. If a stronger contraction is needed, the motor units of fast glycolytic fibers are recruited. And if a maximal contraction is required, motor units of fast oxidative fibers are also called into action. Activation of various motor units is determined in the brain and spinal cord.

Various types of exercises can induce changes in the fibers in a skeletal muscle. Endurance-type exercises, such as running or swimming, cause a gradual transformation of some fast glycolytic (type IIB) fibers into fast oxidative (type IIA) fibers. The transformed muscle fibers show slight increases in the number of mitochondria and strength. Endurance exercises result in cardiovascular and respiratory changes that cause skeletal muscles to receive better supplies of oxygen and nutrients but do not increase muscle mass. On the other hand, exercises that require great strength for short periods of time, such as weight lifting, produce an increase in the size and strength of fast glycolytic fibers. The increase in size is due to increased synthesis of thin and thick filaments. The overall result is muscle enlargement.

The exact way that muscle grows in size with training is not clear. Some studies show increases in size of individual muscle fibers while others claim increases in overall number of muscle fibers. However, the number of muscle fibers is not thought to increase significantly after birth. During childhood, the increase in the size of muscle fibers appears to be at least partially under the control of human growth hormone, which is produced by the anterior pituitary gland. In males, a further increase in the size of muscle fibers is due to the hormone testosterone, produced by the testes.

ANABOLIC STEROIDS

Steroids are lipids derived from cholesterol that serve many useful functions in the body (discussed in later chapters). In recent years, the illegal use of **anabolic steroids** by athletes has received widespread attention. These drugs, which are similar to the male sex hormone testosterone, are taken to increase muscle size and therefore to increase strength and endurance during athletic events.

With an optimal training program, however, there is little evidence that anabolic steroids confer significant additional strength or endurance when taken in moderate doses. The large doses needed to see an effect produce damaging and devastating side effects. These include liver cancer, kidney damage, increased risk of heart disease, stunted growth in young people, increased irritability and aggressive behavior, and wide mood swings. Females may become sterile, develop facial hair, experience deepening of the voice and atrophy of the breasts and uterus, and suffer menstrual irregularities. Males may suffer atrophy of the testes, diminished hormone secretion and sperm production by the testes, and baldness. ■

CARDIAC MUSCLE TISSUE

The principal tissue in the heart wall is **cardiac muscle tissue**. Although it is striated like skeletal muscle, it is involuntary like smooth muscle. Also, certain cardiac muscle fibers (cells) display autorhythmicity, which sets an inherent rhythm for alternating contraction and relaxation (see page 590).

Microscopic Anatomy of Cardiac Muscle

In contrast to skeletal muscle fibers, cardiac muscle fibers are shorter in length, larger in diameter, and squarish rather than circular in transverse section. They also exhibit branching, which gives an individual fiber a Y-shaped appearance. The endomysium (areolar connective tissue) and numerous capillaries occupy the spaces between caridac muscle fibers. A typical cardiac muscle fiber is 50–100 μm long and has a diameter of about 14 μm. Usually there is only one centrally located nucleus (see Exhibit 4.3 on page 117), although an occasional cell may have two nuclei. The sarcolemma of cardiac muscle fibers is similar to that of skeletal muscle, but the sarcoplasm is more abundant and the mitochondria are larger and more numerous. Cardiac muscle fibers have the same arrangement of actin and myosin and the same bands, zones, and Z discs as skeletal muscle fibers (Fig. 10.17). The transverse tubules of mammalian cardiac muscle are wider but less abundant than those of skeletal muscle: there is only one transverse tubule per sarcomere, located at the Z disc. Also, the sarcoplasmic reticulum of cardiac muscle fibers is scanty in comparison with the SR of skeletal muscle fibers. As a result, cardiac muscle has a limited intracellular reserve of Ca^{2+}. During contraction a substantial amount of Ca^{2+} enters cardiac muscle fibers from extracellular fluid.

Although cardiac muscle fibers branch and interconnect with each other, they form two separate networks. The muscular walls and partition of the superior chambers of the heart (atria) compose one network. The muscular walls and partition of the inferior chambers of the heart (ventricles) compose the other network. The ends of each fiber in a network connect to its neighbors by irregular transverse thickenings of the sarcolemma called **intercalated** (in-TER-ka-lāt-ed; *intercalare* = to insert between) **discs**. The discs contain **desmosomes**, which hold the fibers together, and **gap junctions**, which allow muscle action potentials to spread from one muscle fiber to another (see Fig. 4.1). As a consequence, when a single fiber of either network is stimulated, all the other fibers in the network become stimulated as well. Thus each network contracts as a functional unit. When the fibers of the atria contract as a unit, blood moves into the ventricles. Then, when the ventricular fibers contract as a unit, they pump blood out of the heart into arteries.

Physiology of Cardiac Muscle

Under normal resting conditions, cardiac muscle tissue contracts and relaxes about 75 times a minute. This continuous, rhythmic activity is a major physiological difference between cardiac and skeletal muscle tissue. Accordingly, cardiac muscle tissue requires a constant supply of oxygen.

In cardiac muscle fibers, the mitochondria are larger and more numerous than in skeletal muscle fibers. This structural feature correctly suggests that cardiac muscle depends greatly on the aerobic system to generate ATP. Cardiac muscle generates little ATP anaerobically by the

Figure 10.17   Histology of cardiac muscle tissue.

In cardiac muscle, there is only one transverse tubule per sarcomere.

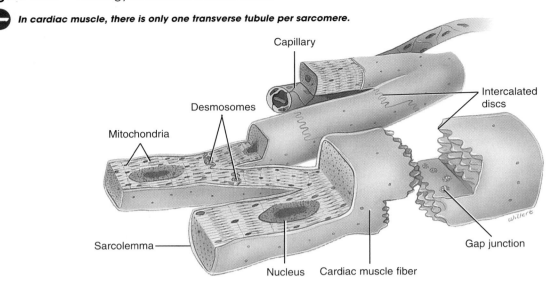

(a) Cardiac muscle fibers

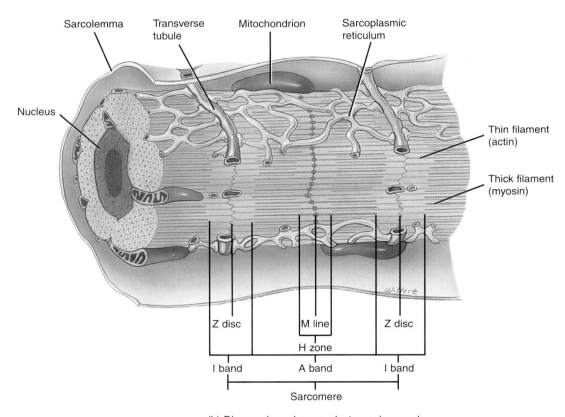

(b) Diagram based on an electron micrograph

Q What are the functions of desmosomes and gap junctions in cardiac muscle cells?

glycogen–lactic acid system. Moreover, cardiac muscle fibers can use lactic acid produced by skeletal muscle fibers to make ATP, a benefit during exercise.

Another difference is the source of stimulation. Skeletal muscle tissue contracts only when stimulated by acetyl-

choline released by an action potential in a motor neuron. In contrast, cardiac muscle tissue can contract without *extrinsic* (outside) nerve or hormonal stimulation. Its source of stimulation is a network of specialized cardiac muscle fibers, called **autorhythmic cells** because they are self-excitable, that is *intrinsic* (within the heart). Neural or hormonal stimulation of the heart causes the autorhythmic fibers to increase or decrease their rate of discharge.

Cardiac muscle tissue remains contracted 10–15 times longer than skeletal muscle tissue. This is due to prolonged delivery of Ca^{2+} into the sarcoplasm. In cardiac muscle fibers, Ca^{2+} enters the sarcoplasm both from the sarcoplasmic reticulum (as in skeletal muscle fibers) and from extracellular fluid (ECF). Since the channels that allow inflow of Ca^{2+} from ECF stay open for a relatively long period of time, cardiac muscle exhibits prolonged depolarization and contraction (see Fig. 20.9).

Cardiac muscle tissue also has a very long refractory period, lasting several tenths of a second, that allows time for the heart chambers to relax and fill with blood between beats. The long refractory period permits heart rate to increase significantly but prevents the heart from undergoing tetanus. If heart muscle could undergo tetanus, blood flow would cease.

SMOOTH MUSCLE TISSUE

There are two types of smooth muscle tissue, visceral and multiunit (Fig. 10.18a,b). The more common type is **visceral (single-unit) smooth muscle tissue**. It is found in wraparound sheets that form part of the walls of small arteries and veins and hollow viscera such as the stomach, intestines, uterus, and urinary bladder. The fibers in visceral smooth muscle form large networks that contract together. Because the fibers connect to one another by gap junctions, muscle action potentials spread throughout the network. When a neurotransmitter, hormone, or autorhythmic signal stimulates one fiber, the muscle action potential spreads to neighboring fibers, which then contract in unison, as a single unit.

The second kind of smooth muscle tissue is **multiunit smooth muscle tissue**. It consists of individual fibers, each with its own motor neuron terminals and with few gap junctions between neighboring fibers. Whereas stimulation of one visceral muscle fiber causes contraction of many adjacent fibers, stimulation of one multiunit fiber causes contraction of only that fiber. Multiunit smooth muscle tissue is found in the walls of large arteries, in large airways to the lungs (bronchioles), in the arrector pili muscles that attach to hair follicles, in the radial and circular muscles of the iris that adjust pupil diameter, and in the ciliary body that adjusts focus of the lens in the eye.

Microscopic Anatomy of Smooth Muscle

Like cardiac muscle tissue, **smooth muscle tissue** is usually involuntary. When examined under the light microscope, no myofibrils are apparent. Smooth muscle fibers are considerably smaller than skeletal muscle fibers. A single smooth muscle fiber is 30–200 μm long, thickest in the middle (3–8 μm), and tapered at each end. Surrounding smooth muscle fibers is

endomysium. Within each fiber is a single, oval, centrally located nucleus (Fig. 10.18; also see Exhibit 4.3 on page 116). The sarcoplasm of smooth muscle fibers contains both **thick filaments** and **thin filaments**, but they are not arranged in orderly sarcomeres as in striated muscle. In smooth muscle fibers, there are 10–15 thin filaments for each thick filament; in skeletal muscle fibers, the ratio is 2:1. Smooth muscle fibers also contain **intermediate filaments**. Since the various filaments have no regular pattern of organization and since there are no A or I bands, smooth muscle fibers do not exhibit cross-striations. This is the reason for the name *smooth*. Smooth muscle fibers also lack transverse tubules and have only scanty sarcoplasmic reticulum for storage of Ca^{2+}.

In smooth muscle fibers, intermediate filaments attach to structures called **dense bodies**, which are similar to Z discs in striated muscle fibers. Some dense bodies are dispersed throughout the sarcoplasm; others are attached to the sarcolemma. Bundles of intermediate filaments stretch from one dense body to another (Fig. 10.18c). During contraction, the sliding filament mechanism involving thick and thin filaments generates tension that is transmitted to intermediate filaments. These, in turn, pull on the dense bodies attached to the sarcolemma, causing a lengthwise shortening of the muscle fiber. Note that shortening of the muscle fiber produces a bubblelike expansion of the sarcolemma. Evidence suggests that a smooth muscle fiber contracts like a corkscrew turns; the fiber twists in a helix as it shortens and rotates in the opposite direction as it lengthens.

Physiology of Smooth Muscle

Although the principles of contraction are similar in skeletal, cardiac, and smooth muscle tissues, smooth muscle tissue exhibits some important physiological differences. In comparison with contraction in a skeletal muscle fiber, contraction in a smooth muscle fiber starts more slowly and lasts much longer. Moreover, smooth muscle can both shorten and stretch to a greater extent than can skeletal or cardiac muscle.

Role of Calmodulin and Myosin Light Chain Kinase

An increase in the concentration of Ca^{2+} in smooth muscle sarcoplasm initiates contraction, just as in striated muscle. Sarcoplasmic reticulum (the reservoir for Ca^{2+} in striated muscle) is scanty in smooth muscle. Calcium ions flow into smooth muscle sarcoplasm from both the extracellular fluid and sarcoplasmic reticulum as also occurs in cardiac, but not in skeletal, muscle fibers. Because there are no transverse tubules in smooth muscle fibers, it takes longer for Ca^{2+} to reach the filaments in the center of the fiber and trigger the contractile process. This accounts, in part, for the slow onset and prolonged contraction of smooth muscle.

Several mechanisms exist to regulate contraction and relaxation of smooth muscle cells. In one, a regulator protein called **calmodulin** binds to Ca^{2+} in the cytosol. (Recall that troponin has this role in striated muscle fibers.) After binding to Ca^{2+}, calmodulin activates an enzyme called **myosin light chain kinase**. This enzyme uses ATP to phosphorylate

Figure 10.18   Histology of smooth muscle tissue.

 Smooth muscle fibers have no transverse tubules and scanty sarcoplasmic reticulum.

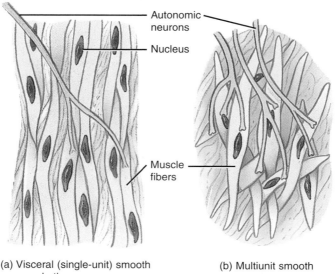

(a) Visceral (single-unit) smooth
muscle tissue

(b) Multiunit smooth
muscle tissue

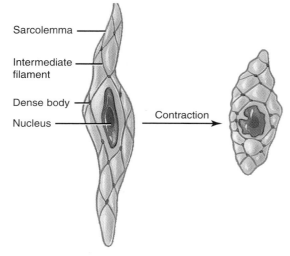

(c) Details of a smooth muscle fiber

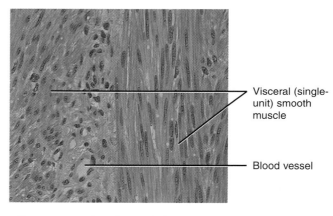

Transverse section   Longitudinal section

(d) Photomicrograph of visceral (single-unit)
smooth muscle in wall of uterus (280x)

Q Which type of smooth muscle is more like cardiac muscle than skeletal muscle, with respect to both its structure and function?

(add a phosphate group to) a portion of the myosin head. Once the phosphate group is attached, the myosin head can bind to actin, and contraction can occur. Myosin light chain kinase works rather slowly, thus contributing to the slowness of smooth muscle contraction.

Smooth Muscle Tone

Not only does Ca^{2+} enter smooth muscle fibers slowly, it also moves slowly out of the muscle fiber when excitation declines. This delays relaxation, and the prolonged presence of Ca^{2+} in the sarcoplasm provides for **smooth muscle tone**, a state of continued partial contraction. Smooth muscle tissue can sustain long-term tone, which is important in the gas-

trointestinal tract where the walls maintain a steady pressure on the contents of the tract and in the walls of blood vessels called arterioles that maintain a steady pressure on blood.

Regulation of Smooth Muscle Contraction

Most smooth muscle fibers contract or relax in response to action potentials from the autonomic (involuntary) nervous system. Thus smooth muscle normally is not under voluntary control. In addition, many smooth muscle fibers contract or relax in response to stretching, hormones, or local factors such as changes in pH, oxygen and carbon dioxide levels, temperature, and ion concentrations. For example, the hormone epinephrine, released by the adrenal medulla, causes relaxation of smooth muscle in the airways and some blood vessel walls (those that have so-called β_2 receptors; see Exhibit 17.3 on page 497).

Unlike striated muscle fibers, smooth muscle fibers can stretch considerably and still maintain their contractile function. When smooth muscle fibers are stretched, they initially contract, developing increased tension. Within a minute or so, the tension decreases. This phenomenon is termed the **stress–relaxation response**. It allows smooth muscle to un-

dergo great changes in length while still retaining the ability to contract effectively. Thus the smooth muscle in the walls of blood vessels and hollow organs such as the stomach, intestines, and urinary bladder can stretch, but the pressure on the contents within increases very little. After the organ empties, on the other hand, the smooth muscle in the wall rebounds, and the wall is firm, not flabby.

Exhibit 10.3 summarizes the principal characteristics of the three types of muscle tissue.

EXHIBIT **10.3** SUMMARY OF THE PRINCIPAL FEATURES OF MUSCLE TISSUE

Characteristic	Skeletal Muscle	Cardiac Muscle	Smooth Muscle
Cell appearance and features	Long cylindrical fiber with many peripherally located nuclei; striated; unbranched.	Branched cylinder usually with one centrally located nucleus; striated; intercalated discs join neighboring fibers.	Spindle-shaped fiber with one, centrally positioned nucleus; no striations.
Location	Attached primarily to bones.	Heart.	Walls of hollow viscera, airways, blood vessels, iris and ciliary body of eye, arrector pili of hair follicles.
Fiber diameter	Very large (10–100 μm).	Large (14 μm).	Small (3–8 μm).
Connective tissue components	Epimysium, perimysium, and endomysium.	Endomysium.	Endomysium.
Fiber length	100 μm to 30 cm.	50–100 μm.	30–200 μm.
Organization of contractile proteins into sarcomeres	Yes.	Yes.	No.
Sarcoplasmic reticulum	Yes.	Yes.	Scanty.
Transverse tubules	Yes, aligned with each A–I band junction.	Yes, aligned with each Z disc.	No.
Gap junctions between fibers	No.	Yes.	Yes in visceral (single-unit) smooth muscle; no in multiunit smooth muscle.
Autorhythmicity	No.	Yes.	Yes (in visceral smooth muscle only).
Source of Ca^{2+} for contraction	Sarcoplasmic reticulum.	Sarcoplasmic reticulum and extracellular fluid.	Sarcoplasmic reticulum and extracellular fluid.
Regulator proteins for contraction	Troponin and tropomyosin.	Troponin and tropomyosin.	Calmodulin and myosin light chain kinase.
Speed of contraction	Fast.	Moderate.	Slow.
Nervous control	Voluntary (somatic nervous system).	Involuntary (autonomic nervous system).	Involuntary (autonomic nervous system).
Contraction regulated by	Acetylcholine released by somatic motor neurons.	Acetylcholine, norepinephrine released by autonomic motor neurons; several hormones.	Acetylcholine, norepinephrine released by autonomic motor neurons; several hormones, local chemical changes (pH, O_2 level, CO_2 level); stretching.
Capacity for regeneration	Limited.	None.	Considerable compared with other muscle tissues but limited compared with tissues such as epithelium.

REGENERATION OF MUSCLE TISSUE

Skeletal muscle fibers have little potential to divide. After the first year of life, growth of skeletal muscle is due to enlargement of existing cells (hypertrophy), rather than an increase in the number of fibers (hyperplasia). Skeletal muscle fibers, however, can be replaced on an individual basis by new cells derived from **satellite cells**, dormant stem cells found in association with skeletal muscle fibers (see Fig. 10.4c). During rapid postnatal growth, satellite cells lengthen existing skeletal muscle fibers by fusing with them. They also persist as a lifelong source of cells that can fuse with each other to form new skeletal muscle fibers. However, the number of new skeletal muscle fibers formed by this mechanism is not sufficient to compensate for any significant skeletal muscle damage. In cases of such damage, skeletal muscle tissue undergoes **fibrosis**, replacement of muscle fibers by fibrous connective (scar) tissue. For this reason, skeletal muscle tissue has only limited powers of regeneration.

Cardiac muscle fibers, like those of skeletal muscle tissue, do not appear to divide in the body. (They have been induced experimentally to undergo division under certain laboratory conditions.) The fibers can undergo hypertrophy, however. Healing of damaged cardiac muscle tissue is by scar formation (fibrosis) because cardiac muscle tissue does not regenerate.

Smooth muscle tissue, like skeletal and cardiac muscle tissues, can undergo hypertrophy. In addition, certain smooth muscle fibers, such as those in the uterus, retain their capacity for division and thus can grow by hyperplasia. Also, new smooth muscle fibers can arise from cells called **pericytes**, stem cells found in association with the endothelium of blood capillaries and small veins. It is also known that smooth muscle fibers can proliferate in certain pathological conditions such as occur in the development of atherosclerosis (described on page 605). Compared with the other two types of muscle tissue, smooth muscle tissue has a considerably higher power of regeneration. It is still limited when compared with other tissues, such as epithelium.

AGING AND MUSCLE TISSUE

Beginning at about 30 years of age, there is a progressive loss of skeletal muscle mass that is replaced largely by fat. In part, this decline is due to increasing inactivity. Accompanying the loss of muscle mass is a decrease in maximal strength and a slowing of muscle reflexes. In some muscles, there may be a selective loss of muscle fibers of a given type. With aging, the relative number of slow oxidative fibers (type I) appears to increase. This could be due to either atrophy of the other fiber types or their conversion into slow oxidative fibers. Whether this is an effect of aging itself or merely reflects the more limited physical activity of older people is still an unresolved question.

DEVELOPMENTAL ANATOMY OF THE MUSCULAR SYSTEM

Smooth muscle fibers of the iris of the eyes and the arrector pili muscles attached to hairs are derived from ectoderm; all other muscles of the body are derived from **mesoderm**. As the mesoderm develops, a portion of it becomes arranged in dense columns on either side of the developing nervous system. These columns of mesoderm undergo segmentation into a series of blocks of cells called **somites** (Fig. 10.19a). The first pair of somites appears on the 20th day of embryologic development. Eventually, 44 pairs of somites are formed by the 30th day.

With the exception of the skeletal muscles of the head and limbs, skeletal muscles develop from the **mesoderm of somites**. Since there are very few somites in the head region

Figure 10.19   Development of the muscular system.

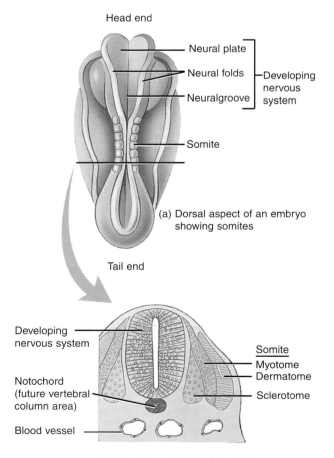

(a) Dorsal aspect of an embryo showing somites

(b) Transverse section of a somite

of the embryo, most of the skeletal muscles there develop from the **general mesoderm** in the head region. The skeletal muscles of the limbs develop from masses of general mesoderm around developing bones in embryonic limb buds (origins of future limbs: see Fig. 6.11).

The cells of a somite have differentiated into three regions: (1) **myotome**, which forms some of the skeletal muscles; (2) **dermatome**, which forms the connective tissues, including the dermis; and (3) **sclerotome**, which gives rise to the vertebrae (Fig. 10.19b).

Cardiac muscle develops from **mesodermal cells** that migrate to and envelop the developing heart while it is still in the form of primitive heart tubes (see Fig. 20.15).

Smooth muscle develops from **mesodermal cells** that migrate to and envelop the developing gastrointestinal tract and viscera.

DISORDERS: HOMEOSTATIC IMBALANCES

FIBROMYALGIA

Fibromyalgia (*algia* = painful condition) refers to a group of common nonarticular rheumatic disorders that are 15 times more common in women than in men and usually appear between the ages of 25 and 50. A striking sign is pain that results from gentle pressure at specific "tender spots." Even without pressure, there is pain, tenderness, and stiffness of muscles, tendons, and surrounding soft tissues. The disorder specifically affects the fibrous connective tissue components of muscles, tendons, and ligaments. Fibromyalgia may be caused or aggravated by physical or mental stress, trauma, exposure to dampness or cold, poor sleep, or a rheumatic condition. Often, a gentle aerobic fitness program is beneficial.

Frequent sites at which fibromyalgia occurs include the lumbar region (**lumbago**), neck, chest, and thighs (**charleyhorse**). Charleyhorse is a slang term that refers to a painful, localized cramp that sometimes is associated with pregnancy or may occur in otherwise healthy persons.

MUSCULAR DYSTROPHIES

Muscular dystrophies (*dystrophy* = degeneration) are inherited muscle-destroying diseases. They are characterized by degeneration of individual muscle fibers, which leads to a progressive atrophy of the skeletal muscle. Usually, the voluntary skeletal muscles are weakened equally on both sides of the body, whereas the internal muscles, such as the diaphragm, are not affected.

The most common form of muscular dystrophy is sex-linked and is called **Duchenne** (doo-SHĀN) **muscular dystrophy**. It strikes boys almost exclusively (1 in 3500), usually appearing between the ages of 3 and 5. Using the techniques of genetic engineering, scientists have discovered that a protein named **dystrophin** is present in the sarcolemma of normal muscle fibers but absent in persons with Duchenne muscular dystrophy. According to one hypothesis, lack of dystrophin may result in leakage of calcium ions into the sarcoplasm. This, in turn, may activate an enzyme (phospholipase A) that causes muscle fibers to degenerate.

MYASTHENIA GRAVIS

Myasthenia gravis ((mī-as-THĒ-nē-a GRAV-is; *myo* = muscle; *asthenia* = weakness; *gravis* = grave or serious) is a weakness of skeletal muscles. It is caused by an abnormality at the neuromuscular junction that partially blocks contraction. Recall that motor neurons stimulate skeletal muscle fibers to contract by releasing acetylcholine (ACh). Myasthenia gravis is an autoimmune disorder caused by antibodies directed against ACh receptors in the motor end plate. The antibodies bind to the receptors and hinder the attachment of ACh (see Fig. 10.3). As the disease progresses, more and more ACh receptors are affected. The muscle becomes increasingly weaker and may eventually cease to function.

Myasthenia gravis, like most autoimmune diseases, is more common in females, occurring most often between the ages of 20 and 50. The muscles of the face and neck are most often affected. Initial symptoms include difficulty in swallowing and a weakness of the eye muscles, which may produce double vision. Later, the individual has difficulty chewing and talking. Eventually, the muscles of the limbs may become involved. Death may result from paralysis of the respiratory muscles, but often the disorder does not progress to this stage.

Anticholinesterase drugs such as neostigmine have been the primary treatment. They act as inhibitors of acetylcholinesterase, the enzyme that breaks down ACh. Thus they raise the level of ACh that is available to bind with still functional receptors. More recently, steroid drugs, such as prednisone, have been used with success to reduce antibody levels. Another treatment is plasmapheresis, a procedure that removes the antibodies from the blood. Sometimes, surgical removal of the thymus gland (thymectomy) is helpful.

ABNORMAL CONTRACTIONS OF SKELETAL MUSCLE

One kind of abnormal muscular contraction is a **spasm**, a sudden involuntary contraction of a single muscle in a large group of muscles. (Cerebral palsy is characterized by generalized spasms.) A painful spasmodic contraction is known as a **cramp**. **Tremor** is a rhythmic, involuntary, purposeless contraction of opposing muscle groups. A **fasciculation** is an involuntary, brief twitch of a muscle visible under the skin. It occurs irregularly and is not associated with movement of the affected muscle. Fasciculations may occur in multiple sclerosis (described on p. 336) or amyotrophic lateral sclerosis (ALS), also called Lou Gehrig's disease. A **fibrillation** is similar to a fasciculation except that it is not visible under the skin. It is recorded by electromyography. A **tic** is a spasmodic twitching made involuntarily by muscles that are ordinarily under voluntary control. Twitching of the eyelid and facial muscles are examples. In general, tics are of psychological origin.

MEDICAL TERMINOLOGY

Gangrene (GANG-rēn; *gangraena* = an eating sore) Death of a soft tissue, such as muscle, that results from interruption of its blood supply.

Myoma (mī-Ō-ma; *myo* = muscle; *oma* = tumor) A tumor consisting of muscle tissue.

Myomalacia (mī′-ō-ma-LĀ-shē-a; *malaco* = soft) Softening of a muscle.

Myopathy (mī-OP-a-thē; *pathos* = disease) Any abnormal condition or disease of muscle tissue.

Myositis (mī′-ō-SĪ-tis; *itis* = inflammation of) Inflammation of muscle fibers (cells).

Myotonia (mī-ō-TŌ-nē-a; *tonia* = tension) Increased muscular excitability and contractility with decreased power of relaxation; tonic spasm of the muscle.

Paralysis (pa-RAL-a-sis; *para* = beyond; *lyein* = to loosen) Loss or impairment of motor (muscular) function resulting from a lesion of nervous or muscular origin.

Volkmann's contracture (FOLK-manz kon-TRAK-tur; *contra* = against) Permanent shortening of a muscle due to replacement of destroyed muscle fibers with fibrous connective tissue that lacks the ability to stretch. Destruction of muscle fibers may occur from interference with circulation caused by a tight bandage, a piece of elastic, or a cast.

STUDY OUTLINE

TYPES OF MUSCLE TISSUE (p. 239)

1. Skeletal muscle tissue is primarily attached to bones. It is striated and voluntary.
2. Cardiac muscle tissue forms the wall of the heart. It is striated and involuntary.
3. Smooth muscle tissue is located in viscera. It is nonstriated (smooth) and involuntary.

FUNCTIONS OF MUSCLE TISSUE (p. 239)

1. Through contraction, muscle tissue performs four important functions.
2. These functions are motion, movement of substances within the body, stabilizing body positions and regulating organ volume, and thermogenesis (heat production).

CHARACTERISTICS OF MUSCLE TISSUE (p. 239)

1. Excitability is the property of responding to stimuli by producing action potentials.
2. Conductivity is the ability of a cell to conduct action potentials along its plasma membrane.
3. Contractility is the ability to shorten and thicken (contract).
4. Extensibility is the ability to be extended (stretched).
5. Elasticity is the ability to return to original shape after contraction or extension.

ANATOMY AND INNERVATION OF SKELETAL MUSCLE TISSUE (p. 240)

Connective Tissue Components (p. 240)

1. Fascia is a sheet or broad band of fibrous connective tissue deep to the skin (superficial fascia) or surrounding muscles and organs of the body (deep fascia).
2. Other connective tissue components are epimysium, covering the entire muscle; perimysium, covering fascicles; and endomysium, covering muscle fibers (cells); all are extensions of deep fascia.
3. Tendons and aponeuroses are extensions of connective tissue beyond muscle fibers that attach the muscle to bone or other muscle.

Nerve and Blood Supply (p. 241)

1. Nerves convey impulses for muscular contraction.
2. Blood provides nutrients and oxygen for muscular contraction.

The Motor Unit (p. 241)

1. A motor neuron and the muscle fibers it stimulates form a motor unit.
2. A single motor unit may contain as few as 2 or as many as 2000 muscle fibers.

The Neuromuscular Junction (p. 241)

1. A motor neuron transmits a nerve impulse (nerve action potential) to a neuromuscular junction.
2. A neuromuscular junction refers to an axon terminal of a motor neuron and the portion of the muscle fiber sarcolemma in close approximation with it (motor end plate).
3. ACh released by a motor neuron diffuses across the synaptic cleft and triggers a muscle action potential.

Microscopic Anatomy of Skeletal Muscle (p. 242)

1. Skeletal muscle consists of fibers (cells) covered by a sarcolemma (plasma membrane). The fibers contain sarcoplasm, nuclei, sarcoplasmic reticulum, and transverse tubules.
2. Each fiber contains myofibrils that consist of thin and thick filaments. The filaments are arranged into sarcomeres.
3. Thin filaments are composed of actin, tropomyosin, and troponin; thick filaments consist mostly of myosin.
4. Projecting myosin heads are called cross bridges; they contain actin-binding and ATP-binding sites.

CONTRACTION OF MUSCLE (p. 247)

1. When a nerve impulse reaches an axon terminal, the synaptic vesicles of the terminal release acetylcholine (ACh), which ultimately initiates a muscle action potential (AP) in the muscle fiber sarcolemma. The muscle AP then travels into the transverse tubules and causes the sarcoplasmic reticulum to release some of its stored Ca^{2+} into the sarcoplasm.
2. The released Ca^{2+} combine with troponin, causing it to pull on tropomyosin to change its orientation, thus exposing the myosin-binding sites on actin.
3. The immediate, direct source of energy for muscle contraction is ATP. ATPase splits ATP into ADP + Ⓟ and the released energy activates (energizes) myosin cross bridges.
4. Activated cross bridges attach to actin and a change in the orientation of the cross bridge occurs (power stroke); this movement results in the sliding of thin filaments.

Figure 11.1 Relationship of skeletal muscles to bones. (a) Skeletal muscles produce movements by pulling on bones. (b) Bones serve as levers, and joints act as fulcrums for the levers. Here the lever–fulcrum principle is illustrated by the movement of the forearm. Note where the resistance and effort are applied in this example.

In the limbs, the origin of a muscle is proximal and the insertion distal.

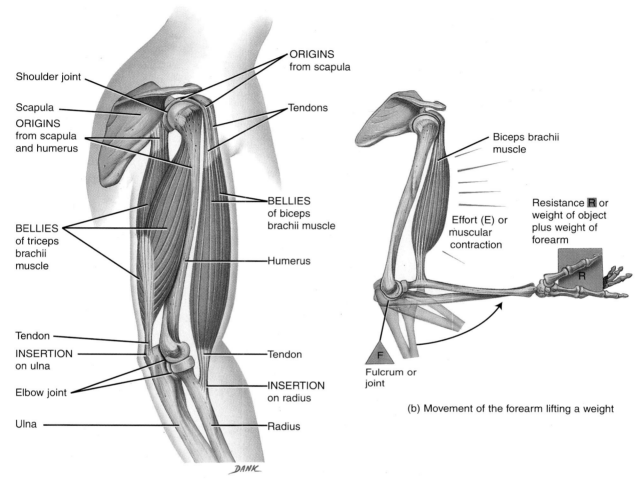

(a) Origin and insertion of a skeletal muscle

(b) Movement of the forearm lifting a weight

Where is the belly of the muscle that extends the forearm located?

ticulating bone toward the other. The two articulating bones usually do not move equally in response to the contraction. One is held nearly in its original position because other muscles contract to pull it in the opposite direction or because its structure makes it less movable. Ordinarily, the attachment of a muscle tendon to the stationary bone is called the **origin** of the muscle. The attachment of the other tendon from the muscle to the movable bone is the **insertion** of the muscle. A good analogy is a spring on a door. In this example, the part of the spring attached to the door represents the insertion; the part attached to the frame is the origin. The fleshy portion of the muscle between the tendons of the origin and insertion is called the **belly**. In the limbs, the origin is proximal and the insertion distal. In addition, muscles that move a body part often do not cover the moving part. For example, Fig. 11.1a

shows that although one of the functions of the biceps brachii muscle is to flex the forearm, the belly of the muscle lies over the humerus, not the forearm. Muscles that cross two joints, such as the rectus femoris and sartorius in the thigh, have more complex actions than muscles that cross only one joint.

Lever Systems and Leverage

In producing a body movement, bones act as levers and joints function as fulcrums of these levers. A **lever** may be defined as a rigid rod that moves about on some fixed point called a **fulcrum**. A fulcrum may be symbolized as △F. A lever is acted on at two different points by two different forces: the **resistance** R and the **effort** E. Whereas resistance is the force that opposes movement, effort is the force exerted to achieve an action. The resistance may be the weight of a part of the body that is to be moved. The effort is the muscular contraction. Motion is produced when the effort is applied to the bone at the insertion and the effort exceeds the resistance (load). Consider the biceps brachii flexing (bending) the forearm at the elbow as a weight is lifted (Fig. 11.1b). When the forearm is raised, the elbow is the fulcrum. The weight of the forearm plus the weight in the hand is the resistance. The shortening due to the force of contraction of the biceps brachii pulling the forearm up is the effort.

Leverage, the mechanical advantage gained by a lever, is largely responsible for a muscle's strength (force of contraction) and range of motion (ROM), that is, the maximum ability to move the bones of a joint through an arc. Consider strength first. Suppose we have two muscles of the same strength crossing and acting on the same joint. Assume also that one is attached farther from the joint and one is nearer.

The muscle attached farther away will produce the more powerful movement. Thus strength of movement depends on the placement of muscle attachments.

In considering ROM, again assume that we have two muscles of the same strength crossing and acting on the same joint and that one is attached farther from the joint than the other. The muscle inserting closer to the joint will produce the greater ROM and speed of movement. Thus ROM also depends on the placement of muscle attachments. Since strength increases with distance from the joint and ROM decreases, maximal strength and maximal range are incompatible; strength and range vary inversely. This means that as strength increases, ROM decreases and as ROM increases, strength decreases.

Levers are categorized into three types according to the positions of the fulcrum, the effort, and the resistance.

1. **First-class levers** have the fulcrum between the effort and resistance (Fig. 11.2a). This is symbolized **EFR**. An

Figure 11.2   Classes of levers.

🔑 *Levers are classified into three classes based on the placement of the fulcrum* ▲F , *effort* E, *and resistance* R .

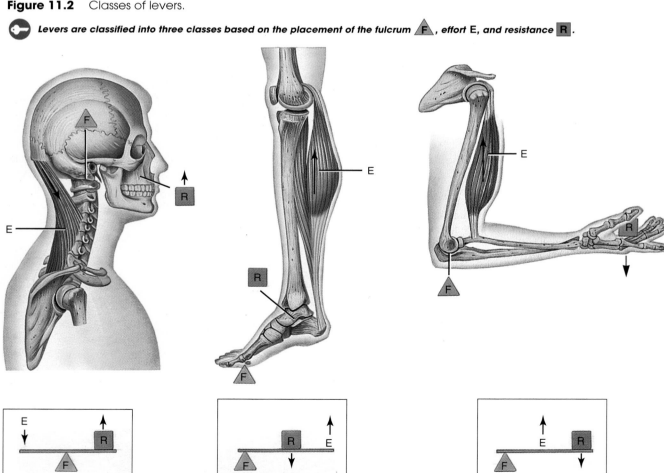

(a) First-class lever   (b) Second-class lever   (c) Third-class lever

ⓠ In the examples shown, bones function as _____, and joints function as _____.

example of a first-class lever is a seesaw. Although some first-class levers provide greater speed and ROM but less strength, others provide less speed and ROM but greater strength. There are not many first-class levers in the body. One example is the head resting on the vertebral column. When the head is raised, the facial portion of the skull is the resistance. The joints between the atlas and occipital bone (atlanto-occipital joints) form the fulcrum. The contraction of the muscles of the upper back is the effort.

2. **Second-class levers** have the fulcrum at one end, the effort at the opposite end, and the resistance between them (Fig. 11.2b). This is symbolized **FRE**. They operate like a wheelbarrow. Second-class levers are levers of strength; however, speed and ROM are sacrificed. Most authorities agree that there are very few examples of second-class levers in the body. One example is raising the body on the toes. The body is the resistance, the ball of the foot is the fulcrum, and the contraction of the calf muscles to pull the heel upward is the effort.

3. **Third-class levers** consist of the fulcrum at one end, the resistance at the opposite end, and the effort between them (Fig. 11.2c). This is symbolized **FER**. They are the most common levers in the body. Third-class levers allow greater speed and ROM that involve relatively little shortening of a muscle. One example is adduction of the thigh, in which the weight of the thigh is the resistance, the hip joint is the fulcrum, and contraction of the adductor muscles is the effort. Another example is flexing the forearm at the elbow. As we have seen, the weight of the forearm is the resistance, the contraction of the biceps brachii is the effort, and the elbow joint is the fulcrum (see Fig. 11.1b).

Group Actions

Most movements require several skeletal muscles acting in groups rather than individually. Also, most skeletal muscles are arranged in opposing (antagonistic) pairs at joints, that is, flexors–extensors, abductors–adductors, and so on. Consider flexing the forearm at the elbow, for example. A muscle that causes a desired action is called the **prime mover** or **agonist** (*agogos* = leader). In this instance, the biceps brachii is the prime mover (see Fig. 11.1a). Simultaneously with the contraction of the biceps brachii, another muscle, called the **antagonist** (*antiagonistes* = opponent), is relaxing. In this movement, the triceps brachii serves as the antagonist (see Fig. 11.1a). The antagonist is usually located on the opposite side of the bone or joint in relation to the prime mover, as is the case for the biceps brachii and triceps brachii. The action of the antagonist is opposite to that of the prime mover; that is, the antagonist stretches and yields to the movement of the prime mover. In moving the arm, the biceps brachii is not always the prime mover and the triceps brachii is not always the an-

tagonist. For example, when extending the forearm at the elbow, the triceps brachii serves as the prime mover and the biceps brachii functions as the antagonist; their roles are reversed. Note that if the prime mover and antagonist contracted simultaneously with equal force, there would be no movement.

In addition to prime movers and antagonists, most movements also involve muscles called **synergists** (SIN-er-jists; *syn* = together; *ergon* = work), which serve to steady a movement, thus preventing unwanted movements and helping the prime mover function more efficiently. For example, flex your hand at the wrist and then make a fist. Note how difficult this is to do. Now, extend your hand at the wrist and then make a fist. Note how much easier it is to clench your fist. In this case, the extensor muscles of the wrist act as synergists in cooperation with the flexor muscles of the fingers acting as prime movers. The extensor muscles of the fingers serve as antagonists (see Fig. 11.17c). Synergists are usually alongside the prime mover.

Some synergistic muscles in a group may also act as **fixators**, which stabilize the origin of the prime mover so that the prime mover can act more efficiently. For example, the scapula is a freely movable bone in the pectoral (shoulder) girdle that serves as a firm origin for several muscles that move the arm. However, for the scapula to do this, it must be held steady. This is accomplished by fixator muscles that hold the scapula firmly against the back of the chest. In abduction of the arm, the deltoid muscle serves as the prime mover, whereas fixators (pectoralis minor, rhomboideus major, rhomboideus minor, trapezius, subclavius, and serratus anterior muscles) hold the scapula firmly (see Fig. 11.14). These fixators stabilize the scapula that serves as the attachment site for the origin of the deltoid muscle, whereas the insertion of the muscle pulls on the humerus to abduct the arm. Under different conditions and depending on the movement and which point is fixed, many muscles act, at various times, as prime movers, antagonists, synergists, or fixators.

Arrangement of Fascicles

Recall from Chapter 10 that skeletal muscle fibers (cells) are arranged within the muscle in bundles called fascicles (fasciculi). The muscle fibers are arranged in a parallel fashion within each bundle, but the arrangement of the fascicles with respect to the tendons has one of several characteristic patterns, as described in Exhibit 11.2.

Fascicular arrangement is correlated with the power of a muscle and its range of motion. When a muscle fiber contracts, it shortens to a length just slightly greater than half of its resting length. Thus the longer the fibers in a muscle, the greater the range of motion it can produce. By contrast, the strength of a muscle depends on the total number of fibers it contains, since a short fiber can contract as

---

EXHIBIT 11.2 ARRANGEMENT OF FASCICLES

Arrangement and Description	Example	Arrangement and Description	Example
PARALLEL Fascicles are parallel with longitudinal axis of muscle and terminate at either end in flat tendons.	Stylohyoid muscle (see Fig. 11.7). 	**PENNATE** Fascicles are short in relation to muscle length and the tendon extends nearly the entire length of the muscle. **Unipennate** Fascicles are arranged on only one side of tendon.	Extensor digitorum longus muscle (see Fig. 11.21c).
FUSIFORM Fascicles are nearly parallel with longitudinal axis of muscle and terminate at either end in flat tendons, but muscle tapers toward tendons where the diameter is less than that of the belly.	Digastric muscle (see Fig. 11.8). 	**Bipennate** Fascicles are arranged on both sides of a centrally positioned tendon.	Rectus femoris muscle (see Fig. 11.19a).
CIRCULAR Fascicles are arranged in a concentric circular pattern to form sphincter muscles that enclose an orifice (opening).	Orbicularis oculi muscle (see Fig. 11.4).	**Multipennate** Fascicles attach obliquely from many directions to several tendons.	Deltoid muscle (see Fig. 11.10a).

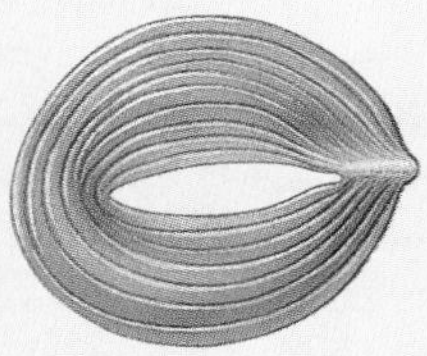

---

forcefully as a long one. Because a given muscle can contain either a small number of long fibers or a large number of short fibers, fascicular arrangement represents a compromise between power and range of motion. Pennate muscles, for example, have a large number of fascicles distributed over their tendons, giving them greater power but a smaller range of motion. Parallel muscles, on the other hand, have comparatively few fascicles that extend the entire length of the muscle. Thus they have a greater range of motion but less power.

PRINCIPAL SKELETAL MUSCLES

Starting on page 277, Exhibits 11.3 through 11.22 describe the principal muscles of the body with their origins, inser-

tions, actions, and innervations. (By no means have all the muscles of the body been included.) An **overview** section in each exhibit provides a general orientation to the muscles under consideration. Refer to Chapters 7 and 8 to review bone markings, since they serve as points of origin and insertion for muscles. Students often have difficulty in pronouncing names of skeletal muscles and understanding how they are named. To make this task easier, we have provided phonetic pronunciations and derivations that indicate how the muscles are named. If you have mastered the naming of the muscles, their actions will have more meaning and be easier to remember.

The muscles are divided into groups according to the part of the body on which they act. Figure 11.3 shows general anterior and posterior views of the muscular system. Do not try to memorize all these muscles yet. As you study groups of muscles in the following exhibits, refer to Fig. 11.3 to see how each group is related to all others.

Figure 11.3 Principal superficial skeletal muscles.

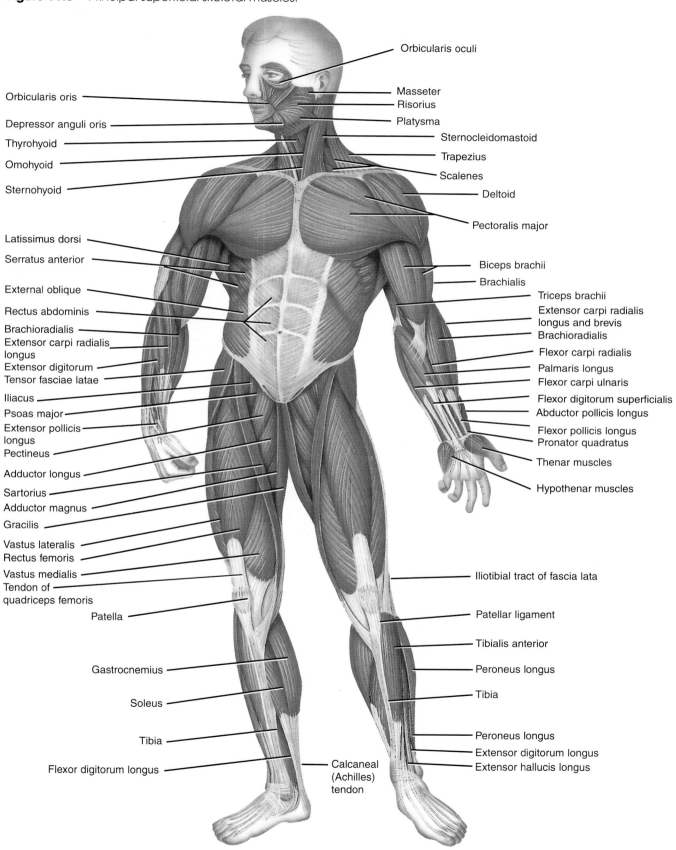

(a) Anterior view

Figure continues

Figure 11.3   (continued)

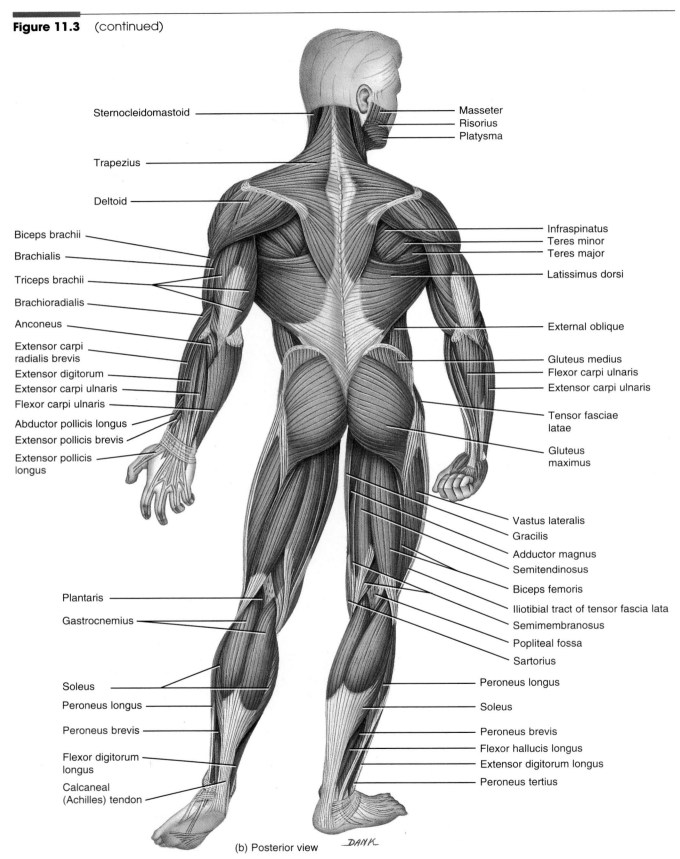

Sternocleidomastoid

Masseter
Risorius
Platysma

Trapezius

Deltoid

Biceps brachii

Brachialis

Triceps brachii

Brachioradialis

Anconeus

Extensor carpi radialis brevis

Extensor digitorum

Extensor carpi ulnaris

Flexor carpi ulnaris

Abductor pollicis longus

Extensor pollicis brevis

Extensor pollicis longus

Infraspinatus
Teres minor
Teres major

Latissimus dorsi

External oblique

Gluteus medius
Flexor carpi ulnaris
Extensor carpi ulnaris

Tensor fasciae latae

Gluteus maximus

Vastus lateralis
Gracilis
Adductor magnus
Semitendinosus
Biceps femoris
Iliotibial tract of tensor fascia lata
Semimembranosus
Popliteal fossa
Sartorius

Plantaris

Gastrocnemius

Soleus

Peroneus longus

Peroneus brevis

Flexor digitorum longus

Calcaneal (Achilles) tendon

Peroneus longus

Soleus

Peroneus brevis

Flexor hallucis longus

Extensor digitorum longus

Peroneus tertius

(b) Posterior view   DANK

Q Select a muscle named for each of the following characteristics: direction of fibers, shape, action, size, origin and insertion, location, and number of origins.

The figures that accompany the exhibits contain superficial and deep, anterior and posterior, or medial and lateral views to show each muscle's position as clearly as possible. The figures also show the relationship of the muscles under consideration to other muscles in the area you are studying.

As a further aid to your learning efforts, the following list of exhibits and figures is provided so you can see the order in which the skeletal muscles will be studied according to region.

EXHIBIT 11.3 MUSCLES OF FACIAL EXPRESSION (Fig. 11.4)

Overview: The muscles in this group provide humans with the ability to express a wide variety of emotions, including surprise, fear, and happiness. The muscles themselves lie within the layers of superficial fascia. They usually originate in the fascia or bones of the skull and insert into the skin. Because of their insertions, the muscles of facial expression move the skin rather than a joint when they contract.

Muscle	Origin	Insertion	Action	Innervation
Epicranius (ep-i-KRĀ-nē-us; *epi* = over; *crani* = skull)	This muscle is divisible into two portions: the frontalis over the frontal bone and the occipitalis over the occipital bone. The two muscles are united by a strong aponeurosis (sheetlike tendon), the **galea aponeurotica** (**epicranial aponeurosis**), which covers the superior and lateral surfaces of the skull.			
Frontalis (fron-TA-lis; *front* = forehead)	Galea aponeurotica.	Skin superior to supraorbital margin.	Draws scalp anteriorly, elevates (raises) eyebrows, and wrinkles skin of forehead horizontally.	Facial (VII) nerve.
Occipitalis (ok-si′-pi-TA-lis; *occipito* = base of skull)	Occipital bone and mastoid process of temporal bone.	Galea aponeurotica.	Draws scalp posteriorly.	Facial (VII) nerve.
Orbicularis oris (or-bi′-kyoo-LAR-is OR-is; *orb* = circular; *or* = mouth)	Muscle fibers surrounding opening of mouth.	Skin at corner of mouth.	Closes lips, compresses lips against teeth, protrudes lips, and shapes lips during speech.	Facial (VII) nerve.
Zygomaticus (zī-go-MA-ti-kus) **major** (*zygomatic* = cheek bone; *major* = greater)	Zygomatic bone.	Skin at angle of mouth and orbicularis oris.	Draws angle of mouth upward and outward as in smiling or laughing.	Facial (VII) nerve.
Levator labii superioris (le-VĀ-tor LA-bē-i soo-per′-ē-OR-is; *levator* = raises or elevates; *labii* = lip; *superioris* = upper)	Superior to infraorbital foramen of maxilla.	Skin at angle of mouth and orbicularis oris.	Elevates upper lip.	Facial (VII) nerve.

Exhibit continues

EXHIBIT **11.3** **MUSCLES OF FACIAL EXPRESSION** (CONTINUED)

Muscle	Origin	Insertion	Action	Innervation
Depressor labii inferioris (de-PRE-sor LĀ-bē-ī in-fer′-ē-OR-is; *depressor* = depresses or lowers; *inferioris* = lower)	Mandible.	Skin of lower lip.	Depresses (lowers) lower lip.	Facial (VII) nerve.
Buccinator (BUK-si-nā′-tor; *bucc* = cheek)	Alveolar processes of maxilla and mandible and pterygomandibular raphe (fibrous band extending from the pterygoid process to the mandible).	Orbicularis oris.	Major cheek muscle; compresses cheek as in blowing air out of mouth and causes cheeks to cave in, producing the action of sucking.	Facial (VII) nerve.
Mentalis (men-TA-lis; *mentum* = chin)	Mandible.	Skin of chin.	Elevates and protrudes lower lip and pulls skin of chin up as in pouting.	Facial (VII) nerve.
Platysma (pla-TIZ-ma; *platy* = flat, broad)	Fascia over deltoid and pectoralis major muscles.	Mandible, muscles around angle of mouth, and skin of lower face.	Draws outer part of lower lip inferiorly and posteriorly as in pouting; depresses mandible.	Facial (VII) nerve.
Risorius (ri-ZOR-ē-us; *risor* = laughter)	Fascia over parotid (salivary) gland.	Skin at angle of mouth.	Draws angle of mouth laterally as in tenseness.	Facial (VII) nerve.
Orbicularis oculi (or-bi′-kyoo-LAR-is OK-yoo-lī; *oculus* = eye)	Medial wall of orbit.	Circular path around orbit.	Closes eye.	Facial (VII) nerve.
Corrugator supercilii (KOR-a-gā′-tor soo-per-SI-lē-ī; *corrugo* = wrinkle; *supercilium* = eyebrow)	Medial end of superciliary arch of frontal bone.	Skin of eyebrow.	Draws eyebrow inferiorly as in frowning.	Facial (VII) nerve.
Levator palpebrae superioris (le-VĀ-tor PAL-pe-brē soo-per′-ē-OR-is; *palpebrae* = eyelids) (see also Fig. 11.6a)	Roof of orbit (lesser wing of sphenoid bone).	Skin of upper eyelid.	Elevates upper eyelid (opens eye).	Oculomotor (III) nerve.

Figure 11.4 Muscles of facial expression.

 When they contract, muscles of facial expression move the skin rather than a joint.

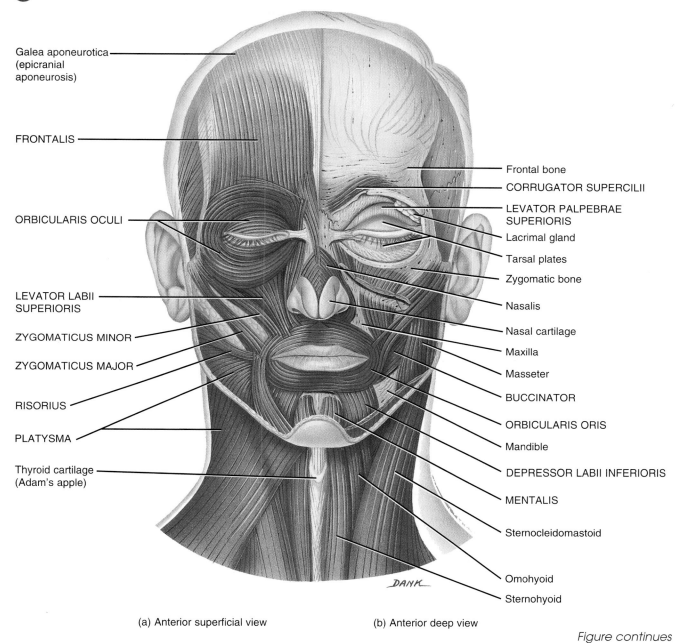

Galea aponeurotica
(epicranial
aponeurosis)

FRONTALIS

ORBICULARIS OCULI

LEVATOR LABII
SUPERIORIS

ZYGOMATICUS MINOR

ZYGOMATICUS MAJOR

RISORIUS

PLATYSMA

Thyroid cartilage
(Adam's apple)

Frontal bone

CORRUGATOR SUPERCILII

LEVATOR PALPEBRAE
SUPERIORIS

Lacrimal gland

Tarsal plates

Zygomatic bone

Nasalis

Nasal cartilage

Maxilla

Masseter

BUCCINATOR

ORBICULARIS ORIS

Mandible

DEPRESSOR LABII INFERIORIS

MENTALIS

Sternocleidomastoid

Omohyoid

Sternohyoid

DANK

(a) Anterior superficial view

(b) Anterior deep view

Figure continues

Figure 11.4 (continued)

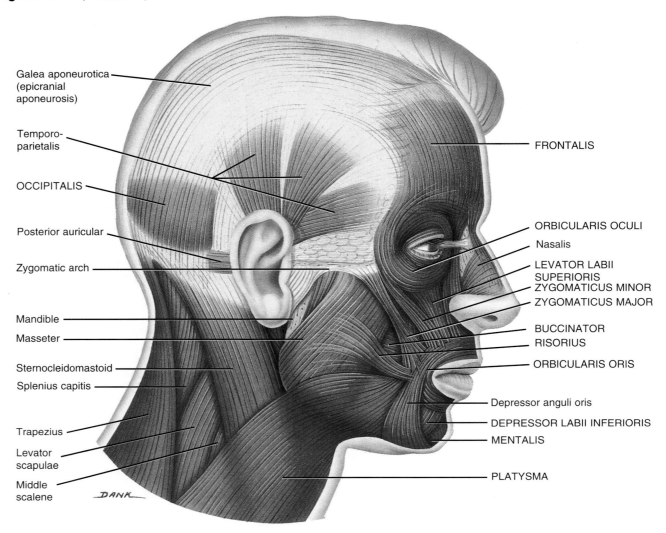

Galea aponeurotica (epicranial aponeurosis)

Temporo-parietalis

OCCIPITALIS

Posterior auricular

Zygomatic arch

Mandible

Masseter

Sternocleidomastoid

Splenius capitis

Trapezius

Levator scapulae

Middle scalene

DANK

FRONTALIS

ORBICULARIS OCULI

Nasalis

LEVATOR LABII SUPERIORIS

ZYGOMATICUS MINOR

ZYGOMATICUS MAJOR

BUCCINATOR

RISORIUS

ORBICULARIS ORIS

Depressor anguli oris

DEPRESSOR LABII INFERIORIS

MENTALIS

PLATYSMA

(c) Right lateral superficial view

Q What major muscle causes frowning? Smiling? Pouting? Squinting?

E X H I B I T **11.4 MUSCLES THAT MOVE THE MANDIBLE (LOWER JAW) (Fig. 11.5)**

Overview: Muscles that move the mandible (lower jaw) are also known as muscles of mastication because they are involved in biting and chewing.

Muscle	Origin	Insertion	Action	Innervation
Masseter (MA-se-ter; *maseter* = chewer)	Maxilla and zygomatic arch.	Angle and ramus of mandible.	Elevates mandible as in closing mouth, assists in side-to-side movement of mandible, and protracts (protrudes) mandible.	Mandibular division of trigeminal (V) nerve.
Temporalis (tem'-por-A-lis; *tempora* = temples)	Temporal and frontal bones.	Coronoid process and ramus of mandible.	Elevates and retracts mandible and assists in side-to-side movement of mandible.	Mandibular division of trigeminal (V) nerve.

Muscle	Origin	Insertion	Action	Innervation
Medial pterygoid (TER-i-goid; *medial* = closer to midline; *pterygoid* = like a wing; pterygoid process of sphenoid bone)	Medial surface of lateral portion of pterygoid process of sphenoid; maxilla.	Angle and ramus of mandible.	Elevates and protracts mandible and moves mandible from side to side.	Mandibular division of trigeminal (V) nerve.
Lateral pterygoid (TER-i-goid; *lateral* = farther from midline)	Greater wing and lateral surface of lateral portion of pterygoid process of sphenoid.	Condyle of mandible; temporomandibular articulation.	Protracts mandible, opens mouth, and moves mandible from side to side.	Mandibular division of trigeminal (V) nerve.

Figure 11.5   Muscles that move the mandible (lower jaw).

🔑 *The muscles that move the mandible are involved in speech and mastication.*

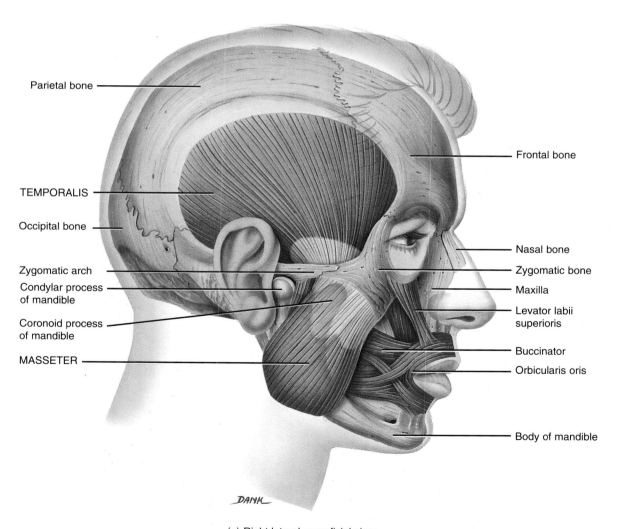

(a) Right lateral superficial view

Figure continues

Figure 11.5   (continued)

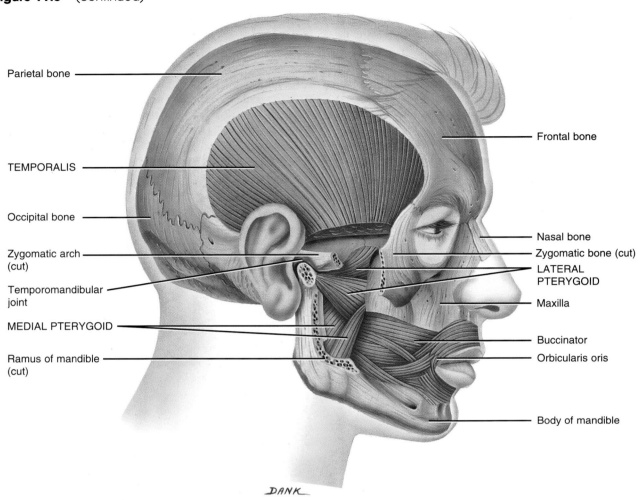

Parietal bone

Frontal bone

TEMPORALIS

Occipital bone

Nasal bone

Zygomatic bone (cut)

Zygomatic arch (cut)

LATERAL PTERYGOID

Temporomandibular joint

Maxilla

MEDIAL PTERYGOID

Buccinator

Ramus of mandible (cut)

Orbicularis oris

Body of mandible

DANK

(b) Right lateral deep view

Q Which muscles protract (protrude) the mandible?

EXHIBIT **11.5** **MUSCLES THAT MOVE THE EYEBALLS—EXTRINSIC MUSCLES (Fig. 11.6)**

Overview: Muscles associated with the eyeball are of two principal types: extrinsic and intrinsic. **Extrinsic muscles** originate outside the eyeballs and insert on their outer surface (sclera). They move the eyeballs in various directions. **Intrinsic muscles** originate and insert entirely within the eyeballs. They move structures within the eyeballs. Movements of the eyeballs are controlled by three pairs of extrinsic muscles. Two pairs of rectus muscles move the eyeballs in the direction indicated by their respective names—superior, inferior, lateral, and medial. One pair of muscles, the oblique muscles—superior and inferior—rotate the eyeballs on their axes. The extrinsic muscles of the eyeballs are among the fastest contracting and most precisely controlled skeletal muscles of the body.

Muscle	Origin	Insertion	Action	Innervation
Superior rectus (REK-tus; *superior* = above; *rectus* = straight, in this case, muscle fibers running parallel to long axis of eyeball)	Tendinous ring attached to bony orbit around optic foramen.	Superior and central part of eyeball.	Rolls eyeball superiorly.	Oculomotor (III) nerve.
Inferior rectus (REK-tus; *inferior* = below)	Same as above.	Inferior and central part of eyeball.	Rolls eyeball inferiorly.	Oculomotor (III) nerve.

Muscle	Origin	Insertion	Action	Innervation
Lateral rectus (REK-tus)	Same as above.	Lateral side of eyeball.	Rolls eyeball laterally.	Abducens (VI) nerve.
Medial rectus (REK-tus)	Same as above.	Medial side of eyeball.	Rolls eyeball medially.	Oculomotor (III) nerve.
Superior oblique (o-BLĒK; *oblique* = slanting, in this case, muscle fibers running diagonally to long axis of eyeball)	Same as above.	Eyeball between superior and lateral recti. The muscle inserts in a round tendon that moves through a ring of fibrocartilaginous tissue called the trochlea (*trochlea* = pulley).	Rotates eyeball on its axis; directs cornea inferiorly and laterally.	Trochlear (IV) nerve.
Inferior oblique (o-BLĒK)	Maxilla (anterior orbital cavity).	Eyeball between inferior and lateral recti.	Rotates eyeball on its axis; directs cornea superiorly and laterally.	Oculomotor (III) nerve.

Figure 11.6   Extrinsic muscles of the eyeballs.

The extrinsic muscles of the eyeballs are among the fastest contracting and most precisely controlled skeletal muscles of the body.

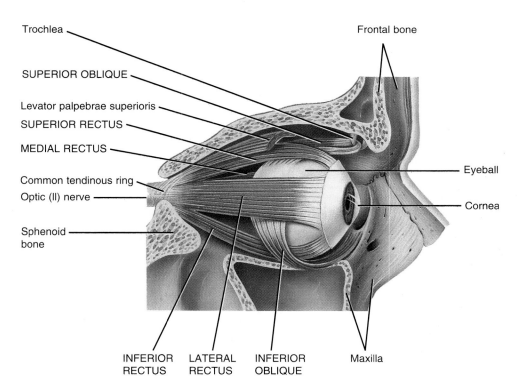

(a) Lateral view of right eyeball

Figure continues

Figure 11.6 (continued)

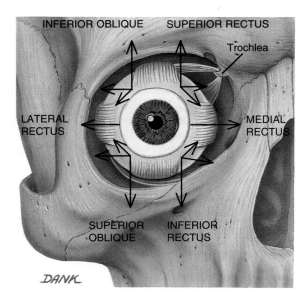

(b) Movements of right eyeball in response to
contraction of its extrinsic muscles

Q Which muscles roll the eyeballs laterally?

E X H I B I T **11.6** **MUSCLES THAT MOVE THE TONGUE—EXTRINSIC MUSCLES (Fig. 11.7)**

Overview: The tongue is divided into lateral halves by a median fibrous septum. The septum extends throughout the length of the tongue and is attached inferiorly to the hyoid bone. Like the muscles of the eyeballs, muscles of the tongue are of two principal types—extrinsic and intrinsic. **Extrinsic muscles** originate outside the tongue and insert into it. They move the entire tongue in various directions, such as anteriorly, posteriorly, and laterally. **Intrinsic muscles** originate and insert within the tongue. These muscles alter the shape of the tongue rather than moving the entire tongue. The extrinsic and intrinsic muscles of the tongue are arranged in both lateral halves of the tongue.

Muscle	Origin	Insertion	Action	Innervation
Genioglossus (jē′-nē-ō-GLOS-us; *geneion* = chin; *glossus* = tongue)	Mandible.	Undersurface of tongue and hyoid bone.	Depresses tongue and thrusts it anteriorly (protraction).	Hypoglossal (XII) nerve.
Styloglossus (stī′-lō-GLOS-us; *stylo* = stake or pole; styloid process of temporal bone)	Styloid process of temporal bone.	Side and undersurface of tongue.	Elevates tongue and draws it posteriorly (retraction).	Hypoglossal (XII) nerve.
Palatoglossus (pal′-a-tō-GLOS-us; *palato* = palate)	Anterior surface of soft palate.	Side of tongue.	Elevates posterior portion of tongue and draws soft palate inferiorly on tongue.	Pharyngeal plexus (pharyngeal branch of vagus (X) nerve.
Hyoglossus (hī′-ō-GLOS-us)	Greater horn and body of hyoid bone.	Side of tongue.	Depresses tongue and draws its sides inferiorly.	Hypoglossal (XII) nerve.

Figure 11.7 Muscles that move the tongue.

 The extrinsic and intrinsic muscles of the tongue are arranged in both lateral halves of the tongue.

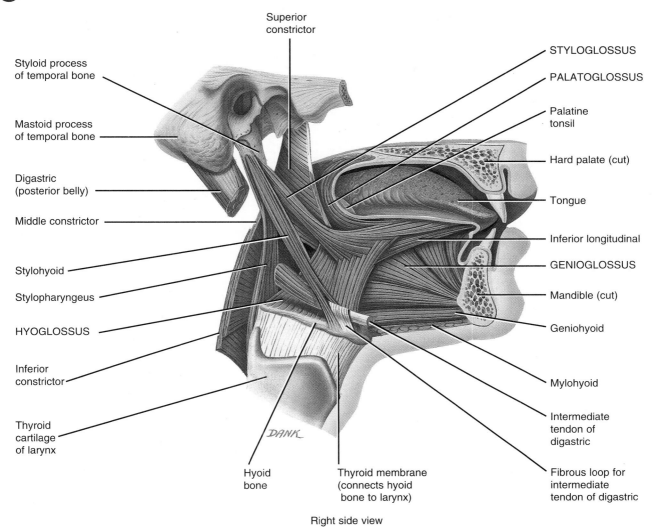

Superior
constrictor

STYLOGLOSSUS

PALATOGLOSSUS

Styloid process
of temporal bone

Palatine
tonsil

Mastoid process
of temporal bone

Hard palate (cut)

Digastric
(posterior belly)

Tongue

Middle constrictor

Inferior longitudinal

Stylohyoid

GENIOGLOSSUS

Stylopharyngeus

Mandible (cut)

HYOGLOSSUS

Geniohyoid

Inferior
constrictor

Mylohyoid

Thyroid
cartilage
of larynx

Intermediate
tendon of
digastric

DANK

Hyoid
bone

Thyroid membrane
(connects hyoid
bone to larynx)

Fibrous loop for
intermediate
tendon of digastric

Right side view

Q Which muscles depress the tongue?

EXHIBIT **11.7**   **MUSCLES OF THE FLOOR OF THE ORAL CAVITY (MOUTH) (Fig. 11.8)**

Overview: As a group, these muscles are referred to as **suprahyoid muscles** because they lie superior to the hyoid bone. The digastric muscle consists of an anterior belly and a posterior belly united by an intermediate tendon that is held in position by a fibrous loop (see also Fig. 11.7).

Muscle	Origin	Insertion	Action	Innervation
Digastric (dī'-GAS-trik; *di* = two; *gaster* = belly)	Anterior belly from inner side of lower border of mandible; posterior belly from mastoid process of temporal bone.	Body of hyoid bone via an intermediate tendon.	Elevates hyoid bone and depresses mandible as in opening the mouth.	Anterior belly from mandibular division of trigeminal (V) nerve; posterior belly from facial (VII) nerve.
Stylohyoid (stī'-lō-HĪ-oid; *stylo* = stake or pole, styloid process of temporal bone; *hyoedes* = U-shaped, pertaining to hyoid bone)	Styloid process of temporal bone.	Body of hyoid bone.	Elevates hyoid bone and draws it posteriorly.	Facial (VII) nerve.
Mylohyoid (mī'-lō-HĪ-oid)	Inner surface of mandible.	Body of hyoid bone.	Elevates hyoid bone and floor of mouth and depresses mandible.	Mandibular division of trigeminal (V) nerve.
Geniohyoid (jē'-nē-ō-HĪ-oid; *geneion* = chin) (see Fig. 11.7)	Inner surface of mandible.	Body of hyoid bone.	Elevates hyoid bone, draws hyoid bone and tongue anteriorly, and depresses mandible.	Cervical nerve C1.

Figure 11.8   Muscles of the floor of the oral cavity (mouth) and anterior neck.

 All suprahyoid muscles insert into the hyoid bone.

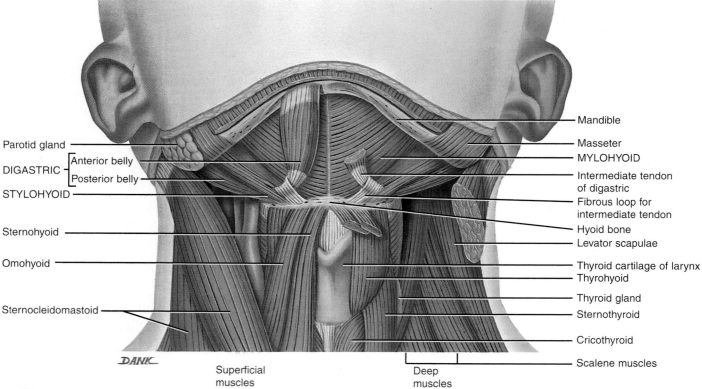

Q   Which muscle does not depress (lower) the mandible?

EXHIBIT **11.8 MUSCLES OF THE LARYNX (VOICE BOX) (Fig. 11.9)**

Overview: The muscles of the larynx (voice box), like those of the eyeballs and tongue, are grouped into extrinsic and intrinsic. The extrinsic muscles of the larynx listed here are together referred to as **infrahyoid (strap) muscles** because they lie inferior to the hyoid bone. The omohyoid muscle, like the digastric muscle, is composed of two bellies and an intermediate tendon. In this case, however, the two bellies are referred to as superior and inferior, rather than anterior and posterior. Extrinsic muscles of the larynx not included here are the **stylopharyngeus** (sti-lō-fa-RIN-jē-us), **inferior constrictor** (kon-STRIK-tor), and **middle constrictor**, which are illustrated in Fig. 11.7, plus the **palatopharyngeus** (pal'-a-tō-fa-RIN-jē-us), which is not illustrated.

Muscle	Origin	Insertion	Action	Innervation
EXTRINSIC				
Omohyoid (ō-mō-HĪ-oid; *omo* = relationship to the shoulder; *hyoedes* = U-shaped; pertaining to hyoid bone)	Superior border of scapula and superior transverse ligament.	Body of hyoid bone.	Depresses hyoid bone.	Branches of ansa cervicalis (C1–C3).
Sternohyoid (ster'-nō-HĪ-oid; *sterno* = sternum)	Medial end of clavicle and manubrium of sternum.	Body of hyoid bone.	Depresses hyoid bone.	Branches of ansa cervicalis (C1–C3).
Sternothyroid (ster'-nō-THĪ-roid; *thyro* = thyroid cartilage of larynx)	Manubrium of sternum.	Thyroid cartilage of larynx.	Depresses thyroid cartilage.	Branches of ansa cervicalis (C1–C3).
Thyrohyoid (thī'-rō-HĪ-oid)	Thyroid cartilage of larynx.	Greater horn of hyoid bone.	Elevates thyroid cartilage and depresses hyoid bone.	Branches of ansa cervicalis (C1–C2) and descending hypoglossal (XII) nerve.
INTRINSIC				
Cricothyroid (kri-kō-THĪ-roid; *crico* = cricoid cartilage of larynx)	Anterior and lateral portion of cricoid cartilage of larynx.	Anterior border of thyroid cartilage of larynx and posterior part of inferior border of thyroid cartilage of larynx.	Elongates and places tension on vocal folds.	External laryngeal branch of vagus (X) nerve.
Posterior cricoarytenoid (kri'-kō-ar'-i-TĒ-noid; *arytaina* = shaped like a jug)	Posterior surface of cricoid cartilage of larynx.	Posterior surface of arytenoid cartilage of larynx.	Opens rima glottidis (space between vocal folds).	Recurrent laryngeal branch of vagus (X) nerve.
Lateral cricoarytenoid (kri'-kō-ar'-i-TĒ-noid)	Superior border of cricoid cartilage of larynx.	Anterior surface of arytenoid cartilage of larynx.	Closes rima glottidis (space between vocal folds).	Recurrent laryngeal branch of vagus (X) nerve.
Arytenoid (ar'-i-TĒ-noid)	Posterior surface and lateral border of one arytenoid cartilage of larynx.	Corresponding parts of opposite arytenoid cartilage of larynx.	Closes rima glottidis (space between vocal folds).	Recurrent laryngeal branch of vagus (X) nerve.
Thyroarytenoid (thī'-rō-ar'-i-TĒ-noid)	Inferior portion of thyroid cartilage of larynx and middle of cricothyroid ligament.	Base and anterior surface of arytenoid cartilage of larynx.	Shortens and relaxes vocal folds.	Recurrent laryngeal branch of vagus (X) nerve.

Figure 11.9   Muscles of the larynx (voice box).

Intrinsic muscles of the larynx adjust tension on the vocal folds and open or close the rima glottidis.

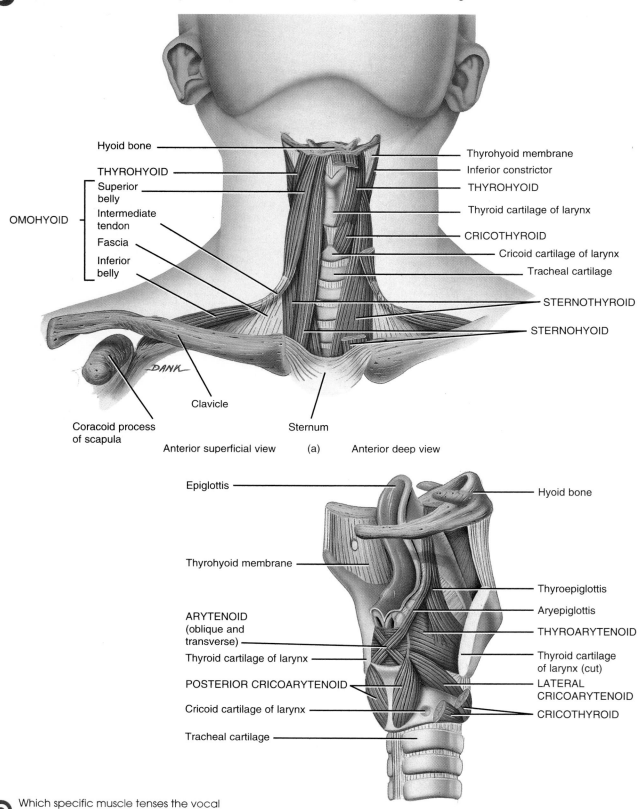

Hyoid bone

THYROHYOID

OMOHYOID
Superior belly
Intermediate tendon
Fascia
Inferior belly

Thyrohyoid membrane
Inferior constrictor
THYROHYOID
Thyroid cartilage of larynx
CRICOTHYROID
Cricoid cartilage of larynx
Tracheal cartilage
STERNOTHYROID
STERNOHYOID

DANK

Coracoid process of scapula

Clavicle

Sternum

Anterior superficial view   (a)   Anterior deep view

Epiglottis

Thyrohyoid membrane

ARYTENOID
(oblique and transverse)

Thyroid cartilage of larynx

POSTERIOR CRICOARYTENOID

Cricoid cartilage of larynx

Tracheal cartilage

Hyoid bone

Thyroepiglottis
Aryepiglottis
THYROARYTENOID
Thyroid cartilage of larynx (cut)
LATERAL CRICOARYTENOID
CRICOTHYROID

Which specific muscle tenses the vocal folds? Relaxes the vocal folds?

(b) Right posterolateral view

E X H I B I T **11.9 MUSCLES THAT MOVE THE HEAD**

Overview: The cervical region is divided by the sternocleidomastoid muscle into two principal triangles—anterior and posterior. The **anterior triangle** is bordered superiorly by the mandible, inferiorly by the sternum, medially by the cervical midline, and laterally by the anterior border of the sternocleidomastoid muscle. The **posterior triangle** is bordered inferiorly by the clavicle, anteriorly by the posterior border of the sternocleidomastoid muscle, and posteriorly by the anterior border of the trapezius muscle. Subsidiary triangles exist within the two principal triangles.

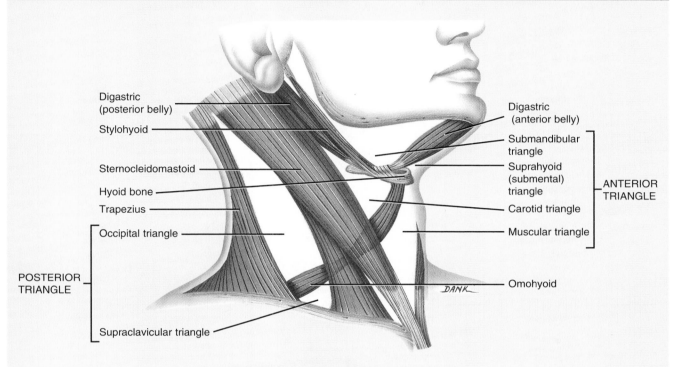

Triangles of the neck viewed from right side

Muscle	Origin	Insertion	Action	Innervation
Sternocleidomastoid (ster′-nō-klī′-dō-MAS-toid; *sternum* = breastbone; *cleido* = clavicle; *mastoid* = mastoid process of temporal bone) (see Fig. 11.14)	Sternum and clavicle.	Mastoid process of temporal bone.	Contraction of both muscles flexes the cervical part of the vertebral column and draws the head forward; contraction of one muscle rotates face toward side opposite contracting muscle.	Accessory (XI) nerve.
Semispinalis capitis (se′-mē-spi-NA-lis KAP-i-tis; *semi* = half; *spine* = spinous process; *caput* = head) (see Fig. 11.18)	Transverse processes of first six or seven thoracic vertebrae and seventh cervical vertebra and articular processes of fourth, fifth, and sixth cervical vertebrae.	Occipital bone.	Both muscles extend head; contraction of one muscle rotates it to side opposite contracting muscle.	Dorsal rami of cervical nerves.
Splenius capitis (SPLĒ-nē-us KAP-i-tis; *splenion* = bandage) (see Fig. 11.18)	Ligamentum nuchae and spinous processes of seventh cervical vertebra and first three or four thoracic vertebrae.	Occipital bone and mastoid process of temporal bone.	Both muscles extend head; contraction of one laterally flexes and rotates it to same side as contracting muscle.	Dorsal rami of middle and lower cervical nerves.
Longissimus capitis (lon-JIS-i-mus KAP-i-tis; *longissimus* = longest) (see Fig. 11.18)	Articular processes of last four cervical vertebrae and transverse processes of upper four thoracic vertebrae.	Mastoid process of temporal bone.	Extends head and rotates it to same side as contracting muscle.	Dorsal rami of middle and lower cervical nerves.

EXHIBIT **11.10**   **MUSCLES THAT ACT ON THE ABDOMINAL WALL (Fig. 11.10)**

Overview: The anterolateral abdominal wall is composed of skin, fascia, and four pairs of flat, sheetlike muscles: rectus abdominis, external oblique, internal oblique, and transversus abdominis. The anterior surfaces of the rectus abdominis muscles are interrupted by three transverse fibrous bands of tissue called **tendinous intersections**, believed to be remnants of septa that separated myotomes during embryological development. The aponeuroses of the external oblique, internal oblique, and transversus abdominis muscles form the **rectus sheath**, which encloses the rectus abdominis muscle, and meet at the midline to form the **linea alba** (white line), a tough, fibrous band that extends from the xiphoid process of the sternum to the pubic symphysis. The inferior free border of the external oblique aponeurosis, plus some collagen fibers, forms the **inguinal ligament**, which runs from the anterior superior iliac spine to the pubic tubercle (see Fig. 11.19a).

Just superior to the medial end of the inguinal ligament is a triangular slit in the aponeurosis referred to as the **superficial inguinal ring**, the outer opening of the **inguinal canal** (see Fig. 28.8). The canal contains the spermatic cord and ilioinguinal nerve in males and round ligament of the uterus and ilioinguinal nerve in females.

The posterior abdominal wall is formed by the lumbar vertebrae, parts of the ilia of the hipbones, psoas major muscle (described in Exhibit 11.20), quadratus lumborum muscle, and iliacus muscle (also described in Exhibit 11.20). Whereas the anterolateral abdominal wall is contractile and distensible, the posterior abdominal wall is bulky and stable by comparison.

Muscle	Origin	Insertion	Action	Innervation
Rectus abdominis (REK-tus ab-DOM-in-is; *rectus* = fibers parallel to midline; *abdomino* = abdomen)	Pubic crest and pubic symphysis.	Cartilage of fifth to seventh ribs and xiphoid process.	Flexes vertebral column and compresses abdomen to aid in defecation, urination, forced expiration, and childbirth.	Branches of thoracic nerves T7–T12.
External oblique (ō-BLĒK; *external* = closer to surface; *oblique* = fibers diagonal to midline).	Inferior eight ribs.	Iliac crest and linea alba.	Contraction of both compresses abdomen; contraction of one side alone bends vertebral column laterally; laterally rotates vertebral column.	Branches of thoracic nerves T7–T12 and iliohypogastric nerve.
Internal oblique (ō-BLĒK; *internal* = farther from surface)	Iliac crest, inguinal ligament, and thoracolumbar fascia.	Cartilage of inferior three or four ribs and linea alba.	Compresses abdomen; contraction of one side alone bends vertebral column laterally; laterally rotates vertebral column.	Branches of thoracic nerves T8–T12, iliohypogastric and ilioinguinal nerves.
Transversus abdominis (tranz-VER-sus ab-DOM-in-is; *transverse* = fibers perpendicular to midline)	Iliac crest, inguinal ligament, lumbar fascia, and cartilages of inferior six ribs.	Xiphoid process, linea alba, and pubis.	Compresses abdomen.	Branches of thoracic nerves T8–T12, iliohypogastric and ilioinguinal nerves.
Quadratus lumborum (kwod-RĀ-tus lum-BOR-um; *quad* = four; *lumbo* = lumbar region) (see Fig. 11.11)	Iliac crest and iliolumbar ligament.	Inferior border of twelfth rib and transverse processes of first four lumbar vertebrae.	During forced expiration, it pulls inferiorly on the twelfth rib; during deep inspiration, it fixes the twelfth rib to prevent its elevation; contraction of one side bends vertebral column laterally.	Branches of thoracic nerve T12 and lumbar nerves L1–L3 or L1–L4.

Figure 11.10 Muscles of the male anterolateral abdominal wall. The arrow in the inset in (c) indicates the direction from which the anterolateral abdominal wall is viewed (superior).

The inguinal ligament separates the thigh from the body wall.

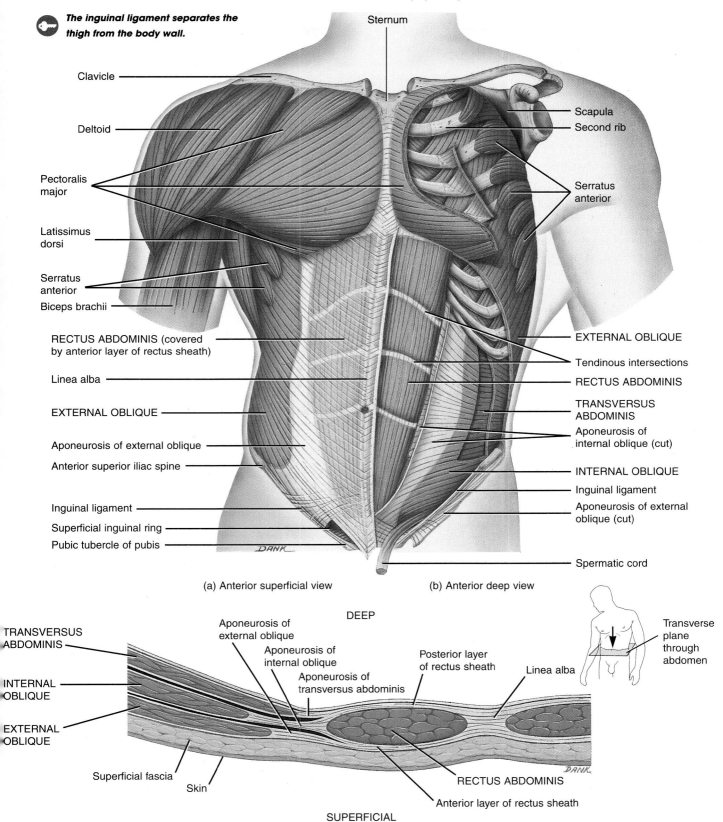

Sternum

Clavicle

Deltoid

Pectoralis major

Latissimus dorsi

Serratus anterior

Biceps brachii

RECTUS ABDOMINIS (covered by anterior layer of rectus sheath)

Linea alba

EXTERNAL OBLIQUE

Aponeurosis of external oblique

Anterior superior iliac spine

Inguinal ligament

Superficial inguinal ring

Pubic tubercle of pubis

DANK

Scapula
Second rib

Serratus anterior

EXTERNAL OBLIQUE

Tendinous intersections

RECTUS ABDOMINIS

TRANSVERSUS ABDOMINIS

Aponeurosis of internal oblique (cut)

INTERNAL OBLIQUE

Inguinal ligament

Aponeurosis of external oblique (cut)

Spermatic cord

(a) Anterior superficial view   (b) Anterior deep view

DEEP

TRANSVERSUS ABDOMINIS

Aponeurosis of external oblique

Aponeurosis of internal oblique

Posterior layer of rectus sheath

Transverse plane through abdomen

INTERNAL OBLIQUE

Aponeurosis of transversus abdominis

Linea alba

EXTERNAL OBLIQUE

Superficial fascia

Skin

RECTUS ABDOMINIS

DANK

Anterior layer of rectus sheath

SUPERFICIAL

(c) Superior view of transverse section of anterior abdominal wall superior to umbilicus (navel)

Which abdominal muscle aids in urination?

EXHIBIT **11.11**   **MUSCLES USED IN BREATHING (Fig. 11.11)**

Overview: The muscles described here are attached to the ribs and by their contraction and relaxation alter the size of the thoracic cavity during breathing. Essentially, inspiration occurs when the thoracic cavity increases in size. Expiration occurs when the thoracic cavity decreases in size. The principal muscles of *inspiration* during normal breathing are the diaphragm and external intercostals. During forced inspiration, accessory muscles, such as the sternocleidomastoid, scalenes, and pectoralis minor are also used. The principal muscles of *expiration* during normal breathing are also the diaphragm and external intercostals. During forced expiration, accessory muscles, such as the internal intercostals and abdominal muscles (external oblique, internal oblique, transversus abdominis, and rectus abdominis) are also used.

The diaphragm, one of the muscles used in breathing, is dome-shaped and has three major openings through which various structures pass between the thorax and abdomen. These structures include the aorta along with the thoracic duct and azygos vein, which pass through the **aortic hiatus**; the esophagus with accompanying vagus (X) nerves, which pass through the **esophageal hiatus**; and the inferior vena cava, which passes through the **foramen for the vena cava**. In a condition called a hiatus hernia, the stomach protrudes superiorly through the esophageal hiatus.

Muscle	Origin	Insertion	Action	Innervation
Diaphragm (DĪ-a-fram; *dia* = across; *phragma* = wall)	Xiphoid process, costal cartilages of inferior six ribs, and lumbar vertebrae.	Central tendon (strong aponeurosis that serves as the tendon of insertion for all muscular fibers of the diaphragm).	Forms floor of thoracic cavity; pulls central tendon inferiorly during inspiration and as dome of diaphragm flattens increases vertical length of thorax.	Phrenic nerve (contains axons from C3, C4, and C5).
External intercostals (in′-ter-KOS-tals; *external* = closer to surface; *inter* = between; *costa* = rib)	Inferior border of rib above.	Superior border of rib below.	Elevate ribs during inspiration and thus increase lateral and anteroposterior dimensions of thorax.	Intercostal nerves.
Internal intercostals (in′-ter-KOS-tals; *internal* = farther from surface)	Superior border of rib below.	Inferior border of rib above.	Draw adjacent ribs together during forced expiration and thus decrease lateral and anteroposterior dimensions of thorax.	Intercostal nerves.

Figure 11.11 Muscles used in breathing as seen in a male.

Openings in the diaphragm transmit the aorta, esophagus, and inferior vena cava.

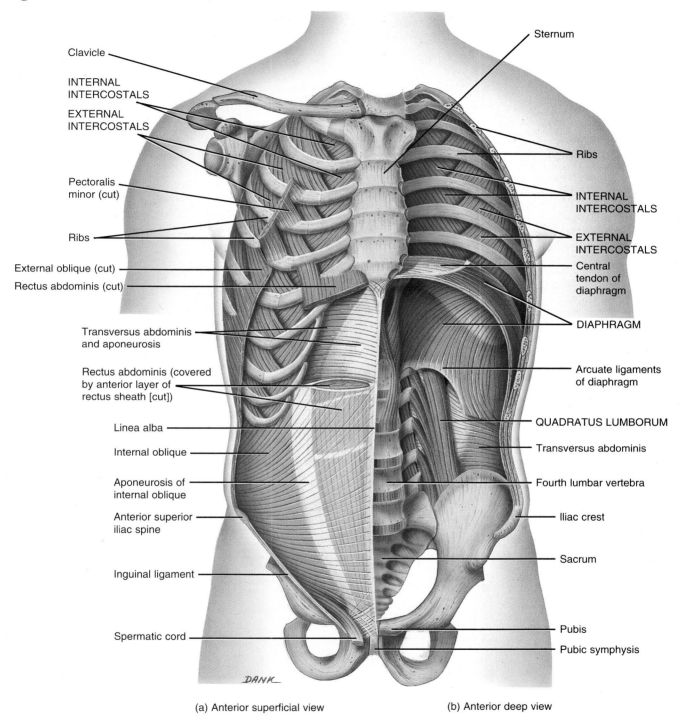

Clavicle

INTERNAL INTERCOSTALS

EXTERNAL INTERCOSTALS

Pectoralis minor (cut)

Ribs

External oblique (cut)

Rectus abdominis (cut)

Transversus abdominis and aponeurosis

Rectus abdominis (covered by anterior layer of rectus sheath [cut])

Linea alba

Internal oblique

Aponeurosis of internal oblique

Anterior superior iliac spine

Inguinal ligament

Spermatic cord

DANK

Sternum

Ribs

INTERNAL INTERCOSTALS

EXTERNAL INTERCOSTALS

Central tendon of diaphragm

DIAPHRAGM

Arcuate ligaments of diaphragm

QUADRATUS LUMBORUM

Transversus abdominis

Fourth lumbar vertebra

Iliac crest

Sacrum

Pubis

Pubic symphysis

(a) Anterior superficial view   (b) Anterior deep view

 Which muscle is innervated by the phrenic nerve?

EXHIBIT **11.12**   **MUSCLES OF THE PELVIC FLOOR (Fig. 11.12)**

Overview: The muscles of the pelvic floor, together with the fasciae covering their external and internal surfaces, are referred to as the **pelvic diaphragm**. The diaphragm is funnel-shaped and forms the floor of the abdominopelvic cavity. It is pierced by the anal canal and urethra in both sexes and also by the vagina in the female.

Muscle	Origin	Insertion	Action	Innervation
Levator ani (le-VĀ-tor Ā-nē; *levator* = raises; *ani* = anus)	This muscle is divisible into two parts, the pubococcygeus muscle and the iliococcygeus muscle.			
Pubococcygeus (pu′-bō-kok-SIJ-ē-us; *pubo* = pubis; *coccygeus* = coccyx)	Pubis.	Coccyx, urethra, anal canal, central tendon of perineum, and anococcygeal raphe (narrow fibrous band that extends from anus to coccyx).	Supports and slightly elevates pelvic floor, resists increased intra-abdominal pressure, and draws anus toward pubis and constricts it.	Sacral nerves S3–S4 or S4 and perineal branch of pudendal nerve.
Iliococcygeus (il′-ē-o-kok-SIJ-ē-us; *ilio* = ilium)	Ischial spine.	Coccyx.	Supports and slightly elevates pelvic floor, resists increased intra-abdominal pressure, and draws anus toward pubis and constricts it.	Sacral nerves S3–S4 or S4 and perineal branch of pudendal nerve.
Coccygeus (kok-SIJ-ē-us)	Ischial spine.	Inferior sacrum and superior coccyx.	Supports and slightly elevates pelvic floor, resists intra-abdominal pressure, and pulls coccyx anteriorly after defecation or parturition (childbirth).	Sacral nerve S3 or S4.

Figure 11.12   Muscles of the pelvic floor seen in the female perineum.

🔑 *The pelvic diaphragm supports the pelvic viscera.*

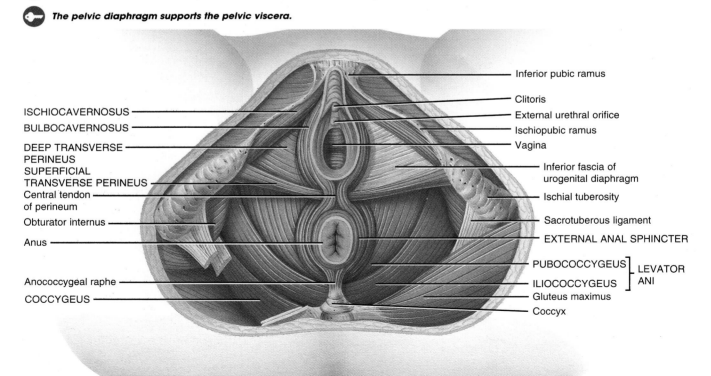

Superficial muscles

Q Which muscle assists in parturition (childbirth)?

EXHIBIT **11.13 MUSCLES OF THE PERINEUM (Fig. 11.13; see also Fig. 11.12)**

Overview: The **perineum** (per-i-NĒ-um) is the entire outlet of the pelvis. It is a diamond-shaped area at the inferior end of the trunk between the thighs and buttocks. It is bordered anteriorly by the pubic symphysis, laterally by the ischial tuberosities, and posteriorly by the coccyx. A transverse line drawn between the ischial tuberosities divides the perineum into an anterior **urogenital triangle** that contains the external genitals and a posterior **anal triangle** that contains the anus (see Fig. 28.19). The deep transverse perineus muscle, the urethral sphincter, and a fibrous membrane constitute the **urogenital diaphragm**.

Muscle	Origin	Insertion	Action	Innervation[a]
Superficial transverse perineus (per-i-NĒ-us; *superficial* = closer to surface; *transverse* = across; *perineus* = perineum)	Ischial tuberosity.	Central tendon of perineum (fibromuscular tissue in the midline between anus and vagina in female and anus and bulb of penis in male).	Helps to stabilize the central tendon of the perineum.	Perineal branch of pudendal nerve.
Bulbospongiosus (bul'-bo-spon'-jē-Ō-sus; *bulbus* = bulb; *spongia* = sponge)	Central tendon of perineum.	Inferior fascia of urogenital diaphragm, corpus spongiosum of penis, and deep fascia on dorsum of penis in male; pubic arch and root and dorsum of clitoris in female.	Helps expel last drops of urine during micturition (urination), helps propel semen along urethra, and assists in erection of the penis in male; decreases size of vaginal orifice and assists in erection of clitoris in female.	Perineal branch of pudendal nerve.
Ischiocavernosus (is'-kē-o-ka'-ver-NŌ-sus; *ischion* = hip; *caverna* = hollow place)	Ischial tuberosity and ischial and pubic rami.	Corpus cavernosum of penis in male and clitoris in female.	Maintains erection of penis in male and clitoris in female.	Perineal branch of pudendal nerve.
Deep transverse perineus (per-i-NĒ-us; *deep* = farther from surface)	Ischial rami.	Central tendon of perineum.	Helps expel last drops of urine and semen in male and urine in female.	Perineal branch of pudendal nerve.
Urethral sphincter (yoo-RĒ-thral SFINGK-ter; *sphincter* = circular muscle that decreases size of an opening; *urethrae* = urethra)	Ischial and pubic rami.	Median raphe in male and vaginal wall in female.	Helps expel last drops of urine and semen in male and urine in female.	Perineal branch of pudendal nerve.
External anal sphincter (Ā-nal)	Anococcygeal raphe.	Central tendon of perineum.	Keeps anal canal and orifice closed.	Sacral nerve S4 and inferior rectal branch of pudendal nerve.

[a]The pudendal nerve is derived from the sacral plexus, which is described and illustrated in Exhibit 13.4 on pages 384–385.

Figure 11.13   Muscles of the male perineum.

 The urogenital diaphragm surrounds the urogenital duct and helps to strengthen the pelvic floor.

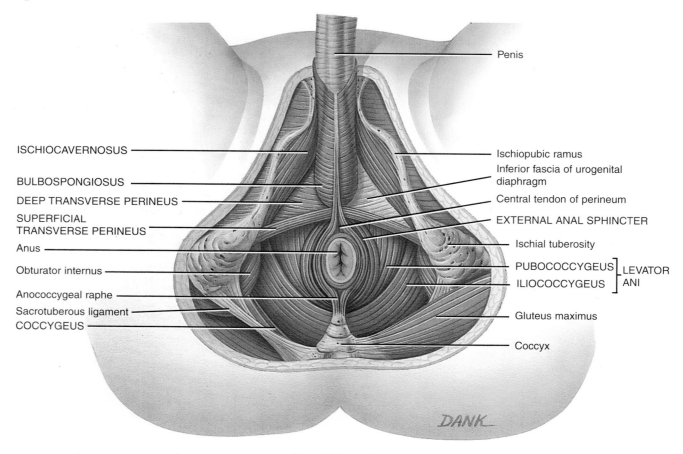

Penis

ISCHIOCAVERNOSUS

Ischiopubic ramus

Inferior fascia of urogenital diaphragm

BULBOSPONGIOSUS

Central tendon of perineum

DEEP TRANSVERSE PERINEUS

EXTERNAL ANAL SPHINCTER

SUPERFICIAL TRANSVERSE PERINEUS

Ischial tuberosity

Anus

PUBOCOCCYGEUS ⎤ LEVATOR

Obturator internus

ILIOCOCCYGEUS ⎦ ANI

Anococcygeal raphe

Gluteus maximus

Sacrotuberous ligament

COCCYGEUS

Coccyx

DANK

Superficial muscles

Which muscles assist in erection of the penis in males and clitoris in females?

EXHIBIT **11.14 MUSCLES THAT MOVE THE PECTORAL (SHOULDER) GIRDLE (Fig. 11.14)**

Overview: Muscles that move the pectoral (shoulder) girdle can be divided into **anterior** and **posterior** groups. The principal action of the muscles is to stabilize the scapula so that it can function as a stable point of origin for most of the muscles that move the humerus (arm).

Muscle	Origin	Insertion	Action	Innervation[a]
ANTERIOR				
Subclavius (sub-KLĀ-vē-us; *sub* = under; *clavius* = clavicle)	First rib.	Clavicle.	Depresses clavicle.	Nerve to subclavius.
Pectoralis (pek′-tor-A-lis) **minor** (*pectus* = breast, chest, thorax; *minor* = lesser)	Third through fifth ribs.	Coracoid process of scapula.	Depresses and moves scapula anteriorly and elevates third through fifth ribs during forced inspiration when scapula is fixed.	Medial pectoral nerve.
Serratus (ser-Ā-tus) **anterior** (*serratus* = saw-toothed; *anterior* = front)	Superior eight or nine ribs.	Vertebral border and inferior angle of scapula.	Rotates scapula superiorly and laterally, protracts scapula, and elevates ribs when scapula is fixed.	Long thoracic nerve.
POSTERIOR				
Trapezius (tra-PĒ-zē-us; *trapezoides* = trapezoid-shaped)	Occipital bone, ligamentum nuchae, and spines of seventh cervical and all thoracic vertebrae.	Clavicle and acromion and spine of scapula.	Elevates clavicle, adducts scapula, rotates scapula superiorly, elevates or depresses scapula, and extends head.	Accessory (XI) nerve and cervical nerves C3–C4.
Levator scapulae (le-VĀ-tor SKA-pyoo-lē; *levator* = raises; *scapulae* = scapula)	Superior four or five cervical vertebrae.	Superior vertebral border of scapula.	Elevates scapula and slightly rotates it inferiorly.	Dorsal scapular nerve and cervical nerves C3–C5.
Rhomboideus (rom-BOID-ē-us) **major** (*rhomboides* = rhomboid or diamond-shaped)	Spines of second to fifth thoracic vertebrae.	Vertebral border of scapula inferior to spine.	Adducts scapula and slightly rotates it inferiorly.	Dorsal scapular nerve.
Rhomboideus (rom-BOID-ē-us) **minor**	Spines of seventh cervical and first thoracic vertebrae.	Vertebral border of scapula superior to spine.	Adducts scapula and slightly rotates it inferiorly.	Dorsal scapular nerve.

[a] Many of the nerves listed here are derived from the brachial plexus, which is described and illustrated in Exhibit 13.2 on pages 379–380.

Figure 11.14   Muscles that move the pectoral (shoulder) girdle.

Muscles that move the pectoral girdle originate on the axial skeleton and insert on the clavicle or scapula.

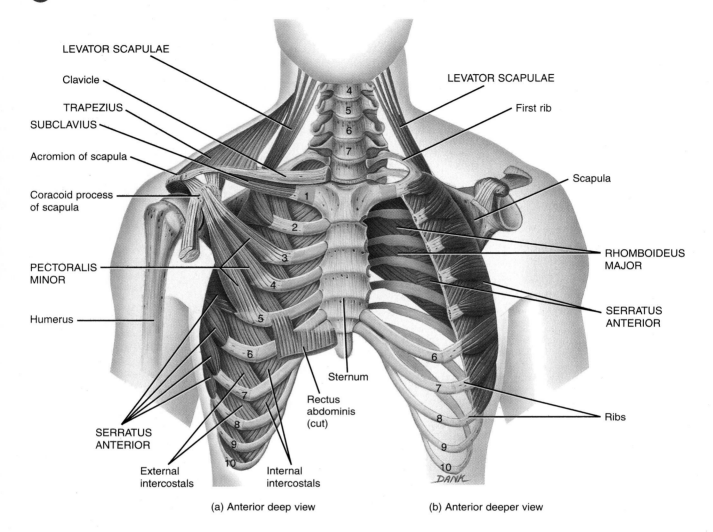

(a) Anterior deep view

(b) Anterior deeper view

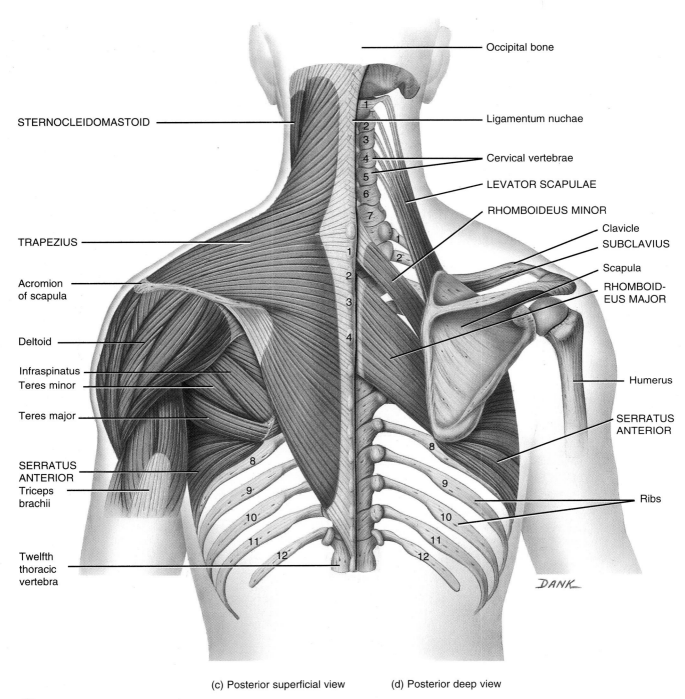

STERNOCLEIDOMASTOID

TRAPEZIUS

Acromion
of scapula

Deltoid

Infraspinatus
Teres minor

Teres major

SERRATUS
ANTERIOR
Triceps
brachii

Twelfth
thoracic
vertebra

Occipital bone

Ligamentum nuchae

Cervical vertebrae

LEVATOR SCAPULAE

RHOMBOIDEUS MINOR

Clavicle
SUBCLAVIUS

Scapula

RHOMBOID-
EUS MAJOR

Humerus

SERRATUS
ANTERIOR

Ribs

DANK

(c) Posterior superficial view   (d) Posterior deep view

Q Which muscles originate on the ribs?

EXHIBIT **11.15   MUSCLES THAT MOVE THE HUMERUS (ARM) (Fig. 11.15)**

Overview: Of the nine muscles that cross the shoulder joint, only two of them (pectoralis major and latissimus dorsi) do not originate on the scapula. These two muscles are thus designated as **axial muscles**, since they originate on the axial skeleton. The remaining seven muscles, the **scapular muscles**, arise from the scapula. The strength and stability of the shoulder joint are not provided by the shape of the articulating bones or its ligaments. Instead, four deep muscles of the shoulder—subscapularis,
supraspinatus, infraspinatus, and teres minor—strengthen and stabilize the shoulder joint. The muscles join the scapula to the
humerus. Their tendons are arranged to form a nearly complete circle around the joint, an arrangement referred to as the **rotator
(musculotendinous) cuff** (see Fig. 9.7).

 After you have studied the muscles in this exhibit, arrange them according to the following actions: flexion, extension, abduction, adduction, medial rotation, and lateral rotation. (The same muscle can be used more than once.)

Muscle	Origin	Insertion	Action	Innervation[a]
AXIAL				
Pectoralis (pek′-tor-A-lis) **major** (see also Fig. 11.10a)	Clavicle, sternum, carti-lages of second to sixth ribs.	Greater tubercle and intertubercular sulcus of humerus.	Flexes, adducts, and rotates arm medially.	Medial and lateral pectoral nerve.
Latissimus dorsi (la-TIS-i-mus DOR-sī; *latissi-mus* = widest; *dorsum* = back)	Spines of inferior six tho-racic vertebrae, lumbar vertebrae, crests of sacrum and ilium, inferior four ribs.	Intertubercular sul-cus of humerus.	Extends, adducts, and rotates arm medially; draws arm infe-riorly and posteriorly.	Thoracodorsal nerve.
SCAPULAR				
Deltoid (DEL-toyd; *deltoides* = triangular)	Acromial extremity of clavicle and acromion and spine of scapula.	Deltoid tuberosity of humerus.	Abducts, flexes, extends, and may alternately medially or laterally rotate arm.	Axillary nerve.
Subscapularis (sub-scap′-yoo-LA-ris; *sub* = below; *scapularis* = scapula)	Subscapular fossa of scapula.	Lesser tubercle of humerus.	Rotates arm medially.	Upper and lower sub-scapular nerves.
Supraspinatus (soo′-pra-spi-NĀ-tus; *supra* = above; *spinatus* = spine of scapula)	Supraspinous fossa of scapula.	Greater tubercle of humerus.	Assists deltoid muscle in abducting arm.	Suprascapular nerve.
Infraspinatus (in′-fra-spi-NĀ-tus; *infra* = below)	Infraspinous fossa of scapula.	Greater tubercle of humerus.	Rotates arm laterally; adducts arm.	Suprascapular nerve.
Teres (TER-ēz) **major** (*teres* = long and round)	Inferior angle of scapula.	Intertubercular sul-cus of humerus.	Extends arm; assists in adduction and medial rotation of arm.	Lower sub-scapular nerve.
Teres (TER-ēz) **minor**	Inferior lateral border of scapula.	Greater tubercle of humerus.	Rotates arm laterally; extends and adducts arm.	Axillary nerve.
Coracobrachialis (kor′-a-ko-BRĀ-kē-a′-lis; *coraco* = coracoid process)	Coracoid process of scapula.	Middle of medial surface of shaft of humerus.	Flexes and adducts arm.	Musculocutan-eous nerve.

[a] The nerves listed here are derived from the brachial plexus, which is described and illustrated in Exhibit 13.2 on pages 379–381.

Figure 11.15 Muscles that move the humerus (arm).

Strength and stability of the shoulder joint are provided by the tendons of the muscles that form the rotator cuff.

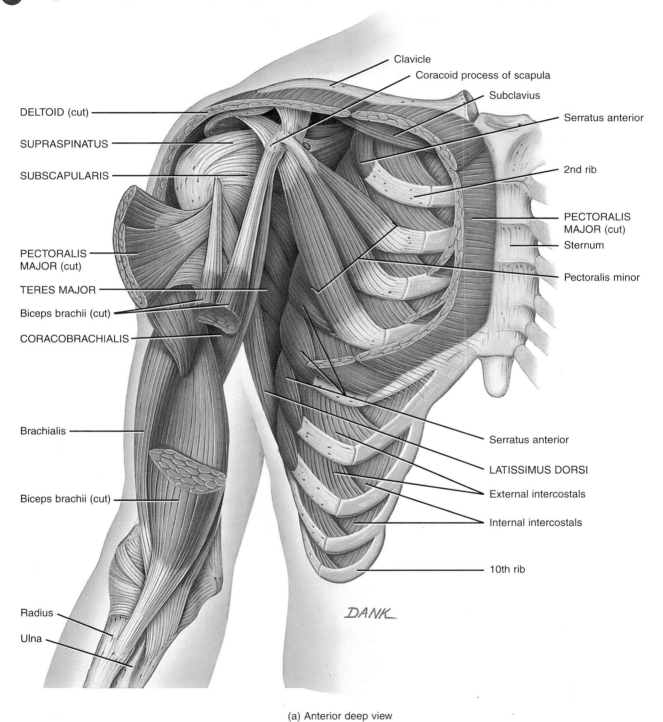

(a) Anterior deep view

Figure continues

Figure 11.15   (continued)

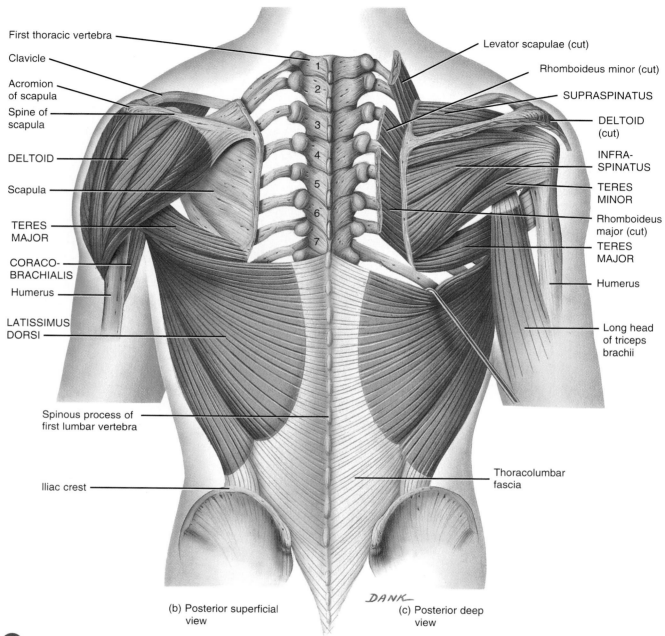

First thoracic vertebra

Clavicle

Acromion
of scapula

Spine of
scapula

DELTOID

Scapula

TERES
MAJOR

CORACO-
BRACHIALIS

Humerus

LATISSIMUS
DORSI

Spinous process of
first lumbar vertebra

Iliac crest

Levator scapulae (cut)

Rhomboideus minor (cut)

SUPRASPINATUS

DELTOID
(cut)

INFRA-
SPINATUS

TERES
MINOR

Rhomboideus
major (cut)

TERES
MAJOR

Humerus

Long head
of triceps
brachii

Thoracolumbar
fascia

DANK

(b) Posterior superficial
view

(c) Posterior deep
view

Q Which muscles flex the arm?

EXHIBIT **11.16 MUSCLES THAT MOVE THE RADIUS AND ULNA (FOREARM) (Fig. 11.16)**

Overview: Most of the muscles that move the radius and ulna (forearm) are divided into **flexors** and **extensors**. Recall that the elbow joint is a hinge joint, capable only of flexion and extension. Whereas the biceps brachii, brachialis, and brachioradialis are flexors of the forearm, the triceps brachii and anconeus are extensors. Other muscles permit pronation and supination of the forearm. (The biceps brachii muscle also permits supination of the forearm).

Muscle	Origin	Insertion	Action	Innervation[a]
FLEXORS				
Biceps brachii (BĪ-ceps BRĀ-kē-ī; *biceps* = two heads of origin; *brachion* = arm)	*Long head* originates from tubercle superior to glenoid cavity; *short head* originates from coracoid process of scapula.	Radial tuberosity and bicipital aponeurosis (broad aponeurosis from tendon of insertion of biceps brachii muscle that descends medially across brachial artery and fuses with deep fascia over forearm flexor muscles).	Flexes and supinates forearm; flexes arm.	Musculocutaneous nerve.
Brachialis (brā′-kē-A-lis)	Distal, anterior surface of humerus.	Ulnar tuberosity and coronoid process of ulna.	Flexes forearm.	Musculocutaneous and radial nerves.
Brachioradialis (brā′-kē-ō-rā′-dē-A-lis; *radialis* = radius) (see also Fig. 11.17)	Medial and lateral borders of distal end of humerus.	Superior to styloid process of radius.	Flexes forearm.	Radial nerve.
EXTENSORS				
Triceps brachii (TRĪ-ceps BRĀ-kē-ī; *triceps* = three heads of origin)	*Long head* originates from a projection inferior to glenoid cavity (infraglenoid tubercle) of scapula; *lateral head* originates from lateral and posterior surface of humerus superior to radial groove; *medial head* originates from entire posterior surface of humerus inferior to a groove for the radial nerve.	Olecranon of ulna.	Extends forearm; extends arm.	Radial nerve.
Anconeus (an-KŌ-nē-us; *anconeal* = pertaining to elbow) (see also Fig. 11.17)	Lateral epicondyle of humerus.	Olecranon and superior portion of shaft of ulna.	Extends forearm.	Radial nerve.
PRONATORS				
Pronator teres (PRŌ-nā′-ter TER-ēz; *pronation* = turning palm inferiorly or posteriorly) (see Fig. 11.17)	Medial epicondyle of humerus and coronoid process of ulna.	Midlateral surface of radius.	Pronates forearm and hand and weakly flexes forearm.	Median nerve.
Pronator quadratus (PRŌ-nā′-ter kwod-RĀ-tus; *quadratus* = squared, four-sided) (see Fig. 11.17a,b)	Distal portion of shaft of ulna.	Distal portion of shaft of radius.	Pronates forearm and hand.	Median nerve.
SUPINATOR				
Supinator (SOO-pi-nā-tor; *supination* = turning palm superiorly or anteriorly) (see Fig. 11.17b)	Lateral epicondyle of humerus and ridge near radial notch of ulna (supinator crest).	Lateral surface of proximal one-third of radius.	Supinates forearm and hand.	Deep radial nerve.

[a] The nerves listed here are derived from the brachial plexus, which is described and illustrated in Exhibit 13.2 on pages 379–381.

Figure 11.16   Muscles that move the radius and ulna (forearm). The arrow in the inset in (c) indicates the direction from which the arm is being viewed (superior).

Whereas anterior arm muscles flex the forearm, posterior arm muscles extend it.

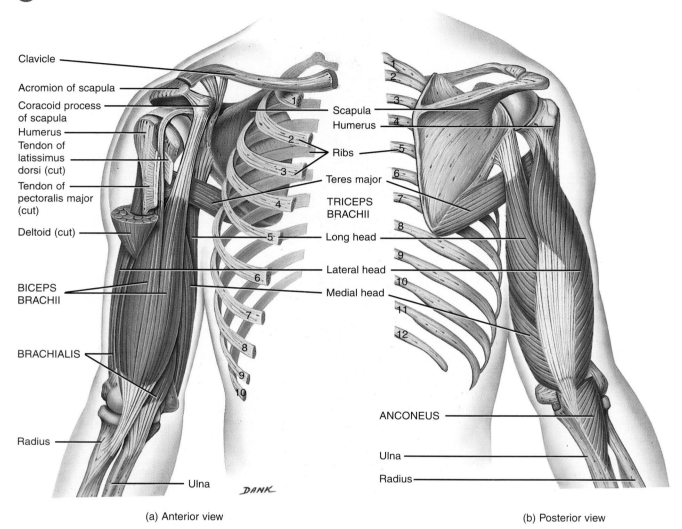

(a) Anterior view

(b) Posterior view

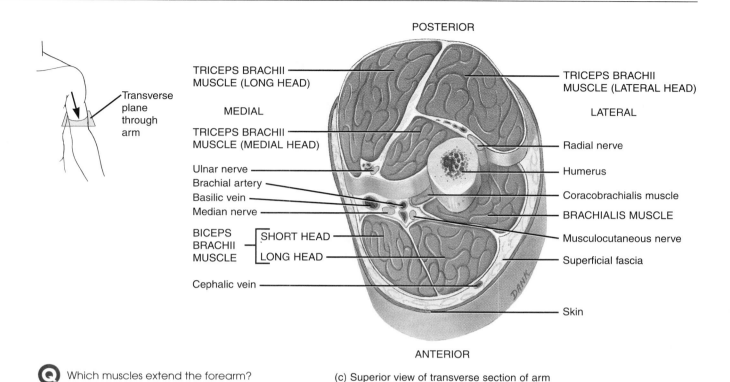

POSTERIOR

TRICEPS BRACHII MUSCLE (LONG HEAD)

TRICEPS BRACHII MUSCLE (LATERAL HEAD)

MEDIAL

LATERAL

TRICEPS BRACHII MUSCLE (MEDIAL HEAD)

Radial nerve

Ulnar nerve

Humerus

Brachial artery

Coracobrachialis muscle

Basilic vein

BRACHIALIS MUSCLE

Median nerve

Musculocutaneous nerve

BICEPS BRACHII MUSCLE — SHORT HEAD / LONG HEAD

Superficial fascia

Cephalic vein

Skin

ANTERIOR

Transverse plane through arm

Q Which muscles extend the forearm?

(c) Superior view of transverse section of arm

EXHIBIT 11.17 MUSCLES THAT MOVE THE WRIST, HAND, AND FINGERS (Fig. 11.17)

Overview: Muscles that move the wrist, hand, and fingers are many and varied. However, as you will see, their names usually give some indication of their origin, insertion, or action. On the basis of location and function, the muscles are divided into two groups—anterior and posterior compartments. The **anterior compartment muscles** originate on the humerus and typically insert on the carpals, metacarpals, and phalanges. The bellies of these muscles form the bulk of the proximal forearm. Anterior compartment muscles are flexors. The **posterior compartment muscles** arise on the humerus and insert on the metacarpals and phalanges. Posterior compartment muscles are extensors. Each of the two principal compartments is also divided into superficial and deep muscles.

The tendons of the muscles of the forearm that attach to the wrist or continue into the hand, along with blood vessels and nerves, are held close to bones by strong fascial structures. The tendons are also surrounded by tendon sheaths. At the wrist, the deep fascia is thickened into fibrous bands called retinacula (*retinere* = retain). The **flexor retinaculum (transverse carpal ligament)** is located over the palmar surface of the carpal bones. Through it pass the long flexor tendons of the digits and wrist and the median nerve. The **extensor retinaculum (dorsal carpal ligament)** is located over the dorsal surface of the carpal bones. Through it pass the extensor tendons of the wrist and digits.

After you have studied the muscles in this exhibit, arrange them according to the following actions: flexion, extension, abduction, adduction, supination, and pronation. (The same muscles can be used more than once.)

Muscle	Origin	Insertion	Action	Innervation[a]
ANTERIOR COMPARTMENT (FLEXORS)				
SUPERFICIAL				
Flexor carpi radialis (FLEK-sor KAR-pē rā'-dē-A-lis; *flexor* = decreases angle at joint; *carpus* = wrist; *radialis* = radius)	Medial epicondyle of humerus.	Second and third metacarpals.	Flexes and abducts wrist.	Median nerve.
Palmaris longus (pal-MA-ris LON-gus; *palma* = palm; *longus* = long	Medial epicondyle of humerus.	Flexor retinaculum and palmar aponeurosis (deep fascia in center of palm).	Weakly flexes wrist. (This muscle is sometimes absent.)	Median nerve.

[a] The nerves listed here are derived from the brachial plexus, which is described and illustrated in Exhibit 13.2 on pages 379–381.

Exhibit continues

EXHIBIT **11.17** **MUSCLES THAT MOVE THE WRIST, HAND, AND FINGERS (Fig. 11.17)** (CONTINUED)

Muscle	Origin	Insertion	Action	Innervation
ANTERIOR COMPARTMENT (FLEXORS), *continued*				
SUPERFICIAL				
Flexor carpi ulnaris (FLEK-sor KAR-pē ul-NAR-is; *ulnaris* = ulna)	Medial epicondyle of humerus and superior posterior border of ulna.	Pisiform, hamate, and fifth metacarpal.	Flexes and adducts wrist.	Ulnar nerve.
Flexor digitorum superficialis (FLEK-sor di′-ji-TOR-um soo′-per-fish′-ē-A-lis; *digit* = finger or toe; *superficialis* = closer to surface)	Medial epicondyle of humerus, coronoid process of ulna, and a ridge along lateral margin of anterior surface (anterior oblique line) of radius.	Middle phalanges.	Flexes middle phalanges of each finger.	Median nerve.
DEEP				
Flexor digitorum profundus (FLEK-sor di′-ji-TOR-um pro-FUN-dus; *profundus* = deep)	Anterior medial surface of body of ulna.	Bases of distal phalanges.	Flexes distal phalanges of each finger.	Median and ulnar nerves.
Flexor pollicis longus (FLEK-sor POL-li-kis LON-gus; *pollex* = thumb)	Anterior surface of radius and interosseous membrane (sheet of fibrous tissue that holds shafts of ulna and radius together).	Base of distal phalanx of thumb.	Flexes thumb.	Median nerve.
POSTERIOR COMPARTMENT (EXTENSORS)				
SUPERFICIAL				
Extensor carpi radialis longus (eks-TEN-sor KAR-pē rā′-dē-A-lis LON-gus; *extensor* = increases angle at joint)	Lateral epicondyle of humerus.	Second metacarpal.	Extends and abducts wrist.	Radial nerve.
Extensor carpi radialis brevis (eks-TEN-sor KAR-pē rā′-dē-A-lis BREV-is; *brevis* = short)	Lateral epicondyle of humerus.	Third metacarpal.	Extends and abducts wrist.	Radial nerve.
Extensor digitorum (eks-TEN-sor di′-ji-TOR-um)	Lateral epicondyle of humerus.	Second through fifth distal and middle phalanges.	Extends phalanges.	Radial nerve.
Extensor digiti minimi (eks-TEN-sor DIJ-i-tē MIN-i-mē; *digiti* = digit; *minimi* = little)	Tendon of extensor digitorum.	Tendon of extensor digitorum on fifth phalanx.	Extends little finger.	Deep radial nerve.
Extensor carpi ulnaris (eks-TEN-sor KAR-pē ul-NAR-is)	Lateral epicondyle of humerus and posterior border of ulna.	Fifth metacarpal.	Extends and adducts wrist.	Deep radial nerve.
DEEP				
Abductor pollicis longus (ab-DUK-tor POL-li-kis LON-gus; *abductor* = moves part away from midline)	Posterior surface of middle of radius and ulna and interosseous membrane.	First metacarpal.	Extends thumb and abducts wrist.	Deep radial nerve.
Extensor pollicis brevis (eks-TEN-sor POL-li-kis BREV-is)	Posterior surface of middle of radius and interosseous membrane.	Base of proximal phalanx of thumb.	Extends thumb and abducts wrist.	Deep radial nerve.
Extensor pollicis longus (eks-TEN-sor POL-li-kis LON-gus)	Posterior surface of middle of ulna and interosseous membrane.	Base of distal phalanx of thumb.	Extends thumb and abducts wrist.	Deep radial nerve.
Extensor indicis (eks-TEN-sor IN-di-kis; *indicis* = index)	Posterior surface of ulna.	Tendon of extensor digitorum of index finger.	Extends index finger.	Deep radial nerve.

Figure 11.17 Muscles that move the wrist, hand, and fingers.

The anterior compartment muscles function as flexors and the posterior compartment muscles function as extensors.

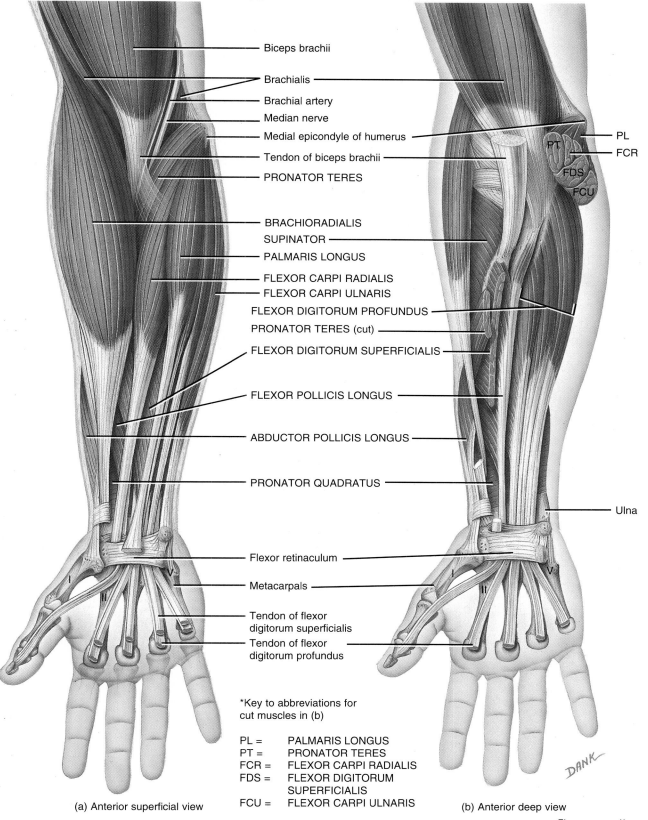

Biceps brachii

Brachialis

Brachial artery

Median nerve

Medial epicondyle of humerus

Tendon of biceps brachii

PRONATOR TERES

BRACHIORADIALIS

SUPINATOR

PALMARIS LONGUS

FLEXOR CARPI RADIALIS

FLEXOR CARPI ULNARIS

FLEXOR DIGITORUM PROFUNDUS

PRONATOR TERES (cut)

FLEXOR DIGITORUM SUPERFICIALIS

FLEXOR POLLICIS LONGUS

ABDUCTOR POLLICIS LONGUS

PRONATOR QUADRATUS

Flexor retinaculum

Metacarpals

Tendon of flexor digitorum superficialis

Tendon of flexor digitorum profundus

PL

FCR

PT

FDS

FCU

Ulna

*Key to abbreviations for cut muscles in (b)

PL =   PALMARIS LONGUS
PT =   PRONATOR TERES
FCR =   FLEXOR CARPI RADIALIS
FDS =   FLEXOR DIGITORUM
             SUPERFICIALIS
FCU =   FLEXOR CARPI ULNARIS

(a) Anterior superficial view

(b) Anterior deep view

Figure continues

Figure 11.17 (continued)

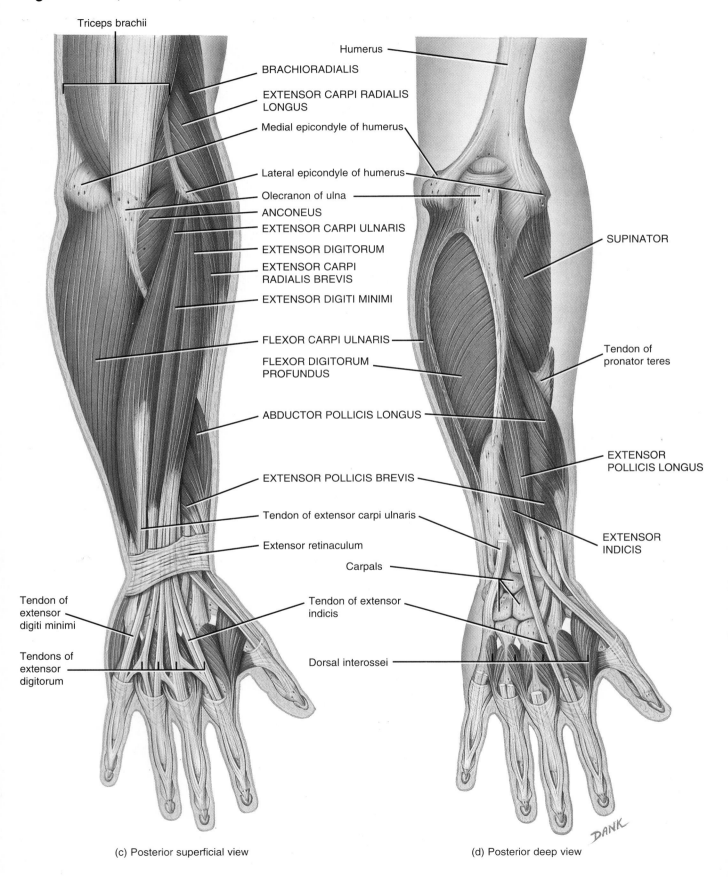

Triceps brachii

BRACHIORADIALIS

EXTENSOR CARPI RADIALIS LONGUS

Medial epicondyle of humerus

Lateral epicondyle of humerus

Olecranon of ulna

ANCONEUS

EXTENSOR CARPI ULNARIS

EXTENSOR DIGITORUM

EXTENSOR CARPI RADIALIS BREVIS

EXTENSOR DIGITI MINIMI

FLEXOR CARPI ULNARIS

FLEXOR DIGITORUM PROFUNDUS

ABDUCTOR POLLICIS LONGUS

EXTENSOR POLLICIS BREVIS

Tendon of extensor carpi ulnaris

Extensor retinaculum

Tendon of extensor digiti minimi

Tendons of extensor digitorum

Humerus

SUPINATOR

Tendon of pronator teres

EXTENSOR POLLICIS LONGUS

EXTENSOR INDICIS

Tendon of extensor indicis

Carpals

Dorsal interossei

(c) Posterior superficial view

(d) Posterior deep view

DANK

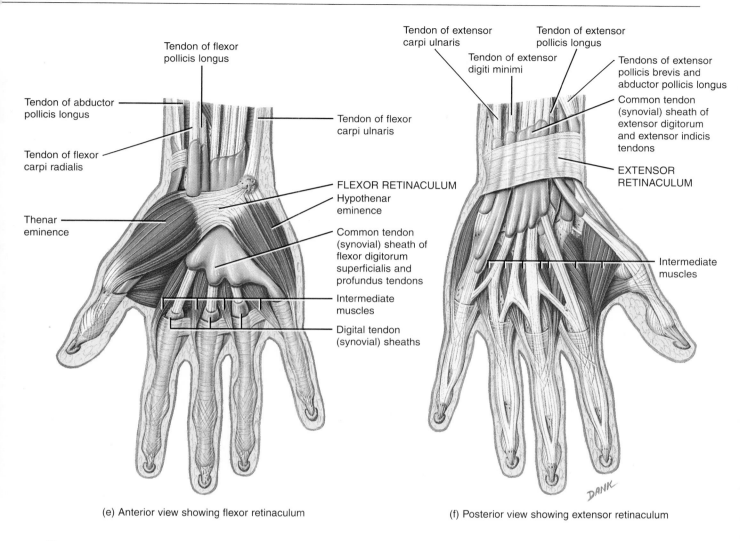

Tendon of flexor pollicis longus

Tendon of abductor pollicis longus

Tendon of flexor carpi radialis

Thenar eminence

Tendon of flexor carpi ulnaris

FLEXOR RETINACULUM
Hypothenar eminence

Common tendon (synovial) sheath of flexor digitorum superficialis and profundus tendons

Intermediate muscles

Digital tendon (synovial) sheaths

Tendon of extensor carpi ulnaris

Tendon of extensor digiti minimi

Tendon of extensor pollicis longus

Tendons of extensor pollicis brevis and abductor pollicis longus

Common tendon (synovial) sheath of extensor digitorum and extensor indicis tendons

EXTENSOR RETINACULUM

Intermediate muscles

DANK

(e) Anterior view showing flexor retinaculum

(f) Posterior view showing extensor retinaculum

Q What passes through the flexor retinaculum?

EXHIBIT **11.18** **INTRINSIC MUSCLES OF THE HAND**

Overview: Several of the muscles discussed in Exhibit 11.17 help to move the digits in various ways. In addition, there are muscles in the palm called **intrinsic muscles** that help to move the digits. These muscles are so named because their origins and insertions are both within the hands. They assist in the intricate and precise movements that are characteristic of the human hand.

The intrinsic muscles of the hand are divided into three principal groups—thenar, hypothenar, and intermediate. The four **thenar muscles** act on the thumb and form the **thenar eminence**. The four **hypothenar muscles** act on the little finger and form the **hypothenar eminence**. The 11 **intermediate (midpalmar) muscles** act on all the digits, except the thumb (see Fig. 11.17e,f).

The functional importance of the hand is readily apparent when one considers that certain hand injuries can result in permanent disability. Most of the dexterity of the hand depends on movements of the thumb. The general activities of the hand are free motion, power grip (forcible movement of the fingers and thumb against the palm, as in squeezing), precision handling (a change in position of a handled object that requires exact control of finger and thumb positions, as in winding a watch or threading a needle), and pinch (compression between the thumb and index finger or between the thumb and first two fingers).

Movement of the thumb is very important in the precise activities of the hand. The five principal movements of the thumb, illustrated below, are flexion (movement of the thumb medially across the palm), extension (movement of the thumb laterally away from the palm), abduction (movement of the thumb in an anteroposterior plane away from the palm), adduction (movement of the thumb in an anteroposterior plane toward the palm), and opposition (movement of the thumb across the palm so that the tip of the thumb meets the tip of a finger). Opposition is the single most distinctive digital movement that gives humans and other primates the ability to precisely grasp and manipulate objects.

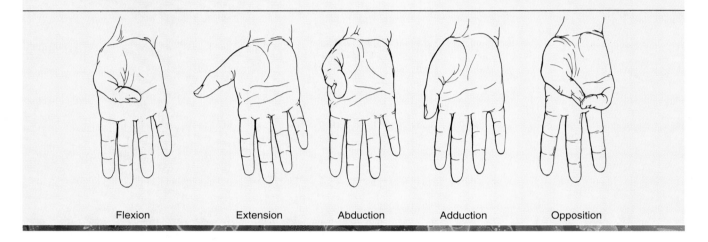

Flexion   Extension   Abduction   Adduction   Opposition

EXHIBIT **11.19** **MUSCLES THAT MOVE THE VERTEBRAL COLUMN (BACKBONE) (Fig. 11.18)**

Overview: The muscles that move the vertebral column (backbone) are quite complex because they have multiple origins and insertions and there is considerable overlap among them. One way to group the muscles is on the basis of the general direction of the muscle bundles and their approximate lengths. For example, the **splenius muscles** arise from the midline and extend laterally and superiorly to their insertions. The **erector spinae (sacrospinalis) muscle** arises from either the midline or more laterally but usually runs almost longitudinally, with neither a significant lateral nor medial direction as it is traced superiorly. The **transversospinalis muscles** arise laterally but extend toward the midline as they are traced superiorly. Deep to these three muscle groups are small **segmental muscles** that extend between spinous processes or transverse processes of vertebrae. Since the scalene muscles also assist in moving the vertebral column, they are included in this exhibit. Note in Exhibit 11.10 that the rectus abdominis, external oblique, internal oblique, and quadratus lumborum muscles also play a role in moving the vertebral column.

Muscle	Origin	Insertion	Action	Innervation
SPLENIUS				
Splenius capitis (SPLĒ-nē-us KAP-i-tis; *splenium* = bandage; *caput* = head)	Ligamentum nuchae and spinous processes of seventh cervical vertebra and first three or four thoracic vertebrae.	Occipital bone and mastoid process of temporal bone.	Acting together, they extend the head and neck; acting singly, each laterally flexes and rotates head to same side.	Dorsal rami of middle cervical nerves.
Splenius cervicis (SPLĒ-nē-us SER-vi-kis; *cervix* = neck)	Spinous processes of third through sixth thoracic vertebrae.	Transverse processes of first two or four cervical vertebrae.	Acting together, they extend the head and neck; acting singly, each laterally flexes and rotates head to same side.	Dorsal rami of lower cervical nerves.
ERECTOR SPINAE (e-REK-tor SPI-nē) (SACROSPINALIS)				
	This is the largest muscular mass of the back and consists of three groups of muscles—iliocostalis, longissimus, and spinalis. These groups, in turn, consist of a series of overlapping muscles. The iliocostalis group is laterally placed, the longissimus group is intermediate in placement, and the spinalis group is medially placed.			
ILIOCOSTALIS (LATERAL) GROUP				
Iliocostalis cervicis (il'-ē-ō-kos-TAL-is SER-vi-kis)	Superior six ribs.	Transverse processes of fourth to sixth cervical vertebrae.	Extends cervical region of vertebral column.	Dorsal rami of cervical nerves.
Iliocostalis thoracis (il'-ē-ō-kos-TAL-is thō-RA-kis; *thorax* = chest)	Inferior six ribs.	Superior six ribs.	Maintains erect position of spine.	Dorsal rami of thoracic (intercostal) nerves.
Iliocostalis lumborum (il'-ē-ō-kos-TAL-is lum-BOR-um; *ilium* = flank; *costa* = rib)	Iliac crest.	Inferior six ribs.	Extends lumbar region of vertebral column.	Dorsal rami of lumbar nerves.
LONGISSIMUS (INTERMEDIATE) GROUP				
Longissimus capitis (lon-JIS-i-mus KAP-i-tis)	Transverse processes of superior four thoracic vertebrae and articular processes of inferior four cervical vertebrae.	Mastoid process of temporal bone.	Extends head and rotates it to same side.	Dorsal rami of middle and lower cervical nerves.
Longissimus cervicis (lon-JIS-i-mus SER-vi-kis)	Transverse processes of fourth and fifth thoracic vertebrae.	Transverse processes of second to sixth cervical vertebrae.	Extends cervical region of vertebral column.	Dorsal rami of spinal nerves.
Longissimus thoracis (lon-JIS-i-mus tho-RA-kis; *longissimus* = longest)	Transverse processes of lumbar vertebrae.	Transverse processes of all thoracic and superior lumbar vertebrae and ninth and tenth ribs.	Extends thoracic region of vertebral column.	Dorsal rami of spinal nerves.

Exhibit continues

EXHIBIT 11.19 MUSCLES THAT MOVE THE VERTEBRAL COLUMN (BACKBONE) (Fig. 11.18) (CONTINUED)

Muscle	Origin	Insertion	Action	Innervation
SPINALIS (MEDIAL) GROUP				
Spinalis capitis (spi-NA-lis KAP-i-tis)	Arises with semispinalis capitis.	Inserts with semispinalis capitis.	Extends vertebral column.	Dorsal rami of spinal nerves.
Spinalis cervicis (spi-NA-lis SER-vi-kis)	Ligamentum nuchae and spinous process of seventh cervical vertebra.	Spinous process of axis.	Extends vertebral column.	Dorsal rami of spinal nerves.
Spinalis thoracis (spi-NA-lis thō-RA-kis; *spinalis* = vertebral column)	Spinous processes of superior lumbar and inferior thoracic vertebrae.	Spinous processes of superior thoracic vertebrae.	Extends vertebral column.	Dorsal rami of spinal nerves.
TRANSVERSOSPINALIS (trans-ver'-sō-spi-NA-lis)				
Semispinalis capitis (sem'-ē-spi-NA-lis KAP-i-tis)	Transverse processes of first six or seven thoracic vertebrae and seventh cervical vertebra and articular processes of fourth, fifth, and sixth cervical vertebrae.	Occipital bone.	Extends vertebral column and rotates it to opposite side.	Dorsal rami of cervical nerves.
Semispinalis cervicis (sem'-ē-spi-NA-lis SER-vi-kis)	Transverse processes of superior five or six thoracic vertebrae.	Spinous processes of first to fifth cervical vertebrae.	Extends vertebral column and rotates it to opposite side.	Dorsal rami of thoracic and cervical spinal nerves.
Semispinalis thoracis (sem'-ē-spi-NA-lis thō-RA-kis; *semi* = partially or one-half)	Transverse processes of sixth to tenth thoracic vertebrae.	Spinous processes of superior four thoracic and last two cervical vertebrae.	Extends vertebral column and rotates it to opposite side.	Dorsal rami of thoracic and cervical spinal nerves.
Multifidus (mul-TIF-i-dus; *multi* = many; *findere* = to split)	Sacrum, ilium, transverse processes of lumbar, thoracic, and inferior four cervical vertebrae.	Spinous process of a more superior vertebra.	Extends vertebral column and rotates it to opposite side.	Dorsal rami of spinal nerves.
Rotatores (rō'-ta-TŌ-rez; *rotare* = to turn)	Transverse processes of all vertebrae.	Spinous process of vertebra superior to the one of origin.	Extend vertebral column and rotate it to opposite side.	Dorsal rami of spinal nerves.
SEGMENTAL (seg-MEN-tal)				
Interspinales (in-ter-SPĪ-nāl-ez; *inter* = between)	Superior surface of all spinous processes.	Inferior surface of spinous process of vertebra superior to the one of origin.	Extends vertebral column.	Dorsal rami of spinal nerves.
Intertransversarii (in'-ter-trans-vers-AR-ē-ī; *inter* = between)	Transverse processes of all vertebrae.	Transverse process of vertebra superior to the one of origin.	Laterally flexes vertebral column.	Dorsal and ventral rami of spinal nerves.
SCALENE (SKĀ-lēn)				
Anterior scalene (SKĀ-lēn; *anterior* = front; *skalenos* = uneven)	Transverse processes of third through sixth cervical vertebrae.	First rib.	Flexes and rotates neck and assists in inspiration.	Ventral rami of fifth and sixth cervical nerves.
Middle scalene (SKĀ-lēn)	Transverse processes of inferior six cervical vertebrae.	First rib.	Flexes and rotates neck and assists in inspiration.	Ventral rami of third through eighth cervical nerves.
Posterior scalene (SKĀ-lēn)	Transverse processes of fourth through sixth cervical vertebrae.	Second rib.	Flexes and rotates neck and assists in inspiration.	Ventral rami of sixth through eighth cervical nerves.

Figure 11.18 Muscles that move the vertebral column (backbone).

 The erector spinae group of muscles is the largest muscular mass of the back.

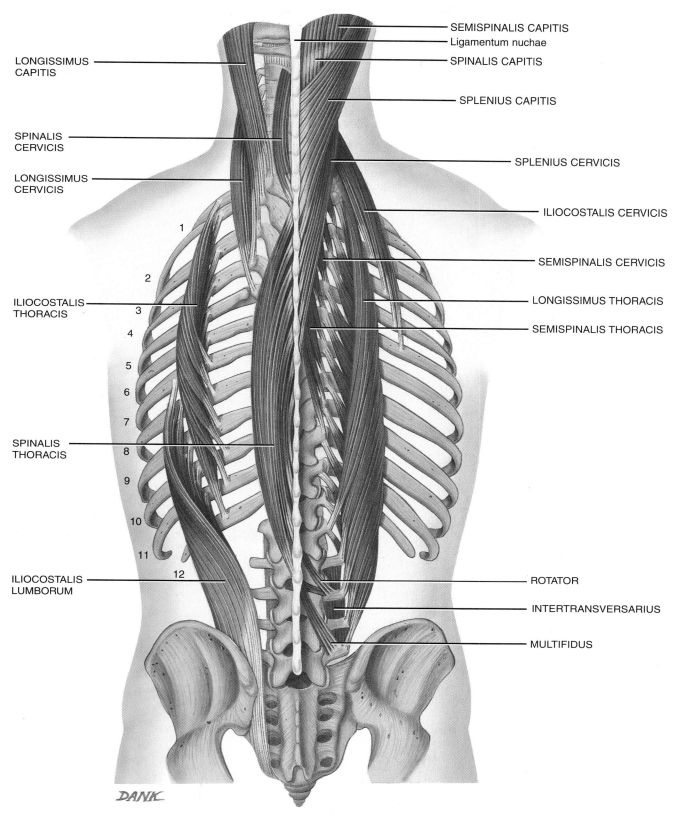

LONGISSIMUS CAPITIS

SPINALIS CERVICIS

LONGISSIMUS CERVICIS

ILIOCOSTALIS THORACIS

SPINALIS THORACIS

ILIOCOSTALIS LUMBORUM

SEMISPINALIS CAPITIS
Ligamentum nuchae
SPINALIS CAPITIS

SPLENIUS CAPITIS

SPLENIUS CERVICIS

ILIOCOSTALIS CERVICIS

SEMISPINALIS CERVICIS

LONGISSIMUS THORACIS

SEMISPINALIS THORACIS

ROTATOR

INTERTRANSVERSARIUS

MULTIFIDUS

DANK

(a) Posterior view

Figure continues

Figure 11.18   (continued)

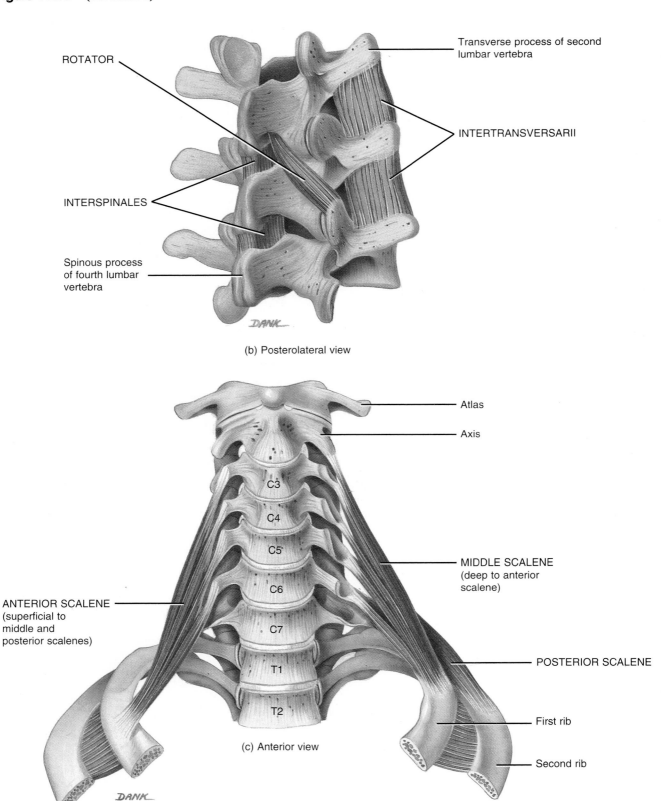

(b) Posterolateral view

(c) Anterior view

Q Which group of muscles originates from the midline and runs laterally and superiorly to its insertions?

EXHIBIT **11.20 MUSCLES THAT MOVE THE FEMUR (THIGH) (Fig. 11.19)**

Overview: As you will see, muscles of the lower limbs are larger and more powerful than those of the upper limbs since lower limb muscles function in stability, locomotion, and maintenance of posture. Upper limb muscles are characterized by versatility of movement. In addition, muscles of the lower limbs often cross two joints and act equally on both.

The anterior muscles are the psoas major and iliacus, together referred to as the iliopsoas (il'-ē-ō-SŌ-as) muscle. The remaining muscles (except for the pectineus, adductors, and tensor fasciae latae) are posterior muscles. Technically, the pectineus and adductors are components of the medial compartment of the thigh, but they are included in this exhibit because they act on the thigh. The tensor fasciae latae muscle is laterally placed. The **fascia lata** is a deep fascia of the thigh that encircles the entire thigh. It is well developed laterally, where together with the tendons of the gluteus maximus and tensor fasciae latae muscles it forms a structure called the **iliotibial tract**. The tract inserts on the margin of the lateral condyle of the tibia.

After you have studied the muscles in this exhibit, arrange them according to the following actions: flexion, extension, abduction, adduction, medial rotation, and lateral rotation. (The same muscles can be used more than once.)

Muscle	Origin	Insertion	Action	Innervation[a]
Psoas (SŌ-as) **major** (*psoa* = muscle of loin)	Transverse processes and bodies of lumbar vertebrae.	Lesser trochanter of femur.	Flexes and rotates thigh laterally; flexes vertebral column.	Lumbar nerves L2–L3.
Iliacus (il'-ē-AK-us; *iliac* = ilium)	Iliac fossa.	Tendon of psoas major.	Flexes and rotates thigh laterally and flexes vertebral column.	Femoral nerve.
Gluteus maximus (GLOO-tē-us MAK-si-mus; *glutos* = buttock; *maximus* = largest; strongest single muscle in body)	Iliac crest, sacrum, coccyx, and aponeurosis of sacrospinalis.	Iliotibial tract of fascia lata and lateral part of linea aspera inferior to greater trochanter (gluteal tuberosity) of femur.	Extends and rotates thigh laterally.	Inferior gluteal nerve.
Gluteus medius (GLOO-tē-us MĒ-dē-us; *media* = middle)	Ilium.	Greater trochanter of femur.	Abducts and rotates thigh medially.	Superior gluteal nerve.
Gluteus minimus (GLOO-tē-us MIN-i-mus; *minimus* = smallest)	Ilium.	Greater trochanter of femur.	Abducts and rotates thigh medially.	Superior gluteal nerve.
Tensor fasciae latae (TEN-sor FA-shē-ē LĀ-tē; *tensor* = makes tense; *fascia* = band; *latus* = wide)	Iliac crest.	Tibia by way of the iliotibial tract.	Flexes and abducts thigh.	Superior gluteal nerve.
Piriformis (pir-i-FOR-mis; *pirum* = pear; *forma* = shape)	Anterior sacrum.	Superior border of greater trochanter of femur.	Rotates thigh laterally and abducts it.	Sacral nerves S1 or S2, mainly S1.
Obturator internus (OB-too-rā'-tor in-TER-nus; *obturator* = obturator foramen; *internus* = inside)	Inner surface of obturator foramen, pubis, and ischium.	Greater trochanter of femur.	Rotates thigh laterally and abducts it.	Nerve to obturator internus.
Obturator externus (OB-too-rā'-tor ex-TER-nus; *externus* = outside)	Outer surface of obturator membrane.	Deep depression inferior to greater trochanter (trochanteric fossa) of femur.	Rotates thigh laterally.	Obturator nerve.
Superior gemellus (jem-EL-lus; *superior* = above; *gemellus* = twins)	Ischial spine.	Greater trochanter of femur.	Rotates thigh laterally and abducts it.	Nerve to obturator internus.
Inferior gemellus (jem-EL-lus; *inferior* = below)	Ischial tuberosity.	Greater trochanter of femur.	Rotates thigh laterally and abducts it.	Nerve to quadratus femoris.

[a] The nerves listed here are derived from the lumbar and sacral plexuses, which are described and illustrated in Exhibits 13.3 and 13.4 on pages 382–385.

Exhibit continues

EXHIBIT **11.20   MUSCLES THAT MOVE THE THIGH (FEMUR) (Fig. 11.19)   (CONTINUED)**

Muscle	Origin	Insertion	Action	Innervation
Quadratus femoris (kwod-RĀ-tus FEM-or-is; *quad* = four; *femoris* = femur)	Ischial tuberosity.	Elevation superior to midportion of intertrochanteric crest (quadrate tubercle) on posterior femur.	Laterally rotates and adducts thigh.	Nerve to quadratus femoris.
Adductor longus (LONG-us; *adductor* = moves part closer to midline; *longus* = long)	Pubic crest and pubic symphysis.	Linea aspera of femur.	Adducts, medially rotates, and flexes thigh.	Obturator nerve.
Adductor brevis (BREV-is; *brevis* = short)	Inferior ramus of pubis.	Superior half of linea aspera of femur.	Adducts, medially rotates, and flexes thigh.	Obturator nerve.
Adductor magnus (MAGnus; *magnus* = large)	Inferior ramus of pubis and ischium to ischial tuberosity.	Linea aspera of femur.	Adducts and medially rotates thigh, anterior part flexes and posterior part extends thigh.	Obturator and sciatic nerves.
Pectineus (pek-TIN-ē-us; *pecten* = comb-shaped)	Superior ramus of pubis.	Pectineal line of femur, between lesser trochanter and linea aspera.	Flexes and adducts thigh.	Femoral nerve.

Figure 11.19 Muscles that move the femur (thigh).

 Most muscles that act on the thigh originate on the pelvic (hip) girdle and insert on the femur.

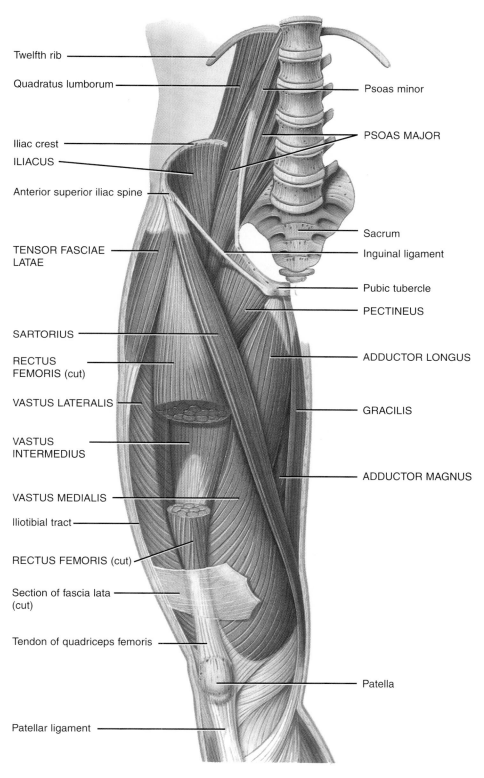

Twelfth rib

Quadratus lumborum

Psoas minor

PSOAS MAJOR

Iliac crest

ILIACUS

Anterior superior iliac spine

Sacrum

TENSOR FASCIAE LATAE

Inguinal ligament

Pubic tubercle

PECTINEUS

SARTORIUS

RECTUS FEMORIS (cut)

ADDUCTOR LONGUS

VASTUS LATERALIS

GRACILIS

VASTUS INTERMEDIUS

ADDUCTOR MAGNUS

VASTUS MEDIALIS

Iliotibial tract

RECTUS FEMORIS (cut)

Section of fascia lata (cut)

Tendon of quadriceps femoris

Patella

Patellar ligament

(a) Anterior superficial view

Figure continues

Figure 11.19   (continued)

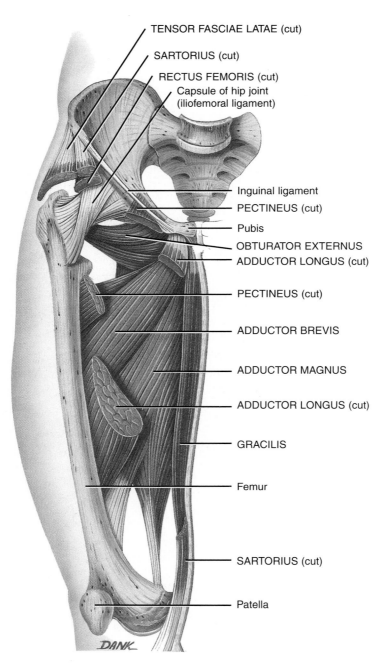

TENSOR FASCIAE LATAE (cut)

SARTORIUS (cut)

RECTUS FEMORIS (cut)

Capsule of hip joint
(iliofemoral ligament)

Inguinal ligament

PECTINEUS (cut)

Pubis

OBTURATOR EXTERNUS

ADDUCTOR LONGUS (cut)

PECTINEUS (cut)

ADDUCTOR BREVIS

ADDUCTOR MAGNUS

ADDUCTOR LONGUS (cut)

GRACILIS

Femur

SARTORIUS (cut)

Patella

DANK

(b) Anterior deep view (femur rotated laterally)

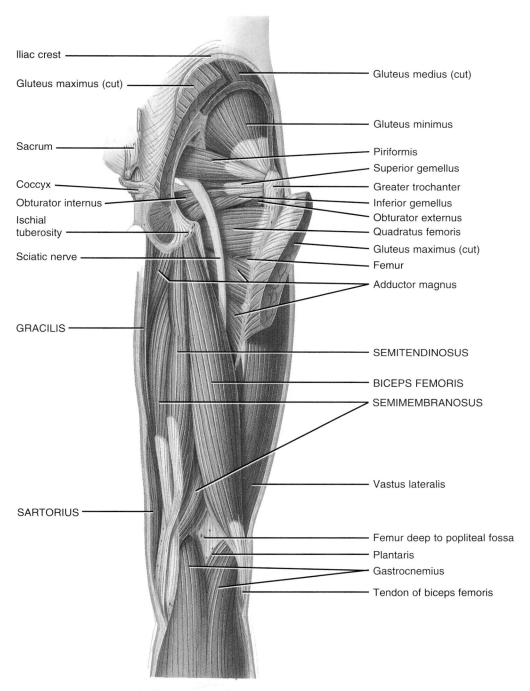

Iliac crest

Gluteus maximus (cut)

Sacrum

Coccyx

Obturator internus

Ischial tuberosity

Sciatic nerve

GRACILIS

SARTORIUS

Gluteus medius (cut)

Gluteus minimus

Piriformis

Superior gemellus

Greater trochanter

Inferior gemellus

Obturator externus

Quadratus femoris

Gluteus maximus (cut)

Femur

Adductor magnus

SEMITENDINOSUS

BICEPS FEMORIS

SEMIMEMBRANOSUS

Vastus lateralis

Femur deep to popliteal fossa

Plantaris

Gastrocnemius

Tendon of biceps femoris

(c) Posterior superficial view

Figure continues

Figure 11.19   (continued)

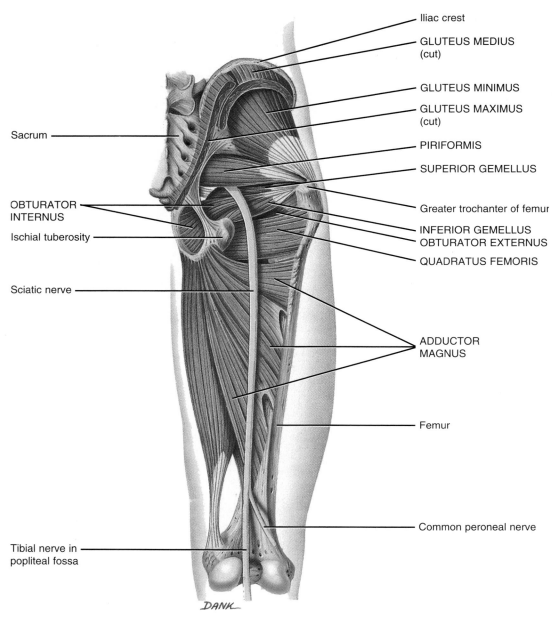

Iliac crest

GLUTEUS MEDIUS (cut)

GLUTEUS MINIMUS

GLUTEUS MAXIMUS (cut)

PIRIFORMIS

SUPERIOR GEMELLUS

Greater trochanter of femur

INFERIOR GEMELLUS
OBTURATOR EXTERNUS

QUADRATUS FEMORIS

ADDUCTOR MAGNUS

Femur

Common peroneal nerve

Sacrum

OBTURATOR INTERNUS

Ischial tuberosity

Sciatic nerve

Tibial nerve in popliteal fossa

DANK

(d) Posterior deeper view

Q   What are the general functions of the medial, anterior, and posterior compartment muscles? (See Overview in Exhibit 11.21.)

EXHIBIT **11.21**   **MUSCLES THAT ACT ON THE TIBIA AND FIBULA (LEG) (Figs. 11.19 and 11.20)**

Overview: The muscles that act on the tibia and fibula (leg) are separated into compartments by deep fascia. The **medial (adductor) compartment** is so named because its muscles adduct the thigh. It is innervated by the obturator nerve. As noted earlier, the adductor magnus, adductor longus, adductor brevis, and pectineus muscles, components of the medial compartment, are included in Exhibit 11.19 because they act on the femur. The gracilis, the other muscle in the medial compartment, not only adducts the thigh but also flexes the leg. For this reason, it is included in this exhibit.

The **anterior (extensor) compartment** is so designated because its muscles act to extend the leg, and some also flex the thigh. It is composed of the quadriceps femoris and sartorius muscles and is innervated by the femoral nerve. The quadriceps femoris muscle is a composite muscle that includes four distinct parts, usually described as four separate muscles (rectus femoris, vastus lateralis, vastus intermedius, and vastus medialis). The common tendon for the four muscles is known as the **quadriceps tendon**, which attaches to the patella. The tendon continues inferior to the patella as the **patellar ligament**, which attaches to the tibial tuberosity. The rectus femoris and sartorius muscles are also flexors of the thigh.

The **posterior (flexor) compartment** is so named because its muscles flex the leg (but also extend the thigh). It is innervated by branches of the sciatic nerve. Included are the hamstrings (biceps femoris, semitendinosus, and semimembranosus). The hamstrings are so named because their tendons are long and stringlike in the popliteal area and from an old practice of butchers in which they hung hams for smoking by these long tendons. The **popliteal fossa** is a diamond-shaped space on the posterior aspect of the knee bordered laterally by the tendons of the biceps femoris and medially by the semitendinosus and semimembranosus muscles.

Muscle	Origin	Insertion	Action	Innervation[a]
MEDIAL (ADDUCTOR) COMPARTMENT				
Adductor magnus (MAG-nus)				
Adductor longus (LONG-us)	See Exhibit 11.20 (page 316)			
Adductor brevis (BREV-is)				
Pectineus (pek-TIN-ē-us)				
Gracilis (gra-SIL-is; *gracilis* = slender)	Pubic symphysis and pubic arch.	Medial surface of body of tibia.	Adducts thigh and flexes leg.	Obturator nerve.
ANTERIOR (EXTENSOR) COMPARTMENT				
Quadriceps femoris (KWOD-ri-ceps FEM-or-is; *quadriceps* = four heads of origin; *femoris* = femur)				
Rectus femoris (REK-tus FEM-or-is; *rectus* = fibers parallel to midline)	Anterior inferior iliac spine.	Patella via quadriceps tendon and then tibial tuberosity via patellar ligament.	All four heads extend leg; rectus portion alone also flexes thigh.	Femoral nerve.
Vastus lateralis (VAS-tus lat'-er-A-lis; *vastus* = large; *lateralis* = lateral)	Greater trochanter and linea aspera of femur.			Femoral nerve.
Vastus medialis (VAS-tus mē'-dē-A-lis; *medialis* = medial)	Linea aspera of femur.			Femoral nerve.
Vastus intermedius (VAS-tus in'-ter-ME̅-dē-us; *intermedius* = middle)	Anterior and lateral surfaces of body of femur.			Femoral nerve.
Sartorius (sar-TOR-ē-us; *sartor* = tailor; contracts when you sit in the cross-legged position of a tailor; longest muscle in body)	Anterior superior iliac spine.	Medial surface of body of tibia.	Flexes leg; flexes thigh and rotates it laterally, thus crossing leg.	Femoral nerve.
POSTERIOR (FLEXOR) COMPARTMENT				
Hamstrings	A collective designation for three separate muscles.			
Biceps femoris (BI̅-ceps FEM-or-is; *biceps* = two heads of origin)	Long head arises from ischial tuberosity; short head arises from linea aspera of femur.	Head of fibula and lateral condyle of tibia.	Flexes leg and extends thigh.	Tibial and common peroneal nerves from sciatic nerve.
Semitendinosus (sem'-ē-TEN-di-nō-sus; *semi* = half; *tendo* = tendon)	Ischial tuberosity.	Proximal part of medial surface of shaft of tibia.	Flexes leg and extends thigh.	Tibial nerve from sciatic nerve.
Semimembranosus (sem'-ē-MEM-bra-nō-sus; *membran* = membrane)	Ischial tuberosity.	Medial condyle of tibia.	Flexes leg and extends thigh.	Tibial nerve from sciatic nerve.

[a] The obturator and femoral nerves are derived from the lumbar plexus (see Exhibit 13.3 on page 382) whereas the sciatic nerve branches from the sacral plexus (see Exhibit 13.4 on page 384).

Figure 11.20   Muscles that act on the tibia and fibula (leg). The arrow in the inset indicates the direction from which the thigh is viewed (superior).

The muscles that act on the leg originate in the hip and thigh and are separated into compartments by deep fascia.

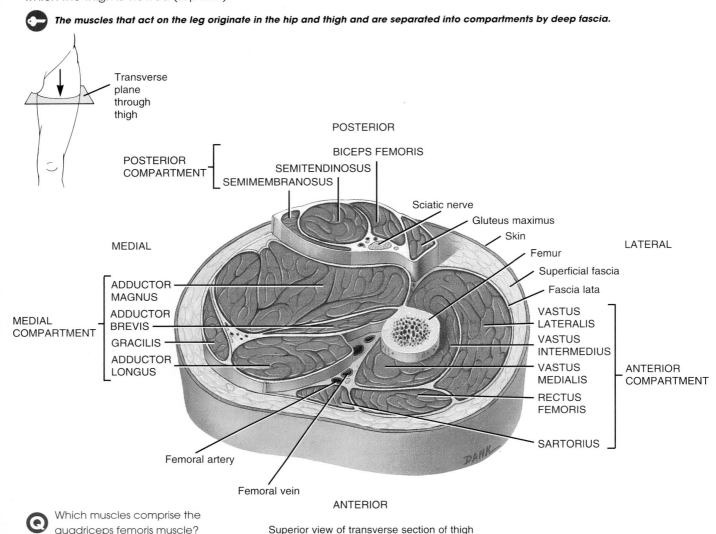

Which muscles comprise the quadriceps femoris muscle?

Superior view of transverse section of thigh

EXHIBIT **11.22 MUSCLES THAT MOVE THE FOOT AND TOES (Fig. 11.21)**

Overview: The musculature of the leg, like that of the thigh, is divided into three compartments by deep fascia. In addition, all the muscles in a given compartment are innervated by the same nerve. The **anterior compartment** consists of muscles that dorsiflex the foot and are innervated by the deep peroneal nerve. In a situation analogous to the wrist, the tendons of the muscles of the anterior compartment are held firmly to the ankle by thickenings of deep fascia called the **superior extensor retinaculum (transverse ligament of the ankle)** and **inferior extensor retinaculum (cruciate ligament of the ankle)**.

The **lateral (peroneal) compartment** contains two muscles that plantar flex and evert the foot. They are supplied by the superficial peroneal nerve.

The **posterior compartment** consists of muscles that are divisible into superficial and deep groups. All are innervated by the tibial nerve. The superficial muscles are plantar flexors of the foot. Of the four deep muscles, three plantar flex the foot.

The calcaneal (Achilles) tendon, the **strongest tendon of the body**, is able to withstand a 1000-pound force without tearing. Despite this, however, the calcaneal tendon ruptures more frequently than any other tendon because of the tremendous pressures placed on it during competitive sports.

Muscle	Origin	Insertion	Action	Innervation[a]
ANTERIOR COMPARTMENT				
Tibialis anterior (tib′-ē-A-lis; *tibialis* = tibia; *anterior* = front)	Lateral condyle and body of tibia and interosseous membrane (sheet of fibrous tissue that holds shafts of tibia and fibula together).	First metatarsal and first (medial) cuneiform.	Dorsiflexes and inverts foot.	Deep peroneal nerve.
Extensor hallucis longus (HAL-a-kis LON-gus; *extensor* = increases angle at joint; *hallucis* = hallux or great toe; *longus* = long)	Anterior surface of fibula and interosseous membrane.	Distal phalanx of great toe.	Dorsiflexes and inverts foot and extends great toe.	Deep peroneal nerve.
Extensor digitorum longus (di′-ji-TOR-um LON-gus)	Lateral condyle of tibia, anterior surface of fibula, and interosseous membrane.	Middle and distal phalanges of four outer toes.	Dorsiflexes and everts foot and extends toes.	Deep peroneal nerve.
Peroneus tertius (per′-ō-NĒ-us TER-shus; *perone* = fibula; *tertius* = third)	Distal third of fibula and interosseous membrane.	Fifth metatarsal.	Dorsiflexes and everts foot.	Deep peroneal nerve.
LATERAL (PERONEAL) COMPARTMENT				
Peroneus longus (per′-ō-NĒ-us LON-gus)	Head and body of fibula and lateral condyle of tibia.	First metatarsal and first cuneiform.	Plantar flexes and everts foot.	Superficial peroneal nerve.
Peroneus brevis (per′-ō-NĒ-us BREV-is; *brevis* = short)	Body of fibula.	Fifth metatarsal.	Plantar flexes and everts foot.	Superficial peroneal nerve.
POSTERIOR COMPARTMENT—SUPERFICIAL				
Gastrocnemius (gas′-trok-NĒ-mē-us; *gaster* = belly; *kneme* = leg)	Lateral and medial condyles of femur and posterior capsule of knee.	Calcaneus by way of calcaneal (Achilles) tendon.	Plantar flexes foot and flexes leg.	Tibial nerve.
Soleus (SŌ-lē-us; *soleus* = sole of foot)	Head of fibula and medial border of tibia.	Calcaneus by way of calcaneal (Achilles) tendon.	Plantar flexes foot.	Tibial nerve.
Plantaris (plan-TA-ris; *plantar* = sole of foot)	Femur superior to lateral condyle.	Calcaneus by way of calcaneal (Achilles) tendon.	Plantar flexes foot.	Tibial nerve.
POSTERIOR COMPARTMENT—DEEP				
Popliteus (pop-LIT-ē-us; *poples* = posterior surface of knee)	Lateral condyle of femur.	Proximal tibia.	Flexes and medially rotates leg.	Tibial nerve.
Flexor hallucis longus (HAL-a-kis LON-gus; *flexor* = decreases angle at joint)	Inferior two-thirds of fibula.	Distal phalanx of great toe.	Plantar flexes and inverts foot and flexes great toe.	Tibial nerve.
Flexor digitorum longus (di′-ji-TOR-um LON-gus; *digitorum* = finger or toe)	Posterior surface of tibia.	Distal phalanges of four outer toes.	Plantar flexes and inverts foot and flexes toes.	Tibial nerve.
Tibialis (tib′-ē-A-lis) **posterior** (*posterior* = back)	Tibia, fibula, and interosseous membrane.	Second, third, and fourth metatarsals; navicular; all three cuneiforms; and cuboid.	Plantar flexes and inverts foot.	Tibial nerve.

[a]The nerves listed here are derived from the sacral plexus, which is described and illustrated in Exhibit 13.4 on page 384.

Figure 11.21   Muscles that move the foot and toes.

The superficial muscles of the posterior compartment share a common tendon of insertion, the calcaneal (Achilles) tendon, that inserts into the calcaneus bone of the ankle.

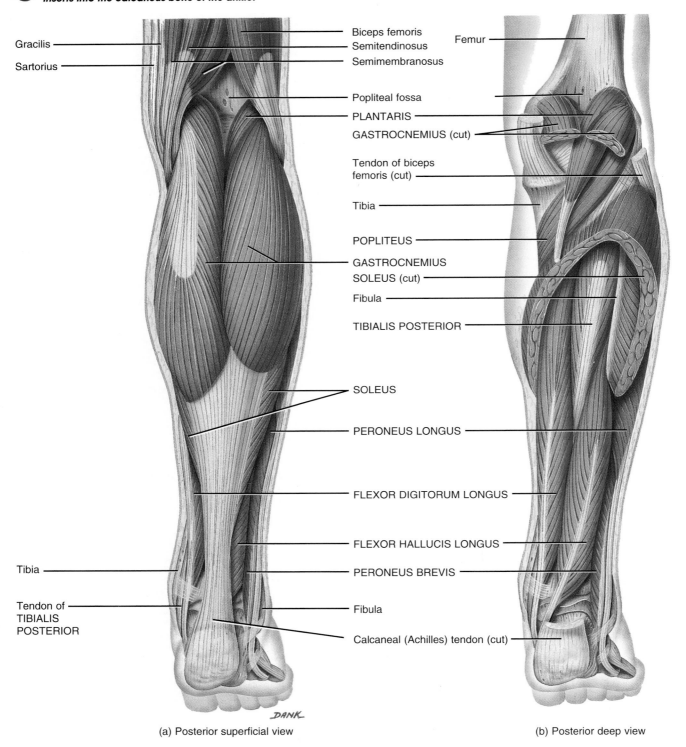

Gracilis

Sartorius

Biceps femoris

Semitendinosus

Semimembranosus

Femur

Popliteal fossa

PLANTARIS

GASTROCNEMIUS (cut)

Tendon of biceps femoris (cut)

Tibia

POPLITEUS

GASTROCNEMIUS

SOLEUS (cut)

Fibula

TIBIALIS POSTERIOR

SOLEUS

PERONEUS LONGUS

FLEXOR DIGITORUM LONGUS

FLEXOR HALLUCIS LONGUS

Tibia

PERONEUS BREVIS

Tendon of TIBIALIS POSTERIOR

Fibula

Calcaneal (Achilles) tendon (cut)

DANK

(a) Posterior superficial view

(b) Posterior deep view

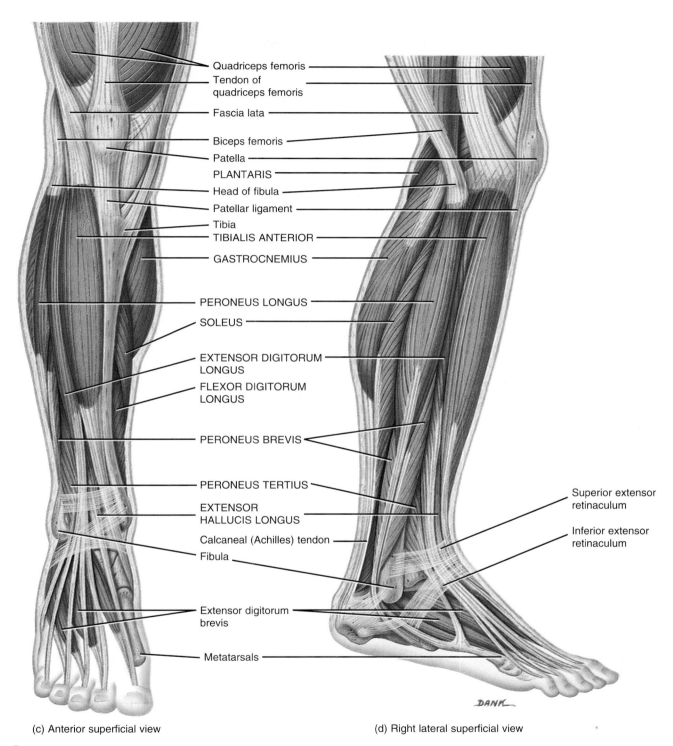

Quadriceps femoris

Tendon of quadriceps femoris

Fascia lata

Biceps femoris

Patella

PLANTARIS

Head of fibula

Patellar ligament

Tibia

TIBIALIS ANTERIOR

GASTROCNEMIUS

PERONEUS LONGUS

SOLEUS

EXTENSOR DIGITORUM LONGUS

FLEXOR DIGITORUM LONGUS

PERONEUS BREVIS

PERONEUS TERTIUS

EXTENSOR HALLUCIS LONGUS

Calcaneal (Achilles) tendon

Fibula

Extensor digitorum brevis

Metatarsals

Superior extensor retinaculum

Inferior extensor retinaculum

DANK

(c) Anterior superficial view

(d) Right lateral superficial view

Q The muscles in which compartment dorsiflex the foot?

INTRAMUSCULAR INJECTIONS

An **intramuscular (IM) injection** penetrates the skin and subcutaneous layer to enter the muscle itself. The common sites for intramuscular injections include the buttock, lateral side of the thigh, and the deltoid region of the arm (Fig. 11.22). Muscles in these areas are fairly thick and absorption is promoted by the extensive blood supply to such large muscles. To avoid injury, intramuscular injections are given deep within the muscle and away from major nerves and blood vessels.

For many intramuscular injections, the preferred site is the **gluteus medius muscle** (ventral gluteal site) of the buttock (Fig. 11.22a). The buttock is divided into quadrants, and the upper outer quadrant is used as the injection site. The iliac crest serves as a landmark for this quadrant. The upper outer quadrant is chosen because the muscle in this area is quite thick and has few nerves. Injection in this area thus reduces the chance of injury to the sciatic nerve, which could cause paralysis of the lower limb. The probability of injecting the drug into a blood vessel is also remote in this area. After the needle is inserted into the gluteus medius muscle, the plunger is pulled up for a few seconds. If the syringe fills with blood, the needle is in a blood vessel, and a different injection site on the opposite buttock is chosen. Injections may also be given in the lateral side of the thigh in the midportion of the **vastus lateralis muscle** (Fig. 11.22b). This site is determined by using the knee and greater trochanter of the femur as landmarks. A **deltoid** injection is given in the midportion of the muscle about two to three fingerbreadths below the acromion of the scapula and lateral to the axilla (Fig. 11.22c). ∎

RUNNING INJURIES

It is estimated that nearly 70% of those who jog or run will sustain some type of running-related injury. Even though most such injuries are minor, such as sprains and strains, some are quite serious; moreover, untreated or inappropriately treated minor injuries may become chronic. Among runners, the knee is the most common site of injury, accounting for about 40% of injuries. Other common sites of injury include the calcaneal (Achilles) tendon, medial aspect of the tibia, hip area, groin area, foot and ankle, and back.

Running injuries are frequently related to faulty training techniques. These include improper or lack of warm-up, running too much, or running too soon. Or it might involve running on hard and/or uneven surfaces. Poorly constructed or worn-out running shoes can also contribute to injury. Any biomechanical problems aggravated by running can also cause injuries.

TREATING SPORTS INJURIES

Most sports injuries should be treated initially with **RICE** therapy, which stands for **R**est, **I**ce, **C**ompression, and **E**levation. Immediately apply ice and rest and elevate the injured part. Then apply an elastic bandage, if possible, to compress the injured tissue. Continue using RICE for 2–3 days, and resist the temptation to apply heat, which may worsen the swelling. Follow-up treatment may include alternating moist heat and ice massage to enhance blood flow in the injured area. Sometimes, nonsteroidal anti-inflammatory drugs (NSAIDs), such as aspirin or ibuprofen, or local injections of corticosteroids, such as cortisone, are needed. During the recovery period, it is important to keep active with an alternate fitness program. And careful exercise is needed to rehabilitate the injured area itself. ∎

Hip, Buttock, and Back Injuries

When back pain occurs in runners, it is often due to a preexisting degenerative condition that is aggravated by an increase in mileage or hill running. Such pain is usually due to an injury in the buttocks, pelvis, or lumbar spine. For example, the pain may be caused by strain of the distal attachments of the abductors (especially the gluteus medius) or the proximal attachment of the abductors to the iliac crest, both of which contribute to a "**pulled groin**" or strain or partial tear of the proximal hamstrings (**hamstring strain** or "**pulled hamstrings**"). Hamstring strains are common sports injuries in individuals who run very hard. Sometimes the violent muscular exertion required to perform a feat tears off part of the tendinous origins of the hamstrings, especially the biceps femoris, from the ischial tuberosity. This is usually accompanied by a contusion (bruising) and tearing of some of the muscle fibers and rupture of blood vessels, producing a hematoma (collection of blood) and pain. Adequate training with good balance between the quadriceps femoris and hamstrings and stretching exercises before running or competing are important in preventing this injury.

Knee Injuries

Patellofemoral stress syndrome ("**runner's knee**") is the single most common problem in runners. During normal flexion and extension of the knee, the patella tracks (glides) up and down in the groove between the femoral condyles. In patellofemoral stress syndrome, normal tracking does not occur; instead, the patella tracks laterally, and the increased pressure of abnormal tracking causes the associated pain. The pain is usually described as an aching or tenderness around or under the patella. The pain typically occurs after a person has been sitting for a while, especially after exercise. A common cause of runner's knee is constantly walking, running, or jogging on the same side of the road. Since roads are high in the middle and slope down on the sides, the slope stresses the knee that is closer to the center of the road.

Besides RICE, treatment of patellofemoral stress syndrome includes cessation of running; avoidance of kneeling, stair climbing, and prolonged sitting; hamstring muscle stretching exercises; short-arc quadriceps strengthening exercises; and use of an orthotic device for the foot, knee wraps, and knee braces.

Figure 11.22   Intramuscular injections. Shown are the three most common sites for intramuscular injections.

 An intramuscular injection penetrates the skin and subcutaneous layer to enter a muscle.

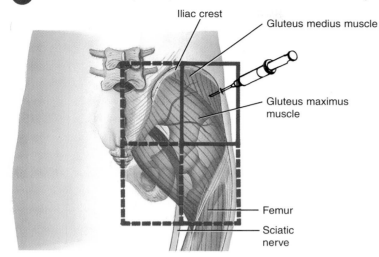

(a) Buttock

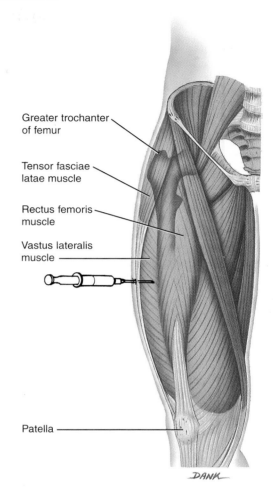

(b) Lateral surface of thigh

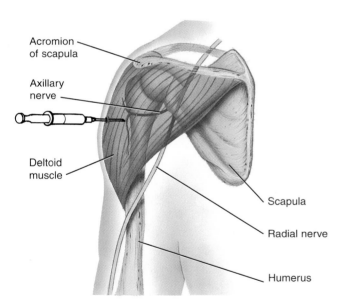

(c) Deltoid region of arm

 What structures should be avoided when giving an IM injection?

Leg and Foot Injuries

Shinsplint Syndrome (Shinsplints)

Shinsplint syndrome, or simply **shinsplints**, refers to pain or soreness along the tibia, specifically the medial, distal two-thirds. It may be caused by tendinitis of the tibialis posterior muscle or toe flexors, inflammation of the periosteum (periostitis) around the tibia, or stress fractures of the tibia. The tendinitis usually occurs when poorly conditioned runners run on hard or banked surfaces with poorly supportive running shoes. The condition may also occur as a result of vigorous activity of the legs following a period of relative inactivity. The muscles in the anterior compartment (mainly the tibialis anterior) can be strengthened to balance the stronger posterior compartment muscles. Patients who do not respond to RICE may be given local injections of corticosteroids or may have to undergo minor surgery to release pressure in the soft tissues around the bone.

Plantar Fasciitis (Painful Heel Syndrome)

Plantar fasciitis (**painful heel syndrome**) is an inflammatory reaction due to chronic irritation of the plantar aponeurosis (fascia) at its origin on the calcaneus (heel bone). The plantar aponeurosis is a deep fascia that extends from the calcaneus to the phalanges. It supports the longitudinal arch of the foot and encloses the flexor tendons of the foot. The condition is the most common cause of heel pain in runners and arises in response to the repeated impact of walking or running. Treatment consists of RICE, heel pads, exercises that gently stretch foot muscles, NSAIDs, and, in some cases, corticosteroid injections.

Stress Fractures

Stress fractures are partial fractures that result from inability to withstand repeated stress owing to a change in training, harder surfaces, longer distances, greater speed, or an existing pathology. Such fractures can occur in the bodies of lumbar vertebrae, sacroiliac joint, pubic symphysis, iliac crest, femoral neck and body, fibula, lateral malleolus, and metatarsals. About 25% of all stress fractures involve the fibula, specifically the distal third. With all stress fractures, running must stop temporarily and immobilization may be needed.

STUDY OUTLINE

NAMING SKELETAL MUSCLES (p. 270)

1. Skeletal muscles are named on the basis of distinctive criteria: direction of fibers, location, size, number of origins (or heads), shape, origin and insertion, and action.

HOW SKELETAL MUSCLES PRODUCE MOVEMENT (p. 270)

1. Skeletal muscles produce movement by pulling on bones.
2. The attachment to the stationary bone is the origin. The attachment to the movable bone is the insertion.
3. Bones serve as levers and joints serve as fulcrums. The lever is acted on by two different forces: resistance and effort.
4. Levers are categorized into three types—first-class, second-class, and third-class (most common)—according to the position of the fulcrum, effort, and resistance on the lever.
5. The prime mover produces the desired action. The antagonist produces an opposite action. The synergist assists the prime mover by reducing unnecessary movement. The fixator stabilizes the origin of the prime mover so that it can act more efficiently.

6. Fascicular arrangements include parallel, fusiform, pennate, and circular. Fascicular arrangement is correlated with the power of a muscle and the range of motion.

PRINCIPAL SKELETAL MUSCLES (p. 274)

1. The principal skeletal muscles of the body are grouped according to region in Exhibits 11.3–11.23.

INTRAMUSCULAR INJECTIONS (p. 326)

1. Advantages of intramuscular injections are prompt absorption, use of larger doses than can be given subcutaneously, and minimal irritation.
2. Common sites for intramuscular injections are the buttock, lateral side of the thigh, and deltoid region of the arm.

RUNNING INJURIES (p. 326)

1. Most running injuries involve the knee. Other commonly injured sites are the calcaneal (Achilles) tendon, medial aspect of tibia, hip, groin, foot, ankle, and back.
2. Running injuries are initially treated by rest, ice, compression, and elevation (RICE). Nonsteroidal anti-inflammatory drugs (NSAIDs), moist heat, and an alternate fitness program may also be included.

REVIEW QUESTIONS

1. What is meant by the muscular system? (p. 270)
2. Select at random several muscles presented in Exhibits 11.3–11.23 and see if you can determine the criterion or criteria employed for naming each. In addition, refer to the prefixes, suffixes, roots, and definitions in each exhibit as a guide. Select as many muscles as you wish, as long as you feel you understand the concept involved.
3. Using the terms origin, insertion, and belly in your discussion, describe how skeletal muscles produce body movements by pulling on bones. (p. 271)
4. What is a lever? Fulcrum? Apply these terms to the body and indicate the nature of the forces that act on levers. Describe the three classes of levers and provide one example for each in the body. (p. 271)
5. Define the role of the prime mover (agonist), antagonist, synergist, and fixator in producing body movements. (p. 273)

6. Describe the various arrangements of fascicles. How is fascicular arrangement correlated with the strength of a muscle and its range of motion? (p. 273)
7. What muscles would you use to do the following: (a) frown, (b) pout, (c) show surprise, (d) show your upper teeth, (e) pucker your lips, (f) squint, (g) blow up a balloon, (h) smile? (p. 277)
8. What are the principal muscles that move the mandible (lower jaw)? Give the function of each. (p. 280)
9. What would happen if you lost tone in the masseter and temporalis muscles? (p. 280)
10. What extrinsic muscles move the eyeballs? In which direction does each muscle move the eyeballs? (p. 282)
11. Describe the action of each of the muscles acting on the tongue. (p. 284)
12. What tongue, facial, and mandibular muscles would you use when chewing food? (pp. 277, 278, 280, 281, 284)

13. Describe the muscles involved, and their actions, in moving the hyoid bone. (p. 286)
14. Describe the actions of the extrinsic and intrinsic muscles of the larynx (voice box). (p. 287)
15. What muscles are responsible for moving the head? And how do they move the head? (p. 289)
16. What muscles would you use to signify "yes" and "no" by moving your head? (p. 289)
17. Describe the composition of the anterolateral and posterior abdominal wall. (p. 290)
18. What muscles accomplish compression of the abdominal wall? (p. 290)
19. What are the principal muscles involved in breathing? What are their actions? (p. 292)
20. Describe the actions of the muscles of the pelvic floor. What is the pelvic diaphragm? (p. 294)
21. Describe the actions of the muscles of the perineum. What is the urogenital diaphragm? (p. 295)
22. In what directions can the pectoral (shoulder) girdle be drawn? What muscles accomplish these movements? (p. 297)
23. What muscles are used to raise your shoulders, lower your shoulders, join your hands behind your back, and join your hands in front of your chest? (p. 297)
24. What movements are possible at the shoulder joint? What muscles accomplish these movements? (p. 297)
25. Distinguish axial and scapular muscles involved in moving the humerus (arm). (p. 300)
26. What muscles move the humerus (arm)? In which directions do these movements occur? (p. 300)
27. Group the muscles that move the ulna and radius (forearm) into flexors and extensors. What muscles move the forearm and what actions are used when striking a match? (p. 303)
28. Discuss the various movements possible at the wrist, hand, and fingers. What muscles accomplish these movements? (p. 305)
29. What muscles and actions of the wrist, hand, and fingers are used when writing? (p. 305)
30. What is the flexor retinaculum? Extensor retinaculum? (p. 305)
31. Discuss the various muscles and movements of the vertebral column (backbone). How are the muscles grouped? (p. 311)
32. What muscles accomplish movements of the femur (thigh)? What actions are produced by these muscles? What is the iliotibial tract? (p. 315)
33. Group the muscles that act on the tibia and fibula (leg) into medial, anterior, and posterior compartments. What is the popliteal fossa? (p. 321)
34. Name the muscles that plantar flex, evert, pronate, and dorsiflex the foot. What is the superior extensor retinaculum? Inferior extensor retinaculum? (p. 323)
35. What are the advantages of intramuscular (IM) injections? Describe how you would locate the sites for an intramuscular injection in the buttock, lateral side of the thigh, and deltoid region of the arm. (p. 326)
36. List the common sites for running injuries. How are most running injuries treated? (p. 326)
37. Define the following: "pulled hamstrings" (p. 326), "pulled groin" (p. 326), patellofemoral stress syndrome (p. 326), shinsplint syndrome (p. 327), plantar fasciitis (p. 328), and stress fracture (p. 328).

ANSWERS TO QUESTIONS WITH FIGURES

11.1 Posterior to the humerus.
11.2 Levers; fulcrums.
11.3 Possible responses: direction of fibers—external oblique; shape—deltoid; action—extensor digitorum; size—gluteus maximus; origin and insertion—sternocleidomastoid; location—tibialis anterior; number of origins—biceps brachii.
11.4 Frowning—frontalis; smiling—zygomaticus major; pouting—platysma; squinting—orbicularis oculi.
11.5 Masseter and lateral pterygoid.
11.6 Lateral rectus, superior oblique, and inferior oblique.
11.7 Genioglossus and hyoglossus.
11.8 Stylohyoid.
11.9 Tenses—cricothyroid; relaxes—thyroarytenoid.
11.10 Rectus abdominis.
11.11 Diaphragm.
11.12 Coccygeus.
11.13 Bulbospongiosus and ischiocavernosus.
11.14 Subclavius, pectoralis minor, and serratus anterior.
11.15 Pectoralis major, deltoid, and coracobrachialis.
11.16 Triceps brachii and anconeus.
11.17 Flexor tendons of the digits and wrist and median nerve.
11.18 Splenius.
11.19 Medial—adduction; anterior—extension; posterior—flexion.
11.20 Rectus femoris, vastus lateralis, vastus medialis, and vastus intermedius.
11.21 Anterior.
11.22 Blood vessels and nerves.

CONTROL SYSTEMS OF THE HUMAN BODY

THIS UNIT WILL EXPLAIN THE SIGNIFICANCE OF the nerve impulse in making rapid adjustments for maintaining homeostasis. You will learn how the nervous system detects changes in the environment, selects a course of action, and responds to the changes. We will also investigate the role of hormones in maintaining long-term homeostasis.

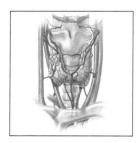

unmyelinated (Fig. 12.2). Electron micrographs reveal that even unmyelinated axons are surrounded by a thin coat of neuroglial plasma membrane.

Two types of neuroglia produce myelin sheaths: neurolemmocytes (in the PNS) and oligodendrocytes (in the CNS). In the PNS, neurolemmocytes (Schwann cells) begin to form myelin sheaths around axons during fetal development. Each neurolemmocyte wraps about 1 millimeter (1 mm = 0.04 in.) of a single axon's length by spiraling many times around the axon (Fig. 12.2b). Up to 500 neurolemmocytes participate in forming a myelin sheath around the longest axons in your body. Eventually, multiple layers of glial plasma membrane surround the axon, with the neurolemmocyte cytoplasm and nucleus forming the outermost layer. The inner portion, con-sisting of up to 100 layers of neurolemmocyte membrane, is the myelin sheath. The outer nucleated cytoplasmic layer of the neurolemmocyte, which encloses the myelin sheath, is called the **neurolemma (sheath of Schwann)**. A neu-rolemma is found only around axons in the PNS. When an axon is injured, the neurolemma aids regeneration by form-ing a regeneration tube that guides and stimulates regrowth of the axon (see Fig. 12.18). At intervals along an axon, the myelin sheath has gaps called **neurofibral nodes** or **nodes of Ranvier** (RON-vē-ā; see Fig. 12.3). Each neurolemmocyte wraps the axon segment between two nodes.

In the CNS an **oligodendrocyte** myelinates parts of many axons in somewhat the same manner as a neurolemmocyte myelinates part of a single PNS axon (see Exhibit 12.1). It

---

Figure 12.2 Myelinated and unmyelinated axons.

Axons of most mammalian neurons are surrounded by a myelin sheath that is produced by neurolemmocytes in the PNS and oligodendrocytes in the CNS.

(a) Diagram of transverse section of unmyelinated axons

(b) Transverse sections of stages in the formation of a myelin sheath

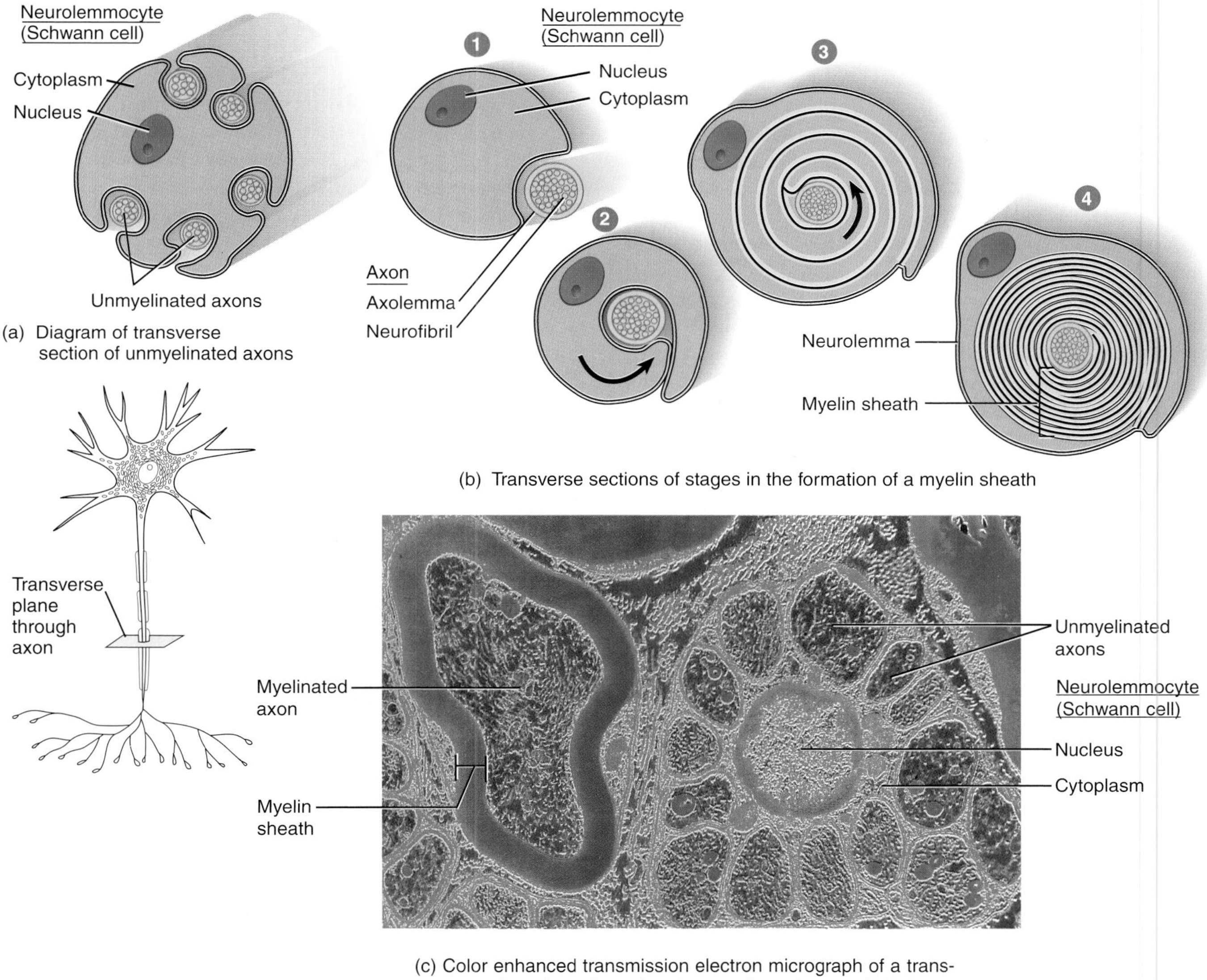

(c) Color enhanced transmission electron micrograph of a transverse section of myelinated and unmyelinated axons (200x)

What is the functional advantage of myelination?

puts forth an average of 15 broad, flat processes that spiral about CNS axons and deposit a myelin sheath. A neurolemma is not formed, however, because the oligodendrocyte cell body and nucleus do not envelop the axon. Neurofibral nodes are present, but they are fewer in number. Axons in the CNS display little regrowth after injury. This is thought to be due, in part, to the absence of a neurolemma and in part to an inhibitory influence exerted by CNS neuroglia.

The amount of myelin increases from birth to maturity, and its presence greatly increases the speed of nerve impulse conduction. Since myelination is still in progress during infancy, an infant's responses to stimuli are not as rapid or coordinated as those of an older child or an adult. Certain diseases such as multiple sclerosis and Tay–Sachs disease cause destruction of myelin sheaths.

MULTIPLE SCLEROSIS (MS)

Multiple sclerosis (MS) is a progressive destruction of myelin sheaths of neurons in the CNS. It is a chronic, often disabling disease that affects 2.6 million people worldwide. The name multiple sclerosis describes the anatomical pathology: myelin sheaths deteriorate to *scleroses,* which are hardened scars or plaques, in *multiple* regions. The destruction of myelin sheaths slows and short-circuits conduction of nerve impulses. Usually, the first symptoms, such as muscular weakness, abnormal sensations, or double vision, occur in early adult life. After an attack there is a period of remission during which the symptoms temporarily disappear. Sometime later a new series of plaques develop, and the victim suffers a second attack. One attack follows another over the years, usually every year or two. The result is a progressive loss of function interspersed with remission periods during which neurons that are undamaged regain their ability to conduct nerve impulses. Although the precise cause of MS is unclear, some evidence implicates a viral infection that precipitates an autoimmune response, specifically the destruction of myelin-producing oligodendrocytes by the body's cytotoxic (killer) cells. In 1993 the Food and Drug Administration licensed treatment of certain patients with MS with injections of Betaseron (a form of interferon beta). This is the first drug approved specifically for treatment of MS and its effectiveness in slowing or preventing progression of MS is still being evaluated. ■

Neurons

Some neurons are tiny and relay signals over a short distance (less than 1 mm) within the CNS. Others are the longest cells in your body. Motor neurons that command muscles to wiggle your toes, for example, extend from the lumbar region of your spinal cord (just above waist level) to your foot. Some sensory neurons are even longer. Those that allow you to feel the position of your wiggling toes stretch all the way from your foot to the lower portion of your brain. Nerve impulses travel these great distances at speeds ranging from less than 1 to more than 100 meters per second (1–280 mi/hr). The site of functional contact be-

tween two neurons or between a neuron and an effector (muscle or gland) cell is called a **synapse.**

Several proteins are **trophic** (TROF-ik; *trophikos* = nourishing) **factors** that regulate the normal growth and development of neurons. These substances are called **neurotropins** (**NTs**). They include **nerve growth factor** (**NGF**), **brain-derived neurotropic factor**, **NT3**, and **NT4/5**. Scientists are just beginning to unravel some of the effects of these crucial molecules. NGF is needed for the normal growth, survival, and development of sympathetic and sensory neurons. Whereas NGF stimulates sensory neurons only during a short period of embryonic development, it stimulates sympathetic neurons into adulthood. NGF also plays a role in the brain, where it increases synthesis of the neurotransmitter acetylcholine (as′-ē-til-KŌ-lēn), stimulates neuronal growth, and may help maintain neuronal function. In mice lacking the gene for NT3, sensory neurons called muscle spindles and Golgi tendon organs, which convey information about joint and muscle position, are absent.

Parts of a Neuron

Most neurons have three parts: (1) cell body, (2) dendrites, and (3) axon (Fig. 12.3). The **cell body** (**soma** or **perikaryon**) contains a nucleus surrounded by cytoplasm that includes typical organelles such as lysosomes, mitochondria, and a Golgi complex. Many neurons also contain cytoplasmic inclusions such as **lipofuscin** pigment that occurs as clumps of yellowish brown granules. Lipofuscin is probably an end-product of lysosomal activity, which collects as one ages but does not seem to harm the neuron.

Other structures in the cytoplasm are characteristic of neurons: chromatophilic substance and neurofibrils. The **chromatophilic substance** (**Nissl bodies**) is an orderly arrangement of rough endoplasmic reticulum, the site of protein synthesis. Newly synthesized proteins replace those being recycled and are used for growth of neurons and regeneration of damaged peripheral nerve axons. **Neurofibrils,** composed of intermediate filaments, form the cytoskeleton, which provides support and shape for the cell.

Neurons have two kinds of processes—dendrites and axons—that differ in their structure and function. **Dendrites** (*dendro* = tree) are the receiving or input portion of a neuron. They usually are short, tapering, and highly branched. Often, the dendrites form a tree-shaped array of processes that emerge from the cell body. Usually, dendrites are not myelinated. Their cytoplasm contains chromatophilic substance, mitochondria, and other organelles.

The second type of process, the **axon** (*axon* = axis), propagates nerve impulses toward another neuron, muscle fiber, or gland cell. An axon is a long, thin, cylindrical projection that often joins the cell body at a cone-shaped elevation called the **axon hillock** (*hilloc* = small hill). The first portion of the axon is called the **initial segment**. Nerve impulses arise at the initial segment, which is called the **trigger zone**, and then conduct along the axon. An axon contains mitochondria, microtubules, and neurofibrils but no rough endoplasmic reticulum; thus it does not synthesize

Figure 12.3 Structure of a typical neuron. Arrows indicate the direction of information flow. The break indicates that the axon actually is longer than shown.

The basic parts of a neuron are (1) dendrites, (2) cell body, and (3) axon.

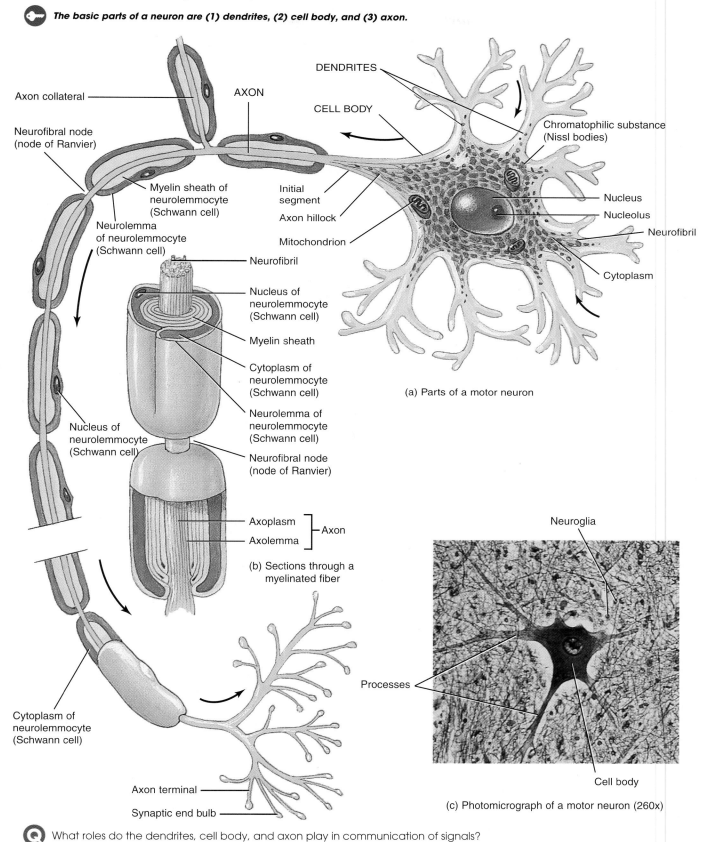

(a) Parts of a motor neuron

(b) Sections through a myelinated fiber

(c) Photomicrograph of a motor neuron (260x)

What roles do the dendrites, cell body, and axon play in communication of signals?

proteins. Its cytoplasm, called **axoplasm,** is surrounded by a plasma membrane known as the **axolemma** (*lemma* = sheath or husk). Along the length of an axon, side branches called **axon collaterals** may branch off, typically at a right angle to the axon. The axon and its collaterals end by dividing into many fine processes called **axon terminals.**

The tips of some axon terminals swell into bulb-shaped structures called **synaptic end bulbs** or **synaptic knobs.** Other axon terminals exhibit a string of swollen bumps called **varicosities.** Both synaptic end bulbs and varicosities contain many minute membrane-enclosed sacs called **synaptic vesicles** that store a chemical substance known as a **neurotransmitter.** Neurons influence the activity of other neurons, muscle fibers, or gland cells by releasing neurotransmitter from synaptic vesicles.

Neurons display great diversity in size and shape. For example, their cell bodies range in diameter from 5 μm (smaller than a red blood cell) up to 135 μm (barely large enough to see with the naked eye). The pattern of dendritic branching is varied and distinctive for neurons in different parts of the nervous system. A few small neurons lack an axon and many others have very short axons. The longest neurons, however, have axons that extend for a meter (3.2 ft) or more.

Nerve fiber is a general term for any neuronal process (dendrite or axon). Most often, it refers to an axon and its sheaths. A **nerve** is a bundle of many nerve fibers that course along the same path in the PNS, such as the ulnar nerve in the arm or the sciatic nerve in the thigh. Most nerves include bundles of both sensory and motor fibers and are surrounded by connective tissue coats. Nerve cell bodies in the PNS generally are clustered together to form **ganglia** (GANG-lē-a; *ganglion* = knot). A **tract** is a bundle of nerve fibers, without connective tissue elements, in the CNS. Tracts may interconnect different regions of the brain or extend long distances up or down the spinal cord and connect with specific regions of the brain.

Axonal Transport

The cell body of a neuron is the site of most synthetic reactions plus recycling of worn-out molecules into new cell parts. However, some substances are needed in the axon or at the axon terminals. Two types of transport systems carry materials from the cell body to the axon terminals and back. The slower one, which moves materials about 1–5 mm per day, is called **slow axonal transport** (**axoplasmic flow**). It conveys axoplasm in one direction only—from the cell body toward the axon terminals. It supplies new axoplasm for developing or regenerating axons and renews axoplasm in growing and mature axons.

The faster system, which is capable of moving materials a distance of 200–400 mm per day, is called **fast axonal transport.** It uses proteins that function as "motors" to move materials in both directions—away from and toward the cell body—along the surfaces of microtubules. Fast axonal transport moves various organelles and materials that form the membranes of the axolemma, synaptic end bulbs,

and synaptic vesicles. Some materials returned to the cell body are degraded or recycled and others influence its growth. Still others, such as certain viruses, may even be harmful to the cell body.

AXONAL TRANSPORT IN DISEASE AND THERAPY

Fast axonal transport is the route by which herpes viruses and rabies viruses make their way from axon terminals near skin cuts to cell bodies of neurons, where they can multiply and cause their damage. The toxin produced by tetanus bacteria reaches the CNS by the same route. The time delay between the release of the toxin and the first appearance of symptoms is, in part, due to the time required for movement of the toxin to the cell body. This is why a tetanus-prone injury of the head or neck, or a bite in this region by a rabid dog, bat, or other animal, is a more serious matter than a similar injury in the leg. The closer to the brain, the shorter the transit time, and the faster must be the treatment to prevent the disease.

Viruses that are able to hop aboard the fast axonal transport system may become a tool for treating genetic disorders of the CNS. Through the techniques of genetic engineering, scientists can insert corrective genes into a messenger virus, and they can also disable potentially harmful viral genes. Current studies will reveal if good copies of defective genes can be shuttled into a person's CNS by viruses without adverse side effects. ■

Classification of Neurons

Both structural and functional features are used to classify the different neurons in the body. *Structural classification* is based on the number of processes extending from the cell body (Fig. 12.4). **Multipolar neurons** usually have several dendrites and one axon (see also Fig. 12.3). Most neurons in the brain and spinal cord are of this type. **Bipolar neurons** have one main dendrite and one axon and are found in the retina of the eye, inner ear, and olfactory area of the brain. **Unipolar neurons** have just one process extending from the cell body and are always sensory neurons. They originate in the embryo as bipolar neurons, but during development the axon and dendrite fuse into a single process that divides into two branches a short distance from the cell body. Both branches have the characteristic structure and function of an axon. They are long, cylindrical processes that may be myelinated, and they propagate action potentials. Dendrites, located at the distal tip of the axon, monitor a sensory stimulus such as touch or pain. The trigger zone for nerve impulses in a unipolar neuron is at the junction of the dendrites and axon (Fig. 12.4c). The impulses then propagate toward the axon terminals, which are in the CNS.

The type of information carried and the direction in which they transmit nerve impulses are used as criteria for the *functional classification* of neurons. **Sensory** or **afferent neurons** transmit sensory nerve impulses from receptors in the skin, sense organs, muscles, joints, and viscera *into the CNS.* Higher order sensory neurons (second-order, third-order) carry sensory signals from lower levels

Figure 12.4 Structural classification of neurons. Breaks indicate that axons are longer than shown.

A multipolar neuron has many processes extending from the cell body whereas a bipolar neuron has two and a unipolar neuron has one.

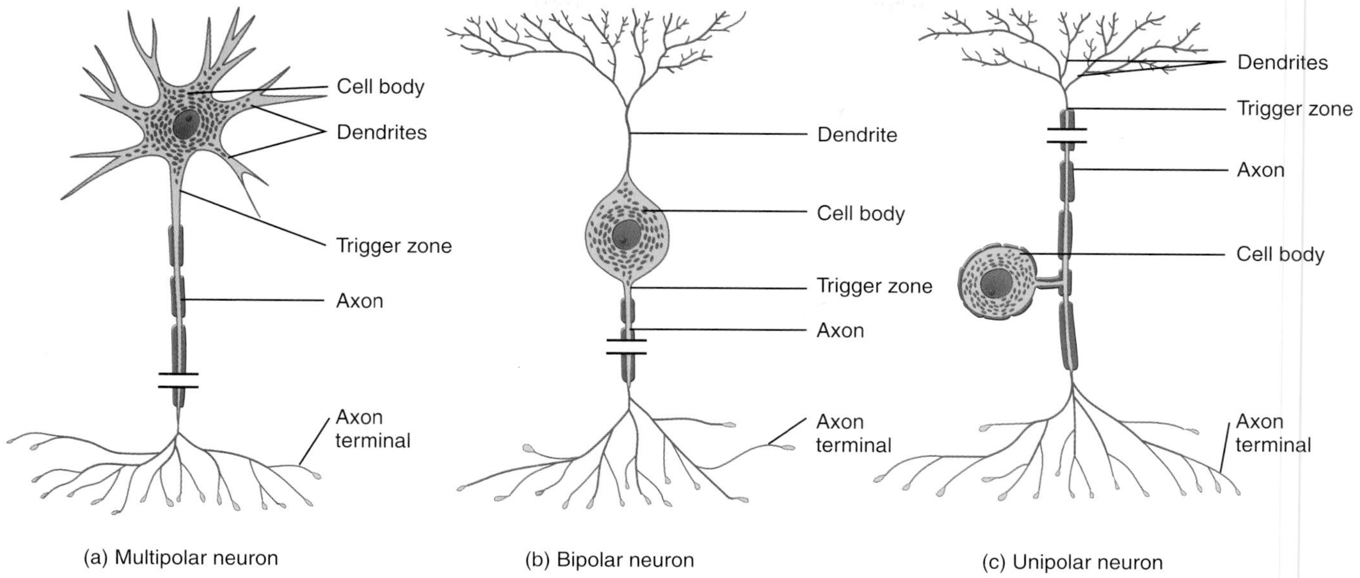

(a) Multipolar neuron (b) Bipolar neuron (c) Unipolar neuron

Q What process occurs at a trigger zone?

of the spinal cord and brain to higher brain regions. **Motor** or **efferent neurons** convey motor nerve impulses *from the CNS to effectors,* which may be either muscles or glands. Neurons that carry commands for movement from the brain into the spinal cord may also be classified as motor neurons. All other neurons that are not specifically sensory or motor neurons are termed **association neurons** or **interneurons**. Most neurons in the body, perhaps 90%, are association neurons.

Spinal and cranial nerves contain fibers in seven functional categories, four sensory and three motor.

1. **General somatic** (*soma* = body) **sensory neurons** convey impulses for the somatic senses of pain, temperature, touch, vibration, and pressure from the skin and for position from joints and muscles via spinal nerves and some cranial nerves.
2. **Special somatic sensory neurons** relay impulses for the special senses of vision, hearing, and balance via cranial nerves.
3. **General visceral sensory neurons** are autonomic neurons that convey information from the viscera, such as distention of organs and chemical conditions within the body, into the CNS via both cranial and spinal nerves.
4. **Special visceral sensory neurons** convey impulses for the special senses of taste and smell (olfaction) via cranial nerves.
5. **General somatic motor neurons** are neurons that conduct impulses to most *skeletal muscles* via spinal nerves and some cranial nerves.

6. **General visceral motor neurons** are neurons that conduct impulses from the CNS to *smooth muscle, cardiac muscle,* and *glands* via cranial and spinal nerves.
7. **Special visceral motor neurons** are neurons that conduct impulses from the CNS to *skeletal muscles* that control facial expression and position of the jaw, neck, larynx, and pharynx via cranial nerves.

There are thousands of different types of association neurons. Neurons often are named for the histologist who first described them, such as **Purkinje** (pur-KIN-jē) **cells** in the cerebellum (Fig. 12.5a) or **Renshaw cells** in the spinal cord. Or, they are named for some aspect of their shape or appearance. For example, **pyramidal** (pi-RAM-i-dal) **cells**, found in the cerebral cortex (the superficial layer of the cerebrum), have a cell body shaped like a pyramid (Fig. 12.5b).

Gray and White Matter

In a freshly dissected section of the brain or spinal cord, some regions look white and glistening whereas others appear gray (Fig. 12.6). **White matter** refers to aggregations of myelinated processes from many neurons. The whitish color of myelin gives white matter its name. The **gray matter** of the nervous system contains either neuron cell bodies, dendrites, and axon terminals or bundles of unmyelinated axons and neuroglia. It looks grayish, rather than white, because there is no myelin in these areas.

In the spinal cord, the white matter surrounds an inner core of gray matter shaped like a butterfly or the letter H

Figure 12.5   Two examples of association neurons (interneurons). Arrows indicate the direction in which information flows.

 Association neurons carry nerve impulses from one neuron to another.

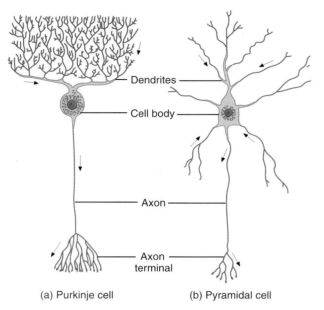

(a) Purkinje cell   (b) Pyramidal cell

Are the dendrites of association neurons receiving input or providing output?

(Fig. 12.6). In the brain, a thin shell of gray matter covers the surface of the largest portions of the brain, the two cerebral hemispheres and the two cerebellar hemispheres (Fig. 12.6). There also are many nuclei of gray matter lying deep

within the brain. A **nucleus** is a mass of cell bodies and dendrites of neurons inside the CNS. (Recall that the same term—nucleus—also refers to the organelle that contains most of a cell's DNA.) Since they consist mainly of cell bodies, nuclei are masses of gray matter. Most nerves in the PNS and all tracts in the CNS are white matter. The arrangement of gray and white matter in the spinal cord and brain is described more extensively in Chapters 13 and 14.

NEUROPHYSIOLOGY

Communication among neurons and from neurons to muscle and gland cells depends on two basic properties of the plasma membrane of excitable cells (neurons and muscle fibers).

1. Like most other cells in the body, the plasma membrane of excitable cells exhibits a **membrane potential**, an electrical voltage difference across the membrane. In excitable cells this voltage is termed the **resting membrane potential**. With proper stimulation, it can change suddenly, producing responses called graded potentials (explained on page 343) and action potentials (explained on page 344). The resting membrane potential is like voltage stored in a battery. If you connect the positive and negative terminals of a battery with a piece of wire, electrons will flow along the wire. Flow of electric charge is called **current**. In living cells, ions (rather than electrons) carry most of the current.

2. Graded potentials and action potentials can occur because the plasma membrane contains a variety of **ion channels** for specific ions that may open or close in response to

Figure 12.6   Distribution of gray and white matter in the spinal cord and brain.

 White matter consists of myelinated processes of many neurons. Gray matter consists of neuron cell bodies, dendrites, axon terminals, bundles of unmyelinated axons, and neuroglia.

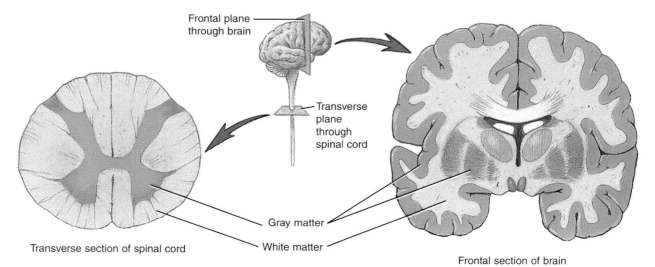

Transverse section of spinal cord   Frontal section of brain

What is responsible for the white appearance of white matter?

specific stimuli. Because the phospholipid bilayer of the plasma membrane is a good electrical insulator, the main path for current to flow across the membrane is through these ion channels. Thus when ion channels open in the plasma membrane, particular ions can enter or leave the cell more rapidly. The result is a change in current flow and this changes the membrane potential.

Excitable cells produce two types of electrical signals—action potentials, which serve for communication over both short and long distances, and graded potentials, which provide short-distance communication. As you will see shortly, production of each type of potential depends on the presence of certain types of ion channels and the establishment of a resting membrane potential.

Ion Channels

Ion channels are of two basic types: leakage and gated. **Leakage (nongated) channels** are always open, like the type of garden hose that has many small holes along its length. **Gated channels**, on the other hand, open and close in response to some sort of stimulus. The plasma membrane of a neuron or muscle fiber has many more potassium ion

(K^+) leakage channels than it has sodium ion (Na^+) leakage channels. This is why membrane permeability to K^+ is much higher. Four categories of stimuli operate gated ion channels: voltage, chemicals, mechanical pressure, and light.

1. **Voltage.** The first type of gated ion channel opens in response to a direct change in the membrane potential (voltage) and is called a **voltage-gated (voltage-regulated) ion channel** (Fig. 12.7a). The presence of voltage-gated ion channels in the plasma membranes of neurons and muscle fibers gives these cells the property of *excitability,* that is, the ability to respond to certain stimuli by producing impulses.

2. **Chemicals.** A **chemically gated ion channel** opens and closes in response to a specific chemical stimulus. A wide variety of chemical ligands, such as neurotransmitters, hormones, and ions such as hydrogen ions (H^+) and calcium ions (Ca^{2+}), regulate chemically gated ion channels. The neurotransmitter acetylcholine, for example, opens cation channels that pass Na^+, K^+, and Ca^{2+} (Fig. 12.7b). Chemically gated ion channels operate in two basic ways. The chemical may *directly* change the membrane permeability to one or more ions, as in the example of acetylcholine. Or it may act

Figure 12.7 Gated ion channels. (a) A change in the membrane potential opens voltage-gated Na^+ channels during an action potential. (b) A chemical stimulus, for example, the neurotransmitter acetylcholine, opens chemically gated ion channels.

 Whereas leakage (nongated) channels are always open, gated channels open and close in response to a stimulus.

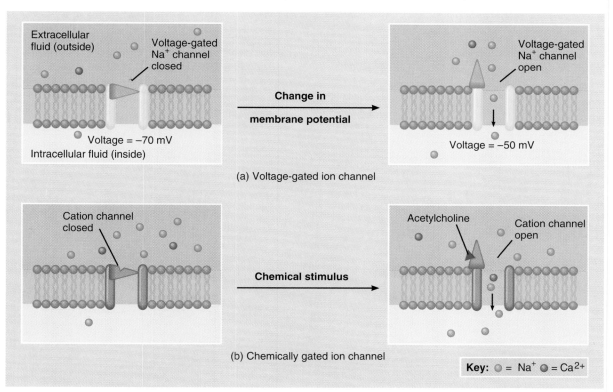

(a) Voltage-gated ion channel

(b) Chemically gated ion channel

Key: ● = Na^+ ● = Ca^{2+}

What type of gated ion channel (not shown here) is activated by a touch on your arm?

indirectly via a type of membrane protein called a G-protein and a second messenger system that uses molecules in the cytosol. Hormones often work by second messenger systems (see Fig. 18.4), and so do some neurotransmitters.

3. **Mechanical pressure. A mechanically gated ion channel** opens or closes in response to mechanical vibration or pressure, such as sound waves or the pressure of a touch. This type of channel is found in sensory receptors that detect mechanical distortions, for example, stretch receptors that monitor stomach stretching or touch receptors in the skin.
4. **Light. A light-gated ion channel** closes in response to light. This type of channel is found in the photoreceptors of the eyes.

Resting Membrane Potential

The resting membrane potential occurs because there is a small buildup of negative charges in the cytosol (intracellular fluid) just inside the membrane and an equal buildup of positive charges in the extracellular fluid next to the outside surface of the membrane (Fig. 12.8a). Such a separation of positive and negative electric charges is a form of potential energy, which is measured in volts or millivolts (1 mV = $\frac{1}{1000}$ V). The greater the difference in charge across the membrane, the larger the membrane potential (voltage). Note in Fig. 12.8a that the buildup of charge is just in fluid that is very close to the membrane. Elsewhere in the cytosol or extracellular fluid, there are equal numbers of positive and negative charges.

In neurons, the resting membrane potential ranges from −40 to −90 mV. A typical value is −70 mV. *The minus sign indicates that the inside is negative relative to the outside.* A cell that exhibits a membrane potential is said to be **polarized**. Most body cells are polarized, with the membrane voltage varying from +5 mV to −100 mV in different types of cells.

Two main factors contribute to the resting membrane potential.

1. **Unequal distribution of ions across the plasma membrane.** Recall that the major anions and cations are different outside and inside cells (Fig. 12.8b). Extracellular fluid is rich in Na$^+$ and chloride ions (Cl$^-$). In cytosol, on the other hand, the main cation is K$^+$ (potassium ions), and the two dominant anions are organic phosphates and amino acids in proteins.
2. **Relative permeability of the plasma membrane to Na$^+$ and K$^+$.** In a resting neuron or muscle fiber, the permeability of the plasma membrane is 50–100 times greater to K$^+$ than to Na$^+$.

To understand how these factors contribute, first consider what would happen if the membrane were permeable only to K$^+$. These positive ions would tend to leak out of the cell into the extracellular fluid down the concentration gradient. But, as more and more positive potassium ions exited, the interior of the membrane would become increasingly negative. The resulting electrical difference (inside negative)

Figure 12.8 Resting membrane potential.

 The resting membrane potential is due to a small buildup of negative ions, mainly phosphates (PO$_4^{3-}$), chloride (Cl$^-$), and proteins, in the cytosol just inside the membrane and an equal buildup of positive ions, mainly sodium ions (Na$^+$) and a much smaller number of potassium ions (K$^+$), in the extracellular fluid on the outside.

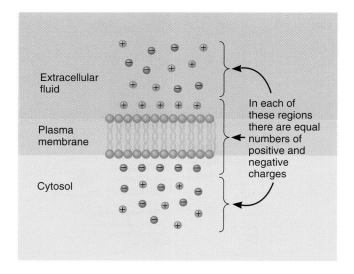

(a) Distribution of charges

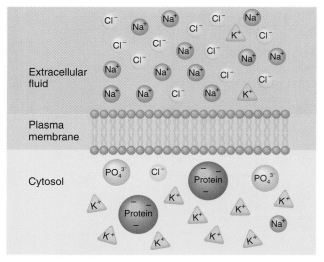

(b) Distribution of ions

 What is a typical value for the resting membrane potential of a neuron?

would then start to pull K⁺ back into the cell. Eventually, just as many K⁺ would be entering because of the electrical difference as would be exiting because of the concentration (chemical) gradient. The membrane potential that just balances the K⁺ concentration difference is −90 mV and is called the **potassium equilibrium potential**. The resting membrane potential (−70 mV) is close to but not exactly at the potassium equilibrium potential, which means that the membrane must be slightly permeable to other ions.

Actually, the membrane is moderately permeable to K⁺ and Cl⁻ and very slightly permeable to Na⁺. One way to balance the electrical effect of K⁺ outflow might be for Na⁺ to flow inward, in effect exchanging one positive particle for another. Because the membrane permeability to Na⁺ is so low, however, inward leakage of Na⁺ is far too slow to keep pace with outward leakage of K⁺. A second way to balance the electrical effect of K⁺ outflow might be the simultaneous exit of anions, but most anions in the cell are not free to leave. They are attached either to large proteins or to other organic molecules, such as phosphates in ATP. Finally, inward leakage of Cl⁻ down its concentration gradient cannot balance the electrical effect of K⁺ outflow. Any chloride ions that enter the cell can only make the inside more negative.

In summary, the result of the low permeability of the membrane to Na⁺ and to anions inside the cell is that fluid just next to the inner surface of the plasma membrane becomes more and more negatively charged as K⁺ leaves.

Note that both the electrical and concentration gradients promote Na⁺ inflow: the negative interior attracts these positive ions and the concentration of Na⁺ is higher outside. Even though membrane permeability to Na⁺ is very low, a slow leak would eventually destroy the electrochemical gradient because there is no gradient to push Na⁺ back out. The small inward Na⁺ leak is taken care of by the sodium pumps (Na⁺/K⁺ ATPase; see Fig. 3.10). They help maintain the resting membrane potential by pumping out Na⁺ as fast as it

leaks in. At the same time, the sodium pumps bring in K⁺. However, the K⁺ merely redistributes according to the electrical and chemical gradients, as previously described. Thus the critical job of the sodium pumps is to expel Na⁺.

Some sodium pumps expel three Na⁺ for each two K⁺ imported. Such pumps are said to be *electrogenic,* which means they contribute to the negativity of the resting membrane potential. The total effect of such pumps, however, is very small, no more than −3 mV of the total −70 mV resting membrane potential in a typical neuron.

Graded Potentials

The presence of chemically, mechanically, or light-gated ion channels in a membrane allows the cell to produce graded potentials. A **graded potential** is a small deviation from the resting membrane potential that is caused by an appropriate stimulus (Fig. 12.9). Graded potentials have different names depending on which type of stimulus causes them. For example, when a neurotransmitter binds to its receptors at a chemically gated ion channel, it produces a graded potential called a postsynaptic potential (explained shortly). On the other hand, sensory receptors produce graded potentials termed receptor potentials and generator potentials (explained in Chapter 15). A graded potential makes the membrane either more polarized (more negative) or less polarized (less negative) than the resting level. Polarization more negative than the resting level is termed **hyperpolarization** (Fig. 12.9a). Polarization less negative than the resting level is termed **depolarization** (Fig. 12.9b).

Graded potentials occur most often in the dendrites and cell body of a neuron and less often in the axon. The region where a graded potential arises contains chemically, mechanically, or light-gated ion channels in the plasma membrane. These electrical signals are *graded,* which means

---

Figure 12.9 Graded potentials.

🔑 *Membrane polarization more negative than the resting level is termed hyperpolarization. Membrane polarization less negative than the resting level is termed depolarization.*

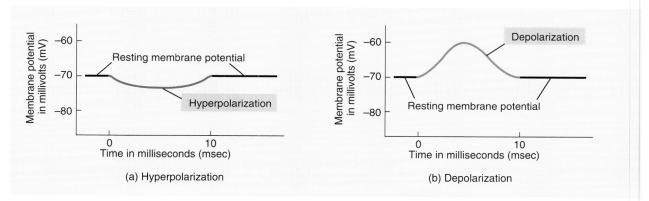

(a) Hyperpolarization

(b) Depolarization

❓ What types of channels produce graded potentials when they open or close?

they vary in amplitude (size), depending on the strength of the stimulus. They are larger or smaller depending on how many gated ion channels have opened (or closed) and how long each one remains open. The opening or closing of gated ion channels alters the flow of specific ions across the membrane, which produces a local flow of current. This current is *localized,* which means that it spreads along the membrane in the extracellular fluid and cytosol for only a few micrometers (μm) before it dies out. Thus graded potentials are useful only for short-distance communication.

Action Potential

When a graded potential is a depolarization that is large enough, an action potential may arise at the trigger zone. An **action potential** (**AP**) or **impulse** is a sequence of rapidly occurring events that decrease and eventually reverse the membrane potential (depolarization) and then restore it to the resting state (repolarization). Because they can travel for long distances without dying out, APs function in both short- and long-distance communication.

During an action potential (Fig. 12.10), two types of voltage-gated ion channels (present in the axon plasma membrane) open and then close. The first channels that open allow Na⁺ to rush into the cell, which causes depolarization. Then K⁺ channels open, allowing K⁺ to flow out, which produces repolarization. Together, the depolarization and repolarization phases last about 1 msec ($\frac{1}{1000}$ sec) in a typical neuron.

Depolarization

Rapid opening of voltage-gated Na⁺ channels brings about depolarization, the loss and then reversal of membrane polarization. If some stimulus causes the membrane to depo-larize to a critical level, called **threshold** (about −55 mV), the inflow of Na⁺ becomes so large that the membrane potential changes from −55 mV toward 0, and then to +30 mV. Throughout depolarization, Na⁺ continues to diffuse inward until the membrane potential *reverses*—the inside becomes 30 mV more positive than the outside (Fig. 12.10). At the beginning of depolarization both the electrical and the concentration (chemical) gradients drive Na⁺ inward.

Each voltage-gated Na⁺ channel has two separate gates, an *activation gate* and an *inactivation gate.* In a resting membrane, the inactivation gate is open, but the activation gate is closed (step ❶ of Fig. 12.11). As a result, Na⁺ cannot diffuse into the cell through these channels. This is the *resting* state of a voltage-gated Na⁺ channel. At threshold, many voltage-gated Na⁺ channels suddenly change from the resting to the *activated* state. In this state, both the activation and inactivation gates in the channel are open, and Na⁺ moves inward (step ❷ of Fig. 12.11). As more channels open, more Na⁺ moves inward, the membrane depolarizes further, and so on. The continued inflow of Na⁺ magnifies the initial stimulus; thus it occurs as part of a positive feedback system. Different neurons may have different thresholds for generation of an action potential, but the threshold in any one neuron usually is constant.

The same depolarization that opens activation gates also closes inactivation gates for Na⁺ (step ❸ of Fig. 12.11). This is called the *inactivated* state of the channel. But the *inactivation gate closes* a few ten-thousandths of a second *after the activation gate opens.* Thus a voltage-gated Na⁺ channel is open for a few ten-thousandths of a second. While the channel is open, about 20,000 Na⁺ flow across the membrane and change the membrane potential considerably. This number represents only one of every million Na⁺ in the fluid just outside the membrane at this site, however, so the change in the Na⁺ concentration is very small.

Figure 12.10   Action potential (impulse).

 An action potential consists of depolarization and repolarization.

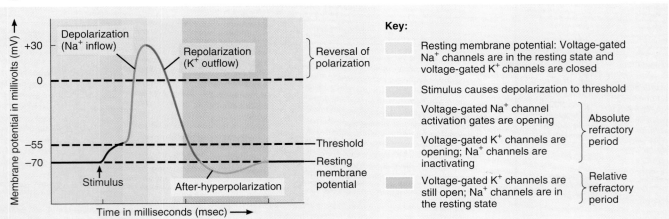

Which channels are open during depolarization? During repolarization?

Because only a few Na$^+$ enter during a single action potential, the sodium pumps easily bail them out and maintain the low concentration of Na$^+$ inside the cell.

Repolarization

A threshold depolarization not only opens voltage-gated Na$^+$ channels but also opens voltage-gated K$^+$ channels (steps ❸ and ❹ of Fig. 12.11). The K$^+$ channels open more slowly, however, so their opening occurs at about the same time the voltage-gated Na$^+$ channels are closing. The slower opening of voltage-gated K$^+$ channels and closing of previously open Na$^+$ channels leads to **repolarization**, the recovery of the resting membrane potential. As the Na$^+$ channels are inactivated, Na$^+$ inflow slows. At the same time the K$^+$

Figure 12.11 Changes in ion flow through voltage-gated channels during action potential depolarization and repolarization. Not illustrated are the leakage channels and the sodium pumps that maintain a low concentration of Na$^+$ inside the cell.

🔑 *Inflow of sodium ions (Na$^+$) causes depolarization and outflow of potassium ions (K$^+$) causes repolarization.*

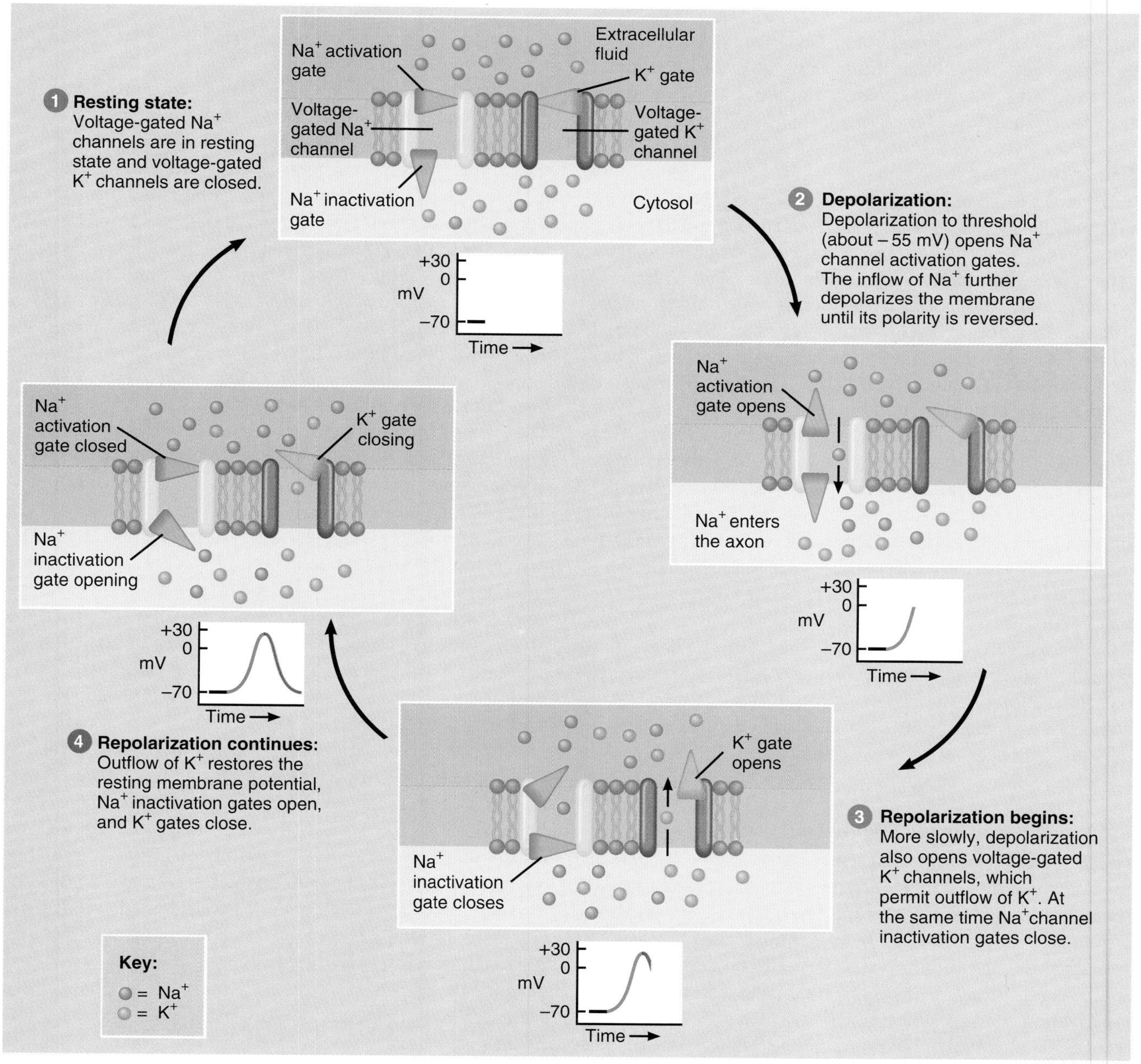

❶ **Resting state:** Voltage-gated Na$^+$ channels are in resting state and voltage-gated K$^+$ channels are closed.

❷ **Depolarization:** Depolarization to threshold (about −55 mV) opens Na$^+$ channel activation gates. The inflow of Na$^+$ further depolarizes the membrane until its polarity is reversed.

❸ **Repolarization begins:** More slowly, depolarization also opens voltage-gated K$^+$ channels, which permit outflow of K$^+$. At the same time Na$^+$ channel inactivation gates close.

❹ **Repolarization continues:** Outflow of K$^+$ restores the resting membrane potential, Na$^+$ inactivation gates open, and K$^+$ gates close.

Key:
● = Na$^+$
● = K$^+$

❓ Recalling that there are leakage channels for both K$^+$ and Na$^+$, could repolarization occur if the voltage-gated K$^+$ channels did not exist?

channels are opening, and K$^+$ outflow accelerates. Slowing of Na$^+$ inflow and acceleration of K$^+$ outflow causes the membrane potential to change from +30 mV to 0 to −70 mV. Repolarization restores the resting membrane potential and allows inactivated Na$^+$ channels to revert to their resting state.

While the voltage-gated K$^+$ channels are open, outflow of K$^+$ may be large enough to cause **after-hyperpolarization** (see Fig. 12.10). After-hyperpolarization is the hyperpolarization that occurs after the repolarizing phase of an action potential. At this point the membrane is even more permeable to K$^+$ than in the resting state, and the membrane potential drifts toward the potassium equilibrium potential (about −90 mV). As the voltage-gated K$^+$ channels close, however, the membrane potential returns to the resting level of −70 mV. In contrast to voltage-gated Na$^+$ channels, most voltage-gated K$^+$ channels do not exhibit an inactivated state. They flip back and forth between closed (resting) and open (activated) states.

Refractory Period

The period of time during which an excitable cell cannot generate another action potential is called the **refractory period** (see Fig. 12.10). The **absolute refractory period** refers to the time period during which a second action potential cannot be initiated, even with a very strong stimulus. It coincides with the period of Na$^+$ channel activation and inactivation. Inactivated Na$^+$ channels cannot reopen. They first must return to the resting state. Graded potentials do not exhibit a refractory period.

Large diameter axons have an absolute refractory period of about 0.4 msec (1/2500 sec). Thus a second nerve impulse can arise 0.4 msec after the first—up to 2500 impulses per second. Small diameter axons, on the other hand, have absolute refractory periods as long as 4 msec (1/250 sec). Thus they can transmit only 250 impulses per second. Under normal body conditions, the frequency of nerve impulses ranges between 10 and 1000 per second.

The **relative refractory period** is the period of time during which a second action potential can be initiated, but only by a suprathreshold (larger than threshold) stimulus. It coincides with the period when the voltage-gated K$^+$ channels are still open after inactivated Na$^+$ channels have returned to their resting state.

Propagation (Conduction) of Action Potentials

Nerve impulses communicate information from one part of the body to another. To do this, they must travel from where they arise at a trigger zone, often the axon hillock, to axon terminals. The special mode of impulse travel is called **propagation (conduction)** and depends on positive feedback. As Na$^+$ flows in, depolarization increases, and the depolarization opens voltage-gated Na$^+$ channels in adjacent patches of membrane. Thus the nerve impulse self-propagates along the membrane. The situation is like toppling a long row of dominoes by pushing on the first one in the line.

Also, since the membrane is refractory behind the leading edge of an action potential, an impulse normally moves only in one direction, from where it arises at the trigger zone toward the axon terminals.

LOCAL ANESTHETICS

Drugs such as procaine (Novocaine) or lidocaine are used to block pain and other sensations, for example, in the skin during suturing of a gash or in the mouth during dental work. They block opening of voltage-gated Na$^+$ channels so nerve impulses cannot pass the obstructed region. Smaller diameter axons, such as pain fibers, are more sensitive than larger diameter fibers to low doses of these anesthetic drugs. ■

The All-or-None Principle

A single neuron, or a single muscle fiber, generates an action potential according to the **all-or-none principle**: if depolarization reaches threshold (about −55 mV in many neurons), voltage-gated channels open and an action potential (impulse) arises. Each time an action potential arises, it has a constant and maximum strength, unless conditions such as toxic materials or fatigue alter the membrane properties. Considering a long row of dominos, when the push on the first domino is strong enough, it falls. Even stronger pushes produce the same effect. Thus toppling the first domino is an all-or-none event; it either falls or remains standing.

Continuous and Saltatory Conduction

The type of impulse conduction considered thus far occurs in muscle fibers and unmyelinated axons. The step-by-step depolarization of each adjacent area of the plasma membrane is called **continuous conduction** (Fig. 12.12a). In myelinated axons, conduction is somewhat different. The myelin sheath acts as an electrical insulator to block ionic currents across the membrane. At intervals, however, are neurofibral nodes (nodes of Ranvier) that interrupt the myelin sheath. The nodes have a very high density of voltage-gated Na$^+$ channels. Here is where membrane depolarization can occur and current carried by Na$^+$ and K$^+$ can flow across the plasma membrane.

When a nerve impulse propagates along a myelinated fiber, current carried by ions flows through the extracellular fluid surrounding the myelin sheath and through the cytosol from one node to the next (Fig. 12.12b). But current flows *across* the membrane only at the nodes. Thus the impulse appears to leap from node to node as each nodal area depolarizes to threshold. This type of impulse conduction, characteristic of myelinated fibers, is called **saltatory** (SAL-ta-tō-rē) **conduction** (*saltare* = leaping).

Since the impulse "leaps" long intervals as current flows from one node to the next in saltatory conduction, it travels much faster than it would by continuous conduction in an unmyelinated fiber of equal diameter. This is especially important in situations where quick responses are necessary.

Figure 12.12   Propagation (conduction) of a nerve impulse after it arises at the trigger zone . Dotted lines indicate ionic current flow. The insets show the path of the current flow. (a) In continuous conduction along an unmyelinated axon, ionic currents flow across each adjacent portion of the membrane. (b) In saltatory conduction along a myelinated axon, the nerve impulse at the first node generates ionic currents in the cytosol and extracellular fluid that open voltage-gated Na^+ channels at the second node. At the second node, the ion flows trigger a nerve impulse. Then the nerve impulse from the second node generates an ionic current that opens voltage-gated Na^+ channels at the third node, and so on. Each node repolarizes after it depolarizes.

🔑 *Unmyelinated axons exhibit continuous conduction whereas myelinated axons exhibit saltatory conduction.*

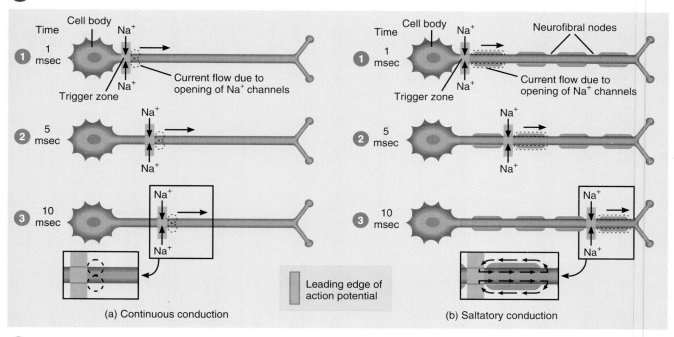

(a) Continuous conduction   (b) Saltatory conduction

Q Where are most of the voltage-gated channels in a myelinated axon?

Saltatory conduction also is more energy efficient. Since only small regions of the membrane depolarize, there is minimal inflow of Na^+ each time a nerve impulse passes by. As a result, less ATP is used by sodium pumps to maintain the low intracellular concentration of Na^+.

Speed of Impulse Propagation

The propagation speed of a nerve impulse is not related to stimulus strength. Rather, the diameter of the fiber and the presence or absence of a myelin sheath are the most important factors that determine the speed of nerve impulse propagation. Also, nerve fibers conduct impulses at higher speeds when warmed and lower speeds when cooled. Localized cooling of a nerve can block impulse conduction. Pain resulting from injured tissue can be reduced by the application of ice because the nerve fibers carrying the pain sensations are partially blocked.

Larger diameter fibers conduct impulses faster than smaller ones. The largest diameter fibers (about 5–20 μm diameter) are called **A fibers** and are all myelinated. The A fibers have a brief absolute refractory period and conduct impulses at speeds from 12 to 130 m/sec (27–280 mi/hr).

The axons of large sensory nerves that relay impulses associated with touch, pressure, position of joints, heat, and cold plus nerve fibers that convey impulses to skeletal muscles are A fibers. Sensory A fibers generally connect the brain and spinal cord with receptors that may detect danger in the outside environment. Motor A fibers innervate the muscles that can do something about the situation. A fibers exist where quick reaction may mean survival.

Smaller fibers, called B and C fibers, conduct impulses more slowly and are generally found where instantaneous response is less urgent. **B fibers** have diameters of 2–3 μm or less and a somewhat longer absolute refractory period than A fibers. They also are myelinated and exhibit saltatory conduction at speeds up to 15 m/sec (32 mi/hr). B fibers are found in nerves that transmit impulses from the viscera to the brain and spinal cord. They also constitute all the axons of the general visceral efferent neurons that extend from the brain and spinal cord to relay stations in the autonomic nervous system called autonomic ganglia.

C fibers have the smallest diameters (½–1½ μm) and the longest absolute refractory periods. Nerve impulse conduction along a C fiber ranges from ½ to 2 m/sec (1–4 mi/hr).

These unmyelinated fibers conduct some impulses for pain, touch, pressure, heat, and cold from the skin and pain impulses from the viscera. Visceral motor fibers that extend from autonomic ganglia to stimulate the heart, smooth muscle, and glands are C fibers. Examples of motor functions of B and C fibers are constricting and dilating the pupils, increasing and decreasing the heart rate, and contracting and relaxing the urinary bladder—actions of the autonomic nervous system.

Coding of Stimulus Intensity

If all action potentials are the same size, then how can your sensory systems detect stimuli of differing intensities? Why does a light touch feel different from firmer pressure? The main factor is the frequency of impulses, that is, how often they are generated at the trigger zone. Thus a light touch generates a low frequency of widely spaced nerve impulses. Firmer pressure, on the other hand, elicits nerve impulses passing down the axon at higher frequency. A second factor is the number of sensory neurons activated by the stimulus. A firm pressure stimulates more pressure-sensitive neurons than does a light touch.

Comparison of Nerve and Muscle Action Potentials

The initiation and conduction of nerve and muscle action potentials are similar, although there are some notable differences. Whereas the typical resting membrane potential of a neuron is −70 mV, it is closer to −90 mV in skeletal and cardiac muscle fibers. The duration of a nerve impulse is ½–2 msec, but a muscle action potential is considerably longer—about 1.0–5.0 msec for skeletal muscle fibers and 10–300 msec for cardiac and smooth muscle fibers. Finally, the velocity of conduction of a nerve impulse can be about 18 times faster than conduction of a muscle action potential.

Comparison of Graded Potentials and Action Potentials

There are six important differences between graded potentials and action potentials (APs).

1. **Amplitude.** Graded potentials have a variable size or amplitude whereas APs are all-or-none (same size in any given neuron or muscle cell). The amplitude of a graded potential varies according to the strength of the stimulus that causes it, from less than 1 mV to more than 50 mV. The amplitude of an action potential is typically about 100 mV.
2. **Duration.** Graded potentials are much longer than nerve impulses. Graded potentials have durations that range from several milliseconds to several minutes whereas the duration of nerve action potentials is between ½ and 2 msec.
3. **Channels.** Different types of channels are used to produce graded potentials and APs. Graded potentials result from opening or closing of chemically gated, mechanically gated, and light-gated ion channels. Action

potentials involve opening and closing of voltage-gated ion channels.
4. **Location.** Graded potentials arise mainly on dendrites and the cell body although some occur in the axon, wherever their channels are located. APs, on the other hand, arise at the trigger zone and propagate along the axon, the site of voltage-gated channels.
5. **Propagation.** Graded potentials are localized whereas APs propagate (conduct). Graded potentials permit communication over short distances of a few micrometers; action potentials can carry messages over much longer distances.
6. **Refractory period.** APs exhibit a refractory period whereas graded potentials do not. Thus two graded potentials may sum if they occur at about the same time (see Fig. 12.16 on page 351).

Transmission at Synapses

In Chapter 10 we described events occurring at one type of synapse—the neuromuscular junction. The focus here is synaptic communication among the billions of neurons in the nervous system. Synapses are essential for homeostasis because they allow information to be integrated and filtered. Certain signals are transmitted while others are blocked. Some diseases and psychiatric disorders result from a disruption of synaptic communication. Synapses also are the sites of action for many drugs that affect the brain, both therapeutic and addictive substances. At a synapse between neurons, the neuron sending the signal is called the **presynaptic neuron**, and the neuron receiving the message is called the **postsynaptic neuron**. Most synapses are **axodendritic** (from axon to dendrite), **axosomatic** (from axon to soma), or **axoaxonic** (from axon to axon).

There are two types of synapses, electrical and chemical, which exhibit both structural and functional differences.

Electrical Synapses

At an **electrical synapse**, ionic current spreads directly from one cell to another through **gap junctions** (see Fig. 4.1). Each gap junction contains a hundred or so tubular protein structures called **connexons** that form tunnels to connect the cytosol of the two cells. This provides a path for ionic current flow. Gap junctions are common in visceral (single-unit) smooth muscle, cardiac muscle, and a developing embryo. They also occur in the CNS.

Electrical synapses have three obvious advantages:

1. They allow *faster communication* than do chemical synapses, since impulses conduct across gap junctions.
2. They can *synchronize* the activity of a group of neurons or muscle fibers. The value of synchronized action potentials in the heart or in visceral smooth muscle is to achieve coordinated contraction of these fibers.
3. They may allow *two-way transmission* of impulses, in contrast to chemical synapses, which function as one-way points of communication.

Chemical Synapses

Although the presynaptic and postsynaptic neurons of a **chemical synapse** are close, their membranes do not touch. They are separated by the **synaptic cleft**, a 20–50-nm space filled with extracellular fluid. Impulses cannot jump the synaptic cleft, so there must be an alternate way for the signal to cross this space. What happens is this (Fig. 12.13): the presynaptic neuron releases a neurotransmitter that diffuses across the synaptic cleft and acts on receptors in the plasma membrane of the postsynaptic neuron to produce a **postsynaptic potential** (a type of graded potential). In essence, the presynaptic electrical signal (nerve impulse) is converted into a chemical signal (liberated neurotransmitter). The postsynaptic neuron receives the chemical signal and, in turn, generates an electrical signal (postsynaptic potential). The time required for these processes at a chemical synapse—the **synaptic delay**—is about ½ msec. This is why chemical synapses relay messages a little more slowly than electrical synapses.

A typical chemical synapse operates as follows:

1 When a nerve impulse arrives at a synaptic end bulb or varicosity of a presynaptic axon, the depolarizing phase opens **voltage-gated Ca²⁺ channels** in addition to the voltage-gated Na⁺ channels normally opened.

2 Because it is more concentrated in the extracellular fluid, Ca²⁺ flows inward.

3 An increase of Ca²⁺ inside the presynaptic neuron triggers exocytosis of synaptic vesicles. As vesicle membranes merge with the plasma membrane, neurotransmitter molecules inside the vesicles enter the synaptic cleft. Each synaptic vesicle may contain several thousand molecules of neurotransmitter.

4 The neurotransmitter molecules then diffuse across the synaptic cleft and bind to **neurotransmitter receptors** in the postsynaptic membrane. The receptor may be a subunit of the ion channel protein, as shown in Fig. 12.13.

At most chemical synapses, there can be only *one-way information transfer*—from a presynaptic neuron to a postsynaptic neuron, muscle fiber, or gland cell. This is because only synaptic end bulbs of presynaptic neurons can release neurotransmitter and only the postsynaptic membrane has the correct receptor proteins to recognize and bind that neurotransmitter. As a result, graded potentials and action potentials must move forward over their pathways. They cannot back up into another presynaptic neuron, a situation that would seriously disrupt homeostasis.

Excitatory and Inhibitory Postsynaptic Potentials

The neurotransmitter causes either an excitatory or an inhibitory graded potential. If it *depolarizes* the postsynaptic membrane, it is excitatory because it brings the membrane closer to threshold (Fig. 12.14a). A depolarizing postsynaptic potential is called an **excitatory postsynaptic potential**

Figure 12.13 Operation of a chemical synapse.

🔑 *At a chemical synapse a presynaptic electrical signal (nerve impulse) is converted into a chemical signal (neurotransmitter release).*

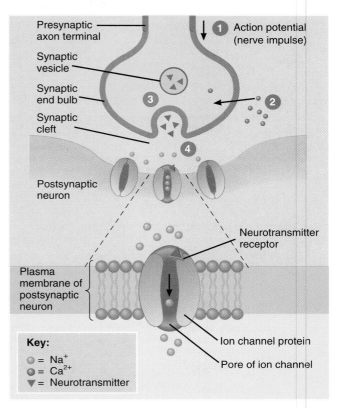

Key:
● = Na⁺
● = Ca²⁺
▼ = Neurotransmitter

❓ Why may electrical synapses work in two directions but chemical synapses transmit information only in one direction?

(**EPSP**). Often, EPSPs result from opening of chemically gated *cation* channels. These channels allow the three most plentiful cations (Na⁺, K⁺, and Ca²⁺) to pass through, but Na⁺ inflow is greater than either Ca²⁺ inflow or K⁺ outflow because both the electrical gradient and a large concentration gradient promote its inward movement. Although a single EPSP normally does not initiate a nerve impulse, the postsynaptic neuron does become more excitable. It is already partially depolarized and thus more likely to reach threshold when the next EPSP occurs.

On the other hand, if a neurotransmitter causes *hyperpolarization* of the postsynaptic membrane, it is inhibitory. It increases the membrane potential by making the inside more negative, and generation of a nerve impulse is more difficult than usual. This is because the membrane potential is even farther from threshold than it was in its resting state. A hyperpolarizing postsynaptic potential is inhibitory and is termed an **inhibitory postsynaptic potential** (**IPSP**) (Fig. 12.14b). IPSPs often result from opening of chemically gated Cl⁻ or K⁺ channels. When Cl⁻ channels open, Cl⁻ tends to diffuse

Figure 12.14 Postsynaptic potentials (PSPs).

Whereas depolarization is excitatory, hyperpolarization is inhibitory.

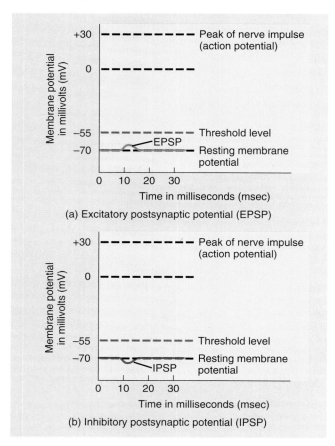

(a) Excitatory postsynaptic potential (EPSP)

(b) Inhibitory postsynaptic potential (IPSP)

Suppose a neurotransmitter binds to a receptor and *closes* K$^+$ channels. Will this produce an EPSP or an IPSP?

inward at a greater pace, but the negativity inside the neuron retards its inflow somewhat. When K$^+$ channels open, the membrane permeability to K$^+$ increases, more K$^+$ diffuses outward, and the inside becomes even more negative.

Removal of Neurotransmitter

Removal of the neurotransmitter from the synaptic cleft is essential for normal synaptic function. If a neurotransmitter could linger in the synaptic cleft, it would influence the postsynaptic neuron, muscle fiber, or gland cell indefinitely. Neurotransmitter is removed in three basic ways:

1. **Diffusion.** Some portion of all neurotransmitters diffuse out of the synaptic cleft.
2. **Enzymatic degradation.** A neurotransmitter may be inactivated through enzymatic degradation. For example, the enzyme acetylcholinesterase breaks down acetylcholine in the synaptic cleft.

3. **Uptake into cells.** Often, neurotransmitters are actively transported back into the neuron that released them (reuptake) or are transported into neighboring neuroglia. Norepinephrine, for example, is rapidly taken up and recycled by the same neurons that release it. The membrane proteins that accomplish such uptake are called **neurotransmitter transporters.** One reason that cocaine produces intense pleasurable feelings (euphoria) is that it blocks transporters for uptake of the neurotransmitter dopamine. This allows dopamine to linger longer in synaptic clefts, producing excessive stimulation of certain brain regions.

Presynaptic Facilitation and Inhibition

Certain synapses can modify the quantity of neurotransmitter released at other synapses. **Presynaptic facilitation** increases the amount of neurotransmitter released whereas **presynaptic inhibition** decreases it. In these situations, a synaptic end bulb of one neuron synapses with the synaptic end bulb of another neuron, making an axoaxonal synapse (Fig. 12.15). If neuron 1 releases excitatory neurotransmitter before an impulse arrives at a synaptic end bulb of neu-

Figure 12.15 Presynaptic facilitation and inhibition. An excitatory or inhibitory neurotransmitter released by neuron 1 can facilitate or inhibit release of neurotransmitter by neuron 2 at its synapse with neuron 3.

Presynaptic facilitation increases the amount of neurotransmitter released whereas presynaptic inhibition decreases it.

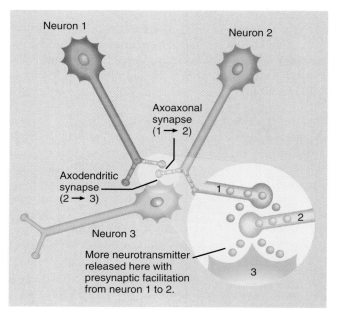

Suppose that neuron 2 transmits pain information. Will more or less pain be felt if neuron 1 provides presynaptic inhibition?

ron 2, more synaptic vesicles undergo exocytosis from neuron 2. Because more neurotransmitter has been released, neuron 3 is then stimulated more strongly. This is presynaptic facilitation. If neuron 1 releases inhibitory neurotransmitter, neuron 2 releases less neurotransmitter at its synapse with neuron 3. This is presynaptic inhibition. Presynaptic facilitation and inhibition may last for several minutes or hours. They are thought to be important in learning and memory.

Spatial and Temporal Summation of PSPs

An EPSP lasts a few milliseconds, and a typical neuron in the CNS receives input from 1000–10,000 synapses. Integration of these inputs is known as **summation** and occurs at the trigger zone. The greater the summation, if it is a depolarization, the greater the chance that a nerve impulse will be initiated.

When summation results from buildup of neurotransmitter released by *several* presynaptic end bulbs, it is called **spatial summation** (Fig. 12.16a). When summation results from buildup of neurotransmitter released by a *single* presynaptic end bulb firing two or more times in rapid succession, it is called **temporal summation** (Fig. 12.16b). Since a typical EPSP lasts about 15 msec, the second firing must occur soon after the first one if temporal summation is to occur.

A single postsynaptic neuron receives input from many presynaptic neurons. Some presynaptic end bulbs produce excitation and some produce inhibition. The sum of all the excitatory and inhibitory effects determines the effect on the postsynaptic neuron. It may respond in the following ways:

1. **EPSP.** If the excitatory effect is greater than the inhibitory effect, but less than the threshold level of stimulation, the result is a small EPSP. Subsequent stimuli can more easily generate a nerve impulse through summation because the neuron is partially depolarized.
2. **Impulse(s).** If the excitatory effect is greater than the inhibitory effect and reaches or surpasses the threshold

Figure 12.16 Summation. (a) When presynaptic neurons a and b separately cause EPSPs (arrows) in postsynaptic neuron c, threshold level is not reached in the postsynaptic neuron. When neurons a and b, plus many others, act simultaneously on the postsynaptic cell, their EPSPs sum to reach the threshold level and trigger a nerve impulse (spatial summation). (b) Stimuli applied to the same axon in rapid succession (arrows) cause overlapping EPSPs that sum (temporal summation). When depolarization reaches the threshold level, a nerve impulse is triggered.

The sum of all excitatory and inhibitory postsynaptic potentials determines whether or not a nerve impulse arises.

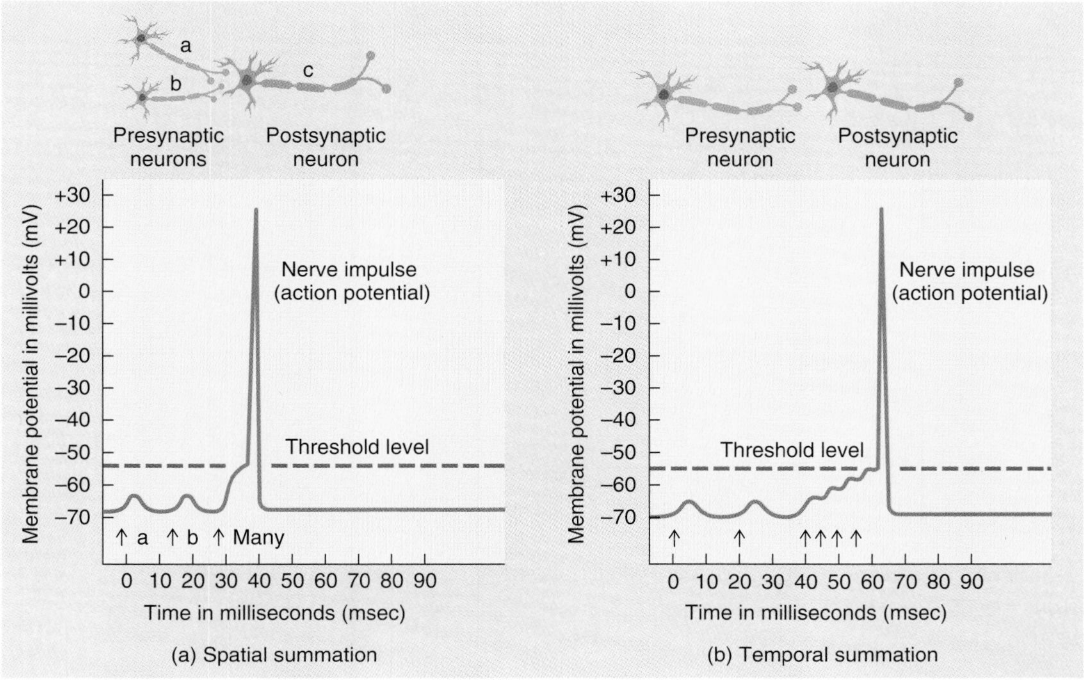

What would be the result if, in addition to the four EPSPs indicated by arrows, an IPSP occurred at time 55 msec in (b)?

level of stimulation, the result is a threshold or suprathreshold EPSP. This spreads to the initial segment of the axon and triggers one or more nerve impulses. Impulses continue to be generated as long as the EPSP stays above the threshold level.

3. **IPSP.** If the inhibitory effect is greater than the excitatory effect, the membrane hyperpolarizes (IPSP). The result is inhibition of the postsynaptic neuron and thus an inability to generate a nerve impulse.

Exhibit 12.2 summarizes the structural and functional elements of a neuron.

Neurotransmitters

Excitatory and inhibitory neurotransmitters are present in both the PNS and CNS. Sometimes the same neurotransmitter is excitatory in one location but inhibitory in another.

Which response occurs depends on the nature of the receptors that bind the neurotransmitter. For many years, neurons were thought to liberate just one type of neurotransmitter at all their synaptic end bulbs. Now we know that two or even three neurotransmitters are present in many neurons.

The best-studied neurotransmitter is **acetylcholine (ACh)**. It is a neurotransmitter released by many PNS neurons and by some CNS neurons. ACh is an excitatory neurotransmitter at the neuromuscular junction, where it acts directly to open chemically gated cation channels. It is also known to be an inhibitory neurotransmitter at other synapses, where its effects on ion channels appear to occur indirectly via receptors that link to a G-protein. One example is parasympathetic fibers of the vagus nerve (cranial nerve X) that innervate the heart; ACh slows heart rate by these inhibitory synapses.

Several amino acids are neurotransmitters in the CNS. **Glutamate** (glutamic acid) and **aspartate** (aspartic acid) have powerful excitatory effects. Two others, **gamma aminobu-**

EXHIBIT 12.2 SUMMARY OF NEURONAL STRUCTURE AND FUNCTION

Diagram	Structure	Functions
	Dendrites	Receive stimuli through activation of chemically or mechanically gated ion channels; in sensory neurons, produce generator or receptor potentials; in motor neurons and association neurons, produce excitatory and inhibitory postsynaptic potentials (EPSPs and IPSPs).
	Cell body	Receives stimuli and produces EPSPs and IPSPs through activation of chemically or mechanically gated ion channels.
	Junction of axon hillock and initial segment of axon	Trigger zone; integrates EPSPs and IPSPs and, if sum is a depolarization that reaches threshold, initiates action potential (nerve impulse).
	Axon	Propagates (conducts) nerve impulses from initial segment (or from dendrites of sensory neurons) to axon terminals in a self-reinforcing manner; impulse amplitude does not change as it propagates along the axon.
	Axon terminals and synaptic end bulbs (or varicosities)	Inflow of Ca^{2+} caused by depolarizing phase of nerve impulse triggers neurotransmitter release by exocytosis of synaptic vesicles.

▬▬▬ Plasma membrane includes chemically gated channels

▬▬▬ Plasma membrane includes voltage-gated Na^+ and K^+ channels

▬▬▬ Plasma membrane includes voltage-gated Ca^{2+} channels

tyric (GAM-ma am-i-nō-byoo-TIR-ik) **acid** (**GABA**) and **glycine** are important inhibitory neurotransmitters. Both cause IPSPs by opening chemically gated Cl⁻ channels. Although GABA is an amino acid, it is not incorporated into proteins in your body. It is found only in the brain, where it is the most common inhibitory neurotransmitter. Perhaps a third of all brain synapses use GABA. Glycine is more prevalent in the spinal cord than in the brain.

The catecholamine neurotransmitters—**norepinephrine**, **epinephrine**, and **dopamine**—are synthesized from the amino acid tyrosine. They are excitatory at some synapses and inhibitory at others.

Even simple gases can function as neurotransmitters. Nitric oxide (NO) has several recognized functions and carbon monoxide (CO) is suspected to be a neurotransmitter as well. Neurotransmitters in the CNS are discussed in more detail on page 414. Certain disorders such as Parkinson's disease, Alzheimer's disease, depression, anxiety, and schizophrenia are caused by problems relating to neurotransmitters.

STRYCHNINE POISONING

The importance of inhibitory neurons can be appreciated by observing what happens when normal inhibition is blocked. Normally, inhibitory neurons in the spinal cord called *Renshaw cells* release glycine at inhibitory synapses with motor neurons. This inhibitory input to motor neurons prevents excessive muscular contraction. Strychnine binds to and blocks glycine receptors, which causes massive tetanic contractions. All skeletal muscles, including the diaphragm, contract fully and remain contracted. Because the diaphragm cannot relax, the person cannot breathe. The normal delicate balance between excitation and inhibition in the CNS is disturbed and motor neurons are firing without restraint. ■

Alteration of Impulse Conduction and Synaptic Transmission

The chemical and physical environment of a neuron influences both impulse conduction and synaptic transmission. *Alkalosis,* an increase in pH above 7.45, increases the excitability of neurons. Impulses may arise inappropriately and cause lightheadedness, numbness around the mouth, tingling in the fingertips, nervousness, muscle spasms, and convulsions. *Acidosis,* a decrease in pH below 7.35, results in a progressive depression of neuronal activity that can produce apathy, weakness, and coma. If a nerve is subjected to excessive or prolonged *pressure,* as when crossing one's legs, nerve impulse conduction is blocked. That part of the body may "go to sleep."

Hypnotics, tranquilizers, and *anesthetics* depress impulse conduction and synaptic transmission by increasing the threshold for excitation of neurons, whereas *caffeine, benzedrine,* and *nicotine* reduce the threshold for excitation of neurons and result in facilitation.

There are several ways to modify chemical synaptic transmission: (1) stimulate or inhibit *neurotransmitter synthesis,* (2) block or enhance *neurotransmitter release,* (3) stimulate or inhibit the *transmitter removal,* and (4) block or activate the *receptor site.* An agent that enhances synaptic transmission or mimics the effect of a natural neurotransmitter is an **agonist**, whereas one that blocks the action of a neurotransmitter is an **antagonist**.

Clostridium botulinum bacteria, which may proliferate in improperly canned food, produce a poison called botulinum toxin. The toxin inhibits the release of acetylcholine, thus weakening muscle contractions. Even a small amount is very poisonous. Yet patients with certain overly strong or uncontrollable muscle contractions may be helped by injections of this toxin. It is used to treat people who have strabismus (crossed eyes) or blepharospasm (uncontrollable winking) and holds promise as a therapy for stuttering.

Neostigmine and *physostigmine* are anticholinesterase agents that inactivate acetylcholinesterase for several hours. The result is very slow removal of acetylcholine from its receptors. As noted on page 265, *myasthenia gravis* results from antibodies that block acetylcholine receptors at neuromuscular junctions and weaken skeletal muscle contraction. Both neostigmine and physostigmine can be used to treat myasthenia gravis. Another anticholinesterase agent is *diisopropyl fluorophosphate,* a very powerful nerve gas that is also the active ingredient of many insecticides. It acts for up to several weeks, making it a particularly lethal poison. It may cause nausea, diarrhea, sweating, bronchial constriction, excess respiratory mucus, and generalized weakness.

Substances that are agonists or antagonists of neurotransmitters have great potential for benefit or harm. The plant derivative *curare* (used by South American Indians on poisoned arrows and blow-gun darts) blocks acetylcholine receptors and thus can cause muscular paralysis. Curare-like drugs, however, often are used in surgery to relax skeletal muscles. Neostigmine is an antidote for curare. It can be used to terminate the effects of curare after surgery or in cases of curare poisoning.

Neuronal Circuits

The CNS contains billions of neurons organized into complicated patterns called **neuronal pools**. Each pool differs from all others and has its own sensory, integrative, or motor functions. A neuronal pool may contain thousands or even millions of neurons.

Neuronal pools in the CNS are arranged in patterns called **circuits** over which the nerve impulses are conducted. In a **simple series circuit** a presynaptic neuron stimulates only a single neuron in a pool. The single neuron then stimulates another, and so on. Most circuits, however, are more complex.

A single presynaptic neuron may synapse with several postsynaptic neurons. Such an arrangement, called **divergence**, permits one presynaptic neuron to influence several postsynaptic neurons or several muscle fibers or gland cells at the same time. In a **diverging circuit**, the nerve impulse from a single presynaptic neuron causes the stimulation of increasing numbers of cells along the circuit (Fig. 12.17a). For example, a small number of neurons in the brain that govern a particular body movement stimulate a much larger number of neurons in the spinal cord. Sensory signals also feed into diverging circuits and are often relayed to several regions of the brain.

In another arrangement, called **convergence**, several presynaptic neurons synapse with a single postsynaptic neuron. This arrangement permits more effective stimulation or inhibition of the postsynaptic neuron. In one type of **converging circuit** (Fig. 12.17b), the postsynaptic neuron receives nerve impulses from several different sources. For example, a single motor neuron that synapses with skeletal muscle fibers at neuromuscular junctions receives input from several pathways that originate in different brain regions.

Some circuits in your body are constructed so that once the presynaptic cell is stimulated, it will cause the postsynaptic cell to transmit a series of nerve impulses. One such circuit is called a **reverberating (oscillatory) circuit** (Fig. 12.17c). In this pattern, the incoming impulse stimulates the first neuron, which stimulates the second, which stimulates the third, and so on. Branches from later neurons synapse with earlier ones, however, sending impulses back through the circuit again and again. The output signal may last from a few seconds to many hours, depending on the number of synapses and arrangement of neurons in the circuit. Inhibitory neurons may turn off a reverberating circuit after a period of time. Among the body responses thought to be the result of output signals from reverberating circuits are breathing, coordinated muscular activities, waking up, sleeping (when reverberation stops), and short-term memory. One form of epilepsy (grand mal) is probably caused by abnormal reverberating circuits.

A fourth type of circuit is the **parallel after-discharge circuit** (Fig. 12.17d). In this circuit, a single presynaptic cell stimulates a group of neurons, each of which synapses with a common postsynaptic cell. A differing number of synapses between the first and last neurons imposes varying synaptic delays so that the last neuron exhibits multiple EPSPs or IPSPs. If the input is excitatory, the postsynaptic neuron then can send out a stream of impulses in quick succession. It is thought that parallel after-discharge circuits may be employed for precise activities such as mathematical calculations.

Figure 12.17 Examples of neuronal circuits.

 Neuronal circuits are groups of neurons arranged in patterns over which nerve impulses are conducted.

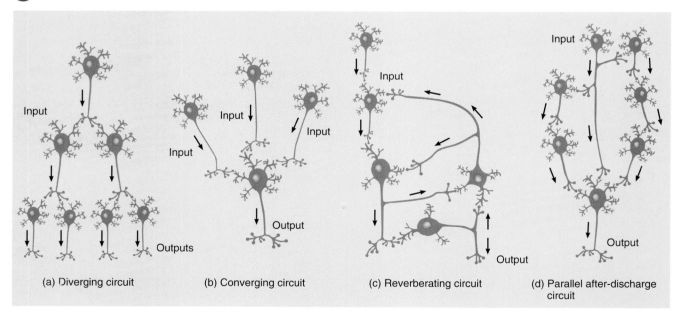

(a) Diverging circuit (b) Converging circuit (c) Reverberating circuit (d) Parallel after-discharge circuit

 A motor neuron in the spinal cord typically receives input from neurons that originate in several different regions of the brain. Is this an example of convergence or divergence?

REGENERATION OF NERVOUS TISSUE

Unlike the cells of epithelial tissue, mammalian neurons have very limited powers of **regeneration**, that is, the capability to replicate or repair themselves. Around 6 months of age, virtually all developing neurons lose their ability to undergo mitosis. Thus when a neuron is damaged or destroyed, it cannot be replaced by daughter cells from other neurons. A neuron destroyed is permanently lost, and only some types of damage may be repaired.

In the PNS, damage to dendrites and myelinated axons may be repaired if the cell body remains intact and if the neurolemmocytes (Schwann cells) that perform the myelination remain active. Substantial regrowth may occur even in a completely severed nerve if it is surgically reattached. The neurolemmocytes form a tube that appears to aid regeneration (see Fig. 12.18d).

In the CNS there is little or no repair of damage to neurons. Axons in the CNS are myelinated by oligodendrocytes that do not form neurolemmas (sheaths of Schwann). An added complication in the CNS is that the neuroglia appear to provide an environment that stops axon regrowth. Perhaps this is the same mechanism that stops axonal growth during development once a target region has been reached. Also, after axonal damage, nearby astrocytes proliferate rapidly, forming a type of scar tissue. The scar tissue is a physical barrier to regeneration. Thus injury of the brain or spinal cord usually is permanent.

ENCOURAGING NEURONAL REGROWTH

Certain types of neuronal tumor cells can grow and replicate in tissue culture. Also, in the brains of some songbirds, new neurons appear and disappear every year. The nearly complete lack of regenerative success in the mammalian CNS seems to result from two factors: (1) inhibitory influences from neuroglia and (2) absence of growth cues that were present during development.

Some developmental cues are electrical in nature. A large research effort is directed at finding a way to promote regrowth of damaged neurons using electrical stimulation. Other developmental cues are chemical in nature, involving both substances that trigger mitosis (mitogenic factors) and those that regulate growth (neurotropins). In 1992 Canadian researchers published their unexpected finding that cells taken from the brains of *adult* mammals could be encouraged to proliferate into both neurons and astrocytes. They cultured small pieces of mouse brain tissue in dishes and added **epidermal growth factor** (**EGF**). Previously, EGF was known to trigger mitosis in a variety of non-neuronal cells and promote wound healing and tissue regeneration (see page 126). These experiments challenge the dogma that stem (progenitor) cells are not present in adult brains. A tantalizing possibility is that scientists may be able to find ways of stimulating these dormant stem cells to replace neurons lost through damage or disease. Also, such tissue-cultured neurons might be useful for transplantation purposes. ■

DISORDERS: HOMEOSTATIC IMBALANCES

DAMAGE AND REPAIR OF PERIPHERAL NEURONS

As we have seen, axons and dendrites that are associated with a neurolemma may undergo repair if the cell body is intact, the neurolemmocytes (Schwann cells) are functional, and scar tissue formation does not occur too rapidly. Most nerves in the PNS consist of processes that are covered with a neurolemma (sheath of Schwann). A person who injures neurons in a nerve in the upper limb, for example, has a good chance of regaining nerve function.

When there is damage to an axon, usually there are changes that occur in the cell body of the affected neuron and in the portion of the axon distal to the site of injury. Changes also may occur in the portion of the axon proximal to the site of injury.

Chromatolysis

About 24–48 hours after injury to a process of a central or peripheral neuron, the chromatophilic substance (Nissl bodies), which normally is arranged in an orderly fashion in an uninjured

cell body, breaks down into finely granular masses (Fig. 12.18b). This alteration is called **chromatolysis** (krō'-ma-TOL-i-sis; *chromo* = color; *lysis* = dissolution). It begins between the axon hillock and nucleus and then spreads throughout the cell body. As a result of chromatolysis, the cell body swells, reaching a maximum size between 10 and 20 days after injury.

Degeneration

The part of the process distal to the damaged region becomes slightly swollen and then breaks up into fragments by the third to fifth day. The myelin sheath also deteriorates (Fig. 12.18c). Degeneration of the distal portion of the neuronal process and myelin sheath is called **Wallerian degeneration**. Following degeneration, macrophages phagocytize the remains.

The changes in the proximal portion of the fiber, called **retrograde degeneration**, are similar to those that occur during Wallerian degeneration. The main difference in retrograde

DISORDERS: HOMEOSTATIC IMBALANCES   CONTINUED

Figure 12.18   Damage and repair of a peripheral neuron.

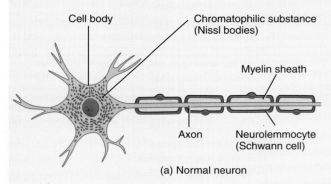

(a) Normal neuron

Labels: Cell body; Chromatophilic substance (Nissl bodies); Myelin sheath; Axon; Neurolemmocyte (Schwann cell)

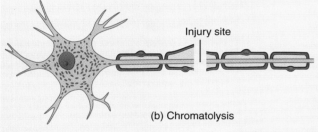

Injury site

(b) Chromatolysis

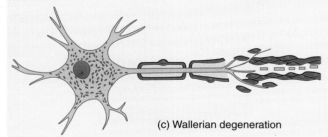

(c) Wallerian degeneration

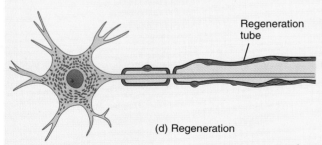

Regeneration tube

(d) Regeneration

Q What is the role of the neurolemma in regeneration?

degeneration is that the changes only extend to the first neurofibral node (node of Ranvier).

Regeneration

Following chromatolysis, there are signs of recovery in the cell body. Synthesis of RNA and protein accelerates, which favors rebuilding or **regeneration** of the axon. Recovery often takes several months.

Even though the neuronal process and myelin sheath degenerate, the neurolemma remains. The neurolemmocytes on either side of the injured site multiply by mitosis, grow toward each other, and may form a regeneration tube across the injured area (Fig. 12.18d). The tube guides growth of new processes from the proximal area across the injured area into the distal area previously occupied by the original nerve fiber. The growth of new axons will not occur if the gap at the site of injury is too large or if the gap becomes filled with collagen fibers.

During the first few days following damage, buds of regenerating axons begin to invade the tube formed by the neurolemmocytes (Fig. 12.18c). Axons from the proximal area grow at the rate of about 1½ mm (0.06 in.) per day across the area of damage, find their way into the distal regeneration tubes, and grow toward the distally located receptors and effectors. Thus sensory and motor connections are reestablished. In time, the neurolemmocytes form a new myelin sheath. However, function is never completely restored after a nerve is severed.

EPILEPSY

Although stroke (rupture or blockage of a brain blood vessel) is the most common neurological disorder, **epilepsy** is the second. It afflicts about 1% of the population and is characterized by short, recurrent, periodic attacks of motor, sensory, or psychological malfunction. The attacks, called **epileptic seizures**, are initiated by abnormal, synchronous electrical discharges from millions of neurons in the brain, perhaps resulting from abnormal reverberating circuits. The discharges stimulate many of the neurons to send nerve impulses over their conduction pathways. As a result, a person undergoing an attack may contract skeletal muscles involuntarily. Lights, noise, or smells may be sensed when the eyes, ears, or nose have not been stimulated. The electrical discharges may also inhibit certain brain centers. For instance, the waking center in the brain (the reticular activating system, or RAS, see page 447) may be depressed so that the person loses consciousness. When RAS activity resumes, the person regains consciousness.

Epilepsy has many causes, including brain damage at birth, the most common cause; metabolic disturbances (hypoglycemia, hypocalcemia, uremia, hypoxia); infections (encephalitis or meningitis); toxins (alcohol, tranquilizers, hallucinogens); vascular disturbances (hemorrhage, hypotension); head injuries; and tumors and abscesses of the brain. Most epileptic seizures, however, are idiopathic; that is, they have no demonstrable cause. Epilepsy almost never affects intelligence.

Epileptic seizures can be eliminated or alleviated by drugs that depress neuronal excitability. One such drug is valproic acid, which increases the quantity of the inhibitory neurotransmitter gamma aminobutyric acid (GABA).

STUDY OUTLINE

INTRODUCTION (p. 332)

1. The nervous system helps regulate homeostasis and integrates all body activities by sensing changes (sensory), interpreting them (integrative), and reacting to them (motor).

NERVOUS SYSTEM DIVISIONS (p. 332)

1. The central nervous system (CNS) consists of the brain and spinal cord.
2. The peripheral nervous system (PNS) consists of cranial and spinal nerves. It has sensory (afferent) and motor (efferent) components.
3. The PNS also is subdivided into somatic (voluntary) and autonomic (involuntary) nervous systems.
4. The somatic nervous system (SNS) consists of neurons that conduct impulses from cutaneous and special sense receptors to the CNS and motor neurons from the CNS to skeletal muscle tissue.
5. The autonomic nervous system (ANS) contains sensory neurons from visceral organs and motor neurons that convey impulses from the CNS to smooth muscle tissue, cardiac muscle tissue, and glands.

HISTOLOGY OF NERVOUS TISSUE (p. 332)

Neuroglia (p. 332)

1. Neuroglia are specialized tissue cells that support neurons, attach neurons to blood vessels, produce the myelin sheath around axons, and carry out phagocytosis.
2. Neuroglia include astrocytes, oligodendrocytes, microglia, ependymal cells, neurolemmocytes (Schwann cells), and satellite cells.
3. Two types of neuroglia produce myelin sheaths: oligodendrocytes myelinate axons in the CNS and neurolemmocytes myelinate axons in the PNS.

Neurons (p. 336)

1. Most neurons, or nerve cells, consist of a cell body (soma), many dendrites, and usually a single axon. The axon conducts nerve impulses from the neuron to the dendrites or cell body of another neuron or to an effector organ of the body (muscle or gland).
2. On the basis of structure, neurons are multipolar, bipolar, and unipolar.
3. On the basis of function, sensory (afferent) neurons conduct impulses from receptors to the CNS; association neurons (interneurons) conduct impulses from one neuron to another within the CNS; and motor (efferent) neurons conduct impulses to effectors.

Gray and White Matter (p. 339)

1. White matter consists of aggregations of myelinated processes whereas gray matter contains neuron cell bodies, dendrites, and axon terminals or bundles of unmyelinated axons and neuroglia.
2. In the spinal cord, gray matter forms an H-shaped inner core, surrounded by white matter. In the brain, a thin superficial shell of gray matter covers the cerebral and cerebellar hemispheres.

NEUROPHYSIOLOGY (p. 340)

Ion Channels (p. 340)

1. The two basic types of ion channels are leakage (nongated) and gated.
2. There are four types of gated channels: voltage-gated, chemically gated, mechanically gated, and light-gated.

Resting Membrane Potential (p. 342)

1. The membrane of a nonconducting neuron is positive outside and negative inside owing to the distribution of different ions across the membrane and the relative permeability of the membrane to Na^+ and K^+.
2. A typical value for the resting membrane potential (RMP) is -70 mV, and the membrane is said to be polarized.
3. Sodium pumps compensate for slow leakage of Na^+ into the cell by pumping it back out.

Graded Potentials (p. 343)

1. A graded potential is a small deviation from the resting membrane potential that is caused by an appropriate stimulus to a region of the neuron (dendrites, cell body, axon) that contains chemically, mechanically, or light-gated channels.
2. In comparison to the resting level of polarization, a graded potential drives the membrane potential more negative (more polarization), termed hyperpolarization, or less negative (less polarization), termed depolarization.
3. The amplitude of a graded potential varies, depending on the strength of the stimulus.
4. During graded potentials, the current (due to ion flow) is localized. Thus they are useful only for short-distance communication.

Action Potential (p. 344)

1. During an action potential (impulse), voltage-gated Na^+ and K^+ channels open in sequence. This results first in depolarization, loss and then reversal of membrane polarization (from -70 to 0 to $+30$ mV), and then in repolarization, recovery of the RMP (from $+30$ to -70 mV).
2. During the refractory period (RP) another impulse cannot be generated at all (absolute RP) or can be triggered only by a suprathreshold stimulus (relative RP).
3. An action potential (impulse) conducts or propagates (travels) from point to point along the membrane and thus is useful for long-distance communication.
4. According to the all-or-none principle, if a stimulus is strong enough to generate an action potential, the impulse travels at a constant and maximum strength for the existing conditions. A stronger stimulus will not cause a larger impulse.
5. Nerve impulse conduction in which the impulse "leaps" from neurofibral node to node is called saltatory conduction.
6. Fibers with larger diameters conduct impulses faster than those with smaller diameters; myelinated fibers conduct impulses faster than unmyelinated fibers.
7. The intensity of a stimulus is coded in the frequency of action potentials.

Transmission at Synapses (p. 348)

1. A synapse is the functional junction between one neuron and another or between a neuron and an effector such as a muscle or gland.
2. Two types of synapses are electrical and chemical.
3. At a chemical synapse, there is only one-way information transfer from a presynaptic neuron to a postsynaptic neuron.
4. An excitatory neurotransmitter is one that can depolarize or make less negative the postsynaptic neuron's membrane, bringing the membrane potential closer to threshold. An inhibitory neurotransmitter hyperpolarizes the membrane of the postsynaptic neuron.
5. Neurotransmitter is removed from the synaptic cleft in three ways: diffusion, enzymatic degradation, and uptake into cells (neurons and neuroglia).
6. Presynaptic facilitation increases the amount of neurotransmitter released by a presynaptic neuron whereas presynaptic inhibition decreases it.
7. If several presynaptic end bulbs release their neurotransmitter at about the same time, the combined effect may generate a nerve impulse, due to summation. Summation may be spatial or temporal.
8. The postsynaptic neuron is an integrator. It receives signals, integrates them, and then responds accordingly.

Neurotransmitters (p. 352)

1. Both excitatory and inhibitory neurotransmitters are present in the CNS and PNS. The same neurotransmitter may be excitatory in some locations and inhibitory in others.

2. Important neurotransmitters include acetylcholine, glutamate, aspartate, gamma aminobutyric acid (GABA), glycine, norepinephrine, epinephrine, dopamine, and nitric oxide.

Alteration of Impulse Conduction and Synaptic Transmission (p. 353)

1. A neuron's chemical and physical environment influences both impulse conduction and synaptic transmission.
2. Chemical synaptic transmission may be stimulated or blocked by affecting neurotransmitter synthesis, release, or removal or by affecting the neurotransmitter receptors.

Neuronal Circuits (p. 353)

1. Neurons in the central nervous system are organized into groupings called neuronal pools. Each pool differs from all others and has its own role.
2. Neuronal pools are organized into circuits. These include simple series, diverging, converging, reverberating (oscillatory), and parallel after-discharge circuits.

REGENERATION OF NERVOUS TISSUE (p. 355)

1. Around 6 months of age, neurons lose the ability to undergo mitosis.
2. Nerve fibers that have a neurolemma are capable of regeneration.

REVIEW QUESTIONS

1. Describe the three basic functions of the nervous system that are necessary to maintain homeostasis. (p. 332)
2. Distinguish between the central and peripheral nervous systems and describe the functions of each subdivision. (p. 332)
3. Relate the terms *voluntary* and *involuntary* to the nervous system. (p. 332)
4. What are neuroglia? List the principal types and their functions. Why are they important clinically? (p. 332)
5. What is a myelin sheath? How is it formed? How does MS affect the nervous system? (p. 334)
6. Define a neuron. Diagram and label a neuron. Next to each part, list its function. (p. 336)
7. Distinguish between slow and fast axonal transport. Why is axonal transport important clinically? (p. 338)
8. Define the neurolemma. Why is it important? (p. 335)
9. Discuss the structural classification of neurons. Give an example of each. (p. 338)
10. What are the functional differences between a typical sensory and motor neuron? Define association neuron. (p. 338)
11. Describe the seven functional types of sensory and motor neurons. (p. 339)
12. Distinguish between gray and white matter. Where is each found? (p. 339)
13. Describe the two basic types of ion channels. (p. 341)
14. What types of stimuli regulate gated channels? (p. 341)
15. Describe the factors that give rise to the resting membrane potential. (p. 342)
16. List and define the charcteristics of graded potentials. (p. 343)
17. Outline the principal steps in the generation and conduction of a nerve impulse. (p. 344)
18. Define the following: resting membrane potential (p. 342), depolarization (p. 344), repolarization (p. 345), nerve impulse (nerve action potential) (p. 344), and refractory period. (p. 346)
19. What is the all-or-none principle? (p. 346)
20. What is saltatory conduction? What factors determine the speed of propagation of nerve impulses? (p. 346)
21. How is the intensity of a stimulus coded in the nervous system? (p. 348)
22. Compare nerve and muscle action potentials. (p. 348)
23. Compare graded potentials and action potentials. (p. 348)
24. Describe the events of chemical synaptic transmission. (p. 349)
25. Distinguish between excitatory and inhibitory postsynaptic potentials. (p. 349)
26. How is neurotransmitter removed from the synaptic cleft? (p. 350)
27. Describe presynaptic facilitation and presynaptic inhibition. (p. 350)
28. Describe spatial and temporal summation. (p. 351)
29. What is a neurotransmitter? List several probable neurotransmitters, indicate their locations in the nervous system and whether or not they may lead to excitation or inhibition. (p. 352)
30. In what ways can impulse conduction or synaptic transmission be altered? (p. 353)
31. What is a neuronal circuit? Distinguish among simple series, diverging, converging, reverberating (oscillatory), and parallel after-discharge circuits. (p. 353)
32. Explain why damaged neurons are not replaced. (p. 355)

ANSWERS TO QUESTIONS WITH FIGURES

12.1 Sensory or afferent; motor or efferent.

12.2 It increases the speed of nerve impulse conduction.

12.3 Dendrites receive (motor or association neurons) or generate (sensory neurons) inputs; cell body also receives input signals; axon conducts action potentials and transmits the message to another neuron or effector cell by releasing neurotransmitter at its synaptic end bulbs.

12.4 Site where nerve impulses arise.

12.5 Receiving input.

12.6 Myelin.

12.7 Mechanically gated.

12.8 -70 mV.

12.9 Chemically, mechanically, or light-gated ion channels.

12.10 Voltage-gated Na^+ channels are open during depolarization and voltage-gated K^+ channels are open during repolarization.

12.11 Yes, because the leakage channels would still allow K^+ to exit more rapidly than Na^+ could enter the axon. In fact,

mammalian myelinated axons have few voltage-gated K^+ channels.

12.12 At the nodes.

12.13 In some electrical synapses (gap junctions) ions flow equally well in either direction, so either neuron may be the presynaptic one. At a chemical synapse, one neuron releases the neurotransmitter and the other neuron has receptors that bind this chemical. Thus the information can flow in only one direction.

12.14 Closing of K^+ channels will slow the outflow of K^+ so the inside will become more positive, thus producing a depolarizing EPSP.

12.15 Less.

12.16 Probably threshold depolarization would not be reached and an impulse would not be generated.

12.17 Convergence.

12.18 Provides a tube to guide regrowth of axon.

CHAPTER 13

THE SPINAL CORD AND SPINAL NERVES

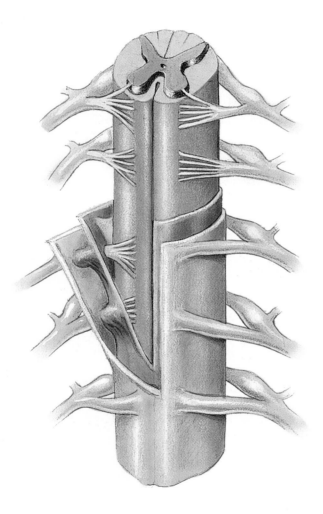

1. Describe the protection and gross anatomical features of the spinal cord. (p. 361)

2. Describe the functions of the principal sensory and motor tracts of the spinal cord. (p. 365)

3. Describe the functional components of a reflex arc and the relationship of reflexes to homeostasis. (p. 368)

4. List and describe several clinically important reflexes. (p. 370)

5. Describe the components and connective tissue coverings of a spinal nerve. (p. 376)

6. Define a plexus and identify the distribution of nerves of the cervical, brachial, lumbar, and sacral plexuses. (p. 378)

7. Describe the clinical significance of dermatomes and myotomes. (p. 385)

8. Explain the causes and symptoms of neuritis, sciatica, shingles, and poliomyelitis. (p. 387)

CHAPTER CONTENTS AT A GLANCE

ogether, the spinal cord and spinal nerves contain neuronal circuits that mediate some of your quickest reactions to environmental changes. If you pick up something hot, for example, the grasping muscles may relax and you may drop it even before the sensation of extreme heat or pain reaches your conscious perception. This is an example of a spinal cord reflex, a quick, automatic response to certain kinds of stimuli that involves neurons (nerve cells) only in the spinal nerves and spinal cord. Besides processing reflexes, the spinal cord also is the site for integration (summing) of nerve impulses that arise locally or arrive from the periphery and brain. Moreover, the spinal cord is the highway traveled by sensory nerve impulses headed for the brain and by motor nerve impulses destined for spinal nerves. Keep in mind that the spinal cord is continuous with the brain and that together they constitute the central nervous system (CNS).

SPINAL CORD ANATOMY

Protection and Coverings

Two types of connective tissue coverings, bony vertebrae and tough meninges, plus a cushion of cerebrospinal fluid (produced in the brain) surround and protect the delicate nervous tissue of the spinal cord and brain.

Vertebral Column

The spinal cord is located within the vertebral (spinal) canal of the vertebral column. The vertebral foramina of all the vertebrae, stacked one on top of the other, form the canal. The surrounding vertebrae provide a sturdy shelter for the enclosed spinal cord (see Fig. 13.1b). The vertebral ligaments, meninges, and cerebrospinal fluid provide additional protection.

Meninges

The **meninges** (me-NIN-jēz; singular, **meninx**, MĒ-ninks) are connective tissue coverings that encircle the spinal cord and brain. They are called, respectively, the **spinal meninges** (Fig. 13.1) and the **cranial meninges** (see Fig. 14.4a). The most superficial of the three spinal meninges is the **dura mater** (DOO-ra MA-ter; *dura* = tough; *mater* = mother), which is composed of dense, irregular connective tissue. It forms a sac from the level of the foramen magnum of the occipital bone, where it is continuous with the dura mater of the brain, to the second sacral vertebra where it is close-ended. The spinal cord is also protected by a cushion of fat and connective tissue located in the **epidural space**, a space between the dura mater and the wall of the vertebral canal.

The middle meninx is an avascular covering called the **arachnoid** (a-RAK-noyd; *arachne* = spider) because of its spider's web arrangement of delicate collagen fibers and some elastic fibers. It is deep to the dura mater and is also continuous with the arachnoid of the brain. Between the

dura mater and the arachnoid is a thin **subdural space**, which contains interstitial fluid.

The innermost meninx is the **pia mater** (PĒ-a MA-ter; *pia* = delicate), a thin transparent connective tissue layer that adheres to the surface of the spinal cord and brain. It consists of interlacing bundles of collagen fibers and some fine elastic fibers and contains many blood vessels that supply nutrients and oxygen to the spinal cord. Between the arachnoid and the pia mater is the **subarachnoid space**, which contains cerebrospinal fluid. Inflammation of the meninges is known as **meningitis**. It may be caused by microbes such as bacteria and viruses.

All three spinal meninges cover the spinal nerves up to the point of exit from the spinal column through the intervertebral foramina. Triangular-shaped membranous extensions of the pia mater suspend the spinal cord in the middle of its dural sheath. These extensions, called **denticulate** (den-TIK-yoo-lāt; *denticulus* = a small tooth) **ligaments**, are thickenings of the pia mater. They project laterally and fuse with the arachnoid and inner surface of the dura mater along the length of the spinal cord between the ventral and dorsal nerve roots of spinal nerves on

Figure 13.1 Spinal meninges. The arrow in the inset in (b) indicates the direction from which the spinal cord is viewed (superior).

 Meninges are connective tissue coverings that surround the spinal cord and brain.

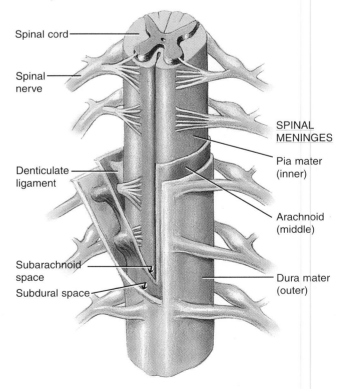

(a) Sections through spinal cord

Figure continues

Figure 13.1 (continued)

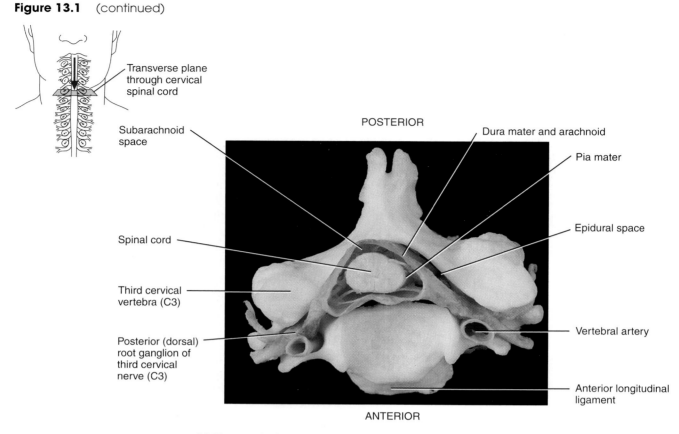

(b) Photograph of a transverse section of the spinal cord within a cervical vertebra

Q What are the superior and inferior boundaries of the spinal dura mater?

either side. They protect the spinal cord against shock and sudden displacement.

External Anatomy of the Spinal Cord

The **spinal cord** is roughly cylindrical but flattened slightly in its anterior–posterior dimension. In the adult, it extends from the medulla oblongata, the most inferior part of the brain, to the superior border of the second lumbar vertebra (Fig. 13.2). In a newborn infant, it extends to the third or fourth lumbar vertebra. During early childhood, both the spinal cord and the vertebral column grow longer as part of overall body growth. Around age 4 or 5, however, elongation of the spinal cord stops. Since the vertebral column continues to elongate, the spinal cord does not extend the entire length of the vertebral column in an adult. The length of the adult spinal cord ranges from 42 to 45 cm (16–18 in.). Its diameter is about 2 cm (¾ in.) in the midthoracic region, somewhat larger in the lower cervical and midlumbar regions, and smallest at the inferior tip.

When the spinal cord is viewed externally, two conspicuous enlargements can be seen. The superior enlargement, the **cervical enlargement**, extends from the fourth cervical to the first thoracic vertebra. Nerves to and from the upper limbs

arise from the cervical enlargement. The inferior enlargement, called the **lumbar enlargement**, extends from the ninth to the twelfth thoracic vertebra. Nerves to and from the lower limbs arise from the lumbar enlargement. Two grooves divide the cord into right and left sides (see Fig. 13.3a). The **anterior median fissure** is a deep, wide groove on the anterior (ventral) side, and the **posterior median sulcus** is a shallower, narrow groove on the posterior (dorsal) surface.

Inferior to the lumbar enlargement, the spinal cord tapers to a conical portion known as the **conus medullaris** (KŌ-nus med-yoo-LAR-is; *konos* = cone), which ends at the level of the intervertebral disc between the first and second lumbar vertebrae in an adult. Arising from the conus medullaris is the **filum terminale** (FĪ-lum ter-mi-NAL-ē; *filum* = filament; *terminale* = terminal), an extension of the pia mater that extends inferiorly and anchors the spinal cord to the coccyx.

Some nerves that arise from the inferior part of the cord do not leave the vertebral column at the same level as they exit from the spinal cord. The roots (points of attachment to the spinal cord) of these nerves angle inferiorly in the vertebral canal from the end of the spinal cord like wisps of hair. Appropriately, the roots of these nerves are collectively named the **cauda equina** (KAW-da ē-KWĪ-na), meaning "horse's tail."

Figure 13.7 Stretch reflex. A monosynaptic reflex arc has only one synapse in the CNS and two neurons—a sensory neuron and a motor neuron. The synapse is between the sensory neuron from the receptor and the motor neuron to the effector. The polysynaptic reflex arc to antagonistic muscles also is illustrated. A polysynaptic reflex arc includes at least two synapses in the CNS and at least one association neuron. Plus signs (+) indicate excitatory synapses; minus sign (–) indicates inhibitory synapse.

 The stretch reflex causes contraction of a muscle that has been stretched.

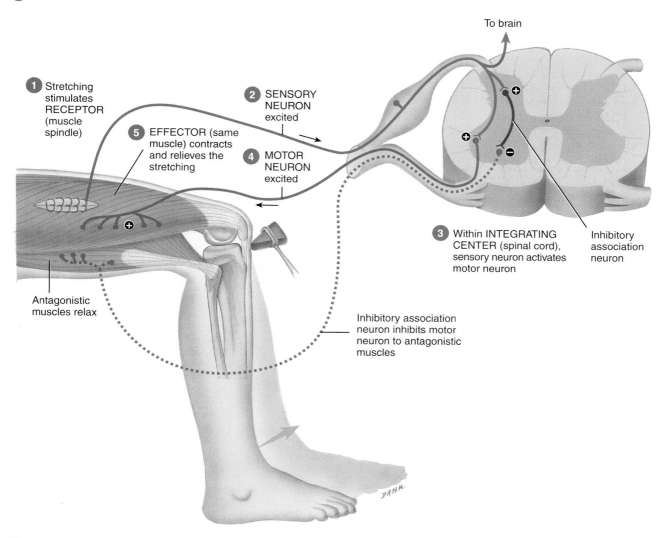

Why is this an ipsilateral reflex?

to changes in muscle tension caused by passive stretch or muscular contraction.

A tendon reflex operates as follows (Fig. 13.8):

1. As the tension applied to a tendon increases, the tendon organ (receptor) is stimulated (depolarized to threshold).

2. Nerve impulses are generated and these propagate into the spinal cord along a sensory neuron.

3. Within the spinal cord (integrating center), the sensory neuron activates an inhibitory association neuron that synapses with a motor neuron.

4. The inhibitory neurotransmitter inhibits (hyperpolarizes) the motor neuron, which then generates fewer nerve impulses.

⑤ The muscle attached to the same tendon relaxes and relieves excess tension.

Thus as tension on the tendon organ increases, the frequency of inhibitory impulses increases, and the inhibition of the motor neurons to the muscle developing excess tension (effector) causes relaxation of the muscle. In this way, the tendon reflex protects the tendon and muscle from damage due to excessive tension.

Note in Fig. 13.8 that the sensory neuron from the tendon organ also synapses with an excitatory association neuron in the spinal cord. The excitatory association neuron, in turn, synapses with motor neurons controlling antagonistic muscles. Thus while the tendon reflex brings about relaxation of the muscle attached to the tendon organ, it also brings about contraction of the antagonists. This is another example of reciprocal innervation. The sensory neuron also relays nerve impulses to the brain by way of sensory tracts, thus informing the brain about the state of muscle tension throughout the body.

Figure 13.8   Tendon reflex. This reflex arc is polysynaptic since there is more than one CNS synapse and more than two different neurons involved in the pathway. The sensory neuron synapses with two association neurons. The inhibitory association neuron causes relaxation of the effector and the stimulatory association neuron causes contraction of the antagonistic muscle. Plus signs (+) indicate excitatory synapses; minus sign (–) indicates inhibitory synapse.

🔑 *The tendon reflex causes relaxation of the muscle attached to the stimulated tendon organ.*

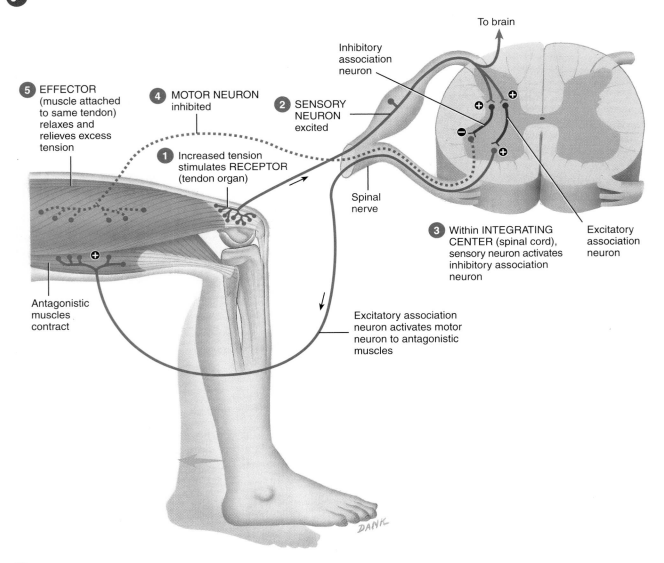

To brain

Inhibitory association neuron

⑤ EFFECTOR (muscle attached to same tendon) relaxes and relieves excess tension

④ MOTOR NEURON inhibited

② SENSORY NEURON excited

① Increased tension stimulates RECEPTOR (tendon organ)

Spinal nerve

③ Within INTEGRATING CENTER (spinal cord), sensory neuron activates inhibitory association neuron

Excitatory association neuron

Antagonistic muscles contract

Excitatory association neuron activates motor neuron to antagonistic muscles

DANK

Q What is reciprocal innervation?

Physiology of the Flexor (Withdrawal) Reflex and Crossed Extensor Reflexes

Another example of a reflex based on a polysynaptic reflex arc is the **flexor (withdrawal) reflex** (Fig. 13.9). Suppose you step on a tack. As a result of the painful stimulus, you immediately withdraw your leg. What has happened?

1 Stepping on a tack stimulates the dendrites of a pain-sensitive neuron (receptor).

2 This sensory neuron then generates nerve impulses, which propagate into the spinal cord.

3 Within the spinal cord (integrating center), the sensory neuron activates association neurons that extend to several spinal cord segments.

Figure 13.9 Flexor (withdrawal) reflex. This reflex arc is polysynaptic and ipsilateral. Plus signs (+) indicate excitatory synapses.

 The flexor reflex causes withdrawal of a part of the body in response to a painful stimulus.

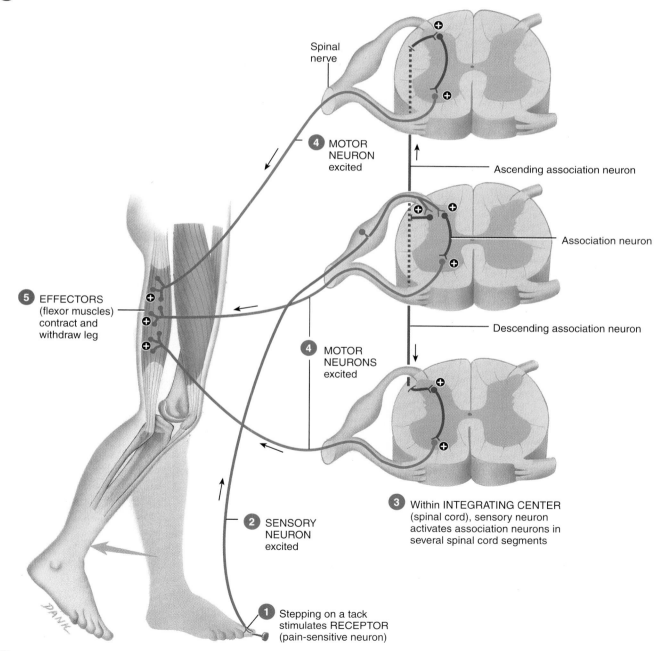

Q Why is the flexor reflex classified as an intersegmental reflex arc?

④ The association neurons activate motor neurons in several spinal cord segments. The motor neurons generate more nerve impulses, which propagate toward the axon terminals.

⑤ Acetylcholine released by the motor neurons causes the flexor muscles in the thigh (effectors) to contract, producing withdrawal of the leg. Again, the reflex is protective because contraction of flexor muscles moves a limb to avoid pain.

The flexor reflex, like the stretch reflex, is ipsilateral. The incoming and outgoing impulses are on the same side of the spinal cord. The flexor reflex also illustrates another feature of polysynaptic reflex arcs. When you withdraw your entire lower or upper limb from a painful stimulus, more than one muscle group is involved. Several motor neurons must simultaneously convey impulses to several upper or lower limb muscles. Because nerve impulses from one sensory neuron ascend and descend in the spinal cord and activate association neurons in different segments of the spinal cord, this type of reflex is called an **intersegmental reflex arc**. Through intersegmental reflex arcs, a single sensory neuron can activate several motor neurons and thereby cause stimulation of more than one effector. The monosynaptic stretch reflex, on the other hand, involves only muscles receiving nerve impulses from one spinal cord segment.

Something else may happen when you step on a tack. You may start to lose your balance as your body weight shifts to the other foot. Automatically, however, reflexes occur to regain balance so you do not fall. The pain impulses from stepping on the tack not only initiate the flexor reflex that causes you to withdraw the limb but also initiate a balance-maintaining **crossed extensor reflex**, as follows (Fig. 13.10).

① Stepping on a tack stimulates a pain-sensitive neuron (receptor) in the right foot.

② This sensory neuron then generates nerve impulses, which propagate into the spinal cord.

③ Within the spinal cord (integrating center), the sensory neuron activates several association neurons that synapse with motor neurons on the left side of the spinal cord in several spinal cord segments. Thus incoming pain signals cross to the opposite side through association neurons at that level and several levels above and below the point of entry into the spinal cord.

④ The association neurons excite motor neurons in several spinal cord segments. The motor neurons, in turn, generate more nerve impulses, which propagate toward the axon terminals.

⑤ Acetylcholine released by the motor neurons causes extensor muscles in the thigh (effectors) of the unstimulated left limb to contract, producing extension of the left leg. This allows weight to be placed on the foot of the limb that must now support the entire body. A comparable reflex occurs with painful stimulation of the left lower limb or either upper limb.

Unlike the flexor reflex, which is an ipsilateral reflex, the crossed extensor reflex involves a **contralateral** (kon′-tra-LAT-er-al) **reflex arc**: sensory impulses enter one side of the spinal cord and motor impulses exit on the opposite side. Thus a crossed extensor reflex synchronizes withdrawal (flexion) of the stimulated limb with extension of the contralateral limbs.

Reciprocal innervation also occurs in both the flexor reflex and the crossed extensor reflex. In the flexor reflex, when the flexor muscles of a painfully stimulated lower limb are contracting, the extensor muscles of the same limb are relaxing to some degree. If both sets of muscles contracted at the same time, the two sets of muscles would pull in opposite directions on the bones. Because of reciprocal innervation, however, one set of muscles contracts while the other relaxes. In the crossed extensor reflex, while the flexor muscles of the limb that has been stimulated by the tack are contracting to produce withdrawal, the extensor muscles of the opposite lower limb are contracting to help maintain balance.

REFLEXES AND NEUROLOGICAL IMPAIRMENT

Several reflexes have clinical significance and are used to assess certain conditions:

1. **Patellar reflex (knee jerk).** This stretch reflex involves extension of the knee joint by contraction of the quadriceps femoris muscle in response to tapping the patellar ligament (see Fig. 13.7). This reflex is blocked by damage of the sensory or motor nerves to the muscle or the integrating centers in the second, third, or fourth lumbar segments of the spinal cord. Often, it is absent in people with chronic diabetes mellitus or neurosyphilis, which cause degeneration of nerves. It is exaggerated in disease or injury involving certain motor tracts descending from the higher centers of the brain to the spinal cord.

2. **Achilles reflex (ankle jerk).** This stretch reflex involves extension (plantar flexion) of the foot by contraction of the gastrocnemius and soleus muscles in response to tapping the calcaneal (Achilles) tendon. Absence of the Achilles reflex indicates damage to the nerves supplying the posterior leg muscles or to neurons in the lumbosacral region of the spinal cord. This reflex may also disappear in people with chronic diabetes, neurosyphilis, alcoholism, and subarachnoid hemorrhages. An exaggerated Achilles reflex indicates cervical cord compression or a lesion of the motor tracts of the first or second sacral segments of the cord.

3. **Babinski sign.** This reflex results from gentle stroking of the lateral outer margin of the sole of the foot. The great toe extends, with or without fanning of the other toes. This phenomenon occurs in normal children under 1½ years of age and is due to incomplete myelination of fibers in the corticospinal tract. A positive Babinski sign after age 1½ is abnormal and indicates an interruption of the corticospinal tract as the result of a lesion of the tract, usually in the upper portion. The normal response after 1½ years of age is the **plantar flexion reflex**, or **negative**

Figure 13.10 Crossed extensor reflex. The flexor reflex arc is shown so that you can correlate it with the crossed extensor reflex arc. Plus signs (+) indicate excitatory synapses.

 A crossed extensor reflex causes contraction of muscles that extend joints in the limb opposite a painful stimulus.

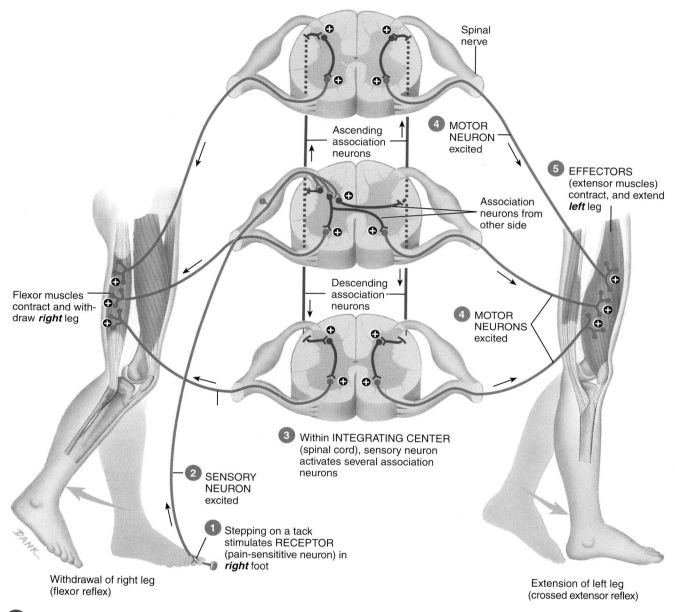

Spinal nerve

Ascending association neurons

4 MOTOR NEURON excited

5 EFFECTORS (extensor muscles) contract, and extend *left* leg

Association neurons from other side

Flexor muscles contract and withdraw *right* leg

Descending association neurons

4 MOTOR NEURONS excited

3 Within INTEGRATING CENTER (spinal cord), sensory neuron activates several association neurons

2 SENSORY NEURON excited

1 Stepping on a tack stimulates RECEPTOR (pain-sensitive neuron) in *right* foot

Withdrawal of right leg (flexor reflex)

Extension of left leg (crossed extensor reflex)

Q Why is the crossed extensor reflex classified as a contralateral reflex arc?

Babinski—a curling under of all the toes, accompanied by a slight turning in and flexion of the anterior part of the foot.

4. **Abdominal reflex.** This reflex involves contraction of the muscles that compress the abdominal wall in response to stroking the side of the abdomen. The response is an abdominal muscle contraction that causes a lateral movement of the umbilicus to the side opposite the stimulus. Absence of this reflex is associated with lesions of the corticospinal tracts. It may also be absent because of lesions of the peripheral nerves, lesions of integrating centers in the thoracic part of the cord, and multiple sclerosis. ■

SPINAL NERVES

Spinal nerves connect the CNS to receptors, muscles, and glands and are part of the peripheral nervous system (PNS). The 31 pairs of spinal nerves (62 total) are named and numbered according to the region and level of the vertebral column from which they emerge (see Fig. 13.2a). The first cervical pair emerges between the atlas (first cervical vertebra) and the occipital bone. All other spinal nerves emerge from the vertebral column through the intervertebral foramina between adjoining vertebrae. Thus there are 8 pairs of cervical nerves, 12 pairs of thoracic, 5 pairs of lumbar, 5 pairs of sacral, and 1 pair of coccygeal nerves.

Not all spinal cord segments are in line with their corresponding vertebrae. Recall that the spinal cord ends near the level of the superior border of the second lumbar vertebra. Thus the roots of the lower lumbar, sacral, and coccygeal nerves descend at an angle to reach their respective foramina before emerging from the vertebral column. This arrangement constitutes the cauda equina (see Fig. 13.2a).

A typical **spinal nerve** has two separate points of attachment to the cord: a posterior (dorsal) root and an anterior (ventral) root (Fig. 13.11). The posterior and anterior roots unite to form a spinal nerve at the intervertebral foramen.

Figure 13.11   Components and connective tissue coverings of a spinal nerve.

Three layers of connective tissue wrappings protect axons: endoneurium surrounds individual axons, perineurium surrounds bundles of axons (fascicles), and epineurium surrounds an entire nerve.

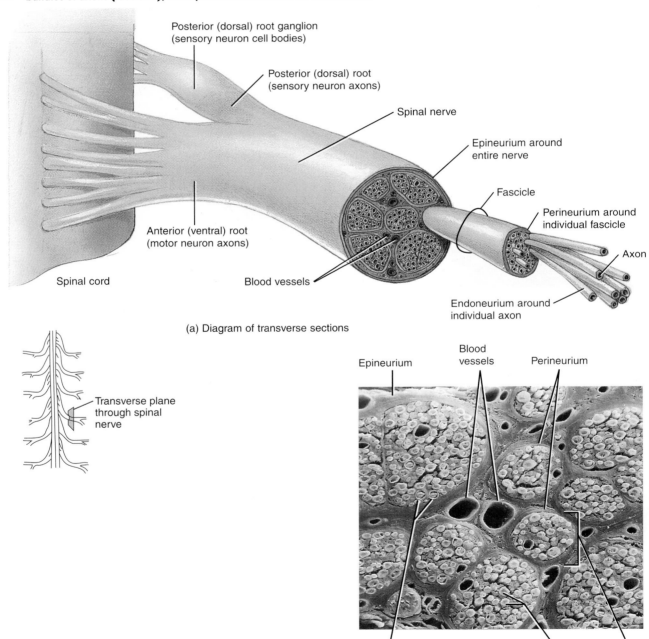

(a) Diagram of transverse sections

(b) Scanning electron micrograph of about ten fascicles of spinal nerves in transverse section (450x)

Why are all spinal nerves classified as mixed nerves?

Since the posterior root contains sensory axons and the anterior root contains motor axons, a spinal nerve is a **mixed nerve**, at least at its origin. The posterior (dorsal) root contains a ganglion in which cell bodies of sensory neurons are located.

Connective Tissue Coverings

Each cranial nerve and spinal nerve is surrounded by protective connective tissue coverings (Fig. 13.11). Individual axons, whether myelinated or unmyelinated, are wrapped in **endoneurium** (en′-dō-NOO-rē-um). Groups of axons with their endoneurium are arranged in bundles called **fascicles**, and each fascicle is wrapped in **perineurium** (per′-i-NOO-rē-um). The superficial covering around the entire nerve is the **epineurium** (ep′-i-NOO-rē-um). The dura mater of the spinal meninges fuses with the epineurium as the nerve passes through the intervertebral foramen. Note the many blood vessels, which nourish nerves, within the coverings (Fig. 13.11b). You may recall from Chapter 10 that the con-

nective tissue coverings of skeletal muscles—endomysium, perimysium, and epimysium—are similar in organization to those of nerves. This organization is common to excitable tissues.

Distribution of Spinal Nerves

Branches

A short distance after passing through its intervertebral foramen, a spinal nerve divides into several branches (Fig. 13.12). These branches are known as **rami** (RĀ-mī; singular is **ramus**). The **dorsal ramus** (RĀ-mus) serves the deep muscles and skin of the dorsal surface of the trunk. The **ventral ramus** serves the muscles and structures of the upper and lower limbs and the lateral and ventral trunk. In addition to dorsal and ventral rami, spinal nerves also give off a **meningeal branch**. This branch reenters the spinal canal through the intervertebral foramen and supplies the vertebrae, vertebral ligaments, blood vessels of the spinal cord, and meninges. Other branches of a spinal nerve are the **rami**

Figure 13.12   Branches of a typical spinal nerve.

 The branches of a spinal nerve are the dorsal ramus, ventral ramus, meningeal branch, and rami communicantes.

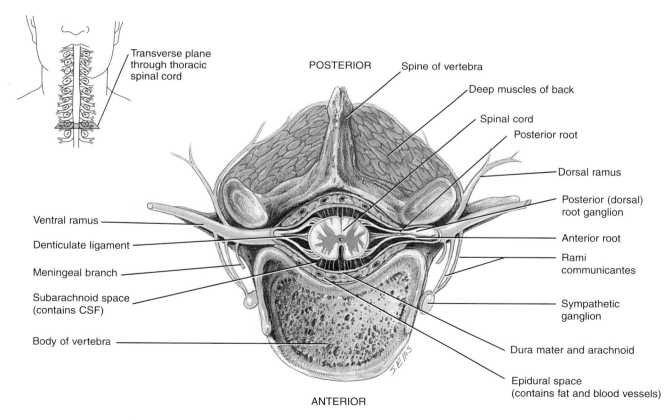

Transverse section

 Which spinal nerve branch serves the upper and lower limbs?

communicantes (kō-myoo-ni-KAN-tēz), components of the autonomic nervous system whose structure and function are discussed in Chapter 17.

Plexuses

The ventral rami of spinal nerves, except for thoracic nerves T2–T12, do not go directly to the body structures they supply. Instead, they form networks on both left and right sides of the body by joining with varying numbers of nerve fibers from ventral rami of adjacent nerves. Such a network is called a **plexus** (*plexus* = braid). The principal plexuses are the **cervical plexus**, **brachial plexus**, **lumbar plexus**, and **sacral plexus**. Look back at Fig. 13.2a to see their relationships to one another. Emerging from the plexuses are nerves bearing names that are often descriptive of the general regions they serve or the course they take. Each of the nerves, in turn, may have several branches named for the specific structures they innervate.

The principal plexuses are summarized in Exhibits 13.1–13.4.

Intercostal (Thoracic) Nerves

The ventral rami of spinal nerves T2–T12 do not enter into the formation of plexuses and are known as **intercostal (thoracic) nerves**. These nerves directly innervate the structures they supply in the intercostal spaces (see Fig. 13.2a). After leaving its intervertebral foramen, the ventral ramus of nerve T2 supplies the intercostal muscles of the second intercostal space and the skin of the axilla and posteromedial aspect of the arm. Nerves T3–T6 pass in the costal grooves of the ribs and then to the intercostal muscles and skin of the anterior and lateral chest wall. Nerves T7–T12 supply the intercostal muscles and the abdominal muscles and overlying skin. The dorsal rami of the intercostal nerves supply the deep back muscles and skin of the dorsal aspect of the thorax.

EXHIBIT **13.1** **CERVICAL PLEXUS**

Overview: The **cervical** (SER-vi-kul) **plexus** is formed by the ventral rami of the first four cervical nerves (C1–C4) with contributions from C5. There is one on each side of the neck alongside the first four cervical vertebrae. The roots of the plexus indicated in the diagram are the ventral rami.

The cervical plexus supplies the skin and muscles of the head, neck, and superior part of the shoulders. Branches of the cervical plexus also run parallel to cranial nerves XI (accessory) and XII (hypoglossal). The phrenic nerves arise from the cervical plexuses and supply motor fibers to the diaphragm.

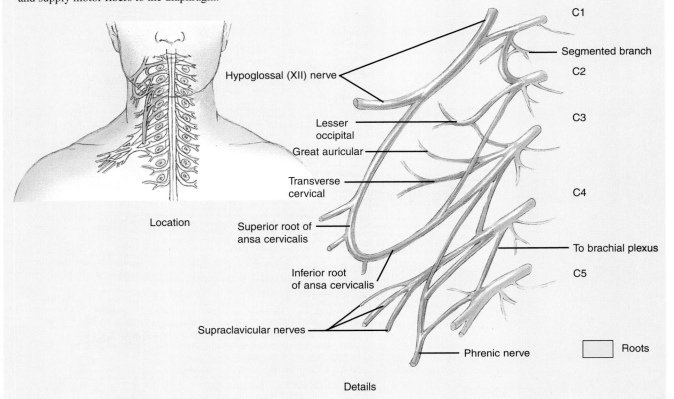

Location

Details

Nerve	Origin	Distribution
SUPERFICIAL (SENSORY) BRANCHES		
Lesser occipital (ok′-SIP-i-tal)	C2.	Skin of scalp posterior and superior to ear.
Great auricular (aw-RIK-yoo-lar)	C2–C3.	Skin anterior, inferior, and over ear and over parotid glands.
Transverse cervical (SER-vi-kul)	C2–C3.	Skin over anterior aspect of neck.
Supraclavicular (soo′-pra-kla-VIK-yoo-lar)	C3–C4.	Skin over superior portion of chest and shoulder.
DEEP (LARGELY MOTOR) BRANCHES		
Ansa cervicalis (AN-sa ser-vi-KAL-is)		This nerve divides into a superior root and an inferior root.
Superior root	C1.	Infrahyoid and geniohyoid muscles of neck.
Inferior root	C2–C3.	Infrahyoid muscles of neck.
Phrenic (FREN-ik)	C3–C5.	Diaphragm between thorax and abdomen.
Segmental (seg-MEN-tal) **branches**	C1–C5.	Prevertebral (deep) muscles of neck, levator scapulae, and middle scalene muscles.

Damage of Cervical Spinal Cord: Damage to the spinal cord above the origin of the phrenic nerves (C3, C4, and C5) causes respiratory arrest. Breathing stops because the phrenic nerves no longer send nerve impulses to the diaphragm.

EXHIBIT **13.2 BRACHIAL PLEXUS**

Overview: The ventral rami of spinal nerves C5–C8 and T1 form the **brachial** (BRĀ-kē-al) **plexus**. The brachial plexus extends inferiorly and laterally on either side of the last four cervical and first thoracic vertebrae. It passes superior to the first rib behind the clavicle and then enters the axilla.

The brachial plexus provides the entire nerve supply of the shoulder and upper limb. Five important nerves arise from the brachial plexus. (1) The *axillary nerve* supplies the deltoid and teres minor muscles. (2) The *musculocutaneous nerve* supplies the flexors of the arm. (3) The *radial nerve* supplies the muscles on the posterior aspect of the arm and forearm. (4) The *median nerve* supplies most of the muscles of the anterior forearm and some of the muscles of the hand. (5) The *ulnar nerve* supplies the anteromedial muscles of the forearm and most of the muscles of the hand.

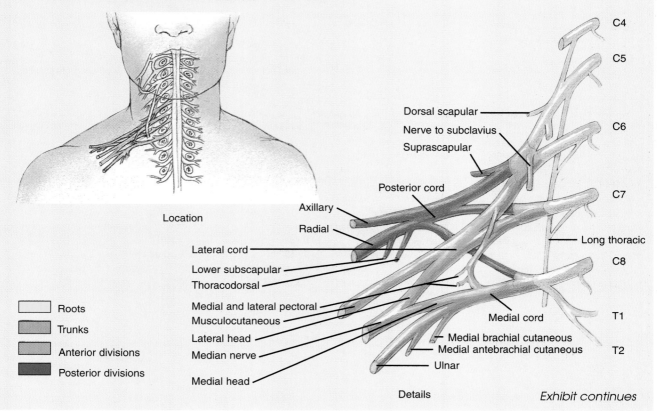

Location

Roots
Trunks
Anterior divisions
Posterior divisions

Dorsal scapular
Nerve to subclavius
Suprascapular
Posterior cord
Axillary
Radial
Lateral cord
Lower subscapular
Thoracodorsal
Medial and lateral pectoral
Musculocutaneous
Lateral head
Median nerve
Medial head

Medial cord
Medial brachial cutaneous
Medial antebrachial cutaneous
Ulnar

C4
C5
C6
C7
Long thoracic
C8
T1
T2

Details   *Exhibit continues*

EXHIBIT **13.2 BRACHIAL PLEXUS** (CONTINUED)

Nerve	Origin	Distribution
Dorsal scapular (SKAP-yoo-lar)	C5.	Levator scapulae, rhomboideus major, and rhomboideus minor muscles.
Long thoracic (tho-RAS-ik)	C5–C7.	Serratus anterior muscle.
Nerve to subclavius (sub-KLĀ-ve-us)	C5–C6.	Subclavius muscle.
Suprascapular (soo′-pra-SKAP-yoo-lar)	C5–C6.	Supraspinatus and infraspinatus muscles.
Musculocutaneous (mus′-kyoo-lo-kyoo-TĀN-ē-us)	C5–C7.	Coracobrachialis, biceps brachii, and brachialis muscles.
Lateral pectoral (PEK-to-ral)	C5–C7.	Pectoralis major muscle.
Upper subscapular (sub-SKAP-yoo-lar)	C5–C6.	Subscapularis muscle.
Thoracodorsal (tho-RA-ko-dor-sal)	C6–C8.	Latissimus dorsi muscle.
Lower subscapular (sub-SKAP-yoo-lar)	C5–C6.	Subscapularis and teres major muscles.
Axillary (AK-si-lar-ē) or **circumflex** (SER-kum-fleks)	C5–C6.	Deltoid and teres minor muscles; skin over deltoid and superior posterior aspect of arm.

Median
    Lateral head
    Medial head
	
	C5–C7.
	C5–C8 and T1.
		Flexors of forearm, except flexor carpi ulnaris and some muscles of the hand (lateral palm); skin of lateral two-thirds of palm of hand and fingers.

Radial (RĀ-dē-al)	C5–C8 and T1.	Triceps brachii and other extensor muscles of arm and extensor muscles of forearm; skin of posterior arm and forearm, lateral two-thirds of dorsum of hand, and fingers over proximal and middle phalanges.
Medial pectoral (PEK-to-ral)	C8–T1.	Pectoralis major and pectoralis minor muscles.
Medial brachial (BRĀ-kē-al) **cutaneous** (kyoo′-TĀ-ne-us)	C8–T1.	Skin of medial and posterior aspects of distal third of arm.
Medial antebrachial cutaneous (an′-tē-BRĀ-kē-al kyoo′-TĀ-ne-us)	C8–T1.	Skin of medial and posterior aspects of forearm.
Ulnar (UL-nar)	C8–T1.	Flexor carpi ulnaris, flexor digitorum profundus, and most muscles of the hand; skin of medial side of hand, little finger, and medial half of ring finger.

Injuries to the Brachial Plexus: Prolonged use of a crutch that presses into the axilla may result in injury to a portion of the brachial plexus. The usual **crutch palsy** involves the posterior cord of the brachial plexus or, more often, just the radial nerve, which, in general, supplies extensors.

    Radial nerve damage is indicated by wrist drop, inability to extend the hand at the wrist. Care must be taken not to injure the radial and axillary nerves during intramuscular injections into the deltoid muscle. The radial nerve may also be injured when a cast is applied too tightly around the midhumerus.

    Median nerve damage is indicated by numbness, tingling, and pain in the palm and fingers; weak thumb movements; inability to pronate the forearm; and difficulty in flexing the wrist properly.

    Carpal tunnel syndrome results from compression of the median nerve inside the carpal tunnel. This narrow passageway is formed anteriorly by the flexor retinaculum (transverse carpal ligament) and posteriorly by the carpal bones (see Fig. 11.17a,b). It may be caused by any condition that aggravates compression of the contents of the carpal tunnel, such as trauma, edema, and repetitive flexion of the wrist as a result of activities such as playing video games, keyboarding at a computer terminal or typing, driving a car, cutting hair, and playing a piano.

    Ulnar nerve damage is indicated by an inability to adduct or abduct the four fingers (not the thumb), weakness in flexing and adducting the wrist, and loss of sensation over the little finger.

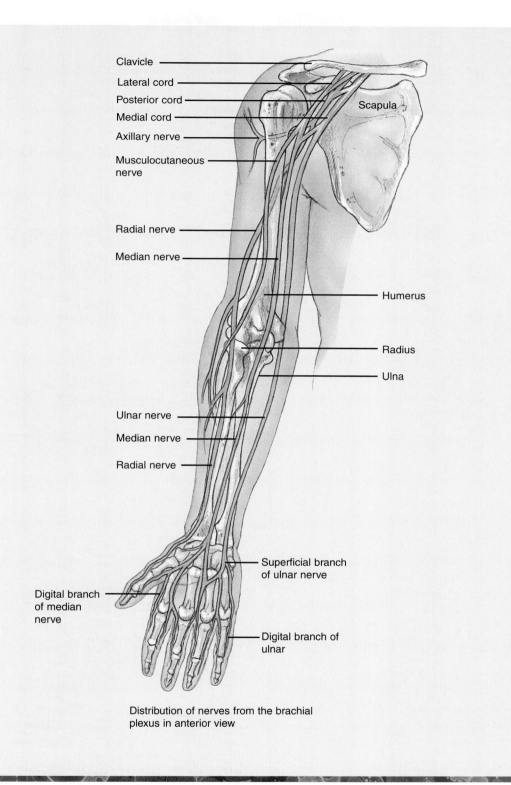

Clavicle

Lateral cord

Posterior cord

Medial cord

Axillary nerve

Musculocutaneous
nerve

Radial nerve

Median nerve

Ulnar nerve

Median nerve

Radial nerve

Digital branch
of median
nerve

Scapula

Humerus

Radius

Ulna

Superficial branch
of ulnar nerve

Digital branch of
ulnar

Distribution of nerves from the brachial
plexus in anterior view

EXHIBIT **13.3 LUMBAR PLEXUS**

Overview: Ventral rami of spinal nerves L1–L4 form the **lumbar** (LUM-bar) **plexus**. It differs from the brachial plexus in that there is no intricate intermingling of fibers. On either side of the first four lumbar vertebrae, the lumbar plexus passes obliquely outward posterior to the psoas major muscle and anterior to the quadratus lumborum muscle. It then gives rise to its peripheral nerves.

The lumbar plexus supplies the anterolateral abdominal wall, external genitals, and part of the lower limb.

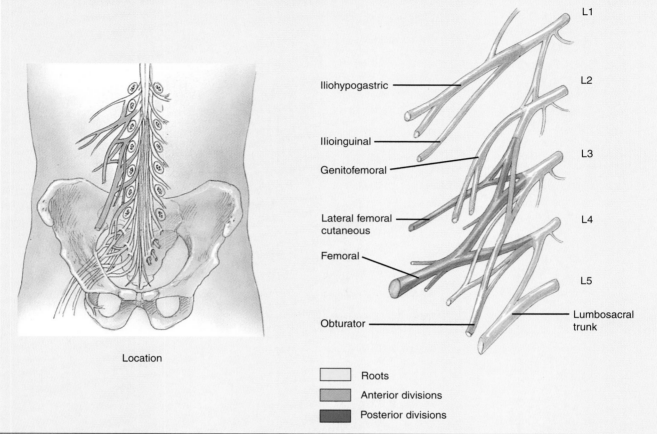

Location

Nerve	Origin	Distribution
Iliohypogastric (il′-ē-o-hī-pō-GAS-trik)	L1.	Muscles of anterolateral abdominal wall; skin of inferior abdomen and buttock.
Ilioinguinal (il′-ē-ō-IN-gwi-nal)	L1.	Muscles of anterolateral abdominal wall; skin of superior medial aspect of thigh, root of penis and scrotum in male, and labia majora and mons pubis in female.
Genitofemoral (jen′-i-tō-FEM-or-al)	L1–L2.	Cremaster muscle; skin over middle anterior surface of thigh, scrotum in male, and labia majora in female.
Lateral femoral cutaneous (FEM-or-al kyoo′-TĀ-nē-us)	L2–L3.	Skin over lateral, anterior, and posterior aspects of thigh.
Femoral (FEM-or-al)	L2–L4.	Flexor muscles of thigh and extensor muscles of leg; skin over anterior and medial aspect of thigh and medial side of leg and foot.
Obturator (OB-too-rā-tor)	L2–L4.	Adductor muscles of leg; skin over medial aspect of thigh.

Femoral Nerve Injury: The largest nerve arising from the lumbar plexus is the femoral nerve. Injury to the femoral nerve is indicated by an inability to extend the leg and by loss of sensation in the skin over the anteromedial aspect of the thigh.

Figure 14.2 The protective coverings of the brain.

Cranial bones and cranial meninges protect the brain.

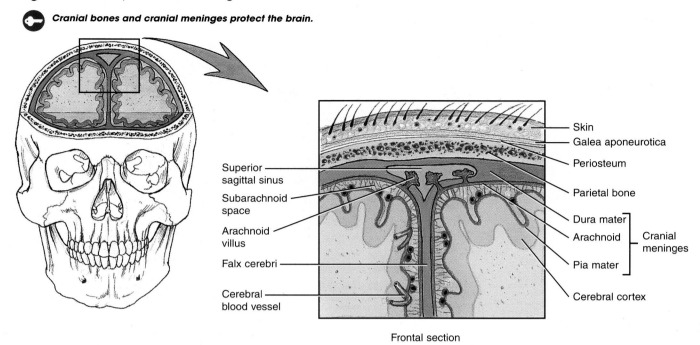

Frontal section

What are the three layers of the cranial meninges, in order from superficial to deep?

around the brain and spinal cord, and through cavities within the brain and spinal cord.

Figure 14.3 shows the four CSF-filled cavities within the brain, which are called **ventricles** (VEN-tri-kuls; *ventriculus* = little belly or cavity). Each of the two **lateral ventricles** is located in a hemisphere of the cerebrum. The **third ventricle** is a narrow cavity at the midline superior to the hypothalamus and between the right and left halves of the thalamus. The **fourth ventricle** lies between the brain stem and the cerebellum.

The entire central nervous system contains between 80–150 ml (3–5 oz) of CSF. It is a clear, colorless liquid that contains glucose, proteins, lactic acid, urea, cations (Na^+, K^+, Ca^{2+}, Mg^{2+}), and anions (Cl^- and HCO_3^-). It also contains some white blood cells.

CSF contributes to homeostasis in three main ways:

1. **Mechanical protection.** The fluid serves as a shock-absorbing medium to protect the delicate tissue of the brain and spinal cord from jolts that would otherwise cause them to crash against the bony walls of the cranial and vertebral cavities. The fluid also buoys the brain so that it "floats" in the cranial cavity.
2. **Chemical protection.** CSF provides an optimal chemical environment for accurate neuronal signaling. Even slight changes in the ionic composition of CSF within the brain could seriously disrupt production of post-synaptic potentials and action potentials.
3. **Circulation.** CSF is a medium for exchange of nutrients and waste products between the blood and nervous tissue.

The **choroid** (KŌ-royd; *chorion* = membrane) **plexuses** (Fig. 14.4) are networks of capillaries (microscopic blood vessels) in the walls of the ventricles. The capillaries are covered by ependymal cells that form cerebrospinal fluid from blood plasma by filtration and secretion. Because the ependymal cells are joined by tight junctions (see Fig. 4.1), materials entering CSF from choroid capillaries cannot leak between these cells. Rather, they must go through the ependymal cells. This **blood–cerebrospinal fluid barrier** permits certain substances to enter the fluid but excludes others. Such a barrier protects the brain and spinal cord from potentially harmful substances in the blood.

The CSF formed in the choroid plexuses of each lateral ventricle flows into the third ventricle through a pair of narrow, oval openings, the **interventricular foramina (foramina of Monro)**. See also Fig. 14.3. More CSF is added by the choroid plexus in the roof of the third ventricle. The fluid then flows through the **cerebral aqueduct (aqueduct of Sylvius)**, which passes through the midbrain, into the fourth ventricle. The choroid plexus of the fourth ventricle contributes more fluid. CSF enters the subarachnoid space through three openings in the roof of the fourth ventricle: a **median aperture (of Magendie)** and the paired **lateral apertures (of Luschka)**, one on each side. It then circulates

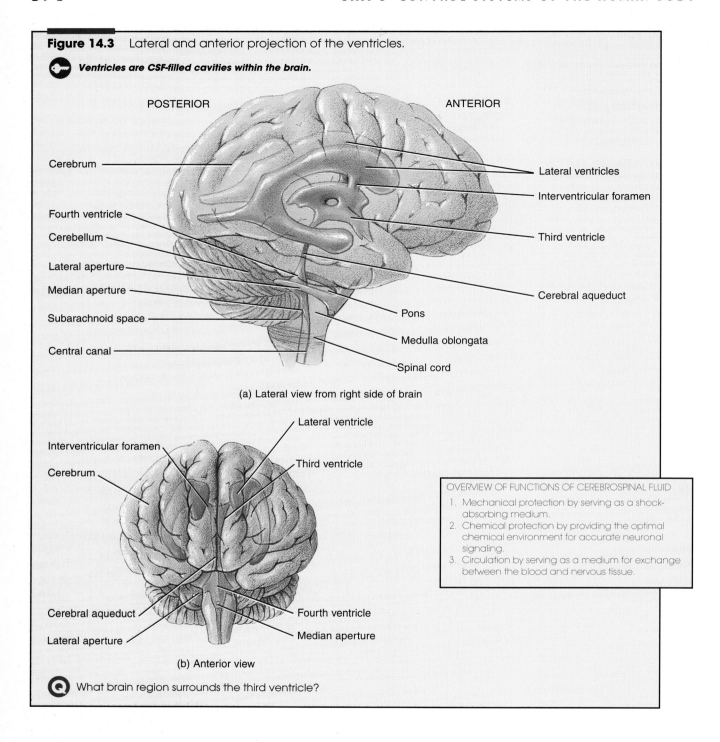

Figure 14.3   Lateral and anterior projection of the ventricles.

Ventricles are CSF-filled cavities within the brain.

(a) Lateral view from right side of brain

(b) Anterior view

Q What brain region surrounds the third ventricle?

OVERVIEW OF FUNCTIONS OF CEREBROSPINAL FLUID

1. Mechanical protection by serving as a shock-absorbing medium.
2. Chemical protection by providing the optimal chemical environment for accurate neuronal signaling.
3. Circulation by serving as a medium for exchange between the blood and nervous tissue.

in the central canal of the spinal cord and in the subarachnoid space around the surface of the brain and spinal cord. CSF is gradually reabsorbed back into the blood through **arachnoid villi**. These are fingerlike extensions of the arachnoid that project into the dural venous sinuses, especially the **superior sagittal sinus** (see Fig. 14.2). Normally, CSF is reabsorbed as rapidly as it is formed by the choroid plexuses, at a rate of about 20 ml/hr (480 ml/day). Because the rates of formation and reabsorption are the same, the pressure of CSF normally is constant.

HYDROCEPHALUS

An obstruction, such as a tumor or a congenital blockage, or an inflammation in the brain can interfere with the drainage of CSF from the ventricles into the subarachnoid space. As fluid accumulates in the ventricles, the CSF pressure rises. This condition is called **hydrocephalus** (hī′-drō-SEF-a-lus; *hydro* = water; *kephale* = head). In a baby, if the fontanels have not yet closed, the head bulges in response to the increased pressure. In time, however, the fluid

Figure 14.4 Meninges and ventricles of the brain. Arrows indicate the direction of flow of cerebrospinal fluid.

CSF is formed by ependymal cells that cover the choroid plexuses of the ventricles.

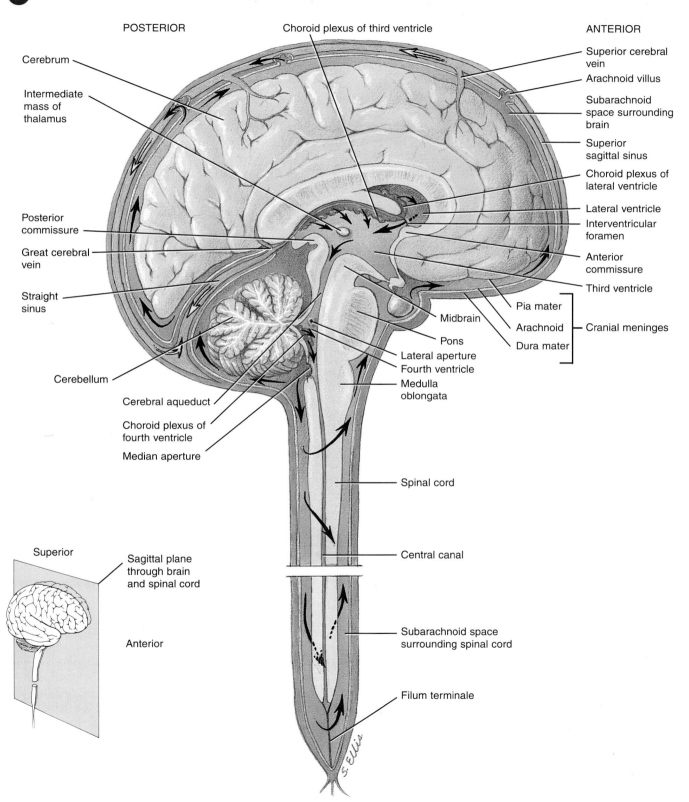

POSTERIOR

Choroid plexus of third ventricle

ANTERIOR

Cerebrum

Intermediate mass of thalamus

Posterior commissure

Great cerebral vein

Straight sinus

Cerebellum

Cerebral aqueduct

Choroid plexus of fourth ventricle

Median aperture

Superior cerebral vein

Arachnoid villus

Subarachnoid space surrounding brain

Superior sagittal sinus

Choroid plexus of lateral ventricle

Lateral ventricle

Interventricular foramen

Anterior commissure

Third ventricle

Pia mater

Arachnoid — Cranial meninges

Dura mater

Midbrain

Pons

Lateral aperture

Fourth ventricle

Medulla oblongata

Spinal cord

Central canal

Subarachnoid space surrounding spinal cord

Filum terminale

Superior

Sagittal plane through brain and spinal cord

Anterior

S. Ellis

(a) Sagittal section of brain, ventricles, spinal cord, and meninges

Figure continues

Figure 14.4   (continued)

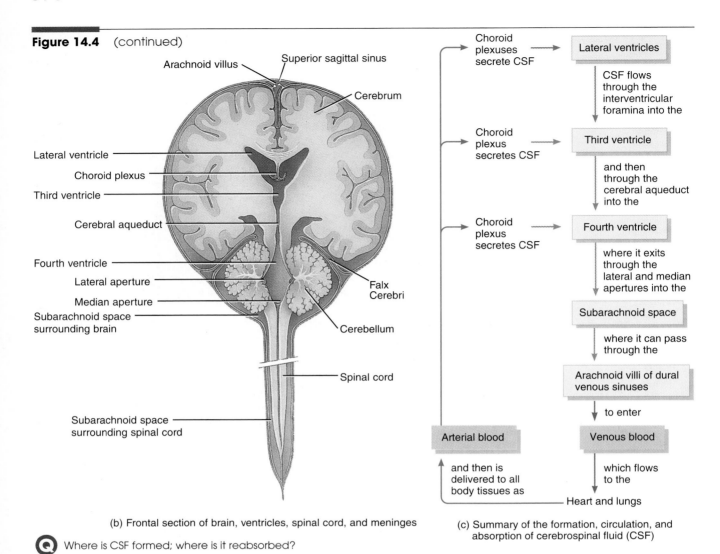

(b) Frontal section of brain, ventricles, spinal cord, and meninges

Ⓠ Where is CSF formed; where is it reabsorbed?

(c) Summary of the formation, circulation, and absorption of cerebrospinal fluid (CSF)

buildup will compress and damage the delicate nervous tissue. To prevent damage, a neurosurgeon may implant a drain line to divert CSF from a brain ventricle into the subclavian vein and then into the right atrium of the heart. In adults, hydrocephalus may occur following head injury, meningitis, or subarachnoid hemorrhage. ∎

Blood Supply

The brain is well supplied with oxygen and nutrients, mainly by blood vessels that form the cerebral arterial circle (circle of Willis) at the base of the brain. Cerebral circulation is shown in Figs. 21.20 (page 640 and 641) and 21.24 (page 651). Blood vessels that enter brain tissue pass along the surface of the brain, and as they penetrate inward, they are surrounded by a loose-fitting layer of pia mater. The amount of oxygen used by a given brain region varies with the degree of activity in that area. When activity of neurons and neuroglia increases in a region of the brain, blood flow to that area also increases.

In an adult, the brain comprises only 2% of total body weight, but it consumes about 20% of the oxygen used at rest. Even a brief interruption of brain blood flow may cause unconsciousness. Lysosomes of brain cells are sensitive to decreased oxygen concentration. If the condition persists long enough, lysosomes break open and release enzymes that destroy neurons and neuroglia. An interruption in blood flow for even 1 or 2 minutes may impair brain cells. If the cells are totally deprived of oxygen for about 4 minutes, many are permanently injured. Occasionally during childbirth, the oxygen supply from the mother's blood is interrupted before the baby leaves the birth canal and can breathe on its own. This can result in the baby being stillborn or suffering permanent brain damage that may result in mental retardation, epilepsy, or paralysis.

Blood supplying the brain also contains glucose, the principal molecule that neurons use to make ATP. Because carbohydrate storage in the brain is limited, the supply of glucose must be continuous. If blood entering the brain has a low level of glucose, mental confusion, dizziness, convulsions, and loss of consciousness may occur.

Glucose, oxygen, carbon dioxide, water, and most lipid-soluble substances, such as alcohol, caffeine, nicotine, heroin, and most anesthetics, pass rapidly from the circulating blood into brain cells. Other substances, such as creatinine, urea, and most ions, such as Na^+, K^+, and Cl^-, enter quite slowly. Still other substances—proteins and most antibiotic drugs—do not pass at all from the blood into brain cells. The different rates of passage of certain materials from the blood into most parts of the brain depend on the **blood–brain barrier** (**BBB**). Brain capillaries are much less permeable than most other body capillaries. Tight junctions seal together the endothelial cells of brain capillaries, which also are surrounded by a continuous basement membrane. Also, processes of large numbers of astrocytes (one type of neuroglial cell) press up against the capillaries. Astrocytes are thought to selectively pass some substances from the blood but inhibit the passage of others. Substances that cross the BBB are either soluble in lipids or are water-soluble substances that receive the assistance of a transporter (carrier) to cross by active transport.

BREACHING THE BLOOD–BRAIN BARRIER

The blood–brain barrier functions as a selective anatomical and physiological barrier to protect brain cells from harmful substances and pathogens. An injury to the brain due to trauma, inflammation, or toxins may cause a breakdown of the BBB, permitting the passage of substances into brain tissue that normally are kept out. On the other hand, the BBB may prevent entry of drugs that could be used as therapy for brain cancer or other CNS disorders. Researchers are finding ways to slip drugs past the BBB. In one method, the drug is injected together with a concentrated sugar solution. Temporarily, the endothelial cells shrink, due to the osmotic effect of the sugar solution, and pull apart to open up gaps in the tight junctions. The result is entry of the drug into the brain tissue. ■

Several small brain regions lying in the walls of the third and fourth ventricles, called **circumventricular organs** (**CVOs**), can monitor chemical changes in the blood because they lack the blood–brain barrier. CVOs include part of the hypothalamus, the pineal gland, the pituitary gland, and a few other nearby structures. Functionally, these regions coordinate homeostatic activities of the endocrine and nervous systems, such as regulation of blood pressure, fluid balance, hunger, and thirst. CVOs are also thought to be the entry site for the virus that causes AIDS (HIV) into the brain. In some cases of HIV infection, dementia (irreversible deterioration of mental state) and other neurologic disorders may appear before other symptoms of AIDS become apparent.

Brain Stem

The brain stem connects the spinal cord to the diencephalon and consists of the medulla oblongata, pons, reticular formation, and midbrain.

Medulla Oblongata

The **medulla oblongata** (me-DULL-la ob′-long-GA-ta), or more simply the **medulla**, develops from the myelencephalon. It is a continuation of the superior portion of the spinal cord and forms the inferior part of the brain stem (see Fig. 14.1). The medulla begins at the foramen magnum and extends upward to the inferior border of the pons, a distance of about 3 cm (1.2 in.).

Within the medulla are all ascending (sensory) and descending (motor) white matter tracts that connect the spinal cord with the brain and many nuclei (gray matter masses of cell bodies and dendrites in the CNS) that regulate various vital body functions. It also contains the nuclei that receive sensory input from or provide motor output to five of the twelve cranial nerves (see Exhibit 14.4 on page 418). Major structural and functional regions of the medulla are as follows.

1. Most sensory and motor tracts cross over from the left to right (or right to left) side as they pass through the medulla. On the anterior aspect of the medulla are two bulges called the **pyramids** (Figs. 14.5 and 14.6). They contain the largest motor tracts that pass from the cerebrum to the spinal cord. Just superior to the junction of the medulla with the spinal cord, most of the axons in the left pyramid cross to the right side, and most of the axons in the right pyramid cross over to the left. This crossing is called the **decussation** (dē′-ku-SĀ-shun) **of pyramids**. Thus neurons in the left cerebral cortex control muscles on the right side of the body, and neurons in the right cerebral cortex control muscles on the left side.

2. The dorsal side of the medulla contains two pairs of prominent nuclei. They are the right and left **nucleus gracilis** (gras-I-lis; *gracilis* = slender) and **nucleus cuneatus** (kyoo-nē-Ā-tus; *cuneus* = wedge). Some ascending sensory axon terminals form synapses in these nuclei and postsynaptic neurons then relay the sensory information to the thalamus on the opposite side (see Fig. 15.4). Most sensory nerve impulses initiated on one side of the body cross to the opposite side either in the spinal cord or in these medullary nuclei and finally are received in the cerebral cortex on the opposite side by way of the thalamus.

3. The **cardiovascular center** regulates the rate and force of the heartbeat and the diameter of blood vessels (see Fig. 21.14).

4. The **medullary rhythmicity area** adjusts the basic rhythm of breathing (see Fig. 23.24).

5. Other centers in the medulla coordinate swallowing, vomiting, coughing, sneezing, and hiccuping.

Figure 14.5   Brain stem. The arrow in the inset in (a) indicates the direction from which the brain is viewed (inferior).

 The brain stem consists of the medulla oblongata, pons, and midbrain.

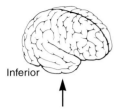

Inferior

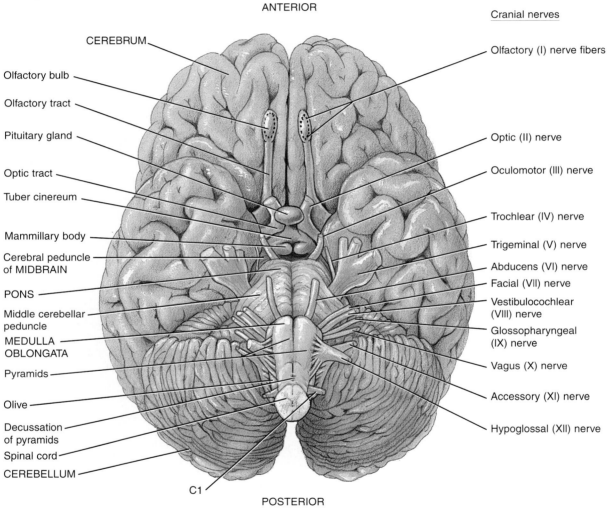

ANTERIOR

Cranial nerves

CEREBRUM

Olfactory (I) nerve fibers

Olfactory bulb

Olfactory tract

Pituitary gland

Optic (II) nerve

Optic tract

Oculomotor (III) nerve

Tuber cinereum

Trochlear (IV) nerve

Mammillary body

Trigeminal (V) nerve

Cerebral peduncle of MIDBRAIN

Abducens (VI) nerve

PONS

Facial (VII) nerve

Middle cerebellar peduncle

Vestibulocochlear (VIII) nerve

MEDULLA OBLONGATA

Glossopharyngeal (IX) nerve

Pyramids

Vagus (X) nerve

Olive

Accessory (XI) nerve

Decussation of pyramids

Hypoglossal (XII) nerve

Spinal cord

CEREBELLUM

C1

POSTERIOR

(a) Inferior aspect of brain showing brain stem in relation to cranial nerves and associated structures

2. **Commissural fibers** transmit impulses from the gyri in one cerebral hemisphere to the corresponding gyri in the opposite cerebral hemisphere. Three important groups of commissural fibers are the **corpus callosum**, **anterior commissure**, and **posterior commissure**.

3. **Projection fibers** form descending and ascending tracts that transmit impulses from the cerebrum and other parts of the brain to the spinal cord or from the spinal cord to the brain. The **internal capsule**, a thick band of sensory and motor axons lying lateral to the thalamus (see Fig. 14.9a), is an example.

Basal Ganglia

The **basal ganglia** are several groups of nuclei in each cerebral hemisphere (Figs. 14.9 and 14.13). The largest nucleus in the basal ganglia is the **corpus striatum** (strī-Ā-tum; *corpus* = body; *striatum* = striped). It consists of the **caudate** (*cauda* = tail) **nucleus** and the **lenticular** (*lenticula* = shaped like a lentil or lens) **nucleus**. The lenticular nucleus, in turn, is subdivided into a lateral portion called the **putamen** (pu-TĀ-men; *putamen* = shell) and a medial portion called the **globus pallidus** (*globus* = ball; *pallid* = pale). The portion of the internal capsule passing between the lenticular nucleus

---

Figure 14.13 Basal ganglia. (a) In this diagram, the basal ganglia have been projected to the surface and are shown in dark green. Refer to Fig. 14.9a for the positions of the basal ganglia in a frontal section of the cerebrum. (b) Left cerebral hemisphere showing portions of the basal ganglia.

 The basal ganglia control large automatic movements of skeletal muscles and muscle tone.

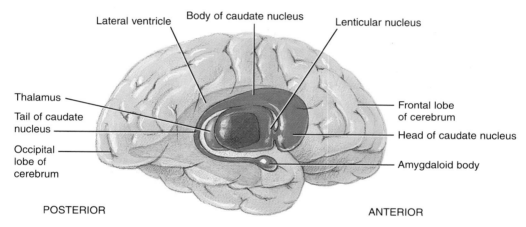

(a) Diagram of lateral view of right side of brain

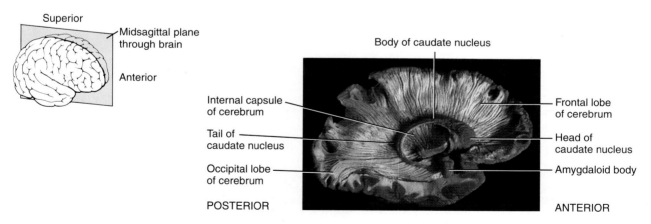

(b) Photograph of medial view of basal ganglia structures after scooping out gray and white matter from a midsagittal section

 What brain regions comprise the basal ganglia? (See Fig. 14.9 also.) In (b), what structure was removed from the midline region to reveal the internal capsule?

and thalamus is sometimes considered part of the corpus striatum (see Fig. 14.9).

Other structures that are functionally linked to and sometimes considered part of the basal ganglia are the **substantia nigra** and **red nuclei** of the midbrain (see Fig. 14.7) and the **subthalamic nuclei** of the diencephalon (see Fig. 14.9a). Axons from the substantia nigra terminate in the caudate nucleus and putamen. The subthalamic nuclei connect with the globus pallidus.

The basal ganglia are interconnected by many nerve fibers. They also receive input from and provide output to the cerebral cortex, thalamus, and hypothalamus. The caudate nucleus and the putamen control large automatic movements of skeletal muscles, such as swinging the arms while walking. The globus pallidus is concerned with the regulation of muscle tone required for specific body movements.

Limbic System

Encircling the brain stem is a ring of structures on the inner border of the cerebrum and floor of the diencephalon that forms the **limbic** (*limbus* = border) **system**. Among its components are the following regions (Fig. 14.14):

1. The **limbic lobe** consists of the **parahippocampal** and **cingulate** (*cingo* = to surround) **gyri**, both gyri of the cerebral hemispheres, and the **hippocampus** (*hippokampos* =

seahorse), a portion of the parahippocampal gyrus that extends into the floor of the lateral ventricle.

2. The **dentate** (*dentatus* = toothed) **gyrus** is between the hippocampus and parahippocampal gyrus.

3. The **amygdaloid** (*amygdale* = almond-shaped) **body (amygdala)** is several groups of neurons located at the tail end of the caudate nucleus.

4. The **septal nuclei** are within the septal area, formed by the regions under the corpus callosum and a cerebral gyrus (paraterminal).

5. The **mammillary bodies of the hypothalamus** are two round masses close to the midline near the cerebral peduncles.

6. The **anterior nucleus of the thalamus** is located in the floor of the lateral ventricle.

7. The **olfactory bulbs** are flattened bodies of the olfactory pathway that rest on the cribriform plate.

8. Bundles of **interconnecting myelinated axons** link various components of the limbic system. They include the fornix, stria terminalis, stria medullaris, medial forebrain bundle, and mammillothalamic tract.

The limbic system governs emotional aspects of behavior. The hippocampus, together with portions of the cerebrum, also functions in memory. Because of this relationship, events that cause a strong emotional response are remembered much more efficiently than those that do not. Memory

---

Figure 14.14   Components of the limbic system and surrounding structures.

 The limbic system governs emotional aspects of behavior.

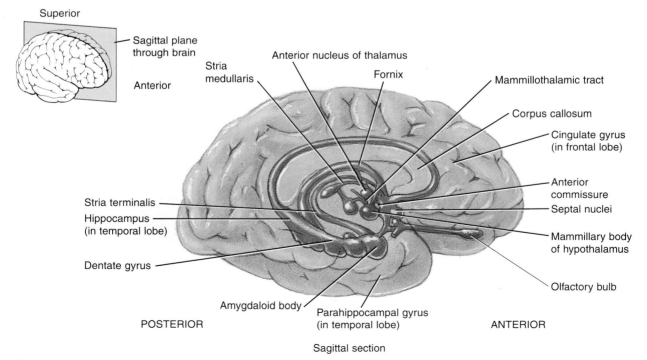

Sagittal section

 Which part of the limbic system functions with the cerebrum in memory?

impairment results from lesions in parts of the limbic system. People with such damage forget recent events and cannot commit anything to memory. Exactly how the limbic system functions in memory is not clear. Although behavior is a function of the entire nervous system, the limbic system controls most of its involuntary aspects.

Experiments have shown that the limbic system is associated with pleasure and pain. When certain areas of the limbic system are stimulated in animals, their reactions indicate they are experiencing intense punishment. When other areas are stimulated, the animals' reactions indicate they are experiencing extreme pleasure. Stimulation of the amygdaloid body or certain nuclei of the hypothalamus in a cat results in a behavioral pattern called rage. The cat assumes a defensive posture—extending its claws, raising its tail, hissing, spitting, and opening its eyes wide. Stimulating other areas of the limbic system results in an opposite behavioral pattern: docility, tameness, and affection. Because the limbic system has a primary function in emotions such as pain, pleasure, anger, rage, fear, sorrow, sexual feelings, docility, and affection, it is sometimes called the "emotional" brain.

BRAIN INJURIES

Brain injuries are commonly associated with head injuries and result in part from displacement and distortion of neuronal tissue at the moment of impact. In addition, some damage occurs when normal blood flow is restored after a period of ischemia (reduced blood flow). The sudden increase in oxygen level produces large numbers of oxygen free radicals (charged oxygen molecules with an unpaired electron). Brain cells recovering from the effects of a stroke or cardiac arrest also release free radicals. Free radicals cause damage by disrupting cellular DNA and enzymes and altering plasma membrane permeability. Various degrees of brain injury are described by the following terms.

Concussion. An abrupt but temporary loss of consciousness following a blow to the head or the sudden stopping of a moving head. A concussion produces no obvious bruising of the brain, but post-traumatic amnesia (memory loss) may occur.

Contusion. A visible bruising of the brain due to trauma and blood leaking from microscopic vessels. The pia mater may be torn, allowing blood to enter the subarachnoid space. A contusion usually results in an extended loss of consciousness, ranging from several minutes to many hours.

Laceration. Tearing of the brain, usually from a skull fracture or gunshot wound. A laceration results in rupture of large blood vessels with bleeding into the brain and subarachnoid space. Consequences include cerebral hematoma (localized pool of blood, usually clotted, that swells against the brain tissue), edema, and increased intracranial pressure. ■

Functional Areas of the Cerebral Cortex

Specific types of sensory, motor, and integrative signals are processed in certain cerebral regions (Fig. 14.15). Gen-

erally, the **sensory areas** receive and interpret sensory impulses, the **motor areas** control muscular movement, and the **association areas** deal with more complex integrative functions such as memory, emotions, reasoning, will, judgment, personality traits, and intelligence.

Sensory Areas

Sensory input to the cerebral cortex flows mainly to the posterior half of the hemispheres, to regions posterior to the central sulci. In the cortex, primary sensory areas have the most direct connections with peripheral sensory receptors. Secondary sensory areas and sensory association areas often are adjacent to the primary areas. Usually, they receive input from the primary areas and from diverse other regions of the brain. They participate in the interpretation of sensory experiences into meaningful patterns of recognition and awareness. For example, a person with damage in the primary visual cortex would be blind in at least part of his visual field. A person with damage to a visual association area, on the other hand, might see normally yet be unable to recognize a friend.

● **Primary Somatosensory Area or General Sensory Area** Located directly posterior to the central sulcus of each cerebral hemisphere in the postcentral gyrus of each parietal lobe, it extends from the longitudinal fissure on the superior aspect of the cerebrum to the lateral cerebral sulcus. In Fig. 14.15, the somatosensory area is designated by the areas numbered 1, 2, and 3.[1]

The primary somatosensory area receives nerve impulses from somatic sensory receptors for touch, proprioception (joint and muscle position), pain, and temperature. Each point within the area receives sensations from a specific part of the body, and essentially the entire body is spatially represented in it. The size of the cortical area receiving impulses from a particular body part depends on the number of receptors present there rather than on the size of the part. For example, a larger portion of the sensory area receives impulses from the lips and fingertips than from the thorax or hip (see Fig. 15.5a). The major function of the primary somatosensory area is to localize exactly the points of the body where the sensations originate. The thalamus is capable of registering sensations in a general way. It receives sensations from large areas of the body but cannot distinguish precisely the specific area of stimulation. This capability depends on the primary somatosensory area of the cortex.

● **Primary Visual Area (Area 17)** Located on the medial surface of the occipital lobe and occasionally extending around to the lateral surface, it receives impulses conveying visual information. Axons of neurons with cell bodies in the eye form the optic nerves (cranial nerve II), which terminate

---

[1]These numbers, as well as most of the others shown, are based on K. Brodmann's map of the cerebral cortex. His map, first published in 1909, was a brilliant and successful attempt to correlate specific brain regions with particular functions.

Figure 14.15   Functional areas of the cerebrum. Broca's area is in the left cerebral hemisphere of most people; it is shown here to indicate its relative location.

 Particular areas of the cerebral cortex process sensory, motor, and integrative signals.

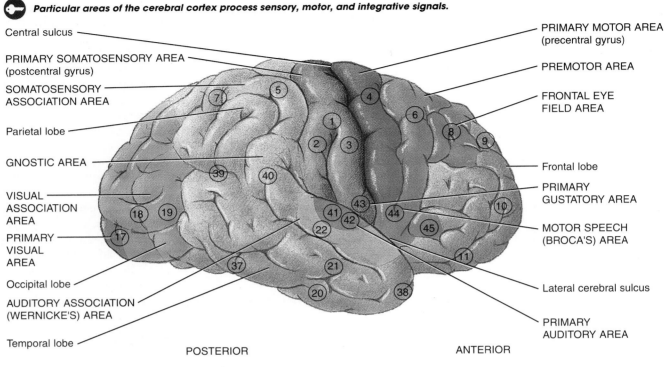

Lateral view of right cerebral hemisphere

What area of the cerebrum integrates interpretation of visual, auditory, and somatic sensations? Translates thoughts into speech? Controls skilled muscular movements? Interprets sensations related to taste? Interprets pitch and rhythm? Interprets shape, color, and movement of objects? Controls voluntary scanning movements of the eyes?

in the thalamus. From the thalamus, neurons project to the primary visual area, with information concerning shape, color, and movement of visual stimuli.

● **Primary Auditory Area (Areas 41 and 42)** Located in the superior part of the temporal lobe near the lateral cerebral sulcus, it interprets the basic characteristics of sound such as pitch and rhythm.

● **Primary Gustatory Area (Area 43)** Located at the base of the postcentral gyrus above the lateral cerebral sulcus in the parietal cortex, it receives impulses related to taste.

● **Primary Olfactory Area** Located in the temporal lobe on the medial aspect, it receives impulses related to smell.

Motor Areas
Motor output from the cerebral cortex flows mainly from the anterior portion of each hemisphere.

● **Primary Motor Area (Area 4)** Located in the precentral gyrus of the frontal lobe (Fig. 14.15), each region in the primary motor area controls voluntary contractions of specific muscles or groups of muscles (see Fig. 15.5b). Electrical stimulation of any point in the primary motor area results in con-

traction of specific skeletal muscle fibers on the opposite side of the body. As is true for the primary somatosensory area, body parts are represented unequally here. More cortical area is devoted to those muscles where skilled, complex, or delicate movement is possible, for example, finger maneuvers.

● **Motor Speech Area (Area 44)** The translation of speech or written words into thought involves both sensory and association areas—primary auditory, auditory association, primary visual, visual association, and gnostic (described shortly). The production of speech occurs in the **motor speech area** (area 44) also called **Broca's** (BRŌ-kaz) **area**. It is located in one frontal lobe, usually the *left* frontal lobe, just superior to the lateral cerebral sulcus.

Association Areas
The **association areas** of the cerebrum consist of motor and sensory areas and large parts of the cortex on the lateral surfaces of the occipital, parietal, and temporal lobes and the frontal lobes anterior to the motor areas. They are connected with one another by association tracts and include the following.

● **Somatosensory Association Area (Areas 5 and 7)** Just posterior to the primary somatosensory area, it receives input from the thalamus, other lower portions of the brain, and the primary somatosensory area. Its role is to integrate and interpret sensations. This area permits you to determine the exact shape and texture of an object without looking at it, to determine the orientation of one object to another as they are felt, and to sense the relationship of one body part to another. Another role of the somatosensory association area is the storage of memories of past sensory experiences. Thus you can compare current sensations with previous experiences.

● **Visual Association Area (Areas 18 and 19)** Located in the occipital lobe, it receives sensory impulses from the primary visual area and the thalamus. It relates present to past visual experiences and is essential for recognizing and evaluating what is seen.

● **Auditory Association (Wernicke's) Area (Area 22)** Located inferior and posterior to the primary auditory area in the temporal cortex, it determines if a sound is speech, music, or noise. It also interprets the meaning of speech by translating words into thoughts.

● **Gnostic** (NOS-tik; *gnosis* = knowledge) **Area (Areas 5, 7, 39, and 40)** This **common integrative area** is located among the somatosensory, visual, and auditory association areas. The gnostic area receives nerve impulses from these areas, as well as from the taste and smell areas, the thalamus, and portions of the brain stem. It integrates sensory interpretations from the association areas and impulses from other areas so that a common thought can be formed from the various sensory inputs. It then transmits signals to other parts of the brain to cause the appropriate response to the interpretation of the sensory signals.

● **Premotor Area (Area 6)** Immediately anterior to the primary motor area is a motor association area. Neurons in this region communicate with the primary motor cortex, sensory association areas in the parietal lobe, the basal ganglia, and the thalamus. The premotor area deals with learned motor activities of a complex and sequential nature. It generates nerve impulses that cause a specific group of muscles to contract in a specific sequence, for example, to write a word. The premotor area controls learned skilled movements and serves as a memory bank for such movements.

● **Frontal Eye Field Area (Area 8)** This area in the frontal cortex is sometimes included in the premotor area. It controls voluntary scanning movements of the eyes—searching for a word in a dictionary, for instance.

APHASIA

Much of what we know about language areas comes from studies of patients with language or speech disturbances that have resulted from brain damage. The motor speech (Broca's) area, auditory association (Wernicke's) area, and other language areas are located in the left cerebral hemisphere of most people, regardless of whether they are left-handed or right-handed. Injury to the association or motor speech areas results in **aphasia** (a-FĀ-zē-a; *a* = without; *phasis* = speech), an inability to speak. Damage to the motor speech area results in nonfluent aphasia, an inability to properly articulate or form words. The person knows what she wishes to say but cannot speak. Damage to the gnostic area or auditory association area (areas 39 and 22) results in fluent aphasia, faulty understanding of spoken or written words. Such a patient may fluently produce strings of words that have no meaning. The deficiency may be **word deafness**, an inability to understand spoken words, or **word blindness**, an inability to understand written words, or both. ■

● **Language Areas** From Broca's area, nerve impulses pass to the premotor regions that control the muscles of the larynx, pharynx, and mouth. The impulses from the premotor area to the muscles result in specific, coordinated contractions that enable you to speak. Simultaneously, impulses are sent from the motor speech area to the primary motor area. From here, impulses reach your breathing muscles to regulate the proper flow of air past the vocal cords. The coordinated contractions of your speech and breathing muscles enable you to speak your thoughts.

Electroencephalogram (EEG)

At any instant, brain cells are generating millions of nerve impulses (nerve action potentials) and graded potentials (excitatory and inhibitory postsynaptic potentials) in individual neurons. These electrical potentials added together are called **brain waves** and reflect electrical activity of the cerebral cortex. Brain waves pass through the skull and can be detected by sensors called electrodes. A record of such waves is called an **electroencephalogram** (**EEG**; e-lek′-trō-en-SEF-a-lō-gram′).

As indicated in Fig. 14.16, four kinds of waves can be recorded from normal individuals.

1. **Alpha waves.** These rhythmic waves occur at a frequency of about 8–13 cycles per second. (The unit commonly used to express frequency is the hertz [Hz]; 1 Hz = 1 cycle per second.) Alpha waves are present in the EEGs of nearly all normal individuals when they are awake and resting with their eyes closed. These waves disappear entirely during sleep.
2. **Beta waves.** The frequency of these waves is between 14 and 30 Hz. Beta waves generally appear when the nervous system is active, that is, during periods of sensory input and mental activity.
3. **Theta waves.** These waves have frequencies of 4–7 Hz. Theta waves normally occur in children and in adults experiencing emotional stress. They also occur in many disorders of the brain.
4. **Delta waves.** The frequency of these waves is 1–5 Hz. In an adult delta waves occur during deep sleep. They are normal, however, in an awake infant. When produced by an awake adult, they indicate brain damage.

Figure 14.16   Types of brain waves recorded in an electroencephalogram (EEG).

 Brain waves indicate electrical activity of the cerebral cortex.

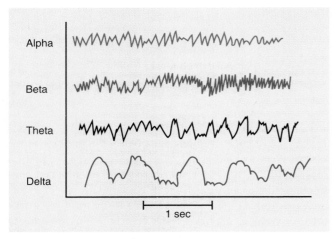

Alpha

Beta

Theta

Delta

1 sec

 Which type of wave indicates emotional stress?

ELECTROENCEPHALOGRAM (EEG)

An **electroencephalogram** (**EEG** or **brain wave test**) is used to diagnose epilepsy and other seizure disorders, infectious diseases, tumors, trauma, hematomas, metabolic abnormalities, degenerative diseases, and periods of unconsciousness and confusion. In some cases, an EEG provides useful information regarding sleep and wakefulness. An EEG may also be one criterion in establishing brain death (complete absence of brain waves in two EEGs taken 24 hours apart). ■

Brain Lateralization

On gross examination, the brain appears the same on both sides. However, detailed examination reveals subtle anatomical differences between the two hemispheres. For example, in left-handed people the parietal and occipital lobes of the right hemisphere are usually narrower than the corresponding lobes of the left hemisphere. In addition, the frontal lobe of the left hemisphere of such individuals is typically narrower than that of the right hemisphere.

Besides the structural differences between the hemispheres, there are important functional differences (Fig. 14.17). The left hemisphere receives sensory signals from and controls the right side of the body, whereas the right hemisphere receives sensory signals from and controls the left side of the body. Also, in most people the left hemisphere is more important for spoken and written language, numerical and scientific skills, ability to use and understand sign language, and reasoning. Conversely, it has been

shown that the right hemisphere is more important for musical and artistic awareness, space and pattern perception, insight, imagination, and generating mental images of sight, sound, touch, taste, and smell to compare relationships.

A summary of the functions of the various parts of the brain is presented in Exhibit 14.1.

NEUROTRANSMITTERS

About 60 substances are either known or suspected neurotransmitters. The list grows longer each year as new techniques confirm the existence of a wide array of possible neurotransmitters. Finding out what each one does, however, is not easy. Dendrites, cell bodies, and axons are tightly packed and closely intermingled in nervous tissue, and the amount of neurotransmitter liberated at a single synapse is tiny. Some neurotransmitters bind to their receptors and act quickly to open or close ion channels in the membrane. Others act more slowly by second messenger systems to influence enzymatic reactions inside the cell.

Figure 14.17   Summary of the principal functional differences between the left and right cerebral hemispheres.

 Brain lateralization means that the two cerebral hemispheres differ slightly in structure and function.

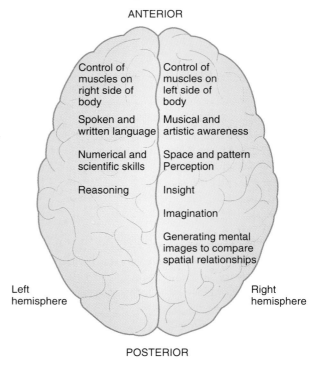

ANTERIOR

Control of muscles on right side of body

Spoken and written language

Numerical and scientific skills

Reasoning

Control of muscles on left side of body

Musical and artistic awareness

Space and pattern Perception

Insight

Imagination

Generating mental images to compare spatial relationships

Left hemisphere

Right hemisphere

POSTERIOR

Superior view of cerebral hemispheres

Which lobes of the cerebrum are usually narrower in left-handed people?

EXHIBIT **14.1** SUMMARY OF FUNCTIONS OF PRINCIPAL PARTS OF THE BRAIN

Part	Function	Part	Function

Brain stem

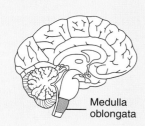
Medulla oblongata

Medulla oblongata: Relays motor and sensory impulses between other parts of the brain and the spinal cord. Reticular formation (also in pons, midbrain, and diencephalon) functions in consciousness and arousal. Vital centers regulate heartbeat, breathing (together with pons), and blood vessel diameter. Other centers coordinate swallowing, vomiting, coughing, sneezing, and hiccuping. Contains nuclei of origin for cranial nerves VIII, IX, X, XI, and XII.

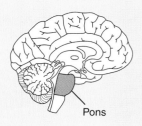
Pons

Pons: Relays impulses from one side of the cerebellum to the other and between the medulla and midbrain. Contains nuclei of origin for cranial nerves V, VI, VII, and VIII. Pneumotaxic area and apneustic area, together with the medulla, help control breathing.

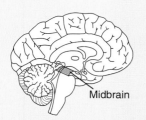
Midbrain

Midbrain: Relays motor impulses from the cerebral cortex to the pons and sensory impulses from the spinal cord to the thalamus. Superior colliculi coordinate movements of the eyeballs in response to visual and other stimuli, and the inferior colliculi coordinate movements of the head and trunk in response to auditory stimuli. Substantia nigra and red nucleus contribute to control of movement. Contains nuclei of origin for cranial nerves III and IV.

Cerebellum

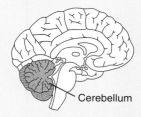
Cerebellum

Compares intended movements with what is actually happening to smooth and coordinate complex, skilled movements. Regulates posture and balance.

Diencephalon

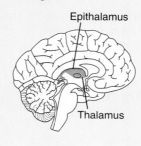
Epithalamus
Thalamus

Epithalamus: Contains pineal gland, which secretes melatonin, habenular nuclei, and the choroid plexus of the third ventricle.

Thalamus: Relays all sensory input to the cerebral cortex. Provides crude appreciation of touch, pressure, pain, and temperature. Includes nuclei involved in voluntary motor actions and arousal; anterior nucleus functions in emotions and memory. Also functions in cognition and awareness.

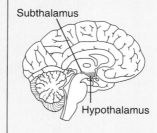
Subthalamus
Hypothalamus

Subthalamus: Contains the subthalamic nuclei and portions of the red nucleus and the substantia nigra, which are positioned mostly lateral to the midline. These regions communicate with the basal ganglia to help control muscle movements.

Hypothalamus: Controls and integrates activities of the autonomic nervous system and pituitary gland. Regulates emotional and behavioral patterns and diurnal rhythms. Controls body temperature and regulates eating and drinking behavior. Helps maintain the waking state and establishes patterns of sleep.

Cerebrum

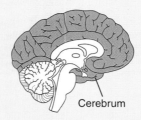
Cerebrum

Sensory areas interpret sensory impulses, motor areas control muscular movement, and association areas function in emotional and intellectual processes. Basal ganglia coordinate gross, automatic muscle movements and regulate muscle tone. Limbic system functions in emotional aspects of behavior related to survival.

The result of either process can be excitation or inhibition of postsynaptic neurons.

Quite a few neurotransmitters also are known to be hormones, released into the blood by endocrine cells in organs throughout the body. Within the brain, certain neurons, called **neurosecretory cells**, also secrete hormones. Still other neurons liberate substances called **neuromodulators** into interstitial or cerebrospinal fluid. Neuromodulators influence signaling of neighboring or distant neurons by intensifying or diminishing their response to other neurotransmitters.

Exhibit 14.2 lists some of the known and suspected neurotransmitters and neuromodulators. Chemically, they are divided into five classes: (1) acetylcholine, (2) amino acids, (3) biogenic amines, (4) neuropeptides, and (5) gases.

Acetylcholine

Depending on the nature of the postsynaptic receptors, acetylcholine (ACh) may be excitatory or inhibitory in the brain. Recall that ACh is the neurotransmitter released at neuromuscular junctions in skeletal muscle (Chapter 10). ACh is inactivated by acetylcholinesterase (AChE), which splits ACh into acetate and choline portions. In the brain, a major center of neurons that liberate ACh is the nucleus basalis, which is inferior to the globus pallidus. Axons of these neurons project widely throughout the cerebral cortex and limbic system and their destruction is a hallmark of Alzheimer's disease (described on page 425). Neurons of the primary motor area in the precentral gyrus that form the pyramids of the medulla and descend in the spinal cord as the corticospinal tract also use ACh.

Amino Acids

The amino acids **glutamate** and **aspartate** are known excitatory neurotransmitters in the brain. The most common inhibitory neurotransmitter in the brain is the amino acid **gamma aminobutyric acid (GABA)**. It is most highly concentrated in the superior and inferior colliculi, thalamus, hypothalamus, and occipital lobes of the cerebrum. Antianxiety drugs such as diazepam (Valium) enhance the action of GABA. The most prevalent inhibitory neurotransmitter in the spinal cord is the amino acid **glycine**.

Biogenic Amines

Certain amino acids are modified and decarboxylated (carboxyl group removed) to produce biogenic amines. Those that are prevalent in the brain include norepinephrine, epinephrine, dopamine, serotonin, and histamine. Depending on the type of receptor (there are three or more different types for each amine), they may cause either excitation or inhibition.

Norepinephrine (NE) is concentrated in a group of neurons in the brain stem that project their axons into the hypothalamus, cerebellum, cerebral cortex, and spinal cord. In these places, NE has been implicated in maintaining arousal (awakening from deep sleep), dreaming, and the regulation of

---

EXHIBIT **14.2 NEUROTRANSMITTERS AND NEUROMODULATORS**
ACETYLCHOLINE (ACH)
AMINO ACIDS
Aspartate
Gamma aminobutyric acid (GABA)
Glutamate
Glycine
BIOGENIC AMINES
Dopamine (DA)
Epinephrine
Histamine
Norepinephrine (NE)
Serotonin (5-HT)
NEUROPEPTIDES[a]
Angiotensin II
Antidiuretic hormone (vasopressin)
Atrial natriuretic peptide (ANP)
Bombesin
Cholecystokinin (CCK)
Corticotropin
Dynorphins
Endorphins
Enkephalins (Enk)
Galanin
Gastrin
Glucagon
Hypothalamic regulating hormones
Inhibin
Melatonin
Motilin
Neuropeptide Y (NPY)
Neurotensin
Oxytocin
Secretin
Substance P
Vasoactive intestinal polypeptide (VIP)
GASES
Nitric oxide (NO)
Carbon monoxide (possibly)

[a] Many also are hormones that are released by tissues outside the brain.

---

mood. A smaller number of neurons in the brain use **epinephrine** as a neurotransmitter. Both epinephrine and norepinephrine also serve as hormones. They are released by the inner portion of the adrenal gland, the adrenal medulla.

Neurons containing the neurotransmitter **dopamine (DA)** are clustered in the substantia nigra of the midbrain (see Fig. 14.7). Some axons projecting from the substantia nigra terminate in the cerebral cortex, where DA is thought to be involved in emotional responses. Other axons project to the corpus striatum (basal ganglia), where the DA is involved in regulating gross, automatic movements of skeletal muscles. Degeneration of these axons occurs in Parkinson's disease (see page 449).

Norepinephrine, dopamine, and epinephrine are *catecholamines*. They all include a catechol ring (*six carbons and two adjacent hydroxyl (OH) groups).* The process for activation of the catecholamines is different from that of ACh. Soon after release from synaptic vesicles, they are actively transported back into the synaptic end bulbs. This is called **reuptake**. Then they are either destroyed by the enzymes **catechol-***O***-methyltransferase** (kat′-e-kōl-ō-meth-il TRANS-fer-ās), or **COMT**, and **monoamine oxidase** (mon-ō-AM-ēn OK-si-dās), or **MAO**, or recycled back into the synaptic vesicles.

Serotonin or **5-hydroxytryptophan (5-HT)** is concentrated in the neurons in a part of the brain stem called the **raphe nucleus**. Axons projecting from the raphe nucleus terminate in the hypothalamus, thalamus, and other parts of the brain and spinal cord. Serotonin is thought to be involved in inducing sleep, sensory perception, temperature regulation, and control of mood. The antidepressant drug Prozac, is a selective inhibitor of serotonin reuptake. Prozac thus makes more serotonin available in the synaptic cleft and may allow signals to pass from one neuron to another more easily.

Histamine is thought to be a neurotransmitter in the hypothalamus. Receptors for certain antihistamine drugs are localized there.

Neuropeptides

Peptide neurotransmiters are called **neuropeptides**. They are numerous and widespread in both the central and peripheral nervous systems. Each one is a chain of 3–40 amino acids and may have excitatory or inhibitory actions. Neuropeptides are formed in the cell body of the neuron,

packaged into vesicles, and transported to axon terminals. Besides their role as neurotransmitters, many of the neuropeptides serve as hormones that regulate physiological responses in other parts of the body, particularly digestion. To give you an idea of their diversity, brief portraits of several important neuropeptides are presented in Exhibit 14.3. We describe the first four here and the rest in later chapters.

In 1974 scientists discovered that certain brain neurons have plasma membrane receptors for opiate drugs such as morphine and heroin. The quest to find the naturally occurring substances that use these receptors brought to light the first neuropeptides. They were two molecules, each a chain of 5 amino acids, named **enkephalins** (en-KEF-a-lins). They have potent analgesic (pain-relieving) effects, 200 times stronger than morphine. Besides the enkephalins, other so-called opioid peptides include the **endorphins** (en-DOR-fins) and **dynorphins** (di-NOR-fins).

Enkephalins are concentrated in the thalamus, hypothalamus, parts of the limbic system (see Fig. 14.14), and those spinal cord pathways that relay impulses for pain. It is thought that opioid peptides are the body's natural painkillers. Acupuncture may produce analgesia (loss of pain sensation) by increasing the release of opioids. They have also been linked to improved memory and learning; feelings of pleasure or euphoria; control of body temperature; regulation of hormones that affect the onset of puberty, sexual drive, and reproduction; and mental illnesses such as depression and schizophrenia.

Another neuropeptide, **substance P**, is found in sensory nerves, spinal cord pathways, and parts of the brain associated

EXHIBIT **14.3** **NEUROPEPTIDES**

Substance	Comment	Substance	Comment
Enkephalins	Concentrated in the thalamus, hypothalamus, parts of limbic system, and spinal cord pathways that relay pain impulses; inhibit pain impulses by suppressing release of substance P.	**Angiotensin II**	Stimulates thirst; improves memory; may regulate blood pressure in the brain; as a hormone, promotes release of aldosterone, which increases the rate of salt and water reabsorption by the kidneys.
Endorphins	Concentrated in the pituitary gland; inhibit pain by blocking release of substance P; may have a role in memory and learning, sexual activity, and control of body temperature; have been linked to depression and schizophrenia.	**Cholecystokinin (CCK)**	Found in the cerebral cortex and small intestine; may regulate feeding as a "stop eating" signal; as a hormone, regulates pancreatic enzyme secretion during digestion and contraction of smooth muscle in the gastrointestinal tract.
Substance P	Found in sensory neurons, spinal cord pathways, and parts of brain associated with pain; stimulates perception of pain.		
Dynorphin	Found in the posterior pituitary gland, hypothalamus, and small intestine; may be related to controlling pain and registering emotions.	**Oxytocin (OT)**	Present in neurons that project to the brain stem and spinal cord; improves memory; as a hormone, stimulates uterine contractions at childbirth and stimulates milk release from mammary glands.
Hypothalamic regulating hormones	Produced by the hypothalamus; regulate the release of hormones by the anterior pituitary gland.	**Antidiuretic hormone (ADH) or vasopressin**	Present in neurons that project to the brain stem and spinal cord; improves memory; as a hormone, regulates rate of water reabsorption by the kidneys and causes constriction of blood vessels.

with pain transmission. When substance P is released by neurons, it transmits pain-related input from peripheral pain receptors into the central nervous system. Acting by presynaptic inhibition (see page 350), enkephalin suppresses the release of substance P, thus decreasing painful input to the CNS. Substance P has also been shown to counter the effects of certain nerve-damaging chemicals, prompting speculation that it might prove useful as a treatment for nerve degeneration.

Gases—Nitric Oxide and Carbon Monoxide

A surprising and important newcomer to the ranks of recognized neurotransmitters is the simple gas **nitric oxide (NO)**. It is not to be confused with nitrous oxide (N_2O, laughing gas), which is sometimes used as an anesthetic during dental procedures. Carbon monoxide (CO) may also function as a neurotransmitter, but the evidence is not yet completely convincing.

NO has widespread effects throughout the body. It is formed from the amino acid arginine by the enzyme **nitric oxide synthase (NOS)**. NO is lipid soluble and thus diffuses out of cells that produce it and into neighboring cells. Its action is brief because it is a highly reactive free radical that lasts less than 10 seconds before it combines with oxygen and water to form inactive nitrates and nitrites.

The first recognition of NO as a regulatory molecule was the discovery in 1987 that EDRF (endothelium-derived relaxing factor) was actually NO. Endothelial cells, which form the lining of blood vessel walls, release NO (EDRF). The NO diffuses to neighboring smooth muscle cells and causes relaxation. The result is vasodilation, an increase in diameter of the blood vessel. The effects of such vasodilation range from a lowering of blood pressure to erection of the penis in males. In larger quantities NO is highly toxic. Phagocytic cells, such as macrophages and certain white blood cells, produce NO to kill microbes and tumor cells.

NO is different from all previously known neurotransmitters because it is not synthesized in advance and packaged into synaptic vesicles. Rather, it is formed on demand and acts immediately. About 2% of the neurons in the cerebral cortex contain NOS, and it is even more abundant in the cerebellum, posterior pituitary gland, and superior and inferior colliculi. NOS is also highly concentrated in autonomic neurons that cause relaxation of smooth muscle in the gut and release of epinephrine and norepinephrine from the adrenal medulla.

The precise functions of NO released by neurons are still unclear. It diffuses into neighboring neurons, where it activates an enzyme for production of a second messenger called cyclic GMP. Some research suggests that NO may play a role in memory and learning.

CRANIAL NERVES

Cranial nerves, like spinal nerves, are part of the peripheral nervous system (PNS). Of the 12 pairs of **cranial nerves**, 10 originate from the brain stem, but all pass through foramina of the skull. The cranial nerves are designated with roman numerals and with names (see Fig. 14.5). The roman numerals indicate the order in which the nerves arise from the brain from anterior to posterior. The names indicate the distribution or function.

Some cranial nerves contain only sensory fibers and thus are called **sensory nerves**. The remainder contain both sensory and motor fibers and are referred to as **mixed nerves**, though a few of them are predominantly motor in function. The cell bodies of sensory neurons are found outside the brain, whereas the cell bodies of motor neurons lie in nuclei within the brain. Motor neurons include both somatic and autonomic efferents.

A summary of cranial nerves and clinical applications related to their dysfunction is presented in Exhibit 14.4.

EXHIBIT **14.4** SUMMARY OF CRANIAL NERVES

Number, Name, Type (Sensory, Motor, or Mixed)	Location	Function and Clinical Application
Cranial nerve I: **olfactory** (*olfacere* = to smell) Sensory Olfactory (I) nerve — Olfactory tract	Arises in olfactory mucosa, passes through olfactory foramina in the cribriform plate of ethmoid bone, and ends in the olfactory bulb. The olfactory tract extends via two pathways to olfactory areas in the temporal lobe of the cerebral cortex.	*Function*: Smell. *Clinical application*: Loss of the sense of smell, called *anosmia*, may result from head injuries in which the cribriform plate of the ethmoid bone is fractured and from lesions along the olfactory pathway.

Number, Name, Type (Sensory, Motor, or Mixed)	Location	Function and Clinical Application
Cranial nerve II: **optic** (*optikos* = vision, eye, or optics) Sensory 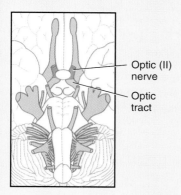 Optic (II) nerve Optic tract	Arises in retina of the eye, passes through optic foramen, forms optic chiasm, passes through optic tracts, and terminates in lateral geniculate nuclei of thalamus. From thalamus, projections extend to visual areas in the occipital lobe of the cerebral cortex.	*Function*: Vision. *Clinical application*: Fractures in the orbit, lesions along the visual pathway, and diseases of the nervous system may result in visual field defects and loss of visual acuity. Loss of vision is called *anopsia*.
Cranial nerve III: **oculomotor** (*oculus* = eye; *motor* = mover) Mixed, primarily motor 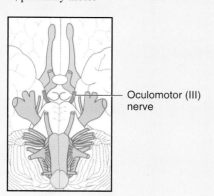 Oculomotor (III) nerve	*Motor portion*: Originates in midbrain, passes through superior orbital fissure, and is distributed to levator palpebrae superioris of upper eyelid and four extrinsic eyeball muscles (superior rectus, medial rectus, inferior rectus, and inferior oblique); parasympathetic innervation to ciliary muscle of eyeball and sphincter muscle of iris. *Sensory portion*: Consists of fibers from proprioceptors in eyeball muscles that pass through superior orbital fissure and terminate in midbrain.	*Motor function*: Movement of eyelid and eyeball, accommodation of lens for near vision, and constriction of pupil. *Sensory function*: Muscle sense (proprioception). *Clinical application*: A lesion in the nerve causes *strabismus* (a deviation of the eye in which both eyes do not fix on the same object), *ptosis* (drooping) of the upper eyelid, pupil dilation, the movement of the eyeball downward and outward on the damaged side, a loss of accommodation for near vision, and *diplopia* (double vision).
Cranial nerve IV: **trochlear** (*trokhileia* = pulley) Mixed, primarily motor 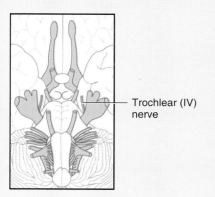 Trochlear (IV) nerve	*Motor portion*: Originates in midbrain, passes through superior orbital fissure, and is distributed to superior oblique muscle, an extrinsic eyeball muscle. *Sensory portion*: Consists of fibers from proprioceptors in superior oblique muscles that pass through superior orbital fissure and terminate in midbrain.	*Motor function*: Movement of eyeball. *Sensory function*: Muscle sense (proprioception). *Clinical application*: In trochlear nerve paralysis, diplopia and strabismus occur.

Exhibit continues

EXHIBIT **14.4** CRANIAL NERVES (CONTINUED)

Number, Name, Type (Sensory, Motor, or Mixed)	Location	Function and Clinical Application
Cranial nerve V: **trigeminal** (*tri* = three; *geminus* = twin; *trigeminus* = threefold, for its three branches) Mixed 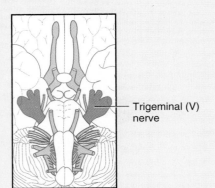 Trigeminal (V) nerve	*Motor portion*: Is part of the mandibular branch; originates in pons, passes through foramen ovale, and ends in muscles of mastication (anterior belly of digastric and mylohyoid muscles). *Sensory portion*: Consists of three branches: **ophthalmic** (*ophthalmos* = eye), which contains fibers from skin over upper eyelid, eyeball, lacrimal glands, nasal cavity, side of nose, forehead, and anterior half of scalp that pass through superior orbital fissure; **maxillary** (*maxilla* = upper jaw bone), which contains fibers from mucosa of nose, palate, parts of pharynx, upper teeth, upper lip, and lower eyelid that pass through foramen rotundum; **mandibular** (*mandibula* = lower jaw bone), which contains somatic sensory fibers (but not special sense of taste) from anterior two-thirds of tongue, lower teeth, skin over mandible, cheek and mucosa deep to it, and side of head in front of ear that pass through foramen ovale. The three branches end in pons. Sensory portion also consists of fibers from proprioceptors in muscles of mastication.	*Motor function*: Chewing. *Sensory function*: Conveys sensations for touch, pain, and temperature from structures supplied; muscle sense (proprioception). *Clinical application*: Injury results in paralysis of the muscles of mastication and a loss of sensation of touch and temperature. *Neuralgia* (pain) of one or more branches of trigeminal nerve is called *trigeminal neuralgia* (*tic douloureux*).
Cranial nerve VI: **abducens** (*ab* = away; *ducere* = to lead) Mixed, primarily motor 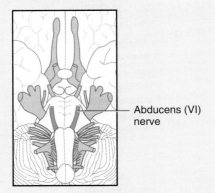 Abducens (VI) nerve	*Motor portion*: Originates in pons, passes through superior orbital fissure, and is distributed to lateral rectus muscle, an extrinsic eyeball muscle. *Sensory portion*: Consists of fibers from proprioceptors in lateral rectus muscle that pass through superior orbital fissure and end in pons.	*Motor function*: Movement of eyeball. *Sensory function*: Muscle sense (proprioception). *Clinical application*: With damage to this nerve, the affected eyeball cannot move laterally beyond the midpoint and the eye is usually directed medially.

Number, Name, Type (Sensory, Motor, or Mixed)	Location	Function and Clinical Application
Cranial nerve VII: **facial** (*facies* = face) Mixed 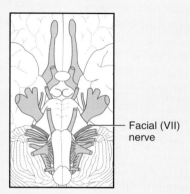 Facial (VII) nerve	*Motor portion*: Originates in pons, passes through stylomastoid foramen, and is distributed to facial, scalp, and neck muscles; parasympathetic fibers are distributed to lacrimal, sublingual, submandibular, nasal, and palatine glands. *Sensory portion*: Arises from taste buds on anterior two-thirds of tongue, passes through stylomastoid foramen, and ends in geniculate ganglion, a nucleus in pons that sends fibers to the thalamus for relay to gustatory areas in the parietal lobe of the cerebral cortex. Also contains fibers from proprioceptors in muscles of face and scalp.	*Motor function*: Facial expression and secretion of saliva and tears. *Sensory function*: Muscle sense (proprioception) and taste. *Clinical application*: Injury produces paralysis of the facial muscles, called *Bell's palsy*, loss of taste, and loss of ability to close the eyes, even during sleep.
Cranial nerve VIII: **vestibulocochlear** (*vestibulum* = vestibule; *kokhlos* = land snail) Sensory 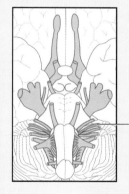 Vestibulocochlear (VIII) nerve	*Cochlear branch*: Arises in spiral organ (organ of Corti), forms spiral ganglion, passes through internal auditory meatus, nuclei in the medulla, and ends in thalamus. Fibers synapse with neurons that relay impulses to auditory areas in the temporal lobe of the cerebral cortex. *Vestibular branch*: Arises in semicircular canals, saccule, and utricle and forms vestibular ganglion; fibers end in pons and cerebellum.	*Cochlear branch function*: Conveys impulses associated with hearing. *Vestibular branch function*: Conveys impulses associated with equilibrium. *Clinical application:* Injury to the cochlear branch may cause *tinnitus* (ringing) or deafness. Injury to the vestibular branch may cause *vertigo* (a subjective feeling of rotation), *ataxia*, and *nystagmus* (involuntary rapid movement of the eyeball).
Cranial nerve IX: **glossopharyngeal** (*glossa* = tongue; *pharynx* = throat) Mixed 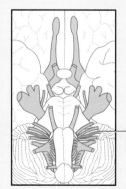 Glossopharyngeal (IX) nerve	*Motor portion*: Originates in medulla, passes through jugular foramen, and is distributed to stylopharyngeus muscle; parasympathetic fibers extend to parotid gland. *Sensory portion*: Arises from taste buds on posterior one-third of tongue and from carotid sinus, passes through jugular foramen, and ends in medulla. Also contains fibers from somatic sensory receptors on posterior one-third of tongue and proprioceptors in swallowing muscles supplied by motor portion.	*Motor function*: Secretion of saliva. *Sensory function*: Taste, regulation of blood pressure, and muscle sense (proprioception). *Clinical application*: Injury results in difficulty during swallowing, reduced secretion of saliva, loss of sensation in the throat, and loss of taste.

Exhibit continues

EXHIBIT **14.4** **CRANIAL NERVES** (CONTINUED)

Number, Name, Type (Sensory, Motor, or Mixed)	Location	Function and Clinical Application
Cranial nerve X: **vagus** (*vagus* = vagrant or wandering) Mixed 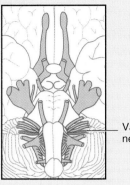 Vagus (X) nerve	*Motor portion*: Originates in medulla, passes through jugular foramen, and terminates in muscles of airways, lungs, esophagus, heart, stomach, small intestine, most of large intestine, and gallbladder; parasympathetic fibers innervate involuntary muscles and glands of the gastrointestinal (GI) tract. *Sensory portion*: Arises from essentially same structures supplied by motor fibers, passes through jugular foramen, and ends in medulla and pons.	*Motor function*: Smooth muscle contraction and relaxation; secretion of digestive fluids. *Sensory function*: Sensations from visceral organs supplied; muscle sense (proprioception). *Clinical application*: Severing of both nerves in the upper body interferes with swallowing, paralyzes vocal cords, and interrupts sensations from many organs.
Cranial nerve XI: **accessory** (*accessorius* = assisting) Mixed, primarily motor 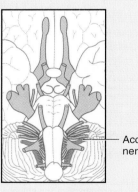 Accessory (XI) nerve	*Motor portion*: Consists of a cranial portion and a spinal portion. *Cranial portion* originates from medulla, passes through jugular foramen, and supplies voluntary muscles of pharynx, larynx, and soft palate. *Spinal portion* originates from anterior gray horn of first five cervical segments of spinal cord, passes through jugular foramen, and supplies sternocleidomastoid and trapezius muscles. *Sensory portion*: Consists of fibers from proprioceptors in muscles supplied by motor portion and passes through jugular foramen.	*Motor function*: Cranial portion mediates swallowing movements; spinal portion mediates movement of head. *Sensory function*: Muscle sense (proprioception). *Clinical application*: If nerves are damaged, the sternocleidomastoid and trapezius muscles become paralyzed, with resulting inability to raise the shoulders and difficulty in turning the head.
Cranial nerve XII: **hypoglossal** (*hypo* = below; *glossa* = tongue) Mixed, primarily motor  Hypoglossal (XII) nerve	*Motor portion*: Originates in medulla, passes through hypoglossal canal, and supplies muscles of tongue. *Sensory portion*: Consists of fibers from proprioceptors in tongue muscles that pass through hypoglossal canal and end in medulla.	*Motor function*: Movement of tongue during speech and swallowing. *Sensory function*: Muscle sense (proprioception). *Clinical application*: Injury results in difficulty in chewing, speaking, and swallowing. The tongue, when protruded, curls toward the affected side and the affected side becomes atrophied, shrunken, and deeply furrowed.

AGING AND THE NERVOUS SYSTEM

One of the effects of aging on the nervous system is loss of neurons. This is a consequence of the aging process, not necessarily because of a disease state. Associated with this decline, there is a decreased capacity for sending nerve impulses to and from the brain so that processing of information diminishes. Conduction velocity decreases, voluntary motor movements slow down, and reflex times increase. Parkinson's disease is the most common movement disorder of the CNS. Degenerative changes and disease states involving the sense organs can alter vision, hearing, taste, smell, and touch. Impaired hearing associated with aging, known as presbycusis, is usually the result of changes in important structures of the inner ear.

DEVELOPMENTAL ANATOMY OF THE NERVOUS SYSTEM

The development of the nervous system begins early in the third week with a thickening of the **ectoderm** called the **neural plate** (Fig. 14.18). The plate folds inward and forms a longitudinal groove, the **neural groove**. The raised edges of the neural plate are called **neural folds**. As development continues, the neural folds increase in height and meet to form a tube called the **neural tube**.

Three types of cells differentiate from the wall that encloses the neural tube. The outer or **marginal layer** develops into the *white matter* of the nervous system; the middle or **mantle layer** develops into the *gray matter*; and the inner or **ependymal layer** eventually forms the *lining of the central canal of the spinal cord and ventricles of the brain*.

The **neural crest** is a mass of tissue between the neural tube and the skin ectoderm (Fig. 14.18b). It differentiates and eventually forms the *posterior (dorsal) root ganglia of spinal nerves, spinal nerves, ganglia of cranial nerves, cranial nerves, ganglia of the autonomic nervous system, adrenal medulla, and meninges.*

When the neural tube forms from the neural plate, its anterior portion develops into three enlarged areas called **primary vesicles** (Fig. 14.19): (1) **prosencephalon (forebrain)**, (2) **mesencephalon (midbrain)**, and (3) **rhombencephalon (hindbrain)**. These are fluid-filled enlargements that develop by the fourth week of the embryonic period. As development progresses, the vesicular region undergoes several flexures (bends), resulting in subdivision of the three primary vesicles, so that by the fifth week of development the embryonic brain consists of five **secondary vesicles**. The prosencephalon divides into an anterior **telencephalon** and a posterior **diencephalon**; the mesencephalon remains unchanged; the rhombencephalon divides into an anterior **metencephalon** and a posterior **myelencephalon**.

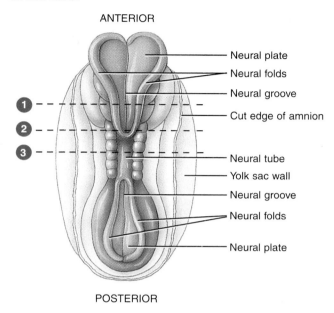

Figure 14.18 Origin of the nervous system. (a) Embryo in which the neural folds have partially united, forming the early neural tube. (b) Transverse sections through the embryo showing the formation of the neural tube.

ANTERIOR

Neural plate
Neural folds
Neural groove
Cut edge of amnion
Neural tube
Yolk sac wall
Neural groove
Neural folds
Neural plate

POSTERIOR

(a) Dorsal view

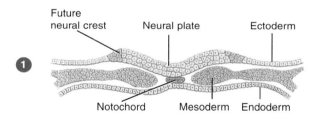

Future neural crest — Neural plate — Ectoderm
Notochord — Mesoderm — Endoderm

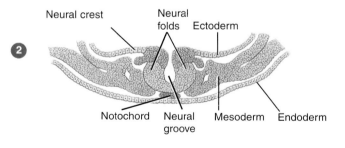

Neural crest — Neural folds — Ectoderm
Notochord — Neural groove — Mesoderm — Endoderm

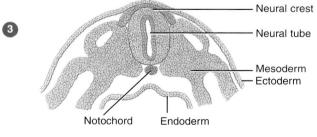

Neural crest
Neural tube
Mesoderm
Ectoderm
Notochord — Endoderm

(b) Transverse sections

Figure 14.19   Development of the brain and spinal cord.

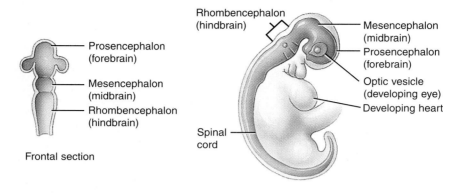

(a) 3-4 week embryo

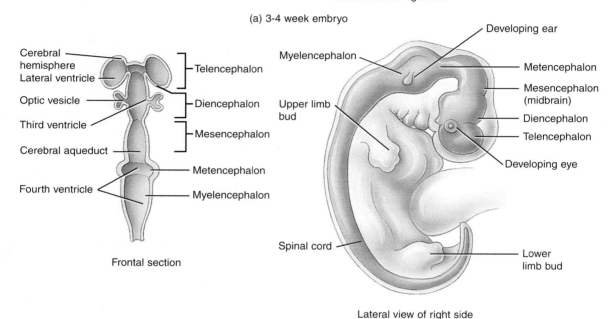

(b) 5-week embryo

Ultimately, the telencephalon develops into the *cerebral hemispheres* and *basal ganglia* and houses the paired *lateral ventricles*; the diencephalon develops into the *thalamus, hypothalamus,* and *pineal gland* and houses the *third ventricle*; the mesencephalon develops into the *midbrain* and houses the *cerebral aqueduct*; the metencephalon develops into the *pons* and *cerebellum*; and the myelencephalon develops into the *medulla oblongata,* each housing a portion of the *fourth*

ventricle. The area of the neural tube inferior to the myelencephalon gives rise to the *spinal cord*.

Two neural tube defects—spina bifida (see page 193) and anencephaly (absence of the skull and cerebral hemispheres)—are associated with low levels of a B vitamin called folic acid. The incidence of both disorders is greatly decreased when women who may become pregnant take folic acid supplements.

DISORDERS: HOMEOSTATIC IMBALANCES

CEREBROVASCULAR ACCIDENT
The most common brain disorder is a **cerebrovascular accident** (**CVA**), also called a **stroke**. CVAs are classified into two principal types: (1) *ischemic,* the most common type, is due to decreased blood flow, and (2) *hemorrhagic* is due to rupture of a blood vessel in the brain. Common causes of

CVAs are intracerebral hemorrhage (from a blood vessel in the pia mater or brain), emboli (blood clots), and atherosclerosis (formation of cholesterol-containing plaques that block blood flow) of the cerebral arteries. A CVA is characterized by abrupt onset of persisting neurological deficits such as paralysis or loss of sensation.

Among the risk factors implicated in CVAs are high blood pressure, high blood cholesterol, heart disease, narrowed carotid arteries, transient ischemic attacks, diabetes, smoking, obesity, and excessive alcohol intake.

TRANSIENT ISCHEMIC ATTACK

A **transient ischemic attack** (**TIA**) is an episode of temporary cerebral dysfunction caused by impaired blood flow to the brain. Symptoms include dizziness, weakness, numbness, or paralysis in a limb or in half of the body; drooping of one side of the face; headache; slurred speech or difficulty understanding speech; or a partial loss of vision or double vision. Sometimes nausea or vomiting also occurs. The onset of symptoms is sudden and reaches maximum intensity almost immediately. A TIA usually persists for a few minutes and only rarely lasts as long as 24 hours; it leaves no persistent neurologic deficits. The causes of the impaired blood flow that lead to TIAs are blood clots, atherosclerosis, and certain blood disorders.

It is estimated that about one-third of patients who experience a TIA will have a CVA within 5 years. Therapy for TIAs includes drugs such as aspirin that block the aggregation of blood cells involved in clotting (platelets) and anticoagulants; cerebral artery bypass grafting; and carotid endarterectomy (removal of the cholesterol-containing plaques and inner lining of an artery).

ALZHEIMER'S DISEASE

Alzheimer's (ALTZ-hī-merz) **disease** (**AD**) is a disabling senile dementia, the loss of reason and ability to care for oneself, that afflicts about 11% of the population over age 65. In the United States, AD afflicts four million people and claims over 100,000 lives a year, making it the fourth leading cause of death among the elderly, after heart disease, cancer, and stroke.

Victims of AD initially have trouble remembering recent events. Next, they become more confused and forgetful, often repeating questions or getting lost while traveling to previously familiar places. Disorientation grows, memories of past events disappear, and there may be episodes of paranoia, hallucination, or violent changes in mood. As their minds continue to deteriorate, they lose their ability to read, write, talk, eat, or walk. Finally, the disease culminates in dementia. A person with AD usually dies of some complication that afflicts bedridden patients, such as pneumonia.

At autopsy, brains of AD victims show three distinct structural abnormalities: (1) great loss of neurons in specific regions, (2) plaques of abnormal proteins deposited outside neurons, and (3) tangled protein filaments within neurons. Some AD brains also have elevated levels of aluminum, an element that serves no known function in the human body.

Neuronal loss is severe in regions such as the hippocampus and cerebral cortex that are important for memory and learning. Many of the lost neurons are thought to use acetylcholine as a neurotransmitter because the level of an enzyme needed to synthesize ACh (choline acetyltransferase) may be greatly reduced in damaged regions. This finding prompted research on treating AD patients with precursors of ACh, which were not helpful. Some success is being seen using drugs that inhibit acetylcholinesterase (AChE), the enzyme that inactivates ACh. AChE blocking drugs such as tacrine (Cognex) and slow-release physostigmine (Synapton) are being tested on AD patients.

Besides neuronal destruction, there is an accumulation of **amyloid plaques**, which consist of degenerating axons and axon terminals and fibrils of an abnormal protein called **beta amyloid**. This 40–42 amino acid fragment is part of a membrane glycoprotein called amyloid precursor protein (APP) that is coded by a gene on chromosome 21. People with Down syndrome, which is caused by an extra chromosome 21 (see page 984), are mentally retarded and their brains also exhibit beta amyloid-containing plaques. This protein is found only rarely in elderly people with normal mental abilities. Beta amyloid is toxic to neurons grown in laboratory cultures because it causes hydrogen peroxide to accumulate.

The third feature of AD is the presence of **neurofibrillary tangles**, which are abnormal bundles of protein filaments inside cells in affected brain regions. These tangles contain a protein called A68 that also appears in the brains of older Down syndrome individuals. A68 appears to be an altered version of a protein called tau that normally associates with microtubules of the cytoskeleton. It also may be involved in neuronal death.

A history of head injury is a risk factor for developing AD, and a similar dementia occurs in boxers, probably caused by repeated blows to the head. Hereditary factors also increase one's AD risk. Three forms (alleles) of a gene on chromosome 19 code for a molecule called apolipoprotein E (apoE) that helps transport cholesterol in the blood. People who have one or two copies of the form that codes for **apolipoprotein E4** (**apoE4**) have a much higher risk of developing AD and an earlier age of onset when compared with people who have genes for the other forms, apoE2 or apoE3. In addition, the brains of people who die shortly after traumatic head injury are more likely to exhibit beta amyloid protein if they have an apoE4 allele. Although the way in which apoE exerts this influence is not clear, some scientists suggest that apoE2 and apoE3 may have a protective effect that apoE4 does not. In a small number of patients, AD is associated with mutation of the chormosome 21 gene that codes for APP or another gene on chromosome 14. Genetic flaws do not explain all cases, however. Even in identical twins, one may have AD while the other does not. Other proposed causes of AD are a slow-acting virus, environmental toxins such as aluminum, genetic defects in mitochondria, and decreased blood flow to the brain that results in inadequate oxygen and glucose.

BRAIN TUMORS

A **brain tumor** refers to any benign or malignant growth within the cranium. Tumors may arise from neuroglial cells in the cerebrum, brain stem, and cerebellum or from supporting or neighboring structures such as cranial nerve coverings, meninges, and the pituitary gland. Pressure in the brain from a growing tumor or edema (tissue swelling) associated with the tumor produces the characteristic signs and symptoms. Among these are headache, altered consciousness, vomiting, seizures, visual problems, cranial nerve abnormalities, hormonal syndromes, personality changes, dementia, and sensory or motor deficits. Treatment of brain tumors involves surgery, radiation therapy, and chemotherapy.

DYSLEXIA

Dyslexia (dis-LEK-sē-a; *dys* = difficulty; *lexis* = words) is a genetic defect that is characterized by an impairment of the brain's ability to translate images received from the eyes into understandable language. The condition is unrelated to basic intellectual

DISORDERS: HOMEOSTATIC IMBALANCES

capacity, but it causes a mysterious difficulty in processing words and symbols. Apparently, some difference in the brain's organizational pattern distorts the ability to read, write, and count. Letters in words seem transposed, reversed, or upside down—*dog* becomes *god* or *bog; b* changes identity with *d;* a sign saying "OIL" inverts into "710." Many dyslectics cannot orient themselves in the three dimensions of space and may exhibit awkward body movements.

Dyslexia is unaccompanied by outward scars of detectable neurological damage, and its symptoms vary. It occurs about three times more often in boys than in girls.

REYE SYNDROME
Reye syndrome (RS), first described in 1963 by the Australian pathologist R. Douglas Reye, seems to occur fol-

lowing a viral infection, particularly chickenpox or influenza. Aspirin at normal doses is a risk factor in the development of RS. Most persons affected are children or teenagers. The disease is characterized by vomiting and brain dysfunction (disorientation, lethargy, and personality changes) and may progress to coma. Also, the liver becomes infiltrated with small lipid droplets and loses some of its ability to detoxify ammonia.

Brain dysfunction and death are typically caused by swelling of brain cells. The pressure not only kills the cells directly but also results in hypoxia that kills them indirectly. The survival rate is about 70%. Swelling may result in irreversible brain damage, including mental retardation, in children who survive. Therapy is directed at controlling the swelling.

MEDICAL TERMINOLOGY

Agnosia (ag-NŌ-zē-a; *a* = without; *gnosis* = knowledge) Inability to recognize the significance of sensory stimuli such as auditory, visual, olfactory, gustatory, and tactile.

Apraxia (a-PRAK-sē-a; *pratto* = to do) Inability to carry out purposeful movements in the absence of paralysis.

Delirium (de-LIR-ē-um; *deliria* = off the track) Also called **acute confusional state (ACS)**. A transient disorder of abnormal cognition and disordered attention that is accompanied by disturbances of the sleep–wake cycle and psychomotor behavior (hyperactivity or hypoactivity of movements and speech).

Dementia (de-MEN-shē-a; *de* = away from; *mens* = mind) A mental disorder that results in permanent or progressive general loss of intellectual abilities such as impairment of memory, judgment, and abstract thinking and changes in personality.

Electroconvulsive therapy (ECT) (e-lek′-trō-con-VUL-siv THER-a-pē) A form of shock therapy in which convulsions are induced by the passage of a brief electric current

through the brain. ECT is regarded as an important therapeutic option in the treatment of severe depression and acute mania in adults. Side effects may include acute confusional states and memory deficits.

Encephalitis (en′-sef-a-LĪ-tis) An acute inflammation of the brain caused by a direct attack by various viruses or by an allergic reaction to any of the many viruses that are normally harmless to the central nervous system. If the virus affects the spinal cord as well, it is called **encephalomyelitis**.

Lethargy (LETH-ar-jē) A condition of functional sluggishness.

Nerve block Loss of sensation in a region, such as in local dental anesthesia, due to injection of a local anesthetic.

Neuralgia (noo-RAL-jē-a; *neur* = nerve) Attacks of pain along the entire course or branch of a peripheral sensory nerve.

Stupor (STOO-por) Unresponsiveness from which a patient can be aroused only briefly and by vigorous and repeated stimulation.

STUDY OUTLINE

BRAIN (p. 391)

Principal Parts (p. 391)
1. The principal parts of the brain are the brain stem, cerebellum, diencephalon, and cerebrum.
2. During embryological development, three primary brain vesicles are formed and serve as forerunners of various parts of the brain. They are the prosencephalon, the mesencephalon, and the rhombencephalon. The prosencephalon develops into the telencephalon and diencephalon; the rhombencephalon develops into the metencephalon and myelencephalon.
3. The telencephalon forms the cerebrum, the diencephalon develops into the thalamus and hypothalamus, the mesencephalon develops into the midbrain, the metencephalon develops into the pons and cerebellum, and the myelencephalon forms the medulla oblongata.

Protection and Coverings (p. 392)
1. The brain is protected by cranial bones, meninges, and cerebrospinal fluid.
2. The cranial meninges are continuous with the spinal meninges and are named dura mater, arachnoid, and pia mater.

Cerebrospinal Fluid (p. 392)
1. Cerebrospinal fluid (CSF) is formed in the choroid plexuses and circulates through the ventricles, subarachnoid space, and central canal. Most of the fluid is absorbed by the arachnoid villi of the superior sagittal blood sinus.
2. Cerebrospinal fluid provides mechanical protection, chemical protection, and circulation.
3. If cerebrospinal fluid accumulates in the ventricles or subarachnoid space, it is called hydrocephalus.

body arrive in a specific region of the cerebral cortex, which interprets the sensation as coming from the stimulated sensory receptors.

Sensory Receptors

The process of sensation begins in a large variety of different types of sensory receptors, each of which is sensitive to a particular type of stimulus. Sensory receptors respond vigorously to one particular kind of stimulus and weakly or not at all to others. This characteristic is termed **selectivity**. The stimulus may be in the form of electromagnetic energy, such as light or heat; mechanical energy, such as pressure; or chemical energy, such as in a molecule of carbon dioxide in body fluids. For instance, auditory receptors in the ears selectivity respond to sound waves but not to light.

Classification of Receptors

Receptors vary in their complexity. The **somatic (general) senses** include touch, pressure, vibration, warmth, cold, and pain plus proprioception (detection of body positions and movements). The anatomically simplest receptors are termed **free nerve endings** because they have no apparent structural specializations. Examples are receptors for pain, temperature, tickle, and itch sensations (some of which are shown in Fig. 15.1). Receptors for other somatic sensations such as touch, pressure, and vibration have distinctive structures, for example, the corpuscles of touch shown in Fig. 15.1. The receptors for the **special senses**—smell, taste, vision, hearing, and equilibrium—are located in sense organs such as the eye and ear. They also have complex and distinctive structures.

Two widely used classifications of receptors are based on the location of the receptors and the type of stimuli they detect.

Figure 15.1   Structure and location of somatic receptors in the skin.

 Somatic senses include touch, pressure, vibration, warmth, cold, and pain plus proprioception.

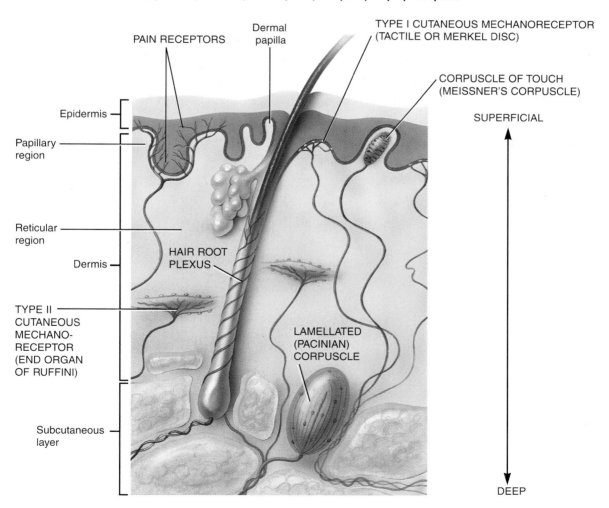

PAIN RECEPTORS

Dermal papilla

TYPE I CUTANEOUS MECHANORECEPTOR (TACTILE OR MERKEL DISC)

CORPUSCLE OF TOUCH (MEISSNER'S CORPUSCLE)

SUPERFICIAL

Epidermis

Papillary region

Reticular region

Dermis

HAIR ROOT PLEXUS

TYPE II CUTANEOUS MECHANO-RECEPTOR (END ORGAN OF RUFFINI)

LAMELLATED (PACINIAN) CORPUSCLE

Subcutaneous layer

DEEP

Section of skin and subcutaneous layer

Which sensations can arise when free nerve endings are stimulated?

● **By Location** One method of classifying receptors is by their location. **Exteroceptors** (eks′-ter-ō-SEP-tors) are located at or near the surface of the body and provide information about the *external* environment. They are sensitive to stimuli outside the body and transmit sensations of hearing, vision, smell, taste, touch, pressure, vibration, temperature, and pain.

Interoceptors or **visceroceptors** (vis′-er-ō-SEP-tors) are located in blood vessels and visceral organs and provide information about the *internal* environment. These sensations arise from within the body and often do not reach conscious perception. Occasionally, they may be felt as pain or pressure.

Proprioceptors (prō′-prē-ō-SEP-tors) are located in muscles, tendons, joints, and the internal ear. They provide information about body position, muscle tension, and the position and activity of our joints.

● **By Type of Stimulus** Another method of classifying receptors is by the type of stimulus they detect. **Mechanoreceptors** detect mechanical pressure or stretching. Stimuli so detected include those related to touch, pressure, vibration, proprioception, hearing, equilibrium, and blood pressure. **Thermoreceptors** detect changes in temperature. **Nociceptors** respond to stimuli that cause physical or chemical damage to tissues. **Photoreceptors** detect light that strikes the retina of the eye. **Chemoreceptors** detect chemicals in the mouth (taste), nose (smell), and body fluids.

Generator Potentials and Receptor Potentials

The electrical response of a sensory receptor to a stimulus is either a **generator potential** or a **receptor potential**. Both types of responses are graded potentials, like postsynaptic potentials, that result when gated ion channels open or close. Recall that graded potentials have a variable amplitude and are not propagated. (See page 348 to review differences between action potentials and graded potentials.) The amplitude of both generator and receptor potentials varies with the intensity of the stimulus. An intense stimulus produces a large generator or receptor potential whereas a weak stimulus elicits a small one.

Receptors that serve the senses of smell (olfactory receptors), touch, pressure, stretching, vibration, temperature, pain, and proprioception produce generator potentials. When a generator potential is large enough to reach threshold, it *generates* one or more nerve impulses in its first-order sensory neuron. A stimulus elicits a generator potential, which then triggers nerve impulses that propagate into the CNS. The resulting nerve impulses propagate along the first-order nerve fiber. Upon arriving at the axon terminals, they trigger exocytosis of synaptic vesicles and release of neurotransmitter molecules.

Receptors that serve the special senses of vision, hearing, equilibrium, and taste produce receptor potentials. These receptors are short cells—just a few micrometers (μm) long—that form synapses with first-order sensory neurons. The re-

ceptor potential spreads to the base of the receptor where synaptic vesicles are located and directly increases or decreases their exocytosis. Neurotransmitter liberated from synaptic vesicles, then, may either stimulate or inhibit production of nerve impulses in the first-order sensory neuron, which extends into the CNS.

Adaptation of Sensory Receptors

A characteristic of many sensations is **adaptation**, which is a change in sensitivity (usually a decrease) during a long-lasting stimulus. The perception of a sensation may even disappear although the stimulus is still being applied. For example, when you first step into a hot shower, the water may feel too hot, but soon the sensation decreases to one of comfortable warmth even though the temperature of the stimulus (hot water) does not change. Receptors vary in their ability to adapt. **Rapidly adapting (phasic) receptors**, for example, those associated with pressure, touch, and smell, adapt very quickly. Such receptors play a major role in signaling changes in a particular sensation. **Slowly adapting (tonic) receptors**, such as those for detecting pain, body position, and chemicals in blood, adapt slowly. These receptors are important in signaling information about steady states of the body. While much adaptation occurs in sensory receptors, further adaptation may occur as sensory signals are processed within the CNS.

SOMATIC SENSES

The somatic (general) senses arise in receptors located in the skin (cutaneous sensations) or embedded in muscles, tendons, joints, and the inner ear (proprioceptive sensations).

Cutaneous Sensations

Cutaneous (kyoo-TĀ-nē-us; *cutis* = skin) **sensations** include tactile sensations (touch, pressure, vibration), thermal sensations (cold and warmth), and pain. The receptors for these sensations are in the skin, connective tissue deep to the skin, mucous membranes, mouth, and anus.

Cutaneous receptors are distributed over the body surface in such a way that some parts of the body are densely populated with receptors and other parts contain only a few. Areas of the body that have few cutaneous receptors are not very sensitive; those containing many are highly sensitive. Areas with high sensitivity include the tip of the tongue, lips, and fingertips.

Cutaneous receptors consist of the dendrites of sensory neurons that may be enclosed in a capsule of epithelial or connective tissue or have no apparent structural specialization (free nerve endings). Nerve impulses generated by cutaneous receptors pass along somatic sensory neurons in spinal and cranial nerves, through the thalamus, to the somatosensory area of the parietal lobe of the cerebral cortex (see Fig. 15.4a).

Tactile Sensations

The **tactile** (TAK-tĭl; *tact* = touch) **sensations** are touch, pressure, and vibration plus itch and tickle. They are all detected by mechanoreceptors.

● **Touch** Sensations of **touch** generally result from stimulation of tactile receptors in the skin or tissues immediately deep to the skin. **Crude touch** refers to the ability to perceive that something has touched the skin, although its exact location, shape, size, or texture cannot be determined. **Discriminative touch** refers to the ability to recognize exactly what point on the body is touched. Tactile receptors include corpuscles of touch, hair root plexuses, and type I and II cutaneous mechanoreceptors (Fig. 15.1).

Corpuscles of touch, or **Meissner's** (MĪS-ners) **corpuscles**, are egg-shaped receptors for discriminative touch. They are a mass of dendrites enclosed by connective tissue and are located in the dermal papillae of the skin. Since they adapt rapidly, they generate impulses most rapidly at the onset of a touch. Corpuscles of touch are most plentiful in the fingertips, palms, and soles. They are also abundant in the eyelids, tip of the tongue, lips, nipples, clitoris, and tip of penis. Other rapidly adapting touch receptors are the **hair root plexuses**, in which dendrites are arranged in networks around hair follicles. Movement of the hair shaft stimulates the dendrites. Hair root plexuses detect movements mainly on the surface of the body when hairs are disturbed.

There are two types of slowly adapting touch receptors. **Type I cutaneous mechanoreceptors**, also called **tactile** or **Merkel** (MER-kel) **discs**, are the flattened portions of dendrites of sensory neurons that make contact with epidermal cells of the stratum basale called Merkel cells (see Fig. 5.2). They are distributed in many of the same locations as corpuscles of touch and also function in discriminative touch. **Type II cutaneous mechanoreceptors**, or **end organs of Ruffini**, are embedded deeply in the dermis and in deeper tissues of the body. They detect heavy and continuous touch sensations.

● **Pressure** Sensations of pressure generally result from stimulation of tactile receptors in deeper tissues. **Pressure** is a sustained sensation that is felt over a larger area than touch. Pressure receptors are type II cutaneous mechanoreceptors and lamellated corpuscles. **Lamellated**, or **Pacinian** (pa-SIN-ē-an), **corpuscles** are large oval structures composed of a connective tissue capsule, layered like an onion, that encloses a dendrite. Like corpuscles of touch, lamellated corpuscles adapt rapidly. They are located in subcutaneous tissues, deep submucosal tissues that lie under mucous membranes, and serous membranes; around joints, tendons, and muscles; and in the mammary glands, external genitalia, and certain viscera, such as the pancreas and urinary bladder.

● **Vibration** Sensations of **vibration** result from rapidly repetitive sensory signals from tactile receptors. The receptors for vibration sensations are corpuscles of touch and lamellated corpuscles. Whereas corpuscles of touch detect lower-frequency vibrations, lamellated corpuscles detect higher-frequency vibrations.

● **Itch and Tickle** The **itch** sensation results from stimulation of free nerve endings by certain chemicals, such as bradykinin, often as a result of a local inflammatory response. Free nerve endings also are thought to mediate the **tickle** sensation. This unusual sensation is the only one that you may not be able to elicit on yourself. Why the tickle sensation may arise only when someone else touches you is still a mystery.

Thermal Sensations

The **thermal** (*therm* = heat) **sensations** are of warmth and coolness. For a long time, the receptors detecting thermal sensations were thought to be the end organs of Ruffini and similar encapsulated structures called Krause's corpuscles. Experiments now confirm, however, that **thermoreceptors** are free nerve endings. Separate thermoreceptors respond to warm and cold stimuli.

Pain Sensations

Pain is indispensable for a normal life. It provides information about noxious, tissue-damaging stimuli and thus often enables us to protect ourselves from greater damage. From a medical standpoint, the subjective description and indication of the location of pain may help pinpoint the underlying cause of disease.

The receptors for **pain**, called **nociceptors** (NŌ-sē-sep′-tors; *noci* = harmful), are free nerve endings (Fig. 15.1). They are found in almost every tissue of the body, and may respond to any type of stimulus if it is strong enough to cause tissue damage. Tissue irritation or injury releases chemicals, for example, prostaglandins and kinins, that stimulate nociceptors. Pain persists even after the initial trauma occurs since these substances linger and nociceptors adapt only slightly or not at all. Stimuli that elicit pain include excessive distension or dilation of a structure, prolonged muscular contractions, muscle spasms, or inadequate blood flow to an organ. Nociceptors perform a protective function by identifying changes that may endanger the body.

● **Types of Pain** Pain is classified as one of two types, based on speed of onset, quality of the sensation, and duration: acute (fast) and chronic (slow). **Acute pain** occurs very rapidly, usually within 0.1 second after a stimulus is applied, and is not felt in deeper tissues of the body. This type of pain is also known as sharp, fast, and pricking pain. The pain felt from a needle puncture or knife cut to the skin are examples of acute pain. Impulses for acute pain conduct along large diameter, myelinated A fibers. **Chronic pain,** by contrast, begins after a second or more and then gradually increases in intensity over a period of several seconds or minutes. This type of pain may be excruciating. It is also referred to as burning, aching, throbbing, and slow pain. Chronic pain can occur both in the skin and deeper tissues or in internal organs. An example is the pain associated with a toothache. Impulses for chronic pain conduct along smaller diameter, unmyelinated C fibers.

Pain that arises from stimulation of receptors in the skin is called **superficial somatic pain**, whereas stimulation of receptors in skeletal muscles, joints, tendons, and fascia, causes **deep somatic pain**. **Visceral pain** results from stimulation of receptors in the visceral organs.

Although receptors for somatic and visceral pain are similar, viscera do not evoke the same pain response as somatic tissues. For example, highly *localized* damage to certain viscera, such as cutting the intestine in two in a patient who is awake, causes very little, if any, pain. But, if stimulation is *diffuse,* involving large areas, visceral pain can be severe. Such stimulation might result from distension, spasms, or ischemia. For example, a kidney stone or a gallstone might obstruct and distend a ureter or the bile duct and cause severe pain.

• **Referred Pain** In most instances of somatic pain and in some instances of visceral pain, the cerebral cortex accurately localizes the pain to the stimulated area. If you burn your finger, you feel the pain in your finger. If the pleural membranes around the lungs are inflamed, you experience pain in the chest. In most instances of visceral pain, however, the pain is felt in or just deep to the skin that overlies the stimulated organ. The pain may also be felt in a surface area far from the stimulated organ. This phenomenon is called **referred pain**. In general, the area to which the pain is referred and the visceral organ involved are served by the same segment of the spinal cord. For example, sensory fibers from the heart and skin over the heart and along the medial aspect of the left arm enter spinal cord segments T1–T5. Thus the pain of a heart attack is typically felt in the skin over the heart and along the left arm. Figure 15.2 illustrates skin regions to which visceral pain may be referred.

Figure 15.2   Referred pain. The colored parts of the diagrams indicate skin areas to which visceral pain is referred.

Nociceptors (receptors for pain) are free nerve endings and are found in almost every tissue of the body.

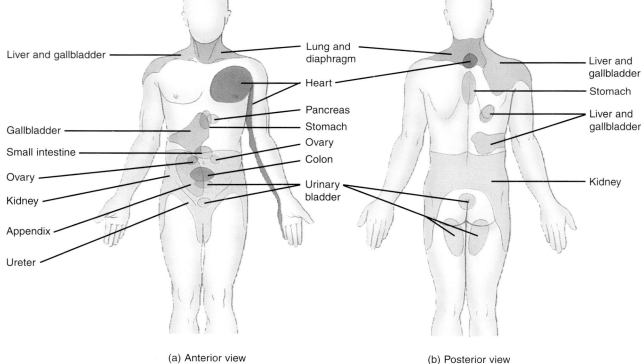

(a) Anterior view   (b) Posterior view

Which visceral organs have the broadest area for referred pain?

• **Phantom Pain** The pain often experienced by patients who have had a limb amputated is called **phantom pain (phantom limb sensation)**. They may still experience sensations such as itching, pressure, tingling, or pain in the extremity as if the limb were still there. A partial explanation for sensations from a phantom limb is that nerve impulses arise in the remaining proximal portions of the sensory neurons that previously received impulses from the limb. The cerebral cortex interprets impulses from these neurons as coming from the nonexistent (phantom) limb. Another theory to explain phantom limb sensations suggests that the brain itself contains networks of neurons that generate sensations of body awareness.

RELIEF FROM PAIN

Some pain sensations are inappropriate; they do not warn of actual or impending damage but rather occur out of proportion to minor damage or chronically for no good reason. In such cases, **analgesia** (*a* = without; *algesis* = sensation of pain) or pain relief is needed. Most pain sensations respond to pain-reducing drugs. Drugs such as aspirin and ibuprofen (Motrin) block formation of prostaglandins, which stimulate nociceptors. Local anesthetics, such as Novocaine, provide short-term pain relief by blocking conduction of nerve impulses in the axons of first-order neurons. Morphine and other opiate drugs alter the quality of pain perception in the brain; pain is still sensed, but it is not perceived as being so noxious.

Sometimes, only surgery can relieve pain. The purpose of surgical treatment is to interrupt the pain pathway somewhere between the receptors and the interpretation centers of the brain. This can be accomplished by severing a sensory nerve, its spinal root, or certain tracts in the spinal cord or brain. Cordotomy is severing a spinal cord tract, usually the lateral spinothalamic; rhizotomy is the cutting of some spinal posterior (sensory) nerve roots. In each instance, a link in the pathway for pain is cut so that pain impulses do not reach the cerebral cortex. ■

Proprioceptive Sensations

An awareness of body position and of movements of parts of the body is provided by the **proprioceptive** (*proprio* = one's own), or **kinesthetic** (kin′-es-THET-ik; *kinesis* = motion), **sense**. It informs us of the degree to which muscles are contracted, the amount of tension created in the tendons, the change of position of a joint, and the orientation of the head relative to the ground and in response to movements. Proprioception tells us the location and rate of movement of one body part in relation to others, so we can walk, type, or dress without using our eyes. It also allows us to estimate the weight of objects and determine the muscular effort necessary to perform a task. For example, as you pick up a bag you quickly realize whether it contains feathers or books, and you exert the correct amount of effort to lift it.

Proprioceptors, the receptors for proprioception, adapt only slightly. This feature allows the brain to be informed continually of the status of different parts of the body so that adjustments can be made to ensure coordination. Proprioceptors include muscle spindles, tendon organs, and joint kinesthetic receptors. They are located within skeletal muscles, tendons, and joint capsules. Also classified as proprioceptors are hair cells of the internal ear, which provide information for maintaining balance (see page 481).

Impulses for conscious proprioception pass along ascending tracts in the spinal cord to the thalamus and from there to the cerebral cortex. The sensation is perceived in the somatosensory area in the parietal lobe of the cerebral cortex, posterior to the central sulcus. At the same time, impulses from proprioceptors also pass to the cerebellum along the spinocerebellar tracts.

Muscle Spindles

Muscle spindles are specialized groupings of muscle fibers interspersed among regular skeletal muscle fibers and oriented parallel to them (Fig. 15.3a). The ends of the spindles are anchored to endomysium and perimysium. A muscle spindle consists of 3–10 specialized muscle fibers called **intrafusal muscle fibers** that are partially enclosed in a spindle-shaped connective tissue capsule. The central region of each intrafusal fiber contains several nuclei but has few or no actin and myosin filaments. Both ends of the intrafusal muscle fibers do contain actin and myosin filaments. They contract when stimulated by small diameter motor neurons called **gamma motor neurons**. Surrounding the muscle spindle are the regular skeletal muscle fibers, which are called **extrafusal muscle fibers**. Large **alpha motor neurons** innervate the extrafusal fibers. Both types of motor neurons arise in the anterior gray horn of the spinal cord.

The central area of an intrafusal fiber cannot contract because it lacks actin and myosin, but it does contain two types of sensory (afferent) fibers. The first are large diameter, rapidly conducting sensory fibers, called **type Ia fibers**. The dendrites of the Ia fiber wrap in a spiral manner around the central area of each intrafusal fiber. Stretching the central part of the spindle stimulates the dendrites, and nerve impulses propagate toward the spinal cord. The central receptive area of some muscle spindles is also served by smaller diameter sensory fibers called **type II fibers**. Their dendrites are located on either side of the type Ia dendrites. Type II dendrites are also stimulated when the central part of the spindle is stretched, and they too send impulses to the spinal cord.

Either sudden or prolonged stretch on the central areas of the intrafusal muscle fibers stimulates the type Ia and type II dendrites. Muscle spindles monitor changes in the length of a skeletal muscle by responding to the rate and degree of change in length. This information is relayed to the cerebrum, which allows conscious perception of limb position (see Fig. 15.4). It also passes to the cerebellum to aid in

Figure 15.3   Proprioceptors.

 Proprioceptors are located mainly in muscles, tendons, and the inner ear; they provide information about movements and position of the body.

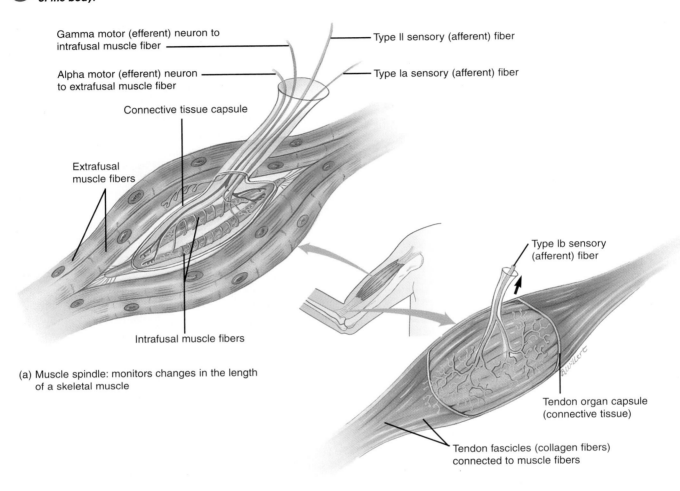

(a) Muscle spindle: monitors changes in the length of a skeletal muscle

(b) Tendon organ: protects muscles and tendons from damage due to excess tension and monitors force of muscle contraction

 Which parts of a muscle spindle are associated with the sensory and motor neurons?

the coordination and efficiency of muscle contraction (see Fig. 15.8).

Tendon Organs

Tendon organs (**Golgi tendon organs**) are proprioceptors found at the junction of a tendon with a muscle. Each tendon organ consists of a thin capsule of connective tissue that encloses a few collagen fibers (Fig. 15.3b). Penetrating the capsule are one or more sensory (afferent) **type Ib fibers** whose dendrites entwine among and around the collagen fibers. When tension is applied to a tendon, tendon organs are stimulated, and nerve impulses are conducted into the CNS (see Fig. 15.4). Tendon organs help protect tendons and their associated muscles from damage due to excessive

tension. Also, they function as contraction receptors: they monitor the force of contraction of associated muscles.

Joint Kinesthetic Receptors

There are several types of **joint kinesthetic receptors** within and around the articular capsules of synovial joints. Encapsulated receptors, similar to type II cutaneous mechanoreceptors (end organs of Ruffini), are present in the capsules of joints and respond to pressure. Small lamellated (Pacinian) corpuscles in the connective tissue outside articular capsules are receptors that respond to acceleration and deceleration of joint movement. Articular ligaments contain receptors similar to tendon organs that adjust reflex inhibition of the adjacent muscles when excessive strain is placed on the joint.

ANESTHESIA
During certain surgical or diagnostic procedures, anesthesia (an′-es-THĒ-zē-a; *an* = without; *aisthesis* = sensation) is used to block sensations. Two commonly used forms of anesthesia are general and spinal. **General anesthesia** removes all sensations, including pain, and also produces unconsciousness and sometimes muscular relaxation. **Spinal anesthesia** involves injection of a drug into the subarachnoid space to block pain and other somatic sensations from that point downward. The procedure is widely used for surgery inferior to the diaphragm such as hernia repair, procedures on the hips and lower limbs, and operations involving the rectum, urinary bladder, prostate gland, and other pelvic structures. ■

SOMATIC SENSORY PATHWAYS

Somatic sensory pathways from receptors to the cerebral cortex involve three-neuron sets. Axon collaterals (branches) of somatic sensory neurons simultaneously carry signals into the cerebellum and the reticular formation of the brain stem.

1. **First-order neurons** carry signals from the somatic receptors into either the brain stem or spinal cord. From the face, mouth, teeth, and eyes, somatic sensory impulses propagate along *cranial nerves* into the brain stem. From the posterior aspect of the head, neck, and body, somatic sensory impulses propagate along *spinal nerves* into the spinal cord.
2. **Second-order neurons** carry signals from the spinal cord and brain stem to the thalamus. Axons of second-order neurons cross over (decussate) to the opposite side in the spinal cord or brain stem before ascending to the thalamus.
3. **Third-order neurons** project from the thalamus to the primary somatosensory area of the cortex (postcentral gyrus; see Fig. 14.15), where conscious perception of the sensations results.

There are two general pathways by which somatic sensory signals entering the spinal cord ascend to the cerebral cortex: the posterior column–medial lemniscus pathway and the anterolateral (spinothalamic) pathways.

Posterior Column–Medial Lemniscus Pathway to the Cortex

Nerve impulses for conscious proprioception and most tactile sensations ascend to the cortex along a common pathway formed by three-neuron sets (Fig. 15.4a). First-order neurons extend from sensory receptors into the spinal cord and up to the medulla oblongata on the same side of the body. The cell bodies of these first-order neurons are in the

posterior (dorsal) root ganglia of spinal nerves. Their axons form the **posterior column: fasciculus gracilis** (fa-SIK-yoo-lus gras-I-lis) and **fasciculus cuneatus** (kyoo-nē-Ā-tus) in the spinal cord (see Exhibit 15.1 on page 443). The axon terminals synapse with second-order neurons in the medulla. The cell body of a second-order neuron is located in the nucleus cuneatus (which receives input conducted along axons in the fasciculus cuneatus from the neck, upper limbs, and upper chest) or nucleus gracilis (which receives input conducted along axons in the fasciculus gracilis from the trunk and lower limbs). The axon of the second-order neuron crosses to the opposite side of the medulla and enters the **medial lemniscus**, a projection tract that extends from the medulla to the thalamus. In the thalamus, the axon terminals of second-order neurons synapse with third-order neurons, which project their axons to the somatosensory area of the cerebral cortex.

Impulses conducted along the posterior column–medial lemniscus pathway give rise to several highly evolved and refined sensations. These are:

1. **Discriminative touch**, the ability to recognize the exact location of a light touch and to make two-point discriminations.
2. **Stereognosis**, the ability to recognize by feel the size, shape, and texture of an object. Examples are reading braille or identifying (with closed eyes) a paperclip put into your hand.
3. **Proprioception**, the awareness of the precise position of body parts, and **kinesthesia**, the awareness of directions of movement.
4. **Weight discrimination**, the ability to assess the weight of an object.
5. **Vibratory sensations**, the ability to sense rapidly fluctuating touch.

Anterolateral (Spinothalamic) Pathways to the Cortex

The **anterolateral** or **spinothalamic** (spī-nō-tha-LAM-ik) **pathways** carry mainly pain and temperature impulses. In addition, they relay the sensations of tickle and itch and some tactile impulses, which give rise to a very crude, poorly localized touch or pressure sensation. Like the posterior column–medial lemniscus pathway, the anterolateral pathways are also composed of three-neuron sets (Fig. 15.4b). The first-order neuron connects a receptor of the neck, trunk, or limbs with the spinal cord. The cell body of the first-order neuron is in the posterior root ganglion. The axon terminals of the first-order neuron synapse with the second-order neuron, which is located in the posterior gray horn of the spinal cord. The axon of the second-order neuron continues to the opposite side of the spinal cord and passes upward to the brain stem in either the **lateral spinothalamic tract** or the **anterior spinothalamic tract** (see Exhibit 15.1 on page 443). The axon from the

Figure 15.4   Somatic sensory pathways.

Nerve impulses are conducted along sets of first-order, second-order, and third-order neurons to the somatosensory area of the cerebral cortex.

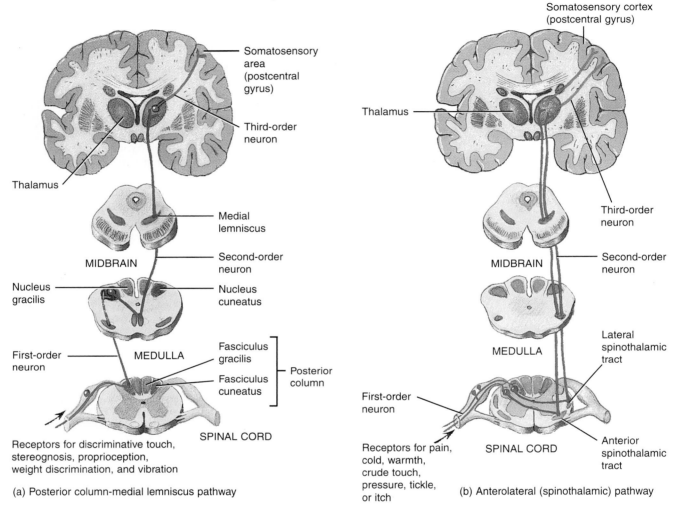

(a) Posterior column-medial lemniscus pathway

Somatosensory area (postcentral gyrus)

Third-order neuron

Thalamus

Medial lemniscus

Second-order neuron

MIDBRAIN

Nucleus gracilis

Nucleus cuneatus

First-order neuron

MEDULLA

Fasciculus gracilis

Fasciculus cuneatus

Posterior column

SPINAL CORD

Receptors for discriminative touch, stereognosis, proprioception, weight discrimination, and vibration

(b) Anterolateral (spinothalamic) pathway

Somatosensory cortex (postcentral gyrus)

Thalamus

Third-order neuron

Second-order neuron

MIDBRAIN

Lateral spinothalamic tract

MEDULLA

First-order neuron

Receptors for pain, cold, warmth, crude touch, pressure, tickle, or itch

SPINAL CORD

Anterior spinothalamic tract

What sorts of sensory deficits could be produced by damage to the right lateral spinothalamic tract?

second-order neuron ends in the thalamus. There, it synapses with the third-order neuron. The axon of the third-order neuron projects to the somatosensory area of the cerebral cortex. The lateral spinothalamic tract conveys sensory impulses for pain and temperature whereas the anterior spinothalamic tract conveys impulses for tickle, itch, crude touch, and pressure.

Somatosensory Cortex

Areas of the somatosensory cortex (postcentral gyrus) that receive sensory information from different parts of the body have been mapped out. Figure 15.5a shows the location and

areas of representation of the somatosensory cortex of the right cerebral hemisphere. The left cerebral hemisphere has a similar somatosensory cortex.

Note that some parts of the body are represented by large areas in the somatosensory cortex. These include the lips, face, tongue, and thumb. Other parts of the body, such as the trunk and lower limbs, are represented by much smaller areas. The relative sizes of the areas in the somatosensory cortex are directly proportional to the number of specialized sensory receptors and thus to the sensitivity in each respective part of the body. For example, there are many receptors in the skin of the lips but few in the skin of the trunk. The size of the cortical area for a particular part of the body re-

Figure 15.5 Primary somatosensory area (postcentral gyrus) and primary motor area (precentral gyrus) of the right cerebral hemisphere. The left hemisphere has similar representation. (after Penfield and Rasmussen)

 Each point on the body surface maps to a specific region in the sensory cortex and motor cortex.

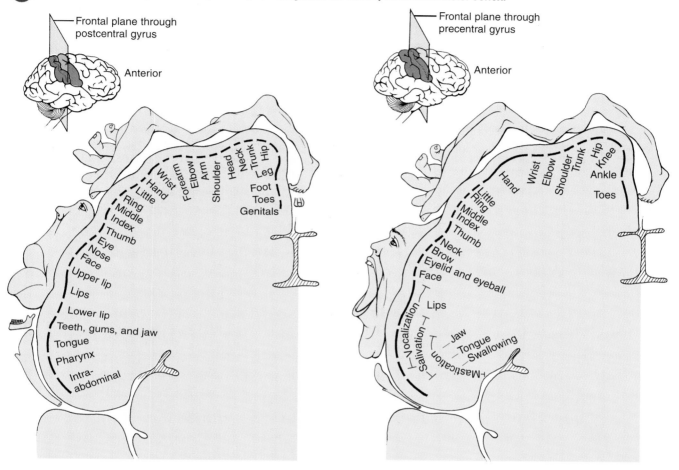

(a) Frontal section of somatosensory cortex in right cerebral hemisphere

(b) Frontal section of motor cortex in right cerebral hemisphere

How do the somatosensory and motor representations compare for the hand and what does this difference imply?

lates directly to the functional importance of high sensitivity for that body part.

Somatic Sensory Pathways to the Cerebellum

Two tracts in the spinal cord, the **posterior spinocerebellar** (spī-nō-ser-e-BEL-ar) **tract** and the **anterior spinocerebellar tract**, are major routes for subconscious proprioceptive input to reach the cerebellum. Sensory input conveyed to the cerebellum along these two pathways is critical for posture, balance, and coordination of skilled movements (described in more detail in Exhibit 15.1 on page 443). Two other tracts also convey proprioceptive input to the cerebellum. These are the **cuneocerebellar tract** and **rostral spinocerebellar tract**. Both tracts transmit nerve impulses from proprioceptors of the trunk and upper limbs.

SYPHILIS

One disabling disorder caused by the syphilis bacterium (*Treponema pallidum*) is progressive degeneration of the posterior portions of the spinal cord, including the posterior columns and posterior roots. This condition is called tabes (*tabes* = wasting or melting) dorsalis. Predictably, these patients experience akinesia and loss of touch and pain sensations. Their gait becomes uncoordinated, resembling patients with cerebellar disease, because proprioceptive input to the cerebellum is diminished. Lacking this proprioceptive input, they must watch their feet while walking to maintain balance. However, visual input does not fully compensate for normal proprioceptive signals relayed via the spinocerebellar tracts, and their movements are uncoordinated and jerky. Syphilis is just one of several sexually transmitted diseases (STDs) that can have a life-threatening prognosis (see page 949). ■

INTEGRATION OF SENSORY INPUT AND MOTOR OUTPUT

Sensory systems provide the input that keeps the central nervous system informed of changes in the external and internal environment. Output from the CNS is then conveyed to motor systems, which enable us to move about, alter glandular secretions, and change our relationship to the world around us. As sensory information reaches the CNS, it becomes part of a large pool of sensory input. Each bit of input the CNS receives does not elicit a response. Rather, the incoming information is integrated with other information arriving from all other operating sensory receptors. The integration process occurs not just once but at many stations along the pathways of the CNS and at both conscious and subconscious levels. It occurs within the spinal cord, brain stem, cerebellum, basal ganglia, and cerebral cortex. As a result, a motor response to make a muscle contract or a gland secrete can be modified and responded to at any of these levels. Motor portions of the cerebral cortex play the major role for initiating and controlling precise, discrete muscular movements. The basal ganglia integrate semivoluntary, automatic movements like walking, swimming, and laughing. The cerebellum assists the motor cortex and basal ganglia by making body movements smooth and coordinated and by contributing significantly to maintaining normal posture and balance.

SOMATIC MOTOR PATHWAYS

The most direct somatic motor pathways extend from the cerebral cortex into the spinal cord and out to skeletal muscles. Other pathways are less direct and include synapses in the basal ganglia, thalamus, reticular formation, and cerebellum.

Motor Cortex

The **primary motor area** (**precentral gyrus**) of the cerebral cortex is the major control region for initiation of voluntary movements. The adjacent **premotor area** and even the **somatosensory area** in the postcentral gyrus (see Fig. 14.15) also contribute fibers to the descending motor pathways. As is true for somatic sensory representation in the somatosensory area, different muscles are represented unequally in the primary motor area (Fig. 15.5b). The degree of representation is proportional to the number of motor units in a particular muscle of the body. For example, the muscles in the thumb, fingers, lips, tongue, and vocal cords have large representations while the trunk has a much smaller representation. By comparing Figs. 15.5a and 15.5b, you will see that somatosensory and motor representations are similar but not identical for the same part of the body.

Direct Pathways

Nerve impulses for voluntary movements propagate from the motor cortex to somatic motor neurons that innervate skeletal muscles via the **direct** or **pyramidal** (pi-RAM-i-dal) **pathways**. The simplest of these pathways consists of sets of two neurons, upper motor neurons and lower motor neurons. About one million pyramidal-shaped cell bodies of direct pathway **upper motor neurons** (**UMNs**) are in the cortex. Their axons descend through the internal capsule of the cerebrum. In the medulla oblongata, the axon bundles form the ventral bulges known as the pyramids. The axons terminate in nuclei of cranial nerves or in the anterior gray horn of the spinal cord. **Lower motor neurons** (**LMNs**) extend from the motor nuclei of cranial nerves to skeletal muscles of the face and head and from the anterior horn of each spinal cord segment to skeletal muscle fibers of the trunk and limbs. Close to their termination point, most upper motor neurons synapse with an association neuron, which, in turn, synapses with a lower motor neuron. A few upper motor neurons synapse directly with lower motor neurons.

PARALYSIS

During a neurological exam, assessment of muscle tone, reflexes, and the ability to perform voluntary movements helps pinpoint certain types of motor system dysfunction. Damage or disease of *lower* motor neurons, either of their cell bodies in the anterior horn or of their axons in the anterior root or spinal nerve, produces a condition called **flaccid paralysis**. There is neither voluntary nor reflex action of the innervated muscle fibers, and the muscle remains limp or flaccid (decreased or lost muscle tone). Injury or disease of *upper* motor neurons causes **spastic paralysis**. This condition is characterized by varying degrees of spasticity (increased muscle tone), exaggerated reflexes, and pathological reflexes such as the Babinski sign (see page 374). ■

The direct pathways convey impulses from the cortex that result in precise, voluntary movements. The main parts of the body governed by the direct pathways are the face, vocal cords (for speech), and hands and feet of the limbs. They channel nerve impulses into three tracts (see Exhibit 15.1 on page 444).

1. **Lateral corticospinal** (kor′-ti-kō-SPI-nal) **tracts.** These pathways begin in the right and left motor cortex and descend through the **internal capsule** of the cerebrum and through the cerebral peduncle of the midbrain and the pons on the same side (Fig. 15.6). About 90% of the axons of upper motor neurons cross over (decussate) to the contralateral (opposite) side in the medulla oblongata. These axons then form the lateral corticospinal tracts in the right and left lateral white columns of the spinal cord. Thus the motor cortex of

the right side of the brain controls muscles on the left side of the body, and vice versa. The lower motor neurons receive input from both upper motor neurons and association neurons. Axons of lower motor neurons

Figure 15.6   Lateral and anterior corticospinal tracts.

 Direct (pyramidal) pathways convey impulses that result in precise, voluntary movements.

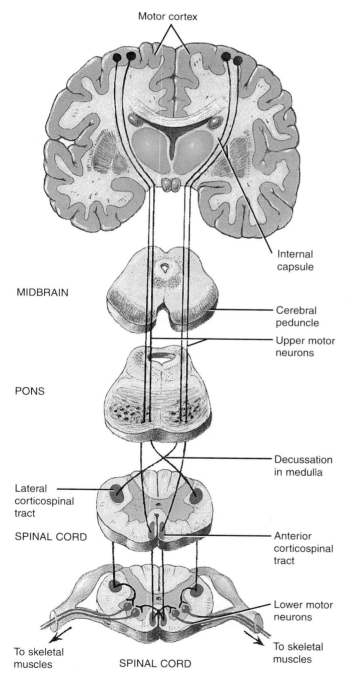

What other tracts (not shown in the illustration) convey impulses that result in precise, voluntary movements?

(somatic motor neurons) emerge from all levels of the spinal cord in the anterior roots of spinal nerves and terminate in skeletal muscles. These motor neurons control skilled movements of the hands and feet.

2. **Corticobulbar** (kor′-ti-kō-BUL-bar) **tracts.** The axons of upper motor neurons of these tracts accompany the corticospinal tracts from the motor cortex through the internal capsule to the brain stem. There some cross whereas others remain uncrossed. They terminate in the nuclei of nine pairs of cranial nerves in the pons and medulla oblongata: the oculomotor (III), trochlear (IV), trigeminal (V), abducens (VI), facial (VII), glossopharyngeal (IX), vagus (X), accessory (XI), and hypoglossal (XII). The lower motor neurons of cranial nerves convey impulses that control precise, voluntary movements of the eyes, tongue, and neck plus chewing, facial expression, and speech.

3. **Anterior corticospinal tracts.** About 10% of the axons of upper motor neurons do not cross over in the medulla oblongata. They pass through the medulla, descend on the ipsilateral (same) side, and form the anterior corticospinal tracts in the right and left anterior white columns (Fig. 15.6). At several spinal cord levels, some of the axons of these upper motor neurons decussate. After crossing over to the opposite side, they synapse with association or lower motor neurons in the anterior gray horn of the spinal cord. Axons of these lower motor neurons exit the cervical and upper thoracic segments of the cord in the anterior roots of spinal nerves. They terminate in skeletal muscles that control movements of the neck and part of the trunk, thus coordinating movements of the axial skeleton.

IMPAIRMENT OF SENSORY AND MOTOR FUNCTIONS

By far the most common cause of impaired sensory and motor function is cerebrovascular disease. When small regions of the brain are damaged, for example, by a cerebrovascular accident (CVA; see page 424), nearby neurons may be able to sprout new dendrites, make new synaptic connections, and take over some of the functions of the lost neurons. After a CVA, the patient's disabilities are related to both the location and the extent of the damage. A common site of cerebral thrombosis (blood clot) is the middle cerebral artery. This artery supplies the lateral portions of the cerebral hemispheres, including the somatic motor and somatosensory areas on either side of the central sulcus. Blockage of the left middle cerebral artery may result in paralysis and loss of sensation on the right side of the body as well as speech difficulties since the main language centers are in the left hemisphere in most people. Cerebral hemorrhage due to rupture of a weakened artery often occurs in the internal capsule. Coursing through the internal capsule are the axons of upper motor neurons of the direct (pyramidal) system. Thus even a small hemorrhage here may produce widespread paralysis on the opposite side of the body. ■

Indirect Pathways

The **indirect (extrapyramidal) pathways** include all descending (motor) tracts other than the corticospinal and corticobulbar tracts. Nerve impulses conducted along the indirect pathways follow complex, polysynaptic circuits that involve the motor cortex, basal ganglia, limbic system, thalamus, cerebellum, reticular formation, and nuclei in the brain stem (Fig. 15.7). Axons of upper motor neurons that carry motor signals from the indirect pathways descend from various nuclei of the brain stem into five major tracts of the spinal cord and terminate on association neurons or lower motor neurons. The five major tracts of the indirect pathways (described in Exhibit 15.1) are the **rubrospinal** (ROO-brō-spī-nal), **tectospinal** (TEK-tō-spī-nal), **vestibulospinal** (ves-TIB-yoo-lō-spī-nal), **lateral reticulospinal** (re-TIK-yoo-lō-spī-nal), and **medial reticulospinal**.

Lower motor neurons receive both excitatory and inhibitory input from many presynaptic neurons in both direct and indirect pathways, an example of convergence. For this reason, lower motor neurons are also called the **final common pathway**. Most nerve impulses from the brain are conveyed to association neurons before being received by lower motor neurons. The sum total of the input from upper motor neurons and association neurons determines the final response of the lower motor neuron. It is not just a simple matter of the brain sending an impulse and the muscle always contracting.

Exhibit 15.1 summarizes the major sensory and motor tracts, their functions, and pathways in the brain.

Basal Ganglia

The basal ganglia have many connections with other parts of the brain. Through these connections, they help to program habitual or automatic movement sequences such as walking or laughing in response to a joke and to set an appropriate level of muscle tone. The basal ganglia also selectively inhibit other motor neuron circuits that are intrinsically active or excitatory. This effect is revealed in certain types of basal ganglia damage, which often are characterized by abnormal movements or tremors caused by loss of inhibitory signals from the basal ganglia.

The caudate nucleus and the putamen receive input from sensory, motor, and association areas of the cortex and the substantia nigra. Output from the basal ganglia comes mainly from the globus pallidus. It sends feedback signals in the form of nerve impulses to the motor cortex by way of the thalamus. This circuit—from cortex to basal ganglia to thalamus to cortex—appears to function in planning and programming movements. The globus pallidus also sends impulses into the reticular formation, which appear to reduce muscle tone. Damage or destruction of certain basal ganglia connections causes a generalized increase in muscle tone that results in abnormal muscle rigidity.

Figure 15.7   Indirect pathways for coordination and control of movement. For comparison, the direct pathway is also shown to the right.

🔑 *Since lower motor neurons receive all input from direct and indirect pathways, and from spinal cord association neurons, they are called the final common pathway.*

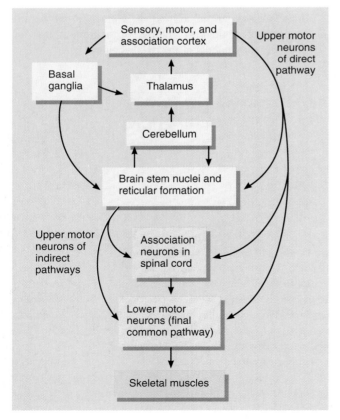

DAMAGE TO BASAL GANGLIA

Much of what we know about the function of the basal ganglia comes from studies of people who have damage or disease in these brain regions. Damage to the basal ganglia results in uncontrollable abnormal body movements, often accompanied by muscle rigidity and tremors (shaking) while at rest.

Parkinson's disease (see page 449) results in a generalized increase in muscle tone and stiffness of the face, arms, and legs. It is associated with deterioration of neural connections between the substantia nigra and the basal ganglia that employ the neurotransmitter dopamine.

Chorea (*choros* = a dance) is quick, purposeless, jerky contractions of the limbs and involuntary facial twitches. Huntington's chorea is a hereditary form of the disease that often does not show symptoms until age 30 or 40. By then, afflicted persons may have passed on this genetic defect to their children. A blood test now can determine if a person has the defective gene. In chorea, degenerative changes occur in which there is (1) loss of acetylcholine-releasing neurons in the caudate nucleus and putamen and (2) loss of GABA-releasing inhibitory neurons that normally project from the caudate nucleus and putamen to the globus pallidus and substantia nigra. ■

Ⓠ Starting with signals from the cortex, describe two circular feedback loops through the thalamus.

EXHIBIT **15.1** **SUMMARY OF MAJOR SENSORY AND MOTOR TRACTS IN THE SPINAL CORD AND PATHWAYS IN THE BRAIN**

Tract*a* and Location	Functions and Pathway in the Brain
SENSORY (ASCENDING) TRACTS **Posterior column** 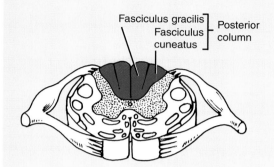Fasciculus gracilis ┐ Posterior Fasciculus ┘ column cuneatus	Conveys nerve impulses for conscious proprioception and tactile sensations from one side of the body to the medulla oblongata on the ipsilateral (same) side. The first-order axons that form the posterior columns convey the sensations of discriminative touch (ability to recognize exact location of light touch stimulus and to make two-point discriminations); stereognosis (ability to recognize an object by feeling it); conscious proprioception (awareness of position of limbs and other body parts); kinesthesia (awareness of direction of movement of limbs and other body parts); weight discrimination (ability to assess weight of an object); and vibration. From the medulla, the axons of second-order neurons form the medial lemniscus, which channels nerve impulses to the thalamus in the opposite cerebral hemisphere. Third-order neurons transmit nerve impulses from the thalamus to the somatosensory area of the cortex, also in the contralateral (opposite) cerebral hemisphere.
Anterolateral (spinothalamic) 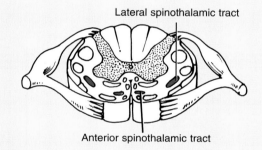Lateral spinothalamic tract Anterior spinothalamic tract	*Lateral spinothalamic*: Conveys nerve impulses for painful and thermal somatic sensations from one side of the body to the thalamus on the opposite side. The second-order axons that form this tract receive input from first-order neurons on the opposite side. From the thalamus, third-order neurons carry nerve impulses to the somatosensory area of the cortex, also in the contralateral cerebral hemisphere. *Anterior spinothalamic*: Conveys nerve impulses for somatic sensations of tickle, itch, crude, poorly localized touch, and pressure from one side of the body to the thalamus on the opposite side. The second-order axons that form this tract receive input from first-order neurons on the opposite side. From the thalamus, third-order neurons carry nerve impulses to the somatosensory area of the cortex, also in the contralateral cerebral hemisphere.
Spinocerebellar 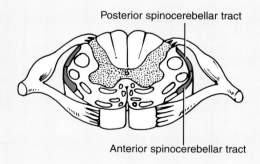Posterior spinocerebellar tract Anterior spinocerebellar tract	*Posterior spinocerebellar*: Conveys nerve impulses for subconscious proprioception from the trunk and lower limb of one side of the body to same side of cerebellum. The second-order axons that form this tract receive input from first-order neurons on the same side, ascend to the medulla, and pass through the inferior cerebellar peduncle into the cerebellum on the same side. This sensory input keeps the cerebellum informed of actual movements and allows it to coordinate, smooth, and refine skilled movements and maintain posture and balance. *Anterior spinocerebellar*: Conveys nerve impulses for subconscious proprioception from the trunk and lower limb of one side of the body to the ipsilateral cerebellum. The second-order axons that form this tract receive input from first-order neurons on the opposite side, ascend to the medulla, and pass through the superior cerebellar peduncle into the cerebellum. Before terminating, however, the axons recross within the cerebellum, thus conveying proprioceptive information from one side of the body to the cerebellum on the same side.

a All tracts are illustrated in the spinal cord except the corticobulbar tract, which extends through the brain stem.

Exhibit continues

EXHIBIT **15.1 SENSORY AND MOTOR PATHWAYS** (CONTINUED)

Tract[a] and Location	Functions and Pathway in the Brain

MOTOR (DESCENDING) TRACTS
Direct (pyramidal)
  Corticospinal

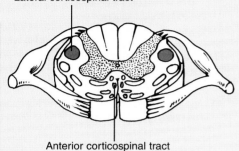

Lateral corticospinal tract

Anterior corticospinal tract

Lateral corticospinal: Conveys nerve impulses from the motor cortex to skeletal muscles on opposite side of body to coordinate precise, discrete, voluntary movements, especially of the hands and feet. Axons of upper motor neurons (UMNs) descend from the cortex into the medulla. Here, 90% cross over to the opposite side (decussate) and then enter the contralateral spinal cord to form this tract. At their level of termination, these UMNs enter the anterior gray horn of the spinal cord on the same side and provide input to lower motor neurons, which provide motor output to skeletal muscles.

Anterior corticospinal: Conveys nerve impulses from the motor cortex to skeletal muscles on opposite side of body to coordinate movements of the axial skeleton. Axons of upper motor neurons (UMNs) descend from the cortex into the medulla. Here, the 10% that do not cross over enter the spinal cord and form this tract. At their level of termination, these UMNs enter the spinal cord anterior gray horn on the opposite side and provide input to lower motor neurons, which provide motor output to skeletal muscles.

  Corticobulbar

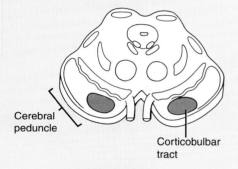

Cerebral
peduncle

Corticobulbar
tract

Midbrain of brain stem

Conveys nerve impulses from the motor cortex to skeletal muscles of the head and neck to coordinate precise, discrete, voluntary contractions. Axons of upper motor neurons descend from the cortex into the brain stem, where some cross to the opposite side and others remain uncrossed. They provide input to lower motor neurons in the nuclei of cranial nerves III, IV, V, VI, VII, IX, X, XI, and XII, which control voluntary movements of the eyes, tongue and neck; chewing; facial expression; and speech.

Indirect (extrapyramidal)

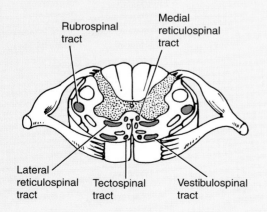

Rubrospinal
tract

Medial
reticulospinal
tract

Lateral
reticulospinal
tract

Tectospinal
tract

Vestibulospinal
tract

Rubrospinal: Conveys motor impulses from the red nucleus, which receives input from the cortex and cerebellum, to skeletal muscles on the opposite side of the body that govern precise, discrete movements of the hands and feet.

Tectospinal: Conveys motor impulses from the superior colliculus to skeletal muscles on the opposite side of the body that move the head and eyes in response to visual stimuli.

Vestibulospinal: Conveys motor impulses from the vestibular nucleus, which receives input about head movements from the vestibular apparatus in the inner ear, to regulate muscle tone for maintaining balance on the same side of the body.

Lateral reticulospinal: Conveys motor impulses from the reticular formation to muscles that facilitate flexor reflexes, inhibit extensor reflexes, and decrease muscle tone in muscles of the axial skeleton and proximal portions of the limbs.

Medial reticulospinal: Conveys motor impulses from the pons that facilitate extensor reflexes, inhibit flexor reflexes, and increase muscle tone in muscles of the axial skeleton and proximal portions of the limbs.

[a] All tracts are illustrated in the spinal cord except the corticobulbar tract, which extends through the brain stem.

Cerebellum

The cerebellum is active in both learning and performing rapid, coordinated, highly skilled movements such as running, speaking, and swimming. It also functions to maintain proper posture and equilibrium (balance). There are four aspects to cerebellar function (Fig. 15.8).

1. **Monitoring intentions.** The cerebellum receives input from the motor cortex and basal ganglia via the pontine nuclei in the pons regarding what movements are planned (red lines).
2. **Monitoring actual movement.** It receives input from proprioceptors in joints and muscles that reveals what actually is happening (blue lines). These nerve impulses travel in the cuneocerebellar and posterior spinocerebellar tracts. The vestibulocerebellar tract transmits impulses from the vestibular (equilibrium-sensing) apparatus in the inner ear to the cerebellum. Nerve impulses from the eyes also enter the cerebellum.
3. **Comparing.** It compares the command signals (intentions for movement) with sensory information (actual performance).
4. **Providing corrective feedback.** It sends out corrective signals, both to nuclei in the brain stem and to the motor cortex via the thalamus (green lines).

In summary, the cerebellum receives information from higher brain centers about what the muscles should be doing and from the peripheral nervous system about what the muscles are doing. If there is a discrepancy between the two, corrective feedback signals are sent from the cerebellum via the thalamus to the cerebrum, where new commands are initiated to decrease the discrepancy and smooth the motion.

Figure 15.8 Input to and output from the cerebellum.

The cerebellum coordinates and smoothes contractions of skeletal muscles during skilled movements and helps maintain posture and balance.

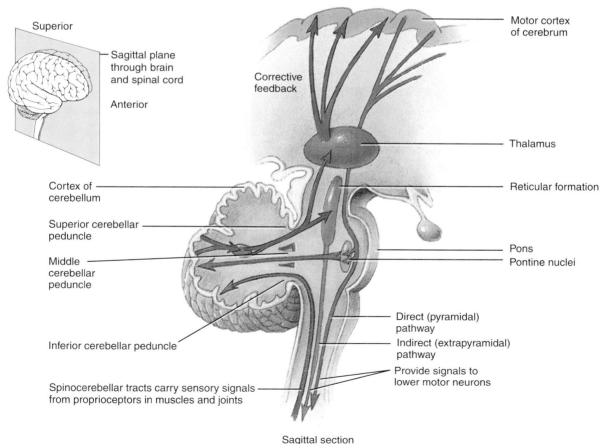

Sagittal section

Which tracts carry information about actual performance of muscles of the trunk and lower limbs to the cerebellum?

Skilled activities such as tennis or volleyball provide good examples of the contribution of the cerebellum to movement. To make a good stop volley or to block a spike, you must bring your racket or arms forward just far enough to make solid contact. How do you stop at exactly the right point? This is where the cerebellum comes in. It receives information about your body status and the position of the approaching ball. Before you even hit the ball, the cerebellum has sent information to the cerebral cortex and basal ganglia informing them where your swing must stop. In response to impulses from the cerebellum, the cortex and basal ganglia transmit motor impulses to opposing body muscles to stop the swing.

DAMAGE TO THE CEREBELLUM

Damage to the cerebellum is characterized by certain symptoms involving skeletal muscles on the same side of the body as the damage. The hallmark of cerebellar trauma or disease is **ataxia** (*a* = without; *taxis* = order), movements that are jerky and uncoordinated. Blindfolded people with ataxia cannot touch the tip of their nose with a finger because they cannot coordinate movement with their sense of where a body part is located. There also is an inability to stop a movement smoothly at the desired point. For example, a person might knock over a glass of water while trying to pick it up. Another sign of ataxia is slurred speech due to a lack of coordination of speech muscles. Cerebellar damage also produces **intention tremor**, a shaking that occurs during deliberate voluntary movement. Chronic alcoholics often suffer degeneration of particular parts of the cerebellum that affects the lower limbs more severely than the upper limbs. Thus they have impaired walking gait with relatively normal control of arm movements. ■

INTEGRATIVE FUNCTIONS

We turn now to a fascinating, though poorly understood, function of the cerebrum: integration. The **integrative functions** include cerebral activities such as memory, sleep and wakefulness, and emotional responses. The role of the limbic system in emotional behavior was discussed in Chapter 14.

Learning and Memory

Without memory, we would repeat mistakes and be unable to learn. Similarly, we would not be able to repeat our successes or accomplishments, except by chance. Although both learning and memory have been studied for many years, there is still no completely satisfactory explanation for how we recall information or how we remember. Some things, however, are known about how information is acquired and stored, and it is clear that there are different categories of memory.

Learning is the ability to acquire new knowledge or skills through instruction or experience. **Memory** is the process in which that knowledge is retained over time. For an experience to become part of memory, it must produce persistent functional changes in the brain that represent the experience. This capability for change with learning is termed **plasticity**. Nervous system plasticity underlies the ability to change behavior in response to stimuli from the external and internal environments. It involves changes in individual neurons, for example, synthesis of different proteins or sprouting of new dendrites, as well as changes in the strengths of synaptic connections among neurons. The portions of the brain known to be associated with memory include the association cortex of the frontal, parietal, occipital, and temporal lobes; parts of the limbic system, especially the hippocampus and amygdaloid nucleus; and the diencephalon.

Memory occurs in stages over a period of time. **Short-term memory** is the temporary ability to recall a few pieces of information. One example is finding an unfamiliar telephone number in a telephone book and then dialing it. If the number has no special significance, it is usually forgotten within a few seconds. Information in short-term memory may later be transformed into a more permanent type of memory, called **long-term memory**, that lasts from days to years. For example, if you use a telephone number often, it becomes part of long-term memory. When information is in long-term memory, it usually can be retrieved for use whenever needed. The reinforcement due to the frequent retrieval of a piece of information is called **memory consolidation**.

Although the brain receives many stimuli, we pay attention to only a few of them at any one time. It has been estimated that of all the information that comes to our consciousness, only about 1% is stored as long-term memory. Moreover, much of what goes into long-term memory is sooner or later forgotten. It is a feature of memory that we can remember short lists easier than long ones. This is to say the obvious, but human memory does not record everything like an endless magnetic tape. Another feature of memory is that even when details are lost, the concept or main idea is retained. Then, interestingly, we can often explain the idea or concept—not like replaying a tape—but with our own selection of words and ways of explanation. Despite several decades of research, the mechanisms of memory are still elusive.

Some evidence supports the notion that short-term memory depends more on electrical and chemical events in the brain than on structural changes, such as the formation of new synapses. Several conditions that inhibit the electrical activity of the brain, such as anesthesia, coma, electroconvulsive shock, and ischemia (reduced blood supply) of the brain, disrupt retention of recently acquired information without altering previously established long-term memories. It is a common experience that people who suffer retrograde amnesia (loss of past memories) cannot remember any events that have occurred for about 30 minutes before the amnesia developed. As the person recovers, the most recent memories return last.

One theory of short-term memory states that memories may be caused by reverberating neuronal circuits—an incoming nerve impulse stimulates the first neuron, which stimulates the second, which stimulates the third, and so on. Branches from the second and third neurons synapse with the first, sending the impulse back through the circuit again and again. Once fired, the output signal may last from a few seconds to many hours, depending on the arrangement of neurons in the circuit. If this pattern is applied to short-term memory, an incoming thought (the phone number) continues in the brain even after the initial stimulus (looking at the number in the phone book) is gone. Thus you can recall the thought only for as long as the reverberation continues.

Most research on long-term memory focuses on anatomical or biochemical changes that might enhance facilitation at synapses. Any of the events in synaptic transmission could be responsible for enhanced communication between neurons. For example, it has been suggested that there could be an increase in the number of receptor molecules in the postsynaptic cell membrane or a decrease in the rate of removal of neurotransmitter.

A phenomenon called **long-term potentiation (LTP)** is believed to underlie some aspects of memory. It occurs at certain synapses within the hippocampus. After a brief period of high-frequency stimulation, transmission at these synapses is facilitated (enhanced) for hours or weeks. The neurotransmitter released is glutamate, which acts on NMDA[1] glutamate receptors on the postsynaptic neurons. Recent studies suggest that induction of LTP depends on the release of nitric oxide (NO) from the postsynaptic neurons after they have been activated by glutamate. The NO, in turn, appears to diffuse into the presynaptic neurons to cause LTP. Acetylcholine (ACh) also seems to play an important role in memory. One of the findings in Alzheimer's disease (see page 425), which wipes out the ability to remember, is depletion in certain brain regions of a key enzyme needed to synthesize ACh.

Anatomical changes also occur in neurons when they are stimulated. For example, electron micrographic studies of presynaptic neurons that have been subjected to prolonged, intense activity exhibit several anatomical changes. These include an increase in the number of presynaptic terminals, enlargement of synaptic end bulbs, and an increase in dendritic branches. Moreover, neurons grow new synaptic end bulbs with increasing age, presumably as a result of increased use. These changes, which are correlated with faster learning, suggest enhancement of facilitation at synapses. Such changes do not occur when neurons are inactive. In fact, in animals that have lost their eyesight, there is thinning of the cerebral cortex in the visual area.

There is also a possible role for nucleic acids in long-term memory. The molecules DNA and RNA store information, and these molecules, especially DNA, persist for the lifetime of the cell. Studies have shown an increase in the RNA content of activated neurons. Also, some evidence shows that long-term memory will not occur to any significant extent when RNA formation is inhibited. Since RNA synthesis precedes protein synthesis, these studies may also imply a relation between synthesis of certain proteins and memory.

Wakefulness and Sleep

Humans sleep and awaken in a fairly constant 24-hour cycle called a **circadian** (ser-KĀ-dē-an) **rhythm** that is established by an area of the hypothalamus called the suprachiasmatic nucleus (see Fig. 14.10). A person who is aroused or awake is in a state of readiness and able to react consciously to various stimuli. Since neuronal fatigue precedes sleep and the signs of fatigue disappear after sleep, fatigue is apparently one cause of sleep. Moreover, EEG recordings show that during wakefulness the cerebral cortex is very active while during sleep fewer impulses arise from the cerebral cortex.

The reticular formation has many ascending connections with the cerebral cortex and descending connections with the spinal cord (Fig. 15.9). Stimulation of portions of the reticular formation results in increased cortical activity. Thus a portion of the reticular formation is known as the **reticular activating system (RAS)**. When this area is active, many nerve impulses pass upward into the thalamus and disperse to widespread areas of the cerebral cortex. The effect is a generalized increase in cortical activity.

Arousal, or awakening from sleep, also involves increased activity in the RAS. For arousal to occur, the RAS must be stimulated by input signals. Many sensory inputs can activate the RAS: painful stimuli detected by nociceptors, touches, proprioceptive signals, bright light, or the buzz of an alarm clock. Once the RAS is activated, the cerebral cortex is also activated and arousal occurs.

The reticular formation also has a feedback system with the spinal cord. Impulses from the reticular formation descend into the spinal cord and then to skeletal muscles over the reticulospinal tracts. Muscle contraction causes proprioceptors to return impulses that activate the RAS. The two feedback systems (from the cortex and skeletal muscles) maintain activation of the RAS, which, in turn, maintains activation of the cerebral cortex. The result is a state of wakefulness called **consciousness**.

ALTERED CONSCIOUSNESS

Consciousness may be altered by various factors. Amphetamines probably activate the RAS to produce a state of wakefulness and alertness. Meditation produces a relaxed, focused consciousness. Anesthetics produce an altered state of consciousness called anesthesia. Drugs such as LSD and alcohol can also alter consciousness.

---

[1]Named after the chemical N-methyl D-aspartate, which is used to detect this type of glutamate receptor.

Figure 15.9 The reticular activating system (RAS) consists of fibers that project from the reticular formation through the thalamus to the cerebral cortex.

 Awakening from sleep (arousal) involves increased activity of the RAS.

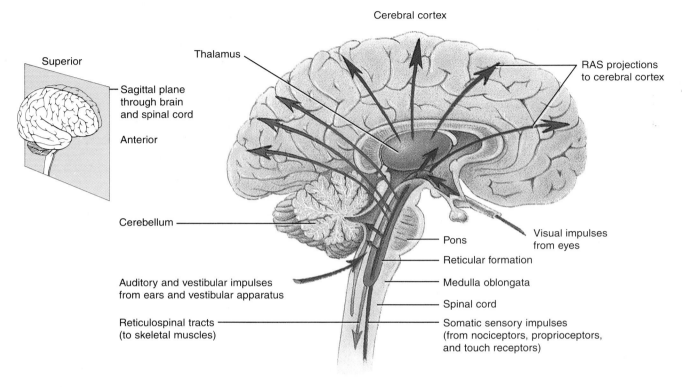

Sagittal section

Why should every dwelling have a smoke alarm? (*Hint:* Think about what a smoke alarm does if you are sleeping and predict the implications for sensory input to the RAS.)

Damage to the RAS or other parts of the brain can produce **coma,** the final stage of brain failure that is characterized by total unresponsiveness to all external stimuli. A comatose patient lies in a sleeplike state with the eyes closed. In the lightest stages of coma, brain stem and spinal cord reflexes persist, but in the deepest stages, these reflexes are lost as are corneal, pupillary, tendon, and plantar reflexes. Finally, respiratory and cardiovascular controls disappear, and the patient usually dies. Coma also may be induced by poisoning, hypoxia, hypoglycemia, ischemia, infection, or acid–base and electrolyte disorders. ■

How does sleep occur if the activating feedback systems are in continual operation? One explanation is that the feedback system slows greatly or is inhibited. During **sleep,** a state of altered consciousness or partial unconsciousness from which an individual can be aroused by different stimuli, activity in the RAS is very low.

Just as there are different levels of awareness when awake, there are different levels of sleep. Normal sleep consists of two types: non–rapid eye movement (NREM) sleep and rapid eye movement (REM) sleep. **NREM sleep,** or **slow wave sleep,** consists of four stages, each of which

gradually merges into the next. Each stage has been identified by EEG recordings.

Stage 1. This is a transition stage between wakefulness and sleep that normally lasts from 1 to 7 minutes. The person is relaxing with eyes closed and has fleeting thoughts. If awakened, the person will often say he has not been sleeping. Alpha waves diminish and theta waves appear on the EEG.

Stage 2. This is the first stage of true sleep, even though the person experiences only light sleep. It is a little harder to awaken the person. Fragments of dreams may be experienced, and the eyes may slowly roll from side to side. The EEG shows *sleep spindles*—sudden, short bursts of sharply pointed waves that occur at 12–14 Hz (cycles per second).

Stage 3. This is a period of moderately deep sleep. The person is very relaxed. Body temperature begins to fall and blood pressure decreases. It is difficult to awaken the person, and the EEG shows a mixture of sleep spindles and delta waves. This stage occurs about 20 minutes after falling asleep.

Stage 4. Deep sleep occurs. The person is very relaxed and responds slowly if awakened. When bed-wetting and sleepwalking occur, they do so during this stage. The EEG is dominated by delta waves.

In a typical 7- or 8-hour sleep period, a person goes from stage 1 to 4 of NREM sleep. Then the person ascends to stages 3 and 2 and then to REM sleep within 50–90 minutes. The cycle normally repeats throughout the sleep period.

In **REM sleep**, the EEG readings are similar to those of stage 1 of NREM sleep. It is during REM sleep that most dreaming occurs. In men, erection of the penis happens during most REM intervals, even when the dream content is not sensual. Presence of penile erections during REM sleep in a man with impotence (inability to attain an erection while awake) indicates a psychogenic, rather than a physical, cause of his impotence. Following REM sleep, the person descends again to stages 3 and 4 of NREM sleep.

REM and NREM sleep alternate throughout the night with approximately 90-minute intervals between REM periods. This cycle repeats from three to five times during the entire sleep period. The REM periods start out lasting from 5 to 10 minutes and gradually lengthen until the final one lasts about 50 minutes. In a normal sleep period, REM sleep totals 90–120 minutes. Most sedatives significantly decrease REM sleep.

As a person ages, the average time spent sleeping decreases. In addition, the percentage of REM sleep decreases. As much as 50% of an infant's sleep is REM, as contrasted with 35% for 2-year olds and 25% for adults. The high percentage of REM sleep in infants and children is thought to be important for the maturation of the brain. Neuronal activity is high during REM sleep; brain oxygen use is higher during REM sleep than during intense mental or physical activity while awake.

SLEEP DISORDERS

Narcolepsy (NAR-kō-lep-sē; *narke* = numbness; *lepsis* = seizure) is a condition of involuntary attacks of sleep that last about 15 minutes and may occur at almost any time of the day. It is an inability, in the waking state, to inhibit REM sleep. **Insomnia** (in-SOM-nē-a; *in* = not; *somnus* = sleep) consists of difficulty in falling asleep and, usually, frequent awakening. It may be related to specific disorders or secondary factors, both medical and psychiatric. **Hypersomnia** refers to an excessively long or deep sleep from which a person can be awakened only by vigorous stimulation. It may be associated with conditions such as with head injury, stroke, and encephalitis. ■

DISORDERS: HOMEOSTATIC IMBALANCES

SPINAL CORD INJURY

The spinal cord may be damaged by compression from a variety of causes. These include a tumor within or adjacent to the spinal cord, herniated intervertebral discs, blood clots, penetrating wounds caused by projectile fragments, or other traumatic events such as automobile accidents. Depending on the location and extent of the injury, paralysis may occur. **Paralysis** is the total loss of voluntary motor function that results from damage to nervous or muscle tissue. Paralysis may be classified as follows: **monoplegia** (*mono* = one; *plege* = stroke), paralysis of one limb only; **diplegia** (*di* = two), paralysis of both upper limbs or both lower limbs; **paraplegia** (*para* = beyond), paralysis of both lower limbs; **hemiplegia** (*hemi* = half), paralysis of the upper limb, trunk, and lower limb on one side of the body; and **quadriplegia** (*quad* = four), paralysis of all four limbs.

Complete transection of the spinal cord means that the cord is severed from one side to the other, thus cutting all ascending and descending tracts. It results in a loss of all sensations and voluntary movement *below* the level of the transection. **Hemisection** is a partial transection of the cord on either the right or left side. Below the level of the hemisection there is a loss of proprioception and discriminative touch sensations on the same side as the injury if the posterior column is cut, paralysis on the same side if the lateral corticospinal tract is cut, and loss of pain, temperature, and crude tactile sensations on the opposite side if the spinothalamic tracts are cut.

Following transection, there is an initial period of **spinal shock** that lasts from a few days to several weeks. During this period, all reflex activity disappears, a condition called

areflexia (a'-rē-FLEK-sē-a). In time, however, reflexes return. The first reflexes to reappear are stretch reflexes (for example, the knee jerk). Their reappearance may take several days. Next, the flexion reflexes return, over a period of up to several months. Then the crossed extensor reflexes return. In some cases, males may not be able to attain erection of the penis or ejaculation of semen. Moreover, urinary bladder and bowel functions are no longer under voluntary control.

Recent studies have shown that patients with spinal cord injury have an improved outcome if they are given an anti-inflammatory corticosteroid drug called methylprednisolone within 8 hours of the injury.

CEREBRAL PALSY

The term **cerebral palsy** (**CP**) refers to a group of motor disorders resulting in muscular incoordination and loss of muscle control. It is caused by damage to the motor areas of the brain during fetal life, birth, or infancy in about 2 of every 1000 children. One cause is infection of the mother by the German measles (rubella) virus during her first 3 months of pregnancy. Radiation during fetal life, temporary lack of oxygen during birth, and hydrocephalus during infancy may also cause cerebral palsy. Cerebral palsy is not a progressive disease; it does not worsen as time elapses. Once the damage is done, however, it is irreversible.

PARKINSON'S DISEASE

Parkinson's disease (**PD**) is a progressive disorder of the CNS that typically affects its victims around age 60. The

DISORDERS: HOMEOSTATIC IMBALANCES ◼CONTINUED

cause is unknown, but toxic environmental factors are suspected. Only 5% of PD patients have a family history of the disease. Pathological changes occur in the substantia nigra and basal ganglia. The substantia nigra contains cell bodies of neurons that extend to the basal ganglia of the cerebrum, where they release the neurotransmitter dopamine (DA). Also in the basal ganglia, in the caudate nucleus, are neurons that liberate acetylcholine (ACh). In Parkinson's disease there is a degeneration of DA-producing neurons in the substantia nigra. The resulting imbalance of neurotransmitter activity—too little DA and too much ACh—is thought to bring about most of the symptoms.

In PD patients, involuntary skeletal muscle contractions often interfere with voluntary movement. For instance, the muscles of the upper limb may alternately contract and relax, causing the hand to shake. This shaking is called **tremor** and is the most common symptom of PD. The tremor may spread to the ipsilateral (same side) lower limb and then to the contralateral (opposite side) limbs. Also, some muscles may contract continuously, causing **rigidity** of the involved body part. Rigidity of the facial muscles gives the face a masklike appearance. The expression is characterized by a wide-eyed, unblinking stare and a slightly open mouth with uncontrolled drooling.

Motor performance is also impaired by **bradykinesia** (*brady* = slow; *kinesis* = motion), in which activities such as shaving, cutting food, and buttoning a blouse take longer and become increasingly more difficult. Muscular movements are performed not only slowly but with decreasing range of motion (**hypokinesia**). For example, handwritten letters get smaller, become poorly formed, and eventually become illegible. Often, walking is impaired; steps become shorter and shuffling, and arm swing diminishes. Even speech may be affected. A spoken sentence may begin intelligibly but become progressively softer spoken and quicker.

Treatment of Parkinson's disease is directed toward increasing levels of DA and decreasing levels of ACh. Although people with Parkinson's disease do not manufacture enough dopamine, taking it is useless because DA cannot cross the blood–brain barrier. However, symptoms are partially relieved by a drug developed in the 1960s called levodopa (L-dopa), a precursor of DA. Administered by itself, levodopa elevates brain levels of DA but also causes undesirable side effects in peripheral tissues, for example, heart arrhythmias and liver dysfunctions. Thus levodopa usually is given in combination with carbidopa, which inhibits the formation of DA outside the brain. The combined drugs diminish the undesirable side effects, but psychiatric symptoms, such as depression and confusion, and disorders of involuntary movements (dyskinesias) are common. Levodopa does not slow the disease progression, however, and as more affected brain cells die, the drug becomes useless. Recently, some patients with PD have been given a drug called deprenyl, which inhibits monoamine oxidase. This is one of the enzymes that degrades catecholamine neurotransmitters. Early results indicate that it can prevent symptoms for nearly a year in newly diagnosed patients. Drugs that reduce ACh levels (anticholinergics) may also provide relief, especially of tremor and rigidity.

Since 1988, physicians in Sweden, Mexico, England, Cuba, China, and the United States have sought to reverse the effects of Parkinson's disease by transplanting fetal nerve tissue (dopamine-rich mesencephalic tissue) into the basal ganglia (usually the putamen) of patients with severe PD. Several patients have shown some degree of improvement after surgery, such as less rigidity and improved quickness of motion.

STUDY OUTLINE ◼

SENSATION (p. 430)

1. Sensation is a conscious or subconscious awareness of external and internal stimuli.
2. The nature of a sensation and the type of reaction generated vary with the level of the CNS at which the sensation is translated.
3. Modality is the property by which one sensation is distinguished from another.
4. Usually, a given sensory neuron serves only one modality.
5. For a sensation to arise, four events typically occur. These are stimulation, transduction, conduction, and translation.
6. Sensory receptors are selective.
7. In terms of simplicity or complexity, simple receptors are associated with the somatic (general) senses, and complex receptors are associated with the special senses.
8. According to their location, receptors are classified as exteroceptors, interoceptors, and proprioceptors.
9. On the basis of type of stimulus detected, receptors are classified as mechanoreceptors, thermoreceptors, nociceptors, photoreceptors, and chemoreceptors.
10. Sensory receptors respond to stimuli by producing receptor or generator potentials.

11. Adaptation is a change in sensitivity (usually a decrease) to a long-lasting stimulus.

SOMATIC SENSES (p. 432)

1. Cutaneous sensations include tactile sensations (touch, pressure, vibration), thermal sensations (warmth and cold), and pain. Receptors for these sensations are located in the skin, connective tissue under the skin, mucous membranes, mouth, and anus.
2. Receptors for touch are hair root plexuses and corpuscles of touch (Meissner's corpuscles), which are rapidly adapting, and type I cutaneous mechanoreceptors (tactile or Merkel discs) and type II cutaneous mechanoreceptors (end organs of Ruffini), which are slowly adapting. Receptors for pressure are type II cutaneous mechanoreceptors and lamellated (Pacinian) corpuscles. Receptors for vibration are corpuscles of touch and lamellated corpuscles. Itch and tickle receptors are free nerve endings.
3. Thermoreceptors are free nerve endings.
4. Pain receptors (nociceptors) are located in nearly every body tissue.
5. Nerve impulses for acute pain propagate along myelinated A fibers whereas those for chronic pain conduct along unmyelinated C fibers.

6. Proprioceptors located in skeletal muscles, tendons, in and around joints, and in the internal ear monitor muscle tone, movement of body parts, and body position.

7. Proprioceptors include muscle spindles, tendon organs (Golgi tendon organs), joint kinesthetic receptors, and hair cells of the inner ear.

SOMATIC SENSORY PATHWAYS (p. 437)

1. Somatic sensory pathways from receptors to the cerebral cortex involve three-neuron sets: first-order, second-order, and third-order neurons.

2. Axon collaterals (branches) of somatic sensory neurons simultaneously carry signals into the cerebellum and the reticular formation of the brain stem.

3. Impulses propagating along the posterior column–medial lemniscus pathway relay discriminative touch, stereognosis, proprioception, weight discrimination, and vibratory sensations.

4. The neural pathway for pain and temperature sensations is the lateral spinothalamic tract.

5. The neural pathway for tickle, itch, crude touch, and pressure sensations is the anterior spinothalamic pathway.

6. Specific areas of the primary somatosensory area (postcentral gyrus) of the cerebral cortex receive different information from different parts of the body.

7. The pathways to the cerebellum are the anterior and posterior spinocerebellar tracts, which are involved in transmitting impulses for subconscious muscle and joint position sense from the trunk and lower limbs.

8. Exhibit 15.1 summarizes the major somatic sensory pathways.

INTEGRATION OF SENSORY INPUT AND MOTOR OUTPUT (p. 440)

1. Sensory input keeps the CNS informed of changes in the environment.

2. Incoming sensory information is integrated at many stations along the CNS at both conscious and subconscious levels.

3. A motor response makes a muscle contract or a gland secrete.

SOMATIC MOTOR PATHWAYS (p. 440)

1. The primary motor area (precentral gyrus) of the cortex is the major control region for initiation of voluntary movement.

2. Impulses governing voluntary movements propagate from the motor cortex to somatic motor neurons that innervate skeletal muscles via the direct (pyramidal) pathways. The simplest pathways consist of upper and lower motor neurons.

3. The direct pathways include the lateral and anterior corticospinal tracts and corticobulbar tracts.

4. Indirect (extrapyramidal) pathways involve the motor cortex, basal ganglia, limbic system, thalamus, cerebellum, reticular formation, and nuclei in the brain stem.

5. Exhibit 15.1 summarizes the major somatic motor pathways.

INTEGRATIVE FUNCTIONS (p. 446)

1. Memory is the ability to store and recall thoughts and involves persistent changes in the brain, a capability called plasticity.

2. Memory is generally classified into two kinds: short-term and long-term memory.

3. Short-term memory is related to electrical and chemical events; long-term memory is related to anatomical and biochemical changes at synapses.

4. Sleep and wakefulness are integrative functions that are controlled by the suprachiasmatic nucleus and the reticular activating system (RAS).

5. Non–rapid eye movement (NREM) sleep consists of four stages identified by characteristic waves in EEG recordings.

6. Most dreaming occurs during rapid eye movement (REM) sleep.

REVIEW QUESTIONS

1. Distinguish between sensation and perception. (p. 430)
2. What is modality? (p. 430)
3. What events take place for a sensation to occur? (p. 430)
4. Classify receptors on the basis of simplicity and complexity, location, and type of stimulus. (p. 431)
5. Distinguish between generator and receptor potentials. (p. 432)
6. What is adaptation? Compare rapidly and slowly adapting receptors. (p. 432)
7. What is a cutaneous sensation? Distinguish tactile, thermal, and pain sensations. (p. 432)
8. For each of the following cutaneous sensations, describe the receptor involved in terms of structure, function, and location: touch, pressure, vibration, itch and tickle, temperature, and pain. (p. 433)
9. How do cutaneous sensations help maintain homeostasis? (p. 433)
10. Why are pain receptors important? Differentiate somatic pain, visceral pain, referred pain, and phantom pain. (p. 433)
11. Why is the concept of referred pain useful to the physician in diagnosing internal disorders? (p. 434)

12. What is the proprioceptive sense? Where are the receptors for this sense located? (p. 435)
13. Describe the structure of muscle spindles, tendon organs (Golgi tendon organs), and joint kinesthetic receptors. (p. 435)
14. Relate proprioception to the maintenance of homeostasis. (p. 436)
15. Distinguish between the posterior column–medial lemniscus pathway and anterolateral (spinothalamic) pathways in terms of location and function. (p. 437)
16. Describe how various parts of the body are represented in the somatosensory cortex. (p. 438)
17. What are the functions of the spinocerebellar tracts? (p. 439)
18. Describe how sensory input and motor output are linked in the central nervous system. (p. 440)
19. Describe how various parts of the body are represented in the motor cortex. (p. 440)
20. Distinguish between direct and indirect motor pathways in terms of location and function. (p. 440)
21. How do the basal ganglia and cerebellum function in body movements? (p. 442)

22. Define memory. What are the two kinds of memory? (p. 446)
23. Define memory consolidation. (p. 446)
24. Describe the mechanisms that are thought to contribute to memory. (p. 446)
25. Describe how sleep and wakefulness are related to the reticular activating system (RAS). (p. 447)

26. What are the four stages of non–rapid eye movement (NREM) sleep? How is NREM sleep distinguished from rapid eye movement (REM) sleep? (p. 448)
27. Define the following: anesthesia (p. 437), paralysis (p. 440), coma (p. 448), narcolepsy, insomnia, and hypersomnia (p. 449).

ANSWERS TO QUESTIONS WITH FIGURES

15.1 Pain, thermal sensations, tickle, and itch.
15.2 The kidneys.
15.3 Type Ia and type II sensory (afferent) fibers wrap around the central region of intrafusal muscle fibers. Gamma motor (efferent) neurons innervate intrafusal muscle fibers.
15.4 Loss of pain and thermal sensations on the left side of the body.
15.5 The hand has a larger representation in the motor than in the sensory cortex, which implies a greater precision in its movement control than discriminative ability in its sensation.

15.6 Corticobulbar and rubrospinal (see Exhibit 15.1 on page 443).
15.7 Cortex to basal ganglia to thalamus to cortex. Cortex to brain stem nuclei to cerebellum to thalamus to cortex.
15.8 Posterior and anterior spinocerebellar tracts.
15.9 A smoke alarm detects smoke and sounds a loud bell or buzzer, which wakes up sleepers by providing auditory input that stimulates the RAS. Since there is little or no olfactory input to the RAS, people may die of toxic substances in smoke without waking up if they do not have a smoke alarm.

CHAPTER 16

THE SPECIAL SENSES

STUDENT OBJECTIVES

1. Locate the receptors for olfaction and describe the neural pathway for smell. (p. 454)

2. Identify the gustatory receptors and describe the neural pathway for taste. (p. 455)

3. List and describe the accessory structures of the eye and the structural divisions of the eyeball. (p. 457)

4. Discuss image formation by describing refraction, accommodation, and constriction of the pupil. (p. 464)

5. Describe how photoreceptors and photopigments function in vision. (p. 467)

6. Describe the retinal processing of visual input and the neural pathway of light impulses to the brain. (p. 469)

7. Describe the anatomical subdivisions of the ear. (p. 472)

8. List the principal events in the physiology of hearing. (p. 478)

9. Identify the receptor organs for equilibrium and how they function. (p. 479)

10. Contrast the causes and symptoms of glaucoma, senile macular degeneration, deafness, Ménière's syndrome, otitis media, and motion sickness. (p. 483)

11. Define medical terminology associated with the sense organs. (p. 484)

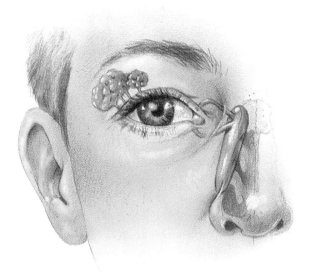

CHAPTER CONTENTS AT A GLANCE

he special senses, like the somatic senses, allow us to detect specific changes in our environment. However, the special senses—smell, taste, vision, hearing, and equilibrium—provide richer sensory experiences. They also have receptor organs that are structurally more complex.

OLFACTORY SENSATIONS: SMELL

Both smell and taste are chemical senses; that is, the sensations arise from the interaction of molecules with smell or taste receptors. Among all sensations, only smell and taste project both to higher cortical areas and to the limbic system. This is probably the reason that certain odors and tastes can evoke strong emotional responses or a flood of memories.

Anatomy of Olfactory Receptors

Between 10 and 100 million receptors for the **olfactory** (ol-FAK-tō-rē; *olfactus* = smell) **sense**, or sense of smell, lie in the nasal epithelium in the superior portion of the nasal cavity. The total area of the olfactory epithelium is 5 cm^2 (a little less than 1 square inch). It occupies the superior portion of the nasal cavity, extending along the superior nasal concha and upper part of the middle nasal concha (Fig. 16.1a). The olfactory epithelium consists of three kinds of cells: olfactory receptors, supporting cells, and basal cells.

Olfactory receptors are the first-order neurons of the olfactory pathway. They are bipolar neurons whose distal (apical) end is a knob-shaped dendrite. Several cilia, called **olfactory hairs**, project from the dendrite. The cilia are the sites of olfactory transduction. Olfactory receptors respond to the chemical stimulation of an odorant molecule by producing a generator potential, thus initiating the olfactory response. From the proximal (basal) part of each olfactory receptor, a single axon projects to the olfactory bulb (Fig. 16.1b).

Supporting (sustentacular) cells are columnar epithelial cells of the mucous membrane lining the nose. **Basal cells** lie between the bases of the supporting cells. They are stem cells that continually produce new olfactory receptors, which live for only a month or so before being replaced. This process is remarkable because olfactory receptors are neurons. It is an exception to the general rule that mature neurons are not replaced in your nervous system.

Within the connective tissue that supports the olfactory epithelium are **olfactory (Bowman's) glands**. They produce mucus, which is carried to the surface of the epithelium by ducts. The secretion moistens the surface of the olfactory epithelium and dissolves odorant gases.

Both supporting cells of the nasal epithelium and olfactory glands are innervated by branches of the facial (VII) nerve. Stimuli such as pepper, ammonia, and chloroform are irritating. They may provoke tears and sniffles by stimulating lacrimal and nasal mucosal receptors in addition to olfactory receptors.

Figure 16.1   Olfactory receptors. (a) Location in nasal cavity. (b) Details.

 The olfactory epithelium consists of olfactory receptors, supporting cells, and basal cells.

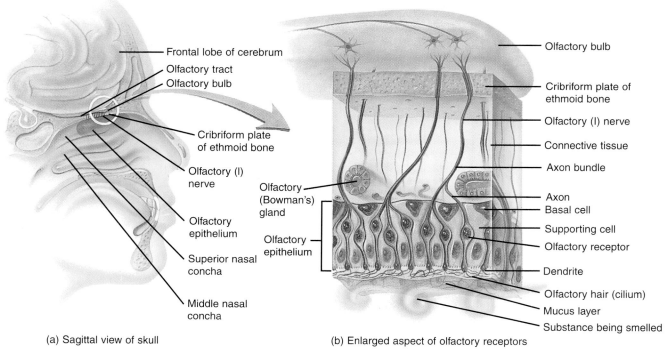

(a) Sagittal view of skull

(b) Enlarged aspect of olfactory receptors

Axons of olfactory bulb neurons form which structure?

Physiology of Olfaction

Many attempts have been made to distinguish and classify "primary" sensations of smell. One older classification included seven primary scents: camphor, musk, floral, peppermint, ether, pungent, and putrid. Genetic evidence now suggests there are many more primary scents (perhaps in the hundreds). In addition, our ability to recognize literally thousands of different scents may depend on patterns of activity in the brain that arise from activation of different combinations of olfactory receptors.

Olfactory receptors react to odorant stimuli in the same way that most sensory receptors react to their specific stimuli: a generator potential (depolarization) develops and triggers one or more nerve impulses. How the generator potential arises is known in some cases. Some odorants bind to receptors that are linked to G-proteins in the plasma membrane and activate the enzyme adenylate cyclase (see page 507). The result is: opening of sodium ion (Na^+) channels → inflow of Na^+ → depolarizing generator potential → nerve impulses.

Adaptation and Odor Thresholds

Adaptation (decreasing sensitivity) to odors is rapid and appears to occur in both olfactory receptors and the central nervous system. Olfactory receptors themselves adapt about 50% in the first second or so after stimulation but adapt very slowly thereafter. Yet we know from common experience that complete insensitivity to certain odors occurs in about a minute after exposure to a strong smell, more than can be explained based on receptor adaptation alone. No doubt other neurons in the olfactory pathway also exhibit adaptation.

Like the other special senses, olfaction has a low threshold. Only a few molecules of certain substances need be present in air to be smelled. A good example is the chemical methyl mercaptan, which can be detected in concentrations as low as $\frac{1}{25,000,000,000}$ mg per milliliter of air. Since natural gas, used for cooking and heating, is odorless but is lethal and potentially explosive if it accumulates, methyl mercaptan is added in small amounts to provide olfactory warning of gas leaks.

Olfactory Pathway

On each side of the nose, bundles of the slender, unmyelinated axons of the olfactory receptors extend through about 20 olfactory foramina in the cribriform plate of the ethmoid bone (Fig. 16.1b). Collectively, these 40 or so bundles of axons are termed the **olfactory (I) nerves**. They terminate in paired masses of gray matter in the brain called the **olfactory bulbs**, which lie inferior to the frontal lobes of the cerebrum and lateral to the crista galli of the ethmoid bone. The first synapse of the olfactory pathway occurs in the olfactory bulbs between the axon terminals of olfactory receptors and dendrites of second-order neurons inside the olfactory bulbs.

Axons of olfactory bulb neurons extend posteriorly and form the **olfactory tract** (Fig. 16.1a). They project to a region called the lateral olfactory area, which is located at the inferior and medial surface of the temporal lobe. This primitive cortical region is a part of the limbic system, and includes some of the amygdaloid body (see Fig. 14.13). Because many olfactory tract axons terminate there, the lateral olfactory area is considered the primary olfactory area, where conscious awareness of smells begins. Connections to other limbic system regions and the hypothalamus probably account for our emotional and memory-evoked responses to odors. Examples include sexual excitement upon smelling a certain perfume, nausea upon smelling a food that once made you violently ill, or an odor-evoked memory flashback to a childhood experience. From the lateral olfactory area, pathways also extend to the frontal lobe both directly and indirectly via the thalamus. An important region for odor identification and discrimination is the orbitofrontal area, corresponding to Brodmann's area 11 (see Fig. 14.15). People who suffer damage there have difficulty identifying different odors. Positron emission tomography (PET) studies suggest a degree of lateralization—there is greater activity in the orbitofrontal area of the *right* hemisphere during olfactory processing.

GUSTATORY SENSATIONS: TASTE

Like olfaction, taste is a chemical sense that requires dissolving of a substance before it can be tasted. Persons with colds or allergies sometimes complain that they cannot taste their food. Although their taste sensations may be operating normally, their olfactory sensations are not. This shows that much of what we think of as taste is actually smell. Odors from foods pass upward into the nasopharynx (portion of the throat behind the nose) and nasal cavity to stimulate olfactory receptors. A given concentration of a substance will stimulate the olfactory system thousands of times more strongly than it stimulates the gustatory system.

Anatomy of Gustatory Receptors

The receptors for **gustatory** (GUS-ta-tō′-rē; *gusto* = taste) **sensations**, or sensations of taste, are located in the taste buds (Fig. 16.2). The nearly 10,000 taste buds of a young adult are mainly on the tongue, but they are also found on the soft palate (posterior portion of roof of mouth), pharynx (throat), and larynx (voice box). The number of taste buds declines with age. Each **taste bud** is an oval body consisting of three kinds of *epithelial* cells: supporting cells, gustatory receptor cells, and basal cells (Fig. 16.2c). The **supporting (sustentacular) cells** form a capsule; inside are about 50 **gustatory (taste) receptor cells**. A single, hairlike **gustatory hair (microvillus)** projects from each gustatory receptor cell to the external surface through an opening in the taste bud called the **taste pore**. The gustatory hairs make contact with taste stimuli through the taste pore. **Basal cells**

Figure 16.2   Tongue papillae and gustatory receptors.

Gustatory (taste) receptors are located in taste buds.

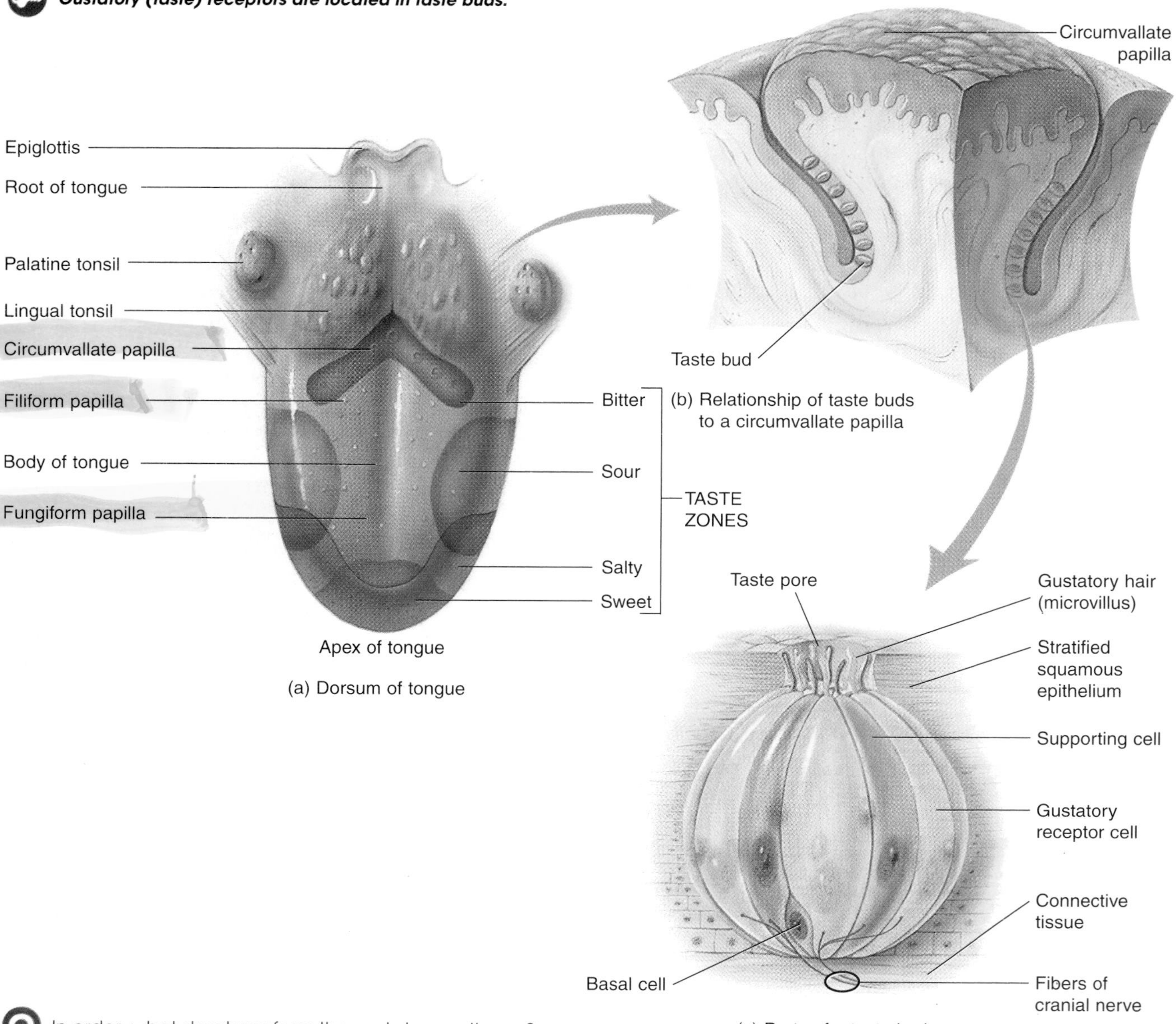

Epiglottis

Root of tongue

Palatine tonsil

Lingual tonsil

Circumvallate papilla

Filiform papilla

Body of tongue

Fungiform papilla

Bitter

Sour

TASTE
ZONES

Salty

Sweet

Apex of tongue

(a) Dorsum of tongue

Circumvallate papilla

Taste bud

(b) Relationship of taste buds to a circumvallate papilla

Taste pore

Gustatory hair (microvillus)

Stratified squamous epithelium

Supporting cell

Gustatory receptor cell

Connective tissue

Basal cell

Fibers of cranial nerve

(c) Parts of a taste bud

In order, what structures form the gustatory pathway?

are found at the periphery of the taste bud near the connective tissue layer. These epithelial cells produce supporting cells, which then develop into gustatory receptor cells that have a life span of about 10 days. At their base, the receptor cells synapse with dendrites of first-order neurons that form the first part of the gustatory pathway. The dendrites of a single fiber branch profusely and contact many receptors in several taste buds.

Taste buds are found in elevations on the tongue called **papillae** (pa-PIL-ē). The papillae give the upper surface of the tongue its rough appearance (Fig. 16.2a,b). **Circumvallate** (ser-kum-VAL-āt) **papillae**, the largest type, are circular and form an inverted V-shaped row at the posterior portion of the tongue. **Fungiform** (FUN-ji-form; meaning mushroom-shaped) **papillae** are knoblike elevations scattered over the entire surface of the tongue. All circumvallate and most fungiform papillae contain taste buds. **Filiform** (FIL-i-form) **papillae** are pointed, threadlike structures that are also distributed over the entire surface of the tongue. They rarely contain taste buds.

Physiology of Gustation

Once a chemical is dissolved in saliva, it can make contact with the plasma membrane of the gustatory hairs, which are the presumed site of taste transduction. The result is a receptor potential (see page 432), which is thought to stimulate the release of neurotransmitter by exocytosis of synaptic vesicles within gustatory receptor cells. Nerve impulses first arise in the first-order sensory neurons that synapse with gustatory receptor cells.

Despite the many substances we seem to taste, there are only four primary taste sensations: sour, salty, bitter, and sweet (Fig. 16.2a). All other "tastes," such as chocolate, pepper, and coffee, are combinations of these four, modified by accompanying olfactory sensations. Individual gustatory receptor cells may respond to more than one of the four primary tastes, but receptors in certain regions of the tongue react more strongly than others to the primary taste sensations. Although the tip of the tongue reacts to all four, it is highly sensitive to sweet and salty substances. The posterior portion of the tongue is highly sensitive to bitter substances. The lateral aspects of the tongue are more sensitive to sour substances.

Adaptation and Taste Thresholds

Complete adaptation to a specific taste can occur in 1–5 minutes of continuous stimulation. As with odors, receptor adaptation alone cannot account for the speed of complete adaptation. Adaptation of receptors to smell contributes to taste adaptation but still does not account for its speed. Adaptation to taste also involves neurons of the taste pathway in the CNS.

The threshold for taste varies for each of the primary tastes. The threshold for bitter substances, as measured by quinine, is lowest. This may have a protective function since poisonous substances often are bitter. The threshold for sour substances, as measured by hydrochloric acid, is somewhat higher. The thresholds for salty substances, as measured by sodium chloride, and sweet substances, as measured by sucrose, are about the same and are higher than those for bitter or sour substances.

Gustatory Pathway

Three cranial nerves include first-order gustatory fibers from taste buds: the facial (VII) serves the anterior two-thirds of the tongue; the glossopharyngeal (IX) nerve serves the posterior one-third of the tongue; and the vagus (X) nerve serves the throat and epiglottis (cartilage lid over the voice box). Taste impulses conduct from the taste buds along these cranial nerves to the medulla oblongata. From the medulla, some taste fibers project to the limbic system and the hypothalamus whereas others project to the thalamus. Fibers that extend from the thalamus to the primary gustatory area in the parietal lobe of the cerebral cortex (see area 43 in Fig. 14.15) are responsible for the conscious perception of taste.

VISUAL SENSATIONS

The study of the structure, function, and diseases of the eye is known as **ophthalmology** (of′-thal-MOL-ō-jē; *ophthalmos* = eye; *logos* = study of). A physician who specializes in the diagnosis and treatment of eye disorders with drugs, surgery, and corrective lenses is known as an **ophthalmologist**. An **optometrist** is a specialist with a degree in optometry who is licensed to examine and test the eyes and treat visual defects by prescribing corrective lenses. An **optician** is a technician who fits, adjusts, and dispenses corrective lenses prescribed by an ophthalmologist or optometrist.

The structures related to vision are the eyeball, the optic (II) nerve, the brain, and several accessory structures of the eye.

Accessory Structures of the Eye

The **accessory structures** of the eye are the eyelids, eyelashes, eyebrows, lacrimal (tearing) apparatus, and extrinsic eye muscles. The upper and lower **eyelids**, or **palpebrae** (PAL-pe-brē), shade the eyes during sleep, protect the eyes from excessive light and foreign objects, and spread lubricating secretions over the eyeballs (Fig. 16.3). From superficial to deep, each eyelid consists of epidermis, dermis, subcutaneous tissue, fibers of the orbicularis oculi muscle, a tarsal plate, tarsal glands, and a conjunctiva. The **tarsal plate** is a thick fold of connective tissue that gives form and support to the eyelids (Fig. 16.4a). Embedded in each tarsal plate is a row of elongated modified sebaceous glands known as **tarsal** or **Meibomian** (mī-BŌ-mē-an) **glands**. Their oily secretion

Figure 16.3 Surface anatomy of the eye.

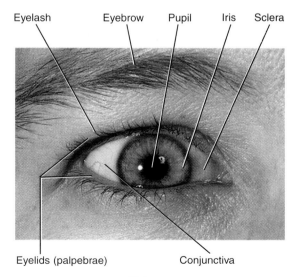

Right eye

Which structure shown here is continuous with the inner lining of the eyelids?

Figure 16.4   Accessory structures of the eye.

🔑 *Accessory structures of the eye are the eyelids, eyelashes, eyebrows, lacrimal apparatus, and extrinsic eye muscles.*

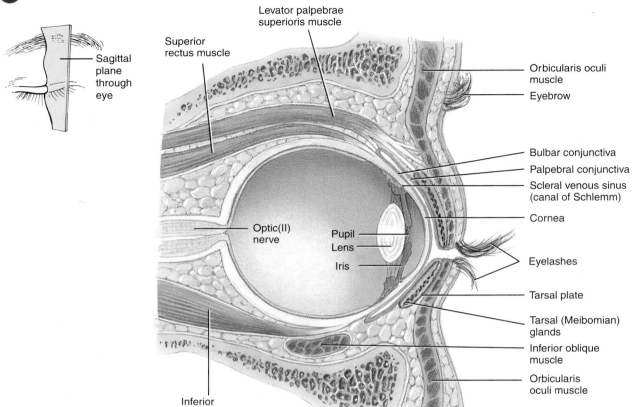

(a) Sagittal section of accessory structures of the eye

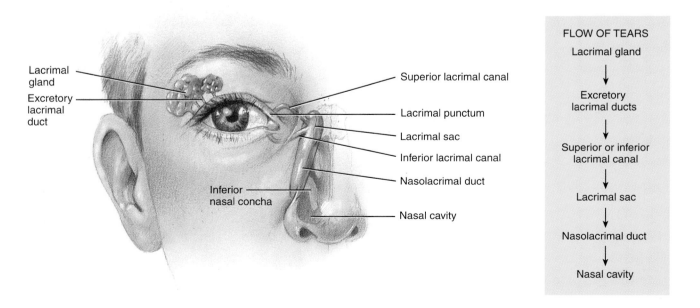

(b) Anterior view of lacrimal apparatus

❓ What are tears, and what are their functions?

helps keep the eyelids from adhering to each other. Infection of the tarsal glands produces a tumor or cyst on the eyelid called a **chalazion** (ka-LĀ-zē-on). The **conjunctiva** (kon'-junk-TĪ-va) is a thin, protective mucous membrane composed mostly of stratified columnar epithelium with numerous goblet cells and areolar connective tissue. The **palpebral conjunctiva** lines the inner aspect of the eyelids and the **bulbar (ocular) conjunctiva** passes from the eyelids onto the anterior surface of the eyeball. When the blood vessels of the bulbar conjunctiva are dilated and congested due to local irritation or infection, the person has bloodshot eyes.

The **eyelashes** project from the border of each eyelid and together with the **eyebrows**, which arch transversely above the upper eyelids, help protect the eyeballs from foreign objects, perspiration, and the direct rays of the sun. Sebaceous glands at the base of the hair follicles of the eyelashes, called **sebaceous ciliary glands** or **glands of Zeis** (ZĪS), pour a lubricating fluid into the follicles. Infection of these glands is called a **sty**.

The **lacrimal** (*lacrima* = tear) **apparatus** is a group of structures that produces and drains **lacrimal fluid** or tears (Fig. 16.4b). The **lacrimal glands**, each about the size and shape of an almond, secrete lacrimal fluid, which drains into 6–12 **excretory lacrimal ducts** that empty tears onto the surface of the conjunctiva of the upper lid. From here the tears pass medially over the anterior surface of the eyeball to enter two small openings called **lacrimal puncta**. Tears then pass into two ducts, the **lacrimal canals**, which lead into the **nasolacrimal sac** and then into the **nasolacrimal duct**. This duct carries the lacrimal fluid into the nasal cavity just inferior to the inferior nasal concha.

Lacrimal fluid is a watery solution containing salts, some mucus, and a bactericidal enzyme called **lysozyme**. The fluid cleans, lubricates, and moistens the eyeball. After being secreted by the lacrimal glands, it is spread medially over the surface of the eyeball by blinking of the eyelids. Each lacrimal gland produces about 1 ml per day.

Normally, tears are cleared away as fast as they are produced by evaporation or by passing into the lacrimal canals and then into the nasal cavity. If an irritating substance contacts the conjunctiva, however, the lacrimal glands are stimulated to oversecrete and tears accumulate (watery eyes). This is a protective mechanism, since the tears dilute and wash away the irritating substance. Watery eyes also occur when an inflammation of the nasal mucosa, such as a cold, obstructs the nasolacrimal ducts and blocks drainage of tears. Humans are unique in expressing emotions, both happiness and sadness, by **crying**. In response to parasympathetic stimulation, the lacrimal glands produce excessive tears that may spill over the edges of the eyelids and even fill the nasal cavity with fluid.

The six extrinsic eye muscles that move each eye (see Fig. 11.6) receive their innervation from cranial nerves III, IV, or VI. In general, the size of motor units is small in these muscles. Some motor neurons serve only two or three muscle fibers, fewer than in any other part of the body except

the larynx (voice box). This permits smooth, precise, and rapid movement of the eyes. As indicated by their names (see Exhibit 11.5 on page 282), each extrinsic eye muscle moves the eyeball in a different direction—laterally, medially, superiorly, or inferiorly. For example, looking to the right requires simultaneous contraction of the right lateral rectus and left medial rectus muscles and relaxation of the left lateral rectus and right medial rectus. The oblique muscles function to preserve rotational stability of the eyeball. Coordinated movement of the two eyes involves circuits in the brain stem and cerebellum.

Anatomy of the Eyeball

The adult **eyeball** measures about 2½ cm (1 in.) in diameter. Of its total surface area, only the anterior one-sixth is exposed. The remainder is recessed and protected by the orbit into which it fits. Anatomically, the wall of the eyeball can be divided into three layers: fibrous tunic, vascular tunic, and retina or nervous tunic (Fig. 16.5).

Fibrous Tunic

The **fibrous tunic** is the superficial coat of the eyeball. It is avascular and consists of the anterior cornea and posterior sclera. The **cornea** (KOR-nē-a) is a nonvascular, transparent coat that covers the colored iris. Because it is curved, the cornea helps focus light. Its outer surface consists of nonkeratinized stratified squamous epithelium. The middle coat of the cornea consists of collagen fibers and fibroblasts, and the inner surface is simple squamous epithelium. The **sclera** (SKLE-ra; *skleros* = hard), the "white" of the eye, is a coat of dense connective tissue made up mostly of collagen fibers and fibroblasts. The sclera covers all the eyeball except the cornea, gives shape to the eyeball, makes it more rigid, and protects its inner parts. At the junction of the sclera and cornea is an opening known as the **scleral venous sinus (canal of Schlemm)**. (Surface features of the eye are shown in Figs. 16.3 and 16.4a.)

CORNEAL TRANSPLANTS

Corneal transplants are the most common organ transplant operation and the most successful type of transplant since rejection rarely occurs. Because the cornea is avascular, blood-borne antibodies that might cause rejection do not enter the transplanted tissue. The defective cornea is removed and a donor cornea of similar diameter is sewn in. The shortage of donated corneas has partially been overcome by the development of artificial corneas made of plastic. ■

Vascular Tunic

The **vascular tunic** or **uvea** (YOO-vē-a) is the middle layer of the eyeball. It has three portions: choroid, ciliary body, and iris. The highly vascularized **choroid** (KŌ-royd) is the

Figure 16.5 Gross structure of the eyeball and responses of the pupil to light. The arrow in the inset in (a) indicates the direction from which the eyeball is viewed (superior).

🔑 *The wall of the eyeball consists of the fibrous tunic, vascular tunic, and retina (nervous tunic).*

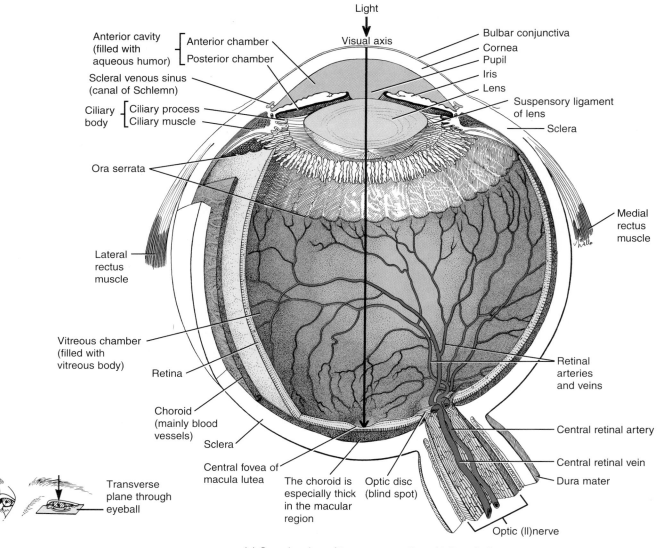

(a) Superior view of transverse section of left eyeball

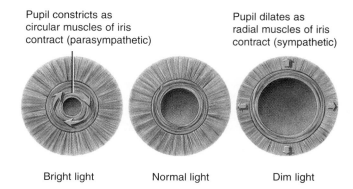

Pupil constricts as circular muscles of iris contract (parasympathetic)

Pupil dilates as radial muscles of iris contract (sympathetic)

Bright light   Normal light   Dim light

(b) Anterior view of responses of the pupil to varying brightness of light

❓ Which structures in the eye do not have their own blood supply and therefore rely on diffusion of oxygen and nutrients from adjacent tissues or fluid? What advantage does this avascularity confer?

posterior portion of the vascular tunic and lines most of the internal surface of the sclera. It provides nutrients to the posterior surface of the retina. Melanocytes, which produce the dark pigment melanin, give the choroid a brown-black appearance.

In the anterior portion of the vascular tunic, the choroid becomes the **ciliary** (SIL-ē-ar′-ē) **body**. It extends from the **ora serrata** (Ō-ra ser-RĀ-ta), the jagged anterior margin of the retina, to a point just posterior to the sclerocorneal junction. The ciliary body consists of the ciliary processes and ciliary muscle. The **ciliary processes** are protrusions or folds on the internal surface of the ciliary body. They contain blood capillaries that secrete a watery fluid called aqueous humor. The **ciliary muscle** is a circular band of smooth muscle that alters the shape of the lens for near or far vision.

The **iris** (*irid* = colored circle) is the colored portion of the eyeball and is shaped like a flattened donut. It is suspended between the cornea and the lens and is attached at its outer margin to the ciliary processes. It consists of circular and radial smooth muscle fibers. The hole in the center of the iris is the **pupil**. A principal function of the iris is to regulate the amount of light entering the vitreous chamber of the eyeball through the pupil (Fig. 16.5b). When bright light stimulates the eye, parasympathetic neurons stimulate the **circular muscles (constrictor** or **sphincter pupillae)** of the iris to contract, causing a decrease in the size of the pupil (constriction). In dim light, sympathetic neurons stimulate the **radial muscles (dilator pupillae)** of the iris to contract, causing an increase in the pupil's size (dilation). These responses are autonomic reflexes.

Retina (Nervous Tunic)

The third and inner coat of the eyeball, the **retina (nervous tunic)**, lines the posterior three-quarters of the eyeball and is the beginning of the visual pathway. By using an ophthalmoscope to peer through the pupil, one can see a magnified image of the retina and the blood vessels that course across its anterior surface. Here blood vessels can be viewed directly and examined for pathological changes such as occur with hypertension or diabetes. Several landmarks are visible (Fig. 16.6). The **optic disc** is the site where the optic nerve exits the eyeball. Bundled together with the optic nerve are the **central retinal artery**, a branch of the ophthalmic artery, and **central retinal vein**. Branches of the central retinal artery fan out to nourish the anterior surface of the retina. The central retinal vein drains blood from the retina through the optic disc.

The retina consists of a pigment epithelium (nonvisual portion) and a neural portion (visual portion). The **pigment epithelium** is a sheet of melanin-containing epithelial cells that lies between the choroid and the neural portion of the retina. Some histologists classify it as part of the choroid rather than the retina. Melanin in the choroid and the pigment epithelium absorbs stray light rays, which prevents reflection and scattering of light within the eyeball. This ensures that the image cast on the retina by the cornea and lens

Figure 16.6 Normal retina as seen through an ophthalmoscope.

 In the retina blood vessels can be viewed directly and examined for pathological changes.

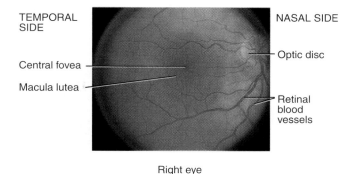

Right eye

Evidence of what diseases may be seen through an ophthalmoscope?

remains sharp and clear. Albinos lack melanin pigment in all parts of the body, including the eye. They often wear sunglasses, even indoors, because moderately bright light is perceived as nothing but glare.

DETACHED RETINA

Detachment of the retina may occur in trauma, such as a blow to the head, or in various eye disorders. The detachment occurs between the neural portion of the retina and the pigment epithelium. Fluid accumulates between these layers, forcing the thin, pliable retina to billow outward. The result is distorted vision and blindness in the corresponding field of vision. The retina may be reattached by laser surgery or cryosurgery (which involves the local application of extreme cold). ■

The **neural portion** of the retina is a multilayered outgrowth of the brain. It processes visual data extensively before transmitting nerve impulses to the thalamus, which then relays nerve impulses to the primary visual cortex. Three distinct layers of retinal neurons are separated by two zones where synaptic contacts are made, the inner and outer synaptic layers. The three layers of retinal neurons, in the order in which they process visual input, are the **photoreceptor layer**, **bipolar cell layer**, and **ganglion cell layer** (Fig. 16.7). Note that light passes through the ganglion and bipolar cell layers before reaching the photoreceptor layer. Two other types of cells present in the retina are called **horizontal cells** and **amacrine cells**. These cells form laterally directed pathways that modify the signals being transmitted along the pathway from photoreceptors to bipolar cells to ganglion cells.

Photoreceptors are specialized to transduce light rays into receptor potentials. The two types of photoreceptors are

Figure 16.7   Microscopic structure of the retina. The downward arrow indicates the direction of the signals passing through the neural portion of the retina. Ultimately, nerve impulses arise in ganglion cells and pass into the optic nerve (cranial nerve II).

The neural portion of the retina processes visual data before axons of ganglion cells propagate nerve impulses to the thalamus via the optic nerve.

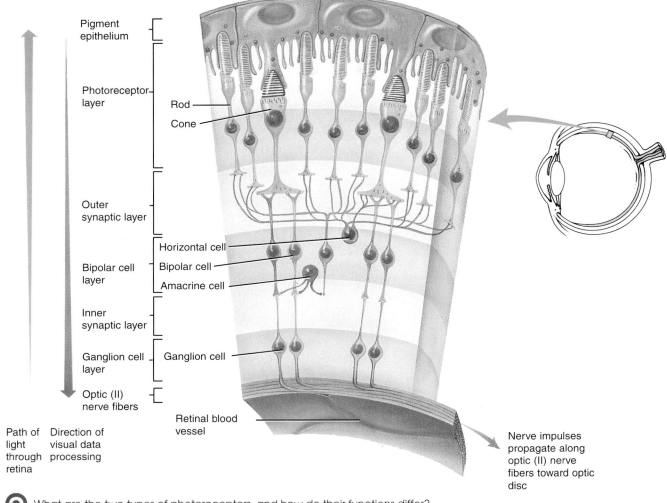

What are the two types of photoreceptors, and how do their functions differ?

rods and cones. Each retina has about 6 million cones and 120 million rods. **Rods** are most important for seeing shades of gray in dim light. They also allow us to see shapes and movement. **Cones** provide color vision in bright light. In moonlight we cannot see colors because only the rods are functioning. Due to the low light leve! cones are not functioning.

The **macula lutea** (MAK-yoo-la LOO-tē-a; *macula* = spot; *lutea* = yellow) is in the exact center of the posterior portion of the retina, at the visual axis of the eye (see Fig. 16.5). The **central fovea**, a small depression in the center of the macula lutea, contains only cone photoreceptors. In addition, the layers of bipolar and ganglion cells, which scatter light to some extent, do not cover the cones here; these layers are displaced to the periphery of the fovea. As a result,

the central fovea is the area of highest **visual acuity** or **resolution** (sharpness of vision). Rods are absent from the fovea and macula and increase in number toward the periphery of the retina. Because rod vision is more sensitive than the cone vision, you can see a faint star on a moonless night better if you gaze to one side of the star rather than directly at it.

From photoreceptors, information flows to bipolar cells through the outer synaptic layer and then from bipolar cells through the inner synaptic layer to ganglion cells. The axons of ganglion cells extend posteriorly to the optic disc and exit the eyeball as the optic nerve. The optic disc is also called the **blind spot**. Since it contains no rods or cones, we cannot see an image that strikes the blind spot. Normally, you are not aware of having a blind spot, but you can easily convince yourself of its presence. Cover your left eye and gaze directly at the cross below. Then increase or decrease the distance between the book and your eye. At some point the square will disappear as its image falls on the blind spot.

Lens

Just posterior to the pupil and iris, within the cavity of the eyeball, is the avascular **lens** (see Fig. 16.5). Proteins called **crystallins**, arranged like the layers of an onion, make up the lens. Normally, the lens is perfectly transparent. It is enclosed by a clear connective tissue capsule and held in position by encircling **suspensory ligaments**. The lens fine-tunes focusing of light rays for clear vision.

CATARACTS

The leading cause of blindness is a loss of transparency of the lens known as a **cataract**. This problem often occurs with aging but may also be caused by injury, exposure to ultraviolet rays, certain medications (such as long-term use of steroids), or complications of other diseases (for example, diabetes). People who smoke also have increased risk of developing cataracts. The lens becomes cloudy or less transparent due to changes in the structure of the lens proteins. Fortunately, sight can usually be restored by surgical removal of the old lens and implantation of an artificial one. ■

Interior of the Eyeball

The interior of the eyeball is a large space divided by the lens into two cavities: the anterior cavity and vitreous chamber. The **anterior cavity**, the space anterior to the lens, is further divided into the **anterior chamber**, which lies behind the cornea and in front of the iris, and the **posterior chamber**, which lies behind the iris and in front of the suspensory ligaments and lens (Fig. 16.8). The anterior cavity is filled with a watery fluid called the **aqueous** (*aqua* = water) **humor** that is continually filtered from blood capillaries in the ciliary processes posterior to the iris. Once the fluid is formed, it flows into the posterior chamber and then forward between the iris and the lens, through the pupil, and into the anterior chamber. From the anterior chamber, aqueous humor drains into the scleral venous sinus (canal of Schlemm) and then into the blood. Normally, aqueous humor is completely replaced about every 90 minutes.

Functionally, aqueous humor helps nourish the lens and cornea. Also, the pressure in the eye, called **intraocular pressure**, is produced mainly by the aqueous humor. The intraocular pressure, along with the vitreous body (described shortly), maintains the shape of the eyeball and keeps the retina smoothly pressed against the choroid so the retina will be an even surface for reception of clear images. Because there is a balance between production and outflow of the aqueous humor, intraocular pressure normally stays about 16 mm Hg. Excessive intraocular pressure is called **glaucoma** (glaw-KŌ-ma). It leads to degeneration of the retina, causing blindness (see page 483).

Figure 16.8   Section through the anterior portion of the eyeball at the sclerocorneal junction. Arrows indicate the flow of aqueous humor.

The lens separates the anterior cavity from the vitreous chamber.

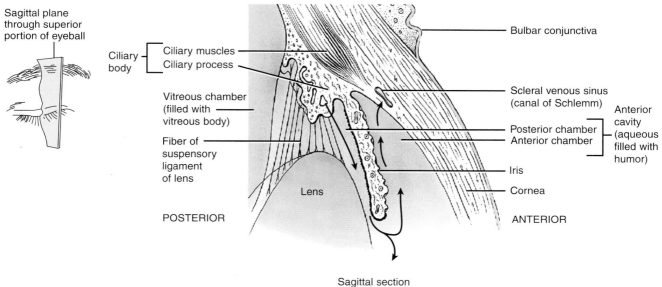

Sagittal section

Where is aqueous humor produced, what is its circulation path, and where does it drain from the eyeball?

The second, and larger, cavity of the eyeball is the **vitreous chamber (posterior cavity)**. It lies between the lens and the retina and contains a jellylike substance called the **vitreous body**. This substance contributes to intraocular pressure, helps prevent the eyeball from collapsing, and holds the retina flush against the internal portions of the eyeball. The vitreous body, unlike the aqueous humor, does not undergo constant replacement. It is formed during embryonic life and is not replaced thereafter.

A summary of structures associated with the eyeball is presented in Exhibit 16.1.

Image Formation

In some respects the eye is like a camera. The cornea and the lens of the eye focus an image of distant objects on a light-sensitive "film"—the retina. Contraction of the ciliary muscle changes the shape of the lens to bring closer objects into focus, in much the same manner as a camera is focused.

Adjustment of the pupil diameter helps to maintain proper light exposure of the retina.

Formation of images on the retina involves three basic processes: (1) refraction of light rays by the cornea and lens, (2) accommodation of the lens, and (3) constriction of the pupil. Accommodation and pupil size are functions of the smooth muscle fibers of the ciliary muscle and iris. These muscles are termed **intrinsic eye muscles** since they are inside the eyeball.

Refraction of Light Rays

When light rays traveling through a transparent substance (such as air) pass into a second transparent substance with a different density (such as water), they bend at the junction between the two. This bending is called **refraction** (Fig. 16.9a). As light rays enter the eye, they are refracted at the anterior and posterior surfaces of the cornea. Both surfaces of the lens of the eye further refract the light rays so they come into exact focus on the retina.

EXHIBIT 16.1 SUMMARY OF STRUCTURES ASSOCIATED WITH THE EYEBALL

Structure	Function	Structure	Function
Fibrous tunic Cornea / Sclera	*Cornea*: Admits and refracts (bends) light. *Sclera*: Provides shape and protects inner parts.	**Lens** Lens	Refracts light.
Vascular tunic Iris / Ciliary body / Choroid	*Iris*: Regulates amount of light that enters eyeball. *Ciliary body*: Secretes aqueous humor and alters shape of lens for near or far vision (accommodation). *Choroid*: Provides blood supply and absorbs scattered light.	**Anterior cavity** Anterior cavity	Contains aqueous humor that helps maintain shape of eyeball and supplies oxygen and nutrients to lens and cornea.
Retina (nervous tunic) Retina	Receives light and converts light into receptor potentials and nerve impulses. Output to brain is via axons of ganglion cells, which form the optic (II) nerve.	**Vitreous chamber** Vitreous chamber	Contains vitreous body that helps maintain shape of eyeball and keeps retina applied to choroid.

Figure 16.9 Refraction of light rays and accommodation.

Refraction is the bending of light rays at the junction of two substances with different densities.

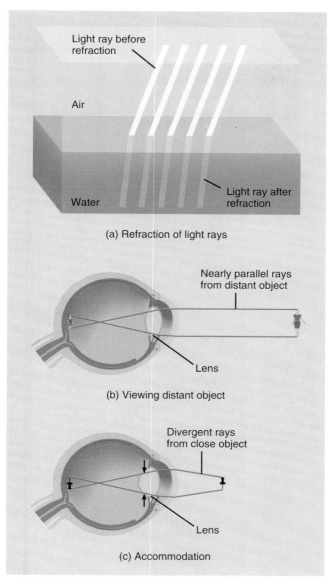

(a) Refraction of light rays

Light ray before refraction

Air

Water

Light ray after refraction

Nearly parallel rays from distant object

Lens

(b) Viewing distant object

Divergent rays from close object

Lens

(c) Accommodation

What is the sequence of changes that occur during accommodation?

Images focused on the retina are inverted; they are upside down (Fig. 16.9b,c). They also undergo right to left reversal; that is, light from the right side of an object strikes the left side of the retina, and vice versa. The reason we do not see an inverted world is that the brain learns early in life to coordinate visual images with the exact locations of objects. The brain stores memories of reaching and touching objects and automatically turns visual images right-side-up and right-side-around.

About 75% of all refraction occurs at the cornea. The lens then is responsible for fine-tuning of image focus and changing the focus for near or distant objects. When an object is 6 m (20 ft) or more away from the viewer, the light rays reflected from the object are nearly parallel to one another (Fig. 16.9b). The parallel rays must be bent sufficiently to fall exactly on the central fovea, where vision is sharpest. Light rays that are reflected from objects closer than 6 m (20 ft) are divergent rather than parallel (Fig. 16.9c). As a result, they must be refracted toward each other to a greater extent.

Accommodation and the Near Point of Vision

If the surface of a lens curves outward, like the surface of a ball, the lens will refract incoming rays toward each other so they eventually intersect. This is a *convex* lens. Conversely, when the surface of a lens curves inward, like the inside of a hollow ball, the rays bend away from each other. This is a *concave* lens. The lens of the eye is convex on both its anterior and posterior surfaces. Furthermore, its focusing power increases as its curvature becomes greater. When the eye is focusing on a close object, the lens curves more to bend the rays toward the central fovea. This increase in the curvature of the lens for near vision is called **accommodation** (Fig. 16.9c).

How does accommodation occur? When you are viewing distant objects, the ciliary muscle is relaxed but the lens is fairly flat because it is stretched in all directions by taut suspensory ligaments. When you view a close object, the ciliary muscle contracts, which pulls the ciliary process and choroid forward toward the lens. This action releases tension on the lens and suspensory ligaments. Since it is elastic, the lens shortens, thickens, and bulges. Now that it is more rounded, its focusing power is greater and the light rays converge more.

The **near point of vision** is the minimum distance from the eye that an object can be clearly focused with maximum effort. This is about 10 cm (4 in.) in a young adult. With aging, the lens loses elasticity and therefore its ability to accommodate. This condition is called **presbyopia** (prez-bē-Ō-pē-a; *presbys* = old man). As a consequence, older people cannot read print at the same close range as can youngsters. By age 40 the near point of vision may have increased to 20 cm (8 in.) and at age 60 to 80 cm (31 in.). Presbyopia usually begins in the midforties and is the reason that those already wearing glasses need bifocals and those who have not previously needed glasses now require reading glasses.

ABNORMALITIES OF REFRACTION

The normal eye, known as an **emmetropic** (em′-e-TROP-ik) **eye**, can sufficiently refract light rays from an object 6 m (20 ft) away to focus a clear image on the retina. Many people, however, do not have this ability because of abnormalities related to improper refraction. Among these abnormalities are **myopia** (mī-Ō-pē-a), or nearsightedness; **hypermetropia** (hī′-per-me-TRŌ-pē-a), or farsightedness, also known as **hyperopia**; and **astigmatism** (a-STIG-ma-tizm), or irregularities in the surface of the lens or cornea. Figure 16.10 illustrates and explains these conditions and how they are corrected.

Most errors of vision can be corrected by eyeglasses or contact lenses. A contact lens floats on a film of tears over the cornea. The anterior outer surface of the contact lens corrects the visual defect, while the posterior surface matches the curvature of the cornea. ■

Constriction of the Pupil

The circular muscle fibers of the iris also have a role in the formation of clear retinal images. Part of the accommodation mechanism consists of the contraction of the circular muscles of the iris to constrict the pupil. **Constriction of the pupil** means narrowing the diameter of the hole through which light enters the eye. This autonomic reflex occurs simultaneously with accommodation of the lens and prevents light rays from entering the eye through the periphery of the lens. Light rays entering at the periphery would not be brought to focus on the retina and would result in blurred vision. The pupil, as noted earlier, also constricts in bright light.

Convergence

Because of the position of their eyes, many animals see a set of objects off to the left through one eye and an entirely different set off to the right through the other. In humans, both eyes focus on only one set of objects—a characteristic called **single binocular vision**.

Single binocular vision occurs when light rays from an object strike corresponding points on the two retinas. When we stare straight ahead at a distant object, the incoming light rays are aimed directly at both pupils and are refracted to comparable spots on the retinas of both eyes. But as we move closer to the object, our eyes must rotate medially for the light rays from the object to strike the same points on both retinas. The term **convergence** refers to this medial movement of the two eyeballs so they are both directed toward the object being viewed. The nearer the object, the greater the degree of convergence needed to maintain single binocular vision. The coordinated action of the extrinsic eye muscles (see Fig. 11.6) brings about convergence.

Figure 16.10   Refraction in the eyeball. (a) Normal (emmetropic) eye. (b) In the nearsighted or myopic eye, the image is focused in front of the retina. The condition may result from an elongated eyeball or thickened lens. (c) Correction is by use of a concave lens that diverges entering light rays so that they come into focus directly on the retina. (d) In the farsighted or hypermetropic eye, the image is focused behind the retina. The condition results from a shortened eyeball or a thin lens. (e) Correction is by a convex lens that converges entering light rays so that they focus directly on the retina. An astigmatism is irregular curvature of the (f) cornea or (g) lens that prevents images from being focused on the retina. This results in blurred or distorted vision. Suitable glasses or contact lenses correct the refraction of an astigmatic eye.

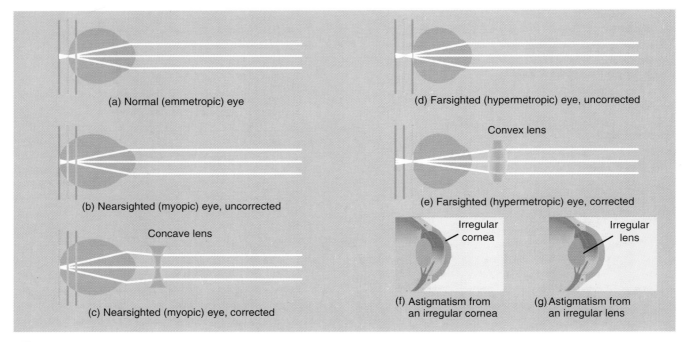

(a) Normal (emmetropic) eye

(b) Nearsighted (myopic) eye, uncorrected

Concave lens

(c) Nearsighted (myopic) eye, corrected

(d) Farsighted (hypermetropic) eye, uncorrected

Convex lens

(e) Farsighted (hypermetropic) eye, corrected

Irregular cornea

Irregular lens

(f) Astigmatism from an irregular cornea

(g) Astigmatism from an irregular lens

Q What is presbyopia?

Physiology of Vision

An image focused on the retina stimulates photoreceptors, which transduce the light stimulus into receptor potentials and pass the information on to bipolar cells. Bipolar cells, in turn, communicate with ganglion cells, which project their axons to the lateral geniculate body of the thalamus. From the thalamus, fibers carrying visual nerve impulses extend to the primary visual cortex in the occipital lobe (see area 17 in Fig. 14.15).

Photoreceptors and Photopigments

Rods and cones were named for the different appearance of their *outer segment,* which is the distal end of the photoreceptor next to the pigment epithelium. The outer segments of cones are tapered or cone-shaped whereas those of rods are cylindrical or rod-shaped (Fig. 16.11). Transduction of light into an electrical signal occurs in the outer segment. The *inner segment* contains the cell nucleus, Golgi complex, and many mitochondria. At its proximal end, the photoreceptor expands into a bulblike synaptic terminal.

The first step in visual transduction is absorption of light by a **photopigment** (**visual pigment**). Photopigments are colored proteins in outer segment membranes that undergo structural changes upon light absorption. They initiate the events that lead to production of a receptor potential. The single type of photopigment in rods is called **rhodopsin** (*rhodo* = rose; *opsis* = vision). A cone contains one of three different kinds of photopigments; thus there are three types of cones.

Photopigments are integral proteins in the plasma membrane of the outer segment, which folds back and forth in a pleated fashion (Fig. 16.11). In rods, the pleats pinch off from the plasma membrane to form discs. The outer segment of each rod contains a stack of about 1000 discs, piled up like coins inside a wrapper. Photoreceptor outer segments are renewed at an astonishingly rapid pace. In rods, one to three new discs are added to the base of the outer segment every hour. Simultaneously, old discs slough off at the tip and are phagocytized by pigment epithelial cells (Fig. 16.11).

All visual photopigments contain two parts: a glycoprotein known as **opsin** and a derivative of vitamin A called **retinal**. Vitamin A derivatives are formed from carotenoids, the plant pigments that give carrots their color. Good vision depends on adequate dietary intake of carotenoid-rich vegetables such as carrots, spinach, broccoli, and yellow squash or meats that contain vitamin A, such as liver. **Night blindness** or **nyctalopia** (nik'-ta-LŌ-pē-a) is an inability to see well at low light levels. It is most often caused by prolonged vitamin A deficiency and the consequent inability to synthesize a normal amount of rhodopsin.

Retinal is the light-absorbing portion of all visual photopigments. In the human retina, there are four different opsins, one for each cone photopigment and one for rhodopsin. Small variations in the amino acid sequences of

Figure 16.11 Rod and cone photoreceptors.

🔑 *Transduction of light into an electrical signal occurs in the outer segments of photoreceptors.*

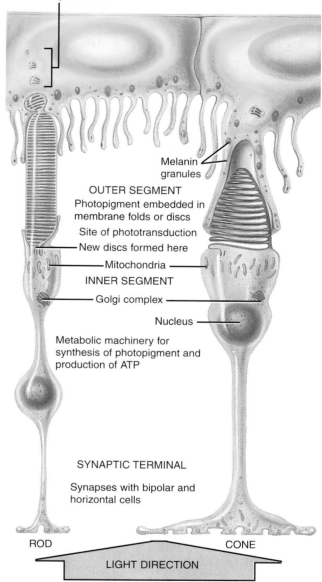

Phagocytosis of old discs by pigment epithelial cell

Melanin granules

OUTER SEGMENT
Photopigment embedded in membrane folds or discs
Site of phototransduction
New discs formed here
Mitochondria
INNER SEGMENT
Golgi complex
Nucleus

Metabolic machinery for synthesis of photopigment and production of ATP

SYNAPTIC TERMINAL

Synapses with bipolar and horizontal cells

ROD   CONE

LIGHT DIRECTION

Structure of rods and cones

❓ What are the functional similarities between rods and cones?

the different opsins permit the rods and cones to absorb different colors (wavelengths) of incoming light. Rhodopsin absorbs blue to green light most effectively whereas the three different cone photopigments most effectively absorb blue, green, or yellow-orange light.

Photopigments are activated and restored as follows (Fig. 16.12):

1 In darkness, retinal has a bent shape, called *cis*-retinal, which fits snugly into the opsin portion of the photopigment. When *cis*-retinal absorbs light, it straightens out to a shape called *trans*-retinal. This *cis* to *trans* conversion is called **isomerization** and it is the first step in visual transduction. Forming a visual image begins with isomerization of photopigments in particular rods and cones. After retinal isomerizes, several unstable intermediates form and disappear.

2 In about a minute, *trans*-retinal completely separates from opsin. The final products look colorless, so the whole process is called **bleaching** of photopigment.

3 In darkness, an enzyme called **retinal isomerase** converts *trans*- back to *cis*-retinal.

4 When *cis*-retinal binds to opsin, it reforms a functional photopigment. Resynthesis of a photopigment is called **regeneration**.

The pigment epithelium, adjacent to the photoreceptors, stores a large quantity of vitamin A and contributes to the

---

Figure 16.12 Bleaching and regeneration of photopigment.

Retinal, a derivative of vitamin A, is the light-absorbing portion of all visual photopigments.

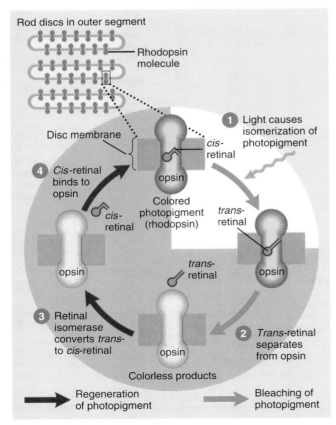

What is the conversion of *cis*- to *trans*-retinal called?

---

regeneration process in rods. The extent of rhodopsin regeneration decreases drastically if the retina detaches from the pigment epithelium. Cone photopigments regenerate much more quickly than does rhodopsin and are less dependent on the pigment epithelium. After complete bleaching, it takes 5 minutes to regenerate half of the rhodopsin but only 1½ minutes to regenerate half of the cone photopigments. Full regeneration of bleached rhodopsin takes 30–40 minutes.

COLOR BLINDNESS

Most forms of **color blindness** (inability to distinguish certain colors) result from absence or deficiency of one of the three cone photopigments. The most common type is **red–green color blindness**, in which a photopigment sensitive to red or green light is missing. As a result, the person has no way to distinguish between red and green. Color blindness is an inherited, sex-linked condition that affects males far more often than females. Its inheritance is discussed on page 982 and illustrated in Fig. 29.19. ■

Light and Dark Adaptation

When you emerge from dark surroundings, such as a tunnel, into the sunshine, your visual system adjusts in seconds to the brighter environment by decreasing its sensitivity. This is called **light adaptation**. On the other hand, when you enter a darkened room, such as a theater, your visual sensitivity increases slowly over many minutes. This is called **dark adaptation**. Bleaching and regeneration of the photopigments account for much but not all of the sensitivity change during light and dark adaptation, respectively.

As the light level increases, more and more photopigment is bleached so more light is needed to stimulate the remaining unbleached photopigment. At the same time that light is bleaching some photopigment molecules, however, others are being regenerated. Regeneration of rhodopsin occurs slowly enough that in daylight it is bleached as fast as it is regenerated. For this reason, rods contribute little to daylight vision. Regeneration of cone photopigments, on the other hand, is fast enough that some of the *cis* form is always present, even in very bright light.

If the light level decreases abruptly, sensitivity increases rapidly at first and then more slowly. In complete darkness, full regeneration of the cone photopigments occurs during the first 8 minutes of dark adaptation. During this time interval, a threshold light flash (barely perceptible) is seen as having color. More slowly, rhodopsin regenerates and our visual sensitivity increases until even a single photon (the smallest unit) of light can be detected. Now, although much dimmer light can be detected, threshold flashes look gray-white, regardless of their color. At very low light levels, such as moonlight or starlight, our visual world appears as shades of gray because only the rods are functioning.

Receptor Potential and Neurotransmitter Release

In darkness, sodium ions (Na^+) flow into photoreceptor outer segments through Na^+ channels that are held open by a molecule called **cyclic GMP (guanosine monophosphate)** (Fig.

16.13). This inflow of Na$^+$, called the "dark current," triggers continual release of neurotransmitter from the synaptic terminals. The neurotransmitter in rods, and perhaps in cones also, is the amino acid glutamate (glutamic acid). Glutamate inhibits (hyperpolarizes) the bipolar cells that synapse with rods. When light strikes the retina and *cis*-retinal undergoes isomerization, the Na$^+$ channels close. Na$^+$ inflow thus decreases, the inside of the rod becomes more negative (hyperpolarization), and release of glutamate decreases (Fig. 16.13). Dim lights cause small and brief hyperpolarizations that partially turn off glutamate release. Brighter lights elicit larger and longer hyperpolarizations that more completely shut down neurotransmitter release. The surprising result is this: light excites the bipolar cells that synapse with rods by turning off release of an inhibitory neurotransmitter!

Several enzymes regulate closing and reopening of the Na$^+$ channels in the outer segment. In light, one enzyme called **transducin** activates another enzyme called **PDE** (**phosphodiesterase**), which breaks down cyclic GMP. This closes the Na$^+$ channels resulting in hyperpolarization of rods and decreased release of glutamate (Fig. 16.13). In darkness, transducin is in an inactive form, and cyclic GMP holds the Na$^+$ channels open. An enzyme called **guanylate cyclase stimulating factor** activates guanylate cyclase, the enzyme that stimulates synthesis of cyclic GMP. As the cyclic GMP level rises, the Na$^+$ channels are again held in the open position and the inflow of Na$^+$ triggers increased release of glutamate (Fig. 16.13).

Visual Pathway

As mentioned earlier, considerable processing of visual input occurs in the retina, at synapses among the various types of cells (see Fig. 16.7). The axons of retinal ganglion cells provide output from the retina to the brain. They exit the eyeball via the **optic (II) nerve**.

Retinal Processing of Visual Input

Within the retina, certain features of visual input are enhanced while other features may be discarded. Input from several cells may converge upon a smaller number of postsynaptic neurons or may diverge to a large number. On the whole, however, convergence predominates since there are only one million ganglion cells that receive input from about 126 million photoreceptors.

Once receptor potentials arise in rods and cones, they spread through the inner segments to the synaptic terminals. Neurotransmitters released by rods and cones induce graded, local potentials in both bipolar cells and horizontal cells. Between 6 and 600 rods synapse with a single bipolar cell in the outer synaptic layer, whereas a cone more often synapses with just one bipolar cell. The convergence of many rods onto a single bipolar cell increases the sensitivity of rod vision but slightly blurs the image that is perceived. Cone vision, although less sensitive, has higher acuity because of the one-to-one synapses between cones and their bipolar cells. Stimulation of rods by light excites their bipo-

Figure 16.13 Operation of rod photoreceptors.

Light causes a hyperpolarizing receptor potential in photoreceptors, which decreases release of neurotransmitter (glutamate in rods).

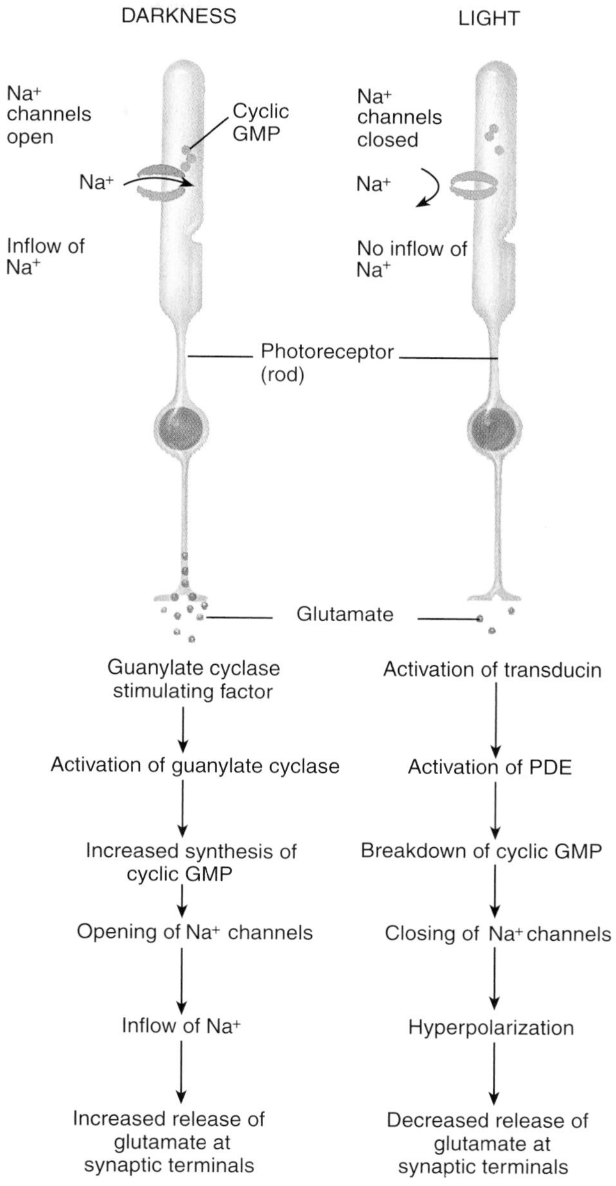

What are the functions of PDE and guanylate cyclase?

lar cells. Cone bipolar cells, on the other hand, may be either excited or inhibited by light.

Horizontal cells transmit inhibitory signals to bipolar cells in the areas lateral to excited rods and cones. This lateral inhibition enhances contrasts in the visual scene between areas of the retina that are strongly stimulated and adjacent areas that are more weakly stimulated. Horizontal cells also assist in the differentiation of various colors. Amacrine cells, which are also excited by bipolar cells, synapse with ganglion cells and transmit information to them that signals a change in the level of illumination of the retina. When bipolar or amacrine cells transmit excitatory signals to ganglion cells, the ganglion cells become depolarized and initiate nerve impulses.

Brain Pathway and Visual Fields

The axons of the optic (II) nerve pass through the **optic chiasm** (kī-AZ-m; *chiasma* = cross over, as in the letter X), a crossing point of the optic nerves (Fig. 16.14). Some fibers cross to the opposite side. Others remain uncrossed. After passing through the optic chiasm, the fibers, now part of the **optic tract**, enter the brain and terminate in the lateral geniculate nucleus of the thalamus. Here the fibers synapse with neurons whose axons form the **optic radiations**. These fibers project to the primary visual areas in the occipital lobes of the cerebral cortex (see area 17 in Fig. 14.15).

Analysis of the visual pathway reveals that the visual field of each eye is divided into two regions: the **nasal (medial) half** and the **temporal (lateral) half**. For each eye, light rays from an object in the nasal half of the visual field fall on the temporal half of the retina. Light rays from an object in the temporal half of the visual field fall on the nasal half of the retina (Fig. 16.15). Also, light rays from objects at the top of the visual field of each eye fall on the inferior portion of the retina, and light rays from objects at the bottom of the visual field fall on the superior portion of the retina.

In the optic chiasm, nerve fibers from the nasal half of each retina cross and continue to the opposite lateral geniculate nucleus of the thalamus. Nerve fibers from the temporal half of each retina do not cross but continue directly to the lateral geniculate nucleus on the same side. As a result, the primary visual area of the cerebral cortex of the right occipital lobe receives visual images from the left side of an object via nerve impulses from the temporal half of the retina of the right eye and the nasal half of the retina of the left eye. The primary visual area of the cerebral cortex of the left occipital lobe receives visual images from the right side of an object via impulses from the nasal half of the right eye and the temporal half of the left eye.

Although we have just described the visual pathway as a more or less single pathway, visual signals are thought to be processed by at least three separate systems in the cerebral cortex, each with its own function. One system processes information related to the shape of objects, another system forms a pathway regarding color of objects, and a third system processes information about movement, location, and spatial organization.

Figure 16.14 Photograph of the visual pathway. The brain is partially dissected to reveal the optic radiations (axons extending from the thalamus to the occipital lobe). The arrow in the inset indicates the direction from which the visual pathway is viewed (inferior).

 The optic chiasm is the crossing point of the optic nerves.

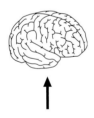

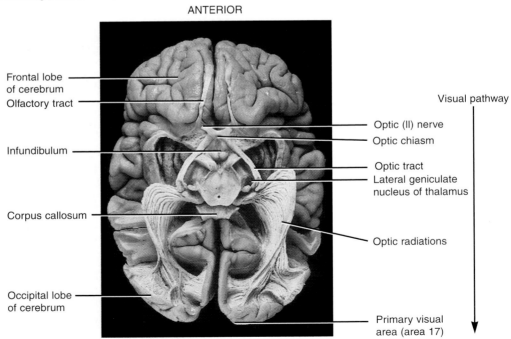

ANTERIOR

Frontal lobe of cerebrum
Olfactory tract

Infundibulum

Corpus callosum

Occipital lobe of cerebrum

Visual pathway

Optic (II) nerve
Optic chiasm

Optic tract
Lateral geniculate nucleus of thalamus

Optic radiations

Primary visual area (area 17)

POSTERIOR

Inferior view

Q What is the correct order of structures that carry nerve impulses from the retina to the occipital lobe of the cerebrum?

Figure 16.15 Visual fields and the visual pathway. The dark circle in the center of the visual fields is the macula lutea. The center of the macula lutea is the central fovea, the area of sharpest vision.

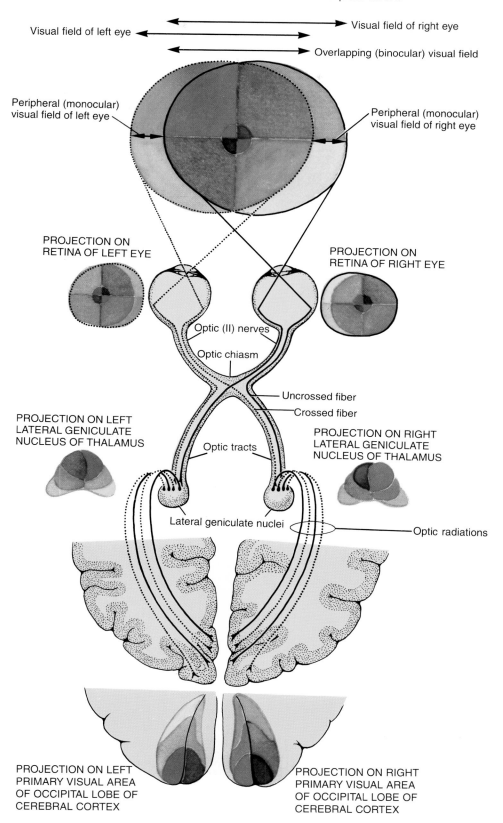

Visual field of left eye

Visual field of right eye

Overlapping (binocular) visual field

Peripheral (monocular) visual field of left eye

Peripheral (monocular) visual field of right eye

PROJECTION ON RETINA OF LEFT EYE

PROJECTION ON RETINA OF RIGHT EYE

Optic (II) nerves

Optic chiasm

Uncrossed fiber

Crossed fiber

PROJECTION ON LEFT LATERAL GENICULATE NUCLEUS OF THALAMUS

PROJECTION ON RIGHT LATERAL GENICULATE NUCLEUS OF THALAMUS

Optic tracts

Lateral geniculate nuclei

Optic radiations

PROJECTION ON LEFT PRIMARY VISUAL AREA OF OCCIPITAL LOBE OF CEREBRAL CORTEX

PROJECTION ON RIGHT PRIMARY VISUAL AREA OF OCCIPITAL LOBE OF CEREBRAL CORTEX

Light rays from an object in the temporal half of the visual field fall on which half of the retina?

AUDITORY SENSATIONS AND EQUILIBRIUM

Besides receptors for sound waves, the ear also contains receptors for equilibrium. Anatomically, the ear is divided into three principal regions: the external (outer) ear, the middle ear, and the internal (inner) ear.

External (Outer) Ear

The **external (outer) ear** collects sound waves and channels them inward (Fig. 16.16). It consists of the auricle, external auditory canal, and eardrum. The **auricle (pinna)** is a flap of elastic cartilage shaped like the flared end of a trumpet and covered by skin. The rim of the auricle is called the **helix**; the inferior portion is the **lobule**. The auricle is attached to the head by ligaments and muscles. The **external auditory** (*audire* = to hear) **canal (meatus)** is a curved tube

about 2.5 cm (1 in.) long that lies in the temporal bone and leads from the auricle to the eardrum. The **eardrum** or **tympanic** (tim-PAN-ik; *tympano* = drum) **membrane** is a thin, semitransparent partition between the external auditory canal and middle ear. The eardrum is covered by epidermis and lined by simple cuboidal epithelium. Between the epithelial layers is connective tissue composed of collagen and elastic fibers and fibroblasts.

Near the exterior opening, the external auditory canal contains a few hairs and specialized sebaceous (oil) glands called **ceruminous** (se-ROO-mi-nus) **glands** that secrete **cerumen** (earwax). The combination of hairs and cerumen (se-ROO-min) helps prevent dust and foreign objects from entering the ear. Usually, cerumen dries up and falls out of the ear canal. Some people, however, produce an abnormal amount of cerumen. It then becomes impacted and muffles incoming sounds. The treatment for **impacted cerumen** is usually periodic ear irrigation or removal of wax with a blunt instrument by trained medical personnel.

Figure 16.16   Structure of the ear illustrated in the right ear.

The ear is divided into three principal regions: external (outer), middle, and internal (inner).

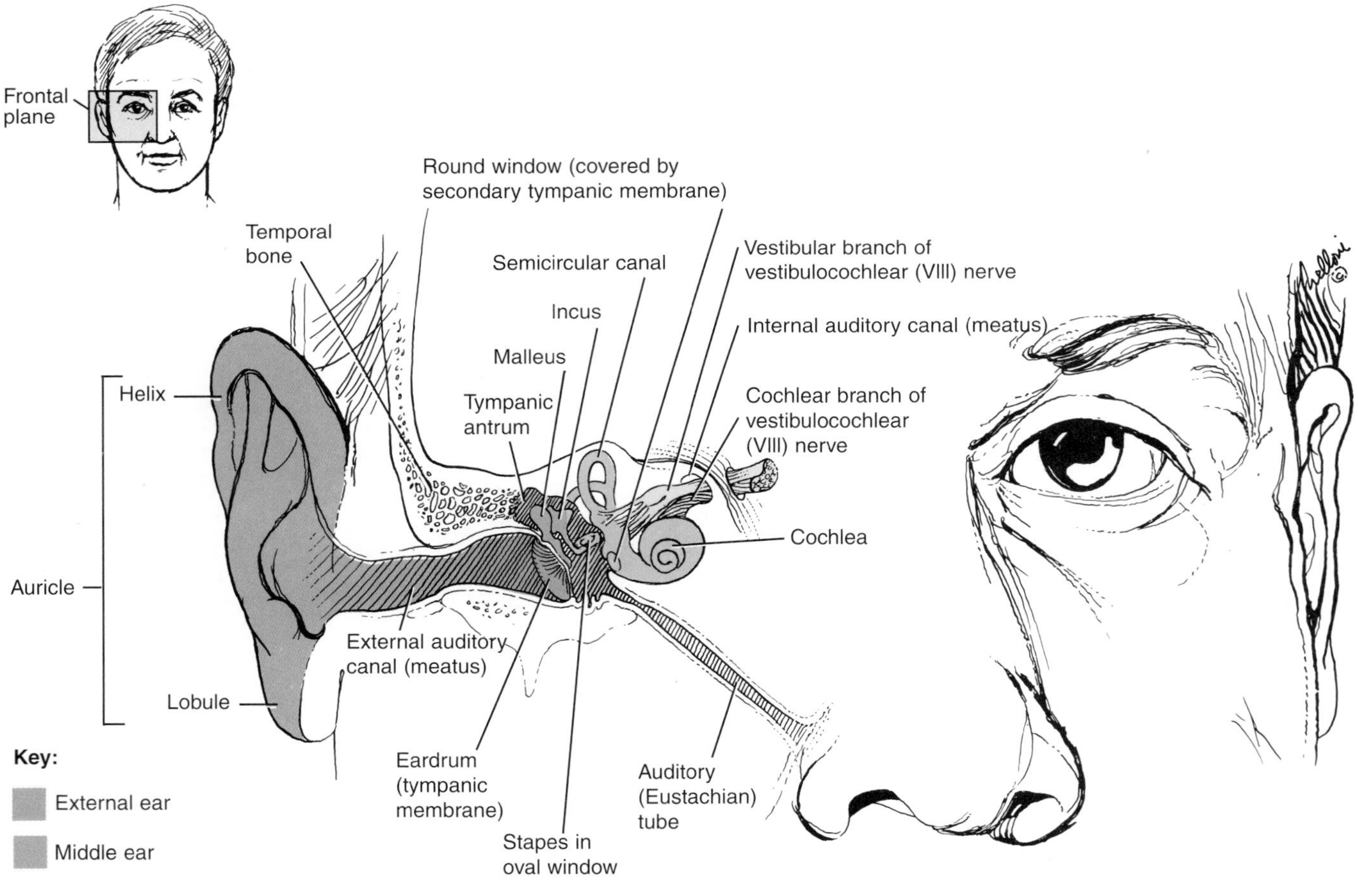

Frontal plane

Temporal bone

Round window (covered by secondary tympanic membrane)

Semicircular canal

Incus

Malleus

Tympanic antrum

Vestibular branch of vestibulocochlear (VIII) nerve

Internal auditory canal (meatus)

Cochlear branch of vestibulocochlear (VIII) nerve

Helix

Auricle

Lobule

Cochlea

External auditory canal (meatus)

Eardrum (tympanic membrane)

Stapes in oval window

Auditory (Eustachian) tube

Key:

External ear

Middle ear

Internal ear

Frontal section through the right side of the skull

Where are the receptors for hearing and equilibrium located?

PERFORATED EARDRUM

A **perforated eardrum** is a hole in the tympanic membrane. The condition is characterized by acute pain initially, ringing or roaring in the affected ear, hearing impairment, and sometimes dizziness. Causes of perforated eardrum include shock waves of compressed air (explosions), scuba diving, trauma (skull fracture or from objects such as ear swabs), or acute middle ear infections. ■

Middle Ear

The **middle ear** (**tympanic cavity**) is a small, air-filled cavity in the temporal bone that is lined by epithelium (Fig. 16.17). It is separated from the external ear by the eardrum and from the internal ear by a thin bony partition that contains two small membrane-covered openings: the oval window and the round window.

The posterior wall of the middle ear communicates with the mastoid air "cells" of the temporal bone through a chamber called the **tympanic antrum** (see also Fig. 16.16). This anatomical feature explains why a middle ear infection may spread to the temporal bone, causing mastoiditis, or even to the brain.

The anterior wall of the middle ear contains an opening that leads directly into the **auditory** (**Eustachian**) **tube**. The auditory tube consists of both bone and hyaline cartilage and connects the middle ear with the nasopharynx (upper portion of the throat). It is normally closed at its medial (pharyngeal) end; during swallowing and yawning, it opens. Then atmospheric air from the throat enters or leaves the middle ear until the internal pressure equals the external pressure. When the pressures are balanced, the eardrum vibrates freely as sound waves strike it. If the pressure is not equalized, intense pain, hearing impairment, ringing in the ears, and vertigo could develop. Sudden pressure changes against the eardrum may be equalized by yawning, swallowing or pinching the nose closed, closing the mouth, and gently forcing air from the lungs into the nasopharynx. The auditory tube also is a route whereby pathogens may travel from the nose and throat to the middle ear.

Extending across the middle ear and attached to it by ligaments are three tiny bones called **auditory ossicles** (OS-si-kuls). The bones, named for their shapes, are the malleus, incus, and stapes, commonly called the hammer, anvil, and stirrup, respectively. They are connected by synovial joints. The "handle" of the **malleus** is attached to the internal surface of the eardrum. Its head articulates with the body of the incus. The **incus** is the intermediate bone in the series and articulates

---

Figure 16.17   Auditory ossicles in the middle ear of the right ear.

🔑 *Common names for the malleus, incus, and stapes are the hammer, anvil, and stirrup, respectively.*

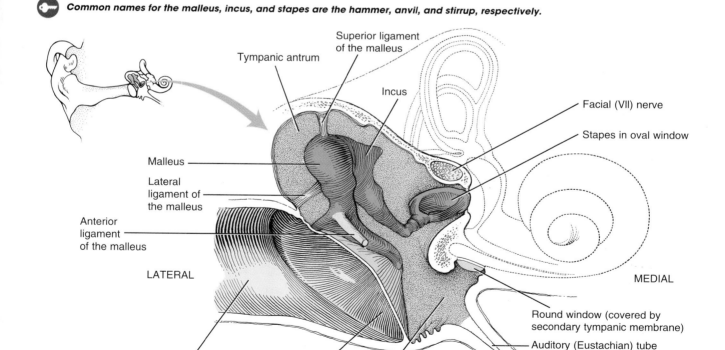

(a) Frontal section showing location of auditory ossicles   *Figure continues*

Figure 16.17   (continued)

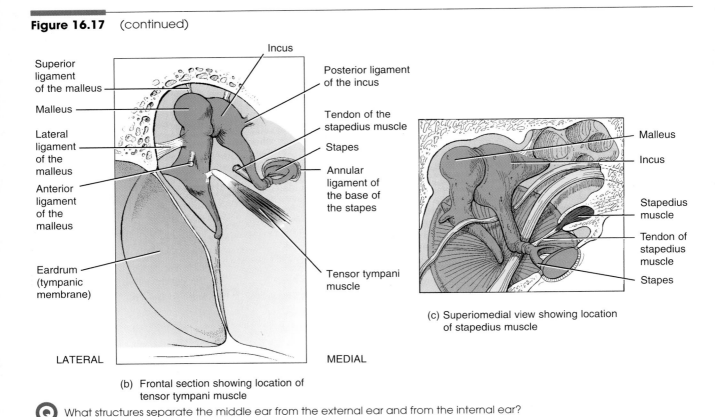

Superior ligament of the malleus

Malleus

Lateral ligament of the malleus

Anterior ligament of the malleus

Eardrum (tympanic membrane)

Incus

Posterior ligament of the incus

Tendon of the stapedius muscle

Stapes

Annular ligament of the base of the stapes

Tensor tympani muscle

LATERAL   MEDIAL

(b) Frontal section showing location of tensor tympani muscle

Malleus

Incus

Stapedius muscle

Tendon of stapedius muscle

Stapes

(c) Superiomedial view showing location of stapedius muscle

Q What structures separate the middle ear from the external ear and from the internal ear?

with the head of the stapes. The base or footplate of the **stapes** fits into a membrane-covered opening in the thin bony partition between the middle and inner ear. The opening is called the **oval window**. Directly below the oval window is another opening, the **round window**. This opening is enclosed by a membrane called the **secondary tympanic membrane**.

Besides the ligaments, two skeletal muscles also attach to the ossicles. The **tensor tympani muscle** limits movement and increases tension on the eardrum to prevent damage to the inner ear from loud noises. It only protects the inner ear from prolonged loud noises, not brief ones, such as a gunshot. The **stapedius muscle** is the smallest of all skeletal muscles and it also has a protective function in that it dampens (checks) large vibrations that result from loud noises. It is for this reason that paralysis of the stapedius muscle is associated with **hyperacusia** (abnormally sensitive hearing).

Internal (Inner) Ear

The **internal (inner) ear** is also called the **labyrinth** (LAB-i-rinth) because of its complicated series of canals (Fig. 16.18). Structurally, it consists of two main divisions: an outer bony labyrinth that encloses an inner membranous labyrinth. The **bony labyrinth** is a series of cavities in the petrous portion of the temporal bone. It can be divided into three areas: (1) the semicircular canals and (2) vestibule, both of which contain receptors for equilibrium, and (3) the cochlea, which contains receptors for hearing. The bony labyrinth is lined with periosteum and contains a fluid called

perilymph. This fluid, which is chemically similar to cerebrospinal fluid, surrounds the **membranous labyrinth**, a series of sacs and tubes inside the bony labyrinth and having the same general form. The membranous labyrinth is lined with epithelium and contains a fluid called **endolymph**, which is chemically similar to intracellular fluid.

The **vestibule** is the oval central portion of the bony labyrinth. The membranous labyrinth in the vestibule consists of two sacs called the **utricle** (YOO-tri-kul; = little bag) and **saccule** (SAK-yool; = little sac). These structures are connected to each other by a small duct. Projecting superiorly and posteriorly from the vestibule are the three bony **semicircular canals**. Each lies at approximately right angles to the other two. Based on their positions, they are called the anterior, posterior, and lateral canals. The anterior and posterior semicircular canals are oriented vertically; the lateral one is oriented horizontally. One end of each canal is a swollen enlargement called the **ampulla** (am-POOL-la; = little jar). The portions of the membranous labyrinth that lie inside the bony semicircular canals are called the **semicircular ducts (membranous semicircular canals)**. These structures communicate with the utricle of the vestibule.

Anterior to the vestibule is the **cochlea** (KŌK-lē-a; = snail's shell). This bony spiral canal (Fig. 16.19) resembles a snail's shell and makes almost three turns around a central bony core called the **modiolus** (Fig. 16.19b). Sections through the cochlea (Fig. 16.19a–c) show that it is divided into three channels. Together, the partitions that separate the channels have a shape like the letter Y. The stem of the Y is

Figure 16.18   The internal ear of the right ear. The outer, blue area is part of the bony labyrinth. The inner, pink-colored area belongs to the membranous labyrinth.

 The bony labyrinth contains perilymph and the membranous labyrinth contains endolymph.

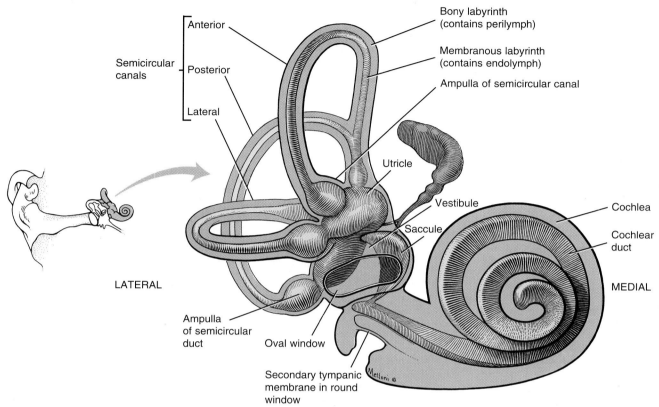

Q   What are the names of the two sacs that lie in the vestibule?

a bony shelf that protrudes into the canal; the wings of the Y are composed mainly of membranous labyrinth. The channel above the bony partition is the **scala vestibuli**, which ends at the oval window; the channel below is the **scala tympani**, which ends at the round window.

The scala vestibuli and scala tympani both contain perilymph and are completely separated except for an opening at the apex of the cochlea called the **helicotrema** (see Fig. 16.20). The cochlea adjoins the wall of the vestibule, into which the scala vestibuli opens. The perilymph of the vestibule is continuous with that of the scala vestibuli. The third channel (between the wings of the Y) is the **cochlear duct (scala media)**. The **vestibular membrane** separates the cochlear duct from the scala vestibuli, and the **basilar membrane** separates the cochlear duct from the scala tympani.

Resting on the basilar membrane is the **spiral organ (organ of Corti)**, the organ of hearing (Fig. 16.19c,d). The spiral organ is a coiled sheet of epithelial cells, including supporting cells and about 16,000 **hair cells**, which are the receptors for auditory sensations. There are two groups of hair cells. The *inner hair cells* are arranged in a single row and extend the entire length of the cochlea. The *outer hair cells* are arranged in several rows. The hair cells have long

hairlike microvilli at their apical ends that extend into the endolymph of the cochlear duct. The basal ends of the hair cells synapse with first-order sensory neurons and motor neurons from the cochlear branch of the vestibulocochlear (VIII) nerve. Projecting over and in contact with the hair cells of the spiral organ is the **tectorial** (*tectum* = cover) **membrane**, a delicate and flexible gelatinous membrane.

Sound Waves

Sound waves result from the alternate compression and decompression of air molecules. They originate from a vibrating object, much the same way that ripples travel over the surface of water when you toss a stone into it. The sounds heard most acutely by human ears are those from sources that vibrate at frequencies between 500 and 5,000 hertz (Hz; 1 Hz = 1 cycle per second). The entire audible range extends from 20 to 20,000 Hz. Sounds of speech contain frequencies mainly between 100 and 3000 Hz, and the "high C" sung by a coloratura soprano has a dominant frequency at 1048 Hz. The sounds from a jet plane several miles away range from 20 to 100 Hz.

The *frequency* of a sound vibration is its *pitch*. The greater the frequency of vibration, the higher the pitch. Also, the greater the *intensity* (size) of the vibration, the *louder* the sound. Sound intensity is measured in units called **decibels (dB)**. Each 10 dB interval represents a tenfold increase in sound intensity. The hearing threshold—that is, the point at which an average young adult can just detect sound from silence—is defined as 0 dB at 1000 Hz. Rustling leaves have a decibel rating of 15, whispered speech 30, normal conversation 60, a vacuum cleaner 75, shouting 80, and a nearby motorcycle or jackhammer 90. Sound becomes uncomfortable to a normal ear at about 120 dB and painful above 140 dB. Because hearing loss results from prolonged noise exposure, employers in the United States must require workers to use hearing protectors when occupational noise levels exceed 90 dB. Rock concerts and even inexpensive headphones can easily produce sounds over 110 dB.

HAIR CELLS, DEAFNESS, AND OTOACOUSTIC EMISSIONS

Exposure to certain antibiotics, such as gentamicin, some anticancer drugs, or high-intensity sounds, such as loud music or the engine roar of jet planes, revved-up motorcycles, lawn mowers, and vacuum cleaners, damages hair cells of the cochlea. Continued exposure to high-intensity sounds causes permanent hearing loss. The louder the sounds, the quicker the loss. Usually, deafness begins with loss of sensitivity for high-pitched sounds. If you are listening to music through headphones and bystanders can hear it, the dB level is in the damaging range. Most people fail to notice their progressive hearing loss until destruction is extensive and they start to have difficulty understanding speech. Wearing earplugs with a noise-reduction rating of 30 dB while engaging in noisy activities can protect the sensitivity of your ears.

Besides its role in detecting sounds, the cochlea has the surprising ability to produce sounds. These inaudible sounds, called **otoacoustic emissions**, were discovered in 1978 and are now thought to come from self-induced vibrations of the outer hair cells. Although outer hair cells outnumber them by 3 to 1, it is mainly the inner hair cells that relay sensory information to the brain. The vibrations of the outer hair cells seem to amplify the responsiveness of the inner hair cells. Clinically, detection of otoacoustic emissions is a fast, inexpensive, noninvasive way to screen newborns for hearing impairment. Since nearly 1 of every 1,000 babies born in the United States is deaf, universal screening of infants using this technique is now recommended by the National Institute on Deafness and Other Communication Disorders. ■

Figure 16.19 Views of the semicircular canals, vestibule, and cochlea of the right ear. Note the cochlea almost makes three complete turns.

 The three channels in the cochlea are the (1) scala vestibuli, (2) scala tympani, and (3) cochlear duct.

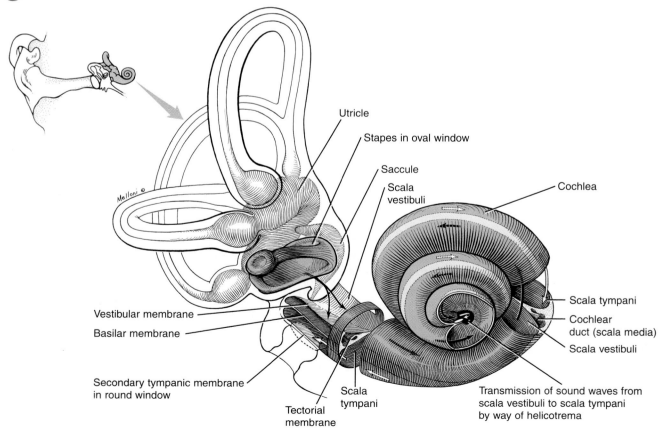

(a) Sections through the cochlea

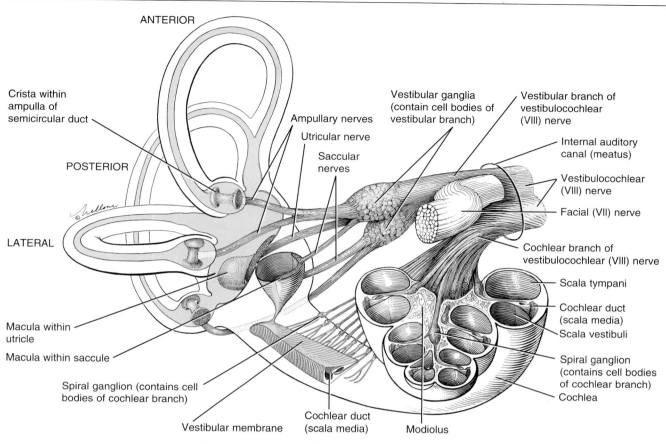

Crista within
ampulla of
semicircular duct

ANTERIOR

POSTERIOR

LATERAL

Macula within
utricle

Macula within saccule

Spiral ganglion (contains cell
bodies of cochlear branch)

Vestibular membrane

Ampullary nerves

Utricular nerve

Saccular
nerves

Vestibular ganglia
(contain cell bodies of
vestibular branch)

Vestibular branch of
vestibulocochlear
(VIII) nerve

Internal auditory
canal (meatus)

Vestibulocochlear
(VIII) nerve

Facial (VII) nerve

Cochlear branch of
vestibulocochlear (VIII) nerve

Scala tympani

Cochlear duct
(scala media)

Scala vestibuli

Spiral ganglion
(contains cell bodies
of cochlear branch)

Cochlea

Cochlear duct
(scala media)

Modiolus

(b) Components of the vestibulocochlear (VIII) nerve

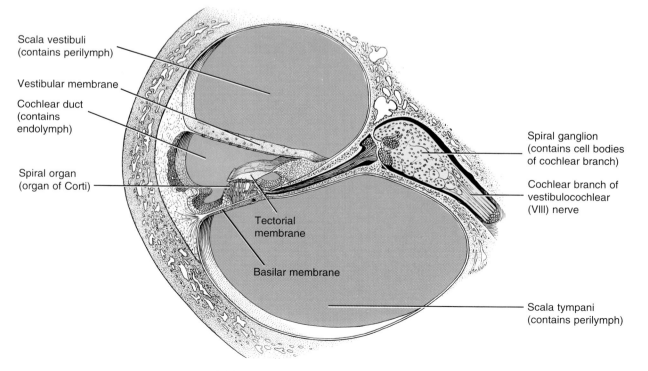

Scala vestibuli
(contains perilymph)

Vestibular membrane

Cochlear duct
(contains
endolymph)

Spiral organ
(organ of Corti)

Tectorial
membrane

Basilar membrane

Spiral ganglion
(contains cell bodies
of cochlear branch)

Cochlear branch of
vestibulocochlear
(VIII) nerve

Scala tympani
(contains perilymph)

(c) Section through one turn of the cochlea

Figure continues

Figure 16.19   (continued)

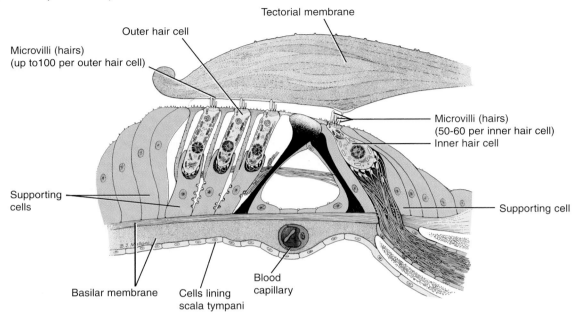

(d) Enlargement of the spiral organ (organ of Corti)

Q What are the three subdivisions of the bony labyrinth?

Physiology of Hearing

The events involved in hearing are as follows (Fig. 16.20):

1 The auricle directs sound waves into the external auditory canal.

2 When sound waves strike the eardrum, the alternate compression and decompression of the air cause the eardrum to vibrate back and forth. The distance it moves is always very small and depends on the intensity and frequency of the sound waves. The eardrum vibrates slowly in response to low-frequency (low-pitch) sounds and rapidly in response to high-frequency (high-pitch) sounds.

3 The central area of the eardrum connects to the malleus, which also starts to vibrate. The vibration is transmitted from the malleus to the incus and then to the stapes.

4 As the stapes moves back and forth, it pushes the membrane of the oval window in and out. The oval window vibrates about 20 times more vigorously than the eardrum because the ossicles efficiently transmit small vibrations spread over a large surface area (eardrum) into larger vibrations of a smaller surface (oval window).

5 The movement of the oval window sets up fluid pressure waves in the perilymph of the cochlea.

6 As the oval window bulges inward, it pushes on the perilymph of the scala vestibuli. Pressure waves are transmitted from the scala vestibuli to the scala tympani and eventually to the round window, causing it to bulge outward into the middle ear. (See number **9** in the illustration.)

7 As the pressure waves deform walls of the scala vestibuli and scala tympani, they also push the vestibular membrane back and forth. As a result, the pressure of the endolymph inside the cochlear duct increases and decreases.

8 The pressure fluctuations of the endolymph move the basilar membrane slightly. When the basilar membrane vibrates, the hair cells of the spiral organ move against the tectorial membrane. The bending of the microvilli produces receptor potentials that ultimately lead to the generation of nerve impulses in cochlear nerve fibers.

Sound waves of various frequencies cause specific regions of the basilar membrane to vibrate more intensely than others. The membrane is narrower but stiffer at the base of the cochlea (portion closer to the oval window); high-frequency (high-pitched) sounds near 20,000 Hz induce maximal vibrations in this region. Toward the apex of the cochlea, the basilar membrane is wider but more flexible; low-frequency (low-pitched) sounds near 20 Hz cause maximal vibration of the basilar membrane there. Loudness is determined by the intensity of sound waves. High-intensity sound waves cause greater vibration of the basilar membrane, which leads to a higher frequency of nerve impulses reaching the brain. More hair cells may also be stimulated by louder sounds.

Hair cells convert a mechanical force (stimulus) into an electrical signal (receptor potential). As the basilar membrane vibrates, the microvilli at the hair cell tip bend back and forth. Mechanical bending in one direction opens ion channels in the microvilli. These mechanically gated channels allow cations in the endolymph, primarily potassium

Figure 16.20 Events in the stimulation of auditory receptors in the right ear. The numbers correspond to the events listed in the text. The cochlea has been uncoiled to more easily visualize transmission of sound waves and their distortion of the vestibular and basilar membranes of the cochlear duct.

 The function of hair cells of the spiral organ (organ of Corti) is to convert a mechanical force (stimulus) into an electrical signal (receptor potential).

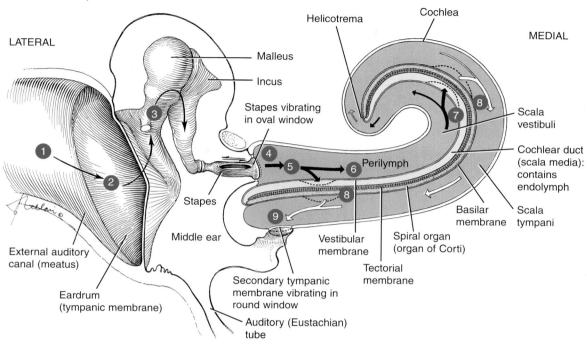

Q Which portion of the basilar membrane vibrates most vigorously in response to high-frequency (high-pitch) sounds?

ions (K^+) and calcium ions (Ca^{2+}), to enter the hair cell cytosol. The result is a depolarizing receptor potential. Depolarization spreads along the plasma membrane and causes voltage-gated Ca^{2+} channels in the base of the hair cell to open. The resulting inflow of Ca^{2+} triggers exocytosis of synaptic vesicles containing neurotransmitter. As more neurotransmitter is released, the rate of nerve impulse firing in the sensory nerve fibers that synapse with the base of the hair cell increases. Bending of microvilli in the opposite direction closes the cation channels, allows repolarization to occur, and reduces neurotransmitter release from the hair cells. This decreases the rate of nerve impulses in sensory nerve fibers. The neurotransmitter is thought to be either glutamate or gamma aminobutyric acid (GABA).

The cochlear branch of the vestibulocochlear (VIII) nerve conducts auditory impulses to the cochlear nuclei in the medulla oblongata. Here, most fibers cross to the opposite side, extend through the midbrain, and terminate in the thalamus. From the thalamus, auditory signals project to the primary auditory area in the temporal lobe of the cerebral cortex (areas 41 and 42 in Fig. 14.15).

COCHLEAR IMPLANTS

Cochlear implants (artificial ears) are devices that translate sounds into electronic signals that can be interpreted by the brain. They take the place of hair cells of the spiral organ (organ of Corti), which normally convert sound waves into nerve impulses. The implants are used for people with deafness due to disease or injury that has destroyed hair cells of the spiral organ.

Sound waves enter a tiny microphone in the ear and travel to a microprocessor pack where they are converted into electrical signals. The signals then travel to electrodes implanted in the cochlea, where they trigger nerve impulses in fibers of the cochlear branch of the vestibulocochlear nerve. These artificially induced nerve impulses then conduct over their normal pathways to the brain. Sounds perceived are crude compared to normal hearing, but they provide a sense of rhythm and loudness as well as information about noises such as telephones, automobiles, and the pitch and cadence of speech. ■

Physiology of Equilibrium

There are two kinds of **equilibrium** (balance). One, called **static equilibrium**, refers to the maintenance of the position of the body (mainly the head) relative to the force of gravity. The second kind, **dynamic equilibrium**, is the maintenance of body position (mainly the head) in response to sudden movements such as rotation, acceleration, and deceleration. Collectively, the receptor organs for equilibrium are called the **vestibular apparatus**, which includes the saccule, utricle, and semicircular ducts.

Otolithic Organs: Saccule and Utricle

The walls of both the utricle and saccule contain a small, thickened region called a **macula** (plural is **maculae**; Fig. 16.21). The maculae are the receptors for static equilibrium and also contribute to some aspects of dynamic equilibrium. For static equilibrium, they provide sensory information on the position of the head in space and are essential for maintaining appropriate posture and balance. For dynamic equi-

---

Figure 16.21   Location and structure of receptors in the maculae of the right ear.

🔑 *The movement of stereocilia initiates depolarizing receptor potentials.*

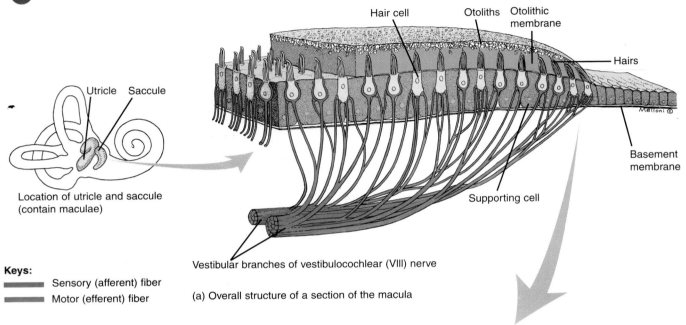

Keys:
▬▬ Sensory (afferent) fiber
▬▬ Motor (efferent) fiber

(a) Overall structure of a section of the macula

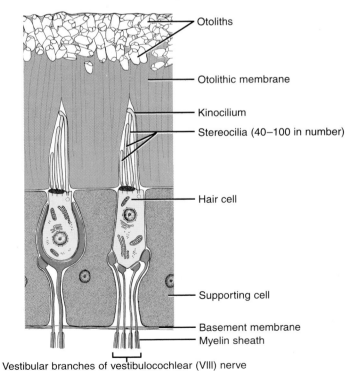

(b) Details of two hair cells

Q With which type of equilibrium are the maculae mainly concerned?

librium, they detect linear acceleration and deceleration, for example, the sensations you feel while in an elevator or a car that is speeding up or slowing down.

The two maculae are perpendicular to one another. They consist of two kinds of cells: **hair (receptor) cells** and **supporting cells**. Hair cells have long extensions of the cell membrane consisting of 70 or more **stereocilia** (they are actually microvilli) and one **kinocilium** (a conventional cilium) anchored firmly to its basal body and extending beyond the longest microvilli. Scattered among the hair cells are columnar supporting cells. They probably secrete the thick, gelatinous, glycoprotein layer, called the **otolithic membrane**, that rests on the hair cells. A layer of heavy calcium carbonate crystals, called **otoliths** (*oto* = ear; *lithos* = stone) or **otoconia**, extends over the entire surface of the otolithic membrane.

Since the heavy otolithic membrane sits on top of the macula, if you tilt your head forward, the otolithic membrane along with the otoliths is pulled by gravity, slides downhill over the hair cells in the direction of the tilt, and stimulates the hair cells. Similarly, if you are sitting upright in a car that suddenly jerks forward, the otolithic membrane, due to its inertia, slides backward and stimulates the hair cells. As the otoliths move, they pull on the gelatinous layer, which pulls on the stereocilia and makes them bend. The movement of the stereocilia initiates depolarizing receptor potentials.

As the hair cells depolarize or repolarize, they release neurotransmitter at a faster or slower rate. The hair cells synapse with first-order sensory neurons in the vestibular branch of the vestibulocochlear (VIII) nerve (Fig. 16.21b). These neurons fire impulses at a slow or rapid pace depending on how much neurotransmitter is present. Motor fibers also synapse with the hair cells and vestibular neurons. Evidently, they regulate the sensitivity of the hair cells and sensory neurons.

Semicircular Ducts

The three semicircular ducts, together with the saccule and utricle, maintain dynamic equilibrium (Fig. 16.22). The ducts lie at right angles to one another in three planes: the two vertical ones are the anterior and posterior semicircular

---

Figure 16.22 Semicircular ducts of the right ear. The ampullary nerves are branches of the vestibular division of the vestibulocochlear (VIII) nerve.

🔑 *The positions of the semicircular ducts permit detection of rotational movements.*

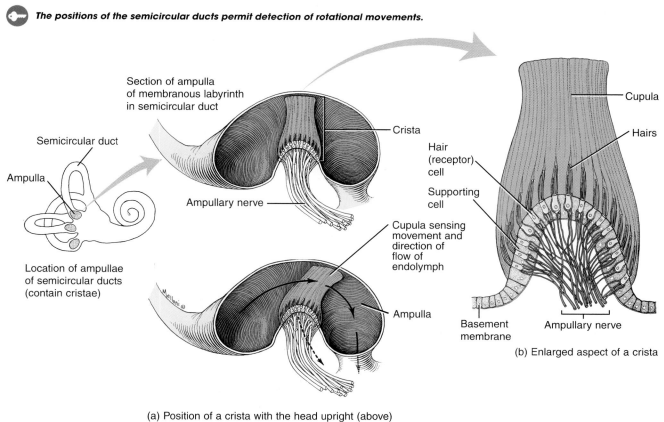

Semicircular duct

Ampulla

Location of ampullae of semicircular ducts (contain cristae)

Section of ampulla of membranous labyrinth in semicircular duct

Crista

Ampullary nerve

Cupula sensing movement and direction of flow of endolymph

Ampulla

Cupula

Hairs

Hair (receptor) cell

Supporting cell

Basement membrane

Ampullary nerve

(b) Enlarged aspect of a crista

(a) Position of a crista with the head upright (above) and when the head moves (below)

❓ With which type of equilibrium are the semicircular ducts (and utricle and saccule) associated?

ducts, and the horizontal one is the lateral semicircular duct. This positioning permits detection of rotational acceleration or deceleration (see Fig. 16.18). In the ampulla, the dilated portion of each duct, there is a small elevation called the **crista**. Each crista contains a group of **hair (receptor) cells** and **supporting cells** covered by a mass of gelatinous material called the **cupula**. When the head moves, the endolymph in the semicircular ducts flows over the hairs and bends them. The movement of the hairs stimulates sensory neurons, and the nerve impulses, evoked by the resulting receptor potentials, pass along the vestibular branch of the vestibulocochlear (VIII) nerve.

Equilibrium Pathways

Most of the vestibular branch fibers of the vestibulocochlear (VIII) nerve enter the brain stem and terminate in the vestibular nuclear complex in the pons. The remaining fibers enter the cerebellum through the inferior cerebellar peduncle (see Fig. 14.5b). Bidirectional pathways connect the vestibular nuclei and cerebellum. Fibers from all the vestibular nuclei extend to the nuclei of cranial nerves that control eye movements—oculomotor (III), trochlear (IV), and abducens (VI)—and to the accessory (XI) nerve nucleus that helps control head and neck movements. In addition, fibers from the lateral vestibular nucleus form the vestibulospinal tract, which conveys impulses to skeletal muscles that regulate muscle tone in response to head movements. Various pathways between the vestibular nuclei, cerebellum, and cerebrum enable the cerebellum to play a key role in maintaining static and dynamic equilibrium. The cerebellum continuously receives updated sensory information from the utricle and saccule. Using this information, the cerebellum monitors and makes corrective adjustments in the motor activities that originate in the cerebral cortex. Essentially, the cerebellum sends continuous nerve impulses to the motor areas of the cerebrum, in response to input from the utricle, saccule, and semicircular ducts. This feedback allows correction of signals from the motor cortex to specific skeletal muscles to maintain equilibrium.

A summary of the structures of the ear related to hearing and equilibrium is presented in Exhibit 16.2.

E X H I B I T **16.2** **SUMMARY OF STRUCTURES OF THE EAR RELATED TO HEARING AND EQUILIBRIUM**

Regions of the Ear and Key Structures	Functions
External (outer) ear 	*Auricle (pinna)*: Collects sound waves. *External auditory canal (meatus)*: Directs sound waves to eardrum. *Eardrum (tympanic membrane)*: Sound waves cause it to vibrate, which, in turn, causes the malleus to vibrate.
Middle ear 	*Auditory ossicles*: Transmit and amplify vibrations from tympanic membrane to oval window. *Auditory (Eustachian) tube*: Equalizes air pressure on both sides of the tympanic membrane.

Regions of the Ear and Key Structures	Functions
Internal (inner) ear	*Cochlea*: Contains a series of fluids, channels, and membranes that transmit vibrations to the spiral organ (organ of Corti), the organ of hearing; hair cells in the spiral organ produce receptor potentials, which elicit nerve impulses in the cochlear branch of the vestibulocochlear (VIII) nerve.
Utricle Saccule	*Semicircular ducts*: Contain cristae, site of hair cells for dynamic equilibrium.
Semicircular ducts Cochlea	*Utricle*: Contains macula, site of hair cells for static and dynamic equilibrium.
	Saccule: Contains macula, site of hair cells for static and dynamic equilibrium.

DISORDERS: HOMEOSTATIC IMBALANCES

Many disorders can alter or damage the organs of special senses. The causes of disorder range from congenital origins to the effects of old age. Here we discuss a few common disorders of the eyes and ears.

GLAUCOMA

Glaucoma is an abnormally high intraocular pressure due to a buildup of aqueous humor inside the anterior chamber. The fluid compresses the lens into the vitreous body and puts pressure on the neurons of the retina. If the pressure continues, there is a progression from mild visual impairment to irreversible destruction of neurons of the retina, degeneration of the optic disc, and blindness. Glaucoma is painless, the other eye compensates to a large extent, and a person may experience considerable retinal damage and visual loss before the condition is diagnosed. Because glaucoma occurs more often with advancing age (affecting 5% of people over age 65), regular measurement of intraocular pressure is an increasingly important part of an eye exam as you age.

SENILE MACULAR DEGENERATION

In **senile macular degeneration (SMD)**, new blood vessels grow over the macula lutea. The effect ranges from distorted vision to blindness. Its cause is unknown, but it occurs mainly in people over age 65.

DEAFNESS

Deafness is significant or total hearing loss. **Sensorineural deafness** is caused by impairment of the cochlea or cochlear branch of the vestibulocochlear (VIII) nerve. **Conduction deafness** is caused by impairment of the external and middle ear mechanisms for transmitting sounds to the cochlea. Among the factors that contribute to deafness are atherosclerosis, which reduces blood supply to the ears; otosclerosis, the deposition of new bone around the oval window; repeated exposure to loud noise, which destroys hair cells of the spiral organ (or-

gan of Corti); certain drugs such as aspirin and streptomycin; impacted cerumen; injury to the eardrum; and aging, which results in thickening of the eardrum, stiffening of the joints of the auditory ossicles, and decreased numbers of hair cells due to diminished cell division.

MÉNIÈRE'S SYNDROME

Ménière's (men′-ē-ĀIRZ) **syndrome** is characterized by an increased amount of endolymph that enlarges the membranous labyrinth. Among the symptoms are fluctuating hearing loss (caused by distortion of the basilar membrane of the cochlea), attacks of vertigo, and roaring tinnitus. Over a period of years, there may be almost total destruction of hearing.

OTITIS MEDIA

Otitis media is an acute infection of the middle ear, caused primarily by bacteria. Symptoms include pain, malaise, fever, and a reddening and outward bulging of the eardrum, which may rupture unless prompt treatment is given (this may involve draining pus from the middle ear). Bacteria from the nasopharynx passing into the auditory (Eustachian) tube are the primary cause of all middle ear infections. Children are more susceptible than adults to middle ear infections because their auditory tubes are shorter, wider, and almost horizontal, which decreases drainage.

MOTION SICKNESS

Motion sickness is nausea and vomiting brought on by repetitive angular, linear, or vertical motion. The cause is excessive stimulation of the vestibular apparatus by motion. Nerve impulses pass from the internal ear to the vomiting center in the medulla. Visual stimuli and emotional factors such as fear or anxiety can also contribute to motion sickness. Susceptible people can take medication (for example, Dramamine) before traveling since prevention is more successful than treatment of symptoms once they have developed.

MEDICAL TERMINOLOGY

Achromatopsia (a-krō′-ma-TOP-sē-a; *a* = without; *chrom* = color) Complete color blindness.

Ametropia (am′-e-TRŌ-pē-a; *ametro* = disproportionate; *ops* = eye) Refractive defect of the eye resulting in an inability to focus images properly on the retina.

Anopsia (an-OP-sē-a; *opsia* = vision) A defect of vision.

Blepharitis (blef-a-RĪ-tis; *blepharo* = eyelid; *itis* = inflammation of) An inflammation of the eyelid.

Conjunctivitis (pinkeye) An inflammation of the conjunctiva, caused by bacteria such as pneumococci, staphylococci, or *Hemophilus influenzae,* that is very contagious and more common in children. Conjunctivitis may also be caused by irritants, such as dust, smoke, or pollutants in the air, in which case it is not contagious.

Eustachitis (yoo′-stā-KĪ-tis) An inflammation or infection of the auditory (Eustachian) tube.

Exotropia (ek′-sō-TRŌ-pē-a; *ex* = out; *tropia* = turning) Turning outward of the eyes.

Keratitis (ker′-a-TĪ-tis; *kerato* = cornea) An inflammation or infection of the cornea.

Labyrinthitis (lab′-i-rin-THĪ-tis) An inflammation of the labyrinth (inner ear).

Mydriasis (mi-DRĪ-a-sis) Dilated pupil.

Myringitis (mir′-in-JĪ-tis; *myringa* = eardrum) An inflammation of the eardrum; also called **tympanitis**.

Nystagmus (nis-TAG-mus; *nystazein* = to nod) A rapid involuntary movement of the eyeballs, possibly caused by a disease of the central nervous system. It is associated with conditions that cause vertigo.

Otalgia (ō-TAL-jē-a; *oto* = ear; *algia* = pain) Earache.

Otosclerosis (ō′-tō-skle-RŌ-sis; *oto* = ear; *sclerosis* = hardening) Pathological process that may be hereditary in which new bone is deposited around the oval window. The result may be immobilization of the stapes, leading to conduction deafness.

Photophobia (fō′-tō-FŌ-bē-a; *photo* = light; *phobia* = fear) Abnormal visual intolerance to light.

Ptosis (TŌ-sis; *ptosis* = fall) Falling or drooping of the eyelid. (This term is also used for the slipping of any organ below its normal position.)

Retinoblastoma (ret′-i-nō-blas-TŌ-ma; *blast* = bud; *oma* = tumor) A tumor arising from immature retinal cells and accounting for 2% of childhood malignancies.

Scotoma (skō-TŌ-ma; *scotoma* = darkness) An area of reduced or lost vision in the visual field. Also called a **blind spot** (other than the normal blind spot at the optic disc).

Strabismus (stra-BIZ-mus) An imbalance in the extrinsic eye muscles that produces a squint (formerly called "cross-eyes"). **Amblyopia** (am′-blē-Ō-pē-a) is the term used to describe the loss of vision in an otherwise normal eye that, because of muscle imbalance, cannot focus in synchrony with the other eye.

Tinnitus (ti-NĪ-tus) A ringing, roaring, or clicking in the ears.

Trachoma (tra-KŌ-ma) A serious form of conjunctivitis, the greatest single cause of blindness in the world. It is caused by a bacterium called *Chlamydia trachomatis*. The disease produces an excessive growth of subconjunctival tissue and invasion of blood vessels into the cornea, which progresses until the entire cornea is opaque, causing blindness.

Vertigo (VER-ti-gō; *vertex* = whorl) A sensation of spinning or movement in which the world is revolving or the person is revolving in space.

STUDY OUTLINE

OLFACTORY SENSATIONS: SMELL (p. 454)

1. The receptors for olfaction, which are bipolar neurons, are in the nasal epithelium.
2. In olfactory reception, a generator potential develops and triggers one or more nerve impulses.
3. Adaptation to odors occurs quickly, and the threshold of smell is low.
4. Axons of olfactory receptors form the olfactory (I) nerves, which convey nerve impulses to the olfactory bulbs, olfactory tracts, limbic system, and cerebral cortex (temporal and frontal lobes).

GUSTATORY SENSATIONS: TASTE (p. 455)

1. The receptors for gustation, the gustatory receptor cells, are located in taste buds.
2. Substances to be tasted must be in solution in saliva.
3. Receptor potentials developed in gustatory receptor cells cause the release of neurotransmitter, which gives rise to nerve impulses.
4. Adaptation to taste occurs quickly, and the threshold varies with the taste involved.

5. Gustatory receptor cells trigger nerve impulses in cranial nerves VII, IX, and X. Taste signals then pass to the medulla oblongata, thalamus, and cerebral cortex (parietal lobe).

VISUAL SENSATIONS (p. 457)

Accessory Structures of the Eye (p. 457)

1. Accessory structures of the eyes include the eyebrows, eyelids, eyelashes, and the lacrimal apparatus.
2. The lacrimal apparatus consists of structures that produce and drain tears.

Anatomy of the Eyeball (p. 459)

1. The eye is constructed of three coats: (a) fibrous tunic (sclera and cornea), (b) vascular tunic (choroid, ciliary body, and iris), and (c) retina (nervous tunic).
2. The retina consists of pigment epithelium and a neural portion (photoreceptor layer, bipolar cell layer, ganglion cell layer, horizontal cells, and amacrine cells).
3. The anterior cavity contains aqueous humor; the vitreous chamber contains the vitreous body.

Image Formation and Convergence (p. 464)

1. Image formation on the retina involves refraction of light rays by the cornea and lens, accommodation of the lens by an increase in its curvature for near vision, and constriction of the pupil to prevent light rays from entering the eye through the periphery of the lens.

2. In convergence, the eyeballs move medially so they are both directed toward an object being viewed.

Physiology of Vision (p. 467)

1. The first step in vision is the absorption of light by photopigments in rods and cones (photoreceptors) and isomerization of *cis*-retinal.

2. Once receptor potentials develop in rods and cones, they decrease the release of inhibitory neurotransmitter, which induces graded potentials in bipolar cells and horizontal cells.

3. Horizontal cells transmit inhibitory signals to bipolar cells; bipolar or amacrine cells transmit excitatory signals to ganglion cells, which depolarize and initiate nerve impulses.

4. Impulses from ganglion cells are conveyed into the optic (II) nerve, through the optic chiasm and optic tract, to the thalamus. From the thalamus visual signals pass to the cerebral cortex (occipital lobe).

AUDITORY SENSATIONS AND EQUILIBRIUM (p. 472)

1. The external or outer ear consists of the auricle, external auditory canal, and eardrum (tympanic membrane).

2. The middle ear consists of the auditory or Eustachian tube, ossicles, oval window, and round window.

3. The internal or inner ear consists of the bony labyrinth and membranous labyrinth. The internal ear contains the spiral organ (organ of Corti), the organ of hearing.

4. Sound waves enter the external auditory canal, strike the eardrum, pass through the ossicles, strike the oval window, set up waves in the perilymph, strike the vestibular membrane and scala tympani, increase pressure in the endolymph, strike the basilar membrane, and stimulate hairs on the spiral organ (organ of Corti).

5. Hair cells convert a mechanical force into a receptor potential.

6. Hair cells release neurotransmitter, which initiates nerve impulses in the first-order sensory neurons.

7. The cochlear branch of the vestibulocochlear (VIII) nerve terminates in the thalamus. From the thalamus auditory signals pass to the temporal lobes of the cerebral cortex.

8. Static equilibrium is the orientation of the body relative to the pull of gravity. The maculae of the utricle and saccule are the sense organs of static equilibrium.

9. Dynamic equilibrium is the maintenance of body position in response to movement. The cristae in the semicircular ducts are the sense organs of dynamic equilibrium.

10. Most vestibular branch fibers of the vestibulocochlear (VIII) nerve enter the brain stem and terminate in the pons; the remaining fibers enter the cerebellum.

REVIEW QUESTIONS

1. Describe the structure of olfactory receptors. What is the function of basal cells? (p. 454)

2. Describe adaptation to odors. What type of threshold does olfaction have? (p. 455)

3. Discuss the origin and path of a nerve impulse that results in olfaction. (p. 455)

4. Describe the structure of gustatory receptors. (p. 455)

5. How are gustatory receptors stimulated? (p. 456)

6. Describe adaptation to taste. What type of threshold does taste have? (p. 457)

7. Discuss how a nerve impulse for gustation travels from a taste bud to the brain. (p. 457)

8. Describe the structure and importance of the following accessory structures of the eye: eyelids, eyelashes, and eyebrows. (p. 457)

9. What is the function of the lacrimal apparatus? Explain how it operates. (p. 459)

10. By means of a labeled diagram, indicate the principal anatomical structures of the eye. (p. 460)

11. Describe the location and contents of the chambers of the eye. What is intraocular pressure (IOP)? How is the scleral venous sinus (canal of Schlemm) related to this pressure? (p. 463)

12. Describe the histology of the neural portion of the retina. (p. 461)

13. Explain how each of the following events is related to the physiology of vision: (a) refraction of light, (b) accommodation of the lens, and (c) constriction of the pupil. (p. 464)

14. Distinguish emmetropia, myopia, hypermetropia, and astigmatism by means of a diagram. (p. 465)

15. What is convergence? How does it occur? (p. 466)

16. Describe the structure of rods and cones. (p. 467)

17. Explain how photopigments respond to light and recover in darkness. (p. 467)

18. How do receptor potentials develop in photoreceptors? (p. 468)

19. Explain how the retina processes visual input. (p. 469)

20. Describe the path of a visual impulse from the optic (II) nerve to the brain. (p. 470)

21. Define visual field. Relate the visual field to image formation on the retina. (p. 470)

22. Diagram the principal parts of the external, middle, and internal ear. Describe the function of each part labeled. (p. 472)

23. What are sound waves? In what units are sound intensities measured? (p. 475)

24. Explain the events involved in the transmission of sound from the auricle to the spiral organ (organ of Corti). (p. 478)

25. What is the sensory pathway for sound impulses from the cochlear branch of the vestibulocochlear (VIII) nerve to the brain? (p. 479)

26. Compare the function of the maculae in the saccule and utricle in maintaining static equilibrium with the role of the cristae in the semicircular ducts in maintaining dynamic equilibrium. (p. 480)

27. Describe the path of a nerve impulse that results in static and dynamic equilibrium. (p. 482)

28. Define the following: corneal transplant (p. 459), detached retina (p. 461), cataract (p. 463), color blindness (p. 468), perforated eardrum (p. 473), otoacoustic emissions (p. 476), and cochlear implant (p. 479).

ANSWERS TO QUESTIONS WITH FIGURES

16.1   Olfactory tract.

16.2   Gustatory receptors → cranial nerves VII, IX, or X → medulla oblongata→ (1) limbic system and hypothalamus or (2) thalamus → primary gustatory area in the parietal lobe of the cerebral cortex.

16.3   The conjunctiva.

16.4   Tears or lacrimal fluid is a watery solution containing salts, some mucus, and lysozyme. It cleans, lubricates, and moistens the eyeball.

16.5   The cornea, lens, and retina are avascular. Since these structures have no blood vessels, they are more transparent to light, an advantage since the photoreceptors lie at the posterior aspect of the eye.

16.6   Hypertension and diabetes mellitus; also cataract and senile macular degeneration.

16.7   Two types of photoreceptors are rods and cones. Rods provide black-and-white vision in dim light whereas cones provide high visual acuity and color vision in bright light.

16.8   Aqueous humor is secreted by the ciliary process, then flows into the posterior chamber, around the iris, into the anterior chamber, and out of the eyeball through the scleral venous sinus.

16.9   The ciliary muscle contracts → suspensory ligaments slacken → lens rounds up, becoming a more powerful lens.

16.10   The loss of elasticity in the lens that occurs with aging.

16.11   Both rods and cones transduce light into receptor potentials, using a photopigment embedded in outer segment membrane folds, and release neurotransmitter at synapses with bipolar cells and horizontal cells.

16.12   Isomerization.

16.13   Breakdown of cyclic GMP by hydrolysis; synthesis of cyclic GMP.

16.14   Retinal ganglion cell axons form the optic nerve (cranial nerve II) → optic chiasm → optic tract → lateral geniculate nucleus of the thalamus → optic radiations → occipital lobe of cerebrum.

16.15   Nasal.

16.16   Inner ear: cochlea (hearing) and semicircular ducts (equilibrium).

16.17   The eardrum (tympanic membrane) separates the external ear from the middle ear. The oval and round windows separate the middle ear from the internal ear.

16.18   The utricle and the saccule.

16.19   Semicircular canals, vestibule, and cochlea.

16.20   Near the base of the cochlea, the region close to the oval and round windows.

16.21   Static.

16.22   Dynamic.

STUDENT OBJECTIVES

1. Compare the structural and functional differences of the somatic and autonomic nervous systems. (p. 488)

2. Identify the principal structural features of the autonomic nervous system. (p. 488)

3. Compare the sympathetic and parasympathetic divisions of the autonomic nervous system in terms of anatomy, physiology, and neurotransmitters released. (p. 492)

4. Describe the various neurotransmitters and receptors involved in autonomic responses. (p. 494)

5. Describe the components of an autonomic reflex. (p. 498)

6. Explain the relationship of the hypothalamus to the autonomic nervous system. (p. 498)

THE AUTONOMIC NERVOUS SYSTEM

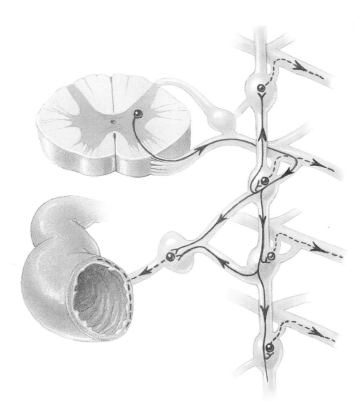

CHAPTER CONTENTS AT A GLANCE

he **autonomic nervous system** (ANS) regulates the activity of smooth muscle, cardiac muscle, and certain glands. Traditionally, the ANS has been described as a specific *motor output* portion of the peripheral nervous system. Operation of the ANS to maintain homeostasis, however, depends on a continual flow of *sensory input* from visceral organs and blood vessels into the CNS. Thus it is reasonable to include these sensory neurons as part of the ANS (see Fig. 12.1). Structurally, then, the ANS includes two main components: general visceral sensory (afferent) neurons and general visceral motor (efferent) neurons.

Functionally, the ANS usually operates without conscious control. The system was originally named *autonomic* because it was thought to function autonomously or in a self-governing manner, without control by the central nervous system (CNS). The ANS, however, is regulated by centers in the brain, mainly the hypothalamus and medulla oblongata, which receive input from the limbic system and other regions of the cerebrum.

COMPARISON OF SOMATIC AND AUTONOMIC NERVOUS SYSTEMS

The somatic nervous system includes both sensory and motor neurons. The sensory neurons convey input from receptors for the special senses (vision, hearing, taste, smell, and equilibrium), proprioceptors (muscle and joint position), and general somatic receptors (pain, temperature, and tactile sensations). All these sensations normally are consciously perceived. In turn, somatic motor neurons innervate skeletal muscle, the effector tissue of the somatic nervous system, and produce conscious, voluntary movements. In the somatic nervous system, the effect of a motor neuron always is *excitation*. When a somatic motor neuron stimulates a skeletal muscle, the muscle contracts. When the neuron ceases to stimulate the muscle, contraction stops.

The input component of the ANS consists of **general visceral sensory (afferent) neurons**. Mostly, these are associated with interoceptors, such as chemoreceptors that monitor carbon dioxide level in the blood and mechanoreceptors that detect stretching of organs or blood vessels. These motor signals are not consciously perceived most of the time, although intense activation of interoceptors may give rise to conscious sensations. Examples of this are the sensations of pain or nausea from damaged viscera, fullness of the urinary bladder, and angina pectoris (chest pain) from inadequate blood flow to the heart.

General visceral motor (efferent) neurons regulate visceral activities by either *exciting* or *inhibiting* their effector tissues, which are cardiac muscle, smooth muscle, and glands. Responses include changes in the size of the pupil, accommodation for near vision, dilation of blood vessels, adjustment of the rate and force of the heartbeat, move-

ments of the gastrointestinal tract, and secretion by most glands. These activities usually lie beyond conscious control. They are automatic. Input from the general somatic and special senses, acting via the limbic system, also may modify responses of autonomic motor neurons. Seeing a bike about to hit you or hearing squealing brakes of a nearby car, or being grabbed by an attacker, for example, would increase the rate and force of your heartbeat.

Autonomic motor pathways consist of sets of two motor neurons in series (one following the other). The first has its cell body in the CNS; its axon extends from the CNS to an **autonomic ganglion**. (Recall that a ganglion is a collection of neuronal cell bodies outside the CNS.) The cell body of the second neuron is in that autonomic ganglion and its axon extends directly from the ganglion to the effector (smooth muscle, cardiac muscle, or gland). In contrast, a single somatic motor neuron extends from the CNS to the effector (skeletal muscle). Also, whereas somatic motor neurons release the neurotransmitter acetylcholine (ACh) at their synapses with skeletal muscle fibers, autonomic motor neurons release either ACh or norepinephrine (NE) at their synapses with cardiac muscle fibers, smooth muscle fibers, or gland cells. Exhibit 17.1 presents a summary of these similarities and differences between the somatic and autonomic nervous systems.

The output (motor) part of the ANS has two principal divisions: **sympathetic** and **parasympathetic**. The many organs that receive impulses from both sympathetic and parasympathetic fibers are said to have **dual innervation**. In general, nerve impulses from one division stimulate the organ to start or increase activity (excitation), whereas impulses from the other division decrease the organ's activity (inhibition). For example, whereas the sympathetic division increases heart rate and force of contraction, the parasympathetic division decreases heart rate and force of contraction. The rest of the chapter focuses on the anatomy and physiology of the sympathetic and parasympathetic divisions and the CNS centers that regulate outflow along these two divisions.

ANATOMY OF AUTONOMIC MOTOR PATHWAYS

Overview

The first of the two autonomic motor neurons is called a **preganglionic neuron** (Exhibit 17.1). Its cell body is in the brain or spinal cord. Its myelinated axon, called a **preganglionic fiber**, passes out of the CNS as part of a cranial or spinal nerve. At some point, the fiber separates from the nerve and extends to an autonomic ganglion. There it synapses with the postganglionic neuron, the second neuron in the autonomic motor pathway. The **postganglionic neuron**

EXHIBIT **17.1** **COMPARISON OF SOMATIC AND AUTONOMIC NERVOUS SYSTEMS**

Sensory Input (Afferent Neurons)	CNS Centers (Process Input and Initiate Output)	Motor Output (Efferent Neurons) and Neurotransmitters	Response of Effectors to Neurotransmitters
SOMATIC Special senses. General somatic senses. Proprioceptors.	Voluntary control from cerebral cortex. Other active regions include basal ganglia, cerebellum, brain stem, and spinal cord.	One-neuron pathway: axon of somatic motor neuron extends from CNS (brain stem or spinal cord) to synapse with effector. At the synapse with effector (the neuromuscular junction), the neurotransmitter is acetylcholine (ACh), which is released from the synaptic end bulbs of the axon terminals.	Excitation of skeletal muscle.

ACh: contraction of skeletal muscle

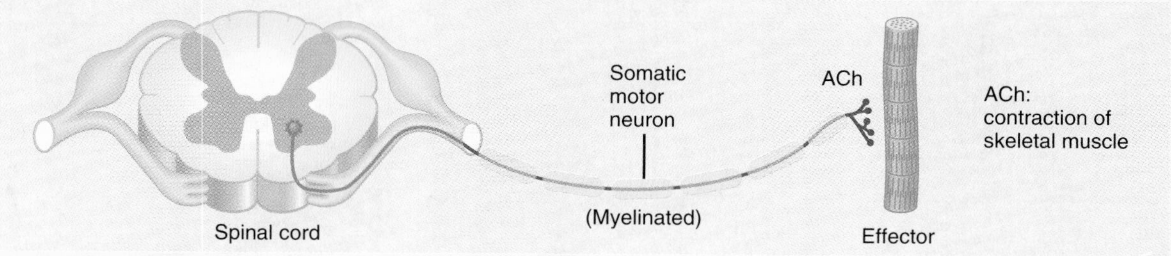

| AUTONOMIC Special senses. General visceral senses (mainly from interoceptors). General somatic senses. | Involuntary control from limbic system, hypothalamus, medulla oblongata, pons, and spinal cord. Cerebral cortex also contributes. | Two-neuron pathway: axon of preganglionic neuron (first autonomic motor neuron) extends from the CNS (brain stem or spinal cord) and synapses with postganglionic neuron (second autonomic motor neuron) in a ganglion. The postganglionic neuron synapses with a visceral effector. Preganglionic axons release acetylcholine (ACh). Postganglionic axons release ACh (parasympathetic division and sympathetic fibers to sweat glands) or norepinephrine (NE; remainder of sympathetic division). | Excitation or inhibition of cardiac muscle, smooth muscle, and glands. |

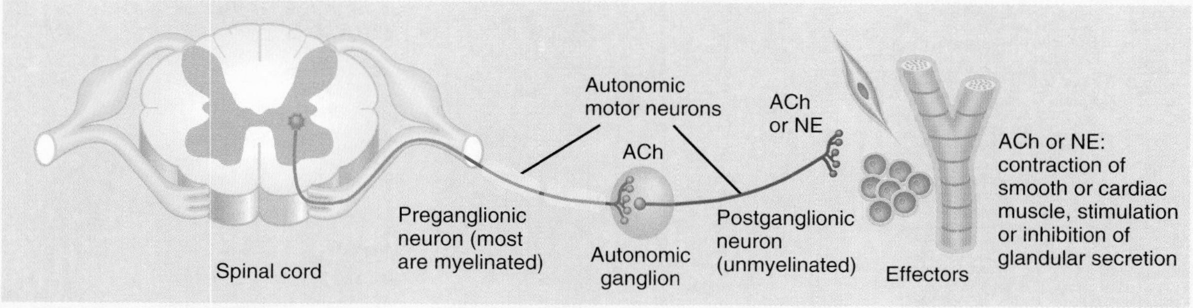

ACh or NE: contraction of smooth or cardiac muscle, stimulation or inhibition of glandular secretion

lies entirely outside the CNS. Its cell body and dendrites are located in an autonomic ganglion, where it makes synapses with one or more preganglionic fibers. The axon of a postganglionic neuron, called a **postganglionic fiber**, is unmyelinated and terminates in a visceral effector. Thus preganglionic neurons convey motor impulses from the CNS to autonomic ganglia, and postganglionic neurons relay the impulses from autonomic ganglia to visceral effectors. Note a major difference between autonomic ganglia and posterior root ganglia: both contain cell bodies but only autonomic ganglia contain synapses.

Preganglionic Neurons

In the sympathetic division, the cell bodies of preganglionic neurons are in the lateral gray horns of the 12 thoracic segments and first two or three lumbar segments of the spinal cord (Fig. 17.1). For this reason, the sympathetic division is also called the **thoracolumbar** (thō′-ra-kō-LUM-bar) **division**, and the fibers of the sympathetic preganglionic neurons are known as the **thoracolumbar outflow**.

The cell bodies of the preganglionic neurons of the parasympathetic division are located in the nuclei of four pairs of cranial nerves—oculomotor (III), facial (VII),

Figure 17.1   Structure of sympathetic and parasympathetic divisions of the autonomic nervous system (ANS).

Sympathetic and parasympathetic stimulation have opposing effects on organs that receive dual innervation.

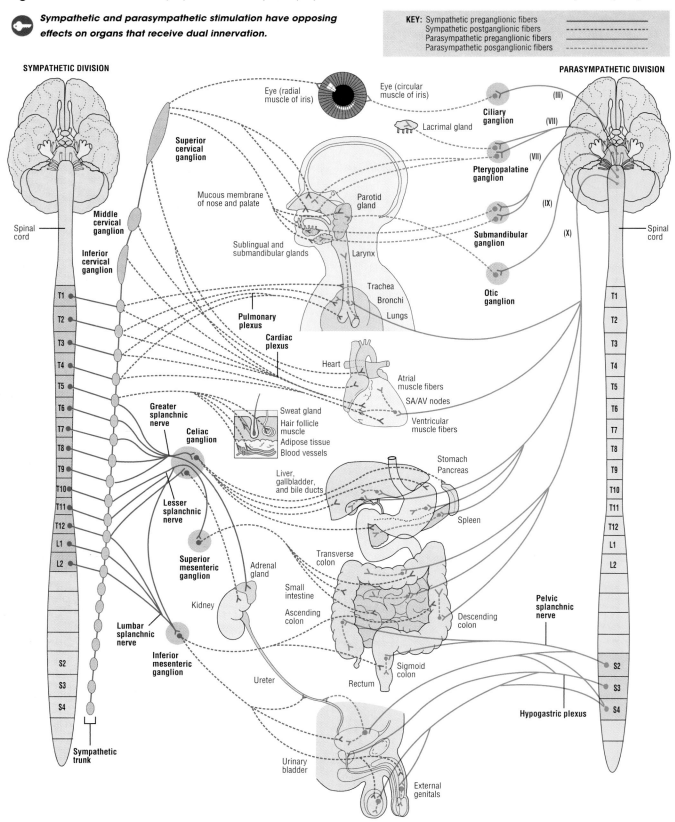

KEY: Sympathetic preganglionic fibers
Sympathetic postganglionic fibers
Parasympathetic preganglionic fibers
Parasympathetic posganglionic fibers

Which division has longer preganglionic fibers? Why?

glossopharyngeal (IX), and vagus (X)—in the brain stem and in the lateral gray horns of the second through fourth sacral segments of the spinal cord. Hence the parasympathetic division is also known as the **craniosacral division**, and the fibers of the parasympathetic preganglionic neurons are referred to as the **craniosacral outflow**.

Autonomic Ganglia

The autonomic ganglia may be divided into three general groups: two of the groups are components of the sympathetic division and the third group is a component of the parasympathetic division.

● **Sympathetic Ganglia** The **sympathetic trunk (vertebral chain) ganglia** are a series of ganglia that lie in a vertical row on either side of the vertebral column, extending from the base of the skull to the coccyx (Figs. 17.1 and 17.2). They are also known as **paravertebral ganglia** and receive preganglionic fibers only from the *sympathetic division*. Because the sympathetic trunk ganglia are so near the spinal cord, sympathetic preganglionic fibers tend to be short.

The second kind of autonomic ganglion also receives preganglionic fibers from the *sympathetic division*. It is called a **prevertebral (collateral) ganglion** (Fig. 17.2). The ganglia

---

Figure 17.2 Ganglia and rami communicantes of the sympathetic division of the ANS.

Sympathetic ganglia lie in two chains on either side of the vertebral column (sympathetic trunk ganglia) and close to large abdominal arteries anterior to the vertebral column (prevertebral ganglia).

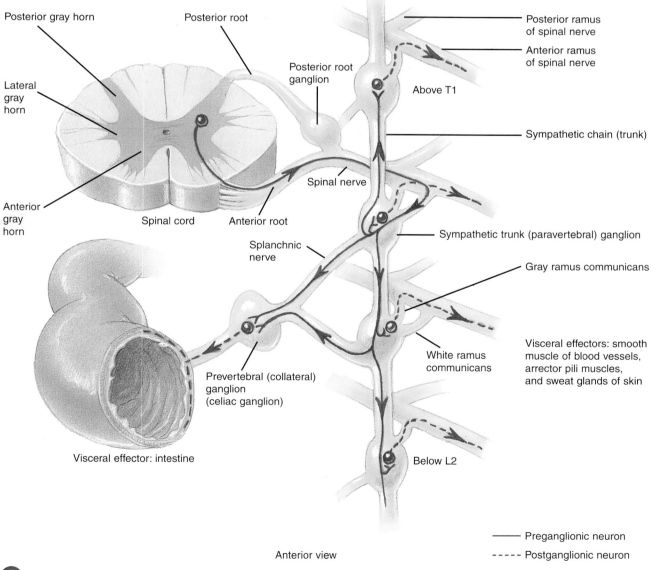

Anterior view

Q What substance gives the white rami their white appearance?

of this group lie anterior to the spinal column and close to the large abdominal arteries. Examples of prevertebral ganglia are the celiac ganglion, on either side of the celiac artery just inferior to the diaphragm; the superior mesenteric ganglion, near the beginning of the superior mesenteric artery in the upper abdomen; and the inferior mesenteric ganglion, located near the beginning of the inferior mesenteric artery in the middle of the abdomen (see Fig. 17.1).

● **Parasympathetic Ganglia** Preganglionic fibers from the *parasympathetic division* make synapses in **terminal (intramural) ganglia**. These ganglia are located at the end of an autonomic motor pathway very close to or actually within the wall of a visceral organ. Since parasympathetic preganglionic fibers extend from the CNS to the terminal ganglion in an innervated organ, they tend to be long.

In addition to autonomic ganglia, the ANS contains **autonomic plexuses**. Slender bundles of preganglionic or postganglionic fibers arranged in a branching network constitute an autonomic plexus.

Postganglionic Neurons

Axons of preganglionic neurons of the sympathetic division pass to ganglia of the sympathetic trunk. Some synapse with postganglionic neurons in the sympathetic trunk ganglia (Fig. 17.2). Others continue, without synapsing, through the sympathetic trunk ganglia to end at a prevertebral ganglion and synapse with the postganglionic neurons there. In either case, a single sympathetic preganglionic fiber may synapse with 20 or more postganglionic fibers. This is an example of divergence (see page 354) and is part of the reason why sympathetic responses tend to be widespread throughout the body. After exiting their ganglia, the postganglionic fibers typically innervate several visceral effectors (see Fig. 17.1).

Axons of preganglionic neurons of the parasympathetic division pass to terminal ganglia near or within a visceral effector (see Fig. 17.1). In the ganglion, the presynaptic neuron usually synapses with only four or five postsynaptic neurons, all of which supply a single visceral effector. Thus parasympathetic effects tend to be localized. With this background in mind, we can now examine some specific structural features of the sympathetic and parasympathetic divisions of the ANS.

Sympathetic Division

Cell bodies of sympathetic preganglionic neurons are part of the lateral gray horns of all thoracic segments and the first two or three lumbar segments of the spinal cord (see Fig. 17.1). The preganglionic axons are myelinated and leave the spinal cord through the anterior root of a spinal nerve along with the somatic motor fibers at the same segmental level. After exiting through the intervertebral foram-

ina, the myelinated preganglionic sympathetic fibers enter a short pathway called a **white ramus** before passing to the nearest sympathetic trunk ganglion on the same side.

Collectively, the white rami are called the **white rami communicantes** (kō-myoo-ni-KAN-tēz). See Fig. 17.2. Their name (white) indicates that they contain myelinated fibers. Only the thoracic and first lumbar nerves have white rami communicantes. The white rami communicantes connect the anterior ramus of the spinal nerve with the ganglia of the sympathetic trunk.

The paired sympathetic trunks lie anterior and lateral to the spinal cord, one on either side. Typically, there are 22 ganglia in each chain: 3 cervical, 11 thoracic, 4 lumbar, and 4 sacral. Although the trunk extends inferiorly from the neck, thorax, and abdomen to the coccyx, it receives preganglionic fibers only from the thoracic and lumbar segments of the spinal cord (see Fig. 17.1).

The cervical portion of each sympathetic trunk is located in the neck anterior to the prevertebral muscles. It is subdivided into superior, middle, and inferior ganglia (see Fig. 17.1). The **superior cervical ganglion** is posterior to the internal carotid artery and anterior to the transverse process of the second cervical vertebra. Postganglionic fibers leaving the ganglion serve the head. They are distributed to sweat glands, smooth muscle of the eye, blood vessels of the face, nasal mucosa, and salivary glands. Gray rami communicantes (described shortly) from the ganglion also pass to the upper 2–4 cervical spinal nerves. The **middle cervical ganglion** lies near the sixth cervical vertebra at the level of the cricoid cartilage, the inferior portion of the larynx (voice box). Postganglionic fibers from it innervate the heart. The **inferior cervical ganglion** is located near the first rib, anterior to the transverse processes of the seventh cervical vertebra. Its postganglionic fibers also supply the heart.

The thoracic portion of each sympathetic trunk usually consists of 11 ganglia, lying anterior to the necks of the ribs. This portion of the sympathetic trunk receives most of the sympathetic preganglionic fibers. Postganglionic fibers from the thoracic sympathetic trunk innervate the heart, lungs, bronchi, and other thoracic viscera. In the skin, they also innervate sweat glands, blood vessels, and arrector pili muscles of hair follicles.

The lumbar portion of each sympathetic trunk lies on either side of the corresponding lumbar vertebrae. The sacral portion of the sympathetic trunk lies in the pelvic cavity on the medial side of the sacral foramina. Unmyelinated postganglionic fibers from the lumbar and sacral sympathetic trunk ganglia enter a short pathway called a **gray ramus** and then either merge with a spinal nerve or join the hypogastric plexus in the pelvis via direct visceral branches. The **gray ramus communicans** (kō-MYOO-ni-kanz; plural is **rami communicantes**) is the structure containing the postganglionic fibers that connect the ganglion of the sympathetic trunk to the spinal nerve (Fig. 17.2). The fibers are unmyelinated. Gray rami communicantes outnumber the white rami, since there is a gray ramus leading to each of the 31 pairs of spinal nerves.

As preganglionic fibers extend from a white ramus communicans into the sympathetic trunk, they give off several axon collaterals (branches). These collateral fibers terminate and synapse in several ways. Some synapse in the first ganglion at the level of entry. Others pass up or down the sympathetic trunk for a variable distance to form the fibers on which the ganglia are strung. These fibers, known as **sympathetic chains** (Fig. 17.2), may not synapse until they reach a ganglion in the cervical or sacral area. Many postganglionic fibers rejoin the spinal nerves through gray rami and supply peripheral visceral effectors such as sweat glands, smooth muscle in walls of blood vessels, and arrector pili muscles of hair follicles.

Some preganglionic fibers pass through the sympathetic trunk without terminating in it. Beyond the trunk, they form nerves known as **splanchnic** (SPLANK-nik) **nerves** (Fig. 17.2), which extend to and terminate in outlying prevertebral ganglia. Splanchnic nerves from the thoracic area terminate in the **celiac** (SĒ-lē-ak) **ganglion** or **solar plexus**. Here, the preganglionic fibers synapse with postganglionic cell bodies. The greater splanchnic nerve passes to the celiac ganglion of the celiac plexus, which is located near the first lumbar vertebra. From here, postganglionic fibers extend to the stomach, spleen, liver, kidney, and small intestine. The lesser splanchnic nerve passes through the celiac plexus to the superior mesenteric ganglion of the superior mesenteric plexus, which is just below the celiac plexus. Postganglionic fibers from this ganglion innervate the small intestine and colon. The lowest splanchnic nerve, not always present, enters the renal plexus near the kidney. Postganglionic fibers supply the renal arterioles and ureter.

The lumbar splanchnic nerve enters the inferior mesenteric plexus, which is inferior to the superior mesenteric plexus. In the plexus, the preganglionic fibers synapse with postganglionic fibers in the inferior mesenteric ganglion. These fibers pass through the hypogastric plexus and supply the distal colon and rectum, urinary bladder, and genital organs. Postganglionic fibers leaving the prevertebral ganglia follow the course of various arteries to abdominal and pelvic visceral effectors.

Sympathetic *preganglionic* fibers also extend to the medullae of the adrenal glands. Developmentally, the adrenal medulla is a modified sympathetic ganglion. Its cells are like sympathetic postganglionic neurons. Rather than extending to another organ, however, these cells release the hormones norepinephrine (20%) and epinephrine (80%) into the blood. This is one exception to the usual pattern of two motor neurons in an autonomic motor pathway.

Parasympathetic Division

Cell bodies of parasympathetic preganglionic neurons are found in nuclei in the brain stem and the lateral gray horn of the second through fourth sacral segments of the spinal cord (see Fig. 17.1). Their fibers emerge as part of a cranial nerve or as part of the anterior root of a spinal nerve. The **cranial parasympathetic outflow** consists of preganglionic fibers that extend from the brain stem in four cranial nerves. The **sacral parasympathetic outflow** consists of preganglionic fibers in anterior roots of the second through fourth sacral nerves. The preganglionic fibers of both the cranial and sacral outflows end in terminal ganglia, where they synapse with postganglionic neurons. We will look first at the cranial outflow.

The cranial outflow has five components: four pairs of ganglia and the plexuses associated with the vagus (X) nerve. The four pairs of cranial parasympathetic ganglia innervate structures in the head and are located close to the organs they innervate. The **ciliary ganglia** lie lateral to each optic (II) nerve near the posterior aspect of the orbit. Preganglionic fibers pass with the oculomotor (III) nerve to the ciliary ganglia. Postganglionic fibers from the ganglion innervate smooth muscle in the eyeball. The **pterygopalatine** (ter'-i-gō-PAL-a-tin) **ganglia** are lateral to the sphenopalatine foramen. They receive preganglionic fibers from the facial (VII) nerve and send postganglionic fibers to the nasal mucosa, palate, pharynx, and lacrimal glands. The **submandibular ganglia** are found near the ducts of the submandibular salivary glands. They receive preganglionic fibers from the facial (VII) nerve and send postganglionic fibers to the submandibular and sublingual salivary glands. The **otic ganglia** are situated just inferior to each foramen ovale. Each otic ganglion receives preganglionic fibers from the glossopharyngeal (IX) nerve and sends postganglionic fibers to the parotid salivary gland.

Preganglionic fibers that leave the brain as part of the vagus (X) nerves are the final components of the cranial outflow. The vagus nerves carry nearly 80% of the parasympathetic outflow. Vagal fibers extend to many terminal ganglia in the thorax and abdomen. Since the terminal ganglia are close to or in the walls of their visceral effectors, postganglionic parasympathetic fibers are very short. By comparison, postganglionic sympathetic fibers are long. As it passes through the thorax, the vagus (X) nerve sends fibers to the heart and the airways of the lungs. In the abdomen, it supplies the liver, gallbladder, stomach, pancreas, small intestine, and part of the large intestine.

The sacral parasympathetic outflow consists of preganglionic fibers from the anterior roots of the second through fourth sacral nerves. Collectively, they form the **pelvic splanchnic nerves**. These nerves synapse with parasympathetic postganglionic neurons located in terminal ganglia in the walls of the innervated viscera. From the ganglia, parasympathetic postganglionic fibers innervate smooth muscle and glands in the walls of the colon, ureters, urinary bladder, and reproductive organs.

Anatomical features of the sympathetic and parasympathetic divisions are compared in Exhibit 17.2.

EXHIBIT **17.2** **ANATOMICAL COMPARISON OF THE SYMPATHETIC AND PARASYMPATHETIC DIVISIONS**

Sympathetic	Parasympathetic
Forms thoracolumbar outflow.	Forms craniosacral outflow.
Contains sympathetic trunk and prevertebral (collateral) ganglia.	Contains terminal ganglia.
Ganglia are close to the CNS and distant from visceral effectors.	Ganglia are near or within the wall of visceral effectors.
Each preganglionic fiber is short and synapses with many postganglionic neurons that pass to many visceral effectors (divergence).	Each preganglionic fiber is long and usually synapses with four or five postganglionic neurons that pass to a single visceral effector.
Distributed throughout the body, including the skin, sweat glands, arrector pili muscles attached to hair follicles, adipose tissue, and smooth muscle of blood vessels.	Distribution limited primarily to head and viscera of thorax, abdomen, and pelvis. No innervation of sweat glands, arrector pili muscles, adipose tissue, the kidneys, and most blood vessels.

PHYSIOLOGICAL EFFECTS OF THE ANS

Most body structures receive dual innervation, that is, fibers from both the sympathetic and parasympathetic divisions. Structures that receive only sympathetic innervation include sweat glands, arrector pili muscles attached to hair follicles in the skin, adipose (fat) cells, the kidneys, and most blood vessels. Lacrimal (tear) glands, on the other hand, receive only parasympathetic fibers. Elsewhere in the body, the two divisions generally have opposing effects on a given organ, one causing excitation and the other inhibition. This is possible because they release different neurotransmitters and their effector tissues have different neurotransmitter receptors. The hypothalamus regulates the balance of sympathetic versus parasympathetic activity or tone. This balance can change from one moment to the next according to demands imposed by the internal and external environments.

ANS Neurotransmitters

Like other neurons, autonomic neurons release neurotransmitters at synapses. Based on the neurotransmitter they produce and liberate, autonomic neurons are classified as either cholinergic or adrenergic (Fig. 17.3).

Cholinergic (kō′-lin-ER-jik) **neurons** release **acetylcholine (ACh)** and include the following: (1) all sympathetic and parasympathetic preganglionic neurons, (2) all

parasympathetic postganglionic neurons, and (3) a few sympathetic postganglionic neurons. The cholinergic sympathetic postganglionic fibers include those to most sweat glands and to a few blood vessels in skeletal muscles.

ACh is stored in synaptic vesicles and is released by exocytosis. It then diffuses the short distance across the synaptic cleft to bind with specific receptors on the postsynaptic membrane. The membrane—of an autonomic postganglionic neuron, a smooth or cardiac muscle fiber, or a glandular cell—either depolarizes and the cell becomes excited or hyperpolarizes and the cell becomes inhibited. (Recall from Chapter 10 that ACh always causes depolarization and excitation of skeletal muscle membranes.) Since acetylcholine is quickly inactivated by the enzyme **acetylcholinesterase (AChE)**, effects triggered by short-lived activity of cholinergic fibers are brief.

Adrenergic (ad′-ren-ER-jik) **neurons** release **norepinephrine (NE)**, also known as **noradrenalin**, or **epinephrine (adrenalin)**. Most sympathetic postganglionic axons are adrenergic; they secrete norepinephrine. Like ACh, NE is synthesized and stored in synaptic vesicles and released by exocytosis. The molecules diffuse across the synaptic cleft and combine with specific receptors on the postsynaptic membrane and trigger either depolarization (excitation) or hyperpolarization (inhibition).

The effects of sympathetic stimulation are longer lasting and more widespread than the effects of parasympathetic stimulation for three reasons. First, there is much more divergence of sympathetic postganglionic fibers. Thus many tissues may be activated simultaneously. Second, acetylcholinesterase quickly inactivates ACh whereas NE lingers in the synaptic cleft for a longer period of time. Slowly, it is both taken up by the axon that released it and enzymatically inactivated by either **catechol-*O*-methyltransferase (COMT)** or **monoamine oxidase (MAO)**. Third, NE and epinephrine secreted into the blood by the adrenal medullae of the two adrenal glands intensify the action of NE liberated from sympathetic postganglionic axons. These hormones in the blood circulate throughout the body, affecting whichever tissues have the appropriate receptors. In time, liver enzymes degrade blood-borne NE and epinephrine.

Cholinergic and Adrenergic Receptors

There are two main categories of both cholinergic and adrenergic receptors plus several subcategories of adrenergic receptors (Fig. 17.3).

The two types of cholinergic (ACh) receptors are nicotinic receptors and muscarinic receptors. **Nicotinic receptors** are found on the dendrites and cell bodies of both sympathetic and parasympathetic postganglionic neurons. They are so named because nicotine mimics the action of ACh on such receptors. Although nicotine is a natural substance in tobacco leaves, it is not present in nonsmoking humans. **Muscarinic receptors** are present on all effectors (muscles and glands) innervated by parasympathetic postganglionic

Figure 17.3 Neurotransmitters and receptors. Cholinergic neurons release acetylcholine whereas adrenergic neurons release norepinephrine.

 Most sympathetic postganglionic neurons are adrenergic (orange); other autonomic neurons and somatic motor neurons are cholinergic (blue).

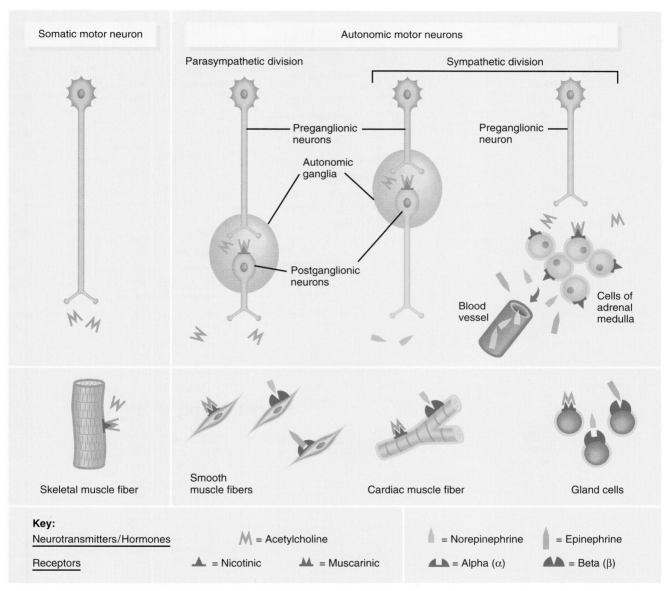

Which neurons are cholinergic and have nicotinic ACh receptors?

axons. Most sweat glands and the smooth muscle in certain blood vessels, which receive their innervation from *cholinergic* sympathetic postganglionic fibers, also display muscarinic receptors. These postsynaptic receptors are so named because a mushroom poison called muscarine mimics the actions of ACh on them. Nicotine does not activate muscarinic receptors, nor does muscarine stimulate nicotinic receptors, but ACh activates both receptor types.

Activation of nicotinic receptors leads to depolarization and thus to excitation of the postsynaptic cell, be it a post-

ganglionic neuron or a visceral effector. Activation of muscarinic receptors sometimes causes depolarization (excitation) and sometimes hyperpolarization (inhibition), depending on which particular cell bears the muscarinic receptors. The binding of ACh to muscarinic receptors inhibits (relaxes) smooth muscle sphincters in the gastrointestinal tract, for example, but excites smooth muscle fibers in the circular muscles of the iris of the eye to contract.

The effects of NE and epinephrine, like those of ACh, also depend on the type of postsynaptic receptor with which they interact. The two types of adrenergic receptors for nor-epinephrine and epinephrine are called **alpha (α) receptors** and **beta (β) receptors** (Fig. 17.3). Such receptors are found on visceral effectors innervated by most sympathetic post-ganglionic axons. Both alpha and beta receptors are, in turn, classified into subtypes—α_1, α_2, β_1, β_2, and β_3. The receptors are distinguished by the specific responses they elicit and by their selective binding of drugs that activate or block them. In general, alpha receptors are excitatory. On the other hand, some beta receptors are excitatory while others are inhibitory. Although cells of most effectors contain either alpha or beta receptors, some visceral effector cells contain both. In general, NE stimulates alpha receptors more vigorously than beta receptors, whereas epinephrine is a potent stimulator of both alpha and beta receptors.

TAKING ADVANTAGE OF RECEPTOR SELECTIVITY

The availability of drugs and natural products that selectively activate or block certain receptors is a cornerstone of modern drug therapy. For example, propranolol (Inderal®) often is prescribed for patients with high blood pressure (hypertension). It is a nonselective **beta blocker**, meaning it binds to all types of beta receptors and prevents their activation by epinephrine and norepinephrine. The desired effects of propranolol are due to its blockade of β_1 receptors, namely, decreased heart rate and contraction force and a consequent decrease in blood pressure. Undesired effects due to blockade of β_2 receptors may include hypoglycemia (low blood glucose) due to decreased glycogen breakdown and decreased gluconeogenesis (the conversion of a noncarbohydrate into glucose in the liver) and mild bronchoconstriction (narrowing of the airways). If these side effects pose a threat to the patient, a selective β_1 blocker such as metoprolol (Lopressor®) can be prescribed instead of propranolol. ■

Parasympathetic and Sympathetic Responses

The parasympathetic division regulates primarily those activities that conserve and restore body energy during times of rest or recovery. It is an **energy conservation–restorative system**. Normally, parasympathetic impulses to the digestive glands and the smooth muscle of the gastrointestinal tract dominate over sympathetic impulses. Thus energy-supplying food can be digested and absorbed.

The acronym "SLUD" is a mental key for remembering many parasympathetic responses. It stands for salivation (S), lacrimation (L), urination (U), and defecation (D). These responses, except for lacrimation, have to do with digestion and absorption of food and elimination of wastes, activities that are stimulated mainly by the parasympathetic

division. Besides the "SLUD" responses, another important parasympathetic response is a decrease in heart rate.

So-called paradoxical fear may cause massive activation of the parasympathetic division. This is the sort of fear that occurs when one is backed into a corner with no way out. It may happen to soldiers in battle, to unprepared students taking an exam, or to athletes before competition. This effect is the basis for loss of control over urination or defecation in such situations.

The sympathetic division, on the other hand, prepares the body for emergency situations. It is primarily concerned with processes involving the expenditure of energy. When the body is in homeostasis, the main function of the sympathetic division is to counteract the parasympathetic effects just enough to carry out normal processes requiring energy. During physical or emotional stress, however, the sympathetic dominates the parasympathetic. Physical exertion stimulates the sympathetic division as do a variety of emotions (such as fear, embarrassment, or rage). Visualizing body changes that occur during "E situations" (emergency, excitement, exercise, embarrassment) will help you remember most of the sympathetic responses. Activation of the sympathetic division and release of hormones by the adrenal medulla set in motion a series of physiological responses collectively called the **alarm reaction** or **fight-or-flight response**, which produces the following effects:

1. The pupils of the eyes dilate.
2. Heart rate and force of contraction and blood pressure increase.
3. The blood vessels of nonessential organs such as the kidneys and gastrointestinal tract constrict.
4. Blood vessels of organs involved in exercise or fighting off danger—skeletal muscles, cardiac muscle, liver, and adipose tissue—dilate to allow faster flow of blood. (The liver splits glycogen to glucose and adipose tissue splits triglycerides to fatty acids, both of which are used by muscle fibers to generate ATP.)
5. The rate and depth of breathing increase and the airways dilate, which allow faster movement of air in and out of the lungs.
6. Blood glucose level rises as liver glycogen is converted to glucose.
7. The medullae of the adrenal glands are stimulated to release epinephrine and norepinephrine. These hormones intensify and prolong the sympathetic effects just described.
8. Processes that are not essential for meeting the stress situation are inhibited. For example, muscular movements of the gastrointestinal tract and digestive secretions slow down or even stop.

Exhibit 17.3 summarizes the responses of glands, smooth muscle, and cardiac muscle to stimulation by the sympathetic and parasympathetic branches of the ANS.

EXHIBIT 17.3   ACTIVITIES OF PARASYMPATHETIC AND SYMPATHETIC DIVISIONS

Visceral Effector	Effect of Sympathetic Stimulation (α or β Receptors, Except as Noted)[a]	Effect of Parasympathetic Stimulation (Muscarinic Receptors)
GLANDS		
Sweat	Increases secretion locally on palms and soles (α); increases secretion in most body regions (muscarinic ACh receptors).	No known functional innervation.
Lacrimal (tear)	No known functional innervation.	Stimulates secretion.
Adrenal medulla	Promotes epinephrine and norepinephrine secretion (nicotinic ACh receptors).	No known functional innervation.
Liver	Promotes conversion of glycogen in the liver into glucose (glycogenolysis), stimulates conversion of noncarbohydrates in the liver into glucose (gluconeogenesis), and decreases bile secretion (α and β_2).	Promotes glycogen synthesis; increases bile secretion.
Adipose (fat) cells[b]	Promotes the breakdown of triglycerides into fatty acids and glycerol (lipolysis) (β_1) and release of fatty acids and glycerol into blood (β_1 and β_3). Promotes thermogenesis (heat production) in brown fat (β_3).	No known functional innervation.
Kidney, juxtaglomerular cells	Stimulates secretion of renin (β_1).	No known functional innervation.
Pancreas	Inhibits secretion of digestive enzymes and the hormone insulin (α); promotes secretion of the hormone glucagon (β_2).	Promotes secretion of digestive enzymes and the hormone insulin.
SMOOTH MUSCLE		
Iris, radial muscle	Contraction → dilation of pupil (α).	No known functional innervation.
Iris, circular muscle	No known functional innervation.	Contraction → constriction of pupil.
Ciliary muscle of eye	Relaxation for far vision (β).	Contraction for near vision.
Salivary gland arterioles	Vasoconstriction, which decreases secretion of saliva (β_2).	Vasodilation, which increases K$^+$ and water secretion.
Gastric gland arterioles	Vasoconstriction, which inhibits secretion (?).	Promotes secretion.
Intestinal gland arterioles	Vasoconstriction, which inhibits secretion (α).	Promotes secretion.
Lungs, bronchial muscle	Relaxation → airway dilation (β_2).	Contraction → airway constriction.
Heart arterioles	Relaxation → dilation of coronary blood vessels (β_1).	Contraction → constriction.
Skin and mucosal arterioles	Contraction → constriction (α).	Dilation, which may not be physiologically significant.
Skeletal muscle arterioles	Contraction → constriction (α); relaxation → dilation (β_2). Relaxation → dilation (muscarinic).	No known innervation.
Abdominal viscera arterioles	Contraction → constriction (α, β).	No known innervation for most.
Brain arterioles	Slight contraction → constriction (α).	No known functional innervation.
Systemic veins	Contraction → constriction (α); relaxation → dilation (β_2).	No known functional innervation.
Gallbladder and ducts	Relaxation (β_2).	Contraction → increased release of bile into small intestine.
Stomach and intestines	Decreases motility and tone (α, β_2); contracts sphincters (α).	Increases motility and tone; relaxes sphincters → enhanced digestive activities and defecation.
Kidney	Constriction of blood vessels → decreased rate of urine production (α).	No known functional innervation.
Ureter	Increases motility.	Decreases motility.
Spleen	Contraction and discharge of stored blood into general circulation (α).	No known functional innervation.

[a] Subcategories of β receptors are listed where known.
[b] Grouped with glands because they release substances into the blood.

Exhibit continues

EXHIBIT **17.3 ACTIVITIES OF PARASYMPATHETIC AND SYMPATHETIC DIVISIONS** (CONTINUED)

Visceral Effector	Effect of Sympathetic Stimulation (α or β Receptors, Except as Noted)[a]	Effect of Parasympathetic Stimulation (Muscarinic Receptors)
Urinary bladder	Relaxation of muscular wall (β); contraction of trigone and sphincter (α).	Contraction of muscular wall; relaxation of trigone and sphincter → urination.
Uterus	Inhibits contraction in nonpregnant woman (β_2); promotes contraction in pregnant woman (α).	Minimal effect.
Sex organs	In male: contraction of smooth muscle of ductus (vas) deferens, seminal vesicle, prostate → ejaculation.	Vasodilation and erection in both sexes.
Hair follicles, arrector pili muscle	Contraction → erection of hairs in skin.	No known functional innervation.
CARDIAC MUSCLE (HEART)	Increases heart rate and force of atrial and ventricular contractions (β_1).	Decreases heart rate; decreases force of atrial contraction.

[a] Subcategories of β receptors are listed where known.

AUTONOMIC REFLEXES

Autonomic (visceral) reflexes adjust the activities of smooth muscle, cardiac muscle, and glands. They cause both contraction and relaxation of smooth and cardiac muscle and changes in the rate of secretion by glands. Autonomic reflexes thus play a key role in activities involved in homeostasis such as regulating heartbeat, blood pressure, respiration, digestion, defecation, and urinary bladder functions.

An autonomic reflex arc consists of the following components:

1. **Receptor.** The receptor is the distal end of a sensory neuron.
2. **Sensory neuron.** This neuron, either a somatic or visceral sensory neuron, conducts nerve impulses to the spinal cord or brain.
3. **Association neurons.** These neurons are found in the central nervous system.
4. **Autonomic motor neurons.**
   Preganglionic neuron. The role of the first (preganglionic) autonomic motor neuron is to convey nerve impulses from the brain or spinal cord to an autonomic ganglion.
   Postganglionic neuron. The second (postganglionic) autonomic motor neuron conducts nerve impulses from an autonomic ganglion to a visceral effector.
5. **Visceral effector.** A visceral effector is smooth muscle, cardiac muscle, or a gland. Alteration of activity in the effector is the response.

Visceral sensations often do not reach the cerebral cortex to give rise to conscious perceptions. Under normal conditions, you are not aware of muscular contractions of the digestive organs, heartbeat, changes in the diameter of blood vessels, and pupil dilation and constriction. Your body adjusts such visceral activities by autonomic reflex arcs whose integrating centers are in the spinal cord or lower regions of the brain. Among such integrating centers are the cardiovascular, respiratory, swallowing, and vomiting centers in the medulla oblongata and the temperature control center in the hypothalamus. Somatic or visceral sensory neurons deliver input to these centers, and autonomic motor neurons provide output that adjusts activity in the visceral effector, usually without conscious recognition.

CONTROL BY HIGHER CENTERS

Axons from many parts of the CNS connect to both the sympathetic and the parasympathetic divisions of the autonomic nervous system and thus exert considerable control over them. The hypothalamus is the major control and integration center of the ANS. Output from the hypothalamus influences autonomic centers in the medulla oblongata and spinal cord.

The hypothalamus receives input from areas of the nervous system concerned with emotions, visceral functions, olfaction (smell), gustation (taste), as well as changes in temperature, osmolarity, and levels of various substances in blood. Anatomically, the hypothalamus is connected to both the sympathetic and the parasympathetic divisions of the ANS by axons of neurons whose dendrites and cell bodies are in various hypothalamic nuclei. The axons form tracts from the hypothalamus to sympathetic and parasympathetic nuclei in the brain stem and spinal cord through relays in the reticular formation.

The posterior and lateral portions of the hypothalamus control the sympathetic division. When these areas are stimulated, there is an increase in heart rate and force of contraction, a rise in blood pressure due to constriction (narrowing) of blood vessels, an increase in body temperature, an increase in the rate and depth of respiration, dilation of the pupils, and inhibition of the gastrointestinal tract. On the other hand, the anterior and medial portions of the hypothalamus control the parasympathetic division. Stimulation of these areas results in a decrease in heart rate, lowering of blood pressure, constriction of the pupils, and increased secretion and motility of the gastrointestinal tract.

Control of the ANS by the cerebral cortex occurs primarily during emotional stress. In extreme anxiety, the cortex can stimulate the hypothalamus as part of the limbic system. This, in turn, stimulates the cardiovascular center of the medulla oblongata, which increases the rate and force of the heartbeat and blood pressure. Stimulation of the cortex by hearing bad news or experiencing an extremely unpleasant sight may cause vasodilation of blood vessels, a lowering of blood pressure, and fainting. While most ANS responses are involuntary, with practice it may be possible to achieve some voluntary control. Biofeedback experiments and the feats of accomplished yogis provide evidence for some degree of learning to control autonomic motor neurons.

STUDY OUTLINE

COMPARISON OF SOMATIC AND AUTONOMIC NERVOUS SYSTEMS (p. 488)

1. The somatic nervous system receives input from the special senses, general somatic senses, and proprioceptors; the ANS receives input from the special senses, general visceral senses, and general somatic senses.
2. The somatic nervous system operates under conscious control; the ANS operates without conscious control.
3. The axons of the motor neurons of the somatic nervous system extend from the CNS, synapse directly with an effector, and release acetylcholine; the axon of the first motor neuron of the ANS (preganglionic) extends from the CNS and synapses in a ganglion with the second motor neuron; the second neuron (postganglionic) synapses on an effector. Preganglionic fibers release acetylcholine and postganglionic fibers release acetylcholine or norepinephrine.
4. Somatic nervous system effectors are skeletal muscles; ANS effectors include cardiac muscle, smooth muscle, and glands.
5. Neurotransmitters released by somatic motor neurons cause excitation; neurotransmitters released by autonomic motor neurons cause excitation or inhibition.

ANATOMY OF AUTONOMIC MOTOR PATHWAYS (p. 488)

1. Preganglionic neurons are myelinated; postganglionic neurons are unmyelinated.
2. The cell bodies of sympathetic preganglionic neurons are in the lateral gray horns of the 12 thoracic and first 2–3 lumbar segments; the cell bodies of parasympathetic preganglionic neurons are in four cranial nerve nuclei (III, VII, IX, and X) in the brain stem and lateral gray horns of the second through fourth sacral segments of the spinal cord.
3. Autonomic ganglia are classified as sympathetic trunk ganglia (on both sides of spinal column), prevertebral ganglia (anterior

to spinal column), and terminal ganglia (near or inside visceral effectors).
4. Sympathetic preganglionic neurons synapse with postganglionic neurons in ganglia of the sympathetic trunk or prevertebral ganglia; parasympathetic preganglionic neurons synapse with postganglionic neurons in terminal ganglia.

PHYSIOLOGICAL EFFECTS OF THE ANS (p. 494)

1. Most body organs receive dual innervation; usually one division causes excitation and the other causes inhibition.
2. Cholinergic neurons release acetylcholine; adrenergic neurons release norepinephrine or epinephrine.
3. The effects of sympathetic stimulation are longer lasting and more widespread than the effects of parasympathetic stimulation.
4. Two types of cholinergic receptors are nicotinic and muscarinic; two types of adrenergic receptors are alpha and beta.
5. The parasympathetic division regulates activities that conserve and restore body energy; the sympathetic division prepares the body for emergency situations (fight-or-flight response).

AUTONOMIC REFLEXES (p. 498)

1. An autonomic (visceral) reflex adjusts the activities of smooth muscle, cardiac muscle, and glands.
2. An autonomic reflex arc consists of a receptor, sensory neuron, association neurons, autonomic motor neurons, and visceral effector.

CONTROL BY HIGHER CENTERS (p. 498)

1. The hypothalamus controls and integrates activities of the autonomic nervous system. It is connected to both the sympathetic and the parasympathetic divisions.
2. Control of the ANS by the cerebral cortex occurs mostly during emotional stress.

REVIEW QUESTIONS

1. What are the principal components of the autonomic nervous system? What is its general function? Why is it called involuntary? (p. 488)
2. What are the principal differences between the somatic nervous system and the autonomic nervous system? (p. 488)

3. Distinguish between preganglionic neurons and postganglionic neurons with respect to location and function. (p. 488)
4. What is an autonomic ganglion? Describe the location and function of the three types of autonomic ganglia. Define white and gray rami communicantes. (p. 491)

5. How do the sympathetic and parasympathetic divisions of the autonomic nervous system differ anatomically and functionally? (p. 492)

6. Discuss the distinction between cholinergic and adrenergic fibers. (p. 494)

7. How is acetylcholine (ACh) related to nicotinic and muscarinic receptors? (p. 494)

8. How are alpha and beta receptors related to norepinephrine (NE) and epinephrine? (p. 496)

9. Give examples of the antagonistic effects of the sympathetic and parasympathetic divisions of the autonomic nervous system. (p. 497)

10. Why is the parasympathetic division of the ANS called an energy conservation–restorative system? (p. 496)

11. Describe what happens during the fight-or-flight response. (p. 496)

12. Give the *sympathetic response* during exercise for each of the following body parts: hair follicles, iris of eye, lungs, spleen, adrenal medullae, urinary bladder, stomach, intestines, gallbladder, liver, heart, arterioles of the abdominal viscera, skeletal muscles, and skin and mucosa. (p. 497)

13. Define an autonomic reflex and give three examples. (p. 498)

14. Describe the components of an autonomic reflex in proper sequence. (p. 498)

15. Describe how the hypothalamus controls and integrates activities of the autonomic nervous system. (p. 498)

ANSWERS TO QUESTIONS WITH FIGURES

17.1 Most parasympathetic preganglionic fibers are longer than most sympathetic preganglionic fibers because most parasympathetic ganglia are in the walls of visceral organs whereas most sympathethic ganglia are close to the spinal cord in the sympathetic trunk.

17.2 Myelin.
17.3 Parasympathetic postganglionic neurons.

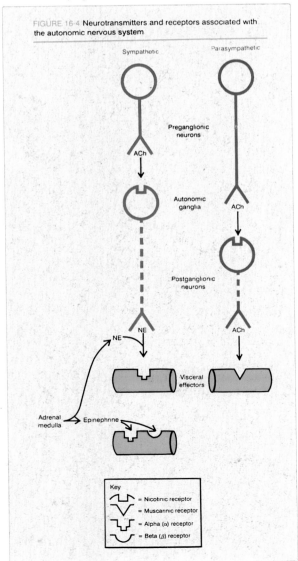

FIGURE 16-4 Neurotransmitters and receptors associated with the autonomic nervous system

STUDENT OBJECTIVES

1. Define the components of the endocrine system and discuss the functions of the endocrine and nervous systems in maintaining homeostasis. (p. 502)

2. Describe how hormones interact with target cell receptors. (p. 503)

3. Compare the four chemical classes of hormones. (p. 504)

4. Explain the two general mechanisms of hormonal action. (p. 506)

5. Describe the control of hormone secretion via feedback cycles and give several examples. (p. 508)

6. Explain why the hypothalamus is considered to be an endocrine gland. (p. 509)

7. Describe the location, histology, hormones, and functions of the following endocrine glands: pituitary, thyroid, parathyroids, adrenals, pancreas, ovaries, testes, pineal, and thymus. (p. 509)

8. Discuss the symptoms of pituitary dwarfism, giantism, and acromegaly (p. 514); diabetes insipidus (p. 519); cretinism, myxedema, Graves' disease, and goiter (p. 524); hypoparathyroidism (p. 527); aldosteronism (p. 531); Addison's disease and Cushing's syndrome (p. 532); congenital adrenal hyperplasia and adrenal tumors (p. 532); pheochromocytomas (p. 533); and diabetes mellitus and hyperinsulinism (p. 537).

9. Describe the effects of aging on the endocrine system. (p. 541)

10. Define the general adaptation syndrome (GAS) and compare homeostatic responses and stress responses. (p. 542)

11. Describe the development of the endocrine system. (p. 545)

CHAPTER 18

THE ENDOCRINE SYSTEM

O'KELLEY

CHAPTER CONTENTS AT A GLANCE

round age 12, as they enter puberty, boys and girls start to develop striking differences in physical appearance and behavior. Perhaps no other period in life so clearly shows the impact of the nervous and endocrine systems in directing development and regulating body functions. Changes in the brain and pituitary gland markedly increase the synthesis of new messenger molecules, the sex hormones, from the gonads. In girls, fatty tissue starts to accumulate in the breasts and hips. At the same time, or a little later, in boys protein synthesis increases; muscle mass builds; and the longer, larger vocal cords produce a lower-pitched voice. These changes provide just a few examples of the powerful influence of secretions from endocrine glands.

ENDOCRINE GLANDS

The body contains two kinds of glands: exocrine and endocrine. **Exocrine** (*ex* = out; *krinein* = to secrete) **glands** secrete their products into ducts, and the ducts carry the secretions into body cavities, into the lumen of an organ, or to the outer surface of the body. Exocrine glands include sudoriferous (sweat), sebaceous (oil), mucous, and digestive glands. **Endocrine** (*endo* = within) **glands**, by contrast, secrete their products (hormones) into the extracellular space around the secretory cells, rather than into ducts. The secretion then diffuses into capillaries and is carried away by the blood. The endocrine glands of the body (Fig. 18.1) consti-

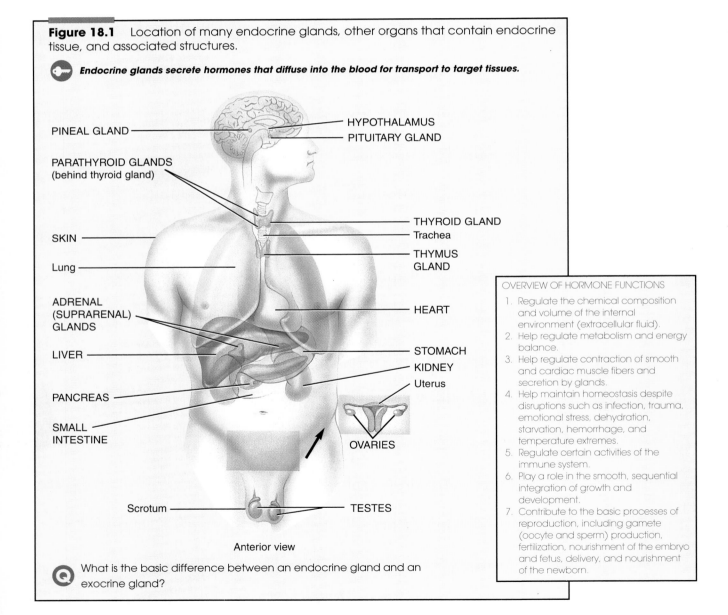

Figure 18.1   Location of many endocrine glands, other organs that contain endocrine tissue, and associated structures.

🔑 *Endocrine glands secrete hormones that diffuse into the blood for transport to target tissues.*

PINEAL GLAND

HYPOTHALAMUS
PITUITARY GLAND

PARATHYROID GLANDS
(behind thyroid gland)

SKIN

Lung

ADRENAL
(SUPRARENAL)
GLANDS

LIVER

PANCREAS

SMALL
INTESTINE

THYROID GLAND
Trachea
THYMUS
GLAND

HEART

STOMACH
KIDNEY
Uterus

OVARIES

Scrotum

TESTES

Anterior view

OVERVIEW OF HORMONE FUNCTIONS

1. Regulate the chemical composition and volume of the internal environment (extracellular fluid).
2. Help regulate metabolism and energy balance.
3. Help regulate contraction of smooth and cardiac muscle fibers and secretion by glands.
4. Help maintain homeostasis despite disruptions such as infection, trauma, emotional stress, dehydration, starvation, hemorrhage, and temperature extremes.
5. Regulate certain activities of the immune system.
6. Play a role in the smooth, sequential integration of growth and development.
7. Contribute to the basic processes of reproduction, including gamete (oocyte and sperm) production, fertilization, nourishment of the embryo and fetus, delivery, and nourishment of the newborn.

❓ What is the basic difference between an endocrine gland and an exocrine gland?

tute the **endocrine system** and include the pituitary, thyroid, parathyroid, adrenal, and pineal glands. In addition, several organs of the body contain cells that secrete hormones but are not endocrine glands exclusively. These include the hypothalamus, thymus, pancreas, ovaries, testes, kidneys, stomach, liver, small intestine, skin, heart, and placenta. Although the functions of hormones are many and varied, their actions can be categorized into seven broad areas. These are listed in Fig. 18.1.

The science concerned with the structure and functions of the endocrine glands and the diagnosis and treatment of disorders of the endocrine system is called **endocrinology** (en′-dō-kri-NOL-ō-jē; *logos* = study of).

The developmental anatomy of the endocrine system is considered at the end of the chapter.

COMPARISON OF NERVOUS AND ENDOCRINE SYSTEMS

Together, the nervous and endocrine systems coordinate functions of all body systems. The nervous system controls homeostasis through nerve impulses (action potentials) conducted along axons of neurons. At axon terminals, impulses trigger release of neurotransmitter molecules. The result is either excitation or inhibition of specific other neurons, muscle fibers (cells), or gland cells. In contrast, the endocrine system releases its messenger molecules, called **hormones** (*hormon* = to urge on), into the bloodstream. The cardiovascular system then delivers hormones to virtually all cells throughout the body.

The nervous and endocrine systems are coordinated as an interlocking supersystem, often referred to as the **neuroendocrine system**. Certain parts of the nervous system stimulate or inhibit the release of hormones. Hormones, in turn, may promote or inhibit the generation of nerve impulses. And certain molecules, for example, norepinephrine, act as hormones in some locations and as neurotransmitters in others.

The nervous system causes muscles to contract and glands to secrete either more or less of their product. The endocrine system alters metabolic activities, regulates growth and development, and guides reproductive processes. Thus it not only helps regulate the activity of smooth and cardiac muscle and some glands, it affects virtually all other tissues as well.

Nerve impulses most often produce their effects within a few milliseconds. While some hormones can act within seconds, others can take several hours or more to bring about their responses. Also, the effects of activating the nervous

EXHIBIT 18.1 COMPARISON OF NERVOUS SYSTEM AND ENDOCRINE SYSTEM REGULATION OF HOMEOSTASIS

Characteristic	Nervous System	Endocrine System
Mechanism of control	Neurotransmitters released in response to nerve impulses.	Hormones delivered to tissues throughout the body by the blood.
Cells affected	Muscle cells, gland cells, other neurons.	Virtually all body cells.
Type of action that results	Muscular contraction or glandular secretion.	Changes in metabolic activities.
Time to onset of action	Typically within milliseconds.	Seconds to hours or days.
Duration of action	Generally briefer.	Generally longer.

system are generally briefer than effects produced by the endocrine system.

A comparison between the nervous and endocrine regulation of homeostasis is presented in Exhibit 18.1

HORMONES

Hormones have powerful effects when present in very low concentration. As a rule, most of the 50 or so hormones affect only a few types of cells. Why is it that some cells respond to a particular hormone and others do not?

Hormone Receptors

Although a given hormone travels throughout the body in the blood, it affects only specific cells called its **target cells**. Hormones, like neurotransmitters, influence their target cells by chemically binding to integral membrane protein or glycoprotein molecules called **receptors**. Only the target cells for a certain hormone have receptors that bind and recognize that hormone. For example, thyroid-stimulating hormone (TSH) binds to receptors on the surface of cells of the thyroid gland, but it does not bind to cells of the ovaries because ovarian cells do not have TSH receptors.

Receptors, like other cellular proteins, are constantly being synthesized and broken down. Generally, a target cell has 2000–100,000 receptors for a particular hormone. When a hormone (or neurotransmitter) is present in excess, the number of target cell receptors may decrease. This effect is called **down-regulation**. For example, when cells of the testes are exposed to a high concentration of luteinizing hormone (LH), the number of LH receptors decreases. Down-regulation thus decreases the responsiveness of target cells to the hormone. On the other hand, when a hormone (or neurotransmitter) is deficient, the number of receptors may increase. This is known as **up-regulation** and makes a target tissue more sensitive to a hormone or neurotransmitter.

BLOCKING HORMONE RECEPTORS

Synthetic hormones that block the receptors for particular naturally occurring hormones can be manufactured. For example, the drug RU486 (mifepristone) binds to the receptors for progesterone (a female sex hormone) and prevents progesterone from exerting its normal effect. When RU486 is given to a pregnant woman, the uterine conditions needed for nurturing an embryo are not maintained and embryonic development stops. This example illustrates an important endocrine principle: if you prevent a hormone from interacting with its receptor, the hormone cannot perform its normal functions. ◼

Circulating and Local Hormones

Hormones that pass into the blood and act on distant target cells are called **circulating hormones** or **endocrines**. Hormones that act locally without first entering the bloodstream are called **local hormones**. Those that act on neighboring cells are called **paracrines** (*para* = beside). Local hormones that act on the same cell that secreted them are termed **autocrines** (*auto* = self). Figure 18.2 compares the sites of action of circulating and local hormones. Local hormones usually are inactivated quickly. Circulating hormones may linger in the blood and exert their effects for a few minutes or occasionally a few hours. In time, circulating hormones are inactivated by the liver and excreted by the kidneys. In cases of kidney or liver failure, excessive buildup of hormones in the blood may cause additional problems.

Chemistry of Hormones

Chemically, there are four principal classes of hormones (Exhibit 18.2): (1) steroids, (2) biogenic amines, (3) pep-

Figure 18.2 Comparison of circulating hormones (endocrines) and local hormones (autocrines and paracrines).

🔑 *Circulating hormones act on distant target cells; paracrines act on neighboring cells; and autocrines act on the same cell that produced them.*

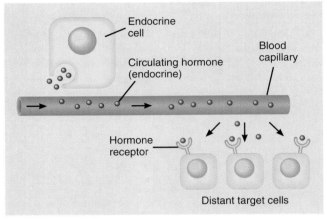

(a) Circulating hormones (endocrines)

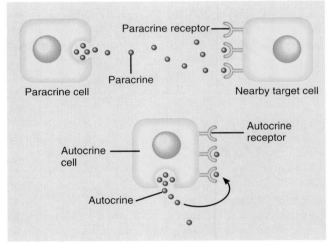

(b) Local hormones (paracrines and autocrines)

 In the stomach, one stimulus for secretion of hydrochloric acid by parietal cells is the release of histamine by neighboring mast cells. Is histamine an endocrine, autocrine, or paracrine in this situation?

tides and proteins, and (4) eicosanoids. In addition, the gas nitric oxide acts as a local hormone in several tissues.

1. **Steroids**. All steroid hormones are lipids that are derived from cholesterol and synthesized on smooth endoplasmic reticulum. The shape of each steroid hor-

mone is subtly different because different side groups are attached at various sites on the four rings. These small differences in side groups allow for a large diversity of functions. Endocrine tissues that secrete steroid hormones all are derived from mesoderm.

2. **Biogenic amines.** Structurally, these are the simplest hormone molecules. Several are synthesized by modifying the amino acid tyrosine. Examples are the two thyroid hormones (T_3 and T_4) and two catecholamines (epinephrine and norepinephrine) secreted by the medulla (inner zone) of the adrenal gland. Histamine is synthesized from the amino acid histidine by mast cells and platelets. Serotonin and melatonin are derived from tryptophan.

3. **Peptides and proteins.** These hormones consist of chains of 3–200 amino acids. Like other cellular pro-

teins, peptide and protein hormones are synthesized on rough endoplasmic reticulum. Some protein hormones, for example, thyroid-stimulating hormone (TSH), have attached carbohydrate groups. Thus they are glycoproteins.

4. **Eicosanoids.** A more recently discovered group of chemical mediators are the eicosanoids (ī-KŌ-sa-noids; *eikosi* = twenty), which are derived from a 20-carbon fatty acid called arachidonic acid. The two major types of eicosanoids are **prostaglandins** and **leukotrienes**. The eicosanoids are important local hormones, and they may also act as circulating hormones.

Exhibit 18.2 contains a summary of the chemical classes of hormones, examples of each, and sites of production.

EXHIBIT **18.2 CHEMICAL CLASSES OF HORMONES**

Chemical Class	Examples	Where Produced
Steroids 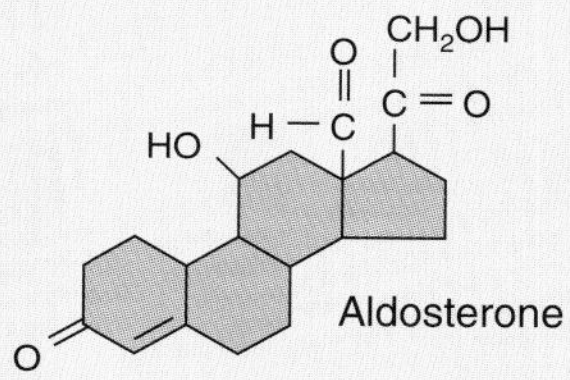 Aldosterone	Aldosterone, cortisol, and androgens (male sex hormones). Calcitriol. Testosterone. Estrogens and progesterone (female sex hormones).	Adrenal cortex. Kidneys. Testes. Ovaries.
Biogenic amines 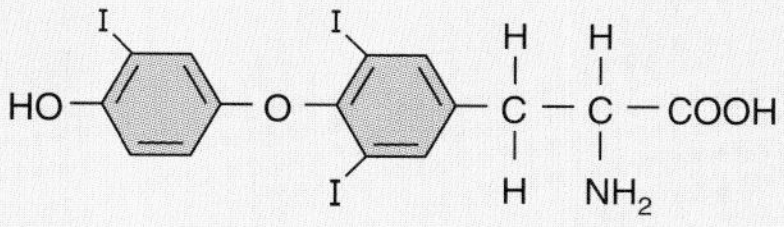 Triiodothyronine (T_3)	T_3 and T_4 (thyroid hormones). Epinephrine and norepinephrine (catecholamines). Histamine. Serotonin. Melatonin.	Thyroid gland (follicular cells). Adrenal medulla. Mast cells in connective tissues. Platelets in blood. Pineal gland.
Peptides and proteins 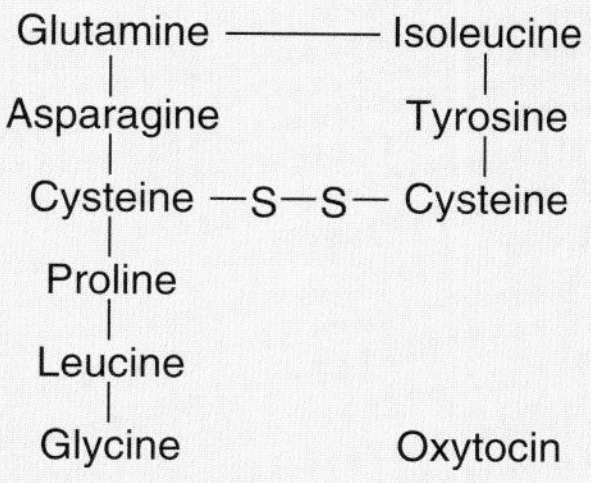 Glutamine — Isoleucine Asparagine — Tyrosine Cysteine —S—S— Cysteine Proline Leucine Glycine — Oxytocin	All hypothalamic releasing and inhibiting hormones (such as thyrotropin releasing hormone). Oxytocin, antidiuretic hormone. All anterior pituitary gland hormones (such as thyroid-stimulating hormone and growth hormone). Insulin, glucagon, somatostatin, and pancreatic polypeptide. Parathyroid hormone. Calcitonin. Hormones that regulate digestion (such as gastrin, secretin, cholecystokinin, and gastric inhibitory peptide). Erythropoietin.	Hypothalamus (neurosecretory cells). Hypothalamus (neurosecretory cells). Anterior pituitary gland. Pancreas. Parathyroid glands. Thyroid gland (parafollicular cells). Stomach and small intestine (enteroendocrine cells). Kidneys.
Eicosanoids 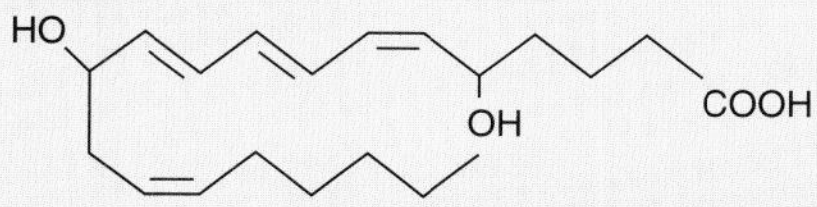 A leukotriene (LTB_4)	Prostaglandins, leukotrienes.	All cells except red blood cells (different cells produce different eicosanoids).

ADMINISTERING HORMONES

The chemical nature of hormones affects how they can be given effectively when people need to take hormone supplements. Steroid hormones (for example, estrogens) and thyroid hormones are effective when taken by mouth. They are not split apart during digestion and easily cross the intestinal lining because they are lipid-soluble. The water-soluble peptide and protein hormones, such as insulin, are not effective oral medications because digestive enzymes destroy them by breaking their peptide bonds. This is why people who need insulin must take it by injection. ■

Hormone Transport in Blood

Endocrine glands are among the most highly vascularized tissues in the body. Catecholamine, peptide, and protein hormones, which by themselves are soluble in watery blood plasma, circulate in free form (not attached to plasma proteins). Upon entering the blood, most steroid and thyroid hormone molecules, however, attach to specific **transport proteins**, which are synthesized by the liver. These hormone "shuttle buses" have three functions. They (1) improve the transportability of the lipid-soluble hormones by making them temporarily water-soluble; (2) retard passage of the small hormone molecules through the filtering mechanism in the kidneys, thus slowing the rate of hormone loss in the urine; and (3) provide a ready reserve of hormone, already present in the bloodstream. In general, 0.1–10% of a lipid-soluble hormone is not bound to a transport protein. It is this **free fraction** that diffuses out of capillaries, binds to receptors, and triggers responses. As free hormone molecules leave the blood and bind to their receptors, transport proteins release new ones to replenish the free fraction.

MECHANISMS OF HORMONE ACTION

The response to a hormone depends on both the hormone and the target cell. Various target cells respond differently to the same hormone. Insulin, for example, stimulates synthesis of glycogen in liver cells but synthesis of triglycerides in adipose cells. Often, the response to a hormone is synthesis of new molecules, as in the examples given for insulin. Other hormone effects include changing the permeability of the plasma membrane, stimulating transport of a substance into or out of the target cells, altering the rate of specific metabolic reactions, or causing contraction of smooth or cardiac muscle. In part, these varied effects of hormones are possible because a single hormone can set in motion several different cellular responses. To begin with, however, a hormone needs to "announce its arrival" to a target cell by binding to receptors. Depending on whether a hormone is lipid-soluble or water-soluble, its receptors are located either inside target cells or at the cell surface in the plasma membrane.

Lipid-Soluble Hormones

Lipid-soluble hormones, including steroid hormones and thyroid hormones, bind to receptors within target cells. Their mechanism of action is as follows (Fig. 18.3):

1 A lipid-soluble hormone diffuses from the blood, through interstitial fluid, and through the phospholipid bilayer of the plasma membrane into a cell.

2 If the cell is a target cell, the hormone will bind to and activate receptors located within the cytosol or nucleus. An activated receptor then alters gene expression: it turns specific genes of the nuclear DNA on or off.

3 As the DNA is transcribed, new messenger RNA (mRNA) forms, leaves the nucleus, and enters the cytosol. There, it directs synthesis of new proteins, usually enzymes, on the ribosomes.

4 The new proteins alter the cell's activity and cause the typical physiological responses of that hormone.

Figure 18.3   Mechanism of action of lipid-soluble hormones (steroid and thyroid hormones).

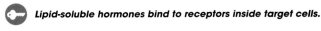

Lipid-soluble hormones bind to receptors inside target cells.

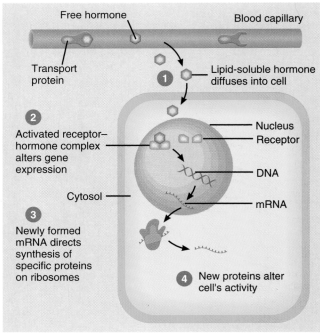

Q How quickly do you think a steroid hormone can start to exert its effects?

Figure 18.6 Hormones secreted by cells in the anterior pituitary gland, target tissues, and general functions.

🔑 *Hormones that influence other endocrine glands are called tropins or trophic hormones.*

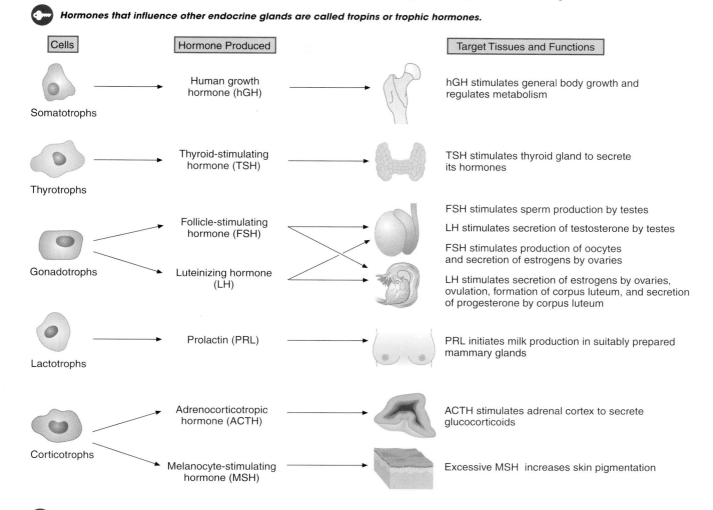

Cells	Hormone Produced	Target Tissues and Functions

Somatotrophs → Human growth hormone (hGH) → hGH stimulates general body growth and regulates metabolism

Thyrotrophs → Thyroid-stimulating hormone (TSH) → TSH stimulates thyroid gland to secrete its hormones

Gonadotrophs → Follicle-stimulating hormone (FSH) / Luteinizing hormone (LH) →
FSH stimulates sperm production by testes
LH stimulates secretion of testosterone by testes
FSH stimulates production of oocytes and secretion of estrogens by ovaries
LH stimulates secretion of estrogens by ovaries, ovulation, formation of corpus luteum, and secretion of progesterone by corpus luteum

Lactotrophs → Prolactin (PRL) → PRL initiates milk production in suitably prepared mammary glands

Corticotrophs → Adrenocorticotropic hormone (ACTH) → ACTH stimulates adrenal cortex to secrete glucocorticoids

Corticotrophs → Melanocyte-stimulating hormone (MSH) → Excessive MSH increases skin pigmentation

Q Which endocrine glands are regulated by hormones released by the anterior pituitary gland?

leasing hormones and two hypothalamic inhibiting hormones are known.

1. **GHRH** (**growth hormone releasing hormone**) stimulates release of human growth hormone (hGH).
2. **GHIH** (**growth hormone inhibiting hormone**) or **somatostatin** inhibits release of both human growth hormone (hGH) and thyroid-stimulating hormone (TSH).
3. **TRH** (**thyrotropin releasing hormone**) stimulates secretion of thyrotropin (TSH) and hGH.
4. **GnRH** (**gonadotropin releasing hormone**) stimulates release of both LH and FSH.
5. **PRH** (**prolactin releasing hormone**) stimulates secretion of prolactin.
6. **PIH** (**prolactin inhibiting hormone**), which is **dopamine**, and TRH both suppress secretion of prolactin.
7. **CRH** (**corticotropin releasing hormone**) stimulates secretion of both corticotropin (ACTH) and MSH.

Negative feedback systems decrease the secretory activity of corticotrophs, thyrotrophs, and gonadotrophs when levels of their target gland hormones rise. As an example, Fig. 18.7 shows how the levels of T_3 and T_4 (thyroid hormones) regulate the secretory activity of anterior pituitary gland and hypothalamus. When blood level of T_3/T_4 (the controlled condition) starts to decrease, the amount of T_3/T_4 bound to receptors in hypothalamic cells that secrete thyrotropin releasing hormone (TRH) and in anterior pituitary gland thyrotrophs that secrete thyroid-stimulating hormone (TSH, or thyrotropin) also decreases. This decrease (input) turns on the genes (control centers) for TRH production in the hypothalamus and TSH production in the anterior pituitary gland. The result is increased secretion of TRH and TSH into the blood (outputs). TRH stimulates thyrotrophs to secrete TSH, and TSH stimulates thyroid gland cells to step up synthesis and secretion of T_3/T_4. Thus thyrotrophs and thyroid cells both are effectors in this negative feedback system. The result is increased T_3/T_4 in the blood

Figure 18.7   Negative feedback system for regulation of T_3 and T_4 concentration in the blood.

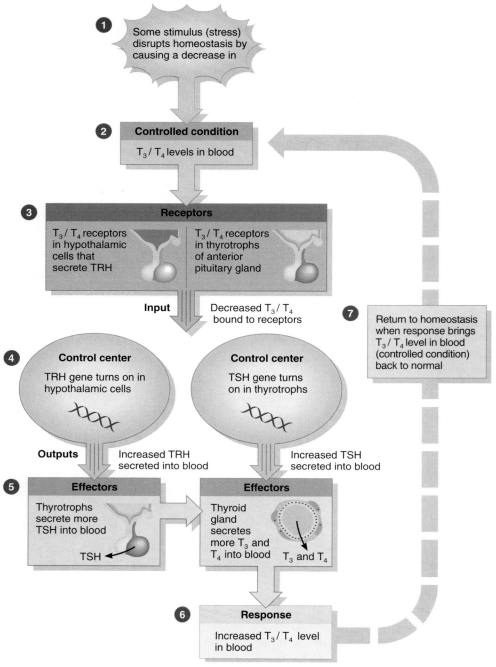

1 Some stimulus (stress) disrupts homeostasis by causing a decrease in

2 **Controlled condition**
T_3 / T_4 levels in blood

3 **Receptors**
T_3 / T_4 receptors in hypothalamic cells that secrete TRH

T_3 / T_4 receptors in thyrotrophs of anterior pituitary gland

Input   Decreased T_3 / T_4 bound to receptors

7 Return to homeostasis when response brings T_3 / T_4 level in blood (controlled condition) back to normal

4 **Control center**
TRH gene turns on in hypothalamic cells

Control center
TSH gene turns on in thyrotrophs

Outputs   Increased TRH secreted into blood

Increased TSH secreted into blood

5 **Effectors**
Thyrotrophs secrete more TSH into blood

TSH

Effectors
Thyroid gland secretes more T_3 and T_4 into blood

T_3 and T_4

6 **Response**
Increased T_3 / T_4 level in blood

Q How can you tell that this is a *negative* feedback system?

(response). When the blood level of T_3/T_4 normalizes, production of TSH by thyrotrophs declines.

Human Growth Hormone

The most abundant anterior pituitary hormone is **human growth hormone** (**hGH** or **GH**), also known as **somatotropin** (sō′-ma-tō-TRŌ-pin). Besides causing body cells to grow, hGH has many effects on metabolism. Generally, hGH (1) stimulates protein synthesis and inhibits protein breakdown, (2) stimulates lipolysis, the breakdown of triglycerides

into fatty acids and glycerol, and (3) retards use of glucose (blood sugar) for ATP production. Most effects of hGH are indirect. The one direct effect of hGH is to promote synthesis and secretion of small protein hormones called **insulin-like growth factors** (**IGFs**), previously called somatomedins. Structurally and functionally IGFs are similar to insulin. However, their growth-promoting effects are even more potent.

In response to hGH, cells in the liver, muscle, cartilage, bone, and other tissues secrete IGFs. They may enter the bloodstream from the liver or act locally in other tissues as autocrine or paracrine substances to enhance growth. IGFs cause cells to grow and multiply by directly increasing the rate at which amino acids enter cells and are used to synthesize proteins. Thus hGH enhances protein anabolism. Also, IGFs decrease the breakdown of proteins and the use of amino acids for ATP production. Due to these effects, hGH increases the growth rate of the skeleton and skeletal muscles during childhood and teenage years. In adults, it helps maintain muscle and bone size and promote healing of injuries and tissue repair.

Human growth hormone, perhaps acting via IGFs, enhances lipid catabolism (breakdown); that is, it causes cells to switch from oxidizing (burning) carbohydrates and proteins to oxidizing fatty acids to produce ATP. It spurs lipolysis in adipose tissue, and it prompts other cells to use the released fatty acids for ATP production. This effect of hGH is most important in periods of fasting or starvation.

Besides affecting protein and fat metabolism, hGH and IGFs influence carbohydrate metabolism. Their effect is to decrease glucose uptake, thus decreasing the use of glucose for ATP production by most body cells. This action spares glucose so neurons may continue to use it for ATP production in times of glucose scarcity. IGFs and hGH may also stimulate liver cells to release glucose into the blood. In this respect, hGH is an insulin antagonist, since insulin lowers blood glucose level by promoting uptake of glucose from the blood into body cells.

DIABETOGENIC EFFECT OF hGH

One symptom of excess hGH is **hyperglycemia** (hī′-per-glī-SĒ-mē-a), high blood glucose concentration. Persistent hyperglycemia, in turn, stimulates the pancreas to continually secrete insulin. Such excessive stimulation, if it lasts for weeks or months, may cause "beta-cell burnout," a greatly decreased capacity of the pancreatic beta cells to synthesize and secrete insulin. Thus, in time, excess secretion of hGH may have a **diabetogenic effect**; that is, it causes diabetes mellitus (lack of insulin activity). ■

Somatotrophs in the anterior pituitary gland release bursts of hGH every few hours. Their secretory activity is controlled mainly by two hypophysiotropic hormones: **growth hormone releasing hormone (GHRH)**, also known as **somatocrinin**, and **growth hormone inhibiting hormone (GHIH)**, also known as **somatostatin**. GHRH promotes whereas GHIH suppresses secretion of hGH. Blood glucose level is a major regulator of GHRH and GHIH secretion as follows (Fig. 18.8):

❶ Low blood glucose level (hypoglycemia) stimulates the hypothalamus to secrete GHRH, which flows toward the anterior pituitary gland in the portal veins.

❷ Upon reaching the anterior pituitary gland, GHRH stimulates somatotrophs to release hGH.

❸ Together, hGH and insulin-like growth factors speed up breakdown of liver glycogen into glucose, which enters the blood.

❹ As a result, blood glucose level rises to the normal level (about 90 mg/100 ml of blood plasma).

❺ An increase in blood glucose above the normal level inhibits release of GHRH.

❻ An abnormally high blood glucose level (hyperglycemia) stimulates the hypothalamus to secrete GHIH (while reducing the secretion of GHRH).

❼ Upon reaching the anterior pituitary gland in portal blood, GHIH inhibits secretion of hGH by somatotrophs.

❽ A low level of hGH slows liver glycogen breakdown and glucose is released into the blood more slowly.

❾ Blood glucose level falls to the normal level.

❿ A decrease in blood glucose below the normal level (hypoglycemia) inhibits release of GHIH.

Figure 18.8 Effects of human growth hormone (hGH) and insulin-like growth factors (IGFs).

🔑 *Secretion of hGH is stimulated by growth hormone releasing hormone (GHRH) and inhibited by growth hormone inhibiting hormone (GHIH).*

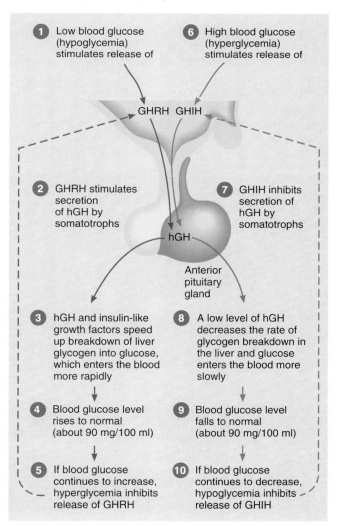

❶ Low blood glucose (hypoglycemia) stimulates release of

❻ High blood glucose (hyperglycemia) stimulates release of

GHRH   GHIH

❷ GHRH stimulates secretion of hGH by somatotrophs

❼ GHIH inhibits secretion of hGH by somatotrophs

hGH

Anterior pituitary gland

❸ hGH and insulin-like growth factors speed up breakdown of liver glycogen into glucose, which enters the blood more rapidly

❽ A low level of hGH decreases the rate of glycogen breakdown in the liver and glucose enters the blood more slowly

❹ Blood glucose level rises to normal (about 90 mg/100 ml)

❾ Blood glucose level falls to normal (about 90 mg/100 ml)

❺ If blood glucose continues to increase, hyperglycemia inhibits release of GHRH

❿ If blood glucose continues to decrease, hypoglycemia inhibits release of GHIH

❓ What condition results from hypersecretion of hGH during childhood?

Other stimuli that promote secretion of hGH include: decreased fatty acids and increased amino acids in the blood; deep sleep (stages 3 and 4 of NREM sleep); increased activity of the sympathetic division of the autonomic nervous system, such as might occur with stress or vigorous physical exercise; and other hormones, for example, glucagon, estrogens, cortisol, and insulin. Factors that inhibit hGH secretion are: increased fatty acids and decreased amino acids in the blood; REM sleep; emotional deprivation; obesity; low level of thyroid hormones; and hGH itself (through negative feedback).

PITUITARY DWARFISM, GIANTISM, AND ACROMEGALY

Disorders of the endocrine system generally involve either **hyposecretion** (underproduction) or **hypersecretion** (overproduction) of hormones. Hyposecretion of hGH during the growth years causes slow bone growth, and the epiphyseal plates close before normal height is reached. This condition is called **pituitary dwarfism**. Other organs of the body also fail to grow, and the pituitary dwarf is childlike in many physical respects. Treatment requires administration of hGH during childhood before the epiphyseal plates close. An ample supply of hGH is now available to treat such children because it is being produced by bacteria using recombinant DNA techniques. Other conditions may also cause dwarfism, in which case hGH treatment is not helpful.

Hypersecretion of hGH during childhood results in **giantism** (**gigantism**), an abnormal increase in the length of long bones. As a result, the person is very tall, but body proportions are about nor-mal. Hypersecretion of hGH during adulthood causes **acromegaly** (ak'-rō-MEG-a-lē) (Fig. 18.9). Further lengthening of the long bones cannot occur because the epiphyseal plates are already closed. Instead, the bones of the hands, feet, cheeks, and jaw thicken. Other tissues also grow. The eyelids, lips, tongue, and nose enlarge, and the skin thickens and develops furrows, especially on the forehead and soles. ■

Thyroid-Stimulating Hormone

Thyroid-stimulating hormone (**TSH**), also called **thyrotropin** (thī-rō-TRŌ-pin), stimulates the synthesis and secretion of two hormones: triiodothyronine (T_3) and thyroxine (T_4), both produced by the thyroid gland. Secretion is controlled by **thyrotropin releasing hormone** (**TRH**) from the hypothalamus. Release of TRH depends on blood levels of TSH, T_3, blood glucose level, and the body's metabolic rate, among other factors, and operates according to a negative feedback system (see Fig. 18.7).

Follicle-Stimulating Hormone

In females, **follicle-stimulating hormone** (**FSH**) is transported from the anterior pituitary gland by the blood to the ovaries. There it initiates the development of follicles each month. Follicles are saclike arrangements of secretory cells that surround a developing oocyte. FSH also stimulates follicular cells to secrete estrogens (female sex hormones). In males, FSH stimulates sperm production in the testes. **Gonadotropin releasing hormone** (**GnRH**) from the hypothalamus stimulates FSH release. GnRH and FSH release is suppressed by estrogens in the female and by testosterone, the principal male sex hormone, through negative feedback systems.

Figure 18.9   Photographs showing characteristic features of acromegaly.

 Acromegaly is caused by hypersecretion of hGH during adulthood.

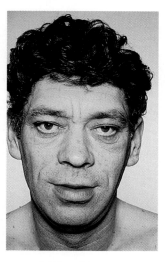

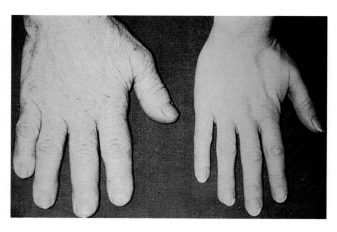

(a) Facial features

Acromegalic   Normal

(b) Hands

 Why would an increase in glove or ring size in an adult be one sign of acromegaly?

Luteinizing Hormone

In females, **luteinizing** (LOO-tē-in′-ī̄z-ing) **hormone (LH)**, together with FSH, stimulates secretion of estrogen by ovarian cells and brings about the release of a secondary oocyte (future ovum) by the ovary, a process called ovulation. LH also stimulates formation of the corpus luteum (structure formed after ovulation) in the ovary and the secretion of progesterone (another female sex hormone) by the corpus luteum. Estrogens and progesterone prepare the uterus for implantation of a fertilized ovum and help prepare the mammary glands for milk secretion. In males, LH stimulates the interstitial cells in the testes to develop and secrete large amounts of testosterone. Secretion of LH, like that of FSH, is controlled by GnRH. GnRH agonists (compounds that mimic GnRH action) are used therapeutically to stimulate the gonads when they are functioning at too low a level. Antagonists of GnRH may be used to inhibit gonadal function when it is thought to be excessive (for example, in precocious puberty) or when other conditions (such as breast or prostate cancer) might be helped by suppressing gonadal secretions. Surprisingly, some GnRH agonists also suppress gonadal functions when they are given continuously for more than 2 weeks.

Prolactin

Prolactin (PRL), together with other hormones, initiates and maintains milk secretion by the mammary glands. Ejection of milk from the mammary glands depends on the hormone oxytocin, which is released from the posterior pituitary gland. Together, milk secretion and ejection are referred to as **lactation**. By itself, prolactin has only a weak effect; the mammary glands require preparation by estrogens, progesterone, glucocorticoids, human growth hormone (hGH), thyroxine, and insulin. When the mammary glands have been primed by these hormones, PRL brings about milk secretion.

The hypothalamus secretes both inhibitory and excitatory hormones that regulate PRL secretion. **Prolactin inhibiting hormone (PIH)**, which is **dopamine**, inhibits the release of PRL from the anterior pituitary gland. As the levels of estrogens and progesterone fall just before menstruation begins, the secretion of PIH diminishes and the blood level of prolactin rises. Breast tenderness just before menstruation may be caused by elevated prolactin. However, the high prolactin level does not last long enough for milk production to start. As the menstrual cycle begins anew, and the level of estrogens rises, PIH is again secreted and the prolactin level drops.

Prolactin level rises during pregnancy, apparently stimulated by a hypothalamic hormone called **prolactin releasing hormone (PRH)**. The sucking action of a nursing infant causes a reduction in hypothalamic secretion of PIH. The function of prolactin is not known in males, but its hypersecretion causes impotence (inability to have an erection of the penis). In females, hypersecretion of prolactin causes absence of menstrual cycles.

Adrenocorticotropic Hormone

Corticotrophs synthesize a large protein precursor molecule called **pro-opiomelanocortin** (prō-ō′-pē-ō-mel-an-ō-KOR-tin) or POMC. POMC is also present in several regions of the brain. Fragments of POMC comprise adrenocorticotropic hormone (ACTH), melanocyte-stimulating hormone (MSH), and beta-endorphin. Corticotrophs secrete mainly **adrenocorticotropic hormone (ACTH)** or **adrenocorticotropin** (ad-rē′-nō-kor′-ti-kō-TRŌ-pin). ACTH controls the production and secretion of hormones called glucocorticoids by the cortex (outer portion) of the adrenal glands. **Corticotropin releasing hormone (CRH)** from the hypothalamus stimulates secretion of ACTH by corticotrophs. Stress-related stimuli, such as low blood glucose or physical trauma, and a substance produced by macrophages called interleukin-1 (IL-1) also stimulate release of ACTH. Glucocorticoids cause negative feedback inhibition of both CRH and ACTH release.

Melanocyte-Stimulating Hormone

Melanocyte-stimulating hormone (MSH) increases skin pigmentation by stimulating the dispersion of melanin granules in melanocytes in amphibians. Its exact role in humans is unknown. However, continued administration of MSH for several days does produce a darkening of the skin, and without MSH the skin may be pallid. Corticotropin releasing hormone (CRH) stimulates MSH release whereas the neurotransmitter dopamine inhibits MSH release.

A summary of anterior pituitary hormones, their principal actions, associated hypothalamic regulating hormones, and selected related disorders is presented in Exhibit 18.3.

Posterior Pituitary Gland (Neurohypophysis)

Although the **posterior pituitary gland (posterior lobe)**, or **neurohypophysis**, does not *synthesize* hormones, it does *store* and *release* two hormones. As noted earlier, it consists of pituicytes and axon terminals of hypothalamic neurosecretory cells (Fig. 18.10). The cell bodies of the neurosecretory cells are in the paraventricular and supraoptic nuclei of the hypothalamus. Their axons form the **supraoptico-hypophyseal** (soo-pra-op′-ti-kō-hī-pō-FIZ-ē-al) **tract**, which extends from the hypothalamus to the posterior pituitary gland and terminates near blood capillaries in the posterior pituitary. Different neurosecretory cells produce two hormones: **oxytocin** (*oxytocia* = rapid childbirth) or **OT** and **antidiuretic hormone (ADH)**, also called **vasopressin**.

After their production in the cell bodies of neurosecretory cells, the hormones are packed into vesicles, which move by

EXHIBIT 18.3   SUMMARY OF ANTERIOR PITUITARY GLAND HORMONES

Hormone and Target Tissues	Principal Actions	Associated Hypothalamic Hormones	Selected Disorders
Human growth hormone (hGH) or somatotropin Liver	Stimulates liver, muscle, cartilage, bone, and other tissues to synthesize and secrete insulin-like growth factors (IGFs); IGFs promote growth of body cells, protein anabolism, tissue repair, lipolysis, and elevation of blood glucose concentration.	Growth hormone releasing hormone (GHRH) and thyrotropin (TRH) promote secretion; growth hormone inhibiting hormone (GHIH), also known as somatostatin, suppresses secretion.	Hyposecretion of hGH during the growth years results in pituitary dwarfism; hypersecretion of hGH during the growth years results in giantism; hypersecretion of hGH during adulthood results in acromegaly.
Thyroid-stimulating hormone (TSH) or thyrotropin Thyroid gland	Stimulates secretion of thyroid hormones by thyroid gland.	Thyrotropin releasing hormone (TRH) stimulates secretion; growth hormone inhibiting hormone (GHIH) suppresses secretion.	Hypersecretion of thyroid hormones stimulated by antibodies (thyroid-stimulating immunoglobulins or TSIs) that mimic the action of TSH causes Graves' disease (see page 524).
Follicle-stimulating hormone (FSH) Ovaries Testes	In females, initiates development of oocytes and induces ovarian secretion of estrogens. In males, stimulates testes to produce sperm.	Gonadotropin releasing hormone (GnRH) promotes secretion.	Lack of FSH causes sterility (inability to produce oocytes or sperm) in both females and males.
Luteinizing hormone (LH) or interstitial cell-stimulating hormone (ICSH) Ovaries Testes	In females, stimulates secretion of estrogens and progesterone, ovulation, and formation of corpus luteum. In males, stimulates interstitial cells in testes to develop and produce testosterone.	Gonadotropin releasing hormone (GnRH) promotes secretion.	Lack of LH causes sterility (inability to produce oocytes or sperm) in both females and males.
Prolactin (PRL) Mammary glands	Together with other hormones promotes milk secretion by the mammary glands.	Prolactin inhibiting hormone (PIH = dopamine) and TRH suppress secretion; prolactin releasing hormone (PRH) promotes secretion.	Hypersecretion causes galactorrhea (inappropriate lactation) and amenorrhea (absence of menstrual cycles) in females and impotence (inability to attain a penile erection) and infertility in males.
Adrenocorticotropic hormone (ACTH) or corticotropin Adrenal cortex	Controls secretion of glucocorticoids (mainly cortisol) by adrenal cortex.	Corticotropin releasing hormone (CRH) stimulates secretion.	Hypersecretion of ACTH leads to excessive release of glucocorticoids from the adrenal gland, which causes Cushing's syndrome (see page 532).
Melanocyte-stimulating hormone (MSH) Skin	Exact role in humans is unknown but can cause darkening of skin; stimulates dispersion of melanin granules in melanocytes of amphibians.	Corticotropin releasing hormone (CRH) stimulates secretion; dopamine inhibits secretion.	

Figure 18.10 Axons of neurosecretory cells form the supraopticohypophyseal tract. Note in the small figure to the right that hormones synthesized by neurosecretory cells in the hypothalamus pass in their axons down to the axon terminals in the posterior pituitary gland. Nerve impulses discharge the hormones, which diffuse into the plexus of the infundibular process and posterior hypophyseal veins for distribution to target cells.

 The posterior pituitary gland stores and releases oxytocin or antidiuretic hormone.

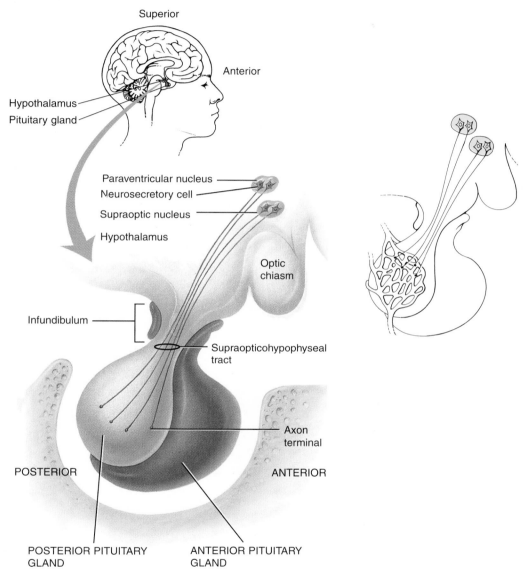

Functionally, how are the supraopticohypophyseal tract and the hypophyseal portal veins similar? Structurally, how are they different?

fast axonal transport (described on page 338) to the axon terminals in the posterior pituitary gland. Nerve impulses that propagate along the axon and reach the axon terminals trigger exocytosis of these secretory vesicles. The released OT or ADH then diffuses into nearby capillaries.

The blood supply to the posterior pituitary gland is from the **inferior hypophyseal arteries** (see Fig. 18.5), which branch from the internal carotid arteries. In the posterior pituitary gland, the inferior hypophyseal arteries form a plexus of capillaries called the **plexus of the infundibular**

process. From this plexus, hormones pass into the **posterior hypophyseal veins** for distribution to tissue cells.

Oxytocin

During and after delivery of a baby, **oxytocin** (ok'-sē-TŌ-sin), or **OT**, has two target tissues—the mother's uterus and breasts. During delivery, it enhances contraction of smooth muscle cells in the wall of the uterus. After delivery, it stimulates milk ejection ("let-down") from the mammary glands in response to the mechanical stimulus provided by a suckling infant. The function of oxytocin in males and in nonpregnant females is not clear. Animal experiments have suggested that it has actions within the brain that foster parental caretaking behavior toward young offspring. It may also be responsible, in part, for the feelings of sexual pleasure during and after intercourse.

During labor and delivery, oxytocin is released in large quantities (Fig. 18.11). When labor contractions begin, the baby's head or body distends (stretches) the cervix (narrow inferior portion) of the uterus. Stretch receptors in the cervix send sensory impulses to the hypothalamus. The nerve impulses cause the posterior pituitary gland to release OT into the blood. It is then carried by the blood to the uterus, where it stimulates muscles of the uterus to contract more forcefully. As the contractions intensify, the resulting sensory impulses stimulate the synthesis and secretion of more OT. Thus a positive feedback cycle is established. With birth of the infant, the cycle is broken because cervical distention suddenly lessens. Note that the input part of the cycle is neural, whereas the output part is hormonal. This is an example of a **neuroendocrine reflex**.

OT affects milk ejection by another neuroendocrine reflex. Milk formed by the glandular cells of the breasts is stored until the baby begins active suckling. Stimulation of touch receptors in the nipple initiates sensory impulses to the hypothalamus. In response, secretion of OT from the posterior pituitary gland quickens. Carried by the bloodstream to the mammary glands, OT stimulates smooth muscle cells around the glandular cells and ducts to contract and eject milk. This sequence is called the **milk ejection (let-down) reflex**. Ejection of milk starts slowly, about 30 seconds to 1 minute after nursing begins. Even in cases of ejection failure, infants can still obtain one-third of the breast's milk. Stimuli other than suckling, such as hearing the baby's cry or touching the genitals, also can trigger OT release and milk ejection. The suckling stimulation that produces the release of OT also inhibits the release of PIH. This results in an increased secretion of prolactin, which maintains lactation.

OXYTOCIN AND CHILDBIRTH

Years before oxytocin was discovered, it was common practice in midwifery to let a first-born twin nurse at the mother's breast to speed the birth of the second child. Now we know why this practice is helpful—it stimulates release of oxytocin. Even after a single birth, nursing promotes expulsion of the placenta (afterbirth) and helps the uterus regain its smaller size. Synthetic OT (Pitocin®) often is given to induce labor or to increase uterine tone and control hemorrhage just after giving birth. ■

Antidiuretic Hormone

An **antidiuretic** is a substance that decreases urine production. A prime action of **antidiuretic hormone (ADH)** is to retain body water. It does this by decreasing water lost by sweating and by causing the kidneys to return more water to the blood, thus decreasing urine volume. In the absence of ADH, urine output increases more than tenfold from the

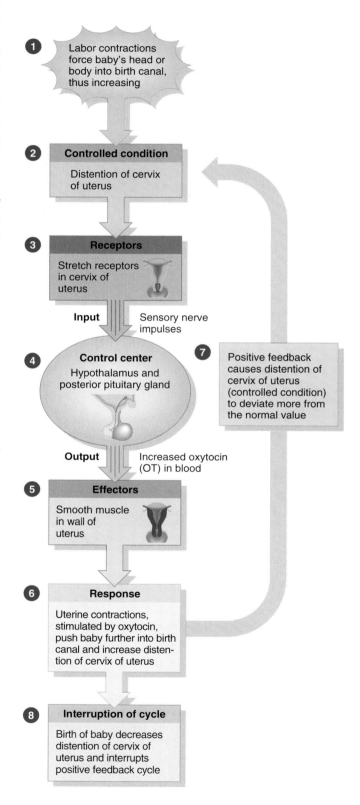

Figure 18.11   Regulation of the secretion of oxytocin (OT) during labor and delivery of a baby.

1 Labor contractions force baby's head or body into birth canal, thus increasing

2 Controlled condition
Distention of cervix of uterus

3 Receptors
Stretch receptors in cervix of uterus

Input   Sensory nerve impulses

4 Control center
Hypothalamus and posterior pituitary gland

7 Positive feedback causes distention of cervix of uterus (controlled condition) to deviate more from the normal value

Output   Increased oxytocin (OT) in blood

5 Effectors
Smooth muscle in wall of uterus

6 Response
Uterine contractions, stimulated by oxytocin, push baby further into birth canal and increase distention of cervix of uterus

8 Interruption of cycle
Birth of baby decreases distention of cervix of uterus and interrupts positive feedback cycle

Q What are the two physiological roles of OT?

Thyroid hormones increase basal metabolic rate or BMR (rate of oxygen consumption at rest after an overnight fast) by stimulating cellular oxygen use to produce ATP. The active transport pumps that continually eject sodium ions (Na^+) from the cytosol into the extracellular fluid a large portion of the ATP produced by most cells. A major effect of the thyroid hormones is to stimulate synthesis of the enzyme that runs the pump, Na^+/K^+ ATPase. As cells use more oxygen to produce ATP, more heat is given off, and body temperature rises. This phenomenon is called the **calorigenic effect** of the thyroid hormones. In this way, they play an important role in the maintenance of normal body temperature (thermoregulation). Normal mammals can survive in freezing temperatures, but those whose thyroid glands have been removed cannot.

In the regulation of metabolism, the thyroid hormones stimulate protein synthesis and increase the use of glucose for ATP production. They also increase lipolysis (triglyceride breakdown) and enhance cholesterol excretion in bile (a substance made in the liver that aids lipid digestion), thus reducing blood cholesterol level.

Together with hGH and insulin, thyroid hormones accelerate body growth, particularly the growth of nervous tissue. Deficiency of thyroid hormones during fetal development can result in fewer and smaller neurons, defective myelination of axons, and mental retardation. During the early years of life, deficiency of thyroid hormones results in small stature and poor development of certain organs such as the brain and reproductive organs.

The thyroid hormones enhance some actions of the catecholamines (norepinephrine and epinephrine) because they up-regulate beta (β) receptors. For this reason, symptoms of hyperthyroidism include increased heart rate and more forceful heartbeats, increased blood pressure, and increased nervousness (see Exhibit 17.3 on page 497).

Control of Thyroid Hormone Secretion

The secretory activity and size of the thyroid gland are controlled in two main ways: (1) by the level of iodine in the thyroid gland and (2) by negative feedback systems involving both the hypothalamus and the anterior pituitary gland. Although needed for synthesis of thyroid hormones, an abnormally high concentration of thyroid iodine suppresses release of thyroid hormones.

Figure 18.15 shows the negative feedback systems that govern synthesis and release of thyroid hormones.

1. Low blood levels of T_3 and T_4 or low metabolic rate stimulate the hypothalamus to secrete thyrotropin releasing hormone (TRH).

2. TRH enters the hypophyseal portal veins and is carried to the anterior pituitary gland, where it stimulates thyrotrophs to secrete thyroid-stimulating hormone (TSH).

3. Then, TSH stimulates virtually all aspects of thyroid follicular cell activity. These include iodide trapping, hormone synthesis and secretion, and growth of the follicular cells.

4. The thyroid follicular cells release T_3 and T_4 into the blood.

5. Blood levels of T_3 and T_4 increase until the metabolic rate returns to normal.

6. An elevated level of T_3 and T_4 inhibits release of TRH and TSH.

Conditions that increase ATP demand—a cold environment, hypoglycemia, high altitude, and pregnancy—also trigger this negative feedback system and increase the secretion of the thyroid hormones.

Figure 18.15 Negative feedback regulation of thyroid hormone secretion.

Thyroid-stimulating hormone (TSH) promotes release of thyroid hormones (T_3 and T_4) by the thyroid gland.

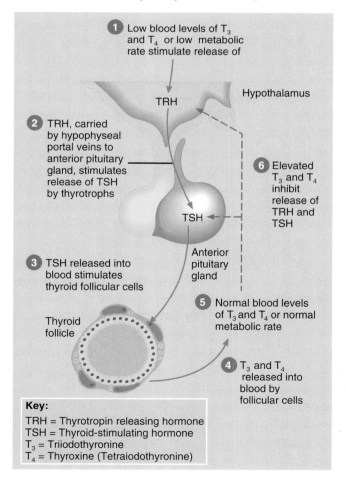

1. Low blood levels of T_3 and T_4 or low metabolic rate stimulate release of

TRH

Hypothalamus

2. TRH, carried by hypophyseal portal veins to anterior pituitary gland, stimulates release of TSH by thyrotrophs

6. Elevated T_3 and T_4 inhibit release of TRH and TSH

TSH

Anterior pituitary gland

3. TSH released into blood stimulates thyroid follicular cells

5. Normal blood levels of T_3 and T_4 or normal metabolic rate

Thyroid follicle

4. T_3 and T_4 released into blood by follicular cells

Key:
TRH = Thyrotropin releasing hormone
TSH = Thyroid-stimulating hormone
T_3 = Triiodothyronine
T_4 = Thyroxine (Tetraiodothyronine)

Q If anterior pituitary gland tumor cells (which are not inhibited by negative feedback) secrete TSH, what will the impact be on the thyroid gland?

CRETINISM, MYXEDEMA, GRAVES' DISEASE, AND GOITER

Hyposecretion of thyroid hormones during fetal life or infancy results in **cretinism** (KRĒ-tin-izm), which is shown in Fig. 18.16a. Cretins exhibit dwarfism because the skeleton fails to grow and mature, and they are severely mentally retarded because the brain fails to develop fully. Although cretinism is not usually apparent in a newborn, because lipid-soluble maternal thyroid hormones pass across the placenta and allow normal development, within months after birth the symptoms start to appear. Most states require testing of all babies to ensure adequate thyroid function. If the hypothyroidism is diagnosed early, cretinism can be prevented by giving oral thyroid hormone.

Hypothyroidism during the adult years produces **myxedema** (mix-e-DĒ-ma). A hallmark of this disorder is an edema (accumulation of interstitial fluid) that causes the facial tissues to swell and look puffy. A person with myxedema has a slow heart rate, low body temperature, sensitivity to cold, hypersensitivity to certain drugs (narcotics, barbiturates, and anesthetics), dry hair and skin, muscular weakness, general lethargy, and a tendency to gain weight easily. Because the brain has already reached maturity, mental retardation does not occur, but mental functions may be dulled so that the person is less alert. Myxedema occurs about five times more often in females than in males. Oral thyroid hormones reduce the symptoms.

Hypersecretion of thyroid hormones increases oxygen use by body cells, elevates heat production, and increases food intake. Common symptoms are increased metabolic rate, heat intolerance, increased sweating, weight loss despite a good appetite, insomnia, tremor of extended fingers, and nervousness. The most common form of hyperthyroidism is **Graves' disease**, which is an autoimmune disorder. The person produces antibodies (thyroid-stimulating immunoglobulins or TSIs) that mimic the action of TSH but are not regulated by the normal negative feedback controls. As a result, the thyroid gland is continually bombarded with stimulation to grow and produce thyroid hormones. A primary sign is an enlarged thyroid, which may be two to three times its normal size. Graves' patients often have a peculiar edema behind the eyes, called **exophthalmos** (ek′-sof-THAL-mos), which causes the eyes to protrude (Fig. 18.16b). This disorder, like myxedema, also occurs more often in females. A variety of treatments are helpful. These include surgical removal of part or all of the thyroid gland (thyroidectomy), using radioactive iodine (^{131}I) to selectively destroy thyroid tissue, and using antithyroid drugs to block synthesis of thyroid hormones.

A **goiter** (GOY-ter; *guttur* = throat) is simply an enlarged thyroid gland, and it is a symptom of many thyroid disorders besides Graves' disease. In some places in the world, dietary iodine intake is inadequate. The resultant low level of thyroid hormone in the blood stimulates secretion of TSH, which causes thyroid gland enlargement (Fig. 18.16c). ■

Calcitonin

The hormone produced by the parafollicular cells of the thyroid gland is **calcitonin** (kal-si-TŌ-nin), or **CT**. Together

Figure 18.16 Abnormalities related to the thyroid gland.

 Cretinism is caused by hyposecretion of thyroid hormones during fetal life or infancy.

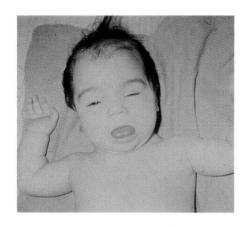

(a) Cretinism

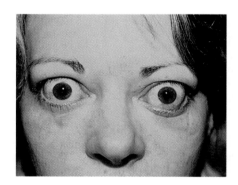

(b) Exophthalmos

(c) Goiter

 How could an iodide-deficient diet lead to goiter?

with parathyroid hormone and calcitriol (described next), calcitonin maintains homeostasis of ionic calcium (Ca^{2+}) and phosphates (PO_4^{3-} and HPO_4^{2-}) in the blood. It lowers the amount of blood calcium and phosphates by inhibiting bone resorption (breakdown of bone matrix) and accelerating uptake of calcium and phosphates into bone matrix. Calcitonin exerts its effect in lowering calcium and phosphate blood levels by inhibiting the action of osteoclasts (bone-destroying cells). Although it acts rapidly in experimental situations, calcitonin's importance in normal physiology is not clear. It can be present in excess or completely absent without causing clinical symptoms. For example, people who have had a complete surgical removal of the thyroid gland maintain calcium homeostasis so long as their parathyroid glands are intact and functional.

A summary of hormones produced by the thyroid gland, their principal actions, control of secretion, and selected disorders is presented in Exhibit 18.5.

PARATHYROID GLANDS

Attached to the posterior surface of the lateral lobes of the thyroid gland are small, round masses of tissue called the **parathyroid** (*para* = beside) **glands**. Usually, there is one superior and one inferior parathyroid gland attached to each lateral thyroid lobe (Fig. 18.17a).

Microscopically, the parathyroids contain two kinds of epithelial cells (Fig. 18.17b,c). The more numerous cells, called **principal (chief) cells**, probably are the major source of **parathyroid hormone (PTH)**, or **parathormone**. The function of the other kind of cell, called an **oxyphil cell**, is not known.

Parathyroid Hormone

PTH increases the number and activity of osteoclasts (bone-destroying cells). The result is enhanced bone resorption, which releases ionic calcium (Ca^{2+}) and phosphates (HPO_4^{2-}) into the blood. PTH also produces two changes in the kidneys. (1) It increases the rate at which the kidneys remove Ca^{2+} and magnesium (Mg^{2+}) from urine that is being formed and returns them to the blood. (2) It inhibits the reabsorption of HPO_4^{2-} filtered by the kidneys so that more of it is excreted in urine. More HPO_4^{2-} is lost through the urine than is gained from the bones. Overall, then, PTH decreases blood HPO_4^{2-} level and increases blood Ca^{2+} and Mg^{2+} levels. With respect to blood Ca^{2+} level, PTH and calcitonin are antagonists; that is, they have opposite actions (see Fig. 18.17).

A third effect of PTH on the kidneys is to promote formation of a hormone called **calcitriol** that is synthesized from vitamin D. Calcitriol is also known as *1,25-dihydroxy cholecalciferol* or *1,25-dihydroxy vitamin D_3*. It increases the rate of Ca^{2+}, HPO_4^{2-}, and Mg^{2+} absorption from the gastrointestinal tract into the blood.

The blood calcium level directly controls the secretion of calcitonin and parathyroid hormone according to negative

---

EXHIBIT **18.5** **SUMMARY OF THYROID GLAND HORMONES**

Hormone	Principal Actions	Control of Secretion	Selected Disorders
T_3 (triiodothyronine) and T_4 (thyroxine) or **thyroid hormones** from follicular cells Thyroid follicle / Folliclular cells	Increase basal metabolic rate, stimulate synthesis of proteins, increase use of glucose for ATP production, increase lipolysis, enhance cholesterol excretion in bile, accelerate body growth, and contribute to normal development of the nervous system.	Secretion increased by thyrotropin releasing hormone (TRH), which stimulates release of thyroid-stimulating hormone (TSH) in response to low thyroid hormone levels, low metabolic rate, cold, pregnancy, and high altitudes; TRH and TSH secretion is inhibited in response to high thyroid hormone levels, high metabolic rate, high levels of estrogens and androgens, and aging; high iodine level suppresses T_3/T_4 secretion.	Hyposecretion of thyroid hormones during infancy or childhood results in cretinism; hypothyroidism during adult years results in myxedema; hypersecretion of thyroid hormones in Graves' disease results in exophthalmos; thyroid enlargement (goiter) may be associated with either thyroid hormone excess or deficiency (see page 524).
Calcitonin (CT) from parafollicular cells Thyroid follicle / Parafollicular cells	Lowers blood levels of ionic calcium and phosphates by inhibiting bone resorption by osteoclasts and accelerating uptake of calcium and phosphates into bone matrix.	High blood Ca^{2+} levels stimulate secretion; low blood Ca^{2+} levels inhibit secretion.	No known disorders result from excess or deficiency of CT secretion.

Figure 18.17   Location, blood supply, and histology of the parathyroid glands.

The parathyroid glands, normally four in number, are embedded in the posterior surface of the thyroid gland.

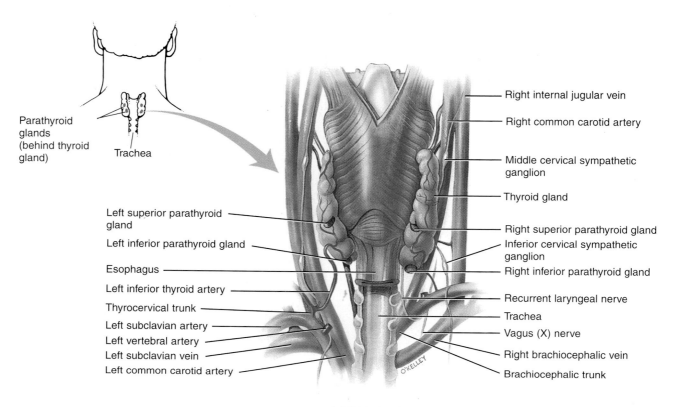

(a) Posterior view

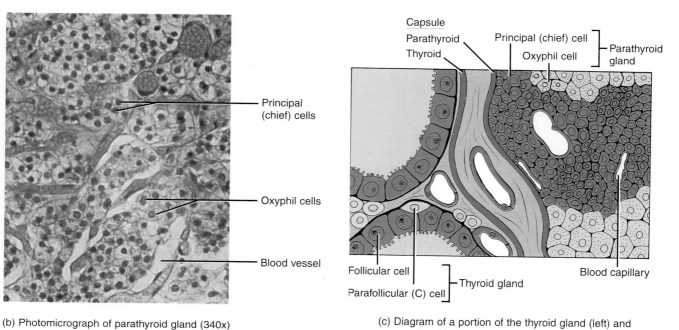

(b) Photomicrograph of parathyroid gland (340x)

(c) Diagram of a portion of the thyroid gland (left) and parathyroid gland (right)

Q   What are the secretory products of parafollicular cells of the thyroid gland and principal cells of the parathyroid glands?

feedback loops that do not involve the pituitary gland. They occur as follows (Fig. 18.18):

① A higher than normal level of calcium ions (Ca^{2+}) in blood stimulates parafollicular cells of the thyroid gland.

② They release more **calcitonin** as blood Ca^{2+} level rises.

③ Calcitonin promotes deposition of blood Ca^{2+} into the matrix of bone tissue. This decreases blood Ca^{2+} level.

④ A lower than normal level of Ca^{2+} in blood stimulates principal cells of the parathyroid gland.

⑤ They release more **parathyroid hormone (PTH)** as blood Ca^{2+} level falls.

⑥ PTH promotes release of Ca^{2+} from bone matrix into the blood and retards loss of Ca^{2+} in the urine. These actions help raise the blood level of Ca^{2+}.

⑦ PTH also stimulates the kidneys to release another hormone called **calcitriol**.

⑧ Calcitriol stimulates increased absorption of Ca^{2+} from foods in the gastrointestinal tract, which helps increase the blood level of Ca^{2+}.

HYPOPARATHYROIDISM

A deficiency of Ca^{2+} caused by **hypoparathyroidism** causes neurons and muscle fibers to depolarize and produce action potentials spontaneously, leading to muscle twitches, spasms, and convulsions. This condition is called **tetany** (*tetanos* = stretched). The leading cause of hypoparathyroidism is accidental damage to the parathyroid glands or their blood supply during thyroidectomy surgery. Other causes include parathyroid disease, infection, or hemorrhage. ■

A summary of the principal actions, control of secretion, and selected disorders related to parathyroid hormone (PTH) is presented in Exhibit 18.6.

ADRENAL GLANDS

The paired **adrenal (suprarenal) glands**, one of which lies superior to each kidney (Fig. 18.19a), are 3–5 cm in height, 2–3 cm in width, and a little less than 1 cm thick; they weigh

---

Figure 18.18 Hormonal regulation of calcium homeostasis by calcitonin (CT), parathyroid hormone (PTH), and calcitriol.

 With respect to regulation of blood Ca^{2+} level, calcitonin and PTH are antagonists.

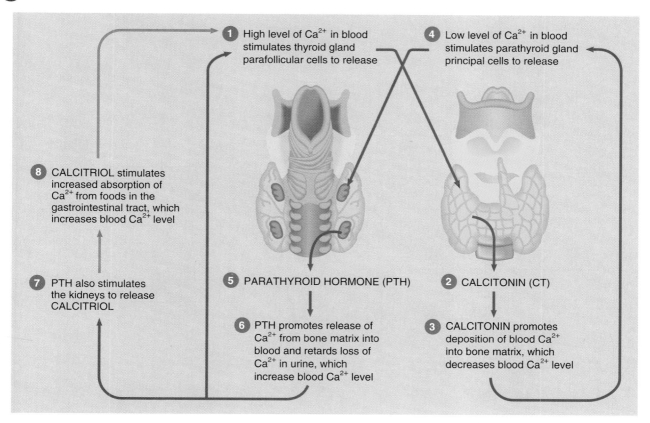

Q What are the primary target tissues for PTH, CT, and calcitriol?

EXHIBIT **18.6** **SUMMARY OF PARATHYROID GLAND HORMONE**

Hormone	Principal Actions	Control of Secretion	Selected Disorders
Parathyroid hormone (PTH) from principal cells Principal (chief) cell	Increases blood Ca^{2+} and Mg^{2+} levels and decreases blood phosphate level by increasing rate of dietary Ca^{2+} and Mg^{2+} absorption; increases number and activity of osteoclasts; increases Ca^{2+} reabsorption by kidneys; increases phosphate excretion by kidneys; and promotes formation of calcitriol (active form of vitamin D).	Low blood Ca^{2+} levels stimulate secretion. High blood Ca^{2+} levels inhibit secretion.	Hypoparathyroidism results in tetany. Hyperparathyroidism, usually due to a tumor of the parathyroid gland, produces osteitis fibrosa cystica, which causes demineralization of bone. The bones become brittle and easily fractured.

Figure 18.19   Location, blood supply, and histology of the adrenal (suprarenal) glands.

The adrenal cortex secretes steroid hormones that are essential for life; the adrenal medulla secretes norepinephrine and epinephrine.

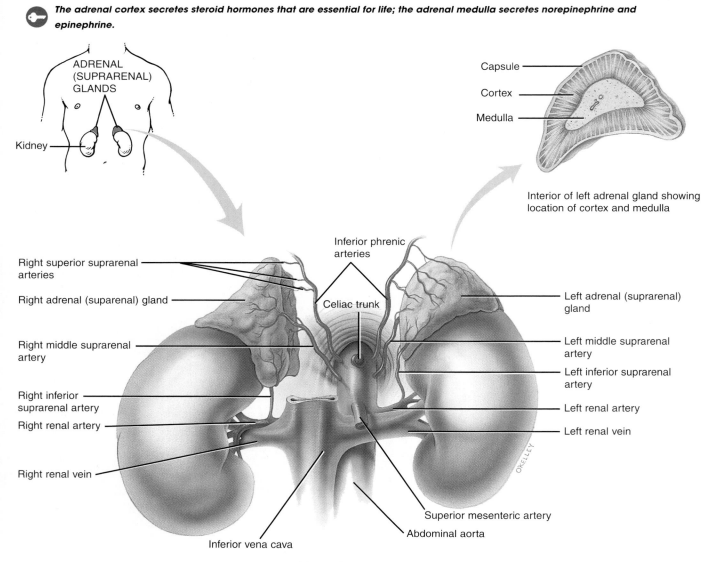

(a) Anterior view

Superficial

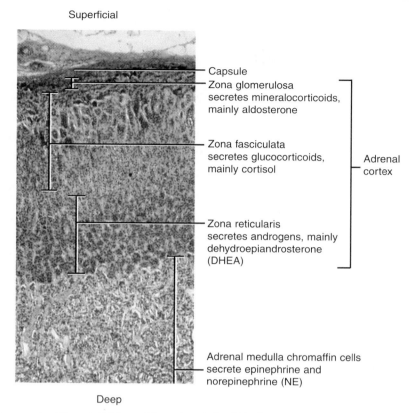

Capsule
Zona glomerulosa
secretes mineralocorticoids,
mainly aldosterone

Zona fasciculata
secretes glucocorticoids,
mainly cortisol

Adrenal
cortex

Zona reticularis
secretes androgens, mainly
dehydroepiandrosterone
(DHEA)

Adrenal medulla chromaffin cells
secrete epinephrine and
norepinephrine (NE)

Deep

(b) Photomicrograph of the subdivisions of the adrenal gland (45x)

Q What is the position of the adrenal glands relative to the pancreas and kidneys?

about 3½–5 g. Structurally and functionally the adrenal glands are differentiated into two regions. The outer **adrenal cortex** makes up the bulk of the gland and surrounds the inner **adrenal medulla** (Fig. 18.19b). The adrenal cortex is derived from mesoderm of a developing embryo and produces steroid hormones that are essential for life. Complete loss of adrenocortical hormones leads to death due to dehydration and electrolyte imbalances in a few days to a week unless hormone replacement therapy begins promptly. The adrenal medulla is derived from ectoderm and produces two catecholamine hormones, norepinephrine and epinephrine. Covering the gland is a connective tissue capsule. The adrenal glands, like the thyroid gland, are highly vascularized.

Adrenal Cortex

The adrenal cortex is subdivided into three zones that secrete different hormones (Fig. 18.19b). The outer zone, just deep to the connective tissue capsule, is called the **zona glomerulosa** (*glomerulus* = little ball). Its cells are arranged in arched loops or round balls. Its primary secretions are a group of hormones called **mineralocorticoids** (min′-er-al-ō-KOR-ti-koyds) because they affect mineral homeostasis.

The middle zone, or **zona fasciculata** (*fasciculus* = little bundle), is the widest of the three zones and consists of cells arranged in long, straight cords. The zona fasciculata secretes mainly **glucocorticoids** (gloo′-kō-KOR-ti-koyds), so named because they affect glucose homeostasis. The inner zone, the **zona reticularis** (*reticular* = net), contains cords of cells that branch freely. This zone synthesizes small amounts of **androgens** (male sex hormones).

Mineralocorticoids

Mineralocorticoids help control water and electrolyte homeostasis, particularly the concentrations of sodium ions (Na^+) and potassium ions (K^+). Although the adrenal cortex secretes at least three different hormones classified as mineralocorticoids, about 95% of the mineralocorticoid activity is due to **aldosterone** (al-DA-ster-ōn). Aldosterone acts on certain tubule cells in the kidneys to increase their reabsorption of Na^+. By stimulating return of Na^+ to the blood, aldosterone prevents depletion of Na^+ from the body. The Na^+ reabsorption leads to reabsorption of Cl^- and HCO_3^- and retention of water. At the same time, aldosterone increases excretion of K^+ in the urine. Aldosterone also promotes excretion of H^+ in the urine, thus removing acids from the body. This action can help prevent acidosis (blood pH below 7.35).

Control of aldosterone secretion involves several mechanisms operating simultaneously:

1 The most important mechanism of control involves the **renin–angiotensin** (an'jē-ō-TEN-sin) **pathway** (Fig. 18.20). Conditions such as dehydration, Na^+ deficiency, or hemorrhage may trigger the renin–angiotensin pathway.

2 These conditions cause a decrease in blood volume.

3 Decreased blood volume causes a decrease in blood pressure.

4 Lowered blood pressure stimulates certain cells of the kidneys, called juxtaglomerular cells, to secrete an enzyme called **renin** (RĒ-nin).

5 This results in increased levels of renin in the blood.

6 Renin converts **angiotensinogen**, a plasma protein produced by the liver, into **angiotensin I**.

7 Increased levels of angiotensin I circulate to the lungs.

8 As blood flows through lung capillaries, an enzyme called **angiotensin converting enzyme** (**ACE**) converts angiotensin I into **angiotensin II**.

9 This results in increased levels of angiotensin II.

10 Angiotensin II is a hormone that has two principal target tissues. One is the adrenal cortex, which it stimulates to secrete aldosterone.

11 Increased levels of aldosterone circulate to the kidneys.

12 In the kidneys, aldosterone increases Na^+ reabsorption, and water follows by osmosis. Aldosterone also leads to increased K^+ secretion by the kidneys into urine.

13 As a result, there is an increase in blood volume.

14 Increased blood volume raises blood volume to normal.

15 As blood volume increases, blood pressure increases to normal.

16 A second target tissue of angiotensin II is smooth muscle in the walls of arterioles, which responds by contracting to produce vasoconstriction.

17 Vasoconstriction of arterioles also increases blood pressure and thus helps raise blood pressure to normal.

18 A second mechanism for the control of aldosterone secretion is the blood K^+ level. An increase in the K^+ concentration of blood and thus of interstitial fluid directly stimulates aldosterone secretion by the adrenal cortex and causes the kidneys to eliminate excess K^+. A decline in the blood K^+ level has the opposite effect.

Figure 18.20   Regulation of aldosterone secretion by the renin–angiotensin pathway.

🔑   *Aldosterone helps regulate blood volume, blood pressure, and levels of Na^+ and K^+ in the body.*

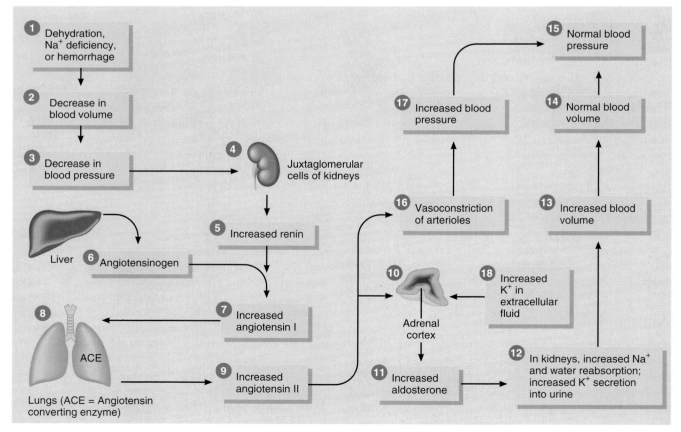

Ⓠ   What are two ways in which angiotensin II can increase blood pressure and what are its target tissues in these cases?

ALDOSTERONISM

Hypersecretion of aldosterone, usually by a tumor of the zona glomerulosa, results in **aldosteronism**, characterized by increased Na$^+$ and decreased K$^+$ levels in the blood. There is excessive retention of Na$^+$ and water. The water increases the volume of the blood and causes hypertension (high blood pressure). If K$^+$ depletion is great, neurons and muscle fibers hyperpolarize, which makes them less responsive to stimulation. Symptoms include muscular weakness, cramps, and paralysis. ■

Glucocorticoids

The glucocorticoids regulate metabolism and resistance to stress. They include **cortisol** (**hydrocortisone**), **corticosterone**, and **cortisone**. Of the three, cortisol is the most abundant and is responsible for about 95% of glucocorticoid activity. The glucocorticoids have the following effects:

1. **Protein catabolism.** Glucocorticoids increase the rate of protein catabolism (breakdown), mainly in muscle fibers, and liberation of amino acids into the bloodstream. The amino acids may be used by liver cells for synthesis of new plasma proteins, such as the enzymes needed for metabolic reactions. Or, they may be used by other body cells for ATP production.
2. **Formation of glucose.** Liver cells also may convert certain amino acids or lactate (lactic acid) to glucose. This conversion of a substance other than glycogen or another monosaccharide into glucose is called **gluconeogenesis** (gloo′-ko-ne′-ō-JEN-e-sis).
3. **Lipolysis.** Glucocorticoids stimulate lipolysis, the breakdown of triglycerides and release of fatty acids from adipose tissue.
4. **Resistance to stress.** Glucocorticoids work in many ways to provide resistance to stress. Added glucose provides tissues with a ready source of ATP to combat a range of stresses such as exercise, fasting, fright, temperature extremes, high altitude, bleeding, infection, surgery, trauma, and disease. Glucocorticoids also make the blood vessels more sensitive to other mediators that cause vasoconstriction. They thereby raise blood pressure. This effect is an advantage if the stress happens to be blood loss, which tends to make blood pressure fall.
5. **Anti-inflammatory effects.** Glucocorticoids are anti-inflammatory compounds that inhibit the cells that participate in inflammatory responses. They: (a) reduce the number of mast cells, thus reducing release of histamine; (b) stabilize lysosomal membranes, thereby slowing release of destructive enzymes; (c) decrease blood capillary permeability; and (d) depress phagocytosis. Unfortunately, they also retard connective tissue repair and are thereby responsible for slow wound healing. Although high doses can cause severe mental disturbances, glucocorticoids are very useful in the treatment of chronic inflammatory disorders, for example, rheumatoid arthritis.

6. **Depression of immune responses.** High doses of glucocorticoids depress immune responses. For this reason, glucocorticoids are taken by organ transplant recipients to retard tissue rejection by the immune system.

The control of glucocorticoid secretion is a typical negative feedback system (Fig. 18.21). Low blood levels of

---

Figure 18.21 Negative feedback regulation of glucocorticoid secretion.

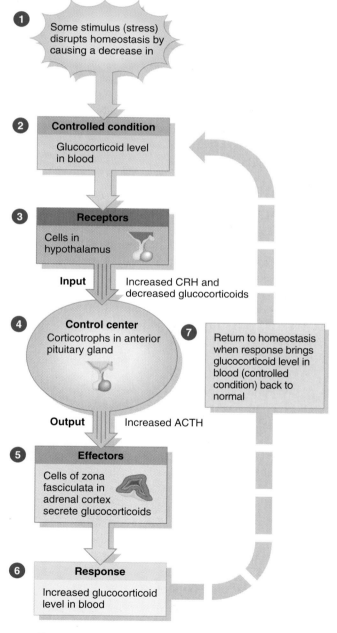

1 Some stimulus (stress) disrupts homeostasis by causing a decrease in

2 **Controlled condition**
Glucocorticoid level in blood

3 **Receptors**
Cells in hypothalamus

Input Increased CRH and decreased glucocorticoids

4 **Control center**
Corticotrophs in anterior pituitary gland

7 Return to homeostasis when response brings glucocorticoid level in blood (controlled condition) back to normal

Output Increased ACTH

5 **Effectors**
Cells of zona fasciculata in adrenal cortex secrete glucocorticoids

6 **Response**
Increased glucocorticoid level in blood

If a person who has had a heart transplant receives prednisone (a glucocorticoid) to help prevent rejection of the transplanted tissue, will blood levels of ACTH and CRH be high or low? Explain.

glucocorticoids (mainly cortisol) stimulate the hypothalamus to secrete **corticotropin releasing hormone** (**CRH**). CRH and a low level of glucocorticoids both promote the release of ACTH from the anterior pituitary. ACTH is carried by the blood to the adrenal cortex, where it stimulates glucocorticoid secretion. As part of the discussion of stress at the end of the chapter, you will see that the hypothalamus also increases CRH release in response to a variety of physical and emotional stresses.

ADDISON'S DISEASE AND CUSHING'S SYNDROME

Hyposecretion of glucocorticoids and aldosterone due to failure of the adrenal cortex results in the condition called **Addison's disease** (**primary adrenocortical insufficiency**). Clinical symptoms include mental lethargy, anorexia, nausea and vomiting, weight loss, hypoglycemia (low blood glucose), and muscular weakness. Loss of aldosterone leads to elevated K^+ and decreased Na^+ in the blood, low blood pressure, dehydration, decreased cardiac output, arrhythmias, and potential cardiac arrest. Blood level of ACTH is high due to loss of negative feedback inhibition by cortisol. At the high concentrations sometimes found in these patients, ACTH can mimic the skin darkening effects of MSH (melanocyte-stimulating hormone). Excessive skin pigmentation, especially in sun-exposed areas and in mucous membranes, also occurs.

Cushing's syndrome is a hypersecretion of glucocorticoids, especially cortisol and cortisone (Fig. 18.22). The condition is characterized by the redistribution of fat. The result is spindly arms and legs, due to catabolism of muscle proteins, accompanied by a rounded "moon face," "buffalo hump" on the back, and pendulous (hanging) abdomen. Facial skin is flushed, and the skin covering the abdomen develops stretch marks (striae). The individual also bruises easily, and

wound healing is poor. Other symptoms include hyperglycemia (high blood glucose), osteoporosis, weakness, hypertension, increased susceptibility to infection, decreased resistance to stress, and mood swings. The most common cause of a cushinoid appearance is the administration of a glucocorticoid such as prednisone to a transplant recipient or to treat asthma or a chronic inflammatory disorder. ■

Androgens

The adrenal cortex secretes small amounts of androgens, male sex hormones that exert masculinizing effects. The major androgen secreted by the adrenal gland is **dehydroepiandrosterone** (dē-hi-drō-ep′-ē-an-DROS-ter-ōn) or **DHEA**. Another important androgen, called testosterone, is produced by the testes. The amount of androgens secreted by normal adult male adrenal gland is usually so low that their effects are insignificant. In females, however, adrenal androgens may contribute to sex drive (libido) and other sexual behavior. They also may be converted into estrogens (female sex hormones) by other body tissues, which is significant when ovarian estrogen secretion diminishes during menopause. Adrenal androgens also assist in the prepubertal growth spurt and early development of axillary and pubic hair in boys and girls.

CONGENITAL ADRENAL HYPERPLASIA AND ADRENAL TUMORS

Congenital adrenal hyperplasia (**CAH**) is a genetic disorder characterized by enlarged adrenal glands. Such people lack one or more enzymes needed for synthesis of cortisol. The low cortisol level stimulates increased secretion of ACTH by the anterior pituitary gland. ACTH, in turn, stimulates growth and secretory activity of the adrenal cortex. Because certain steps leading to synthesis of

---

Figure 18.22 Cushing's syndrome in two different patients.

 Cushing's syndrome is caused by hypersecretion of glucocorticoids.

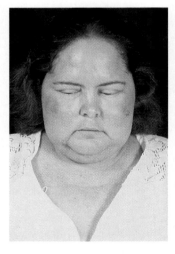

Before treatment   After treatment

(a) Facial features

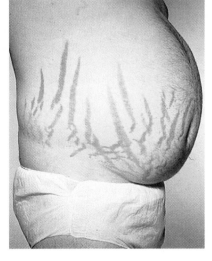

(b) Pendulous abdomen with striae

cortisol are blocked, precursor molecules accumulate, and some of these are converted to androgens. The result is **virilism**, or masculinization. In a female, virile characteristics include growth of a beard, development of a much deeper voice, occasionally the development of baldness, development of a masculine distribution of body hair, growth of the clitoris so it may resemble a penis, atrophy of the breasts, infrequent or absent menstruation, and increased muscularity that produces a male-like physique. In prepubertal males, the syndrome causes the same characteristics as in females, plus rapid development of the male sexual organs and emergence of male sexual desires. In adult males, the virilizing characteristics of CAH may be completely obscured by the normal virilizing characteristics of the testosterone secreted by the testes. As a result, CAH is sometimes difficult to diagnose in adult males.

Virilism may also result from tumors of the adrenal gland called **virilizing adenomas** (*aden* = gland; *oma* = tumor). Occasionally, an adrenal tumor secretes sufficient quantities of feminizing hormones (estrogens) that a male patient develops **gynecomastia** (*gyneca* = woman; *mast* = breast), which means excessive growth (benign) of the male mammary glands. Such a tumor is called a **feminizing adenoma**. ■

Adrenal Medulla

The adrenal medulla consists of hormone-producing cells, called **chromaffin** (krō-MAF-in; *chroma* = color; *affinia* = affinity for) **cells** (see Fig. 18.19b), which surround large blood vessels. Chromaffin cells receive direct innervation from preganglionic neurons of the sympathetic division of the autonomic nervous system (ANS) and develop from the same embryonic tissue as all other sympathetic postganglionic cells. Thus they are sympathetic postganglionic cells that are specialized to secrete hormones (epinephrine and norepinephrine) rather than a neurotransmitter (norepinephrine). Since the ANS controls the chromaffin cells directly, hormone release can occur very quickly.

Epinephrine and Norepinephrine

The two principal hormones synthesized by the adrenal medulla are **epinephrine** and **norepinephrine (NE)**, also called adrenaline and noradrenaline, respectively. Epinephrine constitutes about 80% of the total secretion of the gland. Both hormones are **sympathomimetic** (sim′-pa-thō-mi-MET-ik): their effects mimic those brought about by the sympathetic division of the ANS. To a large extent, they are responsible for the fight-or-flight response. Like the glucocorticoids of the adrenal cortex, these hormones help resist stress. Unlike the hormones of the adrenal cortex, however, the medullary hormones are not essential for life.

Under stress, impulses received by the hypothalamus are conveyed to sympathetic preganglionic neurons, which cause the chromaffin cells to increase their output of epinephrine and norepinephrine. Epinephrine and norepinephrine increase blood pressure by increasing heart rate and force of contraction and constricting blood vessels. They dilate airways to the lungs, decrease the rate of digestion, in-

crease blood glucose level, and stimulate cellular metabolism. Hypoglycemia may also stimulate medullary secretion of epinephrine and norepinephrine.

PHEOCHROMOCYTOMAS

Tumors of the chromaffin cells of the adrenal medulla, called **pheochromocytomas** (fē-ō-krō′-mō-sī-TŌ-mas), cause hypersecretion of the medullary hormones. Such tumors are usually benign. The excess catecholamines cause rapid heart rate, headache, high blood pressure, high levels of glucose in the blood and urine, an elevated basal metabolic rate (BMR), flushing of the face, nervousness, sweating, and decreased gastrointestinal movements. Since the medullary hormones create the same effects as sympathetic nervous stimulation, hypersecretion puts the individual into a prolonged version of the fight-or-flight response (see Fig. 18.26a). Treatment of a pheochromocytoma is surgical removal of the tumor. ■

A summary of the hormones produced by the adrenal glands, their principal actions, control of secretion, and selected disorders is presented in Exhibit 18.7.

PANCREAS

The **pancreas** (*pan* = all; *kreas* = flesh) is both an endocrine and an exocrine gland. We will discuss its endocrine functions here and its exocrine functions with the digestive system (Chapter 24). The pancreas is a flattened organ that measures about 12½–15 cm (4½–6 in.) in length. It is located posterior and slightly inferior to the stomach (Fig. 18.23a) and consists of a head, body, and tail. Roughly 99% of the pancreatic cells are arranged in clusters called **acini**; these cells produce digestive enzymes, which flow into the gastrointestinal tract through a network of ducts. Scattered among the exocrine acini are 1–2 million tiny groups of endocrine tissue called **pancreatic islets** or **islets of Langerhans** (LAHNG-er-hanz; Fig. 18.23b,c). Abundant capillaries serve both the exocrine and endocrine portions of the pancreas.

Cell Types in the Pancreatic Islets

Each pancreatic islet includes four types of hormone-secreting cells: (1) **alpha** or **A cells** constitute about 20% of pancreatic islet cells and secrete **glucagon** (GLOO-ka-gon), (2) **beta** or **B cells** constitute about 70% of pancreatic islet cells and secrete **insulin** (IN-soo-lin), (3) **delta** or **D cells** constitute about 5% of pancreatic islet cells and secrete **somatostatin** (identical to growth hormone inhibiting hormone secreted by the hypothalamus), and (4) **F cells** constitute the remainder of pancreatic islet cells and secrete **pancreatic polypeptide**. The interactions of the four pancreatic hormones is complex and not completely understood. Whereas glucagon raises blood glucose level, insulin lowers it. Somatostatin inhibits

EXHIBIT **18.7** **SUMMARY OF ADRENAL GLAND HORMONES**

Hormone	Principal Actions	Control of Secretion	Selected Disorders
ADRENAL CORTICAL HORMONES			
—Adrenal cortex			
Mineralocorticoids (mainly **aldosterone**) from zona glomerulosa cells	Increase blood levels of Na^+ and water and decrease blood levels of K^+.	Decreased blood volume or Na^+ level initiates renin–angiotensin pathway to stimulate aldosterone secretion; increased blood level of K^+ stimulates aldosterone secretion.	Hypersecretion of aldosterone results in aldosteronism.
Glucocorticoids (mainly **cortisol**) from zona fasciculata cells	Increase rate of protein catabolism (except in liver), stimulate gluconeogenesis and lipolysis, provide resistance to stress, dampen inflammation, and depress immune responses.	ACTH stimulates cortisol release; corticotropin releasing hormone (CRH) promotes ACTH secretion in response to stress and low blood levels of glucocorticoids.	Hyposecretion of glucocorticoids and aldosterone produces Addison's disease; hypersecretion of glucocorticoids results in Cushing's syndrome.
Androgens (mainly **dehydroepiandrosterone** or **DHEA**) from zona reticularis cells	Levels secreted by adult males are so low in comparison to amounts produced by testes that their effects are usually insignificant. In females, adrenal androgens may contribute to sex drive and libido.	Sex hormone secretions by ovaries and testes are discussed in detail in Chapter 28.	In congenital adrenal hyperplasia, synthesis of glucocorticoids is blocked. This leads to excess production of ACTH and androgens, causing virilism. The presence of feminizing hormones in males sometimes causes gynecomastia.
ADRENAL MEDULLARY HORMONES			
—Adrenal medulla			
Epinephrine and **norepinephrine** from chromaffin cells	Sympathomimetic, that is, produce effects that mimic those of the sympathetic division of the autonomic nervous system (ANS) during stress.	Sympathetic preganglionic neurons release acetylcholine, which stimulates secretion by chromaffin cells.	Hypersecretion of medullary hormones, for example, by a pheochromocytoma, results in a prolonged fight-or-flight response.

insulin release and is thought to slow absorption of nutrients from the gastrointestinal tract. Pancreatic polypeptide inhibits secretion of somatostatin, contraction of the gallbladder, and secretion of pancreatic digestive enzymes. Regulation of secretion of somatostatin and pancreatic polypeptide is described in Exhibit 18.8 on page 538.

Figure 18.23 Location, blood supply, and histology of the pancreas.

 Pancreatic hormones regulate blood glucose level.

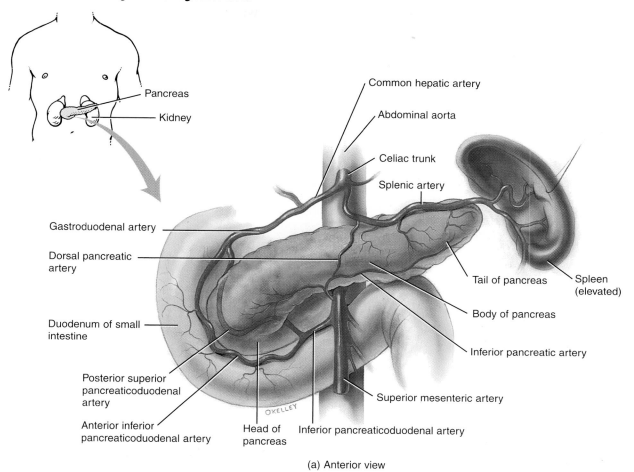

Pancreas

Kidney

Common hepatic artery

Abdominal aorta

Celiac trunk

Splenic artery

Gastroduodenal artery

Dorsal pancreatic artery

Tail of pancreas

Spleen (elevated)

Body of pancreas

Duodenum of small intestine

Inferior pancreatic artery

Posterior superior pancreaticoduodenal artery

Anterior inferior pancreaticoduodenal artery

Head of pancreas

Inferior pancreaticoduodenal artery

Superior mesenteric artery

O'KELLEY

(a) Anterior view

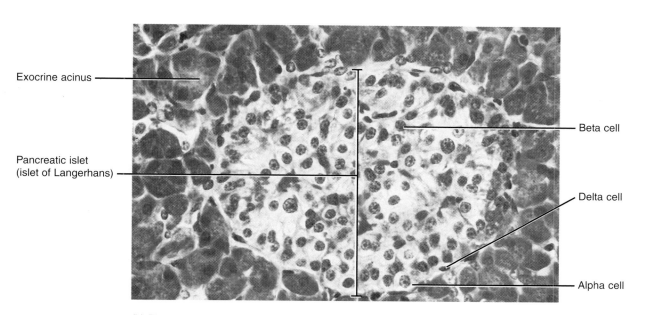

Exocrine acinus

Pancreatic islet (islet of Langerhans)

Beta cell

Delta cell

Alpha cell

(b) Photomicrograph of a pancreatic islet (islet of Langerhans) and surrounding acini (260x)

Figure continues

Figure 18.23   (continued)

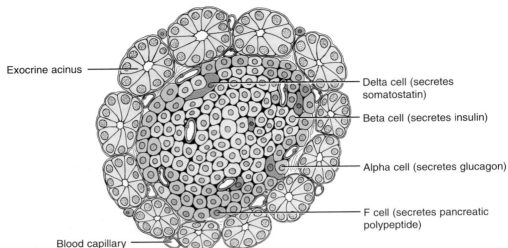

Exocrine acinus

Delta cell (secretes somatostatin)

Beta cell (secretes insulin)

Alpha cell (secretes glucagon)

F cell (secretes pancreatic polypeptide)

Blood capillary

(c) Diagram of a pancreatic islet (islet of Langerhans) and surrounding acini

Q Is the pancreas an exocrine or an endocrine gland?

Regulation of Glucagon and Insulin Secretion

The principal physiological action of glucagon is to increase blood glucose level when it falls below normal. Insulin, on the other hand, helps adjust blood glucose level by decreasing the level if necessary. The level of blood glucose controls secretion of glucagon and insulin via negative feedback systems (Fig. 18.24).

1 Low blood glucose (hypoglycemia) stimulates release of glucagon from alpha cells of the pancreatic islets.

2 Glucagon acts on hepatocytes (liver cells) to:
- accelerate the conversion of glycogen into glucose (glycogenolysis) and
- promote formation of glucose from lactic acid (lactate) and certain amino acids (gluconeogenesis).

3 As a result, hepatocytes release glucose into the blood more rapidly and blood glucose level rises.

4 If blood glucose continues to rise, high blood glucose (hyperglycemia) inhibits release of glucagon (negative feedback).

5 At the same time, however, high blood glucose (hyperglycemia) stimulates release of insulin from beta cells of the pancreatic islets.

6 Insulin acts on various cells in the body to:
- accelerate facilitated diffusion of glucose into cells, especially skeletal muscle fibers
- speed conversion of glucose into glycogen (glycogenesis)
- increase uptake of amino acids by cells and increase protein synthesis

- speed synthesis of fatty acids (lipogenesis)
- slow glycogenolysis, and
- slow gluconeogenesis.

7 As a result, blood glucose level falls.

8 If blood glucose level continues to fall, low blood glucose inhibits release of insulin (negative feedback).

Several hormones and neurotransmitters regulate release of insulin and glucagon. A rising blood glucose level is the most important stimulator of insulin release. In addition, insulin secretion is stimulated by (1) acetylcholine, the neurotransmitter liberated from axon terminals of parasympathetic vagus nerve fibers that innervate the pancreatic islets; (2) the amino acids arginine and leucine; (3) glucagon; and (4) gastric inhibitory peptide (GIP), a hormone released by enteroendocrine cells of the small intestine in response to the presence of glucose in the gastrointestinal tract. Thus digestion and absorption of food containing both carbohydrates and proteins provides strong stimulation for insulin release. Increased activity of the sympathetic division of the ANS, as occurs during exercise, enhances glucagon release. Also, a rise in blood amino acids stimulates glucagon secretion if blood glucose level is low. This could occur after a meal that contained mainly protein. Whereas glucagon stimulates insulin release, insulin suppresses glucagon secretion. Thus as blood levels of glucose decline and less insulin is secreted, the alpha cells are released from the inhibitory effect of insulin and secrete more glucagon. Indirectly, human growth hormone (hGH) and adrenocorticotropic hormone (ACTH) stimulate secretion of insulin because they act to elevate blood glucose level. Somatostatin inhibits the secretion of both insulin and glucagon.

Figure 18.24 Regulation of the secretion of glucagon and insulin.

Low blood glucose level stimulates release of glucagon whereas high blood glucose level stimulates secretion of insulin.

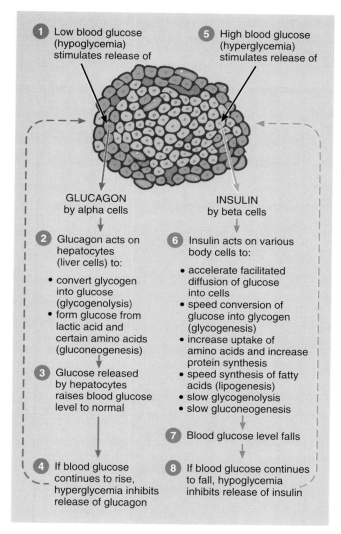

① Low blood glucose (hypoglycemia) stimulates release of

⑤ High blood glucose (hyperglycemia) stimulates release of

GLUCAGON by alpha cells

INSULIN by beta cells

② Glucagon acts on hepatocytes (liver cells) to:
- convert glycogen into glucose (glycogenolysis)
- form glucose from lactic acid and certain amino acids (gluconeogenesis)

③ Glucose released by hepatocytes raises blood glucose level to normal

④ If blood glucose continues to rise, hyperglycemia inhibits release of glucagon

⑥ Insulin acts on various body cells to:
- accelerate facilitated diffusion of glucose into cells
- speed conversion of glucose into glycogen (glycogenesis)
- increase uptake of amino acids and increase protein synthesis
- speed synthesis of fatty acids (lipogenesis)
- slow glycogenolysis
- slow gluconeogenesis

⑦ Blood glucose level falls

⑧ If blood glucose continues to fall, hypoglycemia inhibits release of insulin

Why is glucagon sometimes called an "anti-insulin" hormone?

 ## DIABETES MELLITUS AND HYPERINSULINISM

Diabetes mellitus (MEL-i-tus) is a group of disorders that all lead to an elevation of glucose in the blood (hyperglycemia). As hyperglycemia increases, glucose appears in the urine (glucosuria). Hallmarks of diabetes mellitus are the three "polys": an inability to reabsorb water, resulting in excessive urine production (**polyuria**); excessive thirst (**polydipsia**); and excessive eating (**polyphagia**).

The two major types of diabetes mellitus are type I and type II. In **type I diabetes** there is an absolute deficiency of insulin. Type I

diabetes is called **insulin-dependent diabetes mellitus (IDDM)** because regular injections of insulin are required to prevent death. It previously was known as **juvenile-onset diabetes** because it most commonly develops in people younger than age 20, although it persists throughout life. IDDM appears to be an autoimmune disorder, one in which a person's immune system destroys the pancreatic beta cells, that occurs in genetically susceptible people. The drug cyclosporine, which suppresses the immune system, shows promise of being able to interrupt the destruction of beta cells.

The cellular metabolism of an untreated type I diabetic is similar to that of a starving person. Since insulin is not present to aid the entry of glucose into body cells, most cells use fatty acids to produce ATP. Organic acids called ketones (ketone bodies) are by-products of fatty acid catabolism. As they accumulate, they cause a form of acidosis called **ketoacidosis**, which lowers the pH of the blood and can result in death. The catabolism of stored triglycerides and proteins also causes weight loss. As lipids are transported by the blood from storage depots to cells, lipid particles are deposited on the walls of blood vessels. The deposition leads to atherosclerosis and a multitude of cardiovascular problems including cerebrovascular insufficiency, ischemic heart disease, peripheral vascular disease, and gangrene. One of the major complications of diabetes is loss of vision due to cataracts (excessive glucose attaches to lens proteins, causing cloudiness) or damage to blood vessels of the retina. Severe kidney problems also may result from damage to renal blood vessels.

Considerable research is being done to find ways of preventing IDDM or improving the status of IDDM patients. Work is ongoing to develop an artificial pancreas that monitors blood glucose level and automatically administers insulin from a reservoir. Also, several types of transplants have been tried. Transplantation of the entire pancreas is possible, but the person then needs to take immunosuppressive drugs to prevent rejection. Another approach has been to implant clusters of pancreatic islet cells that have been pretreated to render them incapable of inducing rejection by the recipient. In other cases, pancreatic islet cells have been encapsulated so that the insulin could get out, but elements responsible for rejection could not get in. In still another procedure, patients were injected with fetal pancreatic islet cells.

Type II diabetes represents more than 90% of all cases. Type II diabetes most often occurs in people who are over 40 and overweight. Since type II diabetes usually occurs later in life, it previously was called **maturity-onset diabetes**. Clinical symptoms are mild, and the high glucose levels in the blood usually can be controlled by diet, exercise, and weight loss. Sometimes, an antidiabetic drug such as *glyburide* (DiaBeta®) is needed. This drug stimulates secretion of insulin by beta cells of the pancreas. Many type II diabetics, however, have a sufficient amount or even a surplus of insulin in the blood. For these people, diabetes arises not from a shortage of insulin but because target cells become less sensitive to it, probably through down-regulation of insulin receptors. Type II diabetes is therefore called **non-insulin-dependent diabetes mellitus (NIDDM)**. However, some NIDDM patients do need insulin.

Gestational diabetes refers to diabetes that occurs during pregnancy and then disappears immediately following delivery. It is due to changes in glucose metabolism during pregnancy.

Although the condition may be mild and produce no symptoms in the mother, it presents many of the same hazards to the fetus as other types of diabetes.

Hyperinsulinism most often results when a diabetic injects too much insulin. Rarely, it results from a malignant tumor or hyperplasia of the pancreatic islets. The principal symptom is **hypoglycemia**, decreased blood glucose level, which occurs because the excess insulin stimulates excessive uptake of glucose by many cells of the body. The resulting hypoglycemia stimulates the secretion of epinephrine, glucagon, and hGH. As a consequence, anxiety,

sweating, tremor, increased heart rate, hunger, and weakness occur. When blood glucose level falls, brain cells are deprived of the steady supply of glucose they need to function efficiently. This condition leads to mental disorientation, convulsions, unconsciousness, and shock and is termed **insulin shock**. Death can occur quickly unless blood glucose level is raised. ■

Exhibit 18.8 presents a summary of the hormones produced by the pancreas, their principal actions, control of secretion, and selected disorders.

EXHIBIT **18.8** SUMMARY OF HORMONES PRODUCED BY THE PANCREAS

Hormone	Principal Actions	Control of Secretion	Selected Disorders
Glucagon from alpha or A cells of pancreatic islets Alpha cell	Raises blood glucose level by accelerating breakdown of glycogen into glucose in liver (glycogenolysis) and conversion of other nutrients into glucose in liver (gluconeogenesis) and releasing glucose into blood.	Decreased blood level of glucose, exercise, and mainly protein meals stimulate glucagon secretion; somatostatin and insulin inhibit glucagon secretion.	Hypersecretion of glucagon by a tumor (glucagonoma) causes hyperglycemia if insulin secretion fails to keep pace.
Insulin from beta or B cells of pancreatic islets Beta cell	Lowers blood glucose level by accelerating transport of glucose into cells, converting glucose into glycogen (glycogenesis), and decreasing glycogenolysis and gluconeogenesis; also increases lipogenesis and stimulates protein synthesis.	Increased blood level of glucose, acetylcholine (released by parasympathetic vagus nerve fibers), arginine and leucine (two amino acids), glucagon, GIP, hGH, and ACTH stimulate insulin secretion; somatostatin inhibits insulin secretion.	A deficiency of insulin or defects in insulin receptors produce diabetes mellitus. Hypersecretion of insulin results in hyperinsulinism.
Somatostatin (same as growth hormone inhibiting hormone) from delta or D cells of pancreatic islets Delta cell	Inhibits secretion of insulin and glucagon and slows absorption of nutrients from the gastrointestinal tract.	Pancreatic polypeptide inhibits somatostatin release.	Hypersecretion of somatostatin by a tumor (somatostatinoma) decreases release of insulin (which may lead to diabetes mellitus) and glucagon.
Pancreatic polypeptide from F cells of pancreatic islets F cell	Inhibits secretion of somatostatin, contraction of the gall bladder, and secretion of pancreatic digestive enzymes.	Meals containing protein, fasting, exercise, and acute hypoglycemia stimulate pancreatic polypeptide secretion; somatostatin and elevated blood glucose inhibit pancreatic polypeptide secretion.	

158

OVARIES AND TESTES

The female gonads, called the **ovaries**, are paired oval bodies located in the pelvic cavity. The ovaries produce female sex hormones called **estrogens** and **progesterone**. Along with the gonadotropic hormones of the pituitary gland, the sex hormones regulate the female reproductive cycle, maintain pregnancy, and prepare the mammary glands for lactation. These hormones are also responsible for the development and maintenance of feminine secondary sex characteristics. The ovaries also produce **inhibin**, a protein hormone that inhibits secretion of FSH (and, to a lesser extent, LH). During pregnancy, the ovaries and placenta produce a peptide hormone called **relaxin**, which increases the flexibility of the pubic symphysis during pregnancy and helps dilate the uterine cervix during labor and delivery. These actions help ease the baby's passage by enlarging the birth canal.

The male has two oval gonads, called **testes**, that produce **testosterone**, the primary androgen. Testosterone regulates production of sperm and stimulates the development and maintenance of masculine secondary sex characteristics, for example, beard growth. The testes also produce inhibin, which inhibits secretion of FSH. The specific roles of gonadotropic hormones and sex hormones will be discussed in Chapter 28.

Exhibit 18.9 presents a summary of hormones produced by the ovaries and testes and their principal actions.

PINEAL GLAND

The endocrine gland attached to the roof of the third ventricle is known as the **pineal** (PĪN-ē-al; *pinealis* = shaped like a pine cone) **gland** (see Fig. 18.1). The gland is covered by a capsule formed by the pia mater and consists of masses of **neuroglia** and secretory cells called **pinealocytes** (pin-ē-AL-ō-sits). Sympathetic postganglionic fibers from the superior cervical ganglion terminate in the pineal gland.

Although many anatomical features of the pineal gland have been known for years, its physiological role is still unclear. One hormone secreted by the pineal gland is **melatonin**, a biogenic amine. Less melatonin is liberated in

EXHIBIT 18.9 SUMMARY OF HORMONES OF THE OVARIES AND TESTES

Hormone	Principal Actions	Hormone	Principal Actions
OVARIAN HORMONES		TESTICULAR HORMONES	
Ovary		Testis	
Estrogens and progesterone	Together with gonadotropic hormones of the anterior pituitary gland, they regulate the female reproductive cycle, maintain pregnancy, prepare the mammary glands for lactation, regulate oogenesis, and promote development and maintenance of feminine secondary sex characteristics.	**Testosterone**	Stimulates descent of testes before birth, regulates spermatogenesis, and promotes development and maintenance of masculine secondary sex characteristics.
Relaxin	Increases flexibility of pubic symphysis during pregnancy and helps dilate uterine cervix during labor and delivery.	**Inhibin**	Inhibits secretion of FSH from anterior pituitary gland.
Inhibin	Inhibits secretion of FSH from anterior pituitary gland.		

strong sunlight, according to the following sequence (Fig. 18.25):

1 Light enters the eyes, strikes the retina, and stimulates photoreceptors.

2 Retinal neurons activated by photoreceptors transmit impulses to the suprachiasmatic nucleus of the hypothalamus.

3 From here nerve impulses are transmitted to the superior cervical ganglion.

4 Sympathetic postganglionic fibers from the superior cervical ganglion extend to the pineal gland and form synaptic contacts with cells of the pineal gland.

5 In sunlight, norepinephrine released by the sympathetic fibers inhibits secretion of melatonin by cells of the pineal gland. In darkness, retinal neurons transmit fewer impulses to the pineal gland by way of the suprachiasmatic nucleus and superior cervical ganglion.

6 In sunlight, inhibition of melatonin secretion results in lack of sleepiness. In darkness, lack of norepinephrine stimulates secretion of melatonin by cells of the pineal gland and the result is sleepiness. Thus the release of melatonin is governed by the diurnal (daily) dark–light cycle.

In animals that breed during specific seasons, melatonin alters their capacity for reproduction. An effect of melatonin on human reproductive function, however, has not been demonstrated convincingly. Rather, it seems to contribute to setting the timing of the body's biological clock, which is controlled from the suprachiasmatic nucleus. Research also indicates that small doses of melatonin can reset daily rhythms, which could be a boon to those whose work shift alters between daylight and nighttime hours.

SEASONAL AFFECTIVE DISORDER, JET LAG, AND INSOMNIA

Seasonal affective disorder (**SAD**), is a type of depression that arises during the winter months when day-length is short. It is thought to be due, in part, to overproduction of melatonin. Bright light therapy—repeated doses of several hours exposure to artificial light as bright as sunlight—may provide relief. Three to six hours of bright light exposure also appears to speed recovery from jet lag, the tiredness suffered by travelers who cross several time zones. Some cases of insomnia, on the other hand, may be due to inadequate production of melatonin. Research is under way to see if insomniacs are helped by small doses of melatonin 2 hours before bedtime. ■

Figure 18.25   Pathway whereby light slows release of melatonin from the pineal gland.

 In darkness, secretion of melatonin by the pineal gland promotes sleepiness.

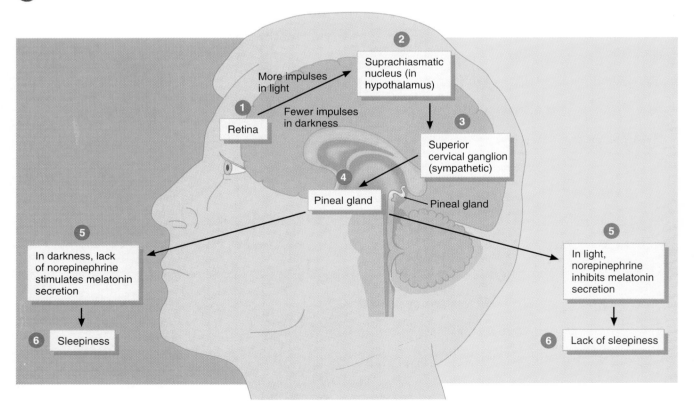

Q To reduce the effects of jet lag by resetting his biological clock, should a New Yorker arriving in London at 9 am (3 am NY time) expose himself to 3 hours of bright light before leaving or after arriving?

THYMUS GLAND

Because of its role in immunity, the details of the structure and functions of the **thymus gland** are discussed in Chapter 22, which deals with the lymphatic system and immunity. Hormones produced by the thymus gland, called **thymosin**, **thymic humoral factor (THF)**, **thymic factor (TF)**, and **thymopoietin**, promote the proliferation and maturation of T cells (a type of white blood cell), which destroy microbes and foreign substances. There is also some evidence that thymic hormones may retard the aging process.

AGING AND THE ENDOCRINE SYSTEM

The endocrine system exhibits a variety of changes, and many researchers look to this system with the hope of finding the key to the aging process. Disorders of the endocrine system most often are related to pathologic changes rather than age. Diabetes mellitus and thyroid disorders, especially hypothyroidism, are relatively common endocrine problems that have a significant effect on health. Sex hormone levels decrease with age in both sexes, but elderly males can still produce active sperm in normal numbers.

OTHER ENDOCRINE TISSUES

There are body tissues other than those normally classified as endocrine glands that contain endocrine cells and thus secrete hormones. These are summarized in Exhibit 18.10.

EICOSANOIDS

Two families of eicosanoid molecules, the **prostaglandins** (pros'-ta-GLAN-dins), or **PGs**, and the **leukotrienes** (loo-kō-TRĪ-ēns), or **LTs**, act as local hormones in most tissues of the body. They are synthesized by clipping a 20-carbon fatty acid called **arachidonic acid** from membrane phospholipid molecules. From arachidonic acid, different enzymatic reactions produce PGs or LTs. **Thromboxane (TX)** is a modified PG that constricts blood vessels and promotes platelet aggregation. Virtually all body cells except red blood cells release these local hormones in response to chemical and mechanical stimuli. They act mainly as potent paracrines and autocrines and appear in blood only in minute quantities. Also, they are present only briefly due to rapid inactivation.

To exert their effects, eicosanoids bind to receptors on target cell plasma membranes and stimulate or inhibit the

EXHIBIT 18.10 SUMMARY OF HORMONES PRODUCED BY TISSUES THAT CONTAIN ENDOCRINE CELLS

Site of Production and Hormones	Action
GASTROINTESTINAL TRACT	
Gastrin	Promotes secretion of gastric juice and increases movements of the gastrointestinal tract.
Gastric inhibitory peptide (GIP)	Inhibits secretion of gastric juice, decreases movements of the gastrointestinal tract, and stimulates release of insulin by pancreas.
Secretin	Stimulates secretion of pancreatic juice and bile.
Cholecystokinin (CCK)	Stimulates secretion of pancreatic juice, regulates release of bile from the gallbladder, and brings about a feeling of fullness after eating.
PLACENTA	
Human chorionic gonadotropin (hCG)	Stimulates the corpus luteum in the ovary to continue the production of estrogens and progesterone to maintain pregnancy.
Estrogens and progesterone	Maintain pregnancy and prepare mammary glands to secrete milk.
Human chorionic somatomammotropin (hCS)	Stimulates the development of the mammary glands for lactation.
KIDNEY	
Erythropoietin (EPO)	Increases rate of red blood cell production.
Calcitriol[a] (active form of Vitamin D)	Aids in the absorption of dietary calcium and phosphorus.
HEART	
Atrial natriuretic peptide (ANP)	Decreases blood pressure.

[a] Synthesis begins in the skin, continues in the liver, and ends in the kidneys

synthesis of second messengers such as cyclic AMP. Leukotrienes stimulate chemotaxis of white blood cells and mediate inflammation. The broad range of biological activities of prostaglandins indicates their importance in both normal physiology and pathology. They alter smooth muscle contraction, glandular secretions, blood flow, reproductive processes, platelet function, respiration, nerve impulse transmission, fat metabolism, and immune responses. They also have roles in inflammation, promoting fever, and intensifying pain. What has intrigued investigators even more than the physiological role of prostaglandins has been their pharmacological effects and their implications for potential therapeutic effects. These effects include lowering or raising blood pressure; reducing gastric secretion; dilation or constriction of airways in the lungs; stimulating or inhibiting blood platelet aggregation; contracting or relaxing intestinal and uterine smooth muscle; mediating inflammation; inducing labor in pregnancy; stimulating steroid production; and promoting diuresis.

NONSTEROIDAL ANTI-INFLAMMATORY DRUGS AND GLUCOCORTICOIDS

In 1971 scientists solved the long-standing puzzle of how aspirin works. Aspirin and related **nonsteroidal anti-inflammatory drugs** (**NSAIDs**), such as ibuprofen (Motrin®), inhibit a key enzyme in prostaglandin synthesis without affecting synthesis of leukotrienes. They are used to treat a wide variety of inflammatory disorders, from rheumatoid arthritis to tennis elbow. The success of NSAIDs in reducing fever, pain, and inflammation points to PGs as contributing to these woes.

Glucocorticoids, such as hydrocortisone (cortisol) and prednisone, slow production of both PGs and LTs by inhibiting the release of arachidonic acid from phospholipids. A major disadvantage of using glucocorticoids to treat chronic inflammation is their tendency to produce effects similar to Cushing's syndrome. ■

GROWTH FACTORS

Besides the well-defined hormones that stimulate cell growth and division (insulin-like growth factor, thymosin, insulin, thyroid hormones, human growth hormone, prolactin, and erythropoietin), several more recently discovered hormones called **growth factors** play important roles. Many growth factors act locally, as autocrines or paracrines. A summary of some known growth factors is presented in Exhibit 18.11.

STRESS AND THE GENERAL ADAPTATION SYNDROME

Homeostatic mechanisms *attempt* to counteract the everyday stresses of living. If they are successful, the internal environment maintains normal physiological limits of chemistry, temperature, and pressure. If a stress is extreme, unusual, or long-lasting, however, the normal mechanisms may not be sufficient. Then, the stress triggers a wide-ranging set of bodily changes called the **stress response** or **general adaptation syndrome** (**GAS**). Hans Selye, a pio-

EXHIBIT 18.11 SUMMARY OF SELECTED GROWTH FACTORS

Growth Factor	Comment
Epidermal growth factor (EGF)	Produced in submaxillary (salivary) glands; stimulates proliferation of epithelial cells, fibroblasts, neurons, and astrocytes; suppresses some cancer cells and secretion of gastric juice by the stomach.
Platelet-derived growth factor (PDGF)	Produced in blood platelets; stimulates proliferation of several cell types, including neuroglia, smooth muscle fibers, and fibroblasts; appears to have a role in wound healing; may contribute to the development of atherosclerosis.
Fibroblast growth factor (FGF)	Found in pituitary gland and brain; stimulates proliferation of many cells derived from embryonic mesoderm (fibroblasts, adrenocortical cells, smooth muscle fibers, chondrocytes, and endothelial cells); also stimulates cell migration and growth and production of the adhesion protein fibronectin.
Nerve growth factor (NGF)	Produced in submaxillary (salivary) glands and hippocampus of brain; stimulates the growth of ganglia in embryonic life, maintains sympathetic nervous system; stimulates hypertrophy and differentiation of neurons.
Tumor angiogenesis factors (TAFs)	Produced by normal and tumor cells; stimulate growth of new capillaries, organ regeneration, and wound healing.
Transforming growth factors (TGFs)	Produced by various cells as separate molecules called TGF-alpha and TGF-beta. TGF-alpha has activities similar to epidermal growth factor (EGF) and TGF-beta inhibits proliferation of many cell types.

neer in stress research, introduced the concept of the GAS. Unlike homeostatic mechanisms, the GAS does not maintain the normal internal environment. It resets the levels of controlled conditions to prepare the body to meet an emergency. For instance, blood pressure and blood glucose level are raised above normal.

It is impossible to remove all stress from our everyday lives. Some stress, called **eustress** (*eu* = true), prepares us to meet certain challenges and thus is productive. Other stress, called **distress**, is harmful. Distress may lower resistance to infection by temporarily inhibiting certain components of the immune system.

Stressors

Any stimulus that produces a stress response is called a **stressor**. A stressor may be almost any disturbance—heat or cold, environmental poisons, toxins given off by bacteria during a raging infection, heavy bleeding from a wound or surgery, or a strong emotional reaction. Stressors may be pleasant or unpleasant and will vary among different people and even in the same person at different times.

When a stressor appears, it stimulates the hypothalamus to initiate the GAS through two pathways. The first pathway produces an immediate set of responses called the alarm reaction. The second pathway, called the resistance reaction, is slower to start, but its effects last longer.

Alarm Reaction

The **alarm reaction** or **fight-or-flight response** is a complex of reactions initiated by hypothalamic stimulation of the sympathetic division of the ANS and the adrenal medulla (summarized in Fig. 18.26a and described in Chapter 17 on page 496). It might occur in a situation in which a person is being attacked by a dog. The responses are immediate, mobilizing the body's resources for immediate physical activity. In essence, the alarm reaction brings huge amounts of glucose and oxygen to the organs that are most active in warding off danger. These are the brain, which must become highly alert; the skeletal muscles, which may have to fight off an attacker; and the heart, which must work furiously to pump enough materials to the brain and muscles. Overall, the stress responses of the alarm stage increase circulation, promote catabolism for ATP production, and decrease nonessential activities. During the alarm reaction, digestive, urinary, and reproductive activities are inhibited. If the stress is great enough, the body mechanisms may not be able to cope and death can result.

Resistance Reaction

The second stage in the stress response is the **resistance reaction** (Fig. 18.26b). Unlike the short-lived alarm reaction, which is initiated by nerve impulses from the hypothalamus, the resistance reaction is initiated in large part by hypothalamic hormones and is a long-term reaction. The hormones are corticotropin releasing hormone (CRH), growth hormone releasing hormone (GHRH), and thyrotropin releasing hormone (TRH).

CRH stimulates the anterior pituitary gland to increase secretion of ACTH, which stimulates the adrenal cortex to secrete more cortisol. Cortisol stimulates the conversion of noncarbohydrates into glucose (gluconeogenesis) and enhances protein catabolism. It makes blood vessels more sensitive to stimuli that bring about their constriction. This response counteracts a drop in blood pressure caused by bleeding. Cortisol also reduces inflammation and prevents it from becoming disruptive rather than protective. Unfortunately, cortisol also discourages formation of new connective tissue. Wound healing is therefore slow during a prolonged resistance stage.

TRH causes the anterior pituitary gland to secrete thyroid-stimulating hormone (TSH); GHRH causes it to secrete human growth hormone (hGH). TSH stimulates the thyroid to secrete T_3 and T_4, which stimulate production of ATP from glucose. hGH stimulates the catabolism of triglycerides and the conversion of glycogen to glucose (glycogenolysis). The combined actions of TSH and hGH increase catabolism and thereby supply additional ATP for metabolically active cells.

Increased secretion of aldosterone by the adrenal cortex acts to conserve Na^+ and eliminate hydrogen ions, which tend to build up as a result of increased catabolism. Thus, during stress, a lowering of body pH is prevented. Na^+ retention also leads to water retention, thus maintaining the high blood pressure of the alarm reaction. Water retention also would help make up for fluid lost through severe bleeding.

The resistance stage of the stress response allows the body to continue fighting a stressor long after the alarm reaction dissipates. It also provides the ATP, enzymes, and circulatory changes needed to meet emotional crises, perform strenuous tasks, or resist the threat of bleeding to death. During the resistance stage, blood chemistry returns to nearly normal. The cells use glucose at the same rate it enters the bloodstream. Thus blood glucose level returns to normal.

Generally, the resistance stage is successful in seeing us through a stressful episode, and our bodies then return to normal. Occasionally, the resistance stage fails to combat the stressor, however, and the GAS moves into the state of exhaustion.

Exhaustion

In the GAS model proposed by Selye, the resources of the body may eventually become depleted such that the resistance stage cannot be sustained and a state termed **exhaustion** occurs. Chronic and severe stressors may deplete the hormonal and other mechanisms of the resistance reaction and lead eventually to failure of mechanisms essential for homeostasis. An alternative explanation involving the body's response to stressors is that pathological changes may occur because resistance mechanisms persist even after the end of the stressor. In summary, stress-related disease may result from failure to initiate appropriate mechanisms of the alarm and resistance reactions of the GAS or from failure to adequately combat the stress (exhaustion).

Figure 18.26 Responses to stressors during the general adaptation syndrome (GAS). Red arrows (hormonal responses) and green arrows (neural responses) indicate immediate alarm ("fight-or-flight") reactions. Black arrows indicate long-term resistance reactions.

🔑 *Stressors stimulate the hypothalamus to initiate the GAS through the alarm reaction and the resistance reaction.*

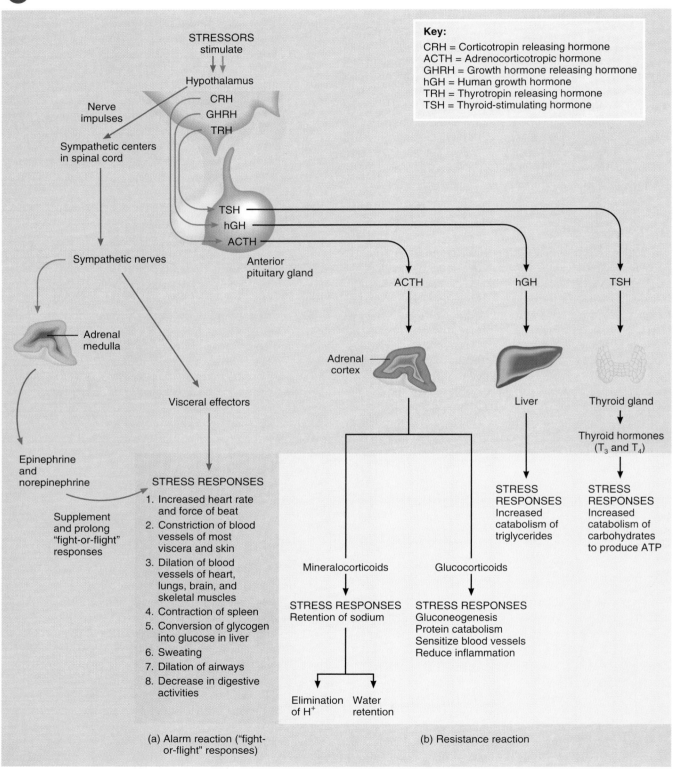

Key:
CRH = Corticotropin releasing hormone
ACTH = Adrenocorticotropic hormone
GHRH = Growth hormone releasing hormone
hGH = Human growth hormone
TRH = Thyrotropin releasing hormone
TSH = Thyroid-stimulating hormone

STRESSORS
stimulate

Hypothalamus
CRH
GHRH
TRH

Nerve impulses

Sympathetic centers in spinal cord

TSH
hGH
ACTH

Anterior pituitary gland

Sympathetic nerves

Adrenal medulla

ACTH

hGH

TSH

Adrenal cortex

Liver

Thyroid gland

Thyroid hormones (T₃ and T₄)

Visceral effectors

Epinephrine and norepinephrine

Supplement and prolong "fight-or-flight" responses

STRESS RESPONSES
1. Increased heart rate and force of beat
2. Constriction of blood vessels of most viscera and skin
3. Dilation of blood vessels of heart, lungs, brain, and skeletal muscles
4. Contraction of spleen
5. Conversion of glycogen into glucose in liver
6. Sweating
7. Dilation of airways
8. Decrease in digestive activities

STRESS RESPONSES
Increased catabolism of triglycerides

STRESS RESPONSES
Increased catabolism of carbohydrates to produce ATP

Mineralocorticoids

Glucocorticoids

STRESS RESPONSES
Retention of sodium

STRESS RESPONSES
Gluconeogenesis
Protein catabolism
Sensitize blood vessels
Reduce inflammation

Elimination of H⁺

Water retention

(a) Alarm reaction ("fight-or-flight" responses)

(b) Resistance reaction

❓ What is the basic difference between the GAS and homeostasis?

Problems may also arise when the stress response doesn't terminate appropriately at the end of stress or if the stress response is activated for too long a period of time. In addition, one's general health determines, to a large degree, the ability to withstand stressors.

Stress and Disease

Although the exact role of stress in human diseases is not known, it is clear that stress can lead to certain diseases by temporarily inhibiting certain components of the immune system. Stress-related disorders include gastritis, ulcerative colitis, irritable bowel syndrome, hypertension, asthma, rheumatoid arthritis (RA), migraine headaches, anxiety, and depression. It has also been shown that people under stress are at a greater risk of developing chronic disease or dying prematurely.

Interleukin-1 (IL-1), a cytokine secreted by macrophages of the immune system (see page 690), is an important link between stress and immunity. In response to infection, inflammation, and other stressors, IL-1 stimulates production of immune substances by the liver, increases the number of circulating neutrophils (white blood cells that are phagocytes), activates cells that participate in immunity, and induces fever. All these responses result in a powerful immune response. However, IL-1 also stimulates the secretion of ACTH. ACTH, in turn, stimulates the production of cortisol, which not only provides resistance to stress and inflammation but also suppresses further production of IL-1. Thus the immune system turns on the stress response. Such a negative feedback system would keep the immune response in check once it has accomplished its goal. Since glucocorticoids, such as cortisol, do suppress certain aspects of the immune response, they are used as immunosuppressive drugs after organ transplantation.

DEVELOPMENTAL ANATOMY OF THE ENDOCRINE SYSTEM

The development of the endocrine system is not as localized as the development of other systems because endocrine organs develop in widely separated parts of the embryo.

The *pituitary gland* (*hypophysis*) originates from two different regions of the **ectoderm**. The *posterior pituitary gland* (*neurohypophysis*) derives from an outgrowth of ectoderm called the **neurohypophyseal bud**, located on the floor of the hypothalamus (Fig. 18.27). The

Figure 18.27 Development of the endocrine system.

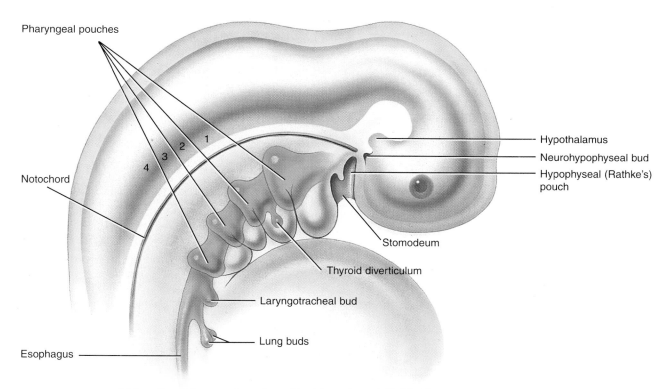

(a) Location of the neurohypophyseal bud, hypophyseal (Rathke's) pouch, and thyroid diverticulum in a 28-day embryo

Figure continues

Figure 18.27   (continued)

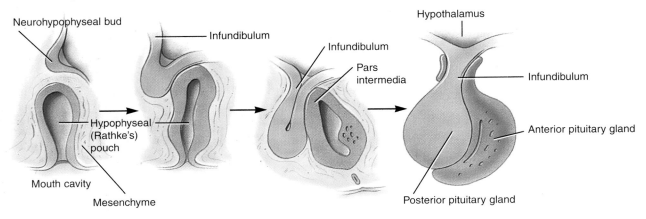

(b) Development of the pituitary gland between 5 and 16 weeks

infundibulum, also an outgrowth of the neurohypophyseal bud, connects the posterior pituitary gland to the hypothalamus. The *anterior pituitary gland (adenohypophysis)* is derived from an outgrowth of ectoderm from the roof of the mouth called the **hypophyseal (Rathke's) pouch**. The pouch grows toward the neurohypophyseal bud and eventually loses its connection with the roof of the mouth.

The *thyroid gland* develops as a midventral outgrowth of **endoderm**, called the **thyroid diverticulum**, from the floor of the pharynx at the level of the second pair of pharyngeal pouches (Fig. 18.27). The outgrowth projects inferiorly and differentiates into the right and left lateral lobes and the isthmus of the gland.

The *parathyroid glands* develop from **endoderm** as outgrowths from the third and fourth **pharyngeal pouches** (Fig. 18.27).

The adrenal cortex and adrenal medulla have completely different embryological origins. The *adrenal cortex* is derived from intermediate **mesoderm** from the same region that produces the gonads. The *adrenal medulla* is **ectodermal** in origin and derives from the **neural crest**, which also gives rise to sympathetic ganglia and other structures of the nervous system (see Fig. 14.18b).

The *pancreas* develops from two outgrowths of **endoderm** from the part of the **foregut** that later becomes the duodenum (see Fig. 24.29). The two outgrowths eventually fuse to form the pancreas. The origin of the ovaries and testes is discussed in the section on the reproductive system.

The *pineal gland* arises as an outgrowth between the thalamus and colliculi from **ectoderm** associated with the **diencephalon** (see Fig. 14.19b).

The *thymus gland* arises from **endoderm** of the third **pharyngeal pouches** (Fig. 18.27a).

STUDY OUTLINE

ENDOCRINE GLANDS (p. 502)

1. Exocrine glands (sudoriferous, sebaceous, digestive) secrete their products through ducts into body cavities or onto body surfaces.
2. Endocrine glands secrete hormones into the blood.
3. The endocrine system consists of endocrine glands and several organs that contain endocrine tissue (see Fig. 18.1).
4. Hormones regulate the internal environment, metabolism, and energy balance.
5. They also help regulate muscular contraction, glandular secretion, and certain immune responses.
6. Hormones play a role in growth, development, and reproduction.

COMPARISON OF NERVOUS AND ENDOCRINE SYSTEMS (p. 503)

1. The nervous system controls homeostasis through nerve impulses; the endocrine system uses hormones.
2. The nervous system causes muscles to contract and glands to secrete; the endocrine system affects virtually all body tissues.

HORMONES (p. 503)

1. Hormones only affect specific target cells that have receptors to recognize (bind) a given hormone.

2. The number of hormone receptors may decrease (down-regulation) or increase (up-regulation).
3. Chemically, hormones are classified as steroids, biogenic amines, peptides or proteins, and eicosanoids. One gas—nitric oxide—is also a hormone.
4. Water-soluble hormones circulate in free form in the blood; lipid-soluble steroid and thyroid hormones are carried attached to transport proteins.

MECHANISMS OF HORMONE ACTION (p. 506)

1. Lipid-soluble steroid hormones and thyroid hormones affect cell function by altering gene expression.
2. Water-soluble hormones alter cell function by activating plasma membrane receptors, which initiate a cascade of events inside the cell.
3. Three hormonal interactions are the permissive effect, synergistic effect, and antagonistic effect.

CONTROL OF HORMONE SECRETION (p. 508)

1. Regulation of secretion is designed to prevent oversecretion or undersecretion.
2. Hormone secretion is controlled by signals from the nervous system, chemical changes in blood, and other hormones.
3. Most often, negative feedback systems regulate hormonal secretions.

HYPOTHALAMUS AND PITUITARY GLAND (p. 509)

1. The hypothalamus is the major integrating link between the nervous and endocrine systems.
2. The hypothalamus and pituitary gland regulate virtually all aspects of growth, development, metabolism, and homeostasis.
3. The pituitary gland is located in the sella turcica and is divided into the anterior pituitary gland and the posterior pituitary gland.
4. Hormones of the anterior pituitary gland are controlled by releasing or inhibiting hormones produced by the hypothalamus.
5. The blood supply to the anterior pituitary gland is from the superior hypophyseal arteries. Hypothalamic releasing and inhibiting hormones enter the primary plexus and are carried to the anterior pituitary gland by the hypophyseal portal veins.
6. Histologically, the anterior pituitary gland consists of somatotrophs that produce human growth hormone (hGH); lactotrophs that produce prolactin (PRL); corticotrophs that secrete adrenocorticotropic hormone (ACTH) and melanocyte-stimulating hormone (MSH); thyrotrophs that secrete thyroid-stimulating hormone (TSH); and gonadotrophs that synthesize follicle-stimulating hormone (FSH) and luteinizing hormone (LH).
7. hGH stimulates body growth through insulin-like growth factors (IGFs). Secretion of hGH is inhibited by GHIH (growth hormone inhibiting hormone, or somatostatin) and promoted by GHRH (growth hormone releasing hormone, or somatocrinin) and TRH (thyrotropin releasing hormone).
8. Disorders associated with abnormal levels of hGH are pituitary dwarfism, giantism, and acromegaly.
9. TSH regulates thyroid gland activities. Its secretion is stimulated by TRH (thyrotropin releasing hormone) and suppressed by GHIH.

10. FSH regulates the activities of the ovaries and testes and is controlled by GnRH (gonadotropin releasing hormone).
11. LH regulates the activities of the ovaries and testes and is controlled by GnRH.
12. PRL helps initiate milk secretion. PIH (prolactin inhibiting hormone) and TRH suppress secretion of PRL whereas PRH (prolactin releasing hormone) stimulates PRL secretion.
13. ACTH regulates the activities of the adrenal cortex and is controlled by CRH (corticotropin releasing hormone).
14. MSH is controlled by corticotropin releasing hormone (CRH).
15. The neural connection between the hypothalamus and posterior pituitary gland is via the supraopticohypophyseal tract.
16. Hormones made by the hypothalamus and stored in the posterior pituitary gland are oxytocin (OT), which stimulates contraction of the uterus and ejection of milk, and antidiuretic hormone (ADH), which stimulates water reabsorption by the kidneys and arteriole constriction. ADH is also known as vasopressin.
17. OT secretion is controlled by uterine distension and nursing; ADH secretion is controlled by osmotic pressure of the blood and blood volume.
18. A disorder associated with lack of ADH or insensitivity of the kidneys to ADH is diabetes insipidus.

THYROID GLAND (p. 520)

1. The thyroid gland is located inferior to the larynx.
2. Histologically, the thyroid gland consists of thyroid follicles composed of follicular cells, which secrete the thyroid hormones thyroxine (T_4) and triiodothyronine (T_3), and parafollicular cells, which secrete calcitonin (CT).
3. Thyroid hormones are synthesized from iodine and tyrosine within thyroglobulin (TGB) and transported in the blood bound to plasma proteins, mostly thyroxine-binding globulin (TBG).
4. Thyroid hormones regulate the rate of metabolism, growth, and development. Their secretion is controlled by TSH.
5. Cretinism, myxedema, Graves' disease, and goiter are disorders associated with the thyroid gland.
6. Calcitonin (CT) lowers the blood level of calcium ions (Ca^{2+}). Secretion of CT is controlled by Ca^{2+} level in the blood.

PARATHYROID GLANDS (p. 525)

1. The parathyroid glands are embedded on the posterior surfaces of the lateral lobes of the thyroid.
2. The parathyroids consist of principal and oxyphil cells.
3. Parathyroid hormone (PTH) regulates the homeostasis of calcium and phosphate ions by increasing blood calcium level and decreasing blood phosphate level. Secretion is controlled by Ca^{2+} level in the blood.
4. Hypoparathyroidism leads to a deficiency of Ca^{2+}, which causes tetany.

ADRENAL GLANDS (p. 527)

1. The adrenal glands are located superior to the kidneys. They consist of an outer adrenal cortex and inner adrenal medulla.
2. Histologically, the adrenal cortex is divided into a zona glomerulosa, zona fasciculata, and zona reticularis; the adrenal medulla consists of chromaffin cells and large blood vessels.

3. Cortical secretions are mineralocorticoids, glucocorticoids, and androgens.

4. Mineralocorticoids (mainly aldosterone) increase sodium and water reabsorption and decrease potassium reabsorption. Secretion is controlled by the renin–angiotensin pathway and K$^+$ level in the blood.

5. A dysfunction related to aldosterone secretion is aldosteronism.

6. Glucocorticoids (mainly cortisol) promote normal organic metabolism, help resist stress, and serve as anti-inflammatory substances. Their secretion is controlled by ACTH.

7. Disorders associated with glucocorticoid secretion are Addison's disease and Cushing's syndrome.

8. Androgens secreted by the adrenal cortex usually have minimal effects. Excessive production occurs in congenital adrenal hyperplasia (CAH).

9. Medullary secretions are epinephrine and norepinephrine (NE), which produce effects similar to sympathetic responses. They are released during stress.

10. Tumors of medullary chromaffin cells are called pheochromocytomas.

PANCREAS (p. 533)

1. The pancreas is posterior and slightly inferior to the stomach.

2. Histologically, it consists of pancreatic islets or islets of Langerhans (endocrine cells) and clusters of enzyme-producing cells (acini). Four types of cells in the endocrine portion are alpha cells, beta cells, delta cells, and F cells.

3. Alpha cells secrete glucagon, beta cells secrete insulin, delta cells secrete somatostatin, and F cells secrete pancreatic polypeptide.

4. Glucagon increases blood glucose level and its secretion is stimulated by a low blood glucose level.

5. Insulin decreases blood glucose level and its secretion is stimulated by a high blood glucose level.

6. Disorders associated with insulin production are diabetes mellitus and hyperinsulinism.

OVARIES AND TESTES (p. 539)

1. The ovaries are located in the pelvic cavity and produce sex hormones related to development and maintenance of feminine secondary sex characteristics, reproductive cycles, pregnancy, lactation, and normal reproductive functions.

2. The testes lie inside the scrotum and produce sex hormones related to the development and maintenance of masculine secondary sex characteristics and normal reproductive functions.

PINEAL GLAND (p. 539)

1. The pineal gland is attached to the roof of the third ventricle.

2. Histologically, it consists of secretory cells called pinealocytes, neuroglial cells, and scattered postganglionic sympathetic fibers.

3. The pineal gland secretes melatonin, which is thought to promote sleepiness. Melatonin is secreted in a diurnal rhythm that is linked to the dark–light cycle, with highest secretion occurring during darkness.

THYMUS GLAND (p. 541)

1. The thymus gland secretes several hormones related to immunity.

2. Thymosin, thymic humoral factor (THF), thymic factor (TF), and thymopoietin promote the maturation of T cells.

AGING AND THE ENDOCRINE SYSTEM (p. 541)

1. Most endocrine disorders are related to pathologies rather than age.

2. Diabetes mellitus and thyroid disorders are common endocrine disorders.

OTHER ENDOCRINE TISSUES (p. 541)

1. The gastrointestinal (GI) tract synthesizes several hormones, including gastrin, gastric inhibitory peptide (GIP), secretin, and cholecystokinin (CCK).

2. The placenta produces human chorionic gonadotropin (hCG), estrogens, progesterone, relaxin, and human chorionic somatomammotropin.

3. The kidneys release erythropoietin, which stimulates red blood cell production, and calcitriol, the active form of vitamin D.

4. The atria of the heart produce atrial natriuretic peptide (ANP).

EICOSANOIDS (p. 541)

1. Prostaglandins and leukotrienes act as paracrines and autocrines in most body tissues by altering the production of second messengers, such as cyclic AMP.

2. Prostaglandins have a wide range of biological activity in normal physiology and pathology.

GROWTH FACTORS (p. 542)

1. Growth factors are local hormones that stimulate cell growth and division.

2. Examples include epidermal growth factor (EGF), platelet-derived growth factor (PDGF), fibroblast growth factor (FGF), nerve growth factor (NGF), tumor angiogenesis factors (TAFs), and transforming growth factors (TGFs).

STRESS AND THE GENERAL ADAPTATION SYNDROME (p. 542)

1. If a stress is extreme or unusual, it triggers a wide-ranging set of bodily changes called the general adaptation syndrome (GAS). The stimuli that produce the general adaptation syndrome are called stressors.

2. Stressors include surgical operations, poisons, infections, fever, and strong emotional responses.

3. Productive stress is termed eustress and harmful, nonproductive stress is termed distress.

4. The alarm reaction is initiated by nerve impulses from the hypothalamus to the sympathetic division of the autonomic nervous system and adrenal medulla.

5. Responses are the immediate and brief fight-or-flight responses that increase circulation, promote catabolism for energy production, and decrease nonessential activities.

6. The resistance reaction is initiated by hormones secreted by the hypothalamus, most importantly CRH, TRH, and GHRH.

7. Resistance reactions are long term and accelerate catabolism to provide energy to counteract stress.

8. Exhaustion results from depletion of body resources during the resistance stage.
9. Stress appears to trigger certain diseases, such as gastritis, ulcerative colitis, irritable bowel syndrome, asthma, anxiety, depression, hypertension, and migraine headaches.
10. A very important link between stress and immunity is interleukin-1 (IL-1) produced by macrophages; it stimulates secretion of ACTH.

DEVELOPMENTAL ANATOMY OF THE ENDOCRINE SYSTEM (p. 545)

1. The development of the endocrine system is not as localized as other systems.
2. The pituitary gland, adrenal medulla, and pineal gland develop from ectoderm; the adrenal cortex develops from mesoderm; and the thyroid gland, parathyroid glands, pancreas, and thymus gland develop from endoderm.

REVIEW QUESTIONS

1. Distinguish between an endocrine gland and an exocrine gland. (p. 502)
2. What are the seven principal effects of hormones? (p. 502)
3. Contrast the nervous and endocrine system control of homeostasis. (p. 503)
4. How are hormones related to receptors? Distinguish down-regulation from up-regulation. (p. 504)
5. Describe the chemical classification of hormones. Give an example of each. (p. 504)
6. How are hormones transported in the blood? (p. 506)
7. Describe the mechanism of hormonal action involving (a) activation of intracellular receptors and (b) interaction with plasma membrane receptors. (p. 506)
8. Distinguish among permissive effects, synergistic effects, and antagonistic effects. (p. 508)
9. How are negative feedback systems related to hormonal control? Discuss the various models of operation. (p. 508)
10. In what respect is the pituitary gland actually two glands? (p. 509)
11. Describe the histology of the anterior pituitary gland. Why does the anterior pituitary gland have such an abundant blood supply? (p. 509)
12. What hormones are produced by the anterior pituitary gland? What are their functions? How are they controlled? (p. 512)
13. Relate the importance of hypothalamic releasing and inhibiting hormones to secretions of the anterior pituitary gland. (p. 513)
14. Describe the clinical symptoms of pituitary dwarfism, giantism, and acromegaly. (p. 514)
15. Discuss the histology of the posterior pituitary gland and the function and regulation of its hormones. Describe the structure and importance of the supraopticohypophyseal tract. (p. 515)
16. What are the clinical symptoms of diabetes insipidus? (p. 519)
17. Describe the location and histology of the thyroid gland. (p. 520)
18. How are the thyroid hormones made, stored, and secreted? (p. 520)
19. Discuss the physiological effects of the thyroid hormones. How is the secretion of these hormones regulated? (p. 522)
20. Discuss the clinical symptoms of cretinism, myxedema, Graves' disease, and goiter. (p. 524)
21. Describe the function and control of calcitonin (CT). (p. 524)
22. Where are the parathyroid glands located? What is their histology? (p. 525)

23. What are the functions of parathyroid hormone (PTH)? (p. 525)
24. Discuss the clinical symptoms of tetany. (p. 527)
25. Compare the adrenal cortex and adrenal medulla with regard to location and histology. (p. 529)
26. Describe the hormones produced by the adrenal cortex in terms of type, normal function, and control. (p. 529)
27. Describe the clinical symptoms of aldosteronism, Addison's disease, Cushing's syndrome, and congenital adrenal hyperplasia. (pp. 531–532)
28. What relationship does the adrenal medulla have to the autonomic nervous system? What is the action of adrenal medullary hormones? (p. 533)
29. What is a pheochromocytoma? (p. 533)
30. Describe the location of the pancreas and the histology of the pancreatic islets. (p. 533)
31. What are the actions of glucagon, insulin, somatostatin, and pancreatic polypeptide? How are blood levels of glucagon and insulin controlled? (p. 536)
32. Describe the clinical symptoms of diabetes mellitus and hyperinsulinism. Distinguish the types of diabetes mellitus. (p. 537)
33. Why are the ovaries and testes considered to be endocrine glands? (p. 539)
34. Where is the pineal gland located? What are its presumed functions? (p. 539)
35. How are hormones of the thymus gland related to immunity? (p. 541)
36. Describe the effects of aging on the endocrine system. (p. 541)
37. List the hormones secreted by the gastrointestinal tract, placenta, kidneys, skin, and heart. (p. 541)
38. Describe the mode of action and function of prostaglandins and leukotrienes. (p. 541)
39. What are growth factors? List several and explain their actions. (p. 542)
40. Define the general adaptation syndrome (GAS). What is a stressor? (p. 542)
41. How do homeostatic responses differ from stress responses? (p. 542)
42. Outline the reactions of the body during the alarm reaction, resistance reaction, and exhaustion when placed under stress. What is the central role of the hypothalamus during stress? (p. 543)
43. Explain how stress and immunity are related. (p. 545)
44. Describe the development of the endocrine system. (p. 545)

ANSWERS TO QUESTIONS WITH FIGURES

18.1 Secretions of endocrine glands diffuse into the blood; exocrine secretions flow into ducts that lead into body cavities or to the body surface.

18.2 It is a paracrine since it acts on nearby parietal cells without entering the blood.

18.3 Synthesis of new proteins takes several minutes to half an hour, due to the time needed for activation of DNA transcription and translation of the mRNA code of nucleotides into a chain of amino acids. The effects of steroid and thyroid hormones thus appear slowly.

18.4 It translates the presence of the first messenger, the water-soluble hormone, into a response inside the cell.

18.5 They carry blood from the median eminence of the hypothalamus, where hypothalamic releasing and inhibiting hormones are secreted, to the anterior pituitary gland, where these hormones act.

18.6 Thyroid, adrenal cortex, ovaries, testes.

18.7 The response is opposite to the stimulus.

18.8 Giantism.

18.9 Elevated hGH in an adult causes bones to grow wider, in which case old gloves and rings are too small.

18.10 Similarity: both carry hypothalamic hormones to the pituitary gland. Difference: the tract is composed of axons of neurons that extend from hypothalamus to posterior pituitary gland; the portal veins are blood vessels that extend to anterior pituitary gland.

18.11 Uterine contraction and milk ejection.

18.12 Absorption of the water in the intestines would decrease the osmotic pressure of your blood plasma, turning off secretion of ADH, and decreasing the ADH level in your blood.

18.13 Follicular cells secrete T_3 and T_4, also known as thyroid hormones. Parafollicular (C) cells secrete calcitonin.

18.14 Thyroglobulin.

18.15 The thyroid will enlarge (goiter formation) because TSH stimulates growth of the follicles.

18.16 Lack of iodine in the diet → diminished production of T_3 and T_4 → increased release of TSH → growth (enlargement) of thyroid gland → goiter.

18.17 Parafollicular cells secrete calcitonin; principal cells secrete PTH.

18.18 PTH: bone and kidneys; CT: bone; calcitriol: gastrointestinal tract.

18.19 The adrenal glands are superior to the kidneys in the retroperitoneal space, at about the same horizontal level as the pancreas.

18.20 It acts to constrict blood vessels (by causing contraction of smooth muscle), and it stimulates secretion of aldosterone (by zona glomerulosa cells of the adrenal cortex), which, in turn, causes the kidneys to conserve water and increase blood volume.

18.21 Low, due to negative feedback suppression.

18.23 Both.

18.24 It has several effects that are opposite to those of insulin.

18.25 Bright light exposure after arriving may help reset the biological clock to earlier times for falling asleep and awakening.

18.26 Homeostasis maintains controlled conditions typical of a normal internal environment, whereas the GAS resets controlled conditions at a different level to cope with various stressors.

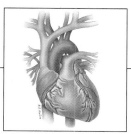

UNIT 4

MAINTENANCE OF THE HUMAN BODY

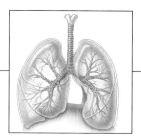

THIS UNIT EXPLAINS HOW THE BODY MAINTAINS homeostasis on a day-to-day basis. In these chapters, you will be studying the interrelations among the cardiovascular, lymphatic, respiratory, digestive, and urinary systems. You will also learn about nonspecific resistance to disease, immunity, metabolism, fluid and electrolyte balance, and acid–base homeostasis.

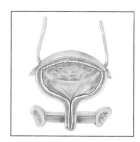

CHAPTER 19

THE CARDIOVASCULAR SYSTEM: THE BLOOD

STUDENT OBJECTIVES

1. Contrast the general roles of blood, lymph, and interstitial fluid in maintaining homeostasis. (p. 553)

2. Define the functions and physical characteristics of the various components of blood. (p. 553)

3. List the components of plasma and explain their importance. (p. 554)

4. Compare the origins, histology, and functions of the formed elements in blood. (p. 554)

5. Describe the mechanisms that contribute to hemostasis. (p. 565)

6. Identify the stages involved in blood clotting and explain the various factors that promote and inhibit blood clotting. (p. 567)

7. Explain the ABO and Rh blood groups. (p. 570)

8. Identify the clinical symptoms of several types of anemia, polycythemia, infectious mononucleosis (IM), and leukemia. (p. 574)

9. Define medical terminology associated with blood. (p. 575)

CHAPTER CONTENTS AT A GLANCE

As cells differentiate and become specialized for particular functions, they also become less capable of an independent existence. They are less able to protect themselves from extreme temperatures, toxic chemicals, and changes in pH. They cannot seek food or devour whole bits of food. And, if they are firmly implanted in a tissue, they cannot move away from their own wastes. The substance that bathes differentiated cells and carries out these vital functions for them is **interstitial fluid** (also known as **intercellular** or **tissue fluid**). The fluid inside lymphatic vessels is called **lymph**. All fluids outside cells, including interstitial fluid, blood plasma (fluid in blood), lymph, and a few others such as aqueous humor in the eyes, comprise **extracellular fluid**.

Because body cells are too specialized to adjust to more than very limited changes in their environment, the internal environment must be kept within normal physiological limits. This is the condition of homeostasis. In preceding chapters, we have discussed how the internal environment is kept in homeostasis. Now we will look at that environment itself, beginning with the blood.

The blood, heart, and blood vessels together make up the **cardiovascular system**. The lymph, lymphatic vessels that transport lymph, and structures and organs containing lymphatic tissue (large clusters of white blood cells called lymphocytes) make up the **lymphatic system** (Chapter 22). The branch of science concerned with the study of blood, blood-forming tissues, and the disorders associated with them is called **hematology** (hēm-a-TOL-ō-jē; *hem* = blood; *logos* = study of).

COMPARISON OF EXTRACELLULAR FLUIDS

The major extracellular fluids—interstitial fluid, blood plasma, and lymph—have similar chemical compositions. Interstitial fluid and lymph, however, contain considerably less protein than is present in plasma. This is because the larger protein molecules are not easily filtered through the endothelial cells that form the walls of capillaries, the smallest blood vessels. The exchange of materials between blood and interstitial fluid occurs by osmosis, diffusion, and bulk flow (filtration and reabsorption) across the endothelial cells. Interstitial fluid and lymph also differ from plasma in that they contain variable numbers of leukocytes (white blood cells). Like plasma, interstitial fluid and lymph lack erythrocytes (red blood cells) and platelets.

Blood plasma and lymph service the interstitial fluid. Blood picks up oxygen from the lungs, nutrients from the gastrointestinal tract, and hormones from endocrine glands. It transports these substances to the tissues, where they diffuse from capillaries into interstitial fluid. From the interstitial fluid, needed substances enter cells and cellular wastes enter the blood. The blood carries carbon dioxide and metabolic wastes to the lungs, kidneys, and sweat glands for elimination from the body. Certain wastes must be detoxified by the liver before they can be excreted. Excess interstitial fluid and the small quantity of plasma proteins that have leaked from blood vessels enter lymphatic capillaries and become part of the lymph. Eventually, lymph drains into the subclavian veins.

Sometimes disease-causing organisms (pathogens) invade the interstitial fluid and blood plasma and thus can be carried throughout the body. Chapter 22 describes how the lymphatic system helps protect the body from such spread of disease.

FUNCTIONS OF BLOOD

Blood is a liquid connective tissue that has three general functions: transportation, regulation, and protection.

1. **Transportation.** Blood transports oxygen from the lungs to the cells of the body and carbon dioxide from the cells to the lungs. It also carries nutrients from the gastrointestinal tract to body cells, heat and waste products away from cells, and hormones from endocrine glands to other body cells.
2. **Regulation.** Blood helps regulate pH through buffers. It also adjusts body temperature through the heat-absorbing and coolant properties of its water content and its variable rate of flow through the skin, where excess heat can be lost from blood to the environment. Blood osmotic pressure also influences the water content of cells, principally through dissolved ions and proteins.
3. **Protection.** The clotting mechanism protects against blood loss, and certain phagocytic white blood cells and plasma proteins (such as antibodies, interferon, and complement) protect against foreign microbes and toxins.

PHYSICAL CHARACTERISTICS OF BLOOD

Blood is heavier, thicker, and more viscous than water. It flows more slowly than water, at least in part because of its viscosity. The adhesive quality of blood, or its stickiness, may be appreciated by touching it. The temperature of blood is about 38°C (100.4°F), which is slightly higher than normal body temperature, and it has a slightly alkaline pH of about 7.40 (normal range 7.35–7.45). Blood constitutes about 8% of the total body weight. The blood volume is 5–6 liters (1.5 gal) in an average-sized adult male and 4–5 liters (1.2 gal) in an average-sized adult female. Several hormonal negative feedback systems ensure that blood volume and osmotic pressure remain relatively constant. Especially important are those involving aldosterone, antidiuretic hormone, and atrial natriuretic peptide, which regulate how much water is excreted in the urine.

WITHDRAWING BLOOD

Blood samples for laboratory testing may be obtained in several ways. The most often used procedure is **venipuncture**, withdrawal of blood from a vein. A commonly used vein is the median cubital vein anterior to the elbow (see Fig. 21.25). A tourniquet is wrapped around the arm to stop blood flow through the veins. This makes the veins distal to the tourniquet stand out. Opening and closing the fist has the same effect.

Another procedure used to withdraw blood is the **finger-stick**. A drop or two of capillary blood is taken from a finger, earlobe, or heel of the foot for evaluation. Diabetics may monitor their blood glucose levels in this way.

Finally, an **arterial stick** may be used to withdraw blood. The sample usually is taken from the radial artery in the wrist or the femoral artery in the groin (see Fig. 21.18). ■

COMPONENTS OF BLOOD

Whole blood is composed of two portions: (1) blood plasma, a watery liquid that contains dissolved substances, and (2) formed elements, which are cells and cell fragments. If a sample of blood is centrifuged (spun) in a small glass tube, the cells sink to the bottom of the tube while the lighter-weight plasma forms a layer on top (Fig. 19.1a). On average, more than 99% of the formed elements are red-colored erythrocytes, also called red blood cells (RBCs). The percentage of the total blood volume occupied by RBCs normally is about 45% and plasma accounts for the remaining 55%. Pale, white-colored leukocytes, or white blood cells (WBCs), and platelets represent less than 1% of the total blood volume. They form a very thin layer, called the buffy coat, between the packed RBCs and plasma. Figure 19.1b shows the composition of blood plasma and the numbers of the various types of formed elements in blood.

Blood Plasma

When the formed elements are removed from blood, a straw-colored liquid called **blood plasma** or simply **plasma** is left. (As you will see later, *serum* is plasma minus its clotting proteins.) Plasma is about 91½% water and 8½% solutes, most of which by weight (7%) are proteins. Some of the proteins in plasma are also found elsewhere in the body, but those confined to blood are called **plasma proteins**. These proteins play a role in maintaining proper

blood osmotic pressure, which is an important factor in exchange of fluid across capillary walls (Chapter 21).

Most plasma proteins are synthesized by hepatocytes (liver cells), including the **albumins** (54% of plasma proteins), most **globulins** (38%), and **fibrinogen** (7%). Plasma cells, which develop from B lymphocytes in lymph nodes and other lymphatic tissues, produce gamma globulins, one of the more important types of globulins. These plasma proteins are also called **antibodies** or **immunoglobulins** because they are produced during certain immune responses. Foreign invaders such as bacteria and viruses stimulate production of millions of different antibodies. An antibody binds to the foreign substance, called an **antigen**, that provoked its production. When the two come together they form an **antigen–antibody complex**, which disables the invading antigen in some way (described on page 696).

Besides proteins, other solutes in plasma include electrolytes; nutrients; regulatory substances such as enzymes and hormones; gases; and waste products, such as urea, uric acid, creatinine, ammonia, and bilirubin.

Exhibit 19.1 on page 556 describes the chemical composition of blood plasma.

Formed Elements

The **formed elements** of the blood are:

> **Erythrocytes (red blood cells)**
> **Leukocytes (white blood cells)**
>   *Granular leukocytes (granulocytes)*
>   Neutrophils
>   Eosinophils
>   Basophils
>   *Agranular leukocytes (agranulocytes)*
>   Lymphocytes (T cells, B cells, and natural killer cells)
>   Monocytes
> **Platelets (thrombocytes)**

FORMATION OF BLOOD CELLS

Although some lymphocytes have a lifetime measured in years, most formed elements of the blood are continually dying and being replaced within hours, days, or weeks. Negative feedback systems regulate the total number of RBCs in circulation and the number normally remains steady. The number of different types of WBCs, however, varies in response to challenges by invading pathogens and antigens.

The process by which formed elements of blood develop is called **hemopoiesis** (hē-mō-poy-Ē-sis; *hemo* = blood; *poiem* = to make) or **hematopoiesis**. About 0.05–0.1% of red bone marrow cells are cells derived from mesenchyme, called **hemopoietic stem cells**, that are able to differentiate and mature into the various formed elements and also

Figure 19.3 Shape of a red blood cell (RBC) and a hemoglobin molecule. In (b) note that each of the four polypeptide chains of one hemoglobin molecule has one heme group, which contains Fe^{2+} (red).

The iron portion of the heme group binds oxygen for transport by hemoglobin.

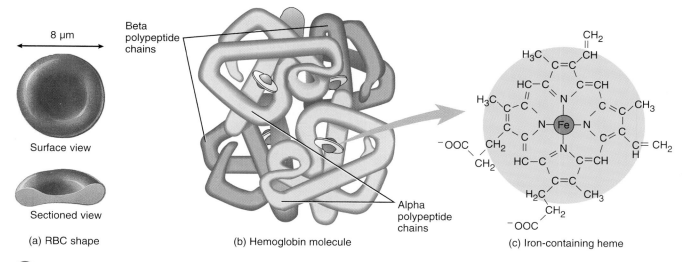

(a) RBC shape

(b) Hemoglobin molecule

(c) Iron-containing heme

How many molecules of O_2 can one hemoglobin molecule transport?

Red blood cells are highly specialized for their oxygen transport function. Each one contains about 280 million hemoglobin molecules. Since RBCs have no nucleus, all their internal space is available for oxygen transport. Moreover, since they lack mitochondria and generate ATP anaerobically (without oxygen), RBCs do not consume any of the oxygen that they transport. Even the shape of a RBC facilitates its function. A biconcave disc has a much greater surface area for its volume than, say, a sphere or a cube. This shape confers two advantages. First, there is a large surface area for the diffusion of gas molecules into or out of the RBC. Second, the biconcave disc is a very flexible shape, which permits RBCs to squeeze through small capillaries, some of which are as narrow as 3 μm in diameter.

A serious disorder called sickle-cell anemia (SCA, described on page 574) is due to a genetic defect that results in substitution of just two out of the 574 amino acids in hemoglobin. Replacement of the polar amino acid glutamate by nonpolar valine at one position in each of the two beta chains greatly decreases the solubility of deoxygenated hemoglobin in water. When this abnormal hemoglobin is exposed to a low level of oxygen, it forms crystals that deform RBCs into a characteristic sickle shape (see Fig. 19.14). Sickled RBCs are more rigid than normal RBCs and may lodge in small capillaries. Interestingly, people who suffer SCA have normal fetal hemoglobin, a slightly different form of hemoglobin that predominates at birth. If a way could be found to reactivate the normal fetal hemoglobin gene, it is thought that the adverse consequences of SCA could be alleviated.

Hemoglobin also transports about 23% of the total carbon dioxide, a waste product of metabolism. Blood flowing through tissue capillaries picks up carbon dioxide, some of which combines with amino acids in the globin portion of hemoglobin to form carbaminohemoglobin. This complex is transported to the lungs, where the carbon dioxide is released and then exhaled.

BLOOD DOPING

In recent years, some athletes have been tempted to try **blood doping**. In this procedure, blood cells are removed from the body, stored for a month or so, and then reinjected a few days before an athletic event. Since delivery of oxygen to muscle is a limiting factor in muscular feats, and red blood cells carry oxygen, it was predicted that increasing the oxygen-carrying capacity of the blood could increase muscular performance. Indeed, blood doping can improve athletic performance in endurance events. The practice is dangerous, however, because it increases the work load of the heart. With increased numbers of RBCs, the viscosity of the blood rises, which makes the blood more difficult for the heart to pump. Moreover, the procedure is banned by the International Olympics Committee. ■

RBC Life Cycle

Red blood cells live only about 120 days because of the wear and tear inflicted on their plasma membranes as they squeeze through blood capillaries. Without a nucleus and other organelles, RBCs cannot synthesize new components to replace damaged ones. The plasma membranes thus become more fragile with age and the cells more likely to burst, especially as they squeeze through narrow channels in the spleen. Worn-out red blood cells are removed from circulation and destroyed by fixed phagocytic macrophages in the spleen and liver and the breakdown products are recycled as follows (Fig. 19.4):

1 Macrophages in the spleen, liver, or red bone marrow phagocytize worn-out red blood cells.

2 The globin and heme portions of hemoglobin are split apart.

3 Globin is broken down into amino acids, which can be

4 reused to synthesize other proteins.

5 Iron removed from the heme portion

6 associates with a plasma protein called **transferrin** (trans-FER-in), which transports iron in the bloodstream.

7 In muscle fibers, liver cells, and macrophages of the spleen and liver, iron detaches from transferrin and attachs to iron-storage proteins called **ferritin** and **hemosiderin** (hē-mō-SID-er-in).

8 Upon release from a storage site or absorption from the gastrointestinal tract, iron attaches to transferrin.

9 It is then transported to bone marrow, where RBC precursors take it up through receptor-mediated endocytosis (see Fig. 3.13)

10 for use in production of new hemoglobin molecules.

11 Erythropoiesis in red bone marrow results in the production of red blood cells, which enter the circulation.

12 At the same time, the non-iron portion of heme is converted to **biliverdin** (bil′-i-VER-din), a green pigment, and then into

13 **bilirubin** (bil′-i-ROO-bin), an orange pigment.

14 Bilirubin enters the blood and is transported to the liver.

15 Within the liver, bilirubin is secreted by liver cells into bile, which passes into the small intestine.

16 In the large intestine bacteria convert bilirubin into **urobilinogen** (yoo′-rō-bi-LIN-ō-jen).

17 Some urobilinogen is absorbed back into the blood, converted to **urobilin** (yoo′-rō-BĪ-lin), a yellow pigment, and excreted in urine.

18 Most urobilinogen is eliminated in feces in the form of a brown pigment called **stercobilin** (ster′-kō-BĪ-lin), which gives feces their characteristic color.

Production of RBCs

The process of erythrocyte formation is called **erythropoiesis** (e-rith′-rō-poy-Ē-sis). It starts in red bone marrow with a proerythroblast (see Fig. 19.2). The **proerythroblast** (**rubriblast**) gives rise to a **basophilic erythroblast** (**prorubricyte**), which then develops into a **polychromatophilic erythroblast** (**rubricyte**), the first cell in the sequence that begins to synthesize hemoglobin. The polychromatophilic erythroblast next develops into an **acidophilic erythroblast** (**normoblast**), in

---

Figure 19.4   Formation and destruction of red blood cells and recycling of hemoglobin components.

🔑 **The rate of RBC formation by red bone marrow equals the rate of RBC destruction by macrophages.**

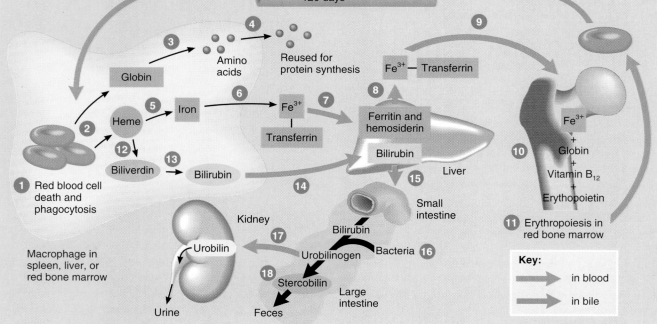

Q What is the function of transferrin?

which hemoglobin synthesis is at a maximum. In the next stage, the acidophilic erythroblast ejects its nucleus and becomes a **reticulocyte**. Loss of the nucleus allows the center of the cell to indent, giving the cell a biconcave shape. Reticulocytes contain about 34% hemoglobin and retain some mitochondria, ribosomes, and endoplasmic reticulum. They pass from red bone marrow into the bloodstream by squeezing between the endothelial cells of blood capillaries. Normally, they develop into **erythrocytes**, or mature red blood cells, 1–2 days after their release from red bone marrow.

Normally, erythropoiesis and red blood cell destruction proceed at the same pace. If the oxygen-carrying capacity of the blood falls because erythropoiesis is not keeping up with RBC destruction, a negative feedback system steps up erythrocyte production (Fig. 19.5). The controlled condition is the amount of oxygen delivered to body tissues. Oxygen delivery may fall due to **anemia**, a lower than normal number of RBCs or quantity of hemoglobin, or to circulatory problems that reduce blood flow to tissues. Cellular oxygen deficiency, called **hypoxia** (hī-POKS-ē-a), may also occur if not enough oxygen enters the blood, for example, when you do not breathe in enough oxygen. This situation often occurs at high altitudes, where the air contains less oxygen. Whatever the cause, hypoxia stimulates the kidneys to step up release of the hormone **erythropoietin** (e-rith′-rō-POY-e-tin). This hormone circulates through the blood to the red bone marrow, where it speeds the development of proerythroblasts into reticulocytes.

Anemia has many causes: lack of iron, lack of certain amino acids, and lack of vitamin B$_{12}$ are but a few. Iron is needed for the heme part of the hemoglobin molecule. The amino acids are needed for the protein, or globin, part. Vitamin B$_{12}$ helps the red bone marrow to produce erythrocytes. This vitamin is obtained from meat, especially liver, but it cannot be absorbed by the lining of the small intestine without the help of another substance—**intrinsic factor** produced by the parietal cells of the stomach mucosa. Intrinsic factor aids the absorption of vitamin B$_{12}$ in the small intestine. Once absorbed, it is stored in the liver. (Several types of anemia are described on page 574.)

RETICULOCYTE COUNT AND HEMATOCRIT

The rate of erythropoiesis is measured by a **reticulocyte** (re-TIK-yoo-lō-sit) **count**. Normally, a little less than 1% of the oldest RBCs are replaced by newcomer reticulocytes on any given day. Then it takes 1–2 days for the reticulocyte to lose the last vestiges of its endoplasmic reticulum and become a mature RBC. Thus reticulocytes account for about ½–1½% of all RBCs in a blood sample. A low "retic" count in a person who is anemic might indicate inability of the red bone marrow to respond to erythropoietin, perhaps because of a nutritional deficiency, lack of intrinsic factor, or leukemia. A high "retic" count might indicate a good red bone marrow response to previous loss of blood or to iron therapy in someone who had been iron-deficient.

Hematocrit (he-MAT-ō-krit), or **Hct**, is the percentage of red blood cells in blood (see Fig. 19.1a). A Hct of 40 means that 40% of the volume of blood is composed of RBCs. The test is used to diag-

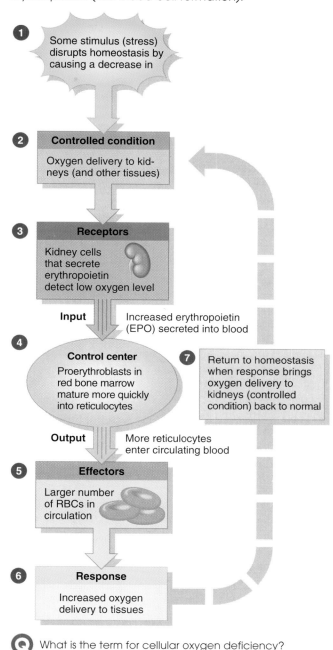

Figure 19.5 Negative feedback regulation of erythropoiesis (red blood cell formation).

1. Some stimulus (stress) disrupts homeostasis by causing a decrease in

2. **Controlled condition**
   Oxygen delivery to kidneys (and other tissues)

3. **Receptors**
   Kidney cells that secrete erythropoietin detect low oxygen level

   Input Increased erythropoietin (EPO) secreted into blood

4. **Control center**
   Proerythroblasts in red bone marrow mature more quickly into reticulocytes

   Output More reticulocytes enter circulating blood

5. **Effectors**
   Larger number of RBCs in circulation

6. **Response**
   Increased oxygen delivery to tissues

7. Return to homeostasis when response brings oxygen delivery to kidneys (controlled condition) back to normal

Q What is the term for cellular oxygen deficiency?

nose anemia and polycythemia (abnormally high percentage of RBCs) and abnormal states of hydration. The normal range of hematocrit for adult females is 38–46% (average 42); for adult males it is 40–54% (average 47). Lower values in women during their reproductive years may be due to loss of menstrual blood. Also, testosterone stimulates synthesis of erythropoietin, which contrbutes to higher hemocrits in men. A significant drop in hematocrit indicates anemia. Premature newborns often exhibit anemia, due in part to inadequate production of erythropoietin. During the first weeks after birth the liver, not the kidneys, produces most EPO. Because the liver is less sensitive than the kidneys to hypoxia, newborns have a smaller

EPO response to anemia than do adults. Polycythemic blood may have a hematocrit of 65% or higher. The average hematocrit of people living in mountainous terrain is greater than that of people living at sea level due to their greater secretion of erythropoietin. ■

LEUKOCYTES (WHITE BLOOD CELLS)

WBC Anatomy and Types

Unlike red blood cells, **leukocytes** (LOO-kō-sīts; *leukos* = white; *cyte* = cell) or **white blood cells** (**WBCs**) have a nucleus and do not contain hemoglobin (see Fig. 19.2). The two major groups of WBCs are granular leukocytes and agranular leukocytes. **Granular leukocytes** (**granulocytes**) develop from myeloblasts (see Fig. 19.2). They have conspicuous granules in the cytoplasm that can be seen under a light microscope. The three types are **eosinophils** (ē-ō-SIN-ō-fils), which are 10–14 μm in diameter and have granules that stain red or orange with acidic dyes; **basophils** (BĀ-sō-fils), which are 8–10 μm in diameter and have granules that stain blue-purple with basic dyes; and **neutrophils** (NOO-trō-fils), which are 10–12 μm in diameter and have granules that are "neutral," staining a pale lilac with a combination of acidic and basic dyes. Common hematology stains, such as Wright's stain are similar to the more familiar hematoxylin and eosin (H&E) stain. Like H&E, they include both acidic eosin dye (pink color) and a basic dye (blue color) as seen in Fig. 19.6.

The nucleus of an eosinophil usually has two lobes connected by a thin or thick strand. Large, uniform-sized granules pack the cytoplasm but usually do not cover or obscure the nucleus. These eosinophilic (= eosin loving) granules stain red-orange. A basophil's nucleus is bilobed or irregular in shape, often in the form of a letter **S**. The cytoplasmic basophilic granules are round, variable in size, stain blue-purple, and commonly obscure the nucleus. The nuclei of neutrophils have two to five lobes, connected by very thin strands of chromatin. As the cells age, the extent of nuclear lobulation increases. Because older neutrophils appear to have many differently shaped nuclei, they are often called **polymorphonuclear leukocytes** (**PMNs**), **polymorphs**, or "**polys**." Younger neutrophils are often called **bands** because their nucleus is more rod-shaped. When stained, the cytoplasm of neutrophils includes fine, evenly distributed pale lilac-colored granules, which may be difficult to see.

Agranular leukocytes (**agranulocytes**) do not have cytoplasmic granules that can be seen under a light microscope, owing to their small size and poor staining qualities. The two kinds of agranular leukocytes are **lymphocytes** (LIM-fō-sīts) and **monocytes** (MON-ō-sīts). Small lymphocytes are 6–9 μm in diameter; large lymphocytes are 10–14 μm in diameter. (Although the functional significance of the size difference between small and large lymphocytes is unclear, the distinction is still useful clinically because an increase in the number of large lymphocytes has diagnostic significance in acute viral infections and some immunodeficiency diseases). Mono-

Figure 19.6 Blood smear and photomicrographs of individual blood cells.

 White blood cells are distinguished from one another by the shape of their nuclei and the presence or absence of granules that are visible with a light microscope.

NEUTROPHIL   EOSINOPHIL   BASOPHIL

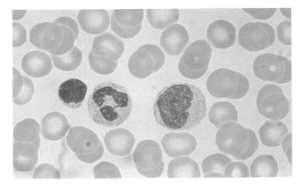

BLOOD SMEAR

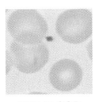

RED BLOOD   MONOCYTE   LARGE
CELLS AND   LYMPHOCYTE
PLATELET

 Which WBCs are called granulocytes? Why?

cytes are 12–20 μm in diameter (Fig. 19.6). Lymphocytes develop from lymphoblasts, and monocytes develop from monoblasts (see Fig. 19.2). The nuclei of lymphocytes are darkly stained and round, or slightly indented. The cytoplasm stains sky blue and forms a rim around the nucleus. The larger the cell, the more cytoplasm is visible. The nucleus of a monocyte is usually kidney-shaped or horseshoe-shaped and the cytoplasm is blue-gray and has a foamy appearance. The blood is merely a conduit for monocytes, which migrate out into the tissues, enlarge, and differentiate into **macrophages** (*macro* = large; *phagein* = to eat). Some are **fixed macrophages**, which means they reside in a particular tissue, for example, alveolar macrophages, spleen macrophages, or stellate reticuloendothelial (Kupffer) cells in the liver. Others are **wandering** (**free**) **macrophages**, which roam the tissues and gather at sites of infection or inflammation.

White blood cells and all other nucleated cells in the body have proteins, called **major histocompatibility** (**MHC**)

antigens protruding from the extracellular surface of their plasma membrane (see Fig. 3.3). These cell identity markers are unique for each person (except for identical twins).

HISTOCOMPATIBILITY TESTING

The success of a proposed organ or tissue transplant depends on **histocompatibility** (his'-tō-kom-pat-i-BIL-i-tē), that is, the tissue compatibility between the donor and the recipient. The more similar the MHC antigens, the greater the histocompatibility, and the higher the probability the transplant will not be rejected. **Tissue typing** (**histocompatibility testing**) is done before any organ transplant. A nationwide computerized registry helps physicians select the most histocompatible and needy organ transplant recipients whenever donor organs become available. Also, in cases of disputed parentage, tissue typing can be used to establish the biological parents. ■

WBC Physiology

In a healthy body, some WBCs, especially lymphocytes, can live for several months or years, but most live only a few days. During a period of infection, phagocytic WBCs may live only a few hours. WBCs are far less numerous than red blood cells, about 5000–10,000 cells per cubic millimeter (mm³) of blood. RBCs therefore outnumber white blood cells about 700:1. The term **leukocytosis** (loo'-kō-sī-TŌ-sis) refers to an increase in the number of WBCs. An abnormally low level of white blood cells (below 5000/mm³) is termed **leukopenia** (loo-kō-PĒ-nē-a).

The skin and mucous membranes of the body are continuously exposed to microbes and their toxins. Some of these microbes can invade deeper tissues to cause disease. Once pathogens enter the body, the general function of white blood cells is to combat them by phagocytosis or immune responses. To accomplish these tasks, many WBCs leave the bloodstream and collect at points of pathogen invasion or inflammation. When granulocytes and monocytes leave the bloodstream to fight injury or infection, they never return. Lymphocytes, on the other hand, continually recirculate from blood to interstitial spaces of tissues to lymphatic fluid and back to blood. Only 2% of the total lymphocyte population is in the blood at any given time. The rest are in lymphatic fluid and organs such as skin, lungs, lymph nodes, and spleen.

WBCs leave the bloodstream by a process termed **emigration**[1] (em'-i-GRĀ-shun; *e* = out; *migrare* = to move) in which they slow down, roll along the endothelium, stop, and then squeeze between endothelial cells (Fig. 19.7). The precise signals that stimulate emigration through a particular blood vessel vary for the different types of WBCs. Molecules known as **adhesion molecules** help WBCs stick to the endothelium. For example, endothelial cells display adhesion molecules called *selectins* in response to nearby injury and inflammation. Selectins stick to carbohydrates on the surface of

---

[1]Synonymous terms are *migration* and *extravasation*. The older term *diapedesis* is no longer applied to this method of WBC exit from blood vessels.

Figure 19.7 Emigration of white blood cells.

Adhesion molecules (selectins and integrins) assist the emigration of WBCs from the bloodstream into interstitial fluid.

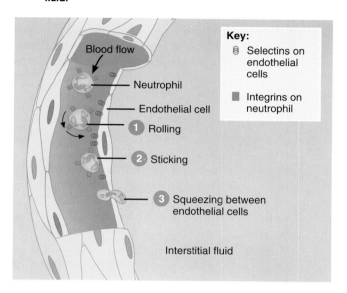

Key:
- Selectins on endothelial cells
- Integrins on neutrophil

Blood flow
Neutrophil
Endothelial cell
1 Rolling
2 Sticking
3 Squeezing between endothelial cells
Interstitial fluid

Q In what way is the traffic of lymphocytes in the body different from that of other WBCs?

neutrophils, causing them to slow down and roll along the endothelial surface. On the neutrophil surface are other adhesion molecules called *integrins,* which then tether neutrophils to the endothelium and assist their movement through the blood vessel wall and into the interstitial fluid of the injured tissue.

Neutrophils and macrophages are active in **phagocytosis**: they can ingest bacteria and dispose of dead matter (see Fig. 3.12). Several different chemicals released by microbes and inflamed tissues attract phagocytes. This phenomenon is called **chemotaxis**. Among the substances that provide stimuli for chemotaxis are toxins produced by microbes, specialized products of damaged tissues called kinins, and some of the colony-stimulating factors. The CSFs also enhance phagocytic activity of neutrophils and macrophages.

Among the WBCs, neutrophils respond to tissue destruction by bacteria most quickly. After engulfing a pathogen during phagocytosis, a neutrophil unleashes several destructive chemicals. These include the enzyme **lysozyme**, which destroys certain bacteria, and **strong oxidants**, such as the superoxide anion (O_2^-), hydrogen peroxide (H_2O_2), and hypochlorite anion (OCl^-), which is similar to household bleach. Neutrophils also contain **defensins**, proteins that exhibit a broad range of antibiotic activity against bacteria, fungi, and viruses. Defensins form peptide spears that poke holes in microbe membranes. The resulting leakiness kills the invader.

Monocytes take longer to reach a site of infection than do neutrophils, but they arrive in larger numbers and destroy more microbes. Upon arrival they enlarge and differentiate into wandering macrophages, which clean up cellular debris and microbes following an infection.

Eosinophils leave the capillaries and enter tissue fluid. They are believed to release enzymes, such as histaminase, that combat the effects of histamine and other mediators of inflammation in allergic reactions. Eosinophils also phagocytize antigen–antibody complexes and are effective against certain parasitic worms. A high eosinophil count often indicates an allergic condition or a parasitic infection.

Basophils are also involved in inflammatory and allergic reactions. They leave capillaries, enter tissues, and develop into mast cells, which can liberate heparin, histamine, and serotonin. These substances intensify the inflammatory reaction and are involved in hypersensitivity (allergic) reactions (Chapter 22).

The major types of lymphocytes are B cells, T cells, and natural killer cells. These cells are the major combatants in defensive responses (described in detail in Chapter 22). B cells are particularly effective in destroying bacteria and inactivating their toxins. T cells attack viruses, fungi, transplanted cells, cancer cells, and some bacteria. Immune responses carried out by B and T cells help combat infection and provide protection against some diseases. They also are responsible for transfusion reactions, allergies, and rejection of transplanted organs. Natural killer cells attack a wide variety of infectious microbes and certain spontaneously arising tumor cells.

An increase in the number of circulating WBCs usually indicates inflammation or infection. To evaluate infection or inflammation, determine the effects of possible poisoning by chemicals or drugs, monitor blood disorders (for example, leukemia) and effects of chemotherapy, or detect allergic reactions and parasitic infections, a physician may order a **differential white blood cell count**. Because each type of white blood cell plays a different role, determining the *percentage* of each type in the blood assists in diagnosing the condition. The percentage of each type of white blood cell in a normal differential white blood cell count is as follows:

	Percentage
Neutrophils	60–70
Lymphocytes	20–25
Monocytes	3–8
Eosinophils	2–4
Basophils	½–1
	100

A high neutrophil count might result from bacterial infections, burns, stress, or inflammation; a low count might be caused by radiation, certain drugs, vitamin B_{12} deficiency, or systemic lupus erythematosus (SLE). A high eosinophil count could indicate allergic reactions, parasitic infections, autoimmune disease, or adrenal insufficiency; a low count could be caused by certain drugs, stress, or Cushing's syndrome. Basophils could be elevated in some types of allergic responses, leukemias, cancers, and hypothyroidism; decreases could occur during pregnancy, ovulation, stress, and hyperthyroidism. High lymphocyte counts could indicate viral infections, immune diseases, and some leukemias; low counts might occur as a result of prolonged severe illness,

high steroid levels, and immunosuppression. Finally, a high monocyte count could result from certain viral or fungal infections, tuberculosis (TB), some leukemias, and chronic diseases; below normal monocyte levels rarely occur.

BONE MARROW TRANSPLANT

A **bone marrow transplant** is the transfer of red bone marrow from a healthy donor to a recipient. The donor marrow must be very closely matched to that of the recipient through histocompatibility testing. In the procedure, donor marrow is withdrawn from the hipbones, mixed with heparin (to prevent blood clotting), and then passed through screens. The suspension of red bone marrow cells is treated to remove T cells and injected into a vein. Stem cells in the suspension pass through the lungs, enter the general circulation, and reseed and grow in the marrow cavities of the recipient's bones. In patients with cancer or certain genetic diseases, the recipient's faulty red bone marrow first must be destroyed by irradiation before transplantation of the healthy marrow.

Bone marrow transplants have been used to treat aplastic anemia, certain types of leukemia, severe combined immunodeficiency disease (SCID), Hodgkin's disease, non-Hodgkin's lymphoma, multiple myeloma, sickle-cell anemia (SCA), breast cancer, ovarian cancer, testicular cancer, and hemolytic anemia. ■

PLATELETS

Besides the immature cell types that develop into erythrocytes and leukocytes, hemopoietic stem cells also differentiate into megakaryoblasts (see Fig. 19.2). Under the influence of a hormone known as **thrombopoietin**, megakaryoblasts transform into metamegakaryocytes, huge cells that splinter into 2000–3000 fragments. Each fragment, enclosed by a piece of the cell membrane, is a **platelet** (**thrombocyte**). Platelets break off from the metamegakaryocytes in red bone marrow and then enter the blood circulation. Between 250,000 and 400,000 platelets are present in each cubic millimeter (mm^3) of blood. They are disc-shaped, 2–4 μm in diameter, and exhibit many granules but no nucleus. Platelets help stop blood loss from damaged blood vessels by forming a platelet plug. Their granules also contain chemicals that upon release promote blood clotting. Platelets have a short life span, normally just 5–9 days. Aged and dead platelets are removed by fixed macrophages in the spleen and liver.

A summary of the formed elements in blood is presented in Exhibit 19.2.

COMPLETE BLOOD COUNT (CBC)

A **complete blood count** is a valuable test that screens for anemia and various infections. Usually included are a determination of red blood cell count, amount of hemoglobin, hematocrit, white blood cell count, differential white blood cell count, and platelet count. ■

EXHIBIT **19.2 SUMMARY OF THE FORMED ELEMENTS IN BLOOD**

Name and Appearance	Number	Characteristics[a]	Functions
Red blood cells (RBCs) or **erythrocytes**	4.8 million/mm^3 in females; 5.4 million/mm^3 in males.	7–8 µm diameter, biconcave discs, without a nucleus; live for about 120 days.	Hemoglobin within RBCs transports most of the oxygen and part of the carbon dioxide in the blood.
White blood cells (WBCs) or **leukocytes**	5000–10,000/mm^3.	Most live for a few hours to a few days.[b]	Combat pathogens and other foreign substances that enter the body.
Granulocytes **Neutrophils**	60–70% of all WBCs.	10–12 µm diameter; nucleus has 2–5 lobes connected by thin strands of chromatin; cytoplasm has very fine, pale lilac granules.	Phagocytosis. Destruction of bacteria with lysozyme, defensins, and strong oxidants, such as superoxide anion, hydrogen peroxide, and hypochlorite anion.
Eosinophils	2–4% of all WBCs.	10–12 µm diameter; nucleus usually has 2 lobes; large, red-orange granules fill the cytoplasm.	Combat the effects of histamine in allergic reactions, phagocytize antigen–antibody complexes, and destroy certain parasitic worms.
Basophils	½–1% of all WBCs.	8–10 µm diameter; nucleus is bilobed or irregular in shape; large cytoplasmic granules appear deep blue-purple.	Liberate heparin, histamine, and serotonin in allergic reactions that intensify the overall inflammatory response.
Agranulocytes **Lymphocytes (T cells, B cells, and natural killer cells)**	20–25% of all WBCs.	Small lymphocytes are 6–9 µm in diameter; large lymphocytes are 10–14 µm in diameter; nucleus is round or slightly indented; cytoplasm forms a rim around the nucleus that looks sky blue; the larger the cell, the more cytoplasm is visible.	Mediate immune responses, including antigen–antibody reactions. B cells develop into plasma cells, which secrete antibodies. T cells attack invading viruses, cancer cells, and transplanted tissue cells. Natural killer cells attack a wide variety of infectious microbes and certain spontaneously arising tumor cells.
Monocytes	3–8% of all WBCs.	12–20 µm diameter; nucleus is oval or kidney-shaped or horseshoe-shaped; cytoplasm is blue-gray and has foamy appearance.	Phagocytosis (after transforming into fixed or wandering macrophages).
Platelets (thrombocytes)	250,000–400,000/mm^3.	2–4 µm diameter cell fragments that live for 5–9 days; contain many granules but no nucleus.	Form platelet plug in hemostasis; release chemicals that promote vascular spasm and blood clotting.

[a] Colors are those seen when using Wright's stain.

[b] Some lymphocytes, called T and B memory cells, can live for many years once they are established. Most white blood cells, however, have life spans ranging from a few hours to a few days.

HEMOSTASIS

Hemostasis (hē-mō-STĀ-sis) refers to the stoppage of bleeding. When blood vessels are damaged or ruptured, the hemostatic response must be quick, localized to the region of damage, and carefully controlled. Three basic mechanisms reduce blood loss: (1) vascular spasm, (2) platelet plug formation, and (3) blood clotting (coagulation). These mechanisms are useful for preventing hemorrhage (loss of blood) in smaller (microcirculation) blood vessels, but extensive hemorrhage from larger vessels usually requires medical intervention.

Vascular Spasm

When arteries or arterioles are damaged, the circularly arranged smooth muscle in their walls contracts immediately. This is called a **vascular spasm** and it reduces blood loss for several minutes to several hours, during which time the other hemostatic mechanisms go into operation. The spasm is probably caused by damage to the smooth muscle and from reflexes initiated by pain receptors.

Platelet Plug Formation

In their unstimulated state, platelets are disc-shaped. Considering their small size, platelets pack an impressive array of chemicals. Two types of granules are present in the cytoplasm: (1) **alpha granules** contain clotting factors and **platelet-derived growth factor** (**PDGF**), which can cause proliferation of vascular endothelial cells, vascular smooth muscle fibers, and fibroblasts to help repair damaged blood vessel walls; and (2) **dense granules** contain ADP, ATP, Ca^{2+}, and serotonin. Also present are enzymes that produce thromboxane A2, a prostaglandin; *fibrin-stabilizing factor,* which helps to strengthen a blood clot; lysosomes; some mitochondria; membrane systems that take up and store calcium and provide channels for release of the contents of granules; and glycogen.

Platelet plug formation occurs as follows (Fig. 19.8):

1 In the first phase of platelet plug formation, platelets contact and stick to parts of a damaged blood vessel, such as collagen underlying the damaged endothelial cells. This process is called **platelet adhesion**.

2 As a result of adhesion, the platelets become activated and their characteristics change dramatically. They extend many projections that enable them to contact one another, and they begin to liberate the contents of their granules. This phase is called the **platelet release reaction**. Liberated ADP and thromboxane A2 play a major role by acting on nearby platelets to activate them as well. Serotonin and thromboxane A2 function as vasoconstrictors, causing contraction of the vascular smooth muscle, which decreases blood flow through the injured vessel.

3 The release of ADP also makes other platelets in the area sticky, and the stickiness of the newly recruited and activated platelets causes them to adhere to the originally activated platelets. This gathering of platelets is called **platelet aggregation**. Eventually, the accumulation and attachment of large numbers of platelets form a mass called a **platelet plug**.

A platelet plug is very effective in preventing blood loss in a small vessel. Although the platelet plug is initially loose, it becomes quite tight when reinforced by fibrin threads formed during clotting. A platelet plug can stop blood loss completely if the hole in a blood vessel is small.

Figure 19.8   Platelet plug formation.

🔑 *A platelet plug can stop blood loss completely if the hole in a blood vessel is small.*

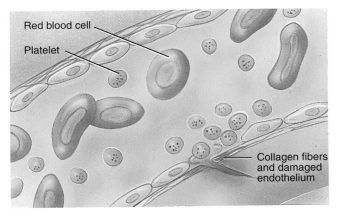

1 Platelet adhesion

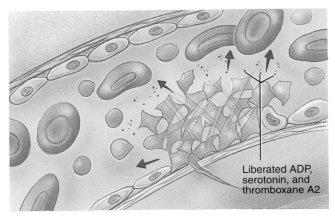

2 Platelet release reaction

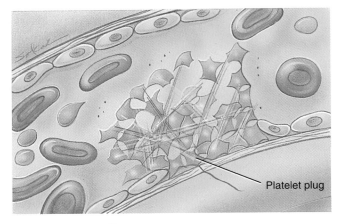

3 Platelet aggregation

❓ Besides platelet plug formation, what are two other mechanisms that contribute to hemostasis?

Clotting (Coagulation)

Normally, blood remains liquid as long as it stays within its vessels. If it is drawn from the body, however, it thickens and forms a gel. Eventually, the gel separates from the liquid. The straw-colored liquid, called **serum**, is simply plasma minus its clotting proteins. The gel is called a **clot** and consists of a network of insoluble protein fibers called fibrin in which the formed elements of blood are trapped (Fig. 19.9).

The process of gel formation is called **clotting** or **coagulation**. If blood clots too easily, the result can be **thrombosis**—clotting in an unbroken blood vessel. If the blood takes too long to clot, a hemorrhage can result.

Clotting involves several substances known as **clotting (coagulation) factors**. These include Ca^{2+}, several inactive enzymes that are synthesized by hepatocytes (liver cells) and released into the bloodstream, and various molecules associated with platelets or released by damaged tissues. Many are identified by Roman numerals which indicate the order of their discovery. The various clotting factors and their synonyms, sources, and pathways of activation are listed in Exhibit 19.3.

Clotting is a complex cascade of reactions in which coagulation factors activate one another. Once the process is initiated, the reactions act in a positive feedback manner to form a large quantity of product. We will describe clotting in three basic stages (see Fig. 19.10):

Stage ❶, the formation of prothrombinase (prothrombin activator), is initiated by either the extrinsic or the in-

Figure 19.9 Blood clot.

 A blood clot is a gel that contains formed elements of the blood entangled in fibrin threads.

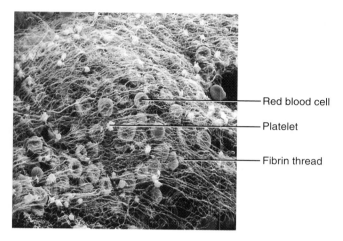

Red blood cell

Platelet

Fibrin thread

Scanning electron micrograph (550x)

Q What is serum?

trinsic pathway or both. Once prothrombinase is formed, the steps involved in the next two stages of clotting are the same for both the extrinsic and intrinsic pathways and the two stages together are referred to as the common pathway.

EXHIBIT **19.3** CLOTTING (COAGULATION) FACTORS AND THEIR SYNONYMS

Clotting Factor[a]	Synonym	Source	Pathway of Activation
I	Fibrinogen.	Liver.	Common.
II	Prothrombin.	Liver.	Common.
III	Tissue factor (thromboplastin).	Damaged tissues and activated platelets.	Extrinsic.
IV	Calcium ions (Ca^{2+}).	Diet, bones, and platelets.	All.
V	Proaccelerin, labile factor, or accelerator globulin (AcG).	Liver and platelets.	Extrinsic and intrinsic.
VII	Serum prothrombin conversion accelerator (SPCA), stable factor, or proconvertin.	Liver.	Extrinsic.
VIII	Antihemophilic factor (AHF), antihemophilic factor A, or antihemophilic globulin (AHG).	Platelets and endothelial cells.	Intrinsic.
IX	Christmas factor, plasma thromboplastin component (PTC), or antihemophilic factor B.	Liver.	Intrinsic.
X	Stuart factor, Prower factor, or thrombokinase.	Liver.	Extrinsic and intrinsic.
XI	Plasma thromboplastin antecedent (PTA) or antihemophilic factor C.	Liver.	Intrinsic.
XII	Hageman factor, glass factor, contact factor, or antihemophiliac factor D.	Liver.	Intrinsic.
XIII	Fibrin-stabilizing factors (FSF).	Liver and platelets.	Common.

[a]There is no factor VI. Prothrombinase (prothrombin activator) is a combination of activated factors V and X.

Stage ❷ is the conversion of prothrombin (a plasma protein formed by the liver) into the enzyme thrombin by prothrombinase.

Stage ❸ is the conversion of soluble fibrinogen (another plasma protein formed by the liver) into insoluble fibrin by thrombin. Fibrin forms the threads of the clot.

Stage ❶ : *Formation of Prothrombinase*

The formation of prothrombinase is initiated by the interplay of two mechanisms: the extrinsic and intrinsic pathways of blood clotting.

● **Extrinsic Pathway** The **extrinsic pathway** of blood clotting has fewer steps than the intrinsic pathway and occurs rapidly, within a matter of seconds if trauma is severe. It is so named because a tissue protein called **tissue factor** (**TF**), also known as **thromboplastin** or **coagulation factor III**, leaks into the blood from cells *outside* (*extrinsic to*) blood vessels and initiates the formation of prothrombinase. TF is a complex mixture of lipoproteins and phospholipids released from the surfaces of damaged cells. It activates clotting factor VII (Fig. 19.10a), which next combines with factor X, thus activating it. Once factor X is activated, it combines with factor V in the presence of calcium ions (Ca^{2+}) to form the active enzyme prothrombinase. This completes the extrinsic pathway.

● **Intrinsic Pathway** The **intrinsic pathway** of blood clotting is more complex than the extrinsic pathway, and it occurs more slowly, usually requiring several minutes. The intrinsic pathway is so named because its activators are in direct contact with blood or contained *within* (*intrinsic to*) the blood; outside tissue damage is not needed. If endothelial cells (cells that line blood vessels) become roughened or damaged, blood can come in contact with collagen in the surrounding basal lamina. In addition, trauma to endothelial cells causes damage to blood platelets, resulting in the release of phospholipids by the platelets. Contact with collagen (or with the slippery glass sides of a blood collection tube) activates clotting factor XII (Fig. 19.10b). In turn, factor XII activates XI, which activates factor IX. Activated factor VII (from the extrinsic pathway) also can activate factor IX. Activated factor IX joins with factor VIII and platelet phospholipids to activate factor X. Once factor X is activated, it combines with factor V to form the active enzyme prothrombinase, just as occurs in the extrinsic pathway. Again, Ca^{2+} is a cofactor in several of the reactions. This completes the intrinsic pathway.

Stages ❷ *and* ❸ : *Common Pathway*

Once prothrombinase is formed, the common pathway follows. In the second stage of blood clotting (Fig. 19.10c), prothrombinase and Ca^{2+} catalyze the conversion of prothrombin to thrombin. In the third stage, thrombin, in the presence of Ca^{2+}, converts fibrinogen, which is soluble, to loose fibrin threads, which are insoluble. Thrombin also activates factor XIII (fibrin stabilizing factor), which strengthens and stabilizes the fibrin

---

Figure 19.10   The blood clotting cascade.

 In clotting, coagulation factors activate each other, resulting in a cascade of reactions that include positive feedback cycles.

Stage ❶

(a) Extrinsic pathway

Tissue trauma → Tissue factor (TF)

TF

VII

Ca^{2+}

Activated VII

Ca^{2+}

X → Activated X

V Ca^{2+}

(b) Intrinsic pathway

Blood trauma

Damaged endothelial cells expose collagen

Damaged platelets

XII → Activated XII

Activated XI ← XI

Ca^{2+}

Activated IX ← IX

Activated platelets ⊕

VIII   Platelet phospholipids

Ca^{2+}

Activated X ← X

V Ca^{2+} ⊕

PROTHROMBINASE

Stage ❷   **(c) Common pathway**

Ca^{2+}

PROTHROMBIN (II) → **THROMBIN**

Ca^{2+} XIII

Stage ❸

FIBRINOGEN (I)   Activated XIII

Loose fibrin threads → **STABILIZED FIBRIN THREADS**

Q   What is the outcome of the first stage of clotting?

---

threads into a sturdy clot. Plasma contains some factor XIII, and it is also released by platelets trapped in the clot.

Thrombin has two positive feedback effects. In one, through factor V, it accelerates the formation of prothrombinase. Prothrombinase, in turn, accelerates the production of more thrombin, and so on. Thrombin also activates platelets, which reinforces their aggregation and release of phospholipids. This is a second positive feedback cycle. If unchecked, a clot would continue to get larger and larger, as a result of the positive feedback cycles. However, fibrin has the ability to absorb and inactivate up to 90% of the thrombin formed from prothrombin. This helps stop the spread of thrombin into the blood and thus limits spread of the clot beyond the site of damage.

HEMOPHILIA

Hemophilia (*hemo* = blood; *philein* = to love) refers to several different hereditary deficiencies of clotting in which bleeding may occur spontaneously or after only minor trauma. The effects of all forms of hemophilia are similar but each is a deficiency of a different blood clotting factor. The most common type (classic hemophilia) is hemophilia A, absence of factor VIII. People with hemophilia B lack factor IX. Hemophilia A and B occur primarily among males because these are sex-linked recessive disorders. A milder form is hemophilia C, a lack of factor XI, which affects both males and females. Recall that factor XI activates factor IX. Hemophilia C is much less severe than hemophilia A or B because an alternate activator of factor IX is present, namely, factor VII.

Hemophilia is characterized by spontaneous or traumatic subcutaneous and intramuscular hemorrhaging, nosebleeds, blood in the urine, and hemorrhages in joints that produce pain and damage. Treatment involves transfusions of fresh plasma or concentrates of the deficient clotting factor to relieve the bleeding tendency. Factor VIII concentrates are made by harvesting the factor from the plasma of a very large number of donors. Sadly, between 1982 and 1985 most such factor VIII preparations were contaminated with HIV, the virus that causes AIDS, and most hemophiliacs who used the products at that time became infected with HIV. Pure factor VIII is now available through recombinant DNA technology. ■

Need for Vitamin K

Normal clotting depends on adequate vitamin K in the body. Although vitamin K is not involved in actual clot formation, it is required for the synthesis of four clotting factors by liver cells: factors II (prothrombin), VII, IX, and X. Vitamin K is normally produced by bacteria that inhabit the large intestine. It is a fat-soluble vitamin and can be absorbed through the lining of the intestine and into the blood only if fat absorption is normal. People suffering from disorders that prevent absorption of fat (for example, inadequate release of bile into the small intestine) often experience uncontrolled bleeding.

Clot Retraction and Repair

Once a clot is formed, it plugs the ruptured area of the blood vessel and thus stops blood loss. **Clot retraction** is the consolidation or tightening of the fibrin clot. The fibrin threads attached to the damaged surfaces of the blood vessel gradually contract owing to platelets pulling on them. As the clot retracts, it pulls the edges of the damaged vessel closer together. Thus the risk of hemorrhage is further decreased.

During retraction, some serum escapes between the fibrin threads, but the formed elements in blood remain trapped in the fibrin threads. Normal clot retraction depends on an adequate number of platelets. Platelets in the clot also release factor XIII, which strengthens and stabilizes the clot, and other factors, which help compress the clot. Permanent repair of the blood vessel can then take place. In time, fibroblasts form connective tissue in the ruptured area, and new endothelial cells repair the lining.

Fibrinolysis

Many times a day little clots start to form, sometimes inappropriately, sometimes at a site of minor roughness or at a developing atherosclerotic plaque inside a blood vessel. Because blood clotting involves several positive feedback cycles, a clot has a tendency to spread, which might block blood flow through undamaged vessels. The **fibrinolytic system** provides checks and balances so that clotting does not get out of hand. It also dissolves clots at a site of damage once the damage is repaired.

Dissolution of a clot is called **fibrinolysis** (fī-brin-OL-i-sis). When a clot is formed, an inactive plasma enzyme called **plasminogen** is incorporated into the clot. Both body tissues and blood contain substances that can activate plasminogen to **plasmin** (**fibrinolysin**), an active plasma enzyme. Among these substances are thrombin, activated factor XII, and tissue plasminogen activator (t-PA), which is synthesized in endothelial cells of most tissues and liberated into the blood. Once plasmin is formed, it can dissolve the clot by digesting fibrin threads and inactivating substances such as fibrinogen, prothrombin, and factors V, VIII, and XII.

THROMBOLYTIC AGENTS

Thrombolytic (**clot-dissolving**) **agents** are chemical substances injected into the body that dissolve blood clots to restore circulation. They either directly or indirectly activate plasminogen. The first thrombolytic agent, approved for use in 1982, was **streptokinase** (Kabikinase, Streptase), produced by streptococcal bacteria. A more recently developed thrombolytic agent, approved in 1988 for dissolving clots in coronary arteries of the heart, is **tissue plasminogen activator** (**t-PA**). This substance, marketed under the brand name Activase®, is a genetically engineered, artificial version of the natural enzyme. ■

Hemostatic Control Mechanisms

It was noted earlier that even though thrombin has a positive feedback effect on blood clotting, clot formation normally occurs locally at the site of damage. It does not extend beyond a wound site into the general circulation, in part because fibrin absorbs the thrombin into the clot. Another reason for localized clot formation is that some of the clotting factors are carried away by the blood so their concentrations are not high enough to bring about widespread clotting.

Several other mechanisms also operate to control blood clotting. For example, both endothelial cells and white blood cells produce a prostaglandin called **prostacyclin** (**PGI₂**). This substance opposes the actions of thromboxane A2. It is a powerful inhibitor of platelet adhesion and release. Also, substances that inhibit clotting are present in blood. Such substances are called **anticoagulants**. These include **antithrombin III** (**AT-III**), which blocks the action of factors XII, XI, IX, X, and II (thrombin); **protein C**, which inactivates factors V and VIII, the two major clotting factors not blocked by AT-III, and enhances activity of plasminogen activators; **alpha-2-macroglobulin**, which inactivates thrombin and plasmin; and **alpha-1-antitrypsin**, which inhibits factor XI.

Heparin is another anticoagulant. It is produced by mast cells and basophils. A similar molecule protrudes from the plasma membrane of endothelial cells into the blood. Heparin's anticoagulant activity is to combine with AT-III and increase its effectiveness in blocking thrombin. Heparin is also a pharmacologic anticoagulant extracted from animal lung tissue and intestinal mucosa. It is often used in open heart surgery and during hemodialysis. Another anticoagulant is the pharmaceutical preparation **warfarin** (**Coumadin**), which may be given to patients who are prone to develop clots. It acts as an antagonist to vitamin K and thus blocks synthesis of four clotting factors (II, VII, IX, and X). Warfarin is slower acting than heparin. To prevent clotting in donated blood, blood banks and laboratories often add a substance, for example, CPD (citrate phosphate dextrose), that removes Ca^{2+}.

Intravascular Clotting

Despite the anticoagulating and fibrinolytic mechanisms, blood clots sometimes form within the cardiovascular system. Such clots may be initiated by roughened endothelial surfaces of a blood vessel as a result of atherosclerosis, trauma, or infection. These conditions induce adhesion of platelets. Intravascular clots may also form when blood flows too slowly (stasis), allowing clotting factors to accumulate locally in high enough concentrations to initiate coagulation. Clotting in an unbroken blood vessel (usually a vein) is called **thrombosis** (*thrombo* = clot). The clot itself is a **thrombus**, and it may dissolve spontaneously. If it remains intact, the thrombus may become dislodged and be swept away in the blood. A blood clot, bubble of air, fat from broken bones, or a piece of debris transported by the bloodstream is called an **embolus** (*em* = in; *bolus* = a mass). When an embolus lodges in the lungs, the condition is called **pulmonary embolism**. An embolus that breaks away from an arterial wall may jam in a smaller diameter artery downstream and block blood flow to a vital organ.

ASPIRIN AND HEMOSTASIS

Although aspirin is used effectively to treat pain and fever, in recent years several studies have shown that low-dose as-

pirin (less than 300 mg/day) has other powerful effects. It reduces the risk of transient ischemic attacks (TIA) and strokes (see page 424), myocardial infarction (see page 589), and blocking of peripheral arteries. The protective effects of aspirin are due to its impact on several events in hemostasis. Normally, injury of a blood vessel wall provokes the response of vasoconstriction, formation of a platelet plug, and coagulation. In patients with heart and blood vessel disease, these events may occur even without vessel injury. The result is formation of a thrombus. At a very low dose, aspirin reduces the chance of thrombus formation. It inhibits vasoconstriction and platelet aggregation by blocking synthesis of thromboxane A2. ■

BLOOD GROUPS AND BLOOD TYPES

The surfaces of erythrocytes contain some glycoproteins and glycolipids that can act as antigens. These antigens, called **isoantigens** or **agglutinogens** (a-gloo-TIN-ō-jens), are normal components of one person's RBC plasma membrane that can trigger damaging antigen–antibody responses in other people. Based on the presence or absence of various isoantigens, blood is categorized into different **blood groups**. Within a given blood group there may be two or more different **blood types**. There are at least 24 blood groups and more than 100 isoantigens that can be detected on the surface of red blood cells. Two major blood groups—ABO and Rh—are the ones we will discuss in detail. Among other blood groups are the Lewis, Kell, Kidd, and Duffy systems.

Because the specific isoantigens are genetically determined, your blood type depends on the genes you inherit from your mother and father. To understand in a general way how blood types are inherited, recall from Chapter 3 that the nuclei of almost all human cells, except gametes (germ cells), contain 23 pairs of chromosomes. In each pair, one chromosome is inherited from the mother, whereas the other is inherited from the father. Paired chromosomes with genes that control the same inherited trait are called homologous chromosomes. For example, if one homologous chromosome contains a gene for height, the other member of the pair will also contain a gene for height. The matched genes that code for the same trait are in the same location on homologous chromosomes and are called **alleles**.

ABO Blood Group

The **ABO blood group** is based on two glycolipid isoantigens called A and B (Fig. 19.11). People whose RBCs display *only antigen A* are said to have **type A** blood. Those who have *only antigen B* are **type B**. Individuals who have *both A and B antigens* are **type AB**, whereas those who have *neither antigen A nor B* are **type O**.

The four ABO blood types result from the inheritance of various combinations of three different alleles of a gene called the *I* gene: (1) *I*^A codes for the A antigen, (2) *I*^B codes for the B antigen, and (3) *i* codes for neither A nor B antigen. Each person inherits two *I*-gene alleles, one from each parent, that give rise to the various blood types. The six possible combinations of the genes produce four blood types as follows:

1. *I*^A*I*^A or *I*^A*i* produces type A blood.
2. *I*^B*I*^B or *I*^B*i* produces type B blood.
3. *I*^A*I*^B produces type AB blood.
4. *ii* produces type O blood.

The incidence of ABO and Rh blood types varies among different populations, as indicated in Exhibit 19.4.

Population	Blood Type (Percentage)				
	O	**A**	**B**	**AB**	**Rh+**
White	45	40	11	4	85
Black	49	27	20	4	95
Korean	32	28	30	10	100
Japanese	31	38	21	10	100
Chinese	42	27	25	6	100
Native American	79	16	4	1	100

EXHIBIT 19.4 INCIDENCE OF BLOOD TYPES IN THE UNITED STATES

In addition to isoantigens on RBCs, blood plasma usually contains naturally occurring **isoantibodies** or **agglutinins** (a-GLOO-ti-nins) that will react with the A or B antigens if the two are mixed. These are **anti-A antibody**, which reacts with antigen A, and **anti-B antibody**, which reacts with antigen B. The antibodies present in each of the four blood types are shown in Fig. 19.11. You do not have antibodies that react with the antigens of your own RBCs, but most likely you do have antibodies for any antigens that your RBCs lack. In an incompatible blood transfusion, isoantibodies in the recipient's plasma bind to the isoantigens on the donated RBCs. This reaction is another example of an antigen–antibody response. When antigen–antibody complexes form in the body, they activate plasma proteins of the complement family (described on page 681). In essence, complement molecules poke holes in the donated RBCs, causing them to burst and release hemoglobin into the plasma. Such a reaction is called **hemolysis**. The liberated hemoglobin may cause kidney damage.

As an example of an incompatible blood transfusion, consider what happens if a person with type A blood receives a transfusion of type B blood. The recipient's blood (type A) contains A antigens on the red blood cells and anti-B antibodies in the plasma. The donor's blood (type B) contains B antigens and anti-A antibodies. Given this situation, two things can happen. First, the anti-B antibodies in the recipient's plasma can bind to the B antigens on the donor's erythrocytes, causing hemolysis of the red blood cells. Second, the anti-A antibodies in the donor's plasma can bind to the A antigens on the recipient's erythrocytes.

Figure 19.11 Antigens and antibodies of the ABO blood types.

The antibodies in your plasma do not react with the antigens on your red blood cells.

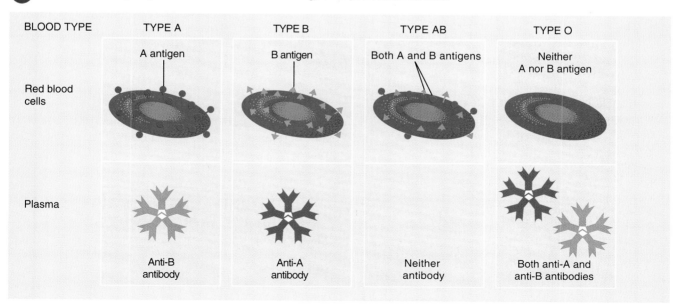

BLOOD TYPE	TYPE A	TYPE B	TYPE AB	TYPE O
Red blood cells	A antigen	B antigen	Both A and B antigens	Neither A nor B antigen
Plasma	Anti-B antibody	Anti-A antibody	Neither antibody	Both anti-A and anti-B antibodies

Q Which antibodies are usually present in type O blood?

However, the second reaction is usually not serious because the donor's anti-A antibodies become so diluted in the recipient's plasma that they do not cause any significant hemolysis of the recipient's RBCs. The interactions of the four blood types of the ABO system are summarized in Exhibit 19.5.

People with type AB blood do not have any anti-A or anti-B antibodies in their plasma. They are sometimes called "universal recipients" because they can theoretically receive blood from donors of all four blood types (Fig. 19.12a). They have no antibodies to attack antigens on donated RBCs. A person with type A blood may receive type A or type O blood but not type B or AB blood because the anti-B antibodies in the recipient's plasma can bind to the B antigens on the donor's RBCs. Likewise, people with type B blood can receive type B or type O but not type A or type AB blood. People with type O blood have no A or B antigens on their RBCs and are sometimes called "universal donors" because they can theoretically donate blood to all four ABO blood types. Type O persons requiring blood may receive only type O blood. In practice, use of the terms "universal recipient" and "universal donor" is misleading and dangerous. There are other antigens and antibodies in blood, besides those associated with the ABO system, that can cause transfusion problems. Thus blood should be carefully cross-matched or screened before transfusion.

Knowledge of blood types is also used in paternity lawsuits, linking suspects to crimes, and as part of anthropolog-

Figure 19.12 Compatible blood types and agglutination due to incompatibility. In (a) arrows indicate possible compatible donors (tail of arrow) and recipients (arrowhead).

 In theory, people having type AB blood could receive all four ABO blood types because their plasma lacks anti-A and anti-B antibodies.

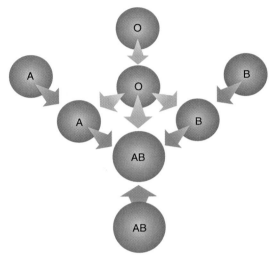

(a) Compatible donors and recipients

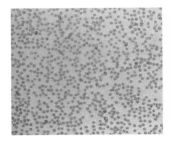 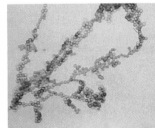

(b) Comparison between normal RBCs (left) and agglutinated RBCs (right) (160x)

 Which blood type is considered the "universal donor" for the ABO blood group system?

ical studies to establish a relationship among races. In about 80% of the population, soluble antigens of the ABO type appear in saliva and other body fluids.

TYPING AND CROSS-MATCHING BLOOD FOR TRANSFUSION

Blood transfusions are most often given to alleviate anemia or when blood volume is low, for example, in cases of circulatory shock after a severe hemorrhage. When blood is transfused, care must be taken to avoid blood type mismatches. This is done by typing the patient's blood and then either cross-matching it to potential donor

EXHIBIT **19.5**	**SUMMARY OF ABO BLOOD GROUP INTERACTIONS**			
Blood Type	**A**	**B**	**AB**	**O**
Antigen on RBCs	A	B	Both A and B	Neither A nor B
Antibody in plasma	anti-B	anti-A	Neither anti-A nor anti-B	Both anti-A and anti-B
Compatible donor blood types (no hemolysis)	A, O	B, O	A, B, AB, O	O
Incompatible donor blood types (hemolysis)	B, AB	A, AB	—	A, B, AB
Alleles (genotype)	$I^A I^A$ or $I^A i$	$I^B I^B$ or $I^B i$	$I^A I^B$	ii
Expressed blood type (phenotype)	A	B	AB	O

blood or screening it for presence of antibodies. Outside the body, at room temperature, mixing of incompatible blood causes **agglutination** (clumping) that is visible to the naked eye (Fig. 19.12b, right) rather than hemolysis. Agglutination is an antigen–antibody response, whereby the cells become cross-linked to one another and form a visible clump. (Note that agglutination is not the same thing as blood clotting.)

In the procedure for ABO blood typing, single drops of blood are mixed with different antisera, solutions that contain antibodies. One drop of blood is mixed with anti-A serum. It contains anti-A antibodies that will agglutinate red blood cells if they contain A antigens. The other drop is mixed with anti-B serum. It contains anti-B antibodies that will agglutinate red blood cells if they contain B antigens. If the red blood cells agglutinate only when mixed with anti-A serum, the blood is type A. If the red blood cells agglutinate only when mixed with anti-B serum, the blood is type B. The sample is type AB if both drops agglutinate; if neither sample agglutinates, the sample is type O.

In the procedure for determining Rh factor (discussed next), a drop of blood is mixed with antiserum containing antibodies that will agglutinate RBCs displaying Rh antigens. If the sample agglutinates, it is Rh$^+$ (Rh positive); no agglutination indicates Rh$^-$ (Rh negative).

Once the patient's blood type is known, donor blood of the same ABO and Rh type is selected. In a **cross-match**, the possible donor RBCs are mixed with the recipient's serum. If agglutination does not occur, the recipient does not have antibodies that will attack the donor RBCs. Alternatively, the recipient's serum can be **screened** against a test panel of RBCs having antigens known to cause blood transfusion reactions to detect any antibodies that may be present. ■

Rh Blood Group

The **Rh blood group** is so named because the antigen was discovered in the blood of the *Rhesus* monkey. The alleles of three genes (C, D, and E) may code for the Rh antigen. People whose RBCs have Rh antigens are designated Rh$^+$. Those who lack Rh antigens are designated Rh$^-$. The incidence of Rh$^+$ and Rh$^-$ individuals in various populations is shown in Exhibit 19.4. Normally, plasma does not contain anti-Rh antibodies. If an Rh$^-$ person receives an Rh$^+$ blood transfusion, however, the immune system starts to make anti-Rh antibodies that will remain in the blood. If a second transfusion of Rh$^+$ blood is given later, the previously formed anti-Rh antibodies will cause hemolysis of the donated blood, and a severe reaction may occur.

HEMOLYTIC DISEASE OF THE NEWBORN

The most common problem with Rh incompatibility may arise during pregnancy (Fig. 19.13). Normally, there is no direct contact between maternal and fetal blood when a woman is pregnant. However, if a small amount of Rh$^+$ blood leaks from the fetus

Figure 19.13 Development of hemolytic disease of the newborn (HDN). (a) At birth, a small quantity of fetal blood usually leaks across the placenta into the maternal bloodstream. (b) A problem can arise when the mother is Rh$^-$ and the baby is Rh$^+$, having inherited an allele for one of the Rh antigens from the father. Upon exposure to Rh antigen, the mother's immune system responds by making anti-Rh antibodies. Because the baby is already born, it suffers no damage. (c) During a subsequent pregnancy, however, the maternal antibodies cross the placenta into the fetal blood. If the second fetus is Rh$^+$, the ensuing antigen–antibody reaction causes hemolysis of fetal RBCs. The result is HDN.

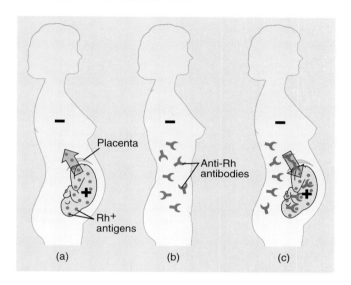

through the placenta into the bloodstream of an Rh$^-$ mother, the mother will start to make anti-Rh antibodies. Since the greatest possibility of fetal blood transfer occurs at delivery, a newborn baby will be unaffected. If the mother becomes pregnant again, however, her anti-Rh antibodies can cross the placenta and enter the bloodstream of the fetus. If the fetus is Rh$^-$, there is no problem, since Rh$^-$ blood does not have the Rh antigen. If the fetus is Rh$^+$, hemolysis may occur in the fetal blood. The hemolysis brought on by fetal–maternal incompatibility is called **hemolytic disease of the newborn (HDN)**, also known as **erythroblastosis fetalis**.

Not too long ago, HDN was a major cause of fetal and newborn deaths, but now its occurrence is very rare. HDN is prevented by giving all Rh$^-$ mothers an injection of anti-Rh antibodies called anti-Rh gamma globulin (RhoGAM) soon after every delivery, miscarriage, or abortion. These antibodies bind to and inactivate the fetal Rh antigens, if they are present, so the mother's immune system does not respond to the foreign antigens by producing antibodies. Thus the fetus of the next pregnancy is protected. Since Rh$^+$ mothers do not make anti-Rh antibodies, their babies are not at risk. ABO incompatibility between mother and fetus rarely causes problems because the anti-A and anti-B antibodies are too large to cross the placenta. ■

DISORDERS: HOMEOSTATIC IMBALANCES

ANEMIA

Anemia is a condition in which the oxygen-carrying capacity of the blood is reduced; it is a sign, not a diagnosis. Many kinds of anemia exist, all characterized by reduced numbers of RBCs or decreased amount of hemoglobin in the blood. These conditions lead to fatigue and intolerance to cold, both of which are related to lack of oxygen needed for ATP and heat production, and to paleness, which is due to low hemoglobin content.

Iron-deficiency Anemia

Iron-deficiency anemia is the most prevalent anemia in the world. It is caused by inadequate absorption or excessive loss of iron. It occurs most frequently in females, young children, and the elderly in undeveloped countries.

Pernicious Anemia

Pernicious anemia is the insufficient hemopoiesis that results from an inability of the stomach to produce intrinsic factor, which is needed for absorption of vitamin B_{12} in the small intestine.

Hemorrhagic Anemia

An excessive loss of RBCs through bleeding is called **hemorrhagic anemia**. Common causes are large wounds, stomach ulcers, and heavy menstrual bleeding. If bleeding is extraordinarily heavy, the anemia is termed *acute*. Excessive blood loss can be fatal. Slow, prolonged bleeding is apt to produce a *chronic* anemia; the chief symptom is fatigue.

Hemolytic Anemia

If RBC plasma membranes rupture prematurely, the cells remain as "ghosts," and their hemoglobin pours out into the plasma. A characteristic sign of this condition, called **hemolytic anemia**, is distortion in the shape of erythrocytes. It may result from inherent defects, such as hemoglobin defects, abnormal red blood cell enzymes, or defects of the red blood cell membrane. Agents that may cause hemolytic anemia are parasites, toxins, and antibodies from incompatible blood (Rh⁻ mother and Rh⁺ fetus, for instance). Hemolytic disease of the newborn (erythroblastosis fetalis) is an example of a hemolytic anemia.

The term **thalassemia** (thal′-a-SE-mē-a) refers to a group of hereditary hemolytic anemias associated with deficient synthesis of hemoglobin. The RBCs are small (microcytic), pale (hypochromic), and short-lived. Thalassemia occurs primarily in populations from countries bordering the Mediterranean Sea. Treatment generally consists of blood transfusions.

Aplastic Anemia

Destruction of the red bone marrow results in **aplastic anemia**. Typically, the marrow is replaced by fatty tissue, fibrous tissue, or tumor cells. Toxins, gamma radiation, and certain medications that inhibit enzymes involved in hemopoiesis are causes. Bone marrow transplants can now be done with a reasonable hope of success in patients with aplastic anemia. Immunosuppressive drugs are given for several days before the transplant and with decreasing frequency afterward.

Sickle-Cell Anemia

The RBCs of a person with **sickle-cell anemia** (SCA) contain an abnormal kind of hemoglobin (Hb-S). When such an erythrocyte gives up its oxygen to the interstitial fluid, the abnormal hemoglobin forms long, stiff, rodlike structures that bend the erythrocyte into a sickle shape (Fig. 19.14). The sickled cells rupture easily. Even though erythropoiesis is stimulated by the loss of the cells, it

Figure 19.14   Sickle-cell anemia.

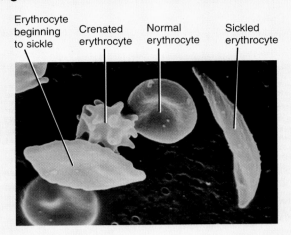

Erythrocyte beginning to sickle   Crenated erythrocyte   Normal erythrocyte   Sickled erythrocyte

Scanning electron micrograph of erythrocytes (3310x)

cannot keep pace with the hemolysis. The individual thus suffers from a hemolytic anemia that reduces the amount of oxygen that can be supplied to the tissues. Prolonged oxygen reduction may eventually cause extensive tissue damage. Furthermore, because of the shape of the sickled cells, they tend to get stuck in blood vessels and can cut off blood supply to an organ altogether.

Sickle-cell anemia is characterized by several symptoms. In young children, hand–foot syndrome is present, in which there is swelling and pain in the wrists and feet. Older patients experience pain in the back and limbs. Complications include neurological disorders (meningitis, seizures, stroke), impaired pulmonary function, orthopedic abnormalities (femoral head necrosis, osteomyelitis), genitourinary tract disorders (involuntary urination, blood in urine, kidney failure), ocular disturbances (hemorrhage, detached retina, blindness), and obstetric complications (convulsions, coma, infection).

Sickle-cell anemia is inherited. People with two SCA genes have severe anemia while those with only one defective gene may have minor problems (see page 979). Sickle-cell genes are found primarily among populations, or descendants of populations, that live in the malaria belt around the world, including parts of Mediterranean Europe, sub-Saharan Africa, and tropical Asia. The gene responsible for the tendency of the RBCs to sickle also alters the permeability of the plasma membranes of sickled cells, causing potassium to leak out. Low levels of potassium kill the malaria parasites that infect sickled cells. Because of this effect, a person with one normal gene and one SCA gene has a high resistance to malaria. The single SCA gene thus confers a survival advantage.

Treatment consists of administration of analgesics to relieve pain, antibiotics to counter infections, and blood transfusions.

POLYCYTHEMIA

The term **polycythemia** (pol′-ē-sī-THĒ-mē-a) refers to a disorder characterized by a hematocrit that is elevated significantly above the normal upper limit of about 55. Blood viscosity rises as hematocrit increases. The increased viscosity elevates blood pressure and contributes to thrombosis and hemorrhage. The thrombosis results from too many red blood cells piling up as they try to enter smaller vessels. The hemorrhage is due to widespread hyperemia (unusually large amount of blood in an organ).

INFECTIOUS MONONUCLEOSIS

Infectious mononucleosis (IM) is a contagious disease primarily affecting lymphatic tissue throughout the body but also affecting the blood. It is caused by the *Epstein–Barr virus* (*EBV*) and occurs mainly in children and young adults. The ratio of affected females to males is 3:1. The virus most commonly enters the body through intimate oral contact, such as kissing, then multiplies in lymphatic tissues, and spreads into the blood, where it infects and multiplies in B lymphocytes, the primary host cells. As a result of this infection, the B cells become enlarged and abnormal in appearance and resemble monocytes, the primary reason for which the disease receives its name, mononucleosis.

Signs and symptoms include an elevated white blood cell count with an abnormally high percentage of lymphocytes, fatigue, headache, dizziness, sore throat, enlarged and tender lymph nodes, and fever. There is no cure for infectious mononucleosis, and treatment consists of watching for and treating complications. Usually the disease runs its course in a few weeks .

LEUKEMIA

Clinically, **leukemia** is classified based on the duration and character of the disease, that is, acute or chronic. Acute leukemia is a malignant disease of blood-forming tissues characterized by uncontrolled production and accumulation of immature leukocytes. In chronic leukemia, there is an accumulation of mature leukocytes in the bloodstream because they do not die at the end of their normal life span. The *human T cell leukemia-lymphoma virus-1* (*HTLV-1*) is strongly associated with some types of leukemia. Leukemia is also classified according to the identity and site of origin of the predominant cell involved, that is, myelocytic, lymphocytic, or monocytic.

In acute leukemia, the anemia and bleeding problems commonly seen result from the crowding out of normal red bone marrow cells by the overproduction of immature cells, preventing normal production of RBCs and platelets. One cause of death from acute leukemia is internal hemorrhaging, especially cerebral hemorrhage that destroys vital centers in the brain. Often, the cause of death is uncontrolled infection due to lack of mature or normal WBCs. The abnormal accumulation of immature leukocytes may be reduced by treatment with x-rays and antileukemic drugs. Partial or complete remissions may be induced, with some lasting as long as 15 years.

MEDICAL TERMINOLOGY

Acute normovolemic (nor-mō-vō-LĒ-mik) **hemodilution** (hē-mō-di-LOO-shun) Removal of blood immediately before surgery and replacing it with a cell-free solution to maintain normal blood volume for adequate circulation. At the end of surgery, when bleeding has been controlled, the collected blood is returned to the body.

Autologous (aw-TOL-o-gus; *auto* = self) **preoperative transfusion** (trans-FYOO-zhun) Donating one's own blood; can be done up to 6 weeks before elective surgery. Also called **predonation**.

Autologous intraoperative transfusion (**AIT**) Procedure in which blood lost during surgery is suctioned from the patient, treated with an anticoagulant, and reinfused into the patient.

Blood bank A stored supply of blood for future use by the donor or others. Because blood banks have now assumed additional and diverse functions (immunohematology reference work, continuing medical education, bone and tissue storage, and clinical consultation), they are more appropriately referred to as **centers of transfusion medicine**.

Citrated (SIT-rā-ted) **whole blood** Whole blood protected from coagulation by CPD (citrate phosphate dextrose) or a similar compound.

Cyanosis (sī-a-NŌ-sis; *cyano* = blue) Slightly bluish, dark purple skin coloration, most easily seen in the nail beds and mucous membranes, due to increased quantity of reduced hemoglobin (hemoglobin not combined with oxygen) in systemic blood.

Exchange transfusion Removing blood from the recipient while simultaneously replacing it with donor blood. This method is used for treating hemolytic disease of the newborn (HDN) and poisoning.

Gamma globulin (GLOB-yoo-lin) Solution of immunoglobulins from blood consisting of antibodies that react with specific pathogens, such as measles, epidemic hepatitis, tetanus, and possibly poliomyelitis viruses. It is prepared by injecting the specific virus into animals, removing blood from the animals after antibodies have accumulated, isolating the antibodies, and injecting them into a human to provide short-term immunity.

Hemochromatosis (hē-mō-krō-ma-TŌ-sis; *heme* = iron; *chroma* = color) Disorder of iron metabolism characterized by excess deposits of iron in tissues, especially the liver and pancreas, that result in bronze coloration of the skin, cirrhosis, diabetes mellitus, and bone and joint abnormalities.

Hemorrhage (HEM-or-ij; *rhegnynai* = bursting forth) Loss of a large amount of blood, either internal (from blood vessels into tissues) or external (from blood vessels directly to the surface of the body).

Multiple myeloma (mī-e-LŌ-ma) Malignant disorder of plasma cells in red bone marrow; symptoms (pain, osteoporosis, hypercalcemia, thrombocytopenia, kidney damage) are caused by the growing tumor cell mass or antibodies produced by malignant cells.

Phlebotomist (fle-BOT-ō-mist; *phlebo* = vein; *tome* = to cut) A technician who specializes in withdrawing blood.

Platelet concentrates A preparation of platelets obtained from freshly drawn whole blood and used for transfusions in platelet-deficiency disorders such as hemophilia.

Porphyria (por-FĒ-rē-a or por-FĪ-rē-a); *porphyra* = purple) Any of a group of inherited disorders caused by the accumulation in the body of substances called porphyrins (molecules formed during the synthesis of hemoglobin

MEDICAL TERMINOLOGY CONTINUED

and other heme-containing molecules). The buildup is due to inherited enzyme deficiencies. Symptoms include a rash or skin blistering brought on by sunlight, abdominal pain, and nervous system disturbances from certain drugs, such as barbiturates and alcohol.

Septicemia (sep′-ti-SĒ-mē-a; *septicus* = decay; *emia* = condition of blood) Toxins or disease-causing bacteria growing in the blood. Also called "blood poisoning."

Thrombocytopenia (throm′-bō-sī′-tō-PĒ-nē-a; *thrombo* = clot; *penia* = poverty) Very low platelet count that results in a tendency to bleed from capillaries.

Transfusion (trans-FYOO-zhun) Transfer of whole blood, blood components (red blood cells only or plasma only), or red bone marrow directly into the bloodstream.

Venesection (vē′-ne-SEK-shun; *veno* = vein) Opening of a vein for withdrawal of blood. Although **phlebotomy** (fle-BOT-ō-me-; *phlebo* = vein; *tome* = to cut) is a synonym for venesection, in clinical practice, phlebotomy refers to therapeutic bloodletting, such as removing some blood to lower the viscosity of blood of a patient with polycythemia.

Whole blood Blood containing all formed elements, plasma, and plasma solutes in natural concentrations.

STUDY OUTLINE

COMPARISON OF EXTRACELLULAR FLUIDS (p. 553)

1. All fluids outside body cells comprise extracellular fluid.
2. Interstitial fluid and lymph are similar in composition to plasma but contain less protein.

FUNCTIONS OF BLOOD (p. 553)

1. Blood transports oxygen, carbon dioxide, nutrients, wastes, and hormones.
2. It helps regulate pH, body temperature, and water content of cells.
3. It prevents blood loss through clotting and combats toxins and microbes through certain phagocytic white blood cells or specialized plasma proteins.

PHYSICAL CHARACTERISTICS OF BLOOD (p. 553)

1. Physical characteristics of blood include a viscosity greater than that of water; a temperature of 38°C (100.4°F); and a pH of 7.35–7.45. Blood constitutes about 8% of body weight and its volume is 4–6 liters in an adult.

COMPONENTS OF BLOOD (p. 554)

1. Blood consists of 55% plasma and 45% formed elements.
2. Plasma consists of 91½% water and 8½% solutes.
3. Principal solutes include proteins (albumins, globulins, fibrinogen), nutrients, vitamins, hormones, respiratory gases, electrolytes, and waste products.
4. The formed elements in blood include erythrocytes (red blood cells), leukocytes (white blood cells), and platelets.

FORMATION OF BLOOD CELLS (p. 554)

1. Blood cells are formed from hemopoietic stem cells in red bone marrow. The process is called hemopoiesis.
2. Several hemopoietic growth factors stimulate differentiation and proliferation of the various blood cells.

ERYTHROCYTES (RED BLOOD CELLS) (p. 558)

1. Mature erythrocytes are biconcave discs without nuclei that contain hemoglobin.
2. The function of the hemoglobin in red blood cells is to transport oxygen and some carbon dioxide.

3. Red blood cells live about 120 days. A healthy male has about 5.4 million/mm³ of blood; a healthy female, about 4.8 million/mm³.
4. After phagocytosis of aged red blood cells by macrophages, hemoglobin is recycled.
5. Erythrocyte formation, called erythropoiesis, occurs in adult red bone marrow of certain bones. It is stimulated by hypoxia, which stimulates release of erythropoietin by the kidneys.
6. A reticulocyte count is a diagnostic test that indicates the rate of erythropoiesis.
7. A hematocrit (Hct) measures the percentage of red blood cells in whole blood.

LEUKOCYTES (WHITE BLOOD CELLS) (p. 562)

1. Leukocytes are nucleated cells. Two principal types are granular (neutrophils, eosinophils, basophils) and agranular (lymphocytes and monocytes).
2. The general function of leukocytes is to combat inflammation and infection. Neutrophils and macrophages (which develop from monocytes) do so through phagocytosis.
3. Eosinophils combat the effects of histamine in allergic reactions, phagocytize antigen–antibody complexes, and combat parasitic worms; basophils develop into mast cells that liberate heparin, histamine, and serotonin in allergic reactions that intensify the inflammatory response.
4. B lymphocytes, in response to the presence of foreign substances called antigens, differentiate into plasma cells that produce antibodies. Antibodies attach to the antigens and render them harmless. This antigen–antibody response combats infection and provides immunity. T lymphocytes destroy foreign invaders directly.
5. Except for lymphocytes, which may live for years, white blood cells usually live for only a few hours or a few days. Normal blood contains 5000–10,000/mm³.

PLATELETS (p. 564)

1. Platelets (thrombocytes) are disc-shaped structures without nuclei.
2. They are fragments derived from metamegakaryocytes and are involved in clotting.
3. Normal blood contains 250,000–400,000 platelets/mm³.

HEMOSTASIS (p. 565)

1. Hemostasis refers to the stoppage of bleeding.
2. It involves vascular spasm, platelet plug formation, and blood clotting (coagulation).

3. In vascular spasm, the smooth muscle of a blood vessel wall contracts, which slows blood loss.

4. Platelet plug formation involves the aggregation of platelets to stop bleeding.

5. A clot is a network of insoluble protein fibers (fibrin) in which formed elements of blood are trapped.

6. The chemicals involved in clotting are known as clotting (coagulation) factors.

7. Blood clotting involves a cascade of reactions that may be divided into three stages: formation of prothrombinase, conversion of prothrombin into thrombin, and conversion of soluble fibrinogen into insoluble fibrin.

8. Clotting is initiated by the interplay of the extrinsic and intrinsic pathways of blood clotting.

9. Normal coagulation requires vitamin K and also involves clot retraction (tightening of the clot) and fibrinolysis (dissolution of the clot).

10. Clotting in an unbroken blood vessel is called thrombosis. A thrombus that moves from its site of origin is called an embolus.

11. Anticoagulants (for example, heparin) prevent clotting.

BLOOD GROUPS AND BLOOD TYPES
(p. 570)

1. ABO and Rh blood groups are genetically determined and based on antigen–antibody responses.

2. In the ABO blood group, A and B antigens on the surface of RBCs determine blood type. Plasma contains anti-A and anti-B antibodies, which react with antigens that are foreign to the individual.

3. In the Rh system, individuals whose RBCs have Rh antigens are classified as Rh$^+$. Those who lack the antigen are Rh$^-$.

REVIEW QUESTIONS

1. How are blood, interstitial fluid, and lymph related to the maintenance of homeostasis? (p. 553)

2. Distinguish between the cardiovascular system and lymphatic system. (p. 553)

3. List the functions of blood and their relationship to other systems of the body. (p. 553)

4. List the principal physical characteristics of blood. (p. 553)

5. How are blood samples obtained for laboratory testing? (p. 554)

6. Distinguish between plasma and formed elements. (p. 554)

7. What are the major constituents of plasma? What do they do? What is the difference between plasma and serum? (p. 554)

8. Describe the origin of blood cells and how hemopoietic growth factors stimulate the process. (p. 554)

9. Describe the microscopic appearance of erythrocytes. What is the function of erythrocytes? What is induced erythrocythemia (blood doping)? (p. 558)

10. Explain how hemoglobin is recycled. (p. 560)

11. Define erythropoiesis. Relate erythropoiesis to red blood cell count. What factors accelerate and slow erythropoiesis? (p. 560)

12. Distinguish between reticulocyte count and hematocrit. (p. 561)

13. Describe the classification of leukocytes and describe their microscopic appearance and functions. (p. 562)

14. What is the importance of emigration, chemotaxis, and phagocytosis in fighting bacterial invaders? (p. 563)

15. What is histocompatibility testing? (p. 563)

16. Distinguish between leukocytosis and leukopenia. (p. 563)

17. What is a differential white blood cell count? (p. 564)

18. What functions are performed by B cells and T cells? (p. 564)

19. How is a bone marrow transplant performed? Why are they used? (p. 564)

20. Describe the structure and function of platelets. (p. 564)

21. Compare erythrocytes, leukocytes, and platelets with respect to size, number per mm^3, and life span. (p. 565)

22. Define hemostasis. Explain the mechanism involved in vascular spasm and platelet plug formation. (p. 565)

23. Briefly describe the stages of clot formation. What is fibrinolysis? Why does blood usually not remain clotted inside blood vessels? (p. 567)

24. How do the extrinsic and intrinsic pathways of blood clotting differ? (p. 568)

25. What is hemophilia? Describe its signs and symptoms. (p. 569)

26. Define the following: thrombus, embolus, anticoagulant. (p. 570)

27. What is the basis for ABO blood grouping? (p. 570)

28. What is the basis for the Rh system? How does hemolytic disease of the newborn (erythroblastosis fetalis) occur? How may it be prevented? (p. 573)

29. Define anemia. Contrast the causes of iron-deficiency, pernicious, hemorrhagic, hemolytic, aplastic, and sickle-cell anemia (SCA). (p. 574)

30. What is infectious mononucleosis (IM)? (p. 575)

31. What is leukemia, and what are the causes of some of its symptoms? (p. 575)

ANSWERS TO QUESTIONS WITH FIGURES

19.1 On average about 6 liters in males and 4–5 liters in females.

19.2 Temperature, 38°C (100.4°F); pH, 7.35–7.45; body weight, 8%.

19.3 Four—one bound to each heme group.

19.4 It is an iron-carrying plasma protein.

19.5 Hypoxia.

19.6 Neutrophils, eosinophils, and basophils are called granulocytes because after staining all have granules in their cytoplasm that can be seen using a light microscope.

19.7 Lymphocytes recirculate back and forth between blood and tissues, whereas other WBCs after leaving the blood remain in the tissues until they die.

19.8 Vascular spasm and blood coagulation.

19.9 Blood plasma minus the clotting proteins.

19.10 Formation of prothrombinase.

19.11 Anti-A and anti-B.

19.12 Type O.

CHAPTER 20

THE CARDIOVASCULAR SYSTEM: THE HEART

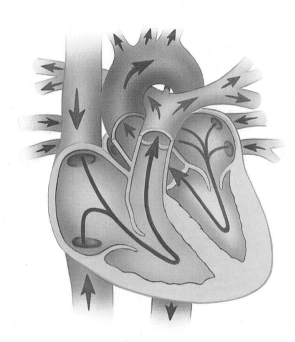

CHAPTER CONTENTS AT A GLANCE

ogether, the heart, blood vessels, and blood make up the **cardiovascular** (*cardio* = heart) **system**. Many-celled organisms need such a system for circulating body fluids because their cells constantly use oxygen and nutrients and produce wastes. Because most cells don't move around, however, they can exchange these materials only with their immediate environment, the interstitial fluid that bathes them. Thus, to maintain homeostasis, this environment must be continually renewed. As blood flows through body tissues, there is a net movement of nutrients and oxygen from the blood into the interstitial fluid and then into cells. At the same time, there is an opposite movement of wastes, carbon dioxide, and heat from cells into interstitial fluid and then into the bloodstream.

The heart is magnificently designed for its task of propelling blood through an estimated 100,000 km (60,000 mi) of blood vessels. Even while you are sleeping, your heart pumps 30 times its own weight each minute, about 5 liters (5.3 qt) to the lungs and the same volume to the rest of the body. At this rate, the heart would pump more than 14,000 liters (3,600 gal) of blood in a day or 10 million liters (2.6 million gal) in a year. You don't spend all your time sleeping, however, and your heart pumps more vigorously when you are active. The actual flow thus is much larger.

This chapter explores the design of the heart and the unique properties of cardiac muscle that permit a lifetime of pumping with never a minute's rest. The study of the normal heart and diseases associated with it is known as **cardiology** (kar-dē-OL-ō-jē).

The developmental anatomy of the heart is considered at the end of the chapter.

OVERVIEW OF THE CIRCULATION

With each beat, the heart pumps blood into two closed circuits—the **systemic circulation** and the **pulmonary** (*pulmonis* = lung) **circulation** (Fig. 20.1). The left side of the heart is the pump for the systemic circulation. It receives freshly oxygenated blood from the lungs and propels it into the largest artery in the body, the **aorta** (*aorte* = to suspend, because the aorta once was believed to suspend the heart). From the aorta, the blood divides into separate streams, entering progressively smaller and smaller **systemic arteries** that carry it to organs throughout the body except the air sacs (alveoli) of the lungs.

Figure 20.1 The systemic and pulmonary circulations.

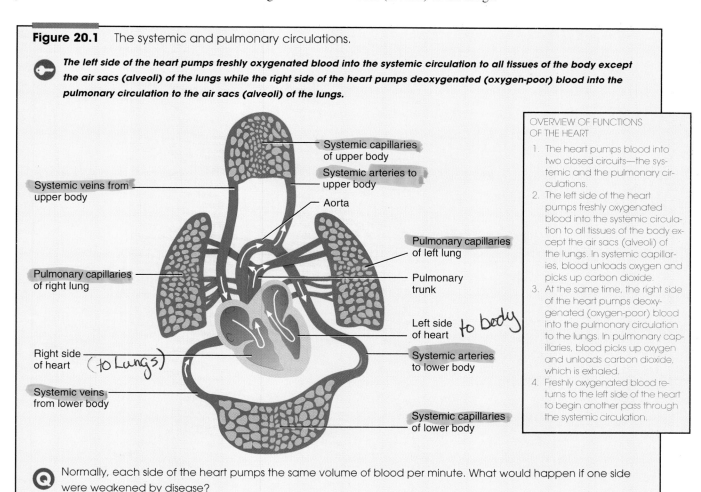

The left side of the heart pumps freshly oxygenated blood into the systemic circulation to all tissues of the body except the air sacs (alveoli) of the lungs while the right side of the heart pumps deoxygenated (oxygen-poor) blood into the pulmonary circulation to the air sacs (alveoli) of the lungs.

Systemic veins from upper body

Pulmonary capillaries of right lung

Right side of heart (to Lungs)

Systemic veins from lower body

Systemic capillaries of upper body

Systemic arteries to upper body

Aorta

Pulmonary capillaries of left lung

Pulmonary trunk

Left side of heart to body

Systemic arteries to lower body

Systemic capillaries of lower body

OVERVIEW OF FUNCTIONS OF THE HEART

1. The heart pumps blood into two closed circuits—the systemic and the pulmonary circulations.
2. The left side of the heart pumps freshly oxygenated blood into the systemic circulation to all tissues of the body except the air sacs (alveoli) of the lungs. In systemic capillaries, blood unloads oxygen and picks up carbon dioxide.
3. At the same time, the right side of the heart pumps deoxygenated (oxygen-poor) blood into the pulmonary circulation to the lungs. In pulmonary capillaries, blood picks up oxygen and unloads carbon dioxide, which is exhaled.
4. Freshly oxygenated blood returns to the left side of the heart to begin another pass through the systemic circulation.

Normally, each side of the heart pumps the same volume of blood per minute. What would happen if one side were weakened by disease?

In the tissues, arteries give rise to smaller diameter **arterioles**, which finally lead into extensive **capillary** beds (Fig. 20.1). Exchange of nutrients and gases occurs across the thin capillary walls. Here, blood unloads oxygen (O_2) and picks up carbon dioxide (CO_2). In most cases, blood flows through only one capillary and then enters a **venule**. Venules carry deoxygenated (oxygen-poor) blood away from tissues and merge to form larger **veins**, and ultimately the blood flows back toward the heart.

The right side of the heart is the pump for the pulmonary circulation. It receives all deoxygenated blood returning from the systemic circulation. Blood ejected from the right side of the heart flows into the **pulmonary trunk**, a large vessel that branches into **pulmonary arteries**, which carry blood to the right and left lungs. In pulmonary capillaries, then, blood picks up oxygen and unloads carbon dioxide, which is exhaled. The freshly oxygenated blood then returns to the left side of the heart to begin another pass through the systemic circulation.

LOCATION AND SIZE OF THE HEART

For all its might, the hollow, cone-shaped heart is relatively small, about the same size as a person's closed fist, and averages only about 300 g (10 oz) in an adult. The heart contains four chambers: two atria and two ventricles. It rests on the diaphragm, near the middle of the thoracic cavity in the **mediastinum** (mē′-dē-a-STĪ-num), a broad median partition, actually a mass of tissues, between the lungs that extends from the sternum to the vertebral column (Fig. 20.2). About two-thirds of the mass of the heart lies to the left of the body's midline. The heart is about 12 cm (5 in.) long, 9 cm (3½ in.) wide at its broadest point, and 6 cm (2½ in.) thick. The pointed inferior end of the heart, the **apex**, is formed by the tip of the left ventricle and tilts obliquely toward the left hip. Opposite the apex, the wide superior and posterior margin of the heart is called the **base**, so named because it is broad and

Figure 20.2   Position of the heart and associated blood vessels in the thoracic cavity. Vessels that carry oxygenated blood are colored red and vessels that carry deoxygenated blood are colored blue throughout the book. The arrow in the inset in (b) indicates the direction from which the section is viewed (inferior).

 The heart is located in the mediastinum with two-thirds of its mass to the left of the midline.

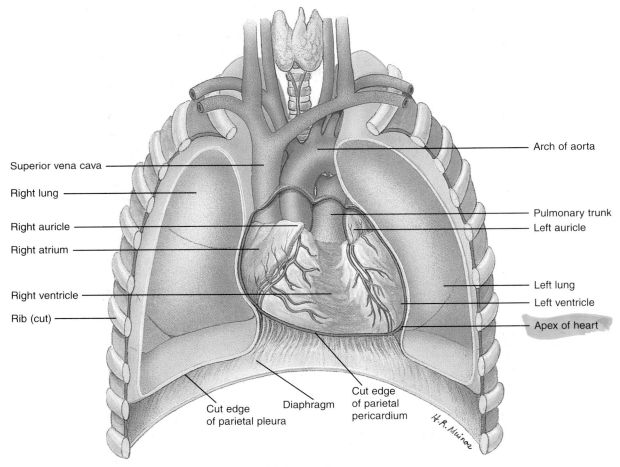

(a) Anterior view

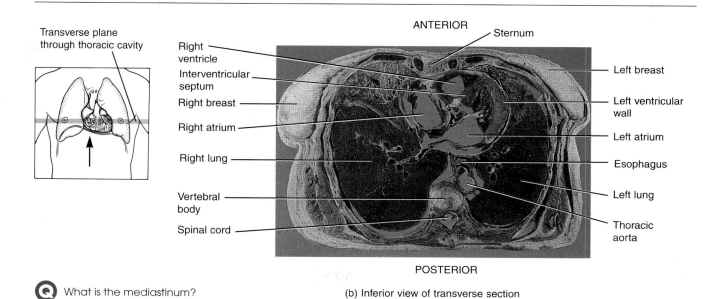

Transverse plane through thoracic cavity

ANTERIOR

Sternum

Right ventricle

Interventricular septum

Right breast

Right atrium

Right lung

Vertebral body

Spinal cord

Left breast

Left ventricular wall

Left atrium

Esophagus

Left lung

Thoracic aorta

POSTERIOR

 What is the mediastinum?

(b) Inferior view of transverse section

rather flat, like the base of a pyramid. The base of the heart is formed by the atria, mostly the left atrium.

CARDIOPULMONARY RESUSCITATION

Because the heart lies between two rigid structures (the vertebral column and the sternum), external pressure (compression) on the chest can be used to force blood out of the heart into the circulation (Fig. 20.2b). In cases where the heart suddenly stops beating, **cardiopulmonary resuscitation** (**CPR**)—properly applied cardiac compressions, performed with artificial ventilation of the lungs—saves lives by keeping oxygenated blood circulating until the heart can be restarted. ■

PERICARDIUM

The **pericardium** (*peri* = around) is a triple-layered sac that surrounds and protects the heart. It confines the heart to its position in the mediastinum, yet allows it sufficient freedom of movement for vigorous and rapid contraction.

The pericardium consists of two principal portions: the fibrous pericardium and the serous pericardium (Fig. 20.3a). The outer **fibrous pericardium** is a tough, inelastic, dense irregular connective tissue. It resembles a bag that rests on and attaches to the diaphragm with its open end fused to the connective tissues of the blood vessels entering and leaving the heart. The fibrous pericardium prevents overstretching of the heart, provides protection, and anchors the heart in the mediastinum.

The inner **serous pericardium** is a thinner, more delicate membrane that forms a double layer around the heart (Fig. 20.3a). The outer **parietal layer** of the serous pericardium

is fused to the fibrous pericardium. The inner **visceral layer** of the serous pericardium, also called the **epicardium** (*epi* = on top), adheres tightly to the surface of the heart. Between the parietal and visceral layers of the serous pericardium is a thin film of serous fluid known as **pericardial fluid**. It is a slippery secretion of the pericardial cells and reduces friction between the membranes as the heart moves. The space that houses the few milliliters of pericardial fluid is called the **pericardial cavity**.

PERICARDITIS AND CARDIAC TAMPONADE

Inflammation of the pericardium is known as **pericarditis**. If production of pericardial fluid diminishes, the result may be painful rubbing together of the parietal and visceral serous pericardial layers. A buildup of pericardial fluid, which may also occur in pericarditis, or extensive bleeding into the pericardium is a life-threatening condition. Because the pericardium cannot stretch, buildup of fluid or blood compresses the heart. This compression, known as **cardiac tamponade** (tam'-pon-ĀD), can stop the beating of the heart. ■

HEART WALL

Three layers form the wall of the heart (Fig. 20.3a): the epicardium (external layer), myocardium (middle layer), and endocardium (inner layer). The outermost **epicardium** (also called the visceral layer of the serous pericardium) is composed of mesothelium and delicate connective tissue that imparts a smooth, slippery texture to the outermost surface of the heart.

The middle **myocardium** (*myo* = muscle), which is cardiac muscle tissue, makes up the bulk of the heart and is responsible for its pumping action. Cardiac muscle fibers (cells)

Figure 20.3   Pericardium, heart wall, and cardiac muscle fibers.

🔑 *The pericardium is a triple-layered sac that surrounds and protects the heart.*

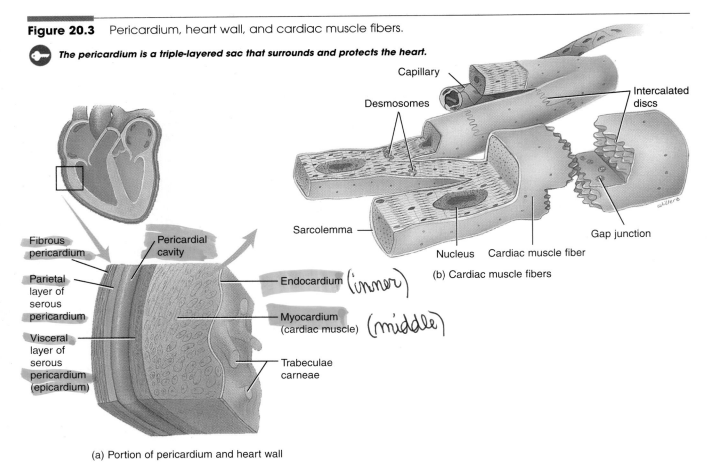

(a) Portion of pericardium and heart wall

(b) Cardiac muscle fibers

❓ Which layer is both a part of the pericardium and a part of the heart wall?

---

are involuntary, striated, and branched. They swirl diagonally around the heart in interlacing bundles and form two large independent networks—one atrial and one ventricular. At its ends, each fiber physically contacts neighboring fibers in its network by transverse thickenings of the sarcolemma called **intercalated discs** (Fig. 20.3b). Within the discs are **gap junctions** (electrical synapses) that allow muscle action potentials to spread from fiber to fiber. As a result, the entire atrial network contracts as one unit and the ventricular network contracts as another. The intercalated discs also contain **desmosomes**. These reinforcing "spot welds" prevent adjacent cardiac fibers from pulling apart during their vigorous contractions.

The innermost **endocardium** (*endo* = within) is a thin layer of endothelium overlying a thin layer of connective tissue. It provides a smooth lining for the inside of the heart and covers the valves of the heart. The endocardium is continuous with the endothelial lining of the large blood vessels associated with the heart and the rest of the cardiovascular system.

CHAMBERS OF THE HEART

The interior of the heart is divided into four compartments called **chambers** (Fig. 20.4). The two superior chambers are

called the **right atrium** and **left atrium** (*atrium* = court or entry hall; plural is **atria**). Whereas the lining of the posterior wall of an atrium has a smooth surface, the anterior wall is rough due to the presence of internal muscular ridges called **pectinate muscles**. Each atrium has an appendage called an **auricle** (OR-i-kul; *auris* = ear), so named because its shape resembles a dog's ear. The auricles increase the size of the atria so that they can hold greater volumes of blood. The two inferior chambers are the **right ventricle** and **left ventricle** (*ventricle* = little belly). Externally, a groove known as the **coronary sulcus** (SUL-kus; plural is **sulci**; SUL-kē) separates the atria from the ventricles (Fig. 20.4a). The **anterior interventricular sulcus** and **posterior interventricular sulcus** separate the right and left ventricles externally (Fig. 20.4a,c). Sulci contain coronary blood vessels and a variable amount of fat.

Connective tissue separates the muscle tissue of the atria from that of the ventricles and effectively divides the myocardium into separate atrial and ventricular muscle masses. The **interatrial septum** (*septum* = partition) is an internal wall that separates the atria. A prominent feature of this septum is an oval depression called the **fossa ovalis** (Fig. 20.4d). This is the remnant of the foramen ovale, an opening in the interatrial septum of the fetal heart that closes soon after birth (see Fig. 21.30a). An internal wall known as the

interventricular septum separates the two ventricles. In the ventricles the irregular surface of ridges and folds of the myocardium covered by endocardium is known as the **trabeculae carneae** (tra-BEK-yoo-lē KAR-nē-ē; *trabecula* = little beam; *carneous* = fleshy). See also Fig. 20.3a.

The thickness of the walls of the four chambers varies according to their functions. The thin walls of the two atria, which contract at the same time, are sufficient to deliver blood into the ventricles (Fig. 20.4d). Although the right and left ventricles act as two separate pumps and expel the same volume of blood at the same time, the left side has a much larger work load. Whereas the right ventricle pumps blood at a fairly low pressure only to the lungs (pulmonary circulation), the left ventricle pumps blood at a much higher pressure to all other parts of the body (systemic circulation). Thus the left ventricle must work harder than the right ventricle to maintain the same rate of blood flow. The anatomy of the two ventricles confirms this functional difference: the muscular wall of the left ventricle is two to four times as thick as the wall of the right ventricle.

VALVES OF THE HEART

As each chamber of the heart contracts, it pushes a portion of blood into a ventricle or out of the heart through an artery. To prevent back flow of blood, the heart has **valves**. These structures are composed of dense connective tissue covered by endocardium. Valves open and close in response to pressure changes as the heart contracts and relaxes.

Atrioventricular Valves

Atrioventricular (AV) valves lie between the atria and ventricles (Fig. 20.4d). The right AV valve between the right atrium and right ventricle is also called the **tricuspid** (tri-KUS-pid) **valve** because it consists of three cusps (flaps). The left AV valve between the left atrium and left ventricle has two cusps and is called the **bicuspid (mitral) valve**. When an AV valve is open, the pointed ends of the cusps project into the ventricle. Tendonlike cords called **chordae tendineae** (KOR-dē ten-DIN-ē-ē; *corda* = cord;

Figure 20.4 Structure of the heart.

 The four chambers of the heart are two superior atria and two inferior ventricles.

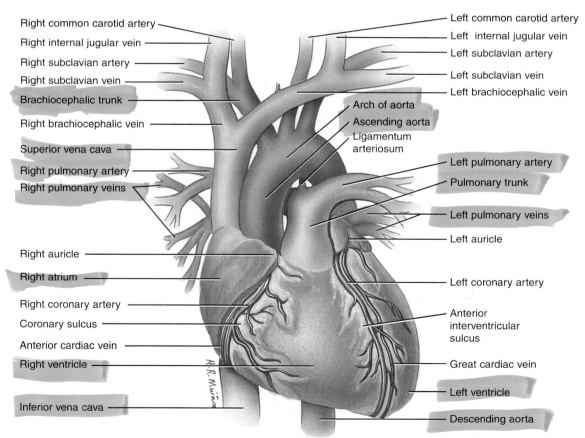

Right common carotid artery
Right internal jugular vein
Right subclavian artery
Right subclavian vein
Brachiocephalic trunk
Right brachiocephalic vein
Superior vena cava
Right pulmonary artery
Right pulmonary veins
Right auricle
Right atrium
Right coronary artery
Coronary sulcus
Anterior cardiac vein
Right ventricle
Inferior vena cava

Left common carotid artery
Left internal jugular vein
Left subclavian artery
Left subclavian vein
Left brachiocephalic vein
Arch of aorta
Ascending aorta
Ligamentum arteriosum
Left pulmonary artery
Pulmonary trunk
Left pulmonary veins
Left auricle
Left coronary artery
Anterior interventricular sulcus
Great cardiac vein
Left ventricle
Descending aorta

(a) Diagram of anterior external view

Figure continues

Figure 20.4   (continued)

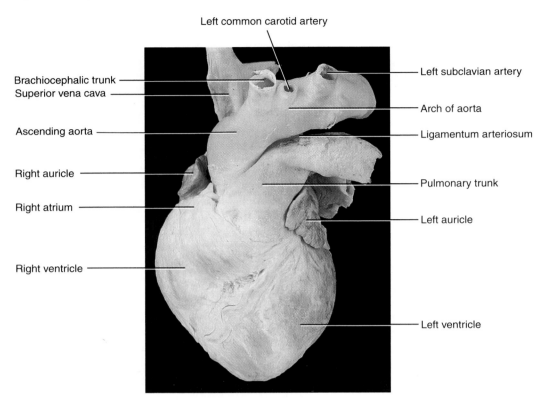

Left common carotid artery

Brachiocephalic trunk

Superior vena cava

Ascending aorta

Right auricle

Right atrium

Right ventricle

Left subclavian artery

Arch of aorta

Ligamentum arteriosum

Pulmonary trunk

Left auricle

Left ventricle

(b) Photograph of anterior external view

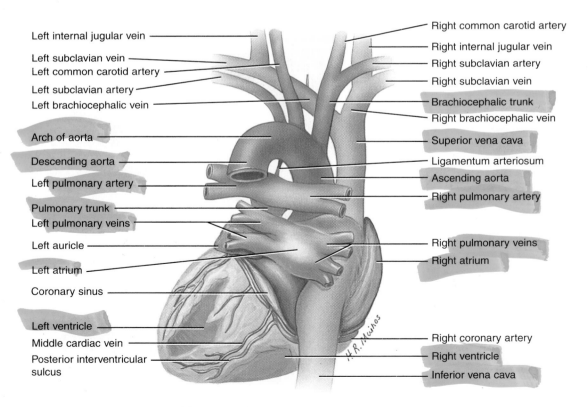

Left internal jugular vein

Left subclavian vein

Left common carotid artery

Left subclavian artery

Left brachiocephalic vein

Arch of aorta

Descending aorta

Left pulmonary artery

Pulmonary trunk

Left pulmonary veins

Left auricle

Left atrium

Coronary sinus

Left ventricle

Middle cardiac vein

Posterior interventricular sulcus

Right common carotid artery

Right internal jugular vein

Right subclavian artery

Right subclavian vein

Brachiocephalic trunk

Right brachiocephalic vein

Superior vena cava

Ligamentum arteriosum

Ascending aorta

Right pulmonary artery

Right pulmonary veins

Right atrium

Right coronary artery

Right ventricle

Inferior vena cava

(c) Diagram of posterior external view

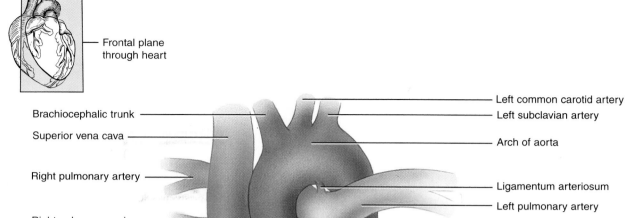

Frontal plane through heart

Brachiocephalic trunk

Superior vena cava

Right pulmonary artery

Right pulmonary veins

Pulmonary semilunar valve

Fossa ovalis

Right atrium

Opening of coronary sinus

Tricuspid valve

Right ventricle

Trabeculae carneae

Inferior vena cava

Left common carotid artery

Left subclavian artery

Arch of aorta

Ligamentum arteriosum

Left pulmonary artery

Pulmonary trunk

Left pulmonary veins

Left atrium

Aortic semilunar valve

Bicuspid valve

Chordae tendineae

Interventricular septum

Papillary muscle

Left ventricle

Descending aorta

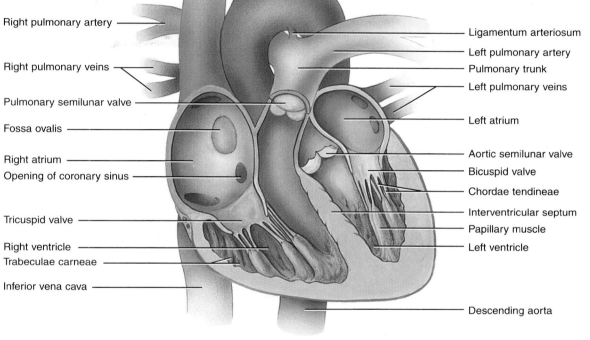

(d) Diagram of anterior view of frontal section

 Which chamber has the thickest wall?

tendo = tendon) connect the pointed ends and undersurfaces to **papillary** (*papilla* = nipple) **muscles** that are located on the inner surface of the ventricles.

Blood moves from the atria into the ventricles through open AV valves when ventricular pressure is lower than atrial pressure (Fig. 20.5a). At this time, the papillary muscles are relaxed, and the chordae tendineae are slack. When the ventricles contract, the pressure of the blood drives the cusps upward until their edges meet and close the opening (Fig. 20.5b). At the same time, the papillary muscles are also contracting, which pulls on and tightens the chordae tendineae. This prevents the valve cusps from everting (swinging back into the atria). If the AV valves or chordae tendineae are damaged, blood may regurgitate (flow back) into the atria when the ventricles contract.

Semilunar Valves

Near the origin of both arteries that emerge from the heart is a heart valve that allows ejection of blood from the heart but prevents backflow of blood into the heart. These are the **semilunar (SL) valves** (see Fig. 20.4d). The **pulmonary semilunar valve** lies in the opening between the pulmonary trunk and the right ventricle. The **aortic semilunar valve** guards the opening between the left ventricle and the aorta.

Both valves consist of three semilunar (half-moon, or crescent-shaped) cusps. Each cusp attaches to the artery wall by its convex outer margin. The free borders of the cusps curve outward and project into the opening inside the blood vessel. When blood starts to flow backward toward

Figure 20.5 Atrioventricular (AV) valves. The bicuspid and tricuspid valves operate in a similar manner. The arrow in the inset in (d) indicates the direction from which the artificial valve is viewed (superior).

 Heart valves prevent backflow of blood.

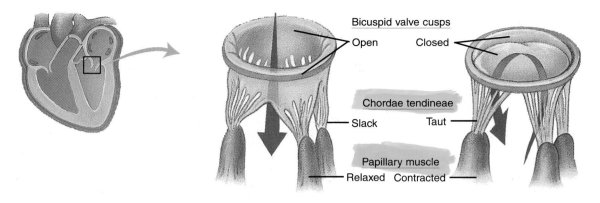

Bicuspid valve cusps

Open   Closed

Chordae tendineae

Slack   Taut

Papillary muscle

Relaxed   Contracted

(a) Bicuspid valve open

(b) Bicuspid valve closed

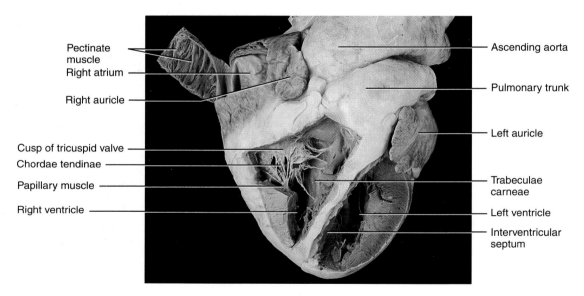

Pectinate muscle

Right atrium

Right auricle

Cusp of tricuspid valve

Chordae tendinae

Papillary muscle

Right ventricle

Ascending aorta

Pulmonary trunk

Left auricle

Trabeculae carneae

Left ventricle

Interventricular septum

(c) Photograph of partially sectioned heart in anterior view

Artificial atrioventricular valve

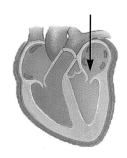

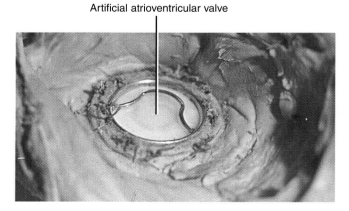

(d) Photograph of a superior view of an artificial atrioventricular valve

Q How do papillary muscles prevent valve cusps from inverting or swinging upward into the atria?

the heart (as the ventricles relax), it fills the cusps and tightly closes the semilunar valves.

RHEUMATIC FEVER

An infection with group A, β-hemolytic *Streptococcus pyogenes* bacteria may result in an inflammation of the heart that damages the valves, particularly the bicuspid and aortic semilunar valves. This condition is called **rheumatic fever** and is usually precipitated by a streptococcal sore throat. The bacteria trigger an immune response. Antibodies that recognize molecules carried by streptococci attack similar molecules in joints, heart valves, and other tissues, which damages the involved structure. Genetic factors may be important in some patients. ■

BLOOD FLOW THROUGH THE PULMONARY AND SYSTEMIC CIRCULATIONS

The right atrium receives *deoxygenated blood* (blood that has given up some of its oxygen to cells) from various parts of the body. Figure 20.6 shows the route of blood flow through the pulmonary and systemic circulations.

From the right atrium (**1**), blood flows into the right ventricle (**2**), which pumps it into the *pulmonary trunk* (**3**). The pulmonary trunk divides into a *right* and *left pulmonary artery*, each of which carries blood to one lung. As blood flows through *pulmonary capillaries*, it loses CO_2 and takes

Figure 20.6   Blood flow through the pulmonary and systemic circulations.

🔑 *Blood flowing through lung capillaries loses CO_2 and gains O_2; blood flowing through systemic capillaries loses O_2 and gains CO_2.*

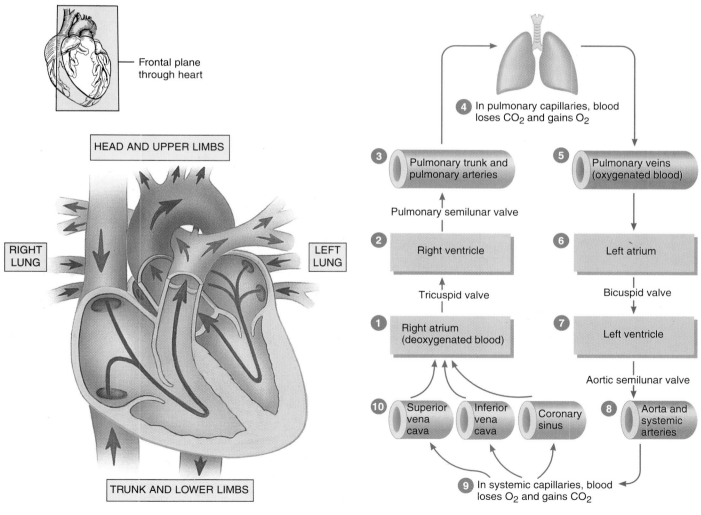

(a) Path of blood flow through the heart (frontal section)   (b) Diagram of blood flow

Q What is the path of blood flow from the heart to the lungs, back to the heart, from the heart to systemic circulation, and back to the heart again? (Be sure to name the vessels, chambers, and valves of the heart, starting in the right atrium and ending in the superior vena cava.)

on O_2 (④). This blood, called *oxygenated blood,* returns to the heart via the *pulmonary veins* (⑤) that empty into the left atrium (⑥). The blood then passes into the *left ventricle* (⑦), which pumps the blood into the *ascending aorta* (⑧). Branches of the arch of the aorta and descending aorta (thoracic aorta and abdominal aorta) deliver blood to *systemic arteries,* which lead into *systemic capillaries* (⑨). In systemic capillaries, blood loses O_2 and gains CO_2. This blood, called *deoxygenated blood,* returns to the right side of the heart through three veins (⑩): the **superior vena cava** (*vena* = vein; *cava* = space) or **SVC** brings blood from most parts of the body superior to the heart, the **inferior vena cava** (**IVC**) brings blood from all parts of the body inferior to the diaphragm (see Fig. 20.4c,d), and the **coronary sinus** drains blood from most of the vessels serving the heart itself (see Fig. 20.7).

During fetal life, a temporary blood vessel, called the ductus arteriosus, connects the pulmonary trunk with the aorta (see Fig. 21.30). It redirects blood so that only a small amount enters the developing but nonfunctioning fetal lungs. The ductus arteriosus normally closes shortly after birth, leaving a remnant known as the **ligamentum arteriosum** (see Fig. 20.4d).

HEART BLOOD SUPPLY

The wall of the heart has its own blood vessels. Nutrients could not possibly diffuse from the chambers of the heart through all the layers of cells that make up the heart tissue. The flow of blood through the many vessels that serve the myocardium is called the **coronary (cardiac) circulation**. The arteries of the heart encircle it like a crown encircles the head (*corona* = crown). While it is contracting, the heart receives little flow of oxygenated blood by way of the **coronary arteries,** which branch from the ascending aorta. When the heart relaxes, however, the high pressure of blood in the aorta propels blood through the coronary arteries, into capillaries, and then into **coronary veins**.

Coronary Arteries

Two coronary arteries—right and left—branch from the ascending aorta (Fig. 20.7a). The **left coronary artery** passes inferior to the left auricle and divides into the anterior interventricular and circumflex branches. The **anterior interventricular branch** or **left anterior descending (LAD) artery** is in the anterior interventricular sulcus and supplies oxygenated blood to the walls of both ventricles and the interventricular septum. The **circumflex branch** lies in the coronary sulcus and distributes oxygenated blood to the walls of the left ventricle and left atrium.

The **right coronary artery** supplies small branches (atrial branches) to the right atrium. It continues inferior to the right auricle and divides into the posterior interventricular and marginal branches. The **posterior interventricular branch** follows the posterior interventricular sulcus and

---

Figure 20.7 Coronary (cardiac) circulation. The heart is viewed from the anterior aspect but is drawn as if it were transparent to reveal blood vessels on the posterior aspect.

 The right and left coronary arteries deliver blood to the heart; the coronary veins drain blood from the heart into the coronary sinus.

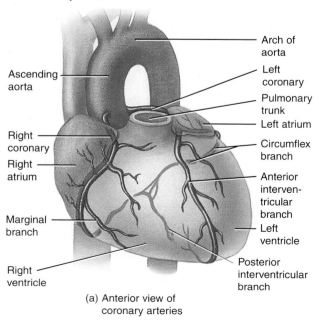

(a) Anterior view of coronary arteries

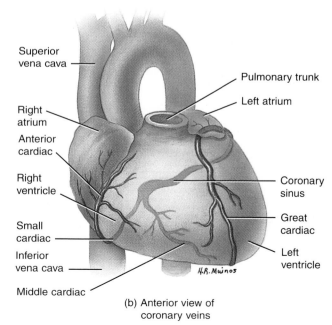

(b) Anterior view of coronary veins

Which blood vessel delivers oxygenated blood to the left atrium and left ventricle?

supplies the walls of the two ventricles and the interventricular septum with oxygenated blood. The **marginal branch** in the coronary sulcus transports oxygenated blood to the myocardium of the right ventricle.

Most parts of the body receive branches from more than one artery, and where two or more arteries supply the same region, they usually connect. The connections, called **anastomoses** (a-nas-tō-MŌ-sēs; singular is **anastomosis**), provide alternate routes for blood to reach a particular organ or tissue. The myocardium contains many anastomoses, connecting branches of one coronary artery or extending between branches of different coronary arteries. In a resting person, heart muscle can remain alive if it receives as little as 10–15% of its normal blood supply, but the person may have little ability to engage in activities.

Coronary Veins

As blood passes through the coronary circulation, it delivers oxygen and nutrients and collects carbon dioxide and wastes. It then drains into a large vascular sinus on the posterior surface of the heart, called the **coronary sinus** (Fig. 20.7b), which empties into the right atrium (see also Fig. 20.4c). A vascular sinus is a venous space with a thin wall that has no smooth muscle to alter its diameter. The principal tributaries carrying blood into the coronary sinus are the **great cardiac vein**, which drains the anterior aspect of the heart, and the **middle cardiac vein**, which drains the posterior aspect of the heart.

ANGINA PECTORIS AND MYOCARDIAL INFARCTION

Most heart problems result from faulty coronary circulation due to blood clots, fatty atherosclerotic plaques, or spasms of the smooth muscle in coronary artery walls. Reduction of blood flow is called **ischemia** (is-KĒ-mē-a; *ischein* = to suppress; *haima* = blood). Usually, ischemia causes **hypoxia** (reduced oxygen supply), which may weaken cells without killing them. **Angina pectoris** (an-JĪ-na, or AN-ji-na, PEK-to-ris), literally meaning "strangled chest," is a severe pain that usually accompanies myocardial ischemia. Typically, sufferers describe it as a tight or squeezing sensation, as though the chest were in a vise. Angina pectoris often occurs during exertion, when the heart demands more oxygen, and disappears with rest. The pain associated with angina pectoris is often referred to the neck, chin, or down the left arm to the elbow (see Fig. 15.2). In some people, ischemic episodes occur without producing pain. This is known as **silent myocardial ischemia** and is particularly dangerous because the person has no forewarning of an impending heart attack.

More serious than ischemia is **myocardial infarction** (in-FARK-shun), or **MI**, commonly called a heart attack. **Infarction** means the death of an area of tissue because of an interrupted blood supply. Myocardial infarction may result from a thrombus (stationary blood clot) or embolus (blood clot transported by blood) in one of the coronary arteries. The tissue beyond the obstruction dies and is replaced by noncontractile scar tissue. Thus the heart muscle loses at least some of its strength. The aftereffects depend partly on the size and location of the infarcted, or dead, area. Besides killing normal heart tissue, the infarction may disrupt the conducting system of the heart (described shortly) and may cause sudden death by triggering ventricular fibrillation. Treatment for a myocardial infarction may involve injection of a thrombolytic (clot-dissolving) agent such as streptokinase or t-PA along with heparin and then performing coronary angioplasty or coronary artery bypass grafting (described on page 605). ■

Reperfusion Damage

Whenever disease or injury deprives any tissue of the body of oxygen, **reperfusion**, reestablishing the blood flow, may damage the tissue further. This surprising effect is due to the formation of oxygen **free radicals** from the reintroduced oxygen. Free radicals are electrically charged molecules that have an unpaired electron (shown in Fig. 3.30). Such molecules are unstable and highly reactive. When an oxygen free radical takes an electron from one molecule, that molecule becomes unstable and borrows an electron from another molecule, which, in turn, becomes unstable. These chain reactions lead to cellular damage and death. Among the molecules attacked by oxygen free radicals are proteins (such as enzymes), neurotransmitters, nucleic acids, and phospholipids of plasma membranes.

Free radicals have been implicated in diseases such as heart disease, cancer, Alzheimer's disease, Parkinson's disease, cataracts, and rheumatoid arthritis. They may also contribute to aging. To counter the effects of oxygen free radicals, the body produces enzymes—for example, *superoxide dismutase, catalase,* and *glutathione peroxidase* —that convert free radicals to less reactive substances. In addition, some nutrients, such as vitamins E and C, beta-carotene, and selenium, are antioxidants that counter oxygen free radicals.

CONDUCTION SYSTEM AND PACEMAKER

An inherent and rhythmical electrical activity is the reason for the heart's continuous beating. Certain cardiac muscle cells repeatedly fire spontaneous action potentials (impulses) that then trigger heart contractions. This is why a heart that has been completely removed from the body—for example, to be transplanted into another person—will continue to beat even though all its nerves have been cut. Signals from the autonomic nervous system and hormones, such as epinephrine, in the blood do modify the heartbeat, but they *do not establish the fundamental rhythm.*

Autorhythmic Cells: The Conduction System

During embryonic development, about 1% of the cardiac muscle fibers become **autorhythmic** (self-excitable) **cells**: they repeatedly and rhythmically generate action potentials (APs). Autorhythmic fibers have two important functions. They act as a **pacemaker**, setting the rhythm for the entire heart, and they form the **conduction system**, the route for conducting APs throughout the heart muscle. The conduction system assures that cardiac chambers become excited to contract in a coordinated manner, which makes the heart an effective pump. Figure 20.8 shows the components of the conduction system:

1 Normally, cardiac excitation begins in the **sinoatrial (SA) node**, which is located in the right atrial wall just inferior to the opening of the superior vena cava. Each AP from the SA node travels throughout both atria via gap junctions in the intercalated discs of atrial fibers. In the wake of the AP, the atria contract.

2 The cardiac AP also spreads from the SA node down to the **atrioventricular (AV) node**, located in the septum between the two atria.

3 From the AV node, the AP enters the **atrioventricular (AV) bundle** (**bundle of His**), the only electrical connection between the atria and the ventricles. (Elsewhere, fibrous rings and sheets of connective tissue act as electrical insulation between the atria and ventricles.)

4 After conducting along the AV bundle, the AP then enters both the **right** and **left bundle branches** that course through the interventricular septum toward the apex of the heart.

5 Finally, large-diameter **conduction myofibers** (**Purkinje fibers**) rapidly conduct the AP first to the apex of the

---

Figure 20.8 Conduction system of the heart. The arrows indicate the conduction of action potentials (impulses) through the atria.

The conduction system ensures that cardiac chambers contract in a coordinated manner.

Frontal plane through heart

Arch of aorta

1 SINOATRIAL (SA) NODE

Left atrium

2 ATRIOVENTRICULAR (AV) NODE

Right atrium

3 ATRIOVENTRICULAR (AV) BUNDLE (BUNDLE OF HIS)

Right ventricle

4 RIGHT AND LEFT BUNDLE BRANCHES

Left ventricle

5 CONDUCTION MYOFIBERS (PURKINJE FIBERS)

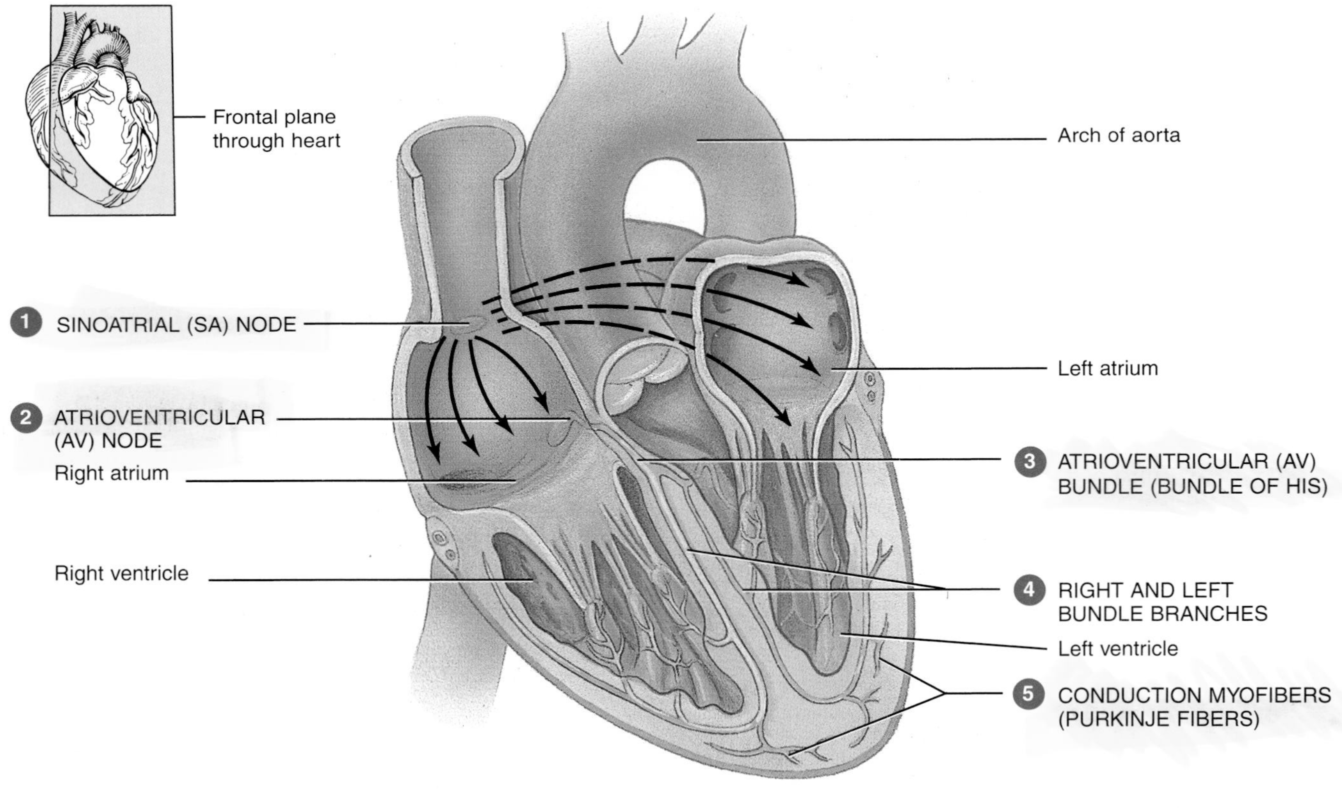

Anterior view of frontal section

Q Which component of the conduction system provides the only electrical connection between the atria and ventricles?

ventricular myocardium and then upward to the remainder of the ventricular myocardium. About 0.20 sec after the atria contract the ventricles contract. Exhibit 20.1 summarizes the conduction system.

On their own, autorhythmic fibers in the SA node initiate action potentials 90–100 times per minute, faster than any other region. As a result, APs from the SA node spread to other areas of the conduction system, stimulating them before they are able to generate an AP at their own slower rate. Thus the normal *pacemaker* of the heart is the SA node. Various hormones and neurotransmitters can speed or slow pacing of the heart by SA node fibers. For example, in a person at rest acetylcholine released by the parasympathetic division of the ANS slows SA node pacing to about 75 impulses per minute.

Sometimes, a site other than the SA node becomes the pacemaker because it develops abnormal self-excitability. Such a site is called an **ectopic** (ek-TOP-ik; *ektopus* = displaced) **pacemaker** or **ectopic focus**. The ectopic focus may operate only occasionally, producing extra beats, or it may pace the heart for some period of time. Triggers of ectopic activity include caffeine and nicotine, electrolyte imbalances, hypoxia, and toxic reactions to drugs such as digitalis.

Timing of Atrial and Ventricular Excitation

From the SA node, a cardiac action potential travels throughout the atrial muscle and down to the AV node in about 0.05 sec (50 milliseconds or msec). The AP slows considerably at the AV node because the fibers there have much smaller diameters. (Recall what happens to automobile traffic when a four-lane highway narrows to just two lanes.) The resulting 0.1-sec (100-msec) delay has an advantage. It gives the atria time to complete their contraction and add to the volume of blood in the ventricles before ventricular contraction begins. After the AP enters the AV bundle, conduction again is rapid; the entire ventricular myocardium undergoes depolarization (loss and then reversal of polarization) about 0.2 sec (200 msec) after the AP arises in the SA node.

If the SA node becomes diseased or damaged, the slower AV node fibers can pick up the pacemaking chores. With pacing by the AV node, however, heart rate is 40–50 beats/min. If the activity of both nodes is suppressed, the heartbeat may still be maintained by autorhythmic fibers in the ventricles—the AV bundle, a bundle branch, or conduction myofibers. These fibers fire APs very slowly, only about 20–40 times per minute. At such a low heart rate, blood flow to the brain is inadequate. In patients with such a condition, normal heart rhythm can be restored and maintained with an **artificial pacemaker**, a device that sends out small electrical currents to stimulate the heart. Many of the newer pacemakers, called activity-adjusted pacemakers, automatically speed up the heartbeat during exercise.

PHYSIOLOGY OF CARDIAC MUSCLE CONTRACTION

The action potential initiated by the SA node travels along the conduction system and spreads out to excite the "working" atrial and ventricular muscle fibers, which are called **contractile fibers**. An AP occurs in a contractile fiber as follows (Fig. 20.9a):

 Contractile fibers have a resting membrane potential close to −90 mV. When they are brought to threshold by excitation in neighboring fibers, certain sodium ion (Na⁺) channels open very rapidly; these are called **voltage-gated fast Na⁺ channels**. When these channels open, the permeability of the sarcolemma (plasma membrane) to sodium ions (P_{Na^+}) increases (Fig. 20.9b). The result is an inflow of Na⁺ along the electrochemical gradient (cytosol more negative than extracellular fluid and Na⁺ concentration higher in extracellular fluid) that produces a **rapid depolarization**.

EXHIBIT 20.1 SUMMARY OF THE CONDUCTION SYSTEM

Structure and Location	Function
Sinoatrial (SA) node in right atrial wall	Autorhythmic fibers initiate cardiac action potentials, which set basic pace for heart rate and conduct throughout both atria.
Atrioventricular (AV) node in septum between atria	Receives action potentials from SA node and passes them to atrioventricular (AV) bundle.
Atrioventricular (AV) bundle (bundle of His) in superior portion of interventricular septum	Receives action potentials from AV node and passes them to right and left bundle branches.
Right and left bundle branches in interventricular septum	Receive action potentials from AV bundle and pass them to conduction myofibers.
Conduction myofibers (Purkinje fibers) in ventricular myocardium	Receives action potentials from bundle branches and passes them to ventricular myocardium.

Figure 20.9   Action potential (impulse) in a ventricular contractile fiber. The resting membrane potential is about –90 mV.

🔑 *The "working" atrial and ventricular muscle fibers are called contractile fibers.*

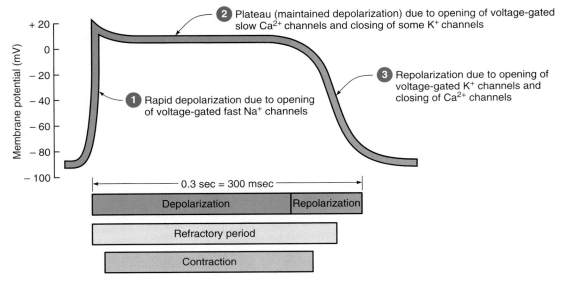

(a) Action potential, refractory period, and contraction

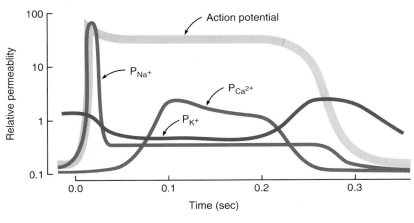

(b) Membrane permeability (P) changes

❓ How does the duration of depolarization and repolarization in a cardiac muscle fiber compare with that in a skeletal muscle fiber?

Within a few milliseconds, the fast Na⁺ channels automatically inactivate and P_{Na^+} decreases.

② During the next phase, called the **plateau**, **voltage-gated slow Ca²⁺ channels** open in the sarcolemma and sarcoplasmic reticulum membrane. This increases permeability to calcium ions ($P_{Ca^{2+}}$), allowing the level of Ca²⁺ to increase in the cytosol. Some Ca²⁺ passes through the sarcolemma from the extracellular fluid (which has a higher Ca²⁺ concentration) while other Ca²⁺ pours out of the sarcoplasmic reticulum within the fiber. At the same time, the membrane permeability to potassium ions (P_{K^+}) decreases due to

closing of K⁺ channels. For about 0.25 sec (250 msec) the membrane potential stays close to 0 mV as a small outflow of K⁺ just balances the inflow of Ca²⁺. By comparison, depolarization in a neuron or skeletal muscle fiber lasts about 1 msec.

③ The recovery of the resting membrane potential during the **repolarization** phase of a cardiac action potential resembles that of other excitable tissues: after a delay (which is particularly prolonged in cardiac muscle), **voltage-gated K⁺ channels** open, thus increasing the membrane permeability to potassium ions. K⁺ then diffuses out more rapidly due to the concentration difference. At the same time, the calcium channels are closing. As more K⁺ leave the fiber and fewer Ca²⁺ enter, the negative resting membrane potential (−90 mV) is restored.

The mechanism of contraction is similar in cardiac and skeletal muscle. The electrical activity (action potential) leads to the mechanical response (contraction) after a short lag time. As its concentration rises inside a contractile fiber, Ca^{2+} binds to the regulator protein troponin, which allows the actin and myosin filaments to begin sliding past one another, and tension starts to develop. Substances that alter the movement of Ca^{2+} through slow Ca^{2+} channels influence the strength of heart contractions. Epinephrine, for example, increases contraction force by enhancing Ca^{2+} inflow. A class of drugs called calcium channel blockers, such as verapamil, reduce Ca^{2+} inflow and diminish the strength of the heartbeat.

In muscle, the **refractory period** is the time interval when a second contraction cannot be triggered. The refractory period of a cardiac fiber is longer than the contraction itself (Fig. 20.9). As a result, another contraction cannot begin until relaxation is well underway. For this reason, tetanus (maintained contraction) cannot occur. The advantage is apparent if you consider how the ventricles work. Their pumping function depends on alternating contraction (when they eject blood) and relaxation (when they refill). If tetanus could occur, blood flow would stop.

ELECTROCARDIOGRAM

Impulse conduction through the heart generates electrical currents that can be detected at the surface of the body. A recording of these electrical changes is called an **electrocardiogram** (e-lek′-trō-KAR-dē-ō-gram), abbreviated either **ECG** or **EKG** (from the German word *elektrokardiogram).* The ECG is a composite of action potentials produced by all the heart muscle fibers during each heartbeat. The instrument used to record the changes is an **electrocardiograph**. In clinical practice, the ECG is recorded by placing electrodes on the arms and legs (the limb leads) and at six positions on the chest. The electrocardiograph amplifies the heart's electrical activity and produces 12 different tracings from different combinations of limb and chest leads. Each limb and chest electrode records slightly different electrical activity because it is in a different position relative to the heart. By comparing these records with one another and with normal records, it is possible to determine (1) if the conduction pathway is abnormal, (2) if the heart is enlarged, and (3) if certain regions are damaged.

In a typical Lead II record (right arm to left leg), three clearly recognizable waves accompany each heartbeat. The first, called the **P wave**, is a small upward wave (Fig. 20.10). It represents **atrial depolarization**, which spreads from the SA node throughout both atria. About 0.1 sec after the P wave begins, the atria contract. The second wave, called the **QRS complex**, begins as a downward deflection,

Figure 20.10   Electrocardiogram or ECG (Lead II).

An ECG is a recording of the electrical activity that accompanies each heartbeat.

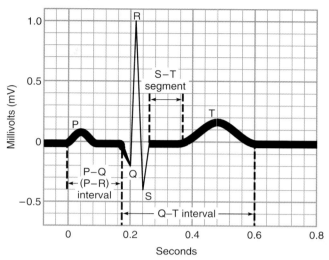

Normal ECG of a single heartbeat

MEANING OF ECG WAVES
P wave = atrial depolarization
QRS complex = onset of ventricular depolarization
T wave = ventricular repolarization

 What is the significance of an enlarged Q wave?

continues as a large, upright, triangular wave, and ends as a downward wave. The QRS complex represents the onset of **ventricular depolarization**, as the wave of electrical excitation spreads through the ventricles. Shortly after the QRS complex begins, the ventricles start to contract. The third wave is a dome-shaped upward deflection called the **T wave**. It indicates **ventricular repolarization** and occurs just before the ventricles start to relax. The T wave is smaller and more spread out than the QRS complex because repolarization occurs more slowly than depolarization. Usually, repolarization of the atria is not evident in an ECG because it is buried in the larger QRS complex.

In reading an ECG, it is important to note the size and timing of the waves. Larger P waves, for example, indicate enlargement of an atrium, as may occur in mitral stenosis. In this condition, the mitral valve narrows, blood backs up into the left atrium, and there is expansion of the atrial wall. An enlarged Q wave may indicate a myocardial infarction (heart attack). An enlarged R wave generally indicates enlarged ventricles. The T wave is flatter than normal when the heart muscle is receiving insufficient oxygen, for example, in coronary artery disease. It may be elevated in hyperkalemia (increased blood K⁺ level).

Analysis of an ECG also involves examination of the time spans between waves, which are called **intervals** or **segments**.

For example, the **P–Q (PR) interval** is measured from the beginning of the P wave to the beginning of the QRS complex. It represents the conduction time from the beginning of atrial excitation to the beginning of ventricular excitation. The P–Q interval is the time required for an impulse to travel through the atria, atrioventricular node, and the remaining fibers of the conduction system. In coronary artery disease and rheumatic fever, scar tissue may form in the heart. As the impulse detours around scar tissue, the P–Q interval lengthens.

The **S–T segment** begins at the end of the S wave and ends at the beginning of the T wave. It represents the time when the ventricular contractile fibers are fully depolarized, during the plateau phase of the action potential. The S–T segment is elevated (above the baseline) in acute myocardial infarction and depressed (below the baseline) when the heart muscle receives insufficient oxygen. The **Q–T interval** extends from the start of the QRS complex to the end of the T wave. It is the time from the beginning of ventricular depolarization to the end of ventricular repolarization. The Q–T interval may be lengthened by myocardial damage, coronary ischemia, or conduction abnormalities.

Sometimes it is necessary to evaluate the heart's response to the stress of physical exercise. Such a test is called a **stress electrocardiogram**, or **stress test**. It is based on the principle that narrowed coronary arteries may carry adequate oxygenated blood while a person is at rest, but during exercise will be unable to meet the heart's increased need for oxygen, creating changes that can be noted on an electrocardiogram.

CARDIAC CYCLE

Since heartbeats automatically follow one another throughout life, you can appreciate much of cardiac physiology by understanding a single **cardiac cycle**, that is, all the events associated with one heartbeat. In each cardiac cycle, pressure changes occur as the atria and ventricles alternately contract and relax, and blood flows from areas of higher blood pressure to areas of lower blood pressure. As a chamber of the heart contracts, pressure of the fluid within it increases. Figure 20.11 shows the relation between the heart's electrical signals (electrocardiogram) and mechanical events (contraction and relaxation) and the consequent changes in atrial pressure, ventricular pressure, ventricular volume, and aortic pressure during the cardiac cycle. The pressures given in Fig. 20.11 apply to the left side of the heart; on the right side, pressures are considerably lower because the wall of the right ventricle is thinner than that of the left. Each ventricle, however, expels the same volume of blood per beat, and the same pattern exists for both pumping chambers.

In a normal cardiac cycle, the two atria contract while the two ventricles relax. Then, while the two ventricles contract, the two atria relax. The term **systole** (SIS-tō-lē; *systole* = contraction) refers to the phase of contraction; the phase of relaxation is **diastole** (dī-AS-tō-lē; *diastole* = expansion). A cardiac cycle consists of a systole and diastole of both atria plus a systole and diastole of both ventricles.

Phases of the Cardiac Cycle

For the purposes of our discussion, we divide the cardiac cycle of a resting adult into three main phases. Follow along on Fig. 20.11 as you read the descriptions.

①   **Relaxation period.** At the end of a heartbeat when the ventricles start to relax, all four chambers are in diastole. This is the beginning of the **relaxation** or **quiescent** (kwī-ES-ent) **period**. Repolarization of the ventricular muscle fibers (T wave in the ECG) initiates relaxation. As the ventricles relax, pressure within the chambers drops, and blood starts to flow from the pulmonary trunk and aorta back toward the ventricles. As this blood becomes trapped in the semilunar cusps, however, the valves close. Rebound of blood off the closed cusps produces a bump called the *dicrotic wave* on the aortic pressure curve.

With closing of the semilunar valves, there is a brief interval when ventricular blood volume does not change because both semilunar and AV valves are closed. This period is called **isovolumetric relaxation**. As the ventricles continue to relax, the space inside expands, and the pressure falls quickly. When ventricular pressure drops below atrial pressure, the AV valves open and ventricular filling begins.

②   **Ventricular filling.** The major part of ventricular filling occurs just after the AV valves open. Blood that has been flowing into the atria and building up while the ventricles were contracting now rushes into the ventricles. The first third of ventricular filling time thus is known as the period of **rapid ventricular filling**. During the middle third, called **diastasis**, a much smaller volume of blood flows into the ventricles.

Firing of the SA node results in atrial depolarization, noted as the P wave in the ECG. Atrial contraction follows the P wave, which also marks the end of the quiescent period. Atrial systole occurs in the last third of the ventricular filling period and accounts for the final 20–25 ml of the blood that fills the ventricles. At the end of ventricular diastole, there is about 130 ml in each ventricle. This volume of blood is called **end-diastolic volume (EDV)**. Since atrial systole contributes only 20–30% of the total blood volume in the ventricles, atrial contraction is not absolutely necessary for adequate blood flow at normal heart rates. Throughout the period of ventricular filling, the AV valves are open and the semilunar valves are closed.

③   **Ventricular systole (contraction).** Near the end of atrial systole, the impulse from the SA node has passed through the AV node and into the ventricles, causing them to depolarize. This is represented by the QRS complex in the ECG. Then, ventricular contraction begins, and blood is pushed up against the AV valves, forcing them shut. For about 0.05 sec (50 msec), all four valves are closed again. This period is called **isovolumetric contraction**. During this interval cardiac muscle fibers are contracting and exerting

Figure 20.11 Cardiac cycle. (a) ECG. (b) Left atrial, left ventricular, and aortic pressure changes along with the opening and closing of valves. (c) Left ventricular volume changes. (d) Heart sounds. (e) Phases of the cardiac cycle.

🔑 *A cardiac cycle is composed of all the events associated with one heartbeat.*

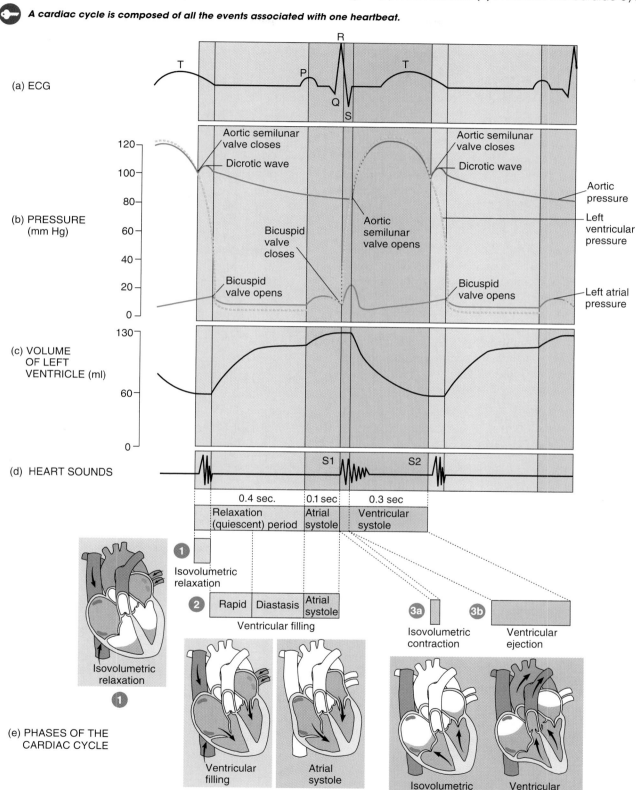

(a) ECG

(b) PRESSURE (mm Hg)

(c) VOLUME OF LEFT VENTRICLE (ml)

(d) HEART SOUNDS

(e) PHASES OF THE CARDIAC CYCLE

❓ How much blood is in each ventricle at the end of ventricular diastole? What is this volume called?

force, but are not yet shortening because it is very difficult to compress any liquid, including blood. Thus the muscle contraction is isometric (same length). Moreover, since there is no escape route for the blood, ventricular volume remains the same (isovolumic).

As ventricular contraction continues, pressure inside the chambers rises sharply. When left ventricular pressure surpasses aortic pressure (about 80 mm Hg) and right ventricular pressure rises above the pressure in the pulmonary trunk (about 15–20 mm Hg), both semilunar valves open, and ejection of blood from the heart begins. This period is called **ventricular ejection** and lasts for about 0.25 sec (250 msec), until the ventricles start to relax. Then, the semilunar valves close and another relaxation period begins. The volume of blood still left in a ventricle after its systole is called **end-systolic volume (ESV)**. At rest, it is about 60 ml.

As mentioned earlier, the different pressures developed by the two ventricles are a reflection of their differing wall thickness. During contraction, the pressure in the left ventricle rises to about 120 mm Hg, while the pressure in the right ventricle climbs to about 30 mm Hg. At rest, the **stroke volume**, the volume ejected per beat from each ventricle, is about 70 ml (a little more than 2 oz). This is a little more than half of the total volume in the ventricle at the end of diastole; during ejection, ventricular volume decreases from about 130 ml (end-diastolic volume) to 60 ml (end-systolic volume).

Timing of Systole and Diastole

Because resting heart rate (HR) is about 75 beats/min, each cardiac cycle lasts about 0.8 sec (see Fig. 20.11). During the first 0.4 sec of the cycle, the relaxation period, all four chambers are in diastole. At first, all valves are closed; then, the atrioventricular valves open and blood starts draining into the ventricles. During the next 0.1 sec, the atria contract and the atrioventricular valves are open, but the ventricles are still relaxed and the semilunar valves are closed. For the next 0.3 sec, the atria are relaxing and the ventricles are contracting. During the first part of this period, all valves are closed (isovolumetric contraction); during the second part, the semilunar valves are open (ventricular ejection). As the heart beats faster, the relaxation period (diastole) becomes shorter and shorter whereas the duration of the contraction period (systole) shortens only slightly.

Heart Sounds

The act of listening to sounds within the body is called **auscultation** (aws-kul-TĀ-shun; *auscultare* = to listen), and it is usually done with a stethoscope. The sound of the heartbeat comes primarily from blood turbulence caused by closing of the heart valves. During each cardiac cycle, four **heart sounds** are generated. In a normal heart, however,

only the first two (first and second heart sounds) are loud enough to be heard by listening through a stethoscope.

The first sound (S1), which can be described as a **lubb** sound, is louder and a bit longer than the second sound. The lubb is the sound created by blood turbulence associated with closure of the AV valves soon after ventricular systole begins. The second sound (S2), which is shorter and not as loud as the first, can be described as a **dupp** sound. Dupp is the sound created by blood turbulence associated with closure of the semilunar valves at the beginning of ventricular diastole. Although these heart sounds are related to blood turbulence associated with the closure of valves, they are not necessarily heard best over these valves. Each sound tends to be clearest in a slightly different location on the chest surface (Fig. 20.12). Normally the third and fourth heart sounds (S3 and S4) are not loud enough to be heard. They are associated with rapid ventricular filling (S3) and atrial contraction (S4). Figure 20.11 shows the timing of S1 and S2 relative to other changes in the cardiac cycle.

In a person at rest, the time between the S2 and the next S1 is about two times longer than the time between S1 and S2 within a cycle. Thus the rhythm is lubb, dupp, pause; lubb, dupp, pause; lubb, dupp, pause. As heart rate increases, the pause interval shortens.

HEART MURMURS

Just as the ECG gives important information about the electrical operation of the heart, heart sounds provide valuable information about the mechanical operation of the heart. A **heart**

Figure 20.12 Heart sounds. The red circles indicate where heart sounds associated with closure of the respective valves are best heard.

Listening to sounds within the body is called auscultation and is usually done with a stethoscope.

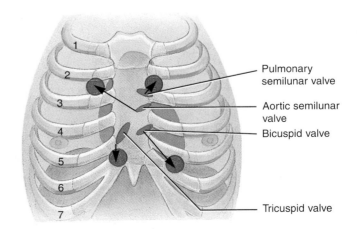

Pulmonary semilunar valve

Aortic semilunar valve

Bicuspid valve

Tricuspid valve

 Which heart sound is related to blood turbulence associated with closure of the AV valves?

murmur is an abnormal sound that consists of a flow noise that is heard before, between, or after the normal heart sounds or that may mask the normal heart sounds. Although some heart murmurs are "innocent," meaning they do not suggest a heart problem, most often a murmur indicates a valve disorder.

Among the valvular abnormalities that may contribute to murmurs are **mitral stenosis** (narrowing of the mitral valve by scar formation or a congenital defect), **mitral insufficiency** (backflow or regurgitation of blood from the left ventricle into the left atrium due to a damaged mitral valve or ruptured chordae tendineae), **aortic stenosis** (narrowing of the aortic semilunar valve), and **aortic insufficiency** (backflow of blood from the aorta into the left ventricle). Another cause of a heart murmur is **mitral valve prolapse** (**MVP**), an inherited disorder in which a portion of a mitral valve is pushed back too far (prolapsed) during ventricular contraction. Although a small volume of blood may flow back into the left atrium during ventricular systole, mitral valve prolapse does not always pose a serious threat. Up to 10% of otherwise healthy youngsters may have MVP. ■

CARDIAC OUTPUT

Although the heart has autorhythmic fibers that enable it to beat independently, its operation is governed by events occurring in the rest of the body. All body cells must receive a certain amount of oxygenated blood each minute to maintain health and life. When cells are very active, as during exercise, they take up more oxygen from the blood. During rest periods, cellular need is reduced, and the work load of the heart decreases.

Cardiac output (**CO**) is the amount of blood ejected from the left ventricle (or the right ventricle) into the aorta (or pulmonary trunk) each minute. Cardiac output equals the **stroke volume** (**SV**), the volume of blood ejected by the ventricle with each contraction, times the **heart rate** (**HR**), the number of heartbeats per minute. Thus CO = SV × HR. In a typical resting adult, stroke volume averages 70 ml/beat and heart rate is about 75 beats/min. This gives an average CO of:

$$
\begin{aligned}
\text{CO} \quad &= \quad \text{SV} \quad \times \quad \text{HR} \\
\text{(ml/min)} \quad &\quad \text{(ml/beat)} \qquad \text{(beats/min)} \\
&= \quad \text{(70 ml/beat)} \ \times \ \text{(75 beats/min)} \\
&= \quad \text{5250 ml/min or 5.25 liters/min}
\end{aligned}
$$

This volume is close to the total blood volume, which is about 5 liters in a typical adult male. The entire blood volume thus flows through the pulmonary and systemic circulations about once a minute. When the demands of the body for oxygen increase or decrease, cardiac output changes to meet the need. Factors that increase stroke volume or heart rate tend to increase CO. During mild exercise, for example, stroke volume may increase to 100 ml/beat and heart rate to 100 beats/min. Cardiac output then would be 10 liters/min. During intense (but still not maximal) exercise,

the heart rate may accelerate to 150 beats/min and stroke volume may rise to 130 ml/beat, providing a cardiac output of 19.5 liters/min.

Cardiac reserve is the ratio between the maximum cardiac output a person can achieve and the cardiac output at rest. An average figure is four or five times the resting value. Top athletes may have a cardiac reserve seven or eight times their resting CO. People with severe heart disease may have little or no cardiac reserve, which limits their ability to carry out even simple daily tasks.

The next sections explain how large increases in cardiac output are possible by increasing stroke volume and heart rate. As we shall see, these two factors are not completely separate but rather depend somewhat on one another.

Regulation of Stroke Volume

A healthy heart pumps out all the blood that has entered its chambers during the previous diastole. At rest this is 50–60% of the total volume because 40–50% remains in the ventricles after each contraction (end-systolic volume). Thus stroke volume (SV) equals end-diastolic volume (EDV) minus end-systolic volume (ESV): SV = EDV − ESV. Three important factors regulate stroke volume in different circumstances and ensure that the left and right ventricles pump equal volumes of blood: (1) **preload**, the *stretch* on the heart before it contracts, (2) **contractility**, the *forcefulness of contraction* of individual ventricular muscle fibers, and (3) **afterload**, the *pressure that must be exceeded* before ejection of blood from the ventricles can begin.

Preload: Effect of Stretching

A greater preload (stretch) on cardiac muscle fibers just before they contract increases their force of contraction. Within limits, the more the heart is filled during diastole, the greater the force of contraction during systole. This is known as the **Frank–Starling law of the heart**. In the body, the preload is the volume of blood that fills the ventricles at the end of diastole, the end-diastolic volume (EDV). The greater the EDV (preload), within limits, the more forceful the contraction.

The duration of ventricular diastole and venous pressure are the two key factors that determine EDV. When heart rate increases, the duration of diastole is shorter. Less filling time means a smaller EDV, and the ventricles may contract before they are adequately filled. On the other hand, when venous pressure increases, a greater volume of blood is forced into the ventricles, and the EDV is increased.

When heart rate exceeds about 160 beats/min, stroke volume usually declines. At such rapid heart rates, the ventricular filling time is severely shortened, EDV is less, and the preload thus is lower. People who have slow resting heart rates, on the other hand, usually have large stroke volumes because filling time is prolonged and preload thus is larger.

The Frank–Starling law of the heart equalizes the output of the right and left ventricles and keeps the same volume of blood flowing to both the systemic and the pulmonary circulations. If the left side of the heart pumps a little more blood than the right side, for example, the volume of blood returning to the right ventricle (venous return) increases. With increased EDV, then, the right ventricle contracts more forcefully on the next beat, and the two sides are again in balance.

Contractility

The second factor that influences stroke volume is myocardial **contractility**, which is the strength of contraction at any given preload. Substances that increase contractility are called **positive inotropic agents** whereas those that decrease contractility are called **negative inotropic agents**. Thus, for a constant preload, the stroke volume is larger when a positive inotropic substance is present. Positive inotropic agents often promote Ca^{2+} inflow during cardiac action potentials, which strengthens the force of the subsequent muscle fiber contraction. They include stimulation of the sympathetic division of the autonomic nervous system (ANS), hormones such as epinephrine and norepinephrine, increased Ca^{2+} level in the extracellular fluid, and the drug digitalis. Digitalis is often used to treat people who have congestive heart failure (described shortly). Negative inotropic agents include inhibition of the sympathetic division of the ANS, anoxia, acidosis, some anesthetics (for example, halothane), and increased K^+ level in the extracellular fluid.

Afterload

Ejection of blood from the heart begins when pressure in the right ventricle exceeds the pressure in the pulmonary trunk (about 20 mm Hg) and the pressure in the left ventricle exceeds the pressure in the aorta (about 80 mm Hg). At this point, the higher pressure in the ventricles causes blood to press against the semilunar valves and push them open. The pressure that must be overcome before the semilunar valves can open is termed the **afterload**. When the afterload increases, for example, when blood pressure is elevated or arteries are narrowed by atherosclerosis (described on page 605), stroke volume decreases, and more blood remains in the ventricles at the end of systole.

CONGESTIVE HEART FAILURE

In **congestive heart failure** (**CHF**), the heart is a failing pump. Causes include coronary artery disease (described on page 605), congenital defects, long-term high blood pressure (which increases the afterload), myocardial infarcts (regions of dead heart tissue due to a prior heart attack), and valve disorders.

Congestive heart failure is an example of a positive feedback cycle in a pathological process. As the pump becomes less effective, more blood remains in the ventricles at the end of each cycle, and gradually the end-diastolic volume (preload) increases. Initially,

this increased preload may promote increased force of contraction (the Frank–Starling law of the heart). As the preload increases further, however, the heart is overstretched and contracts less forcefully. Now there is a positive feedback situation: less effective pumping leads to even lower pumping capability.

Often, one side of the heart starts to fail before the other. If the left ventricle fails first, it can't pump out all the blood it receives. As a result, blood backs up in the lungs. The result is **pulmonary edema**, fluid accumulation in the lungs, which can suffocate an untreated person. If the right ventricle fails first, blood backs up in the systemic vessels. In this case, the resulting **peripheral edema** is usually most noticeable in the feet and ankles. ■

Regulation of Heart Rate

Cardiac output depends on heart rate as well as stroke volume. Changing heart rate is important in the short-term control of cardiac output and blood pressure. The sinoatrial (SA) node initiates contraction and, left to itself, would set a constant heart rate of 90–100 beats/min. However, tissues require a varying blood supply under different conditions. For example, during exercise cardiac output rises to supply working tissues with increased amounts of oxygen and nutrients. Stroke volume may fall if the ventricular myocardium is damaged or if blood volume is reduced by bleeding. In these cases, homeostatic mechanisms strive to maintain an adequate cardiac output by increasing the rate and strength of contraction.

Several factors contribute to regulation of heart rate. The most important ones are the autonomic nervous system and hormones released by the adrenal medulla (epinephrine and norepinephrine).

Autonomic Control of Heart Rate

Nervous system control of the heart stems from the **cardiovascular center** in the medulla oblongata. This region of the brain stem receives input from a variety of sensory receptors and from higher brain centers, such as the limbic system and cerebral cortex. The cardiovascular center then provides appropriate output via both the sympathetic and parasympathetic branches of the ANS (Fig. 20.13).

Consider what happens during exercise. Even before a physical activity begins, especially in competitive situations, heart rate may climb. This anticipatory increase occurs because the limbic system in the brain sends signals to the cardiovascular center in the medulla. Then, as movements start, **proprioceptors**, which monitor the position of limbs and muscles, send increased input to the cardiovascular center. Proprioceptor input is a major stimulus for the quick rise in heart rate at the onset of physical activity.

Other sensory receptors that provide input to the cardiovascular center include **chemoreceptors** that monitor

Figure 20.13 Nervous system control of the heart.

The cardiovascular center in the medulla oblongata controls both sympathetic and parasympathetic nerves that innervate the heart.

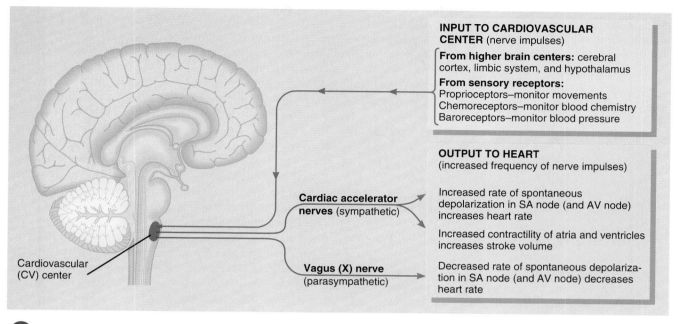

INPUT TO CARDIOVASCULAR CENTER (nerve impulses)

From higher brain centers: cerebral cortex, limbic system, and hypothalamus

From sensory receptors:
Proprioceptors–monitor movements
Chemoreceptors–monitor blood chemistry
Baroreceptors–monitor blood pressure

OUTPUT TO HEART
(increased frequency of nerve impulses)

Increased rate of spontaneous depolarization in SA node (and AV node) increases heart rate

Increased contractility of atria and ventricles increases stroke volume

Decreased rate of spontaneous depolarization in SA node (and AV node) decreases heart rate

Cardiac accelerator nerves (sympathetic)

Vagus (X) nerve (parasympathetic)

Cardiovascular (CV) center

Q What region of the heart is innervated by the sympathetic but not by the parasympathetic division?

chemical changes in the blood and **baroreceptors** that monitor blood pressure in major arteries and veins. Important baroreceptors are located in the arch of the aorta and carotid arteries (see Fig. 21.13). They detect changes in blood pressure and relay the information to the cardiovascular center. From there, nerve impulses propagate along sympathetic and parasympathetic nerves to the heart. These reflexes are important for the regulation of blood pressure, as well as heart rate, and are discussed in detail in Chapter 21. Here we focus specifically on the innervation of the heart by the sympathetic and parasympathetic branches of the ANS.

Sympathetic fibers extend from the medulla oblongata into the spinal cord. From the thoracic region of the spinal cord, **cardiac accelerator nerves** extend out to the SA node, AV node, and most portions of the myocardium (Fig. 20.13). Impulses in the cardiac accelerator nerves release norepinephrine, which binds to beta-1 (β_1) receptors on cardiac muscle fibers. This interaction has two separate effects. First, in SA (and AV) node fibers it speeds the rate of spontaneous depolarization. These pacemakers thus fire impulses more rapidly and heart rate increases. Second, in contractile fibers throughout the atria and ventricles norepinephrine enhances Ca^{2+} entry through the voltage-gated slow Ca^{2+} channels and thus increases contractility. This results in greater ejection of blood during systole. Thus, with a moderate increase in heart rate, stroke volume does not decline. With maximal sympathetic stimulation, how-

ever, heart rate may reach 250 beats/min. In this case, stroke volume is lower than at rest due to the very short filling time. The highest cardiac output usually occurs at a heart rate between 160 and 200 beats/min.

Parasympathetic nerve impulses reach the heart via the right and left **vagus (X) nerves**. These fibers innervate the SA node, AV node, and atrial myocardium. They release acetylcholine, which decreases heart rate by causing hyperpolarization and slowing the rate of spontaneous depolarization in autorhythmic fibers. Since only a few vagal fibers innervate ventricular muscle, changes in parasympathetic activity have little or no effect on contractility of the ventricles.

There always is a balance between sympathetic and parasympathetic stimulation of the heart. At rest the parasympathetic effects predominate. The resting heart rate, about 75 beats/min, usually is lower than the autorhythmic rate of the SA node (90–100 beats/min). With maximal stimulation by the parasympathetic division, the heart can slow to 20 or 30 beats/min or even stop momentarily.

Chemical Regulation of Heart Rate

Certain chemicals influence both the basic physiology of cardiac muscle and heart rate. For example, hypoxia (lowered oxygen level), acidosis (low pH), and alkalosis (high pH) all depress cardiac activity. Two types of chemicals—hormones and ions—have major effects on the heart.

1. **Hormones.** Epinephrine and norepinephrine (from the adrenal medulla) enhance the heart's pumping efficiency. These hormones affect cardiac muscle fibers in much the same way as does norepinephrine released by cardiac accelerator nerves; they increase both heart rate and contractility. Exercise, stress, and excitement cause the adrenal medulla to release more of its hormones. Thyroid hormones also enhance cardiac contractility and increase heart rate. One sign of hyperthyroidism (excessive thyroid hormone) is tachycardia (elevated heart rate at rest).

2. **Ions.** Differences between intracellular and extracellular concentrations of several ions, for example, Na^+ and K^+, are critical to the production of action potentials in all nerve and muscle fibers. So it is not surprising that ion imbalances can quickly compromise the pumping effectiveness of the heart. In particular, the relative concentrations of three cations—K^+, Ca^{2+}, and Na^+—have a large impact on cardiac function. Elevated blood levels of K^+ or Na^+ decrease heart rate and contractility. Excess Na^+ blocks Ca^{2+} inflow during cardiac action potentials and thus decreases the force of contraction, whereas excess K^+ blocks generation of action potentials. A moderate increase in extracellular Ca^{2+} speeds heart rate and strengthens the heartbeat.

Other Factors

Age, gender, physical fitness, and body temperature also influence heart rate. A newborn baby is likely to have a heart rate over 120 beats/min. Heart rate then declines through childhood and much of the adult life span. Senior citizens may develop a more rapid heartbeat, however. Females generally have slightly higher heart rates than males. In both sexes, regular exercise tends to bring resting heart rate down. A physically fit person may even exhibit bradycardia, a resting heart rate under 60 beats/min. This is a beneficial effect of endurance-type training because a slowly beating heart is more energy efficient than one that beats more rapidly.

Increased body temperature, such as occurs during fever or strenuous exercise, causes the SA node to discharge impulses faster and thereby increases heart rate. Decreased body temperature decreases heart rate and strength of contraction. During surgical repair of certain heart abnormalities, it is helpful to slow a patient's heart rate. One method is called **hypothermia** (hī-pō-THER-mē-a), in which the person is deliberately cooled to a low body temperature. Hypothermia also slows metabolism and reduces the oxygen needs of the tissues. Thus the heart and brain can withstand short periods of interrupted or reduced blood flow.

Figure 20.14 summarizes the influences on both stroke volume and heart rate in the overall regulation of cardiac output.

Help for Failing Hearts

Millions of people have inadequate cardiac output because their heart is diseased or damaged. Throughout the industrialized world, heart disease is the number one cause of premature death. One in every five persons who reach age 60 will have a myocardial infarction (MI, heart attack). Cardiomyopathies (degenerative disorders of the heart) also afflict thousands. Although a variety of drugs are helpful in the earlier stages of heart disease, at some point they are no longer effective because there is too little functional cardiac muscle left.

Researchers are investigating a wide variety of devices and techniques that might aid a failing heart. Even a 10% increase in cardiac output can get a bedridden patient out of bed. One possibility is a **heart transplant**, but the availability of donor hearts is very limited. In the United States there are about 50 potential candidates for each available heart. A second possibility is an **artificial heart**. During the 1980s several patients received a Jarvik-7 artificial heart, which used an external power source to drive a mechanical pump inside the body by compressed air. Persistent problems with blood clotting (which caused strokes and failure of other organs) and infection (due to the chest tube for compressed air) led the U.S. Food and Drug Administration to ban use of the device in 1990. Research continues, however, on other types of artificial hearts, which are being used as a "bridge to transplant," to keep patients alive while they wait for suitable donor hearts. A third approach is to develop **cardiac assist devices**, which supplement cardiac output without removing the heart. These devices may be made from artificial materials or fashioned from a person's own skeletal muscle. Exhibit 20.2 on page 602 describes several cardiac assist devices.

RISK FACTORS IN HEART DISEASE

About 1½ million people suffer a myocardial infarction every year in the United States and of these more than 500,000 die suddenly before reaching a hospital. However, the prevalence of heart disease has diminished in recent years, due in part to changes in life-style. Some of the causes of heart disease can be foreseen and prevented. People who develop combinations of certain risk factors are more likely to have heart attacks. **Risk factors** are characteristics, symptoms, or signs present in a person free of disease that are statistically associated with a greater chance of developing a disease. The major risk factors in heart disease are:

1. High blood cholesterol level.
2. High blood pressure.
3. Cigarette smoking.
4. Obesity.
5. Lack of regular exercise.
6. Diabetes mellitus.
7. Genetic predisposition (family history of heart disease at an early age).
8. Male gender. (But after age 70, the risk of heart attack is similar in males and females.)

Figure 20.14 Factors that increase cardiac output.

🔑 *Cardiac output equals stroke volume multiplied by heart rate.*

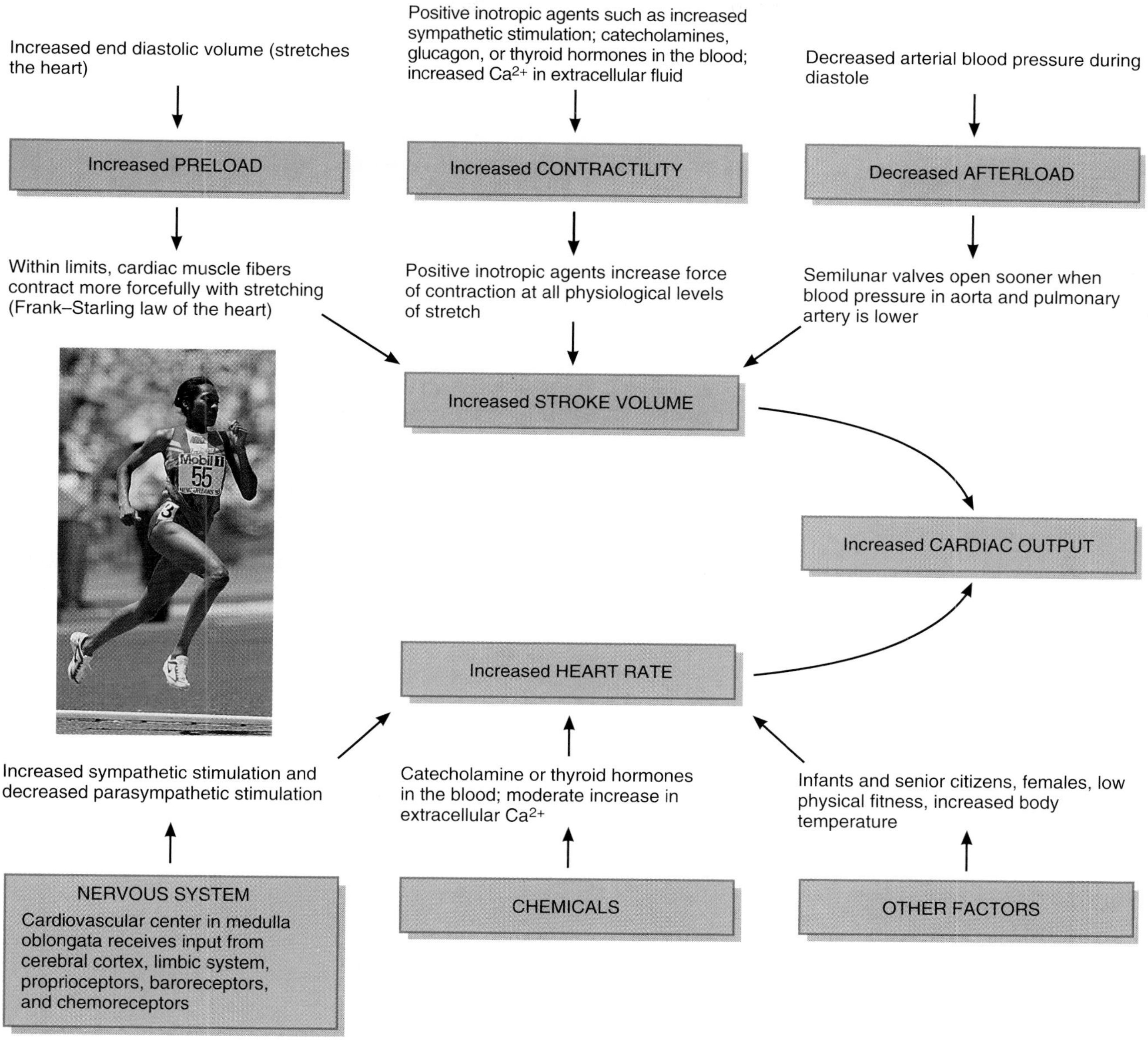

Increased end diastolic volume (stretches the heart)

Positive inotropic agents such as increased sympathetic stimulation; catecholamines, glucagon, or thyroid hormones in the blood; increased Ca^{2+} in extracellular fluid

Decreased arterial blood pressure during diastole

Increased PRELOAD

Increased CONTRACTILITY

Decreased AFTERLOAD

Within limits, cardiac muscle fibers contract more forcefully with stretching (Frank–Starling law of the heart)

Positive inotropic agents increase force of contraction at all physiological levels of stretch

Semilunar valves open sooner when blood pressure in aorta and pulmonary artery is lower

Increased STROKE VOLUME

Increased CARDIAC OUTPUT

Increased HEART RATE

Increased sympathetic stimulation and decreased parasympathetic stimulation

Catecholamine or thyroid hormones in the blood; moderate increase in extracellular Ca^{2+}

Infants and senior citizens, females, low physical fitness, increased body temperature

NERVOUS SYSTEM
Cardiovascular center in medulla oblongata receives input from cerebral cortex, limbic system, proprioceptors, baroreceptors, and chemoreceptors

CHEMICALS

OTHER FACTORS

❓ When you are exercising, contraction of skeletal muscles helps return blood to the heart more rapidly. Would this effect tend to increase or decrease stroke volume?

The first five risk factors can all be modified to reduce one's risk, for example, quitting smoking, losing weight, and becoming more physically active. High blood cholesterol is discussed shortly, and high blood pressure is discussed in the next chapter. Nicotine in cigarette smoke enters the bloodstream and constricts small blood vessels. It also stimulates the adrenal gland to oversecrete epinephrine and norepinephrine, which elevate heart rate and blood pressure. Obese people develop extra capillaries to nourish adipose tissue. An estimate is an additional 300 km (about 200 mi) of blood vessels

for each pound of fat. The heart has to work harder to pump blood this extra distance. With lack of regular exercise, the heart's stroke volume declines and heart rate must increase to achieve the needed cardiac output. As resting heart rate increases, the pumping efficiency of the heart decreases.

Other factors may also contribute to development of heart disease. Alcoholism (which damages cardiac and skeletal muscle), a high blood level of fibrinogen (which promotes blood clot formation), renin (which increases blood pressure), and uric acid all increase the risk of myocardial

EXHIBIT **20.2 CARDIAC ASSIST DEVICES**

Device	Description
Intra-aortic balloon pump (IABP)	A 40-ml polyurethane balloon mounted on a catheter is inserted into an artery in the groin and threaded into the thoracic aorta. An external pump inflates the balloon with gas at the beginning of ventricular diastole. As the balloon inflates, it pushes blood both backward, toward the heart, which improves coronary blood flow, and forward toward peripheral tissues. The balloon then is rapidly deflated just before the next ventricular systole. This decreases afterload, making it easier for the left ventricle to eject blood. Because the balloon is inflated between heartbeats, this technique is called intra-aortic balloon counterpulsation.
Hemopump	This propeller-like pump is threaded through an artery in the groin and then into the left ventricle. There, the blades of the pump whirl at about 25,000 revolutions per minute, pulling blood out of the left ventricle and pushing it into the aorta.
Left ventricular assist device (LVAD)	The LVAD, first used in 1991, is designed to be a completely portable assist device. It is implanted within the abdomen and powered by a battery pack worn in a shoulder holster. The LVAD is connected to the patient's weakened left ventricle and pumps blood into the aorta. The pumping rate increases automatically during exercise.
Cardiomyoplasty	A large piece of the patient's own skeletal muscle (left latissimus dorsi) is partially freed from its connective tissue attachments and wrapped around the heart, leaving the blood and nerve supply intact. An implanted pacemaker stimulates the skeletal muscle's motor neurons to cause contraction 10–20 times per minute, in synchrony with some of the heartbeats.
Skeletal muscle assist device	A piece of the patient's own skeletal muscle is used to fashion a pouch that is inserted between the heart and the aorta, functioning as a booster heart. A pacemaker stimulates the muscle's motor neurons to elicit contraction.

infarction. Enlargement (hypertrophy) of the left ventricle, associated with both high blood pressure and obesity, also is a risk factor for myocardial infarction.

PLASMA LIPIDS AND HEART DISEASE

A strong risk factor for developing heart disease is high blood cholesterol level. The reason is that high blood cholesterol promotes growth of fatty plaques that build up in the walls of arteries (see Fig. 20.16). As a plaque enlarges, the passageway for blood progressively narrows. Not only does the narrowed opening restrict blood flow, the roughened plaque tends to promote blood clotting. If a blood clot forms at the site of a plaque or lodges there, it may suddenly cut off blood flow. If the blocked vessel is in the brain, the result may be a fatal stroke. Blockage of a coronary artery may cause a heart attack.

Lipoproteins in Blood

Most lipids, such as cholesterol and triglycerides, are nonpolar and therefore very hydrophobic molecules. To be transported in watery blood, such molecules first must be dissolved. They are made water-soluble by combining them with proteins produced by the liver and intestine (apopro-

teins). The combinations thus formed are called **lipoproteins**, which vary in size, weight, and density. There are several types of lipoproteins, each having different functions, but all essentially are transport vehicles. They provide delivery and pick-up service so the various types of lipids can be available to cells that need them or removed from circulation if not needed. For example, all cells need cholesterol because it is a major building block of plasma membranes. It also is a key compound for the synthesis of steroid hormones and bile salts. Three classes of lipoproteins are called **low-density lipoproteins (LDLs)**, **high-density lipoproteins (HDLs)**, and **very low-density lipoproteins (VLDLs)**.

LDLs contain 25% proteins, 20% triglycerides, and 55% cholesterol. They deliver cholesterol to body cells that need it. Under abnormal conditions, however, LDLs also deposit cholesterol in and around smooth muscle fibers in arteries. Most cells of the body contain LDL receptors. Once LDL attaches to its receptor, it is taken into the cell by receptor-mediated endocytosis (described on page 65). Within the cell, the LDL is broken down, and the cholesterol is released to serve the cell's needs. Once a cell has sufficient cholesterol for its activities, a negative feedback system inhibits the cell from synthesizing new LDL receptors. Some people have too few LDL receptors, owing to various environmental and genetic factors. Since their cells cannot remove LDL from the blood as effectively, their plasma LDL level is abnormally high, and they are thus more likely to develop fatty plaques.

HDLs, which contain 50% proteins, 37% triglycerides, and only 13% cholesterol, remove excess cholesterol from body cells and transport it to the liver for elimination. This pick-up service prevents accumulation of cholesterol in the blood. Thus a high HDL level is associated with decreased risk of heart disease caused by plaque formation.

VLDLs contain about 10% proteins, 65% triglycerides, and 25% cholesterol. They transport triglycerides synthesized by liver cells to adipose cells for storage. A high fat diet promotes production of VLDLs. After depositing some of their triglycerides in adipose cells, however, VLDLs are converted to LDLs. This is one way that a fatty diet is believed to increase fatty plaque formation.

Blood Cholesterol

There are two sources of cholesterol in the body. Some is present in foods (eggs, dairy products, organ meats, beef, pork, and processed luncheon meats), but most is synthesized by the liver. Fatty foods that don't contain any cholesterol at all can still dramatically increase blood cholesterol level in two ways. First, a high intake of dietary fats stimulates reabsorption of cholesterol-containing bile back into the blood so less cholesterol is lost in the feces. Second, when saturated fats (described on page 41) are broken down in the body, the liver uses some of the breakdown products to produce cholesterol.

The total cholesterol (as part of HDLs, LDLs, and VLDLs) in blood plasma is one commonly used indicator of risk for coronary artery disease (described on page 605). A lipid profile test usually measures total cholesterol (TC), HDL-cholesterol, and triglycerides (VLDLs). LDL-cholesterol then is calculated by using the following formula: $LDL = TC - HDL - (triglycerides/5)$. In the United States, blood cholesterol is usually measured in milligrams per deciliter (mg/dl); a deciliter is $1/10$ of a liter. As the total cholesterol level increases above 150 mg/dl (3.9 mmol/liter), the risk of coronary artery disease slowly begins to rise. Above 200 mg/dl (5.2 mmol/liter), the risk increases even more. The chance of a heart attack doubles with every 50 mg/dl (1.3 mmol/liter) increase in total cholesterol once the level goes over 200 mg/dl.

For adults, desirable levels of blood cholesterol are total cholesterol under 200 mg/dl, LDL under 130 mg/dl, and HDL over 40 mg/dl. Normally, triglycerides are in the range of 10–190 mg/dl. TC of 200–239 mg/dl and LDL of 130–159 mg/dl are borderline-high while TC above 239 mg/dl and LDL above 159 are classified as high blood cholesterol. The risk of developing coronary artery disease may be predicted by determining the ratio of total cholesterol to HDL cholesterol. For example, a person with a total cholesterol of 180 and a HDL of 60 has a risk ratio of 3. Ratios above 4 are considered undesirable; the higher the ratio, the greater the risk of developing coronary artery disease.

Among the therapies used to reduce blood cholesterol level are exercise, diet, and drugs. Regular physical activity at aerobic and nearly aerobic levels tends to raise HDL level. Dietary changes are aimed at reducing the intake of total fat, saturated fats, and cholesterol. Among the drugs used to treat high blood cholesterol levels are cholestyramine (Questran®) and colestipol (Colestid®), which promote excretion of bile in the feces; nicotinic acid (Lipo-nicin®); and lovastatin (Mevacor®) and simvastatin (Zocor®), which block synthesis of cholesterol by liver cells.

EXERCISE AND THE HEART

No matter what a person's level of fitness, it can be improved at any age with regular exercise. Of the various types of exercise, some are more effective than others for improving the health of the cardiovascular system. **Aerobics**, or any activity that works large body muscles for at least 20 minutes, elevates cardiac output and accelerates metabolic rate. Three to five such sessions a week are usually recommended for improving the health of the cardiovascular system. Brisk walking, running, bicycling, cross-country skiing, and swimming are examples of aerobic exercises.

Sustained exercise increases the oxygen demand of the muscles. Whether the demand is met depends primarily on the adequacy of cardiac output and proper functioning of the respiratory system. After several weeks of training, a healthy person increases maximal cardiac output and thereby increases the maximal rate of oxygen delivery to the tissues. Oxygen delivery also rises because hemoglobin level increases and skeletal muscles develop more capillary networks in response to long-term training.

During strenuous activity, a well-trained athlete can achieve a cardiac output double or triple that of a sedentary person because training causes hypertrophy (enlargement) of the heart. Even though the heart of a well-trained athlete is larger, *resting* cardiac output is about the same as in a healthy untrained person. This is because stroke volume is increased while heart rate is decreased. The resting heart rate of a trained athlete often is only 40–60 beats per minute.

Other benefits of physical conditioning are an increase in high-density lipoprotein (HDL), a decrease in triglyceride levels, and improved lung function. Exercise also helps to reduce blood pressure, anxiety, and depression; control weight; and increase the body's ability to dissolve blood clots by increasing fibrinolytic activity. Intense exercise increases levels of endorphins, the body's natural painkillers. This may explain the psychological "high" that athletes experience with strenuous training and the "low" they feel when they miss regular workouts. Exercise also helps make bones stronger and thus may be a factor in inhibiting and treating osteoporosis. Some research indicates that exercise may even offer protection against cancer and diabetes.

DEVELOPMENTAL ANATOMY OF THE HEART

The *heart*, a derivative of **mesoderm**, begins to develop before the end of the third week of gestation. It begins its development in the ventral region of the embryo inferior to the foregut (shown in Fig. 24.29). The first step in its development is the formation of a pair of tubes, the **endothelial (endocardial) tubes**. They develop from mesodermal cells (Fig. 20.15). These tubes then unite to form a common tube, referred to as the **primitive heart tube**. Next, the primitive heart tube develops into five distinct regions: (1) **ventricle**, (2) **bulbus cordis**, (3) **atrium**, (4) **sinus venosus**, and (5) **truncus arteriosus**. Because the bulbus cordis and ventricle grow most rapidly, and the heart enlarges more rapidly than its superior and inferior attachments, the heart first assumes a U-shape and then an S-shape. The flexures of the heart re-

orient the regions so that the atrium and sinus venosus eventually come to lie superior to the bulbus cordis, ventricle, and truncus arteriosus. Contractions of the primitive heart begin by day 22. They begin in the sinus venosus and force blood through the tubular heart.

At about the seventh week of development, a partition called the **interatrial septum** forms in the atrial region. This septum divides the atrial region into a *right atrium* and *left atrium*. The opening in the partition is the **foramen ovale**, which normally closes at birth and later forms a depression called the *fossa ovalis*. An **interventricular septum** also develops and partitions the ventricular region into a *right ventricle* and *left ventricle*. The bulbus cordis and truncus arteriosus divide into two vessels, the *aorta* (arising from the left ventricle) and the *pulmonary trunk* (arising from the right ventricle). The great veins of the heart, *superior vena cava* and *inferior vena cava*, develop from the venous end of the primitive heart tube.

Figure 20.15   Development of the heart. Arrows within the structures show the direction of blood flow.

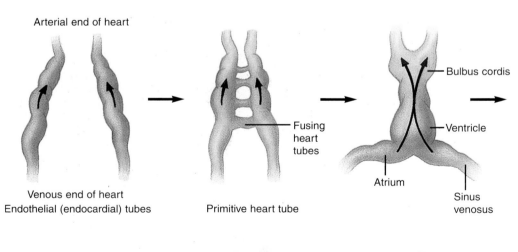

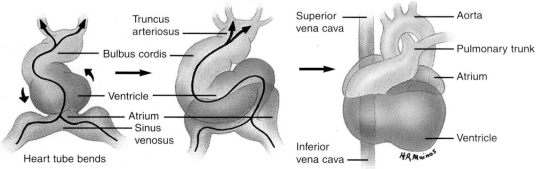

DISORDERS: HOMEOSTATIC IMBALANCES

CORONARY ARTERY DISEASE

In **coronary artery disease** (**CAD**) the coronary arteries have become narrowed so that blood flow to the heart is reduced. This results in coronary heart disease (CHD), a condition in which the heart muscle is damaged because of an inadequate amount of blood due to obstruction of its blood supply. CAD is the leading cause of death in the United States. Depending on the degree of obstruction, symptoms can range from mild angina to a full-scale heart attack. Generally, symptoms start when there is about a 75% narrowing of a coronary artery. The principal causes of narrowing are atherosclerosis, coronary artery spasm, or a clot in a coronary artery.

Atherosclerosis

Thickening of the walls of arteries and loss of elasticity are the main characteristics of a group of diseases called **arteriosclerosis** (ar-tē'-rē-ō'-skle-RŌ-sis; *sclerosis* = hardening). One form of arteriosclerosis is **atherosclerosis** (ath'-er-ō-skle-RŌ-sis), a process in which smooth muscle cells proliferate and fatty substances, especially cholesterol and triglycerides, accumulate in the walls of medium and large diameter arteries. The first event in atherosclerosis is thought to be damage to the endothelial lining of the artery. One theory is that a common virus, perhaps cytomegalovirus (a member of the herpes family), triggers endothelial damage after being harbored in a dormant state for some time. Other theories hold that prolonged high blood pressure, carbon monoxide in cigarettes, and diabetes mellitus can produce endothelial damage. Whatever the causes, two events follow endothelial damage. Within the artery wall, (1) smooth muscle fibers proliferate and (2) lipids build up, both within cells and in the interstitial spaces. The accumulated cholesterol, triglycerides, and cells form a lesion called an **atherosclerotic plaque** (Fig. 20.16). As it grows, a plaque obstructs blood flow in the affected artery, and tissues supplied by the artery suffer damage. The condition is reversible to some extent. With reduction of a high blood cholesterol level, atherosclerotic plaques tend to shrink.

An additional danger is that the plaque provides a roughened surface that causes blood platelets to release *platelet-derived growth factor* (*PDGF*). PDGF is a hormone that promotes the proliferation of smooth muscle fibers. Macrophages and endothelial cells also produce PDGF. This worsens the atherosclerosis because additional smooth muscle cells enlarge the size of the plaque. Platelets in the area of the plaque also release clot-forming chemicals. Thus a thrombus may form. If the clot breaks off and forms an embolus (blood clot transported by blood), it may obstruct smaller arteries and capillaries downstream from the site of formation.

Coronary Artery Spasm

An atherosclerotic plaque is a fixed obstruction to blood flow. Obstruction can also be caused intermittently by **coronary artery spasm**, in which the smooth muscle of a coronary artery undergoes a sudden contraction that narrows the lumen of a blood vessel. Although the causes of coronary artery spasm are unknown, several factors are being investigated. These include smoking, stress, and a vasoconstrictor chemical released by platelets.

Diagnosis and Treatment

Diagnosis and treatment of coronary artery disease vary with the nature and urgency of symptoms. Among the treatment options

Figure 20.16 Photomicrograph of an artery partially obstructed by an atherosclerotic plaque.

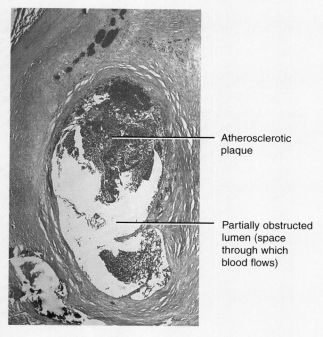

Atherosclerotic plaque

Partially obstructed lumen (space through which blood flows)

Transverse section

Q Which type of lipoprotein contributes most to the development of atherosclerosis?

are drug therapy (nitroglycerine, beta blockers, and clot-dissolving agents) and various surgical and nonsurgical procedures.

Cardiac catheterization (kath'-e-ter-i-ZĀ-shun) is an invasive procedure that is used to visualize the heart's coronary arteries, chambers, valves, and great vessels. It may also be used to measure pressure in the heart and blood vessels; to assess left ventricular function, cardiac output, and diastolic properties of the left ventricle; to measure the flow of blood through the heart and blood vessels, the oxygen content of blood, status of heart valves and conduction system; and to identify the exact location of septal and valvular defects. The basic procedure involves inserting a long, flexible, radiopaque **catheter** (plastic tube) into a peripheral vein (for right heart catheterization) or artery (for left heart catheterization) and guiding it under fluoroscopy (x-ray observation).

Cardiac angiography (an'-jē-OG-ra-fē) is also an invasive procedure in which a cardiac catheter is used to inject a radiopaque contrast medium into blood vessels or heart chambers. The procedure may be used to visualize coronary arteries, the aorta, pulmonary blood vessels, and the ventricles to assess structural abnormalities in blood vessels such as atherosclerotic plaques and emboli and ventricular volume, wall thickness, and wall motion. Angiography can also be used to inject clot-dissolving drugs, such as streptokinase or tissue plasminogen activator (t-PA), into a coronary artery to dissolve an obstructing thrombus.

Coronary artery bypass grafting (**CABG**) is one way of increasing the blood supply to the heart. It is a surgical procedure in which a blood vessel from another part of the body is

DISORDERS: HOMEOSTATIC IMBALANCES CONTINUED

used to bypass the blocked region of a coronary artery. The two vessels used most often are the saphenous vein, which first is removed from the thigh and leg, and the internal mammary artery. The distal end of a vein segment is sutured into the aorta and its proximal end into a coronary artery beyond the obstructed area (Fig. 20.17a). When the internal mammary artery is used, its distal end is cut and then sutured to the coronary artery, distal to the blocked area. If more than one artery is clogged, additional bypasses may be made.

A nonsurgical procedure used to treat CAD is termed **percutaneous transluminal coronary angioplasty** (**PTCA**) (*percutaneous* = through the skin; *trans* = across; *lumen* = channel in a tube; *angio* = blood vessel; *plasty* = to mold or shape). Like coronary artery bypass grafting, it is an attempt to increase the blood supply to the heart muscle. A balloon catheter is inserted into an artery of an arm or leg. Using fluoroscopy, it is gently guided through the arterial system (Fig. 20.17b) until it is threaded into a coronary artery. Then, while dye is being released, angiograms (x-rays of blood vessels) are taken to localize the plaques. Next, the catheter is advanced to the point of obstruction and the balloon is inflated with air. This stretches the arterial wall and squashes the plaque. If successful, PTCA increases the inside diameter of the vessel, and blood flow improves. PTCA is most often used to relieve angina pectoris. Because about 30% of PTCA-opened arteries fail due to restenosis (renarrowing), a special device called a **stent** may be inserted via the catheter to keep the artery patent (open). A stent is a stainless steel device, resembling a spring coil, that is permanently placed in an artery to maintain patency, permitting blood to circulate (Fig. 20.17c).

Another technique for opening clogged arteries is a procedure called **laser angioplasty**. In one variation of this procedure, a laser vaporizes the atherosclerotic plaque and makes a channel through the blood vessel obstruction. Then a balloon catheter is inserted, and the balloon is inflated to widen the vessel. Two of the latest techniques for clearing arteries are balloon–laser welding and catheter artherectomy. In **balloon–laser welding**, an artery is first widened by PTCA. On the last balloon inflation, a laser heats the surrounding tissue sufficiently to stretch and weld the arterial wall into a smooth surface. In **catheter artherectomy**, a rotating drill shaves off plaque. Shavings are trapped and suctioned out.

CONGENITAL DEFECTS

A defect that exists at birth, and usually before, is called a **congenital defect**. Many defects are not serious and may go unnoticed for a lifetime. Others heal themselves. But some are life-threatening and must be mended by surgical techniques ranging from simple suturing to replacement of malfunctioning parts with synthetic materials.

One congenital defect that can occur is **coarctation** (kō′-ark-TĀ-shun) **of the aorta** (Fig. 20.18a). In this condition, a segment of the aorta is too narrow. As a result, the flow of oxygenated blood to the body is reduced, the left ventricle is forced to pump harder, and high blood pressure develops.

Another common congenital defect is **patent ductus arteriosus** (Fig. 20.18b). The ductus arteriosus, a temporary blood vessel between the aorta and the pulmonary trunk, normally closes shortly after birth. In some babies, the ductus

Figure 20.17 Several procedures for reestablishing blood flow in occluded coronary arteries.

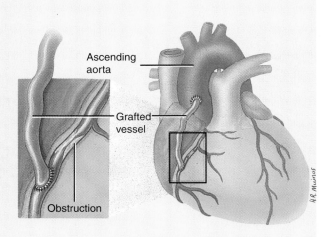

Ascending aorta

Grafted vessel

Obstruction

(a) Coronary artery bypass grafting (CABG)

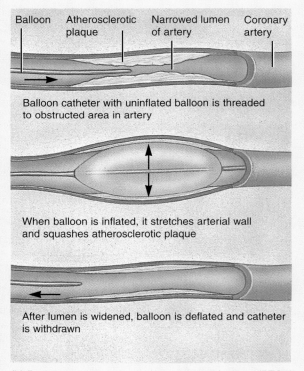

Balloon   Atherosclerotic plaque   Narrowed lumen of artery   Coronary artery

Balloon catheter with uninflated balloon is threaded to obstructed area in artery

When balloon is inflated, it stretches arterial wall and squashes atherosclerotic plaque

After lumen is widened, balloon is deflated and catheter is withdrawn

(b) Percutaneous transluminal coronary angioplasty (PTCA)

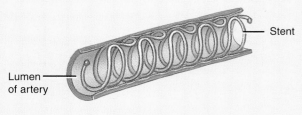

Stent

Lumen of artery

(c) Stent in an artery

Figure 20.18   Some common congenital heart defects.

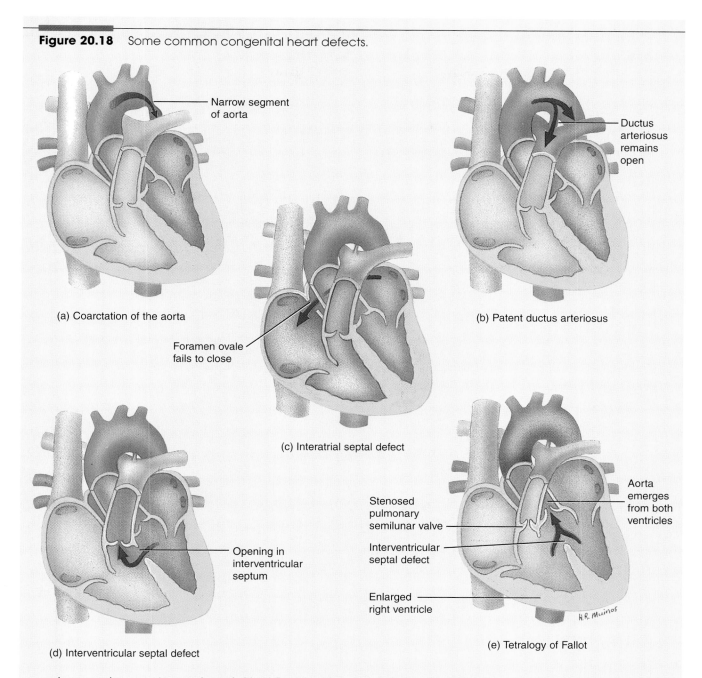

(a) Coarctation of the aorta

— Narrow segment of aorta

Foramen ovale fails to close

(c) Interatrial septal defect

(b) Patent ductus arteriosus

Ductus arteriosus remains open

(d) Interventricular septal defect

Opening in interventricular septum

(e) Tetralogy of Fallot

Stenosed pulmonary semilunar valve

Interventricular septal defect

Enlarged right ventricle

Aorta emerges from both ventricles

H.R. Muinos

arteriosus remains open. As a result, aortic blood flows into the lower-pressure pulmonary trunk, thus increasing the pulmonary trunk blood pressure and overworking both ventricles.

A **septal defect** is an opening in the septum that separates the interior of the heart into left and right sides. In one type of **interatrial septal defect** the fetal foramen ovale between the two atria fails to close after birth (Fig. 20.18c). Because pressure in the right atrium is lower, blood flows from the left atrium to the right without going through the systemic circulation. This defect overloads the pulmonary circulation. **Interventricular septal defect** is caused by incomplete closure of the interventricular septum (Fig. 20.18d). It permits oxygenated blood to flow directly from the left ventricle into the right ventricle where it mixes with deoxygenated blood.

Valvular stenosis is a narrowing of one of the valves regulating blood flow in the heart. All stenoses are serious because they place a severe burden on the heart by making it work harder to push blood through the abnormally narrow valve openings. As a result of mitral valve stenosis, systemic blood pressure is increased. Most stenosed valves are totally replaced with artificial valves.

Tetralogy of Fallot (tet-RAL-ō-jē of fal-Ō) is a combination of four defects: an interventricular septal defect, an aorta that emerges from both ventricles instead of from the left ventricle only, a stenosed pulmonary semilunar valve, and an enlarged right ventricle (Fig. 20.18e). Because there is stenosis of the pulmonary semilunar valve, the increased right ventricular pressure forces deoxygenated blood from the right ventricle to enter the left ventricle through the interventricular septum. As a result,

DISORDERS: HOMEOSTATIC IMBALANCES CONTINUED

deoxygenated blood mixes with oxygenated blood and is pumped into the systemic circulation. Also, because the aorta emerges from the right ventricle and the pulmonary trunk is stenosed, very little blood reaches the lungs, and pulmonary circulation is bypassed almost completely. This causes cyanosis, the blue or dark purple discoloration that is most easily seen in nail beds and mucous membranes when deoxygenated hemoglobin level is high. For this reason, tetralogy of Fallot is one of the conditions that cause a "blue baby."

ARRHYTHMIAS

Arrhythmia (a-RITH-mē-a) is a general term that refers to an abnormality or irregularity in the heart rhythm. Some physicians use the term **dysrhythmia** since that implies an abnormal rhythm, whereas arrhythmia implies no rhythm. An arrhythmia results when there is a disturbance in the conduction system of the heart. It may be due to either faulty production of electrical impulses or poor conduction of impulses as they pass through the system.

Many different types of arrhythmias can occur, some normal and some quite serious. Arrhythmias may be caused by factors such as caffeine, nicotine, alcohol, anxiety, certain drugs, hyperthyroidism, potassium deficiency, and certain heart diseases. Serious arrhythmias can result in cardiac arrest (stopping of the heartbeat) if the heart cannot supply its own oxygen demands.

Heart Block

One serious arrhythmia is called a **heart block**. Perhaps the most common blockage is in the atrioventricular (AV) node, which is the only path for impulses in the atria to reach the ventricles. This disturbance is called **atrioventricular (AV) block**. In first-degree AV block, the P–Q (PR) interval (see Fig. 20.10) is prolonged, usually because conduction through the AV node is slower than normal. In second-degree AV block, some of the impulses from the SA node are not conducted through the AV node. This results in "dropped" beats because excitation doesn't reach the ventricles. In third-degree (complete) AV block, none of the SA node impulses get through the AV node. Autorhythmic cells in the atria and ventricles pace the upper and lower chambers independently. With complete AV block, the ventricular contraction rate is less than 40 beats/min. Due to decreased cardiac output and diminished brain blood flow, patients may experience dizziness, unconsciousness, or convulsions.

Flutter and Fibrillation

Two other abnormal rhythms are **flutter** and **fibrillation**. In **atrial flutter** the atrial rhythm is 240–360 beats/min. The condition is basically rapid atrial contractions accompanied by a second-degree AV block. Flutter may result from rheumatic heart disease, coronary artery disease, or certain congenital heart diseases. **Atrial fibrillation** is asynchronous contraction of the atrial muscle fibers so that atrial pumping ceases altogether. Atrial fibrillation may occur in myocardial infarction, acute and chronic rheumatic heart disease, and hyperthyroidism. In a strong heart, atrial fibrillation reduces the pumping effectiveness of the heart by only 20–30%.

Ventricular fibrillation (VF) is the most ominous arrhythmia. It almost always indicates imminent death unless corrected quickly. It is characterized by asynchronous, haphazard, ventricular muscle contractions. Because part of the ventricles are contracting while other parts are relaxing, ventricular pumping ceases, blood is not ejected, and circulatory failure and death occur.

A strong, brief electrical current passed across the chest can stop ventricular fibrillation. This is called **defibrillation**. It is accomplished by giving the electric shock through large paddle-shaped electrodes pressed against the skin of the chest. Patients who face a high risk of dying from heart rhythm disorders now can have a device implanted that monitors their heart rhythm and delivers a smaller shock directly to the heart when a life-threatening rhythm disturbance occurs. Such an **automatic implantable cardioverter defibrillator** (AICD) has been used in thousands of patients around the world.

Ventricular Premature Contraction (VPC)

Another form of arrhythmia arises when a small region of the heart outside the pacemaker (an ectopic focus) becomes more excitable than normal, causing an occasional abnormal impulse to arise between normal impulses. As a wave of depolarization spreads outward from the ectopic focus, it causes a **ventricular premature contraction** (VPC) or **premature ventricular contraction** (PVC). The contraction occurs early in diastole before the SA node is normally scheduled to discharge its impulse. VPCs may be relatively benign and may be caused by emotional stress, excessive intake of stimulants such as caffeine or nicotine, and lack of sleep. In other cases, the contractions may indicate an underlying pathology.

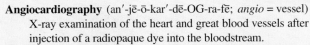

MEDICAL TERMINOLOGY

Angiocardiography (an′-jē-ō-kar′-dē-OG-ra-fē; *angio* = vessel) X-ray examination of the heart and great blood vessels after injection of a radiopaque dye into the bloodstream.

Cardiac arrest (KAR-dē-ak a-REST) A clinical term meaning cessation of an effective heartbeat. The heart may be completely stopped (cardiac standstill) or quivering ineffectively (ventricular fibrillation).

Cardiomegaly (kar′-dē-ō-MEG-a-lē; *mega* = large) Heart enlargement, hypertrophy.

Compliance (kom-PLĪ-ans) The passive or diastolic stiffness properties of the left ventricle. A hypertrophied or fibrosed heart with a stiff wall, for example, has decreased compliance. Also, the stiffness properties of the lungs or major arteries.

Constrictive pericarditis (kon-STRIK-tiv per′-i-kar-DĪ-tis) A shrinking and thickening of the pericardium that prevent heart muscle from expanding and contracting normally.

Cor pulmonale (CP) (kor pul-mōn-ALE; *cor* = heart; *pulmon* = lung) Right ventricular hypertrophy associated with hypertension (high blood pressure) in the pulmonary circulation.

Palpitation (pal′-pi-TA-shun) A fluttering of the heart or abnormal rate or rhythm of the heart.

Paroxysmal tachycardia (par′-ok-SIZ-mal tak′-e-KAR-dē-a) A period of rapid heartbeats that begins and ends suddenly.

STUDY OUTLINE

LOCATION AND SIZE OF THE HEART
(p. 580)

1. The heart is situated between the lungs in the mediastinum.
2. About two-thirds of its mass is to the left of the midline.
3. The heart is about 12 cm long, 9 cm wide, and 6 cm thick.

PERICARDIUM (p. 581)

1. The pericardium consists of an outer fibrous layer and an inner serous pericardium.
2. The serous pericardium is composed of a parietal and a visceral layer.
3. Between the parietal and visceral layers of the serous pericardium is the pericardial cavity, a potential space filled with a few milliliters of pericardial fluid that reduces friction between the two membranes.

HEART WALL (p. 581)

1. The wall of the heart has three layers: epicardium (visceral layer of the serous pericardium), myocardium, and endocardium.
2. The epicardium consists of mesothelium and connective tissue, the myocardium is composed of cardiac muscle tissue, and the endocardium consists of endothelium and connective tissue.

CHAMBERS OF THE HEART (p. 582)

1. The chambers include two superior atria and two inferior ventricles.
2. An interatrial septum separates the atria and an interventricular septum separates the ventricles.

VALVES OF THE HEART (p. 583)

1. Valves prevent backflow of blood in the heart.
2. Atrioventricular (AV) valves, between the atrium and ventricle on the same side, are the tricuspid valve on the right side of the heart and the bicuspid (mitral) valve on the left.
3. The chordae tendineae and papillary muscles stabilize the valve cusps and stop blood from backing into the atria.
4. The two arteries that receive blood ejected by the ventricles each have a semilunar valve (aortic and pulmonary).

BLOOD FLOW THROUGH THE PULMONARY AND SYSTEMIC CIRCULATIONS (p. 587)

1. Blood flows into the heart from the superior and inferior venae cavae and the coronary sinus to the right atrium, through the tricuspid valve to the right ventricle, through the pulmonary semilunar valve into the pulmonary trunk and pulmonary arteries to the lungs, through the pulmonary veins into the left atrium, through the bicuspid valve to the left ventricle, and through the aortic semilunar valve into the aorta.
2. Divisions of the aorta are the ascending aorta, arch of the aorta, and descending aorta (thoracic aorta and abdominal aorta).

HEART BLOOD SUPPLY (p. 588)

1. The flow of blood through the heart is called the coronary (cardiac) circulation.
2. The principal arteries are left and right coronary arteries; the principal veins are the cardiac vein and the coronary sinus.
3. Reperfusion damage is caused by oxygen free radicals.

CONDUCTION SYSTEM AND PACEMAKER
(p. 589)

1. The conduction system consists of tissue specialized for generation and conduction of action potentials.
2. Components of this system are the sinoatrial (SA) node (pacemaker), atrioventricular (AV) node, atrioventricular (AV) bundle (bundle of His), bundle branches, and conduction myofibers (Purkinje fibers).

PHYSIOLOGY OF CARDIAC MUSCLE CONTRACTION (p. 591)

1. An impulse in a ventricular contractile fiber is characterized by rapid depolarization, plateau, and repolarization.
2. The refractory period of a cardiac muscle fiber lasts longer than its contraction.

ELECTROCARDIOGRAM (p. 593)

1. The record of electrical changes during each cardiac cycle is called an electrocardiogram (ECG).
2. A normal ECG consists of a P wave (atrial depolarization), QRS complex (onset of ventricular depolarization), and T wave (ventricular repolarization).
3. The P–Q (PR) interval represents the conduction time from the beginning of atrial excitation to the beginning of ventricular excitation. The S–T segment represents the time when ventricular contractile fibers are fully depolarized.

CARDIAC CYCLE (p. 594)

1. A cardiac cycle consists of the systole (contraction) and diastole (relaxation) of both atria, plus the systole and diastole of both ventricles.
2. The phases of the cardiac cycle are (a) the relaxation period, (b) ventricular filling, and (c) ventricular systole.
3. With an average heartbeat of 75 beats/min, a complete cardiac cycle requires 0.8 sec.
4. The first heart sound (lubb) is created by blood turbulence associated with the closing of the atrioventricular valves. The second sound (dupp) is created by blood turbulence associated with the closing of semilunar valves.

CARDIAC OUTPUT (p. 597)

1. Cardiac output (CO) is the amount of blood ejected by the left ventricle (or right ventricle) into the aorta (or pulmonary trunk) per minute. It is calculated as follows: CO (ml/min) = stroke volume (SV) in ml/beat × heart rate (HR) in beats per minute.
2. Stroke volume (SV) is the amount of blood ejected by a ventricle during each systole.
3. Cardiac reserve is the ratio between the maximum cardiac output a person can achieve and the cardiac output at rest.
4. Stroke volume is related to preload (stretch on the heart before it contracts), contractility (forcefulness of contraction), and afterload (pressure that must be exceeded before ventricular ejection can begin).
5. According to the Frank–Starling law of the heart, a greater preload (stretch) on cardiac muscle fibers just before they contract increases their force of contraction until the stretching becomes excessive.
6. Nervous control of the cardiovascular system stems from the cardiovascular center in the medulla oblongata.
7. Sympathetic impulses increase heart rate and force of contraction; parasympathetic impulses decrease heart rate.

8. Heart rate is affected by hormones (epinephrine, norepinephrine, thyroid hormones), ions (Na^+, K^+, Ca^{2+}), age, gender, physical fitness, and temperature.

RISK FACTORS IN HEART DISEASE (p. 600)

1. Risk factors in heart disease that can be modified include high blood cholesterol, high blood pressure, cigarette smoking, obesity, and lack of regular exercise.
2. Other factors include diabetes mellitus; genetic predisposition; male gender; high levels of fibrinogen, renin, and uric acid; and left ventricular hypertrophy.

PLASMA LIPIDS AND HEART DISEASE (p. 602)

1. High blood cholesterol promotes growth of fatty plaques in the walls of arteries.

2. Whereas HDLs remove excess cholesterol from circulation, high levels of LDLs are associated with the formation of fatty plaques in arteries.

EXERCISE AND THE HEART (p. 603)

1. Sustained exercise increases oxygen demand on muscles.
2. Among the benefits of aerobic exercise are increased cardiac output, increased HDLs, decreased triglycerides, improved lung function, decreased blood pressure, and weight control.

DEVELOPMENTAL ANATOMY OF THE HEART (p. 604)

1. The heart develops from mesoderm.
2. The endothelial tubes develop into the four-chambered heart and great vessels of the heart.

REVIEW QUESTIONS

1. Describe the position of the heart in the mediastinum. (p. 580)
2. Distinguish the subdivisions of the pericardium. What is the purpose of this structure? (p. 581)
3. Compare the three layers of the heart wall according to composition, location, and function. (p. 581)
4. Define atria and ventricles. What vessels enter or exit the atria and ventricles? (p. 582)
5. Describe the principal valves in the heart and how they operate. (p. 583)
6. Trace the path of blood flow through the pulmonary and systemic circulations. (p. 587)
7. Describe the route of blood flow in coronary (cardiac) circulation. (p. 588)
8. Describe how reperfusion damage occurs. (p. 589)
9. Describe the structure and function of the heart's conduction system. What are autorhythmic fibers? (p. 589)
10. Describe the phases of an impulse in ventricular contractile fibers. (p. 591)
11. Define and label the deflection waves of a normal electrocardiogram (ECG). (p. 593) What is the significance of the P–Q (PR) interval and S–T segment? (p. 594)
12. Define cardiac cycle and list the principal events of the relaxation period, ventricular filling, and ventricular systole. (p. 594)
13. By means of a labeled diagram, relate the events of the cardiac cycle to time. (p. 595) What is the relaxation period? (p. 594)
14. Describe the source and significance of the heart sounds. (p. 596)
15. What is cardiac output (CO)? How is it calculated? (p. 597)
16. Define stroke volume (SV). Explain the factors that regulate stroke volume. (p. 597)
17. What is the Frank–Starling law of the heart? What is its significance? (p. 597)
18. Define cardiac reserve. Why is it important? (p. 597)
19. Explain how the sympathetic and parasympathetic divisions of the autonomic nervous system adjust heart rate. (p. 598)
20. Explain how each of the following affects heart rate: hormones, ions, age, gender, physical fitness, and temperature. (p. 600)
21. Describe the various devices and techniques that are used to help failing hearts. (p. 600)
22. Describe the risk factors involved in heart disease. (p. 600)
23. How are plasma lipids related to heart disease? (p. 602)
24. What are some of the cardiovascular benefits of regular exercise? (p. 603)
25. Describe how the heart develops. (p. 604)

ANSWERS TO QUESTIONS WITH FIGURES

20.1   If one side of the heart pumps less blood than the other for some time, blood backs up and fluid accumulates in the tissues (edema) served by the other side of the heart.

20.2   The broad median partition (mass of tissue) between the lungs and between the sternum and backbone.

20.3   Visceral layer of the serous pericardium (epicardium).

20.4   Left ventricle.

20.5   They contract, pulling on the chordae tendineae.

20.6   Right atrium, tricuspid valve, right ventricle, pulmonary semilunar valve, pulmonary trunk, pulmonary arteries, pulmonary capillaries, pulmonary veins, left atrium, bicuspid valve, left ventricle, aortic semilunar valve, aorta, systemic capillaries, systemic veins, and superior vena cava.

20.7   Circumflex.

20.8   Atrioventricular (AV) bundle.

20.9   Cardiac: about 0.3 sec (300 msec). Skeletal: 1–2 msec.

20.10   May indicate a myocardial infarction (heart attack).

20.11   About 130 ml; end-diastolic volume.

20.12   First sound (S1).

20.13   The ventricular myocardium.

20.14   The skeletal muscle "pump" increases stroke volume by increasing preload (end-diastolic volume).

20.16   LDL.

1. Contrast the structure and function of the various types of blood vessels. (p. 612)

2. Discuss the various pressures involved in the movement of fluids between capillaries and interstitial spaces. (p. 617)

3. Explain the factors that regulate the velocity and volume of blood flow. (p. 620)

4. Explain how blood pressure changes throughout the cardiovascular system and describe the factors that determine mean arterial blood pressure. (p. 621)

5. Describe the factors that determine systemic vascular resistance and explain how the return of venous blood to the heart is accomplished. (p. 622)

6. Describe how blood pressure is regulated. (p. 623)

7. Define the three stages of shock. (p. 629)

8. Define pulse and blood pressure (BP) and contrast the clinical significance of systolic, diastolic, and pulse pressures. (p. 631)

9. Identify the principal arteries and veins and describe the flow of blood through the systemic, hepatic portal, pulmonary, and fetal circulations. (p. 632)

10. Explain the effects of aging on the cardiovascular system. (p. 664)

11. Describe the development of blood vessels and blood. (p. 664)

12. List the causes and symptoms of hypertension, aneurysm, coronary artery disease (CAD), and deep-venous thrombosis (DVT). (p. 665)

13. Define medical terminology associated with blood vessels. (p. 666)

CHAPTER 21

THE CARDIOVASCULAR SYSTEM: BLOOD VESSELS AND HEMODYNAMICS

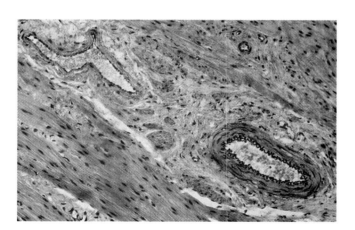

CHAPTER CONTENTS AT A GLANCE

The focus of this chapter is a study of the major blood vessels and **hemodynamics** (hē-mō-dī-NAM-ics; *hemo* = blood; *dynamis* = power), that is, a study of the forces involved in circulating blood throughout the body. The developmental anatomy of blood vessels and blood will be considered later in the chapter.

ANATOMY OF BLOOD VESSELS

Blood vessels form a closed system of tubes that carries blood away from the heart, transports it to the tissues of the body, and then returns it to the heart. **Arteries** are vessels that carry blood from the heart to the tissues. Large, elastic arteries leave the heart and divide into medium-sized, muscular arteries that branch out into the various regions of the body. Medium-sized arteries then divide into small arteries, which, in turn, divide into still smaller arteries called **arterioles** (ar-TER-ē-ōls). As the arterioles enter a tissue, they branch into countless microscopic vessels called **capillaries** (KAP-i-lar'-ēs). Substances are exchanged between the blood and body tissues through the thin walls of capillaries. Before leaving the tissue, groups of capillaries unite to form small veins called **venules** (VEN-yools). These, in turn, merge to form progressively larger blood vessels called veins. **Veins** then convey blood from the tissues back to the heart. Because blood vessels require oxygen (O_2) and nutrients just like other tissues of the body, larger blood vessels especially also have their own blood vessels, called **vasa vasorum** (literally, vasculature of vessels), in their walls.

Arteries

In ancient times, **arteries** (*aer* = air; *tereo* = to carry), found empty at death, were thought to contain only air. The hollow center through which blood flows is called the **lumen** (Fig. 21.1). The surrounding arterial wall has three coats or tunics. The inner coat, the **tunica interna** (**intima**), is composed of a lining of *endothelium* (simple squamous epithelium) that is in contact with the blood, a *basement membrane,* and a layer of elastic tissue called the *internal elastic lamina.* As you will discover, the endothelium is a continuous layer of cells that line the inner surface of the entire cardiovascular system (heart and all blood vessels). The middle coat, or **tunica media**, is usually the thickest layer. It consists of elastic fibers and smooth muscle fibers (cells). The outer coat, the **tunica externa** (**adventitia**), is composed principally of elastic and collagen fibers. In muscular arteries (described shortly), an *external elastic lamina* composed of elastic tissue separates the tunica externa from the tunica media.

The smooth muscle of arteries is arranged circularly around the lumen. Sympathetic fibers of the autonomic nervous system innervate vascular smooth muscle. Usually, when there is an increase in sympathetic stimulation, the smooth muscle contracts, squeezes the wall around the lumen, and narrows the vessel. Such a decrease in the size of

the lumen of a blood vessel is called **vasoconstriction**. Conversely, when sympathetic stimulation decreases, the smooth muscle fibers relax and the size of the lumen increases. This increase is called **vasodilation**.

The smooth muscle layer of blood vessels, especially of arteries and arterioles (described shortly), also helps limit bleeding from wounds. When an artery or arteriole is cut, the smooth muscle contracts, producing vascular spasm of the vessel. This is one of the three mechanisms involved in hemostasis (described on page 566). However, there is a limit to how much vascular spasm can prevent hemorrhaging since the heart's pumping action causes blood to flow through arteries under great pressure.

Elastic (Conducting) Arteries

Large arteries are referred to as **elastic** (**conducting**) **arteries**. Examples include the aorta and the brachiocephalic, common carotid, subclavian, vertebral, pulmonary, and common iliac arteries. The wall of an elastic artery is relatively thin in proportion to its diameter, and its tunica media contains more elastic fibers and less smooth muscle. Elastic arteries are called conducting arteries because they **conduct** blood from the heart to medium-sized muscular arteries.

As the heart alternately contracts and relaxes, blood flow speeds up and slows down accordingly. When the heart contracts and ejects blood, the walls of elastic arteries stretch to accommodate the surge of blood (Fig. 21.2a). The stretched elastic fibers momentarily store some of the energy. For this reason, the elastic arteries function as a **pressure reservoir**. During relaxation of the heart they recoil, converting stored (potential) energy into kinetic energy of the blood. Thus blood moves forward in a more-or-less continuous flow (Fig. 21.2b).

Muscular (Distributing) Arteries

Medium-sized arteries are called **muscular** (**distributing**) **arteries**. Examples include the axillary, brachial, radial, intercostal, splenic, mesenteric, femoral, popliteal, and tibial arteries. In comparison with an elastic artery, the tunica media contains more smooth muscle and fewer elastic fibers. Thus they are capable of greater vasoconstriction and vasodilation to adjust the rate of blood flow to suit the needs of the structure supplied. The walls of muscular arteries are relatively thick, due mainly to the large amount of smooth muscle. Muscular arteries also are called distributing arteries because they *distribute* blood to various parts of the body.

Anastomoses

Most tissues of the body receive blood from more than one artery. The union of the branches of two or more arteries supplying the same body region is called an **anastomosis** (a-nas-tō-MŌ-sis; = coming together; plural is **anastomoses**). Fig. 21.21c on page 644 illustrates several anastomoses between branches of the superior mesenteric artery as they approach the jejunum of the small intestine. Anastomoses may also occur between veins and between arterioles and venules. Anastomoses between arteries provide alternate routes for blood to reach a tissue or organ.

Figure 21.1 Comparative structure of blood vessels. The relative size of the capillary in (c) is enlarged.

Arteries carry blood from the heart to tissues; veins carry blood from tissues to the heart.

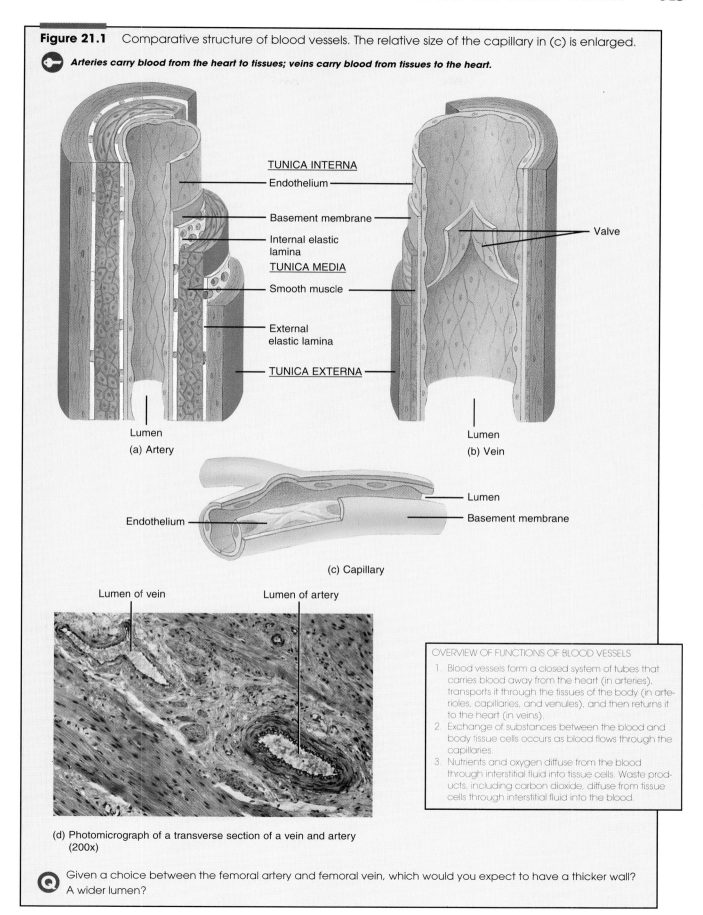

TUNICA INTERNA
Endothelium

Basement membrane

Internal elastic lamina

TUNICA MEDIA

Smooth muscle

External elastic lamina

TUNICA EXTERNA

Valve

Lumen

(a) Artery

Lumen

(b) Vein

Endothelium

Lumen

Basement membrane

(c) Capillary

Lumen of vein

Lumen of artery

OVERVIEW OF FUNCTIONS OF BLOOD VESSELS

1. Blood vessels form a closed system of tubes that carries blood away from the heart (in arteries), transports it through the tissues of the body (in arterioles, capillaries, and venules), and then returns it to the heart (in veins).
2. Exchange of substances between the blood and body tissue cells occurs as blood flows through the capillaries.
3. Nutrients and oxygen diffuse from the blood through interstitial fluid into tissue cells. Waste products, including carbon dioxide, diffuse from tissue cells through interstitial fluid into the blood.

(d) Photomicrograph of a transverse section of a vein and artery (200x)

Given a choice between the femoral artery and femoral vein, which would you expect to have a thicker wall? A wider lumen?

Figure 21.2   Pressure reservoir function of elastic arteries.

Recoil of elastic arteries keeps blood moving during ventricular relaxation (diastole).

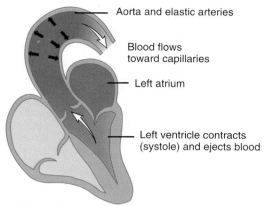

(a) Elastic aorta and arteries stretch during ventricular contraction

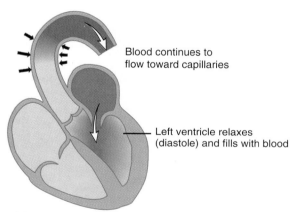

(b) Elastic aorta and arteries recoil during ventricular relaxation

 In atherosclerosis, the walls of elastic arteries become stiffer and lose some of their elasticity. How do you think this affects the pressure reservoir function of the arteries?

Thus if blood flow stops momentarily as normal movements compress a vessel or if a vessel is blocked by disease, injury, or surgery, circulation to a part of the body is not necessarily stopped. The alternate route of blood flow to a body part through an anastomosis is known as **collateral circulation**. An alternate blood route may also be from nonanastomosing vessels that supply the same region of the body.

Arteries that do not anastomose are known as **end arteries**. Obstruction of an end artery interrupts the blood supply to a whole segment of an organ, producing necrosis (death) of that segment.

Arterioles

An **arteriole** (ar-TIR-rē-ol; *arteriola* = small artery) is a very small, almost microscopic artery that delivers blood to

capillaries. Arterioles closer to the arteries from which they branch have a tunica interna like that of arteries, a tunica media composed of smooth muscle and very few elastic fibers, and a tunica externa composed mostly of elastic and collagen fibers. In the smallest diameter arterioles, which are closest to capillaries, the tunics consist of little more than a ring of endothelial cells surrounded by a few scattered smooth muscle fibers (Fig. 21.3).

Arterioles play a key role in regulating blood flow from arteries into capillaries. When the smooth muscle of arterioles contracts, causing vasoconstriction, blood flow into capillaries decreases. When the smooth muscle relaxes, the arterioles vasodilate, and blood flow into capillaries increases. A change in diameter of arterioles can also significantly affect blood pressure.

Capillaries

Capillaries (KAP-i-lar′-ēs; *capillaris* = hairlike) are microscopic vessels that usually connect arterioles and venules (Fig. 21.3). The flow of blood from arterioles to venules through capillaries is called the **microcirculation**. Capillaries are found near almost every cell in the body, but their distribution varies with the metabolic activity of the tissue. Body tissues with high metabolic activity, for example, muscles, the liver, kidneys, lungs, and nervous system, use more O_2 and nutrients. Accordingly, they have extensive capillary networks. In areas where activity is lower, such as tendons and ligaments, there are fewer capillaries. A few tissues have no capillaries—all covering and lining epithelia, the cornea and lens of the eye, and cartilage.

The primary function of capillaries is to permit the exchange of nutrients and wastes between the blood and tissue cells through interstitial fluid. The structure of capillaries is admirably suited to this purpose. Capillary walls are composed of only a single layer of epithelial cells (endothelium) and a basement membrane (see Fig. 21.1c). They have no tunica media or tunica externa. Thus a substance in the blood passes through just one cell layer into interstitial fluid before reaching tissue cells. Exchange of materials occurs only through capillary walls; the walls of arteries and veins present too thick a barrier.

Capillaries form extensive branching networks that increase the surface area for diffusion and filtration and thereby allow rapid exchange of large quantities of materials. In most tissues, blood flows through only a small portion of the capillary network when metabolic needs are low. But when a tissue is active, such as contracting muscle, the entire capillary network fills with blood.

The flow of blood through capillaries is regulated by vessels with smooth muscle in their walls. A **metarteriole** (*met* = beyond) is a vessel that emerges from an arteriole, passes through the capillary network, and empties into a venule (Fig. 21.3). The proximal portion of a metarteriole is surrounded by scattered smooth muscle fibers whose contraction and relaxation help regulate blood flow and pressure. The distal portion of a metarteriole, which empties into a

Figure 21.3 Arteriole, capillaries, and venule.

Arterioles regulate blood flow into capillaries, where nutrients, gases, and wastes are exchanged between blood and tissue cells.

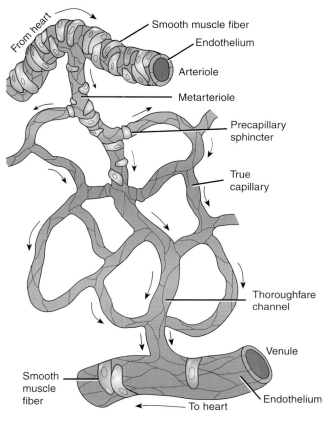

(a) Details of a capillary network

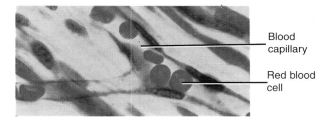

(b) Photomicrograph showing red blood cells squeezing through capillaries

Why do metabolically active tissues have extensive capillary networks?

venule, has no smooth muscle fibers and is called a **thoroughfare channel**. It serves as a low-resistance pathway that opens when constriction of precapillary sphincters (described shortly) reduces blood flow through the capillary network. Thoroughfare channels thus bypass the capillary bed and sustain blood flow through a region when the capillaries are not being utilized.

True capillaries emerge from arterioles or metarterioles and are not on the direct flow route from arteriole to venule. At their sites of origin, there is a ring of smooth muscle fibers called a **precapillary sphincter** that controls the flow of blood entering a true capillary. Blood usually does not flow in a continuous manner through capillary networks. Rather, it flows intermittently, because of contraction and relaxation of the smooth muscle of metarterioles and the precapillary sphincters of true capillaries. This intermittent contraction and relaxation may occur 5–10 times per minute and is called **vasomotion**. In part, vasomotion is due to certain chemicals released by the endothelium. The various factors that regulate the contraction of smooth muscle fibers of metarterioles and precapillary sphincters are discussed later.

Many capillaries of the body are said to be **continuous capillaries**. Except for **intercellular clefts**, which are gaps between neighboring endothelial cells, the plasma membranes of the cells form a continuous, uninterrupted ring around the capillary (Fig. 21.4a). Continuous capillaries are found in skeletal and smooth muscle, connective tissues, and the lungs. Other capillaries of the body are called **fenestrated** (*fenestra* = window) **capillaries**. They differ from continuous capillaries in that their endothelial cells have many fenestrations (pores) in the plasma membrane (Fig. 21.4b). These range from 70 to 100 nm in diameter. Fenestrated capillaries are found in the kidneys, villi of the small intestine, choroid plexuses of the ventricles in the brain, ciliary processes of the eyes, and endocrine glands.

Blood capillaries in certain parts of the body, such as the liver, are termed sinusoids. They are wider than other capillaries and more winding. Instead of the usual endothelial lining, sinusoids contain spaces between endothelial cells, and the basement membrane is incomplete or absent (Fig. 21.4c). In addition, sinusoids contain specialized lining cells that are adapted to the function of the tissue. For example, sinusoids in the liver contain hepatocytes (liver cells) and phagocytic cells called **stellate reticuloendothelial (Kupffer's) cells** that remove bacteria and other debris from the blood (see Fig. 24.20b). Sinusoids are also present in the spleen, anterior pituitary gland, parathyroid glands, and bone marrow.

Materials can cross the blood capillary walls through four basic routes: through intercellular clefts, via pinocytic vesicles, directly across endothelial membranes, and through fenestrations (Fig. 21.4).

Venules

When several capillaries unite, they form small veins called **venules** (VEN-yools; *venula* = little vein). Venules collect blood from capillaries and drain it into veins. The venules closest to the capillaries consist of a tunica interna of endothelium and a tunica media that has only a few scattered smooth muscle fibers (see Fig. 21.3a). The venules become larger as they approach the veins; here they also contain the tunica externa characteristic of veins.

Figure 21.4　Types of capillaries shown in transverse sections.

 Capillaries are microscopic blood vessels that connect arterioles and venules.

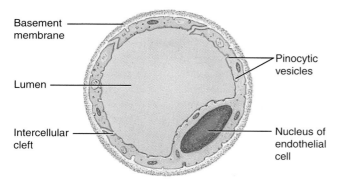

(a) Continuous capillary formed by endothelial cells

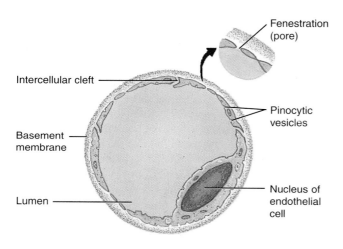

(b) Fenestrated capillary

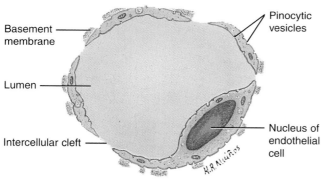

(c) Sinusoid

 What are the ways that materials can cross capillary walls?

Veins

Veins (VĀNZ) are composed of essentially the same three coats as arteries, but there are variations in their relative thickness. The tunica interna of veins is thinner than that of their companion arteries and the tunica media of veins is much thinner with relatively little smooth muscle and elastic fibers. The tunica externa of veins is the thickest layer, consisting of collagen and elastic fibers. The tunica externa of the inferior vena cava also contains longitudinal fibers of smooth muscle. Veins do not contain the external or internal elastic laminae found in arteries (see Fig. 21.1b). Despite these differences, veins are still distensible enough to adapt to variations in the volume and pressure of blood passing through them. Also, the lumen of a vein is larger than that of a comparable artery and a vein frequently appears collapsed (flattened) when sectioned (see Fig. 21.1d).

The average blood pressure in veins is considerably lower than in arteries. The difference in pressure can be noticed when blood flows from a cut vessel. Blood leaves a cut vein in an even, slow flow but spurts rapidly from a cut artery. Most of the structural differences between arteries and veins reflect this pressure difference. For example, the walls of veins are not as strong as those of arteries.

Many veins, especially those in the limbs, also feature abundant **valves** (Fig. 21.5, see also Figs. 21.1b and 21.10), which are needed because venous blood pressure is so low. When you stand, the pressure pushing blood up the veins in your legs is barely enough to overcome the force of gravity pulling it back down. Each valve is composed of thin folds of tunica interna that form flaplike cusps. The cusps project into the lumen of the veins, pointing toward the heart. Veins pass between groups of skeletal muscles and when the muscles contract, venous pressure is increased and the valve opens as the cusps are pushed against the wall of the vein by the blood as it flows through the valve toward the heart. When the muscles relax, the blood tends to move back toward the feet but is prevented from doing so by the coming together of the cusps that close the lumen of the vein. This prevents the flow of blood away from the heart. In this way, valves prevent backflow of blood and aid in moving blood in one direction only—toward the heart.

A **vascular** (**venous**) **sinus** is a vein with a thin endothelial wall that has no smooth muscle to alter its diameter. Surrounding dense connective tissue replaces the tunica media and tunica externa to provide support. Intracranial vascular sinuses, which are supported by the dura mater, return cerebrospinal fluid and deoxygenated blood from the brain to the heart. Another example of a vascular sinus is the coronary sinus of the heart.

Blood Distribution

The largest portion of your blood volume at rest, about 60%, is in systemic veins and venules (Fig. 21.6). Systemic

Figure 21.5 Photographs of a valve in a vein.

🔑 *Valves in veins allow blood to flow in one direction only—toward the heart.*

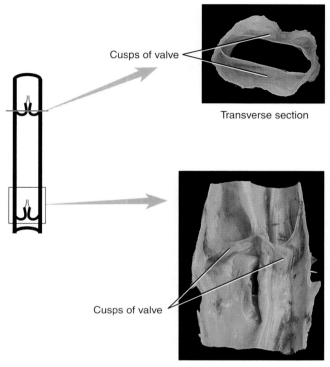

Cusps of valve

Transverse section

Cusps of valve

Longitudinal section

❓ Why is it more important for arm and leg veins to have valves than it is for neck veins to have valves?

capillaries hold only about 5% of the blood volume, and arteries and arterioles about 15%. Since systemic veins and venules contain so much of the blood, they are called **blood reservoirs**. They serve as storage depots for blood, which can be diverted quickly to other vessels if the need arises. For example, when there is increased muscular activity, an area in the brain stem called the vasomotor center sends more sympathetic impulses to veins that serve as blood reservoirs. The result is vasoconstriction, which reduces the volume of blood in venous reservoirs and allows a greater blood volume to flow to skeletal muscles, where it is needed most. A similar mechanism operates in cases of hemorrhage, when blood volume and pressure decrease. Vasoconstriction of veins in venous reservoirs helps to compensate for the blood loss. Among the principal blood reservoirs are the veins of the abdominal organs (especially the liver and spleen) and the veins of the skin.

Figure 21.6 Blood distribution in the heart and various types of blood vessels at rest.

🔑 *Since systemic veins and venules contain so much blood, they are called blood reservoirs.*

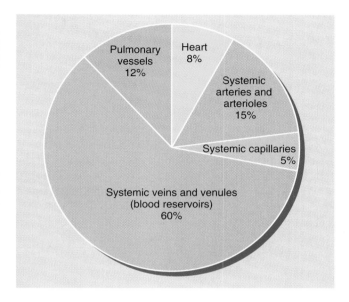

❓ If your total blood volume is 5 liters, what volume is in your venules and veins right now? In your capillaries?

CAPILLARY EXCHANGE

The 5% of the blood in systemic capillaries represents the bulk of blood that exchanges materials with systemic tissue cells. Nutrients, oxygen, and other substances in the blood pass through thin capillary walls, into interstitial fluid, and then into tissue cells. Wastes and substances being secreted by tissue cells move in the opposite direction. Substances enter and leave capillaries in three basic ways: (1) diffusion, (2) vesicular transport (endocytosis and exocytosis), and (3) bulk flow (filtration and reabsorption).

Diffusion

The most important method of capillary exchange is by simple diffusion. Substances such as O_2, carbon dioxide (CO_2), glucose, amino acids, hormones, and others diffuse through capillary walls down their concentration gradients. All plasma solutes, except larger proteins, pass freely across most capillary walls. Lipid-soluble materials, such as O_2, CO_2, and steroid hormones, may pass directly through the phospholipid bilayer of the endothelial cell plasma membranes. Water-soluble substances, such as glucose and amino acids, pass either through fenestrations or intercellular clefts (see Fig. 21.4). In the liver, intercellular clefts between endothelial cells that line the sinusoids are so large that even proteins may pass through. This provides an easy route for proteins synthesized by liver cells, for example, fibrinogen and albumin, to enter the plasma. The prime exception to diffusion of water-soluble materials across capillary walls is in the brain, where the endothelial cells in most regions are nonfenestrated and sealed together by tight junctions. (The blood–brain barrier is described on page 397.)

Vesicular Transport

A small quantity of material crosses capillary membranes by **vesicular transport** (**transcytosis**). Substances in blood plasma become enclosed within tiny vesicles that enter endothelial cells by endocytosis and then exit on the other side by exocytosis (see Fig. 21.4). This method of transport is important mainly for large, lipid-insoluble molecules that cannot cross capillary walls in any other way. For example, certain antibody proteins pass from maternal into fetal circulation by vesicular transport.

Bulk Flow (Filtration and Reabsorption)

Whereas diffusion is more important for *solute exchange* between plasma and interstitial fluid, bulk flow is more important for regulation of the *relative volumes of blood and interstitial fluid.* **Bulk flow** is a passive process. It involves the movement in the same direction of *large* numbers of ions, molecules, or particles that are dissolved or carried in a medium such as fluid or air. The substances move in unison in response to forces such as hydrostatic (water) pressure or air pressure, and they move at rates far greater than can be accounted for by diffusion or osmosis alone. Only 20–25% of the extracellular fluid is confined within blood vessels. But if the need arises, for example, if you lose blood through hemorrhage, interstitial fluid can move into capillaries, expand the blood volume, and help maintain blood pressure.

Bulk flow occurs because some pressures promote *filtration* of water and solutes *from capillaries into the surrounding interstitial (tissue) spaces.* Two pressures promote filtration: blood hydrostatic pressure and interstitial fluid colloid osmotic pressure. Fluid doesn't build up in interstitial spaces because opposing forces promote *reabsorption* of water and solutes *from interstitial fluid into blood capillaries.* The main pressure promoting reabsorption of fluid is blood colloid osmotic pressure. The balance of these pressures, called **net filtration pressure** (**NFP**), determines whether blood volume remains steady or changes. Overall, the volume reabsorbed normally is almost as large as the volume filtered. This near equilibrium is known as **Starling's law of the capillaries**. Let us now see how the pressures operate.

First we will consider the hydrostatic pressures. These pressures are due to the pressure of water in the fluids against blood vessel walls. Blood pressure in capillaries, called **blood hydrostatic pressure** (**BHP**), tends to push fluid out of capillaries into interstitial fluid. BHP is about 35 mm Hg at the arterial end of a capillary and about 16 mm Hg at the venous end (Fig. 21.7). The pressure of the interstitial fluid, called **interstitial fluid hydrostatic pressure** (**IFHP**), is close to zero. It is difficult to measure, and its reported values vary from small positive to small negative values. For purposes of our discussion, we will assume that IFHP is 0 mm Hg all along the capillaries. Regardless of its exact value, however, the basic principles of fluid movement still apply.

Now let us consider the osmotic pressures involved in fluid movement. The difference in osmotic pressures across a capillary wall is due almost entirely to the presence of plasma proteins, which are too large to pass through either fenestrations or gaps between endothelial cells. **Blood colloid osmotic pressure** (**BCOP**), also called **oncotic pressure**, is a force caused by the colloidal suspension of these large plasma proteins. The effect of BCOP is to pull fluid from interstitial spaces into capillaries. It averages about 26 mm Hg in capillaries. Opposing BCOP is **interstitial fluid osmotic pressure** (**IFOP**), which tends to move fluid out of capillaries into interstitial fluid. Normally, IFOP is very small, only 0.1–5 mm Hg. The small amount of protein that leaks from plasma into interstitial fluid does not accumulate there because it enters lymphatic fluid and is returned to the blood. For discussion, we will use a value of 1 mm Hg for IFOP.

Whether fluids leave or enter capillaries depends on how the pressures relate to one another. If the pressures that push fluid out of capillaries are greater than the pressures that pull fluid into capillaries, fluid will move from capillaries into interstitial spaces (filtration). If, on the other hand, the pressures that move fluid out of interstitial spaces into capillaries are greater than the pressures that move fluid out of capillaries, then fluid will move from interstitial spaces into capillaries (reabsorption).

The **net filtration pressure** (**NFP**) shows the direction of fluid movement. It is calculated as follows:

$$NFP = (BHP + IFOP) - (BCOP + IFHP)$$

At the arterial end of a capillary,

$$NFP = (35 + 1) - (26 + 0) = (36) - (26) = 10 \text{ mm Hg}$$

At the venous end of a capillary,

$$NFP = (16 + 1) - (26 + 0) = (17) - (26) = -9 \text{ mm Hg}$$

Thus at the arterial end of a capillary, there is a *net outward pressure* of 10 mm Hg, and fluid moves out of the capillary into interstitial spaces (filtration). At the venous end of a capillary, the negative value (−9 mm Hg) represents a *net inward pressure,* and fluid moves into the capillary from tissue spaces (reabsorption).

Figure 21.7 Dynamics of capillary exchange (Starling's law of the capillaries).

🔑 *Blood hydrostatic pressure pushes fluid out of capillaries (filtration) whereas blood colloid osmotic pressure pulls fluid into capillaries (reabsorption).*

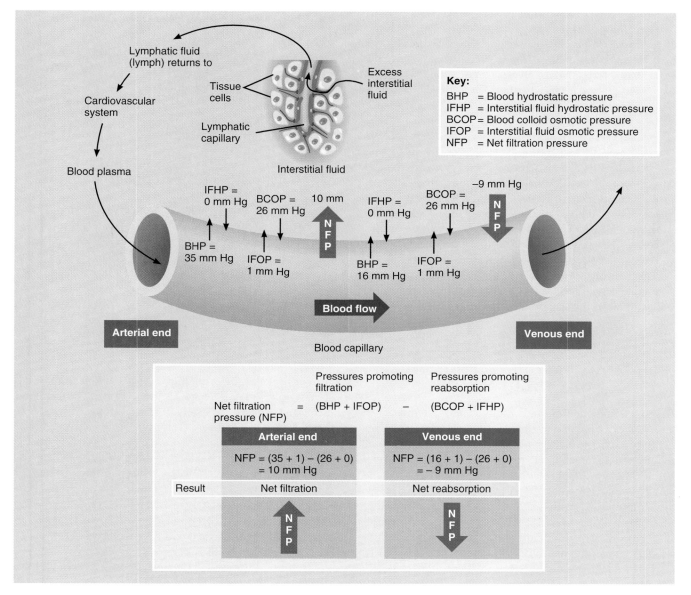

❓ A person who has liver failure cannot synthesize the normal amount of plasma proteins. How will this affect blood colloid osmotic pressure and what will be the impact on capillary filtration and reabsorption?

On average, about 85% of the fluid filtered at the arteriolar ends of capillaries is reabsorbed at their venular ends. Some of the filtered fluid and any proteins that escape from blood into interstitial fluid flow into lymphatic capillaries and return to the blood via the lymphatic system. On a daily basis, about 20 liters of fluid filters out of capillaries, 17 liters is reabsorbed, and 3 liters enters lymphatic capillaries. Not considered in these numbers is the fluid that filters from capillaries in the kidneys during formation of urine (described on page 861).

EDEMA

Occasionally, the balance of filtration and reabsorption between interstitial fluid and plasma is disrupted. If filtration greatly exceeds reabsorption, there is an abnormal increase in interstitial fluid volume, which is termed **edema** (*edemas* = swelling). Usually, edema is not detectable in tissues until interstitial fluid volume has increased to about 30% above normal.

Edema may result from several main causes:

1. **Increased blood hydrostatic pressure in capillaries** due to an increase in venous pressure. This may result from poor blood flow back to the heart due to cardiac failure or blood clots.

2. **Decreased concentration of plasma proteins**, which lowers blood colloid osmotic pressure. Protein loss may result from burns, malnutrition, liver disease, and kidney disease.

3. **Increased permeability of capillaries**, which raises interstitial fluid osmotic pressure by allowing greater amounts of plasma proteins to leave the blood and enter tissue fluid. This may be caused by chemical, bacterial, thermal, or mechanical agents.

4. **Increased extracellular fluid volume** as a result of fluid retention. When a person has difficulty excreting fluids, for whatever reason, but continues to drink normal amounts of water, extracellular fluid in the body increases. Some of the fluid enters blood and increases blood hydrostatic pressure.

5. **Blockage of lymphatic vessels** as often occurs after a radical mastectomy (breast removal, usually because of cancer) or infection by filariasis roundworms. In a radical mastectomy, nearby lymph nodes that appear cancerous are removed with the breast tissue. Edema occurs in the arm on the same side because lymph drainage is blocked. The larvae of the tropical filariasis parasite invade and block lymphatic channels, causing the grossly disfiguring type of edema known as elephantiasis. ■

HEMODYNAMICS: PHYSIOLOGY OF CIRCULATION

The overall function of the cardiovascular system is to ensure adequate circulation of blood to all body tissues and capillary exchange between blood plasma, interstitial fluid, and tissue cells. Each tissue requires a minimum number of milliliters of blood per minute to sustain its metabolic activities and remove its wastes. In Chapter 20 you saw that total cardiac output depends on heart rate and stroke volume. Now you will discover that the distribution of the cardiac output to various tissues depends on the interplay of (1) the *pressure difference* that drives the blood flow and (2) the *resistance* to blood flow, which is the opposition to flow of blood through specific blood vessels.

Velocity of Blood Flow

The *volume* of blood that flows through any tissue in a given period of time (in milliliters per minute) is called **blood flow**. The *velocity* (speed) of blood flow (in centimeters per second) is inversely related to the cross-sectional area available. This means that blood flows slowest where the cross-sectional area is greatest (Fig. 21.8), just as a river flows more slowly as it becomes broader. Each time an artery branches, the total cross-sectional area of all its branches is greater than that of the original vessel. On the other hand, when branches combine, for example, as venules merge to form veins, the total cross-sectional area

Figure 21.8   Relationship between velocity of blood flow and cross-sectional area in different types of blood vessels.

 Velocity of blood flow is slowest in the capillaries because altogether they have the largest cross-sectional area.

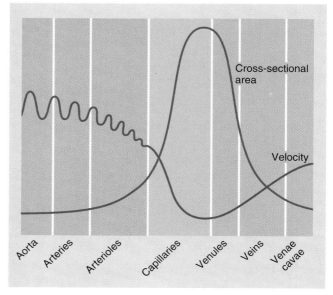

Q In which blood vessels is the velocity of blood flow fastest?

becomes smaller. In an adult, the cross-sectional area of the aorta is only 3–5 cm², and the average velocity of the blood there is 40 cm/sec. In capillaries, the estimated cross-sectional area is 4500–6000 cm², and the velocity of blood flow is less than 0.1 cm/sec. In the two venae cavae combined, the cross-sectional area is about 14 cm², and the velocity is 5–20 cm/sec. Thus the velocity of blood flow decreases as it flows from the aorta to arteries to arterioles to capillaries and increases as it leaves capillaries and returns to the heart. Because blood moves most slowly through the capillaries, there is adequate time for the exchange of materials between the capillaries and adjacent tissues.

Circulation time is the time required for a drop of blood to pass from the right atrium, through the pulmonary circulation, back to the left atrium, through systemic circulation down to the foot, and back again to the right atrium. In a resting person, such a trip normally takes about 1 min.

Volume of Blood Flow

In an adult, cardiac output (CO) is about 5.25 liters/min. This is the volume of blood that circulates through systemic (or pulmonary) blood vessels each minute. In Chapter 20 we noted that CO equals stroke volume (SV) multiplied by heart rate (HR).

$$\text{CO} = \text{SV} \times \text{HR}$$
$$= 70 \text{ ml/beat} \times 75 \text{ beats/minute}$$
$$= 5250 \text{ ml/min} = 5.25 \text{ liters/min}$$

Besides heart rate and stroke volume, two other factors influence cardiac output: (1) blood pressure and (2) resistance, which is due mainly to friction between blood and blood vessel walls. Blood flows from regions of higher to lower pressure; the greater the pressure difference, the greater the blood flow. The higher the resistance, on the other hand, the lower the blood flow.

Blood Pressure

Blood pressure (**BP**) is the hydrostatic pressure exerted by blood on the walls of a blood vessel. Blood pressure is highest in the aorta and large systemic arteries and is generated by contraction of the ventricles. In the aorta and large systemic arteries of a resting, young adult, BP rises to about 120 mm Hg during systole (contraction) and drops to about 80 mm Hg during diastole (relaxation). Because the blood pressure curve has a rather triangular shape (see Fig. 21.16), the mean (average) arterial blood pressure (MABP) is closer to diastolic than to systolic blood pressure. MABP is approximately one-third of the way between diastolic and systolic blood pressure: MABP = diastolic BP + ⅓ (systolic BP − diastolic BP). Thus in a person whose BP is 120/80 mm Hg, MABP is about 93 mm Hg.

Cardiac output equals mean arterial blood pressure (MABP) divided by resistance (R): CO = MABP ÷ R. If cardiac output rises due to an increase in stroke volume or heart rate, then blood pressure rises so long as resistance remains steady. Likewise, a decrease in cardiac output causes a decrease in blood pressure if resistance does not change.

As blood leaves the aorta and flows through the systemic circulation, its pressure falls progressively as the distance from the pump (left ventricle) increases (Fig. 21.9). Blood pressure decreases from 93 mm Hg to about 35 mm Hg as blood passes into the arteriolar end of a capillary. At the venous end of a capillary, blood pressure has dropped to about 16 mm Hg. Blood pressure continues to drop as blood enters venules and then veins because these vessels are far from the pressure source (the left ventricle). Finally blood pressure reaches 0 mm Hg as blood flows into the right ventricle. Blood (or any fluid) always flows along a tube down a pressure gradient (difference). If there is no pressure difference, there is no flow.

Blood pressure also depends on the total volume of blood in the cardiovascular system. The normal volume of blood in an adult is about 5 liters (5.3 qt). Any decrease in this volume, as from hemorrhage, decreases the amount of blood that is circulated through the arteries each minute. A modest decrease can be compensated for by homeostatic mechanisms (described on page 629 in the section on Compensated Shock), but if the decrease in blood volume is

Figure 21.9 Blood pressures in various parts of the cardiovascular system. The dashed line is the mean (average) blood pressure.

 Blood pressure rises and falls with each heartbeat in blood vessels leading to capillaries.

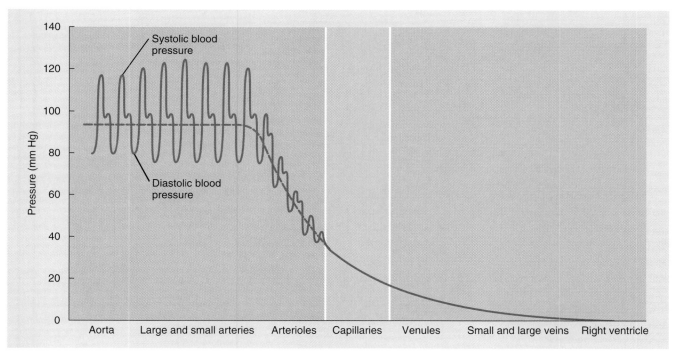

What is the mean blood pressure in the aorta?

significant (greater than 10% of total blood volume), blood pressure drops. On the other hand, anything that increases blood volume, such as water retention in the body, tends to increase blood pressure.

Resistance

As noted earlier, **resistance** refers to the opposition to blood flow principally as a result of friction between blood and the walls of blood vessels. The friction, and thus the resistance, depends on (1) blood viscosity, (2) total blood vessel length, and (3) average blood vessel radius. Normally, the first two are steady and only changes in blood vessel radius contribute to changing resistance.

1. **Blood viscosity.** The viscosity ("thickness") of blood depends largely on the ratio of red blood cells to plasma (fluid) volume and to a smaller extent on the concentration of proteins in plasma. Resistance to blood flow is directly proportional to the viscosity of blood. Any condition that increases the viscosity of blood, such as dehydration, an unusually high number of red blood cells (polycythemia), or severe burns, increases resistance and thus blood pressure. A depletion of plasma proteins or red blood cells, as a result of anemia or hemorrhage, decreases resistance and thus blood pressure.

2. **Total blood vessel length.** Resistance to blood flow through a vessel is directly proportional to the length of the blood vessel. The longer a blood vessel, the greater the resistance as blood flows through it. An obese person may have hypertension (elevated blood pressure) due to increase in total blood vessel length caused by the additional blood vessels in adipose tissue. An estimated 300 km (about 200 miles) of additional blood vessels develop for each extra pound of fat.

3. **Average blood vessel radius.** Resistance is inversely proportional to the fourth power of the radius of the blood vessel ($R \propto 1/r^4$). The smaller the radius of the blood vessel, the greater the resistance it offers to blood flow. As an example, if the radius of a blood vessel decreases by one-half, its resistance to blood flow increases 16 times $[1 \div (\frac{1}{2})^4 = 2^4 = 2 \times 2 \times 2 \times 2 = 16]$. For the same reason, water flows more quickly through a large diameter fire hose than through a small diameter garden hose.

Systemic vascular resistance (**SVR**) (also known as **total peripheral resistance**) refers to all the vascular resistances offered by systemic blood vessels. Most resistance is in arterioles, capillaries, and venules. The diameter of arteries and veins is large and thus their resistance is very small because most of the blood does not come into physical contact with the walls of the blood vessel. A major function of arterioles is to control SVR—and therefore blood pressure and blood flow to particular tissues—by changing their diameters. Arterioles need to vasodilate or vasoconstrict only slightly to have a large effect on SVR. The principal center for regulation of SVR is the vasomotor center in the brain stem (described shortly).

Venous Return

Venous return, the volume of blood flowing back to the heart from the systemic veins, depends on the pressure difference from venules (averaging about 16 mm Hg) to the right ventricle (0 mm Hg). Although this pressure difference is small, venous return to the right atrium keeps pace with output from the left ventricle because resistance of veins also is low. If pressure increases in the right atrium, however, venous return will decrease. One cause of increased pressure in the right atrium is an incompetent (leaky) tricuspid valve, which lets blood flow backwards as the ventricles contract. The result is buildup of blood on the venous side of the systemic circulation.

Besides the heart, two other mechanisms act as pumps to boost venous return: (1) contraction of skeletal muscles in the lower limbs and (2) the pressure changes in the thorax and abdomen during respiration (breathing). The presence of valves in veins allows both of these pumps to contribute to venous return.

1. **Skeletal muscle pump.** When skeletal muscles contract, they tighten around the vein passing through them, which increases the venous blood pressure, and the proximal valve opens. This pressure drives the blood toward the heart: the action is called *milking* (Fig. 21.10). When the muscles relax, this valve closes and prevents the backflow of blood away from the heart. People who are immobilized through injury or disease lack these contractions. As a result, the return of venous blood to the heart is slower, and the heart has to work harder.

Figure 21.10   Role of the skeletal muscle pump and venous valves in returning blood to the heart.

 Milking refers to skeletal muscle contractions that drive venous blood toward the heart.

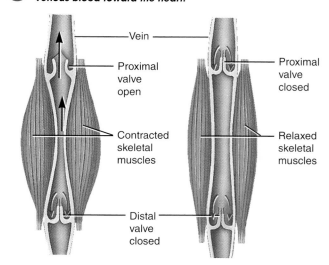

(a) Contracted skeletal muscles   (b) Relaxed skeletal muscles

 What mechanisms, beside cardiac contractions, act as pumps to boost venous return?

2. **Respiratory pump.** During inspiration (breathing in), the diaphragm moves inferiorly. This causes a decrease in pressure in the thoracic (chest) cavity and an increase in pressure in the abdominal cavity. As a result, a greater volume of blood moves from the compressed abdominal veins into the decompressed thoracic veins. When the pressures reverse during expiration (breathing out), the valves in the veins prevent backflow of blood.

A summary of factors that affect blood pressure is presented in Fig. 21.11.

VARICOSE VEINS

In people with weak venous valves, gravity forces large quantities of blood back down into distal parts of the vein. The resulting back-pressure overloads the vein and pushes its wall outward. After repeated overloading, the walls lose their elasticity and become stretched and flabby. Such dilated and tortuous veins caused by leaky valves are called **varicose** (*varicosus* = swollen, knotted) **veins**. They may be due to heredity, mechanical factors (prolonged standing and pregnancy), or aging. Because a varicosed wall is not able to exert a firm resistance against the blood, blood tends to accumulate in the pouched-out area of the vein. This causes it to bulge and also forces fluid into the surrounding tissue. Veins close to the surface of the lower limbs, especially the saphenous veins, are highly susceptible to varicosities whereas deeper veins are not as vulnerable because surrounding skeletal muscles prevent their walls from excessive stretching. ■

CONTROL OF BLOOD PRESSURE AND BLOOD FLOW

From moment to moment and day to day, several interconnected negative feedback systems control blood pressure by adjusting heart rate, stroke volume, systemic vascular resistance, and blood volume. Some systems allow rapid adjustment of blood pressure to cope with sudden changes such as the drop in brain blood pressure that occurs when you get out of bed. Others act more slowly to provide long-term regulation of blood pressure. Even if blood pressure is steady,

Figure 21.11 Summary of factors that affect blood pressure.

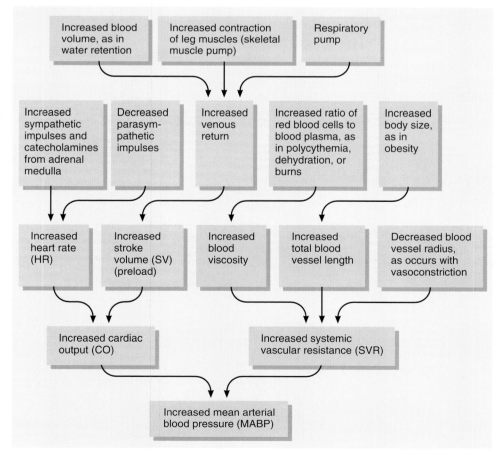

Which type of blood vessel exerts the major control of systemic vascular resistance on a moment-to-moment basis, and how does it achieve this?

there may be a need to change the distribution of blood flow, which is accomplished mainly by altering the diameter of arterioles. For example, during exercise a greater percentage of blood is diverted to organs directly involved in exercise. In strenuous exercise, blood flow to skeletal muscles may increase tenfold and blood flow to the heart and skin may triple. At the same time, blood flow to the digestive tract and kidneys may decrease to half the resting value. No matter what the level of exercise, total blood flow to the brain remains nearly constant. Blood flow to different regions of the brain, however, depends on neuronal metabolic activity—the higher the activity, the greater the blood flow.

In Chapter 20 we noted how the cardiovascular center in the medulla oblongata contributes to regulation of heart rate and stroke volume. Now we will complete that account by describing the neural, hormonal, and local negative feedback systems that regulate blood pressure and blood flow to specific tissues.

Cardiovascular Center

Groups of neurons scattered within the medulla oblongata of the brain stem regulate heart rate, contractility (force of contraction) of the ventricles, and blood vessel diameter (vasoconstriction versus vasodilation). As a whole, this region is known as the **cardiovascular (CV) center**. Some of its neurons stimulate the heart (cardiostimulatory center) whereas others inhibit the heart (cardioinhibitory center). Still others control blood vessel diameter (vasomotor center), either by causing constriction (vasoconstrictor center) or dilation (vasodilator center). Since these clusters of neurons communicate with one another, function together, and are not clearly separated anatomically, we will discuss them all as a group.

Input to the Cardiovascular Center

The CV center receives input both from higher brain regions and from sensory receptors (Fig. 21.12). Nerve impulses descend from higher brain regions including the cerebral cortex, limbic system, and hypothalamus to affect the CV center. For example, even before you start to run a race, your heart rate may increase due to nerve impulses conveyed from the limbic system to the CV center. If your body temperature rises during a race, the thermoregulatory center of the hypothalamus sends nerve impulses to the CV center of the medulla oblongata. The result is vasodilation of skin blood vessels, which allows heat to dissipate more rapidly. The two main types of sensory receptors that provide input to the cardiovascular center are baroreceptors and chemoreceptors. Baroreceptors are important pressure-sensitive sensory neurons that monitor stretching of the walls of blood vessels and the atria. Chemoreceptors monitor blood acidity, CO_2 level, and O_2 level.

Output from the Cardiovascular Center

Output from the CV center flows along sympathetic and parasympathetic fibers of the ANS (Fig. 21.13; see also Fig. 21.12). Sympathetic impulses reach the heart via the **cardiac accelerator nerves**. Sympathetic stimulation of the heart in-

---

Figure 21.12   The cardiovascular (CV) center in the medulla oblongata receives input from higher brain centers, proprioceptors, baroreceptors, and chemoreceptors. It provides output to both the sympathetic and parasympathetic divisions of the autonomic nervous system.

🔑 *The cardiovascular center is the main region for nervous system regulation of the heart and blood vessels.*

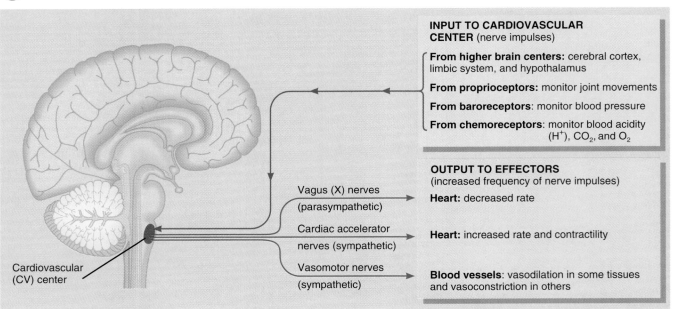

INPUT TO CARDIOVASCULAR CENTER (nerve impulses)

From higher brain centers: cerebral cortex, limbic system, and hypothalamus

From proprioceptors: monitor joint movements

From baroreceptors: monitor blood pressure

From chemoreceptors: monitor blood acidity (H^+), CO_2, and O_2

OUTPUT TO EFFECTORS (increased frequency of nerve impulses)

Heart: decreased rate

Heart: increased rate and contractility

Blood vessels: vasodilation in some tissues and vasoconstriction in others

Vagus (X) nerves (parasympathetic)

Cardiac accelerator nerves (sympathetic)

Vasomotor nerves (sympathetic)

Cardiovascular (CV) center

❓ What types of effector tissue are regulated by the CV center?

creases heart rate and contractility. Parasympathetic stimulation, conveyed along the **vagus (X) nerves**, decreases heart rate. The CV center also continually sends impulses to smooth muscle in blood vessel walls via sympathetic fibers called **vasomotor nerves**. Thus autonomic control of the heart is the result of opposing sympathetic (stimulatory) and

parasympathetic (inhibitory) influences. Autonomic control of blood vessel diameter, on the other hand, is mostly by the sympathetic division. Sympathetic vasomotor nerve fibers exit the spinal cord through all thoracic and the first one or two lumbar spinal nerves and pass into the sympathetic trunk ganglia (see Fig. 17.2). From here, impulses propagate along

Figure 21.13 Innervation of the heart by the autonomic nervous system and baroreceptor reflexes that help regulate bood pressure.

 Baroreceptors are pressure-sensitive neurons that monitor stretching.

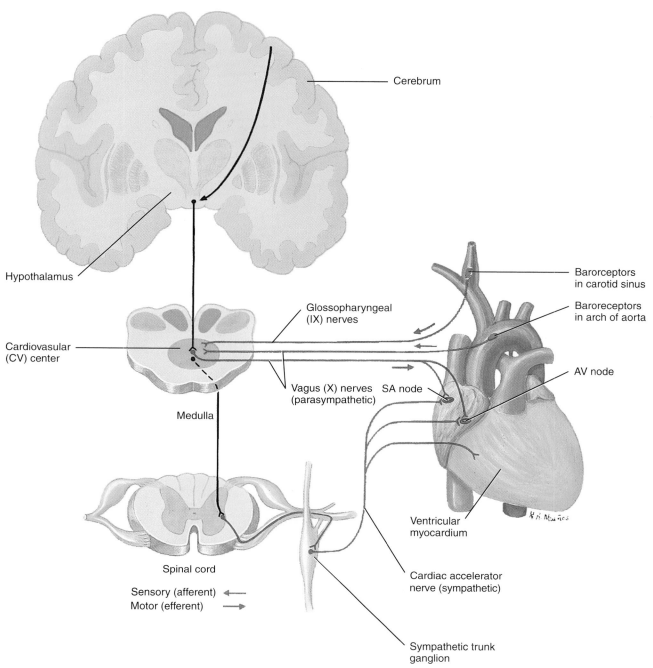

What are the paths taken by nerve impulses from baroreceptors in carotid sinuses and the arch of the aorta to the CV center?

sympathetic nerves that innervate blood vessels in viscera and peripheral areas. Over these routes, the CV center (specifically, the vasomotor center) continually sends impulses to arterioles throughout the body but especially in the skin and abdominal viscera. The result is a moderate state of tonic contraction or vasoconstriction, called **vasomotor tone**, which sets the resting level of systemic vascular resistance.

In the smooth muscle of most small arteries and arterioles, sympathetic stimulation causes vasoconstriction, which raises blood pressure and restricts blood flow to a tissue. This is due to activation of alpha (α) adrenergic receptors for norepinephrine (and epinephrine) in the vascular smooth muscle. In skeletal muscles and the heart, the smooth muscle of blood vessels displays beta (β) adrenergic receptors instead, and sympathetic stimulation causes vasodilation rather than vasoconstriction. In addition, some of the sympathetic fibers to blood vessels in skeletal muscle are cholinergic; they release acetylcholine, which causes vasodilation. (Refer to Exhibit 17.3 on page 497 for a listing of receptors in vascular smooth muscle.) When sympathetic stimulation increases, for example, during exercise, both vasoconstriction and vasodilation occur, but in different tissues. As a result, systemic vascular resistance may increase, decrease, or stay the same. The tissues that have dilated arterioles, however, will receive a larger share of the cardiac output. Sympathetic stimulation of most veins results in constriction that moves blood from reservoirs and increases blood pressure.

Neural Regulation of Blood Pressure

Regulation of blood pressure by the nervous system depends on receptors in the periphery that monitor blood pressure (baroreceptors) and blood chemistry (chemoreceptors) and provide input to the cardiovascular center. Several neural reflexes contribute to blood pressure regulation by negative feedback systems.

Baroreceptors

Nerve cells capable of responding to changes in pressure or stretch are called **baroreceptors**. Baroreceptors in the walls of the arteries, veins, and right atrium monitor blood pressure. The three most important negative feedback systems that baroreceptors participate in are the aortic reflex, carotid sinus reflex, and right heart reflex.

The **carotid sinus reflex** helps maintain normal blood pressure in the brain and is initiated by baroreceptors in the wall of the carotid sinus (Fig. 21.13). The **carotid sinus** is a small widening of the internal carotid artery just above the point where it branches from the common carotid artery. Any increase in blood pressure stretches the wall of the aorta and carotid sinus, and the stretching stimulates the baroreceptors. For the carotid sinus reflex, the impulses travel from the baroreceptors over sensory (afferent) fibers in the glossopharyngeal (IX) nerves to the CV center of the medulla oblongata. The **aortic reflex** governs general systemic blood pressure and is initiated by baroreceptors in the wall of the arch of the aorta. Impulses from baroreceptors in the arch of the aorta reach the CV center via sensory (afferent) fibers of the vagus (X) nerves.

When an increase in aortic and carotid artery pressures is detected in this manner, the CV center responds by putting out more parasympathetic impulses via motor (efferent) fibers of the vagus (X) nerves to the heart and fewer sympathetic impulses via cardiac accelerator nerves to the heart. The resulting decreases in heart rate and force of contraction lower cardiac output. Also, the CV center sends out decreased sympathetic impulses along vasomotor fibers that normally cause vasoconstriction. The result is vasodilation, which lowers systemic vascular resistance (SVR). Decreased cardiac output and SVR both lower systemic arterial blood pressure.

If blood pressure (controlled condition) falls, on the other hand, the baroreceptors (receptors) are stretched less (Fig. 21.14). They send nerve impulses (input) at a slower rate to the cardiovascular center (control center). In response, the CV center calls for increased sympathetic impulses, decreased parasympathetic impulses, and increased secretion of epinephrine and norepinephrine (NE) by the adrenal medulla (outputs). The effects on the heart and blood vessels (effectors) are to accelerate heart rate, increase force of contraction, and promote vasoconstriction. As the heart beats faster and more forcefully and SVR increases, blood pressure increases (response) and there is a return to homeostasis when blood pressure returns to the normal level. This relationship between heart rate and blood pressure is called **Marey's law of the heart**. The ability of the aortic and carotid sinus reflexes to correct a drop in blood pressure is very important when a person sits or stands from a lying position. Moving from a prone to an erect position decreases blood pressure in the head and upper part of the body. The drop in pressure, however, is quickly counteracted by the reflexes. If the pressure were to fall markedly, unconsciousness could occur.

The **right heart (atrial) reflex** responds to increases in venous blood pressure. It is initiated by baroreceptors in the right atrium and venae cavae. When venous pressure increases, the baroreceptors send impulses through the vagus (X) nerves to the CV center. Returning impulses via sympathetic nerves increase heart rate and force of contraction. This mechanism is called the **Bainbridge reflex**.

Chemoreceptors

Receptors sensitive to chemicals are called **chemoreceptors**. Chemoreceptors that monitor blood chemicals are located close to the baroreceptors of the carotid sinus and arch of the aorta in small structures called **carotid bodies** and **aortic bodies**, respectively. These chemoreceptors detect changes in blood level of oxygen, carbon dioxide, and hydrogen ions. If there is a severe deficiency of oxygen (hypoxia), an increase in hydrogen ion concentration (increased acidity or acidosis), or an excess of carbon dioxide (hypercapnia), the chemoreceptors are stimulated and send impulses to the CV center. In response, the CV center increases sympathetic stimulation to arterioles and veins. This brings about vasoconstriction and an

Figure 21.14 Negative feedback regulation of blood pressure via baroreceptor reflexes.

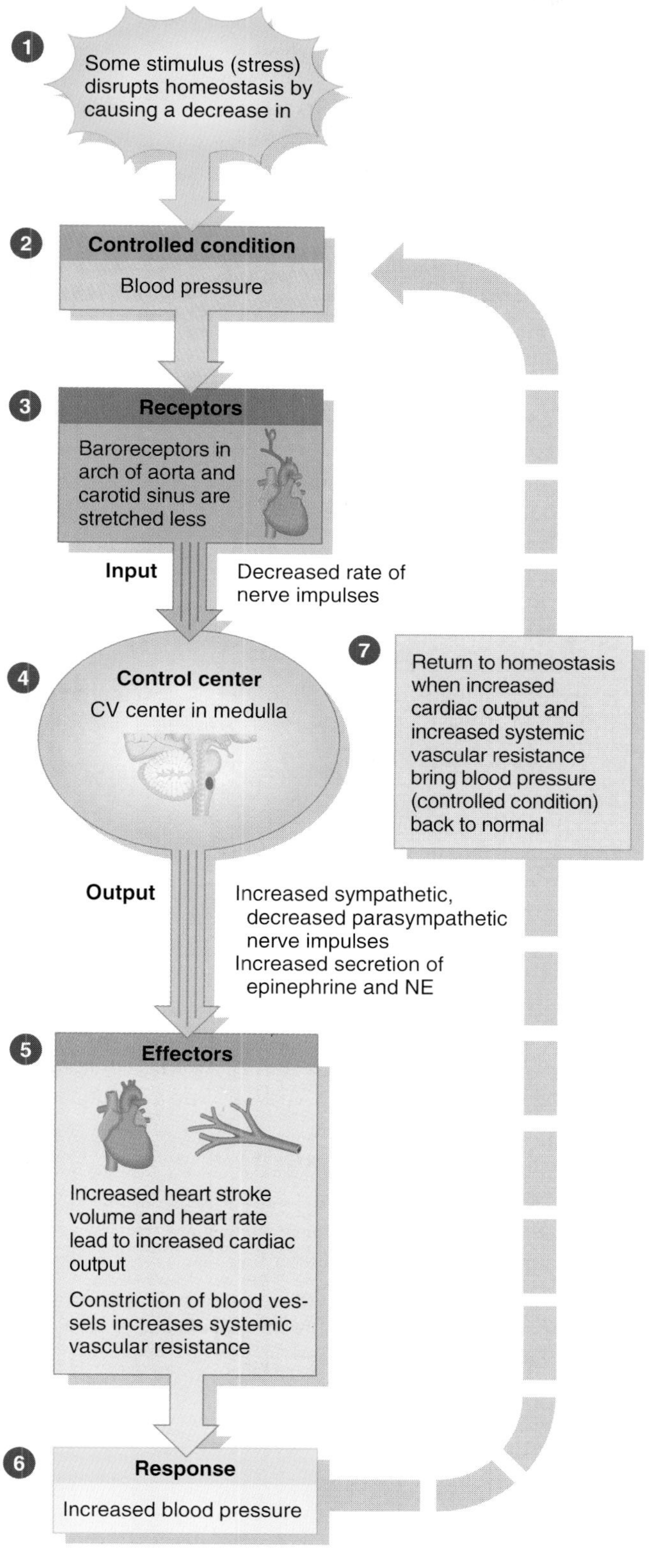

1 Some stimulus (stress) disrupts homeostasis by causing a decrease in

2 Controlled condition

Blood pressure

3 Receptors

Baroreceptors in arch of aorta and carotid sinus are stretched less

Input Decreased rate of nerve impulses

4 Control center

CV center in medulla

7 Return to homeostasis when increased cardiac output and increased systemic vascular resistance bring blood pressure (controlled condition) back to normal

Output Increased sympathetic, decreased parasympathetic nerve impulses
Increased secretion of epinephrine and NE

5 Effectors

Increased heart stroke volume and heart rate lead to increased cardiac output

Constriction of blood vessels increases systemic vascular resistance

6 Response

Increased blood pressure

Q Does this negative feedback cycle represent what happens when you lie down or when you stand up?

increase in blood pressure. As you will see in Chapter 23, these chemoreceptors also stimulate respiratory neurons in the medulla oblongata to adjust the rate of breathing.

Hormonal Regulation of Blood Pressure

Several hormones affect blood pressure and blood flow by three mechanisms: (1) altering cardiac output, (2) changing systemic vascular resistance, or (3) adjusting the total blood volume.

1. **Renin–angiotensin–aldosterone (RAA) system.** When blood volume falls or blood flow to the kidneys decreases, juxtaglomerular cells in the kidneys release increased amounts of an enzyme called **renin** into the bloodstream. Renin acts on angiotensinogen to form **angiotensin I**. As this molecule passes through capillaries in the lungs, **angiotensin converting enzyme** (**ACE**) changes it into **angiotensin II**. Angiotensin II helps to raise blood pressure in two ways. (1) It is a potent vasoconstrictor and thus raises total systemic resistance. (2) It stimulates secretion of **aldosterone**, which increases sodium ion (Na^+) and water reabsorption by the kidneys. This action increases total blood volume and thus increases blood pressure.

2. **Epinephrine and norepinephrine (NE).** Produced by the adrenal medulla, these hormones increase cardiac output (by increasing the rate and force of heart contractions) and bring about vasoconstriction of arterioles and veins in the skin and in abdominal organs. In addition, they also vasodilate arterioles in cardiac and skeletal muscle.

3. **Antidiuretic hormone (ADH).** ADH is produced by the hypothalamus and released from the posterior pituitary gland. One of the functions of ADH is to cause vasoconstriction if there is a severe loss of blood due to hemorrhage. For this reason, ADH is also called **vasopressin**. Alcohol inhibits release of ADH and has an inhibitory effect on the vasomotor center of the medulla oblongata. These effects bring about vasodilation, which lowers blood pressure.

4. **Atrial natriuretic peptide (ANP).** Released by cells in the atria of the heart, ANP lowers blood pressure by causing vasodilation and by promoting loss of salt and water in the urine, which reduces blood volume.

5. **Parathyroid hormone (PTH) and calcitriol.** These two hormones, which regulate the circulating levels of calcium ions (Ca^{2+}) and phosphate ions (HPO_4^{2-}) in the blood, also influence vascular smooth muscle. PTH causes vasodilation, which tends to decrease blood pressure. On the other hand, calcitriol, the active form of vitamin D, causes vasoconstriction, which increases blood pressure.

Exhibit 21.1 summarizes the relationship between hormones and blood pressure regulation.

EXHIBIT 21.1 HORMONAL REGULATION OF BLOOD PRESSURE

Factor Influencing Blood Pressure	Hormone	Effect on Blood Pressure
CARDIAC OUTPUT		
Increased heart rate and force of contraction	Norepinephrine Epinephrine	Increase
SYSTEMIC VASCULAR RESISTANCE		
Vasoconstriction	Angiotensin II Antidiuretic hormone (vasopressin) Norepinephrine[a] Epinephrine[a] Calcitriol (active form of vitamin D)	Increase
Vasodilation	Atrial natriuretic peptide Epinephrine[b] Parathyroid hormone	Decrease
BLOOD VOLUME		
Blood volume increase	Aldosterone Antidiuretic hormone	Increase
Blood volume decrease	Atrial natriuretic peptide	Decrease

[a] Acts at α receptors in arterioles of abdomen and skin.
[b] Acts at β receptors in arterioles of cardiac and skeletal muscle; norepinephrine has a much smaller vasodilating effect.

Autoregulation of Blood Pressure

Autoregulation refers to a local, automatic adjustment of blood flow in a given region of the body to match the particular needs of the tissue. In most body tissues, O_2 is the principal, though not direct, stimulus for autoregulation. Autoregulation is important in meeting the O_2 and nutritional demands of active tissues, such as heart and muscle tissue, where the demand might increase as much as 10-fold. It also is the major regulator of regional brain blood flow. Total blood flow to the brain remains almost constant, independent of level of exercise, but distribution to various parts of the brain changes dramatically, depending on your mental and physical activities. For example, blood flow increases to the motor speech areas when you talk, whereas it increases to the auditory areas when you listen. There are two general types of stimuli that cause autoregulatory changes in blood flow—physical and chemical.

1. **Physical changes.** Warming promotes vasodilation whereas cooling causes vasoconstriction. Smooth mus-

cle in arteriole walls exhibits a **myogenic response**; that is, it contracts more forcefully when it is stretched and relaxes when stretching is less. In an arteriole, the amount it is stretched depends on its blood flow. If blood flow decreases, stretch decreases, the smooth muscle relaxes, and vasodilation occurs. With vasodilation, blood flow increases.

2. **Chemical mediators.** Cells in the blood, such as white blood cells and platelets, and cells near blood vessels, including smooth muscle fibers, macrophages, and endothelial cells, synthesize and release a wide variety of **vasoactive factors**. These are chemicals that alter blood vessel diameter. One of the most important is **endothelium-derived relaxation factor** (**EDRF**), now known to be **nitric oxide**. Other vasodilators, which widen blood vessels, include certain ions (K^+ and H^+), metabolic products such as lactic acid (lactate), and adenosine (from ATP). Vasoconstrictors, which narrow blood vessels, include certain eicosanoids such as thromboxane A_2 and prostaglandin $F_{2\alpha}$, superoxide radicals, angiotensins, and endothelins. Once released, vasodilators produce a local dilation of arterioles and relaxation of precapillary sphincters. The result is an increased flow of blood into the tissue, which restores O_2 level to normal. Vasoconstrictors have opposite effects. Stimuli that promote release of vasoactive factors include changes in tissue CO_2 and O_2 levels, mechanical stretch of the tissue, hormones in the blood, and local hormones (autocrines and paracrines).

An important difference between pulmonary and systemic circulations is their autoregulatory response to changes in O_2 level. In systemic circulation, blood vessels *dilate* in response to low O_2 concentration. In pulmonary circulation, blood vessels *constrict* in response to low levels of O_2. This mechanism is very important in distributing blood to areas of the lungs where it can pick up the most O_2. For example, if some alveoli (air sacs) are not well ventilated by fresh air, the blood vessels in the affected area constrict and blood will largely bypass the poorly functioning areas. As a result, most of the blood flows to other areas of the lung that are well ventilated and have plenty of O_2.

SYNCOPE

Syncope (SIN-kō-pē), or faint, refers to a sudden, temporary loss of consciousness followed by spontaneous recovery. It is most commonly due to cerebral ischemia (lack of sufficient blood flow) and may be preceded by uneasiness, malaise, light-headedness, nausea, vertigo, confusion, disturbances in vision, weakness, sweating, or tinnitus. Among the common causes of syncope are sudden emotional stress or real, threatened, or fantasized injury (vasodepressor syncope); pressure stress associated with urination, defecation, or severe coughing (situational syncope); drugs such as antihypertensives, diuretics, vasodilators, and tranquilizers (drug-induced syncope); an excessive decrease in blood pressure that occurs upon standing up (orthostatic hypotension); situations that stretch the carotid sinus, such as hyperextension of the head, tight collars, or carrying shoulder loads (carotid sinus syncope); and reduced cardiac output. ■

SHOCK AND HOMEOSTASIS

Shock is an inadequate cardiac output that results in a failure of the cardiovascular system to deliver enough O$_2$ and nutrients to meet the metabolic needs of body cells. As a result, cellular membranes dysfunction, cellular metabolism is abnormal, and, without proper treatment, cellular death may eventually occur.

Signs and Symptoms of Shock

The signs and symptoms of shock vary with the severity of the condition, including the following:

1. Clammy, cool, pale skin due to vasoconstriction of skin blood vessels.
2. Tachycardia due to sympathetic stimulation and increased levels of epinephrine.
3. Weak, rapid pulse due to generalized vasodilation and reduced cardiac output.
4. Sweating due to sympathetic stimulation.
5. Hypotension in which the systolic blood pressure is lower than 90 mm Hg as a result of generalized vasodilation and decreased cardiac output.
6. Altered mental status due to cerebral ischemia.
7. Reduced urine formation due to hypotension and increased levels of aldosterone and antidiuretic hormone (ADH).
8. Thirst due to loss of extracellular fluid.
9. Acidosis due to buildup of lactic acid.
10. Nausea due to impaired circulation to the digestive system.

Stages of Shock

The causes of shock are many and varied, but all are characterized by inadequate perfusion (blood flow) to sustain body tissues. One type of shock, called hypovolemic (hī-pō-vō-LĒ-mik) **shock**, refers to decreased blood volume resulting from loss of blood or plasma. Situations that may lead to hypovolemic shock are *acute hemorrhage* due to trauma, gastrointestinal bleeding, and hematomas; and *excessive fluid loss* as a result of excess vomiting, diarrhea, sweating, dehydration, excessive urine production, and burns. Since hypovolemic shock has been studied so extensively, we will describe the stages of shock as they apply to this type of shock. The development of shock occurs in three principal stages, which merge with one another.

Stage I: Compensated (Nonprogressive) Shock

During stage I, when symptoms and signs are minimal, certain homeostatic mechanisms of the cardiovascular system compensate for the shock so that no serious damage results. If the initiating cause does not get any worse, a full recovery follows. The major mechanisms of compensation are negative feedback systems that attempt to return cardiac output (CO) and arterial blood pressure to normal. These compensatory adjustments are mediated through the sympathetic nervous system and the release of various substances. In an otherwise healthy person, acute loss of as much as 10% of the total blood volume can be dealt with by compensatory mechanisms. Among the adjustments are the following:

1. **Activation of the sympathetic division of the ANS.** A decrease in blood pressure, detected by baroreceptors, quickly initiates powerful sympathetic responses throughout most of the body. One result is marked vasoconstriction of arterioles and veins of the skin (which produces coolness and paleness), kidneys, and other abdominal viscera. (Vasoconstriction does not occur in the brain or heart.) The vasoconstriction increases systemic vascular resistance (SVR), which helps maintain adequate venous return. Sympathetic stimulation also increases heart rate and force of contraction and increases secretion of epinephrine and norepinephrine (NE) by the adrenal medulla. These hormones intensify vasoconstriction and increase heart rate and contractility, which all help raise blood pressure (Fig. 21.15).
2. **Renin–angiotensin pathway.** Decreased blood flow to the kidneys causes the kidneys to secrete renin and initiates the renin–angiotensin pathway (see Fig. 18.20). Recall that angiotensin II is a potent vasoconstrictor and also stimulates the adrenal cortex to secrete aldosterone—a hormone that increases reabsorption of Na$^+$ and, indirectly, water by the kidneys. The increase in systemic vascular resistance and blood volume help raise blood pressure (Fig. 21.15).
3. **Antidiuretic hormone (vasopressin).** In response to decreased blood pressure, the posterior pituitary gland releases more antidiuretic hormone (ADH). This hormone brings about water conservation by the kidneys and vasoconstriction (Fig. 21.15).
4. **Hypoxia.** In response to hypoxia (lowered O$_2$ availability), affected cells liberate vasodilators that dilate arterioles and relax precapillary sphincters. This increases regional blood flow and restores O$_2$ level to normal. However, vasodilation also has the potentially harmful effect of decreasing systemic vascular resistance and thus lowering blood pressure.

The various compensatory mechanisms may require from 30 seconds to 48 hours. Eventually, recovery occurs, provided the shock does not intensify and enter the second stage.

Stage II: Decompensated (Progressive) Shock

If blood volume drops more than 15–25%, the shock becomes steadily worse as compensatory mechanisms are no longer able to maintain adequate perfusion. This is stage II of shock. As the cardiovascular system progressively

Figure 21.15   Negative feedback systems that elevate blood pressure during the compensated (nonprogressive) stage of hypovolemic shock.

🔑 *During the compensated stage of shock, homeostatic mechanisms prevent serious damage to body tissues.*

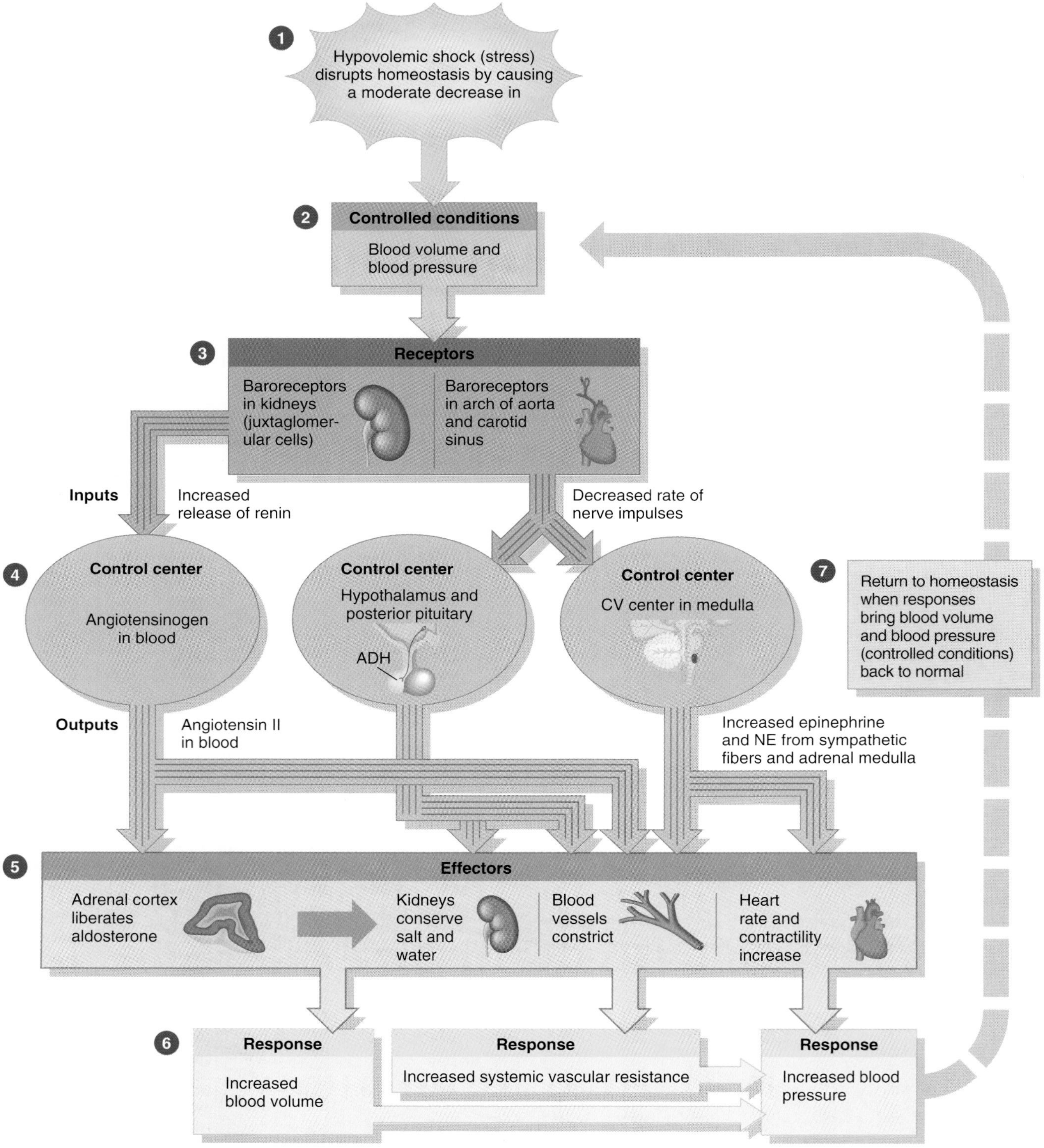

1 Hypovolemic shock (stress) disrupts homeostasis by causing a moderate decrease in

2 **Controlled conditions**
Blood volume and blood pressure

3 **Receptors**
Baroreceptors in kidneys (juxtaglomerular cells)
Baroreceptors in arch of aorta and carotid sinus

Inputs Increased release of renin

Decreased rate of nerve impulses

4 **Control center**
Angiotensinogen in blood

Control center
Hypothalamus and posterior pituitary
ADH

Control center
CV center in medulla

7 Return to homeostasis when responses bring blood volume and blood pressure (controlled conditions) back to normal

Outputs Angiotensin II in blood

Increased epinephrine and NE from sympathetic fibers and adrenal medulla

5 **Effectors**
Adrenal cortex liberates aldosterone
Kidneys conserve salt and water
Blood vessels constrict
Heart rate and contractility increase

6 **Response**
Increased blood volume

Response
Increased systemic vascular resistance

Response
Increased blood pressure

❓ If blood pressure is almost normal in a person who has lost blood, does that mean that his tissues are receiving adequate perfusion (blood flow)?

deteriorates, cardiac output falls dramatically. Several positive feedback cycles characterize stage II. They lead to further reductions in blood pressure and cardiac output and even more cardiovascular deterioration as follows:

1. **Depression of cardiac activity.** When mean blood pressure falls below 60 mm Hg, the pressure is no longer adequate to force blood through coronary arteries and the myocardium becomes ischemic. This weakens the heart muscle, decreases cardiac output further, and depresses blood pressure even more. The decreased cardiac output and blood pressure produce additional ischemia and even more severe depression of cardiac output and blood pressure.
2. **Depression of vasoconstriction.** Decreased blood pressure in the vasomotor center depresses its activity. Progressively less activity of the center results in lower blood pressure due to generalized vasodilation and even more depression of the center. This positive feedback cycle occurs when mean blood pressure falls below 40–50 mm Hg.
3. **Increased permeability of capillaries.** In very late stages of prolonged shock, hypoxia causes an increase in blood capillary permeability. As more blood plasma components move from the capillaries into tissue spaces, blood volume decreases. This decreases cardiac output, the reduced cardiac output intensifies the hypoxia, and the cycle leads to progressively worsening shock.
4. **Intravascular clotting.** With decreased cardiac output, blood velocity slows. Sluggish circulation increases the risk of platelet aggregation and blood clot formation. The resulting obstructions further reduce cardiac output.
5. **Cellular destruction.** In response to shock, cellular destruction occurs throughout the body, including the heart, which pumps less effectively as a consequence. The cellular changes include lysosomal rupture, depressed mitochondrial activity, diminished active transport, and decreased metabolism.
6. **Acidosis.** As a result of metabolic dysfunction, cells produce excess lactic acid, which results in acidosis, a condition in which the pH of blood drops below 7.35. (Recall that the pH of blood is normally 7.35–7.45.) Acidosis depresses the central nervous system, including the vasomotor center in the medulla oblongata.

To reverse the changes that occur during the decompensated stage, immediate medical intervention is required. If this fails, shock progresses to a third stage.

Stage III: Irreversible Shock

In stage III, there is rapid deterioration of the cardiovascular system that cannot be helped by compensatory mechanisms or medical intervention. As the shock cycle perpetuates itself, there are life-threatening reductions in cardiac output, blood pressure, and tissue perfusion. Ultimately, ATP reserves are depleted, especially in the heart and liver, and the heart deteriorates so much that it can no longer pump blood.

CHECKING CIRCULATION

Pulse

The alternate expansion and recoil of elastic arteries after each systole of the left ventricle create a traveling pressure wave that is called the **pulse**. Pulse is strongest in the arteries closest to the heart. It becomes weaker in the arterioles and disappears altogether in the capillaries. The pulse may be felt in any artery that lies near the surface of the body and over a bone or other firm tissue. The radial artery at the wrist is most commonly used to feel the pulse (see Fig. 21.20a). Others include the:

1. **Temporal artery,** lateral to the orbit of the eye.
2. **Facial artery,** at the mandible (lower jawbone) on a line with the corners of the mouth.
3. **Common carotid artery,** lateral to the larynx (voice box). See Fig. 21.18a,b.
4. **Brachial artery,** along the medial side of the biceps brachii muscle (see Fig. 21.20a).
5. **Femoral artery,** inferior to the inguinal ligament (see Fig. 21.22).
6. **Popliteal artery,** posterior to the knee (see Fig. 21.22).
7. **Posterior tibial artery,** posterior to the medial malleolus of the tibia (see Fig. 21.22).
8. **Dorsalis pedis artery,** superior to the instep of the foot (see Fig. 21.22a).

The pulse rate normally is the same as the heart rate. Resting pulse rate in a normal person is between 70 and 80 beats per minute. The term **tachycardia** (tak′-i-KAR-dē-a; *tachy* = fast) means a rapid resting heart or pulse rate (over 100/min). The term **bradycardia** (brād′-i-KAR-dē-a; *brady* = slow) indicates a slow resting heart or pulse rate (under 60/min).

Other characteristics of the pulse may give additional information about circulation. For example, the intervals between beats should be equal in length. If a pulse is missed at intervals, the pulse is said to be irregular. Also, each pulse beat should be of equal strength. Irregularities in strength may indicate a lack of smooth muscle tone in the arteries or varying force of cardiac contraction.

Measurement of Blood Pressure (BP)

Blood pressure is usually measured in the left brachial artery using a **sphygmomanometer** (sfig′-mō-ma-NOM-e-ter; *sphygmo* = pulse). A commonly used sphygmomanometer consists of a rubber cuff attached by a rubber tube to a compressible hand pump or bulb for inflating the cuff. Another tube attaches to the cuff and to a column of mercury or pressure dial marked off in millimeters of mercury (mm Hg) to measure the pressure. The cuff is wrapped around the arm over the brachial artery and inflated, which creates a pressure on the artery. Inflation is continued until the pressure in the cuff exceeds the pressure in the artery. At this point, the walls of the brachial artery are compressed tightly against each other, and no blood can flow through. Two signs can confirm that the artery is occluded. First, if a stethoscope is placed over the artery below the cuff, no sounds can be heard. Second, no pulse can be felt by placing the fingers over the radial artery at the wrist.

Then the cuff is deflated gradually until the pressure in the cuff is slightly less than the maximal pressure in the brachial artery. At this point, the artery opens, a spurt of blood passes through, and a sound may be heard through the stethoscope due to the turbulence of the blood flow. When this first sound is heard, a reading on the mercury column is made. This sound corresponds to **systolic blood pressure (SBP)**—the highest force with which blood pushes against arterial walls as a result of ventricular contraction (Fig. 21.16). As cuff pressure is further reduced, the sounds suddenly become faint as the blood turbulence reduces significantly. The pressure recorded on the mercury column when the sounds suddenly become faint is called **diastolic blood pressure (DBP)**. It corresponds to the lowest force of blood in arteries during ventricular relaxation. Whereas systolic pressure reflects the force of left ventricular contraction, diastolic pressure provides information about systemic vascular resistance. Below diastolic blood pressure, the sounds disappear altogether. The various sounds that are heard while taking blood pressure are called **Korotkoff** (kō-ROT-kof) **sounds**.

Although some people may have a lower or higher blood pressure, the normal blood pressure of a young adult male is about 120 mm Hg systolic and 80 mm Hg diastolic, reported as "120 over 80" and written as 120/80. In young adult females, the pressures are 8–10 mm Hg less. People who exercise regularly and are in good physical condition also tend to have lower blood pressures. Thus blood pressure slightly lower than 120/80 may be a sign of good health and fitness.

Figure 21.16   Relationship of blood pressure changes to cuff pressure.

🔑 *As the cuff is deflated, the first sound occurs at the systolic blood pressure; the sounds suddenly become faint at the diastolic blood pressure.*

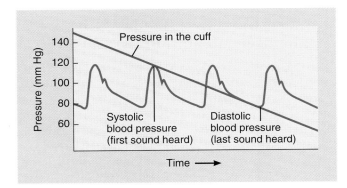

Ⓠ If a blood pressure is reported as "142 over 95," what are the diastolic, systolic, and pulse pressures? Does this person have hypertension (see page 665)?

The difference between systolic and diastolic pressure is called **pulse pressure** (**PP**). This pressure, which averages 40 mm Hg, provides information about the condition of the cardiovascular system. For example, conditions such as atherosclerosis and patent (open) ductus arteriosus greatly increase pulse pressure. The normal ratio of systolic pressure to diastolic pressure to pulse pressure is about 3:2:1.

CIRCULATORY ROUTES

Arteries, arterioles, capillaries, venules, and veins are organized into routes that deliver blood throughout the body. We can now look at the basic routes the blood takes as it is transported through its vessels.

Figure 21.17 shows the **circulatory routes** for blood flow. The two basic postnatal (after birth) routes are systemic and pulmonary. The **systemic circulation** includes all the arteries and arterioles that carry oxygenated blood from the left ventricle to systemic capillaries plus the veins and venules that carry deoxygenated blood returning to the right atrium after flowing through body organs. The nutrient arteries to the lungs, such as the bronchial arteries, also are part of the systemic circulation. As blood leaves the aorta and flows through the systemic circulation, its pressure falls progressively to 0 mm Hg by the time it reaches the right ventricle (see Fig. 21.9).

Blood leaving the aorta and flowing through the systemic arteries is a bright red color. As it moves through capillaries, it loses some of its oxygen and picks up carbon dioxide, so that blood in systemic veins is a dark red color. When blood returns to the heart from the systemic route, it is pumped out of the right ventricle through the **pulmonary circulation** to the lungs (shown in Fig. 21.29). In pulmonary capillaries of the alveoli of the lungs, it loses some of its CO_2 and takes on O_2. Bright red again, it returns to the left atrium of the heart and reenters the systemic circulation as it is ejected by the left ventricle.

Another major route—**fetal circulation**—exists only in the fetus and contains special structures that allow the developing fetus to exchange materials with its mother (shown in Fig. 21.30).

Systemic Circulation

The systemic circulation carries O_2 and nutrients to body tissues and removes CO_2 and other wastes and heat from the tissues. All systemic arteries branch from the aorta. The portion of the aorta that passes superiorly and then posteriorly to the pulmonary trunk as it emerges from the left ventricle is called the **ascending aorta**. It gives off two coronary artery branches to the myocardium of the heart. Then it turns to the left, forming the **arch of the aorta**, which descends to the level of the fourth thoracic vertebra. The **descending aorta** begins at this point. It lies close to the

Figure 21.17 Circulatory routes. Heavy black arrows indicate systemic circulation, thin black arrows pulmonary circulation, and thin red arrows hepatic portal circulation. Refer to Fig. 20.7 for details of coronary circulation and to Fig. 21.30 for details of fetal circulation.

 Blood vessels are organized into various routes that deliver blood to tissues of the body.

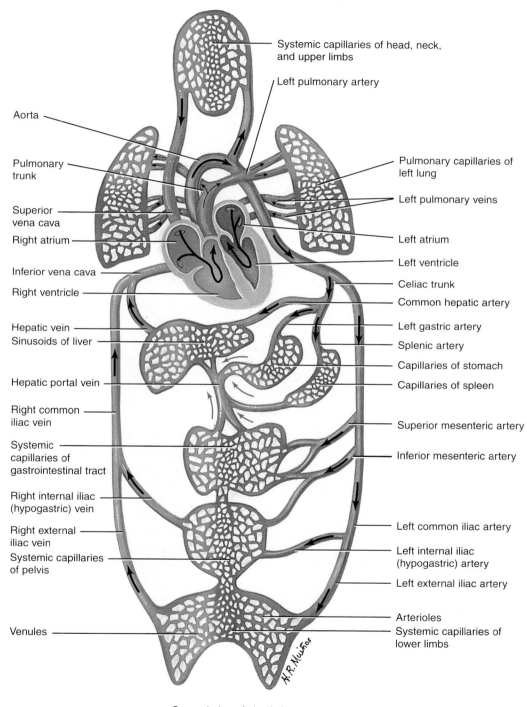

General plan of circulation

 What are the two principal postnatal circulatory routes?

vertebral bodies, passes through the diaphragm, and divides at the level of the fourth lumbar vertebra into two **common iliac arteries**, which carry blood to the lower limbs. The section of the descending aorta between the arch of the aorta and the diaphragm is called the **thoracic aorta**. The section between the diaphragm and the common iliac arteries is termed the **abdominal aorta**. Each section of the aorta gives off arteries that continue to branch into distributing arteries leading to organs and finally into the arterioles and

capillaries that service the systemic tissues (all tissues except the alveoli of the lungs).

Deoxygenated blood returns to the heart through the systemic veins. All the veins of the systemic circulation drain into the **superior vena cava**, **inferior vena cava**, or **coronary sinus**, which, in turn, empty into the right atrium. The principal arteries and veins of systemic circulation are described and illustrated in Exhibits 21.2–21.13 and Figs. 21.18–21.27.

EXHIBIT **21.2** AORTA AND ITS BRANCHES (FIG. 21.18)

Overview: The **aorta** is the largest artery of the body and has a diameter of 2–3 cm (about 1 in.). It begins at the left ventricle and contains a valve at its origin, called the aortic semilunar valve (see Fig. 20.4d), which prevents backflow of blood into the left ventricle during its diastole (relaxation). The principal divisions of the aorta are the ascending aorta, arch of the aorta, thoracic aorta, and abdominal aorta.

Division of Aorta	Arterial Branch	Region Supplied
Ascending aorta (ā-OR-ta)	Right and left coronary	Heart.
Arch of aorta	Brachiocephalic (brā′-kē-ō-se-FAL-ik) trunk → Right common carotid (ka-ROT-id)	Right side of head and neck.
	→ Right subclavian (sub-KLĀ-vē-an)	Right upper limb.
	Left common carotid	Left side of head and neck.
	Left subclavian	Left upper limb.
Thoracic (thō-RAS-ik) **aorta**	Intercostals (in′-ter-KOS-tals)	Intercostal and chest muscles and pleurae.
	Superior phrenics (FREN-iks)	Posterior and superior surfaces of diaphragm.
	Bronchials (BRONG-kē-als)	Bronchi of lungs.
	Esophageals (e-sof′a-JE-als)	Esophagus.
Abdominal (ab-DOM-i-nal) **aorta**	Inferior phrenics (FREN-iks)	Inferior surface of diaphragm.
	Celiac → Common hepatic (he-PAT-ik)	Liver.
	→ Left gastric (GAS-trik)	Stomach and esophagus.
	→ Splenic (SPLEN-ik)	Spleen, pancreas, and stomach.
	Superior mesenteric (MES-en-ter′-ik)	Small intestine, cecum, ascending and transverse colons, and pancreas.
	Suprarenals (soo′-pra-RE-nals)	Adrenal (suprarenal) glands.
	Renals (RE-nals)	Kidneys.
	Gonadals (gō-NAD-als) → Testiculars (tes-TIK-yoo-lars)	Testes.
	or → Ovarians (ō-VA-rē-ans)	Ovaries.
	Inferior mesenteric (MES-en-ter′-ik)	Transverse, descending, and sigmoid colons and rectum.
	Common iliacs (IL-ē-aks) → External iliacs	Lower limbs.
	→ Internal iliacs (hypogastrics)	Uterus, prostate gland, muscles of buttocks, and urinary bladder.

Figure 21.18 Aorta and its principal branches.

The aorta is the largest artery in the body.

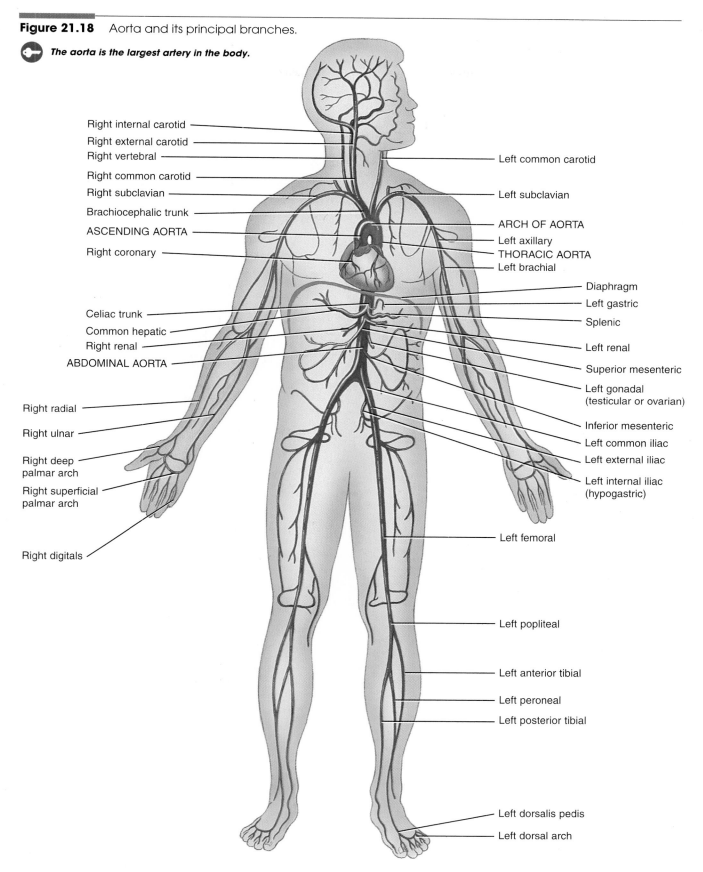

Right internal carotid

Right external carotid

Right vertebral

Right common carotid

Right subclavian

Brachiocephalic trunk

ASCENDING AORTA

Right coronary

Celiac trunk

Common hepatic

Right renal

ABDOMINAL AORTA

Right radial

Right ulnar

Right deep palmar arch

Right superficial palmar arch

Right digitals

Left common carotid

Left subclavian

ARCH OF AORTA

Left axillary

THORACIC AORTA

Left brachial

Diaphragm

Left gastric

Splenic

Left renal

Superior mesenteric

Left gonadal (testicular or ovarian)

Inferior mesenteric

Left common iliac

Left external iliac

Left internal iliac (hypogastric)

Left femoral

Left popliteal

Left anterior tibial

Left peroneal

Left posterior tibial

Left dorsalis pedis

Left dorsal arch

(a) Overall anterior view

Figure continues

Figure 21.18   (continued)

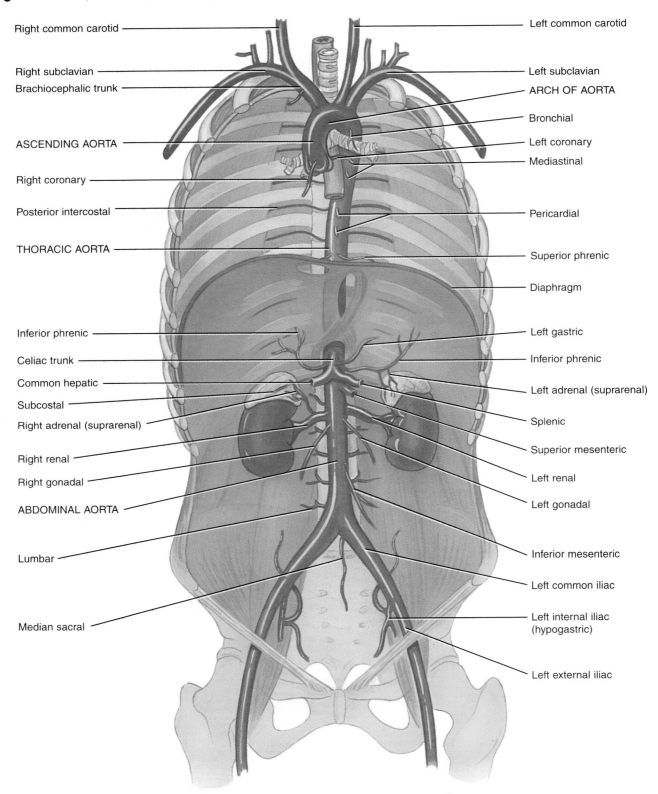

Right common carotid

Right subclavian

Brachiocephalic trunk

ASCENDING AORTA

Right coronary

Posterior intercostal

THORACIC AORTA

Inferior phrenic

Celiac trunk

Common hepatic

Subcostal

Right adrenal (suprarenal)

Right renal

Right gonadal

ABDOMINAL AORTA

Lumbar

Median sacral

Left common carotid

Left subclavian

ARCH OF AORTA

Bronchial

Left coronary

Mediastinal

Pericardial

Superior phrenic

Diaphragm

Left gastric

Inferior phrenic

Left adrenal (suprarenal)

Splenic

Superior mesenteric

Left renal

Left gonadal

Inferior mesenteric

Left common iliac

Left internal iliac (hypogastric)

Left external iliac

(b) Detailed anterior view

Q   What are the four principal subdivisions of the aorta?

EXHIBIT **21.3** **ASCENDING AORTA (FIG. 21.19)**

Overview: The **ascending aorta** is the first division of the aorta, about 5 cm (2 in.) in length. It is directed superiorly, slightly anteriorly, and to the right and ends at the level of the sternal angle, where it becomes the arch of the aorta. The beginning of the ascending aorta is hidden by the pulmonary trunk and right auricle; the right pulmonary artery is posterior to it. At its origin, the ascending aorta contains three dilations, called aortic sinuses. Two of these, the right and left sinuses, give rise to the right and left coronary arteries, respectively.

Branch	Description and Region Supplied
Coronary (KOR-ō-nar-ē) **arteries**	The right and left branches arise from the ascending aorta just superior to the aortic semilunar valve. They form a crown-like ring around the heart, giving off branches to the atrial and ventricular myocardium. The **posterior interventricular** (in'-ter-ven-TRIK-yoo-lar) **branch** of the right coronary artery supplies both ventricles and the **marginal branch** supplies the right ventricle. The **anterior interventricular branch (left anterior descending)** of the left coronary artery supplies both ventricles and the **circumflex branch** supplies the left atrium and left ventricle.

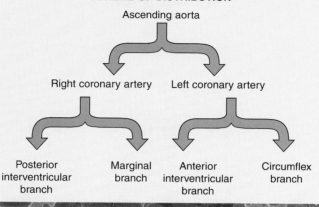

SCHEME OF DISTRIBUTION

Ascending aorta

Right coronary artery → Left coronary artery

Posterior interventricular branch — Marginal branch — Anterior interventricular branch — Circumflex branch

---

Figure 21.19 Ascending aorta and its branches.

🔑 *The ascending aorta is the first division of the aorta.*

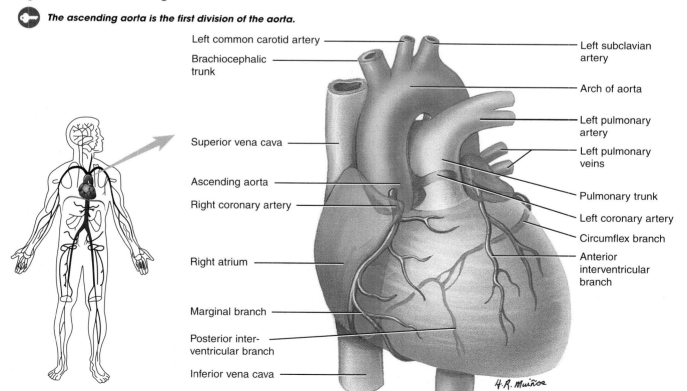

Left common carotid artery

Brachiocephalic trunk

Superior vena cava

Ascending aorta

Right coronary artery

Right atrium

Marginal branch

Posterior inter-ventricular branch

Inferior vena cava

Left subclavian artery

Arch of aorta

Left pulmonary artery

Left pulmonary veins

Pulmonary trunk

Left coronary artery

Circumflex branch

Anterior interventricular branch

H. R. Muiños

❓ Which arteries arise from the ascending aorta?

Anterior view

EXHIBIT **21.4 ARCH OF THE AORTA (FIG. 21.20)**

Overview: The **arch of the aorta** is about $4\frac{1}{2}$ cm (almost 2 in.) in length and is the continuation of the ascending aorta. It emerges from the pericardium posterior to the sternum at the level of the sternal angle. Initially, the arch is directed superiorly, posteriorly, and to the left, and then inferiorly on the left side of the body of the fourth thoracic vertebra. Actually, the arch is directed not only from right to left, but from anterior to posterior as well. The arch of the aorta ends at the level of the intervertebral disc between the fourth and fifth thoracic vertebrae, where it becomes the thoracic aorta. The thymus gland lies anterior to the arch of the aorta, while the trachea lies posterior to it.

Three major arteries branch from the arch of the aorta. In order of their origination, they are the brachiocephalic trunk, left common carotid artery, and left subclavian artery.

Branch	Description and Region Supplied
Brachiocephalic (brā′-kē-ō-se-FAL-ik)	The **brachiocephalic trunk**, which is found only on the right side, is the first and largest branch off the arch of the aorta. (There is no left brachiocephalic artery.) It bifurcates (divides) to form the right subclavian artery and right common carotid artery. The **right subclavian** (sub-KLĀ-vē-an) **artery** extends from the brachiocephalic to the first rib and then passes into the armpit (axilla) and supplies the arm, forearm, and hand. Continuation of the right subclavian into the axilla is called the **axillary** (AK-si-ler′-ē) **artery**.[a] From here, it continues into the arm as the **brachial** (BRĀ-kē-al) **artery**. At the bend of the elbow, the brachial artery divides into the medial **ulnar** (UL-nar) and lateral **radial** (RĀ-dē-al) **arteries**. These vessels pass inferiorly to the palm, one on each side of forearm. In the palm, branches of the two arteries anastomose to form two palmar arches—the **superficial palmar** (PAL-mar) **arch** and the **deep palmar arch**. From these arches arise the **digital** (DIJ-i-tal) **arteries**, which supply the fingers and thumb.
	Before passing into the axilla, the right subclavian gives off a major branch to the brain called the **right vertebral** (VER-te-bral) **artery**. The right vertebral artery passes through the foramina of transverse processes of the cervical vertebrae and enters the skull through the foramen magnum to reach the inferior surface of the brain. Here it unites with the left vertebral artery to form the **basilar** (BAS-i-lar) **artery**.
	The **right common carotid artery** passes superiorly in the neck. At the superior border of the larynx, it divides into the **right external** and **right internal carotid** (ka-ROT-id) **arteries**. The external carotid supplies the right side of the thyroid gland, tongue, throat, face, ear, scalp, and dura mater. The internal carotid supplies the brain, right eye, and right sides of the forehead and nose.
	Inside the cranium, anastomoses of the left and right internal carotids along with the basilar artery form an arrangement of blood vessels at the base of the brain near the sella turcica called the **cerebral** (se-RĒ-bral) **arterial circle (circle of Willis)**. From this circle arise arteries supplying most of the brain. Essentially the cerebral arterial circle is formed by union of the **anterior cerebral arteries** (branches of internal carotids) and **posterior cerebral arteries** (branches of basilar artery). The posterior cerebral arteries are connected with the internal carotids by the **posterior communicating** (ko-MYOO-ni-kā′-ting) **arteries**. The anterior cerebral arteries are connected by the **anterior communicating arteries**. The **internal carotid** (ka-ROT-id) **arteries** are also considered part of the cerebral arterial circle. The function of the cerebral arterial circle is to equalize blood pressure to the brain and provide alternate routes for blood to the brain, should the arteries become damaged.

[a] The right subclavian artery, which passes deep to the clavicle, is a good example of the practice of giving the same vessel different names as it passes through different regions.

Branch	Description and Region Supplied
Left common carotid (ka-ROT-id)	The **left common carotid** is the second branch off the arch of the aorta (see Fig. 21.21a). Corresponding to the right common carotid, it divides into basically the same branches with the same names, except that the arteries are now labeled "left" instead of "right."
Left subclavian (sub-KLĀ-vē-an)	The **left subclavian artery** is the third branch off the arch of the aorta (see Fig. 21.21a). It distributes blood to the left vertebral artery and vessels of the left upper limb. Arteries branching from the left subclavian are named like those of the right subclavian.

SCHEME OF DISTRIBUTION

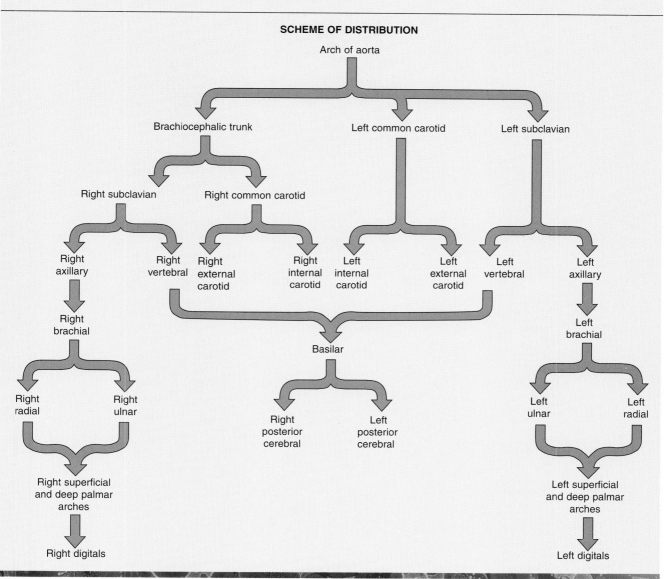

Figure 21.20   Arch of the aorta and its branches. Note in (c) the arteries that comprise the cerebral arterial circle (circle of Willis).

🔑 *The arch of the aorta ends at the level of the intervertebral disc between the fourth and fifth thoracic vertebrae.*

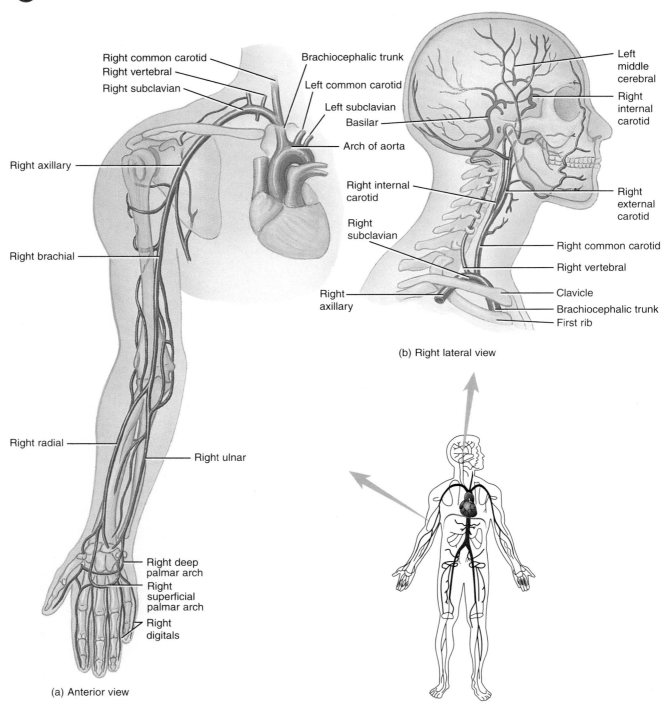

(a) Anterior view

(b) Right lateral view

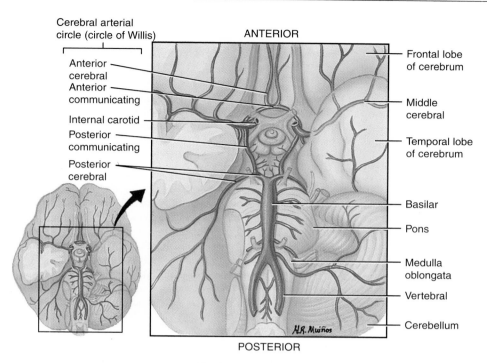

(c) Inferior view of base of brain

Q What are the three major branches of the arch of the aorta in order of their origination?

EXHIBIT **21.5 THORACIC AORTA (FIG. 21.21)**

Overview: The **thoracic aorta** is about 20 cm (8 in.) long and is a continuation of the arch of the aorta. It begins at the level of the intervertebral disc between the fourth and fifth thoracic vertebrae, where it lies to the left of the vertebral column. As it descends, it moves closer to the midline and ends at an opening in the diaphragm (aortic hiatus) anterior to the vertebral column at the level of the intervertebral disc between the twelfth thoracic and first lumbar vertebrae.

Along its course, the thoracic aorta sends off numerous small arteries to viscera (**visceral branches**) and body wall structures (**parietal branches**).

Branch	Description and Region Supplied
VISCERAL	
Pericardial (per′-i-KAR-dē-al)	Several minute **pericardial arteries** supply blood to the posterior aspect of the pericardium.
Bronchial (BRONG-kē-al)	One right and two left **bronchial arteries** supply the bronchial tubes, visceral pleurae, bronchial lymph nodes, and esophagus. (Whereas the right bronchial artery arises from the third posterior intercostal artery, the two left bronchial arteries arise from the thoracic aorta.)
Esophageal (e-sof′-a-JĒ-al)	Four or five **esophageal arteries** supply the esophagus.
Mediastinal (mē′-dē-as-TĪ-nal)	Numerous small **mediastinal arteries** supply blood to structures in the posterior mediastinum.
PARIETAL	
Posterior intercostal (in′-ter-KOS-tal)	Nine pairs of **posterior intercostal arteries** supply the intercostal, pectoral, and abdominal muscles; overlying subcutaneous tissue and skin; mammary glands; and vertebral canal and its contents.
Subcostal (sub-KOS-tal)	The left and right **subcostal arteries** have a distribution similar to that of the posterior intercostals.
Superior phrenic (FREN-ik)	Small **superior phrenic arteries** supply the posterior and superior surfaces of the diaphragm.

EXHIBIT **21.6 ABDOMINAL AORTA (FIG. 21.21)**

Overview: The **abdominal** (ab-DOM-i-nal) **aorta** is the continuation of the thoracic aorta. It begins at the aortic hiatus in the diaphragm and ends at about the level of the fourth lumbar vertebra, where it divides into right and left common iliac arteries. The abdominal aorta lies anterior to the vertebral column.

As with the thoracic aorta, the abdominal aorta gives off visceral and parietal branches. The unpaired visceral branches arise from the anterior surface of the aorta and include the celiac, superior mesenteric, and inferior mesenteric arteries. The paired visceral branches arise from the lateral surfaces of the aorta and include the suprarenal, renal, and gonadal arteries. The paired parietal branches arise from the posterolateral surfaces of the aorta and include the inferior phrenic and lumbar arteries. The unpaired parietal artery is the median sacral.

Branch	Description and Region Supplied
VISCERAL **Celiac** (SĒ-lē-ak)	The **celiac artery** (trunk) is the first visceral aortic branch inferior to the diaphragm. It has three branches: (1) **common hepatic** (he-PAT-ik) **artery,** (2) **left gastric** (GAS-trik) **artery,** and (3) **splenic** (SPLĒN-ik) **artery.** The common hepatic artery has three main branches: (1) **hepatic artery proper,** a continuation of the common hepatic artery, which supplies the liver and gallbladder; (2) **right gastric artery,** which supplies the stomach and duodenum; and (3) **gastroduodenal** (gas′-trō-doo′-ō-DĒ-nal) **artery,** which supplies the stomach, duodenum, and pancreas. The left gastric artery supplies the stomach, and its **esophageal** (e-sof′-a-JĒ-al) **branch** supplies the esophagus. The splenic artery supplies the spleen and has three main branches: (1) **pancreatic** (pan′-krē-AT-ik) **arteries,** which supply the pancreas; (2) **left gastroepiploic** (gas′-trō-ep′-i-PLO-ik) **artery,** which supplies the stomach and greater omentum; and (3) **short gastric** (GAS-trik) **arteries,** which supply the stomach.
Superior mesenteric (MES-en-ter′-ik)	The **superior mesenteric artery** anastomoses extensively and has several principal branches: (1) **inferior pancreaticoduodenal** (pan′-krē-at′-i-kō-doo′-ō-DĒ-nal) **artery,** which supplies the pancreas and duodenum; (2) **jejunal** (je-JOO-nal) and **ileal** (IL-ē-al) **arteries,** which supply the jejunum and ileum, respectively; (3) **ileocolic** (il′-ē-ō-KŌL-ik) **artery,** which supplies the ileum and ascending colon; (4) **right colic** (KŌL-ik) **artery,** which supplies the ascending colon; and (5) **middle colic artery,** which supplies the transverse colon.
Suprarenals (soo′-pra-RĒ-nals)	Right and left **suprarenal arteries** supply blood to the adrenal (suprarenal) glands. The glands are also supplied by branches of the renal and inferior phrenic arteries.
Renals (RĒ-nals)	Right and left **renal arteries** carry blood to the kidneys and adrenal (suprarenal) glands.
Gonadals (gō-NAD-als) [**testiculars** (tes-TIK-yoo-lars) **or ovarians** (ō-VA-rē-ans)]	Right and left **testicular arteries** extend into the scrotum and terminate in the testes; right and left **ovarian arteries** are distributed to the ovaries.
Inferior mesenteric (MES-en-ter′-ik)	The principal branches of the **inferior mesenteric artery,** which also anastomose, are the (1) **left colic** (KŌL-ik) **artery,** which supplies the transverse and descending colons; (2) **sigmoid** (SIG-moyd) **arteries,** which supply the descending and sigmoid colons; and (3) **superior rectal** (REK-tal) **artery,** which supplies the rectum.
PARIETAL **Inferior phrenics** (FREN-iks)	The **inferior phrenic arteries** are distributed to the inferior surface of the diaphragm and adrenal (suprarenal) glands.
Lumbars (LUM-bars)	The **lumbar arteries** supply the spinal cord and its meninges and the muscles and skin of the lumbar region of the back.
Median sacral (SĀ-kral)	The **median sacral artery** supplies the sacrum, coccyx, and rectum.

Figure 21.21 Thoracic and abdominal aorta and their principal branches.

The thoracic aorta ends at the level of the intervertebral disc between the twelfth thoracic and first lumbar vertebrae.

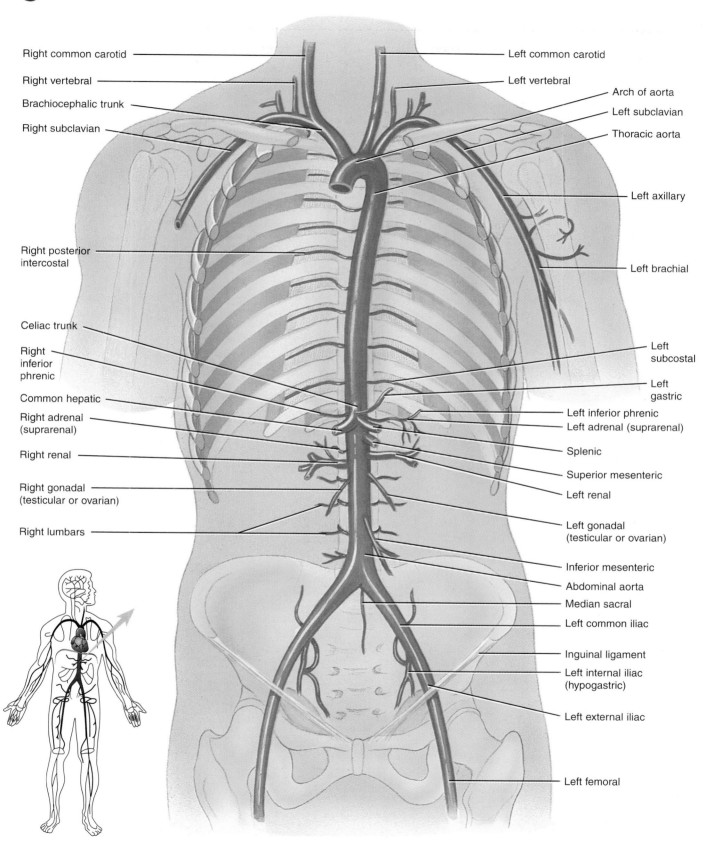

Right common carotid

Right vertebral

Brachiocephalic trunk

Right subclavian

Right posterior intercostal

Celiac trunk

Right inferior phrenic

Common hepatic

Right adrenal (suprarenal)

Right renal

Right gonadal (testicular or ovarian)

Right lumbars

Left common carotid

Left vertebral

Arch of aorta

Left subclavian

Thoracic aorta

Left axillary

Left brachial

Left subcostal

Left gastric

Left inferior phrenic

Left adrenal (suprarenal)

Splenic

Superior mesenteric

Left renal

Left gonadal (testicular or ovarian)

Inferior mesenteric

Abdominal aorta

Median sacral

Left common iliac

Inguinal ligament

Left internal iliac (hypogastric)

Left external iliac

Left femoral

(a) Overview in anterior view

Figure continues

Figure 21.21   (continued)

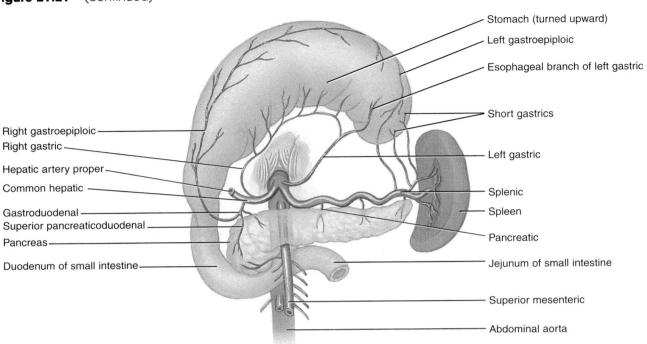

(b) Anterior view of celiac trunk and its branches

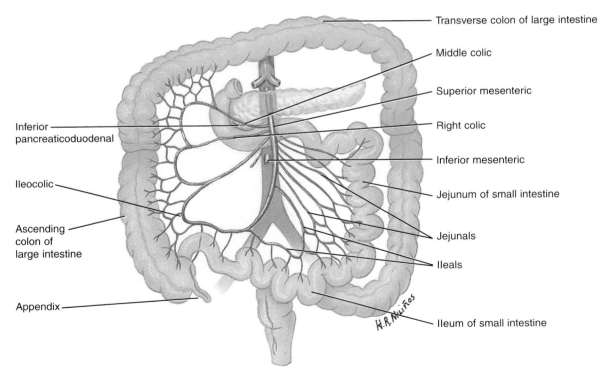

(c) Anterior view of superior mesenteric artery and its branches

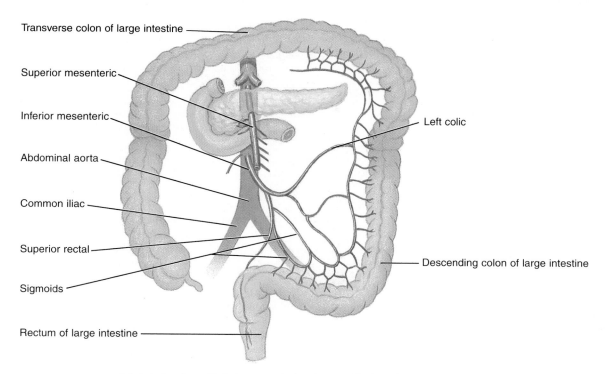

Transverse colon of large intestine

Superior mesenteric

Inferior mesenteric

Abdominal aorta

Common iliac

Superior rectal

Sigmoids

Rectum of large intestine

Left colic

Descending colon of large intestine

(d) Anterior view of inferior mesenteric artery and its branches

Q Where does the thoracic aorta begin?

EXHIBIT 21.7 ARTERIES OF THE PELVIS AND LOWER LIMBS (FIG. 21.22)

Overview: The **internal iliac arteries** enter the pelvic cavity anterior to the sacroiliac joint and supply most of the blood to the pelvic viscera and wall. The **external iliac arteries** travel along the brim of the lesser (true) pelvis. Posterior to the midportion of the inguinal ligament, each external iliac artery enters the thigh, where its name changes to the femoral artery.

Branch	Description and Region Supplied
Common iliacs (IL-ē-aks)	At about the level of the fourth lumbar vertebra, the abdominal aorta divides into the right and left **common iliac arteries**. Each passes inferiorly about 5 cm (2 in.) and gives rise to two branches: internal iliac and external iliac.
Internal iliacs	The **internal iliac (hypogastric) arteries** form branches that supply psoas major, gluteal muscles, quadratus lumborum, medial side of each thigh, urinary bladder, rectum, prostate gland, ductus (vas) deferens, uterus, and vagina.
External iliacs	The **external iliac arteries** diverge through the greater (false) pelvis and enter the thighs to become the right and left **femoral** (FEM-ō-ral) **arteries**. Both femorals send branches superiorly to the genitals and the wall of the abdomen. Other branches run to the muscles of the thigh. The femoral continues down the medial and posterior side of the thigh posterior to the knee joint, where it becomes the **popliteal** (pop′-li-TĒ-al) **artery**. Between the knee and ankle, the popliteal runs down on the posterior aspect of the leg and is called the **posterior tibial** (TIB-ē-al) **artery**. Inferior to the knee, the **peroneal** (per′-ō-NĒ-al) **artery** branches off the posterior tibial to supply structures on the medial side of the fibula and calcaneus. In the calf, the **anterior tibial artery** branches off the popliteal and runs along the anterior surface of the leg. At the ankle, it becomes the **dorsalis pedis** (PED-is) **artery**. At the ankle, the posterior tibial divides into the **medial and lateral plantar** (PLAN-tar) **arteries**. The lateral plantar artery and the dorsalis pedis artery unite to form the **plantar arch**. From this arch, **digital arteries** supply the toes.

SCHEME OF DISTRIBUTION

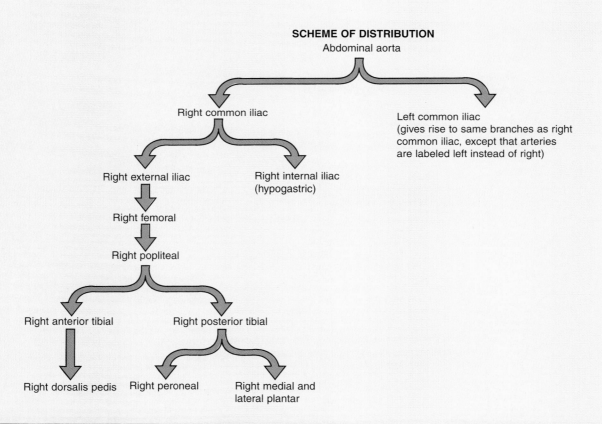

Figure 21.22   Arteries of the pelvis and right lower limb.

 The internal iliac arteries supply most blood to the pelvic viscera and wall.

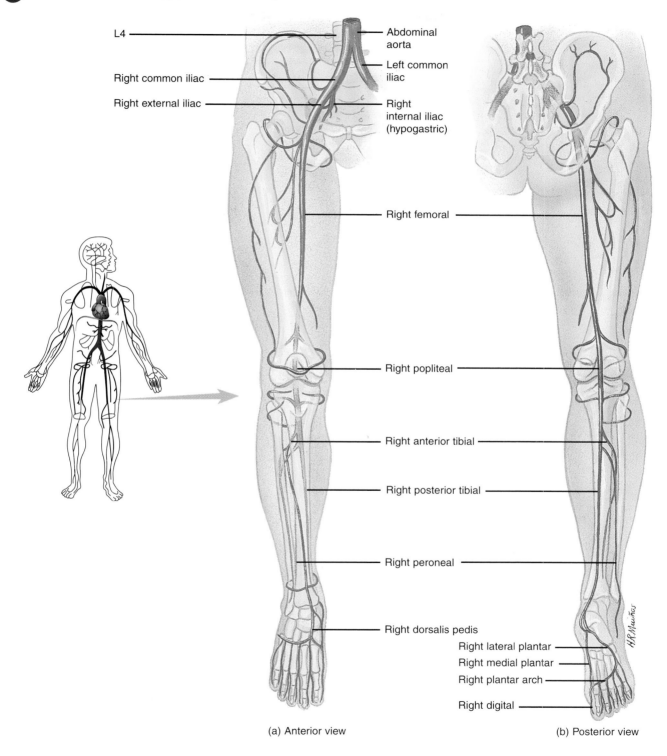

L4

Abdominal aorta

Left common iliac

Right common iliac

Right internal iliac (hypogastric)

Right external iliac

Right femoral

Right popliteal

Right anterior tibial

Right posterior tibial

Right peroneal

Right dorsalis pedis

Right lateral plantar

Right medial plantar

Right plantar arch

Right digital

(a) Anterior view

(b) Posterior view

At what point does the abdominal aorta divide into the common iliac arteries?

EXHIBIT **21.8 VEINS OF SYSTEMIC CIRCULATION (FIG. 21.23)**

Overview: Deoxygenated blood returns to the right atrium from three veins: **coronary sinus, superior vena cava**, and **inferior vena cava**. The coronary sinus receives blood from the cardiac veins; the superior vena cava receives blood from other veins superior to the diaphragm, except the air sacs (alveoli) of the lungs. This includes the head, neck, upper limbs, and thoracic wall. The inferior vena cava receives blood from veins inferior to the diaphragm. This includes the lower limbs, most of the abdominal walls, and abdominal viscera.

Vein	Description and Region Drained
Coronary (KOR-ō-nar-ē) sinus	The **coronary sinus** receives almost all venous blood from the myocardium. It is located in the coronary sulcus (see Fig. 20.4c) and opens into the right atrium between the orifice of the inferior vena cava and the tricuspid valve.
Superior vena cava (VĒ-na CA-va) (SVC)	The **SVC** is about $7^1/_2$ cm (3 in.) long and empties its blood into the superior part of the right atrium. It begins posterior to the right first costal cartilage by the union of the right and left brachiocephalic veins and ends at the level of the right third costal cartilage where it enters the right atrium.
Inferior vena cava (IVC)	The **IVC** is the largest vein in the body, about $3^1/_2$ cm ($1^1/_2$ in.) in diameter. It begins anterior to the fifth lumbar vertebra by the union of the common iliac veins, ascends behind the peritoneum to the right of the midline, pierces the costal tendon of the diaphragm at the level of the eighth thoracic vertebra, and enters the inferior part of the right atrium. The inferior vena cava is commonly compressed during the later stages of pregnancy by the enlarging uterus. This produces edema of the ankles and feet and temporary varicose veins.

Figure 21.23 Principal veins.

 Deoxygenated blood returns to the heart via the superior and inferior venae cavae and the coronary sinus.

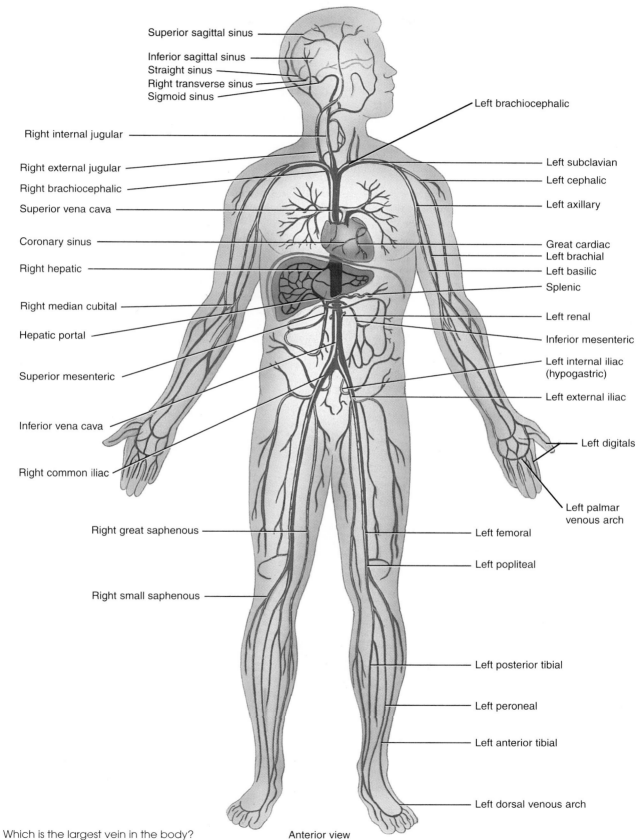

Superior sagittal sinus

Inferior sagittal sinus
Straight sinus
Right transverse sinus
Sigmoid sinus

Left brachiocephalic

Right internal jugular

Right external jugular

Left subclavian

Right brachiocephalic

Left cephalic

Superior vena cava

Left axillary

Coronary sinus

Great cardiac
Left brachial

Right hepatic

Left basilic
Splenic

Right median cubital

Left renal

Hepatic portal

Inferior mesenteric

Superior mesenteric

Left internal iliac
(hypogastric)

Left external iliac

Inferior vena cava

Left digitals

Right common iliac

Left palmar
venous arch

Right great saphenous

Left femoral

Left popliteal

Right small saphenous

Left posterior tibial

Left peroneal

Left anterior tibial

Left dorsal venous arch

Q Which is the largest vein in the body?

Anterior view

EXHIBIT **21.9** **VEINS OF THE HEAD AND NECK (FIG. 21.24)**

Overview: The majority of blood draining from the head passes into three pairs of veins: **internal jugular**, **external jugular**, and **vertebral**. Within the brain, all veins drain into the internal jugular veins.

Vein	Description and Region Drained
Internal jugulars (JUG-yoo-lars)	Right and left **internal jugular veins** receive blood from the face, brain, and neck. They arise as a continuation of the **sigmoid** (SIG-moyd) **sinuses** at the base of the skull. Intracranial vascular sinuses are located between layers of the dura mater and receive blood from the brain. Other sinuses that drain into the internal jugulars include the **superior sagittal** (SAJ-i-tal) **sinus, inferior sagittal sinus**, **straight sinus**, and **transverse (lateral) sinuses**. The internal jugulars descend on either side of the neck and pass behind the clavicles, where they join with the right and left subclavian veins. Unions of the internal jugulars and subclavians form the right and left **brachiocephalic** (brā′-kē-ō-se-FAL-ik) **veins**. From here blood flows into the superior vena cava.
External jugulars	Right and left **external jugular veins** run inferiorly in the neck along the outside of the internal jugulars. They drain blood from the parotid (salivary) glands, facial muscles, scalp, and other superficial structures into the subclavian veins.
Vertebrals (VER-te-brals)	Right and left **vertebral veins** descend through the transverse foramina of the cervical vertebrae and enter the subclavian veins. They drain deep structures of the neck such as the vertebrae and muscles. In cases of heart failure, the venous pressure in the right atrium may rise. In such patients the pressure in the column of blood in the external jugular vein rises so that, even with the patient at rest and sitting in a chair, the external jugular vein will be visibly distended. Temporary distention of the vein is often seen in healthy adults when the intrathoracic pressure is raised during coughing and physical exertion.

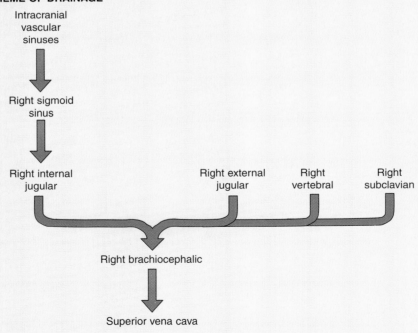

SCHEME OF DRAINAGE

blood via the hepatic artery, a branch of the celiac artery. Ultimately, all blood leaves the liver through the **hepatic veins**, which drain into the inferior vena cava.

The other portal veins in the body, the hypophyseal portal veins that carry blood from the hypothalamus to the anterior pituitary gland, may be reviewed in Fig. 18.5.

Figure 21.28 Hepatic portal circulation. The scheme of blood flow through the liver, including arterial circulation, is shown in (b). Deoxygenated blood is indicated in blue, oxygenated blood in red.

🔑 *The hepatic portal circulation delivers venous blood from the gastrointestinal organs and spleen to the liver.*

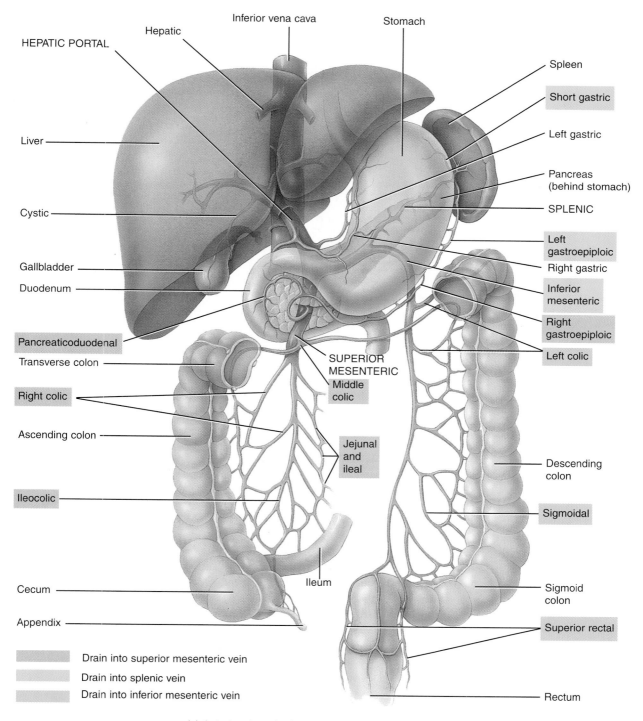

Drain into superior mesenteric vein
Drain into splenic vein
Drain into inferior mesenteric vein

(a) Anterior view of veins draining into the hepatic portal vein

Figure continues

Figure 21.28 (continued)

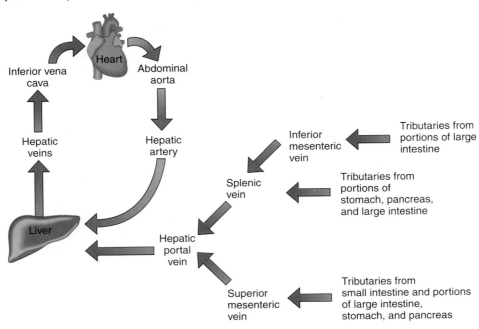

(b) Scheme of principal blood vessels of hepatic portal circulation and arterial supply and venous drainage of liver

 Which veins carry blood away from the liver?

Pulmonary Circulation

The **pulmonary** (*pulmo* = lung) **circulation** carries deoxygenated blood from the right ventricle to the air sacs (alveoli) within the lungs and returns oxygenated blood from the air sacs within the lungs to the left atrium (Fig. 21.29). The **pulmonary trunk** emerges from the right ventricle and passes superiorly, posteriorly, and to the left. It then divides into two branches: the **right pulmonary artery** extends to the right lung; the **left pulmonary artery** goes to the left lung. The pulmonary arteries are the only postnatal arteries that carry deoxygenated blood. On entering the lungs, the branches divide and subdivide until finally they form capillaries around the air sacs within the lungs. CO_2 passes from the blood into the air sacs within the lungs and is exhaled. Inhaled O_2 passes from the air within the lungs into the blood. The pulmonary capillaries unite, form venules, and eventually **pulmonary veins** which exit from the lungs and transport the oxygenated blood to the left atrium. Two left and two right pulmonary veins enter the left atrium. The pulmonary veins are the only postnatal veins that carry oxygenated blood. Contractions of the left ventricle then send the blood into the systemic circulation.

The pulmonary and systemic circulations are different in several ways. For one thing, blood in pulmonary circulation is not pumped so far as in systemic circulation. Also, in comparison to systemic arteries, pulmonary arteries have larger diameters, thinner walls, and less elastic tissue. As a result, the resistance to blood flow is very low. This means that less pressure is needed to move blood through the lungs. The peak systolic pressure achieved by the right ventricle is only about one-fifth that reached by the left ventricle.

Because resistance in the pulmonary circulation is low, normal *pulmonary* capillary hydrostatic pressure, the principal force that moves fluid out of capillaries into interstitial fluid, is only 10 mm Hg. This compares to the average *systemic* capillary pressure of about 25 mm Hg. The relatively low capillary hydrostatic pressure tends to prevent pulmonary edema. If capillary blood pressure increases in the lungs (due to increased left atrial pressure as may occur in mitral valve stenosis) or capillary permeability increases (as may occur from bacterial toxins), however, edema may develop. Pulmonary edema reduces the rate of diffusion of O_2 and CO_2 and thus slows the exchange of these gases in the lungs.

Fetal Circulation

The circulatory system of a fetus, called **fetal circulation**, differs from the postnatal (after birth) circulation because the lungs, kidneys, and gastrointestinal organs begin to function at birth. The fetus obtains its O_2 and nutrients by diffusion from the maternal blood and eliminates its CO_2 and wastes by diffusion into the maternal blood.

The exchange of materials between fetal and maternal circulation occurs through a structure called the **placenta** (pla-SEN-ta). It is attached to the umbilicus (navel) of the fetus by the umbilical (um-BIL-i-kal) cord, and it communicates with the mother through countless small blood vessels that emerge from the uterine wall. The umbilical cord con-

Figure 21.29 Pulmonary circulation.

 The pulmonary circulation brings deoxygenated blood from the right ventricle to the lungs and returns oxygenated blood from the lungs to the left atrium.

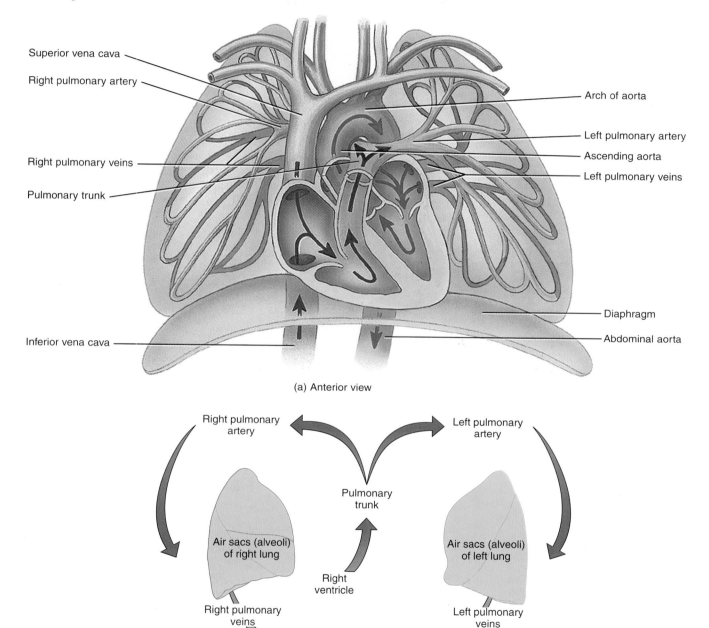

(a) Anterior view

(b) Scheme of pulmonary circulation

Which are the only postnatal arteries that carry deoxygenated blood?

tains blood vessels that branch into capillaries in the placenta. Wastes from the fetal blood diffuse out of the capillaries, into spaces containing maternal blood (intervillous spaces) in the placenta, and finally into the mother's uterine veins (see Fig. 29.7). Nutrients travel the opposite route—

from the maternal blood vessels to the intervillous spaces to the fetal capillaries. Normally, there is no actual mixing of maternal and fetal blood.

Blood passes from the fetus to the placenta via two **umbilical arteries** (Fig. 21.30a,c). These branches of the

Figure 21.30   Fetal circulation and changes at birth. The boxed areas indicate the fate of certain fetal structures once postnatal circulation is established.

The lungs, kidneys, and gastrointestinal organs begin to function at birth.

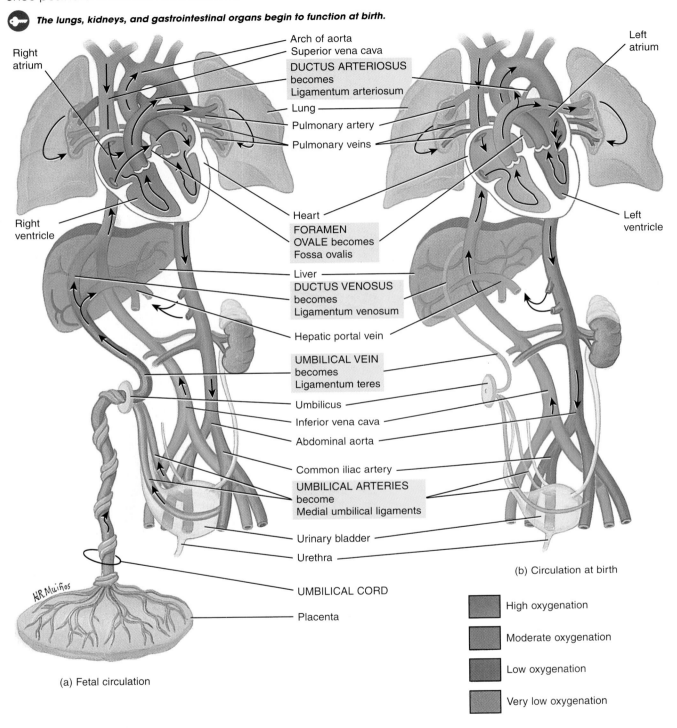

(a) Fetal circulation

(b) Circulation at birth

High oxygenation

Moderate oxygenation

Low oxygenation

Very low oxygenation

internal iliac (hypogastric) arteries are within the umbilical cord. At the placenta, fetal blood picks up O_2 and nutrients and eliminates CO_2 and wastes. The oxygenated blood returns from the placenta via a single **umbilical vein**. This vein ascends to the liver of the fetus, where it divides into two branches. Some blood flows through the branch that joins the hepatic portal vein and enters the liver. Most of the blood flows into the second branch, the **ductus venosus** (DUK-tus ve-NŌ-sus), which drains into the inferior vena cava.

Deoxygenated blood returning from the inferior regions mingles with oxygenated blood from the ductus venosus in

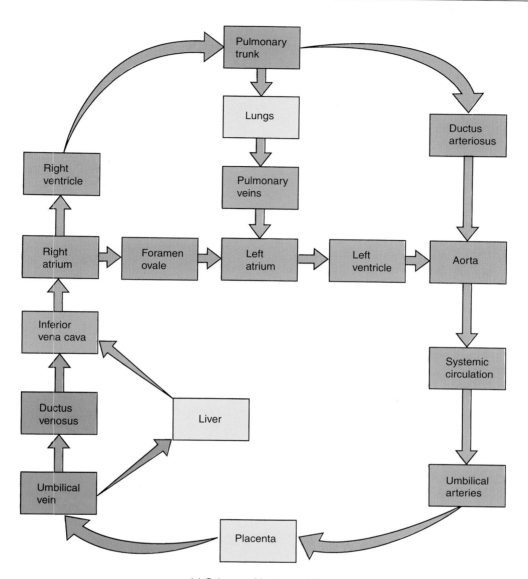

(c) Scheme of fetal circulation

Q Through which structure does the exchange of materials between mother and fetus occur?

the inferior vena cava. This mixed blood then enters the right atrium. Deoxygenated blood returning from the superior regions of the fetus enters the superior vena cava and passes into the right atrium.

Most of the fetal blood does not pass from the right ventricle to the lungs, as it does in postnatal circulation because an opening called the **foramen ovale** (fō-RĀ-men ō-VAL-ē) exists in the septum between the right and left atria. About one third of the blood passes through the foramen ovale directly into the systemic circulation. The blood that does pass into the right ventricle is pumped into the pulmonary trunk, but little of this blood reaches the nonfunctioning fetal lungs. Most is sent through the **ductus arteriosus** (ar-tē-rē-Ō-sus). This vessel connects the pulmonary trunk with the aorta and allows most blood to bypass the fetal

lungs. The blood in the aorta is carried to all parts of the fetus through the systemic circulation. When the common iliac arteries branch into the external and internal iliacs, part of the blood flows into the internal iliacs. It then goes to the umbilical arteries and back to the placenta for another exchange of materials. The only fetal vessel that carries fully oxygenated blood is the umbilical vein.

At birth, when pulmonary (lung), renal, and digestive functions begin, the following vascular changes occur (Fig. 21.30b).

1. When the umbilical cord is tied off, no blood flows through the umbilical arteries, they fill with connective tissue, and the distal portions of the umbilical arteries become the **medial umbilical ligaments**.

2. Tying off the umbilical cord results in the conversion of the umbilical vein into the **ligamentum teres (round ligament)**, a structure that attaches the umbilicus to the liver.
3. The placenta is expelled as the "**afterbirth**."
4. The ductus venosus collapses as blood stops flowing through the umbilical vein and becomes the **ligamentum venosum**, a fibrous cord on the inferior surface of the liver.
5. The foramen ovale normally closes shortly after birth to become the **fossa ovalis**, a depression in the interatrial septum. When an infant takes its first breath, the lungs expand and blood flow to the lungs increases. Blood returning from the lungs to the heart increases pressure in the left atrium. This closes the foramen ovale by pushing the valve that guards it against the interatrial septum.
6. The ductus arteriosus closes by vasoconstriction, atrophies, and becomes the **ligamentum arteriosum**.

Anatomical defects resulting from failure of these changes to occur are illustrated in Fig. 20.18 on page 607.

AGING AND THE CARDIOVASCULAR SYSTEM

General changes associated with aging and the cardiovascular system include loss of compliance (extensibility) of the aorta, reduction in cardiac muscle fiber size, progressive loss of cardiac muscular strength, reduced cardiac output, a decline in maximum heart rate, and an increase in systolic blood pressure. Total blood cholesterol tends to increase with age, as does low-density lipoprotein (LDL); high-density lipoprotein (HDL) tends to decrease. There is an increase in the incidence of coronary artery disease (CAD), the major cause of heart disease and death in older Americans. Congestive heart failure (CHF), a set of symptoms associated with impaired pumping of the heart, also occurs. Changes in blood vessels that serve brain tissue, such as atherosclerosis, reduce nourishment to the brain and result in the malfunction or death of brain cells. By age 80, cerebral blood flow is 20% less and renal blood flow is 50% less than in the same person at age 30.

 ## DEVELOPMENTAL ANATOMY OF BLOOD VESSELS AND BLOOD

The human yolk sac has little yolk to nourish the developing embryo. Blood and blood vessel formation starts as early as 15–16 days in the **mesoderm** of the yolk sac, chorion, and body stalk (Fig. 21.31).

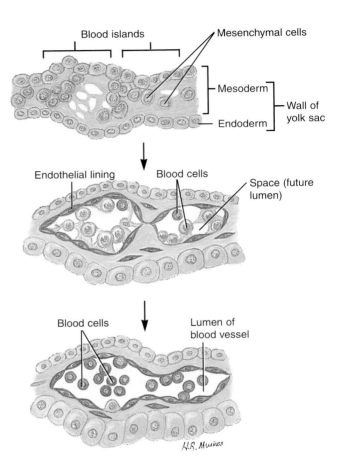

Figure 21.31 Development of blood vessels and blood cells from blood islands.

Blood vessels develop from isolated masses and cords of mesenchyme in the mesoderm called **blood islands**. Spaces soon appear in the islands and become the lumens of the blood vessels. Some of the mesenchymal cells immediately around the spaces give rise to the *endothelial lining of the blood vessels.* Mesenchyme around the endothelium forms the *tunics* (intima, media, externa) of the larger blood vessels. Growth and fusion of blood islands form an extensive network of blood vessels throughout the embryo.

Blood plasma and *blood cells* are produced by the endothelial cells and appear in the blood vessels of the yolk sac and allantois quite early. Blood formation in the embryo itself begins at about the second month in the liver and spleen, a little later in red bone marrow, and much later in lymph nodes.

DISORDERS: HOMEOSTATIC IMBALANCES

HYPERTENSION

Hypertension, or persistently high blood pressure, is defined as systolic blood pressure of 140 mm Hg or greater and diastolic blood pressure of 90 mm Hg or greater. Recall that a blood pressure of 120/80 is normal and desirable in a healthy adult. In industrialized societies, hypertension is the most common disorder affecting the heart and blood vessels; it is a major cause of heart failure, kidney disease, and stroke. The classification system, first used in 1993, ranks blood pressure values for adults as follows:

Normal	Systolic less than 130 mm Hg; diastolic less than 85 mm Hg
High-normal	Systolic 130–139; diastolic 85–89
Hypertension	Systolic 140 or greater; diastolic 90 or greater
Stage 1	Systolic 140–159; diastolic 90–99
Stage 2	Systolic 160–179; diastolic 100–109
Stage 3	Systolic 180–209; diastolic 110–119
Stage 4	Systolic 210 or higher; diastolic 120 or higher

Types and Causes of Hypertension

Primary (**essential**) **hypertension** is a persistently elevated blood pressure that cannot be attributed to any identifiable cause. Approximately 90–95% of all hypertension cases fit this definition. It is suspected that several factors combine to predispose a person to hypertension, including diet, lack of exercise, metabolic defects, stress, and heredity. The remaining 5–10% of cases are **secondary hypertension**. Secondary hypertension has an identifiable underlying cause such as:

1. **Aldosteronism.** The hypersecretion of aldosterone causes hypertension by stimulating excess reabsorption of salt and water by the kidneys. Elevated aldosterone secretion most often is due to a tumor of the adrenal cortex.
2. **Kidney disease.** Obstruction of renal blood flow or disorders that damage renal tissue may cause the kidneys to release excessive amounts of renin into the blood. This enzyme promotes the formation of angiotensin II, which is a powerful vasoconstrictor and also stimulates aldosterone release.
3. **Pheochromocytoma** (fē-ō-krō′-mō-sī-TŌ-ma). This type of tumor in the adrenal medulla produces and releases large quantities of norepinephrine and epinephrine into the blood. Both hormones raise blood pressure. Epinephrine causes an increase in heart rate and contractility and norepinephrine increases systemic vascular resistance.

Damaging Effects of Untreated Hypertension

High blood pressure is known as the "silent killer" because it can cause considerable damage to the heart, brain, and kidneys before a person notices symptoms. The heart is most commonly affected by hypertension. When pressure is high, the heart uses more energy to pump against the increased afterload caused by the elevated arterial blood pressure. If the heart cannot meet the demands put on it, angina pectoris or even myocardial infarction may develop. Hypertension is also a factor in the development of

atherosclerosis. Continued high blood pressure may produce a cerebral vascular accident (CVA), or stroke. Brain arteries are usually less protected by surrounding tissues than are the major arteries in other parts of the body. With prolonged hypertension, they may finally rupture, and a brain hemorrhage follows.

Hypertension also compromises kidney function. The principal site of damage is in the arterioles, where the continual high blood pressure pushing against the walls causes them to thicken, which narrows the lumen. The blood supply to the kidneys is thereby gradually reduced. In response, the kidneys may secrete more renin, which elevates the blood pressure even more and complicates the problem. Eventually, the reduced blood flow may lead to the death of the kidney cells.

Lifestyle Changes to Reduce Hypertension

Although there are several categories of drugs (described next) that can reduce elevated blood pressure, the following lifestyle changes also are effective in managing hypertension.

1. **Lose weight.** This is the best treatment for high blood pressure short of using drugs. Loss of even a few pounds helps reduce blood pressure in overweight hypertensives.
2. **Limit alcohol intake.** Drink less than 2 ounces of 100-proof liquor a day, or avoid alcohol altogether.
3. **Exercise.** Becoming more physically fit by engaging in moderate activity, such as brisk walking, several times a week for 30–45 minutes can lower systolic blood pressure by about 10 mm Hg.
4. **Reduce intake of sodium (salt).** Roughly half of the people with hypertension are "salt sensitive." For them, a high-salt diet appears to promote hypertension whereas a low-salt diet can lower blood pressure.
5. **Maintain recommended dietary intake of potassium, calcium, and magnesium.** Higher levels of potassium, calcium, and magnesium in the diet are associated with a lower risk of hypertension.
6. **Don't smoke.** Although smoking is not thought to cause hypertension, it has devastating effects on the heart and can augment the damaging effects of high blood pressure.
7. **Manage stress.** Various meditation and biofeedback techniques help some people reduce high blood pressure. These methods may work by decreasing the daily release of epinephrine and norepinephrine by the adrenal medulla.

Drug Treatment

Although there is presently no cure for primary hypertension, certain forms of secondary hypertension can be cured by removing the underlying cause. For example, an aldosterone-secreting tumor could be removed. And even in some cases of primary hypertension, certain measures may result in a dramatic reduction in blood pressure. For those people who need medication, several types of drugs are available. Most drugs will be even more effective if combined with appropriate lifestyle modifications. Drugs having several different mechanisms of action

DISORDERS: HOMEOSTATIC IMBALANCES  CONTINUED

are effective in lowering blood pressure. Many people are successfully treated with *diuretics*, which decrease blood pressure by decreasing blood volume. They increase elimination of water and salt in the urine. *ACE (angiotensin converting enzyme) inhibitors* block formation of angiotensin II and thereby promote vasodilation and decrease liberation of aldosterone. *Beta blockers* reduce blood pressure by inhibiting secretion of renin and decreasing heart rate and contractility. *Vasodilators* relax the smooth muscle in arterial walls, causing vasodilation and thus lowering blood pressure by lowering systemic vascular resistance. An important category of vasodilators are the *calcium channel blockers.* They slow inflow of Ca^{2+} into vascular smooth muscle cells. They also reduce the heart's work load by slowing Ca^{2+} entry into myocardial fibers.

ANEURYSM

An **aneurysm** (AN-yoo-rizm) is a thin, weakened section of the wall of an artery or a vein that bulges outward, forming a balloonlike sac. Common causes of aneurysms are atherosclerosis, syphilis, congenital blood vessel defects, and trauma. If an aneurysm goes untreated, it grows larger and larger until the blood vessel wall becomes so thin that it bursts. The result is massive hemorrhage with shock, severe pain, stroke, or death, depending on which vessel is involved. Even an unruptured aneurysm can lead to damage by interrupting blood flow or putting pressure on adjacent blood vessels, organs, or bones.

Surgical repair consists of temporarily clamping the damaged artery on both sides of the aneurysm and surgically removing it. A graft, usually of Dacron, is then sutured to healthy segments of the artery to reestablish normal blood flow.

CORONARY ARTERY DISEASE

Coronary artery disease (**CAD**) is a condition in which the heart muscle receives inadequate O_2 due to blockage of its blood supply. Depending on the extent of blockage, symptoms can range from mild chest pain (angina pectoris) to a full-scale heart attack. The underlying causes of CAD are many and varied. Two of the principal ones are atherosclerosis and coronary artery spasm (see page 605).

DEEP-VENOUS THROMBOSIS

Venous thrombosis, the presence of a thrombus (blood clot) in a vein, usually occurs in deep veins of the lower limbs. This condition is called **deep-venous thrombosis** (**DVT**) and has two serious complications: (1) *pulmonary embolism,* in which the thrombus dislodges and finds its way into the pulmonary arterial blood flow, and (2) *postphlebitic syndrome,* which consists of edema, pain, and skin changes due to destruction of venous valves. Diagnosis is by ultrasound and treatment consists of anticoagulant therapy, elevation of the limb, fibrinolytic therapy (streptokinase or t-PA), and sometimes thrombectomy.

MEDICAL TERMINOLOGY

Angiogenesis (an′-jē-ō-JEN-e-sis) Formation of new blood vessels.

Aortography (ā′-or-TOG-ra-fē) X-ray examination of the aorta and its main branches after injection of a radiopaque dye.

Arteritis (ar′-te-RĪ-tis; *itis* = inflammation of) Inflammation of an artery, probably due to an autoimmune response.

Carotid endarterectomy (ka-ROT-id end′-ar-ter-EK-tō-mē) The removal of atherosclerotic plaque from the carotid artery to restore greater blood flow to the brain.

Claudication (klaw′-di-KĀ-shun) Pain and lameness or limping caused by defective circulation of the blood in the vessels of the limbs.

Hypotension (hī-pō-TEN-shun) Low blood pressure; most commonly used to describe an acute drop in blood pressure, as occurs during excessive blood loss.

Normotensive (nor′-mō-TEN-siv) Characterized by normal blood pressure.

Occlusion (ō-KLOO-shun) The closure or obstruction of the lumen of a structure such as a blood vessel. An example is an atherosclerotic plaque in an artery.

Orthostatic (or′-thō-STAT-ik) **hypotension** (*ortho* = straight; *statikos* = causing to stand) An excessive lowering of sys-

temic blood pressure with the assumption of an erect or semierect posture; it is usually a sign of a disease. May be caused by excessive fluid loss, certain drugs (antihypertensives), and cardiovascular or neurogenic factors. Also called **postural hypotension**.

Phlebitis (fle-BĪ-tis; *phleb* = vein) Inflammation of a vein, often in a leg.

Raynaud's (rā-NOZ) **disease** A vascular disorder, primarily of females, characterized by bilateral attacks of ischemia, usually of the fingers and toes, in which the skin becomes pale and exhibits burning and pain. It is brought on by cold temperatures or emotional stimuli.

Thrombectomy (throm-BEK-tō-mē; *thrombo* = clot) An operation to remove a blood clot from a blood vessel.

Thrombophlebitis (throm′-bō-fle-BĪ-tis) Inflammation of a vein with clot formation. Superficial thrombophlebitis occurs in veins under the skin, especially the calf.

White coat (office) hypertension A stress-induced syndrome found in patients who have elevated blood pressures while being examined by health-care personnel, but otherwise have normal blood pressure.

STUDY OUTLINE

ANATOMY OF BLOOD VESSELS (p. 612)

Arteries (p. 612)

1. Arteries carry blood away from the heart. The wall of an artery consists of a tunica interna, tunica media (which maintains elasticity and contractility), and tunica externa.
2. Large arteries are termed elastic (conducting) arteries, and medium-sized arteries are called muscular (distributing) arteries.
3. Many arteries anastomose: the distal ends of two or more vessels unite. An alternate blood route from an anastomosis is called collateral circulation. Arteries that do not anastomose are called end arteries.

Arterioles (p. 614)

1. Arterioles are small arteries that deliver blood to capillaries.
2. Through constriction and dilation, they assume a key role in regulating blood flow from arteries into capillaries and in altering arterial blood pressure.

Capillaries (p. 614)

1. Capillaries are microscopic blood vessels through which materials are exchanged between blood and tissue cells; some capillaries are continuous, whereas others are fenestrated.
2. Capillaries branch to form an extensive network throughout the tissue. This network increases the surface area, allowing a rapid exchange of large quantities of materials.
3. Precapillary sphincters regulate blood flow through capillaries.
4. Microscopic blood vessels in the liver are called sinusoids.

Venules (p. 615)

1. Venules are small vessels that continue from capillaries and merge to form veins.
2. They drain blood from capillaries into veins.

Veins (p. 616)

1. Veins consist of the same three tunics as arteries but have a thinner tunica interna and media. The lumen of a vein is also larger than that of a comparable artery.
2. They contain valves to prevent backflow of blood.
3. Weak valves can lead to varicose veins.
4. Vascular (venous) sinuses are veins with very thin walls.

Blood Distribution (p. 616)

1. Systemic veins are collectively called blood reservoirs because they hold a large volume of blood.
2. If the need arises, this blood can be shifted into other blood vessels through vasoconstriction of veins.
3. The principal reservoirs are the veins of the abdominal organs (liver and spleen) and skin.

CAPILLARY EXCHANGE (p. 617)

1. Substances enter and leave capillaries by diffusion, vesicular transport, and bulk flow.
2. The movement of water and solutes (except proteins) through capillary walls depends on hydrostatic and osmotic pressures.
3. The near equilibrium between filtration and reabsorption in capillaries is called Starling's law of the capillaries.
4. Edema is an abnormal increase in interstitial fluid.

HEMODYNAMICS: PHYSIOLOGY OF CIRCULATION (p. 620)

1. The velocity of blood flow is inversely related to the cross-sectional area of blood vessels; blood flows slowest where cross-sectional area is greatest.
2. The velocity of blood flow decreases from the aorta to arteries to capillaries and increases as blood returns to the heart.
3. Blood flow is determined by blood pressure and resistance.
4. Blood flows from regions of higher to lower pressure; the higher the resistance the lower the blood flow.
5. Cardiac output equals the mean arterial blood pressure divided by total resistance ($CO = MABP \div R$).
6. Blood pressure is the pressure exerted on the walls of a blood vessel.
7. Factors that affect blood pressure are cardiac output, blood volume, viscosity, resistance, and the elasticity of arteries.
8. As blood leaves the aorta and flows through the systemic circulation, its pressure progressively falls to 0 mm Hg by the time it reaches the right ventricle.
9. Resistance depends on blood viscosity, blood vessel length, and blood vessel radius.
10. Venous return depends on pressure differences between the venules and the right ventricle.
11. Blood return to the heart is maintained by several factors including skeletal muscular contractions, valves in veins (especially in the limbs), and pressure changes associated with breathing.

CONTROL OF BLOOD PRESSURE AND BLOOD FLOW (p. 623)

1. The cardiovascular (CV) center is a group of neurons in the medulla oblongata that regulates heart rate, contractility, and blood vessel diameter.
2. The CV center receives input from higher brain regions and sensory receptors (baroreceptors and chemoreceptors).
3. Output from the CV center flows along sympathetic and parasympathetic fibers. Sympathetic impulses along cardioaccelerator nerves increase heart rate and contractility and parasympathetic impulses along vagus nerves decrease heart rate.
4. Baroreceptors monitor blood pressure and chemoreceptors monitor blood levels of O_2, CO_2, and hydrogen ions.
5. The carotid sinus reflex helps regulate blood pressure in the brain.
6. The aortic reflex is concerned with general systemic blood pressure.
7. The right heart reflex responds to increases in venous blood pressure.
8. Hormones that help regulate blood pressure are epinephrine, norepinephrine, ADH (vasopressin), angiotensin II, and ANP.
9. Autoregulation refers to local, automatic adjustments of blood flow in a given region to meet a particular tissue's need.
10. O_2 level is the principal stimulus for autoregulation.

SHOCK AND HOMEOSTASIS (p. 629)

1. Shock is a failure of the cardiovascular system to deliver enough O_2 and nutrients to meet the metabolic needs of cells.
2. Signs and symptoms include clammy, cool, pale skin; tachycardia; weak, rapid pulse; sweating; hypotension; altered mental state; decreased urinary output; thirst; and acidosis.

3. Stages of shock are (1) compensated, in which negative feedback cycles restore homeostasis; (2) decompensated, in which positive feedback cycles intensify the shock and immediate medical intervention is required; and (3) irreversible, in which the cardiovascular system collapses and death results.

CHECKING CIRCULATION (p. 631)

Pulse (p. 631)

1. Pulse is the alternate expansion and elastic recoil of an artery wall with each heartbeat. It may be felt in any artery that lies near the surface or over a hard tissue.
2. A normal resting pulse (heart) rate is 70–80 beats/min.

Measurement of Blood Pressure (BP) (p. 631)

1. Blood pressure is the pressure exerted by blood on the wall of an artery when the left ventricle undergoes systole and then diastole. It is measured by the use of a sphygmomanometer.
2. Systolic blood pressure (SBP) is the force of blood recorded during ventricular contraction. Diastolic blood pressure (DBP) is the force of blood recorded during ventricular relaxation. Normal blood pressure is 120/80 mm Hg.
3. Pulse pressure is the difference between systolic and diastolic pressure. It normally is about 40 mm Hg.

CIRCULATORY ROUTES (p. 632)

1. The largest circulatory route is the systemic circulation.
2. Two of the several subdivisions of the systemic circulation are coronary (cardiac) circulation and hepatic portal circulation.
3. Other routes are the cerebral, pulmonary, and fetal circulations.

Systemic Circulation (p. 632)

1. The systemic circulation carries oxygenated blood from the left ventricle through the aorta to all parts of the body, including some lung tissue, but does *not* supply the air sacs (alveoli) of the lungs, and returns the deoxygenated blood to the right atrium.
2. The aorta is divided into the ascending aorta, the arch of the aorta, and the descending aorta. Each section gives off arteries that branch to supply the whole body.
3. Blood returns to the heart through the systemic veins. All veins of the systemic circulation drain into the superior or inferior venae cavae or the coronary sinus, which, in turn, empty into the right atrium.

Hepatic Portal Circulation (p. 658)

1. The hepatic portal circulation detours venous blood from the gastrointestinal organs and spleen and directs it into the hepatic portal vein of the liver before it is returned to the heart.
2. This circulation enables the liver to utilize nutrients and detoxify harmful substances in the blood.

Pulmonary Circulation (p. 660)

1. The pulmonary circulation takes deoxygenated blood from the right ventricle to the alveoli within the lungs and returns oxygenated blood from the alveoli to the left atrium.
2. It allows blood to be oxygenated for systemic circulation.

Fetal Circulation (p. 660)

1. The fetal circulation involves the exchange of materials between fetus and mother.
2. The fetus derives O_2 and nutrients and eliminates CO_2 and wastes through the maternal blood supply by means of a structure called the placenta.
3. At birth, when pulmonary (lung), digestive, and liver functions begin, the special structures of fetal circulation are no longer needed.

AGING AND THE CARDIOVASCULAR SYSTEM (p. 664)

1. General changes include loss of elasticity of blood vessels, reduction in cardiac muscle size, reduced cardiac output, and increased systolic blood pressure.
2. The incidence of coronary artery disease (CAD), congestive heart failure (CHF), and atherosclerosis increases with age.

DEVELOPMENTAL ANATOMY OF BLOOD VESSELS AND BLOOD (p. 664)

1. Blood vessels develop from isolated masses of mesenchyme in mesoderm called blood islands.
2. Blood is produced by the endothelium of blood vessels.

REVIEW QUESTIONS

1. Describe the structural and functional differences among arteries, arterioles, capillaries, venules, and veins. (p. 612)
2. Discuss the importance of smooth muscle in the tunica media of arteries. (p. 612)
3. Distinguish between elastic and muscular arteries in terms of location, histology, and function. What is an anastomosis? What is collateral circulation? (p. 612)
4. Describe how capillaries are structurally adapted for exchanging materials between blood and body cells. (p. 614)
5. Define varicose veins. (p. 623)
6. What are blood reservoirs? Why are they important? (p. 617)
7. Describe how substances enter and leave capillaries. (p. 617)
8. Describe how hydrostatic and osmotic pressures determine fluid movement through capillaries. Set up the equation for net filtration pressure (NFP) to substantiate your response. How does edema develop? (p. 618)
9. What is Starling's law of the capillaries? (p. 618)
10. What is velocity of blood flow? Why does blood flow faster in arteries and veins than in capillaries? (p. 620)
11. Explain how blood pressure and resistance determine volume of blood flow. (p. 621)
12. Summarize the factors that affect blood pressure. (p. 623)
13. Define resistance and explain the factors that contribute to it. (p. 622)
14. Describe the factors that assist the return of systemic venous blood to the heart. (p. 622)
15. What is the cardiovascular center? What are the principal inputs and outputs? (p. 624)
16. Describe the operation of the carotid sinus, aortic, and right heart reflexes. (p. 626)
17. Explain the role of chemoreceptors in the regulation of blood pressure. (p. 626)
18. Describe the hormonal regulation of blood pressure. (p. 627)
19. What is autoregulation? Explain how it occurs through physical and chemical changes in blood. (p. 628)

20. Define shock. What are its signs and symptoms? (p. 629)
21. Describe the three stages of shock. (p. 629)
22. Define pulse. Where may pulse be felt? Contrast tachycardia and bradycardia. (p. 631)
23. What is blood pressure (BP)? How are systolic and diastolic blood pressures measured with a sphygmomanometer? (p. 631)
24. Compare the clinical significance of systolic and diastolic pressures. How are these pressures written? (p. 632)
25. What is meant by a circulatory route? Define systemic circulation. (p. 632)
26. Diagram the major divisions of the aorta, their principal arterial branches, and the regions supplied. (p. 634)
27. Trace a drop of blood from the arch of the aorta through its systemic circulatory route to the tip of the big toe on your left foot and back to the heart again. Be sure to also indicate which veins return the blood to the heart. (pp. 634–656)
28. What is the cerebral arterial circle (circle of Willis)? Why is it important? (p. 638)
29. Distinguish between visceral and parietal branches of an artery. What major organs are supplied by branches of the thoracic aorta? How is blood returned from these organs to the heart? (p. 641)
30. What organs are supplied by the celiac, superior mesenteric, renal, inferior mesenteric, inferior phrenic, and median sacral arteries? How is blood returned to the heart? (p. 642)
31. Trace a drop of blood from the brachiocephalic artery into the digits of the right upper limb and back again to the right atrium. (pp. 638–651)
32. What are the three major groups of systemic veins? (p. 648)
33. What is hepatic portal circulation? Describe the route by means of a diagram. Why is this route significant? (p. 658)
34. Define pulmonary circulation. Prepare a diagram to indicate the route. What is the purpose of the route? (p. 660)
35. Discuss the anatomy and physiology of fetal circulation. Indicate the function of umbilical arteries, umbilical vein, ductus venosus, foramen ovale, and ductus arteriosus. (p. 660)
36. How does aging affect the cardiovascular system? (p. 664)
37. Describe the development of blood vessels and blood. (p. 664)

ANSWERS TO QUESTIONS WITH FIGURES

21.1 Artery; vein.
21.2 Because less energy is stored in the elastic arteries during systole, the heart has to pump harder to maintain the same rate of blood flow.
21.3 They use O_2 and produce wastes more rapidly than inactive tissues.
21.4 Through intercellular clefts, via pinocytic vesicles, directly across endothelial membranes, and through fenestrations.
21.5 When you are standing, gravity tends to cause pooling of blood in leg veins. The valves prevent backflow as the blood is pushed back toward the right atrium after each heartbeat. When you are erect, gravity aids the flow of blood in neck veins back toward the heart.
21.6 Volume in veins is about $60\% \times 5$ liters = 3 liters; volume in capillaries is about $5\% \times 5$ liters = 250 ml.
21.7 Blood colloid osmotic pressure will be lower than normal and therefore capillary reabsorption will be low. The result will be edema (see Clinical Application on page 619).
21.8 Aorta and arteries.
21.9 About 93 mm Hg.
21.10 Skeletal muscle and respiratory pumps.
21.11 Arteriole by vasodilation and vasoconstriction.
21.12 Cardiac muscle in the heart and smooth muscle in blood vessel walls.
21.13 Impulses pass from baroreceptors in the carotid sinuses via the glossopharyngeal (IX) nerves and from the arch of the aorta via the vagus (X) nerves to the CV center.

21.14 It represents a change from lying down to standing because the gravity causes increased pooling of blood in leg veins when you stand up, and this decreases the blood pressure in your upper body.
21.15 Not necessarily. If systemic vascular resistance has increased greatly, flow may be inadequate.
21.16 Diastolic = 95 mm Hg; systolic = 142 mm Hg; pulse pressure = 47 mm Hg. This person has stage I hypertension, because the systolic pressure is greater than 140 mm Hg and the diastolic pressure is greater than 90 mm Hg.
21.17 Systemic and pulmonary.
21.18 Ascending aorta, arch of aorta, thoracic aorta, and abdominal aorta.
21.19 Coronary.
21.20 Brachiocephalic trunk, left common carotid artery, and left subclavian artery.
21.21 At the level of the intervertebral disc between T4 and T5.
21.22 About the level of L4.
21.23 Inferior vena cava.
21.24 Internal jugulars.
21.25 Median cubital.
21.26 Inferior vena cava.
21.27 Dorsal venous arch, great saphenous, and small saphenous.
21.28 Hepatic.
21.29 Pulmonary.
21.30 Placenta.

CHAPTER 22

THE LYMPHATIC SYSTEM, NONSPECIFIC RESISTANCE TO DISEASE, AND IMMUNITY

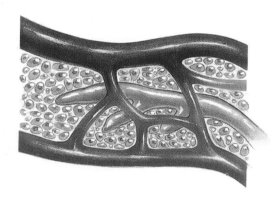

rom hour to hour, survival and good health depend on fending off attacks by disease-producing organisms, termed **pathogens**, neutralizing toxic products of pathogens, repairing tissues damaged by bumps, cuts, burns, ultraviolet light, and caustic chemicals, and eliminating cancerous body cells. This chapter describes the many mechanisms that provide defenses against intruders from without and within and promote repair of body tissues.

The ability to ward off disease through our defenses is called **resistance**. Vulnerability or lack of resistance is termed **susceptibility**. Resistance to disease may be grouped into two broad areas: nonspecific resistance and immunity. **Nonspecific resistance** to disease includes defense mechanisms that provide general protection against invasion by a wide range of pathogens, such as the many different kinds of bacteria and viruses. These include mechanical barriers provided by the skin and mucous membranes, antimicrobial chemicals, phagocytosis, inflammation, and fever. For example, the acidity of the stomach contents kills many bacteria ingested in food. **Immunity** involves activation of specific lymphocytes that combat a particular pathogen or other foreign substance. The body system that carries out immune responses is the lymphatic system.

The **lymphatic** (lim-FAT-ik) **system** consists of a fluid called lymph flowing within lymphatic vessels (lymphatics), several structures and organs that contain lymphatic tissue, and red bone marrow, which houses stem cells that develop into lymphocytes (Fig. 22.1). Interstitial (tissue) fluid and lymph are basically the same. The major difference between the two is location. After fluid passes from interstitial spaces into lymphatic vessels, it is called **lymph** (*lympha* = clear water). Lymphatic tissue is a specialized form of reticular connective tissue that contains large numbers of lymphocytes.

The lymphatic system has several functions:

1. **Draining interstitial fluid.** Lymphatic vessels drain tissue spaces of excess interstitial fluid.
2. **Transporting dietary lipids.** Lymphatic vessels carry lipids and lipid-soluble vitamins (A, D, E, and K) absorbed by the gastrointestinal tract to the blood.
3. **Protecting against invasion.** Lymphatic tissue carries out **immune responses**. These are highly specific responses targeted to particular invaders or abnormal cells. Lymphocytes, aided by macrophages, recognize foreign cells and substances, microbes (bacteria, viruses, and so on), and cancer cells and respond to them in two basic ways. Some lymphocytes (called T cells) destroy the intruders by causing them to rupture or by releasing cytotoxic (cell-killing) substances. Other lymphocytes (called B cells) differentiate into plasma cells that secrete antibodies. These are proteins that combine with and cause destruction of specific foreign substances. In carrying out immune responses, the lymphatic system concentrates foreign substances in certain lymphatic organs, circulates lymphocytes through the organs to make contact with the foreign substances, destroys the foreign substances, and eliminates them from the body.

The developmental anatomy of the lymphatic system is considered later in the chapter.

LYMPHATIC VESSELS AND LYMPH CIRCULATION

Lymphatic vessels begin as closed-ended vessels called **lymphatic capillaries** in spaces between cells (Fig. 22.2a on page 673). Just as blood capillaries converge to form venules and veins, lymphatic capillaries unite to form larger tubes called **lymphatic vessels** (see Fig. 22.1). Lymphatic vessels resemble veins in structure but have thinner walls and more valves. At intervals along the lymphatic vessels, lymph flows through lymphatic tissue structures called lymph nodes. In the skin, lymphatic vessels lie in subcutaneous tissue and generally follow veins. Lymphatic vessels of the viscera generally follow arteries, forming plexuses around them.

Lymphatic Capillaries

Capillaries containing lymph are found throughout the body, except in (1) avascular tissues, (2) the central nervous system, (3) splenic pulp, and (4) bone marrow. Lymphatic capillaries have a slightly larger diameter than blood capillaries and have a unique structure that permits interstitial fluid to flow into them but not out. The ends of endothelial cells that make up the wall of a lymphatic capillary overlap. When pressure is greater in the interstitial fluid than in lymph, the cells separate slightly, like opening of a one-way valve, and fluid enters the lymphatic capillary. When pressure is greater inside the lymphatic capillary, the cells adhere more closely so lymph cannot flow back into interstitial fluid.

At right angles to the lymphatic capillary are structures called **anchoring filaments**, which are partly composed of fine collagen fibrils and attach lymphatic endothelial cells to surrounding tissues (Fig. 22.2b). During edema, excess interstitial fluid accumulates and causes tissue swelling. This swelling pulls on the anchoring filaments, making the openings between cells even larger so that more fluid can flow into the lymphatic capillary.

As you will see in Chapter 24, the lining of the small intestine contains fingerlike projections called villi. Each contains blood capillaries and a specialized lymphatic capillary called a **lacteal** (LAK-tē-al; *lacteus* = milky). Lacteals transport lipids absorbed from the small intestine into lymphatic vessels and ultimately into blood. Due to the

Figure 22.1   The lymphatic system.

The lymphatic system consists of lymph, lymphatic vessels, lymphatic tissues, and red bone marrow.

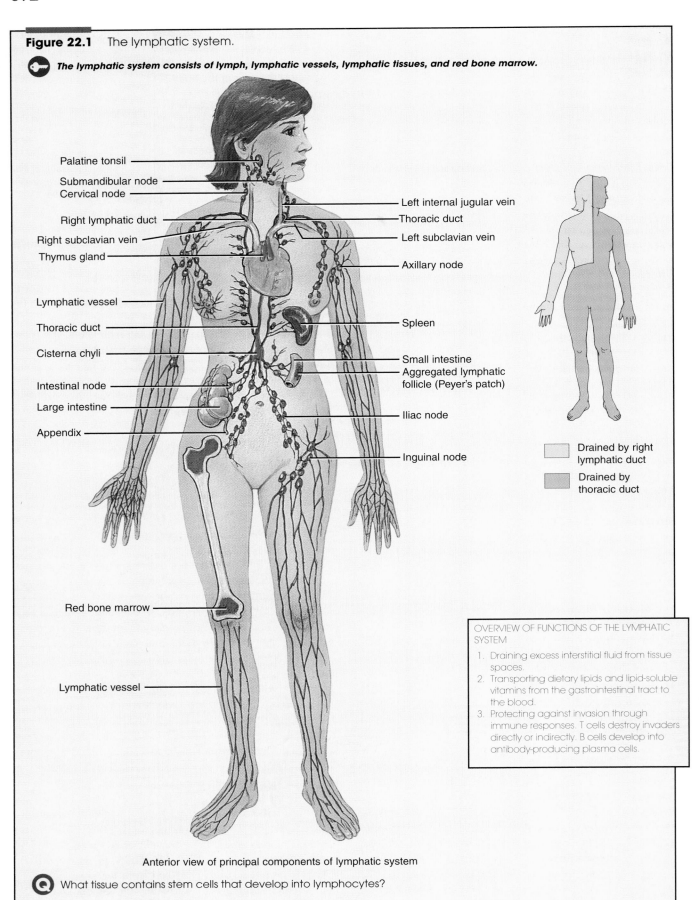

Palatine tonsil
Submandibular node
Cervical node
Right lymphatic duct
Right subclavian vein
Thymus gland

Lymphatic vessel
Thoracic duct
Cisterna chyli
Intestinal node
Large intestine
Appendix
Red bone marrow

Lymphatic vessel

Left internal jugular vein
Thoracic duct
Left subclavian vein
Axillary node

Spleen

Small intestine
Aggregated lymphatic
follicle (Peyer's patch)
Iliac node

Inguinal node

Drained by right
lymphatic duct
Drained by
thoracic duct

OVERVIEW OF FUNCTIONS OF THE LYMPHATIC
SYSTEM

1. Draining excess interstitial fluid from tissue
   spaces.
2. Transporting dietary lipids and lipid-soluble
   vitamins from the gastrointestinal tract to
   the blood.
3. Protecting against invasion through
   immune responses. T cells destroy invaders
   directly or indirectly. B cells develop into
   antibody-producing plasma cells.

Anterior view of principal components of lymphatic system

Q What tissue contains stem cells that develop into lymphocytes?

Figure 22.2 Lymphatic capillaries.

🔑 *Lymphatic capillaries are found throughout the body except in avascular tissues, the central nervous system, splenic pulp, and bone marrow.*

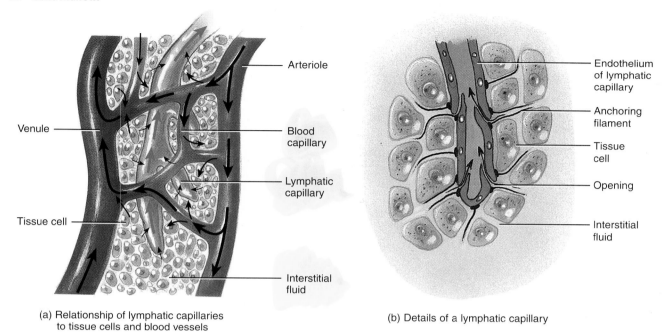

(a) Relationship of lymphatic capillaries to tissue cells and blood vessels

(b) Details of a lymphatic capillary

Q Is lymph more similar to blood plasma or to interstitial fluid? Why?

presence of lipids, the fluid within lacteals is creamy white and may be called **chyle** (KĪL; *chylus* = juice).

Formation and Flow of Lymph

Most components of blood plasma freely filter through the capillary walls to form interstitial fluid. More fluid seeps out of blood capillaries by filtration than returns to them by reabsorption. The excess fluid, about 3 liters per day, drains into lymphatic vessels and becomes lymph. Ultimately, lymph drains into venous blood through the right lymphatic duct and thoracic duct (left lymphatic duct) at the junction of the internal jugular and subclavian veins (see Fig. 22.4). Thus the sequence of fluid flow is (Fig. 22.3): arteries (blood plasma) → blood capillaries (blood plasma) → interstitial spaces (interstitial fluid) → lymphatic capillaries (lymph) → lymphatic vessels (lymph) → lymphatic ducts (lymph) → subclavian veins (blood plasma).

Since most plasma proteins are too large to leave blood vessels, interstitial fluid contains only small amounts of proteins. Any proteins that do escape, however, cannot return to the blood by diffusion. The concentration gradient (high level of proteins inside blood capillaries, low level outside) prevents this. Thus an important function of lymphatic vessels is to return leaked plasma proteins to the blood.

The flow of lymph from tissue spaces to the large lymphatic ducts to the subclavian veins is maintained primarily by the contraction of skeletal muscles (milking action). Skeletal muscle contractions compress lymphatic vessels and force lymph toward the subclavian veins. One-way valves (similar to those found in veins) within the lymphatic vessels prevent backflow of lymph.

Another factor that maintains lymph flow is breathing movements. These movements create a pressure gradient between the two ends of the lymphatic system. With each inhalation, lymph flows from the abdominal region, where the pressure is higher, toward the thoracic region, where it is lower. In addition, when a lymphatic vessel distends, the smooth muscle in its wall contracts. This helps move lymph from one segment of a lymphatic vessel to another.

Lymph Trunks and Ducts

Lymph passes from lymphatic capillaries into lymphatic vessels and through lymph nodes. Lymphatic vessels exiting lymph nodes pass lymph toward another node of the same group or on to another group of nodes. From the most proximal group of each chain of nodes, the exiting vessels unite to form **lymph trunks**. The principal trunks are the **lumbar**, **intestinal**, **bronchomediastinal**,

Figure 22.3   Schematic diagram showing the relationship of the lymphatic system to the cardiovascular system.

 The flow of fluid is from arteries (blood plasma) to blood capillaries (blood plasma) to interstitial spaces (interstitial fluid) to lymphatic capillaries (lymph) to lymphatic vessels (lymph) to lymphatic ducts (lymph) to subclavian veins (blood plasma).

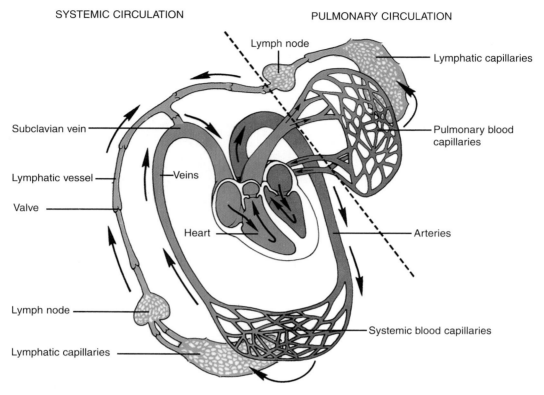

Arrows show direction of flow of lymph and blood

 Does inhalation promote or hinder the flow of lymphatic fluid?

subclavian, and **jugular trunks** (see Fig. 22.4). The principal trunks pass their lymph into two main channels, the thoracic duct and the right lymphatic duct. From these ducts, lymph passes into venous blood.

Thoracic (Left Lymphatic) Duct

The **thoracic** (**left lymphatic**) **duct** (Fig. 22.4) is about 38–45 cm (15–18 in.) in length and begins as a dilation called the **cisterna chyli** (sis-TER-na KĪ-lē), which is anterior to the second lumbar vertebra. The thoracic duct is the main collecting duct of the lymphatic system. It receives lymph from the left side of the head, neck, and chest, the left upper limb, and the entire body inferior to the ribs (see Fig. 22.1).

The cisterna chyli receives lymph from the right and left lumbar trunks and from the intestinal trunk. The lumbar trunks drain lymph from the lower limbs, wall and viscera of the pelvis, kidneys, adrenal (suprarenal) glands, and the deep lymphatics from most of the abdominal wall. The intestinal trunk drains lymph from the stomach, intestines, pancreas, spleen, and part of the liver.

In the neck, the thoracic duct also receives lymph from the left jugular, left subclavian, and left bronchomediastinal trunks. The left jugular trunk drains lymph from the left side of the head and neck, and the left subclavian trunk drains lymph from the left upper limb. The left bronchomediastinal trunk drains lymph from the left side of the deeper parts of the anterior thoracic wall, superior part of the anterior abdominal wall, anterior part of the diaphragm, left lung, and left side of the heart.

Right Lymphatic Duct

The **right lymphatic duct** (Fig. 22.4) is about 1.25 cm (½ in.) long and drains lymph from the upper right side of the body (see Fig. 22.1). The right lymphatic duct collects lymph from the right jugular trunk, which drains the right side of the head and neck; the right subclavian trunk, which drains the right upper limb; and the right bronchomediastinal trunk, which drains the right side of the thorax, right lung, right side of the heart, and part of the liver.

Figure 22.4 Routes for drainage of lymph from lymph trunks into thoracic and right lymphatic ducts.

All lymph returns to the bloodstream through the thoracic (left) lymphatic duct and right lymphatic duct.

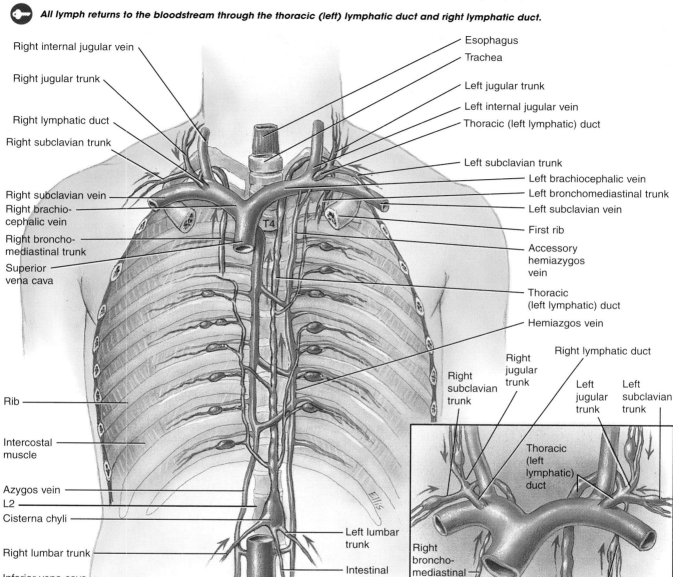

Overall anterior view

Detailed anterior view

Which lymphatic vessels empty into the cisterna chyli and which duct receives lymph from the cisterna chyli?

LYMPHATIC TISSUES

The **primary lymphatic (lymphoid) organs** of the body are the **red bone marrow** (in flat bones and the epiphyses of long bones) and the **thymus gland**. They are termed primary lymphatic organs because they produce B and T cells, the lymphocytes that carry out immune responses. Hemopoietic

stem cells in red bone marrow give rise to B cells and pre-T cells. The pre-T cells then migrate to the thymus gland. The major **secondary lymphatic organs** are the **lymph nodes** and **spleen**. Also included among the secondary lymphatic organs, although strictly speaking they are not discrete organs because they are not surrounded by a capsule, are the **lymphatic nodules**. Lymphatic nodules are clusters of lymphocytes that stand guard in all mucous membranes, where invaders might try to enter the body. Mucous membranes line the gastrointestinal tract, respiratory passageways, urinary tract, and reproductive tract. Most immune responses occur in secondary lymphatic organs.

Thymus Gland

Usually a bilobed lymphatic organ, the **thymus gland** is located in the mediastinum, posterior to the sternum and between the lungs (Fig. 22.5a). An enveloping layer of connective tissue holds the two **thymic lobes** closely together but a connective tissue **capsule** encloses each lobe. The capsule gives off extensions into the lobes called **trabeculae**, which divide the lobes into **lobules** (Fig. 22.5b).

Each lobule consists of a deeply staining outer **cortex** and a lighter-staining central **medulla**. The cortex is composed of tightly packed lymphocytes, epithelial cells, and macrophages. Pre-T cells migrate (via the blood) from red bone marrow to the thymus, where they proliferate and develop into mature T cells. The medulla consists mostly of epithelial cells and more widely scattered lymphocytes. The epithelial cells produce thymic hormones, which are thought to aid in maturation of T cells. Exactly what they do, however, is not known. In addition, the medulla contains characteristic **thymic (Hassall's) corpuscles,** concentric layers of epithelial cells. Possibly, they are remnants of dying cells.

The thymus gland is large in the infant, and it reaches its maximum size of about 40 g (about 1.4 oz) at 10–12 years of age. After puberty, adipose and areolar connective tissue begin to replace the thymic tissue. By the time a person reaches maturity, the gland has atrophied considerably (involution with age). Although most T cells arise before puberty, some continue to mature throughout life.

Lymph Nodes

The oval or bean-shaped structures located along the length of lymphatic vessels are called **lymph nodes** (Fig. 22.6). They range from 1 to 25 mm (0.04 to 1 in.) in length. Lymph nodes are scattered throughout the body, usually in groups (see Fig. 22.1). They are heavily concentrated in areas such as the mammary glands, axillae, and groin.

Each node is covered by a **capsule** of dense connective tissue that extends strands into the node. The capsular extensions are called **trabeculae** (tra-BEK-yoo-lē; *trabecula* = little beam) that divide the node into compartments, provide support, and convey blood vessels into the interior of a node. Internal to the capsule is a supporting network of reticular fibers and fibroblasts. The capsule, trabeculae, reticular fibers, and fibroblasts constitute the stroma (framework) of a lymph node. The parenchyma of a lymph node is specialized into two regions: cortex and medulla. The outer **cortex** contains many **follicles,** which are regions of densely packed lymphocytes arranged in masses that resemble lymphatic nodules. The outer rim of each follicle contains **T cells (T lymphocytes)** plus **macrophages** and **follicular dendritic cells,** which participate in activation of T cells. The follicles contain lighter-staining central areas, the **germinal centers,** where **B cells (B lymphocytes)** proliferate into antibody-secreting plasma cells. The inner region of a lymph node is called the **medulla.** In the medulla, the lymphocytes are tightly packed in strands called **medullary cords.** These cords also contain macrophages and plasma cells.

Lymph flows through a node in one direction. It enters through **afferent** (*ad* = to; *ferre* = to carry) **lymphatic vessels,** which penetrate the convex surface of the node at sev-

---

Figure 22.5   Thymus gland.

 The bilobed thymus gland is largest at puberty; with age it atrophies.

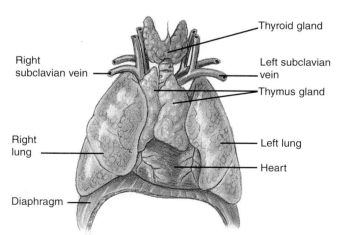

(a) Anterior view of thymus gland of a young child

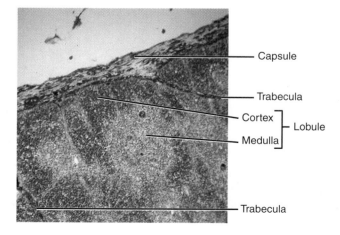

(b) Photomicrograph of several thymus lobules (37x)

 Which lymphocytes mature in the thymus?

Figure 22.6 Structure of a lymph node. Arrows indicate direction of lymph flow through a lymph node.

🔑 *Lymph nodes are present throughout the body, usually clustered in groups.*

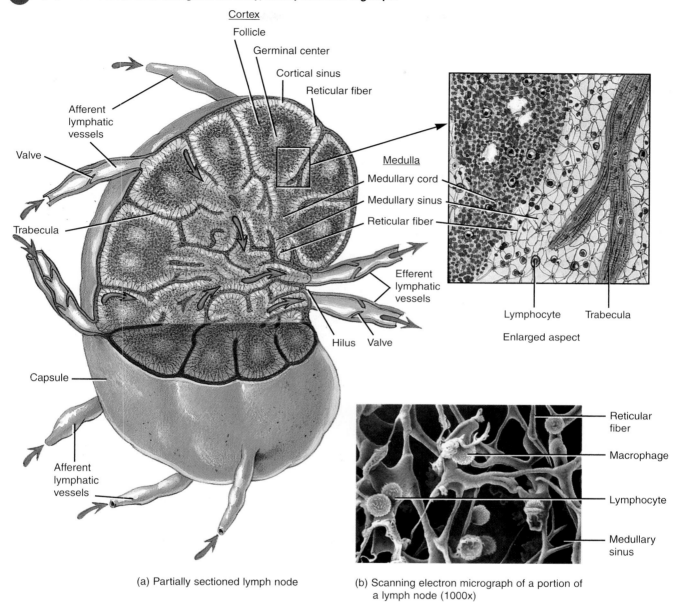

(a) Partially sectioned lymph node

(b) Scanning electron micrograph of a portion of a lymph node (1000x)

❓ What happens to foreign substances in lymph that enter a lymph node?

eral points. They contain valves that open toward the node so that the lymph is directed *inward.* Inside the node, lymph enters the **sinuses**, which are a series of irregular channels between the medullary cords. Lymph flows through sinuses in the cortex (**cortical sinuses**) and then in the medulla (**medullary sinuses**) and exits the lymph node via one or two **efferent** (*ex* = away) **lymphatic vessels**. Efferent lymphatic vessels are wider than the afferent vessels and are fewer in number. They contain valves that open away from the node to convey lymph *out* of the node. Efferent lymphatic vessels emerge from one side of the lymph node at a

slight depression called a **hilus** (HĪ-lus). Blood vessels also enter and leave the node at the hilus.

Among lymphatic tissues, only lymph nodes filter lymph by having it enter at one end and exit at another. The lymph nodes filter foreign substances from lymph as it passes back toward the bloodstream. These substances are trapped by the reticular fibers within the node. Then macrophages destroy some foreign substances by phagocytosis and lymphocytes bring about destruction of others by immune responses. Plasma cells and T cells that have proliferated within a lymph node also can leave and circulate to other parts of the body.

METASTASIS THROUGH THE LYMPHATIC SYSTEM

Knowledge of the location of the lymph nodes and the direction of lymph flow is important in the diagnosis and prognosis of the spread of cancer by **metastasis** (me-TAS-ta-sis). Cancer cells may travel via the lymphatic system and produce clusters of tumor cells where they lodge. Such secondary tumor sites are predictable by the direction of lymph flow from the organ primarily involved. (Cancer may also spread by extending locally or being carried by the cardio-

vascular system.) Cancerous lymph nodes feel enlarged, firm, and nontender. Most lymph nodes that enlarge during an infection, by contrast, are not firm and are very tender. ■

Spleen

The oval **spleen** is the largest single mass of lymphatic tissue in the body. It measures about 12 cm (5 in.) in length (Fig. 22.7a) and is situated in the left hypochondriac region between the stomach and diaphragm lateral to the liver (see Fig. 1.11). Like lymph nodes, the spleen has a hilus, where

---

Figure 22.7   Structure of the spleen.

 The spleen is the largest single mass of lymphatic tissue in the body.

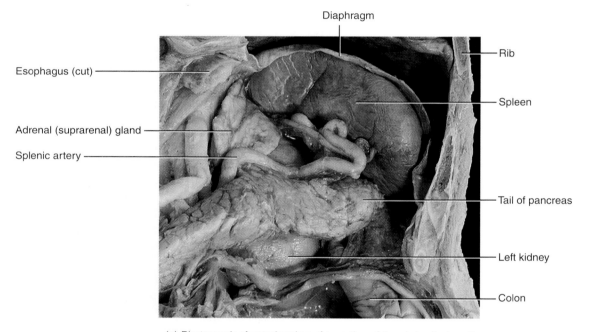

(a) Photograph of anterior view of a portion of the abdominal cavity

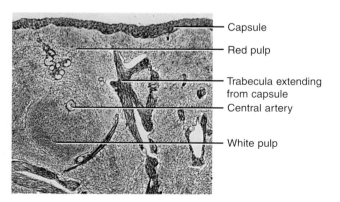

(b) Photomicrograph of portion of the spleen (approx. 23x)

Q After birth, what is the main function of the spleen?

Mechanism of Phagocytosis

Phagocytosis has three phases: chemotaxis, adherence, and ingestion. After it ingests a particle, a phagocyte unleashes chemicals that kill any microbes and digest the remains (Fig. 22.10).

1. **Chemotaxis.** The chemical attraction of phagocytes to a particular location is termed **chemotaxis** (kē′-mō-TAK-sis; *chemo* = chemicals; *taxis* = arrangement). Among the chemotactic chemicals that attract phagocytes are microbial products, components of white blood cells and damaged tissue cells, and activated complement proteins.

2. **Adherence.** The *attachment* of the plasma membrane of a phagocyte to the surface of a microorganism or other foreign material is called **adherence**. In some instances, adherence occurs easily, and the microorganism is readily phagocytized. In other cases, adherence is more difficult because of certain microbial defenses. Opsonization of microorganisms by complement proteins facilitates phagocytosis.

3. **Ingestion.** Following adherence, **ingestion** occurs. The cell membrane of the phagocyte extends projections, called pseudopods, that engulf the microorganism. Once the microorganism is surrounded, the pseudopods meet and fuse, enclosing the microorganism within a sac called a **phagocytic vesicle** (**phagosome**).

Digestion and Killing

After phagocytosis has been accomplished, several deadly mechanisms come into play. The phagocytic vesicle, which forms when a portion of the membrane completely pinches off, enters the cytoplasm (Fig. 22.10). Within the cytoplasm, it merges with lysosomes to form a single, larger structure called a **phagolysosome**. The phagolysosome contains lysozyme, which breaks down microbial cell walls, and digestive enzymes, which degrade carbohydrates, proteins, lipids, and nucleic acids. The phagocyte also forms lethal oxidants, such as superoxide anion (O_2^-), hypochlorite anion (OCl^-), and hydrogen peroxide (H_2O_2), in a process called the **respiratory** (**oxidative**) **burst**. Finally, phagocytes produce other bactericidal substances, such as **defensins**, so named because of their apparent role in preventing and overcoming infections. Defensins are active against bacteria, fungi, and viruses.

The chemical onslaught kills many types of microbes in only 10–30 minutes. Any materials that cannot be degraded further remain in structures called **residual bodies**. The cell disposes of residual bodies by exocytosis, a process in

---

Figure 22.10 Phagocytosis, killing, and digestion.

The major types of phagocytes are neutrophils and macrophages.

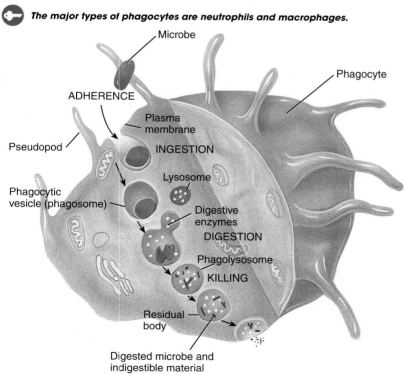

(a) Phases of phagocytosis

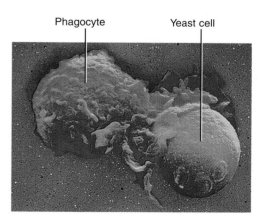

(b) Scanning electron micrograph of a phagocyte engulfing a yeast cell (about 2500x)

 What chemicals are responsible for killing microbes?

which the residual body migrates to the plasma membrane, fuses with it, ruptures, and releases its contents.

Some microbes, such as toxin-producing staphylococci, may be ingested but are not necessarily killed. Rather, their toxins may kill the phagocytes. Other microbes, such as the tubercle bacillus, may multiply within the phagolysosome and eventually destroy the phagocyte. Still other microbes, such as the causative agents of tularemia and brucellosis, may remain dormant in phagocytes for months or years at a time.

Inflammation

When cells are damaged by microbes, physical agents, or chemical agents, the injury is a form of stress. The response to the stress of tissue damage is called **inflammation**. It is a defensive response that is usually characterized by four symptoms: **redness**, **pain**, **heat**, and **swelling**. A fifth symptom can be the **loss of function** in the injured area. Whether loss of function occurs depends on the site and extent of the injury. Inflammation aids disposal of microbes, toxins, or foreign material at the site of injury, prevents their spread to other organs, and prepares the site for tissue repair. Thus it helps restore tissue homeostasis.

Since inflammation is one of the body's nonspecific defenses, the response of a tissue to an accidental cut is similar to the response that results from other types of tissue damage, caused by burns, radiation, or bacterial or viral invasion. In each case there are three basic stages of inflammation: (1) vasodilation and increased permeability of blood vessels, (2) phagocyte migration, and (3) tissue repair.

Vasodilation and Increased Permeability of Blood Vessels

Immediately after tissue damage, blood vessels in the area of the injury dilate and become more permeable. **Vasodilation** is an increase in diameter of the blood vessels. **Increased permeability** means that an increased amount of material, including some proteins normally retained in blood, is allowed to pass out of the blood vessels. Vasodilation allows more blood to flow through the damaged area, and increased permeability permits defensive materials in the blood to enter the injured area (Fig. 22.11). Such defensive mediators include antibodies, phagocytes, and clot-forming chemicals. The increased blood flow also helps remove toxic products released by the invading micro-organisms and dead cells.

Among the substances that contribute to vasodilation, increased permeability, and other aspects of the inflammatory response are the following:

1. **Histamine.** This substance is found in many body cells, especially mast cells in connective tissue and basophils and platelets in blood. In response to injury, cells that

Figure 22.11   Inflammation.

The three stages of inflammation are (1) vasodilation and increased permeability of blood vessels, (2) phagocyte migration, and (3) tissue repair.

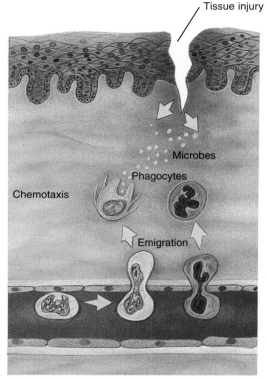

Phagocytes migrate from blood to site of tissue injury

What causes each of the following symptoms of inflammation: redness, pain, heat, and swelling?

contain histamine release it. Phagocytes (neutrophils and macrophages) attracted to the site of injury also stimulate release of histamine. Histamine causes vasodilation and increased permeability of blood vessels.

2. **Kinins.** These polypeptides are formed in blood from inactive precursors called kininogens. They induce vasodilation and increased permeability and serve as chemotactic agents for phagocytes.

3. **Prostaglandins (PGs).** These lipids, especially of the E series, are released by damaged cells and intensify the effects of histamine and kinins. PGs also may stimulate the emigration of phagocytes through capillary walls.

4. **Leukotrienes (LTs).** Produced by basophils and mast cells by breakdown of membrane phospholipids, LTs cause increased permeability and also function in adherence of phagocytes to pathogens and as chemotactic agents that attract phagocytes.

5. **Complement.** Different members of the complement system stimulate histamine release, attract neutrophils

by chemotaxis, and promote phagocytosis (see Fig. 22.9). They can also destroy bacteria.

Within minutes after an injury, dilation of arterioles and increased permeability of capillaries produce heat, redness, and edema (swelling) in the affected area. The large amount of warm blood flowing through the area produces both heat and redness (erythema). As the local temperature rises slightly, metabolic reactions proceed more rapidly and release additional heat. Edema results from increased permeability of blood vessels, which permits more fluid to move from blood into tissue spaces. Pain, whether immediate or delayed, is a cardinal symptom of inflammation. It can result from injury of nerve fibers or from irritation by toxic chemicals from microorganisms. Kinins affect some nerve endings, causing much of the pain associated with inflammation. Prostaglandins intensify and prolong the pain associated with inflammation. Pain may also be due to increased pressure from edema.

The increased permeability of capillaries allows leakage of clotting factors into tissues. The clotting cascade (shown in Fig. 19.10) is set into motion and fibrinogen is ultimately converted to an insoluble, thick network of fibrin threads, which localizes and traps invading microbes and blocks their spread.

Phagocyte Migration

Within an hour after the inflammatory process starts, phagocytes appear on the scene, first neutrophils and then monocytes (Fig. 22.11). They leave the bloodstream at an inflamed site by a process termed **emigration** (see Fig. 19.7). Neutrophil emigration depends on chemotaxis. They are attracted by microbes, kinins, complement, and other neutrophils and attempt to destroy the invaders by phagocytosis. A steady stream of neutrophils is ensured by the production and release of additional cells from bone marrow. This increase in white blood cells in the blood is termed **leukocytosis**.

As the inflammatory response continues, monocytes follow the neutrophils into the infected area. Once in the tissue, monocytes transform into wandering macrophages that augment the phagocytic activity of fixed macrophages. Neutrophils predominate in the early stages of inflammation but die off rapidly. Macrophages arrive on the scene hours later. They are more powerful phagocytes than neutrophils and engulf damaged tissue, worn-out neutrophils, and invading microbes.

Eventually phagocytes die. Within a few days, a pocket of dead phagocytes and damaged tissue forms. This collection of dead cells and fluid is called **pus**. Pus formation usually continues until the infection subsides. At times, the pus pushes to the surface of the body or into an internal cavity for dispersal. On other occasions, the pus remains even after the infection is terminated. In this case, the pus is gradually destroyed over a period of days and is absorbed.

ABSCESS AND ULCERS

If the pus cannot drain out of an inflamed region, an abscess develops. An **abscess** is simply an excessive accumulation of pus in a confined space. Common examples are pimples and boils. When inflamed tissue is shed many times, it produces an open sore, called an **ulcer**, on the surface of an organ or tissue. People with poor circulation are susceptible to ulcers in the tissues of their legs. The ulcers, called stasis ulcers, develop because of poor oxygen and nutrient supply to tissues that then become very susceptible to even a very mild injury or infectious process. ■

Fever

Fever is an abnormally high body temperature. Although its significance is still not understood, fever commonly occurs during infection and inflammation. Many bacterial toxins elevate body temperature, sometimes by triggering release of fever-causing cytokines such as interleukin-1. Elevated body temperature is thought to intensify the effects of interferons, inhibit the growth of some microbes, and speed up body reactions that aid repair. (Fever is discussed in more detail on page 812.)

A summary of some of the components of nonspecific resistance is presented in Exhibit 22.1.

IMMUNITY

The nonspecific defenses all have one thing in common. They offer immediate protection against a *variety* of pathogens or other foreign substances. They are not specifically directed against one particular invader. The ability of the body to defend itself against specific invading agents such as bacteria, toxins, viruses, and foreign tissues is called **immunity**. Substances that are recognized as foreign by the immune system and provoke immune responses are called **antigens** (**Ags**). Two properties distinguish immunity from the nonspecific defenses: (1) *specificity* for particular foreign molecules (antigens), which also involves distinguishing self from nonself molecules, and (2) *memory* for most previously encountered antigens such that a second encounter prompts an even more rapid and vigorous response. The branch of science that deals with the responses of the body when challenged by antigens is called **immunology** (im'-yoo-NOL-ō-jē; *immunis* = free).

Formation of T Cells and B Cells

Lymphocytes that develop **immunocompetence**, the ability to carry out immune responses if properly stimulated, are

EXHIBIT 22.1 SUMMARY OF NONSPECIFIC RESISTANCE

Component	Functions
SKIN AND MUCOUS MEMBRANES	
MECHANICAL FACTORS	
Epidermis of skin	Forms a physical barrier to the entrance of microbes.
Mucous membranes	Inhibit the entrance of many microbes, but not as effective as intact skin.
Mucus	Traps microbes in respiratory and gastrointestinal tracts.
Hairs	Filter out microbes and dust in nose.
Cilia	Together with mucus, trap and remove microbes and dust from upper respiratory tract.
Lacrimal apparatus	Tears dilute and wash away irritating substances and microbes.
Saliva	Washes microbes from surfaces of teeth and mucous membranes of mouth.
Urine	Washes microbes from urethra.
Defecation and vomiting	Expel microbes from body.
CHEMICAL FACTORS	
Acid pH of skin	Discourages growth of many microbes.
Unsaturated fatty acids	Antibacterial substance in sebum.
Lysozyme	Antimicrobial substance in perspiration, tears, saliva, nasal secretions, and tissue fluids.
Hyaluronic acid	Prevents spread of noxious agents in localized infection.
Gastric juice	Destroys bacteria and most toxins in stomach.
ANTIMICROBIAL SUBSTANCES	
Interferons (IFNs)	Protect uninfected host cells from viral infection.
Complement system	Causes cytolysis of microbes, promotes phagocytosis, and contributes to inflammation.
NATURAL KILLER (NK) CELLS	Kill a wide variety of microbes and certain tumor cells.
PHAGOCYTOSIS	Ingestion of foreign particulate matter by neutrophils, eosinophils, and macrophages.
INFLAMMATION	Confines and destroys microbes and initiates tissue repair.
FEVER	Intensifies the effects of interferons, inhibits growth of some microbes, and speeds up body reactions that aid repair.

the B cells and T cells. Both develop from hemopoietic stem cells that originate in red bone marrow (see Fig. 19.2). B cells complete their development into mature, immunocompetent cells in bone marrow, a process that continues throughout one's lifetime (Fig. 22.12). T cells develop from pre-T cells that migrate from bone marrow into the thymus.

Before T cells leave the thymus or B cells leave bone marrow, they acquire several distinctive surface proteins. Some function as **antigen receptors**, that is, molecules capable of recognizing specific antigens (Fig. 22.12). In addition, T cells exit the thymus as either CD4+ or CD8+ cells, which means they display either a protein called CD4 or one called CD8 on their plasma membrane. These two types of T cells, called T4 and T8 cells, have very different functions.

Types of Immune Responses

Immunity consists of two kinds of closely allied immune responses, both triggered by antigens. In the first kind, called **cell-mediated (cellular) immune (CMI) responses**, CD8+ T cells proliferate into "killer" T cells and directly attack the invading antigen. In the second kind, called **antibody-mediated (humoral) immune (AMI) responses**, B cells transform into plasma cells, which synthesize and secrete specific proteins called **antibodies (Abs)** or **immunoglobulins** (im′-yoo-nō-GLOB-yoo-lins). Antibodies bind to and inactivate a particular antigen. Most CD4+ T cells become helper T cells that aid both CMI and AMI responses. The AIDS virus uses the CD4 molecule to enter and then destroy helper T cells (see page 701).

To some extent, each type of immune response specializes in dealing with certain invaders. Cell-mediated immunity is particularly effective against (1) intracellular pathogens, such as fungi, parasites, and viruses; (2) some cancer cells; and (3) foreign tissue transplants. Thus CMI always involves cells attacking cells. Antibody-mediated immunity works mainly against (1) antigens dissolved in body fluids and (2) extracellular pathogens, primarily bacteria, that multiply in body fluids but rarely enter body cells. Often, however, a pathogen provokes both types of immune responses.

Antigens

Antigens have two important characteristics. The first is **immunogenicity** (im′-yoo-nō-jen-IS-it-ē). This means the ability to provoke an immune response, that is, to stimulate production of specific antibodies or proliferation of specific T cells, or both. The second is **reactivity**, the ability of the antigen to react specifically with the produced antibodies or cells. Strictly speaking, immunologists define an antigen as a substance that has reactivity, whereas they call a substance with both of these characteristics a **complete antigen** or **immunogen**. Commonly, however, the term antigen im-

Figure 22.12 Lymphocyte maturation and types of immune responses.

 B cells and pre-T cells develop from hemopoietic stem cells in red bone marrow.

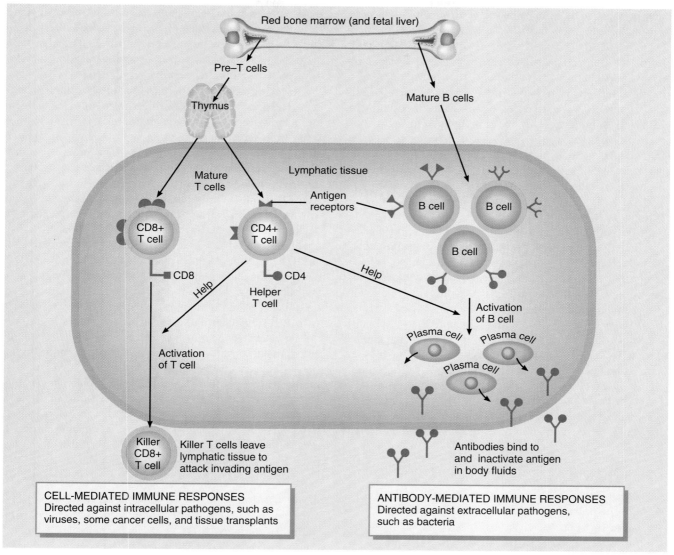

CELL-MEDIATED IMMUNE RESPONSES
Directed against intracellular pathogens, such as viruses, some cancer cells, and tissue transplants

ANTIBODY-MEDIATED IMMUNE RESPONSES
Directed against extracellular pathogens, such as bacteria

Q Which type of T cell participates in both cell-mediated and antibody-mediated immune responses?

plies both immunogenicity and reactivity, and we use the word in this way.

Entire microbes, such as bacteria or viruses, or parts of microbes may act as antigens. Bacterial structures such as flagella, capsules, and cell walls are antigenic, as are bacterial toxins. Nonmicrobial examples of antigens include pollen, egg white, incompatible blood cells, and transplanted tissues and organs. The myriad of antigens in the environment provides many opportunities for provoking immune responses.

An antigen that gets past the nonspecific defenses generally follows one of three routes into lymphatic tissue: (1) most antigens in the bloodstream, for example, having en-

tered an injured blood vessel, are deposited in the spleen; (2) antigens that penetrate the skin enter lymphatic vessels and end up in lymph nodes; and (3) those that penetrate mucous membranes lodge in mucosa-associated lymphoid tissue (MALT).

Chemical Nature of Antigens

Antigens are large, complex molecules. Most often they are proteins. However, nucleic acids, nucleoproteins (nucleic acid + protein), lipoproteins (lipid + protein), glycoproteins (carbohydrate + protein), and certain large polysaccharides may also act as antigens. T cells respond only to antigens that include protein. B cells respond to antigenic proteins, certain lipids, carbohydrates, and nucleic acids. Complete antigens usually have molecular weights of 10,000 daltons or more. Large molecules that have simple, repeating subunits—for example, cellulose and most plastics—usually are not antigenic. This is why plastics can be used in artificial heart valves or joints.

Smaller substances that have reactivity but lack immunogenicity are called **partial antigens** or **haptens** (HAP-tenz; *haptein* = to grasp). A hapten can stimulate an immune response if it is attached to a larger carrier molecule. An example is the small lipid toxin in poison ivy, which triggers an immune response after combining with a body protein. Likewise, some drugs, for example, penicillin, may combine with proteins in the body, and the complex becomes immunogenic. Such hapten-stimulated immune responses are responsible for allergic reactions to drugs and other chemicals in our environment (see page 703).

Antigenic Determinants

Specific portions of antigen molecules trigger immune responses. These regions are called **antigenic determinants** or **epitopes** (Fig. 22.13). Most antigens have many antigenic determinants, each of which induces production of a specific antibody or activates a specific T cell. As a rule, antigens are foreign substances. They are not usually part of body tissues. However, sometimes the immune system fails to distinguish friend (self) from foe (nonself). The result is an autoimmune disorder (see page 702), in which self molecules or cells are attacked as though they were foreign.

Diversity of Antigen Receptors

An amazing feature of the human immune system is its ability to recognize and bind to at least a billion (10^9) different antigenic determinants. Before a particular antigen ever enters the body, T and B lymphocytes that can recognize and respond to that intruder are ready and waiting. Cells of the immune system can even recognize artificially made molecules that do not exist in nature. The basis for the ability to recognize so many antigenic determinants is an equally large diversity of antigen receptors. Since human cells contain only about 100,000 genes, however, it was puzzling how a billion or more different antigen receptors could arise.

The answer to this puzzle turned out to be simple in concept. The diversity of antigen receptors in both B and T cells and the diversity of antibodies produced by B cell progeny result from shuffling and rearranging of a few hundred versions of several small gene segments. This process is called **genetic recombination**. The mini-genes are put together in different combinations as the lymphocytes are developing from stem cells in red bone marrow and the thymus gland. The situation is like shuffling a deck of 52 cards and then dealing out three cards. If you did this over and over, you could generate many more than 52 different sets of three cards. As a result of genetic recombination, each B or T cell has a unique set of mini-genes that codes for its unique antigen receptor. After transcription and translation, the receptor molecules are inserted into the plasma membrane.

In B cells another mechanism also contributes to receptor diversity. The gene segments that code for antigen receptors and antibodies are unusually susceptible to **somatic mutations**. These are changes in the nucleotides of DNA that arise in somatic (body) cells. The somatic mutations occur in particular gene segments after antigenic stimulation of B cells. Some of these mutations give rise to antibodies that bind more tightly to (have a higher affinity for) the stimulating antigen. If the same antigen appears again, it preferentially activates those B cells that make higher affinity antibodies.

Major Histocompatibility Complex Antigens

Unless you have an identical twin, your **major histocompatibility complex** (**MHC**) **antigens** are unique. These glycoproteins are also called human leukocyte associated (HLA) antigens because they were first identified on white blood cells. In some form, however, several hundred thousand MHC molecules mark the surface of all your body cells (except red blood cells). Although MHC antigens are the reason that tissues are rejected when they are transplanted from one person to another, their normal function is to help T cells recognize foreign invaders. In other words, these self-antigens aid in the detection of foreign antigens, which is an important first step in any immune response.

There are two types of MHC antigens: class I and class II. Class I MHC (MHC-I) molecules are built into the plasma membranes of all body cells except red blood cells. Class II MHC (MHC-II) molecules appear only on the surface of antigen-presenting cells (described next), cells of the thymus, and T cells that have been activated by exposure to an antigen.

 Figure 22.13   Antigenic determinants (epitopes).

Most antigens have several antigenic determinants that induce production of different antibodies or activate different T cells.

Antigenic determinants

Antigen

 What is the difference between an antigenic determinant and a hapten?

Pathways of Antigen Processing

For an immune response to occur, B and T cells must recognize that a foreign antigen is present. B cells can recognize and bind to antigens in extracellular fluid. T cells, however, can only recognize fragments of antigenic proteins that first have been processed and presented in association with MHC self-antigens (see Fig. 22.14).

Proteins inside body cells are continually being broken down and their amino acids are recycled into other proteins. Some of these peptide fragments, however, associate with a peptide-binding groove of newly synthesized MHC molecules. This association appears to stabilize the MHC molecule and aid its proper folding. Then it can move to and be inserted in the plasma membrane. MHC-I molecules pick up peptide fragments that are 8–9 amino acids long whereas MHC-II molecules pick up peptides having 13–17 amino acids. When a peptide fragment from a *self-protein* is associated with an MHC antigen on the surface of a cell, T cells ignore it. When the fragment is from a *foreign protein,* however, a few T cells recognize an intruder and an immune response ensues.

● **Processing of Exogenous Antigens** Foreign antigens present in fluids outside body cells are termed **exogenous antigens**. They include intruders such as bacteria, bacterial toxins, inhaled pollen or dust, and viruses that have not yet infected a body cell. A special class of cells called **antigen-presenting cells** (**APCs**) process and present exogenous antigens. APCs include **macrophages**, **B cells**, and **dendritic** (*dendro* = tree) **cells**, which are named for their long, branch-

like projections. APCs are strategically located in places where antigens are likely to penetrate nonspecific defenses and enter the body. These are the epidermis and dermis of the skin (Langerhans cells are a type of dendritic cell); mucous membranes that line the respiratory, gastrointestinal, urinary, and reproductive tracts; and lymph nodes. APCs can migrate from tissues via lymphatic vessels to lymph nodes.

The steps in processing and presenting an exogenous antigen by an APC are shown in Fig. 22.14.

① **Ingestion of the antigen.** APCs ingest antigens by phagocytosis or endocytosis. This could happen almost anywhere in the body that invaders, such as microbes, have penetrated the nonspecific defenses.

② **Digestion of antigen into peptide fragments.** Within the phagosome or endosome, digestive enzymes split large antigens into short peptide fragments. At the same time, the APC is synthesizing MHC-II molecules on its endoplasmic reticulum and packing them into secretory vesicles in the Golgi apparatus.

③ **Fusion of vesicles.** The vesicles containing peptide fragments and MHC-II molecules merge and fuse.

④ **Binding of peptide fragments to MHC-II molecules.** After fusion of the two types of vesicles, antigen peptide fragments bind to MHC-II molecules.

⑤ **Insertion of antigen–MHC-II complex into the plasma membrane.** The combined vesicle containing antigen–MHC-II complexes undergoes exocytosis. As a result, the antigen–MHC-II complexes are inserted into the plasma membrane.

Figure 22.14 Steps in processing and presenting of exogenous antigen by an antigen-presenting cell (APC).

🔑 *Except for identical twins, MHC self-antigens are unique in each person. They help T cells recognize foreign invaders.*

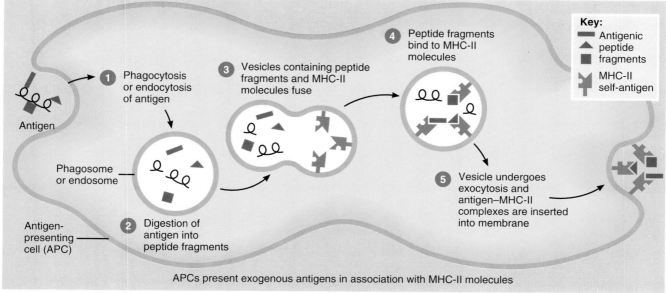

APCs present exogenous antigens in association with MHC-II molecules

❓ What cells are APCs and where in the body are they found?

After processing an antigen, the APC migrates to lymphatic tissue to present the antigen to T cells. There the complex of antigen fragment and MHC is recognized by a small number of T cells that have correctly shaped receptors. The interaction of antigen and antigen receptor then triggers either a cell-mediated immune response or an antibody-mediated immune response.

● **Processing of Endogenous Antigens** Foreign antigens that have been synthesized in a body cell are termed **endogenous antigens**. Usually, these are viral proteins produced after a virus infects the cell and takes over its metabolic machinery. Fragments of endogenous antigens associate with MHC-I molecules and the complex of antigen fragment and MHC-I then moves to the plasma membrane, where it is displayed at the surface of the cell. Most cells of the body, including APCs, can process and present foreign antigens that were synthesized in a body cell.

Cytokines

Cytokines are small protein hormones that stimulate or inhibit many normal cell functions such as cell growth and differentiation. Lymphocytes and APCs both secrete cytokines, as do fibroblasts, endothelial cells, and monocytes. Most cytokines act locally as autocrines (on the cell that secreted them) or as paracrines (on neighboring cells). A few, for example, erythropoietin, are transported in the blood to distant targets (endocrine action). Several cytokines, called colony-stimulating factors and interleukins, stimulate proliferation of progenitor blood cells in bone marrow.

Other cytokines regulate activities of the cells that carry out specific and nonspecific defense responses. When secreted by lymphocytes, cytokines are termed **lymphokines**. When secreted by monocytes or macrophages, they are called **monokines**. The more general term cytokine includes both. Exhibit 22.2 describes some of the cytokines that participate in immune responses.

Antibodies

An **antibody** (**Ab**) can combine specifically with the antigenic determinant on the antigen that triggered its production. It binds to its antigen just as a key fits into a specific lock. In

---

EXHIBIT **22.2** **SUMMARY OF CYTOKINES PARTICIPATING IN IMMUNE RESPONSES**

Cytokine	Comments
Interleukin-1 (IL-1)	Produced by monocytes and macrophages; costimulator of T cell and B cell proliferation; acts on hypothalamus to cause fever.
Interleukin-2 (IL-2) (T cell growth factor)	Secreted by helper T cells to costimulate the proliferation of helper T cells, cytotoxic T cells, and B cells; activates natural killer cells.
Interleukin-4 (IL-4) (B cell-stimulating factor 1)	Produced by activated helper T cells; costimulator for B cells; causes plasma cells to secrete IgE antibodies (see Exhibit 22.3); promotes growth of T cells.
Interleukin-5 (IL-5)	Produced by certain activated CD4+ T cells and activated mast cells; costimulator for B cells and causes plasma cells to secrete IgA antibodies.
Tumor necrosis factor (TNF)	Produced mainly by macrophages; stimulates accumulation of neutrophils and macrophages at sites of inflammation and stimulates their killing of microbes; stimulates macrophages to produce IL-1; induces synthesis of colony-stimulating factors by endothelial cells and fibroblasts; exerts an interferon-like protective effect against viruses; and functions as an endogenous pyrogen to induce fever (see page 812).
Transforming growth factor beta (TGF-β)	Secreted by T cells and macrophages; has some positive effects but is thought to be important for turning off immune responses; inhibits proliferation of T cells and activation of macrophages.
Gamma-interferon (γ-IFN)	Secreted by helper and cytotoxic T cells and NK cells; strongly stimulates phagocytosis by neutrophils and macrophages (formerly called macrophage activating factor [MAF] for this action); activates NK cells; enhances both cellular and antibody-mediated immune responses.
Alpha- and beta-interferons (α-IFN and β-IFN)	Produced by virus-infected cells to inhibit viral replication in uninfected cells; produced by antigen-stimulated macrophages to stimulate T cell growth; activate natural killer cells; inhibit cell growth and suppress formation of some tumors.
Lymphotoxin (LT)	Secreted by cytotoxic T cells; kills cells by causing fragmentation of DNA.
Perforin	Secreted by cytotoxic T cells and perhaps by natural killer cells; perforates cell membranes of target cells, which causes cytolysis.
Macrophage migration inhibiting factor	Produced by T cells; prevents macrophages from migrating away from site of infection.

theory, one could produce as many different antibodies as there are antigen receptors on B cells. Recall that the same recombined genes code for both the antigen receptors on B cells and the antibodies eventually secreted by plasma cells.

Antibodies belong to a group of glycoproteins called globulins, and for this reason they are also known as **immunoglobulins (Igs)**. Most antibodies contain four polypeptide chains (Fig. 22.15). Two of the chains are identical to each other and are called **heavy (H) chains**. Each consists of about 450 amino acids. Short carbohydrate chains are attached to each heavy chain. The other two chains, also identical to each other, are called **light (L) chains**, and each consists of about 220 amino acids. A disulfide bond (S—S) holds each light chain to a heavy chain. Two disulfide bonds also link the midregion of the two heavy chains. This part of the antibody displays considerable flexibility and is called the **hinge region**. Since the antibody "arms" can move somewhat as the hinge region bends, an antibody can assume either a T shape (Fig. 22.15a) or a Y shape (Fig. 22.15b). Disulfide bonds also form loop-shaped **domains** within the light and heavy chains (Fig. 22.15b). Each domain contains about 110 amino acids and folds into a separate unit.

Within each H and L chain are two distinct regions. The tips of the H and L chains, called the **variable (V) regions**, contain the **antigen binding site**. The variable region is different for each kind of antibody. This is the part of the antibody that recognizes and attaches specifically to a particular antigen. Since most antibodies have two antigen binding sites, they are said to be bivalent. Flexibility at the hinge allows the antibody to simultaneously bind to two antigenic determinants that are some distance apart, for example, on the surface of a microbe.

The remainder of each H and L chain is called the **constant (C) region**. The constant region is nearly the same in all antibodies of the same class and is responsible for the type of antigen–antibody reaction that occurs. However, the constant region of the H chain differs from one class of antibody to another, and its structure serves as a basis for distinguishing five different classes, as shown in Exhibit 22.3. These are designated as IgG, IgA, IgM, IgD, and IgE. Each has a distinct chemical structure and a specific biological role. Because they appear first and are relatively short-lived, the presence of IgM antibodies indicates a recent invasion. In a sick patient, the responsible pathogen may be suggested by finding high levels of IgM to a particular organism. Exhibit 22.3 summarizes the structures and functions of the five classes of antibodies.

Cell-Mediated Immunity

A cell-mediated immune response begins with *activation* of a small number of T cells (lymphocytes) by a particular antigen. Once a T cell has been activated, it can undergo *proliferation* and *differentiation* into a clone of **effector cells**, a population of identical cells that can recognize the same antigen and carry out some aspect of the immune attack. Finally, the immune response results in *elimination* of the intruder.

Figure 22.15 Chemical structure of the immunoglobulin G (IgG) class of antibody. It is composed of four polypeptide chains (two heavy and two light) plus a short carbohydrate chain attached to each heavy chain. In (a) each circle represents one amino acid. In (b) V_L = variable region of light chain; C_L = constant region of light chain; V_H = variable region of heavy chain; C_H = constant region of heavy chain.

🔑 *An antibody can combine specifically with the antigenic determinant on the antigen that triggered its production.*

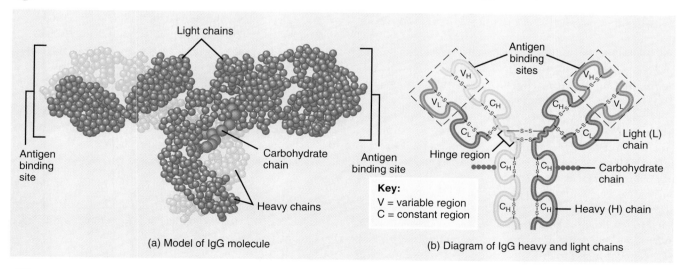

(a) Model of IgG molecule

(b) Diagram of IgG heavy and light chains

Key:
V = variable region
C = constant region

❓ What is the function of the variable regions?

EXHIBIT **22.3**   **CLASSES OF IMMUNOGLOBULINS (Igs)**

Name and Structure	Characteristics and Functions
IgG	Most abundant, about 75% of all antibodies in the blood; found in blood, lymph, and the intestines; monomer (one unit) structure. Protect against bacteria and viruses by enhancing phagocytosis, neutralizing toxins, and triggering the complement system. They are the only class of antibody to pass the placenta from mother to fetus and thereby confer considerable immune protection in newborns.
IgA	Make up about 15% of all antibodies in the blood; occur as monomers and dimers (two units). Found mainly in sweat, tears, saliva, mucus, milk, and gastrointestinal secretions. Smaller quantities present in blood and lymph. Levels decrease during stress, lowering resistance to infection. Provide localized protection on mucous membranes against bacteria and viruses.
IgM	About 5–10% of all antibodies in the blood; occur as pentamers (five units); first antibodies to be secreted by plasma cells after an initial exposure to any antigen; found in blood and lymph. Activate complement and cause agglutination and lysis of microbes. Also present as monomers on the surfaces of B cells, where they serve as antigen receptors. In blood plasma the anti-A and anti-B antibodies of the ABO blood group, which bind to A and B antigens during incompatible blood tranfusions, are also IgM antibodies (see Fig. 19.11 on page 571).
IgD	Less than 1% of all antibodies in the blood; occur as monomers; found in blood, in lymph, and on the surfaces of B cells as antigen receptors. Involved in activation of B cells.
IgE	Less than 0.1% of all antibodies in the blood; occur as monomers; located on mast cells and basophils. Involved in allergic and hypersensitivity reactions; provide protection against parasitic worms.

Activation, Proliferation, and Differentiation of T Cells

Antigen receptors on the surface of T cells are called **T cell receptors (TCRs)**. They recognize and bind to specific foreign antigen fragments that are presented together with self-MHC molecules (see Fig. 22.16). There are literally millions of different T cells, each with its own unique TCRs that can recognize a specific antigen–MHC combination. At any given time, most T cells are inactive. When an antigen enters the body, only a few T cells have TCRs that can recognize and bind to the antigen. Antigen recognition by a TCR is the *first signal* in activation of a T cell.

A *second signal*, called a **costimulator**, also is needed. More than 20 such costimulators are known. Some are secreted cytokines, for example, **interleukin-1 (IL-1)** and **interleukin-2 (IL-2)**. Other costimulators include pairs of plasma membrane molecules, one on the surface of the T cell and a second on the surface of an APC, that cause the two cells to adhere to one another for a period of time. The need for two signals is a little like starting and driving a car. When you insert the correct key (antigen) in the ignition (TCR) and turn it, the car starts (recognition of specific antigen). But the car does not move forward until you move the gear shift into drive (costimulation). The need for costimu-

lation may prevent an immune response from occurring accidentally. It is thought that different costimulators affect the activated T cell in different ways, just as shifting a car into reverse has a different effect than shifting it into drive. Moreover, recognition (antigen binding to receptor) without costimulation is thought to lead to a prolonged *state of inactivity* called **anergy** in both T cells and B cells. (Imagine a car in neutral with its engine running until it's out of gas!)

When a T cell has received two signals (antigen recognition and costimulation), it is said to be **activated** or **sensitized**. It enlarges and begins to **proliferate** (divide several times) and **differentiate** (form more highly specialized cells). The result is a **clone**, or population of cells that can recognize the same antigen. Before the first exposure to a certain antigen, only a handful of T cells might be able to recognize it. But after an immune response has occurred, there are thousands.

Types of T Cells

Several different types of differentiated T cells appear: helper T cells, cytotoxic (killer) T cells, suppressor T cells, and memory T cells.

● **Helper T (T$_H$) Cells**   Most T cells that display CD4 develop into **helper T (T$_H$) cells** or **T4 cells**. Resting (inactive) T_H cells recognize antigen fragments associated with MHC-II molecules and are costimulated by interleukin-1, which is secreted by macrophages (Fig. 22.16a). This means they are activated mainly by antigen-presenting cells.

Within hours after costimulation, helper T cells start secreting a variety of cytokines (see Exhibit 22.2). Different subsets of helper T cells specialize in production of particular cytokines. In mice, which are a popular animal for

Figure 22.16 Activation, proliferation, and differentiation of T cells.

The binding of CD4 to MHC-II and CD8 to MHC-I helps anchor the TCR–antigen interaction so that antigen recognition can occur.

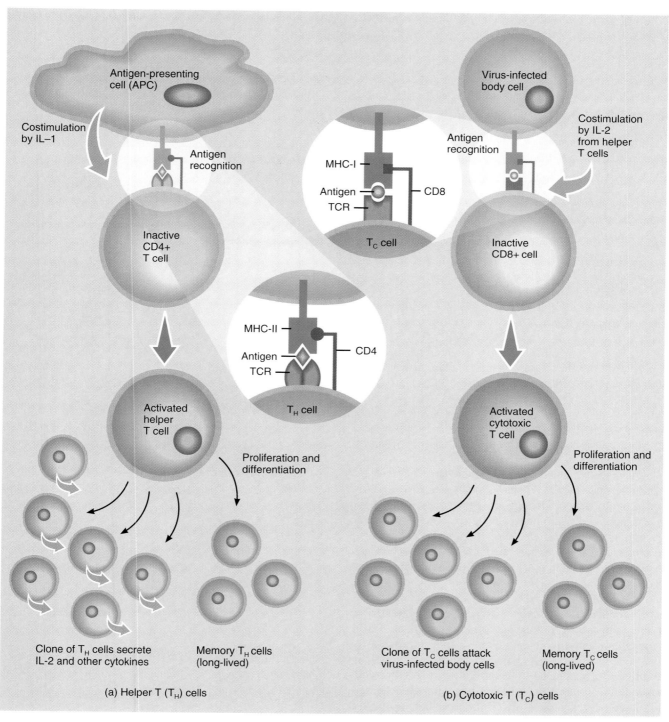

(a) Helper T (T_H) cells

(b) Cytotoxic T (T_C) cells

What are the first and second signals in activation of a T cell?

immune system studies, there are two main subsets, called type 1 and type 2 helper T cells (T_{H1} and T_{H2}). In humans the variety of T cells is much more complex. Moreover, particular subtypes appear in certain diseases, such as asthma, multiple sclerosis, and Lyme arthritis. One very important cytokine produced by helper T cells is interleukin-2 (IL-2). IL-2 is needed for virtually all immune responses and it is the prime substance that triggers T cell proliferation. It can act as a costimulator for resting helper T or cytotoxic T cells, and it enhances activation and proliferation of T cells, B cells, and natural killer cells. Other cytokines secreted by helper T cells are gamma-interferon, interleukin-4 (IL-4), and transforming growth factor beta (TGF-β).

Some actions of interleukin-2 provide a good example of a positive feedback system that makes a beneficial contribution. As noted earlier, activation of a helper T cell stimulates it to start secreting IL-2. IL-2 acts in an autocrine manner by binding to IL-2 receptors on the plasma membrane of the cell that secreted it. One effect is stimulation of cell division. As the T_H cells proliferate, they secrete more and more IL-2, a positive feedback effect. IL-2 may also act in a paracrine manner by binding to IL-2 receptors on neighboring T_H, T_C, or B cells. If any of these cells have already bound an antigen, IL-2 serves as a costimulator to activate them.

● **Cytotoxic T (T_C) cells** T cells that display CD8 develop into **cytotoxic T (T_C) cells** or **T8 cells**. They also are known as **cytolytic** or **killer T cells**. T_C cells recognize foreign antigens combined with MHC-I molecules on the surfaces of (1) body cells infected by viruses, (2) some tumor cells, and (3) cells of a tissue transplant (Fig. 22.16b). However, to become cytolytic (able to lyse cells) they need costimulation by IL-2 or other cytokines produced by helper T cells. (Recall that T_H cells are activated by antigen associated with MHC-II molecules.) Thus maximal activation of T_C cells requires presentation of antigen associated with both MHC-I and MHC-II molecules.

● **Other Types of T Cells** T cells that mediate a class of allergic response called **delayed type hypersensitivity** (see page 703) may display either CD4 or CD8. In response to activation by antigen, these T cells secrete cytokines, especially gamma-interferon, that activate macrophages. The activated macrophages, in turn, eliminate the antigen. **Suppressor T (T_S) cells** are thought to be a class of T cells distinct from T_H cells and T_C cells. However, their presence is difficult to demonstrate, the nature of their receptors for antigen is not known, and their very existence is controversial. They may down-regulate or dampen parts of the immune response by producing cytokines such as TGF-β, which inhibits proliferation of T cells. Another possibility is that they directly destroy activated lymphocytes. **Memory T cells** are programmed to recognize the original invading antigen. Should the same type of pathogen invade the body at a later date, thousands of memory cells are available to initiate a far swifter reaction than occurred during the first invasion. The second response usually is so fast that the

pathogens are destroyed before any signs or symptoms of disease occur.

Elimination of Invaders

Cytotoxic T cells are the army that marches forth to do battle with foreign invaders in cell-mediated immune responses. They leave lymphatic tissues and migrate to the site of invasion, infection, or tumor formation. They recognize and attach to the target cell that bears the same antigen as stimulated activation and proliferation of their progenitor cells. Then they deliver a "lethal hit" that kills the target cell without damaging the T_C itself. After detaching from the target cell, the T_C can seek out and destroy another invader that displays the same antigen (Fig. 22.17).

Two killing mechanisms are used by T_C cells. In the first, granules containing perforin undergo exocytosis from the

Figure 22.17 *Activity of cytotoxic T cell. After delivering a lethal hit, a cytotoxic T cell can detach and attack another target cell displaying the same antigen.*

Cytotoxic T cells kill microbes directly by secreting perforin and lymphotoxin.

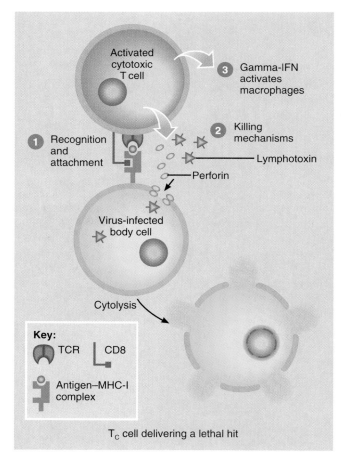

T_C cell delivering a lethal hit

Besides cells infected by viruses, what other types of target cells are attacked by T_C cells?

T_C cell. Perforin forms holes in the plasma membrane of the target cell, which allow extracellular fluid to flow in, and the cell bursts (cytolysis). In the second, the T_C cell secretes a toxic molecule known as lymphotoxin (LT) that activates damaging enzymes within the target cell. These enzymes cause the DNA to fragment, and the target cell dies. Cytotoxic T cells also secrete gamma-interferon, which activates neutrophils and macrophages, greatly increasing their phagocytic activity.

In summary, cytotoxic T cells can destroy antigens directly by killing the cells that bear them and indirectly by secreting gamma-interferon to activate phagocytic cells at the scene of the battle. T_C cells are especially effective against slowly developing bacterial diseases (such as tuberculosis and brucellosis), some viruses, fungi, cancer cells associated with viral infection, and transplanted cells.

Antibody-Mediated (Humoral) Immunity

The body contains not only millions of different T cells but also millions of different B cells, each capable of responding to a specific antigen. Whereas cytotoxic T cells leave lymphatic tissue to seek out and destroy a foreign antigen, B cells stay put. In the presence of a foreign antigen, specific B cells in lymph nodes, the spleen, or lymphatic tissue in the gastrointestinal tract become activated. They differentiate into plasma cells that secrete specific antibodies, which then circulate in the lymph and blood to reach the site of invasion.

Activation, Proliferation, and Differentiation of B Cells

During activation of a B cell, an antigen binds to antigen receptors on the cell surface (Fig. 22.18). B cell antigen receptors are chemically similar to the antibodies that will eventually be secreted by their progeny. Although B cells can respond to unprocessed antigen dissolved in lymph or interstitial fluid, their response is much more intense when nearby follicular dendritic cells also process and present antigen to them. Some antigen is then taken into the B cell, broken down into peptide fragments and combined with MHC-II self-antigen, and moved to the B cell surface. Helper T cells recognize the antigen–MHC-II combination and deliver the costimulation needed for B cell proliferation and differentiation. The T_H cell produces interleukin-2 and other cytokines that act as costimulators to activate B cells. Interleukin-1, secreted by macrophages, also enhances B cell proliferation and differentiation into plasma cells.

Some of the activated B cells enlarge and divide and differentiate into a clone of **plasma cells**. The phenomenal rate of antibody secretion by plasma cells is about 2000 molecules per second for each cell, and it occurs for 4 or 5 days until the plasma cell dies. The activated B cells that do not differentiate into plasma cells remain as **memory B cells**,

Figure 22.18 Activation, proliferation, and differentiation of B cells.

🔑 *Plasma cells are the progeny of activated B cells that produce antibodies.*

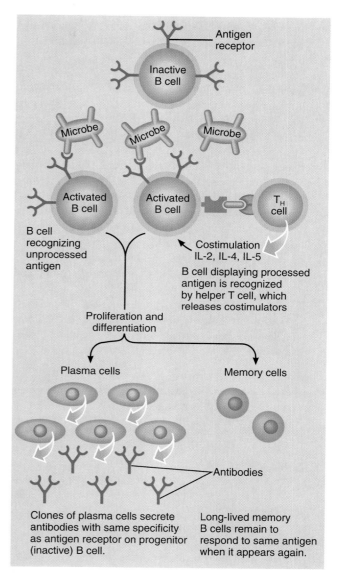

❓ How many different kinds of antibodies will be secreted by the plasma cells in the clone shown here?

ready to respond more rapidly and forcefully should the same antigen reappear at a future time.

Different antigens stimulate different B cells to develop into plasma cells and their accompanying memory B cells. The B cells of a particular clone are capable of secreting only one kind of antibody (immunoglobulin). The secreted antibody is identical in its specificity to the antigen receptor displayed by the progenitor B cell that responded to the antigen in the first place. Each specific antigen activates only those B cells that are predestined (by the combination of mini-genes they carry) to secrete antibody specific to that antigen. Antibodies produced by a clone of plasma cells enter the circulation and form antigen–antibody complexes with the antigen that initiated their production.

Functions of Antibodies

The five classes of immunoglobulins have functions that differ somewhat (see Exhibit 22.3), but all attack antigens in several ways:

1. **Neutralizing antigen.** The reaction of antibody with antigen blocks or neutralizes the damaging effect of some bacterial toxins and prevents attachment of some viruses to body cells.
2. **Immobilization of bacteria.** If antibodies form against cilia or flagella of motile bacteria, the antigen–antibody reaction may cause the bacteria to lose their motility, which limits their spread into nearby tissues.
3. **Agglutination and precipitation of antigen.** Because antibodies have two or more sites for binding to antigen, the antigen–antibody reaction may cross-link pathogens to one another, causing agglutination (clumping together). Likewise, soluble antigens may come out of solution and form a precipitate when cross-linked by antibodies. Phagocytic cells more readily ingest insoluble materials.
4. **Activation of complement.** Antigen–antibody complexes initiate the classical pathway of the complement system by activating C1 (see Fig. 22.9).

5. **Enhancing phagocytosis.** Antibodies enhance the activity of phagocytes by causing agglutination and precipitation, by activating complement, and by coating microbes so that they are more susceptible to phagocytosis, a process known as **opsonization**.
6. **Providing fetal and newborn immunity.** Resistance of the fetus and newborn baby to infection stems mainly from maternal IgG antibodies that pass across the placenta before birth and IgA antibodies that are absorbed from breast milk after birth.

A summary of the various cells involved in immune responses is presented in Exhibit 22.4.

MONOCLONAL ANTIBODIES

Scientists have known for many years how to stimulate laboratory animals (or humans) to produce antibodies. After injecting a particular antigen, antibodies are produced against the antigen by plasma cells and can be harvested from the blood. However, since an antigen typically has many antigenic determinants, the antibodies are produced by many different clones of plasma cells. These antibodies vary physically and chemically; they are not pure. If a single plasma cell could be isolated and induced to proliferate into a clone of identical cells, then the antibodies produced would all be the same. But lymphocytes and plasma cells are difficult to grow in culture. Scientists sidestepped this difficulty by fusing B

EXHIBIT **22.4** **SUMMARY OF FUNCTIONS OF CELLS PARTICIPATING IN IMMUNE RESPONSES**

Cell	Functions
ANTIGEN-PRESENTING CELLS (APCS)	
Macrophage	Phagocytosis; processing and presentation of foreign antigens to T cells; secretion of interleukin-1, which stimulates secretion of interleukin-2 by helper T cells and induces proliferation of B cells; secretion of interferons that stimulate T cell growth.
Dendritic cell	Processes and presents antigen to T and B cells; found in mucous membranes, skin, and lymph nodes.
B cell	Processes and presents antigen to helper T cells.
LYMPHOCYTES	
Cytotoxic (cytolytic or killer) T cell (T_C or T8 cell)	Causes lysis and death of foreign cells by releasing perforin and lymphotoxin; releases other cytokines that attract macrophages and increase their phagocytic activity (gamma-IFN) and prevent macrophage migration from site of action (macrophage migration inhibition factor).
Helper T cell (T_H or T4 cell)	Cooperates with B cells to amplify antibody production by plasma cells and secretes interleukin-2, which stimulates proliferation of T and B cells. May secrete gamma-IFN and tumor necrosis factor (TNF), which stimulate inflammatory response.
Suppressor T cell (T_S cell)	Thought to down-regulate immune responses by producing cytokines such as transforming growth factor beta, which inhibits proliferation of T cells. May also directly destroy activated lymphocytes.
Memory T cell	Remains in lymphoid tissue and recognizes original invading antigens, even years after the first encounter.
B cell	Differentiates into antibody-producing plasma cell. May process and present antigen to helper T cell.
Plasma cell	Descendant of B cell that produces and secretes antibodies.
Memory B cell	Ready to respond more rapidly and forcefully than initially should the same antigen enter the body in the future.

cells with tumor cells that grow easily and proliferate endlessly. The resulting hybrid cell is called a **hybridoma** (hī-bri-DŌ-ma). Hybridoma cells are a long-term source of large quantities of pure antibodies called **monoclonal antibodies (MAbs)** because they come from a single clone of identical cells. Such antibodies combine with just one antigenic determinant.

One clinical use of monoclonal antibodies is for measuring levels of a drug in a patient's blood. Other uses include the diagnosis of strep throat, pregnancy, allergies, and diseases such as hepatitis, rabies, and some sexually transmitted diseases. MAbs have also been used to detect cancer at an early stage and to determine the extent of metastasis. They may also be useful in preparing vaccines to counteract the rejection associated with transplants, to treat autoimmune diseases, and perhaps to treat AIDS. ■

Immunological Memory

A hallmark of immune responses is memory for specific antigens that have triggered immune responses in the past. Immunological memory is due to the presence of long-lived antibodies and very long-lived lymphocytes that arise during proliferation and differentiation of antigen-stimulated B and T cells.

Immune responses, whether cell-mediated or antibody-mediated, are much quicker and more intense after a second or subsequent exposure to an antigen than after the first exposure. Initially, only a few cells have the correct specificity to respond, and the immune response may take several days to build to maximum intensity. Because thousands of memory cells exist after an encounter with an antigen, the next time the same antigen appears they can proliferate and differentiate into plasma cells or cytotoxic T cells within hours.

Immunological memory can be demonstrated by measuring the amount of antibody in serum, called the *antibody titer* (TĪ-ter). After an initial contact with an antigen, there is a period of several days during which no antibody is present. Then there is a slow rise in the antibody titer, first IgM and then IgG, followed by a gradual decline (Fig. 22.19). This is called the **primary response**.

Memory cells may remain for decades. Every time the same antigen is encountered again, there is a rapid proliferation of memory cells. The antibody titer is far greater than during a primary response and is mainly IgG antibodies. This accelerated, more intense response is called the **secondary response**. Antibodies produced during a secondary response have an even higher affinity for the antigen than those secreted during a primary response and are thus more successful in disposing of it.

Primary and secondary responses occur during microbial infection. When you recover from an infection without taking antibiotic drugs, it is usually because of the primary response. If, at a later time, you are infected by the same microbe, the secondary response could be so swift that the microbes are quickly destroyed and you do not exhibit any signs or symptoms.

Figure 22.19 Secretion of antibodies in the primary (after first exposure) and secondary (after second exposure) responses to the same antigen.

 Immunological memory is the basis for successful immunization by vaccination.

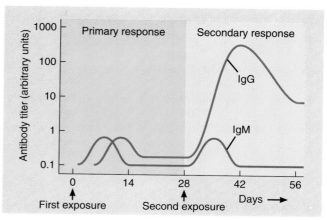

 According to this graph, how much more IgG is secreted into the blood in the secondary than in the primary response?

Immunological memory provides the basis for immunization by vaccination against certain diseases, for example, polio. When you receive the vaccine, your B and T cells are activated. Should you encounter the same pathogen again as an infecting microbe, your body experiences the secondary response. Exhibit 22.5 summarizes the various types of naturally and artificially acquired immunity.

EXHIBIT **22.5** TYPES OF IMMUNITY

Type of Immunity	How Acquired
Naturally acquired active immunity	Antigen recognition by B cells and T cells and costimulation lead to antibody-secreting plasma cells, cytotoxic T cells, and B and T memory cells.
Naturally acquired passive immunity	Transfer of IgG antibodies from mother to fetus across placenta or IgA antibodies from mother to baby in milk during breast-feeding.
Artificially acquired active immunity	Antigens introduced in a vaccination stimulate cell-mediated and antibody-mediated immune responses, leading to production of memory cells. The antigens are pretreated to be immunogenic but not pathogenic; that is, they will trigger an immune response but not cause significant illness.
Artificially acquired passive immunity	Intravenous injection of immunoglobulins (antibodies).

Self-Recognition and Immunological Tolerance

To function properly, all T cells must have two traits. Each of your T cells must (1) be able to recognize your own MHC molecules, a process known as **self-recognition**, and (2) lack reactivity to peptide fragments from your own proteins, a condition known as **immunological tolerance**. B cells also display immunological tolerance. Loss of immunological tolerance leads to autoimmune diseases (see page 702).

While residing in the thymus gland, immature T cells that become capable of recognizing self-MHC molecules survive while those that do not undergo apoptosis (programmed cell death). This aspect of development of immunocompetence is termed **positive selection** (Fig. 22.20a). The T cells selected to survive *can recognize* the MHC part of an antigen–MHC complex.

The development of immunological tolerance, on the other hand, occurs by a weeding out process called **negative selection**. Cells with TCRs that recognize peptide fragments from self-proteins are eliminated or inactivated (Fig. 22.20a). The T cells selected to survive *do not respond* to fragments of molecules that are normally present in the body. Negative selection occurs in two ways: **deletion** and **anergy**. In deletion, self-reactive T cells undergo apoptosis and die whereas in anergy they remain alive but are unresponsive to antigenic stimulation. It is estimated that only 1 in 100 immature T cells in the thymus receives the proper signals to survive apoptosis during both positive and negative selection and emerges as a mature, immunocompetent T cell.

Even after T cells emerge from the thymus, it is possible they may come in contact with an unfamiliar self-protein. In such cases, they may also become anergic if there is no co-stimulator (Fig. 22.20b). Some evidence suggests that dele-

Figure 22.20   Development of self-recognition and immunological tolerance.

Positive selection allows recognition of self-MHC-I and self-MHC-II; negative selection provides immunological tolerance of self-peptides.

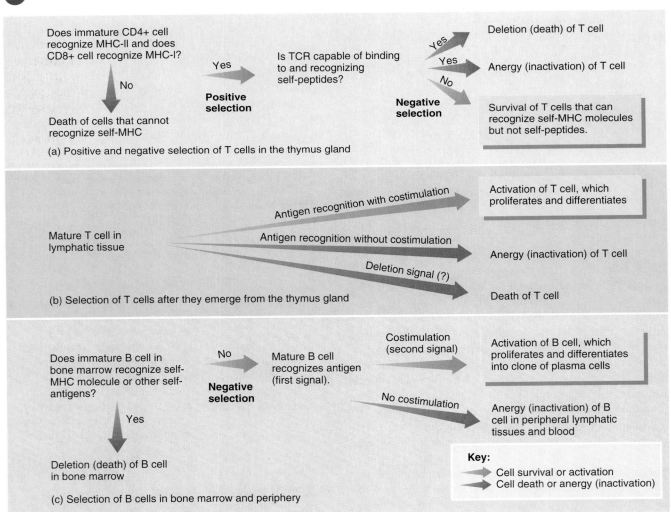

(a) Positive and negative selection of T cells in the thymus gland

(b) Selection of T cells after they emerge from the thymus gland

(c) Selection of B cells in bone marrow and periphery

How does deletion differ from anergy?

tion of self-reactive T cells may also occur after they leave the thymus. B cells also develop tolerance through deletion and anergy (Fig. 22.20c). While B cells are developing in bone marrow, those cells exhibiting antigen receptors that recognize common self-antigens such as MHC antigens or blood group antigens are deleted. Once B cells are released into the blood, however, anergy appears to be the main mechanism for preventing responses to self-proteins. When B cells encounter antigen not associated with an antigen-presenting cell (APC), the necessary costimulation signal often is missing. In this case, the B cell is likely to become anergic (inactivated) rather than activated.

Immunology and Cancer

When a normal cell transforms into a cancer cell, it may display cell surface components called **tumor antigens**. These are molecules that are rarely, if ever, displayed on the surface of normal cells. If the immune system can recognize tumor antigens as nonself, it can destroy the cancer cells carrying them. Such an immune response is called **immunological surveillance** and is carried out by cytotoxic T cells, macrophages, and natural killer cells. It appears to be most effective in eliminating tumor cells that arise due to a cancer-causing virus. In part, evidence for immunological surveillance comes from observations in people whose immune system is depressed, for whatever reason. For example, patients who are taking immunosuppressive drugs to prevent rejection of a tissue transplant (see page 704) do not have a higher than normal incidence of most cancers. They do have a greatly increased incidence, however, of virus-associated cancers.

IMMUNOTHERAPY

For many years, researchers have been trying to induce the immune system to mount an attack against cancer. This approach is called **tumor immunotherapy** and is effective against some types of cancer.

In adoptive cellular immunotherapy, cells that have antitumor activity are injected into the blood of a cancer patient. The hope is that these "adopted" cells will seek out and destroy tumor cells. One method uses a patient's own inactive cytotoxic T cells and natural killer cells. They are removed in a blood sample and cultured with IL-2, which activates them. Such cells are called lymphokine-activated killer (LAK) cells. The LAK cells are then transfused back into the patient's blood. Although LAK cells cause some tumor regression, severe complications affect most patients.

Cytokine therapy uses cytokines as therapeutic agents. The first cytokines that were shown to be effective against any human cancer were interferons. Of the interleukins, the one most widely employed to fight cancer is interleukin-2, either alone or with LAK cells. The treatment is effective in causing tumor regression in some patients, presumably by stimulating cytotoxic T cells and natural killer cells. It also can be very toxic. Among the adverse effects are high fever, severe weakness, difficult breathing due to pulmonary edema, and hypotension leading to shock. Interleukin-4 is being tested because it can also activate T_C cells and may have fewer adverse side effects.

Monoclonal antibodies, either alone or in combination with radioactive isotopes, toxins, or drugs, are being used in trial studies of **antibody therapy** to treat cancer. This approach has the advantage of destroying diseased tissues only, while sparing healthy tissues, thus overcoming some of the major adverse effects of chemotherapy and radiation treatment. ■

AGING AND THE IMMUNE SYSTEM

Elderly people are more susceptible to all types of infections and malignancies because the immune system functions less effectively. With age, there are progressive declines in both cell-mediated and antibody-mediated immune responses. The response to vaccines is decreased, production of antibodies against self-proteins increases, and the number of helper T cells decreases.

DISORDERS: HOMEOSTATIC IMBALANCES

AIDS: ACQUIRED IMMUNODEFICIENCY SYNDROME

Never before have humans been confronted with an epidemic in which the primary disease only lowers the victim's immunity, and then one or more unrelated diseases produce the fatal symptoms. **Acquired immunodeficiency syndrome (AIDS)** is caused by the **human immunodeficiency virus (HIV)**. The initial response to HIV invasion is a modest decline in the number of circulating T4 cells. Infected people experience a brief flu-like illness, with chills and fever, but the immune system fights back by making antibodies against HIV and the number of circulating T4 cells recovers nearly to normal. Although infected people test positive for HIV antibodies, they typically have few clinical signs or symptoms and do not yet have AIDS.

Over the next 2–10 years, the virus slowly destroys the T4 cell population in lymphatic tissues throughout the body. As immune responses weaken, people develop certain **indicator diseases** (diseases that are rare in the general population but common in AIDS patients). At this point, the diagnosis of AIDS is made. A simplified definition of AIDS includes anyone infected with HIV and having a T4 lymphocyte count under 200/mm³ of blood. (Normally, the T4 count would be about 1200/mm³.)

Two indicator diseases provided the original clues that a new disorder had appeared. AIDS was first recognized in June 1981 as a result of reports from the Los Angeles area to the Centers for Disease Control and Prevention (CDC) of several

DISORDERS: HOMEOSTATIC IMBALANCES   CONTINUED

cases of a very rare type of pneumonia caused by a fungus. The pneumonia, called *Pneumocystis carinii* (noo-mō-SIS-tis kar-RIN-ē-ī) pneumonia (PCP), occurred among homosexual males. At about the same time, the CDC also received reports from New York and Los Angeles concerning an increase in the incidence of Kaposi's sarcoma (KS) among homosexual males. Until the 1980s, this had been a rare, generally benign, skin cancer usually found in elderly Jewish or Italian men. In AIDS patients it is aggressive and rapidly fatal. KS produces painless purple or brownish skin lesions that resemble bruises.

Scientists first isolated the virus that causes AIDS in 1983. There have been several theories about the origin of HIV. It is believed that it arose by mutation of a virus found in monkeys that had been endemic in some areas of central Africa for many years. The virus has been found in blood samples preserved from as early as 1959 in several African nations and in Great Britain and as early as 1969 in the United States.

In the United States AIDS is present in all 50 states. The primary victims are homosexual men, intravenous drug users, and patients who received blood products before 1985, when testing of all donated blood for HIV antibodies began. Others at high risk are the heterosexual partners of HIV-infected individuals. Worldwide, however, an estimated 75% of those who have AIDS were infected through heterosexual contacts. At present, the average incubation period (time interval from infection with HIV to full-blown AIDS) is about 10 years. About 5–8% of those infected with HIV have now survived 13–16 years, however, without developing AIDS. It is hoped that studies of these

people, termed **nonprogressives**, will reveal the mechanism of their successful resistance to destruction of the immune system.

The World Health Organization (WHO) projects that 40 million people will be infected with HIV throughout the world by the year 2000. More than half will be women and a quarter will be children. The current number of infected Americans is estimated to be 1–1½ million, and most do not even know they carry the virus. All carriers of the virus are assumed to be infected for life and are capable of transmitting the virus to others.

HIV: Structure and Pathogenesis

Viruses consist of a core of DNA or RNA surrounded by a protein coat (capsid). Some viruses, including HIV, also contain an envelope (outer layer) composed of a phospholipid bilayer penetrated by glycoproteins (Fig. 22.21). Outside a living host cell a virus is unable to replicate. However, once a virus enters a cell, the viral nucleic acid uses the host cell's enzymes, ribosomes, nutrients, and other resources to make copies of itself. As these components accumulate, they are assembled into a large number of viruses that can then leave the host cell to infect other cells.

As viruses go through their cycle of replication, they can damage or kill host cells in various ways. These include shutting down the synthesis of host cell proteins, RNA, and DNA; inhibiting cell division; damaging DNA; and rupturing lysosomes that can lead to autolysis. Moreover, the body's own defenses will attack the infected cells, killing them as well as the viruses they harbor.

Figure 22.21   Human immunodeficiency virus (HIV), the cause of AIDS. (a) The core contains RNA and reverse transcriptase, plus several other enzymes. The protein coat (capsid) around the core consists of a protein called P24. Outside the protein coat is a layer composed of another protein called P17. The envelope consists of a phospholipid bilayer studded with gycoproteins (GP120 and GP41). These glycoproteins play a vital role when HIV binds to and enters certain target cells.

HIV enters helper T cells by receptor-mediated endocytosis in which GP120 binds to CD4.

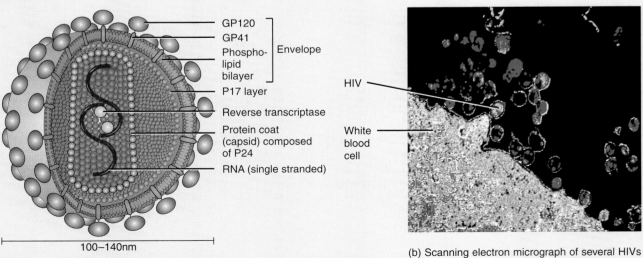

(a) Structure of HIV

GP120
GP41
Phospholipid bilayer ⎤ Envelope
P17 layer
Reverse transcriptase
Protein coat (capsid) composed of P24
RNA (single stranded)

100–140nm

HIV
White blood cell

(b) Scanning electron micrograph of several HIVs targeting a white blood cell (93,500x)

 What is the function of reverse transcriptase?

HIV is a **retrovirus**. Retroviruses carry their genetic material in RNA and copy this genetic code into DNA by using an enzyme in the core called **reverse transcriptase**. Once HIV produces DNA from its RNA, the DNA is integrated into the host cell's DNA. There it can remain dormant, giving no sign of its presence, or it can take over the host cell's genetic machinery to produce more viruses.

HIV enters body cells by receptor-mediated endocytosis (see Fig. 3.13). The receptor or docking protein that permits HIV entry is the CD4 molecule on the surface of T4 cells, although other cellular factors, not yet understood, must also contribute. With time, the number of T4 cells, mainly helper T cells, declines due to death of infected cells. The result is progressive collapse of the immune system. Since cytokines secreted by helper T cells normally stimulate the activity of monocytes, neutrophils, and macrophages, nonspecific defense mechanisms also are depressed. The person becomes susceptible to **opportunistic infections**, invasion of normally harmless microorganisms that now proliferate wildly because of the defective immune system.

Besides *Pneumocystis carinii* pneumonia, AIDS victims have persistent diarrhea and are especially susceptible to tuberculosis, leukoplakia (whitish patches on mucous membranes primarily due to yeast infections), cytomegalovirus (leading to blindness and dementia), herpes simplex, and severe shingles, among many other bacterial and fungal infections. The AIDS virus infects macrophages, dendritic cells, brain cells (where HIV-infected cells may release toxins that disrupt and kill other brain cells), as well as T4 cells. Not all infected cells display CD4, so there are probably other receptors that permit HIV entry. Also, the virus membrane appears able to directly fuse with the host cell plasma membrane in some cases. Dendritic cells and macrophages are pivotal cells in the development of AIDS. They serve as reservoirs for the virus, since viral infection does not seem to harm them, and they spread the virus throughout the body.

Some weeks after infection with HIV, the host develops antibodies against several proteins in the virus. Usually, antibodies appear in blood within 6 weeks to 6 months after exposure. Antibodies normally are protective because they help eliminate an intruder. In the case of the AIDS virus, this is not necessarily the case because HIV can remain hidden inside body cells, unavailable to form antigen–antibody complexes. HIV may also escape detection by cytotoxic T cells, natural killer cells, and phagocytes. The virus further evades immune defenses by undergoing rapid antigenic changes in its surface proteins. Moreover, infected cells displaying viral antigens can fuse to uninfected cells and spread the virus that way.

In rare cases a person may harbor HIV without forming antibodies against it. Thus a standard blood test that detects antibodies would be negative. The presence of nucleic acids from HIV can still be detected using a method called the polymerase chain reaction (PCR). A PCR test is difficult and expensive, however, which prevents its use for general screening.

Transmission and Protection

Although HIV has been isolated from several body fluids, the only documented transmissions are from blood, semen, or vaginal secretions or by way of breast milk from a nursing mother to her baby. The virus is found free and in macrophages in these fluids.

HIV is transmitted by sexual contact between males and females through vaginal, oral, or anal intercourse. Because she receives a large quantity of infected fluid (semen), the female's risk of becoming infected is much higher than the male's. Male homosexuals transmit HIV to their partner by anal or oral intercourse. Dendritic cells, which are present in mucosal membranes and are efficient antigen presenting cells, are the suspected route for HIV entry into the lymphatic system. HIV is also effectively transmitted through exchanges of blood, for example, by contaminated hypodermic needles, contact with open wounds, or using the same razor blade for shaving. Infected mothers may transmit the virus to their infants before or during birth. It does not appear that people become infected as a result of routine, nonsexual contacts.

No evidence exists that AIDS can be spread through kissing although "deep kissing" with exchange of saliva poses a theoretical danger since dendritic cells are present in the oral mucosa. There is no known case of transmission from a mosquito bite. It also appears that health-care personnel who take proper routine barrier precautions when dealing with body fluids (gloves, masks, safety glasses) are not at risk unless the barriers fail.

Outside the body, HIV is fragile and can easily be eliminated. For example, dishwashing and clotheswashing, by exposing the virus to 135°F (56°C) for 10 minutes, will kill HIV. Chemicals such as hydrogen peroxide (H_2O_2), rubbing alcohol, Lysol, household bleach, and germicidal skin cleaner (such as Betadine and Hibiclens) are also very effective, as is standard chlorination in swimming pools and hot tubs.

Drugs Against HIV

Medical scientists are engaged in an immense effort to find a cure for AIDS. One of the problems in treating AIDS is that HIV can lie undetected in body cells. In addition, HIV can infect a variety of cells, including those in the central nervous system that are protected by the blood–brain barrier. Added to this is the problem of opportunistic infections, which may be very difficult to treat. Any therapy must overcome the problem that antiviral agents may also harm host cells. Thus scientists are trying to devise strategies for disrupting specific viral activities. Some research centers on preventing binding of the virus to the host CD4 protein. Other strategies are to prevent conversion of RNA to DNA, which is catalyzed by reverse transcriptase, block processing of viral proteins by specific viral enzymes, inhibit assembly of viruses within the host cell, and thwart release of new viruses.

To date, four drugs with similar action are used to inhibit HIV replication and slow the progression of AIDS. All are nucleoside analogs, substances that are similar to the naturally occurring nucleosides in RNA and DNA. They block conversion of retroviral RNA into DNA. The first and still most commonly used drug to treat AIDS is AZT (azidothymidine) or Retrovir. Among patients taking AZT, there is a slowing in the progression of symptoms. The main side effects are red bone marrow damage and anemia. Eventually, the virus develops resistance to the drug. Other drugs are DDI (dideoxyinosine), DDC (dideoxycytidine), and D4T (stavudine) or Zerit, which may be used in patients who do not respond to AZT or have become resistant to it. Although patients taking these nucleoside analogs show improvement in immunologic functions, for example,

DISORDERS: HOMEOSTATIC IMBALANCES CONTINUED

increased T4 counts, there are serious side effects such as pancreatitis and inflammation of peripheral nerves.

Alpha-interferon is believed to inhibit the final stage of virus production. It reduces the spread of Kaposi's sarcoma and is being tried both alone and in combination with other drugs. A host of other drugs are also being tested. It is quite possible that AIDS, like cancer, will have to be treated with a variety of drugs.

Vaccines Against HIV

Considerable effort has been expended to develop a vaccine against HIV, with little success. A vaccine would stimulate the production of antibodies to block the virus before it could infect body cells. Development of an effective vaccine has been impeded by the ability of HIV to mutate so quickly and the lack of a suitable experimental animal model. Most animals are not susceptible to HIV infection. And although chimpanzees can be infected, they are in short supply and very expensive to maintain. Moreover, there may be a shortage of volunteers when a potential vaccine is ready for testing in humans.

Prevention of Transmission

At present, the only means of preventing AIDS is to block transmission of HIV. Sexual transmission of HIV can be prevented by abstinence from vaginal, oral, and anal intercourse with infected persons. Sexual transmission can be reduced by the use of effective barrier methods (condoms and spermicides such as nonoxynol 9) during intercourse. Infection from donated blood and blood products is now very rare. It is important to note that no one can acquire AIDS by donating blood. In the United States all blood has been tested for HIV antibodies since 1985. HIV transmission via contaminated hypodermic needles could be avoided by sterilization of the needles with chlorox before each use. In 1994 researchers found that AZT taken during pregnancy dramatically decreases the risk of transmitting the virus to a fetus. If these measures are to be effective, however, they must be part of an overall program involving health education, counseling, screening individuals who engage in high-risk behaviors, tracing contacts, and modifying behavior. Until there is effective drug therapy or an effective vaccine, blocking the spread of HIV depends on education and safer sexual practices.

AUTOIMMUNE DISEASES

Normally, the immune system displays self-tolerance and does not attack the body's own components. At times, however, self-tolerance breaks down, and this leads to an **autoimmune disease (autoimmunity)**. The immune system fails to recognize self-antigens and mounts a misguided immune attack against them. Currently, there are two theories as to why autoimmune diseases occur.

First, T cells that react dangerously with self-proteins may escape from the thymus. Recall that normally such T cells are weeded out by negative selection. Second, T cells that were anergized (turned off) because they did respond to a self-antigen may inappropriately get turned back on. Further research is needed to confirm these theories.

Among human autoimmune diseases are rheumatoid arthritis (RA), systemic lupus erythematosus (SLE), thyroiditis, rheumatic fever, encephalomyelitis, hemolytic and pernicious ane-

mias, Addison's disease, Graves' disease, insulin-dependent (type I) diabetes mellitus, myasthenia gravis, multiple sclerosis (MS), and ulcerative colitis.

Therapy for autoimmune diseases typically involves giving drugs to suppress immune responses. This, of course, leaves the person prone to opportunistic infections. An experimental form of treatment for autoimmune disorders is **oral tolerization (oral antigen therapy)**, in which the patient takes daily doses of the same protein antigens that are thought to have caused the disease in the first place. Animal experiments indicate that such antigen feeding acts like a vaccine to suppress some autoimmune responses. Clinical trials have shown benefits for some patients with multiple sclerosis and rheumatoid arthritis.

SYSTEMIC LUPUS ERYTHEMATOSUS

Systemic lupus erythematosus (er-e'-thēm-a-TŌ-sus), **SLE**, or **lupus** (*lupus* = wolf) is an autoimmune, noncontagious, inflammatory disease of connective tissue, occurring mostly in young women. In SLE, damage to blood vessel walls results in the release of chemicals that mediate inflammation. The blood vessel damage can be associated with virtually every body system.

The cause of SLE is not known, and its onset may be abrupt or gradual. There seems to be a strong incidence of other connective tissue disorders—especially rheumatoid arthritis and rheumatic fever—in relatives of SLE victims. The disease may be triggered by drugs, such as penicillin, sulfa, or tetracycline, exposure to excessive sunlight, injury, emotional upset, infection, or other stress.

Symptoms of SLE include painful joints, slight fever, fatigue, mouth ulcers, weight loss, enlarged lymph nodes and spleen, photosensitivity, rapid loss of large amounts of scalp hair, and sometimes an eruption across the bridge of the nose and cheeks called a "butterfly rash." Other skin lesions may occur with blistering and ulceration. The erosive nature of some of the SLE skin lesions was thought to resemble the damage inflicted by the bite of a wolf—thus the term lupus. The most serious complications of the disease involve inflammation of the kidneys, liver, spleen, lungs, heart, and the central nervous system.

CHRONIC FATIGUE SYNDROME

Chronic fatigue syndrome (**CFS**) usually occurs in young adults, primarily females. It is characterized by extreme fatigue that impairs normal activities for at least 6 months and by the absence of known diseases (cancer, infections, drug abuse, toxicity, or psychiatric disorders) that might produce similar symptoms. The cause of CFS is not known. Among the possible causes of CFS are lowered levels of corticotropin-releasing hormone (CRH) and cortisol, emotional factors such as depression and excess stress, and viral infection. Diagnostic guidelines specify that at least 8 of the following 11 indications persist or recur over 6 months: mild fever or chills, sore throat, painful lymph nodes, general muscle weakness, muscle pain, fatigue for more than 24 hours after mild exercise, headaches that differ in type and severity from past ones, joint pain without swelling, neuropsychological complaints (irritability, memory loss, confusion, depression), sleep disturbances, and development of the initial symptoms over a few hours to a few days. The diagnosis can also be made if the patient reports 6 of the 11 symptoms just listed plus observation by a physician of two of these three physical signs: low-grade fever, inflamed throat, and enlarged lymph nodes in the neck or axilla.

SEVERE COMBINED IMMUNODEFICIENCY

Severe combined immunodeficiency (SCID) is a rare inherited disorder in which both B cells and T cells are missing or inactive in providing immunity. Scientists have now identified mutations in two genes on different chromosomes that are responsible for some of the five known types of SCID. One gene codes for an enzyme known as adenosine deaminase (ADA) whereas the other codes for one of the three polypeptide chains (called the gamma chain) that together form the interleukin-2 receptor. In some cases, an infusion of red bone marrow cells from a sibling having very similar MHC (HLA) antigens can provide normal stem cells that give rise to normal B and T cells. The result can be a complete cure. Less than 30% of afflicted patients, however, have a compatible sibling who could be the donor. Another promising approach for treatment of SCID is to provide good copies of the defective gene through gene therapy (described on page 983). Several different clinical trials are in progress to evaluate the risks and benefits of this approach. In 1990 and 1991, two girls with ADA deficiency were the first persons in the U.S. to receive gene therapy for a life-threatening disorder, and they have continued to improve while receiving infusions of gene-corrected T cells. Because the procedure involves infusion of mature T cells, however, the girls must be retreated every couple of months.

Until recently, the most famous patient with SCID was David, the "bubble boy." He was placed in a sterile chamber shortly after birth in the 1970s to protect him from microbes that his body could not fight. At age 12, David underwent a bone marrow transplant in an effort to correct his disorder. Eighty days after the transplant and still in a germ-free environment, David developed some of the symptoms of infectious mononucleosis, a condition caused by the Epstein–Barr virus. He was brought out of isolation for easier treatment, with the hope that the transplant would provide the same protection as his sterile plastic chamber. It probably was successful but, unfortunately, David died 4 months later from cancer associated with Epstein–Barr virus. David's legacy has continued into the 1990s, however, because some of his cells were used in the experiments that localized the genetic defect to the gene that codes for the gamma chain of the interleukin-2 receptor.

HYPERSENSITIVITY (ALLERGY)

A person who is overly reactive to an antigen that is tolerated by most others is said to be **hypersensitive (allergic)**. Whenever an allergic reaction occurs, there is tissue injury. The antigens that induce an allergic reaction are called **allergens**. Common allergens include certain foods (milk, peanuts, shellfish, eggs), antibiotics (penicillin, tetracycline), vitamins (thiamine, folic acid), drugs (insulin, ACTH, estradiol), vaccines (pertussis, typhoid), venoms (honeybee, wasp, snake), cosmetics, chemicals in plants such as poison ivy, pollens, dust, molds, iodine-containing dyes used in certain x-ray procedures, and even microbes.

There are four basic types of hypersensitivity reactions: type I (anaphylaxis), type II (cytotoxic), type III (immune complex),

and type IV (cell-mediated). The first three involve antibodies; the last involves T cells.

Type I (anaphylaxis) reactions are the most common and occur within a few minutes after a person sensitized to an allergen is reexposed to it. **Anaphylaxis** (an′-a-fi-LAK-sis) literally means "against protection" and results from the interaction of allergens with IgE antibodies on the surface of mast cells and basophils. Basophils circulate in blood; mast cells are especially numerous in connective tissue of the skin and respiratory system and endothelium of blood vessels. In response to certain allergens, some people produce IgE antibodies that bind to the surface of mast cells and basophils.

The next time the same allergen enters the body, it attaches to the IgE antibodies already present on the surface of mast cells and basophils. In response, the cells release chemicals called **mediators of anaphylaxis**, among which are histamine, prostaglandins, leukotrienes, and kinin. Collectively, the mediators cause vasodilation, increased blood capillary permeability, increased smooth muscle contraction in the airways of the lungs, and increased mucus secretion. As a result, a person may experience inflammatory responses, difficulty in breathing through the constricted bronchial tubes, and a "runny" nose from excess mucus secretion.

Some anaphylactic reactions, such as hives, eczema, swelling of the lips or tongue, abdominal cramps, and diarrhea, are **localized** (affecting one part or a limited area). Other anaphylactic reactions are **systemic** (affecting several parts or the entire body). An example is acute anaphylaxis (anaphylactic shock), which may occur in a susceptible individual who has just received a triggering drug or been stung by a wasp. The person develops respiratory symptoms (wheezing and shortness of breath) as bronchioles constrict, usually accompanied by cardiovascular failure and collapse due to vasodilation and fluid loss from blood. This life-threatening emergency is usually treated by injecting epinephrine to dilate the airways and strengthen the heartbeat.

Type II (cytotoxic) reactions are caused by antibodies (IgG or IgM) directed against antigens on a person's blood cells (red blood cells, lymphocytes, or platelets) or tissue cells. The reaction of antibodies and antigens usually leads to activation of complement. Type II reactions, which may occur in incompatible transfusion reactions, damage cells by causing lysis.

Type III (immune complex) reactions involve antigens (not part of a host tissue cell), antibodies (IgA or IgM), and complement. When certain ratios of antigen to antibody occur, the complexes are small and escape phagocytosis. The complexes become trapped in the basement membrane under the endothelium of blood vessels, activate complement, and cause an inflammation. Conditions that so arise include glomerulonephritis, systemic lupus erythematosus (SLE), and rheumatoid arthritis (RA).

Type IV (cell-mediated) reactions or **delayed type hypersensitivity (DTH) reactions** are carried out by macrophages that have become activated by T cells. They usually appear 12–72 hours after exposure to an allergen. Type IV reactions occur when allergens are taken up by antigen-presenting cells, such as Langerhans cells in the skin, which then migrate to lymph nodes and present the allergen to T cells. This results in sensitization and proliferation of T cells, some of which migrate via the lymph

DISORDERS: HOMEOSTATIC IMBALANCES   CONTINUED

and blood to the site of allergen entry into the body. There they secrete cytokines, such as gamma-interferon, which activates macrophages, and tumor necrosis factor (TNF), which stimulates an inflammatory response. Intracellular bacteria, such as *Listeria monocytogenes* and *Mycobacterium tuberculosis,* trigger this type of cell-mediated immunity, as do certain haptens, such as poison ivy toxin. The skin test for tuberculosis also is a delayed hypersensitivity type reaction.

TISSUE REJECTION

Transplantation involves the replacement of an injured or diseased tissue or organ. Usually, the immune system recognizes the proteins in the transplanted tissue or organ as foreign and mounts both CMI and AMI responses against them. This phenomenon is known as **tissue rejection**. The more closely matched the MHC (HLA) antigens between donor and recipient, the weaker the tissue rejection response.

The most successful transplants are **autografts**, transplants in which one's own tissue is grafted to another part of the body (such as skin grafts for burn treatment or plastic surgery) and **isografts**, transplants in which the donor and recipient are genetically identical. An **allograft** is a transplant between individuals of the same species but with different genetic backgrounds. The success of this type of transplant has been moderate. Often, it is used as a temporary measure until the damaged or diseased tissue is able to repair itself. Skin transplants from other people and blood transfusions might properly

be considered allografts. A **xenograft** is a transplant between animals of different species. This type of transplantation is used primarily as a physiological dressing over severe burns.

Until recently, **immunosuppressive drugs** suppressed not only the recipient's immune rejection of the donor tissue but also the immune response to all antigens as well. This causes patients to become very susceptible to infectious diseases. A drug called *cyclosporine,* derived from a fungus, has largely overcome this problem with regard to kidney, heart, and liver transplants. Cyclosporine inhibits secretion of IL-2 by helper T cells but has only a minimal impact on B cells. Thus rejection is avoided and resistance against some diseases is still maintained.

HODGKIN'S DISEASE

Hodgkin's disease (HD) is a form of cancer, usually arising in lymph nodes, the cause of which is unknown. It may, however, arise from a combination of genetic predisposition, disturbance of the immune system, and the Epstein–Barr virus. Initially, the disease is characterized by a painless, nontender enlargement of one or more lymph nodes, most commonly in the neck but occasionally in the axilla, inguinal, or femoral region. Some patients also have an unexplained and persistent fever and/or night sweats. Fatigue and weight loss are also associated complaints, as is pruritus (itching). Treatment consists of radiation therapy, chemotherapy, and bone marrrow transplants. Hodgkin's disease is considered to be a curable malignancy.

MEDICAL TERMINOLOGY

Adenitis (ad′-e-NĪ-tis; *adeno* = gland; *itis* = inflammation of) Enlarged, tender, and inflamed lymph nodes resulting from an infection.

Hypersplenism (hī′-per-SPLĒN-izm; *hyper* = over) Abnormal splenic activity due to splenic enlargement and associated with an increased rate of destruction of normal blood cells.

Lymphadenectomy (lim-fad′-e-NEK-tō-mē; *ectomy* = removal) Removal of a lymph node.

Lymphadenopathy (lim-fad′-e-NOP-a-thē; *patho* = disease) Enlarged, sometimes tender lymph nodes.

Lymphangioma (lim-fan′-jē-Ō-ma; *angio* = vessel; *oma* = tumor) A benign tumor of the lymphatic vessels.

Lymphangitis (lim′-fan-JĪ-tis) Inflammation of the lymphatic vessels.

Lymphedema (lim′-fe-DĒ-ma; *edema* = swelling) Accumulation of lymph producing subcutaneous tissue swelling.

Lymphoma (lim′-FŌ-ma) Any tumor composed of lymphatic tissue.

Lymphostasis (lim-FŌ-stā-sis; *stasis* = halt) A lymph flow stoppage.

Splenomegaly (splē′-nō-MEG-a-lē; *mega* = large) Enlarged spleen.

STUDY OUTLINE

INTRODUCTION (p. 671)

1. The ability to ward off disease is called resistance. Lack of resistance is called susceptibility.
2. Nonspecific resistance refers to a wide variety of body responses against a wide range of pathogens. Immunity involves activation of specific lymphocytes to combat a particular foreign substance.
3. The lymphatic system carries out immune responses and consists of lymph, lymphatic vessels, and structures and organs

that contain lymphatic tissue (specialized reticular tissue containing many lymphocytes).

4. The lymphatic system functions to drain interstitial fluid, transport dietary lipids, and protect against invasion through immune responses.

LYMPHATIC VESSELS AND LYMPH CIRCULATION (p. 671)

1. Lymphatic vessels begin as closed-ended lymph capillaries in tissue spaces between cells.

2. Interstitial fluid drains into lymphatic capillaries, thus forming lymph.

3. Lymph capillaries merge to form larger vessels, called lymphatic vessels, which convey lymph into and out of structures called lymph nodes.

4. The passage of lymph is from lymph capillaries to lymphatic vessels to lymph trunks to the thoracic duct or right lymphatic duct to the subclavian veins.

5. Lymph flows as a result of skeletal muscle contractions and respiratory movements. It is also aided by valves in the lymphatic vessels.

LYMPHATIC TISSUES (p. 675)

1. The primary lymphatic organs are red bone marrow and the thymus gland. Secondary lymphatic organs are lymph nodes, spleen, and lymphatic nodules.

2. The thymus gland lies between the sternum and the large blood vessels above the heart. It is the site of T cell maturation.

3. Lymph nodes are encapsulated, oval structures located along lymphatic vessels.

4. Lymph enters nodes through afferent lymphatic vessels, is filtered, and exits through efferent lymphatic vessels.

5. Lymph nodes are the site of proliferation of plasma cells and T cells.

6. The spleen is the largest mass of lymphatic tissue in the body. It is a site of B cell proliferation into plasma cells and phagocytosis of bacteria and worn-out red blood cells.

7. Lymphatic nodules are scattered throughout the mucosa of the gastrointestinal, respiratory, urinary, and reproductive tracts. This lymphatic tissue is referred to as mucosa-associated lymphoid tissue (MALT).

DEVELOPMENTAL ANATOMY OF THE LYMPHATIC SYSTEM (p. 679)

1. Lymphatic vessels develop from lymph sacs, which develop from veins. Thus they are derived from mesoderm.

2. Lymph nodes develop from lymph sacs that become invaded by mesenchymal cells.

NONSPECIFIC RESISTANCE TO DISEASE (p. 680)

1. Mechanisms of nonspecific resistance include mechanical factors, chemical factors, antimicrobial substances, natural killer cells, phagocytosis, inflammation, and fever.

2. The skin and mucous membranes are the first line of defense against entry of pathogens.

3. Antimicrobial substances include transferrins, interferons, and the complement system.

4. Natural killer cells and phagocytes attack and kill pathogens and defective cells in the body.

5. Inflammation aids disposal of microbes, toxins, or foreign material at the site of an injury and prepares the site for tissue repair.

IMMUNITY (p. 685)

1. Specific resistance to disease involves the production of a specific lymphocyte or antibody against a specific antigen and is called immunity.

Formation of T Cells and B Cells (p. 685)

1. Both B and T cells derive from stem cells in red bone marrow.

2. T cells complete their maturation and develop immunocompetence in the thymus gland.

Types of Immune Responses (p. 686)

1. Cell-mediated immunity refers to destruction of antigens by T cells.

2. Antibody-mediated (humoral) immunity refers to destruction of antigens by antibodies.

Antigens (p. 686)

1. Antigens (Ags) are chemical substances that are recognized as foreign by the immune system.

2. Antigen receptors exhibit great diversity due to genetic recombination.

3. Major histocompatibility complex (MHC) antigens are unique to each person's body cells. All cells except red blood cells display MHC-I antigens. Some cells also display MHC-II antigens.

4. Peptide fragments help stabilize MHC molecules.

5. Cells called antigen-presenting cells (APCs) process exogenous antigens (formed outside the body) and present them together with MHC class II molecules to T cells. APCs include macrophages, B cells, and dendritic cells.

Cytokines (p. 690)

1. Cytokines are small protein hormones needed for many normal cell functions. Some of them regulate immune responses (see Exhibit 22.2).

Antibodies (p. 690)

1. An antibody (Ab) is a protein that combines specifically with the antigen that triggered its production.

2. Antibodies consist of heavy and light chains and variable and constant regions.

3. Based on chemistry and structure, antibodies are grouped into five principal classes, each with specific biological roles (IgG, IgA, IgM, IgD, and IgE).

Cell-Mediated Immunity (p. 691)

1. In a cell-mediated immune response, an antigen is recognized, specific T cells proliferate and differentiate into effector cells, and the antigen is eliminated.

2. T cell receptors (TCRs) recognize antigen fragments associated with MHC molecules on the surface of a body cell.

3. Proliferation of T cells requires costimulation, by cytokines such as interleukin-1 (IL-1) and IL-2 or by pairs of plasma membrane molecules.

4. T cells consist of several subpopulations. Helper T cells display CD4 protein, recognize antigen fragments associated with MHC-II molecules, and secrete several cytokines, most importantly interleukin-2, which acts as a costimulator for other helper T cells, cytotoxic T cells, and B cells. Cytotoxic (killer) T cells display CD8 protein and recognize antigen fragments associated with MHC-I molecules. Delayed hypersensitivity T cells produce cytokines and are important in hypersensitivity (allergic) responses. Suppressor T cells down-regulate immune responses.

5. Cytotoxic T cells eliminate invaders by secreting lymphotoxin, which causes fragmentation of the DNA of a target cell, and perforin, which causes cytolysis. They also secrete gamma-interferon.

Antibody-Mediated (Humoral) Immunity (p. 695)

1. B cells can respond to unprocessed antigens, but their response is more intense when dendritic cells present antigen to them. Interleukin-2 and other cytokines secreted by helper T cells provide costimulation for proliferation of B cells.

2. An activated B cell develops into a clone of antibody-producing plasma cells.

Immunological Memory (p. 697)

1. Immunization against certain microbes is possible because memory B cells and memory T cells remain after the primary response to an antigen. The secondary response provides protection should the same microbe enter the body again.

Self-Recognition and Immunological Tolerance (p. 698)

1. T cells undergo positive selection to ensure that they can recognize self-MHC antigens (self-recognition) and negative selec-

tion to ensure that they do not react to other self-proteins (tolerance). Negative selection involves both deletion and anergy.

2. B cells develop tolerance through deletion and anergy.

Immunology and Cancer (p. 699)

1. Cancer cells may display tumor-specific antigens and are often destroyed by the body's immune system (immunological surveillance).

AGING AND THE IMMUNE SYSTEM (p. 699)

1. With advancing age, the immune system functions less effectively.

REVIEW QUESTIONS

1. Identify the components and functions of the lymphatic system. (p. 671)
2. Compare veins and lymphatic vessels with regard to structure. (p. 671)
3. Construct a diagram to show the route of lymph circulation. (p. 675)
4. Describe the role of the thymus gland in immunity. (p. 676)
5. Describe the structure of a lymph node. What functions do lymph nodes serve? (p. 676)
6. Describe the location, gross anatomy, histology, and functions of the spleen. (p. 678)
7. Identify the tonsils by location. (p. 679)
8. Describe how the lymphatic system develops. (p. 679)
9. Describe the various mechanical and chemical factors involved in nonspecific resistance. (p. 680)
10. Outline the role of the following antimicrobial substances: interferons (IFNs), complement, and transferrin. (p. 681)

11. What are natural killer (NK) cells? (p. 681)
12. Describe the three phases of phagocytosis. (p. 683)
13. Define inflammation. Describe the principal symptoms associated with inflammation and outline its stages. (p. 684)
14. Define immunity and summarize the two types of immune responses. Where do B cells and T cells form? (p. 685)
15. List the various characteristics of antigens and explain the function of antigen presenting cells. (p. 686)
16. Describe the functions of cytokines. (p. 690)
17. Describe the chemical characteristics of antibodies. (p. 690)
18. Compare and contrast cell-mediated and antibody-mediated immune responses. (p. 691)
19. Discuss the importance of the secondary response to an antigen. (p. 697)
20. How are monoclonal antibodies (MAbs) produced? (p. 696)
21. Explain how immunology is related to cancer. (p. 699)
22. Describe the effects of aging on the immune system. (p. 699)

ANSWERS TO QUESTIONS WITH FIGURES

22.1 Red bone marrow.
22.2 Interstitial fluid, because the protein content is low.
22.3 Inhalation promotes the flow of lymphatic fluid from abdominal lymphatic vessels toward the thoracic region.
22.4 Left and right lumbar trunks and intestinal trunk; thoracic.
22.5 T cells.
22.6 Macrophages may phagocytize them or lymphocytes attack them via immune responses.
22.7 Phagocytosis of bacteria and aged red blood cells.
22.9 Classical pathway: Ag–Ab complexes activate C1.
22.10 Digestive enzymes, oxidants, and defensins.
22.11 *Redness,* increased blood flow due to vasodilation; *pain,* injury of nerve fibers, irritation by toxins from microbes, kinins, prostaglandins, pressure from edema; *heat,* increased blood flow, heat release by locally increased metabolic reactions; *swelling,* leakage of fluid from capillaries due to increased permeability.

22.12 Helper T cells.
22.13 Antigenic determinant = small immunogenic part of large foreign antigen; hapten = small molecule that becomes immunogenic when it attaches to a body protein.
22.14 Macrophages in tissues throughout the body; B cells in blood and lymphatic tissue; dendritic cells in mucous membranes and the skin.
22.15 It recognizes and binds to antigen.
22.16 First signal, antigen binding to TCR; second signal, costimulator, such as a cytokine or another pair of plasma membrane molecules.
22.17 Some tumor cells, transplanted tissue cells.
22.18 Just one.
22.19 At peak secretion, 1000 times more.
22.20 Deletion—self-reactive T or B cells die; anergy—T or B cells are alive but are unresponsive to antigenic stimulation.
22.21 It catalyzes the formation of DNA from viral RNA.

STUDENT OBJECTIVES

1. Identify the structures of the respiratory system and describe their functions. (p. 708)

2. Explain the structure of the alveolar–capillary (respiratory) membrane and describe its function in the diffusion of respiratory gases. (p. 722)

3. Describe the events involved in inspiration and expiration. (p. 723)

4. Explain how respiratory gases are transported by blood. (p. 732)

5. Describe the various factors that control the rate of respiration. (p. 736)

6. Describe the responses of the respiratory system to exercise. (p. 742)

7. Describe the effects of aging on the respiratory system. (p. 743)

8. Describe the development of the respiratory system. (p. 743)

9. Define asthma, bronchitis, emphysema, bronchogenic carcinoma, pneumonia, tuberculosis, respiratory distress syndrome, respiratory failure, sudden infant death syndrome, coryza, influenza, pulmonary embolism, pulmonary edema, cystic fibrosis, and smoke inhalation injury as disorders of the respiratory system. (p. 744)

10. Define medical terminology associated with the respiratory system. (p. 748)

CHAPTER 23

THE RESPIRATORY SYSTEM

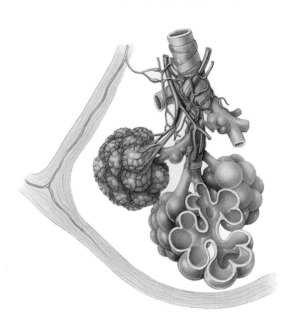

CHAPTER CONTENTS AT A GLANCE

ells continually use oxygen (O_2) for the metabolic reactions that release energy from nutrient molecules and produce ATP. At the same time, these reactions release carbon dioxide (CO_2). Since an excessive amount of CO_2 produces acidity that is toxic to cells, the excess CO_2 must be eliminated quickly and efficiently. The two systems that cooperate to supply O_2 and eliminate CO_2 are the cardiovascular system and the respiratory system. The respiratory system provides for gas exchange, intake of O_2 and elimination of CO_2, whereas the cardiovascular system transports the gases in the blood between the lungs and body cells. Failure of either system has the same effect on the body: disruption of homeostasis and rapid death of cells from oxygen starvation and buildup of waste products. In addition to functioning in gas exchange, the respiratory system also contains receptors for the sense of smell, filters inspired air, produces sounds, and helps eliminate wastes.

Respiration is the exchange of gases between the atmosphere, blood, and cells. It takes place in three basic steps:

1. **Pulmonary ventilation.** The first process, **pulmonary** (*pulmo* = lung) **ventilation**, or breathing, is the inspiration (inflow) and expiration (outflow) of air between the atmosphere and the lungs.
2. **External (pulmonary) respiration.** This is the exchange of gases between the air spaces of the lungs and blood in pulmonary capillaries. The blood gains O_2 and loses CO_2.
3. **Internal (tissue) respiration.** The exchange of gases between blood in systemic capillaries and tissue cells is known as internal (tissue) respiration. The blood loses O_2 and gains CO_2. Within cells, the metabolic reactions that consume O_2 and give off CO_2 during production of ATP are termed **cellular respiration** (discussed in Chapter 25).

The developmental anatomy of the respiratory system is considered at the end of the chapter.

STRUCTURES OF THE RESPIRATORY SYSTEM

The **respiratory system** consists of the nose, pharynx (throat), larynx (voice box), trachea (windpipe), bronchi, and lungs (Fig. 23.1). Structurally, the respiratory system consists of two portions. (1) The term **upper respiratory system** refers to the nose, pharynx, and associated structures. (2) The **lower respiratory system** refers to the larynx, trachea, bronchi, and lungs. Functionally, the respiratory system also consists of two portions. (1) The **conducting portion** consists of a series of interconnecting cavities and tubes—nose, pharynx, larynx, trachea, bronchi,

bronchioles, and terminal bronchioles—that conduct air into the lungs. (2) The **respiratory portion** consists of those portions of the respiratory system where the exchange of gases occurs—respiratory bronchioles, alveolar ducts, alveolar sacs, and alveoli.

The branch of medicine that deals with the diagnosis and treatment of diseases of the ears, nose, and throat is called **otorhinolaryngology** (ō′-tō-rī′-nō-lar′-in-GOL-ō-jē; *oto* = ear; *rhino* = nose).

Nose

The **nose** has an external portion and an internal portion inside the skull (Fig. 23.2). The external portion consists of a supporting framework of bone and hyaline cartilage covered with muscle and skin and lined by mucous membrane. The bridge of the nose is formed by the nasal bones, which hold it in a fixed position. Because it has a framework of pliable hyaline cartilage, the rest of the external nose is somewhat flexible. On the undersurface of the external nose are two openings called the **external nares** (NA-rēz; singular is **naris**), or **nostrils**. The interior structures of the nose are specialized for three functions: (1) incoming air is warmed, moistened, and filtered; (2) olfactory stimuli are received; and (3) large, hollow resonating chambers modify speech sounds.

The internal portion of the nose is a large cavity in the anterior aspect of the skull that lies inferior to the nasal bone and superior to the mouth. Anteriorly, the internal nose merges with the external nose, and posteriorly it communicates with the pharynx through two openings called the **internal nares (choanae).** Ducts from the paranasal sinuses (frontal, sphenoidal, maxillary, and ethmoidal) and the nasolacrimal ducts also open into the internal nose. The lateral walls of the internal nose are formed by the ethmoid, maxillae, lacrimal, palatine, and inferior nasal conchae bones. The ethmoid also forms the roof. The floor of the internal nose is formed mostly by the palatine bones and palatine processes of the maxillae, which together comprise the hard palate.

The space inside the internal nose is called the **nasal cavity.** It is divided into right and left sides by a vertical partition called the **nasal septum.** The anterior portion of the septum consists primarily of hyaline cartilage. The remainder is formed by the vomer, perpendicular plate of the ethmoid, maxillae, and palatine bones (see Fig. 7.10a). The anterior portion of the nasal cavity, just inside the nostrils, is called the **vestibule** and is surrounded by cartilage. The superior part of the nasal cavity is surrounded by bone.

When air enters the nostrils, it passes first through the vestibule. The vestibule is lined by skin containing coarse hairs that filter out large dust particles. Three shelves formed by projections of the superior, middle, and inferior nasal conchae extend out of each lateral wall of the cavity. The conchae, almost reaching the septum, subdivide each

Figure 23.1 Structures of the respiratory system.

The upper respiratory system includes the nose, pharynx, and associated structures. The lower respiratory system incudes the larynx, trachea, bronchi, and lungs.

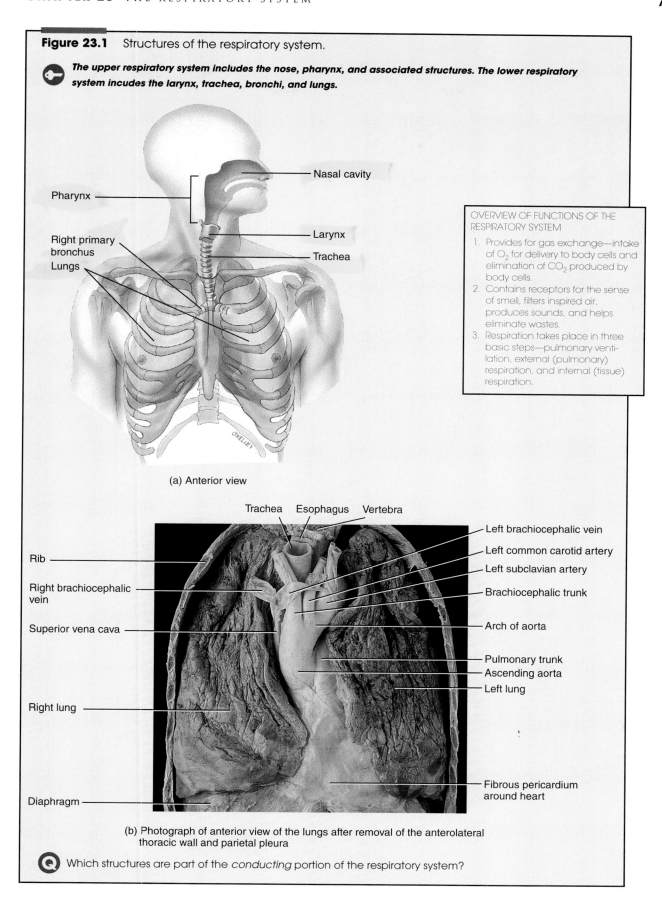

OVERVIEW OF FUNCTIONS OF THE RESPIRATORY SYSTEM

1. Provides for gas exchange—intake of O_2 for delivery to body cells and elimination of CO_2 produced by body cells.
2. Contains receptors for the sense of smell, filters inspired air, produces sounds, and helps eliminate wastes.
3. Respiration takes place in three basic steps—pulmonary ventilation, external (pulmonary) respiration, and internal (tissue) respiration.

(a) Anterior view

(b) Photograph of anterior view of the lungs after removal of the anterolateral thoracic wall and parietal pleura

Q Which structures are part of the *conducting* portion of the respiratory system?

Figure 23.2   Respiratory organs in the head and neck.

 As air passes through the nose, it is warmed, filtered, and moistened.

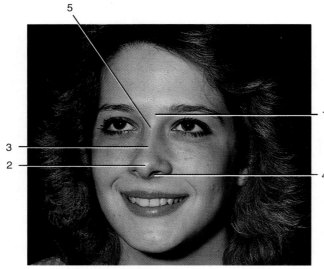

(a) Anterior view of surface anatomy of the nose

1. **Root.** Superior attachment of nose at forehead between eyes.
2. **Apex.** Tip of nose.
3. **Dorsum nasi.** Rounded anterior border connecting root and apex; in profile, may be straight, convex, concave, or wavy.
4. **External naris.** External opening into nose.
5. **Bridge.** Superior portion of dorsum nasi, superficial to nasal bones.

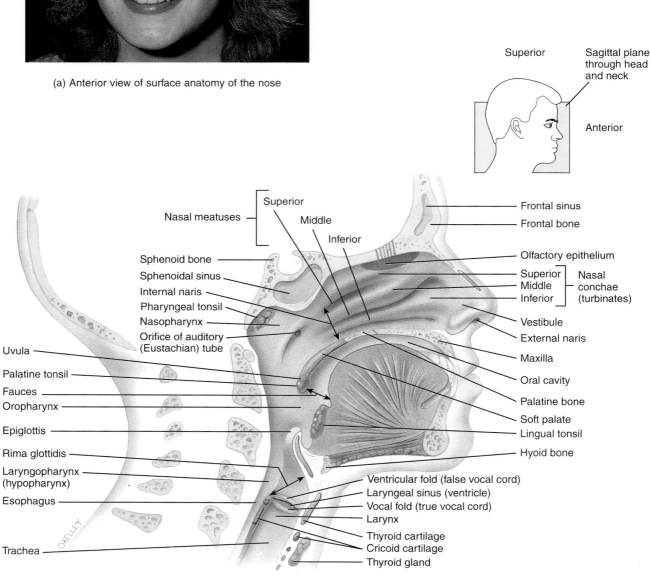

(b) Diagram of sagittal section of the left side of the head and neck

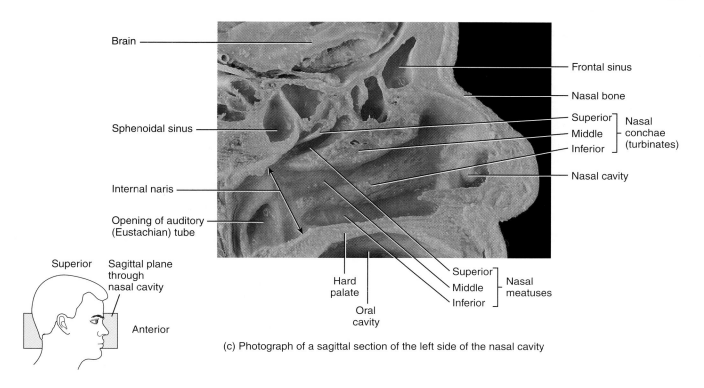

Brain

Frontal sinus

Nasal bone

Superior ⎤
Middle ⎬ Nasal conchae (turbinates)
Inferior ⎦

Sphenoidal sinus

Nasal cavity

Internal naris

Opening of auditory (Eustachian) tube

Superior Sagittal plane through nasal cavity

Anterior

Hard palate

Oral cavity

Superior ⎤
Middle ⎬ Nasal meatuses
Inferior ⎦

(c) Photograph of a sagittal section of the left side of the nasal cavity

Q What is the path taken by air molecules into and through the nose?

side of the nasal cavity into a series of groovelike passage-ways—the **superior**, **middle**, and **inferior meatuses** (mē-Ā-tes-ez; *meatus* = passage; singular is **meatus**). Mucous membrane lines the cavity and its shelves.

The olfactory receptors lie in the membrane lining the su-perior nasal conchae and adjacent septum. This region is called the **olfactory epithelium**. Inferior to the olfactory ep-ithelium, the mucous membrane contains capillaries and pseudostratified ciliated columnar epithelium with many goblet cells. As the air whirls around the conchae and mea-tuses, it is warmed by blood in the capillaries. Mucus se-creted by the goblet cells moistens the air and traps dust par-ticles. Drainage from the nasolacrimal ducts and perhaps secretions from the paranasal sinuses also help moisten the air. The cilia move the mucus–dust particles toward the pharynx so they can be eliminated from the respiratory tract by swallowing or expectoration (spitting). Substances in cigarette smoke inhibit movement of cilia. When this hap-pens, only coughing can remove mucus–dust particles from the airways. This is one reason that smokers cough often.

RHINOPLASTY

Rhinoplasty (RĪ-nō-plas′-tē; *rhino* = nose; *plassein* = to form), commonly called a "nose job," is a surgical procedure in which the structure of the external nose is altered. Although it is

frequently done for cosmetic reasons, it is sometimes performed to repair a fractured nose or deviated nasal septum. In the procedure, the nasal bones are fractured and repositioned to achieve the de-sired shape. Any extra bone fragments or cartilage are shaved down and removed through the nostrils. Both an internal packing and an external splint keep the nose in the desired position while it heals. In some cases, only the cartilage has to be reshaped to achieve the desired appearance. ■

Pharynx

The **pharynx** (FAIR-inks), or throat, is a somewhat funnel-shaped tube about 13 cm (5 in.) long that starts at the inter-nal nares and extends to the level of the cricoid cartilage, the most inferior cartilage of the larynx (voice box) (Fig. 23.3). It lies just posterior to the nasal cavity and oral cavity, supe-rior to the larynx, and just anterior to the cervical vertebrae. Its wall is composed of skeletal muscles and lined with mu-cous membrane. The pharynx functions as a passageway for air and food, provides a resonating chamber for speech sounds, and houses the tonsils, which help eliminate foreign invaders through immunological reactions.

The superior portion of the pharynx, called the **naso-pharynx**, lies posterior to the nasal cavity and extends to the plane of the soft palate. There are five openings in its wall: two internal nares, two openings that lead into the auditory

Figure 23.3   Portion of the left side of the head and neck with the nasal septum removed.

🔑 *The three subdivisions of the pharynx are the (1) nasopharynx, (2) oropharynx, and (3) laryngopharynx.*

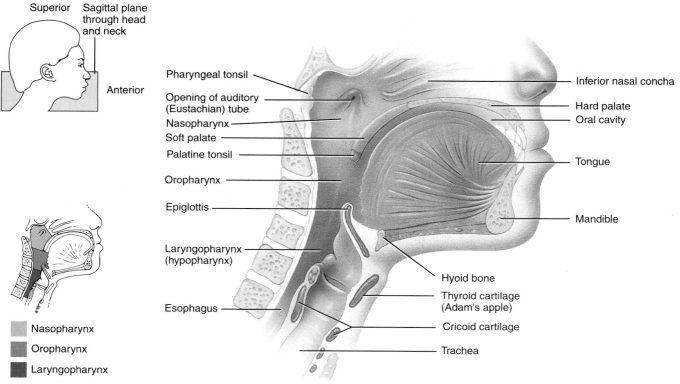

Sagittal section showing left side of the head and neck

❓ In this figure, can you identify the superior and inferior borders of the pharynx?

(Eustachian) tubes, and the opening into the oropharynx. The posterior wall also contains the pharyngeal tonsil. Through the internal nares, the nasopharynx receives air from the nasal cavity and receives packages of dust-laden mucus. It is lined with pseudostratified ciliated columnar epithelium, and the cilia move the mucus down toward the most inferior part of the pharynx. The nasopharynx also exchanges small amounts of air with the auditory (Eustachian) tubes to equalize air pressure between the pharynx and middle ear.

The intermediate portion of the pharynx, the **oropharynx**, lies posterior to the oral cavity and extends from the soft palate inferiorly to the level of the hyoid bone. It has only one opening, the **fauces** (FAW-sēz; *fauces* = throat), the opening from the mouth. This portion of the pharynx is both respiratory and digestive in function, because it is a common passageway for air, food, and drink. Because the oropharynx is subject to abrasion by food particles, it is lined with nonkeratinized stratified squamous epithelium. Two pairs of tonsils, the palatine and lingual tonsils, are found in the oropharynx.

The inferior portion of the pharynx, the **laryngopharynx** (la-rin′-gō-FAIR-inks), or **hypopharynx**, begins at the level

of the hyoid bone and connects the esophagus (food tube) with the larynx (voice box). Like the oropharynx, the laryngopharynx is both a respiratory and a digestive pathway and is lined by nonkeratinized stratified squamous epithelium.

Larynx

The **larynx** (LAIR-inks), or voice box, is a short passageway that connects the laryngopharynx with the trachea. It lies in the midline of the neck anterior to the fourth through sixth cervical vertebrae (C4–C6).

The wall of the larynx is composed of nine pieces of cartilage (Fig. 23.4). Three are single (thyroid cartilage, epiglottis or epiglottic cartilage, and cricoid cartilage) and three are paired (arytenoid, cuneiform, and corniculate cartilages). Of the paired cartilages, the arytenoid cartilages are the most important since they influence the positions and tensions of the vocal folds (true vocal cords).

The **thyroid cartilage** (**Adam's apple**) consists of two fused plates of hyaline cartilage that form the anterior wall

Figure 23.4 Larynx.

 The larynx is composed of nine pieces of cartilage.

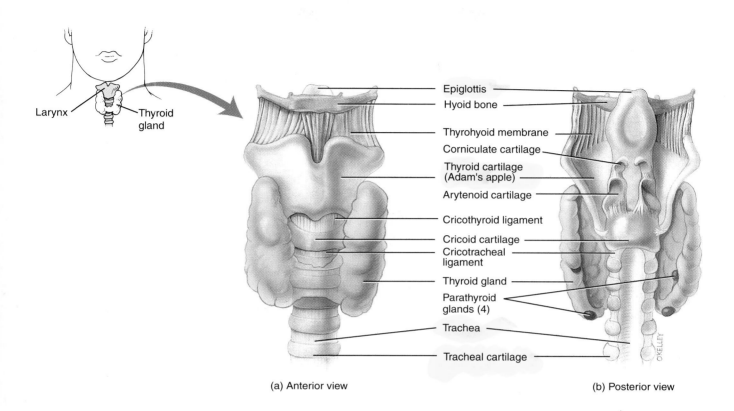

(a) Anterior view

(b) Posterior view

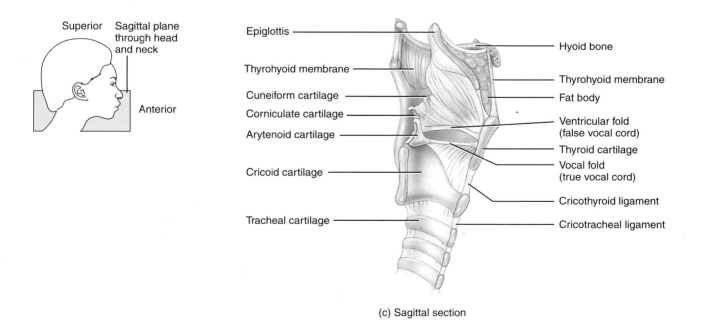

(c) Sagittal section

Q How does the epiglottis prevent aspiration of foods and liquids?

of the larynx and give it its triangular shape. It is usually larger in males than in females due to the influence of male sex hormones during puberty. The ligament that connects the thyroid cartilage to the hyoid bone is called the **thyrohyoid membrane**.

The **epiglottis** (*epi* = above; *glotta* = tongue) is a large, leaf-shaped piece of elastic cartilage that is covered with epithelium (see also Fig. 23.3). The "stem" of the epiglottis is attached to the anterior rim of the thyroid cartilage, but the "leaf" portion is unattached and free to move up and down like a trap door. During swallowing, the larynx rises. This causes the free edge of the epiglottis to move down and form a lid over the glottis, closing it off. The **glottis** consists of a pair of folds of mucous membrane, the vocal folds (true vocal cords) in the larynx, and the space between them called the **rima glottidis** (RĪ-ma GLOT-ti-dis). In this way, the larynx is closed off, and liquids and foods are routed into the esophagus and kept out of the larynx and airways inferior to it. When small particles, such as dust, smoke, food, or liquids pass into the larynx, a cough reflex occurs to expel the material.

The **cricoid** (KRĪ-koyd; *krikos* = ring) **cartilage** is a ring of hyaline cartilage that forms the inferior wall of the larynx. It is attached to the first ring of cartilage of the trachea by the **cricotracheal ligament**. The thyroid cartilage is connected to the cricoid cartilage by the **cricothyroid ligament**. Clinically, the cricoid cartilage is the landmark for making an emergency airway (a tracheostomy, see page 717).

The paired **arytenoid** (ar′-i-TĒ-noyd; *arytaina* = ladle) **cartilages** are triangular pieces of mostly hyaline cartilage and located at the posterior, superior border of the cricoid cartilage. They attach to the vocal folds and intrinsic pharyngeal muscles. Supported by the arytenoid cartilages, the intrinsic pharyngeal muscles contract and thus move the vocal folds.

The paired **corniculate** (kor-NIK-yoo-lāt; *corniculate* = shaped like a small horn) **cartilages** are horn-shaped pieces of elastic cartilage. One is located at the apex of each arytenoid cartilage. The paired **cuneiform** (kyoo-NĒ-i-form; *cuneus* = wedge) **cartilages** are club-shaped elastic cartilages anterior to the corniculate cartilages (see Fig. 23.5). The cuneiform cartilages support the vocal folds and lateral aspects of the epiglottis.

The lining of the larynx inferior to the vocal folds is pseudostratified ciliated columnar epithelium. It consists of ciliated columnar cells, goblet cells, and basal cells, and its mucous covering helps trap dust not removed in the upper passages. Whereas the cilia in the upper respiratory tract move mucus and trapped particles *down* toward the pharynx, the cilia in the lower respiratory tract move them *up* toward the pharynx.

Voice Production

The mucous membrane of the larynx forms two pairs of folds (Fig. 23.5): a superior pair called the **ventricular folds (false vocal cords)** and an inferior pair called simply the **vocal folds (true vocal cords)**. The space between the ventricular folds is known as the **rima vestibuli**. The **laryngeal sinus (ventricle)** is a lateral expansion of the middle portion of the laryngeal cavity between the ventricular folds above and the vocal folds below (see Fig. 23.2b).

When the ventricular folds are brought together, they function in holding the breath against pressure in the tho-

Figure 23.5　Vocal folds. The arrow in the inset indicates the direction from which the vocal folds are viewed (superior).

Sound originates from vibrations of the vocal folds.

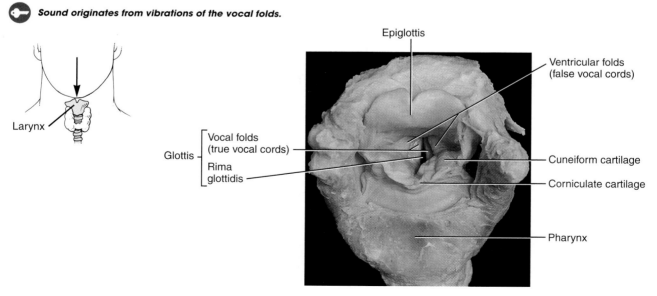

Superior view

What is the main function of the vocal folds?

racic cavity, such as might occur when a person strains to lift a heavy object. The mucous membrane of the vocal folds is lined by nonkeratinized stratified squamous epithelium. Deep to the membrane are bands of elastic ligaments stretched between pieces of rigid cartilage like the strings on a guitar. Skeletal muscles of the larynx, called intrinsic muscles, attach to both the rigid cartilage and the vocal folds themselves. When the muscles contract, they pull the elastic ligaments tight and stretch the vocal folds out into the airways so that the rima glottidis is narrowed. If air is directed against the vocal folds, they vibrate and set up sound waves in the column of air in the pharynx, nose, and mouth. The greater the pressure of air, the louder the sound.

Pitch is controlled by the tension on the vocal folds. If they are pulled taut by the muscles, they vibrate more rapidly, and a higher pitch results. Lower sounds are produced by decreasing the muscular tension on the vocal folds. Due to the influence of androgens (male sex hormones), vocal folds are usually thicker and longer in males than in females, and therefore they vibrate more slowly. Thus men generally have a lower range of pitch than women.

Sound originates from the vibration of the vocal folds, but other structures are necessary for converting the sound into recognizable speech. The pharynx, mouth, nasal cavity, and paranasal sinuses all act as resonating chambers that give the voice its human and individual quality. By constricting and relaxing the muscles in the wall of the pharynx, we produce the vowel sounds. Muscles of the face, tongue, and lips help us enunciate words.

LARYNGITIS AND CANCER OF THE LARYNX

Laryngitis is an inflammation of the larynx that is most often caused by a respiratory infection or irritants such as cigarette smoke. Inflammation of the vocal folds causes hoarseness or loss of voice by interfering with the contraction of the folds or by causing them to swell to the point where they cannot vibrate freely. Many long-term smokers acquire a permanent hoarseness from the damage done by chronic inflammation. Cancer of the larynx is found almost exclusively in individuals who smoke. The condition is characterized by hoarseness, pain on swallowing, or pain radiating to an ear. Treatment consists of radiation therapy and/or surgery. ■

Trachea

The **trachea** (TRĀ-kē-a; *tracheia* = sturdy), or windpipe, is a tubular passageway for air about 12 cm (5 in.) in length and 2½ cm (1 in.) in diameter. It is located anterior to the esophagus (Fig. 23.6a) and extends from the larynx to the superior border of the fifth thoracic vertebra (T5), where it divides into right and left primary bronchi (see Fig. 23.7).

The layers of the trachea, from deep to superficial, are (1) a mucosa, (2) submucosa, (3) hyaline cartilage, and (4) adventitia, composed of areolar connective tissue (Fig. 23.6b). The mucosa of the trachea consists of an epithelial layer of pseudostratified ciliated columnar epithelium (Fig. 23.6c)

Figure 23.6   Histology of the trachea.

 The trachea is anterior to the esophagus and extends from the larynx to the superior border of the fifth thoracic vertebra.

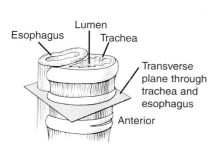

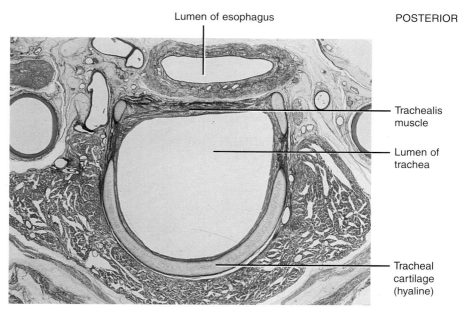

(a) Photomicrograph of a transverse section of the trachea in relation to the esophagus (2.6x)

Figure continues

Figure 23.6 (continued)

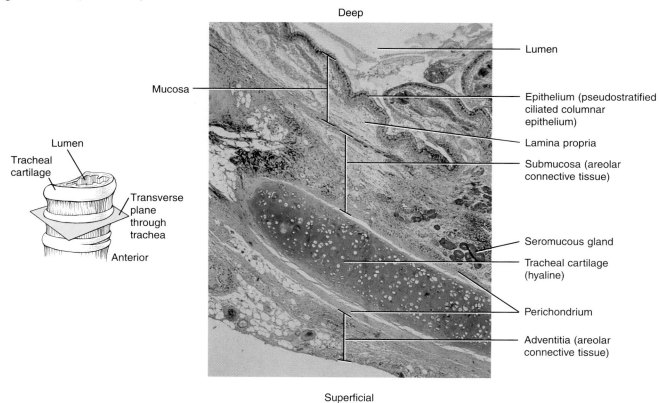

(b) Photomicrograph of a transverse section of part of the tracheal wall (80x)

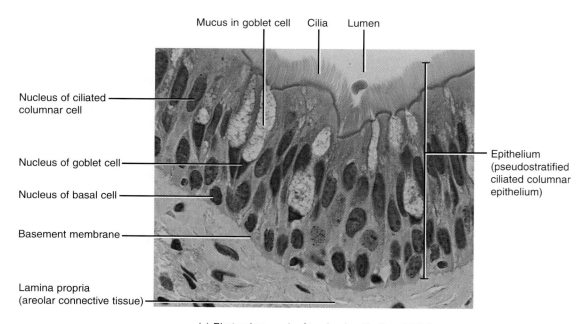

(c) Photomicrograph of tracheal epithelium (850x)

What is the benefit of not having cartilage between the trachea and esophagus?

and an underlying layer of lamina propria that contains elastic and reticular fibers. The epithelium consists of ciliated columnar cells and goblet cells that reach the luminal surface plus basal cells that do not reach the luminal surface. The epithelium provides the same protection against dust as the membrane lining the nasal cavity and larynx. The submucosa consists of areolar connective tissue and contains seromucous glands and their ducts. The 16–20 incomplete rings of hyaline cartilage look like letter Cs and are arranged horizontally and stacked one on top of another. They may be felt through the skin inferior to the larynx. The open part of each C-shaped cartilage ring faces the esophagus. This accommodates slight expansion of the esophagus into the trachea during swallowing (Fig. 23.6a). Transverse smooth muscle fibers, called the **trachealis muscle,** and elastic connective tissue hold the open ends of the cartilage rings together. The solid C-shaped cartilage rings provide a semirigid support so the tracheal wall does not collapse inward (especially during inspiration) and obstruct the air passageway. The adventitia of the trachea consists of areolar connective tissue which joins the trachea to surrounding tissues.

At the point where the trachea divides into right and left primary bronchi, there is an internal ridge called the **carina** (ka-RĪ-na; *karina* = keel of a boat). It is formed by a posterior and somewhat inferior projection of the last tracheal cartilage. The mucous membrane of the carina is one of the most sensitive areas of the entire larynx and trachea for triggering a cough reflex.

TRACHEOSTOMY AND INTUBATION

In some situations, for example, a crushing injury to the chest, the rings of cartilage may not be strong enough to overcome collapse and obstruction of the trachea. Or the mucous membrane may become inflamed and swell so much that it closes off the airways; inflamed membranes secrete a great deal of mucus that may clog the lower airways. A large object may be breathed in (aspirated) while the rima glottidis is open; or an aspirated foreign object may cause spasm of the muscles of the larynx. If the obstruction is superior to the level of the larynx, a **tracheostomy** (trā-kē-OS-tō-mē) may be performed. A skin incision is made, followed by a short longitudinal incision into the trachea inferior to the cricoid cartilage. The patient breathes through a metal or plastic tracheal tube inserted through the incision. Another method is **intubation.** A tube is inserted into the mouth or nose and passed inferiorly through the larynx and trachea. The firm wall of the tube pushes back any flexible obstruction, and the inside of the tube provides a passageway for air. If mucus is clogging the trachea, it can be suctioned out through the tube. ■

Bronchi

At the superior border of the fifth thoracic vertebra, the trachea divides into a **right primary bronchus** (BRON-kus; *bronchos* = windpipe), which goes into the right lung, and a **left primary bronchus,** which goes into the left lung (Fig.

23.7). The right primary bronchus is more vertical, shorter, and wider than the left. As a result, an aspirated object is more likely to enter and lodge in the right primary bronchus than the left. Like the trachea, the primary bronchi (BRON-kē) contain incomplete rings of cartilage and are lined by pseudostratified ciliated columnar epithelium.

BRONCHOSCOPY

Bronchoscopy (bron-KOS-kō-pē) is the visual examination of the bronchi through a **bronchoscope,** an illuminated, tubular instrument that is passed through the trachea into the bronchi. The examiner can view the interior of the trachea and bronchi to biopsy a tumor, clear an obstructing object or secretions from an airway, take cultures or smears for microscopic examination, stop bleeding, or deliver drugs. ■

On entering the lungs, the primary bronchi divide to form smaller bronchi—the **secondary (lobar) bronchi,** one for each lobe of the lung. (The right lung has three lobes; the left lung has two.) The secondary bronchi continue to branch, forming still smaller bronchi, called **tertiary (segmental) bronchi,** that divide into **bronchioles.** Bronchioles, in turn, branch repeatedly and the smallest bronchioles branch into even smaller tubes called **terminal bronchioles.** This extensive branching from the trachea resembles a tree trunk with its branches and is commonly referred to as the **bronchial tree.**

Bronchography (bron-KOG-ra-fē) is an imaging technique to visualize the bronchial tree using x-rays. An intratracheal catheter is passed into the right or left bronchus through the mouth or nose. Then an opaque contrast medium, usually containing iodine, is inhaled, which distributes it throughout the bronchioles. Radiographs of the chest in various positions are taken, and the developed film, a **bronchogram** (BRON-kō-gram), provides a picture of the bronchial tree (Fig. 23.7b).

As the branching becomes more extensive in the bronchial tree, several structural changes may be noted. First, the epithelium gradually changes from pseudostratified ciliated columnar epithelium in the bronchi to nonciliated simple cuboidal epithelium in the terminal bronchioles. (In regions where nonciliated cuboidal epithelium is present, inhaled particles are removed by macrophages.) Second, incomplete rings of cartilage in primary bronchi are gradually replaced by plates of cartilage that finally disappear in the distal bronchioles. Third, as the cartilage decreases, the amount of smooth muscle increases. Smooth muscle encircles the lumen in spiral bands and its contraction is affected by both the autonomic nervous system (ANS) and various chemicals.

During exercise, activity in the sympathetic division of the ANS increases and the adrenal medulla releases the hormones epinephrine and norepinephrine. Both changes relax smooth muscle in the bronchioles, thus dilating the airways. This improves lung ventilation because more air reaches the alveoli. The parasympathetic division of the ANS and mediators of allergic reactions, such as histamine, cause constriction of distal bronchioles. During an **asthma attack** the smooth muscle of

Figure 23.7   Bronchial tree in relation to the lungs.

 The bronchial tree begins at the trachea and ends at the terminal bronchioles.

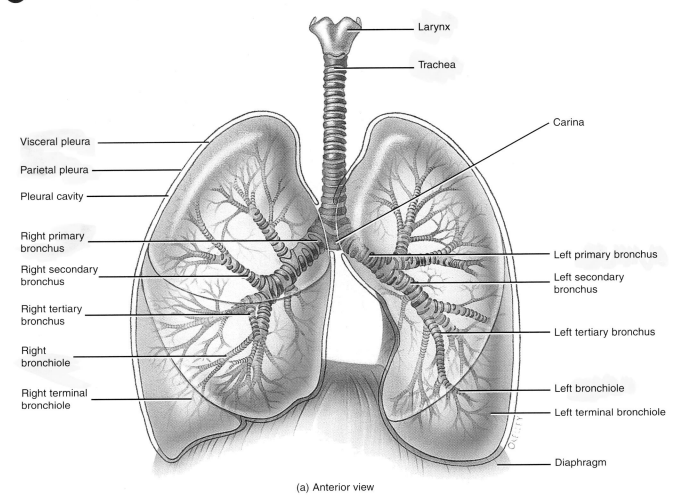

Larynx

Trachea

Carina

Visceral pleura

Parietal pleura

Pleural cavity

Right primary bronchus

Right secondary bronchus

Right tertiary bronchus

Right bronchiole

Right terminal bronchiole

Left primary bronchus

Left secondary bronchus

Left tertiary bronchus

Left bronchiole

Left terminal bronchiole

Diaphragm

(a) Anterior view

BRANCHING OF BRONCHIAL TREE

Trachea

↓

Primary bronchi

↓

Secondary bronchi

↓

Tertiary bronchi

↓

Bronchioles

↓

Terminal bronchioles

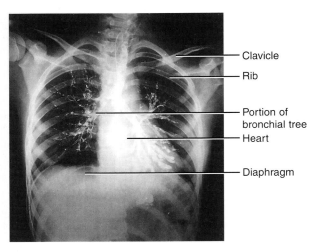

Clavicle

Rib

Portion of bronchial tree

Heart

Diaphragm

(b) Bronchogram

Q   How many lobes and secondary bronchi are present in each lung?

bronchioles contracts, reducing the diameter of the airways. Because there is no supporting cartilage, the muscle spasms can even close off the airways, a life-threatening situation. Epinephrine, administered as an inhaled mist (nebulization), often is used to relax the smooth muscle and open the airways.

NEBULIZATION

Many respiratory disorders are treated by means of **nebulization** (neb-yoo-li-ZA-shun). This procedure consists of administering medication in the form of droplets that are suspended in air into the respiratory tract. The patient inhales the medication as a fine mist. Nebulization therapy can be used with many different types of drugs, such as chemicals that relax the smooth muscle of the airways, chemicals that reduce the thickness of mucus, and antibiotics. ■

Lungs

The **lungs** (*lunge* = lightweight, since the lungs float) are paired cone-shaped organs lying in the thoracic cavity. They are separated from each other by the heart and other structures in the mediastinum. The mediastinum separates the thoracic cavity into two anatomically distinct chambers. Accordingly, if trauma causes one lung to collapse, the other may remain expanded. Two layers of serous membrane, collectively called the **pleural** (*pleura* = side) **membrane**, enclose and protect each lung (Fig. 23.8). The superficial layer lines the wall of the thoracic cavity and is called the **parietal pleura**. The deep layer, the **visceral pleura**, covers the lungs themselves. Between the visceral and parietal pleurae is a small potential space, the **pleural cavity**, which contains a lubricating fluid secreted by the

Figure 23.8   Relationship of the pleural membranes to the lungs. The arrow in the inset indicates the direction from which the lungs are viewed (inferior).

The parietal pleura lines the thoracic cavity, whereas the visceral pleura covers the lungs.

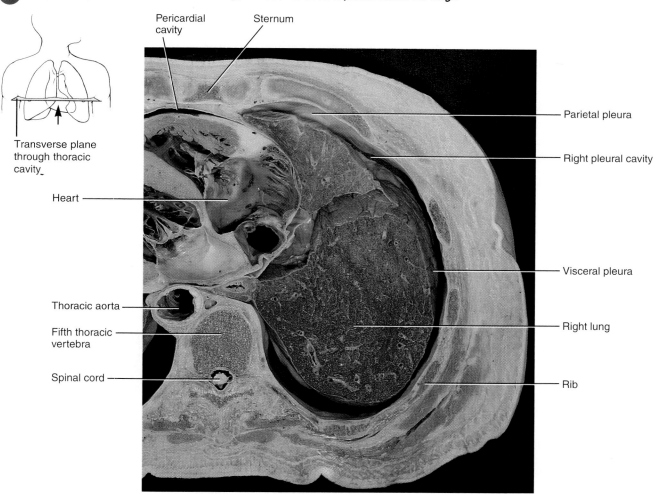

Photograph of an inferior view of a transverse section through the thoracic cavity

What type of membrane is the pleural membrane?

membranes. This fluid reduces friction between the membranes and allows them to slide easily on one another during breathing.

PNEUMOTHORAX, HEMOTHORAX, AND PLEURISY

In certain conditions, the pleural cavity may fill with air (**pneumothorax**; *pneumo* = air or breath), blood (**hemothorax**), or pus. Air in the pleural cavity, most commonly introduced in a surgical opening of the chest or as a result of a stab or gunshot wound, may cause the lung to collapse (atelectasis). Fluid can be drained from the pleural cavity by inserting a needle, usually posteriorly through the seventh intercostal space. The needle is passed along the superior border of the lower rib to avoid damage to the intercostal nerves and blood vessels. Inferior to the seventh intercostal space there is danger of penetrating the diaphragm.

Inflammation of the pleural membrane, or **pleurisy**, may in its early stages cause pain due to friction between the parietal and visceral layers of the pleura. If the inflammation persists, fluid accumulates in the pleural space, a condition known as **pleural effusion**. One cause of pleural effusion is lung cancer. ■

Gross Anatomy

The lungs extend from the diaphragm to just slightly superior to the clavicles and lie against the ribs anteriorly and posteriorly. The broad inferior portion of the lung, the **base**, is concave and fits over the convex area of the diaphragm (Fig. 23.9). The narrow superior portion of the lung is termed the **apex** (**cupula**). The surface of the lung lying against the ribs, the **costal surface**, is rounded to match the curvature of the ribs. The **mediastinal** (**medial**) **surface** of each lung contains a region, the **hilus**, through which bronchi, pulmonary blood vessels, lymphatic vessels, and nerves enter and exit. These structures are held together by the pleura and connective tissue and constitute the **root** of the lung. Medially, the left lung also contains a concavity, the **cardiac notch**, in which the heart lies.

The right lung is thicker and broader than the left. It is also somewhat shorter than the left because the diaphragm is higher on the right side to accommodate the liver that lies inferior to it.

Lobes and Fissures

Each lung is divided into lobes by one or more fissures (Fig. 23.9). Both lungs have an **oblique fissure**, which extends inferiorly and anteriorly. The right lung also has a **horizontal fissure**. The oblique fissure in the left lung separates the **superior lobe** from the **inferior lobe**. The superior part of the oblique fissure of the right lung separates the superior lobe from the inferior lobe, whereas the inferior part of the oblique fissure separates the inferior lobe from the **middle lobe**. The horizontal fissure of the right lung subdivides the superior lobe, thus forming a middle lobe.

Each lobe receives its own secondary (lobar) bronchus. Thus the right primary bronchus gives rise to three secondary (lobar) bronchi called the **superior**, **middle**, and **inferior secondary (lobar) bronchi**. The left primary bronchus gives rise to **superior** and **inferior secondary** (**lobar**) **bronchi**. Within the substance of the lung, the secondary bronchi give rise to the **tertiary (segmental) bronchi**, which are constant in both origin and distribution. There are ten tertiary bronchi in each lung. The segment of lung tissue that each supplies is called a **bronchopulmonary segment**. Bronchial and pulmonary disorders, such as tumors or abscesses, may be localized in a bronchopulmonary segment and may be surgically removed without seriously disrupting surrounding lung tissue.

Figure 23.9   Lungs. The arrows in the inset indicate the directions from which the lungs are viewed (lateral and medial).

🔑 *The subdivisions of the lungs are lobes → bronchopulmonary segments → lobules.*

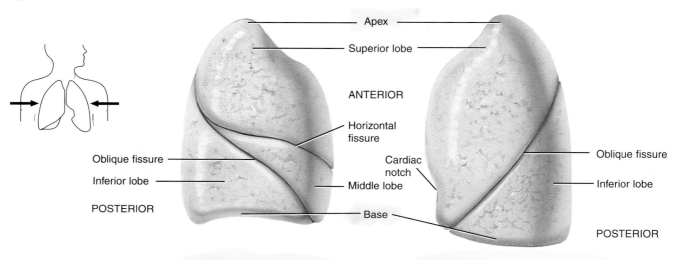

(a) Lateral view of right lung   (b) Lateral view of left lung

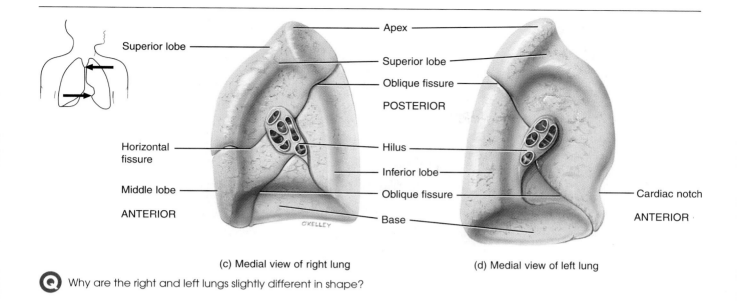

(c) Medial view of right lung

(d) Medial view of left lung

 Why are the right and left lungs slightly different in shape?

Lobules

Each bronchopulmonary segment of the lungs has many small compartments called **lobules**. Each lobule is wrapped in elastic connective tissue and contains a lymphatic vessel, an arteriole, a venule, and a branch from a terminal bronchiole (Fig. 23.10a). Terminal bronchioles subdivide into microscopic branches called **respiratory bronchioles**. As the respiratory bronchioles penetrate more deeply into the lungs, the epithelial lining changes from simple cuboidal to simple squamous. Respiratory bronchioles, in turn, subdivide into several (2–11) **alveolar ducts**. From the trachea to the alveolar ducts, there are about 25 orders of branching of the respiratory passageways. That is, the trachea divides into primary bronchi (first order), the primary bronchi divide into secondary bronchi (second order), and so on.

Around the circumference of the alveolar ducts are numerous alveoli and alveolar sacs (Fig. 23.10b). An **alveolus** (al-VĒ-ō-lus) is a cup-shaped outpouching lined by simple

Figure 23.10   Histology of the lungs.

Alveolar sacs are two or more alveoli that share a common opening.

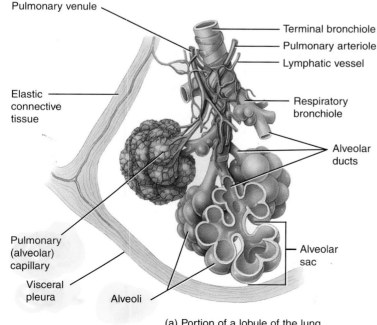

(a) Portion of a lobule of the lung

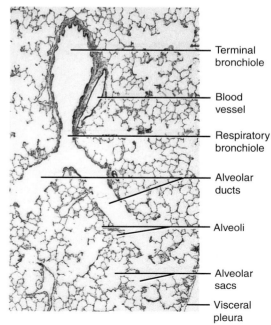

(b) Photomicrograph of the lung (approx. 30x)

 What are the roles of goblet cells and cilia?

squamous epithelium and supported by a thin elastic basement membrane. **Alveolar sacs** are two or more alveoli that share a common opening (Fig. 23.10a,b). The alveolar walls consist of two types of alveolar epithelial cells or **pneumocytes** (Fig. 23.11). **Type I alveolar (squamous pulmonary epithelial) cells** are simple squamous epithelial cells that form a continuous lining of the alveolar wall, interrupted by occasional **type II alveolar (septal) cells**. The thin type I alveolar cells are the main site where gas exchange takes place. Type II alveolar cells are rounded or cuboidal epithelial cells whose free surfaces contain microvilli. They secrete alveolar fluid, which keeps the alveolar cells moist. Associated with the alveolar wall are **alveolar macrophages (dust cells)**. These wandering phagocytes remove fine dust particles and other debris in the alveolar spaces. Also present are fibroblasts that produce reticular and elastic fibers. Superficial to the layer of type I alveolar cells is an elastic basement membrane. Around the alveoli, the lobule's arteriole and venule disperse into a capillary net-work. The blood capillaries consist of a single layer of endothelial cells and basement membrane.

Alveolar–Capillary Membrane

The exchange of respiratory gases (O_2 and CO_2) between the lungs and blood takes place by diffusion across alveolar and capillary walls. Collectively, the layers through which the respiratory gases diffuse are known as the **alveolar–capillary (respiratory) membrane** (Fig. 23.11b). It consists of:

1. A layer of type I and type II alveolar cells with wandering alveolar macrophages that constitute the **alveolar (epithelial) wall**.
2. An **epithelial basement membrane** underlying the alveolar wall.
3. A **capillary basement membrane** that is often fused to the epithelial basement membrane.
4. The **endothelial cells** of the capillary.

Figure 23.11   Structure of an alveolus.

The exchange of respiratory gases occurs by diffusion across the alveolar-capillary (respiratory) membrane.

- Monocyte
- Reticular fiber
- Elastic fiber
- Type II alveolar (septal) cell
- Alveolar-capillary (respiratory) membrane
- Alveolus
- Type I alveolar (squamous pulmonary epithelial) cell
- Alveolar macrophage
- Red blood cell
- Diffusion of O_2
- Diffusion of CO_2
- Aveolus
- Red blood cell
- Capillary endothelium
- Capillary basement membrane
- Epithelial basement membrane
- Type I alveolar (squamous pulmonary epithelial) cell
- Surfactant layer (alveolar fluid and surfactant)
- Interstitial space

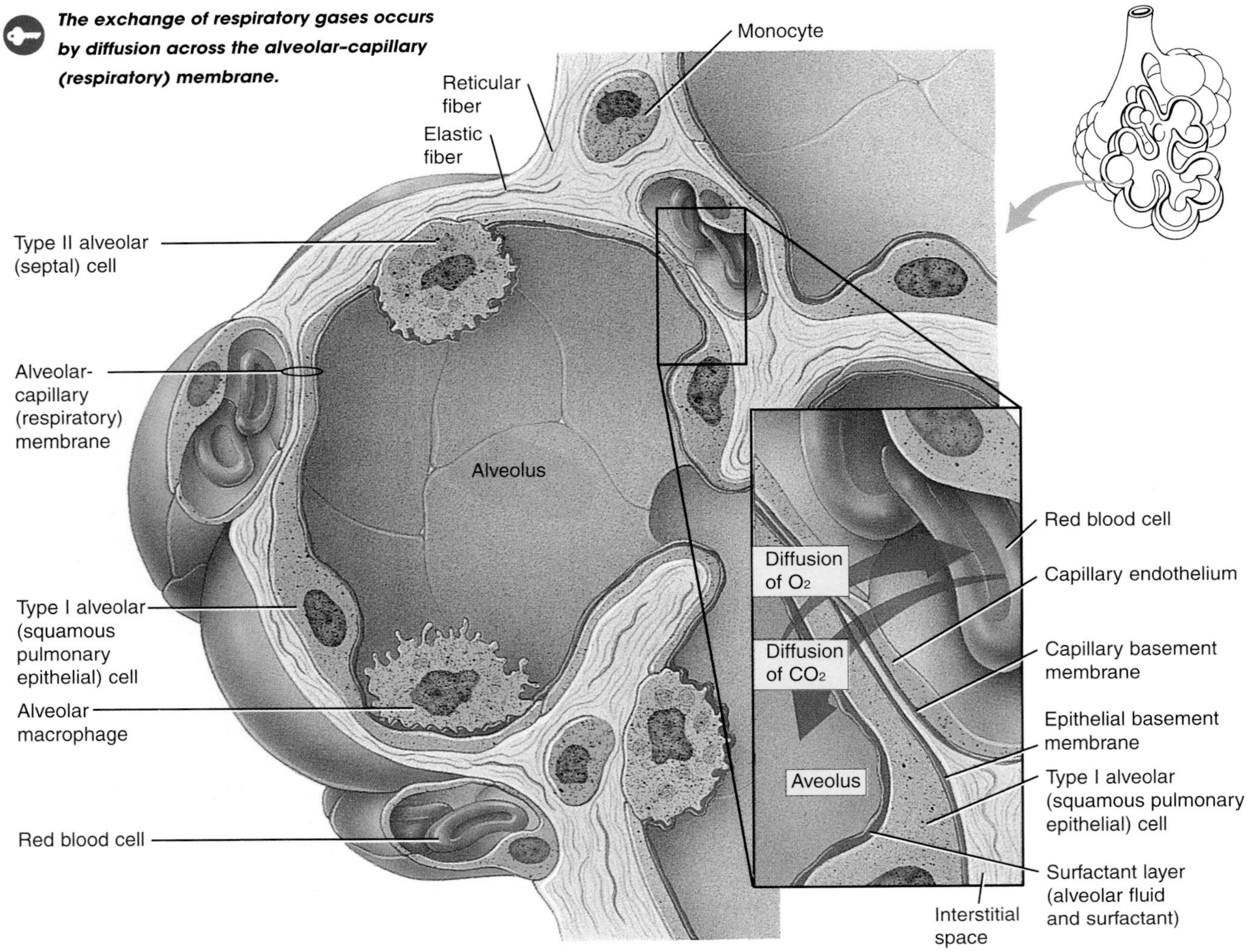

(a) Transverse section of an alveolus

(b) Details of alveolar–capillary (respiratory) membrane

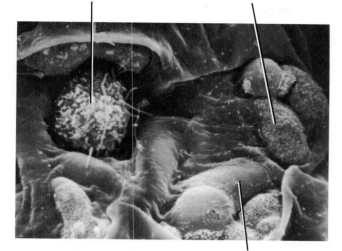

Alveolar macrophage (functions as a phagocyte)

Type II alveolar (septal) cell (secretes alveolar fluid, which contains surfactant)

Type I alveolar (squamous pulmonary epithelial) cell (main site for gas exchange)

(c) Scanning electron micrograph of an alveolus (3430x)

 What is the thickness of the alveolar-capillary membrane?

Despite having several layers, the alveolar–capillary membrane averages only ½ μm in thickness, about 1/16 the diameter of a red blood cell. This allows rapid diffusion of respiratory gases. Moreover, it has been estimated that the lungs contain 300 million alveoli, providing an immense surface area of 70 m² (750 ft²), about the size of a handball court, for the exchange of gases.

Blood Supply to the Lungs

There is a double blood supply to the lungs. Deoxygenated blood passes through the pulmonary trunk, which divides into a left pulmonary artery that enters the left lung and a right pulmonary artery that enters the right lung. The return of the oxygenated blood to the heart is by way of the pulmonary veins, which drain into the left atrium (see Fig. 21.29).

Bronchial arteries, which branch from the aorta, deliver oxygenated blood to the lungs. This blood mainly perfuses the walls of the bronchi and bronchioles. Connections exist between branches of the bronchial arteries and branches of the pulmonary arteries, however, and most blood returns to the heart via pulmonary veins. Some blood, however, drains into bronchial veins, branches of the azygos system, and returns to the heart via the superior vena cava.

Recall from Chapter 21 that pulmonary circulation differs from systemic circulation in that pulmonary blood vessels provide less resistance to blood flow; less pressure is required to move blood through the pulmonary circulation. In addition, pulmonary blood vessels constrict in response to hypoxia (low O_2 level) so that pulmonary blood bypasses poorly aerated areas. In all other body tissues, hypoxia causes dilation of blood vessels, which serves to increase blood flow to a tissue that is not receiving adequate O_2.

PULMONARY VENTILATION

Pulmonary ventilation (**breathing**) is the process by which gases are exchanged between the atmosphere and lung alveoli. The bulk flow of air between the atmosphere and lungs occurs for the same reason that blood flows through the body: a pressure gradient exists. Air moves into the lungs when the pressure inside the lungs is less than the air pressure in the atmosphere. Air moves out of the lungs when the pressure inside the lungs is greater than the pressure in the atmosphere. Let us examine the mechanics of pulmonary ventilation by first looking at inspiration.

Inspiration

Breathing in is called **inspiration** (**inhalation**). Just before each inspiration, the air pressure inside the lungs equals the pressure of the atmosphere, which is about 760 mm Hg, or 1 atmosphere (atm), at sea level. For air to flow into the lungs, the pressure inside the alveoli must become lower than the pressure in the atmosphere. This condition is achieved by increasing the volume (size) of the lungs.

The pressure of a gas in a closed container is inversely proportional to the volume of the container. This means that if the size of a closed container is increased, the pressure of the gas inside the container decreases. If the size of the container is decreased, then the pressure inside it increases. This relationship is called **Boyle's law** and may be demonstrated as follows. Suppose we place a gas in a cylinder that has a movable piston and a pressure gauge, and the initial pressure is 1 atm (Fig. 23.12). This pressure is created by the gas molecules striking the wall of the container. If the piston is pushed down, the gas is compressed into a smaller volume. The same number of gas molecules strike less wall space. The gauge shows that the pressure doubles as the gas is compressed to half its volume. In other words, the same

Figure 23.12 Boyle's law.

 The volume of a gas varies inversely with the pressure.

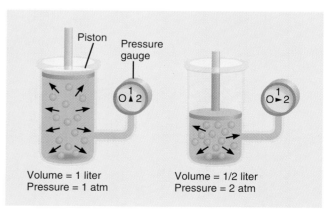

Volume = 1 liter
Pressure = 1 atm

Volume = 1/2 liter
Pressure = 2 atm

 If the volume is decreased to ¼ liter, how does the pressure change?

number of molecules in half the space produce twice the pressure. Conversely, if the piston is raised to increase the volume, the pressure decreases. Thus the volume of a gas varies inversely with pressure (if temperature is constant). Boyle's law also applies to everyday activities such as the operation of a bicycle pump or the blowing up of a balloon.

Differences in pressure, caused by changes in lung volume, force air into our lungs when we inhale and out when we exhale. For inspiration to occur, the lungs must expand. This increases lung volume and thus decreases the pressure in the lungs below atmospheric pressure. The first step in expanding the lungs involves contraction of the principal inspiratory muscles—the diaphragm and external intercostals (Fig. 23.13; see also Fig. 11.11).

The diaphragm, the most important muscle of inspiration, is a dome-shaped skeletal muscle that forms the floor of the

---

Figure 23.13   Pulmonary ventilation: muscles of inspiration and expiration. The pectoralis minor muscle, an accessory inspiratory muscle, is not shown here but is illustrated in Fig. 11.15a.

🔑 *During deep, labored inspiration, accessory muscles of inspiration (sternocleidomastoids, scalenes, and pectoralis minors) participate.*

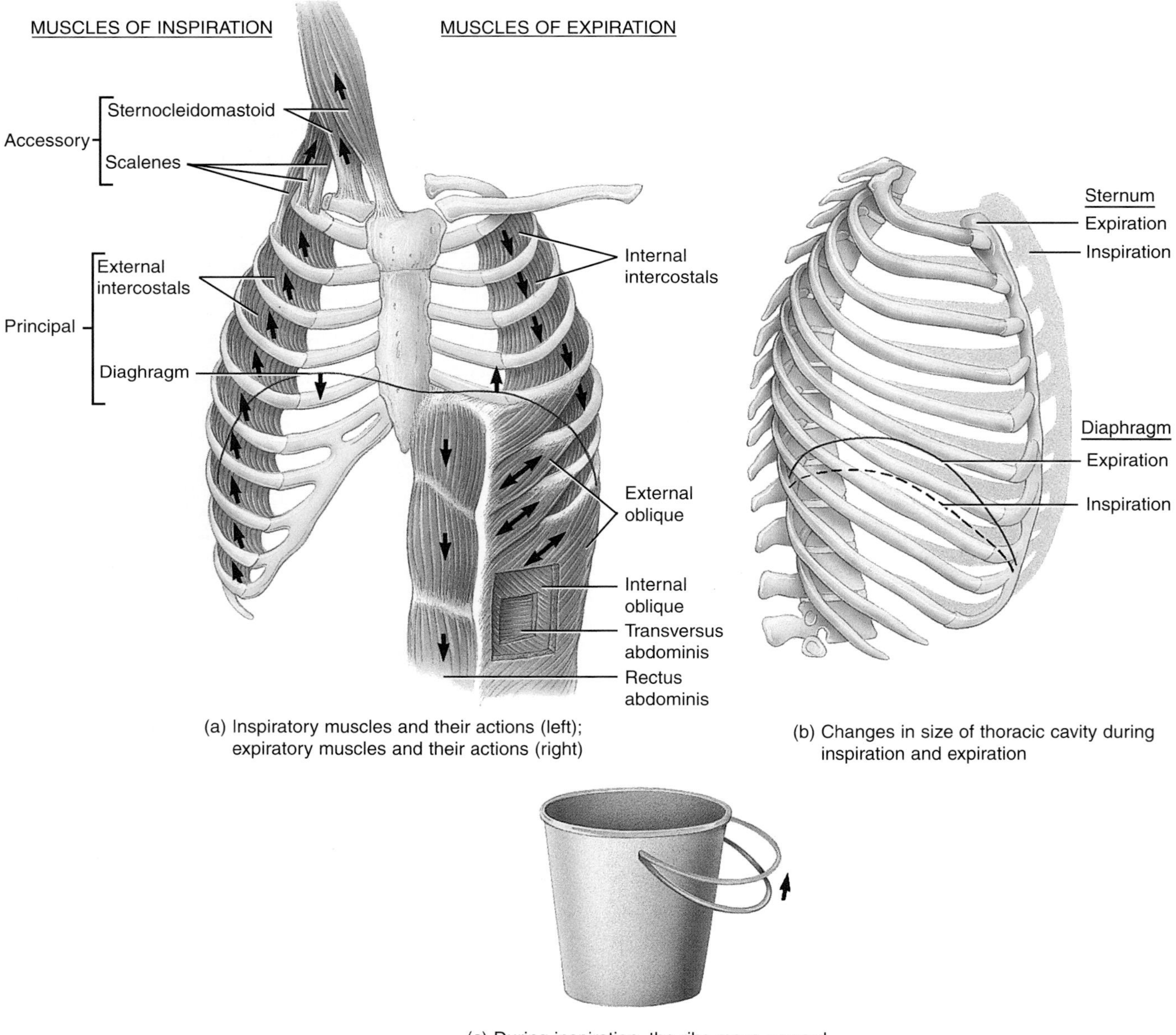

MUSCLES OF INSPIRATION   MUSCLES OF EXPIRATION

Accessory —
- Sternocleidomastoid
- Scalenes

Principal —
- External intercostals
- Diaghragm

Internal intercostals

External oblique

Internal oblique

Transversus abdominis

Rectus abdominis

(a) Inspiratory muscles and their actions (left); expiratory muscles and their actions (right)

Sternum
- Expiration
- Inspiration

Diaphragm
- Expiration
- Inspiration

(b) Changes in size of thoracic cavity during inspiration and expiration

(c) During inspiration, the ribs move upward and outward like the handle on a bucket

 Right now, what muscle is mainly responsible for your breathing?

thoracic cavity. It is innervated by fibers of the phrenic nerves, which emerge laterally from the spinal cord at cervical levels 3, 4, and 5. Contraction of the diaphragm causes it to flatten, lowering its dome. This increases the vertical dimension of the thoracic cavity and accounts for the movement of about 75% of the air that enters the lungs during inspiration. During normal quiet breathing, the diaphragm descends about 1 cm (0.4 in.), which produces a pressure difference of 1–3 mm Hg and inhalation of about ½ liter of air. In strenuous breathing, the diaphragm may descend 10 cm (4 in.), which produces a pressure difference of 100 mm Hg and inhalation of 2–3 liters.

At the same time the diaphragm contracts, the external intercostals contract. These skeletal muscles run obliquely in-

feriorly and anteriorly between adjacent ribs, and when these muscles contract, the ribs are pulled superiorly and the sternum is pushed anteriorly. This increases the anterior–posterior dimension of the thoracic cavity.

During normal breathing, the pressure between the two pleural layers, called **intrapleural (intrathoracic) pressure**, is always subatmospheric (lower than atmospheric pressure). Just before inspiration, it is about 4 mm Hg less than the atmospheric pressure, or 756 mm Hg if the atmospheric pressure is 760 mm Hg (Fig. 23.14). As the diaphragm contracts and the overall size of the thoracic cavity increases, the intrapleural pressure falls to about 754 mm Hg. The parietal and visceral pleurae normally adhere strongly to each other because of the subatmospheric pressure between them and

Figure 23.14 Pulmonary ventilation: pressure changes. At the beginning of inspiration, the diaphragm contracts, the chest expands, the lungs are pulled outward, and alveolar pressure decreases. As the diaphragm relaxes, the lungs recoil inward. Alveolar pressure rises, forcing air out until alveolar pressure equals atmospheric pressure.

🔑 *Air moves into the lungs when alveolar pressure is less than atmospheric pressure and out of the lungs when alveolar pressure is greater than atmospheric pressure.*

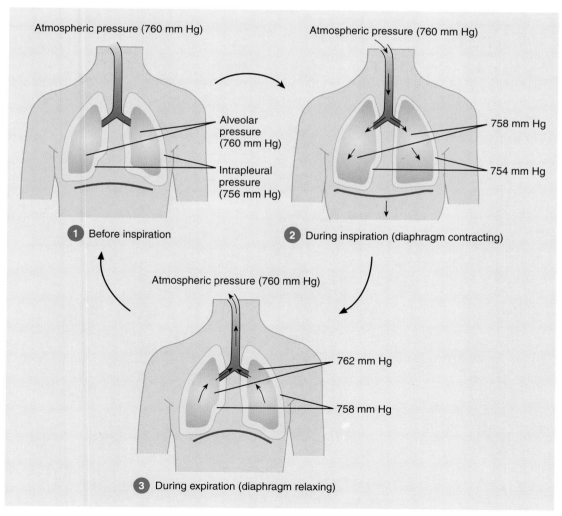

1 Before inspiration

2 During inspiration (diaphragm contracting)

3 During expiration (diaphragm relaxing)

Atmospheric pressure (760 mm Hg)

Alveolar pressure (760 mm Hg)

Intrapleural pressure (756 mm Hg)

758 mm Hg
754 mm Hg
762 mm Hg
758 mm Hg

❓ How does the intrapleural pressure change during a normal, quiet breath?

because of the surface tension created by their moist adjoining surfaces. As the thoracic cavity expands, the parietal pleura lining the cavity is pulled outward in all directions, and the visceral pleura and lungs are pulled along with it.

As the volume of the lungs increases in this way, the pressure inside the lungs, called the **alveolar (intrapulmonic) pressure**, drops from 760 to 758 mm Hg. A pressure difference is thus established between the atmosphere and the alveoli. Air always flows from a region of higher pressure to a region of lower pressure and inspiration takes place. Air continues to flow into the lungs as long as the pressure difference exists.

A summary of inspiration is presented in Fig. 23.15a.

Expiration

Breathing out, called **expiration (exhalation)**, is also achieved by a pressure gradient, but in this case the gradient is reversed: the pressure in the lungs is greater than the pressure of the atmosphere. Normal expiration during quiet breathing, unlike inspiration, is a *passive process* because no muscular contractions are involved. It results from **elastic recoil** of the chest wall and lungs: they have a natural tendency to spring back after they have been stretched. Two inwardly directed forces contribute to elastic recoil: (1) the recoil of elastic fibers that were stretched during inspiration and (2) the inward pull of surface tension due to the film of alveolar fluid (see page 727).

Expiration starts when the inspiratory muscles relax. As the external intercostals relax, the ribs move inferiorly, and as the diaphragm relaxes, its dome moves superiorly owing to its elasticity. These movements decrease the vertical and anterior–posterior dimensions of the thoracic cavity. Also, surface tension exerts an inward pull and the elastic basement membranes of the alveoli and elastic fibers in bronchioles and alveolar ducts recoil. As a result, lung volume decreases and the alveolar pressure increases to 762 mm Hg. Air then flows from the area of higher pressure in the alveoli to the area of lower pressure in the atmosphere (see Fig. 23.14, part ❸).

Expiration becomes active during labored breathing and when air movement out of the lungs is impeded. During these times, muscles of expiration—the abdominals and internal intercostals—contract. Contraction of the abdominal muscles moves the inferior ribs downward and compresses the abdominal viscera, thus forcing the diaphragm superiorly. Contraction of the internal intercostals, which extend inferiorly and posteriorly between adjacent ribs, pulls the ribs inferiorly.

A summary of expiration is presented in Fig. 23.15b.

Breathing Patterns

The term for normal quiet breathing is **eupnea** (yoop-NĒ-a; *eu* = normal; *pnoia* = breath). Eupnea involves shallow, deep, or combined shallow and deep breathing. **Apnea** (AP-nē-a) refers to a temporary cessation of breathing. **Dyspnea**

Figure 23.15   Summary of inspiration and expiration.

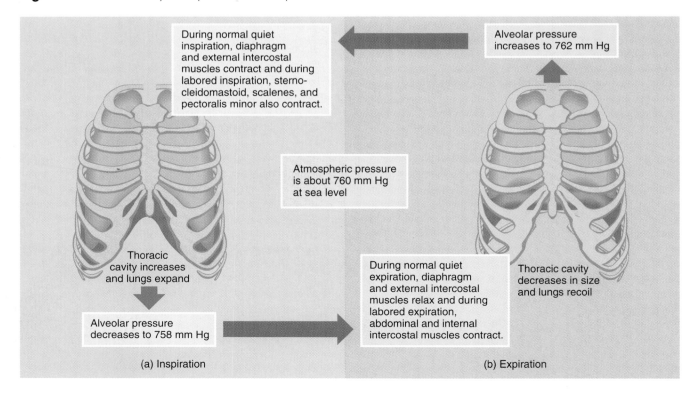

During normal quiet inspiration, diaphragm and external intercostal muscles contract and during labored inspiration, sternocleidomastoid, scalenes, and pectoralis minor also contract.

Alveolar pressure increases to 762 mm Hg

Atmospheric pressure is about 760 mm Hg at sea level

Thoracic cavity increases and lungs expand

During normal quiet expiration, diaphragm and external intercostal muscles relax and during labored expiration, abdominal and internal intercostal muscles contract.

Thoracic cavity decreases in size and lungs recoil

Alveolar pressure decreases to 758 mm Hg

(a) Inspiration

(b) Expiration

(DISP-nē-a; *dys* = painful, difficult) means painful or labored breathing whereas **tachypnea** (tak´-ip-NĒ-a; *tachy* = rapid) means rapid breathing. Shallow (chest) breathing is called **costal breathing**. It consists of an upward and outward movement of the chest as a result of contraction of the external intercostal muscles. The intercostal muscles receive innervation from the intercostal nerves, which emerge from thoracic segments of the spinal cord. Deep (abdominal) breathing is called **diaphragmatic breathing**. It consists of the outward movement of the abdomen as a result of the contraction and descent of the diaphragm. During deep, labored inspiration, accessory muscles of inspiration also participate in increasing the size of the thoracic cavity (see Fig. 23.13a). These include the sternocleidomastoids, which elevate the sternum, the scalenes, which elevate the superior two ribs, and the pectoralis minors, which elevate the third through fifth ribs. Although normally subatmospheric, intrapleural pressure may briefly exceed atmospheric pressure during any type of forced expiration, such as coughing or straining during defecation.

Alveolar Surface Tension

A thin layer of alveolar fluid lies next to air in the alveoli and exerts a force known as **surface tension**. Surface tension arises at all air–water junctions because the polar water molecules are more strongly attracted to each other than they are to gas molecules in the air. When liquid surrounds a sphere of air, as in an alveolus or a soap bubble, surface tension produces an inwardly directed force. Soap bubbles "burst" because they collapse inward due to surface tension. In the lungs, surface tension causes the alveoli to assume the smallest possible diameter. During breathing, surface tension must be overcome to expand the lungs during each inspiration. It is also the major component of lung elastic recoil, which acts to decrease the size of alveoli during expiration.

The surface tension of alveolar fluid is not as great as that of pure water due to the presence of a detergent-like substance called **surfactant** (sur-FAK-tant). Produced by type II alveolar cells, surfactant is a complex mixture of phospholipids and lipoproteins. It lowers the surface tension of alveolar fluid and thus reduces the tendency of alveoli to collapse completely. A deficiency of surfactant in premature infants greatly increases the surface tension and causes collapse of the alveoli at the end of each exhalation. Great effort is then needed at the next inhalation to reopen the collapsed alveoli (see respiratory distress syndrome, page 746).

ATELECTASIS

It was noted earlier that intrapleural pressure normally is lower than atmospheric pressure. The pleural cavities are sealed off from the outside environment and cannot equalize their pressure with that of the atmosphere. If an injury of the chest wall allows air to enter the intrapleural space, either from the outside or from the alveoli (pneumothorax), surface tension and recoil of elastic fibers cause the lung on that side to collapse. Collapse or incomplete expansion of lung tissue is called **atelectasis** (at´-ē-LEK-ta-sis; *ateles* = incomplete; *ektasis* = expansion). ■

Compliance

Compliance refers to the ease with which the lungs and thoracic wall can be expanded. When a small increase in alveolar pressure produces a large increase in lung volume, compliance is high. Higher compliance means that the lungs and thoracic wall expand easily; lower compliance means that they resist expansion. By analogy, a thin balloon that is easy to inflate has high compliance whereas a heavy and stiff balloon that takes a lot of effort to inflate has low compliance. In the lungs, compliance is related to two principal factors: elasticity and surface tension. The lungs normally have high compliance and expand easily because elastic fibers in lung tissue are easily stretched and surfactant in alveolar fluid reduces surface tension. If surface tension within lung tissue were high, the tissues would resist expansion (lower compliance). Compliance decreases in conditions that (1) scar lung tissue, for example, tuberculosis (see page 745), (2) cause it to become filled with fluid (pulmonary edema), (3) produce a deficiency in surfactant, or (4) impede lung expansion in any way, for example, paralysis of the intercostal muscles. Emphysema (see page 744) increases compliance.

Airway Resistance

As is true for blood flow through blood vessels, the flow of air through the airways depends on both the pressure difference and the resistance: airflow equals the pressure difference between the alveoli and the atmosphere divided by the resistance. The walls of the airways, especially the bronchi and bronchioles, offer some resistance to the normal flow of air into and out of the lungs. During inhalation, the increase in size of the thoracic cavity decreases resistance to airflow by increasing the diameter of bronchi and bronchioles. Resistance increases, however, during exhalation because the airways narrow somewhat. Also, the degree of contraction or relaxation of smooth muscle in the walls of the airways regulates airway diameter and thus resistance. Increased signals from the sympathetic division of the autonomic nervous system cause relaxation of this smooth muscle, which results in bronchodilation and decreased resistance.

Any condition that narrows or obstructs the airways increases resistance, and more pressure is required to force air through. During a forced expiration, as in coughing or playing a wind instrument, intrapleural pressure may increase. This compresses the airways, which greatly increases airway resistance. People with chronic obstructive pulmonary disease (COPD), such as bronchial asthma and emphysema, have some degree of obstruction or collapse of airways, which increases resistance.

Modified Respiratory Movements

Respirations also provide humans with methods for expressing emotions such as laughing, sighing, and sobbing. Moreover, respiratory air can be used to expel foreign matter from the lower air passages through actions such as sneezing and coughing. Respiratory movements are also modified and controlled during talking and singing. Some of the modified respiratory movements that express emotion or clear the airways are listed in Exhibit 23.1. All these movements are reflexes, but some of them also can be initiated voluntarily.

LUNG VOLUMES AND CAPACITIES

In clinical practice, the word **respiration** (**ventilation**) means one inspiration plus one expiration. The healthy adult averages 12 respirations a minute and moves about 6 liters of air into and out of the lungs while at rest. A lower-than-normal volume of air exchange is usually a sign of pulmonary malfunction. The apparatus commonly used to measure the volume of air exchanged during breathing and the rate of ventilation is a **spirometer** (*spiro* = breathe) or **respirometer**. The record is called a **spirogram**. Inspiration is recorded as an upward deflection and expiration is recorded as a downward deflection, and the recording pen usually moves from right to left (Fig. 23.16).

About 500 ml of air moves into and then out of the airways with each inspiration and expiration during normal quiet breathing (Fig. 23.16). This is the volume of one breath and is called the **tidal volume** (V_T). Tidal volume varies considerably from one person to another and in the same person at different times. In an average adult, about 70% (350 ml) of the tidal volume actually reaches respiratory bronchioles, alveolar ducts, alveolar sacs, and alveoli (respiratory portion of the respiratory system) and participates in respiration. The other 30% (150 ml) remains in air spaces of the nose, pharynx, larynx, trachea, bronchi, bronchioles, and terminal bronchioles (conducting portion of the repiratory system). These areas are known as the **anatomic dead space** (V_D). An easy rule of thumb is that the volume of your anatomic dead space in milliliters is about the same as your weight in pounds.

The total volume of air taken in during 1 minute is called the **minute volume of respiration** (**MVR**) or **minute ventilation**. It is calculated by multiplying the tidal volume by the normal breathing rate per minute. An average MVR would be 500 ml times 12 respirations per minute or 6000 ml/min. Not all of the MVR can be used in gas exchange, however, because some of it remains in the anatomic dead space. The **alveolar ventilation rate** (**AVR**) is the volume of air per minute that reaches the alveoli. In the example just given, AVR would be 350 ml times 12 respirations per minute or 4200 ml/min.

Several other lung volumes result when one engages in strenuous breathing. In general, these volumes are larger in males, taller persons, and younger adults and smaller in females, shorter persons, and the elderly. Various disorders also may be diagnosed by comparison of actual with predicted normal values for one's sex, height, and age. The values given here are averages for young adults.

EXHIBIT 23.1 MODIFIED RESPIRATORY MOVEMENTS

Movement	Comment
Coughing	A long-drawn and deep inspiration followed by a complete closure of the rima glottidis, which results in a strong expiration that suddenly pushes the rima glottidis open and sends a blast of air through the upper respiratory passages. Stimulus for this reflex act may be a foreign body lodged in the larynx, trachea, or epiglottis.
Sneezing	Spasmodic contraction of muscles of expiration that forcefully expels air through the nose and mouth. Stimulus may be an irritation of the nasal mucosa.
Sighing	A long-drawn and deep inspiration immediately followed by a shorter but forceful expiration.
Yawning	A deep inspiration through the widely opened mouth producing an exaggerated depression of the mandible. It may be stimulated by drowsiness, fatigue, or someone else's yawning, but precise cause is unknown.
Sobbing	A series of convulsive inspirations followed by a single prolonged expiration. The rima glottidis closes earlier than normal after each inspiration so only a little air enters the lungs with each inspiration.
Crying	An inspiration followed by many short convulsive expirations, during which the rima glottidis remains open and the vocal folds vibrate; accompanied by characteristic facial expressions and tears.
Laughing	The same basic movements as crying, but the rhythm of the movements and the facial expressions usually differ from those of crying. Laughing and crying are sometimes indistinguishable.
Hiccuping	Spasmodic contraction of the diaphragm followed by a spasmodic closure of the rima glottidis to produce a sharp inspiratory sound. Stimulus is usually irritation of the sensory nerve endings of the gastrointestinal tract.
Valsalva maneuver	Forced expiration against a closed rima glottidis as during periods of straining.

Figure 23.16   Spirogram of lung volumes and capacities (average values for a healthy adult).

 Lung capacities are combinations of various lung volumes.

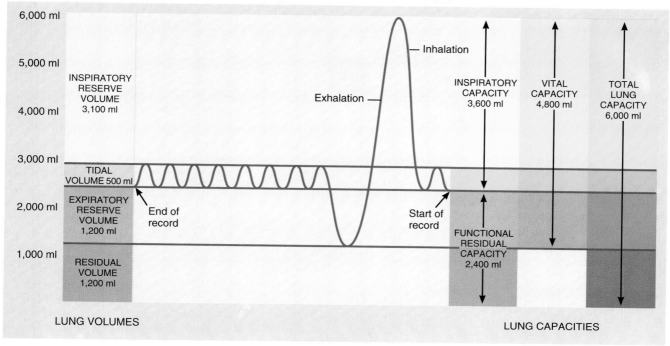

Q Breathe in as deeply as possible and then exhale as much air as you can. Which lung capacity have you demonstrated?

By taking a very deep breath, you can inspire a good deal more than 500 ml. This additional inhaled air, called the **inspiratory reserve volume**, is about 3100 ml above the 500 ml of tidal volume for a total of 3600 ml. Even more air can be inhaled if inspiration follows forced expiration. If you inhale normally and then exhale as forcibly as possible, you should be able to push out 1200 ml of air in addition to the 500 ml of tidal volume. The extra 1200 ml is called the **expiratory reserve volume**. The **FEV₁** is the **forced expiratory volume in 1 second**, which is the volume of air that can be expelled from the lungs in 1 second with maximal effort following a maximal inhalation. Disorders such as asthma and emphysema (see page 744) can greatly reduce FEV_1.

Even after the expiratory reserve volume is expelled, considerable air remains in the lungs because the lower intrapleural pressure keeps the alveoli slightly inflated, and some air also remains in the noncollapsible airways. This volume, which cannot be measured by spirometry, is called the **residual volume** and amounts to about 1200 ml.

Opening the thoracic cavity allows the intrapleural pressure to equal the atmospheric pressure, forcing out some of the residual volume. The air remaining is called the **minimal volume**. Minimal volume provides a medical and legal tool for determining whether a baby was born dead or died after birth. The presence of minimal volume can be demonstrated by placing a piece of lung in water and watch-

ing it float. Fetal lungs contain no air, and so the lung of a stillborn baby will not float in water.

Lung capacities are combinations of specific lung volumes (Fig. 23.16). **Inspiratory capacity**, the total inspiratory ability of the lungs, is the sum of tidal volume plus inspiratory reserve volume (3600 ml). **Functional residual capacity** is the sum of residual volume plus expiratory reserve volume (2400 ml). **Vital capacity** is the sum of inspiratory reserve volume, tidal volume, and expiratory reserve volume (4800 ml). Finally, **total lung capacity** is the sum of all volumes (6000 ml).

EXCHANGE OF OXYGEN AND CARBON DIOXIDE

At birth, as soon as the lungs fill with air, O_2 starts to diffuse from the alveoli into the blood, through the interstitial fluid, and finally into the cells. Carbon dioxide diffuses in the opposite direction—from the cells, through interstitial fluid to the blood, and into alveolar air.

Gas Laws

The exchanges of O_2 and CO_2 are purely passive—no ATP is consumed—and they depend on the behavior of gases described in Dalton's law and Henry's law.

Dalton's Law

According to **Dalton's law**, each gas in a mixture of gases exerts its own pressure as if all the other gases were not present. The pressure of a specific gas in a mixture is called its *partial pressure* and is denoted as *p*. The total pressure of the mixture is calculated by simply adding all the partial pressures. Atmospheric air is a mixture of several gases—oxygen, carbon dioxide, nitrogen (N_2), water vapor (H_2O), and several other gases present in such small quantities that we will ignore them. Atmospheric pressure is the sum of the pressures of all these gases:

$$\text{Atmospheric pressure (760 mm Hg)} = pO_2 + pCO_2 + pN_2 + pH_2O$$

We can determine the partial pressure exerted by each component in the mixture by multiplying the percentage of the gas in the mixture by the total pressure of the mixture. For example, to find the partial pressure of O_2 in the atmosphere, multiply the percentage of atmospheric air composed of O_2 (21%) by the total atmospheric pressure (760 mm Hg):

$$\text{Atmospheric } pO_2 = 0.21 \times 760 \text{ mm Hg}$$
$$= 159.6 \text{ or } 160 \text{ mm Hg}$$

Since the percentage of CO_2 in the atmosphere is 0.04%,

$$\text{Atmospheric } pCO_2 = 0.0004 \times 760 \text{ mm Hg}$$
$$= 0.3 \text{ mm Hg}$$

These partial pressures are important in determining the movement of O_2 and CO_2 between the atmosphere and lungs, the lungs and blood, and the blood and body cells. When a mixture of gases diffuses across a permeable membrane, each gas diffuses from the area where its partial pressure is greater to the area where its partial pressure is less. Every gas is on its own and behaves as if the other gases in the mixture did not exist. The partial pressures of O_2 and CO_2 in mm Hg in air, blood, and tissue cells are as follows:

Atmosphere (sea level)	$pO_2 = 160$; $pCO_2 = 0.3$
Alveoli	$pO_2 = 105$; $pCO_2 = 40$
Oxygenated blood	$pO_2 = 100$; $pCO_2 = 40$
Tissue cells (average at rest)	$pO_2 = 40$; $pCO_2 = 45$
Deoxygenated blood	$pO_2 = 40$; $pCO_2 = 45$

The relative amounts of O_2 and CO_2 differ in inspired (atmospheric) air, alveolar air, and expired air as follows:

Inspired air—21% O_2, 0.04% CO_2
Alveolar air—14% O_2, 5.2% CO_2
Expired air—16% O_2, 4.5% CO_2

Compared with inspired air, alveolar air has less O_2 (21% versus 14%) and more CO_2 (0.04% versus 5.2%) because gas exchange is occurring in the alveoli. On the other hand, expired air contains more O_2 than alveolar air (16% versus 14%) and less CO_2 (4.5% versus 5.2%) because some of the expired air was in the anatomic dead space and did not participate in gas exchange. Expired air is a mixture of alveolar air and inspired air that was in the anatomic dead space. Also, expired air and alveolar air have higher water vapor content than inspired air because the moist mucosal linings humidify air as it is inhaled.

Henry's Law

You have probably noticed that a soft drink makes a hissing sound when the top of the container is removed, and bubbles rise to the surface for some time afterward. The gas dissolved in carbonated beverages is CO_2. The ability of a gas to stay in solution depends on its partial pressure and solubility coefficient, that is, its physical or chemical attraction for water. The solubility coefficient of CO_2 is high (0.57), that of O_2 is low (0.024), and that of nitrogen is still lower (0.012). The higher the partial pressure of a gas over a liquid and the higher the solubility coefficient, the more gas will stay in solution. Since the soft drink is bottled or canned under pressure and capped, the CO_2 remains dissolved as long as the container is unopened. Once you remove the cap, the pressure is released and the gas begins to bubble out. This phenomenon is explained by **Henry's law**: the quantity of a gas that will dissolve in a liquid is proportional to the partial pressure of the gas and its solubility coefficient, when the temperature remains constant.

Henry's law explains two conditions resulting from changes in the solubility of nitrogen in body fluids. Even though the air we breathe contains about 79% N_2, this gas has no known effect on bodily functions, and very little of it dissolves in blood plasma because of its low solubility coefficient at sea level pressure. But when a **scuba (self-contained underwater breathing apparatus) diver** breathes air under high pressure, the N_2 in the mixture can have serious negative effects. Partial pressure is a function of total pressure, and therefore the partial pressures of all the components of a mixture increase as the total increases. Since the partial pressure of N_2 is higher in a mixture of compressed air than in air at sea level pressure, a considerable amount of N_2 dissolves in plasma and interstitial fluid. Excessive amounts of dissolved N_2 may produce giddiness and other symptoms similar to alcohol intoxication. The condition is called **nitrogen narcosis** or "rapture of the depths." The greater the depth, the more severe the condition.

If a diver comes to the surface slowly, the dissolved N_2 can be eliminated through the lungs. However, if the ascent is too rapid, N_2 comes out of solution too quickly to be eliminated by exhalation. Instead, it forms gas bubbles in the tissues, resulting in **decompression sickness** (the **bends**). The effects of decompression sickness typically result from bubbles in nervous tissue and can be mild or severe, depending on the number of bubbles formed. Symptoms include joint pain, especially in the arms and legs, dizziness, shortness of breath, extreme fatigue, paralysis, and unconsciousness. Decompression sickness can be prevented by a slow ascent or by the use of a special decompression tank within 5 minutes after arriving at the surface. The use of helium–oxygen mixtures instead of air containing N_2 may reduce the dangers of decompression sickness since helium is only about 40% as soluble as N_2 in blood.

HYPERBARIC OXYGENATION

A major clinical application of Henry's law is **hyperbaric** (*hyper* = over; *baros* = pressure) **oxygenation** (**HBO**). Using pressure to cause more O_2 to dissolve in the blood is an effective technique in treating patients infected by anaerobic bacteria, such as those that cause tetanus and gangrene. (Anaerobic bacteria cannot live in the presence of free O_2.) A person undergoing hyperbaric oxygenation is placed in a hyperbaric chamber, which contains O_2 at a pressure of 3–4 atm (2280–3040 mm Hg). The body tissues pick up the O_2, and the bacteria are killed. Hyperbaric chambers may also be used for treating certain heart disorders, carbon monoxide poisoning, gas embolisms, crush injuries, cerebral edema, certain hard-to-treat bone infections, smoke inhalation, near-drowning, asphyxia, vascular insufficiencies, and burns. ■

External Respiration

External (pulmonary) respiration is the exchange of O_2 and CO_2 between air in the alveoli of the lungs and blood in pulmonary capillaries (Fig. 23.17a). It results in the conversion of **deoxygenated blood** (depleted of some O_2) coming from the heart to **oxygenated blood** (saturated with O_2) returning to the heart (see Fig. 20.1). During inspiration, atmospheric air containing O_2 enters the alveoli and during expiration CO_2 is exhaled into the atmosphere. Deoxygenated blood is pumped from the right ventricle through the pulmonary arteries into the pulmonary capillaries surrounding the alveoli. The number of capillaries next to alveoli in the lungs is very large and blood flows rather slowly through these capillaries even during vigorous exercise. As a result, the partial pressure of gases in the bloodstream has time to come into equilibrium with the partial pressure of gases in the alveolar air.

The pO_2 of alveolar air is 105 mm Hg. If you are at rest, the pO_2 of deoxygenated blood entering your pulmonary capillaries is about 40 mm Hg. If you have been exercising, it will be even lower. As a result of the difference in pO_2, there is a net diffusion of O_2 from the alveoli into the deoxygenated blood until equilibrium is reached. The pO_2 of the now oxygenated blood increases to 105 mm Hg. Because blood leaving capillaries near alveoli mixes with a small volume of blood that has not flowed close to alveoli, the pO_2 of blood in the pulmonary veins is about 100 mm Hg.

While O_2 diffuses from the alveoli into deoxygenated blood, there is a net diffusion of CO_2 in the opposite direction. The pCO_2 of deoxygenated blood is 45 mm Hg in a resting person, whereas that of the alveolar air is 40 mm Hg. Because of this difference in pCO_2, carbon dioxide diffuses from de-

Figure 23.17 Changes in partial pressures (in mm Hg) during external and internal respiration.

Gases diffuse from areas of higher partial pressure to areas of lower partial pressure.

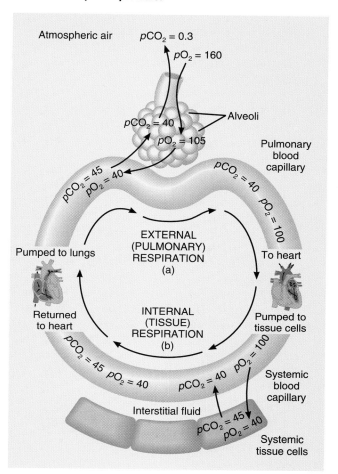

What causes oxygen to enter pulmonary capillaries from alveoli and to enter tissue cells from systemic capillaries?

oxygenated blood into the alveoli until the pCO_2 of the blood decreases to 40 mm Hg. This is the pCO_2 of fully oxygenated blood.

The rate of external respiration (diffusion of gases) depends on several factors:

1. **Partial pressure difference of the gases.** As long as alveolar pO_2 is higher than pO_2 in pulmonary capillaries, O_2 diffuses from the alveoli into the blood. When a person ascends to a higher altitude, the total atmospheric pressure decreases although O_2 still comprises 21% of the total. As the pO_2 of inhaled air decreases, alveolar pO_2 decreases correspondingly, and less O_2 diffuses into the blood. For example, at sea level, pO_2 is 160 mm Hg. At 10,000 ft, it decreases to 110 mm Hg; at 20,000 ft, to 73 mm Hg; and at 50,000 ft, to 18 mm Hg. The common

symptoms of **high altitude sickness (acute mountain sickness)**—shortness of breath, headache, fatigue, insomnia, nausea, dizziness—are due to the low O_2 content of the blood. The partial pressures of O_2 and CO_2 in the alveoli also depend on the rate of airflow into and out of the lungs. Certain drugs, such as morphine, slow ventilation, thereby decreasing the amount of O_2 and CO_2 that can be exchanged between the alveoli and the blood.

2. **Surface area for gas exchange.** The surface area over which diffusion may occur is large (about 70 m² or 750 ft²). Any pulmonary disorder that decreases the functional surface area formed by the alveolar–capillary membranes decreases the rate of external respiration. In emphysema, for example, alveolar walls disintegrate, and surface area declines.

3. **Diffusion distance.** The total thickness of the alveolar–capillary (respiratory) membranes is only 0.5 μm. Thicker membranes would slow the rate of diffusion. Also, the capillaries are so narrow that the red blood cells must pass through them in single file. This minimizes the diffusion distance from an alveolar air space to hemoglobin inside red blood cells. Buildup of fluid, as occurs in pulmonary edema, slows the rate of gas exchange because it increases diffusion distance.

4. **Solubility and molecular weight of the gases.** Because O_2 has a lower molecular weight than CO_2, it should diffuse across the alveolar–capillary membrane about 1.2 times faster. However, the solubility of CO_2 is about 24 times greater than the solubility of O_2 in the fluid portions of the alveolar–capillary membrane. Taking into account both of these factors, net outward CO_2 diffusion occurs 20 times more rapidly than net inward O_2 diffusion. When diffusion is impaired, for example, in emphysema or pulmonary edema, O_2 insufficiency (hypoxia) typically occurs before there is significant retention of CO_2 (hypercapnia).

Internal Respiration

Even when cardiac output increases greatly during strenuous exercise so that blood flows rapidly through pulmonary capillaries, the blood picks up a maximal amount of O_2 and returns to the heart fully oxygenated. The left ventricle pumps oxygenated blood into the aorta and through the systemic arteries to capillaries to tissue cells. The exchange of O_2 and CO_2 between tissue blood capillaries and tissue cells is called **internal (tissue) respiration** (Fig. 23.17b). It results in the conversion of oxygenated blood into deoxygenated blood. Oxygenated blood entering tissue capillaries has a pO_2 of 100 mm Hg, whereas tissue cells have an average pO_2 of 40 mm Hg. Because of this difference in pO_2, oxygen diffuses from the oxygenated blood through interstitial fluid and into tissue cells until the pO_2 in the blood decreases to 40 mm Hg. This is the average pO_2 of deoxygenated blood entering tissue venules when you are at rest.

At rest, only about 25% of the available O_2 in oxygenated blood actually enters tissue cells. This amount is sufficient to support the needs of resting cells. Thus deoxygenated blood still retains considerable O_2. During exercise, more O_2 diffuses from the blood into active cells.

While O_2 diffuses from the tissue blood capillaries into tissue cells, CO_2 diffuses in the opposite direction. The average pCO_2 of tissue cells is 45 mm Hg, whereas that of tissue capillary oxygenated blood is 40 mm Hg. As a result, CO_2 diffuses from tissue cells through interstitial fluid into the oxygenated blood until the pCO_2 in the blood increases to 45 mm Hg, the pCO_2 of tissue capillary deoxygenated blood. The deoxygenated blood now returns to the heart. From here it is pumped to the lungs for another cycle of external respiration.

TRANSPORT OF OXYGEN AND CARBON DIOXIDE

The transport of gases between the lungs and body tissues is a function of the blood. When O_2 and CO_2 enter the blood, certain physical and chemical changes occur that aid in gas transport and exchange.

Oxygen Transport

Oxygen does not dissolve easily in water, and therefore very little O_2, only about 1.5%, is carried in the dissolved state in watery blood plasma. The remainder of the O_2, about 98.5%, is transported in chemical combination with hemoglobin inside red blood cells (Fig. 23.18). Each 100 ml of oxygenated blood contains about 20 ml of O_2; 0.3 ml dissolved in the plasma and 19.7 ml bound to hemoglobin.

Hemoglobin consists of a protein portion called globin and an iron-containing pigment portion called heme (see Fig. 19.3b). Each hemoglobin molecule has four heme groups, and each heme group can combine with one molecule of O_2. Oxygen and hemoglobin combine in an easily reversible reaction to form **oxyhemoglobin** as follows:

$$\text{Hb} + \text{O}_2 \underset{\text{Dissociation of O}_2}{\overset{\text{Binding of O}_2}{\rightleftharpoons}} \text{HbO}_2$$

Reduced hemoglobin   Oxygen   Oxyhemoglobin
(deoxyhemoglobin)

Since 98.5% of the O_2 is bound to hemoglobin and is trapped inside RBCs, only the dissolved O_2 (1.5%) can diffuse out of tissue capillaries into tissue cells. Thus it is important to understand the factors that promote O_2 binding to and dissociation (separation) from hemoglobin.

Hemoglobin and Oxygen Partial Pressure

The most important factor that determines how much O_2 combines with hemoglobin is the O_2 partial pressure (pO_2). When reduced hemoglobin (deoxyhemoglobin) is com-

Figure 23.18 Transport of oxygen (O_2) and carbon dioxide (CO_2) in the blood.

Most O_2 is transported by hemoglobin as oxyhemoglobin within red blood cells; most CO_2 is transported in blood plasma as bicarbonate ions.

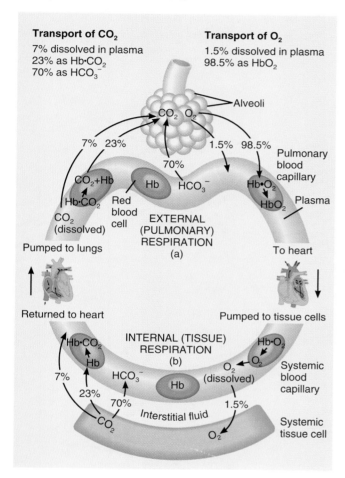

Transport of CO_2
7% dissolved in plasma
23% as $Hb \cdot CO_2$
70% as HCO_3^-

Transport of O_2
1.5% dissolved in plasma
98.5% as HbO_2

How many milliliters of CO_2 are in 100 ml of deoxygenated blood at rest? How many milliliters of O_2 are in 100 ml of oxygenated blood during strenuous exercise?

pletely converted to HbO_2, it is said to be **fully saturated**. When hemoglobin consists of a mixture of Hb and HbO_2, it is **partially saturated**. The **percent saturation of hemoglobin** is the percentage of HbO_2 in total hemoglobin. The relation between the percent saturation of hemoglobin and pO_2 is illustrated in Fig. 23.19, the oxygen–hemoglobin dissociation curve. Note that when the pO_2 is high, hemoglobin binds with large amounts of O_2 and is almost fully saturated. When pO_2 is low, hemoglobin is only partially saturated and O_2 is released from hemoglobin. In other words, the greater the pO_2, the more O_2 will combine with hemoglobin, until the available hemoglobin molecules are saturated. Therefore in pulmonary capillaries, where pO_2 is

high, a lot of O_2 binds with hemoglobin. In tissue capillaries, where the pO_2 is lower, hemoglobin does not hold as much O_2, and the O_2 is released for diffusion into tissue cells. Note that hemoglobin is still 75% saturated with O_2 at a pO_2 of 40 mm Hg, the average pO_2 of tissue cells when you are at rest. This is the basis for our earlier statement that only 25% of the available O_2 splits from hemoglobin and is used by tissue cells under resting conditions.

When the pO_2 is between 60 and 100 mm Hg, hemoglobin is 90% or more saturated with O_2 (Fig. 23.19). Thus blood picks up a nearly full load of O_2 from the lungs even when the alveolar pO_2 is as low as 60 mm Hg. This explains why people can still perform well at high altitudes or when they have certain cardiac and pulmonary diseases, even though pO_2 may drop as low as 60 mm Hg. Note also in the curve that at a considerably lower pO_2 of 40 mm Hg, hemoglobin is still 75% saturated with O_2. However, oxygen saturation drops to 35% at 20 mm Hg. This means that large amounts of O_2 are released from hemoglobin in response to only small decreases in pO_2 between 40 and 20 mm Hg. In active tissues such as contracting muscles, pO_2 may drop well below 40 mm Hg. Then, a large percentage of the O_2 is released from hemoglobin. This provides more O_2 to tissues that are using it more rapidly.

Figure 23.19 Oxygen–hemoglobin dissociation curve at normal body temperature showing the relationship between hemoglobin saturation and pO_2.

As pO_2 increases, more O_2 combines with hemoglobin.

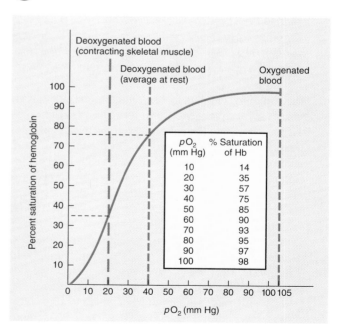

pO_2 (mm Hg)	% Saturation of Hb
10	14
20	35
30	57
40	75
50	85
60	90
70	93
80	95
90	97
100	98

What point on the curve would represent blood in your pulmonary veins right now? In your pulmonary veins if you were jogging?

Hemoglobin and Other Factors

Although pO_2 is the most important factor that determines the percent O_2 saturation of hemoglobin, several other factors influence the tightness or **affinity** with which hemoglobin binds O_2. In effect, these factors shift the entire curve to the left (higher affinity) or the right (lower affinity). The changing affinity of hemoglobin for O_2 is another example of how homeostatic mechanisms adjust body activities to cellular needs. Each one makes sense if you keep in mind that metabolically active tissue cells need O_2 and produce acids, CO_2, and heat.

1. **Acidity (pH).** As acidity increases (pH decreases), the affinity of hemoglobin for O_2 decreases and O_2 dissociates more readily from hemoglobin (Fig. 23.20a). This change shifts the oxygen–hemoglobin dissociation curve to the right and is referred to as the **Bohr effect**. The explanation for the Bohr effect is that hydrogen ions (H^+) bind to certain amino acids in hemoglobin, slightly alter its structure, and thereby decrease its oxygen-carrying capacity. Thus lowered pH drives O_2 off hemoglobin, making more O_2 available for tissue cells. By contrast, elevated pH increases the affinity of hemoglobin for O_2 and shifts the curve to the left.

2. **Partial pressure of carbon dioxide.** CO_2 also can bind to hemoglobin and the effect is similar to that of H^+ (shifting the curve to the right). As pCO_2 rises, hemoglobin releases O_2 more readily (Fig. 23.20b). pCO_2 and pH are related factors because low blood pH (acidity) results from high pCO_2. As CO_2 enters the blood, much of it is temporarily converted to carbonic acid (H_2CO_3). This conversion is catalyzed by an enzyme in red blood cells called *carbonic anhydrase* (CA).

$$CO_2 + H_2O \overset{CA}{\rightleftharpoons} H_2CO_3 \rightleftharpoons H^+ + HCO_3^-$$

Carbon dioxide   Water   Carbonic acid   Hydrogen ion   Bicarbonate ion

The carbonic acid thus formed in red blood cells dissociates into hydrogen ions and bicarbonate ions (HCO_3^-). As the H^+ concentration increases, pH decreases. Thus an increased pCO_2 produces a more acidic environment, which helps release O_2 from hemoglobin. During exercise, lactic acid, a by-product of anaerobic metabolism within muscles, also decreases the blood pH. Decreased pCO_2 (and elevated pH) shifts the saturation curve to the left.

3. **Temperature.** Within limits, as temperature increases, so does the amount of O_2 released from hemoglobin (Fig. 23.21). Heat is a by-product of the metabolic reactions of all cells, and contracting muscle fibers release an especially large amount of heat, which tends to raise body temperature. Metabolically active cells require more O_2 and liberate more acids and heat. The acids and heat, in turn, promote release of O_2 from oxyhemoglobin. Fever produces a similar result. On the other

Figure 23.20 Oxygen–hemoglobin dissociation curve at normal body temperature showing the relationship between pH, pCO_2, and hemoglobin saturation. As pH increases or pCO_2 decreases, O_2 combines more tightly with hemoglobin and less is available to tissues. These relationships are emphasized by the broken lines.

 As pH decreases or pCO_2 increases, the affinity of hemoglobin for O_2 is less, so less O_2 combines with hemoglobin and more is available to tissues.

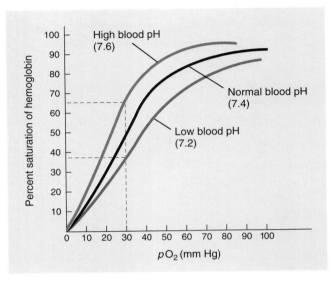

(a) Effect of pH on affinity of hemoglobin for oxygen

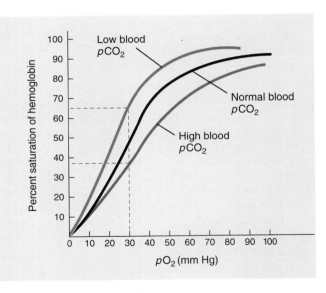

(b) Effect of pCO_2 on affinity of hemoglobin for oxygen

Q In comparison to the value when you are sitting, is the affinity of your hemoglobin for O_2 higher or lower when you are jogging? What is the benefit of this change to you?

Figure 23.21   Oxygen–hemoglobin dissociation curve showing the relationship between temperature and hemoglobin saturation with O₂.

 As temperature increases, the affinity of hemoglobin for O₂ decreases.

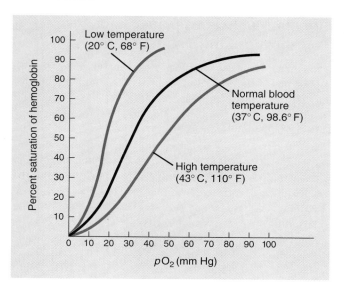

 Is O₂ more available or less available to tissue cells when you have a fever? Why?

Figure 23.22   Oxygen–hemoglobin dissociation curves comparing fetal and maternal hemoglobin.

 Fetal hemoglobin has a higher affinity for O₂ than does adult hemoglobin.

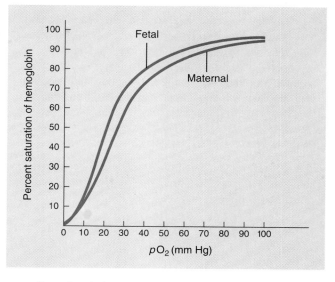

 The pO_2 of placental blood is about 40 mm Hg. What are the O₂ saturations of maternal and fetal hemoglobin at this pO_2?

hand, during hypothermia (lowered body temperature) cellular metabolism slows, the need for O₂ is less, and more O₂ remains bound to hemoglobin (a shift to the left in the saturation curve).

4. **BPG.** A substance found in red blood cells called **2,3-bisphosphoglycerate** (**BPG**), previously called diphosphoglycerate (DPG), decreases the affinity of hemoglobin for O₂ and thus helps to release O₂ from hemoglobin. BPG is formed in red blood cells when they break down glucose for energy in a process called glycolysis (described on page 816). When BPG combines with hemoglobin, the hemoglobin binds O₂ less tightly. The greater the level of BPG, the more O₂ is released from hemoglobin. Certain hormones, such as thyroxine, human growth hormone, epinephrine, norepinephrine, and testosterone, increase the formation of BPG. The level of BPG also is higher in people living at higher altitudes.

Fetal Hemoglobin

Fetal hemoglobin (Hb-F) differs from adult hemoglobin (Hb-A) in structure and in its affinity for O₂. Hb-F has a higher affinity for O₂ because it binds BPG less strongly. Thus when pO_2 is low, Hb-F can carry up to 30% more O₂ than maternal Hb-A (Fig. 23.22). As the maternal blood enters the placenta, O₂ is readily transferred to fetal blood. This is very important since the O₂ saturation in maternal

blood in the placenta is quite low and the fetus might suffer hypoxia were it not for the greater affinity of fetal hemoglobin for O₂.

 ## CARBON MONOXIDE POISONING

Carbon monoxide (CO) is a colorless and odorless gas found in exhaust fumes from automobiles and in tobacco smoke. It is a by-product of burning carbon-containing materials such as coal, gas, and wood. CO combines with the heme group of hemoglobin, just as O₂ does, except that the binding of carbon monoxide to hemoglobin is over 200 times as strong as the binding of O₂ to hemoglobin. Thus at a concentration as small as 0.1% ($pCO = 0.5$ mm Hg), CO will combine with half the hemoglobin molecules and reduce the oxygen-carrying capacity of the blood by 50%. Elevated CO leads to hypoxia, and the result is **carbon monoxide poisoning**. The condition may be treated by administering pure O₂, which hastens the separation of carbon monoxide from hemoglobin. ■

Hypoxia

Hypoxia (hī-POK-sē-a; *hypo* = below or under) is a deficiency of O₂ at the tissue level. Based on the cause, we can classify types of hypoxia as follows:

1. **Hypoxic hypoxia.** This is caused by a low pO_2 in arterial blood and may be the result of high altitude, obstructions in airways, or fluid in the lungs.

2. **Anemic hypoxia.** In this case, there is too little functioning hemoglobin in the blood. Among the causes are hemorrhage, anemia, or failure of hemoglobin to carry its normal complement of O_2, as in carbon monoxide poisoning.

3. **Stagnant (ischemic) hypoxia.** In this condition, blood flow to a tissue is so low that adequate O_2 is not delivered to it even though pO_2 and oxyhemoglobin are normal.

4. **Histotoxic hypoxia.** In this condition, the blood delivers adequate O_2 to tissues, but the tissues are unable to use it properly because of the action of a toxic agent. One cause is cyanide poisoning, in which cyanide blocks the metabolic machinery of cells related to O_2 utilization.

Carbon Dioxide Transport

Under normal resting conditions, each 100 ml of deoxygenated blood contains about 55 ml of CO_2, which is carried by the blood in three main forms (see Fig. 23.18):

1. **Dissolved CO_2.** The smallest percentage, about 7%, is dissolved in plasma. Upon reaching the lungs, it diffuses into the alveoli.

2. **Carbaminohemoglobin.** A somewhat higher percentage, about 23%, combines with the globin portion of hemoglobin to form **carbaminohemoglobin (Hb·CO_2)**:

$$Hb \;+\; CO_2 \;\rightleftharpoons\; Hb \cdot CO_2$$

Hemoglobin   Carbon dioxide   Carbaminohemoglobin

The formation of carbaminohemoglobin is greatly influenced by pCO_2. For example, in tissue capillaries pCO_2 is relatively high, and this promotes formation of carbaminohemoglobin. But in pulmonary capillaries, pCO_2 is relatively low, and the CO_2 readily splits apart from globin and enters the alveoli by diffusion.

3. **Bicarbonate ions.** The greatest percentage of CO_2, about 70%, is transported in plasma as bicarbonate ions (HCO_3^-). The reaction that brings about this method of transport of CO_2 is the same one noted earlier:

$$CO_2 + H_2O \;\overset{CA}{\rightleftharpoons}\; H_2CO_3 \;\rightleftharpoons\; H^+ \;+\; HCO_3^-$$

Carbon   Water   Carbonic   Hydrogen Bicarbonate
dioxide   acid   ion   ion

As CO_2 diffuses into tissue capillaries and enters the red blood cells, it reacts with water, in the presence of the enzyme carbonic anhydrase (CA), to form carbonic acid (Fig. 23.23a). The carbonic acid dissociates into H^+ and HCO_3^-. Many of the H^+ combine with hemoglobin (H·Hb) as well as other buffers. Recall that oxygen dissociates from hemoglobin more readily when hemoglobin is buffering more H^+ (Bohr effect). As HCO_3^- accumulates inside the RBC, some diffuses into the plasma, down the concentration gradient. In exchange, chloride ions (Cl^-) diffuse from plasma into the RBCs. This exchange of negative ions maintains the

electrical balance between plasma and RBCs and is known as the **chloride shift**. The net effect of these reactions is that CO_2 is carried from tissue cells as HCO_3^- in plasma.

Summary of Gas Exchange and Transport in Lungs and Tissues

Deoxygenated blood returning to the lungs (Fig. 23.23a) contains CO_2 dissolved in plasma, CO_2 combined with globin as carbaminohemoglobin, and CO_2 incorporated in HCO_3^-. The blood has also picked up H^+, some of which is buffered by hemoglobin (H·Hb).

In the pulmonary capillaries, the chemical reactions reverse. CO_2 dissolved in plasma diffuses into the alveoli and is exhaled. CO_2 combined with hemoglobin splits from the globlin, diffuses into alveoli, and is exhaled. Carbon dioxide in HCO_3^- is released when H^+ combines with HCO_3^- inside RBCs to form H_2CO_3, which splits into CO_2 and H_2O (Fig. 23.23b). As the concentration of HCO_3^- declines inside RBCs, HCO_3^- diffuses in from the plasma, again in exchange for Cl^-. Carbon dioxide diffuses out of RBCs, into alveoli, and is exhaled. At the same time, inhaled O_2 is diffusing from alveoli into RBCs and is attaching to hemoglobin. Thus oxygenated blood leaving the lungs has increased O_2 content and decreased CO_2 and H^+.

Just as an increase in CO_2 in blood causes O_2 to split from hemoglobin, the binding of O_2 to hemoglobin causes the release of CO_2 from blood. In the presence of O_2, less CO_2 binds to hemoglobin. This reaction, the reverse of the Bohr effect, is called the **Haldane effect**. It occurs because when O_2 combines with hemoglobin, the hemoglobin becomes a stronger acid. In this state, hemoglobin combines with less CO_2. Also, the more acidic hemoglobin releases more H^+ that bind to HCO_3^- to form H_2CO_3; the H_2CO_3 breaks down into $H_2O + CO_2$, and the CO_2 diffuses from blood into alveoli. The direction of the carbonic acid reaction depends mostly on pCO_2. In tissue capillaries, where pCO_2 is high, H^+ and HCO_3^- are formed. In pulmonary capillaries, where pCO_2 is low, CO_2 and H_2O are formed.

CONTROL OF RESPIRATION

At rest about 200 ml of O_2 are used each minute. During strenuous exercise, however, O_2 use can increase as much as 30-fold. Thus mechanisms must exist to match respiratory effort to metabolic demand. The basic rhythm of respiration is controlled by portions of the nervous system in the medulla oblongata and pons. We will first examine the principal mechanisms involved in the nervous control of the rhythm of respiration.

Figure 23.23 Summary of gas transport. As CO_2 leaves tissue cells and enters blood cells, it causes more O_2 to dissociate from hemoglobin (Bohr effect) and thus more CO_2 combines with hemoglobin and more bicarbonate ions (HCO_3^-) are produced. As O_2 passes from alveoli into blood cells, hemoglobin becomes saturated with O_2 and becomes a stronger acid. The more acidic hemoglobin releases more hydrogen ions (H^+), which bind to HCO_3^- to form carbonic acid (H_2CO_3). The H_2CO_3 dissociates into $H_2O + CO_2$ and the CO_2 diffuses from blood into alveoli (Haldane effect).

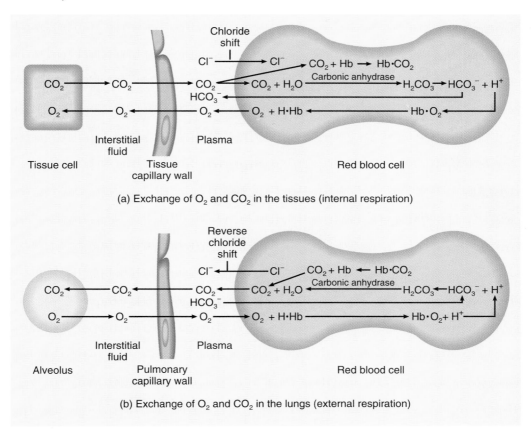

(a) Exchange of O_2 and CO_2 in the tissues (internal respiration)

(b) Exchange of O_2 and CO_2 in the lungs (external respiration)

Q Would you expect the concentration of HCO_3^- to be greater in plasma taken from an artery or vein of an arm?

Respiratory Center

The size of the thorax is affected by the action of the respiratory muscles. These muscles contract and relax as a result of nerve impulses transmitted to them from centers in the brain. The area from which nerve impulses are sent to respiratory muscles is located bilaterally in the medulla oblongata and pons of the brain stem. This area is called the **respiratory center** and consists of a widely dispersed group of neurons that is functionally divided into three areas: (1) the medullary rhythmicity (rith-MIS-i-tē) area in the medulla oblongata; (2) the pneumotaxic (noo-mō-TAK-sik) area in the pons; and (3) the apneustic (ap-NOO-stik) area, also in the pons (Fig. 23.24)

Medullary Rhythmicity Area
The function of the **medullary rhythmicity area** is to control the basic rhythm of respiration. In the normal rest-

ing state, inspiration usually lasts for about 2 seconds and expiration for about 3 seconds. This is the basic rhythm of respiration. Within the medullary rhythmicity area are both inspiratory and expiratory neurons that comprise inspiratory and expiratory areas, respectively. Let us first consider the proposed role of the inspiratory neurons in respiration.

The basic rhythm of respiration is determined by nerve impulses generated in the inspiratory area (Fig. 23.25a). At the beginning of expiration, the inspiratory area is inactive, but after 3 seconds it automatically becomes active. This activity seems to result from neurons that are autorhythmic cells. Even when all incoming nerve connections to the inspiratory area are cut or blocked, neurons in this area still rhythmically discharge impulses that result in inspiration. Nerve impulses from the active inspiratory area last for about 2 seconds and travel to the muscles of inspiration. The impulses reach the diaphragm by the phrenic nerves and the

Figure 23.24   Location of areas of the respiratory center. The respiratory center is composed of neurons in the medullary rhythmicity center in the medulla oblongata plus the pneumotaxic area and apneustic area in the pons.

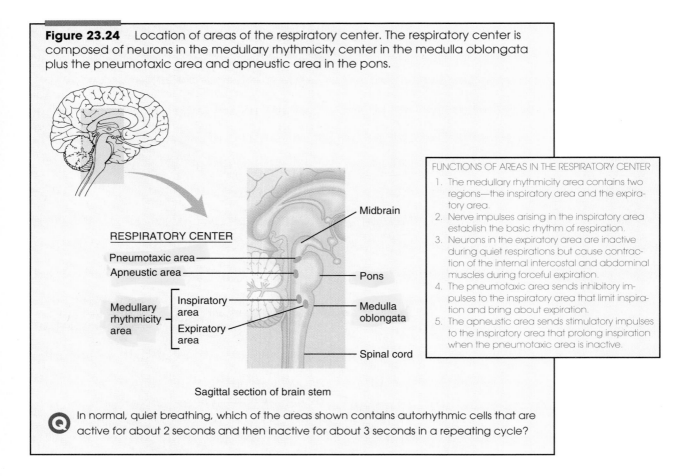

FUNCTIONS OF AREAS IN THE RESPIRATORY CENTER

1. The medullary rhythmicity area contains two regions—the inspiratory area and the expiratory area.
2. Nerve impulses arising in the inspiratory area establish the basic rhythm of respiration.
3. Neurons in the expiratory area are inactive during quiet respirations but cause contraction of the internal intercostal and abdominal muscles during forceful expiration.
4. The pneumotaxic area sends inhibitory impulses to the inspiratory area that limit inspiration and bring about expiration.
5. The apneustic area sends stimulatory impulses to the inspiratory area that prolong inspiration when the pneumotaxic area is inactive.

Sagittal section of brain stem

Q In normal, quiet breathing, which of the areas shown contains autorhythmic cells that are active for about 2 seconds and then inactive for about 3 seconds in a repeating cycle?

Figure 23.25   Proposed role of the medullary rhythmicity area in controlling the basic rhythm of respiration.

🔑 *Autorhythmic neurons in the inspiratory area generate the basic rhythm of respiration.*

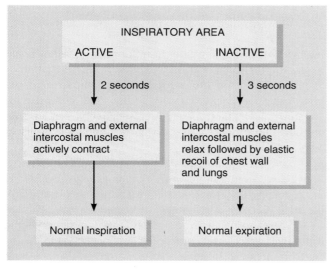

(a) During normal quiet breathing

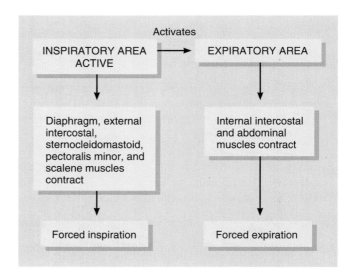

(b) During labored breathing

Q Which nerves convey impulses from the respiratory center to the diaphragm? To the external intercostal muscles?

external intercostal muscles by the intercostal nerves. When the impulses reach the inspiratory muscles, the muscles contract and inspiration occurs. At the end of 2 seconds, the inspiratory muscles relax and the cycle repeats itself over and over.

The expiratory neurons remain inactive during most normal, quiet respirations. During quiet respiration, inspiration is accomplished by active contraction of the inspiratory muscles, and expiration results from passive elastic recoil of the lungs and thoracic wall as the inspiratory muscles relax. However, during high levels of ventilation, it is believed that impulses from the inspiratory area activate the expiratory area (Fig. 23.25b). Impulses discharged from the expiratory area cause contraction of the internal intercostals and abdominal muscles that decrease the size of the thoracic cavity to cause forced (labored) expiration.

Pneumotaxic Area

Although the medullary rhythmicity area controls the basic rhythm of respiration, other parts of the brain stem help coordinate the transition between inspiration and expiration. One of these is the **pneumotaxic** (*pneuma* = breath; *taxis* = arrangement) **area** in the superior portion of the pons (see Fig. 23.24). It transmits inhibitory impulses to the inspiratory area. The major effect of these impulses is to help turn off the inspiratory area before the lungs become too full of air. In other words, the impulses limit the duration of inspiration and thus facilitate the onset of expiration. When the pneumotaxic area is more active, the breathing rate is quicker.

Apneustic Area

Another part of the brain stem that coordinates the transition between inspiration and expiration is the **apneustic area** in the inferior portions of the pons (see Fig. 23.24). It sends stimulatory impulses to the inspiratory area that activate it and prolong inspiration, thus inhibiting expiration. This occurs when the pneumotaxic area is inactive. When the pneumotaxic area is active, it overrides the apneustic area.

Regulation of the Respiratory Center

Although the basic rhythm of respiration is set and coordinated by the inspiratory area, the rhythm can be modified in response to inputs from other brain regions and receptors in the peripheral nervous system.

Cortical Influences

The cerebral cortex has connections with the respiratory center, which means we can voluntarily alter our pattern of breathing. We can even refuse to breathe at all for a short time. Voluntary control is protective because it enables us to prevent water or irritating gases from entering the lungs. The ability to not breathe is limited by the buildup of CO_2 and H^+ in the blood, however. When pCO_2 and the concentration of H^+ increase to a certain level, the inspiratory area is strongly stimulated, nerve impulses are sent along the phrenic and intercostal nerves to inspiratory muscles, and breathing resumes whether or not the person wishes. It is impossible for people to kill themselves by holding their breath. Even if a person faints, breathing resumes when consciousness is lost. Nerve impulses from the hypothalamus and limbic system also stimulate the respiratory center. These pathways permit emotional stimuli to alter respirations, for example, in crying.

Chemical Regulation

Certain chemical stimuli determine how fast and deeply we breathe. The respiratory system functions to maintain proper levels of CO_2 and O_2, and the system is highly responsive to changes in the blood levels of either. Chemoreceptors in two locations monitor levels of CO_2 and O_2 and provide input to the respiratory center. **Central chemoreceptors** are located in the medulla oblongata (*central* nervous system) whereas **peripheral chemoreceptors** are located in the walls of systemic arteries and relay impulses to the respiratory center over two cranial nerves of the *peripheral* nervous system.

Central chemoreceptors respond to changes in H^+ concentration or pCO_2, or both, in cerebrospinal fluid. Peripheral chemoreceptors are sensitive to changes in H^+, pCO_2, and pO_2 in the blood. They are located in the **aortic body**, a cluster of chemoreceptors located in the wall of the arch of the aorta, and in the **carotid bodies**, which are oval nodules in the wall of the left and right common carotid arteries where they bifurcate (divide) into internal and external carotid arteries. Sensory fibers from the aortic body join the vagus (X) nerve whereas those from the carotid bodies join the right and left glossopharyngeal (IX) nerves.

Because CO_2 is lipid-soluble, it easily diffuses across plasma membranes, including those that form the blood–brain barrier. Inside cells, including red blood cells and neurons, CO_2 combines with water (H_2O) to form carbonic acid (H_2CO_3). But the carbonic acid quickly breaks down into H^+ and HCO_3^-. Any increase in CO_2 will thus cause an increase in H^+, and any decrease in CO_2 will cause a decrease in H^+.

Under normal circumstances, the pCO_2 in arterial blood is 40 mm Hg. If there is even a slight increase in pCO_2—a condition called **hypercapnia**—the central chemoreceptors are stimulated. They respond vigorously to the increase in H^+ concentration that accompanies hypercapnia (Fig. 23.26). H^+ and CO_2 concentrations fluctuate more readily in cerebrospinal fluid (CSF) than in blood plasma because

Figure 23.26   Negative feedback control of breathing by changes in blood pCO_2, pO_2, and pH (H^+ concentration).

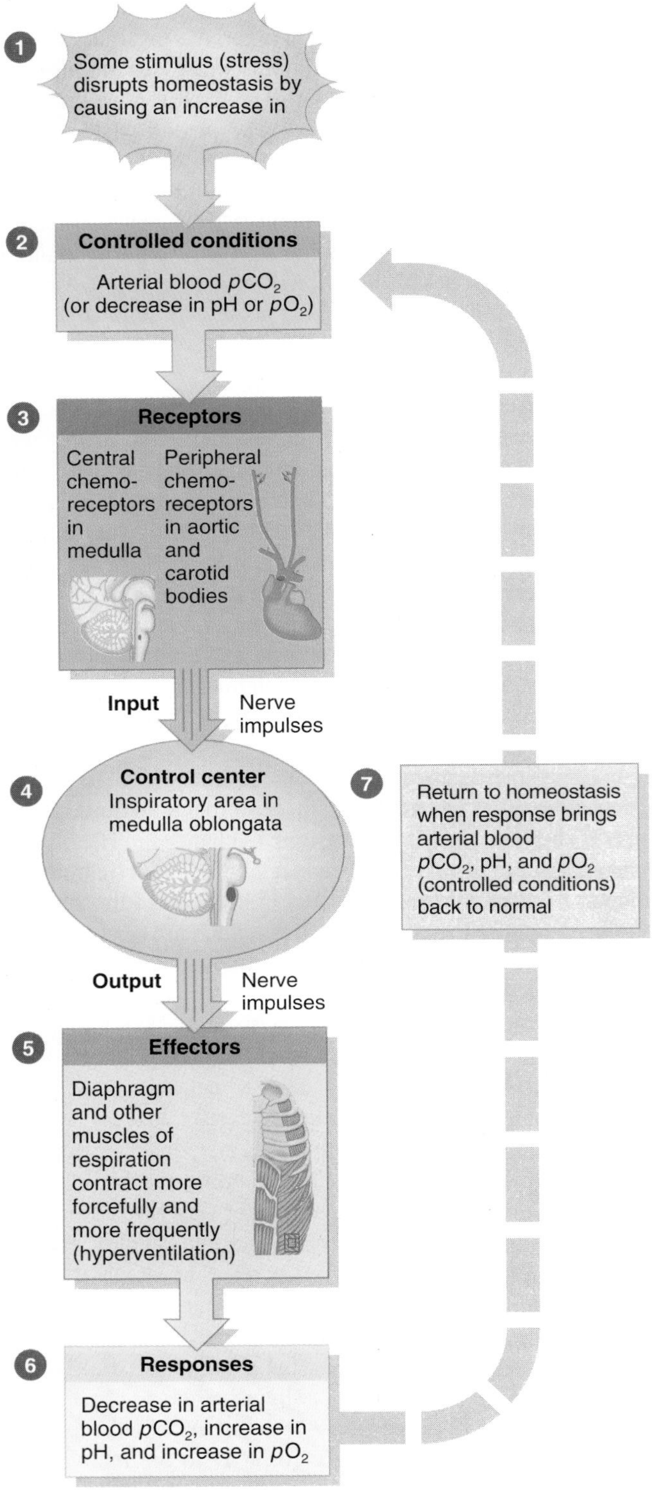

1 Some stimulus (stress) disrupts homeostasis by causing an increase in

2 Controlled conditions

Arterial blood pCO_2 (or decrease in pH or pO_2)

3 Receptors

Central chemo-receptors in medulla

Peripheral chemo-receptors in aortic and carotid bodies

Input   Nerve impulses

4 Control center
Inspiratory area in medulla oblongata

7 Return to homeostasis when response brings arterial blood pCO_2, pH, and pO_2 (controlled conditions) back to normal

Output   Nerve impulses

5 Effectors

Diaphragm and other muscles of respiration contract more forcefully and more frequently (hyperventilation)

6 Responses

Decrease in arterial blood pCO_2, increase in pH, and increase in pO_2

Q What is the normal arterial blood pCO_2?

CSF has fewer buffers than blood. The peripheral chemoreceptors in the carotid and aortic bodies also are stimulated by both the high pCO_2 and the rise in H^+ concentration. In addition, the peripheral chemoreceptors respond to hypoxia (oxygen starvation). If arterial pO_2 falls from a normal of 100 mm Hg to about 50 mm Hg, the peripheral chemoreceptors are strongly stimulated.

As a result of increased pCO_2, increased H^+ concentration, and decreased pO_2, input from the central and peripheral chemoreceptors causes the inspiratory area to become highly active, and the rate and depth of breathing increase (Fig. 23.26). Rapid and deep breathing is called **hyperventilation** and allows exhalation of more CO_2 until pCO_2 and H^+ are lowered to normal. Slow and shallow breathing is called **hypoventilation**.

If arterial pCO_2 is lower than 40 mm Hg, a condition called **hypocapnia**, the central and peripheral chemoreceptors are not stimulated and stimulatory impulses are not sent to the inspiratory area. Consequently, the area sets its own moderate pace until CO_2 accumulates and the pCO_2 rises to 40 mm Hg. If a person hyperventilates voluntarily and causes hypocapnia, then the breath can be held for an unusually long period of time. At one time swimmers were encouraged to hyperventilate just before diving in to compete. This is risky, however, because the O_2 level may fall dangerously low and cause fainting before the pCO_2 rises high enough to stimulate inspiration. A person who faints on dry land may suffer bumps and bruises but one who faints in the water may drown.

Severe hypoxia depresses activity of the central chemoreceptors and respiratory center, which then do not respond well to any inputs (Fig. 23.27). They send fewer impulses to the muscles of respiration. As the respiration rate decreases or breathing ceases altogether, pO_2 falls lower and lower, thus establishing a positive feedback cycle.

Neural Changes Due to Movement

As soon as you start exercising, your rate and depth of breathing increase, even before there are changes in pO_2, pCO_2, or H^+ concentration. The main stimulus for these quick changes in respiratory effort is thought to be input from proprioceptors, which monitor movement of joints and muscles. Nerve impulses from the proprioceptors stimulate the inspiratory area of the medulla oblongata. At the same time, axon collaterals (branches) of upper motor neurons that originate in the primary motor cortex (precentral gyrus) also feed excitatory impulses into the inspiratory area.

Inflation Reflex

Located in the walls of bronchi and bronchioles within the lungs are receptors sensitive to stretch called **barorecep-**

Figure 23.27 Positive feedback reduction of partial pressure of O_2 in blood.

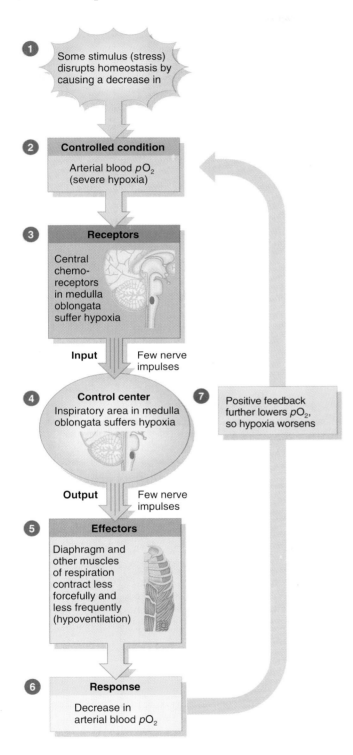

1. Some stimulus (stress) disrupts homeostasis by causing a decrease in

2. **Controlled condition**
Arterial blood pO_2 (severe hypoxia)

3. **Receptors**
Central chemo-receptors in medulla oblongata suffer hypoxia

Input Few nerve impulses

4. **Control center**
Inspiratory area in medulla oblongata suffers hypoxia

7. Positive feedback further lowers pO_2, so hypoxia worsens

Output Few nerve impulses

5. **Effectors**
Diaphragm and other muscles of respiration contract less forcefully and less frequently (hypoventilation)

6. **Response**
Decrease in arterial blood pO_2

How could you intervene to break this positive feedback loop?

tors or **stretch receptors**. When the receptors become stretched during overinflation of the lungs, nerve impulses are sent along the vagus (X) nerves to the inspiratory area and apneustic area. In response, the inspiratory area is inhibited and the apneustic area is inhibited from activating the inspiratory area. The result is that expiration begins. As air leaves the lungs during expiration, the lungs deflate and the stretch receptors are no longer stimulated. Thus the inspiratory and apneustic areas are no longer inhibited, and a new inspiration begins. This reflex is referred to as the **inflation (Hering–Breuer) reflex**. Some evidence suggests that the reflex is mainly a protective mechanism for preventing excessive inflation of the lungs rather than a key component in the normal regulation of respiration.

Other Influences

Many other factors contribute to regulation of respiration as follows:

1. **Blood pressure.** The carotid and aortic sinuses, which are close to the carotid and aortic bodies, contain baroreceptors (pressure receptors) that detect changes in blood pressure. Although these baroreceptors are concerned mainly with the control of circulation (shown in Fig. 21.12), they also affect respiration. For example, a sudden rise in blood pressure decreases the rate of respiration, and a drop in blood pressure increases the respiratory rate.
2. **Limbic system.** Anticipation of activity or emotional anxiety both may stimulate the limbic system, which then sends excitatory input to the inspiratory area to increase rate and depth of ventilation.
3. **Temperature.** An increase in body temperature, as during a fever or vigorous muscular exercise, increases the rate of respiration. A decrease in body temperature decreases respiratory rate. A sudden cold stimulus such as plunging into cold water causes apnea.
4. **Pain.** A sudden, severe pain brings about apnea, but a prolonged pain triggers the general adaptation syndrome (shown in Fig. 18.27) and increases respiratory rate.
5. **Stretching the anal sphincter muscle.** This increases the respiratory rate and is sometimes employed to stimulate respiration in a person who has stopped breathing.
6. **Irritation of airways.** Mechanical or chemical irritation of the pharynx or larynx brings about an immediate cessation of breathing followed by coughing or sneezing.

Exhibit 23.2 summarizes the changes that increase or decrease ventilation rate and depth.

EXHIBIT 23.2 SUMMARY OF REGULATION OF VENTILATION RATE AND DEPTH

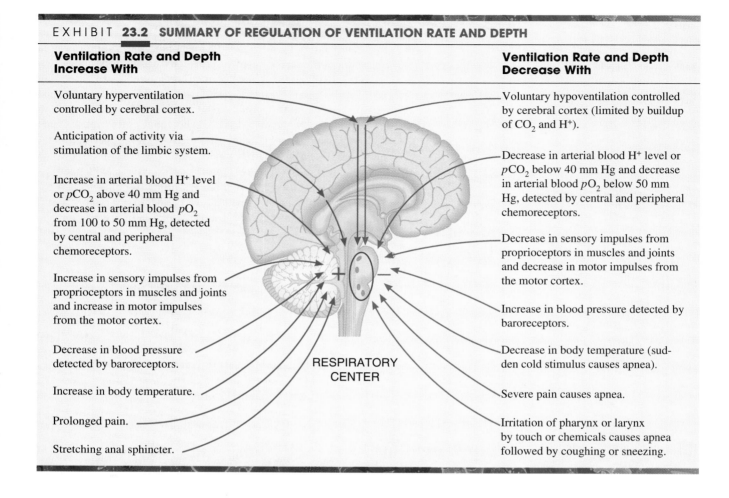

Ventilation Rate and Depth Increase With

Voluntary hyperventilation controlled by cerebral cortex.

Anticipation of activity via stimulation of the limbic system.

Increase in arterial blood H$^+$ level or pCO$_2$ above 40 mm Hg and decrease in arterial blood pO$_2$ from 100 to 50 mm Hg, detected by central and peripheral chemoreceptors.

Increase in sensory impulses from proprioceptors in muscles and joints and increase in motor impulses from the motor cortex.

Decrease in blood pressure detected by baroreceptors.

Increase in body temperature.

Prolonged pain.

Stretching anal sphincter.

Ventilation Rate and Depth Decrease With

Voluntary hypoventilation controlled by cerebral cortex (limited by buildup of CO$_2$ and H$^+$).

Decrease in arterial blood H$^+$ level or pCO$_2$ below 40 mm Hg and decrease in arterial blood pO$_2$ below 50 mm Hg, detected by central and peripheral chemoreceptors.

Decrease in sensory impulses from proprioceptors in muscles and joints and decrease in motor impulses from the motor cortex.

Increase in blood pressure detected by baroreceptors.

Decrease in body temperature (sudden cold stimulus causes apnea).

Severe pain causes apnea.

Irritation of pharynx or larynx by touch or chemicals causes apnea followed by coughing or sneezing.

RESPIRATORY CENTER

EXERCISE AND THE RESPIRATORY SYSTEM

During exercise, the respiratory and cardiovascular systems make adjustments in response to both the intensity and duration of the exercise. Since the effects of exercise on the heart have been discussed in Chapter 20, here we will focus on how exercise affects the respiratory system.

Recall that the heart pumps the same amount of blood to the lungs as to all the rest of the body. Thus as cardiac output rises, the blood flow to the lungs, termed **pulmonary perfusion**, also rises by as much as five times. The **O$_2$ diffusing capacity** is a measure of the rate at which O$_2$ can diffuse from alveolar air into the blood. This capacity may increase threefold during maximal exercise because more pulmonary capillaries become perfused to the maximum level. This provides a greater surface area for diffusion of O$_2$ into pulmonary blood capillaries.

When muscles contract during exercise, they consume large amounts of O$_2$ and produce large amounts of CO$_2$. During vigorous exercise, both O$_2$ consumption and pul-

monary ventilation can increase up to 30-fold from the resting state. At the onset of exercise, there is an abrupt increase in pulmonary ventilation followed by a more gradual increase. With moderate exercise, the increase is due mostly to an increase in the depth of ventilation rather than an increased breathing rate. When exercise is more strenuous, the rate of pulmonary ventilation also increases.

The abrupt increase in ventilation at the start of exercise is due to *neural* changes that feed excitatory impulses into the inspiratory area in the medulla oblongata. These include (1) anticipation of the activity, which stimulates the limbic system, (2) sensory impulses from proprioceptors in muscles, tendons, and joints, and (3) motor impulses from the primary motor cortex (precentral gyrus). The more gradual increase in ventilation during moderate exercise is due to *chemical* and *physical* changes in the bloodstream, including (1) decreased pO$_2$, due to increased O$_2$ consumption, (2) increased pCO$_2$, due to increased CO$_2$ production by contracting muscle fibers, and (3) increased temperature due to liberation of more heat as more O$_2$ is utilized. Moreover, during strenuous exercise, HCO$_3^-$

buffers H^+ released by lactic acid. This reaction liberates CO_2, which further increases pCO_2.

At the end of a bout of exercise, there is an abrupt decrease in pulmonary ventilation followed by a more gradual decline to the resting level. The initial decrease is due mainly to changes in the neural factors when movement stops or slows whereas the slower phase reflects the slower return of blood chemistry and temperature to the resting state.

WHY SMOKERS HAVE LOWERED RESPIRATORY EFFICIENCY

It is commonly observed that smoking may cause a person to become easily "winded" with even moderate exercise. Several factors decrease respiratory efficiency. (1) Nicotine constricts terminal bronchioles and this decreases airflow into and out of the lungs. (2) Carbon monoxide in smoke binds to hemoglobin and reduces its oxygen-carrying capability. (3) Irritants in smoke cause increased fluid secretion by the mucosa of the bronchial tree and swelling of the mucosal lining, which both impede airflow into and out of the lungs. (4) Irritants in smoke also inhibit the movement of cilia in the lining of the respiratory system. Thus excess fluids and foreign debris are not easily removed, which adds further to the difficulty in breathing. (5) With time, smoking leads to destruction of elastic fibers in the lungs and emphysema (described on page 744). These changes cause collapse of small bronchioles and trapping of air in alveoli at the end of exhalation. As a result, gas exchange is less efficient. ■

AGING AND THE RESPIRATORY SYSTEM

With advancing age, the airways and tissues of the respiratory tract, including the alveoli, become less elastic and more rigid. In addition to the lungs becoming less elastic, the chest wall also becomes more rigid. As a result, there is a decrease in lung capacity. In fact, vital capacity (the maximum amount of air that can be expired after maximal inspiration) can decrease as much as 35% by age 70. Also, there is a decrease in blood levels of O_2, decreased activity of alveolar macrophages, and diminished ciliary action of the epithelium lining the respiratory tract. Owing to all these age-related factors, elderly people are more susceptible to pneumonia, bronchitis, emphysema, and other pulmonary disorders.

DEVELOPMENTAL ANATOMY OF THE RESPIRATORY SYSTEM

The development of the mouth and pharynx is considered in Chapter 24 on the digestive system. Here we consider the remainder of the respiratory system. At about 4 weeks of fetal development, the respiratory system begins as an outgrowth of the **endoderm** of the foregut (precursor of some digestive organs) just posterior to the pharynx. This outgrowth is called the **laryngotracheal bud** (see Fig. 18.26). As the bud grows, it elongates and differentiates into the future epithelial lining of the *larynx* and other structures as well. Its proximal end maintains a slitlike opening into the pharynx called the *glottis*. The middle portion of the bud gives rise to the epithelial lining of the *trachea*. The distal portion divides into two **lung buds**, which grow into the epithelial lining of the *bronchi* and *lungs* (Fig. 23.28).

As the lung buds develop, they branch and rebranch and give rise to all the *bronchial tubes*. After the sixth month, the closed terminal portions of the tubes dilate and become the *alveoli* of the lungs. The smooth muscle, cartilage, and connective tissues of the bronchial tubes and the pleural sacs of the lungs are contributed by **mesenchymal (mesodermal) cells**.

Figure 23.28 Development of the bronchial tubes and lungs.

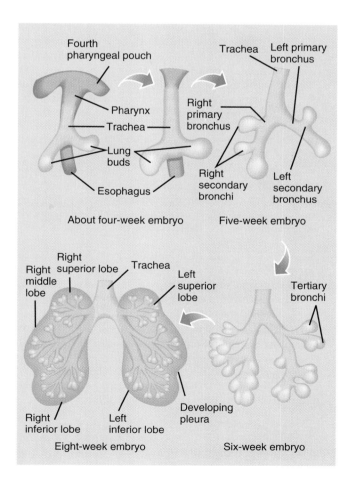

DISORDERS: HOMEOSTATIC IMBALANCES

Diseases such as asthma, bronchitis, and emphysema have in common some degree of obstruction of the airways. The term **chronic obstructive pulmonary disease** (**COPD**) is used to refer to these disorders. Among the symptoms that might indicate significant airflow obstruction are coughing, wheezing, and dyspnea. Spirometry reveals decreased FEV_1.

ASTHMA

Asthma is a chronic, inflammatory disorder that produces sporadic narrowing of airways. Attacks are brought on by spasms of the smooth muscle in the walls of the smaller bronchi and bronchioles, causing these airways to close partially or completely (bronchoconstriction). Symptoms include periods of coughing, difficult breathing, and wheezing that may abate spontaneously or with treatment. The patient has trouble exhaling, and air may be trapped in the alveoli during expiration.

In the early phase (acute) response, besides smooth muscle spasm, there is excessive secretion of mucus that may clog the bronchi and bronchioles and worsen the attack. In the late phase (chronic) response, inflammation continues, accompanied by fibrosis, edema, and necrosis (death) of bronchial epithelial cells. A host of mediator chemicals, including leukotrienes, prostaglandins, thromboxane, platelet-activating factor, and histamine, take part.

The airways of people with asthma are hyperreactive to a variety of stimuli that normally do not trigger bronchoconstriction. Sometimes the trigger is an allergen such as pollen, house dust mites, molds, or a particular food. Other common triggers of asthma attacks are emotional upset, aspirin, sulfiting agents (used in wine and beer and to keep greens fresh on salad bars), exercise, and breathing cold air or cigarette smoke. An acute attack is treated by giving a beta-adrenergic agonist such as epinephrine to help open up the airways. However, long-term therapy of asthma strives to suppress the underlying inflammation. The anti-inflammatory drugs used most often are inhaled glucocorticoids (corticosteroids) and cromolyn sodium.

BRONCHITIS

Bronchitis is inflammation of the bronchi characterized by hypertrophy and hyperplasia of glands and goblet cells lining the bronchial airways. The typical symptom is a cough in which sputum is raised (productive cough). This secretion signifies the presence of the underlying inflammation that is causing excessive secretion of mucus. Cigarette smoking remains the leading cause of chronic bronchitis, that is, bronchitis that lasts for at least 3 months of the year for two successive years.

EMPHYSEMA

In **emphysema** (em'-fi-SĒ-ma), alveolar walls disintegrate, producing abnormally large air spaces that remain filled with air during expiration (see Fig. 23.30b). The name means "blown up" or "full of air." With less surface area for gas exchange, O_2 diffusion across the damaged alveolar–capillary membrane is reduced. Blood O_2 level is somewhat lowered, and any mild exercise that raises the O_2 requirements of the cells leaves the patient breathless. As increasing numbers of alveolar walls are damaged, lung compliance increases due to loss of elastic fibers. Although inhalation is easy as a result, the patient has to

work voluntarily to exhale and air becomes trapped in the lungs. Over several years, the added exertion increases the size of the chest cage, resulting in a "barrel chest."

Emphysema is generally caused by a long-term irritation. Cigarette smoke, air pollution, and occupational exposure to industrial dust are the most common irritants. Some destruction of alveolar sacs may be caused by an imbalance between enzymes called proteases, such as elastase, and a molecule (alpha-1-antitrypsin) that inhibits proteases. When there is decreased production by the liver of the plasma protein alpha-1-antitrypsin, elastase is not inhibited and is free to attack the connective tissue in the walls of alveolar sacs. Cigarette smoke not only deactivates a protein that is seemingly crucial in preventing emphysema but also prevents the repair of affected lung tissue.

BRONCHOGENIC CARCINOMA

A common lung cancer, **bronchogenic carcinoma**, starts in the walls of the bronchi (Fig. 23.29). The constant irritation by inhaled smoke and other pollutants causes the goblet cells of the bronchial epithelium to enlarge. They respond by secreting excessive mucus. The basal cells also respond to the stress by undergoing cell division so fast that they push into the area occupied by the goblet and columnar cells. Many researchers believe that if the stress is removed at this point, the epithelium can return to normal.

If the stress persists, however, more and more mucus is secreted, and the cilia become less effective. As a result, mucus is not carried toward the throat but remains trapped in the bronchial tubes. The person then develops a "smoker's cough." Moreover, the constant irritation from the pollutant slowly destroys the alveoli, which are replaced with thick, inelastic connective tissue. Alveoli are destroyed by protein-digesting enzymes produced by white blood cells and macrophages in response to the stress. Mucus that has accumulated becomes trapped in the alveoli. Millions of alveoli rupture, reducing the diffusion surface for the exchange of O_2 and CO_2. The person has now developed emphysema. If the stress is removed at this point, there is little chance for improvement. Alveolar tissue that has been destroyed cannot be repaired. But removal of the stress can stop further destruction of lung tissue.

If the stress still continues, the emphysema progressively worsens, and the basal cells of the bronchial tubes continue to divide and break through the basement membrane (Fig. 23.29c). At this point the stage is set for bronchogenic carcinoma. Columnar and goblet cells disappear and may be replaced with squamous cancer cells. If this happens, the malignant growth spreads throughout the lung (Fig. 23.30) and may block a bronchial tube. Treatment involves surgical removal of the diseased lung. However, metastasis (spreading) of the growth through the lymphatic or blood vessels may result in new growths in other parts of the body such as the brain and liver.

Other factors may be associated with lung cancer. For instance, malignant cells from breast, stomach, and prostate cancer can metastasize to the lungs. Radioactive radon gas, which may seep from the earth into buildings, causes lung cancer in some nonsmokers. Smoking raises a person's radon-related

Figure 23.29 Effects of smoking on the respiratory epithelium. (a) Microscopic view of the normal epithelium of a bronchial tube. (b) Initial response of the bronchial epithelium to irritation by pollutants. (c) Advanced response of the bronchial epithelium.

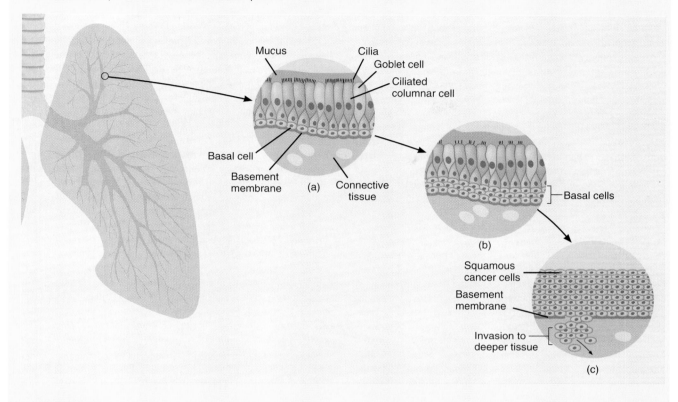

Figure 23.30 Lung cancer and emphysema.

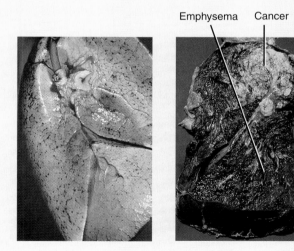

(a) Normal lung

(b) Cancerous and emphysematous lung

cancer risk to a level 15 times higher than that of a nonsmoker. Asbestos is also associated with lung cancer, as well as with other pulmonary diseases. The occurrence of bronchogenic carcinoma is over 20 times higher in heavy cigarette smokers than it is in nonsmokers. Exposure to environmental tobacco smoke (second-hand smoke) is also associated with lung cancer and heart disease. In the United States second-hand smoke causes an estimated 4000 deaths a year from lung cancer and nearly 40,000 deaths a year from heart disease.

PNEUMONIA

Pneumonia refers to an acute infection or inflammation of the alveoli. It is the most common infectious cause of death in the United States. The alveolar sacs fill up with fluid and dead white blood cells, reducing the amount of air space in the lungs. Oxygen has difficulty diffusing through the inflamed alveoli, and the blood pO_2 may be drastically reduced. Blood pCO_2 usually remains normal because CO_2 diffuses through the alveoli more easily than O_2.

The most common cause of pneumonia is the pneumococcal bacterium *Streptococcus pneumoniae*, but other bacteria or fungi, protozoans, or viruses may also cause pneumonia. Those who are most susceptible to pneumonia are the elderly, infants, immunocompromised individuals (persons with AIDS, malignancy, or those taking immunosuppressive drugs), cigarette smokers, and persons with obstructive lung disease.

TUBERCULOSIS

The bacterium *Mycobacterium tuberculosis* produces an infectious, communicable disease called **tuberculosis (TB)**. It most often affects the lungs and the pleurae. The bacteria destroy parts of the lung tissue, and the tissue is replaced by fibrous connective tissue. Because the connective tissue is inelastic and

DISORDERS: HOMEOSTATIC IMBALANCES CONTINUED

thick, the affected areas of the lungs do not recoil well during expiration and air is retained. O_2 and CO_2 no longer diffuse easily through the fibrous tissue.

Tuberculosis bacteria are spread by inhalation and exhalation. Although they can withstand exposure to many disinfectants, they die quickly in sunlight. Thus tuberculosis is sometimes associated with crowded, poorly lighted housing. Many drugs, such as isoniazid and rifampin, are successful in treating tuberculosis. Rest, sunlight, and good diet are vital parts of treatment.

During the past decade, the number of cases of tuberculosis reported annually in the United States has risen dramatically. This rise followed an average annual decline of 6% during the prior 30 years. Perhaps the single most important factor related to the increase in the number of tuberculosis patients is the presence of the human immunodeficiency virus (HIV). People infected with HIV are much more likely to develop tuberculosis because their immune systems are impaired. Among the other factors that have contributed to the increased number of cases are homelessness, increased drug abuse, increased immigration from countries with a high prevalence of tuberculosis, increased crowding in housing among the poor, and airborne transmission of tuberculosis in prisons and shelters. In addition, there have been recent outbreaks of tuberculosis involving multi-drug-resistant strains of *Mycobacterium tuberculosis* because patients fail to complete their drug and other treatments. Drug resistance is due to an adaptive response of microbes that allows them to tolerate a certain amount of an antibiotic that would normally inhibit them. For example, the microbes may produce enzymes that inactivate antibiotics, prevent an antibiotic from passing into their interior, or prevent attachment of an antibiotic to their surface.

RESPIRATORY DISTRESS SYNDROME

Adult respiratory distress syndrome (ARDS) is characterized by excessive leakiness of the alveolar–capillary membranes. Although the exact pathogenic mechanism is not clear, several situations can cause ARDS. These include near-drowning, aspiration of acidic gastric juice, drug reactions, inhalation of an irritating gas such as ammonia, an allergic reaction, various lung infections such as pneumonia or TB, and pulmonary hypertension. ARDS strikes an estimated 150,000 people a year in the United States and about 50% die despite intensive medical care. New studies are underway to test the effectiveness of inhaled nitric oxide (NO) for ARDS. Inhaled NO may be better than oral medications because it can improve respiratory function by dilating pulmonary arterioles without similarly dilating systemic arterioles, which would cause blood pressure to drop.

Respiratory distress syndrome (RDS) of the newborn, also called **hyaline membrane disease**, affects 65,000 of the premature babies born in the United States each year. Before birth, the respiratory passages are filled with fluid. At birth, these fluid-filled passageways must become air-filled, and the collapsed alveoli must expand to function in gas exchange. The success of this transition depends largely on the presence of surfactant, which lowers surface tension in the fluid layer and eases the expansion of the lungs. RDS is more common in premature babies delivered by cesarean section than by the normal vaginal route. It is thought that the stress of being born vaginally elevates cor-

tisol levels. Cortisol, in turn, stimulates surfactant production by type II alveolar cells. Pregnant mothers at risk of delivering a baby prematurely may be given cortisol to enhance development of the fetal type II alveolar cells.

Since 1989, when commercially prepared surfactants first became widely available, the death rate from RDS of the newborn has decreased dramatically. Besides receiving surfactant, which is squirted down the trachea into their lungs, babies with RDS are put on a ventilator to assist breathing and help distribute the surfactant.

RESPIRATORY FAILURE

Respiratory failure refers to a condition in which the respiratory system cannot supply sufficient O_2 to maintain metabolism or cannot eliminate enough CO_2 to prevent respiratory acidosis, a lower-than-normal pH in extracellular fluid. Specifically, respiratory failure occurs when the O_2 level of arterial blood is too low or when the CO_2 level of arterial blood is too high. Among the causes of respiratory failure are lung disorders; mechanical disorders that disturb the chest wall or neuromusculature (anesthetic blocking agents, cervical spinal cord injuries, myasthenia gravis, and amyotrophic lateral sclerosis); depression of the respiratory center by drugs, strokes, or trauma; and carbon monoxide poisoning. Symptoms of respiratory failure include disorientation, malaise, headache, muscular weakness, difficulty with sleep, palpitations, breathlessness, coughing, tachycardia, dysrhythmia, cyanosis, hypertension, edema, stupor, and coma. Treatment consists of providing adequate oxygenation and reversing the respiratory acidosis.

SUDDEN INFANT DEATH SYNDROME

Sudden infant death syndrome (SIDS), also called crib death, kills about 10,000 U.S. babies every year. It kills more infants between the ages of 1 week and 12 months than any other disease. SIDS occurs without warning. Although about half of its victims had an upper respiratory infection within 2 weeks of death, the babies tended to be otherwise remarkably healthy.

Many hypotheses have been proposed as explanations for SIDS but the precise pathogenic mechanisms are not completely clear. A main cause is thought to be hypoxia in infants who sleep in a prone position (on their stomach) and rebreathe exhaled air trapped in a depression of the mattress. For this reason, it is suggested that normal newborns be placed on their backs for sleeping. The seasonal distribution of SIDS incidence, being lowest during the summer months and highest in the late fall and winter, suggests an infectious agent, viruses being the most likely possibility. Most cases occur at a time in life when antibody levels are low and the infant is at a critical period of susceptibility. Alternatively, infants may be more likely to die from hyperthermia due to being dressed too warmly in the winter months.

CORYZA AND INFLUENZA

Hundreds of viruses are responsible for **coryza (common cold)**. A group of viruses, called rhinoviruses, is responsible for about 40% of all colds in adults. Typical symptoms include sneezing, excessive nasal secretion, dry cough, and congestion. The uncomplicated common cold is not usually accompanied by a fever. Complications include sinusitis, asthma, bronchitis, ear infections, and laryngitis. Recent investigations suggest an association between emotional stress and the common cold: the higher the stress level, the greater the frequency and duration of colds.

Influenza (flu) is also caused by a virus. Its symptoms include chills, fever (usually higher than 101°F = 39°C), headache, and muscular aches. Coldlike symptoms appear as the fever subsides.

PULMONARY EMBOLISM

A **pulmonary embolism** is a blood clot or other foreign substance in a pulmonary blood vessel that obstructs circulation to lung tissue. Most pulmonary emboli originate from thrombi in the proximal deep veins of the lower limbs. The immediate effect of pulmonary embolism is partial obstruction of arterial blood flow to the lung, resulting in dysfunction of the affected lung tissue. A large embolus can produce death in a few minutes. Pulmonary embolism usually develops in patients with one or more risk factors. Among these are immobilization in bed for more than 3 days; heart disease (especially with congestive heart failure); trauma, such as fractures; malignant disease; obesity; pregnancy; polycythemia; and the use of oral contraceptives.

Among the clinical findings associated with pulmonary embolism are sudden and unexplained dyspnea, chest pain, apprehension, rapid breathing, rales (bubbling or rattling sounds), cough, deep-venous thrombosis, pulmonary infarction as evidenced by expectoration of blood or blood-stained sputum, and syncope (fainting).

Treatment of pulmonary embolism consists of intravenous (IV) injections of heparin (an anticoagulant); administration of oral anticoagulants, such as warfarin; administration of clot-dissolving enzymes, such as t-PA or streptokinase; bed rest; analgesia; and O_2 administration. In some cases, surgery (embolectomy) is needed to prevent death due to a massive pulmonary embolism.

PULMONARY EDEMA

Pulmonary edema refers to an abnormal accumulation of interstitial fluid in the interstitial spaces and alveoli of the lungs. The edema may arise from increased pulmonary capillary permeability (pulmonary origin) or increased pulmonary capillary pressure (cardiac origin). The latter cause may coincide with congestive heart failure. The most common symptom is dyspnea. Others include wheezing, tachypnea (rapid respirations), restlessness, feeling of suffocation, cyanosis, pallor (paleness), and diaphoresis (excessive perspiration). Treatment consists of administering O_2, drugs that dilate the bronchioles and lower blood pressure, and drugs that correct acid–base imbalance; suctioning of airways; and mechanical ventilation.

HANTAVIRUS PULMONARY SYNDROME (HPS)

In May 1993, an outbreak of a severe respiratory illness occurred in the remote Four Corners area where Arizona, Colorado, Utah, and New Mexico meet. The most common initial symptoms are fever, muscle aches, headache, cough, nausea and vomiting, chills, malaise, and diarrhea. Then there is a rapid progression to acute respiratory distress due to leakage of fluid from lung blood capillaries into alveolar sacs (pulmonary edema), cardiac arrhythmias, and ultimately complete respiratory failure.

An intensive investigation by the Centers for Disease Control and Prevention (CDC) identified the cause of the disorder as a rodent-borne (deer mice, primarily) virus called a **hantavirus**, named for the Hantaan River in South Korea. In Asia and Europe, hantavirus illnesses have been known since the 1930s and have been characterized by periodic flu-like symptoms, fever, and kidney problems. The recently discovered strain of hantavirus, called **Muerto Canyon virus**, is a newly altered form of the virus that emerged in the Four Corners and causes a fast-moving deadly pneumonia called **hantavirus pulmonary syndrome** (HPS). Whereas the mortality rate of the original virus is 5–30%, it is about 60% for the newly discovered strain.

Victims of HPS apparently contract the disease through exposure to rodent saliva, feces, or urine. Domestic, occupational, or leisure activities (planting or harvesting crops, camping out or occupying cabins that have been vacant, and cleaning barns) are likely to bring potential victims in contact with infected rodents. If caught soon enough, HPS is treatable. Ribavirin, an antiviral drug, is being used experimentally and may be of some benefit. Treatment is supportive and consists of having patients hospitalized early, monitored carefully, and treated with life-sustaining fluids and medications that help normalize heart rate and breathing.

CYSTIC FIBROSIS

Cystic fibrosis (CF) is an inherited disease of secretory epithelia that affects the airways, pancreas, salivary glands, and sweat glands. It is the most common lethal genetic disease of Caucasians: 5% of the population are thought to be genetic carriers. The cause of cystic fibrosis is a genetic mutation affecting a transporter protein that carries chloride ions (Cl^-) across the plasma membranes of many epithelial cells.

Among the most common signs and symptoms are breathing difficulty, pancreatic insufficiency, and cirrhosis of the liver. CF is characterized by the production of thick mucous secretions that do not drain easily from the airways. Buildup of these secretions leads to inflammation and replacement of injured cells with connective tissue that blocks the airways. Another prominent feature is blockage of the pancreatic ducts so that the digestive enzymes cannot reach the intestine. Since pancreatic juice contains the main fat-digesting enzyme, the person fails to absorb fats or fat-soluble vitamins and thus suffers from vitamin A, D, and K deficiency diseases.

A child suffering from cystic fibrosis is given pancreatic extract and large doses of vitamins A, D, and K. The therapeutic diet is low, but not lacking, in fats and high in carbohydrates and proteins that can be used for energy and can also be converted by the liver into the lipids essential for life processes. Studies are in progress to evaluate the ease and effectiveness of gene therapy for CF patients.

SMOKE INHALATION INJURY

When smoke is inhaled, the lungs are injured directly by heat from flames and substances in fumes. **Smoke inhalation injury** has three components that occur in sequence: (1) inhibition of O_2 delivery and utilization, (2) upper airway injury from heat, and (3) lung damage from acids and aldehydes in smoke.

Inhibition of O_2 delivery and utilization is due to inhalation of mainly carbon monoxide (CO) and, to a lesser extent, cyanide. CO binds to hemoglobin and components of the cytochrome system (required for cellular respiration) and thus

DISORDERS: HOMEOSTATIC IMBALANCES  CONTINUED

impairs O_2 delivery and utilization and also causes acidosis (low blood pH). The onset of symptoms (confusion and disorientation with moderate toxicity, and coma and cardiac arrest with severe toxicity) is immediate.

Upper airway injury from heat may cause edema of the laryngotracheal mucosa and obstruction of upper airways with atelectasis (collapsed lung). Symptoms, such as labored breathing and hypoxia, occur after 18–24 hours.

The acids and aldehydes in inhaled smoke result in bronchoconstriction, atelectasis, bronchiolar edema, impairment of mucociliary action, and destruction of surfactant. The onset of symptoms (labored breathing, hypovolemia, hypoxemia, and bronchopneumonia) is both early and delayed.

Treatment consists of administering 90–100% O_2, inserting an endotracheal tube, performing pulmonary suction, administering bronchodilators, and restoring and maintaining fluid balance.

MEDICAL TERMINOLOGY

Asphyxia (as-FIK-sē-a; *sphyxis* = pulse) Oxygen starvation due to low atmospheric oxygen or interference with ventilation, external respiration, or internal respiration.

Aspiration (as′-pi-RA-shun; *spirare* = to breathe) Inhalation of a foreign substance such as water, food, or foreign body into the bronchial tree; drawing of a substance in or out by suction.

Bronchiectasis (bron′-kē-EK-ta-sis; *ektasis* = expansion) A chronic dilation of the bronchi or bronchioles.

Cardiopulmonary resuscitation (kar′-dē-ō-PUL-mō-ner-ē re-sus′-i-TA-shun) (**CPR**) The artificial establishment of normal or near-normal respiration and circulation. The **A**, **B**, **C's** of cardiopulmonary resuscitation are **Airway**, **Breathing**, and **Circulation**, meaning the rescuer must establish an airway, provide artificial ventilation if breathing has stopped, and reestablish circulation if there is inadequate cardiac action. The procedure should be performed in that order (A, B, C).

Cheyne–Stokes respiration (CHAN STOKS res′-pi-RA-shun) A repeated cycle of irregular breathing beginning with shallow breaths that increase in depth and rapidity, then decrease and cease altogether for 15–20 seconds. Cheyne–Stokes is normal in infants. It is also often seen just before death from pulmonary, cerebral, cardiac, and kidney disease.

Epistaxis (ep′-i-STAK-sis) Loss of blood from the nose due to trauma, infection, allergy, malignant growths, and bleeding

disorders. It can be arrested by cautery with silver nitrate, electrocautery, and firm packing. Also called **nosebleed**.

Heimlich (abdominal thrust) maneuver (HĪM-lik ma-NOO-ver) First-aid procedure designed to clear the airways of obstructing objects. It is performed by applying a quick upward thrust that causes sudden elevation of the diaphragm and forceful, rapid expulsion of air in the lungs; this action forces air out the trachea to eject the obstructing object. The Heimlich maneuver is also used to expel water from the lungs of near-drowning victims before resuscitation is begun.

Hemoptysis (hē-MOP-ti-sis; *hemo* = blood; *ptein* = to spit) Spitting of blood from the respiratory tract.

Pneumonectomy (noo′-mō-NEK-tō-mē; *pneumo* = lung; *tome* = cutting) Surgical removal of a lung.

Rales (RALS) Sounds sometimes heard in the lungs that resemble bubbling or rattling. Rales are to the lungs what murmurs are to the heart. Different types are due to the presence of an abnormal amount or type of fluid or mucus inside the bronchi or alveoli, or to bronchoconstriction that causes turbulent airflow.

Respirator (RES-pi-rā′-tor) An apparatus fitted to a mask over the nose and mouth, or hooked directly to an endotracheal or tracheotomy tube, that is used to assist or support ventilation or to provide nebulized medication to the air passages under pressure.

Rhinitis (ri-NĪ-tis; *rhino* = nose) Chronic or acute inflammation of the mucous membrane of the nose.

STUDY OUTLINE

STRUCTURES OF THE RESPIRATORY SYSTEM (p. 708)

1. Respiratory organs include the nose, pharynx, larynx, trachea, bronchi, and lungs.
2. They act with the cardiovascular system to supply oxygen (O_2) and remove carbon dioxide (CO_2) from the blood.

Nose (p. 708)

1. The external portion of the nose is made of cartilage and skin and is lined with mucous membrane. Openings to the exterior are the external nares.
2. The internal portion of the nose communicates with the paranasal sinuses and nasopharynx through the internal nares.

3. The nasal cavity is divided by a septum. The anterior portion of the cavity is called the vestibule.
4. The nose warms, moistens, and filters air and functions in olfaction and speech.

Pharynx (p. 711)

1. The pharynx (throat) is a muscular tube lined by a mucous membrane.
2. The anatomic regions are the nasopharynx, oropharynx, and laryngopharynx.
3. The nasopharynx functions in respiration. The oropharynx and laryngopharynx function both in digestion and in respiration.

Larynx (p. 712)

1. The larynx (voice box) is a passageway that connects the pharynx with the trachea.
2. It contains the thyroid cartilage (Adam's apple); the epiglottis, which prevents food from entering the larynx; the cricoid cartilage, which connects the larynx and trachea; and the paired arytenoid, corniculate, and cuneiform cartilages.
3. The larynx contains vocal folds, which produce sound as they vibrate. Taut folds produce high pitches, and relaxed ones produce low pitches.

Trachea (p. 715)

1. The trachea (windpipe) extends from the larynx to the primary bronchi.
2. It is composed of C-shaped rings of cartilage and smooth muscle and is lined with pseudostratified ciliated columnar epithelium.
3. Two methods of bypassing obstructions in the airways are tracheostomy and intubation.

Bronchi (p. 717)

1. The bronchial tree consists of the trachea, primary bronchi, secondary bronchi, tertiary bronchi, bronchioles, and terminal bronchioles. Walls of bronchi contain rings of cartilage; walls of bronchioles contain plates of decreasing cartilage and increasing smooth muscle.
2. Bronchoscopy is the visual examination of the bronchus through a bronchoscope.
3. A bronchogram is a radiograph of the bronchial tree after introduction of an opaque contrast medium usually containing iodine.

Lungs (p. 719)

1. Lungs are paired organs in the thoracic cavity. They are enclosed by the pleural membrane. The parietal pleura is the superficial layer that lines the thoracic cavity; the visceral pleura is the deep layer that covers the lungs.
2. The right lung has three lobes separated by two fissures; the left lung has two lobes separated by one fissure and a depression, the cardiac notch.
3. Secondary bronchi give rise to branches called segmental bronchi, which supply segments of lung tissue called bronchopulmonary segments.
4. Each bronchopulmonary segment consists of lobules, which contain lymphatics, arterioles, venules, terminal bronchioles, respiratory bronchioles, alveolar ducts, alveolar sacs, and alveoli.
5. Alveolar walls consist of type I alveolar cells, type II alveolar cells, and alveolar macrophages.
6. Gas exchange occurs across the alveolar–capillary (respiratory) membranes.

PULMONARY VENTILATION (p. 723)

1. Pulmonary ventilation, or breathing, consists of inspiration and expiration.
2. The movement of air into and out of the lungs depends on pressure changes governed in part by Boyle's law, which states that the volume of a gas varies inversely with pressure, assuming that temperature is constant.
3. Inspiration occurs when alveolar pressure falls below atmospheric pressure. Contraction of the diaphragm and external intercostal muscles increases the size of the thorax, thus decreasing the intrapleural pressure so that the lungs expand. Expansion of the lungs decreases alveolar pressure so that air moves along the pressure gradient from the atmosphere into the lungs.
4. During forced inspiration, accessory muscles of inspiration (sternocleidomastoids, scalenes, and pectoralis minors) are also used.
5. Expiration occurs when alveolar pressure is higher than atmospheric pressure. Relaxation of the diaphragm and external intercostal muscles results in elastic recoil of the chest wall and lungs, which increases intrapleural pressure, lung volume decreases, and alveolar pressure increases so that air moves from the lungs to the atmosphere.
6. Forced expiration employs contraction of the internal intercostal and abdominal muscles.
7. Normal quiet breathing is termed eupnea. Other patterns are costal breathing and diaphragmatic breathing.
8. The surface tension exerted by alveolar fluid is decreased by the presence of surfactant.
9. A collapsed lung or portion of a lung is called atelectasis.
10. Compliance is the ease with which the lungs and thoracic wall expand.
11. The walls of the airways offer some resistance to breathing.
12. Modified respiratory movements, such as coughing, sneezing, sighing, yawning, sobbing, crying, laughing, and hiccuping, are used to express emotions and to clear the airways.

LUNG VOLUMES AND CAPACITIES (p. 728)

1. Lung volumes exchanged during breathing and rate of respiration are measured with a spirometer.
2. Lung volumes measured by spirometry include tidal volume, minute volume of respiration, alveolar ventilation rate, inspiratory reserve volume, expiratory reserve volume, and FEV_1. Other lung volumes are anatomic dead space, residual volume, and minimal volume.
3. Lung capacities, the sum of two or more volumes, include inspiratory, functional residual, vital, and total lung capacity.

EXCHANGE OF OXYGEN AND CARBON DIOXIDE (p. 729)

Gas Laws (p. 729)

1. The partial pressure of a gas is the pressure exerted by that gas in a mixture of gases. It is symbolized by *p*.
2. According to Dalton's law, each gas in a mixture of gases exerts its own pressure as if all the other gases were not present.
3. Henry's law states that the quantity of a gas that will dissolve in a liquid is proportional to the partial pressure of the gas and its solubility coefficient, when the temperature remains constant.
4. A major clinical application of Henry's law is hyperbaric oxygenation (HBO).

External Respiration and Internal Respiration (p. 731)

1. In internal and external respiration, O_2 and CO_2 diffuse from areas of their higher partial pressures to areas of their lower partial pressures.
2. External (pulmonary) respiration is the exchange of gases between alveoli and pulmonary blood capillaries. It depends on partial pressure differences, a large surface area for gas exchange, a small diffusion distance across the alveolar–capillary (respiratory) membrane, and the rate of airflow into and out of the lungs.

3. Internal (tissue) respiration is the exchange of gases between tissue blood capillaries and tissue cells.

TRANSPORT OF OXYGEN AND CARBON DIOXIDE (p. 732)

1. In each 100 ml of oxygenated blood, 1.5% of the O_2 is dissolved in plasma and 98.5% is bound to hemoglobin as oxyhemoglobin (HbO_2).
2. The association of O_2 and hemoglobin is affected by pO_2, acidity (pH), pCO_2, temperature, and BPG.
3. Fetal hemoglobin differs from adult hemoglobin in structure and has a higher affinity for O_2.
4. Hypoxia refers to O_2 deficiency at the tissue level and is classified as hypoxic, anemic, stagnant, or histotoxic.
5. In each 100 ml of deoxygenated blood, 7% of CO_2 is dissolved in plasma, 23% combines with hemoglobin as carbaminohemoglobin ($Hb \cdot CO_2$), and 70% is converted to bicarbonate ions (HCO_3^-).
6. Carbon monoxide poisoning occurs when CO combines with hemoglobin. The result is hypoxia.
7. In an acidic environment, hemoglobin's affinity for O_2 is lower and O_2 dissociates more readily from it (Bohr effect).
8. In the presence of O_2, less CO_2 binds to hemoglobin (Haldane effect).

CONTROL OF RESPIRATION (p. 736)

1. The respiratory center consists of a medullary rhythmicity area, pneumotaxic area, and apneustic area.
2. The inspiratory area has an intrinsic excitability (autorhythmicity) that sets the basic rhythm of respiration.
3. The pneumotaxic and apneustic areas coordinate the transition between inspiration and expiration.
4. Respirations may be modified by a number of factors, including cortical influences, the inflation reflex, chemical stimuli, such as O_2 and CO_2, and H^+ levels, neural changes due to movement, blood pressure changes, the limbic system, temperature, pain, and irritation to the airways.

EXERCISE AND THE RESPIRATORY SYSTEM (p. 742)

1. The rate and depth of ventilation change in response to both the intensity and duration of the exercise.
2. There is an increase in pulmonary perfusion and O_2 diffusing capacity during exercise.
3. The abrupt increase in ventilation at the start of exercise is due to neural changes that feed excitatory impulses into the inspiratory area in the medulla oblongata. The more gradual increase in ventilation during moderate exercise is due to chemical and physical changes in the bloodstream.
4. Several factors decrease respiratory efficiency in people who smoke.

AGING AND THE RESPIRATORY SYSTEM (p. 743)

1. Aging results in decreased vital capacity, decreased blood level of O_2, and diminished alveolar macrophage activity.
2. Elderly people are more susceptible to pneumonia, emphysema, bronchitis, and other pulmonary disorders.

DEVELOPMENTAL ANATOMY OF THE RESPIRATORY SYSTEM (p. 743)

1. The respiratory system begins as an outgrowth of endoderm called the laryngotracheal bud.
2. Smooth muscle, cartilage, and connective tissue of the bronchial tubes and pleural sacs develop from mesoderm.

REVIEW QUESTIONS

1. What organs make up the respiratory system? Distinguish between the upper and lower respiratory system. What functions do the respiratory and cardiovascular systems have in common? (p. 708)
2. Describe the structure of the external and internal nose, and describe their functions in filtering, warming, and moistening air. (p. 708)
3. What is the pharynx? Differentiate the three anatomical regions of the pharynx and indicate their roles in respiration. (p. 711)
4. Describe the structure of the larynx and explain how it functions in respiration and voice production. (p. 712)
5. Describe the location and structure of the trachea. (p. 715)
6. What is the bronchial tree? Describe its structure. What is a bronchogram? (p. 717)
7. Where are the lungs located? Distinguish the parietal pleura from the visceral pleura. (p. 719)
8. Define each of the following parts of a lung: base, apex, costal surface, medial surface, hilus, root, cardiac notch, and lobe. (p. 720)
9. What is a bronchopulmonary segment? (p. 720)
10. What is a lobule of the lung? Describe its composition and function in respiration. (p. 721)
11. Describe the histology and function of the alveolar–capillary (respiratory) membrane. (p. 722)
12. What are the basic differences among pulmonary ventilation, external respiration, and internal respiration? (p. 723)
13. Discuss the basic steps involved in inspiration and expiration. Be sure to include values for all pressures involved. (p. 724)
14. Distinguish between quiet and forced inspiration and expiration. (p. 725)
15. Describe how alveolar surface tension, compliance, and airway resistance affect pulmonary ventilation. (p. 727)
16. Define the various kinds of modified respiratory movements. (p. 728)
17. What is a spirometer? Define the various lung volumes and capacities. How is the minute volume of respiration (MVR) calculated? What are alveolar ventilation rate (AVR) and FEV_1? (p. 728)
18. Define the partial pressure of a gas. How is it calculated? (p. 730)
19. Define Boyle's law (p. 723), Dalton's law (p. 730), and Henry's law (p. 730).
20. How does decompression sickness occur? (p. 731)

21. Construct a diagram to illustrate how and why the respiratory gases move during external and internal respiration. (p. 731)
22. What factors affect external respiration? (p. 731)
23. Describe the relationship between hemoglobin and pO_2, acidity, pCO_2, temperature, and BPG. (p. 732)
24. Define hypoxia and distinguish the principal types. (p. 735)
25. Explain how CO_2 is picked up by tissue capillary blood and then discharged into the alveoli. (p. 736)
26. How does the medullary rhythmicity area function in controlling respiration? How are the apneustic area and pneumotaxic area related to the control of respiration? (p. 737)
27. Explain how each of the following modifies respiration: cerebral cortex, inflation reflex, CO_2, O_2, proprioceptors, temperature, pain, and irritations of the respiratory mucosa. (p.739)

28. How does the control of respiration demonstrate the principle of homeostasis? (p. 736)
29. Describe the effects of exercise on the respiratory system. (p. 742)
30. Describe the effects of aging on the respiratory system. (p. 743)
31. Describe the development of the respiratory system. (p. 743)
32. Define the following terms: rhinoplasty (p. 711), laryngitis (p. 715), cancer of the larynx (p. 715), tracheostomy (p. 717), intubation (p. 717), nebulization (p. 719), pneumothorax (p. 720), hemothorax (p. 720), pleurisy (p. 720), hyperbaric oxygenation (p. 731), and carbon monoxide poisoning (p. 735).
33. Explain why smokers have breathing difficulties. (p. 743)

ANSWERS TO QUESTIONS WITH FIGURES

23.1 Nose, pharynx, larynx, trachea, bronchi, and bronchioles, except the respiratory bronchioles.

23.2 External nares → vestibule → nasal cavity → internal nares.

23.3 Superior, internal nares; inferior, cricoid cartilage.

23.4 During swallowing, the epiglottis closes over the rima glottidis, the entrance to the trachea.

23.5 Voice production.

23.6 Since the tissues between the esophagus and trachea are soft, the esophagus can bulge into the trachea when you swallow food.

23.7 Left lung two of each; right lung three of each.

23.8 Serous membrane.

23.9 Since two-thirds of the heart lies to the left of the midline, the left lung contains a cardiac notch to accommodate the position of the heart.

23.10 Goblet cells produce and secrete mucus, and cilia move mucus laden with foreign particles toward the throat.

23.11 It averages 0.5 μm in thickness.

23.12 The pressure would increase to 4 atm.

23.13 If you are at rest while reading, your diaphragm.

23.14 At the start of inspiration, intrapleural pressure is about 756 mm Hg. With contraction of the diaphragm, it decreases to about 754 mm Hg, as the volume of the space between the two pleural layers expands. With relaxation of the diaphragm, it increases back to 756 mm Hg.

23.16 Vital capacity.

23.17 A difference in pO_2 that promotes diffusion in the indicated direction.

23.18 55 ml; 20 ml (the same as at rest!).

23.19 In both cases, hemoglobin in your pulmonary veins would be fully saturated with O_2 (upper right-most point on curve).

23.20 Since lactic acid (lactate) and CO_2 are produced by active skeletal muscles, blood pH decreases a little and pCO_2 increases when you are actively working out. The result is lowered affinity of hemoglobin for O_2. Thus more O_2 is available to the working muscles.

23.21 More available because the affinity of hemoglobin for O_2 decreases with increasing temperature.

23.22 Fetal Hb is 80% saturated while maternal Hb is about 75% saturated with O_2.

23.23 Vein.

23.24 The medullary inspiratory area.

23.25 Diaphragm—phrenic nerves; external intercostal muscles—intercostal nerves.

23.26 40 mm Hg.

23.27 Administer air enriched in O_2, with artificial ventilation if the person has stopped breathing. Even mouth-to-mouth resuscitation might save the person.

CHAPTER 24

THE DIGESTIVE SYSTEM

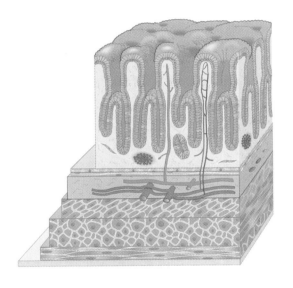

STUDENT OBJECTIVES

1. Identify the organs of the gastrointestinal (GI) tract and the accessory organs of digestion and their functions in digestion. (p. 753)

2. Describe the mechanical movements of the gastrointestinal tract. (p. 754)

3. Explain how salivary secretion, gastric secretion, gastric emptying, pancreatic secretion, bile secretion, and small intestinal secretion are regulated. (pp. 761, 771, 773, 776, 781, and 788)

4. Define absorption and explain how the end products of digestion are absorbed. (p. 788)

5. Define the processes involved in the formation of feces and defecation. (p. 796)

6. Describe the effects of aging on the digestive system. (p. 797)

7. Describe the development of the digestive system. (p. 798)

8. Describe the clinical symptoms of the following disorders: dental caries, periodontal disease, peptic ulcer disease (PUD), gastrointestinal tumors, cirrhosis, hepatitis, gallstones, anorexia nervosa, and bulimia. (p. 799)

9. Define medical terminology associated with the digestive system. (p. 801)

CHAPTER CONTENTS AT A GLANCE

ood contains a variety of nutrients—molecules needed for building new and repairing damaged body tissues and sustaining needed chemical reactions. Food also is vital for life because it is the source of energy that drives the chemical reactions occurring in every cell. Energy is needed for muscle contraction, conduction of nerve impulses, and secretory and absorptive activities of many cells. Food as it is consumed, however, is not in a state suitable for use as an energy source by any cell. First, it must be broken down into molecules small enough to cross the plasma (cell) membranes. The breaking down of larger food particles into molecules small enough to enter body cells is called **digestion**. The passage of these smaller molecules into blood and lymph is termed **absorption**. The organs that collectively perform these functions compose the **digestive system**.

The organs of digestion are divided into two main groups. First is the **gastrointestinal (GI) tract**, or **alimentary** (*alimentum* = nourishment) **canal**, a continuous tube that extends from the mouth to the anus and opens to the outside at each end (Fig. 24.1). Dietary materials in the lumen (inside cavity) of this tube have not yet truly entered the body. Organs composing the gastrointestinal tract include the mouth, pharynx, esophagus, stomach, small intestine, and large intestine.

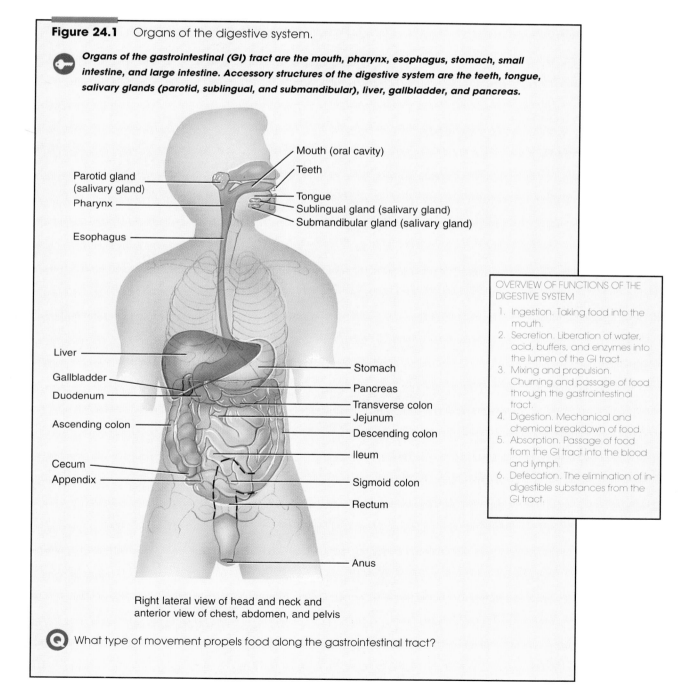

Figure 24.1 Organs of the digestive system.

Organs of the gastrointestinal (GI) tract are the mouth, pharynx, esophagus, stomach, small intestine, and large intestine. Accessory structures of the digestive system are the teeth, tongue, salivary glands (parotid, sublingual, and submandibular), liver, gallbladder, and pancreas.

Parotid gland (salivary gland)

Pharynx

Esophagus

Mouth (oral cavity)

Teeth

Tongue

Sublingual gland (salivary gland)

Submandibular gland (salivary gland)

Liver

Gallbladder

Duodenum

Ascending colon

Cecum

Appendix

Stomach

Pancreas

Transverse colon

Jejunum

Descending colon

Ileum

Sigmoid colon

Rectum

Anus

OVERVIEW OF FUNCTIONS OF THE DIGESTIVE SYSTEM

1. Ingestion. Taking food into the mouth.
2. Secretion. Liberation of water, acid, buffers, and enzymes into the lumen of the GI tract.
3. Mixing and propulsion. Churning and passage of food through the gastrointestinal tract.
4. Digestion. Mechanical and chemical breakdown of food.
5. Absorption. Passage of food from the GI tract into the blood and lymph.
6. Defecation. The elimination of indigestible substances from the GI tract.

Right lateral view of head and neck and anterior view of chest, abdomen, and pelvis

What type of movement propels food along the gastrointestinal tract?

The second group of organs composing the digestive system are the **accessory structures**—the teeth, tongue, salivary glands, liver, gallbladder, and pancreas. Teeth aid in the physical breakdown of food. The tongue assists in chewing and swallowing. The other accessory structures never come into direct contact with food. They produce or store secretions that aid in the chemical breakdown of food. These secretions flow into the tract through ducts.

The medical specialty that deals with the structure, function, diagnosis, and treatment of diseases of the stomach and intestines is called **gastroenterology** (gas′-trō-en′-ter-OL-ō-jē; *gastro* = stomach; *enteron* = intestines; *logos* = study of). The medical specialty that deals with the diagnosis and treatment of disorders of the rectum and anus is called **proctology** (prok-TOL-ō-jē; *proct* = rectum). The developmental anatomy of the digestive system is considered at the end of the chapter.

OVERVIEW OF DIGESTIVE PROCESSES

The GI tract contains food from the time it is eaten until it is digested and absorbed or eliminated. Muscular contractions in the wall of the GI tract break down the food physically by churning it. The contractions also help to dissolve foods by mixing them with the fluids secreted into the tract. Enzymes secreted by accessory structures and cells that line the tract break down the food chemically. Peristalsis, wavelike contraction of the smooth muscle in the wall of the GI tract, propels the food along the tract, from the esophagus to the anus. Overall, digestion includes six basic processes.

1. **Ingestion.** Taking food into the mouth (eating) is termed ingestion.
2. **Secretion.** Cells within the walls of the GI tract and accessory organs secrete a total of about 9 liters per day of water, acid, buffers, and enzymes into the lumen of the tract.
3. **Mixing and propulsion.** Alternating contraction and relaxation of smooth muscle in the walls of the GI tract mix food and secretions and propel them toward the anus.
4. **Digestion.** Both mechanical and chemical processes mix secreted fluids with ingested food and break down food molecules into smaller fragments. **Mechanical digestion** consists of various movements of the gastrointestinal tract. The teeth cut and grind food before it is swallowed. Then the smooth muscles of the stomach and small intestine churn the food so it is thoroughly mixed with enzymes that digest foods. This aids in dissolving food molecules in the liquid. **Chemical digestion** is a series of catabolic (hydrolysis) reactions. Enzymes break down large carbohydrate, lipid, protein, and nucleic acid molecules that we eat into smaller molecules.

5. **Absorption.** Most secreted fluids, small molecules, and ions that are products of digestion enter the epithelial cells lining the lumen of the GI tract either by active transport or by passive diffusion. This is absorption. Then, materials pass into blood and lymph and circulate to cells throughout the body.
6. **Defecation.** The elimination of variable amounts of indigestible substances and bacteria from the gastrointestinal tract through the anus is termed defecation.

LAYERS OF THE GI TRACT

The wall of the GI tract, from the stomach to the anal canal, has the same basic arrangement of tissues. The four layers or tunics of the tract from deep to superficial are the mucosa, submucosa, muscularis, and serosa (Fig. 24.2).

Mucosa

The **mucosa**, or deep lining of the tract, is a mucous membrane. Three layers compose the mucosa in the GI tract: (1) a lining layer of **epithelium** in direct contact with the contents of the GI tract, (2) an underlying layer of areolar connective tissue called the **lamina** (*lamina* = thin, flat plate) **propria**, and (3) a thin layer of smooth muscle called the **muscularis mucosae**.

The epithelial layer is mainly nonkeratinized stratified squamous epithelium, which has a protective function, in the mouth, esophagus, and anal canal. The stomach and intestines are lined by simple columnar epithelium, which functions in secretion and absorption. Among the epithelial cells are exocrine cells that secrete mucus and fluid into the lumen of the tract and several types of endocrine cells, collectively called **enteroendocrine** (*enteron* = intestine) **cells**, that secrete hormones into the bloodstream.

The lamina propria is areolar connective tissue containing many blood and lymphatic vessels and scattered lymphatic nodules. This layer supports the epithelium, binds it to the muscularis mucosae, and provides it with a blood and lymph supply. The blood and lymphatic vessels are the avenues by which absorbed molecules reach other tissues of the body.

The lamina propria also contains most of the cells of the **mucosa-associated lymphoid tissue (MALT)**. These patches of lymphatic tissues contain cells of the immune system that protect against disease. They are prevalent all along the GI tract, especially in the tonsils, small intestine, appendix, and large intestine. It is estimated that there are as many immune cells associated with the GI tract as in all the rest of the body. This makes sense when you realize that the GI tract is in contact with the outside environment and contains food that often carries harmful pathogens. The lymphocytes and macrophages in MALT mount immune responses against pathogens, such as bacteria, that penetrate the mucous membrane. The muscularis mucosae contains smooth muscle fibers typically arranged into an inner circular layer

Figure 24.2 Composite of various sections of the gastrointestinal tract seen in a three-dimensional drawing depicting the various layers and related structures.

 The four layers of the GI tract from deep to superficial are the mucosa, submucosa, muscularis, and serosa.

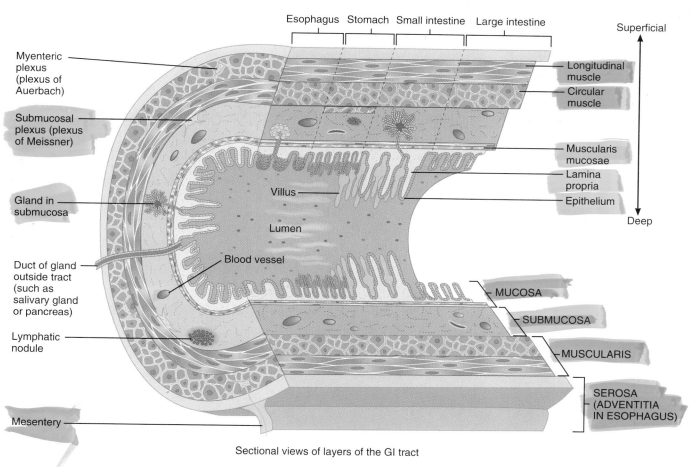

Sectional views of layers of the GI tract

Q What is the function of the nerve plexuses in the wall of the gastrointestinal tract?

and an outer longitudinal layer that throw the mucous membrane of the stomach and small intestine into folds. They also cause local movements of these folds, which increase the surface area for digestion and absorption.

Submucosa

The **submucosa** consists of areolar connective tissue that binds the mucosa to the third layer, the muscularis. It is highly vascular and contains a portion of the **submucosal plexus (plexus of Meissner)**, which is part of the autonomic nerve supply to the smooth muscle cells of the muscularis mucosae and blood vessels. Accordingly, the plexus regulates movements of the mucosa and vasoconstriction of blood vessels. The plexus also innervates secretory cells of mucosal glands and thus is important in controlling secretions by the GI tract. The submucosa may also contain glands and lymphatic tissue.

Muscularis

The **muscularis** of the mouth, pharynx, and superior part of the esophagus contains *skeletal muscle* that produces voluntary swallowing. Skeletal muscle also forms the external anal sphincter, which permits voluntary control of defecation. Throughout the rest of the tract, the muscularis consists of *smooth muscle* that is generally found in two sheets: an inner sheet of circular fibers and an outer sheet of longitudinal fibers. Involuntary contractions of the smooth muscles help break down food physically, mix it with digestive secretions, and propel it along the tract. The muscularis also contains the major nerve supply to the gastrointestinal tract—the **myenteric** (*myo* = muscle; *enteron* = intestine) **plexus (plexus of Auerbach)**, which consists of fibers from both autonomic divisions. This plexus mostly controls GI tract **motility**—the frequency and strength of contraction of the muscularis. Together, the submucosal and myenteric plexuses have as many neurons as does the spinal cord.

Serosa

The **serosa** is the superficial layer of those portions of the GI tract that are suspended in the abdominopelvic cavity. It is a serous membrane composed of connective tissue and simple squamous epithelium. As you will see shortly, the esophagus, which passes through the mediastinum, has a superficial layer called the adventitia composed of areolar connective tissue. Inferior to the diaphragm, the serosa is also called the **visceral peritoneum** and forms a portion of the peritoneum, which we shall now describe in detail.

PERITONEUM

The **peritoneum** (per′-i-tō-NĒ-um; *peri* = around; *tonos* = tension) is the largest serous membrane of the body. Serous membranes are also associated with the heart (pericardium) and lungs (pleurae). Serous membranes consist of a layer of simple squamous epithelium (called mesothelium) and an underlying supporting layer of connective tissue. Whereas the **parietal peritoneum** lines the wall of the abdominopelvic cavity, the **visceral peritoneum** covers some of the organs in the cavity and is their serosa. The potential space between the parietal and visceral portions of the peritoneum is called the **peritoneal cavity** and contains serous fluid (Fig. 24.3a). In certain diseases, the peritoneal cavity may become distended by several liters of fluid so that it forms an actual space. Such an accumulation of serous fluid is called **ascites** (a-SĪ-tēz).

As you will see later, some organs lie on the posterior abdominal wall and are covered by peritoneum on their anterior surfaces only. Such organs, including the kidneys and pancreas, are said to be **retroperitoneal** (*retro* = backward or located behind).

Unlike the pericardium and pleurae, which smoothly cover the heart and lungs, the peritoneum contains large folds that weave between the viscera. The folds bind the organs to each other and to the walls of the abdominal cavity and contain blood and lymphatic vessels and nerves that supply the abdominal organs. One extension of the peritoneum is called the **mesentery** (MEZ-en-ter′-ē; *meso* = middle; *enteron* = intestine). It is an outward fold of the

---

Figure 24.3   Relationship of the peritoneal extensions to each other and to organs of the digestive system.

🔑 *The peritoneum is the largest serous membrane in the body.*

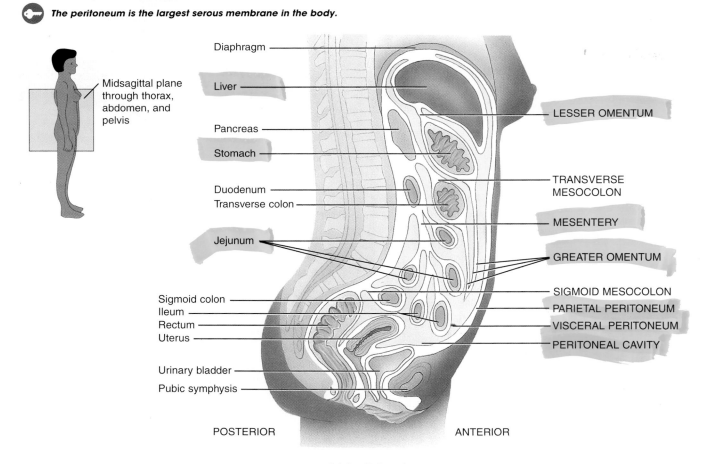

(a) Sagittal section

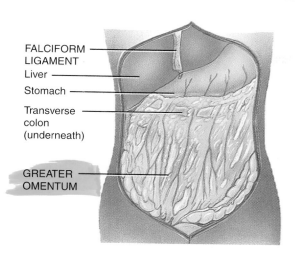

FALCIFORM LIGAMENT
Liver
Stomach
Transverse colon (underneath)
GREATER OMENTUM

(b) Greater omentum, anterior view

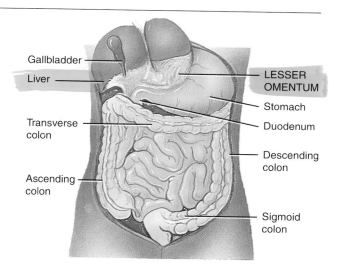

Gallbladder
Liver
LESSER OMENTUM
Stomach
Duodenum
Transverse colon
Descending colon
Ascending colon
Sigmoid colon

(c) Lesser omentum, anterior view
(liver and gallbladder lifted)

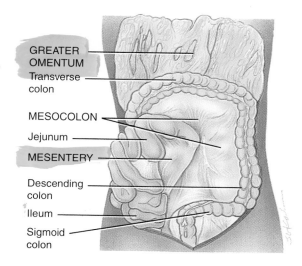

GREATER OMENTUM
Transverse colon
MESOCOLON
Jejunum
MESENTERY
Descending colon
Ileum
Sigmoid colon

(d) Mesentery and mesocolon, anterior view
(greater omentum lifted and small
intestine reflected to right side)

Q Which peritoneal extension binds the small intestine to the posterior abdominal wall?

serous coat of the small intestine (Fig. 24.3a,d). The tip of the fold is attached to the posterior abdominal wall. The mesentery binds the small intestine to the wall. A fold of peritoneum, called the **mesocolon** (mez′-ō-KŌ-lon), binds the large intestine to the posterior abdominal wall. It also carries blood and lymphatic vessels to the intestines. The mesentery and mesocolon hold the intestines loosely in place and also allow for a great amount of movement as the muscular contractions mix and move the luminal contents along the GI tract.

Other important peritoneal folds are the falciform ligament, lesser omentum, and greater omentum. The **falciform** (FAL-si-form; *falx* = sickle-shaped) **ligament** attaches the

liver to the anterior abdominal wall and diaphragm (Fig. 24.3b). The liver is the only digestive organ that is attached to the anterior abdominal wall. The **lesser omentum** (ō-MENT-um; = fat skin) arises as two folds in the serosa of the stomach and duodenum suspending the stomach and duodenum from the liver (Fig. 24.3c). The **greater omentum** is the largest peritoneal fold that drapes over the transverse colon and coils of the small intestine (Fig. 24.3b,d).

Because the greater omentum contains large quantities of adipose tissue, it looks like a "fatty apron" draped over the viscera. It is a double sheet that folds upon itself and thus is a four-layered structure. It attaches along the stomach and duodenum, extends downward anterior to the small intestine, then turns and extends upward, and attaches to the transverse colon. The greater omentum contains many lymph nodes. If an infection occurs in the intestines, plasma cells formed in the lymph nodes may produce antibodies to combat the infection and help prevent it from spreading to the peritoneum.

PERITONITIS

Peritonitis is an acute inflammation of the peritoneum. As in pleurisy and pericarditis, rubbing together of the inflamed peritoneal surfaces can cause considerable pain. One possible cause is contamination of the peritoneum by pathogenic bacteria from the external environment. This contamination could result from accidental or surgical wounds in the abdominal wall or from perforation or rupture of abdominal organs. For example, the large intestine contains colonies of bacteria that live on undigested nutrients and break them down so they can be eliminated. But if the bacteria enter the peritoneal cavity through an intestinal perforation or rupture of the appendix, they produce acute infection. Such an infection can quickly become life-threatening. ■

MOUTH

The **mouth**, also referred to as the **oral** or **buccal** (BUK-al; *bucca* = cheeks) **cavity**, is formed by the cheeks, hard and soft palates, and tongue (Fig. 24.4). Forming the lateral walls of the oral cavity are the **cheeks**—muscular structures covered externally by skin and internally by nonkeratinized stratified squamous epithelium. The anterior portions of the cheeks end at the lips.

The **lips** (**labia**) are fleshy folds surrounding the opening of the mouth. They are covered externally by skin and internally by a mucous membrane. The transition zone where the two kinds of covering tissue meet is called the **vermilion** (ver-MIL-yon). This portion of the lips is nonkeratinized, and the color of the blood in the underlying blood vessels is visible through the transparent surface layer. The inner surface of each lip is attached to its corresponding gum by a

midline fold of mucous membrane called the **labial frenulum** (LĀ-bē-al FREN-yoo-lum; *labium* = fleshy border; *frenulum* = small bridle).

The orbicularis oris muscle and connective tissue lie between the external integumentary covering and the internal mucosal lining. During chewing, contraction of the buccinator muscles in the cheeks and orbicularis oris muscle in the lips help keep food between the upper and lower teeth. They also assist in speech.

The **vestibule** (= entrance to a canal) of the oral cavity is a space bounded externally by the cheeks and lips and internally by the gums and teeth. The **oral cavity proper** is a space that extends from the gums and teeth to the **fauces** (FAW-sēs; *fauces* = passages), the opening between the oral cavity and the pharynx or throat.

The **hard palate**, the anterior portion of the roof of the mouth, is formed by the maxillae and palatine bones, is covered by mucous membrane, and forms a bony partition be-

Figure 24.4 Structures of the mouth (oral cavity).

 The mouth is formed by the cheeks, hard and soft palates, and tongue.

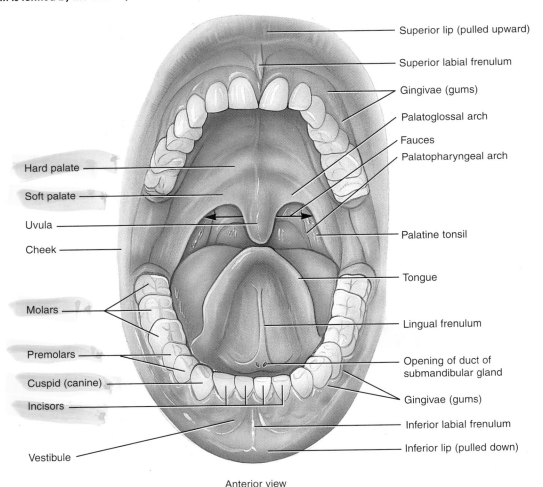

Superior lip (pulled upward)

Superior labial frenulum

Gingivae (gums)

Palatoglossal arch

Fauces

Palatopharyngeal arch

Palatine tonsil

Tongue

Lingual frenulum

Opening of duct of submandibular gland

Gingivae (gums)

Inferior labial frenulum

Inferior lip (pulled down)

Hard palate

Soft palate

Uvula

Cheek

Molars

Premolars

Cuspid (canine)

Incisors

Vestibule

Anterior view

Q Which structure of the mouth contains intrinsic and extrinsic muscles and what are their functions?

tween the oral and nasal cavities. The **soft palate** forms the posterior portion of the roof of the mouth. It is an arch-shaped muscular partition between the oropharynx and nasopharynx and is lined by mucous membrane.

Hanging from the free border of the soft palate is a conical muscular process called the **uvula** (YOU-vyoo-la; = little grape). During swallowing, the soft palate and uvula are drawn superiorly, closing off the nasopharynx. This prevents swallowed foods and liquids from entering the nasal cavity. Lateral to the base of the uvula are two muscular folds that run down the lateral sides of the soft palate. Anteriorly, the **palatoglossal arch (anterior pillar)** extends to the side of the base of the tongue. Posteriorly, the **palatopharyngeal** (PAL-a-tō-fa-rin′-jē-al) **arch (posterior pillar)** extends to the side of the pharynx. The palatine tonsils are situated between the arches, and the lingual tonsils are situated at the base of the tongue. At the posterior border of the soft palate, the mouth opens into the oropharynx through the fauces (Fig. 24.4).

Tongue

The **tongue**, together with its associated muscles, forms the floor of the oral cavity. It is composed of skeletal muscle covered with mucous membrane. The tongue is divided into symmetrical lateral halves by a median septum that extends throughout its entire length and is attached inferiorly to the hyoid bone, styloid process of the temporal bone, and mandible. Each half of the tongue consists of an identical complement of extrinsic and intrinsic muscles.

The **extrinsic muscles** of the tongue originate outside the tongue (to bones in the area) and insert into connective tissues in the tongue. They include the hyoglossus, genioglossus, and styloglossus (see Fig. 11.7). The extrinsic muscles move the tongue from side to side and in and out. These movements maneuver food for chewing, shape the food into a rounded mass, and force the food to the back of the mouth for swallowing. They also form the floor of the mouth and hold the tongue in position. The **intrinsic muscles** originate and insert into connective tissue within the tongue and alter the shape and size of the tongue for speech and swallowing. The intrinsic muscles include the longitudinalis superior, longitudinalis inferior, transversus linguae, and verticalis linguae. The **lingual** (*lingua* = tongue) **frenulum**, a fold of mucous membrane in the midline of the undersurface of the tongue, is attached to the floor of the mouth and aids in limiting the movement of the tongue posteriorly (Fig. 24.4).

The dorsum (upper surface) and lateral surfaces of the tongue are covered with **papillae** (pa-PIL-ē; = nipple-shaped projections; singular is **papilla**), projections of the lamina propria covered with epithelium (shown in Fig. 16.2a). **Filiform** (= threadlike) **papillae** are conical projections distributed in parallel rows over the anterior two-thirds of the tongue. They are whitish and contain no taste buds. **Fungiform** (= shaped like a mushroom) **papillae** are mush-

roomlike elevations distributed among the filiform papillae and are more numerous near the tip of the tongue. They appear as red dots on the surface of the tongue, and most of them contain taste buds. **Circumvallate** (*circum* = around; *vallum* = wall) **papillae** are arranged in the form of an inverted V on the posterior surface of the tongue and all of them contain taste buds. Ducts of lingual (von Ebner's) glands, which are serous glands, surround the circumvallate papillae. On the dorsum of the tongue are glands that secrete a digestive enzyme called **lingual lipase**, which initiates digestion of triglycerides into fatty acids and monoglycerides.

Salivary Glands

Saliva is a fluid that is continuously secreted into the mouth. It helps cleanse the mouth and teeth. Ordinarily, just enough saliva is secreted to keep the mucous membranes of the mouth and pharynx moist. When food enters the mouth, however, secretion of saliva increases. It lubricates, dissolves, and begins the chemical breakdown of the food. The mucous membrane lining the mouth contains many small glands. The **buccal glands** and minor salivary glands secrete small amounts of saliva. However, most saliva is secreted by the **salivary glands (major salivary glands)**, accessory structures that lie outside the mouth. Their secretions pour into ducts that empty into the oral cavity. There are three pairs of salivary glands: parotid, submandibular, and sublingual glands (Fig. 24.5a). The **parotid** (*para* = near; *otia* = ear) **glands** are located inferior and anterior to the ears between the skin and the masseter muscle. Each secretes into the oral cavity vestibule via a duct, called the **parotid (Stensen's) duct**, that pierces the buccinator muscle to open into the vestibule opposite the second maxillary (upper) molar tooth. The **submandibular glands** are found beneath the base of the tongue in the posterior part of the floor of the mouth (Fig. 24.5a). Their ducts, the **submandibular (Wharton's) ducts**, run under the mucosa on either side of the midline of the floor of mouth and enter the oral cavity proper lateral to the lingual frenulum. The **sublingual glands** are superior to the submandibular glands. Their ducts, the **lesser sublingual (Rivinus') ducts**, open into the floor of the mouth in the oral cavity proper.

MUMPS

Although any of the salivary glands may be the target of a nasopharyngeal infection, the mumps virus (myxovirus) typically attacks the parotid glands. **Mumps** is an inflammation and enlargement of the parotid glands accompanied by moderate fever, malaise, and extreme pain in the throat, especially when swallowing sour foods or acidic juices. Swelling occurs on one or both sides of the face, just anterior to the ramus of the mandible. In about 20–35% of males past puberty, the testes may also become inflamed, and, although it rarely occurs, sterility is a possible consequence. ■

Figure 24.5   Salivary glands. The submandibular gland shown in the photomicrograph in (b) consists mostly of serous acini (serous fluid-secreting exocrine portion of gland) and a few mucous acini (mucus-secreting exocrine portion of gland). The parotid glands consist of all serous acini, and the sublingual glands consist of mostly mucous acini and a few serous acini.

 Saliva lubricates and dissolves foods and begins the chemical breakdown of carbohydrates and lipids.

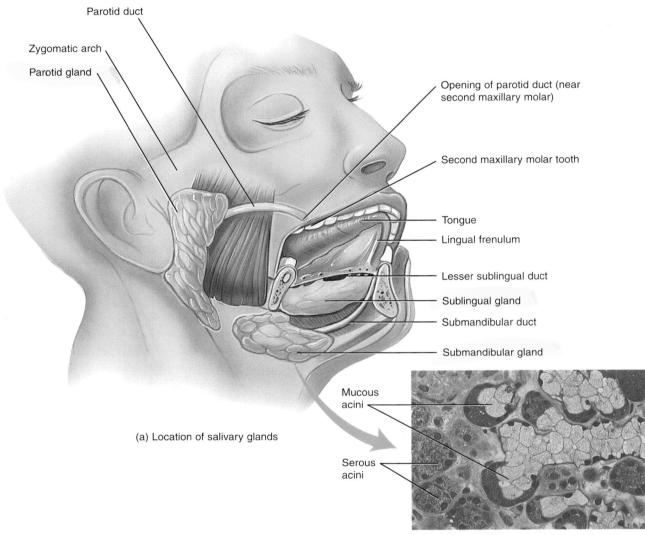

(a) Location of salivary glands

What is the function of the chloride ions in saliva?

(b) Photomicrograph of submandibular gland (350x)

Composition of Saliva

Chemically, **saliva** is 99½% water and ½% solutes. Among the solutes are ions, such as sodium, potassium, chloride, bicarbonate, and phosphates. Also present are some dissolved gases and various organic substances including urea and uric acid, serum albumin and globulin, mucus, the bacteriolytic enzyme lysozyme, and the digestive enzyme salivary amylase, which begins the digestion of carbohydrates.

Each salivary gland supplies different proportions of ingredients to saliva. The parotid glands contain cells that secrete a watery serous liquid containing salivary amylase. The submandibular glands contain cells similar to those found in the parotid glands as well as some mucous cells. Therefore they secrete a fluid that is thickened with mucus but still contains quite a bit of enzyme. The sublingual glands contain mostly mucous cells, so they secrete a much thicker fluid that contributes only a small amount of amylase to the saliva.

The water in saliva provides a medium for dissolving foods so they can be tasted and for initiating digestive reactions. Chloride ions in the saliva activate salivary amylase. Bicarbonate and phosphate ions buffer acidic foods that enter the mouth. As a result, saliva is only slightly acidic (pH of 6.35–6.85). Urea and uric acid are found in saliva because the saliva-producing glands (like the sweat glands of the skin) help remove waste molecules from the body. Mucus lubricates the food so it can easily be moved about in the mouth, formed into a ball, and swallowed. The enzyme lysozyme, which is found in small quantities, helps destroy bacteria. This contributes to protecting the mucous membrane from infection and the teeth from decay. However, it is not present in large enough quantities to eliminate all oral bacteria.

Salivation

Secretion of saliva or **salivation** (sal-i-VĀ-shun) is controlled by the nervous system. Amounts of saliva secreted daily vary considerably but average 1000–1500 ml (1–1.6 qt). Normally, parasympathetic stimulation promotes continuous secretion of a moderate amount of saliva. It keeps the mucous membranes moist and lubricates the movements of the tongue and lips during speech. The saliva is then swallowed and helps to moisten the esophagus. Eventually, most components of saliva are reabsorbed, which prevents fluid loss. Sympathetic stimulation dominates during stress, resulting in dryness of the mouth. During dehydration, the salivary and buccal glands stop secreting saliva to conserve water. The resulting dryness in the mouth contributes to the sensation of thirst. Drinking will then not only moisten the mouth but also restore the homeostasis of body water.

The touch and taste of food also are potent stimulators of salivary gland secretions. Chemicals in the food stimulate receptors in taste buds on the tongue. Impulses are conveyed from the receptors to two salivary nuclei in the brain stem. Returning parasympathetic impulses in fibers of the facial (VII) and glossopharyngeal (IX) nerves stimulate the secretion of saliva. Saliva continues to be secreted heavily some time after food is swallowed. This flow of saliva washes out the mouth and dilutes and buffers the chemical remnants of irritating substances.

The smell, sight, sound, or memory of food may also stimulate secretion of saliva. These stimuli constitute psychological activation and involve learned behavior. Memories that associate the stimuli with food are activated. Nerve impulses propagate from the cortex to the nuclei in the brain stem and the salivary glands are activated. Psychological activation of the glands has some benefit to the body because it allows chemical digestion to start in the mouth as soon as the food is ingested.

Salivation also occurs in response to swallowing irritating foods or during nausea. Reflexes originating in the stomach and upper small intestine stimulate salivation. This mechanism presumably helps to dilute or neutralize the irritating substance.

Teeth

The **teeth** (**dentes**) are accessory structures of the digestive system located in sockets of the alveolar processes of the mandible and maxillae. The alveolar processes are covered by the **gingivae** (jin-JI-vē; singular is **gingiva**), or gums (Fig. 24.6), which extend slightly into each socket forming the gingival sulcus. The sockets are lined by the **periodontal** (*peri* = around; *odous* = tooth) **ligament**, which consists of dense fibrous connective tissue and is attached to the socket walls and the cemental surface of the roots. Thus it anchors the teeth in position and also acts as a shock absorber during chewing.

A typical tooth consists of three principal regions. The **crown** is the visible portion above the level of the gums. Embedded in the socket are one to three **roots**. The **neck** is the constricted junction line of the crown and the root near the gum line.

Teeth are composed primarily of **dentin**, a calcified connective tissue layer that gives the tooth its basic shape and rigidity. The dentin encloses a cavity. The enlarged part of the cavity, the **pulp cavity**, lies in the crown and is filled with **pulp**, a connective tissue containing blood vessels, nerves, and lymphatic vessels. Narrow extensions of the pulp cavity run through the root of the tooth and are called **root canals**. Each root canal has an opening at its base, the **apical foramen**. Extending through the foramen are blood vessels, lymphatic vessels, and nerves.

The dentin of the crown is covered by a layer of **enamel** that consists primarily of calcium phosphate and calcium carbonate. Enamel is the hardest substance in the body and protects the tooth from the wear of chewing. It is also a barrier against acids that could easily dissolve the dentin. The dentin of the root is covered by **cementum**, another bonelike substance, which attaches the root to the periodontal ligament.

The branch of dentistry that is concerned with the prevention, diagnosis, and treatment of diseases that affect the pulp, root, periodontal ligament, and alveolar bone is known as **endodontics** (en′-dō-DON-tiks; *endo* = within). **Orthodontics** (or′-thō-DON-tiks; *ortho* = straight), by contrast, is a branch of dentistry that is concerned with the prevention and correction of abnormally aligned teeth.

ROOT CANAL THERAPY

Root canal therapy refers to a procedure, accomplished in several phases, in which all traces of pulp tissue are removed from the pulp cavity and root canals of a badly decayed tooth. After a hole is made in the tooth, the root canals are filed out and irrigated to remove bacteria. Then the canals are treated with medication and sealed tightly. The damaged crown is then repaired. ∎

Humans have two **dentitions**, or sets of teeth, deciduous and permanent. The first of these—the **deciduous** (*deciduus* = falling out) **teeth**, also called **primary teeth**, **milk teeth**, or **baby teeth**—begin to erupt at about 6 months of age, and one pair appears at about each month thereafter until

Figure 24.6   Parts of a typical tooth.

 Teeth are anchored in sockets of the alveolar processes of the mandible and maxilla.

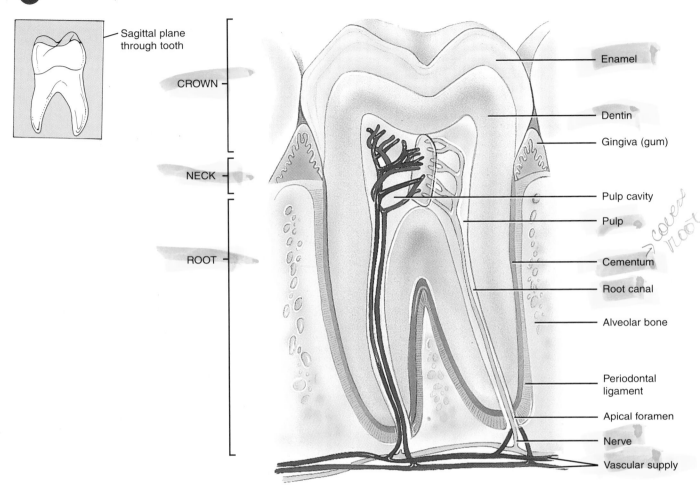

Sagittal plane through tooth

CROWN

NECK

ROOT

Enamel

Dentin

Gingiva (gum)

Pulp cavity

Pulp

Cementum

Root canal

Alveolar bone

Periodontal ligament

Apical foramen

Nerve

Vascular supply

Sagittal section of a mandibular (lower) molar

Q   What type of tissue is the main component of teeth?

all 20 are present. Figure 24.7 illustrates the deciduous teeth. The incisors, which are closest to the midline, are chisel-shaped and adapted for cutting into food. They are referred to as either **central** or **lateral incisors** on the basis of their position. Next to the incisors, moving posteriorly, are the **cuspids** (**canines**), which have a pointed surface called a cusp. Cuspids are used to tear and shred food. The incisors and cuspids have only one root apiece. Posterior to them lie the **first** and **second molars**, which have four cusps. Maxillary (upper) molars have three roots; mandibular (lower) molars have two roots. The molars crush and grind food.

All the deciduous teeth are lost—generally between 6 and 12 years of age—and are replaced by the **permanent** (**secondary**) **teeth** (Fig. 24.7b). The permanent dentition contains 32 teeth that erupt between age 6 and adulthood. The pattern resembles the deciduous dentition with the fol-

lowing exceptions. The deciduous molars are replaced with the **first** and **second premolars** (**bicuspids**), which have two cusps and one root (upper first bicuspids have two roots) and are used for crushing and grinding. The permanent molars erupt into the mouth posterior to the bicuspids. They do not replace any deciduous teeth and erupt as the jaw grows to accommodate them—the **first molars** at age 6, the **second molars** at age 12, the **third molars** (**wisdom teeth**) after age 17.

The human jaw often does not afford enough room posterior to the second molars for the eruption of the third molars. In this case, the third molars remain embedded in the alveolar bone and are said to be "impacted." Often they cause pressure and pain and must be removed surgically. In some people, third molars may be dwarfed in size or may not develop at all.

the stomach. When the rate of gastric emptying is slower, proportionally more alcohol will be absorbed and converted to acetaldehyde in the stomach and thus less alcohol will reach the bloodstream. After drinking the same amount of alcohol, females often experience higher blood alcohol levels and therefore become more drunk than males of comparable size for two reasons: (1) females typically have a smaller total body fluid volume (see Fig. 27.1) and (2) the activity of gastric alcohol dehydrogenase is up to 60% lower in females than in males. Asian males may also have lower levels of this gastric enzyme. ■

PANCREAS

The next organ of the GI tract involved in digestion and absorption of food is the small intestine. Chemical digestion there depends on activities of three accessory structures of the

digestive system: the pancreas, liver, and gallbladder. We first consider the activities of these accessory structures and then examine their contributions to digestion in the small intestine.

Anatomy

The **pancreas** (*pan* = all; *kreas* = flesh) is a retroperitoneal gland about 12–15 cm (5–6 in.) long and 2½ cm (1 in.) thick. It lies posterior to the greater curvature of the stomach and is connected, usually by two ducts, to the duodenum. The pancreas is divided into a head, body, and tail. The **head** is the expanded portion near the C-shaped curve of the duodenum. Located superior to and to the left of the head are the central **body** and the tapering **tail** (Fig. 24.16).

Pancreatic secretions pass from the secreting cells in the pancreas into small ducts. The small ducts ultimately unite to

Figure 24.16   Relation of pancreas to liver, gallbladder, and duodenum. The inset shows details of the common bile duct and pancreatic duct forming the hepatopancreatic ampulla (of Vater) and emptying into the duodenum.

🔑 *Pancreatic enzymes are involved in digestion of starches (polysaccharides), proteins, triglycerides, and nucleic acids.*

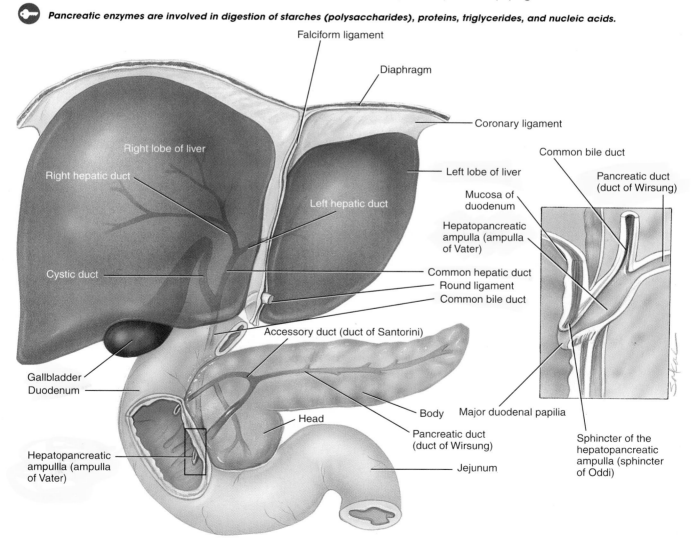

Anterior view

❓ What type of fluid would you find in the pancreatic duct? The common bile duct? The hepatopancreatic ampulla?

form two larger ducts that convey the secretions into the small intestine. The larger of the two ducts is called the **pancreatic duct** (**duct of Wirsung**). In most people, the pancreatic duct joins the common bile duct from the liver and gallbladder and enters the duodenum as a common duct called the **hepatopancreatic ampulla** (**ampulla of Vater**). The ampulla opens on an elevation of the duodenal mucosa known as the **major duodenal papilla**, about 10 cm (4 in.) inferior to the pyloric sphincter of the stomach. The smaller of the two ducts is the **accessory duct** (**duct of Santorini**). It leads from the pancreas and empties into the duodenum about 2½ cm (1 in.) superior to the hepatopancreatic ampulla.

Histology

The pancreas is made up of small clusters of glandular epithelial cells. About 99% of the cells are arranged in clusters called **acini** (AS-i-nī) and constitute the *exocrine* portion of the organ (see Fig. 18.23b,c). The cells within acini secrete a mixture of fluid and digestive enzymes called **pancreatic juice**. The remaining 1% of the cells are organized into clusters called **pancreatic islets** (**islets of Langerhans**). They form the *endocrine* portion of the pancreas and consist of cells that secrete the hormones glucagon, insulin, somatostatin, and pancreatic polypeptide. The functions of these hormones may be reviewed in Chapter 18.

Pancreatic Juice

Each day the pancreas produces 1200–1500 ml (about 1.3–1.6 qt) of pancreatic juice, which is a clear, colorless liquid. It consists mostly of water, some salts, sodium bicarbonate, and several enzymes. The sodium bicarbonate gives pancreatic juice a slightly alkaline pH (7.1–8.2) that buffers acidic gastric juice in chyme, stops the action of pepsin from the stomach, and creates the proper pH for action of digestive enzymes in the small intestine. The enzymes in pancreatic juice include a carbohydrate-digesting enzyme called **pancreatic amylase**; several protein-digesting enzymes called **trypsin** (TRIP-sin), **chymotrypsin** (kī′-mō-TRIP-sin), **carboxypeptidase** (kar-bok′-sē-PEP-ti-dās), and **elastase** (ē-LAS-tās); the principal triglyceride-digesting enzyme in an adult, called **pancreatic lipase**; and nucleic acid-digesting enzymes called **ribonuclease** and **deoxyribonuclease**.

Just as pepsin is produced in the stomach in an inactive form (pepsinogen), so too are the protein-digesting enzymes of the pancreas. This prevents the enzymes from digesting cells of the pancreas itself. Trypsin is secreted in an inactive form called **trypsinogen** (trip-SIN-ō-jen). Pancreatic acinar cells also secrete a protein called **trypsin inhibitor** that combines with any trypsin formed accidently in the pancreas or pancreatic juice and blocks its enzymatic activity. When trypsinogen reaches the lumen of the small intestine, it comes in contact with an activating brush border enzyme called **enterokinase** (en′-ter-ō-KĪ-nās), which splits off part of the trypsinogen molecule to form trypsin. In turn, trypsin acts on the inactive precursors (**chymotrypsinogen**,

procarboxypeptidase, and **proelastase**) to produce chymotrypsin, carboxypeptidase, and elastase, respectively.

PANCREATITIS

Inflammation of the pancreas, as may occur in association with mumps, is called **pancreatitis** (pan′-krē-a-TĪ-tis). In a more severe condition, known as **acute pancreatitis**, which is associated with heavy alcohol intake or biliary tract obstruction, the pancreatic cells may release trypsin instead of trypsinogen or insufficient amounts of trypsin inhibitor, and the trypsin begins to digest the pancreatic cells. The patient with acute pancreatitis usually responds to treatment, but recurrent attacks are the rule. ■

Regulation of Pancreatic Secretions

Pancreatic secretion, like gastric secretion, is regulated by both neural and hormonal mechanisms (Fig. 24.17).

Figure 24.17 Neural and hormonal factors that enhance secretion of pancreatic juice.

Parasympathetic stimulation (vagus nerves) and acidic chyme in the small intestine stimulate secretin release into the blood. Vagal stimulation and fatty acids and amino acids in chyme of the small intestine stimulate cholecystokinin (CCK) release into the blood.

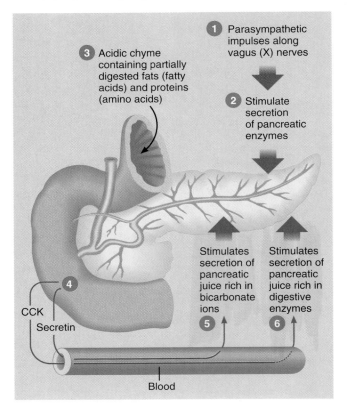

Why is secretion of pancreatic juice rich in bicarbonate ions helpful at this point in digestion?

1. During the cephalic and gastric phases of gastric digestion, parasympathetic impulses are transmitted along the vagus (X) nerves to the pancreas.
2. These nerve impulses stimulate increased secretion of pancreatic enzymes.
3. Acidic chyme containing partially digested fats and proteins enters the small intestine.
4. In response to fatty acids and amino acids, some enteroendocrine cells secrete cholecystokinin (CCK) into the blood. In response to acidic chyme, other enteroendocrine cells in the small intestinal mucosa liberate secretin into the blood.
5. Secretin stimulates the flow of pancreatic juice that is rich in bicarbonate ions.
6. Cholecystokinin (CCK) stimulates a pancreatic secretion rich in digestive enzymes (see Exhibit 24.4 on page 783).

LIVER

The **liver** is the heaviest gland of the body, weighing about 1.4 kg (about 3 lb) in an average adult, and after the skin is the second largest organ of the body. It is inferior to the diaphragm and occupies most of the right hypochondriac and part of the epigastric regions of the abdominopelvic cavity (see Fig. 1.11b).

Anatomy

The liver is almost completely covered by visceral peritoneum as well as completely covered by a dense irregular connective tissue layer that lies deep to the peritoneum. It is divided into two principal lobes—a large **right lobe** and a smaller **left lobe**—separated by the **falciform ligament** (Fig. 24.18). The right lobe is considered by many anatomists to include an inferior **quadrate lobe** and a posterior **caudate lobe**. However, on the basis of internal morphology, primarily the distribution of blood, the quadrate and caudate lobes more appropriately belong to the left lobe. The falciform ligament is a reflection of the parietal peritoneum. It extends from the undersurface of the diaphragm to the superior surface of the liver, between the two principal lobes of the liver, helping to suspend the liver. In the free border of the falciform ligament is the

Figure 24.18 External anatomy of the liver. The arrows in the insets in (a) and (c) indicate the directions from which the liver is viewed (posterior and inferior, respectively). The anterior view is illustrated in Fig. 24.16.

The liver is the heaviest gland of the body.

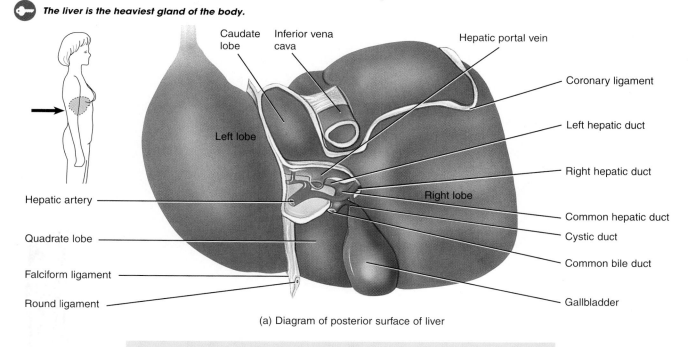

(a) Diagram of posterior surface of liver

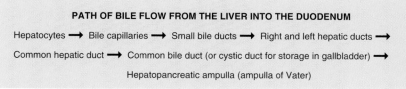

PATH OF BILE FLOW FROM THE LIVER INTO THE DUODENUM

Hepatocytes ➔ Bile capillaries ➔ Small bile ducts ➔ Right and left hepatic ducts ➔

Common hepatic duct ➔ Common bile duct (or cystic duct for storage in gallbladder) ➔

Hepatopancreatic ampulla (ampulla of Vater)

Figure continues

Figure 24.18 (continued)

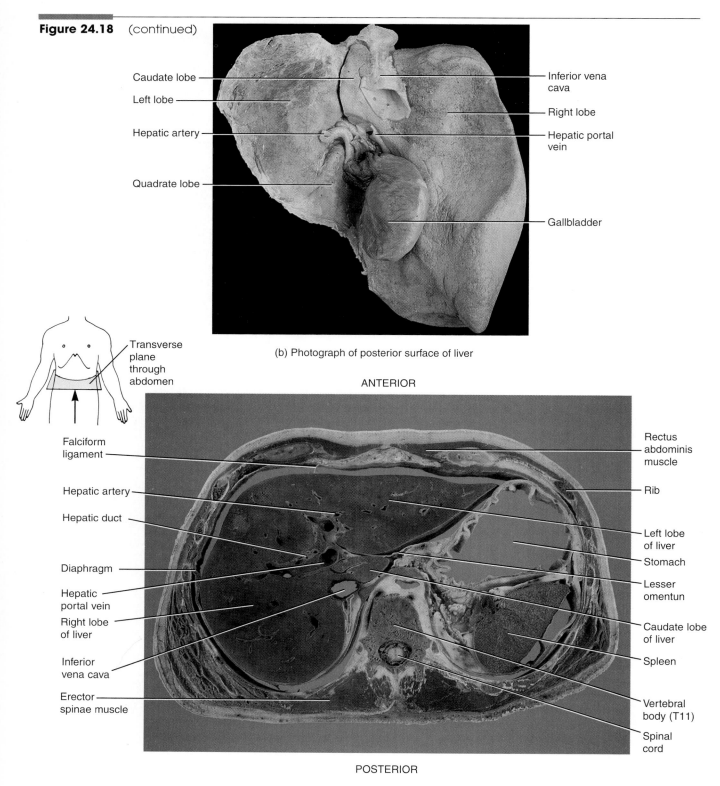

(b) Photograph of posterior surface of liver

(c) Photograph of an inferior view of a transverse section of the abdomen

Within which abdominopelvic region (see Fig. 1.11a) could you palpate (feel) most of the liver to decide if it is enlarged?

ligamentum teres (**round ligament**). It extends from the liver to the umbilicus. The ligamentum teres is a fibrous cord that is a remnant of the umbilical vein of the fetus (see Fig. 21.30a,b). The right and left **coronary ligaments** are narrow reflections of the parietal peritoneum that suspend the liver from the diaphragm.

Histology

The lobes of the liver are made up of many functional units called **lobules** (Fig. 24.19). A lobule consists of specialized epithelial cells, called **hepatocytes** (*hepatikos* = liver) arranged in irregular, branching, interconnected plates around a **central vein**. Rather than capillaries, the liver has larger spaces lined by endothelium called **sinusoids**, through which blood passes.

Figure 24.19 Histology of the liver.

🔑 *A lobule consists of hepatocytes arranged around a central vein.*

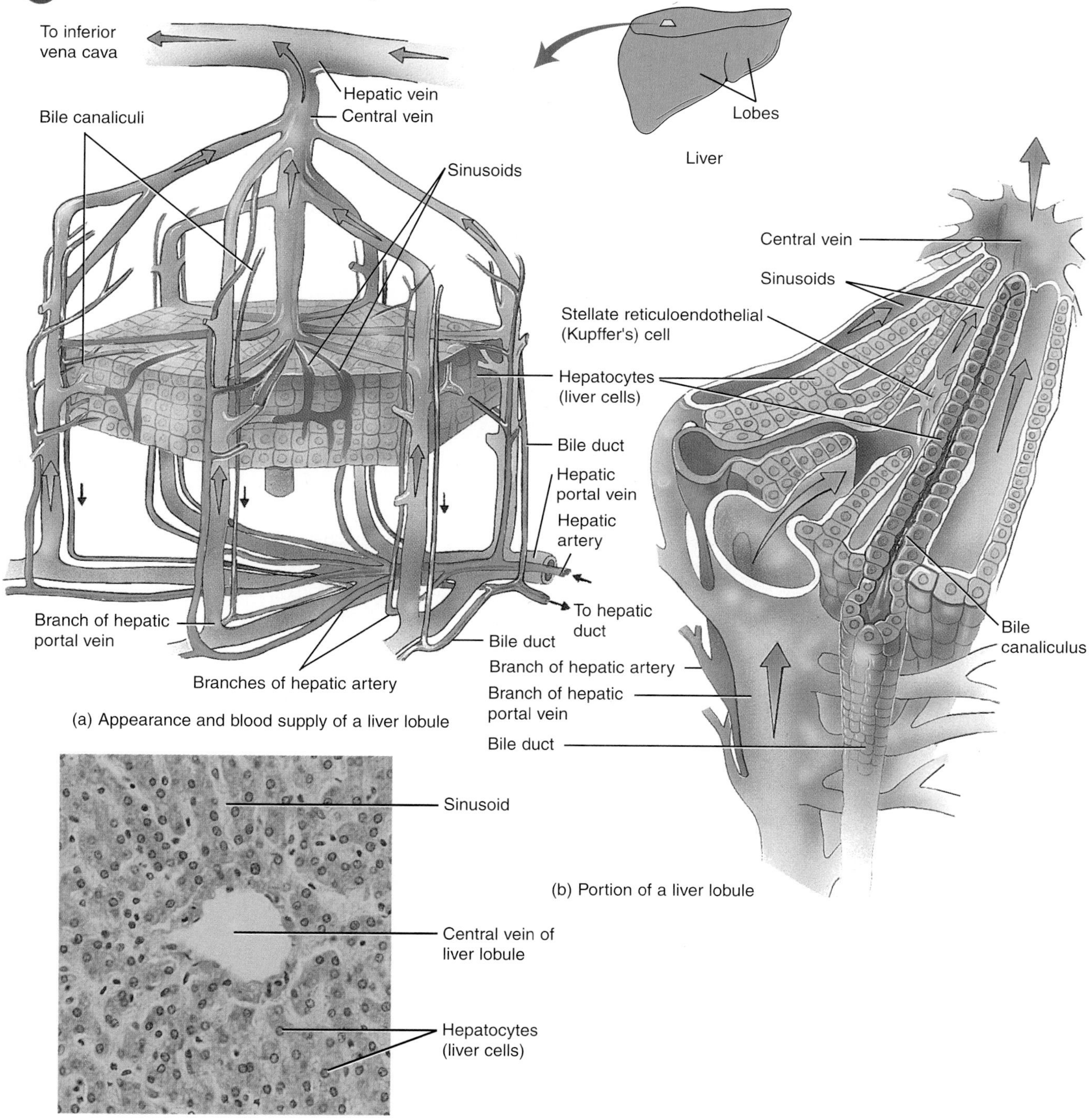

To inferior vena cava

Hepatic vein
Central vein

Bile canaliculi

Sinusoids

Lobes

Liver

Central vein

Sinusoids

Stellate reticuloendothelial (Kupffer's) cell

Hepatocytes (liver cells)

Bile duct

Hepatic portal vein

Hepatic artery

To hepatic duct

Branch of hepatic portal vein

Bile duct

Branches of hepatic artery

Bile canaliculus

Branch of hepatic artery

Branch of hepatic portal vein

Bile duct

(a) Appearance and blood supply of a liver lobule

(b) Portion of a liver lobule

Sinusoid

Central vein of liver lobule

Hepatocytes (liver cells)

(c) Photomicrograph of a portion of a liver lobule (300x)

❓ Which cells in the liver are phagocytic?

The sinusoids are also partly lined with **stellate reticulo-endothelial (Kupffer's) cells**. These phagocytes destroy worn-out white and red blood cells, bacteria, and other foreign matter in the blood draining the gastrointestinal tract.

Bile, secreted by hepatocytes, enters **bile canaliculi** (kan'-a-LIK-yoo-lī; = small canals), which are narrow intercellular canals that empty into small bile ducts (Fig. 24.19a). These small ducts eventually merge to form the larger **right** and **left hepatic ducts**, which unite and exit the liver as the **common hepatic duct** (see Fig. 24.16). Further on, the common hepatic duct joins the **cystic duct** from the gallbladder to form the **common bile duct**. Bile enters the cystic duct and is temporarily stored in the gallbladder. After a meal, various stimuli cause contraction of the gallbladder, which releases stored bile into the common bile duct.

Blood Supply

The liver receives blood from two sources. From the hepatic artery it obtains oxygenated blood, and from the hepatic portal vein it receives deoxygenated blood containing newly absorbed nutrients, drugs, and possibly microbes and toxins from the gastrointestinal tract (see Figs. 21.28 and 24.20). Branches of both the hepatic artery and the hepatic portal vein carry blood into liver sinusoids, where oxygen, most of the nutrients, and certain poisons are extracted by the hepatocytes. Products manufactured by the hepatocytes and nutrients needed by other cells are secreted back into the blood. The blood then drains into the central vein and eventually passes into a hepatic vein. Branches of the hepatic portal vein, hepatic artery, and bile duct typically accompany each other in their distribution through the liver. Collectively, these three structures are called a **portal triad** (Fig. 24.20).

LIVING-DONOR LIVER TRANSPLANT

In November 1989, surgeons at the University of Chicago Medical Center performed the first **living-donor liver transplant** in the United States. In the 14-hour procedure, a 21-month-old child received part of the left lobe of her mother's liver. The child suffered from biliary atresia, the closure or absence of some or all of the major bile ducts. Since the liver is capable of regeneration, the mother's liver returned to its normal size in about 2 months and the child's liver is growing as she does. Living-donor transplants have also been done for the lungs, pancreas, kidneys, and bone marrow. ■

Bile

Each day, hepatocytes secrete 800–1000 ml (about 1 qt) of **bile**, a yellow, brownish, or olive-green liquid. It has a pH of 7.6–8.6 and consists mostly of water and bile acids, bile salts, cholesterol, a phospholipid called lecithin, bile pigments, and several ions.

 Figure 24.20 Blood flow through the liver and return to the heart.

The liver receives oxygen-rich blood via the hepatic artery and nutrient-rich blood via the hepatic portal vein.

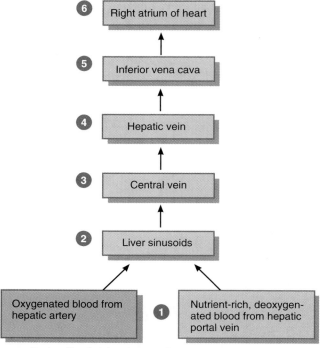

During the first few hours after a meal, how does the chemical composition of blood change as it flows through the liver sinusoids?

Bile is partially an excretory product and partially a digestive secretion. Bile salts, which are sodium and potassium salts of bile acids (mostly cholic acid and chenodeoxycholic acid), play a role in **emulsification**, the breakdown of large lipid globules into a suspension of droplets about 1 mm in diameter, and absorption of lipids following their digestion. The tiny lipid droplets present a very large surface area for the action of the pancreatic lipase necessary for rapid lipid digestion. Cholesterol is made soluble in bile by bile salts and lecithin.

The principal bile pigment is **bilirubin**. When worn-out red blood cells are broken down, iron, globin, and bilirubin (derived from heme) are released. The iron and globin are recycled, but some of the bilirubin is excreted into the bile ducts. Bilirubin eventually is broken down in the intestine, and one of its breakdown products—**stercobilin**—gives feces their normal brown color.

JAUNDICE

Jaundice (*jaune* = yellow) is a yellowish coloration of the sclerae of the eyes, skin, and mucous membranes due to a buildup of bilirubin (which is yellow in color) in the body. After bilirubin is formed from the breakdown of the heme pigment in aged red blood cells, it is transported to the liver where it is processed and eventually excreted into bile. The three main categories of jaundice are (1) *prehepatic jaundice,* due to excess production of bilirubin; (2) *hepatic jaundice,* due to congenital liver diseases, cirrhosis of the liver, or hepatitis; and (3) *extrahepatic jaundice,* due to blockage of bile drainage by gallstones or cancer of the bowel or the pancreas. Surgery must usually be performed to relieve the obstruction.

Because the liver of a newborn functions poorly for the first week or so, many babies experience a mild form of jaundice called *neonatal (physiological) jaundice* that disappears as the liver matures. Usually, it is treated by exposing the infant to blue light, which converts bilirubin into substances the kidneys can excrete. ■

Regulation of Bile Secretion

Both neural and hormonal factors regulate bile secretion. Following are some of the principal factors (Fig. 24.21):

① Parasympathetic impulses along the vagus (X) nerves can increase bile production to more than twice the baseline rate.

② Fatty acids and amino acids in chyme entering the duodenum stimulate enteroendocrine cells of the duodenum to secrete the hormone cholecystokinin (CCK) into the blood. Acidic chyme entering the duodenum stimulates enteroendocrine cells of the duodenum to secrete the hormone secretin into the blood.

③ CCK causes contraction of the wall of the gallbladder (and stimulates the production of pancreatic juice rich in digestive enzymes). The first action of CCK squeezes stored bile out of the gallbladder into the common bile duct. CCK also causes relaxation of the sphincter of the hepatopancreatic ampulla (see Fig. 24.16), which allows bile to flow into the duodenum.

④ Secretin stimulates the secretion of bile by hepatocytes that is rich in bicarbonate ions (HCO_3^-) (and the secretion of pancreatic juice that also is rich in HCO_3^-).

Within limits, as blood flow through the liver increases, so does the secretion of bile. Finally, the presence of large amounts of bile salts in the blood also increases the rate of bile production.

Functions of the Liver

The liver performs many vital functions, many of which are related to metabolism and are discussed in Chapter 25. Among the functions of the liver are the following:

1. **Carbohydrate metabolism.** The liver is especially important in maintaining a normal blood glucose level. When blood glucose level is low, the liver can break down glycogen to glucose (glycogenolysis) and release glucose into the bloodstream. The liver can also convert certain amino acids and lactic acid to glucose (gluconeogenesis), and convert other sugars, such as

---

Figure 24.21   Neural and hormonal stimuli that promote production and release of bile.

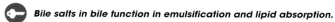

Bile salts in bile function in emulsification and lipid absorption.

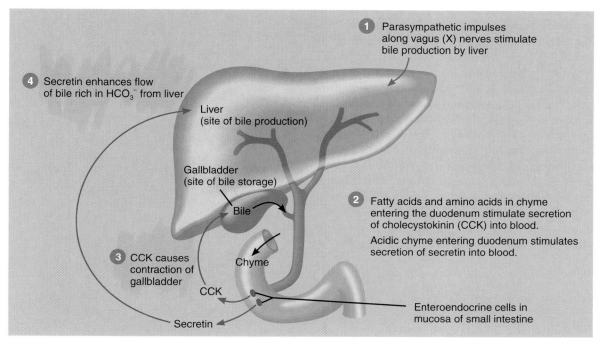

In what respect are the effects of secretin on the liver and on the pancreas similar?

fructose and galactose, into glucose. When blood glucose level is high, for example, just after eating a meal, the liver converts glucose to glycogen (glycogenesis) and to triglycerides (lipogenesis) for storage.

2. **Lipid metabolism.** The liver stores some triglycerides, breaks down fatty acids into acetyl coenzyme A, a process called beta oxidation, and converts excess acetyl coenzyme A into ketone bodies (ketogenesis). It synthesizes lipoproteins, which transport fatty acids, triglycerides, and cholesterol to and from body cells. Hepatocytes synthesize cholesterol and use cholesterol to make bile salts.

3. **Protein metabolism.** Without the role of the liver in protein metabolism, death would occur in a few days. Hepatocytes synthesize most plasma proteins, such as alpha and beta globulins, albumin, prothrombin, and fibrinogen. Also, liver enzymes can perform transamination, the transfer of an amino group from an amino acid to another substance to convert one amino acid into another. The liver deaminates (removes the amino group, NH_2, from) amino acids so that they can be used for ATP production or converted to carbohydrates or fats. It converts the resulting toxic ammonia (NH_3) into the much less toxic urea for excretion in urine. (Ammonia also is produced by bacteria in the GI tract.)

4. **Removal of drugs and hormones.** The liver can detoxify substances such as alcohol or excrete drugs such as penicillin, erythromycin, and sulfonamides into bile. It can also chemically alter or excrete thyroid hormones and steroid hormones, such as estrogens and aldosterone.

5. **Excretion of bilirubin.** As noted earlier, bilirubin, derived from the heme of worn-out red blood cells, is absorbed by the liver from the blood and secreted into bile. Most of the bilirubin in bile is metabolized in the intestine by bacteria and eliminated in feces.

6. **Synthesis of bile salts.** Bile salts are used in the small intestine for the emulsification and absorption of lipids, cholesterol, phospholipids, and lipoproteins.

7. **Storage.** In addition to glycogen, the liver stores vitamins (A, B_{12}, D, E, and K) and minerals (iron and copper). Hepatocytes contain a protein called **apoferritin** that combines with iron to form **ferritin**, the form in which iron is stored in the liver (see Fig. 19.4). The iron is released from the liver when needed elsewhere in the body.

8. **Phagocytosis.** The stellate reticuloendothelial (Kupffer's) cells of the liver phagocytize worn-out red and white blood cells and some bacteria.

9. **Activation of vitamin D.** The skin, liver, and kidneys participate in activating vitamin D (see page 124).

GALLBLADDER

The **gallbladder** (*galla* = bile) is a pear-shaped sac about 7–10 cm (3–4 in.) long. It is located in a depression on the posterior surface of the liver and usually hangs from the an-

terior inferior margin of the liver (see Figs. 24.16 and 24.18a).

Histology

The mucosa of the gallbladder consists of simple columnar epithelium arranged in rugae resembling those of the stomach. The gallbladder lacks a submucosa. The middle, muscular coat of the wall consists of smooth muscle fibers. Contraction of these fibers ejects the contents of the gallbladder into the **cystic** (*kystis* = bladder) **duct**. The outer coat is the visceral peritoneum.

Physiology

The functions of the gallbladder are to store and concentrate bile (up to tenfold) until it is needed in the small intestine. In the concentration process, water and ions are absorbed by the mucosa of the gallbladder. When the level of cholecystokinin (CCK) increases, the smooth muscle in the wall of the gallbladder contracts and forces bile into the cystic duct, through the common bile duct and into the small intestine. When the small intestine is empty, a valve around the hepatopancreatic ampulla (ampulla of Vater) called the **sphincter of the hepatopancreatic ampulla (sphincter of Oddi)** closes, and the backed-up bile flows into the cystic duct to the gallbladder for storage (see Fig. 24.16).

SUMMARY: DIGESTIVE HORMONES

The effects of the four major digestive hormones—gastrin, secretin, CCK, and GIP—and the stimuli that promote release of each are summarized in Exhibit 24.4. All four are secreted into the blood by enteroendocrine cells located in the GI tract mucosa. Gastrin comes from cells in the stomach and the other three from cells in the small intestine. Gastrin and gastric inhibitory peptide both exert their major effects on the stomach, whereas secretin and cholecystokinin affect the pancreas, liver, and gallbladder most strongly.

Stretching of the stomach as it receives food and buffering of gastric acids by proteins in food are stimuli that trigger release of gastrin. Gastrin, in turn, promotes secretion of gastric juice and increases gastric motility so that ingested food becomes well-mixed into a thick, soupy chyme. Reflux of the acidic chyme into the esophagus is prevented by tonic contraction of the lower esophageal sphincter, which is enhanced by gastrin.

Fatty acids (digested from triglycerides) and glucose (digested from disaccharides and polysaccharides) are dissolved in a thin layer of fluid next to mucosal epithelial cells in the first part of the small intestine. These products of digestion trigger release of GIP. In turn, GIP inhibits secretion

EXHIBIT **24.4** MAJOR HORMONES THAT CONTROL DIGESTIVE PROCESSES

Hormone	Stimulus and Site of Secretion	Actions
Gastrin	Distension of stomach, partially digested proteins and caffeine in stomach, and high pH of stomach chyme stimulate gastrin secretion by enteroendocrine G cells, located mainly in the mucosa of pyloric antrum.	*Major effects*: Promotes secretion of gastric juice, increases gastric motility, and promotes growth of gastric mucosa. *Minor effects*: Constricts lower esophageal sphincter; relaxes pyloric sphincter and ileocecal sphincter.
Gastric inhibitory peptide (GIP)	Fatty acids and glucose that enter the small intestine stimulate secretion of GIP by enteroendocrine K cells in the mucosa of the small intestine.	Stimulates release of insulin by pancreatic beta cells, inhibits secretion of gastric juice, and slows gastric emptying.
Secretin	Acidic (high H$^+$ level) chyme that enters the small intestine stimulates secretion of secretin by enteroendocrine S cells in the mucosa of the small intestine.	*Major effects*: Stimulates secretion of pancreatic juice and bile that are rich in HCO$_3^-$ (bicarbonate ions). *Minor effects*: Inhibits secretion of gastric juice, promotes normal growth and maintenance of the pancreas, and enhances effects of CCK.
Cholecystokinin (CCK)	Partially digested proteins (amino acids) and triglycerides (fatty acids) that enter the small intestine stimulate secretion of CCK by enteroendocrine CCK cells in the mucosa of the small intestine; CCK is also released in the brain.	*Major effects*: Stimulates secretion of pancreatic juice rich in digestive enzymes; causes ejection of bile from the gallbladder and opening of the sphincter of the hepatopancreatic ampulla (sphincter of Oddi); and induces satiety (feeling full to satisfaction). *Minor effects*: Inhibits gastric emptying, promotes normal growth and maintenance of the pancreas, and enhances effects of secretin.

of gastric juice and slows emptying of the stomach. GIP also stimulates the pancreas to secrete insulin, as if to "get ready" for the glucose that is about to be absorbed.

The major stimulus for secretion of secretin is acidic chyme (high concentration of H$^+$) entering the small intestine. In turn, secretin promotes secretion of bicarbonate ions (HCO$_3^-$) into pancreatic juice and bile. HCO$_3^-$ acts to buffer or soak up excess H$^+$ (see Figs. 24.17 and 24.20). Besides these major effects, secretin inhibits secretion of gastric juice and promotes normal growth and maintenance of the pancreas. It also enhances the effects of CCK. Overall, secretin causes buffering of acid in chyme that reaches the duodenum and slows production of acid in the stomach.

Amino acids from partially digested proteins and fatty acids from partially digested triglycerides stimulate secretion of cholecystokinin by enteroendocrine cells in the mucosa of the small intestine. CCK stimulates secretion of pancreatic juice that is rich in digestive enzymes (see Fig. 24.17) and ejection of bile into the duodenum (see Fig. 24.21). It also slows gastric emptying by promoting contraction of the pyloric sphincter and produces satiety (feeling full to satisfaction) by acting on the hypothalamus in the brain. Like secretin, CCK promotes normal growth and maintenance of the pancreas. It also enhances the effects of secretin.

In addition to the "big four," there are at least ten other so-called gut hormones, that is, hormones secreted by and having effects on the GI tract. A few of these are: motilin, substance P, and bombesin, which stimulate motility of the intestines; vasoactive intestinal polypeptide (VIP), which stimulates secretion of ions and water by the intestines and inhibits gastric acid secretion; gastrin-releasing peptide, which stimulates release of gastrin; and somatostatin, which inhibits gastrin release. Some are thought to act as local hormones (paracrines), while others are secreted into the blood or even into the lumen of the GI tract. The physiological roles of these and other gut hormones are still under investigation.

SMALL INTESTINE

The major events of digestion and absorption occur in a long tube called the **small intestine**. Since almost all the digestion and absorption of nutrients occur in the small intestine, its structure is specially adapted for this function. Its length alone provides a large surface area for digestion and absorption, and that area is further increased by circular folds, villi, and microvilli. The small intestine begins at the pyloric sphincter of the stomach, coils through the central and inferior part of the abdominal cavity, and eventually opens into the large intestine. It averages 2½ cm (1 in.) in diameter. The length is about 3 m (10 ft) in a living person and about 6½ m (21 ft) in a cadaver due to loss of smooth muscle tone after death.

Anatomy

The small intestine is divided into three segments (Fig. 24.22). The **duodenum** (doo'-ō-DĒ-num) is the shortest segment and is retroperitoneal. It starts at the pyloric sphincter of the stomach and extends about 25 cm (10 in.) until it merges with the jejunum. *Duodenum* means "12"; the structure is 12 fingers' breadth in length. The **jejunum** (jē-JOO-num) is about 1 m (3 ft) long and extends to the ileum. *Jejunum* means "empty," since at death it is found empty. The final portion of the small intestine, the **ileum** (IL-ē-um; *eileos* = twisted) measures about 2 m (6 ft) and joins the large intestine at the **ileocecal** (il'-ē-ō-SĒ-kal) **sphincter** (**valve**).

Projections called **circular folds**, or **plicae circulares** (PLĪ-kē SER-kyoo-lar-es), are permanent ridges in the mucosa, about 10 mm (0.4 in.) high (see Fig. 24.23c). Some extend all the way around the circumference of the intestine, and others extend only part of the way around. The circular folds begin near the proximal portion of the duodenum and end at about the midportion of the ileum. They enhance absorption by increasing surface area and causing the chyme to spiral, rather than to move in a straight line, as it passes through the small intestine.

Histology

The wall of the small intestine is composed of the same four coats that make up most of the GI tract. However, special features of both the mucosa and the submucosa facilitate the processes of digestion and absorption. The mucosa forms a se-

ries of fingerlike **villi** (= tuft of hair; singular is **villus**). These projections are ½–1 mm long and give the intestinal mucosa a velvety appearance (Fig. 24.23a). The large number of villi (20–40 per square millimeter) vastly increases the surface area of the epithelium available for absorption and digestion. Each villus has a core of lamina propria (areolar connective tissue). Embedded in this connective tissue are an arteriole, a venule, a capillary network, and a **lacteal** (LAK-tē-al), which is a lymphatic capillary. Nutrients absorbed by the epithelial cells covering the villus pass through the wall of a capillary or a lacteal to enter blood or lymph, respectively.

The epithelium of the mucosa consists of simple columnar epithelium that contains absorptive cells, goblet cells, enteroendocrine cells, and Paneth cells (Fig. 24.23b). The apical (free) membrane of the absorptive cells features **microvilli** (mī'-krō-VIL-ī). Each microvillus is a 1 μm-long cylindrical, membrane-covered projection that contains a bundle of 20–30 actin filaments (see Fig. 24.24c). In a photomicrograph taken through a light microscope, the microvilli are too small to be seen individually. They form a fuzzy line, called the **brush border**, at the apical surface of the absorptive cells, next to the lumen of the small intestine (see Fig. 24.24b). Larger amounts of digested nutrients can diffuse into the absorptive cells of the intestinal wall because the microvilli greatly increase the surface area of the plasma membrane. There are an estimated 200 million microvilli per square millimeter of small intestine. The brush border also contains several digestive enzymes.

The mucosa contains many cavities lined with glandular epithelium. Cells lining the cavities form the **intestinal glands** (**crypts of Lieberkühn**) and secrete intestinal juice. **Paneth**

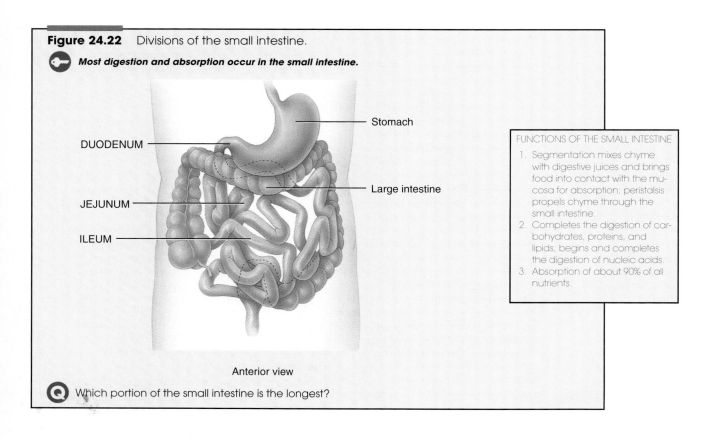

Figure 24.22   Divisions of the small intestine.

🔑 *Most digestion and absorption occur in the small intestine.*

Stomach

DUODENUM

Large intestine

JEJUNUM

ILEUM

FUNCTIONS OF THE SMALL INTESTINE

1. Segmentation mixes chyme with digestive juices and brings food into contact with the mucosa for absorption; peristalsis propels chyme through the small intestine.
2. Completes the digestion of carbohydrates, proteins, and lipids; begins and completes the digestion of nucleic acids.
3. Absorption of about 90% of all nutrients.

Anterior view

❓ Which portion of the small intestine is the longest?

Figure 24.23 Small intestine.

🔑 *Circular folds, villi, and microvilli increase the surface area of the small intestine for digestion and absorption.*

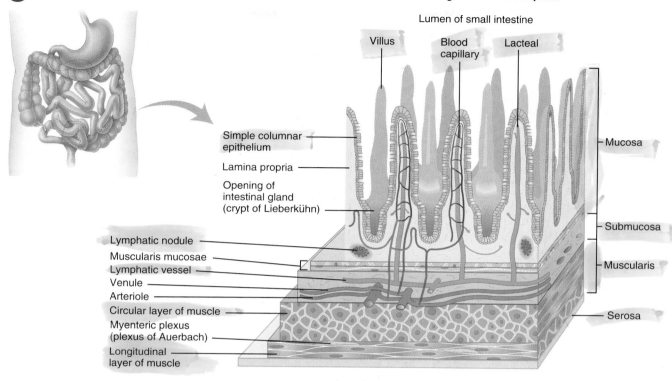

(a) Three-dimensional view of layers of the small intestine showing villi

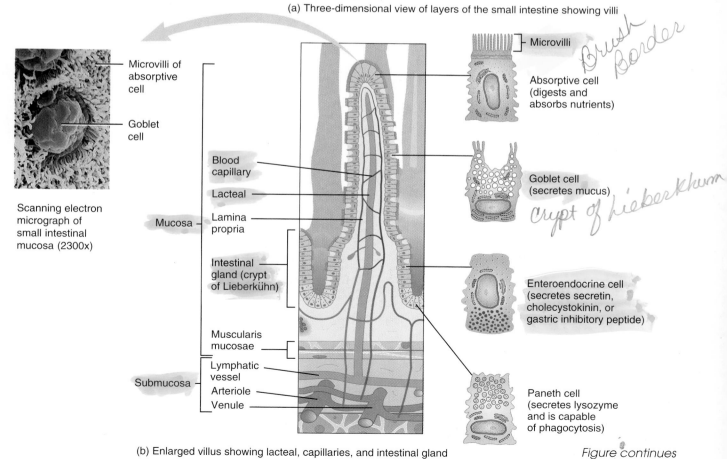

(b) Enlarged villus showing lacteal, capillaries, and intestinal gland

Figure continues

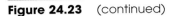

Figure 24.23 (continued)

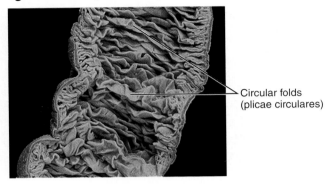

Circular folds
(plicae circulares)

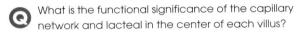

(c) Photograph of jejunum cut open to expose
the circular folds

Q What is the functional significance of the capillary
network and lacteal in the center of each villus?

cells are found in the deepest parts of the intestinal glands. They secrete lysozyme, a bactericidal enzyme, and are also capable of phagocytosis. They may have a role in regulating the microbial population in the intestines. Enteroendocrine cells, also in the deepest part of the intestinal glands, secrete three hormones: secretin (S cells), cholecystokinin (CCK cells), and gastric inhibitory peptide (K cells). Their functions are described in Exhibit 24.4 on page 783. The submucosa of the duodenum contains **duodenal** (**Brunner's**) **glands** (Fig. 24.24a). They secrete an alkaline mucus that helps neutralize gastric acid in the chyme. Many of the epithelial cells in the mucosa are goblet cells, which secrete additional mucus.

The lamina propria of the small intestine has an abundance of mucosa-associated lymphoid tissue (MALT). **Solitary lymphatic nodules** are most numerous in the distal part of the ileum. Groups of lymphatic nodules, referred to as **aggregated lymphatic follicles** (**Peyer's patches**), are numerous in the ileum. The muscularis mucosae consists of smooth muscle.

The **muscularis** of the small intestine consists of two layers of smooth muscle. The outer, thinner layer contains longitudinally arranged fibers. The inner, thicker layer contains circularly arranged fibers. Except for a major portion of the duodenum, the serosa (or visceral peritoneum) completely surrounds the small intestine.

Intestinal Juice and Brush Border Enzymes

Intestinal juice is a clear yellow fluid secreted in amounts of 1–2 liters (about 1–2 qt) a day. It has a slightly alkaline pH of 7.6 and contains water and mucus. Together, pancreatic and intestinal juices provide a vehicle for the absorption of substances from chyme as they come in contact with the microvilli. The absorptive epithelial cells synthesize several digestive enzymes, called **brush border enzymes**, and insert them in the plasma membrane of the microvilli. Thus some enzymatic digestion occurs at the surface of the epithelial cells that line the villi, rather than in the lumen exclusively, as in other parts of the GI tract. Among the brush border en-

zymes are four carbohydrate-digesting enzymes called α-**dextrinase**, **maltase**, **sucrase**, and **lactase**; protein-digesting enzymes called **peptidases** (**aminopeptidase** and **dipeptidase**); and two types of nucleotide-digesting enzymes, **nucleosidases** and **phosphatases**. Also, as small intestinal cells slough off into the lumen of the intestine, they break apart and release enzymes that digest nutrients in the chyme.

Physiology of Digestion in the Small Intestine

Mechanical Digestion

The movements of the small intestine are divided into two types: segmentation and peristalsis. **Segmentation** is the major movement of the small intestine. It is strictly a localized contraction in areas containing food. It mixes chyme with the digestive juices and brings the particles of food into contact with the mucosa for absorption. Segmentation starts with the contractions of circular muscle fibers in a portion of the small intestine, an action that constricts the intestine into segments. Next, muscle fibers that encircle the middle of each segment also contract, dividing each segment again. Finally, the fibers that contracted first relax, and each small segment unites with an adjoining small segment so that large segments are formed. This sequence of events is repeated 12–16 times a minute, sloshing the chyme back and forth. This movement is similar to alternately squeezing opposite ends of a tube of toothpaste.

Peristalsis propels the chyme onward through the intestinal tract. Peristaltic contractions in the small intestine are normally very weak compared with those in the esophagus or stomach, and chyme remains in the small intestine for 3–5 hours. Peristalsis, like segmentation, is controlled by the autonomic nervous system.

Chemical Digestion

In the mouth, salivary amylase converts starch (polysaccharide) to maltose (a disaccharide), maltotriose (a trisaccharide), and α-dextrins (short-chain, branched fragments of starch with five to ten glucose units). In the stomach, pepsin converts proteins to peptides (small fragments of proteins), and lingual and gastric lipases convert some triglycerides into fatty acids and monoglycerides. Thus chyme entering the small intestine contains partially digested carbohydrates, proteins, and lipids. The completion of the digestion of carbohydrates, proteins, and lipids is a collective effort of pancreatic juice, bile, and intestinal juice in the small intestine.

● **Carbohydrate Digestion** Even though the action of **salivary amylase** may continue in the stomach for awhile, the acidic pH of the stomach destroys salivary amylase and blocks its activity. Thus few starches are reduced to maltose by the time chyme leaves the stomach. Those starches not already broken down into maltose, maltotriose, and α-dextrins are cleaved by **pancreatic amylase**, an enzyme in pancreatic juice that acts in the small intestine. Although amylase acts on both glycogen and starches, it does not act on another polysaccharide called cellulose, an indigestible plant fiber. After amylase

Figure 24.24 Histology of the small intestine.

 Microvilli greatly increase the surface area of the small intestine for digestion and absorption.

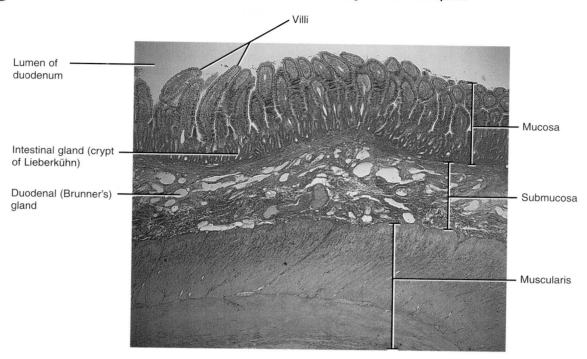

(a) Photomicrograph of a portion of the wall of the duodenum (160x)

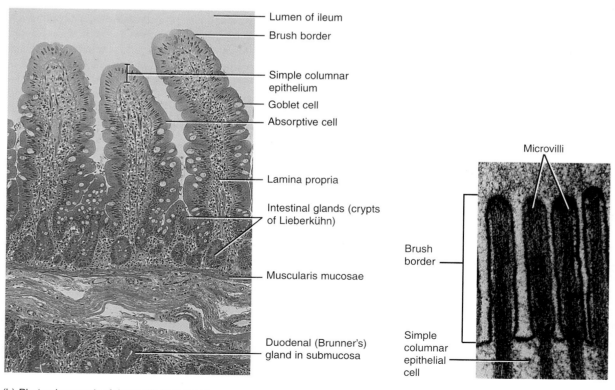

(b) Photomicrograph of three villi from the ileum of the small intestine (90x)

(c) Transmission electron micrograph of several microvilli from the duodenum (46,800x)

What is the function of the fluid secreted by duodenal (Brunner's) glands?

(either salivary or pancreatic) has split starch into smaller fragments, a brush border enzyme called α-**dextrinase** acts on the resulting α-dextrins, clipping off one glucose unit at a time.

Sucrose and lactose, two disaccharides, are ingested as such and are not acted on until they reach the small intestine. Three brush border enzymes digest the disaccharides into monosaccharides. **Maltase** splits maltose and maltotriose into two or three molecules of glucose, respectively. **Sucrase** breaks sucrose into a molecule of glucose and a molecule of fructose. **Lactase** digests lactose into a molecule of glucose and a molecule of galactose. This completes the digestion of carbohydrates since these monosaccharides are small enough to be absorbed.

LACTOSE INTOLERANCE

In some people, especially senior citizens and blacks, the mucosal cells of the small intestine fail to produce enough lactase, which is essential for the digestion of lactose. This results in a condition called **lactose intolerance**. Undigested lactose in chyme retains fluid, and bacterial fermentation of lactose results in the production of gases. Symptoms of lactose intolerance include diarrhea, gas, bloating, and abdominal cramps after consumption of milk and other dairy products. The severity of symptoms varies, from relatively minor to sufficiently serious to require medical attention. ■

● **Protein Digestion** Protein digestion starts in the stomach, where proteins are fragmented by the action of **pepsin** into peptides. Enzymes in pancreatic juice, **trypsin**, **chymotrypsin**, **carboxypeptidase** and **elastase**, continue to break down proteins into peptides. Although these enzymes all convert whole proteins into peptides, their actions differ somewhat because each splits peptide bonds between different amino acids. Pancreatic **carboxypeptidase** acts on peptides and breaks the peptide bond that attaches the terminal amino acid to the carboxyl (acid) end of the peptide. Protein digestion is completed by the **peptidases** in the brush border: aminopeptidase and dipeptidase. **Aminopeptidase** acts on peptides and breaks the peptide bond that attaches the terminal amino acid to the amino end of the peptide. **Dipeptidase** splits dipeptides (two amino acids joined by a peptide bond) into single amino acids.

Because trypsin, chymotrypsin, and elastase act on interior peptide bonds of a protein molecule, they are termed **endopeptidases**. Carboxypeptidases and aminopeptidases, which hydrolyze amino acids at the carboxyl end and amino end of proteins, respectively, are known as **exopeptidases**.

● **Lipid Digestion** The most abundant lipids in the diet are the triglycerides, which consist of a molecule of glycerol bonded to three molecules of fatty acid (see Fig. 2.9). Enzymes that split triglycerides and phospholipids are called **lipases**. In an adult, most lipid digestion occurs in the small intestine, although some occurs in the stomach through the action of **lingual** and **gastric lipases**. When chyme enters the small intestine, bile salts break the globules of triglycerides into droplets about 1 mm in diameter. This process is called **emulsification**. It increases the surface area exposed to **pancreatic lipase**, an enzyme found in pancreatic juice. The enzyme thus

can hydrolyze more triglycerides into fatty acids and monoglycerides, the end products of triglyceride digestion. Lipases remove two of the three fatty acids from glycerol; the third remains attached to the glycerol, thus forming a monoglyceride.

● **Nucleic Acid Digestion** Pancreatic juice contains two nucleases called **ribonuclease**, which digests RNA, and **deoxyribonuclease**, which digests DNA. The nucleotides that result from the action of the two nucleases are further digested by brush border enzymes called **nucleosidases** and **phosphatases** into pentoses, phosphates, and nitrogenous bases. These products are absorbed via active transport.

A summary of digestive enzymes in terms of source, substrate acted on, and product is presented in Exhibit 24.5.

Regulation of Intestinal Secretion and Motility

The most important means for regulating small intestinal secretion and motility are local neural reflexes that respond to the presence of chyme. Also, the hormone known as vasoactive intestinal polypeptide (VIP) stimulates the production of intestinal juice. Segmentation movements depend mainly on intestinal distension, which initiates nerve impulses to the submucosal and myenteric plexuses and the central nervous system. Local reflexes and returning parasympathetic impulses from the CNS increase motility. Sympathetic impulses decrease intestinal motility. Peristalsis increases when most nutrients and water have been absorbed, that is, when the walls of the small intestine are stretched less. With more vigorous peristalsis, the chyme moves along toward the large intestine as fast as 10 cm/sec. The first remnants of a meal reach the beginning of the large intestine in about 4 hours.

Physiology of Absorption in the Small Intestine

All the chemical and mechanical phases of digestion from the mouth through the small intestine are directed toward changing food into forms that can pass through the epithelial cells lining the mucosa into the underlying blood and lymphatic vessels. These forms are monosaccharides (glucose, fructose, and galactose) from carbohydrates; single amino acids, dipeptides, and tripeptides from proteins; and fatty acids, glycerol, and monoglycerides from triglycerides. Passage of these digested nutrients from the gastrointestinal tract into the blood or lymph is called **absorption**.

About 90% of all absorption of nutrients takes place in the small intestine. The other 10% occurs in the stomach and large intestine. Any undigested or unabsorbed material left in the small intestine passes on to the large intestine. Absorption of materials occurs by diffusion, facilitated diffusion, osmosis, and active transport.

Carbohydrate Absorption

Essentially all carbohydrates are absorbed as monosaccharides. They pass through the apical (free) surface by

EXHIBIT 24.5 SUMMARY OF DIGESTIVE ENZYMES

Enzyme	Source	Substrate	Product
SALIVA			
Salivary amylase	Salivary glands.	Starches (polysaccharides).	Maltose (disaccharide), maltotriose (trisaccharide), and α-dextrins.
Lingual lipase	Glands in the tongue.	Triglycerides (fats and oils) and other lipids.	Fatty acids and monoglycerides.
GASTRIC JUICE			
Pepsin (activated from pepsinogen by pepsin and hydrochloric acid)	Stomach chief (zymogenic) cells.	Proteins.	Peptides.
Gastric lipase	Stomach chief (zymogenic) cells.	Short-chain triglycerides (fats and oils) in butterfat molecules in milk.	Fatty acids and monoglycerides.
PANCREATIC JUICE			
Pancreatic amylase	Pancreatic acinar cells.	Starches (polysaccharides).	Maltose (disaccharide), maltotriose (trisaccharide), and α-dextrins.
Trypsin (activated from trypsinogen by enterokinase)	Pancreatic acinar cells.	Proteins.	Peptides.
Chymotrypsin (activated from chymotrypsinogen by trypsin)	Pancreatic acinar cells.	Proteins.	Peptides.
Elastase (activated from proelastase by trypsin)	Pancreatic acinar cells.	Proteins.	Peptides.
Carboxypeptidase (activated from procarboxypeptidase by trypsin)	Pancreatic acinar cells.	Terminal amino acid at carboxyl (acid) end of peptides.	Peptides and amino acids.
Pancreatic lipase	Pancreatic acinar cells	Triglycerides (fats and oils) that have been emulsified by bile salts.	Fatty acids and monoglycerides.
Nucleases			
Ribonuclease	Pancreatic acinar cells.	Ribonucleic acid.	Nucleotides.
Deoxyribonuclease	Pancreatic acinar cells.	Deoxyribonucleic acid.	Nucleotides.
BRUSH BORDER			
α-Dextrinase	Small intestine.	α-Dextrins.	Glucose.
Maltase	Small intestine.	Maltose.	Glucose.
Sucrase	Small intestine.	Sucrose.	Glucose and fructose.
Lactase	Small intestine.	Lactose.	Glucose and galactose.
Enterokinase	Small intestine.	Trypsinogen.	Trypsin.
Peptidases			
Aminopeptidase	Small intestine.	Terminal amino acid at amino end of peptides.	Peptides and amino acids.
Dipeptidase	Small intestine.	Dipeptides.	Amino acids.
Nucleosidases and phosphatases	Small intestine.	Nucleotides.	Nitrogenous bases, pentoses, and phosphates.

facilitated diffusion or *active transport*. Fructose, a monosaccharide found in fruits, is transported by *facilitated diffusion*. Glucose and galactose are transported into epithelial cells of the villi by *secondary active transport* that is coupled to the active transport of Na⁺. The transporter has binding sites for glucose and Na⁺. Unless both sites are filled, neither substance is transported. Galactose competes with glucose to ride the same transporter. Since both Na⁺ and

glucose or galactose move in the same direction, this is a *symporter*. Monosaccharides then move out of the epithelial cells through their sides and basal surface by *facilitated diffusion* and enter the capillaries of the villi (Fig. 24.25a,b).

Protein Absorption
Most proteins are absorbed as amino acids by *active transport* processes that occur mainly in the duodenum and jejunum.

About half of the absorbed amino acids are present in food, whereas the other half comes from proteins in digestive juices and dead cells that slough off the mucosal surface. There are several transporters that carry different types of amino acids. Some amino acids enter epithelial cells of the villi by Na^+-dependent secondary active transport processes that are similar to the glucose transporter. Other amino acids are actively transported by themselves. At least one secondary active transport process brings in dipeptides and tripeptides together with H^+. The peptides then are hydrolyzed to single amino acids inside the epithelial cells. Amino acids move out of the epithelial cells by diffusion and enter capillaries of the villus (Fig. 24.25a,b). Both monosaccharides and amino acids are transported in the blood to the liver by way of the hepatic portal system. If not removed by hepatocytes, they will then enter the general circulation.

Lipid Absorption

Dietary lipids are all absorbed by *simple diffusion.* As a result of their emulsification and digestion, triglycerides are broken down into monoglycerides and fatty acids. Recall that lipase removes two of the three fatty acids from glycerol during digestion of a triglyceride; the other fatty acid remains attached to glycerol, thus forming a monoglyceride. Short-chain fatty acids having fewer than 10–12 carbon atoms pass into the ep-

ithelial cells by simple diffusion and follow the same route taken by monosaccharides and amino acids into a blood capillary of a villus (Fig. 24.25a,b).

Most fatty acids, however, are long-chain fatty acids. They and the monoglycerides reach the bloodstream by a different route and require bile for adequate absorption. Bile salts are amphipathic; they have both polar (hydrophilic) and nonpolar (hydrophobic) portions. Thus they can form tiny spheres called **micelles** (mi-SELZ), which are 2–10 nm in diameter and include 20–50 bile salt molecules. Because they are small and have the polar portions of bile salt molecules at their surface, micelles can dissolve in the water of intestinal fluid. On the other hand, dietary triglycerides can dissolve in the nonpolar central core of micelles. It is in this form that fatty acids and monoglycerides reach the epithelial cells of the villi.

At the apical surface of the epithelial cells, fatty acids and monoglycerides diffuse into the cells, leaving the micelles behind in chyme. The micelles continually repeat this ferrying function. When chyme reaches the ileum, 90–95% of the bile salts are reabsorbed and returned by the blood to the liver through the hepatic portal system for resecretion. This cycle is called the **enterohepatic circulation.** Insufficient bile salts, due to obstruction of the bile ducts or removal of the gallbladder, can result in the loss of up to 40% of dietary

---

Figure 24.25   Absorption of digested nutrients in the small intestine. For simplicity, all digested foods are shown in the lumen of the small intestine, even though some nutrients are digested by brush border enzymes.

🔑 *Long-chain fatty acids and monoglycerides are absorbed into lacteals; other products of digestion enter blood capillaries.*

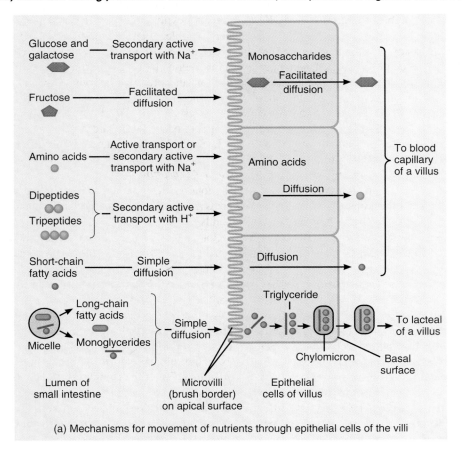

(a) Mechanisms for movement of nutrients through epithelial cells of the villi

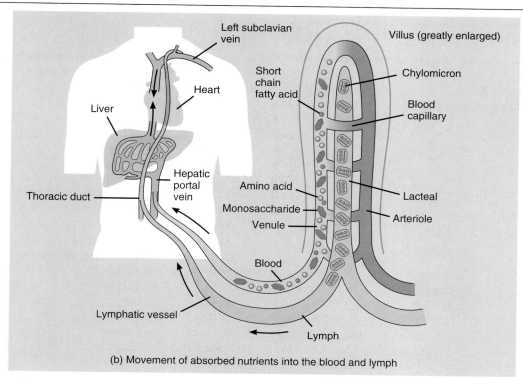

(b) Movement of absorbed nutrients into the blood and lymph

Q A monoglyceride may be larger than an amino acid. Why can monoglycerides be absorbed by simple diffusion whereas amino acids cannot?

lipids due to diminished absorption. Also, when lipids are not absorbed properly, the fat-soluble vitamins (A, D, E, K) are not adequately absorbed.

Within the epithelial cells, many monoglycerides are further digested by lipase in the cells to glycerol and fatty acids. Then the fatty acids and glycerol are recombined to form triglycerides, aggregate into globules along with phospholipids and cholesterol, and become coated with proteins. These large spherical masses, about 80 nm in diameter, are called **chylomicrons**. The protein coat keeps the chylomicrons suspended and prevents their sticking to each other. Chylomicrons leave the epithelial cell by exocytosis. Because they are so large and bulky, chylomicrons cannot enter blood capillaries in the small intestine. Rather they enter the much more leaky lacteals. From here, they are transported by way of lymphatic vessels to the thoracic duct and enter the blood at the left subclavian vein.

Within 10 minutes after their absorption, half the chylomicrons have already been removed from the blood as they pass through blood capillaries in the liver and adipose tissue. This is accomplished by an enzyme called **lipoprotein lipase** found in capillary endothelial cells. The enzyme breaks down triglycerides in chylomicrons and other lipoproteins into fatty acids and glycerol. The fatty acids diffuse into hepatocytes and adipose cells and combine with glycerol during resynthesis of triglycerides. Two or three hours after a meal, few chylomicrons remain in the blood.

The plasma lipids—fatty acids, triglycerides, and cholesterol—are insoluble in water and body fluids. To be transported in blood and utilized by body cells, the lipids must be combined with protein carriers, called **apoproteins**, to make them soluble. The combination of lipid and protein is referred to as a **lipoprotein**. Most lipoproteins are synthesized in the liver and all contain triglycerides, phospholipids, cholesterol, and protein in varying proportions. In addition to chylomicrons, there are several other types of lipoproteins, known as **high-density lipoproteins (HDLs)**, **low-density lipoproteins (LDLs)**, and **very low-density lipoproteins (VLDLs)**. Their functions may be reviewed on page 602.

Water Absorption

The total volume of fluid that enters the small intestine each day is about 9.3 liters (about 9.8 qt). It comes from ingestion of liquids (about 2.3 liters) and from various gastrointestinal secretions (about 7.0 liters). Figure 24.26 reviews the fluid input to the GI tract. The small intestine absorbs about 8.3 liters of the fluid; the remainder passes into the large intestine. There, most of the rest of it, about 0.9 liter, is also absorbed. Only 0.1 liter (100 ml) per day of water is excreted in the feces.

All water absorption in the GI tract occurs by *osmosis* from the lumen of the intestines through epithelial cells and into blood capillaries. Water can move across the intestinal mucosa in both directions. The absorption of water from the small intestine depends on the absorption of electrolytes and nutrients to maintain an osmotic balance with the blood. The absorbed electrolytes, monosaccharides, and amino acids establish a concentration gradient for water that promotes water absorption by osmosis.

Figure 24.26   Daily volumes of fluid ingested, secreted, absorbed, and excreted from the GI tract.

🔑 *All water absorption in the GI tract occurs by osmosis.*

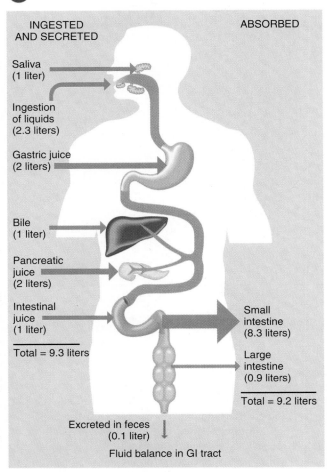

Which two organs of the digestive system secrete the most fluid?

Electrolyte Absorption

Many of the electrolytes absorbed by the small intestine come from gastrointestinal secretions and some are part of ingested foods and liquids. Sodium ions are actively transported out of intestinal epithelial cells by sodium (Na^+/K^+ ATPase) pumps after they have moved into epithelial cells by diffusion and secondary active transport. Thus most of the Na^+ in gastrointestinal secretions is reclaimed and not lost in the feces. Chloride, iodide, and nitrate ions can passively follow sodium ions or be actively transported. Calcium ions also are absorbed actively, and their movement depends on parathyroid hormone (PTH) and vitamin D. Other electrolytes such as iron, potassium, magnesium, and phosphate ions also are absorbed by active transport.

Vitamin Absorption

Fat-soluble vitamins (A, D, E, and K) are included along with ingested dietary lipids in micelles and are absorbed by simple diffusion. Most water-soluble vitamins, such as most

B vitamins and vitamin C, also are absorbed by simple diffusion. Vitamin B_{12}, however, combines with intrinsic factor produced by the stomach and the combination is absorbed in the ileum by receptor-mediated endocytosis.

A summary of the digestive and absorptive activities of the small intestine and associated accessory structures is presented in Exhibit 24.6.

EXHIBIT 24.6	SUMMARY OF DIGESTIVE ACTIVITIES IN THE PANCREAS, LIVER, GALLBLADDER, AND SMALL INTESTINE
Structure	**Activity**
PANCREAS	Delivers pancreatic juice into the duodenum via the pancreatic duct (see Exhibit 24.5 for pancreatic enzymes and their functions).
LIVER	Produces bile (bile salts), necessary for emulsification and absorption of fats.
GALLBLADDER	Stores, concentrates, and delivers bile into the duodenum via the common bile duct.
SMALL INTESTINE	Major site of digestion and absorption of nutrients and water in the gastrointestinal tract.
Mucosa/submucosa	
Intestinal glands	Secrete intestinal juice.
Duodenal (Brunner's) glands	Secrete alkaline fluid to buffer stomach acids and mucus for protection and lubrication.
Microvilli	Microscopic, membrane-covered projections of epithelial cells that contain brush border enzymes (listed in Exhibit 24.5) and increase surface area for absorption and digestion.
Villi	Fingerlike projections of mucosa that are the sites of absorption of digested food and also increase the surface area for digestion and absorption.
Circular folds	Folds of mucosa and submucosa that increase surface area for absorption and digestion.
Muscularis	
Segmentation	Consists of alternating contractions of circular fibers that produce segmentation and resegmentation of portions of the small intestine; mixes chyme with digestive juices and brings food into contact with the mucosa for absorption.
Peristalsis	Consists of mild waves of contraction and relaxation of circular and longitudinal muscles passing the length of the small intestine; moves chyme toward ileocecal sphincter.

LARGE INTESTINE

The overall functions of the large intestine are the completion of absorption, the manufacture of certain vitamins, the formation of feces, and the expulsion of feces from the body.

Anatomy

The **large intestine** is about 1½ m (5 ft) long and 6½ cm (2½ in.) in diameter. It extends from the ileum to the anus and is attached to the posterior abdominal wall by its **mesocolon**, which is a double layer of peritoneum. Structurally, the large intestine is divided into four principal regions: cecum, colon, rectum, and anal canal (Fig. 24.27a).

The opening from the ileum into the large intestine is guarded by a fold of mucous membrane called the **ileocecal sphincter** (**valve**). This structure allows materials from the small intestine to pass into the large intestine. Hanging infe-

rior to the ileocecal valve is the **cecum**, a blind pouch about 6 cm (2½ in.) long. Attached to the cecum is a twisted, coiled tube, measuring about 8 cm (3 in.) in length, called the **appendix** or **vermiform appendix** (*vermis* = worm; *appendix* = appendage). The mesentery of the appendix, called the **mesoappendix**, attaches the appendix to the inferior part of the ileum and adjacent part of the posterior abdominal wall.

APPENDICITIS

Inflammation of the appendix is termed **appendicitis**. It is preceded by obstruction of the lumen of the appendix by fecal material, inflammation, a foreign body, carcinoma of the cecum, stenosis, or kinking of the organ. The infection that follows may result in edema, ischemia, gangrene, and perforation. Rupture of the appendix develops into peritonitis. Typically, appendicitis begins with referred pain in the umbilical region of the abdomen, followed by

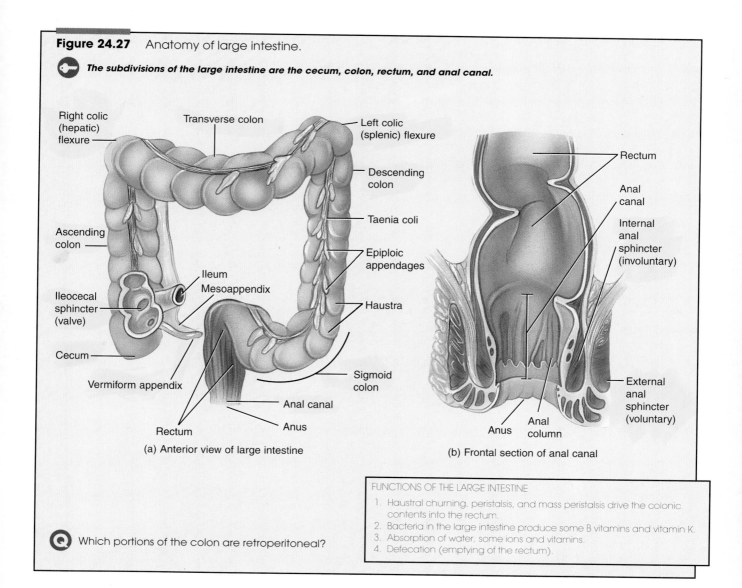

Figure 24.27 Anatomy of large intestine.

The subdivisions of the large intestine are the cecum, colon, rectum, and anal canal.

Right colic (hepatic) flexure
Transverse colon
Left colic (splenic) flexure
Descending colon
Taenia coli
Epiploic appendages
Ascending colon
Ileum
Mesoappendix
Haustra
Ileocecal sphincter (valve)
Cecum
Vermiform appendix
Sigmoid colon
Rectum
Anal canal
Anus

(a) Anterior view of large intestine

Rectum
Anal canal
Internal anal sphincter (involuntary)
Anal column
Anus
External anal sphincter (voluntary)

(b) Frontal section of anal canal

Which portions of the colon are retroperitoneal?

FUNCTIONS OF THE LARGE INTESTINE
1. Haustral churning, peristalsis, and mass peristalsis drive the colonic contents into the rectum.
2. Bacteria in the large intestine produce some B vitamins and vitamin K.
3. Absorption of water, some ions and vitamins.
4. Defecation (emptying of the rectum).

anorexia (lack or loss of appetite for food), nausea, and vomiting. After several hours, the pain localizes in the right lower quadrant (RLQ) and is continuous, dull or severe, and intensified by coughing, sneezing, or body movements. Early appendectomy (removal of the appendix) is recommended in all suspected cases because it is safer to operate than to risk gangrene, rupture, and peritonitis. ■

The open end of the cecum merges with a long tube called the **colon** (*kolon* = food passage). The colon is divided into ascending, transverse, descending, and sigmoid portions. The ascending and descending colon are retroperitoneal whereas the transverse and sigmoid colon are not. The **ascending colon** ascends on the right side of the abdomen, reaches the inferior surface of the liver, and turns abruptly to the left. Here it forms the **right colic (hepatic) flexure**. The colon continues across the abdomen to the left side as the **transverse colon**. It curves beneath the inferior end of the spleen on the left side as the **left colic (splenic) flexure** and passes inferiorly to the level of the iliac crest as the **descending colon**. The **sigmoid colon** begins near the left iliac crest, projects medially to the midline, and terminates as the rectum at about the level of the third sacral vertebra.

The **rectum**, the last 20 cm (8 in.) of the GI tract, lies anterior to the sacrum and coccyx. The terminal 2–3 cm (1 in.) of the rectum is called the **anal canal** (Fig. 24.27b). The mucous membrane of the anal canal is arranged in longitudinal folds called **anal columns** that contain a network of arteries and veins. The opening of the anal canal to the exterior is called the **anus**. It is guarded by an internal sphincter of smooth muscle (involuntary) and an external sphincter of skeletal muscle (voluntary). Normally the anus is closed except during the elimination of feces.

HEMORRHOIDS

Varicosities in veins are regions that are enlarged and inflamed. In the rectal veins, varicosities are known as **hemorrhoids (piles)**. Hemorrhoids develop when the veins are put under pressure and become engorged with blood. If the pressure continues, the wall of the vein stretches. Such a distended vessel oozes blood, and bleeding or itching are usually the first signs that a hemorrhoid has developed. Stretching of a vein also favors clot formation, further aggravating swelling and pain. Hemorrhoids may be caused by constipation, which may be brought on by low-fiber diets. Also, repeated straining during defecation forces blood into the rectal veins, increasing pressure in these veins and possibly causing hemorrhoids. ■

Histology

The wall of the large intestine differs from that of the small intestine in several respects. No villi or permanent circular folds are found in the mucosa. The **mucosa** consists of simple columnar epithelium, lamina propria (areolar connective tissue), and muscularis mucosae (smooth muscle). The epithelium contains mostly absorptive and goblet cells (Fig. 24.28). The absorptive cells function primarily in water absorption. The goblet cells secrete mucus that lubricates the colonic contents as they pass through. Both absorptive and goblet cells are located in long, straight, tubular intestinal glands that extend the full thickness of the mucosa. Solitary lymphatic nodules are also found in the mucosa. The **submucosa** of the large intestine is similar to that found in the rest of the GI tract. The **muscularis** consists of an external layer of longitudinal muscles and an internal layer of circular muscles. Unlike other parts of the GI tract, portions of the longitudinal muscles are thickened, forming three conspicuous longitudinal bands called **taeniae coli** (TĒ-nē-ē KŌ-lī; *taenia* = flat band), alternating with a wall section with less or no longitudinal muscle. Each band runs the length of most of the large intestine (see Fig. 24.27a). Tonic contractions of the bands gather the colon into a series of pouches called **haustra** (HAWS-tra; singular is **haustrum** = shaped like a pouch), which give the colon a puckered appearance. There is a single layer of circular muscle between taeniae coli. The **serosa** of the large intestine is part of the visceral peritoneum. Small pouches of visceral peritoneum filled with fat are attached to taeniae coli and are called **epiploic appendages**.

DIVERTICULITIS

Diverticula are saclike outpouchings of the wall of the colon in places where the muscularis has become weak. The development of diverticula is called **diverticulosis**. Many people who develop diverticulosis are asymptomatic and experience no complications. About 15% of people with diverticulosis will eventually develop an inflammation within the diverticula, a condition known as **diverticulitis**. Because diets low in fiber contribute to development of diverticulitis, patients who change to high-fiber diets show marked relief of symptoms. In severe cases, affected portions of the colon may require surgical removal. ■

Physiology of Digestion in the Large Intestine

Mechanical Digestion

The passage of chyme from the ileum into the cecum is regulated by the action of the ileocecal sphincter. Normally, the valve remains partially closed so that the passage of chyme into the cecum is usually a slow process. Immediately after a meal, there is a **gastroileal reflex** in which ileal peristalsis intensifies and forces any chyme in the ileum into the cecum. The hormone gastrin also relaxes the sphincter. Whenever the cecum is distended, the degree of contraction of the ileocecal sphincter intensifies.

Movements of the colon begin when substances pass the ileocecal sphincter. Since chyme moves through the small

Figure 24.28 Histology of the large intestine.

Intestinal glands formed by simple columnar and goblet cells extend the full thickness of the mucosa.

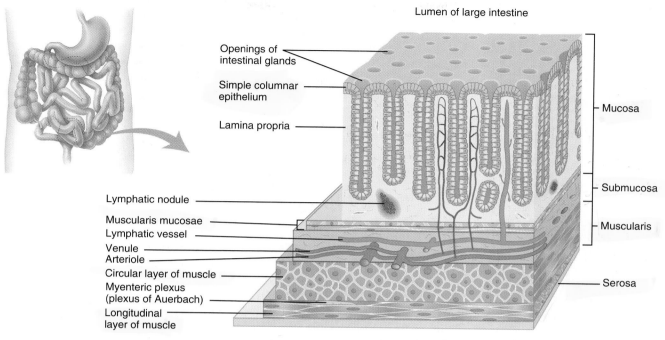

(a) Three-dimensional view of layers of the large intestine

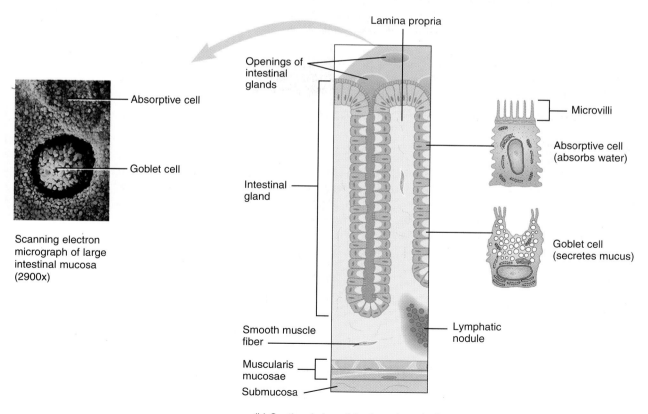

(b) Sectional view of the large intestinal mucosa
showing intestinal glands

Figure continues

Figure 24.28   (continued)

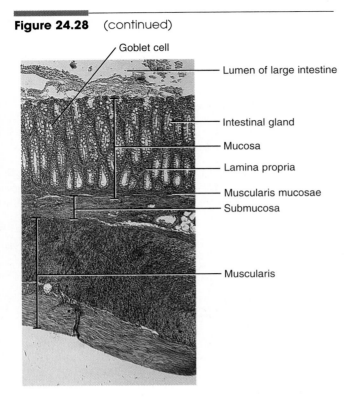

- Goblet cell
- Lumen of large intestine
- Intestinal gland
- Mucosa
- Lamina propria
- Muscularis mucosae
- Submucosa
- Muscularis

(c) Photomicrograph of a portion of the
wall of the large intestine (90x)

Q   What is the function of the goblet cells?

intestine at a fairly constant rate, the time required for a meal to pass into the colon is determined by gastric emptying time. As food passes through the ileocecal sphincter, it fills the cecum and accumulates in the ascending colon.

One movement characteristic of the large intestine is **haustral churning**. In this process, the haustra remain relaxed and distended while they fill up. When the distension reaches a certain point, the walls contract and squeeze the contents into the next haustrum. **Peristalsis** also occurs, although at a slower rate (3–12 contractions per minute) than in more proximal portions of the tract. A final type of movement is **mass peristalsis**, a strong peristaltic wave that begins at about the middle of the transverse colon and quickly drives the colonic contents into the rectum. Food in the stomach initiates this **gastrocolic reflex** in the colon. Thus mass peristalsis usually takes place three or four times a day, during or immediately after a meal.

Chemical Digestion

The last stage of digestion occurs in the colon through the activity of bacteria that live in the lumen. Mucus is secreted by the glands of the large intestine, but no enzymes are secreted. Chyme is prepared for elimination by the action of bacteria, which ferment any remaining carbohydrates and

release hydrogen, carbon dioxide, and methane gas. These gases contribute to flatus (gas) in the colon. Bacteria also convert remaining proteins to amino acids and break down the amino acids into simpler substances: indole, skatole, hydrogen sulfide, and fatty acids. Some of the indole and skatole is carried off in the feces and contributes to their odor. The rest are absorbed and transported to the liver, where they are converted to less toxic compounds and excreted in the urine. Bacteria also decompose bilirubin to simpler pigments (such as stercobilin), which give feces their brown color. Several vitamins needed for normal metabolism are bacterial products that are absorbed in the colon. These include some B vitamins and vitamin K.

Absorption and Feces Formation in the Large Intestine

By the time the chyme has remained in the large intestine 3–10 hours, it has become solid or semisolid as a result of water absorption and is now known as **feces**. Chemically, feces consist of water, inorganic salts, sloughed-off epithelial cells from the mucosa of the gastrointestinal tract, bacteria, products of bacterial decomposition, and undigested parts of food.

Although most water absorption occurs in the small intestine, the large intestine absorbs enough to make it an important organ in maintaining the body's water balance. Of the ½–1.0 liter of water that enters the large intestine, all but about 100–200 ml is absorbed by osmosis. The large intestine also absorbs electrolytes, including sodium and chloride, and some vitamins.

Physiology of Defecation

Mass peristaltic movements push fecal material from the sigmoid colon into the rectum. The resulting distension of the rectal wall stimulates stretch receptors, which initiates a **defecation reflex** that empties the rectum. The defecation reflex occurs as follows. In response to distension of the rectal wall, the receptors send sensory nerve impulses to the sacral spinal cord. Motor impulses from the cord travel along parasympathetic nerves back to the descending colon, sigmoid colon, rectum, and anus. Contraction of the longitudinal rectal muscles shortens the rectum, thereby increasing the pressure inside it. The pressure, along with voluntary contractions of the diaphragm and abdominal muscles, and parasympathetic stimulation open the internal sphincter, and the feces are expelled through the anus.

The external sphincter is voluntarily controlled. If it is voluntarily relaxed, defecation occurs; if it is voluntarily constricted, defecation can be postponed. Voluntary contractions of the diaphragm and abdominal muscles aid defecation by increasing the pressure inside the abdomen, which pushes the walls of the sigmoid colon and rectum inward. If defecation does not occur, the feces back into the sigmoid colon until the next wave of mass peristalsis again stimulates the stretch receptors, creating the desire to defecate. In infants, the defecation reflex causes automatic emptying of the rectum because voluntary control of the external anal sphincter has not yet developed.

Diarrhea refers to defecation of liquid feces. It is caused by increased motility of and decreased absorption by the intestines. When chyme passes too quickly through the small intestine and feces pass too quickly through the large intestine, there is not enough time for absorption. Frequent diarrhea can result in dehydration and electrolyte imbalances. It may be caused by lactose intolerance, stress, and microbes that irritate the gastrointestinal mucosa.

Constipation refers to infrequent or difficult defecation and is caused by decreased motility of the intestines. Because the feces remain in the colon for prolonged periods of time, there is excessive water absorption, and the feces become dry and hard. Constipation may be caused by improper bowel habits, spasms of the colon, insufficient fiber in the diet, inadequate fluid intake, lack of exercise, and emotions. The usual treatment is a mild laxative, such as milk of magnesia, that induces defecation. However, many physicians maintain that laxatives are habit-forming, and that adding fiber to the diet, increasing one's amount of exercise, and improving fluid intake are safer ways of controlling this common problem.

Digestive activities in the large intestine are summarized in Exhibit 24.7.

DIETARY FIBER

Adequate fiber (bulk or roughage) in the diet has several beneficial effects. Dietary fiber consists of indigestible plant substances, such as cellulose, lignin, and pectin, found in fruits, vegetables, grains, and beans. Fiber may be classified as **insoluble**, which does not dissolve in water, and **soluble**, which does dissolve in water. Insoluble fiber includes the woody or structural parts of plants such as fruit and vegetable skins and the bran coating around wheat and corn kernels. Insoluble fiber passes through the GI tract largely unchanged and speeds up the passage of material through the tract. Soluble fiber is found in abundance in beans, oats, barley, broccoli, prunes, apples, and citrus fruits. It has the consistency of a gel and tends to slow the passage of material through the tract. Both types of fiber contribute to normal digestive functions.

People who choose a fiber-rich, unrefined diet may reduce their risk of developing obesity, diabetes, atherosclerosis, gallstones, hemorrhoids, diverticulitis, appendicitis, and colon cancer. Each of these conditions is directly related to the digestion and metabolism of food and the operation of the digestive system. Insoluble fiber also helps protect against colon cancer and soluble fiber may help lower blood cholesterol level. One possible explanation for the relationship of soluble fiber to cholesterol level is that the fiber binds bile salts and prevents their reabsorp-

tion. Since cholesterol is a precursor to bile salt formation, this leads to use of more cholesterol to replace the bile salts lost by the binding action of soluble fiber. ■

EXHIBIT 24.7 SUMMARY OF DIGESTIVE ACTIVITIES IN THE LARGE INTESTINE

Structure	Action	Function
Mucosa	Secretes mucus.	Lubricates colon and protects mucosa.
	Absorbs water and other soluble compounds.	Maintains water balance; solidifies feces. Absorbs vitamins and some ions.
Lumen	Bacterial activity.	Breaks down undigested carbohydrates, proteins, and amino acids into products that can be expelled in feces or absorbed and detoxified by liver. Synthesizes certain B vitamins and vitamin K.
Muscularis	Haustral churning.	Moves contents from haustrum to haustrum by muscular contractions.
	Peristalsis.	Moves contents along length of colon by contractions of circular and longitudinal muscles.
	Mass peristalsis.	Forces contents into sigmoid colon and rectum.
	Defecation reflex.	Eliminates feces by contractions in sigmoid colon and rectum.

AGING AND THE DIGESTIVE SYSTEM

Overall changes associated with aging of the digestive system include decreased secretory mechanisms, decreased motility of the digestive organs, loss of strength and tone of the muscular tissue and its supporting structures, changes in neurosensory feedback regarding enzyme and hormone release, and diminished response to pain and internal sensations. In the upper portion of the GI tract, common changes include reduced sensitivity to mouth irritations and sores, loss of taste, periodontal disease, difficulty in swallowing, hiatal hernia, gastritis, and peptic ulcer disease (see page 799). Changes that may appear in the small intestine include duodenal ulcers, appendicitis, malabsorption, and maldigestion. Other pathologies that increase in incidence with age are gallbladder problems, jaundice, cirrhosis, and acute pancreatitis. Large intestinal changes such as constipation, hemorrhoids, and diverticular disease may also occur. Cancer of the colon or rectum (colorectal cancer, see page 800) is quite common.

DEVELOPMENTAL ANATOMY OF THE DIGESTIVE SYSTEM

About the fourteenth day after fertilization, the cells of the endoderm form a cavity referred to as the **primitive gut** (Fig. 24.29). Soon after the mesoderm forms and splits into two layers (somatic and splanchnic), the splanchnic mesoderm associates with the endoderm of the primitive gut. Thus the primitive gut has a double-layered wall. The **endodermal layer** gives rise to the *epithelial lining and glands* of most of the gastrointestinal tract. The **mesodermal layer** produces the *smooth muscle* and *connective tissue* of the tract.

The primitive gut elongates, and during the third week, it differentiates into an anterior **foregut**, an intermediate **midgut**, and a posterior **hindgut**. Until the fifth week of development, the midgut opens into the yolk sac. After that time, the yolk sac constricts, detaches from the midgut, and

the midgut seals. In the region of the foregut, a depression consisting of **ectoderm**, the **stomodeum**, appears. This develops into the *oral cavity.* The **oral membrane** that separates the foregut from the stomodeum ruptures during the fourth week of development, so that the foregut is continuous with the outside of the embryo through the oral cavity. Another depression consisting of **ectoderm**, the **proctodeum**, forms in the hindgut and goes on to develop into the *anus.* The **cloacal membrane**, which separates the hindgut from the proctodeum, ruptures, so that the hindgut is continuous with the outside of the embryo through the anus. Thus the GI tract forms a continuous tube from mouth to anus.

The foregut develops into the *pharynx, esophagus, stomach,* and a *portion of the duodenum.* The midgut is transformed into the *remainder of the duodenum,* the *jejunum,* the *ileum,* and *portions of the large intestine* (cecum, appendix, ascending colon, and most of the transverse colon). The hindgut develops into the *remainder of the large intestine,* except for a portion of the anal canal that is derived from the proctodeum.

As development progresses, the endoderm at various places along the foregut develops into hollow buds that grow into the mesoderm. These buds will develop into the *salivary glands, liver, gallbladder,* and *pancreas.* Each of the glands retains a connection with the gastrointestinal tract through ducts.

Figure 24.29   Development of the digestive system.

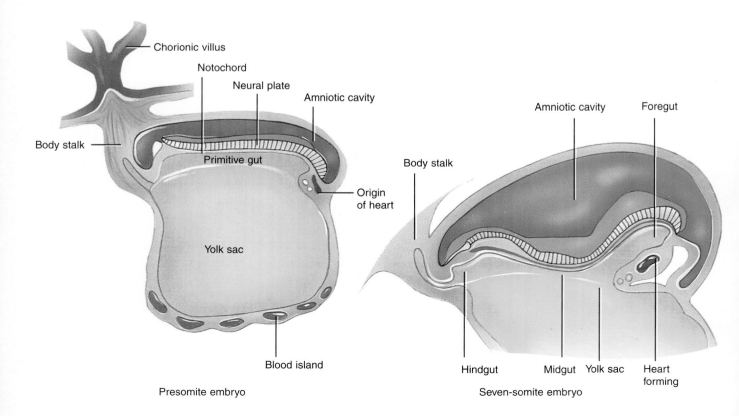

Presomite embryo

Seven-somite embryo

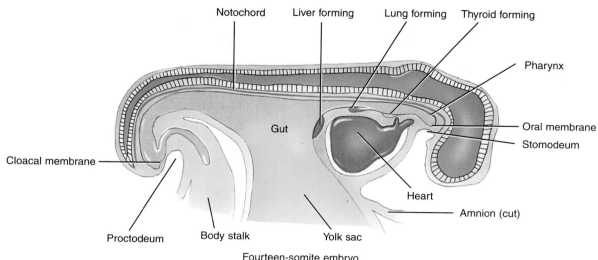

Fourteen-somite embryo

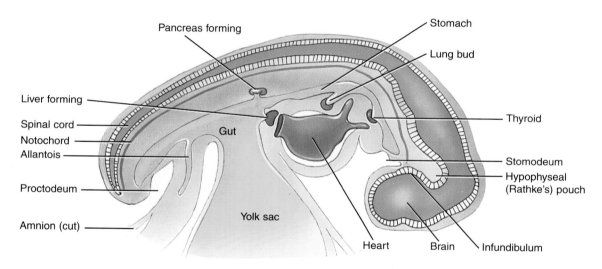

Embryo at end of first month

DISORDERS: HOMEOSTATIC IMBALANCES

DENTAL CARIES

Dental caries, or tooth decay, involve a gradual demineralization (softening) of the enamel and dentin. If untreated, microorganisms may invade the pulp, causing inflammation and infection with subsequent death of the pulp and abscess of the alveolar bone surrounding the root's apex. Such teeth are treated by root canal therapy.

Dental caries begin when bacteria, acting on sugars, give off acids that demineralize the enamel. **Dextran**, a sticky polysaccharide produced from sucrose, causes the bacteria to stick to the teeth. Masses of bacterial cells, dextran, and other debris adhering to teeth constitute **dental plaque**. Saliva cannot reach the tooth surface to buffer the acid because the plaque covers the teeth. Brushing the teeth immediately after eating removes the plaque from flat surfaces before the bacteria have a chance to go to work. Dentists also suggest that the plaque between the teeth be removed every 24 hours with dental floss.

PERIODONTAL DISEASE

Periodontal disease is a collective term for a variety of conditions characterized by inflammation and degeneration of the gingivae, alveolar bone, periodontal ligament, and cementum. One such condition is called **pyorrhea**. The initial symptoms are enlargement and inflammation of the soft tissue and bleeding of the gums. Without treatment, the soft tissue may deteriorate and the alveolar bone may be resorbed, causing loosening of the teeth and recession of the gums. Periodontal diseases are frequently caused by poor oral hygiene; by local irritants, such as bacteria, impacted food, and cigarette smoke; or by a poor "bite."

PEPTIC ULCER DISEASE

In the United States about 5–10% of the population develops **peptic ulcer disease** (**PUD**). An **ulcer** is a craterlike lesion in a

DISORDERS: HOMEOSTATIC IMBALANCES   CONTINUED

membrane. Ulcers that develop in areas of the GI tract exposed to acidic gastric juice are called **peptic ulcers**. Most occur in the pylorus or first part of the duodenum, where they are called **duodenal ulcers**. A smaller number occur in the body of the stomach, where they are called **gastric ulcers**. The most common complication of peptic ulcers is bleeding, which can lead to anemia if blood loss is serious.

At present, there are three well-identified causes of PUD: (1) the bacterium *Helicobacter pylori*; (2) nonsteroidal anti-inflammatory drugs (NSAIDs), such as aspirin; and (3) hypersecretion of HCl, as occurs in Zollinger–Ellison syndrome, a gastrin-producing tumor usually of the pancreas. At this point, we will concentrate on *Helicobacter pylori*, the most frequent cause .

In causing PUD, *Helicobacter pylori* (previously named *Campylobacter pylori*) produces an enzyme called urease, which splits urea into ammonia and other substances. The bacterium surrounds itself in the ammonia which shields the bacterium from the acidity of the stomach and also damages the protective mucus layer of the stomach and the underlying gastric cells. *H. pylori* also produces an enzyme (catalase) that may protect the microbe from phagocytosis by neutrophils, another enzyme that may inhibit gastric acid, and several adhesion proteins that allow the bacterium to attach itself to gastric cells.

Mucosal defensive mechanisms are a major factor in the development of PUD. These include bicarbonate ions in pancreatic juice which buffer HCl, prostaglandin synthesis which stimulates gastric mucus production, gastric mucosal blood flow, and epithelial cell regeneration in response to mucosal injury. Stress, cigarette smoking, alcohol, caffeine, and NSAIDs can impair mucosal defensive mechanisms, thereby increasing mucosal susceptibility to the damaging effects of HCl.

Several therapeutic approaches are helpful to treat PUD. In cases associated with *Helicobacter pylori,* treatment with an antibiotic drug often resolves the problem. Oral antacids, such as Tums® or Maalox®, can help temporarily by buffering gastric acid. In cases where hypersecretion of HCl is the cause of PUD, H_2 blockers, such as Tagamet®, may be used (see page 773).

TUMORS

Both benign and malignant **tumors** can occur in all parts of the gastrointestinal tract. Although benign growths are much more common than malignant ones, cancers of the gastrointestinal tract are responsible for 30% of all deaths from cancer in the United States.

Colorectal cancer is one of the most deadly and common malignant diseases, ranking second to lung cancer in males and third after lung and breast cancer in females. Development of colorectal cancer is a multistep process, involving both environmental and genetic factors (see page 89). Dietary fiber, retinoids, calcium, and selenium may be protective, whereas intake of alcohol, animal fat, and protein may cause an increase in the disease. Genetics plays a very important role in that an inherited predisposition contributes to more than half of all cases of colorectal cancer. Signs and symptoms of colorectal cancer include changes in the normal pattern of bowel habits (diarrhea, constipation), cramping, abdominal pain, and rectal bleeding, either visible or occult. Screening for colorectal cancer includes testing for blood in the feces, digital rectal examination, sigmoidoscopy, colonoscopy, and barium enema. Gastrointestinal tumors may be removed endoscopically or surgically.

CIRRHOSIS

Cirrhosis refers to a distorted or scarred liver as a result of chronic inflammation. The parenchymal (functional) hepatocytes are replaced by fibrous or adipose connective tissue. The symptoms of cirrhosis include jaundice, edema in the legs, uncontrolled bleeding, and increased sensitivity to drugs. Cirrhosis may be caused by hepatitis (inflammation of the liver), certain chemicals that destroy hepatocytes, parasites that infect the liver, and alcoholism.

HEPATITIS

Hepatitis refers to inflammation of the liver and can be caused by viruses, drugs, and chemicals, including alcohol. Clinically, several viral types are recognized. **Hepatitis A** (**infectious hepatitis**) is caused by hepatitis A virus and is spread by fecal contamination of food, clothing, toys, eating utensils, and so forth (fecal–oral route). It is generally a mild disease of children and young adults characterized by anorexia (loss of appetite), malaise, nausea, diarrhea, fever, and chills. Eventually, jaundice appears. It does not cause lasting liver damage. Most people recover in 4–6 weeks.

Hepatitis B (**serum hepatitis**) is caused by hepatitis B virus and is spread primarily by sexual contact and contaminated syringes and transfusion equipment. It can also be spread by saliva and tears. Hepatitis B virus can be present for years or even a lifetime and can produce cirrhosis and possibly cancer of the liver. Persons who harbor the active hepatitis B virus are at risk for cirrhosis and also become carriers. Vaccines produced through recombinant DNA technology (for example, Recombivax HB) are available to prevent hepatitis B infection.

Hepatitis C (**non-A**, **non-B hepatitis**) is caused by hepatitis C virus. It is clinically similar to hepatitis B and is often spread by blood transfusions. Hepatitis C can cause cirrhosis and possibly liver cancer.

Hepatitis D (**delta hepatitis**) is caused by hepatitis D virus. It is transmitted like hepatitis B and, in fact, a person must be coinfected with hepatitis B before contracting hepatitis D. Hepatitis D results in severe liver damage and has a fatality rate higher than that of people infected with hepatitis B virus alone.

Hepatitis E (**infectious NANB hepatitis**) is caused by hepatitis E virus and is spread like hepatitis A. Although it does not cause chronic liver disease, hepatitis E virus is responsible for a very high mortality rate in pregnant women.

GALLSTONES

Most **gallstones** (**biliary calculi**) stem from the fusion of crystals of cholesterol in bile. The presence of gallstones is called **cholelithiasis** (kō′-lē-li-THĪ-a-sis; *lithos* = stone). Following their formation, gallstones gradually grow in size and number and may cause minimal, intermittent, or complete obstruction to the flow of bile from the gallbladder into the duct system. If obstruction of the outlet occurs and the gallbladder cannot empty as it normally does after eating, the pressure within it increases, and the person may have intense pain or discomfort (**biliary colic**). Jaundice, due to the inability to secrete bilirubin into the intestine, will accompany complete biliary obstruction.

Treatment of gallstones consists of using gallstone-dissolving drugs, lithotripsy (shock-wave therapy), or surgery. For people who have had recurrent gallstones or in whom drugs or lithotripsy is not indicated, removal of the gallbladder and its contents (cholecystectomy) is necessary. More than half a million cholecystectomies are performed each year in the United States.

ANOREXIA NERVOSA

Anorexia nervosa is a chronic disorder characterized by self-induced weight loss, negative perception of body image, and physiological changes that result from nutritional depletion. Patients with anorexia nervosa have a fixation on weight control and, often, the insistence of having a bowel movement every day despite a lack of adequate food intake. They abuse laxatives, which worsens the fluid, electrolyte, and nutrient deficiencies. The disorder is found predominantly in young, single females and may be inherited. Abnormal patterns of menstruation, amenorrhea (absence of menstruation), and a lowered basal metabolic rate reflect the depressant effects of the starvation. Individuals may become emaciated and may ultimately die of starvation or one of its complications. Also associated with the disorder are osteoporosis, depression, and brain abnormalities coupled with impaired mental performance. Treatment consists of psychotherapy and dietary regulation.

BULIMIA

A disorder that typically affects single, middle-class, young, white females is known as **bulimia** (*bous* = ox; *limos* = hunger), or **binge–purge syndrome**. It is characterized by overeating at least twice a week followed by purging by self-induced vomiting, strict dieting or fasting, vigorous exercise, or use of laxatives or diuretics. This binge–purge cycle occurs in response to fears of being overweight, stress, depression, and physiological disorders such as hypothalamic tumors.

Bulimia can seriously upset the body's electrolyte balance. In addition, it can increase susceptibility to flu, salivary gland infections that result in bilateral parotid gland enlargement, pharyngeal scratches from self-induced gagging, dry skin, acne, muscle spasms, loss of hair, kidney and liver diseases, erosion of dental enamel from stomach acids, ulcers, hernias, constipation, and hormone imbalances. Although the exact cause of bulimia is unknown, some evidence suggests that it may be related to impaired release of cholecystokinin (CCK), a hormone that induces satiety, which is the sensation of being full to satisfaction. Treatment of bulimia combines nutrition counseling, psychotherapy, and medical treatment.

MEDICAL TERMINOLOGY

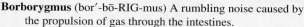

Borborygmus (bor′-bō-RIG-mus) A rumbling noise caused by the propulsion of gas through the intestines.

Botulism (BOCH-yoo-lism; *botulus* = sausage) A type of food poisoning caused by a toxin produced by *Clostridium botulinum*. The bacterium is ingested when improperly cooked or preserved foods are eaten. The toxin causes paralysis of skeletal muscles by inhibiting the release of acetylcholine. Symptoms include paralysis, nausea, vomiting, blurred or double vision, difficulty in speech and swallowing, dryness of the mouth, and general weakness.

Canker (KANG-ker) **sore** Painful ulcer on the mucous membrane of the mouth that affects females more often than males and usually occurs between ages 10–40; may be an autoimmune reaction or a food allergy.

Cholecystitis (kō′-lē-sis-TĪ-tis; *chole* = bile; *kystis* = bladder; *itis* = inflammation of) Some cases are caused by an autoimmune inflammation of the gallbladder. Other cases are caused by obstruction of the cystic duct with bile stones.

Colitis (ko-LĪ-tis) Inflammation of the mucosa of the colon and rectum in which absorption of water and salts is reduced, producing watery, bloody feces and, in severe cases, dehydration and salt depletion. Spasms of the irritated muscularis produce cramps. It is thought to be an autoimmune condition.

Colostomy (kō-LOS-tō-mē; *stomoun* = provide an opening) The diversion of the fecal stream through an opening in the colon, creating a surgical "stoma" (artificial opening) that is affixed to the exterior of the abdominal wall. This opening serves as a substitute anus through which feces are eliminated into a bag worn on the abdomen.

Dysphagia (dis-FĀ-jē-a; *dys* = abnormal; *phagein* = to eat) Difficulty in swallowing that may be caused by inflammation, paralysis, obstruction, or trauma.

Enteritis (en′-ter-Ī-tis; *enteron* = intestine) An inflammation of the intestine, particularly the small intestine.

Flatus (FLĀ-tus) Air (gas) in the stomach or intestine, usually expelled through the anus. If the gas is expelled through the mouth, it is called **eructation** or **belching** (burping). Flatus may result from gas released during the breakdown of foods in the stomach or from swallowing air or gas-containing substances such as carbonated drinks.

Gastrectomy (gas-TREK-tō-mē; *gastro* = stomach; *tome* = excision) Removal of a portion of or the entire stomach.

Hernia (HER-nē-a) Protrusion of an organ or part of an organ through a membrane or cavity wall, usually the abdominal cavity. *Diaphragmatic (hiatal) hernia* is the protrusion of the lower esophagus, stomach, or intestine into the thoracic cavity through the opening in the diaphragm (esophageal hiatus) that allows passage of the esophagus. *Umbilical hernia* is usually a mild defect that contains the protrusion of a portion of peritoneum through the navel area of the abdominal wall. *Inguinal hernia* is the protrusion of the hernial sac into the inguinal opening. It may contain a portion of the bowel in an advanced stage and may extend into the scrotal compartment in males, causing strangulation of the herniated part.

Inflammatory bowel (in-FLAM-a-tō′-rē BOW-el) **disease** Disorder that exists in two forms: (1) Crohn's disease (inflammation of the gastrointestinal tract, especially the distal ileum and proximal colon, in which the inflammation may extend from the mucosa through the serosa) and (2) ulcerative colitis (inflammation of the mucosa of the gastrointestinal tract, usually limited to the large intestine and usually accompanied by rectal bleeding).

Irritable bowel (IR-i-ta-bul BOW-el) **syndrome (IBS)** Disease of the entire gastrointestinal tract in which a person reacts to stress by developing symptoms such as cramping and abdominal pain associated with alternating patterns of diarrhea and constipation. Excessive amounts of mucus may appear in the stools, and other symptoms include flatulence,

MEDICAL TERMINOLOGY CONTINUED

nausea, and loss of appetite. The condition is also known as **irritable colon** or **spastic colitis**.

Malocclusion (mal'-ō-KLOO-zhun; *mal* = disease; *occlusio* = to fit together) Condition in which the maxillary (upper) and mandibular (lower) teeth do not close together.

Nausea (NAW-sē-a; *nausia* = seasickness) Discomfort characterized by a loss of appetite and the sensation of impending vomiting. Its causes include local irritation of the gastrointestinal tract, a systemic disease, brain disease or injury, overexertion, or the effects of medication or drug overdosage.

Traveler's diarrhea Infectious disease of the gastrointestinal tract that results in loose, urgent bowel movements, cramping, abdominal pain, malaise, nausea, and occasionally fever and dehydration. It is acquired through ingestion of food or water that has become contaminated with fecal material containing bacteria (especially *Escherichia coli*). Viruses or protozoan parasites are less frequently involved.

STUDY OUTLINE

INTRODUCTION (p. 753)

1. The breaking down of larger food molecules into smaller molecules is called digestion; the passage of these smaller molecules into blood and lymph is termed absorption.
2. The organs that collectively perform digestion and absorption compose the digestive system and are usually divided into two main groups: those composing the gastrointestinal (GI) tract and accessory structures.
3. The GI tract is a continuous tube extending from the mouth to the anus.
4. The accessory structures include the teeth, tongue, salivary glands, liver, gallbladder, and pancreas.

OVERVIEW OF DIGESTIVE PROCESSES (p. 754)

1. Digestion includes six basic processes: ingestion, secretion, mixing and propulsion, mechanical and chemical digestion, absorption, and defecation.
2. Mechanical digestion consists of movements of the gastrointestinal tract that aid chemical digestion.
3. Chemical digestion is a series of catabolic (hydrolysis) reactions that break down large carbohydrate, lipid, and protein food molecules into smaller molecules that are usable by body cells.

LAYERS OF THE GI TRACT (p. 754)

1. The basic arrangement of layers in most of the gastrointestinal tract from deep to superficial is the mucosa, submucosa, muscularis, and serosa.
2. Associated with the lamina propria of the mucosa are extensive patches of lymphatic tissue called mucosa-associated lymphoid tissue (MALT).

PERITONEUM (p. 756)

1. The peritoneum is the largest serous membrane of the body; it lines the wall of the abdominal cavity and covers some abdominal organs.
2. Extensions of the peritoneum include the mesentery, mesocolon, falciform ligament, lesser omentum, and greater omentum.

MOUTH (p. 758)

1. The mouth is formed by the cheeks, hard and soft palates, lips, and tongue, which aid mechanical digestion.

2. The vestibule is the space between the cheeks and lips and teeth and gums.
3. The oral cavity proper extends from the vestibule to the fauces.

Tongue (p. 759)

1. The tongue, together with its associated muscles, forms the floor of the oral cavity. It is composed of skeletal muscle covered with mucous membrane.
2. The upper surface and sides of the tongue are covered with papillae. Some papillae contain taste buds.

Salivary Glands (p. 759)

1. The major portion of saliva is secreted by the salivary glands, which lie outside the mouth and pour their contents into ducts that empty into the oral cavity.
2. There are three pairs of salivary glands: parotid, submandibular (submaxillary), and sublingual glands.
3. Saliva lubricates food and starts the chemical digestion of carbohydrates.
4. Salivation is controlled by the nervous system.

Teeth (p. 761)

1. The teeth (dentes) project into the mouth and are adapted for mechanical digestion.
2. A typical tooth consists of three principal regions: crown, root, and neck.
3. Teeth are composed primarily of dentin and are covered by enamel, the hardest substance in the body.
4. There are two dentitions—deciduous and permanent.

Physiology of Digestion in the Mouth (p. 763)

1. Through mastication, food is mixed with saliva and shaped into a bolus.
2. Salivary amylase converts polysaccharides (starches) to disaccharides (maltose) and lingual lipase acts on triglycerides.

Physiology of Deglutition (p. 764)

1. Deglutition, or swallowing, moves a bolus from the mouth to the stomach.
2. It consists of a voluntary stage, pharyngeal stage (involuntary), and esophageal stage (involuntary).

ESOPHAGUS (p. 764)

1. The esophagus is a collapsible, muscular tube that connects the pharynx to the stomach.

2. It passes a bolus into the stomach by peristalsis.

3. It contains an upper and a lower esophageal sphincter.

STOMACH (p. 767)

Anatomy; Histology (p. 767)

1. The stomach connects the esophagus to the duodenum.

2. The principal anatomic subdivisions of the stomach are the cardia, fundus, body, and pylorus.

3. Adaptations of the stomach for digestion include rugae; glands that produce mucus, hydrochloric acid, pepsin, gastric lipase, and intrinsic factor; and a three-layered muscularis.

Physiology of Digestion and Absorption in the Stomach (p. 770)

1. Mechanical digestion consists of mixing waves.

2. Chemical digestion consists mostly of the conversion of proteins into peptides by pepsin.

3. The stomach wall is impermeable to most substances.

4. Among the substances absorbed are water, certain ions, drugs, and alcohol.

Regulation of Gastric Secretion and Motility (p. 771)

1. Gastric secretion is regulated by neural and hormonal mechanisms.

2. Stimulation of gastric secretion occurs in three phases: cephalic, gastric, and intestinal.

3. During the cephalic and gastric phases, peristalsis is stimulated; during the intestinal phase, motility is inhibited.

Regulation of Gastric Emptying (p. 773)

1. Gastric emptying is stimulated in response to distension and gastrin is released in response to the presence of certain types of foods.

2. Gastric emptying is inhibited by the enterogastric reflex and by hormones (CCK and GIP).

PANCREAS (p. 775)

1. The pancreas is divisible into a head, body, and tail and is connected to the duodenum via the pancreatic duct and accessory duct.

2. Endocrine pancreatic islets (islets of Langerhans) secrete hormones, and exocrine acini secrete pancreatic juice.

3. Pancreatic juice contains enzymes that digest starch (pancreatic amylase), proteins (trypsin, chymotrypsin, carboxypeptidase, and elastase), triglycerides (pancreatic lipase), and nucleic acids (ribonuclease and deoxyribonuclease).

4. Pancreatic secretion is regulated by neural and hormonal mechanisms.

LIVER (p. 777)

1. The liver has left and right lobes; associated with the right lobe are the caudate and quadrate lobes.

2. The lobes of the liver are made up of lobules that contain hepatocytes (liver cells), sinusoids, stellate reticuloendothelial (Kupffer's) cells, and a central vein.

3. Hepatocytes produce bile that is carried by a duct system to the gallbladder for concentration and temporary storage.

4. Bile's contribution to digestion is the emulsification of dietary lipids.

5. The liver also functions in carbohydrate, lipid, and protein metabolism; removal of drugs and hormones; excretion of bilirubin; synthesis of bile salts; storage of vitamins and minerals; phagocytosis; and activation of vitamin D.

6. Bile secretion is regulated by neural and hormonal mechanisms.

GALLBLADDER (p. 782)

1. The gallbladder is a sac located in a depression on the posterior surface of the liver.

2. The gallbladder stores and concentrates bile.

3. Cholecystokinin (CCK) stimulates ejection of bile into the common bile duct.

SUMMARY: DIGESTIVE HORMONES (p. 782)

1. The four major hormones that regulate digestive processes are gastrin, gastric inhibitory peptide (GIP), secretin, and cholecystokinin (CCK).

2. Gastrin and GIP exert their major effects on the stomach, whereas secretin and CCK affect the pancreas, liver, and gallbladder most strongly.

SMALL INTESTINE (p. 783)

Anatomy; Histology; Brush Border Enzymes (p. 784)

1. The small intestine extends from the pyloric sphincter to the ileocecal sphincter.

2. It is divided into duodenum, jejunum, and ileum.

3. Its glands secrete fluid and mucus, and the circular folds, villi, and microvilli of its wall provide a large surface area for digestion and absorption.

4. Brush border enzymes digest α-dextrins, maltose, sucrose, lactose, peptides, and nucleotides at the surface of mucosal epithelial cells.

Physiology of Digestion in the Small Intestine (p. 786)

1. Pancreatic and intestinal brush border enzymes break down starches into maltose, maltotriose, and α-dextrins (pancreatic amylase), α-dextrins into glucose (dextrinase), maltose to glucose (maltase), sucrose to glucose and fructose (sucrase), lactose to glucose and galactose (lactase), and proteins into peptides (trypsin, chymotrypsin, and elastase). Also, enzymes break peptide bonds that attach terminal amino acids to carboxyl ends of peptides (carboxypeptidases) and peptide bonds that attach terminal amino acids to amino ends of peptides (aminopeptidases). Finally, enzymes split dipeptides to amino acids (dipeptidase), triglycerides to fatty acids and monoglycerides (lipases), and nucleotides to pentoses and nitrogenous bases (nucleosidases and phosphatases).

2. Mechanical digestion in the small intestine involves segmentation and peristalsis.

Regulation of Intestinal Secretion and Motility (p. 788)

1. The most important regulators are local reflexes.

2. Hormones also assume a role.

3. Parasympathetic impulses increase motility; sympathetic impulses decrease motility.

Physiology of Absorption in the Small Intestine (p. 788)

1. Absorption occurs by diffusion, facilitated diffusion, osmosis, and active transport; most occurs in the small intestine.

2. Monosaccharides, amino acids, and short-chain fatty acids pass into the blood capillaries.

3. Long-chain fatty acids and monoglycerides are absorbed from micelles, resynthesized to triglycerides, and formed into chylomicrons.

4. Chylomicrons move into lymph in the lacteal of a villus.

5. The small intestine also absorbs water, electrolytes, and vitamins.

LARGE INTESTINE (p. 793)

Anatomy; Histology (p. 793)

1. The large intestine extends from the ileocecal sphincter to the anus.
2. Its subdivisions include the cecum, colon, rectum, and anal canal.
3. The mucosa contains many goblet cells, and the muscularis consists of taeniae coli and haustra.

Physiology of Digestion in the Large Intestine (p. 794)

1. Mechanical movements of the large intestine include haustral churning, peristalsis, and mass peristalsis.
2. The last stages of chemical digestion occur in the large intestine through bacterial action. Substances are further broken down and some vitamins are synthesized.

Absorption and Feces Formation in the Large Intestine (p. 796)

1. The large intestine absorbs water, electrolytes, and vitamins.
2. Feces consist of water, inorganic salts, epithelial cells, bacteria, and undigested foods.

Physiology of Defecation (p. 796)

1. The elimination of feces from the rectum is called defecation.
2. Defecation is a reflex action aided by voluntary contractions of the diaphragm and abdominal muscles and relaxation of the external anal sphincter.

AGING AND THE DIGESTIVE SYSTEM (p. 797)

1. General changes include decreased secretory mechanisms, decreased motility, and loss of tone.
2. Specific changes may include loss of taste, pyorrhea, hernias, peptic ulcer disease, constipation, hemorrhoids, and diverticular diseases.

DEVELOPMENTAL ANATOMY OF THE DIGESTIVE SYSTEM (p. 798)

1. The endoderm of the primitive gut forms the epithelium and glands of most of the gastrointestinal tract.
2. The mesoderm of the primitive gut forms the smooth muscle and connective tissue of the gastrointestinal tract.

REVIEW QUESTIONS

1. Define digestion and describe the six processes involved. Distinguish between chemical and mechanical digestion. (p. 754)
2. Identify, in sequence, the organs of the gastrointestinal (GI) tract. How does the gastrointestinal tract differ from the accessory structures of digestion? (p. 753)
3. Describe the structure and function of each of the four layers of the gastrointestinal tract and how they vary in different organs. (p. 754)
4. What is the peritoneum? Describe the location and function of the mesentery, mesocolon, falciform ligament, lesser omentum, and greater omentum. (p. 756)
5. What structures form the mouth (oral cavity)? (p. 758)
6. Make a simple diagram of the tongue. Indicate the location of the three types of papillae and the four taste zones. (p. 756)
7. Distinguish buccal from salivary glands. Describe the location of the salivary glands and their ducts. What is mumps? (p. 759)
8. How are salivary glands distinguished histologically? (p. 760)
9. Describe the composition of saliva and the role of each of its components in digestion. What is the pH of saliva? (p. 760)
10. How is salivary secretion regulated? (p. 761)
11. What are the principal regions of a typical tooth? What are the functions of each region? (p. 761)
12. Compare deciduous and permanent dentitions with regard to number of teeth and time of eruption. (p.761)
13. Contrast the functions of incisors, cuspids, premolars, and molars. (p. 761)
14. What is a bolus? How is it formed? (p. 763)
15. Define deglutition. List the sequence of events involved in passing a bolus from the mouth to the stomach. Be sure to discuss the voluntary, pharyngeal, and esophageal stages of swallowing. (p. 764)
16. Describe the location and histology of the esophagus. What is its role in digestion? (p. 764)
17. Explain the operation of the upper and lower esophageal sphincters. (p. 766)
18. Describe the location of the stomach. List and briefly explain the anatomical features of the stomach. (p. 767)
19. What is the importance of rugae, mucous surface cells, mucous neck cells, chief cells, parietal cells, and enteroendocrine (G) cells in the stomach? (p. 768)
20. Describe mechanical digestion in the stomach. (p. 770)
21. What is the role of pepsin? Why is it secreted in an inactive form? (p. 771)
22. What are the functions of gastric lipase and lingual lipase in the stomach? (p. 771)
23. Outline the factors that stimulate and inhibit gastric secretion. Be sure to discuss the cephalic, gastric, and intestinal phases. (p. 771)
24. How is gastric emptying stimulated and inhibited? (p. 773)
25. Describe the role of the stomach in absorption. (p. 774)
26. Where is the pancreas located? Describe the duct system connecting the pancreas to the duodenum. (p. 775)
27. What are pancreatic acini? Contrast their functions with those of the pancreatic islets (islets of Langerhans). (p. 776)
28. Describe the composition of pancreatic juice and the digestive functions of each component. (p. 776)
29. How is pancreatic juice secretion regulated? (p. 776)
30. Where is the liver located? What are its principal functions? (p. 777)
31. Describe the anatomy of the liver. Draw a labeled diagram of a liver lobule. (p. 777)
32. How is blood carried to and from the liver? (p. 780)
33. Once bile has been formed by the liver, how is it collected and transported to the gallbladder for storage? (p. 780)
34. What is the function of bile? (p. 780)
35. How is bile secretion regulated? (p. 781)
36. Where is the gallbladder located? How is it connected to the duodenum? (p. 782)
37. Describe the function of the gallbladder. How is emptying of the gallbladder regulated? (p. 782)
38. What are the subdivisions of the small intestine? How are the mucosa and submucosa of the small intestine adapted for digestion and absorption? (p. 784)

39. Describe the movements in the small intestine. (p. 786)
40. Explain the function of each enzyme in intestinal juice. (p. 786)
41. How is small intestinal secretion regulated? (p. 788)
42. Define absorption. How are the end products of carbohydrate and protein digestion absorbed? How are the end products of lipid digestion absorbed? (p. 788)
43. What routes are taken by absorbed nutrients to reach the liver? (p. 789)
44. Describe the absorption of water, electrolytes, and vitamins by the small intestine. (p. 791)

45. What are the principal subdivisions of the large intestine? How does the muscularis of the large intestine differ from that of the rest of the gastrointestinal tract? What are haustra? (p. 794)
46. Describe the mechanical movements that occur in the large intestine. (p. 794)
47. Explain the activities of the large intestine that change its contents into feces. (p. 796)
48. Define defecation. How does it occur? (p. 796)
49. Describe the effects of aging on the digestive system. (p. 797)
50. Describe the development of the digestive system. (p. 798)

ANSWERS TO QUESTIONS WITH FIGURES

24.1 Peristalsis.
24.2 They help regulate secretions and motility of the tract.
24.3 Mesentery.
24.4 The tongue. Extrinsic muscles move the tongue during eating and swallowing. Intrinsic muscles alter the shape of the tongue for speech production and swallowing.
24.5 They activate salivary amylase.
24.6 Connective tissue, specifically dentin.
24.7 First, second, and third molars.
24.8 Both. Initiation of swallowing is voluntary and the action is carried out by skeletal muscles. Completion of swallowing—moving a bolus along the esophagus and into the stomach—is involuntary and involves peristalsis in smooth muscle.
24.9 Mucosa and submucosa.
24.10 Food is pushed along by contraction of smooth muscle behind the bolus and relaxation of smooth muscle in front of it.
24.11 Probably not, since as the stomach fills, they stretch out.
24.12 Mucous surface and neck cells secrete mucus; chief cells secrete pepsinogen and gastric lipase; parietal cells secrete HCl and intrinsic factor; enteroendocrine (G) cells secrete gastrin.
24.13 Secretion of gastric juice and increased gastric motility.

24.14 Because all parts of the cycle are within the stomach. Impulses from the brain or spinal cord are not needed.
24.15 Inhibition.
24.16 Pancreatic juice (fluid and digestive enzymes); bile; pancreatic juice plus bile.
24.17 The HCO_3^- helps buffer gastric acid and raises the pH of chyme so that pancreatic enzymes can work effectively.
24.18 Epigastric.
24.19 Stellate reticuloendothelial (Kupffer's) cells.
24.20 While a meal is being absorbed, hepatocytes remove nutrients from blood flowing through liver sinusoids. They also remove O_2 and extract certain toxic substances.
24.21 Secretin promotes liberation of fluid rich in HCO_3^- in both organs.
24.22 Ileum.
24.23 Nutrients being absorbed enter blood or lymph there.
24.24 Their alkaline mucus neutralizes gastric acid and protects the mucosal lining of the duodenum.
24.25 They are nonpolar (hydrophobic) molecules and thus can dissolve in and diffuse through the phospholipid bilayer of the plasma membrane.
24.26 Stomach and pancreas.
24.27 Ascending and descending portions of the colon.
24.28 Secrete mucus to lubricate colonic contents.

CHAPTER 25

METABOLISM

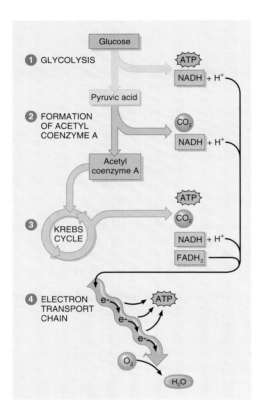

① GLYCOLYSIS

Glucose

ATP

NADH + H⁺

Pyruvic acid

② FORMATION OF ACETYL COENZYME A

CO₂

NADH + H⁺

Acetyl coenzyme A

③ KREBS CYCLE

ATP

CO₂

NADH + H⁺

FADH₂

④ ELECTRON TRANSPORT CHAIN

e⁻

ATP

e⁻

e⁻

O₂

H₂O

STUDENT OBJECTIVES

1. Explain how food intake is regulated. (p. 807)

2. Define a nutrient and list the functions of the six principal classes of nutrients. (p. 807)

3. Define metabolism and explain the role of ATP in anabolism and catabolism. (p. 808)

4. Define basal metabolic rate (BMR) and explain several factors that affect it. (p. 809)

5. Explain how normal body temperature is maintained by the hypothalamic thermostat, heat production and conservation, and heat loss. (p. 809)

6. Describe fever and hypothermia as abnormalities of temperature regulation. (p. 812)

7. Describe oxidation–reduction reactions and explain the role of ATP in metabolism. (p. 813)

8. Describe the metabolism of carbohydrates, lipids, and proteins. (p. 814)

9. Compare the metabolic reactions occurring during the absorptive (fed) and postabsorptive (fasting) states and explain their hormonal regulation. (p. 832)

10. Describe how to select foods to include in a healthy diet. (p. 835)

11. Compare the sources, functions, and importance of minerals and vitamins in metabolism. (p. 837)

12. Define heat cramps, heat exhaustion, heatstroke, obesity, vitamin or mineral overdose, malnutrition, phenylketonuria (PKU), and celiac disease. (p. 842)

CHAPTER CONTENTS AT A GLANCE

lants use the green pigment chlorophyll to trap energy in sunlight. Because we don't have chlorophyll, the food we eat is our only source of energy for performing biological work such as synthesis of proteins, contraction of muscle, mitosis, and active transport. In addition, food provides building blocks, such as amino acids, needed for repair and growth of body tissues. Many molecules needed to maintain cells and tissues can be made from simpler precursors by metabolic reactions within the body. Others—the essential amino acids, essential fatty acids, vitamins, and minerals—must be present in our food because we cannot synthesize them. Food molecules absorbed by the gastrointestinal (GI) tract have three fates:

1. Most are used to supply energy for sustaining life processes, for example, active transport, DNA replication, synthesis of proteins and other molecules, and muscle contraction.
2. Some are used as building blocks during synthesis of structural or functional molecules such as muscle proteins, hormones, and enzymes.
3. Others are stored for future use, for example, glycogen stored in liver cells and triglycerides stored in adipose cells.

Any molecule that serves one or more of these functions is called a **nutrient**. There are six principal classes of nutrients: carbohydrates, lipids, proteins, minerals, vitamins, and water. Carbohydrates, lipids, and proteins are digested by enzymes in the GI tract. The end products of digestion that reach body cells are monosaccharides, fatty acids, glycerol, monoglycerides, and amino acids. Some minerals and many vitamins are part of enzyme systems that catalyze the reactions of carbohydrates, lipids, and proteins.

In this chapter, we will discuss the regulation of food intake, the reactions that harvest the chemical energy in nutrients to power cellular activities, and how each group of nutrients contributes to the body's growth, repair, and energy needs.

REGULATION OF FOOD INTAKE

When the energy content of your food balances the energy needs of all the cells of your body, you stay at the same weight. (This is true for adults unless there is a gain or loss of water, which is a separate topic that is discussed on page 892.) In the more affluent countries of the world, however, a significant fraction of the population is obese (fat). Although we often think of weight gain as "middle age spread," the problem of obesity among children has risen dramatically in the U.S. Being grossly overweight increases one's risk of dying from a variety of cardiovascular and metabolic disorders, for example, diabetes mellitus.

Most mature animals and many men and women maintain a stable body weight for long periods of time. Stability may persist despite large day-to-day variations in activity level and food intake. But there are no sensory receptors that monitor one's weight or size. How then is food intake regulated? The answer to this question is still not well understood. It seems to depend on many factors, including levels of certain nutrients in the blood, particular hormones, psychological elements such as stress or depression, signals from the GI tract and the special senses, and neural connections between the hypothalamus and other parts of the brain.

Within the hypothalamus are nuclei that play key roles in regulating food intake (see Fig. 14.10). One is a cluster of neurons (nerve cells) in the lateral hypothalamic nuclei called the **feeding (hunger) center**. When this area is stimulated in animals, they begin to eat heartily, even if they are already full. The second center is a cluster of neurons in the ventromedial nuclei of the hypothalamus called the **satiety center**. (**Satiety** is the opposite of appetite; it is a sensation of fullness with lack of desire to eat.) Stimulation of this center causes animals to stop eating, even if they have been starved for days. Apparently the feeding center is constantly active, but it is inhibited by the satiety center. The hypothalamus functions as a complex integrating and relay center rather than the absolute controller of eating behavior. It receives nerve impulses from other parts of the brain, such as the brain stem, limbic system, and cerebral cortex, and from a variety of sensory receptors. These include chemoreceptors, photoreceptors, olfactory receptors, taste receptors, and stretch receptors in the GI tract.

If neurons in the hypothalamus regulate food intake, how do they sense whether a person is well-fed or in need of food? One possibility is that changes in the chemical composition of the blood after eating and fasting might alert the hypothalamic neurons. According to the **glucostatic theory**, when blood glucose levels are low, feeding increases. It is thought that hypoglycemia (low blood glucose level) decreases the activity of the neurons in the satiety center to the point where they no longer inhibit the feeding center, and the person eats.

A second theory relates changes in lipid metabolism to control of food intake. As the amount of adipose tissue increases in the body, the rate of feeding usually decreases. The **lipostatic theory** holds that some substance or substances, perhaps fatty acids, are released from adipose stores in proportion to the total fat content of the body. The released substances then activate neurons of the satiety center, which, in turn, inhibit the feeding center.

Another factor that affects food intake is body temperature. Whereas a cold environment enhances eating, a warm environment depresses it. Food intake is also regulated by distension of the GI tract, particularly the stomach and duodenum. When these organs are stretched, a reflex is initiated that activates the satiety center and depresses the feeding center. It has also been shown that the hormone cholecystokinin (CCK), secreted when triglycerides enter the small intestine, inhibits eating. For this reason, CCK has been called the satiety hormone. Psychological factors may override the usual intake mechanisms, for example, in anorexia nervosa, bulimia, and obesity (see pages 801 and 842).

METABOLISM

Metabolism (me-TAB-ō-lizm; *metabole* = change) refers to all the chemical reactions of the body. It may be thought of as an energy-balancing act between catabolic (degradative) reactions and anabolic (synthesis) reactions. Overall, catabolic reactions are exergonic; they produce more energy than they consume. On the other hand, anabolic reactions are endergonic; they consume more energy than they produce. The molecule that participates most often in energy exchanges of living cells is **ATP (adenosine triphosphate)**. It couples energy-releasing catabolic reactions and energy-requiring anabolic reactions.

Which reactions occur depend on which enzymes are active in a particular cell at a particular time. (You might want to review the characteristics of enzymes on page 46.) Often, catabolic reactions occur in one compartment of a cell, for example, the mitochondria, whereas synthetic reactions take place in another location such as the cytosol or endoplasmic reticulum.

Once a molecule is synthesized, it has a limited lifetime. With few exceptions, it will be broken down, and its component atoms recycled into other molecules or excreted from the body. Recycling is a continual and ongoing process in living tissues, occurring rapidly in some and very slowly in others. Individual cells may be refurbished, molecule-by-molecule or a whole tissue may be rebuilt cell-by-cell.

Catabolism

The chemical reactions that break down complex organic molecules and polymers into simpler ones are collectively known as **catabolism** (ka-TAB-ō-lizm; *cata* = downward; *ballein* = to throw). Catabolic reactions release the available chemical energy in organic molecules. This chemical energy is captured in an easy-to-use form by phosphorylating ADP (adenosine diphosphate) to create ATP (Fig. 25.1). Important sets of catabolic reactions are those occurring in glycolysis, the Krebs cycle, and the electron transport chain (see Fig. 25.2). Glycolysis consists of the breakdown of a glucose molecule, which has six carbon atoms, into two pyruvic acid molecules, with three carbons each. It consumes two ATP but produces four ATP, for a net gain of two ATP. The Krebs cycle and ensuing reactions of the electron transport chain provide even more ATP. The breakdown of glycogen into glucose, proteins into amino acids, and triglycerides into fatty acids and glycerol are other examples of catabolism.

Anabolism

Chemical reactions that combine simple molecules and monomers to make more complex ones that form the body's structural and functional components are collectively known as **anabolism** (a-NAB-ō-lizm; *ana* = upward). One example of an anabolic process is the formation of peptide bonds between amino acids, thereby building the amino acids into proteins.

Figure 25.1   Role of ATP in linking anabolic and catabolic reactions. When complex molecules and polymers are split apart (catabolism), some of the energy is transferred to and trapped in ATP and the rest is given off as heat. When simple molecules and monomers are combined to form complex molecules (anabolism), ATP provides the energy for synthesis and again some energy is given off as heat.

The coupling of energy-releasing and energy-requiring reactions is achieved through ATP.

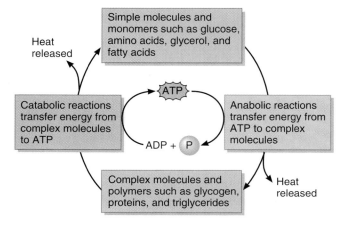

In a cell of the pancreas that is making digestive enzymes, does anabolism or catabolism dominate?

Also, fatty acids can be built into phospholipids that form the plasma membrane bilayer, and glucose monomers can be put together to form glycogen through anabolic reactions.

Coupling of Catabolism and Anabolism by ATP

The chemical reactions of living systems depend on efficiently transferring manageable amounts of energy from one molecule to another. ATP is the molecule that accomplished this task. It is the "energy currency" of a living cell. Like money, it is readily available to "buy" cellular activities; it is spent and remade over and over. A typical cell has about a billion molecules of ATP, which usually last for less than a minute before being used. So it is not a long-term storage form of currency (like gold in a vault), but rather convenient cash for moment-to-moment transactions.

Recall from Chapter 2 that a molecule of ATP consists of an adenine molecule, a ribose molecule, and three phosphate groups bonded to each other (see Fig. 2.17). Figure 25.1 shows how ATP links anabolic and catabolic reactions. When the terminal phosphate group is split from ATP, adenosine diphosphate (ADP) and a phosphate group (symbolized Ⓟ) are formed. The energy released is used to drive anabolic reactions such as formation of glycogen from glucose. Then the energy from catabolic reactions is used to combine ADP and a phosphate group to resynthesize ATP (ADP + Ⓟ + energy → ATP).

About 40% of the energy released in catabolism is available for cellular functions. The rest of the energy is converted to heat, which helps maintain normal body temperature. Excess heat is lost to the environment. Compared with man-made machines, which typically operate at 10–20% efficiency, a 40% energy yield is quite large. Still, there is a continuous need for new external sources of energy for cells to synthesize enough ATP to sustain life.

METABOLISM AND BODY TEMPERATURE

Depending on the rates of metabolic reactions, more or less heat is produced. Homeostasis of body temperature can be maintained only if the rate of body heat loss equals the rate of body heat production. Thus it is important to understand the mechanisms of heat gain, heat conservation, and heat loss.

Basal Metabolic Rate

The overall rate at which heat is produced is termed the **metabolic rate**. Since many factors affect metabolic rate, it is measured under standard conditions designed to reduce or eliminate those factors as much as possible. These conditions of the body are called the **basal state**, and the measurement obtained is the **basal metabolic rate** (**BMR**). BMR is a measure of the rate at which the quiet, resting, fasting body breaks down nutrients to liberate energy. Some of the energy is used to form ATP and some is released as heat. BMR also is an indicator of how much thyroid hormone is present because this hormone is the main regulator of the basal rate of ATP use.

Heat is a form of kinetic energy that can be measured as **temperature** and expressed in units called calories. A **calorie**, spelled with a little c, is the amount of heat energy required to raise the temperature of 1 gram of water from 14°C to 15°C. Since the calorie is a small unit relative to the large amount of energy stored in foods, the **kilocalorie** (**kcal**) or **Calorie** (**Cal**), spelled with a big C, is used instead. A kilocalorie or Calorie is equal to 1000 calories. The kilocalorie is the unit used to express the heating value of foods and to measure the body's metabolic rate.

Basal metabolic rate is most often measured indirectly by measuring oxygen consumption using a spirometer. To release a given amount of heat energy when it is oxidized, a nutrient must combine with a given amount of oxygen. Thus by measuring the amount of oxygen needed for the metabolism of foods, we can determine how many kilocalories are produced. The amount of heat energy released when 1 liter of oxygen combines with carbohydrates is 5.05 kcal; with triglycerides, the heat released is 4.70 kcal; with proteins, the heat released is 4.60 kcal. On a typical diet, uptake of 1 liter of oxygen indicates about 4.9 kcal of heat produced.

The usual way to express basal metabolic rate is in kilocalories per square meter of body surface area per hour ($kcal/m^2/hr$). Suppose you use 1.8 liters of oxygen in 6 minutes as recorded on a spirometer. This means that your oxygen consumption in an hour would be 18 liters:

$$(1.8 \text{ liters}/6 \text{ min}) \times 60 \text{ min/hr} = 18 \text{ liters/hr}$$

Your basal metabolic rate would be about 18 liters/hr × 4.9 kcal/liter or 88 kcal/hr. To express the kilocalories per square meter of body surface, a standardized chart is used. Such a chart shows square meters of body surface relative to height in centimeters and weight in kilograms. If you weigh 75 kg (165 lb) and are 190 cm (75 in.) tall, your body surface area is 2 m^2. Your basal metabolic rate is equal to 88 kcal/hr divided by 2m^2, or about 44 $kcal/m^2/hr$.

The normal BMRs for various age groups by sex are also listed in standardized charts. Values 15% above or below the standard may indicate an excess or deficiency of thyroid hormone. When the thyroid gland is secreting extremely large quantities of thyroid hormone, BMR can double. If, on the other hand, it is secreting very little thyroid hormone, BMR may be half the normal value.

Homeostasis of Body Temperature

Even though there are wide fluctuations in environmental temperature, homeostatic mechanisms can maintain a normal range for the internal body temperature. If your heat production equals heat loss, you maintain a constant core temperature near 37°C (98.6°F). **Core temperature** refers to the body's temperature in body structures below the skin and subcutaneous tissue. **Shell temperature** refers to the body's temperature at the surface, that is, the skin and subcutaneous tissue. Core temperature is usually a little higher than shell temperature. If your heat-producing mechanisms generate more heat than is lost by your heat-losing mechanisms, your core temperature rises. If your heat-losing mechanisms give off more heat than is generated by heat-producing mechanisms, your core temperature falls. Too high a core temperature kills by denaturing body proteins, while too low a core temperature causes cardiac arrhythmias that result in death.

Hypothalamic Thermostat

Body temperature is regulated by mechanisms that attempt to keep heat production and heat loss in balance. A center of control for those mechanisms that are reflex in nature is found in the hypothalamus in a group of neurons in the anterior portion referred to as the **preoptic area**. This area receives input from temperature receptors in the skin and mucous membranes (peripheral thermoreceptors) and in internal structures (central thermoreceptors), including the hypothalamus. If blood temperature increases, the neurons of the preoptic area fire nerve impulses more rapidly. If something causes the blood's temperature to decrease, these neurons fire nerve impulses less rapidly. The preoptic area maintains normal body temperature and thus serves as your thermostat.

Nerve impulses from the preoptic area are sent to other portions of the hypothalamus known as the heat-losing center and the heat-promoting center. The **heat-losing center**, when stimulated by the preoptic area, sets into operation a series of responses that lower body temperature. The **heat-promoting center**, when stimulated by the preoptic area, sets into operation a series of responses that raise body temperature. The heat-losing center is mainly parasympathetic in function; the heat-promoting center is primarily sympathetic.

Heat Production and Conservation

The production and conservation of body heat are influenced by metabolic rate and by responses that occur when body temperature starts to fall. Among the factors that affect metabolic rate and thus production of body heat are the following:

1. **Exercise.** During strenuous exercise, the metabolic rate may increase to as much as 15 times the basal rate. In well-trained athletes, the rate may increase up to 20 times.
2. **Hormones.** Thyroid hormones (thyroxine and triiodothyronine) are the main regulators of BMR, which increases as the blood levels of thyroid hormones rise. Increased secretions of testosterone and human growth hormone also increase the metabolic rate.
3. **Nervous system.** In a stressful situation, the sympathetic division of the autonomic nervous system is stimulated and the nerves release norepinephrine (NE). The sympathetic division also stimulates release of the hormones epinephrine and norepinephrine by the adrenal medulla in stressful situations. Both epinephrine and norepinephrine increase the metabolic rate of body cells.
4. **Body temperature.** The higher the body temperature, the higher the metabolic rate. Each 1°C rise in temperature increases the rate of biochemical reactions by about 10%. Thus metabolic rate may be substantially increased during fever.
5. **Ingestion of food.** The ingestion of food can raise metabolic rate by as much as 10–20%. This effect is called **specific dynamic action** (**SDA**) and is greatest with proteins and less with carbohydrates and lipids.
6. **Age.** The metabolic rate of a child, in relation to its size, is about double that of an elderly person because the high rates of reactions related to growth.
7. **Others.** Other factors that affect metabolic rate are gender (lower in females, except during pregnancy and lactation), climate (lower in tropical regions), sleep (lower), and malnutrition (lower).

If body temperature starts to decrease, changes occur that help conserve heat and produce heat at a quicker pace. The changes are part of a negative feedback system that attempts to raise body temperature (controlled condition) to normal. It works as follows (Fig. 25.2). Thermoreceptors in the skin and hypothalamus (receptors) send nerve impulses (input) to the preoptic area and heat-promoting center in the hypothalamus (control centers). In response, the hypothalamus discharges nerve impulses and secretes thyrotropin-releasing hormone or TRH (output), which activate several effectors. Each effector responds in a way that helps increase body temperature to the normal value.

- **Vasoconstriction** Nerve impulses from the heat-promoting center stimulate sympathetic nerves that cause blood vessels of the skin to constrict. The net effect of vasoconstriction is to decrease the flow of warm blood from the internal organs to the skin, thus decreasing the transfer of heat from the internal organs to the skin. This reduction in heat loss helps raise the internal body temperature.

- **Sympathetic Stimulation** Another response triggered by the heat-promoting center is the sympathetic stimulation of metabolism. The heat-promoting center stimulates sympathetic nerves leading to the adrenal medulla. This stimulation causes the medulla to secrete epinephrine and norepinephrine into the blood. The hormones, in turn, bring about an increase in cellular metabolism, a reaction that also increases heat production. This effect is called **chemical thermogenesis**.

- **Skeletal Muscles** Heat production is also increased by responses of skeletal muscles. For example, stimulation of the heat-promoting center causes stimulation of parts of the brain that increase muscle tone and hence heat production. As muscle tone increases, the stretching of the agonist muscle initiates the stretch reflex and the muscle contracts. This contraction causes the antagonist muscle to stretch, and it too develops a stretch reflex. The repetitive cycle—called **shivering** (**involuntary thermogenesis**)—increases the rate of heat production. During maximal shivering, body heat production can rise to about four times the basal rate in just a few minutes.

- **Thyroid Hormones** Another body response that increases heat production is increased production of thyroid hormones. A cold environmental temperature increases the secretion of TRH, which, in turn, stimulates the anterior pituitary gland to secrete thyroid-stimulating hormone (TSH). The thyroid gland responds to TSH by releasing more thyroid hormones into the blood. As increased levels of thyroid hormones increase the metabolic rate, body temperature rises.

Heat Loss

Maintaining normal body temperature depends on the ability to lose heat just as fast as it is produced. Heat is lost from the body by radiation, evaporation, conduction, and convection.

- **Radiation** The transfer of heat as infrared heat rays from a warmer object to a cooler one without physical contact is called **radiation**. Your body loses heat by the radiation of heat waves to cooler objects nearby such as ceilings, floors, and walls. If these objects are at a higher temperature, you absorb heat by radiation. Incidentally, the air temperature has no relationship to the radiation of heat to and from objects. Skiers can remove their shirts in bright sunshine even though the air temperature is very low because the radiant heat from the sun is adequate to warm them. In a room at 21°C (70°F), about 60% of heat loss is by radiation in a resting person.

- **Evaporation** The conversion of a liquid to a vapor is called **evaporation**. Water has a *high heat of evaporation,*

Figure 25.2   Negative feedback mechanisms that conserve heat and increase heat production.

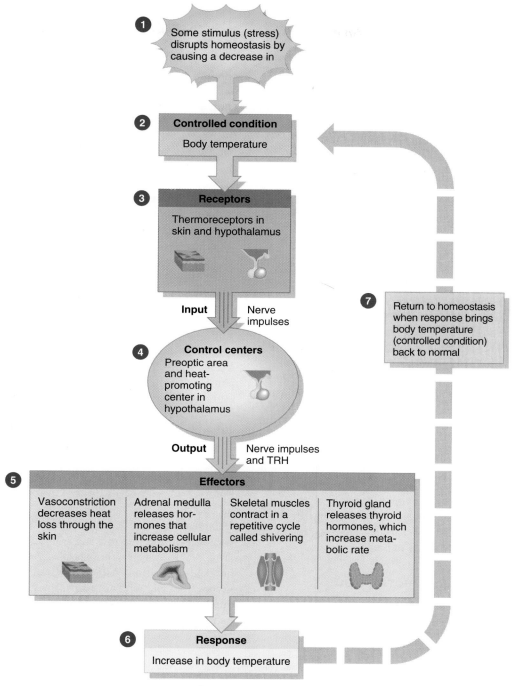

1   Some stimulus (stress) disrupts homeostasis by causing a decrease in

2   **Controlled condition**
Body temperature

3   **Receptors**
Thermoreceptors in skin and hypothalamus

Input   Nerve impulses

4   **Control centers**
Preoptic area and heat-promoting center in hypothalamus

Output   Nerve impulses and TRH

7   Return to homeostasis when response brings body temperature (controlled condition) back to normal

5   **Effectors**

| Vasoconstriction decreases heat loss through the skin | Adrenal medulla releases hormones that increase cellular metabolism | Skeletal muscles contract in a repetitive cycle called shivering | Thyroid gland releases thyroid hormones, which increase metabolic rate |

6   **Response**
Increase in body temperature

Q   What factors can increase your metabolic rate and thus increase your rate of heat production?

which is the amount of heat needed to evaporate 1 gram of water at 30°C (86°F). Because of water's high heat of evaporation, every gram of water evaporating from the skin takes with it a great deal of heat—about 0.58 kcal per gram of water. Under normal resting conditions, about 22% of heat loss occurs through evaporation. Under extreme conditions, about 4 liters (1 gal) of perspiration are produced each hour, and this volume can remove more than 2000 kcal of heat from the body. This is approximately 32 times the basal level of heat production. The rate of evaporation is inversely related to **relative humidity**, the ratio of the actual amount of moisture in the air to the greatest amount it can hold at a given temperature. The higher the relative humidity, the lower the rate of evaporation due to a reduced diffusion rate for water leaving the surface of the body.

● **Conduction** The transfer of heat to a substance or object in contact with the body, such as chairs, clothing, jewelry, air, or water is termed **conduction**. At rest, about 3% of body heat is lost via conduction to solid objects.

● **Convection** The transfer of heat by the movement of a liquid or gas between areas of different temperature is called **convection**. The contact of air or water with your body results in heat transfer by both conduction and convection. When cool air makes contact with the body, it becomes warmed and therefore less dense and is carried away by convection currents as the less dense air rises. Then the cycle repeats as more cool air makes contact with the body and is carried away as it warms by conduction and becomes less dense. The faster the air moves, for example, by a breeze or a fan, the faster the rate of convection. At rest, about 15% of body heat is lost to the air by conduction and convection.

If some stress raises body temperature above normal, a negative feedback loop opposite to the one shown in Fig. 25.2 swings into action. The higher temperature of the blood stimulates thermoreceptors that send nerve impulses to the preoptic area, which, in turn, stimulate the heat-losing center and inhibit the heat-promoting center. The heat-losing center discharges nerve impulses to blood vessels in the skin, causing them to dilate. The skin becomes warm, and the excess heat is lost to the environment as an increased volume of blood flows from the core of the body into the skin. At the same time, metabolic rate decreases and shivering does not occur. The high temperature of the blood, by way of hypothalamic activation of sympathetic nerves, stimulates sweat glands of the skin to produce perspiration. As the water of the perspiration evaporates from the surface of the skin, the skin is cooled. All these responses reverse the heat-promoting effects and decrease body temperature to normal.

Voluntary activities, such as putting on or taking off clothing, along with the negative feedback responses discussed above, help achieve homeostasis of body temperature.

Body Temperature Abnormalities

Fever

A **fever** is an abnormally high body temperature. The most frequent cause of fever is a viral or bacterial infection (or bacterial toxins). Other causes are heart attacks, tumors, tissue destruction by x-rays, surgery or trauma, and reactions to vaccines. A fever-producing substance is called a **pyrogen** (PĪ-rō-gen; *pyr* = fire). The mechanism of fever production is believed to occur as follows. When phagocytes—namely, monocytes and macrophages—ingest certain bacteria, a portion of the cell wall of the bacteria is released, causing the phagocytes to secrete interleukin-1. Interleukin-1 acts as a pyrogen; it circulates to the anterior hypothalamus and induces neurons of the preoptic area to secrete prostaglandins, particularly of the E series. Prostaglandins reset the hypothalamic thermostat at a higher temperature, and temperature-regulating reflex mechanisms will then act to bring the core body temperature up to this new setting. Aspirin, acetaminophen

(for example, Tylenol®), and ibuprofen (for example, Advil®) reduce fever by inhibiting synthesis of prostaglandins.

Suppose that as a result of pyrogens the thermostat is set at 39°C (103°F). Now the heat-promoting mechanisms (vasoconstriction, increased metabolism, shivering) are operating at full force. Thus even though body temperature is climbing higher than normal—say, 38°C (101°F)—the skin remains cold, and shivering occurs. This condition, called a **chill**, is a definite sign that body temperature is rising. After several hours, body temperature reaches the setting of the thermostat and the chills disappear. But the body will continue to regulate temperature at 39°C (103°F) until the stress is removed. When the stress is removed, the thermostat is reset at normal—37.0°C (98.6°F). Since body temperature remains high in the beginning, the heat-losing mechanisms (vasodilation and sweating) go into operation to decrease body temperature. The skin becomes warm and the person begins to sweat. This phase of the fever is called the **crisis** and indicates that body temperature is falling.

Up to a point, fever is beneficial. Interleukin-1 helps step up production of T cells. Higher body temperature intensifies the effect of interferon and the phagocytic activities of macrophages while hindering replication of some pathogens. Fever also increases heart rate so that white blood cells are delivered to sites of infection more rapidly and their secretions are increased. In addition, antibody production and T cell proliferation increase. Moreover, heat speeds up the rate of chemical reactions. This increase may help body cells repair themselves more quickly during a disease. Among the complications of fever are dehydration, acidosis, and permanent brain damage. As a rule, death results if body temperature rises above 44–46°C (112–114°F). On the other end of the scale, death usually results when body temperature falls below 21–24°C (70–75°F).

Hypothermia

Hypothermia refers to a lowering of body temperature to 35°C (95°F) or below. It may be caused by an overwhelming cold stress (immersion in icy water), metabolic diseases (hypoglycemia, adrenal insufficiency, or hypothyroidism), drugs (alcohol, antidepressants, sedatives, or tranquilizers), burns, malnutrition, transection of the cervical spinal cord, and lowering of body temperature for surgery. Hypothermia is characterized by the following as body temperature falls: sensation of cold, shivering, confusion, vasoconstriction, muscle rigidity, bradycardia, acidosis, hypoventilation, hypotension, ventricular fibrillation, no reflexes and loss of spontaneous movement, coma, and possibly death usually caused by cardiac arrhythmias. The elderly have lessened metabolic protection against a cold environment coupled with reduced perception of cold. As a result, they are at greater risk for developing hypothermia.

In 1986, a 2½-year-old girl was submerged in icy water for 66 minutes and suffered no neurological damage. Part of the reason for a successful outcome was the use of a heart–lung bypass machine that rewarmed the blood of the child and brought body temperature back to normal.

ENERGY PRODUCTION

Nutrient molecules, like all molecules, have energy stored in the bonds between their atoms. Various reactions in catabolic pathways concentrate the energy as it is released into the high-energy phosphate bonds of ATP. Before discussing metabolic pathways, we will first consider two important aspects of energy production—oxidation–reduction reactions and mechanisms of ATP generation.

Oxidation–Reduction Reactions

Within a cell, oxidation and reduction reactions are always coupled; that is, whenever one substance is oxidized, another is reduced almost simultaneously. Such coupled reactions are referred to as **oxidation–reduction (redox) reactions**.

Oxidation is the *removal of electrons* from a molecule and results in a decrease in the energy content of the molecule. In many cellular oxidations, a hydrogen ion or proton (H^+, a hydrogen nucleus with no orbiting electrons) and a hydride ion (H^-, a hydrogen nucleus with two orbiting electrons) are removed at the same time; this is equivalent to the removal of two hydrogen atoms ($H^+ + H^- = 2 H$). Because most biological oxidations involve the loss of hydrogen atoms, they are called *dehydrogenation reactions*. An example of an oxidation is the conversion of lactic acid into pyruvic acid.

$$\underset{\substack{\text{Lactic acid}}}{\begin{array}{c} COOH \\ | \\ HC-OH \\ | \\ CH_3 \end{array}} \xrightarrow[\text{(oxidation)}]{\text{remove 2 H (H}^+ + \text{ H}^-)} \underset{\substack{\text{Pyruvic acid}}}{\begin{array}{c} COOH \\ | \\ C=O \\ | \\ CH_3 \end{array}}$$

Reduction is the opposite of oxidation; it is the *addition of electrons* to a molecule. Reduction results in an increase in the energy content of the molecule. An example of reduction is the conversion of pyruvic acid into lactic acid.

$$\underset{\substack{\text{Pyruvic acid}}}{\begin{array}{c} COOH \\ | \\ C=O \\ | \\ CH_3 \end{array}} \xrightarrow[\text{(reduction)}]{\text{add 2 H (H}^+ + \text{ H}^-)} \underset{\substack{\text{Lactic acid}}}{\begin{array}{c} COOH \\ | \\ HC-OH \\ | \\ CH_3 \end{array}}$$

When a substance is oxidized, the liberated hydride ions do not remain free in the cell but are transferred immediately by coenzymes to another compound. Two coenzymes are commonly used by animal cells to carry hydrogen atoms. They are **nicotinamide adenine dinucleotide (NAD^+)**, a derivative of the B vitamin and niacin, **flavin adenine dinucleotide (FAD)**, a derivative of vitamin B_2 (riboflavin). The oxidation and reduction states of NAD^+ and FAD can be represented as follows:

$$\underset{\text{Oxidized}}{NAD^+} \underset{-2 \text{ H (H}^+ + \text{ H}^-)}{\overset{+2 \text{ H (H}^+ + \text{ H}^-)}{\rightleftharpoons}} \underset{\text{Reduced}}{NADH + H^+}$$

$$\underset{\text{Oxidized}}{FAD} \underset{-2 \text{ H (H}^+ + \text{ H}^-)}{\overset{+2 \text{ H (H}^+ + \text{ H}^-)}{\rightleftharpoons}} \underset{\text{Reduced}}{FADH_2}$$

In the preceding equations, 2 H ($H^+ + H^-$) means that two neutral hydrogen atoms (2 H) are equivalent to one hydrogen ion (H^+) plus one hydride ion (H^-). When NAD^+ is reduced to $NADH + H^+$, the NAD^+ gains a hydride ion (H^-). The H^+ ion is released into the surrounding solution. The addition of a hydride ion to NAD^+ neutralizes the charge on NAD^+ and adds a hydrogen atom so that the reduced form is NADH. On the other hand, when NADH is oxidized to NAD^+, a hydride ion is lost from NADH, which results in one less hydrogen atom and an additional positive charge. Thus the oxidized form is NAD^+. FAD is reduced to $FADH_2$ when it gains a hydrogen ion and a hydride ion, and $FADH_2$ is oxidized to FAD when it loses the same two ions.

When lactic acid is *oxidized* to form pyruvic acid, the two hydrogen atoms removed in the reaction are used to *reduce* NAD. This coupled oxidation–reduction reaction may be written as follows:

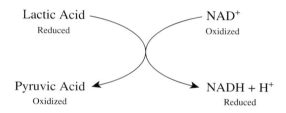

An important point to remember about oxidation–reduction reactions is that oxidation is usually an energy-producing reaction. Cells degrade nutrients that serve as energy sources in many biochemical steps from energy-rich, highly reduced compounds (with many hydrogen atoms) to lower energy, highly oxidized compounds (with many oxygen atoms or multiple bonds). For example, when a cell oxidizes a molecule of glucose ($C_6H_{12}O_6$), the energy in the glucose molecule is removed in a stepwise manner. Ultimately, some of the energy is trapped by transferring it to ATP, which then serves as an energy source for energy-requiring reactions. Compounds such as glucose that have many hydrogen atoms are highly reduced compounds. They contain more chemical potential energy than oxidized compounds. For this reason, glucose is a valuable nutrient for organisms.

Generation of ATP

Some of the energy released during oxidation reactions is trapped within a cell as ATP is formed. To summarize, a phosphate group (Ⓟ) is added to ADP, with an input of energy, to form ATP. The two high-energy phosphate bonds that can be used to transfer energy are indicated by "squiggles" ($\sim$).

$$\text{Adenosine–}\textcircled{P}\sim\textcircled{P}+\text{Energy}\rightarrow\text{Adenosine–}\textcircled{P}\sim\textcircled{P}\sim\textcircled{P}$$
$$\underset{\text{ADP}}{}\qquad\qquad\qquad\underset{\text{ATP}}{}$$

The high-energy phosphate bond that attaches the third phosphate group contains the energy stored in this reaction. The addition of a phosphate group to a chemical compound is called **phosphorylation** (fos′-for-i-LĀ-shun) and raises the energy level of the molecule. Enzymes that catalyze phosphorylation reactions are called **kinases**. Organisms use three mechanisms of phosphorylation to generate ATP.

1. In **substrate-level phosphorylation**, ATP is generated when a high-energy phosphate group is transferred directly from an intermediate phosphorylated metabolic compound to ADP. In cells of the human body, this occurs in the cytosol. For example, substrate-level phosphorylation occurs twice in glycolysis (see steps seven and ten in Fig. 25.4).
2. In **oxidative phosphorylation**, electrons are removed from organic compounds (usually by NAD$^+$) and passed through a series of electron acceptors to molecules of oxygen (O$_2$). This process occurs in the inner mitochondrial membrane of cells. The series of electron acceptors used in oxidative phosphorylation is called the **electron transport chain** (see Fig. 25.8). The transfer of electrons from one electron acceptor to the next releases energy, which is used to generate ATP from ADP and $\textcircled{P}$ through a process called **chemiosmosis** (see Fig. 25.7).
3. The third type of phosphorylation, **photophosphorylation**, which will not be discussed here, occurs only in photosynthetic cells that contain a light-absorbing pigment such as chlorophyll.

We will now consider the metabolism of carbohydrates, lipids, and proteins in body cells.

CARBOHYDRATE METABOLISM

During digestion, polysaccharides and disaccharides are hydrolyzed into the monosaccharides glucose (about 80%), fructose, and galactose. Some fructose is converted into glucose as it is absorbed through the intestinal epithelial cells. The three monosaccharides are absorbed into the capillaries of the villi of the small intestine and then are carried through the hepatic portal vein to the liver. Liver cells convert much of the remaining fructose and practically all the galactose to glucose. Thus the story of carbohydrate metabolism is really the story of glucose metabolism.

Fate of Carbohydrates

Since glucose is the body's preferred source for synthesizing ATP, the fate of absorbed glucose depends on the energy needs of body cells.

1. **ATP production.** If the cells require immediate energy, glucose is oxidized by the cells. Each gram of carbohydrate produces about 4 kilocalories (kcal). (The kilocalorie content of a food is a measure of the heat it releases upon oxidation. Determination of caloric value is described on page 835.) Glucose not needed for immediate ATP production can enter one of several metabolic pathways.
2. **Amino acid synthesis.** Glucose can be used to form several amino acids, which then can be incorporated into proteins.
3. **Glycogenesis.** The liver can store a small amount of excess glucose by converting it to glycogen (*glycogenesis*). Later, when blood glucose starts to decrease, hepatocytes (liver cells) can convert glycogen back to glucose (*glycogenolysis*) and release it into the blood. In this way the liver provides glucose for other cells to oxidize. Skeletal muscle fibers (cells) can also store glycogen and oxidize it to provide ATP for their own use. However, they lack the enzyme needed to release glucose into the blood.
4. **Lipogenesis.** If the glycogen storage areas are filled up, hepatocytes can transform the glucose to glycerol and fatty acids that can be used for synthesis of triglycerides (*lipogenesis*). Triglycerides then are deposited in adipose tissue, which has virtually unlimited storage capacity.
5. **Excretion in urine.** Excess glucose occasionally is excreted in the urine. When blood glucose level is very high, the kidneys may not recover all glucose molecules that are filtered and some glucose may be lost in the urine. Normally, this happens only when a meal containing mostly carbohydrates and no triglycerides is eaten. Without the inhibiting effect of triglycerides, the stomach empties its contents quickly, and the carbohydrates are all digested at the same time. As a result, a large amount of glucose suddenly floods into the bloodstream. Diabetics who experience high blood glucose levels often, however, may regularly lose glucose in the urine.

Glucose Movement into Cells

Before glucose can be used by body cells, it must first pass through the plasma membrane and enter the cytosol. Whereas glucose absorption in the GI tract (and kidney tubules) is accomplished by secondary active transport (Na$^+$–glucose symporters), glucose movement from blood into most other body cells occurs by facilitated diffusion (see page 61). Insulin increases the rate of facilitated diffusion of glucose into most cells. In neurons and hepatocytes, however, glucose entry is always "turned on" and insulin does not regulate glucose uptake. Immediately upon entry into cells, glucose is phosphorylated. It combines with a phosphate group, produced by the breakdown of ATP, to form glucose 6-phosphate (see Fig. 25.11). Phosphorylation traps glucose in the cell so that it cannot diffuse back out. Hepatocytes, kidney tubule cells, and intestinal epithelial cells have the necessary enzyme (phosphatase) to remove the phosphate group, which enables glucose to diffuse out of these cells and into the bloodstream (see Fig. 25.11).

Glucose Catabolism

The **oxidation** of glucose is also known as **cellular respiration** (Fig. 25.3). It involves **1** glycolysis, **2** formation of acetyl coenzyme A, **3** the Krebs cycle, and **4** the electron transport chain. Glycolysis is the oxidation of glucose to pyruvic acid and occurs in most cells in the body. It pro-

vides some ATP and energy-containing $NADH + H^+$. Because glycolysis does not require oxygen, it is a way to produce ATP anaerobically (without oxygen) and is known as **anaerobic cellular respiration**. The formation of acetyl coenzyme A from pyruvic acid is a transition step between glycolysis and the Krebs cycle that prepares pyruvic acid for entrance into the cycle. In this step, energy-containing $NADH + H^+$ plus carbon dioxide (CO_2) are produced.

The Krebs cycle and the electron transport chain together require oxygen to produce ATP and are known as **aerobic cellular respiration**. The Krebs cycle oxidizes acetyl coenzyme A and produces ATP, energy-containing $NADH + H^+$, and $FADH_2$ plus CO_2. In the electron transport chain, $NADH + H^+$ and $FADH_2$ are oxidized, contributing their electrons to a series of electron carriers. The various electron transfers generate considerable amounts of ATP. Let us now examine each of the steps in the oxidation of glucose in some detail.

Figure 25.3 Overview of cellular respiration (oxidation of glucose). A small version of this figure will be used in several places in this chapter to indicate the relationships of different reactions to the overall process of cellular respiration.

 The oxidation of glucose involves glycolysis, formation of acetyl coenzyme A, the Krebs cycle, and the electron transport chain.

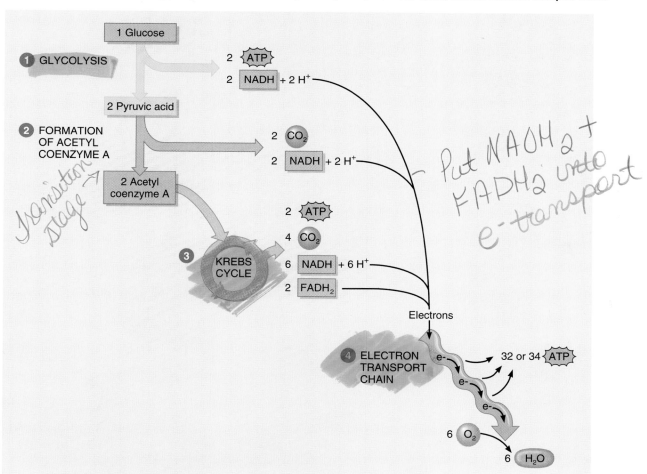

Which of the four processes shown here is also called anaerobic cellular respiration?

Glycolysis

The term **glycolysis** (gli-KOL-i-sis; *glyco* = sugar; *lysis* = breakdown) refers to ten chemical reactions that split a six-carbon molecule of glucose into two three-carbon molecules of pyruvic acid. These reactions occur in the cytosol. (Caution: The word *glycolysis* sounds a lot like *glycogenolysis*, which is the breakdown of glycogen to glucose.)

● **Steps in Glycolysis**   Figure 25.4a shows the steps of glycolysis. Each of the ten reactions is catalyzed by a specific enzyme. The reactions of glycolysis use two ATP molecules, but produce four, a net gain of two (Fig. 25.4b). The essential features of the process are:

Steps ❶ , ❷ , and ❸ . The first three reactions involve the addition of a phosphate group (phosphorylation) to

Figure 25.4   Glycolysis. Part (a) shows each of the 10 steps of glycolysis, and part (b) is a simplified summary.

As a result of glycolysis, there is a net gain of two ATP and two NADH + 2 H⁺.

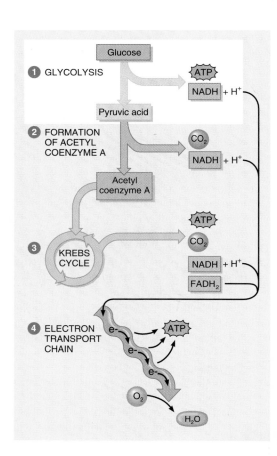

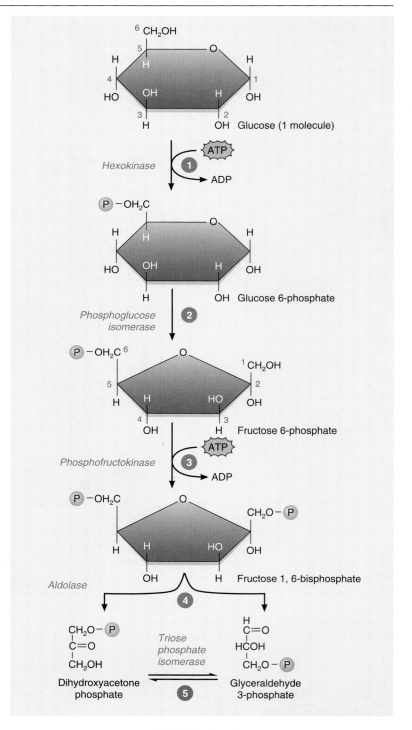

(a) Detailed version

Q   Why is the enzyme that catalyzes step ❶ named hexokinase?

glucose, its conversion to fructose, and the addition of another phosphate group to fructose. This involves an input of energy in which two molecules of ATP are converted to ADP. Phosphofructokinase, the enzyme that catalyzes step ③, is the key regulator of the rate of glycolysis. When ADP concentration is high, this enzyme has a high activity. Then both pyruvic acid and ATP are produced rapidly. When ATP is plentiful, on the other hand, the enzyme activity is low, and most glucose 6-phosphate is converted to glycogen for storage rather than catabolized to produce ATP.

Steps ④ and ⑤. The doubly phosphorylated molecule of fructose splits into two three-carbon compounds, glyceraldehyde 3-phosphate (G 3-P) and dihydroxyacetone phosphate. These two compounds are interconvertible, but it is G 3-P that undergoes further reactions to pyruvic acid.

Step ⑥. Oxidation occurs here as two molecules of NAD^+ accept two pairs of electrons and hydrogen ions from two molecules of G 3-P, forming two molecules each of NADH and 1,3-bisphosphoglyceric acid (BPG). From the two NADH produced here, many body cells generate 4 ATPs in the electron transport chain. A few types of cells, such as hepatocytes, kidney cells, and cardiac muscle fibers can generate 6 ATPs.

Steps ⑦ through ⑩. These reactions generate four molecules of ATP and produce two molecules of pyruvic acid (pyruvate[1]).

---

[1]The carboxyl groups (—COOH) of intermediates in glycolysis and the Krebs cycle are mostly ionized at the pH of body fluids. The ending "-ate" indicates the ionized form, that is, —COO^-. The suffix "-ic acid" indicates the nonionized form, that is, —COOH. Although the "-ate" names are more correct, we will use the "acid" names because these terms are more familiar.

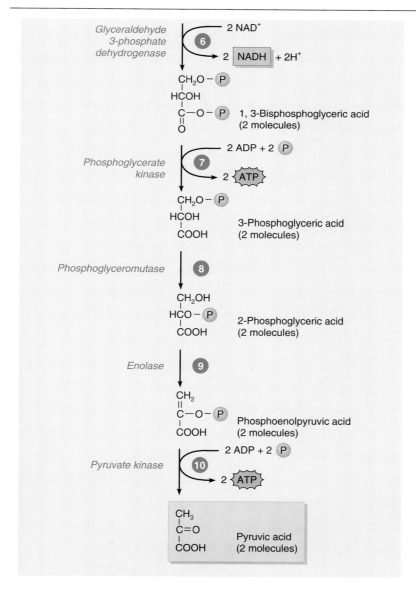

(a) Detailed version (continued)

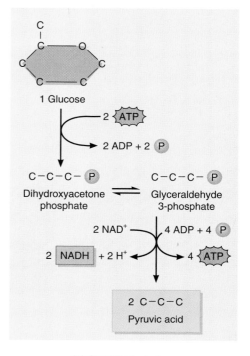

(b) Simplified version

● **Fate of Pyruvic Acid** The fate of pyruvic acid produced during glycolysis depends on the availability of oxygen (Fig. 25.5). If oxygen is scarce (anaerobic conditions), for example, in skeletal muscle fibers during strenuous exercise, pyruvic acid is reduced by the addition of two hydrogen atoms to form lactic acid (lactate). The reaction is:

$$2 \text{ Pyruvic acid} + 2 \text{ NADH} + 2 \text{ H}^+ \rightarrow 2 \text{ Lactic acid} + 2 \text{ NAD}^+$$
(oxidized)   (reduced)

This reaction regenerates the NAD^+ that was used in the oxidation of glyceraldehyde 3-phosphate (step ❻ in glycolysis) and thus allows glycolysis to continue. As lactic acid is produced, it rapidly diffuses out of the cell, enters the blood, and is transported to the liver, where it can be converted back to pyruvic acid. (Neurons do not produce lactic acid.)

When oxygen is plentiful (aerobic conditions), most cells convert pyruvic acid to acetyl coenzyme A. This molecule links glycolysis, which occurs in the cytosol, with the Krebs cycle, which occurs in the matrix of mitochondria. Pyruvic acid enters the mitochondrial matrix with the help of a special transporter protein. Since they lack mitochondria, red blood cells can produce ATP only through glycolysis.

Formation of Acetyl Coenzyme A

Each step in the oxidation of glucose requires a different enzyme and often a coenzyme as well. We are interested in only one coenzyme at this point: **coenzyme A (CoA)**. This important coenzyme is derived from pantothenic acid, another B vitamin.

During the transitional step between glycolysis and the Krebs cycle, pyruvic acid is prepared for entrance into the cycle. It is converted to a two-carbon fragment by removing a molecule of carbon dioxide (Fig. 25.5). The loss of a molecule of CO_2 by a substance is called **decarboxylation** (dē-kar-bok′-si-LĀ-shun). During this reaction, NAD^+ is reduced to $NADH + H^+$. Recall that the oxidation of one glucose molecule produces two molecules of pyruvic acid, so for each molecule of glucose two molecules of carbon dioxide are lost, and two $NADH + H^+$ are produced. The two-carbon fragment, called an **acetyl group**, attaches to coenzyme A, and the whole complex is called **acetyl coenzyme A (acetyl CoA)**. Once the pyruvic acid has undergone decarboxylation and the remaining acetyl group has attached to CoA, the resulting compound (acetyl CoA) is ready to enter the Krebs cycle.

Krebs (Citric Acid) Cycle

The **Krebs cycle** is also called the **citric acid cycle**, or the **tricarboxylic acid (TCA) cycle**. It is a series of nine biochemical reactions that occur in the matrix of mitochondria (Fig. 25.6). In step ❶, coenzyme A carries the two-carbon acetyl unit into the Krebs cycle, which is called a "cycle" because the starting substance (oxaloacetic acid) is formed again at the end. The acetyl group combines with oxaloacetic acid to form citric acid. From this point, the Krebs cycle consists mainly of a series of decarboxylation and oxidation–reduction reactions, each catalyzed by a different enzyme.

Figure 25.5 Fate of pyruvic acid.

🗝 **When oxygen is plentiful, pyruvic acid enters mitochondria, is converted to acetyl coenzyme A, and enters into the Krebs cycle (aerobic pathway). When oxygen is scarce, most pyruvic acid is converted to lactic acid (anaerobic pathway).**

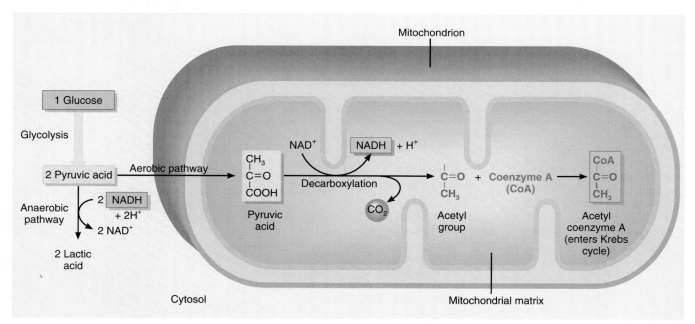

In which cellular compartment does glycolysis occur? The Krebs cycle?

The large amount of chemical potential energy stored in intermediate substances derived from pyruvic acid is released step-by-step. In the Krebs cycle, a series of oxidations and reductions transfers the chemical energy, in the form of electrons, to several coenzymes. The pyruvic acid derivatives are oxidized, whereas the coenzymes are reduced. In addition, step ❻ generates one molecule of guanosine triphosphate (GTP). In turn, GTP can transfer a high-energy phosphate group to ADP to form ATP. Several of the intermediates in the Krebs cycle also are building blocks used to synthesize cellular components, for example amino acids (discussed later in this chapter).

Figure 25.6   Krebs cycle. Part (a) shows each of the 9 steps of the cycle, and part (b) is a simplified summary.

The net results of the Krebs cycle are (1) the production of reduced coenzymes (NADH + H⁺ and FADH₂), which contain stored energy; (2) the generation of GTP, a high-energy compound that is used to produce ATP; and (3) the formation of CO₂, which is transported to the lungs for exhalation.

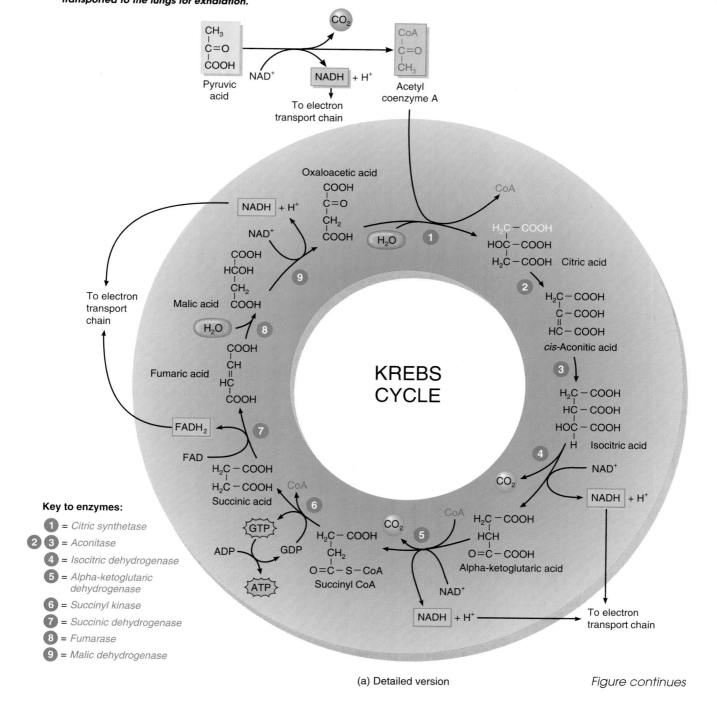

(a) Detailed version

Figure continues

Figure 25.6 (continued)

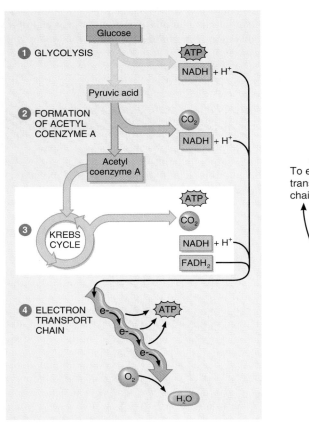

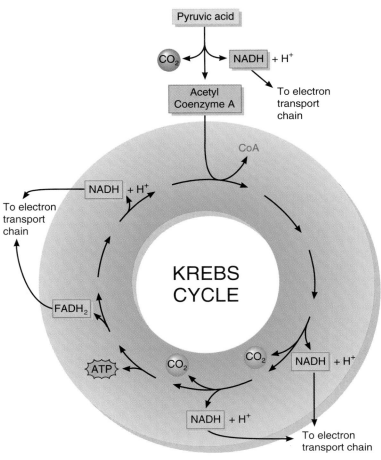

(b) Simplified version

Q Why is the production of reduced coenzymes important in the Krebs cycle?

Let us look at the decarboxylation reactions first. Isocitric acid, a six-carbon compound, loses a molecule of CO_2 to form a five-carbon compound called alpha-ketoglutaric acid (step **4**). Then alpha-ketoglutaric acid is decarboxylated and picks up a molecule of CoA to form succinyl CoA, a four-carbon compound (step **5**). Overall, each time acetyl CoA enters the Krebs cycle, two molecules of CO_2 are liberated by decarboxylation (Fig. 25.6b). Because each molecule of glucose generates two molecules of pyruvic acid, however, six molecules of CO_2 are liberated from each original glucose molecule catabolized along this pathway. The molecules of CO_2 leave the mitochondria, diffuse through the cytosol to the plasma membrane, and then diffuse into the blood. Eventually, the CO_2 is transported by the blood to the lungs and is exhaled.

Now let us look at the oxidation–reduction reactions (Fig. 25.6a). Recall, when a molecule is oxidized, it loses hydrogen atoms and their associated electrons. When a molecule is re-

duced, it gains hydrogen atoms and their associated electrons. Every oxidation is coupled with a reduction. In the oxidation of pyruvic acid to acetyl CoA, each pyruvic acid loses two hydrogen atoms in the form of one hydride ion (H^-) plus one hydrogen ion (H^+). At the same time, the coenzyme NAD^+ is reduced as it picks up the hydride ion from pyruvic acid. (The hydrogen ion is released into the mitochondrial matrix.) The reduction of NAD^+ to NADH + H^+ is indicated in the diagram by the curved arrow entering and then leaving the reaction.

Note that steps **4**, **5**, and **9** of the Krebs cycle also involve transfer of hydride ions to NAD^+. In each case, the molecule that donates the hydride ion is oxidized and NAD^+ is reduced to NADH. In step **7**, succinic acid is oxidized to fumaric acid and the coenzyme flavin adenine dinucleotide (FAD) picks up two hydrogen atoms to become $FADH_2$.

Overall, for every two molecules of acetyl CoA that enter the Krebs cycle, 6 NADH, 6 H^+, and 2 $FADH_2$ are produced by oxidation–reduction reactions, and two molecules of ATP are generated by substrate-level phosphorylation (Fig. 25.6b).

In the electron transport chain, the 6 NADH and 6 H^+ will later yield 18 ATP molecules and the 2 $FADH_2$ will later yield 4 ATP molecules. The reduced coenzymes (NADH and $FADH_2$) are the most important outcome of the Krebs cycle because they contain the energy originally stored in glucose and then in pyruvic acid. During the next phase of aerobic respiration, a series of reductions transfers the energy stored in the coenzymes to ADP + ⓟ to form ATP. These reactions involve the electron transport chain and occur on the cristae of the inner mitochondrial membrane.

Electron Transport Chain

The **electron transport chain** involves a sequence of **electron carrier molecules** on the inner mitochondrial membrane that are capable of oxidation and reduction. As electrons pass through the chain, there is a stepwise release of energy that is used to form ATP. In aerobic cellular respiration, the last electron acceptor of the chain is oxygen. Because this mechanism of ATP generation links chemical reactions (electrons passing along the electron chain) with a pumping process, it is called **chemiosmosis** (kem'-ē-oz-MŌ-sis; *chemi* = chemical; *osmos* = push). Briefly, it occurs as follows (Fig. 25.7):

❶ Energy from NADH + H^+ passes along the electron transport chain and is used to pump H^+ (protons) from the matrix of the mitochondrion into the space between the inner and outer mitochondrial membranes.

❷ A high concentration of H^+ accumulates between the inner and outer mitochondrial membranes.

❸ ATP synthesis then occurs as H^+ diffuse back into the mitochondrial matrix through a special type of H^+ channel in the inner membrane.

We will first examine the electron carriers and then describe chemiosmosis in more detail.

● **Electron Carriers** The electron transport chain involves carrier molecules in the inner mitochondrial membrane that are alternately oxidized and reduced. There are several types of carriers.

1. **Flavin mononucleotide (FMN)**, like FAD (flavin adenine dinucleotide), is a flavoprotein derived from riboflavin (vitamin B_2).

2. **Cytochromes** (SĪ-tō-krōmz) are proteins with an iron-containing group (heme) capable of existing alternately in a reduced form (Fe^{2+}) and an oxidized form (Fe^{3+}). The several cytochromes involved in the electron transport chain are cytochrome *b* (cyt *b*), cytochrome c_1 (cyt c_1), cytochrome *c* (cyt *c*), cytochrome *a* (cyt *a*), and cytochrome a_3 (cyt a_3).

Figure 25.7 Chemiosmosis.

In chemiosmosis, ATP is produced when H^+ diffuses back into the mitochondrial matrix.

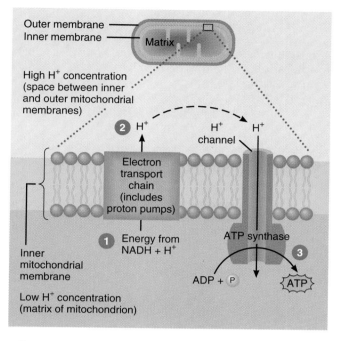

Q What is the energy source that runs the proton pumps?

3. **Iron–sulfur (Fe–S) centers** contain either two or four iron atoms bound to sulfur atoms that form an electron transfer center within a protein.

4. **Copper (Cu) atoms** bound to two proteins in the chain also participate in electron transfer.

5. **Ubiquinones** are nonprotein carriers of low molecular weight that are mobile in the phospholipid bilayer of the membrane. They are also called **coenzyme Q**, symbolized **Q**.

● **Steps in Electron Transport** Figure 25.8 shows the steps in electron transport. The first step is the transfer of high-energy electrons from NADH + H^+ to FMN, the first carrier in the chain. From each molecule of glucose, two NADH + 2 H^+ are generated from glycolysis, two from the formation of acetyl CoA, and six from the Krebs cycle. In this transfer, a hydride ion (H^-) passes to FMN, which then picks up an additional H^+ from the surrounding aqueous medium. As a result, NADH + H^+ is oxidized to NAD^+, and FMN is reduced to $FMNH_2$.

In the second step in the electron transport chain, $FMNH_2$ passes electrons to several iron–sulfur centers and then to Q, which picks up an additional H^+ from the surrounding aqueous medium. As a result, $FMNH_2$ is oxidized to FMN.

The next sequence in the electron transport chain involves cytochromes, iron–sulfur centers, and copper atoms located between Q and molecular oxygen. Electrons are passed successively from Q to cyt b, to Fe–S, to cyt c_1, to cyt c, to Cu, to cyt a, and finally to cyt a_3. Each carrier in the chain is reduced as it picks up electrons and is oxidized as it gives up electrons. The last cytochrome, cyt a_3, passes its electrons to one-half of a molecule of oxygen (O_2),

Figure 25.8   Steps in the electron transport chain. The energy drop for electrons passing through the chain occurs in a stepwise fashion. Examine Fig. 25.9 to see where ATP is formed. The inset here highlights the portion of cellular respiration shown in both Fig. 25.8 and 25.9.

🔑 *Each carrier in the chain is reduced as it picks up electrons and oxidized as it gives up electrons.*

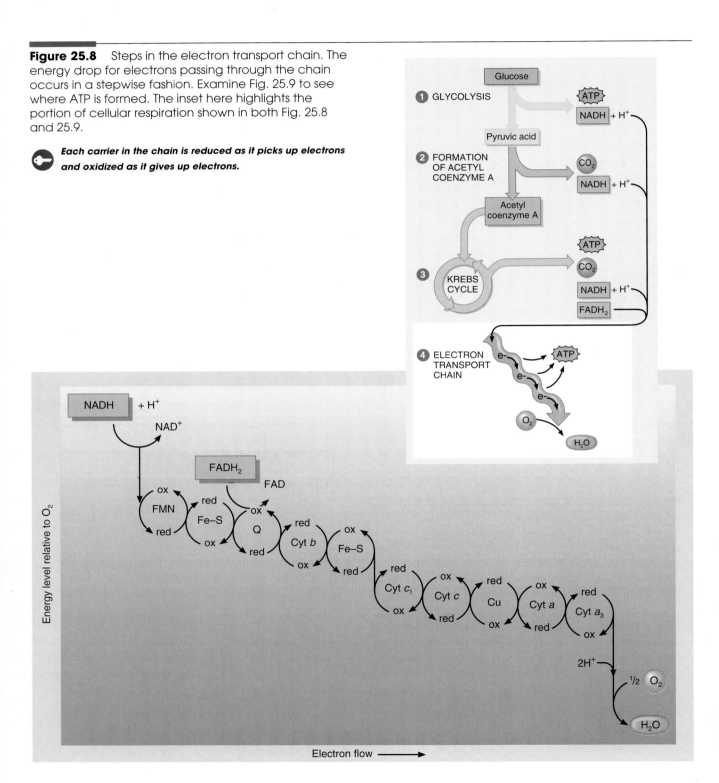

Ⓠ Why is each electron carrier drawn on a lower level than the carrier from which it received electrons and on a higher level than the carrier to which it passes electrons?

which becomes negatively charged and then picks up 2 H$^+$ from the surrounding medium to form H$_2$O. This is the only point in aerobic cellular respiration where O$_2$ is consumed. The deadly poison cyanide binds to the enzyme cytochrome oxidase and blocks the last step in electron transport.

Note in Fig. 25.8 that FADH$_2$, derived from the Krebs cycle, is another source of electrons. However, FADH$_2$ adds its electrons to the electron transport chain at a lower energy level than does NADH + H$^+$. Because of this, the electron transport chain produces about one-third less energy for ATP generation when FADH$_2$ donates electrons as compared with NADH + H$^+$.

● **Chemiosmotic Mechanism of ATP Generation** Within the inner mitochondrial membrane, the carriers of the electron transport chain cluster into three complexes. Each complex acts as a **proton pump** that expels H$^+$ from the mitochondrial matrix and helps create an electrochemical gradient of H$^+$ (Fig. 25.9). Each proton pump complex includes three or more electron carriers:

① **NADH dehydrogenase complex** contains flavin mononucleotide (FMN) and at least five Fe–S centers. Its name stems from its action, to remove hydrogen from NADH.

② **Cytochrome b–c$_1$ complex** contains cytochromes b and c$_1$ and an Fe–S center.

③ **Cytochrome oxidase complex** contains cytochromes a and a$_3$ and two copper atoms.

One electron carrier (Q) ferries electrons from the first to the second complex and a second electron carrier (cytochrome c) ferries electrons from the second to the third.

Because the inner mitochondrial membrane is nearly impermeable to H$^+$, the pumping of H$^+$ leads to both a concentration gradient of protons and an electrical gradient. The buildup of H$^+$ on one side of the membrane confers a positive charge there, whereas the fluid on the other side is left with a negative charge. This proton gradient has potential energy and is called the **proton motive force**. In regions where specific H$^+$ channels exist, H$^+$ can diffuse back across the membrane, driven by the proton motive force. As H$^+$ diffuse back, they generate ATP because the H$^+$ channels also include an enzyme called **ATP synthase**. The enzyme uses the proton motive force to synthesize ATP from ADP and ℗. This process of chemiosmosis is responsible for most of the ATP produced during cellular respiration.

Summary of Aerobic Cellular Respiration
The various electron transfers in the electron transport chain generate 32 or 34 ATP molecules from each molecule of glucose that is oxidized: 28 or 30[1] from the 10 molecules of

---

[1]The two NADH produced in the cytosol during glycolysis cannot themselves enter mitochondria. Rather, they donate their electrons to one of two transfer molecules, known as the malate shuttle and the glycerol phosphate shuttle. In some body cells, such as in the liver, kidneys, and heart, use of the malate shuttle yields three ATP for each NADH. In other body cells, such as skeletal muscle fibers and neurons, use of the glycerol phosphate shuttle yields two ATP for each NADH.

Figure 25.9 Proton pumps. Each pump is a complex of three or more electron carriers.

🔑 *The three proton pumps move H$^+$ from the matrix into the space between inner and outer mitochondrial membranes.*

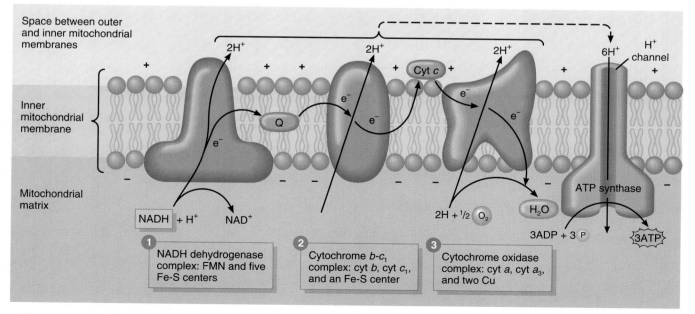

🔍 Where is the concentration of H$^+$ highest?

NADH + H⁺ and 2 from each of the 2 molecules of FADH₂ (4 total). Thus, during aerobic respiration, 36 or 38 ATPs can be generated from one molecule of glucose. Note that two of those ATPs come from substrate-level phosphorylation in glycolysis and two come from substrate-level phosphorylation in the Krebs cycle. Exhibit 25.1 summarizes the ATP yield during aerobic respiration.

The overall reaction for aerobic respiration is:

$$C_6H_{12}O_6 + 6 O_2 + 36 \text{ or } 38 \text{ ADPs} + 36 \text{ or } 38 \, \textcircled{P}$$

Glucose   Oxygen

$$\longrightarrow 6 CO_2 + 6 H_2O + 36 \text{ or } 38 \text{ ATPs}$$

Carbon   Water
dioxide

A summary of the sites of the principal events of the various stages of cellular respiration is presented in Fig. 25.10. The actual ATP yield may be lower than 36 or 38 ATPs per glucose. One uncertainty is the exact number of H⁺ that must be pumped out to generate one ATP during chemiosmosis. Also, the ATP generated in mitochondria must be transported out of these organelles into the cytoplasm for use elsewhere in a cell. This process uses up a portion of the proton motive force to drive out ATP in exchange for the inward movement of ADP that is formed from metabolic reactions in the cytosol.

Figure 25.10   Sites of principal events of cellular respiration. ETC = electron transport chain and chemiosmosis.

 Except for glycolysis, which occurs in the cytosol, all other reactions of cellular respiration occur within mitochondria.

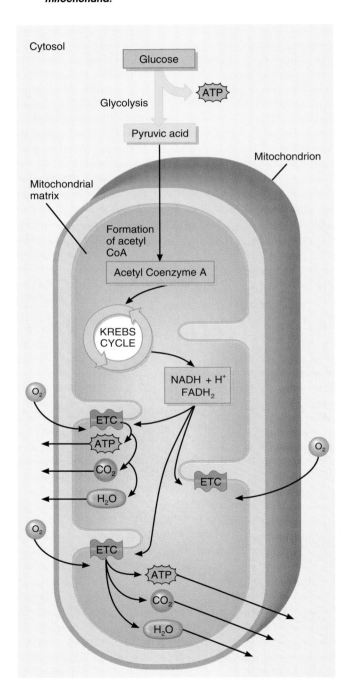

Q How many molecules of O_2 are used during the complete oxidation of one glucose molecule? How many molecules of CO_2 are produced?

EXHIBIT **25.1**   **SUMMARY OF ATP PRODUCED IN AEROBIC RESPIRATION**

Source	ATP Yield (Method)
GLYCOLYSIS	
Oxidation of one glucose molecule to two pyruvic acid molecules	2 ATPs (substrate-level phosphorylation)
Production of 2 NADH + 2 H⁺	4 or 6 ATPs (oxidative phosphorylation in electron transport chain)
FORMATION OF ACETYL COENZYME A	
2 NADH + 2 H⁺	6 ATPs (oxidative phosphorylation in electron transport chain)
KREBS CYCLE AND ELECTRON TRANSPORT CHAIN	
Oxidation of succinyl CoA to succinic acid	2 GTPs that are converted to ATP (substrate-level phosphorylation)
Production of 6 NADH + 6 H⁺	18 ATPs (oxidative phosphorylation in electron transport chain)
Production of 2 FADH₂	4 ATPs (oxidative phosphorylation in electron transport chain)
Total:	36 or 38 ATPs per glucose molecule (theoretical maximum)

Glycolysis, the Krebs cycle, and especially the electron transport chain provide all the ATP for cellular activities. And because the Krebs cycle and electron transport chain are aerobic processes, cells cannot carry on their activities for long without sufficient oxygen.

CARBOHYDRATE LOADING

During exercise, the body preferentially uses carbohydrates to produce ATP. Carbohydrates are stored as glycogen in the liver and skeletal muscles. Since glycogen stores are limited and can be completely exhausted after running more than 32 km (20 mi), many marathon runners follow a precise dietary regimen that includes a large amount of carbohydrates before the event. This practice, called **carbohydrate loading**, is designed to maximize body stores of carbohydrates to provide additional energy for an athletic event. ■

Glucose Anabolism

Most of the glucose in the body is catabolized to generate ATP. However, glucose may take part in several anabolic reactions. One is the synthesis of glycogen. Another is the synthesis of new glucose molecules from the breakdown products of proteins and lipids.

Glucose Storage: Glycogenesis

If glucose is not needed immediately for ATP production, it is combined with many other molecules of glucose to form a long-chain molecule called glycogen. This process is termed **glycogenesis** (glī´-kō-JEN-e-sis; *glyco* = sweet; *genesis* = to generate). About 500 g (1.1 lb) of glycogen can be stored, 25% in the liver and 75% in skeletal muscle fibers.

In the process of glycogenesis (Fig. 25.11), glucose that enters cells is first phosphorylated to glucose 6-phosphate. This is then converted to glucose 1-phosphate, then to uridine diphosphate glucose, and finally to glycogen. Glycogenesis is stimulated by insulin from the pancreas.

Glucose Release: Glycogenolysis

When body activities need ATP, glycogen stored in hepatocytes is broken down into glucose and released into the blood to be transported to cells, where it will be catabolized. The process of converting glycogen back to glucose is called **glycogenolysis** (glī´-kō-je-NOL-e-sis; *lysis* = breakdown). Glycogenolysis usually occurs in between meals and is stimulated by glucagon and epinephrine.

Glycogenolysis is not a simple reversal of the steps of glycogenesis (Fig. 25.11). It begins by splitting glucose molecules from the branched glycogen molecule by phosphorylation to form glucose 1-phosphate. Phosphorylase, the enzyme that catalyzes this reaction, is activated by the hormones glucagon from the pancreas and epinephrine from the adrenal medulla. Glucose 1-phosphate is then converted to glucose 6-phosphate and finally to glucose. Phosphatase, the enzyme

Figure 25.11 Glycogenesis and glycogenolysis. The glycogenesis pathway converts glucose into glycogen whereas the glycogenolysis pathway breaks down glycogen into glucose.

 About 500 grams of glycogen are stored in the liver and skeletal muscle fibers.

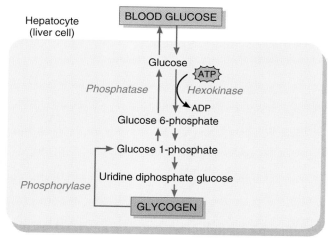

Key:
→ Reactions of glycogenesis (stimulated by insulin)   → Reactions of glycogenolysis (stimulated by glucagon and epinephrine)

Q Besides hepatocytes, which cells can synthesize glycogen? Why can't they release glucose into the blood?

that converts glucose 6-phosphate into glucose, is present in hepatocytes but absent in skeletal muscle cells. Thus hepatocytes can release glucose derived from glycogen to the bloodstream whereas skeletal muscle cells cannot. In skeletal muscle cells, glycogen is broken down into glucose 1-phosphate, which is then catabolized for ATP production via glycolysis and the Krebs cycle. However, the lactic acid produced by glycolysis in muscle cells can be converted to glucose in the liver. In this way, muscle glycogen can be an indirect source of blood glucose.

Formation of Glucose from Proteins and Fats: Gluconeogenesis

When the supply of liver glycogen runs low, it is time to eat. If you do not eat, your body starts catabolizing more triglycerides and proteins. Although the body normally catabolizes some of its triglycerides and a few of its proteins, large-scale triglyceride and protein catabolism does not occur unless you are fasting, starving, eating meals that contain very few carbohydrates, or suffering from an endocrine disorder.

Certain molecules may be broken down and converted to glucose in the liver. The process by which new glucose is formed from noncarbohydrate sources is called

gluconeogenesis (gloo′-kō-nē′-ō-JEN-e-sis; *neo* = new). Lactic acid, certain amino acids, and the glycerol portion of triglyceride molecules can all be used to form new glucose molecules through gluconeogenesis (Fig. 25.12). About 60% of the amino acids in the body can undergo this conversion. Amino acids such as alanine, cysteine, glycine, serine, and threonine are converted to pyruvic acid. The pyruvic acid may be synthesized into glucose or enter the Krebs cycle. Glycerol may be converted into glyceraldehyde 3-phosphate, which may form pyruvic acid or be used to synthesize glucose. Figure 25.12 also shows how gluconeogenesis is related to other metabolic reactions.

Gluconeogenesis is stimulated by cortisol, the main glucocorticoid hormone of the adrenal cortex, and glucagon from the pancreas. In addition, cortisol stimulates breakdown of proteins into amino acids, thus expanding the pool of amino acids for gluconeogenesis. Thyroid hormones (thyroxin and triiodothyronine) also mobilize proteins and may mobilize triglycerides from adipose tissue, thus making glycerol available for gluconeogenesis.

LIPID METABOLISM

When triglycerides are eaten, they are digested into fatty acids and monoglycerides. Short-chain fatty acids diffuse into epithelial cells of the intestinal villi and then into the blood capillaries. Long-chain fatty acids and monoglyc-erides are carried in micelles to epithelial cells of the villi for entrance. Once inside, they are further digested to glycerol and fatty acids and then recombined to form triglyc-erides. They leave intestinal cells in chylomicrons and enter lymph in lacteals of villi. Finally, the chylomicrons enter the blood through the thoracic duct. Since lipids dissolve poorly in water, they are transported in the bloodstream in several types of lipoprotein particles. The major ones are chylomicrons, very low-density lipoproteins (VLDLs), low-density lipoproteins (LDLs), and high-density lipoproteins (HDLs), which are described on page 602.

Fate of Lipids

Lipids, like carbohydrates, may be oxidized to produce ATP. Each gram of triglyceride produces about 9 kilocalo-ries (= 9 Cal), about twice that of carbohydrates and proteins. If the body has no immediate need to use lipids this way, they are stored in adipose tissue (fat depots) throughout the body and in the liver. A few lipids are used as structural molecules or to synthesize other essential substances. For example, phospholipids are constituents of plasma membranes, lipoproteins are used to transport cholesterol throughout the body, thromboplastin is needed for blood clotting, and myelin sheaths speed up nerve impulse conduction. Cholesterol, another lipid, is used in the synthesis of bile salts and steroid hormones (adrenocortical hormones and sex hormones). The various functions of lipids in the body may be reviewed in Exhibit 2.5 on page 40.

Figure 25.12   Gluconeogenesis. This process involves the conversion of noncarbohydrate molecules (amino acids, lactic acid, and glycerol) into glucose.

About 60% of the amino acids in the body can undergo gluconeogenesis.

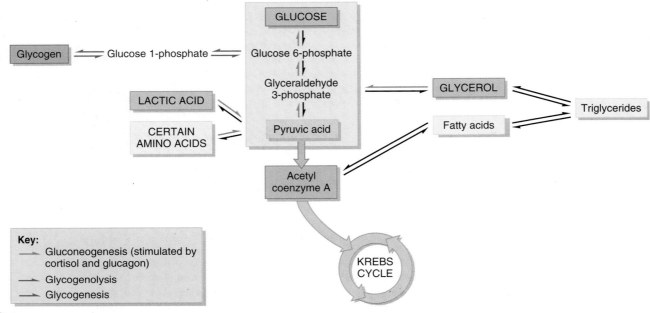

What cells can carry out gluconeogenesis and glycogenesis?

Triglyceride Storage

A major function of adipose tissue is to remove triglycerides from chylomicrons and VLDLs and store them until they are needed for ATP production in other parts of the body. Subcutaneous adipose tissue contains about 50% of the stored triglycerides, with other adipose tissues accounting for the other half—about 12% around the kidneys, 10–15% in the omenta, 15% in genital areas, 5–8% between muscles, and 5% behind the eyes, in the sulci of the heart, and in the folds of the large intestine. Triglycerides in adipose tissue are continually broken down and resynthesized. Thus the triglycerides stored in your adipose tissue today are not the same ones that were present last month. Also, they are continually released from storage, transported in the blood, and redeposited in other adipose tissue cells.

Lipid Catabolism: Lipolysis

Triglycerides stored in adipose tissue constitute 98% of all body energy reserves. They are stored more readily than is glycogen. One reason is that triglycerides are hydrophobic and do not exert osmotic pressure on cell membranes. Also, the energy yield of triglycerides is more than twice that of carbohydrates (9 Cal/g versus 4 Cal/g). Several tissues, such as muscle, liver, and adipose tissue, routinely oxidize fatty acids derived from triglycerides to produce ATP. Before triglycerides can be catabolized to produce ATP, they must be split into glycerol and fatty acids, a process called **lipolysis** (li-POL-i-sis) that is catalyzed by enzymes called **lipases**. Hormones that enhance triglyceride breakdown into fatty acids and glycerol include epinephrine, norepinephrine, cortisol, thyroid hormones, and human growth hormone. Glycerol and fatty acids are then catabolized separately (Fig. 25.13).

Glycerol
Glycerol is converted easily by many cells of the body to glyceraldehyde 3-phosphate, one of the compounds also formed during the catabolism of glucose. If ATP supply is high in a cell, glyceraldehyde 3-phosphate is converted into glucose. This is one example of gluconeogenesis. When a cell needs to produce more ATP, however, glyceraldehyde 3-phosphate enters the catabolic sequence to pyruvic acid.

Fatty Acids
Fatty acids are catabolized differently than glycerol and yield more ATP. The process occurs in the matrix of mitochondria. The first stage in fatty acid catabolism involves a series of reactions called **beta oxidation**. By dehydration, hydration, and cleavage reactions, enzymes remove one pair of carbon atoms at a time from the long chain of carbon atoms composing a fatty acid. The resulting two-carbon fragment then is attached to coenzyme A, forming **acetyl coenzyme A (acetyl CoA)**. In the second stage of fatty acid catabolism, the acetyl CoA formed as a result of beta oxidation enters the Krebs cycle (Fig. 25.13). A 16-carbon fatty

Figure 25.13   Metabolism of lipids. Glycerol may be converted to glyceraldehyde 3-phosphate, which can then be converted to glucose or enter the Krebs cycle for oxidation. Fatty acids undergo beta oxidation and enter the Krebs cycle via acetyl coenzyme A. Lipogenesis is the synthesis of lipids from glucose or amino acids.

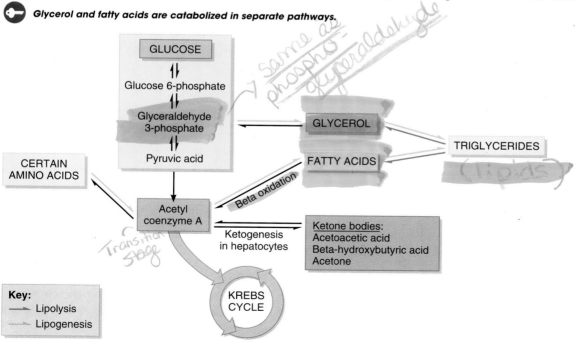

Glycerol and fatty acids are catabolized in separate pathways.

What types of cells can carry out lipogenesis, beta oxidation, and lipolysis? Ketogenesis?

acid, such as palmitic acid, can yield a net of 129 ATPs upon its complete oxidation via beta oxidation, the Krebs cycle, and the electron transport chain.

As part of normal fatty acid catabolism, hepatocytes can take two acetyl CoA molecules at a time and condense them to form a substance called **acetoacetic acid**. This reaction liberates the bulky CoA portion, which cannot diffuse out of cells. Some acetoacetic acid is converted into **beta-hydroxybutyric acid** and **acetone**. These three substances collectively are known as **ketone bodies**. The formation of ketone bodies is called **ketogenesis** (Fig. 25.13). Since ketone bodies freely diffuse through plasma membranes, they leave the hepatocytes and enter the bloodstream.

Other cells take up acetoacetic acid, attach its four carbons to two coenzyme A molecules, and form two acetyl CoA molecules. These then can enter the Krebs cycle for oxidation. Heart muscle and the cortex (outer part) of the kidneys use acetoacetic acid in preference to glucose to generate ATP. Although brain cells normally use glucose almost exclusively for ATP production, during starvation they become able to use acetoacetic acid. Hepatocytes, which make acetoacetic acid, cannot use it themselves for ATP production because they lack the enzyme that transfers acetoacetic acid back to coenzyme A.

KETOSIS

The level of ketone bodies in the blood normally is very low because other tissues use them for ATP production as fast as the liver generates them. During periods of excessive beta oxidation, however, production of ketone bodies exceeds their uptake and use by body cells. This might occur after a meal rich in triglycerides, or during fasting or starvation, because few carbohydrates are available for catabolism. It may also occur in poorly controlled or untreated diabetes mellitus (described on page 537) because adequate glucose cannot get into cells. When the concentration of ketone bodies in the blood rises above normal—a condition called **ketosis**—the ketone bodies, most of which are acids, must be buffered. If too many accumulate, they use up the body's buffers, such as bicarbonate ions, and the blood pH falls. Thus extreme or prolonged ketosis can lead to **acidosis** (**ketoacidosis**), or abnormally low blood pH. When a diabetic becomes seriously insulin deficient, one of the telltale signs is a sweet smell of acetone on the breath. ∎

Lipid Anabolism: Lipogenesis

Liver cells and adipose cells can synthesize lipids from glucose or amino acids through a process called **lipogenesis** (lip′-ō-JEN-ē-sis), which is stimulated by insulin (Fig. 25.13). Lipogenesis occurs when you consume more calories than are needed to satisfy your ATP needs. Excess dietary carbohydrates, proteins, and fats all have the same fate—they are converted into triglycerides. Many amino acids can be converted into acetyl CoA, which can then be converted into triglycerides (Fig. 25.13). The steps in the conversion of glucose to lipids involve the formation of glyceraldehyde 3-phosphate, which can be converted to glycerol, and acetyl

CoA, which can be converted to fatty acids. The resulting glycerol and fatty acids can undergo anabolic reactions to become triglycerides that can be stored or can go through a series of anabolic reactions to produce other lipids such as lipoproteins, phospholipids, and cholesterol.

PROTEIN METABOLISM

During digestion, proteins are broken down into their constituent amino acids. The amino acids are then absorbed into blood capillaries in villi and transported to the liver via the hepatic portal vein. Unlike carbohydrates and triglycerides, which are stored, proteins are not warehoused for future use. Amino acids may be oxidized to produce ATP or used to synthesize new proteins for body growth and repair. Excess dietary amino acids are not excreted in the urine or feces but rather are converted into glucose (gluconeogenesis) or triglycerides (lipogenesis).

Fate of Proteins

Amino acids enter body cells by active transport. This process is stimulated by human growth hormone and insulin. Almost immediately after entrance, they are incorporated into proteins. Many proteins function as enzymes. Other proteins are involved in transportation (hemoglobin) or serve as antibodies, clotting chemicals (fibrinogen), hormones (insulin), and contractile elements in muscle fibers (actin and myosin). Several proteins serve as structural components of the body (collagen, elastin, and keratin). When proteins are oxidized, each gram produces about 4 Cal. The various functions of proteins in the body may be reviewed in Exhibit 2.6 on page 43.

Protein Catabolism

A certain amount of protein catabolism occurs in the body each day. Proteins are extracted from worn-out cells, such as red blood cells, and broken down into free amino acids. Some amino acids are converted into other amino acids, peptide bonds are reformed, and new proteins are made as part of the constant state of turnover in all cells. A significant fraction of the amino acids absorbed by the GI tract are from proteins in worn-out cells that have sloughed off the mucosa into the lumen.

Proteins being recycled are first broken down into amino acids. Hepatocytes, then, can convert amino acids to fatty acids, ketone bodies, or glucose or oxidize them to carbon dioxide and water. However, before amino acids can be catabolized, they must first be converted to various substances that can enter the Krebs cycle. One such conversion consists of removing the amino group (NH_2) from the amino acid, a process called **deamination** (dē-am′-i-NĀ-shun), and converting it to ammonia (NH_3). Hepatocytes then convert ammonia to urea, which is excreted in the urine. Other conversions are decarboxylation and dehydrogenation. The fate of the remainder of the amino acid depends on which amino acid it is. Figure 25.14 shows that particular amino acids enter the Krebs cycle at different points.

The conversion of amino acids into glucose (gluconeogenesis) may be reviewed in Fig. 25.12. The conversion of amino acids into fatty acids (lipogenesis) or ketone bodies (ketogenesis) is shown in Fig. 25.13.

Protein Anabolism

Protein anabolism involves the formation of peptide bonds between amino acids to produce new proteins. Protein anabolism, or synthesis, occurs on the ribosomes of almost every cell in the body, directed by the cells' DNA and RNA (see Fig. 3.22). Human growth hormone, thyroid hormones, insulin, estrogen, and testosterone stimulate protein synthesis. Because proteins are a main component of most cell structures, adequate dietary protein is especially essential during the growth years, during pregnancy, and when tissue has been damaged by disease or injury. Once dietary intake of protein is adequate, however, eating more protein will not increase bone or muscle mass. Only a regular program of forceful, weight-bearing muscular activity accomplishes that goal.

Figure 25.14   Various points at which amino acids, shown in yellow boxes, enter the Krebs cycle for oxidation.

 Before amino acids can be catabolized, they must first be converted to various substances that can enter the Krebs cycle.

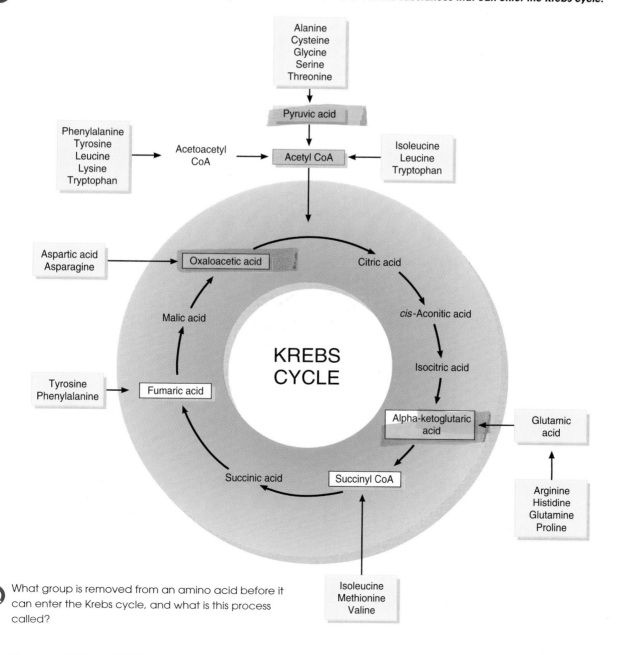

What group is removed from an amino acid before it can enter the Krebs cycle, and what is this process called?

Of the 20 amino acids in your body, 10 are called **essential amino acids**. We are unable to synthesize eight (isoleucine, leucine, lysine, methionine, phenylalanine, threonine, tryptophan, and valine) and synthesize two others in inadequate amounts, especially during childhood (arginine and histidine). These amino acids are synthesized by plants or bacteria, and so foods containing these amino acids are "essential" for human growth and must be part of the diet. **Nonessential amino acids** can be synthesized by a process called **transamination**, the transfer of an amino group from an amino acid to a substance such as pyruvic acid or an acid of the citric acid cycle. Once the appropriate essential and nonessential amino acids are present in cells, protein synthesis occurs rapidly.

SUMMARY OF KEY MOLECULES IN METABOLISM

Although there are thousands of different chemicals in your cells, three molecules play key roles in metabolism. They are glucose 6-phosphate (G 6-P), pyruvic acid, and acetyl coenzyme A (acetyl CoA). Figure 25.15 shows the various options open to these molecules. Double-headed arrows indicate the reactions between two molecules may proceed in either direction, if the appropriate enzymes are present and the conditions are favorable. Single-headed arrows signify the presence of an irreversible step or steps.

Glucose 6-Phosphate

Upon entering any cell in the body, glucose is phosphorylated to glucose 6-phosphate. If a given cell has the enzymes needed to catalyze the steps, four possible fates await this molecule:

1. If the enzyme glucose 6-phosphatase is present and active, G 6-P may be dephosphorylated; that is, it loses a phosphate group and is converted to glucose. Glucose 6-phosphatase allows glucose to leave a cell.
2. G 6-P may be used to synthesize glycogen. In turn, when glycogen is broken down, G 6-P reforms.
3. Although it hasn't been discussed here, G 6-P may be used to generate ribose 5-phosphate, a five-carbon sugar that is needed for synthesis of RNA (ribonucleic acid) and DNA (deoxyribonucleic acid). The same sequence of reactions that produces ribose 5-phosphate

Figure 25.15   Summary of key molecules and pathways in metabolism.

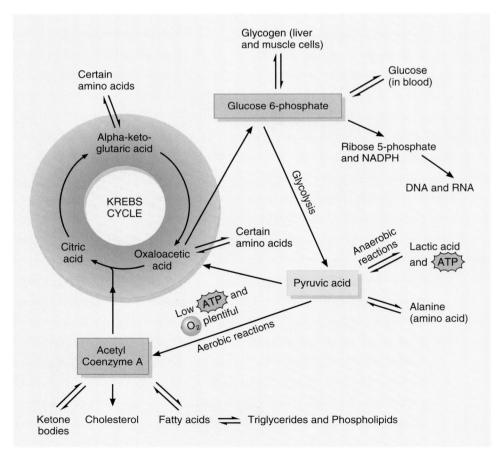

Q   Which substance is the gateway into the Krebs cycle for fuel molecules that are being oxidized to generate ATP?

also produces NADPH, which is needed for reduction reactions.

4. Some ATP is produced anaerobically via glycolysis, in which G 6-P is converted to pyruvic acid, another key molecule in metabolism.

Pyruvic Acid

Pyruvic acid marks a decision point in anaerobic versus aerobic cellular respiration. When the ATP level in a cell is low but oxygen is plentiful, most pyruvic acid streams toward ATP-producing reactions (the Krebs cycle and electron transport chain) via conversion to acetyl CoA. On the other hand, when oxygen is in short supply in a tissue, as in actively contracting skeletal or cardiac muscle, most pyruvic acid is changed to lactic acid. This conversion generates a small amount of ATP. The lactic acid then diffuses into the bloodstream and is taken up by hepatocytes, which sooner or later convert it back into pyruvic acid.

One link between carbohydrate and protein metabolism occurs at pyruvic acid. An amino group can be added to pyruvic acid (a carbohydrate) to produce the amino acid alanine or removed from alanine to generate pyruvic acid. Pyruvic acid and certain amino acids also can be converted to oxaloacetic acid, one of the Krebs cycle intermediates, which, in turn, can generate glucose 6-phosphate. This sequence of reactions bypasses certain one-way reactions of glycolysis and is called gluconeogenesis.

Acetyl Coenzyme A

Acetyl coenzyme A is the gateway into the Krebs cycle for fuel molecules that are being oxidized to generate ATP. It also is used to synthesize fatty acids, ketone bodies, and cholesterol. Since pyruvic acid can be converted to acetyl coenzyme A, carbohydrates can be turned into triglycerides; this is the metabolic path for storage of most excess calories as triglycerides. Mammals, including humans, cannot reconvert acetyl coenzyme A to pyruvic acid, however, so fatty acids cannot be used to generate glucose or other carbohydrate molecules.

A summary of carbohydrate, lipid, and protein metabolism is presented in Exhibit 25.2.

EXHIBIT 25.2 SUMMARY OF METABOLISM

Process	Comment
CARBOHYDRATES	
Glucose catabolism	Complete oxidation of glucose, also referred to as cellular respiration, is the chief source of ATP in cells. The process requires glycolysis, Krebs cycle, and electron transport chain. The complete oxidation of 1 molecule of glucose yields a maximum of 36 or 38 molecules of ATP.
Glycolysis	Conversion of glucose into pyruvic acid results in the production of some ATP. Reactions do not require oxygen (anaerobic cellular respiration).
Krebs cycle	Cycle includes series of oxidation–reduction reactions in which coenzymes (NAD^+ and FAD) pick up hydrogen ions and hydride ions from oxidized organic acids, and some ATP is produced. CO_2 and H_2O are by-products. Reactions are aerobic.
Electron transport chain	Third set of reactions in glucose catabolism is another series of oxidation–reduction reactions, in which electrons are passed from one carrier to the next, and most of the ATP is produced. Reactions require oxygen (aerobic cellular respiration).
Glucose anabolism	Some glucose is converted into glycogen (glycogenesis) for storage if not needed immediately for ATP production. Glycogen can be reconverted to glucose (glycogenolysis). The conversion of amino acids, glycerol, and lactic acid into glucose is called gluconeogenesis.
LIPIDS	
Triglyceride catabolism	Triglycerides are broken down into glycerol and fatty acids. Glycerol may be converted into glucose (gluconeogenesis) or catabolized via glycolysis. Fatty acids are catabolized via beta oxidation into acetyl coenzyme A that can enter the Krebs cycle for ATP production or be converted into ketone bodies (ketogenesis).
Triglyceride anabolism	The synthesis of triglycerides from glucose and fatty acids is called lipogenesis. Triglycerides are stored in adipose tissue.
PROTEINS	
Protein catabolism	Amino acids are oxidized via the Krebs cycle after deamination. Ammonia resulting from deamination is converted into urea in the liver, passed into blood, and excreted in urine. Amino acids may be converted into glucose (gluconeogenesis), fatty acids, or ketone bodies.
Protein anabolism	Protein synthesis is directed by DNA and utilizes the cells' RNA and ribosomes.

ABSORPTIVE AND POSTABSORPTIVE STATES

Metabolic reactions depend on how recently you have eaten. During the **absorptive (fed) state**, ingested nutrients are entering the bloodstream and glucose is readily available for ATP production. During the **postabsorptive (fasting) state**, absorption of nutrients from the GI tract is complete, and the energy needs of the body must be satisfied by nutrients already in the body. Hormones are the major regulators of reactions occurring in each state. The effects of insulin dominate in the absorptive state whereas several hormones contribute to regulation of metabolic reactions in the postabsorptive state. By analyzing the major events of both states, we can gain a better understanding of the interrelations of metabolic pathways.

Absorptive (Fed) State

An average meal requires about 4 hours for complete absorption, and given three meals a day, the absorptive state exists for about 12 hours each day. The other 12 hours, in the late morning, late afternoon, and most of the night, are spent in the postabsorptive state unless one eats between-meal snacks.

Reactions of the Absorptive State

Figure 25.16 presents the reactions that dominate during the absorptive state.

Figure 25.16   Principal pathways during the absorptive (fed) state.

🔑 *During the absorptive state, most body cells produce ATP by oxidizing glucose to $CO_2 + H_2O$.*

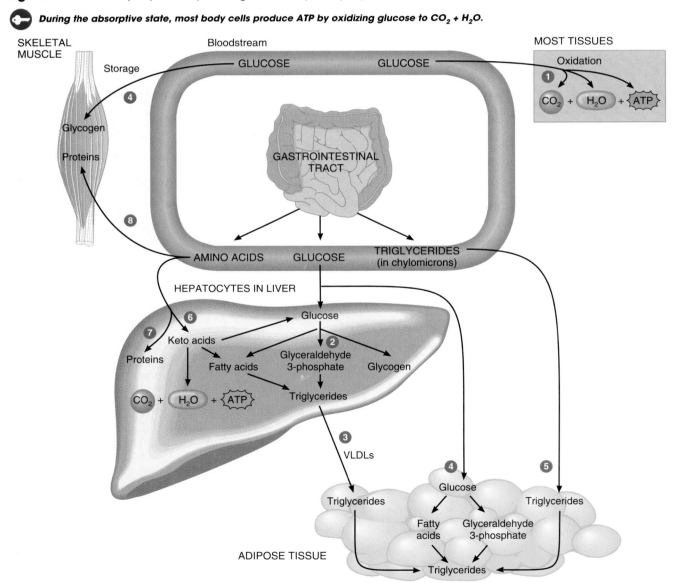

❓ Are the reactions shown here mainly anabolic or catabolic?

1 During the absorptive state, most body cells produce ATP by oxidizing glucose to carbon dioxide and water. This occurs through glycolysis, the Krebs cycle, and the electron transport chain.

2 Glucose that enters hepatocytes is mostly converted to triglycerides or glycogen; little is oxidized for energy.

3 Some fatty acids and triglycerides synthesized in the liver remain there, but hepatocytes package most into a type of transport lipoprotein called a very low-density lipoprotein (VLDL). VLDLs enter the blood and ferry lipids to adipose tissue for storage. As adipocytes remove lipids, VLDLs become smaller and eventually become LDLs (low-density lipoproteins).

4 Adipose tissue cells also take up glucose not picked up by the liver and convert it into triglycerides for storage. Also, some blood glucose is stored as glycogen in skeletal muscles.

5 During the absorptive state, most dietary lipids (mainly triglycerides and fatty acids) are stored in adipose tissue; only a small portion is used for synthesis reactions. Adipose cells obtain the lipids from chylomicrons, from VLDLs, and from their own synthesis reactions.

6 Many absorbed amino acids that enter hepatocytes are deaminated to keto acids. These then can enter the Krebs cycle for ATP production or be used to synthesize glucose or fatty acids.

7 Some amino acids that enter hepatocytes are used to synthesize proteins, for example, plasma proteins.

8 Amino acids not taken up by hepatocytes enter other cells of the body, such as muscle cells, for synthesis of proteins or regulatory chemicals such as hormones or enzymes.

Hormonal Regulation of Absorptive State Reactions

Soon after eating, gastric inhibitory peptide (GIP) and the rise in blood glucose concentration both stimulate insulin release from pancreatic beta cells. In several ways, insulin stimulates absorptive state metabolism. Insulin promotes entry of glucose and amino acids into cells of many tissues. It stimulates phosphorylation of glucose in hepatocytes and conversion of glucose 6-phosphate to glycogen in both liver and muscle. In liver and in adipose tissue, insulin enhances synthesis of triglycerides. (See page 536 to review the effects of insulin.) Human growth hormone and the thyroid hormones (T_3 and T_4) also stimulate some absorptive state reactions.

Exhibit 25.3 summarizes hormonal regulation of reactions occurring in the absorptive (fed) state.

Postabsorptive (Fasting) State

The principal metabolic challenge during the postabsorptive state is to maintain the normal blood glucose level of 70–110 mg/100 ml (3.9–5.6 mmol/liter). Cells are continually removing glucose from the blood, but none is being absorbed by the GI tract. The maintenance of this level is especially important for the nervous system because neurons use only glucose (or ketone bodies during starvation) to produce ATP.

Reactions of the Postabsorptive State

Usually, blood glucose level starts to drop about 4 hours after a large meal. Blood glucose concentration is then maintained in two ways: (1) by producing new glucose molecules and (2) by switching to alternative fuels for ATP

EXHIBIT **25.3 HORMONAL REGULATION IN THE ABSORPTIVE (FED) STATE**

Process	Location	Main Stimulating Hormones
Facilitated diffusion of glucose into cells	Most cells.	Insulin.[a]
Active transport of amino acids into cells	Most cells.	Insulin.
Glycogenesis (glycogen synthesis)	Hepatocytes and muscle fibers.	Insulin.
Protein synthesis	All body cells.	Insulin, thyroid hormone, and human growth hormone.
Lipogenesis (triglyceride synthesis)	Adipose cells and hepatocytes.	Insulin.

[a] Facilitated diffusion of glucose into hepatocytes (liver cells) and neurons is always "turned on" and does not require insulin.

production to conserve scarce glucose. Figure 25.17 shows the major reactions of the postabsorptive state. Reactions that produce glucose are:

1 **Breakdown of liver glycogen.** During fasting a major source of blood glucose is liver glycogen, which can provide about a 4-hour supply of glucose. Liver glycogen is continually being formed and broken down as needed.

2 **Lipolysis.** Glycerol, produced by breakdown of triglycerides in adipose tissue, is also used to form glucose.

3 **Gluconeogenesis using lactic acid.** During exercise, skeletal muscle tissue breaks down stored glycogen (see step **9**) and produces some ATP anaerobically by glycolysis. The pyruvic acid that results is converted partially to acetyl CoA and partially to lactic acid, which diffuses into the blood. In the liver, lactic acid can be used for gluconeogenesis and the resulting glucose released into the blood.

4 **Gluconeogenesis using amino acids.** During fasting or starvation, breakdown of proteins in skeletal muscle and other tissues releases large amounts of amino

acids, which then can be converted to glucose by gluconeogenesis in the liver. Prolonged use of amino acids in this way destroys tissues, especially muscle.

Despite these ways to produce glucose, blood glucose level cannot be maintained for very long without further metabolic changes. Thus a major adjustment must be made during the postabsorptive state to produce ATP while conserving glucose. Reactions that produce ATP without using glucose are:

5 **Oxidation of fatty acids.** The fatty acids released by lipolysis of triglycerides cannot be used for glucose production because acetyl CoA cannot readily be converted to pyruvic acid. But most cells can oxidize the fatty acids directly, feed them into the Krebs cycle as acetyl CoA, and produce ATP through the electron transport chain.

6 **Oxidation of lactic acid.** Cardiac muscle can produce ATP aerobically from lactic acid.

7 **Oxidation of amino acids.** In hepatocytes, amino acids may be oxidized directly to produce ATP.

Figure 25.17   Principal pathways during the postabsorptive (fasting) state.

🔑 *The principal function of the postabsorptive state is to maintain normal blood glucose level.*

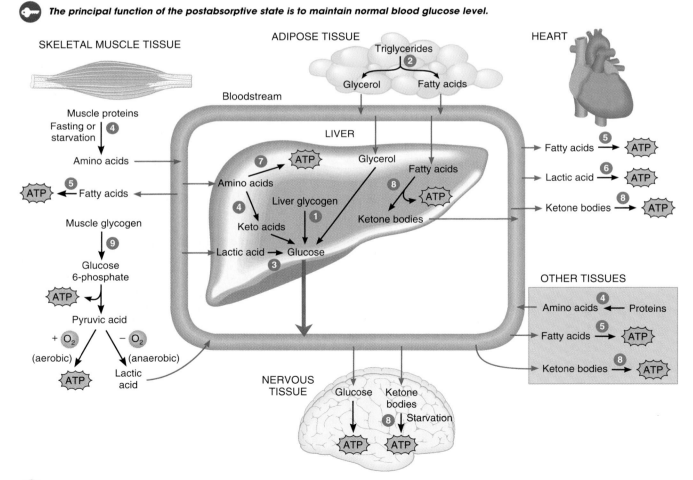

❓ What processes *directly* elevate blood glucose level during this state and where are they taking place?

Vitamins

Organic nutrients required in minute amounts to maintain growth and normal metabolism are called **vitamins**. Unlike carbohydrates, lipids, or proteins, vitamins do not provide energy or serve as building materials. The essential function of vitamins is the regulation of physiological processes. Of the vitamins whose functions are known, most serve as coenzymes.

Most vitamins cannot be synthesized by the body; they must be ingested. Other vitamins, such as vitamin K, are produced by bacteria in the GI tract and then absorbed. The body can assemble some vitamins if the raw materials called **provitamins** are provided. For example, vitamin A is produced by the body from the provitamin beta-carotene, a chemical present in yellow vegetables such as carrots and dark green vegetables such as spinach. No single food contains all the required vitamins—one of the best reasons to eat a varied diet.

Based on solubility, vitamins are divided into two principal groups: fat-soluble and water-soluble. **Fat-soluble vitamins** are emulsified into micelles and absorbed along with other dietary lipids by the small intestine. They cannot be absorbed in adequate quantity unless they are ingested with other lipids. Fat-soluble vitamins may be stored in cells, particularly hepatocytes. The fat-soluble vitamins include vitamins A, D, E, and K. **Water-soluble vitamins**, by contrast, are absorbed along with water in the GI tract and dissolve in the body fluids. Excess quantities of these vitamins are not stored but rather are excreted in the urine. Examples of water-soluble vitamins are the B vitamins and vitamin C.

Besides their other functions, three vitamins—C, E, and beta-carotene (a provitamin)—are termed **antioxidant vitamins** because they inactivate oxygen free radicals. Free radi-cals are highly reactive particles that carry an unpaired electron (see Fig. 3.30). They damage cell membranes, DNA, and other cellular structures and contribute to formation of artery-narrowing atherosclerotic plaques. Some free radicals arise naturally in the body and others derive from environmental hazards such as tobacco smoke and radiation. Antioxidant vitamins are thought to play a role in protecting against some kinds of cancer, reducing the buildup of atherosclerotic plaque, delaying some effects of aging, and decreasing the chance of cataract formation in the lens of the eyes.

VITAMIN AND MINERAL SUPPLEMENTS

Most physicians recommend eating a balanced diet that includes a variety of foods rather than taking vitamin or mineral supplements, except in special circumstances. Common situations that may warrant using supplements include: iron for women who have excessive menstrual bleeding; iron and calcium for women who are pregnant or breast-feeding; folic acid (folate) for all women who may become pregnant to reduce the risk of fetal neural tube defects (see page 424); calcium for adults who do not receive the recommended amount in their diets; and vitamin B_{12} for strict vegetarians who eat no meat. Because most Americans do not ingest in their food the high levels of antioxidant vitamins thought to have beneficial effects, some experts recommend supplementing these vitamins. More is not always better, however; megadoses of vitamins or minerals can be very harmful. The dangers of vitamin or mineral overdose are described on page 842. ■

Exhibit 25.6 lists the principal vitamins, their sources, functions, and related deficiency disorders.

EXHIBIT **25.6** THE PRINCIPAL VITAMINS

Vitamin	Comment and Source	Functions	Deficiency Symptoms and Disorders
FAT-SOLUBLE	All require bile salts and some dietary lipids for adequate absorption.		
A	Formed from provitamin beta-carotene (and other provitamins) in GI tract. Stored in liver. Sources of carotene and other provitamins include yellow and green vegetables; sources of vitamin A include liver and milk.	Maintains general health and vigor of epithelial cells. Beta-carotene acts as an antioxidant to inactivate free radicals. Its potential role in cancer prevention is currently under investigation.	Deficiency results in atrophy and keratinization of epithelium, leading to dry skin and hair, increased incidence of ear, sinus, respiratory, urinary, and digestive system infections, inability to gain weight, drying of cornea and ulceration (**xerophthalmia**), nervous disorders, and skin sores.
		Essential for formation of photopigments, light-sensitive chemicals in photoreceptors of retina.	**Night blindness** or decreased ability for dark adaptation.
		Aids in growth of bones and teeth apparently by helping to regulate activity of osteoblasts and osteoclasts.	Slow and faulty development of bones and teeth.

Exhibit continues

EXHIBIT **25.6** **THE PRINCIPAL VITAMINS** (CONTINUED)

Vitamin	Comment and Source	Functions	Deficiency Symptoms and Disorders
D	Sunlight converts 7-dehydrocholesterol in the skin to cholecalciferol (vitamin D_3). A liver enzyme then converts cholecalciferol to 25-hydroxycholecalciferol. A second enzyme in the kidneys converts 25-hydroxycholecalciferol to calcitriol (1,25-dihydroxycalciferol), which is the active form of vitamin D. Stored in tissues to slight extent. Most excreted via bile. Dietary sources include fish-liver oils, egg yolk, fortified milk.	Essential for absorption and utilization of calcium and phosphorus from GI tract. Works with parathyroid hormone (PTH) to maintain Ca^{2+} homeostasis.	Defective utilization of calcium by bones leads to **rickets** in children and **osteomalacia** in adults. Possible loss of muscle tone.
E (tocopherols)	Stored in liver, adipose tissue, and muscles. Sources include fresh nuts and wheat germ, seed oils, green leafy vegetables.	Believed to inhibit catabolism of certain fatty acids that help form cell structures, especially membranes. Involved in formation of DNA, RNA, and red blood cells. May promote wound healing, contribute to the normal structure and functioning of the nervous system, and prevent scarring. Believed to help protect liver from toxic chemicals like carbon tetrachloride. Acts as an antioxidant to inactivate free radicals.	May cause the oxidation of monounsaturated fats, resulting in abnormal structure and function of mitochondria, lysosomes, and plasma membranes. A possible consequence is hemolytic anemia. Deficiency also causes muscular dystrophy in monkeys and sterility in rats.
K	Produced by intestinal bacteria. Stored in liver and spleen. Dietary sources include spinach, cauliflower, cabbage, liver.	Coenzyme essential for synthesis of several clotting factors by liver, including prothrombin.	Delayed clotting time results in excessive bleeding.
WATER-SOLUBLE	Absorbed along with water in GI tract and dissolved in body fluids.		
B_1 (thiamine)	Rapidly destroyed by heat. Not stored in body. Excessive intake eliminated in urine. Sources include whole-grain products, eggs, pork, nuts, liver, yeast.	Acts as coenzyme for many different enzymes that break carbon-to-carbon bonds and are involved in carbohydrate metabolism of pyruvic acid to CO_2 and H_2O. Essential for synthesis of acetylcholine.	Improper carbohydrate metabolism leads to buildup of pyruvic and lactic acids and insufficient production of ATP for muscle and nerve cells. Deficiency leads to: (1) **beriberi**—partial paralysis of smooth muscle of GI tract, causing digestive disturbances; skeletal muscle paralysis; atrophy of limbs; (2) **polyneuritis**—due to degeneration of myelin sheaths; impaired reflexes related to kinesthesia, impaired sense of touch, stunted growth in children, and poor appetite.
B_2 (riboflavin)	Not stored in large amounts in tissues. Most is excreted in urine. Small amounts supplied by bacteria of GI tract. Dietary sources include yeast, liver, beef, veal, lamb, eggs, whole-grain products, asparagus, peas, beets, peanuts.	Component of certain coenzymes (for example, FAD and FMN) in carbohydrate and protein metabolism, especially in cells of eye, integument, mucosa of intestine, blood.	Deficiency may lead to improper utilization of oxygen resulting in blurred vision, cataracts, and corneal ulcerations. Also dermatitis and cracking of skin, lesions of intestinal mucosa, and development of one type of anemia.

Vitamin	Comment and Source	Functions	Deficiency Symptoms and Disorders
Niacin (nicotinamide)	Derived from amino acid tryptophan. Sources include yeast, meats, liver, fish, whole-grain products, peas, beans, nuts.	Essential component of NAD and NADP, coenzymes in oxidation–reduction reactions. In lipid metabolism, inhibits production of cholesterol and assists in triglyceride breakdown.	Principal deficiency is **pellagra,** characterized by dermatitis, diarrhea, and psychological disturbances.
B$_6$ (pyridoxine)	Synthesized by bacteria of GI tract. Stored in liver, muscle, brain. Other sources include salmon, yeast, tomatoes, yellow corn, spinach, whole-grain products, liver, yogurt.	Essential coenzyme for normal amino acid metabolism. Assists production of circulating antibodies. May function as coenzyme in triglyceride metabolism.	Most common deficiency symptom is dermatitis of eyes, nose, and mouth. Other symptoms are retarded growth and nausea.
B$_{12}$ (cyanocobalamin)	Only B vitamin not found in vegetables; only vitamin containing cobalt. Absorption from GI tract depends on HCl and intrinsic factor secreted by gastric mucosa. Sources include liver, kidney, milk, eggs, cheese, meat.	Coenzyme necessary for red blood cell formation, formation of amino acid methionine, entrance of some amino acids into Krebs cycle, and manufacture of choline (used to synthesize acetylcholine).	Pernicious anemia, neuropsychiatric abnormalities (ataxia, memory loss, weakness, personality and mood changes, and abnormal sensations), and impaired osteoblast activity.
Pantothenic acid	Stored primarily in liver and kidneys. Some produced by bacteria of GI tract. Other sources include kidney, liver, yeast, green vegetables, cereal.	Constituent of coenzyme A essential for transfer of pyruvic acid into Krebs cycle, conversion of lipids and amino acids into glucose, and synthesis of cholesterol and steroid hormones.	Experimental deficiency tests indicate fatigue, muscle spasms, neuromuscular degeneration, insufficient production of adrenal steroid hormones.
Folic acid (folate, folacin)	Synthesized by bacteria of GI tract. Dietary sources include green leafy vegetables and liver.	Component of enzyme systems synthesizing purines and pyrimidines built into DNA and RNA. Essential for normal production of red and white blood cells.	Production of abnormally large red blood cells (macrocytic anemia). Higher risk of neural tube defects in babies born to folate-deficient mothers.
Biotin	Synthesized by bacteria of GI tract. Dietary sources include yeast, liver, egg yolk, kidneys.	Essential coenzyme for conversion of pyruvic acid to oxaloacetic acid and synthesis of fatty acids and purines.	Mental depression, muscular pain, dermatitis, fatigue, nausea.
C (ascorbic acid)	Rapidly destroyed by heat. Some stored in glandular tissue and plasma. Sources include citrus fruits, tomatoes, green vegetables.	Exact role not understood. Promotes many metabolic reactions, particularly protein metabolism, including laying down of collagen in formation of connective tissue. As coenzyme, may combine with poisons, rendering them harmless until excreted. Works with antibodies, promotes wound healing, and functions as an antioxidant. Role in cancer prevention is under investigation.	Scurvy; anemia; many symptoms related to poor connective tissue growth and repair including tender swollen gums, loosening of teeth (alveolar processes also deteriorate), poor wound healing, bleeding (vessel walls fragile because of connective tissue degeneration), and retardation of growth.

DISORDERS: HOMEOSTATIC IMBALANCES

HEAT CRAMPS, HEAT EXHAUSTION, AND HEATSTROKE

Heat cramps occur as a result of profuse sweating that removes water and salt (NaCl) from the body. The salt loss causes painful contractions of muscles called heat cramps. The cramps tend to occur in muscles used while working but do not appear until the person relaxes after work. Salted liquids usually lead to rapid improvement.

In **heat exhaustion** (**heat prostration**), the body temperature is generally normal, or a little below, and the skin is cool and clammy (moist) due to profuse perspiration. Heat exhaustion is normally characterized by fluid and electrolyte loss, especially salt. The salt loss results in muscle cramps, dizziness, vomiting, and fainting. Fluid loss may cause low blood pressure. Complete rest and salt tablets are recommended.

Heatstroke (**sunstroke**) occurs when the temperature and relative humidity are high, making it difficult for the body to lose heat by radiation, conduction, or evaporation. There is a decreased flow of blood to the skin, perspiration is greatly reduced, and body temperature rises sharply. The skin is thus dry and hot—the temperature may reach 43°C (110°F). Brain cells are quickly affected and may be destroyed permanently. As a result, the body temperature-regulating reflexes fail to operate. Treatment must be undertaken immediately and consists of cooling the body by immersing the victim in cool water and by administering fluids and electrolytes.

OBESITY

Obesity is defined as a body weight more than 20% above a desirable standard due to an excessive accumulation of adipose tissue. (An athlete may be overweight due to higher-than-normal amounts of muscle tissue without being obese.) Even moderate obesity is hazardous to health. It is implicated as a risk factor in cardiovascular disease, hypertension, pulmonary disease, non-insulin-dependent diabetes mellitus (type II), arthritis, certain cancers (breast, uterus, and colon), varicose veins, and gallbladder disease. Also, loss of body fat in obese persons has been shown to elevate HDL cholesterol, the type associated with prevention of cardiovascular disease.

In a few cases, obesity may result from trauma or tumors in the food-regulating centers in the hypothalamus. In most cases of obesity, no specific cause can be identified. Contributing factors include eating habits taught early in life, overeating to relieve tension, and social customs. Recently, obesity has been linked to genetic factors. Studies indicate that some people inherit a low metabolic rate and that they become obese not because they eat too much but because they burn calories too slowly.

Morbid obesity refers to obese people who weigh twice their ideal weight or more. The condition is so named because it is associated with serious and life-threatening conditions such as hypertension, diabetes mellitus, and atherosclerosis. For those with morbid obesity, a surgical procedure may be considered. Most operations for morbid obesity either produce a certain amount of malabsorption, such as the intestinal by-

pass, or reduce the size of the stomach so that it will hold less food, such as the gastric bypass or gastroplasty.

VITAMIN OR MINERAL OVERDOSE

Hypervitaminosis refers to an excess of one or more vitamins. High doses of certain vitamins and minerals can be deadly. Rarely is vitamin or mineral overdose the result of eating foods. Rather, it stems from consumption of supplements. (But bear liver supposedly contains enough vitamin A to kill a person. So if you like to eat liver, stick to chicken or beef.) The fat-soluble vitamins (A, E, D, and K) are particularly toxic because they tend to be stored and accumulate in the body rather than being excreted in the urine. Some adverse effects of high doses of minerals and vitamins are as follows.

Minerals

Calcium: Depresses nerve function, causes drowsiness, extreme lethargy, calcium deposits, kidney stones.

Iron: Damage to liver, heart, and pancreas.

Zinc: Masklike fixed expression, difficulty in walking, slurred speech, hand tremor, involuntary laughter.

Cobalt: Goiter, polycythemia, and heart damage.

Selenium: Nausea, vomiting, fatigue, irritability, and loss of fingernails and toenails.

Vitamins

A: Blurred vision, dizziness, ringing in the ears, headache, insomnia, irritability, apathy, ataxia, stupor, skin rash, nausea, vomiting, diarrhea, hair loss, joint pain, menstrual irregularities, fatigue, liver damage, abnormal bone growth, and damage to the nervous system.

D: Calcium deposits, deafness, nausea, fatigue, headache, loss of appetite, kidney stones, weak bones, hypertension, and high cholesterol.

E: Thrombophlebitis, pulmonary embolism, hypertension, muscular weakness, severe fatigue, breast tenderness, and slow wound healing.

Niacin: Acute flushing, peptic ulcers, liver dysfunction, gout, faintness, dizziness, tingling of fingertips, arrhythmias, and hyperglycemia.

B_6: Impaired sense of position and vibration, diminished tendon reflexes, numbness and loss of sensations in hands and feet, difficulty in walking, impaired memory, depression, headache, and fatigue.

C: Dependence on megadoses may lead to scurvy when withdrawn, kidney stones, diarrhea, hemolysis, hot flashes, headache, fatigue, and insomnia.

MALNUTRITION

Whereas **nutrition** refers to the intake of adequate essential nutrients and calories to maintain health, **malnutrition** refers to a state of poor nutrition. One cause of malnutrition is undernutrition, that is, inadequate food intake. It can result from conditions such as fasting, anorexia nervosa, deprivation, cancer, gastrointestinal obstructions, inability to swallow, renal disease, and poor dentition. Other causes of malnutrition are imbalance of nutrients, malabsorption of nutrients, improper distribution of nutrients, inability to use nutrients (for example, diabetes mellitus), increased nutrient requirements (due to fever, infections, burns, fractures, stress, and exposure to heat or cold), increased nutrient losses (diarrhea, bleeding, or glycosuria), and overnutrition (excess vitamins, minerals, and calories).

One of the major types of undernutrition is known as protein–calorie undernutrition, which occurs when there is inadequate intake of protein and/or calories to meet a person's nutritional requirements. Protein–calorie undernutrition may be classified into two types based on which factor is lacking in the diet. In one type, called **kwashiorkor** (kwash-ē-OR-kor), protein intake is deficient despite normal or nearly normal calorie intake. Some dietary proteins are called complete proteins; that is, they contain adequate amounts of essential amino acids. Sources of complete proteins are primarily animal products such as milk, meat, fish, poultry, and eggs. Incomplete proteins lack certain essential amino acids. An example is zein, the protein in corn, which lacks the essential amino acids tryptophan and lysine. The diet of many Africans consists largely of cornmeal. As a result, many African children develop kwashiorkor. It is characterized by edema of the abdomen, enlarged liver, decreased blood pressure, bradycardia, hypothermia, anorexia, lethargy, dry and hyperpigmented skin, easily pluckable hair, and sometimes mental retardation.

Another type of protein–calorie undernutrition is called **marasmus** (mar-AZ-mus). It results from inadequate intake of both protein and calories. Its characteristics include retarded growth, low weight, muscle wasting, emaciation, dry skin, and thin, dry, dull hair.

PHENYLKETONURIA

Phenylketonuria (fen′-il-kē′-tō-NOO-rē-a) or **PKU** is a genetic error of metabolism characterized by an elevation of the amino acid phenylalanine in the blood. The DNA of children with phenylketonuria lacks the gene that normally programs the manufacture of the enzyme phenylalanine hydroxylase. This enzyme is necessary for converting phenylalanine into the amino acid tyrosine, an amino acid that enters the Krebs cycle. As a result, phenylalanine cannot be metabolized, and what is not used in protein synthesis builds up in the blood. High levels of phenylalanine are toxic to the brain during the early years of life when the brain is developing, leading eventually to mental retardation. When the condition is detected early, mental retardation can be prevented by restricting the child to a diet that supplies only the amount of phenylalanine necessary for growth. The artificial sweetener aspartame contains phenylalanine, and its consumption should be restricted in children with PKU. PKU may be detected by blood or urine tests.

CELIAC DISEASE

Celiac disease results in malabsorption by the intestinal mucosa due to the ingestion of gluten. **Gluten** is the water-insoluble protein fraction of wheat, rye, barley, and oats. In susceptible persons, ingestion of gluten induces destruction of villi and inhibition of enzyme secretion accompanied by a variable amount of malabsorption. The remedy is to eat a diet that excludes all cereal grains except rice and corn.

STUDY OUTLINE

INTRODUCTION (p. 807)

1. Our only source of energy for performing biological work is the food we eat. Food also provides building blocks needed for tissue growth and repair and essential substances that we cannot synthesize.
2. Nutrients are chemical substances in food that provide energy, act as building blocks in forming new body components, serve as storage molecules, or assist in the functioning of various body processes.
3. There are six major classes of nutrients: carbohydrates, lipids, proteins, minerals, vitamins, and water.

REGULATION OF FOOD INTAKE (p. 807)

1. Two centers in the hypothalamus related to regulation of food intake are the feeding center and satiety center; the feeding center is constantly active but may be inhibited by the satiety center.
2. Among the stimuli that affect the feeding and satiety centers are glucose, amino acids, lipids, body temperature, distension of the GI tract, and cholecystokinin (CCK).

METABOLISM (p. 809)

1. Metabolism refers to all chemical reactions of the body and has two phases: catabolism and anabolism.
2. Catabolism is the term for reactions that break down complex organic compounds into simple ones. Overall, catabolic reactions produce more energy than they consume.
3. Chemical reactions that combine simple molecules to form more complex ones that form the body's structural and functional components are collectively known as anabolism. Overall, anabolic reactions consume more energy than they produce.
4. The coupling of anabolism and catabolism is via ATP.

METABOLISM AND BODY TEMPERATURE (p. 809)

Basal Metabolic Rate (p. 809)

1. Measurement of the metabolic rate under basal conditions is called the basal metabolic rate (BMR).
2. A kilocalorie (kcal) or Calorie is the amount of energy required to raise the temperature of 1000 g of water from 14 to 15°C.

3. BMR is expressed in kilocalories per square meter of body area per hour ($kcal/m^2/hr$).

Homeostasis of Body Temperature (p. 809)

1. A normal body temperature is maintained by a delicate balance between heat-producing and heat-losing mechanisms.
2. The hypothalamic thermostat is in the preoptic area.
3. Metabolic rate is affected by exercise, the nervous system, hormones, body temperature, ingestion of food, age, sex, climate, sleep, and malnutrition.
4. Responses that produce, conserve, or retain heat when body temperature falls are vasoconstriction, sympathetic stimulation, skeletal muscle contraction, and thyroid hormone production.
5. Mechanisms of heat loss are radiation, evaporation, conduction, and convection.
6. Radiation is the transfer of heat from a warmer object to a cooler object without physical contact.
7. Evaporation is the conversion of a liquid to a vapor. In the process, heat is lost.
8. Conduction is the transfer of body heat to a substance or object in contact with the body.
9. Convection is the transfer of body heat by a liquid or gas between areas of different temperatures.
10. Responses that increase heat loss when body temperature increases include vasodilation, decreased metabolic rate, decreased skeletal muscle contraction, and perspiration.

Body Temperature Abnormalities (p. 812)

1. Fever is an abnormally high body temperature most commonly caused by bacteria (and their toxins) and viruses. A fever-producing substance is called a pyrogen.
2. Hypothermia refers to a lowering of body temperature.

ENERGY PRODUCTION (p. 813)

1. Oxidation is the removal of electrons from a substance; reduction is the addition of electrons to a substance; oxidation–reduction reactions are coupled.
2. ATP can be generated by substrate-level phosphorylation, oxidative phosphorylation, and photophosphorylation (if chlorophyll is present).

CARBOHYDRATE METABOLISM (p. 814)

1. During digestion, polysaccharides and disaccharides are converted to monosaccharides, which are absorbed through capillaries in villi and transported to the liver via the hepatic portal vein.
2. Carbohydrate metabolism is primarily concerned with glucose metabolism.

Fate of Carbohydrates (p. 814)

1. Glucose is the body's preferred source for ATP production; it is oxidized by cells to provide energy.
2. Excess glucose can be used to form amino acids, stored by the liver and skeletal muscles as glycogen, converted to triglycerides, or occasionally excreted in the urine.

Glucose Movement into Cells (p. 814)

1. Absorption of glucose into cells of the GI tract and kidneys is by secondary active transport with Na^+.
2. Glucose moves into most cells by facilitated diffusion and becomes phosphorylated to glucose 6-phosphate; this process is stimulated by insulin.

3. Glucose entry into neurons and hepatocytes is always "turned on."

Glucose Catabolism (p. 815)

1. Glucose oxidation is also called cellular respiration.
2. The complete oxidation of glucose to CO_2 and H_2O involves glycolysis, the Krebs cycle, and the electron transport chain.
3. Glycolysis refers to the breakdown of glucose into two molecules of pyruvic acid; there is a net production of two molecules of ATP.
4. When oxygen is in short supply, pyruvic acid is reduced to lactic acid; under aerobic conditions, pyruvic acid enters the Krebs cycle.
5. Pyruvic acid is prepared for entrance into the Krebs cycle by conversion to a two-carbon acetyl group followed by the addition of coenzyme A to form acetyl coenzyme A.
6. The Krebs cycle involves decarboxylations and oxidations and reductions of various organic acids.
7. Each molecule of pyruvic acid that enters the Krebs cycle produces three molecules of CO_2, four molecules of NADH and 4 H^+, one molecule of $FADH_2$, and one molecule of ATP.
8. The energy originally in glucose and then pyruvic acid is transferred primarily to reduced coenzymes NADH and $FADH_2$.
9. The electron transport chain involves a series of oxidation–reduction reactions in which the energy in NADH and $FADH_2$ is liberated and transferred to ATP.
10. The electron carriers include FMN, cytochromes, iron–sulfur centers, copper atoms, and ubiquinones.
11. The electron transport chain yields 32 or 34 molecules of ATP and six molecules of H_2O.
12. The complete oxidation of glucose can be represented as follows:

$$C_6H_{12}O_6 + 6\ O_2 + 36 \text{ or } 38 \text{ ADPs} + 36 \text{ or } 38 \text{ } \textcircled{P} \longrightarrow$$
$$6\ CO_2 + 6\ H_2O + 36 \text{ or } 38 \text{ ATPs}$$

Glucose Anabolism (p. 825)

1. The conversion of glucose to glycogen for storage in the liver and skeletal muscle is called glycogenesis. It is stimulated by insulin.
2. The body can store about 500 g (1.1 lb) of glycogen.
3. The conversion of glycogen back to glucose is called glycogenolysis. It occurs between meals and is stimulated by glucagon and epinephrine.
4. Gluconeogenesis is the conversion of noncarbohydrate molecules into glucose. It is stimulated by cortisol and glucagon.

LIPID METABOLISM (p. 826)

1. During digestion, triglycerides are ultimately broken down into fatty acids and monoglycerides.
2. Long-chain fatty acids and monoglycerides are carried in micelles, digested to glycerol and fatty acids in intestinal epithelial cells, recombined to form triglycerides, and transported by chylomicrons through the lacteals of villi into the thoracic duct.

Fate of Lipids (p. 826)

1. Lipids may be oxidized to produce ATP or stored in adipose tissue as triglycerides.

2. A few lipids are used as structural molecules or to synthesize essential molecules. Examples include phospholipids of plasma membranes, lipoproteins that transport cholesterol, thromboplastin for blood clotting, and cholesterol used to synthesize bile salts and steroid hormones.

Triglyceride Storage (p. 827)

1. Triglycerides are stored in adipose tissue, mostly in the subcutaneous layer.
2. Adipose tissue contains lipases that catalyze the deposition of triglycerides from chylomicrons and hydrolyze triglycerides into fatty acids and glycerol.

Lipid Catabolism: Lipolysis (p. 827)

1. Triglycerides are split into fatty acids and glycerol and released from adipose tissue under the influence of epinephrine, norepinephrine, cortisol, thyroid hormones, and human growth hormone.
2. Glycerol can be converted into glucose by conversion into glyceraldehyde 3-phosphate.
3. In beta oxidation of fatty acids, carbon atoms are removed in pairs from fatty acid chains; the resulting molecules of acetyl coenzyme A enter the Krebs cycle.
4. The formation of ketone bodies by the liver is a normal phase of fatty acid catabolism, but an excess of ketone bodies, called ketosis, may cause acidosis.

Lipid Anabolism: Lipogenesis (p. 828)

1. The conversion of glucose or amino acids into lipids is called lipogenesis; it is stimulated by insulin.
2. The intermediary links in lipogenesis are glyceraldehyde 3-phosphate and acetyl coenzyme A.

PROTEIN METABOLISM (p. 828)

1. During digestion, proteins are hydrolyzed into amino acids.
2. Amino acids are absorbed by the capillaries of villi and enter the liver via the hepatic portal vein.

Fate of Proteins (p. 828)

1. Amino acids, under the influence of human growth hormone (hGH) and insulin, enter body cells by active transport.
2. Inside cells, amino acids are synthesized into proteins that function as enzymes, hormones, structural elements, and so forth; stored as fat or glycogen; or used for energy.

Protein Catabolism (p. 828)

1. Before amino acids can be catabolized, they must be converted to substances that can enter the Krebs cycle; these conversions involve deamination, decarboxylation, and hydrogenation.
2. Amino acids may also be converted into glucose, fatty acids, and ketone bodies.

Protein Anabolism (p. 829)

1. Protein synthesis is stimulated by human growth hormone (hGH), thyroxine, and insulin.
2. The process is directed by DNA and RNA and carried out on the ribosomes of cells.

SUMMARY OF KEY MOLECULES IN METABOLISM (p. 830)

1. Three molecules play a key role in metabolism: glucose 6-phosphate, pyruvic acid, and acetyl coenzyme A.

2. Glucose 6-phosphate may be converted to glucose, glycogen, ribose 5-phosphate, and pyruvic acid.
3. When ATP is low and oxygen is plentiful, pyruvic acid is converted to acetyl coenzyme A; when oxygen supply is low, pyruvic acid is converted to lactic acid. One link between carbohydrate and protein metabolism is via pyruvic acid.
4. Acetyl coenzyme A is the molecule that enters the Krebs cycle and is also used to synthesize fatty acids, ketone bodies, and cholesterol.

ABSORPTIVE AND POSTABSORPTIVE STATES (p. 832)

1. During the absorptive (fed) state, ingested nutrients enter the blood and lymph from the GI tract.
2. During the absorptive state, blood glucose is oxidized by body cells to form ATP. Glucose transported to the liver also is converted to glycogen or triglycerides. Most triglycerides are stored in adipose tissue. Amino acids in hepatocytes are converted to carbohydrates, fats, and proteins. Exhibit 25.3 summarizes hormonal regulation of metabolism during the absorptive state.
3. During the postabsorptive (fasting) state, absorption is complete and the ATP needs of the body are satisfied by nutrients already present in the body.
4. The major task during the postabsorptive state is to maintain normal blood glucose level. This involves conversion of liver and skeletal muscle glycogen into glucose, conversion of glycerol into glucose, conversion of amino acids into glucose, and oxidation of fatty acids, ketone bodies, and amino acids to supply ATP. Exhibit 25.4 summarizes hormonal regulation of metabolism during the postabsorptive state.

CALORIES AND NUTRITION (p. 835)

1. The apparatus used to determine the caloric value of foods is called a calorimeter. Most teens and adults need between 1600 and 2800 Calories per day.
2. Nutrition experts suggest dietary calories be 50–60% from carbohydrates, 30% or less from fats, and 12–15% from proteins, although the optimal levels of these nutrients are not known for sure.
3. The Food Guide Pyramid shows how many servings of five food groups to eat each day to attain the number of calories and variety of nutrients needed for wellness.

Minerals (p. 837)

1. Minerals are inorganic substances that help regulate body processes.
2. Minerals known to perform essential functions are calcium, phosphorus, sodium, chlorine, potassium, magnesium, iron, sulfur, iodine, manganese, cobalt, copper, zinc, selenium, and chromium. Their functions are summarized in Exhibit 25.5.

Vitamins (p. 839)

1. Vitamins are organic nutrients that maintain growth and normal metabolism. Many function in enzyme systems.
2. Fat-soluble vitamins are absorbed with fats and include A, D, E, and K.
3. Water-soluble vitamins are absorbed with water and include the B vitamins and vitamin C.
4. The functions and deficiency disorders of the principal vitamins are summarized in Exhibit 25.6.

REVIEW QUESTIONS

1. Define a nutrient. List the six classes of nutrients and indicate the function of each. (p. 807)
2. Discuss how food intake is regulated. (p. 807)
3. What is metabolism? Distinguish between anabolism and catabolism and give examples of each. (p. 808)
4. How does ATP couple anabolism and catabolism? (p. 808)
5. Define a kilocalorie (kcal). How is the unit used? (p. 809)
6. Define basal metabolic rate (BMR). How is it measured? (p. 809)
7. Explain how body temperature is regulated by describing the mechanisms of heat production, conservation, and loss. (p. 810)
8. Contrast fever and hypothermia. What causes fever? What are its benefits? (p. 812)
9. Define oxidation and reduction and give an example of each. (p. 813)
10. Describe the three ways that ATP can be generated. (p. 814)
11. How are carbohydrates absorbed and what are their fates in the body? How does glucose move into body cells? (p. 814)
12. Define glycolysis. Describe its principal events and outcome. What is the fate of pyruvic acid? (p. 816)
13. Describe how acetyl coenzyme A is formed. (p. 818)
14. Outline the principal events and outcomes of the Krebs cycle. (p. 818)
15. Explain what happens in the electron transport chain. (p. 821)
16. What is chemiosmosis? (p. 823)
17. Summarize the outcomes of the complete oxidation of a molecule of glucose. (p. 823)
18. Define glycogenesis and glycogenolysis. Under what circumstances does each occur? (p. 825)
19. Why is gluconeogenesis important? Give specific examples to substantiate your answer. (p. 825)
20. How are triglycerides absorbed and what are their fates in the body? Where are triglycerides stored in the body? (p. 826)
21. Explain the principal events of the catabolism of glycerol and fatty acids. (p. 827)
22. What are ketone bodies? What is ketosis? (p. 828)
23. Define lipogenesis and explain its importance. (p. 828)
24. How are proteins absorbed and what are their fates in the body? (p. 828)
25. Relate deamination to amino acid catabolism. (p. 828)
26. Summarize the major steps involved in protein synthesis. (p. 79)
27. Distinguish between essential and nonessential amino acids. (p. 830)
28. Explain the importance of glucose 6-phosphate, pyruvic acid, and acetyl coenzyme A in metabolism. (p. 830)
29. What is the absorptive (fed) state? Outline its principal events. (p. 832)
30. What is the postabsorptive (fasting) state? Outline its principal events. (p. 833)
31. Indicate the roles of the following hormones in the regulation of metabolism: insulin, glucagon, epinephrine, human growth hormone (hGH), thyroxine, cortisol, estrogen, and testosterone. (p. 835)
32. How is the caloric value of foods determined? (p. 835)
33. List the guidelines for choosing a healthy diet. (p. 835)
34. Describe the Food Guide Pyramid and give examples of foods from each food group. (p. 836)
35. What is a mineral? Briefly describe the functions of the following minerals: calcium, phosphorus, iron, iodine, copper, sodium, potassium, chlorine, magnesium, sulfur, zinc, fluorine, manganese, cobalt, chromium, and selenium. (p. 837)
36. Define a vitamin. Explain how we obtain vitamins. Distinguish between a fat-soluble and a water-soluble vitamin. (p. 839)
37. For each of the following vitamins, indicate its principal function and effect of deficiency: A, D, E, K, B_1, B2, niacin, B_6, B_{12}, pantothenic acid, folic acid, biotin, and C. (p. 839)
38. Describe abnormalities associated with megadoses of several minerals and vitamins. (p. 842)

ANSWERS TO QUESTIONS WITH FIGURES

25.1 Anabolism dominates because the cells are synthesizing complex molecules.

25.2 Exercise, sympathetic nervous system, hormones (epinephrine, norepinephrine, thyroxine, testosterone, human growth hormone), elevated body temperature, and ingestion of food.

25.3 Glycolysis.

25.4 The substrate is glucose, a hexose. Kinases are enzymes that phosphorylate (add phosphate to) their substrate.

25.5 Cytosol; mitochondria.

25.6 They will later yield ATP in the electron transport chain.

25.7 Electrons provided by NADH + H^+.

25.8 Each carrier in the series has a lower energy than the carrier it received electrons from because some of the energy in the electrons is used to pump protons.

25.9 In the space between the outer and inner mitochondrial membranes.

25.10 Six; six.

25.11 Muscle fibers; they lack the enzyme phosphatase.

25.12 Hepatocytes.

25.13 Hepatocytes and adipose cells; hepatocytes.

25.14 The amino group is removed by deamination.

25.15 Acetyl coenzyme A.

25.16 Anabolic.

25.17 Lipolysis (adipose and hepatocytes); gluconeogenesis (hepatocytes); glycogenolysis (hepatocytes).

25.18 Milk, yogurt, cheeses, and meats.

STUDENT OBJECTIVES

1. List the functions of the kidneys. (p. 848)

2. Identify the external and internal gross anatomical features of the kidneys. (p. 849)

3. Trace the path of blood flow through the kidneys. (p. 853)

4. Discuss how the structure of a nephron contributes to regulating the volume, composition, and pressure of blood. (p. 854)

5. Discuss the processes of glomerular filtration, tubular reabsorption, and tubular secretion. (p. 861)

6. Describe how the kidneys produce dilute and concentrated urine. (p. 873)

7. Explain how dialysis is performed and what it accomplishes. (p. 877)

8. Discuss the anatomy, histology, and physiology of the ureters, urinary bladder, and urethra. (p. 878)

9. List and describe the physical characteristics, normal constituents, and abnormal constituents of urine. (p. 882)

10. Describe the effects of aging on the urinary system. (p. 883)

11. Describe the development of the urinary system. (p. 883)

12. Discuss the causes of urinary tract infections, glomerulonephritis, nephrotic syndrome, renal failure, polycystic kidney disease, and diabetes insipidus. (p. 885)

13. Define medical terminology associated with the urinary system. (p. 886)

CHAPTER 26

THE URINARY SYSTEM

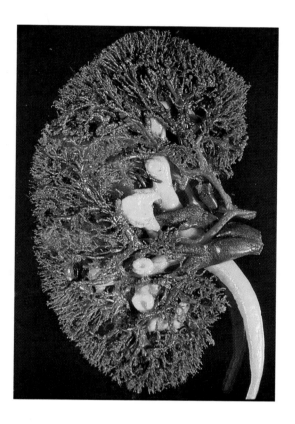

CHAPTER CONTENTS AT A GLANCE

his chapter describes the anatomy and physiology of the urinary system. Just one of its many functions is to help rid the body of waste materials. In homes and communities we use many methods for temporary confinement of wastes, for recycling some materials, and for eventually eliminating those that have no use. Each bedroom has a wastebasket, which temporarily can hold a limited amount of paper wastes. The kitchen sink may have a waste disposer, which grinds food garbage and sends it down the drain into the sewer or septic system. Likewise, toilet and bathtub waste water drains into the sewers. Newspaper, glass, and plastics may be picked up for recycling, whereas garbage usually is taken to a landfill. Just as there are many solutions for waste management in a home or community, several body systems cooperate to meet this need in the human body.

A human waste product is any substance that has no function in the body, for example, excess carbon dioxide (CO_2) from cellular respiration; toxic nitrogen-containing molecules from the catabolism of proteins, such as ammonia and urea; bilirubin from the breakdown of hemoglobin; and uric acid from the catabolism of nucleic acids. Also, essential ions such as sodium (Na^+), chloride (Cl^-), sulfate (SO_4^{2-}), phosphate (HPO_4^{2-}), and hydrogen (H^+) can build up in excess quantity. Even excess water and heat need disposal. Toxic materials and excess essential materials must be excreted (eliminated) from the body. Several tissues, organs, and processes contribute to the temporary confinement of wastes, transport of waste materials for disposal, recycling of materials, and excretion of excess or toxic substances in the body. These include:

1. **Body buffers.** Buffers bind excess hydrogen ions (H^+), which prevents an increase in the acidity of body fluids. Buffers are like wastebaskets because they have a limited capacity and eventually the H^+, like the paper in a wastebasket, must be eliminated from the body by excretion. The contribution of body buffers to regulation of acid–base balance is described in Chapter 27 on page 900.
2. **Blood.** The bloodstream provides pick-up and delivery services for the transport of wastes, just as garbage trucks and sewer lines supply the same services for a community.
3. **Liver.** The liver is the primary site for metabolic recycling, for example, conversion of amino acids into glucose or glucose into fatty acids. The liver also changes toxic substances into less toxic ones, such as ammonia into urea. These functions of the liver were described in Chapters 24 and 25.
4. **Lungs.** With each exhalation, the lungs excrete CO_2, heat, and a little water vapor (see Chapter 23).
5. **Sudoriferous (sweat) glands in the skin.** Especially during exercise, sudoriferous glands in the skin help eliminate excess heat, water, and CO_2 plus small quantities of salts and urea.
6. **Gastrointestinal tract.** Through defecation, the gastrointestinal tract eliminates solid, undigested foods, wastes, some CO_2, water, salts, and heat.
7. **Kidneys.** The kidneys eliminate excess water, ammonia, urea, bilirubin, uric acid, some bacterial toxins, H^+ and other ions, plus some heat and CO_2.

The major function of the **urinary system** is to help maintain homeostasis by controlling the composition, volume, and pressure of the blood. Two kidneys, two ureters, one urinary bladder, and one urethra make up the urinary system (Fig. 26.1). The kidneys filter blood and restore selected amounts of water and solutes to the bloodstream. The remaining water and solutes constitute **urine.** Urine is excreted from each kidney through its ureter and is stored in the urinary bladder until it is expelled from the body through the urethra. (In the male, the urethra is also the route for semen to exit the body.)

Although the kidneys play a key role in waste management, their other functions are equally important. These include:

1. **Regulation of blood volume and composition.** The kidneys adjust the composition and volume of the blood and remove wastes from it. In the process, urine is formed. By regulating blood volume and composition, the kidneys in effect also regulate the volume and composition of interstitial fluid, the internal environment of the body.
2. **Regulation of blood pH.** The kidneys excrete a variable amount of H^+, which helps control blood pH.
3. **Regulation of blood pressure.** The kidneys help regulate blood pressure by secreting the enzyme renin, which activates the renin–angiotensin pathway (shown in Fig. 18.20). The response to an increase in release of renin is an increase in blood pressure and blood volume.
4. **Contributions to metabolism.** The kidneys (1) help synthesize the hormone calcitriol, the active form of vitamin D; (2) secrete erythropoietin, the hormone that stimulates production of red blood cells; and (3) perform gluconeogenesis (synthesis of new glucose molecules) during periods of fasting or starvation.

The specialized branch of medicine that deals with the structure, function, and diseases of the male and female urinary systems and the male reproductive system is known as **nephrology** (nef-ROL-ō-jē; *nephros* = kidney; *logos* = study of). The branch of medicine related to the male and female urinary systems and the male reproductive system is called **urology** (yoo-ROL-ō-jē; *urina* = urine).

The developmental anatomy of the urinary system is considered at the end of the chapter.

Figure 26.1 Organs of the female urinary system in relation to surrounding structures.

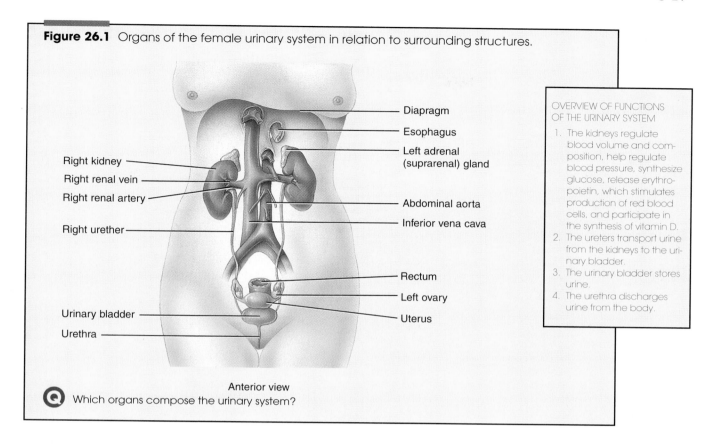

Right kidney

Right renal vein

Right renal artery

Right urether

Urinary bladder

Urethra

Diapragm

Esophagus

Left adrenal (suprarenal) gland

Abdominal aorta

Inferior vena cava

Rectum

Left ovary

Uterus

OVERVIEW OF FUNCTIONS OF THE URINARY SYSTEM

1. The kidneys regulate blood volume and composition, help regulate blood pressure, synthesize glucose, release erythropoietin, which stimulates production of red blood cells, and participate in the synthesis of vitamin D.
2. The ureters transport urine from the kidneys to the urinary bladder.
3. The urinary bladder stores urine.
4. The urethra discharges urine from the body.

Anterior view

Which organs compose the urinary system?

KIDNEYS

The paired **kidneys** are reddish organs shaped like kidney beans. They are located just above the waist between the peritoneum and the posterior wall of the abdomen (Fig. 26.2). Since their position is behind the peritoneum of the abdominal cavity, they are said to be **retroperitoneal** (re'-trō-per-i-tō-NĒ-al; *retro* = behind) organs. Other retroperitoneal structures include the ureters and adrenal glands. The kidneys are located between the levels of the last thoracic and third lumbar vertebrae. They are partially protected by the eleventh and twelfth pairs of ribs. The right kidney is slightly lower than the left because the liver occupies a large area on the right side superior to the kidney.

Three layers of tissue surround each kidney. The deep layer, the **renal** (*renalis* = kidney) **capsule**, is a smooth, transparent, fibrous membrane that is continuous with the outer coat of the ureter. It serves as a barrier against trauma and helps to maintain the shape of the kidney. The intermediate layer, the **adipose capsule (perirenal fat)**, is a mass of fatty tissue surrounding the renal capsule. It also protects the kidney from trauma and holds it firmly in place within the abdominal cavity. The superficial layer, the **renal fascia**, is a thin layer of dense, irregular connective tissue that anchors the kidney to its surrounding structures and to the abdominal wall. On the anterior surface of the kidneys, the renal fascia is deep to the peritoneum.

NEPHROPTOSIS

Nephroptosis (nef'-rō-TŌ-sis; *ptosis* = falling), or **floating kidney**, is an inferior displacement or dropping of the kidney. It occurs when the kidney slips from its normal position because it is not securely held in place. Nephroptosis develops most often in very thin people whose adipose capsule or renal fascia is deficient. It is dangerous because the ureter may kink and block urine flow. The resulting backup of urine puts pressure on the kidney, which damages the renal tissue. Twisting of the ureter also causes pain. ■

External and Internal Anatomy

An average adult kidney is 10–12 cm (4–5 in.) long, 5–7½ cm (2–3 in.) wide, and 2½ cm (1 in.) thick. Its concave medial border faces the vertebral column (Fig. 26.3 on page 851). Near the center of the concave border is a deep vertical fissure called the **renal hilus**, through which the ureter leaves the kidney. Blood and lymphatic vessels and nerves also enter and exit the kidney through the renal hilus. The renal hilus is the entrance to a cavity within the kidney called the **renal sinus** (see Fig. 26.4).

A frontal section through a kidney reveals two distinct regions: a superficial reddish area called the **renal cortex** (*cortex* = rind or bark) and a deep reddish-brown region called the

Figure 26.2 Position and coverings of the kidneys. The arrow in the inset indicates the direction from which the abdomen is viewed (inferior).

 The kidneys are surrounded by a renal capsule, adipose capsule, and renal fascia.

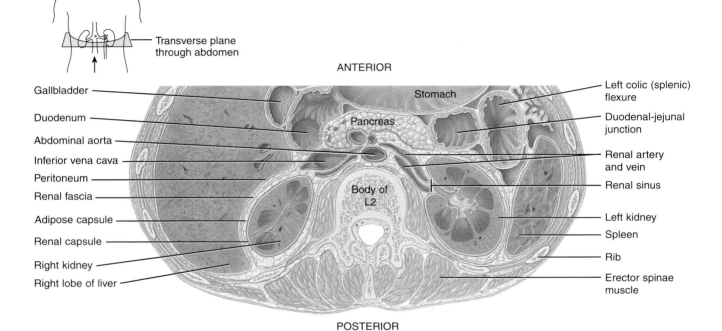

(a) Diagram of inferior view of transverse section of abdomen (L2)

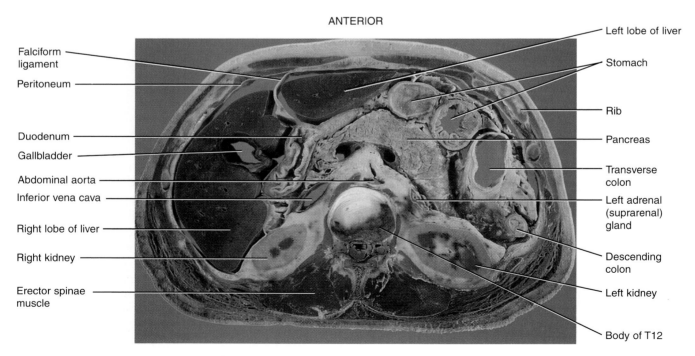

(b) Photograph of inferior view of transverse section of abdomen (level T12)

 Why are the kidneys said to be retroperitoneal?

Figure 26.3 External and internal anatomy of the kidneys.

 The kidneys are located between the levels of the last thoracic and third lumbar vertebrae.

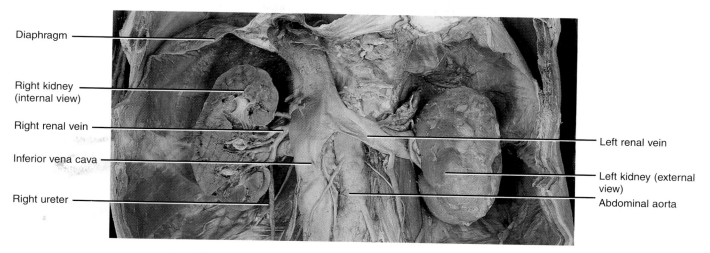

Diaphragm

Right kidney (internal view)

Right renal vein

Inferior vena cava

Right ureter

Left renal vein

Left kidney (external view)

Abdominal aorta

(a) Photograph of anterior view of kidneys and associated structures

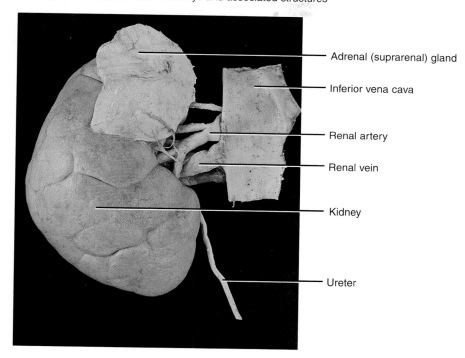

Adrenal (suprarenal) gland

Inferior vena cava

Renal artery

Renal vein

Kidney

Ureter

(b) Photograph of anterior view of right kidney and adrenal (suprarenal) gland

 What structures pass through the renal hilus?

renal medulla (*medulla* = inner portion; Fig. 26.4 on page 852). Within the medulla are 8–18 cone-shaped structures termed **renal (medullary) pyramids**. The base of each pyramid faces the renal cortex, and its apex, called a **renal papilla** (plural is **papillae**), points toward the center of the kidney. The renal cortex is the smooth-textured area extending from the renal capsule to the bases of the renal pyramids and into the spaces between them. Those portions of the renal cortex that extend between renal pyramids are called the **renal columns**.

Together, the renal cortex and renal pyramids constitute the functional portion or **parenchyma** of the kidney. Within the parenchyma are about 1 million microscopic structures called **nephrons** (NEF-rons), which are the functional units of the kidney. The number of nephrons is constant from birth; an increase in kidney size is due solely to the growth of individual nephrons. If nephrons are injured or become diseased, new ones cannot form. The remaining functional nephrons, however, may gradually be able to

Figure 26.4 Internal anatomy of the kidneys.

The two main regions of the kidney parenchyma are the renal cortex and the renal medulla.

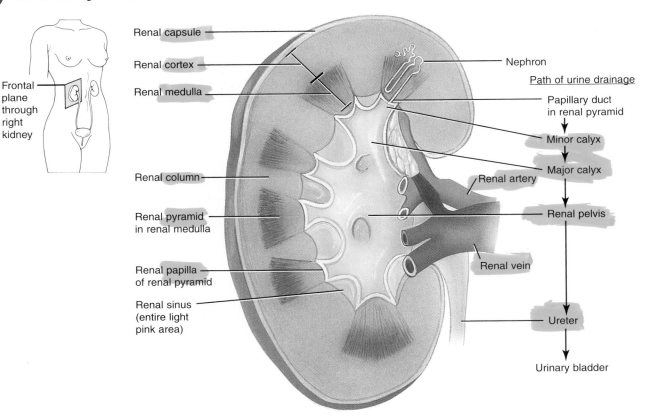

(a) Diagram of frontal section of right kidney

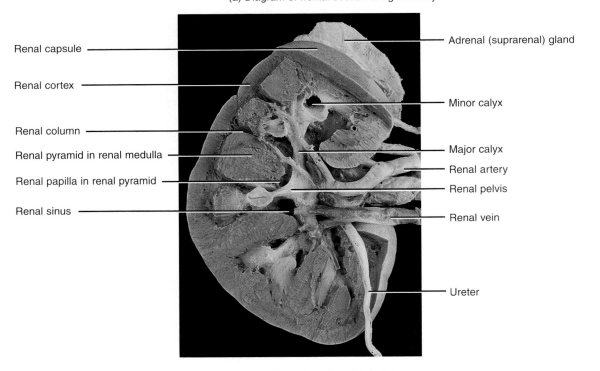

(b) Photograph of frontal section of right kidney

 What structure is considered to be the functional unit of the kidneys?

handle a larger than normal load. For example, surgical removal of one kidney results in hypertrophy (enlargement) of the remaining kidney. Eventually, the one remaining kidney is able to filter blood at 80% of the rate of two normal kidneys.

Urine formed by the nephrons ultimately drains into large ducts called **papillary ducts**. They lead to cuplike structures called **minor** and **major calyces** (KĀ-li-sēz; *calyx* = cup). Each kidney has 8–18 minor calyces and 2–3 major calyces. A minor calyx receives urine from the papillary ducts of one renal pyramid and delivers urine to a major calyx. From the major calyces, the urine drains into a large cavity called the **renal pelvis** (*pelvis* = basin) and then out through the ureter to the urinary bladder.

Blood and Nerve Supply

Because the kidneys remove wastes from the blood and regulate its fluid and electrolyte content, it is not surprising that they are abundantly supplied with blood vessels. Although the kidneys make up just 1% of the total body mass, they receive 20–25% of the resting cardiac output via the right and left **renal arteries** (Fig. 26.5). This amounts to about 1200 ml of blood per minute.

Figure 26.5 Blood supply of the kidneys.

The renal arteries deliver 20–25% of the resting cardiac output to the kidneys.

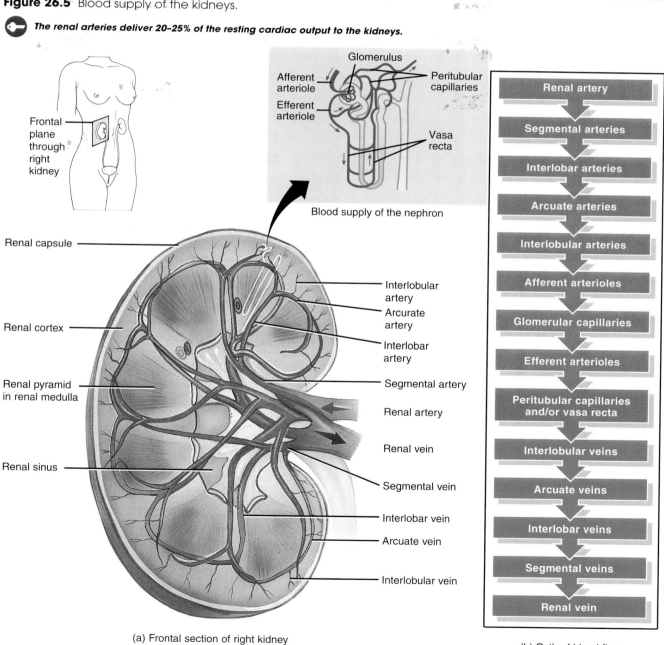

(a) Frontal section of right kidney

(b) Path of blood flow

Figure continues

Figure 26.5 (continued)

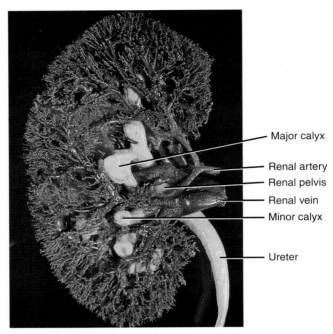

— Major calyx

— Renal artery
— Renal pelvis
— Renal vein
— Minor calyx

— Ureter

(c) Photograph of a cast of the blood vessels and
urine-draining structures of the right kidney

 What volume of blood enters the renal arteries per minute?

Within the kidney, the renal artery divides into several **segmental arteries**, each of which supplies one renal segment (region). Each segmental artery gives off several branches that enter the parenchyma and pass as the **interlobar arteries** in the renal columns between the renal pyramids. At the bases of the renal pyramids, the interlobar arteries arch between the renal medulla and cortex; here they are known as the **arcuate** (*arcuatus* = shaped like a bow) **arteries**. Divisions of the arcuate arteries produce a series of **interlobular arteries**, which enter the renal cortex and give off branches called **afferent** (*ad* = toward; *ferre* = to carry) **arterioles**.

Each nephron receives one afferent arteriole, which divides into a tangled, ball-shaped capillary network called the **glomerulus** (glō-MER-yoo-lus; *glomus* = ball; *ulus* = small; plural is **glomeruli**). The glomerular capillaries then reunite to form an **efferent** (*efferens* = to bring out) **arteriole** that drains blood out of the glomerulus. The afferent–efferent arteriole situation is unique because blood usually flows out of capillaries into venules and not into other arterioles. Since they are capillary networks, the glomeruli are part of the cardiovascular system as well as the urinary system.

The efferent arterioles divide to form a network of capillaries, called the **peritubular** (*peri* = around) **capillaries**, which surround tubular portions of the nephron in the *renal cortex*. Extending from some efferent arterioles are long loop-shaped capillaries called **vasa recta** (VĀ-sa REK-ta; *vasa* = vessels; *recta* = straight) that supply tubular portions of the nephron in the *renal medulla* (see Fig. 26.8b).

The peritubular capillaries eventually reunite to form **peritubular venules** and then **interlobular veins**. The interlobular veins also receive blood from the vasa recta. Then the blood drains through the **arcuate veins** to the **interlobar veins** running between the renal pyramids, and on to the **segmental veins**. Blood leaves the kidney through a single **renal vein** that exits at the renal hilus.

The nerve supply to the kidneys is derived from the **renal plexus** of the sympathetic division of the autonomic nervous system. Nerves from this plexus accompany the renal arteries and their branches and are distributed to the blood vessels. Because these are vasomotor nerves, they regulate the flow of blood through the kidney by regulating the diameters of the arterioles.

Nephron

A nephron consists of two portions: a **renal corpuscle** (KŌR-pus-sul; *corpus* = body; *cle* = tiny) where plasma is filtered and a **renal tubule** into which the filtered fluid (the filtrate) passes. Nephrons perform three basic functions—glomerular filtration, tubular secretion, and tubular reabsorption (Fig. 26.6).

Figure 26.6 Overview of nephron structure and three basic functions: filtration, reabsorption, and secretion.

Filtration occurs in the renal corpuscle whereas reabsorption and secretion occur all along the renal tubule. Excreted substances are those that remain in the urine and leave the body.

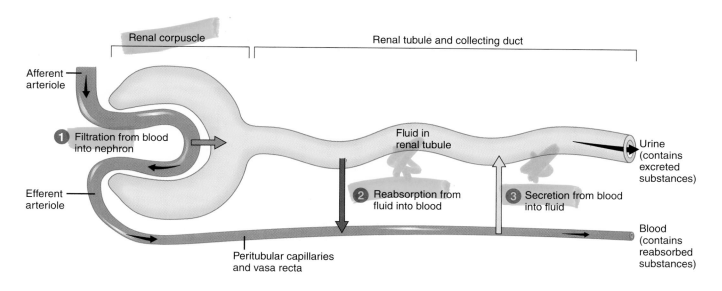

In tubular secretion, are substances entering or leaving the bloodstream?

❶ In *glomerular filtration,* or simply *filtration,* substances in the blood that are small enough pass across the wall of the glomerular capillaries into the renal tubule.

❷ Then, as the fluid moves along the renal tubule, many useful materials are returned to the blood in peritubular capillaries and vasa recta; this is *tubular reabsorption,* or simply *reabsorption.*

❸ As the fluid passes along the tubule, it also gains some additional materials (wastes and excess substances) from tubule cells and blood capillaries; this is *tubular secretion,* or simply *secretion.*

By filtering, reabsorbing, and secreting, nephrons maintain the homeostasis of the blood. Needed materials are returned to the bloodstream and wastes are excreted in the urine. The situation is analogous to a recycling center. Garbage trucks dump refuse into an input hopper, where the smaller refuse passes onto a conveyor belt (filtration of plasma). As the conveyor belt carries the garbage along, workers remove useful and recyclable items, such as aluminum cans, plastics, and glass containers (reabsorption). Other workers place additional garbage left at the center and larger items onto the conveyor belt (secretion). At the end of the belt, all remain-

ing garbage falls into a truck for transport to the landfill (excretion of wastes in urine).

Renal Corpuscle

Renal corpuscles all lie in the renal cortex. Each corpuscle has two components—the **glomerulus,** which is a capillary network, and the **glomerular (Bowman's) capsule,** a double-walled epithelial cup that surrounds the glomerulus (Fig. 26.7). Their arrangement is analogous to a fist (glomerulus) punched into a limp balloon (glomerular capsule) until the fist is covered by two layers of balloon with a space in between. The space is called the **capsular (Bowman's) space.** The parietal layer of the glomerular capsule consists of simple squamous epithelium and forms the outer wall of the cup. The visceral layer consists of modified simple squamous epithelial cells called podocytes (described shortly) that form the inner wall of the cup and adhere closely to the endothelial cells of the glomerular capillaries. Together, the endothelial cells and podocytes form an **endothelial–capsular (filtration) membrane** that acts as a filter. As blood flows through the glomerular capillaries, water and most solutes filter from blood plasma into the capsular space. Large plasma proteins and the formed elements in blood do not normally pass through. Filtered substances pass through the three layers of this membrane in the following order:

Figure 26.7 Endothelial–capsular (filtration) membrane. The size of the endothelial fenestrations and filtration slits in (b) have been exaggerated for emphasis.

🔑 *The endothelial–capsular membrane permits some materials to pass from the blood into the glomerular filtrate but prevents filtration of others.*

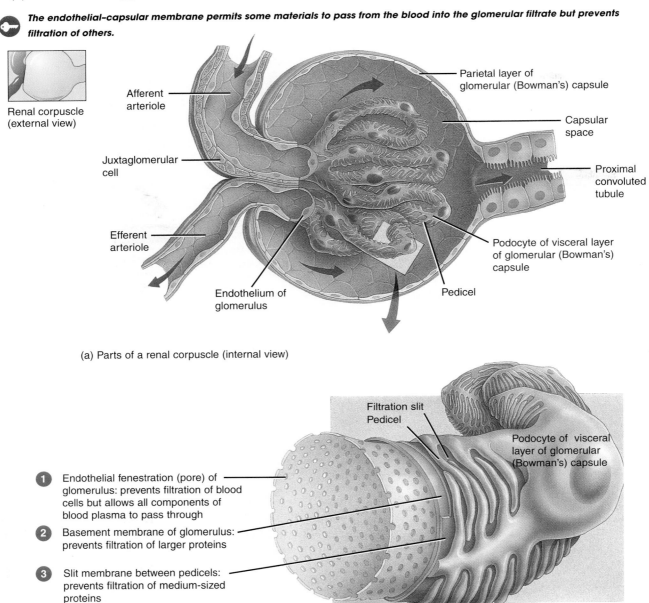

(a) Parts of a renal corpuscle (internal view)

(b) Details of endothelial–capsular (filtration) membrane

1 **Endothelial fenestrations (pores) of the glomerulus.** The single layer of endothelial cells has large fenestrations (pores) that prevent filtration of blood cells but allow all components of blood plasma to pass through.

2 **Basement membrane of the glomerulus.** This layer of extracellular material lies between the endothelium and the visceral layer of the glomerular capsule. It consists of fibrils in a glycoprotein matrix and prevents filtration of larger proteins.

3 **Slit membranes between pedicels.** The specialized epithelial cells that cover the glomerular capillaries are called **podocytes** (*podos* = foot). Extending from each podocyte are thousands of footlike structures called **pedicels** (PED-i-sels; *pediculus* = little foot). The pedicels cover the basement membrane, except for spaces between them, which are called **filtration slits**. A thin membrane, the **slit membrane**, extends across filtration slits and prevents filtration of medium-sized proteins.

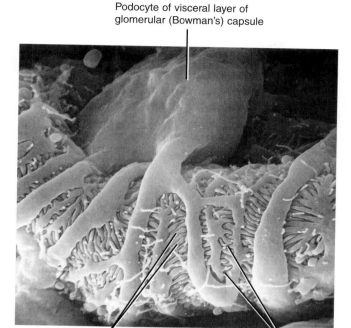

Podocyte of visceral layer of glomerular (Bowman's) capsule

Pedicels

Filtration slits

(c) Scanning electron micrograph of a podocyte (5400x)

Slit membrane   Filtration slit   Pedicel of podocyte

Lumen of glomerulus

Endothelial fenestration (pore) of glomerulus

Basement membrane of glomerulus

(d) Transmission electron micrograph of the endothelial–capsular membrane (42,700x)

Q Which part of the endothelial–capsular membrane prevents red blood cells from entering the capsular space?

Renal Tubule and Ducts

From the capsular space, filtered fluid passes into the renal tubule, which has three main sections (Fig. 26.8). In the order that fluid passes through them, the renal tubule consists of a (1) **proximal convoluted tubule (PCT)**, (2) **loop of Henle (nephron loop)**, and (3) **distal convoluted tubule (DCT)**. *Convoluted* means the tubule is coiled rather than straight. *Proximal* denotes the tubule portion attached to the glomerular capsule and *distal* the portion that is further away. The renal corpuscle and both convoluted tubules lie in the renal cortex of the kidney, whereas the loop of Henle extends into the renal medulla, makes a hairpin turn, and then returns to the renal cortex.

The distal convoluted tubules of several nephrons empty into a single **collecting duct**. Collecting ducts then unite and converge until eventually there are only several hundred large **papillary ducts** at the apices of the renal pyramids, which drain into the minor calyces. The collecting ducts and papillary ducts extend from the renal cortex through the renal medulla to the renal pelvis. On average, there are about 30 papillary ducts per renal papilla. Although a kidney has about 1 million nephrons (each consisting of a renal corpuscle, PCT, loop of Henle, and DCT), it has a much smaller number of collecting ducts and even fewer papillary ducts.

Cortical and Juxtamedullary Nephrons

In a nephron, the loop of Henle connects the proximal and distal convoluted tubules. The first portion of the loop of Henle dips into the renal medulla, where it is called the **descending limb of the loop of Henle**. It then bends in a U-shape and returns to the renal cortex as the **ascending limb of the loop of Henle**. About 80–85% of the nephrons have short loops of Henle that penetrate only into the superficial region of the renal medulla (Fig. 26.8a). These nephrons usually have glomeruli in the superficial region of the renal cortex and are termed **cortical nephrons**. They receive their blood supply from peritubular capillaries that arise from efferent arterioles. The remaining 15–20% are **juxtamedullary nephrons** (*juxta* = beside). They have glomeruli deep in the renal cortex close to the renal medulla and long loops of Henle that stretch through the renal medulla, almost to the renal papilla. They receive their blood supply from peritubular capillaries and vasa recta that arise from efferent arterioles. In addition, the ascending limb of the loop of Henle of juxtamedullary nephrons consists of two portions; the first part is the **thin ascending limb** and the second part is the **thick ascending limb** (Fig. 26.8b). These long-loop nephrons enable the kidneys to excrete a very dilute or very concentrated urine (described shortly).

Figure 26.8 Nephrons (colored gold).

 Nephrons are the functional units of the kidneys.

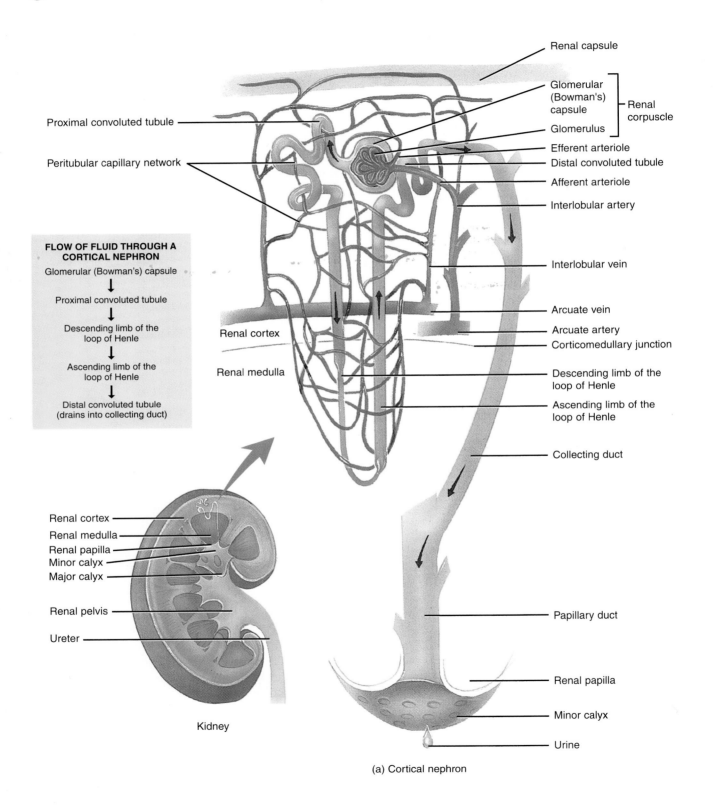

FLOW OF FLUID THROUGH A CORTICAL NEPHRON

Glomerular (Bowman's) capsule
↓
Proximal convoluted tubule
↓
Descending limb of the loop of Henle
↓
Ascending limb of the loop of Henle
↓
Distal convoluted tubule (drains into collecting duct)

Renal capsule

Glomerular (Bowman's) capsule ⎤
Glomerulus ⎦ Renal corpuscle

Efferent arteriole
Distal convoluted tubule
Afferent arteriole
Interlobular artery

Interlobular vein

Arcuate vein
Arcuate artery
Corticomedullary junction

Descending limb of the loop of Henle
Ascending limb of the loop of Henle

Collecting duct

Proximal convoluted tubule

Peritubular capillary network

Renal cortex

Renal medulla

Papillary duct

Renal papilla

Minor calyx

Urine

Renal cortex
Renal medulla
Renal papilla
Minor calyx
Major calyx

Renal pelvis

Ureter

Kidney

(a) Cortical nephron

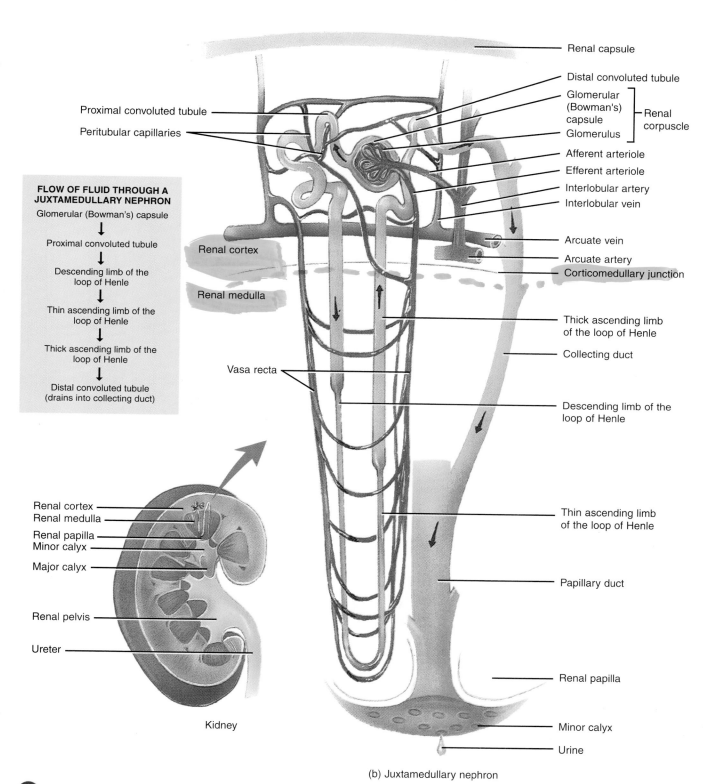

FLOW OF FLUID THROUGH A JUXTAMEDULLARY NEPHRON

Glomerular (Bowman's) capsule
↓
Proximal convoluted tubule
↓
Descending limb of the loop of Henle
↓
Thin ascending limb of the loop of Henle
↓
Thick ascending limb of the loop of Henle
↓
Distal convoluted tubule (drains into collecting duct)

Renal capsule

Distal convoluted tubule

Glomerular (Bowman's) capsule ⎤
⎥ Renal corpuscle
Glomerulus ⎦

Afferent arteriole

Efferent arteriole

Interlobular artery

Interlobular vein

Arcuate vein

Arcuate artery

Corticomedullary junction

Thick ascending limb of the loop of Henle

Collecting duct

Descending limb of the loop of Henle

Thin ascending limb of the loop of Henle

Papillary duct

Renal papilla

Minor calyx

Urine

Proximal convoluted tubule

Peritubular capillaries

Renal cortex

Renal medulla

Vasa recta

Renal cortex
Renal medulla
Renal papilla
Minor calyx
Major calyx
Renal pelvis
Ureter

Kidney

(b) Juxtamedullary nephron

Q What are the basic differences between cortical and juxtamedullary nephrons?

Histology of a Nephron and Associated Structures

A single layer of epithelial cells forms the wall of the renal corpuscle, renal tubule, collecting duct, and papillary duct. Each portion, however, has distinctive histological features that reflect its particular functions (Fig. 26.9).

1. The parietal layer of the glomerular (Bowman's) capsule consists of simple squamous epithelium, whereas the visceral layer consists of modified simple squamous epithelial cells called podocytes.

2. In the proximal convoluted tubule (PCT), the cells are cuboidal and have a prominent brush border of microvilli on their apical surface (surface facing the lumen). These microvilli, like those of the small intestine, increase the surface area for reabsorption and secretion. About 65% of the filtered water and up to 100% of some filtered solutes return to the bloodstream from the PCT.

3. The descending limb of the loop of Henle and the first part of the ascending limb of the loop of Henle, the

---

Figure 26.9 Histology of a nephron, juxtaglomerular apparatus, and collecting duct.

🔑 *A single layer of epithelial cells forms the two layers of the renal corpuscle and the entire renal tubule.*

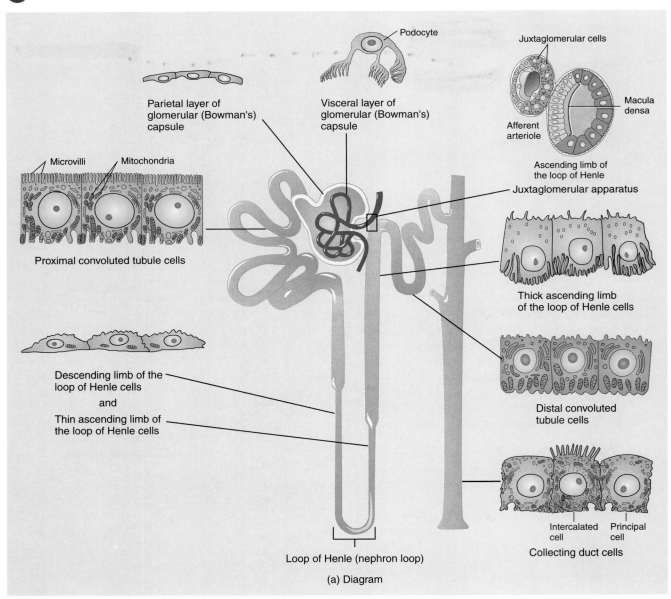

(a) Diagram

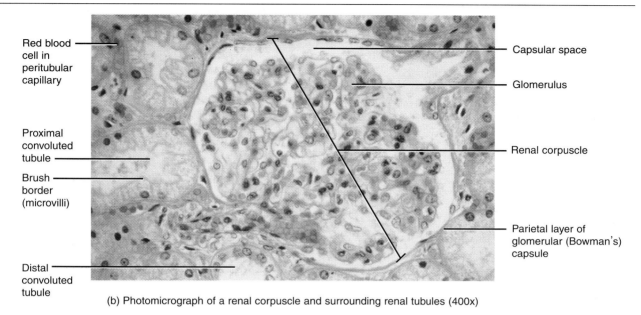

Red blood cell in peritubular capillary

Proximal convoluted tubule

Brush border (microvilli)

Distal convoluted tubule

Capsular space

Glomerulus

Renal corpuscle

Parietal layer of glomerular (Bowman's) capsule

(b) Photomicrograph of a renal corpuscle and surrounding renal tubules (400x)

Q What type of epithelium forms the parietal layer of the glomerular capsule? The proximal convoluted tubule? The distal convoluted tubule?

thin ascending limb, is simple squamous epithelium. (Cortical nephrons lack the thin ascending limb).

4. The thick ascending limb of the loop of Henle is composed of cuboidal to low columnar epithelium.

5. In each nephron, the final portion of the ascending limb of the loop of Henle makes contact with the afferent arteriole serving its own renal corpuscle. The cells of the renal tubule in this region are tall and crowded together. Collectively, they are known as the **macula densa** (*macula* = spot; *densa* = dense). These cells monitor the Na^+ and Cl^- concentration of fluid in the tubule lumen. Next to the macula densa, the wall of the afferent arteriole (and sometimes efferent arteriole) contains modified smooth muscle fibers called **juxtaglomerular (JG) cells**. Together with the macula densa, they constitute the **juxtaglomerular apparatus**, or **JGA**. The JGA helps regulate arterial blood pressure and the rate of blood filtration by the kidneys. The distal convoluted tubule begins a short distance past the macula densa.

6. The cells of the distal convoluted tubule (DCT) and collecting ducts are cuboidal and have a few microvilli. Up to the DCT, the cells of a given tubule segment are all alike. Beginning in the last portion of the DCT and continuing into the collecting ducts, however, two different cell types are present. Most are **principal cells**, which respond to antidiuretic hormone (ADH) and aldosterone, two hormones that regulate kidney functions (described shortly). A few are **intercalated cells**, which secrete H^+ and thereby rid the body of excess acids.

7. Cells of the large papillary ducts are simple columnar epithelium.

RENAL PHYSIOLOGY

The major work of the urinary system is done by the nephrons and the collecting ducts within the kidneys. The other parts of the system are primarily passageways and storage areas.

Glomerular Filtration

The first step in renal regulation of blood composition and volume is **glomerular filtration**. In filtration, pressure forces fluid and all solutes smaller than a certain size through a membrane. The principle is the same in glomerular capillaries as in capillaries elsewhere in the body (see Starling's law of the capillaries, discussed on page 618). Filtration occurs in the renal corpuscles of the kidneys across the endothelial–capsular membranes. Blood pressure forces water and dissolved blood components through the endothelial fenestrations (pores) of the capillaries, basement membrane, and on through the slit membranes between pedicels (see Fig. 26.7b). The resulting fluid that enters the capsular space is called **glomerular filtrate**.

The fraction of plasma in the afferent arterioles of the kidneys that becomes glomerular filtrate is termed the **filtration fraction**. Although the filtration fraction normally is 16–20%, the value varies considerably in both health and disease. Typically, about 180 liters (48 gal) of filtrate enter the capsular spaces each day. This represents 65 times the entire blood plasma volume. However, 178–179 liters return to the bloodstream by tubular reabsorption so only 1 to 2 liters (about 1–2 qt) are excreted as urine. In a healthy person, glomerular filtrate contains all the materials present in the blood except the formed elements and most plasma proteins, which are too large to pass through the endothelial–capsular membranes. Several structural features of the renal corpuscles enhance their blood-filtering capacity.

1. **The endothelial–capsular membrane is thin and porous.** Glomerular capillaries are about 50 times more permeable than capillaries elsewhere in the body. The diameter of the endothelial fenestrations (pores) is quite large (50–100 nm or 0.05–0.1 μm) and generally they do not prevent filtration of solutes in plasma. The basement membrane and slit membranes, however, only permit the passage of molecules smaller than 6–7 nm (0.006–0.007 μm) in diameter. These include water, glucose, vitamins, amino acids, small proteins, nitrogenous wastes, and ions, which easily pass into the capsular space. The major plasma protein, albumin, has a diameter of 7.1 nm and less than 1% of it passes the filter.

2. **Glomerular capillaries present a large surface area.** Altogether, the glomerular capillaries are very extensive, presenting a vast total surface area for filtration.

3. **Capillary blood pressure is high.** The efferent arteriole is smaller in diameter than the afferent arteriole, so there is high resistance to the outflow of blood from the glomerulus. Thus blood pressure is higher in the glomerular capillaries than in capillaries elsewhere in the body. A higher pressure produces more filtrate.

Net Filtration Pressure

In the glomerulus, blood filtering depends on three main pressures, one that promotes filtration and two that oppose filtration (Fig. 26.10).

① **Glomerular blood hydrostatic pressure** (**GBHP**) promotes filtration—it pushes water and solutes in blood plasma through the glomerular filter. *Hydrostatic (hydro = water) pressure* is the force that a fluid under pressure exerts against the walls of its container. Glomerular blood hydrostatic pressure is the blood pressure in glomerular capillaries, which is about 55 mm Hg.

② **Capsular hydrostatic pressure** (**CHP**) is a back-pressure that opposes filtration. It develops in the following way. As the filtrate is forced into the capsular space between the walls of the glomerular capsule, it meets two forms of resistance: the wall of the capsule and the fluid that has already filled the renal tubule. As a result, some filtrate is pushed back into the capillary. The amount of back-pressure is the capsular hydrostatic pressure, about 15 mm Hg.

③ The second force opposing filtration is the **blood colloid osmotic pressure** (**BCOP**), which is mainly due to the

---

Figure 26.10 Pressures that determine net filtration pressure (NFP).

 Glomerular filtration is promoted by GBHP and opposed by CHP and BCOP.

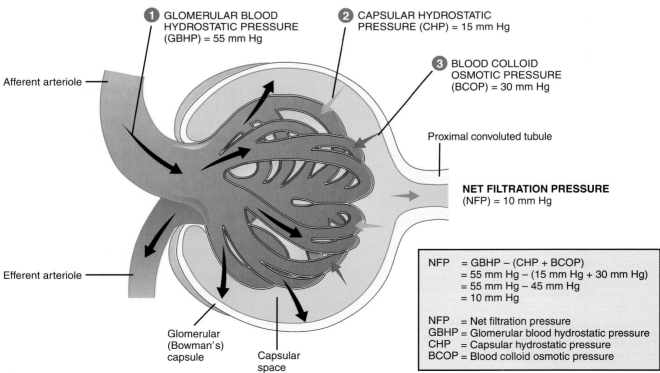

① GLOMERULAR BLOOD HYDROSTATIC PRESSURE (GBHP) = 55 mm Hg

② CAPSULAR HYDROSTATIC PRESSURE (CHP) = 15 mm Hg

③ BLOOD COLLOID OSMOTIC PRESSURE (BCOP) = 30 mm Hg

Afferent arteriole

Proximal convoluted tubule

NET FILTRATION PRESSURE (NFP) = 10 mm Hg

Efferent arteriole

Glomerular (Bowman's) capsule

Capsular space

NFP = GBHP – (CHP + BCOP)
    = 55 mm Hg – (15 mm Hg + 30 mm Hg)
    = 55 mm Hg – 45 mm Hg
    = 10 mm Hg

NFP = Net filtration pressure
GBHP = Glomerular blood hydrostatic pressure
CHP = Capsular hydrostatic pressure
BCOP = Blood colloid osmotic pressure

Q Suppose a large stone (calculus) is blocking a major calyx. What effect might this have on CHP and thus on NFP?

presence of proteins (such as albumin, globulins, and fibrinogen) in blood plasma. These proteins normally cannot pass through the endothelial–capsular membrane and so remain within the glomerular capillaries. *Osmotic pressure* is the pressure required to prevent the net movement of water into a solution containing solutes when the solutions are separated by a selectively permeable membrane. The greater the solute concentration of a solution, the greater its osmotic pressure. Because blood contains a much higher concentration of proteins than glomerular filtrate, some water moves from the filtrate back into the glomerular capillaries.

Blood entering glomerular capillaries has a BCOP close to 25 mm Hg. As water and other solutes leave plasma and filter into the capsular space, however, the BCOP steadily increases. As a result, blood entering the efferent arterioles has a BCOP close to 35 mm Hg. The *average* BCOP in glomerular capillaries is about 30 mm Hg, which is larger than the average of 26 mm Hg in capillaries elsewhere in the body.

Net filtration pressure (NFP) is the total pressure that promotes filtration. To calculate NFP, we subtract the forces that oppose filtration from the glomerular blood hydrostatic pressure.

$$NFP = GBHP - (CHP + BCOP)$$

Promotes	Oppose
filtration	filtration

By substituting the values just given, a normal NFP may be calculated as follows:

$$
\begin{aligned}
NFP &= (55 \text{ mm Hg}) - (15 \text{ mm Hg} + 30 \text{ mm Hg}) \\
&= (55 \text{ mg Hg}) - (45 \text{ mm Hg}) \\
&= 10 \text{ mm Hg}
\end{aligned}
$$

This means that a pressure of only 10 mm Hg causes a normal amount of plasma (minus plasma proteins) to filter from the glomerulus into the capsular space.

LOSS OF PLASMA PROTEINS

In some kidney diseases, such as nephrotic syndrome (see page 885), damaged glomerular capillaries become so permeable that plasma proteins enter the filtrate. As a result, the filtrate exerts a colloid osmotic pressure that draws water out of the blood. In this situation, NFP increases, which means more fluid is filtered. At the same time, blood colloid osmotic pressure decreases because plasma proteins are being lost in the urine. The result is a potentially life-threatening decrease in blood volume and buildup of interstitial fluid (edema). ■

Glomerular Filtration Rate

The amount of filtrate formed in all the renal corpuscles of both kidneys each minute is called the **glomerular filtra-** tion rate (GFR). In a normal adult, GFR is about 125 ml/min, which amounts to 180 liters (48 gal) a day.

GFR is directly related to the pressures that determine NFP. Any factor that alters NFP will affect GFR. For example, severe blood loss reduces systemic blood pressure, which also decreases the glomerular blood hydrostatic pressure. If GBHP drops to 45 mm Hg, filtration stops because this is the magnitude of the opposing forces. Such a condition causes **anuria** (a-NOO-rē-a), which is defined as a daily urine output of less than 50 ml. It may be caused by insufficient pressure for filtration or by inflammation of the glomeruli that blocks filtration of plasma into the capsular space.

Homeostasis of body fluids requires that the kidneys maintain a relatively constant GFR. If the GFR is too high, needed substances may pass so quickly through the renal tubules that they are not reabsorbed and instead are lost in the urine. On the other hand, if the GFR is too low, nearly all the filtrate may be reabsorbed and certain waste products may not be adequately excreted.

Regulation of GFR

When blood flows into the glomerular capillaries more rapidly than normal, glomerular filtration rate increases. Glomerular blood flow, in turn, depends on the systemic blood pressure and the diameter of the afferent and efferent arterioles. Three principal mechanisms control systemic blood pressure and blood vessel diameter: renal autoregulation, hormonal regulation, and neural regulation.

● **Renal Autoregulation of GFR** The ability of the kidneys to maintain a constant blood pressure and GFR despite changes in systemic arterial pressure is called **renal autoregulation**. Renal autoregulation is intrinsic, which means that it operates completely *within* the kidneys (even if they are removed from the body, for example, in a transplant operation). The kidneys have a built-in system to compensate for moderate changes in systemic arterial pressure for short periods of time.

Renal autoregulation operates by a negative feedback system that involves the juxtaglomerular apparatus or JGA (Fig. 26.11). When NFP and GFR (controlled conditions) are low due to low blood pressure, the proximal convoluted tubules and loop of Henle reabsorb more than the normal fraction of Na$^+$, Cl$^-$, and water. At the end of the loop of Henle, macula densa cells in the JGA (receptors) detect decreased delivery of Na$^+$, Cl$^-$, and water. By an unknown mechanism (input), cells in the JGA (control center) are inhibited and decrease their release of a vasoconstrictor substance (output). (Which JGA cells are responsible for this secretion and the chemical nature of the vasoconstrictor are not yet known.) With less vasoconstrictor substance present, the afferent arteriole (effector) dilates, allowing more blood to flow into the glomerulus. This increases NFP and GFR (responses) and brings about a return to homeostasis. If blood pressure is elevated, on the other hand, NFP and

GFR are higher than normal and the opposite sequence of events occurs.

● **Hormonal Regulation of GFR** Two hormones—angiotensin II and atrial natriuretic peptide (ANP)—contribute

Figure 26.11 Negative feedback regulation of glomerular filtration rate by the juxtaglomerular apparatus (JGA).

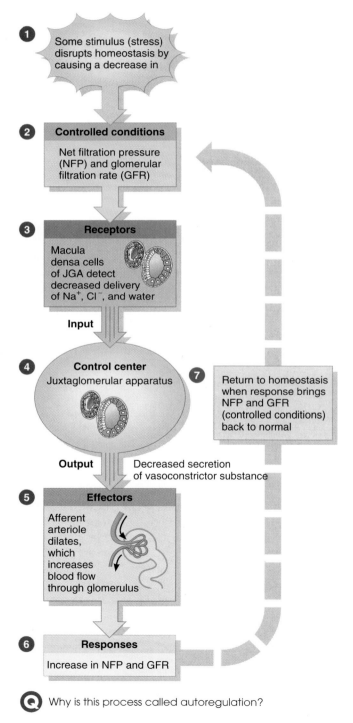

Q Why is this process called autoregulation?

to the regulation of blood pressure and therefore GFR. Figure 26.12 describes the renin–angiotensin system. When blood pressure (BP) and therefore GFR (controlled conditions) decrease, juxtaglomerular cells and macula densa cells of the JGA (receptors) detect decreased stretch and decreased delivery of Na^+, Cl^-, and water, respectively. The juxtaglomerular cells then secrete an enzyme called **renin** (input) into the blood. In the blood, renin acts on a large plasma protein produced by the liver called **angiotensinogen** (control center) and clips off a small fragment called angiotensin I. As angiotensin I passes through the lungs, a second enzyme called **angiotensin converting enzyme** (**ACE**) converts it to **angiotensin II**, which is an active hormone.

Angiotensin II (output) is transported in the blood and has important actions on several effectors:

1. **Vasoconstriction of arterioles.** Angiotensin II constricts efferent arterioles (effectors), which increases glomerular blood hydrostatic pressure (response) and raises GFR back to normal. It also constricts arterioles elsewhere in the body. This increases systemic vascular resistance and raises mean arterial blood pressure.

2. **Stimulation of aldosterone secretion by the adrenal cortex.** Angiotensin II stimulates the adrenal cortex (effector) to secrete aldosterone, which increases retention of Na^+, Cl^-, and water by the kidneys. Water retention increases blood volume (response), which restores blood pressure and GFR to normal.

3. **Stimulation of thirst center in the hypothalamus.** Angiotensin II acts on the thirst center in the hypothalamus (effector) to increase water intake. This results in an increase in blood volume (response), which restores blood pressure and GFR to normal.

4. **Stimulation of ADH secretion from the posterior pituitary gland.** Angiotensin II stimulates the release of antidiuretic hormone (ADH) from the posterior pituitary gland (effector). ADH promotes water retention by the kidneys and increases blood volume (response), which restores blood pressure and GFR to normal.

In all its actions, angiotensin II helps restore normal systemic and renal blood pressure, which normalize GFR and bring a return to homeostasis.

A second hormone that influences glomerular filtration and other renal processes is **atrial natriuretic peptide** (**ANP**). As its name suggests, it is secreted by cells in the atria (superior chambers) of the heart. It was identified in 1983 after a long search for the elusive "third factor" (aldosterone and ADH were the other two) that could explain certain clinical and research results. ANP promotes both excretion of water (diuresis) and excretion of sodium (natriuresis). Secretion of ANP is stimulated by increased stretching of the atria, as occurs when blood volume increases. ANP increases GFR, perhaps by increasing the permeability of the filter or by dilating the afferent arterioles. It also suppresses secretion of ADH, aldosterone, and

Figure 26.12 Renin–angiotensin system in regulation of blood pressure and glomerular filtration rate.

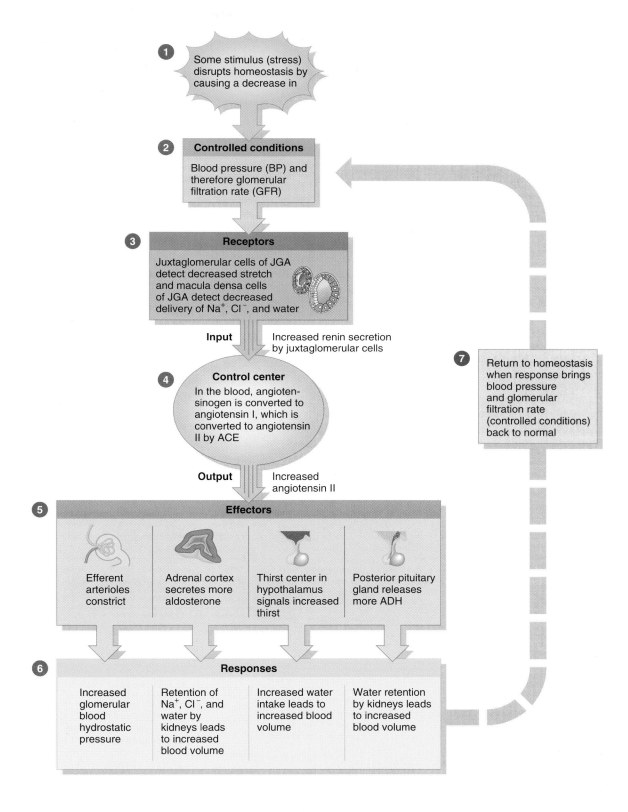

1 Some stimulus (stress) disrupts homeostasis by causing a decrease in

2 **Controlled conditions**

Blood pressure (BP) and therefore glomerular filtration rate (GFR)

3 **Receptors**

Juxtaglomerular cells of JGA detect decreased stretch and macula densa cells of JGA detect decreased delivery of Na$^+$, Cl$^-$, and water

Input Increased renin secretion by juxtaglomerular cells

4 **Control center**

In the blood, angiotensinogen is converted to angiotensin I, which is converted to angiotensin II by ACE

Output Increased angiotensin II

7 Return to homeostasis when response brings blood pressure and glomerular filtration rate (controlled conditions) back to normal

5 **Effectors**

| Efferent arterioles constrict | Adrenal cortex secretes more aldosterone | Thirst center in hypothalamus signals increased thirst | Posterior pituitary gland releases more ADH |

6 **Responses**

| Increased glomerular blood hydrostatic pressure | Retention of Na$^+$, Cl$^-$, and water by kidneys leads to increased blood volume | Increased water intake leads to increased blood volume | Water retention by kidneys leads to increased blood volume |

Q What enzymes and hormones are involved in this system?

renin. Studies are underway to see if ANP can benefit renal failure patients by increasing their GFR and hypertensive patients by lowering their blood pressure and reducing water retention (edema).

● **Neural Regulation of GFR** Like most blood vessels of the body, those of the kidneys are supplied by vasoconstrictor fibers from the sympathetic division of the autonomic nervous system. At rest, sympathetic stimulation is minimal and renal blood vessels are maximally dilated. With moderate sympathetic stimulation, both afferent and efferent arterioles constrict to the same degree. Blood flow into and out of the glomerulus is restricted to the same extent, which decreases GFR only slightly. With greater sympathetic stimulation, however, as might occur during exercise, hemorrhage, or a fight-or-flight response, vasoconstriction of the afferent arterioles predominates. This greatly decreases blood flow into glomerular capillaries and thus reduces GFR. By reducing renal blood flow in this way, more blood flow is made available for other body tissues. Strong sympathetic stimulation also causes the adrenal medulla to secrete epinephrine, which also decreases GFR by bringing about vasoconstriction of the afferent arteriole.

Tubular Reabsorption

The normal rate of glomerular filtration is so high that the volume of fluid entering the proximal convoluted tubules in half an hour is greater than the total plasma volume. Returning most of the filtered water and many of the filtered solutes to the bloodstream is a critical kidney function. This selective reclamation process is called **tubular reabsorp-** **tion** or simply **reabsorption**. As the filtrate passes through the renal tubules and collecting ducts, about 99% of it is reabsorbed and only 1% leaves the body as urine (1–2 liters a day). In tubular reabsorption, water and solutes move from the tubule lumen back into the blood within a peritubular capillary or vasa recta. Solutes that are reabsorbed by both active and passive processes include glucose, amino acids, urea, and ions such as Na^+ (sodium), K^+ (potassium), Ca^{2+} (calcium), Cl^- (chloride), HCO_3^- (bicarbonate), and HPO_4^{2-} (phosphate). Water reabsorption accompanies solute reabsorption by osmosis. Most small proteins and peptides that pass through the filter also are reabsorbed, usually by the process of pinocytosis.

Reabsorption of sodium ions is especially important because more of them pass the glomerular filter than any other substance except water. Sodium ions are reabsorbed in each portion of the renal tubule and collecting duct by several different types of transport systems. These mechanisms recover not only filtered Na^+ but also water, anions (negatively charged ions), and nutrients. Sometimes they also permit secretion of unneeded substances such as excess H^+ and K^+.

Epithelial cells all along the renal tubule carry out tubular reabsorption. The proximal convoluted tubules, where the epithelial cells have many microvilli that increase the surface area for reabsorption, make the largest contribution. They reabsorb 100% of the filtered glucose and amino acids; 80–90% of the HCO_3^-; 65% of the water, Na^+, and K^+; and 50% of the Cl^- and urea. More distal portions of the nephron and collecting ducts are responsible for fine-tuning the reabsorption processes to maintain homeostatic balances. Exhibit 26.1 compares the values for the substances in the filtrate immediately after passing into the glomerular (Bowman's) capsule with those reabsorbed from the filtrate.

EXHIBIT **26.1**   **SUBSTANCES IN PLASMA AND AMOUNTS FILTERED, REABSORBED, AND EXCRETED IN URINE**

Substance	Plasma (Total Amount)	Filtered (Enters Glomerular Capsule per Day)	Reabsorbed (Returned to Blood per Day)	Urine (Excreted per Day)
Water	3000 ml	180,000 ml	178,500 ml	1500 ml
Proteins	200 g	2 g	1.9 g	0.1 g
Sodium ions (Na^+)	9.7 g (420 mmol)	579.6 g (25,200 mmol)	575.0 g (25,000 mmol)	4.6 g (200 mmol)
Chloride ions (Cl^-)	10.7 g (300 mmol)	639.0 g (18,000 mmol)	633.7 g (17,850 mmol)	6.3 g (150 mmol)
Bicarbonate ions (HCO_3^-)	4.6 g (75 mmol)	274.5 g (4500 mmol)	274.5 g (4500 mmol)	0
Glucose	3 g	180 g	180 g	0
Urea	4.8 g	53 g	28 g	25 g[a]
Potassium ions (K^+)	0.5 g (12.6 mmol)	29.6 g (756 mmol)	29.6 g (756 mmol)	2.0 g (50 mmol)[b]
Uric acid	0.15 g	8.5 g	7.7 g	0.8 g
Creatinine	0.03 g	1.6 g	0	1.6 g

[a] Urea is secreted in addition to being filtered and reabsorbed.
[b] After being 100% reabsorbed in the PCT, loop of Henle, and DCT, a variable amount of K^+ is secreted in the collecting ducts.

Reabsorption of Na+ in the PCT

Figure 26.13a shows a major mechanism for Na+ reabsorption that occurs in the proximal convoluted tubule (PCT) as well as the distal convoluted tubule (DCT) and collecting ducts. The concentration of Na+ inside the tubule cells is low and the interior of the cell is negatively charged with respect to the exterior. As a result of these differences, Na+ passively diffuses from the fluid in the tubule lumen through leakage channels in the brush border into the tubule cells. At the same time, sodium pumps actively expel Na+ from the base and sides of the cell into interstitial fluid. Na+ then diffuses through interstitial fluid into the peritubular capillaries.

Recall from Chapter 3 that in **primary active transport** the energy derived from splitting ATP *directly* moves or "pumps" a substance across a membrane. Because the sodium pump uses ATP in this manner, it is a primary active

Figure 26.13 Reabsorption of sodium ions in the proximal convoluted tubule by primary active transport. Na+ = sodium ion; K+ = potassium ion; H_2O = water; HCO_3^- = bicarbonate ion; Cl^- = chloride ion.

Active transport of Na+ leads to passive reabsorption of water by osmosis and passive reabsorption of urea and anions, such as Cl^- and HCO_3^-, by simple diffusion.

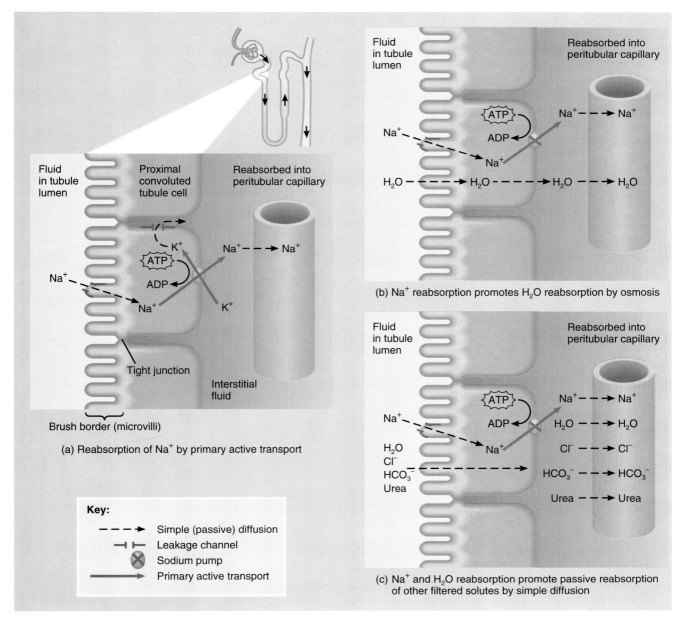

(a) Reabsorption of Na+ by primary active transport

Key:
- – – – ▸ Simple (passive) diffusion
- ⊣ ⊢ Leakage channel
- ⊗ Sodium pump
- ——▸ Primary active transport

(b) Na+ reabsorption promotes H_2O reabsorption by osmosis

(c) Na+ and H_2O reabsorption promote passive reabsorption of other filtered solutes by simple diffusion

Q Which step in Na+ movement in (a) is promoted by the electrical and chemical differences?

transport mechanism. Although the sodium pump imports K$^+$ at the same time it expels Na$^+$, K$^+$ can diffuse back out through K$^+$ leakage channels because membranes typically have many more leakage channels for K$^+$ than for Na$^+$. Thus reabsorption of Na$^+$ is the main effect of the sodium pump. The total ATP used by sodium pumps in the renal tubules is large, an estimated 6% of the total resting energy budget. For comparison, this is about the same as the energy needed for contraction of the diaphragm in quiet breathing.

Reabsorption of Na$^+$ also promotes the reabsorption of water by osmosis (Fig. 26.13b). Each reabsorbed Na$^+$ increases the osmotic pressure, first inside the tubule cell, then in the interstitial fluid, and finally in the blood. Water thus moves rapidly from the filtrate into the peritubular capillaries and this movement restores the osmotic balance. As water leaves the filtered fluid, the concentration of the remaining filtered solutes increases. This creates a concentration difference for some substances, such as K$^+$, Cl$^-$, HCO$_3^-$, and urea, that drives their reabsorption by simple diffusion (Fig. 26.13c). Also, the reabsorption of positively charged Na$^+$ into the peritubular capillaries makes the blood more positive than the fluid in the tubular lumen. This electrical difference drives reabsorption of negatively charged ions, such as Cl$^-$ and HCO$_3^-$. Thus the reabsorption of Na$^+$ by active transport also promotes the reabsorption of other solutes by simple diffusion.

Reabsorption of Nutrients in the PCT

Normally, the PCTs reabsorb 100% of the filtered glucose, amino acids, lactic acid, and other possibly useful nutrients. These substances are reabsorbed by secondary active transport. Again recall from Chapter 3 that in **secondary active transport** the energy stored in an ionic concentration difference (gradient), for example, the high concentration of Na$^+$ in extracellular fluid versus its low concentration in cytosol, drives substances across a membrane. Since the ionic concentration difference is established by primary active transport pumps, secondary active transport *indirectly* uses the energy obtained from splitting ATP. The membrane proteins that perform secondary active transport are called **symporters** when they move two substances in the same direction across a membrane.

Figure 26.14 shows how glucose is reabsorbed by a symporter that uses the Na$^+$ concentration difference. Na$^+$ and a molecule of glucose both attach to a Na$^+$–glucose symporter that transports the two substances from the fluid in the tubule lumen into a tubule cell. Because the symporter cannot bring Na$^+$ into the cell against a concentration gradient, it depends on the sodium pumps to keep the concentration of Na$^+$ low inside the cell.

Substances brought into PCT cells by symporters generally leave the cells by facilitated diffusion and then enter peritubular capillaries by simple diffusion. Several different Na$^+$ symporters reclaim filtered glucose, various amino acids, and other nutrients. As was true for Na$^+$ reabsorption, the reabsorption of these nutrients leads to the reabsorption of water by osmosis.

Figure 26.14 Reabsorption of glucose by Na$^+$–glucose symporters in the proximal convoluted tubule (PCT).

🔑 *Normally, all filtered glucose is reabsorbed in the PCT.*

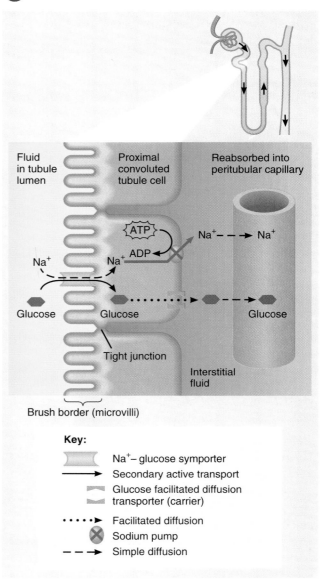

🔍 How does filtered glucose enter and leave a PCT cell?

Normally, all the filtered glucose and amino acids are reabsorbed in the PCT. But each type of symporter has a limit on how fast it can work, just as an escalator has a limit on how many people it can carry in a given period of time. This limit is called the **transport maximum (T$_m$)** and is *measured in mg/min*. When blood concentration of one of these substances is abnormally high, it passes the endothelial–capsular membrane more rapidly than usual, and the T$_m$ may be exceeded. As a result, the substance spills over into the urine. The **renal threshold**, *measured in mg/ml*, is the plasma concentration at which a substance begins to spill into the urine because its T_m has been exceeded.

Whenever excess solutes, for example, glucose molecules, spill into the urine, they cause urine volume to be larger than normal. The reason is simple: when solutes are not reabsorbed, water molecules also remain in the urine because they are not reabsorbed by osmosis.

GLUCOSURIA

The Na^+–glucose symporters can reclaim about 300 mg of glucose/min (T_m). If glomerular filtration is normal, the renal threshold for glucose is a plasma concentration of about 200 mg/100 ml. When the blood concentration of glucose is above this level, the symporters cannot work fast enough to reabsorb it all, and the excess glucose remains in the urine. This condition is called **glucosuria** (gloo′-kō-SOO-rē-a) or **glycosuria**. The most common cause of glucosuria is diabetes mellitus, in which the blood glucose level may rise far above normal because insulin activity is deficient. Another cause of glucosuria is a rare genetic defect in the Na^+–glucose symporter that greatly reduces its T_m. In these cases, glucose appears in the urine even though the blood glucose level is normal. ■

Reabsorption in the Loop of Henle

Because the PCTs reabsorb about 65% of the filtered water, 40–45 ml/min of fluid remains and enters the loop of Henle. The chemical composition of this fluid is different from that of blood plasma (and glomerular filtrate) because glucose, amino acids, and other nutrients are no longer present. The tonicity of the tubular fluid is still close to the tonicity of blood, however, because reabsorption of water by osmosis keeps pace with reabsorption of solutes all along the PCT.

The loop of Henle reabsorbs about 30% of the filtered K^+, 20% of the filtered Na^+, 35% of the filtered Cl^-, and 15% of the filtered water. Here, for the first time, reabsorption of water by osmosis is *not* automatically coupled to the reabsorption of filtered solutes. This feature allows production of either dilute or concentrated urine. Thus there can be independent regulation of both your total body water and the tonicity of your body fluids (described shortly).

Figure 26.15 shows the main reabsorption mechanism in the thick ascending limb of the loop of Henle. Cells in this part of the nephron have symporters that simultaneously reclaim one Na^+, one K^+, and two Cl^- from the fluid in the tubular lumen. As is true for other symporters, these depend on sodium pumps to maintain a low concentration of Na^+ inside the thick ascending limb cells. Na^+ that is actively transported into interstitial fluid at the base and sides of the cell diffuses into the vasa recta. K^+ diffuses down its concentration gradient into the blood, and negatively charged Cl^- follows Na^+ and K^+.

Some water is reabsorbed in the descending limb of the loop of Henle. Little or no water is reabsorbed in the as-

cending limb of the loop of Henle, however, because the apical surfaces (surfaces facing the fluid in the lumen) of these cells are quite impermeable to water.

Reabsorption in the DCT and Collecting Ducts

Fluid enters the distal convoluted tubules at a rate of about 25 ml/min because 80% of the filtered water has now been reabsorbed. As fluid flows along the DCT, Na^+–Cl^- symporters in the apical membranes reabsorb Na^+ and Cl^-. By

---

Figure 26.15 Na^+–K^+–$2Cl^-$ symporters in the thick ascending limb of the loop of Henle.

Cells in the thick ascending limb have symporters that simultaneously reabsorb one Na^+, one K^+ and two Cl^-.

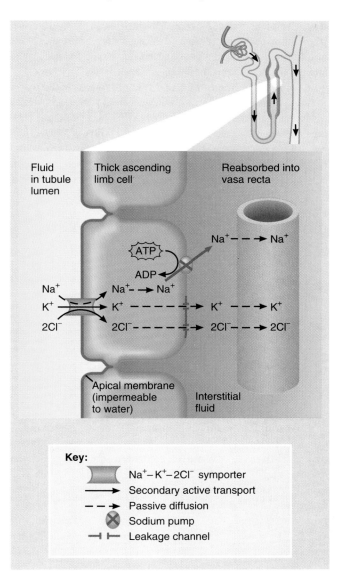

Key:
	Na^+–K^+–$2Cl^-$ symporter
→	Secondary active transport
– – →	Passive diffusion
⊗	Sodium pump
⊣⊢	Leakage channel

 Why is this process called secondary active transport? Does water reabsorption accompany ion reabsorption in this region of the nephron?

the time fluid reaches the end of the DCT, about 90% of the filtered solutes and water have been returned to the bloodstream. Still, in one day almost 20 liters of fluid enters the collecting ducts whereas only 1–2 liters appear as urine. Two hormones—aldosterone and antidiuretic hormone—help maintain homeostasis of blood volume, composition, and pressure by fine-tuning reabsorption. They act on the **principal cells**, which are located in the final portion of the DCT and throughout the collecting duct.

Aldosterone, secreted by the cortex of the adrenal gland, increases Na$^+$ and water reabsorption by principal cells. This happens because the steroid hormone aldosterone stimulates principal cells to synthesize more sodium pump proteins. As more sodium pumps are inserted into the plasma membrane, the principal cells reabsorb more and more Na$^+$. Aldosterone also stimulates secretion of K$^+$ from principal cells into the tubular fluid. When the level of aldosterone is low, the principal cells reabsorb little Na$^+$. It thus passes into urine and is excreted. At the same time, increased amounts of water are also excreted due to the osmotic pressure of the Na$^+$ that remains in the urine.

Antidiuretic hormone (**ADH**) is produced by the hypothalamus and released into the bloodstream by the posterior pituitary gland. In the absence of ADH, the principal cells have a very low permeability to water. ADH acts to stimulate insertion of water-channel proteins into the apical membranes of the principal cells. As a result, water molecules can pass more rapidly into the cells and then into blood. When ADH is absent, a condition known as diabetes insipidus (described on page 886), the kidneys may excrete up to 20 liters per day of very dilute urine. When ADH concentration is maximal, the kidneys excrete as little as 400 to 500 ml per day of very concentrated urine.

Water Reabsorption

About 90% of the water reabsorption, a total of 160 liters per day, occurs by osmosis together with the reabsorption of solutes such as Na$^+$, Cl$^-$, and glucose. Water reabsorbed by osmosis is termed **obligatory** (ob-LIG-a-tor′-ē) **water reabsorption** because the water is "obliged" to follow the solutes. This type of water reabsorption occurs in the PCT, descending limb of the loop of Henle, and early DCT because these structures are always permeable to water.

Reabsorption of the remaining 10% of the water in the tubular fluid, a total of 10–20 liters per day, is termed **facultative** (FAK-ul-tā′-tiv) **water reabsorption**. The word *facultative* means capable of adapting to a need. It is this portion of renal water reabsorption that is regulated in response to the needs of the body to maintain homeostasis. The main regulator of facultative water reabsorption is the hormone ADH. As shown in Fig. 26.16, a negative feedback system regulates ADH-stimulated water reabsorption.

When the water concentration of blood is low—that is, when its osmotic pressure is high (controlled condition)—osmoreceptors in the hypothalamus (receptors) detect the change. They send nerve impulses (input) that cause the hypothalamus and posterior pituitary gland (control center)

to release ADH into the blood. ADH then acts on principal cells (effectors), which are present in the final portion of the DCT and throughout the collecting duct. Here, ADH makes the principal cells more permeable to water and water moves through them into the blood. This is facultative water reabsorption. As more water is reabsorbed, the blood's water concentration increases—that is, its osmotic

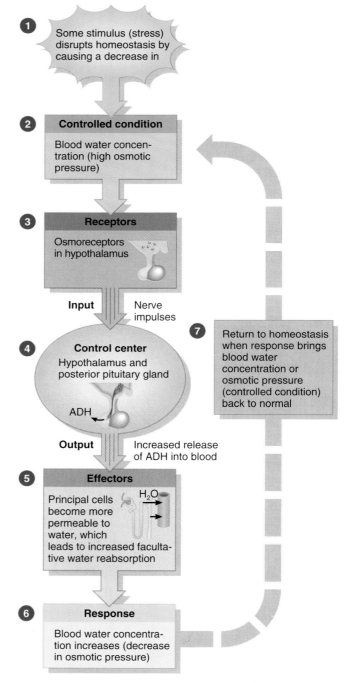

Figure 26.16 Negative feedback regulation of facultative water reabsorption by ADH.

1. Some stimulus (stress) disrupts homeostasis by causing a decrease in

2. **Controlled condition**
Blood water concentration (high osmotic pressure)

3. **Receptors**
Osmoreceptors in hypothalamus

Input   Nerve impulses

4. **Control center**
Hypothalamus and posterior pituitary gland

ADH

Output   Increased release of ADH into blood

5. **Effectors**
Principal cells become more permeable to water, which leads to increased facultative water reabsorption

H$_2$O

6. **Response**
Blood water concentration increases (decrease in osmotic pressure)

7. Return to homeostasis when response brings blood water concentration or osmotic pressure (controlled condition) back to normal

Q   Besides ADH, which other hormones contribute to the regulation of water reabsorption?

pressure decreases (response). Thus the controlled condition is brought back to normal.

Tubular Secretion

The third function of nephrons is **tubular secretion**, which is the movement of materials from the blood into tubular fluid. Whereas tubular reabsorption *returns filtered substances* to the blood, tubular secretion *removes materials* from the blood. Secreted substances include hydrogen ions (H^+), potassium ions (K^+), ammonium ions (NH_4^+), creatinine, and some drugs, for example, penicillin. Tubular secretion has two principal effects: (1) the secretion of H^+ helps control blood pH and (2) the secretion of other substances helps eliminate them from the body.

Secretion of H+

Arterial blood pH remains between 7.35 and 7.45 despite the continual production of far more acids than bases by metabolic reactions. The cells of the renal tubule and col-

lecting duct can *raise blood pH* (make it less acidic) in three ways: (1) by secreting H^+ into the fluid in the lumen, which rids the blood of acid and makes the urine more acidic; (2) by reabsorbing filtered bicarbonate ions (HCO_3^-), which are the most important buffers of H^+ in extracellular fluids; and (3) by producing new HCO_3^-, to provide more buffering capacity. Amazingly, one process—the secretion of H^+ by tubular epithelial cells—also accomplishes both the reabsorption of filtered HCO_3^- and generation of new HCO_3^-. Figure 26.17 shows how this occurs.

Secretion of H^+ takes place in the cells of the PCTs and collecting ducts. It begins when carbon dioxide (CO_2) diffuses from peritubular blood or tubular fluid into the cells or is produced by metabolic reactions within the cells (Fig. 26.17a). Here, in the presence of the enzyme *carbonic anhydrase* (CA), CO_2 combines with water (H_2O) to form carbonic acid (H_2CO_3). The H_2CO_3 then dissociates into H^+ and HCO_3^-, and the H^+ is secreted into the tubular fluid. In the proximal convoluted tubule, secretion of H^+ is linked to reabsorption of Na^+. This is because PCT cells have *Na+/H+*

Figure 26.17 Secretion of hydrogen ions (H^+) in the proximal convoluted tubule. Note that reabsorption of sodium ions (Na^+) and bicarbonate ions (HCO_3^-) accompany H^+ secretion. CO_2 = carbon dioxide; H_2O = water; H_2CO_3 = carbonic acid.

The kidneys help maintain blood pH by secreting H^+ and reabsorbing HCO_3^-.

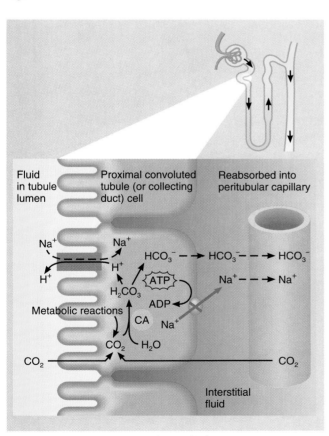

(a) Secretion of H+ by Na+/H+ antiporter

(b) Reabsorption of HCO_3^-

What effects would a drug have that blocks the activity of carbonic anhydrase?

antiporters. Recall that an **antiporter** is a membrane protein that moves an ion (usually Na^+) and another substance in *opposite* directions across a membrane. In this case, an antiporter moves Na^+ into the cell and H^+ out, using the energy stored in the Na^+ concentration difference.

An important feature of H^+ secretion is that HCO_3^-, formed during the dissociation of H_2CO_3 inside tubule cells, can diffuse from the cell into peritubular blood. For every H^+ secreted into tubular fluid, one HCO_3^- is returned to the blood. To see exactly how this occurs, we will follow the fate of one H^+.

As H^+ is secreted into the fluid within the lumen of the PCT, it combines with filtered HCO_3^-. This reaction, catalyzed by carbonic anhydrase in the brush border membrane, forms H_2CO_3 which in turn dissociates into CO_2 and H_2O (Fig. 26.17b). CO_2 diffuses into the tubule cells and joins with H_2O to form H_2CO_3, which dissociates into H^+ and HCO_3^-. The HCO_3^- is reabsorbed into the blood together with Na^+. Thus, when secreted H^+ reacts with filtered HCO_3^-, the end result is reabsorption of Na^+ and HCO_3^-.

In the collecting ducts, the secretion of H^+ is a little different. Intercalated cells in the ducts secrete H^+ by primary active transport in which the H^+ pump itself uses ATP (Fig. 26.18). Intercalated cells can secrete H^+ so effectively that urine can be up to 1000 times more acidic than blood. Inside the intercalated cell, HCO_3^- produced by dissociation of H_2CO_3 may enter the interstitial fluid by way of a HCO_3^-/Cl^- antiporter. The HCO_3^- that enters the blood in this way is *new* (not filtered). For this reason, blood leaving the kidney in the renal vein may have a higher concentration of HCO_3^- than blood entering the kidney in the renal artery.

Some H^+ secreted into the tubular fluid of the collecting duct is buffered. By now, most of the filtered HCO_3^- has been reabsorbed so that little is left in the tubule lumen to combine with and buffer secreted H^+. Two other buffers, however, are present and available to combine with H^+ (Fig. 26.18). The most plentiful buffer is HPO_4^{2-} (monohydrogen phosphate ion); a small amount of NH_3 (ammonia) also is present. H^+ combines with HPO_4^{2-} to form $H_2PO_4^-$ (dihydrogen phosphate ion) and with NH_3 to form NH_4^+ (ammonium ion). These ionized products cannot diffuse back into tubule cells and so are excreted in the urine.

Secretion of K^+

Normally, close to 100% of the filtered K^+ is reabsorbed in the PCT, loop of Henle, and DCT. To adjust for varying dietary intake of K^+ and to maintain the homeostasis of K^+ concentration in body fluids, the principal cells in the final portion of the DCT and collecting duct secrete a variable amount of K^+. Secretion of K^+ is controlled by:

1. **Aldosterone.** When the level of aldosterone in the blood rises, more K^+ is secreted into tubular fluid.
2. **K^+ concentration in plasma.** When plasma K^+ concentration is high, K^+ secretion increases.

Figure 26.18 Secretion of hydrogen ions (H^+) by intercalated cells in the collecting duct. HCO_3^- = bicarbonate ion; CO_2 = carbon dioxide; H_2O = water; H_2CO_3 = carbonic acid; Cl^- = chloride ion; NH_3 = ammonia; NH_4^+ = ammonium ion; HPO_4^{2-} = monohydrogen phosphate ion; $H_2PO_4^-$ = dihydrogen phosphate ion.

 Urine can be up to 1000 times more acidic than blood due to operation of these H^+ primary active transport pumps.

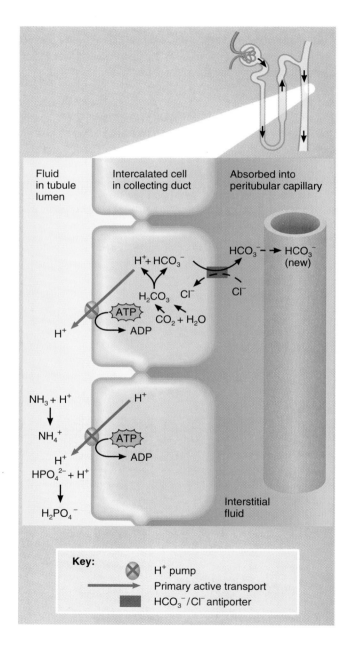

 What substances can combine with and buffer H^+ that is secreted into urine?

3. **Na⁺ concentration in distal convoluted tubules.** A high level of Na^+ in the fluid within the DCTs increases the rate of Na^+ absorption and K^+ secretion.

Secretion of K^+ is very important. As the K^+ concentration in plasma rises, disturbances in cardiac rhythm may develop, and at higher concentrations, cardiac arrest may occur.

Secretion of NH₃ and NH₄⁺

Ammonia (NH_3) is a poisonous waste product derived from the deamination of amino acids (removal of an amino group) by liver cells. The liver converts much of the ammonia to a less toxic compound called urea. Urea and ammonia in blood both are excreted in urine. Proximal convoluted tubule cells produce additional ammonia by deaminating the amino acid glutamine. This reaction generates both NH_3 and HCO_3^-. At pH 7.4 most NH_3 quickly picks up a H^+ and becomes an ammonium ion (NH_4^+). NH_4^+ can substitute for H^+ aboard Na^+/H^+ antiporters and be secreted into the proximal tubular fluid. When NH_4^+ is secreted in this way, the newly formed HCO_3^- can enter the bloodstream.

Ammonia production in the PCT increases in acidosis (blood pH below 7.35), and the resulting increase in blood HCO_3^- helps correct the acidic condition. On the other hand, ammonia formation decreases during alkalosis (blood pH above 7.45). Acid–base balance is discussed more fully in Chapter 27.

PRODUCTION OF DILUTE AND CONCENTRATED URINE

Despite the large intake of fluids at some times and little or no intake at others, the total volume of fluid in the body remains fairly stable. Homeostasis of body fluid volume depends in large part on the ability of the kidneys to regulate the rate at which water is lost in the urine. Normally functioning kidneys produce a large volume of dilute urine when fluid intake is excessive and a small volume of concentrated urine when fluid intake is meager or fluid loss is large. The hormone that regulates this function of the kidneys is antidiuretic hormone (ADH). As noted earlier, ADH controls the water permeability of the principal cells in the last portion of the distal convoluted tubules and collecting ducts. In the absence of ADH, the ducts are virtually impermeable to water, and urine contains a large volume of water (dilute urine). However, in the presence of ADH, the collecting ducts become quite permeable to water. Much water is reabsorbed back into blood, leaving less water in the urine (concentrated urine).

Mechanism of Urine Dilution

For the kidneys to produce dilute urine, they must form urine that contains fewer solutes in a given volume than

does blood. This occurs when the renal tubules allow excretion of an increased amount of water. The mechanism of urine dilution is as follows.

Glomerular filtrate has the same ratio of water and solute particles as blood; its osmotic concentration is about 300 mOsm/liter.[1] As noted earlier, fluid leaving the proximal convoluted tubule is still isotonic to plasma. In the loop of Henle, the osmolarity of the fluid in the tubular lumen *increases* as fluid flows down the descending limb and *decreases* as fluid flows up the ascending limb (Fig. 26.19). One reason for these changes in osmolarity is the reabsorption of Na^+, K^+, and Cl^- from the tubular fluid by symporters in cells of the thick ascending limb (see Fig. 26.15). The ions pass from the tubular fluid into thick ascending limb cells, then into interstitial fluid, and finally into the vasa recta capillaries. Although ions are being reabsorbed here, the water permeability of this portion of the nephron is quite low so water cannot follow by osmosis. Since ions but not water molecules are leaving the tubular fluid, its concentration drops to about 150 mOsm/liter. The fluid leaving the ascending limb is thus more dilute than plasma.

As the tubular fluid continues through the DCT and collecting duct, reabsorption of ions continues and the fluid becomes even more dilute. By the time the dilute fluid passes into the papillary ducts, its concentration can be as low as 65–70 mOsm/liter. Thus urine may be up to four times more dilute than blood plasma and glomerular filtrate (300 mOsm/liter).

Mechanism of Urine Concentration

During times of low water intake or excessive water loss, for example, through heavy perspiration, the kidneys must conserve water while still eliminating wastes and excess ions. They accomplish this by increasing the volume of water that is reabsorbed into blood and thereby producing a small volume of highly concentrated urine. It is primarily the long-loop juxtamedullary nephrons that establish the conditions for producing concentrated urine. Urine can be four times more concentrated (up to 1200 mOsm/liter) than blood plasma and glomerular filtrate (300 mOsm/liter).

The ability of ADH to cause excretion of concentrated urine depends on the presence of a high concentration of solutes in the interstitial fluid of the renal medulla. In Fig. 26.20 on page 875 you can see that the solute concentration of the interstitial fluid in the kidney increases from about 300 mOsm/liter in the renal cortex to about 1200 mOsm/liter deep in the renal medulla. The major solutes that contribute to this high osmolarity are Na^+, Cl^-, and urea. Two main factors

---

[1] The osmotic concentration of a solution, called **osmolarity**, is the total number of dissolved particles per liter of solution. Because fluids in living systems are dilute, the units used most often are **milliosmoles per liter (mOsm/liter)**.

Figure 26.19 Mechanism of urine dilution. Numbers indicate concentrations in milliosmoles per liter (mOsm/liter). The portions of the renal tubule indicated by heavy brown lines represent areas that are impermeable to water.

 When ADH level is low, urine is dilute and has an osmolarity less than the osmolarity of blood.

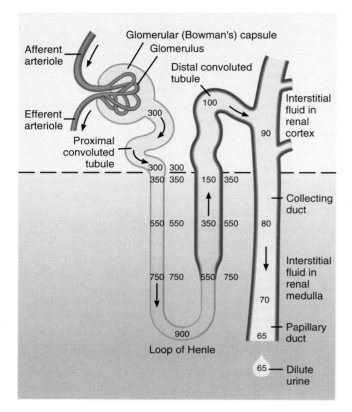

 Which portions of the renal tubule and collecting duct reabsorb more solutes than water to produce dilute urine?

contribute to building and maintaining this osmotic gradient: (1) differences in solute and water reabsorption in different sections of the loop of Henle and collecting duct and (2) the countercurrent flow of fluid and blood that occurs due to the anatomical arrangement of the loop of Henle and vasa recta.

Solute and Water Reabsorption

Production of concentrated urine occurs as follows:

1 **Thick ascending limb cells of the loop of Henle establish the ionic osmotic gradient in the renal medulla.** In the thick ascending limb of the loop of Henle, Na⁺–K⁺–2Cl⁻ symporters reabsorb these solutes from the tubular fluid (Fig. 26.20a). As the ions pass into the interstitial fluid of the outer renal medulla, they become concentrated in the fluid and are carried deep into the inner renal medulla by the blood flowing in the vasa recta (Fig.26.20b).

2 **The collecting duct reabsorbs more water.** When ADH increases the water permeability of the principal cells, water quickly moves by osmosis from the collecting duct tubular fluid, through the principal cells, into the interstitial fluid of the inner renal medulla, and into the blood of the vasa recta.

3 **Urea recycling causes a buildup of urea in the renal medulla.** With the loss of water from fluid in the collecting and papillary ducts, urea and other solutes left behind become more concentrated. Because duct cells deep in the renal medulla are permeable to urea, it diffuses from the fluid in the duct into the interstitial fluid of the renal medulla (Fig. 26.20a). As urea builds up in the interstitial fluid, some diffuses into the tubular fluid in the descending and thin ascending limbs of the long loops of Henle. However, while the fluid flows through the thick ascending limb, DCT, and cortical portion of the collecting duct, urea remains in the lumen because cells there are quite impermeable to it. As fluid flows along the collecting and papillary ducts, water continues to move out by osmosis because ADH is present. This *further* increases the concentration of urea in the tubular fluid, *more* urea diffuses into the interstitial fluid of the inner renal medulla, and the cycle repeats itself. Thus reabsorption of water from the tubular fluid of the ducts promotes the buildup of urea in the interstitial fluid of the renal medulla, which in turn promotes water reabsorption. The solutes left behind in the tubular fluid thus become very concentrated and a small volume of concentrated urine is excreted.

Countercurrent Mechanism

The second factor that maintains the osmotic gradient in the renal medulla is the **countercurrent mechanism**. It is based on the anatomical arrangement of the long loops of Henle of juxtamedullary nephrons and the vasa recta.

You can see in Fig. 26.20a that the descending limb of the loop of Henle carries tubular fluid from the renal cortex deep into the renal medulla and the ascending limb carries it in the opposite direction. Thus fluid flowing in one tube runs counter (opposite) to fluid flowing in a nearby parallel tube, an arrangement called *countercurrent* flow.

The descending limb is relatively permeable to water and relatively impermeable to solutes except urea. Because the interstitial fluid outside the descending limb is more concentrated than the tubular fluid within it, water moves out of the descending limb by osmosis. This water movement causes the concentration of the tubular fluid to increase. As the fluid continues down the descending limb, more water leaves it by osmosis, and the concentration of the tubular fluid increases even more. At the hairpin turn of the loop, its concentration is as high as 1200 mOsm/liter.

As noted earlier, the ascending limb is relatively impermeable to water, but its symporters reabsorb Na⁺, K⁺, and Cl⁻ from the tubular fluid into the interstitial fluid of the renal medulla. As the fluid flows through the ascending limb and the ions move out, the concentration of tubular fluid in the ascending limb progressively decreases. Near the renal cortex, the concentration has fallen to 100 mOsm/liter. The overall effect of the countercurrent flow is that tubular fluid becomes progressively more concentrated as it flows down the descending limb and progressively more dilute as it moves up the ascending limb.

Look again at Fig. 26.20 and note that the vasa recta also consists of descending and ascending portions that are parallel to each other and to the loop of Henle. Just as tubular fluid flows in opposite directions in the loop of Henle, blood flows in opposite directions in the ascending and descending portions of the vasa recta. This countercurrent flow of blood helps

maintain the osmotic gradient in the renal medulla while providing oxygen and nutrients to cells in the renal medulla.

Let's see how this happens. Blood entering the vasa recta has a solute concentration of about 300 mOsm/liter. As it flows down the descending portion into the renal medulla, where the interstitial fluid becomes more and more concentrated, Na^+, Cl^-, and urea diffuse from interstitial fluid into the blood. But after the blood becomes more concentrated, it flows into the ascending portion of the vasa recta. Then the interstitial fluid becomes less and less concentrated. As a result, ions and urea diffuse out of the blood into the interstitial fluid so that blood leaving the vasa recta has a solute

Figure 26.20 Mechanism of urine concentration in long-loop juxtamedullary nephrons. The green line indicates the presence of symporters in the thick ascending limb of the loop of Henle that simultaneously reabsorb a sodium ion (Na^+), a potassium ion (K^+), and two chloride ions ($2Cl^-$) into the interstitial fluid of the renal medulla. Also, this portion of the nephron is relatively impermeable to water and urea. All concentrations are in milliosmoles per liter (mOsm/liter).

The formation of concentrated urine depends on high concentrations of solutes in interstitial fluid in the renal medulla.

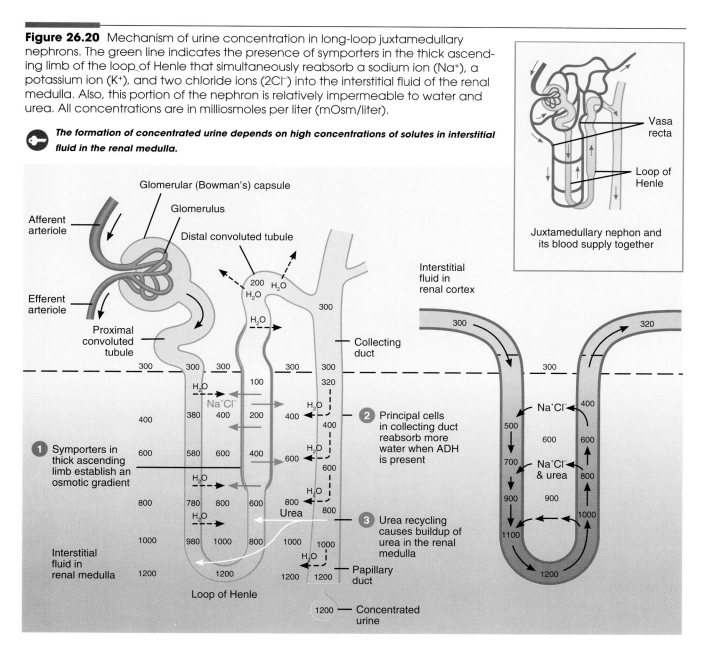

(a) Reabsorption of Na^+, Cl^-, and water in a long-loop juxtamedullary nephron

(b) Recycling of salts and urea in the vasa recta

 What solutes contribute to the high osmotic pressure of interstitial fluid in the renal medulla?

concentration only slightly higher than blood entering it. Thus blood flows through the renal medulla without disturbing the osmotic gradient because it carries away very few extra solutes.

Figure 26.21 summarizes the processes of filtration, reabsorption, and secretion in each segment of the nephron and collecting duct. The effects of aldosterone, ADH and ANP also are noted.

Figure 26.21 Summary of filtration, reabsorption, and secretion in the nephron and collecting duct.

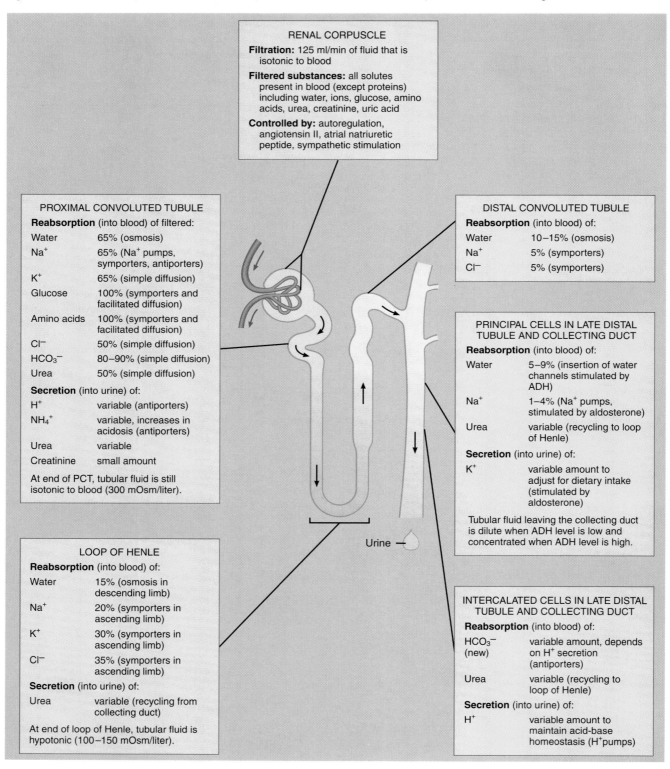

RENAL CORPUSCLE

Filtration: 125 ml/min of fluid that is isotonic to blood

Filtered substances: all solutes present in blood (except proteins) including water, ions, glucose, amino acids, urea, creatinine, uric acid

Controlled by: autoregulation, angiotensin II, atrial natriuretic peptide, sympathetic stimulation

PROXIMAL CONVOLUTED TUBULE

Reabsorption (into blood) of filtered:

Water	65% (osmosis)
Na^+	65% (Na^+ pumps, symporters, antiporters)
K^+	65% (simple diffusion)
Glucose	100% (symporters and facilitated diffusion)
Amino acids	100% (symporters and facilitated diffusion)
Cl^-	50% (simple diffusion)
HCO_3^-	80–90% (simple diffusion)
Urea	50% (simple diffusion)

Secretion (into urine) of:

H^+	variable (antiporters)
NH_4^+	variable, increases in acidosis (antiporters)
Urea	variable
Creatinine	small amount

At end of PCT, tubular fluid is still isotonic to blood (300 mOsm/liter).

DISTAL CONVOLUTED TUBULE

Reabsorption (into blood) of:

Water	10–15% (osmosis)
Na^+	5% (symporters)
Cl^-	5% (symporters)

PRINCIPAL CELLS IN LATE DISTAL TUBULE AND COLLECTING DUCT

Reabsorption (into blood) of:

Water	5–9% (insertion of water channels stimulated by ADH)
Na^+	1–4% (Na^+ pumps, stimulated by aldosterone)
Urea	variable (recycling to loop of Henle)

Secretion (into urine) of:

K^+	variable amount to adjust for dietary intake (stimulated by aldosterone)

Tubular fluid leaving the collecting duct is dilute when ADH level is low and concentrated when ADH level is high.

LOOP OF HENLE

Reabsorption (into blood) of:

Water	15% (osmosis in descending limb)
Na^+	20% (symporters in ascending limb)
K^+	30% (symporters in ascending limb)
Cl^-	35% (symporters in ascending limb)

Secretion (into urine) of:

Urea	variable (recycling from collecting duct)

At end of loop of Henle, tubular fluid is hypotonic (100–150 mOsm/liter).

Urine

INTERCALATED CELLS IN LATE DISTAL TUBULE AND COLLECTING DUCT

Reabsorption (into blood) of:

HCO_3^- (new)	variable amount, depends on H^+ secretion (antiporters)
Urea	variable (recycling to loop of Henle)

Secretion (into urine) of:

H^+	variable amount to maintain acid-base homeostasis (H^+ pumps)

DIURETICS

Diuretics are drugs that increase the rate of urine flow, which speeds the loss of water from the body, usually by interfering with reabsorption of filtered Na⁺. Since there are so many mechanisms for Na⁺ reabsorption, there is a great diversity of diuretic drugs. One of the earliest known classes of diuretics slows Na⁺ reabsorption by inhibiting carbonic anhydrase in the brush border of the proximal convoluted tubule cells. Acetazolamide (Diamox®) is an example of a carbonic anhydrase inhibitor. It is rarely used today because an adverse side effect is loss of filtered HCO_3^- in the urine. (Reexamine Fig. 26.17 to see why.) Loop diuretics, such as furosemide (Lasix®), are the most potent diuretic drugs. They selectively inhibit the Na⁺–K⁺–2Cl⁻ symporters in the thick ascending limb of the loop of Henle (see Fig. 26.15). The thiazide diuretics, such as chlorothiazide (Diuril®), are thought to inhibit Na⁺–Cl⁻ symporters in the distal convoluted tubule.

Most diuretic drugs have the adverse side effect of causing excessive loss of K⁺ in the urine. One exception is the class of diuretics that inhibits the action of aldosterone, such as spironolactone (Aldactone®). These are called potassium-sparing diuretics. They promote mild diuresis by slowing reabsorption of Na⁺ and water in the collecting duct, and at the same time they also decrease urinary loss of K⁺ because they inhibit K⁺ secretion.

Naturally occurring diuretics include caffeine in coffee, tea, and cola sodas, which inhibit Na⁺ reabsorption, and alcohol in beer, wine, and mixed drinks, which inhibit secretion of ADH. Diuretic drugs often are prescribed to treat congestive heart failure, high blood pressure, and edema. ■

EVALUATION OF KIDNEY FUNCTION

Several screening tests can provide information about kidney function. One is the **blood urea nitrogen (BUN)** test. It measures the nitrogen in blood that is part of urea, which results from the catabolism of amino acids. When glomerular filtration rate decreases severely, as may occur with renal disease or obstruction of the urinary tract, BUN rises steeply. One strategy in treating such patients is to minimize their protein intake, thus reducing the rate of urea production.

Another test often used to evaluate kidney function is measurement of **plasma creatinine**. Creatinine is the end product of the catabolism of creatine phosphate in skeletal muscle. Normally, the blood creatinine level is steady because creatinine is continually lost in the urine as fast as it is discharged from muscle. When the creatinine level rises above 1.5 mg/dl (about 135 mmol/liter), this usually is an indication of poor renal function.

Even more useful than BUN and blood creatinine values in the diagnosis of kidney problems are renal plasma clearance values for specific substances. **Renal plasma clearance** expresses how effectively the kidneys remove a sub-

stance from blood plasma. High renal clearance indicates efficient removal of a substance from the blood into the urine; low renal clearance indicates a low rate of excretion in the urine. For example, the clearance of glucose normally would be zero because it is not excreted at all but rather is returned to the blood by reabsorption. In drug therapy, knowing the clearance is essential for determining the correct dosage. If clearance is high (an example is penicillin), the dose must also be high, and the drug must be given several times a day to maintain an adequate therapeutic level in the blood.

Clearance is usually expressed in units of *milliliters per minute* and is calculated from the following equation.

$$\text{Renal clearance} = \frac{UV}{P}$$

U is the concentration of the substance in urine and P is the concentration of the same substance in plasma, in mg/ml. V is the urine flow rate in ml/min.

The clearance of a solute depends on the three basic processes that occur in a nephron: glomerular filtration, tubular reabsorption, and tubular secretion. If a substance is filtered but not reabsorbed at all and not secreted, then its clearance equals the glomerular filtration rate. In other words, all the molecules that pass the filter appear in the urine. This situation is almost true for creatinine; it easily passes the filter, it is not reabsorbed, and it is secreted to a very small extent. Measuring the creatinine clearance, which normally is about 140 ml/min, is the easiest way to evaluate glomerular filtration rate. A polysaccharide called **inulin**, which can be introduced into the blood by a steady intravenous infusion, gives an even more accurate determination of GFR because it is not reabsorbed or secreted at all. *The clearance of inulin equals the GFR,* normally about 125 ml/min. The waste product urea is filtered, reabsorbed, and secreted to varying extents. Its clearance typically is less than the GFR, about 70 ml/min.

Often it is helpful to know the rate of blood flow through the kidneys. For example, a low GFR might result from inadequate renal blood flow or from some other problem. Renal plasma flow can be determined by measuring the clearance of a substance called **para-aminohippuric acid (PAH)**. Because PAH is both filtered and very vigorously secreted, virtually all the PAH entering the kidneys in arterial blood plasma is gone by the time the blood enters the renal veins. Clearance of PAH, and thus renal plasma flow, normally is about 600 ml/min. Since PAH is not a naturally occurring substance, it must be injected intravenously to measure its clearance.

DIALYSIS THERAPY

If the kidneys are so impaired by disease or injury that they are unable to excrete nitrogenous wastes, regulate pH, and adjust the concentrations of various ions in blood plasma, then blood must be cleansed artificially by dialysis. *Dialysis* means the separation of large particles from smaller ones through use of a selectively permeable membrane.

One of the best-known devices for accomplishing dialysis is the artificial kidney machine (Fig. 26.22), which performs **hemodialysis** (*hemo* = blood) because it connects directly to the blood. The patient's blood flows through tubing made of selectively permeable dialysis membrane. As blood flows through the tubing, waste products diffuse from the blood into the dialysis solution or *dialysate* (dī -AL-i-sāt) surrounding the dialysis membrane. Also, if nutrients are provided in the solution, they can diffuse from the dialysate into the blood. The dialysate is continuously replaced to maintain favorable concentration gradients between the solution and the blood. After passing through the dialysis tubing, the blood flows back to the body. In removing wastes from the blood, the dialysis membrane performs one of the kidney's principal functions. Hemodialysis typically is performed three times a week, each session lasting for several hours. There are serious drawbacks to hemodialysis. Blood cells can be damaged and anticoagulants must be added, which can lead to bleeding problems. The slow rate at which the blood can be processed makes the treatment time-consuming.

Continuous ambulatory peritoneal dialysis (**CAPD**) is more convenient and less time-consuming for many patients because it can be done at home or while working. CAPD uses the peritoneum as the dialysis membrane. Since the peritoneum is a selectively permeable membrane, it permits rapid two-directional transfer of substances. The tip of a catheter is placed in the patient's peritoneal cavity and connected to a supply of dialysate. The dialysate flows into the cavity from its plastic container by gravity. When the process is complete, the solution is drained from the cavity into the plastic container and then is discarded. Although CAPD is more convenient than hemodialysis, the risk of infection is greater due to the indwelling catheter.

URETERS

Urine drains through papillary ducts into the minor calyces. They join to become major calyces that unite to form the renal pelvis. From the renal pelvis, urine drains into the ureters and then into the urinary bladder (see Fig. 26.4a). From the urinary bladder, urine is discharged from the body through the single urethra.

Anatomy

There are two **ureters** (YOO-re-ters; *oureter* = ureter)—one for each kidney. Each ureter connects the renal pelvis of one kidney to the urinary bladder (see Fig. 26.1). The ureters have a length of 25–30 cm (10–12 in.) and, as they descend, their thick walls increase in diameter to a maximum of about 1.7 cm (0.7 in.). Like the kidneys, the ureters are retroperitoneal. At the base of the urinary bladder, the ureters curve medially and enter the posterior aspect of the urinary bladder (Fig. 26.23a).

Although there is no anatomical valve at the opening of each ureter into the urinary bladder, there is a functional one that is quite effective. The ureters pass obliquely through the wall of the urinary bladder. As the urinary bladder fills with urine, pressure inside the urinary bladder compresses the openings and prevents backup of urine into the ureters. When this physiological valve is not operating, it is possible for cystitis (urinary bladder inflammation) that is caused by a microbial infection to spread into a kidney infection.

Histology

Three coats of tissue form the wall of the ureters. The deepest coat, or **mucosa**, is a mucous membrane with **transitional epithelium** (see Exhibit 4.1, transitional epithelium, on page 100) and an underlying **lamina propria** of areolar connective tissue with considerable collagen and elastic fibers and lymphatic tissue. Transitional epithelium is able to stretch—a marked advantage for any organ that must continually inflate and deflate. The solute concentration and

---

Figure 26.22 Operation of an artificial kidney machine. The blood route is indicated in red and blue. The route of the dialysate is indicated in gold.

🔑 *Dialysis is the separation of large particles from smaller ones through a selectively permeable membrane.*

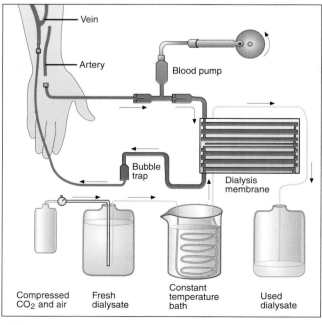

Q What happens to plasma proteins during hemodialysis?

Figure 26.23 Ureters, urinary bladder, and urethra (female).

🔑 *Urine is stored in the urinary bladder until it is expelled by an act called micturition.*

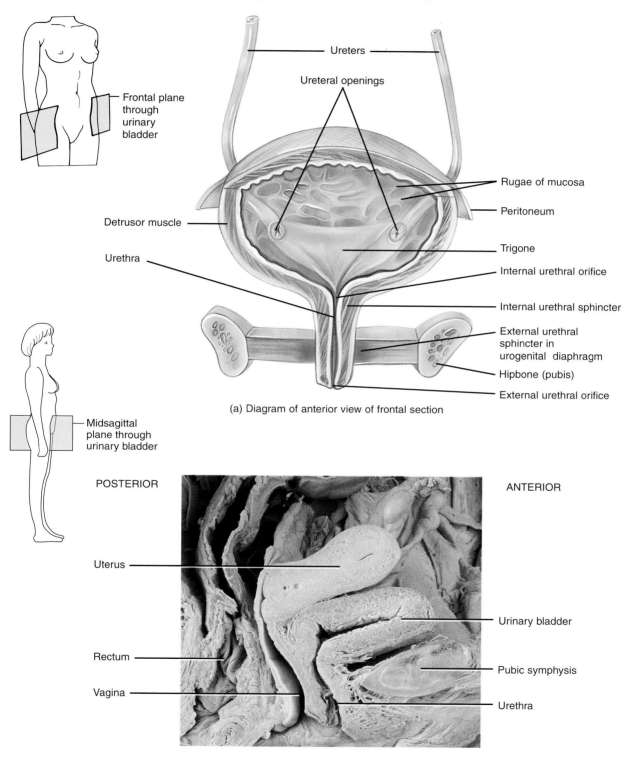

Frontal plane through urinary bladder

Ureters

Ureteral openings

Rugae of mucosa

Peritoneum

Detrusor muscle

Trigone

Urethra

Internal urethral orifice

Internal urethral sphincter

External urethral sphincter in urogenital diaphragm

Hipbone (pubis)

External urethral orifice

(a) Diagram of anterior view of frontal section

Midsagittal plane through urinary bladder

POSTERIOR

ANTERIOR

Uterus

Rectum

Vagina

Urinary bladder

Pubic symphysis

Urethra

(b) Photograph of midsagittal section

❓ What is a lack of voluntary control over micturition called?

pH of urine may differ drastically from the cytosol of cells that form the wall of the ureters. Mucus secreted by the mucosa prevents the cells from coming in contact with urine. Throughout most of the length of the ureters, the intermediate coat, the **muscularis**, is composed of inner longitudinal and outer circular layers of smooth muscle fibers. The muscularis of the distal third of the ureters also contains an outer layer of longitudinal muscle fibers. Peristalsis is the major function of the muscularis. The superficial coat of the ureters is the **adventitia**, a layer of areolar connective tissue containing blood vessels, lymphatic vessels, and nerves that serve the muscularis and mucosa. The adventitia blends in with surrounding connective tissue and anchors the ureters in place.

Physiology

The ureters transport urine from the renal pelvis into the urinary bladder. Peristaltic contractions of the muscular walls of the ureters push urine toward the urinary bladder, but hydrostatic pressure and gravity also contribute. Peristaltic waves pass from the renal pelvis to the urinary bladder, varying in rate from one to five per minute, depending on how fast urine is being formed.

RENAL CALCULI

Occasionally, the crystals of salts present in urine may solidify into insoluble stones called **renal calculi** (singular is *calculus* = pebble) or **kidney stones**. They may form in any portion of the urinary tract. Conditions leading to calculus formation include the ingestion of excessive calcium, a decrease in water intake, abnormally alkaline or acidic urine, and overactivity of the parathyroid glands. Common components of kidney stones are calcium oxalate, uric acid, and calcium phosphate crystals. When a stone gets stuck in a narrow opening, for example, a ureter, the pain can be excruciating. Recovery from surgery to remove a calculus in a kidney or a ureter takes weeks.

A technique called **shock wave lithotripsy** (LITH-ō-trip′-sē; *litho* = stone; *tribein* = to rub) offers an easier solution. A device, called a lithotripter, delivers brief, high-intensity sound shock waves through a water bath or water-filled cushion. Over a period of 30–60 minutes, 1000 or more hydraulic shock waves pulverize the stone until the fragments are small enough to wash out in the urine. Recovery time is minimal because no incision is made, and the cost is much less. ■

URINARY BLADDER

The **urinary bladder** is a hollow muscular organ situated in the pelvic cavity posterior to the pubic symphysis. In the male, it is directly anterior to the rectum. In the female, it is anterior to the vagina and inferior to the uterus (Fig. 26.23b). It is a freely movable organ held in position by folds of the peritoneum. The shape of the urinary bladder depends on how much urine it contains. Empty, it is collapsed. It becomes spherical when slightly distended. As urine volume increases, it becomes pear-shaped and rises into the abdominal cavity. In general, urinary bladder capacity is smaller in females because the uterus occupies the space just superior to the bladder.

Anatomy

In the floor of the urinary bladder is a small triangular area called the **trigone** (TRĪ-gōn; *trigonium* = triangle). The two posterior corners of the trigone contain the two ureteral openings whereas the opening into the urethra, the **internal urethral orifice**, lies in the anterior corner (Fig. 26.23a). Because its mucosa is firmly bound to the muscularis, the trigone has a smooth appearance.

Histology

Three coats make up the wall of the urinary bladder. The deepest is the **mucosa**, a mucous membrane composed of **transitional epithelium** and an underlying **lamina propria** similar to that of the ureters. Rugae (folds in the mucosa) are also present. Surrounding the mucosa is the intermediate **muscularis**, called the **detrusor** (de-TROO-ser; *detrudere* = to push down) **muscle**. It consists of three layers of smooth muscle fibers: inner longitudinal, middle circular, and outer longitudinal. Around the opening to the urethra, the circular fibers form an **internal urethral sphincter**. Inferior to the internal sphincter is the **external urethral sphincter**, which is composed of skeletal muscle and is a modification of the urogenital diaphragm muscle (see Fig. 11.12). The most superficial coat of the urinary bladder on the posterior and inferior surfaces is the **adventitia**, a layer of areolar connective tissue that is continuous with that of the ureters. Over the superior surface of the urinary bladder is a layer of visceral peritoneum called the **serosa**.

Physiology

Expulsion of urine from the urinary bladder is termed **micturition** (mik′-too-RISH-un; *micturire* = to urinate), commonly known as urination or voiding. This response is brought about by a combination of involuntary and voluntary activities. The average capacity of the urinary bladder is 700–800 ml. When the amount of urine in the urinary bladder exceeds 200–400 ml, stretch receptors in the wall transmit sensory impulses to the spinal cord. These impulses, by way of sensory tracts to the cortex, initiate a conscious desire to expel urine and, by way of a center in the sacral spinal cord, trigger a reflex called the **micturition** (**urination**) **reflex**. In this reflex arc, parasympathetic fibers from the micturition reflex center of the spinal cord (spinal segments S2 and S3) conduct motor impulses to the urinary bladder wall and internal urethral sphincter. The nerve impulses cause contraction of the detrusor muscle and relaxation of the internal urethral sphincter. Urination does not yet occur, however, because the external urethral sphincter, which is skeletal muscle, remains closed. When nerve impulses from the cerebral cortex of the brain inhibit activity in motor neurons to the external urethral sphincter (skeletal muscle), there is voluntary relaxation of this muscle and urination takes place. Although emptying the urinary bladder is a reflex, it may be initiated voluntarily and stopped at will because of cerebral cortical control of the external urethral sphincter and certain muscles of the urogenital (pelvic) diaphragm.

INCONTINENCE AND RETENTION

A lack of voluntary control over micturition is referred to as **incontinence**. In infants less than 2 years old, incontinence is normal because neurons to the external urethral sphincter are not completely developed. Infants void whenever the urinary bladder is sufficiently distended to stimulate the reflex. Involuntary micturition in adults may occur as a result of unconsciousness, injury to the spinal cord or spinal nerves controlling the urinary bladder, irritation due to abnormal constituents in urine, disease of the urinary bladder, damage to the external urethral sphincter, and inability of the detrusor muscle to relax due to emotional stress.

Retention, a failure to completely or normally void urine, may be due to an obstruction in the urethra or neck of the urinary bladder, nervous contraction of the urethra, or lack of sensation to urinate. ■

URETHRA

The **urethra** is a small tube leading from the internal urethral orifice in the floor of the urinary bladder to the exterior of the body (see Fig. 26.1).

Anatomy

In females, the urethra lies directly posterior to the pubic symphysis (Fig. 26.23b). Its length is approximately 4 cm (1.5 in.). The female urethra is directed obliquely inferiorly and anteriorly. The opening of the urethra to the exterior, the **external urethral orifice**, is located between the clitoris and the vaginal opening (see Fig. 28.18).

In males, the urethra also extends from the internal urethral orifice to the exterior, but its length and passage through the body are considerably different from those in females. The male urethra is about 15–20 cm (6–8 in.) long (see Figs. 28.1 and 28.10). From its origin, it first passes through the prostate gland, then through the urogenital diaphragm, and finally through the penis.

Histology

The wall of the female urethra consists of a deep **mucosa** and superficial **muscularis**. The mucosa is a mucous membrane composed of **epithelium** and **lamina propria** (areolar connective tissue with elastic fibers and a plexus of veins). The muscularis consists of circularly arranged smooth muscle fibers and is continuous with that of the urinary bladder. Near the urinary bladder, the mucosa contains transitional epithelium that is continuous with that of the urinary bladder. Near the external urethral orifice, the epithelium is nonkeratinized stratified squamous epithelium. Between these areas, the mucosa contains stratified columnar or pseudostratified columnar epithelium.

The male urethra also consists of a deep **mucosa** and a superficial **muscularis.** It is subdivided into three anatomical regions: (1) the **prostatic urethra** passes through the prostate gland; (2) the **membranous urethra**, the shortest portion, passes through the urogenital diaphragm; and (3) the **spongy urethra**, the longest portion, passes through the penis. The epithelium of the prostatic urethra is continuous with that of the urinary bladder and consists of transitional epithelium that becomes stratified columnar or pseudostratified columnar epithelium more distally. The mucosa of the membranous urethra contains stratified columnar or pseudostratified columnar epithelium. The epithelium of the spongy urethra is stratified columnar or pseudostratified columnar epithelium, except near the external urethral orifice, which is nonkeratinized stratified squamous epithelium. The **lamina propria** of the male urethra, like that of the female, is areolar connective tissue with elastic fibers and a plexus of veins.

The muscularis of the prostatic urethra is composed of wisps of mostly circular smooth muscle fibers superficial to the lamina propria that help form the internal urethral sphincter of the urinary bladder. The muscularis of the membranous urethra consists of circularly arranged skeletal muscle fibers of the urogenital diaphragm that help form the external urethral sphincter of the urinary bladder.

Physiology

In both males and females, the urethra is the terminal portion of the urinary system. It serves as the passageway for discharging urine from the body. The male urethra also serves as the duct through which various reproductive secretions are discharged from the body (Chapter 28). The prostatic urethra contains the openings of ducts that transport secretions from the **prostate gland** and from the **seminal vesicles** and **ductus (vas) deferens** that deliver sperm into the urethra, provide secretions that neutralize the acidity of the female reproductive tract, and contribute to sperm motility and viability. The openings of the ducts of the **bulbourethral (Cowper's)**

glands empty into the spongy urethra. They deliver an alkaline substance prior to ejaculation that neutralizes the acidity of the urethra. The glands also secrete mucus, which lubricates the end of the penis during sexual arousal. Throughout the urethra, but especially in the spongy urethra, the openings of the ducts of **urethral (Littré) glands** discharge mucus during sexual arousal or ejaculation.

URINALYSIS

An analysis of the volume and physical, chemical, and microscopic properties of urine, called a **urinalysis**, tells us much about the state of the body. The principal physical characteristics of urine are summarized in Exhibit 26.2.

The volume of urine eliminated per day in a normal adult is 1–2 liters (about 1–2 qt). Urine volume is influenced by fluid intake, blood pressure, blood osmotic pressure, diet, temperature, diuretics, mental state, and general health. Low blood pressure triggers the renin–angiotensin pathway, which increases reabsorption of water and salts in the renal

tubules and decreases urine volume. On the other hand, when blood osmotic pressure decreases—for example, after drinking a large volume of water—secretion of ADH is inhibited and a larger volume of urine is excreted. Opposite effects occur with high blood pressure and increased blood osmotic pressure.

Water accounts for about 95% of the total volume of urine. The remaining 5% consists of solutes derived from cellular metabolism and outside sources such as drugs. Typical solutes normally present in urine are described in Exhibit 26.3.

EXHIBIT 26.2　PHYSICAL CHARACTERISTICS OF NORMAL URINE

Characteristic	Description
Volume	One to two liters in 24 hours but varies considerably.
Color	Yellow or amber but varies with urine concentration and diet. Color is due to urochrome (pigment produced from breakdown of bile). Concentrated urine is darker in color. Diet (reddish colored urine from beets), medications, and certain diseases affect color. Kidney stones may produce blood in urine.
Turbidity	Transparent when freshly voided but becomes turbid (cloudy) upon standing.
Odor	Mildly aromatic but becomes ammonia-like upon standing. Some people inherit the ability to form methylmercaptan from digested asparagus that gives urine a characteristic odor. Urine of diabetics has a fruity odor due to presence of ketone bodies.
pH	Ranges between 4.6 and 8.0; average 6.0; varies considerably with diet. High-protein diets increase acidity; vegetarian diets increase alkalinity.
Specific gravity	Specific gravity (density) is the ratio of the weight of a volume of a substance to the weight of an equal volume of distilled water. It ranges from 1.001 to 1.035. The higher the concentration of solutes, the higher the specific gravity.

EXHIBIT 26.3　PRINCIPAL SOLUTES IN NORMAL URINE

Solute	Comments
ORGANIC	
Urea	Composes 60–90% of all nitrogen-containing material in urine; derived from ammonia produced by deamination of amino acids, which combines with carbon dioxide to form urea; amount excreted increases with increased dietary protein intake.
Creatinine	Normal constituent of blood. Derived primarily from breakdown of creatine phosphate in muscle tissue.
Uric acid	Product of catabolism of nucleic acids (DNA and RNA) derived from food or cellular destruction. Because of its insolubility, uric acid tends to crystallize and is a common component of kidney stones.
Urobilinogen	Bile pigment derived from breakdown of hemoglobin.
Other substances	May be present in small quantities, depending on diet and general health. Include carbohydrates, pigments, fatty acids, mucin, enzymes, and hormones.
INORGANIC	
Na^+, K^+	Amount excreted varies with dietary intake and level of aldosterone.
Cl^-, Mg^{2+}	Amount excreted varies with dietary intake.
SO_4^{2-}	Derived from amino acids. Amount excreted varies with dietary protein intake.
$H_2PO_4^-$, HPO_4^{2-}, PO_4^{3-}	Serve as buffers in blood and urine. Parathyroid hormone increases urinary excretion.
NH_4^+	Derived from protein catabolism and from deamination of the amino acid glutamine in kidneys. Amount produced by kidneys may vary with need to produce HCO_3^- to offset acidity of blood and tissue fluids.
Ca^{2+}	Amount excreted varies with dietary intake. Parathyroid hormone increases urinary excretion.

If the body's chemical processes are not operating efficiently, traces of substances not normally present may appear in the urine, or normal constituents may appear in abnormal amounts. Exhibit 26.4 lists several abnormal constituents in urine that may be detected as part of a urinalysis. Normal values of urine components and the clinical implications of deviations from normal are listed in Appendix B.3 on page A–10.

AGING AND THE URINARY SYSTEM

As one grows older, the kidneys become less effective, and by age 70 the filtering mechanism is only about half as effective as it was at age 40. Renal blood flow, glomerular filtration rate, and clearance of urea all decline with age. Urinary incontinence and urinary tract infections are two common problems associated with aging. Other pathologies include **polyuria** (excessive urine production), **nocturia** (excessive urination at night), increased frequency of urination, **dysuria** (painful urination), **retention** (failure to release urine from the urinary bladder), and **hematuria** (blood in the urine). Changes and diseases in the kidney include acute and chronic kidney inflammations and renal calculi (kidney stones). Since water balance and thirst may be altered, elderly persons are susceptible to dehydration. The prostate gland is often implicated in various disorders of the urinary tract, and cancer of the prostate (described on page 920) is the most common malignancy in older males. Since the prostate gland encircles part of the male urethra (prostatic urethra), an enlarged prostate may cause retention and difficulty in urination.

DEVELOPMENTAL ANATOMY OF THE URINARY SYSTEM

Starting in the third week of fetal development, a portion of the mesoderm along the posterior aspect of the embryo, the **intermediate mesoderm**, differentiates into the kidneys.

EXHIBIT **26.4** **SUMMARY OF ABNORMAL CONSTITUENTS IN URINE**

Abnormal Constituent	Comments
Albumin	Normal constituent of plasma, but it usually appears in only very small amounts in urine because it is too large to pass through the pores in capillary walls. The presence of excessive albumin in the urine—**albuminuria** (al′-byoo-mi-NOO-rē-a)—indicates an increase in the permeability of endothelial–capsular membranes due to injury or disease, increased blood pressure, or irritation of kidney cells by substances such as bacterial toxins, ether, or heavy metals.
Glucose	The presence of glucose in the urine is called **glucosuria** (gloo-kō-SOO-rē-a) and usually indicates diabetes mellitus. Occasionally, it may be caused by stress, which can cause excessive amounts of epinephrine to be secreted. Epinephrine stimulates the breakdown of glycogen and liberation of glucose from the liver.
Red blood cells (erythrocytes)	The presence of red blood cells in the urine is called **hematuria** (hēm-a-TOO-rē-a) and generally indicates a pathological condition. One cause is acute inflammation of the urinary organs as a result of disease or irritation from kidney stones. Other causes include tumors, trauma, and kidney disease. One should make sure the urine sample was not contaminated with menstrual blood from the vagina.
White blood cells (leukocytes)	The presence of white blood cells and other components of pus in the urine, referred to as **pyuria** (pī-YOO-rē-a), indicates infection in the kidney or other urinary organs.
Ketone bodies	High levels of ketone bodies, called **ketosis** (kē-TŌ-sis), may indicate diabetes mellitus, anorexia, starvation, or simply too little carbohydrate in the diet.
Bilirubin	When red blood cells are destroyed by macrophages, the globin portion of hemoglobin is split off and the heme is converted to biliverdin. Most of the biliverdin is converted to bilirubin, which gives bile its major pigmentation. An above-normal level of bilirubin in urine is called **bilirubinuria** (bil′-ē-roo-bi-NOO-rē-a).
Urobilinogen	The presence of urobilinogen (breakdown product of hemoglobin) in urine is called **urobilinogenuria** (yoo′-rō-bi-lin′-ō-je-NOO-rē-a). Traces are normal, but increased urobilinogen may be due to hemolytic or pernicious anemia, infectious hepatitis, biliary obstruction, jaundice, cirrhosis, congestive heart failure, or infectious mononucleosis.
Casts	**Casts** are tiny masses of material that have hardened and assumed the shape of the lumen of a tubule in which they formed. They are then flushed out of the tubule when filtrate builds up behind them. Casts are named after the cells or substances that compose them or on the basis of their appearance. For example, there are white blood cell casts, red blood cell casts, and epithelial cell casts that contain cells from the walls of the tubules.
Microbes	The number and type of bacteria vary with specific infections in the urinary tract. The most common fungus to appear in urine is *Candida albicans*, a cause of vaginitis. The most frequent protozoan seen is *Trichomonas vaginalis*, a cause of vaginitis in females and urethritis in males.

Three pairs of kidneys form within the intermediate meso-derm in successive time periods: pronephros, mesonephros, and metanephros (Fig. 26.24). Only the last pair remains as the functional kidneys of the newborn.

The first kidney to form, the **pronephros**, is the most su-perior of the three. Associated with its formation is a tube termed the **pronephric duct**. This duct empties into the **cloaca**, which functions as a common outlet from the uri-nary, digestive, and reproductive ducts. The pronephros be-gins to degenerate during the fourth week and is completely gone by the sixth week. The pronephric ducts, however, remain.

The second kidney, the **mesonephros**, replaces the pro-nephros. The retained portion of the pronephric duct, which connects to the mesonephros, becomes known as the **mesonephric duct**. The mesonephros begins to degen-erate by the sixth week and is almost gone by the eighth week.

At about the fifth week, a hollow tube, called the **ureteric bud**, grows out from the distal end of the meso-nephric duct near the cloaca. This bud is the developing **metanephros**. The ureteric bud progressively lengthens and branches to form the *pelvis* of the kidney, the *major* and *minor calyces,* the *papillary ducts,* and the *collecting ducts.* The stalk of the ureteric bud, the **metanephric duct**, becomes the *ureter.* The *nephrons* arise from a mass of intermediate mesoderm around each ureteric bud. By the third month, the fetal kidneys begin excreting urine into the surrounding amniotic fluid. In fact, fetal urine makes up most of the amniotic fluid.

During development, the cloaca divides into a **urogeni-tal sinus**, into which urinary and genital ducts empty, and a *rectum* that discharges into the anal canal. The *urinary bladder* develops from the urogenital sinus. In the female, the *urethra* develops from lengthening of the short duct that extends from the urinary bladder to the urogenital si-nus. The *vestibule,* into which the urinary and genital ducts empty, is also derived from the urogenital sinus. In the male, the urethra is considerably longer and more compli-cated but is also derived from the urogenital sinus.

Figure 26.24 Development of the urinary system.

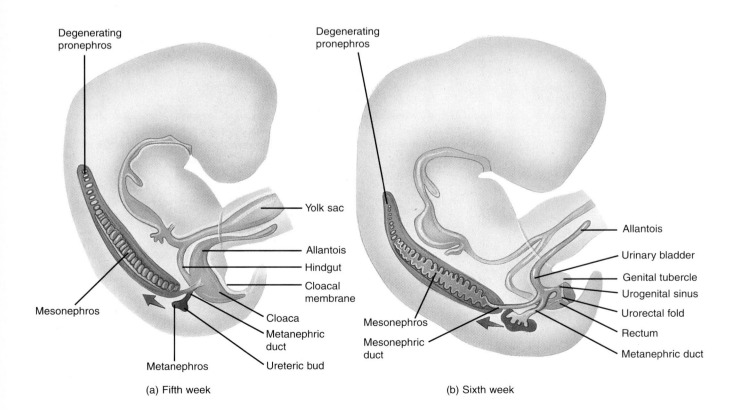

(a) Fifth week

(b) Sixth week

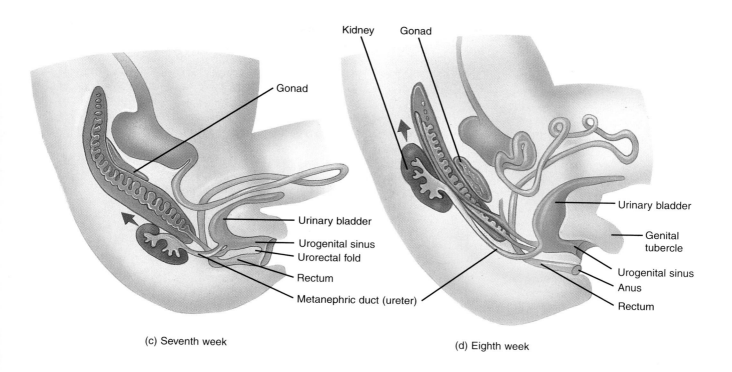

(c) Seventh week

(d) Eighth week

DISORDERS: HOMEOSTATIC IMBALANCES

URINARY TRACT INFECTIONS

The term **urinary tract infection (UTI)** is used to describe either an infection of a part of the urinary system or the presence of large numbers of microbes in urine. Symptoms include burning on urination or painful urination, urgency and frequent urination, and low back pain. Bed-wetting may also occur. UTIs include **urethritis** (inflammation of the urethra), **cystitis** (inflammation of the urinary bladder), **pyelonephritis** (inflammation of the kidneys), and **pyelitis** (inflammation of the renal pelvis and its calyces). If pyelonephritis becomes chronic, scar tissue forms in the kidneys and severely impairs their function.

GLOMERULONEPHRITIS

Glomerulonephritis (Bright's disease) is an inflammation of the kidney that involves the glomeruli. One of the most common causes is an allergic reaction to the toxins given off by streptococci bacteria that have recently infected another part of the body, especially the throat. The glomeruli become so inflamed, swollen, and engorged with blood that the endothelial–capsular membranes become highly permeable and allow blood cells and proteins to enter the filtrate. Thus the urine contains many erythrocytes (hematuria) and much protein (protein-

uria). The glomeruli may be permanently damaged, leading to acute or chronic renal failure.

NEPHROTIC SYNDROME

Nephrotic syndrome refers to protein in the urine (proteinuria), primarily albumin, that results in edema and hyperlipidemia (high blood levels of cholesterol, phospholipids, and triglycerides). The proteinuria is due to an increased permeability of the endothelial–capsular membrane, which permits proteins to escape from blood into urine. Loss of albumin results in a decline in the blood level of albumin as soon as liver production of albumin fails to meet increased urinary losses. The edema associated with nephrotic syndrome, usually seen around the eyes, ankles, feet, and abdomen, occurs because loss of albumin from the blood decreases blood colloid osmotic pressure. This leads to water movement from blood into interstitial spaces. Among the causes of nephrotic syndrome are diabetes mellitus, rheumatoid arthritis, systemic lupus erythematosus, lymphoma, leukemia, bacterial and viral infections, certain drugs (nonsteroidal anti-inflammatory drugs and street heroin), heavy metals (gold or mercury), hypertension, and sickle-cell anemia. Treatment is aimed at alleviating the proteinuria.

DISORDERS: HOMEOSTATIC IMBALANCES CONTINUED

RENAL FAILURE

Renal failure is a decrease or cessation of glomerular filtration. In **acute renal failure** (**ARF**) the kidneys abruptly stop working entirely or almost entirely. The main feature of ARF is suppression of urine flow, usually characterized by *oliguria* (*olig* = scanty), daily urine output less than 250 ml, or *anuria*, daily urine output less than 50 ml. Causes include low blood volume (for example, due to hemorrhage), decreased cardiac output, damaged renal tubules, and kidney stones.

Chronic renal failure (**CRF**) refers to a progressive and usually irreversible decline in glomerular filtration rate (GFR). CRF may result from chronic glomerulonephritis, pyelonephritis, polycystic kidney disease, or traumatic loss of kidney tissue. CRF develops in three stages. In the first stage, *diminished renal reserve*, nephrons are destroyed until about 75% of the functioning nephrons are lost. At this stage, the person may show no symptoms since the remaining nephrons enlarge and take over the function of those that have been lost. When more nephrons are lost, the balance between glomerular filtration and tubular reabsorption is upset, and any change in diet or fluid intake brings on the symptoms. Once 75% of the nephrons are lost, the person enters the second stage, called *renal insufficiency*. In this stage, there is a decrease in GFR and increased blood levels of nitrogen-containing wastes and creatinine. Also, the kidneys cannot effectively concentrate or dilute the urine. The final stage, called *end-stage renal failure*, occurs when about 90% of the nephrons have been lost. At this stage, GFR diminishes to about 10% of normal, and blood levels of nitrogen-containing wastes and creatinine increase further. The low GFR results in oliguria. People with CRF need dialysis therapy and are possible candidates for a kidney transplant operation.

Among the effects of renal failure are edema from salt and water retention, acidosis due to inability of the kidneys to excrete acidic substances, increased blood urea nitrogen (BUN), and elevated blood potassium levels that can lead to cardiac arrest. Anemia usually occurs since the kidneys no longer produce enough erythropoietin (EPO) for adequate red blood cell production. The availability of recombinant EPO has greatly aided renal failure patients. Osteomalacia is another common problem since the kidneys are no longer able to convert vitamin D to the active form (calcitriol) that aids calcium absorption from the small intestine.

POLYCYSTIC KIDNEY DISEASE

Polycystic kidney disease (**PKD**) is the most common inherited disorder of the kidneys. It occurs about once in every thousand births and affects 500,000 people in the United States (about 5 million worldwide). The main cause is a break in chromosome 16 that affects a gene named *PKD1*. The kidney tubules become riddled with hundreds or thousands of cysts (fluid-filled cavities). In addition, inappropriate apoptosis (programmed cell death) of cells in noncystic tubules leads to progressive impairment of renal function and eventually to endstage renal failure.

People with PKD also may exhibit cysts and apoptosis in the liver, pancreas, spleen, and gonads; increased risk of cerebral aneurysms; heart valve defects; and diverticuli in the colon. Typically, symptoms are not noticed until adulthood, when patients may have back pain, urinary tract infections, blood in the urine, hypertension, and large abdominal masses. Aside from renal transplant surgery, there is no specific treatment for PKD. Progression to renal failure may be slowed by using drugs to restore normal blood pressure, restricting protein in the diet, and controlling urinary tract infections.

DIABETES INSIPIDUS

Diabetes insipidus (**DI**) is characterized by excretion of a large volume (5–20 liters per day) of very dilute urine. As might be expected, patients with the disorder exhibit extreme thirst and drink huge volumes of water (polydipsia). The cause of DI is either a defect in production of ADH (central DI) or an insensitivity of the principal cells in distal convoluted tubules and collecting ducts to stimulation by ADH (nephrogenic DI).

Central DI most commonly results from surgical or traumatic injury of the hypothalamic nuclei that synthesize ADH. For example, a fracture at the base of the skull may damage this region. Most cases of nephrogenic DI are associated with advanced kidney disease, but there is also a rare hereditary form. Interestingly, all North American patients with congenital nephrogenic DI are thought to be descendants of a man from Scotland who settled in Nova Scotia in 1761.

Treatment of central DI consists of taking a modified form of ADH as a nasal spray. Patients with nephrogenic DI do not respond to such hormone replacement therapy.

MEDICAL TERMINOLOGY

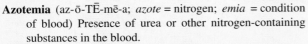

Azotemia (az-ō-TĒ-mē-a; *azote* = nitrogen; *emia* = condition of blood) Presence of urea or other nitrogen-containing substances in the blood.

Cystocele (SIS-tō-sēl; *cyst* = bladder; *cele* = cyst) Hernia of the urinary bladder.

Diuresis (dī-yoo-RĒ-sis; *dia* = through; *urina* = urine) Increased excretion of urine.

Dysuria (dis-YOO-rē-a; *dys* = painful) Painful urination.

Enuresis (en′-yoo-RĒ-sis; *enourein* = to urinate) Bed-wetting; may be due to faulty toilet training, to some psychological or emotional disturbance, or rarely to some physical disor-

der such as diabetes insipidus (lack of ADH). Also referred to as **nocturia**.

Intravenous pyelogram (in′-tra-VĒ-nus PĪ-e-lō-gram′; *intra* = within; *veno* = vein; *pyelos* = pelvis of kidney; *gramma* = record), or **IVP** Radiograph (x-ray) of the kidneys after venous injection of a dye.

Polyuria (pol′-ē-YOO-rē-a; *poly* = much) Excessive urine formation.

Stricture (STRIK-chur) Narrowing of the lumen of a canal or hollow organ, as may occur in the ureter, urethra, or any other tubular structure in the body.

STUDY OUTLINE

INTRODUCTION (p. 848)

1. Substances, organs, and tissues that temporarily confine wastes, transport waste materials for disposal, recycle materials, and excrete excess or toxic substances and wastes are the body buffers, blood, liver, kidneys, lungs, skin, and gastrointestinal tract.
2. The organs of the urinary system are the kidneys, ureters, urinary bladder, and urethra.
3. The kidneys regulate the volume and composition of blood, blood pressure, and some aspects of metabolism.

KIDNEYS (p. 849)

1. The kidneys are retroperitoneal organs attached to the posterior abdominal wall.
2. Three layers of tissue surround the kidneys: renal capsule, adipose capsule, and renal fascia.
3. Internally, the kidneys consist of a renal cortex, renal medulla, renal pyramids, renal papillae, renal columns, calyces, and a renal pelvis.
4. Blood flows into the kidney through the renal artery and successively into segmental, interlobar, arcuate, and interlobular arteries, afferent arterioles, glomerular capillaries, efferent arterioles, peritubular capillaries and vasa recta, and interlobular, arcuate, interlobar, and segmental veins before flowing out of the kidney through the renal vein.
5. Vasomotor nerves from the sympathetic branch of the autonomic nervous system supply kidney blood vessels and help regulate renal blood pressure.
6. The nephron is the functional unit of the kidneys. A nephron consists of a renal corpuscle (glomerulus and glomerular or Bowman's capsule) and a renal tubule.
7. A renal tubule consists of a proximal convoluted tubule, loop of Henle, and distal convoluted tubule, which drains into a collecting duct (shared by several nephrons). The loop of Henle consists of a descending limb of the loop of Henle and ascending limb of the loop of Henle.
8. The filtering unit of a nephron is the endothelial–capsular (filtration) membrane. It consists of the glomerular endothelium, glomerular basement membrane, and slit membranes between pedicels of podocytes.
9. A cortical nephron has its glomerulus in the superficial third of the renal cortex and a short loop that dips only into the superficial region of the renal medulla; a juxtamedullary nephron has its glomerulus deep in the renal cortex near the renal medulla and a long loop of Henle that stretches through the renal medulla almost to the renal papilla.
10. The wall of the entire renal tubule consists of a single layer of epithelial cells and a basement membrane. The epithelium is modified in different portions of the tubule.
11. The juxtaglomerular apparatus (JGA) consists of the juxtaglomerular cells of an afferent arteriole (and sometimes efferent arteriole) and the macula densa of the final portion of the ascending limb of the loop of Henle. It monitors blood pressure and secretes renin when blood pressure falls.

RENAL PHYSIOLOGY (p. 861)

1. Nephrons regulate blood volume and composition and form urine by glomerular filtration, tubular reabsorption, and tubular secretion.

2. Most substances in plasma easily pass through the glomerular filter. However, blood cells and most proteins normally are not filtered.
3. Glomerular filtrate amounts to about 180 liters of fluid per day. This large amount of fluid is filtered because the filter is porous and thin, the glomerular capillaries are long, and the capillary blood pressure is high.
4. One force promotes glomerular filtration: glomerular blood hydrostatic pressure (GBHP). Two forces oppose glomerular filtration: capsular hydrostatic pressure (CHP) and blood colloid osmotic pressure (BCOP). Net filtration pressure (NFP) = GBHP – (CHP + BCOP). NFP is about 10 mm Hg.
5. Filtration fraction is the percentage of plasma entering the nephrons that becomes glomerular filtrate, normally 16–20%.
6. Glomerular filtration rate (GFR) is the amount of filtrate formed in both kidneys per minute; it is normally about 125 ml/min.
7. Glomerular blood flow depends on renal autoregulation, hormonal regulation, and neural regulation.
8. Tubular reabsorption is a selective process that reclaims materials from tubular fluid and returns them to the bloodstream. Reabsorbed substances include water, glucose, amino acids, and ions, such as sodium, chloride, potassium, and bicarbonate. The maximum amount of a substance that can be reabsorbed per unit time is called the transport maximum (T_m).
9. Sodium ions are reabsorbed throughout the renal tubule by primary active transport. This then promotes reabsorption of water by osmosis and other substances by passive diffusion.
10. Nutrients such as glucose and amino acids are reabsorbed in the proximal convoluted tubule by secondary active transport (symporters).
11. The loop of Henle reabsorbs about 30% of the filtered K^+, 20% of the filtered Na^+, 35% of the filtered Cl^-, and 15% of the filtered water.
12. Antidiuretic hormone (ADH) and aldosterone regulate solute and water reabsorption in the final portion of the distal convoluted tubule and the collecting duct.
13. About 90% of the water reabsorption occurs by osmosis, together with reabsorption of solutes such as ions and glucose; this is termed obligatory water reabsorption. The remaining 10% varies according to body needs and is regulated by ADH; this component is termed facultative water reabsorption.
14. Some chemicals not needed by the body are removed from the blood and discharged into the urine by tubular secretion. Included are ions (potassium, hydrogen, ammonium), nitrogen-containing wastes (urea, creatinine), and certain drugs.
15. The kidneys help maintain blood pH by secreting H^+ and increasing or decreasing production of ammonia and bicarbonate ions.

PRODUCTION OF DILUTE AND CONCENTRATED URINE (p. 873)

1. In the absence of ADH, the kidneys produce dilute urine; renal tubules absorb more solutes than water.
2. In the presence of ADH, the kidneys produce concentrated urine; large amounts of water are reabsorbed from the tubular fluid into interstitial fluid, increasing solute concentration of the urine.
3. The countercurrent mechanism establishes an osmotic concentration gradient in the interstitial fluid of the renal

medulla. This enables production of concentrated urine when ADH is present.

EVALUATION OF KIDNEY FUNCTION (p. 877)

1. Renal clearance refers to the ability of the kidneys to clear (remove) a specific substance from blood.
2. In diagnostic testing, the clearance of inulin is equal to the glomerular filtration rate, whereas the clearance of para-aminohippuric acid (PAH) is equal to the renal plasma flow.

DIALYSIS THERAPY (p. 877)

1. Filtering blood through a kidney machine that cleanses the blood of wastes and adds nutrients is called hemodialysis.
2. A portable method of dialysis is called continuous ambulatory peritoneal dialysis (CAPD).

URETERS (p. 878)

1. The ureters are retroperitoneal and consist of a mucosa, muscularis, and adventitia.
2. The ureters transport urine from the renal pelvis to the urinary bladder, primarily by peristalsis.

URINARY BLADDER (p. 880)

1. The urinary bladder is located in the pelvic cavity posterior to the pubic symphysis. Its function is to store urine prior to micturition.
2. Histologically, the urinary bladder consists of a mucosa (with rugae), a muscularis (detrusor muscle), and an adventitia (serosa over superior surface).
3. A lack of control over micturition is called incontinence; failure to void urine completely or normally is termed retention.

URETHRA (p. 881)

1. The urethra is a tube leading from the floor of the urinary bladder to the exterior. Its anatomy and histology differ between females and males.
2. In both sexes the urethra functions to discharge urine from the body; in males it discharges semen as well.

URINALYSIS (p. 882)

1. Urine volume is influenced by blood pressure, blood concentration (blood osmotic pressure), temperature, diuretics, and emotions.
2. The physical characteristics of urine evaluated in a urinalysis are volume, color, odor, turbidity, pH, and specific gravity. See Exhibit 26.2.
3. Chemically, normal urine contains about 95% water and 5% solutes. The solutes normally include urea, creatinine, uric acid, urobilinogen, and various ions. See Exhibit 26.3.
4. Abnormal constituents that can be detected in a urinalysis include albumin, glucose, erythrocytes, leukocytes, ketone bodies, bilirubin, excessive urobilinogen, casts, and microbes. See Exhibit 26.4.

AGING AND THE URINARY SYSTEM (p. 883)

1. After age 40, kidney function decreases.
2. Common problems related to aging include incontinence, urinary tract infections, prostate disorders, and renal calculi.

DEVELOPMENTAL ANATOMY OF THE URINARY SYSTEM (p. 883)

1. The kidneys develop from intermediate mesoderm.
2. They develop in the following sequence: pronephros, mesonephros, metanephros. Only the metanephros remains and develops into a functional kidney.

REVIEW QUESTIONS

1. What organs contribute to the elimination of wastes from the body? (p. 848)
2. What are the functions of the kidneys? What organs compose the urinary system? (p. 848)
3. Describe the location of the kidneys. Why are they said to be retroperitoneal? (p. 849)
4. Prepare a labeled diagram that illustrates the principal external and internal features of the kidney. What is nephroptosis? (p. 849)
5. How are nephrons supplied with blood? Which nerves innervate renal blood vessels? (p. 854)
6. What is a nephron? List and describe the parts of a nephron. (p. 854)
7. Describe the structure of the endothelial–capsular membrane. How is the membrane adapted for filtration? (p. 855)
8. Distinguish between cortical and juxtamedullary nephrons. (p. 857)
9. Describe the histology of the various portions of a nephron. (p. 860)
10. Describe the structure and importance of the juxtaglomerular apparatus (JGA). (p. 861)
11. What is glomerular filtration? Define glomerular filtrate. (p. 861)
12. How are renal corpuscles structurally adapted for filtration? (p. 862)
13. Set up an equation to indicate how net filtration pressure (NFP) is calculated. (p. 863)
14. What are the major chemical differences among plasma, glomerular filtrate, and urine? (p. 866)
15. Define tubular reabsorption. Why is the process physiologically important? (p. 866)
16. What chemical substances are normally reabsorbed by the kidneys? Define transport maximum (T_m). (p. 866)
17. Describe the mechanisms for reabsorption of sodium ions, nutrients, and water by the kidneys. Where does each process occur? (p. 867)
18. Define tubular secretion. Why is it important? List some substances that are secreted. (p. 871)
19. Explain the mechanisms by which the kidneys help control body pH. (p. 871)
20. Describe how the kidneys produce a dilute or concentrated urine. (p. 873)

21. Describe the countercurrent mechanism. Why is it important? (p. 874)
22. Define renal clearance. Why is it important? (p. 877)
23. What is dialysis? Briefly describe hemodialysis using an artificial kidney machine. What is continuous ambulatory peritoneal dialysis (CAPD)? (p. 877)
24. Describe the structure, histology, and function of the ureters. (p. 878)
25. How is the urinary bladder adapted to its storage function? (p. 880)
26. What is micturition? Describe the micturition reflex. (p. 880)
27. Contrast the causes of incontinence and retention. (p. 881)
28. Compare the location, length, and histology of the urethra in the male and female. (p. 881)
29. Describe the effects of blood pressure, blood concentration, and diuretics, on the volume of urine formed. (p. 882)
30. Describe the following physical characteristics of normal urine: color, turbidity, odor, pH, and specific gravity. (p. 882)
31. Describe the chemical composition of normal urine. (p. 882)
32. Define each of the following: albuminuria, glucosuria, hematuria, pyuria, ketosis, bilirubinuria, urobilinogenuria, casts, and renal calculi. (p. 883)
33. Describe the effects of aging on the urinary system. (p. 883)
34. Describe the development of the urinary system. (p. 883)

ANSWERS TO QUESTIONS WITH FIGURES

26.1 Kidneys, ureters, urinary bladder, and urethra.
26.2 They are posterior to the peritoneum.
26.3 Blood and lymphatic vessels, nerves, and ureter.
26.4 The nephron is the functional unit of the kidneys.
26.5 About 1200 ml.
26.6 Leaving.
26.7 Endothelial fenestrations (pores) of the glomerulus.
26.8 Cortical nephrons have glomeruli in the superficial renal cortex and their short loops of Henle penetrate only into the superficial renal medulla; juxtamedullary nephrons have glomeruli deep in the renal cortex and their long loops of Henle extend through the renal medulla nearly to the renal papilla.
26.9 Simple squamous; simple cuboidal with numerous microvilli; simple cuboidal with few microvilli.
26.10 It would tend to decrease NFP by increasing CHP.
26.11 *Auto* means self; it takes place entirely *within* the kidneys.
26.12 Enzymes—renin and ACE; hormones—angiotensin II, aldosterone, and ADH.

26.13 Diffusion through the brush border into the tubule cell.
26.14 Enters via a Na^+ symporter; leaves via facilitated diffusion.
26.15 Because the symporter uses the energy stored in the concentration difference of Na^+ in extracellular fluid versus cytosol. No, because the thick ascending limb of the loop of Henle is almost impermeable to water.
26.16 Aldosterone and atrial natriuretic peptide.
26.17 It would reduce secretion of H^+ into the urine and reduce reabsorption of Na^+ and HCO_3^- into the blood. See Diuretics on page 877.
26.18 HCO_3^-, NH_3, and HPO_4^{2-}.
26.19 Thick ascending limb of the loop of Henle, distal convoluted tubule, and collecting duct.
26.20 Mainly Na^+, Cl^-, and urea.
26.22 They remain in the blood because they are too large to pass through the pores in the dialysis membrane.
26.23 Incontinence.

CHAPTER 27

FLUID, ELECTROLYTE, AND ACID–BASE HOMEOSTASIS

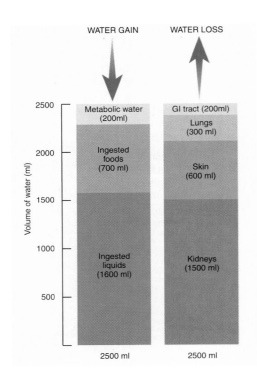

WATER GAIN   WATER LOSS

CHAPTER CONTENTS AT A GLANCE

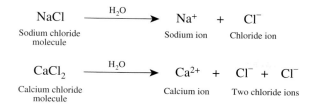

Because glucose does not dissociate when dissolved in water, a molecule of glucose contributes only one particle to the solution. Sodium chloride, on the other hand, contributes two ions, or particles, and calcium chloride contributes three. For example, a 5 mmol/liter solution of $CaCl_2$ has an osmolarity of 15 mOsm/liter (if the salt dissociates completely). On the other hand, a 5 mmol/liter solution of glucose has an osmolarity of only 5 mOsm/liter. Thus each mole of calcium chloride has three times as great an osmotic effect as each mole of glucose. Osmotic pressure is usually expressed in mm Hg. Each 1 mOsm/liter exerts a pressure equal to 19.3 mm Hg.

ELECTROLYTES IN BODY FLUIDS

The ions formed when electrolytes dissolve serve four general functions in the body. (1) Because they are largely confined to particular body fluid compartments and are more numerous than nonelectrolyes, certain ions control the osmosis of water between body compartments. (2) Ions help maintain the acid–base balance required for normal cellular activities. (3) Ions carry electrical current, which allows production of action potentials and graded potentials and controls secretion of some hormones and neurotransmitters. (4) Several ions are cofactors needed for optimal activity of enzymes.

Concentrations of Electrolytes in Body Fluids

Figure 27.4 compares the concentrations of electrolytes and protein anions in plasma, interstitial fluid, and intracellular

Figure 27.4 Comparison of electrolyte and protein anion concentrations in the body fluid compartments—plasma, interstitial fluid, and intracellular fluid. The height of each column represents the milliequivalents/liter (mEq/liter).

🔑 *In each body fluid compartment, the total milliequivalents (charges) of positive and negative electrolytes balance.*

Key to symbols:
Na^+ Sodium
K^+ Potassium
Ca^{2+} Calcium
Mg^{2+} Magnesium
HCO_3^- Bicarbonate
Cl^- Chloride
SO_4^{2-} Sulfate
HPO_4^{2-} Phosphate
H_2CO_3 Carbonic acid

❓ What is the major cation and the two major anions in ECF? In ICF?

fluid. The chief difference between plasma and interstitial fluid is that plasma contains quite a few protein anions, whereas interstitial fluid has hardly any. Since normal capillary membranes are virtually impermeable to protein, only a few plasma proteins leak out of blood vessels into the interstitial fluid. Plasma also contains slightly more Na^+ but less Cl^- than the interstitial fluid. In other respects the two fluids are similar.

The electrolyte content of intracellular fluid, however, differs considerably from that of extracellular fluid. In extracellular fluid, the most abundant cation is Na^+, and the most abundant anion is Cl^-. In intracellular fluid, the most abundant cation is K^+, and the most abundant anions are proteins and phosphates (HPO_4^{2-}).

Sodium

Sodium (Na^+), the most abundant extracellular ion, represents about 90% of extracellular cations. Normal plasma (serum) Na^+ concentration is 136–142 mEq/liter. Na^+ has a pivotal role in fluid and electrolyte balance because it accounts for almost half of the osmolarity (140 of about 300 mOsm/liter) of extracellular fluid (ECF). The flow of Na^+ through voltage-gated channels in the plasma membrane also is necessary for the generation and conduction of action potentials (impulses) in nervous and muscle tissue. The average daily intake of Na^+ far exceeds the body's normal daily requirements due largely to excess dietary salt. The kidneys excrete excess Na^+ but also can conserve it during periods of shortage.

The Na^+ level in the blood is controlled by aldosterone, antidiuretic hormone (ADH), and atrial natriuretic peptide (ANP). Aldosterone, secreted by the cortex of the adrenal glands, acts on the distal convoluted tubules and collecting ducts of the nephrons of the kidneys and causes them to increase their reabsorption of Na^+. As Na^+ moves from the filtrate back into the blood, it establishes an osmotic gradient. This causes water to follow Na^+ from the filtrate back into the blood. Aldosterone is secreted in response to reduced blood volume or cardiac output, decreased extracellular Na^+ concentration, and increased extracellular K^+ concentration. When the blood concentration of Na^+ drops below about 135 mEq/liter, ADH release from the posterior pituitary gland ceases. The lack of ADH, in turn, permits greater excretion of water in urine and restoration of the normal Na^+ level in ECF. The atria of the heart produce the hormone atrial natriuretic peptide (ANP), which increases Na^+ and water excretion by the kidneys when Na^+ level is too high.

HYPONATREMIA AND HYPERNATREMIA

Sodium loss from the body may occur through excessive perspiration, vomiting, or diarrhea; taking certain diuretics; and burns. Such a loss can result in **hyponatremia** (hī-pō-na-TRĒ-mē-a; *natrium* =

sodium), a lower-than-normal blood Na^+ level. Signs of hyponatremia include muscular weakness, dizziness, headache, hypotension, tachycardia, and shock. Severe sodium loss can result in mental confusion, stupor, and coma. Osmosis of water from the ECF into body cells, including those of the nervous system, causes these symptoms (see Fig. 27.5). **Hypernatremia** is a higher-than-normal blood sodium level. It may occur with water loss, water deprivation, or sodium gain. Since sodium is the major determinant of the osmotic pressure of ECF, hypernatremia causes hypertonicity of ECF. As a result, water moves out of body cells into ECF, resulting in cellular dehydration. Symptoms include intense thirst, fatigue, hypertension, restlessness, agitation, and coma. ■

Chloride

Chloride ions (Cl^-) are the most prevalent extracellular anions. Normal plasma (serum) Cl^- concentration is 95–103 mEq/liter. Cl^- diffuses relatively easily between the extracellular and intracellular compartments because most plasma membranes contain quite a few Cl^- leakage channels. For this reason, Cl^- can help balance the level of anions in different body fluid compartments. One example is the chloride shift that occurs between the intracellular fluid of red blood cells and plasma as the blood level of carbon dioxide either increases or decreases (see Fig. 23.23). In this case, the exchange of Cl^- for HCO_3^- maintains the correct balance of anions in ECF and ICF. The gastric mucosal glands secrete both Cl^- and H^+ to form hydrochloric acid.

Aldosterone indirectly adjusts Cl^- balance in body fluids because it regulates reabsorption of Na^+ in distal portions of the renal tubules. In many cases, Cl^- passively follows Na^+ due to the electrical attraction.

HYPOCHLOREMIA AND HYPERCHLOREMIA

An abnormally low level of Cl^- in the blood, called **hypochloremia** (hī-pō-klō-RĒ-mē-a), may be caused by excessive vomiting, dehydration, and therapy with certain diuretics such as furosemide (Lasix®). Symptoms include muscle spasms, alkalosis, depressed respirations, and even coma. Elevated blood Cl^- level is termed **hyperchloremia** and may result from either dehydration, due to water loss or water deprivation, or chloride gain. Severe renal failure, hyperaldosteronism, certain types of acidosis (elevated blood concentration of H^+), and some drugs can cause hyperchloremia. ■

Potassium

Potassium ions (K^+) are the most abundant cations in intracellular fluid. K^+ plays a key role in establishing the resting membrane potential and in the repolarization phase of action potentials in nervous and muscle tissue. K^+ also helps maintain fluid volume in cells. When K^+ moves in or out of cells, it tends to be exchanged for H^+. This shift of H^+ helps regulate pH.

Normal plasma (serum) K^+ concentration is 3.8–5.0 mEq/liter. The plasma level of K^+ is controlled mainly by aldosterone. When plasma K^+ concentration is high, more aldosterone is secreted into the blood. Aldosterone then stimulates secretion of K^+ into the urine so that more K^+ is lost from the body. When plasma K^+ concentration is low, on the other hand, aldosterone secretion decreases and less K^+ is excreted in urine. Abnormal plasma K^+ levels adversely affect neuromuscular and cardiac function.

HYPOKALEMIA AND HYPERKALEMIA

A lower-than-normal level of K^+, called **hypokalemia** (hī-pō-ka-LĒ-mē-a; *kalium* = potassium), may result from vomiting, diarrhea, high sodium intake, kidney disease, and therapy with some diuretics. Symptoms include cramps and fatigue, flaccid paralysis, nausea, vomiting, mental confusion, increased urine output, shallow respirations, and changes in the electrocardiogram, including a lengthening of the Q–T interval and flattening of the T wave.

A higher-than-normal blood K^+ level, called **hyperkalemia**, is characterized by irritability, anxiety, abdominal cramping, diarrhea, weakness (especially of the lower limbs), and paresthesia (abnormal sensation, such as burning or prickling). Hyperkalemia can cause death by inducing fibrillation of the heart. ■

Bicarbonate

Bicarbonate ions (HCO_3^-) are the second most prevalent extracellular anions. Normal plasma (serum) HCO_3^- concentration is 22–26 mEq/liter in arterial blood and 19–24 mEq/liter in venous blood. HCO_3^- concentration decreases as blood flows through systemic capillaries because it is the major buffer of H^+ in plasma. Some of the HCO_3^- is used up as it reacts with H^+, released by tissue cells, to form carbonic acid, a reaction that is catalyzed by carbonic anhydrase. As blood flows through pulmonary capillaries, however, the concentration of HCO_3^- increases again (Fig. 23.23 shows these reactions). Intracellular fluid also contains a small amount of HCO_3^-. As mentioned earlier, the exchange of Cl^- for HCO_3^- helps maintain the correct balance of anions in ECF and ICF.

The kidneys are the main regulators of blood HCO_3^- concentration. They can form HCO_3^- and release it into the blood when the level is low (see Fig. 26.18) and excrete excess HCO_3^- in the urine when the blood level is too high. Changes in the blood level of HCO_3^- are considered later in the chapter in the section on acid–base balance.

Calcium

Because such a large amount is stored in bone, calcium is the most abundant mineral in the body. About 98% of the calcium in an adult is in the skeleton (and teeth), combined with phosphates to form a crystal lattice of mineral salts. In body fluids, calcium is principally an extracellular cation (Ca^{2+}). The normal concentration of free or unattached Ca^{2+} in plasma is 4.6–5.5 mEq/liter. A roughly equal amount of Ca^{2+} is attached to various plasma proteins. Besides contributing to the hardness of bones and teeth, Ca^{2+} plays important roles in blood clotting, neurotransmitter release, maintenance of muscle tone, and excitability of nervous and muscle tissue.

The level of Ca^{2+} in plasma is regulated principally by parathyroid hormone (PTH) and calcitonin (CT) (see Fig. 18.17). A low plasma Ca^{2+} level promotes release of more PTH. PTH stimulates osteoclasts in bone tissue to release calcium (and phosphate) from mineral salts of bone matrix. Thus PTH increases bone *resorption*. PTH also increases Ca^{2+} *absorption* from the gastrointestinal tract and enhances *reabsorption* of Ca^{2+} from glomerular filtrate through renal tubule cells and back into blood. Calcitonin, produced by the thyroid gland, is released in increased quantities when plasma Ca^{2+} level is high. It decreases blood Ca^{2+} by stimulating osteoblasts and inhibiting osteoclasts. In the presence of calcitonin, osteoblasts remove calcium (and phosphate) ions from plasma and deposit them in bone matrix as calcium phosphate salts.

HYPOCALCEMIA AND HYPERCALCEMIA

An abnormally low level of Ca^{2+} is called **hypocalcemia** (hī-pō-kal-SĒ-mē-a). It may be due to increased calcium loss, reduced calcium intake, elevated levels of phosphate (as one goes up, the other goes down), or altered regulation as might occur in hypoparathyroidism. Hypocalcemia is characterized by numbness and tingling of the fingers, hyperactive reflexes, muscle cramps, tetany, and convulsions. Bone fractures may also occur. In addition, hypocalcemia may cause spasms of laryngeal muscles that can cause death by asphyxiation.

Characteristics of **hypercalcemia**, an abnormally high level of Ca^{2+}, are lethargy, weakness, anorexia, nausea, vomiting, polyuria, itching, bone pain, depression, confusion, paresthesia, stupor, and coma. ■

Phosphate

About 85% of the phosphate in an adult is present as calcium phosphate salts, which are structural components of bone and teeth. The remaining 15% is ionized. Mostly, phosphate ions are combined with lipids (phospholipids), proteins, carbohydrates, nucleic acids (DNA and RNA), and adenosine triphosphate (ATP) inside cells. The three phosphate ions ($H_2PO_4^-$, HPO_4^{2-}, and PO_4^{3-}) are important intracellular anions. At the normal pH of body fluids, however, HPO_4^{2-} is the most prevalent form. Also, $H_2PO_4^-$ and HPO_4^{2-} play an important role in buffering reactions (the phosphate buffer system is discussed later in the chapter).

The normal plasma (serum) concentration of ionized phosphate is only 1.7–2.6 mEq/liter. The level of HPO_4^{2-} in blood plasma is regulated by PTH and calcitonin. PTH stimulates osteoclasts to release phosphate from mineral salts of bone matrix and causes renal tubular cells to excrete phosphate ions. Calcitonin lowers blood phosphate level by stimulating osteoblasts and inhibiting osteoclasts. In the presence of calcitonin, osteoblasts remove phosphate ions from blood and deposit them in bone matrix, where they combine with calcium ions to form the mineral salts of bone matrix.

HYPOPHOSPHATEMIA AND HYPERPHOSPHATEMIA

An abnormally low level of phosphate, called **hypophosphatemia** (hī′-pō-fos′-fa-TĒ-mē-a), may occur through increased urinary losses, decreased intestinal absorption, or increased utilization. Hypophosphatemia is characterized by confusion, seizures, coma, chest and muscle pain, increased susceptibility to infection, numbness and tingling of the fingers, uncoordination, memory loss, and lethargy.

Hyperphosphatemia occurs most often when the kidneys fail to excrete excess phosphate, as may occur in renal failure. It also can result from increased intake of phosphates or destruction of cells, which releases phosphate into the blood. The primary complication of hyperphosphatemia is the precipitation of calcium phosphate in soft tissues, joints, and arteries. Symptoms of hyperphosphatemia include anorexia, nausea, vomiting, muscular weakness, hyperactive reflexes, tetany, and tachycardia. ■

Magnesium

In an adult, about 54% of the total body magnesium is deposited in bone matrix as magnesium salts. The remaining 46% occurs as magnesium ions (Mg^{2+}) in intracellular fluid (45%) and extracellular fluid (1%). Mg^{2+} is the second most common intracellular cation (after K^+). Functionally, Mg^{2+} is a cofactor for enzymes involved in the metabolism of carbohydrates and proteins and Na^+/K^+ ATPase (the sodium pump enzyme). Mg^{2+} is also important in neuromuscular activity, impulse transmission, and myocardial functioning.

Normal plasma (serum) Mg^{2+} concentration is low, only 1.3–2.1 mEq/liter. Several factors regulate the blood level of Mg^{2+} by varying the rate at which it is excreted in the urine. The kidneys increase urinary excretion of Mg^{2+} in hypercalcemia, hypermagnesemia, an increase in extracellular fluid volume, a decrease in parathyroid hormone (PTH), and acidosis. The opposite conditions decrease renal excretion of Mg^{2+}.

HYPOMAGNESEMIA AND HYPERMAGNESEMIA

Magnesium deficiency, called **hypomagnesemia** (hī′-pō-mag′-ne-SĒ-mē-a), may be caused by inadequate absorption, diarrhea, alco-holism, malnutrition, excessive lactation, diabetes mellitus, and diuretic therapy. The condition is characterized by weakness, irritability, tetany, delirium, convulsions, confusion, anorexia, nausea, vomiting, paresthesia, and cardiac arrhythmias.

Hypermagnesemia, or magnesium excess, occurs mainly in persons with renal failure or those who have an increased intake of magnesium, for example, magnesium-containing antacids. Other causes include Addison's disease, acute diabetic acidosis, severe dehydration, and hypothermia. The condition is characterized by hypotension, muscular weakness or paralysis, nausea, vomiting, and altered mental functioning. ■

MOVEMENT OF BODY FLUIDS

Blood is the vehicle for transport and exchange of materials between body cells and the outside world. Nutrients in food enter the blood for distribution to tissues throughout the body. Oxygen enters the lungs and then the blood. At the same time, waste products generated by cellular metabolism diffuse from the cells that produce them into the bloodstream. From blood, wastes may be excreted into urine, exhaled by the lungs, or follow some other route out of the body. Interstitial fluid, on the other hand, is the go-between for exchanges between intracellular fluid and blood plasma.

Exchange Between Plasma and Interstitial Fluid

The movement of substances between plasma and interstitial fluid occurs across capillary walls. Capillary exchange was discussed in detail in Chapter 21 on page 617, but we will recap the important points here. Substances enter and leave capillaries in three ways: (1) vesicular transport, (2) diffusion, and (3) bulk flow (filtration and reabsorption).

In vesicular transport or transcytosis; substances in blood plasma cross the capillary wall first by endocytosis into an endothelial cell and then by exocytosis into interstitial fluid. This method accounts for only a tiny fraction of the exchange between plasma and interstitial fluid.

Most substances in blood or interstitial fluid can cross capillary walls by diffusion. This process accounts for the largest part of capillary exchange in most body tissues. One exception is in the brain, where the blood–brain barrier blocks diffusion of many substances, especially those that are not lipid-soluble.

Bulk flow consists of both filtration (net movement of materials from blood into interstitial fluid) and reabsorption (net movement of materials from interstitial fluid into blood). Filtration predominates at the arteriolar ends of capillaries whereas reabsorption predominates at the venular ends of capillaries. In most capillaries slightly more fluid is filtered than is reabsorbed. (To review the forces that govern filtration and reabsorption, see page 618 and Fig. 21.7.)

The fluid not reabsorbed (about 3 liters/day from all capillaries excluding those in the glomeruli of the kidneys) and the small quantity of proteins that escape from plasma pass into lymphatic capillaries. From here, the fluid (lymph) moves through lymphatic vessels to the thoracic duct or right lymphatic duct, where lymph drains into the subclavian veins. Occasionally, the near balance of filtration and reabsorption between plasma and interstitial fluid is disrupted and interstitial fluid volume increases. An abnormal increase in the volume of interstitial fluid is called **edema**.

Exchange Between Interstitial and Intracellular Fluids

Because intracellular fluid and interstitial fluid normally have the same osmotic pressures, cells neither shrink nor swell. A fluid imbalance between these two compartments can be caused by a change in their osmotic pressures. Most often an osmotic pressure change is due to a change in the concentration of Na^+, which is the principal cation outside cells, or the concentration of K^+, which is the principal cation inside cells. Recall that aldosterone, ANP, and ADH all regulate Na^+ balance in the body whereas aldosterone is the main regulator of K^+ balance.

A decrease in the osmotic pressure of interstitial fluid normally shuts down secretion of ADH. Normally functioning kidneys then excrete excess water in the urine. When renal function is poor, however, a decrease in the osmotic pressure of interstitial fluid can produce two very serious results—water intoxication or circulatory (hypovolemic) shock (Fig. 27.5):

1. Both body water and Na^+ are lost during excessive sweating, vomiting, or diarrhea. If the lost fluid is replaced by drinking plain water, body fluids become more dilute.
2. This dilution can cause the Na^+ concentration of plasma and then of interstitial fluid to fall below the normal range (hyponatremia).
3. When its Na^+ concentration decreases, the interstitial fluid's osmotic pressure also falls; it becomes hypotonic.
4. The result is net osmosis of water from the hypotonic interstitial fluid into intracellular fluid.
5. This osmosis-driven water movement has two serious consequences. The first effect is to increase tonicity of the interstitial fluid as its water content and volume decreases. The second effect is to decrease the tonicity of intracellular fluid.
6. Increased tonicity of interstitial fluid causes water to move from plasma into interstitial fluid. As water enters cells, they become hypotonic and swell, a condition called **water intoxication**.

Figure 27.5 Relation between Na^+ imbalance and water imbalance.

🔑 ***Na^+ loss results in water loss.***

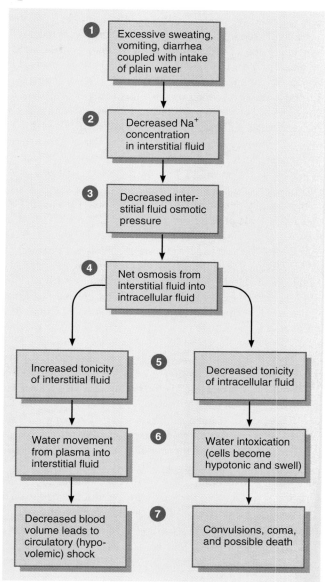

❓ What effect will step ③ have on secretion of ADH and thus on urine output?

7. As water moves out of plasma, blood volume decreases, which may lead to circulatory (hypovolemic) shock. Among the symptoms of water intoxication are convulsions, coma, and possible death.

To prevent this sequence of events in cases of severe electrolyte and water loss, solutions given for intravenous or oral rehydration therapy (ORT) include a small amount of table salt (NaCl).

ACID–BASE BALANCE

From our discussion thus far, you can see that various ions play different roles in helping to maintain homeostasis. A very important ion in terms of the body's acid–base balance is the hydrogen ion (H^+). A major challenge to homeostasis is keeping the H^+ concentration at an appropriate level to maintain proper acid–base balance of body fluids. This is of critical importance because the three-dimensional shape of all body proteins, which enables them to perform specific functions, is very sensitive to changes in concentration of H^+. Although some H^+ enter the body in ingested foods, most are produced as a result of the cellular metabolism of substances such as glucose, fatty acids, and especially amino acids. (You might want to review the discussion of acids, bases, and pH in Chapter 2.)

In a healthy person, the pH of ECF remains between 7.35 and 7.45. A pH of 7.4 corresponds to a H^+ concentration of 0.04 mEq/liter. Metabolic reactions often produce a huge excess of H^+. If there were no mechanisms for disposal of H^+, its rising concentration in body fluids would quickly lead to death. Homeostasis of H^+ concentration within a narrow pH range is essential to survival and depends on three major mechanisms that cooperate to remove H^+ from body fluids and then eliminate them from the body. The mechanisms are:

1. **Buffer systems.** Buffers act quickly to temporarily bind H^+, which removes the highly reactive, excess H^+ from solution but not from the body.
2. **Exhalation of carbon dioxide.** By increasing the rate and depth of breathing, more carbon dioxide can be exhaled. In minutes, this reduces the level of carbonic acid, which raises blood pH (reduces blood H^+ level).
3. **Kidney excretion.** The slowest mechanism, taking hours or days, but the only way to eliminate acids other than carbonic acid is through their passage into urine and their excretion by the kidneys.

Buffer Systems

Most buffer systems of the body consist of a weak acid and an anion of that acid, which functions as a weak base. Buffers function to prevent rapid, drastic changes in the pH of a body fluid by changing strong acids and bases into weak acids and bases. Buffers work within fractions of a second. A strong acid dissociates into H^+ more easily than does a weak acid. Strong acids therefore lower pH more than weak ones because strong acids contribute more H^+. Similarly, strong bases raise pH more than weak ones because strong bases dissociate more easily into hydroxide ions (OH^-). The principal buffer systems of the body fluids are the protein buffer system, the carbonic acid–bicarbonate system, and the phosphate system.

Protein Buffer System

The **protein buffer system** is the most abundant buffer in body cells and plasma. The protein hemoglobin is an especially good buffer within red blood cells. Proteins are composed of amino acids. An amino acid is an organic compound that contains at least one carboxyl group (—COOH) and at least one amino group (—NH_2). The components of the protein buffer system are the *carboxyl group(s)* and *amino group(s)*. The free carboxyl group at one end of a protein acts like an acid by releasing H^+ when pH rises and can dissociate in this way:

$$
\underset{\underset{H}{|}}{\overset{\overset{R}{|}}{NH_2-C-COOH}} \longrightarrow \underset{\underset{H}{|}}{\overset{\overset{R}{|}}{NH_2-C-COO^-}} + H^+
$$

The H^+ is then able to react with any excess OH^- in the solution to form water.

The free amino group at the other end of a protein can act as a base by combining with hydrogen ions when pH falls as follows:

$$
\underset{\underset{H}{|}}{\overset{\overset{R}{|}}{NH_2-C-COOH}} + H^+ \longrightarrow \underset{\underset{H}{|}}{\overset{\overset{R}{|}}{{}^+NH_3-C-COOH}}
$$

Thus proteins can buffer both acids and bases. Besides the terminal carboxyl and amino groups, side chains that can buffer H^+ are present on seven of the twenty amino acids.

At blood pH (7.4) the two most important amino acid buffers are histidine and cysteine. Proteins that are rich in these two amino acids are especially effective buffers. For example, each molecule of hemoglobin, a good buffer, contains 37 histidines. Hemoglobin is very effective in buffering H^+ in red blood cells. As blood flows through the systemic capillaries, carbon dioxide (CO_2) passes from tissue cells into red blood cells where it combines with water (H_2O) to form carbonic acid. Once formed, H_2CO_3 dissociates into H^+ and HCO_3^-. At the same time CO_2 is entering red blood cells, oxyhemoglobin ($Hb \cdot O_2$) is giving up its oxygen to tissue cells. Reduced hemoglobin (deoxyhemoglobin) is an excellent buffer of H^+, so it picks up most of the H^+. For this reason, reduced hemoglobin usually is written as $Hb \cdot H$. The following reactions summarize these relations:

$$
\underset{\text{Water}}{H_2O} + \underset{\substack{\text{Carbon dioxide} \\ \text{(entering RBCs)}}}{CO_2} \longrightarrow \underset{\text{Carbonic acid}}{H_2CO_3}
$$

$$
\underset{\text{Carbonic acid}}{H_2CO_3} \longrightarrow \underset{\text{Hydrogen ion}}{H^+} + \underset{\text{Bicarbonate ion}}{HCO_3^-}
$$

$$
\underset{\substack{\text{Oxyhemoglobin} \\ \text{(in RBCs)}}}{Hb \cdot O_2} + \underset{\substack{\text{Hydrogen ion} \\ \text{(from carbonic acid)}}}{H^+} \longrightarrow \underset{\substack{\text{Reduced} \\ \text{hemoglobin}}}{Hb \cdot H} + \underset{\substack{\text{Oxygen} \\ \text{(released to} \\ \text{tissue cells)}}}{O_2}
$$

Carbonic Acid–Bicarbonate Buffer System

The **carbonic acid–bicarbonate buffer system** is based on the *bicarbonate ion* (HCO_3^-), which can act as a weak base, and *carbonic acid* (H_2CO_3), which can act as a weak acid. HCO_3^- is a significant anion in both intracellular and extracellular fluids (see Fig. 27.4) and the constant release of CO_2 during cellular respiration produces H_2CO_3. If there is an excess of H^+, HCO_3^- can function as a weak base and remove the excess H^+ as follows:

$$H^+ \quad + \quad HCO_3^- \quad \longrightarrow \quad H_2CO_3$$

Hydrogen ion   Bicarbonate ion   Carbonic acid
             (weak base)

In the lungs, then, H_2CO_3 dissociates into water and carbon dioxide and the CO_2 is exhaled.

On the other hand, if there is a shortage of H^+, H_2CO_3 can function as a weak acid and provide H^+ as follows:

$$H_2CO_3 \quad \longrightarrow \quad H^+ \quad + \quad HCO_3^-$$

Carbonic acid   Hydrogen ion   Bicarbonate ion
(weak acid)

When the diet contains a large amount of protein, as is typical in North America, normal metabolism produces more acids than bases and thus tends to acidify the blood rather than make it more alkaline. Accordingly, there must be more HCO_3^- than H_2CO_3. When extracellular pH is normal (7.4), HCO_3^- concentration is about 24 mEq/liter whereas H_2CO_3 concentration is about 1.2 mEq/liter. Thus bicarbonate ions outnumber carbonic acid molecules 20:1. Because CO_2 and H_2O combine to form H_2CO_3, this buffer system cannot protect against pH changes due to respiratory problems in which there is an excess or shortage of CO_2.

Phosphate Buffer System

The **phosphate buffer system** acts in essentially the same manner as the carbonic acid–bicarbonate buffer system. The components of the phosphate buffer system are the *dihydrogen phosphate ion* ($H_2PO_4^-$) and the *monohydrogen phosphate ion* (HPO_4^{2-}). Recall that phosphates are major anions in intracellular fluid and minor ones in extracellular fluids (see Fig. 27.4). The dihydrogen phosphate ion acts as a weak acid and is capable of buffering strong bases such as OH^-.

$$OH^- \quad + \quad H_2PO_4^- \quad \longrightarrow \quad H_2O \quad + \quad HPO_4^{2-}$$

Hydroxide ion   Dihydrogen   Water   Monohydrogen
(strong base)   phosphate   phosphate
         (weak acid)   (weak base)

On the other hand, the monohydrogen phosphate ion acts as a weak base and is capable of buffering the H^+ released by a strong acid such as hydrochloric acid (HCl).

$$H^+ \quad + \quad HPO_4^{2-} \quad \longrightarrow \quad H_2PO_4^-$$

Hydrogen ion   Monohydrogen   Dihydrogen
(strong acid)   phospate   phosphate
         (weak base)   (weak acid)

Since the concentration of phosphates is highest in intracellular fluid, the phosphate buffer system is an important regulator of pH in the cytosol. It also is present at a lower level in extracellular fluids and acts to buffer acids in urine. $H_2PO_4^-$ is formed when excess H^+ in the kidney tubule fluid combines with HPO_4^{2-} (see Fig. 26.18). The H^+ that becomes part of the $H_2PO_4^-$ passes into the urine. This reaction is one way in which the kidneys help maintain blood pH by excreting H^+ in the urine.

Exhalation of Carbon Dioxide

Breathing also plays a role in maintaining the pH of body fluids. An increase in the carbon dioxide (CO_2) concentration in body fluids increases H^+ concentration and thus lowers the pH (makes it more acidic). Because H_2CO_3 can be eliminated by exhaling CO_2, it is called a **volatile acid.** Conversely, a decrease in the CO_2 concentration of body fluids raises the pH (makes it more basic). This is illustrated by the following reactions:

$$CO_2 \; + \; H_2O \; \rightleftharpoons \; H_2CO_3 \; \rightleftharpoons \; H^+ \; + \; HCO_3^-$$

Carbon   Water   Carbonic   Hydrogen   Bicarbonate
dioxide   acid   ion   ion

The pH of body fluids may be adjusted, usually in 1–3 minutes, by a change in the rate and depth of breathing. With increased ventilation, more CO_2 is exhaled, the reaction just given is driven to the left, H^+ concentration falls, and the blood pH rises. Doubling the ventilation increases the pH by about 0.23, from 7.4 to 7.63. If ventilation slows, less carbon dioxide is exhaled, and the blood pH falls. Reducing ventilation to one-quarter of normal lowers the pH by 0.4, from 7.4 to 7.0. These examples show the powerful effect of alterations in breathing on pH of body fluids.

The pH of body fluids, in turn, affects the rate and depth of breathing (Fig. 27.6). If, for example, the blood becomes more acidic, the decrease in pH or increase in concentration of H^+ (controlled condition) is detected by central chemoreceptors in the medulla oblongata and peripheral chemoreceptors in the aortic and carotid bodies (receptors) that stimulate the inspiratory area in the medulla oblongata (control center). As a result, the diaphragm and other muscles of respiration (effectors) contract more forcefully and frequently so more CO_2 is exhaled. As less H_2CO_3 forms and fewer H^+ are present, blood pH increases (response). When the response brings blood pH (H^+ concentration) back to normal, there is a return to homeostasis. The same negative feedback loop occurs if the blood level of CO_2 increases. Ventilation increases, which removes more CO_2 from blood and reduces the H^+ concentration; thus blood pH increases.

On the other hand, if the pH of the blood increases, the respiratory center is inhibited and ventilation decreases. A decrease in the CO_2 concentration of blood has the same effect. When ventilation decreases, CO_2 accumulates in blood

Figure 27.6 Negative feedback regulation of blood pH by the respiratory system.

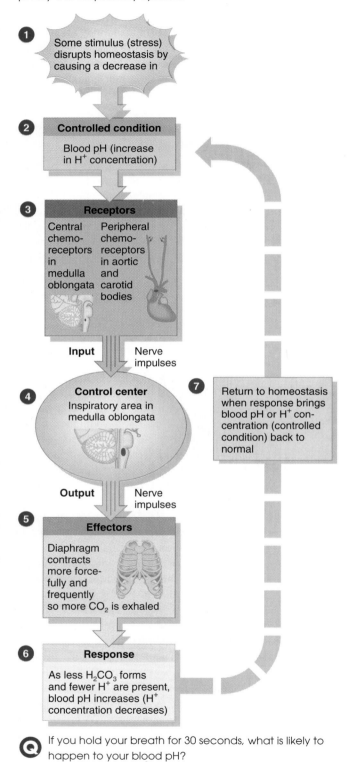

1. Some stimulus (stress) disrupts homeostasis by causing a decrease in

2. **Controlled condition**

   Blood pH (increase in H⁺ concentration)

3. **Receptors**

   Central chemo-receptors in medulla oblongata

   Peripheral chemo-receptors in aortic and carotid bodies

   Input — Nerve impulses

4. **Control center**

   Inspiratory area in medulla oblongata

   Output — Nerve impulses

5. **Effectors**

   Diaphragm contracts more force-fully and frequently so more CO_2 is exhaled

6. **Response**

   As less H_2CO_3 forms and fewer H⁺ are present, blood pH increases (H⁺ concentration decreases)

7. Return to homeostasis when response brings blood pH or H⁺ con-centration (controlled condition) back to normal

Q If you hold your breath for 30 seconds, what is likely to happen to your blood pH?

and the H⁺ concentration increases. This respiratory mechanism is a powerful eliminator of acid, but it can get rid of only the single volatile acid—carbonic acid.

Kidney Excretion of H⁺

On a daily basis metabolic reactions produce nonvolatile or **fixed acids** at a rate of about 1 mEq/liter of H⁺ for every kilogram of body weight. The normal concentration of H⁺ in body fluids is only 0.04 mEq/liter. The only way to eliminate this huge fixed acid load is to excrete H⁺ in the urine. The kidneys also synthesize new HCO_3^- and reabsorb filtered HCO_3^- so this important buffer is not lost in the urine. Given these contributions to acid–base balance, it's not surprising that renal failure can quickly cause death. Since the role of the kidneys in maintaining pH has already been discussed in Chapter 26, we simply refer you to that chapter. Carefully review Figs. 26.17 and 26.18.

Exhibit 27.2 presents a summary of the mechanisms that maintain pH of body fluids.

ACID–BASE IMBALANCES

The normal pH range of systemic arterial blood is between 7.35 (= 0.045 mEq of H⁺/liter) and 7.45 (= 0.035 mEq of H⁺/liter). **Acidosis** (or **acidemia**) is a condition in which blood pH is below 7.35. **Alkalosis** (or **alkalemia**) is a condition in which blood pH is higher than 7.45.

EXHIBIT **27.2** **MECHANISMS THAT MAINTAIN pH OF BODY FLUIDS**

Mechanism	Comments
BUFFER SYSTEMS	Most consist of a weak acid and the salt of that acid, which functions as a weak base. They prevent drastic changes in body fluid pH.
Proteins	Most abundant buffers in body cells and blood. Histidine and cysteine are the two amino acids that contribute most of the buffering capacity of proteins. Hemoglobin inside red blood cells is a good buffer.
Carbonic acid– bicarbonate	Important regulator of blood pH. Most abundant buffers in extracellular fluid (ECF).
Phosphates	Important buffers inside cells and in urine.
EXHALATION OF CO_2	With increased exhalation of CO_2, pH rises (less H⁺). With decreased exhalation of CO_2, pH falls (more H⁺).
KIDNEYS	Renal tubules secrete H⁺ into the urine and reabsorb HCO_3^- so it is not lost in the urine.

A change in blood pH that leads to acidosis or alkalosis may be compensated. **Compensation** is the physiological response to an acid–base imbalance that attempts to normalize arterial blood pH. It may be either *complete,* if pH indeed is brought within the normal range, or *partial,* if pH is still lower than 7.35 or higher than 7.45. If a person has an altered pH due to metabolic causes, hyperventilation or hypoventilation can help bring blood pH back toward the normal range. Because it involves the respiratory system, this form of compensation is termed **respiratory compensation.** Respiratory compensation occurs within minutes and reaches its maximum within hours. On the other hand, if a person has an altered pH due to respiratory causes, renal mechanisms (mainly changes in secretion of H^+ and reabsorption of HCO_3^-), can help compensate for the alteration. This form of compensation is termed **metabolic compensation** because it doesn't involve the respiratory system. Metabolic compensation may begin in minutes but takes days to reach maximum effectiveness.

Physiological Effects of Acidosis and Alkalosis

The principal physiological effect of acidosis is depression of the central nervous system through depression of synaptic transmission. If the blood pH falls below 7, depression of the nervous system is so severe that the individual becomes disoriented and comatose and dies. Patients with severe acidosis usually die in a state of coma. On the other hand, a major physiological effect of alkalosis is overexcitability in both the central nervous system and peripheral nerves. Nerves conduct impulses repetitively, even when not stimulated by normal stimuli, resulting in nervousness, muscle spasms, and even convulsions and death.

In the discussion that follows, note that both respiratory acidosis and respiratory alkalosis are disorders due to changes in the partial pressure of CO_2 (pCO_2) in blood (normal range is 35–45 mm Hg). On the other hand, both metabolic acidosis and metabolic alkalosis are disorders due to changes in HCO_3^- concentration (normal range is 22–26 mEq/liter in arterial blood).

Respiratory Acidosis

The hallmark of **respiratory acidosis** is elevation of pCO_2 of arterial blood above 45 mm Hg. Inadequate exhalation of CO_2 causes the blood pH to drop. Any condition that decreases the movement of CO_2 from the blood to the alveoli of the lungs to the atmosphere causes a buildup of CO_2, H_2CO_3, and H^+. Such conditions include emphysema, pulmonary edema, injury to the respiratory center of the medulla oblongata, airway obstruction, or disorders of the muscles involved in breathing. If the respiratory problem is not too severe, the kidneys can help raise the blood pH into the normal range by increasing their excretion of H^+ and reabsorption of HCO_3^- (metabolic compensation). The goal in treatment of respiratory acidosis is to increase the exhalation of CO_2. Excessive secretions may be suctioned out of the respiratory tract, and artificial respiration may be given. In addition, intravenous administration of HCO_3^- and ventilation therapy to remove excessive CO_2 may be used.

Respiratory Alkalosis

In **respiratory alkalosis** arterial blood pCO_2 falls below 35 mm Hg. Hyperventilation causes the pH to increase. It occurs in conditions that stimulate the respiratory center. Such conditions include oxygen deficiency due to high altitude or pulmonary disease, cerebrovascular accident (CVA), severe anxiety, and aspirin overdose. Again, metabolic compensation may bring blood pH into the normal range if the kidneys decrease excretion of H^+ and reabsorption of HCO_3^-. Treatment of respiratory alkalosis is aimed at increasing the level of CO_2 in the body. One simple measure is to have the person breathe into a paper bag and then rebreathe the exhaled mixture of CO_2 and oxygen from the bag for a short period of time.

Metabolic Acidosis

In **metabolic acidosis** the blood HCO_3^- concentration drops below 22 mEq/liter. A decline in the level of this important buffer causes the blood pH to decrease (become more acidic). Three situations may lower the blood level of HCO_3^-: (1) actual loss of HCO_3^-, such as may occur with severe diarrhea or renal dysfunction; (2) accumulation of an acid, other than carbonic acid, as may occur in ketosis (described on page 828); or (3) failure of the kidneys to excrete H^+ from metabolism of dietary proteins. If the metabolic problem is not too severe, hyperventilation can help bring blood pH into the normal range (respiratory compensation). Treatment of metabolic acidosis consists of intravenous solutions of sodium bicarbonate and correcting the cause of acidosis.

Metabolic Alkalosis

In **metabolic alkalosis** the blood HCO_3^- concentration is above 26 mEq/liter. A nonrespiratory loss of acid by the body or excessive intake of alkaline drugs causes the pH to increase above 7.45. Excessive vomiting of gastric contents results in a substantial loss of hydrochloric acid and is probably the most frequent cause of metabolic alkalosis. Other causes of metabolic alkalosis include gastric suctioning, use of certain diuretics, endocrine disorders, administration of alkali, and severe dehydration. Respiratory compensation, through hypoventilation, may bring blood pH into the normal range. Treatment of metabolic alkalosis consists of fluid therapy to correct chloride, potassium, and other electrolyte deficiencies and correcting the cause of alkalosis.

A summary of acidosis and alkalosis is presented in Exhibit 27.3.

EXHIBIT **27.3** **SUMMARY OF ACIDOSIS AND ALKALOSIS**

Condition	Definition	Common Cause	Compensatory Mechanism
Respiratory acidosis	Increased pCO_2 (above 45 mm Hg) and decreased pH (below 7.35) if there is no compensation.	Hypoventilation due to emphysema, pulmonary edema, trauma to respiratory center, airway obstructions, dysfunction of muscles of respiration.	*Renal*: increased excretion of H^+; increased reabsorption of HCO_3^-. If compensation is complete, pH will be within the normal range but pCO_2 will be high.
Respiratory alkalosis	Decreased pCO_2 (below 35 mm Hg) and increased pH (above 7.45) if there is no compensation.	Hyperventilation due to oxygen deficiency, pulmonary disease, cerebrovascular accident (CVA), anxiety, or aspirin overdose.	*Renal*: decreased excretion of H^+; decreased reabsorption of HCO_3^-. If compensation is complete, pH will be within the normal range but pCO_2 will be low.
Metabolic acidosis	Decreased HCO_3^- (below 22 mEq/liter) and decreased pH (below 7.35) if there is no compensation.	Loss of bicarbonate ions due to diarrhea, accumulation of acid (ketosis), renal dysfunction.	*Respiratory*: hyperventilation, which increases loss of CO_2. If compensation is complete, pH will be within the normal range but HCO_3^- will be low.
Metabolic alkalosis	Increased HCO_3^- (above 26 mEq/liter) and increased pH (above 7.45) if there is no compensation.	Loss of acid due to vomiting, gastric suctioning, or use of certain diuretics; excessive intake of alkaline drugs.	*Respiratory*: hypoventilation, which slows loss of CO_2. If compensation is complete, pH will be within the normal range but HCO_3^- will be high.

Diagnosis of Acid-Base Imbalances

One can often pinpoint the cause of an acid–base imbalance by careful evaluation of three factors in a sample of systemic arterial blood—pH, concentration of HCO_3^-, and pCO_2. These three blood chemistry values are examined in a four-step sequence.

1. Note whether the pH is high (alkalosis) or low (acidosis).
2. Then decide which value—pCO_2 or HCO_3^-—is out of the normal range and could be the *cause* of the pH change. For example, *elevated pH* could be caused by *low pCO_2* or *high HCO_3^-*.
3. If the cause is a *change in pCO_2*, the problem is *respiratory*. If the cause is a *change in HCO_3^-*, the problem is *metabolic*.
4. Now look at the value that doesn't correspond with the observed pH change. If it is within its normal range, there is no compensation. If it is outside the normal range, compensation is occurring and partially correcting the pH imbalance.

Here's a case to practice your diagnostic skills. The patient is in the intensive care unit because he suffered a severe myocardial infarction 3 days ago. Make a diagnosis and decide whether or not compensation is occurring. The lab reports the following values from an arterial blood sample: pH = 7.30, HCO_3^- = 20 mEq/liter, pCO_2 = 32 mm Hg. (The answer is given on page 906.)

FLUID, ELECTROLYTE, AND ACID–BASE BALANCE IN INFANTS

There are significant differences between an adult and an infant, especially a premature infant, with respect to fluid distribution, regulation of fluid and electrolyte balance, and acid–base homeostasis. Accordingly, infants experience more problems than adults in these areas. The differences are related to the following conditions:

1. **Proportion and distribution of water.** Whereas a newborn's total body weight is about 75% water and can be as high as 90% in a premature infant, an adult's is about 55–60%. The adult percentage is not achieved until about 2 years of age. Also, whereas an adult has twice as much water in ICF as ECF, the opposite is true in a premature infant. Because ECF is subject to more changes than ICF, rapid losses or gains of body water are much greater in infants. As the rate of fluid intake and output is about seven times higher in infants than in adults, the slightest changes in fluid balance can result in severe abnormalities.
2. **Metabolic rate.** The metabolic rate of an infant is about double that of an adult. This results in the production of more metabolic wastes and acids, which can lead to the development of acidosis in infants.
3. **Functional development of the kidneys.** In infants, the kidneys are only about half as efficient as those of an adult in concentrating urine. (Functional development is not complete until about the end of the first month after birth.) As a result, the kidneys of a newborn can neither concentrate urine nor rid the body of excess acids produced as a result of the high metabolic rate as effectively as those of an adult.
4. **Body surface area.** The body surface area of an infant relative to body volume is about three times greater than that of an adult. This significantly increases water loss through the skin.
5. **Breathing rate.** The higher breathing rate of an infant (about 30–80 times a minute) causes greater water loss from the lungs. Also, respiratory alkalosis may occur because greater ventilation eliminates more CO_2 and lowers the pCO_2.
6. **Ion concentrations.** The newborn has higher K^+ and Cl^- concentrations than an adult. This causes a tendency toward metabolic acidosis. ■

STUDY OUTLINE

FLUID COMPARTMENTS AND FLUID BALANCE (p. 891)

1. Body fluid is water and its dissolved substances.
2. About two-thirds of the body's fluid is located within cells and is called intracellular fluid (ICF).
3. The other one-third is called extracellular fluid (ECF). It includes interstitial fluid; plasma and lymph; cerebrospinal fluid; gastrointestinal tract fluids; synovial fluid; fluids of the eyes and ears; pleural, pericardial, and peritoneal fluids; and glomerular filtrate.
4. Fluid balance means that the various body compartments contain the required amount of water.
5. Fluid balance and electrolyte balance are inseparable.
6. Water is the largest single constituent in the body, 45–75% of total body weight depending on age and the amount of adipose tissue (fat) present.
7. Sources of water gain are ingested liquids and foods (preformed water) and water produced by cellular respiration and dehydration synthesis reactions (metabolic water).
8. Avenues of water loss are the kidneys, skin, lungs, and gastrointestinal tract.
9. Normally, water loss equals water gain, so the body fluid volume is constant.
10. The stimulus for fluid intake is dehydration resulting in thirst sensations. Under normal conditions, water loss is regulated by antidiuretic hormone (ADH), atrial natriuretic peptide (ANP), and aldosterone.

CONCENTRATION OF SOLUTIONS (p. 893)

1. An inorganic substance that dissociates into ions in solution is called an electrolyte. Cations are positively charged ions and anions are negatively charged ions.
2. The total amount of solute in a solution is expressed as a percentage.
3. The total concentration of cations and anions is expressed as milliequivalents/liter (mEq/liter).
4. The total concentration of particles in solution is expressed as milliosmoles/liter (mOsm/liter).
5. Electrolytes have a greater effect on osmosis than do nonelectrolytes.

ELECTROLYTES IN BODY FLUIDS (p. 895)

1. Ions formed when electrolytes dissolve in body fluids control the osmosis of water between body fluid compartments, help maintain acid–base balance, and carry electrical current.
2. Plasma, interstitial fluid, and intracellular fluid contain varying kinds and amounts of ions.
3. Sodium ions (Na^+) are the most abundant extracellular ions. They are involved in impulse transmission, muscle contraction, and fluid and electrolyte balance. Na^+ level is controlled by aldosterone.
4. Chloride ions (Cl^-) are the major extracellular anions. They play a role in regulating osmotic pressure and forming HCl. Cl^- level is controlled indirectly by aldosterone.
5. Potassium ions (K^+) are the most abundant cations in intracellular fluid. They are involved in maintaining fluid volume, impulse conduction, muscle contraction, and regulating pH. K^+ level is controlled by aldosterone.

6. Bicarbonate ions (HCO_3^-) are the second most abundant anions in extracellular fluid. They are the most important buffer in plasma.
7. Calcium is the most abundant mineral in the body. Calcium salts are structural components of bones and teeth. Ca^{2+} is principally an extracellular cation. It also functions in blood clotting, neurotransmitter release, and contraction of muscle. Ca^{2+} level is controlled by parathyroid hormone (PTH) and calcitonin (CT).
8. Phosphate ions ($H_2PO_4^-$, HPO_4^{2-}, and PO_4^{3-}) are principally intracellular anions and their salts are structural components of bones and teeth. They are also required for the synthesis of nucleic acids and ATP and participate in buffer reactions. Their level is controlled by PTH and CT.
9. Magnesium ions (Mg^{2+}) are primarily intracellular cations. They act as cofactors in several enzyme systems.

MOVEMENT OF BODY FLUIDS (p. 898)

1. At the arterial end of a capillary, fluid moves from plasma into interstitial fluid (filtration). At the venous end, fluid moves in the opposite direction (reabsorption).
2. A net shift of about 3 liters/day of fluid into interstitial spaces is returned to blood plasma by the lymphatic system.
3. Fluid movement between interstitial and intracellular compartments depends on the concentrations of Na^+ and K^+ and the secretion of aldosterone, antidiuretic hormone (ADH), and atrial natriuretic peptide (ANP).
4. Fluid imbalance may lead to circulatory (hypovolemic) shock and water intoxication.

ACID–BASE BALANCE (p. 900)

1. The overall acid–base balance of the body is maintained by controlling the H^+ concentration of body fluids, especially extracellular fluid.
2. The normal pH of extracellular fluid is 7.35 (= 0.045 mEq H^+/liter) to 7.45 (= 0.035 mEq H^+/liter).
3. Homeostasis of pH is maintained by buffer systems, exhalation of carbon dioxide, and kidney excretion of H^+ and reabsorption of HCO_3^-.
4. The important buffer systems include proteins, carbonic acid–bicarbonate, and phosphates.
5. An increase in exhalation of carbon dioxide (hyperventilation) increases pH; a decrease in exhalation (hypoventilation) decreases pH.
6. The kidneys excrete H^+ and reabsorb HCO_3^-.

ACID–BASE IMBALANCES (p. 902)

1. Acidosis is a systemic arterial blood pH below 7.35. Its principal effect is depression of the central nervous system (CNS).
2. Alkalosis is a systemic arterial blood pH above 7.45. Its principal effect is overexcitability of the CNS.
3. Respiratory acidosis is characterized by an elevated pCO_2 and is caused by hypoventilation; metabolic acidosis is characterized by a decreased level of HCO_3^- and results from an abnormal increase in acidic metabolic products (other than CO_2) and loss of HCO_3^-.
4. Respiratory alkalosis is characterized by a decreased pCO_2 and is caused by hyperventilation; metabolic alkalosis is

characterized by increased HCO_3^- and results from nonrespiratory loss of acid or excess intake of alkaline substances.

5. Metabolic acidosis or alkalosis can be compensated by respiratory mechanisms (respiratory compensation); respiratory aci-

dosis or alkalosis can be compensated by renal mechanisms (metabolic compensation).

6. By examining pH, HCO_3^-, and pCO_2 values, it is possible to pinpoint the cause of an acid–base imbalance.

REVIEW QUESTIONS

1. Define body fluid. List the principal body fluid compartments and describe how they are separated. (p. 891)
2. What is meant by fluid balance? How are fluid balance and electrolyte balance related? (p. 892)
3. Describe the normal avenues of fluid gain and loss. Be sure to indicate volumes in each case. (p. 892)
4. Discuss the role of thirst in regulating fluid intake. How is thirst stimulated and satisfied? (p. 892)
5. Explain how aldosterone, atrial natriuretic peptide (ANP), and antidiuretic hormone (ADH) adjust normal fluid output. What are some abnormal routes of fluid output? (p. 893)
6. How are concentrations of chemicals in body fluids expressed? (p. 894)
7. Define a nonelectrolyte and an electrolyte. Give specific examples of each. (p. 894)
8. Describe the functions of electrolytes in the body. (p. 895)
9. Distinguish between a cation and an anion. Give several examples of each. (p. 894)
10. Describe some of the major differences in the ion concentrations of the three major fluid compartments in the body. (p. 895)
11. Name three important extracellular electrolytes and three important intracellular electrolytes. (p. 896)
12. Indicate the function and regulation of each of the following ions: Na^+, Cl^-, K^+, HCO_3^-, Ca^{2+}, phosphate ions, and Mg^{2+}. (p. 896)

13. Describe the physiological effects of hyponatremia (p. 896), hypernatremia (p. 896), hypochloremia (p. 896), hyperchloremia (p. 896), hypokalemia (p. 897), hyperkalemia (p. 897), hypocalcemia (p. 897), hypercalcemia (p. 897), hypophosphatemia (p. 898), hyperphosphatemia (p. 898), hypomagnesemia (p. 898), and hypermagnesemia (p. 898).
14. Explain the factors involved in fluid movement between the interstitial fluid and the intracellular fluid. (p. 898)
15. Explain how the following buffer systems help maintain the pH of body fluids: proteins, carbonic acid–bicarbonate, and phosphates. (p. 900)
16. Describe how exhalation of carbon dioxide is related to the maintenance of pH. (p. 901)
17. Briefly discuss the role of the kidneys in maintaining pH. (p. 902)
18. Define acidosis and alkalosis. Distinguish between respiratory and metabolic acidosis and alkalosis. (p. 902)
19. What are the principal physiological effects of acidosis and alkalosis? (p. 903)
20. How are acidosis and alkalosis compensated and treated? (p. 903)
21. If you know pH, HCO_3^-, and pCO_2 values, how can this information help you determine the cause of an acid–base imbalance? (p. 904)

ANSWERS TO QUESTIONS WITH FIGURES

27.1 Male—60 kg total body weight $\times$ 60% fluid $\times$ 1 liter/kg $\times$ 1/3 ECF $\times$ 20% plasma = 2.4 liters. Female—2.2 liters.
27.2 All would increase fluid loss.
27.3 Negative because the result (an increase in fluid intake) is opposite to the initiating stimulus (dehydration).
27.4 ECF: major cation = Na^+; major anions = Cl^- and HCO_3^-. ICF: major cation = K^+; major anions = proteins and phosphates, for example, ATP.
27.5 It will tend to depress ADH secretion and thus increase urine output, which could make the progression to circulatory shock more rapid.

27.6 It will decrease slightly as CO_2 and H^+ accumulate.

Problem on page 904. (1) pH = 7.30 indicates slight acidosis, which could be caused by elevated pCO_2 or lowered HCO_3^-. (2) The HCO_3^- is lower than normal (20 mEq/liter), so the cause is metabolic. (3) The pCO_2 is lower than normal (32 mm Hg), so hyperventilation is providing some compensation. Diagnosis: partially compensated metabolic acidosis. A possible cause is kidney damage that resulted from loss of blood flow during the heart attack.

CONTINUITY

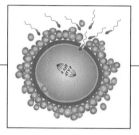

THIS UNIT EXPLAINS HOW THE HUMAN ORGANISM is adapted for reproduction, the process by which new individuals of a species are produced and genetic material is passed from one generation to the next. It also traces the sequence of embryonic and fetal development during pregnancy and discusses the basic principles of inheritance.

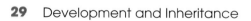

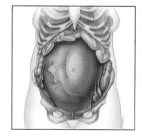

CHAPTER 28

THE REPRODUCTIVE SYSTEMS

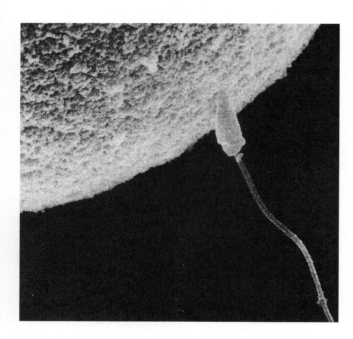

STUDENT OBJECTIVES

1. Define reproduction and classify the organs of reproduction by function. (p. 909)

2. Explain the structure, histology, and functions of the testes, ducts, accessory sex glands, and penis. (p. 910)

3. Describe the location, histology, and functions of the ovaries, uterine (Fallopian) tubes, uterus, vagina, vulva, and mammary glands. (p. 923)

4. Compare the principal events of the ovarian and uterine cycles. (p. 937)

5. Describe the similarities and differences in the male and female sexual response. (p. 942)

6. Contrast the various kinds of birth control (BC) and their effectiveness. (p. 943)

7. Describe the effects of aging on the reproductive systems. (p. 946)

8. Describe the development of the reproductive systems. (p. 946)

9. Explain the symptoms and causes of sexually transmitted diseases (STDs) such as chlamydia, gonorrhea, syphilis, genital herpes, trichomoniasis, and genital warts. (p. 949)

10. Describe the symptoms and causes of male disorders (testicular cancer, prostate disorders, impotence, and infertility) and female disorders (amenorrhea, dysmenorrhea, premenstrual syndrome [PMS], toxic shock syndrome [TSS], ovarian cysts, endometriosis, infertility, breast cysts and benign tumors, cervical cancer, pelvic inflammatory disease [PID], and vulvovaginal candidiasis). (p. 950)

11. Define medical terminology associated with the reproductive systems. (p. 953)

CHAPTER CONTENTS AT A GLANCE

Reproduction is the process by which new individuals of a species are produced and the genetic material is passed from generation to generation. This maintains continuation of the species. Cell division in a multicellular organism is necessary for growth as well as repair and it involves passing of genetic material from parent cells to daughter cells. Recall from Chapter 3 that in somatic cell division a parent cell produces two identical daughter cells. This process is involved in replacing cells and growth. In reproductive cell division, sperm and egg cells are produced for continuity of the species.

The organs of the male and female reproductive systems may be grouped by function. (1) The testes and ovaries, also called **gonads** (*gonos* = seed), function in the production of gametes—sperm and oocytes, respectively. The gonads also

secrete hormones. (2) The **ducts** of the reproductive systems transport and store gametes. (3) Still other reproductive organs, called **accessory sex glands**, produce materials that support gametes. (4) Finally, several **supporting structures**, including the penis, have various roles in reproduction.

The developmental anatomy of the reproductive systems is considered later in the chapter.

MALE REPRODUCTIVE SYSTEM

The organs of the male reproductive system are the testes, a system of ducts, accessory sex glands, and several supporting structures, including the penis (Fig. 28.1). The testes

Figure 28.1 Male organs of reproduction and surrounding structures.

Reproductive organs are adapted to produce new individuals and pass on genetic material from one generation to the next.

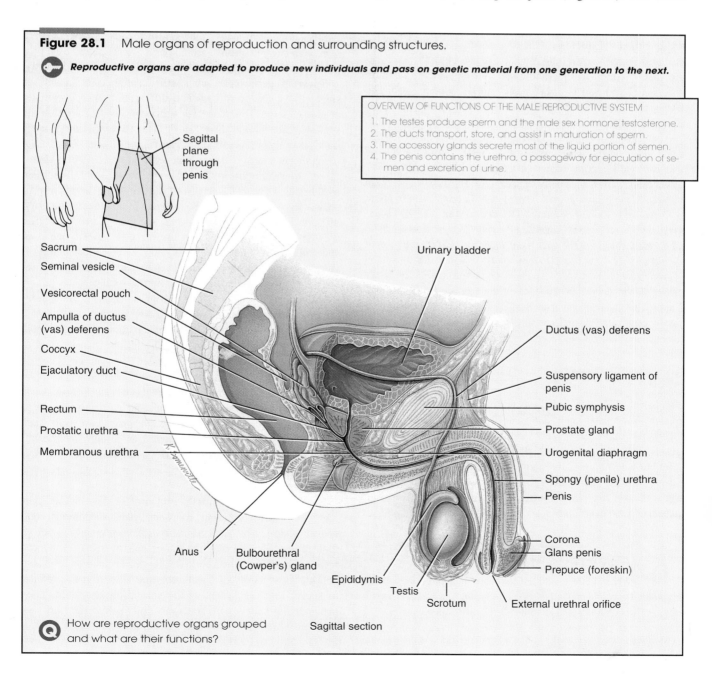

OVERVIEW OF FUNCTIONS OF THE MALE REPRODUCTIVE SYSTEM
1. The testes produce sperm and the male sex hormone testosterone.
2. The ducts transport, store, and assist in maturation of sperm.
3. The accessory glands secrete most of the liquid portion of semen.
4. The penis contains the urethra, a passageway for ejaculation of semen and excretion of urine.

Sagittal section

How are reproductive organs grouped and what are their functions?

(male gonads) produce sperm and also secrete hormones. A system of ducts transports and stores sperm, assists in their maturation, and conveys them to the exterior. Together with the sperm, secretions provided by accessory sex glands constitute semen. As noted in Chapter 26, a **urologist** (yoo-ROL-ō-gist′; *uro* = urine or urinary tract) is a physician who treats diseases related to the male reproductive system and the urinary system of both sexes.

Scrotum

The **scrotum** (SKRŌ-tum; *scrotum* = bag) is a sac that hangs from the root (attached portion) of the penis and consists of loose skin and superficial fascia (Fig. 28.1). It is the supporting structure for the testes. Internally, a vertical septum divides it into two sacs, each containing a single testis (see Fig. 28.8). The septum consists of superficial fascia and muscle tissue called the **dartos** (DAR-tōs; *dartos* = skinned), which contains bundles of smooth muscle fibers (cells). Dartos muscle is also found in the subcutaneous tissue of the scrotum (see Fig. 28.8) and is directly continuous with the subcutaneous tissue of the abdominal wall. When it contracts, the dartos muscle causes wrinkling of the skin of the scrotum.

The location of the scrotum and contraction of its muscle fibers regulate the temperature of the testes. Both production and survival of sperm require a temperature that is about 3°C lower than normal core body temperature. Because the scrotum is outside the body cavities, it provides an environment about 3°C below body temperature. The **cremaster** (krē-MAS-ter; *kremaster* = suspender) **muscle** (see Fig. 28.8) is a small band of skeletal muscle in the spermatic cord that is a continuation of the internal oblique muscle. It elevates the testes during sexual arousal and on exposure to cold. This action moves the testes closer to the pelvic cavity where they can absorb body heat. Exposure to warmth reverses the process. The dartos also is reflexly controlled to help assure that the temperature of the testes is maintained below core body temperature.

Testes

The **testes** (TES-tēz; singular is **testis**), or **testicles**, are paired oval glands measuring about 5 cm (2 in.) in length and 2½ cm (1 in.) in diameter (Fig. 28.2). Each testis weighs 10–15 grams. The testes develop high on the embryo's posterior abdominal wall and usually begin their descent into the scrotum through the inguinal canals (passageways in anterior abdominal wall, see Fig. 28.8) during the latter half of the seventh month of fetal development.

CRYPTORCHIDISM

When the testes do not descend, the condition is called cryptorchidism (krip-TOR-ki-dizm; *kryptos* = hidden; *orchis* = testis). The condition occurs in about 3% of full-term infants and about 30% of premature infants. Untreated cryptorchidism on both sides results in sterility because the cells involved in the initial de-velopment of sperm are destroyed by the higher temperature of the pelvic cavity. The chance of testicular cancer is 30–50 times greater in cryptorchid testes. The testes of about 80% of boys with cryptorchidism will descend spontaneously during the first year of life. When the testes remain undescended, injections of human chorionic gonadotropin (hCG) given at 2–5 years of age may stimulate descent. If such hormonal treatment is unsuccessful, the condition can be corrected surgically at about age 5. ■

The testes are partially covered by a serous membrane called the **tunica** (*tunica* = sheath) **vaginalis**, which is derived from the peritoneum and forms during the descent of the testes. Internal to the tunica vaginalis is a dense white fibrous capsule, the **tunica albuginea** (al′-byoo-JIN-ē-a; *albus* = white). It extends inward forming septa that divide each testis into a series of internal compartments called **lobules**. Each of the 200–300 lobules contains one to three tightly coiled tubules, the **seminiferous** (*semen* = seed; *ferre* = to carry) **tubules**. Here sperm are produced by a process called **spermatogenesis**, which will be considered shortly.

Spermatogenic cells are sperm-forming cells in various stages that undergo mitosis and differentiation to eventually produce sperm. Together with supporting cells, they line the seminiferous tubules (Fig. 28.3). The most immature spermatogenic cells are called **spermatogonia** (sper′-ma-tō-GŌ-nē-a; *sperm* = seed; *gonium* = generation or offspring; singular is **spermatogonium**). They lie next to the basement membrane. Toward the lumen of the tubule are layers of progressively more mature cells. In order of advancing maturity, these are primary spermatocytes, secondary spermatocytes, spermatids, and sperm. By the time a **sperm cell**, or **spermatozoon** (sper′-ma-tō-ZŌ-on; *zoon* = life; plural is **sperm** or **spermatozoa**), has nearly reached maturity, it is released into the lumen of the seminiferous tubule.

Embedded among the spermatogenic cells in the tubules are large **sustentacular** (sus′-ten-TAK-yoo-lar; *sustentare* = to support) **cells** (**Sertoli cells**) that extend from the basement membrane to the lumen of the tubule. Just internal to the basement membrane, tight junctions join neighboring sustentacular cells to one another. These tight junctions form the **blood–testis barrier**. To reach the developing gametes, substances must first pass through the sustentacular cells. This barrier is important because spermatogenic cells have surface antigens that are recognized as foreign by the immune system. The barrier prevents an immune response against the surface antigens by isolating the spermatogenic cells from the blood.

Sustentacular cells support and protect developing spermatogenic cells; nourish spermatocytes, spermatids, and sperm; phagocytize excess spermatid cytoplasm as development proceeds; and mediate the effects of testosterone and follicle-stimulating hormone (FSH). Sustentacular cells also control movements of spermatogenic cells and the release of sperm into the lumen of the seminiferous tubule. They produce fluid for sperm transport and secrete the hormone inhibin, which helps regulate sperm production by inhibiting secretion of FSH.

transrectal **ultrasonography**, in which a rectal probe is used to image the prostate gland, can detect tumors as small as a grain of rice. Treatment for prostate cancer may involve surgery, radiation, hormonal therapy, and chemotherapy. Because many prostate cancers grow very slowly, some urologists recommend "watchful waiting" before treating small tumors in men over age 70. ■

The paired **bulbourethral** (bul'-bō-yoo-RĒ-thral), or **Cowper's, glands** are each about the size of a pea. They lie inferior to the prostate gland on either side of the membranous urethra within the urogenital diaphragm and have ducts that open into the spongy urethra (Fig. 28.9a). During sexual arousal, the bulbourethral glands pave the way for arrival of sperm by secreting an alkaline substance that protects sperm by neutralizing acids in the urethra. At the same time, they also secrete mucus that lubricates the end of the penis and the lining of the urethra, which decreases the number of sperm injured during ejaculation.

Semen

Semen (*semen* = seed) is a mixture of sperm and **seminal fluid**, which is the liquid portion of semen that consists of the secretions of the seminiferous tubules, seminal vesicles, prostate gland, and bulbourethral glands. The average volume of semen in an ejaculation is 2½–5 ml, with a sperm count (concentration) of 50–150 million sperm per milliliter. When the number of sperm falls below 20 million/ml, the male is likely to be infertile. The very large number is required because only a tiny fraction ever reach the secondary oocyte. And although only a single sperm cell fertilizes a secondary oocyte, its penetration requires the combined action of a large number of sperm. Their acrosomes release hyaluronidase and proteinases to digest the materials surrounding the secondary oocyte. A single sperm does not produce enough of these enzymes to dissolve the barrier, but the combined action of many permits the entrance of one.

Despite the slight acidity of prostatic fluid, semen has a slightly alkaline pH of 7.2–7.7 due to the higher pH and larger volume of fluid from the seminal vesicles. The prostatic secretion gives semen a milky appearance, whereas fluids from the seminal vesicles and bulbourethral glands give it a sticky consistency. Semen provides sperm with a transportation medium and nutrients. It neutralizes the hostile acidic environment of the male urethra and the female vagina.

Semen contains an antibiotic, *seminalplasmin,* that has the ability to destroy certain bacteria. Since both semen and the lower female reproductive tract contain bacteria, the antibiotic activity of seminalplasmin may keep these bacteria under control.

Once ejaculated, liquid semen coagulates within 5 minutes due to the presence of clotting proteins from the the seminal vesicles. The functional role of semen coagulation is not known, but the proteins involved are different from those that cause blood coagulation. After about 10–20 minutes, semen reliquefies because PSA and other proteolytic enzymes produced by the prostate gland break down the clot. Abnormal or delayed liquefaction of clotted semen may cause complete or partial immobilization of sperm, thus inhibiting their movement through the cervix of the uterus.

SEMEN ANALYSIS

Semen analysis is a valuable test for evaluating male fertility. It is used to determine if sterility is related to sperm production and, after vasectomy, to check that sperm are absent. Among the factors analyzed are the following:

1. **Volume.** A low volume might suggest an anatomical or functional defect or inflammation.
2. **Motility.** In a normal semen sample at least 60% of the sperm should show good forward motility within the first 3 hours after collecting the specimen.
3. **Count.** Sperm counts below 20 million/ml could indicate sterility. Counts of 20–40 million/ml are borderline normal.
4. **Liquefaction.** Delayed liquefaction of more than 2 hours suggests inflammation of accessory sex glands or enzyme defects in the secretory products of the glands.
5. **Morphology.** No more than 30–35% of sperm should have abnormal shapes, such as a poorly formed head or tail.
6. **pH.** Normal is pH 7.2–7.7. A pH below 7.0 indicates semen that contains primarily prostatic fluid, which may be due to congenital lack of seminal vesicle formation.
7. **Fructose.** This sugar is present in a normal ejaculate. Its absence indicates obstruction or congenital absence of the ejaculatory ducts or seminal vesicles.

A normal semen analysis does not guarantee fertility. On the other hand, the absence of sperm and zero motility are the only definitive signs of sterility. ■

Penis

The **penis** contains the urethra, a passageway for ejaculation of semen and for excretion of urine. It is cylindrical in shape and consists of a body, root, and glans penis (Fig. 28.10). The **body** of the penis is composed of three cylindrical masses of tissue, each bound by fibrous tissue called the **tunica albuginea.** The paired dorsolateral masses are called the **corpora cavernosa penis** (*corpus* = body; *caverna* = hollow). The smaller midventral mass, the **corpus spongiosum penis**, contains the spongy urethra and functions in keeping the spongy urethra open during ejaculation. All three masses are enclosed by fascia and skin and consist of erectile tissue permeated by blood sinuses. With sexual stimulation, which may be visual, tactile, auditory, olfactory, or from the imagination, the arteries supplying the penis dilate, and large quantities of blood enter the blood sinuses. Expansion of these spaces compresses the veins draining the penis, so more blood that enters is trapped. These vascular changes result in an **erection**, a parasympathetic reflex. The penis returns to its flaccid state when the arteries constrict and pressure on the veins is relieved.

Ejaculation is a sympathetic reflex. As part of the reflex, the smooth muscle sphincter at the base of the urinary bladder closes. Thus urine is not expelled during ejaculation,

Figure 28.10   Internal structure of the penis. The inset in (b) shows details of the skin and fascia.

🔑 *The penis contains a pathway for ejaculation of semen and excretion of urine.*

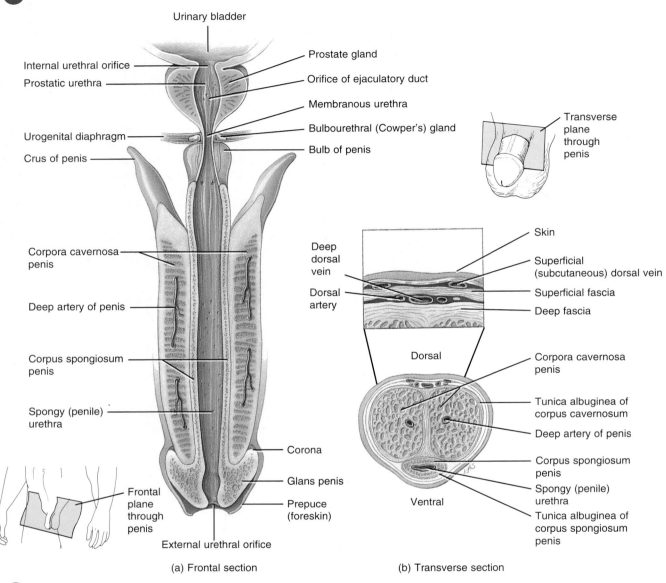

(a) Frontal section

(b) Transverse section

❓ Which tissue masses form the erectile tissue in the penis and why do they become rigid during sexual arousal?

and semen does not enter the urinary bladder. Even before ejaculation occurs, peristaltic contractions in the ampulla of the ductus deferens, seminal vesicles, ejaculatory ducts, and prostate gland propel semen into the penile portion of the urethra (spongy urethra). Typically, this leads to **emission** (ē-MISH-un), the discharge of a small volume of semen before ejaculation. Emission may also occur during sleep (nocturnal emission).

The **root** of the penis is the attached portion (proximal portion) and consists of the **bulb of the penis**, the expanded portion of the base of the corpus spongiosum penis, and the **crura** (*crus* = leg; singular is **crus**) **of the penis**, the two separated and tapered portions of the corpora cavernosa pe-

nis. The bulb of the penis is attached to the inferior surface of the urogenital diaphragm and enclosed by the bulbospongiosus muscle. Each crus of the penis is attached to the ischial and inferior pubic rami and surrounded by the ischiocavernosus muscle (see Fig. 11.13). Contraction of these skeletal muscles aids ejaculation.

The distal end of the corpus spongiosum penis is a slightly enlarged, acorn-shaped region called the **glans penis** (*glandes* = acorn). The margin of the glans penis is termed the **corona**. The distal urethra enlarges within the glans penis and forms a terminal slitlike opening, the **external urethral orifice**. Covering the glans in an uncircumsized penis is the loosely fitting **prepuce** (PRĒ-pyoos), or **foreskin**. The weight of the penis is supported by two ligaments that are continuous with the fascia of the penis. These are the **fundiform ligament**, which arises from the inferior part of the linea alba, and the **suspensory ligament of the penis**, which arises from the pubic symphysis.

CIRCUMCISION

Circumcision (*circumcido* = to cut around) is a surgical procedure in which part or all of the prepuce is removed. It is usually performed just after delivery, 3–4 days after birth, or on the eighth day as part of a Jewish religious rite. Some health care professionals think there is no medical justification for circumcision. Others feel that it has benefits such as a lower risk of urinary tract infections, protection against penile cancer, and possibly a lower risk for sexually transmitted diseases. ■

FEMALE REPRODUCTIVE SYSTEM

The female organs of reproduction (Fig. 28.11) include the ovaries, which produce secondary oocytes and hormones such

---

Figure 28.11   Female organs of reproduction and surrounding structures.

The female organs of reproduction include the ovaries, uterine (Fallopian) tubes, uterus, vagina, vulva, and mammary glands.

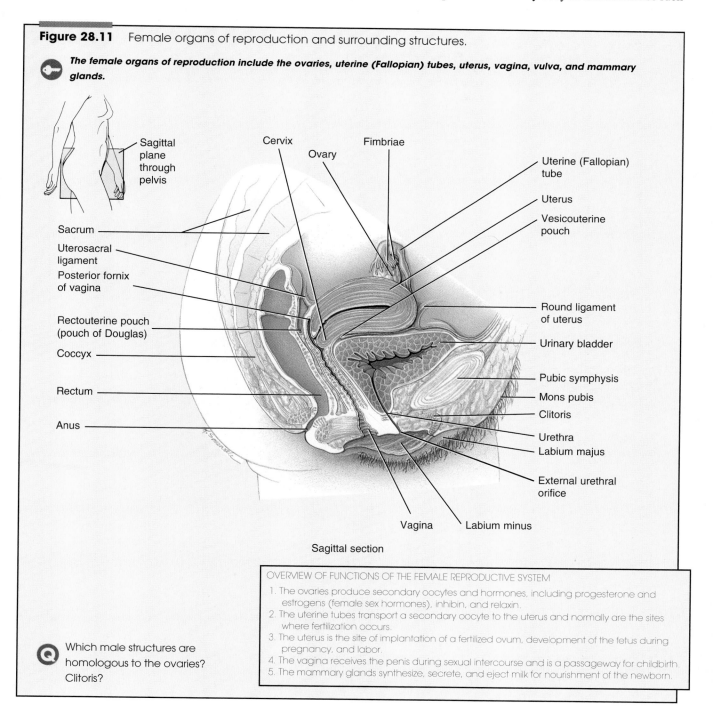

Sagittal section

OVERVIEW OF FUNCTIONS OF THE FEMALE REPRODUCTIVE SYSTEM

1. The ovaries produce secondary oocytes and hormones, including progesterone and estrogens (female sex hormones), inhibin, and relaxin.
2. The uterine tubes transport a secondary oocyte to the uterus and normally are the sites where fertilization occurs.
3. The uterus is the site of implantation of a fertilized ovum, development of the fetus during pregnancy, and labor.
4. The vagina receives the penis during sexual intercourse and is a passageway for childbirth.
5. The mammary glands synthesize, secrete, and eject milk for nourishment of the newborn.

Q Which male structures are homologous to the ovaries? Clitoris?

as progesterone and estrogens (the female sex hormones), inhibin, and relaxin; the uterine (Fallopian) tubes, or oviducts, which transport secondary oocytes and fertilized ova to the uterus; the uterus, in which embryonic and fetal development occur; the vagina; and external organs that constitute the vulva, or pudendum. The mammary glands also are considered part of the female reproductive system.

The specialized branch of medicine that deals with the diagnosis and treatment of diseases of the female reproductive system is called **gynecology** (gī'-ne-KOL-ō-jē; *gyneco* = woman). **Obstetrics** (ob-STET-riks; *obstetrix* = midwife), on the other hand, is the branch of medicine that deals with the management of pregnancy, labor, and the neonatal period (about 42 days) following childbirth.

Ovaries

The **ovaries** (*ovarium* = egg receptacle), or female gonads, are paired glands that resemble unshelled almonds in size and shape. Because they have the same embryonic origin, ovaries are said to be *homologous* to the testes. The ovaries lie in the superior portion of the pelvic cavity, one on each side of the uterus. A series of ligaments holds them in position (Fig. 28.12). The **broad ligament** of the uterus, which is

---

Figure 28.12 Position of the ovaries relative to the uterus and the ligaments that hold the ovaries in position. The arrow in the inset indicates the viewing direction (superior).

 Ligaments holding the ovaries in position are the mesovarium, the ovarian ligament, and the suspensory ligament.

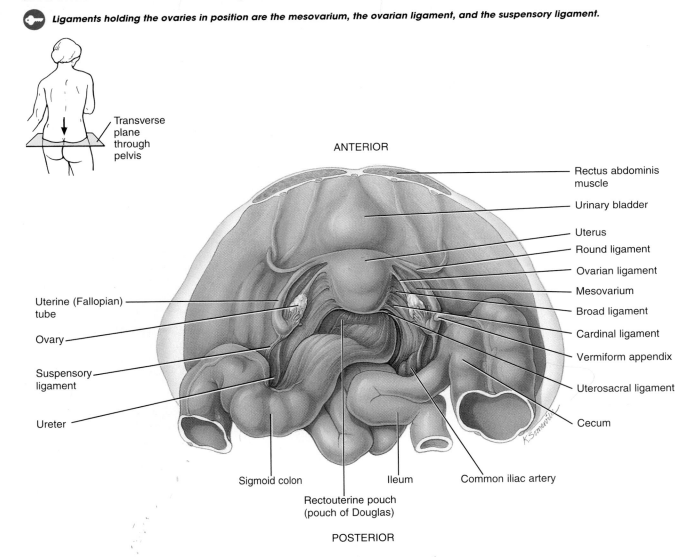

Superior view of transverse section

To which structures do the mesovarium, ovarian ligament, and suspensory ligament anchor the ovary?

itself part of the parietal peritoneum, attaches to the ovaries by a double-layered fold of peritoneum called the **mesovarium**. The **ovarian ligament** anchors the ovaries to the uterus and the **suspensory ligament** attaches them to the pelvic wall. Each ovary contains a **hilus**, the point of entrance and exit for blood vessels and nerves and along which the mesovarium is attached.

Each ovary consists of the following parts (Fig. 28.13):

1. The **germinal epithelium** is a layer of simple epithelium (low cuboidal or squamous) that covers the surface of the ovary and is continuous with the mesothelium that covers the mesovarium. The term *germinal epithelium* is a misnomer because it does not give rise to oocytes. At one time people believed that it did.
2. The **tunica albuginea** is a whiteish capsule of dense, irregular connective tissue immediately deep to the germinal epithelium.
3. The **stroma** is a region of connective tissue deep to the tunica albuginea and composed of a superficial, dense layer called the **cortex** and a deep, loose layer known as the **medulla**.
4. **Ovarian follicles** (*folliculus* = little bag) lie in the cortex and consist of **oocytes** in various stages of development and their surrounding cells. When the surrounding cells form a single layer, they are called **follicular cells**. Later in development, when they form several layers, they are referred to as **granulosa cells**. The surrounding cells nourish the developing oocyte and begin to secrete estrogens as the follicle grows larger.
5. A **mature (Graafian) follicle** is a large, fluid-filled follicle that soon will rupture and expel a secondary oocyte, a process called **ovulation**.
6. A **corpus luteum** (= yellow body) contains the remnants of an ovulated mature follicle. The corpus luteum produces progesterone, estrogens, relaxin, and inhibin

Figure 28.13 Histology of the ovary. The arrows indicate the sequence of developmental stages that occur as part of the ovarian cycle.

 The ovaries are the female gonads; they produce haploid oocytes.

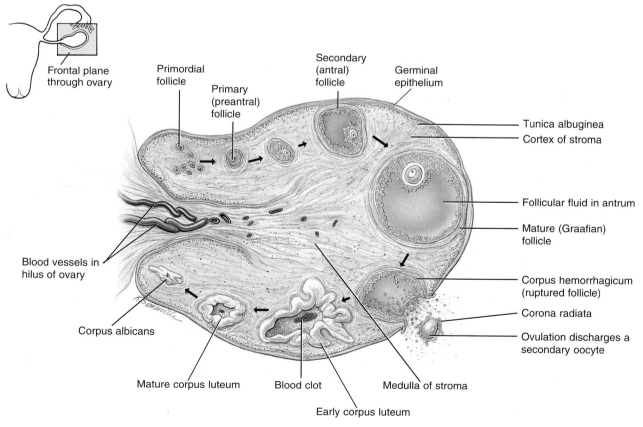

Frontal section

Q What structures in the ovary contain endocrine tissue and what hormones do they secrete?

until it degenerates and turns into fibrous tissue called a **corpus albicans** (= white body).

OVARIAN CANCER

Ovarian cancer is the sixth most common form of cancer in females and is the leading cause of death from all gynecological malignancies (excluding breast cancer). This is because it is difficult to detect before it metastasizes (spreads) beyond the ovaries. Risk factors associated with ovarian cancer include age (usually over age 50); race (Caucasians are at highest risk); family history of ovarian cancer; more than 40 years of active ovulation; nulliparity (never having had children) or first pregnancy after age 30; high-fat, low-fiber, vitamin A-deficient diet; and prolonged exposure to asbestos and talc. Early ovarian cancer has no symptoms or only mild ones associated with other common problems, such as abdominal discomfort, heartburn, nausea, loss of appetite, bloating, and flatulence. Later-stage signs and symptoms include an enlarged abdomen, abdominal and/or pelvic pain, persistent gastrointestinal disturbances, urinary complications, menstrual irregularities, and heavy menstrual bleeding. ■

Oogenesis

The formation of haploid (*n*) secondary oocytes in the ovaries involves several phases, including meiosis, and is called

oogenesis (ō′-ō-JEN-e-sis; *oon* = egg; *genesis* = production). Like spermatogenesis, oogenesis involves meiosis.

● **Reduction Division (Meiosis I)** During early fetal development, primordial (primitive) germ cells migrate from the endoderm of the yolk sac to the ovaries. There, germ cells differentiate within the ovaries into **oogonia** (ō′-o-GŌ-nē-a; singular is **oogonium**; ō′-o-GŌ-nē-um). Oogonia are diploid (2*n*) cells that divide mitotically to produce millions of germ cells. Even before birth, many of these germ cells degenerate, a process known as **atresia**. A few develop into larger cells called **primary** (*primus* = first) **oocytes** (Ō-ō-sītz) that enter prophase of reduction division (meiosis I) during fetal development but do not complete it until after puberty. At birth 200,000–2,000,000 oogonia and primary oocytes remain in each ovary. Of these, about 400 will mature and ovulate during a woman's reproductive lifetime; the remainder undergo atresia.

Each primary oocyte is surrounded by a single layer of follicular cells, and the entire structure is called a **primordial follicle** (Fig. 28.14a). Although the stimulating mechanism is unclear, a few primordial follicles periodically start to grow, even during childhood. They become **primary (preantral) follicles**, which are surrounded first by one layer of cuboidal-shaped follicular cells and then by six to seven layers of

---

Figure 28.14　Primordial, primary (preantral), and secondary (antral) follicles in the ovary.

🔑 ***As an ovarian follicle enlarges, follicular fluid accumulates in a cavity called the antrum.***

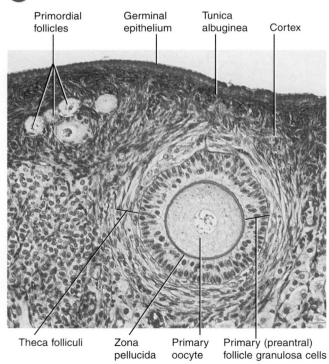

(a) Photomicrograph of the cortex of the ovary (about 200x)

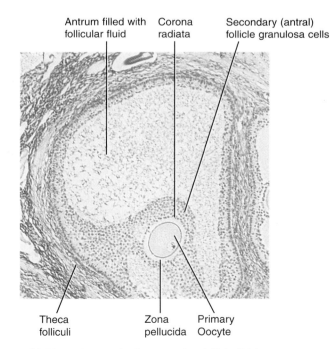

(b) Photomicrograph of a secondary follicle (60x)

 What happens to most ovarian follicles?

cuboidal and low-columnar cells called **granulosa cells**. As a follicle grows, it forms a clear glycoprotein layer, called the **zona pellucida** (pe-LOO-si-da) between the primary oocyte and the granulosa cells. The innermost layer of granulosa cells becomes firmly attached to the zona pellucida and is called the **corona radiata** (*corona* = crown; *radiata* = radiation). The outermost granulosa cells rest on a basement membrane that separates them from the surrounding ovarian stroma. This outer region is called the **theca folliculi**. As the primary follicle continues to grow, the theca differentiates into two layers: (1) the **theca interna**, a vascularized internal layer of secretory cells, and (2) the **theca externa**, an outer layer of connective tissue cells. The granulosa cells begin to secrete follicular fluid, which builds up in a cavity called the **antrum** in the center of the follicle. The follicle is now termed a **secondary (antral) follicle** (Fig. 28.14b). During childhood, primordial and developing follicles continue to undergo atresia.

After puberty, under the influence of the gonadotropin hormones secreted by the anterior pituitary gland, each month meiosis resumes in one secondary follicle (Fig. 28.15). The diploid primary oocyte completes reduction division (meiosis I), and two haploid cells of unequal size, both with 23 chromosomes (*n*) of two chromatids each, are produced. The smaller cell produced by meiosis I, called the **first polar body**, is essentially a packet of discarded nuclear material. The larger cell, known as the **secondary oocyte**, receives most of the cytoplasm. Once a secondary oocyte is formed, it proceeds to the metaphase of equatorial division (meiosis II) and then stops at this stage. The follicle in which these events are taking place, termed the **mature (Graafian) follicle** (also called a **vesicular ovarian follicle**) will soon rupture and release its secondary oocyte.

● **Equatorial Division (Meiosis II)** At ovulation, usually one secondary oocyte (with the first polar body and corona radiata) is expelled into the pelvic cavity. Normally, the cells are swept into the uterine (Fallopian) tube. If fertilization does not occur, the secondary oocyte degenerates. If sperm are present in the uterine tube and one penetrates the secondary oocyte (fertilization), however, equatorial division (meiosis II) resumes. The secondary oocyte splits into two haploid (*n*) cells of unequal size. The larger cell is the **ovum**, or mature egg; the smaller one is the **second polar body**. The nuclei of the sperm cell and the ovum then unite, forming a diploid (2*n*) **zygote**. The first polar body may also undergo another division to produce two polar bodies. If it does, the primary oocyte ultimately gives rise to a single haploid (*n*) ovum and three haploid (*n*) polar bodies, which all degenerate. Thus one oogonium gives rise to a single gamete (ovum), whereas one spermatogonium produces four gametes (sperm).

Uterine (Fallopian) Tubes

Females have two **uterine (Fallopian) tubes**, also called **oviducts**, that extend laterally from the uterus. They trans-

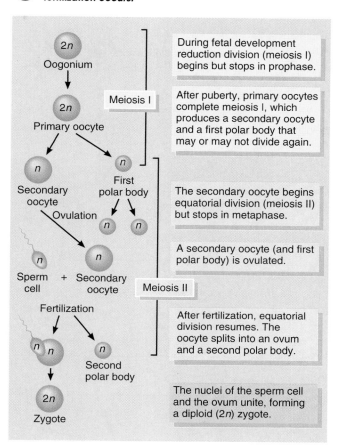

Figure 28.15 Oogenesis. 2*n* indicates diploid (46 chromosomes); *n* indicates haploid (23 chromosomes).

 In an oocyte, equatorial division is completed only if fertilization occurs.

During fetal development reduction division (meiosis I) begins but stops in prophase.

After puberty, primary oocytes complete meiosis I, which produces a secondary oocyte and a first polar body that may or may not divide again.

The secondary oocyte begins equatorial division (meiosis II) but stops in metaphase.

A secondary oocyte (and first polar body) is ovulated.

After fertilization, equatorial division resumes. The oocyte splits into an ovum and a second polar body.

The nuclei of the sperm cell and the ovum unite, forming a diploid (2*n*) zygote.

 How does the age of a primary oocyte in a female compare with the age of a primary spermatocyte in a male?

port secondary oocytes and fertilized ova from the ovaries to the uterus (Fig. 28.16). Measuring about 10 cm (4 in.) long, the tubes lie between the folds of the broad ligaments of the uterus. The open, funnel-shaped portion of each tube, called the **infundibulum**, is close to the ovary. It ends in a fringe of fingerlike projections called **fimbriae** (FIM-brē-ē; *fimbrae* = fringe; singular is **fimbria**), one of which is attached to the lateral end of the ovary. From the infundibulum, the uterine tube extends medially and inferiorly and attaches to the superior lateral angle of the uterus. The **ampulla** of the uterine tube is the widest, longest portion, making up about the lateral two-thirds of its length. The **isthmus** of the uterine tube is the more medial, short, narrow, thick-walled portion that joins the uterus.

Histologically, the uterine tubes are composed of three layers. The internal **mucosa** contains ciliated columnar epithelial cells, which help move the fertilized ovum (or secondary oocyte) along the tube, and secretory cells, which

Figure 28.16 Ovaries and uterus with associated structures. In (a), the left side of the uterine (Fallopian) tube and uterus has been sectioned to show internal structures. In (b), part of the posterior wall of the uterus has been removed.

After ovulation, a secondary oocyte and its corona radiata move from the pelvic cavity into the infundibulum of the uterine tube. The uterus is the site of menstruation, implantation of a fertilized ovum, development of the fetus, and labor.

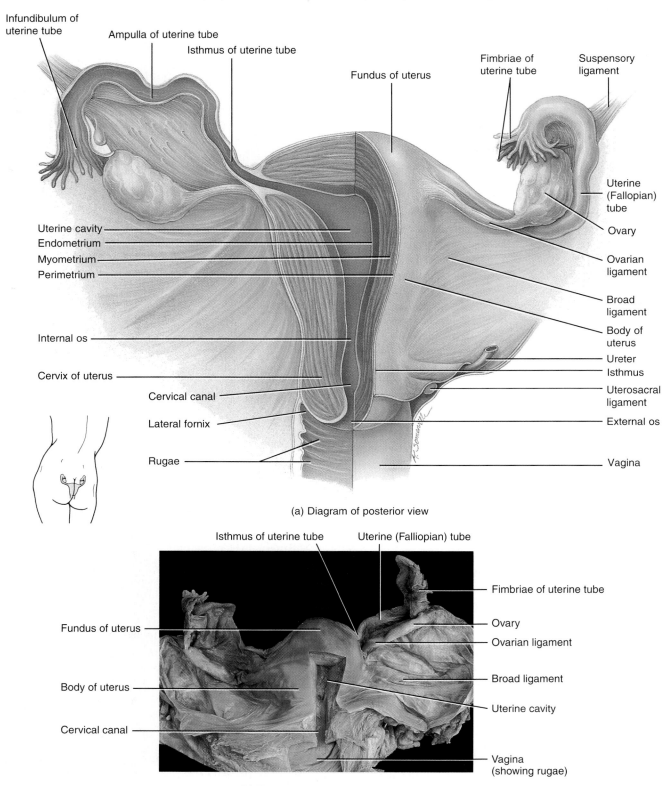

(a) Diagram of posterior view

(b) Photograph of posterior view

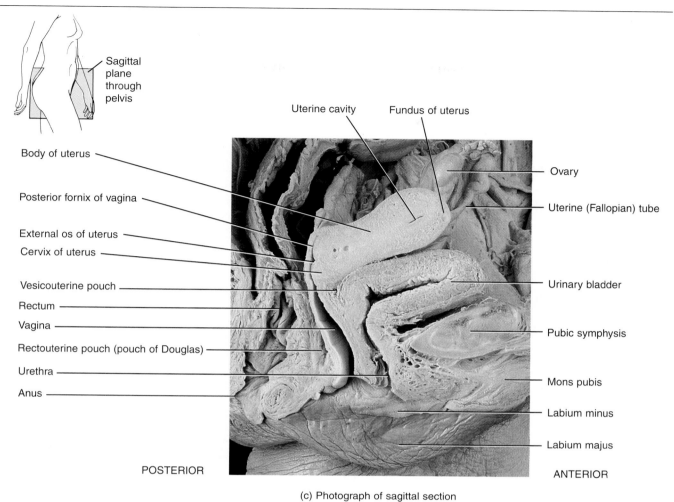

Sagittal plane through pelvis

Uterine cavity

Fundus of uterus

Body of uterus

Ovary

Posterior fornix of vagina

Uterine (Fallopian) tube

External os of uterus

Cervix of uterus

Urinary bladder

Vesicouterine pouch

Rectum

Pubic symphysis

Vagina

Rectouterine pouch (pouch of Douglas)

Urethra

Mons pubis

Anus

Labium minus

Labium majus

POSTERIOR

ANTERIOR

(c) Photograph of sagittal section

Q Where does fertilization usually occur?

have microvilli and may provide nutrition for the ovum (Fig. 28.17). The middle layer, the **muscularis**, is composed of an inner, thick, circular region of smooth muscle and an outer, thin, longitudinal region of smooth muscle. Peristaltic contractions of the muscularis and the ciliary action of the mucosa help move the oocyte or fertilized ovum toward the uterus. The outer layer of the uterine tubes is a serous membrane, the **serosa**.

After ovulation, local currents produced by movements of the fimbriae, which surround the surface of the mature follicle just before ovulation occurs, sweep the secondary oocyte into the uterine tube. A sperm cell usually encounters and fertilizes a secondary oocyte in the ampulla of the uterine tube, although fertilization in the abdominopelvic cavity is not uncommon. Fertilization may occur within the tube at any time up to about 24 hours after ovulation. Some hours after fertilization, the nuclear materials of the haploid ovum and sperm unite; the diploid fertilized ovum is now called a **zygote**. After several cell divisions, it should arrive at the uterus about 7 days after ovulation. At this point it is called a blastocyst.

Uterus

The **uterus** (womb) serves as part of the pathway for sperm to reach the uterine tubes (see Fig. 28.16). It is also the site of menstruation, implantation of a fertilized ovum, development of the fetus during pregnancy, and labor. Situated between the urinary bladder and the rectum, the uterus is the size and shape of an inverted pear. In a female who has never been pregnant, it is about 7½ cm (3 in.) long, 5 cm (2 in.) wide, and 2½ cm (1 in.) thick. It is larger in females who have recently been pregnant and smaller (atrophied) when female sex hormone levels are low, as occurs while taking birth control pills (see page 943) or after menopause (see page 946).

Anatomical subdivisions of the uterus include: (1) the dome-shaped portion superior to the uterine tubes called the **fundus**, (2) the major tapering central portion called the **body**, and (3) the inferior narrow portion opening into the vagina called the **cervix**.

The secretory cells of the mucosa of the cervix produce a secretion called **cervical mucus**, a mixture of water, glycoprotein, serum-type proteins, lipids, enzymes, and inorganic salts.

Figure 28.17 Histology of the uterine (Fallopian) tube.

Peristaltic contractions of the muscularis and ciliary action of the mucosa help move the oocyte or fertilized ovum toward the uterus.

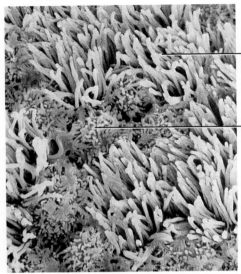

Cilia of ciliated columnar epithelial cell

Secretory cell with microvilli

Scanning electron micrograph (4000x)

Q What types of cells compose the mucosa of the uterine tubes?

Females of reproductive age secrete 20–60 ml of cervical mucus per day. Cervical mucus is more receptive to sperm at or near the time of ovulation because it is then less viscous and more alkaline (pH 8.5). At other times, viscous mucus forms a plug (cervical plug) that physically impedes sperm penetration. The mucus also supplements the energy needs of sperm. Both the cervix and mucus serve as a sperm reservoir, protect sperm from the hostile environment of the vagina, and protect sperm from phagocytes. They may also play a role in capacitation—a functional change that sperm undergo in the female reproductive tract before they can fertilize a secondary oocyte.

PAP SMEAR

Early diagnosis of cancer of the cervix of the uterus is accomplished by a **Pap smear** (Papanicolaou test). A few cells are removed from the part of the vagina surrounding the cervix and the cervix itself and examined microscopically. Malignant cells have a characteristic appearance and indicate an early stage of cancer, even before symptoms occur. ■

Between the body of the uterus and the cervix is the **isthmus** (IS-mus), a constricted region about 1 cm (½ in.) long. The interior of the body of the uterus is called the **uterine cavity**, and the interior of the narrow cervix is called the **cervical canal**. The cervical canal opens into the uterine cavity at the **internal os** (*os* = mouth) and into the vagina at the **external os**.

Normally, there is a flexure (bend) between the body of the uterus and the cervix. In this position, called **anteflexion** (*ante* = before), the body of the uterus projects anteriorly and superiorly over the urinary bladder. The cervix projects inferiorly and posteriorly and enters the anterior wall of the vagina at nearly a right angle (see Fig. 28.11). Several structures that are either extensions of the parietal peritoneum or fibromuscular cords, referred to as ligaments, maintain the position of the uterus (see Fig. 28.12).

The paired **broad ligaments** are double folds of peritoneum attaching the uterus to either side of the pelvic cavity. The paired **uterosacral ligaments**, also peritoneal extensions, lie on either side of the rectum and connect the uterus to the sacrum. The **cardinal** (**lateral cervical**) **ligaments** extend inferior to the bases of the broad ligaments between the pelvic wall and the cervix and vagina. The **round ligaments** are bands of fibrous connective tissue between the layers of the broad ligament. They extend from a point on the uterus just inferior to the uterine tubes to a portion of the labia majora of the external genitalia. Although the ligaments normally maintain the anteflexed position of the uterus, they also afford the uterine body some movement. As a result, the uterus may become malpositioned. A posterior tilting of the uterus is called **retroflexion** (*retro* = backward).

UTERINE PROLAPSE

A condition called **uterine prolapse** (*prolapses* = falling down or downward displacement) may result from weakening of supporting ligaments and pelvic musculature associated with age or disease, traumatic vaginal delivery, chronic straining from coughing or difficult bowel movements, or pelvic tumors. The prolapse may be characterized as *first degree* (*mild*), in which the cervix remains within the vagina; *second degree* (*marked*), in which the cervix protrudes to the exterior through the vagina; and *third degree* (*complete*), in which the entire uterus is outside the vagina. Depending on the degree of prolapse, treatment may involve pelvic exercises, dieting if a patient is overweight, stool softeners to minimize straining during defecation, pessary therapy (placement of a rubber device around the uterine cervix that helps prop up the uterus), and surgery. ■

Histologically, the uterus consists of three layers of tissue: the perimetrium, myometrium, and endometrium (Fig. 28.18). The outer layer, the **perimetrium** (*peri* = around; *metron* = uterus) or **serosa**, is part of the visceral peritoneum. It is composed of simple squamous epithelium and areolar connective tissue. Laterally, it becomes the broad ligament. Anteriorly, it covers the urinary bladder and forms a shallow pouch, the **vesicouterine** (ves′-i-kō-YOO-ter-in; *vesico* = bladder) **pouch** (see Fig. 28.11). Posteriorly, it covers the rectum and forms a deep pouch, the **rectouterine** (rek-tō-YOO-ter-in; *recto* = rectum) **pouch** (**pouch of Douglas**)—the most inferior point in the pelvic cavity.

The middle layer of the uterus, the **myometrium** (*myo* = muscle), forms the bulk of the uterine wall. This layer consists of three layers of smooth muscle fibers and is thickest in the

Figure 28.18 Histology of the uterus.

 The three layers of the uterus from superficial to deep are the perimetrium (serosa), myometrium, and endometrium.

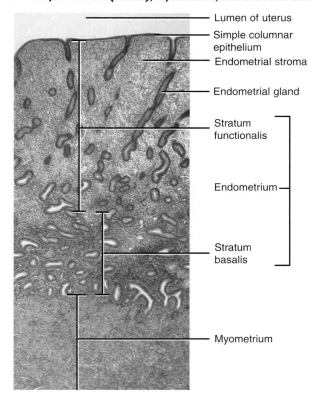

Photomicrograph of portion of endometrium and myometrium (25x)

What structural features of the endometrium and myometrium contribute to their functions?

fundus and thinnest in the cervix. The thicker middle layer is circular, whereas the inner and outer layers are longitudinal or oblique. During childbirth, coordinated contractions of the muscles (labor) in response to oxytocin from the posterior pituitary gland help expel the fetus from the body of the uterus.

The inner layer of the uterus, the **endometrium** (*endo* = within), is highly vascular and is composed of an innermost layer of simple columnar epithelium (ciliated and secretory cells) that lines the lumen; an underlying endometrial stroma that is a very thick region of lamina propria (areolar connective tissue); and endometrial (uterine) glands that develop as invaginations of the luminal epithelium and extend almost to the myometrium. The endometrium is divided into two layers. (1) The **stratum functionalis (functional layer)**, the layer closer to the uterine cavity, is shed during menstruation. (2) The deeper layer, the **stratum basalis (basal layer)**, is permanent. It gives rise to a new stratum functionalis after each menstruation.

Blood is supplied to the uterus by branches of the internal iliac artery called **uterine arteries** (Fig. 28.19). Branches called **arcuate** (*arcuatus* = shaped like a bow) **arteries** are arranged

in a circular fashion in the myometrium and give off **radial arteries** that penetrate deeply into the myometrium. Just before the branches enter the endometrium, they divide into two kinds of arterioles. The **straight arterioles** supply the basalis with the materials needed to regenerate the functionalis. The **spiral arterioles** supply the functionalis and change markedly during the menstrual cycle. The uterus is drained by the **uterine veins** into the internal iliac veins. The extensive blood supply of the uterus is essential to support regrowth of a new stratum functionalis after menstruation, implantation of a fertilized ovum, and development of the placenta.

HYSTERECTOMY

Hysterectomy (hiss-te-RECT-tō-mē; *hyster* = uterus; *ectomy* = surgical removal) refers to surgical removal of the uterus and is the most common gynecological operation. It may be indicated in conditions such as fibroid tumors, endometriosus, pelvic inflammatory disease, recurrent ovarian cysts, excessive uterine bleeding, uterine prolapse, and cancer of the cervix, uterus, or ovaries. The traditional operation to remove the uterus was through an abdominal incision (*abdominal hysterectomy*). Recently, many have been perfomed by inserting instruments into the vagina and pulling excised sections out through the vagina, without the need for an abdominal incision (*vaginal hysterectomy*). A *complete hysterectomy* is the removal of the body and cervix of the uterus. In a *partial* or *subtotal hysterectomy,* the body of the uterus is removed but the cervix is left in place. A *radical hysterectomy* includes removal of the body and cervix of the uterus, uterine tubes, possibly the ovaries, superior portion of the vagina, pelvic lymph nodes, and supporting structures, such as ligaments. ∎

Vagina

The **vagina** (*vagina* = sheath) serves as a passageway for the menstrual flow and childbirth. It also receives semen from the penis during sexual intercourse. It is a tubular, fibromuscular organ lined with mucous membrane and measures about 10 cm (4 in.) in length (see Figs. 28.11 and 28.16). Situated between the urinary bladder and the rectum, the vagina is directed superiorly and posteriorly, where it attaches to the uterus. A recess, called the **fornix** (*fornix* = arch or vault), surrounds the vaginal attachment to the cervix.

The **mucosa** of the vagina is continuous with that of the uterus. Histologically, it consists of nonkeratinized stratified squamous epithelium and areolar connective tissue that lies in a series of transverse folds called **rugae**. Dendritic cells in the mucosa are APCs (antigen-presenting cells; described on page 689). They are thought to participate in the transmission of HIV (the virus that causes AIDS) to a female during intercourse with an infected male.

The mucosa of the vagina contains large stores of glycogen, which upon decomposition produce organic acids. These acids create a low pH environment that retards microbial growth. However, the acidity is also harmful to sperm. Alkaline components of semen, mainly from the seminal

Labels on figure:
Lumen of uterus
Simple columnar epithelium
Endometrial stroma
Endometrial gland
Stratum functionalis
Endometrium
Stratum basalis
Myometrium

Figure 28.19   Blood supply of the uterus. The inset to the right shows the histological details of the blood vessels of the endometrium.

 Straight arterioles supply the necessary materials for regeneration of the stratum functionalis.

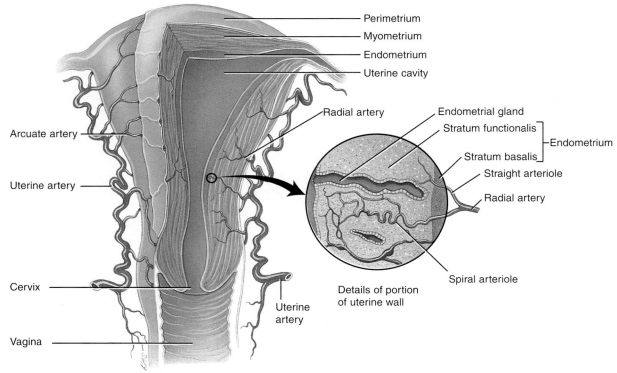

Anterior view with left side partially sectioned

Q   What is the functional significance of the stratum basalis of the endometrium?

vesicles, neutralize the acidity of the vagina and increase viability of the sperm.

The **muscularis** is composed of an outer circular layer and an inner longitudinal layer of smooth muscle that can stretch considerably to receive the penis during sexual intercourse and allow for birth of a fetus.

The **adventitia** is the superficial layer of the vagina. It consists of areolar connective tissue and anchors the vagina to adjacent organs such as the urethra and urinary bladder anteriorly and the rectum and anal canal posteriorly.

At the inferior end of the vaginal opening to the exterior, the **vaginal orifice**, there may be a thin fold of vascularized mucous membrane called the **hymen** (*hymen* = membrane). It forms a border around the orifice, partially closing it (see Fig. 28.20). Sometimes the hymen completely covers the orifice, a condition called **imperforate** (im-PER-fō-rāt) **hymen**, which may require surgery to open the orifice and permit the discharge of the menstrual flow.

COLPOSCOPY

Colposcopy (kol-POS-kō-pē; *kolpos* = vagina) is a procedure used to evaluate the status of the mucosa of the vagina and

cervix. It is often the first test done after an abnormal Pap smear. Colposcopy is the direct examination of the vaginal and cervical mucosa with a low-power binocular microscope called a colposcope that magnifies the mucous membrane 6–40 times its actual size. The application of a 3% solution of acetic acid removes mucus and enhances the appearance of mucosal epithelium. ◼

Vulva

The term **vulva** (VUL-va; *volvere* = to wrap around), or **pudendum** (pyoo-DEN-dum), refers to the external genitalia of the female (Fig. 28.20). Its components are as follows:

1. **Mons pubis.** Anterior to the vaginal and urethral openings is the **mons pubis** (MONZ PŪ-bis; *mons* = mountain). It is an elevation of adipose tissue covered by skin and coarse pubic hair that cushions the pubic symphysis.

2. **Labia majora.** From the mons pubis, two longitudinal folds of skin, the **labia majora** (LĀ-bē-a ma-JŌ-ra; *labium* = lip; singular is **labium majus**), extend inferiorly and posteriorly. The labia majora are homologous to the scrotum and are covered by pubic hair. They contain an abundance of adipose tissue and sebaceous (oil) and apocrine sudoriferous (sweat) glands.

Figure 28.20 Components of the vulva (pudendum).

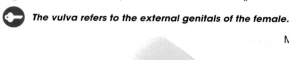 *The vulva refers to the external genitals of the female.*

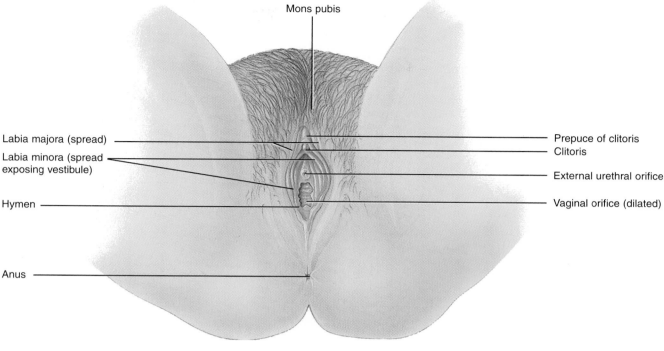

Mons pubis

Labia majora (spread)

Labia minora (spread exposing vestibule)

Hymen

Anus

Prepuce of clitoris

Clitoris

External urethral orifice

Vaginal orifice (dilated)

Inferior view

What surface structures are anterior to the vaginal opening? Lateral to it?

3. **Labia minora.** Medial to the labia majora are two smaller folds of skin called the **labia minora** (mī-NŌ-ra; singular is **labium minus**). Unlike the labia majora, the labia minora are devoid of pubic hair and fat and have few sudoriferous (sweat) glands. They do, however, contain many sebaceous (oil) glands.

4. **Clitoris.** The **clitoris** (KLĪ-to-ris) is a small, cylindrical mass of erectile tissue and nerves. It is located at the anterior junction of the labia minora. A layer of skin called the **prepuce** (foreskin) is formed at the point where the labia minora unite and covers the body of the clitoris. The exposed portion of the clitoris is the **glans**. The clitoris is homologous to the penis of the male and like the penis is capable of enlargement upon tactile stimulation and has a role in sexual excitement of the female.

5. **Vestibule.** The region between the labia minora is called the **vestibule**. Within the vestibule are the hymen (if still present), vaginal orifice, external urethral orifice, and the openings of several ducts. The **vaginal orifice**, the opening of the vagina to the exterior, occupies the greater portion of the vestibule and is bordered by the hymen. The bulb of the vestibule (see Fig. 28.21) consists of two elongated masses of erectile tissue just deep to the labia on either side of the vaginal orifice. The bulb becomes engorged with blood during sexual arousal, narrowing the vaginal orifice and placing pressure on the penis during intercourse.

Anterior to the vaginal orifice and posterior to the clitoris is the **external urethral orifice**, the opening of the urethra to the exterior. On either side of the external urethral orifice are the openings of the ducts of the **paraurethral (Skene's) glands**. They are embedded in the wall of the urethra and secrete mucus. The paraurethral glands are homologous to the male prostate gland. On either side of the vaginal orifice itself are the **greater vestibular (Bartholin's) glands** (see Fig. 28.21). These glands open by ducts into a groove between the hymen and labia minora and produce a small quantity of mucus during sexual arousal and intercourse that adds to cervical mucus and provides lubrication. The greater vestibular glands are homologous to the male bulbourethral (Cowper's) glands. Several **lesser vestibular glands** also open into the vestibule.

Perineum

The **perineum** (per'-i-NĒ-um) is the diamond-shaped area medial to the thighs and buttocks of both males and females that contains the external genitals and anus (Fig. 28.21). It is bounded anteriorly by the pubic symphysis, laterally by the ischial tuberosities, and posteriorly by the coccyx. A transverse line drawn between the ischial tuberosities divides the perineum into an anterior **urogenital** (yoo'-rō-JEN-i-tal) **triangle** that contains the external genitalia and a posterior **anal triangle** that contains the anus.

Figure 28.21 Female perineum. (Figure 11.13 shows the male perineum.)

🔑 *The perineum is a diamond-shaped area medial to the thighs and buttocks that contains the external genitals and anus.*

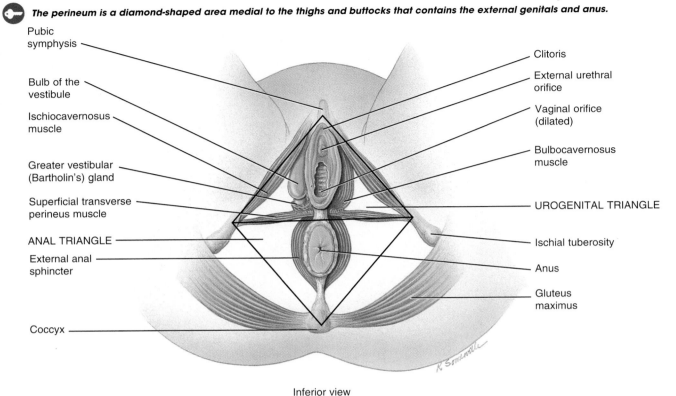

Inferior view

❓ Why is the anterior portion of the perineum called the urogenital triangle?

EPISIOTOMY

During childbirth, the emerging fetus stretches the perineal region. To prevent undue stretching and even tearing of this region, a physician sometimes performs an **episiotomy** (e-piz'-ē-OT-ō-mē; *epision* = pubic region; *tome* = incision), a perineal cut made with surgical scissors. This cut enlarges the vaginal opening to make room for the fetus to pass. In effect, a controlled cut is substituted for a jagged, uncontrolled tear. The incision is closed in layers with a continuous suture that is absorbed within a few weeks, so that stitches do not have to be removed. ■

Mammary Glands

The two **mammary** (*mamma* = breast) **glands** are modified sudoriferous (sweat) glands that produce milk. They lie over the pectoralis major and serratus anterior muscles and are attached to them by a layer of deep fascia (dense irregular connective tissue) (Fig. 28.22).

Anatomy and Histology

Each breast has one pigmented projection, the **nipple**, that has a series of closely spaced openings. These are the external openings of ducts called **lactiferous ducts** where milk

emerges. The circular pigmented area of skin surrounding the nipple is called the **areola** (a-RĒ-ō-la; *areola* = small space). It appears rough because it contains modified sebaceous (oil) glands. Strands of connective tissue called the **suspensory ligaments of the breast** (**Cooper's ligaments**) run between the skin and deep fascia and support the breast. These ligaments become looser with age or with undue stress, as occurs in long-term jogging or high-impact aerobics. Wearing a supportive bra slows the appearance of "Cooper's droop."

Within each breast, the mammary gland consists of 15–20 **lobes**, or compartments, separated by adipose tissue. The amount of adipose tissue, not the amount of milk produced, determines the size of the breasts. In each lobe are several smaller compartments called **lobules**, composed of grapelike clusters of milk-secreting glands termed **alveoli** (*alveolus* = small cavity) embedded in connective tissue. Surrounding the alveoli are spindle-shaped cells called **myoepithelial cells**, whose contraction helps propel milk toward the nipples. When milk is being produced, it passes from the alveoli into a series of **secondary tubules**. From here the milk enters the **mammary ducts**. Close to the nipple the mammary ducts expand to form sinuses called **lactiferous** (*lact* = milk; *ferre* = to carry) **sinuses**, where some milk may be stored before draining into a lactiferous duct. Each lactiferous duct carries milk from one of the lobes to the exterior, although some may join before reaching the surface.

Figure 28.22 Mammary glands

 The mammary glands function in the synthesis, secretion, and ejection of milk (lactation).

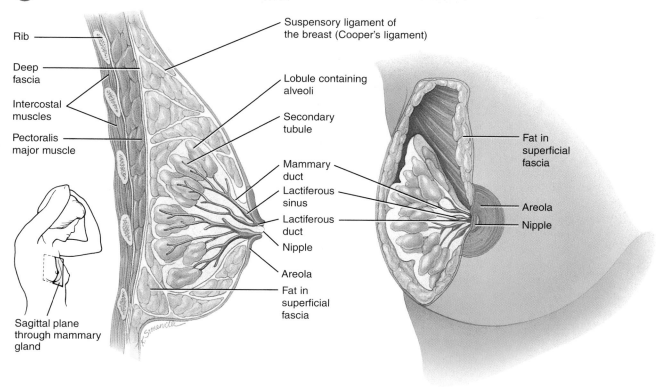

Rib

Deep fascia

Intercostal muscles

Pectoralis major muscle

Suspensory ligament of the breast (Cooper's ligament)

Lobule containing alveoli

Secondary tubule

Mammary duct

Lactiferous sinus

Lactiferous duct

Nipple

Areola

Fat in superficial fascia

Fat in superficial fascia

Areola

Nipple

Sagittal plane through mammary gland

(a) Sagittal section

(b) Anterior view, partially sectioned

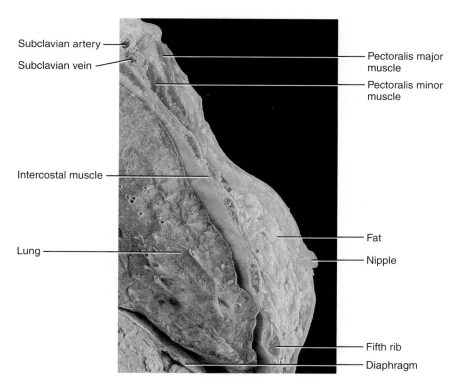

Subclavian artery

Subclavian vein

Intercostal muscle

Lung

Pectoralis major muscle

Pectoralis minor muscle

Fat

Nipple

Fifth rib

Diaphragm

(c) Photograph of sagittal section

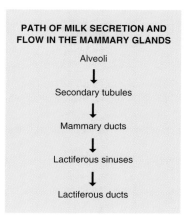

PATH OF MILK SECRETION AND FLOW IN THE MAMMARY GLANDS

Alveoli
↓
Secondary tubules
↓
Mammary ducts
↓
Lactiferous sinuses
↓
Lactiferous ducts

Q What hormones regulate synthesis and release of milk?

Physiology

The essential functions of the mammary glands are synthesis, secretion, and ejection of milk. These functions are associated with pregnancy and childbirth and together are called **lactation**. Milk production is stimulated largely by the hormone prolactin, with contributions from progesterone and estrogens. The ejection of milk occurs in the presence of oxytocin, which is released from the posterior pituitary gland in response to the sucking action of an infant on the mother's nipple (suckling). Lactation is discussed in more detail on page 976.

Breast Cancer

One in nine American women faces the prospect of **breast cancer**. After lung cancer, it is the second-leading cause of death from cancer in U.S. women but seldom occurs in men. In females, breast cancer is rarely seen before age 30, and its occurrence rises rapidly after menopause. Breast cancer is generally not painful until it becomes quite advanced, so often it is not discovered early or, if noted, is ignored. Any lump, no matter how small, should be reported to a physician at once. Early detection—by breast self-examination and mammograms—is the best way to increase the chance of survival.

● **Risk Factors** Among the factors that increase the risk of breast cancer development are (1) a family history of breast cancer, especially in a mother or sister; (2) never having a child or having a first child after age 34; (3) previous cancer in one breast; (4) exposure to ionizing radiation, such as x-rays; (5) excessive fat and alcohol intake; and (6) cigarette smoking. Recent studies in the United States show that modern, low-dose birth control pills do not increase a woman's risk of developing breast cancer.

An estimated 5% of the 180,000 cases diagnosed each year in the United States, particularly those that arise in younger women, stem from inherited genetic mutations (changes in the DNA). Two genes that increase susceptibility to breast cancer now have been identified—*BRCA1* (*br*east *ca*ncer *1*), mapped to chromosome 17 in 1990 and *BRCA2*, mapped to chromosome 13 in 1994. Mutation of *BRCA1* (but not *BRCA2*) also confers high risk for ovarian cancer. In addition, mutations of the *p53* gene increase the risk of breast cancer in both males and females and mutations of the androgen receptor gene are associated with occurrence of breast cancer in some males.

● **Detection** The most effective technique for detecting tumors less than 1 cm (about ½ in.) in diameter is **mammography** (mam-OG-ra-fē; *graphein* = to record). It is a type of radiography using very sensitive x-ray film. The image of the breast, called a **mammogram**, is obtained by placing the breasts, one at a time, on a flat surface and using a flat plate to compress the breast for better imaging (Fig. 28.23). A supplementary procedure for evaluating breast abnormalities is **ultrasound** (described on page 22). Although ultrasound cannot detect tumors less than 1 cm in diameter, it can be used to determine whether a lump is a benign, fluid-filled cyst or a solid and therefore possibly malignant tumor.

Figure 28.23 Mammograms.

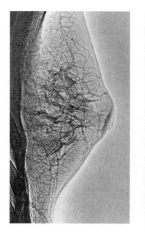 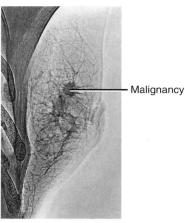

— Malignancy

(a) Normal breast　　(b) Breast with malignant tumor

The American Cancer Society recommends the following steps to help detect breast cancer as early as possible:

1. A mammogram should be taken between the ages of 35 and 39, to be used later for comparison (baseline mammogram).
2. A physician should examine the breasts every 3 years when a female is between the ages of 20 and 40, and every year after age 40.
3. Females with no symptoms should have a mammogram every year or two between ages 40 and 49, and every year after 50.
4. Females of any age with a history of breast cancer, a strong family history of the disease, or other risk factors such as smoking should consult a physician to determine a schedule for mammography.
5. All females over age 20 should develop the habit of monthly breast self-examination. Each month after the menstrual period, or on the same day each month in women who are not menstruating, the breasts should be thoroughly examined for lumps, puckering of the skin, and nipple retraction or discharge.

● **Treatment** Treatment for breast cancer may involve hormone therapy, chemotherapy, radiation therapy, **lumpectomy** (removal of just the tumor and immediate surrounding tissue), a modified or radical mastectomy, or a combination of these. A **radical mastectomy** (*mastos* = Greek for breast; *ektome* = excision) involves removal of the affected breast along with the underlying pectoral muscles and the axillary lymph nodes. Lymph nodes are removed because metastasis of cancerous cells is usually through lymphatic or blood vessels. Radiation treatment and chemotherapy may follow the surgery to ensure the destruction of any stray cancer cells. By using artificial implants, skin, fat, and muscles from other parts of the body, the breast can be reconstructed after a radical mastectomy. Using these techniques, it is possible to reconstruct a natural-looking breast.

FEMALE REPRODUCTIVE CYCLE

During their reproductive years, nonpregnant females normally experience a cyclical sequence of changes in the ovaries and uterus. Each cycle takes about a month and involves both oogenesis and preparation of the uterus to receive a fertilized ovum. Hormones secreted by the hypothalamus, anterior pituitary gland, and ovaries control the principal events. The **ovarian cycle** is a series of events associated with the maturation of an oocyte. The **uterine (menstrual) cycle** is a series of changes in the endometrium of the uterus. Each month, the endometrium is prepared for arrival of a fertilized ovum that will develop in the uterus until birth. If fertilization does not occur, the stratum functionalis of the endometrium is shed. The general term **female reproductive cycle** encompasses the ovarian and uterine cycles, the hormonal changes that regulate them, and cyclical changes in the breasts and cervix.

Hormonal Regulation

The uterine cycle and ovarian cycle are controlled by gonadotropin releasing hormone (GnRH) from the hypothalamus (Fig. 28.24). GnRH stimulates the release of follicle-stimulating hormone (FSH) and luteinizing hormone (LH) from the anterior pituitary gland. FSH stimulates the initial secretion of estrogens by growing follicles. LH stimulates the further development of ovarian follicles and their full secretion of estrogen, brings about ovulation, promotes formation of the corpus luteum, and stimulates the production of estrogens, progesterone, relaxin, and inhibin by the corpus luteum.

At least six different estrogens have been isolated from the plasma of human females. However, only three are present in significant quantities: *beta (β)-estradiol, estrone,* and *estriol.* In a nonpregnant woman, the principal estrogen is β-estradiol. It is synthesized from cholesterol in the ovaries.

Estrogens secreted by follicular cells have four important functions. (1) They promote development and maintenance

---

Figure 28.24 Secretion and physiological effects of estrogens, progesterone, relaxin, and inhibin.

 The uterine and ovarian cycles are controlled by GnRH and ovarian hormones.

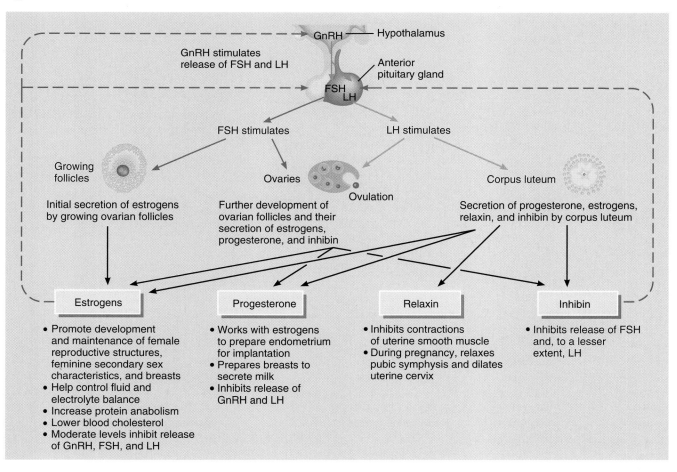

of female reproductive structures, secondary sex characteristics, and the breasts. The secondary sex characteristics include distribution of adipose tissue in the breasts, abdomen, mons pubis, and hips; voice pitch; broad pelvis; and pattern of hair growth on the head and body. (2) They help control fluid and electrolyte balance. (3) They increase protein anabolism. In this regard, estrogens are synergistic with human growth hormone (hGH). (4) They lower blood cholesterol level, which is probably the reason that women under age 50 have a much lower risk of coronary artery disease than do comparably aged men. *Moderate* levels of estrogens in the blood inhibit the release of GnRH by the hypothalamus and secretion of LH and FSH by the anterior pituitary gland. This inhibition provides the basis for the action of contraceptive (birth control) pills.

Progesterone, secreted mainly by cells of the corpus luteum, acts synergistically with estrogens to prepare the endometrium for implantation of a fertilized ovum and the mammary glands for milk secretion. High levels of progesterone also inhibit secretion of GnRH and LH.

A small quantity of **relaxin** is produced by the corpus luteum in each monthly cycle. It relaxes the uterus by inhibiting contractions; presumably implantation of a fertilized ovum occurs more readily in a relaxed uterus. During pregnancy the placenta produces much more relaxin, and it continues to relax uterine smooth muscle. It also relaxes the pubic symphysis and may help dilate the uterine cervix, both of which ease delivery of the baby.

Inhibin is secreted by granulosa cells of growing follicles and by the corpus luteum of the ovary. It inhibits secretion of FSH and, to a lesser extent, LH.

Phases of the Female Reproductive Cycle

The duration of the female reproductive cycle typically is 24–35 days. For this discussion, we shall assume a duration of 28 days, divided into three phases: the menstrual phase, preovulatory phase, and postovulatory phase (Fig. 28.25).

Figure 28.25 Correlation of ovarian and uterine cycles with the hypothalamic and anterior pituitary gland hormones. In the cycle shown, fertilization and implantation have not occurred.

The length of the female reproductive cycle typically is 24–36 days; the preovulatory phase is more variable in length than the other phases.

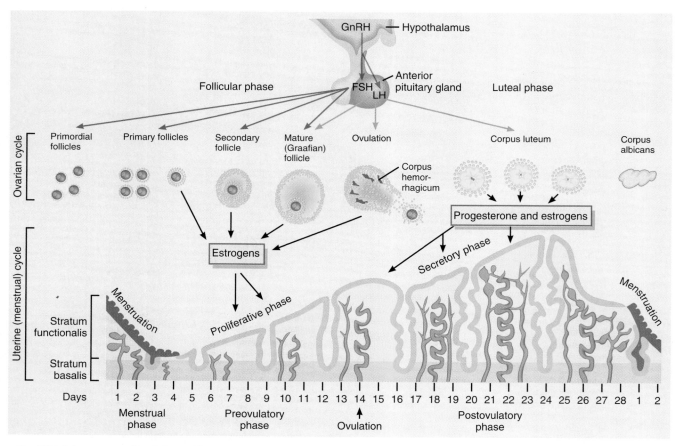

Which hormones stimulate proliferation of the endometrium? Ovulation? Growth of the corpus luteum? The surge of LH at midcycle?

Menstrual Phase (Menstruation)

The **menstrual** (MEN-stroo-al) **phase**, also called **menstruation** (men′-stroo-Ā-shun) or **menses** (*mensis* = month), lasts for roughly the first 5 days of the cycle. (By convention, the first day of menstruation marks the first day of a new cycle.)

● **Events in the Ovaries** During the menstrual phase, 20 or so small secondary (antral) follicles, some in each ovary, begin to enlarge. Follicular fluid, secreted by the granulosa cells and oozing from blood capillaries, accumulates in the enlarging antrum while the oocyte remains near the edge of the follicle (see Fig. 28.14b).

● **Events in the Uterus** Menstrual flow from the uterus consists of 50–150 ml of blood, tissue fluid, mucus, and epithelial cells derived from the endometrium. This discharge occurs because the declining level of estrogens and progesterone causes the uterine spiral arteries to constrict. As a result the cells they supply become ischemic (deficient in blood) and start to die. Eventually, the entire stratum functionalis sloughs off. At this time the endometrium is very thin because only the stratum basalis remains. The menstrual flow passes from the uterine cavity to the cervix and through the vagina to the exterior.

Preovulatory Phase

The **preovulatory phase**, the second phase of the female reproductive cycle, is the time between menstruation and ovulation. The preovulatory phase of the cycle is more variable in length than the other phases and accounts for most of the difference when cycles are shorter or longer than 28 days. It lasts from days 6 to 13 in a 28-day cycle.

● **Events in the Ovaries** Under the influence of FSH, the group of about 20 secondary follicles continues to grow and begins to secrete estrogen and inhibin. By about day 6, one follicle in one ovary has outgrown all the others and is thus the dominant follicle. Estrogen and inhibin secreted by the dominant follicle decrease the secretion of FSH, which causes the other less well-developed follicles to stop growing and undergo atresia.

The one dominant follicle becomes the **mature (Graafian) follicle** that continues to enlarge until it is more than 20 mm in diameter and ready for ovulation (see Fig. 28.13). This follicle forms a blisterlike bulge on the surface of the ovary. Fraternal (nonidentical) twins may result if two secondary follicles achieve codominance and both ovulate. During the final maturation process, the dominant follicle continues to increase its estrogen production under the influence of an increasing level of LH (Fig. 28.26). Estrogens are the primary ovarian hormones before ovulation, but small amounts of progesterone are produced by the mature follicle a day or two before ovulation.

With reference to the ovaries, the menstrual phase and preovulatory phase together are termed the **follicular** (fō-

Figure 28.26 Relative concentrations of anterior pituitary gland hormones (FSH and LH) and ovarian hormones (estrogens and progesterone) during a normal female reproductive cycle. Note the relationship of the hormones to the ovarian and uterine cycles.

 Estrogens are the primary ovarian hormones before ovulation; after ovulation, both progesterone and estrogens are secreted by the corpus luteum.

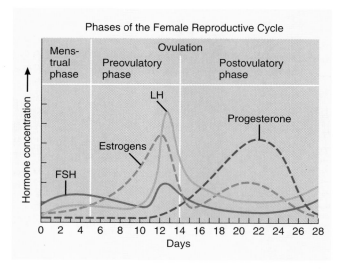

Q Which hormones stimulate rebuilding of the stratum functionalis?

LIK-yoo-lar) **phase** because ovarian follicles are growing and developing.

● **Events in the Uterus** Estrogens liberated into the blood by growing ovarian follicles (described next) stimulate the repair of the endometrium. Cells of the stratum basalis undergo mitosis and produce a new stratum functionalis. As the endometrium thickens, the short, straight endometrial glands develop and the arterioles coil and lengthen as they penetrate the stratum functionalis. The thickness of the endometrium approximately doubles to about 4–6 mm. With reference to the uterus, the preovulatory phase is also termed the **proliferative phase** because the endometrium is proliferating.

Ovulation

Ovulation, the rupture of the mature (Graafian) follicle with release of the secondary oocyte into the pelvic cavity, usually occurs on day 14 in a 28-day cycle. During ovulation, the secondary oocyte remains surrounded by its zona pellucida and corona radiata. It generally takes a total of about 20 days (spanning the last 6 days of the previous cycle and the first 14 days of the current cycle) for a secondary follicle to develop into a fully mature follicle. During this time the primary oocyte completes reduction division (meiosis I) to become a secondary oocyte, which begins equatorial division (meiosis II) but then halts in metaphase.

The *high* levels of estrogens during the last part of the preovulatory phase exert a *positive feedback* effect on both LH and GnRH (Fig. 28.27) and cause ovulation as follows:

1 When estrogens are present in high enough concentration, they stimulate the hypothalamus to release more GnRH and the anterior pituitary gland to produce more LH.

2 GnRH promotes release of FSH and more LH by the anterior pituitary gland.

3 The LH surge brings about rupture of the fully mature, dominant follicle and expulsion of a secondary oocyte. The ovulated oocyte and its corona radiata cells are usually swept into the uterine tube, but some are lost into the pelvic cavity and disintegrate.

An over-the-counter home test that detects the LH surge associated with ovulation is now available. The test predicts ovulation a day in advance. FSH also increases at this time, but not as dramatically as LH because FSH is stimulated only by the increase in GnRH. The positive feedback effect of estrogens on the hypothalamus and anterior pituitary gland does not occur if progesterone is present at the same time.

After ovulation, the mature follicle collapses and blood within it forms a clot due to minor bleeding during rupture

---

Figure 28.27 Positive feedback effect of *high* levels of estrogens on secretion of GnRH and LH, which triggers ovulation.

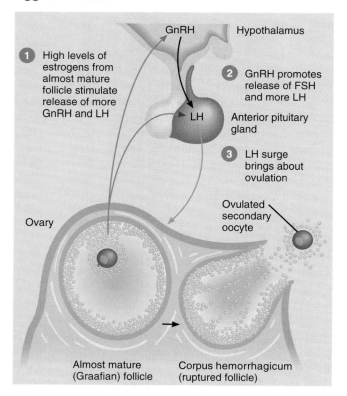

What is the effect of rising but still moderate levels of estrogens on secretion of GnRH, LH, and FSH?

---

of the follicle to become the **corpus hemorrhagicum** (*hemo* = blood; *rhegnynai* = to burst forth). See Fig. 28.13. The clot is absorbed by the remaining follicular cells, which enlarge, change character, and form the corpus luteum under the influence of LH. Stimulated by LH, the corpus luteum secretes progesterone, estrogen, relaxin, and inhibin.

SIGNS OF OVULATION

One **sign of ovulation** is an increase in **basal temperature** (body temperature at rest). A 0.4–0.6°F increase in temperature typically occurs about 14 days after the start of the menstrual cycle and is due to a small increase in progesterone just before ovulation. The 24 hours following this rise in temperature is the period immediately after ovulation and is the best time to have intercourse if pregnancy is desired.

Another sign of ovulation is the amount and consistency of **cervical mucus**. Secretion of cervical mucus is regulated by estrogens and progesterone. At midcycle, near the time of ovulation, increasing levels of estrogens cause secretory cells of the cervix to produce large amounts of cervical mucus. As ovulation approaches, the mucus becomes clear and very stretchy. This type of mucus indicates the time of greatest fertility.

The cervix also exhibits signs of ovulation. The external os dilates slightly, the cervix rises, and the cervix becomes softer. Some women also experience a pain in the area of one or both ovaries around the time of ovulation. Such pain is called **mittelschmerz** (MIT-el-shmarts), meaning "pain in the middle," and may last from several hours to a day or two. ■

Postovulatory Phase

The **postovulatory phase** of the female reproductive cycle is the most constant in duration and lasts for 14 days, from days 15 to 28 in a 28-day cycle. It represents the time between ovulation and the onset of the next menses. After ovulation, LH secretion stimulates the remnants of the mature follicle to develop into the corpus luteum. During its 2-week lifespan, the corpus luteum secretes increasing quantities of progesterone and some estrogen.

● **Events in One Ovary** If the secondary oocyte is fertilized and begins to divide, the corpus luteum persists past its normal 2-week lifespan. It is maintained by **human chorionic** (kō-rē-ON-ik) **gonadotropin** (**hCG**), a hormone produced by the chorion of the embryo as early as 8–12 days after fertilization. The chorion eventually develops into part of the placenta and the presence of hCG in maternal blood or urine is an indication of pregnancy. As the pregnancy progresses, the placenta itself begins to secrete estrogens to support pregnancy and progesterone to support pregnancy and breast development for lactation. Once the placenta begins its secretion, the role of the corpus luteum becomes minor. With reference to the ovaries, this phase of the cycle is also called the **luteal phase**.

If hCG does not rescue the corpus luteum, after 2 weeks its secretions decline and it degenerates into a scar called the corpus albicans (see Fig. 28.13). The lack of progesterone and estrogens due to degeneration of the corpus luteum then causes menstruation. In addition, the decreased levels of progesterone, estrogens, and inhibin promote the release of GnRH, FSH, and LH, which stimulate follicular growth and a new ovarian cycle begins. A summary of these hormonal interactions is presented in Fig. 28.28.

• **Events in the Uterus** Progesterone and estrogens produced by the corpus luteum promote growth and coiling of the endometrial glands, which begin to secrete glycogen, vascularization of the superficial endometrium, thickening of the endometrium, and an increase in the amount of tissue fluid. These preparatory changes are maximal about one week after ovulation, corresponding to the time of possible arrival of a fertilized ovum. With reference to the uterus, this phase of the cycle is called the **secretory phase** because of the secretory activity of the endometrial glands.

DONOR INSEMINATION

Donor (artificial) insemination (*inseminatus* = sown seed) is the deposition of semen into the vagina by a physician at a time during the female reproductive cycle when pregnancy is most likely to occur. One risk of artificial insemination is transmission of the AIDS virus and the hepatitis B and C viruses from semen donors to inseminated females. To decrease the risk of passing on sexually transmitted diseases, guidelines have been established for screening semen donors. ■

Figure 28.28   Summary of hormonal interactions of the uterine and ovarian cycles.

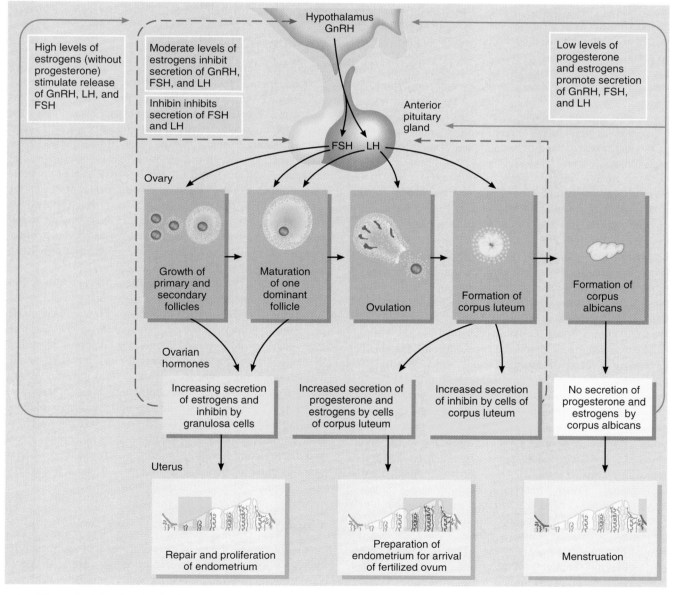

When declining levels of estrogens and progesterone stimulate secretion of GnRH, is this a positive or negative feedback effect? Why?

THE HUMAN SEXUAL RESPONSE

During **sexual intercourse**, also called **coitus** (KŌ-i-tus; *coitio* = a coming together) when it is between a male and a female, sperm are ejaculated from the male urethra into the vagina. The similar sequence of physiological and emotional changes experienced by both males and females before, during, and after intercourse is termed the **human sexual response**. William Masters and Virginia Johnson, who began their pioneering research on the human sexual response in the late 1950s, described these changes as occurring in four stages—excitement, plateau, orgasm, and resolution.

Stages of the Human Sexual Response

During sexual **excitement**, also known as **arousal**, various physical and psychological stimuli trigger *parasympathetic* reflexes in which nerve impulses propagate from the gray matter of the second, third, and fourth sacral segments of the spinal cord along the pelvic splanchnic nerves. Some of these parasympathetic postganglionic fibers produce relaxation of vascular smooth muscle, which allows **vasocongestion**, engorgement with blood, of genital tissues. Parasympathetic impulses also stimulate secretion of lubricating fluids, most of which is contributed by the female. Without satisfactory lubrication, the sexual act is difficult and painful for both partners and inhibits orgasm.

Other changes include increases in heart rate and blood pressure, increased tone in skeletal muscles throughout the body, and hyperventilation. Excitement can be induced by direct physical contact, as in kissing or tactile stimulation of the penis, especially the glans penis, clitoris, nipples of the breasts, earlobes, and other body areas. However, anticipation or fear, memories, visual, olfactory, and auditory sensations, and fantasies can enhance or diminish the likelihood that excitement will occur. Due to the tactile stimulation of the breast while nursing an infant, feelings of sexual arousal are fairly common and should not be cause for concern.

The changes that begin during excitement are sustained at an intense level in the **plateau** stage, which may last for only a few seconds or for many minutes. At this stage, many females and some males display a sex flush—a rashlike redness of the face and chest. Generally the briefest stage is **orgasm** (**climax**), during which the male ejaculates and both sexes experience several rhythmic muscular contractions about 0.8 sec apart, accompanied by intense pleasurable sensations. The sex flush also is most prominent at this time. During orgasm bursts of *sympathetic* nerve impulses leave the spinal cord at the first and second lumbar levels and cause rhythmic contractions of smooth muscle in genital organs. At the same time, *somatic motor neurons* from lumbar and sacral segments of the spinal cord trigger powerful, rhythmic contractions of perineal skeletal muscles, particularly the bulbocavernosus muscles, ischiocavernosus muscles, and anal sphincter. These muscles are shown in Figs 11.13 (male) and 28.21 (female).

In males, orgasm usually accompanies ejaculation. Receiving the ejaculate, on the other hand, provides little stimulus for the female, especially if she is not already at the plateau stage. This is why a female does not automatically experience orgasm simultaneously with her partner. On the other hand, females may experience two or more orgasms in rapid succession whereas males enter a **refractory period**, a recovery time during which a second ejaculation and orgasm is physiologically impossible. In some males the refractory period lasts only a few minutes; in others it lasts for several hours.

In the final stage—**resolution**, which begins with a sense of profound relaxation—genital tissues, heart rate, blood pressure, breathing, and muscle tone return to the unaroused state. If sexual excitement has been intense but orgasm has not occurred, resolution takes place more slowly. The four phases of the human sexual response are not always clearly separated from one another and may vary considerably among different people and even in the same person at different times.

Changes in the Male

Most of the time, the penis is flaccid or limp because sympathetic impulses cause vasoconstriction of its arteries, which limits the blood inflow. The first sign of sexual excitement is **erection**, the enlargement and stiffening of the penis. Parasympathetic impulses cause release of several neurotransmitters, including the gas nitric oxide, which relaxes vascular smooth muscle in the penis. The arteries of the penis dilate and blood fills the blood sinuses of the three corpora. Expansion of these erectile tissues compresses the superficial veins that normally drain the penis. The result is enlargement and rigidity (erection). Parasympathetic impulses also cause the bulbourethral (Cowper's) glands to secrete mucus, which flows through the urethra and provides a small amount of lubrication for intercourse.

During the plateau stage, the head of the penis increases in diameter and vasocongestion causes the testes to swell. As orgasm begins, rhythmic sympathetic impulses cause peristaltic contractions of smooth muscle in the ducts of each testis, epididymis, and ductus (vas) deferens as well as in the walls of the seminal vesicles and prostate gland. These contractions propel sperm and fluid into the urethra (emission). During this time, contraction of urethral sphincter muscles prevents flow of semen into the urinary bladder or urine into the urethra. Ejaculation quickly follows emission. During ejaculation, peristaltic contractions in the ducts and urethra combine with rhythmic contractions of skeletal muscles in the perineum and at the base of the penis to propel the semen from the urethra to the exterior.

Changes in the Female

Just as blood engorgement causes erection of the penis, the first signs of sexual excitement in females are also due to vasocongestion. Within a few seconds to a minute, parasympathetic impulses stimulate release of fluids that lubricate the walls of the vagina. Although the vaginal mucosa lacks glands, engorgement of its connective tissue with blood during sexual excitement causes lubricating fluid to ooze from capillaries and seep through the epithelial layer through a process called **transudation** (trans'-ū-DĀ-shun; *trans* = across or through; *sudare* = to sweat). Glands within the cervical mucosa and the greater vestibular (Bartholin's) glands contribute a small quantity of lubricating mucus. During excitement, parasympathetic impulses also trigger erection of the clitoris, engorgement of the labia, and relaxation of vaginal smooth muscle. The breasts also may swell due to vasocongestion, and erection may occur in the nipples of the breasts.

Late in the plateau stage, pronounced vasocongestion of the distal third of the vagina swells the tissue and narrows the opening. Because of this response, the vagina grips the penis more firmly. If effective sexual stimulation continues, orgasm may occur, associated with 3–15 rhythmic contractions of the vagina, uterus, and perineal muscles. In both males and females, orgasm is a total body response that may produce milder sensations on some occasions and more intense, explosive sensations at other times.

BIRTH CONTROL METHODS

Although there is no single, ideal method of **birth control**, several methods are available, each with advantages and disadvantages. The methods discussed here are sterilization, hormonal, intrauterine, barrier, chemical, physiological, coitus interruptus (withdrawal), and induced abortion.

Sterilization

One means of **sterilization** of males is **vasectomy** (discussed on page 917). Sterilization in females most often is achieved by performing a **tubal ligation** (lī-GĀ-shun), in which the uterine tubes are tied closed and then cut. Thus the secondary oocyte cannot pass to the uterus and sperm cannot reach the oocyte.

Hormonal Methods

By adjusting hormone levels, it is possible to interfere with production of gametes (sperm and ova) or implantation of a fertilized ovum in the uterus. This may be accomplished by use of **oral contraceptives** ("**the pill**"). The pills used most often contain a higher concentration of a progestin (similar to progesterone) and a lower concentration of estrogens (combination pill). These two hormones act via negative feedback on the anterior pituitary gland to decrease the secretion of FSH and LH and on the hypothalamus to inhibit secretion of GnRH. The low levels of FSH and LH usually prevent both follicular development and ovulation; thus pregnancy cannot occur because there's no secondary oocyte to fertilize. Even if ovulation does occur, as it does in some cases, oral contraceptives also alter cervical mucus so that it is more hostile to sperm.

Among the noncontraceptive benefits of oral contraceptives are regulation of the length of menstrual cycles, decreased menstrual flow (and therefore decreased risk of anemia), and prevention of ovarian cysts. The pill also provides protection against endometrial and ovarian cancers. Oral contraceptives may not be advised for women with a history of thromboembolic disorders (predisposition to blood clotting), cerebral blood vessel damage, hypertension, liver malfunction, or heart disease. Women who take the pill and smoke face far higher odds of having a heart attack or stroke than do nonsmoking pill users. Smokers should quit or use an alternative method of birth control. Oral contraceptives do not provide any protection against sexually transmitted diseases.

If daily pill-taking is not desired, a woman may opt for **Norplant** or **Depo-provera**, two other hormonal methods of contraception. Norplant is six slender hormone-containing capsules that are surgically implanted under the skin of the arm using local anesthesia. They slowly and continually release progestin, which inhibits ovulation and thickens the cervical mucus. The effects last for 5 years, and Norplant is as reliable as sterilization (less than 1% failure rate). Removing the Norplant capsules restores fertility. Over a 5-year period Norplant is less expensive than most birth control pills. Depo-provera is given as an intramuscular injection once every 3 months. It contains a hormone similar to progesterone that prevents maturation of the ovum and causes changes in the uterine lining that make it less likely for pregnancy to occur.

The quest for an efficient male oral contraceptive has been disappointing. The challenge is to find substances that will block production of functional sperm without disrupting the ability to have an erection.

Intrauterine Devices

An **intrauterine device** (**IUD**) is a small object made of plastic, copper, or stainless steel that is inserted into the cavity of the uterus. IUDs cause changes in the uterine lining that block implantation of a fertilized ovum. The dangers associated with the use of IUDs in some females include pelvic inflammatory disease (described on page 952), infertility, excessive menstrual bleeding, and pain. Females in monogamous (single-partner) relationships have a lower risk of developing PID and recent research discounts IUDs as a cause of PID.

Barrier Methods

Barrier methods are designed to prevent sperm from gaining access to the uterine cavity and uterine tubes. Among the barrier methods are use of a condom, vaginal pouch, diaphragm, or cervical cap.

The **condom** is a nonporous, elastic (latex or similar material) covering placed over the penis that prevents deposition of sperm in the female reproductive tract. The **vaginal pouch**, sometimes called a **female condom**, is made of two flexible rings connected by a polyurethane sheath. One ring lies inside the sheath and is inserted to fit over the cervix. The other ring remains outside the vagina and covers the female external genitals. Proper use of condoms with each act of sexual intercourse, especially when used with a spermicide (sperm-killing chemical), is a fairly reliable method of birth control. Condom use also greatly reduces, but does not eliminate, the risk of acquiring a sexually transmitted disease (STD) such as AIDS (see page 699) or gonorrhea (see page 949). Even when used properly, male or female condoms fail to protect against pregnancy and disease transmission 10–20% of the time.

The **diaphragm** is a rubber dome-shaped structure that fits over the cervix and is used together with a spermicide. The diaphragm stops the sperm from passing into the cervix. The chemical kills the sperm cells. Toxic shock syndrome (TSS) and recurrent urinary tract infections are associated with diaphragm use in some females, and a diaphragm does not protect against STDs.

The **cervical cap** is a thimble-shaped contraceptive device made of latex or plastic that measures about 4 cm (1½ in.) in diameter. It fits snugly over the cervix of the uterus and is held in position by suction. Like the diaphragm, the cervical cap is used with a spermicide. Both the diaphragm and the cervical cap must be fitted initially by a health-care provider. Advantages of the cervical cap in comparison to the diaphragm are (1) the cap can be worn up to 48 hours versus 24 hours for the diaphragm, and (2) since the cap fits tightly and rarely leaks, it is not necessary to reintroduce spermicide before intercourse. The cervical cap is not recommended for females with known or suspected cervical or uterine malignancies and current vaginal or cervical infections, and it also does not protect against STDs.

Chemical Methods

Chemical methods of contraception are spermicidal agents. Various foams, creams, jellies, suppositories, and douches that contain spermicidal agents make the vagina and cervix unfavorable for sperm survival and are available without prescription. The most widely used spermicide is nonoxynol-9, which kills sperm by disrupting the plasma membrane. It also kills the AIDS virus and decreases the incidence of chlamydia and gonorrhea (described on page 949). A spermicide is most effective when used with a diaphragm or condom.

Physiological Methods

Physiological methods are based on knowledge of certain physiological changes that occur during the menstrual cycle. In females with normal and regular menstrual cycles these events help to predict on which day ovulation is likely to occur. Physiological methods are used both for birth control and for increasing the chance of becoming pregnant.

The first physiological method, developed in the 1930s, is known as the **rhythm method**. It takes advantage of the fact that a secondary oocyte is fertilizable for only 24 hours and is

available for only 3–5 days in each menstrual cycle. During this time, the couple refrains from intercourse (3 days before ovulation, the day of ovulation, and 3 days after ovulation). The effectiveness of the rhythm method for birth control is poor because few women have absolutely regular cycles.

Another natural family planning system, developed during the 1950s and 1960s, is the **sympto-thermal method**. According to this method, couples are instructed to know and understand certain signs of fertility and infertility. Recall that the signs of ovulation include increased basal body temperature; the production of clear, stretchy cervical mucus; abundant cervical mucus; and pain associated with ovulation (mittelschmerz). If the couple refrains from sexual intercourse when the signs of ovulation are present, the chance of pregnancy is decreased.

Coitus Interruptus (Withdrawal)

Coitus interruptus refers to withdrawal of the penis from the vagina just before ejaculation. Failures with this method are due to either failure to withdraw before ejaculation or preejaculatory emission of sperm-containing fluid from the urethra. In addition, this method offers no protection against transmission of STDs.

Induced Abortion

Abortion refers to the premature expulsion from the uterus of the products of conception, usually before the 20th week of pregnancy. An abortion may be spontaneous (naturally occurring), sometimes called a miscarriage, or induced (intentionally performed). When birth control methods are not used or fail to prevent an unwanted pregnancy, **induced abortion** may be performed. Induced abortions may involve vacuum aspiration (suction), infusion of a saline solution, or surgical evacuation (scraping).

Certain drugs, most notably the French drug RU 486, can induce abortion, a so-called nonsurgical abortion. RU 486 (**mifepristone**) is an antiprogestin; it blocks the action of progesterone. Progesterone prepares the uterine endometrium for implantation and then maintains the uterine lining after implantation. If progesterone levels fall during pregnancy or if the action of the hormone is blocked, menstruation occurs, and the embryo is sloughed off along with the uterine lining. RU 486 occupies the endometrial receptor sites for progesterone. In effect, it blocks the action of progesterone on the endometrium. Within 12 hours the endometrium starts to degenerate and then begins to slough off within 72 hours. Prostaglandin, which stimulates uterine contractions, is given after RU 486 to aid in expulsion of the endometrium. RU 486 can be taken up to 5 weeks after conception. One side effect of the drug is uterine bleeding. RU 486 is being tested in clinical trials in the United States and has been used for several years in France, Sweden, the United Kingdom, and China.

A summary of birth control methods is presented in Exhibit 28.1.

EXHIBIT **28.1** **SUMMARY OF BIRTH CONTROL (BC) METHODS**

Method	Comments
Sterilization	In males, *vasectomy*, in which each ductus (vas) deferens is cut and tied. In females, *tubal ligation* (lī-GĀ-shun), in which both uterine (Fallopian) tubes are cut and tied. Failure rate: 0.1% (0.4%).[a]
Hormonal	Except for total abstinence or surgical sterilization, hormonal methods are the most effective means of birth control. *Oral contraceptives*, commonly known as "the pill," usually include both an estrogen and a progestin. Side effects include nausea, occasional light bleeding between periods, breast tenderness or enlargement, fluid retention, and weight gain. Pill users may have an increased risk of infertility. Failure rate: 0.1% (3%). *Norplant* is a set of implantable, progestin-loaded cylinders that slowly release hormone for 5 years. Fertility is restored by removing the cylinders. Failure rate: 0.3% (0.3%). *Depo-provera* is given as an intramuscular injection once every 3 months. It contains a hormone similar to progesterone that prevents maturation of the ovum and causes changes in the uterine lining that make it less likely for pregnancy to occur. Failure rate: 0.3% (0.3%).
Intrauterine device (IUD)	Small object made of plastic, copper, or stainless steel and inserted into uterus by physician. May be left in place for long periods of time. (Copper T is approved for 8 years of use.) Some women cannot use them because of expulsion, bleeding, or discomfort. Not recommended for women who have not had children because uterus is too small and cervical canal too narrow. Failure rate: 0.8% (3%).
Barrier	A *condom* is a thin, strong sheath of latex or similar material worn by male to prevent sperm from entering the vagina. A similar device for use by a woman is called a *vaginal pouch*. Failures are caused by the condom or pouch tearing or slipping off after climax or not putting it on soon enough. Failure rate: 2% (12%). A *diaphragm* is a flexible rubber dome inserted into vagina to cover the cervix, providing a barrier to sperm. Usually used with spermicidal cream or jelly. Must be left in place at least 6 hours after intercourse and may be left in place as long as 24 hours. Must be fitted by a health-care professional and refitted every two years and after each pregnancy. Offers high level of protection if used with spermicide. Occasional failures are caused by improper insertion or displacement during sexual intercourse. Diaphragm alone failure rate: 6% (18%). A *cervical cap* is a thimble-shaped latex device that fits snugly over the cervix of the uterus. Used with a spermicide, must be fitted by a health-care professional. May be left in place for up to 48 hours, and it is not necessary to reintroduce spermicide before sexual intercourse. Failure rate: 6% (18%).
Chemical	Sperm-killing chemicals inserted into vagina to coat vaginal surfaces and cervical opening. Provide protection for about 1 hour. Effective when used alone but significantly more effective when used with diaphragm or condom. Spermicide alone failure rate: 3% (21%).
Physiological	In the *rhythm method*, sexual intercourse is avoided just before and just after ovulation (about 7 days). Failure rate is about 20% even in females with regular menses. In the *sympto-thermal method*, signs of ovulation are noted (increased basal body temperature, clear and stretchy cervical mucus, opening of the external os, elevation and softening of the cervix, abundant cervical mucus, and pain associated with ovulation), and sexual intercourse is avoided. Failure rate: 1–9% (20%).
Coitus interruptus	Withdrawal of penis from vagina before ejaculation occurs. Failure rate: 4% (18%).
Induced abortion	Surgical or drug-induced removal of products of conception at an early stage from uterus or from uterine tube in cases of tubal pregnancy. Surgical removal from the uterus may involve vacuum aspiration (suction), saline solution, or surgical evacuation (scraping). Drug-induced abortions make use of RU 486, which blocks the action of progesterone.

[a] Failure rates: The first number is the expected percentage of women experiencing an unintended pregnancy in the first year of continuous and proper use. In parentheses is the typical percentage of women who become pregnant, including those who forgot to use their birth control. With no method of birth control, the failure rate is 85%.

AGING AND THE REPRODUCTIVE SYSTEMS

During the first decade of life, the reproductive system is in a juvenile state. At about age 10, hormone-directed changes start to occur in both sexes. **Puberty** (PŪ-ber-tē; *puber* = marriageable age) refers to the period of time when secondary sexual characteristics begin to develop and the potential for sexual reproduction is reached. The factors that initiate puberty are poorly understood, but the sequence of events is well established.

Male Puberty

Male puberty begins about age 10–11 years and ends at age 15–17. During the prepubertal years, plasma levels of LH, FSH, and testosterone are low. Around age six or seven, a prepubertal growth spurt occurs that is probably related to secretion of adrenal androgens and human growth hormone (hGH).

Sleep-associated increases in LH and, to a lesser extent, FSH signal the onset of puberty. As puberty advances, elevated LH and FSH levels are present throughout the day and are accompanied by increased levels of testosterone. The rises in LH and FSH are believed to result from increased GnRH secretion and enhanced responsiveness of the anterior pituitary gland to GnRH. With sexual maturity, the hypothalamic–pituitary system becomes less sensitive to the feedback inhibition of testosterone on LH and FSH secretion.

The changes in the testes that occur during puberty include maturation of sustentacular (Sertoli) cells and initiation of spermatogenesis. The anatomical and functional changes associated with puberty are the result of increased testosterone secretion. Usually, the first sign is enlargement of the testes. About a year later, the penis increases in size. The prostate gland, seminal vesicles, bulbourethral glands, and epididymides increase in size over a period of several years. Development of masculine secondary sex characteristics occurs and a growth spurt takes place as elevated testosterone levels increase both bone and muscle growth.

Female Puberty and Menarche

In girls, prepubertal levels of LH, FSH, and estrogens are low. Around age seven or eight, girls experience an increase in the secretion of adrenal androgens (adrenarche), which are responsible for the eventual growth of pubic and axillary hair. The onset of puberty is signaled by sleep-associated increases in LH and FSH. As puberty progresses, the increases in LH and FSH are sustained throughout the day. The rising levels stimulate the ovaries to secrete estrogens, which are responsible for the development of feminine secondary sexual characteristics.

At birth, both male and female mammary glands are poorly developed and appear as slight elevations on the chest. With the onset of puberty, under the influence of estrogens and progesterone, the female breasts begin to develop. Budding of the breasts is the first outward sign of puberty. The duct system matures, fat deposition occurs, and the areola and nipple grow and become more pigmented. Further mammary gland development occurs at reproductive maturity with the onset of ovulation and the formation of the corpus luteum, which lead to higher levels of estrogens and progesterone.

Estrogens and progesterone also stimulate the growth of the uterine tubes, uterus, and vagina. **Menarche** (me-NAR-kē; *arche* = beginning), the first menses, occurs at an average of 12 years of age. As you will see in Chapter 29, a female must have a minimum amount of body fat to begin and maintain a normal menstrual cycle.

Age-Related Changes in Older Males

Healthy men often retain reproductive capacity into their 80s or 90s. At about age 55 a decline in testosterone synthesis leads to less muscle strength, fewer viable sperm, and decreased sexual desire. However, abundant sperm may be present even in old age. A common problem of older men is enlargement (hypertrophy) of the prostate gland. As the prostate enlarges, it squeezes the urethra and causes difficulty in urinating. The enlargement often is benign, but prostate cancer (see page 920) is quite common.

Menopause and Age-Related Changes in Older Females

Menopause (*pausis* = to cease) refers to the permanent cessation of menses. Between the ages of 40 and 50 the ovaries become less responsive to the stimulation of gonadotropic hormones from the anterior pituitary gland. As a result, estrogen and progesterone production decline, and follicles do not undergo normal development. Changes in GnRH release patterns and decreased responsiveness to it by cells of the anterior pituitary gland that secrete LH also contribute to the onset of menopause. Some women experience hot flashes, copious sweating, headache, hair loss, muscular pains, vaginal dryness, insomnia, depression, weight gain, and mood swings. In the postmenopausal woman there will be some atrophy of the ovaries, uterine tubes, uterus, vagina, external genitalia, and breasts. Osteoporosis is also associated with a diminished level of estrogens. Sexual desire (libido) does not show a parallel decline due to continued production of adrenal adrogens.

The female reproductive system has a time-limited span of fertility between menarche and menopause. Fertility declines with age, possibly as a result of less frequent ovulation and the declining ability of the uterine (Fallopian) tubes and uterus to support the young embryo. Uterine cancer peaks at about 65 years of age, but cervical cancer is more common in younger women.

 # DEVELOPMENTAL ANATOMY OF THE REPRODUCTIVE SYSTEMS

The *gonads* develop from the **intermediate mesoderm**. By the sixth week, they appear as bulges that protrude into the ventral body cavity (Fig. 28.29). The gonads develop near the

mesonephric (Wolffian) ducts. A second pair of ducts, the **paramesonephric (Müllerian) ducts**, develop lateral to the mesonephric ducts. Both sets of ducts empty into the urogenital sinus. An early embryo contains primative gonads that have the potential to differentiate into either testes or ovaries. The male pattern of differentiation depends on the presence of a master gene on the Y chromosome called *SRY*, which stands for *S*ex-determining *R*egion of the *Y* chromosome, and the release of testosterone. During the seventh week, primitive sustentacular (Sertoli) cells start to appear in the gonadal tissues of male embryos, which have one X and one Y chromosome (sex chromosomes, see page 981). Stimulated by

Figure 28.29 Development of the internal reproductive systems.

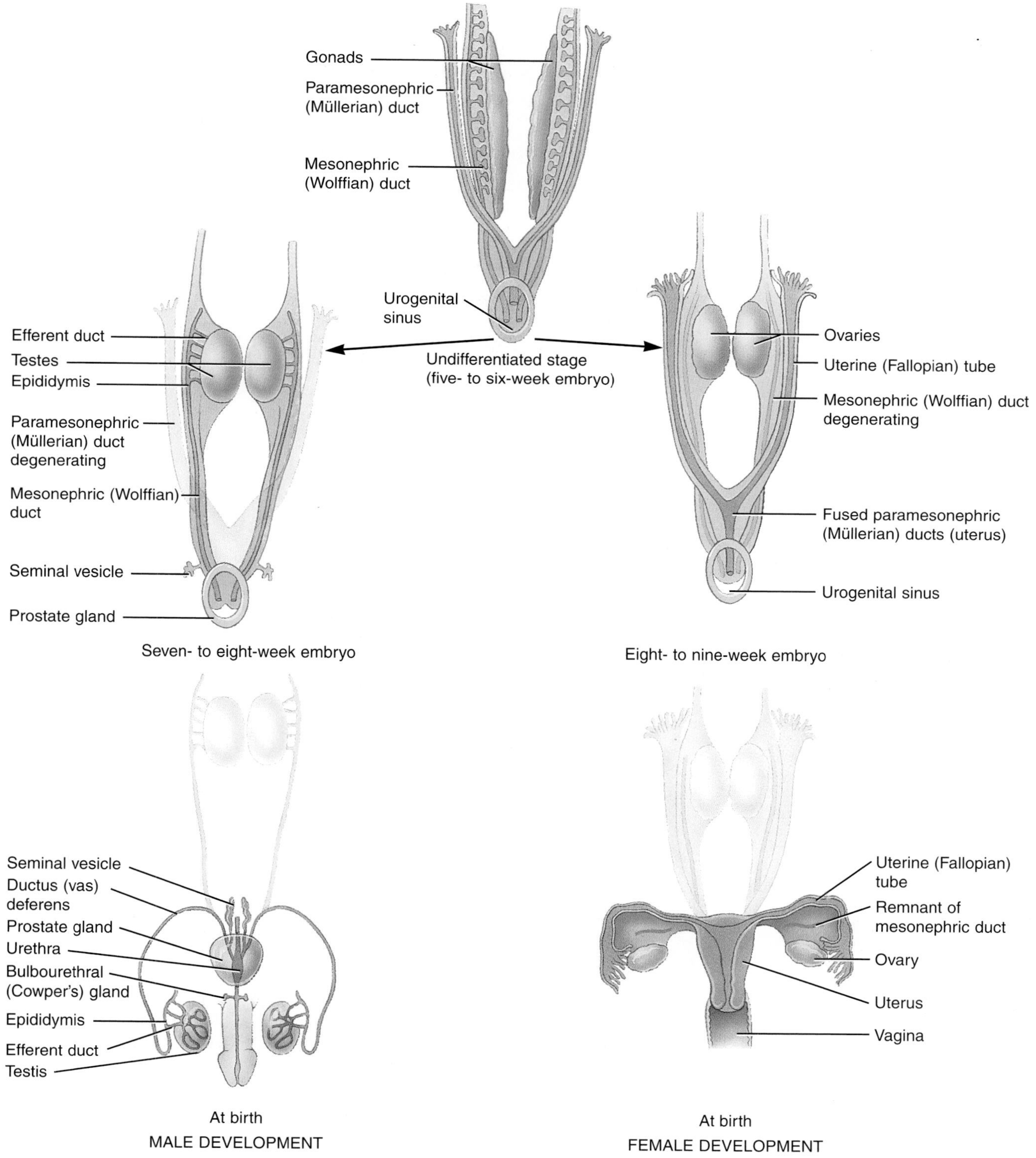

Gonads

Paramesonephric (Müllerian) duct

Mesonephric (Wolffian) duct

Urogenital sinus

Undifferentiated stage (five- to six-week embryo)

Efferent duct
Testes
Epididymis
Paramesonephric (Müllerian) duct degenerating
Mesonephric (Wolffian) duct
Seminal vesicle
Prostate gland

Seven- to eight-week embryo

Ovaries
Uterine (Fallopian) tube
Mesonephric (Wolffian) duct degenerating
Fused paramesonephric (Müllerian) ducts (uterus)
Urogenital sinus

Eight- to nine-week embryo

Seminal vesicle
Ductus (vas) deferens
Prostate gland
Urethra
Bulbourethral (Cowper's) gland
Epididymis
Efferent duct
Testis

At birth
MALE DEVELOPMENT

Uterine (Fallopian) tube
Remnant of mesonephric duct
Ovary
Uterus
Vagina

At birth
FEMALE DEVELOPMENT

hCG, primitive interstitial endocrinocytes (Leydig cells) begin to secrete testosterone. Differentiation of gonadal tissue into ovaries in embryos with two X chromosomes depends on the absence of *SRY* and the absence of testosterone.

In the male embryo, the *testes* connect to the mesonephric duct through a series of tubules. These tubules become the *seminiferous tubules.* Stimulated by testosterone, the mesonephric duct on each side continues to develop and gives rise to the *epididymis, ductus (vas) deferens, ejaculatory duct,* and *seminal vesicle.* The developing sustentacular cells also secrete a second hormone called Müllerian-inhibiting substance (MIS), which causes the death of cells within the paramesonephric (Müllerian) ducts. As a result, those cells do not contribute any functional structures to the male reproductive system. The *prostate* and *bulbourethral (Cowper's) glands* are **endodermal** outgrowths of the urethra.

In the female embryo, the gonads develop into *ovaries.* In the absence of MIS, the paramesonephric ducts flourish. Their distal ends fuse to form the *uterus* and *vagina.* The unfused proximal portions become the *uterine (Fallopian) tubes.* The mesonephric ducts in the female degenerate without contributing any functional structures to the female reproductive system. The *greater (Bartholin's)* and *lesser vestibular glands* develop from **endodermal** outgrowths of the vestibule.

The *external genitals* of both male and female embryos also remain undifferentiated until about the eighth week. Before differentiation, all embryos have an elevated region, the **genital tubercle.** This is a point between the tail (future coccyx) and the umbilical cord where the mesonephric and paramesonephric ducts open to the exterior (Fig. 28.30). The tubercle consists of the **urethral groove** (opening into

Figure 28.30   Development of the external genitals.

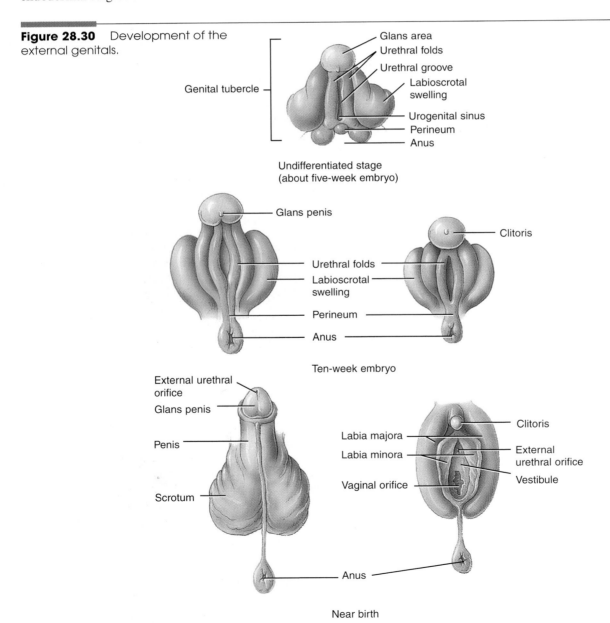

Glans area
Urethral folds
Urethral groove
Labioscrotal swelling
Urogenital sinus
Perineum
Anus

Genital tubercle

Undifferentiated stage
(about five-week embryo)

Glans penis

Clitoris

Urethral folds
Labioscrotal swelling
Perineum
Anus

Ten-week embryo

External urethral orifice
Glans penis

Penis

Scrotum

Labia majora
Labia minora

Vaginal orifice

Clitoris

External urethral orifice

Vestibule

Anus

Near birth

MALE DEVELOPMENT   FEMALE DEVELOPMENT

the urogenital sinus), paired **urethral folds**, and paired **labioscrotal swellings**.

In the male embryo, DHT stimulates development of the urethra, prostate gland, and external genitals (scrotum and penis). A portion of the genital tubercle elongates and develops into a *penis*. Fusion of the urethral folds forms the *spongy* (*penile*) *urethra* and leaves an opening to the exterior only at the distal end of the penis, the *external urethral orifice*. The labioscrotal swellings develop into the *scrotum*. In the absence of DHT, the genital tubercle gives rise to the *clitoris* in a female embryo. The urethral folds remain open as the *labia minora,* and the labioscrotal swellings become the *labia majora*. The urethral groove becomes the *vestibule*. After birth, androgen levels decline because hCG is no longer present to stimulate secretion of testosterone.

DEFICIENCY OF 5 ALPHA-REDUCTASE

A rare genetic defect leads to a deficiency in 5 alpha-reductase, the enzyme that converts testosterone to DHT. It is caused by mutation of the gene on chromosome 2 that codes for this enzyme. At birth such a baby has a female external appearance, due to absence of DHT during development. At puberty, however, testosterone level rises and masculine characteristics, such as pubic and axillary hair, start to appear. In addition, the breasts fail to develop. Internal examination reveals testes and structures that normally develop from the mesonephric duct (epididymis, ductus deferens, seminal vesicle, and ejaculatory duct) rather than ovaries and uterus. ■

DISORDERS: HOMEOSTATIC IMBALANCES

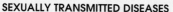

SEXUALLY TRANSMITTED DISEASES

The general term **sexually transmitted disease (STD)** is applied to any of the large group of diseases that can be spread by sexual contact. The group includes conditions traditionally specified as **venereal** (ve-NIR-rē-al; from Venus, goddess of love) **diseases** (**VD**), such as chlamydia, gonorrhea, syphilis, and genital herpes. In most developed countries of the world, such as those of the European Community, Japan, Australia, and New Zealand, the incidence of STDs has declined markedly during the past 20 years. In the United States by contrast, STDs have been rising to near epidemic proportions, especially among urban populations. AIDS and hepatitis, which are sexually transmitted diseases that also may be contracted in other ways, are discussed on pages 699 and 800, respectively.

Chlamydia

Chlamydia (kla-MID-ē-a) is a sexually transmitted disease caused by the bacterium *Chlamydia trachomatis* (*chlamys* = cloak). This unusual bacterium cannot reproduce outside body cells; it "cloaks" itself inside cells to divide. At present, chlamydia is the most prevalent sexually transmitted disease, affecting 3–5 million persons annually in the United States in all socioeconomic groups. For example, it is found in up to 5% of female college students and 10% of young men in the military. On an annual basis, more than 20,000 young men and women in the United States are rendered sterile by chlamydia.

In males, urethritis is the principal result. Symptoms of urethritis include a thick discharge, burning on urination, frequent urination, and painful urination. Without treatment, the epididymides may also become inflamed, leading to sterility. In females, the most common site of infection is the cervix, resulting in cervicitis and production of a thick mucus and pus discharge. Symptoms include pelvic soreness, lower back pain, and abdominal pain. Due to its proximity to the vagina, the female urethra may also become infected, leading to urethritis and symptoms similar to those that occur in males. Moreover, the uterine tubes may also become inflamed, which increases the risk of ectopic pregnancy (implantation of a fertilized ovum outside the uterus) and sterility due to formation of scar tissue in the tubes. Chlamydia may be passed from mother to infant during childbirth, infecting the eyes. Treatment consists of the administration of tetracycline or doxycycline.

Gonorrhea

Gonorrhea (gon'-ō-RĒ-a; *gonas* = seed; *rhein* = to flow) or "**clap**" is an infectious sexually transmitted disease caused by the bacterium *Neisseria gonorrhoeae*. In the United States 1–2 million new cases of gonorrhea appear each year. Most cases occur in those aged 15–24 years. Discharges from infected mucous membranes are the source of transmission of the bacteria during sexual contact or during passage of a newborn through the birth canal. The site of infection relates to the type of sexual contact, occurring in the mouth and throat after oral-genital contact, vagina and penis after genital intercourse, or rectum after recto-genital contact.

Males usually suffer inflammation of the urethra with pus and painful urination. The prostate gland and epididymis may also become infected. In females, infection typically occurs in the vagina, often with a discharge of pus. Both infected males and females may harbor the disease without any symptoms, however, until it has progressed to a more advanced stage. In females the infection and consequent inflammation can proceed from the vagina into the uterus, uterine tubes, and pelvic cavity. Peritonitis, or inflammation of the peritoneum, is a life-threatening disorder. Although antibiotics have greatly reduced the mortality rate of acute peritonitis, it is estimated that 50,000–80,000 women in the United States are made sterile by gonorrhea every year as a result of scar tissue formation that closes the uterine tubes. If the bacteria are transmitted to the eyes of a newborn in the birth canal, blindness can result.

Administration of a 1% silver nitrate solution in the infant's eyes prevents infection. For many years, penicillin and tetracycline were the drugs of choice for the treatment of gonorrhea in adults. However, bacterial strains resistant to these antibiotics have become very prevalent since the mid-1980s. Currently, ceftriaxone is the antibiotic that most effectively attacks the majority of gonorrhea bacteria.

Syphilis

Syphilis is a sexually transmitted disease caused by the bacterium *Treponema pallidum*. It is transmitted through sexual contact or exchange of blood, or through the placenta to a fetus. The disease progresses through several stages. During the **primary stage**, the chief symptom is a painless open sore, called a

DISORDERS: HOMEOSTATIC IMBALANCES CONTINUED

chancre (SHANG-ker), at the point of contact. The chancre heals within 1–5 weeks. From 6 to 24 weeks later, symptoms such as a skin rash, fever, and aches in the joints and muscles usher in the **secondary stage**. These symptoms also eventually disappear (in about 4–12 weeks), and the disease ceases to be infectious, but a blood test for the presence of the bacteria generally remains positive. During this "symptomless" period, called the **latent stage**, which may last up to 20 years, the bacteria may invade body organs. When signs of organ degeneration appear, the disease is said to be in the **tertiary stage**.

If the organs of the nervous system become involved, the tertiary stage is called **neurosyphilis**. Neurosyphilis may take different forms, depending on the tissue involved. Cerebellar damage is manifested by uncoordinated movements in such activities as writing. As the motor areas become extensively damaged, victims may be unable to control urine and bowel movements. Eventually, they may become bedridden, unable even to feed themselves. Damage to the cerebral cortex produces memory loss and personality changes that range from irritability to hallucinations. AIDS and other disorders that compromise the immune system may speed the progression of neurosyphilis, possibly by impairing macrophages and antibody production.

Syphilis can be treated with antibiotics (penicillin) during the primary, secondary, and latent periods. Certain forms of neurosyphilis may also be successfully treated, but the prognosis for others is very poor.

Genital Herpes

Another sexually transmitted disease, **genital herpes**, is common in the United States. Each year, 400,000–600,000 new cases join an estimated 30 million who are already infected. Type II herpes simplex virus (HSV-2) causes genital infections, such as painful genital blisters on the prepuce, glans penis, and penile shaft in males and on the vulva or sometimes high up in the vagina in females. The blisters disappear and reappear in most patients, but the virus itself remains in the body. A related virus, type I herpes simplex virus (HSV-1), causes cold sores on the mouth and lips.

Infected persons typically experience recurrences of symptoms several times a year. Treatment of the symptoms involves pain medication, saline compresses, sexual abstinence for the duration of the eruption, and use of a salve or oral drug called acyclovir (Zovirax). This drug interferes with viral DNA replication but not with host cell DNA replication. Acyclovir speeds the healing and sometimes reduces the pain of initial genital herpes infections. It also shortens the duration of lesions in patients with recurrent genital herpes and reduces the number of flareups. A topically applied ointment that contains Inter Vir-A (Immuvir), an antiviral substance, is another drug used to treat genital herpes. Inter Vir-A provides rapid relief for the pain, itching, and burning associated with genital herpes. An experimental genital herpes vaccine is being tested.

Trichomoniasis

The microorganism *Trichomonas vaginalis,* a flagellated protozoan, causes **trichomoniasis** (trik′-ō-mō-NI-a-sis), an inflammation of the mucous membrane of the vagina in females and the urethra in males where it is a common inhabitant. If the normal acidity of the vagina is disrupted, the protozoan may overgrow the normal microbial population and cause trichomonia-

sis. Symptoms include a yellow vaginal discharge with a particularly offensive odor and severe vaginal itch. Men can have it without overt symptoms but can nevertheless transmit it to women. Sexual partners must be treated simultaneously. The drug of choice is metronidazole.

Genital Warts

Warts are an infectious disease caused by viruses. Sexual transmission of **genital warts** is common and is caused by the *human papillomavirus* (*HPV*). It is estimated that nearly one million persons a year develop genital warts in the United States. Patients with a history of genital warts may be at increased risk for cervical, vaginal, anal, vulval, and penile cancers. There is no cure for genital warts. Treatment consists of cryotherapy with liquid nitrogen, electrocautery, excision, laser surgery, and topical application of podophyllin in tincture of benzoin. Alpha-interferon is also used to treat genital warts.

MALE DISORDERS

Testicular Cancer

Testicular cancer occurs most often between the ages of 15 and 34 and is one of the most common cancers seen in young males. Although the cause is unknown, the condition is more prevalent in males who have a history of cryptorchidism (described on page 910). Most testicular cancers arise from the sperm-producing cells. An early sign of testicular cancer is a mass in the testis, often associated with pain or discomfort. All males should perform regular testicular self-exams. Treatment involves removal of the affected testis.

Prostate Disorders

Because the prostate surrounds a portion of the urethra, any infection, enlargement, or tumor can obstruct the flow of urine. Prolonged obstruction may result in serious changes in the urinary bladder, ureters, and kidneys and may perpetuate urinary tract infections. One treatment consists of widening a narrowed urethra with a balloon catheter (balloon urethroplasty). If the obstruction cannot be relieved by other means, the gland may be partially or completely removed. The surgical procedure is called **prostatectomy** (pros′-ta-TEK-tō-mē).

Acute and chronic infections of the prostate gland are common in postpubescent males, often in association with inflammation of the urethra. In **acute prostatitis**, the prostate gland becomes swollen and tender. Appropriate antibiotic therapy, bed rest, and above-normal fluid intake are effective treatment. **Chronic prostatitis** is one of the most common chronic infections in men of the middle and later years. On examination, the prostate gland feels enlarged, soft, and very tender, and its surface outline is irregular. When bacteria are present, treatment is with long-term antibiotics. Often, however, bacteria cannot be detected in the urinary tract, and the symptoms do not resolve with antibiotic therapy. Such cases may be an autoimmune disorder.

An enlarged prostate gland, two to four times the normal size, occurs in approximately one-third of all males over age 60. The condition is called **benign prostatic hyperplasia** (**BPH**) and is characterized by nocturia (bed-wetting), hesitancy in urination, decreased force of urinary stream, postvoiding dribbling, and a sensation of incomplete emptying. Surgical correction is possible by a procedure called **transurethral resection of the prostate**

(TURP), in which pieces of the gland are removed using a special cystoscope inserted into the urethra.

Both benign and malignant growths are common in elderly men. Both types of tumors put pressure on the urethra, making urination painful and difficult. Therefore, even when the tumor is benign, surgery may be needed.

Abnormalities of Sexual Function

Impotence (*impotenia* = lack of strength) is the inability of an adult male to ejaculate or to attain or hold an erection long enough for sexual intercourse. Many cases of impotence are thought to be caused by insufficient release of the appropriate neurotransmitter, perhaps nitric oxide, which relaxes the smooth muscle of the penile arteries. Other causes include diabetes mellitus, physical abnormalities of the penis, systemic disorders such as syphilis, vascular disturbances (arterial or venous obstructions), neurological disorders, testosterone deficiency, or drugs (alcohol, antidepressants, antihistamines, antihypertensives, narcotics, nicotine, and tranquilizers). Psychic factors such as fear of causing pregnancy, fear of sexually transmitted diseases, religious inhibitions, and emotional immaturity may also cause impotence.

In some impotent men, penile implants may be helpful. Penile injections of papaverine (Pavabid), a vasodilator, and phentolamine mesylate (Regitine), an alpha-adrenergic blocker, can produce excellent effects in overcoming both physical and psychological impotence.

Male infertility (**sterility**) is an inability to fertilize a secondary oocyte. It does not imply impotence. Male fertility requires production of adequate quantities of viable, normal sperm by the testes, unobstructed transport of sperm through the ducts, and satisfactory deposition in the vagina. The tubules of the testes are sensitive to many factors—x-rays, infections, toxins, malnutrition, and significantly higher-than-normal scrotal temperatures—that may cause degenerative changes and produce male sterility. If inadequate sperm production is suspected, a sperm analysis should be performed.

FEMALE DISORDERS

Menstrual Abnormalities

Because menstruation reflects not only the health of the uterus but also the health of the endocrine glands that control it, the ovaries, anterior pituitary, and hypothalamus, disorders of the female reproductive system often involve menstrual disorders.

Amenorrhea (ā-men′-ō-RĒ-a; *a* = without; *men* = month; *rhein* = to flow) is the absence of menstruation. If a woman has never menstruated, the condition is called **primary amenorrhea**. Primary amenorrhea can be caused by endocrine disorders, most often in the pituitary gland and hypothalamus, or by a genetically caused abnormal development of the ovaries or uterus. **Secondary amenorrhea**, the skipping of one or more periods, is commonly experienced by women at some time during their lives. Changes in body weight often cause amenorrhea. Either obesity or extreme weight loss, such as occurs in anorexia nervosa, can disturb ovarian function and cause amenorrhea. Amenorrhea may also be associated with very low body fat level, as sometimes occurs during rigorous athletic training. When amenorrhea is unrelated to weight, analysis of levels of estrogens often reveals deficiencies of pituitary and ovarian hormones.

Dysmenorrhea (dis′-men-ō-RĒ-a; *dys* = difficult) refers to pain associated with menstruation and is usually reserved to describe an individual with menstrual symptoms that are severe enough to prevent her from functioning normally for one or more days each month. **Primary dysmenorrhea** is painful menstruation with no detectable organic disease. Besides pain, other signs and symptoms may include headache, nausea, diarrhea or constipation, and urinary frequency. The pain of primary dysmenorrhea is thought to result from uterine contractions, probably associated with uterine muscle ischemia and prostaglandins produced by the uterus. Primary dysmenorrhea is less of a problem after pregnancy and vaginal delivery, perhaps because of enlargement of the cervical canal. Drugs that inhibit prostaglandin synthesis (naproxen and ibuprofen) are used to treat primary dysmenorrhea.

Secondary dysmenorrhea is painful menstruation that is frequently associated with a pelvic pathology. Some cases are caused by uterine tumors, ovarian cysts, pelvic inflammatory disease (described on page 952), endometriosis, and intrauterine devices (IUDs). Treatment is aimed at correction of the underlying cause.

Abnormal uterine bleeding includes menstruation of excessive duration or excessive amount, diminished menstrual flow, too frequent menstruation, intermenstrual bleeding, and postmenopausal bleeding. These abnormalities may be caused by disordered hormonal regulation, emotional factors, fibroid tumors of the uterus, and systemic diseases.

Premenstrual syndrome (**PMS**) refers to severe physical and emotional distress that occurs late in the postovulatory phase of the menstrual cycle and sometimes extends into the menstrual phase. Signs and symptoms usually increase in severity until the onset of menstruation and then dramatically disappear. Among the signs and symptoms are edema, weight gain, breast swelling and tenderness, abdominal distension, backache, joint pain, constipation, skin eruptions, fatigue and lethargy, greater need for sleep, depression or anxiety, irritability, mood swings, headache, poor coordination and clumsiness, and cravings for sweet or salty foods. The basic cause of PMS is unknown. Although PMS is related to the cyclic production of ovarian hormones, the symptoms are not directly due to changes in the levels of these hormones. Treatment may include dietary changes, exercise, over-the-counter drugs for pain relief (ibuprofen or acetaminophen), psychoactive drugs (sedatives, tranquilizers, and antidepressants), prostaglandins, diuretics, and vitamin B_6. Management of PMS involves medical, psychological, and social support that may include education of the patient and her family; elimination of fears or inappropriate beliefs regarding the menstrual cycle; alteration in coping style; change in life-style, occupation, or family relationships; and use of appropriate medications.

Toxic Shock Syndrome

Toxic shock syndrome (**TSS**), first described in 1978, is primarily a disease of previously healthy, young, menstruating females who use tampons. It is also recognized in males, children, and nonmenstruating females. Clinically, TSS is characterized by high fever up to 40.6°C (105°F), sore throat or very tender mouth, headache, fatigue, lethargy, memory loss, hypotension, irritability, muscle soreness and tenderness, conjunctivitis, diarrhea and vomiting, abdominal pain, vaginal irritation, and erythematous rash.

DISORDERS: HOMEOSTATIC IMBALANCES CONTINUED

Toxin-producing strains of the bacterium *Staphylococcus aureus* are necessary for development of the disease. The risk is greatest in females who use highly absorbent tampons. These tampons absorb magnesium that is normally present in the vagina. When the amount of magnesium in the vagina is reduced, *S. aureus* produces large amounts of toxin that cause the disease. TSS can also occur as a complication of influenza and influenza-like illness. Initial therapy is directed at correcting all homeostatic imbalances as quickly as possible. Anti-staphylococcal antibiotics, such as penicillin or clindamycin, are also administered.

Ovarian Cysts

Ovarian cysts are fluid-containing sacs within the ovary. Follicular cysts may occur in the ovaries of elderly women, in ovaries that have inflammatory diseases, and in menstruating females. They have thin walls and contain a serous albuminous material. Cysts may also arise in the corpus luteum or the endometrium.

Endometriosis

Endometriosis (en'-dō-mē-trē-Ō-sis; *endo* = within; *metri* = uterus; *osis* = condition) is characterized by the growth of endometrial tissue outside the uterus. The tissue enters the pelvic cavity via the open uterine tubes and may be found in any of several sites—on the ovaries, rectouterine pouch, outer surface of the uterus, sigmoid colon, pelvic and abdominal lymph nodes, cervix, abdominal wall, kidneys, and urinary bladder. One theory for the development of endometriosis is that there is regurgitation of menstrual flow through the uterine tubes. Endometriosis is common in women 25–40 years of age who have not had children. Symptoms include premenstrual pain or unusual menstrual pain. The unusual pain is caused by the displaced tissue sloughing off at the same time the normal uterine endometrium is being shed during menstruation. Infertility can be a consequence. Treatment usually consists of hormone therapy, modified GnRH (nafarelin), videolaseroscopy (laparoscope with camera and laser), or conventional surgery. Endometriosis disappears at menopause or when the ovaries are removed.

Female Infertility

Female infertility, or the inability to conceive, may be caused by ovarian disease, tubal obstruction, and certain conditions of the uterus. An upset in hormone balance, so that the endometrium is not adequately prepared to receive the fertilized ovum, may also be the problem. Some research suggests that an autoimmune disease might underlie many cases of infertility. Infertility treatment may involve the use of fertility drugs, donor (artificial) insemination, or surgery. Gynecologists are now using a procedure called **transcervical balloon tuboplasty** to clear obstructions in the uterine tubes. The technique, borrowed from the cardiology procedure to unclog coronary arteries, consists of inserting a catheter through the cervix of the uterus and into the uterine tube. Then a balloon is inflated, compressing the obstruction.

Breast Cysts and Benign Tumors

The breasts of females are highly susceptible to cysts and tumors. In females, **fibrocystic disease** is the most common cause of a breast lump in which one or more cysts (fluid-filled sacs) and thickening of alveoli (clusters of milk-secreting cells) develop. The condition occurs mainly in females between the ages of 30 and 50 and is probably due to a hormonal imbalance; a relative excess of estrogens or deficiency of progesterone in the postovulatory (luteal) phase of the reproductive cycle may be responsible. Fibrocystic disease usually causes one or both breasts to become lumpy, swollen, and tender about a week or so before menstruation begins. The cysts may be aspirated to relieve the pain. Medical management may involve administration of progesterone, antiestrogens, prolactin inhibitors, and pituitary gonadotropin-inhibiting agents.

Benign **fibroadenoma** is a common tumor of the breast. It occurs most often in young women. Fibroadenomas have a firm rubbery consistency and are easily moved about within the breast tissue. The usual treatment is excision of the growth. The breast itself is not removed.

Cervical Cancer

Another common disorder of the female reproductive tract is **cervical cancer**, carcinoma of the cervix of the uterus. The condition starts with **cervical dysplasia** (dis-PLA-sē-a), a change in the shape, growth, and number of the cervical cells. If the condition is minimal, the cells may regress to normal. If it is severe, it may progress to cancer. Cervical cancer may be detected in most cases in its earliest stages by a Pap smear. There is some evidence linking cervical cancer to penile virus (papillomavirus) infections of male sexual partners. Depending on the progress of the disease, treatment may consist of excision of lesions, radiotherapy, chemotherapy, and hysterectomy (removal of the uterus).

Pelvic Inflammatory Disease

Pelvic inflammatory disease (**PID**) is a collective term for any extensive bacterial infection (primarily involving *Chlamydia trachomatis, Neisseria gonorrhoeae, Bacteroides, Peptostreptococcus,* and *Gardnerella vaginalis*) of the pelvic organs, especially the uterus, uterine tubes, or ovaries. A vaginal or uterine infection may spread into the uterine tube (**salpingitis**) or even farther into the abdominal cavity, where it infects the peritoneum (**peritonitis**). Diagnosis of PID depends on three findings: abdominal tenderness; cervical tenderness; and ovarian, uterine tube, and uterine ligament tenderness. In addition, diagnosis is based on at least one of the following: fever, leukocytosis, pelvic abscess or inflammation, purulent cervical discharge, and the presence of certain bacteria. Early treatment with bed rest and antibiotics (cefoxitin, penicillin, tetracycline, doxycycline) can stop the spread of PID.

Vulvovaginal Candidiasis

Candida albicans is a yeastlike fungus that commonly grows on mucous membranes of the gastrointestinal and genitourinary tracts. The organism is responsible for **vulvovaginal candidiasis** (vul'-vō-VAJ-i-nal can'-di-DI-a-sis), the most common form of vaginitis. It is characterized by severe itching; a thick, yellow, cheesy discharge; a yeasty odor; and pain. The disorder, experienced at least once by about 75% of females, is usually a result of proliferation of the fungus following antibiotic therapy for another condition. Predisposing conditions include use of oral contraceptives, cortisone-like medications, pregnancy, and diabetes. Treatment is by topical (clostrimazole) or oral (ketoconazole) drugs.

MEDICAL TERMINOLOGY

Castration (kas-TRĀ-shun; *castrare* = to prune) Removal, inactivation, or destruction of the gonads.

Colpotomy (kol-POT-ō-mē; *colp* = vagina; *tome* = cutting) Incision of the vagina.

Culdoscopy (kul-DOS-kō-pē; *skopein* = to examine) A procedure in which a culdoscope (endoscope) is used to view the female pelvic cavity. The approach is through the vagina.

Endocervical curettage (ku-re-TAZH; *curette* = scraper) The cervix is dilated and the endometrium of the uterus is scraped with a spoon-shaped instrument called a curette. This procedure is commonly called a **D and C** (**dilation and curettage**).

Hermaphroditism (her-MAF-rō-di-tizm′) Presence of both male and female sex organs in one individual.

Hypospadias (hī′-pō-SPĀ-dē-as; *hypo* = below; *span* = to draw) A displaced urethral opening. In the male, the

opening may be on the underside of the penis, at the penoscrotal junction, between the scrotal folds, or in the perineum. In the female, the urethra opens into the vagina.

Leukorrhea (loo′-kō-RĒ-a; *leuco* = white; *rrhea* = discharge) A nonbloody vaginal discharge that may occur at any age and affects most women at some time.

Oophorectomy (ō′-of-ō-REK-tō-mē; *oophoro* = bearing eggs; *ektome* = excision) Removal of the ovaries.

Salpingectomy (sal′-pin-JEK-tō-mē; *salpingo* = tube) Removal of a uterine (Fallopian) tube.

Smegma (SMEG-ma; *smegma* = soap) The secretion, consisting principally of desquamated epithelial cells, found chiefly about the external genitalia and especially under the foreskin of the male.

Vaginitis (vaj′-i′-NĪ-tis) Inflammation of the vagina.

STUDY OUTLINE

MALE REPRODUCTIVE SYSTEM (p. 909)

1. Reproduction is the process by which new individuals of a species are produced and the genetic material is passed from generation to generation.

2. The organs of reproduction are grouped as gonads (produce gametes), ducts (transport and store gametes), accessory sex glands (produce materials that support gametes), and supporting structures (have various roles in reproduction).

3. The male structures of reproduction include the testes, ductus epididymis, ductus (vas) deferens, ejaculatory duct, urethra, seminal vesicles, prostate gland, bulbourethral (Cowper's) glands, and penis.

Scrotum (p. 910)

1. The scrotum is a sac that hangs from the root of the penis and consists of loose skin and superficial fascia. It supports the testes.

2. It regulates the temperature of the testes by contraction of the cremaster muscle and dartos, which elevates them and brings them closer to the pelvic cavity or relaxes causing testes to move farther from the pelvic cavity.

Testes (p. 910)

1. The testes are oval-shaped glands (gonads) in the scrotum containing seminiferous tubules, in which sperm cells are made; sustentacular (Sertoli) cells, which nourish sperm cells and secrete inhibin; and interstitial endocrinocytes (Leydig cells), which produce the male sex hormone testosterone.

2. The testes descend into the scrotum through the inguinal canals during the seventh month of fetal development. Failure of the testes to descend is called cryptorchidism.

3. Secondary oocytes and sperm are collectively called gametes and are produced in gonads.

4. Somatic cells divide by mitosis, the process in which each daughter cell is identical and receives the full complement of 23 chromosome pairs (46 chromosomes). Somatic cells are said to be diploid (2*n*).

5. Germ cells divide by meiosis to produce nonidentical gametes, in which the pairs of chromosomes are split so that the gamete has only 23 chromosomes. It is said to be haploid (*n*).

6. Spermatogenesis occurs in the testes. It results in the formation of four haploid sperm (spermatozoa) from each primary spermatocyte.

7. Spermatogenesis is a process in which immature spermatogonia develop into mature sperm. The spermatogenesis sequence includes reduction division (meiosis I), equatorial division (meiosis II), and spermiogenesis.

8. Mature sperm consist of a head, midpiece, and tail. Their function is to fertilize a secondary oocyte.

9. At puberty, gonadotropin releasing hormone (GnRH) stimulates anterior pituitary gland secretion of FSH and LH. LH stimulates production of testosterone. FSH and testosterone stimulate spermatogenesis. Sustentacular (Sertoli) cells secrete androgen-binding protein (ABP), which binds to testosterone and keeps its concentration high in the seminiferous tubule.

10. Testosterone controls the growth, development, and maintenance of sex organs; stimulates bone growth, protein anabolism, and sperm maturation; and stimulates development of masculine secondary sex characteristics.

11. Inhibin is produced by sustentacular (Sertoli) cells. Its inhibition of FSH helps regulate the rate of spermatogenesis.

Ducts (p. 916)

1. The duct system of the testes includes the seminiferous tubules, straight tubules, and rete testis.

2. Sperm flow out of the testes through the efferent ducts.

3. The ductus epididymis is the site of sperm maturation and storage.

4. The ductus (vas) deferens stores sperm and propels them toward the urethra during ejaculation.

5. Cutting and tying-off of the ductus (vas) deferens to achieve sterility by blocking entry of sperm into semen is called vasectomy.

6. Each ejaculatory duct is formed by the union of the duct from the seminal vesicle and ductus (vas) deferens. It is the passageway

for ejection of sperm and secretions of the seminal vesicles into the first portion of the urethra, the prostatic urethra.

7. The male urethra is subdivided regionally into three portions: prostatic, membranous, and spongy (penile).

Accessory Sex Glands (p. 918)

1. The seminal vesicles secrete an alkaline, viscous fluid that constitutes about 60% of the volume of semen and contributes to sperm viability.
2. The prostate gland secretes a slightly acidic fluid that constitutes about 25% of the volume of semen and contributes to sperm motility. Prostate cancer is the leading cause of death from cancer in U.S. men.
3. The bulbourethral (Cowper's) glands secrete mucus for lubrication and an alkaline substance that neutralizes acid.
4. Semen is a mixture of sperm and accessory sex gland secretions that provides the fluid in which sperm are transported, provides nutrients, and neutralizes the acidity of the male urethra and female vagina.

Penis (p. 921)

1. The penis consists of a root, body, and glans penis.
2. Expansion of its blood sinuses under the influence of sexual excitation is called erection.

FEMALE REPRODUCTIVE SYSTEM (p. 923)

1. The female organs of reproduction include the ovaries (gonads), uterine (Fallopian) tubes or oviducts, uterus, vagina, and vulva.
2. The mammary glands are considered part of the reproductive system.

Ovaries (p. 924)

1. The ovaries are female gonads located in the superior portion of the pelvic cavity, lateral to the uterus.
2. They produce secondary oocytes, discharge secondary oocytes (ovulation), and secrete estrogens, progesterone, relaxin, and inhibin.
3. Oogenesis (production of haploid secondary oocytes) begins in the ovaries. The oogenesis sequence includes reduction division (meiosis I) and equatorial division (meiosis II), which goes to completion only after an ovulated secondary oocyte is fertilized by a sperm cell.

Uterine (Fallopian) Tubes (p. 927)

1. The uterine (Fallopian) tubes transport secondary oocytes from the ovaries to the uterus and are the normal sites of fertilization.
2. Ciliated cells and peristaltic contractions help move a secondary oocyte or fertilized ovum toward the uterus.

Uterus (p. 929)

1. The uterus is an organ the size and shape of an inverted pear that functions in menstruation, implantation of a fertilized ovum, development of a fetus during pregnancy, and labor. It also is part of the pathway for sperm to reach the uterine tubes to fertilize a secondary oocyte.
2. The uterus is normally held in position by a series of ligaments.
3. Histologically, the layers of the uterus are an outer perimetrium (serosa), middle myometrium, and inner endometrium.

Vagina (p. 931)

1. The vagina is a passageway for sperm and the menstrual flow, the receptacle of the penis during sexual intercourse, and the inferior portion of the birth canal.
2. It is capable of considerable distension.

Vulva (p. 932)

1. The vulva is a collective term for the external genitals of the female.
2. It consists of the mons pubis, labia majora, labia minora, clitoris, vestibule, vaginal and urethral orifices, hymen, bulb of the vestibule, and the paraurethral (Skene's), greater vestibular (Bartholin's), and lesser vestibular glands.

Perineum (p. 933)

1. The perineum is a diamond-shaped area at the inferior end of the trunk medial to the thighs and buttocks.
2. An incision in the female perineum before delivery is called an episiotomy.

Mammary Glands (p. 934)

1. The mammary glands are modified sweat glands lying superficial to the pectoralis major muscles. Their function is to synthesize, secrete, and eject milk (lactation).
2. Mammary gland development depends on estrogens and progesterone.
3. Milk production is stimulated by prolactin, estrogen, and progesterone. Milk ejection is stimulated by oxytocin.
4. In females, breast cancer is rarely seen before age 30, its occurrence rises rapidly after menopause, and it has a high fatality rate. Early detection—by breast self-examination and mammography—is the best way to increase the chance of survival.

FEMALE REPRODUCTIVE CYCLE (p. 937)

1. The function of the ovarian cycle is to develop a secondary oocyte. The function of the uterine (menstrual) cycle is to prepare the endometrium each month to receive a fertilized egg.
2. The uterine and ovarian cycles are controlled by GnRH from the hypothalamus, which stimulates the release of FSH and LH by the anterior pituitary gland.
3. FSH stimulates development of secondary follicles and initiates secretion of estrogens by the follicles. LH stimulates further development of the follicles, secretion of estrogens by follicular cells, ovulation, formation of the corpus luteum, and the secretion of progesterone and estrogens by the corpus luteum.
4. Estrogens stimulate the growth, development, and maintenance of female reproductive structures; stimulate the development of secondary sex characteristics; regulate fluid and electrolyte balance; and stimulate protein synthesis.
5. Progesterone works with estrogens to prepare the endometrium for implantation and the mammary glands for milk synthesis.
6. Relaxin relaxes the pubic symphysis and helps dilate the uterine cervix to facilitate delivery.
7. During the menstrual phase (days 1–5), the stratum functionalis of the endometrium is shed, discharging blood, tissue fluid, mucus, and epithelial cells.
8. During the preovulatory phase, a group of follicles in the ovaries begin to undergo final maturation. One follicle outgrows the others and becomes dominant while the others degenerate. At the same time endometrial repair occurs in the uterus. Estrogens are the dominant ovarian hormones during the preovulatory phase.
9. Ovulation is the rupture of the dominant mature (Graafian) follicle and the release of a secondary oocyte into the pelvic cavity. It is brought about by a surge of LH. Signs of ovulation include increased basal body temperature; clear, stretchy cervical mucus; changes in the uterine cervix; and ovarian pain.

10. During the postovulatory phase, both progesterone and estrogens are secreted in large quantity by the corpus luteum of the ovary and the uterine endometrium thickens in readiness for implantation.

11. If fertilization and implantation do not occur, the corpus luteum degenerates, and the resulting low levels of estrogens and progesterone allow discharge of the endometrium followed by initiation of another uterine and ovarian cycle.

12. If fertilization and implantation do occur, the corpus luteum is maintained by placental hCG, and the corpus luteum and later the placenta secrete progesterone and estrogens to support pregnancy and breast development for lactation.

THE HUMAN SEXUAL RESPONSE (p. 942)

1. The similar sequence of changes experienced by both males and females before, during, and after intercourse is termed the human sexual response. Masters and Johnson described it as occurring in 4 stages—excitement (arousal), plateau, orgasm, and resolution.

2. During excitement and plateau, parasympathetic nerve impulses produce genital vasocongestion, engorgement of tissues with blood, and secretion of lubricating fluids. Heart rate, blood pressure, breathing rate, and muscle tone increase.

3. During orgasm sympathetic and somatic motor nerve impulses cause rhythmical contractions of smooth and skeletal muscles.

4. During resolution, there is relaxation and return of the body to the unaroused state.

BIRTH CONTROL METHODS (p. 943)

1. Methods include sterilization (vasectomy, tubal ligation), hormonal, intrauterine devices, barriers (condom, vaginal pouch, diaphragm, cervical cap), chemicals (spermicides), physiological (rhythm, sympto-thermal method), coitus interruptus, and induced abortion. See Exhibit 28.1 on page 945.

2. Contraceptive pills of the combination type contain estrogens and progestins in concentrations that decrease the secretion of FSH and LH and thereby inhibit development of ovarian follicles and ovulation.

AGING AND THE REPRODUCTIVE SYSTEMS (p. 946)

1. Puberty refers to the period of time when secondary sex characteristics begin to develop and the potential for sexual reproduction is reached.

2. The onset of male puberty is signaled by increased levels of LH, FSH, and testosterone.

3. The onset of female puberty is signaled by increased levels of LH, FSH, and estrogens.

4. In older males, decreased levels of testosterone are associated with decreased muscle strength, sexual desire, and viable sperm; prostate disorders are common.

5. In older females, levels of progesterone and estrogens decrease, resulting in changes in menstruation and then menopause; uterine and breast cancer increase in incidence.

DEVELOPMENTAL ANATOMY OF THE REPRODUCTIVE SYSTEMS (p. 946)

1. The gonads develop from intermediate mesoderm. In the presence of the *SRY* gene and the hormone testosterone, the gonads begin to differentiate into testes during the seventh week of fetal development. The gonads differentiate into ovaries when testosterone is absent.

2. The external genitals develop from the genital tubercle and are stimulated to develop into typical male structures by the hormone dihydrotestosterone (DHT). The external genitals appear female when testosterone is not produced (the normal situation in female embryos) or when 5 alpha-reductase, which converts testosterone to DHT, is deficient due to a genetic defect.

REVIEW QUESTIONS

1. Define reproduction. Describe how the reproductive organs are classified and list the male and female organs of reproduction. (p. 909)

2. Describe the function of the scrotum in protecting the testes from temperature fluctuations. What is cryptorchidism? (p. 910)

3. Describe the internal structure of a testis. Where are the sperm cells made? What are the functions of sustentacular (Sertoli) cells and interstitial endocrinocytes (Leydig cells)? (p. 910)

4. Describe the principal events of spermatogenesis. Why is meiosis important? Distinguish between haploid (*n*) and diploid (2*n*) cells. (p. 911)

5. Identify the principal parts of a sperm cell (spermatozoon). List the functions of each. (p. 914)

6. Explain the effects of FSH and LH on the male reproductive system. How are these hormones controlled by GnRH? (p. 915)

7. Describe the physiological effects of testosterone and inhibin on the male reproductive system. How is testosterone level controlled? (p. 915)

8. Which ducts are involved in transporting sperm *within* the testes? (p. 916)

9. Describe the location, structure, and functions of the ductus epididymis, ductus (vas) deferens, and ejaculatory duct. (p. 917)

10. What is a vasectomy? (p. 917)

11. What is the spermatic cord? What is an inguinal hernia? (p. 917)

12. Give the location of the three subdivisions of the male urethra. (p. 918)

13. Trace the course of sperm through the male system of ducts from the seminiferous tubules through the urethra. (p. 917)

14. Briefly explain the locations and functions of the seminal vesicles, prostate gland, and bulbourethral (Cowper's) glands. (p. 920)

15. What is semen? What is its function? What is a semen analysis? (p. 921)

16. How is the penis structurally adapted as an organ of sexual intercourse? How does an erection occur? What is circumcision? (p. 921)

17. How are the ovaries held in position in the pelvic cavity? (p. 924)

18. Describe the microscopic structure and function of an ovary. What are the early and late symptoms of ovarian cancer? (p. 925)

19. Describe the principal events of oogenesis. (p. 926)

20. Where are the uterine (Fallopian) tubes located? What is their function? (p. 927)

21. Diagram the principal parts of the uterus. What is a Pap smear? Uterine prolapse? Hysterectomy? (p. 929)

22. Describe the arrangement of ligaments that hold the uterus in its normal position. What is retroflexion? (p. 931)

23. Describe the histology of the uterus. (p. 930)
24. Discuss the blood supply to the uterus. Why is an abundant blood supply important? (p. 931)
25. What is the function of the vagina? Describe its histology. What is colposcopy? (p. 931)
26. List the parts of the vulva and the functions of each part. What is an episiotomy? (p. 932)
27. Describe the structure of the mammary glands. How are they supported? (p. 934)
28. Describe the passage of milk from the alveoli of the mammary gland to the nipple. (p. 934)
29. Define lactation. How is it controlled? (p. 936)
30. Explain the roles of estrogens and progesterone in the development of the mammary glands. (p. 937)
31. How is breast cancer detected? How is breast cancer treated? (p. 936)
32. What is the function of each of the following in the uterine and ovarian cycles: GnRH, FSH, LH, estrogens, progesterone, and inhibin? (p. 937)

33. Briefly outline the major events of each phase of the uterine cycle and correlate them with the events of the ovarian cycle. (p. 939)
34. Prepare a labeled diagram of the principal hormonal interactions involved in the uterine and ovarian cycles. (p. 941)
35. What are the signs that ovulation has occurred? (p. 940)
36. Describe the stages of the human sexual response. (p. 942)
37. How are the physiological changes similar and different in males and females during the human sexual response? (p. 942)
38. Briefly describe the various methods of birth control (BC) and the effectiveness of each. (p. 943)
39. Describe the events involved in male and female puberty. (p. 946)
40. Explain the effects of aging on the reproductive systems. (p. 946)
41. Distinguish between menarche and menopause. (p. 946)
42. Describe the development of the reproductive systems. (p. 946)

ANSWERS TO QUESTIONS WITH FIGURES

28.1 Gonads: produce gametes and hormones; ducts: transport, store, and receive gametes; accessory sex glands: produce materials that support gametes.

28.2 Tunica vaginalis and tunica albuginea.

28.3 Interstitial endocrinocytes (Leydig cells).

28.4 Because the number of chromosomes in each cell is reduced by half. It is necessary because otherwise the normal chromosome number would be doubled after fertilization occurs.

28.5 Head contains DNA and enzymes for penetration of secondary oocyte; midpiece contains mitochondria for ATP production; tail consists of a flagellum that provides propulsion.

28.6 Sustentacular (Sertoli) cells.

28.7 Testosterone inhibits secretion of LH and inhibin inhibits secretion of FSH.

28.8 Ilioinguinal nerve and spermatic cord, which consists of the testicular artery and veins, lymphatic vessels, autonomic nerves, ductus deferens, and the cremaster muscle.

28.9 Seminal vesicles.

28.10 Two corpora cavernosa penis and one corpus spongiosum penis contain blood sinuses that fill with blood that cannot flow out of the penis as quickly as it flows in. The trapped blood stiffens the tissue. The corpus spongiosum penis keeps the spongy urethra open so that ejaculation can occur.

28.11 Testes; penis.

28.12 Mesovarium, to broad ligament of the uterus and the uterine tube; ovarian ligament, to uterus; suspensory ligament, to pelvic wall.

28.13 Ovarian follicles secrete estrogens and a corpus luteum secretes progesterone, estrogens, relaxin, and inhibin.

28.14 They undergo atresia (degeneration).

28.15 Primary oocytes are present in the ovary at birth, so they are as old as the woman is. In males, primary spermatocytes are continually being formed from stem cells (spermatogonia) and thus are only a few days old.

28.16 In the ampulla of the uterine tube.

28.17 Ciliated columnar epithelial cells and secretory cells with microvilli.

28.18 Endometrium: highly vascular, secretory epithelium that provides oxygen and nutrients to sustain a fertilized egg; myometrium: thick smooth muscle layer that supports the uterine wall during pregnancy and contracts to expel fetus at birth.

28.19 The stratum basalis provides cells to replace those shed (stratum functionalis) during each menstruation.

28.20 Anterior: mons pubis, clitoris, and prepuce. Lateral: labia minora and labia majora.

28.21 Its borders form a triangle that encloses the urethral (uro-) and vaginal (-genital) orifices.

28.22 Synthesis: prolactin, estrogens, progesterone. Release: oxytocin.

28.23 β-estradiol.

28.24 Estrogens; LH; LH; estrogens.

28.25 Estrogens.

28.26 Negative feedback inhibition of secretion of these hormones.

28.27 Negative, because the response is opposite to the stimulus. Decreasing estrogens and progesterone stimulate release of GnRH, which, in turn, increases production and release of FSH and LH, which ultimately stimulate secretion of estrogens.

DEVELOPMENT AND INHERITANCE

CHAPTER CONTENTS AT A GLANCE

evelopmental anatomy is the study of the sequence of events from the fertilization of a secondary oocyte to the formation of an adult organism. As we look at the sequence from fertilization to birth, we will consider fertilization, implantation, placental development, embryonic development, fetal growth, gestation, labor, and parturition (birth). The development of an embryo and fetus is a wonderfully complex and precisely coordinated series of events.

DEVELOPMENT DURING PREGNANCY

Once sperm and a secondary oocyte have developed through meiosis and maturation, and the sperm have been deposited in the vagina, pregnancy can occur. **Pregnancy** is a sequence of events that normally includes fertilization, implantation, embryonic growth, and fetal growth that ends with birth about 38 weeks later.

Fertilization and Implantation

Fertilization

During **fertilization** (fer-til-i-ZĀ-shun; *fertilis* = fruitful) the genetic material from a sperm cell (spermatozoon) and secondary oocyte merges into a single nucleus. Of the 300–500 million sperm introduced into the vagina, less than 1% reach the secondary oocyte. Fertilization normally occurs in the uterine (Fallopian) tube about 12–24 hours after ovulation. Since ejaculated sperm remain viable for about 48 hours and a secondary oocyte is viable for about 24 hours after ovulation, there typically is a 3-day window during which pregnancy can occur—from 2 days before to 1 day after ovulation. Peristaltic contractions and the action of cilia transport the oocyte through the uterine tube. Sperm swim up the female tract by whiplike movements of their tail (flagellum). The acrosome of sperm produces an enzyme called **acrosin** that stimulates sperm motility and migration within the female reproductive tract. Also, muscular contractions of the uterus, stimulated by prostaglandins in semen, probably aid sperm movement toward the uterine tube. Finally, the oocyte is thought to secrete a chemical substance that attracts sperm.

Besides contributing to sperm movement, the female reproductive tract also confers on sperm the capacity to fertilize a secondary oocyte. Although sperm undergo maturation in the epididymis, they are still not able to fertilize an oocyte until they have been in the female reproductive tract for several hours.

Capacitation (ka-pas'-i-TĀ-shun) refers to the functional changes that sperm undergo in the female reproductive tract that allow them to fertilize a secondary oocyte. During this process, the membrane around the acrosome becomes fragile so that several destructive enzymes—hyaluronidase, acrosin,

and neuraminidase—are released from the acrosomes. It requires the collective action of many sperm for just one to penetrate the secondary oocyte. The enzymes help penetrate the **corona radiata**, a ring of cells that surrounds the oocyte, and a clear glycoprotein layer internal to the corona radiata called the **zona pellucida** (pe-LOO-si-da; *pellucida* = translucent) (Fig. 29.1a). Normally only one sperm cell penetrates and enters a secondary oocyte. This event is called **syngamy** (*syn* = together; *gamos* = marriage). Syngamy causes depolarization, which triggers the release of calcium ions inside the cell. Calcium ions stimulate the release of granules by the oocyte that, in turn, promote changes in the zona pellucida to block

Figure 29.1 Fertilization. (a) Sperm cell penetrating the corona radiata and zona pellucida around a secondary oocyte. (b) Female and male pronuclei. Figure 28.5b shows a sperm cell in contact with the surface of a secondary oocyte.

 During fertilization, genetic material from a sperm cell merges with that of an oocyte to form a single nucleus.

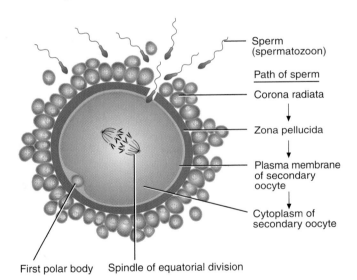

First polar body   Spindle of equatorial division

(a) Diagram

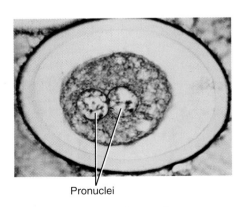

Pronuclei

(b) Photomicrograph

 What is capacitation?

entry of other sperm. This prevents **polyspermy**, fertilization by more than one sperm cell. Once a sperm cell has entered a secondary oocyte, the oocyte completes equatorial division (meiosis II). It divides into a larger ovum (mature egg) and a smaller second polar body that fragments and disintegrates (see Fig. 28.15).

When a sperm cell has entered a secondary oocyte, the tail is shed and the nucleus in the head develops into a structure called the **male pronucleus**. The nucleus of the fertilized ovum develops into a **female pronucleus** (Fig. 29.1b). After the pronuclei are formed, they fuse to produce a **segmentation nucleus**. The segmentation nucleus is diploid since it contains 23 chromosomes (n) from the male pronucleus and 23 chromosomes (n) from the female pronucleus. Thus the fusion of the haploid (n) pronuclei restores the diploid number ($2n$). The fertilized ovum, consisting of a segmentation nucleus, cytoplasm, and zona pellucida, is called a **zygote** (ZĪ-gōt; *zygosis* = a joining).

Dizygotic (fraternal) twins are produced from the independent release of two secondary oocytes and the subsequent fertilization of each by different sperm. They are the same age and are in the uterus at the same time, but they are genetically as dissimilar as any other siblings. Dizygotic twins may or may not be the same sex. **Monozygotic (identical) twins** develop from a single fertilized ovum that splits at an early stage in development and so they contain exactly the same genetic material and are always the same sex. On rare occasions, monozygotic twins may be joined in varying degrees, from slight skin fusion to sharing of limbs, trunks, and viscera. Such twins are called **conjoined twins**.

Formation of the Morula

After fertilization, rapid mitotic cell divisions of the zygote take place. These early divisions of the zygote are called **cleavage**. Although cleavage increases the number of cells, it does not increase the size of the embryo, which is still contained within the zona pellucida.

The first cleavage begins about 24 hours after fertilization and is completed about 30 hours after fertilization. Each succeeding division takes slightly less time (Fig. 29.2a,b). By the second day after fertilization, the second cleavage is completed. By the end of the third day, there are 16 cells. The progressively smaller cells produced by cleavage are called **blastomeres** (BLAS-tō-mērz; *blast* = germ, sprout; *meros* = part). Successive cleavages produce a solid sphere of cells, still surrounded by the zona pellucida, called the **morula** (MOR-yoo-la; *morula* = mulberry). See Fig. 29.2c. The morula is about the same size as the original zygote.

Development of the Blastocyst

By the end of the fourth day, the number of cells in the morula increases and it continues to move through the uterine (Fallopian) tube toward the uterine cavity. At 4½–5 days, the dense cluster of cells has developed into a hollow ball of cells and enters the uterine cavity; it is now called a **blastocyst** (*kystis* = bag) (Fig. 29.2d,e).

Figure 29.2 Cleavage and formation of the morula and blastocyst.

Cleavage refers to the early, rapid mitotic divisions in a zygote.

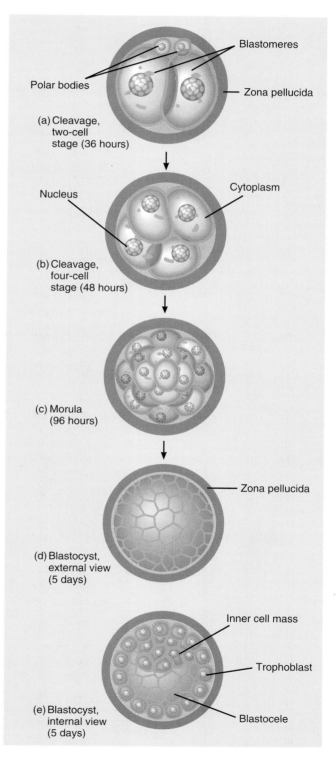

(a) Cleavage, two-cell stage (36 hours)

Blastomeres

Polar bodies

Zona pellucida

Nucleus

Cytoplasm

(b) Cleavage, four-cell stage (48 hours)

(c) Morula (96 hours)

Zona pellucida

(d) Blastocyst, external view (5 days)

Inner cell mass

Trophoblast

(e) Blastocyst, internal view (5 days)

Blastocele

 What is the histological difference between a morula and a blastocyst?

The blastocyst has an outer covering of cells called the **trophoblast** (TRŌF-ō-blast; *troph* = nourish), an **inner cell mass** (**embryoblast**), and an internal fluid-filled cavity called the **blastocele** (BLAS-tō-sēl; *koilos* = hollow). The trophoblast and part of the inner cell mass ultimately form the membranes composing the fetal portion of the placenta; the rest of the inner cell mass develops into the embryo.

ECTOPIC PREGNANCY

Ectopic (*ektos* = outside; *topos* = place) **pregnancy** refers to the development of an embryo or fetus outside the uterine cavity. Most occur in the uterine (Fallopian) tube, usually in the ampullar and infundibular portions. Some occur in the abdominal cavity or uterine cervix. Tubal pregnancies usually occur because passage of the fertilized ovum through the uterine tube is impaired. The cause may be decreased motility of the uterine tube smooth muscle or abnormal anatomy. In comparison with nonsmokers, women who smoke and become pregnant are twice as likely to have an ectopic pregnancy. It is thought that the nicotine in cigarette smoke paralyzes the cilia in the lining of the uterine tube as it does in the respiratory passageways. Scars from pelvic inflammatory disease (PID), previous uterine tube surgery, and previous ectopic pregnancy may hinder movement of the fertilized ovum. Other causes include repeated elective abortions, pelvic tumors, and developmental abnormalities.

Ectopic pregnancy may be characterized by one or two missed menstrual cycles, followed by bleeding and acute abdominal and pelvic pain. Unless removed, the developing embryo can rupture the tube, often resulting in death of the mother. ■

Implantation

The blastocyst remains free within the cavity of the uterus for a short period of time before it attaches to the uterine wall. During this time, the zona pellucida disintegrates and the blastocyst enlarges. The blastocyst receives nourishment from glycogen-rich secretions of endometrial (uterine) glands, sometimes called uterine milk. About 6 days after fertilization the blastocyst attaches to the endometrium, a process called **implantation** (Fig. 29.3). At this time, the endometrium is in its secretory phase (see Fig. 28.28).

As the blastocyst implants, usually on the posterior wall of the fundus or body of the uterus, it is oriented so that the inner cell mass is toward the endometrium. The trophoblast develops two layers in the region of contact between the blastocyst and endometrium. These layers are a **syncytiotrophoblast** (sin-sit'-ē-ō-TRŌF-ō-blast; *syn* = joined; *cyto* = cell) that contains no cell boundaries and a **cytotrophoblast** (sī-tō-TRŌF-ō-blast) between the inner cell mass and syncytiotrophoblast that is composed of distinct cells (Fig. 29.3c). These two layers of trophoblast become part of the chorion (one of the fetal membranes) as they undergo further growth (see Fig. 29.5). During implantation, the syncytiotrophoblast secretes enzymes that enable the blastocyst to penetrate the uterine lining. The enzymes digest and liquefy the endometrial cells. The fluid and nutrients further nourish the burrowing blastocyst for about a week after implantation. Eventually, the blastocyst becomes buried in the endometrium. The trophoblast also secretes human chorionic gonadotropin (hCG), which has actions similar to LH. It rescues the corpus luteum from degeneration and sustains its secretion of progesterone and estrogens (see Fig. 29.9). Thus menstruation does not begin.

Since a developing embryo, and later a fetus, contains genes from the father as well as the mother, it is essentially a foreign graft. Thus it is surprising that it is not rejected by the mother's immune system. The trophoblast is the only tissue of the developing organism that contacts the mother's uterus. Even though trophoblast cells have paternal antigens that could provoke a rejection response, some mechanism in the uterus prevents this to permit development of the fetus to term. One possibility is that the mother makes antibodies that mask paternal antigens so that other components of her immune system cannot recognize and attack the antigens.

A summary of the principal events associated with fertilization and implantation is shown in Fig. 29.4 on page 962.

Alternative Methods of Fertilization

On July 12, 1978, Louise Joy Brown was born near Manchester, England. Her birth was the first recorded case of **in vitro fertilization** (**IVF**)—fertilization in a laboratory dish. In this procedure for IVF, the mother-to-be is given follicle-stimulating hormone (FSH) soon after menstruation, so that several secondary oocytes, rather than the typical single one, will be produced (superovulation). Administration of luteinizing hormone (LH) may also ensure the maturation of the secondary oocytes. Next, a small incision is made near the umbilicus, and the secondary oocytes are aspirated from the stimulated follicles and then transferred to a solution of the male's sperm.

Once fertilization has taken place in a lab dish, the fertilized ovum is put in another medium and observed for cleavage. When the zygote reaches the 8-cell or 16-cell stage, it is introduced into the uterus for implantation and subsequent growth. It is also possible to freeze embryos (cryopreservation) to permit parents another pregnancy several years later or allow a second attempt at implantation if the first attempt is unsuccessful.

In **embryo transfer**, a husband's semen is used to artificially inseminate a fertile secondary oocyte donor. After fertilization in the donor's uterine (Fallopian) tube, the morula or blastocyst is transferred from the donor to the infertile wife who carries it to term. Embryo transfer is indicated for females who are infertile or who do not want to pass on their own genes because they are carriers of a serious genetic disorder.

Figure 29.3 Implantation. Shown is the blastocyst in relation to the endometrium of the uterus at various time intervals after fertilization.

Implantation refers to the attachment of a blastocyst to the endometrium, which occurs about 6 days after fertilization.

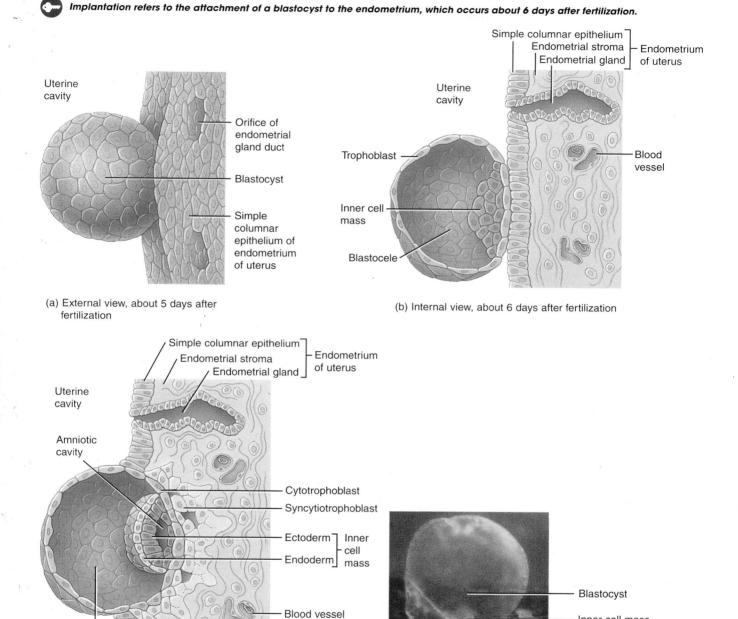

(a) External view, about 5 days after fertilization

(b) Internal view, about 6 days after fertilization

(c) Internal view, about 7 days after fertilization

(d) Photomicrograph

How does the blastocyst merge with and burrow into the endometrium?

In **gamete intrafallopian transfer (GIFT)** the goal is to mimic the normal process of conception by uniting sperm and secondary oocyte in the prospective mother's uterine tubes. It is an attempt to bypass conditions in the female reproductive tract that might prevent fertilization such as high acidity or inappropriate mucus. In the procedure, the female is given FSH and LH to stimulate the production of several secondary oocytes. The secondary oocytes are aspirated

Figure 29.4   Summary of events associated with fertilization and implantation.

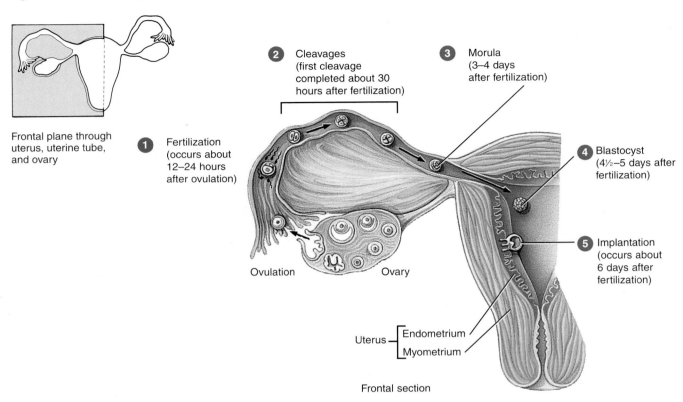

Frontal plane through uterus, uterine tube, and ovary

1 Fertilization (occurs about 12–24 hours after ovulation)

2 Cleavages (first cleavage completed about 30 hours after fertilization)

3 Morula (3–4 days after fertilization)

4 Blastocyst (4½–5 days after fertilization)

5 Implantation (occurs about 6 days after fertilization)

Ovulation   Ovary

Uterus —⌈ Endometrium
          ⌊ Myometrium

Frontal section

Q In which phase of the uterine cycle does implantation occur?

from the mature follicles with a laparoscope fitted with a suction device, mixed with a solution of the male's sperm outside the body, and then immediately inserted into the uterine (Fallopian) tubes.

FERTILITY AND BODY FAT

Body fat has a regulatory role in reproduction, especially in females. To begin and maintain a normal reproductive cycle, a female must have a minimum amount of body fat. A moderate loss of fat, from 10–15% below normal weight for height, may delay the onset of menstruation (menarche), inhibit ovulation during the reproductive cycle, or induce the cessation of menstruation (amenorrhea). Both dieting and intensive exercise may reduce body fat below the minimum amount and lead to infertility. The resulting infertility is reversible following weight gain or reduction of intensive exercise or both.

It appears that in underweight or very lean females, the secretion of gonadotropin releasing hormone (GnRH) by the hypothalamus is abnormal in quantity and timing. The result is abnormally low levels of follicle-stimulating hormone (FSH) and luteinizing hormone (LH). Since these hormones are deficient, ovarian follicles fail to develop normally and there is little secretion of progesterone and estrogens. The same hormonal deficiencies that cause infertility in underweight or athletic females also provide a degree of protection against cancers that are sensitive to estrogens, such as breast cancer, although the reason is not known.

Studies of very obese females also indicate that they, like very lean ones, experience problems with amenorrhea and infertility. Males also experience problems related to reproduction in response to undernutrition and weight loss. For example, they produce less prostatic fluid and reduced numbers of sperm with decreased motility. ■

EMBRYONIC DEVELOPMENT

The first 2 months of development are generally considered the **embryonic period**. During this period the developing human is called an **embryo** (*bryein* = grow). The study of development from the fertilized egg through the eighth week *in utero* is termed **embryology** (em-brē-OL-ō-jē). The months of development after the second month are considered the **fetal period**, and during this time the developing human is called a **fetus** (*feo* = to bring forth). By the end of the embryonic period, the rudiments of all the principal adult organs are present and the embryonic membranes are developed. By the end of the third month, the placenta, which is the site of exchange of nutrients and wastes between the mother and the fetus, is functioning.

Beginnings of Organ Systems

After implantation, the first major event of the embryonic period occurs. The inner cell mass of the blastocyst begins to differentiate into the three **primary germ layers**: ectoderm, endoderm, and mesoderm (Fig. 29.5a). These are the major embryonic tissues from which all tissues and organs of the body will develop. The process by which the two-layered inner cell mass is converted into a structure composed of the primary germ layers is called **gastrulation** (gas'-troo-LĀ-shun; *gastrula* = little belly).

Within 8 days after fertilization, the cells of the inner cytotrophoblast proliferate and form the amnion (a fetal membrane) and a space, the **amniotic** (am-nē-OT-ik; *amnion* = lamb) or **amniotic cavity**, adjacent to the inner cell mass. The layer of cells of the inner cell mass that is closer to the amniotic cavity develops into the **ectoderm** (*ecto* = outside; *derm* = skin). The layer of the inner cell mass that borders the blastocele develops into the **endoderm** (*endo* = inside). As the amniotic cavity forms, the inner cell mass at this stage is called the **embryonic disc**. It will form the embryo. At this stage, the embryonic disc contains ectodermal and endodermal cells; the mesodermal cells are scattered external to the disc.

About the 12th day after fertilization, striking changes appear (Fig. 29.5a). The cells of the endodermal layer have been dividing rapidly, so that groups of them now extend around in a circle, forming the yolk sac, another fetal membrane (described shortly). The cells of the **mesoderm** (*meso* = middle), which develop between the ectodermal and endodermal layers, also have been dividing, and many have left the area of the embryonic disc and can be seen around the structures that are becoming fetal membranes.

About the 14th day, the cells of the embryonic disc differentiate into three distinct layers: the ectoderm, mesoderm, and endoderm (Fig. 29.5b). The mesoderm in the disc soon splits into two layers, and the space between the layers becomes the **extraembryonic coelom** (SĒ-lōm; *koiloma* = cavity), the future ventral body cavity.

As the embryo develops (Fig. 29.5c), the endoderm becomes the epithelial lining of the gastrointestinal tract, respiratory tract, and several other organs. The mesoderm forms muscle, bone and other connective tissues, and the peritoneum. The ectoderm develops into the epidermis of the skin and the nervous system. Exhibit 29.1 on page 965 provides more details about the fates of these primary germ layers.

Embryonic Membranes

A second major event that occurs during the embryonic period is the formation of the **embryonic (extraembryonic) membranes**. These membranes lie outside the embryo and protect and nourish the embryo and, later, the fetus. The membranes are the yolk sac, amnion, chorion, and allantois.

In many species such as birds whose young develop inside a shelled egg, the **yolk sac** is a membrane that is the primary source of nourishment for the embryo (Fig. 29.5c). However, human embryo receives nutrients from the endometrium. The yolk sac remains small and functions as an early site of blood formation. The yolk sac also contains cells that migrate into the gonads and differentiate into the primitive germ cells (spermatogonia and oogonia).

The **amnion** is a thin, protective membrane that forms by the eighth day after fertilization and initially overlies the embryonic disc (Fig. 29.5a,b). As the embryo grows, the

---

Figure 29.5 Formation of the primary germ layers and associated structures.

🔑 *The primary germ layers (ectoderm, mesoderm, and endoderm) are the embryonic tissues from which all tissues and organs develop.*

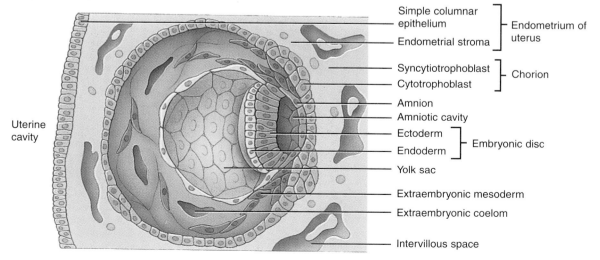

(a) Internal view, about 12 days after fertilization

Figure continues

Figure 29.5 (continued)

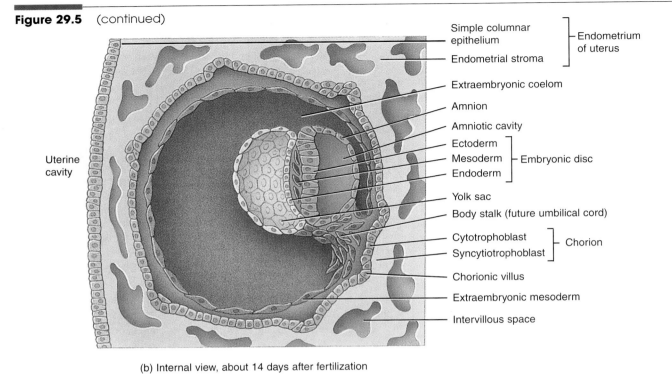

Simple columnar epithelium
Endometrial stroma
} Endometrium of uterus

Extraembryonic coelom

Amnion

Amniotic cavity

Ectoderm
Mesoderm
Endoderm
} Embryonic disc

Yolk sac

Body stalk (future umbilical cord)

Cytotrophoblast
Syncytiotrophoblast
} Chorion

Chorionic villus

Extraembryonic mesoderm

Intervillous space

Uterine cavity

(b) Internal view, about 14 days after fertilization

Embryo

Head   Heart   Tail

Extraembryonic coelom

Amniotic cavity
Chorion

Body stalk (future umbilical cord)

Chorionic villi

Yolk sac

Intervillous space

Uterine cavity

(c) External view, about 25 days after fertilization

Which cells of the blastocyst give rise to the embryonic disc?

amnion entirely surrounds the embryo, creating a cavity that becomes filled with **amniotic fluid** (Fig. 29.6). Most amniotic fluid is initially derived from a filtrate of maternal blood. Later, the fetus makes daily contributions to the fluid by excreting urine into the amniotic cavity. Amniotic fluid serves as a shock absorber for the fetus, helps regulate fetal body temperature, and prevents adhesions between the skin of the fetus and surrounding tissues. Embryonic cells are sloughed off into amniotic fluid; they can be examined in the procedure called **amniocentesis** (am′-nē-ō-sen-TĒ-sis), which is described on page 973. The amnion usually ruptures just before birth and with its fluid constitutes the "bag of waters."

The **chorion** (KŌR-ē-on) is derived from the trophoblast of the blastocyst and the mesoderm that lines the trophoblast. It surrounds the embryo and, later, the fetus.

EXHIBIT 29.1 STRUCTURES PRODUCED BY THE THREE PRIMARY GERM LAYERS

Endoderm	Mesoderm	Ectoderm
Epithelial lining of gastrointestinal tract (except the oral cavity and anal canal) and the epithelium of its glands.	All skeletal, most smooth, and all cardiac muscle.	All nervous tissue.
	Cartilage, bone, and other connective tissues.	Epidermis of skin.
Epithelial lining of urinary bladder, gall-bladder, and liver.	Blood, bone marrow, and lymphatic tissue.	Hair follicles, arrector pili muscles, nails, and epithelium of skin glands (sebaceous and sudoriferous).
Epithelial lining of pharynx, auditory (Eustachian) tubes, tonsils, larynx, tra-chea, bronchi, and lungs.	Endothelium of blood vessels and lym-phatic vessels.	Lens, cornea, and internal eye muscles.
Epithelium of thyroid, parathyroid, pan-creas, and thymus glands.	Dermis of skin.	Internal and external ear.
	Fibrous tunic and vascular tunic of eye.	Neuroepithelium of sense organs.
Epithelial lining of prostate and bul-bourethral (Cowper's) glands, vagina, vestibule, urethra, and associated glands such as the greater (Bartholin's) vestibu-lar and lesser vestibular glands.	Middle ear.	Epithelium of oral cavity, nasal cavity, paranasal sinuses, salivary glands, and anal canal.
	Mesothelium of ventral body cavity.	
	Epithelium of kidneys and ureters.	Epithelium of pineal gland, pituitary gland (hypophysis), and adrenal medulla.
	Epithelium of adrenal cortex.	
	Epithelium of gonads and genital ducts.	

Figure 29.6 Embryonic membranes.

 Embryonic membranes are outside the embryo; they protect and nourish the embryo and, later, the fetus.

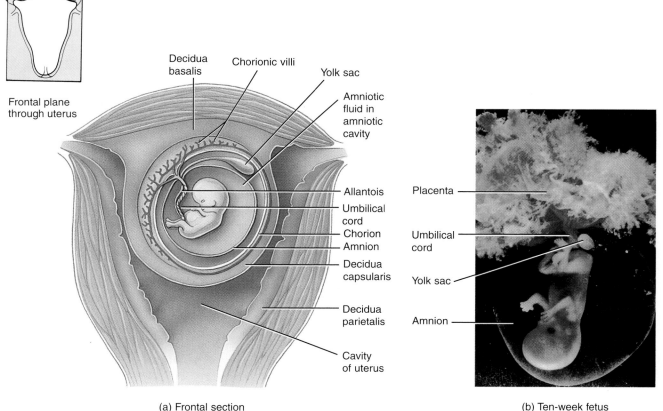

(a) Frontal section

(b) Ten-week fetus

Q How do the amnion and chorion differ in function?

Eventually, the chorion becomes the principal embryonic part of the placenta, the structure for exchange of materials between mother and fetus. It also produces human chorionic gonadotropin (hCG). The amnion, which also surrounds the fetus, eventually fuses to the inner layer of the chorion.

The **allantois** (a-LAN-tō-is; *allas* = sausage) is a small, vascularized structure that serves as an early site of blood formation. Later its blood vessels serve as the umbilical connection in the placenta between mother and fetus. This connection is the umbilical cord.

Placenta and Umbilical Cord

Development of the **placenta** (pla-SEN-ta; *placenta* = flat cake), the third major event of the embryonic period, is accomplished by the third month of pregnancy. The placenta has the shape of a flat cake when fully developed and is formed by the chorion of the embryo and a portion of the endometrium of the mother (see Fig. 29.8b). Functionally, the placenta allows oxygen and nutrients to diffuse into fetal blood from maternal blood. Simultaneously, carbon dioxide and wastes diffuse from fetal blood into maternal blood at the placenta.

The placenta also is a protective barrier since most microorganisms cannot cross it. However, certain viruses, such as those that cause AIDS, German measles, chickenpox, measles, encephalitis, and poliomyelitis, may pass through the placenta. The placenta also stores nutrients such as carbohydrates, proteins, calcium, and iron, which are released into fetal circulation as required. Finally, the placenta produces several hormones that are necessary to maintain pregnancy (see page 970). Almost all drugs, including alcohol, and many substances that can cause birth defects pass freely through the placenta.

If implantation occurs, a portion of the endometrium becomes modified and is known as the **decidua** (dē-SID-yoo-a; *deciduus* = falling off). The decidua includes all but the stratum basalis layer of the endometrium and separates from the endometrium after the fetus is delivered as it does in normal menstruation. Different regions of the decidua, which are all areas of the stratum functionalis, are named based on their positions relative to the site of the implanted blastocyst (Fig. 29.7). The **decidua basalis** is the portion of the endometrium between the chorion and the stratum basalis of the uterus. It becomes the maternal part of the placenta. The **decidua capsularis** is the portion of the endometrium that covers the embryo and is located between the embryo and the uterine cavity. The **decidua parietalis** (par-rī-e-TAL-is) is the remaining modified endometrium that lines the noninvolved areas of the entire pregnant uterus. As the embryo and later the fetus enlarges, the decidua capsularis bulges into the uterine cavity and initially fuses with the decidua parietalis, thus obliterating the uterine cavity. By about 27 weeks, the decidua capsularis degenerates and disappears.

Figure 29.7   Regions of the decidua.

The decidua is a modified endometrium that develops after implantation.

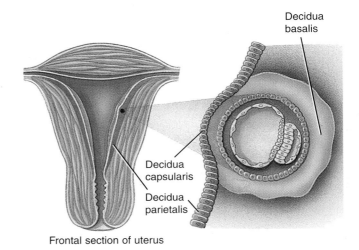

Frontal section of uterus

Which part of the decidua helps form the maternal part of the placenta?

During embryonic life, fingerlike projections of the chorion, called **chorionic villi** (kō′-rē-ON-ik VIL-ī), grow into the decidua basalis of the endometrium (Fig. 29.8a). These will contain fetal blood vessels of the allantois. They continue growing until they are bathed in maternal blood sinuses called **intervillous** (in-ter-VIL-us) **spaces**. Thus maternal and fetal blood vessels are brought into proximity. It should be noted, however, that maternal and fetal blood do not normally mix. Rather, oxygen and nutrients in the blood of the mother's intervillous spaces diffuse across the cell membranes into the capillaries of the villi while waste products diffuse in the opposite direction. From the capillaries of the villi, nutrients and oxygen enter the fetus through the umbilical vein. Wastes leave the fetus through the umbilical arteries, pass into the capillaries of the villi, and diffuse into the maternal blood. A few materials, such as IgG antibodies, pass from the blood of the mother into the capillaries of the villi by vesicular transport (described on page 618), in which the endocytosis step is receptor-mediated.

The **umbilical** (um-BIL-i-kul) **cord** is a vascular connection between mother and fetus. It consists of two umbilical arteries that carry deoxygenated fetal blood to the placenta, one umbilical vein that carries oxygenated blood into the fetus, and supporting mucous connective tissue called Wharton's jelly from the allantois. The entire umbilical cord is surrounded by a layer of amnion (Fig. 29.8a).

After the birth of the baby, the placenta detaches from the uterus and is termed the **afterbirth**. At this time, the umbilical

Figure 29.8 Placenta and umbilical cord.

The placenta is formed by the chorion of the embryo and part of the endometrium of the mother.

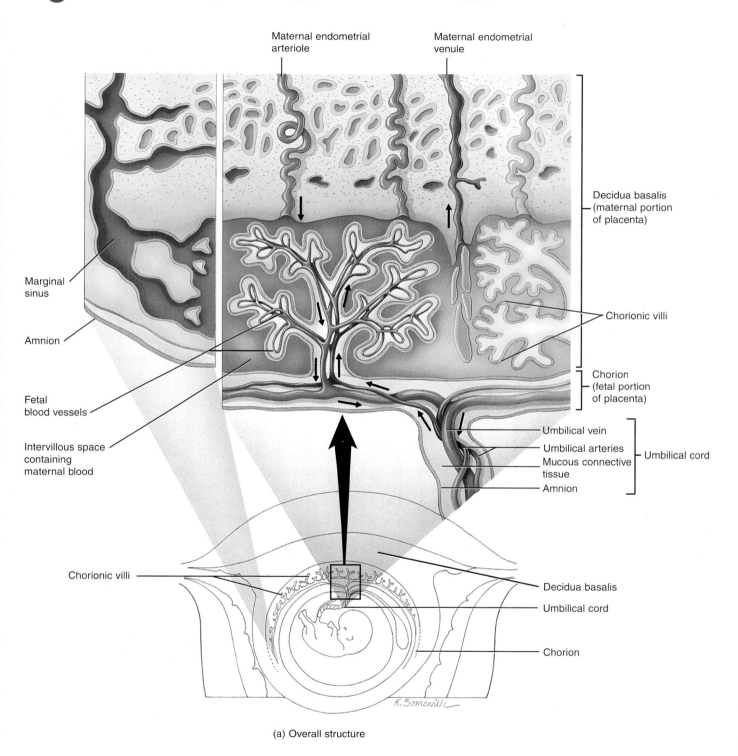

(a) Overall structure

Figure continues

Figure 29.8   (continued)

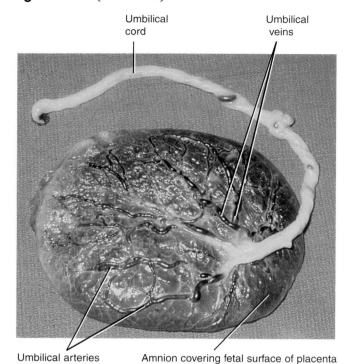

Umbilical cord   Umbilical veins

Umbilical arteries   Amnion covering fetal surface of placenta

(b) Photograph of fetal aspect of placenta

What is the function of the placenta?

cord is severed, leaving the baby on its own. The small portion (about an inch) of the cord that remains still attached to the infant begins to wither and falls off, usually within 12–15 days after birth. The area where the cord was attached becomes covered by a thin layer of skin and scar tissue forms. The scar is the **umbilicus** (**navel**).

Pharmaceutical companies use human placentas to harvest hormones, drugs, and blood. Portions of placentas are also used for burn coverage. The placental and umbilical cord veins can also be used in blood vessel grafts.

PLACENTA PREVIA AND UMBILICAL CORD ACCIDENTS

In some cases, part or all of the placenta becomes implanted in the inferior portion of the uterus, near or over the internal os of the cervix. This condition is called **placenta previa** (PRĒ-vē-a; *previa* = before or in front of). It may lead to spontaneous abortion and occurs in approximately 1 in 250 live births. It is dangerous to the fetus because it may cause premature birth and intrauterine hypoxia due to maternal bleeding. Maternal mortality is increased due to hemorrhage and infection. It is also associated with fetal abnormalities, twin gestation, and multiple uterine curettages. The most important symptom is sudden, painless, bright red vaginal bleeding in the third trimester. Cesarean section is the preferred method of delivery in placenta previa.

Several problems involving the umbilical cord are termed **umbilical cord accidents**. In **prolapse**, the umbilical cord descends in advance of the fetus during delivery. In **entrapment**, circulation through the cord is endangered because of pressure between the fetus and uterine wall. These, plus other conditions, such as knots in the cord, strictures (narrowings), and thromboses (blood clots), can reduce the oxygen supply to the fetus. In such cases the fetus may die or suffer brain damage. ■

FETAL GROWTH

During the **fetal period**, organs established by the primary germ layers grow rapidly and the fetus takes on a human appearance. Throughout the text, we have discussed developmental anatomy of the various body systems in their respective chapters. A listing of these sections is presented here, for your review.

Integumentary System (page 136)
Skeletal System (page 157)
Muscular System (page 264)
Nervous System (page 423)
Endocrine System (page 545)
Heart (page 604)
Blood and Blood Vessels (page 664)
Lymphatic System (page 679)
Respiratory System (page 743)
Digestive System (page 798)
Urinary System (page 883)
Reproductive Systems (page 946)

A summary of changes associated with the embryonic and fetal periods is presented in Exhibit 29.2.

FETAL SURGERY

Fetal surgery is a new medical field that had its beginnings in 1985. In a pioneering operation, a team of surgeons removed a 23-week-old fetus from its mother's uterus, operated to correct a blocked urinary tract, and then returned the fetus to the uterus. Nine weeks later, a healthy baby was born. Since then, surgeons have performed procedures to repair diaphragmatic hernias and are experimenting on animals to try to correct spina bifida (see page 193) and hydrocephalus (see page 394). It has been observed that surgery on fetuses does not leave any scars, although the reason is not known. It is hoped that surgeons can perform craniofacial surgery before birth to correct conditions such as cleft lip without leaving scars. ■

EXHIBIT 29.2 CHANGES ASSOCIATED WITH EMBRYONIC AND FETAL GROWTH

End of Month	Approximate Size and Weight	Representative Changes
1	0.6 cm (3/16 in.)	Eyes, nose, and ears not yet visible. Vertebral column and vertebral canal form. Small buds that will develop into limbs form. Heart forms and starts beating. Body systems begin to form. The central nervous system appears at the start of the third week.
2	3 cm (1¼ in.) 1 g (1/30 oz)	Eyes far apart, eyelids fused, nose flat. Ossification begins. Limbs become distinct and digits are well formed. Major blood vessels form. Many internal organs continue to develop.
3	7½ cm (3 in.) 30 g (1 oz)	Eyes almost fully developed but eyelids still fused, nose develops a bridge, and external ears are present. Ossification continues. Limbs are fully formed and nails develop. Heartbeat can be detected. Urine starts to form. Fetus begins to move, but it cannot be felt by mother. Body systems continue to develop.
4	18 cm (6½–7 in.) 100 g (4 oz)	Head large in proportion to rest of body. Face takes on human features and hair appears on head. Many bones ossified, and joints begin to form. Rapid development of body systems.
5	25–30 cm (10–12 in.) 200–450 g (½–1 lb)	Head less disproportionate to rest of body. Fine hair (lanugo) covers body. Brown fat forms and is the site of heat production. Fetal movements commonly felt by mother (quickening). Rapid development of body systems.
6	27–35 cm (11–14 in.) 550–800 g (1¼–1½ lb)	Head becomes even less disproportionate to rest of body. Eyelids separate and eyelashes form. Substantial weight gain. Skin wrinkled. Type II alveolar cells begin to produce surfactant.
7	32–42 cm (13–17 in.) 1100–1350 g (2½–3 lb)	Head and body more proportionate. Skin wrinkled. Seven-month fetus (premature baby) is capable of survival. Fetus assumes an upside-down position. Testes start to descend into scrotum.
8	41–45 cm (16½–18 in.) 2000–2300 g (4½–5 lb)	Subcutaneous fat deposited. Skin less wrinkled. Chances of survival much greater at end of eighth month.
9	50 cm (20 in.) 3200–3400 g (7–7½ lb)	Additional subcutaneous fat accumulates. Lanugo shed. Nails extend to tips of fingers and maybe even beyond.

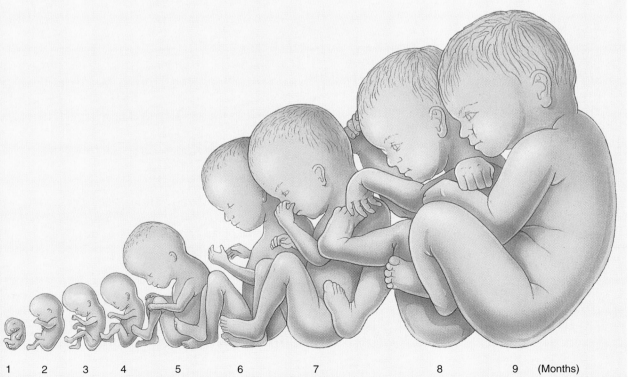

1   2   3   4   5   6   7   8   9   (Months)

HORMONES OF PREGNANCY

During the first 3–4 months of pregnancy, the corpus luteum continues to secrete **progesterone** and **estrogens**. These hormones maintain the lining of the uterus during pregnancy and prepare the mammary glands to secrete milk. The amounts secreted by the corpus luteum, however, are only slightly more than that produced after ovulation in a normal menstrual cycle. From the third month through the rest of the pregnancy the placenta itself provides the high levels of progesterone and estrogens needed to maintain pregnancy and develop the mammary glands for lactation. As noted earlier, the chorion of the placenta secretes **human chorionic gonadotropin (hCG)**. It mimics LH and its primary role is to rescue the corpus luteum from degeneration and to stimulate its continued production of progesterone and estrogens—an activity necessary to prevent menstruation and for the continued attachment of the embryo and fetus to the lining of the uterus (Fig. 29.9). By the eighth day after fertilization, hCG can be detected in the blood of a pregnant woman. Peak secretion of hCG occurs at about the ninth week of pregnancy. The hCG level decreases sharply during the fourth and fifth months and then levels off until childbirth.

EARLY PREGNANCY TESTS

Early pregnancy tests detect tiny amounts of human chorionic gonadotropin (hCG) in the urine. hCG starts to be released about 8 days after fertilization. The test kits can detect pregnancy as early as the first day of a missed menstrual period, that is, at about 14 days after fertilization. All kits include antibodies to hCG and other chemicals that produce a color change if there is a reaction between hCG in the urine and the hCG antibody in the test kit.

Several of the test kits available at pharmacies are as sensitive and accurate as test methods used in many hospitals. Still, false-negative and false-positive results can occur. A false-negative result (test is negative, but the female is pregnant) may occur from testing too soon or from an ectopic pregnancy. A false-positive result (test is positive, but the female is not pregnant) may be due to excess protein or blood in urine or hCG production due to a rare type of uterine cancer. Thiazide diuretics, hormones, steroids, and thyroid drugs may also affect the outcome of an early pregnancy test. ■

The chorion of the placenta begins to secrete estrogens after the first 3 or 4 weeks and progesterone by the sixth week of pregnancy. These hormones are secreted in increasing quantities until the time of birth. By the fourth month, when the placenta is fully established, the secretion of hCG is greatly reduced because the secretions of the corpus luteum are no longer essential. Thus from the third to ninth month, the placenta supplies the levels of estrogens and progesterone needed to maintain the pregnancy. The placental hormones take over management of the mother's body in preparation

Figure 29.9 Hormones of pregnancy.

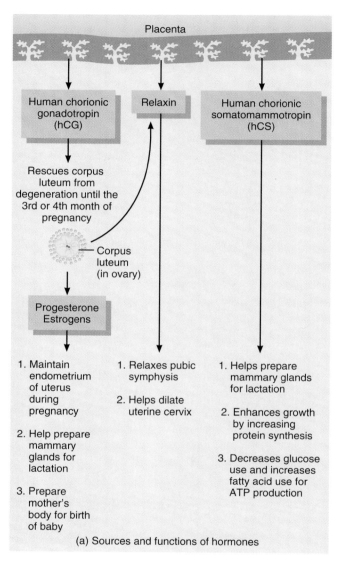

Whereas the corpus luteum produces progesterone and estrogens during the first 3–4 months of pregnancy, the placenta assumes this function from the third month on.

(a) Sources and functions of hormones

for parturition (birth) and lactation. After delivery, estrogens and progesterone in the blood decrease to normal levels.

Relaxin is a hormone produced first by the corpus luteum of the ovary and later by the placenta. It increases the flexibility of the pubic symphysis and ligaments of the sacroiliac and sacrococcygeal joints and helps dilate the uterine cervix during labor. Both of these actions ease delivery of the baby.

A third hormone produced by the chorion of the placenta is **human chorionic somatomammotropin (hCS)**, also known as **human placental lactogen (hPL)**. The rate of secretion of hCS increases in proportion to placental mass, reaching maximum levels after 32 weeks and remaining rel-

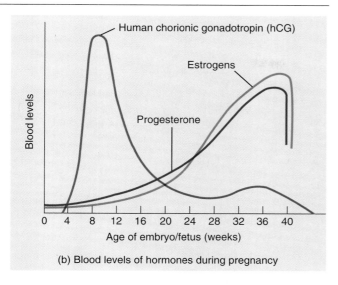

(b) Blood levels of hormones during pregnancy

Ⓠ Which hormone serves as the basis for early pregnancy tests?

atively constant after that. It is thought to help prepare the mammary glands for lactation, enhance growth by increasing protein synthesis, and regulate certain aspects of metabolism. For example, hCS causes decreased use of glucose by the mother, thus making more available for the fetus. Also, hCS promotes the release of fatty acids from adipose tissue, providing an alternative to glucose for the mother's ATP production.

MORNING SICKNESS

In the early months of pregnancy, **morning sickness** (**emesis gravidarum**) may occur, characterized by episodes of nausea and possibly vomiting that are most likely to occur in the morning. The cause is unknown, but the high levels of human chorionic gonadotropin (hCG) secreted by the placenta and progesterone secreted by the ovaries have been implicated. In some women the severity of these symptoms requires hospitalization for intravenous feeding, and the condition is then known as **hyperemesis gravidarum**. ∎

GESTATION

The time a zygote, embryo, or fetus is carried in the female reproductive tract is called **gestation** (jes-TĀ-shun; *gestare* = to bear). The human gestation period is about 38 weeks, counted from the estimated day of fertilization. The specialized branch of medicine that deals with pregnancy, labor, and the period of time immediately following delivery (about 42 days) is called **obstetrics** (ob-STET-riks; *obstetrix* = midwife).

Anatomical and Physiological Changes

By about the end of the third month of gestation, the uterus occupies most of the pelvic cavity. As the fetus continues to grow, the uterus extends higher and higher into the abdominal cavity. Toward the end of a full-term pregnancy, the uterus fills nearly all the abdominal cavity, reaching above the costal margin nearly to the xiphoid process of the sternum (Fig. 29.10). It pushes the maternal intestines, liver, and stomach superiorly, elevates the diaphragm, and widens the thoracic cavity. Pressure on the stomach may force the stomach contents superiorly into the esophagus, resulting in heartburn. In the pelvic cavity, there is compression of the ureters and urinary bladder.

Besides the anatomical changes associated with pregnancy, there are also pregnancy-induced physiological changes. General changes include weight gain due to the fetus, amniotic fluid, placenta, uterine enlargement, and increased total body water. Also, there is increased storage of proteins, triglycerides, and minerals; marked breast enlargement in preparation for lactation; and lower back pain due to lordosis (swayback).

With respect to the maternal cardiovascular system, there is an increase in stroke volume by about 30%; a rise in cardiac output of 20–30% due to increased maternal blood flow to the placenta and increased metabolism; an increase in heart rate of about 10–15%; and an increase in blood volume of 30–50%, mostly during the second half of pregnancy. These increases are necessary to meet the additional demands of the fetus for nutrients and oxygen. When a pregnant female is lying on her back, the enlarged uterus may compress the aorta, resulting in diminished blood flow to the uterus. Compression of the inferior vena cava also decreases venous return, which leads to edema in the lower limbs and may produce varicose veins. Compression of the renal artery can lead to renal hypertension.

Pulmonary function is also altered during pregnancy to meet the added oxygen demands of the fetus. Tidal volume can increase 30–40%, expiratory reserve volume can decrease up to 40%, functional residual capacity can decrease up to 25%, minute volume of respiration (MVR) can increase up to 40%, and airway resistance in the bronchial tree can decrease up to 36%. There is also an increase in total body oxygen consumption by about 10–20%. Dyspnea (difficult breathing) also occurs.

With regard to the gastrointestinal tract, there is an increase in appetite and a general decrease in motility that can result in constipation and a delay in gastric emptying time. Nausea, vomiting, and heartburn can also occur.

Pressure on the urinary bladder by the enlarging uterus can produce urinary symptoms, such as frequency, urgency, and stress incontinence. An increase in renal plasma flow up to 35% and an increase in glomerular filtration rate (GFR) up to 40% increase the renal filtering capacity, which allows faster elimination of the extra wastes produced by the fetus.

Figure 29.10   Normal fetal position at end of full-term pregnancy.

 The gestation period is the time interval (about 38 weeks) from fertilization to birth.

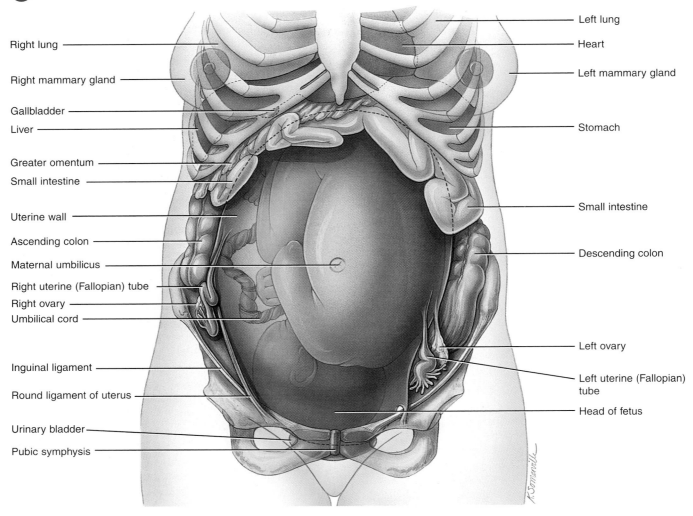

Right lung

Right mammary gland

Gallbladder

Liver

Greater omentum

Small intestine

Uterine wall

Ascending colon

Maternal umbilicus

Right uterine (Fallopian) tube

Right ovary

Umbilical cord

Inguinal ligament

Round ligament of uterus

Urinary bladder

Pubic symphysis

Left lung

Heart

Left mammary gland

Stomach

Small intestine

Descending colon

Left ovary

Left uterine (Fallopian) tube

Head of fetus

Anterior view

Q What hormone increases the flexibility of the pubic symphysis and helps dilate the cervix of the uterus to ease delivery of the baby?

Changes in the skin during pregnancy are more apparent in some women than others. Included are increased pigmentation around the eyes and cheekbones in a masklike pattern (chloasma), in the areolae of the breasts, and in the linea alba of the lower abdomen (linea nigra). Striae (stretch marks) over the abdomen can occur as the uterus enlarges, and hair loss also increases.

Changes in the reproductive system include edema and increased vascularity of the vulva and increased pliability and vascularity of the vagina. The uterus increases in weight from its nonpregnant state of 60–80 g to 900–1200 g at term. This increase is due to hyperplasia of muscle fibers (cells) in the myometrium in early pregnancy and hypertrophy of muscle fibers during the second and third trimesters.

PREGNANCY-INDUCED HYPERTENSION

About 10–15% of all pregnant women in the United States experience **pregnancy-induced hypertension** (**PIH**), elevated blood pressure associated with pregnancy. The major cause is **preeclampsia** (prē′-e-KLAMP-sē-a), in which the hypertension seems to result from impaired renal function. It typically appears after the 20th week of gestation and there are large amounts of protein in urine. Other signs and symptoms are generalized edema, blurred vision, and headaches. It might be related to an autoimmune or allergic reaction due to the presence of a fetus. When the condition is also associated with convulsions and coma, it is termed **eclampsia**. Other forms of PIH are not associated with protein in the urine. ■

Exercise and Pregnancy

Since pregnancy results in so many major body changes, it has an impact on a female's ability to exercise. In early pregnancy, there are only a few changes that affect exercise. A pregnant woman may tire earlier than usual, or she may become lethargic. Morning sickness may also curtail regular exercise. As the pregnancy progresses, weight is gained and posture changes. As a result, more energy is needed to perform activities, and certain maneuvers (sudden stopping, changes in direction, rapid movements) are difficult to execute. In addition, certain joints, especially the pubic symphysis, become less stable in response to the increased level of the hormone relaxin. As compensation, many mothers-to-be walk with widely spread legs and a shuffling motion.

Although during exercise blood shifts from viscera (including the uterus) to the muscles and skin, there is no evidence of inadequate blood flow to the placenta. The heat generated during exercise may cause dehydration and further increase body temperature. During early pregnancy, especially, excessive exercise and heat buildup should be avoided since elevated body temperature has been implicated in neural tube defects. Exercise has no known effect on lactation, provided the female remains hydrated and wears a bra with good support. Moderate physical activity does not endanger the fetuses of healthy females who have a normal pregnancy.

Among the benefits of exercise during pregnancy are improvement in oxygen capacity, greater sense of well-being, and fewer minor complaints.

PRENATAL DIAGNOSTIC TESTS

Several tests are available to detect genetic disorders and assess fetal well-being. Here we will describe fetal ultrasonography, amniocentesis, and chorionic villi sampling (CVS).

Fetal Ultrasonography

If there is a question about the normal progress of a pregnancy, **fetal ultrasonography** (ul′-tra-son-OG-ra-fē) may be performed. By far the most common use of diagnostic ultrasound is to determine true fetal age when the date of conception is uncertain. It is also used to evaluate fetal viability and growth, determine fetal position, ascertain multiple pregnancies, identify fetal–maternal abnormalities, and serve as an adjunct to special procedures such as amniocentesis. Ultrasound is not used routinely to determine the sex of a fetus; it is performed only for a specific medical indication.

An instrument (transducer) that emits high-frequency sound waves is passed back and forth over the abdomen. The reflected sound waves from the developing fetus are picked up by the transducer and converted to an image on a screen. This image is called a **sonogram** (see Exhibit 1.4 on page 22). Since the urinary bladder serves as a landmark during the procedure, the patient needs to drink liquids and not void to maintain a full bladder.

Amniocentesis

Amniocentesis (am′-nē-ō-sen-TĒ-sis; *amnio* = amnion; *kentesis* = puncture) involves withdrawing some of the amniotic fluid that bathes the developing fetus and analyzing the fetal cells and dissolved substances. It is used to test for the presence of certain genetic disorders, such as Down syndrome (DS), spina bifida, hemophilia, Tay–Sachs disease, sickle-cell anemia, and certain muscular dystrophies, or to determine fetal maturity and well-being near the time of delivery. To detect suspected genetic abnormalities, the test is usually done at 14–16 weeks of gestation. To assess fetal maturity, it is usually done after the 35th week of gestation. About 300 chromosomal disorders and over 50 biochemical defects can be detected through amniocentesis. It can also reveal gender. This information is important for diagnosis of sex-linked disorders, in which an abnormal gene is carried by the mother but affects only her male offspring. If the fetus is female, it will not be afflicted unless the father also carries the defective gene.

During amniocentesis, the position of the fetus and placenta is first determined using ultrasound and palpation. After the skin is prepared with an antiseptic, a local anesthetic is given, a hypodermic needle is inserted through the mother's abdominal wall and uterus into the amniotic cavity, and about 10 ml of fluid are aspirated (Fig. 29.11a). The fluid and suspended cells are subjected to microscopic examination and biochemical testing. Elevated levels of alpha-fetoprotein (AFP) and acetylcholinesterase may indicate failure of the nervous system to develop properly, for example, anencephaly (absence of the cerebrum) or spina bifida. Chromosome studies, which require growing the cells for 2–4 weeks in a culture medium, may reveal rearranged, missing, or extra chromosomes. There is about a 0.5% chance of spontaneous abortion after the test.

Chorionic Villi Sampling

Chorionic (ko-rē-ON-ik) **villi** (VIL-ī) **sampling** (**CVS**) can determine the same defects as amniocentesis because chorion cells and fetal cells contain the same genome. Moreover, CVS offers several advantages over amniocentesis. It can be performed as early as 8 weeks of gestation and test results are available in a few days, which permit an earlier decision on whether or not to continue the pregnancy. In addition, the procedure does not require penetration of the abdomen, uterus, or amniotic cavity by a needle. However, the procedure is slightly more risky than amniocentesis; there is a 1–2% chance of spontaneous abortion after the test.

During CVS, a catheter is placed through the vagina and cervix of the uterus and then advanced to the chorionic villi under ultrasound guidance (Fig. 29.11b). About 30 mg of tissue are suctioned out and prepared for chromosomal analysis.

Figure 29.11 Amniocentesis and chorionic villi sampling.

To detect genetic abnormalities, amniocentesis is performed at 14–16 weeks of gestation, whereas chorionic villi sampling may be performed as early as 8 weeks of gestation.

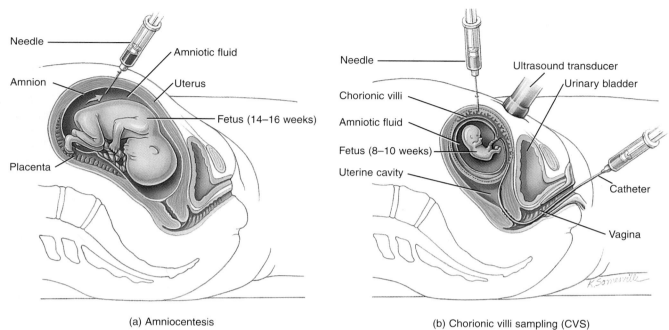

(a) Amniocentesis

(b) Chorionic villi sampling (CVS)

Q What kind of information can be provided by amniocentesis?

Alternatively, the chorionic villi can be sampled by inserting a needle through the abdominal cavity, as for amniocentesis.

LABOR

Labor (*labor* = toil, suffering) is the process by which the fetus is expelled from the uterus through the vagina to the outside. The term **parturition** (par′-too-RISH-un; *parturitio* = childbirth) also means giving birth. The onset of labor is related to a complex interaction of many factors. Just before birth, the muscles of the uterus contract rhythmically and forcefully. Both placental and ovarian hormones seem to play a role in these contractions. Since progesterone inhibits uterine contractions, labor cannot take place until its effects are diminished. At the end of gestation, progesterone level falls, the level of estrogens in the mother's blood is sufficient to overcome the inhibiting effects of progesterone, and labor commences. It has been suggested that cortisol released by the fetus overcomes the inhibiting effects of progesterone so that estrogens can exert their effect. Prostaglandins may also play a role in labor. Oxytocin (OT) from the posterior pituitary gland stimulates uterine contractions, and relaxin assists by relaxing the pubic symphysis and helping to dilate the uterine cervix.

Uterine contractions occur in waves, quite similar to peristaltic waves, that start at the top of the uterus and move downward. These waves expel the fetus. **True labor** begins when uterine contractions occur at regular intervals, usually producing pain. As the interval between contractions shortens, the contractions intensify. Another sign of true labor in some females is localization of pain in the back, which is intensified by walking. A reliable indication of true labor is the "show" and dilation of the cervix. The "show" is a discharge of a blood-containing mucus that accumulates in the cervical canal during labor. In **false labor**, pain is felt in the abdomen at irregular intervals. The pain does not intensify and is not altered significantly by walking. There is no "show" and no cervical dilation. True labor can be divided into three stages (Fig. 29.12).

① **Stage of dilation.** The time from the onset of labor to the complete dilation of the cervix is the **stage of dilation.** During this stage, which typically lasts 6–12 hours, there are regular contractions of the uterus, usually a rupturing of the amniotic sac, and complete dilation (10 cm) of the cervix. If the amniotic sac does not rupture spontaneously, it is done deliberately.

② **Stage of expulsion.** The time (10 minutes to several hours) from complete cervical dilation to delivery of the baby is the **stage of expulsion**.

③ **Placental stage.** The time (5–30 minutes or more) after delivery until the placenta or "afterbirth" is expelled by powerful uterine contractions is the **placental stage**. These contractions also constrict blood vessels that were torn during delivery. In this way, the chance of hemorrhage is reduced.

Figure 29.12 Stages of true labor.

 Parturition refers to birth.

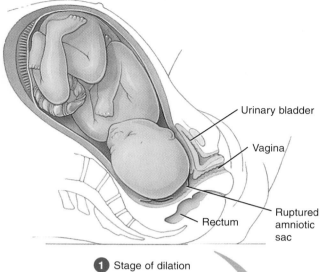

- Urinary bladder
- Vagina
- Rectum
- Ruptured amniotic sac

1 Stage of dilation

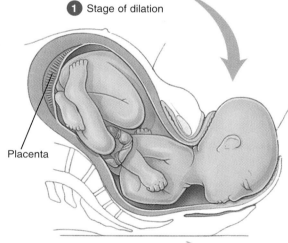

Placenta

2 Stage of expulsion

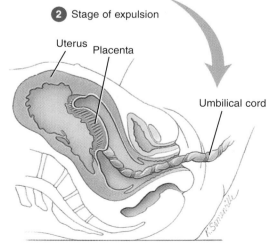

Uterus   Placenta

Umbilical cord

3 Placental stage

 What event marks the beginning of the stage of expulsion?

As a rule, labor lasts longer with first babies (typically 14 hours). In a female who has previously given birth, an average time for labor is 8 hours. During labor, the fetus may be squeezed through the birth canal (cervix and vagina) for up to several hours. As a result, the fetal head is compressed, and there is some degree of intermittent hypoxia due to compression of the umbilical cord and placenta during uterine contractions. Thus the fetus is stressed during and by the birth process. In response to this stress, the fetal adrenal medullae secrete very high levels of epinephrine and norepinephrine, the "fight-or-flight" hormones. Much of the protection afforded against the stresses of the birth process and preparation of the infant to survive extrauterine life are provided by the adrenal medullary hormones. Among other functions, the hormones clear the lungs and alter their physiology for breathing outside the uterus, mobilize readily usable nutrients for cellular metabolism, and promote an increased blood flow to the brain and heart.

About 7% of pregnant females have not delivered by 2 weeks after their due date. In such cases, there is increased risk of brain damage to the fetus and even fetal death. This is due to inadequate supply of oxygen and nutrients from an aging placenta. Post-term deliveries may be facilitated by induced labor (initiated by artificial means such as the administration of oxytocin) or surgical delivery.

After delivery of the baby and placenta, there is a period of time that lasts about 6 weeks during which the maternal reproductive organs and physiology return to the prepregnancy state. This is called the **puerperium** (pyoo'-er-PE-rē-um). Through a process of tissue catabolism, the uterus undergoes a remarkable reduction in size (especially in lactating females), called **involution** (in'-vō-LOO-shun). It is related primarily to a decrease in cytoplasm and cell size of myometrial cells. The cervix loses its elasticity and regains its prepregnancy firmness. For 2–4 weeks after delivery, there is a uterine discharge called **lochia** (LŌ-kē-a), which consists initially of blood and later serous fluid derived from the former placental site.

DYSTOCIA AND CESAREAN SECTION

Dystocia (dis-TŌ-sē-a; *dys* = difficult; *tokos* = birth), or difficult labor, may result from impaired uterine forces, an abnormal position (presentation) of the fetus, or a birth canal of inadequate size to permit vaginal birth. For example, in a **breech presentation** the fetal buttocks or lower extremities, rather than the head, enter the birth canal first; this occurs most often in premature births. If fetal or maternal distress prevents a vaginal birth, the baby may be delivered surgically through an abdominal incision. A low, horizontal cut is made through the abdominal wall and lower portion of the uterus, through which the baby and placenta are removed. Although popularly associated with the birth of Julius Caesar, the procedure for performing this operation is termed a **cesarean section** (**C-section**) because it was described in Roman Law, *lex cesarea,* 600 years before Julius Caesar was born. Even a history of multiple C-sections need not exclude a pregnant woman from attempting a vaginal delivery. ■

ADJUSTMENTS OF THE INFANT AT BIRTH

During pregnancy, the embryo and later the fetus is totally dependent on the mother for its existence. The mother supplies the fetus with oxygen and nutrients, eliminates its carbon dioxide and other wastes, and protects it against shocks, temperature changes, and certain harmful microbes. At birth, a physiologically mature baby becomes self-supporting, and the newborn's body systems must make various adjustments. Following are some changes that occur in the respiratory and cardiovascular systems.

Respiratory System

The fetus depends entirely on the mother for obtaining oxygen and eliminating carbon dioxide. The fetal lungs are either collapsed or partially filled with amniotic fluid, which is absorbed at birth. The production of surfactant begins by the end of the sixth month of development, and the respiratory system is fairly well developed at least 2 months before birth. Premature babies delivered at 7 months are able to breathe and cry. After delivery, the baby's supply of oxygen from the mother ceases. Circulation in the baby continues, and as the blood level of carbon dioxide increases, the respiratory center in the medulla oblongata is stimulated. This causes the respiratory muscles to contract, and the baby draws its first breath. Since the first inspiration is unusually deep because the lungs contain no air, the baby exhales vigorously and naturally cries. A full-term baby may breathe 45 times a minute for the first 2 weeks after birth. Breathing rate gradually declines until it approaches a normal rate.

Cardiovascular System

After the baby's first inspiration, the cardiovascular system must make several adjustments (see Fig. 21.30). The foramen ovale between the atria of the fetal heart closes at the moment of birth. This diverts deoxygenated blood to the lungs for the first time. The foramen ovale is closed by two flaps of septal heart tissue that fold together and permanently fuse. The remnant of the foramen ovale is the fossa ovalis. Once the lungs begin to function, the ductus arteriosus is shut off by contractions of the muscles in its wall and becomes the ligamentum arteriosum. This is probably mediated by the polypeptide bradykinin, released from the lungs during their initial inflation. The ductus arteriosus generally does not completely and irreversibly close for about 3 months after birth. Incomplete closing results in a condition called **patent ductus arteriosus**.

When the umbilical cord is tied off, no blood flows through the umbilical arteries, they fill with connective tissue, and their distal portions become the medial umbilical ligaments. Also, after tying off the umbilical cord, the umbilical vein becomes the ligamentum teres (round ligament) of the liver.

The ductus venosus connects the umbilical vein directly with the inferior vena cava. It allows blood from the placenta to bypass the fetal liver. When the umbilical cord is severed, blood from the viscera of the fetus goes directly to the fetal heart via the inferior vena cava. This shunting of blood usually occurs within minutes after birth but may take a week or two to complete. The ligamentum venosum, the remnant of the ductus venosus, is well established by the eighth week after birth.

At birth, the infant's pulse may be from 120 to 160 beats per minute and may go as high as 180 with excitation. After birth, oxygen use increases, which stimulates an increase in the rate of red blood cell and hemoglobin production. Moreover, the white blood cell count at birth is very high, sometimes as much as 45,000 cells per cubic millimeter, but this decreases rapidly by the seventh day.

Finally, the infant's liver may not be adjusted at birth to control the production of bile pigment. As a result of this and other complicating factors, a temporary jaundice may result in as many as 50% of normal newborns by the third or fourth day after birth.

PREMATURE INFANTS

Delivery of a physiologically immature baby carries certain risks. A **premature infant** or "preemie" is generally considered to be one who weighs less than 2500 g (5 lb, 8 oz) at birth. Poor prenatal care, drug abuse, a history of an earlier premature delivery, and mother's age below 16 or above 35 increase the chance of premature delivery. The problems related to survival of premature infants are due to the fact that they are not yet ready to take over functions the mother's body normally is performing. The major problem with delivery of an infant under 36 weeks of gestation is infant respiratory distress syndrome (RDS) due to insufficient surfactant (see page 746). This condition can be helped by use of artificial surfactant and a ventilator that delivers oxygen until the lungs can operate on their own. ■

PHYSIOLOGY OF LACTATION

The term **lactation** (lak'-TĀ-shun) refers to the secretion and ejection of milk by the mammary glands. A principal hormone in promoting lactation is **prolactin (PRL)** from the anterior pituitary gland. It is released in response to prolactin releasing hormone (PRH) secreted by the hypothalamus. Even though prolactin levels increase as the pregnancy progresses, there is no milk secretion because progesterone inhibits the prolactin from being effective. After delivery, the levels of estrogens and progesterone in the mother's blood decrease, and the inhibition is removed.

The principal stimulus in maintaining prolactin secretion during lactation is the sucking action of the infant. Sucking initiates nerve impulses from receptors in the nipples to the hypothalamus. The impulses decrease release of prolactin inhibiting hormone (PIH) and/or increase release of prolactin releasing hormone (PRH), so more prolactin is released by the anterior pituitary gland.

The sucking action also initiates nerve impulses to the posterior pituitary gland via the hypothalamus. These impulses stimulate the release of the hormone **oxytocin (OT)** by the posterior pituitary gland. Oxytocin induces myoepithelial cells surrounding the outer walls of the alveoli to contract, thereby compressing the alveoli and ejecting milk. The compression moves milk from the alveoli of the mammary glands into the ducts, where it can be suckled. This process is termed **milk ejection (let-down)**. Although the actual ejection of milk does not occur until 30–60 seconds after nursing begins (the latent period), some milk is stored in lactiferous sinuses near the nipple. Thus some milk is available during the latent period.

Recall from Chapter 18 (see Fig. 18.11) that oxytocin, as part of a positive feedback cycle, also stimulates contraction of smooth muscle in the pregnant uterus during labor and delivery. This action of oxytocin in nursing mothers results in a more rapid return of the uterus to its prepregnant size than occurs in nonnursing mothers. OT also compresses torn placental vessels at delivery and thus reduces blood loss by the mother.

During late pregnancy and the first few days after birth, the mammary glands secrete a cloudy fluid called **colostrum**. Although it is not as nutritious as milk, since it contains less lactose and virtually no fat, it serves adequately until the appearance of true milk on about the fourth day. Colostrum and maternal milk contain antibodies that protect the infant during the first few months of life.

Following birth of the infant, the prolactin level starts to return to the nonpregnant level. However, each time the mother nurses the infant, nerve impulses from the nipples to the hypothalamus increase the release of PRH (and/or decrease the release of PIH) resulting in a tenfold increase in prolactin secretion by the anterior pituitary gland that lasts about an hour. Prolactin acts on the mammary glands to provide milk for the next nursing period. If this surge of prolactin is blocked by injury or disease, or if nursing is discontinued, the mammary glands lose their ability to secrete milk in a few days. Milk secretion normally decreases considerably within 7–9 months. However, it can carry on for several years if nursing continues. A female who maintains lactation by nursing other women's infants is called a wet-nurse.

Lactation often prevents the occurrence of female ovarian cycles for the first few months following delivery, if the frequency of sucking is about 8–10 times a day. This effect is inconsistent, however, and ovulation will normally precede the first menstrual period after delivery of a baby. So the mother can never be certain she is not fertile. Breast-feeding is therefore not a very effective birth control measure. The suppression of ovulation during lactation is believed to occur as follows. During breast-feeding, neural input from the nipple reaches the hypothalamus and causes it to produce neurotransmitters that suppress the release of gonadotropin releasing hormone (GnRH). Thus there is a decreased production of LH and FSH, and ovulation is inhibited.

Breast-feeding offers the following advantages to the infant:

1. Fats and iron in human milk are more easily absorbed than those in cow's milk, and the amino acids in human milk are more readily metabolized. Also, the lower sodium content of human milk is more suited to the infant's needs.

2. Breast-feeding provides important antibodies that prevent gastroenteritis. Immunity to respiratory infections and meningitis is also greater.

3. Premature infants benefit from breast-feeding because the milk produced by mothers of premature infants seems to be specially adapted to the infant's needs by having a higher protein content than the milk of mothers of full-term infants.

4. There is less likelihood of an allergic reaction in the baby to the milk of the mother and several proteins in breast milk may stimulate the infant's immune system by increasing B lymphocyte maturation.

5. It establishes early and prolonged contact between mother and infant.

6. The infant is more in control of intake.

INHERITANCE

As indicated earlier, fertilization involves the coming together of genetic material of the father with that of the mother when a sperm cell merges with a secondary oocyte to form a zygote. Through development, the zygote becomes an embryo and later a fetus. The newborn child resembles its parents because of the inheritance of traits passed down from both parents. We will now examine some of the principles involved in inheritance.

Inheritance is the passage of hereditary traits from one generation to another. It is the process by which you acquired your characteristics from your parents and may transmit some of your traits to your children. The branch of biology that deals with inheritance is called **genetics** (je-NET-iks). The area of health care that offers advice on genetic problems is called **genetic counseling**.

MAPPING THE HUMAN GENOME

A **genome** (JĒ-nōm) is the complete genetic makeup of an organism. Of the 70,000–100,000 genes in a human cell, about 3000 have been identified and roughly located on various chromosomes. In 1988 the Human Genome Project, an international research effort to map and sequence the entire human genome, got underway. A major goal is the application of molecular genetics to medical research to identify and find cures for certain diseases. Equally important is the expectation that this knowledge will lead to more preventative measures for avoiding disease in genetically susceptible people. ■

Genotype and Phenotype

The nuclei of all human cells except gametes contain 23 pairs of chromosomes—the diploid number. One chromosome in each pair came from the mother, and the other came from the father. Homologues, the two chromosomes in a pair, contain genes that control the same traits. If a chromosome contains a gene for height, its homologue will contain a gene for height. The two alternative forms of a gene that code for the same trait and are at the same location (locus) on homologous chromosomes are termed **alleles** (ah-LĒLZ). A **mutation** (myoo-TĀ-shun; *mutare* = change) is a permanent heritable change in a gene that causes it to have a different effect than it had previously.

The relationship of genes to heredity is illustrated by examining the alleles involved in a disorder called **phenylketonuria** or **PKU** (Fig. 29.13). People with PKU (described on page 843) are unable to manufacture the enzyme phenylalanine hydroxylase. PKU is brought about by an abnormal allele, which we symbolize as *p*. The normal allele, which most people have, is symbolized as *P*. The maternal chromosome concerned with directions for phenylalanine hydroxylase production includes either the *p* or *P* allele. Its paternal homologue will also have *p* or *P*. Thus every individual will have one of the following genetic makeups, or **genotypes** (JĒ-nō-tīps): *PP*, *Pp*, or *pp*. People who inherit *PP* or *Pp* genotypes do not have PKU. Although people with a *Pp* genotype have one abnormal allele, only those with a *pp* genotype suffer from the disorder. When one normal allele (*P*) is present, it dominates the abnormal one (*p*). An allele that dominates or masks the presence of another allele and is fully expressed (*P* in this example) is said to be **dominant**, and the trait expressed is a dominant trait. The allele whose presence is completely masked (*p* in this example) is said to be **recessive**, and the trait it controls is a recessive trait.

By tradition, the symbol for the dominant allele is a capital letter and for the recessive one is a lowercase letter. A person with the same alleles on homologous chromosomes (for example, *PP* or *pp*) is said to be **homozygous** for the trait. Whereas *PP* is homozygous dominant, *pp* is homozygous recessive. An individual with different alleles on homologous chromosomes (for example, *Pp*) is said to be **heterozygous** for the trait.

Phenotype (FĒ-nō-tīp; *pheno* = showing) refers to how the genetic makeup is expressed in the body. It is the physical or outward expression of a gene. A person with *Pp* (heterozygous) has a different genotype from one with *PP* (homozygous), but both have the same phenotype—normal production of phenylalanine hydroxylase. Individuals who carry a recessive gene but do not express it (*Pp*) can pass the gene on to their offspring. Such individuals are called **carriers**.

Most genes give rise to the same phenotype whether they are inherited from the mother or the father. In a few cases, however, the phenotype is dramatically different, depending on the parental origin. This surprising phenomenon, first appreciated in the 1980s, is called **genomic imprinting**. In humans the abnormalities most clearly associated with mutation of an imprinted gene are *Angelman syndrome*, which results when the abnormal gene is inherited from the mother, and *Prader–Willi syndrome*, which results when the abnormality is inherited from the father.

To determine the possible ways that haploid gametes can unite to form diploid fertilized eggs, special charts called **Punnett squares** are used. Usually, the possible paternal alleles in sperm are placed at the side of the chart, and the possible maternal alleles in secondary oocytes are at the top (as in Fig. 29.13). The four spaces on the chart represent the possible genotypes for that trait in fertilized ova formed by the union of these male and female gametes.

Normal traits do not always dominate over abnormal ones, but dominant genes for severe disorders usually are lethal; they cause death of the embryo or fetus. One exception is Huntington's disease (HD), which is caused by a dominant gene. Homozygous dominant and heterozygous people both exhibit the disease whereas homozygous recessive people are normal. HD causes progressive degeneration

---

Figure 29.13 Inheritance of phenylketonuria (PKU).

🔑 *Whereas genotype refers to genetic makeup, phenotype refers to the physical or outward expression of a gene.*

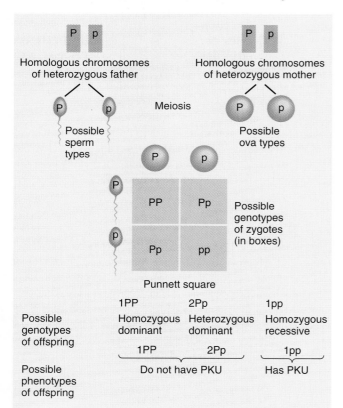

❓ If a couple has the genotypes shown here, what is the percent chance that their first child will have PKU? What is the chance of PKU occurring in their second child?

of the nervous system and eventual death but symptoms typically do not appear until after age 30 or 40. By this time many afflicted persons have passed the defective allele on to their children.

Occasionally an error in meiosis called **nondisjunction** results in an abnormal number of chromosomes. In this situation, homologous chromosomes fail to separate properly during anaphase of reduction division of meiosis. A cell that has one or more chromosomes of a set added or deleted is called an **aneuploid** (an'-yoo-PLOID). A monosomic cell $(2n - 1)$ is missing a chromosome; a trisomic cell $(2n + 1)$ has an added chromosome. Down syndrome (described on page 984) is an example of an aneuploid disorder in which there is trisomy of chromosome 21.

Exhibit 29.3 lists some dominant–recessive inherited structural and functional traits in humans.

EXHIBIT 29.3 SELECTED HEREDITARY TRAITS IN HUMANS

Dominant	Recessive
Coarse body hair	Fine body hair
Male pattern baldness	Baldness
Normal skin pigmentation	Albinism
Freckles	Absence of freckles
Astigmatism	Normal vision
Near- or farsightedness	Normal vision
Normal hearing	Deafness
Broad lips	Thin lips
Tongue roller	Inability to roll tongue into a U shape
PTC taster	PTC nontaster
Large eyes	Small eyes
Polydactylism (extra digits)	Normal digits
Brachydactylism (short digits)	Normal digits
Syndactylism (webbed digits)	Normal digits
Feet with normal arches	Flat feet
Hypertension	Normal blood pressure
Diabetes insipidus	Normal excretion
Huntington's disease	Normal nervous system
Normal mentality	Schizophrenia
Migraine headaches	Normal
Widow's peak	Straight hairline
Curved (hyperextended) thumb	Straight thumb
Normal Cl⁻ transport	Cystic fibrosis
Hypercholesterolemia (familial)	Normal cholesterol level

Variations on Dominant–Recessive Inheritance

Most patterns of inheritance are not simple **dominant–recessive inheritance** in which only dominant and recessive genes interact. In fact, the phenotypic expression of a particular gene may be influenced not only by which alleles are present, but also by other genes and the environment. Moreover, most inherited traits are influenced by more than one gene and most genes can influence more than a single trait. Following are some examples.

Incomplete Dominance

In **incomplete dominance**, neither member of an allelic pair is dominant over the other and the heterozygote has a phenotype intermediate between the homozygous dominant and homozygous recessive. An example of incomplete dominance in humans is the inheritance of **sickle-cell anemia (SCA)** (Fig. 29.14). Individuals with the homozygous dominant genotype $Hb^A Hb^A$ form normal hemoglobin. Those with the homozygous recessive genotype $Hb^S Hb^S$ have the disease called sickle-cell anemia and have severe anemia. Although they are usually healthy, those with the heterozygous genotype $Hb^A Hb^S$ have minor problems with anemia since they produce both normal and sickle-cell hemoglobin. Since people in this last category are carriers, they are sometimes referred to as having the "sickle-cell" trait.

Figure 29.14 Inheritance of sickle-cell anemia.

Sickle-cell anemia is an example of incomplete dominance.

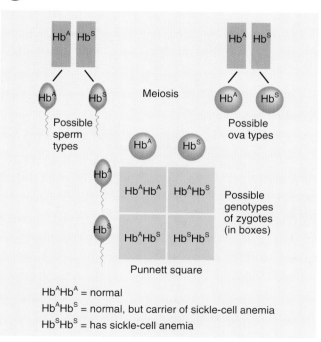

$Hb^A Hb^A$ = normal
$Hb^A Hb^S$ = normal, but carrier of sickle-cell anemia
$Hb^S Hb^S$ = has sickle-cell anemia

 What are the distinguishing features of incomplete dominance?

Another example of incomplete dominance in humans is the inheritance of hair texture. In this example, *HH* represents the homozygous genotype for curly hair; H^lH^l represents the homozygous genotype for straight hair; and HH^l represents the heterozygous genotype for wavy hair.

Multiple-Allele Inheritance

Although a single individual inherits only two alleles for each gene, some genes may have more than two alternate forms and this is the basis for **multiple-allele inheritance**.

One example of multiple-allele inheritance in humans is the inheritance of the ABO blood group. The four blood types (phenotypes) of the ABO group—A, B, AB, and O—result from the inheritance of six combinations of three different alleles of a single gene called the *I* gene: (1) allele I^A produces the A antigen, (2) allele I^B produces the B antigen, and (3) allele *i* produces neither A nor B antigen. Each person inherits two *I*-gene alleles, one from each parent, that give rise to the various phenotypes. The six possible genotypes produce four blood types as follows:

Genotype	Blood type
I^AI^A or I^Ai	A
I^BI^B or I^Bi	B
I^AI^B	AB
ii	O

Notice that both I^A and I^B are inherited as dominant traits but *i* is inherited as a recessive trait. Since an individual with type AB blood has characteristics of both type A and type B red blood cells expressed in the phenotype, alleles I^A and I^B are said to be **codominant**. In other words, both genes are expressed in the heterozygote equally. Depending on the parental blood types, different offspring may have one, two, three, or four different blood types. Figure 29.15 shows which blood types offspring could or could not inherit relative to their parent's blood types.

Polygenic Inheritance

Most inherited traits are not controlled by one gene but rather by the combined effects of many genes. This is referred to as **polygenic** (*poly* = many) **inheritance**. A polygenic trait shows a continuous gradation of small differences between extremes among individuals. Examples of polygenic traits include skin color, hair color, eye color, height, and body build. Whereas it is relatively easy to predict the risk of passing on an undesirable trait that is due to a single dominant or recessive gene, it is very difficult to make this prediction when the trait is polygenic. Such traits are difficult to follow in a family because the range of variation is so large and even the number of different genes involved usually is not known.

As an example of polygenic inheritance, we will consider skin color. Suppose that skin color is controlled by three separate genes, each having two alleles: *A, a*; *B, b*; and *C, c*. Whereas a person with the genotype *AABBCC* is very dark

Figure 29.15 The ten possible combinations of parental ABO blood types showing which blood types their offspring could or could not inherit. For each set of parents the blue letters show the blood types their offspring could inherit whereas the red letters show blood types their offspring could not inherit.

 Inheritance of ABO blood types is an example of multiple-allele inheritance.

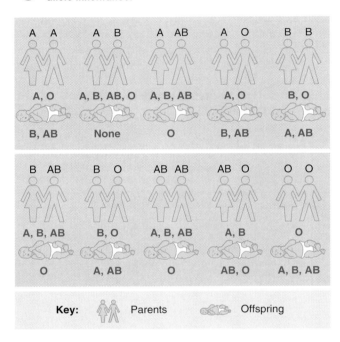

Is it possible for a baby to have type O blood if neither parent is type O?

skinned, an individual with the genotype *aabbcc* is very light skinned; and a person with the genotype *AaBbCc* has an intermediate skin color. Parents having an intermediate skin color may have children with very light, very dark, or intermediate skin color. Figure 29.16 illustrates the inheritance of skin color and the various gradations of skin tone. Note that the **P generation** (parental generation) is the starting generation, the F_1 **generation** (first filial generation) is produced from the P generation, and the F_2 **generation** (second filial generation) is produced from the F_1 generation.

Autosomes, Sex Chromosomes, and Inheritance of Gender

When viewed under a microscope, each of the 46 human chromosomes in a normal somatic cell can be identified by its size, shape, and staining pattern to be one member of 23 different pairs of chromosomes. In 22 of the pairs, the homologous chromosomes look alike and have the same appearance in both males and females; these 22 pairs are called **autosomes**. The two members of the 23rd pair are

Figure 29.16 Polygenic inheritance of skin color.

 In polygenic inheritance, the trait is controlled by the combined effects of several genes.

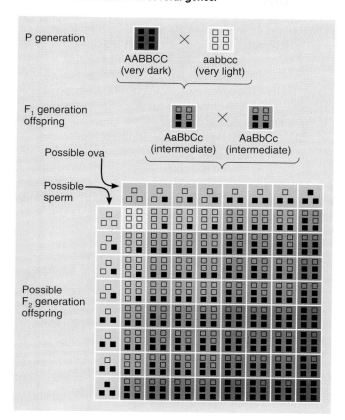

 What other traits are transmitted by polygenic inheritance?

Figure 29.17 Gender (sex) determination.

 Gender is determined at the time of fertilization.

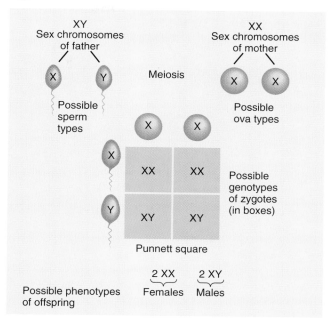

 What are chromosomes other than sex chromosomes called?

termed the **sex chromosomes**; they look different in males and females (see Fig. 29.18). In females, the pair consists of two chromosomes called X chromosomes. One X chromosome is also present in males, but its mate is a much smaller chromosome called a Y chromosome.

When a spermatocyte undergoes meiosis to reduce its chromosome number, one daughter cell will contain the X chromosome, and the other will contain the Y chromosome. Oocytes have no Y chromosomes and produce only X-containing ova. If the secondary oocyte is fertilized by an X-bearing sperm, the offspring normally will be female (XX). Fertilization by a Y-bearing sperm normally produces a male (XY). Thus gender (sex) is determined by the father's chromosomes (Fig. 29.17).

Both female and male embryos develop identically until about 7 weeks after fertilization. At that point, one or more genes set into motion a cascade of events that leads to the development of a male. In the absence of the gene or genes, the female pattern of development occurs. Since 1959 it has been known that the Y chromosome is needed to initiate male development. Experiments published in 1991 established that the prime male-determining gene is one called *SRY* (sex-

determining region of the Y chromosome). When a small DNA fragment containing this gene was inserted into 11 female mouse embryos, three of them developed as males. (The researchers suspect that the gene failed to be integrated into the genetic material in the other eight.) *SRY* apparently acts as a molecular switch to turn on the male pattern of development. Only if the *SRY* gene is present and functional in a fertilized ovum, will the fetus develop testes and differentiate into a male. In the absence of *SRY,* the fetus will develop ovaries and differentiate into a female.

An experiment of nature provided confirming evidence for humans. In two cases, phenotypic females with an XY genotype were found to have mutated *SRY* genes. In other words, they failed to develop normally as males because their *SRY* gene was defective.

KARYOTYPING

A **karyotype** (KAR-ē-ō-tīp; *karyon* = nucleus) is an arrangement of chromosomes from a cell based on their shape, size, and position of their centromeres. A karyotype is prepared by photographing the chromosomes, usually in a white blood cell or a cell obtained during amniocentesis, cutting them out of a printed photograph or a computer disc file, and then arranging them in standard order (Fig. 29.18). Karyotyping is done when a chromosomal abnormality is suspected that might be responsible for a disease or developmental problem. It is often done to investigate birth defects, abnormal growth, mental retardation, delayed puberty, abnormal sexual development, infertility, or certain inherited disorders. When

Figure 29.18   Normal human karyotype. The bent appearance of some chromosomes results from their position at the time they were photographed.

🔑 *A karyotype is an arrangement of chromosomes based on their size, shape, and centromere position.*

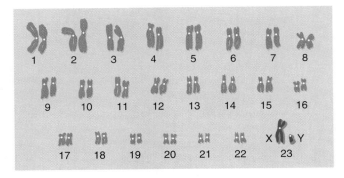

Ⓠ Was this karyotype from a male or a female? How did you decide?

nondisjunction occurs during meiosis, the abnormal segregation of chromosomes results in a gamete containing two or no copies of a particular parental chromosome. In nondisjunction of sex chromosomes, oocytes can contain two X's or no X chromosome (symbolized as O), and sperm can contain both an X and a Y, two X's or two Y's, or no sex chromosome at all. Turner's and Klinefelter's syndromes (see page 984) are examples of aneuploidy caused by nondisjunction of sex chromosomes. ∎

Sex-Linked Inheritance

The sex chromosomes also are responsible for the transmission of several nonsexual traits. Genes for these traits appear on X chromosomes but are absent from Y chromosomes. This feature produces a pattern of heredity that is different from the pattern described earlier.

Red–Green Color Blindness

Let us consider the most common type of color blindness, called **red–green color blindness**. In this condition, there is a deficiency in either red or green cones and red and green are seen as the same color, either red or green, depending on which cone is present. The gene for red–green color blindness is a recessive one designated c. Normal color vision, designated C, dominates. The C/c genes are located on the X chromosome. The Y chromosome does not contain these genes. Thus the ability to see colors depends entirely on the X chromosomes. The possible combinations are:

Genotype	Phenotype
$X^C X^C$	Normal female
$X^C X^c$	Normal female (carrying the recessive gene)
$X^c X^c$	Red–green color-blind female
$X^C Y$	Normal male
$X^c Y$	Red–green color-blind male

Only females who have two X^c genes are red–green color blind. This rare situation can result only from the mating of a color-blind male and a color-blind or carrier female. In $X^C X^c$ females the trait is masked by the normal, dominant gene. Males, on the other hand, do not have a second X chromosome that would mask the trait. Therefore all males with an X^c gene will be red–green color blind. The inheritance of red–green color blindness is illustrated in Fig. 29.19.

Traits inherited in the manner just described are called **sex-linked traits**. The most common type of **hemophilia**—a condition in which the blood fails to clot or clots very slowly after an injury (described on page 569)—is a sex-linked trait. Like the trait for red–green color blindness, hemophilia is caused by a recessive gene. Other sex-linked traits in humans are fragile X syndrome (described on page 984), nonfunctional sweat glands, certain forms of diabetes, some types of deafness, uncontrollable rolling of the eyeballs, absence of central incisors, night blindness, one form of cataract, juvenile glaucoma, and juvenile muscular dystrophy.

X-Chromosome Inactivation

Whereas females have two X chromosomes in every cell (except developing oocytes), males have only one. Females thus have a double dose of all genes on the X chromosome. A mechanism termed **X-chromosome inactivation** or **lyonization** balances this gender inequality. In each cell of

Figure 29.19   Inheritance of red–green color blindness.

🔑 *Red-green color blindness and hemophilia are examples of sex-linked traits.*

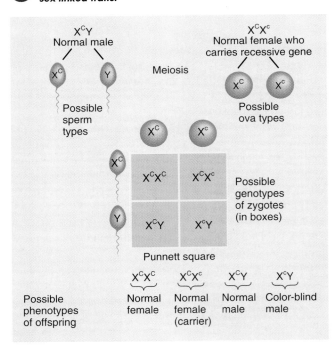

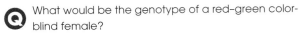

Ⓠ What would be the genotype of a red–green color-blind female?

the female's body, one X chromosome is randomly and permanently inactivated early in development. The genes of an inactivated X chromosome are not expressed (transcribed and translated). The nuclei of cells in female mammals contain a dark-staining body, called a **Barr body**, that is not present in the nuclei of cells in males. Geneticist Mary Lyon correctly predicted in 1961 that the Barr body is the inactivated X chromosome. During inactivation, chemical groups that prevent transcription into RNA are added to the X chromosome's DNA. Thus an inactivated X chromosome reacts differently to histological stains and has a different appearance than the rest of the DNA. In nondividing (interphase) cells, it remains tightly coiled and can be seen as a dark-staining body within the nucleus. In a blood smear, the Barr body of neutrophils is termed a "drumstick" because is looks like a tiny drumstick-shaped projection of the nucleus.

GENE THERAPY

A very exciting and important outcome of recombinant DNA technology is **gene therapy**—inserting a missing gene or replacing a defective one in human cells. This technique typically uses a harmless virus or liposomes to carry the missing or new gene into certain host cells, where the gene is picked up and inserted into the appropriate chromosome. Since 1990, gene therapy has been used to treat patients with ADA (adenosine deaminase) deficiency, a cause of severe combined immunodeficiency disease (SCID); Duchenne's muscular dystrophy; cystic fibrosis (CF); and a genetic type of hypercholesterolemia, an elevated blood cholesterol level. Other genetic disorders, such as hemophilia, diabetes, sickle-cell anemia, leukemia, malignant melanoma, phenylketonuria (PKU), and Huntington's disease, may one day also be treatable by gene therapy. ■

Genes and the Environment

A given phenotype often is the result of both genotype and environment. Factors in the environment seem to be particularly influential in the case of polygenic traits. For example, even though a person inherits several genes for tallness, full potential may not be reached due to environmental factors, such as disease or malnutrition during the growth years. Also, the risk of having a child with a neural tube defect is greater in pregnant women who lack adequate folic acid in their diet (environmental factor) and is more prevalent in some families than in others, which suggests a genetic component.

Exposure of a developing embryo or fetus to certain environmental factors can damage the developing organism or even cause death. A **teratogen** (*terato* = monster) is any agent or influence that causes developmental defects in the embryo. Following are several examples.

Chemicals and Drugs

Since the placenta is a porous barrier between the maternal and fetal circulations, any drug or chemical dangerous to an infant may be considered potentially dangerous to the fetus when given to the mother. Alcohol is by far the number one fetal teratogen. It has been a suspected teratogen for centuries, but only recently has a relationship been recognized between maternal alcohol intake and the characteristic pattern of malformations in the fetus. Intrauterine exposure to even a small amount of alcohol may result in **fetal alcohol syndrome** (**FAS**). The symptoms of FAS may include slow growth before and after birth, small head, characteristic facial features (narrow eye slits and sunken nasal bridge), defective heart and other organs, malformed limbs, genital abnormalities, and central nervous system damage. Behavioral problems, such as hyperactivity, extreme nervousness, a poor attention span, and an inability to appreciate cause-and-effect relationships, are common. Acetaldehyde is one of the toxic products of alcohol metabolism and may cross the placenta, thereby causing fetal damage.

Other fetal teratogens include pesticides; defoliants (chemicals that cause plants to shed their leaves prematurely); industrial chemicals; some hormones; antibiotics; oral anticoagulants, anticonvulsants, antitumor agents, thyroid drugs, thalidomide, diethylstilbestrol (DES), and numerous other prescription drugs; LSD; marijuana; and cocaine. For example, a pregnant female who uses cocaine subjects the fetus to higher risk of retarded growth, attention and orientation problems, hyperirritability, a tendency to stop breathing, malformed or missing organs, strokes, and seizures. The risks of spontaneous abortion (miscarriage), premature birth, and stillbirth also increase from fetal exposure to cocaine.

Cigarette Smoking

Strong evidence implicates cigarette smoking during pregnancy as a cause of low infant birth weight. There also is a strong association between smoking and a higher fetal and infant mortality rate. Women who smoke have a much higher risk of an ectopic pregnancy. Cigarette smoke may be teratogenic and cause cardiac abnormalities and anencephaly (a developmental anomaly with absence of a cerebrum). Maternal smoking also appears to be a significant factor in the development of cleft lip and palate and has tentatively been linked with sudden infant death syndrome (SIDS). Infants nursing from smoking mothers have also been found to have an increased incidence of gastrointestinal disturbances. Even exposure to secondhand cigarette smoke (in the air they breathe) predisposes infants to increased incidence of respiratory problems during the first year of life, including bronchitis and pneumonia.

Irradiation

Ionizing radiations are potent teratogens. Exposure of pregnant mothers to x-rays or radioactive isotopes during the embryo's susceptible period of development may cause microcephaly (small size of head in relation to the rest of the body), mental retardation, and skeletal malformations. Caution is advised, especially during the first trimester of pregnancy.

DISORDERS: HOMEOSTATIC IMBALANCES

DOWN SYNDROME

Down syndrome (DS) is a disorder that results from nondis-junction of chromosome 21. As a result, an extra chromosome passes to one of the daughter cells (gametes). For this reason DS is also called trisomy 21. Most of the time, the extra chromosome comes from the mother. This is not too surprising when you consider that her oocytes all began meiosis when she was a fetus herself. For years, they may have been exposed to chromosome-damaging chemicals and radiation. Undoubtedly, the kinetochore microtubules responsible for pulling sister chromatids to opposite poles of the cell (see Fig. 3.27) sustain increasing damage with the passing of years. Sperm, on the other hand, may be less than 10 weeks old at the time they fertilize a secondary oocyte. Overall, 1 in 800 infants is born with Down syndrome. However, older women are more likely to have a Down baby. The chance of conceiving a baby with this syndrome, which is less than 1 in 3000 for women under age 30 increases to 1 in 300 in the age 35–39 group and 1 in 9 at age 48.

Down syndrome is characterized by mental retardation, retarded physical development (short stature and stubby fingers), distinctive facial structures (large tongue, flat profile, broad skull, slanting eyes, and round head), and malformation of the heart, ears, hands, and feet (Fig. 29.20). Sexual maturity is rarely attained.

Figure 29.20 Photograph of a boy with Down syndrome.

SEX CHROMOSOME ANEUPLOIDS

Turner's syndrome, caused by the presence of only one X chromosome (XO), is an example of a sex chromosome aneuploid. Recall that an aneuploid is a cell that has one or more chromosomes of a set added or deleted. Such females are sterile with virtually no ovaries and limited development of secondary sex characteristics. Other features include short stature, webbed neck, underdeveloped breasts, and widely spaced nipples. There is usually no mental retardation. Another sex chromosome aneuploid condition is called **metafemale syndrome**, characterized by at least three X chromosomes (XXX). These females have underdeveloped genital organs and limited fertility and are generally mentally retarded.

One male in 1000 has an extra Y chromosome (XYY). These so-called "supermales" are normal in most respects, although they may be taller and larger than average and may have reading or speech problems. **Klinefelter's syndrome** is usually due to trisomy XXY. Such individuals are sterile males. They have undeveloped testes, scant body hair, and enlarged breasts and are characteristically somewhat mentally disadvantaged.

FRAGILE X SYNDROME

Fragile X syndrome is a recently recognized disorder due to a defective gene on the X chromosome. It is so named because a small portion of the tip of the X chromosome seems susceptible to breakage. It is the leading cause of mental retardation among newborns. Fragile X syndrome affects males more than females and results in learning difficulties, mental retardation, and physical abnormalities such as oversized ears, elongated forehead, enlarged testes, and double jointedness. The syndrome may also be involved in autism in which the individual exhibits extreme withdrawal and refusal to communicate.

Fragile X syndrome can be diagnosed by amniocentesis. Although the syndrome is largely sex-linked, 20–50% of males who inherit the trait are unaffected, but they can pass it on to their daughters. Although the daughters also are unaffected, their children, both male and female, may suffer mental retardation. Researchers have suggested that this pattern of inheritance could be explained by maternal imprinting of the gene. According to this theory, the father's fragile X gene becomes chemically changed as it passes through the daughter so that the gene is expressed in the daughter's offspring.

STUDY OUTLINE

DEVELOPMENT DURING PREGNANCY
(p. 958)

1. Pregnancy is a sequence of events that starts with fertilization.
2. Its various events are hormonally controlled.
3. Fertilization refers to the penetration of a secondary oocyte by a sperm cell and the subsequent union of their pronuclei to form a zygote.
4. Penetration is facilitated by enzymes produced by sperm acrosomes.

5. Normally, only one sperm cell fertilizes a secondary oocyte.
6. Early rapid cell division of a zygote is called cleavage, and the cells produced by cleavage are called blastomeres.
7. The solid mass of cells produced by cleavage is a morula.
8. The morula develops into a blastocyst, a hollow ball of cells differentiated into a trophoblast and inner cell mass.
9. The attachment of a blastocyst to the endometrium is called implantation; it occurs by enzymatic degradation of the endometrium.

10. In vitro fertilization (IVF) refers to the fertilization of a secondary oocyte outside the body and the subsequent implantation of the zygote. Other alternatives to IVF include embryo transfer and gamete intrafallopian transfer (GIFT).

EMBRYONIC DEVELOPMENT (p. 962)

1. During embryonic growth, the primary germ layers and embryonic membranes are formed.
2. The primary germ layers—ectoderm, mesoderm, and endoderm—form all tissues of the developing organism.
3. Embryonic membranes include the yolk sac, amnion, chorion, and allantois.
4. Fetal and maternal materials are exchanged through the placenta.

FETAL GROWTH (p. 968)

1. During the fetal period, organs established by the primary germ layers grow rapidly.
2. The principal changes associated with fetal growth are summarized in Exhibit 29.2.

HORMONES OF PREGNANCY (p. 970)

1. Pregnancy is maintained by human chorionic gonadotropin (hCG), estrogens, and progesterone.
2. Human chorionic somatomammotropin (hCS) assumes a role in breast development, protein anabolism, and glucose and fatty acid catabolism.
3. Relaxin relaxes the pubic symphysis and helps dilate the uterine cervix toward the end of pregnancy.

GESTATION (p. 971)

1. The time an embryo or fetus is carried in the uterus is called gestation.
2. Human gestation lasts about 38 weeks after fertilization.
3. During gestation, several anatomical and physiological changes occur in the mother.

PRENATAL DIAGNOSTIC TESTS (p. 973)

1. In fetal ultrasonography, an image of a fetus is displayed on a screen.
2. Amniocentesis is the withdrawal of amniotic fluid. It can be used to diagnose inherited biochemical defects and chromosomal disorders, such as hemophilia, Tay–Sachs disease, sickle-cell anemia, and Down syndrome.
3. Chorionic villi sampling (CVS) involves withdrawal of chorionic villi tissue for chromosomal analysis. CVS can be done earlier than amniocentesis, and the results are available more quickly, but it is also slightly more risky.

LABOR (p. 974)

1. True labor involves dilation of the cervix, expulsion of the fetus, and delivery of the placenta.
2. Parturition refers to birth.

ADJUSTMENTS OF THE INFANT AT BIRTH (p. 976)

1. The fetus depends on the mother for oxygen and nutrients, removal of wastes, and protection.

2. Following birth, the respiratory and cardiovascular systems undergo changes in adjusting to self-supporting postnatal life.

PHYSIOLOGY OF LACTATION (p. 976)

1. Lactation refers to the production and ejection of milk by the mammary glands.
2. Milk production is influenced by prolactin (PRL), estrogens, and progesterone.
3. Ejection is stimulated by oxytocin (OT).

INHERITANCE (p. 977)

Genotype and Phenotype (p. 977)

1. Inheritance is the passage of hereditary traits from one generation to another.
2. The genetic makeup of an organism is called its genotype. The traits expressed are called its phenotype.
3. Dominant genes control a particular trait; expression of recessive genes is inhibited by dominant genes.

Variations on Dominant–Recessive Inheritance (p. 979)

1. Most patterns of inheritance do not conform to the simple dominant–recessive patterns.
2. In incomplete dominance, neither member of an allelic pair dominates; phenotypically the heterozygote is intermediate between the homozygous dominant and homozygous recessive. An example is sickle-cell anemia.
3. In multiple-allele inheritance, genes have more than two alternate forms. An example is the inheritance of ABO blood groups.
4. In polygenic inheritance, an inherited trait is controlled by the combined effects of many genes. An example is skin color.

Autosomes, Sex Chromosomes, and Inheritance of Gender (p. 980)

1. Each somatic cell has 46 chromosomes—22 pairs of autosomes and 1 pair of sex chromosomes.
2. In females, the sex chromosomes are two X chromosomes; in males, they are one X chromosome and a much smaller Y chromosome, which normally includes the prime male-determining gene, called *SRY*.
3. If the *SRY* gene is present and functional in a fertilized ovum, the fetus will develop testes and differentiate into a male. In the absence of *SRY,* the fetus will develop ovaries and differentiate into a female.

Sex-Linked Inheritance (p. 982)

1. Red–green color blindness and hemophilia result from recessive genes located on the X chromosome and occur primarily in males because there are no counterbalancing dominant genes on the Y chromosome.
2. A mechanism termed X-chromosome inactivation or lyonization balances the gender inequality between males (one X) and females (two Xs). In each cell of the female's body, one X chromosome is randomly and permanently inactivated early in development and is called a Barr body.

Genes and the Environment (p. 983)

1. A given phenotype is the result of the interactions of genotype and the environment.
2. Teratogens are agents that cause physical defects in developing embryos. They include chemicals and drugs, alcohol, nicotine, and ionizing radiation.

REVIEW QUESTIONS

1. Define developmental anatomy. (p. 958)
2. Define fertilization. Where does it normally occur? How is a morula formed? (p. 958)
3. Explain how dizygotic (fraternal) and monozygotic (identical) twins are produced. (p. 959)
4. Describe the components of a blastocyst. (p. 959)
5. What is implantation and how does it occur? Why doesn't the mother reject the implanted blastocyst? (p. 960)
6. Describe in vitro fertilization (IVF), embryo transfer, and gamete intrafallopian transfer (GIFT). (p. 960)
7. Define the embryonic period and the fetal period. (p. 962)
8. List several body structures or layers formed by the endoderm, mesoderm, and ectoderm. (p. 965)
9. What is an embryonic membrane? Describe the functions of the four embryonic membranes. (p. 963)
10. Explain the importance of the placenta and umbilical cord to fetal growth. (p. 966)
11. Outline some of the major developmental changes during fetal growth. (p. 969)
12. List the hormones involved in pregnancy and describe the functions of each. (p. 970)
13. Define gestation. (p. 971)
14. Describe several anatomical and physiological changes that occur during gestation. (p. 971)
15. Explain the effects of pregnancy on exercise and exercise on pregnancy. (p. 973)
16. Explain the procedure and diagnostic value of the following: fetal ultrasonography (p. 973), amniocentesis (p. 973), and chorionic villi sampling (CVS) (p. 973).
17. Distinguish between false and true labor. Describe what happens during the stage of dilation, the stage of expulsion, and the placental stage of delivery. (p. 974)
18. Discuss the principal respiratory and cardiovascular adjustments made by an infant at birth. (p. 976)
19. What is lactation? Name the hormones involved and their functions. (p. 976)
20. Define inheritance. What is genetics? (p. 977)
21. Define the following terms: genotype, phenotype, dominant, recessive, homozygous, and heterozygous. (p. 978)
22. What is a Punnett square? (p. 978)
23. Describe genomic imprinting and nondisjunction. (p. 978)
24. Define incomplete dominance. Give an example. (p. 979)
25. What is multiple-allele inheritance? Give an example. (p. 980)
26. Define polygenic inheritance and give an example. (p. 980)
27. Set up Punnett squares to show the inheritance of the following: gender, red–green color blindness, and hemophilia. (p. 981)
28. What is sex-linked inheritance? Why does X-chromosome inactivation occur? (p. 982)
29. How do genes and the environment determine phenotype? Give several examples. (p. 983)
30. Define the following: ectopic pregnancy (p. 960), placenta previa (p. 968), umbilical cord accident (p. 968), fetal surgery (p. 968), morning sickness (p. 971), dystocia (p. 975), cesarean section (p. 975), premature infant (p. 976), genome (p. 977), karyotyping (p. 981), and gene therapy (p. 983).

ANSWERS TO QUESTIONS WITH FIGURES

29.1 Functional changes experienced by sperm after they have been deposited in the female vagina that allow them to fertilize a secondary oocyte.
29.2 Morula is a solid ball of cells. Blastocyst is a rim of cells (trophoblast) surrounding a cavity (blastocele) plus an inner cell mass.
29.3 It secretes digestive enzymes that eat away the endometrial lining at the site of implantation.
29.4 Postovulatory.
29.5 Inner cell mass.
29.6 The amnion functions as a shock absorber, whereas the chorion forms the placenta.
29.7 Exchange of materials between fetus and mother.
29.8 Decidua basalis.
29.9 Human chorionic gonadotropin (hCG).
29.10 Relaxin.
29.11 It is used mainly to detect genetic disorders but also gives evidence of the maturity (and survivability) of the fetus.
29.12 Complete dilation of cervix.
29.13 Twenty-five percent. The odds will be the same for each child—25%.
29.14 Neither member of an allelic pair is dominant; the heterozygote has a phenotype intermediate between the homozygous dominant and homozygous recessive.
29.15 Yes, if each parent has one i allele and passes it on to the offspring.
29.16 Hair color, height, and body build, among others.
29.17 Autosomes.
29.18 Male because of the presence of a Y chromosome.
29.19 $X^c X^c$.

MEASUREMENTS

UNITS OF MEASUREMENT

When you measure something, you are comparing it with some standard scale to determine its *magnitude.* How long is it? How much does it weigh? How fast is it going? Some measurements are made directly by comparing the unknown quantity with the known unit of the same kind, for example, weighing a patient on a scale and taking the reading directly in pounds. Other measurements are indirect and are done by calculation, for example, counting a person's blood cells in a certain number of squares on a microscope slide and then calculating the total blood count.

Regardless of how a measurement is taken, it always requires two things: a *number* and a *unit.* When recording the weight of a patient, you would not just say 145. You have to give both the number (145) and the unit (pounds). When you count blood cells, you report the measurement as 10,000 (number) white blood cells per cubic millimeter of blood (unit).

All the units in use can be expressed in terms of one of three special units called **fundamental units**. These fundamental units have been established arbitrarily as length, mass, and time. Mass is perhaps an unfamiliar term to you. **Mass** is the amount of matter an object contains. The mass of this textbook is the same whether it is measured in a laboratory, under the sea, on top of a mountain, or even on the moon. No matter where you take it, it still has the same quantity of matter. **Weight**, on the other hand, is determined by the pull of gravity on an object. This textbook will not have the same weight on earth as on the moon because of the differences in gravitation. However, as long as we are dealing only with earthbound objects, weight and mass may be considered synonymous terms because the force of gravity on the surface of the earth is nearly constant. Thus, weight remains nearly the same regardless of where the measurements are taken.

All units other than the fundamental ones are **derived units**— they can always be written as some combination of the three fundamental units. For example, units of volume are derived from units of length (the volume of a cube = length × width × height). Units of speed are combinations of distance and time (miles per hour).

Units are grouped into systems of measurement. The two principal systems of measurement commonly used are the U.S. and the metric systems. The apothecary system is used by physicians and pharmacists.

U.S. SYSTEM

The *U.S. system* of measurement is used in everyday household work, industry, and some fields of engineering. The fundamental units in the U.S. system are the foot (length), the pound (mass), and the second (time).

The basic problem with the U.S. system is that there is no *uniform* progression from one unit to another. If you want to convert a measurement of 2½ yd to feet, you have to multiply it by 3 because there are 3 ft in a yard. If you want to convert the same length to inches, you have to multiply by 3 and then by 12 (or by 36) because there are 12 in. in a foot. In other words, to convert one unit of length to another, it is necessary to use *different* numbers each time. Conversions in the metric system are much easier since they are based on progressions of the number 10.

Exhibit A.1 lists U.S. units of measurement.

EXHIBIT A.1 U.S. UNITS OF MEASUREMENT

Fundamental or Derived Unit	Units and U.S. Equivalents
Length	1 inch (in.) = 0.083 foot
	1 foot (ft) = 12 in.
	= 0.333 yard
	1 yard (yd) = 3 ft = 36 in.
	1 mile (mi) = 1,760 yd = 5,280 ft
Mass	1 grain (gr) = 0.002285 ounce
	1 dram (dr) = 27.34 gr
	= 0.063 ounce
	1 ounce (oz) = 16 dr = 437.5 gr
	1 pound (lb) = 16 oz = 7,000 gr
	1 ton = 2,000 lb
Time	second (sec) = $\frac{1}{86,400}$ of a day
	1 minute (min) = 60 sec
	1 hour (hr) = 60 min = 3,600 sec
	1 day = 24 hr = 1,440 min
	= 86,400 sec
Volume	1 fluidram (fl dr) = 0.125 fluidounce
	1 fluidounce (fl oz) = 8 fl dr
	= 0.0625 quart
	= 0.008 gallon
	1 pint (pt) = 16 fl oz = 128 fl dr
	1 quart (qt) = 2 pt = 32 fl oz
	= 256 fl dr
	= 0.25 gallon
	1 gallon (gal) = 4 qt = 8 pt
	= 128 fl oz
	= 1,024 fl dr

Figure A.1   Metric and U.S. units of length.

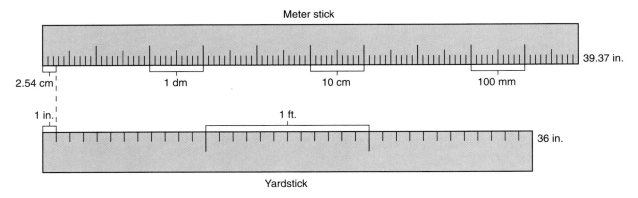

METRIC SYSTEM

The *metric system*, introduced in France in 1790, is now used by all major countries except the United States. Scientific observations are almost universally expressed in metric units.

LENGTH

The standard of length in the metric system is the *meter* (m). It was originally defined in 1790 as one ten-millionth of the distance from the North Pole to the Equator. In 1889 it was redefined as the distance measured at 0°C between two lines on a bar of platinum-iridium kept at the International Bureau of Weights and Measures in France. In 1960 the meter was redefined as the length of 1,650,763.73 light waves emitted by atoms of the gas krypton under strictly specified conditions. The meter is now defined as the distance light travels in $^1/_{299,792,458}$ second. The meter is equal to 39.37 inches.

A major advantage of the metric system is that units are related to one another by factors of 10. Thus 1 m is 10 decimeters (dm) or

100 centimeters (cm) or 1,000 millimeters (mm). Conversion from one unit to another is simple. Figure A.1 illustrates the differences between metric and U.S. conversions by comparing the meter stick and the yardstick. Exhibit A.2 lists the metric units of length with U.S. equivalents.

Since numbers with many zeros (very large numbers or very small fractions) are cumbersome to work with, they are expressed in *exponential form,* that is, as powers of 10. The form of exponential notation is

$$M \times 10^n$$

You can determine M and n in two steps. For example, how is 0.0000000001 written in exponential form? First, determine M by moving the decimal point so that only one nonzero digit is to the left of it:

EXHIBIT **A.2   METRIC UNITS OF LENGTH AND SOME U.S. EQUIVALENTS**

Metric Unit	Meaning of Prefix	Metric Equivalent	U.S. Equivalent
1 kilometer (km)	kilo = 1,000	1,000 m = 10^3 m	3,280.84 ft = 0.62 mi 1 mi = 1.61 km
1 hectometer (hm)	hecto = 100	100 m = 10^2 m	328 ft
1 dekameter (dam)	deka = 10	10 m = 10^1 m	32.8 ft
1 meter (m)		Standard unit of length	39.37 in. = 3.28 ft = 1.09 yd
1 decimeter (dm)	deci = $^1/_{10}$	0.1 m = 10^{-1} m	3.94 in.
1 centimeter (cm)	centi = $^1/_{100}$	0.01 m = 10^{-2} m	0.394 in. 1 in. = 2.54 cm
1 millimeter (mm)	milli = $^1/_{1,000}$	0.001 m = $^1/_{10}$ cm = 10^{-3} m	0.0394 in.
1 micrometer (μm) [formerly micron (μ)]	micro = $^1/_{1,000,000}$	0.000,001 m = $^1/_{10,000}$ cm = 10^{-6} m	3.94 x 10^{-5} in.
1 nanometer (nm) [formerly millimicron (mμ)]	nano = $^1/_{1,000,000,000}$	0.000,000,001 m = $^1/_{10,000,000}$ cm = 10^{-9} m	3.94 x 10^{-8} in.

The digit to the left of the decimal is 1; therefore $M = 1$. Second, determine n by counting the number of places you moved the decimal point. If you moved the point to the left, make the number positive; if you moved it to the right, it is negative. Since you moved the decimal point 10 places to the right, $n = -10$. Thus

$$0.0000000001 = 1 \times 10^{-10}$$

Now do a problem on your own. The wavelength of yellow light is about 0.000059 cm. Convert the centimeters to exponential form. If your answer is 5.9×10^{-5} cm, you are ready to continue. If you got the wrong answer, reread the discussion.

When we are working with a very large number, the same rules apply, but our exponential value will be positive rather than negative. Refer to Exhibit A.2. Note that 1 km equals 1,000 m. Even though 1,000 is not a cumbersome number, we can still convert it into exponential form. First, move the decimal point so there is only one nonzero digit to the left of it to determine M:

1.000.

Now, because the decimal has been moved three places to the left, n equals +3 or simply 3. Thus

$$1 \text{ km} = 1 \times 10^3 \text{ m}$$

Do another problem on your own. The speed of light is about 30,000,000,000 cm/sec. Convert the centimeters to exponential form. Your answer should be 3×10^{10} cm.

Review Exhibit A.2 and note some common metric and U.S. equivalents. Note also the exponential forms.

MASS

Now let us look at the second fundamental unit of the metric system: mass. The standard unit of mass is the *kilogram* (kg). A kilogram is defined as the mass of a platinum-iridium cylinder kept at the International Bureau of Weights and Measures in France. The standard pound is defined in terms of standard kilogram: 1 lb equals 0.4536 kg.

Exhibit A.3 lists metric units of mass and some U.S. equivalents.

TIME

The third fundamental unit of both the metric and the U.S. systems is time. The standard of time is the *second*. Formerly, the second was defined as $\frac{1}{86,400}$ of a mean solar day. (A mean solar day is the average of the lengths of all days throughout the year.) Currently, the second is defined as the time required for 9,192,631,770 vibrations of cesium atoms when they are vibrating in a specific manner. Units of time are used in measuring pulse and heart rate, metabolic rate, x-ray exposure, and intervals between medications.

Exhibit A.1 lists the units of time.

VOLUME

Units of volume, or capacity, are derived units based on length. *Volume* in the U.S. system may be expressed as cubic feet (ft³), cubic inches (in.³), and cubic yards (yd³), or as a unit of volume such as the quart. Volume in the metric system may be expressed in cu-

bic units of length such as cubic centimeters (cm³) or in terms of the basic unit of volume, the *liter*. A liter is defined as the volume occupied by 1,000 g of pure water at 4°C. Since 1 cm³ of water at this temperature weighs 1 g, then 1,000 g of water occupies a volume of 1,000 cm³. This means that a liter is equal to 1,000 cm³ and 1 milliliter (ml) is equal to 1 cm³. Because of this relationship, many volume-measuring devices, such as hypodermic needles, may be graduated in either milliliters or cubic centimeters.

Exhibit A.4 lists metric units of volume and some U.S. equivalents.

APOTHECARY SYSTEM

In addition to the U.S. and metric systems, there is the *apothecary system*. This system is commonly used by physicians prescribing

EXHIBIT A.3 METRIC UNITS OF MASS AND SOME U.S. EQUIVALENTS

Metric Unit	Metric Equivalent	U.S. Equivalent
1 kilogram (kg)	1,000 g	2.205 lb
		1 ton = 907 kg
1 hectogram (hg)	100 g	0.353 oz
1 dekagram (dag)	10 g	0.0353 oz
1 gram (g)	1 g	0.0035 oz
		1 lb = 453.6 g
		1 oz = 28.35 g
1 decigram (dg)	0.1 g	
1 centigram (cg)	0.01 g	
1 milligram (mg)	0.001 g	
1 microgram (μg)	0.000,001 g	
1 nanogram (ng)	0.000,000,001 g	
1 picogram (pg)	0.000,000,000,001 g	

EXHIBIT A.4 METRIC UNITS OF VOLUME AND SOME U.S. EQUIVALENTS

Metric Unit	Metric Equivalent	U.S. Equivalent
1 liter (l)	1,000 ml	33.81 fl oz = 1.057 qt
		946 ml = 1 qt
1 milliliter (ml)	0.001 liter	0.0338 fl oz
		30 ml = 1 fl oz
		5 ml = 1 teaspoon
1 cubic centimeter (cm³)	1.0 ml	0.0338 fl oz

medications and by pharmacists preparing them. Exhibit A.5 lists the important units and equivalents of the apothecary system. Note that the units of mass have the same names as in the U.S. system, but they are not equivalent (1 oz = 28.35 g; 1 oz ap = 30 g) and they do not have the same relationship to one another (1 oz = 16 dr; 1 oz ap = 8 dr ap). The units of volume are the same in both systems.

EXHIBIT **A.5 APOTHECARY SYSTEM OF MASS AND VOLUME WITH METRIC EQUIVALENTS**

Fundamental Unit	Apothecary Unit and Conversion	Metric Equivalent
Mass	1 grain (gr) = 0.002083 ounce	1 g = 15 gr
	1 dram (dr ap) = 60 gr	4 g = 1 dr
	1 ounce (oz ap) = 8 dr ap	30 g = 1 oz ap
	1 pound (lb ap) = 12 oz ap	1 kg = 32 oz
Volume	1 fluidram (fl dr) = 60 minims (min)	1 ml (or cm^3) = 15 min
		4 ml (or cm^3) = 1 fl dr
	1 fluidounce (fl oz) = 8 fl dr	30 ml (or cm^3) = 1 fl oz
	1 pint (pt) = 16 fl oz	500 ml (or cm^3) = 1 pt
		1,000 ml (or cm^3) = 1 qt

FREQUENTLY USED CONVERSIONS BASED ON THE MILLIGRAM

1,000 mg (1 g) = 15 grains

600 mg (0.6 g) = 10 grains

300 mg (0.3 g) = 5 grains

60 mg (0.06 g) = 1 grain

30 mg (0.03 g) = 0.50 (½) grain

20 mg (0.02 g) = 0.33 (⅓) grain

10 mg (0.01 g) = 0.166 (⅙) grain

5 mg (0.005 g) = 0.083 (1/12) grain

4 mg (0.004 g) = 0.66 (1/15) grain

1 mg (0.001 g) = 0.016 (1/60) grain

0.5 mg (0.0005 g) = 0.0083 (1/120) grain

0.1 mg (0.0001 g) = 0.0016 (1/600) grain

NORMAL VALUES FOR SELECTED BLOOD AND URINE TESTS

The system of international (SI) units (Système Internationale d'Unités) is used in most countries and in many medical and scientific journals.[*] Clinical laboratories in the United States, however, usually report values for blood and urine tests in conventional (traditional) units, which are also used in most medical and biological textbooks. To encompass both systems, the laboratory values in this Appendix give conventional units and, in parentheses, their SI equivalents.

A major difference between SI and conventional units is in expressing concentration as number of molecules or particles per volume (millimoles/liter) rather than mass per volume (grams/100 milliliters) or chemical activity per volume (milliequivalents per liter). In certain cases, for example, proteins that have a variable molecular weight, it is not practical to change the values to SI units. Therefore, protein concentration may be expressed in grams per liter rather than moles per liter. Electrolyte (ion) values have traditionally been given in milliequivalents per liter. In SI they are expressed in millimoles/liter. For monovalent (singly charged) ions such as Na^+, K^+, and Cl^-, the values will be numerically the same. For example, a Na^+ concentration of 135 mEq/liter is equal to 135 mmol/liter. For divalent ions such as Ca^{2+} and Mg^{2+}, milliequivalents are divided by the valence of 2 to obtain millimoles. Thus a Ca^{2+} concentration of 3.0 mEq/liter is equal to 1.5 mmol/liter.

In some laboratory and medical measurements, SI units offer little advantage. Their use might require replacement or revision of current instruments, for example the sphygmomanometer for measuring blood pressure or the pH meter for measuring hydrogen ion concentration. In these cases, traditional units are commonly retained.

It is important to note that values listed for various laboratory tests should be viewed as reference values rather than absolute "normal" values for all well people. A single reference range may be inadequate for some measurements. Values may vary due to age, gender, body build, diet, and environment of the subject or the equipment, methods, and standards of the lab performing the measurement. Also, in most pathological processes, there is a gradual transition between normal and abnormal values. Thus the values given here are representative reference ranges. They are grouped into three categories: **blood chemistry tests** (mainly blood plasma and serum values), **hematology tests** (blood clotting parameters and counts of blood formed elements), and **urine tests**.

KEY TO SYMBOLS
mm^3 = cubic millimeter
dl = deciliter = 10^{-1} liter
g = gram
> = greater than
hr = hour
IU = international unit
kg = kilogram = 10^3 grams
< = less than
% = percent
μg = microgram = 10^{-6} gram
μmol/l = micromole per liter
mEq/l = milliequivalent per liter
mg = milligram = 10^{-3} gram
ml = milliliter = 10^{-3} liter
mm = millimeter = 10^{-3} meter
mm Hg = millimeter of mercury
mmol/l = millimole per liter
mOsm = milliosmole
ng = nanogram = 10^{-9} gram
nmol/l = nanomole per liter
U = unit

[*]Introductory paragraphs and SI units provided by Henry Ruschin, Humber College, Toronto, Canada. Reference ranges previously researched by John Lo Russo, Bergen Community College, New Jersey.

EXHIBIT B.1 BLOOD CHEMISTRY TESTS

(WB) = WHOLE BLOOD   (S) = SERUM   (P) = PLASMA

Test (Specimen)	Reference Values: Conventional U.S. Units (SI Units)	Clinical Implications
Acid phosphatase (ACP) (S)	0.1–5.0 U/dl by King–Armstrong method (0–0.8 IU/liter)	Values increase in prostatic cancer (especially if it has spread beyond the prostate gland), some liver diseases, hyperparathyroidism, hemolytic anemia, and sickle cell crisis; values are decreased in Down syndrome.
Alkaline phosphatase (ALP) (S)	4–13 U/dl by King–Armstrong method (30–120 U/liter)	Values increase in some liver and bone diseases, hyperparathyroidism, and pregnancy; values decrease in cretinism, growth retardation, scurvy, and achondroplasia.
Alphafetoprotein (AFP) (WB or amniotic fluid)	Nonpregnant adult: <25 ng/ml (<25 µg/liter)	Major plasma protein synthesized by fetal liver during first 3 months of development. In amniotic fluid and maternal blood, values increase with faulty development of the fetal nervous system, in particular neural tube defects such as spina bifida. In nonpregnant adults, values increase in liver cancer, cirrhosis, or chronic active hepatitis.
Aminotransferases (S) **Alanine amino-transferase (ALT)**; formerly serum glutamic–pyruvic transaminase (SGPT)	10–30 IU/ml; 5–25 Reitman–Frankel units (10–30 U/liter)	Values increase in liver disease or liver damage due to toxic drugs.
Aspartate amino-transferase (AST); formerly serum glutamic–oxaloacetic transaminase (SGOT)	5–24 IU/liter; 5–35 Reitman–Frankel units (5–30 U/liter)	Values increase in myocardial infarction, liver disease, trauma to skeletal muscles, and severe burns; values decrease in beriberi and uncontrolled diabetes mellitus with acidosis.
Ammonia (P)	20–120 mg/dl (12–55 µmol/liter)	Values increase in liver disease, heart failure, emphysema, pneumonia, cor pulmonale, and hemolytic disease of the newborn (erythroblastosis fetalis).
Amylase (S)	60–160 Somogyi U/dl (25–125 U/liter)	Values increase in acute pancreatitis, mumps, and obstruction of pancreatic duct; values decrease in hepatitis, cirrhosis, burns, and toxemia of pregnancy.
Bilirubin (S)	Conjugated: <0.5 mg/dl (<5.0 µmol/liter) Unconjugated: 0.2–1.0 mg/dl (18–20 µmol/liter) Newborn: 1.0–12.0 mg/dl (<200 µmol/liter)	An increase in conjugated bilirubin probably results from liver dysfunction or biliary obstruction; an increase in unconjugated bilirubin probably results from excessive hemolysis of red blood cells.
Blood urea nitrogen (BUN) (S)	8–26 mg/dl (2.9–9.3 mmol/liter)	Values increase in kidney disease, shock, dehydration, diabetes, and acute myocardial infarction (MI); values decrease in liver failure, impaired absorption, and overhydration.
Calcium (Ca and Ca^{2+}) (S)	Total: 9–11 mg/dl (2.3–2.7 mmol/liter) Ionized (50% of total): 4.5–5.5 mg/dl (1.15–1.35 mmol/liter)	Values increase in cancer, hyperparathyroidism, Addison's disease, hyperthyroidism, and Paget's disease; values decrease in hypopara-thyroidism, chronic renal failure, osteomalacia, rickets, and diarrhea.
Carbon dioxide (CO_2), content (WB)	Arterial: 19–24 mEq/l (19–24 mmol/liter) Venous: 22–26 mEq/l (22–26 mmol/liter)	Values increase in severe vomiting, emphysema, and aldosteronism; values decrease in severe diarrhea, starvation, and acute renal failure.
Carbon dioxide, partial pressure (pCO_2) (WB)	Arterial: 35–40 mm Hg (same) Venous: 45 mm Hg (same)	Values increase in hypoventilation, obstructive lung disease, and emphysema; values decrease in hyperventilation, hypoxia, and pregnancy.

Test (Specimen)	Reference Values: Conventional U.S. Units (SI Units)	Clinical Implications
Carcinoembryonic antigen (CEA) (P)	<3ng/ml (<3 µg/liter)	Values increase in carcinoma of the colon, rectum, breast, ovary, liver, and pancreas; inflammatory bowel disease (IBD); cirrhosis; and chronic cigarette smoking.
Carotene, beta (S)	40–200 mg/dl (0.4–2.0 µg/liter)	Value varies with diet but increases in myxedema, diabetes mellitus, and excessive dietary intake; values decrease in fat malabsorption, liver disease, and poor dietary intake.
Chloride ion (Cl⁻) (S)	95–103 mEq/liter (95–103 mmol/liter)	Values increase in dehydration, Cushing's syndrome, and anemia; values decrease in severe vomiting, severe burns, diabetic acidosis, and fever.
Cholesterol, total (S)	<200 mg/dl (<5.2 mmol/liter) is desirable	Value varies with diet, gender, and age. Values increase in diabetes mellitus, cardiovascular disease, nephrosis, and hypothyroidism; values decrease in liver disease, hyperthyroidism, fat malabsorption, pernicious anemia, severe infections, and terminal stages of cancer.
HDL cholesterol (P)	>40 mg/dl (>1.0 mmol/liter) is desirable	
LDL cholesterol (P)	<130 mg/dl (<3.2 mmol/liter) is desirable	
Cortisol (hydrocortisone) (P)	8 A.M.–10 A.M.: 5–23 µg/dl (270–700 nmol/liter) 4 P.M.–6 P.M.: 3–13 µg/dl (135–350 nmol/liter)	Values increase in hyperthyroidism, stress, obesity, and Cushing's syndrome; values decrease in hypothyroidism, liver disease, and Addison's disease.
Creatine (S or P)	Male: 0.1–0.4 mg/dl (1–4 g/liter) Female: 0.2–0.7 mg/dl (2–7 g/liter)	Values increase in muscular dystrophy, damage to muscle tissue, nephritis, and pregnancy.
Creatine kinase (CK); formerly creatine phosphokinase (CPK) (S)	Male: 55–170 U/liter (same) Female: 30–135 U/liter (same)	Values increase in myocardial infarction, progressive muscular dystrophy, myxedema, convulsions, hypothyroidism, and pulmonary edema.
Creatinine (S)	0.5–1.2 mg/dl (45–105 µmol/liter)	Values increase in impaired renal function, giantism, and acromegaly; values decrease in muscular dystrophy.
Fetal hemoglobin (WB)	Newborns: 60–90% Before age 2: 0–4% Adults: 0–2%	Values increase in thalassemia, sickle-cell anemia, and leakage of fetal blood into maternal bloodstream.
Gamma-glutamyl transferase (GGT) (S)	5–40 IU/liter (5–40 U/liter)	Values increase in obstruction of bile duct, cirrhosis of the liver, metastatic cancer of the liver, cholelithiasis, congestive heart failure (CHF), and alcoholism.
Glucose (S)	70–110 mg/dl (3.9–6.1 mmol/liter)	Values increase in diabetes mellitus, acute stress, hyperthyroidism, chronic liver disease, and nephritis; values decrease in Addison's disease, hypothyroidism, and cancer of the pancreas.
Immunoglobulins (S)		
IgG	800–1,801 mg/dl (8.0–18.0 g/liter)	IgG values increase in infections of all types, liver disease, and severe malnutrition.
IgA	113–563 mg/dl (1.1–5.6 g/liter)	IgA values increase in cirrhosis of the liver, chronic infections, and autoimmune disorders and decrease in immunologic deficiency states.
IgM	54–222 mg/dl (0.5–2.2 g/liter)	IgM values increase in trypanosomiasis and decrease in lymphoid aplasia.
IgD	0.5–3.0 mg/dl (5–30 mg/liter)	IgD values increase in chronic infections and myelomas.
IgE	0.01–0.04 mg/dl (0.1–0.4 mg/liter)	IgE values increase in hay fever, asthma, and anaphylactic shock.

Exhibit continues

EXHIBIT **B.1**   **BLOOD CHEMISTRY TESTS** (CONTINUED)

(WB) = WHOLE BLOOD   (S) = SERUM   (P) = PLASMA

Test (Specimen)	Reference Values: Conventional U.S. Units (SI Units)	Clinical Implications
Iron, total (S)	50–170 µg/dl (13–30 µmol/liter)	Values higher in males than females; values increase in liver disease and various anemias; values decrease in iron-deficiency anemia.
Iron-binding capacity, total (TIBC)	300–420 µg/dl (50–70 µmol/liter)	Values increase in iron-deficiency anemia, blood loss, and pregnancy and in women taking birth control pills; values decrease in many chronic diseases, widespread cancer, malnutrition, and nephrotic syndrome.
Ketone bodies (acetone, acetoacetic acid, and β-hydroxy-butyric acid) (S or P)	Negative Toxic level: 20 mg/dl (0.2 g/liter)	Values increase in ketoacidosis, fever, anorexia, fasting, starvation, high-fat diet, low-carbohydrate diet, and following vomiting.
Lactic acid (lactate) (WB)	Arterial: 3–7 mg/dl (0.3–0.7 mmol/liter) Venous: 5–20 mg/dl (0.5–2.0 mmol/liter)	Values increase during muscular activity, congestive heart failure (CHF), shock, and severe hemorrhage.
Lactic dehydrogenase (LDH) (S)	71–207 IU/liter (70–207 U/liter)	Values increase in myocardial infarction, liver disease, skeletal muscle necrosis, and extensive cancer.
Lipids (S)		
Total	400–800 mg/dl (4.0–8.0 g/liter)	Values increase in hyperlipidemia, diabetes mellitus, and hypothyroidism; values decrease in fat malabsorption.
Cholesterol	150–220 mg/dl (3.9–5.7 mmol/liter)	
Triglycerides	10–190 mg/dl (0.1–1.9 g/liter)	
Osmolality (S)	285–295 mOsm/kg water (285–295 mmol/kg water)	Values increase in cirrhosis, congestive heart failure (CHF), and high-protein diets; values decrease in aldosteronism, diabetes insipidus, and hypercalcemia.
Oxygen (O_2) content (WB)	Arterial: 15–23 volume % (same)	Values increase in polycythemia; values decrease in chronic obstructive lung disease.
Oxygen, partial pressure (pO_2) (WB)	Arterial: 80–105 mm Hg (same)	Values increase in polycythemia and hyperventilation; values decrease in anemias, insufficient atmospheric oxygen, and hypoventilation.
pH (WB)	Arterial: 7.35–7.45 (same) [corresponds to H^+ concentrations of 45–35 nmol/liter = 0.045–0.035 mmol/liter]	Values increase in vomiting, hyperventilation, excessive bicarbonate, and lack of oxygen; values decrease in renal failure, diabetic ketoacidosis, hypoxia, airway obstruction, and shock.
Phosphorous, inorganic (P); also reported as phosphate (S)	Adults: 2.5–4.5 mg/dl (0.8–1.5 mmol/liter) Children: 4–7 mg/dl (1.3–2.3 mmol/liter)	Values increase in renal failure, hypoparathyroidism, hypocalcemia, bone tumors, diabetic ketoacidosis, and acromegaly; values decrease in hyperparathyroidism, alcoholism, rickets, and osteomalacia.
Protein (S)		Total protein values increase in dehydration, shock, systemic lupus erythematosus (SLE), rheumatoid arthritis (RA), chronic infections, and chronic liver disease; total protein values decrease in insufficient protein intake, hemorrhage, malabsorption, diarrhea, chronic renal failure, and severe burns; reversed A/G ratio (A lowered, G elevated) may indicate chronic liver disease, leukemia, Hodgkin's disease, tuberculosis, or chronic hepatitis.
Total	6.0–7.8 g/dl (60–78 g/liter)	
Albumin	3.5–5.0 g/dl (35–50 g/l)	
Globulin	2.3–3.5 g/dl (23–35 g/liter)	
A/G ratio	1.5:1 to 2.5:1	
Sodium (Na^+) (S)	136–142 mEq/liter (136–142 mmol/liter)	Values increase in dehydration, aldosteronism, coma, Cushing's disease, and diabetes insipidus; values decrease in severe burns, vomiting, diarrhea, Addison's disease, nephritis, excessive sweating, and edema.

Test (Specimen)	Reference Values: Conventional U.S. Units (SI Units)	Clinical Implications
Thyroid hormones (S)		Values increase in hyperthyroidism; values decrease in hypothyroidism.
T_3 (triiodothyronine)	80–200 ng/dl by radioimmunoassay (RIA) (1.2–3.1 nmol/liter)	
T_4 (thyroxine)	4–11 mg/dl by RIA (52–141 nmol/liter)	
Thyroxine-binding globulin (TBG) (S)	10–26 µg/dl (130–335 nmol/liter)	Values increase in hypothyroidism; values decrease in hyperthyroidism.
Uric acid (urate) (S)	Male: 4.0–8.5 mg/dl (240–510 µmol/liter) Female: 2.7–7.3 mg/dl (160–430 µmol/liter)	Values increase in impaired renal function, gout, metastatic cancer, shock, and starvation; values decrease in persons treated with uricosuric drugs.

EXHIBIT B.2 HEMATOLOGY TESTS

(WB) = WHOLE BLOOD   (S) = SERUM   (P) = PLASMA

Test (Specimen)	Reference Values: Conventional U.S. Units (SI Units)	Clinical Implications
Bleeding time (WB)	4–8 minutes using Simplate (same)	Values increase in thrombocytopenia, severe liver disease, leukemia, and aplastic anemia.
Erythrocyte sedimentation rate (ESR) (WB)	(Westergren) Female, under 50 years: <20 mm/hr (same) Female, over 50 years: 30 mm/hr (same) Male, under 50 years: <15 mm/hr (same) Male, over 50 years: <20 mm/hr (same)	Values increase in pregnancy, infection, carcinoma, tissue destruction, and nephritis; values decrease in sickle-cell anemia and congestive heart failure (CHF).
Hemoglobin (S or P)	Male: 13.5–18 g/100 ml (135–180 g/liter) Female: 12–16 g/100 ml (120–160 g/liter) Newborn: 14–20 g/100 ml (140–200 g/liter)	Values increase in polycythemia, congestive heart failure (CHF), chronic obstructive pulmonary disease, and at high altitudes; values decrease in anemia, hyperthyroidism, cirrhosis of the liver, and severe hemorrhage.
Hematocrit (WB)	Male: 40–54%; average 47% (same) Female: 38–47%; average 42% (same)	Values increase in polycythemia, severe dehydration, and shock; values decrease in anemia, leukemia, cirrhosis, and hyperthyroidism.
Platelet count (WB)	150,000–400,000/mm³ (150–400 × 10⁹/liter)	Values increase in cancer, trauma, heart disease, and cirrhosis; values decrease in anemias, allergic conditions, and during cancer chemotherapy.
Prothrombin time (PT) (WB)	11–15 seconds (same)	Values increase in prothrombin and vitamin K deficiency, liver disease, and hypervitaminosis A.
Red blood cell count (WB)	Male: 4.5–6.5 million/mm³ (4.5–6.5 × 10¹²/liter) Female: 3.9–5.6 million/mm³ (3.9–5.6 × 10¹²/liter)	Values increase in polycythemia, dehydration, and following hemorrhaging; values decrease in systemic lupus erythematosus (SLE), anemias, and Addison's disease.

Exhibit continues

EXHIBIT **B.2** **HEMATOLOGY TESTS** (CONTINUED)

(WB) = WHOLE BLOOD   (S) = SERUM   (P) = PLASMA

Test (Specimen)	Reference Values: Conventional U.S. Units (SI Units)	Clinical Implications
Reticulocyte count (WB)	0.5–2.0% (same)	Values increase in hemolytic anemia, metastatic carcinoma, and leukemia; values decrease in iron-deficiency and pernicious anemia, radiation therapy, and kidney disease in which kidney cells do not make erythropoietin.
White blood cell count, differential (WB)		Neutrophils increase in acute infections; eosinophils and basophils increase in allergic reactions; lymphocytes increase during antigen-antibody reactions; monocytes increase in chronic infections.
Neutrophils	60–70% (same)	
Eosinophils	2–4% (same)	
Basophils	0.5–1% (same)	
Lymphocytes	20–25% (same)	
Monocytes	3–8% (same)	
White blood cell count, total (WB)	5,000–10,000/mm^3 (5–10 × 10^9/liter)	Values increase in acute infections, trauma, malignant diseases, and cardiovascular diseases; values decrease in diabetes mellitus, anemias, and following cancer chemotherapy.

EXHIBIT **B.3** **URINE TESTS**

Test (Sample)	Reference Values: Conventional U.S. Units (SI Units)	Clinical Implications
Amylase (2 hour)	35–260 Somogyi units/hr (6.5–48.1 units/hr)	Values increase in inflammation of the pancreas (pancreatitis) or salivary glands, obstruction of the pancreatic duct, and perforated peptic ulcer.
Bilirubin[a] (random)	Negative	Values increase in liver disease and obstructive biliary disease.
Blood[a] (random)	Negative	Values increase in renal disease, extensive burns, transfusion reactions, and hemolytic anemia.
Calcium (Ca^{2+}) (random)	10 mg/dl (2.5 mmol/liter); up to 300 mg/24 hr (7.5 mmol/24 hr)	Amount depends on dietary intake; values increase in hyperparathyroidism, metastatic malignancies, and primary cancer of breasts and lungs; values decrease in hypoparathyroidism and vitamin D deficiency.
Casts (24 hour)		
Epithelial	Occasional	Values increase in nephrosis and heavy metal poisoning.
Granular	Occasional	Values increase in nephritis and pyelonephritis.
Hyaline	Occasional	Values increase in glomerular membrane damage and fever.
Red blood cell	Occasional	Values increase in pyelonephritis, kidney stones, and cystitis.
White blood cell	Occasional	Values increase in kidney infections.
Chloride (Cl$^-$) (24 hour)	140–250 mEq/24 hr (140–250 mmol/24 hr)	Amount depends on dietary salt intake; values increase in Addison's disease, dehydration, and starvation; values decrease in pyloric obstruction, diarrhea, and emphysema.
Color (random)	Yellow, straw, amber	Varies with many disease states, hydration, and diet.
Creatinine (24 hour)	Male: 1.0–2.0 g/24 hr (9–18 mmol/24 hr) Female: 0.8–1.8 g/24 hr (7–16 mmol/24 hr)	Values increase in infections; values decrease in muscular atrophy, anemia, and kidney diseases.

Test (Sample)	Reference Values: Conventional U.S. Units (SI Units)	Clinical Implications
Glucose[a] (random)	Negative	Values increase in diabetes mellitus, brain injury, and myocardial infarction.
Hydroxycorticosteroids (17-hydroxysteroids) (24 hour)	Male: 5–15 mg/24 hr (13–41 μmol/24 hr) Female: 2–13 mg/24 hr (5–36 μmol/24 hr)	Values increase in Cushing's syndrome, burns, and infections; values decrease in Addison's disease.
Ketone bodies[a] (random)	Negative	Values increase in diabetic acidosis, fever, anorexia, fasting, and starvation.
17-ketosteroids (KS) (24 hour)	Male: 8–25 mg/24 hr (28–87 μmol/24 hr) Female: 5–15 mg/24 hr (17–53 μmol/24 hr)	Values decrease in surgery, burns, infections, adrenogenital syndrome, and Cushing's syndrome.
Odor (random)	Aromatic	Becomes acetonelike in diabetic ketosis.
Osmolality (24 hour)	500–1400 mOsm/kg water (500–1400 mmol/kg water)	Values increase in cirrhosis, congestive heart failure (CHF), and high-protein diets; values decrease in aldosteronism, diabetes insipidus, and hypokalemia.
pH[a] (random)	4.6–8.0	Values increase in urinary tract infections and severe alkalosis; values decrease in acidosis, emphysema, starvation, and dehydration.
Phenylpyruvic acid (random)	Negative	Values increase in phenylketonuria (PKU).
Potassium (K^+) (24 hour)	40–80 mEq/24 hr (40–80 mmol/24 hr)	Values increase in chronic renal failure, dehydration, starvation, and Cushing's syndrome; values decrease in diarrhea, malabsorption syndrome, and adrenal cortical insufficiency.
Protein[a] (albumin) (random)	Negative	Values increase in nephritis, fever, severe anemias, trauma, and hyperthyroidism.
Sodium (Na^+) (24 hour)	75–200 mg/24 hr (75–200 mmol/24 hr)	Amount depends on dietary salt intake; values increase in dehydration, starvation, and diabetic acidosis; values decrease in diarrhea, acute renal failure, emphysema, and Cushing's syndrome.
Specific gravity[a] (random)	1.001–1.035 (same)	Values increase in diabetes mellitus and excessive water loss; values decrease in absence of antidiuretic hormone (ADH) and severe renal damage.
Urea (random)	25–35 g/24 hr (420–580 mmol/24 hr)	Values increase in response to increased protein intake; values decrease in impaired renal function.
Uric acid (24 hour)	0.4–1.0 g/24 hr (1.5–4.0 mmol/24 hr)	Values increase in gout, leukemia, and liver disease; values decrease in kidney disease.
Urobilinogen[a] (2 hour)	0.3–1.0 Ehrlich units (1.7–6.0 μmol/24 hr)	Values increase in anemias, hepatitis A (infectious), biliary disease, and cirrhosis; values decrease in cholelithiasis and renal insufficiency.
Volume, total (24 hour)	1000–2000 ml/24 hr (1.0–2.0 liters/24 hr)	Varies with many factors.

[a]Test often performed using a **dipstick**, a plastic strip impregnated with chemicals that is dipped into a urine specimen to detect particular substances. Certain colors indicate the presence or absence of a substance and sometimes give a rough estimate of the amount(s) present.

PERIODIC TABLE

Key:

1	← Atomic number
H	← Chemical symbol
1.00797	← Atomic weight

1 **H** 1.00794																	**2** **He** 4.00260
3 **Li** 6.941	**4** **Be** 9.01218											**5** **B** 10.811	**6** **C** 12.011	**7** **N** 14.0067	**8** **O** 15.9994	**9** **F** 18.9984	**10** **Ne** 20.1797
11 **Na** 22.9898	**12** **Mg** 24.3050											**13** **Al** 26.9815	**14** **Si** 28.0855	**15** **P** 30.9738	**16** **S** 32.066	**17** **Cl** 35.4527	**18** **Ar** 39.948
19 **K** 39.0983	**20** **Ca** 40.078	**21** **Sc** 44.9559	**22** **Ti** 47.88	**23** **V** 50.9415	**24** **Cr** 51.9961	**25** **Mn** 54.9381	**26** **Fe** 55.847	**27** **Co** 58.9332	**28** **Ni** 58.69	**29** **Cu** 63.546	**30** **Zn** 65.38	**31** **Ga** 69.723	**32** **Ge** 72.61	**33** **As** 74.9216	**34** **Se** 78.96	**35** **Br** 79.904	**36** **Kr** 83.80
37 **Rb** 85.4678	**38** **Sr** 87.62	**39** **Y** 88.9059	**40** **Zr** 91.224	**41** **Nb** 92.9064	**42** **Mo** 95.94	**43** **Tc** (98)	**44** **Ru** 101.07	**45** **Rh** 102.906	**46** **Pd** 106.42	**47** **Ag** 107.868	**48** **Cd** 112.41	**49** **In** 114.82	**50** **Sn** 118.710	**51** **Sb** 121.75	**52** **Te** 127.60	**53** **I** 126.904	**54** **Xe** 131.29
55 **Cs** 132.905	**56** **Ba** 137.327	**57*** **La** 138.906	**72** **Hf** 178.49	**73** **Ta** 180.9479	**74** **W** 183.85	**75** **Re** 186.2	**76** **Os** 190.2	**77** **Ir** 192.22	**78** **Pt** 195.08	**79** **Au** 196.967	**80** **Hg** 200.59	**81** **Tl** 204.383	**82** **Pb** 207.2	**83** **Bi** 208.980	**84** **Po** (209)	**85** **At** (210)	**86** **Rn** (222)
87 **Fr** (223)	**88** **Ra** 226.025	**89**† **Ac** 227.028	**104** **Unq** (261)	**105** **Unp** (262)	**106** **Unh** (263)	**107** **Uns** (262)	**108** **Uno** (265)	**109** **Une** (266)									

*Lanthanides	**58** **Ce** 140.115	**59** **Pr** 140.908	**60** **Nd** 144.24	**61** **Pm** (145)	**62** **Sm** 150.36	**63** **Eu** 151.965	**64** **Gd** 157.25	**65** **Tb** 158.925	**66** **Dy** 162.50	**67** **Ho** 164.930	**68** **Er** 167.26	**69** **Tm** 168.934	**70** **Yb** 173.04	**71** **Lu** 174.967
†Actinides	**90** **Th** 232.038	**91** **Pa** 231.036	**92** **U** 238.029	**93** **Np** 237.048	**94** **Pu** (244)	**95** **Am** (243)	**96** **Cm** (247)	**97** **Bk** (247)	**98** **Cf** (251)	**99** **Es** (252)	**100** **Fm** (257)	**101** **Md** (258)	**102** **No** (259)	**103** **Lr** (260)

GLOSSARY OF COMBINING FORMS, WORD ROOTS, PREFIXES, AND SUFFIXES

PRONUNCIATION KEY

1. The most strongly accented syllable appears in capital letters, for example, bilateral (bī-LAT-er-al) and diagnosis (dī-ag-NŌ-sis).
2. If there is a secondary accent, it is noted by a prime ('), for example, constitution (kon'-sti-TOO-shun) and physiology (fiz'-ē-OL-ō-jē). Any additional secondary accents are also noted by a prime, for example, decarboxylation (dē'-kar-bok'-si-LĀ-shun).
3. Vowels marked with a line above the letter are pronounced with the long sound as in the following common words:

    ā as in *māke*
    ē as in *bē*
    ī as in *īvy*
    ō as in *pōle*

4. Vowels not so marked are pronounced with the short sound as in the following words:

    a as in *above*
    e as in *bet*
    i as in *sip*
    o as in *not*
    u as in *bud*

5. Other phonetic symbols are used to indicate the following sounds:

    oo as in *sue*
    yoo as in *cute*
    oy as in *oil*

Many medical terms are "compound" words; that is, they are made up of one or more word roots or combining forms of word roots with prefixes or suffixes. For example, *leukocyte* (white blood cell) is a combination of *leuko,* the combining form for the word root meaning "white," and *cyt,* the word root meaning "cell." Learning the medical meanings of the fundamental word roots will enable you to analyze many long, complicated terms.

The following list includes the most commonly used combining forms, word roots, prefixes, and suffixes used in making medical terms and an example for each.

COMBINING FORMS AND WORD ROOTS

Acou-, Acu- hearing Acoustics (A-KOO-stiks), the science of sounds or hearing.

Acr-, Acro- extremity Acromegaly (ak'-rō-MEG-a-lē), hyperplasia of the nose, jaws, fingers, and toes.

Aden-, Adeno- gland Adenoma (ad-en'-Ō-ma), a tumor with a glandlike structure.

Alg-, Algia- pain Neuralgia (nyoo-RAL-ja), pain along the course of a nerve.

Angi- vessel Angiocardiography (an'-jē-ō-kard-ē-OG-ra-fē), x-ray of the great blood vessels and heart after intravenous injection of radiopaque fluid.

Arthr-, Arthro- joint Arthropathy (ar-THROP-a-thē), disease of a joint.

Aut-, Auto- self Autolysis (aw-TOL-i-sis), destruction of cells of the body by their own enzymes after death.

Bio- life, living Biopsy (BĪ-op-sē), examination of tissue removed from a living body.

Blast- germ, bud Blastocyte (BLAS-tō-sīt), an embryonic or undifferentiated cell.

Blephar- eyelid Blepharitis (blef-a-RĪT-is), inflammation of the eyelids.

Brachi- arm Brachialis (brā-kē-AL-is), muscle that flexes the forearm.

Bronch- trachea, windpipe Bronchoscopy (bron-KOS-kō-pē), direct visual examination of the bronchi.

Bucc- cheek Buccocervical (bū-kō-SER-vi-kal), pertaining to the cheek and neck.

Capit- head Decapitate (dē-KAP-i-tāt), to remove the head.

Carcin- cancer Carcinogenic (kar-sin-ō-JENK-ik), causing cancer.

Cardi-, Cardia-, Cardio- heart Cardiogram (KARD-ē-o-gram), a recording of the force and form of the heart's movements.

Cephal- head Hydrocephalus (hī-drō-SEF-a-lus), enlargement of the head due to an abnormal accumulation of fluid.

Cerebro- brain   Cerebrospinal (se-rē-brō-SPĪN-al) fluid, fluid contained within the cranium and spinal canal.

Cheil- lip   Cheilosis (kī-LŌ-sis), dry scaling of the lips.

Chole- bile, gall   Cholecystogram (kō-lē-SIS-tō-gram), x-ray image of the gallbladder.

Chondr-, Chondri-, Chondrio- cartilage   Chondrocyte (KON-drō-sīt), a cartilage cell.

Chrom-, Chromat-, Chromato- color   Hyperchromic (hī-per-KRŌ-mik), highly colored.

Cili- eyelash   Supercilia (soo′-per-SIL-ē-a), eyebrow (hairs above eyelash).

Colpo- vagina   Colpotomy (kol-POT-ō-mē), incision into the wall of the vagina.

Cor-, coron- heart   Coronary (KOR-ō-na-rē), arteries supplying blood to the heart muscle.

Cost- rib   Costal (KOS-tal), pertaining to a rib.

Crani- skull   Craniotomy (krā-ne-OT-ō-mē), surgical opening of the skull.

Cry-, Cryo- cold   Cryosurgery (kri-ō-SERJ-e-rē), surgical procedure using a very cold liquid nitrogen probe.

Cut- skin   Subcutaneous (sub-kyoo-TĀ-ne-us), under the skin.

Cysti-, Cysto- sac, bladder   Cystoscope (SIS-tō-skōp), instrument for internal examination of the urinary bladder.

Cyt-, Cyto-, Cyte- cell   Cytology (sī-TOL-ō-jē), the study of cells.

Dactyl-, Dactylo- digits   (usually fingers, but sometimes toes) Polydactylism (pol-ē-DAK-til-ism), more than the normal number of fingers or toes.

Derma-, Dermato- skin   Dermatosis (der-ma-TŌ-sis), any skin disease.

Dura- hard   Dura mater (DYOO-ra MĀ-ter), outer membrane covering brain and spinal cord.

Entero- intestine   Enteritis (ent-e-RĪT-is), inflammation of the intestine.

Erythro- red   Erythrocyte (e-RITH-rō-sīt), red blood cell.

Galacto- milk   Galactose (ga-LAK-tōse), a milk sugar.

Gastr- stomach   Gastrointestinal (gas′-trō-in-TES-tin-al), pertaining to the stomach and intestine.

Gloss-, Glosso- tongue   Hypoglossal (hī′-pō-GLOS-al), located under the tongue.

Gluco- sugar   Glucosuria (gloo′-kō-SUR-ē-a), sugar in the urine.

Gravid- pregnant   Gravidity (gra-VID-i-tē), condition of being pregnant.

Gyn-, Gyne-, Gynec- female, women   Gynecology (gīn′-e-KOL-ō-jē), the medical specialty dealing with disorders of the female reproductive system.

Hem-, Hemat- blood   Hematoma (hē′-ma-TO-ma), a tumor or swelling filled with blood.

Hepar-, Hepato- liver   Hepatitis (hep-a-TĪT-is), inflammation of the liver.

Hist-, Histio- tissue   Histology (his-TOL-ō-jē), the study of tissues.

Hydr- water   Hydrocele (HĪ-drō-sēl), accumulation of fluid in a saclike cavity.

Hyster- uterus   Hysterectomy (his′-te-REK-tō-mē), surgical removal of the uterus.

Ileo- ileum   Ileocecal (il′-ē-ō-SE-kal) valve, folds at the opening between ileum and cecum.

Ilio- ilium   Iliosacral (il′-ē-ō-SĀ-kral), pertaining to ilium and sacrum.

Kines- motion   Kinesiology (ki-ne-sē-OL-ō-jē), study of movement of body parts.

Labi- lip   Labial (LĀ-bē-al), pertaining to a lip.

Lachry-, Lacri- tears   Nasolacrimal (nā-zō-LAK-rim-al), pertaining to the nose and lacrimal apparatus.

Laparo- loin, flank, abdomen   Laparoscopy (lap′-a-ROS-kōpē), examination of the interior of the abdomen by means of a laparoscope.

Leuco-, Leuko- white   Leukocyte (LYOO-kō-sīt), white blood cell.

Lingua- tongue   Lingual (LIN-gwal), pertaining to the tongue.

Lip-, Lipo fat   Lipoma (lī-PŌ-ma), a fatty tumor.

Lith- stone   Lithiasis (li-THĒ-a-sis), the formation of stones.

Lumbo- lower back, loin   Lumbar (LUM-bar), pertaining to the loin.

Macul- spot, blotch   Macula (MAK-yoo-la), spot or blotch.

Malign- bad, harmful   Malignant (ma-LIG-nant), condition that gets worse and results in death.

Mamm- breast   Mammography (ma-MOG-ra-fē), x-ray of the mammary gland.

Mast- breast   Mastitis (ma-STĪT-is), inflammation of the mammary gland.

Meningo- membrane   Meningitis (men-in-JĪT-is), inflammation of the membranes of spinal cord and brain.

Metro- uterus   Endometrium (en′-dō-MĒ-trē-um), lining of the uterus.

Morpho- form, shape   Morphology (mor-FOL-o-jē), the study of form and structure of things.

Myelo- marrow, spinal cord   Poliomyelitis (pō-lē-ō-mī′-a-LĪT-is), inflammation of the gray matter of the spinal cord.

Myo- muscle   Myocardium (mī-ō-KARD-ē-um), heart muscle.

Necro- corpse, dead   Necrosis (ne-KRŌ-sis), death of areas of tissue surrounded by healthy tissue.

Nephro- kidney   Nephrosis (ne-FRŌ-sis), degeneration of kidney tissue.

Neuro- nerve   Neuroblastoma (nyoor′-ō-blas-TŌ-ma), malignant tumor of the nervous system composed of embryonic nerve cells.

Oculo- eye   Binocular (bī-NOK-yoo-lar), pertaining to the two eyes.

Odont- tooth   Orthodontic (or-thō-DONT-ik), pertaining to the proper positioning and relationship of the teeth.

Onco- mass, tumor   Oncology (ong-KOL-ō-jē), study of tumors.

Oo- egg   Oocyte (Ō-ō-sīt), original egg cell.

Oophor- ovary, egg carrier   Oophorectomy (ō′-of-o-REK-tō-mē), surgical removal of ovaries.

Ophthalm- eye   Ophthalmology (of-thal-MOL-ō-jē), the study of the eye and its diseases.

Or- mouth   Oral (Ō-ral), pertaining to the mouth.

Orchido- testicle   Orchidectomy (or′-ki-DEK-tō-mē), surgical removal of a testicle.

Osmo- odor, sense of smell   Anosmia (an-OZ-mē-a), absence of sense of smell.

Oss-, Osseo-, Osteo- bone   Osteoma (os-tē-Ō-ma), bone tumor.

Oto- ear   Otosclerosis (ō′-tō-skle-RŌ-sis), formation of bone in the labyrinth of the ear.

Palpebr- eyelid   Palpebra (PAL-pe-bra), eyelid.

Part- birth, delivery, labor   Parturition (par′-too-RISH-un), act of giving birth.

Patho- disease   Pathogenic (path′-ō-JEN-ik), causing disease.

Ped- children   Pediatrician (pēd-ē-a-TRISH-an), medical specialist in the treatment of children.

Peps- digest   Peptic (PEP-tik), pertaining to digestion.

Phag-, Phago- to eat   Phagocytosis (fag′-ō-sī-TŌ-sis), the process by which cells ingest particulate matter.

Philic-, Philo- to like, have an affinity for   Hydrophilic (hī-drō-FIL-ik), having an affinity for water.

Phleb- vein   Phlebitis (fle-BĪT-is), inflammation of the veins.

Phon- voice, sound   Phonogram (FŌ-nō-gram), record made of sound.

Phren- diaphragm   Phrenic (FREN-ik), pertaining to the diaphragm.

Pilo- hair   Depilatory (de-PIL-a-tō-rē), hair remover.

Pneumo- lung, air   Pneumothorax (nyoo-mō-THŌR-aks), air in the thoracic cavity.

Pod- foot   Podiatry (po-DĪ-a-trē), the diagnosis and treatment of foot disorders.

Procto- anus, rectum   Proctoscopy (prok-TOS-kō-pē), instrumental examination of the rectum.

Psycho- soul, mind   Psychiatry (sī-KĪ-a-trē), treatment of mental disorders.

Pulmon- lung   Pulmonary (PUL-mō-ner′-ē), pertaining to the lungs.

Pyle-, Pyloro opening, passage   Pyloric (pī-LOR-ik), pertaining to the pylorus of the stomach.

Pyo- pus   Pyuria (pī-YOOR-ē-a), pus in the urine.

Ren- kidneys   Renal (RĒ-nal), pertaining to the kidney.

Rhin- nose   Rhinitis (rī-NĪT-is), inflammation of nasal mucosa.

Salpingo- uterine (Fallopian) tube   Salpingitis (sal′-pin-JĪ-tis), inflammation of the uterine (Fallopian) tubes.

Scler-, Sclero- hard   Atherosclerosis (ath′-er-ō-skle-RŌ-sis), hardening of the arteries.

Sep-, Septic- toxic condition due to microorganisms   Septicemia (sep′-ti-SĒ-mē-a), presence of bacterial toxins in the blood (blood poisoning).

Soma-, Somato- body   Somatotropic (sō-mat-ō-TRŌ-pik), having a stimulating effect on body growth.

Somni- sleep   Insomnia (in-SOM-nē-a), inability to sleep.

Stasis-, Stat- stand still   Homeostasis (hō′-mē-ō-STĀ-sis), achievement of a steady state.

Sten- narrow   Stenosis (ste-NŌ-sis), narrowing of a duct or canal.

Tegument- skin, covering   Integumentary (in-teg-yoo-MEN-ta-rē), pertaining to the skin.

Therm- heat   Thermometer (ther-MOM-et-er), instrument used to measure and record heat.

Thromb- clot, lump   Thrombus (THROM-bus), clot in a blood vessel or heart.

Tox-, Toxic- poison   Toxemia (tok-SĒ-mē-a), poisonous substances in the blood.

Trich- hair   Trichosis (trik-Ō-sis), disease of the hair.

Tympan- eardrum   Tympanic (tim-PAN-ik) membrane, eardrum.

Vas- vessel, duct   Cerebrovascular (se-rē-brō-VAS-kyoo-lar), pertaining to the blood vessels of the cerebrum of the brain.

Viscer- organ   Visceral (VIS-e-ral), pertaining to the abdominal organs.

Zoo- animal   Zoology (zō-OL-ō-jē), the study of animals.

Zyg(o)- joined   Zygote (ZĪ-gōt), cell resulting from fertilization of an ovum by a sperm cell.

PREFIXES

A-, An- without, lack of, deficient   Anesthesia (an′-es-THĒ-zha), without sensation.

Ab- away from, from   Abnormal (ab-NOR-mal), away from normal.

Ad- to, near, toward   Adduction (a-DUK-shun), movement of a limb toward the axis of the body.

Alb- white   Albino (al-BĪ-no), person whose skin, hair, and eyes lack the pigment melanin.

Alveol- cavity, socket   Alveolus (al-VĒ-o-lus), air sac in the lung.

Ambi- both sides   Ambidextrous (am′-bi-DEK-strus), able to use either hand.

Ambly- dull   Amblyaphia (am-blē-A-fē-a), dull sense of touch.

Andro- male, masculine   Androgen (AN-drō-jen), male sex hormone.

Ankyl(o)- bent, fusion   Ankylosed (ANG-ki-lōsd), fused joint.

Ante- before   Antepartum (ant-ē-PAR-tum), before delivery of a baby.

Anti- against   Anticoagulant (an-tī-kō-AG-yoo-lant), a substance that prevents coagulation (clotting) of blood.

Basi- base, foundation   Basal (BĀ-sal), located near the base.

Bi- two, double, both   Biceps (BĪ-seps), a muscle with two heads of origin.

Bili- bile, gall   Biliary (BIL-ē-er-ē), pertaining to bile, bile ducts, or gallbladder.

Brachy- short   Brachyesophagus (brā-kē-e-SOF-a-gus), short esophagus.

Brady- slow   Bradycardia (brād′-ē-KARD-ē-a), abnormally slow resting heart rate

Cata- down, lower, under, against   Catabolism (ka-TAB-a-lizm), metabolic breakdown into simpler substances.

Circum- around   Circumrenal (ser-kum-RĒN-al), around the kidney.

Cirrh- yellow   Cirrhosis (si-RŌ-sis), liver disorder that causes yellowing of skin.

Co-, Con-, Com- with, together   Congenital (kon-JEN-i-tal), existing at birth.

Contra- against, opposite   Contraception (kon-tra-SEP-shun), the prevention of conception.

Crypt- hidden, concealed   Cryptorchidism (krip-TOR-ka-dizm′), undescended or hidden testes.

Cyano- blue   Cyanosis (sī-a-NŌ-sis), bluish discoloration due to inadequate oxygen.

De- down, from   Decay (de-KA), waste away from normal.

Demi-, hemi- half   Hemiplegia (hem′-ē-PLĒ-jē-a), paralysis on one side of the body.

Di-, Diplo- two   Diploid (DIP-loyd), having double the haploid number of chromosomes.

Dis- separation, apart, away from   Disarticulate (dis′-ar-TIK-yoo-lāt′), to separate at a joint.

Dys- painful, difficult   Dyspnea (disp-NĒ-a), difficult breathing.

E-, Ec-, Ex- out from, out of   Eccentric (ek-SEN-trik), not located at the center.

Ecto-, Exo- outside   Ectopic (ek-TOP-ik) pregnancy, gestation outside the uterine cavity.

Em-, En- in, on   Empyema (em′-pī-Ē-ma), pus in a body cavity.

End-, Endo- inside   Endocardium (en′-dō-KARD-ē-um), membrane lining the inner surface of the heart.

Epi- upon, on, above   Epidermis (ep′-i-DER-mis), outermost layer of skin.

Eu- well   Eupnea (YOOP-nē-a), normal breathing.

Ex-, Exo- out, away from   Exocrine (EK-sō-krin), excreting outwardly or away from.

Extra- outside, beyond, in addition to   Extracellular (ek′-stra-SEL-yoo-lar), outside the cell.

Fore- before, in front of   Forehead (FOR-hed), anterior part of head.

Gen- originate, produce, form   Genetics (gen-ET-iks), the study of heredity.

Gingiv- gum   Gingivitis (jin′-je-VĪ-tus), inflammation of the gums.

Hemi- half   Hemiplegia (hem-ē-PLĒ-jē-a), paralysis of only half of the body.

Heter-, Hetero- other, different   Heterogeneous (het′-e-rō-JEN-ē-us), composed of different substances.

Homeo-, Homo- unchanging, the same, steady   Homeostasis (hō′-mē-ō-STĀ-sis), achievement of a steady state.

Hyper- beyond, excessive   Hyperglycemia (hī-per-glī-SĒ-mē-a), excessive amount of glucose in the blood.

Hypo- under, below, deficient   Hypodermic (hī-pō-DER-mik), below the skin or dermis.

Idio- self, one's own, separate   Idiopathic (id′-ē-o-PATH-ik), a disease without recognizable cause.

In-, Im- in, inside, not   Incontinent (in-KON-ti-nent), not able to retain urine or feces.

Infra- beneath   Infraorbital (in′-fra-OR-bi-tal), beneath the orbit.

Inter- among, between   Intercostal (int′-er-KOS-tal), between the ribs.

Intra- within, inside   Intracellular (in′-tra-SEL-yoo-lar), inside the cell.

Iso- equal, like   Isotonic (ī-sō-TON-ik), equal tension or tone.

Later- side   Lateral (LAT-er-al), pertaining to a side or farther from the midline.

Lepto- small, slender, thin   Leptodermic (lep′-tō-DER-mik), having thin skin.

Macro- large, great   Macrophage (MAK-rō-fāj), large phagocytic cell.

Mal- bad, abnormal   Malnutrition (mal′-noo-TRISH-un), lack of necessary food substances.

Medi-, Meso- middle   Medial (MĒD-ē-al), nearer to midline.

Mega-, Megalo- great, large   Megakaryocyte (meg′-a-KAR-ē-ō-sīt), giant cell of bone marrow.

Melan- black   Melanin (MEL-a-nin), black or dark brown pigment found in skin and hair.

Meta- after, beyond   Metacarpus (met-a-KAR-pus), the part of the hand between the wrist and fingers.

Micro- small   Microtome (MĪ-krō-tōm), instrument for preparing very thin slices of tissue for microscopic examination.

Mono- one   Monorchid (mon-OR-kid), having one testicle.

Neo- new   Neonatal (nē-ō-NĀT-al), pertaining to the first weeks after birth.

Noct(i)- night   Nocturia (nok-TOO-rē-a), involuntary urination occurring at night during sleep.

Null(i)- none   Nullipara (nu-LIP-a-ra), woman with no children.

Nyct- night   Nyctalopia (nik′-ta-LŌ-pē-a), night blindness.

Oligo- small, deficient   Oliguria (ol-ig-YOO-rē-a), abnormally small amount of urine.

Ortho- straight, normal   Orthopnea (or-THOP-nē-a), inability to breathe in any position except when straight or erect.

Pan- all   Pancarditis (pan-kar-DĪ-tis), inflammation of the entire heart.

Para- near, beyond, apart from, beside   Paranasal (par-a-NĀ-zal), near the nose.

Per- through   Percutaneous (per′-kyoo-TĀ-nē-us), through the skin.

Peri- around   Pericardium (per′-i-KARD-ē-um), membrane or sac around the heart.

Poly- much, many   Polycythemia (pol′-i-sī-THĒ-mē-a), an excess of red blood cells.

Post- after, beyond   Postnatal (pōst-NĀT-al), after birth.

Pre-, Pro- before, in front of   Prenatal (prē-NĀ-T-al), before birth.

Prim- first   Primary (PRĪ-me-rē), first in time or order.
Proto- first   Protocol (PRŌ-to-kol), clinical report made from first notes taken.
Pseud-, Pseudo- false   Pseudoangina (soo′-dō-an-JĪ-na), false angina.

Retro- backward, located behind   Retroperitoneal (re′-trō-per′-it-on-Ē-al), located behind the peritoneum.

Schizo- split, divide   Schizophrenia (skiz′-ō-FRE-nē-a), split personality mental disorder.
Semi- half   Semicircular (sem′-i-SER-kyoo-lar) canals, canals in the shape of a half circle in the ears.
Sub- under, beneath, below   Submucosa (sub′-myoo-KŌ-sa), tissue layer under a mucous membrane.
Super- above, beyond   Superficial (soo-per-FISH-al), confined to the surface.
Supra- above, over   Suprarenal (soo-pra-RĒN-al), adrenal gland above the kidney.
Sym-, Syn- with, together, jointed   Syndrome (SIN-drōm), all the symptoms of a disease considered as a whole.

Tachy- rapid   Tachycardia (tak′-i-KARD-ē-a), abnormally rapid resting heart rate.
Terat(o)- malformed fetus   Teratogen (ter-AT-ō-jen), an agent that causes development of a malformed fetus.
Tetra-, quadra- four   Tetrad (TET-rad), group of four with something in common.
Trans- across, through, beyond   Transudation (trans-yoo-DĀ-shun), oozing of a fluid through a membrane or tissue surface.
Tri- three   Trigone (TRĪ-gon), a triangular space, as at the base of the bladder.

SUFFIXES

-able capable of, having ability to   Viable (VĪ-a-bal), capable of living.
-ac, -al pertaining to   Cardiac (KARD-ē-ak), pertaining to the heart.
-algia painful condition   Myalgia (mī-AL-jē-a), pain in a muscle.
-an, -ian pertaining to   Circadian (ser-KĀ-dē-an), pertaining to a daily (24 hour) cycle.
-ant having the characteristic of   Malignant (ma-LIG-nant), having the characteristic of badness.
-ary connected with   Ciliary (SIL-ē-ar-ē), resembling any hairlike structure.
-asis, -asia, -esis, -osis condition or state of   Hemostasis (hē-mō-STĀ-sis), stopping of bleeding.
-asthenia weakness   Myasthenia (mi-as-THĒ-nē-a), weakness of skeletal muscles.
-ation process, action, condition   Inspiration (in-spi-RĀ-shun), process of drawing air into lungs.

-cel, -cele swelling, an enlarged space or cavity   Meningocele (men-IN-gō-sēl), enlargement of the meninges.

-centesis puncture, usually for drainage   Amniocentesis (am′-nē-ō-sen-TĒ-sis), withdrawal of amniotic fluid.
-cid, -cide, -cis cut, kill, destroy   Germicide (jer-mi-SĪD), a substance that kills germs.

-ectasia, -ectasis stretching, dilation   Bronchiectasis (bron-kē-EK-ta-sis), dilation of a bronchus or bronchi.
-ectomize, ectomy excision of, removal of   Thyroidectomy (thi-royd-EK-tō-mē), surgical removal of a thyroid gland.
-ema swelling, distension   Emphysema (em′-fi-SĒ-ma), swelling of air sacs in lungs.
-emia condition of blood   Lipemia (lip-Ē-mē-a), abnormally high concentration of fat in the blood.
-esis condition, process   Enuresis (en′-yoo-RĒ-sis), condition of involuntary urination.
-esthesia sensation, feeling   Anesthesia (an′-es-THĒ-zē-a), total or partial loss of feeling.

-ferent carry   Efferent (EF-e-rent), carrying away from a center.
-form shape   Fusiform (FYOO-zi-form), spindle-shaped.

-gen agent that produces or originates   Pathogen (PATH-ō-jen), microorganism or substance capable of producing a disease.
-genic produced from, producing   Pyogenic (pī-ō-JEN-ik), producing pus.
-gram record, that which is recorded   Electrocardiogram (e-lek′-trō-KARD-ē-ō-gram), record of heart action.
-graph instrument for recording   Electroencephalograph (e-lek′-trō-en-SEF-a-lō-graf), instrument for recording electrical activity of the brain.

-ia state, condition   Hypermetropia (hī′-per-me-TRŌ-pē-a), condition of farsightedness.
-iatrics, iatry medical practice specialities   Pediatrics (pēd-ē-A-triks), medical science relating to care of children and treatment of their diseases.
-ician person associated with   Technician (tek-NISH-an), person skilled in a technical field.
-ics art of, science of   Optics (OP-tiks), science of light and vision.
-ion action, condition resulting from action   Incision (in-SIZH-un), act or result of cutting into flesh.
-ism condition, state   Rheumatism (ROO-ma-tizm), inflammation, especially of muscles and joints.
-ist one who practices   Internist (in-TER-nist), one who practices internal medicine.
-itis inflammation   Neuritis (nyoo-RĪT-is), inflammation of a nerve or nerves.
-ive relating to   Sedative (SED-a-tive), relating to a pain or tension reliever.

-logy, -ology the study or science of   Physiology (fiz-ē-OL-ō-jē), the study of function of body parts.
-lyso, -lysis dissolution, loosening, destruction   Hemolysis (hē-MOL-i-sis), destruction of red blood cell membranes, causing release of hemoglobin.

-malacia softening   Osteomalacia (os'-tē-ō-ma-LĀ-shē-a), softening of bone.

-megaly enlarged   Cardiomegaly (kar'-dē-ō-MEG-a-lē), enlarged heart.

-oid resembling   Lipoid (li-POYD), resembling fat.

-ologist specialist   Dermatologist (der-ma-TOL-ō-gist), specialist in the study of the skin.

-oma tumor   Fibroma (fi-BRŌ-ma), tumor composed mostly of fibrous tissue.

-ory pertaining to   Sensory (SENS-o-rē), pertaining to sensation.

-ose full of   Adipose (AD-i-pōz), characterized by presence of fat.

-osis condition, disease   Necrosis (ne-KRŌ-sis), condition of death of cells.

-ostomy create an opening   Colostomy (kō-LOS-tō-me), surgical creation of an opening between the colon and body surface.

-otomy surgical incision   Tracheotomy (trā-kē-OT-ō-me), surgical incision of the trachea.

-pathy disease   Neuropathy (nyoo-ROP-a-thē), disease of the peripheral nervous system.

-penia deficiency   Thrombocytopenia (throm'-bō-sīt'-ō-PĒ-nē-a), deficiency of thrombocytes in the blood.

-phobe, -phobia fear of, aversion to   Hydrophobia (hī-drō-FŌ-bē-a), fear of water.

-plasia, -plasty development, formation   Rhinoplasty (RĪ-nō-plas-tē), surgical reconstruction of the nose.

-plegia, -plexy stroke, paralysis   Apoplexy (AP-ō-plek-sē), sudden loss of consciousness and paralysis.

-pnea to breathe   Apnea (AP-nē-a), temporary absence of respiration, following a period of overbreathing.

-poiesis production   Hemopoiesis (hēm'-ō-poy-Ē-sis), formation and development of blood cells.

-ptosis falling, sagging   Blepharoptosis (blef'-a-rō-TŌ-sis), drooping of upper eyelid.

-rrhage bursting forth, abnormal discharge   Hemorrhage (HEM-or-rij), bursting forth of blood.

-rrhea flow, discharge   Diarrhea (dī-a-RĒ-a), abnormal frequency of bowel evacuation, the stools with a more or less fluid consistency.

-scope instrument for viewing   Bronchoscope (BRON-kō-skōp), instrument used to examine the interior of a bronchus.

-stomy creation of a mouth or artificial opening   Tracheostomy (trā-kē-OST-ō-mē), creation of an opening in the trachea.

-tic pertaining to   Diagnostic (dī'-ag-NOS-tik), pertaining to diagnosis.

-tomy cutting into, incision into   Laparotomy (lap-a-ROT-ō-mē), an abdominal incision to gain access to the peritoneal cavity.

-tripsy crushing   Lithotripsy (LITH-ō-trip-sē), crushing of a calculus (stone).

-trophy state relating to nutrition or growth   Hypertrophy (hī-PER-trō-fē), excessive growth of an organ or part.

-tropic turning toward, influencing, changing   Gonadotropic (gō-nad-ō-TRŌ-pic), influencing the gonads.

-uria urine   Polyuria (pol-ē-YOOR-ē-a), excessive excretion of urine.

GLOSSARY OF TERMS

Abdomen (ab-DŌ-men or AB-dō-men)   The area between the diaphragm and pelvis.

Abdominal (ab-DOM-i-nal) **cavity**   Superior portion of the abdominopelvic cavity that contains the stomach, spleen, liver, gallbladder, pancreas, small intestine, and most of the large intestine.

Abdominal thrust maneuver   A first-aid procedure for choking. Employs a quick, upward thrust against the diaphragm that forces air out of the lungs with sufficient force to eject any lodged material. Also called the **Heimlich** (HĪM-lik) **maneuver**.

Abdominopelvic (ab-dom′-i-nō-PEL-vic) **cavity**   Inferior component of the ventral body cavity that is subdivided into a superior abdominal cavity and an inferior pelvic cavity.

Abduction (ab-DUK-shun)   Movement away from the axis or midline of the body.

Abortion (a-BOR-shun)   The premature loss (spontaneous) or removal (induced) of the embryo or nonviable fetus; any failure in the normal process of developing or maturing.

Abrasion (a-BRĀ-shun)   A portion of skin that has been scraped away.

Abscess (AB-ses)   A localized collection of pus and liquefied tissue in a cavity.

Absorption (ab-SORP-shun)   The taking up of liquids by solids or of gases by solids or liquids; intake of fluids or other substances by cells of the skin or mucous membranes; the passage of digested foods from the gastrointestinal tract into blood or lymph.

Absorptive (fed) state   Metabolic state during which ingested nutrients are being absorbed by the blood or lymph from the gastrointestinal tract.

Accessory duct   A duct of the pancreas that empties into the duodenum about 2.5 cm (1 in.) superior to the ampulla of Vater (hepatopancreatic ampulla). Also called the **duct of Santorini** (san′-tō-RE-nē).

Accommodation (a-kom-ō-DĀ-shun)   A change in the curvature of the eye lens to adjust for vision at various distances.

Acetabulum (as′-e-TAB-yoo-lum)   The rounded cavity on the external surface of the hipbone that receives the head of the femur.

Acetylcholine (as′-ē-til-KŌ-lēn) **(ACh)**   A neurotransmitter liberated by many peripheral nervous system neurons and some central nervous system neurons. It is excitatory at neuromuscular junctions but inhibitory at some other synapses (slows heart rate). The enzyme **acetylcholinesterase** breaks down ACh, thus terminating its action.

Achilles′ tendon   *See* **Calcaneal tendon**.

Acid (AS-id)   A proton donor, or substance that dissociates into hydrogen ions (H⁺) and anions; characterized by an excess of hydrogen ions and a pH less than 7.

Acidosis (as-i-DŌ-sis)   A condition in which blood pH is below 7.35. Also known as **acidemia**.

Acini (AS-i-nē)   Masses of cells in the pancreas that secrete digestive enzymes.

Acne (AK-nē)   Inflammation of sebaceous (oil) glands that usually begins at puberty; the basic acne lesions in order of increasing severity are comedones, papules, pustules, and cysts.

Acoustic (a-KOOS-tik)   Pertaining to sound or the sense of hearing.

Acquired immunodeficiency syndrome (AIDS)   Caused by a virus called human immunodeficiency virus (HIV). A disorder characterized by a positive HIV-antibody test, low T4 (helper) cell count, and certain indicator diseases (Kaposi's sarcoma, *Pneumocystis carinii* pneumonia, tuberculosis, fungus diseases, etc.). Other symptoms include fever or night sweats, coughing, sore throat, fatigue, body aches, weight loss, and enlarged lymph nodes.

Acromegaly (ak′-rō-MEG-a-lē)   Condition caused by hypersecretion of human growth hormone (hGH) during adulthood characterized by thickened bones and enlargement of other tissues.

Acrosome (AK-rō-sōm)   A dense lysosome-like body in the head of a sperm cell that contains enzymes that facilitate the penetration of a sperm cell into a secondary oocyte.

Actin (AK-tin)   The contractile protein that makes up thin filaments in muscle fiber (cell).

Action potential   A wave of negativity that self-propagates along the membrane of a neuron or muscle fiber (cell); a rapid change in membrane potential that involves a depolarization followed by a repolarization. Also called a **nerve action potential** or **nerve impulse** as it relates to a neuron and a **muscle action potential** as it relates to a muscle fiber (cell).

Activation (ak′-ti-VĀ-shun) **energy**   The minimum amount of energy required for a chemical reaction to occur.

Active transport   The movement of substances across cell membranes against a concentration gradient, requiring the expenditure of energy (ATP).

Acuity (a-KYOO-i-tē)   Clearness or sharpness, usually of vision.

Acupuncture (AK-yoo-punk′-chur)   The insertion of a needle into a tissue for the purpose of drawing fluid or relieving pain. It is also an ancient Chinese practice employed to cure illnesses by inserting needles into specific locations of the skin.

Acute (a-KYOOT)   Having rapid onset, severe symptoms, and a short course; not chronic.

Adam's apple   *See* **Thyroid cartilage**.

Adaptation (ad′-ap-TĀ-shun)   The adjustment of the pupil of the eye to light variations. The property by which a neuron relays a decreased frequency of action potentials from a receptor even though the strength of the stimulus remains constant; the decrease in perception of a sensation over time while the stimulus is still present.

Addison's (AD-i-sonz) **disease**   Disorder caused by hyposecretion of glucocorticoids (and aldosterone) characterized by muscular weakness, hypoglycemia, mental lethargy, anorexia, nausea and vomiting, weight loss, low blood pressure, dehydration, and excessive skin and mucous membrane pigmentation.

Adduction (ad-DUK-shun)   Movement toward the axis or midline of the body.

Adenoids (AD-e-noyds)   Inflamed and enlarged pharyngeal tonsils.

Adenosine triphosphate (a-DEN-ō-sēn tri-FOS-fāt) **(ATP)**   The universal energy-carrying molecule manufactured in all living cells as a means of capturing and storing energy. It consists of the purine base *adenine* and the five-carbon sugar *ribose,* to which are added, in linear array, three *phosphate* groups.

Adenylate cyclase (a-DEN-i-lāt SĪ-klās)   An enzyme in the postsynaptic membrane that is activated when certain neurotransmitters (or hormones) bind to their receptors; the enzyme that converts ATP into cyclic AMP.

Adherence (ad-HĒR-ens)   Firm contact between the plasma membrane of a phagocyte and an antigen or other foreign substance.

Adhesion (ad-HĒ-zhun)   Abnormal joining of parts to each other.

Adipocyte (AD-i-pō-sit)   Fat cell, derived from a fibroblast.

Adipose (AD-i-poz) **tissue**   Tissue composed of adipocytes specialized for triglyceride storage and present in the form of soft pads between various organs for support, protection, and insulation.

Adrenal cortex (a-DRĒ-nal KOR-teks)   The outer portion of an adrenal gland, divided into three zones: the zona glomerulosa secretes mainly aldosterone, the zona fasciculata secretes mainly cortisol, and the zona reticularis secretes mainly weak androgens.

Adrenal (a-DRĒ-nal) **glands**   Two glands located superior to each kidney. Also called the **suprarenal** (soo′-pra-RĒ-nal) **glands.**

Adrenal medulla (me-DUL-a)   The inner portion of an adrenal gland, consisting of cells that secrete epinephrine and norepinephrine (NE) in response to the stimulation of preganglionic sympathetic neurons.

Adrenergic (ad′-ren-ER-jik) **fiber**   A nerve fiber that when stimulated releases epinephrine (adrenaline) or norepinephrine (noradrenaline) at a synapse.

Adrenocorticotropic (ad-rē′-nō-kor-ti-kō-TRŌP-ik) **hormone (ACTH)**   A hormone produced by the anterior pituitary gland that influences the production and secretion of certain hormones of the adrenal cortex.

Adventitia (ad-ven-TISH-ya)   The outermost covering of a structure or organ.

Aerobic (air-Ō-bik)   Requiring molecular oxygen.

Afferent arteriole (AF-er-ent ar-TĒ-rē-ōl)   A blood vessel of a kidney that breaks up into the capillary network called a glomerulus; there is one afferent arteriole for each glomerulus.

Agglutination (a-gloo′-ti-NĀ-shun)   Clumping of microorganisms or blood cells; typically an antigen–antibody reaction.

Aggregated lymphatic follicles   Aggregated lymph nodules that are most numerous in the ileum. Also called **Peyer's** (PĪ-erz) **patches.**

Agnosia (ag-NŌ-zē-a)   A loss of the ability to recognize the meaning of stimuli from the various senses (visual, auditory, touch).

Agraphia (a-GRAF-ē-a)   An inability to write.

AIDS   *See* **Acquired immunodeficiency syndrome.**

Albinism (AL-bin-izm)   Abnormal, nonpathological, partial or total absence of pigment in skin, hair, and eyes.

Albumin (al-BYOO-min)   The most abundant (60%) and smallest of the plasma proteins, which functions primarily to regulate osmotic pressure of plasma.

Albuminuria (al-byoo′-min-UR-ē-a)   Presence of albumin in the urine.

Aldosterone (al-do-STER-ōn)   A mineralocorticoid produced by the adrenal cortex that brings about sodium and water reabsorption and potassium excretion.

Aldosteronism (al′-do-STER-ōn-izm′)   Condition caused by hypersecretion of aldosterone that results in increased sodium concentration and decreased potassium concentration in blood and characterized by muscular paralysis, high blood pressure, and edema.

Alimentary (al-i-MEN-ta-rē)   Pertaining to nutrition.

Alkaline (AL-ka-lin)   Containing more hydroxide ions (OH⁻) than hydrogen ions (H⁺) to produce a pH higher than 7.

Alkalosis (al-ka-LŌ-sis)   A condition in which blood pH is higher than 7.45. Also known as **alkalemia.**

Allantois (a-LAN-tō-is)   A small, vascularized membrane between the chorion and amnion of the fetus that serves as an early site for blood formation.

Alleles (a-LĒZ)   Alternate forms of a single gene that control the same inherited trait (such as height or eye color) and are located at the same position (locus) on homologous chromosomes.

Allergen (AL-er-jen)   An antigen that evokes a hypersensitivity reaction.

Allergic (a-LER-jik)   Pertaining to or sensitive to an allergen.

All-or-none principle   In muscle physiology, individual muscle fibers (cells) contract to their fullest extent or not at all. In neuron physiology, if a stimulus is strong enough to initiate an action potential, a nerve impulse is propagated along the entire neuron at a constant strength.

Alpha (AL-fa) **cell**   A cell in the pancreatic islets (islets of Langerhans) in the pancreas that secretes glucagon. Also termed an **A cell.**

Alpha receptor   Receptor found on visceral effectors innervated by most sympathetic postganglionic axons.

Alveolar–capillary (al-VĒ-ō-lar) **membrane**   Structure in the lungs consisting of the alveolar wall and basement membrane and a capillary endothelium and basement membrane through which the diffusion of respiratory gases occurs. Also called the **respiratory membrane.**

Alveolar duct   Branch of a respiratory bronchiole around which alveoli and alveolar sacs are arranged.

Alveolar macrophage (MAK-rō-fāj)   Cell found in the alveolar walls of the lungs that is highly phagocytic. Also called a **dust cell.**

Alveolar (al-VĒ-ō-lar) **pressure**   Air pressure within the lungs. Also called **intrapulmonic pressure.**

Alveolar sac   A collection or cluster of alveoli that share a common opening.

Alveolus (al-VĒ-ō-lus)   A small hollow or cavity; an air sac in the lungs; milk-secreting portion of a mammary gland. *Plural is* **alveoli** (al-VĒ-ol-ī).

Alzheimer's (ALTZ-hi-merz) **disease (AD)**   Disabling, progressive neurological disorder characterized by dysfunction and death of specific cerebral neurons resulting in widespread intellectual impairment and personality changes.

Ambulatory (AM-byoo-la-tō′-rē)   Capable of walking.

Amenorrhea (ā-men-ō-RĒ-a)   Absence of menstruation.

Amino acid An organic acid, containing an acid carboxyl group (COOH) and a basic amino group (NH$_2$), that is the building unit from which proteins are formed.

Amnesia (am-NĒ-zē-a) A lack or loss of memory.

Amniocentesis (am′-nē-ō-sen-TĒ-sis) Removal of amniotic fluid by inserting a needle transabdominally into the amniotic cavity.

Amnion (AM-nē-on) The deepest fetal membrane; a thin transparent sac that holds the fetus suspended in amniotic fluid. Also called the "**bag of waters**."

Amniotic (am′-nē-OT-ik) **fluid** Fluid in the amniotic cavity, the space between the developing embryo (or fetus) and amnion; the fluid is initially produced as a filtrate from maternal blood and later from fetal urine.

Amphiarthrosis (am′-fē-ar-THRŌ-sis) A slightly movable articulation midway between diarthrosis and synarthrosis, in which the articulating bony surfaces are separated by fibrous connective tissue or fibrocartilage to which both are attached; types are syndesmosis and symphysis.

Ampulla (am-POOL-la) A saclike dilation of a canal.

Ampulla of Vater *See* **Hepatopancreatic ampulla**.

Amyotrophic (a-mē-ō-TROF-ik) **lateral sclerosis (ALS)** Progressive neuromuscular disease characterized by degeneration of motor neurons in the cerebral cortex, brain stem, and spinal cord that leads to muscular weakness. Also called **Lou Gehrig's disease**.

Anabolism (a-NAB-ō-lizm) Synthetic energy-requiring reactions whereby small molecules are built up into larger ones.

Anaerobic (an-air-Ō-bik) Not requiring molecular oxygen.

Anal (Ā-nal) **canal** The terminal 2 or 3 cm (1 in.) of the rectum; opens to the exterior through the anus.

Anal column A longitudinal fold in the mucous membrane of the anal canal that contains a network of arteries and veins.

Analgesia (an-al-JĒ-zē-a) Pain relief.

Anal triangle The subdivision of the female or male perineum that contains the anus.

Anaphase (AN-a-fāz) The third stage of mitosis in which the chromatids that have separated at the centromeres move to opposite poles of the cell.

Anaphylaxis (an′-a-fi-LAK-sis) Against protection; a hypersensitivity (allergic) reaction in which IgE antibodies attach to mast cells and basophils, causing them to produce mediators of anaphylaxis (histamine, leukotrienes, kinins, and prostaglandins) that bring about increased permeability of blood vessels, increased smooth muscle contraction, and increased mucus production. Examples are hay fever, hives, and anaphylactic shock.

Anastomosis (a-nas-tō-MŌ-sis) An end-to-end union or joining together of blood vessels, lymphatic vessels, or nerves.

Anatomical (an′-a-TOM-i-kal) **position** A position of the body universally used in anatomical descriptions in which the body is erect, facing the observer, the upper limbs are at the sides, the palms are facing forward, and the feet are on the floor.

Anatomic dead space Spaces of the nose, pharynx, larynx, trachea, bronchi, and bronchioles that contain 150 ml of tidal volume; the air does not reach the alveoli to participate in gas exchange.

Anatomy (a-NAT-ō-mē) The structure or study of structure of the body and the relation of its parts to each other.

Androgen (AN-drō-jen) Substance producing or stimulating masculine characteristics, such as the male hormone testosterone.

Anemia (a-NĒ-mē-a) Condition of the blood in which the number of functional red blood cells or their hemoglobin content is below normal.

Anesthesia (an′-es-THĒ-zē-a) A total or partial loss of feeling or sensation, usually defined with respect to loss of pain sensation; may be general or local.

Aneuploid (an′-yoo-PLOID) A cell that has one or more chromosomes of a set added or deleted.

Aneurysm (AN-yoo-rizm) A saclike enlargement of a blood vessel caused by a weakening of its wall.

Angina pectoris (an-JĪ-na *or* AN-ji-na PEK-tō-ris) A pain in the chest related to reduced coronary circulation that may or may not involve heart or artery disease.

Angiography (an-jē-OG-ra-fē) X-ray examination of blood vessels after injection of a radiopaque substance.

Angiotensin (an-jē-ō-TEN-sin) Either of two forms of a protein associated with regulation of blood pressure. Angiotensin I is produced by the action of renin on angiotensinogen and is converted by the action of ACE (angiotensin converting enzyme) into angiotensin II, which stimulates aldosterone secretion by the adrenal cortex, stimulates the sensation of thirst, and causes vasoconstriction with resulting increase in systemic vascular resistance.

Anion (AN-ī-on) A negatively charged ion. An example is the chloride ion (Cl$^-$).

Ankylosis (ang′-ki-LŌ-sus) Severe or complete loss of movement at a joint.

Anomaly (a-NOM-a-lē) An abnormality that may be a developmental (congenital) defect; a variant from the usual standard.

Anopsia (an-OP-sē-a) A defect of vision.

Anorexia nervosa (an-ō-REK-sē-a ner-VŌ-sa) A chronic disorder characterized by self-induced weight loss, body-image and other perceptual disturbances, and physiologic changes that result from nutritional depletion.

Anosmia (an-OZ-mē-a) Loss of the sense of smell.

Anoxia (an-OK-sē-a) Deficiency of oxygen.

Antagonist (an-TAG-ō-nist) A muscle that has an action opposite that of the prime mover (agonist) and yields to the movement of the prime mover.

Antagonistic (an-tag-ō-NIST-ik) **effect** A hormonal interaction in which the effect of one hormone on a target cell is opposed by another hormone. For example, calcitonin (CT) lowers blood calcium level, whereas parathormone (PTH) raises it.

Antepartum (an-tē-PAR-tum) Before delivery of the child; occurring (to the mother) before childbirth.

Anterior (an-TER-ē-or) Nearer to or at the front of the body. Also called **ventral**.

Anterior pituitary gland Anterior portion of the pituitary gland. Also called the **adenohypophysis** (ad′-e-nō-hī-POF-i-sis).

Anterior root The structure composed of axons of motor (efferent) fibers that emerges from the anterior aspect of the spinal cord and extends laterally to join a posterior root, forming a spinal nerve. Also called a **ventral root**.

Antibiotic (an′-ti-bī-OT-ik) Literally, "antilife"; a chemical produced by a microorganism that is able to inhibit the growth of or kill other microorganisms.

Antibody (AN-ti-bod′-ē) A protein produced by certain cells in response to a specific antigen; the antibody combines with that antigen to neutralize, inhibit, or destroy it. Also called an **immunoglobulin** (im-yoo-nō-GLOB-yoo-lin) or **Ig**.

Antibody-mediated immunity That component of immunity in which lymphocytes (B cells) develop into plasma cells that produce antibodies that destroy antigens. Also called **humoral** (YOO-mor-al) **immunity**.

Anticoagulant (an-tī-cō-AG-yoo-lant)   A substance that is able to delay, suppress, or prevent the clotting of blood.

Antidiuretic (an′-ti-dī-yoo-RET-ik)   A substance that inhibits urine formation.

Antidiuretic hormone (ADH)   Hormone produced by neurosecretory cells in the paraventricular and supraoptic nuclei of the hypothalamus that stimulates water reabsorption from kidney cells into the blood and vasoconstriction of arterioles. Also called **vasopressin** (Vāz-ō-PRES-in).

Antigen (AN-ti-jen)   A substance that has immunogenicity—the ability to provoke an immune response—and reactivity—the ability to react with the antibodies or cells that result from the immune response. Also termed complete antigen or immunogen.

Antigen-presenting cell (APC)   Special class of migratory cells that process and present antigens to T cells during an immune response; APCs include macrophages, B cells, and dendritic cells, which are present in the skin and in mucous membranes.

Antiport   Process by which two substances, often Na⁺ and another substance, move in opposite directions across a plasma membrane. Also called **countertransport**.

Anulus fibrosus (AN-yoo-lus fi-BRŌ-sus)   A ring of fibrous tissue and fibrocartilage that encircles the pulpy substance (nucleus pulposus) of an intervertebral disc.

Anuria (a-NOO-rē-a)   A daily urine output of less than 50 ml.

Anus (Ā-nus)   The distal end and outlet of the rectum.

Aorta (ā-OR-ta)   The main systemic trunk of the arterial system of the body that emerges from the left ventricle.

Aortic (ā-OR-tik) **body**   Receptor on or near the arch of the aorta that responds to alterations in blood levels of oxygen, carbon dioxide, and hydrogen ions (H⁺).

Aortic reflex   A reflex concerned with maintaining normal general systemic blood pressure.

Aperture (AP-er-chur)   An opening or orifice.

Apex (Ā-peks)   The pointed end of a conical structure, such as the apex of the heart.

Aphasia (a-FĀ-zē-a)   Loss of ability to express oneself properly through speech or loss of verbal comprehension.

Apnea (AP-nē-a)   Temporary cessation of breathing.

Apneustic (ap-NOO-stik) **area**   Portion of the respiratory center in the pons that sends stimulatory nerve impulses to the inspiratory area that activate and prolong inspiration and inhibit expiration.

Apocrine (AP-ō-krin) **gland**   A type of gland in which the secretory products gather at the free end of the secreting cell and are pinched off, along with some of the cytoplasm, to become the secretion, as in mammary glands.

Aponeurosis (ap′-ō-noo-RŌ-sis)   A sheetlike tendon joining one muscle with another or with bone.

Apoptosis (ap-ō-TŌ-sis)   A normal type of cell death that removes unneeded cells during embryological development, regulates the number of cells in tissues, and eliminates many potentially dangerous cells such as cancer cells. During apoptosis, the DNA fragments, the nucleus condenses, mitochondria cease to function, and the cytoplasm shrinks, but the plasma membrane remains intact. Phagocytes engulf and digest the apoptotic cells, and an inflammatory response does not occur.

Appendage (a-PEN-dij)   A structure attached to the body.

Appendicitis (a-pen-di-SĪ-tis)   Inflammation of the vermiform appendix.

Appositional (ap′-ō-ZISH-o-nal) **growth**   Growth due to surface deposition of material, as in the growth in diameter of cartilage and bone. Also called **exogenous** (eks-OJ-e-nus) **growth**.

Aqueous humor (AK-wē-us HYOO-mor)   The watery fluid, similar in composition to cerebrospinal fluid, that fills the anterior cavity of the eye.

Arachnoid (a-RAK-noyd)   The middle of the three coverings (meninges) of the brain or spinal cord.

Arachnoid villus (VIL-us)   Berrylike tuft of arachnoid that protrudes into the superior sagittal sinus and through which cerebrospinal fluid is reabsorbed into the bloodstream.

Arbor vitae (AR-bor VĪ-tē)   The treelike appearance of the white matter tracts of the cerebellum when seen in midsagittal section.

Arch of the aorta   The most superior portion of the aorta, lying between the ascending and descending segments of the aorta.

Areflexia (a′-rē-FLEK-sē-a)   Absence of reflexes.

Areola (a-RĒ-ō-la)   Any tiny space in a tissue. The pigmented ring around the nipple of the breast.

Arm   The portion of the upper limb from the shoulder to the elbow.

Arousal (a-ROW-zal)   Awakening from sleep, a response due to stimulation of the reticular activating system (RAS).

Arrector pili (a-REK-tor PI-lē)   Smooth muscles attached to hairs; contraction pulls the hairs into a more vertical position, resulting in "goose bumps."

Arrhythmia (a-RITH-mē-a)   Irregular heart rhythm. Also called a **dysrhythmia**.

Arteriogram (ar-TER-ē-ō-gram)   An x-ray image of an artery after injection of a radiopaque substance into the blood.

Arteriole (ar-TE-rē-ōl)   A small, almost microscopic, artery that delivers blood to a capillary.

Arteriosclerosis (ar-te′-rē-ō-skle-RŌ-sis)   Group of diseases characterized by thickening of the walls of arteries and loss of elasticity.

Artery (AR-ter-ē)   A blood vessel that carries blood away from the heart.

Arthritis (ar-THRĪ-tis)   Inflammation of a joint.

Arthrology (ar-THROL-ō-jē)   The study or description of joints.

Arthroscopy (ar-THROS-co-pē)   A procedure for examining the interior of a joint, usually the knee, by inserting an arthroscope into a small incision; used to determine extent of damage, remove torn cartilage, repair cruciate ligaments, and obtain samples for analysis.

Arthrosis (ar-THRŌ-sis)   A joint or articulation.

Articular (ar-TIK-yoo-lar) **capsule**   Sleevelike structure around a synovial joint composed of a fibrous capsule and a synovial membrane.

Articular cartilage (KAR-ti-lij)   Hyaline cartilage attached to articular bone surfaces.

Articular disc   Fibrocartilage pad between articular surfaces of bones of some synovial joints. Also called a **meniscus** (men-IS-cus).

Articulate (ar-TIK-yoo-lāt)   To join together as a joint to permit motion between parts.

Articulation (ar-tik′-yoo-LĀ-shun)   A joint; a point of contact between bones, cartilage and bones, or teeth and bones.

Artificial pacemaker   A device that generates and delivers electrical signals to the heart to maintain a regular heart rhythm.

Arytenoid (ar′-i-TĒ-noyd) **cartilages**   A pair of small, pyramidal cartilages of the larynx that attach to the vocal folds and intrinsic pharyngeal muscles and can move the vocal folds.

Ascending colon (KŌ-lon)   The portion of the large intestine that passes superiorly from the cecum to the inferior edge of the liver where it bends at the right colic (hepatic) flexure to become the transverse colon.

Ascites (as-SĪ-tēz)   Abnormal accumulation of serous fluid in the peritoneal cavity.

Aseptic (ā-SEP-tik)   Free from any infectious or septic material.

Asphyxia (as-FIX-ē-a)   Unconsciousness due to interference with the oxygen supply of the blood.

Aspiration (as′-pi-RA-shun)   Inhalation of a foreign substance (water, food, or foreign body) into the bronchial tree; drainage of a substance in or out by suction.

Association area   A portion of the cerebral cortex connected by many motor and sensory fibers to other parts of the cortex. The association areas are concerned with motor patterns, memory, concepts of word-hearing and word-seeing, reasoning, will, judgment, and personality traits.

Association neuron (NOO-ron)   A nerve cell lying completely within the central nervous system. Also called an **interneuron**.

Astereognosis (as-ter′-ē-ōg-NO-sis)   Inability to recognize objects or forms by touch.

Asthenia (as-THE-nē-a)   Lack or loss of strength.

Asthma (AZ-ma)   Usually allergic reaction characterized by smooth muscle spasms in bronchi resulting in wheezing and difficult breathing. Also called **bronchial asthma**.

Astigmatism (a-STIG-ma-tizm)   An irregularity of the lens or cornea of the eye causing the image to be out of focus and producing faulty vision.

Astrocyte (AS-trō-sit)   A neuroglial cell having a star shape that participates in brain development and the metabolism of neurotransmitters, helps form the blood–brain barrier and maintain the proper balance of K$^+$ for generation of nerve impulses, and provides a link between neurons and blood vessels.

Ataxia (a-TAK-sē-a)   A lack of muscular coordination, lack of precision, usually due to cerebellar damage.

Atelectasis (at′-ē-LEK-ta-sis)   A collapsed or airless state of all or part of the lung, which may be acute or chronic.

Atherosclerosis (ath′-er-ō-skle-RO-sis)   A process in which fatty substances (cholesterol and triglycerides) are deposited in the walls of medium and large arteries in response to certain stimuli (hypertension, carbon monoxide, dietary cholesterol). Following endothelial damage, monocytes stick to the tunica interna, develop into macrophages, and take up cholesterol and low-density lipoproteins. Smooth muscle fibers (cells) in the tunica media ingest cholesterol. This results in the formation of an atherosclerotic plaque that decreases the size of the arterial lumen.

Atherosclerotic (ath′-er-ō-skle-RO-tic) **plaque** (PLAK)   A lesion that results from accumulated cholesterol and smooth muscle fibers (cells) of the tunica media of an artery; may become obstructive.

Atom   Unit of matter that comprises a chemical element; consists of a nucleus and electrons.

Atomic mass (weight)   Average mass of all stable atoms of an element, reflecting the relative proportion of atoms with different mass numbers.

Atomic number   Number of protons in an atom.

Atresia (a-TRE-zē-a)   Degeneration and reabsorption of an ovarian follicle before it fully matures and ruptures; abnormal closure of a passage, or absence of a normal body opening.

Atrial fibrillation (A-trē-al fib-ri-LA-shun)   Asynchronous contraction of the atria that results in the cessation of atrial pumping.

Atrial natriuretic (na′-trē-yoo-RET-ik) **peptide (ANP)**   Peptide hormone produced by the atria of the heart in response to stretching that inhibits aldosterone production and thus lowers blood pressure.

Atrioventricular (AV) (ā′-trē-ō-ven-TRIK-yoo-lar) **bundle**   The portion of the conduction system of the heart that begins at the atrioventricular (AV) node, passes through the cardiac skeleton separating the atria and the ventricles, then extends a short distance down the interventricular septum before splitting into right and left bundle branches. Also called the **bundle of His** (HISS).

Atrioventricular (AV) node   The portion of the conduction system of the heart made up of a compact mass of conducting cells located near the orifice of the coronary sinus in the right atrial wall.

Atrioventricular (AV) valve   A structure made up of membranous flaps or cusps that allows blood to flow in one direction only, from an atrium into a ventricle.

Atrium (A-trē-um)   A superior chamber of the heart.

Atrophy (AT-rō-fē)   Wasting away or decrease in size of a part, due to a failure, abnormality of nutrition, or lack of use.

Auditory ossicle (AW-di-tō-rē OS-si-kul)   One of the three small bones of the middle ear called the malleus, incus, and stapes.

Auditory tube   The tube that connects the middle ear with the nose and nasopharynx region of the throat. Also called the **Eustachian** (yoo-STA-kē-an) **tube**.

Aura (OR-a)   A feeling or sensation that precedes an epileptic seizure or any paroxysmal attack (like those of bronchial asthma).

Auscultation (aws-kul-TA-shun)   Examination by listening to sounds in the body.

Autocrine (AW-tō-krin)   Local hormone, such as interleukin-2, that acts on the same cell that secreted it.

Autograft (AW-tō-graft)   A graft of tissue from a donor site to a recipient site of the same individual.

Autoimmunity   An immunologic response against a person's own tissue antigens.

Autologous (aw-TOL-ō-gus) **preoperative transfusion**   Donating one's own blood up to 6 weeks before elective surgery to ensure an abundant supply and reduce transfusion complications such as those that may be associated with diseases such as AIDS and hepatitis. Also called **predonation**.

Autolysis (aw-TOL-i-sis)   Self-destruction of cells by their own lysosomal digestive enzymes after death or in a pathological process.

Autonomic ganglion (aw′-tō-NOM-ik GANG-lē-on)   A cluster of sympathetic or parasympathetic cell bodies located outside the central nervous system.

Autonomic nervous system (ANS)   Visceral sensory (afferent) and motor (efferent) neurons, both sympathetic and parasympathetic. Motor neurons transmit nerve impulses from the central nervous system to smooth muscle, cardiac muscle, and glands; so named because this portion of the nervous system was thought to be self-governing or spontaneous.

Autonomic plexus (PLEK-sus)   An extensive network of sympathetic and parasympathetic fibers; the cardiac, celiac, and pelvic plexuses are located in the thorax, abdomen, and pelvis, respectively.

Autophagy (aw-TOF-a-jē)   Process by which worn-out organelles are digested within lysosomes.

Autopsy (AW-top-sē)   An examination of the body after death to determine or confirm the cause of death.

Autoregulation (aw-tō-reg-yoo-LA-shun)   A local, automatic adjustment of blood flow in a given region of the body in response to tissue needs.

Autorhythmic cells   Cardiac or smooth muscle fibers that are self-excitable (generate impulses without an external stimulus); act as the heart's pacemaker and conduct the pacing impulse through the conduction system of the heart. Self-excitable neurons in the central nervous system, as in the inspiratory area of the brain stem.

Autosome (AW-tō-sōm)   Any chromosome other than the pair of sex chromosomes.

Axilla (ak-SIL-a)   The small hollow beneath the arm where it joins the body at the shoulders. Also called the **armpit**.

Axon (AK-son)   The usually single, long process of a nerve cell that propagates a nerve impulse toward the axon terminals.

Axon terminal   Terminal branch of an axon.

Azygos (AZ-i-gos)   An anatomical structure that is not paired; occurring singly.

Babinski (ba-BIN-skē) **sign**   Extension of the great toe, with or without fanning of the other toes, in response to stimulation of the outer margin of the sole of the foot; a normal reflex up to 1½ years of age; abnormal sign thereafter.

Back   The posterior part of the body; the dorsum.

Bainbridge (BĀN-bridge) **reflex**   The increased heart rate that follows increased pressure or distension of the right atrium.

Ball-and-socket joint   A synovial joint in which the rounded surface of one bone moves within a cup-shaped depression or fossa of another bone, as in the shoulder or hip joint. Also called a **spheroid** (SFĒ-roid) **joint**.

Barium (BA-rē-um) **swallow**   X-ray examination of the upper gastrointestinal tract to evaluate for ulcers, tumors, and bleeding.

Baroreceptor (bar'-ō-re-SEP-tor)   Nerve cell capable of responding to changes in blood, air, or fluid pressure. Also called a **pressoreceptor**.

Bartholin's glands   *See* **Greater vestibular glands**.

Basal ganglia (GANG-glē-a)   Paired clusters of cell bodies that make up the central gray matter in each cerebral hemisphere, including the caudate nucleus, lentiform nucleus, claustrum, and amygdaloid body. Also called **cerebral nuclei** (SER-e-bral NOO-klē-ī).

Basal metabolic (BĀ-sal met'-a-BOL-ik) **rate (BMR)**   The rate of metabolism measured under standard or basal conditions (awake, at rest, fasting).

Base   The broadest part of a pyramidal structure. A nonacid or a proton acceptor, characterized by excess of hydroxide ions (OH⁻) and a pH greater than 7. A ring-shaped, nitrogen-containing organic molecule that is one of the components of a nucleotide, namely, adenine, guanine, cytosine, thymine, and uracil.

Basement membrane   Thin, extracellular layer consisting of a basal lamina secreted by epithelial cells and a reticular lamina secreted by connective tissue cells.

Basilar (BĀ-S-i-lar) **membrane**   A membrane in the cochlea of the inner ear that separates the cochlear duct from the scala tympani and on which the spiral organ (organ of Corti) rests.

Basophil (BĀ-sō-fil)   A type of white blood cell characterized by a pale nucleus and large granules that stain blue-purple with basic dyes.

B cell   A lymphocyte that can develop into an antibody-producing plasma cell or a memory cell.

Belly   The abdomen. The gaster or prominent, fleshy part of a skeletal muscle.

Benign (be-NĪN)   Not malignant; favorable for recovery; a mild disease.

Beta (BĀ-ta) **cell**   A cell in the pancreatic islets (islets of Langerhans) in the pancreas that secretes insulin. Also termed a **B cell**.

Beta receptor   Receptor found on some visceral effectors innervated by sympathetic postganglionic axons.

Bicuspid (bi-KUS-pid) **valve**   Atrioventricular (AV) valve on the left side of the heart. Also called the **mitral valve**.

Bifurcate (bī-FUR-kāt)   Having two branches or divisions; forked.

Bilateral (bī-LAT-er-al)   Pertaining to two sides of the body.

Bile (BĪL)   A secretion of the liver consisting of water, bile salts, bile pigments, cholesterol, lecithin, and several ions; it emulsifies lipids prior to their digestion.

Biliary (BIL-ē-er-ē)   Relating to bile, the gallbladder, or the bile ducts.

Biliary calculi (CAL-kyoo-lē)   Gallstones formed by the crystallization of cholesterol in bile.

Bilirubin (bil-ē-ROO-bin)   An orange pigment that is one of the end products of hemoglobin breakdown in the hepatocytes and is excreted as a waste material in the bile.

Bilirubinuria (bil-ē-roo-bi-NOO-rē-a)   The presence of above-normal levels of bilirubin in urine.

Biliverdin (bil-ē-VER-din)   A green pigment that is one of the first products of hemoglobin breakdown in the hepatocytes and is converted to bilirubin or excreted as a waste material in bile.

Biopsy (BĪ-op-sē)   Removal of tissue or other material from the living body for examination, usually microscopic.

Blastocele (BLAS-tō-sēl)   The fluid-filled cavity within the blastocyst.

Blastocyst (BLAS-tō-sist)   In the development of an embryo, a hollow ball of cells that consists of a blastocele (the internal cavity), trophoblast (outer cells), and inner cell mass.

Blastomere (BLAS-tō-mēr)   One of the cells resulting from the cleavage of a fertilized ovum.

Blastula (BLAS-tyoo-la)   An early stage in the development of a zygote.

Bleeding time   The time required for the cessation of bleeding from a small skin puncture; identifies platelet function defects and integrity of small blood vessels; ranges from 4 to 8 minutes.

Blind spot   Area in the retina at the end of the optic (II) nerve in which there are no photoreceptors.

Blood   The fluid that circulates through the heart, arteries, capillaries, and veins and that constitutes the chief means of transport within the body.

Blood–brain barrier (BBB)   A barrier consisting of specialized brain capillaries and astrocytes that prevents the passage of materials from the blood to the cerebrospinal fluid and brain.

Blood pressure (BP)   Force exerted by blood against the walls of blood vessels, due to contraction of the heart and influenced by the elasticity of the vessel walls; clinically, a measure of the pressure in arteries during ventricular systole and ventricular diastole. *See also* **Mean arterial blood pressure**.

Blood reservoir (REZ-er-vwar)   Systemic veins that contain large amounts of blood that can be moved quickly to parts of the body requiring the blood.

Blood–testis barrier   A barrier formed by sustentacular (Sertoli) cells that prevents an immune response against antigens produced by sperm and developing gametes by isolating the cells from the blood.

Body cavity   A space within the body that contains various internal organs.

Body fluid   Body water and its dissolved substances; comprises about 60% of total body weight.

Bohr (BOR) **effect**   In an acid environment, oxygen splits more readily from hemoglobin because when hydrogen ions (H⁺) bind to hemoglobin, they alter the structure of hemoglobin and this reduces its oxygen-carrying capacity.

Bolus (BŌ-lus)   A soft, rounded mass, usually food, that is swallowed.

Bone scan   Procedure in which a radioisotope is injected and the radiation emitted from bone is measured.

Bony labyrinth (LAB-i-rinth)   A series of cavities within the petrous portion of the temporal bone forming the vestibule, cochlea, and semicircular canals of the inner ear.

Bowman's capsule   *See* **Glomerular capsule**.

Brachial plexus (BRĀ-kē-al PLEK-sus)   A network of nerve fibers of the ventral rami of spinal nerves C5, C6, C7, C8, and T1. The nerves that emerge from the brachial plexus supply the upper limb.

Bradycardia (brād′-i-KAR-de-a)   A slow resting heart or pulse rate (under 60/min).

Brain   A mass of nervous tissue located in the cranial cavity.

Brain stem   The portion of the brain immediately superior to the spinal cord, made up of the medulla oblongata, pons, and midbrain.

Brain waves   Electrical activity produced as a result of action potentials of brain cells.

Broad ligament   A double fold of parietal peritoneum attaching the uterus to the side of the pelvic cavity.

Broca's (BRŌ-kaz) **area**   Motor area of the brain in the frontal lobe that translates thoughts into speech. Also called the **motor speech area**.

Bronchi (BRONG-kē)   Branches of the respiratory passageway including primary bronchi (the two divisions of the trachea), secondary or lobar bronchi (divisions of the primary that are distributed to the lobes of the lung), and tertiary or segmental bronchi (divisions of the secondary that are distributed to bronchopulmonary segments of the lung). *Singular is* **bronchus**.

Bronchial tree   The trachea, bronchi, and their branching structures up to and including the terminal bronchioles.

Bronchiectasis (brong′-kē-EK-ta-sis)   A chronic disorder in which there is a loss of the normal tissue and expansion of lung air passages; characterized by difficult breathing, coughing, expectoration of pus, and foul breath.

Bronchiole (BRONG-kē-ōl)   Branch of a tertiary bronchus further dividing into terminal bronchioles (distributed to lobules of the lung), which divide into respiratory bronchioles (distributed to alveolar sacs).

Bronchitis (brong-KĪ-tis)   Inflammation of the bronchi characterized by hypertrophy and hyperplasia of seromucous glands and goblet cells that line the bronchi and results in a productive cough.

Bronchogenic carcinoma (brong′-kō-JEN-ik kar′-si-NŌ-ma)   Cancer originating in the bronchi.

Bronchogram (BRONG-kō-gram)   An x-ray image of the bronchial tree.

Bronchography (bron-KOG-ra-fē)   Technique for examining the bronchial tree in which an opaque contrast medium is introduced into the trachea for distribution to the bronchial branches. The x-ray image produced is called a bronchogram.

Bronchopulmonary (brong′-kō-PUL-mō-ner-ē) **segment**   One of the smaller divisions of a lobe of a lung supplied by its own branches of a bronchus.

Bronchoscope (BRONG-kō-skōp)   An instrument used to examine the interior of the bronchi of the lungs.

Bronchoscopy (brong-KOS-kō-pē)   Visual examination of the interior of the trachea and bronchi with a bronchoscope to biopsy a tumor, clear an obstruction, take cultures, stop bleeding, or deliver drugs.

Brunner's gland   *See* **Duodenal gland**.

Buccal (BUK-al)   Pertaining to the cheek or mouth.

Buffer (BUF-er) **system**   A pair of chemicals, one a weak acid and one the salt of the weak acid, which functions as a weak base, that resists changes in pH.

Bulb of penis   Expanded portion of the base of the corpus spongiosum penis.

Bulbourethral (bul′-bō-yoo-RĒ-thral) **gland**   One of a pair of glands located inferior to the prostate gland on either side of the urethra that secretes an alkaline fluid into the cavernous urethra. Also called a **Cowper's** (KOW-perz) **gland**.

Bulimia (boo-LIM-ē-a)   A disorder characterized by overeating, followed by purging by self-induced vomiting, strict dieting or fasting, vigorous exercise, or use of laxatives.

Bulk flow   The movement of large numbers of ions, molecules, or particles in the same direction as a result of pressure differences (osmotic, hydrostatic, or air pressure).

Bundle branch   One of the two branches of the atrioventricular (AV) bundle made up of specialized muscle fibers (cells) that transmit electrical impulses to the ventricles.

Bundle of His   *See* **Atrioventricular (AV) bundle**.

Bunion (BUN-yun)   Lateral deviation of the great toe that produces inflammation and thickening of the bursa, bone spurs, and calluses.

Burn   An injury in which tissue proteins are destroyed (denatured) as a result of heat (fire, steam), chemicals, electricity, or the ultraviolet rays of the sun.

Bursa (BUR-sa)   A sac or pouch of synovial fluid located at friction points, especially about joints.

Bursitis (bur-SĪ-tis)   Inflammation of a bursa.

Buttocks (BUT-oks)   The two fleshy masses on the posterior aspect of the inferior trunk, formed by the gluteal muscles.

Cachexia (kah-KEK-sē-ah)   A state of ill health, malnutrition, and wasting.

Calcaneal tendon   The tendon of the soleus, gastrocnemius, and plantaris muscles at the back of the heel. Also called the **Achilles'** (a-KIL-ēz) **tendon**.

Calcification (kal-si-fi-KĀ-shun)   Deposition of mineral salts, primarily hydroxyapatite, in a framework formed by collagen fibers in which the tissue hardens. Also called **mineralization** (min′-e-ral-i-ZĀ-shun).

Calcitonin (kal-si-TŌ-nin) **(CT)**   A hormone produced by the thyroid gland that lowers the calcium and phosphate levels of the blood by inhibiting bone breakdown and accelerating calcium absorption by bones.

Calculus (KAL-kyoo-lus)   A stone, or insoluble mass of crystallized salts or other material, formed within the body, as in the gallbladder, kidney, or urinary bladder.

Callus (KAL-lus)   A growth of new bone tissue in and around a fractured area, ultimately replaced by mature bone. An acquired, localized thickening.

Calmodulin (kal-MOD-yoo-lin)   An intracellular protein that binds with calcium ions and activates or inhibits enzymes, many of which are protein kinases, to elicit physiological responses of hormones.

Calorie (KAL-ō-rē)   A unit of heat. A calorie (cal) is the standard unit and is the amount of heat necessary to raise 1 g of water from 14 to 15°C. The kilocalorie (kcal), used in metabolic and nutrition studies, is equal to 1000 cal.

Calyx (KĀL-iks) Any cuplike division of the kidney pelvis. *Plural is* **calyces** (KĀ-li-sēz).

Canal (ka-NAL) A narrow tube, channel, or passageway.

Canaliculus (kan′-a-LIK-yoo-lus) A small channel or canal, as in bones, where they connect lacunae. *Plural is* **canaliculi** (kan′-a-LIK-yoo-lī).

Canal of Schlemm *See* **Scleral venous sinus**.

Cancellous (KAN-sel-us) Having a reticular or latticework structure, as in spongy tissue of bone.

Cancer (KAN-ser) A malignant tumor of epithelial origin tending to infiltrate and give rise to new growths or metastases. Also called **carcinoma** (kar′-si-NŌ-ma).

Canker (KANG-ker) **sore** Painful ulcer on the mucous membrane of the mouth that may result from an autoimmune response.

Capacitation (ka′-pas-i-TĀ-shun) The functional changes that sperm undergo in the female reproductive tract that allow them to fertilize a secondary oocyte.

Capillary (KAP-i-lar′-ē) A microscopic blood vessel located between an arteriole and venule through which materials are exchanged between blood and body cells.

Carbohydrate (kar′-bō-HĪ-drāt) An organic compound containing carbon, hydrogen, and oxygen in a particular amount and arrangement and composed of sugar subunits; usually has the formula $(CH_2O)_n$.

Carbon monoxide (CO) poisoning Hypoxia due to increased levels of carbon monoxide as a result of its preferential and tenacious combination with hemoglobin rather than with oxygen.

Carcinoembryonic (car′-sin-ō-em-brē-ON-ik) **antigen (CEA)** A glycoprotein secreted by normally developing fetal tissue during the first or second trimester, after birth, and in certain malignant and benign conditions.

Carcinogen (kar-SIN-ō-jen) Any substance that causes cancer.

Carcinoma (kar′-si-NŌ-ma) A malignant tumor consisting of epithelial cells.

Cardiac (KAR-dē-ak) **arrest** Cessation of an effective heartbeat in which the heart is completely stopped or in ventricular fibrillation.

Cardiac catheterization (KAR-dē-ak kath′-e-ter-i-ZĀ-shun) Introduction of a catheter into the heart and/or its blood vessels to measure pressure; assess left ventricular function and cardiac output; measure blood flow, oxygen content of blood, and the status of valves and conduction system; and identify valvular and septal defects.

Cardiac (KAR-dē-ak) **cycle** A complete heartbeat consisting of systole (contraction) and diastole (relaxation) of both atria plus systole and diastole of both ventricles.

Cardiac muscle Striated muscle fibers (cells) that form the wall of the heart; stimulated by an intrinsic conduction system and autonomic motor neurons.

Cardiac notch An angular notch in the anterior border of the left lung into which a portion of the heart fits.

Cardiac output (CO) The volume of blood pumped from one ventricle of the heart (usually measured from the left ventricle) in 1 min; about 5.2 liters/min under normal resting conditions.

Cardiac reserve The maximum percentage that cardiac output can increase above normal.

Cardiac tamponade (tam′-pon-ĀD) Compression of the heart due to excessive fluid or blood in the pericardial sac that could result in cardiac failure.

Cardinal ligament A ligament of the uterus, extending laterally from the cervix and vagina as a continuation of the broad ligament.

Cardiology (kar-dē-OL-ō-jē) The study of the heart and diseases associated with it.

Cardiopulmonary resuscitation (rē-sus-i-TĀ-shun) **(CPR)** A technique employed to restore life or consciousness to a person apparently dead or dying; includes external respiration (exhaled air respiration) and external cardiac massage.

Cardiovascular (kar-dē-ō-VAS-kyoo-lar) **center** Groups of neurons scattered within the medulla oblongata that regulate heart rate, force of contraction, and blood vessel diameter.

Carotene (KAR-o-tēn) Antioxidant vitamin; yellow-orange pigment present in the stratum corneum of the epidermis. Accounts for the yellowish coloration of skin. Also termed **beta carotene**.

Carotid (ka-ROT-id) **body** Receptor on or near the carotid sinus that responds to alterations in blood levels of oxygen, carbon dioxide, and hydrogen ions.

Carotid sinus A dilated region of the internal carotid artery immediately superior to the branching of the common carotid artery that contains receptors that monitor blood pressure.

Carotid sinus reflex A reflex concerned with maintaining normal blood pressure in the brain.

Carpus (KAR-pus) A collective term for the eight bones of the wrist.

Cartilage (KAR-ti-lij) A type of connective tissue consisting of chondrocytes in lacunae embedded in a dense network of collagen and elastic fibers and a matrix of chondroitin sulfate.

Cartilaginous (kar′-ti-LAJ-i-nus) **joint** A joint without a synovial (joint) cavity where the articulating bones are held tightly together by cartilage, allowing little or no movement.

Cast A small mass of hardened material formed within a cavity in the body and then discharged from the body; can originate in different areas and can be composed of various materials.

Castration (kas-TRĀ-shun) Removal of the gonads (testes or ovaries).

Catabolism (ka-TAB-ō-lizm) Chemical reactions that break down complex organic compounds into simple ones with the net release of energy.

Catalyst (KAT-a-list) A substance that speeds up a chemical reaction without itself being altered; enzyme.

Cataract (KAT-a-rakt) Loss of transparency of the lens of the eye or its capsule or both.

Catheter (KATH-i-ter) A tube that can be inserted into a body cavity through a canal or into a blood vessel; used to remove fluids, such as urine and blood, and to introduce diagnostic materials or medication.

Cation (KAT-ī-on) A positively charged ion. An example is a sodium ion (Na^+).

Cauda equina (KAW-da ē-KWĪ-na) A tail-like collection of roots of spinal nerves at the inferior end of the spinal canal.

Caudal (KAW-dal) Pertaining to any tail-like structure; inferior in position.

Cecum (SĒ-kum) A blind pouch at the proximal end of the large intestine to which the ileum is attached.

Celiac plexus (PLEK-sus) A large mass of ganglia and nerve fibers located at the level of the superior part of the first lumbar vertebra. Also called the **solar plexus**.

Cell The basic structural and functional unit of all organisms; the smallest structure capable of performing all the activities vital to life.

Cell cycle Growth and division of a single cell into daughter cells, consisting of interphase and cell division.

Cell division Process by which a cell reproduces itself that consists of a nuclear division (mitosis) and a cytoplasmic division (cytokinesis); types include somatic and reproductive cell division.

Cell inclusion Principally organic substance produced by a cell that is not enclosed by a membrane and may appear or disappear at various times in the life of a cell, such as glycogen.

Cell-mediated immunity That component of immunity in which specially sensitized lymphocytes (T cells) attach to antigens to destroy them. Also called **cellular immunity**.

Cellular respiration *See* **Oxidation**.

Cementum (se-MEN-tum) Calcified tissue covering the root of a tooth.

Center of ossification (os′-i-fi-KĀ-shun) An area in the cartilage model of a future bone where the cartilage cells hypertrophy and then secrete enzymes that result in the calcification of their matrix, resulting in the death of the cartilage cells, followed by the invasion of the area by osteoblasts that then lay down bone.

Central canal A microscopic tube running the length of the spinal cord in the gray commissure. A circular channel running longitudinally in the center of an osteon (Haversian system) of mature compact bone, containing blood and lymphatic vessels and nerves. Also called a **Haversian** (ha-VER-shun) **canal**.

Central nervous system (CNS) That portion of the nervous system that consists of the brain and spinal cord.

Centrioles (SEN-trē-ōlz) Paired, cylindrical structures within a centrosome, each consisting of a ring of microtubules and arranged at right angles to each other.

Centromere (SEN-trō-mēr) The clear, constricted portion of a chromosome where the two chromatids are joined; serves as the point of attachment for the chromosomal microtubules.

Centrosome (SEN-trō-sōm) A rather dense area of cytoplasm, near the nucleus of a cell, containing a pair of centrioles. During prophase, it forms the mitotic spindle.

Cephalic (se-FAL-ik) Pertaining to the head; superior in position.

Cerebellar peduncle (ser-e-BEL-ar pe-DUNG-kul) A bundle of nerve fibers connecting the cerebellum with the brain stem.

Cerebellum (ser-e-BEL-um) The portion of the brain lying posterior to the medulla oblongata and pons, concerned with coordination of movements.

Cerebral aqueduct (SER-ē-bral AK-we-dukt) A channel through the midbrain connecting the third and fourth ventricles and containing cerebrospinal fluid. Also termed the **aqueduct of Sylvius**.

Cerebral arterial circle A ring of arteries forming an anastomosis at the base of the brain between the internal carotid and basilar arteries and arteries supplying the brain. Also called the **circle of Willis**.

Cerebral cortex The surface of the cerebral hemispheres, 2–4 mm thick, consisting of six layers of nerve cell bodies (gray matter) in most areas.

Cerebral palsy (PAL-zē) A group of motor disorders resulting in muscular uncoordination and loss of muscle control and caused by damage to motor areas of the brain (cerebral cortex, basal ganglia, and cerebellum) during fetal life, birth, or infancy.

Cerebral peduncle One of a pair of nerve fiber bundles located on the anterior surface of the midbrain, conducting nerve impulses between the pons and the cerebral hemispheres.

Cerebrospinal (se-rē′-brō-SPĪ-nal) **fluid (CSF)** A fluid produced by ependymal cells that cover choroid plexuses in the ventricles of the brain; the fluid circulates in the ventricles and the subarachnoid space around the brain and spinal cord.

Cerebrovascular (se-rē′-brō-VAS-kyoo-lar) **accident (CVA)** Destruction of brain tissue (infarction) resulting from disorders of blood vessels that supply the brain. Also called a **stroke**.

Cerebrum (SER-ē-brum *or* ser-Ē-brum) The two hemispheres of the forebrain, making up the largest part of the brain.

Ceruminous (se-ROO-mi-nus) **gland** A modified sudoriferous (sweat) gland in the external auditory meatus that secretes cerumen (ear wax).

Cerumen (se-ROO-men) Waxlike secretion produced by ceruminous glands in the external auditory meatus (ear canal).

Cervical dysplasia (dis-PLĀ-sē-a) A change in the shape, growth, and number of cervical cells of the uterus that, if severe, may progress to cancer.

Cervical ganglion (SER-vi-kul GANG-glē-on) A cluster of nerve cell bodies of postganglionic sympathetic neurons located in the neck, near the vertebral column.

Cervical mucus A mixture of water, glycoproteins, proteins, lipids, enzymes, and inorganic salts produced by secreting cells of the mucosa of the cervix of the uterus.

Cervical plexus (PLEK-sus) A network of nerve fibers formed by the ventral rami of the first four cervical nerves.

Cervix (SER-viks) Neck; any constricted portion of an organ, such as the inferior cylindrical part of the uterus.

Cesarean (se-SA-rē-an) **section** Procedure in which a low, horizontal incision is made through the abdominal wall and uterus for removal of the baby and placenta. Also called a **C-section**.

Chalazion (ka-LĀ-zē-on) A small tumor of the eyelid.

Chemical bond Force of attraction in a molecule or compound that holds its atoms together. Examples include ionic and covalent bonds.

Chemical element Unit of matter that cannot be decomposed into a simpler substance by ordinary chemical reactions. Examples include hydrogen (H), carbon (C), and oxygen (O).

Chemically gated ion channel An ion channel that opens and closes in response to a specific chemical stimulus, such as a neurotransmitter, hormone, or certain ions, for example, H^+ or Ca^{2+}.

Chemical reaction The combination or breaking apart of atoms in which chemical bonds are formed or broken and new products with different properties are produced.

Chemiosmosis (kem′-ē-oz-MŌ-sis) Mechanism for ATP generation that links chemical reactions (electrons passing along the electron transport chain) with pumping of H^+ out of the mitochondrial matrix. ATP synthesis occurs as H^+ diffuse back into the mitochondrial matrix through special H^+ channels in the membrane.

Chemoreceptor (kē′-mō-rē-SEP-tor) Receptor that detects the presence of chemicals.

Chemotaxis (kē-mō-TAK-sis) Attraction of phagocytes to microbes by a chemical stimulus.

Chemotherapy (kē-mō-THER-a-pē) The treatment of illness or disease by chemicals.

Chiasm (KĪ-azm) A crossing; especially the crossing of the optic (II) nerve fibers.

Chief cell The secreting cell of a gastric gland that produces pepsinogen, the precursor of the enzyme pepsin, and the enzyme gastric lipase. Also called a **zymogenic** (zī′-mō-JEN-ik) **cell**.

Chiropractic (kī′-rō-PRAK-tik) A system of treating disease by using one's hands to manipulate body parts, mostly the vertebral column.

Chlamydia (kla-MID-ē-a)   Prevalent sexually transmitted disease; characterized by burning on urination, frequent and painful urination, and low back pain; may spread to uterine (Fallopian) tubes in females.

Chloride shift   Exchange of bicarbonate ions (HCO_3^-) for chloride ions (Cl^-) between red blood cells and plasma; maintains electrical balance inside red blood cells as bicarbonate ions are produced or eliminated during respiration.

Cholecystogram (kō-lē-SIS-tō-gram)   X-ray of the gallbladder to evaluate for the presence of gallstones, inflammations, and tumors.

Cholecystectomy (kō′-lē-sis-TEK-tō-mē)   Surgical removal of the gallbladder.

Cholesterol (kō-LES-terol)   Classified as a lipid, the most abundant steroid in animal tissues; located in cell membranes and used for the synthesis of steroid hormones and bile salts.

Cholinergic (kō′-lin-ER-jik) **fiber**   A nerve ending that liberates acetylcholine at a synapse.

Chondrocyte (KON-drō-sīt)   Cell of mature cartilage.

Chondroitin (kon-DROY-tin) **sulfate**   An amorphous matrix material found outside connective tissue cells.

Chordae tendineae (KOR-dē TEN-di-nē-ē)   Tendonlike, fibrous cords that connect the heart valves with the papillary muscles.

Chorion (KŌ-rē-on)   The most superficial fetal membrane that becomes the principal embryonic portion of the placenta; serves a protective and nutritive function.

Chorionic villi (kō′-rē-ON-ik VIL-lī)   Fingerlike projections of the chorion that grow into the decidua basalis of the endometrium and contain fetal blood vessels.

Chorionic villi sampling (CVS)   The removal of a sample of chorionic villus tissue by means of a catheter to analyze the tissue for prenatal genetic defects.

Choroid (KŌ-royd)   One of the vascular coats of the eyeball.

Choroid plexus (PLEK-sus)   A network of capillaries located in the roof of each of the four ventricles of the brain; ependymal cells around choroid plexuses produce cerebrospinal fluid.

Chromaffin (krō-MAF-in) **cell**   Cell that has an affinity for chrome salts, due in part to the presence of the precursors of the neurotransmitter epinephrine; found, among other places, in the adrenal medulla.

Chromatid (KRŌ-ma-tid)   One of a pair of identical connected nucleoprotein strands that are joined at the centromere and separate during cell division, each becoming a chromosome of one of the two daughter cells.

Chromatin (KRŌ-ma-tin)   The threadlike mass of the genetic material consisting principally of DNA, which is present in the nucleus of a nondividing or interphase cell.

Chromatolysis (krō′-ma-TOL-i-sis)   The breakdown of chromatophilic substance (Nissl bodies) into finely granular masses in the cell body of a central or peripheral neuron whose process (axon or dendrite) has been damaged.

Chromatophilic substance   Rough endoplasmic reticulum in the cell bodies of neurons that functions in protein synthesis. Also called **Nissl bodies**.

Chromosomal microtubule (mī-krō-TOOB-yool)   Formed during prophase of mitosis; originates from centromeres, extends from a centromere to a pole of the cell, and assists in chromosomal movement; part of the mitotic spindle.

Chromosome (KRŌ-mō-sōm)   One of the 46 small, dark-staining bodies that appear in the nucleus of a human diploid ($2n$) cell during cell division.

Chronic (KRON-ik)   Long term or frequently recurring; applied to a disease that is not acute.

Chronic obstructive pulmonary disease (COPD)   Any disease, such as asthma, bronchitis, or emphysema, in which there is some degree of obstruction of air passageways.

Chyle (KĪL)   The milky fluid found in the lacteals of the small intestine after digestion.

Chylomicron (kī-lō-MĪK-ron)   Protein-coated spherical structure that contains triglycerides, phospholipids, and cholesterol and is absorbed into the lacteal of a villus in the small intestine.

Chyme (KĪM)   The semifluid mixture of partly digested food and digestive secretions found in the stomach and small intestine during digestion of a meal.

Ciliary (SIL-ē-ar′-ē) **body**   One of the three portions of the vascular tunic of the eyeball, the others being the choroid and the iris; includes the ciliary muscle and the ciliary processes.

Ciliary ganglion (GANG-glē-on)   A very small parasympathetic ganglion whose preganglionic fibers come from the oculomotor (III) nerve and whose postganglionic fibers carry nerve impulses to the ciliary muscle and the sphincter muscle of the iris.

Cilium (SIL-ē-um)   A hair or hairlike process projecting from a cell that may be used to move the entire cell or to move substances along the surface of the cell. *Plural is* **cilia**.

Circadian (ser-KĀ-dē-an) **rhythm**   A cycle of active and nonactive periods in organisms determined by internal mechanisms and repeating about every 24 hours.

Circle of Willis   *See* **Cerebral arterial circle**.

Circular folds   Permanent, deep, transverse folds in the mucosa and submucosa of the small intestine that increase the surface area for absorption. Also called **plicae circulares** (PLĪ-kē SER-kyoo-lar-ēs).

Circulation time   Time required for blood to pass from the right atrium, through pulmonary circulation, back to the left ventricle, through systemic circulation to the foot, and back again to the right atrium; normally about 1 min.

Circumcision (ser′-kum-SIZH-un)   Surgical removal of the foreskin (prepuce), the fold of skin over the glans penis.

Circumduction (ser′-kum-DUK-shun)   A movement at a synovial joint in which the distal end of a bone moves in a circle while the proximal end remains relatively stable.

Circumvallate papilla (ser′-kum-VAL-āt pa-PIL-a)   One of the circular projections that is arranged in an inverted V-shaped row at the back of the tongue; the largest of the elevations on the upper surface of the tongue containing taste buds.

Cirrhosis (si-RŌ-sis)   A liver disorder in which the parenchymal cells are destroyed and replaced by connective tissue.

Cisterna chyli (sis-TER-na KĪ-lē)   The origin of the thoracic duct.

Cleavage   The rapid mitotic divisions following the fertilization of a secondary oocyte, resulting in an increased number of progressively smaller cells, called blastomeres.

Cleft palate   Condition in which the palatine processes of the maxillae do not unite before birth; cleft lip, a split in the upper lip, is often associated with cleft palate.

Climacteric (klī-mak-TER-ik)   Cessation of the reproductive function in the female or diminution of testicular activity in the male.

Climax   The peak period or moments of greatest intensity during sexual excitement.

Clitoris (KLI-to-ris)   An erectile organ of the female located at the anterior junction of the labia minora that is homologous to the male penis.

Clone (KLŌN)   A population of identical cells.

Clot   The end result of a series of biochemical reactions that changes liquid plasma into a gelatinous mass; specifically, the conversion of fibrinogen into a tangle of polymerized fibrin molecules.

Clot retraction (rē-TRAK-shun)   The consolidation of a fibrin clot to pull damaged tissue together.

Coagulation (cō-ag-yoo-LĀ-shun)   Process by which a blood clot is formed.

Coarctation (kō′-ark-TĀ-shun) **of the aorta**   Congenital condition in which the aorta is too narrow and results in reduced blood supply, increased ventricular pumping, and high blood pressure.

Coccyx (KOK-six)   The fused bones at the inferior end of the vertebral column.

Cochlea (KŌK-lē-a)   A winding, cone-shaped tube forming a portion of the inner ear and containing the spiral organ (organ of Corti).

Cochlear duct   The membranous cochlea consisting of a spirally arranged tube enclosed in the bony cochlea and lying along its outer wall. Also called the **scala media** (SCA-la MĒ-dē-a).

Coenzyme   A type of cofactor; a nonprotein organic molecule that is associated with and activates an enzyme; many are derived from vitamins. An example is nicotinamide adenine dinucleotide (NAD), derived from the B vitamin niacin.

Coitus (KŌ-i-tus)   Sexual intercourse. Also called **copulation** (cop-yoo-LĀ-shun).

Colitis (ko-LĪ-tis)   Inflammation of the mucosa of the colon and rectum in which absorption of water and salts is reduced, producing watery, bloody feces, and, in severe cases, dehydration and salt depletion. Spasms of the irritated muscularis produce cramps.

Collagen (KOL-a-jen)   A protein that is the main organic constituent of connective tissue.

Collateral circulation   The alternate route taken by blood through an anastomosis.

Colliculus (ko-LIK-yoo-lus)   A small elevation.

Colon   The division of the large intestine consisting of ascending, transverse, descending, and sigmoid portions.

Colony-stimulating factor (CSF)   One of a group of hematopoietins that stimulates development of white blood cells. Examples are macrophage CSF and granulocyte CSF.

Color blindness   Any deviation in the normal perception of colors, usually resulting from the lack of one or more of the photopigments of the cones.

Colostomy (kō-LOS-tō-mē)   The diversion of feces through an opening in the colon, creating a surgical opening at the exterior of the abdominal wall.

Colostrum (kō-LOS-trum)   A thin, cloudy fluid secreted by the mammary glands a few days prior to or after delivery before true milk is produced.

Colposcopy (kol-POS-kō-pē)   Direct examination of the vaginal and cervical mucosa using a magnifying device; frequently the first procedure performed following an abnormal Pap smear.

Column (KOL-um)   Group of white matter tracts in the spinal cord.

Coma (KŌ-ma)   Final stage of brain failure that is characterized by total unresponsiveness to all external stimuli.

Common bile duct   A tube formed by the union of the common hepatic duct and the cystic duct that empties bile into the duodenum at the hepatopancreatic ampulla (ampulla of Vater).

Compact (dense) bone tissue   Bone tissue that contains few spaces between osteons (Haversian systems); forms the external portion of all bones and the bulk of the diaphysis (shaft) of long bones; is found immediately deep to the periosteum and external to spongy bone.

Complement (KOM-ple-ment)   A group of at least 20 normally inactive proteins found in serum that forms a component of nonspecific resistance and immunity by bringing about cytolysis, inflammation, and opsonization.

Complete blood count (CBC)   Hematology test that usually includes hemoglobin determination, hematocrit, red and white blood cell counts, differential white blood cell count, and platelet count.

Compliance   The ease with which the lungs and thoracic wall or blood vessels can be expanded.

Compound   A substance that can be broken down into two or more other substances by chemical means.

Computed tomography (tō-MOG-ra-fē) **(CT)**   X-ray technique that provides a cross-sectional image of any area of the body. Also called **computed axial tomography (CAT)**.

Concha (KONG-ka)   A scroll-like bone found in the skull. *Plural is* **conchae** (KONG-kē).

Concussion (kon-KUSH-un)   Traumatic injury to the brain that produces no visible bruising but may result in abrupt, temporary loss of consciousness.

Conduction myofiber   Muscle fiber (cell) in the ventricular tissue of the heart specialized for conducting an action potential to the myocardium; part of the conduction system of the heart. Also called a **Purkinje** (pur-KIN-jē) **fiber.**

Conduction system   A series of autorhythmic cardiac muscle fibers that generates and distributes electrical impulses to stimulate coordinated contraction of the heart chambers; includes the sinoatrial (SA) node, the atrioventricular (AV) node, the atrioventricular (AV) bundle, the right and left bundle branches, and the conduction myofibers (Purkinje fibers).

Conductivity (kon′-duk-TIV-i-tē)   The ability of a cell to propagate (conduct) action potentials along its plasma membrane; highly developed in nerve and muscle fibers (cells).

Condyloid (KON-di-loid) **joint**   A synovial joint structured so that an oval-shaped condyle of one bone fits into an elliptical cavity of another bone, permitting side-to-side and back-and-forth movements, such as the joint at the wrist between the radius and carpals. Also called an **ellipsoidal** (e-lip-soy-dal) **joint**.

Cone   The type of photoreceptor in the retina that is specialized for highly acute, color vision in bright light.

Congenital (kon-JEN-i-tal)   Present at the time of birth.

Congestive heart failure (CHF)   Chronic or acute state that results when the heart is not capable of supplying the oxygen demands of the body.

Conjunctiva (kon′-junk-TĪ-va)   The delicate membrane covering the eyeball and lining the eyes.

Conjunctivitis (kon-junk′-ti-VĪ-tis)   Inflammation of the conjunctiva, the delicate membrane covering the eyeball and lining the eyelids.

Connective tissue   The most abundant of the four basic tissue types in the body, performing the functions of binding and supporting; consists of relatively few cells in a generous matrix (the ground substance and fibers between the cells).

Consciousness (KON-shus-nes)   A state of wakefulness in which an individual is fully alert, aware, and oriented, partly as a result of feedback between the cerebral cortex and reticular activating system.

Constipation (con-sti-PĀ-shun)   Infrequent or difficult defecation caused by decreased motility of the intestines.

Contact inhibition Phenomenon by which migration of a growing cell is stopped when it makes contact with another cell of its own kind.

Continuous conduction (kon-DUK-shun) Propagation of an action potential (nerve impulse) in a step-by-step depolarization of each adjacent area of an axon membrane.

Contraception (kon′-tra-SEP-shun) The prevention of fertilization or impregnation without destroying fertility.

Contractility (kon′-trak-TIL-i-tē) The ability of cells or parts of cells to actively generate force to undergo shortening and change form for purposeful movements. Muscle fibers (cells) exhibit a high degree of contractility.

Contralateral (kon′-tra-LAT-er-al) On the opposite side; affecting the opposite side of the body.

Control center The component of a feedback system, such as the brain, that determines the point at which a controlled condition, such as body temperature, is maintained.

Contusion (kon-TOO-shun) Condition in which tissue below the skin is damaged, but the skin is not broken.

Conus medullaris (KŌ-nus med-yoo-LAR-is) The tapered portion of the spinal cord inferior to the lumbar enlargement.

Convergence (con-VER-jens) An anatomical arrangement in which the synaptic end bulbs of several presynaptic neurons terminate on one postsynaptic neuron. The medial movement of the two eyeballs so that both are directed toward a near object being viewed in order to produce a single image.

Convulsion (con-VUL-shun) Violent, involuntary, tetanic contractions of an entire group of muscles.

Cornea (KOR-nē-a) The nonvascular, transparent fibrous coat through which the iris can be seen.

Coronal (kō-RŌ-nal) **plane** A plane that runs vertical to the ground and divides the body into anterior and posterior portions. Also called **frontal plane**.

Corona radiata The innermost layer of granulosa cells that is firmly attached to the zona pellucida.

Coronary angiography (KOR-ō-na-rē an′-jē-OG-ra-fē) Procedure in which the severity and location of blocked coronary arteries are visualized by injection of contrast dyes or in which clot-dissolving drugs may be injected into coronary arteries.

Coronary artery bypass grafting (CABG) Surgical procedure in which a portion of a blood vessel is removed from another part of the body and grafted onto a coronary artery distal to the obstruction in the coronary artery.

Coronary artery disease (CAD) A condition such as atherosclerosis that causes narrowing of coronary arteries so that blood flow to the heart is reduced. This results in **coronary heart disease (CHD)** in which the heart muscle receives inadequate blood flow due to an interruption of its blood supply.

Coronary artery spasm A condition in which the smooth muscle of a coronary artery undergoes a sudden contraction, resulting in vasoconstriction.

Coronary circulation The pathway followed by the blood from the ascending aorta through the blood vessels supplying the heart and returning to the right atrium. Also called **cardiac circulation**.

Coronary sinus (SĪ-nus) A wide venous channel on the posterior surface of the heart that collects the blood from the coronary circulation and returns it to the right atrium.

Corpora quadrigemina (KOR-por-a kwad-ri-JEM-in-a) Four small elevations (superior and inferior colliculi) in the posterior portion of the midbrain concerned with visual and auditory reflexes.

Cor pulmonale (kor pul-mōn-AL-ē) **(CP)** Right ventricular hypertrophy from disorders that bring about hypertension in pulmonary circulation.

Corpus (KOR-pus) The principal part of any organ; any mass or body.

Corpus albicans (KOR-pus AL-bi-kanz) A white fibrous patch in the ovary that forms after the corpus luteum regresses.

Corpus callosum (kal-LŌ-sum) The great commissure of the brain between the cerebral hemispheres.

Corpuscle of touch The sensory receptor for the sensation of touch; found in the dermal papillae, especially in palms and soles. Also called a **Meissner's** (MĪS-nerz) **corpuscle**.

Corpus luteum (LOO-tē-um) A yellow endocrine gland in the ovary formed when a follicle has discharged its secondary oocyte; secretes estrogens, progesterone, relaxin, and inhibin.

Corpus striatum (strī-Ā-tum) An area in the interior of each cerebral hemisphere composed of the caudate and lentiform nuclei of the basal ganglia and white matter of the internal capsule, arranged in a striated manner.

Cortex (KOR-teks) An outer layer of an organ. The convoluted layer of gray matter covering each cerebral hemisphere.

Costal (KOS-tal) Pertaining to a rib.

Costal cartilage (KAR-ti-lij) Hyaline cartilage that attaches a rib to the sternum.

Countercurrent mechanism One mechanism involved in the ability of the kidneys to produce a hypertonic urine.

Cowper's gland *See* **Bulbourethral gland**.

Cramp A spasmodic, especially a tonic, contraction of one of many muscles, usually painful.

Cranial (KRĀ-nē-al) **cavity** A subdivision of the dorsal body cavity formed by the cranial bones and containing the brain.

Cranial nerve One of 12 pairs of nerves that leave the brain, pass through foramina in the skull, and supply the head, neck, and part of the trunk; each is designated by a Roman numeral and a name.

Craniosacral (krā-nē-ō-SĀ-kral) **outflow** The fibers of parasympathetic preganglionic neurons, which have their cell bodies located in nuclei in the brain stem and in the lateral gray matter of the sacral portion of the spinal cord.

Craniotomy (krā′-nē-OT-ō-mē) Any operation on the skull, as for surgery on the brain or decompression of the fetal head in difficult labor.

Cranium (KRĀ-nē-um) The skeleton of the skull that protects the brain and the organs of sight, hearing, and balance; includes the frontal, parietal, temporal, occipital, sphenoid, and ethmoid bones.

Creatine phosphate (KRĒ-a-tin FOS-fāt) Molecule in skeletal muscle fibers that contains high-energy phosphate bonds; used to generate ATP rapidly from ADP by transfer of phosphate group. Also called **phosphocreatine** (fos′-fō-KRĒ-a-tin).

Crenation (krē-NĀ-shun) The shrinkage of red blood cells into knobbed, starry forms when they are placed in a hypertonic solution.

Cretinism (KRĒ-tin-izm) Severe congenital thyroid deficiency during childhood leading to physical and mental retardation.

Crista (KRIS-ta) A crest or ridged structure. A small elevation in the ampulla of each semicircular duct that contains receptors for dynamic equilibrium.

Crossed extensor reflex A reflex in which extension of the joints in one limb occurs in conjunction with contraction of the flexor muscles of the opposite limb.

Crossing-over The exchange of a portion of one chromatid with another in a tetrad during meiosis. It permits an exchange of genes among chromatids and is one factor that results in genetic variation of progeny.

Crus (KRUS) **of penis** Separated, tapered portion of the corpora cavernosa penis. *Plural is* **crura** (KROO-ra).

Cryosurgery (KRĪ-ō-ser-jer-ē) The destruction of tissue by application of extreme cold.

Crypt of Lieberkühn *See* **Intestinal gland**.

Cryptorchidism (krip-TOR-ki-dizm) The condition of undescended testes.

Cupula (KUP-yoo-la) A mass of gelatinous material covering the hair cells of a crista; a receptor in the ampulla of a semicircular canal stimulated when the head moves.

Cushing's syndrome Condition caused by a hypersecretion of glucocorticoids characterized by spindly legs, "moon face," "buffalo hump," pendulous abdomen, flushed facial skin, poor wound healing, hyperglycemia, osteoporosis, weakness, hypertension, and increased susceptibility to disease.

Cutaneous (kyoo-TĀ-nē-us) Pertaining to the skin.

Cyanosis (si-a-NŌ-sis) Reduced hemoglobin (deoxygenated) concentration of blood of more than 5 g/dl that results in a blue or dark purple discoloration that is most easily seen in nail beds and mucous membranes.

Cyclic AMP (cyclic adenosine-3′,5′-monophosphate) Molecule formed from ATP by the action of the enzyme adenylate cyclase; serves as an intracellular messenger (second messenger) for some hormones.

Cyst (SIST) A sac with a distinct connective tissue wall, containing a fluid or other material.

Cystic (SIS-tik) **duct** The duct that transports bile from the gallbladder to the common bile duct.

Cystic (SIS-tik) **fibrosis** (fī-BRŌ-sis) Inherited disease of secretory epithelia that affects the respiratory passageways, pancreas, salivary glands, and sweat glands; the most common lethal genetic disease among whites.

Cystitis (sis-TĪ-tis) Inflammation of the urinary bladder.

Cytolysis (si-TOL-i-sis) The rupture of living cells in which the contents leak out.

Cystoscope (SIS-to-skōp) An instrument used to examine the inside of the urinary bladder.

Cystoscopy (sis-TOS-kō-pē) Direct visual examination of the urinary tract (and prostate gland in males as well) using a cystoscope to evaluate urinary tract disorders and remove tissue for biopsy, kidney stones, urinary bladder tumors, and urine samples.

Cytochrome (SĪ-tō-krōm) A protein with an iron-containing group (heme) capable of alternating between a reduced form (Fe^{2+}) and an oxidized form (Fe^{3+}).

Cytokines (SI-to-kīns) Small protein hormones produced by lymphocytes, fibroblasts, endothelial cells, and antigen-presenting cells that act as autocrine or paracrine substances to stimulate or inhibit cell growth and differentiation, regulate immune responses, or aid nonspecific defenses.

Cytokinesis (si′-tō-ki-NĒ-sis) Distribution of the cytoplasm into two separate cells during cell division; coordinated with nuclear division (mitosis).

Cytology (si-TOL-ō-jē) The study of cells.

Cytoplasm (SĪ-tō-plazm) Cytosol, all organelles (except the nucleus), and inclusions.

Cytoskeleton Complex internal structure of cytoplasm consisting of microfilaments, microtubules, and intermediate filaments.

Cytosol (SĪ-tō-sol) Semifluid portion of cytoplasm in which organelles and inclusions are suspended and solutes are dissolved. Also called **intracellular fluid**.

Dartos (DAR-tōs) The contractile tissue deep to the skin of the scrotum.

Deafness Lack of the sense of hearing or a significant hearing loss.

Debility (dē-BIL-i-tē) Weakness of tonicity in functions or organs of the body.

Decibel (DES-i-bel) **(dB)** A unit that measures relative sound intensity (loudness).

Decidua (dē-SID-yoo-a) That portion of the endometrium of the uterus (all but the deepest layer) that is modified during pregnancy and shed after childbirth.

Deciduous (dē-SID-yoo-us) Falling off or being shed seasonally or at a particular stage of development. In the body, referring to the first set of teeth.

Decompression sickness A condition characterized by joint pains and neurologic symptoms; occurs after a too-rapid reduction of environmental pressure or decompression, so that nitrogen dissolved in body fluids comes out of solution as bubbles; the bubbles form air emboli and occlude blood vessels. Also called **caisson** (KĀ-son) **disease** or **bends**.

Decussation (dē′-ku-SĀ-shun) A crossing-over; usually refers to the crossing of 90% of the fibers in the large motor tracts to opposite sides in the medullary pyramids.

Deep Away from the surface of the body or an organ.

Deep fascia (FASH-ē-a) A sheet of connective tissue wrapped around a muscle to hold it in place.

Deep inguinal (IN-gwi-nal) **ring** A slitlike opening in the aponeurosis of the transversus abdominis muscle that represents the origin of the inguinal canal.

Deep-venous thrombosis (DVT) The presence of a thrombus (clot) in a vein, usually a deep vein of the lower limbs.

Defecation (def-e-KĀ-shun) The discharge of feces from the rectum.

Defibrillation (dē-fib-ri-LĀ-shun) Delivery of a very strong electrical current to the heart in an attempt to stop ventricular fibrillation.

Deglutition (dē-gloo-TISH-un) The act of swallowing.

Dehydration (dē-hi-DRĀ-shun) Excessive loss of water from the body or its parts.

Delirium (de-LIR-ē-um) A transient disorder of abnormal cognition (perception, thinking, and memory) and disordered attention that is accompanied by disturbances of the sleep–wake cycle and psychomotor behavior (hyperactivity or hypoactivity of movements and speech). Also called **acute confusional state (ACS)**.

Delta cell A cell in the pancreatic islets (islets of Langerhans) in the pancreas that secretes somatostatin. Also termed a **D cell**.

Dementia (de-MEN-shē-a) An organic mental disorder that results in permanent or progressive general loss of intellectual abilities such as impairment of memory, judgment, and abstract thinking and changes in personality; most common cause is Alzheimer's disease.

Demineralization (de-min′-er-al-i-ZĀ-shun) Loss of calcium and phosphorus from bones.

Denaturation (de-nā-chur-Ā-shun) Disruption of the tertiary structure of a protein by agents, such as heat, changes in pH, or other physical or chemical methods, in which the protein loses its physical properties and biological activity.

Dendrite (DEN-drīt)   A neuronal process that carries a nerve impulse toward the cell body.

Dendritic (den-DRIT-ik) **cell**   One type of antigen-presenting cell with long branchlike projections that commonly is present in epithelial linings, such as the vagina, and in the skin, for example, Langerhans cells in the epidermis.

Dental caries (KA-rēz)   Gradual demineralization of the enamel and dentin of a tooth that may invade the pulp and alveolar bone. Also called **tooth decay**.

Denticulate (den-TIK-yoo-lāt)   Finely toothed or serrated; characterized by a series of small, pointed projections.

Dentin (DEN-tin)   The bony tissues of a tooth enclosing the pulp cavity.

Dentition (den-TI-shun)   The eruption of teeth. The number, shape, and arrangement of teeth.

Deoxyribonucleic (dē-ok′-sē-ri′-bō-nyoo-KLĒ-ik) **acid (DNA)**   A nucleic acid constructed of nucleotides consisting of one of four nitrogenous bases (adenine, cytosine, guanine, or thymine), deoxyribose, and a phosphate group; encoded in the nucleotides is genetic information.

Depolarization (dē-pō-lar-i-ZA-shun)   Used in neurophysiology to describe the reduction of voltage across a plasma membrane; expressed as a change toward less negative (more positive) voltages on the interior surface of the plasma membrane.

Depression (de-PRESS-shun)   Movement in which a part of the body moves inferiorly.

Dermal papilla (pa-PILL-a)   Fingerlike projection of the papillary region of the dermis that may contain blood capillaries or corpuscles of touch (Meissner's corpuscles).

Dermatology (der-ma-TOL-ō-jē)   The medical specialty dealing with diseases of the skin.

Dermatome (DER-ma-tōm)   The cutaneous area developed from one embryonic spinal cord segment and receiving most of its sensory innervation from one spinal nerve. An instrument for incising the skin or cutting thin transplants of skin.

Dermis (DER-mis)   A layer of dense irregular connective tissue lying deep to the epidermis.

Descending colon (KŌ-lon)   The part of the large intestine descending from the left colic (splenic) flexure to the level of the left iliac crest.

Detritus (de-TRI-tus)   Particulate matter produced by or remaining after the wearing away or disintegration of a substance or tissue; scales, crusts, or loosened skin.

Detrusor (de-TROO-ser) **muscle**   Muscle in the wall of the urinary bladder.

Developmental anatomy   The study of development from the fertilized egg to the adult form. The branch of anatomy called embryology is generally restricted to the study of development from the fertilized egg through the eighth week in utero.

Diabetes insipidus (dī-a-BĒ-tēz in-SIP-i-dus)   Condition caused by hyposecretion of antidiuretic hormone (ADH) or nonfunctional ADH receptors; characterized by excretion of large amounts of urine and extreme thirst.

Diabetes mellitus (MEL-i-tus)   Condition caused by hyposecretion of insulin; characterized by hyperglycemia, increased urine production, excessive thirst, and excessive eating.

Diagnosis (dī-ag-NŌ-sis)   Distinguishing one disease from another or determining the nature of a disease from signs and symptoms by inspection, palpation, laboratory tests, and other means.

Dialysis (dī-AL-i-sis)   The process of separating small from large molecules by the difference in their rates of diffusion through a selectively permeable membrane.

Diapedesis (di-a-pe-DĒ-sis)   Term formerly used for the passage of white blood cells through intact blood vessel walls. *See* **Emigration.**

Diaphragm (DĪ-a-fram)   Any partition that separates one area from another, especially the dome-shaped skeletal muscle between the thoracic and abdominal cavities. Also a dome-shaped device that is placed over the cervix, usually with a spermicide, to prevent conception.

Diaphysis (di-AF-i-sis)   The shaft of a long bone.

Diarrhea (di-a-RĒ-a)   Frequent defecation of liquid feces caused by increased motility of the intestines.

Diarthrosis (di-ar-THRO-sis)   A freely movable joint; types are gliding, hinge, pivot, condyloid, saddle, and ball-and-socket.

Diastole (di-AS-tō-lē)   In the cardiac cycle, the phase of relaxation or dilation of the heart muscle, especially of the ventricles.

Diastolic (di-as-TOL-ik) **blood pressure**   The force exerted by blood on arterial walls during ventricular relaxation; the lowest blood pressure measured in the large arteries, about 80 mm Hg under normal conditions for a young adult.

Diencephalon (di′-en-SEF-a-lon)   A part of the brain consisting primarily of the thalamus and the hypothalamus.

Differential (dif-fer-EN-shal) **white blood cell count**   Determination of the number of each kind of white blood cell in a sample of 100 cells for diagnostic purposes.

Differentiation (dif′-e-ren′-shē-A-shun)   Acquisition of specific functions different from those of the original general type.

Digestion (di-JES-chun)   The mechanical and chemical breakdown of food to simple molecules that can be absorbed and used by body cells.

Digital subtraction angiography (an-jē-OG-ra-fē) **(DSA)**   A medical imaging technique that compares an x-ray image of the same artery of the body before and after a contrast substance containing iodine has been introduced intravenously.

Dilate (DĪ-lāt)   To expand or swell.

Dilation (di-LA-shun) **and curettage** (ku-re-TAZH)   Following dilation of the uterine cervix, the uterine endometrium is scraped with a curette (spoon-shaped instrument). Also called a **D and C.**

Diploid (DIP-loyd)   Having the number of chromosomes characteristically found in the somatic cells of an organism. Symbolized 2*n*.

Diplopia (di-PLO-pē-a)   Double vision.

Direct (motor) pathways   Collections of upper motor neurons with cell bodies in the motor cortex that project axons into the spinal cord, where they synapse with lower motor neurons or association neurons in the anterior horns. Also called the **pyramidal pathways.**

Disease   Any change from a state of health.

Dislocation (dis-lō-KA-shun)   Displacement of a bone from a joint with tearing of ligaments, tendons, and articular capsules. Also called **luxation** (luks-A-shun).

Dissect (di-SEKT)   To separate tissues and parts of a cadaver (corpse) or an organ for anatomical study.

Dissociation (dis′-sō-sē-A-shun)   *See* **Ionization.**

Distal (DIS-tal)   Farther from the attachment of a limb to the trunk; farther from the point of origin or attachment.

Diuretic (di-yoo-RET-ik)   A chemical that inhibits sodium reabsorption, reduces antidiuretic hormone (ADH) concentration, and increases urine volume by inhibiting facultative reabsorption of water.

Diurnal (di-UR-nal)   Daily.

Divergence (di-VER-jens) An anatomical arrangement in which the synaptic end bulbs of one presynaptic neuron terminate on several postsynaptic neurons.

Diverticulitis (di-ver-tik-yoo-LĪ-tis) Inflammation of diverticula, saclike outpouchings of the colonic wall, when the muscularis becomes weak.

Diverticulum (di-ver-TIK-yoo-lum) A sac or pouch in the wall of a canal or organ, especially in the colon.

Dominant gene A gene that is able to override the influence of the complementary gene on the homologous chromosome; the gene that is expressed.

Donor insemination (in-sem′-i-NĀ-shun) The deposition of semen within the vagina or cervix at a time during the menstrual cycle when pregnancy is most likely to occur. Also called **artificial insemination.**

Dorsal body cavity Cavity near the dorsal (posterior) surface of the body that consists of a cranial cavity and vertebral canal.

Dorsal ramus (RĀ-mus) A branch of a spinal nerve containing motor and sensory fibers supplying the muscles, skin, and bones of the posterior part of the head, neck, and trunk.

Dorsiflexion (dor′-si-FLEK-shun) Bending the foot in the direction of the dorsum (upper surface).

Down-regulation Phenomenon in which there is a decrease in the number of receptors in response to an excess of a hormone or neurotransmitter.

Down syndrome (DS) An inherited defect due to an extra copy of chromosome 21. Symptoms include mental retardation; a small skull, flattened from front to back; a short, flat nose; short fingers; and a widened space between the first two digits of the hand and foot. Also called **trisomy 21.**

Duct of Santorini *See* **Accessory duct.**

Duct of Wirsung *See* **Pancreatic duct.**

Ductus arteriosus (DUK-tus ar-tē-rē-Ō-sus) A small vessel connecting the pulmonary trunk with the aorta; found only in the fetus.

Ductus (vas) deferens (DEF-er-ens) The duct that carries sperm from the epididymis to the ejaculatory duct. Also called the **seminal duct.**

Ductus epididymis (ep′-i-DID-i-mis) A tightly coiled tube inside the epididymis, distinguished into a head, body, and tail, in which sperm undergo maturation.

Ductus venosus (ve-NO-sus) A small vessel in the fetus that helps the circulation bypass the liver.

Duodenal (doo-ō-DĒ-nal) **gland** Gland in the submucosa of the duodenum that secretes an alkaline mucus to protect the lining of the small intestine from the action of enzymes and to help neutralize the acid in chyme. Also called **Brunner's** (BRUN-erz) **gland.**

Duodenum (doo′-ō-DĒ-num) The first 25 cm (10 in.) of the small intestine, which connects the stomach and the ileum.

Dura mater (DYOO-ra MĀ-ter) The outer membrane (meninx) covering the brain and spinal cord.

Dynamic equilibrium (ē-kwi-LIB-rē-um) The maintenance of body position, mainly the head, in response to sudden movements such as rotation.

Dynamic spatial reconstruction (DSR) A technique that has the ability to construct moving, three-dimensional, life-size images of all or part of an internal organ from any view desired.

Dysfunction (dis-FUNK-shun) Absence of completely normal function.

Dyslexia (dis-LEX-sē-a) Impairment of the ability to read.

Dysmenorrhea (dis′-men-ō-RĒ-a) Painful menstruation.

Dysphagia (dis-FĀ-jē-a) Difficulty in swallowing.

Dysplasia (dis-PLA-zē-a) Change in the size, shape, and organization of cells due to chronic irritation or inflammation; may revert to normal if stress is removed or progress to neoplasia.

Dyspnea (DISP-nē-a) Shortness of breath.

Dystocia (dis-TO-sē-a) Difficult labor due to factors such as pelvic deformities, malpositioned fetus, and premature rupture of fetal membranes.

Dystrophia (dis-TRO-fē-a) Progressive weakening of a muscle.

Dysuria (dis-YOO-rē-a) Painful urination.

Eardrum A thin, semitransparent partition of fibrous connective tissue between the external auditory meatus and the middle ear. Also called the **tympanic membrane.**

Echocardiogram (ek-ō-KAR-dē-ō-gram) A procedure in which high-frequency sound waves directed at the heart are bounced back and the echoes are picked up by a transducer and converted to an image.

Ectoderm The primary germ layer that gives rise to the nervous system and the epidermis of skin and its derivatives.

Ectopic (ek-TOP-ik) Out of the normal location, as in ectopic pregnancy.

Eczema (EK-ze-ma) A skin rash characterized by itching, swelling, blistering, oozing, and scaling of the skin.

Edema (e-DĒ-ma) An abnormal accumulation of interstitial fluid.

Effector (e-FEK-tor) An organ of the body, either a muscle or a gland, that responds to a motor neuron impulse.

Efferent arteriole (EF-er-ent ar-TE-rē-ōl) A vessel of the renal vascular system that transports blood from a glomerulus to a peritubular capillary.

Efferent (EF-er-ent) **ducts** A series of coiled tubes that transport sperm cells from the rete testis to the epididymis.

Effusion (e-FYOO-zhun) The escape of fluid from the lymphatic vessels or blood vessels into a cavity or into tissues.

Eicosanoids (ī-KŌ-sa-noyds) Local hormones derived from a 20-carbon fatty acid (arachidonic acid); two important types are prostaglandins and leukotrienes.

Ejaculation (e-jak-yoo-LĀ-shun) The reflex ejection or expulsion of semen from the penis.

Ejaculatory (e-JAK-yoo-la-tō-rē) **duct** A tube that transports sperm cells from the ductus (vas) deferens to the prostatic urethra.

Elasticity (e-las-TIS-i-tē) The ability of tissue to return to its original shape after contraction or extension.

Electrocardiogram (e-lek′-trō-KAR-dē-ō-gram) (**ECG** *or* **EKG**) A recording of the electrical changes that accompany the cardiac cycle that can be detected at the surface of the body; may be resting, stress, or ambulatory.

Electroencephalogram (e-lek′-trō-en-SEF-a-lō-gram) (**EEG**) A recording of the electrical impulses of the brain from the scalp surface; used to diagnose certain diseases (such as epilepsy), furnish information regarding sleep and wakefulness, and confirm brain death.

Electrolyte (ē-LEK-trō-līt) Any compound that separates into ions when dissolved in water and is able to conduct electricity.

Electromyography (e-lek′-trō-mi-OG-ra-fē) Evaluation of the electrical activity of resting and contracting muscle to ascertain causes of muscular weakness, paralysis, involuntary twitching, and abnormal levels of muscle enzymes; also used as part of biofeedback studies.

Electron transport chain   A sequence of electron carrier molecules on the inner mitochondrial membrane that undergo oxidation and reduction as they pump hydrogen ions (H⁺) through the membrane. ATP synthesis then occurs as H⁺ diffuse back into the mitochondrial matrix through special H⁺ channels. *See also* **Chemiosmosis.**

Elevation (el-e-VĀ-shun)   Movement in which a part of the body moves superiorly.

Ellipsoidal (e-lip-SOY-dal) **joint**   *See* **Condyloid joint.**

Embolism (EM-bō-lizm)   Obstruction or closure of a vessel by an embolus.

Embolus (EM-bō-lus)   A blood clot, bubble of air or fat from broken bones, mass of bacteria, or other debris or foreign material transported by the blood.

Embryo (EM-brē-ō)   The young of any organism in an early stage of development; in humans, the developing organism from fertilization to the end of the eighth week in utero.

Embryology (em′-brē-OL-ō-jē)   The study of development from the fertilized egg to the end of the eighth week in utero.

Embryo transfer   A procedure in which semen is used to artificially inseminate a fertile secondary oocyte donor and the morula or blastocyst is then transferred from the donor to the infertile woman, who then carries it to term.

Emesis (EM-e-sis)   Vomiting.

Emigration (em′-e-GRĀ-shun)   Process whereby white blood cells (WBCs) leave the bloodstream by slowing down, rolling along the endothelium, and squeezing between the endothelial cells. Adhesion molecules help WBCs stick to the endothelium. Also known as **migration** or **extravasation.**

Emission (ē-MISH-un)   Propulsion of sperm into the urethra in response to peristaltic contractions of the ducts of the testes, epididymides, and ductus (vas) deferens as a result of sympathetic stimulation.

Emmetropia (em′-e-TRŌ-pē-a)   The ideal optical condition of the eyes.

Emphysema (em′-fi-SĒ-ma)   A swelling or inflation of air passages due to loss of elasticity in the alveoli.

Emulsification (ē-mul′-si-fi-KĀ-shun)   The dispersion of large lipid globules to smaller, uniformly distributed particles in the presence of bile.

Enamel (e-NAM-el)   The hard, white substance covering the crown of a tooth.

End-diastolic (dī-as-TO-lik) **volume (EDV)**   The volume of blood, about 130 ml, remaining in a ventricle at the end of its diastole (relaxation).

Endergonic (end′-er-GON-ik) **reaction**   Type of chemical reaction in which the energy released as new bonds form is less than the energy needed to break apart old bonds; an energy-requiring reaction.

Endocardium (en-dō-KAR-dē-um)   The layer of the heart wall, composed of endothelium and smooth muscle, that lines the inside of the heart and covers the valves and tendons that hold the valves open.

Endochondral ossification (en′-dō-KON-dral os′-i-fi-KĀ-shun)   The replacement of cartilage by bone. Also called **intracartilaginous** (in′-tra-kar′-ti-LAJ-i-nus) **ossification.**

Endocrine (EN-dō-krin) **gland**   A gland that secretes hormones into the blood; a ductless gland.

Endocrinology (en′-dō-kri-NOL-ō-jē)   The science concerned with the structure and functions of endocrine glands and the diagnosis and treatment of disorders of the endocrine system.

Endocytosis (en′-dō-si-TŌ-sis)   The uptake into a cell of large molecules and particles in which a segment of plasma membrane surrounds the substance, encloses it, and brings it in; includes phagocytosis, pinocytosis, and receptor-mediated endocytosis.

Endoderm (EN-dō-derm)   The primary germ layer of the developing embryo that gives rise to the gastrointestinal tract, urinary bladder and urethra, and respiratory tract.

Endodontics (en′-dō-DON-tiks)   The branch of dentistry concerned with the prevention, diagnosis, and treatment of diseases that affect the pulp, root, periodontal ligament, and alveolar bone.

Endogenous (en-DOJ-e-nus)   Growing from or beginning within the organism.

Endolymph (EN-dō-lymf′)   The fluid within the membranous labyrinth of the inner ear.

Endometriosis (en′-dō-MĒ-trē-ō′-sis)   The growth of endometrial tissue outside the uterus.

Endometrium (en′-dō-MĒ-trē-um)   The mucous membrane lining the uterus.

Endomysium (en′-dō-MIZ-ē-um)   Invagination of the perimysium separating each individual muscle fiber (cell).

Endoneurium (en′-dō-NYOO-rē-um)   Connective tissue wrapping around individual nerve fibers (cells).

Endoplasmic reticulum (en′-do-PLAZ-mik re-TIK-yoo-lum) **(ER)**   A network of channels running through the cytoplasm of a cell that serves in intracellular transportation, support, storage, synthesis, and packaging of molecules. Portions of ER where ribosomes are attached to the outer surface are called **rough (granular) reticulum;** portions that have no ribosomes are called **smooth (agranular) reticulum.**

End organ of Ruffini   *See* **Type II cutaneous mechanoreceptor.**

Endorphin (en-DOR-fin)   A neuropeptide in the central nervous system that acts as a painkiller.

Endoscope (EN-dō-skōp′)   An illuminated tube with lenses used to look inside hollow organs such as the stomach (gastroscope) or urinary bladder (cystoscope).

Endoscopy (en-DOS-kō-pē)   The visual examination of any cavity of the body using an endoscope, an illuminated tube with lenses.

Endosteum (en-DOS-tē-um)   The membrane that lines the medullary (marrow) cavity of bones, consisting of osteoprogenitor cells and scattered osteoclasts.

Endothelial–capsular membrane   A filtration membrane in a nephron of a kidney consisting of the endothelium and basement membrane of the glomerulus and the epithelium of the visceral layer of the glomerular (Bowman's) capsule.

Endothelium (en′-dō-THĒ-lē-um)   The layer of simple squamous epithelium that lines the cavities of the heart, blood vessels, and lymphatic vessels.

End-systolic (sis-TO-lik) **volume (ESV)**   The volume of blood, about 60 ml, remaining in a ventricle after its systole (contraction).

Energy   The capacity to do work.

Enkephalin (en-KEF-a-lin)   A peptide found in the central nervous system that acts as a painkiller.

Enteroendocrine (en-ter-ō-EN-dō-krin) **cell**   A cell of the mucosa of the gastrointestinal tract that secretes the hormones gastrin, cholecystokinin, gastric inhibitory peptide, or secretin.

Enterogastric (en-te-rō-GAS-trik) **reflex**   A reflex that inhibits gastric secretion; initiated by food in the small intestine.

Enuresis (en′-yoo-RĒ-sis)   Involuntary discharge of urine, complete or partial, after age 3.

Enzyme (EN-zim)   A substance that affects the speed of chemical changes; an organic catalyst, usually a protein.

Eosinophil (ē′-ō-SIN-ō-fil)　A type of white blood cell characterized by granules that stain red or pink with acid dyes.

Ependymal (e-PEN-de-mal) **cells**　Neuroglial cells that cover choroid plexuses and produce cerebrospinal fluid (CSF); they also line the ventricles of the brain and probably assist in the circulation of CSF.

Epicardium (ep′-i-KAR-dē-um)　The thin outer layer of the heart wall, composed of serous tissue and mesothelium. Also called the **visceral pericardium.**

Epidemic (ep′-i-DEM-ik)　An infectious disease that spreads rapidly and extensively within a population.

Epidemiology (ep′-i-dē-mē-OL-ō-jē)　Medical science concerned with the occurrence and distribution of diseases and disorders in human populations.

Epidermis (ep-i-DERM-is)　The superficial, thinner layer of skin, composed of keratinized stratified squamous epithelium.

Epididymis (ep′-i-DID-i-mis)　A comma-shaped organ that lies along the posterior border of the testis and contains the ductus epididymis, in which sperm undergo maturation. *Plural is* **epididymides** (ep′-i-DID-i-mi-dēz).

Epidural (ep′-i-DOO-ral) **space**　A space between the spinal dura mater and the vertebral canal, containing areolar connective tissue and a plexus of veins.

Epiglottis (ep′-i-GLOT-is)　A large, leaf-shaped piece of cartilage lying on top of the larynx, with its "stem" attached to the thyroid cartilage and its "leaf" portion unattached and free to move up and down to cover the glottis (vocal folds and rima glottidis).

Epilepsy (EP-i-lep′-sē)　Neurological disorder characterized by short, periodic attacks of motor, sensory, or psychological malfunction.

Epimysium (ep′-i-MĪZ-ē-um)　Fibrous connective tissue around muscles.

Epinephrine (ep-ē-NEF-rin)　Hormone secreted by the adrenal medulla that produces actions similar to those that result from sympathetic stimulation. Also called **adrenaline** (a-DREN-a-lin).

Epineurium (ep′-i-NYOO-rē-um)　The superficial covering around the entire nerve.

Epiphyseal (ep′-i-FIZ-ē-al) **line**　The remnant of the epiphyseal plate in a long bone.

Epiphyseal (ep′-i-FIZ-ē-al) **plate**　The hyaline cartilage plate between the epiphysis and diaphysis that is responsible for the lengthwise growth of long bones.

Epiphysis (ē-PIF-i-sis)　The end of a long bone, usually larger in diameter than the shaft (diaphysis).

Epiphysis cerebri (se-RĒ-brē)　*See* **Pineal gland.**

Episiotomy (e-piz′-ē-OT-ō-mē)　A cut made with surgical scissors to avoid tearing of the perineum during birth of a baby.

Epistaxis (ep′-i-STAK-sis)　Loss of blood from the nose due to trauma, infection, allergy, neoplasm, and bleeding disorders. Also called **nosebleed.**

Epithelial (ep′-i-THĒ-lē-al) **tissue**　The tissue that forms glands or the superficial part of the skin and lines blood vessels, hollow organs, and passages that lead externally from the body.

Eponychium (ep′-ō-NIK-ē-um)　Narrow band of stratum corneum at the proximal border of a nail that extends from the margin of the nail wall. Also called the **cuticle.**

Erection (ē-REK-shun)　The enlarged and stiff state of the penis or clitoris resulting from the engorgement of the spongy erectile tissue with blood.

Eructation (e-ruk′-TA-shun)　The forceful expulsion of gas from the stomach. Also called **belching.**

Erythema (er′-e-THĒ-ma)　Skin redness usually caused by engorgement of the capillaries in the deeper layers of the skin.

Erythematosus (er-i′-them-a-TŌ-sus)　Pertaining to redness.

Erythrocyte (e-RITH-rō-sit)　Red blood cell.

Erythrocyte sedimentation rate (ESR)　A test that measures the distances, in millimeters (mm), that red blood cells fall in 1 hour when a sample of blood is placed in a vertical tube; used as a screening test for infections, inflammations, and cancers.

Erythropoiesis (e-rith′-rō-poy-Ē-sis)　The process by which erythrocytes (red blood cells) are formed.

Erythropoietin (e-rith′-rō-POY-ē-tin)　A hormone released by the kidneys that stimulates erythrocyte (red blood cell) production.

Esophagus (e-SOF-a-gus)　A hollow muscular tube connecting the pharynx and the stomach.

Essential amino acids　Those ten amino acids that cannot be synthesized by the human body at an adequate rate to meet its needs and therefore must be obtained from the diet.

Estrogens (ES-tro-jens)　Female sex hormones produced by the ovaries concerned with the development and maintenance of female reproductive structures and secondary sex characteristics, fluid and electrolyte balance, and protein anabolism. Examples are β-estradiol, estrone, and estriol.

Etiology (ē′-tē-OL-ō-jē)　The study of the causes of disease, including theories of origin and the organisms, if any, involved.

Euphoria (yoo-FOR-ē-a)　A subjectively pleasant feeling of well-being marked by confidence and assurance.

Eupnea (YOOP-nē-a)　Normal quiet breathing.

Eustachian tube　*See* **Auditory tube.**

Euthanasia (yoo′-tha-NĀ-zē-a)　The practice of deliberately ending a life in case of incurable disease.

Eversion (ē-VER-zhun)　The movement of the sole laterally at the ankle joint or of an atrioventricular valve into an atrium during ventricular contraction.

Exacerbation (eg-zas′-er-BĀ-shun)　An increase in the severity of symptoms or of disease.

Excitability (ek-sit′-a-BIL-i-tē)　The ability of muscle tissue to receive and respond to stimuli; the ability of nerve cells to respond to stimuli and convert them into nerve impulses.

Excitatory postsynaptic potential (EPSP)　A small depolarization of the postsynaptic membrane when it is stimulated by an excitatory neurotransmitter. The EPSP is a localized event that decreases in strength from the point of excitation.

Excrement (EKS-kre-ment)　Material eliminated from the body as waste, especially fecal matter.

Excretion (eks-KRĒ-shun)　The process of eliminating waste products from the body or the products excreted.

Exergonic (ek′-er-GON-ik) **reaction**　Type of chemical reaction in which the energy released as new bonds form is greater than the energy needed to break apart old bonds; an energy-releasing reaction.

Exocrine (EK-sō-krin) **gland**　A gland that secretes substances into ducts that empty at covering or lining epithelium or directly onto a free surface.

Exocytosis (ex′-ō-si-TŌ-sis)　A process of discharging large cellular products through the membrane. Particles for export are enclosed by Golgi membranes when they are synthesized. Vesicles pinch off from the Golgi complex and carry the enclosed particles to the interior surface of the cell membrane, where the vesicle membrane and plasma membrane fuse and the contents of the vesicle are discharged.

Exogenous (ex-SOJ-e-nus)　Originating outside an organ or part.

Exon (EX-on)   A region of DNA that codes for synthesis of a protein.

Exophthalmic goiter (ek′-sof-THAL-mik GOY-ter)   An autoimmune disease that may result in hypersecretion of thyroid hormones characterized by protrusion of the eyeballs (exophthalmos) and an enlarged thyroid (goiter). Also called **Graves disease**.

Exophthalmos (ek′-sof-THAL-mus)   An abnormal protrusion or bulging of the eyeball.

Expiration (ek-spi-RA-shun)   Breathing out; expelling air from the lungs into the atmosphere. Also called **exhalation**.

Expiratory (eks-PI-ra-tō-rē) **reserve volume**   The volume of air in excess of tidal volume that can be exhaled forcibly; about 1200 ml.

Extensibility (ek-sten′-si-BIL-i-tē)   The ability of muscle tissue to be stretched when pulled.

Extension (ek-STEN-shun)   An increase in the angle between two bones; restoring a body part to its anatomical position after flexion.

External   Located on or near the surface.

External auditory (AW-di-tōr-ē) **canal** or **meatus** (mē-A-tus)   A curved tube in the temporal bone that leads to the middle ear.

External ear   The outer ear, consisting of the pinna, external auditory canal, and tympanic membrane (eardrum).

External nares (NA-rēz)   The external nostrils, or the openings into the nasal cavity on the exterior of the body.

External respiration   The exchange of respiratory gases between the lungs and blood. Also called **pulmonary respiration**.

Exteroceptor (eks′-ter-ō-SEP-tor)   A receptor adapted for the reception of stimuli from outside the body.

Extracellular fluid (ECF)   Fluid outside body cells, such as interstitial fluid and plasma.

Extrapyramidal pathways.   *See* **Indirect pathways**.

Extravasation (eks-trav-a-SA-shun)   The escape of fluid, especially blood, lymph, or serum, from a vessel into the tissues.

Extrinsic (ek-STRIN-sik)   Of external origin.

Extrinsic pathway (of blood clotting)   Sequence of reactions leading to blood clotting that is initiated by the release of tissue factor (TF), also known as thromboplastin, that leaks into the blood from damaged cells *outside* the blood vessels.

Exudate (EKS-yoo-dāt)   Escaping fluid or semifluid material that oozes from a space that may contain serum, pus, and cellular debris.

Eyebrow   The hairy ridge superior to the eye.

Face   The anterior aspect of the head.

Facilitated diffusion (fa-SIL-i-tā-ted dif-YOO-zhun)   Diffusion in which a substance not soluble by itself in lipids diffuses across a selectively permeable membrane with the help of a transporter (carrier) protein.

Falciform ligament (FAL-si-form LIG-a-ment)   A sheet of parietal peritoneum between the two principal lobes of the liver. The ligamentum teres, or remnant of the umbilical vein, lies within its fold.

Fallopian tube   *See* **Uterine tube**.

Falx cerebelli (FALKS ser′-e-BEL-lē)   A small triangular process of the dura mater attached to the occipital bone in the posterior cranial fossa and projecting inward between the two cerebellar hemispheres.

Falx cerebri (FALKS SER-e-brē)   A fold of the dura mater extending deep into the longitudinal fissure between the two cerebral hemispheres.

Fascia (FASH-ē-a)   A fibrous membrane covering, supporting, and separating muscles.

Fascicle (FAS-i-kul)   A small bundle or cluster, especially of nerve or muscle fibers (cells). Also called a **fasciculus** (fa-SIK-yoo-lus). *Plural is* **fasciculi** (fa-SIK-yoo-lī).

Fasciculation (fa-sik′-yoo-LA-shun)   Abnormal, spontaneous twitch of all skeletal muscle fibers in one motor unit that is visible at the skin surface; not associated with movement of the affected muscle; present in progressive diseases of motor neurons, for example, poliomyelitis and amyotrophic lateral sclerosis (ALS).

Fat   *See* **Triglyceride**.

Fauces (FAW-sēz)   The opening from the mouth into the pharynx.

F cell   A cell in the pancreatic islets (islets of Langerhans) that secretes pancreatic polypeptide.

Febrile (FE-bril)   Feverish; pertaining to a fever.

Feces (FE-sēz)   Material discharged from the rectum and made up of bacteria, excretions, and food residue. Also called **stool**.

Feedback system   A sequence of events in which information about the status of a situation is continually reported (fed back) to a control center.

Feeding (hunger) center   A cluster of neurons in the lateral nuclei of the hypothalamus that, when stimulated, brings about feeding.

Female reproductive cycle   General term for the ovarian and uterine cycles, the hormonal changes that accompany them, and cyclic changes in the breasts and cervix; includes changes in the endometrium of a nonpregnant female that prepares the lining of the uterus to receive a fertilized ovum. Less correctly termed **menstrual cycle**.

Fertilization (fer′-ti-li-ZA-shun)   Penetration of a secondary oocyte by a sperm cell and subsequent union of the nuclei of the gametes.

Fetal (FE-tal) **alcohol syndrome (FAS)**   Term applied to the effects of intrauterine exposure to alcohol, such as slow growth, defective organs, and mental retardation.

Fetal circulation   The cardiovascular system of the fetus, including the placenta and special blood vessels involved in the exchange of materials between fetus and mother.

Fetus (FE-tus)   The latter stages of the developing young of an animal; in humans, the developing organism *in utero* from the beginning of the third month to birth.

Fever   An elevation in body temperature above the normal temperature of 37°C (98.6°F).

Fibrillation (fi-bri-LA-shun)   Abnormal, spontaneous twitch of a single skeletal muscle fiber (cell) that can be detected with electromyography but is not visible at the skin surface; not associated with movement of the affected muscle; present in certain disorders of motor neurons, for example, amyotrophic lateral sclerosis (ALS). With reference to cardiac muscle, *see* **Atrial fibrillation** and **Ventricular fibrillation**.

Fibrin (FI-brin)   An insoluble protein that is essential to blood clotting; formed from fibrinogen by the action of thrombin.

Fibrinogen (fi-BRIN-ō-jen)   A clotting factor in the blood plasma that by the action of thrombin is converted to fibrin.

Fibrinolysis (fi-bri-NOL-i-sis)   Dissolution of a blood clot by the action of a proteolytic enzyme, such as plasmin (fibrinolysin), that dissolves fibrin threads and inactivates fibrinogen and other blood clotting factors.

Fibroblast (FI-brō-blast)   A large, flat cell that secretes most of the matrix (extracellular) material of areolar and dense connective tissues.

Fibromyalgia (fi-brō-mi-AL-jē-a)   Groups of common nonarticular rheumatic disorders characterized by pain, tenderness, and stiffness of muscles, tendons, and surrounding tissues. Examples of fibromyalgia are lumbago and charley horse.

Fibrosis (fī-BRŌ-sis) Abnormal formation of fibrous tissue.

Fibrous (FĪ-brus) **joint** A joint that allows little or no movement, such as a suture and syndesmosis.

Fibrous tunic (TOO-nik) The superficial coat of the eyeball, made up of the posterior sclera and the anterior cornea.

Fight-or-flight response The effects produced upon stimulation of the sympathetic division of the autonomic nervous system.

Filiform papilla (FIL-i-form pa-PIL-a) One of the conical projections that are distributed in parallel rows over the anterior two-thirds of the tongue and contain no taste buds.

Filtration (fil-TRĀ-shun) The passage of a liquid through a filter or membrane that acts like a filter.

Filtration fraction The percentage of plasma entering the kidneys that becomes glomerular filtrate.

Filum terminale (FĪ-lum ter-mi-NAL-ē) Nonnervous fibrous tissue of the spinal cord that extends inferiorly from the conus medullaris to the coccyx.

Fimbriae (FIM-brē-ē) Fingerlike structures, especially the lateral ends of the uterine (Fallopian) tubes.

Fissure (FISH-ur) A groove, fold, or slit that may be normal or abnormal.

Fistula (FIS-choo-la) An abnormal passage between two organs or between an organ cavity and the outside.

Fixator A muscle that stabilizes the origin of the prime mover so that the prime mover can act more efficiently.

Fixed macrophage (MAK-rō-fāj) Stationary phagocytic cell found in the liver, lungs, brain, spleen, lymph nodes, subcutaneous tissue, and red bone marrow. Also called a **histiocyte** (HIS-tē-ō-sit).

Flaccid (FLAS-sid) Relaxed, flabby, or soft; lacking muscle tone.

Flagellum (fla-JEL-um) A hairlike, motile process on the extremity of a bacterium, protozoan, or sperm cell. *Plural is* **flagella** (fla-JEL-a).

Flatfoot A condition in which the ligaments and tendons of the arches of the foot are weakened and the height of the longitudinal arch decreases.

Flatus (FLĀ-tus) Gas in the stomach or intestines, commonly used to denote expulsion of gas through the rectum.

Flexion (FLEK-shun) Movement in which there is a decrease in the angle between two bones.

Flexor reflex A protective reflex in which flexor muscles are stimulated while extensor muscles are inhibited.

Fluid mosaic (mō-ZĀ-ik) **model** Model of plasma membrane structure that describes the molecular arrangement of the plasma membrane and other membranes in living organisms.

Fluoroscope (FLOOR-ō-skōp) An instrument for visual observation of the body using x-rays.

Follicle (FOL-i-kul) A small secretory sac or cavity.

Follicle-stimulating hormone (FSH) Hormone secreted by the anterior pituitary gland that initiates development of ova and stimulates the ovaries to secrete estrogens in females and initiates sperm production in males.

Fontanel (fon′-ta-NEL) A membrane-covered spot where bone formation is not yet complete, especially between the cranial bones of an infant's skull.

Foot The terminal part of the lower limb.

Foramen (fō-RĀ-men) A passage or opening; a communication between two cavities of an organ or a hole in a bone for passage of vessels or nerves. *Plural is* **foramina** (fō-RAM-i-nah).

Foramen ovale (fō-RĀ-men ō-VAL-ē) An opening in the fetal heart in the septum between the right and left atria. A hole in the greater wing of the sphenoid bone that transmits the mandibular branch of the trigeminal (V) nerve.

Forearm (FOR-arm) The part of the upper limb between the elbow and the wrist.

Fornix (FOR-niks) An arch or fold; a tract in the brain made up of association fibers, connecting the hippocampus with the mammillary bodies; a recess around the cervix of the uterus where it protrudes into the vagina.

Fossa (FOS-a) A furrow or shallow depression.

Fourth ventricle (VEN-tri-kul) A cavity filled with cerebrospinal fluid within the brain lying between the cerebellum and the medulla oblongata and pons.

Fovea (FŌ-vē-ah) A cuplike depression in the center of the macula lutea of the retina, containing cones only; the area of clearest vision.

Fracture (FRAK-chur) Any break in a bone.

Fragile X syndrome Inherited disorder characterized by learning difficulties, mental retardation, and physical abnormalities; due to a defective gene on the X chromosome.

Frenulum (FREN-yoo-lum) A small fold of mucous membrane that connects two parts and limits movement.

Frontal plane A plane at a right angle to a midsagittal plane that divides the body or organs into anterior and posterior portions. Also called a **coronal** (kō-RŌ-nal) **plane**.

Functional residual (re-ZID-yoo-al) **capacity** The sum of residual volume plus expiratory reserve volume; about 2400 ml.

Fundus (FUN-dus) The part of a hollow organ farthest from the opening.

Fungiform papilla (FUN-ji-form pa-PIL-a) A mushroomlike elevation on the upper surface of the tongue appearing as a red dot; most contain taste buds.

Gallbladder A small pouch that stores bile, located inferior to the liver, which fills with bile and empties via the cystic duct.

Gallstone A solid mass, usually containing cholesterol, in the gallbladder or a bile-containing duct; formed anywhere between bile canaliculi in the liver and the hepatopancreatic ampulla (ampulla of Vater), where bile enters the duodenum. Also called a **biliary calculus**.

Gamete (GAM-ēt) A male or female reproductive cell; a sperm cell or secondary oocyte.

Gamete intrafallopian transfer (GIFT) A procedure in which aspirated secondary oocytes are mixed with a solution containing sperm outside the body and the mixture is then immediately inserted into the uterine (Fallopian) tubes.

Ganglion (GANG-glē-on) Usually, a group of nerve cell bodies lying outside the central nervous system (CNS); also used for one group of nerve cell bodies within the CNS—the basal ganglia. *Plural is* **ganglia** (GANG-glē-a).

Gangrene (GANG-rēn) Death and rotting of a considerable mass of tissue that usually is caused by interruption of blood supply followed by bacterial (*Clostridium*) invasion.

Gastric (GAS-trik) **glands** Glands in the mucosa of the stomach composed of cells that empty their secretions into narrow channels called gastric pits. Types of cells are chief cells (secrete pepsinogen), parietal cells (secrete hydrochloric acid and intrinsic factor), mucous surface and mucous neck cells (secrete mucus), and G cells (secrete gastrin).

Gastroenterology (gas′-trōen′-ter-OL-ō-jē) The medical specialty that deals with the structure, function, diagnosis, and treatment of diseases of the stomach and intestines.

Gastrointestinal (gas-trō-in-TES-ti-nal) **(GI) tract** A continuous tube running through the ventral body cavity extending from the mouth to the anus. Also called the **alimentary** (al'-i-MEN-tar-ē) **canal**.

Gastroscopy (gas-TROS-kō-pē) Diagnostic procedure in which the interior of the stomach is examined with a gastroscope to detect and biopsy lesions, stop bleeding, and remove foreign objects.

Gastrulation (gas'-troo-LA-shun) The migration of groups of cells that transform a blastula into a gastrula and establish the primary germ layers.

Gene (jēn) Biological unit of heredity; a segment of DNA located in a definite position on a particular chromosome; a sequence of DNA that codes for a particular mRNA, rRNA, or tRNA.

General adaptation syndrome (GAS) Wide-ranging set of bodily changes triggered by a stressor that gears the body to meet an emergency.

Generator potential The graded depolarization that results in a change in the resting membrane potential in a receptor (specialized neuronal ending); may trigger a nerve action potential (nerve impulse) if depolarization reaches threshold.

Genetic engineering The manufacture and manipulation of genetic material.

Genetics The study of genes and heredity.

Genital herpes (JEN-i-tal HER-pēz) A sexually transmitted disease caused by type II herpes simplex virus.

Genitalia (jen'-i-TAL-ya) Reproductive organs.

Genome (JE-nōm) The complete gene complement of an organism.

Genotype (JE-nō-tip) The total hereditary information carried by an individual; the genetic makeup of an organism.

Geriatrics (jer'-ē-AT-riks) The branch of medicine devoted to the medical problems and care of elderly persons.

Gestation (jes-TA-shun) The period of development from fertilization to birth.

Giantism (GI-an-tizm) Condition caused by hypersecretion of human growth hormone (hGH) during childhood; characterized by excessive bone growth and body size. Also called **gigantism**.

Gingivae (jin-JI-vē) Gums. They cover the alveolar processes of the mandible and maxilla and extend slightly into each socket.

Gingivitis (jin'-je-VI-tis) Inflammation of the gums.

Gland Single or group of specialized epithelial cells that secrete substances.

Glans penis (glanz PE-nis) The slightly enlarged region at the distal end of the penis.

Glaucoma (glaw-KO-ma) An eye disorder in which there is increased intraocular pressure due to an excess of aqueous humor.

Gliding joint A synovial joint having articulating surfaces that are usually flat, permitting only side-to-side and back-and-forth movements, as between carpal bones, tarsal bones, and the scapula and clavicle. Also called an **arthrodial** (ar-THRO-de-al) **joint**.

Glomerular (glō-MER-yoo-lar) **capsule** A double-walled globe at the proximal end of a nephron that encloses the glomerular capillaries. Also called **Bowman's** (BO-manz) **capsule**.

Glomerular filtrate (glō-MER-yoo-lar FIL-trāt) The fluid produced when blood is filtered by the endothelial–capsular membrane in the glomeruli of the kidneys.

Glomerular filtration The first step in urine formation in which substances in blood are filtered at the endothelial–capsular membrane and the filtrate enters the proximal convoluted tubule of a nephron.

Glomerular filtration rate (GFR) The total volume of fluid that enters all the glomerular (Bowman's) capsules of the kidneys in 1 min; about 125 ml/min.

Glomerulonephritis (glō-mer-yoo-lō-nef-RI-tis) Inflammation of the glomeruli of the kidney that increases the permeability of the endothelial–capsular membrane and permits blood cells and proteins to enter the filtrate. Also called **Bright's disease**.

Glomerulus (glō-MER-yoo-lus) A rounded mass of nerves or blood vessels, especially the microscopic tuft of capillaries that is surrounded by the glomerular (Bowman's) capsule of each kidney tubule.

Glottis (GLOT-is) The vocal folds (true vocal cords) in the larynx plus the space between them (rima glottidis).

Glucagon (GLOO-ka-gon) A hormone produced by the alpha cells of the pancreatic islets (islets of Langerhans) that increases blood glucose level.

Glucocorticoids (gloo-kō-KOR-ti-koyds) Hormones secreted by the cortex of the adrenal gland, especially cortisol, that influence glucose metabolism.

Gluconeogenesis (gloo'-kō-nē-ō-JEN-e-sis) The conversion of certain amino acids or lactic acid into glucose.

Glucose (GLOO-kōs) A six-carbon sugar, $C_6H_{12}O_6$; the major energy source for every cell type in the body for the production of ATP.

Glucosuria (gloo'-kō-SOO-rē-a) The presence of glucose in the urine; may be temporary or pathological. Also called **glycosuria**.

Glycogen (GLI-kō-jen) A highly branched polymer of glucose containing thousands of subunits; functions as a compact store of glucose molecules in liver and muscle fibers (cells).

Glycogenesis (gli'-kō-JEN-e-sis) The process by which many molecules of glucose combine to form a molecule called glycogen.

Glycogenolysis (gli-kō-je-NOL-i-sis) The breakdown of glycogen into glucose.

Glycolysis (gli-KOL-i-sis) Series of chemical reactions in the cytosol of a cell in which a molecule of glucose is split into two molecules of pyruvic acid with production of two ATPs.

Gnostic (NOS-tik) Pertaining to the faculties of perceiving and recognizing.

Gnostic area Sensory area of the cerebral cortex that receives and integrates sensory input from various parts of the brain so that a common thought can be formed.

Goblet cell A goblet-shaped unicellular gland that secretes mucus; present in epithelium of airways and intestines.

Goiter (GOY-ter) An enlargement of the thyroid gland.

Golgi (GOL-jē) **complex** An organelle in the cytoplasm of cells consisting of four to six flattened sacs (cisterns), stacked on one another, with expanded areas at their ends; functions in processing, sorting, packaging, and delivering proteins and lipids to the plasma membrane, lysosomes, and secretory vesicles.

Golgi tendon organ *See* **Tendon organ**.

Gomphosis (gom-FO-sis) A fibrous joint in which a cone-shaped peg fits into a socket.

Gonad (GO-nad) A gland that produces gametes and hormones; the ovary in the female and the testis in the male.

Gonorrhea (gon'-ō-RE-a) Infectious, sexually transmitted disease caused by the bacterium *Neisseria gonorrhoeae* and characterized by inflammation of the urogenital mucosa, discharge of pus, and painful urination.

Gout (gowt) Hereditary condition associated with excessive uric acid in the blood; the acid crystallizes and deposits in joints, kidneys, and soft tissue.

Graafian follicle *See* **Mature follicle.**

Gray commissure (KOM-i-shur) A narrow strip of gray matter connecting the two lateral gray masses within the spinal cord.

Gray matter Area in the central nervous system and ganglia consisting of nonmyelinated nerve tissue.

Gray ramus communicans (RĀ-mus kō-MYOO-ni-kans) A short nerve containing postganglionic sympathetic fibers; the cell bodies of the fibers are in a sympathetic chain ganglion, and the nonmyelinated axons extend by way of the gray ramus to a spinal nerve and then to the periphery to supply smooth muscle in blood vessels, arrector pili muscles, and sweat glands. *Plural is* **rami communicantes** (RĀ-mē kō-myoo-ni-KAN-tēz).

Greater omentum (ō-MEN-tum) A large fold in the serosa of the stomach that hangs down like an apron anterior to the intestines.

Greater vestibular (ves-TIB-yoo-lar) **glands** A pair of glands on either side of the vaginal orifice that open by a duct into the space between the hymen and the labia minora. Also called **Bartholin's** (BAR-to-linz) **glands.**

Groin (GROYN) The depression between the thigh and the trunk; the inguinal region.

Gross anatomy The branch of anatomy that deals with structures that can be studied without using a microscope. Also called **macroscopic anatomy.**

Growth An increase in size due to an increase in the number of cells or an increase in the size of existing cells as internal components increase in size or an increase in the size of intercellular substances.

Gustatory (GUS-ta-tō′-rē) Pertaining to taste.

Gynecology (gī′-ne-KOL-ō-jē) The branch of medicine dealing with the study and treatment of disorders of the female reproductive system.

Gynecomastia (gīn′-e-kō-MAS-tē-a) Excessive growth (benign) of the male mammary glands; may be due to secretion of estrogens by an adrenal gland tumor (feminizing adenoma).

Gyrus (JĪ-rus) One of the folds of the cerebral cortex of the brain. *Plural is* **gyri** (JĪ-rī). Also called a **convolution.**

Hair A threadlike structure produced by hair follicles that develops in the dermis. Also called **pilus** (PĪ-lus).

Hair follicle (FOL-li-kul) Structure composed of epithelium surrounding the root of a hair from which hair develops.

Hair root plexus (PLEK-sus) A network of dendrites arranged around the root of a hair as free or naked nerve endings that are stimulated when a hair shaft is moved.

Haldane (HAWL-dān) **effect** In the presence of oxygen, less carbon dioxide binds in the blood because when oxygen combines with hemoglobin, the hemoglobin becomes a stronger acid, which combines with less carbon dioxide.

Hallucination (ha-loo′-si-NĀ-shun) A sensory perception of something that does not really exist in the world, that is, a sensory experience created from within the brain.

Hand The terminal portion of an upper limb, including the carpus, metacarpus, and phalanges.

Haploid (HAP-loyd) Having half the number of chromosomes characteristically found in the somatic cells of an organism; characteristic of mature gametes. Symbolized *n.*

Hard palate (PAL-at) The anterior portion of the roof of the mouth, formed by the maxillae and palatine bones and lined by mucous membrane.

Haustra (HAWS-tra) The sacculated elevations of the colon. *Singular is* **haustrum.**

Haversian canal *See* **Central canal.**

Haversian system *See* **Osteon.**

Head The superior part of a human, cephalic to the neck. The superior or proximal part of a structure.

Heart A hollow muscular organ lying slightly to the left of the midline of the chest that pumps the blood through the cardiovascular system.

Heart block An arrhythmia (dysrhythmia) of the heart in which the atria and ventricles contract independently because of a blocking of electrical impulses through the heart at some point in the conduction system.

Heartburn Burning sensation in the esophagus due to reflux of hydrochloric acid (HCl) from the stomach.

Heart–lung machine A device that pumps blood, functioning as a heart, and adds oxygen to and removes carbon dioxide from blood, functioning as lungs; used during heart transplantation, open-heart surgery, and coronary artery bypass grafting.

Heart murmur (MER-mer) An abnormal sound that consists of a flow noise that is heard before, between, or after the normal lubb–dupp or that may mask normal heart sounds.

Heat exhaustion Condition characterized by cool, clammy skin, profuse perspiration, and fluid and electrolyte (especially salt) loss that results in muscle cramps, dizziness, vomiting, and fainting. Also called **heat prostration.**

Heatstroke Condition produced when the body cannot easily lose heat and characterized by reduced perspiration and elevated body temperature. Also called **sunstroke.**

Heimlich maneuver *See* **Abdominal thrust maneuver.**

Hematocrit (hē-MAT-ō-krit) **(Hct)** The percentage of blood made up of red blood cells. Usually calculated by centrifuging a blood sample in a graduated tube and then reading off the volume of red blood cells and total blood.

Hematology (hē′-ma-TOL-ō-jē) The study of blood.

Hematoma (hē′-ma-TŌ-ma) A localized pool of blood, usually clotted, that produces swelling in an organ, tissue, or space.

Hematuria (hē′-ma-TOOR-ē-a) Blood in the urine.

Hemiballismus (hem′-i-ba-LIZ-mus) Violent muscular restlessness of half of the body, especially of the upper limb.

Hemiplegia (hem-i-PLĒ-jē-a) Paralysis of the upper limb, trunk, and lower limb on one side of the body.

Hemodialysis (hē′-mō-di-AL-i-sis) Filtering of the blood by means of an artificial device so that certain substances are removed from the blood as a result of the difference in rates of their diffusion through a selectively permeable membrane while the blood is being circulated outside the body.

Hemodynamics (hē-mō-di-NA-miks) The study of factors and forces that govern the flow of blood through blood vessels.

Hemoglobin (hē′-mō-GLO-bin) **(Hb)** A substance in red blood cells consisting of the protein globin and the iron-containing red pigment heme and constituting about 33% of the cell volume; involved in the transport of oxygen and carbon dioxide.

Hemolysis (hē-MOL-i-sis) The escape of hemoglobin from the interior of an erythrocyte (red blood cell) into the surrounding medium; results from disruption of the cell membrane by toxins or drugs, freezing or thawing, or hypotonic solutions.

Hemolytic disease of the newborn A hemolytic anemia of a newborn child that results from the destruction of the infant's erythrocytes (red blood cells) by antibodies produced by the mother; usually the antibodies are due to an Rh blood type incompatibility. Also called **erythroblastosis fetalis** (e-rith′-rō-blas-TŌ-sis fe-TAL-is).

Hemophilia (hē'-mō-FĒL-ē-a)   A hereditary blood disorder where there is a deficient production of certain factors involved in blood clotting, resulting in excessive bleeding into joints, deep tissues, and elsewhere.

Hemopoiesis (hē-mō-poy-Ē-sis)   Blood cell production occurring in red bone marrow. Also called **hematopoiesis** (hem'-a-tō-poy-Ē-sis).

Hemopoietic (hem'-ō-poy-Ē-tic) **stem cell**   Immature stem cell in red bone marrow that gives rise to precursors of all the different mature blood cells. Previously called a **hemocytoblast** (hē'-mō-SĪ-tō-blast) or pluripotent hematopoietic stem cell.

Hemoptysis (hē-MOP-ti-sis)   Spitting of blood from the respiratory tract.

Hemorrhage (HEM-or-rij)   Bleeding; the escape of blood from blood vessels, especially when it is profuse.

Hemorrhoids (HEM-ō-royds)   Dilated or varicosed blood vessels (usually veins) in the anal region. Also called **piles**.

Hemostasis (hē-MOS-tā-sis)   The stoppage of bleeding.

Hemostat (HĒ-mō-stat)   An agent or instrument used to prevent the flow or escape of blood.

Heparin (HEP-a-rin)   An anticoagulant given to slow the conversion of prothrombin to thrombin, thus reducing the risk of blood clot formation; also found naturally in basophils and most cells.

Hepatic (he-PAT-ik)   Refers to the liver.

Hepatic duct   A duct that receives bile from the bile capillaries. Small hepatic ducts merge to form the larger right and left hepatic ducts that unite to leave the liver as the common hepatic duct.

Hepatic portal circulation   The flow of blood from the gastrointestinal organs to the liver before returning to the heart.

Hepatitis (hep-a-TĪ-tis)   Inflammation of the liver due to a virus, drugs, and chemicals.

Hepatocyte (he-PAT-ō-cyte)   A liver cell.

Hepatopancreatic (hep'-a-tō-pan'-krē-A-tik) **ampulla**   A small, raised area in the duodenum where the combined common bile duct and main pancreatic duct empty into the duodenum. Also called the **ampulla of Vater** (VA-ter).

Hering–Breuer reflex   See **Inflation reflex**.

Hernia (HER-nē-a)   The protrusion or projection of an organ or part of an organ through a membrane or cavity wall, usually the abdominal cavity.

Herniated (HER-nē-ā'-ted) **disc**   A rupture of an intervertebral disc so that the nucleus pulposus protrudes into the vertebral cavity. Also called a **slipped disc**.

Heterozygous (he-ter-ō-ZĪ-gus)   Possessing a pair of different genes on homologous chromosomes for a particular hereditary characteristic.

Hiatus (hi-A-tus)   An opening; a foramen.

High altitude sickness   Disorder caused by decreased levels of alveolar pO_2 as altitude increases and characterized by headache, fatigue, insomnia, shortness of breath, nausea, and dizziness. Also called **acute mountain sickness**.

Hilus (HĪ-lus)   An area, depression, or pit where blood vessels and nerves enter or leave an organ. Also called a **hilum**.

Hinge joint   A synovial joint in which a convex surface of one bone fits into a concave surface of another bone, such as the elbow, knee, ankle, and interphalangeal joints. Also called a **ginglymus** (JIN-gli-mus) **joint**.

Hirsutism (HER-soot-izm)   An excessive growth of hair in females and children, with a distribution similar to that in adult males, due to the conversion of vellus hairs into large terminal hairs in response to higher-than-normal levels of androgens.

Histamine (HISS-ta-mēn)   Substance found in many cells, especially mast cells, basophils, and platelets, released when the cells are injured; results in vasodilation, increased permeability of blood vessels, and constriction of bronchioles.

Histocompatibility (hiss'-tō-kom-pat-i-BIL-i-tē) **testing**   Comparison of human leukocyte associated (HLA) antigens between donor and recipient to determine histocompatibility, the degree of compatibility between the two. Also called **HLA antigen typing** or **tissue typing**.

Histology (hiss-TOL-ō-jē)   Microscopic study of the structure of tissues.

Hives (HĪVZ)   Condition of the skin marked by reddened elevated patches that are often itchy; may be caused by infections, trauma, medications, emotional stress, food additives, and certain foods.

Hodgkin's disease (HD)   A malignant disorder, usually arising in lymph nodes.

Holocrine (HOL-ō-krin) **gland**   A type of gland in which entire secretory cells, along with their accumulated secretions, make up the secretory product of the gland, as in the sebaceous (oil) glands.

Holter monitor   Electrocardiograph worn by a person while going about everyday routines.

Homeostasis (hō'-mē-o-STA-sis)   The condition in which the body's internal environment remains relatively constant, within physiological limits.

Homologous (hō-MOL-ō-gus)   Correspondence of two organs in structure, position, and origin.

Homologous chromosomes   Two chromosomes that belong to a pair. Also called **homologues**.

Homozygous (hō-mō-ZĪ-gus)   Possessing a pair of similar genes on homologous chromosomes for a particular hereditary characteristic.

Hormone (HOR-mōn)   A secretion of endocrine cells that alters the physiological activity of target cells of the body.

Horn   Principal area of gray matter in the spinal cord.

Human chorionic gonadotropin (hCG) (kō-rē-ON-ik gō-nad-ō-TRO-pin)   A hormone produced by the developing placenta that maintains the corpus luteum.

Human chorionic somatomammotropin (sō-mat-ō-mam-ō-TRO-pin) **(hCS)**   A hormone produced by the chorion of the placenta that may stimulate breast tissue for lactation, enhance body growth, and regulate metabolism. Also called **human placental lactogen (hPL)**.

Human growth hormone (hGH)   Hormone secreted by the anterior pituitary gland that stimulates growth of body tissues, especially skeletal and muscular. Also known as **somatotropin** and **somatotropic hormone (STH)**.

Human leukocyte associated (HLA) antigens   See **major histocompatibility antigens**.

Hyaluronic (hi'-a-loo-RON-ik) **acid**   A viscous, amorphous extracellular material that binds cells together, lubricates joints, and maintains the shape of the eyeballs.

Hyaluronidase (hi'-a-loo-RON-i-dās)   An enzyme that breaks down hyaluronic acid, increasing the permeability of connective tissues by dissolving the substances that hold body cells together.

Hydride ion (HĪ-drid Ī-on)   A hydrogen nucleus with two orbiting electrons (H^-), in contrast to the more common **hydrogen ion**, which has no orbiting electrons (H^+).

Hydrocele (HĪ-drō-sēl)   A fluid-containing sac or tumor. Specifically, a collection of fluid formed in the space along the spermatic cord and in the scrotum.

Hydrocephalus (hī-drō-SEF-a-lus) Abnormal accumulation of cerebrospinal fluid on the brain.

Hymen (HĪ-men) A thin fold of vascularized mucous membrane at the vaginal orifice.

Hyperbaric oxygenation (hī′-per-BA-rik ok′-sē-je-NA-shun) **(HBO)** Use of pressure supplied by a hyperbaric chamber to cause more oxygen to dissolve in blood to treat patients infected with anaerobic bacteria (tetanus and gangrene bacteria). Also used to treat carbon monoxide poisoning, asphyxia, smoke inhalation, and certain heart disorders.

Hypercalcemia (hī′-per-kal-SE-mē-a) An excess of calcium in the blood.

Hypercapnia (hī′-per-KAP-nē-a) An abnormal increase in the amount of carbon dioxide in the blood.

Hyperemia (hī′-per-E-mē-a) An excess of blood in an area or part of the body.

Hyperextension (hī′-per-ek-STEN-shun) Continuation of extension beyond the anatomical position, as in bending the head backward.

Hyperglycemia (hī′-per-glī-SE-mē-a) An elevated blood glucose level.

Hyperkalemia (hī′-per-kā-LE-mē-a) An excess of potassium ions in the blood.

Hypermagnesemia (hī′-per-mag′-ne-SE-mē-a) An excess of magnesium ions in the blood.

Hypermetropia (hī′-per-mē-TRO-pē-a) A condition in which visual images are focused behind the retina with resulting defective vision of near objects; farsightedness.

Hyperphosphatemia (hī-per-fos′-fa-TE-mē-a) An abnormally high level of phosphates in the blood.

Hyperplasia (hī′-per-PLA-zē-a) An abnormal increase in the number of normal cells in a tissue or organ, increasing its size.

Hyperpolarization (hī′-per-POL-a-ri-zā′-shun) Increase in the internal negativity across a cell membrane, thus increasing the voltage and moving it farther away from the threshold value.

Hypersecretion (hī′-per-se-KRE-shun) Overactivity of glands resulting in excessive secretion.

Hypersensitivity (hī′-per-sen-si-TI-vi-tē) Overreaction to an allergen that results in pathological changes in tissues. Also called **allergy**.

Hypertension (hī′-per-TEN-shun) High blood pressure.

Hyperthermia (hī′-per-THERM-ē-a) An elevated body temperature.

Hypertonia (hī-per-TO-nē-a) Increased muscle tone that is expressed as spasticity or rigidity.

Hypertonic (hī′-per-TON-ik) Solution that causes cells to shrink due to loss of water by osmosis.

Hypertrophy (hī-PER-trō-fē) An excessive enlargement or overgrowth of tissue without cell division.

Hyperventilation (hī′-per-ven-ti-LA-shun) A rate of respiration higher than that required to maintain a normal level of plasma pCO_2.

Hypervitaminosis (hī′-per-vī′-ta-min-O-sis) An excess of one or more vitamins in the body.

Hypocalcemia (hī′-pō-kal-SE-mē-a) A below normal level of calcium in the blood.

Hypochloremia (hī′-pō-klō-RE-mē-a) Deficiency of chloride ions in the blood.

Hypoglycemia (hī′-pō-glī-SE-mē-a) An abnormally low concentration of glucose in the blood; can result from excess insulin (injected or secreted).

Hypokalemia (hī′-pō-ka-LE-mē-a) Deficiency of potassium ions in the blood.

Hypomagnesemia (hī′-pō-mag′-ne-SE-mē-a) Deficiency of magnesium ions in the blood.

Hyponatremia (hī′-pō-na-TRE-mē-a) Deficiency of sodium ions in the blood.

Hyponychium (hī′-pō-NIK-ē-um) Free edge of the fingernail.

Hypophosphatemia (hi-pō-fos′-fa-TE-mē-a) An abnormally low level of phosphates in the blood.

Hypophyseal (hī′-pō-FIZ-ē-al) **pouch** An outgrowth of ectoderm from the roof of the mouth from which the anterior pituitary gland develops.

Hypophysis (hi-POF-i-sis) Pituitary gland.

Hypoplasia (hi-pō-PLA-zē-a) Defective development of tissue.

Hyposecretion (hī′-pō-se-KRE-shun) Underactivity of glands resulting in diminished secretion.

Hypospadias (hī′-pō-SPA-dē-as) A displaced urethral opening. In the male, the opening may be on the underside of the penis, at the penoscrotal junction, between the scrotal folds, or in the perineum. In the female, the urethra opens into the vagina.

Hypothalamus (hī-pō-THAL-a-mus) A portion of the diencephalon, lying beneath the thalamus and forming the floor and part of the wall of the third ventricle.

Hypothermia (hi-pō-THER-mē-a) Lowering of body temperature below 35°C (95°F); in surgical procedures, it refers to deliberate cooling of the body to slow down metabolism and reduce oxygen needs of tissues.

Hypotonia (hī′-pō-TO-nē-a) Decreased or lost muscle tone in which muscles appear flaccid.

Hypotonic (hī′-pō-TON-ik) Solution that causes cells to swell and perhaps rupture due to gain of water by osmosis.

Hypoventilation (hi-pō-ven-ti-LA-shun) A rate of respiration lower than that required to maintain a normal level of plasma pCO_2.

Hypovolemic (hi-pō-vō-LE-mik) **shock** A type of shock characterized by decreased intravascular volume resulting from blood loss; may be caused by acute hemorrhage or excessive fluid loss.

Hypoxia (hi-POKS-ē-a) Lack of adequate oxygen at the tissue level.

Hysterectomy (hiss-te-REK-tō-mē) The surgical removal of the uterus.

Ileocecal (il′-ē-ō-SE-kal) **sphincter** A fold of mucous membrane that guards the opening from the ileum into the large intestine. Also called the **ileocecal valve**.

Ileum (IL-ē-um) The terminal portion of the small intestine.

Immunity (im-YOO-ni-tē) The state of being resistant to injury, particularly by poisons, foreign proteins, and invading pathogens.

Immunogenicity (im-yoo-nō-jen-IS-it-ē) Ability of an antigen to provoke an immune response.

Immunoglobulin (im-yoo-nō-GLOB-yoo-lin) **(Ig)** An antibody synthesized by plasma cells derived from B lymphocytes in response to the introduction of antigen. Immunoglobulins are divided into five kinds (IgG, IgM, IgA, IgD, IgE) based primarily on the larger protein component present in the immunoglobulin.

Immunology (im′-yoo-NOL-ō-jē) The branch of science that deals with the responses of the body when challenged by antigens.

Immunosuppression (im′-yoo-nō-su-PRESH-un) Inhibition of immune responses.

Immunotherapy (im-yoo-nō-THER-a-pē) Attempt to induce the immune system to mount an attack against cancer cells.

Impetigo (im′-pe-TĪ-go) A contagious skin disorder characterized by pustular eruptions.

Implantation (im-plan-TĀ-shun) The insertion of a tissue or a part into the body. The attachment of the blastocyst to the lining of the uterus 7–8 days after fertilization.

Impotence (IM-pō-tens) Weakness; inability to copulate; failure to maintain an erection long enough for sexual intercourse.

Incontinence (in-KON-ti-nens) Inability to retain urine, semen, or feces, through loss of sphincter control.

Indirect (motor) pathways Motor tracts that convey information from the brain down the spinal cord for automatic movements, coordination of body movements with visual stimuli, skeletal muscle tone and posture, and balance. Also known as **extrapyramidal pathways**.

Infarction (in-FARK-shun) The presence of a localized area of necrotic tissue, produced by inadequate oxygenation of the tissue.

Infection (in-FEK-shun) Invasion and multiplication of microorganisms in body tissues, which may be inapparent or characterized by cellular injury.

Infectious mononucleosis (mon-ō-nook′-lē-Ō-sis) **(IM)** Contagious disease caused by the Epstein–Barr virus (EBV) and characterized by an elevated mononucleocyte and lymphocyte count, fever, sore throat, stiff neck, cough, and malaise.

Inferior (in-FĒR-ē-or) Away from the head or toward the lower part of a structure. Also called **caudad** (KAW-dad).

Inferior vena cava (VĒ-na CĀ-va) **(IVC)** Large vein that collects blood from parts of the body inferior to the heart and returns it to the right atrium.

Infertility Inability to conceive or to cause conception. Also called **sterility**.

Inflammation (in′-fla-MĀ-shun) Localized, protective response to tissue injury designed to destroy, dilute, or wall off the infecting agent or injured tissue; characterized by redness, pain, heat, swelling, and sometimes loss of function.

Inflammatory bowel (in-FLAM-a-tō′-re BOW-el) **disease** Disorder that exists in two forms: (1) Crohn's disease (inflammation of the gastrointestinal tract, especially the distal ileum and proximal colon, in which the inflammation may extend from the mucosa through the serosa); and (2) ulcerative colitis (inflammation of the mucosa of the gastrointestinal tract, usually limited to the large intestine and usually accompanied by rectal bleeding).

Inflation reflex Reflex that prevents overinflation of the lungs. Also called **Hering–Breuer reflex**.

Infundibulum (in′-fun-DIB-yoo-lum) The stalklike structure that attaches the pituitary gland to the hypothalamus of the brain. The funnel-shaped, open, distal end of the uterine (Fallopian) tube.

Ingestion (in-JES-chun) The taking in of food, liquids, or drugs, by mouth.

Inguinal (IN-gwi-nal) Pertaining to the groin.

Inguinal canal An oblique passageway in the anterior abdominal wall just superior and parallel to the medial half of the inguinal ligament that transmits the spermatic cord and ilioinguinal nerve in the male and round ligament of the uterus and ilioinguinal nerve in the female.

Inheritance The acquisition of body characteristics and qualities by transmission of genetic information from parents to offspring.

Inhibin A hormone secreted by the gonads that inhibits release of follicle-stimulating hormone (FSH) by the anterior pituitary gland.

Inhibiting hormone Chemical secretion of the hypothalamus that can suppress secretion of hormones by the anterior pituitary gland.

Inhibitory postsynaptic potential (IPSP) A small hyperpolarization caused by neurotransmitter at a synapse in which the membrane potential becomes more negative.

Inner cell mass A region of cells of a blastocyst that differentiates into the three primary germ layers—ectoderm, mesoderm, and endoderm—from which all tissues and organs develop; also called an **embryoblast**.

Inorganic (in′-or-GAN-ik) **compound** Compound that usually lacks carbon, usually small, and often contains ionic bonds. Examples include water and many acids, bases, and salts.

Insertion (in-SER-shun) The attachment of a muscle tendon to a movable bone or the end opposite the origin.

Insomnia (in-SOM-nē-a) Difficulty in falling asleep and, usually, frequent awakening.

Inspiration (in-spi-RĀ-shun) The act of drawing air into the lungs. Also termed **inhalation**.

Inspiratory (in-SPĪ-ra-tor-ē) **capacity** Total inspiratory capacity of the lungs; the total of tidal volume plus inspiratory reserve volume; averages 3600 ml.

Inspiratory (in-SPĪ-ra-tor-ē) **reserve volume** Additional inspired air over and above tidal volume; averages 3100 ml.

Insula (IN-su-la) A triangular area of cerebral cortex that lies deep within the lateral cerebral fissue, under the parietal, frontal, and temporal lobes, and cannot be seen in an external view of the brain. Also called the **island** or **isle of Reil** (RĪL).

Insulin (IN-su-lin) A hormone produced by the beta cells of a pancreatic islet (islet of Langerhans) that decreases the blood glucose level.

Insulin-like growth factor (IGF) Small protein produced by the liver and other tissues in response to stimulation by human growth hormone (hGH) that mediates most of the effects of human growth hormone. Previously called **somatomedin** (sō′-ma-tō-ME-din).

Integrin (IN-te-grin) Receptor on a plasma membrane that interacts with an adhesion protein found in intercellular material and blood.

Integumentary (in-teg′-yoo-MEN-tar-ē) Relating to the skin.

Intercalated (in-TER-ka-lāt-ed) **disc** An irregular transverse thickening of sarcolemma that contains desmosomes that hold cardiac muscle fibers (cells) together and gap junctions that aid in conduction of muscle action potentials.

Intercostal (in′-ter-KOS-tal) **nerve** A nerve supplying a muscle located between the ribs.

Interferons (in′-ter-FĒR-ons) **(IFNs)** Three principal types of protein (alpha, beta, gamma) naturally produced by virus-infected host cells that induce uninfected cells to synthesize antiviral proteins that inhibit intracellular viral replication in uninfected host cells.

Intermediate Between two structures, one of which is medial and one of which is lateral.

Intermediate filament Protein filament, ranging from 8 to 12 nm in diameter, that may provide structural reinforcement, hold organelles in place, and give shape to a cell.

Internal Away from the surface of the body.

Internal capsule A tract of projection fibers connecting various parts of the cerebral cortex and lying between the thalamus and the caudate and lentiform nuclei of the basal ganglia.

Internal ear The inner ear or labyrinth, lying inside the temporal bone, containing the organs of hearing and balance.

Internal nares (NA-rēz) The two openings posterior to the nasal cavities opening into the nasopharynx. Also called the **choanae** (kō-A-nē).

Internal respiration The exchange of respiratory gases between blood and body cells. Also called **tissue respiration**.

Interoceptor (in′-ter-ō-SEP-tor) Receptor located in blood vessels and viscera that provides information about the body's internal environment. Also termed **visceroceptor** (vis′-er-ō-SEP-tor).

Interphase (IN-ter-fāz) The period of the cell cycle between cell divisions, consisting of the G_1- (gap or growth) phase when the cell is engaged in growth, metabolism, and production of substances required for division, S- (synthesis) phase during which chromosomes are replicated, and G_2-phase.

Interstitial cell of Leydig *See* **Interstitial endocrinocyte**.

Interstitial (in′-ter-STISH-al) **endocrinocyte** A type of cell that secretes testosterone; located in the connective tissue between seminiferous tubules in a mature testis. Also called an **interstitial cell of Leydig** (LĪ-dig).

Interstitial (in′-ter-STISH-al) **fluid** The portion of extracellular fluid that fills the microscopic spaces between the cells of tissues; the internal environment of the body. Also called **intercellular** or **tissue fluid**.

Interstitial growth Growth from within, as in the growth of cartilage. Also called **endogenous** (en-DOJ-e-nus) **growth**.

Interventricular (in′-ter-ven-TRIK-yoo-lar) **foramen** A narrow, oval opening through which the lateral ventricles of the brain communicate with the third ventricle. Also called the **foramen of Monro**.

Intervertebral (in′-ter-VER-te-bral) **disc** A pad of fibrocartilage located between the bodies of two vertebrae.

Intestinal gland A gland that opens onto the surface of the intestinal mucosa and secretes digestive enzymes. Also called a **crypt of Lieberkühn** (LĒ-ber-kyoon).

Intracellular (in′-tra-SEL-yoo-lar) **fluid (ICF)** Fluid located within cells. Also called **cytosol** (SĪ-tō-sol).

Intrafusal (in′-tra-FYOO-zal) **fibers** Three to ten specialized muscle fibers (cells), partially enclosed in a spindle-shaped connective tissue capsule; these fibers make up a muscle spindle.

Intramembranous ossification (in′-tra-MEM-bra-nus os′-i′-fi-KĀ-shun) The method of bone formation in which the bone is formed directly in membranous tissue.

Intraocular (in-tra-OK-yoo-lar) **pressure (IOP)** Pressure in the eyeball, produced mainly by aqueous humor.

Intrapleural pressure Air pressure between the two pleural layers of the lungs, usually subatmospheric. Also called **intrathoracic pressure**.

Intrauterine device (IUD) A small metal or plastic object inserted into the uterus for the purpose of preventing pregnancy.

Intrinsic (in-TRIN-sik) Of internal origin; for example, intrinsic factor, a glycoprotein formed by the gastric mucosa that is necessary for the absorption of vitamin B_{12} in the small intestine.

Intrinsic pathway (of blood clotting) Sequence of reactions leading to blood clotting that is initiated by damage to blood vessel endothelium or platelets; activators of this pathway are contained *within* blood itself or are in direct contact with blood.

Intrinsic factor (IF) A glycoprotein synthesized and secreted by the parietal cells of the gastric mucosa that facilitates vitamin B_{12} absorption in the small intestine.

Intron (IN-tron) A region of DNA that does not code for the synthesis of a protein.

Intubation (in′-too-BA-shun) Insertion of a tube through the nose or mouth into the larynx and trachea for entrance of air or to dilate a stricture.

In utero (YOO-ter-ō) Within the uterus.

Invagination (in-vaj′-i-NA-shun) The pushing of the wall of a cavity into the cavity itself.

Inversion (in-VER-zhun) The movement of the sole medially at the ankle joint.

In vitro (VĒ-trō) Literally, in glass; outside the living body and in artificial environment such as a laboratory test tube.

In vivo (VĒ-vō) In the living body.

Ion (Ī-on) Any charged particle or group of particles; usually formed when a substance, such as a salt, dissolves and dissociates.

Ionization (ī′-on-i-ZĀ-shun) Separation of inorganic acids, bases, and salts into ions when dissolved in water. Also called **dissociation**.

Ipsilateral (ip′-si-LAT-er-al) On the same side, affecting the same side of the body.

Iris The colored portion of the eyeball seen through the cornea that consists of circular and radial smooth muscle; the hole in the center of the iris is the pupil.

Irritable bowel (IR-i-ta-bul BOW-el) **syndrome (IBS)** Disease of the entire gastrointestinal tract in which persons with the condition may react to stress by developing symptoms such as cramping and abdominal pain associated with alternating patterns of diarrhea and constipation. Excessive amounts of mucus may appear in the stools, and other symptoms include flatulence, nausea, and loss of appetite. The condition is also known as **irritable colon** or **Spastic colitis**.

Ischemia (is-KĒ-mē-a) A lack of sufficient blood to a part due to obstruction or constriction of a blood vessel.

Island of Reil *See* **Insula**.

Islet of Langerhans *See* **Pancreatic islet**.

Isoantibody A specific antibody in blood serum that reacts with specific isoantigens and causes the clumping of bacteria, red blood cells, or particles. Also called an **agglutinin**.

Isoantigen A genetically determined antigen located on the surface of red blood cells; basis for the ABO grouping and Rh system of blood classification. Also called an **agglutinogen**.

Isometric contraction A muscle contraction in which tension on the muscle increases, but there is only minimal muscle shortening so that no movement is produced.

Isotonic (ī′-sō-TON-ik) Having equal tension or tone. A solution having the same concentration of impermeable solutes as cytosol.

Isotonic (i-so-TON-ik) **contraction** Contraction in which the tension remains the same; occurs when a constant load is moved through the range of motions possible at a joint.

Isotopes (Ī-sō-tōpes′) Chemical elements that have the same number of protons but different numbers of neutrons. Radioactive isotopes change into other elements with the emission of alpha or beta particles or gamma rays.

Isovolumetric (ī-sō-vol-yoo′-MET-rik) **contraction** The period of time, about 0.05 sec, between the start of ventricular systole and the opening of the semilunar valves; there is contraction of the ventricles, but no emptying, and there is a rapid rise in ventricular pressure.

Isovolumetric relaxation The period of time, about 0.05 sec, between the closing of the semilunar valves and the opening of the atrioventricular (AV) valves; there is a drastic decrease in ventricular pressure without a change in ventricular volume.

Isthmus (IS-mus)   A narrow strip of tissue or narrow passage connecting two larger parts.

Jaundice (JAWN-dis)   A condition characterized by yellowness of skin, white of eyes, mucous membranes, and body fluids because of a buildup of bilirubin.

Jejunum (jē-JOO-num)   The middle portion of the small intestine.

Joint kinesthetic (kin′-es-THET-ik) **receptor**   A proprioceptive receptor located in a joint, stimulated by joint movement.

Juxtaglomerular (juks-ta-glō-MER-yoo-lar) **apparatus (JGA)**   Consists of the macula densa (cells of the distal convoluted tubule adjacent to the afferent and efferent arteriole) and juxtaglomerular cells (modified cells of the afferent and sometimes efferent arteriole); secretes renin when blood pressure starts to fall.

Karyotype (KAR-ē-ō-tip)   An arrangement of chromosomes based on shape, size, and position of centromeres.

Keratin (KER-a-tin)   An insoluble protein found in the hair, nails, and other keratinized tissues of the epidermis.

Keratinocyte (ke-RAT-in′-ō-sit)   The most numerous of the epidermal cells that function in the production of keratin.

Keratosis (ker′-a-TO-sis)   Formation of a hardened growth of tissue.

Ketone (KE-tōn) **bodies**   Substances produced primarily during excessive triglyceride catabolism, such as acetone, acetoacetic acid, and β-hydroxybutyric acid.

Ketosis (kē-TO-sis)   Abnormal condition marked by excessive production of ketone bodies.

Kidney (KID-nē)   One of the paired reddish organs located in the lumbar region that regulates the composition, volume, and pressure of blood and produces urine.

Kidney stone   A solid mass, usually consisting of calcium oxalate, uric acid, or calcium phosphate crystals, that may form in any portion of the urinary tract. Also called a **renal calculus**.

Kilocalorie (KIL-ō-kal′-ō-rē) **(kcal)**   The amount of heat required to raise the temperature of 1000 g of water 1°C from 14° to 15°C; the unit used to express the heating value of foods and to measure metabolic rate.

Kinesiology (ki-nē′-sē-OL-ō-jē)   The study of the movement of body parts.

Kinesthesia (kin-is-THE-szē-a)   Ability to perceive extent, direction, or weight of movement; muscle sense.

Korotkoff (kō-ROT-kof) **sounds**   The various sounds that are heard while taking blood pressure.

Krebs cycle   A series of biochemical reactions that occurs in the matrix of mitochondria in which electrons are transferred to coenzymes and carbon dioxide is formed. The electrons carried by the coenzymes then enter the electron transport chain, which generates a large quantity of ATP. Also called the **citric acid cycle** or **tricarboxylic acid (TCA) cycle**.

Kupffer's cell   *See* **Stellate reticuloendothelial cell.**

Kyphosis (ki-FO-sis)   An exaggeration of the thoracic curve of the vertebral column, resulting in a "round-shouldered" appearance. Also called **hunchback.**

Labial frenulum (LA-bē-al FREN-yoo-lum)   A medial fold of mucous membrane between the inner surface of the lip and the gums.

Labia majora (LA-bē-a ma-JO-ra)   Two longitudinal folds of skin extending downward and backward from the mons pubis of the female.

Labia minora (min-OR-a)   Two small folds of mucous membrane lying medial to the labia majora of the female.

Labium (LA-bē-um)   A lip. A liplike structure. *Plural is* **labia** (LA-bē-a).

Labor   The process by which a fetus is expelled from the uterus through the vagina.

Labyrinth (LAB-i-rinth)   Intricate communicating passageway, especially in the internal ear.

Labyrinthine (lab-i-RIN-thēn) **disease**   Malfunction of the internal ear characterized by deafness, tinnitus, vertigo, nausea, and vomiting.

Laceration (las′-er-A-shun)   Wound or irregular tear of the skin.

Lacrimal (LAK-ri-mal)   Pertaining to tears.

Lacrimal canal   A duct, one on each eyelid, beginning at the punctum at the medial margin of an eyelid and conveying tears medially into the nasolacrimal sac.

Lacrimal gland   Secretory cells located at the superior anterolateral portion of each orbit that secrete tears into excretory ducts that open onto the surface of the conjunctiva.

Lacrimal sac   The superior expanded portion of the nasolacrimal duct that receives the tears from a lacrimal canal.

Lactation (lak-TA-shun)   The secretion and ejection of milk by the mammary glands.

Lacteal (LAK-tē-al)   One of many intestinal lymphatic vessels in villi of the intestines that absorb triglycerides and other lipids from digested food.

Lactose intolerance   Inability to digest lactose because of failure of small intestinal mucosal cells to produce lactase.

Lacuna (la-KOO-na)   A small, hollow space, such as that found in bones in which the osteocytes lie. *Plural is* **lacunae** (la- KOO-nē).

Lambdoid (lam-DOYD) **suture**   The joint in the skull between the parietal bones and the occipital bone; sometimes contains sutural (Wormian) bones.

Lamellae (la-MEL-ē)   Concentric rings of hard, calcified matrix found in compact bone.

Lamellated corpuscle   Oval-shaped pressure receptor located in subcutaneous tissue and consisting of concentric layers of connective tissue wrapped around a sensory nerve fiber. Also called a **Pacinian** (pa-SIN-ē-an) **corpuscle**.

Lamina (LAM-i-na)   A thin, flat layer or membrane, as the flattened part of either side of the arch of a vertebra. *Plural is* **laminae** (LAM-i-nē).

Lamina propria (PRO-prē-a)   The connective tissue layer of a mucous membrane.

Langerhans (LANG-er-hans) **cell**   Epidermal dendritic cell that functions as an antigen-presenting cell (APC) during an immune response.

Lanugo (lan-YOO-gō)   Fine downy hairs that cover the fetus.

Laparoscopy (lap′-a-ROS-kō-pē)   A procedure in which a laparoscope is inserted through an incision in the abdominal wall to view abdominal and pelvic viscera, remove fluids and tissues for biopsy, drain ovarian cysts, cut adhesions, stop bleeding, and perform tubal ligation.

Large intestine   The portion of the gastrointestinal tract extending from the ileum of the small intestine to the anus, divided structurally into the cecum, colon, rectum, and anal canal.

Laryngitis (la-rin-JI-tis)   Inflammation of the mucous membrane lining the larynx.

Laryngopharynx (la-rin′-gō-FAR-inks)   The inferior portion of the pharynx, extending downward from the level of the hyoid bone that divides posteriorly into the esophagus and anteriorly into the larynx. Also called the **hypopharynx.**

Laryngoscope (la-RIN-gō-skōp) An instrument for examining the larynx.

Laryngotracheal (la-rin'-gō-TRA-kē-al) **bud** An outgrowth of endoderm of the foregut from which the respiratory system develops.

Larynx (LAR-inks) The voice box, a short passageway that connects the pharynx with the trachea.

Lateral (LAT-er-al) Farther from the midline of the body or a structure.

Lateral ventricle (VEN-tri-kul) A cavity within a cerebral hemisphere that communicates with the lateral ventricle in the other cerebral hemisphere and with the third ventricle by way of the interventricular foramen.

Learning The ability to acquire knowledge or a skill through instruction or experience.

Leg The part of the lower limb between the knee and the ankle.

Lens A transparent organ constructed of proteins (crystallins) lying posterior to the pupil and iris of the eyeball and anterior to the vitreous body.

Lesion (LE-zhun) Any localized, abnormal change in tissue formation.

Lesser omentum (ō-MEN-tum) A fold of the peritoneum that extends from the liver to the lesser curvature of the stomach and the first part of the duodenum.

Lesser vestibular (ves-TIB-yoo-lar) **gland** One of the paired mucus-secreting glands with ducts that open on either side of the urethral orifice in the vestibule of the female.

Lethargy (LETH-ar-jē) A condition of drowsiness or indifference.

Leukemia (loo-KE-mē-a) A malignant disease of the blood-forming tissues characterized by either uncontrolled production and accumulation of immature leukocytes in which many cells fail to reach maturity (acute) or an accumulation of mature leukocytes in the blood because they do not die at the end of their normal life span (chronic).

Leukocyte (LOO-kō-sīt) A white blood cell.

Leukocytosis (loo'-kō-si-TO-sis) An increase in the number of leukocytes (white blood cells), characteristic of many infections and other disorders.

Leukopenia (loo-kō-PE-nē-a) A decrease in the number of leukocytes (white blood cells) below 5000/mm³.

Leukoplakia (loo-kō-PLA-kē-a) A disorder in which there are white patches in the mucous membranes of the tongue, gums, and cheeks.

Leukotriene (loo'-kō-TRI-ēn) A type of eicosanoid produced by basophils and mast cells; acts as a local hormone; produces increased vascular permeability and acts as a chemotactic agent for phagocytes in tissue inflammation.

Leydig cell _See_ **Interstitial endocrinocyte**.

Libido (li-BE-dō) Sexual desire.

Ligament (LIG-a-ment) Dense, regular, connective tissue that attaches bone to bone.

Ligand (LI-gand) A chemical substance that binds to a specific receptor.

Limbic system A portion of the forebrain, sometimes termed the visceral brain, concerned with various aspects of emotion and behavior, that includes the limbic lobe, dentate gyrus, amygdaloid body, septal nuclei, mammillary bodies, anterior thalamic nucleus, olfactory bulbs, and bundles of myelinated axons.

Lingual frenulum (LIN-gwal FREN-yoo-lum) A fold of mucous membrane that connects the tongue to the floor of the mouth.

Lingual lipase (LI-pās) Digestive enzyme secreted by glands on the dorsum of the tongue that digests triglycerides.

Lipase An enzyme that splits fatty acids from triglycerides and phospholipids.

Lipid An organic compound composed of carbon, hydrogen, and oxygen that is usually insoluble in water, but soluble in alcohol, ether, and chloroform; examples include triglycerides (fats and oils), phospholipids, steroids, and eicosanoids.

Lipid profile Blood test that measures total cholesterol, high-density lipoprotein, low-density lipoprotein, and triglycerides, to assess risk for cardiovascular disease.

Lipogenesis (li-pō-GEN-e-sis) The synthesis of triglycerides from glucose or amino acids by hepatocytes.

Lipolysis (lip-OL-i-sis) The splitting of fatty acids from a triglyceride (fat) or phospholipid molecule.

Lipoma (li-PO-ma) A fatty tissue tumor, usually benign.

Lipoprotein (lip'-ō-PRO-tēn) One of several types of particles containing lipids (cholesterol and triglycerides) and proteins, which make it water-soluble for transport in the blood; high levels of **low-density lipoproteins (LDLs)** are associated with increased risk of atherosclerosis, while high levels of **high-density lipoproteins (HDLs)** are associated with decreased risk of atherosclerosis.

Lithotripsy (LITH-ō-trip'-sē) A noninvasive procedure in which mechanical shock waves generated by a lithotriptor are used to pulverize kidney stones or gallstones.

Liver Large gland under the diaphragm that occupies most of the right hypochondriac region and part of the epigastric region; functionally, it produces bile and synthesizes most plasma proteins; converts one nutrient into another; detoxifies substances; stores glycogen, minerals, and vitamins; carries on phagocytosis of worn-out blood cells and bacteria; and helps synthesize the active form of vitamin D.

Lobe (LOB) A curved or rounded projection.

Long-term potentiation (LTP) Prolonged, enhanced synaptic transmission that occurs at certain synapses within the hippocampus of the brain; believed to underlie some aspects of memory.

Lordosis (lor-DO-sis) An exaggeration of the lumbar curve of the vertebral column. Also called **swayback**.

Lou Gehrig's disease _See_ **Amyotrophic lateral sclerosis**.

Lower limb The appendage attached at the pelvic (hip) girdle, consisting of the thigh, knee, leg, ankle, foot, and toes. Also called **lower extremity**.

Lumbar (LUM-bar) Region of the back and side between the ribs and pelvis; loin.

Lumbar plexus (PLEK-sus) A network formed by the ventral branches of spinal nerves L1 through L4.

Lumen (LOO-men) The space within an artery, vein, intestine, renal tubule, or other tubelike structure.

Lungs Main organs of respiration, lying on either side of the heart in the thoracic cavity.

Lung scan A diagnostic test in which a radioactive substance is detected in the lungs by a scanning camera; used to evaluate for pulmonary embolism, pneumonia, or cancer.

Lunula (LOO-nyoo-la) The moon-shaped white area at the base of a nail.

Luteinizing (LOO-tē-in'-īz-ing) **hormone (LH)** A hormone secreted by the anterior pituitary gland that stimulates ovulation and progesterone secretion by the corpus luteum, and readies the mammary glands for milk secretion in females and stimulates testosterone secretion by the testes in males.

Lyme (LIM) **disease** Disease caused by the bacterium _Borrelia burgdorferi_ that is transmitted to humans by ticks (mainly deer ticks) and is often characterized by a bull's eye rash. Symptoms include joint stiffness, fever and chills, headache, stiff neck, nausea, and low back pain. Later stages may involve cardiac and neurological problems and arthritis.

Lymph (LIMF)   Fluid confined in lymphatic vessels and flowing through the lymphatic system until it is returned to the blood.

Lymphangiography (lim-fan′-jē-OG-ra-fē)   A procedure by which lymphatic vessels and lymph organs are filled with a radiopaque substance in order to be radiographed.

Lymphatic (lim-FAT-ik) **capillary**   Closed-ended microscopic lymphatic vessel that begins in spaces between cells and converges with other lymphatic capillaries to form lymphatic vessels.

Lymphatic tissue   A specialized form of reticular tissue that contains large numbers of lymphocytes.

Lymphatic vessel   A large vessel that collects lymph from lymphatic capillaries and converges with other lymphatic vessels to form the thoracic and right lymphatic ducts.

Lymph node   An oval or bean-shaped structure located along lymphatic vessels.

Lymphocyte (LIM-fō-sit)   A type of white blood cell, found in lymph nodes, associated with the immune system.

Lymphokines (LIM-fō-kins)   Powerful proteins secreted by T cells that endow T cells with their ability to assist in immunity.

Lymphoma (lim′-FŌ-ma)   Any tumor composed of lymphatic tissue.

Lysosome (LĪ-sō-sōm)   An organelle in the cytoplasm of a cell, enclosed by a single membrane and containing powerful digestive enzymes.

Lysozyme (LĪ-sō-zim)   A bactericidal enzyme found in tears, saliva, and perspiration.

Macrophage (MAK-rō-fāj)   Phagocytic cell derived from a monocyte. May be fixed or wandering.

Macula (MAK-yoo-la)   A discolored spot or a colored area. A small, thickened region on the wall of the utricle and saccule that contains receptors for static equilibrium.

Macula lutea (LOO-tē-a)   The yellow spot in the center of the retina.

Magnetic resonance imaging (MRI)   A diagnostic procedure that focuses on the nuclei of atoms of a single element in a tissue, usually hydrogen, to determine if they behave normally in the presence of an external magnetic force; used to indicate the biochemical activity of a tissue. Formerly called **nuclear magnetic resonance (NMR)**.

Major histocompatibility (MHC) antigens   Surface proteins on white blood cells and other nucleated cells that are unique for each person (except for identical siblings) and are used to type tissues and help prevent rejection. Also known as **human leukocyte associated (HLA) antigens**.

Malaise (ma-LĀYZ)   Discomfort, uneasiness, and indisposition, often indicative of infection.

Malignant (ma-LIG-nant)   Referring to diseases that tend to become worse and cause death; especially the invasion and spreading of cancer.

Malignant melanoma (mel′-a-NŌ-ma)   A melanin-containing, usually dark, highly malignant tumor of the skin.

Malnutrition (mal′-noo-TRISH-un)   State of bad or poor nutrition that may be due to inadequate food intake, imbalance of nutrients, malabsorption of nutrients, improper distribution of nutrients, increased nutrient requirements, increased nutrient losses, or overnutrition.

MALT   *See* **Mucosa-associated lymphatic tissue**.

Mammary (MAM-ar-ē) **gland**   Modified sudoriferous (sweat) gland of the female that produces milk for the nourishment of the young.

Mammillary (MAM-i-ler-ē) **bodies**   Two small rounded bodies posterior to the tuber cinereum that are involved in reflexes related to the sense of smell.

Mammography (mam-OG-ra-fē)   Procedure using x-rays for imaging the breasts to evaluate for breast disease or screen for breast cancer.

Marfan (MAR-fan) **syndrome**   Inherited defect of the fibrillin-gene that results in abnormal elastic fibers in connective tissues.

Marrow (MAR-ō)   Soft, spongelike material in the cavities of bone. Red bone marrow produces blood cells; yellow bone marrow, formed mainly of adipose tissue, has no blood-producing function.

Mass number   The total number of protons and neutrons in an atom of a chemical element.

Mast cell   A cell found in areolar connective tissue along blood vessels that produces histamine, a dilator of small blood vessels during inflammation.

Mastectomy (mas-TEK-tō-mē)   Surgical removal of breast tissue.

Mastication (mas′-ti-KĀ-shun)   Chewing.

Matrix (MĀ-triks)   The ground substance and fibers between cells in a connective tissue.

Matter   Anything that occupies space and has mass.

Mature follicle   A relatively large, fluid-filled follicle containing a secondary oocyte and surrounding granulosa cells that secrete estrogens. Also called a **Graafian** (GRAF-ē-an) **follicle**.

Maximal oxygen uptake   Maximum rate of oxygen consumption during aerobic catabolism of pyruvic acid that is determined by age, sex, and body size.

Mean arterial blood pressure (MABP)   The average force of blood pressure exerted against the walls of arteries; approximately equal to diastolic pressure plus one-third of pulse pressure, for example, 93 mm Hg when systolic pressure is 120 mm Hg and diastolic pressure is 80 mm Hg.

Meatus (mē-Ā-tus)   A passage or opening, especially the external portion of a canal.

Mechanoreceptor (me-KAN-ō-rē-sep-tor)   Receptor that detects mechanical deformation of the receptor itself or adjacent cells; stimuli so detected include those related to touch, pressure, vibration, proprioception, hearing, equilibrium, and blood pressure.

Medial (MĒ-dē-al)   Nearer the midline of the body or a structure.

Medial lemniscus (lem-NIS-kus)   A band of myelinated nerve fibers extending from the medulla oblongata to the thalamus on the same side. Sensory neurons in this tract transmit impulses for the sensations of proprioception, discriminative touch, hearing, equilibrium, and vibration.

Median aperture (AP-er-choor)   One of the three openings in the roof of the fourth ventricle through which cerebrospinal fluid enters the subarachnoid space of the brain and cord. Also called the **foramen of Magendie**.

Median plane   A vertical plane dividing the body into right and left halves. Situated in the middle.

Mediastinum (mē′-dē-as-TĪ-num)   A broad, median partition, actually a mass of tissue, found between the pleurae of the lungs, that extends from the sternum to the vertebral column.

Medulla (me-DULL-la)   An inner layer of an organ, such as the medulla of the kidneys.

Medulla oblongata (ob′-long-GA-ta) The most inferior part of the brain stem.

Medullary (MED-yoo-lar′-ē) **cavity** The space within the diaphysis of a bone that contains yellow bone marrow. Also called the **marrow cavity**.

Medullary rhythmicity (rith-MIS-i-tē) **area** Portion of the respiratory center in the medulla oblongata that controls the basic rhythm of respiration.

Meibomian gland *See* **Tarsal gland.**

Meiosis (mē-O-sis) A type of cell division that occurs during production of gametes, involving two successive nuclear divisions that result in daughter cells with the haploid (*n*) number of chromosomes.

Meissner's corpuscle *See* **Corpuscle of touch.**

Melanin (MEL-a-nin) A dark black, brown, or yellow pigment found in some parts of the body such as the skin and hair.

Melanoblast (MEL-a-nō-blast) Precursor cell in the epidermis that gives rise to melanocytes, cells that produce melanin.

Melanocyte (MEL-a-nō-sit′) A pigmented cell located between or beneath cells of the deepest layer of the epidermis that synthesizes melanin.

Melanocyte-stimulating hormone (MSH) A hormone secreted by the anterior pituitary gland that stimulates the dispersion of melanin granules in melanocytes in amphibians; continued administration produces darkening of skin in humans.

Melatonin (mel-a-TON-in) A hormone secreted by the pineal gland that helps set the timing of the body's biological clock.

Membrane A thin, flexible sheet of tissue composed of an epithelial layer and an underlying connective tissue layer, as in an epithelial membrane, or of areolar connective tissue only, as in a synovial membrane.

Membranous labyrinth (mem-BRA-nus LAB-i-rinth) The portion of the labyrinth of the inner ear that is located inside the bony labyrinth and separated from it by the perilymph; made up of the membranous semicircular canals, the saccule and utricle, and the cochlear duct.

Memory The ability to recall thoughts; commonly classified as short-term and long-term.

Menarche (me-NAR-kē) The first menses (menstrual flow) and beginning of ovarian and uterine cycles.

Ménière's (men-YAIRZ) **syndrome** A disease characterized by fluctuating loss of hearing, vertigo, and tinnitus due to an increased amount of endolymph that enlarges the labyrinth.

Meninges (me-NIN-jēz) Three membranes covering the brain and spinal cord, called the dura mater, arachnoid, and pia mater. *Singular is* **meninx** (MEN-inks).

Meningitis (men-in-JI-tis) Inflammation of the meninges, most commonly the pia mater and arachnoid.

Menopause (MEN-ō-pawz) The termination of the menstrual cycles.

Menstrual (MEN-stroo-al) **cycle** *See* **Female reproductive cycle.**

Menstruation (men′-stroo-A-shun) Periodic discharge of blood, tissue fluid, mucus, and epithelial cells that usually lasts for 5 days; caused by a sudden reduction in estrogens and progesterone. Also called the **menstrual phase** or **menses.**

Merkel (MER-kel) **cell** Type of cell in the epidermis of hairless skin that makes contact with a tactile (Merkel) disc, which functions in touch.

Merocrine (MER-ō-krin) **gland** Gland made up of secretory cells that remain intact throughout the process of formation and discharge of the secretory product, as in the salivary and pancreatic glands.

Mesenchyme (MEZ-en-kim) An embryonic connective tissue from which all other connective tissues arise.

Mesentery (MEZ-en-ter′-ē) A fold of peritoneum attaching the small intestine to the posterior abdominal wall.

Mesocolon (mez′-ō-KO-lon) A fold of peritoneum attaching the colon to the posterior abdominal wall.

Mesoderm The middle primary germ layer that gives rise to connective tissues, blood and blood vessels, and muscles.

Mesothelium (mez′-ō-THE-lē-um) The layer of simple squamous epithelium that lines serous membranes.

Mesovarium (mez′-ō-VAR-ē-um) A short fold of peritoneum that attaches an ovary to the broad ligament of the uterus.

Metabolism (me-TAB-ō-lizm) All the biochemical reactions that occur within an organism, including the synthetic (anabolic) reactions and decomposition (catabolic) reactions.

Metacarpus (met′-a-KAR-pus) A collective term for the five bones that make up the palm.

Metaphase (MET-a-phāz) The second stage of mitosis in which chromatid pairs line up on the metaphase plate (equatorial plane) of the cell.

Metaphysis (me-TAF-i-sis) Growing portion of a bone.

Metaplasia (met′-a-PLA-zē-a) The abnormal change of one type of adult, differentiated cell into another, for example, as occurs in bronchogenic carcinoma.

Metarteriole (met′-ar-TE-rē-ōl) A blood vessel that emerges from an arteriole, traverses a capillary network, and empties into a venule.

Metastasis (me-TAS-ta-sis) The spread of cancer to surrounding tissues (local) or to other body sites (distant).

Metatarsus (met′-a-TAR-sus) A collective term for the five bones located in the foot between the tarsals and the phalanges.

Micelle (mi-SEL) A spherical aggregate of bile salts that dissolves fatty acids and monoglycerides so that they can be absorbed into small intestinal epithelial cells.

Microcephalus (mi-krō-SEF-a-lus) An abnormally small head; premature closing of the anterior fontanel so that the brain has insufficient room for growth, resulting in mental retardation.

Microfilament (mi-krō-FIL-a-ment) Rodlike, protein filament about 6 nm in diameter; comprises contractile units in muscle fibers (cells) and provides support, shape, and movement in nonmuscle cells.

Microglia (mi-krō-GLE-a) Neuroglial cells that carry on phagocytosis. Also called **brain macrophages** (MAK-rō-fāj-ez).

Microphage (MIK-rō-fāj) Granular leukocyte that carries on phagocytosis, especially neutrophils and eosinophils.

Microtubule (mi-krō-TOOB-yool′) Cylindrical protein filament, ranging in diameter from 18 to 30 nm, consisting of the protein tubulin; provides support, structure, and transportation.

Microvilli (mi′-krō-VIL-ē) Microscopic, fingerlike projections of the plasma membranes of cells that increase surface area for absorption, especially in the small intestine and proximal convoluted tubules of the kidneys.

Micturition (mik′-too-RISH-un) The act of expelling urine from the urinary bladder. Also called **urination** (yoo-ri-NA-shun).

Midbrain The part of the brain between the pons and the diencephalon. Also called the **mesencephalon** (mes′-en-SEF-a-lon).

Middle ear A small, epithelial-lined cavity hollowed out of the temporal bone, separated from the external ear by the eardrum and from the internal ear by a thin bony partition containing the oval and round windows; extending across the middle ear are the three auditory ossicles. Also called the **tympanic** (tim-PAN-ik) **cavity.**

Midline An imaginary vertical line that divides the body into equal left and right sides.

Midsagittal plane A vertical plane through the midline of the body that divides the body or organs into *equal* right and left sides. Also called a **median plane**.

Milk ejection reflex Contraction of alveolar cells to force milk into ducts of mammary glands, stimulated by oxytocin (OT), which is released from the posterior pituitary gland in response to suckling action. Also called the **milk let-down reflex**.

Mineral Inorganic, homogeneous solid substance that may perform a function vital to life; examples include calcium and phosphorus.

Mineralocorticoids (min′-er-al-ō-KOR-ti-koyds) A group of hormones of the adrenal cortex that help regulate sodium and potassium balance.

Minimal volume The volume of air in the lungs even after the thoracic cavity has been opened forcing out some of the residual volume.

Minute volume of respiration (MVR) Total volume of air taken into the lungs per minute; about 6000/ml.

Mitochondrion (mī′-tō-KON-drē-on) A double-membraned organelle that plays a central role in the production of ATP; known as the "powerhouse" of the cell.

Mitosis (mi-TŌ-sis) The orderly division of the nucleus of a cell that ensures that each new daughter nucleus has the same number and kind of chromosomes as the original parent nucleus. The process includes the replication of chromosomes and the distribution of the two sets of chromosomes into two separate and equal nuclei.

Mitotic spindle Collective term for a football-shaped assembly of microtubules (nonkinetochore, kinetochore, and aster) that is responsible for the movement of chromosomes during cell division.

Mitral (MĪ-tral) **insufficiency** Backflow of blood from the left ventricle into the left atrium due to a damaged mitral valve or ruptured chordae tendineae.

Mitral stenosis (ste-NŌ-sis) Narrowing of the mitral valve by scar formation or a congenital defect.

Mitral valve prolapse (PRŌ-laps) **(MVP)** An inherited disorder in which a portion of a mitral valve is pushed back too far (prolapsed) during ejection due to expansion of the cusps and elongation of the chordae tendineae.

Mittelschmerz (MIT-el-shmerz) Abdominopelvic pain that supposedly indicates the release of a secondary oocyte from the ovary.

Modality (mō-DAL-i-tē) Any of the specific sensory entities, such as vision, smell, taste, or touch.

Modiolus (mō-DI-ō′-lus) The central pillar or column of the cochlea.

Mole The weight, in grams, of the combined atomic weights of the atoms that comprise a molecule of a substance.

Molecule (MOL-e-kyool) The chemical combination of two or more atoms covalently bonded together.

Monoclonal antibody (MAb) Antibody produced by *in vitro* clones of B cells hybridized with cancerous cells.

Monocyte (MON-ō-sit′) A type of white blood cell characterized by agranular cytoplasm; the largest of the leukocytes.

Monounsaturated fat A fatty acid that contains one double covalent bond between its carbon atoms; it is not completely saturated with hydrogen atoms. Plentiful in triglycerides of olive and peanut oils.

Mons pubis (MONZ PYOO-bis) The rounded, fatty prominence over the pubic symphysis, covered by coarse pubic hair.

Morbid (MOR-bid) Diseased; pertaining to disease.

Morning sickness Sensation of nausea during pregnancy that is thought to be due to high levels of the hormone human chorionic gonadotropin (hCG). Also called **emesis gravidarum**.

Morula (MOR-yoo-la) A solid sphere of cells produced by successive cleavages of a fertilized ovum about four days after fertilization.

Motor area The region of the cerebral cortex that governs muscular movement, particularly the precentral gyrus of the frontal lobe.

Motor end plate Portion of the sarcolemma of a muscle fiber (cell) that receives neurotransmitter liberated by an axon terminal.

Motor neuron (NOO-ron) A neuron that conducts nerve impulses from the brain and spinal cord to effectors that may be either muscles or glands. Also called an **efferent neuron**.

Motor unit A motor neuron together with the muscle fibers (cells) it stimulates.

Mucin (MYOO-sin) A protein found in mucus.

Mucosa-associated lymphatic tissue Lymphatic nodules scattered throughout the lamina propria (connective tissue) of mucous membranes lining the gastrointestinal tract, respiratory airways, urinary tract, and reproductive tract.

Mucous (MYOO-kus) **cell** A unicellular gland that secretes mucus. Two types are mucous neck and mucous surface cells in the stomach.

Mucous membrane A membrane that lines a body cavity that opens to the exterior. Also called the **mucosa** (myoo-KŌ-sa).

Mucus The thick fluid secretion of goblet cells, mucous cells, mucous glands, and mucous membranes.

Multiple motor unit summation Type of summation in which stimuli occur at the same time but at different locations (different motor units).

Multiple sclerosis (skler-Ō-sis) Progressive destruction of myelin sheaths of neurons in the central nervous system, short-circuiting conduction pathways.

Mumps Inflammation and enlargement of the parotid glands accompanied by fever and extreme pain during swallowing.

Muscarinic (mus′-ka-RIN-ik) **receptor** Receptor found on all effectors innervated by parasympathetic postganglionic axons and some effectors innervated by sympathetic postganglionic axons; so named because the actions of acetylcholine (ACh) on such receptors are similar to those produced by muscarine.

Muscle An organ composed of one of three types of muscle tissue (skeletal, cardiac, or smooth), specialized for contraction to produce voluntary or involuntary movement of parts of the body.

Muscle action potential A stimulating impulse that propagates along a sarcolemma and then into transverse tubules; in skeletal muscle, it is generated by acetylcholine, which alters permeability of the sarcolemma to cations, especially sodium ions (Na^+).

Muscle fatigue (fa-TĒG) Inability of a muscle to maintain its strength of contraction or tension; may be related to insufficient oxygen, depletion of glycogen, and/or lactic acid buildup.

Muscle spindle An encapsulated proprioceptor in a skeletal muscle, consisting of specialized intrafusal muscle fibers and nerve endings; stimulated by changes in length or tension of muscle fibers.

Muscle tissue A tissue specialized to produce motion in response to muscle action potentials by its qualities of contractility, extensibility, elasticity, and excitability. Types include skeletal, cardiac, and smooth.

Muscle tone A sustained, partial contraction of portions of a skeletal or smooth muscle in response to activation of stretch receptors or a baseline level of action potentials in the innervating motor neurons.

Muscular dystrophies (DIS-trō-fēz′) Inherited muscle-destroying diseases, characterized by degeneration of the individual muscle fibers (cells), which leads to progressive atrophy of the skeletal muscle.

Muscularis (MUS-kyoo-la′-ris) A muscular layer (coat or tunic) of an organ.

Muscularis mucosae (myoo-KŌ-sē) A thin layer of smooth muscle fibers (cells) located in the most superficial layer of the mucosa of the gastrointestinal tract, underlying the lamina propria of the mucosa.

Mutation (myoo-TĀ-shun) Any change in the sequence of bases in the DNA molecule resulting in a permanent alteration in some inheritable characteristic.

Myasthenia (mi-as-THĒ-nē-a) **gravis** Weakness of skeletal muscles caused by antibodies that block acetylcholine receptors and thus diminish muscle contractions.

Myelin (MĪ-e-lin) **sheath** Multilayered lipid and protein covering, formed by neurolemmocytes (Schwann cells) and oligodendrocytes around axons of many peripheral and central nervous system neurons.

Myelography (mi-e-LOG-ra-fē) Introduction of a contrast medium into the subarachnoid space of the spinal cord to demonstrate tumors or herniated (slipped) discs within or near the spinal cord.

Myenteric plexus A network of nerve fibers from both autonomic divisions located in the muscularis coat of the small intestine. Also called the **plexus of Auerbach** (OW-er-bak).

Myocardial infarction (mi′-ō-KAR-dē-al in-FARK-shun) **(MI)** Gross necrosis of myocardial tissue due to interrupted blood supply. Also called a **heart attack**.

Myocardium (mi′-ō-KAR-dē-um) The middle layer of the heart wall, made up of cardiac muscle tissue, comprising the bulk of the heart, and lying between the epicardium and the endocardium.

Myofibril (mi′-ō-FĪ-bril) A threadlike structure, running longitudinally through a muscle fiber (cell) consisting mainly of thick filaments (myosin) and thin filaments (actin, troponin, and tropomyosin).

Myoglobin (mi-ō-GLŌ-bin) The oxygen-binding, iron-containing conjugated protein complex present in the sarcoplasm of muscle fibers; contributes the red color to muscle.

Myogram (MĪ-ō-gram) The record or tracing produced by a myograph, an apparatus that measures and records the effects of muscular contractions.

Myology (mi-OL-ō-jē) The study of the muscles.

Myometrium (mi′-ō-MĒ-trē-um) The smooth muscle layer of the uterus.

Myopathy (mi-OP-a-thē) Any abnormal condition or disease of muscle tissue.

Myopia (mi-Ō-pē-a) Defect in vision in which objects can be seen distinctly only when very close to the eyes; nearsightedness.

Myosin (MĪ-ō-sin) The contractile protein that makes up the thick filaments of muscle fibers (cells).

Myotome (MĪ-ō-tōm) A group of muscles innervated by the motor neurons of a single spinal segment; in embryos, that portion of a somite that developes into some skeletal muscle.

Myotonia (mi-ō-TŌ-nē-a) A continuous spasm of muscle; increased muscular irritability and tendency to contract, and less ability to relax.

Myxedema (mix-e-DĒ-ma) Condition caused by hypothyroidism during the adult years characterized by swelling of facial tissues.

Nail A hard plate, composed largely of keratin, that develops from the epidermis of the skin to form a protective covering on the dorsal surface of the distal phalanges of the fingers and toes.

Nail matrix (MĀ-triks) The part of the nail beneath the body and root from which the nail is produced.

Nasal (NĀ-zal) **cavity** A mucosa-lined cavity on either side of the nasal septum that opens onto the face at external nares and into the nasopharynx at internal nares.

Nasal septum (SEP-tum) A vertical partition composed of bone (perpendicular plate of ethmoid and vomer) and cartilage, covered with a mucous membrane, separating the nasal cavity into left and right sides.

Nasolacrimal (nā′-zō-LAK-ri-mal) **duct** A canal that transports the lacrimal secretion (tears) from the nasolacrimal sac into the nose.

Nasopharynx (nā′-zō-FAR-inks) The superior portion of the pharynx, lying posterior to the nose and extending inferiorly to the soft palate.

Nausea (NAW-sē-a) Discomfort characterized by loss of appetite and sensation of impending vomiting.

Nebulization (neb′-yoo-li-ZĀ-shun) Administration of medication to selected portions of the respiratory tract by droplets suspended in air.

Neck The part of the body connecting the head and the trunk. A constricted portion of an organ such as the neck of the femur or uterus.

Necrosis (ne-KRŌ-sis) A pathological type of cell death that results from disease, injury, or lack of blood supply in which many adjacent cells swell, burst, and spill their contents into the interstitial fluid, triggering an inflammatory response.

Negative feedback The principle governing most control systems; a mechanism of response in which a stimulus initiates actions that reverse or reduce the stimulus.

Neonatal (nē′-ō-NĀ-tal) Pertaining to the first 4 weeks after birth.

Neoplasm (NE-ō-plazm) A new growth that may be benign or malignant.

Nephritis (ne-FRĪT-is) Inflammation of the kidney.

Nephron (NEF-ron) The functional unit of the kidney.

Nephrotic (ne-FROT-ik) **syndrome** A condition in which the endothelial–capsular membrane leaks, allowing large amounts of protein to escape into urine.

Nerve A cordlike bundle of nerve fibers (axons and/or dendrites) and its associated connective tissue coursing together outside the central nervous system.

Nerve fiber General term for any process (axon or dendrite) projecting from the cell body of a neuron.

Nerve impulse A wave of depolarization and repolarization that self-propagates along the plasma membrane of a neuron; also called a **nerve action potential**.

Nervous tissue Tissue that initiates and transmits nerve impulses to coordinate homeostasis.

Net filtration pressure (NFP) Net pressure that promotes fluid outflow at the arterial end of a capillary and fluid inflow at the venous end of a capillary; net pressure that promotes glomerular filtration in the kidneys.

Neuralgia (noo-RAL-jē-a) Attacks of pain along the entire course or branch of a peripheral nerve.

Neural plate A thickening of ectoderm that forms early in the third week of development and represents the beginning of the development of the nervous system.

Neuritis (noo-RĪ-tis)   Inflammation of a single nerve, two or more nerves in separate areas, or many nerves simultaneously.

Neuroeffector (noo-rō-e-FEK-tor) **junction**   Collective term for neuromuscular and neuroglandular junctions.

Neurofibral (noo-rō-FĪ-bral) **node**   A space, along a myelinated nerve fiber, between the individual neurolemmocytes (Schwann cells) that form the myelin sheath and the neurolemma. Also called **node of Ranvier** (ron-vē-Ā).

Neurofibril (noo-rō-FĪ-bril)   One of the delicate threads that forms a complicated network in the cytoplasm of the cell body and processes of a neuron.

Neuroglandular (noo-rō-GLAND-yoo-lar) **junction**   Area of contact between a motor neuron and a gland.

Neuroglia (noo-RŌG-lē-a)   Cells of the nervous system that perform various supportive functions. The neuroglia of the central nervous system are the astrocytes, oligodendrocytes, microglia, and ependymal cells; neuroglia of the peripheral nervous system include the neurolemmocytes (Schwann cells) and the satellite cells. Also called **glial** (GLĒ-al) **cells**.

Neurohypophyseal (noo´-rō-hī´-pō-FIZ-ē-al) **bud**   An outgrowth of ectoderm located on the floor of the hypothalamus that gives rise to the posterior pituitary gland.

Neurolemma (noo-rō-LEM-ma)   The peripheral, nucleated cytoplasmic layer of the neurolemmocyte (Schwann cell). Also called **sheath of Schwann** (SCHVON).

Neurolemmocyte   A neuroglial cell of the peripheral nervous system that forms the myelin sheath and neurolemma of a nerve fiber by wrapping around a nerve fiber in a jelly-roll fashion. Also called a **Schwann** (SCHVON) **cell**.

Neurology (noo-ROL-ō-jē)   The branch of science that deals with the normal functioning and disorders of the nervous system.

Neuromuscular (noo-rō-MUS-kyoo-lar) **junction**   The area of contact between the axon terminal of a motor neuron and a portion of the sarcolemma of a muscle fiber (cell). Also called a **myoneural** (mī-ō-NOO-ral) **junction**.

Neuron (NOO-ron)   A nerve cell, consisting of a cell body, dendrites, and an axon.

Neuropeptide (noo-rō-PEP-tīd)   Chain of 3 to about 40 amino acids that occurs naturally in the nervous system, and that acts primarily to modulate the response of or to a neurotransmitter. Examples are enkephalins and endorphins.

Neurosecretory (noo-rō-SĒC-re-tō-rē) **cell**   A cell in a nucleus (paraventricular and supraoptic) in the hypothalamus that produces oxytocin (OT) or antidiuretic hormone (ADH), hormones released in the posterior pituitary gland.

Neurosyphilis (noo-rō-SIF-i-lis)   A form of the tertiary stage of syphilis in which various types of nervous tissue are attacked by bacteria and degenerate.

Neurotransmitter   One of a variety of molecules within axon terminals that are released into the synaptic cleft in response to a nerve impulse and affect the membrane potential of the postsynaptic neuron. Also called a **transmitter substance**.

Neutrophil (NOO-trō-fil)   A type of white blood cell characterized by granules that stain pale lilac with a combination of acidic and basic dyes.

Nicotinic (nik´-ō-TIN-ik) **receptor**   Receptor found on both sympathetic and parasympathetic postganglionic neurons, so named because the actions of acetylcholine (ACh) on such receptors are similar to those produced by nicotine.

Night blindness   Poor or no vision in dim light or at night, although good vision is present during bright illumination; often

caused by a deficiency of vitamin A. Also referred to as **nyctalopia** (nik´-ta-LŌ-pē-a).

Nipple   A pigmented, wrinkled projection on the surface of the breast that is the location of the openings of the lactiferous ducts for milk release.

Nissl bodies   *See* **Chromatophilic substance**.

Nociceptor (nō´-sē-SEP-tor)   A free (naked) nerve ending that detects painful stimuli.

Node of Ranvier   *See* **Neurofibral node**.

Nondisjunction (non´-dis-JUNGK-shun)   Failure of sister chromatids to separate properly during anaphase of mitosis (or equatorial division of meiosis) or failure of homologous chromosomes to separate properly during reduction division of meiosis in which chromatids or chromosomes pass into the same daughter cell; the result is too many copies of that chromosome in the daughter cell or gamete.

Nonessential amino acid   An amino acid that can be synthesized by body cells through transamination, the transfer of an amino group from an amino acid to another substance.

Norepinephrine (nor´-ep-ē-NEF-rin) **(NE)**   A hormone secreted by the adrenal medulla that produces actions similar to those that result from sympathetic stimulation. Also called **noradrenaline** (nor-a-DREN-a-lin).

Notochord (NŌ-tō-cord)   A flexible rod of embryonic tissue that lies where the future vertebral column will develop.

Nuclear medicine   The branch of medicine concerned with the use of radioisotopes in the diagnosis of disease and therapy.

Nuclease (NOO-klē-ās)   An enzyme that breaks nucleotides into pentoses and nitrogenous bases; examples are ribonuclease and deoxyribonuclease.

Nucleic (noo-KLĒ-ic) **acid**   An organic compound that is a long polymer of nucleotides, with each nucleotide containing a pentose sugar, a phosphate group, and one of four possible nitrogenous bases (adenine, cytosine, guanine, and thymine or uracil).

Nucleolus (noo-KLĒ-ō-lus)   Nonmembranous spherical body within the nucleus composed of protein, DNA, and RNA that functions in the synthesis and storage of ribosomal RNA.

Nucleosome (NOO-klē-ō-sōm)   Elementary structural subunit of a chromosome consisting of histones and DNA.

Nucleus (NOO-klē-us)   A spherical or oval organelle of a cell that contains the hereditary factors of the cell, called genes. A cluster of unmyelinated nerve cell bodies in the central nervous system. The central portion of an atom made up of protons and neutrons.

Nucleus cuneatus (kyoo-nē-Ā-tus)   A group of nerve cells in the inferior portion of the medulla oblongata in which fibers of the fasciculus cuneatus terminate.

Nucleus gracilis (gras-I-lis)   A group of nerve cells in the inferior portion of the medulla oblongata in which fibers of the fasciculus gracilis terminate.

Nucleus pulposus (pul-PŌ-sus)   A soft, pulpy, highly elastic substance in the center of an intervertebral disc, a remnant of the notochord.

Nutrient   A chemical substance in food that provides energy, forms new body components, or assists in the functioning of various body processes.

Nystagmus (nis-TAG-mus)   Rapid, involuntary, rhythmic movement of the eyeballs; horizontal, rotary, or vertical.

Obesity (ō-BĒS-i-tē)   Body weight more than 20% above a desirable standard due to excessive accumulation of fat.

Oblique (ō-BLĒK) **plane**   A plane that passes through the body or an organ at an angle between the transverse plane and either the midsagittal, parasagittal, or frontal plane.

Obstetrics (ob-STET-riks)   The specialized branch of medicine that deals with pregnancy, labor, and the period of time immediately following delivery (about 42 days).

Occlusion (ō-KLOO-zhun)   The act of closure or state of being closed.

Occult (o-KULT)   Obscured or hidden from view, as, for example, occult blood in stools or urine.

Olfactory (ōl-FAK-tō-rē)   Pertaining to smell.

Olfactory bulb   A mass of gray matter containing cell bodies of neurons that form synapses with neurons of the olfactory (I) nerve, lying inferior to the frontal lobe of the cerebrum on either side of the crista galli of the ethmoid bone.

Olfactory receptor   A bipolar neuron with its cell body lying between supporting cells located in the mucous membrane lining the superior portion of each nasal cavity; transduces odors into neural signals.

Olfactory tract   A bundle of axons that extends from the olfactory bulb posteriorly to the olfactory portion of the cerebral cortex.

Oligodendrocyte (o-lig-ō-DEN-drō-sīt)   A neuroglial cell that supports neurons and produces a myelin sheath around axons of neurons of the central nervous system.

Oligospermia (ol′-i-gō-SPER-mē-a)   A deficiency of sperm cells in the semen.

Oliguria (ol′-i-GYOO-rē-a)   Daily urinary output usually less than 250 ml.

Olive   A prominent oval mass on each lateral surface of the superior part of the medulla oblongata.

Oncogene (ONG-kō-jēn)   Gene that has the ability to transform a normal cell into a cancerous cell when it is inappropriately activated.

Oncology (ong-KOL-ō-jē)   The study of tumors.

Oogenesis (ō′-ō-JEN-e-sis)   Formation and development of the ovum.

Oophorectomy (ō′-of-ō-REK-tō-mē)   The surgical removal of the ovaries.

Ophthalmic (of-THAL-mik)   Pertaining to the eye.

Ophthalmologist (of′-thal-MOL-ō-jist)   A physician who specializes in the diagnosis and treatment of eye disorders with drugs, surgery, and corrective lenses.

Ophthalmology (of′-thal-MOL-ō-jē)   The study of the structure, function, and diseases of the eye.

Ophthalmoscopy (of′-thal-MOS-co-pē)   Examination of the interior fundus of the eyeball to detect retinal changes associated with hypertension, diabetes mellitus, atherosclerosis, and increased intracranial pressure.

Opsin (OP-sin)   The glycoprotein portion of a photopigment.

Opsonization (op-sō-ni-ZA-shun)   The action of some antibodies that renders bacteria and other foreign cells more susceptible to phagocytosis. Also called **immune adherence**.

Optic (OP-tik)   Refers to the eye, vision, or properties of light.

Optic chiasm (KĪ-azm)   A crossing point of the optic (II) nerves, anterior to the pituitary gland. Also called **optic chiasma**.

Optic disc   A small area of the retina containing openings through which the axons of the ganglion cells emerge as the optic (II) nerve. Also called the **blind spot**.

Optician (op-TISH-an)   A technician who fits, adjusts, and dispenses corrective lenses on prescription of an ophthalmologist or optometrist.

Optic tract   A bundle of axons that transmits nerve impulses from the retina of the eye between the optic chiasm and the thalamus.

Optometrist (op-TOM-e-trist)   Specialist with a doctorate degree in optometry who is licensed to examine and test the eyes and treat visual defects by prescribing corrective lenses.

Oral contraceptive (OC)   A hormone compound, usually a high concentration of progesterone and a low concentration of estrogens, that is swallowed and prevents ovulation, and thus pregnancy. Also called "**the pill.**"

Ora serrata (Ō-ra ser-RA-ta)   The irregular margin of the retina lying internal and slightly posterior to the junction of the choroid and ciliary body.

Orbit (OR-bit)   The bony, pyramidal-shaped cavity of the skull that holds the eyeball.

Organ   A structure composed of two or more different kinds of tissues with a specific function and usually a recognizable shape.

Organelle (or-gan-EL)   A permanent structure within a cell with characteristic morphology that is specialized to serve a specific function in cellular activities.

Organic (or-GAN-ik) **compound**   Compound that always contains carbon in which the atoms are held together by covalent bonds. Examples include carbohydrates, lipids, proteins, and nucleic acids (DNA and RNA).

Organism (OR-ga-nizm)   A total living form; one individual.

Orgasm (OR-gazm)   Sensory and motor events involved in ejaculation for the male and involuntary contraction of the perineal muscles in the female at the climax of sexual intercourse.

Orifice (OR-i-fis)   Any aperture or opening.

Origin (OR-i-jin)   The attachment of a muscle tendon to a stationary bone or the end opposite the insertion.

Oropharynx (or′-ō-FAR-inks)   The intermediate portion of the pharynx, lying posterior to the mouth and extending from the soft palate to the hyoid bone.

Orthopedics (or′-thō-PĒ-diks)   The branch of medicine that deals with the preservation and restoration of the skeletal system, articulations, and associated structures.

Orthopnea (or′-THOP-nē-a)   Dyspnea that occurs in the horizontal position.

Osmoreceptor (oz′-mō-re-CEP-tor)   Receptor in the hypothalamus that is sensitive to changes in blood osmotic pressure and, in response to high osmotic pressure (low water concentration), causes synthesis and release of antidiuretic hormone (ADH).

Osmosis (os-MŌ-sis)   The net movement of water molecules through a selectively permeable membrane from an area of higher water concentration to an area of lower water concentration until an equilibrium is reached.

Osmotic pressure   The pressure required to prevent the movement of pure water into a solution containing solutes when the solutions are separated by a selectively permeable membrane.

Osseous (OS-ē-us)   Bony.

Ossicle (OS-si-kul)   Small bone, as in the middle ear (malleus, incus, stapes).

Ossification (os′-i-fi-KA-shun)   Formation of bone. Also called **osteogenesis**.

Osteoblast (OS-tē-ō-blast)   Cell formed from an osteoprogenitor cell that participates in bone formation by secreting some organic components and inorganic salts.

Osteoclast (OS-tē-ō-clast′)   A large multinuclear cell that destroys or resorbs bone tissue.

Osteocyte (OS-tē-ō-sīt′)   A mature bone cell that maintains the daily activities of bone tissue.

Osteogenic (os′-tē-ō-JEN-ik) **layer**   The inner layer of the periosteum that contains cells responsible for forming new bone during growth and repair.

Osteology (os′-tē-OL-ō-jē)   The study of bones.

Osteomalacia (os′-tē-ō-ma-LĀ-shē-a)   A deficiency of vitamin D in adults causing demineralization and softening of bone.

Osteomyelitis (os′-tē-ō-mī-i-LĪ-tis)   Inflammation of bone marrow or of the bone and its marrow.

Osteon (OS-tē-on)   The basic unit of structure in adult compact bone, consisting of a central (Haversian) canal with its concentrically arranged lamellae, lacunae, osteocytes, and canaliculi. Also called a **Haversian** (ha-VER-shun) **system**.

Osteoporosis (os′-tē-ō-pō-RŌ-sis)   Age-related disorder characterized by decreased bone mass and increased susceptibility to fractures, often as a result of decreased levels of estrogens.

Osteoprogenitor (os′-tē-ō-prō-JEN-i-tor) **cell**   Stem cell derived from mesenchyme that has mitotic potential and the ability to differentiate into an osteoblast.

Otalgia (ō-TAL-jē-a)   Pain in the ear; earache.

Otic (Ō-tik)   Pertaining to the ear.

Otitis media (ō-ti-tus MĒ-dē-a)   Acute infection of the middle ear characterized by pain, malaise, fever, and an inflamed tympanic membrane, subject to rupture.

Otolith (Ō-tō-lith)   A particle of calcium carbonate embedded in the otolithic membrane that functions in maintaining static equilibrium.

Otolithic (ō-tō-LITH-ik) **membrane**   Thick, gelatinous, glycoprotein layer located directly over hair cells of the macula in the saccule and utricle of the inner ear.

Otorhinolaryngology (ō′-tō-rī-nō-lar′-in-GOL-ō-jē)   The branch of medicine that deals with the diagnosis and treatment of diseases of the ears, nose, and throat.

Oval window   A small, membrane-covered opening between the middle ear and inner ear into which the footplate of the stapes fits. Also called the **fenestra vestibuli** (fe-NES-tra ves-TIB-yoo-lē).

Ovarian (ō-VAR-ē-an) **cycle**   A monthly series of events in the ovary associated with the maturation of an ovum.

Ovarian follicle (FOL-i-kul)   A general name for oocytes (immature ova) in any stage of development, along with their surrounding epithelial cells.

Ovarian ligament (LIG-a-ment)   A rounded cord of connective tissue that attaches the ovary to the uterus.

Ovary (Ō-var-ē)   Female gonad that produces ova and the hormones estrogens, progesterone, and relaxin.

Ovulation (ō-vyoo-LĀ-shun)   The rupture of a mature ovarian (Graafian) follicle with discharge of a secondary oocyte into the pelvic cavity.

Ovum (Ō-vum)   The female reproductive or germ cell; an egg cell.

Oxidation (ok-si-DĀ-shun)   The removal of electrons from a molecule or, less commonly, the addition of oxygen to a molecule that results in a decrease in the energy content of the molecule. The oxidation of glucose in the body is called **cellular respiration**.

Oxygen debt   *See* **Recovery oxygen consumption**.

Oxyhemoglobin (ok′-sē-HĒ-mō-glō-bin) (**Hb·O₂**)   Hemoglobin combined with oxygen.

Oxyphil cell   A cell found in the parathyroid gland that secretes parathyroid hormone (PTH).

Oxytocin (ok′-sē-TŌ-sin) (**OT**)   A hormone secreted by neurosecretory cells in the paraventricular and supraoptic nuclei of the hypothalamus that stimulates contraction of the smooth muscle fibers in the pregnant uterus and contractile cells around the ducts of mammary glands.

Pacinian corpuscle   *See* **Lamellated corpuscle**.

Paget's (PAJ-ets) **disease**   A disorder characterized by a greatly accelerated remodeling process in which osteoclastic resorption is massive and new bone formation by osteoblasts is extensive. As a result, there is an irregular thickening and softening of the bones.

Palate (PAL-at)   The horizontal structure separating the oral and the nasal cavities; the roof of the mouth.

Palliative (PAL-ē-a-tiv)   Serving to relieve or alleviate without curing.

Palpate (PAL-pāt)   To examine by touch; to feel.

Palpitation (pal′-pi-TĀ-shun)   A fluttering of the heart or abnormal rate or rhythm of the heart.

Pancreas (PAN-krē-as)   A soft, oblong organ lying along the greater curvature of the stomach and connected by a duct to the duodenum. It is both exocrine (secreting pancreatic juice) and endocrine (secreting insulin, glucagon, somatostatin, and pancreatic polypeptide).

Pancreatic (pan′-krē-AT-ik) **duct**   A single, large tube that unites with the common bile duct from the liver and gallbladder and drains pancreatic juice into the duodenum at the hepatopancreatic ampulla (ampulla of Vater). Also called the **duct of Wirsung**.

Pancreatic islet   A cluster of endocrine gland cells in the pancreas that secretes insulin, glucagon, somatostatin, and pancreatic polypeptide. Also called an **islet of Langerhans** (LANG-er-hanz).

Pancreatic polypeptide   Hormone secreted by the F cells of pancreatic islets (islets of Langerhans) that regulates release of pancreatic digestive enzymes.

Papanicolaou (pap′-a-NIK-ō-la-oo) **test**   A cytological staining test for the detection and diagnosis of premalignant and malignant conditions of the female genital tract. Cells scraped from the genital epithelium are smeared, fixed, stained, and examined microscopically. Also called a **Pap smear**.

Papilla (pa-PIL-a)   A small nipple-shaped projection or elevation.

Paracrine (pa-RA-krin)   Local hormone, such as histamine, that acts on neighboring cells without entering the bloodstream.

Paralysis (pa-RAL-a-sis)   Loss or impairment of motor function due to a lesion of nervous or muscular origin.

Paranasal sinus (par′-a-NĀ-zal SĪ-nus)   A mucus-lined air cavity in a skull bone that communicates with the nasal cavity. Paranasal sinuses are located in the frontal, maxillary, ethmoid, and sphenoid bones.

Paraplegia (par-a-PLĒ-jē-a)   Paralysis of both lower limbs.

Parasagittal plane   A vertical plane that does not pass through the midline and that divides the body or organs into *unequal* left and right portions.

Parasympathetic (par′-a-sim-pa-THET-ik) **division**   One of the two subdivisions of the autonomic nervous system, having cell bodies of preganglionic neurons in nuclei in the brain stem and in the lateral gray matter of the sacral portion of the spinal cord; primarily concerned with activities that conserve and restore body energy. Also called the **craniosacral** (krā-nē-ō-SĀ-kral) **division**.

Parathyroid (par′-a-THĪ-royd) **gland**   One of four (usually) small endocrine glands embedded in the posterior surfaces of the lateral lobes of the thyroid gland.

Parathyroid hormone (PTH)   A hormone secreted by the parathyroid glands that increases blood calcium level and decreases blood phosphate level.

Paraurethral (par′-a-yoo-RĒ-thral) **gland**   Gland embedded in the wall of the urethra whose duct opens on either side of the urethral orifice and secretes mucus. Also called **Skene's** (SKĒNZ) **gland**.

Parenchyma (par-EN-ki-ma) The functional parts of any organ, as opposed to tissue that forms its stroma or framework.

Parenteral (par-EN-ter-al) Situated or occurring outside the intestines; referring to introduction of substances into the body other than by way of the intestines such as intradermal, subcutaneous, intramuscular, intravenous, or intraspinal.

Parietal (pa-RĪ-e-tal) Pertaining to or forming the outer wall of a body cavity.

Parietal cell A type of secretory cell in gastric glands that produces hydrochloric acid and intrinsic factor. Also called an **oxyntic cell**.

Parietal pleura (PLOO-ra) The outer layer of the serous pleural membrane that encloses and protects the lungs; the layer that is attached to the wall of the pleural cavity.

Parkinson's disease Progressive degeneration of the basal ganglia and substantia nigra of the cerebrum resulting in decreased production of dopamine (DA) that leads to tremor, slowing of voluntary movements, and muscle weakness.

Parotid (pa-ROT-id) **gland** One of the paired salivary glands located inferior and anterior to the ears connected to the oral cavity via a duct (Stensen's) that opens into the inside of the cheek opposite the maxillary (upper) second molar tooth.

Paroxysm (PAR-ok-sizm) A sudden periodic attack or recurrence of symptoms of a disease.

Parturition (par´-too-RISH-un) Act of giving birth to young; childbirth, delivery.

Patellar (pa-TELL-ar) **reflex** Extension of the leg by contraction of the quadriceps femoris muscle in response to tapping the patellar ligament. Also called the **knee jerk reflex**.

Patent ductus arteriosus Congenital anatomical heart defect in which the fetal connection between the aorta and pulmonary trunk remains open instead of closing completely after birth.

Pathogen (PATH-ō-jen) A disease-producing organism.

Pathogenesis (path´-ō-JEN-e-sis) The development of disease or a morbid or pathological state.

Pathological (path´-ō-LOJ-i-kal) Pertaining to or caused by disease.

Pathological (path´-ō-LOJ-i-kal) **anatomy** The study of structural changes caused by disease.

Pectinate (PEK-ti-nāt) **muscles** Projecting muscle bundles of the anterior atrial walls and the lining of the auricles.

Pectoral (PEK-tō-ral) Pertaining to the chest or breast.

Pediatrician (pē´dē-a-TRISH-un) A physician who specializes in the care and treatment of children.

Pedicel (PED-i-sel) Footlike structure, as on podocytes of a glomerulus.

Pelvic (PEL-vik) **cavity** Inferior portion of the abdominopelvic cavity that contains the urinary bladder, sigmoid colon, rectum, and internal female and male reproductive structures.

Pelvic inflammatory disease (PID) Collective term for any extensive bacterial infection of the pelvic organs, especially the uterus, uterine (Fallopian) tubes, and ovaries.

Pelvic splanchnic (PEL-vik SPLANGK-nik) **nerves** Preganglionic parasympathetic fibers from the levels of S2, S3, and S4 that supply the urinary bladder, reproductive organs, and the descending and sigmoid colon and rectum.

Pelvimetry (pel-VIM-e-trē) Measurement of the size of the inlet and outlet of the birth canal.

Pelvis The basinlike structure formed by the two hipbones, the sacrum, and the coccyx. The expanded, proximal portion of the

ureter, lying within the kidney and into which the major calyces open.

Penis (PĒ-nis) The male copulatory organ, used to introduce semen into the female vagina.

Pepsin Protein-digesting enzyme secreted by chief (zymogenic) cells of the stomach as the inactive form pepsinogen, which is converted to active pepsin by hydrochloric acid.

Peptic ulcer An ulcer that develops in areas of the gastrointestinal tract exposed to hydrochloric acid; classified as a gastric ulcer if in the lesser curvature of the stomach and as a duodenal ulcer if in the first part of the duodenum.

Percussion (per-KUSH-un) The act of striking (percussing) an underlying part of the body with short, sharp blows as an aid in diagnosing the part by the quality of the sound produced.

Perforating canal A minute passageway by means of which blood vessels and nerves from the periosteum penetrate into compact bone. Also called **Volkmann's** (FŌLK-manz) **canal**.

Pericardial (per´-i-KAR-dē-al) **cavity** Small potential space between the visceral and parietal layers of the serous pericardium that contains pericardial fluid.

Pericarditis (per´-i-KAR-dī-tis) Inflammation of the pericardium.

Pericardium (per´-i-KAR-dē-um) A loose-fitting membrane that encloses the heart, consisting of a superficial fibrous layer and a deep serous layer.

Perichondrium (per´-i-KON-drē-um) The membrane that covers cartilage.

Perilymph (PER-i-lymf) The fluid contained between the bony and membranous labyrinths of the inner ear.

Perimetrium (per-i-MĒ-trē-um) The serosa of the uterus.

Perimysium (per´-i-MĪZ-ē-um) Invagination of the epimysium that divides muscles into bundles.

Perineum (per´-i-NĒ-um) The pelvic floor; the space between the anus and the scrotum in the male and between the anus and the vulva in the female.

Perineurium (per´-i-NYOO-rē-um) Connective tissue wrapping around fascicles in a nerve.

Periodontal (per-ē-ō-DON-tal) **disease** A collective term for conditions characterized by degeneration of gingivae, alveolar bone, periodontal ligament, and cementum.

Periodontal ligament The periosteum lining the alveoli (sockets) for the teeth in the alveolar processes of the mandible and maxillae.

Periosteum (per´-ē-OS-tē-um) The membrane that covers bone and consists of connective tissue, osteoprogenitor cells, and osteoblasts and is essential for bone growth, repair, and nutrition.

Peripheral (pe-RIF-er-al) Located on the outer part or a surface of the body.

Peripheral nervous system (PNS) The part of the nervous system that lies outside the central nervous system—nerves and ganglia.

Periphery (pe-RIF-er-ē) Outer part or a surface of the body; part away from the center.

Peristalsis (per´-i-STAL-sis) Successive muscular contractions along the wall of a hollow muscular structure.

Peritoneum (per´-i-tō-NĒ-um) The largest serous membrane of the body that lines the abdominal cavity and covers the viscera.

Peritonitis (per´-i-tō-NĪ-tis) Inflammation of the peritoneum.

Permissive (per-MIS-sive) **effect** A hormonal interaction in which the effect of one hormone on a target cell requires previous or simultaneous exposure to another hormone(s) to enhance the response of a target cell or increase the activity of another hormone.

Pernicious (per-NISH-us)   Tending to cause death.

Peroxisome (pe-ROKS-i-sōm)   Organelle similar in structure to a lysosome that contains enzymes that use molecular oxygen to oxidize various organic compounds. Such reactions produce hydrogen peroxide; abundant in liver cells.

Perspiration   Sweat; produced by sudoriferous (sweat) glands and containing water, salts, urea, uric acid, amino acids, ammonia, sugar, lactic acid, and ascorbic acid; helps maintain body temperature and eliminate wastes.

Peyer's patches   *See* **Aggregated lymphatic follicles**.

pH   A symbol of the measure of the concentration of hydrogen ions (H^+) in a solution. The pH scale extends from 0 to 14, with a value of 7 expressing neutrality, values lower than 7 expressing increasing acidity, and values higher than 7 expressing increasing alkalinity.

Phagocytosis (fag'-ō-si-TŌ-sis)   The process by which phagocytes ingest particulate matter; especially the ingestion and destruction of microbes, cell debris, and other foreign matter.

Phalanx (FĀ-lanks)   The bone of a finger or toe. *Plural is* **phalanges** (fa-LAN-jēz).

Phantom pain   A sensation of pain as originating in a limb that has been amputated.

Pharmacology (far'-ma-KOL-ō-jē)   The science that deals with the effects and uses of drugs in the treatment of disease.

Pharynx (FAR-inks)   The throat; a tube that starts at the internal nares and runs partway down the neck where it opens into the esophagus posteriorly and the larynx anteriorly.

Phenotype (FĒ-nō-tip)   The observable expression of genotype; physical characteristics of an organism determined by genetic makeup and influenced by interaction between genes and internal and external environmental factors.

Phenylketonuria (fen'-il-kē'-tō-NOO-rē-a) **(PKU)**   A disorder characterized by an elevation of the amino acid phenylalanine in the blood.

Pheochromocytoma (fē-ō-krō'-mō-si-TŌ-ma)   Tumor of the chromaffin cells of the adrenal medulla that results in hypersecretion of epinephrine and norepinephrine.

Phlebitis (fle-BĪ-tis)   Inflammation of a vein, usually in a lower limb.

Phospholipid (fos'-fō-LIP-id) **bilayer**   Arrangement of phospholipid molecules in two parallel sheets in which the hydrophilic "heads" face outward and the hydrophobic "tails" face inward.

Phosphorylation (fos'-for-i-LĀ-shun)   The addition of a phosphate group to a chemical compound; types include substrate-level, oxidative, and photophosphorylation.

Photopigment   A substance that can absorb light and undergo structural changes that can lead to the development of a receptor potential. An example is rhodopsin. Also called **visual pigment**.

Photoreceptor   Receptor that detects light shining on the retina of the eye.

Physiology (fiz'-ē-OL-ō-jē)   Science that deals with the functions of an organism or its parts.

Pia mater (PĪ-a MĀ-ter)   The deep membrane (meninx) covering the brain and spinal cord.

Pineal (PĪN-ē-al) **gland**   The cone-shaped gland located in the roof of the third ventricle. Also called the **epiphysis cerebri** (ē-PIF-i-sis se-RĒ-brē).

Pinealocyte (pin-ē-AL-ō-sīt)   Secretory cell of the pineal gland that releases melatonin.

Pinna (PIN-na)   The projecting part of the external ear composed of elastic cartilage and covered by skin and shaped like the flared end of a trumpet. Also called the **auricle** (OR-i-kul).

Pinocytosis (pi'-nō-si-TŌ-sis)   The process by which cells ingest liquid.

Pituicyte (pi-TOO-i-sit)   Supporting cell of the posterior pituitary gland.

Pituitary (pi-TOO-i-tar'-ē) **dwarfism**   Condition caused by hyposecretion of human growth hormone (hGH) during the growth years and characterized by childlike physical traits in an adult.

Pituitary gland   A small endocrine gland lying in the sella turcica of the sphenoid bone and attached to the hypothalamus by the infundibulum. Also called the **hypophysis** (hi-POF-i-sis).

Pivot joint   A synovial joint in which a rounded, pointed, or conical surface of one bone articulates with a ring formed partly by another bone and partly by a ligament, as in the joint between the atlas and axis and between the proximal ends of the radius and ulna. Also called a **trochoid** (TRŌ-koid) **joint**.

Placenta (pla-SEN-ta)   The special structure through which the exchange of materials between fetal and maternal circulations occurs. Also called the **afterbirth**.

Plantar flexion (PLAN-tar FLEK-shun)   Bending the foot in the direction of the plantar surface (sole).

Plaque (PLAK)   A cholesterol-containing mass in the tunica media of arteries. A mass of bacterial cells, dextran (polysaccharide), and other debris that adheres to teeth.

Plasma (PLAZ-ma)   The extracellular fluid found in blood vessels; blood minus the formed elements.

Plasma cell   Cell that produces antibodies and develops from a B cell (lymphocyte).

Plasma (cell) membrane   Outer, limiting membrane that separates the cell's internal parts from extracellular fluid or the external environment.

Plasmapheresis (plaz'-ma-fe-RĒ-sis)   A procedure in which blood is withdrawn from the body, its components are selectively separated, the undesirable component causing disease is removed, and the remainder is returned to the body. Among the substances removed are toxins, metabolic substances, and antibodies. Also called **therapeutic plasma exchange (TPE)**.

Platelet (PLĀT-let)   A fragment of cytoplasm enclosed in a cell membrane and lacking a nucleus; found in the circulating blood; plays a role in blood clotting. Also called a **thrombocyte** (THROM-bō-sīt).

Platelet plug   Aggregation of thrombocytes at a damaged blood vessel to prevent blood loss.

Pleura (PLOOR-a)   The serous membrane that covers the lungs and lines the walls of the chest and the diaphragm.

Pleural cavity   Small potential space between the visceral and parietal pleurae.

Plexus (PLEK-sus)   A network of nerves, veins, or lymphatic vessels.

Plexus of Auerbach   *See* **Myenteric plexus**.

Plexus of Meissner   *See* **Submucosal plexus**.

Pneumonia (noo-MŌ-nē-a)   Acute infection or inflammation of the alveoli of the lungs.

Pneumotaxic (noo-mō-TAK-sik) **area**   Portion of the respiratory center in the pons that continually sends inhibitory nerve impulses to the inspiratory area that limit inspiration and facilitate expiration.

Podiatry (pō-DĪ-a-trē)   The diagnosis and treatment of foot disorders.

Polar body   The smaller cell resulting from the unequal division of cytoplasm during the meiotic divisions of an oocyte. The polar body has no function and degenerates.

Polarized A condition in which opposite effects or states exist at the same time. In electrical contexts, having one portion negative and another positive; for example, a polarized nerve cell membrane has the outer surface positively charged and the inner surface negatively charged.

Poliomyelitis (pō'-lē-ō-mi-e-LĪ-tis) Viral infection marked by fever, headache, stiff neck and back, deep muscle pain and weakness, and loss of certain somatic reflexes; a serious form of the disease, **bulbar polio**, results in destruction of motor neurons in anterior horns of spinal nerves that leads to paralysis.

Polycythemia (pol'-ē-si-THĒ-mē-a) Disorder characterized by a hematocrit above the normal level of 55 in which hypertension, thrombosis, and hemorrhage occur.

Polyp (POL-ip) A tumor on a stem found especially on a mucous membrane.

Polysaccharide (pol'-ē-SAK-a-rīd) A carbohydrate in which three or more monosaccharides are joined chemically.

Polyunsaturated fat A fatty acid that contains more than one double covalent bond between its carbon atoms; abundant in triglycerides of corn oil, safflower oil, and cottonseed oil.

Polyuria (pol'-ē-YOO-rē-a) An excessive production of urine.

Pons (PONZ) The portion of the brain stem that forms a "bridge" between the medulla oblongata and the midbrain, anterior to the cerebellum.

Positive feedback A feedback mechanism in which the response enhances the original stimulus.

Positron emission tomography (PET) A type of radioactive scanning based on the release of gamma rays when positrons collide with negatively charged electrons in body tissues; it indicates where radioisotopes are used in the body.

Postabsorptive (fasting) state Metabolic state during which absorption is complete and energy needs of the body must be satisfied.

Postcentral gyrus *See* **Primary somatosensory area**.

Posterior (pos-TĒR-ē-or) Nearer to or at the back of the body. Also called **dorsal**.

Posterior column–medial lemniscus pathways Sensory pathways that carry information related to proprioception, discriminitive touch, two-point discrimination, pressure, and vibration. First-order neurons project from the spinal cord to the ipsilateral medulla in the posterior columns (fasciculus gracilis and fasciculus cuneatus). Second-order neurons project from the medulla to the contralateral thalamus in the medial lemniscus. Third-order neurons project from the thalamus to the somatosensory cortex (postcentral gyrus) on the same side.

Posterior pituitary gland Posterior portion of the pituitary gland. Also called the **neurohypophysis** (noo-rō-hī-POF-i-sis).

Posterior root The structure composed of sensory fibers lying between a spinal nerve and the dorsolateral aspect of the spinal cord. Also called the **dorsal (sensory) root**.

Posterior root ganglion (GANG-glē-on) A group of cell bodies of sensory neurons and their supporting cells located along the posterior root of a spinal nerve. Also called a **dorsal (sensory) root ganglion**.

Postganglionic neuron (pōst'-gang-lē-ON-ik NOO-ron) The second visceral motor neuron in an autonomic pathway, having its cell body and dendrites located in an autonomic ganglion and its unmyelinated axon ending at cardiac muscle, smooth muscle, or a gland.

Postpartum (pōst-PAR-tum) After parturition; occurring after the delivery of a baby.

Postsynaptic (pōst-sin-AP-tik) **neuron** The nerve cell that is activated by the release of a neurotransmitter substance from another neuron and carries nerve impulses away from the synapse.

Pouch of Douglas *See* **Rectouterine pouch**.

Precapillary sphincter (SFINGK-ter) A ring of smooth muscle fibers (cells) at the site of origin of true capillaries that regulate blood flow into true capillaries.

Precentral gyrus *See* **Primary motor area**.

Preeclampsia (prē'-ē-KLAMP-sē-a) An abnormal condition of pregnancy characterized by sudden hypertension, large amounts of protein in urine, and generalized edema.

Preganglionic (prē'-gang-lē-ON-ik) **neuron** The first visceral motor neuron in an autonomic pathway, with its cell body and dendrites in the brain or spinal cord and its myelinated axon ending at an autonomic ganglion, where it synapses with a postganglionic neuron.

Pregnancy Sequence of events that normally includes fertilization, implantation, embryonic growth, and fetal growth that terminates in birth.

Premenstrual syndrome (PMS) Severe physical and emotional stress occurring late in the postovulatory phase of the menstrual cycle and sometimes overlapping with menstruation.

Prepuce (PRĒ-pyoos) The loose-fitting skin covering the glans of the penis and clitoris. Also called the **foreskin**.

Presbyopia (prez-bē-Ō-pē-a) A loss of elasticity of the lens of the eye due to advancing age with resulting inability to focus clearly on near objects.

Pressure sore Tissue destruction due to a constant deficiency of blood to tissues overlying a bony projection that has been subjected to prolonged pressure against an object such as a bed, cast, or splint. Also called **bedsore**, **decubitus** (dē-KYOO-bi-tus) **ulcer**, or **trophic ulcer**.

Presynaptic (prē-sin-AP-tik) **inhibition** Type of inhibition in which neurotransmitter released by an inhibitory neuron depresses the release of neurotransmitter by a presynaptic neuron.

Presynaptic (prē-sin-AP-tik) **neuron** A neuron that propagates nerve impulses toward a synapse.

Prevertebral ganglion (prē-VERT-e-bral GANG-lē-on) A cluster of cell bodies of postganglionic sympathetic neurons anterior to the spinal column and close to large abdominal arteries. Also called a **collateral ganglion**.

Primary germ layer One of three layers of embryonic tissue, called ectoderm, mesoderm, and endoderm, that give rise to all tissues and organs of the organism.

Primary motor area A region of the cerebral cortex in the precentral gyrus of the frontal lobe of the cerebrum that controls specific muscles or groups of muscles.

Primary somatosensory area A region of the cerebral cortex posterior to the central sulcus in the postcentral gyrus of the parietal lobe of the cerebrum that localizes exactly the points of the body where somatic sensations originate.

Prime mover The muscle directly responsible for producing the desired motion. Also called an **agonist** (AG-ō-nist).

Primigravida (pri-mi-GRAV-i-da) A woman pregnant for the first time.

Primitive gut Embryonic structure composed of endoderm and mesoderm that gives rise to most of the gastrointestinal tract.

Primordial (pri-MOR-dē-al) Existing first; especially primordial follicles in the ovary.

Principal cell Cell found in the parathyroid glands that secretes parathyroid hormone (PTH); cell type in the distal convoluted tubule and collecting duct of nephron that is stimulated by aldosterone and antidiuretic hormone.

Proctology (prok-TOL-ō-jē) The branch of medicine that treats the rectum and its disorders.

Progeny (PROJ-e-nē) Refers to offspring or descendants.

Progesterone (prō-JES-te-rō-n) **(PROG)** A female sex hormone produced by the ovaries that helps prepare the endometrium for implantation of a fertilized ovum and the mammary glands for milk secretion.

Prognosis (prog-NŌ-sis) A forecast of the probable results of a disorder; the outlook for recovery.

Prolactin (prō-LAK-tin) **(PRL)** A hormone secreted by the anterior pituitary gland that initiates and maintains milk secretion by the mammary glands.

Prolapse (PRŌ-laps) A dropping or falling down of an organ, especially the uterus or rectum.

Proliferation (pro-lif′-er-Ā-shun) Rapid and repeated reproduction of new parts, especially cells.

Pronation (prō-NĀ-shun) A movement of the forearm in which the palm of the hand is turned posteriorly or inferiorly.

Prophase (PRŌ-fāz) The first stage of mitosis during which chromatid pairs are formed and aggregate around the metaphase plate of the cell.

Proprioception (prō-prē-ō-SEP-shun) The receipt of information from muscles, tendons, and the labyrinth that enables the brain to determine movements and position of the body and its parts. Also called **kinesthesia** (kin′-es-THĒ-zē-a).

Proprioceptor (prō′-prē-ō-SEP-tor) A receptor located in muscles, tendons, or joints that provides information about body position and movements.

Prostaglandin (pros′-ta-GLAN-din) **(PG)** A membrane-associated lipid; released in small quantities and acts as a local hormone.

Prostatectomy (pros′-ta-TEK-tō-mē) The surgical removal of part of or the entire prostate gland.

Prostate (PROS-tāt) **gland** A doughnut-shaped gland inferior to the urinary bladder that surrounds the superior portion of the male urethra and secretes a slightly acidic solution that contributes to sperm motility and viability.

Prosthesis (pros-THĒ-sis) An artificial device to replace a missing body part.

Protein An organic compound consisting of carbon, hydrogen, oxygen, nitrogen, and sometimes sulfur and phosphorus, and made up of amino acids linked by peptide bonds.

Prothrombin (prō-THROM-bin) An inactive blood-clotting factor synthesized by the liver, released into the blood, and converted to active thrombin in the process of blood clotting by the activated enzyme prothrombinase.

Proto-oncogene (prō′-tō-ONG-kō-jēn) Gene responsible for some aspect of normal growth and development; it may transform into an oncogene, a gene capable of causing cancer.

Protraction (prō-TRAK-shun) The movement of the mandible or shoulder girdle forward on a plane parallel with the ground.

Proximal (PROK-si-mal) Nearer the attachment of a limb to the trunk; nearer to the point of origin or attachment.

Pruritus (proo′-RĪ-tus) Itching.

Pseudopods (SOO-dō-pods) Temporary, protruding projections of cytoplasm in a migrating cell.

Psoriasis (sō-RĪ-a-sis) Chronic skin disease characterized by reddish plaques or papules covered with scales.

Psychosomatic (si′-kō-sō-MAT-ik) Pertaining to the relation between mind and body. Commonly used to refer to those physio-

logical disorders thought to be caused entirely or partly by emotional disturbances.

Pterygopalatine ganglion (ter′-i-gō-PAL-a-tin GANG-glē-on) A cluster of cell bodies of parasympathetic postganglionic neurons ending at the lacrimal and nasal glands.

Ptosis (TŌ-sis) Drooping, as of the eyelid or the kidney (nephroptosis).

Puberty (PYOO-ber-tē) The time of life during which the secondary sex characteristics begin to appear and the capability for sexual reproduction is possible; usually between the ages of 10 and 17.

Pubic symphysis A slightly movable cartilaginous joint between the anterior surfaces of the hipbones.

Puerperium (pyoo′-er-PER-ē-um) The state immediately after childbirth, usually 4–6 weeks.

Pulmonary (PUL-mo-ner′-ē) Concerning or affected by the lungs.

Pulmonary circulation The flow of deoxygenated blood from the right ventricle to the lungs and the return of oxygenated blood from the lungs to the left atrium.

Pulmonary edema (e-DĒ-ma) An abnormal accumulation of interstitial fluid in the tissue spaces and alveoli of the lungs due to increased pulmonary capillary permeability or increased pulmonary capillary pressure.

Pulmonary embolism (EM-bō-lizm) **(PE)** The presence of a blood clot or other foreign substance in a pulmonary arterial blood vessel that obstructs circulation to lung tissue.

Pulmonary (PUL-mo-ner-ē) **function tests** Any number of tests (forced vital capacity, forced expiratory volume in one second, maximum midrespiratory flow, maximum voluntary ventilation) designed to evaluate lung disease and measure pulmonary impairment.

Pulmonary ventilation The inflow (inspiration) and outflow (expiration) of air between the atmosphere and the lungs. Also called **breathing.**

Pulp cavity A cavity within the crown and neck of a tooth, filled with pulp, a connective tissue containing blood vessels, nerves, and lymphatic vessels.

Pulse (PULS) The rhythmic expansion and elastic recoil of a systemic artery after each contraction of the left ventricle.

Pulse pressure The difference between the maximum (systolic) and minimum (diastolic) pressures; normally about 40 mm Hg.

Pupil The hole in the center of the iris, the area through which light enters the posterior cavity of the eyeball.

Purkinje fiber *See* **Conduction myofiber**.

Pus The liquid product of inflammation containing leukocytes or their remains and debris of dead cells.

P wave The deflection wave of an electrocardiogram that reflects atrial depolarization.

Pyelitis (pi′-e-LĪ-tis) Inflammation of the kidney pelvis and its calyces.

Pyelonephritis (pī′-e-lō-ne-FRĪ-tis) Inflammation of the nephrons and renal pelvis of one or both kidneys.

Pyemia (pi-Ē-mēa) Infection of the blood, with multiple abscesses, caused by pus-forming microorganisms.

Pyloric (pi-LOR-ik) **sphincter** A thickened ring of smooth muscle through which the pylorus of the stomach communicates with the duodenum. Also called the **pyloric valve**.

Pyogenesis (pi′-ō-JEN-e-sis) Formation of pus.

Pyorrhea (pi-ō-RĒ-a) A discharge or flow of pus, especially in the alveoli (sockets) and the tissues of the gums.

Pyramid (PIR-a-mid) A pointed or cone-shaped structure; one of two roughly triangular structures on the ventral side of the medulla oblongata composed of the largest motor tracts that run from the cerebral cortex to the spinal cord; a triangular-shaped structure in the renal medulla.

Pyramidal (pi-RAM-i-dal) **tracts (pathways)** *See* **Direct (motor) pathways**.

Pyuria (pī-YOO-rē-a) The presence of leukocytes and other components of pus in urine.

QRS wave The deflection waves of an electrocardiogram that represent onset of ventricular depolarization.

Quadrant (KWOD-rant) One of four parts.

Quadriplegia (kwod′-ri-PLĒ-jē-a) Paralysis of four limbs: two upper and two lower.

Radiographic (rā′-dē-ō-GRAF-ic) **anatomy** Diagnostic branch of anatomy that includes the use of x-rays.

Rales (RALS) Sounds sometimes heard in the lungs that resemble bubbling or rattling due to the presence of an abnormal amount or type of fluid or mucus inside the bronchi or alveoli, or to bronchoconstriction so that air cannot enter or leave the lungs normally.

Rami communicantes (RĀ-mē ko-myoo-ni-KAN-tēz) Branches of a spinal nerve. *Singular is* **ramus communicans** (RĀ-mus ko-MYOO-ni-kans).

Rapid eye movement (REM) sleep Stage of sleep in which dreaming occurs, lasting for 5–10 minutes several times during a sleep cycle; characterized by rapid movements of the eyes beneath the eyelids.

Rathke's pouch *See* **Hypophyseal pouch**.

Raynaud's (rā-NOZ) **disease** A vascular disorder, primarily of females, characterized by bilateral attacks of ischemia, usually of the fingers and toes, in which the skin becomes pale and exhibits burning and pain; it is brought on by cold temperatures or emotional stimuli.

Reactivity (rē-ak-TI-vi-tē) Ability of an antigen to react specifically with the antibody whose formation it induced.

Receptor A specialized cell or a distal portion of a neuron that responds to a specific sensory modality, such as touch, pressure, cold, light, or sound, and converts it to an electrical signal (generator or receptor potential). A specific molecule or cluster of molecules that recognizes and binds a particular ligand.

Receptor-mediated endocytosis A highly selective process whereby cells take up specific ligands, which usually are large molecules or particles, by enveloping them within a sac of plasma membrane. Ligands are eventually broken down by enzymes in lysosomes.

Receptor potential Depolarization or hyperpolarization of the plasma membrane of a receptor that alters release of neurotransmitter from the cell; if the neuron that synapses with the receptor cell becomes depolarized to threshold, a nerve impulse is triggered.

Recessive gene A gene that is not expressed in the presence of a dominant gene on the homologous chromosome.

Reciprocal innervation (re-SIP-rō-kal in-ner-VĀ-shun) The phenomenon by which action potentials stimulate contraction of one muscle and simultaneously inhibit contraction of antagonistic muscles.

Recombinant DNA Synthetic DNA, formed by joining a fragment of DNA from one source to a portion of DNA from another.

Recovery oxygen consumption Elevated oxygen use after exercise ends due to metabolic changes that starts during exercise and continues after exercise. Previously called **oxygen debt**.

Recruitment (rē-KROOT-ment) The process of increasing the number of active motor units. Also called **motor unit summation**.

Rectouterine pouch A pocket formed by the parietal peritoneum as it moves posteriorly from the surface of the uterus and is reflected onto the rectum; the most inferior point in the pelvic cavity. Also called the **pouch** or **cul de sac of Douglas**.

Rectum (REK-tum) The last 20 cm (7 in.) of the gastrointestinal tract, from the sigmoid colon to the anus.

Red nucleus A cluster of cell bodies in the midbrain, occupying a large portion of the tectum and sending fibers into the rubroreticular and rubrospinal tracts.

Red pulp That portion of the spleen that consists of venous sinuses filled with blood and thin plates of splenic tissue called splenic (Billroth's) cords.

Reduction The addition of electrons to a molecule or, less commonly, the removal of oxygen from a molecule that results in an increase in the energy content of the molecule.

Referred pain Pain that is felt at a site remote from the place of origin.

Reflex Fast response to a change (stimulus) in the internal or external environment that attempts to restore homeostasis; passes over a reflex arc.

Reflex arc The most basic conduction pathway through the nervous system, connecting a receptor and an effector and consisting of a receptor, a sensory neuron, an integrating center in the central nervous system, a motor neuron, and an effector.

Refraction (rē-FRAK-shun) The bending of light as it passes from one medium to another.

Refractory (re-FRAK-to-rē) **period** A time during which an excitable cell (neuron or muscle fiber) cannot respond to a stimulus that is usually adequate to evoke an action potential.

Regeneration (rē-jen′-er-Ā-shun) The natural renewal of a structure.

Regional anatomy The division of anatomy dealing with a specific region of the body, such as the head, neck, chest, or abdomen.

Regurgitation (rē-gur′-ji-TĀ-shun) Return of solids or fluids to the mouth from the stomach; flowing backward of blood through incompletely closed heart valves.

Relapse (RĒ-laps) The return of a disease weeks or months after its apparent cessation.

Relaxin (RLX) A female hormone produced by the ovaries that increases flexibility of the pubic symphysis and helps dilate the uterine cervix to ease delivery of a baby.

Releasing hormone Chemical secretion of the hypothalamus that can stimulate secretion of hormones of the anterior pituitary gland.

Remodeling Replacement of old bone by new bone tissue.

Renal (RĒ-nal) Pertaining to the kidney.

Renal corpuscle (KOR-pus′-el) A glomerular (Bowman's) capsule and its enclosed glomerulus.

Renal failure Inability of the kidneys to function properly, due to abrupt failure (acute) or progressive failure (chronic).

Renal pelvis A cavity in the center of the kidney formed by the expanded, proximal portion of the ureter, lying within the kidney, and into which the major calyces open.

Renal pyramid A triangular structure in the renal medulla containing the straight segments of renal tubules and the vasa recta.

Renin (RĒ-nin) An enzyme released by the kidney into the plasma where it converts angiotensinogen into angiotensin I.

Renin–angiotensin (an′-jē-ō-TEN-sin) **pathway** A mechanism for the control of aldosterone secretion by angiotensin II, initiated by the secretion of renin by the kidney in response to low blood pressure.

Repolarization (rē-pō′-lar-i-ZĀ-shun) Restoration of a resting membrane potential following depolarization.

Reproduction (rē′-prō-DUK-shun) Either the formation of new cells for growth, repair, or replacement, or the production of a new individual.

Reproductive cell division Type of cell division in which gametes (sperm and egg cells) are produced; consists of meiosis and cytokinesis.

Residual (re-ZID-yoo-al) **volume** The volume of air still contained in the lungs after a maximal expiration; about 1200 ml.

Resistance (re-ZIS-tans) Hindrance (impedance) to blood flow as a result of higher viscosity, longer total blood vessel length, and smaller blood vessel radius. Ability to ward off disease. The hindrance encountered by an electrical charge as it moves through a substance from one point to another. The hindrance encountered by air as it moves through the respiratory passageways.

Respiration (res-pi-RĀ-shun) Overall exchange of gases between the atmosphere, blood, and body cells consisting of pulmonary ventilation, external respiration, and internal respiration.

Respirator (RES-pi-rā′-tor) An apparatus fitted to a mask over the nose and mouth, or hooked directly to an endotracheal or tracheotomy tube, that is used to assist or support ventilation or to provide nebulized medication to the air passages under positive pressure.

Respiratory center Neurons in the reticular formation of the brain stem that regulate the rate and depth of respiration.

Respiratory distress syndrome (RDS) of the newborn A disease of newborn infants, especially premature ones, in which insufficient amounts of surfactant are produced and breathing is labored. Also called **hyaline** (HĪ-a-lin) **membrane disease (HMD)**.

Respiratory failure Condition in which the respiratory system cannot supply sufficient oxygen to maintain metabolism or eliminate enough carbon dioxide to prevent respiratory acidosis.

Resting membrane potential The voltage difference between the inside and outside of a cell membrane when the cell is not responding to a stimulus; in many neurons and muscle fibers it is −70 to −90 mV, with the inside of the cell negative relative to the outside.

Resuscitation (rē-sus′-i-TĀ-shun) Act of bringing a person back to full consciousness.

Retention (rē-TEN-shun) A failure to void urine due to obstruction, nervous contraction of the urethra, or absence of sensation of desire to urinate.

Rete (RĒ-tē) **testis** The network of ducts in the testes.

Reticular (re-TIK-yoo-lar) **activating system (RAS)** A portion of the reticular formation that has many ascending connections with the cerebral cortex; when this area of the brain stem is active, nerve impulses pass to the thalamus and widespread areas of the cerebral cortex, resulting in generalized alertness or arousal from sleep.

Reticular formation A network of small groups of neuron cell bodies scattered among bundles of axons (mixed gray and white matter) beginning in the medulla oblongata and extending superiorly through the central part of the brain stem.

Reticulocyte (re-TIK-yoo-lō-sīt) An immature red blood cell.

Reticulocyte count Examination of a stained sample of blood to determine the percentage of reticulocytes in the total number of red blood cells; used to evaluate rate of erythropoiesis and monitor treatment for anemia.

Reticulum (re-TIK-yoo-lum) A network.

Retina (RET-i-na) The deep coat of the posterior portion of the eyeball and consisting of nervous tissue (where the process of vision begins) and a pigmented layer of epithelial cells that contact the choroid. Also called the **nervous tunic** (TOO-nik).

Retinal (RE-ti-nal) A derivative of vitamin A that functions as the light-absorbing portion of the photopigment rhodopsin.

Retraction (rē-TRAK-shun) The movement of a protracted part of the body posteriorly on a plane parallel to the ground, as in pulling the lower jaw back in line with the upper jaw.

Retroflexion (re-trō-FLEK-shun) A malposition of the uterus in which it is tilted posteriorly.

Retrograde degeneration (RE-trō-grād dē-jen-er-Ā-shun) Changes that occur in the proximal portion of a damaged axon only as far as the first neurofibral node (node of Ranvier); similar to changes that occur during Wallerian degeneration.

Retroperitoneal (re′-trō-per-i-tō-NĒ-al) External to the peritoneal lining of the abdominal cavity.

Reye (RĪ) **syndrome** Disorder of unknown cause, primarily of children but occasionally of adults, that is characterized by vomiting, liver dysfunction, and brain dysfunction; may progress to coma and death; usually occurs after a viral infection, particularly chickenpox or influenza, and is linked to intake of salicylates (aspirin).

Rheumatism (ROO-ma-tizm′) Any painful state of the supporting structures of the body—bones, ligaments, joints, tendons, or muscles.

Rh factor An inherited antigen on the surface of erythrocytes (red blood cells).

Rhinology (rī-NOL-ō-jē) The study of the nose and its disorders.

Rhinoplasty (RĪ-nō-plas′-tē) Surgical procedure in which the structure of the nose is altered.

Rhodopsin (rō-DOP-sin) The photopigment in rods of the retina, consisting of a glycoprotein called opsin and a derivative of vitamin A called retinal.

Ribonucleic (rī-bō-nyoo-KLĒ-ik) **acid (RNA)** A single-stranded nucleic acid constructed of nucleotides, each consisting of a nitrogenous base (adenine, cytosine, guanine, or uracil), ribose, and a phosphate group; three types are messenger RNA (mRNA), transfer RNA (tRNA), and ribosomal RNA (rRNA), each of which has a specific role during protein synthesis.

Ribosome (RĪ-bō-sōm) An organelle in the cytoplasm of cells, composed of ribosomal RNA and ribosomal proteins, that synthesizes proteins; nicknamed the "protein factory."

Rickets (RIK-ets) Condition affecting children, characterized by soft and deformed bones resulting from inadequate calcium metabolism due to a vitamin D deficiency.

Right heart (atrial) reflex A reflex concerned with maintaining normal venous blood pressure.

Right lymphatic (lim-FAT-ik) **duct** A vessel of the lymphatic system that drains lymph from the upper right side of the body and empties it into the right subclavian vein.

Rigidity (ri-JID-i-tē) Hypertonia characterized by increased muscle tone, but reflexes are not affected.

Rigor mortis State of partial contraction of muscles following death due to lack of ATP; myosin heads (cross bridges) remain attached to actin, thus preventing relaxation.

Rod One of two types of photoreceptor in the retina of the eye; specialized for vision in dim light.

Roentgen (RENT-gen) The international unit of radiation; a standard quantity of x-rays or gamma radiation.

Root canal A narrow extension of the pulp cavity lying within the root of a tooth.

Root of penis Attached portion of penis that consists of the bulb and crura.

Rotation (rō-TĀ-shun) Moving a bone around its own axis, with no other movement.

Round ligament (LIG-a-ment) A band of fibrous connective tissue enclosed between the folds of the broad ligament of the uterus, emerging from the uterus just inferior to the uterine tube, extending laterally along the pelvic wall and through the deep inguinal ring to end in the labia majora.

Round window A small opening between the middle and inner ear, directly inferior to the oval window, covered by the secondary tympanic membrane.

Rugae (ROO-jē) Large folds in the mucosa of an empty hollow organ, such as the stomach and vagina.

Saccule (SAK-yool) The inferior and smaller of the two chambers in the membranous labyrinth inside the vestibule of the inner ear containing a receptor organ for static equilibrium.

Sacral plexus (PLEK-sus) A network formed by the ventral branches of spinal nerves L4 through S3.

Sacral promontory (PROM-on-tor′-ē) The superior surface of the body of the first sacral vertebra that projects anteriorly into the pelvic cavity; a line from the sacral promontory to the superior border of the pubic symphysis divides the abdominal and pelvic cavities.

Saddle joint A synovial joint in which the articular surface of one bone is saddle shaped and the articular surface of the other bone is shaped like the legs of the rider sitting in the saddle, as in the joint between the trapezium and the metacarpal of the thumb.

Sagittal (SAJ-i-tal) **plane** A vertical plane that divides the body or organs into left and right portions. Such a plane may be **midsagittal (median)**, in which the divisions are equal, or **parasagittal**, in which the divisions are unequal.

Saliva (sa-LĪ-va) A clear, alkaline, somewhat viscous secretion produced mostly by the three pairs of salivary glands; contains various salts, mucin, lysozyme, salivary amylase, and lingual lipase (produced by glands in the tongue).

Salivary amylase (SAL-i-ver-ē AM-i-lās) An enzyme in saliva that initiates the chemical breakdown of starch, mostly in the mouth.

Salivary gland One of three pairs of glands that lie external to the mouth and pour their secretory product (called saliva) into ducts that empty into the oral cavity; the parotid, submandibular, and sublingual glands.

Salt A substance that, when dissolved in water, ionizes into cations and anions, neither of which are hydrogen ions (H^+) or hydroxide ions (OH^-).

Saltatory (sal-ta-TŌ-rē) **conduction** The propagation of an action potential (nerve impulse) along the exposed portions of a myelinated nerve fiber. The action potential appears at successive neurofibral nodes (nodes of Ranvier) and therefore seems to jump or leap from node to node.

Sarcolemma (sar′-kō-LEM-ma) The cell membrane of a muscle fiber (cell), especially of a skeletal muscle fiber.

Sarcoma (sar-KŌ-ma) A connective tissue tumor, often highly malignant.

Sarcomere (SAR-kō-mēr) A contractile unit in a striated muscle fiber extending from one Z disc to the next Z disc.

Sarcoplasm (SAR-kō-plazm) The cytoplasm of a muscle fiber (cell).

Sarcoplasmic reticulum (sar′-kō-PLAZ-mik re-TIK-yoo-lum) A network of saccules and tubes surrounding myofibrils of a muscle fiber (cell), comparable to endoplasmic reticulum; functions to reabsorb calcium ions during relaxation and to release them to cause contraction.

Satiety (sa-TĪ-e-tē) Fullness or gratification, as of hunger or thirst.

Satiety center A collection of neurons located in the ventromedial nuclei of the hypothalamus that, when stimulated, bring about the cessation of eating.

Saturated fat A fatty acid that contains only single bonds (no double bonds) between its carbon atoms; all carbon atoms are bonded to the maximum number of hydrogen atoms; prevalent in triglycerides of animal products such as meat, milk, milk products, and eggs.

Scala tympani (SKA-la TIM-pan-ē) The inferior spiral-shaped channel of the bony cochlea, filled with perilymph.

Scala vestibuli (ves-TIB-yoo-lē) The superior spiral-shaped channel of the bony cochlea, filled with perilymph.

Schwann cell See **Neurolemmocyte.**

Sciatica (si-AT-i-ka) Inflammation and pain along the sciatic nerve; felt along the posterior aspect of the thigh extending down the inside of the leg.

Sclera (SKLE-ra) The white coat of fibrous tissue that forms the superficial protective covering over the eyeball except in the most anterior portion; the posterior portion of the fibrous tunic.

Scleral venous sinus A circular venous sinus located at the junction of the sclera and the cornea through which aqueous humor drains from the anterior chamber of the eyeball into the blood. Also called the **canal of Schlemm** (SHLEM).

Sclerosis (skle-RŌ-sis) A hardening with loss of elasticity of tissues.

Scoliosis (skō′-lē-Ō-sis) An abnormal lateral curvature from the normal vertical line of the backbone.

Scotoma (skō-TŌ-ma) An area of depressed or lost vision within the visual field.

Scrotum (SKRŌ-tum) A skin-covered pouch that contains the testes and their accessory structures.

Sebaceous (se-BĀ-shus) **gland** An exocrine gland in the dermis of the skin, almost always associated with a hair follicle, that secretes sebum. Also called an **oil gland.**

Sebum (SĒ-bum) Secretion of sebaceous (oil) glands.

Secondary response Accelerated, more intense cell-mediated or antibody-mediated immune response upon a subsequent exposure to an antigen after the primary response.

Secondary sex characteristic A characteristic of the male or female body that develops at puberty under the influence of sex hormones but is not directly involved in sexual reproduction; examples are distribution of body hair, voice pitch, body shape, and muscle development.

Secretion (se-KRĒ-shun) Production and release from a gland cell of a fluid, especially a functionally useful product as opposed to a waste product.

Selective permeability (per′-mē-a-BIL-i-tē) The property of a membrane by which it permits the passage of certain substances but restricts the passage of others.

Sella turcica (SEL-a TUR-si-ka) A depression on the superior surface of the sphenoid bone that houses the pituitary gland.

Semen (SĒ-men)   A fluid discharged at ejaculation by a male that consists of a mixture of sperm and the secretions of the seminiferous tubules, seminal vesicles, prostate gland, and bulbourethral (Cowper's) glands.

Semicircular canals   Three bony channels (anterior, posterior, lateral), filled with perilymph, in which lie the membranous semicircular canals filled with endolymph. They contain receptors for equilibrium.

Semicircular ducts   The membranous semicircular canals filled with endolymph and floating in the perilymph of the bony semicircular canals. They contain cristae that are concerned with dynamic equilibrium.

Semilunar (sem'-ē-LOO-nar) **valve**   A valve between the aorta or the pulmonary trunk and a ventricle of the heart.

Seminal vesicle (SEM-i-nal VES-i-kul)   One of a pair of convoluted, pouchlike structures, lying posterior and inferior to the urinary bladder and anterior to the rectum, that secrete a component of semen into the ejaculatory ducts.

Seminiferous tubule (sem'-i-NI-fer-us TOO-byool)   A tightly coiled duct, located in the testis, where sperm are produced.

Senescence (se-NES-ens)   The process of growing old; the period of old age.

Senile macular (MAK-yoo-lar) **degeneration (SMD)**   A disease in which blood vessels grow over the macula lutea.

Senility (se-NIL-i-tē)   A loss of mental or physical ability due to old age.

Sensation   A state of awareness of external or internal conditions of the body.

Sensory area   A region of the cerebral cortex concerned with the interpretation of sensory impulses.

Sensory neuron (NOO-ron)   A neuron that conducts nerve impulses into the central nervous system. Also called an **afferent neuron**.

Sepsis (SEP-sis)   A morbid condition that results from the presence in the blood or other body tissues of pathogenic bacteria and their products.

Septal defect   An opening in the septum (interatrial or interventricular) between the left and right sides of the heart.

Septicemia (sep'-ti-SĒ-mē-a)   Toxins or disease-causing bacteria in blood. Also called "**blood poisoning**."

Septum (SEP-tum)   A wall dividing two cavities.

Serosa (ser-Ō-sa)   Any serous membrane. The external layer of an organ formed by a serous membrane. The membrane that lines the pleural, pericardial, and peritoneal cavities.

Serous (SIR-us) **membrane**   A membrane that lines a body cavity that does not open to the exterior. Also called the **serosa** (se-RŌ-sa).

Serum   Blood plasma minus its clotting proteins.

Serum enzyme studies   Evaluation of levels of certain enzymes in blood (creatine phosphokinase, serum glutamic oxaloacetic transaminase, lactic dehydrogenase) to diagnose and monitor heart attacks, liver damage, or other tissue damage.

Sesamoid (SES-a-moyd) **bones**   Small bones usually found in tendons.

Sex chromosomes   The twenty-third pair of chromosomes, designated X and Y, which determine the genetic sex of an individual; in males, the pair is XY; in females, XX.

Sexual intercourse   The insertion of the erect penis of a male into the vagina of a female. Also called **coitus** (KŌ-i-tus) or **copulation**.

Sexually transmitted disease (STD)   General term for any of a large number of diseases spread by sexual contact. Also called a **venereal disease (VD)**.

Sheath of Schwann   *See* **Neurolemma**.

Shingles   Acute infection of the peripheral nervous system caused by reactivation of latent varicella zoster virus. Also termed **herpes zoster**.

Shinsplints   Soreness or pain along the tibia probably caused by inflammation of the periosteum brought on by repeated tugging of the muscles and tendons attached to the periosteum. Also called **tibia stress syndrome**.

Shivering   Involuntary contraction of skeletal muscles that generates heat. Also called **involuntary thermogenesis**.

Shock   Failure of the cardiovascular system to deliver adequate amounts of oxygen and nutrients to meet the metabolic needs of the body due to inadequate cardiac output. It is characterized by hypotension; clammy, cool, and pale skin; sweating; reduced urine formation; altered mental state; acidosis; tachycardia; weak, rapid pulse; and thirst. Types include hypovolemic, cardiogenic, obstructive, neurogenic, and septic.

Shoulder joint   A synovial joint where the humerus articulates with the scapula.

SIDS   *See* **Sudden infant death syndrome**.

Sigmoid colon (SIG-moyd KŌ-lon)   The S-shaped portion of the large intestine that begins at the level of the left iliac crest, projects medially, and terminates at the rectum at about the level of the third sacral vertebra.

Sigmoidoscopy (sig'-moy-DOS-kō-pē)   Visualization of the anal canal, rectum, and colon with a fiber optics instrument to screen for colorectal cancer, collect biopsy samples, remove polyps, or gather specimens for culture.

Sign   Any objective evidence of disease that can be observed or measured such as a lesion, swelling, or fever.

Simple diffusion (dif-YOO-zhun)   A passive process in which there is a net or greater movement of molecules or ions from a region of high concentration to a region of low concentration until equilibrium is reached.

Sinoatrial (si-nō-Ā-trē-al) **(SA) node**   A compact mass of cardiac muscle fibers (cells) specialized for conduction, located in the right atrium inferior to the opening of the superior vena cava. Also called the **pacemaker**.

Sinus (SĪ-nus)   A hollow in a bone (paranasal sinus) or other tissue; a channel for blood (vascular sinus); any cavity having a narrow opening.

Sinusitis (sin-yoo-SĪT-is)   Inflammation of the mucous membrane of a paranasal sinus.

Sinusoid (SĪN-yoo-soyd)   A microscopic space or passage for blood in certain organs such as the liver or spleen.

Skeletal muscle   An organ specialized for contraction, composed of striated muscle fibers (cells), supported by connective tissue, attached to a bone by a tendon or an aponeurosis, and stimulated by somatic motor neurons.

Skene's gland   *See* **Paraurethral gland**.

Skin   The external covering of the body that consists of a superficial, thinner epidermis (epithelial tissue) and a deep, thicker dermis (connective tissue) that is anchored to the subcutaneous layer.

Skin cancer   Any one of several types of malignant tumors that arise from epidermal cells. Types include basal cell carcinoma, squamous cell carcinoma, and malignant melanoma.

Skull The skeleton of the head consisting of the cranial and facial bones.

Sleep A state of partial unconsciousness from which a person can be aroused; associated with a low level of activity in the reticular activating system.

Sliding-filament mechanism The explanation of how thick and thin filaments slide relative to one another during striated muscle contraction to decrease sarcomere length.

Small intestine A long tube of the gastrointestinal tract that begins at the pyloric sphincter of the stomach, coils through the central and inferior part of the abdominal cavity, and ends at the large intestine; divided into three segments: duodenum, jejunum, and ileum.

Smooth muscle A tissue specialized for contraction, composed of smooth muscle fibers (cells), located in the walls of hollow internal organs, and innervated by autonomic motor neurons.

Snellen (SNEL-en) **test** Test used to evaluate any problem or changes in vision by measuring visual acuity (sharpness).

Sodium–potassium ATPase An active transport pump located in the cell membrane that transports sodium ions out of the cell and potassium ions into the cell at the expense of cellular ATP. It functions to keep the ionic concentrations of these elements at physiological levels. Also called the **sodium pump**.

Soft palate (PAL-at) The posterior portion of the roof of the mouth, extending from the palatine bones to the uvula. It is a muscular partition lined with mucous membrane.

Solution A homogeneous molecular or ionic dispersion of one or more substances (solutes) in a dissolving medium (solvent) that is usually liquid.

Somatic (sō-MAT-ik) **cell division** Type of cell division in which a single starting cell (parent cell) duplicates itself to produce two identical cells (daughter cells); consists of mitosis and cytokinesis.

Somatic nervous system (SNS) The portion of the peripheral nervous system consisting of somatic sensory (afferent) neurons and somatic motor (efferent) neurons.

Somite (SŌ-mit) Block of mesodermal cells in a developing embryo that is distinguished into a myotome (which forms most of the skeletal muscles), dermatome (which forms connective tissues), and sclerotome (which forms the vertebrae).

Spasm (SPAZM) A sudden, involuntary contraction of large groups of muscles.

Spastic (SPAS-tik) An increase in muscle tone (stiffness) associated with an increase in tendon reflexes and abnormal reflexes (Babinski sign).

Spasticity (spas-TIS-i-tē) Hypertonia characterized by increased muscle tone, increased tendon reflexes, and pathological reflexes (Babinski sign).

Spermatic (sper-MAT-ik) **cord** A supporting structure of the male reproductive system, extending from a testis to the deep inguinal ring, that includes the ductus (vas) deferens, arteries, veins, lymphatic vessels, nerves, cremaster muscle, and connective tissue.

Spermatogenesis (sper′-ma-tō-JEN-e-sis) The formation and development of sperm in the seminiferous tubules of the testes.

Spermatozoon (sper′-ma-tō-ZŌ-on) *See* **Sperm cell**.

Sperm cell A mature male gamete. Also termed **spermatozoon**.

Spermicide (SPER-mi-sīd′) An agent that kills sperm.

Spermiogenesis (sper′-mē-ō-JEN-e-sis) The maturation of spermatids into sperm.

Sphincter (SFINGK-ter) A circular muscle that constricts an opening.

Sphincter of Oddi *See* **Sphincter of the hepatopancreatic ampulla**.

Sphincter of the hepatopancreatic ampulla A circular muscle at the opening of the common bile and main pancreatic ducts in the duodenum. Also called the **sphincter of Oddi** (OD-ē).

Sphygmomanometer (sfig′-mō-ma-NOM-e-ter) An instrument for measuring arterial blood pressure.

Spina bifida (SPĪ-na BIF-i-da) A congenital defect of the vertebral column in which the halves of the neural arch of a vertebra fail to fuse in the midline; a neural tube defect.

Spinal (SPĪ-nal) **cord** A mass of nerve tissue located in the vertebral canal from which 31 pairs of spinal nerves originate.

Spinal nerve One of the 31 pairs of nerves that originate on the spinal cord from posterior and anterior roots.

Spinal shock A period of time, from several days to several weeks, following transection of the spinal cord and characterized by the abolition of all reflex activity.

Spinal tap Withdrawal of some of the cerebrospinal fluid from the subarachnoid space in the lumbar region for diagnostic purposes, introduction of various substances, and evaluation of the effects of treatment. Also called **lumbar puncture**.

Spinothalamic (spi-no-THAL-am-ik) **tracts** Sensory (ascending) tracts that convey information up the spinal cord to the thalamus for sensations of pain, temperature, crude touch, and deep pressure.

Spinous (SPĪ-nus) **process** A sharp or thornlike process or projection. Also called a **spine**. A sharp ridge running diagonally across the posterior surface of the scapula.

Spiral organ The organ of hearing, consisting of supporting cells and hair cells that rest on the basilar membrane and extend into the endolymph of the cochlear duct. Also called the **organ of Corti** (KOR-tē).

Spirometer (spi-ROM-e-ter) An apparatus used to measure lung volumes and capacities.

Splanchnic (SPLANK-nik) Pertaining to the viscera.

Spleen (SPLĒN) Large mass of lymphatic tissue between the fundus of the stomach and the diaphragm that functions in formation of blood cells during early fetal development, phagocytosis of worn-out blood cells, and proliferation of B cells during immune responses.

Spongy (cancellous) bone tissue Bone tissue that consists of an irregular latticework of thin plates of bone called trabeculae; spaces between trabeculae of some bones are filled with red bone marrow; found inside short, flat, and irregular bones and in the epiphyses (ends) of long bones.

Sprain Forcible wrenching or twisting of a joint with partial rupture or other injury to its attachments without dislocation.

Sputum (SPYOO-tum) Substance ejected from the mouth containing saliva and mucus.

Squamous (SKWĀ-mus) Flat.

Starling's law of the capillaries The movement of fluid between plasma and interstitial fluid is in a state of near equilibrium at the arterial and venous ends of a capillary; that is, filtered fluid and absorbed fluid plus that returned to the lymphatic system are nearly equal.

Starling's law of the heart The force of muscular contraction is determined by the length of the cardiac muscle fibers just before they contract; within limits, the greater the length of stretched fibers, the stronger the contraction.

Starvation (star-VĀ-shun)   The loss of energy stores in the form of glycogen, triglycerides, and proteins due to inadequate intake of nutrients or inability to digest, absorb, or metabolize ingested nutrients.

Stasis (STĀ-sis)   Stagnation or halt of normal flow of fluids, as blood or urine, or of the intestinal contents.

Static equilibrium (ē-kwi-LIB-rē-um)   The maintenance of posture in response to changes in the orientation of the body, mainly the head, relative to the ground.

Stellate reticuloendothelial (STEL-āte re-tik′-yoo-lō-en′-dō-THĒ-lē-al) **cell**   Phagocytic cell bordering a sinusoid of the liver. Also called a **Kupffer's** (KOOP-ferz) **cell**.

Stenosis (sten-Ō-sis)   An abnormal narrowing or constriction of a duct or opening.

Stereocilia (ste′-rē-ō-SIL-ē-a)   Groups of extremely long, slender, nonmotile microvilli projecting from epithelial cells lining the epididymis.

Stereognosis (ste′-rē-og-NŌ-sis)   The ability to recognize the size, shape, and texture of an object by touching it.

Sterile (STE-ril)   Free from any living microorganisms. Unable to conceive or produce offspring.

Sterilization (ster′-i-li-ZĀ-shun)   Elimination of all living microorganisms. The rendering of an individual incapable of reproduction (e.g., castration, vasectomy, hysterectomy).

Sternal puncture   Introduction of a wide-bore needle into the spongy bone of the sternum for aspiration of a sample of red bone marrow.

Stimulus   Any stress that changes a controlled condition; any change in the internal or external environment that excites a receptor, a neuron, or a muscle fiber.

Stomach   The J-shaped enlargement of the gastrointestinal tract directly inferior to the diaphragm in the epigastric, umbilical, and left hypochondriac regions of the abdomen, between the esophagus and small intestine.

Strabismus (stra-BIZ-mus)   A condition in which the visual axes of the two eyes differ, so that they do not fix on the same object.

Straight tubule (TOO-byool)   A duct in a testis leading from a convoluted seminiferous tubule to the rete testis.

Stratum (STRĀ-tum)   A layer.

Stratum basalis (ba-SAL-is)   The superficial layer of the endometrium (next to the myometrium) that is maintained during menstruation and gestation and produces a new stratum functionalis following menstruation or parturition.

Stratum functionalis (funk′-shun-AL-is)   The deep layer of the endometrium (next to the uterine cavity) that is shed during menstruation and that forms the maternal portion of the placenta during gestation.

Stress   Any stimulus that creates an imbalance in the internal environment.

Stressor   A stress that is extreme, unusual, or long-lasting and triggers the general adaptation syndrome.

Stretch receptor   Receptor in the walls of blood vessels, airways, or organs that monitors the amount of stretching. Also termed **baroreceptor**.

Stretch reflex   A monosynaptic reflex triggered by sudden stretching of muscle spindles within a muscle that elicits contraction of that same muscle. Also called a **tendon jerk**.

Stricture (STRIK-cher)   A local constriction of a tubular structure.

Stroke   *See* **Cerebrovascular accident**.

Stroke volume   The volume of blood ejected by either ventricle in one systole; about 70 ml at rest.

Stroma (STRŌ-ma)   The tissue that forms the ground substance, foundation, or framework of an organ, as opposed to its functional parts (parenchyma).

Stupor (STOO-por)   Unresponsiveness from which a patient can be aroused only briefly and by vigorous and repeated stimulation.

Subarachnoid (sub′-a-RAK-noyd) **space**   A space between the arachnoid and the pia mater that surrounds the brain and spinal cord and through which cerebrospinal fluid circulates.

Subcutaneous (sub′-kyoo-TĀ-nē-us)   Beneath the skin. Also called **hypodermic** (hi-pō-DER-mik).

Subcutaneous layer   A continuous sheet of areolar connective tissue and adipose tissue between the dermis of the skin and the deep fascia of the muscles. Also called the **superficial fascia** (FASH-ē-a).

Subdural (sub-DOO-ral) **space**   A space between the dura mater and the arachnoid of the brain and spinal cord that contains a small amount of fluid.

Sublingual (sub-LING-gwal) **gland**   One of a pair of salivary glands situated in the floor of the mouth deep to the mucous membrane and to the side of the lingual frenulum, with a duct (Rivinus') that opens into the floor of the mouth.

Submandibular (sub′-man-DIB-yoo-lar) **gland**   One of a pair of salivary glands found inferior to the base of the tongue under the mucous membrane in the posterior part of the floor of the mouth, posterior to the sublingual glands, with a duct (Wharton's) situated to the side of the lingual frenulum. Also called the **submaxillary** (sub′-MAK-si-ler-ē) **gland**.

Submucosa (sub-myoo-KŌ-sa)   A layer of connective tissue located deep to a mucous membrane, as in the gastrointestinal tract or the urinary bladder; the submucosa connects the mucosa to the muscularis layer.

Submucosal plexus   A network of autonomic nerve fibers located in the superficial portion of the submucous layer of the small intestine. Also called the **plexus of Meissner** (MIS-ner).

Substrate   A molecule upon which an enzyme acts.

Subthreshold stimulus   A stimulus of such weak intensity that it cannot initiate an action potential (nerve impulse).

Sudden infant death syndrome (SIDS)   Unexpected and unexplained death of an apparently well infant; death usually occurs during sleep.

Sudoriferous (soo′-dor-IF-er-us) **gland**   An apocrine or eccrine exocrine gland in the dermis or subcutaneous layer that produces perspiration. Also called a **sweat gland**.

Sulcus (SUL-kus)   A groove or depression between parts, especially between the convolutions of the brain. *Plural is* **sulci** (SUL-sī).

Summation (sum-MĀ-shun)   The addition of the excitatory and inhibitory effects of many stimuli applied to a neuron. The increased strength of muscle contraction that results when stimuli follow one another in rapid succession.

Superficial (soo′-per-FISH-al)   Located on or near the surface of the body or an organ.

Superficial fascia (FASH-ē-a)   A continuous sheet of fibrous connective tissue between the dermis of the skin and the deep fascia of the muscles. Also called **subcutaneous** (sub′-kyoo-TĀ-nē-us) **layer**.

Superficial inguinal (IN-gwi-nal) **ring**   A triangular opening in the aponeurosis of the external oblique muscle that represents the termination of the inguinal canal.

Superior (soo-PĒR-ē-or)   Toward the head or upper part of a structure. Also called **cephalad** (SEF-a-lad) or **craniad**.

Superior vena cava (VĒ-na CĀ-va) **(SVC)** Large vein that collects blood from parts of the body superior to the heart and returns it to the right atrium.

Supination (soo-pi-NĀ-shun) A movement of the forearm in which the palm is turned anteriorly or superiorly.

Suppuration (sup'-yoo-RĀ-shun) Pus formation and discharge.

Supraopticohypophyseal (soo'-pra-op'-tik-ō-hi-pō-FIZ-ē-al) **tract** A bundle of axons that have their cell bodies in the hypothalamus but release their neurosecretions in the posterior pituitary gland.

Surface anatomy The study of the structures that can be identified from the outside of the body.

Surfactant (sur-FAK-tant) Complex mixture of phospholipids and lipoproteins produced by type II alveolar (septal) cells in the lungs that decreases surface tension.

Susceptibility (sus-sep'-ti-BIL-i-tē) Lack of resistance of a body to the deleterious or other effects of an agent such as pathogenic microorganisms.

Suspensory ligament (sus-PEN-so-rē LIG-a-ment) A fold of peritoneum extending laterally from the surface of the ovary to the pelvic wall.

Sustentacular (sus'-ten-TAK-yoo-lar) **cell** A supporting cell of seminiferous tubules that produces secretions for supplying nutrients to sperm and the hormone inhibin. Also called a **Sertoli** (ser-TŌ-lē) **cell**.

Sutural (SOO-cher-al) **bone** A small bone located within a suture between certain cranial bones. Also called **Wormian** (WER-mē-an) **bone**.

Suture (SOO-cher) An immovable fibrous joint that joins skull bones.

Sympathetic (sim'-pa-THET-ik) **division** One of the two subdivisions of the autonomic nervous system, having cell bodies of preganglionic neurons in the lateral gray columns of the thoracic segment and first two or three lumbar segments of the spinal cord; primarily concerned with processes involving the expenditure of energy. Also called the **thoracolumbar** (thō'-ra-kō-LUM-bar) **division**.

Sympathetic trunk ganglion (GANG-glē-on) A cluster of cell bodies of postganglionic sympathetic neurons lateral to the vertebral column, close to the body of a vertebra. These ganglia extend inferiorly through the neck, thorax, and abdomen to the coccyx on both sides of the vertebral column and are connected to one another to form a chain on each side of the vertebral column. Also called **lateral**, or **sympathetic**, **chain** or **vertebral chain ganglia**.

Sympathomimetic (sim'-pa-thō-mi-MET-ik) Producing effects that mimic those brought about by the sympathetic division of the autonomic nervous system.

Symphysis (SIM-fi-sis) A line of union. A slightly movable cartilaginous joint such as the pubic symphysis.

Symport Process by which two substances, often Na$^+$ and another substance, move in the same direction across a cell membrane. Also called **cotransport**.

Symptom (SIMP-tum) A subjective change in body function not apparent to an observer, such as fever or nausea, that indicates the presence of a disease or disorder of the body.

Synapse (SYN-aps) The functional junction between two neurons or between a neuron and an effector, such as a muscle or gland; may be electrical or chemical.

Synapsis (sin-AP-sis) The pairing of homologous chromosomes during prophase I of meiosis.

Synaptic (sin-AP-tik) **cleft** The narrow gap that separates the axon terminal of one neuron from another neuron or muscle fiber (cell) and across which a neurotransmitter diffuses to affect the postsynaptic cell.

Synaptic delay The length of time between the arrival of the action potential at a presynaptic axon terminal and the membrane potential (IPSP or EPSP) change on the postsynaptic membrane; usually about 0.5 msec.

Synaptic end bulb Expanded distal end of an axon terminal that contains synaptic vesicles. Also called a **synaptic knob** or **end foot**.

Synaptic vesicle Membrane-enclosed sac in a synaptic end bulb that stores neurotransmitters.

Synarthrosis (sin'-ar-THRO-sis) An immovable joint; types are suture, gomphosis, and synchondrosis.

Synchondrosis (sin'-kon-DRO-sis) A cartilaginous joint in which the connecting material is hyaline cartilage.

Syncope (SIN-kō-pē) Faint; a sudden, temporary loss of consciousness with loss of postural tone and followed by spontaneous recovery; commonly caused by cerebral ischemia.

Syndesmosis (sin'-dez-MO-sis) A slightly movable joint in which articulating bones are united by fibrous connective tissue.

Syndrome (SIN-drōm) A group of signs and symptoms that occur together in a pattern that is characteristic of a particular disease or abnormal condition.

Syneresis (si-NER-e-sis) The process of clot retraction.

Synergist (SIN-er-jist) A muscle that assists the prime mover by reducing undesired action or unnecessary movement.

Synergistic (syn-er-GIS-tik) **effect** A hormonal interaction in which the effects of two or more hormones acting together is greater or more extensive than the sum of each hormone acting alone.

Synostosis (sin'-os-TŌ-sis) A joint in which the dense fibrous connective tissue that unites bones at a suture has been replaced by bone, resulting in a complete fusion across the suture line.

Synovial (si-NO-vē-al) **cavity** The space between the articulating bones of a diarthrotic joint, filled with synovial fluid. Also called a **joint cavity**.

Synovial fluid Secretion of synovial membranes that lubricates joints and nourishes articular cartilage.

Synovial joint A fully movable or diarthrotic joint in which a synovial (joint) cavity is present between the two articulating bones.

Synovial membrane The deeper of the two layers of the articular capsule of a synovial joint, composed of areolar connective tissue that secretes synovial fluid into the synovial (joint) cavity.

Syphilis (SIF-i-lis) A sexually transmitted disease caused by the bacterium *Treponema pallidum*.

System An association of organs that have a common function.

Systemic (sis-TEM-ik) Affecting the whole body; generalized.

Systemic anatomy The anatomic study of particular systems of the body, such as the skeletal, muscular, nervous, cardiovascular, or urinary systems.

Systemic circulation The routes through which oxygenated blood flows from the left ventricle through the aorta to all the organs of the body and deoxygenated blood returns to the right atrium.

Systemic lupus erythematosus (er-i-them-a-TO-sus) **(SLE)** An autoimmune, inflammatory disease that may affect every tissue of the body.

Systemic vascular resistance (SVR) All the vascular resistances offered by systemic blood vessels. Also called **total peripheral resistance**.

Systole (SIS-tō-lē)   In the cardiac cycle, the phase of contraction of the heart muscle, especially of the ventricles.

Systolic (sis-TO-lik) **blood pressure**   The force exerted by blood on arterial walls during ventricular contraction; the highest pressure measured in the large arteries, about 120 mm Hg under normal conditions for a young adult.

Tachycardia (tak′-i-KAR-dē-a)   An abnormally rapid resting heartbeat or pulse rate (over 100/min).

Tactile (TAK-til)   Pertaining to the sense of touch.

Tactile disc   Modified epidermal cell in the stratum basale of hairless skin that functions as a cutaneous receptor for discriminative touch. Also called a **Merkel** (MER-kel) **disc**.

Taenia coli (TĒ-nē-a KŌ-li)   One of three flat bands of thickened, longitudinal smooth muscle running the length of the large intestine.

Target cell   A cell whose activity is affected by a particular hormone.

Tarsal gland   Sebaceous (oil) gland that opens on the edge of each eyelid. Also called a **Meibomian** (mi-BŌ-mē-an) **gland**.

Tarsal plate   A thin, elongated sheet of connective tissue, one in each eyelid, giving the eyelid form and support. The aponeurosis of the levator palpebrae superioris is attached to the tarsal plate of the superior eyelid.

Tarsus (TAR-sus)   A collective term for the seven bones of the ankle.

Tay–Sachs (TĀ SAKS) **disease**   Inherited, progressive neuronal degeneration of the central nervous system due to a deficient lysosomal enzyme that causes excessive accumulations of a lipid called ganglioside.

T cell   A lymphocyte that becomes immunocompetent in the thymus gland and can differentiate into one of several types of effector cells that function in cell-mediated immunity.

Tectorial (tek-TO-rē-al) **membrane**   A gelatinous membrane projecting over and in contact with the hair cells of the spiral organ (organ of Corti) in the cochlear duct.

Teeth (TĒTH)   Accessory structures of digestion composed of calcified connective tissue and embedded in bony sockets of the mandible and maxilla that cut, shred, crush, and grind food. Also called **dentes** (DEN-tēz).

Telophase (TEL-ō-fāz)   The final stage of mitosis in which the daughter nuclei become established.

Temporomandibular joint (TMJ) syndrome   A disorder of the temporomandibular joint (TMJ) characterized by dull pain around the ear, tenderness of jaw muscles, a clicking or popping noise when opening or closing the mouth, limited or abnormal opening of the mouth, headache, tooth sensitivity, and abnormal wearing of the teeth.

Tendon (TEN-don)   A white fibrous cord of dense, regularly arranged connective tissue that attaches muscle to bone.

Tendon organ   A proprioceptive receptor, sensitive to changes in muscle tension and force of contraction, found chiefly near the junctions of tendons and muscles. Also called a **Golgi** (GOL-jē) **tendon organ**.

Tendon reflex   A polysynaptic, ipsilateral reflex that is designed to protect tendons and their associated muscles from damage that might be brought about by excessive tension. The receptors involved are called tendon organs (Golgi tendon organs).

Tenosynovitis (ten′-ō-sin-ō-VI-tis)   Inflammation of a tendon sheath and synovial membrane at a joint.

Tentorium cerebelli (ten-TŌ-rē-um ser′-e-BEL-ē)   A transverse shelf of dura mater that forms a partition between the occipital lobe of the cerebral hemispheres and the cerebellum and that covers the cerebellum.

Teratogen (TER-a-tō-jen)   Any agent or factor that causes physical defects in a developing embryo.

Terminal ganglion (TER-min-al GANG-glē-on)   A cluster of cell bodies of postganglionic parasympathetic neurons either lying very close to the visceral effectors or located within the walls of the visceral effectors supplied by the postganglionic fibers.

Testis (TES-tis)   Male gonad that produces sperm and the hormones testosterone and inhibin. Also called a **testicle**.

Testosterone (tes-TOS-te-rō-n)   A male sex hormone (androgen) secreted by interstitial endocrinocytes (cells of Leydig) of a mature testis; needed for development of sperm; together with a second androgen termed **dihydrotestosterone (DHT)**, controls the growth and development of male reproductive organs, secondary sex characteristics, and body growth.

Tetanus (TET-a-nus)   An infectious disease caused by the toxin produced by *Clostridium tetani* bacteria, characterized by tonic muscle spasms and exaggerated reflexes, lockjaw, and arching of the back. A smooth, sustained contraction produced by a series of very rapid stimuli to a muscle.

Tetany (TET-a-nē)   Hyperexcitability of neurons and muscle fibers caused by hypocalcemia and characterized by intermittent or continuous tonic muscular contractions; may be due to hypoparathyroidism.

Tetralogy of Fallot (tet-RAL-ō-jē of fal-Ō)   A combination of four congenital heart defects: (1) a constricted pulmonary semilunar valve, (2) an interventricular septal opening, (3) emergence of the aorta from both ventricles instead of from the left only, and (4) an enlarged right ventricle.

Thalamus (THAL-a-mus)   A large, oval structure located superior to the midbrain, consisting of two masses of gray matter covered by a thin layer of white matter.

Thalassemia (thal′-a-SĒ-mē-a)   A group of hereditary hemolytic anemias.

Thallium (THAL-ē-um) **imaging**   Diagnostic procedure used to evaluate blood flow through coronary arteries, cardiac disorders, and effectiveness of drug therapy; thallium shows up in healthy myocardial tissue.

Therapy (THER-a-pē)   The treatment of a disease or disorder.

Thermoreceptor (THER-mō-rē-sep-tor)   Receptor that detects changes in temperature.

Thigh   The portion of the lower limb between the hip and the knee.

Third ventricle (VEN-tri-kul)   A slitlike cavity between the right and left halves of the thalamus and between the lateral ventricles of the brain.

Thirst center   A cluster of neurons in the hypothalamus that is sensitive to the osmotic pressure of extracellular fluid and brings about the sensation of thirst.

Thoracic (thō-RAS-ik) **cavity**   Superior portion of the ventral body cavity that contains two pleural cavities, the mediastinum, and the pericardial cavity.

Thoracic duct   A lymphatic vessel that begins as a dilation called the cisterna chyli, receives lymph from the left side of the head, neck, and chest, the left arm, and the entire body below the ribs, and empties into the left subclavian vein. Also called the **left lymphatic** (lim-FAT-ik) **duct**.

Thoracolumbar (thō'-ra-kō-LUM-bar) **outflow** The fibers of the sympathetic preganglionic neurons, which have their cell bodies in the lateral gray columns of the thoracic segments and first two or three lumbar segments of the spinal cord.

Thorax (THŌ-raks) The chest.

Threshold potential The membrane voltage that must be reached to trigger an action potential.

Threshold stimulus Any stimulus strong enough to initiate an action potential or activate a sensory receptor.

Thrombin (THROM-bin) The active enzyme formed from prothrombin that acts to convert fibrinogen to fibrin during formation of a blood clot.

Thrombolytic (throm-bō-LIT-ik) **agent** Chemical substance injected into the body to dissolve blood clots and restore circulation; mechanism of action is direct or indirect activation of plasminogen; examples include tissue plasminogen activator (t-PA), streptokinase, and urokinase.

Thrombophlebitis (throm'-bō-fle-BĪ-tis) A disorder in which inflammation of the wall of a vein is followed by the formation of a blood clot (thrombus).

Thrombosis (throm-BŌ-sis) The formation of a clot in an unbroken blood vessel, usually a vein.

Thrombus A stationary clot formed in an unbroken blood vessel, usually a vein.

Thymus (THĪ-mus) **gland** A bilobed organ, located in the superior mediastinum posterior to the sternum and between the lungs, that plays an essential role in immune responses.

Thyroglobulin (thi-rō-GLŌ-byoo-lin) **(TGB)** A large glycoprotein molecule produced by follicle cells of the thyroid gland in which iodine is combined with tyrosine to form thyroid hormones.

Thyroid cartilage (THĪ-royd KAR-ti-lij) The largest single cartilage of the larynx, consisting of two fused plates that form the anterior wall of the larynx. Also called the **Adam's apple**.

Thyroid colloid (KOL-loyd) A complex in thyroid follicles consisting of thyroglobulin and stored thyroid hormones.

Thyroid follicle (FOL-i-kul) Spherical sac that forms the parenchyma of the thyroid gland and consists of follicular cells that produce thyroxine (T_4) and triiodothyronine (T_3).

Thyroid function tests Tests used to evaluate a swelling or lump in the thyroid gland, ascertain symptoms of abnormal thyroxine levels, monitor responses of thyroid diseases to therapy, and screen newborns for cretinism. Examples are serum T_4 concentration, serum T_3 concentration, and serum TSH concentration tests.

Thyroid gland An endocrine gland with right and left lateral lobes on either side of the trachea connected by an isthmus located anterior to the trachea just inferior to the cricoid cartilage; secretes thyroxine (T_4), triiodothyronine (T_3), and calcitonin.

Thyroid-stimulating hormone (TSH) A hormone secreted by the anterior pituitary gland that stimulates the synthesis and secretion of thyroxine (T_4) and triiodothyronine (T_3).

Thyroxine (thi-ROK-sēn) **(T_4)** A hormone secreted by the thyroid gland that regulates organic metabolism, growth and development, and the activity of the nervous system.

Tic Spasmodic twitching made involuntarily by muscles that are ordinarily under voluntary control.

Tidal volume The volume of air breathed in and out in any one breath; about 500 ml in quiet, resting conditions.

Tinnitus (ti-NĪ-tus) A ringing, roaring, or clicking in the ears.

Tissue A group of similar cells and their intercellular substance joined together to perform a specific function.

Tissue factor (TF) A factor, or collection of factors, whose appearance initiates the blood clotting process. Also called **thromboplastin** (throm-bō-PLAS-tin).

Tissue plasminogen activator (t-PA) An enzyme that dissolves small blood clots by initiating a process that converts plasminogen to plasmin, which degrades the fibrin of a clot.

Tissue rejection Phenomenon by which the immune system recognizes the protein (MHC antigens) in transplanted tissues or organs as foreign and produces antibodies against them.

Tongue A large skeletal muscle covered by a mucous membrane located on the floor of the oral cavity.

Tonicity (tō-NIS-i-tē) A measure of the concentration of impermeable solute particles in a solution relative to cytosol. When cells are bathed in an **isotonic solution**, they neither shrink nor swell.

Tonometry (tō-NOM-e-trē) A procedure for measuring intraocular pressure to screen for glaucoma.

Tonsil (TON-sil) An aggregation of large lymphatic nodules embedded in the mucous membrane of the throat.

Topical (TOP-i-kal) Applied to the surface rather than ingested or injected.

Torn cartilage A tearing of an articular disc (meniscus) in the knee.

Torpor (TOR-por) State of lethargy and sluggishness; may progress to stupor, semicoma, and finally coma.

Total lung capacity The sum of tidal volume, inspiratory reserve volume, expiratory reserve volume, and residual volume; about 6000 ml in an average adult.

Toxic (TOK-sik) Pertaining to poison; poisonous.

Toxic shock syndrome (TSS) A disease caused by the bacterium *Staphylococcus aureus,* occurring in menstruating females who use tampons and characterized by high fever, sore throat, headache, fatigue, irritability, and abdominal pain.

Trabecula (tra-BEK-yoo-la) Irregular latticework of thin plate of spongy bone. Fibrous cord of connective tissue serving as supporting fiber by forming a septum extending into an organ from its wall or capsule. *Plural is* **trabeculae** (tra-BEK-yoo-lē).

Trabeculae carneae (KAR-nē-ē) Ridges and folds of the myocardium in the ventricles.

Trachea (TRĀ-kē-a) Tubular air passageway extending from the larynx to the fifth thoracic vertebra. Also called the **windpipe**.

Tracheostomy (trā-kē-OS-tō-mē) Surgical opening into the trachea through the neck (inferior to the cricoid cartilage) to faciliate passage of air or evacuation of secretions.

Trachoma (tra-KŌ-ma) A chronic infectious disease of the conjunctiva and cornea of the eye caused by *Chlamydia trachomatis.*

Tract A bundle of nerve fibers in the central nervous system.

Transcription (trans-KRIP-shun) The first step in the transfer of genetic information in which a single strand of DNA serves as a template for the formation of an RNA molecule.

Transfusion (trans-FYOO-shun) Transfer of whole blood, blood components, or red bone marrow directly into the bloodstream.

Transient ischemic (is-KĒ-mik) **attack (TIA)** Episode of temporary focal, nonconvulsive cerebral dysfunction caused by interference of the blood supply to a region of the brain.

Translation (trans-LĀ-shun) The synthesis of a new protein on the ribosome of a cell as dictated by the sequence of codons in messenger RNA.

Transplantation (trans-plan-TĀ-shun) The replacement of injured or diseased tissues or organs with natural ones.

Transport maximum (T$_m$) The maximum amount of a substance that can be reabsorbed, especially by renal tubules, under any condition.

Transverse colon (trans-VERS KŌ-lon) The portion of the large intestine extending across the abdomen from right colic (hepatic) flexure to the left colic (splenic) flexure.

Transverse fissure (FISH-er) The deep cleft that separates the cerebrum from the cerebellum.

Transverse plane A plane that runs parallel to the ground and divides the body or organs into superior and inferior portions. Also called a **horizontal plane**.

Transverse tubules (TOO-byools) **(T tubules)** Small, cylindrical invaginations of the sarcolemma of striated muscle fibers (cells) that conduct muscle action potentials toward the center of the muscle fiber.

Trauma (TRAW-ma) An injury, either a physical wound or psychic disorder, caused by an external agent or force, such as a physical blow or emotional shock; the agent or force that causes the injury.

Traveler's diarrhea Infectious disease of the gastrointestinal tract that results in loose, urgent bowel movements, cramping, abdominal pain, malaise, nausea, and occasionally fever and dehydration. It is acquired through ingestion of food or water that has become contaminated with fecal material containing mostly bacteria (especially *Escherichia coli*). Also called **Montezuma's revenge**, **turista**, or **Tut's tummy**.

Tremor (TREM-or) Rhythmic, involuntary, purposeless contraction of opposing muscle groups.

Treppe (TREP-eh) The gradual increase in the force of contraction of a muscle caused by repetitive stimuli of the same strength.

Triad (TRĪ-ad) A complex of three units in a muscle fiber composed of a transverse tubule and the sarcoplasmic reticulum terminal cisterns on both sides of it.

Tricuspid (tri-KUS-pid) **valve** Atrioventricular (AV) valve on the right side of the heart.

Trigeminal neuralgia (tri-JEM-i-nal noo-RAL-jē-a) Pain in one or more of the branches of the trigeminal (V) nerve. Also called **tic douloureux** (doo-loo-ROO).

Triglyceride (tri-GLI-cer-īde) A lipid formed from one molecule of glycerol and three molecules of fatty acids that may be either solid (fats) or liquid (oils) at room temperature; the body's most highly concentrated source of chemical potential energy. Found mainly within adipocytes. Also called a **neutral fat**.

Trigone (TRĪ-gon) A triangular area at the base of the urinary bladder.

Triiodothyronine (tri-i-ō-dō-THĪ-rō-nēn) **(T$_3$)** A hormone produced by the thyroid gland that regulates organic metabolism, growth and development, and the activity of the nervous system.

Trochlea (TROK-lē-a) A pulleylike surface.

Trophoblast (TRŌF-ō-blast) The superficial covering of cells of the blastocyst.

Tropic (TRŌ-pik) **hormone** A hormone whose target is another endocrine gland.

True vocal cords Pair of mucous membrane folds below the ventricular folds that function in voice production. Also called **vocal folds**.

Trunk The part of the body to which the upper and lower limbs are attached.

Tubal ligation (lī-GA-shun) A sterilization procedure in which the uterine (Fallopian) tubes are tied and cut.

Tuberculosis (too-berk-yoo-LŌ-sis) An infection of the lungs and pleurae caused by *Mycobacterium tuberculosis;* results in destruction of lung tissue, which is replaced by fibrous connective tissue.

Tubular reabsorption The movement of filtrate from renal tubules back into blood in response to the body's specific needs.

Tubular secretion The movement of substances in blood into renal tubular fluid in response to the body's specific needs.

Tumor (TOO-mor) A growth of excess tissue due to an unusually rapid division of cells.

Tumor suppressor gene A gene coding for a protein that normally inhibits cell division; loss or alteration of a tumor suppressor gene called *p53* is the most common genetic change in a wide variety of cancer cells.

Tunica albuginea (TOO-ni-ka al′-byoo-JIN-ē-a) A dense white fibrous capsule covering a testis or deep to the surface of an ovary.

Tunica externa (eks-TER-na) The superficial coat of an artery or vein, composed mostly of elastic and collagen fibers. Also called the **adventitia**.

Tunica interna (in-TER-na) The deep coat of an artery or vein, consisting of a lining of endothelium, basement membrane, and internal elastic lamina. Also called the **tunica intima** (IN-ti-ma).

Tunica media (MĒ-dē-a) The intermediate coat of an artery or vein, composed of smooth muscle and elastic fibers.

T wave The deflection wave of an electrocardiogram that represents ventricular repolarization.

Twitch contraction Brief contraction of all muscle fibers in a motor unit triggered by a single action potential in its motor neuron.

Tympanic antrum (tim-PAN-ik AN-trum) An air space in the middle ear that leads into the mastoid air cells or sinus.

Type II cutaneous mechanoreceptor A receptor embedded deeply in the dermis and deeper tissues that detects heavy and continuous touch sensations. Also called an **end organ of Ruffini**.

Ulcer (UL-ser) An open lesion of the skin or a mucous membrane of the body with necrosis of the tissue.

Ultrasound (US) Medical imaging technique that utilizes high-frequency sound waves to produce an image called a **sonogram**.

Umbilical (um-BIL-i-kal) Pertaining to the umbilicus or navel.

Umbilical cord The long, ropelike structure containing the umbilical arteries and vein that connect the fetus to the placenta.

Umbilicus (um-BIL-i-kus *or* um-bil-Ī-kus) A small scar on the abdomen that marks the former attachment of the umbilical cord to the fetus. Also called the **navel**.

Upper limb The appendage attached at the shoulder girdle, consisting of the arm, forearm, wrist, hand, and fingers. Also called **upper extremity**.

Up-regulation Phenomenon in which there is an increase in the number of receptors in response to a deficiency of a hormone or neurotransmitter.

Uremia (yoo-RE-mē-a) Accumulation of toxic levels of urea and other nitrogenous waste products in the blood, usually resulting from severe kidney malfunction.

Ureter (YOO-re-ter) One of two tubes that connect the kidney with the urinary bladder.

Urethra (yoo-RĒ-thra) The duct from the urinary bladder to the exterior of the body that conveys urine in females and urine and semen in males.

Urinalysis The physical, chemical, and microscopic analysis or examination of urine.

Urinary (YOO-ri-ner-ē) **bladder**   A hollow, muscular organ situated in the pelvic cavity posterior to the pubic symphysis.

Urinary tract infection (UTI)   An infection of a part of the urinary tract or the presence of many microbes in urine.

Urine   The fluid produced by the kidneys that contains wastes or excess materials and is excreted from the body through the urethra.

Urobilinogenuria (yoo-rō-bi′-lin-ō-je-NOOR-ē-a)   The presence of urobilinogen in urine.

Urogenital (yoo′-rō-JEN-i-tal) **triangle**   The region of the pelvic floor inferior to the pubic symphysis, bounded by the pubic symphysis and the ischial tuberosities and containing the external genitalia.

Urology (yoo-ROL-ō-je-)   The specialized branch of medicine that deals with the structure, function, and diseases of the male and female urinary systems and the male reproductive system.

Urticaria (ur′-ti-KĀ-rē-a)   A skin reaction to certain foods, drugs, or other allergy-causing substances; hives.

Uterine (YOO-ter-in) **tube**   Duct that transports ova from the ovary to the uterus. Also called the **Fallopian** (fal-LŌ-pē-an) **tube** or **oviduct**.

Uterosacral ligament (yoo′-ter-ō-SĀ-kral LIG-a-ment)   A fibrous band of tissue extending from the cervix of the uterus laterally to the sacrum.

Uterovesical (yoo′-ter-ō-VES-i-kal) **pouch**   A shallow pouch formed by the reflection of the peritoneum from the anterior surface of the uterus, at the junction of the cervix and the body, to the posterior surface of the urinary bladder.

Uterus (YOO-te-rus)   The hollow, muscular organ in females that is the site of menstruation, implantation, development of the fetus, and labor. Also called the **womb**.

Utricle (YOO-tri-kul)   The larger of the two divisions of the membranous labyrinth located inside the vestibule of the inner ear, containing a receptor organ for static equilibrium.

Uvea (YOO-vē-a)   The three structures that together make up the vascular tunic of the eye.

Uvula (YOO-vyoo-la)   A soft, fleshy mass, especially the V-shaped pendant part, descending from the soft palate.

Vagina (va-JĪ-na)   A muscular, tubular organ that leads from the uterus to the vestibule, situated between the urinary bladder and the rectum of the female.

Valence (VĀ-lens)   The combining capacity of an atom; the number of deficit or extra electrons in the outermost electron shell of an atom.

Valvular stenosis (VAL-vyoo-lar sten-Ō-sis)   A narrowing of the opening of a heart valve, most often the bicuspid (mitral) valve.

Varicocele (VAR-i-kō-sēl)   A twisted vein; especially, the accumulation of blood in the veins of the spermatic cord.

Varicose (VAR-i-kōs)   Pertaining to an unnatural swelling, as in the case of a varicose vein.

Vas   A vessel or duct.

Vasa recta (VĀ-sa REK-ta)   Extensions of the efferent arteriole of a juxtamedullary nephron that run alongside the loop of the nephron (Henle) in the medullary region of the kidney.

Vasa vasorum (va-SŌ-rum)   Blood vessels that supply nutrients to the larger arteries and veins.

Vascular (VAS-kyoo-lar)   Pertaining to or containing many blood vessels.

Vascular spasm   Contraction of the smooth muscle in the wall of a damaged blood vessel to prevent blood loss.

Vascular (venous) sinus   A vein with a thin endothelial wall that lacks a tunica media and externa and is supported by surrounding tissue.

Vascular tunic (TOO-nik)   The middle layer of the eyeball, composed of the choroid, ciliary body, and iris. Also called the **uvea** (YOO-vē-a).

Vasectomy (va-SEK-tō-mē)   A means of sterilization of males in which a portion of each ductus (vas) deferens is removed.

Vasoconstriction (vāz-ō-kon-STRIK-shun)   A decrease in the size of the lumen of a blood vessel caused by contraction of the smooth muscle in the wall of the vessel.

Vasodilation (vāz′-ō-DĪ-lā-shun)   An increase in the size of the lumen of a blood vessel caused by relaxation of the smooth muscle in the wall of the vessel.

Vasomotion (vāz-ō-MO-shun)   Intermittent contraction and relaxation of the smooth muscle of the metarterioles and precapillary sphincters that result in an intermittent blood flow.

Vasopressin   *See* **Antidiuretic hormone.**

Vein   A blood vessel that conveys blood from tissues back to the heart.

Vena cava (VĒ-na KĀ-va)   One of two large veins that open into the right atrium, returning to the heart all of the deoxygenated blood from the systemic circulation except from the coronary circulation.

Venereal (ve-NĒ-rē-al) **disease (VD)**   *See* **Sexually transmitted disease**.

Venesection (ven′-e-SEK-shun)   Cutting or puncturing a vein for withdrawal of blood. Also termed **phlebotomy**.

Ventral (VEN-tral)   Pertaining to the anterior or front side of the body; opposite of dorsal.

Ventral body cavity   Cavity near the ventral aspect of the body that contains viscera and consists of a superior thoracic cavity and an inferior abdominopelvic cavity.

Ventral ramus (RĀ-mus)   The anterior branch of a spinal nerve, containing sensory and motor fibers to the muscles and skin of the anterior surface of the head, neck, trunk, and the limbs.

Ventricle (VEN-tri-kul)   A cavity in the brain or an inferior chamber of the heart.

Ventricular fibrillation (ven-TRIK-yoo-lar fib-ri-LĀ-shun)   Asynchronous ventricular contractions; unless reversed by defibrillation results in heart failure.

Venule (VEN-yool)   A small vein that collects blood from capillaries and delivers it to a vein.

Vermiform appendix (VER-mi-form a-PEN-diks)   A twisted, coiled tube attached to the cecum.

Vermilion (ver-MIL-yon)   The area of the mouth where the skin on the outside meets the mucous membrane on the inside.

Vermis (VER-mis)   The central constricted area of the cerebellum that separates the two cerebellar hemispheres.

Vertebral (VER-te-bral) **canal**   A cavity within the vertebral column formed by the vertebral foramina of all the vertebrae and containing the spinal cord. Also called the **spinal canal**.

Vertebral column   The 26 vertebrae of an adult and 33 vertebrae of a child; encloses and protects the spinal cord and serves as a point of attachment for the ribs and back muscles. Also called the **backbone, spine,** or **spinal column**.

Vertigo (VER-ti-go)   Sensation of spinning or movement.

Vesicle (VES-i-kul)   A small bladder or sac containing liquid.

Vestibular (ves-TIB-yoo-lar) **apparatus**   Collective term for the organs of equilibrium, which includes the saccule, utricle, and semicircular ducts.

Vestibular membrane   The membrane that separates the cochlear duct from the scala vestibuli.

Vestibule (VES-ti-byool)   A small space or cavity at the beginning of a canal, especially the inner ear, larynx, mouth, nose, and vagina.

Villus (VIL-lus)   A projection of the intestinal mucosal cells containing connective tissue, blood vessels, and a lymphatic vessel; functions in the absorption of the end products of digestion. *Plural is* **villi** (VIL-Ī).

Viscera (VIS-er-a)   The organs inside the ventral body cavity. *Singular is* **viscus** (VIS-kus).

Visceral (VIS-er-al)   Pertaining to the organs or to the covering of an organ.

Visceral effector (e-FEK-tor)   Cardiac muscle, smooth muscle, and glandular epithelium.

Visceral pleura (PLOO-ra)   The deep layer of the serous membrane that covers the lungs.

Viscosity (vis-KOS-i-tē)   The state of being sticky or thick.

Vital capacity   The sum of inspiratory reserve volume, tidal volume, and expiratory reserve volume; about 4800 ml.

Vital signs   Signs necessary to life that include temperature (T), pulse (P), respiratory rate (RR), and blood pressure (BP).

Vitamin   An organic molecule necessary in trace amounts that acts as a catalyst in normal metabolic processes in the body.

Vitiligo (vit-i-LĪ-go)   Patchy, white spots on the skin due to partial or complete loss of melanocytes.

Vitreous (VIT-rē-us) **body**   A soft, jellylike substance that fills the vitreous chamber of the eyeball, lying between the lens and the retina.

Vocal cords   *See* **True vocal cords**.

Volkmann's canal   *See* **Perforating canal**.

Voltage-gated channel   An ion channel in a plasma membrane composed of integral proteins that functions like a gate to permit or restrict the movement of ions across the membrane in response to changes in the voltage.

Vomiting   Forcible expulsion of the contents of the upper gastrointestinal tract through the mouth.

Vulva (VUL-va)   Collective designation for the external genitalia of the female. Also called the **pudendum** (poo-DEN-dum).

Wallerian (wal-LE-rē-an) **degeneration**   Degeneration of the portion of the axon and myelin sheath of a neuron distal to the site of injury.

Wandering macrophage (MAK-rō-fāj)   Phagocytic cell that develops from a monocyte, leaves the blood, and migrates to infected tissues.

Wart   Generally benign tumor of epithelial skin cells caused by a virus.

Wave summation (sum-MA-shun)   The increased strength of muscle contraction that results when stimuli follow one another in rapid succession.

White matter   Aggregations or bundles of myelinated axons located in the brain and spinal cord.

White pulp   The portion of the spleen composed of lymphatic tissue, mostly lymphocytes (B cells).

White ramus communicans (RĀ-mus kō-MYOO-ni-kans)   The portion of a preganglionic sympathetic nerve fiber that branches from the anterior ramus of a spinal nerve to enter the nearest sympathetic trunk ganglion.

X-chromosome inactivation   The random and permanent inactivation of one X chromosome in each cell of a developing female embryo. Also called **lyonization**.

Xiphoid (ZĪ-foyd)   Sword-shaped. The inferior portion of the sternum is the **xiphoid process**.

Yolk sac   An extraembryonic membrane that connects with the midgut during early embryonic development but is nonfunctional in humans.

Zona fasciculata (ZŌ-na fa-sik′-yoo-LA-ta)   The middle zone of the adrenal cortex consisting of cells arranged in long, straight cords that secrete glucocorticoid hormones.

Zona glomerulosa (glo-mer′-yoo-LŌ-sa)   The external zone of the adrenal cortex, directly under the connective tissue covering, consisting of cells arranged in arched loops or round balls that secrete mineralocorticoid hormones.

Zona pellucida (pe-LOO-si-da)   Clear glycoprotein layer that surrounds an oocyte.

Zona reticularis (ret-ik′-yoo-LAR-is)   The inner zone of the adrenal cortex, consisting of cords of branching cells that secrete sex hormones, chiefly androgens.

Zygote (ZĪ-gōt)   The single cell resulting from the union of male and female gametes; the fertilized ovum.

CREDITS

PHOTOS

CHAPTER 1

Figure 1.8a: Stephen A. Kieffer and E. Robert Heitzman, *An Atlas of Cross-Sectional Anatomy.* Harper & Row, Publishers, Inc., New York, 1979.

Figure 1.8b: Lester Bergman & Associates.

Figure 1.8c: © 1988 Martin Rotker.

Exhibit 1.4 (conventional radiography): © Biophoto/Photo Researchers.

Exhibit 1.4 (computed tomography scanning): Simon Fraser/SPL/Photo Researchers.

Exhibit 1.4 (dynamic spatial reconstruction): Robb, Mayo Foundation.

Exhibit 1.4 (digital subtraction angiography): The Bayer Company.

Exhibit 1.4 (positron emission tomography): © Howard Sochurek, Medical Images, Inc.

Exhibit 1.4 (magnetic resonance imaging): Technicare Corp.

Exhibit 1.4 (ultrasound): Anthony Gerard Tortora.

CHAPTER 3

Figure 3.5: © Scott, Foresman Photo.

Figure 3.12b: Abbott Laboratories.

Figure 3.14b: Lennart Nilsson, *The Body Victorious*, Delacorte Press, © Boehringer Ingelheim International GmbH.

Figure 3.17b: Lennart Nilsson, *The Body Victorious*, Delacorte Press, © Boehringer Ingelheim International GmbH.

Figure 3.18b: Lennart Nilsson, *The Body Victorious*, Delacorte Press, © Boehringer Ingelheim International GmbH.

Figure 3.20: Lennart Nilsson, *The Body Victorious*, Delacorte Press, © Boehringer Ingelheim International GmbH.

Figure 3.26a–f: CABISCO/Phototake NY.

CHAPTER 4

Chapter 4 opening photo: Biophoto/Photo Researchers.

Exhibit 4.1 (simple squamous epithelium; left): Biophoto/Photo Researchers.

Exhibit 4.1 (simple squamous epithelium; right): Douglas Merrill.

Exhibit 4.1 (simple cuboidal epithelium): © Ed Reschke.

Exhibit 4.1 (nonciliated simple columnar epithelium): © Ed Reschke.

Exhibit 4.1 (ciliated simple columnar epithelium): Douglas Merrill.

Exhibit 4.1 (stratified squamous epithelium; stratified cuboidal epithelium): Biophoto/Photo Researchers.

Exhibit 4.1 (stratified columnar epithelium): Biophoto/Photo Researchers.

Exhibit 4.1 (transitional epithelium): Andrew J. Kuntzman.

Exhibit 4.1 (pseudostratified columnar epithelium): © Ed Reschke.

Exhibit 4.1 (exocrine glands): Bruce Iverson.

Exhibit 4.1 (endocrine glands): © Lester Bergman & Associates.

Exhibit 4.2 (mesenchyme): © R. Kessel/VU.

Exhibit 4.2 (mucous connective tissue): © Lester Bergman & Associates.

Exhibit 4.2 (areolar connective tissue): Biophoto/Photo Researchers.

Exhibit 4.2 (adipose tissue): © Ed Reschke.

Exhibit 4.2 (reticular connective tissue): Biophoto/Photo Researchers.

Exhibit 4.2 (dense regular connective tissue): Andrew J. Kuntzman.

Exhibit 4.2 (dense irregular connective tissue): © Ed Reschke.

Exhibit 4.2 (elastic connective tissue; hyaline cartilage; fibrocartilage; elastic cartilage): Biophoto/Photo Researchers.

Exhibit 4.2 (bone tissue): © Ed Reschke.

Exhibit 4.2 (blood): © Ed Reschke.

Exhibit 4.3 (skeletal muscle tissue; cardiac muscle tissue): © Ed Reschke.

Exhibit 4.3 (smooth muscle tissue): Andrew J. Kuntzman.

Exhibit 4.4: © Ed Reschke.

CHAPTER 5

Figure 5.2a: Lester Bergman & Associates.

Figure 5.3b: Brain/SPL/Photo Researchers.

Figure 5.3c: Shi/Kessel/VU.

CHAPTER 6

Figure 6.1b: Lester Bergman & Associates.

Figure 6.3b: © Ed Reschke.

Figure 6.6 (top): Arthur Provost, R.T.

Figure 6.6 (bottom): Biophoto/Photo Researchers.

Figure 6.9g: Lester Bergman & Associates.

CHAPTER 9

Figure 9.1b: Mark Nielsen, University of Utah.

Figure 9.3a–d: © 1994 Evan J. Collela.

Figure 9.3e–f: © 1983 by Gerard J. Tortora. Courtesy of Lynn and James Borghesi.

Figure 9.4: Evan J. Collela.

Figure 9.5a,b: © 1994 Evan J. Collela.

Figure 9.5c: © 1983 by Gerard J. Tortora. Courtesy of Lynn and James Borghesi.
Figure 9.5d: © 1994 Evan J. Collela.
Figure 9.6a,b: © 1983 by Gerard J. Tortora. Courtesy of Lynn and James Borghesi.
Figure 9.6c,d: © 1992 Scott, Foresman photo by John Moore.
Figure 9.6e,f: © 1983 by Gerard J. Tortora. Courtesy of Lynn and James Borghesi.
Figure 9.6g: Evan J. Collela.
Figure 9.6h: © 1983 by Gerard J. Tortora. Courtesy of Lynn and James Borghesi.
Figure 9.7d: Mark Nielsen, University of Utah.
Figure 9.8d: Mark Nielsen, University of Utah.
Figure 9.9: © 1988 Camazine/Photo Researchers.

CHAPTER 10

Figure 10.3a: Fujija.
Figure 10.5: H. E. Huxley.
Figure 10.15a–c: © 1994; Scott, Foresman Photo.
Exhibit 10.2: Biophoto/Photo Researchers.
Figure 10.18d: Douglas Merrill.

CHAPTER 12

Chapter 12 opening photo: Martin Rotker/CNRI/Phototake NY.
Figure 12.2b: Martin Rotker/CNRI/Phototake NY.
Figure 12.3c: © 1991 Jan Leesma, M.D./Custom Medical Stock Photo.

CHAPTER 13

Figure 13.1b: N. Gluhbegovic and T. H. Williams, *The Human Brain: A Photographic Guide*, Harper & Row, Publishers, Inc. Hagerstown, MD, 1980.
Figure 13.2b,c: Mark Nielsen, University of Utah.
Figure 13.3b: Mark Nielsen, University of Utah.
Figure 13.11b: Richard G. Kessel and Randy H. Kardon, *from Tissues and Organs: A Text–Atlas of Scanning Electron Microscopy.* Copyright © 1979 by W. H. Freeman and Company. Reprinted by permission.

CHAPTER 14

Figure 14.1b: Mark Nielsen, University of Utah.
Figure 14.11b: Mark Nielsen, University of Utah.
Figure 14.12: N. Gluhbegovic and T. H. Williams, *The Human Brain: A Photographic Guide*, Harper & Row, Publishers, Inc. Hagerstown, MD, 1980.
Figure 14.13b: N. Gluhbegovic and T. H. Williams, *The Human Brain: A Photographic Guide*, Harper & Row, Publishers, Inc. Hagerstown, MD, 1980.

CHAPTER 16

Figure 16.3: © 1992 Scott, Foresman photo by John Moore.
Figure 16.6: © 1991 Eagle/Photo Researchers.
Figure 16.14: N. Gluhbegovic and T. H. Williams, *The Human Brain: A Photographic Guide*, Harper & Row, Publishers, Inc. Hagerstown, MD, 1980.

CHAPTER 18

Figure 18.9a,b: Hall/Evered, *A Colour Atlas of Endocrinology.* Mosby-Year Book/Wolfe Publishing.
Figure 18.13b: Lester Bergman & Associates.
Figure 18.16a,b: Lester Bergman & Associates.
Figure 18.16c: Kay/Peter Arnold.
Figure 18.18b: Biophoto/Photo Researchers.
Figure 18.19b: Andrew J. Kuntzman.
Figure 18.22a: Beverly M. K. Biller, Massachusetts General Hospital.
Figure 18.22b: Hall/Evered, *A Colour Atlas of Endocrinology.* Mosby-Year Book/Wolfe Publishing.
Figure 18.23b: Biophoto/Photo Researchers.

CHAPTER 19

Figure 19.2b: Lennart Nilsson, *Our Body Victorious.* © Boehringer Ingelheim International GmbH.
Figure 19.6 (neutrophil): © Ed Reschke.
Figure 19.6 (eosinophil; basophil; monocyte): Biophoto/Photo Researchers.
Figure 19.6 (blood smear): Douglas Merrill.
Figure 19.6 (large lymphocyte; red blood cells and platelet): Lester Bergman & Associates.
Figure 19.9: Lennart Nilsson, *Our Body Victorious.* © Boehringer Ingelheim International GmbH.
Figure 19.12b (left and right): Lester Bergman & Associates.
Figure 19.14: Lewin/Royal Free Hospital/SPL/Photo Researchers.

CHAPTER 20

Figure 20.2b: Stephen A. Kieffer and E. Robert Heitzman. *An Atlas of Cross-Sectional Anatomy.* Harper & Row, Publishers, Inc., New York, 1979.
Figure 20.4b: Mark Nielsen, University of Utah.
Figure 20.5c: Mark Nielsen, University of Utah.
Figure 20.5d: John Eads.
Figure 20.14 inset: © 1992 David Madison.
Figure 20.16: Sklar/Photo Researchers.

CHAPTER 21

Chapter 21 opening photo: Andrew J. Kuntzman.
Figure 21.1d: Andrew J. Kuntzman.
Figure 21.3b: Lennart Nilsson. *Behold Man.* Dell Publishing Company. © Boehringer Ingelheim International GmbH.
Figure 21.5a,b: Mark Nielsen, University of Utah.

CHAPTER 22

Figure 22.5b: Andrew J. Kuntzman.
Figure 22.6b: Leroy/Biocosmos/Photo Researchers.
Figure 22.7a: Mark Nielsen, University of Utah.
Figure 22.7b: Biophoto/Photo Researchers.
Figure 22.10b: © National Cancer Institute/Photo Researchers.
Figure 22.21b: Wagner Herbert Stock/Phototake NY.

CHAPTER 23

Figure 23.1b: Mark Nielsen, University of Utah.
Figure 23.2a: © 1982 by Gerard J. Tortora. Courtesy of Lynn Borghesi.
Figure 23.2c: Mark Nielsen, University of Utah.
Figure 23.5: Mark Nielsen, University of Utah.
Figure 23.6a: Cunningham/VU.
Figure 23.6b,c: Andrew J. Kuntzman.
Figure 23.7b: © 1992 Lester Bergman & Associates.
Figure 23.8: Mark Nielsen, University of Utah.
Figure 23.10b: Biophoto/Photo Researchers.
Figure 23.11c: Richard G. Kessel and Randy H. Kardon, from *Tissues and Organs: A Text–Atlas of Scanning Electron Microscopy.* Copyright © 1979 by W. H. Freeman and Company. Reprinted by permission.
Figure 23.30a,b: Gordon T. Hewett.

CHAPTER 24

Figure 24.5b: Douglas Merrill.
Figure 24.9: John D. Cunningham/VU.
Figure 24.11b: Mark Nielsen, University of Utah.
Figure 24.12b: Hessler/VU.
Figure 24.12c: © 1988 Ed Reschke.
Figure 24.18b: Mark Nielsen, University of Utah.
Figure 24.18c: Stephen A. Kieffer and E. Robert Heitzman, *An Atlas of Cross-Sectional Anatomy.* Harper & Row, Publishers, Inc., New York, 1979.
Figure 24.19c: Bruce Iverson.
Figure 24.23b: P. Motta and A. Familiari/University La Sapienza, Rome/SPL/Photo Researchers.
Figure 24.23c: Mark Nielsen, University of Utah.
Figure 24.24a: Birke, Peter Arnold.
Figure 24.24b: Willis/BPS.
Figure 24.24c: CNRI/Phototake NY.
Figure 24.28b: P. Motta/University La Sapienza, Rome/SPL/Photo Researchers.
Figure 24.28c: Birke, Peter Arnold.

CHAPTER 26

Chapter 26 opening photo: Lester Bergman & Associates.
Figure 26.2b: Stephen A. Kieffer and E. Robert Heitzman, *An Atlas of Cross-Sectional Anatomy.* Harper & Row, Publishers, Inc., New York, 1979.
Figure 26.3a,b: Mark Nielsen, University of Utah.
Figure 26.4b: Mark Nielsen, University of Utah.
Figure 26.5c: Lester Bergman & Associates.
Figure 26.7c,d: Richard K. Kessel and Randy H. Kardon, *Tissues and Organs: A Text–Atlas of Scanning Electron Microscopy.* © 1979 W. H. Freeman and Company. Reprinted by permission.
Figure 26.9b: Andrew J. Kuntzman.
Figure 26.23b: Mark Nielsen, University of Utah.

CHAPTER 28

Chapter 28 opening photo: Fawcett/Phillips/Photo Researchers.
Figure 28.2a,b: Mark Nielsen, University of Utah.
Figure 28.3a: © Ed Reschke.
Figure 28.5b: Fawcett/Phillips/Photo Researchers.
Figure 28.9b: Mark Nielsen, University of Utah.
Figure 28.14a,b: Biophoto/Photo Researchers.
Figure 28.16b,c: Mark Nielsen, University of Utah.
Figure 28.17: P. Motta/University La Sapienza, Rome/SPL/Photo Researchers.
Figure 28.18: Andrew J. Kuntzman.
Figure 28.21c: Mark Nielsen, University of Utah.
Figure 28.23a,b: Xerox Medical Systems.

CHAPTER 29

Chapter 29 opening photo: Roberts Rugh, Landrum B. Shettles, with Richard Einhorn: *Conception to Birth: The Drama of Life's Beginnings.* Copyright © 1971 by Roberts Rugh and Landrum B. Shettles. By permission of Harper & Row, Publishers, Inc.
Figure 29.1b: CABISCO/Phototake NY.
Figure 29.3d: Roberts Rugh, Landrum B. Shettles, with Richard Einhorn: *Conception to Birth: The Drama of Life's Beginnings.* Copyright © 1971 by Roberts Rugh and Landrum B. Shettles. By permission of Harper & Row, Publishers, Inc.
Figure 29.6b: Roberts Rugh, Landrum B. Shettles, with Richard Einhorn: *Conception to Birth: The Drama of Life's Beginnings.* Copyright © 1971 by Roberts Rugh and Landrum B. Shettles. By permission of Harper & Row, Publishers, Inc.
Figure 29.7b: © 1991 Siu/Biomedical Communication/CMS.
Figure 29.20: Michael Mohan, Lifetouch Studios.

LINE ART

CHARLES BRIDGMAN

16.15

BURMAR TECHNICAL CORPORATION

20.9a,b, 20.10

LEONARD DANK

5.4 a,b, 6.1a, 6.3a, 6.3c, 6.4, 6.5, 6.7, 6.8a–f, 6.11a–d, 7.1a,b, 7.2, 7.3, Exhibit 7.3, 7.4 inset, 7.5, 7.6a,b, 7.7, 7.7 inset, 7.8a, 7.8a inset, 7.8b, 7.8b inset, 7.9a–d, 7.9 inset, 7.10, 7.11a, 7.11a inset, 7.11b, 7.12a, 7.12b, 7.12b inset, 7.13, 7.14, 7.14 inset, 7.15a,b, 7.16a–d, 7.17a,b, 7.18a–d, 7.18b inset, 7.19a–c, 7.19a inset, 7.20a–c, 7.20a inset, 7.21a,b, 7.21 inset, 7.22, 7.23a–c, 7.24, 8.1a,b, 8.2, 8.3a–c, 8.4a,b, 8.5a–e, 8.6a,b, 8.7a–c, 8.7a inset, 8.8a,b, 8.9a,b, 8.10a–c, 8.11a,b, 8.12a–c, 8.13, 8.14, 8.15a,b, 9.1a, 9.2a–f, 9.7a–c, 9.7b inset, 9.7c inset, 9.7d inset, 9.8a–c, 9.8e, 11.1, 11.2a–c, 11.3a,b, 11.4a–c, 11.5a,b, 11.6a,b, 11.7, 11.8, Exhibit 11.9, 11.9a,b, 11.10a–c, 11.10b inset, 11.11a,b, 11.12, 11.13, 11.14a–d, 11.15a–c, 11.16a–c, 11.16c inset, 11.17a–f, 11.18a–c, 11.19a–d, 11.20 inset, 11.20, 11.21a–d, 11.22a–c, 13.7, 13.8, 13.9 , 13.10, 21.10a,b.

SHARON ELLIS

12.3a,b, 12.6, 13.1a, 13.2a, 13.3a, 13.4, 13.5, 13.11a, 13.12, Exhibit 13.1, Exhibit 13.2, Exhibit 13.3, Exhibit 13.4 , 13.13a,b, 14.1a, 14.2, 14.3a,b, 14.4a,b, 14.5a,b, 14.6, 14.7, 14.8a–c, 14.9a,b, 14.10, 14.11a, 14.11c, 14.13a, 14.14, 14.15, 15.4a,b, 15.6, 15.8, 15.9, 16.4a,b, 16.16 inset, 17.2, 22.1, 22.1 inset, 22.2a,b, 22.4, 22.4 inset, 22.5a, 22.6, 22.8.

JEAN JACKSON

19.11, Exhibit 22.3.

LAUREN KESWICK

3.1, 3.2, 3.2 inset, Exhibit 3.2, 3.14 inset, 3.14a, 3.16 inset, 3.17 inset, 3.17a, 3.18 inset, 3.18a, 3.19, 3.20a, 3.20a inset, 3.21a,b, 3.21a inset, 3.27a–h, 3.28a–c, 3.29a,b, 4.1, 4.2, Exhibit 4.2, Exhibit 4.3, 4.3a–c, 5.1, 5.2b, 5.3 inset, 5.3a, 5.5 a,b, 5.6a,b, 6.2, Exhibit 10.3.

BIAGIO JOHN MELLONI, Ph.D.

16.5a, 16.16, 16.17a, 16.18, 16.19a–d, 16.20, 16.21a,b, 16.22a,b.

HILDA MUINOS

3.11a,b, 10.3c–d, 10.6a–c, 10.7a–c, 10.8a,b, 10.9, 10.10, 12.1, Exhibit 12.2, 12.9 a,b, 14.4c, Exhibit 14.4, 14.16, 14.17, 17.1, 18.25, 19.1a, 19.7, 20.2a, 20.2b inset, 20.3a, 20.3 inset, 20.4a, 20.4c–d, 20.4d inset, 20.5a inset, 20.5a,b, 20.5c inset, 20.5d inset, 20.6a, 20.6a inset, 20.6b, 20.7a,b, 20.8, 20.8 inset, 20.11, 20.12, 20.14, 20.15, 20.17a–c, 20.18a–e, 21.1a–c, 21.2, 21.4a–c, 21.13, 21.17, 21.18a,b, 21.19, 21.20a,b, 21.20c, 21.21a–d, 21.22a,b, 21.23, 21.24, 21.25, 21.26, 21.27a,b, 21.28a, 21.29a, 21.30a,b, 21.31, 23.20a,b, 23.21, 23.22, 24.2, 24.12a,b, 24.23a,b, 24.28a,b, 28.4, 28.6, 28.15, 29.9a,b.

LYNN O'KELLEY

1.7, 15.1, 15.5a,b, 16.1a,b, 16.2a–c, 16.4a inset, 16.5a inset, 16.7, 16.8, 16.11, 16.13, 16.17b, 16.21a inset, 16.22a inset, Exhibit 16.2 inset, 18.1, 18.5, 18.5 insets 1,2, 18.10, 18.13a, 18.13a inset, 18.13c, 18.17a, 18.17a inset, 18.17c, 18.19a, 18.19a inset 1, 18.19a inset 2, 18.23a inset, 18.23a,c, 18.27a,b, 23.1a,b, 23.3, 23.3 inset 2, 23.4a inset, 23.4a–c, 23.5 inset, 23.6a,b insets, 23.7a, 23.9a–d.

PAGE TWO ASSOCIATES

25.1, 25.3, 25.4a,b, 25.4a inset, 25.5, 25.5 inset, 25.6a,b, 25.6b inset, 25.7, 25.8, 25.8 inset, 25.9, 25.9 inset, 25.10, 25.11, 25.12, 25.13, 25.14, 25.15, 25.16, 25.17, 25.18.

JARED SCHNEIDMAN DESIGN

1.3, 1.4, 2.1, 2.2, 2.3a–c, 2.4a–d, 2.5, 2.6a–c, 2.7, 2.8a,b, 2.9a–c, 2.10a–c, 2.11a–d, 2.12a–c, 2.13, 2.14a–d, 2.15, 2.16a,b, 2.17, 3.3, 3.4a,b, 3.6a,b, 3.7a,b, 3.8a–c, 3.9, 3.10, 3.15, 3.16, 3.22, 3.23, 3.24, 3.25, 3.30a,b, 5.7, 5.8a,b, 6.10, 10.11, 10.12, 10.13a–c, 10.14, 10.16a–c, 12.7a,b, 12.8a,b, 12.10, 12.11, 12.12a,b, 12.13, 12.14a,b, 12.15, 12.16a,b, 12.17a–d, 13.6, 15.7, 16.9a–c, 16.10a–g, 16.12, Exhibit 17.1, 17.3, 18.2, Exhibit 18.2, 18.3, 18.4, 18.7, Exhibit 18.7, 18.8, 18.11, 18.12, 18.14, 18.15, 18.18, 18.20, 18.21, 18.24, 18.26a,b, 19.1b, Exhibit 19.2, 19.4, 19.5, 19.10, 20.13, 21.6, 21.7, 21.8, 21.9, 21.11, 21.12, 21.14, 21.15, 21.16, 22.9, 22.12, 22.13, 22.14, 22.15a,b, 22.16, 22.17, 22.18, 22.19, 22.20, Exhibit 23.2, 23.12, 23.15, 23.16, 23.17, 23.18, 23.19, 23.23, 23.24, 23.25a,b, 23.26, 23.27, 23.28, 23.29a–c, 24.13, 24.14, 24.15, 24.17, 24.21, 24.25a,b, 24.26, 25.2, 26.5a inset, 26.5b, 26.11, 26.12, 26.13, 26.14, 26.15, 26.16, 26.17, 26.18, 26.19, 26.20, 27.1, 27.3, 27.4, 27.5, 27.6, 28.7, 28.24, 28.25, 28.26, 28.27, 28.28, 29.13, 29.14, 29.15, 29.16, 29.17, 29.18, 29.19.

NADINE SOKOL

1.2, Exhibit 4.1, Exhibit 7.2, 7.22 inset, 7.23a inset, 7.23c inset, 8.2 inset, 8.3a inset, 8.4a inset, 8.5a inset, 8.5c–d inset, 8.6a inset, 8.8a inset, 8.9a inset, 8.10a inset, 8.10c inset, 8.11a inset, 8.12a inset, 18.12c inset, 8.13 inset, Exhibit 9.3, 10.1, 12.2a,b, 12.2c inset, 12.4a–c, 12.5a,b, 12.18a–d, 13.1b inset, 13.3a inset, 13.11a inset, 13.12 inset, Exhibit 14.1, 15.3 inset, 16.8 inset, 16.14 inset, Exhibit 18.3, Exhibit 18.4, Exhibit 18.5, 18.6, Exhibit 18.6, Exhibit 18.8, 19.2a, 19.3a–c, 19.8, 19.12a, 19.13, 20.1, 21.3a, 21.5a,b inset, 21.19 inset, 21.20a,b inset, Exhibit 21.3, Exhibit 21.4, Exhibit 21.7, Exhibit 21.9, Exhibit 21.10, Exhibit 21.13, 21.21a inset, 21.22a,b inset, 21.24 inset, 21.25 inset, 21.26 inset, 21.27a,b inset, 21.28b, 21.29b, 22.7a inset, 22.10a, 22.11, 22.21a, 23.2b inset, 23.2c inset, 23.3 inset, 23.4c inset, 23.8 inset, 23.9a–d inset, 23.14, 23.24 inset, 24.1, 24.3a inset, 24.3a–d, 24.4, 24.5a, 24.6, 24.6 inset, 24.7, 24.8a,b, 24.10, 24.11a, 24.16, 24.18a, 24.18a inset, 24.18c inset, 24.19a inset, 24.19a,b, 24.27 a,b, 24.29, 26.2a inset, 26.4a, 26.4a inset, 26.5a inset, 26.6, 26.7a inset, 26.8a,b, 26.9a, 26.10, 26.21, 26.22, 26.23a, 26.23a,b inset, 26.24a–d, 27.2, 28.9a inset, 28.9b inset, 29.1a.

KEVIN SOMERVILLE

1.1, Exhibit 1.4 insets, 1.5a,b, 1.6, 1.8a–c, 1.9a,b, 1.10, 1.10 inset, 1.11a–c, 1.12, 5.4b inset, 8.15a,b inset, 9.1a,b inset, 9.8d inset, Exhibit 11.2, Exhibit 11.18, Exhibit 12.1, 14.1a inset, 14.4a inset, 14.5a inset, 14.6 inset, 14.7 inset, 14.8a–c inset, 14.9a inset, 14.12, 14.12 inset, 14.13b inset, 14.14 inset, 14.18a,b, 14.19a,b, Exhibit 15.1, 15.2, 15.8 inset, 15.9 inset, 16.4b inset, Exhibit 16.1, 16.17a inset, 16.17c, 16.18 inset, 16.19a inset, Exhibit 16.2 (excluding inset), 23.10a, 23.11a,b, 23.11a,b inset, 23.13a–c, 24.12a inset, 24.22, 24.23a inset, 24.28a inset, 26.1, 26.2a, 26.7a,b, 28.1, 28.1 inset, 28.2c, 28.2c inset, 28.3a inset, 28.3b, 28.5a, 28.8, 28.9a, 28.10a inset, 28.10a,b, 28.10b inset, 28.11, 28.11 inset, 28.12, 28.12 inset, 28.13, 28.13 inset, 28.16a,b inset, 28.16a, 28.16c inset, 28.19, 28.10, 28.21, 28.22a inset, 28.22a,b, 28.29, 28.30, 29.3a–c, 29.4, 29.4 inset, 29.5a–c, 29.6a, 29.6a inset, 29.7, 29.8a, Exhibit 29.2, 29.10, 29.11a,b, 29.12.

BETH WILLERT

10.2, 10.3b, 10.4a–e, 10.17a,b, 10.18a,b, 10.19a,b, 15.3a,b, 20.3b.

INDEX

---

* Those page numbers followed by the letter *E* indicate terms to be found in exhibits, and those followed by the letter *f* indicate terms to be found in figures.

INDEX

EPONYMS USED IN THIS TEXT

Eponymous terms are those named after a person. In general, eponyms should be avoided where possible, since they are totally nondescriptive, often vague, and do not necessarily indicate that the person whose name is used actually contributed anything very original. However, since eponyms are still in frequent use, this glossary has been prepared to indicate which current terms have been used to replace eponyms in this book. In the body of the text eponyms are cited in parentheses, immediately following the current terms where they are used for the first time in a chapter or later in the book. In addition, although eponyms are included in the index, they have been cross-referenced to their current terminology.

EPONYM	CURRENT TERMINOLOGY
Achilles tendon	calcaneal tendon
Adam's apple	thyroid cartilage
ampulla of Vater (VA-ter)	hepatopancreatic ampulla
Bartholin's (BAR-tō-linz) gland	greater vestibular gland
Billroth's (BIL-rōtz) cord	splenic cord
Bowman's (BŌ-manz) capsule	glomerular capsule
Bowman's (BŌ-manz) gland	olfactory gland
Broca's (BRŌ-kaz) area	motor speech area
Brunner's (BRUN-erz) gland	duodenal gland
bundle of His (HISS)	atrioventricular (AV) bundle
canal of Schlemm (SHLEM)	scleral venous sinus
circle of Willis (WIL-is)	cerebral arterial circle
Cooper's (KOO-perz) ligament	suspensory ligament of the breast
Cowper's (KOW-perz) gland	bulbourethral gland
crypt of Lieberkuühn (LĒ-ber-kyoon)	intestinal gland
duct of Rivinus (re-VĒ-nus)	lesser sublingual duct
duct of Santorini (san'-tō-RĒ-nē)	accessory duct
duct of Wirsung (VĒR-sung)	pancreatic duct
end organ of Ruffini (roo-FĒ-nē)	type II cutaneous mechanoreceptor
Eustachian (yoo-STĀ-kē-an) tube	auditory tube
Fallopian (fal-LŌ-pē-an) tube	uterine tube
gland of Littré (LĒ-tra)	urethral gland
gland of Zeis (ZĪS)	sebaceous ciliary gland
Golgi (GOL-jē) tendon organ	tendon organ
Graafian (GRAF-ē-an) follicle	vesicular ovarian follicle
Granstein (GRAN-stēn) cell	nonpigmented granular dendrocyte
Hassall's (HAS-alz) corpuscle	thymic corpuscle
Haversian (ha-VĒR-shun) canal	central canal
Haversian (ha-VĒR-shun) system	osteon
Heimlich (HĪM-lik) maneuver	abdominal thrust maneuver
interstitial cell of Leydig (LĪ-dig)	interstitial endocrinocyte
islet of Langerhans (LANG-er-hanz)	pancreatic islet
Kupffer's (KOOP-ferz) cells	stellate reticuloendothelial cell
Langerhans (LANG-er-hanz) cell	nonpigmented granular dendrocyte
loop of Henle (HEN-lē)	nephron loop
Luschka's (LUSH-kaz) aperture	lateral aperture
Magendie's (ma-JEN-dēz) aperture	median aperture
Malpighian (mal-PIG-ē-an) corpuscle	splenic nodule
Meibomian (mi-BŌ-mē-an) gland	tarsal gland
Meissner's (MĪS-nerz) corpuscle	corpuscle of touch
Merkel's (MER-kelz) disc	tactile disc
Müller's (MIL-erz) duct	paramesonephric duct
Nissl (NIS-l) bodies	chromatophilic substance
node of Ranvier (ron-VĒ-ā)	neurofibral node
organ of Corti (KOR-tē)	spiral organ
Pacinian (pa-SIN-ē-an) corpuscle	lamellated corpuscle
Peyer's (PĪ-erz) patch	aggregated lymphatic follicle
plexus of Auerbach (OW-er-bak)	myenteric plexus
plexus of Meissner (MĪS-ner)	submucosal plexus
pouch of Douglas	rectouterine pouch
Purkinje (pur-KIN-jē) fiber	conduction myofiber
Rathke's (rath-KĒZ) pouch	hypophyseal pouch
Schwann (SCHVON) cell	neurolemmocyte
Sertoli (ser-TŌ-lē) cell	sustentacular cell
Skene's (SKĒNZ) gland	paraurethral gland
sphincter of Oddi (OD-dē)	sphincter of the hepatopancreatic ampulla
Stensen's (STEN-senz) duct	parotid duct
Volkmann's (FŌLK-manz) canal	perforating canal
Wernicke's (VER-ni-kēz) area	auditory association area
Wharton's (HWAR-tunz) duct	submandibular duct
Wharton's (HWAR-tunz) jelly	mucous connective tissue
Wormian (WER-mē-an) bone	sutural bone